46th International Conference on Computers & Industrial Engineering 2016 (CIE46)

Tianjin, China
29 – 31 October 2016

Volume 1 of 2

ISBN: 978-1-5108-3478-1

Printed from e-media with permission by:

Curran Associates, Inc.
57 Morehouse Lane
Red Hook, NY 12571

Some format issues inherent in the e-media version may also appear in this print version.

Printed by Curran Associates, Inc. (2017)

Computers & Industrial Engineering
c/o Mohamed Dessouky
3715 McClintock Avenue, GER 240
Los Angeles CA 90089-0193

Phone: (949) 425-1686
Fax: (949) 425-1696

dessouky@usc.edu

Additional copies of this publication are available from:

Curran Associates, Inc.
57 Morehouse Lane
Red Hook, NY 12571 USA
Phone: 845-758-0400
Fax: 845-758-2633
Email: curran@proceedings.com
Web: www.proceedings.com

TABLE OF CONTENTS

VOLUME 1

VOLUME 2

STUDY OF CREATIVE MANAGEMENT PATH BASED ON PRODUCT INNOVATION DESIGN PROCESS

Wang Jin-Feng[1,*] and Yue Jun-Ju[2], Feng Li-Jie[2]
[1] School of Economics and Management
Shanghai Maritime University ,Shanghai, China
jinfengwang@zzu.edu.cn

[2]School of Management Engineering
Zhengzhou University, Zhengzhou, China
yuejunju@163.com

ABSTRACT

Numerous enterprise practice shows that if the creative management can effectively participate in product innovation design process, the product innovation design efficiency and quality can improve significantly. But the related research is not can come up with suitable reference model and strategy for enterprise applications. This paper put forward the demand oriented and able to coupling creative management process of product innovation design process, then design the design creative process management correspondence enterprise innovation path, combined with the path studied the enterprise creative group strategy and creative management ability as creative management support system of the path. In this paper, the research makes up for the deficiency of the innovative design ideas in the process of enterprise management and provides an important theoretical basis promoting the ability of the enterprise innovation.

Keywords: Product Innovation Design, Creative Management, Creative Ability, Creative Group

1 INTRODUCTION

The starting point of the innovation is originality. The enterprise must get the new originality, accumulate and apply the existing originality continuously if it wants to innovate quickly. Whether the enterprise has the ability to create, accumulate, arrange and assimilate and apply the originality or not will determine the competitive ability of the enterprise directly. It is performed like the following. Firstly, if the enterprise does not attach importance to the acquisition of the originality, then the enterprise can only rely on the knowledge left before, new product's protrusive speed will be very slow and the quality and function of the product will not be improved. The enterprise will face the danger of being

* *Corresponding author. Present address: Shanghai Maritime University School of Economics & Management, NO.1550, Haigang Road, Pudong New District, Shanghai, China*
E-mail address: jinfengwang@zzu.edu.cn

eliminated when the marketing environment gets worse or there are swarming into more market competitors(Mráček and Mucha(2015)). Secondly, if the enterprise does not attach importance to the accumulation of the originality, then it can not draw lesson of the previous experience, thus cause the quality problems' repeat. If the enterprise does not attach importance to the accumulation of the originality, then the product which costs a large amount of manpower, material and finical resources to invent will be lost, and what can only do is to spend a lot to invent repeatedly again(Kok and Biemans(2009)). If the enterprise does not attach importance to the accumulation of the knowledge, then the key originality which is important to the enterprise will be flowed away when the experts retire or leave office. The above examples illustrate that if the enterprise does not implement the originality management, then the enterprise will be hard to walk. And it will be surely reduce the market competitiveness of the enterprise(Murmann and Frenken(2006)).

Nowadays with the economic globalization and rapid development of science and technology, the market competence that the enterprise faces is getting more and more intensive. Due to the experts' leaving office cause the loss of the enterprise's originality. Due to the lack of accumulating and arranging the design knowledge cause the repeated innovation and making the costly mistakes repeatedly. Due to the lack of fully using the existing originality and randomly generating the new originality cause the parts number's largely increasing. These phenomena weaken the enterprise's competitive power heavily. To solve these problems, the enterprise must implement originality management. But the existing researches and practices for originality management can not provide a strong operational solution of the originality management. This paper mainly study the designing process, originality ability and originality grouping strategy that support the product originality management based on constructing the process of product innovative design that can mix with the process of the enterprise's originality management.

2 BACKGROUND

Enterprise is generally accepted that pay more attention to product quality and customer demand can guarantee enterprise survive in the competitive business world. In order to achieve this goal, QFD, life cycle assessment, invented the theory of problem solving, particle swarm optimization algorithm and other tools are widely used in product innovation design process. Younesi and Roghanian(2015)put forward the front end product innovation design model integrated with environment oriented, fuzzy decision making trial and evaluation laboratory quality function deployment (QFD), fuzzy network analysis method. Lin et al.(2015)based on the multidisciplinary cross characteristics of product design, put forward the design scheme selection system by using of taguchi method design the product innovation process and analyzing though the integration of neural network and genetic algorithm. Jiang et al.(2015)ready solve the problem of customer satisfaction is difficult to determine, introduced in the adaptive neural fuzzy inference system to the process of rough set and particle swarm optimization, then constructing customer satisfaction evaluation model. Fernandez-Mesa and Alegre-Vidal(2013) analyze the influence mechanism between organizational learning ability and product innovation design, studied the way of influence dynamic capability of product innovation design by design management and dynamic capability. Warell(2015)provides a means for identifying and developing strategies for brand representation through visual product design, in order to illustrate how product and brand identity is communicated through visual references in product design; Pommier et al.(2016) propose a Life Cycle Assessment (LCA) in order to validate the use of the vacuum process with green veneers and set the directions for research towards high efficiency sustainable innovation; Kır and Yazgan(2016) use Fuzzy Axiomatic Design (FAD) to determine earliness and tardiness penalty costs, solves a scheduling problem on a single machine producing dairy products subject; Moldovan(2014)present an integrated approach about translating the voic of customer in QFD for the development of a new product in a mineral water company by means of knowledge management; Renev and Chechurin(2016) focused on literature review in the area of TRIZ application in building industry, reveals that TRIZ usage in construction is still quite limited. The further research will adapt classic TRIZ tools for construction engineering and management and provide a number of specific case studies.

The Innovation elements value and information in the process of product innovation design have become the decisive factors can influence the competitive of new products and many companies are trying to improve product innovation to gain core competitiveness. And creativity as a source of innovation and starting point, product innovation design in fact can be regarded as a process of proposing creative, screening creative, protecting creative, realizing creative and creating value. But it is worth noting that companies know little about the creative management in the process of product innovation design, the lack of creative management directly led to the innovation management blindly and efficient.

In the study of existing creative management, some scholars study the formation of organizational creativity from the entrepreneurial spirit, organization atmosphere, individual ability. Some scholars study the relationship between creativity and innovation from the aspects of organization structure, enterprise innovation strategy. However, when study the concrete application of creative management in the process of product innovation design often require in virtue of the relevant theories of knowledge management. For example, Moreno and Cavazotte(2015)]take the creative management as one of the dimensions of knowledge management system, study the correlation between working characteristics of the individual and the knowledge innovation process. Gemino et al.(2015)construct the conceptual model of knowledge management, study the promoting for project come from the creative produce in the project practice. Kim et al.(2014) referring to the methods of knowledge view study the implementation method of the knowledge management in virtual organization and the innovation of creative management model from the point of view of one dimensional universal. Sivri and Krallmann(2015)study the mechanism from the point of view of organization and employee personal knowledge management, creative management strategy and corporate performance, then give implementation advice of creative management. Based on this, this article hopes to build product innovation design process fusion with enterprise creative management , design the process suitable for the creative management system and study the creative ability and creative clustering strategy fitting this system operation to support the implementation of the management system.

3 PRODUCT INNOVATION DESIGN PROCESS BUILDING

Most of the above models study the technology fuzzy front end and innovative process of single phase in developing new product and they are lack of the methods that can be used in the fuzzy front end and the whole process of product development. As a result, there are not enough methods supporting the process of innovating design. It also causes the disjoint of the conceptual design and product detailed design. Aiming at the problem, this article constructs the process of the product innovative design facing requirement like figure 1.

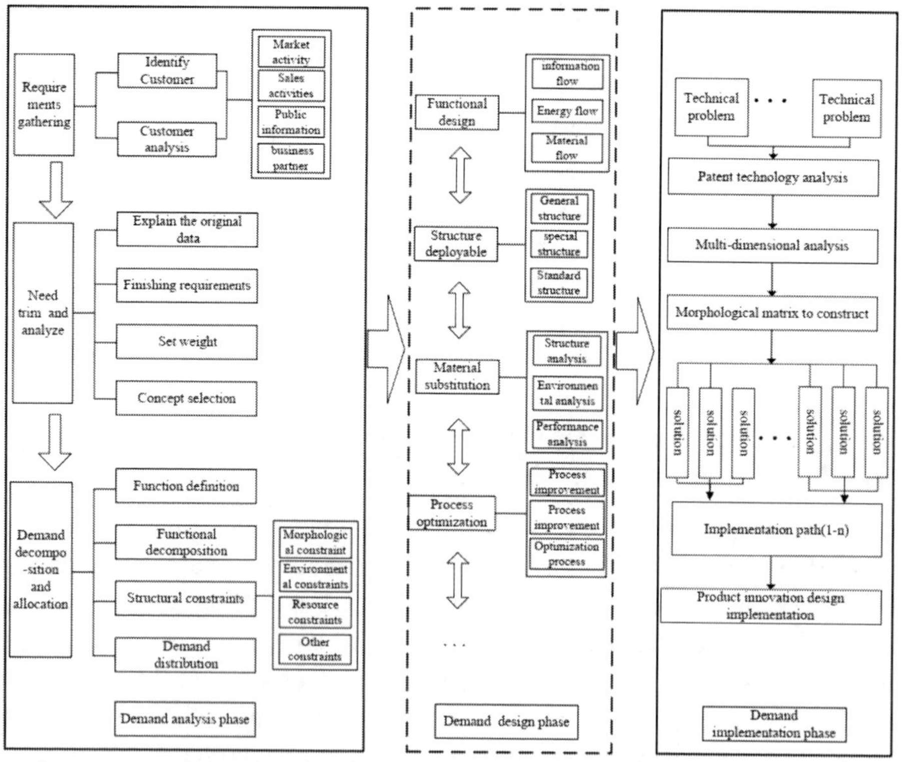

Figure 1: Process of Product Innovative Design

4 CREATIVE MANAGEMENT SYSTEM BUILDING

4.1 Creative Management Process

Based on the creative commons in every stage of the product innovation design can build a creative management system oriented market, then constantly tracking found new ideas and implement design knowledge in the process of innovation diffusion and accumulation. Based on the above, we can form creative management system adapt to the enterprise product innovation design process by integrating the fragmentation obtain. At the same time ,by establishing the feedback mechanism, we can not only make creative commons in different stages of innovation, but also optimize the product innovation design process. Specifically, creative management system including creative acquisition and processing, creative integration and accumulation, creative dissemination and sharing, creative curing and evaluation, creative implementation and application, as shown in figure 2.

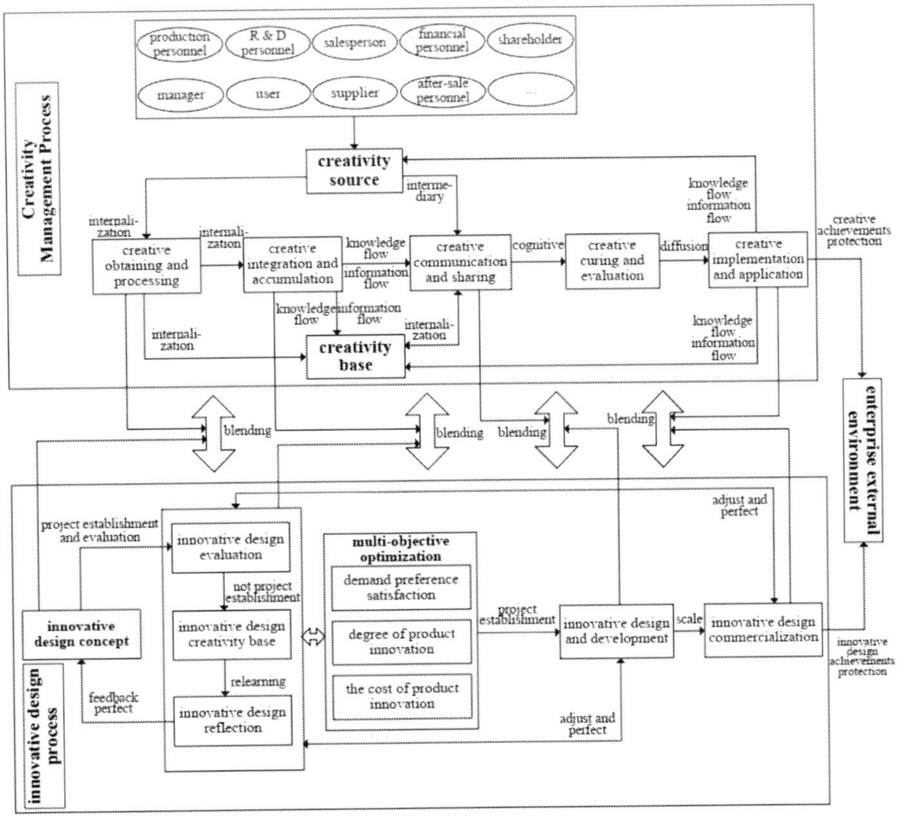

Figure 2: Creative Management System

(1) creative acquisition and processing

Creative acquisition and processing is the process of organizations use creative source for creative collection and processing. It includes internal ideas get, external access, creative process. In-house ideas get refers to get ideas from employees inside the company. External ideas get from customers, competitors, suppliers, partners.

(2) creative integration and accumulation

Creative integration is analyzing the creative acquired and possessed by enterprises, Through a variety of creative integration mechanism to guarantee the correctness of the enterprise existing ideas, coordination between creativity and creativity and enterprise facing environmental compatibility. Creative integration mechanisms need to get ideas for correctness and validity checking, ensure the correctness of the enterprise management ideas and practical.

(3)Creative communication and sharing

Creative dissemination and sharing is used to solve the problem the effective transfer of different individuals and internal organization, different between organizations and individuals. Creative communication and sharing in fact is to break the barriers between different creative owner, to realize the free flow and free use to the ideas within a certain range.

(4) creative curing and evaluation

Creativity is a subject to new ideas or new plan into practice which do not like innovation brings considerable benefits for the enterprise to quickly return. Therefore, in the fierce market competition environment it is particularly important to prevent the enterprise not enough attention resulting in loss of creative.

(5) creative implementation and applications

Creative application is a process that use the knowledge to solve the problem that when companies face a problem can with the aid of enterprises have the creativity to solve practical problems and create new value for the enterprise. In the process of using existing creative solving problem and with the deepening of the solution, the enterprise usually generate new ideas.

4.2 Creative Ability

As shown in figure 3, because creativity as the basis of innovation, the, etc. After the external environment contain information, resources, technology, personnel, demand entering into the enterprise internal, the formation of innovation efficiency is closely related to creativity. Concrete embodiment in the following aspects:

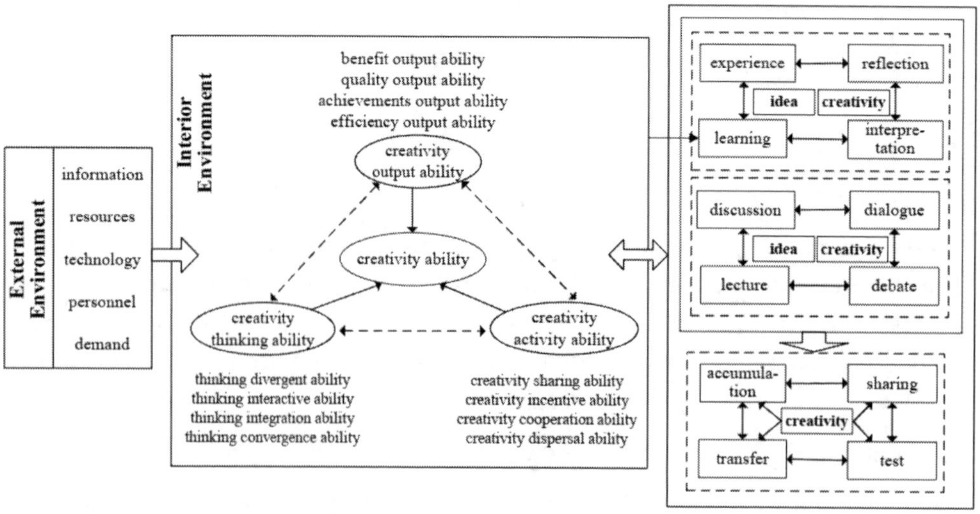

Figure 3:Creative Ability

(1) Rapid response to customer demand. After the external environment contain information, resources, technology, personnel, demand entering into the enterprise internal, we can through to the ideas library for reuse that could reduce the time of the repeated work and with the help of the creative management system greatly shorten the idea of time required to obtain, which laid a foundation for rapid response customer needs through discussion of creativity, learning, sharing.

(2) Continue to improve the quality of the products. By reusing existing prevailing ideas interpret and debate ideas, draw lessons from past experience and lessons which can guarantee the quality of the product and repeat to avoid mistakes.

(3) Reduce the cost of the products. By reusing existing ideas and view of the existing teaching, avoid the repeated inventions and expensive mistakes repeat price.

(4) Improve enterprise innovation ability. Through the cultivation of creative ability, the employees have more time to engage in creative work. Good creative Commons atmosphere will promote the rapid growth of the employees while promoting creativity sharing between enterprises.

(5)Avoid relying too much on some experts, to achieve the creative inheritance. In the process of creative management, we can by sharing scattered in various experts in creative mind implement the valuable experience of expert knowledge creativity especially recessive creative inheritance, avoid due to retirement or resignation of individual experts project delay and even paralysis of the company.

5 CONCLUSION

Creative management applied to product innovation design process can effectively promote for users, enterprise and society have put forward the quality requirements of the best solution. It is great significance to build a creative laws, systematic integration of management and product innovation design process model, narrating for designers to understand the creative and design rule, quickly get both meet the requirements of the function of product innovation.

Based on this, this paper put forward product innovation design process suitable for the creative management, then propose innovation management implementation path. At the same time, to effectively support the implementation of the path study the creative ability and put forward the creative clustering strategy to form the system of creative management system and effectively support the enterprise innovation activities.

REFERENCE

1. **Mráček P, Mucha M**. The Use of Knowledge Management in Marketing Communication of Small and Medium-sized Companies. Procedia - Social and Behavioral Sciences. **2015**, 175: 185-92.

2. **Kok RAW, Biemans WG**. Creating a market-oriented product innovation process: A contingency approach. Technovation. **2009**, 29(8): 517-26.

3. **Murmann JP, Frenken K.** Toward a systematic framework for research on dominant designs, technological innovations, and industrial change. Research Policy. **2006**, 35(7): 925-52.

4. **Younesi M, Roghanian E.** A framework for sustainable product design: a hybrid fuzzy approach based on Quality Function Deployment for Environment. Journal of Cleaner Production. **2015**, 108: 385-94.

5. **Lin M-C, Lin Y-H, Lin C-C, Chen M-S, Hung Y-C.** An integrated neuro-genetic approach incorporating the Taguchi method for product design. Advanced Engineering Informatics. **2015**, 29(1): 47-58.

6. **Jiang H, Kwong CK, Siu KWM, Liu Y.** Rough set and PSO-based ANFIS approaches to modeling customer satisfaction for affective product design. Advanced Engineering Informatics. **2015**, 29(3): 727-38.

7. **Fernandez-Mesa A, Alegre-Vidal J.** Design management capability and product innovation in SMEs. Emerald Insight. **2013**, 51(3): 547-64.

8. **Warell A.** Identity References in Product Design: An Approach for Inter-relating Visual Product Experience and Brand Value Representation. Procedia Manufacturing. **2015**, 3: 2118-25.

9. **Pommier R, Grimaud G, Prinçaud M, Perry N, Sonnemann G.** LCA (Life Cycle Assessment) of EVP - engineering veneer product: plywood glued using a vacuum moulding technology from green veneers. Journal of Cleaner Production. **2016**, 124: 383-94.

10. **Kır S, Yazgan HR.** A sequence dependent single machine scheduling problem with fuzzy axiomatic design for the penalty costs. Computers & Industrial Engineering. **2016**, 92: 95-104.

11. **Moldovan L.** QFD Employment for a New Product Design in a Mineral Water Company. Procedia Technology. **2014**, 12: 462-8.

12. **Renev IA, Chechurin LS.** Application of TRIZ in Building Industry: Study of Current Situation. Procedia CIRP. **2016**, 39: 209-15.

13. **Moreno V, Cavazotte F.** Using Information Systems to Leverage Knowledge Management Processes: The Role of Work Context, Job Characteristics and Task-Technology Fit. Procedia Computer Science. **2015**, 55: 360-9.

14. **Gemino A, Reich BH, Sauer C.** Plans versus people: Comparing knowledge management approaches in IT-enabled business projects. International Journal of Project Management. **2015**, 33(2): 299-310.

15. **Kim TH, Lee J-N, Chun JU, Benbasat I.** Understanding the effect of knowledge management strategies on knowledge management performance: A contingency perspective. Information & Management. **2014**, 51(4): 398-416.

16. **Sivri SD, Krallmann H.** Process-oriented Knowledge Management within the Product Change Systems of the Automotive Industry. Procedia Engineering. **2015**, 100: 1032-9.

DESIGN OF THE ENVIRONMENTAL SERVICE CLOUD PLATFORM CONSIDERING THE CROWDSOURCING MODE

Shengren Piao1* and Jian Li[1,2]
[1]School of Management, Tianjin University of Technology, Tianjin 300384
piaoshengren@163.com

[2]College of Management and Economics, Tianjin University, Tianjin 300391, China
185523348@qq.com

ABSTRACT

Under the Background of Internet+,aiming at the asymmetry in the process of service supply and demand information and the low public participation in the environmental protection, a cloud service platform of crowdsourcing was presented. The problem of environmental service industry was analysed based on the analysis of process. After describing the general idea and technology system of platform, the whole architecture of cloud platform was constructed on the basis of technology supporting system. Finally, taking Tianjin as an example, based on the analysis of operation process of platform information, a preliminary application was discussed from the business centre, cloud management centre and trading process. This study provides some support for promoting efficiency of environmental protection service and reference in crowdsourcing cloud service platform.

Key words: Environmental service, Crowdsourcing mode, Cloud platform

1 INTRODUCTION

The environmental service has become the mainstream concept of environmental policy [1]. It has unique evolution path and is different from other services industry. Some scholars showed that the environmental and ecological could use equivalently[2]. It has been widely discussed over the world and utilized in various research areas. Connolly et al.[3] presented the structure of the network for managing environmental services in NYC. Primmer et al.[4] provided a comprehensive framework for empirical analysis of environmental services governance, which took into account the people and organizations making decisions. Ingram et al.[5] explored the utility of environmental services for supporting livelihoods by drawing from quantitative data derived from relative initiatives. Some studies focused on the payment system of environmental service. Such as, Vijay K et al.[6] suggested environmental services billing system was more effective than traditional government imperative to solve environmental problems. Leander Raes[7] examined the development of American environmental services market mechanism and found that services related with low carbon were given priority to use the market mechanism, however, sewage treatment was depended on government organizations. Shamah[8] established supply chain of environmental service in order to explore the important role of value innovation for environmental services. Although scholars have analyzed environmental service mode from different angles, but most of them focused on discussing the revolves between "government—environmental service company", mainly relying on top-down system. However, very few studies research on the down-top mechanism for public participation in environmental service. With the development of cloud computing and the crowdsourcing mode, many industries begin to apply crowdsourcing cloud platform to improve service efficiency and increase the participation of public, such as the application in computer, transportation, logistics and other fields, it also provides reference for the development of environmental service mode.

Crowdsourcing was introduced by Jeff HoWe who was a Wired magazine writer[9]. On the platform, the seekers put out some tasks, and a crowd of contributors from all over the world would solve the problem. The one who finishes the task would obtain remuneration. Such as Amazon Mechanical Turk has become an efficient and low cost means of obtaining a great quantity of labeled data for supervised learning. Academically, the crowdsourcing mode has been widely studied. Someone used crowdsourcing model to monitor the energy information, and supply help to developing country[10]. Paloheimo et al.[11] put the crowdsourcing model in Book Express in Finland in order to increase the efficiency of public services and reduce transport carbon emissions. Other scholars analyzed the way of crowdsourcing to enhance the competitiveness of enterprises (Prpić J et al.[12]; Kohler T.[13]).

The crowdsourcing mode has been applied to many different aspects in last decade, mainly in computer field included data collection and social network. Few of literatures analyzed the crowdsourcing model in environmental service, but it is not comprehensive. These systems are not flexible enough to support a dynamic service environment. Therefore, it is necessary for us to use the crowdsourcing model idea on environmental service, constructing the cloud platform in crowdsourcing environmental service which can increase the efficiency of service and the public participation.

2 THE DESIGN OF CROWDSOURCING MODEL ENVIRONMENTAL SERVICE CLOUD PLATFORM

2.1 Design concept

Firstly, the convenient cloud platform can be constructed by putting out environmental service information and maintaining the user information and systems. It can achieve the communication between supply and demand sides in the process of environmental service, trading and management, meeting the needs of different enterprises and individual environmental service.Secondly, the category is planned reasonably by cloud computer technology and considers the service time and ways and so on. In this way, it can reduce the costs of trading , transportation and labor and play a major role in the environmental service decision-making process.Finally, the environmental service information will be selected by cloud platform in order to achieve the effective flow of environmental services, mitigate the press of environmental resources and provide comprehensive support for decision making of businesses and individuals.Based on this, the general idea is showed in Figure 1. As shows from the graph, the environmental service information are classified, filtered and pushed by the cloud for enterprises or persons according to their environmental service demand. The solver can get task after being checked qualification and present results to the cloud platform, then the completed program will be marked and the solvers get paid by the seekers. So, the main model of the crowdsourcing cloud platform is “putting out tasks-solicitation-screening program-determining the program-completing the task-issues reward”.

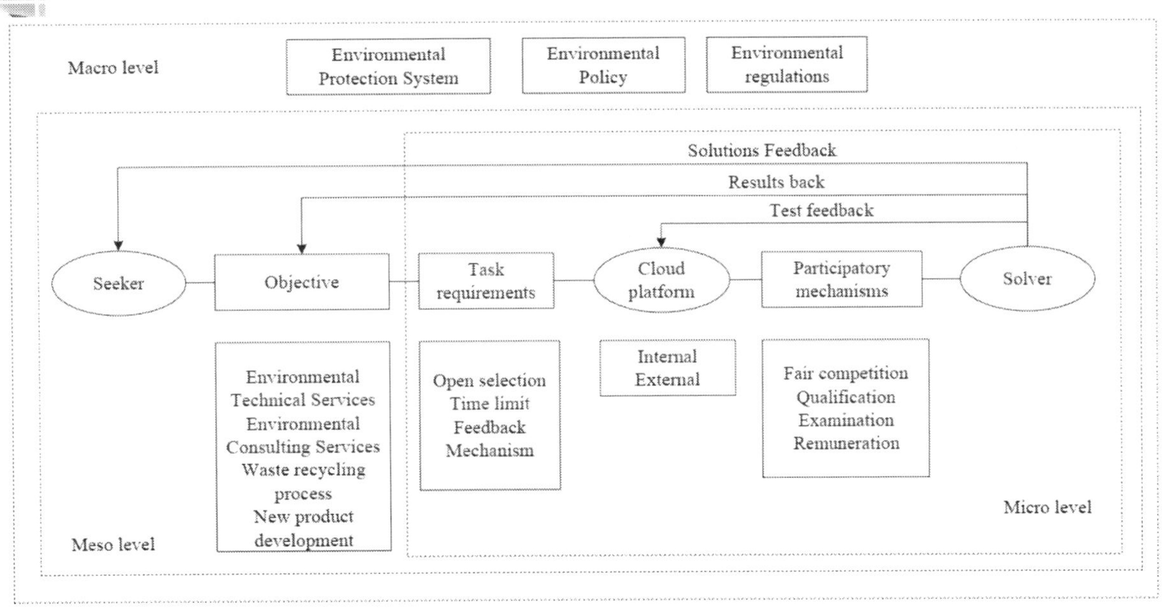

Figure 1: General Idea

2.2 Supporting system

Environmental service cloud platform has multilevel retrieval, task releasing, task receiving ,information browsing, searching, online transactions and payment functions involving dynamic pages, networking, security, electronic payment, cloud security, virtualization technology and other technical support. All that presented is mature and reliable technology, and it can ensure the stability and security of the cloud platform. Technical architecture of the cloud platform includes crowdsourcing technical system and supporting system of cloud platform. The supporting system is constituted by supporting of standard system, shared switching technology and data integrating. The environmental service is sorted as particle size, simple task, macro task and micro-tasks according to the complexity, difficulty and granularity. Firstly, the cloud platform uses the algorithms called DBSCAN, DENCLUE to cluster analysis and match the environmental service data. At the designing stage of environmental service program, all kinds of talents attend to analyze the specific group with reference the algorithms of Native Bayes, GSP, BIRCH et al, then seek the balance among "Cost - feasibility - satisfaction". In the end, the quality of crowdsourcing service is estimated by gold standard data method, the EM algorithm and kappa consistency test method.

2.3 Overall architecture

Application as a service (SaaS), platform as a service (PaaS) and infrastructure as a service(IaaS) are considered to be the most common cloud delivery model.

1.Infrastructure as a Service is the foundation of the whole service, it can supply the service of coordinating the hardware resources of the server and storage by basic information integration and sharing. Its service foundation is CPU, memory and storage and so on. In addition to providing computing and storage services, it can also supply mode-demand by using the technology of configuring the server and virtualization for building a variety of resource pools. According to the diversification of demand, different categories and difficulty, the crowdsourcing programs will be processed by exchange service of computing clusters and virtual resources layer.

2.Platform as a Service provides the supporting technology of collaborative applications, deployment and usage patterns for the large innovation network. At present, the PaaS is usually used as a middle platform of public cloud, private cloud and hybrid cloud, which can handle the customers' supporting system integration, environmental service, publishing tasks and other aspects of integration programs. It also has an effect on developing the applications and services layer.

3.Software as a Service is proposed to deal with these issues in monitoring management, system management, environmental services and metadata management for remote users with timely information.. While this may be suitable for operating space objects on a graphical interface, it plays a key role in developing crowdsourcing model and enhancing the velocity of processing crowdsourcing program and shortening the period of environmental service.

3 APPLICATION ANALYSIS

At present, TianJin has set up the environmental technologies supermarkets with reserving more than 300 kinds of technology. So far, a few demand and docking technology promotion activities have been carried out with various forms.Facing with the development of "Internet +"and crowdsourcing models ,Tianjin environmental technology supermarket is necessary to combine the theory of crowdsourcing and cloud computing technology for the design and optimization platform.According to the foundation of the foregoing, we build the Tianjin environmental service cloud platform which is composed by business centers, cloud management center and transaction process as shown in figure 2. Enterprises and public users can enter the cloud platform by computers,intercoms,etc.

1.The business functions of the platform include the searching module, payment module, helping module, case module, service provider module, tendering and administrative module , etc. And the service mode is composed by B2C, B2B, C2B, C2C and O2O. Referring to the platform Witkey and the actual situation, the environmental services business model include three types: cash reward mode, tendering mode and solver map mode.

2.Cloud management center is divided into base server, IDC Center(Web server and authentication server constitution), Hadoop platform and switches. It can establish and improve the components of user information, take needs of uers as the main service, access the database with parallel computing, grid computing and utility computing and so on. The platform can operate the cloud network by problem management, incident management, configuration management, release management and knowledge management.

3.The transaction process of environmental service could be developed by combining Witkey model. Firstly, the providers submit an application, then when completing the task he will ask for remuneration in the cloud platform with the certification verification by the third party agency. In this process, the bill centre should attend to handle the transaction. Finally, the special department should supervise and manage the entire transaction process.

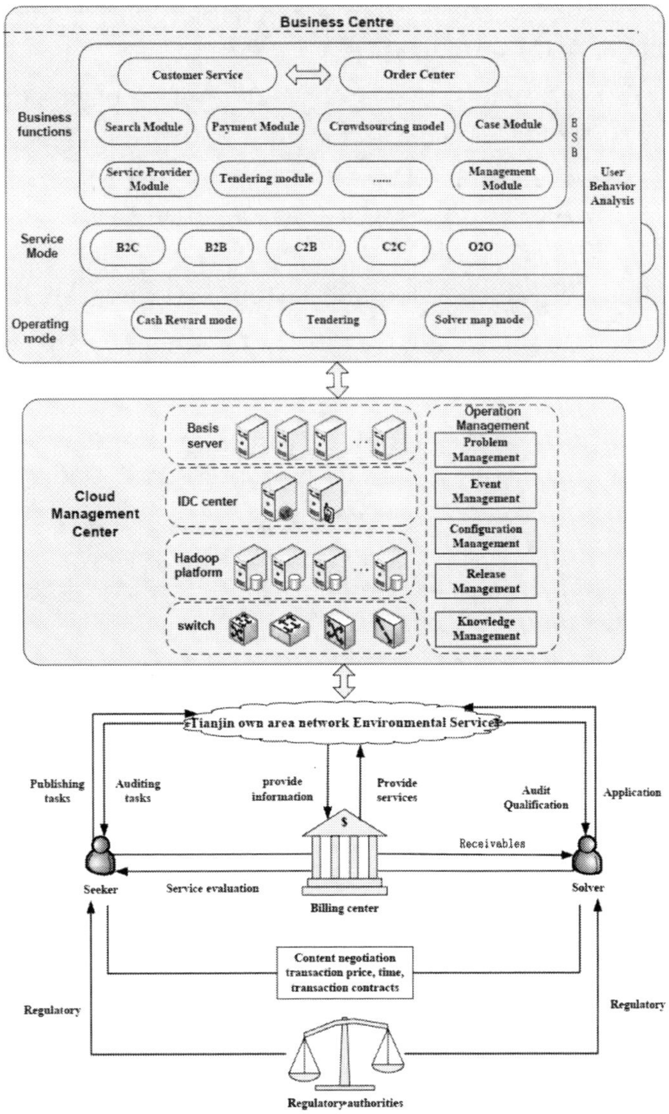

Figure 2: Environmental Services Cloud Platform Application Schematic

4 CONCLUSION

In this paper, we proposed crowdsourcing environmental service cloud platform for the enterprises and persons. In order to prevent the asymmetric information and motivate more public participation, the cloud platform interlocks the crowdsourcing model and cloud computing technology, supplies the real-time environmental service information and encourage broad public participation, so that the public and businesses are in the "win-win". At the same time it can provide the effective protection of natural resources. Finally, the application analysis is provided by Tianjin District environmental service platform. Through it, constructing the crowdsourcing environmental service platform is feasibility. It seems that a rapid development of "internet and cloud computer" for environmental service will be the trend of environmental industry.

5 REFERENCES

[1] **Gómez-Baggethun E, De Groot R, Lomas P L, et al.,2010.**The history of ecosystem services in economic theory and practice: from early notions to markets and payment schemes, Ecological economics, 69(6),pp 1209-1218.

[2] **Wunder,Sven.2015.**Revisiting the concept of payments for environmental services,Ecological Economics,117,pp 234-243.

[3] **Connolly J J T, Svendsen E S, Fisher D R, et al.,2014.**Networked governance and the management of ecosystem services: The case of urban environmental stewardship in New York City,Ecosystem Services, 10,pp 187-194.

[4] **Primmer, E., Jokinen, P., Blicharska, M., Barton, D.N., Bugter, R., Potschin, M., 2015.**

Governance of ecosystem services: a framework for empirical analysis,Ecosyst. Serv.,

16, pp158-166.

[5] **Ingram J C, Wilkie D, Clements T, et al.,2014.**Evidence of Payments for Ecosystem Services as a mechanism for supporting biodiversity conservation and rural livelihoods,Ecosystem Services, 7,pp 10-21.

[6] **Kolinjivadi, Vijay K., and Terry Sunderland.2012.**A review of two payment schemes for watershed services from China and Vietnam: the interface of government control and PES theory,Ecology and Society,17(4),pp10-25.

[7] **Raes, Leander, et al.,2016.**Towards market-or command-based governance? The evolution of payments for environmental service schemes in Andean and Mesoamerican countries,Ecosystem Services,18,pp 20-32.

[8] **Shamah R A M.2012.**Innovation within green service supply chains for a value creation,Journal of Modelling in Management, 7(3),pp 357-374.

[9] **Howe J.2006.**The rise of crowdsourcing,Wired magazine,14(6),pp 1-4.

[10] **Bazilian M, Rice A, Rotich J, et al.,2012.**Open source software and crowdsourcing for energy analysis,Energy Policy,49(10),pp149-153.

[11] **Paloheimo, H., et al.,2015.**Transport reduction by crowdsourced deliveries e-a library case in Finland,Journal of Cleaner Production.http://dx.doi.org/10.1016/j.jclepro.2015.04.103.

[12] **Prpić J, Shukla P P, Kietzmann J H, et al.,2015.** How to work a crowd: Developing crowd capital through crowdsourcing ,Business Horizons,58(1),pp 77-85.

[13] **Kohler T.2015.**Crowdsourcing-Based Business Models,California Management Review, 57(4),pp 63-84.

FDI OUTFLOW AND LIFE SATISFACTION: A RESEARCH OF CHINESE FDI OUTFLOW TO OECD COUNTRIES IN A COMPLEX NETWORK BASED ON TABU SEARCH

Da Huo[1*], Ken Hung[2], and Haibo Wang[3*]

[1] School of International Trade and Economics
Central University of Finance and Economics,China
dhuo@cufe.edu.cn

[2]A.R. Sanchez School of Business
Texas A&M International University, USA
ken.hung@tamiu.edu

[3]A.R. Sanchez School of Business
Texas A&M International University, USA
hwang@tamiu.edu

ABSTRACT

With development of the Chinese economy, Chinese FDI outflow has been enlarged, including investment flows leading to a number of OECD countries. This research studies the investment efficiency of Chinese FDI outflow to OECD destination countries based on life satisfaction by the DEA model. It further reveals key locations in the complex network based on Chinese FDI outflow to OECD countries, life satisfaction of OECD destination countries, and their cultural distances, through a Tabu Search algorithm. This research offers decision support to global managers and policy makers by identifying key locations of Chinese FDI in OECD countries based on life satisfaction.

Keywords: Chinese FDI Outflow, Life Satisfaction, OECD Countries, Cultural Distance, Complex Network, DEA Model, Tabu Search

1 INTRODUCTION

China has made considerable effort to transition from a planned economy to a market-oriented economy (Li et al. [1]). Lower entry barriers and more relaxed regulation of foreign direct investment (FDI) have allowed foreign investors to initiate merger and acquisitions in China. This has resulted in an increase in production activity (Reid and Walsh [2]). With the deregulation of foreign investment in China and the relaxation of the constraints on foreign products, the open-door policy of China has increased the level of competition for both Chinese companies and foreign investors. Increasing competition has driven down the margins of Chinese companies in the domestic market, forcing Chinese companies to learn to compete in the global marketplace as the Chinese economy is further globally integrated. As most of the FDI comes from developed countries, the Chinese government has encouraged FDI through Sino-foreign joint ventures and wholly owned foreign enterprises. The success of the Chinese open door economic policy has in turn further accelerated the outward FDI of the country (Robert [3]). Chinese international companies have purchased foreign companies in hopes of restoring them to their former profitability (Lin & Lin [4]).

In both co-operation and competition with foreign companies, Chinese companies have developed management and technical skills to meet the international standards required for entry to higher value-added markets. Chinese companies have also increased their global expansion with increased trade and serving as the manufacturing plant of the world market. Domestic constraints and the pressures of global competition have provided the dynamic for Chinese companies to operate in foreign markets (Child & Rodrigues [5]). As OECD countries are important destination countries of Chinese FDI outflows, the increase of Chinese FDI can have important effect on the life satisfaction of OECD destination countries, and cultural distances between China and OECD countries are also important to the life satisfaction of FDI destinations. This research reveals the efficiency of FDI outflow from China to OECD countries, based on the life satisfaction of OECD destination countries, and also measures cultural distances between China and FDI destination countries. This research further reveals the key locations of Chinese FDI outflows to OECD countries by a heuristic method, based on Chinese FDI outflows to different OECD countries, investment efficiency based on life satisfaction, and cultural distance between China and OFDI destinations. This research can offer support to global managers and policy makers in determining future location of FDI outflows.

* Corresponding Author

2 LITERATURE REVIEW

2.1 Enhanced Involvement of Chinese Companies in OFDI

According to international product life cycle theory, developed countries will shift to import from the developing countries adversely, while the market in the developing countries increases to a sufficient size and competition is increased based on lower cost of production in the developing countries (Shenkar and Luo [6]). These companies become international companies by creating and exploiting innovations through global expansion. Chinese companies are often started to focus on the domestic market with development of new product markets. In cooperation with foreign partners, Chinese companies learned the necessary management skills and technology to compete in the international marketplace and then they start to make efforts to launch their products in foreign markets. The open-door policy helped Chinese companies to complete the process of internationalization and compete for a larger market share in the global market.

Chinese companies have become international players with a long-term strategic plan and have developed the capacity to organize their resources systematically in pursuit of global expansion. According to the Ministry of Commerce of China, the FDI out flow from China to global market has increased from 2.7 billion $US in 2002 to 107.84 billion $US in 2013 (see Figure 1). Chinese companies have become strategic asset-seeking investors and seek to acquire information and knowledge on how to operate internationally. Chinese companies have started to use mergers and acquisitions to acquire strategic assets such as advanced technology and intangible strategic assets as their shortcut to establish their companies as international players. Examples include Lenovo acquiring IBM PC division and recent acquisition of Volvo by Zhejiang Geely Holding Group in 2010 (Gugler & Fetscherin [7]). Buckley et al. [8] pointed out that the advantages of Chinese companies in capital markets can offer a great support to FDI outflow activities. Chinese companies with OFDI activities can benefit from soft budget constraints from banking systems, and internal capital market based on conglomeration and family owned businesses. The advantages of Chinese companies in FDI outflow activities can be transformed to ownership advantages that allow them to operate investment activities in foreign countries more effectively. Government support can further offer Chinese companies better access to financial and institutional resources that offset ownership and location restraints abroad.

2.2 FDI Outflow and Life Satisfaction of Destination Countries

FDI flows are important in improving the life relevant indexes of destination countries. Bjornskov et al. [9] pointed out that investment price level is an important cross country determinant of life satisfaction. Barclay & Gary [10]suggest that while the MNEs initially used the natural assets of the country, their activities ultimately resulted in the quality of the workforce being upgraded towards higher value-added, skill-intensive activities. Better institutional quality and human capital development are associated with reducing poverty. Shiralashetti & Hugar [11] also pointed out that FDI can motivate GDP growth rate, growth in industry and service sector, reduction in unemployment, reduction in poverty, improvement in standard of living, increase in foreign exchange reserves, increase in exports, and improvement in stock market. Fowowe & Shuaibu [12] pointed out that FDI inflows have significantly contributed to poverty reduction in African countries, and better institutional quality and human capital development as well as better functioning financial systems with FDI are associated with reducing poverty. Yang et al. [13] pointed out that the location of FDI has a great importance to firm performance of multinational companies. Li et al. [14] pointed out that the major motive of China's outflow FDI to Thailand is marketing seeking. But if the Chinese invested firm lacks specific advantages, it tends to have a poor performance. Filatotchev et al. [15] pointed out that high-commitment entry is found to be positively associated with the affiliate being located in areas with strong economic, cultural and historic links with the parent company. Sun et al. [16] pointed out that R&D investment and patents are significantly positively related to the high tech export of BRICS countries in the U.S. market.

2.3 Cultural Distances and Effect of FDI Outflows on Life Satisfaction of Destinations

Cultural distance is an important factor to be considered in determining the optimal location and performance of FDI outflows. Salavrakos [17] pointed out that culture plays a key role in determining the success of cross-national investment and management practices by developed countries in China. Tang (2012) [18] pointed out that difference in individualism between two countries encourages FDI from OECD countries to non-OECD countries while difference in power distance discourages it. Lu et al. [19] pointed out that FDI has significant effects on the degree of future orientation, performance orientation and in-group collectivism in changes of Chinese culture. Zeng et al. [20] pointed out that a subsequent FDI located in a country that is culturally similar to an MNE's home country benefits from the MNE's prior double-proximate experience.

Furthermore, the life satisfaction of different countries is also associated with their cultural differences. Ungerer & Strasheim [21] pointed out that life satisfaction has also been found to be related to income at aggregate cultural levels, and living standards have a moderating effect on the relationships between higher-order individual cultural values and life satisfaction. Yetim [22] (2003) pointed out that life satisfaction is positively correlated with individualism, self-esteem and mastery and negatively correlated with collectivism. Tong & Tak [23] pointed out that organizational culture significantly influences knowledge sharing and job satisfaction, and knowledge sharing plays an important mediating role between organizational culture and job satisfaction. Therefore, life satisfaction of OFDI destination countries is associated with the cultural distances across different countries.

Cultural distances can have an impact on the location selection of Chinese FDI outflows. Qi & Yang [24] found that Chinese OFDI prefers to choose locations with smaller cultural gaps. Qi & Yang [25] also found that cultural distance has an important effect on entry mode choices of Chinese OFDI. Huang et al. [26] pointed out that cultural differences and conflicts can influence greenfield or M&A decisions of Chinese OFDI. Different locations of FDI outflows can have varying cultural traits referring to conceptions and values in work-life balance, job satisfaction, financial conditions and health sectors, which also influences life satisfaction. Akram & Hassaan [27] pointed out that conflicts between work and life have a negative relationship with job satisfaction. Loewe et al. [28] pointed out that life satisfaction is positively associated with t financial situation, and family, work, and health sectors. Su & Ma [29] pointed out that shorter cultural distances can facilitate the Chinese FDI outflows to host countries. Zhang & Yi [30] also pointed out that larger cultural distances can reduce the positive effect of technological innovations on Chinese OFDI.

3 RESEARCH METHOD

3.1 Investment Efficiency of Chinese OFDI to OECD Countries based on Life Satisfaction

The DEA (data envelopment analysis) model is used to identify investment efficiencies of Chinese FDI outflows based on life satisfaction in OECD destination countries. Charnes et al. [31] applied the DEA model to measure the efficiency based on constant return of scale. Banker et al. [32] applied the variable return of scale to measure the efficiency in DEA model. In this research, China's outward FDI flow to OECD countries in 2013 is set as the short-term input in cross-border investment activities, and China's outward FDI stock in OECD countries is set as the long-term input in cross-border investment activities from China to OECD countries, using the OFDI data from the foreign direct investment bulletin. Life satisfaction of 30 OECD countries is set as the output of Chinese FDI outflows according to life satisfaction data from OECD countries, including Australia, Austria, Belgium, Brazil, Canada, Chile, Czech Republic, Denmark, Finland, Germany, Hungary, Ireland, Italy, Japan, South Korea, Luxembourg, Mexico, Netherlands, New Zealand, Norway, Poland, Portugal, Russia, Slovak Republic, Sweden, Switzerland, Turkey, U.K., and U.S. Life relevant indexes of OECD countries are also set as output variables for life satisfaction of OECD countries, including housing, jobs, community, education, environment, civic engagement, health, safety, and work-life balance.

The life satisfaction and life relevant indexes are given on the scale of 1-10, and both Chinese FDI outflow and Chinese FDI stock in OECD countries by the unit of US$1000 are transformed by logarithm for scaling to a 1~10 scale. The scale assumption is variable return of scale, and the objective is to maximize the profit in investment to OECD countries. The function is input oriented and based on a two stage slack model. The constant return of scale technological efficiency is decomposed to the variable return of scale technological efficiency and scale efficiency, which is equal to the ratio of the constant return of scale technological efficiency to the variable return of scale technological efficiency. The function for technological efficiencies is as following:

$$\begin{aligned} &\min_{\theta,\lambda}\ \theta \\ &s.t.\ \ -q_i + Q\lambda \geq 0 \\ &\qquad \theta x_i - X\lambda \geq 0 \\ &\qquad\quad I'_{Ix1}\lambda \leq 1 \\ &\qquad\qquad \lambda \geq 0 \end{aligned} \qquad (1)$$

q_i is the output of life satisfaction and life relevant variables in OECD destination countries. xi is the input of Chinese FDI outflows to OECD countries in 2013 and Chinese OFDI stock in OECD countries. is the technological efficiency of Chinese FDI outflow investment based on life satisfaction of OECD destination countries, and efficiency values are also scaled to a 1~10 scale.

3.2 Cultural Distance between China and OECD Destination Countries of Chinese FDI Outflow

Cultural distances between Chinese and OECD destination countries are measured based on Hofstede cultural dimensions. Kogut & Singh [33] pointed out that the cultural distance can be measured on the basis of four Hofstede cultural dimensions, including individualism vs. collectivism, power distance, masculinity vs. femininity, and uncertainty avoidance. The function is as following:

$$CD_j = \sum_{i=1}^{n} \frac{(C_{ij} - C_{ih})^2 / V_i}{n} \quad (2)$$

CD_j is the cultural distance of country j to country h, and C_{ij} is the cultural dimension i of country j. C_{ih} is the cultural dimension i of country h. V_i is the variance of cultural dimension i. Groot et al. [34] further measured the cultural distances of different countries based on five Hofstede cultural dimensions, including the traditional four dimensions and long-term vs. short-term orientation. This research studies the cultural distances between China and OECD destination countries of Chinese FDI outflows based on six cultural dimensions according to Hofstede cultural indexes, including individualism, power distance, masculinity, uncertainty avoidance, pragmatism, and indulgence. And cultural distances are also in the scale of 1~10.

3.3 Key Locations of OECD Countries in Complex Network based on Chinese FDI Outflows, Life Satisfaction of FDI Destinations and Cultural Distances

Our research applies the heuristic method to identify the key locations of OECD countries in the complex network based on Chinese FDI outflows, life satisfaction of FDI destinations and cultural distances. A Tabu search is performed to determine the key-node OECD countries in the complex network. Glover [35] initiated the use of the Tabu search method in artificial intelligence and operational research. This research reveals the Euclidean distances of OECD destination countries by visualization of the complex network based on six dimensions, including Chinese FDI outflows in 2013, Chinese FDI outflow stocks, constant return of scale technological efficiency (CRSTE), variable return of scale technological efficiency (VRSTE), and scale return as investment efficiency based on life satisfaction of FDI destinations, as well as cultural distances between Chinese and OECD destination countries. The function of Euclidean distance is as following:

$$d_{ij} = \sqrt{\sum_{k=1}^{p} (x_{ik} - x_{jk})^2} \quad (3)$$

d_{ij} is the Euclidean distance between country i and country j. x_{ik} is dimension k of country i, and x_{jk} is dimension k of country j. The key locations of OECD destination countries is identified by Tabu search method that targets at the shortest Euclidean distance between key node countries and other connected OECD destination countries of Chinese FDI outflows in the complex network.

3.4 Tabu Search algorithm

Tabu search is an efficient method to obtain global optimal solution for complexity combinatorial optimal problem. Let be a feasible solution set. Each has an associated neighbourhood, and each solution is reached from by an operation called a move. Tabu search is an iterative method and begins in the same way as ordinary local or neighbourhood search. The general steps of an iterative procedure start with (1) choosing, and then (2) find such that, (3) is the local optimum (minimal), denoted with, if no such can be found and the method stops, (4) otherwise, designate to be the new and go to step (2). However, algorithms for ordinary local or neighbourhood search often face the risk of being trapped in local optimal instead of global optimal. Tabu neighbourhood search method has been designed to avoid being trapped in a local minimum or maximum by using memory structure which can be considered as modifying the neighbourhood of the current solution. The modified neighbourhood denoted by may be expanded to include solutions not ordinarily found in. Therefore, Tabu search can be viewed as a dynamic neighbourhood method where the neighbourhood of is not a static set, but rather a set that can change according to the history of the search. The dynamic neighbourhood is achieved via the memory strategy. The selected attributes that occur in solutions recently visited are labelled 'Tabu-active', and solutions that contain Tabu-active elements are those become Tabu. Current moves are taken in Tabu list which record the Tabu-active attributes and identify their current status. These Tabu-active attributes are prohibited in following number of iterative that defined as Tabu tenure. This prevents certain solutions from the recent past from belonging to and hence from being revisited, and consequently exploiting solutions in a modified neighbourhood. This method has proven to be highly effective for solving the problems on clustering and minimum spanning tree based on the distance measures (Wang et al. [36]; Kochenberger et al. [37]; Pereira et al. [38]).

4 RESULT

Table 1 shows the descriptive analysis of Chinese FDI outflow to OECD Countries. It is found that the average level of Chinese FDI to OECD countries in 2013 is 4.98, and the standard deviation is 1.03. The average level of Chinese OFDI stock in OECD countries is 5.98, and standard deviation is 0.85. It is found that the average level of life satisfaction in OECD destination countries is 6.65, and standard deviation is 2.81. In addition, the average level of power distance in OECD destination countries of Chinese FDI outflow is 49.13, and the standard deviation is 21.57. The average level of individualism is 61.2, with a standard deviation of 20.6. The average level of masculinity is 52.3, with a standard deviation of 24.42. The average level of uncertainty avoidance is 66.53, and the standard deviation is 21.63. The average level of pragmatism is 53.03, with a standard deviation of 22.19. The average level of indulgence is 53.97, and the standard deviation is 19.03.

The Chinese FDI outflow to OECD countries in 2013 is shown in Figure 2. U.S., Australia, U.K., Luxembourg, Russia, Canada, Germany, Japan, Brazil, and France are OECD destination countries with higher level of Chinese FDI outflow in 2013. The Chinese OFDI stock in OECD countries is shown in Figure 3. Australia, U.S., U.K., Luxembourg, Russia, Canada, Norway, France, Germany, and Netherlands are OECD destination countries with higher level of Chinese OFDI stock. Table 2 shows the investment efficiencies of Chinese FDI outflows to OECD countries based on life satisfaction and life relevant indexes by DEA model. It is found that Austria, Belgium, Denmark, Finland, New Zealand, Portugal, Slovak Republic, Switzerland, and U.S. are at the frontier of investment efficiency, as the constant return of scale technological efficiencies, variable return of scale technological efficiencies and return of scale technological efficiency are equal to 1.

Furthermore, Table 3 shows the cultural distances between China and OECD destination countries of Chinese FDI outflows. OECD destination countries such as Brazil, Czech Republic Russia, Slovak Republic, South Korea, Turkey, Germany, and Japan have shorter cultural distances to China. Figure 4 shows the attributes of OECD destination countries, including investment efficiencies of Chinese FDI outflows based on life satisfaction of OECD destination countries, Chinese FDI outflows to OECD countries in 2013, Chinese FDI stock in OECD countries, and cultural distances between China and OECD destination countries.

The complex network based on Chinese FDI outflow investment efficiencies of life satisfaction, Chinese FDI outflows to OECD countries, and cultural distances between China and OECD destination countries can be visualized in Figure 5. Table 4 shows the connection sequence of key locations in OECD destination countries by Tabu search method that targets the shortest Euclidean distances between key node countries and other connected OECD destination countries of Chinese FDI outflows in the complex network. It is found that Switzerland, Belgium, U.S., Czech Republic, Italy, Mexico, Hungary, Turkey, France and Brazil are in the front group of key location connections in the complex network. Russia, Chile, U.K., Germany, Netherlands, Australia, Japan, Norway, Luxembourg and Canada are in the middle group of key location connections in the complex network. Poland, South Korea, Ireland, Sweden, New Zealand, Finland, Austria, Denmark, Portugal, and Slovak Republic are in the final group of key location connections in the complex network.

5 CONCLUSION

With the development of Chinese economy and further implementation of the open-door economic policy, China has enhanced its international involvement through increased foreign direction investment (FDI) activities to a number of destinations, including OECD countries. The Chinese FDI outflow can be important factor in life satisfaction of OECD destination countries. This research identifies the investment efficiencies of Chinese FDI outflows based on life satisfaction of OECD destination countries using DEA model. It is found that Austria, Belgium, Denmark, Finland, New Zealand, Portugal, Slovak Republic, Switzerland, and U.S. are at the frontier of investment efficiency. This research also measures the cultural distances between China and the OECD destination countries based on Hofstede cultural dimensions. It is also found that OECD destination countries such as Brazil, Czech Republic Russia, Slovak Republic, South Korea, Turkey, Germany, and Japan have shorter cultural distances to China.

This research further identifies the key locations of OECD countries in the complex network based on investment efficiencies of life satisfaction, Chinese FDI outflows, and cultural distances by the Tabu search heuristic method. This method targets the shortest Euclidean distances between key node countries and other connected OECD destination countries of Chinese FDI outflows in the complex network. The complex network based on Chinese FDI outflow investment efficiencies of life satisfaction, Chinese FDI outflows to OECD countries, and cultural distances between China and OECD destination countries is visualized in this research. It is found that Switzerland, Belgium, U.S., Czech Republic, Italy, Mexico, Hungary, Turkey, France and Brazil are in the front group of key location connections in the complex network. Russia, Chile, U.K., Germany, Netherlands, Australia, Japan, Norway, Luxembourg and Canada are in the middle group of connections in the complex network. Poland, South Korea, Ireland, Sweden, New Zealand, Finland, Austria, Denmark, Portugal, and Slovak Republic are in the final group of connections in the complex network. This research offers decision support to global managers and policy makers by identifying key locations in complex network based on Chinese FDI, life satisfaction, and cultural distances.

6 ACKNOWLEDGEMENT

This research was supported in part by Central University of Finance and Economics Graduate Program Development Project (KC20150003) and National Natural Science Foundation of China General Project (71573023). The authors thank Professor Joseph Brennan for editing this paper.

7 APPENDIX

Table 1: Descriptive Analysis of Chinese FDI Outflow to OECD Countries

LIFE	FDI2013	FDI Stock	Life Satisfy	Housing	Income	Jobs	Communit
Mean	4.98	5.98	6.56	5.68	3.73	7.01	7.71
Std	1.03	0.85	2.81	1.54	2.33	1.49	1.99
Max	6.59	8.07	10.00	7.80	10.00	9.60	10.00
Min	2.18	4.63	0.50	2.20	0.10	4.30	2.00
Median	5.17	5.78	7.40	6.20	4.00	7.50	8.15
LIFE	Education	Envrion	Civic Engag	Health	Safety	Work-life	
Mean	6.42	7.30	5.42	6.92	8.24	6.64	
Std	1.94	1.83	1.66	2.01	2.09	1.97	
Max	9.10	9.80	9.50	9.40	9.90	9.80	
Min	1.20	2.90	2.10	0.60	0.40	0.00	
Median	7.10	7.85	5.30	7.60	8.80	7.15	
CULTURE	Power Dist	Individ	Masculine	Unctn Avd	Pragmat	Indulgenc	
Mean	49.13	61.20	52.30	66.53	53.03	53.97	
Std	21.57	20.60	24.42	21.63	22.19	19.03	
Max	100.00	91.00	100.00	99.00	100.00	97.00	
Min	11.00	18.00	5.00	23.00	21.00	20.00	
Median	43.00	67.50	55.50	70.00	52.00	57.00	

Table 2: Investment Efficiency of Chinese FDI to OECD Countries based on Life Satisfaction in DEA Model

Country	CRSTE	VRSTE	Scale	Country	CRSTE	VRSTE	Scale
Australia	0.85	1.00	0.85	South Korea	0.92	1.00	0.92
Austria	1.00	1.00	1.00	Luxembourg	0.87	1.00	0.87
Belgium	1.00	1.00	1.00	Mexico	0.83	0.83	1.00
Brazil	0.68	0.74	0.91	Netherlands	0.84	1.00	0.84
Canada	0.87	1.00	0.87	New Zealand	1.00	1.00	1.00
Chile	0.68	0.89	0.77	Norway	0.86	1.00	0.86
Czech Republic	0.88	0.88	1.00	Poland	0.91	1.00	0.91
Denmark	1.00	1.00	1.00	Portugal	1.00	1.00	1.00
Finland	1.00	1.00	1.00	Russia	0.60	0.68	0.88
France	0.76	0.77	0.99	Slovak Republic	1.00	1.00	1.00
Germany	0.78	1.00	0.78	Sweden	0.98	1.00	0.98
Hungary	0.81	0.83	0.98	Switzerland	1.00	1.00	1.00
Ireland	0.97	1.00	0.97	Turkey	0.75	0.81	0.92
Italy	0.84	0.84	1.00	UK	0.76	1.00	0.76
Japan	0.86	1.00	0.86	USA	1.00	1.00	1.00

Table 3: Cultural Distance between China and OECD Destination Countries based on Hofstede Cultural Dimensions

Country	Cultural Distance	Country	Cultural Distance	Country	Cultural Distance
Australia	4.97	Germany	2.02	Norway	4.09
Austria	3.58	Hungary	3.03	Poland	2.75
Belgium	3.07	Ireland	3.87	Portugal	3.24
Brazil	2.09	Italy	2.36	Russia	1.89
Canada	3.77	Japan	2.15	Slovak Republic	1.02
Chile	3.44	South Korea	1.43	Sweden	4.49
Czech Republic	1.48	Luxembourg	2.36	Switzerland	2.66
Denmark	4.92	Mexico	4.58	Turkey	2.14
Finland	3.43	Netherlands	3.84	UK	3.73
France	2.63	New Zealand	4.65	USA	4.51

Table 4: Key Node Destinations of OECD Countries for Chinese FDI Outflow based on Tabu Search

Country	Key Location	Country	Key Location	Country	Key Location
Australia	16	Germany	14	Norway	18
Austria	27	Hungary	7	Poland	21
Belgium	2	Ireland	23	Portugal	29
Brazil	10	Italy	5	Russia	11
Canada	20	Japan	17	Slovak Republic	30
Chile	12	South Korea	22	Sweden	24
Czech Republic	4	Luxembourg	19	Switzerland	1
Denmark	28	Mexico	6	Turkey	8
Finland	26	Netherlands	15	UK	13
France	9	New Zealand	25	USA	3

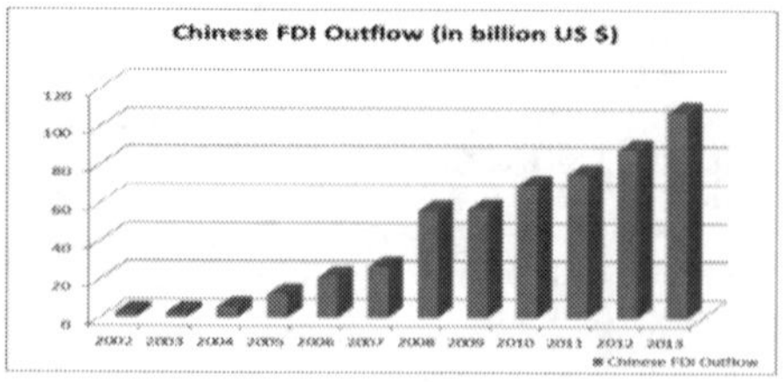

Figure 1: Chinese FDI Outflow from 2002 to 2013

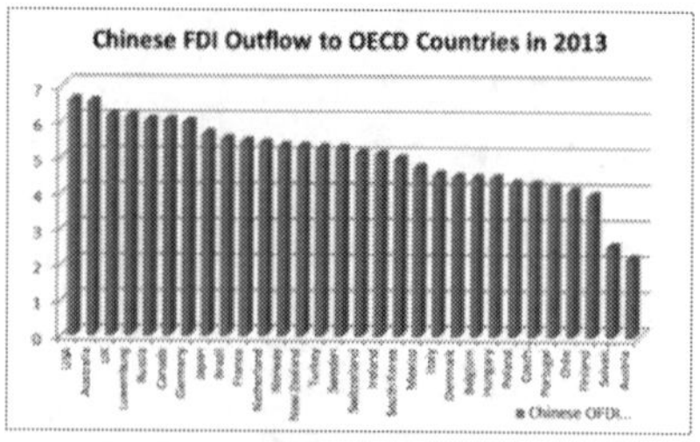

Figure 2: Chinese FDI Outflow to OECD Countries in 2013

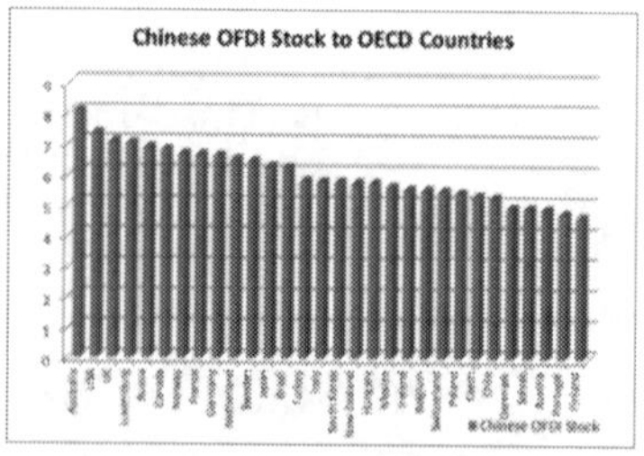

Figure 3: Chinese OFDI Stock to OECD Countries

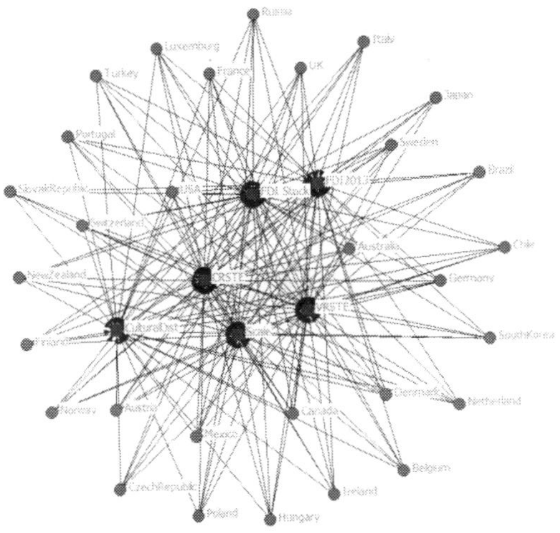

Figure 4: Attributes of OECD Countries based on Chinese FDI Outflows, investment efficiency for life satisfaction, and Cultural Distances

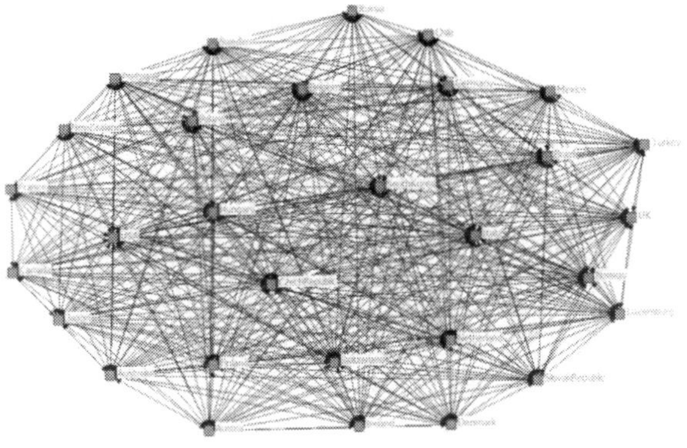

Figure 5: Complex Network based on Chinese FDI Outflows, Life Satisfaction of OECD Destination Countries, and Cultural Distances

8 REFERENCE

[1]**Li, Shaomin & Li, Mingfang & Tan, J. J.** 1998. Understanding Diversification in a Transition Economy: A Theoretical Exploration, *Journal of Applied Management Studies*, 7(1), 77-92

[2]**Reid, David McHardy & Walsh, John.** 2003. Market Entry Decisions in China, *Thunderbird International Business Review*, 45(3), 289-312

[3]**Robert, T.** 2002. Globalization Strategies of Chinese Companies: Current Developments and Future Prospects, *Asian Business & Management*, 1(2), 209-225

[4]**Lin, Chia Chia & Lin, Jason.** 2008. Capitalism in Contemporary China: Globalization Strategies, Opportunities, Threats, and Cultural Issues, *Journal of Global Business Issues*, 2(1), 31-40

[5]**Child, John & Rodrigues, Suzana B.** 2005. The Internationalization of Chinese Firms: A Case for Theoretical Extension, *Management and Organization Review*, Volume 1, Issue 3, 381-410

[6]**Shenkar, Oded and Luo, Yadong.** 2004. International Business (Chapter 2. International Trade Theory and Application): John Wiley & Sons, Inc, 23-25

[7]**Gugler, Philippe & Fetscherin, Marc.** 2010. The Role and Importance of the Chinese Government for Chinese Outward Foreign Direct Investments, *Academy of International Business Insights*, Vol.10, No. 4, 12-15

[8]**Buckley, Peter J. & Clegg, L. Jeremy & Cross, Adam R., & Liu, Xin.** 2007. The Determinants of Chinese Outward Foreign Direct Investment, *Journal of International Business Studies*, 38, 499-518

[9]**Bjornskov, Christian & Dreher, Axel & Fischer, Justina A. V.** 2008. Cross-Country Determinants of Life Satisfaction: Exploring Different Determinants across Groups in Society, *Social Choice and Welfare*, 30(1), 119-173

[10]**Barclay, Lou Anne & Gray, Sidney J.** 2001. Upgrading the Diamond of Developing Countries through Inward FDI: The Case of Four MNEs in the Information Service Industry of Barbados, *Management International Review*, 41(4), 333-357

[11]**Shiralashetti, A. S. & Hugar, S. S.** 2009. Foreign Direct Investment and Economic Development of India: A Diagnostic Study, *IUP Journal of Managerial Economics*, 7(1), 54-69

[12]**Fowowe, Babajide & Shuaibu, Mohammed I.** 2014. Is Foreign Direct Investment Good for the Poor? New Evidence from African Countries, *Economic Change and Restructuring*, 47(4), 321-339

[13]**Yang, Yong & Martins, Pedro S. & Driffield, Nigel.** 2013. Multinational Performance and the Geography of FDI: Evidence from 46 Countries, *Management International Review*, 53(6), 763-794

[14]**Li, Mengying & Ruangkanjanases, Athapol & Chen, Chenin.** 2014. China"s Foreign Direct Investment in Thailand — Current Status and Future Prospects, International Journal of Trade, *Economics and Finance*, 5(4), 296-304

[15]**Filatotchev, Igor & Strange, Roger & Piesse, Jenifer & Lien, Yung-Chih.** 2007. FDI by Firms from Newly Industrialized Economies in Emerging Markets: Corporate Governance, Entry Mode and Location, *Journal of International Business Studies*, 38(4), 556-572

[16]**Sun, Ying & Li, Miao & Chen, Yibo.** 2014. High-Tech Products Export Competitiveness, BRIC Countries in U.S. Markets: A Comparative Analysis, *Journal of Developing Areas*, 48(3), 195-218

[17]**Salavrakos, Ioannis-Dionysios.** 2010. Business History and Conflicting Entrepreneurial Culture as Explanatory Factors of Success and Failure of FDI in Chinese Economy, *Studies in Economics and Finance*, 27(4), 276-298
[18]**Tang, Linghui.** 2012. The Direction of Cultural Distance on FDI: Attractiveness or Incongruity? *Cross Cultural Management*, 19(2), 233-256
[19]**Lu, Jiang & Chen, Qiangbing & Liu, Yali.** 2010. FDI and the Change of the Chinese Culture, International *Journal of Social Economics*, 37(2), 101-118
[20]**Zeng, Yuping & Shenkar, Oded & Song, Sangcheol & Lee, Seung-Hyun.** 2013. FDI Experience Location and Subsidiary Mortality Differences in National Culture and the Expansion of Korean MNEs, *Management International Review*, 53, 477-509
[21]**Ungerer, Leona M & Strasheim, Arien.** 2011. The Moderating Effect of Living Standards on the Relationship between Individual-level Culture and Life Satisfaction: A Value Segmentation Perspective, *Management Dynamics*, 20(3), 25-50
[22]**Yetim, Unsal.** 2003. The Impacts of Individualism/Collectivism, Self-esteem, and Feeling of Mastery on Life Satisfaction among the Turkish University Students and Academicians, *Social Indicators Research*, 61(3), 297-317
[23]**Tong, Canon & Tak, Walder Ip Wah.** 2013. The Impact of Knowledge Sharing on the Relationship between Organizational Culture and Job Satisfaction: the Perception of Information Communication and Technology (ICT) Practitioners in Hong Kong, *International Journal of Human Resource Studies*, 3(1), 9-37
[24] **Qi, Jianhong & Yang, Li.** 2012. Determinants of Chinese OFDI Location: Based on the Geographical Distance and Cultural Distance Test, *Economic Geography*, 32(12), 40-46
[25]**Qi, Jianhong & Yang, Li.** 2014. Cultural Distance and OFDI Entry Mode Choice of Chinese Enterprises: A Study based on the Test of Large Enterprises' Micro-Data, *World Economy Study*, 6, 55-61
[26]**Huang, Lingyun & Yang, Na & Wang, Yu.** 2014. Influences of Cultural Traits and Conflicts on OFDI *Strategies of Chinese Companies, Economic Science*, 3, 114-128
[27]**Akram, Annam & Hassaan, Muhammad.** 2013. Impact of Work Life Conflict on Job Satisfaction, *Interdisciplinary Journal of Contemporary Research In Business*, 5(8), 434-448
[28]**Loewe, Nicolas & Bagherzadeh, Mehdi & Araya-castillo, Luis & Thieme, Claudio & Batista-foguet, Joan Manuel.** 2014. Life Domain Satisfactions as Predictors of Overall Life Satisfaction Among Workers: Evidence from Chile, *Social Indicators Research*, 118(1), 71-86
[29]**Su, Yang & Ma, Lijun.** 2013. Impact of Host's System on Enterprise's Overseas M&A, *Technology Economics*, 32(12), 104-110
[30]**Zhang, Jipeng & Yi, Changjun.** 2014. Technology Endowment and Location Choice of Chinese OFDI- The Moderating Effect of Cultural Distance, Journal of Industrial Technological Economics, 4, 90-97
[31]**Charnes, A. & Cooper, W. W. & Rhodes, E.** 1978. Measuring the Efficiency of Decision Making Units, *European Journal of Operational Research*, 2, 429-444
[32]**Banker, R. D. & Charnes, A. & Cooper W. W.** 1984. Some Models for Estimating Technical and Scale Inefficiencies in Data Envelopment Analysis, *Management Science*, 30, 1078-1092
[33]**Kogut, Bruce & Singh, Harbir.** 1988. The Effect of National Culture on the Choice of Entry Mode, Journal of International Business Studies, 19(3), 411-432
[34]**Groot L.T., David A.G., Raig J. R.** 2005. The Effect of Cultural Distance on Entry Mode Choice, International Diversification, and MNE Performance: a Meta-Analysis, *Journal of International Business Studies*, 36, 270-283
[35]**Glover, Fred. W.** 1986. Future Paths for Integer Programming and Links to Artificial Intelligence, *Computers and Operations Research*, 13(5), 533-549
[36]**Wang, H. & Obremski, Tom & Alidaee, Bahram & Kochenberger, Gary.** 2007. Clique Partitioning Model for Clustering: A Comparison with K-means and Latent Class Analysis, *Communications in Statistics- Simulation and Computation*, 37(1), 1-13
[37]**Kochenberger, G., F. Glover, and H. Wang.** 2013. The Binary Unconstrained Quadratic Optimization Problem: A Unified Framework for Combinatorial Optimization, in Handbook on Combinatorial Optimization, P. Pardalos, D.-Z. Du, and R. Graham, Editors, 2013, Springer.
[38]**Pereira, D.L. & Gendreau, M. & Cunha, A. Salles da.** 2013. Stronger Lower Bounds for the Quadratic Minimum Spanning Tree Problem with Adjacency Costs, *Electronic Notes in Discrete Mathematics*, 2013, 41, 229-236.

APPLICATION AND RESEARCH OF INDUSTRIAL BIG DATA IN EQUIPMENT PRODUCTION MANAGEMENT

Xuying Wang[1*] and Xiaotong Hu[2]
[1,2]Department of Computer science and information engineering
Tianjin University of Science and Technology, Tianjin, China
Katherine_Ashin@163.com

ABSTRACT

With the industry 4.0 be proposed and popularized, the use of the big data in industrial has been more and more important in the change of industrial and the restructure of economic. With network technology's rising, the new service company is occupying the market, which has made people pay more attention to the storage, management and analysis of data. This paper describes the development and application prospects of industrial big data, and analysis the role of equipment management from the perspective of production management to combine big data analytic with the OEE, and study speed of the device operating in printing factory and the regulation of the equipment abrasion and consumption of the resource, illustrating the significant role to improve the effective management of equipment and the industrial production by industrial big data

Keywords: Big Data, Industrial Big Data, Big Data Analysis, Equipment Management, Industry 4.0

1 INTRODUCTION:

With the development of cloud computing memory, the internet of things and other technology, the applications based on network is more and more flourishing. Germany first proposed the national strategy - Industry 4.0, which have leaded people advance into the period of big data, and caused the academia and industry's attention.

The United States has jointly issued "Challenges and Opportunities with Big Data", which described the generation and handle of big data from the perspective of professional learning, presented opportunities and challenges the future [1]. At the 2012 Davos World Economic Forum, the big data was one of the titles, and published "Big Data, big impact: New possibilities for international development", put forward the idea of using the data to create social benefits.

This paper explains the development and applications of industrial big data technology from the perspective of industrial production efficiency and device management. And combined with the Overall Equipment Efficiency (OEE), deeply mine the data, complete production mechanism, improve production efficiency and maximize the use of the value of data.

2 THE DEFINITION AND CHARACTERISTICS OF INSDUSTRY 4.0:

Industry 4.0 is proposed and advocated by Germany, which use Cyber-Physical System(CPS) as the core technology thinking , make equipment automatic and system can be connected by the internet,

---

*Xuying Wang(1991-), female, Tianjin, Tianjin University of Science and Technology graduate student, mainly engaged in intelligent information processing, production management, machine learning,E-mail:Katherine_Ashin@163.com;

the digitization of information as the leading, to realize the zero risk, zero pollution, zero failure green production target for industrial intelligence reform in the whole industrial manufacturing. It is a new normality in the future, also represents a kind of deep integration between network and the industrial production.

Industry 4.0 speeds up process of the integration between China and the world information, the production mode is based on the big data analytic [3], and special in two aspects, the first is interconnection, the second is a high degree of integration [4].The integration of information technology and industrial production aims to make the data into useful information timely. This way can make the production activity data, equipment data, industrial environmental data, market research data, as well as chain data displayed in the same platform, which can create a good source and environment for the big data management. Big data platform comes out and also makes the data from a simple processed object into valuable information resources [5].

3 THE DEVELOPMENT OF INDUSTRIAL BIG DATA:

Now with the generation of huge amounts of data, we need more specific management ways to classify them into systematization and make them have valuable information resources, means big data [2].After cloud computing, big data has become another information industry growth point in the information technology area [2]. the intelligent analysis advocated by Industry 4.0 can better timely stored the data in the industrial equipment generating operation[14], also can through the analysis of historical data, a lot of information can be refined into useful information intelligently [6].

3.1 The Rise of Industrial Big Data:

Big Data promoted the development of artificial intelligence [1], the reason why the big data is gradually rising in the industrial area as follows:

1. The equipment automation which claimed by Industry 4.0, leaving a large amount of data on the device controller, but the useful information and potential value hidden behind the industrial data have not been fully tapped;

2. As the development of Information Communications Technology (ITC) and sensing devices, online access to real-time data has been able to achieve basically.

3. The application of other advanced Internet technology such as the widely using of cloud computing and the embedded systems development can greatly improve the computing capacity of the equipment, and lay a solid foundation for the big data handling.

The human's capacity about analysing and handling data has been unable to satisfy the complicated manufacturing process and the management of active commercial projects.

3.2 Characteristics of Big Data:

Industrial big data has 6V features [14]:Volume - Large scale unstructured data grows rapidly , and has no doubt about the number; Velocity - If we want to achieve real-time collection and non-batch analysis about the data, there must ensure the speed ;Variety - isomer and diversity is to ensure the value of analysis of big data ;Veracity [7]- collecting and refining data must be based on the quality of data is real and effective; Visibility - the meaning of Big data analysis is mining and analysing the invisible factors and information hidden behind the data; Value [7]- Big Data analysis focused on the discovery of the value hidden behind the information.

3.3 The Application of Industrial Big Data:

The High degree integration of information technology and industrial manufacturing makes the ICT technology permeate every aspect of industrial production chain, so the production line with high speed must generate a wealth of industrial data which hidden value in use, the data generated by the equipment mostly belongs to huge unstructured type, he acquisition speed must have more high demands in order to ensure the timeliness of data stream. Application of industrial big data would make different industrial sectors embark on a fast and efficient innovation.

(1) Application of Industrial Thing in production line

With the development of the intelligent industry, more and more wireless mobile sensing devices are installed in the production line. These devices can be used to obtain data for device diagnostics, productivity analysis, energy consumption statistics and quality assessment. We use these data to know the trend of entire production process, and resolve the implementation conflict between ERP and MES and also can reduce the loss. Through analysing the historical data, we can establish the virtual production system, and use simulation technology to optimize existing production processes, observe the production efficiency curve to improve production.

(2) Diagnosis and prediction of fault about equipment

In Modern large enterprises ,equipment is stable or not, can continuous operate or not are closely related to the business interests[12].In industrial production, the key factors that determine the productivity is the equipment, from the perspective of the equipment operating speed, analyse the factors which impact velocity peaking can increase productivity; from the perspective of equipment wear, analyse the best use duration of equipment use and maintenance can make equipment utilized mostly; from the perspective of the frequency of equipment fault, we can make a clear fine troubleshooting instructions and effectively make predictions.

(3) Production Planning and Scheduling

A typical feature of many varieties and small batch production is the widely kinds of products, the objects of production always change, if using a common production equipment to package process for operating, will result in a low degree of specialization and decrease the productivity. By analyzing historical data, it is possible to visually obtain the degree of deviation between planned production and the actual production, select the appropriately optimization algorithm, formulate more suitable production planning about this factory's equipment, real-time to correct the deviations caused by production execution, dynamically adjusting schedule.

4 EQUIPMENT MANAGEMENT AND INDUSTRIAL BIG DATA:

4.1 Significance of Equipment Management:

"We must first of its profits." Equipment is an essential integral part of the production system, also is the most basic guarantee of production. With advances in production technology and the development of equipment automation, how to manage equipment to mostly play the effectiveness in the man-machine combination is the primary task.

Now business management departs is not isolated from each other, and each work are inseparable from equipment management. The degree of equipment management is directly related to the quality and output of products, which directly affect the cost of production's inputs. If we can build an advanced equipment management platform to improve equipment management level [8], the industry funds can be better allocated, also can bring more economic benefits.

4.2 Equipment Comprehensive Benefits:

In the early stage, Japan has presented Total Productive Maintenance (TPM) as a standard of production efficiency with absorbing Europe research results. TPM standard [15] is a comprehensive enterprise management mechanisms, the Overall Equipment Effectiveness(OEE) is the most important part, which from the time starting rate, performance starting rate and rate of qualified products three angles, make the equipment productivity visualization, enables the operator to fully assess the actual production capacity of equipment from the perspective of production system .Sometimes there will be a variety of inevitable activities losses during industrial production, and OEE summarizes six losses related to equipment: equipment failure, setting and adjustment, idling or short downtime, speed loss, quality defects losses and pre-production trial.

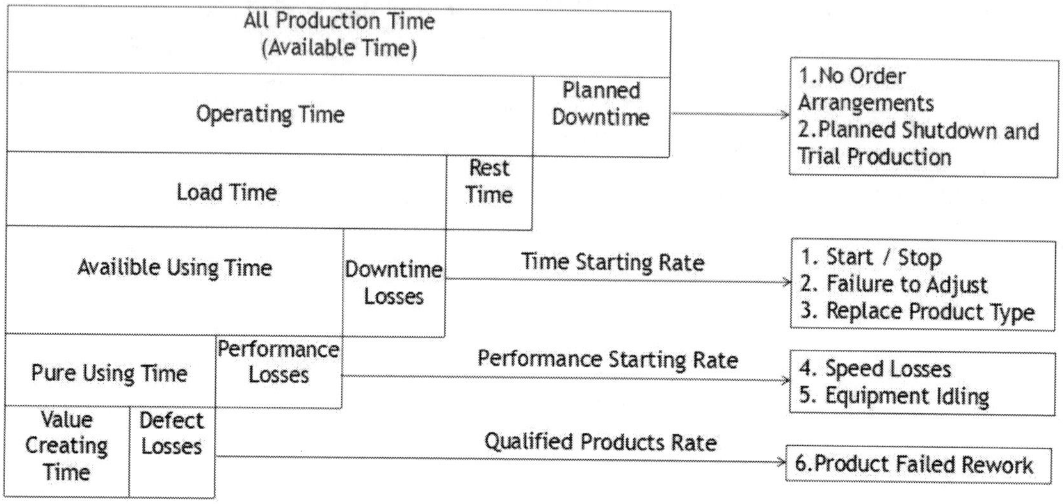

Figure 1: Relation in Equipment 6 Losses with OEE

We can see from Figure 1 the downtime losses is corresponding to the time starting rate in OEE, which needs us to improve the equipment from effectiveness of operating time; performance losses is corresponding to the performance starting rate in OEE, and it reflects the operating state of the equipment; defect losses refers to the qualified products rate, it directly reflects the equipment effective job situation [9] with quality losses of nonconforming product. OEE reflects the work efficiency of equipment value [10].

4.3 Applications of Industrial Big Data in Equipment Management:

The important thing in Production systems is the potential factors can influence on production efficiency, such as speed curve of equipment, wear curve of equipment, equipment resources waste of different processes, decrease of equipment precision. These problems which are easily overlooked may cause downtime, quality deviations and other industrial loss problem. Therefore, we should pay more attention to mining, summary, prediction and management these invisible factors, and the use of industrial big data analysis technology can help us understand the health of machinery and equipment, even the production efficiency problem better.

Next, use the production status of a printing production line as an example to illustrate the application of industrial big data and OEE in Equipment Management. At first, through the historical data we establish a new database entity model, which mainly includes state look-up table, product management table, the equipment look-up table, as follows.

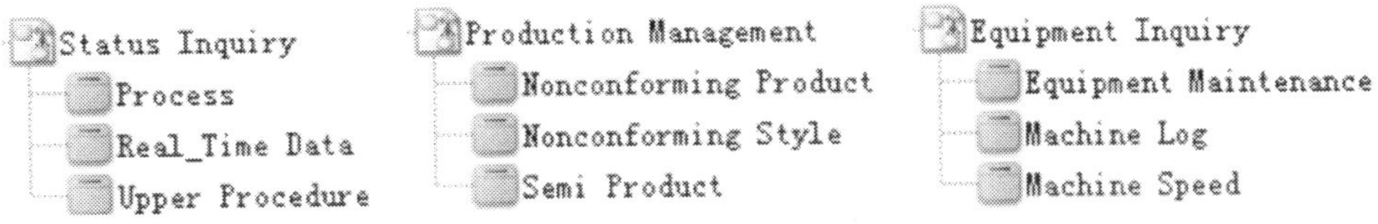

Figure2: Database Entity Mode

The state look-up table saves the current production schedule and deliveries real-time transmission data, and data collection interval can be adjusted by the production. The sensor will collect the printing number and time and save to database, the data is huge. Product management table statistics the unqualified number and variety. Equipment look-up table can read the speed records of the equipment, maintenance records and production logs.

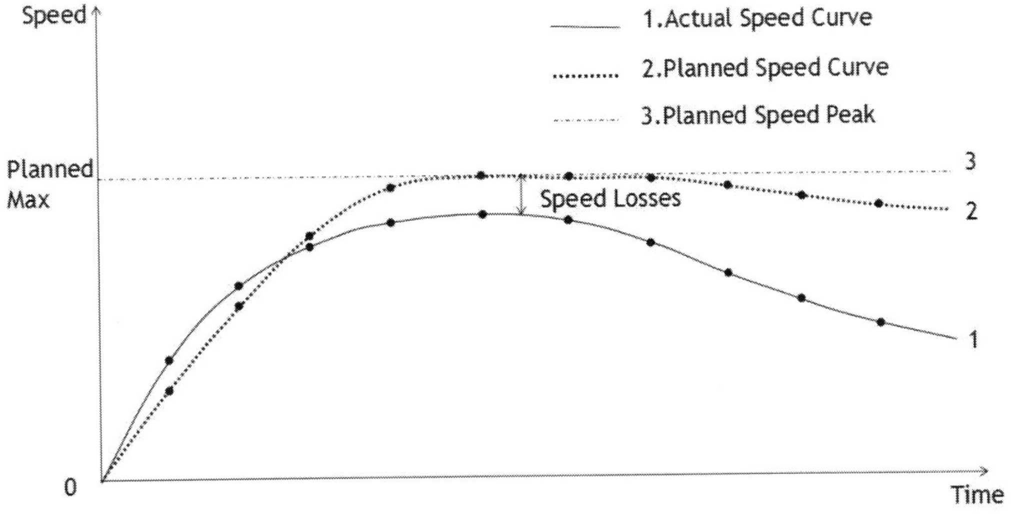

Figure 3: Equipment Speed Curve

In this experiment, everyday can draw the device speed graph, and Figure 3 fitted the operation speed profile in these five months, curve 1 represents the actual speed curvature of the device, curve 2 shows planned speed curve. From figure we can see the deviation between actual production and the plan. Equipment firstly run through the trial production stage, and gradually adapt to the pace of production and then the speed reached the maximum into the production phase, but through the graph we found that the equipment did not achieve the desired maximum speed (line 3), the latter period production velocity was gradually decreasing and forming the speed loss. Combining OEE knowledge and big data analysis, we found that loss of speed and equipment itself abrasion has a close relationship. Because of the equipment wear and the operating life's shorten, the speed of equipment operating will be gradually reduced, they both belong to the performance loss.

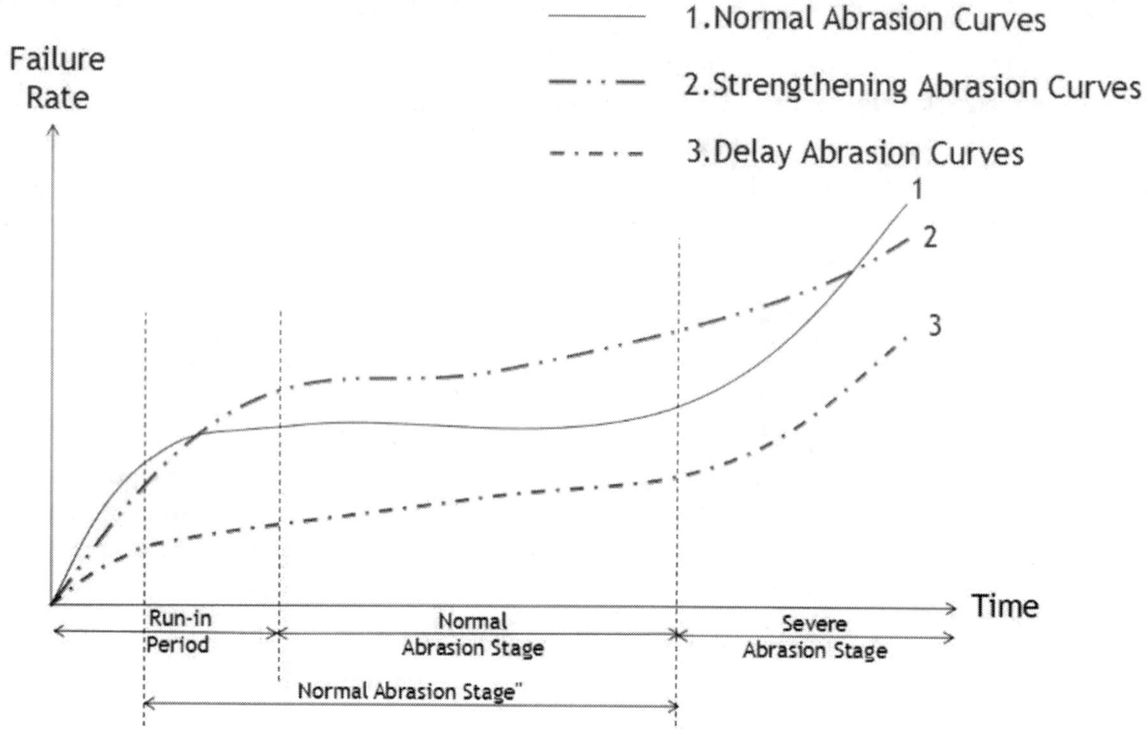

Figure4: Equipment Abrasion Curve

Figure 4, curve 1 represents the wear ordinary regularity of equipment, which reflects the equipment abrasion caused by the force during the operation in the run-in period, this stage is very short. In the second stage of normal abrasion, it shows the normal use lifetime of the equipment, then enter the final of severe abrasion stage, due to the increased wear speed, equipment precision degree deteriorate rapidly; curve 2 represents the regularity of equipment in strengthening wear stage, the figure shows, the second stage and third stage have no obvious turning point, and due to the device overload work or poor lubrication, the wear often is exacerbated; curve 3 shows a delay wear regularity, equipment wear rate has slow down, the second stage significantly prolonged. The reason why is this state of affairs is reduce of the use intensity of the device and no fully use the equipment, so it is not ideal production state. In conclusion, any device needs to have a reasonable and moderate operating standards, and also should focus on production equipment maintenance to protect the accuracy of the equipment.

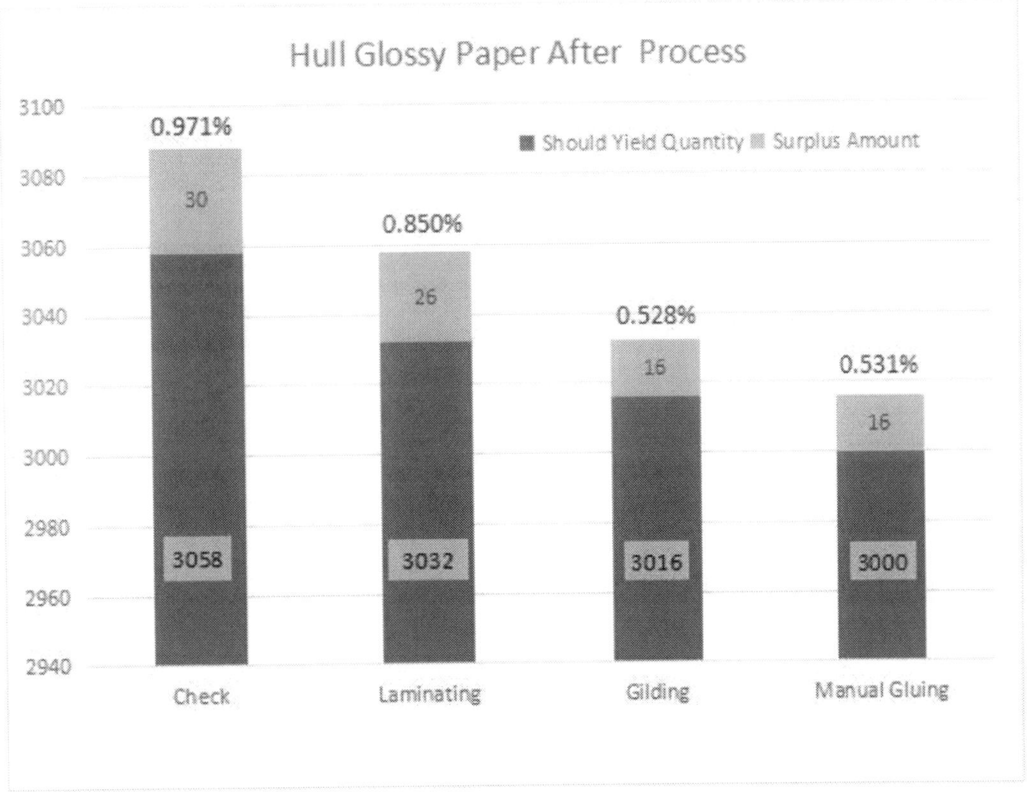

Figure 5: Different Process Equipment Resource Consumption

Figure 5 shows that in a printing order form the proportion of equipment resource consumption when the same part through different processes. Through the display and analysis of data, we can easily find that more complex the process operations is, more easily lead to failure, so we not only needs to ensure the equipment use intensity ,but also further improve the accuracy of the device to reduce the waste of resources.

5 CONCLUSION:

Through experiments, we understand the meaning of OEE and we can better manage the equipment, then it is more targeted when collect the production data from the equipment.

Big Data era under the industrial 4.0 will play a powerful role in promoting productivity, reducing production costs and management [11]. Smart Plant proposed to make the equipment has the self-learning awareness and self- forecasting capabilities, at the same time, pay attention to the integrated data analysis and deep mining big data from the whole industry chain perspective to achieve industrial optimization and decision-making can help us realize further implement intelligent services. The core competitiveness of enterprise is not the industrial data, and the true meaning of its existence is mining the use value of data.

Find the hidden factors which result equipment production capacity decrease, deep excavate the various losses through industrial big data analytic technology can improve the productivity of factories and enterprises and open up a broader prospects for modern automated production line control system.

6 REFERENCES:

[1] **Liang Jiye. 2016.** The Challenge of Big Data Mining and Considerations, Computer Science, 7(43), pp 3-6.

[2] **Feng Dengguo, Zhang Min, Li Hao. 2014.** Big Data Security and Privacy Protection, CHINESE JOURNAL OF COMPUTERS, 1(37), pp 243-258.

[3] **Yang Guohua, Liu Yong, Huang Xiaoli, Luo Jialong, Huang Ping, Jiang Ying. 2015.** The Roles of Big Data Analysis in "Industry 4.0"Times and in the Innovation & Development of Traditional Liquor-making Industry. Liquor-Making Science &Technology, 11, pp 1-8.

[4] **Zhong Hai. 2015.** Application and research of big data in industrial manufacturing. Technological Development of Enterprise, 34(13), pp 104-105.

[5] **Meng Xiaofeng, Ci Xiang. 2013.** Big Data Management: Concepts, Techniques and Challenges. Journal of Computer Research and Development, 50(1), pp 146-169.

[6] **Cheng Xiaolei. 2015.** Demand, Environment and Service of Industrial Big Data under the Industry 4.0 Structure. Journal of Chifeng University (Natural Science Edition), 31(4), pp 14-15.

[7] **Li Jie. 2014.** Predictive Manufacturing under Industry 4.0 Environment, World Science, (5), PP 13-16.

[8] **Wang Shuangli, Wang Taiyong, Wang Zhengying, Zeng Hui, Yuanfeng. 2006.** Construction of equipment management integration platform for condition monitoring and diagnosis, Computer Applications, 9(26), pp 2254-2260.

[9] **Liang Yue. 2014.** Design and Application of OEE Data Analysis, Electronic Design Engineering, 22 (12), pp 53-65.

[10] **Li Chengyu.2006.** Application of OEE in manufacturing management. SCI-Tech Information Development &Economy, 16(20), pp 98-200.

[11] **Hao Yuanlong.2015.** The Opportunity, Challenge and Route Choice of Government Governance in the Era of Big Data, Brand, (3), pp 117.

[12] **Tang Sheng, Hu Jie, Zhao Jinghu. 2012.** An Equipment Condition Monitoring Algorithm Based on Massive Data Mining, Computer Science, 6(39), pp 319-321.

[13] **Wu Xindong. 2016.** From Big Data to Big Knowledge: HACE + BigKe, Computer Science, 7(43), pp 3-6.

[14] **Su Xin, Wu Yingya, Pei Huajian, Lan Xingying, Gao Jinsen. 2016.** Recent development of the application of big data technology in process industries, CHEMICAL INDUSTRY AND ENGINEERING PROGRESS, 6(35), pp 1652-1659.

[15] **Wang Chen, Guo Xuemeng. 2012.** Research on the Application of Total Production Maintenance in Enterprise, The 4th International Conference on Engineering and Business, pp 508-511.

COLUMN GENERATION ALGORITHM FOR TEMPERED GLASS CUTTING PROBLEM

Ming Liu[1],Xuenan Yang[1*], Shijin Wang[1], Jiantong Zhang[1]
[1]Department of Management Science andEngineering, School of Economics and Management, University of Tongji, Shanghai, P.R. China
mingliu@tongji.edu.cn
yangxuenan@yeah.net
shijinwang@tongji.edu.cn
zhangjiant@163.com

ABSTRACT

With growing demand for tempered glass and rapid shrinkage of raw material. To save raw material while meet customer's demand, a tempered glass cutting problem originated from real-world tempered glass manufacturer is considered in this paper, which aims to minimizing the cost of materials. An integer programming model is formulated, with the originated problem used as the master problem, and corresponding dual problem considered as the subproblem. Then a modified column generation algorithm is proposed to solve the problem. A real-world case is conducted, the computed results illustrated the effectiveness of the proposed method.

Keywords: Tempered glass cutting, Integer programming, Column generation algorithm

1 INTRODUCTION

With the development of the technology and the elevation of human safety consciousness, more and more common glasses have been replaced by tempered glasses, thus market demand for tempered glass has also grown rapidly. However, rapidly decreasing of natural resources has lead to material's price rising dramatically, which erodes the tempered glass manufacture's margin profit further. Thus to improve profit, manufactures have to consider how to save raw material while meet customer's demand simultaneously. As a completed tempered glass's process contains cutting from float glass first, and then edging, printing and tempering. Of the whole process, cutting is more likely to generate waste of raw material if following irrational cutting scheme, while waste generated in other procedures mainly due to scrap glasses. Therefore, cutting procedure is a significant segment to save cost of raw materials.

The tempered glass manufacturer investigated in this paper is a famous German company, whose tempered glass has supplied to customers all over the world, and has concentrated on developing and producing special glasses since 19th century. Rich experience and many advanced technology as the manufacturer has, but the significant cutting scheme in cutting procedure mainly assigned by workers manually. Since those workers doesn't receive much education and have little sense of optimization, easy to operate is their main principle, thus usually 1 piece of material cut into a type(size) of glass. Apparently, the current principle is easy to lead to waste, thus an optimization method is designed here to optimize the cutting procedure.

*Corresponding Author

Generally speaking, in the cutting procedure, a piece of lager float has to be cut into several pieces of demand size of glass, while the storage of float glasses in stock is limited. Thus cutting feature of tempered glass makes its cutting problem can be classified into typical cutting stock problem.

1.1 Literature review

The cutting stock problem(CSP) is to cut large pieces in stock into smaller pieces to fulfill customers' requirements while optimizing one or more objective, for instance, minimizing the total amount of scrap generated or total number of the stock out. These problems are highly relevant in the production planning of industries such as furniture (Ragsdale and Zobel, 2004), metal coils (Cui, 2014), paper (Rinaldi and Franz, 2007) industry. Hence, since Gilmore and Gomory (1961), much interest from both industry and academia are continuously put on CSP and its variants.

Ragsdale and Zobel(2004) considered the ordered one-dimension CSP and its applications in manufacturing of a wooden windows and doors. Furini and Malaguti(2013) studied a two-dimensional two-stage guillotine constrained cutting stock problem(2D-CSP) with different sizes of stocks. Three mixed integer programming (MIP) models, namely compact formulation, pseudo-polynomial formulation and exponential formulation, are proposed and compared with the benchmark problems. Belov and Scheithauer(2006) proposed a branch-and-cut-and-price algorithm for both the one-dimension CSP, and the two-dimensional two-stage guillotine constrained cutting problem.

According to the reviews above, many researchers have studied on CSP or its extended problem, and many solution methods have also proposed correspondingly. But as far as we know, most of this researchers were basing on theory, and did not apply its theory to real situation. Therefore, this paper investigate a practical case, and propose an optimization method to solve it.

1.2 Contribution

In this paper, we consider a tempered glass cutting problem originated from a real-world tempered glass manufacturer, the contribution may include:

(i) We formulated an integer programming model for the cutting problem, with the goal of minimizing the cost of materials.

(ii) We design a modified column generation algorithm, which consist of simplex method to solve the master problem and dynamic programming method to solve the subproblem.

(iii) A real-world cutting case is conducted to illustrate the effectiveness of the proposed method.

1.3 Organization

The reminders are: Problem description will depict in Section2. Mathematical model of the researched problem is formulated in Section3. Section4 mainly describe the proposed column generation algorithm. Detail of the case is depicted in Section5. Conclusion is presented in Section6.

2 PROBLEM DESCRIBTION

For the tempered glass manufacturer, there are n orders should be processed. Each order $j \in \{1, \dots, n\}$ has n_j jobs to be cut from larger piece of rectangular float glass (object) in stock. Each jobof order j, J_{ij}, $i \in \{1,..,n_j\}, j \in \{1,..,n\}$ requires cutting a smaller piece of rectangular float glass(item) with the width w_{ij} and the length l_{ij}, and $l_{ij} \gg w_{ij}$. It would be better if the jobs of an order can be finished at the same time. However, jobs of each order is allowable for splitting.

There are M larger pieces of float glasses(objects) with different sizes available in stock, each k is with width W_k and lengthL_k, $k \in \{1, \dots, M\}$.And $w_{ij} < W_k$, $l_{ij} > W_k$. It can be assumed that w_{ij}, l_{ij},

W_k, and $L_k$are integers. As illustrated in Figure 1, each larger piece of float glass can be slitted(cut horizontally) for a single item or for a combination of multiple items with different size of widths. Trim allowances should be considered during the cutting process, that is, the leftovers of each side of the objective should at least have the allowable range of about 5cm. It is possible that L_k is less than l_{ij}, which means that another larger piece of float glass should be cut for J_{ij}. If the remnant is kept in stock for other jobs; Otherwise, the remnant is discarded as a scrap(a kind of waste), which is not what the manufacturer wants.

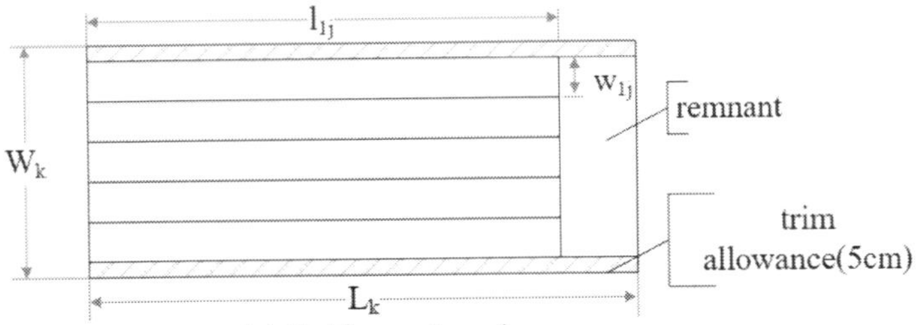

(a) Cutting pattern 1

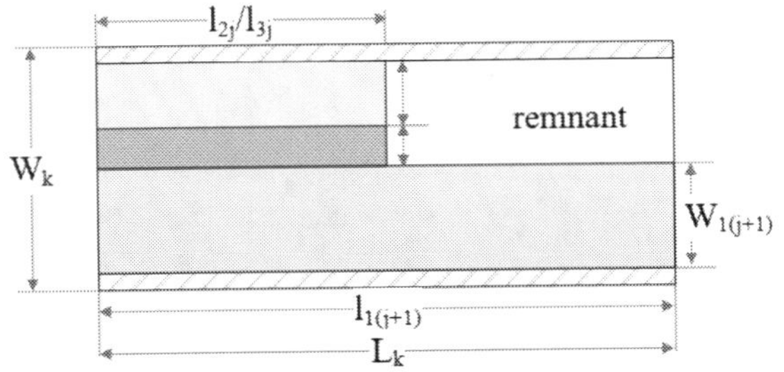

(b) Cutting pattern 2

Figure 1: The illustration of the cutting patterns

For each larger float glass, there are at most 10 slitting can be cut due to the number of cutters in the cutting machine. Thus there are at most 10 items can be cut at a time.The problem is to specify how to cut the items to meet customers' requirements with the minimum cost(measured by the total area of used stock of float glasses), while subject to production constraints. This problem is a variant of the 2D-CSP where stock of float glasses with different sizes is available.

3 MATHEMATICAL FORMULATION

For the glass cutting stock problem in the tempered glass manufacture, an integer programming model is formulated in this section. Before introducing the integer programming model, relevant indices, parameters and variables contained in the model will denote first.

Indices:

k: index of materials;

j: index of cutting patterns;

i: index of various types of demand;

Parameters:

K: set of materials;

P^k: set of feasible patterns for material $k \in K$;

D: set of types of demand;

s^k: number of materials of type k in storage;

W^k: width of material k;

T^k: thickness of material k;

d_i: number of items(also called item i) in demand type$i \in D$;

a_{ij}^k: number of item i in pattern $j \in P^k$ for material k. Note that if $t_i \neq T^k$, then $a_{ij}^k = 0$ (i.e., if the thickness does not match, then it is impossible to produce item I from material k);

w_i: width of item i;

t^k: thickness of item i;

Variables:

x_j^k: number of pattern $j \in P^k$ is produced;

For the concerned problem, the master problem can be described as follows:

$$\textbf{\textit{Master Problem}}\text{: } min \sum_{k \in K} \sum_{j \in P^k} c^k x_j^k \quad (1)$$

$$s.t. \sum_{k \in K} \sum_{j \in P^k} a_{ij}^k x_j^k \geq d_i, \quad i \in D \quad (2)$$

$$\sum_{i \in P^k} x_j^k \leq s^k, \quad k \in K \quad (3)$$

$$x_j^k \in Z_+, \quad j \in P^k, k \in K \quad (4)$$

To solve the CSP by column generation, we start with a subproblem below. We make decision on x_j on the basis of their reduced cost. That is, if the reduced cost of a variable (or column) is negative, then we increase the value of this variable will decrease the minimization objective. In light of this, we use u_i as the dual variable of constraint (2), and v_k as the dual variable of constraint (3).

For a column for material$k \in K$, its reduced cost can be obtained by the following subproblem, i.e., function$f(c^k, W^k, \vec{u}, \vec{v}, T^k, \vec{t})$:

$$\textbf{\textit{Subproblem}}: \quad min\ c^k - \sum_{i \in D} u_i y_i - v_k \quad (5)$$

$$s.t. \quad \sum_{i \in D} w_i y_i \leq W^k \quad (6)$$

$$\sum_{i \in D} y_i \leq 10 \quad (7)$$

$$(t_i - T^K) y_i = 0, \quad i \in D \quad (8)$$

$$y_i \in Z_+, \quad i \in D \quad (9)$$

In the subproblem, we design a pattern by determining y_i, i.e., number of item i in this pattern. Constraint(6) ensures that the total width of material k could not be exceeded. Constraint (7) guarantees that maximum number of items in a single material could not be exceeded. Constraint(8) enforces that if the thickness of item i and material k does not match, then item $i$can be processed on material k. To generate all potential columns, we need to run subproblem for each$k \in K$.

4 COLUMN GENERATION ALGORITHM

Column generation algorithm is a kind of exact algorithm, and has often been used to solve large-scale optimization problems(Du et al. 1999), such as tactical operating room planning(Fei et al. 2009). Parallel machine scheduling(Mellouli et al. 2013). Endoscopy scheduling(Fei et al. 2009). Column generation shows excellent performance in solving such optimization problems, thus a column generation algorithm is proposed for the concerned problem. In this algorithm, the set P of feasible patterns for each material are dynamically enlarged. Recall that each pattern corresponds to a column in the master problem's coefficient matrix. Namely, coefficient matrix of the master problem is dynamically widened. As known that in column generation algorithm, all columns are not explicitly described in the coefficient matrix, and only promising ones appear there. The procedure of proposed algorithm is as below:

Step 0: Input problem parameters K, D, s, W, d, w.
Step 1: (Initialize the master problem) Set material index k = 1.
Step1.1: If $k > |K|$, go to Step2; otherwise, set demand type index i = 1.
Step1.2: If $i > |D|$, set material index$k = k + 1$ and go to Step 1.1.
Step 1.2.1: (Add the first pattern for material $k$in P^k) Initialize a $(|D| + |K|) \times 1$ zero column (or m-atrix) A_i^k.
Step 1.2.2: Compute an integer $a_{i1}^k = [W^k/w_i]$(i.e., the maximum integer not greater than W^k/w_i).
Step 1.2.3: Update column A_i^k, by setting the ith row of equal to a_{i1}^k and the $(|D| + k)$th row equal to 1 (i.e., material k can be exclusively used to produce a number ofa_{i1}^k of demand type i in the first pattern in P^k). Set demand type index$i = i + 1$ and go to Step 1.2.
Step 2: Set material index k=1.
Step 2.1: If $k > |K|$, go to Step 3; otherwise, set cutting pattern index $j = 1$.
Step 2.2: Initialize a $(|D| + |K|) \times 1$ zero column (or matrix) P_0^k.
Step 2.3: Compute the reduced cost under current scheme (solution of Master problem in Step 1), and generate new pattern P_0^k by setting the jth row equal to x_{i+1}^k, and the $(|D| + k)$th row equal to 1.
Step 2.4: If the computed reduced cost is non-negative, then set $k = k + 1$, and go to Step 2.1; oth-erwise set $j = j + 1$, and go to Step 2.2.
Step 3: Output the optimal cutting scheme x_j^k and corresponding cost.

As depict in the procedure, an initial solution of the master problem is calculated through Simplex method, which provides the first relaxation for the subproblem. Then feasibility cuts are added in the subproblem to tighten the solution space until a feasible solution of the subproblem is found.

5 CASE STUDY

In this section, a 4 (K=4) types of materials, and 7 (D=7) types of demand float glass cutting case is conducted. The case will be solved through the proposed algorithm, the algorithm is coded in Matlab R2013b, all the computer tests ran on Inter Core i5-2450M 2.5G Hz personal computer.

The quantity and size information of the 7 types of demand $(d_1, d_2, d_3, d_4, d_5, d_6, d_7)$ float glasses are listed in Table1. The detailed storage, width and cost information of the 4 types of materials (k=1, k=2, k=3, k=4) are listed in Table2. In this case, all 7 types of tempered glass share the same thickness ($T^k = t^k = 10\text{mm}$) with the materials. Table3 shows the cutting scheme manually assigned by workers. Of which, P_j^k is the jth pattern of the kth material. Take P_1^1 as an example, P_1^1 = (5,0,0,0,0,0,0) indicates that 1 piece of the first (k=1) material can be cut into 5 pieces of $d_1$under pattern P_1^1. And x_j^k indicates how many pieces of kth material should be cut according to pattern P_j^k. Thus according to manual assignment, altogether 68 pieces of the first (k=1) material is need. Of which, 10 pieces should be cut according to pattern P_1^1; 12 pieces should be cut according to pattern P_2^1; 12 pieces should be cut according to pattern P_3^1; 4 pieces should be cut according to pattern P_4^1, 30 pieces should be cut according to pattern P_5^1, and the total material cost is 6500yuan.

Table 1: Quantity and size information of demands

	d_1	d_2	d_3	d_4	d_5	d_6	d_7
Quantity	50	22	36	26	24	8	30
Size(mm)	250	300	400	450	500	550	700

Table 2: Detailed information of materials in storage

	k=1	k=2	k=3	k=4
Width(mm)	1250	1880	2100	2440
Storage	70	70	70	70
Cost(yuan)	76	100	115	140

Table 3: Patterns of each material in the manual assignment

		d_1	d_2	d_3	d_4	d_5	d_6	d_7
k=1	P_1^1	5						
	P_2^1		4					
	P_3^1			3				
	P_4^1				2			
	P_5^1					2		
	P_6^1						2	
	P_7^1							1
k =2	P_1^2	7						
	P_2^2		6					
	P_3^2			4				
	P_4^2				4			
	P_5^2					3		
	P_6^2						3	
	P_7^2							2
k =3	P_1^3	8						
	P_2^3		7					
	P_3^3			5				
	P_4^3				4			
	P_5^3					4		
	P_6^3						3	
	P_7^3							3

Table3: Patterns of each material in the manual assignment

		d_1	d_2	d_3	d_4	d_5	d_6	d_7
k =4	P_1^4	9						
	P_2^4		8					
	P_3^4			6				
	P_4^4				5			
	P_5^4					4		
	P_6^4						4	
	P_7^4							3

Table 4: Cutting scheme in the manual assignment

	x_1^2	x_4^2	x_6^2	x_2^3	x_3^3	x_5^3	x_7^3
Quantity	8	7	3	4	8	6	10

Table 5: Newly generated patterns in final solution

		d_1	d_2	d_3	d_4	d_5	d_6	d_7
k=2	P_8^2	1				1	2	
	P_9^2	1		4				
	P_{10}^2	1	2			2		
	P_{11}^2	5	2					
	P_{12}^3	4		1	1			
	P_{13}^2				3	1		
	P_{14}^2				1			2

Table 5 and Table 6 illustrate the newly generated patterns information and cutting scheme in the final solution calculated from the proposed method. P_j^ksin Table 5 share the same mean as in Table 3. As the newly generated 7 patterns are combination of different type of demand tempered glasses, they can lead to less waste of materials. Take P_8^2 as an example, 1 piece of second (k=2) material can simultaneously be cut into 1 piece of d_1, 1 piece of d_5 and 2 pieces of d_6, thus the waste width is 30mm. Table 6 is calculated cutting scheme, and corresponding material cost is 4700yuan, thus 1800yuan (about 27.69%) is saved of the total material cost. As the manufacturer receives lot of order every day, if the proposed algorithm is adopted, then the saved cost will be very substantial.

Table 6: Cutting scheme in final solution

	x_8^2	x_9^2	x_{10}^2	x_{11}^2	x_{12}^2	x_{13}^2	x_{14}^2
Quantity	4	8	9	3	5	3	15

6 CONCLUSIONS

This paper investigate a glass cutting problem in a tempered glass manufacturer, which can be classfied into a typical two-dimensional cutting stock problem. MIP model is formulated, which takes the originated problem as the master problem and corresponding dual problem as the subproblem. A column generation algorithm is proposed to solve the problem. To test the effectiveness of the proposed method, raw material cost is compared with the original manual assignment.

7 ACKNOWLEDGMENT

This work was supported by the National Science Fundation of China (NSFC) under Grants 7110106, 71428002, 71571134 and 71571135.

8 REFERENCES

[1] **Belov G, Scheithauer G.2006.** A branch-and-cut-and-price algorithm for one-dimensional stock cutting and two-dimensional two-stage cutting[J]. European journal of operational research,171(1): 85-106.

[2] **Cherri A C, Arenales M N, Yanasse H H. 2009.**The one-dimensional cutting stock problem with usable leftover-A heuristic approach[J]. European Journal of Operational Research, 196(3): 897-908.

[3] **Cui Y. 2014.** Heuristic for the cutting and purchasing decisions of multiple metal coils[J]. Omega, 46: 117-125.

[4] **Du Merle O, Villeneuve D, Desrosiers J, et al. 1999.**Stabilized column generation[J]. Discrete Mathematics, 194(1): 229-237.

[5] **Fei H, Chu C, 2009.** Meskens N. Solving a tactical operating room planning problem by a column-generation-based heuristic procedure with four criteria[J]. Annals of Operations Research, 166(1): 91-108.

[6] **Fei H, Meskens N, Combes C, et al. 2009.**The endoscopy scheduling problem: a case study with two specialised operating rooms[J]. International Journal of Production Economics, 120(2): 452-462.

[7] **Furini F, Malaguti E. 2013.** Models for the two-dimensional two-stage cutting stock problem with multiple stock size[J]. Computers & Operations Research, 40(8): 1953-1962.

[8] **Gilmore P C, Gomory R E. 1961.**A linear programming approach to the cutting-stock problem[J]. Operations research, 9(6): 849-859.

[9] **Mellouli R, Kacem I, Sadfi C, et al. 2013.**Lagrangian relaxation and column generation-based lower bounds for the Pm, hj1 ‖ Σ wiCi scheduling problem[J]. Applied Mathematics and Computation, 219(22): 10783-10805.

[10] **Ragsdale C T, Zobel C W. 2004.**The ordered cutting stock problem[J]. Decision Sciences, 35(1): 83-100.

[11] **Rinaldi F, Franz A. 2007.**A two-dimensional strip cutting problem with sequencing constraint[J]. European Journal of Operational Research, 183(3): 1371-1384.

ALGORITHM FOR THE TACTICAL BERTH ALLOCATION OPTIMIZATION WITH THE CONSIDERATION OF TIME-VARYING WATER DEPTHS

Ming Liu [1*], Yujian Song[1], Jiantong Zhang[1], and Feng Chu[2]
[1] School of Economics & Management, Tongji University, Shanghai 200092, P.R. China
mingliu@tongji.edu.cn, soonvict@163.com, zhangjiant@163.com

[2]Laboratory IBISC, University of Evry-Val dÉssonne, Evry 91020, France
feng.chu@ufrst.univ-evry.fr

ABSTRACT

In this paper, we study a tactical berth allocation problem with time-variant water depths. Confronting the challenges provided by mega ships, managers of world-class seaports must employ water depths related constraints when making vessel berthing decisions. When the water depths of high tides satisfy the requirement of jumbo vessels' drafts, these vessels are allowed to be served at berth. We propose an integer programming formulation for the tactical berth allocation problem, i.e., the joint optimization of berth allocation and quay crane assignment. The aim is to minimize the total weighted turn-around times of vessels. We will present an efficient heuristic.

Keywords: Maritime logistics, Berth allocation, Quay crane allocation, Water depth

1 INTRODUCTION

The emergence of containerized sea-freight transportation provides cheaper, faster and safer transport of cargo all over the world. With these merits, containerized transport has witnessed a drastic growth in the last decades, which is also attributed to rapid increase in international trade. The fast growth and the fierce competition among terminals impel container terminals to strengthen their technological competence and management capability in pursuit of enhancing their productivity to compete with other counterparts. In this respect, inefficient utilization of scarce resources such as berths and quay cranes (QCs) could be translated into a stumbling block for fast ship handling operations. For this reason, the optimization of container terminal operations is of great importance in boosting efficiency, reducing the overall trip duration and attaining higher customer satisfaction, which has been gaining more and more attention from academics and practitioners. From a practical point of view, berths and QCs, serving as two kinds of scarce and costly resource, have a great impact on terminal operational efficiency. This naturally gives rise to two important interrelated planning problems known as berth allocation problem and quay crane assignment problem. Generally, berth allocation problem aims to determine the berthing time and position for each incoming vessel within a given planning horizon, and quay crane assignment problem deals with assigning available quay cranes to vessels for loading and unloading operations. The number of quay cranes is a direct determinant of the dwell time on berth for a vessel. Concurrently, when and where the vessel is berth determines the number of available quay cranes. Given this important interrelationship, integrating planning of berths and QCs from a holistic perspective conforms

* Corresponding Author

to a more practical scenario and becomes a meaningful research topic,i.e., B&CAP (or called the tactical berth allocation problem). Its solutions might potentially yield significant improvement in efficiency and reduction in spending for the container terminal.

The original designed berth water depth capacities cannot accommodate these Mega container ships, during a low-tidal period. To guarantee the safe sailing of such jumbo vessels with numerous containers, these vessels are only allowed to access terminal berths during a period with high tide. In the execution of operational schedules, a big challenge confronting container terminals and shipping lines is that vessels are exposed to tide effects. For example, in Shanghai Yangshan Port's practice, some large container vessels can access the berths only during some particular times. The fluctuation in water depth of tide-sensitive areas, i.e., navigation channel, has a great impact on some vessels' sailing in and departure from port basin. This is primarily due to the fact that the water depth should meet the requirement of vessels in term of draft which is the distance between waterline and the vessels' keel, to protect jumbo vessels from being stranded. In other words, if a vessel arrive at a time when water depth is not adequate to transit through the navigation channel, then it should wait at the anchorage ground for the next incoming tide. Seaports like Shanghai port, Antwerp port and Hamburg Port are still plagued with the time-varying water depth constraints and endures its resultant inefficiency. In this regard, it is worthwhile to take into account the tide-varying (time-varying) water depth. Though there exist some works considering the tide effects, they roughly divide the planning horizon into low-water and high-water period (Xu et al. [1]; Eduardo et al. [2]), irrespective of time-variant observation and vessel-dependent requirement, which is not consistent with the practical situations. In fact, the variation of water depth follows a certain pattern with respect to time and the draft is also vessel-specific.

We formulate it as an integer programming problem and will develop a differential evolutionary algorithm (DE for short hereafter) with local search scheme to solve it. The paper is organized as follows. Section 2 gives a brief literature review. Problem description and integer linear programming formulation are presented in Section 3.

2 LITERATURE REVIEW

The importance of optimal resource scheduling in container terminal has long been recognized by academia, and there have been a growing number of works examining berth allocation problem, quay crane assignment problem and their integration. Several articles comprehensively summarize related literature and provide exhaustive information on optimization models and algorithms for a wide variety of specific problems (Stahlbock et al. [3]; Bierwirth et al. [4]). In this section, we briefly review the integration studies of B&CAP, which is intimately relevant to our research scope.

2.1 B&CAP Literature

The pioneering research on integration of the BAP and the QCAP was conducted by Park and Kim. [5]. This problem is addressed in a hierarchical optimization framework. The first level problem deals with berth allocation problem and the second level problem concerns quay crane assignment problem. For the resolution, a two-phase solution procedure is developed to solve this problem, with each phase being exclusive to one problem. However, this strategy may not yield best-performing operation schedule, which has been proven by subsequent relevant works, such as Vacca et al. [6].

Liu et al. [7] propose a two-level framework for the integrated problem. The vessel-level problem centers on each individual vessel with the objective of determining optimal handing time for a given number of QCs assigned to it. Afterwards, a berth-level problem considers all vessels together and assigns QCs among the vessels based on the result of berth allocation and vessel-level optimization. Meisel et al. [9] regard the B&CAP as a multi-mode resource constrained project scheduling problem where each vessel corresponds to an activity and a specific QC-vessel assignment to a mode. Imai et al. [10] consider the B&CAP with QCs' movement restriction, as QCs are all mounted on the same track and cannot bypass each other.

Meisel et al, [11] study the integration problem. They take into consideration some real-world aspect such as possible decrease in marginal productivity of quay cranes resulting from interference effect and increase in processing time due to deviation from preferred berth position. Giallombardo et al. [12] extend previous works by introducing a new concept-QC profiles which captures real-world issues. Raa et al. [13] present a enriched model for B&CAP which takes into account many real-life operational constraints such as vessel priorities, preferred berthing locations and handling time.

Vacca et al. [6] make the fist attempt to solve the integration problem to optimality using branch and price algorithm. Computational results reveal that the proposed approach outperforms commercial solvers. Turkogullari et al. [14] also make their endeavor to solve this problem to optimality. According to whether the quay crane is time-invariant or not, they consider two different versions with distinct complexity: BACAP and BACASP. Cagatay et al. [15] propose a new model based on set partitioning for the integrated planning.

Despite the presence of numerous works mentioned above, research on B&CAP with the consideration of time-varying water depth is still in its infancy. Kelareva et al. [16] investigate the vessel scheduling problem that can take time-varying draft constraints into consideration so as to optimize throughput of ports. Both mixed-integer programming and constraint programming are proposed. Sheikholeslam et al. [17] deal with the continuous BAP in conjunction with historical tidal data of Shahid Rajaee Port. They adopt a hybrid of genetic algorithm and pattern search algorithm to achieve a satisfactory solution for small-scale instances. Du et al. [18] identify the tide effect on entrance/departure of vessels into/from ports. They model the tide effect by introducing tidal time windows. However, this increases the difficulty of implementation because extra efforts to eliminate variables from conditions of constraints are required, to make the formulation suitable to commercial solver (CPLEX). Other works related to tide effects on berth at a river port can be found in literature (see Xu et al.[1], Lalla-Ruiz et al. [2], Qin et al. [19]).

3 PROBLEM DESCRIPTION AND MATHEMATICAL FORMULATION

Within a given planning horizon T, a set of incoming vessels V dynamically arrive at a tide-sensitive seaport and are to be berthed on a quay for loading/uploading operations which are performed by quay cranes. The quay is partitioned into a number of sections M, referred to as berth positions. Only one vessel can be served at each single berth at a time. Each vessel v, $v \in V$ has an arrival time a_v; a duration draft p^v, i.e., the time needed to sail through the channel; l_v, i.e., distance between the waterline and the vessel' s keel which defines the water requirement and processing time P_q^v, i.e., the time taken for the loading/uploading which depends on the number of quay cranes assigned to it. The number of quay cranes serving a vessel simultaneously is also restricted by some technical and physical constraints. Additionally, each vessel is also associated with a hard time window $[EST^v, LCT^v]$, i.e., an earliest arrival time and a latest departure time. This is because a vessel can be speeded up in order to meet a berthing time earlier than the expected arrival time. In order to keep liner schedules, the latest departure times of the vessel can be preset.

Before berthed, each vessel should sail through the navigation channel, a shallow water area where the water depth periodically change due to tide effect. Figure 1 shows a schematic representation of offshore topography at a seaport. In this context, when optimizing the B&CAP, the terminal operator must take this factor into account to ensure that the water depth can satisfy the draft requirement and grantee a safe sailing with a duration of p^v for each vessel v. Reference data is available based on historical information, which records the water depth at every hour over a day. Figure 2 schematically shows the varying pattern we considered. If a vessel with a draft of 13.2m arrives at 1:00 when the water depth is 12.6m, it cannot sail through the channel until the clock strikes 5:00. A similar situation might confront the vessel when it departs for next destination.

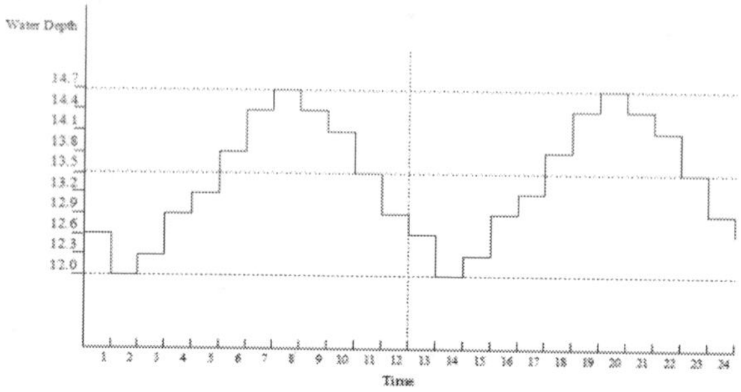

Figure 1. Water depth varying pattern

The assumptions contemplated in this work are enumerated as follows: (1) Anchorage area in port basin is sufficiently large to accommodate potential incoming vessels; (2) Quay cranes can be rearranged among multiple vessels, and the relocation time is ignored; (3) The number of quay cranes assigned for each vessel keeps unchanged during its stay at berth; (4) Each berth can only serve one vessel at a time; (5) Berth shift is not allowed; (6) Stevedoring operation for each vessel should be finished without interruption. To model this integrated problem, following parameters and decision variables are used:

V	set of vessels, i.e., $V = \{1, 2, ..., \lvert V \rvert\}$
M	set of berths, i.e., $M = \{1, 2, ..., \lvert M \rvert\}$
T	set of time steps in planning horizon, i.e., $T = \{1, 2, ..., \lvert T \rvert\}$. One hour per time
a^v	step arrival at anchorage area of vessel v
P_q^v	processing time of vessel v if q QCs simultaneously serve it (output of QCSP)
$Dept_t$	the water depth at time t
l^v	the draft for each vessel v
p^v	the navigation duration of sailing through the channel for each vessel v
Q	the number of all QCs
R^v	the range of assignable number of QCs for vessel v, i.e., $R^v = [R_{min}^v, R_{max}^v]$
$[EST^v, LCT^v]$	hard time window of vessel v, $EST^v < LCT^v$
w^v	weight of vessel v
c_t^v	equal to 1, if vessel v starts its sailing through the channel at time t
A^v	the arrival time at port basin for vessel v
d_t^v	equal to 1, if vessel v sail out of the port basin at time t
D^v	the departure time for vessel v
s^v	berthing time of vessel v
e^v	departure time of vessel v
x_q^v	equal to 1, if q QCs are assigned to vessel v
r_{qt}^v	equal to 1, if vessel is processed by QCs at time

r_t^v equal to 1, if vessel v is processed at time t
b^v berth position of vessel v
y^{ij} equal to 1, if berthing position of vessel i is smaller than berthing position of vessel j, i.e., $b^i < b^j$
z^{ij} equal to 1, if vessel i completes not later than vessel j starts

We formulate the problem as integer programming model as follows.

$$\min \sum_{v \in V} w^v (D^v - a^v)$$

Each vessel can start to sail into the navigation channel after its arrival at the port.

$$c_t^v = 0, \quad v \in V, t \leq a^v - 1. \tag{1}$$

Each vessel passes through the navigation channel exactly once.

$$\sum_{t \in T} c_t^v = 1, \quad v \in V. \tag{2}$$

Adequate water depth guarantee a safe sailing with a specific duration for each vessel.

$$\sum_{\hat{t} \in [max\{t - p^v + 1, 0\}, t]} l^v c_{\hat{t}}^v \leq Dept_t, \quad v \in V, t \in T. \tag{3}$$

The time of arrival at port basin is calculated for each vessel as follows.

$$\sum_{t \in T} (t + p^v) c_t^v = A^v, \quad v \in V, t \in T. \tag{4}$$

No interruption between processing of vessel:

$$\sum_{q \in R^v} x_q^v P_q^v = e^v - s^v, \quad v \in V. \tag{5}$$

Relation between r_t^v and $[s^v, e^v]$:

$$\sum_{t \in T} r_t^v = e^v - s^v, \quad v \in V. \tag{6}$$

$$(t + 1) r_t^v \leq e^v, \quad v \in V, t \in T. \tag{7}$$

$$t \cdot r_t^v + |T|(1 - r_t^v) \geq s^v, \quad v \in V, t \in T. \tag{8}$$

Each vessel must be served by exact q QCs.

$$\sum_{q \in R^v} x_q^v = 1, \quad v \in V. \tag{9}$$

Relation between r_{qt}^v and r_t^v:

$$\sum_{q \in R^v} r_{qt}^v = r_t^v, \quad v \in V, t \in T. \tag{10}$$

Definition of y^{ij}:

$$b^j + |M|(1 - y^{ij}) \geq b^i + 1, \quad i,j \in V. \tag{11}$$

Definition of z^{ij}:

$$s_j + |T|(1 - z^{ij}) \geq e_i, \quad i,j \in V. \tag{12}$$

One berth can serve one vessel at a time:

$$y^{ij} + y^{ji} + z^{ij} + z^{ji} \geq 1, \quad i,j \in V. \tag{13}$$

Total number of QCs (constraint):

$$\sum_{v \in V} \sum_{q \in R^v} q \cdot r^v_{qt} \leq Q, \quad t \in T. \tag{14}$$

Hard time window (constraint):

$$s^v \geq EST^v, \quad v \in V. \tag{15}$$

$$e^v \leq LDT^v, \quad v \in V. \tag{16}$$

Each outgoing vessel can only start to sail into the navigation channel after its completion.

$$\sum_{t \in T} t d^v_t \geq e^v, \quad v \in V, t \in T. \tag{17}$$

Each vessel sails out of the port basin exactly once.

$$\sum_{t \in T} d^v_t = 1, \quad v \in V. \tag{18}$$

Water depth requirement should be met for each vessel's navigation.

$$\sum_{\tilde{t} \in [max\{t - p^v + 1, 0\}, t]} l^v d^v_{\tilde{t}} \leq Dept_t, \quad v \in V, t \in T. \tag{19}$$

The departure time from port basin is computed for each vessel as follows.

$$\sum_{t \in T} (t + p^v) d^v_t = D^v, \quad v \in V, t \in T. \tag{20}$$

$$s^v \in [EST^v, |T|], \quad v \in V. \tag{21}$$

$$e^v \in [EST^v, LDT^v], \quad v \in V. \tag{22}$$

$$A^v, D^v \in T, \quad v \in V. \tag{23}$$

$$1 \leq b^v \leq |M|, \quad v \in V. \tag{24}$$

$$c^v_t, d^v_t, x^v_q, r^v_{qt}, r^v_t \in \{0, 1\}, \quad v \in V, q \in R^v, t \in T. \tag{25}$$

We consider a small size instance with $|V|$=6, $|M|$=3, $|T|$=18, Q=5. Other information is depicted in Table 1 and Table 2. A feasible solution is graphically represented on a space-time diagram.

Table 1. Problem information

vessel v	1	2	3	4	5	6
a^v	0	2	3	5	5	7
l^v	12.4	12.5	13.8	13.2	14.4	14
p^v	1	1	1	1	1	1
$[EST^v, LCT^v]$	[0, 4]	[3, 10]	[5, 11]	[6, 15]	[9, 17]	[9, 16]

Table 2. Handling time

QCs\vessel v	1	2	3	4	5	6
1	-	6	-	8	-	-
2	4	3	5	4	6	5
3	3	2	4	3	4	4
4	-	2	3	2	3	3

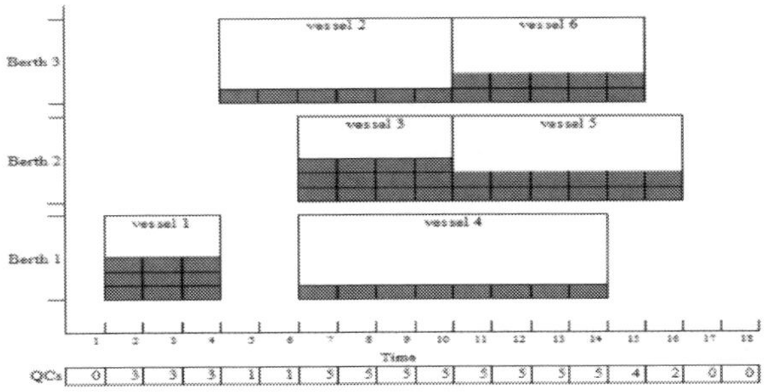

Figure 2. A feasible solution for example

4 ACKNOWLEDGEMENT

This work was supported by the National Science Foundation of China (NSFC) under Grants 71571134, 71571135, 71101106, 71171149 and 71428002.

5 REFERENCES

[1] **Xu D, Li C L, Leung Y T. 2012.** Berth allocation with time-dependent physical limitations on vessels. *European Journal of Operational Research*, 216(1), pp 47-56.

[2] **Lalla-Ruiz E, Expsito-Izquierdo C, Meliscn-Batista B, et al. 2015.** A Set-Partitioning-based model for the Berth Allocation Problem under Time-Dependent Limitations. *European Journal of Operational Research*, 250(3), pp 1001-1012.

[3] **Stahlbock, R., Voss, S. 2008.** Operations research at container terminals: A literature update, OR Spectrum, 30(1),pp 1-52.

[4] **Bierwirth C, Meisel F.2015.** A survey of berth allocation and quay crane scheduling problems in container terminals. *European Journal of Operational Research*, 244(202),pp 615-627.

[5] **Park, Y.M., Kim, K.H., 2003.** A scheduling method for berth and quay cranes, OR Spectrum, 25 (1),pp 1-23.

[6] **Vacca, I., Salani, M., Bierlaire, M., 2013.** An exact algorithm for the integrated planning of berth allocation and quay crane assignment. *Transportation Science*, 47(2),pp 148-161.

[7] **Liu, J., Wan, Y. W., Wang, L., 2006.** Quay crane scheduling at container terminals to minimize the maximum relative tardiness of vessel departures. *Naval Research Logistics*, 53 (1),pp 60-74.

[8] **Chen J H, Lee D H, Cao J X.2012.** A Combinatorial Benders' Cuts Algorithm for the Quayside Operation Problem at Container Terminals. *Transportation Research Part E Logistics & Transportation Review*, 48(1), pp 266-275.

[9] **Meisel, F., Bierwirth, C., 2006.** Integration of berth allocation and crane assignment to improve the resource utilization at a seaport container terminal. *Operations Research Proceedings 2005*, 105-C110.

[10] **Imai, A., Chen, H., Nishimura, E., Papadimitriou, S., 2008.** The simultaneous berth and quay crane allocation problem.*Transportation Research Part E*, 44(5), pp 900-920.

[11] **Meisel, F., Bierwirth, C., 2009.** Heuristics for the integration of crane productivity in the berth allocation problem. *Transportation Research Part E*, 45(1), pp 196-209.

[12] **Giallombardo, G., Moccia, L., Salani, M., Vacca, I. 2010.** Modeling and solving the tactical berth allocation problem. *Transportation Research Part B*, 44(2), pp 232-245.

[13] **Raa B, Dullaert W, Schaeren R V. 2011.** An enriched model for the integrated berth allocation and quay crane assignment problem. *Expert Systems with Applications*, 38(11), pp 14136-14147.

[14] **Turkogullar Y B, Taskin Z C, Aras N, et al.2014.** Optimal berth allocation and time-invariant quay crane assignment in container terminals. *European Journal of Operational Research*, 235(1),pp 88-101.

[15] **Cagatay Iris, Pacino D, Ropke S, et al. 2015.** Integrated Berth Allocation and Quay Crane Assignment Problem: Set partitioning models and computational results. *Transportation Research Part E Logistics & Transportation Review*, 81,pp 75-97.

[16] **Kelareva E, Brand S, Kilby P, et al. 2012.** CP and MIP Methods for Ship Scheduling with Time-Varying Draft. pp 110-118.

[17] **Sheikholeslami, A., Ilati, G., Kobari, M., 2014.** The continuous dynamic berth allocation problem at a marine container terminal with tidal constraints in the access channel. *International Journal of Civil Engineering*, 12(3), pp 344-353.

[18] **Du, Y., Chen, Q., Lam, J.S.L., Xu, Y., Cao, J.X., 2015.** Modeling the impacts of tides and the virtual arrival policy in berth allocation. *Transportaion Science*, 49(4),pp 939-956.

[19] **Qin T, Du Y, Sha M. 2016.** Evaluating the solution performance of IP and CP for berth allocation with time-varying water depth. *Transportation Research Part E Logistics & Transportation Review*, 87, pp 167-185.

ENTERPRISE R&D, EQUIPMENT AND PRODUCT SERVICE SYNERGY INNOVATION SUPPORT SYSTEM AND ITS SYSTEM DYNAMICS ANALYSIS

J.F. Wang[1],H.Z. Wu[1], L.J. Feng[1],J.J. Yue[1]
[1]Zhengzhou university of engineering management, Henan Zhengzhou, China
alwhz@163.com

ABSTRACT

Innovation ability as one of the core elements of value creation of enterprises to improve business efficiency. Howecer, the development level and innovation ability of enterprises in our country is still at low level compared with the world-famous enterprises, it's an urgent need to build a scientific enterprise innovation support system, to promote the enterprise innovation ability rapidly. Therefore, this article summarize and put forward the enterprise R&D, equipment and product service synergy innovation support system, then using system dynamics model to simulate the supporting system. The result shows that under the premise of ensuring the production and R&D equipment investment, increasing the investment in R&D and product service can make enterprise's innovation ability improving more rapidly.

Keywords: Enterprise innovation, Synergy innovation, Support system, System dynamics

1 INTRODUCTION

Under the new situation of economic globalization, science and technology independent innovation has replaced natural resources become the decisive factor of national competitiveness. Under the construction of national innovation system, the first thing should take action is to make enterprises as the main body of independent innovation promoting, because in the increasingly fierce market competition, as well as the dramatically increasing of social demand, enterprises desperately needed an efficiently innovation system to push them forward, as innovation ability is an inexhaustible motive force for the survival of enterprises.

However, compared with developed countries, the innovation capability of Chinese enterprises is generally low. Lack of systematic and comprehensive innovation support system to promote the innovation capability and development of enterprise. Therefore, how to build the scientific enterprise innovation support system, to reasonably allocate enterprise innovation resources, to constantly improve the innovation capability and meet the growing individual needs of users, thus constantly promote the social and economic benefits of the enterprise, become the key problems of Chinese enterprises[1].

This paper structure is as follows: In chapter 2, we summarize and introduces the predecessors' research results on enterprise innovation support system. In chapter 3, we analysis the synergy relationship of enterprise R&D, equipment and product service, then put forward the enterprise R&D, equipment and product service synergy innovation support system. In chapter 4, we make a further study on enterprise innovation system based on causal relationship analysis, and build the enterprise R&D, equipment and product service synergy innovation support system using System Dynamics(SD) model. In chapter 5, we use the built SD model to do the enterprise cooperative innovation support system dynamic simulation research. In chapter 6, we made the conclusion of this article.

2 LITERATURE REVIEW

For business innovation support system problems, Rosenberg[2] had already studied the positive correlation between the level of enterprise independent innovation and economic benefit through empirical research in 1992. Since then, experts and scholars have carried out a lot of research on the components of enterprise innovation support system. Polder et al.[3] based on the investigation and study of manufacturing service companies, put forward three factors to support enterprise innovation process, namely product innovation, process innovation and organizational innovation. Hassink[4] carried out research on enterprise innovation support system from the perspective of Korean industry and economic value chain, and put forward the regional enterprise innovation support system with the core of infrastructure, resource investment and social environment as the core. Großmann et al.[5] through the research of enterprise innovation achieving process, build a enterprise innovation network model with social science and technology level, economic environment and patent technology as the index, in order to support enterprise service innovation. Robert[6] through the analysis of the regional innovation environment, believes that in the imperfect competitive market environment, the innovation performance of enterprises will be affected by the double effects of incentive policies and financing environment. Radicic et al.[7] studied the manufacture of small and medium-sized enterprise innovation environment , and summed up the 4 elements of the enterprise innovation environment, including financial environment, national policy, technology environment and social environment. Koziol[8] and Oh[9]consider that the level of innovation of enterprises depends on the company's investment in product marketing and customer service, through marketing and services can fully tap the potential value of the user to enhance the innovation ability of the enterprise.

Based on the enterprise innovation data, Cabrer-Borrás and Serrano-Domingo[10] through the analysis of 17 regional innovation data, obtains the local enterprise innovation level is mainly affected by R&D investment, human capital and innovation culture. Song et al.[11] based on the analysis of 269 private listed companies, that without fully competitive market environment, innovation performance will be influenced by both incentives and financing environment. After studying the relationship between investment and innovation performance of small and medium enterprises, Kaufmann and Todtling[12] consider that R&D investment and non R&D investment can both improve the innovation performance. Samara et al.[13] based on the system dynamics model, from the national policy level to the enterprise's original innovation support system to carry out the research, think that the improvement of enterprise innovation performance is closely linked with national economy, institution and social environment.

In summary, innovation plays an important supporting role in the development of enterprise and community economic. In recent years, experts and scholars have made a lot of research on the elements of enterprise innovation support system, which provides an important reference for the construction of enterprise synergy innovation support system in this paper. However, most of the above studies are from perspective of national policy, technology and financial environment, innovative services, social environment and other external environment to carry on the enterprise innovation support system analysis, just a small amount of literature related to the business-led R&D funding, infrastructure, marketing, product service, non R&D investment and other enterprise internal innovation support system. But, compared with the external environment of the enterprise, the rational integration and coordination of enterprise internal innovation resources can be more efficient and system to improve the performance of enterprise innovation[14-16].

In view of this, based on product innovation process, this paper from the view of enterprise internal innovation resources synergistic utilization, plans to construct enterprise R&D, equipment and product service synergy innovation support system, and use the SD model of the synergy innovation support system to verify its effectiveness. At the same time, use the model to do a dynamic simulation study of China's enterprise synergy innovation support system, in order to provide the theoretical support for the construction and optimization of China's enterprise synergy innovation support system.

3 ENTERPRISE R&D, EQUIPMENT AND PRODUCT SERVICE SYNERGY INNOVATION SUPPORT SYSTEM

3.1 Analysis on the synergy relationship between enterprise R&D, equipment and product service

R&D in the enterprise innovation value chain is to explore the potential demand of customers, using the knowledge of science and technology (such as literature, patent, knack) for innovation, which is a activity achieving innovation through the function, structure or product improving[17]. Equipment is the basis of enterprise R&D and manufacturing, mainly through processing and utilization the manufacturing resources, such as material, energy, technology and information , to produce new products or improving existing products. Product service refers to the marketing department provide products and services to customers through advertising, brand, packaging, sales and after service[18]. Enterprise R&D, equipment and product service are interdependent and promoting in each other in the process of enterprise innovation, they have a close logical relationship and synergy.

3.1.1 Synergy effect between enterprise R&D and production equipment investment

The synergy effect between enterprise R&D and production equipment investment refers to R&D department and production department through make full use of the production equipment to achieve mutually beneficial effect in the production chain. The engineering and technical personnel of R&D departments carry on scientific and technological knowledge with digestion and innovation, allowing production department to achieve the optimization and efficient use of the production equipment resources. At the same time, the production department provide the necessary R&D equipment support for R&D department in order to improve the efficiency of product development.

3.1.2 Synergy effect between production equipment and product service investment

Synergy effect between enterprise production equipment and product service investment is refers to the production department and service department built up mutual benefit relationship through the circulation of products. The production department through improved or input production equipment to produce customers needed products and improve their satisfaction, thereby to improve the level of service products. At the same time, according to the customer feedback information from product side, can assist the production department reasonable arrange production equipment to maximize improve the production efficiency.

3.1.3 Synergy effect between product service and enterprise R&D investment

Synergy effect between service investment and R&D investment of enterprise product refers to apply customer feedback information to carry out precise manufacturing in product innovation and to achieve mutually beneficial effect. The enterprise R&D department develop and design new products to meet the needs of customers through technological innovation. At the same time, R&D department to grasp the direction of R&D by statistic the customer feedback information from production service department, to enhance the success rate of new product development.

3.2 The structure of enterprise R&D, equipment and product service synergy innovation support system

Based on the relationship of enterprise R&D, equipment and product service, the synergy effect between them has a significant impact on the improvement of enterprise innovation ability. Therefore, this paper integrated enterprise R&D, equipment and product service, and put forward the enterprise R&D, equipment and product service synergy innovation support system, as shown in Fig. 1.

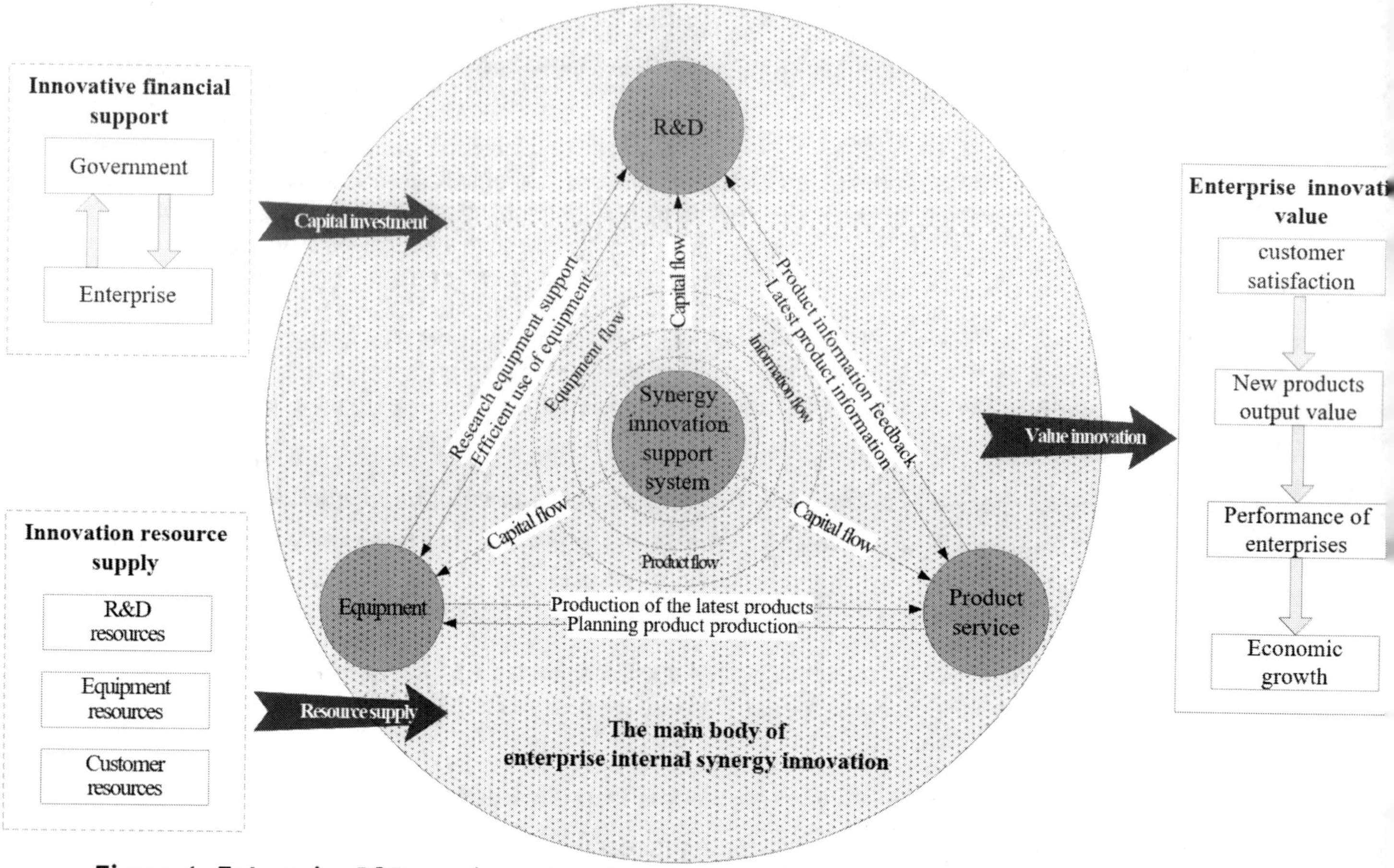

Figure 1: Enterprise R&D, equipment and product service synergy innovation support system

It can be seen from Fig. 1 that enterprise R&D, equipment and product service synergy innovation support system is an enterprise innovation development model in the allocation of resources and cooperative operation, in the foundation of government and enterprise financing investment, and with the the final output of increasing the enterprise overall efficiency as well as social economy[19-23].

4 ENTERPRISE R&D, EQUIPMENT AND PRODUCT SERVICE SYNERGY INNOVATION SUPPORT SYSTEM SD MODEL BUILD

SD model is a set of system theory, cybernetics and information theory concept, which is applied to the analysis of complex nonlinear system and a method of comprehensive analysis of the interaction of feedback information. we can use the theory of SD to build the system model from the microstructure of internal system, and use the flow chart to express the internal structure of the interaction and feedback information interaction relations, and use computer software system simulation to simulate the reality of the state[24-25]. Due to the enterprise R&D, equipment and product service synergy innovation support system is an internal structure of complex non-linear system, this paper use SD model to analysis the system of enterprise R&D, equipment and products service synergy innovation support system and reveal the synergy relationship between the elements.

4.1 Causality model and path analysis

Based on the above construction of enterprise R&D, equipment and products service support synergy innovation system, the capital investment of Chinese enterprises in product innovation mainly includes the government funds and the enterprise funds, and the final output of the innovation achievement is the enterprise sales income and the overall social benefits. Therefore, we can build enterprise R&D, equipment and product service synergy innovation support system causal relation model based on the synergy relationship between enterprise R&D, equipment and products services about synergy innovation support system. which is shown in Fig. 2.

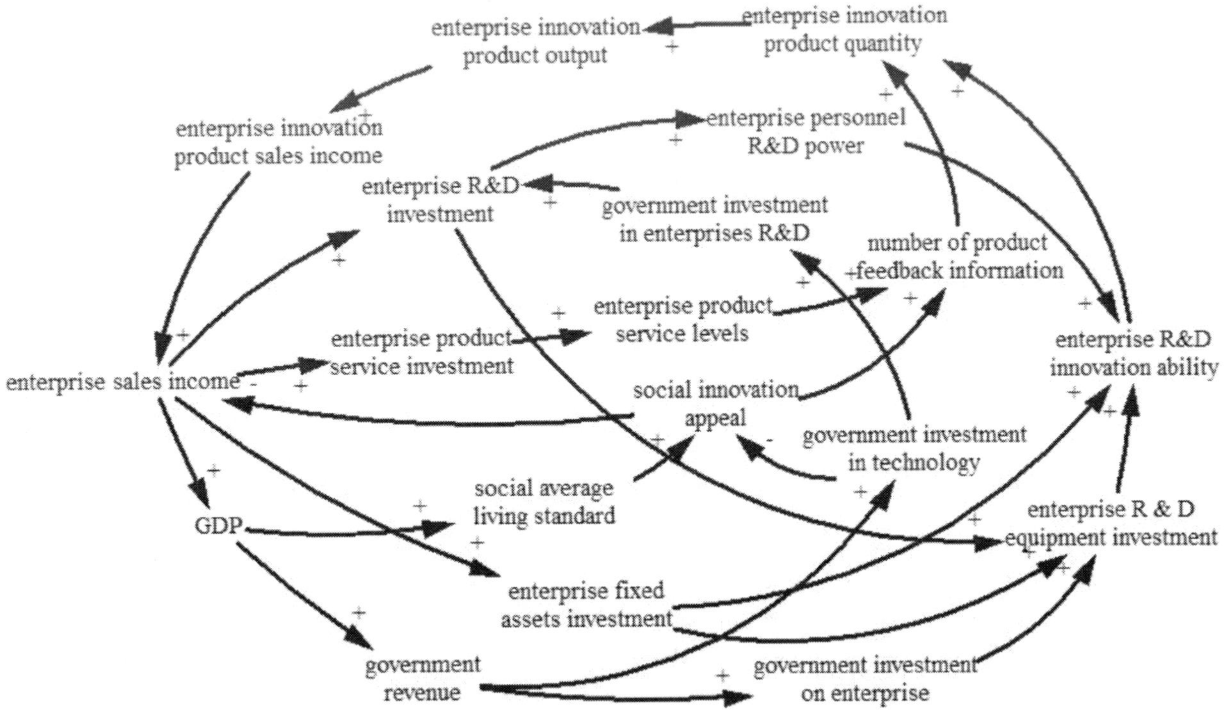

Figure 2: The causality model of enterprise R&D, equipment and product service synergy innovation support system

The model mainly contains four causal loop:

(1)firm sales income → enterprise R&D investment → enterprise R&D power → enterprise R&D innovation ability → enterprise innovation product quantity → enterprise innovation product output → enterprise innovation product sales income → firm sales income.

This loop is from the perspective of enterprise R&D investment, by the increasing investment in product R&D we can enhance the level and power of R&D and, further increase the additional value of the innovative products, improve the sales income of enterprises, and the increase in sales revenue and its further increase investment in R&D link, thus forms the enterprise innovation ability continuously enhanced dynamic closed loop. The loop emphasizes the importance of the investment of R&D on the innovation ability.

(2) firm sales income → enterprise fixed assets investment → enterprise R&D equipment investment → enterprise R&D innovation ability → enterprise innovation product quantity → enterprise innovation product output → enterprise innovation product sales income → firm sales income.

This loop is from the perspective of enterprise equipment investment, by the increasing investment of production and R&D equipment we can optimize the basic conditions in R&D innovation activities of enterprises so that to improve the innovation ability of them,.using advanced equipment to speed up the product R&D of enterprises in order to create more sales and constantly expand the scale.The expansion of enterprise scale will make enterprises purchase more R&D equipment, leading to the promotion of enterprise innovation ability.The loop emphasizes the importance of the investment in production and R&D on the equipment to improve the innovation ability.

(3) firm sales income → enterprise product service investment → enterprise product service levels → enterprise product feedback information quantity → enterprise innovation product quantity → enterprise innovation product output → enterprise innovation product sales income → firm sales income.

This loop is from the perspective of enterprise product service investment, by intensifying investment in services link to improve the level of enterprise product service,which can also promote the interactions between service personnel and users and increase the effective feedback so as to provide the basis for new product R&D, and make the number of innovative products and sales continue to increase, It can promote the enterprise investment in product service so that we can exert the guiding role of users in the enterprise innovation and make the enterprise R&D level's continuous improvement. The loop emphasizes the importance of the investment in product service to improve the innovation ability.

(4)firm sales income → GDP → government revenue → government investment on enterprise → enterprise R&D innovation ability → enterprise innovation product quantity → enterprise innovation product output → enterprise innovation product sales → firm sales income.

This loop is from the perspective of government investment and the model mainly aimed at new high-tech enterprise. The increase of sales revenue about new high-tech enterprise will drive the growth of GDP, so that government revenue input for the new high-tech enterprise increased, and the innovation ability of enterprises improved too,which make the sales revenue increased and lead to the increase of GDP. With this circulation, the innovation ability of enterprises continue to improve. The loop emphasizes the importance of government revenue input in construction enterprise to improve the innovation ability.

Through the above analysis can indicate that, the innovation ability of the enterprise is positively related to the investment of internal research, equipment ,and products service ,meanwhile, government investment in the construction of new high-tech enterprises is also conducive to the continuous improvement of enterprise innovation capability

4.2 flow chart analysis

Considering the data availability and calculation, combined with our country enterprise reality, the causality model of synergy innovation support system about enterprise R&D, equipment and products service showed in Fig. 3 can be simplified as SD flow chart of enterprise R&D, equipment and product service input synergy innovation support system[26], as is shown in Fig. 3.

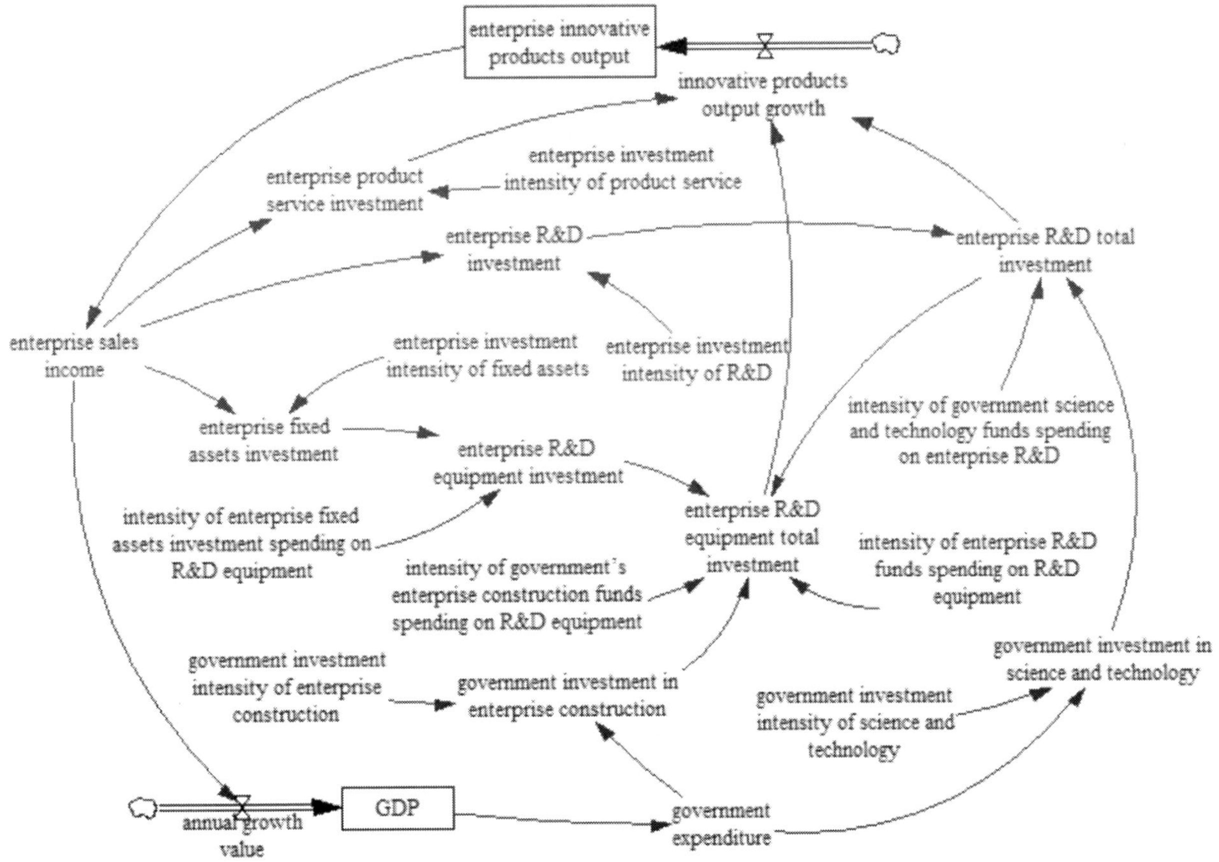

Figure 3: SD flow chart of enterprise R&D, equipment and product service synergy innovation support system

The variable data in the model mainly from *China Statistical Yearbook of 2005-2014* and *China High Technology Industries Statistical Yearbook*, and the function relation among variables are obtained by SPSS software calculation of function regression analysis and fitting tools.

4.3 Output of model

The above SD model is used to verify the promotion effect of enterprise R&D, equipment and product service synergy innovation support system on enterprise innovation ability, and the intuitive expression of enterprise innovation ability is the output value of innovative products. This paper chooses enterprise innovation product output value as the first output value of the model to reflect the change of enterprise innovation ability. Further, in order to verify the scientific and accuracy of the SD model, this paper chooses the GDP as the second output value of the model to reflect the level of enterprise's innovation ability[27].

5 SYNAMIC SIMULATION RESEARCH OF THE ENTERPRISE R&D, EQUIPMENT AND PRODUCT SERVICE SYNERGY INNOVATION SUPPORT SYSTEM

5.1 Model validity test

SD model is a kind of method to dynamic simulate the real system through the process building and mathematical calculation. Although it can get good simulation results under the premise of comprehensive system flow and accurate mathematical formula, but it still exists deviation between the fitting value and real value[28]. To verifiy the effectiveness of the SD model of the enterprise R&D, equipment and product service synergy innovation support system, we carried out a consistency test on it through following ways.

The first and second output values of the enterprise R&D, equipment and product service synergy innovation support system constructed above is shown in Table 1.

Table 1: Comparison table of the model output value

Time	Enterprise innovative product output value			GDP		
	Output value	Ture value	Error rate	Output value	Ture value	Error rate
2005	7034.82	7034.82	0.00	185896	185896	0.00
2006	8466.11	8493.26	0.00	228412	217657	0.05
2007	10173.5	10671.3	-0.05	271814	268019	0.01
2008	12203.5	13018.3	-0.06	316274	316752	0.00
2009	14610.4	12501.2	0.17	361992	345629	0.05
2010	17457.6	16502.6	0.06	409200	408903	0.00
2011	20819.3	20384.5	0.02	458173	484124	-0.05
2012	24782.2	23765.3	0.04	509228	534123	-0.05
2013	29447.6	29028.8	0.01	562738	588019	-0.04
2014	34933.9	35012.7	0.00	619139	636139	-0.03

The data shown in Table 1 indicate that, the error rate between the two output values and the true value of the enterprise innovation product output and the GDP is less than 0.06, in addition to the output value of enterprise innovation product in 2009. In 2009, the output value of the enterprise innovative product was larger, is because of the huge impact to enterprises by the international financial crisis in 2008, most enterprises were faced with great economic difficulties, therefore, the output value of enterprise innovative products were affected seriously.

According to the analysis above, the fitting degree of the output value and the true value of the enterprise R&D, equipment and product service synergy innovation support system SD model is good, which can objectively reflect the real operating conditions of chinese enterprises from 2005 to 2014. On the other hand, it also proves that the enterprise R&D, equipment and product service synergy innovation support system is scientific and reasonable.

5.2 Simulation and result analysis

The enterprise R&D, equipment and product service synergy innovation support system is based on the enterprise investment. Consequently, the investment intensity of the enterprise funds in R&D, equipment and product service directly impact the enterprise's innovation capability. In order to guide the rational allocation of enterprise investment, and make the enterprise R&D, equipment and product service synergy innovation support system better help enterprises to improve the ability of innovation, this paper carry on a simulation study to the output value of innovative products by changing the intensity of enterprise investment in R&D, equipment and product service.

Assuming that the intensity of enterprise investment in the 3 aspects respectively increased by 1%, and when 1 of the 3 intensity changes, the other parameters keep invariant, the simulation results of the model are shown in Fig. 4, and the specific data is shown in Table 2.

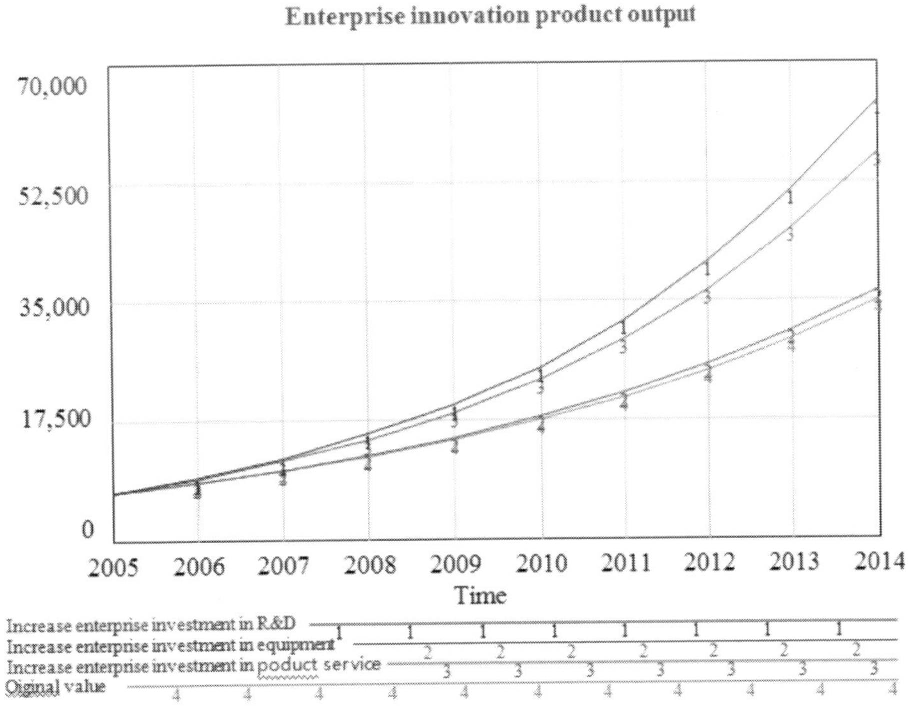

Figure 4: The impact on the value of innovative products when the investment strength changed

Table 2: The change rate of the enterprise innovation product output value

Time	Original value	Increase R&D investment strength		Increase equipment investment strength		Increase service input strength	
		Output value	Change rate	Output value	Change rate	Output value	Change rate
2005	7034.82	7034.82	0	7034.82	0	7034.82	0
2006	8466.11	9237.78	0.09	8519.95	0.01	9069.39	0.07
2007	10173.5	12008.5	0.18	10297.8	0.01	11598.3	0.14
2008	12203.5	15484.4	0.27	12419.2	0.02	14733.2	0.21
2009	14610.4	19836.3	0.36	14943.6	0.02	18611.0	0.27
2010	17457.6	25276.0	0.45	17941.1	0.03	23399.5	0.34

2011	20819.3	32066.9	0.54	21493.7	0.03	29304.7	0.41
2012	24782.2	40536.3	0.64	25697.8	0.04	36578.7	0.48
2013	29447.6	51090.4	0.73	30666.4	0.04	45531.1	0.55
2014	34933.9	64234.1	0.84	36532.4	0.05	56541.0	0.62

The data from Fig. 4 and table 2 indicate that, the enterprise innovative product output were increased by 0.84, 0.05 and 0.62 when the intensity of enterprise investment in R&D, equipment and products service respectively increase 1%, each of them played a role in promoting the enterprise innovation capability. Among them, increase the enterprise investment in R&D and product service, has a significant influence on promoting the enterprise innovation capability. However, increase the enterprise investment in equipment also promotes the enterprise innovation capability, but the effect is not obvious. What's more, the actual investments on enterprise are generally focused on R&D funding and equipment procurement, it is easy to ignore the important role of product service for the promotion of innovation capability[29].

The result indicates that equipment as the basis for the development of enterprises, it needs to purchase but do not need excessive purchasing, or blind introduce the advanced equipment beyond the level of production and R&D. At the same time, due to the important role of product service investment on enterprise innovation capability promotion, the enterprise should increase the investment in product service appropriately on the basis of ensuring the necessary R&D investment, in order to improve the efficiency of innovation.

6 CONCLUSION AND DISCUSSION

According to innovation development present situation of enterprise, this paper summarize and put forward a synergy innovation support system based on the enterprise R&D, equipment and product service, and the relationship of 3 elements in the synergy innovation support system are described in this paper. In order to provide theoretical reference for China's manufacturing enterprises improving its own R&D and innovation ability. At the same time, this paper uses SD model to structure system modeling based on the proposed enterprise R&D, equipment and product service synergy innovation support system, and the scientific and reasonable of the system are proved by the validity test of SD model. Then, we apply the SD model in a simulation research for enterprise's R&D, equipment and product service investment synergy innovation support system.

The results show that the R&D, equipment and product service investment 3 elements has a different degree of promote effection on improving the innovation ability. Under the premise of ensuring the enterprise necessary production and R&D equipment investment, by increasing the investment in R&D and product service, is conducive to the rapid upgrading of enterprise innovation ability.

However, due to the limit of research time and data coverage, the enterprise R&D, equipment and product service synergy innovation support system we put forward can be well applied in high-tech enterprises innovation practice, but hard to cover all the enterprise field. So, further study can be made in future on the universal innovation support system, which can overall promote the construction of national innovation system.

7 REFERENCES

[1] **Cohen W M, Nelson R R, Walsh J P. 2003.** Links and Impacts: The Influence of Public Research on Industrial R&D, *Management Science*, 48(1), pp 1-23.

[2] **Rosenberg N, Landau R, Mowery D C. 1992.** *Technology and the wealth of nations*, Stanford: Stanford University Press.

[3] **Polder M, Leeuwen G V, Mohnen P, et al. 2010.** Product, process and organizational innovation: drivers, complementarity and productivity effects, *Merit Working Papers*, 35, pp 1-54.

[4] **Hassink R. 2001.** Towards Regionally Embedded Innovation Support Systems in South Korea? Case Studies from Kyongbuk-Taegu and Kyonggi, *Urban Studies*, 38(8), pp 1373-1395.

[5] **Großmann D J, Kasperek D, Stahl B, et al. 2015.** Supporting PSS Innovation Processes by an Integrating Model Grid, *Procedia Cirp*, 30, pp 227-232.

[6] **Robert H. 2002.** Regional Innovation Support Systems: Recent Trends in Germany and East Asia, *European Planning Studies*, 10(10), pp 153-164.

[7] **Radicic D, Pugh G T, Hollanders H, et al. 2014.** The impact of innovation support programmes on SME innovation in traditional manufacturing industries: an evaluation for seven EU regions, *Merit Working Papers*, 33, pp 1-44.

[8] **Kozioł L, Kozioł W, Wojtowicz A, et al. 2014.** Relationship Marketing - A Tool for Supporting the Company's Innovation Process, *Procedia - Social and Behavioral Sciences*, 148, pp 324-329.

[9] **Oh E T, Chen K M, Wang L M, et al. 2015.** Value creation in regional innovation systems: The case of Taiwan's machine tool enterprises, *Technological Forecasting & Social Change*, 100, pp 118-129.

[10] **Cabrer-Borrás B, Serrano-Domingo G. 2007.** Innovation and R&D spillover effects in Spanish regions: A spatial approach, *Research Policy*, 36(9), pp 1357-1371.

[11] **Song M, Ai H, Li X. 2015.** Political connections, financing constraints, and the optimization of innovation efficiency among China's private enterprises, Journal of Child Psychology & Psychiatry, 92(3), pp 290-299.

[12] **Kaufmann A, Tödtling F. 2002.** How effective is innovation support for SMEs? An analysis of the region of Upper Austria, *Technovation*, 22(3), pp 147-159.

[13] **Samara E, Georgiadis P, Bakouros I. 2012.** The impact of innovation policies on the performance of national innovation systems: A system dynamics analysis, *Technovation*, 32(11), pp 624-638.

[14] **Tidd, Joe, Pavitt, Keith. 1997.** Managing Innovation: Integrating Technological, Market and Organizational Change, *International Journal of Entrepreneurship & Innovation*, 18(98), pp 369-370.

[15] **Oke A, Prajogo D I, Jayaram J. 2013.** Strengthening the Innovation Chain: The Role of Internal Innovation Climate and Strategic Relationships with Supply Chain Partners, *Journal of Supply Chain Management*, 49(49), pp 43-58.

[16] **Wu A H, Su J Q, Wang H. 2013.** Internal innovation or external innovation? An organizational context-based analysis in China, *Journal of High Technology Management Research*, 24(2), pp 118-129.

[17] **Becker W, Dietz J. 2004.** R&D cooperation and innovation activities of firms—evidence for the German manufacturing industry, *Research Policy*, 33(2), pp 209-223.

[18] **Zhong X, Yang X. 2007.** Science and technology policy reform and its impact on China's national innovation system, *Technology in Society*, 29(3), pp 317-325.

[19] **Wonglimpiyarat J. 2013.** Innovation financing policies for entrepreneurial development — Cases of Singapore and Taiwan as newly industrializing economies in Asia, *Journal of High Technology Management Research*, 24(2), pp 109-117.

[20] **Herath D, Cranfield J, Henson S. 2010.** Understanding the financing of innovation and commercialization: the case of the Canadian functional food and nutraceutical sector, *Applied*

Economics, 42(21), pp 2667-2682.

[21] **Doloreux D, Parto S, Doloreux D, et al. 2005.** Regional innovation systems: Current discourse and unresolved issues, *Technology in Society*, 27(2), pp 133-153.

[22] **Guo Y, Yu F, Le-Nguyen K, et al. 2015.** The Impact of Government Subsidies on Enterprises' R&D Investment: A Panel Data Study From Renewable Energy in China, *Energy Policy*, 89, pp 106-113.

[23] **Lukas B A, Ferrell O C. 2000.** The Effect of Market Orientation on Product Innovation, *Journal of the Academy of Marketing Science*, 28(2), pp 239-247.

[24] **Wang Q P. 1988.** *System Dynamics*, Beijing: Tsinghua University Press.

[25] **Xing R, Wang G, Tang L. 2013.** The support system of regional industry integrative innovation based on system dynamics, *Science Research Management*, 34(1), pp 19-27.

[26] **Wu D D, Xie K, Liu H, et al. 2010.** Modeling technological innovation risks of an entrepreneurial team using system dynamics: An agent-based perspective, *Technological Forecasting & Social Change*, 77(6), pp 857-869.

[27] **Hu J Y, Zhu G L, Ma Y Y. 2011.** System Dynamic Analysis on Influencing Factors of Industry-University-Research Cooperation in Open Innovation Context, *Science of Science & Management of S & T*, 32(8), pp 49-57.

[28] **Sterman J. 2002.** Business dynamics: systems thinking and modeling for a complex world[J]. Journal of the Operational Research Society, 53(4), pp 324-326.

[29] **Haryanto A T, Haryono T. 2015.** The influence of market orientation on innovation type and enterprise performance[J]. Polish Journal of Management Studies, 11(1), pp 68-78.

USING A NEW HYBRID MULTI-OBJECTIVE META-HEURISTIC ALGORITHM FOR A BI-OBJECTIVE TRANSPORTATION VEHICLE ROUTING PROBLEM IN WASTE COLLECTION

R. Tavakkoli-Moghaddam[1], H. Farrokhi-Asl[2] and H. Mirzahosseinian[3*]
[1]School of Industrial Engineering, College of Engineering
University of Tehran, Tehran, Iran
tavakoli@ut.ac.ir

[2]School of Industrial Engineering
Iran University of Science & Technology, Tehran, Iran
hamed.farrokhi@ut.ac.ir

[3]National Elites Foundation of Iran and School of Industrial Engineering, College of Engineering
University of Tehran, Tehran, Iran
mirzahosseinian@ut.ac.ir

ABSTRACT

This paper is an extension for the well-known operations research (OR) problem, called vehicle routing problem (VRP), in which two stages are considered. The first stage deals with the VRP and the second one is dedicated to transportation problems. Also, the application of the presented problem in waste collection is investigated. This problem aims to minimize the waste collection cost and decreases the risk posed to the environment for hazardous wastes transportation, simultaneously. Given the NP-hard nature of the problem, a new hybrid multi-objective meta-heuristic algorithm combining genetic and cultural algorithms is proposed to obtain Pareto solutions. A straightforward solution representation is applied to help us in reducing the computational time. To validate the proposed algorithm, a number of test problems are conducted and the results of the algorithm are compared with the results obtained by non-dominated sorting genetic algorithm (NSGA-II) with respect to some comparison metrics. Finally, the conclusion remarks and future research directions are provided.

Keywords: Waste collection problem, Vehicle routing problem, Hazardous waste, Evolutionary algorithm.

1 INTRODUCTION

A wastes collection problem has been categorized as an important part of reverse logistics management recently. Markov et al. [1] applied a vehicle routing problem (VRP) in waste management, called a waste collection problem. Sbihi and Eglese [2] introduced the waste collection problem as the one part of green logistics and divided waste collection into two segments including hazardous waste and rollon-rolloff problems. Also, Lin et al. [3] reviewed a waste collection problem and categorized it as a subset of the VRP in reverse logistics (VRPRL). Reverse logistics has attracted attention of researchers in recent years. Dekker et.al [4] defined the reverse logistics as: "the process of planning, implementing and controlling backward flows of raw materials

* Corresponding Author.

in process inventory, packaging and finished goods, from a manufacturing, or distribution or use point, to a point of recovery of point of proper disposal". Wy et al. [5] divided the waste collection problem into three major business lines (i.e., residential, commercial and industrial). In the residential waste collection problem, the vehicles pass from streets to gather homemade wastes. Bonomo et al. [6] transformed an arc routing waste collection problem into a traveling salesman problem (TSP) and used it in the real case in Buenos Aires. In a commercial waste collection problem, a goal is to find logical routes in order to gather wastes from commercial sites, such as shopping centers. Duzgun et al. [7] applied a heuristic approach and geographical information system (GIS) for solving the commercial waste collection problem. The rollon-rolloff waste collection problem regards the collection of wastes from industrial sites and this problem has a specific limitations such as each vehicle should serve only one customer at a time [8]. Hazardous waste can be defined as the useless materials which are harmful to environment and human's life. Some kinds of these materials cause immediate or long-term risk for humans, plants, animals and in total for our environment. Hence, proper planning for collection of hazardous waste can be advantageous for the environment and can enhance the lives of people. According to Alumur and Kara [9], waste can be characterized as hazardous if it has any of the following attributes: ignitability, corrosiveness, reactivity or toxicity. In the current research, we investigate collection of hazardous wastes from commercial location, such as hospitals, shopping centers and industrial companies.

Two risk measures are commonly used in the literature of waste collection, namely societal risk and population exposure [10]. Population exposure is the number of populations influenced from hazardous wastes. Caballero et al. [11] considered three objective functions for transportation risk. The first one takes into account rejection of towns that trucks cross from it in their way. The second one is equity distribution of damage between the towns. The last objective function is about towns that are near to incineration plants, called collective disutility. In this paper, we utilize the risk type 2 meaning that population exposure and dissatisfaction of people in transportation routes are considered as one of criteria to evaluate the solutions.

Martinez-Salazar et al. [12] considered a transportation location-routing problem and presented a new bi-objective mathematical model for this two-stage location-routing problem. They implemented two meta-heuristic algorithms based on scatter and tabu search for non-linear multi-objective optimization (SSPMO) and NSGA-II. Two objectives are considered as decrease of distribution cost and the equitable distribution of workloads for drivers. The transportation vehicle routing problem belongs to the class of NP-hard problems [13]. This paper considers a bi-objective transportation VRP, so we use the NSGA-II and proposed a hybrid multi-objective algorithm based on cultural and genetic algorithms. Several experiments are conducted for this problem and the results are compared with each other with the respect to some assessment measures.

The remainder of the paper is structured as follows. In Section 2, the problem description is provided. Section 3 explains the details of the proposed algorithm. Comparisons and discussion on the experimental results are presented in Section 4. The study is finally ended by conclusions and future research in Section 5.

2 PROBLEM DEFINITION

The study concentrates on social and economic issues related to commercial hazardous wastes collection, treatment of wastes and disposal of wastes residues. There are predefined number of customers and each customer generates several kinds of waste. Customers require collection of wastes, treatment and disposal of these wastes. There exist number of treatment facilities and disposal centers with different characteristics. Treatment facilities aim to reduce hazard of wastes collected from customers' locations and decrease the destructive effects of hazardous wastes. Each treatment facility has its own technology (i.e., each treatment facility can treat only specific kinds of wastes those are compatible with its implemented technology). In addition to these nodes, there exits several depots node and various disposal centers. Trucks start their routes from depots to serve customers and each vehicle must return to the depot, which departs from it. The capacity of each depot is limited and only certain number of trucks can leave the related depot to serve customers. Trucks are considered multi-compartment meaning that each vehicle has specific space for each type of waste; therefore, the capacity of trucks for each type of waste is different. Also the traveling distance must be limited for each truck. Trucks start their routes from depots and move to customers' locations and collect the customers' wastes with regard to capacity and traveling distance limitations. After these actions, each truck must move to treatment facility that has compatible technology with the loads of truck. In other words, trucks must try to find the treatment facilities that can treat all hazardous wastes shipped with them and move to one of these appropriate treatment facilities. Of course, we must regard to capacity limitation of each treatment facility. And the total amount of all types of waste are treated in each treatment, in which facility must not exceed from capacity of treatment facility. After unloading of wastes in location of a treatment facility, trucks must return to its own depot. Thus, the first stage of the problem (i.e., routing phase) is ended at this point. Treatment facilities eradicate the potential risks of hazardous wastes and each treatment facility according to applied technology can decrease the ratio of wastes' volume. For the reason of eradication of potential risk and reduction of wastes' volume, residue wastes can be transported from treatment facilities' location to the disposal centers by means of general vehicles. Each disposal center has specific capacity limitations. The second phase is about transportation of waste residues from treatment facilities to disposal centers.

In this research, we consider two objective functions minimizing the economic cost and societal risk. The economic cost includes the routes cost, treatment cost, transportation cost of residue wastes and disposal cost of these residue wastes. The societal risk encompasses the reduction of populations suffering from transportation of hazardous wastes. Trucks pass from towns in their routes to servicing customers and each town has certain population. Hence, we intend to choose the routes that pass from thin towns. Figure 1 depicts an example of a solution to the problem.

The problem can be seen as two distribution stages linked with each other. The first stage also called a vehicle routing problem, which corresponds to the collection of hazardous wastes from customers' locations. The second stage is transportation stage, which corresponds to the transportation of waste residues from treatment facilities to the disposal centers.

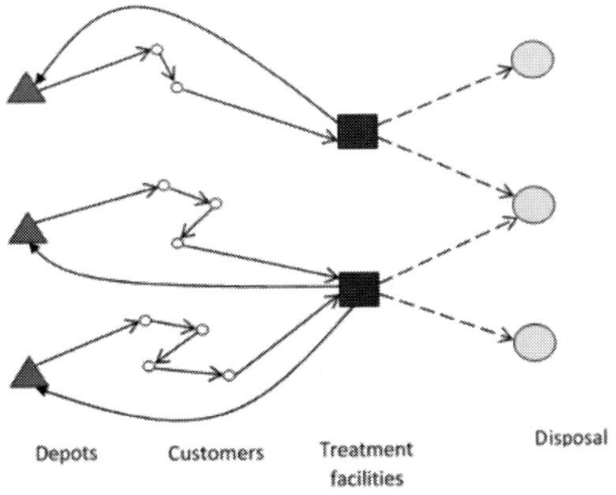

Figure 1: Schematic Figure of the Problem

3 METHODOLOGY

We utilize two solution approaches to find approximated Pareto solutions, in which the first one is based on multi-objective hybrid cultural and genetic algorithm and another one is based on the non-dominated sorting genetic algorithm (NSGA-II) [14]. Additionally, we present a straight-forward representation to reduce the complexity and decrease of computational time. A new representation, which will be explained in the next subsection, considers a feasibility of generated solutions and some constraints caused the infeasibility will be managed by introducing penalty function as $f3$. It is obvious that we must try to minimize the penalty objective function.

3.1 Solution Representation

This representation is based on order of problem's component. We specify four major components for the problem and the solutions can be extractable by determining the situation of these components. These main components of the problem include customers order, depots order, treatment facilities order and disposal centers order. Therefore, the representation proposed for this problem consists of order of customers, order of depots, order of treatment facilities, and order of disposal centers. The methodologies, in which we specify the order of the components of the problem, are partly different for the solution approaches investigated in this paper and this difference is related to distinction in nature of these two solution approaches. The nature of NSGA-II is discrete and the nature of the proposed algorithm is continuous. In NSGA-II, we produce a permutation for the integer numbers in the range (1-n) and this permutation specifies the priorities of customers. Order of the depots is a permutation for integer numbers in range (1-d), order of the treatment facilities is a permutation for integer numbers in range (1-t) and the order of disposal centers is a permutation for integer numbers in range (1-f). In proposed algorithm the manner is partly different with NSGA-II. In this method, we generate random real numbers in range (0-1) for all components of the problem (i.e. customers, depots, treatment facilities and disposal centers). For example in customers ordering, each random real number corresponds to one customer. We can specify the order of customers by sorting of corresponding real numbers. Table 1 demonstrates a simple example for this type of ordering based on random real numbers between 0 and 1.

Table 1: Example of Customers Ordering in the Proposed Algorithm

Customers	1	2	3	4	5	6
Random real numbers	0.123	0.222	0.809	0.437	0.011	0.943
Numbers rank	5	4	2	3	6	1
Resulted order of customers	6	3	4	2	1	5

3.2 Proposed Hybrid Algorithm

A cultural algorithm (CA) introduced by Reynolds [15] is an augmentation of the classical genetic algorithm (GA). In a single objective problems, it considers interactions between individuals (solutions); however, in the GA, these interactions are neglected. Individuals communicate with each other to generate a belief space (i.e., culture) and the belief space saves knowledge about the problem within itself and applies this knowledge to improve generated solutions. Experiences of individuals which are selected from the population space with respect to some criteria is used to produce problem solving knowledge influencing on the belief space. The belief space stores and manipulates the knowledge and information to impact on the evolution and improvement of the population space. The steps of the proposed algorithm are summarized as follows:

Step 1. The initial population of solutions is produced by means of solution structures described in the previous subsection.

Step 2. The solutions are evaluated and a rank is assigned to each solution regarding to non-domination concept. Level 1 is the best level, level 2 is the next best level, and so forth. For the solutions have a same rank, crowding distance is calculated. By applying this, the selection process goes toward a uniformly spread Pareto front at different stages of the algorithm.

Step 3. Belief space is adjusted by means of specific operators. One operator updates the situational knowledge and another one updates the normative knowledge.

Step 4. Population is affected by cultural space. Knowledge about normative solutions and situational solutions influences on population components. Population is updated and new population is available to continue algorithm.

Step 5. GA's operators including crossover and mutation operators are utilized to create the next generation called the offspring population from the parents.

Step 6. The new generated population and the old population are merged with each other.

Step 7. The merged populations are sorted according to the rank and crowding distance criteria.

Step 8. The solutions with the ranked 1 are reported as approximated Pareto solution.

Step 9. The stop criterion in this algorithm is considered as maximum number of iterations. If the criterion is satisfied the algorithm is stopped; otherwise, Steps 3 to 9 are repeated.

4 EXPERIMENTAL RESULTS

The performances of the proposed algorithm and NSGA-II, which is one of the most well-known meta-heuristics for multi-objective optimization problems are compared with each other and the associated results are analyzed. The algorithms are coded in MATLAB R2013a and run on Intel Core i5 2.27 GHz personal computer with 4 GB RAM.

4.1 Test Problems

To the best of our knowledge, there are no benchmark instances in the literature for this problem, so the instances are generated by a defined instance generator in Matlab software. The instances have six main components. The generated instances are named ($d \times t \times f \times c \times w \times q$). These characters denote a number of depots, number of treatment facilities, number of disposal centers, number of customers, number of waste types and number of available technologies respectively. Instances can be divided into two sections including small instances, and medium and large-scale instances. This results in a set of 20 instance types for two sizes of problems. Each instance runs five times and the results are collected. Instances are available for download from:

https://www.dropbox.com/sh/770opj0ut22jyho/AACsvMzKCHJeCSD2ZaSuYU8ka?dl=0

4.2 Comparison Metrics

Five runs for each instance are performed to gain results. In order to compare the efficiency of the proposed algorithms, we use three performance metrics including a number of Pareto solutions (NPS), computational time and spacing metric (SM). The number of Pareto solutions measures the algorithms ability to find an efficient point. Tables 2 and 3 are summarized the results of NPS and computational time for small and medium/large-sized problems, respectively.

In small-sized problems, computational times for both algorithms is in the same range. In large-scale problems, the computational time for NSGA-II is almost better than the proposed algorithm; however, this difference is not notable. In most examples, the proposed algorithm is better than NSGA-II in producing Pareto numbers and the average values in small and large-scale problems.

Spacing metric provides us details about the uniformity of the distribution of the solutions obtained by way of each algorithm. This metric is computed by:

$$SM = \sqrt{\frac{1}{N-1} \times \sum_{i=1}^{n} (d_i - \bar{d})^2} \quad (1)$$

where d_i is the Euclidean distance between solution i and the nearest solution is belongs to Pareto sets of solutions. $\bar{d}$ is the average value of all d_i. In some instances, the proposed algorithm operates better and in other instances NSGA-II operates better. This measure does not show a superiority of none of them; however, the average value for NSGA-II is better than the proposed algorithm. The results are shown in Tables 4 and 5.

Table 2: Quantity of Pareto Solutions and Computational Time for Small-sized Problems

Problem characters	Quantity of Pareto solution		Computational time(s)	
	NSGA-II	MOHCG	NSGA-II	MOHCG
4-4-4-5-2-2	7.2	10.8	24.2	25.8
5-5-4-6-4-3	10.2	14	25	26
5-6-5-20-4-3	9.4	10	24.8	25.4
6-6-6-8-3-4	11.2	8.6	22	24
7-8-8-10-4-3	12.4	12.2	24	24.8
8-7-9-12-2-3	9.8	14.2	24.6	23.4
10-10-10-10-4-4	8	10	25	26.2
10-10-10-5-4-3	8.8	12.4	31	29.8
11-11-11-15-2-2	11.4	11.6	26	27
12-12-10-15-3-4	8.6	10.2	26	27.4
Average	9.7	11.4	25.26	25.98

Table 3: Quantity of Pareto Solutions and Computational Time for Medium/Large-sized Problems

Problem characters	Quantity of Pareto solution		Computational time (s)	
	NSGA-II	MOHCG	NSGA-II	MOHCG
15-15-15-25-4-4	7	5	26.2	30
15-15-15-30-5-4	14.8	13.2	27	30.4
20-18-20-30-5-4	11.2	15.4	29.4	32.4
20-20-18-25-5-5	16	15.2	27.4	29.2
20-20-20-40-4-4	7.2	8.6	31	33.4
20-20-20-40-5-5	5.8	9	30.4	33.6
20-25-25-50-3-3	12	14	33.6	35
20-25-25-100-3-3	12.4	11.4	45.8	48.8
30-25-25-100-2-2	14.8	11.8	49	51
30-30-30-150-3-5	13	13.2	58	64.6
Average	11.42	11.68	35.78	38.84

Table 4: SM Measure for Small-sized Instances

Problem characters	NSGA-II	MOHCG
4-4-4-5-2-2	2.0596e4	1.1687e4
5-5-4-6-4-3	1.5370e4	1.2737e4
5-6-5-20-4-3	2.5077e4	2.3223e4
6-6-6-8-3-4	1.2326e4	1.5078e4
7-8-8-10-4-3	1.8156e5	1.5560e5
8-7-9-12-2-3	0.7814e4	2.3286e4
10-10-10-10-4-4	3.9840e4	5.5144e4
10-10-10-50-4-3	9.0894e4	6.9924e4
11-11-11-15-2-2	1.8823e4	1.4247e4
12-12-10-15-3-4	1.8119e3	2.4534e4
Average	4.3695e4	6.1064e4

Table 5: SM Measure for Medium/Large-sized Instances

Problem characters	NSGA-II	MOHCG
15-15-15-25-4-4	1.445e5	3.1505e4
15-15-15-30-5-4	2.0391e4	3.3161e4
20-18-20-30-5-4	1.8133e4	2.8232e5
20-20-18-25-5-5	1.5975e4	7.1180e4
20-20-20-40-4-4	1.2129e6	2.9243e6
20-20-20-40-5-5	2.1604e5	1.2046e5
20-25-25-50-3-3	3.1683e4	5.135e5
20-25-25-100-3-3	4.4345e4	3.5291e4
30-25-25-100-2-2	5.6636e6	2.4809e7
30-30-30-150-3-5	1.5787e8	1.4095e8
Average	16.5339e6	16.9819e6

5 CONCLUSION

In this work, a bi-objective problem for a waste collection transportation vehicle routing problem was addressed. An efficient order based solution representation for this problem was proposed to decrease the computational time and complexity as much as possible. All constraints were addressed in a construction algorithm. For traveling distance constraints that were not addressed in the construction algorithm, a penalty function was defined and one solution penalized if traveling distance limitation was exceed. In this representation, four main components of the problem were sorted in the chromosomes formation, and operators generating new solutions or updating culture were applied on these chromosomes. When all solutions were considered as feasible solutions applying of some operators, such as crossover and mutation, would be easy and executable. This new representation was implemented in NSGA-II and another meta-heuristic algorithm presented in this paper. In order to show the effectiveness of the proposed algorithm, several test instances were conducted and the results were compared with each other with respect to three comparison metrics. This experiments showed the effectiveness of the proposed algorithm compared with NSGA-II, which is one of most frequent and effective multi-objective algorithms. For the future research direction, we suggest to combine the available constraints with other real world constraints. For the reason of importance of time in waste collection problems, time is a critical factor and the time windows constraint can be added to the considered problem.

6 ACKNOWLEGEMENTS

This work has been partially supported by the National Elites Foundation and the Science & Technology Vice President of Iran as well as the University of Tehran.

7 REFERENCES

[1] **Markov, I., Maknoon, Y., de Lapparent, M., Cordeau, J.F., Varone, S., Bierlaire, M.** 2015. November. Vehicle routing and demand forecasting in recyclable waste collection, In: *CIRRELT Seminar* (No. EPFL-TALK-215058).

[2] **Sbihi A., Eglese R.W.** 2007. Combinatorial optimization and Green Logistics, *4OR*, 5, 99-116.

[3] **Lin, C., Choy, K.L., Ho, G.T., Chung, S.H., Lam, H.Y.** 2014. Survey of green vehicle routing problem: Past and future trends. *Expert Systems with Appl.*, 41(4), pp. 1118-1138.

[4] **Dekker, R., Fleischmann, M., Inderfurth, K., Van Wassenhov, L.N.** 2004. *Reverse Logistics: quantitative models for closed loop supply chain*. Berlin: Springer.

[5] **Wy, J., Kim, B.I., Kim, S.**, 2013. The rollon-rolloff waste collection vehicle routing problem with time windows, *European Journal of Operational Research*, 224(3), pp 466-476.

[6] **Bonomo, F., Duaran, G., Larumbe, F., Marenco, J.** 2012. A method for optimizing waste collection using mathematical programming: A Buenos Aires case study, *Waste Management & Research*, 30(3), pp 311-324.

[7] **Düzgün, H.Ş., Uşkay, S.O., Aksoy, A.** 2015. Parallel hybrid genetic algorithm and gis-based optimization for municipal solid waste collection routing, *Journal of Computing in Civil Engineering*, *30*(3), pp 04015037.

[8] **Kim B.I., Kim S., Sahoo S.**, 2006. Waste collection vehicle routing problem with time windows, *Computers & Operations Research*, 33, pp 3624-3642.

[9] **Alumur, S., Kara, B. Y.**, 2007. A new model for the hazardous waste location-routing problem, *Computers & Operations Research*, 34(5), pp 1406-1423.

[10] **Verter, V., Erkut, E.**, 1996. *Hazardous materials logistics: An annotated bibliography* (pp. 221-267). Springer Netherlands.

[11] **Caballero R., Gonzalez M., Guerrero F.M., Molina J., Paralera C.** 2007. Solving a multi-objective location routing problem with a metaheuristic based on tabu search. Application to real case in Andalusia, *European Journal of Operational Research*, 177, pp 1751-1763.

[12] **Martínez-Salazar, I. A., Molina, J., Ángel-Bello, F., Gómez, T., Caballero, R.** 2014. Solving a bi-objective transportation location routing problem by metaheuristic algorithms, *European Journal of Operational Research*, 234(1), pp 25-36.

[13] **Rabbani, M., Farrokhi-asl, H., Rafiei, H.**, 2016. A hybrid genetic algorithm for waste collection problem by heterogeneous fleet of vehicles with multiple separated compartments, *Journal of Intelligent & Fuzzy Systems*, *30*(3), pp1817-1830.

[14] **Deb, K., Agrawal, S., Pratab, A., Meyunvian, T.**, 2002. A fast elitist non-dominated sorting algorithm for multi-objective optimization NSGA-II, *IEEE Transactions on Evolution Computation*, 6(2), pp 2182-2197.

[15] **Reynolds, R.G.** 1994. An introduction to cultural algorithms. In: *Proceedings of the third annual conference on evolutionary programming*, Singapore, pp. 131-139.

DEVELOPMENT AND EXPERIMENT OF ENGINE INJECTOR DEPOSITS FORMATION SIMULATING APPARATUS

Li Wang[1], Wanzhong Li[2*], Jianwei Zhou[3], Yanhong Du[4]
[1]School of Engineering Technology
Tianjin Agricultural University, Tianjin, China
wanglimay@163.com

[2]School of Mechanical Engineering
Tianjin University, Tianjin, China
lwzhong@tju.edu.cn

[3]School of Mechanical Engineering
Tianjin University, Tianjin, China
zhoujianwei@tju.edu.cn

[4]School of Engineering Technology
Tianjin Agricultural University, Tianjin, China
yanhong_du@163.com

ABSTRACT

In order to study on the mechanism of injector deposits formation in a direct injection gasoline engine, a device to simulate the injector deposits formation environment and its control system were developed. The apparatus is mainly composed of a MAX6675 and thermocouple based injector head heating control module, a driving current based injection timing and dosing control module and a spray tank. The contrast experiments on injector deposits formation were carried out on an engine dyno and the injector deposits formation simulating apparatus, and the FTIR analysis on the composition of the injector deposits derived from the two methods shows that the injector deposits formation simulating apparatus in the lab is of the same property compare with the actual injector deposit in the engine, this engine injector deposits formation simulating system can well simulate the injector working environment inside the engine.

Keywords: Direct injection gasoline engine, Injector deposit, Simulating apparatus, Fourier transform infrared spectroscopy

1 INTRODUCTION

Since gasoline direct injection technology can significantly improve fuel economy and meet future more stringent emission standards, it has become an important development direction of automotive gasoline engine. Gasoline Direct Injection (GDI) of the excellent performance is mainly derived from its direct precise control to the internal air-fuel ratio distribution of the engine cylinder, which requires a set of precisely-controlled fuel supply system for the direct injection

* Corresponding Author

gasoline engine. Injector, as a terminal and performer of the fuel supply system, is one of the most critical parts of direct injection gasoline engine, and the injector performance has a direct impact on gasoline atomization quality, thus affecting the engine's performance and emissions. However, the changes in the structural parameter of injector due to carbon deposit has always influences the spray shape and atomization of the fuel, thus affecting the performance of GDI gasoline engine[1-3] .

The injector deposit phenomenon in GDI engine can be formed in a very short time after the engine is running, and it is not a linear relationship with the running time[4,5]. In order to study the formation mechanism and the process of injector deposit, it is required to make quantitative research on the influence to *injector* deposit by various factors in the course of engine operation. Due to injectors mounted on the engine, it is more difficult to precisely control the various factors affecting the deposit, resulted in the extreme observation difficulty in the formation of deposit in the engine. This paper develops a set of fuel injector deposit formation simulating apparatus and control system, and the real test experiments on fuel injector deposit has proven that it is feasible to study injector deposit in the laboratory.

2 SYSTEM ARCHITECTURE

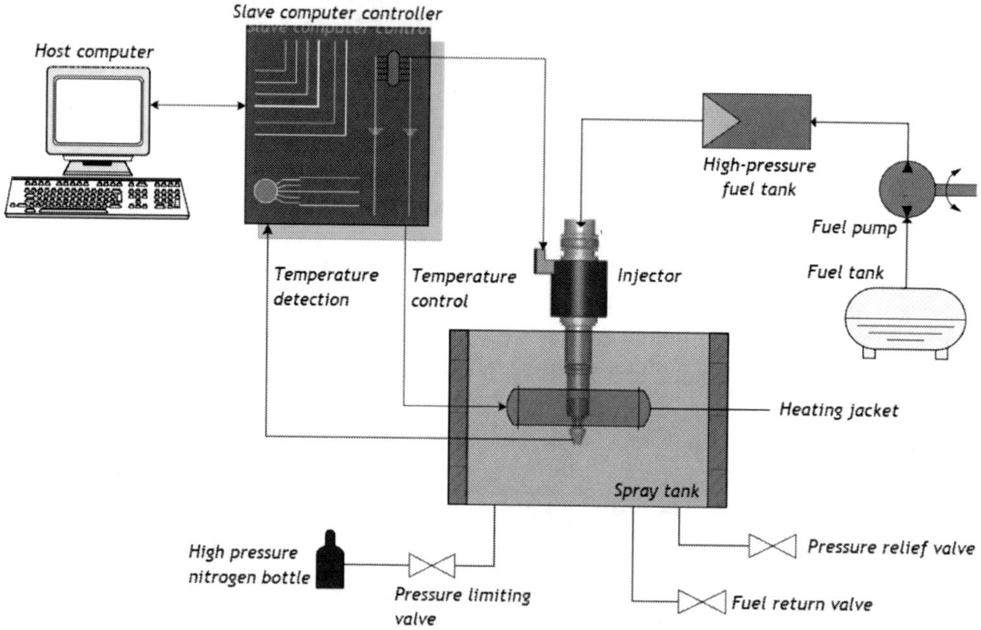

Figure 1: Overview of Injector Deposits Formation Simulating Apparatus and Control System

The injector deposit formation simulating experiment apparatus and control system is composed of host computer, injector control module, injector drive module, injector head temperature control module, injector, spray tank, high-pressure fuel tank and high pressure nitrogen bottle, and the structure is shown in Figure 1. Wherein the injector control module, injector drive module, injector head temperature control module are integrated in the slave computer controller, the injector is placed in a pressure-controllable spray tank, the host computer via the serial communication send the start time, peak current holding time and time intervals of adjacent injections, etc. control parameters to the injector controller, which based on the fuel injection control signal sent by the host computer generates injector drive control signal (PWM signal). The injector drive circuit drives the injector open and sprays the high-pressure fuel into the spray tank. The head of the injector is mounted with the heating jacket for 20-250 ℃ heating, on which a heating wire is equipped as heating element, using K-type thermocouple for temperature acquisition, and the external heating jacket with asbestos material for insulation, the host computer via the serial port communication sends the set value of injector head heating temperature to the controller for the closed-loop control of injector head temperature. The high-pressure fuel driven by the high pressure nitrogen gas is of pressure range 0-15MPa; N2 and CO2 gas mixture in the spray tank, back pressure adjustable, the maximum pressure 1.0MPa.

2.1 Hardware of Control System

The slave computer controller of the control system applied MSP430F2112 microcontroller as the main control chip[6], and its function modules include injector head heating control, fuel injection control and communication modules, etc. Wherein the communication module applied the RS-232 serial port communication based on MAX232 chip[7], to achieve communication between host computer and slave controller, and the control parameters can be modified in real time.

2.1.1 Injector Head Heating Control

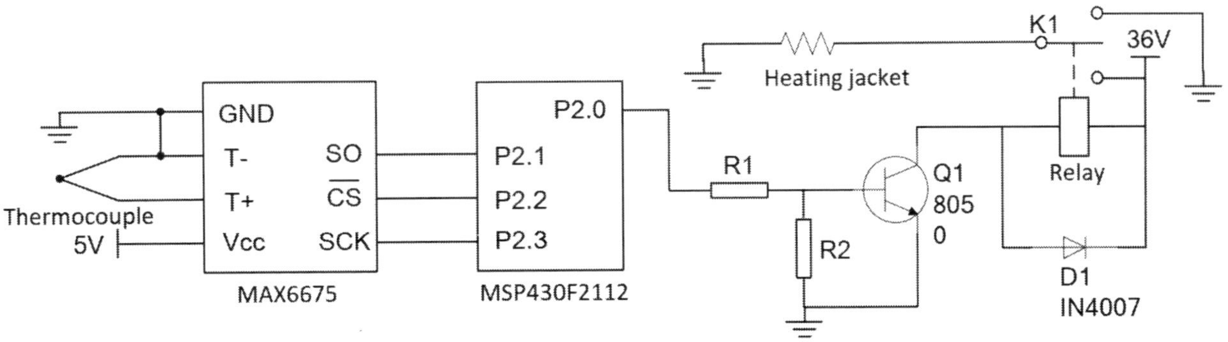

Figure 2: Interface between Injector Head Heating Control Module and Microcontroller

The interface between injector head heating control module and *microcontroller* is shown in Figure 2, the temperature control of heating jacket is in the range of 20-250 ℃, the sensor applied K-type thermocouple, the thermocouple measuring circuit used MAX6675. The MAX6675 produced by MAXIM, America is a K-type thermocouple measuring conversion circuit with cold junction temperature compensation, linearity correction, thermocouple line break detection, etc. functions, of output 12-bit binary digital[8]. MAX6675 will convert the thermocouple output signal directly into a digital signal to the microcontroller MSP430F2112, and the cold junction compensation can be solved by the MAX6675 itself. Microcontroller will communicate with the host computer via the serial port, transmit the measurement data, receive the temperature given value, and analyze the measured values of the temperature and the given value, send the control signal, through the triode amplification to control relay in heating operation, to make the heating jacket temperature stabilized around the given point.

2.1.2 Fuel Injection Control

Fuel injection control module applied the dual power bilateral drive single segment held PWM (Pulse Width Modulation) wave driving method[9], as shown in Figure 3, in which: O1 , O2 are optocoupler 4N25 for isolation, Q3, Q4 are field effect transistor (FET) IRFH58N20 as high-speed switch in the circuit, M1, M2 Schottky diode MUR1660CT, isolators in the circuit to prevent the excessive instantaneous induced electromotive force of injector electromagnetic coil when injector closes damaging the driving circuit. IR2101 is a high and low side driver chip, for receiving the driver instruction issued by the controller and controlling the high speed switching of the FET[10].

When the driving circuit powers on, +12V power supply through the Schottky diode MUR1660CT connects to the driving circuit output terminal, and at this moment, Q3 is not switched on. At the opening phase of injector, the high *voltage* output signal from ECU is received by optocoupler O1 and O2 simultaneously, and O1 switches on the Q4 through IR2101, O2 switches on Q3. +65V high-voltage from boost circuit opens the injector by switched on the Q4, while +12V voltage is temporarily not involved in the driving. At the holding phase, ECU output low voltage to O1 making Q4 off and output PWM wave signal to O2, which control Q3 on/off according to the PWM waveform. The injector solenoid coil current during the off-period of Q3 through resistor R40 and diode D5 to form discharge circuit, will control holding phase current at a certain level. In the closing phase, ECU will at the same issue a low *voltage* signal to O1 and O2, which switches FET Q3 and Q4 off, and then the injector will close quickly.

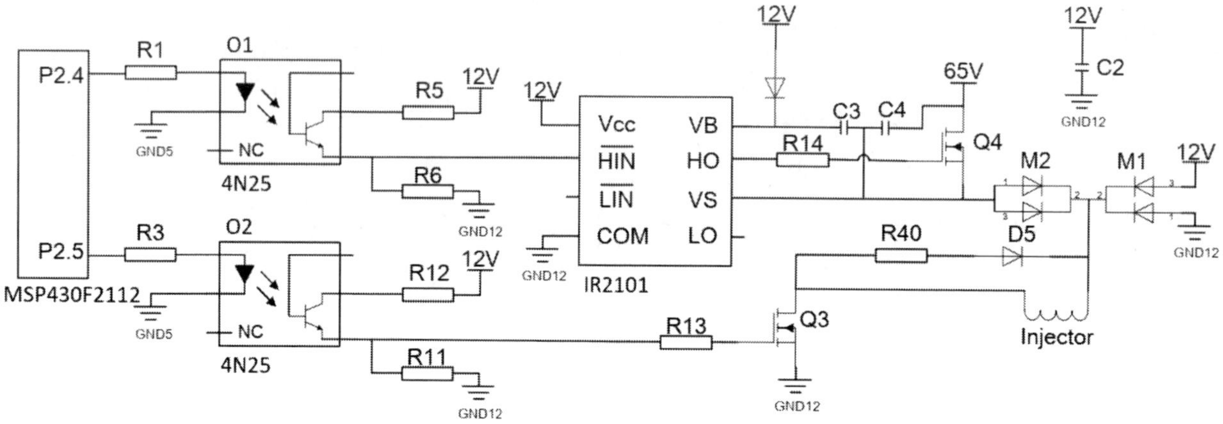

Figure 3: Diagram of Injector Driver Circuit

2.2 Software of Control System

Control System Software Design must realize the injector head heating jacket temperature acquisition and processing, heating jacket temperature control, injector opening time and keeping duration control, the injection frequency *control,* the experiment duration control, and the communications between the controller, host computer and driver chips, etc. functions. The designed injector software control concept includes a controller initialization, controller communication, heating jacket control parameter update and heating control, fuel injection control parameter update and fuel injection control, etc. function modules, with the *slave* computer running flowchart shown in Figure 4.

The slave controller working process is: Upon receiving the temperature setting and the injection frequency and duration of experiment setting, firstly to control the heating jacket's temperature, upload the current temperature and injection data after the temperature meets the goal, and then to set the injection duration and the interval between two injections according to the injection frequency received, and to set the number of injection times in accordance with the duration of the experiment. Both the high voltage signal and the PWM signal shall be set simultaneously upon each injection, to enable the quick open of the injector. After the start-up phase completed, reset the high voltage signal and output PWM signal to make the injector running in the keeping phase, and then reset PWM signal after the completion of the spray. After each injection completed, check the current temperature and upload, then the next injection will be carried on after the interval. Finally, the slave *controller working sequence* finished and the data will be uploaded after the injecting times meets the goal.

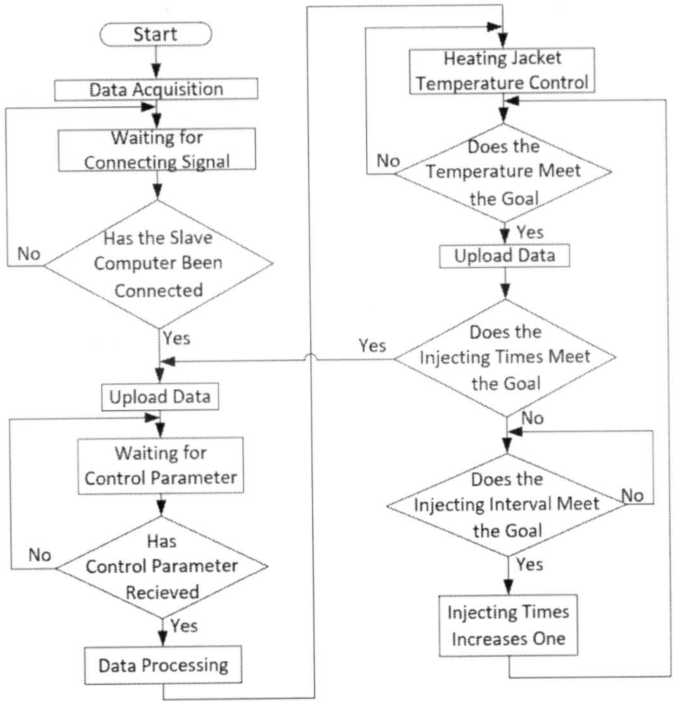

Figure 4: Operation Flowchart of Lower Computer Controller Software

3 EXPERIMENTAL RESULTS

In order to verify whether the simulation of deposit in the laboratory is of the same physical and chemical characteristics as the deposit in the injector of actual engine, the engine experiment to form injector deposit is carried out. To carry out the test of injector deposit on a GDI engine, a 2-liter four-cylinder TDI with EGR gasoline engine bench was set up in the laboratory. The engine was equipped with dual VVT system and a developing ECU which can realize the control of injection pressure, injection pulse width, ignition advance angle and EGR rate, pressure ratio, etc.. Applied the commercial 97 # gasoline, the engine continuously ran at low speed and high load for 8 hours, and there is obvious carbon deposits in the injectors.

Comparative experiment is performed on the *injector deposit formation simulating experiment apparatus*, using the same injector as in the above experiment. Since the carbon deposition will be more serious when the temperature reaches T90, that is, around 180℃, the test apparatus temperature control system will control the injector temperature at 180 ± 5℃. Furthermore, due to the carbon deposit is susceptible to be eroded, the frequency of injection in the test is set at 1Hz, the injection pressure is set at 1Mpa, fuel injection pulse width is set at 1ms. The experiments lasts 120 hours, during which every 30h take off the injector, shoot injector choking of each nozzle hole using scanning electron microscopy. At the end of the injector deposit simulation test, the injector head deposit sample was collected for FTIR spectroscopy analysis.

Figure 5: (a) Photo of injector head deposit in the time course of 60h, (b) Photo of injector head deposit in the time course of 90h and (c) Photo of injector head deposit in the time course of 120h

Figure 5 shows the injector deposit accumulation at the nozzle head during heating under different time histories. It's very obvious that in the laboratory simulation apparatus, by heating the injector could produce carbon deposit at a clean injector head, and the amount of carbon deposit increased along with the addition of heating time course. Meanwhile observed that, in the course of time, the carbon deposit expanded from the central region of *injector* head to the surrounding, and finally covered the entire area of the injector head. The reason may be a part of the fuel left over after its leaving from injection hole, and after oxidation at high temperature, as the increase of time, the carbon deposits were formed, another part of the subsequent injected fuel absorbed by the carbon deposit formed previously and gradually oxidized into carbon deposit, and the carbon deposition region also will gradually increase.

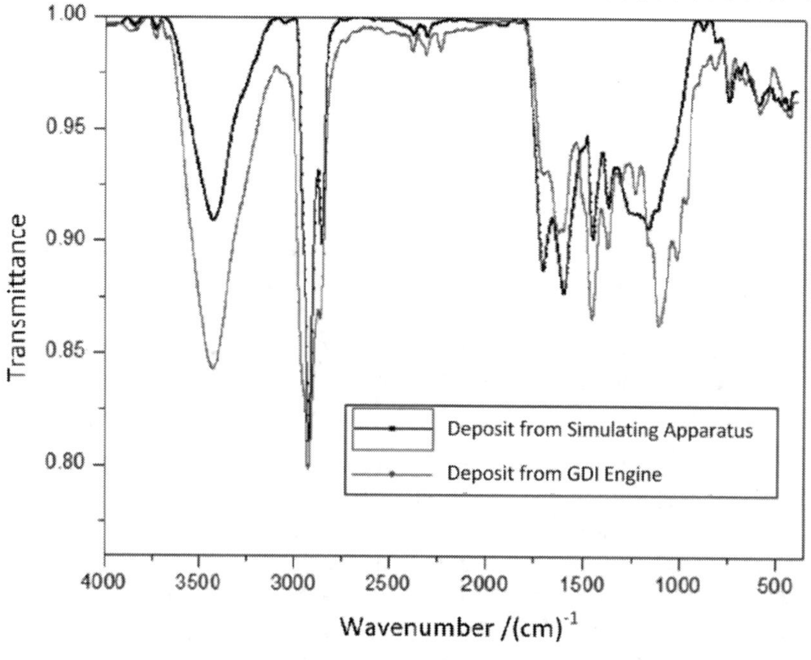

Figure 6: FT-IR Spectra Comparison between Simulating Apparatus Formed and GDI Engine Produced Deposit

The FT-IR spectra of the injector deposits were recorded with a VECTOR 22 infrared spectrometer. Measurements were done for wavenumbers between 300 cm^{-1} and 4000 cm^{-1} at a resolution of 1cm^{-1}. Sample preparation using the compression method, with comparative analysis of the results as shown in Figure 6, in which the abscissa is for the wave number, and the ordinate is for the transmittance.

When the gasoline on the surface of injector at high temperatures formed carbon deposit, it experienced the volatile of the light component, part of the long chain structure constituent splitting into olefins, the concentration of heavy constituent, high temperature oxidation of the heavy constituent producing a large number of hydroxy and oxygen compounds, and the growth and accumulation of carbon deposits. By comparison and analysis, the simulated carbon deposit had the same trend of infrared spectrum as the actual engine injector carbon deposit, and there are peak of wave that represent the same functional group at the same wavelength, that is, hydroxyl functional groups in the vicinity of 3500 cm^{-1}, methylene functional groups in the range of 2800-3000cm^{-1}, methyl functional group in the vicinity of 1600 cm^{-1}, olefin functional group in the vicinity of 1680-1620 cm^{-1}, sulfur oxides and other functional groups in the range of 1100-1250 cm^{-1}, etc., but there are differences on the content of each functional group, which may be caused by the accumulation of carbon deposition in the two different environments. However, the simulating of carbon deposition in the laboratory can still be used in the study of carbon deposit formation mechanism instead of the actual carbon deposit formation, which will reduce the test time and costs in the study of carbon deposition, meanwhile it is convenient to study in the lab scale the influence to injector deposit formation by these single influence factors such as fuel injection temperature, injection pressure, injection frequency, constituent contents of gasoline, etc..

4 CONCLUSIONS

This paper designed a simulating apparatus on the formation of GDI engine injector deposit, carried out the test experiment on injector carbon deposit in a typical operating conditions, and came to the following conclusions:

1) The measurement and control structure of the injector head heating control module based on MAX6675 is simple, easy to control, and able to achieve the control requirement on injector head heating temperature control in the experiment.

2) The fuel injection control module based on the drive current control, can achieve precise control of the injection timing and injection quantity, the fuel injector response rapidly and running reliably, the use of single-chip microcomputer control on the number of injections, can make the injector carbon deposit formation simulating experiment easier and save the test workload.

3) The injector carbon deposit component analysis and comparison between the actual engine aging and applied simulating apparatus shows that the formed carbon deposits are of similar component. The engine injector carbon deposit formation simulating system can well simulate the working environment of the injector inside the engine.

5 ACKNOWLEDGEMENT

This work was supported by the National Natural Science Foundation of China (No. 51306132).

6 REFERENCES

[1] **Smith, S. 2013.** Measurement and Control of Fuel Injector Deposits in Direct Injection Gasoline Vehicles, *SAE Technical Paper 2013-01-2616.*

[2] **Wang, C. 2014.** Impact of fuel and injection system on particle emissions from a gdi engine, Applied Energy, 132(11), pp 178-191.

[3] **Pei, Y.Q. 2015.** Effect of GDI Engine Injector Coking on Spray, Journal of Tianjin University , 48(4), pp 334-340.

[4] **William, I. 2012.** A DOE Approach to Engine Deposit Testing used to Optimize the Design of a Gasoline Direct Injector Seat and Orifice, SAE Technical Paper 2012-01-1642.

[5] **Von Bacho, P. 2009.** Engine Test for Accelerated Fuel Deposit Formation on Injectors Used in Gasoline Direct Injection Engines, SAE Technical Paper 2009-01-1495.

[6] **Hong, L. 2010.** Detailed Annotation on Principle and Application of MSP430 Single Chip Microcomputer, Beihang University Press.

[7] **Liang, T.T. 2015.** Computer and CPU Communication Based on MAX232, New Technology and New Process, 2015(8), pp 34-36

[8] **Yu, Z. 2002.** Principle and Application of MAX6675, International Electronic Elements, 2002(12), pp 41-43.

[9] **Hu, C.M. 2014.** Development and Optimization of Injector Driving Module for GDI Engine, Chinese Internal Combustion Engine Engineering, 35(3), pp 83-88.

[10] **Jiang, L. 2010.** Design and Realization of Maximum Power Tracking Inverter based on IR2101, Electronic Design and Engineering, 2010 (11), pp 27-29.

FINITE ELEMENT ANALYSIS ON THE RING DIE STRUCTUREMOLD HOLE OF RING DIE PELLET MILL

Yang Lei[1*], Li Yancong, Song Xin, Wang li
[1]College of Engineering and Technology
Tianjin Agricultural University, Tianjin, China
yanglei16@tjau.edu.cn

ABSTRACT

The ring die structure mold hole of the ring die pellet mill is one of the main parameters affecting the quality of the ring die. By analyzing the working conditions of ring die and the press roll, carrying out the stress analysis of the material, and using Pro/E and ANSYS software to complete the modeling and finite element static analysis on the three types of ring die structure mold holes whose chamfer are 30°、45°、60°, obtained the stress and deformation distribution of the ring die mold hole in the axial direction, found out the influence relationship between the stress and deformation of ring die chamfer and ring die hole in the axial path, and determined the mold hole structure with the chamfer of 30° is preferably, which provides a basis for structural optimization of ring die.

Keywords: Finite Element, Ring Die, Stress, Mold Hole, Pellet Mill

1 INTRODUCTION

Ring die pellet mill is one of the main particle type feed processing equipment, plays an important role in feed production, and is widely used in breeding industry. Ring die as the core component of the pellet mill, its performance directly affects the quality of processed feed, and the yield and service life of pellet mill [1], thereby affects the production cost of feed. In addition, the ring die is the main wear parts of the pellet mill, with the wear mainly in the die hole and the inner surface. Wherein the shape and size of the die hole is directly related to the rate of feed out of the die, but also have a significant impact on the extent of the wear on the structure mold hole and its surface[2]. Therefore, in order to improve the production efficiency and service life of ring die pellet mill, the analysis on the structure mold hole has important and practical significance for improving the quality of the ring die.

2 STRUCTURE AND ITSANALYSIS ON RING DIE PELLET MILL

2.1 The Ring Die Structure in the Ring Die Pellet Mill

Figure 1 is the assembly of pellet mill, containing a ring die, press roll (including the roll axis), the bottom support shaft and the front panel, one end of the support shaft fixed via screw-thread onto the bed plate and the other end through the center hole of the front panel screwed up for fixing the press roll[3].

---

*Yang Lei

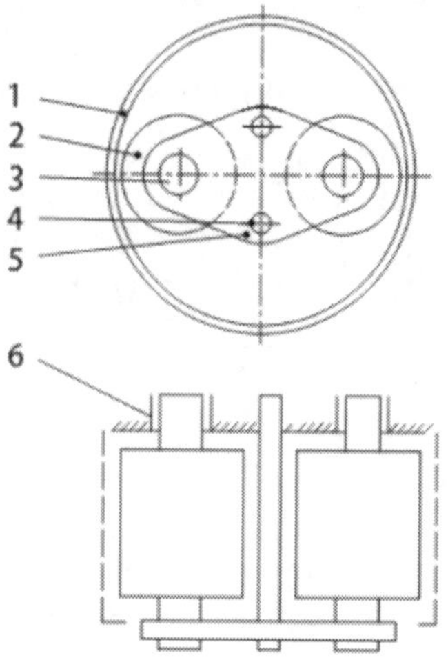

1. ring die,2.press roll,3. roll axis,4. bottom support shaft,5. front panel,6. Roll sleeve

Figure 1: Assembly of Pellet Mill

2.2 Work Conditions of Ring Die and Press Roll

Figure 2 is the working schematic of ring die and the press roll[4], in which the ring die is driven by motor at a constant rotational speed, with the material rotated into the extrusion zone where the press roll under friction will rotate along with the ring die in the same direction. The rotation of the ring die and the press roll make the material continuously enter into the extrusion zone, in which the extrusion force and the density of the material will escalate[5-6]. When the extrusion force increased up to overcome the friction of the hole wall, the material of a certain density will be extruded into the hole. With the input of material, the material in the ring hole is extruded out of the die hole to form a cylindrical shape.

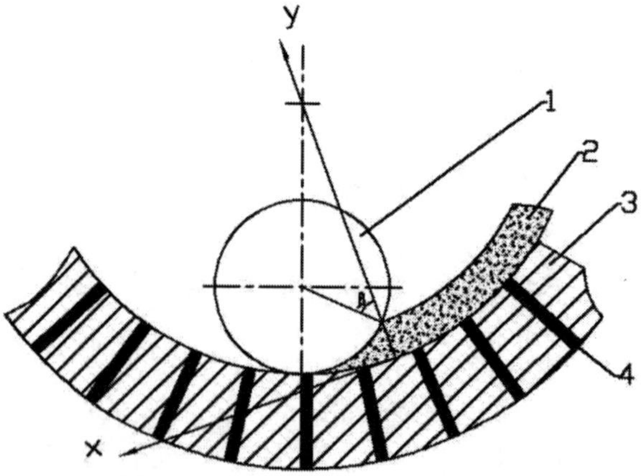

1.Press Roll,2.Feed,3.Ring Die,4. Formed Feed

Figure 2:Working Conditions of Ring Die and Press Roll

3 FEA OF RING DIE STRUCTURE MOLD HOLE

3.1 Design of Mold Hole Structure

Ring mold is a key part of the pellet mill, is the main wearing parts of the pellet mill. According to statistics, the ring mold loss costs accounted for more than 25% of the maintenance costs of the whole production plant, at the same time, has a direct impact on the quality of extruded pellets. Therefore, understanding the characteristics of the ring die and its proper selection, rational use and effective maintenance is crucial for the particle producers. In this study, the ring die material is 42CrMo, the surface hardness after quenching is HRC54-58, except the thickness of the ring die, the specification of extrusion hole shall be considered during selection, such as: inlet tapered hole, tapered hole depth, the length of tapered extruded hole, the angle and depth of outlet tapered hole, etc., they all must be applicable to the particular materials to pellet mill. According to the properties of the particles, and the final hole shape structure determined is as shown in Figure 3.

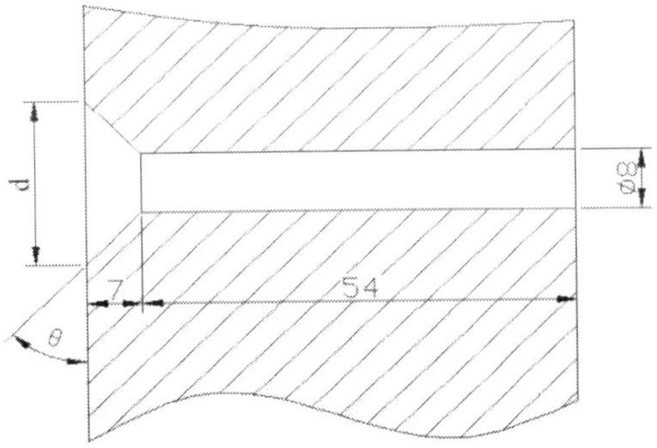

Figure 3: Structure of Mold Hole

Roughness is a contradiction with the compression ratio, where the higher compression ratio is, the more difficult in roughness processing. It is harder for ring die to yield the material, so the roughness is also an important indicator of the quality of the ring die. Therefore, the die hole roughness value is determined as 0.18.

3.2 Structure Model of Ring Die Mold Hole and Finite Element Meshing

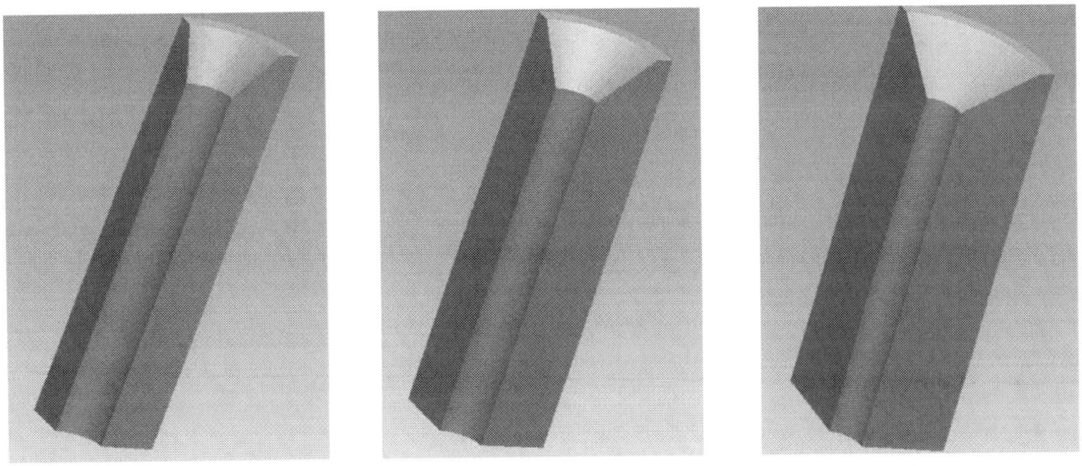

(a) model of 30° (b) model of 45° (c) model of 60°

Figure 4: Schematic of Ring Die Mold Hole

In the process of using ANSYS for finite element analysis, geometric modeling is an essential process. In terms of modeling function, ANSYS is not as powerful as Pro/E and UG, but ANSYS has powerful interface and structure computing capability. The ring die mold hole adopts Pro/E to establish three-dimensional solid model, by IGES as intermediate data conversion platform using ANSYS interface technology, realized the import of the model [7-8]. For symmetrical structure, in order to save analysis time and computer resources as far as possible, the size of the analytical model can be reduced. Since the ring hole model is axial symmetric structure, so when creating the model, establish only 1/4 of the mold hole. Figure 4 is the model for 30°、45°、60°chamfered ring structure mold hole.

After import the model into ANSYS software, define its cell attribute. Select Solid 45 as an analysis unit of ring die mold hole, with the material parameters shown in Table 1, the elastic modulus $E = 2.06e^{+11}$ and density $\rho = 7800$. And distribute the cell attribute to the solid model.

Table 1: Parameters of Materials

Name of Parts	Material	Yield Strength (MPa)	Limit of Strength (MPa)	Poisson Ratio
Mold Hole	42CrMo	930	1080	0.3

ANSYS software provides two kinds of meshing methods: free meshing and mapped meshing. Free meshing is reflected in the absence of specific guidelines, unlimited cell shape, and the generated cell is irregular[9]. While the mapped meshing is required to meet certain rules, in which the generated cell shape is more regular, fit for the body or face of regular shapes. In order to ensure the accuracy of analysis, the mapped meshing is applied. For 30°chamfered hole, after meshing generation, the number of cellsis 1104 and the number of nodes is 1890, 45° chamfered hole, after meshing generation, the number of cellsis 728 and the number of nodes is 1080, 60°chamfered hole, after meshing generation, the number of cellsis 486 and the number of nodes is 756.

3.3 Finite Element Model Load

In order to carry out the FEA on the ring die mold hole, the relevant assumptions shall be made in advance[10]:

Take $\mu_1 = \mu_2 = c$, after checking from the relevant files, $c = 0.18$.

Assume the force F at each position of the ring die mold hole is the same. Take any one of the mold hole as the research object, to carry out the force analysis, refer to Figure 5 for the analysis[11].

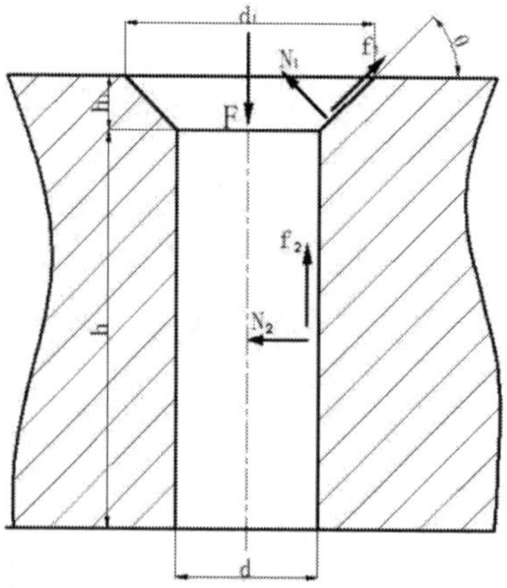

Figure 5: Force Analysis on Ring Die Mold Hole

When material at the place of chamfer, the following condition must be satisfied for material to be extruded into the mold hole.

$$F \geq f_1 \sin\theta + N_1 \cos\theta = N_1(\mu_1 \sin\theta + \cos\theta) \quad (1)$$

When the material not at the place of chamfer, the following condition must be satisfied for material to be extruded out and mold the shape.

$$F \geq f_2 = \mu_2 N_2 \quad (2)$$

In which: F ——Extrusion Force; f_1 ——Friction of Chamfer Face ; N_1 ——Normal Pressure of Chamfer Face ; f_2 ——Friction of Inner Surface of Mold Hole; N_2 ——Normal Pressure of Inner Surface of Mold Hole; μ_1 ——Friction Efficient of Chamfer Face; μ_2 ——Friction Efficient of Inner Surface of Mold Hole.

Take the critical condition between the two: $F = N_1(\mu_1 \sin\theta + \cos\theta) = \mu_2 N_2$, then

$$N_1 = \frac{F}{\mu_1 \sin\theta + \cos\theta} \quad (3)$$

$$f_1 = \mu_1 N_1 \quad (4)$$

$$N_2 = \frac{F}{\mu_2} \quad (5)$$

$$f_2 = F \quad (6)$$

In which, the extrusion force F, according to its relationship with the density of material, that is, subject to different density of the material, there is the following empirical formula[12].

$$F = 39.7e^{-\frac{2.352}{\rho}} \quad (7)$$

When the material density specified is $\rho = 1.306 g/cm^3$, it is concluded $F = 6556.34N$

Therefore, according to the formula, the f_1 、 N_1 and f_2 、 N_2 under different chamfer θ will be obtained, as listed in the table 2.

Table 2: Load of the Ring Die Structure Mold Hole

Category	$f_1(N)$	$N_1(N)$	$f_2(N)$	$N_2(N)$
30°	1234.42	6857.91	6556.34	36424.11
45°	1414.38	7857.68	6556.34	36424.11
60°	1799.31	9996.18	6556.34	36424.11

3.4 Constraints of Finite Element Model

Since the ring die mold hole is of axis symmetrical structure, when set up the model, only 1/4 of ring die mold hole is built. Therefore, symmetry constraint shall be applied on the two sides of the model, and full constraint on the bottom of the model.

4 SOLUTION.

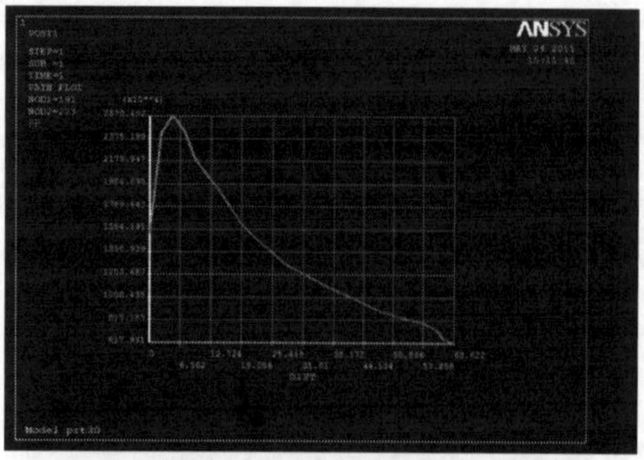

Figure 6:Stress distribution of 30°

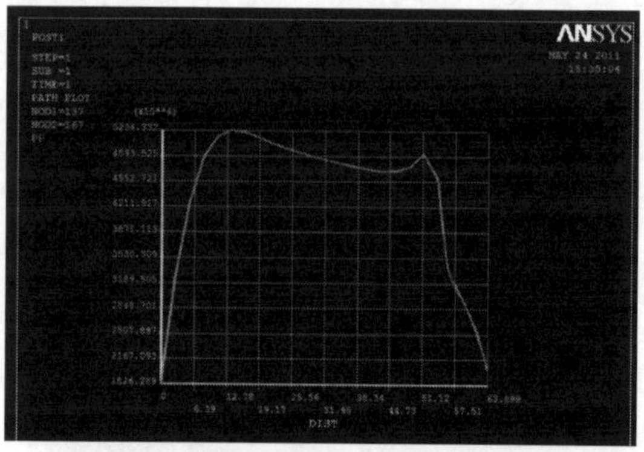

Figure 7:Stress distribution of 45°

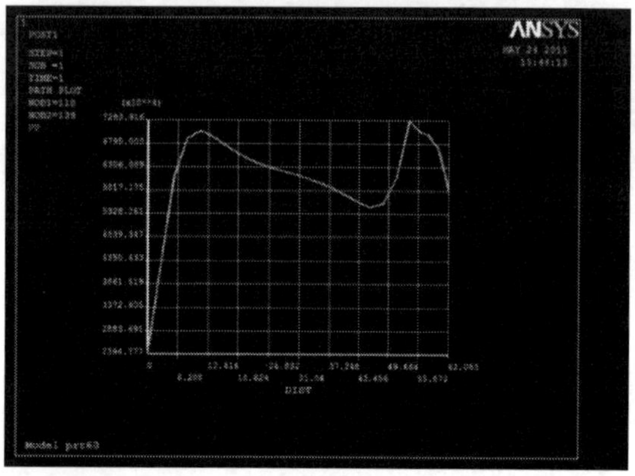

Figure 8:Stress distribution of 60°

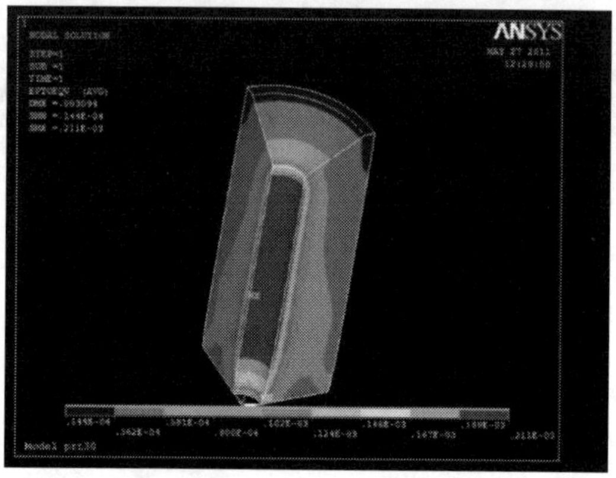

Figure9: Node Displacement Nephogram of 30°

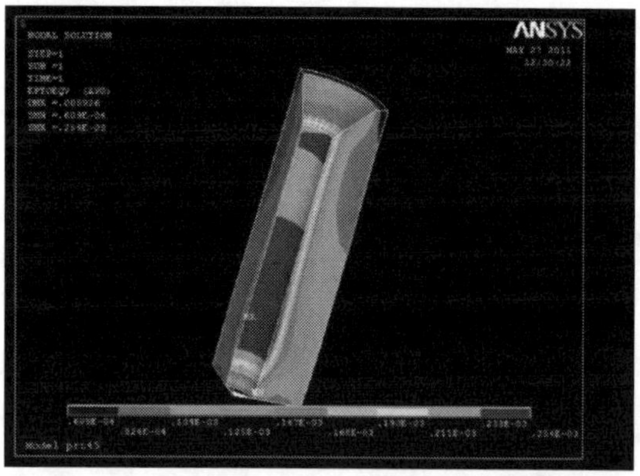

Figure10: Node Displacement Nephogram of 45°

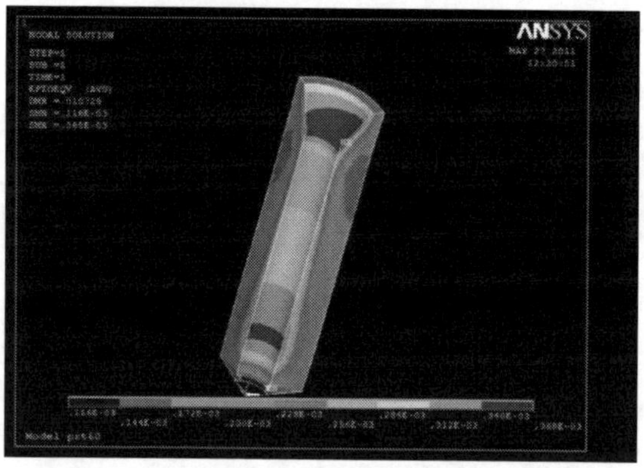

Figure11: Node Displacement Nephogram of 60°

Run the Solution-Analysis, type New Analysis-static, select linear static analysis on finite element, run Solution-Solve-Current LS, for the static structure finite element calculation. Refer to Figure 6-11 for the 3 kinds of stress distribution and deformation arrangement of ring die mold hole along the axial direction respectively.

5 CONCLUSION

(i) From the node displacement nephogram of the three kinds of chamfered ring die structure mold hole, it shows that the nodal displacement of ring die mold hole increases along with the increase of the chamfer, the smaller the chamfer is, the smaller the nodal displacement will be, then its stress and deformation will be smaller, and its structural stiffness will be higher, therefore, the 30° chamfered ring die mold hole structure and shape is better.

(ii) From the stress distribution in the axial direction of the three kinds of chamfered ring die structure mold hole, it shows that the axial stress increases along with the increase of the chamfer, the smaller chamfer is, the smaller the axial stress will be, then its load during work will be smaller, and its structural strength will be higher, therefore, the 30° chamfered ring die mold hole structure and shape is better.

(iii) From the maximum deformation and maximum stress table of the three kinds of chamfered ring die structure mold hole, it shows that the smaller the chamfer is, the smaller the maximum deformation and the maximum stress are, therefore, the 30° chamfered ring die mold hole structure and shape is better.

(iv) In the actual production, the 30° chamfered ring die mold hole can make the material pelleting in the hole for longer time, thereby improving the quality of pelleting, and matching with the actual.

6 ACKNOWLEDGMENT

In this paper, the research was sponsored by Tianjin Science and Technology Correspondent Project No.15JCTPJC57700 and National Spark Plan Project of China (The design and demonstration of efficiency ring die pellet mill based on the feed quality requirements, 2013).

7 REFERENCES

[1] **Chen Bingwei.2009.** High Efficient Pellet Production Principle and Performance Analysis of Ring Die Pellet Mill. Nanjing, Nanjing University of Science and Technology

[2] **LI Yancong, Wan Zhisheng, Li Shuhuang, etc. 2011.** Concept Design of Ring Die of Feed Pellet Mill. Journal of ANHUI Agricultural Sciences, 39(13), pp 8100_8101

[3] **Zhu Binfeng, Miao Jian, Chen Zhen. 2009.** Analysis and Study on the Ring Mould of Pelleter. Feed Industry, 30(11), pp 1-3

[4] **Li Shuhuan, Li Yancong, Song Xin, etc. 2011.** Force Analysis on Ring Die Structure of Feed Pellet Mill. Feed Industry, 32 （7）, pp 8-10

[5] **Shi Shuijuan, Wu Kai, Jiang Aijun, etc. 2011.**The Establishment of Mechanical Model of Ring Die and its FEA in the Pellet Production. Mechanical Design and Manufacture, （1）, pp 38-40

[6] **Jens K. Holm,Ulrik B.Henriksen,Johan E.Hustad,Lasse H.Sorensen. 2006.** Toward an Understanding of Controlling Parameters in Softwood and Hardwood Pellets Production. Energy & Fuels, (20), pp 2686-2694

[7] **Prepared by Saeed Moaveni; translated by Wang Song, Liu Lijuan, etc. 2008.** Finite Element Analysis- ANSYSY theory and application (3rd Edition). Beijing: Electronic Press Industry

[8] **Wang Jing, Ding Wangcai, etc. 2007.** FEA on Axis Shaft Based on Pro/E and ANSYS. Mechanical Engineering and Automation, (1), pp 54-55

[9] **Zhang Wei,Wu Jingfeng. 2009.** The Finite Element Analysis of Structure Mould Hole of Circular Mould Pelleting Machine.Chinese Agricultural Mechaniztion, (2), pp 83-85

[10] **Li Yancong, Song xin, Li Shuhuang, etc. 2011.**Study on the design method of ring die in the feed pellet mill based on equivalent elastic modulus. Journal of Agricultural Mechanization Research, (11), pp 79-82

[11] **Wu Kai, Shi Shuijuan, Peng Binbin, etc. 2012.** Modeling and Analysis on Extruding Force in Pelleting Process.Transactions of the CSAE, 26(12), pp 142-147

[12] **Wu Jinfeng, Huang Jianlong, Zhang Weiguo etc. 2007.** Simulation Experiment on Pellet Density and Extrusion Force of Alfalfa Meal. Agricultural Mechanics Journal, 38(1), pp 68-71.

DESIGN OF REMOTE ALARM SYSTEM FOR WAREHOUSE FIRE

Yanhong Du*, Li Wang, Yong Wei, Panlong Yan
College of Engineering and Technology
Tianjin Agricultural University, Tianjin, China
yanhong_du@163.com

ABSTRACT

In order to solve the disadvantages of traditional warehouse fire alarm system being too expensive and lack of flexibility, where the wired system is arranged due to the dispersed location of the monitoring points, this paper, taking STC89C51 series MCU as the controller, by fuzzy control algorithm to set the threshold of temperature change rate, and using the transparent transmission function of GPRS DTU, designed the remote warehouse fire alarm monitoring system. This system has local LCD display of ambient temperature, smoke concentration and time, local voice alarm and remote fire monitoring, etc. functions. The experiments show that the system can collect the real-time ambient temperature, smoke concentration, early warning of fire, and is of remote monitoring and control as well as good stability, suitable for work in the warehouse environment.

Keywords: Warehouse Fire, STC89C51, Remote Alarm, Fuzzy Control

1 INTRODUCTION

With the rapid development of the logistics and express company, more and more goods are packaged for distribution, and more and more goods are hoarded in the warehouse to be delivered, where the categories are more and more diverse, therefore the security management of the goods becomes particularly important, the automatic fire alarm system construction is also increasing. Viewing from the routine fire inspection condition at the fire fighting department, there are various problems of the automatic fire alarm system in the building, such as system failure or more false alarm, and even system crashes or even shut down. One reason is the existing warehouse fire alarm network to exchange data via a dedicated communication line or bus, but the communication line or bus network of the cabling, transmission distance, anti-interference ability, etc. deficiencies; the other is the difficulty in the line checking, simple and single function, not applicable in the volatile storage environment, meantime generally lack of monitoring system, in the event of a fire, not able to fully act the pre-set functions [1,2]. At present the rise and popularity of the GPRS network makes the microcontroller have practical significance for remote data transmission over the wireless network. The system relying on China Mobile's GPRS network and STC89C51 series MCU, combines the fire alarm equipment and the monitoring system, has designed a remote fire alarm system, extended the fire alarm mode, reduced the false alarm rate, and enhanced the reliability of remote fire alarm monitoring system.

*Corresponding Author

2 GENERAL DESIGN OF THE SYSTEM

The overall design of the system is as shown in Figure 1, mainly consists of local monitoring, communication platform and remote monitoring three parts. The local monitoring is composed by fire detection, alarm controller and alarm display three parts, in which the fire information is detected by the temperature sensor and smoke concentration sensors, the controller applied STC89C51 series MCU, and the display and alarm part mainly composed by the ISD4004 chip voice module and LCD display; The communication platform by means of GPRS networks, internet network and the I/O device drivers equipped with configuration software, CommBridge components, via GPRS DTU, to realize the wireless data transmission between the site data and remote control center; The remote monitoring is a monitoring function of the host PC machine. In the warning processing, the system uses fuzzy algorithm [3], taking the rate of temperature rise as the fire prediction scale; when the rate of temperature rise reaches the set value, identify that the fire has occurred, and then send the fire event to the control center, for early action.

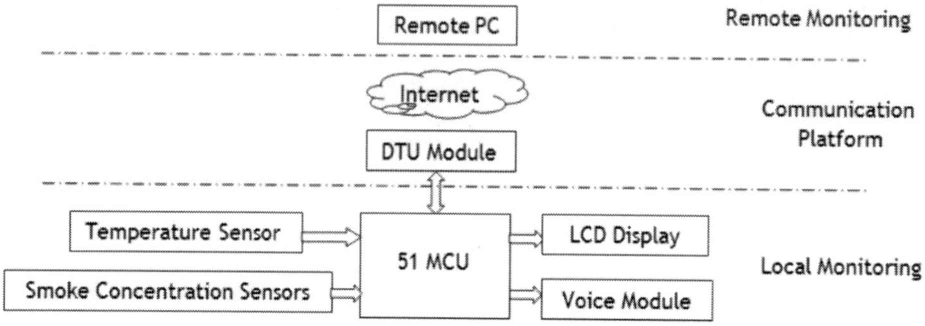

Figure 1: Block Diagram of the System General Design

3 DESIGN OF COMMUNICATION PLATFORM

3.1 Wireless Communication Module

Taking into account that the domestic GPRS network has also become increasingly mature and stable, and has become a good way to remote data transmission, and the current market has a variety of GPRS data transmission terminals with TCP/IP protocol conversion and affordable price [4]. The system selected the GPRS DTU equipment of Guangzhou Zhiyuan Electronics Co., Ltd., as the core module for wireless data communication.

3.2 Network Structure

Monitoring centre server is applied a fixed IP networking, in which the IP address is fixed, of more convenient addressing, easier data networking with terminal unit DTU but hard to break, more reliable working , suitable for multi-node and high volume data transmission.

3.3 Data Transmission Process

Data transmission mainly adopted the transparent transmission function of GPRS DTU equipment. Based on the above network scheme, the data uplink transmission process is ① lower PC through RS232 serial port transmits data to DTU; ②DTU makes data into TCP/IP packet, then transmitted over the wireless network to the remote PC; ③ remote PC processes and display after receiving the data [5]. The downlink data transmission is of reversed process compared with the above, and no more repeated explanation.

4 DESIGN OF SITE MONITORING SYSTEM

4.1 Fire Information Detection Module

4.1.1 Temperature Detection Module

DS18b20 temperature sensor has advantages of being small and faster conversion of the temperature data format, in addition, the temperature alarm threshold when fire occurred is at about 60℃, not exceeding the maximum detected temperature of DS18b20, therefore suitable for detecting the temperature of the fire scene. The sensor is capable of measuring maximum up to 125℃ and the measurement accuracy up to 0.1℃, whose internal structure is mainly composed of 64-bit lithography ROM, temperature-sensitive element, nonvolatile temperature alarm triggers TH and TL and configuration register four parts [6] It is designed to connect the I/O port of DS18b20 temperature sensor and the programmable I/O port of microcontroller, via a single bus in serial mode to transmit data to the CPU. Figure 2 is the pin wiring diagram of DS18b20.

4.1.2 Smoke Detection Module

MQ-2 smoke sensor is a new type of widely used combustible gas and smoke concentration detection apparatus, whenever there is flammable gas or smoke in the environment, MQ-2 conductivity will increase along with the increase of concentration of flammable gases or smoke. The concentration detected ranges from 300-10000ppm, and the higher precision when work at a humidity less than 70% of ambient temperature environment. It is designed to obtain the concentration values of MQ-2 smoke sensor by means of read and write control of the single-chip microcomputer on SCL clock signal line and SDA data line, and then the MQ-2 smoke sensor through A/D conversion chip converts the voltage value into digital data to the microcontroller, Figure 3 illustrated the A/D conversion circuit diagram of MQ-2 smoke sensor, which PCF8591T is a four-circuit 8-digit A/D conversion chip.

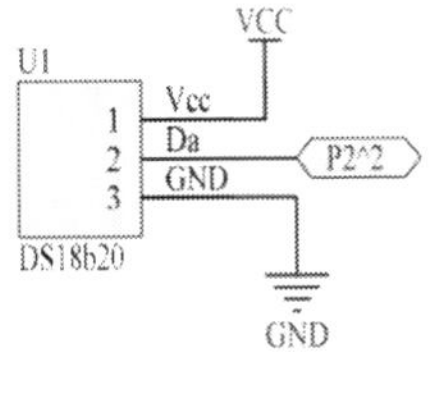

Figure 2: Pin Wiring Diagram of DS18b20

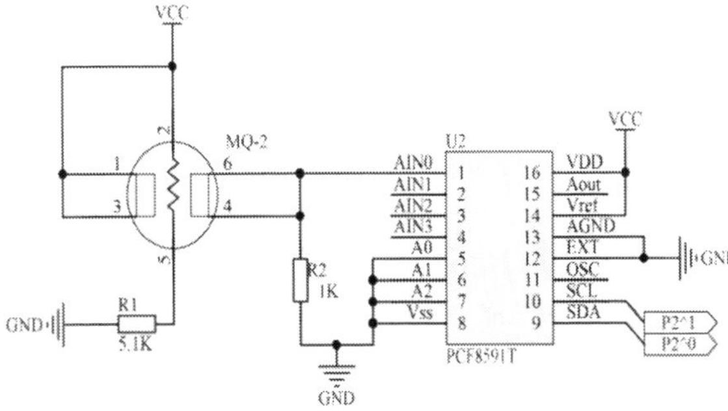

Figure 3: A/D Conversion Circuit Diagram of MQ-2 Smoke Sensor

4.2 Display and Alarm Module

Display module selected the LCD12864 with word stock, and the clock chip selected DS1302. The system transmits data through MCU P0 port and the 8-digit data lines of CD12864, showing the time on site, temperature and smoke concentration value. The alarm module took ISD4004 chip with stronger anti-jamming capability as the core, where the ISD4004 chip transmits data with the microcontroller via SPI serial interface, during transmission, SS must be kept low level, and high level must be kept between the two instructions [7]. Voice alarm circuit is as shown in Figure 4, which can record and store 8 minutes for 1200 paragraphs of voice, and control the voice speaker output via an external amplifier circuit. During recording, the data of MOSI pin at the serial data input terminal of the ISD4004 chip will be latched at the rising edge of the serial synchronous clock SCLK; while during playback, the data of MISO pin at the serial data output terminal will be removed at the falling edge of SCLK.

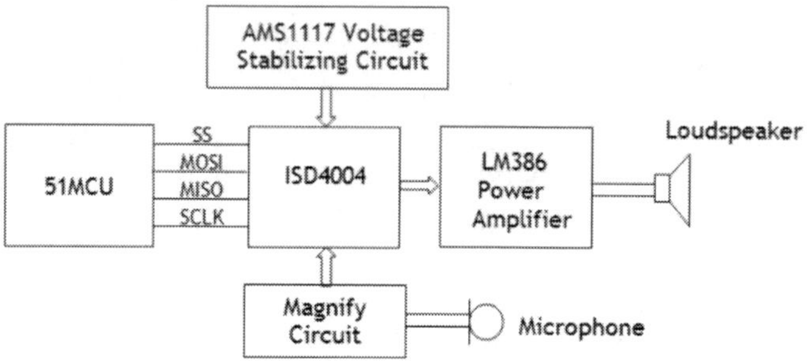

Figure 4: Circuit Diagram of Alarming Part in the System

4.3 Control Module of Lower PC

The lower PC took STC89C51 series microcontroller as the core, through a single bus protocol mode to control the collection of ambient temperature by temperature sensor DS18b20, and using I2C protocol to control PCF8591 chip to carry out AD conversion of the data collected by MQ-2 smoke sensor, to obtain standard smoke concentration value. To achieve bi-directional data transmission between lower PC and host PC and to reduce the workload of wiring, Apply GPRS DTU communication module, only requiring the connection of TXD and RXD of the communication module with the RXD and the TXD of lower PC.

5 DESIGN OF SYSTEM SOFTWARE

5.1 Setting of Temperature Threshold

Fire alarm system focuses on forecasting; early warning correspondingly increases the processing time of the fire. Because the fuzzy algorithm has strong anti-jamming capability and fast response, etc. advantages, the system according to the fuzzy algorithm has been designed a program of fire forecast, where the early fire diffusion rate is simulated under the natural state [8,9]:

$$\mathrm{v} = \frac{2k}{\rho c_o} \cdot \frac{\beta_o}{\beta} \qquad (1)$$

Where c_o is the concentration of oxygen in space, ρ is material concentration; β/β_o is combustible ash layer proportion.

When the concentration of the material is constant, the fire diffusion rate is proportional to the concentration of oxygen but inversely proportional to the combustible ash layer. In a closed environment, along with the fire, the oxygen concentration will decrease, while the combustion ash layer will get thicker and thicker. Thus it is concluded that the speed of fire will certainly turn slow. However, in the real life, the scope of fire will not reduce, but will continue to expand, even difficult to control. This is because the abundance of burning material makes the fire reaction time become longer. This design is mainly to make the system work at the earlier stage where the fire is rapidly spreading, since any later stage fire condition has no meaning to the alarm system [10]. According to the formula (1), to take polyurethane foam as an example, it is concluded the diffusion rate of fire at earlier stage V=0.075mm/s.

Analyzing from the fuzzy algorithms and the experimental data, to conclude the subordinate function between the fire temperature rising speed (V_T) and the fire diffusion rate (V), the fire distance (L) and the fire center temperature (T), etc. influencing factors:

$$v_T = \frac{k_1}{L} \cdot k_2 \cdot V \cdot k_3 \cdot T \qquad (2)$$

Where in k_1, k_2, k_3 for the membership of each factor, k1+k2+k3=1.

5.2 Design of Single Chip Microcomputer Software

Take the combustion of packaging cartons as an example, where k_1=0.5, k_2=0.4, k_3=0.1, according to the formula (2), it is concluded that the temperature rise velocity at 3 meters high at the fire scene is: V_T= 0.031℃/s. That is when the change rate of temperature collected by the temperature sensor is greater than 0.031℃/s, it can be considered a fire will occur. However, whether the rate of temperature rise of other combustibles is 0.031℃/s or not shall be further studied. This design used the data as a critical reference point for the fire occurrence. with the main flow chart of the software design as shown in Figure 5.

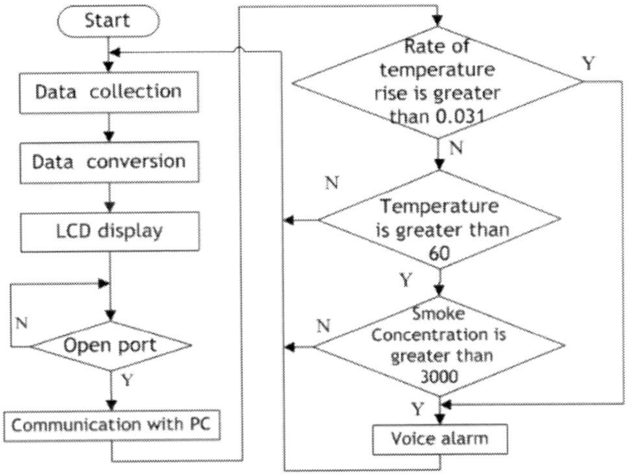

Figure 5: Main Program Flow Chart of Lower PC

5.3 Design of Remote Configuration

The remote PC will be installed with three dimensional force control configuration software, using the extension component CommBridge provided thereby to achieve GRPS / CDMA mode acquisition of warehouse site temperature, and the smoke concentration. The key of technology application is when defining the I/O device microcontroller, it is required to switch the "communication" mode into "bridge" mode. In addition, by setting the device address to determine which equipment in the network is connected with the force control configuration software.

6 REMOTE ALARM SYSTEM TEST

To simulate a fire at the scene under the experimental conditions, when the system is in operation, the remote monitoring interface is as shown in Figure 6, in which the displayed fire temperature is 23℃, the smoke concentration is 4635ppm, exceeding the predetermined concentration of 3000ppm, and the concentration indicator lights in red indicates the abnormal concentration, consistent with the on-site environmental parameters.

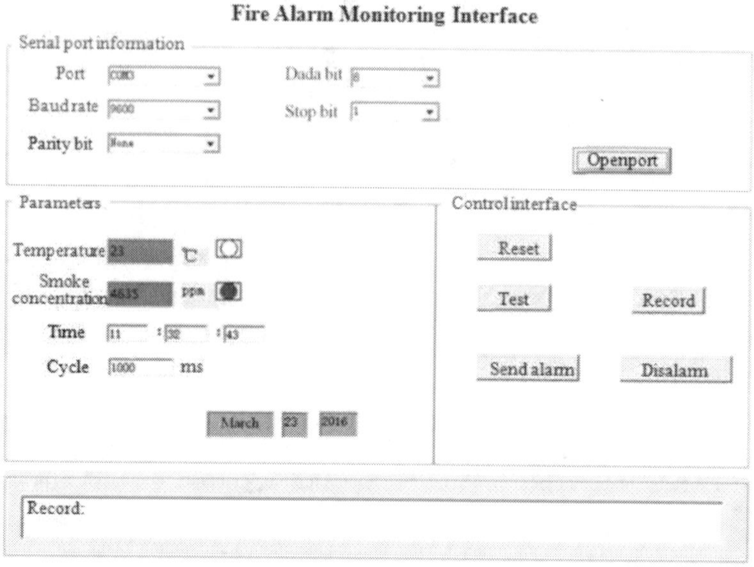

Figure 6: Monitoring Interface of Remote PC

7 CONCLUSION

This warehouse remote fire alarm system designed on basis of the single-chip microcontroller STC89C51 is capable of the accurate and real-time acquisition of warehouse temperature, smoke concentration value, and to reasonably set the upper limit on the rate of temperature rise by fuzzy algorithm, whereby early warning of fire, by GPRS RTU to send the fire conditions to the remote monitoring centre, for an early fire information issued to the staff. The microcontroller applied in the system has expandable functions, such as video surveillance, fire sprinklers control, etc., to facilitate future system enhancement and upgrade.

8 ACKNOWLEDGEMENT

This work was supported by the National Natural Science Foundation of China (No.51306132).

9 REFERENCES

[1] **Sun Zhangjun. 2011.** The development of fire alarm technology and system. Science and Technology Innovation Herald, (8), pp 123.

[2] **Gao Xiaoming, Wan Minkai. 2015.** Design of fire alarm system in cigarettes factory. modern building electrical,6(7), pp 51-54,65.

[3] **Ren Guosheng Wang Changle. 2013.** Present situation and countermeasures of application efficiency of automatic fire alarm products. China CIO News. (3),pp 82,84.

[4] **Hu Wei, Fu Xingwu. 2004.** Intelligent fire report and control system based on fuzzy neural network. Microcomputer Information, 20(3), pp 20-21.

[5] www.embedcontrol.com/products/wuxian/GPRS/imgaes/dtu.pdf

[6] Shi **Luchuan, Liang Guodong. 2009.** Remote monitoring of the waste water treatment based on the GPRS network. Techniques of Automation & Applications, 28(3), pp 121-122,125.

[7] **Meng Xianglian. 2010.** principle and application of embedded system. Beijing: Tsinghua university press.

[8] **Gao Yan. 2011.** The application of ISD4004 voice chip in the telephone remote control system . Science and Technology Innovation Herald, (9), pp 11.

[9] **Peng Lei, Zou Yanghui, Zhou Jianjun. 2004.** A model of transition from smoldering to flaming with natural diffusion. Fire Safety Science, 13(1), pp 28-34.

[10] **Guo Zhen, Yuan Yingshu. 2013.** Numerical model of hot-smoke flow in residential fire. Architectural & Environmental Engineering, 35(2), pp 73-78.

CUSTOMER SATISFACTION OF BIKE SHARING SYSTEMS AND AN ARTIFICIAL BEE COLONY ALGORITHM FOR BIKE REPOSITIONING PROBLEM

X.H. Feng[1*] Y. Chen[2] Y.Z. Wu[1] and S.Z. Zhang[3]

[1]Ocean College
Zhejiang University, Zhejiang, China
fengxuehao@zju.edu.cn, yezhouwu@zju.edu.cn

[2]Department of Transportation Management
Zhejiang Institute of Communication, Zhejiang, China
cyath@sina.com

[3]Department of Industrial and Systems Engineering
The Hongkong Polytechnic University, Hong Kong, China
shuzhu.zhang@connect.polyu.hk

ABSTRACT

The paper studies the bike-sharing system that is popular in many cities all around the world. It is also a useful way of relieving greenhouse gas emission and traffic congestion. However, the customer satisfaction (CS) of such systems is rarely explored. Therefore, it would be interesting to discuss the factors that have significant impacts on the CS of bike-sharing systems. First, we develop a model to investigate the CS based on the Cognitive-Affective System Theory of Personality that has been used in analyzing the CS in several industries. We design encoding process and customer behavior generation process to analyze how factors influence the CS. The factors that influence the CS significantly are revealed. Second, based on the CS analysis, we develop a model of the bike repositioning problem to maximize the CS of bike-sharing systems. Because it is an NP-hard problem, an artificial bee colony (ABC) algorithm, derived from the intelligent collective behavior of bee colony, is employed to solve the bike repositioning problem. Three unique features of ABC algorithm, i.e., self-organization, decentralization and labor division, guarantee the effectiveness and efficiency of the ABC algorithm. The advantage of the proposed model and algorithm can be further examined and validated through numerical experiments.

Keywords: Bike sharing, Customer satisfaction, Bike repositioning, Artificial bee colony algorithm

* Corresponding Author: X.H. Feng

1 INTRODUCTION

With the rapid development of populations of metropolises all around the world, traffic congestion has been a critical problem in modern cities. The case can be worse when we consider the greenhouse gas emission resulting from the bad traffic conditions. In recent years, public bicycle sharing systems have been a popular approach of handling with the above two challenges simultaneously. Under such a system, customers can rent bicycles from stations located in different areas of the city. After the utilization of the bicycle, customers can return the bicycles to any a station in the system. Until the April 2013, around 535 bicycle sharing (BS) systems have been implemented in the world, equipped with about 517,000 bicycles [1]. Many cities in China have implemented their BS systems, such as Hangzhou, Zhoushan, Yangzhou, and so on. There are also a large number of BS systems in other countries in the world, such as France, Italy, and Singapore.

However, these cities implement the BS systems with different purposes. Take Hangzhou and Zhoushan from the same province as examples. As reported in 2013, Hangzhou had a population of 8.84 million, while Zhoushan's population is slightly larger than 1 million. In addition, Hangzhou's traffic congestion is the worse three cases in China, but Zhoushan is almost free of traffic congestion. Therefore, Hangzhou implemented her BS system in order to relief the traffic congestion; while Zhoushan's BS system is mainly used a tool of tour and exercise. The advantages of BS systems in these aspects are straightforward, including less traffic congestion, less greenhouse gas emission, higher flexibility of parking, and so on. However, these advantages can be achieved only when many customers are willing to use the BS systems instead of using cars, and the BS systems face fierce competition with private cars, taxis, and buses. In the light of this statement, the rental prices in BS systems in the world are usually much lower than those of other transportation modes. Because of the development of economy and citizens' income, the impact of price on the attractiveness of BS systems is not that significant. As a consequence, we need to learn more about customer satisfaction (CS) of the BS systems and find out the key elements that have significant positive/negative impacts on the CS of the BS systems.

Practitioners in marketing have paid much attention to improving the CS of their products, service, or even image. In the short-run, high CS can bring the repeated consumption and word of mouth. In the long-run, CS can convert into customer loyalty that makes the customers adhere to the brand. Researchers conduct a great number of studies on CS from the view of marketing. In this paper, we develop a model to investigate the CS based on the Cognitive-Affective System Theory of Personality (CASP Theory) that has been used in analyzing the CS in several industries. We design encoding process and customer behavior generation process to analyze how factors influence the CS. The factors that influence the CS significantly are revealed. Next, based on the CS analysis, we discuss the development of a model of the bike repositioning problem to maximize the CS of bike-sharing systems.

This article is organized as follows. Section 2 proposes a literature survey of recent studies on BS system management. Section 3 shows the model of the CS of BS systems based on the CASP Theory. Section 4 discusses the bicycle repositioning problem under the consideration of the CS of BS systems. We conclude the article and suggest future avenues in Section 5.

2 LITERATURE SURVEY

In this section, we briefly review the relevant studies on the BS system management.

There are two main streams of literature in BS system management. The first direction focuses on the operation mode of BS systems. Ji et al. [2] analysed the operations mode of electronic bicycle sharing systems in the US by using simulation. Zhang et al. [3] discussed the characteristics of BS systems in several Chinese cities (e.g., Beijing, Hangzhou, and Shanghai) and pointed out that the design of BS systems should consider it as a product-service system. Borgnat [4] proposed a model of the traffic flow of citizens based on the demand of bicycles at stations and the operation of BS systems, which is conducted under the analysis of bicycle flow in BS systems. Vogel [5] used data mining to analyse the demand of bicycles in the BS system in Vienna, Austria. The results in that study show that bicycle repositioning among stations is a main approach of balancing the demand and inventory of bicycles at stations, which may have an impact on the acceptance of BS systems to citizens. Shaheen et al. [6] collected the citizens' demand for the BS system in Hangzhou with questionnaires and proposed several suggestions to improve the performance of the system, including increasing the number of stations, improving the customer satisfaction rate, maintaining better qualities of bicycles, and extending the operation time. Raviv and Kolka [7] discussed the bicycle inventory problem of a single station in BS systems with random demand. Nair et al. [8] proposed suggestions on the functions of BS systems and involved policies, based on the analysis of the BS system in Paris, France.

The second direction of literature in BS system management focuses on the operational decision in the static bicycle repositioning problem in BS systems. In the problems, the objective is to minimize the total repositioning cost by satisfying some certain constraints on the satisfaction of customers' order. In most of studies in this area, only one positioning is considered in models. Lin and Yang [10] discussed the optimal station location of BS systems and the customers' optimal routing of riding bicycles with the purpose of minimizing the total riding distance of customers in BS systems. Nair and Miller-Hooks [10] developed a mixed integer program (MIP) model for the bicycle repositioning problem in BS systems by considering the uncertainty of demands of bicycles. Romero et al. [11] analysed the optimal location of stations in BS systems to maximize the number of satisfied customers with simulation models. Chemla et al. [12] proposed a static bicycle repositioning model of BS systems and discussed the feasible solutions and the lower bound on the optimal solution. Lin et al. [13] considered the BS system network design problem as a warehouse location problem and developed an MIP model. In this model, the bicycle repositioning among stations is not involved. Chow and Sayarshad [14] proposed a multi-objective programming model with existence of multiple networks for the design of BS systems. Ho and Szeto [15] developed an iterated tabu search algorithm for the static bicycle repositioning problem. Dell'Amico et al. [16] proposed four mathematical model to present the static bicycle repositioning problem. That studied compared the computational time and quality of solutions of the four models, and figure out the optimal MIP model. Forma et al. [17] proposed a three-stage heuristic algorithm for the static bicycle repositioning problem. Li et al. [18] developed a hybrid genetic algorithm for multi-type bicycle repositioning problem.

3 THEORETICAL MODEL AND HYPOTHESIS

In this section, we propose our theoretical model of CS on the BS systems based on the CASP Theory. We measure this relationship between the BS system and the CS via perceived relative advantage and perceived emotional attachment. Existing studies show that that perceived relative advantage (PRA) and perceived emotional attachment (PEA) are have positive impacts on the CS of products and service. We summarize the features of BS systems into three parts, hardware quality (HQ), service quality (SQ), and pricing issue (PI). The HQ consists of several hardware characteristics of BS systems, such as the quality of bicycles, the quality of dock and locks, and the quality of payment system. For some BS systems, the image or decoration of bicycles is also an important element of the HQ group. Obviously, the HQ is the basic element that influences the CS on BS systems. The SQ presents the features that are related to service of BS systems, such as the satisfaction ratio of customers, the convenience of reaching a station, waiting time, and traffic information service. Price is another factor that influences the CS. A higher price usually makes the customers expect higher level of HQ and SQ. Consequently, we consider PI as a factor in our model, which not only presents rental price, but also contains the compensation price for loss or damage of bicycles due to customers' mistake. Moreover, we add customers' mental factors as moderators of the model, such as the idea of environment protection and reputation. Figure 1 illustrates our theoretical model.

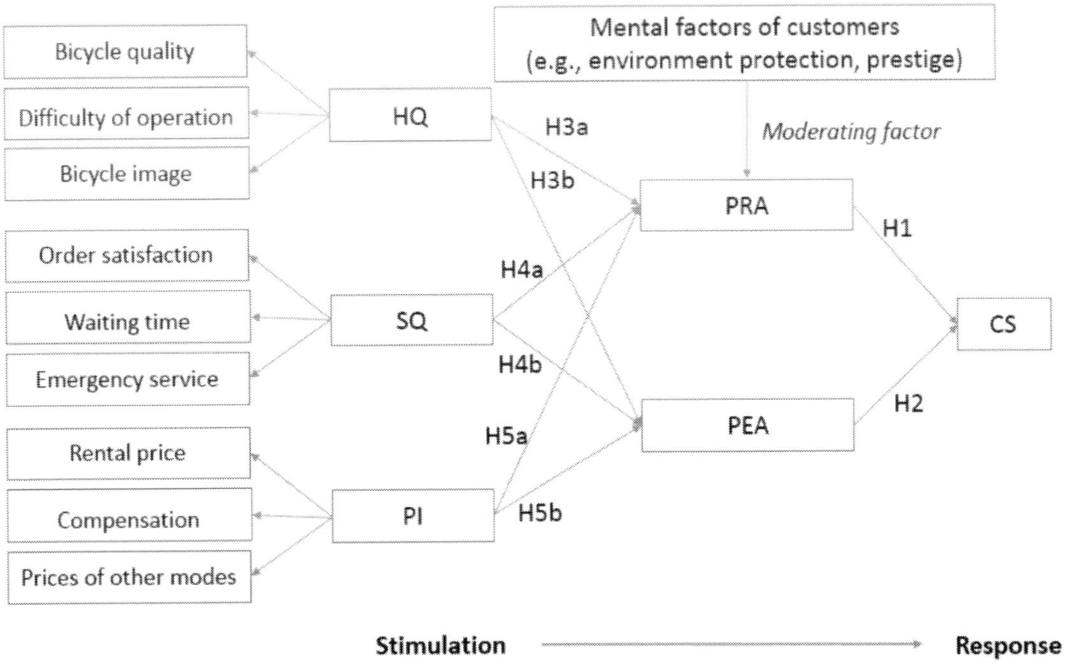

Note: PRA presents perceived relative advantage and PEA presents perceived emotional attachment.

Figure 1: Theoretical model

3.1 Perceived relative advantages

Relative advantage of a product or service has been defined as the degree to which a customer finds that a product is better at some function(s) than other products [19]. Existing definitions imply that relative advantage is perceived by customers who have experienced different products and those products are substitutable to each other. As we can observe from the practice, citizens have multiple options for their traffic, including private cars, taxi, buses, private bicycles, public bicycles, and foot. These traffic modes have quite different features in HQ, SQ, and PI. Inevitably, customers compare these modes and make decisions on their traffic mode selection. In the present article, perceived relative advantage is selected as the customer's cognitive evaluation to explain CS. We conjecture that customers would obtain a high level of satisfaction on BS systems coupled with a higher perceived relative advantage.

Hypothesis 1 (H1). Perceived relative advantage is positively associated with the CS on BS systems.

3.2 Perceived relative advantages

According to existing studies, some products can keep an internal connection with customers, which can result in a type of attachment in the long run [20]. Many companies attempt to create customers' attachments to their brands and the emotional attachment implies the customers' feeling the products. In our model, perceived emotional attachment is used as the customer's affective evaluation.

Hypothesis 2 (H2). Perceived emotional attachment is positively associated with the CS on BS systems.

3.3 Hardware quality

Hardware of BS systems refer to physical components of the system, such as such as bicycle quality, image, difficulty of operation, reliability of payment system, and so on. Therefore, the HQ plays an importation part in the performance of BS systems. Suppose a BS system with bicycles that are of low quality. It is difficult to imagine that customers are willing to use that system for their traffic. Hence, the BS systems with higher HQ would enjoy a relative advantage compared with other traffic modes.

Hypothesis 3a (H3a). Hardware quality is positively associated with the perceived relative advantage to BS systems.

Hypothesis 3b (H3b). Hardware quality is positively associated with the perceived emotional attachment to BS systems.

3.4 Service quality

Service of BS systems refer to the additional service provided by the system, such as order satisfaction ratio, waiting time, and emergency service. For instance, if customers arrive in a station and find no available bicycles there, they may leave the system and switch to other traffic mode. They may also wait at the stations for some time before they obtain a bicycle or leave the queue. These two factors are influenced by the bicycle repositioning among stations. Emergency service is another important factor of system service. Some customers, for instance, may find their bicycles broken during their utilization. Then, emergency service is required to help customer to handle with the problem. As a consequence, the SQ may have a significant impact on the performance of BS systems.

Hypothesis 4a (H4a). Service quality is positively associated with the perceived relative advantage to BS systems.

Hypothesis 4b (H4b). Service quality is positively associated with the perceived emotional attachment to BS systems.

3.5 Pricing issue

Pricing strategy is one of the core decision of marketing and operation management. The price issue of BS systems may influence the CS significantly. It contains rental price of bicycles and the compensation price for loss or damage of bicycles resulting from customers' mistake. It is related to both certain price and risk issue.

Hypothesis 5a (H5a). Price is negatively associated with the perceived relative advantage to BS systems.

Hypothesis 5b (H5b). Price is negatively associated with the perceived emotional attachment to BS systems.

4 DISCUSSION AND ARTIFICIAL BEE COLONY ALGORITHM

The hypotheses can be tested with data collected from the real world. With the results, we can tell the key factors influencing the CS on BS systems. Then, new models with the considerations of these factors in objective functions or constraints can be developed. There are many different algorithms to solve the bicycle repositioning problem and as far as we know, the artificial bee colony (ABC) algorithm has not be explored in this problem. Three unique features of the ABC algorithm, i.e., self-organization, decentralization and labor division, guarantee the effectiveness and efficiency of the ABC algorithm. The ABC algorithm exemplifies the two inherent features of swarm intelligence, which are decentralization and collective intelligence. In addition, a unique feature of ABC algorithm is the labor division. ABC algorithm consists of three types of bees, i.e., scout bees, employed bees and onlooker bees, and each type of bees plays different role, which balances the exploration and exploitation of ABC algorithm.

The operations of a bee colony can be described as follows. At the very beginning, scout bees are sent out to search for food sources randomly. Once the food sources are located, employed bees are sent out to exploit them. One employed bee matches with one food source. During the exploitation process, employed bees also search around for better food sources. After employed bees return to the hive, they share information with onlooker bees waiting in the hive. Onlooker bees then determine to exploit some food sources based on the information they acquired. Once a food source is exhausted, it would be abandoned and replaced by another food source found by scout bee. Correspondingly, the design of ABC algorithm consists of four distinct phases, which are initialization phase, employed bee phase, onlooker bee phase and scout bee phase. Different types of bees are involved in different phases with different purposes.

5 CONCLUSION

In this article, we develop a model of CS on BS systems with the CASP Theory. We selected PRA and PEA as the two factors that have direct impact on the CS. We use HQ, SQ, and PI to explain how the features of BS systems influence the CS via PRA and PEA. Several hypotheses are proposed for further evaluation of our model. With the data analysis, we can check whether these hypotheses hold or not. It could also be possible that some of them hold under some certain cases in some countries. The numerical results would provide managerial insights for managers about how to invest their money into the BS systems to improve the CS. The contribution can be more significant when some managers have budget constraints that is a common phenomenon in current economy [21].

For further research, we would like to suggest a couple of directions. First, with the results of the CS on BS systems, a model of optimal pricing and investment in hardware and service under budget constraints can be a fruitful direction. Mathematical models with multiple objectives can be proposed and some algorithms may need to be developed to obtain the optimal solution. Second, the ABC algorithm could be extended and tailored to the bicycle positioning problem in BS systems with new and realistic objective functions and constraints, such as order satisfaction rate, waiting time, and emergency service time.

ACKNOWLEDGEMENT

This work is supported by the Fundamental Research Funds for the Central Universities.

6 REFERENCES

[1] **Shu, J., Chou, M.C., Liu, Q.Z., Teo, C.P., Wang, I.P. 2013.** Models for effective deployment and redistribution of bicycles within public bicycle-sharing systems, *Operations Research*, 61(6), pp 1346-1359.

[2] **Ji, S.G., Cherry, C.R., Han, L.D., Jordan, D.A. 2014.** Electric bike sharing: simulation of user demand and system availability, *Journal of Cleaner Production*, 85, pp 250-257.

[3] **Zhang, L.H., Zhang, J., Duan Z.Y., Bryde, D. 2015.** Sustainable bike-sharing systems: characteristics and commonalities across cases in urban China, *Journal of Cleaner Production*, 97(15), pp 124-133.

[4] **Borgnat, P., Abry, P., Flandrin, P., Robardet, C., Rouquier, J. B., Fleury, E. 2011.** Shared bicycles in a city: A signal processing and data analysis perspective, *Advance in Complex Systems*, 14 (3), pp 415-438.

[5] **Vogel, P., Greiser, T., Mattfeld, DC. 2011.** Understanding bike-sharing systems using data mining: Exploring activity patterns, *Procedia-Social and Behavioral Sciences*, 20, pp 514-523.

[6] **Shaheen, S.A., Zhang, H., Martin, E., Guzman, S., 2011.** China's Hangzhou public bicycle: Understanding early adoption and behavioural, *Transportation Research Record*, 2247, pp 33-41.

[7] **Raviv, T., Kolka, O. 2013.** Optimal inventory management of a bike-sharing station, *IIE Transactions*, 45, pp 1077-1093.

[8] **Nair, R., Miller-Hooks, E., Hampshire, R.C., Busic, A. 2013.** Large-scale vehicle sharing systems: Analysis of Vélib', *International Journal of Sustainable Transportation*, 7(1), pp 85-106.

[9] **Lin, J.R., Yang, T.H. 2011.** Strategic design of public bicycle sharing systems with service level constraints, *Transportation Research Part E*, 47(2), pp 284-294.

[10] **Nair, R., Miller-Hooks, E. 2011.** Fleet management for vehicle sharing operations, *Transportation Science*, 45(4), pp 524-540.

[11] **Romero, J.P., Ibeas, A., Moura, J.L., Benavente, J., Alonso, B. 2012.** A simulation-optimization approach to design efficient systems of bike-sharing, *Procedia-Social and Behavioral Sciences*, 54 (4), pp 646-655.

[12] **Chemla, D., Meunier, F., Wolfler-Calvo, R. 2013.** Bike sharing systems: solving the static rebalancing problem, *Discrete Optimization*, 10(2), pp 120-146.

[13] **Lin, J.R., Yang, T.L., Chang, Y.C. 2013.** A hub location inventory model for bicycle sharing system design: Formulation and solution, *Computers & Industrial Engineering*, 65, pp 77-86.

[14] **Chow, J.Y.J., Sayarshad, H.R. 2014.** Symbiotic network design strategies in the presence of coexisting transportation networks, *Transportation Research Part B*, 62(1), pp 13-34.

[15] **Ho, S.C., Szeto, W.Y. 2014.** Solving a static repositioning problem in bike-sharing systems using iterated tabu search, *Transportation Research Part E*, 69, pp 180-198.

[16] **Dell'Amico, M., Hadjicostantinou, E., Iori, M., Novellani, S. 2014.** The bike sharing rebalancing problem: Mathematical formulations and benchmark instances, *Omega*, 45, pp 7-19.

[17] **Forma, I.A., Raviv, T., Tzur, M. 2015.** A 3-step math heuristic for the static repositioning problem in bike-sharing systems, *Transportation Research Part B*, 71, pp 230-247.

[18] **Li, Y.F., Szeto, W.Y., Long, J.C., Shui, C.S. 2016.** A multiple type bike repositioning problem, *Transportation Research Part B*, 90, pp 263-278.

[19] **Rogers, E. 2003.** *Diffusion of Innovations.* 5th Ed... New York, Free Press.

[20] **Thomson, M., MacInnis, D.J., Park, C.W. 2005.** The ties that bind: Measuring the strength of consumers' emotional attachments to brands, *Journal of Consumer Psychology*, 15(1), pp 77-91.

[21] **Feng, X.H., Moon, I.K., Ryu, K.Y. 2015.** Supply chain coordination under budget constraints, *Computers & Industrial Engineering*, 88, pp 487-500.

DEVELOPMENT AND APPLICATION OF STRIP COATING WEIGHT CONTROL SYSTEM*

Z.S. Pan[1†], X.H. Zhou[2] and P. Chen[2]
[1]Institute of Cyber-Systems and Control, Zhejiang University, Hangzhou, China
panzs@zju.edu.cn

[2] Zhejiang SUPCON Research CO.,LTD., 309, Liuhe Road, Hangzhou, China
zhouxuanhao@supcon.com chenpeng@supcon.com

ABSTRACT

This study proposes a novel neural-network-based control system for the hot dip coating line (HDCL) in steel-making where several control difficulties such as large time-varying delay, nonlinear characteristic, and strong disturbance are addressed. The proposed system is successfully applied to a real plant at VaLin LY Steel Co., Loudi, China and industrial application results show great performance improvement, both the variance of coating weight and the transition time are significantly reduced.

Keywords: Neural network, HDCL, Coating weight control

1 INTRODUCTION

Hot dip coating is a popular production process after cold mill in steelmakers, which produces high value-added products such as galvanized sheet widely used in automotive and household electrical appliances manufacturing.

The hot dip coating process shows some complicated characteristics such as large time-varying delay, nonlinear relationship, as well as strong disturbance. All of above will lead to the difficulties in the coating weight control. Nowadays, manual control is still widely used in HDCLs, while only a few model-based control approaches including first principle Adams[1], regression Liu[2], Yao[3], Guelton[4], Shin[5] and neural network Lu[6][7], are incorporated. However, the first principle model[1] contains several parameters hardly derived through physical analysis leading to the model precision problem; the regression models[2][3][4][5] are weak in description of nonlinear characteristic of the system results in the accuracy problem coving all the operating conditions; the conventional neural network[6][7] is strong in nonlinear modeling but the prediction performance will be poor under varying operating conditions. Thus, these methods still have the limit of control ability, resulting in fluctuation of coating weight, long transition time, and large zinc cost in practical production.

In order to cope with the control difficulties of HDCL, and further enhance the quality of coating weight as well as reduce the waste of zinc, this paper presents a novel neural-network-based coating weight control system, which is composed by a feed forward control (FFC) and a feedback control (FBC). More specially, FFC works when drastic change happens to the line speed or target coating weight changes, while FBC runs under stable working status to handle disturbance. In

---

* This work is supported by National High Technology Research and Development Program of China (863:2012AA041703)

† Corresponding Author

addition, the neural network embedded in both FFC and FBC is not only strong in reflecting the nonlinear characteristic of real plant, but also addresses the time-variant large measurement delay problem with a bias update module.

In the following, the problem statement, the framework of the novel control system and the introduction of each component will be addressed. Finally, the application results as well as the conclusion will be presented.

2 PROBLEM STATEMENT

2.1 Introduction of the line

Generally, the schematic diagram of a typical hot dip coating line is shown as follows:

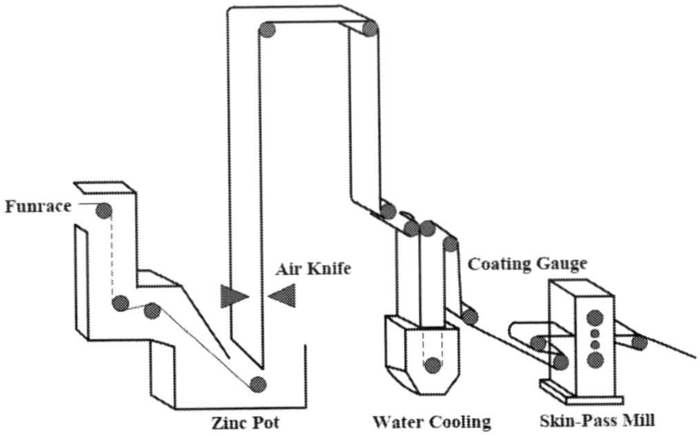

Figure 1: The Schematic Diagram of A Typical Hot Dip Coating Line

As shown in Figure.1, after heat treatment in the furnace, the strip enters the zinc pot where liquid hot zinc is attached to the steel's surface, and the extra zinc is carried out when passing the air knife subsequently. This work takes emphasis on the coating weight control system for an HDCL which aims to make the thickness of zinc that putting on the strip meeting the specifications of customer orders.

2.2 System analysis

Fundamentally, the galvanizing process can be summarized as the block diagram in Figure. 2.

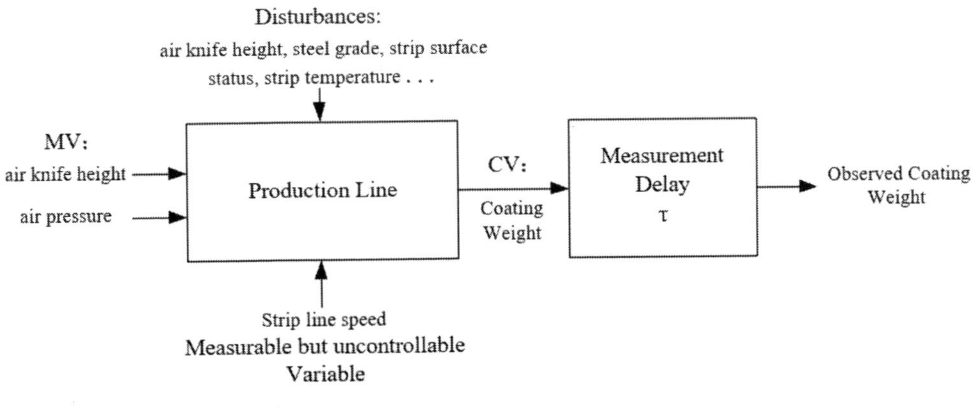

Figure 2: Block Diagram of HDCL in Terms of Coating Weight Control

As shown in Figure.2, the coating weight of coil is affected by several factors which can be classified into three categories, namely MV, line speed, and disturbance. Besides the strip line speed, the setting of air knife including the air pressure applied to a tapered gap and the distance between the knives (air knife gap) are three dominant factors. However, only the latter two are selected as manipulated variables (MV) while the line speed is measurable but not controllable which plays important role in feed forward control. Except for the above variables, other factors such as air knife height, grade, strip temperature and so on are all regarded as disturbance to the control system.

Although the basic mechanism of system is clear shown above, the HDCL system shows some complicated characteristics, which lead to the difficulty in coating weight control, can be concluded as follows:

- The coating weight (CV) is determined by all the factors forced when it passing the air knife, the value of that can't be observed until reaching the X-ray coating gauge which installed about 100 meters behind the air knife, introducing a large measurement delay and increasing the difficulty of closed-loop feedback control system where the oscillation may be caused.
- Various disturbance exists in the control system may influence the model precision and impose negative effect on control performance.
- Strong nonlinear relationship exists between the controlled variable and the manipulated variables, as well as other influential factors so that it's quite difficult to establish an accurate model coving all the operating points.

2.3 Control target

The HDCL is a typical order-driven discrete-event production system, in which the target coating weight of each strip is determined by the customer demand and varies within a wide range. Thus, the first objective of coating weight control system is to minimize the error between the target and actual coating weight. In particular, the negative error leads to off-specification products while the positive error means the waste of raw material (zinc ingots).

On the other hand, the order scheduling is determined by the inherent properties of strips such as the target coating weight, the steel grade, as well as the strip width. It can't be guaranteed that the strips with same target coating weight are processed successively leading to the frequent change in target coating weight. Thus, it is important to minimize the coating weight transition time which equals to interval from the beginning of a strip with new target to the time when that target is reached. Obviously, the shorter the transition time, the less the waste of raw material or the less the yield of off-specification products.

3 DESIGN OF CONTROL SYSTEM

3.1 Framework of the control system

To deal with the above control puzzle, this work proposed a novel coating weight control system based on neural network with the framework shown in Figure.3.

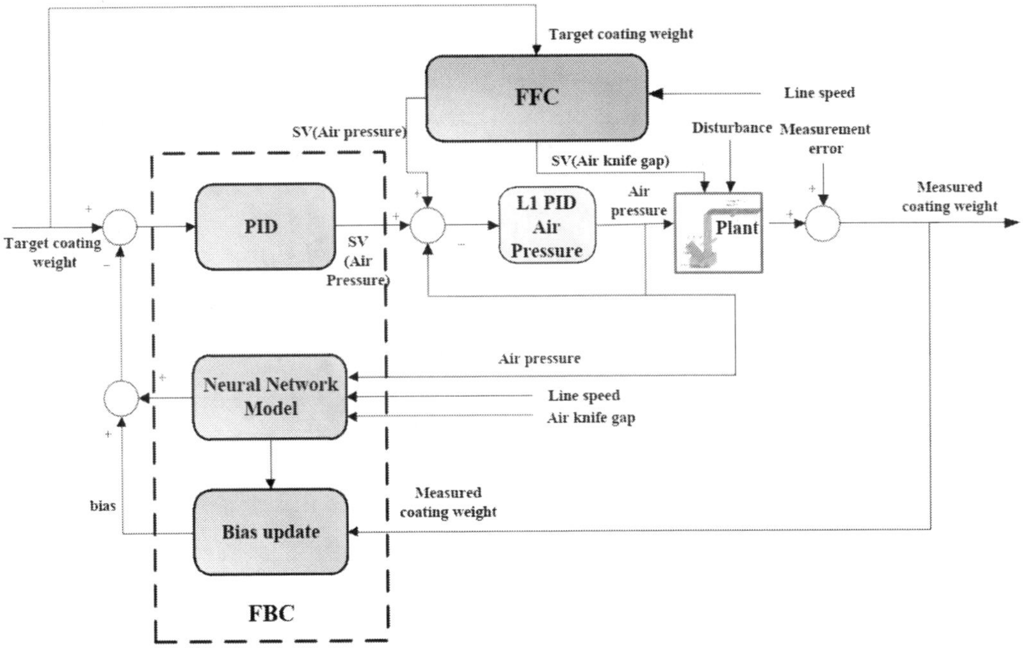

Figure 3: Framework of the Novel Control System

As shown in the figure above, the control system is mainly composed by two modes, namely feed forward control (FFC) and feedback control (FBC), aims to improve the control performance under different operating conditions. It should be noted that to assure the stability of control system, a basic constraint should be satisfied that two control modes do not run simultaneously and a well-designed logical rule is implemented to determine the switching time between these two modes. In the following, the detail demonstration of each component of the control system will be presented.

3.2 Neural network predictive model

The neural network predictive model Basheer[8] is developed as the base of FFC and FBC, describing the steady-state relationship between coating weight (CWp) and the manipulated variable (air pressure P, air knife gap D), with additional measurable terms - the line speed S. Thus, the nonlinear mapping of neural network can be described as follows:

$$NNp(P,S,D) = CWp \qquad (1)$$

Fundamentally, the neural network is in a BP form Basheer[8], whose topology and architecture are determined in advance and synaptic weights (sw) are first optimized through off-line BP learning. The target function that drives the training process is the error between the actual coating weight *CW* and CWp in the training space Ω, which is made up of four dimensions (air pressure, air knife gap, line speed, actual coating weight), is as follows:

$$\min_{sw} \sum\nolimits_{s\in\Omega} \|CW(s) - CWp(s)\| \qquad (2)$$

Due to the strong ability in nonlinear mapping, this neural network can generate accurate predictive value of coating weight under almost normal operating status with less prediction error, which is quite important for control both in FFC and FBC. Meanwhile, this model can be further updated on-line easily when the real data with information of new operating status is gathered. Thus, the neural network model is "clever" enough to adjust to the change of operational status.

3.3 FFC control

3.3.1 Outline

In this work, FFC consists of two modules, namely preview control and speed-FFC. In particular, preview control copes with the situation that a new strip with distinct target coating weight approaching while speed-FFC handles the change of the line speed correspondingly. The preview control calculates the point of air pressure $P_{\mathrm{pre}}(t)$, as well as air knife gap $D_{\mathrm{pre}}(t)$ based on the predetermined production order sent from L3 (MES) system. On the other hand, the speed-FFC belongs to the typical feed forward approach to perceive the drastic speed changes, and adjust the air pressure as well as air knife gap in advance to maintain the stability of coating weight.

3.3.2 Preview control

When a strip with new target coating weight approaching, a preview control is used to regulate the air pressure and air knife gap simultaneously in order to adjust the coating weight to a new set point as fast as possible and keeping it above the set point strictly throughout the transition process as required by the contract.

Thus, the preview control problem can be transformed to an optimization problems P1 with decision variables air pressure $P_{\mathrm{pre}}(t)$, air knife gap $D_{\mathrm{pre}}(t)$ as follows.

P1:

Obviously, the major objective of preview control is to minimize the error between the future target coating weight $Rfcw(t)$ and the predictive coating weight, which depends on current line speed $S(t)$, set point of air knife gap $D_{\mathrm{pre}}(t)$, as well as air pressure $P_{\mathrm{pre}}(t)$. i.e.

$$\min_{D_{\mathrm{pre}}(t),P_{\mathrm{pre}}(t)} \left|NNp\left(P_{\mathrm{pre}}(t),S(t),D_{\mathrm{pre}}(t)\right)-Rfcw(t)\right| \quad (3)$$

Meanwhile, the other objective is to minimize the sensitive of the coating weight to gas pressure under the line speed $S(t)$, set point of air knife gap $D_{\mathrm{pre}}(t)$, air pressure $P_{\mathrm{pre}}(t)$, which is beneficial to the stable of production, i.e.

$$\min_{D_{\mathrm{pre}}(t),P_{\mathrm{pre}}(t)} \left|\frac{NNp\left(P_{\mathrm{pre}}(t)+\Delta P,S(t),D_{\mathrm{pre}}(t)\right)-NNp\left(P_{\mathrm{pre}}(t),S(t),D_{\mathrm{pre}}(t)\right)}{\Delta P}\right| \quad (4)$$

Here, ΔP is a predefined small disturbance term.

By employing two weights w_1, w_2, the above objectives can be combined to a single one as follow:

$$\min_{D_{\mathrm{pre}}(t),P_{\mathrm{pre}}(t)} \left(\begin{array}{c} w_1\left|NNp\left(P_{\mathrm{pre}}(t),S(t),D_{\mathrm{pre}}(t)\right)-Rfcw(t)\right| \\ +w_2\left|\frac{NNp\left(P_{\mathrm{pre}}(t)+\Delta P,S(t),D_{\mathrm{pre}}(t)\right)-NNp\left(P_{\mathrm{pre}}(t),S(t),D_{\mathrm{pre}}(t)\right)}{\Delta P}\right| \end{array}\right) \quad (5)$$

In addition, the listed below should be respected:

$$D_{pre}(t) \in \left[D_{Lower}, D_{Upper}\right]$$

$$P_{\text{pre}}(t) \in [P_{\text{Lower}}, P_{\text{Upper}}] \quad (6)$$

Here, the values of $D_{\text{Lower}}, D_{\text{Upper}}, P_{\text{Lower}}, P_{\text{Upper}}$ are the upper and lower limits of air knife gap and air pressure correspondingly given in advance.

The above optimization problem can be easily solved by commonly used computational intelligence algorithm such as GA, SA or PSO etc. The result will be sent to PLC as reference.

3.3.3 *Speed-FFC*

If drastic change happens to the line speed $S(t)$, it is too late to regulate the MVs when the fluctuations is observed by coating gauge several minutes later. Thus, both the air pressure and air knife gap should be regulated immediately when change happens, to prevent big fluctuations in coating weight in the manner of feed forward control. On the other hand, if small change happens to the line speed, it is sufficient to just regulate the air pressure alone by FBC demonstrated below.

Similar to the preview control approach, the Speed-FFC solve an optimization problem with decision variables air pressure $P_{\text{FFC}}(t)$, air knife gap $D_{\text{FFC}}(t)$ on the basis of the current line speed $S(t)$ and current target coating weight $Rcw(t)$. Since the only different between the P1 and this is that the future target coating weight $Rfcw(t)$ is replaced by the current target coating weight $Rcw(t)$, the demonstration in detail is neglected here for brevity.

3.4 Feedback control with Neural Network

3.4.1 *Outline*

Different from the conventional feedback control approaches such as PID, which using the observed coating weight as feedback information, this study proposed a novel FBC framework utilizing a bias update Alvarez-Ramirez[9] predictive coating weight value rather than the actual one that with a large time-variant delay.

In particular, besides the conventional PID controller, FBC is composed by a predictive neural network module, an on-line bias update module. Hereinto, the neural network used for generating a primitive predictive coating weight has been described above while the on-line bias update module are demonstrated in detail in the following.

3.4.2 *Bias update*

The workflow of on-line bias update module is as follows.

Step 1: Calculate the primitive predictive coating weight $CWp(t)$ in equation (1) with current air pressure $P(t)$, air knife gap $D(t)$, and line speed $S(t)$

$$CWp(t) = NNp(P(t), S(t), D(t)) \quad (7)$$

Step 2: Calculate the time delay corresponding to current observed coating weight $CW(t)$. Fundamentally, this delay equals to the travel time of strip from air knife to coating gauge. Consider the simple relationship between velocity, time and distance in first principle, the following equation can be established:

$$L = \sum_{k=0}^{\tau(t)} \Delta T \cdot S(t-k) \quad (8)$$

Here ΔT is the sample cycle, $S(t), S(t-1)$,... are a series of line speed in recent sample cycles.

Based on it, a simple iterative algorithm is used to compute the time delay $\tau(t)$ as follows:

Initialization: define variable $L_{\text{SUM}} = 0, k = 0$;

Step 2.1: Calculate $L_{\text{SUM}} = L_{\text{SUM}} + \Delta T \cdot S(t-k)$;

Step 2.2: if $L_{\text{SUM}} < L$, let $k = k+1$, return to Step 2.1, otherwise go to Step 2.3;

Step 2.3: let $\tau(t) = k$, terminate the calculation.

Step 3: Shift the MVs and line speed $\tau(t)$ in time horizon, and find the value of $P(t-\tau(t))$, $S(t-\tau(t))$, $D(t-\tau(t))$ in history database. Fundamentally, the current observed coating weight $CWm(t)$ is the result of these shifted variable.

Step 4: Calculate the model predictive coating weight $CWd(t)$ corresponding to the current observed coating weight $CWm(t)$, using the past terms $P(t-\tau(t))$, $S(t-\tau(t))$, $D(t-\tau(t))$.i.e.

$$CWd(t) = NNp\,(P(t-\tau(t)),\ S(t-\tau(t)),\ D(t-\tau(t))) \qquad (9)$$

Restore the coating-weight-tuple $(CWm(t), CWd(t))$ into database.

Step 5: Bias calculation. Based on the series of coating-weight-tuples in recent sample cycles such as $(CWm(t), CWd(t))$,$(CWm(t-1), CWd(t-1))$, …, $(CWm(t-N+1), CWd(t-N+1))$, a moving average method is used to calculate the adjusted model predictive coating weight as follows:

$$CW_P'(\mathrm{t}) = CWp(t) + \sum_{k=0}^{N-1} \beta_k \cdot \big(CWd(t-k) - CWm(t-k)\big) \qquad (10)$$

Here, $\beta_k, k = 0,1,\ldots,N-1$ is a series of weight coefficients given in advance.

Different from the conventional approach that deriving the delay parameter approximately with identification method as in Smith Predictor ASTROM[10], in the above algorithm, the measurement delay of coating gauge is derived precisely, so that the predictive coating weight is matched with the plant value through shifting the process variables. On the basis of these coating-weight-tuple, the updated predictive coating weight $CW_P'(t)$ is derived and inputted into the PID controller to calculate the set point of air pressure. Through this method, the dynamic error with a trend between the model and the real system, caused by inaccurate model, working condition change, or other disturbance, can be significantly reduced.

4 APPLICATION TO A REAL PLANT

According to the mechanism of the coating weight control introduced above, a real HDCL control system is implemented at VaLin LY Steel Co., Loudi, Hunan, P.R.China.

The long-term performance of the control system is evaluated based on one month's actual data (Nov, 2015) after putting the system into operation. The rate of the “auto mode” is about 60% where totally 354 strips in different specifications are produced. The statistical result is compared with the original performance of manual operation before the implementation of the control system and presented in Table.1.

As shown in Table.1, after this system being put into operation, the variance of coating weight reduce 41%、12% and 29.3% in terms of Z80、Z120、Z180 respectively. It indicates that the fluctuation of actual coating weight in the production line is obviously decreased so that the quality of the products is improved and more zinc is saved as the coating weight is controlled more precisely. Meanwhile, the average transition time is reduced by 33.3%, indicating that the production is more stable and more qualified product is produced.

Table 1: Performance Comparison Based on Long-term Statistic Date

	Variance of coating weight (g/m^2)			Tran. Time avg. (s)
	Z80	Z120	Z180	
Manual	6.44	4.19	10.42	580
Auto	3.80	3.69	7.37	387

*Z80, Z120, Z180 represent three different kinds of strips with target coating weight 80g/m^2, 120g/m^2, 180g/m^2, respectively

5 CONCLUSION

In this paper, we proposed a novel neural-network-based coating weight control approach for HDCL. The system consists of an FFC and an FBC, which is composed by a neural network predictive model, a bias update module. Through this framework, three control puzzles, including strong nonlinear, large time-variant delay, strong disturbance are largely solved.

The industrial application results have shown that both the variance of coating weight and the transition time are significantly reduced, which is beneficial to enhance the quality of coating weight with less zinc cost.

6 REFERENCES

[1] **Adams, J., et al. 1996.** Coating mass control on no. 2 galvanizing line at LTV steel's Indiana Harbor works, *Iron and Steel Eng.*, 73(1), pp 22-27.

[2] **Liu, H.L. 2006.** Hot galvanized line zinc coat thickness close-loop control introduction, *Sichuan Metallurgy*, 28(6), pp 34-37.

[3] **Yao, L., Yang, J.H., Xu, J.W., Wang, Z. 2007.** Quality monitoring method of strip hot-dip galvanizing based on partial least squares regression, *Journal of University of Science and Technology Beijing*, 29(6), pp 627-631.

[4] **Guelton, N., Lerouge, A. 2010.** Coating weight control on ArcelorMittal's galvanizing line at Florange Works, *Control Engineering Practice*, 18(10), pp 1220-1229.

[5] **Shin, K., Park, H.D., Chung, W.K. 2006.** Synthesis method for the modelling and robust control of coating weight at galvanizing process, *ISIJ International*, 46(10), pp 1442-1451.

[6] **Lu, Y.Z., Markward, S. W. 1997.** Development and Application of an Integrated Neural System for an HDCL, *IEEE Transactions on Neural Networks*, 8(6), pp 146-156.

[7] **Lu, Y.Z., Markward, S. W. 1995.** Development and application of an integrated neural system for a HDCL, Proceedings of the American Control Conference, Westin Hotel, Seattle, Washington, June 21-23, pp 2302-2306.

[8] **Basheer, IA, Hajmeer, M. 2000.** Artificial neural networks: fundamentals, computing, design, and application, *Journal of Microbiological Methods*, 43(1), pp 3-31.

[9] **Alvarez-Ramirez, Jose. 2008.** Optimization of crude oil blending with neural networks and bias update scheme, *Engineering Intelligent Systems*, 16(1), pp 7-36.

[10] **ASTROM, KJ, HANG, CC, LIM, BC. 1994.** A new smith predictor for controlling a process with an integrator and long dead-time, *IEEE Transactions on Automatic Control*, 39(2), pp 343-345.

Railway Network Blocking Problem: Modeling Formulation andBranch-and-cut Algorithm

Xuecheng Yan

School of Traffic and Transportation, Beijing Jiaotong University,

No.3 Shang Yuan Cun, Hai Dian District, Beijing, China. 100044
Phone: 8618813099094 Fax: 86-10-51683124

13251273@bjtu.edu.cn

Yixiang Yue*

School of Traffic and Transportation, Beijing Jiaotong University,
No.3 Shang Yuan Cun, Hai Dian District, Beijing, China. 100044
Phone: 8613511054454 Fax: 86-10-51683124
yixiangyue@yahoo.com

Qunxing Yue

Institute of Space System Engineering (ISSE) of China Academy of Space Technology (CAST)
No.31 South Street, Zhong Guan Cun, Hai Dian District, Beijing, China. 100044
Phone: 8618910263978 Fax: 86- 010-68745759
icywater@163.com

Hongpeng Ma

School of Traffic and Transportation, Beijing Jiaotong University,

No.3 Shang Yuan Cun, Hai Dian District, Beijing, China. 100044
Phone: 8613810439221 Fax: 86-10-51683124

11252013@bjtu.edu.cn

Congli Hao

School of Traffic and Transportation, Beijing Jiaotong University,

No.3 Shang Yuan Cun, Hai Dian District, Beijing, China. 100044
Phone: 8617888843575 Fax: 86-10-51683124

14120828@bjtu.edu.cn

* Correspond author

ABSTRACT

Railway Blocking Problem (RBP) is the core issue of the railway; in this paper we propose a novel network based mathematic model and branch-and-cut algorithms for RBP. Firstly we analyse the complexity of RBP for network compared with a single railway line. Then we propose model formulations and the details of the branch-and-cutalgorithm for RBP. The objective function is to minimize the costs of flow traveling and delay for the trains in marshalling stations, by deciding which block is built and specifying the assignment of commodities to these blocks, while observing limits on the break-up/make-up capacity at each terminal. The model is solved using branch-and-cut algorithm. The algorithm has very fast computation speed due to the principle of cutting. The model and algorithm is tested on a real-world railway network located in North of China, the computation results show that the model and algorithm have the potential to apply and can yield the dramatic railroad's operating costs saving.

Keywords: RBP, branch-and-cut algorithm, optimization, railway network

1 INTRODUCTIORN

1.1 Railway Blocking Problem

The Railway Blocking Problem (RBP) determines how to aggregate a large number of shipments into blocks of shipments as they travel from origins to destinations [1]. In other words, RBP determines which blocks should be built at each yard and what shipments should be placed in the block. RBP is one of the most important decision in freight railroads. A good solution of RBP can save railway operation costs of delivering all commodities by deciding which inter-terminal blocks are built and specifying the assignment of commodities to these blocks, while observing limits on the number and aggregate volume of the blocks assembled at each terminal. And these costs are usually broken down into car-handling costs associated with handling (or blocking) a car and car-miles costs associated with the movement of a car. And we use the Mixed Integer Programming (MIP) to build the model in this paper.

1.2 Literature review

In the study about blocking problem of single railway line, Xu [7] and Yang [8] proposed a 0-1 programming with the target of minimum balanced using of adaptation capacity and hour of freight train in marshalling station. In connection with these models, Wu [9] and Zhuo [10] proposed Genetic Algorithms for a problem defined on a line network.

At present, in spite of blocking problem of technical station single freight train having mature arithmetic, the arithmetic can't solve extensive blocking problem. So for complicated RBP, there are very few study in the field. Li [11]proposed Chance Constrained Programming with considering flow, assembly time and volatility of vehicle adaptation extra time consuming. And they developed Branch-and-cut algorithms to solve it. Lin [12] proposed Nonlinear Mixed Integer Programming Model with the minimum of car detention time under accumulation and marshalling time. They developed Simulated Annealing to solve it. Lin [13] proposed Bilevel Programming Model with the minimum of flow organization cost and the maximum of flow adapted distance. Wang [14] proposed stochastic optimization model. They developed Simulated Annealing to solve it.

Newton [15] and Newton et al. [16] modeled the blocking problem as a network-design model and formulated it as a MIP.They developed column generation and branch-and-price algorithms to solve it. Barnhart et al. [4] used the same formulation as Newton et al. and proposed the Lagrangian Relaxation technique to decompose the problem into two sub-problems. Bodin [17] established Nonlinear Mixed Integer Programming model with the target of minimum total cost of adaptation and transport, and proposed heuristic decomposition method. Newton established Linear Mixed

Integer Programming model with the target of minimum transport cost, and proposed Dantzig-Wolfe decomposition method in literature [16].

2 PROBLEM DESCRIPTION

The target of RBP is to determine which route should be chosen and which blocks should be built at each yard. Due to some constraints, such as passing capacity of sections, receiving/departure capacity and break-up/make-up train capacity of marshalling stations, the physical shortest route may not be the optimal solution of RBP.

We give a simple example of comparison between RBP of railway single line and network. The two networks are shown as Figure 1.

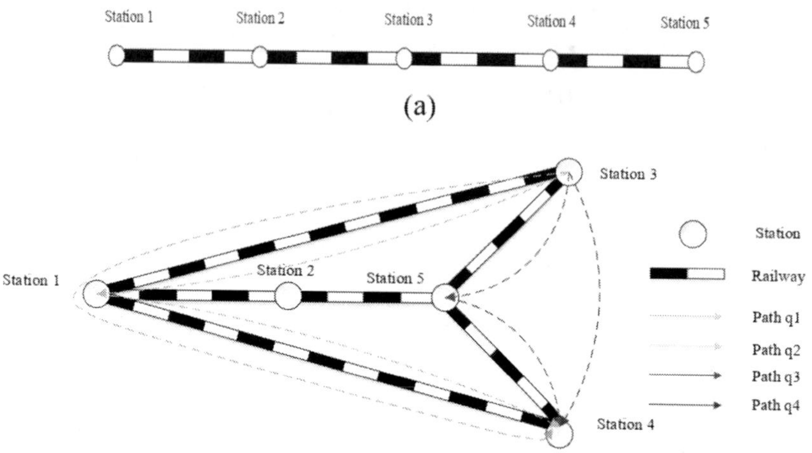

(b)

Figure 1: Railway Network Structure

Figure 1(a) is a single railway line including five marshalling stations, with only one route between two marshalling stations. To every traffic flow, we need to determine whether it is necessary to break up and then make up a new train at each marshalling station in its journey.

To other traffic OD flows, the RBP solutions can be enumerated. Suppose there are enough transport demand for any two marshalling stations, the number of transport plans is shown in Table 1calculating by Equation (1).The equation about the number of grouping methods of passing through the pathway for the shipment is as follow.

$$M = \sum_{i=0}^{N} C_N^i = 2^N \qquad (1)$$

Figure 1(b) is a railway network including five marshalling stations. Compared with single railway line, there may be more than one route between two marshalling stations.

And there is only one marshalling station on every route. So according to Equation (1), the number of grouping methods of passing through every route is 2. Because there are many routes in railway network, each route can be considered as a single line, the growth numbers of RBP solutions for network are explosive. We give the number of transport plan between any two marshalling stations is shown in Table 1.

Table 1: The Number of Possible Solutions (Single Railway Line/ Railway Network)

arrival leave	1	2	3	4	5
1	-	1/1	2/9	4/9	8/6
2	1/1	-	1/4	2/4	4/1
3	2/9	1/4	-	1/4	2/1
4	4/9	2/4	1/4	-	1/1
5	8/6	4/1	2/1	1/1	-

Based on transport plan between any two marshalling stations, we can give the formulation of total solution number of RBP as Equation (2). Suppose station *i* is originating station and station *j* is final station, and number of possible solution between station *i* and station *j* is $q_{i,j}$. So total number of solutions for RBP in network is:

$$n = \prod_{i,j} q_{i,j} = \prod_{i,j} 2^{N_{i,j}} \qquad (2)$$

Obviously, RBP is more complicated than single railway line blocking problem. The number of combinations for routing shipments on railroad network grows exponentially with the number of terminals. And when number of terminals increases to 50, or 500, even 5000, there will be huger combinations.

3 MIP MODEL FOR BRP OF RAILWAY NETWORK

3.1 Influence factors of RBP

There are 6 main influence factors of RBP: 1.Delay on marshalling with reclassification operation. Every shipment will be reclassified at each yard on its itinerary in this situation. Thus, there are the delays on marshalling with reclassification operation. Furthermore, the increase in reclassification of cars may cause congestion at some yards [18].2. Accumulation delay. This is because that there are not enough classification tracks for storing large numbers of outbound trains in many yard sand the daily demand of a single OD pair may be too small to guarantee dispatching a direct train service. [19]. 3.Flow transport cost. 4. Passing capacity of sections. At the same time, in the same section, on the same line can only be occupied by a train, and double tracks railway must ensure the separation distance of train operation [20]. 5. Receiving/departure capacity [21]. 6. Break-up/make-up Capacity. Break-up/make-up capacity refers to the number of freight trains that the station shunting device within a day and night can break up of trains and organize into groups under the circumstance of using legitimately technical equipment.

3.2 Formulation of MIP model

Due to so many factors and interdependent constraints, when railway network become large enough, RBP will become much complicated. In order to construct mathematical models, we try to simplify RBP. For passing capacity of sections, with the rapid development of railway and continuous improvement of road network, the influence of passing capacity of sections is getting smaller and smaller. Receiving/departure capacity in connection with station yard layout, track allocation schemes and flow organization, can increase by adjusting station yard layout and organizing legitimately flow. So when establishing models, we will not consider passing capacity of sections and receiving/departure capacity.

3.2.1 Notation

The notation for the proposed formulation is first introduced in Tables 2-4, followed by a few illustrative examples. For the convenience of using algebraic modeling language, we use these tables to list general subscripts, input parameters and decision variables, respectively.

Table 2: General subscripts

Symbol	Description
i, j, m, n	macroscopic node index, corresponding to physical station in a railway network, i, j ,m ,n∈ N, N is the set of nodes
k, q	macroscopic node index, corresponding to origin and destination station of commodity in a railway network

Table 3: Input parameters

Symbol	Description
$c_{i,j}$	One car transport cost from station *i* to station *j*.
p	Conversion factor that changing car-hour to cost, where *p*=80.
$m_{i,j}$	Car detention time under accumulation from station *i* to station *j*.
t_i	The save time of transit car passing station *i* without resorting.
$u_{i,j}$	Blocking capacity from station *i* to direction *j*.
$d_{k,q}$	The number of cars from station *k* to station *q*.
$b_{i,j}$	Through trains judgment factor, $b_{i,j}=1$ if there is through trains from station i to station j, and otherwise $b_{i,j}=0$.
M	The great number

Table 4: Decision variables

Symbol	Description
$y_{i,j}$	0-1 binary variables, $y_{i,j}$=1, if there is train from station *i* to station *j*, and otherwise $y_{i,j}$=1.
$x_{i,j,k,q}$	The number of cars from station i to station j carried the goods from station *k* to station *q*.

3.2.2 MIP model

$$
\begin{aligned}
z &= min\sum_{i,j}\sum_{k,q} x_{i,j,k,q}c_{i,j} + p[\sum_{i,j} m_{i,j}y_{i,j}b_{i,j} + \sum_{j} t_j(\sum_{i,k,q} x_{i,j,k,q} - \sum_{k} d_{k,j})] \\
&= min\sum_{i,j,k,q} x_{i,j,k,q}(c_{i,j} + pt_j) + p\sum_{i,j} m_{i,j}y_{i,j}b_{i,j} - p\sum_{j,k} d_{k,j}
\end{aligned} \quad (3)
$$

$$
s.t. \sum_{i,q} x_{i,j,k,q} - d_{k,j} \le u_{j,k} \qquad \forall j,k \quad (4)
$$

$$\sum_j x_{i,j,k,q} = d_{k,q} \qquad \forall i,k,q \text{ and } i = k \tag{5}$$

$$\sum_i x_{i,j,k,q} = d_{k,q} \qquad \forall j,k,q \text{ and } j = q \tag{6}$$

$$\sum_j x_{i,j,k,q} - \sum_j x_{j,i,k,q} = 0 \qquad \forall i,k,q \quad i \neq q \text{ and } i \neq k \tag{7}$$

$$x_{i,j,k,q} \leq M y_{i,j} \qquad \forall i,j,k,q \tag{8}$$

$$x_{i,j,k,q} \geq 0 \qquad \forall i,j,k,q \tag{9}$$

Constraint (4) is the forcing constraint requiring that the number of blocking cars satisfied break-up/make-up capacity. Constraints (5), (6) and (7) are flow balance constrains.

4 BRANCH-AND-CUT ALGORITHM FOR MIP MODEL

The model is the Mixed Integer Programming Model, so branch-and-cut is chosen to solve MIP model [21]-[23]. Branch-and-bound is a technique to solve a discrete variable problem by solving a sequence of simpler problems derived from the original problem. The search is organized via a branch-and-bound tree. The solution of each problem on the tree gives a lower bound on the solutions of the problems that are descendants, of that problem, on the tree. The leaves of a tree correspond to all the feasible solutions.

In Figure 2 case, suppose there are enough transport demand for any two marshalling stations, steps by branch-and-cut to solve are as follows:

Step 1. Determine the upper bound of the original objective function.

Step 2. Set A is the set of flows without determined transport plan.

Step 3. Branch. We need change blocking problem of anyone flow which doesn't exist in set A, with blocking problem of other flows fixedly. So we get a branch.

Step 4. Judge. If one branch doesn't satisfied constraints of break-up/make-up capacity, revising the branch until satisfying break-up/make-up capacity constraints. The new upper bound can be calculated.

Step 5. Compare. The upper bound is compared with the minimum objective function value of all branch. If the upper bound is bigger, we have got feasible solution of the branch corresponding to the minimum objective function value. Continuing Step 6. If the upper bound is smaller, algorithm is end and feasible solution has been got.

Step 6. Returning to Step 3.

Flow chart is shown in Figure 2.

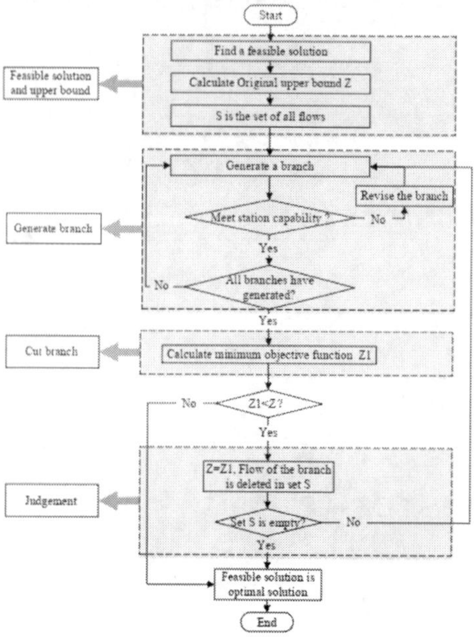

Figure 2: Flow Chart of Branch-and-cut

5 NUMERICAL EXAMPLE

5.1 Introduction of the network

Based on eight marshalling stations, the railroad network of North China is constructed to calculate as follows in Figure 3.

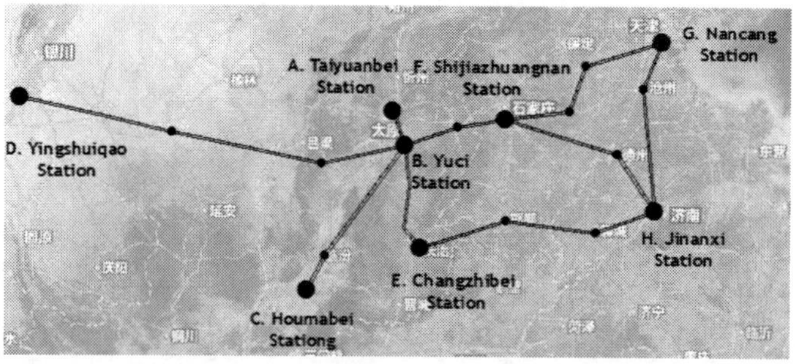

Figure 3: Case Railroad Network

There are eight marshalling stations including in station A, station B, station C, station D, station E, station F, station G, station H. In spite of the rail network simply, blocking problem is complicated. And we will use station B as the centre of the network for analysing.

5.2 Descriptions of experimental settings

We get the number of transit car with resorting and transit car without resorting of station B. And according to the number, flow transport cost between any two nodes can be calculated as Table 5.

Table 5: Demand between Each Pair OD

Arrival leave	A	B	C	D	E	F	G	H
A	0	190	67	125	106	13	9	34

B	161	0	231	22	24	315	153	396
C	72	442	0	161	17	131	160	29
D	89	401	293	0	9	132	49	185
E	59	135	4	3	0	4	3	0
F	15	36	231	360	40	0	50	32
G	11	21	282	134	26	30	0	0
H	38	30	50	505	4	59	0	0

The mileage between the pair OD and flow transport cost between each pair OD are in Table 6.

Table 6: The Mileage between Each Pair OD/Flow Transport Costs between Each Pair OD

arrival leave	A	B	C	D	E	F	G	H
A	0/0	34/84	339/832	782/832	274/324	242/292	643/693	559/609
B	34/84	0/0	305/355	748/798	240/290	208/258	609/659	525/575
C	339/389	305/355	0/0	1053/1103	415/465	513/563	914/964	824/874
D	782/832	748/798	1053/1103	0/0	988/1038	956/1006	1357/1407	1273/1323
E	274/324	240/290	415/465	988/1038	0/0	380/430	781/831	486/536
F	242/292	208/258	513/563	956/1006	380/430	0/0	401/451	317/367
G	654/704	609/659	914/964	1487/1537	781/831	401/451	0/0	352/402
H	559/609	525/575	824/874	1273/1323	486/536	317/367	352/402	0/0

Accumulation-time from station B to other station is in Table 7.

Table 7: Accumulation Delay from Station B to Other Stations

departure	A	C	D	E	F	G	H
Number of cars of one train of station B	49.3	49.3	52.6	42	41.8	50.5	52.3
Accumulation Parameter of station B	9.1	10.1	8.2	9	9.3	10.6	4.1
Accumulation delay of station B	449	498	431	378	389	535	215

Accumulation delay of other stations is in Table 8.

Table 8: Accumulation Delay of Other Stations

station	A	C	D	E	F	G	H
Accumulation delay	550	550	636	600	500	530	530

There are transit car times with reclassification, transit car time without resorting and accumulation delay in "Station Operation Detailed Rules". So the save time of transit car without resorting can be calculated by the formula as Table 9.

Table 9: the Save Time of Transit Car without Reclassification

station	A	B	C	D	E	F	G	H
t_i	3	2.9	4	4.7	3	3	4	4

Break-up/make-up capacity can be calculated by station type as Table 10.

Table 10: Break-up/make-up Capacity

station	A	B	C	D	E	F	G	H
Break-up/make-up capacity of B	321	-	651	1110	39	1125	1125	1125
Break-up/make-up capacity of F	1125	1125	1125	1125	88	-	700	1100
Break-up/make-up capacity of E	500	500	500	500	-	500	500	30

5.3 Solution and Analysis

In order to verify the results of Branch-and-cut for MIP model, we use General Algebraic Modeling System (GAMS) to solve MIP model of RBP. And we got the result as follows. To illustrate the optimal solution clearly, the rail network is abstracted as two railway lines in Figure 4.

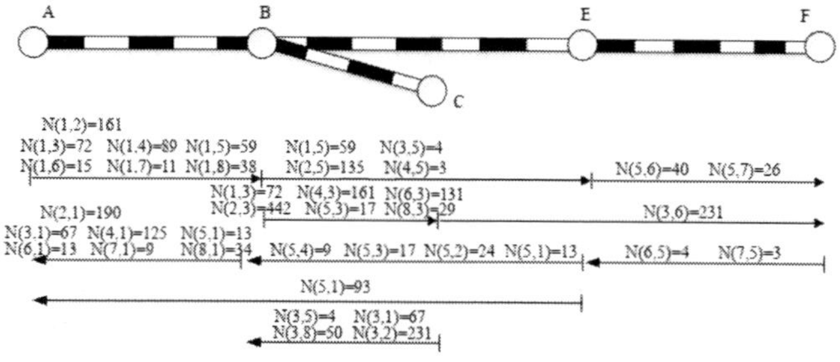

(a)

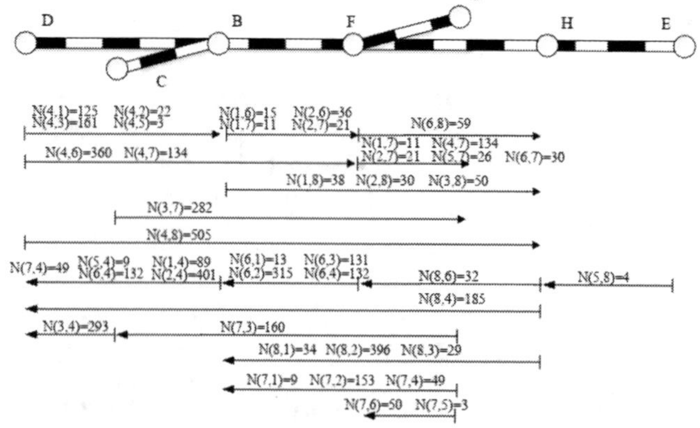

(b)

Figure 4: The solution of variable$x_{i,j,k,q}$

And we get that the numerical example's objective function value is : $z = 5496920$ .And we used the data of the flow of transit car with resorting and transit car without resorting of station B to get that the objective function value of real-world in the condition of the same parameters is $z_{actual} = 6527430$.

Comparison with real-world RBP: Compared with the actual blocking problem, the feasibility of the model can be verified from the following aspects: First, Some directed blocks are canceled, for example, there are 13 cars from station F to station A. Because of a directed block for the flow, there are 500 car-hours about car detention time under accumulation and 38 car-hours of the save time because of transit car without resorting. The directed block causes the loss of 400 car-hours, so the directed block is canceled. Second, some directed blocks are added, for example, there are 293 cars from station C to station D. Because of a directed block for the flow, there are 550 car-hours about car detention time under accumulation and 850 car-hours of the save time because of transit car without resorting. The directed block causes the loss of 300 car-hours, so the directed block is canceled. Third, the cost is smaller, there are two pathways between station E and station G. One is from station E to station B, from station B to station F and from station F to station G. Other is from station E to station F, from station F to station G. The mileage of the second pathway is shorter than the first, so the solution chooses the second. And the results show that the blocking plans reduce the railroad's operating costs by millions of dollars annually by comparing objective function value of RBP by MIP model and Branch-and-cut and the objective function value of real-world BRP. Fourth, the solution meets blocking capability, the solution meets blocking capability of every station, and some flow routes are changed because of the limit of blocking capability. Fifth, the line capability is taken full advantage; the solution considers the balance of railway line. And some flows in busy railway line are adjusted to other rail lines to improve transportation efficiency.

6 CONCLUSIONS

The paper mainly study on modelling formulation and branch-and-cut algorithm of railway network blocking problem. The objective function of RBP model is to minimize the costs of flow traveling and hours of freight train in marshalling station by deciding which inter-terminal blocks to build and specifying the assignment of commodities to these blocks, while observing limits on break-up/make-up capacity at each terminal. And the paper introduces branch-and-cut algorithm to solve this problem, and verify feasibility of Branch-and-cut for MIP model by GAMS. The model is tested on blocking problems from the railway network based on the actual data. By analysing the solution, it can be concluded that the model and algorithm can optimize blocking problem. The results show that the blocking plans generated have the potential to reduce the railroad's operating costs by millions of dollars annually.

7 REFERENCE

[1] **Yaghini M, et.al.2011**. "Solving railroad blocking problem using ant colony optimization algorithm, " *Applied Mathematical Modelling*, vol. 35, pp. 5579-5591.

[2] **R.K. Ahuja, et.al.2007**. "Solving Real-Life Railroad Blocking Problems, " *Interfaces*, vol. 37 pp. 404-419.

[3] **Masoud Yaghinia and Rahim Akhavan,2012**. "Multicommodity Network Design Problem in Rail FreightTransportation Planning, " *Procedia - Social and Behavioral Sciences*, vol. 43,pp. 728-739.

[4] **Barnhart C, et.al.*2000*.**"Railroad Blocking: A Network Design Application, " *Operations Research*, vol. 48, pp. 603-614.

[5] **Ahuja,et.al.2005**. "Network Models in Railroad planning and scheduling," *Operation Research*, vol. 1, pp. 54-101.

[6] **Balakrishnan, A, et.al.1989**. "A Dual-ascent Procedure for Large-scale Uncapacitated Network Design," *Operations Research*, vol. 73, pp. 716-740.

[7] **Xu Hong, et al.2006**. "Study on the Model and Algorithm of the Formation Plan of Single Group Trains at Technical Service Stations," *Jourmal of the China Railway Society*, vol. 28, pp. 12-17.

[8] **Yang Shi-gang, et al.2002**."An Artificial Neural Network Method for Marshalling Plan," *Journal of Changsha Railway University*, vol. 20, pp. 79-84.

[9] **Zuo Wu**. "Study of Genetic Algorithms for Marshalling Plan of Group Freight Train," *Central South University*.

[10] **Zuo Wu, et.al.2011**. "The Study on Genetic Algorithms of Freight Train Blocking Problem," *journal of Shijiazhuang Railway Institute(Nature Science)*, vol. 24, pp. 84-90.

[11] **Li Xiao-yin,2002**."Study on Optimization of Marshalling Plan and Flow Path Based on Uncertain Parameters," *Southwest Jiaotong University*.

[12] **Lin Bo-liang,1997**." A Simulated Annealing Approach to Railroad Operating Problem with Limits on Yard and Line Capacity," *Journal of Beijing Jiaotong University*, vol. 21, pp. 264-272.

[13] **Lin Bo-liang, et.al.2011**. "Double Programming Optimizition Model of Blocking Problem Based on Farthest Station Law," *Journal of the China Railway Society*, vol. 32, pp. 108-113.

[14] **Wang Bao-hua and He Shi-wei,2009**. "Stochastic Optimization Model and Algorithms of Flow Group Problems," *CHINA RAILWAY SCIENCE*, vol. 30, pp.104-108.

[15] **H.N. Newton,1996**. "Network Design Under Budget Constraints With Application to The Railroad BlockingProblem, " *Auburn University*.

[16] **H.N. Newton, et.al.1998**. "Constructing Railroad Blocking Plans to Minimize HandlingCosts, " *Transportation Science*, vol. 32, pp. 330-345.

[17] **Bodin L, et.al.1980**. "A Model for the Blocking of Trains," *Transportation Research Part B Methodological*, vol. 14, pp. 115-120.

[18] **Lin B L, *et.al.*2012**."Optimizing the freight train connection service network of a large-scale rail system, " *Transportation Research Part B Methodological*, vol. 46, pp. 649-667.

[19] **Yang Hao,2006**."Railway Transport Organization.".

[20] **Li Hai-ying,2014**. "Railway Yards and hub".

[21] **Yaghini M, et.al.2011**. "Solving Railroad Blocking Problem Using Ant Colony Optimization Algorithm," *Applied Mathematical Modelling*, vol. 35, pp. 5579-5591.

[22] **Yue Y, et.al.2011**. "Multi-route Railroad Blocking Problem by Improved Model and Ant Colony Algorithm in Real World," *Computers & Industrial Engineering*, vol. 60, pp. 34-42.

[23] **Zhu E, et.al.2014**. "Scheduled Service Network Design for Freight Rail Transportation," *Operations Research*, vol. 62, pp. 383-400.

GENERAL OPTIMIZING ALGORITHM FOR URBAN RAIL DIAGRAM USING PASSENGER FLOW AND TRAIN FLOW

Mingxuan Zhong
School of Traffic and Transportation, Beijing Jiaotong University,
No.3 Shang Yuan Cun, Hai Dian District, Beijing, China. 100044
Phone: 8618813096344 Fax: 86-10-51683124
13251152@bjtu.edu.cn

Yixiang Yue*
School of Traffic and Transportation, Beijing Jiaotong University,
No.31 Shang Yuan Cun, Hai Dian District, Beijing, China. 100044
Phone: 8613511054454 Fax: 86-10-51683124
yixiangyue@yahoo.com

Qunxing Yue
Institute of Space System Engineering (ISSE) of China Academy of Space Technology (CAST)
No.31 South Street, Zhong Guan Cun, Hai Dian District, Beijing, China. 100044
Phone: 86 18910263978 Fax: 86- 010-68745759
icywater@163.com

Shifeng Wang
School of Traffic and Transportation, Beijing Jiaotong University,
No.3 Shang Yuan Cun, Hai Dian District, Beijing, China. 100044
Phone: 8613810439221 Fax: 86-10-51683124
13120904@bjtu.edu.cn

Leishan Zhou
School of Traffic and Transportation, Beijing Jiaotong University,
No.3 Shang Yuan Cun, Hai Dian District, Beijing, China. 100044
Phone: 8613511054454 Fax: 86-10-51683124
lshzhou@bjtu.edu.cn

ABSTRACT

We focus on a general optimizing algorithm for urban rail diagram using passenger flow and train flow under different types of transit lines with given time-varying origin-to-destination flows. We use passenger volumes in section to replace origin-to-destination flows and construct a general form for all types of transit lines. Based on this, a nonlinear integer programing model with linear constraints is developed to minimize waiting times for passengers; and an efficient local improvement algorithm is presented to solve it. We take three cases in Beijing Urban Rail Transit (URT) network as real-world examples; the computation results prove that the optimization model and algorithm are effective and efficient and have potential for practical implementation.

Keywords: Urban rail transit, Train working diagram, Rolling stock using plan, Passenger waiting time, Time-varying demand

* Corresponding author

1 INTRODUCTION

Train Working Diagram (TWD) of an URT is important to urban public transport system. A high quality TWD can decrease passengers' waiting times and improve train operation efficiency. The scheduling task and challenge for public transit management is optimizing operation and rolling stock usage depending on passenger flow. Over the years, many scholars dedicated to study the optimization of TWD for urban rail diagram, and many useful related studies have been observed.

Many algorithms and models have been put forward to optimize coordinated timetables design and train operation with time-dependent demand, such as the real-time control model of Chien et al. [1], the multi-commodity flow model of Borndörfer et al. [2], and the two-level approach (computation of transport offer, timetable design) of Albrecht [3]. Niu and Zhou use accumulative input-output diagrams and a local improvement algorithm to find optimal timetables for individual station cases [4]. Another paper of them focuses on the demand-oriented passenger train-timetable optimization for a rail corridor [5]. Canca et al. tackle a problem of scheduling in networks by means of a non-linear integer-programming model [6]. Min et al. consider the train-conflict resolution problem for which the optimal conflict-free timetable and use a column-generation method to compute it [7]. Corman et al. considers a bi-objective problem of minimizing train delays and missed connections in order to provide a set of feasible non-dominated schedules to support this decisional process [8]. Kecman et al. propose underlying algorithms automatically identify route conflicts with conflicting trains, and reconstruct the train paths on track section and blocking time level [9]. Meanwhile, urban railway train usage planning has important influence on operating costs and transportation service quality. Tong et al. [10], Liu and Yue [11], Corman et al. [12] respectively use ant colony algorithm, Rolling Time and Space Assignment Model, and the rolling horizon scheme to deal the urban vehicle-use problem. However, they compute train usage planning not considering passenger flows.

However, some proposed models, including input parameters, variables and constraints, are complicated to solve and compute. Simplification for parameters and variables is necessary in order to improve computing efficiency. Some papers focus on arranging trains depending on conflicts and use some algorithms to solve it; but they can only compute under some established condition. Further, there are few studies on TWD considering both passenger flows and train flows.This paper will focus on rail TWD problem with the main challenge of connection of passenger flow and train flow. We are intended to use passenger volumes and train flows to propose a general model which can find the minimum waiting times for passengers, and give a local improve algorithm to solve it.

2 PROBLEM DESCRIPTION

We consider train operation on bi-directional urban rail lines. Each train departs from depot when start to serve passengers, and return to depot when end to serve passengers. In our study, all trains' running times in section and dwell times at stations are given and trains need stop at every station they pass by. Specially, stations connected to depot are more important, we call them **"D-station"**.

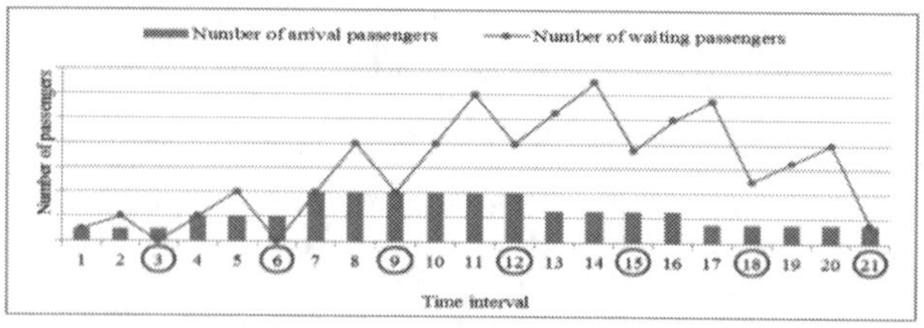

Figure 1: Illustration for Connection of Passenger Flows and Train Flows

We use Figure 1 to introduce the connection of passenger flows and train flows. In Figure 1, we give a time span with 21 time intervals. Given that, trains depart at time 3, 6, 9, and so on. The bur chart is the number of arrival passengers and the line chart is the number of waiting passengers. Passenger flows is time varying. Time span before "time 6" is in case of under-saturation and the other is in case of oversaturation. In case of oversaturation, urban rail operator has to put as more train us into operation as possible. In case of under-saturation, urban rail operator does not need to put so many trains into operation.

From the users' perspective, service coverage and frequency level is much important. From the operator's perspective, the objective is for the system to make as much profit as possible. It is the main challenge in transit planning to find equilibrium between these conflicting objectives [13].

2.1 The Description of Passenger Flow and Train Flows

For rail transit operator, he may more concerns passenger volume in section (the total number of passengers from one station to next station). We can give a formula to calculate the number of passengers from one station to next station by using related passenger flows. When not considering trains capacity, some passenger volumes in different section can always get on the same train. In this way, we can get that passenger volumes that in pane will always get on the same train. We regard these passenger volumes as a set. The number of passengers in a set may very different from each other. To make the number of passengers in a set is equal, we can define a reference value to replace initial values.

A real-world rail transit line is constructed depending on urban planning and urban land use. For a linear transit line, up trains and down trains are running separately, there is no conflicts between two directions in intermediate stations. Therefore, we can regard tracks have different stations in different directions, and these stations form a loop line. In this way, we can modify any transit line as a loop line with one or multiple depots. It is a general form to describe transit line. In a one-depot transit line, trains can only depart from the depot and return to the depot (Scenario 1). In a multiple-depot transit line, there are two management patterns. One is that trains only depart from a depot and return to the origin depot (Scenario 2). The other is that trains depart from a depot and enter into any depot (Scenario 3). As Figure 2 shows, in Scenario 1 and Scenario 2, trains' trajectory can be regarded as a loop, in Scenario 3, trains trajectory can be regarded as a loop or a part of loop. Please note that, a loop line needs to consider trains' trajectory both clockwise direction and anti-clockwise direction, while a linear line only needs to consider one direction.

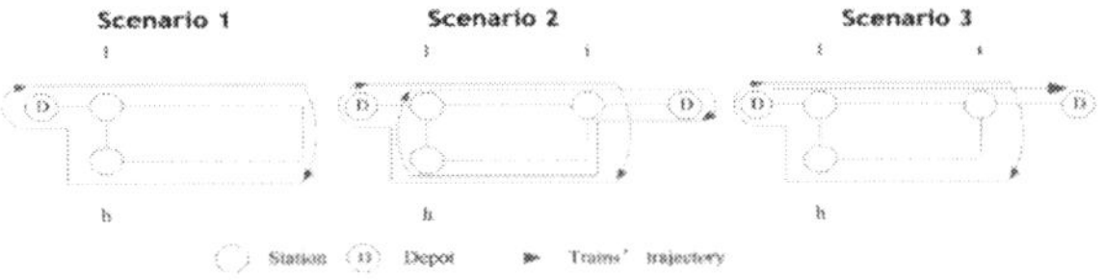

Figure 2: Illustration of Trains' Trajectories

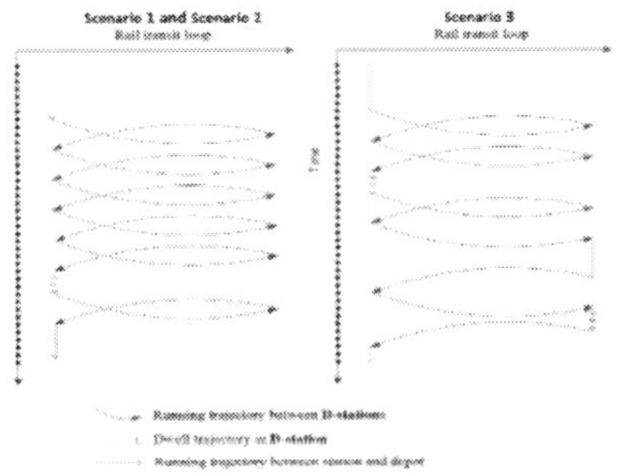

Figure 3: Illustration for Different Types of Trains' Trajectories

A complete trajectory can be composed of three parts: 1) running trajectories between stations, 2) dwell trajectories in stations, 3) running trajectories between station and depot (including go to, depart from and dwell at depot). We give a simpler schematic of trains' trajectory as shown in Figure 3. Running times between **D-stations** for all trains are constant, and trajectories between **D-stations** are pictured in helical line, In **Scenario 1** and **Scenario 2**, the trajectory is a complete loop, while in **Scenario 3**, the trajectory is a complete loop or a part of a loop. In our study, we only need to consider trains departure times in **D-stations** and the set of passenger volumes is between a **D-station** and the next **D-station**. In previous model, the number of parameters for passenger flows is $r \times h \times h \times 2$ and the number of trajectories is $r \times h \times 2 \times g$. In our study, the numbers decrease to $\mathrm{r} \times \mathrm{d}$ and $\mathrm{r} \times \mathrm{d} \times \mathrm{g}$. If a transit line has 10 stations and 2 depots, the number of parameters for passenger flow in our study is only 1% compared to previous model, and the number of decision variables is 10%.

3 TWD MODEL FOR URBAN RAIL TRANSIT

In our paper, we consider how to optimize the TWD for URT according to waiting times for passengers in a loop and ensure operation efficiency for trains. By describing and simplifying passenger flows and train flows, we use a simpler and general model to describe TWD. Meanwhile, we minimize waiting times in objective function and ensure operation efficiency for trains in constraints.

3.1 Notations

The notations, input parameters and variables for the optimization problems under consideration are listed as follows.

General subscripts used in mathematical formulations are summarized as follows:

G	set of trains, the maximum number of trains is g
H	set of stations, $\|H\| = h$
D	set of D-stations, $D \in H, \|D\| = d$
R	set of time, $[1, r]$
f	index of trains, $f \in G$
i, j, k	index of stations and depots, $i, j, k \in H$
t, s	index of time, $t, s \in R$

Parameters used in mathematical formulations are summarized as follows:

$p_{i,j}^{t}$	number of passengers who arrive at station i towards station j at time t
$\overline{p}_{k,k+1}^{t}$	passenger volume in section k (between station k and station $k+1$) at time t
G_{capa}	rated capacity for loading people for a train
q	minimum loading-rate for a train
$o(f)$	depot of train f dwelling at in origin time, $o(f) \in D$
c	minimum departure headway (time interval) between two consequent trains at a same station
M	maximum dwell time for trains stop at D-station
N	maximum running time for a train in one day
$R_{i,j}$	running time between station i and station j
R_{loop}	running time of a loop or a round trip

Variables are as follows:

$v_{i,j}^{t}$	passenger volume between station i and station j at time t
w_{i}^{t}	number of passengers waiting in station i at time t, it is an accumulative variable
$G_{oper}^{t}(f)$	chance for train f to be selected to start to operate at time t

Decision variables are as follows:

$x_{i,j}^{t,s}(f)$	1, if train f departs from station i at time t and arrives at station j at time s 0, otherwise $\forall f \in G, i, j \in H, t, s \in R$

$y_i^{t,s}(f)$	1, if train f dwells at D-station i from time t to time s 0, otherwise $\forall f \in G, i \in D, t, s \in R, \lvert t-s \rvert = 1$
$z_i^{t,s}(f)$	1, if train f departs from, return to or dwells at depot i from time t to time s 0, otherwise $\forall f \in G, i \in D, t, s \in R$

$x_{i,j}^{t,s}(f)$, $y_i^{t,s}(f)$ and $z_i^{t,s}(f)$ represent operation status of train f.These three variables decide w_i^t.

3.2 Mathematical Formulation

1) Objection Function

In our model, we replace waiting time of each passenger with the number of waiting passengers in each time interval. The objection function is :

$$min\, L = \sum_{t \in R} \sum_{k \in D} w_k^t \tag{1}$$

In the model, $x_{i,j}^{t,s}$ can be rewritten as $x_{k,k+1}^{t,t+R_{k,k+1}}$, and $v_{i,j}^t$ can be rewritten as $v_{k,k+1}^t$,$(\forall I, j \in H, k \in D)$.

1) Timetable continuity constraint

For a train, the number of departure times in a time interval is no more than one, and equal to the number of arrival times

$$\left(x_{k,k+1}^{t,t+R_{k,k+1}}(f) + \sum_{s \in R} y_k^{t,s}(f) + \sum_{s \in R} z_k^{t,s}(f)\right) \leq 1 \; \forall f \in G, k \in D, t \in R \tag{2}$$

$$\left(x_{k,k+1}^{t,t+R_{k,k+1}}(f) + \sum_{s \in R} y_k^{t,s}(f) + \sum_{s \in R} z_k^{t,s}(f)\right) - \left(x_{k-1,k}^{t-R_{k-1,k},t}(f) + \sum_{s \in R} y_k^{s,t}(f) + \sum_{s \in R} z_k^{s,t}(f)\right) = \begin{cases} 1 & t=1 \\ -1 & t=r \\ 0 & otherwise \end{cases} \quad \forall f \in G, k \in D, t \in R \tag{3}$$

2) Passenger flow constraints

The number of waiting passengers in time t is the number of waiting passengers in time $t-1$ adds the number of arrival passengers in time t, and subtracts the number of departure passengers in time t. In case of oversaturation, the number of departure passengers is equal to G_{capa} and in case of under-saturation, the number of departure passengers is less than G_{capa}. Considering cases of oversaturation and under-saturation, we use “greater than” in formula (7) and formula (8).

$$w_k^t \geq 0 \tag{4}$$

$$w_k^1 \geq v_{k,k+1}^1 - \sum_{f \in G} x_{k,k+1}^{1,1+R_{k,k+1}}(f) \times G_{capa} \quad \forall k \in D \tag{5}$$

$$w_k^t \geq w_k^{t-1} + v_{k,k+1}^t - \sum_{f \in G} x_{k,k+1}^{1,1+R_{k,k+1}}(f) \times G_{capa} \quad \forall k \in D, t \in R/\{1\} \tag{6}$$

The number of real departure passengers need more than $G_{capa} \times q$ to ensure operation efficiency.

$$w_k^{t-1} + v_k^t - w_k^t \geq \sum_{f \in G} x_{k,k+1}^{1,1+R_{k,k+1}}(f) \times G_{capa} \times q \quad \forall k \in D, t \in R/\{1\} \tag{7}$$

(3) Train flow constraints

Two trains in one direction cannot depart from the same station, only one train can stay in the station at a time interval. Besides, a train cannot dwell in a D-station for too long in a loop.

$$\sum_{t-c}^{t+c} \sum_{f \in G} x_{k,k+1}^{t,t+R_{k,k+1}}(f) \leq 1 \quad \forall k \in D, t \in R \tag{8}$$

$$\sum_{f \in G} \sum_{s \in R} y_k^{t,s}(f) \leq 1 \quad \forall k \in D, t \in R \tag{9}$$

$$\sum_{t}^{t+R_{loop}} \sum_{s \in R} \sum_{k \in D} y_k^{t,s}(f) \leq M \quad \forall f \in G, t \in [1, r - R_{loop}] \tag{10}$$

$$\sum_{t\in R}\sum_{k\in D} x_{k,k+1}^{t,t+R_{k,k+1}}(f) \times R_{k,k+1} \le N \ \forall f \in G \tag{11}$$

(4) Constraints of depot

A running trajectory is a complete loop. These constraints are only active in **Scenario 2**.

$$z_k^{t,s}(f) = 0 \quad \forall t,s \in R, f \in G, k \in D/\{o(f)\} \tag{12}$$

The number of trains in each depot is constant in a day. This constraint is only active in **Scenario 3**.

$$\sum_{t\in R}\sum_{f\in G} x_{k,k+1}^{t,t+R_{k,k+1}}(f) - \sum_{t\in R}\sum_{f\in G} x_{k-1,k}^{t-R_{k-1,k},t}(f) = 0 \ \forall k \in D \tag{13}$$

In **Scenario 1**, constraints are constraints (2)-(11); In **Scenario 2**, constraints are constraints (2)-(12); In **Scenario 3**, constraints are constraints (2)-(11) and constraint (13). These three scenarios are a nonlinear integer programing with linear constraints.

3.3 Local Improvement Algorithm

In our paper, we use local improvement algorithm to solve our proposed models. We judge w_k^t at each time interval and improve it by changing $x_{i,j}^{t,s}(f)$. For a one-depot transit line (**Scenario 1**), it just considers passenger flow and train flow for a **D-station.** For multi-depot transit lines, it need consider influences between **D-stations**. For **Scenario 2** and **Scenario 3**, when $x_{i,i+1}^{t,t+R_{i,i+1}}(f) = 1$, $x_{i+1,i+2}^{t,t+R_{i+1,i+2}}(f)$may equal to "1". Meanwhile, we give a dynamic parameter, $G_{oper}^t(f)$ to represent chance of train f selected to start to operate at time t. $G_{oper}^t(f)$ is calculated depending on its trajectory and accumulative running time in a day.

Firstly, we give a roulette algorithm for selecting and arranging trains. Please note that, it may output null, due to Train flow constraints.

Algorithm 1. Roulette algorithm for selecting and arranging trains

```
Step 1 (Initialization)
    sum=0
Step 2 (Summation)
    For train index f=1 to g
        sum = sum + G^t_oper(f)
    End for //f
Step 3 (Generate random number)
    Generate a random number: u (u ∈ (0,sum])
Step 4 (Get train f)
    sum=0
    For train index f=1 to g
        sum = sum + G^t_oper(f)
        If sum ≥ u
            Output f
            Update x^{t,s}_{i,j}(f), y^{t,s}_i(f), z^{t,s}_i(f)
            Update G^t_oper(f)
            Break
        End if
    End for //f
```

The following is a developed local improvement algorithm for solving proposed models.

Algorithm 2. Developed local improvement algorithm for proposed models

Step 1: (Initialization)

  Calculate $v^t_{k,k+1}$ $(k \in D)$ using $p^t_{i,j}$ $(\forall i, j \in H)$.

Step 2: (Iterative computation)

  For time index $t = 1$ to r

    For **D-station** index $k = 1$ to d

      If $\sum_{f \in G} x^{f,t,t+R_{k,k+1}}_{k,k+1} = 0$

        Calculate $w^{t-1}_k + v^t_{k,k+1}$ (or w^1_k)

        If $w^{t-1}_k + v^t_{k,k+1} \geq G_{capa} \times q$

          Select and arrange trains using **Algorithm 1**

        End if

      End if

      Calculate w^t_k

    End for // k

  End for // t

Step 3: (Termination)

  Check and let the diagram appropriate

4 NUMERICAL EXAMPLES

In this study, we take three real-world instances (Line 2, Line 10 and Line 13) from the Beijing urban rail transit network to test our proposed method.

4.1 Introduction to Cases

Line 2 is a loop line with one depot (**Scenario 1**); Line 10 is a loop line with three depots (**Scenario 2**); and Line 13 is a linear line with one depot (**Scenario 3**). These three lines can serve 3.5 millions of passengers in a day. Figure 4 shows their scales.

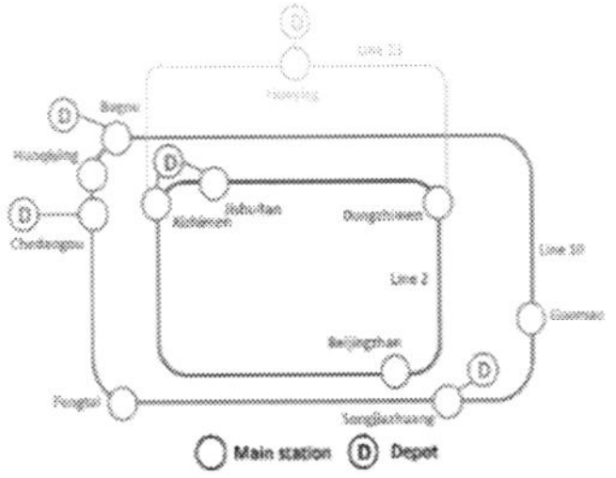

Figure 4: Schematic of Line 2, Line 10 and Line 13 in Beijing Railway Transit Network

4.2 Input Data and Parameter Settings

Table 1 is some detailed descriptions about three lines (Line 2, Line 10 and Line 13). Figure 5 is their number of arrival passengers per one minute in a day. The data in Table 1 and Figure 5 originate from Wikipedia and Baidu Encyclopedia. Our time span is from 5:30 A.M. to 23:30 P.M..

Table 1: Overview of Line 2, Line 10 and Line 13

Name	Transit line type	Number of depots	Number of stations	Number of trains	Running time in a loop(min)	Distance of a loop(km)	Average number of passengers in a day (million)
Line 2	5	1	18	42	40	23.1	1.2
Line 10	6	3	45	116	107	57.1	1.5
Line 13	3	1	16	72	140	91	0.8

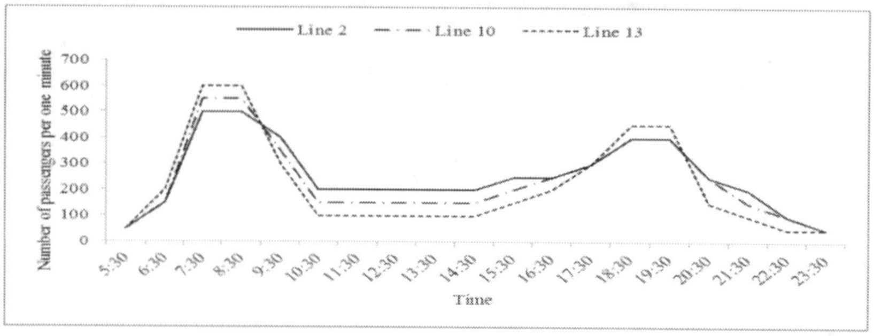

Figure 5: Schematic of Number of Passengers per One Minute

In our study, related parameters are defined as Table 2 shows. In order to improve balance of trains' running times, we define several cases to calculate chances selected to start operate.

Table 2: Parameters' Definitions

Definition of Parameters of Trains		Definition of $G^{t}_{oper}(f)$		
Symbol	Value	Case	Definition	Value
G_{capa}	1600(person)	Case 1	Dwelling at D-station (accumulative running time < 0.6×N)	100
		Case 2	Dwelling at D-station (accumulative running time ≥ 0.6×N)	1
q	1.6	Case3	Stopping in depot(accumulative running time < 0.6×N)	10
c	2 (min)	Case 4	Stopping in depot(accumulative running time≥ 0.6×V and <N)	1
M	5 (min)	Case 5	Stopping in depot (accumulative running time ≥ N)	0
N	720 (min)	Case 6	In operation	0

4.3 Results and Analyses

We use Line 2(**Scenario 1**). The depot of Line 2 is connected to "Xizhimen" and "Jishuitan". We add a virtual station, "V-Station", between them and regard it as an only **D-station**. The complete loop is "V-Station→Xizhimen→Beijingzhan→Jishuitan→V-Station". Figure 6(a) is the obtained diagram for Line 2 (just anti-clockwise direction) and Figure 6(b) is the obtained usage of trains for Line 2. In Figure 6(a), the horizontal axis is time and vertical axis is station. In Figure 6(b), the horizontal axis is time and vertical axis is index of trains. Gray rectangle means this train is running.

We use Line 10 to test (**Scenario 2**). Line 10 is a loop line with three depots, which connected to "Bagou", "Chedaogou" and "Songjiazhuang". Figure 6(c) is the obtained diagram for Line 10 (just clockwise direction) and Figure 6(d) is the obtained usage of trains for Line 10.

We use Line 13 to test (**Scenario 3**). Line 13 is a linear line with one depot, connected to "Huoying". Its terminal stations are "Xizhimen" and "Dongzhimen". We split "Huoying" into two **D-stations**, "Huoying-E" and "Huoying-W". The complete loop for Huoying-E is "Huoying-E→Dongzhimen→Huoying-W→Xizhimen→Huoying-E". The complete loop for Huoying-W is "Huoying-W→Xizhimen→Huoying-E→Dongzhimen→Huoying-W". However, trains can run in a part of loop. Figure 6(e) is the obtained diagram for Line 13 and Figure 6(f) is the obtained usage of trains for Line 13.

Figure 6 shows our rail diagram is feasible and train usage is appropriate in all three types of models.

(a) The Obtained Diagram for the Line 2 (Anti-clockwise Direction)

(b) The Obtained Usage of Trains in Line 2

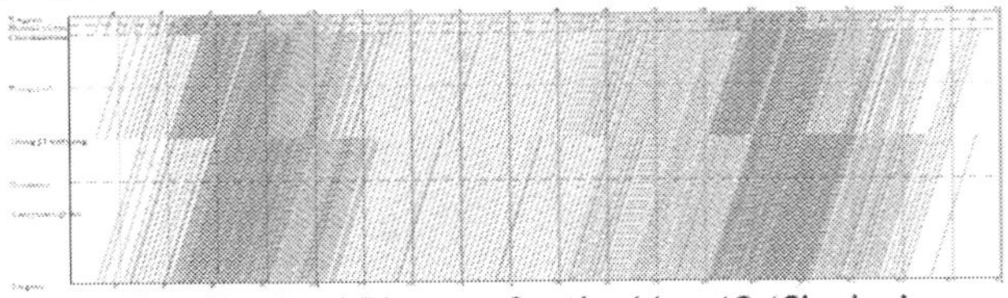

(c) The Obtained Diagram for the Line 10 (Clockwise Direction)

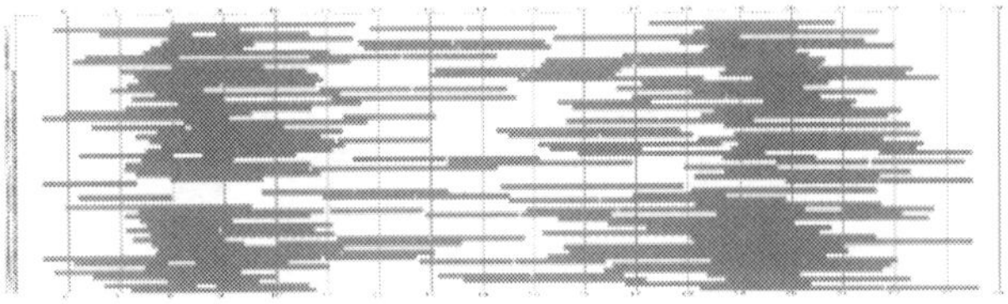

(d) The Obtained Usage of Trains in Line 10

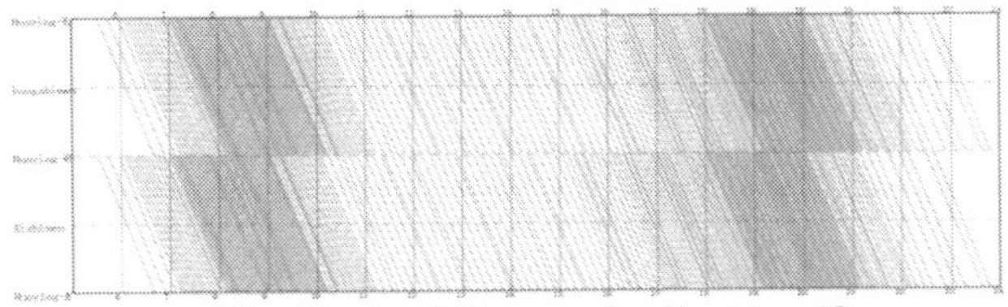

(e) The Obtained Diagram for the Line 13

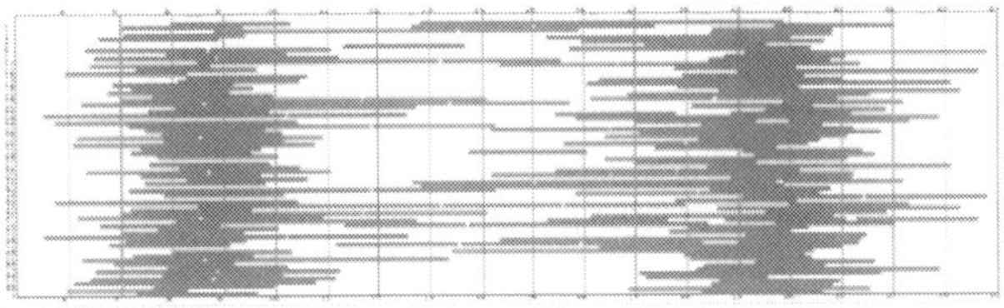

(f) The Obtained Usage of Trains in Line 13

Figure 6: The Obtained Diagram and Train Usage by Our Model

After that, we will give some comparisons between our optimal diagrams and regular diagrams.

We record the number of waiting passengers in each time interval at each **D-station** in our optimal diagrams and calculate an average number of all **D-stations** per 30 min. Figure 7(a) shows the results. It shows that number of waiting passengers is stable and in a low range.

In the regular diagram commonly used in practice, the headways of the even-spacing schedule are 2 min from 7:00A.M. to 10:00A.M. and from 5:00P.M. to 8:00P.M., 8 min for other time spans. We also calculate an average number of all **D-stations** per 30 min. Figure 7(b) shows the results. It shows that number of waiting passengers is not stable and in a very high range in some time span.

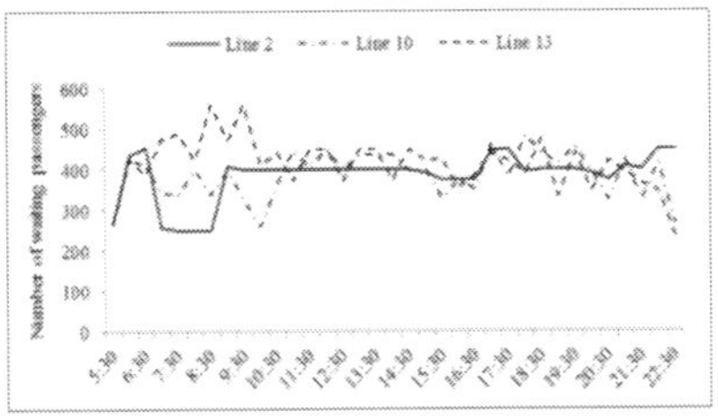

(a) Waiting Passengers in Diagram in Our Diagram

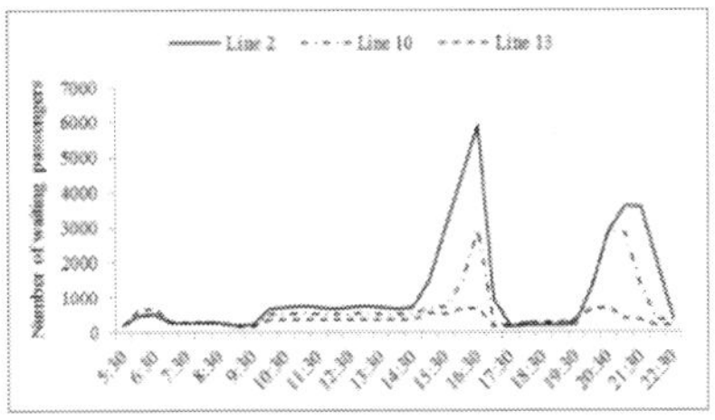

(b) Waiting Passengers in Regular Diagram

Figure 72: Schematic of Waiting Passengers in Different Diagrams

Table 3: Comparison of Different Diagrams

Line	Type	Average of arrival passengers per minute	Average of waiting passengers per minute	Average of waiting time(min)	Ratio of waiting time	Number of loops	Ratio of loops
Line 2	Our study	264	385	1.46	34%	538	99%
	Regular diagram		1121	4.25	100%	542	100%
Line 10	Our study	256	400	1.56	53%	508	98%
	Regular diagram		757	2.96	100%	518	100%
Line 13	Our study	245	413	1.69	102%	227	89%
	Regular diagram		405	1.65	100%	256	100%

Finally, we present a comparison of different diagrams for three lines in Table 3. For Line 2, we decrease waiting time for passengers for about 65%; for Line 10, we decrease waiting time for passengers for about 47%, and number of loops (running times of trains) in these two cases remain unchanged. For Line 13, we decrease loops (running times of trains) over 10%, with waiting time remains unchanged. Thus, we can draw the conclusion that our proposed models and algorithm provides satisfactory performance when applied to urban train diagram problem.

5 CONCLUSIONS

This paper focuses on a general optimizing algorithm for urban rail diagram using passenger flows and train flows. We replace origin-to-destination flows with passenger volume in section and propose a general form for train flows. Based on this, we develop a nonlinear integer programing model with linear constraints, whose objective function is waiting times for passengers, and design a local improvement algorithm to solve it. We give three cases of Beijing rail transit network as real-world examples. In two of them, we decrease waiting times for passengers for about 50%, and in another case, we decrease running times of trains over 10%. It is sure that our proposed models and algorithm are effective, efficient and useful for application in the real world.

6 REFERENCES

[1] **Y.Ding and C. Steven, 2001**. Improving transit service quality and headway regularity with real-time control, Transportation Research Record: Journal of the Transportation Research Board, vol.1760, pp. 161-170,

[2] **R.Borndörfer, et al.**, 2007. A column-generation approach to line planning in public transport, Transportation Science, vol. 41, pp.123-132,.

[3] **T. Albrecht,** 2008. Automated timetable design for demand-oriented service on suburban railways, Public Transport, vol. 1, pp. 5-20,

[4] **H.Niu and X. Zhou,** 2013. Optimizing urban rail timetable under time-dependent demand and oversaturated conditions, Transportation Research Part C Emerging Technologies, vol. 36, pp. 212-230,

[5] **H Niu, X Zhou, R Gao,** 2015. Train scheduling for minimizing passenger waiting time with time-dependent demand and skip-stop patterns: Nonlinear integer programming models with linear constraints, Transportation Research Part B Methodological, vol. 76, pp. 117-135,

[6] **D.Canca, et al.**, 2011. Confrontation of Different Objectives in the determination of train scheduling, Procedia-Social and Behavioral Sciences, vol.20, pp.302-312,

[7] **Y H.Min, M J Park,** S P Hong, S H Hong et al., 2011. An appraisal of a column-generation-based algorithm for centralized train-conflict resolution on a metropolitan railway network, Transportation Research Part B Methodological, vol.45, pp. 409-429,

[8] **F.Corman, et al.**, 2012. Bi-objective conflict detection and resolution in railway traffic management, Transportation Research Part C Emerging Technologies, vol. 20, pp. 79-94,

[9] **P.Kecman and RMP. Goverde**, 2013. Process Mining Approach for Recovery of Realized Train Paths and Route Conflict Identification, Transportation Research Board 92nd Annual Meeting, America, Washington DC, Jan 13-17, p14.

[10] **L.Tong, et al.**, 2009. Application of Ant Colony Algorithm in Train-Set Scheduling Problem, Journal of Transportation Systems Engineering\s&\sinformation Technology, vol. 9, pp.161-167,

[11] **C.Liu and Y. Yue,** 2012. "Optimization model and method for application modes of long and short routing in urban rail transit," Beijing jiaotong university.

[12] **F.Corman and E. Quaglietta,** 2014. Closing the Loop in Railway Traffic Management, Transportation Research Board 93rd Annual Meeting, America, Washington DC, Jan 12-16 p17

[13] **V.Guihaire and J K. Hao**, 2008. Transit network design and scheduling: A global review, Transportation Research Part A Policy & Practice, vol. 42, pp. 1251-1273,.

A KNOWLEDGE-BASED FRAMEWORK FOR DESIGN AUTOMATION OF AGRICULTURAL VEHICLE TRANSMISSION SYSTEM

Xiaoping Jin[1*]
[1]College of Engineering
China Agricultural University, Beijing, China
jinxp@cau.edu.cn

ABSTRACT

This paper explores methodologies for design automation of agricultural vehicle transmission system. To realize the application, the knowledge-based engineering is adopted aiming at achieving knowledge reuse and fast design by shifting from manual modeling to computer aided design automation. The proposed framework is based on four separate models: component model, structure model, process model and mesh model. Particularly the product architecture is represented by a tree structure while the design process by a list of activities with possible iterations and data and information transference. Finally, application examples are presented in which the proposed framework is assessed.

Keywords: Agricultural vehicle, Knowledge-based engineering, Design automation, Transmission system, Design process

1 INTRODUCTION

As one of the most important and largely-planted food crops in China, wheat requires higher mechanization in harvesting. Agricultural vehicle manufacturers in China are faced with difficulties in fast designing new product and promoting competitiveness in marketing because of lacking of advanced design methods and techniques [1]. Especially for combine harvesting machinery, as working in very intense environment, they need to be designed with more power, higher speed and better efficiency. As a key part of the machines, the transmission system influencing the whole set performance is a basic part to guarantee the whole operation quality.

Design Automation (DA) systems are important part of manufacturing automation, using IT technology as basis and CAD technology as platform to integrate domain knowledge and automatically generate a solution to specific design problem [2]. Design activities include choosing proper parts and components, the dimensioning, the validation or the optimization of the parts and assemblies. DA refers to various tasks of the product development, from automatic modification of parametric geometric models and automatic definition of parts in the embodiment design to the automatic detailed design of parts or even the generation of new conceptual design [3]. And it is based on the representation of the product and of the design process after having acquired and formalized the associated knowledge. Generally, three main issues should be considered: product structure, characteristics of the design process and the evaluation of product configuration result [4].

* Corresponding Author

Currently, a typical mechanical final product is often obtained assembling instances of a fixed set of components and these components interact with each other in a predefined way. This requires definition of general product architecture and its characterization according to design specifications. For the purpose of fast design, it usually needs to reconstruct the structure of a product (with different variants) from the same set of data. It represents not only the hierarchical organization of the product with all possible components, but also the knowledge/rules how they assembled. Colombo et al [4] proposed a knowledge based system for chemical plant design using this idea, setting a good example for this study. However, it did not concern the detailed design (e.g. CAE analysis) other than embodiment design.

For the above data processing and flow of the customized product, it easily causes management difficulties using traditional solutions. Computer aided techniques can be used to support design problems which spent much time on iterations or data exchange, assuring the best possible solutions and less time for the design targets [5-6]. For instance, the KBE (Knowledge Based Engineering) systems can well help to solve this problem by simplifying and automating the design process [7-9].

This paper discusses a framework of a knowledge based application to automate design process by taking crawler combine harvester transmission system as example, which is a typical representative of agricultural vehicles.

2 KNOWLEDGE MODELLING

Knowledge in design automation is sharable among all designing sections. It consists in replacing design tasks that require human intervention by procedures and automatically generating the desired solutions. So it needs a deep analysis of the acquisition and representation of the knowledge in design process. Two types of knowledge are introduced, that is, product knowledge and design process knowledge.

2.1 Classification of transmission system

A transmission mechanism can change running style, directions and velocity of power as required. Three types of transmission systems are usually used in crawler combine harvester (Figure 1), (A) Mechanical-hydraulic mixed type, (B) Mechanical type, and (C) Hydraulic type.

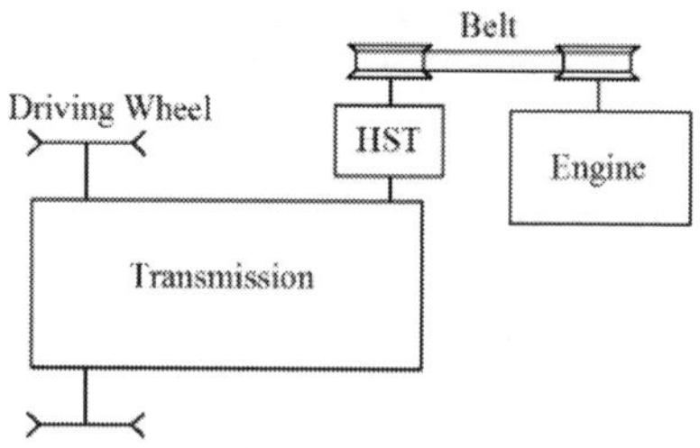

(A)

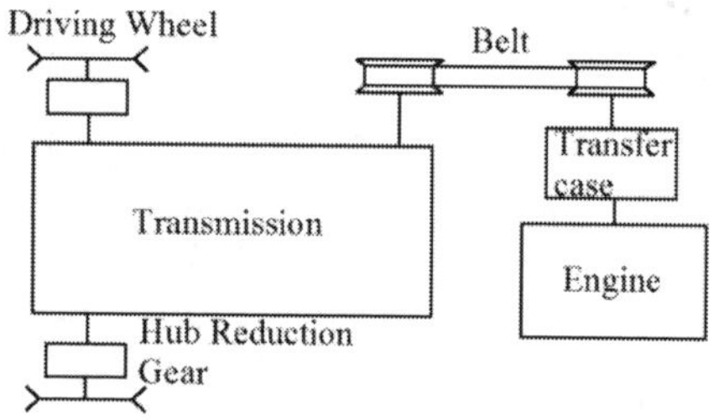

(B)

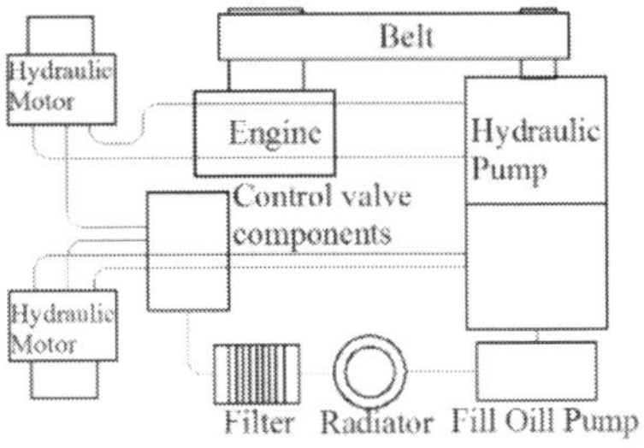

(C)

Figure 1: Three types of transmission system: (A) Mechanical-hydraulic (B) Mechanical (C) Hydraulic

2.2 Product knowledge

The product knowledge here refers to both component and structure knowledge. Some methods and tools are available to represent product related knowledge [1]. Here the object-oriented method is adopted. Firstly, the product architecture was analysed, consisting of three main functional subsystems (taking the mechanical-hydraulic mixed type for example): hydraulic transmission subsystem, mechanical gearbox and control mechanisms. Figure 2 showed the product configuration represented in a tree structure with the possibility of illustrating the subsystems and their primary sub-components. It contained not only the information of components, but also the information of structure, e.g. design rules and constraints linking to properties of all components.

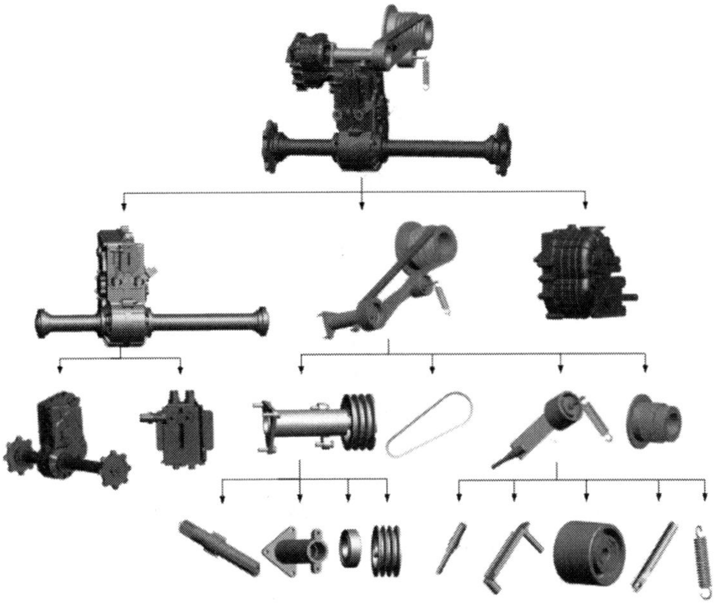

Figure 2: Tree structure of a transmission system

The tree structure was translated into an object-oriented pattern where the subsystems, components and relationships were well represented. Figure 3 illustrated the object-oriented diagram for the subsystems. It contained the attributes of each subsystem, e.g. the power efficiency, speed, drive ratio and number of gear shift, etc.

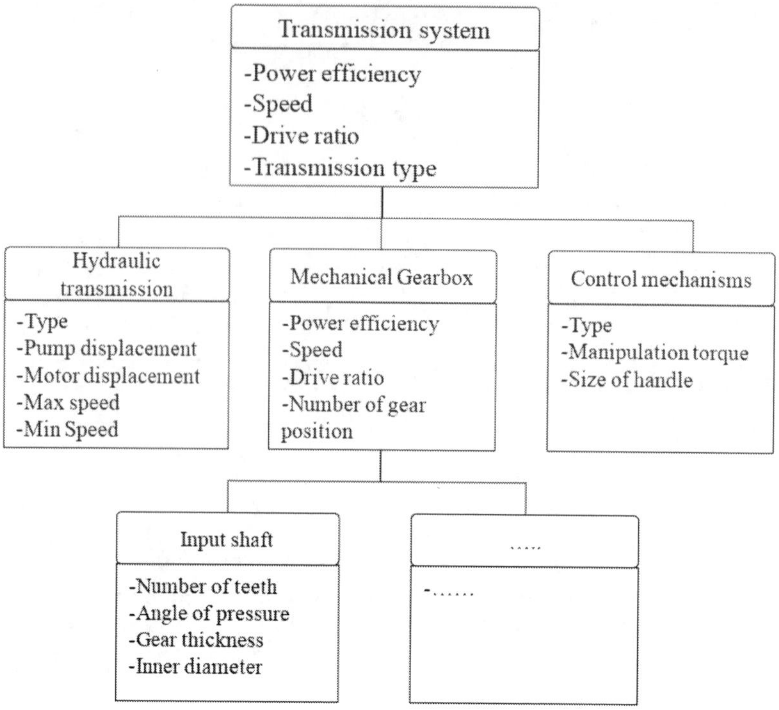

Figure 3: Component properties representation in object-oriented mode

2.3 Design process knowledge

The design process of a transmission system configuration mainly depended on three interrelated design activities: determination of fundamental dimensions / parameters of primary parts using general mechanical design rules, then selection of parts and assemblies, and verification of their values. In detail the following activities were performed in parallel but not in sequence so as to save time: firstly, selection of transmission layout and calculation of drive ratio according to design requirements; next, gearbox and belt drive model selection; next, parts parameters calculation and determination; finally, the check of design results.

2.4 Knowledge modelling

The following activities were characterized as knowledge-based procedure for design automation of agricultural vehicles:

- To build the hierarchic product structure representing all possible configuration of the transmission system;
- To define the properties of each part and assembly;
- Design rules representation;
- To build the parametric geometric model of each part in three-dimensional CAD software platform;
- Representation of the design process flow;
- To develop a user interface as design guidance.

The application was developed based on four separate models: component model, structure model, process model and mesh model. The component model and structure model (Figure 4a) mainly represented product parts and assembling information. In the process model (Figure 4b), there were four functional modules, input, design, analysis and output. The input module needed designers to provide design requirements and initial parameters for mechanical calculation, e.g. drive ratio. In the design module, the detailed design and type selections were conducted, such as hydraulic transmission type and gearbox type selection, belt parts detailed design, etc. After that, check of parts strength, stiffness and other performances would be done in the analysis module using the mesh model for CAE analysis. Finally, the output module would produce three-dimensional CAD models and related documents.

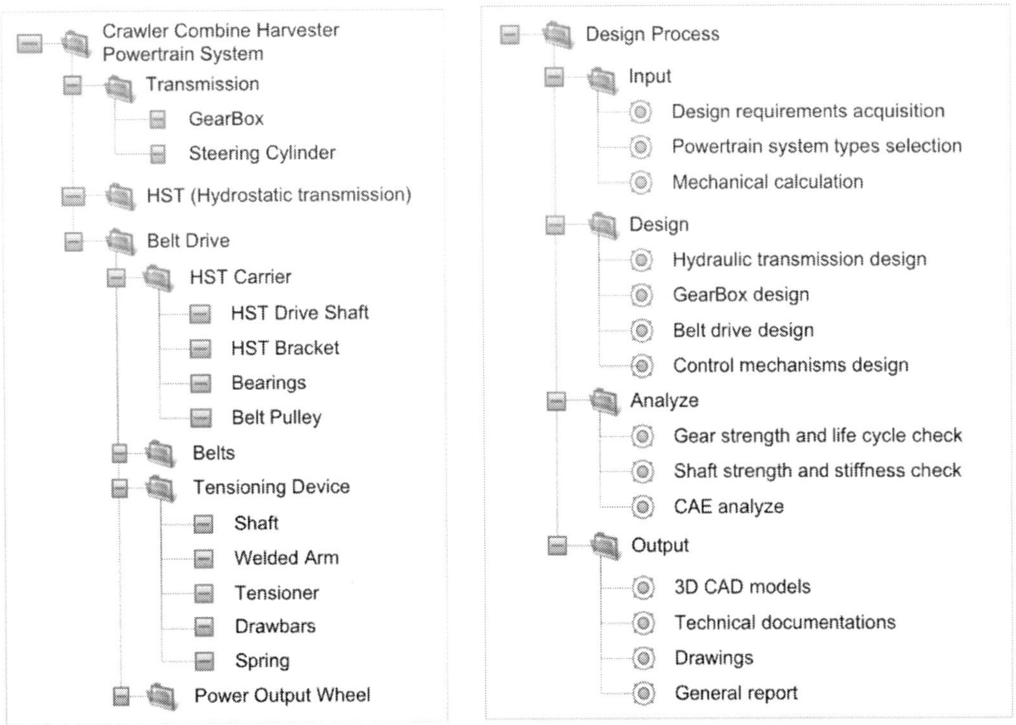

(a) Structure model **(b) Design process model**

Figure 4: Models of product structure and design process

The communication of data and information transference within the design process was shown in Figure 5. It was a procedure consisting of automatic iteration and sharable design knowledge as guideline.

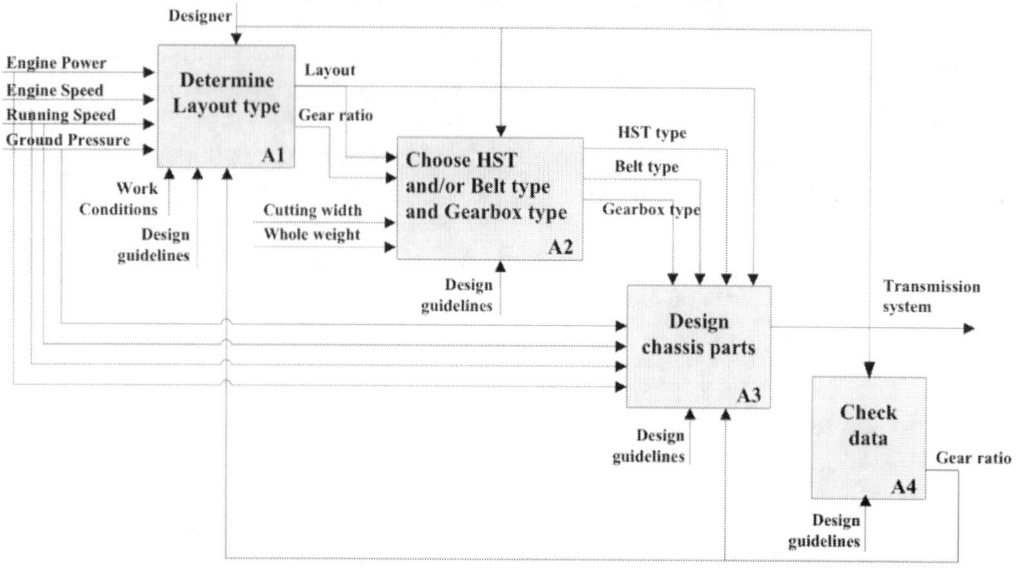

Figure 5: Data and information flow in design process

3 CASE STUDY AND CONCLUSION

Framework of the models in knowledge-based design automation application for agricultural vehicle transmission system was described in Figure 6. And Figure 7 pictured the user interface of the application system. Entering the input data, such as the whole weight, engine speed and engine power, etc., a solution was worked out for crawler combine harvester transmission system design. The obtained result had the accordance with those obtained by traditional process [10], and could improve design efficiency indeed. The results of the tests confirmed the potential of such application for complex products design. The possibility to follow task sequence and the simultaneous control of design parameters and rules also allow designers to share knowledge within the domain easily.

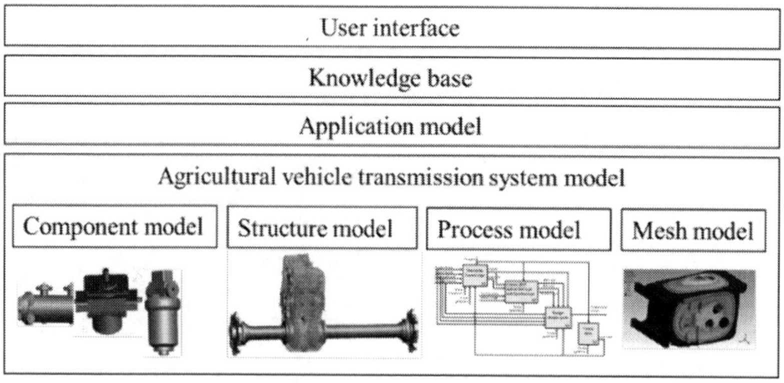

Figure 6: Framework of the models in knowledge-based design automation system

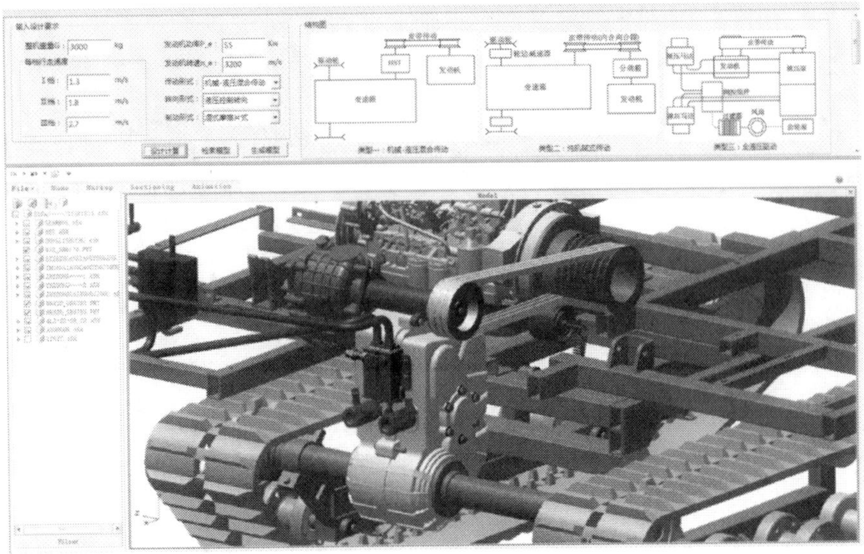

Figure 7: User interface of the knowledge-based application

4 REFERENCES

[1] **Bi, S.Q., Song Z.H., Jin X.P., et al. 2013.** Design of Knowledge Base for Crawler Combine Harvester Transmission System, *Agricultural Engineering*, 03, pp 94-98.

[2] **Elgh F. 2008.** Supporting management and maintenance of manufacturing knowledge in design automation systems, *Advanced Engineering Informatics*, 22(4), pp 445-456.

[3] **Lin Y.S., Shea K., Johnson A., et al. 2009.** A Method and Software Tool for Automated Gearbox Synthesis. ASME 2009 International Design Engineering Technical Conferences and Computers and Information in Engineering Conference, San Diego, CA, USA, pp 111-121.

[4] **Rizzi C., Colombo G., Scotto M. 2011.** A knowledge based application to enhance fired heater design. Research in Interactive Design. Vol. 3, Virtual, Interactive and Integrated Product Design and Manufacturing for Industrial Innovation.

[5] **Amadori K., Tarkian M., Ölvander J., et al. 2012.** Flexible and robust CAD models for design automation, *Advanced Engineering Informatics*, 26(2), pp 180-195.

[6] **Sandberg M, Tyapin I, Kokkolaras M, et al. 2011.** A Knowledge-based Master-model Approach with Application to Rotating Machinery Design. *Concurrent Engineering*, 19(4), pp 295-305.

[7] **Tosserams S., Kokkolaras M., Etman L.F.P. and Rooda, J.E. 2010.** A Non-hierarchical Formulation of Analytical Target Cascading. *Transaction of the ASME Journal of Mechanical Design*, 132(5), 051002.

[8] **Rasmussen C.C., Canfield R.A. and Blair M. 2009.** Optimization Process for Configuration of Flexible Joinedwing. *Structural Multidisciplinary Optimization*, 37,pp 265-277.

[9] **Kesseler, E. 2006.** Advancing the State-of-the-art in the Civil Aircraft Design: A Knowledge-based Multidisciplinary Engineering Approach, Proceedings of the European Conference on Computational Fluid Dynamics, Egmond aan Zee, The Netherlands, pp. 1-14.

[10] **Song Z.H., Bi S.Q., Jin X.P., et al. 2013.** Rapid Design Reasoning Method for Crawler Harvester Transmission System, *Transactions of the Chinese Society for Agricultural Machinery*, S2, pp 268~272.

CIE46 Proceedings, 29-31 October 2016, Tianjin / China

SYSTEM DYNAMICS BASED IOT CONTROL STRATEGY ASSESSMENT

FOR TYPICAL CUSTOMIZED PRODUCTION SYSTEM

B.Q. Tan[1], T. Qu[2*], C.D. Li[2], G.Q. Huang[3]
[1] Guangdong CIMS Provincial Key Lab
Guangdong University of Technology, Guangzhou, China
tan_bqing@163.com
[2] Institute of Physical Internet
Jinan University (Zhuhai Campus), Zhuhai, China
quting@jnu.edu.cn
[3]Department of Industrial and Manufacturing Systems Engineering
The University of Hong Kong, Hong Kong, China
gqhuang@hku.hk

ABSTRACT

Customized Production (CP) is uncertain and complicated in the production process caused by customers' dynamic demands. This paper presents the analysis of two IoT-based control strategies in CP systems. These approaches combined production system layout policies with an IoT deployment. In order to analyze control strategies, we propose to use System Dynamics as a structural modeling and simulation approach to achieve the statistical optimization of the operation by controlling the structure and parameters of the CP system. The results show improvements of quality on the part of all these production system layout policies. Sensitivity analysis demonstrates that quality related decisions, supported by the proposed control strategies, bring important performance and quality improvements. Moreover, the research can enhance the use of system dynamics, present a quantitative IoT system analysis approach and provide guidance on IoT implementation in CP.

Keywords: Customized Production (CP), Internet of things (IoT), System Dynamics (SD)

1 INTRODUCTION

Customized production (CP) is a powerful manufacturing mode for utilizing flexibility of production system to response small-batch, high-variety and frequently-changing personalized customer requirement by sharing the resources such as a series of standard unit promptly. CP is similar to mass-customization production (MCP), both of which are customers' demand-oriented and make-to-order (MTO) production mode of producing diverse products by using a large-scale mass production strategy to satisfy different customer requirements and to gain benefits both from mass production. Unlike MCP, customers play a leading roles in CP and obtain their own exclusive products by designing personalized attribute, which are produced by manufacturers. However, in typical CP mode, large-scale mass production strategy is hardly satisfied with multiple personalized attributes, which results in increasing production cost. On the other hand, highly stochastic events (e.g. the arrival of customer orders and operators maintenance) lead to dynamic disturbances in production system. Furthermore, these real-time and highly frequent disturbances cause snow-ball effects on production efficiency due

* Corresponding Author

to hard-to-assess process time. Therefore, how to overcome contradiction between large-scale mass production strategy and multiple personalized attributes is of interest.

In order to guarantee personalized customer requirements, there are three approaches to solving contradiction. The first approach is to introduce management information system (MIS) such as ERP and MES systems. ERP systems and MES systems achieve integration of business functions inside and outside an organization and management of shop-floor operation respectively, while paper-based data collection that is always incomplete, inaccurate and untimely makes low efficiency of production system. The second approach is to improve flexibility of automatic machines. Nevertheless, it takes large amounts of money to continue it with high flexibility of them. The third approach is to reconfiguration manufacturing layout. In old-brand companies, the manufacturing layout has already been done. The large-scale reconfiguration is limited with finite space and budget constraint. Therefore, the fourth approach to realize CP is utilizing existing resource.

Concomitant with the rapid development of Internet-of-Things (IoT), IoT has been applied widely in production system. Early research on IoT application has been demonstrated by Brewer et al[1] and further more and more scholars widely explore IoT application in different projects. Huang et al,[2-4] shows the three typical IoT application in a product assembly line, a fixed-position walking-worker product assembly shop and a part fabrication job shop. Furthermore, each individual stage of production process is tracked and traced by deploying IoT device, e.g. IoT-based dynamic resource management system[5], IoT-based manufacturing execution system[6] and warehouse operation management system[7]. The author's team has evaluated the improvement after deploying IoT devices by SD simulation method[8, 9]. Finally, our team provides production logistics system with the best IoT solution[9].

This paper proposed two IoT-based information control strategy based on previous research. Those control strategies is applied in production system, which integrate customized requirement and characteristics of production system. A simulation environment is established according to SD. How dynamics caused by personalized requirement have effect on production system is analysed. Furthermore, an appropriate simulation result is obtained for enterprises.

The remainder of this paper is organized as follow. Section 2 introduces SD method. Section 3 introduces two IoT-based control strategies and SD simulation model. Section 4 is experiment simulation and analysis. Conclusion are drawn in Section 5.

2 SYSTEM DYNAMICS

System Dynamics (SD) established by Forrester is a simulation method which integrates qualitative analysis with quantitative research. Based on Feedback Control Theory, SD is a competitive theory, perspective and method to deal with sophisticated dynamic problems. SD can analyse a complex system by Causal Loop Diagram as well as Stock and Flow Diagram. Then its performance can be evaluated by adjusting a series of variables, such as state/level, rate and auxiliary variables. At present, SD is widely applied to supply chain management, management and operation of enterprise, etc. Its steps are as follows:

Step 1: System analysis. Apply theory and principle to analyse research objective as a system.

Step 2: Structure analysis. Process system information, partition system level and identify global and local feedback mechanism.

Step 3: Establish standard mathematical model. Identify a series of initial equations about state/level, rate, auxiliary and constant variables.

Step 4: Simulation and modification. Run the model and discover error. Then modify equations of model and acquire relative information

Step 5: Adjustment and Analysis. Adjust the overall structure and analyse the result.

3 PRODUCTION SYSTEM LAYOUT AND IOT DEPLOYMENT

In this section, production system layout with IoT deployment is presented. Firstly, according to the reconfiguration manufacturing system (RMS) proposed by Yoram[10], production system layout can be designed for CM. Secondly, combining with the best IoT device deployment introduced by Qu[9], we propose two IoT-based information control strategies, namely time-triggered and event-triggered.

3.1 Production System Layout Description

With the hypothesis above, this paper reaches the solution through the production system layout planning of professor Yoram. The theoretical layout planning might be 56, but the real final layout planning is 6, see figure 1. Simulation environment defines as follows. A shoes industrial manufacturer provides with customized sneakers service. In this case, 5 customized attributes are provided with customers, namely sole inch, sole technology, vamp width, vamp colour and personalized ID. Shown as Figure 1, processing time is different for different customized attributes. We assumed that expected yield of shoes is 500 pairs daily and net processing time for each shoes is 12.2 minutes. And workshop runs 1000 minutes every day.

sole		upper		
inch	technology	breath	color	Personality ID
3.7 min	1.7 min	1.7 min	1.7 min	1.7 min

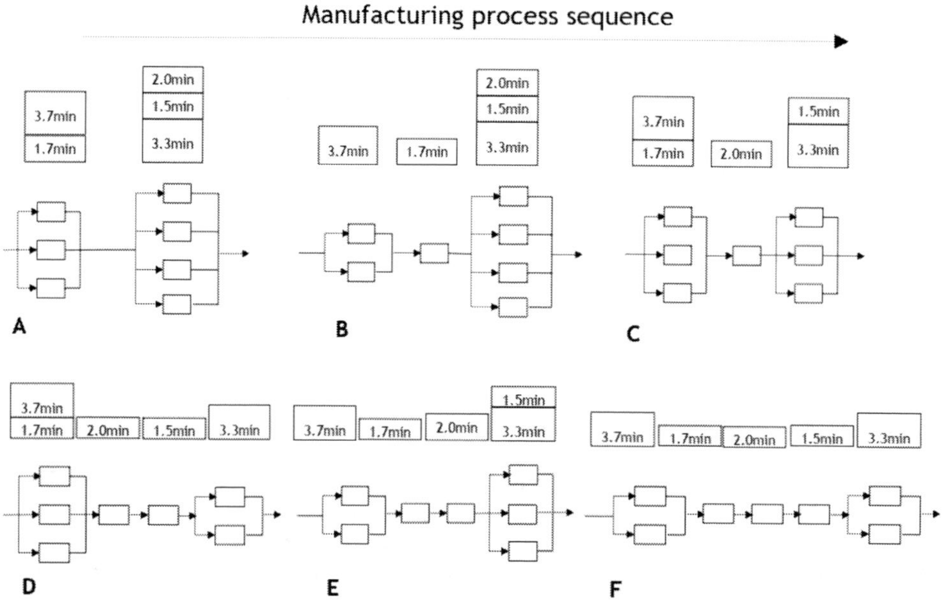

Figure 1: CM System Layout

3.2 IOT-based Deployment Description

Two IoT-based information control strategies proposed in this paper are to monitor real-time production status in every manufacturing unit and acquire the quantity of material in buffer. The information obtained by IoT is used to make decision. The first strategy is time-triggered, which is triggered at fixed time not fixed quantities. The second strategy is event-triggered, which is triggered at fixed quantities not fixed time. For the sake of simplicity the strategies are named as LOE, LOT, and are described as follows figure 2.

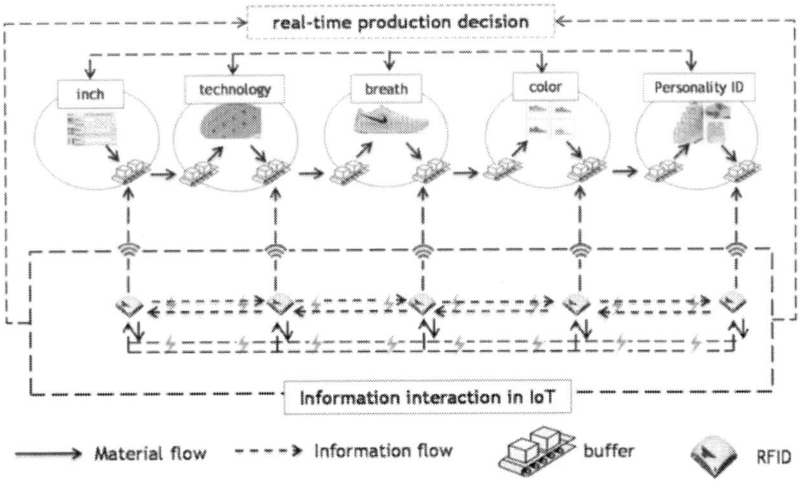

Figure 2: IoT-based Deployment

3.3 IOT-based Control Strategies

3.3.1 L-O-T

Quantity of material in Out-buffer equals to quantity of material produced by this machine, namely reduced quantity of material in In-buffer. According to pre-set frequency of RFID devices in Out-buffer, the quantity of cumulate material is detected. The information is delivered to IoT-based production logistics system. Then the system makes material plan, and the information is delivered to manager of previous manufacturing unit. According to the information, manager arrange distribution batch which is delivered to this manufacturing unit. Finally, quantity of cumulate material is under control by this means according to fixed time not quantity.

3.3.2 L-O-E

When quantity of material reaching in Out-buffer which is detected by RFID devices equals to pre-set quantity, the information is delivered to IoT-based production logistics system. Then, the system makes material plan and delivers it to managers. According to the information, managers arrange distribution task to deliver the cumulate material to next unit in time. Finally, quantity of cumulate material is under control by this means according to fixed quantity not time.

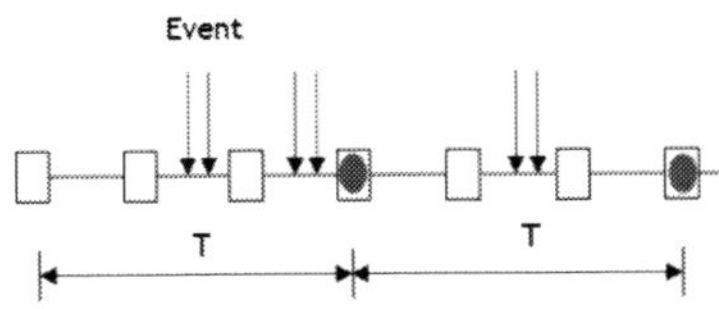

Figure 3: Time-Triggered Control Strategy

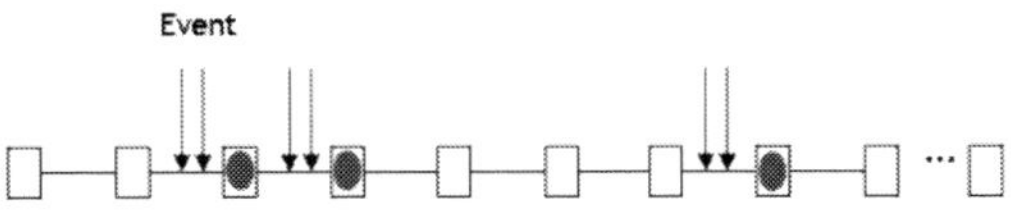

Figure 4: Event-Triggered Control Strategy

3.4 SD Simulation for IoT-based System

3.4.1 Symbol Explanation

Parameter notations in the model and definition are given in Table 1.

Table 1: Parameter notations and Definition

SYMBOL	EXPLAINATION	SYMBOL	EXPLAINATION
M-a	manufacturing Unit(e.g. M-a M-b M-c)	RM DB	Distribution Batch of raw material
M-a1	parallel machine in the manufacturing Unit(e.g. M-a1 M-a2)	RM DR	Distribution Rate of raw material
RW	raw material	RM DC	Distribution Cycle of raw material
SPT a(b/c)	Standard Processing Time of M-a(b/c)	M-a(b/c) RR	Replenishment Rate of M-a(b/c)
FPD	finished product delay	M-a(b/c) RD	Replenishment Delay of M-a(b/c)
FPER	finished product warehousinig rate	M-a(b/c) DR	Distribution Rate of M-a(b/c)
FPTN	total number of finished product	M-a(b/c) DC	Distribution Cycle of M-a(b/c)
In-a/Out-a	Buffer (e.g. In-Buffer, Out-Buffer) of M-a(b/c)	M-a(b/c) DD	Distribution Delay of M-a(b/c)
In-(b/c) N	In-buffer Number of M-a(b/c)	M-a(b/c) DB	Distribution Batch of M-a(b/c)
In-a(b/c) CR	In-buffer Consumption Rate of M-a(b/c)	M-a(b/c) BD	Bufffer Delay of M-a(b/c)
Out-a(b/c) N	Out-buffer Number of M-a(b/c)	M-a(b/c) PD	Production Delay of M-a(b/c)
Out- a (b/c) RR	Replenishment Rate of Out-a(b/c)	M-a1(a2/b1)WIP	the inventory of work of prodyct of M-a1 in period t
Out-a RM INV	raw material of Out-buffer inventory	FPD	finished product delay
IPR a(b/c)	Initial Production Rate	FPWR	finished product ware-housinig rate
EIPR a(b/c)	Expected Initial Production Rate of M-a(b/c)	FPTN	total number of finished product
PCR a(b/c)	Production Completion Rate of M-a(b/c)	EPCR a(b/c)	Expected Production CompletionRate of M-a(b/c)

3.4.2 Loop Diagram

Loop Diagram mainly describes logic relationship among system elements. Since the loop diagrams are alike in essence, only the four stages selected is analysed for illustration in detail. There are eight feedback loops, half of which are positive loops, namely loop 1, 3, 5 and 7. And rest of them are negative loops. Balanced relationship exists in feedback loop among every variables. Aforementioned six typical production system owns 2, 3, 4 and 5 manufacturing units. Loop diagram is established based on program E in this section.

Loop 1 shows Causal loop between raw material warehouse and ship/replenishment of product unit and explores the relationship of influence of interaction between production and Replenishment in product, as well as Loop3, 5 and 7.And Loop2 shows Causal loop of self-product of manufacturing unit and explores interaction among all kinds of factors in production, as well as Loop4, 6, 8.

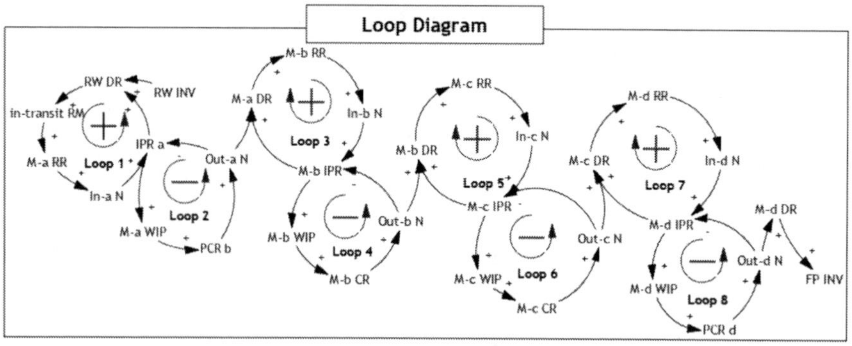

Figure 5: Loop Diagram

3.4.3 Rate and Level Diagram

As shown in Fig.6, loop diagram is converted to rate and level diagram by SD method. Material flow and production logistics is included, ranging from replenishment, production initialization, production completion, distribution stages. Pipeline inventory exists in manufacturing unit and buffer, and total number of finished products is equal to material which is out of warehouse plus finished products which is in warehouse.

In this paper, RFID device is deployed in out-buffer as key control point for CP system and SD model is established. We assume that material supplement and customer order are sufficient, which means production process doesn't stop because of lack of order and material. The work time is equal to 1000 min and simulation time is one day, namely 1000 min, and interval is one minute. Online time of material in buffer, offline time and the number of products can be obtained by IoT system. Each point in IoT system can deliver information and feedback. When the number of information increases continually, it is impractical to share information for every information collection point. Only by sharing information selectively, delivery rate of information can be accelerated and redundant information can be reduced.

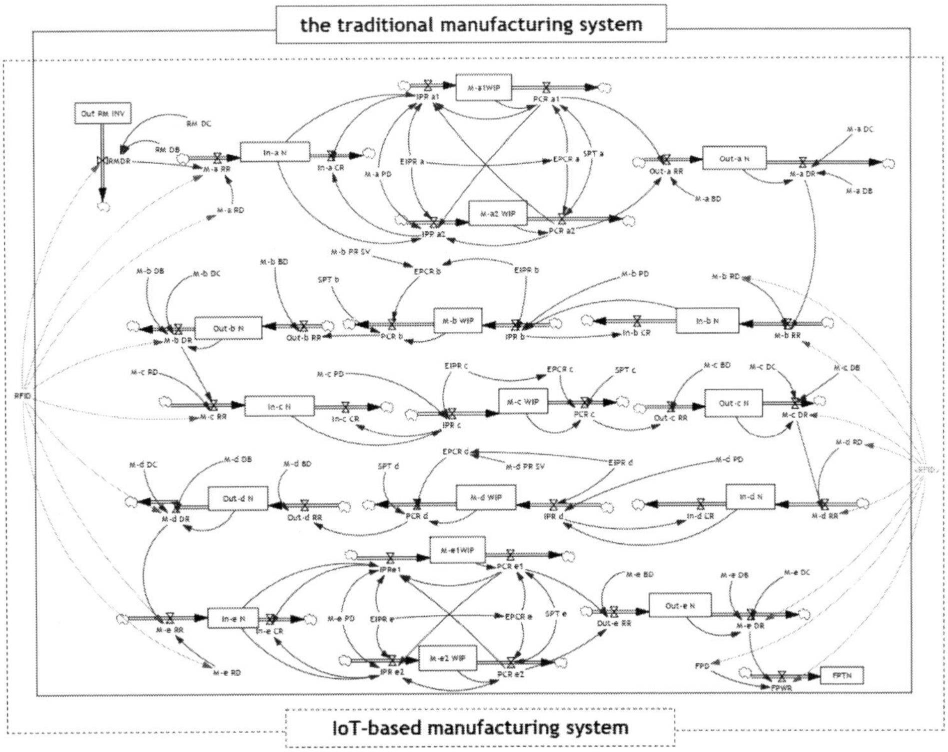

Figure 6: Rate and Level Diagram

3.4.4 Mathematical Model

The operation process of manufacturing system is not changed by IoT device. Under IoT environment, the parameters and their relationship are different from the traditional manufacturing system. Therefore, the equation explanation for model is described in this paragraph and the quantitative relationship between parameters is defined in detail in Table 2.

Table 2: Model Equation

SYMBOL	Traditional	LOE	LOT
RW DC	IF THEN ELSE (PULSE TRAIN(0, 1, RW DC, 1000), RW DC, 0)	IF THEN ELSE (Out N>RM DB, Out N, 0)	IF THEN ELSE (PULSE TRAIN(0, 1, RW DC, 1000), Out N, 0)
PD	RANDOM NORMAL(1, 3, 2, 0.5 , 1)	RANDOM NORMAL(1, 2, 1.5, 0.5 , 1)	RANDOM NORMAL(1, 2, 1.5, 0.5 , 1)
BD	RANDOM NORMAL(1, 3, 2, 0.5 , 1)	RANDOM NORMAL(1, 3, 2, 0.5 , 1)	RANDOM NORMAL(1, 3, 2, 0.5 , 1)
M DR	IF THEN ELSE (PULSE TRAIN(0, 1, RW DC, 1000), M DB0)	IF THEN ELSE (Out N> M DB, Out N,0)	IF THEN ELSE (PULSE TRAIN(0, 1, M DC, 1000), Out-a N, 0)
DD	RANDOM NORMAL(1, 3 , 2, 0.5 , 1)	RANDOM NORMAL(0.5, 3 , 1.5, 0.5 , 1)	RANDOM NORMAL(0.5, 3 , 1.5, 0.5 , 1)
FPWR	DELAY1I(M-b DR, FPD, 0)	IF THEN ELSE (Out N>M DB,Out N, 0)	DELAY1I(M-b DR, FPD, 0)

4 EXPERIMENT SIMULATION AND ANALYSIS

Simulation results indicates that the difference of production layout policy is that the number of manufacturing unit is different. At the same period, the performance of different production layout policy is different, such as production rate and yield. In this paper, finished production and pipeline inventory are chosen as key parameters for demonstration. The results are shown in Fig.7 by combination different production layout and two control strategies.

1. Before deploying IoT device: pipeline inventory of system A is least, while pipeline inventory of system F is most. It is easy to influence system which owns more phase. Pipeline inventory refers to the number of WIP in production logistics system, which shows directly cost in production and distribution stage. The more the number of pipeline inventory is, the more invisible cost is in system except production and warehouse cost.
2. After deploying IoT device: production performance in all system increases clearly. Event-triggered control strategy has a remarkable impact on system which owns less manufacturing phase. However, time-triggered control strategy does well in system which owns more phase.

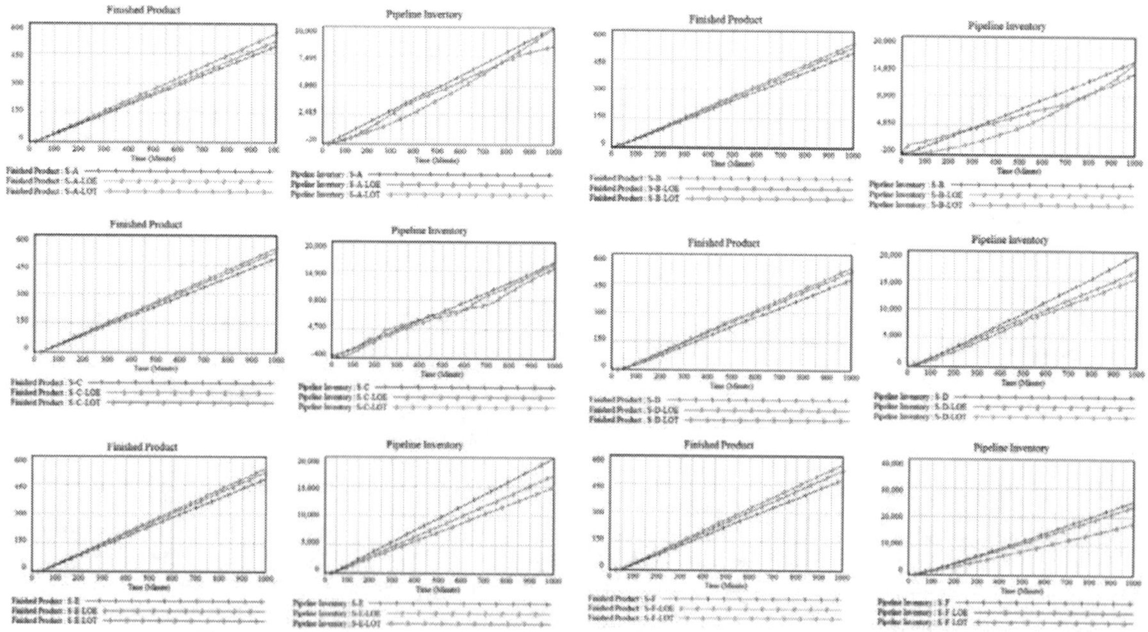

Figure 7: Simulation of Pipeline Inventory and Finished Product

5 CONCLUSION

This paper proposes a SD methodology for studying two control strategies of CM with IoT application. The application in the CM system deals with dynamic contexts that can not only simplify the manual data collection operations, but also improve the performance of the whole process by presented two control strategies. This provides a reference basis for manufacturers to select and implement IoT systems. Previously, CM and IoT technology proposals are investigated separately. Using SD simulation and sensitivity analysis, the optimal IoT solutions are identified.

This research does not only enhance the use of system dynamics, but also present two control strategies to introduce a quantitative IoT system analysis approach. The main contributions can be summarized as follows: (1) This study used the dynamic information captured from the IoT system to analyse the operations of CM systems; (2) Making a quantitative analysis on the IoT system, and provides two control strategies to analyse and evaluate the IoT-based system from quantitative perspective. (3) Focusing on the research of the CM system under IoT environment. Apart from analysing the simulation of systems to provide the best IoT solution to them independently, we also analyse the simulation of two control strategies, and provide the best solution of model for the manager.

6 REFERENCES

[1] Brewer A., et al.1999. Intelligent tracking in manufacturing, Journal of Intelligent Manufacturing, 10(3), pp 245-250.

[2] Huang G. Q., et al.2008. RFID-based wireless manufacturing for real-time management of job shop WIP inventories, The International Journal of Advanced Manufacturing Technology, 36(7-8), pp: 752-764.

[3] Huang G. Q., et al.2008. RFID-enabled real-time wireless manufacturing for adaptive assembly planning and control. Journal of Intelligent Manufacturing, 19(6), pp: 701-713.

[4] Huang G. Q., et al.2007. RFID-based wireless manufacturing for walking-worker assembly islands with fixed-position layouts, Robotics and Computer-Integrated Manufacturing, 23(4), pp: 469-477.

[5] Qu T., et al.2012. A case of implementing RFID-based real-time shop-floor material management for household electrical appliance manufacturers, Journal of Intelligent Manufacturing, 23(6), pp: 2343-2356.

[6] Zhong R. Y., et al.2013. RFID-enabled real-time manufacturing execution system for mass-customization production, Robotics and Computer-Integrated Manufacturing, 29(2), pp: 283-292.

[7] Chow H., et al.2006. Design of a RFID case-based resource management system for warehouse operations, Expert Systems with Applications, 30(4), pp: 561-576.

[8] Qu T., et al.2016. System dynamics analysis for an Internet-of-Things-enabled production logistics system, International Journal of Production Research, pp: 1-28.

[9] Qu T., et al.2015. System Dynamics for Internet-of-Things-based Real-Time Manufacture Systems, Industrial Engineering Journal.

[10] Koren Y., et al.2010. Design of reconfigurable manufacturing systems, Journal of manufacturing systems, 29(4), pp: 130-141.

BIOMECHANICS OF HUMAN LOWER LIMB JOINTS UNDER LOAD DURING SUPPORT PHASE

CHEN Jian-hua[1*]MU Xi-hui[1,2†]DU Feng-po[2]GUO Hao-liang[1]

[1]Shijiazhuang Mechanical Engineering College, Shijiazhuang, China
cai4728qi@yeah.net

[2] Shijiazhuang New Technology Application Institute, Shijiazhuang, China

ABSTRACT

The multi-camerasmotioncapturesystem and force measuring system are constructed for study the biomechanics of human lower limb's walk gait during support phase, which consists of three sub-phases. The articlefocuses onissues about the kinematics and dynamics of three lower limbjoints, namely hip, knee and ankle, which will bemeasured and analyzed under different loads (5 kg, 10 kg, 15 kg,). Firstly,angle ranges of the three joints during support phaseare obtained under different loads, by multi-rigid-body analysis, after building joint angle equations. Secondly, the ranges of torque and power are studied by Newton-Euler formulation. Finally, the knee joint stiffness is investigated. The result shows that the measure of knee joint stiffness is about 15.5~18.6 N.m/° during sub-phase I and 4.65~6.2 N.m/° during sub-phase III, while which is as large as beyond 27.9 N.m/° during sub-phase II. The conclusion of the experiment and analysis contributes to the bionics design of exoskeleton.

Keywords: Gait; Support phase; Joint Stiffness; Biomechanical; Load

1 INTRODUCTION

Lower limb exoskeleton, considered as wearable robots, can help restore movement to the paralyzed and improving the capability of workers or soldiers [1-4]. As it fits with human body, clinical gait biomechanics analysis has been the foundation of exoskeleton design [5]. Most researchers would study the biomechanics of human before design exoskeleton robot. They made the exoskeleton imitating human body to improve the man-machine coupling when human wears the exoskeleton on the outside of the body.

The human walking gait is divided into support phase and swing phase.Support phase is the process when supporting leg contacts with the ground, supporting body and leaves the ground,which is a quite complexprogess. Therefore, the article utilizes motion capture system to find out the movement rule of human lower limb under different load.

2 MOTION CAPTURE AND FOOT FORCE MEASURING SYSTEM

*CHEN Jian-hua, Shijiazhuang Mechanical Engineering College,050000, Shijiazhuang, Hebei, China. Email: chen_jhualei@163.com

†MU Xihui, Shijiazhuang New Technology Application Institute,050000, Shijiazhuang, Hebei, China.

The human body kinematics parameters include time, space and space-time parameter. The main methods of measurement of human body movement are high-speed camera capture system and body position sensor system. The former takes some sets of high-speed cameras to capture the body's instantaneous movement and reconstructthree-dimensional movement process. The latter shows the human movement by installing potentiometer or inertial navigation equipment on human body. Each has both advantages and disadvantages. For example, the former has a smaller cumulative error and high precision but only used in specific places—indoor spaces. In contrast, the latter could be used outdoor, but unable to get the space displacement.

In this article, we utilize multi-cameras motion capture system and foot force measuring system to analyse the lower limb's joint movement, as showed in Figure 1. There are six high-speed cameras decorated in the scene. Participant dressed in special black clothes on which there were many reflective markers. Each marker would be captured in all of the six cameras. After internal correlation algorithm and anatomy empirical formula, three dimensional trajectory of each marker could be achieved and human movement would be reconstructed.

Figure 1: Multi-cameras capture and force acquisitionsystem

3 KINEMATICS ANALYSIS BASED ON MULTIPLE RIGID BODY

3.1 Motion Equations

Incase of knee joint, the triangle, whose three vertices are *p3, p4 and p5*, constitutes the shank, as illustrated in Figure 2. Based on the triangle, a *vuw* coordinate is defined, wherev is an unit vector from *p3* to *p5, u* is the unit normal vector perpendicular to the triangle and *w* is determined by right-hand rule. The equations are as equation (1)~(3).

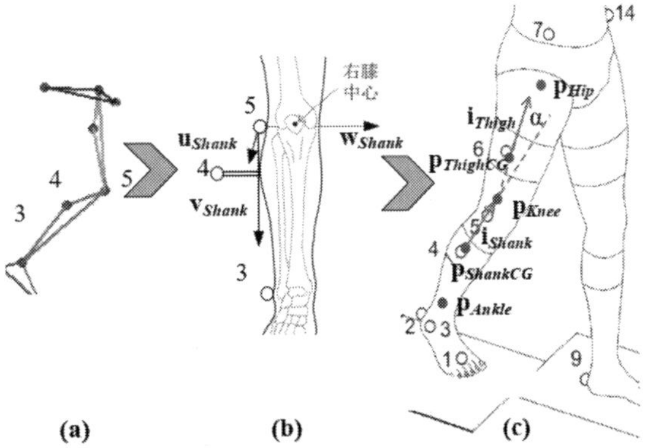

Figure 2: 3D coordinate reconstruction

$$v_{Shank} = \frac{p_3 - p_5}{|p_3 - p_5|} \quad (1)$$

$$u_{Shank} = \frac{(p_4 - p_5) \times (p_3 - p_5)}{|(p_4 - p_5) \times (p_3 - p_5)|} \quad (2)$$

$$w_{Shank} = u_{Shank} \times v_{Shank} \quad (3)$$

According *to* empirical formula, thecentre's coordinate of knee joint is as follows:

$$p_{Knee} = p_5 + 0.5 A_{11} u_{Shank} w_{Shank} \quad (4)$$

Here, A_{11} is the width of the knee joint.

In the same way, the centre coordinates of ankle joint P_{hip} and hip joint P_{ankle} is achieved.

The i shaft unit vectors fixed in the centre of gravity of shank and thigh are as Eq.(5) and (6), respectively.

$$i_{Thigh} = \frac{p_{Hip} - p_{Knee}}{|p_{Hip} - p_{Knee}|} \quad (5)$$

$$i_{Shank} = \frac{p_{Knee} - p_{Ankle}}{|p_{Knee} - p_{Ankle}|} \quad (6)$$

Then the angle of the two unit vectors of knee joint is as follows:

$$\alpha_{RKnee} = \arcsin(i_{Shank} \times i_{Thigh}) \quad (7)$$

3.2 Experiment Design

The subject (25 years, male, healthy) walked naturally through the camera capture area carrying three different loadsof 5 kg, 10 kg, 15 kg. Six cameras will take photos in 100 Hz. Each group of walking is repeated for three times, 15 minutes for intervals rest.

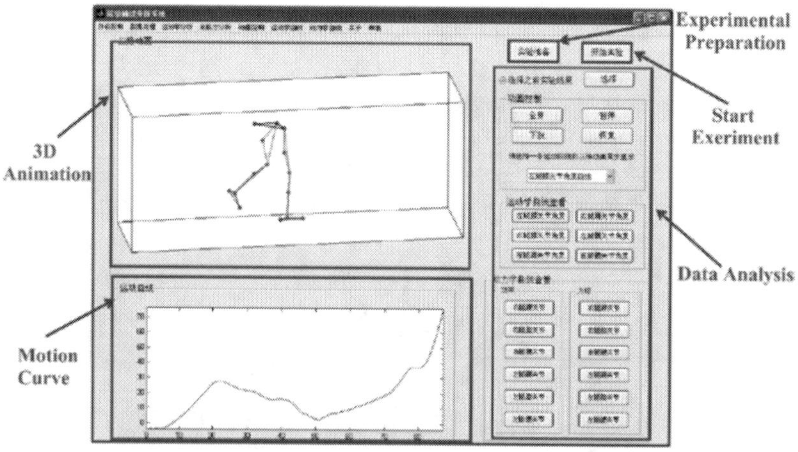

Figure 3: Gui of MotionCaptureSystem

The MotionCaptureSystem is show as Figuire 3. The experiment steps are divided into three in totals. The first step is *Experiment Preparation*, which is for a series of preparation before the experiment, mainly includes *adjust the field of view*, *shooting calibration*, *the calibration calculation* , *set markers* , *shooting body template* and *calculating*. Among them, *set markers* need 22 reflective markers installed on human left hand, right hand, left leg and right leg according Helen HayesMethod. After the completion of *Experiment Preparation*, we could conduct human motion capture and analysis of experiment. Click the second step button *START EXPERIMENT* to start wizard interface to begin capture and analysis of human movement experiment. The third step is to analyze results and curve display

3.3 Experiment result

Human walking gait is dividing into two, namely support phase and swing phase. The support phase consists of phase I, phase II and phase III according to different process. In phase I, support leg's heel touches the ground andtwo feet support on the ground at the same time, the centre of gravity of body shifting from the original support leg to the new support legs gradually. In Phase II, only one leg support on the ground with another leg swing in the air. In pphase III, the swing leg touches the ground and two feet support on the ground at the same time again. The results are showed as Figure 4.

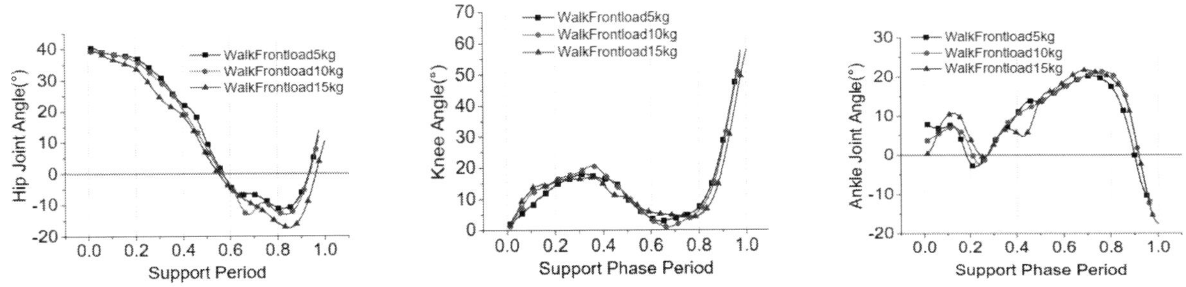

(a)Hip (b) Knee (c) Ankle

Figure 4: Angle range of three joints

Table 1: Angle range of joint under different sub-phase

Phase	Joint	Period	Range of Angle(°)		
			5kg	10kg	15kg
I	Hip	0~0.2	36~40	35~39	33~39
	Knee		2~14	1~16	2~15
	Ankle		-3~8	0~7	0~11
II	Hip	0.2~0.8	-10~36	-12~35	-14~33
	Knee		7~18	1~21	4~17
	Ankle		-2~20	-2~21	0~21
III	Hip	0.8~1	-10~13	-12~12	-12~10
	Knee		7~58	7~56	4~58
	Ankle		-12~18	-15~20	-17~20

The following conclusions can be drawn from Figure 4 and Table 1, the human body has a quite good ability for regulation. Under different loads, human joints always maintain stability. 1) Hip's whole range of angle is from 40°to -15°. In phase I, as the load increases, hip joint has a larger range of angle, however in phase II load has little effect on hip because of a large enough stiffness. 2) Knee's whole range of angle is from 0°to 60°. A larger load causes a smaller range. In whole its behaviour is like a sine function, but in support phase III, the angle increases linearly. 3) Ankle shows a three-stage variation, there is a small fluctuation of about 10 in phase I, however in phase II and phase III there is a trend of respectively monotone linear rise and fall. And a larger load causes a larger angle variation.

4 DYNAMICS ANALYSIS BASED ON NEWTON EULER METHOD

Based on the theory of multi-body, the body is divided into thighsegment,shank segment and foot segment (Figure 5). They are connected with each otherthrough the knee and ankle joint, where the force and reaction force are in a same size but opposite direction.

For ankle joint, connecting the foot and shank, according to Newton's second law:

$$F_{Ankle} + F_{plate} + m_{Foot} g\vec{k} = m_{Foot}\vec{a} \quad (8)$$

$$M_{Ankle} + M_{plate} + F_{plate}(p_{plate} - p_{FootCG}) + F_{Ankle}(p_{Ankle} - p_{FootCG}) = \dot{L}_{Ankle} \quad (9)$$

For knee joint:

$$F_{Knee} + F_{Ankle} + m_{Shank} g\vec{k} = m_{Shank}\vec{a} \quad (10)$$

$$M_{Knee} + M_{Ankle} + F_{Knee}(p_{Knee} - p_{ShankCG}) + F_{Ankle}(p_{Ankle} - p_{ShankCG}) = \dot{L}_{Ankle} \quad (11)$$

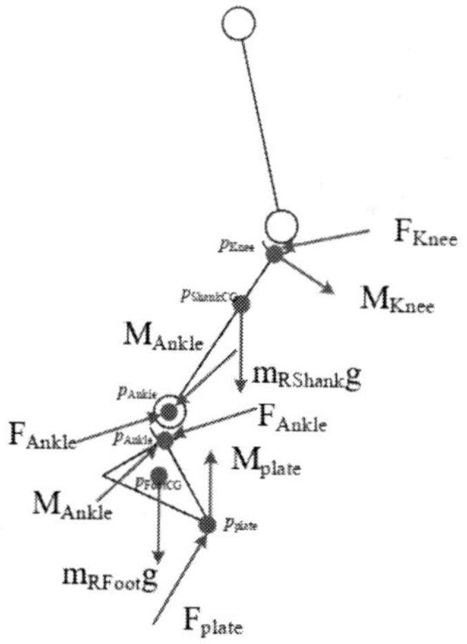

Figure 5: Multi-body dynamics

4.1 Torque

The following conclusions can be drawn from Figure 6 and Table 2. 1) The torque curves of three joints are approximate sine function. 2) It is found that the valueof the maximum torque at the ankle is largest, which is equal to the sum of knee and hip approximately.3) With the increase of load weight, the torque of joints also increases. In hip joint the value is approximately 5kg / 62 kg≈0.05.

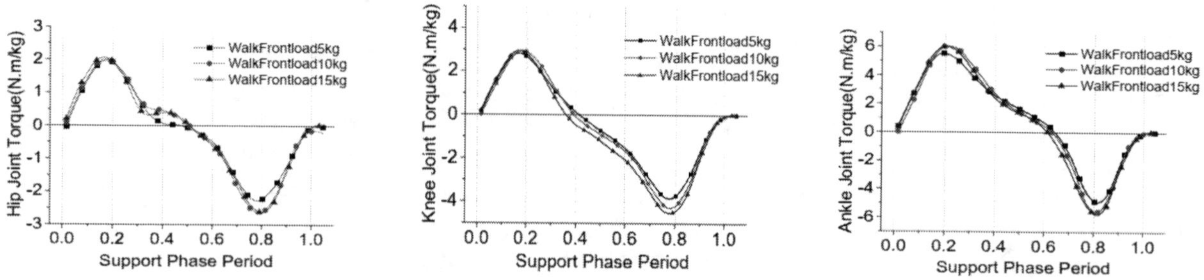

(a) Hip (b) Knee (c) Ankle

Figure 6:Torque range of three joints

4.2 Power

It should be noted that the negative stage of power means human body does against the external force passively when the external force acting on human body. 1) For hip, in phase I, the drive power needed is the largest, as much as 5W/kg and in phase II, it needs little power. For exoskeleton design, this potential energy storage could be reused when the power of electrical machine is cut offto improve energy efficiency (e.g. place a storage spring). There is a largest negative power in the initial support phase ofankle joint, due to compression of body weight on it. In this section, ankle muscles do negative work passively, resulting in a large waste of power. The power curves of hip and knee are illustrated as Figure 7.

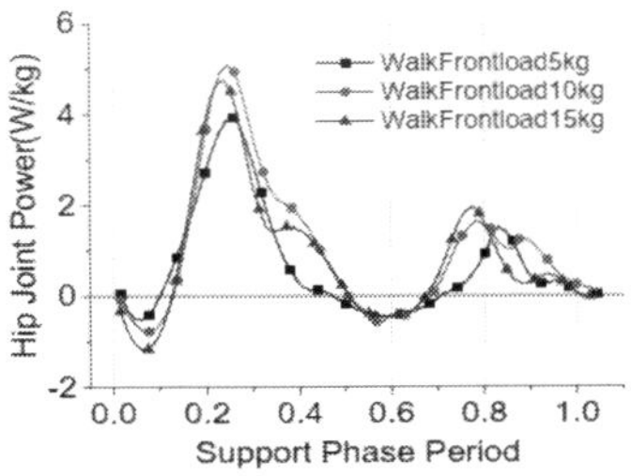

(a) Hip

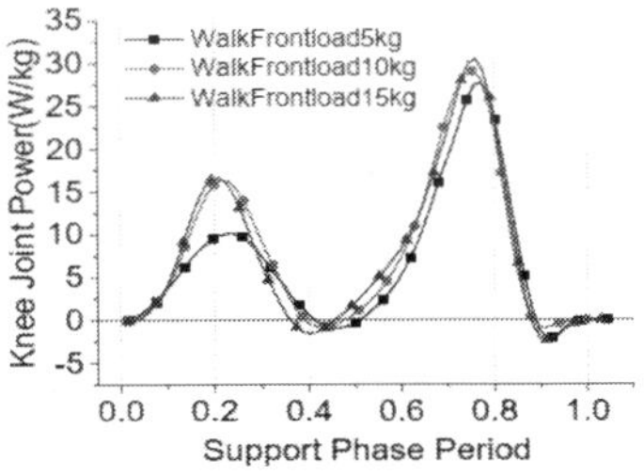

(b) K nee

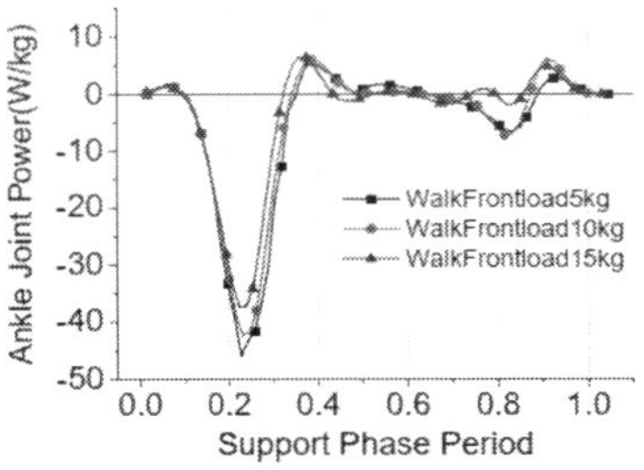

(c) Ankle

Figure 7:Torque range of three joints

4.3 The Variation of Torque-Angle of Knee

When human walks, leg stiffness is variable, enhancing environmental adaptability and energy efficiency. The term of stiffness belongs to material mechanics. So it shall be cautions to introduce it to describebody's property.

1) In material area, stiffness is passive, whose expression is as follows:

$$K = \Delta F / \Delta x \tag{12}$$

2) If the formula is introduced directly intohuman lower limb musculoskeletal system, there will produce a mistake.Because human lower limb musculoskeletal system includes passive components, such as bones, tendons, etc., and active ingredients, such as muscle, whose spring-mass model and differential formula derivation formula is illustrated as Equation (13)[6].As can be seen from the formulas, dF/dx does not mean stiffness, because it also includes other complex elements. However, since dF/dx is often used as a description of the characteristics of an organism, it used to be referred to as "quasi-stiffness" in the field of biomechanics[7]

$$F(t) = m(t)d^2x / dt^2 + b(t)dx / dt + k(t)x(t) \tag{13}$$

$$\frac{dF}{dx} = \frac{dm}{dt}\frac{a}{v} + m(t)\frac{j}{v} + \frac{db}{dt} + b(t)\frac{a}{v} + \frac{dk}{dt}\frac{x}{v} + k(t) \tag{14}$$

Here, v=dx/dt, a=d2x/dt2, j=d3x/dt3。

So in this case the lower limb stiffness refers toquasi-stiffness, also known as apparent stiffness [8]. Based on this understanding, the expression of quasi-stiffness of lower limbs can be written into three ways: vertical stiffness, leg stiffness and joint stiffness [9, 10].

The knee joint involves only one degree of freedom and has a simple structure but obvious role. So among the three joints, variation analysis of knee joint's torque-angle is particularly important. In phase I, the torquemonotonically increases with angle, so the stiffness can be equivalent to a constant, whose range is 15.5~18.6 Nm/° .However, lesserload causes a larger value. Phase II is the most complex sub-phase due to the process of the foot touches the ground. The stiffness increases if given a larger load , whose value is above 27.9 N.m/° . In phase III, the torque and angle has a monotonically increasing relationship. There's no matter between stiffness and load. Under different load conditions, although the required maximum torque is different, they all has a similar stiffness from 4.65 to 6.2 Nm / ° .

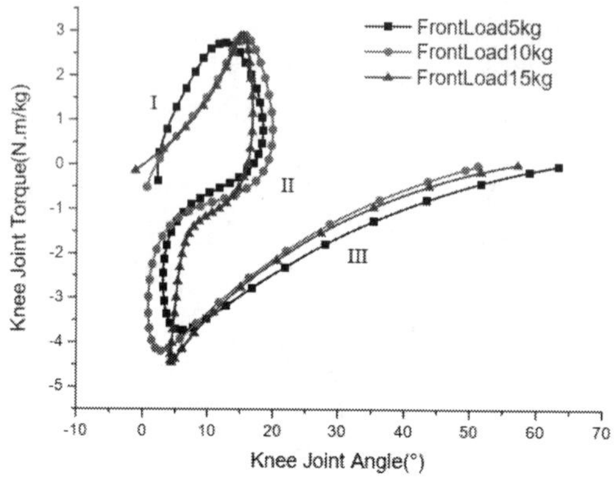

Figure 8:Torque-angle of knee

5 SUMMARY

In order to analyse the body's natural gait during the support phase, multi-cameras motion capture system and force measuring system are built. Though the acquisition of data, kinematic and dynamic performance of three joint are obtained under different body weight during the support phase, which is useful in the design of bionic

Firstly, curves of joints' angle variation under different weight-bearing are obtained by the multi-body motion analysis. In overall, the three joints all have an obvious change in phase I, however in phase II, III the effect of increasing load on joint angle is quite small.

Secondly, the kinetic equation about joints' torque and power is derived by Newton-Euler method. 1) In support phase, the torque curves of the three joints are approximate sine function; 2) Torque of ankle is approximate the sum of hip and knee; 3) With the increase of weight-bearing, the torque on joints also has an increase of approximately 0.05.

Finally, considering the particularity of the knee joint, the variation analysis of knee joint's torque-angle is particularly noted. In phase I, the torque monotonically increases with angle, so the stiffness can be equivalent to a constant, which is relatively larger value as a lesser load, whose range is 15.5 ~ 18.6 Nm / °. Phase II is the most complex sub-phase due to the process of foot touches the ground. The stiffness increases if given a larger load, whose value is above 27.9 N.m / °. In phase III, the torque and angle has a monotonically increasing relationship. Under different load conditions, although the required maximum torque is different, they all has a similar stiffness from 4.65 to 6.2 Nm / °

In conclusion, the article studies the kinematics and dynamics of human lower limb during support phase, based on the gait experiments, to get the regulation of angle, torque and power in three sub-phases, for bionics design of exoskeleton.

6 REFERENCES

[1] **Hayashi Y, Kiguchi K**. 2011. A Lower-Limb Power-Assist Robot with Perception-Assist, 2011 IEEE International Conference on Rehabilitation Robotics, Rehab Week Zurich, ETH Zurich Science City, Switzerland,June 29 - July 1.

[2] **Wehner M, Quinlivan B, Aubin P M**. 2013. A Lightweight Soft Exosuit for Gait Assistance, 2013 IEEE International Conference on Robotics and Automation (ICRA), Karlsruhe, Germany, May 6-10, pp 3362-3369.

[3] **Spencer A. MurrayKevin H. HaClareHartiganMichael Goldfarb**. 2015. An Assistive Control Approach for a Lower-Limb Exoskeleton to Facilitate Recovery of Walking Following Stroke, Ieee Transactions on Neural Systems And Rehabilitation Engineering, 23(3), pp 441-449

[4] **Steven H. CollinsM. Bruce WigginGregory S. Sawicki**. 2015. Reducing the energy cost of human walking using an unpowered exoskeleton, NATURE, 522(7555), pp 212-215

[5] **Lee H, Kim W, Han J**. 2012. The Technical Trend of the Exoskeleton Robot System for Human Power Assistance, *International Journal of Precision Engineering and Manufacturing*, 13(8), pp 1491-1497.

[6] **Farley C T, González O**. 1996. Leg stiffness and stride frequency in human running, *Journal of Biomechanics*, 29(2), pp 181-6.

[7] **Latash M L, Zatsiorsky V M**. 1993. Joint stiffness Myth or reality, *Human Movement Science*, 12, pp 653-692.

[8] **Zatsiorsky, V. M**. 2002. Kinetics of human motion. Human Kinetics.

[9] **Rapoport S, Mizrahi J, Kimmel E**. 2003. Constant and Variable Stiffness and Damping of the Leg Joints in Human Hopping, *Journal of Biomechanical Engineering*, 125(4), pp 507-516.

[10] **Stefanyshyn D J, Nigg B M**. 1998. Dynamic Angular Stiffness of the Ankle Joint During Running and Sprinting, *Journal of Applied Biomechanics*, 14,pp 292-299.

MULTI-OBJECITVE SLOTTING OPTIMIZATION PROBLEM OF WAREHOUSING SYSTEM

Yuzhen Hu[1*] Song Zhang[1] Jiajia Chen[1]

[1]School of Economics and Management
Harbin Engineering University, Heilongjiang Harbin, China
yuzhenhu@hrbeu.edu.cn

ABSTRACT

In this paper, a multi-objective slotting optimization problem of stereoscopic warehousing system based on different demand frequencies and the principle of shelf stability is studied. The objectives are to minimize the total routing with the weight of turnover rate of the goods, and the shelf center of gravity, at which a multi objective 0-1 integer programming model is established aimed. Dimensionless method and weighted sum method are first applied to transfer the multiple objective model to a single objective model formulation. The Hungarian method is then used to solve the problem and the solution has been proved to get optimal. Finally, a data test is provided to illustrate the effectiveness of the method. This study can provide theoretical and technical support for warehousing slotting optimization system.

Keywords: Warehousing system, Slotting optimization, Multi-objective, Hungarian algorithm, Optimization

1 INTRODUCTION

As a part of logistics system, storage plays an important role in the process of logistics integration. However, with the rapid development of modern industry, the traditional storage methods cannot meet the needs of production and circulation, and automated warehouse has been applied more and more widely. The current management of warehouse still follows the habit and experience of many years ago to store goods. It does not only waste storage space, but also degrades working efficiency of warehouse. Therefore, the location allocation problem is a very important issue in storage management system and slotting optimization has been emerged under this background.

In the face of thousands of ware location and cargo space, effective storage location assignment scheme by warehouse slotting optimization methods has become the key factor to improve storage efficiency and cut down the warehouse cost. Furthermore, it can also significantly improve the efficiency of space using and the goods picking, removal of cargo and overcome the peak load and polish up the storage space layout. The main factors we need to consider for solving location allocation problem include picking strategy, resource feasibility and warehouse maintenance and so on.

* Corresponding Author

supported by the project of Chinese Postdoctoral Science Foundation (2016M590276), the project of Central University Basic Research Fund (HEUCF150903), the project of Heilongjiang province postdoctoral Fund (LBH-Z15047), the Natural Science Foundation of Heilongjiang Province (QC2016095).

The available literature on warehouse strategic and performance evaluation/optimization may be classified based on three main issues(see Johnson [1] and Gu [2]: (i) contributions to address warehouse design decisions, typically concentrating on order picking routing or product location strategies; (ii) papers to propose analytic or simulation models and (iii) publications on benchmarking and performance evaluation. Slotting optimization belongs to the first issue. De Koster et al [3] thinks that order picking accounts for over 40% of the total picking time and over 65% of the total warehousing operating cost. Therefore, reasonable slotting optimization based on picking efficiencies plays an important role in improving the efficiency of order picking in warehousing system. During space distribution, the major factors that are considered by Yan et al [4] include picking strategy, the feasibility of resource, warehouse maintenance and so on.

As stated in Brynzér et al [5] and Muppani et al [6], traditional studies of slotting optimization focus on storage location for the highest picking efficiency. For small-scale problems, the branch and bound algorithm is generally used (Muppani et al [7]), and for large-scale problems, many intelligent algorithms, such as simulated annealing algorithm (Wang et al [8]), and genetic algorithm (Poulos et al [9], Liu et al [10], Lin et al [11], Zhang et al [12] and Ma et al [13]), are often considered by many studies. Exploiting the SKU correlations during location assignment is to assign them to the adjacent position. Slotting optimization problem was solved by use of historical statistical correlation in much previous researches. Kim and Smith[15] found a way to generate SKUs correlation data, and proposed a slotting optimization algorithm which achieved well results. Cartonization optimization is to pack SKUs into carton. It aims to minimize the cartons and picking-walking time (Koster et al [3] and Kim et al [14]). When requiring multiple cartons, assigning SKUs which are in the same carton to the adjacent location would reduce picking-walking time in possible (Henn et al [15] and Bonassa et al [16]).

We can find that most of studies use heuristic algorithm to solve slotting optimization problem. But few studies solve the integrated optimization problem of the goods picking path and stability of shelf simultaneously by polynomial time algorithms. In this paper, we follow the principles that the high turnover rate of the goods should be stored close to the entrance or exits and that heavy goods should be kept in the bottom of shelf, establish 0-1 integer programming model, use the Hungarian method to solve the problem and prove that the optimal solution can be obtained.

This paper is organized as follows. In the next section, a formal description of the problem, including a statement of assumptions and constraints is given. This is followed in Section 3 by a full mathematical model. Solution methods are presented in Section 4. Section 5 presents the computational results. Conclusions are drawn in Section 6.

2 PROBLEM STATEMENT

Warehouse location optimization is to reasonably determine the proper storage allocation solution of each type of goods. First, during different period, different demand quantities and the demand frequencies will lead to different path and time of access goods in different location. It can help reduce storage costs and enable enterprises to obtain higher profits to improve the work efficiency and to reduce the useless movement of the warehousing as more as possible by optimizing the storage system. At this point we should adopt the principle "something commonly used should be easy to get" to optimize the goods location, in order to improve access efficiency. Second, due to frequent access on certain goods, the weight of shelf may vary. If shelves appear serious imbalance phenomenon, such as "top heavy and bottom light", it may lead to deformation and even overturning of shelves. So we should follow the "weight goods should be in the bottom of shelves" to ensure the shelf stability.

In this paper, we first calculate the time to deliver goods to the shipping port by stacker in warehouse, place the goods of higher frequency of the inbound and outbound operations in the cargo space near an exit, and store the goods of lower frequency of the inbound and outbound operations in the cargo space far away from an exit. We minimize the total time of the inbound and outbound operations of all goods, in order to improve the efficiency of warehousing system.

Given a stereoscopic warehouse, there are I rows of shelves, each shelf has J columns and each column has K layers. We denote the row nearest the exits as row 1, the column nearest the exits as column 1, and the layer nearest the exits as layer 1. The coordinate (i, j, k) illustrates the cargo space of row i, column j and layer k. We assume that the length, width and height of each cargo space are the same as L. The shelf layout is shown in Figure 1.

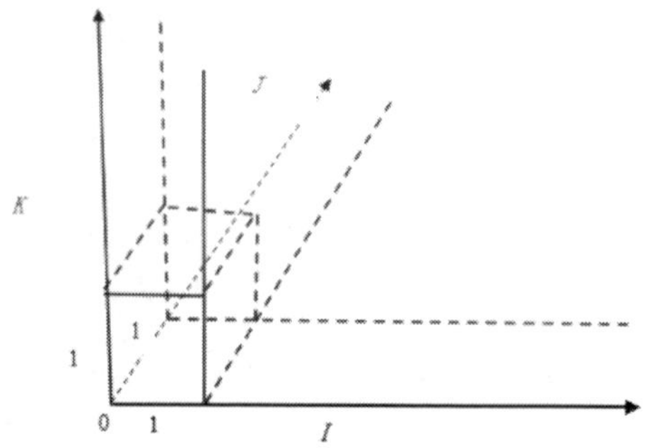

Figure1: Shelf Layout

It is assumed that there are N types of goods to be stored in the warehouse, and the turnover rate P is used to measure the inbound and outbound frequency of the goods, and the turnover rate of the goods n is P_n, and the mass of the goods n is m_n.

Stacking machine is needed to use for the goods access procedure. It is assumed that the stacking machine can only move along the I axis direction with the rapid v_i, along the J axis direction of the movement speed of v_j and along the K axis direction of the movement speed of v_k.

Other assumptions should also be given as follows.

1. The number of goods types N has been given as a constant, and N is no more than the number of cargo spaces.
2. The volume of each good is smaller than the capacity of the cargo space.
3. Each location can only access a piece of goods, each kind of goods can only be stored to one location.
4. The time of accessing goods from the location is negligible, and we only consider the stacker transportation time.
5. Single exit for goods transportation.

3 MATHEMATICAL FORMULATIONS

A mathematical formulation model of slotting optimization is as follows, according to the assumptions and constraints in Section 2. The objective should be considered in two aspects. One objective is the minimization of picking time and the other is minimization of cargo center of gravity

3.1 Picking efficiency

In order to improve the efficiency of inbound and outbound storage, it is necessary to shorten the time of storage of the goods, and it is the key to shorten the route of picking operation, that is the distance of the stacking movement for storage (the total path of the stacking machine).

In the plane warehouse, we set that the coordinates of a cargo space is (i,j,k), the coordinates of the exit is (0,0,0). As the stacker carrying goods cannot move along the straight line according to the shortest path, the distance of movement of the stacker from the cargo space to exits can be computed as $(i+j+k)\times L$.

According to the corresponding speed of the stacker along the axis, the time the stacker spending from cargo space (i,j,k) to exit can be computed as

$$t_{ijk} = \left(\frac{i}{v_i} + \frac{j}{v_j} + \frac{k}{v_k}\right) \tag{1}$$

According to the goal of location optimization, the goods with larger turnover rate should be moved to the place near the exits and the ones with smaller turnover rate should be moved to the place far away from the exits. If we set binary variables x_{ijkn}, then

$$x_{ijkn} = \begin{cases} 1, n\ in\ the\ location\ (i,j,k) \\ 0, otherwise \end{cases} \tag{2}$$

Therefore, the objective function of the problem is as follows.

$$min\, f_1 = \sum_{n=1}^{N}\sum_{i=1}^{I}\sum_{j=1}^{J}\sum_{k=1}^{K} t_{ijk}\, P_n x_{ijkn} \tag{3}$$

3.2 Stability of shelf

In actual situation, shelf stability is related to center of gravity of the shelf. To keep the shelf stable, so the placement principle of goods should be that heavier goods are placed on the bottom of shelf and lighter goods are placed on the up of shelf. And in accordance with the principle "bottom heavier and up lighter", the shelf center of gravity can be reduced apparently, so as to enhance the stability of the shelf. We first calculate the shelf center of gravity and then minimizing it.

The height of cargo space (i,j,k) is

$$h_{ijk} = kL \tag{4}$$

It is assumed that height of N types goods are $h_1, h_2, \dots, h_N$ respectively, the weight are $m_1, m_2, \dots, m_N$ respectively, then gravity center is stated as follows.

$$M = \frac{m_1h_1 + m_2h_2 + \cdots + m_nh_n}{m_1 + m_2 + \cdots + m_n} \tag{5}$$

Then the second objective can be

$$minf_2 = \frac{\left(\sum_{n=1}^{N}\sum_{i=1}^{I}\sum_{j=1}^{J}\sum_{k=1}^{K} h_{ijk}m_nx_{ijkn}\right)}{\sum_{n=1}^{N} m_i} \tag{6}$$

From equation (4) and equation (6), we get:

$$minf_2 = \frac{\left(\sum_{n=1}^{N}\sum_{i=1}^{I}\sum_{j=1}^{J}\sum_{k=1}^{K} kLm_nx_{ijkn}\right)}{\sum_{n=1}^{N} m_i} \tag{7}$$

3.3 Constraints

The constraints are as follows as well. Constraints (6) presents one piece of goods can only be located one cargo space, and Constraints (7) shows that one cargo space can accommodate at most one piece of goods.

$$\sum_{i=1}^{I}\sum_{j=1}^{J}\sum_{k=1}^{K} x_{ijkn} = 1 \qquad \forall n \in \{1,2,\dots,N\} \tag{8}$$

$$\sum_{n=1}^{N} x_{ijkn} \leq 1 \quad \forall i \in \{1,2,\dots,I\}, j \in \{1,2,\dots,J\}, k \in \{1,2,\dots,K\} \tag{9}$$

The decision variables are defined as follows.

$$x_{ijkn} = \{0,1\}, \forall i \in \{1,2,\dots,I\}, j \in \{1,2,\dots,J\}, k \in \{1,2,\dots,K\}, n \in \{1,2,\dots,N\} \tag{10}$$

The model formulations are established with two objective functions through the above analysis. The objectives are both interrelated and mutually conflict. It needs to combine them by weighted sum method stated in next section.

4 SOLUTION METHOD

In this section, we solve the model formulation by weighted sum method transferring the multiple objective problem to single objective problem. As both objective functions have different dimensions, therefore, dimensionless method will be applied respectively for each objective function.

Let $f_1' = f_1/T$ and $f_2' = f_2/H$, where $T = \sum_{i=1}^{I}\sum_{j=1}^{J}\sum_{k=1}^{K} t_{ijk}$ and $H = \sum_{i=1}^{I}\sum_{j=1}^{J}\sum_{k=1}^{K} h_{ijk}$, and given two weights $\omega_1$and ω_2, where $\omega_1 + \omega_2 = 1$, for the model formulation, then the single model can be formulated as follows.

$$min\, F = \omega_1 f_1' + \omega_2 f_2' = \sum_{n=1}^{N}\sum_{i=1}^{I}\sum_{j=1}^{J}\sum_{k=1}^{K} x_{ijkn}\left[\frac{w_1}{T}\left(\frac{i}{v_i}+\frac{j}{v_j}+\frac{k}{v_k}\right) L\, P_n + \frac{w_2}{H}\frac{kLm_n}{\sum_{n=1}^{N} m_n}\right] \tag{11}$$

$I \times J \times K - N$ dummy goods are added into the shelf with $p_n = 0$, $m_n = 0$, $n = N+1$, $N+2$, ... $I \times J \times K$. Then the problem can be described as the linear sum assignment problem.

Theorem 1. Given a pair of parameters ω_1 and ω_2, the optimal solution stays unchanged by adding dummy goods for slotting optimization problem.

Proof. As $p_n = 0$, $m_n = 0$, $n = N+1$, $N+2$, ... $I \times J \times K$, addition of the $I \times J \times K - N$ dummy goods into the objective function has no any impact on the value of the model formulation. So the conclusion promises.

Kuhn [17] has given a polynomial optimal method (Hungarian algorithm) for the linear sum assignment problem. Similarly with linear sum assignment problem, the slotting optimization problem of stereoscopic warehousing system can also be solved by Hungarian algorithm.

Theorem 2. The solution of the model formulation (8-11) got from the Hungarian algorithm is optimal.

Proof. From Theorem 1, we get that the slotting optimization problem of stereoscopic warehousing system can be transferred to the linear sum assignment problem by adding some coefficients. From Kuhn [17], we know that solution of linear sum assignment problem can be obtained by Hungarian algorithm. So the conclusion promises.

5 COMPUTATION RESULTS

We extract a part of the cargo area of the warehouse in a Business Company as sampled data, and $I = 2, J = 2, K = 2, L = 1m, v_i = v_j = v_k = 1m/s$, and the time of picking goods is Negligible. 6 types of goods are located in the cargo spaces. We give the information of goods including the item numbers "Item No.", turnover rate P_n, the mass m_n and the coordinate of initial storage position in Table 1. Table 2 gives the moving time of a stacker from each cargo space to the exit and Table 3 gives the height of each cargo.

Table 1: Information of goods for sampled data

Item No.	m_n/kg	P_n	Initial coordinate
1	5	0.8	(2,2,2)
2	3	0.4	(1,1,2)
3	4	0.2	(1,1,1)
4	1	0.3	(2,1,1)
5	2	0.5	(2,1,2)
6	5	0.6	(1,2,2)

Table 2: Moving time to exit

Cargo spaces	(1,1,1)	(1,2,1)	(2,1,1)	(2,2,1)	(1,1,2)	(1,2,2)	(2,1,2)	(2,2,2)
t_{ijk}	4	5	5	6	6	7	7	8

Table 3: Height of each cargo

Cargo spaces	(1,1,1)	(1,2,1)	(2,1,1)	(2,2,1)	(1,1,2)	(1,2,2)	(2,1,2)	(2,2,2)
h_{ijk}	1	1	1	1	2	2	2	2

According to practical situation of the warehouse system, the values of two weights are $\omega_1 = 0.8$, $\omega_2 = 0.2$. The optimal solution derived from Hungarian algorithm is illustrated in Table 4. Table 5 shows the comparison of objective values between initial solution and the optimal solution given in table 4. The sum of both objectives is displayed in the first row, the first objective value is shown in the second row and the second objective value is represented in the last row. It is displayed that the objective value in the optimal solution is lower by 21.96% than initial solutions.

Table 4: Optimal solution

Item No.	m_n/kg	P_n	Coordinate space
1	5	0,8	(1,1,1)
2	3	0.4	(2,1,1)
3	4	0.2	(1,2,2)
4	1	0.3	(2,1,2)

5	2	0.5	(2,2,1)
6	5	0.6	(1,2,1)

Table 5: Comparison of objective values

objective values	intial solution(is)	optimal solution(os)	(is-os)/os
weighted sum	15.39	12.01	21.96%
picking time	18.8	14.7	21.81%
weighted center	1.75	1.25	28.57%

6 CONCLUSION

This paper focuses on slotting optimization problem of stereoscope warehousing system, based on the multi-objective of minimizing the picking time and weighted centre of shelves. A detailed formal description for the problem is first given, and some factors affecting on cargo space assignment of stereoscope warehouse system are then analyzed. Finally an integer programming model is constructed.

A logical equivalent transformation is adapted from an integer programming model to the model that can be solved by the Hungarian algorithm. A dataset test illustrates the effectiveness of the method presented in this paper.

In the future, it would be worthwhile to examine several related issues. Firstly, the volumes of different types of goods vary greatly, therefore volume difference should be considered when optimizing the warehousing storage. Secondly, if it were possible to develop an expanded model that describes the three-dimensional warehouse, improved results would likely follow. Finally, other solution approaches, such as meta-heuristics or robust optimization might prove advantageous for some large scale slotting optimization problem. Although we were able to find optimal solutions for the instance examined, stable solutions are more often preferred by practitioners.

7 REFERENCES

[1] **Johnson, A., Mcginnis, L. 2011.** Performance measurement in the warehousing industry. *IIE Transactions*, 43(3),pp 220-230.

[2] **Gu, J.X., Goetschalckx. J.X., McGinnis, L.F. 2010.** Research on warehouse design and performance evaluation. *a comprehensive review European Journal of Operational Research*, 203 (3) : 539-549.

[3] **Koster, D., René, Le-Duc, T ,and Roodbergen, K.J. 2007.** Design and control of warehouse order picking: A literature review. *European Journal of Operational Research*, 182(2):481-501.

[4] **Yan, Y.Z., Sun, X.M., Pan, E.S. 2000.** Some discussion on the relation between shelves number and stacker operating efficiency in Stereo Warehouse. *Industrial Engineering and Management*, 5(6):29-33.

[5] **Brynzér, Henrik, and Johansson, M.I. 1996.** Storage location assignment: Using the product structure to reduce order picking times. *International Journal of Production Economics*, 46:595-603.

[6] **Muppani, Reddy, V., and Adil, G.K. 2008.** A branch and bound algorithm for class based storage location assignment. *European Journal of Operational Research* , 189(2):492-507.

[7] **Muppani, V.R., Adil, GK.A. 2007.** branch and bound algorithm for class based storage location assignment. *European Journal of Operational Research*, 1-15.

[8] **Wang, Z., Lin, J.H., Liu, C.Y., Li, X.L. 1996.** Order picking optimization for a multi-carousel-single-server system solved by improved annealing procedure. *Control and Decision*, 11(1):182-187.

[9] **Poulos, P.N., Rigatos, G.G., Tzafestas, S.G. 2001.** A Pareto-optimal genetic algorithm for warehouse mu1ti-objective optimization. *Engineering Applications of Artificial Intelligence*, 14(6):737-749.

[10] **Liu, S.N., Ke, Y.L., Li, J.X., Lv, Z. 2006.** Optimization for automated warehousebased on scheduling policy. *Computer Integrated Manufacturing System*, 12(9): 1438-1443.

[11] **Lin, J.H., Li, G.F., Li, J.X. 1997.** Multi-carousel with two servers. *Journal of Shandong University of Technology*, 27(3): 236-239.

[12] **Zhang, P., Tian, G.H., Jia, L., Li, X.L., Lu, F. 2004.** New hybrid Genetic algorithm solving the order-picking optimization problem of a multi-carousel system. *Chinese Journal of Mechanical Engineering*, 40(6): 34-38.

[13] **Ma, Y., Li, Z., Yun, W. 2015.** Fast Genetic Algorithm for Pick-up Path Optimization in the Large Warehouse System. *Lecture Notes in Computer Science*, 9142: 38-44.

[14] **Kim, B. S., and Smith. J. S. 2008.** Dynamic slotting for zone based distribution center picking operation. 10th International material handling research colloquium, Dortmund, Germany.

[15] **Henn, Sebastian, and Schmid, V. 2013.** Metaheuristics for order batching and sequencing in manual order picking systems. *Computers & Industrial Engineering* , 66(2):338-351.

[16] **Bonassa, Carlos, A., and Cunha, C.B. 2011.** The order-picking routing problem for low-level order picker in a warehouse. *Gestão & Produção* , 18(1):105-118.

[17] **Kuhn, H. W. 1955.** The Hungarian method for the assignment problem. *Naval Research Logistics Quarterly*, 2: 83-97.

THE PREDICTION MODEL OF ONLINE CONSUMER REVIEWS PERFORMANCE BASED ON SENTIMENT MINING AND TOBIT REGRESSION

SunQing[1, 2], ZhaoTeng[2], YaoZhong[1*], XuXi[1]
[1]School of Economics and Management, Beihang University, China
iszhyao@buaa.edu.cn
[2]School of Computer Science and Engineering, Beihang University, China

ABSTRACT

This paper proposes a research model that examines the effect of text content and social context on the performance of online reviews. To validate the model, a Tobit regression analysis was conducted in the Amazon.cn online shopping context. Hypotheses were tested on a data set of 2384 reviews over six different product categories. The results suggest that review sentiment, review quality, review perspective and the source credibility have positive impacts on perceived helpfulness, further influencing the online reviews performance. Moreover, the results confirm that product type is an influential variable to mediate the impacts. The findings also indicate that ratings does not play a behoved role in deciding whether the review is helpful as traditionally expected, nor does the date when reivews are published. This work applies sentiment mining method to understand the review content. The findings contribute to relevant literature by providing a deeper understanding of quantitative and qualitative predictors of online review performance. The model we suggest can be applied in predicting the adoption of online reviews in the online shopping context, also proposes another perspective of review ranking.

Keywords: Online review, Sentiment mining, Review sentiment, Helpfulness

1 INTRODUCTION

Nowadays, an important trend has been a growth in consumers sharing their experiences, opinions, and feedback regarding products or services in the form of online reviews for other consumers. According to a research report from the China Internet Network Information Centre, 82.1 percent of online shoppers read online reviews, and 41.1 percent of them indicate that they use online reviews every time (CNNIC, 2014). The interesting issue in this area is why most online shoppers read online reviews, but only half of them adopt online reviews every time, in other words, among the enormous amount of online reviews only half of them make a good performance.

Related to areas, significant academic effort has been made to understand the performance of online reviews and how they affect the information adoption process of consumers. Most previous studies use helpfulness to observe online review performance. Among them numerical rating, length of reviews or other formal influential factors are evaluated [1-5], where textual information contained in online reviews is relatively less discussed. Even some studies referred to the concept of sentiment, most of them use ratings to represent sentiment [6], while star ratings and sentiments may have different proximities to the final choice. Moreover, as a typical format of User Generated Content (UGC), the personal characteristic of the reviewer provide important clues for the

* Corresponding Author

perceived helpfulness. Nonetheless, not many researches have dove into this issue. Therefore, we accordingly put forward the following two questions:

RQ1: What kind of online comments will get more selective attention and comprehension（perceived helpful）?

RQ2: Does the online comments performance differs by product type?

To find the answers to the two questions above, we use sentiment mining approach to analyse the text of online reviews. Then we try to apply the results to performance prediction model of online reviews across different product types. The most similar work to ours in previous is Mohammad and Dan [7], however, they did not investigate into the product type, which has been discussed to be contributing in related researches [5, 8, 9]. Meanwhile, social context, such as the reviewer's background and behaviour of history commenting have not been exploited. Finally, the sentiment analysis to big data in their study was limited to sentence subjectivity and sentiment classification, the more in-depth sentiment mining of feature extraction has not been considered.

The remainder of this paper is structured as follows. Section 2 provides a theoretical background and presents the literature review. Section 3 discusses the influential factors and proposes the research model. Section 3 and Section 4 describes the experiments on Amazon online review dataset, including details on the performance evaluation, model validation and hypotheses testing. Detailed analysis and explanation for the results are presented. The discussion part proposes the theoretical and practical contribution, together with the limitations and future work.

2 THEORETICAL BACKGROUND

2.1 Needs Theory

According to the Needs Theory, when deficiencies of a need exist, individuals are motivated to take action to remove them in order to satisfy the need (Steers and Porter, 1991)[10]. McClelland (1965) [11] suggested that individuals may have varied levels of needs and that this variation motivates them to pursue one goal instead of another. This research stream points out that individuals perceive less value in one goal when some of their needs are fulfilled in another goal (Hsu, 2014) [12]. This suggests that customers who explore the properties of products or services should be able to derive the most value from online reviews and hence their needs.

2.2 Information Diagnosticity Theory

Information Diagnosticity Theory was originally put forward by the Feldman [13]. Diagnosticity refers to how easily the information could be extracted in human brain. That vivid, shocking information will be easily recorded. Information diagnostic expresses how much the information can be referred to the product discrimination. If the information can help consumers to classify a product into unique category, this information is considered to be highly diagnostic. On the contrary, if the information is too vaguely expressed, so that consumers can divide products into different categories from different perspective, the information is considered to be lack of diagnosticity.

2.3 Product Type: Experience Goods versus Search Goods

Most products available on Internet retail sites can be classified into either search goods or experience goods. Search goods refer to the products whose quality can be easily estimated based on product-related information even before purchase [14], such as calculators, computer printers and vitamin pills. Experience goods refer to the products whose quality is difficult to assess before direct experience, such as video games, mattresses, wine, computer games and bicycles [14-19].

There are intuitional differences between perceived helpfulness of search and experience goods. For search goods, consumers can obtain information about product quality prior to purchase while evaluation of experience goods requires sampling or purchase. Product type moderates the relationship between the number of positive and negative statements with the helpfulness of a review [20].

3 RESEARCH MODEL

Integrating concepts and findings from prior research, we propose that perception of online review performance can be understood in terms of the interactive effects of four factors:

3.1 Review Sentiment

Review sentiment is factor used by previous studies to predict the performance of computer-mediated communications (CMC) [7]. Selective attention refers to the process of selectively pay attention, contact or accept to information stimulus, namely in a particular period of time, only a part of information is allowed to be input, large amount of information will be filtered or blocked. According to this general rule, people tend to choose information according to their interest. Therefore, intensely emotional comments may attract people's attention.

H1: The exhibited total amount of sentiment of the review text has a positive effect on review helpfulness.

According to previous study, low involvement of customer is habitual behaviour, while high involvement should be attributed to enthusiasm and passion. Negative expression of practice for experience goods tend to enhance source credibility than search attributes [21], the effect of review sentiment should vary by product type. Thus, we predict:

H1a: Product type moderates the effect of review sentiment on the helpfulness of the review. The effect will be more profound for experience goods than the search goods.

3.2 Review Quality

In the literature on the effects of eWOM, review quality has often been defined as the persuasive strength of the message [22]. High-quality reviews should meet the expectations of users, fundamentally could be perceived as accurate, reliable and complete. Length of a review has been proved to be an effective substitution variable to predict the performance of online reviews [23], since there is more chance for the online reviews that consist of more words to provide consumers with greater amount of information than those who contains fewer words.

H2: Length of the review has a positive effect on the perceived helpfulness of the review.

The effect of review quality might vary depending on the product type. Considering that consumers rely more on others' product reviews for experience than search goods, they may elaborate more on review messages when evaluating experience goods. According to the elaboration likelihood model (ELM), heightened motivation leads people to examine the message more carefully (Petty, Cacioppo, & Schumann, 1983)[24].

H2a: On condition of search goods, the impact of review length on perceived helpfulness is greater than on experience goods.

3.3 Review Perspective

We next extend our study by incorporating the concept of argument perspective. This is a relatively new concept that is rarely involved in past studies. Consumers comment on different points of a product from different perspectives. For example, online reviews about mobile phone can focus on appearance, usage, after-sales, and many other attributes. The ways in which consumers seek out and evaluate online reviews are radically different due to several unique perspectives of argument (King et al., 2014) [25]. That means argument perspective could be important in the perceived helpfulness of online reviews. When a consumer thinks that the argument perspective is wide, there is a higher chance that such online reviews will be adopted. We use the listing of product features to represent argument perspective. Considering that there is more chance for search goods to acquire information of product features from the detailed introduction on website, we thus propose:

H3. The breadth of review perspective (Listing product features) is positively associated with the perceived helpfulness of online Reviews for experience goods.

H3a: This effect of review perspective on perceived helpfulness is attenuated for experience goods.

3.4 Source Credibility

Credibility is usually defined as how expert the communicator is perceived to be, and also as how trusted by the individual receiving the communication. Attribution theory suggests that consumers, when presented with a message, will make an effort to assess whether the message provides an accurate representation and/or whether the source of the message lacks credibility. Therefore, we focus on review characteristics that should contribute to the consumer goals of reducing uncertainty and equivocality and also provide information on reviewer expertise and trustworthiness.

H4: Source credibility is positively associated with the perceived helpfulness of online reviews.

The Attribution theory also proposes that uninvolved recipients respond with more negative thoughts to a low credibility communicator than to a high credibility communicator. Involved recipients respond with more negative thoughts to a high credibility communicator.

H4a: This effect of source credibility on perceived helpfulness is attenuated for search goods.

3.5 Control Variables

In order to explain the differences among individuals or situations, we also include several control variables known in the literature to affect behaviour intention. These variables include social context and demographic variables which has mostly been viewed in prior researches, including rating and the longevity of a review.

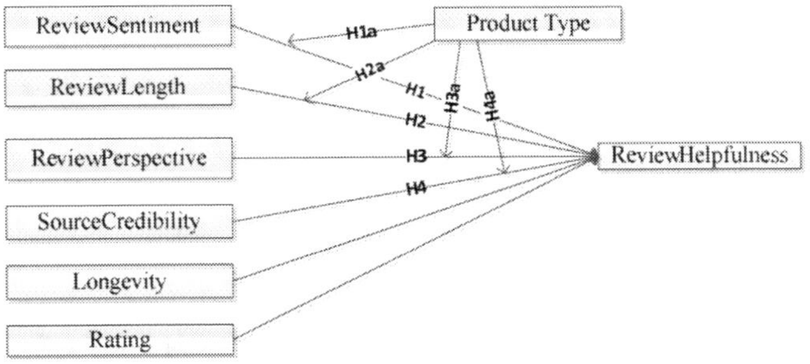

Figure 1: The Research Model

4 METHOD

4.1 Data Collection

To evaluate our research model empirically, we focus on product reviews that have been published on amazon.com.cn. Thereby, we select reviews related to six product categories as applied in previous study and select the best selling products according to the TOP Sales issued by amazon for each category. These goods could be categorized into search goods (Mobile, PC, Camera) and experience goods (Book, Cosmetic, Food). For each product, we acquire the corresponding product reviews, including review text, review title, star rating, review date, product price, the number of people rating the review as helpful and the total number of people rating the review. Finally, a panel of data on 94 different products over 11800 items from Amazon.com.cn review web pages were collected using the crawler software developed by the authors in Python IDE. We eliminate the reviews that had no helpfulness vote. The final sample consisted of 2384 reviews. We then developed a comprehensive model to extracted sentiments and explore listing of product features from the product reviews dataset and approach to the above research questions.

4.2 Variable Calculation

4.2.1 Sentiment Analysis

Sentiment mining module was developed using sentiment strength analysis API (SentiStrength) provided by Tencent, which is free for academic research and has already been tested and validated in business apps such as (Weixin、QQ、myapp、TecentSocialAds). It has been proved to be a successful API in sentiment strength analysis for SNS language environment. The reason why we choose this API attributes to the result of sentiment analysis. It reports two separate numbers for positive and negative emotions. Both the two numbers range from 0 (not positive or negative) to 1 (extremely positive or negative) to represent a relative emotional feature. Relative sentiment strength could better describe the sentiment of the reviewer, because each reviewer has different habit of understanding and applying emotional words. We could not simply use the total sentiment and make a comparison between reviews from different reviewers. What we can infer is the reviewer's sentiment expressing habit keep the same when he produce the current comment. Thus, we calculate the review balance of each statement by creating the following measure, which determines the strength of the sentiment:

$$ReviewSentiment = |positive\ sentiment - negative\ sentiment| \qquad (1)$$

4.2.2 Features Extraction

We use the Listing features to describe the argument perspective. The extraction of product features is executed by association rules and mutual information. First of all, we use the LTP （Language Technology Platform） tools to analysis sentences in online reviews. The output of this step is word vectors with standard labels, adjective, noun, verb, adverb or other types. Then we use association rules to extract nouns and noun phrases from the Amazon online reviews dataset we crawled. Those who satisfy the rule of Minimal Support will be chosen as the candidates (i.e., "mobile" and "screen" appear together frequently means the minimal support for them is high). During experiment, we found that product features were preferentially rhetoric by the same or similar adjectives, these gave us more clues of association rules for better coverage. The next step we use mutual information for extract product features. What we noticed is when people comment on a certain category of products, they always expect to concern particular features of the products. In our collected dataset, when people comment on "camera", they always comment details on "focusing", "aperture", "pixel", and this kind of words with appear much probability in certain category, but relatively seldom appear in other category.

Mutual information is described as a statistic index for connection of two random variables. In our study, we use mutual information to represent the relationship between words and product categories. We design a module of word frequency filtering for pre-processing, those words with few occurrences are disregarded. Then mutual information is calculated according to the following model,

$$I(W, C) = \log_2(P(W/C)/P(W)) \quad (2)$$

where W represents the targeting word, C represents the product category, P(W) is the frequency of W occurs in the whole review dataset, $P(W/C)$is the probability W appears in C. For word W, larger mutual information value indicates more probability of being a product feature.

Consequently, we ranked the mutual information for each category and merged the result with the candidate words in step2. Artificial screening was carried out for a more accurate result. We finally chose the top ranking words in each category for further experiment. The breath of review perspective was calculated by the coverage scale of these feature words. Synonym Thesaurus was used when calculate the index of similarity to these top features. Then we ranked the reviews from 0-10 using coverage statistical zonation method by category. This number reflects the relative perspective breadth for a product among all the products in the category.

Table 1: The extracted features

Category	**TOP ranking features**
Mobile	screen,battery,price,cost,performance,quality,voice,packaging,software,headset, logistics, system, speed, game, service, appearance, pixel, feel, function, the phone, camera
PC	Speed, fans, batteries, screen, colour, price, appearance, freezing, logistics, system, heat, sound, memory, keyboard, hard disk, after-sales, charger
Camera	Focus, voice, memory, functional, batteries, size, weight, quality, delivery, attitude, appearance, quality goods, the aperture, the colour, display, accessories, packaging, overall, screen, performance
Book	Plot, design, content, story, emotional statement, text, logistics, paper, printing, type, setting, feelings, translation, quality, cover, packaging, illustration, logic
Cosmetic	Taste, prices, overall, quality, goods, logistics, packaging, feeling, effect of use, fake, ,label production date, supermarket, counters, quality product
Food	Quality product, taste, flavour, ingredients, production date, logistics, packaging, delivery, price, cheap, shelf life, taste, carton, weight, effect

4.2.3 Other Variables Measurement

Longevity was measured by counting the number of the days since the review was created. Length of review was measured by counting the number of words in the text and their squared. Source credibility is represented by Reviewer Rank. We use number 1-5 to respectively represent reviewers who rank in “best 50 reviewers”, “best 100 reviewers”, “best 1000 reviewers”, “best 1000 reviewers” and ordinary reviewers. Helpfulness of a review was measured as the ratio of “helpful votes” to “total votes” cast by readers in each review.

4.3 Tobit Regression Model

To investigate the impact of the independent variables on review helpfulness, we follow previous research and do the Tobit regression analysis (Croissant, 2008) [26]. Tobit model belongs to the standard model of censored variable, which is always applied in the regression analysis when the observed value of dependent variable is deficient due to some limitations. In this research we adopt the Tobit model mainly due to the sample selection basis of dataset. In the online shopping environment, not all consumers who read the online reviews will participate in the helpfulness voting. The panel data we got could only reflect part of the consumers' response. Meanwhile, the dependent variable review helpfulness never becomes negative. Thereby, we include the different hypothesized independent variables as well as the control variables into the model. Meanwhile, in order to test the hypothesis H1a-H4a, the cross-term results of the above factors were added to our model for verifying the corresponding interaction effects.

$$\frac{HelpfulVote}{TotalVote}\% = Constant + \beta_1 ReviewSentiment + \beta_2 ReviewLength + \beta_3 ReviewPerspetive + \beta_4 SourceCredibility + \beta_5(ReviewSentiment \times ProductType) + \beta_6(ReviewLength \times ProductType) + \beta_7(ReviewPerspetive \times ProductType) + \beta_8(SourceCredibility \times ProductType) + \beta_9(Controls) + \varepsilon)$$

5 RESULT

In order to provide first sights into the difference between search and experience goods, we test whether both product types differ. Specific statistics are aggregated in the table below.

Table 2: Descriptive statistics of dataset

	Search Product		Experience Product	
Variable	Mean	SD	Mean	SD
ListOfFeature	2.670406	1.797642	2.305306	1.660236
Rating	4.014668	1.478911	3.947755	1.462677
ReviewSentiment	0.5141813	0.3107421	0.443988	0.305826
WordCount	99.98016	161.7957	97.84653	109.6452
Longevity	199.2166	231.1338	573.1755	526.5889
ReviewerRank	1.103538	0.4529461	1.109388	0.419444
HelpfulVote	9.33132	43.2916	5.766531	27.05887
TotalVote	54.4478	32.53889	95.47837	57.40156

According to the statistics, there is no obvious variance for WordCount, Ratings, ListOfFeature and ReviewerRank between search products and experience products. However, the longevity of experience products is much larger (573.1755) than search products (199.2166), which reflect the matter of fact that search products have more frequently online reviews updating. This is corresponding to the product replacement and popularizing in search products (mobile, pc, camera, etc.).Meanwhile, the average total votes for search products are obviously less than experience products. It could be owned to the dataset we collected. Among the items we collected, books are typical experienced good where people are easily willing to engage in comment.

In our study, we use StataSE12.0 for regression analysis. As the regression results shows, the fitting result is satisfactory, likelihood ratio test (1194.55), which has a statistical significance ($p > \chi^2$=0.000). Significant interaction between the focal independent variables and the product type variable lend strong support to the proposed framework.

Table 3: Tobit Regression Result (n=2384,$Prob > chi2 = 0.000$)

Helpfulness	Coef.	Std. Err.	t	P>t	[95% Conf.	Interval]
ReviewSentiment	0.061	0.018	3.340	0.001	0.025	0.096
ListOfFeature	0.051	0.004	14.260	0.000	0.044	0.058
logWordCount	0.030	0.005	5.570	0.000	0.020	0.041
ReviewerRank	0.099	0.014	7.250	0.000	0.072	0.126
RSP	-0.076	0.022	-3.420	0.001	-0.120	-0.033
LOFP	0.030	0.005	5.850	0.000	0.020	0.040
RLP	0.009	0.006	1.570	0.116	-0.002	0.020
RRP	-0.005	0.019	-0.250	0.801	-0.043	0.033
_cons	-0.289	0.023	-12.530	0.000	-0.334	-0.244

$(RSP: ReviewSentiment \times ProductType, LOFP: ListOfFeature \times ProductType,$

$RLP: ReviewLength \times ProductType, RRP: ReviewerRank \times ProductType)$

In order to test H1, we related to the impact of review sentiment on review helpfulness, the regression coefficient is 0.061, which means if reviews express a striking contrast in sentiment, review helpfulness increases. However, this impact could be moderated by product type. H1a could be accepted since the moderating effect $ReviewSentiment \times ProductType$ is significant with a regression coefficient of -0.076. This can be explained as for experience goods, the impact of review sentiment on perceived helpfulness is more obvious.

Related to H2, we find the length of review increase the perceived helpfulness of online review, which is significant at 1% level. This conclusion agrees with most previous research [1]. Related to H2a, the regression result is not significant and the hypothesis should be rejected, which means there is no difference between search goods and experience goods when discussing the review length impact on perceived helpfulness.

As shown in Table 3, the breadth of review perspective positively impacts the perceived helpfulness of online reviews with a regression coefficient of 0.051. Thus, in support of H3, listing product features significantly increases the odds of a review being perceived as helpful by a factor of

1.051(5.1%). Meanwhile, $ListOfFeature \times ProductType$ is significantly positive, the manifestations included that H3a could be accepted with a significance of $P>t=0.000$.This finding reveal that for search goods, the expected consumer concerned how much the review can eliminate product uncertainty through feature listing in online reviews.

The significance level of reviewer rank is much higher than the significance levels of other explanatory variables. This indicates that reviewer's community status influence review helpfulness more evidently than other variables. In contrast, H4a has to be rejected since the moderating effect $ReviewerRank \times ProductType$ is not significant. Thus, product type does not moderate the positive relationship between reviewer rank and perceived helpfulness.

Finally, we test the relationship between perceived helpfulness and control variables. There is a significant, negative interaction between the control variable Longevity and Rating. The significance is 0.944 with Rating, which cannot be accepted. The impact of review longevity on perceived helpfulness also cannot be accepted with a significance of 0.706.

6 DISCUSSIONS

6.1 Implications

The purpose of this paper is to examine several highly possible factors and their joint effects on review helpfulness. Built on relevant online review literatures, eight hypotheses were proposed and tested on the dataset from the most famous e-commerce community (Amazon.cn). The regression results show that the reviewer rank is a significant predictor of the perceived helpfulness. While trust is a critical issue in online shopping, the profile of the reviewer needs to be evaluated when processing the reviews. The review sentiment is another significant predictor. The result demonstrates that the perceived helpfulness increased with the review sentimental strength augment. Our study has achieved a different conclusion compared with prior studies [7]. This probably can be explained by the different method we used in calculating sentiment strength. The review sentiment strength we define is relative strength index, which we suggest is more reasonable and practical. Furthermore, the regression result can be explained by the related theory of information selection. Reviews with greater strength of sentiment could be easier for consumer's selective attention and selective processing. Nevertheless, the impact is attenuated for search goods. For experience goods, emotional resonance are more common among consumers, the most typical representative is book comments.

Listing features has a positive impact on perceived helpfulness of online reviews. Listing features also enhanced helpfulness more for search goods than for experience goods. Reviews that list product features many eliminate the uncertainty by disclosing product information that the shopper missed or deliberately concealed when provided product introductions on the online community. On the other hand, listing features could provide the readers with a more convenient way to get an accurate product profile or focus on the features that interest them most. The effect will be more obvious for search goods, because the features of search goods are easily recognized and always with common concern.

Previous literature has suggested that word count positively affects review helpfulness. Our work also confirms existing findings. Previous studies have also shown that reviews with extreme rating received more YES votes on helpfulness, however, the finding of our study is different. Ratings, as a relatively obscure evaluation of certain products, has not affect the perceived helpfulness as expected. The reason can be ascribed to the new review ranking mechanism carried out by Amazon. Totally different from the traditional ranking by comment date, Amazon now provides consumer with two different ways of review ranking, by comment date or by popularity. This produces huge shocks to the impact of rating. The same situation exists in perceive the impact of review longevity on review helpfulness.

6.2 Conclusions and Future Work

The results of this study has contributed to relevant literature by providing a deeper understanding of quantitative and qualitative predictors of online review helpfulness, it sheds light on the empirical relationship between a group of typical variables and review helpfulness. More specifically, the paper apply the method of sentiment analysis to deeply mining the review content, compared with questionnaire, this method can achieve more objective result and reproduce the comment behaviour to maximum extent. Additionally, the findings have extended the results found in existing research by looking at also the review context (i.e., reviewer's cumulative ranking, review content sentiment) to see whether each of those aspects influences online review helpfulness.

This study is also practically meaningful. Most previous studies just use numerical rating and length of reviews to investigate their performance. Few studies have tried to analyse the textual information contained in online reviews by questionnaires or deep interviews, this kind of method can barely provide a practical solution for online vendors because of the vigorous growth of online reviews. Instead, we use sentiment mining to analyse the text of online reviews. Based on the online product reviews in the six product categories, we get a reasonable model from panel data, which could be applied in the prediction of future data. In addition, this study suggests a probable effective way of review ranking. The results can assist online vendors in developing better decision support system for consumers.

Although this article has important theoretical and practical significance, there are still some limitations. The sample selection included 6 kinds of brand products, these products mainly comes from Amazon who has a mature system of reviewer ranking, therefore whether the conclusion is applicable to other products and other kind of community website also, still need further validation, future research can increase the sample size and diverse data sources. Meanwhile, as a text-driven method, especially for Chinese text processing, our future work will focus on improve the result of sentiment mining and continually improve the model with better degree of fitting.

7 ACKNOWLEDGEMENTS

This work is partially supported by the National Natural Science Foundation of China

(Project No.:71271012, 71302002 and 71332003).

8 REFERENCES

[1] **Mudambi, S.M. and D. Schuff. 2010.** What makes a helpful online review? a study of customer reviews on amazon.com. Mis Quarterly, 34(1), pp 185-200.

[2] **Ghose, A. and P.G. Ipeirotis. 2011.** Estimating the Helpfulness and Economic Impact of Product Reviews: Mining Text and Reviewer Characteristics. IEEE Transactions on Knowledge & Data Engineering, 23(10), pp 1498-1512.

[3] **Korfiatis, N., E. Garcia-Bariocanal and S. Sinchez-Alonso. 2012.** Evaluating content quality and helpfulness of online product reviews: The interplay of review helpfulness vs. review content. Electronic Commerce Research & Applications, 11(3), pp 205-217.

[4] **Baek, H., J. Ahn and Y. Choi. 2012.** Helpfulness of Online Consumer Reviews: Readers' Objectives and Review Cues. International Journal of Electronic Commerce, 17(2), pp 99-126.

[5] **Racherla, P. and W. Friske. 2013.** Perceived 'usefulness' of online consumer reviews: An exploratory investigation across three services categories. Electronic Commerce Research & Applications, 11(6), pp 548-559.

[6] **Sun, M. 2012.** How Does the Variance of Product Ratings Matter? Management Science, 58(4), pp 696-707.

[7] **Salehan, M. and D.J. Kim. 2016.** Predicting the performance of online consumer reviews: A sentiment mining approach to big data analytics. Decision Support Systems, 81(4), pp 30 - 40.

[8] **Connors, L., S.M. Mudambi and D. Schuff.** 2011. Is It the Review or the Reviewer? a Multi-Method Approach to Determine the Antecedents of Online Review Helpfulness. Hicss.

[9] **Muntermann, M.S.J.** 2013. What Drives the Helpfulness of Online Product Reviews? From Stars to Facts and Emotions, Wirtschafts informatik Proceedings.

[10] **Steers, R.M. and L.W. Porter.** 2004. Motivation and work behavior: McGraw-Hill/Irwin.

[11] **Mcclella, D. 1965.** TOWARD A THEORY OF MOTIVE ACQUISITION. American Psychologist, 5(20), pp 321-333.

[12] **Hsu, S.C. 2013.** Understanding the role of satisfaction in the formation of perceived switching value. Decision Support Systems, 59(1), pp 152-162.

[13] **Feldman, J.M. and J.G. Lynch. 1988.** Self-Generated Validity and Other Effects of Measurement on Belief, Attitude, Intention and Behavior. Journal of Applied Psychology, 73(3), pp 421-435.

[14] **Nelson, P. 1974.** Advertising as Information. Journal of Political Economy, 82(4): p. 729-754.

[15] **Senecal, S. and J. Nantel. 2004.** The influence of online product recommendations on consumers' online choices. Journal of Retailing, 80(2), pp 159-169.

[16] **Weathers, D., S. Sharma and S.L. Wood. 2007.** Effects of online communication practices on consumer perceptions of performance uncertainty for search and experience goods. Journal of Retailing, 83(4), pp 393-401.

[17] **Suwelack, T., J. Hogreve and W.D. Hoyer. 2011.** Understanding Money-Back Guarantees: Cognitive, Affective, and Behavioral Outcomes. Journal of Retailing, 87(4), pp 462-478.

[18] **Mudambi, B.S.M. and D. Schuff. 2010.** What Makes a Helpful Online Review? A. in Study of Customer Reviews on Amazon.com. MIS Quarterly, 34(1), pp 185-200.

[19] **Weathers, D., S.D. Swain and V. Grover. 2015.** Can online product reviews be more helpful? Examining characteristics of information content by product type. Decision Support Systems, 79(C), pp 12-23.

[20] **Sen, S. and D. Lerman. 2007.** Why are you telling me this? An examination into negative consumer reviews on the Web. Journal of Interactive Marketing, 21(4), pp 76-94.

[21] **Thelwall, M., et al. 2010.** Sentiment strength detection in short informal text. Journal of the American Society for Information Science & Technology, 62(2), pp 200-208.

[22] **Cheung, C.M.K. and D.R. Thadani. 2012.** The impact of electronic word-of-mouth communication: A literature analysis and integrative model. Decision Support Systems, 54(1), pp 461-470.

[23] **Chevalier, J.A. and D. Mayzlin. 2004.** The Effect of Word of Mouth on Sales: Online Book Reviews. Social Science Electronic Publishing, 43(3), pp 345-354.

[24] **Petty, R.E., J.T. Cacioppo and D. Schumann. 1983.** Central and Peripheral Routes to Advertising Effectiveness: The Moderating Role of Involvement. Current Oncology, 21(1), pp 35-42.

[25] **King, R.A., P. Racherla and V.D. Bush. 2014.** What We Know and Don't Know About Online Word-of-Mouth: A Review and Synthesis of the Literature. Journal of Interactive Marketing, 28(3),pp 167-183.

[26] **Stieglitz, S. and L. Dang-Xuan. 2013.** Emotions and Information Diffusion in Social Media-Sentiment of Microblogs and Sharing Behavior. Journal of Management Information Systems,29(4), pp 217-248.

A NOVEL HIERARCHICAL TIME DELAY ESTIMATION METHOD FOR WI-FI INDOOR POSITIONING

Ke Li[1*], Song Chen[2], Yulei Zong[3]
Zhengzhou Information Science and Technology Institute, Zhengzhou, China
[1]happylike8023@163.com, [2]wirelesscs@163.com, [3]zongyulei1992@163.com

ABSTRACT

Wi-Fi based indoor positioning techniques have advantages over other schemes, and positioning algorithms based on time delay estimation (TDE) can provide high accuracy. Aiming at the problem that existing kinds of TDE algorithms either perform poorly in accuracy or need high complexity, a novel hierarchical TDE method combines Leading-edge Detection based Cross-Correlation and Phase Difference Fit (LDCC-PDF) is proposed. Firstly, perform cross-correlation to the transmitted and received signal, and get the coarse time of arrival (TOA) estimate (sample level) by detecting the leading-edge of FAP step by step. Then, the fine estimate value (sub-sample level) can be obtained by getting the linear fit to the phase difference between the transmitted and received data sub-carriers. The Long Training Sequence (LTS) of IEEE 802.11n is used to test the performance of the algorithm. Simulation tests show that the proposed method provides high accuracy of TDE with a good anti-noise performance. It is found to achieve the ranging error of less than 0.3m in a single path and less than 1m in the multipath non-line-of-sight (NLOS) environment.

Keywords: Wi-Fi, TDE, Cross-Correlation, Sub-carrier phase difference, Linear Fit, Multipath

1 INTRODUCTION

The demand for indoor location based service is increasingly growing. Techniques widely used for outdoor localization, such as GPS can't work well indoors because of the severe signal attenuation by building blocks[1]. As a result, seeking effective indoor positioning techniques with high accuracy is of great importance. Thanks to the low cost and wide popularity, Wi-Fi network using IEEE 802.11 standard has application advantages in high accuracy indoor positioning field over existing techniques using UWB[2], RFID[3], Bluetooth[4], etc. Previous studies on Wi-Fi positioning mostly focused on the received signal strength indicator (RSSI) according to the radio propagation loss model[5][6]. Then RSSI fingerprinting methods have been proposed and improved[7], with accuracy reported to be 1.6-5m[8] to date. Fingerprinting needs a series of heavy data collection, pre-processing and post-maintenance work. Tiny real-time environment changes will cause severe error or even failure. Therefore, time-based Wi-Fi positioning has been considered to be an alternative with good robustness.

Conventional baseband TDE methods mostly employ cross-correlation of the received signal with a reference signal, getting the TOA value by extracting the correlation peak[9]. It is easy to implement but limited to sampling frequency of the system, as for 802.11n ADC (20MHz), the time resolution (15m) fails to meet the accuracy requirement. Another method is detecting the phase difference of the continuous radio wave since the phase difference is proportional to the time delay, which provides high accuracy but can only be applied in a small range for fear of causing phase ambiguity[10]. Makki, et al[11] combined the cross-correlation with phase difference method, reaching

*Corresponding Author

an actual measurement error of 49cm under the propagation environment with only a single line of sight (LOS) path. Oversampling schemes[12] can achieve high-resolution (sub-meter) accuracy by using expensive A/D chips and special signal pre-processing algorithms, but additional hardware support and cost are required. Frequency super-resolution algorithms have been proposed to enhance TDE performance in the multipath existed environment, such as MUSIC[13], ESPRIT[14] and their improvements[15][16]. Currently, ranging accuracies of 1-5m[16] in real world have been provided at the expense of high signal-to-noise ratio (SNR) and computational complexity.

Herein, taking the real-world application needs into consideration, we propose a novel TDE method based on Wi-Fi baseband signal. It is a hierarchical TDE scheme combines Leading-edge Detection based Cross-Correlation and Phase Difference Fit (LDCC-PDF) according to the sample and sub-sample level TOA estimates. The LTS of IEEE 802.11n is used to be the processed signal. The method absorbs advantages of both cross-correlation and phase detection. It needs lower complexity and less measurements, keeping high accuracy when applied in the multipath environment.

2 SIGNAL MODEL

The time domain channel model in the multipath environment can be expressed as

$$h(t)=\sum_{i=0}^{L_p-1}\alpha_i\delta(t-\tau_i)=\alpha_0\delta(t-\tau_0)+\sum_{i=1}^{L_p-1}\alpha_i\delta(t-\tau_i) \tag{1}$$

where L_p is the multipath number, $\alpha_i=|\alpha_i(t)|e^{j\theta_i(t)}$, τ_i are the complex fading coefficient and propagation delay of the *i-th* path respectively. τ_0 is the propagation delay of the first arrive path (FAP), which refers to the expected TOA estimate.

OFDM modulated Wi-Fi signals have been widely used indoors. We use 802.11n physical layer baseband signal as the target signal. Time domain form of the LTS from the transmitting antenna can be expressed as

$$s(t)=\frac{1}{\sqrt{12}}\sum_{k=-26}^{26}S_k\exp\left(j2\pi\Delta_F t\right) \tag{2}$$

$$\begin{aligned}S_{-26,26}=\{&1,1,-1,-1,1,1,-1,1,-1,1,1,1,1,1,1,-1,-1,1,1,-1,1,-1,1,1,1,1,0,\\&1,-1,-1,1,1,-1,1,-1,1,-1,-1,-1,-1,-1,1,1,-1,-1,1,-1,1,-1,1,1,1,1\}\end{aligned} \tag{3}$$

where S_k stands for the frequency domain data of LTS in 53 sub-carriers(including 52 non-zero data sub-carriers and a direct-current(DC) carrier). Δ_F is the sub-carrier spacing (312.5kHZ). Received signal propagated by the multipath channel can then be given by

$$\begin{aligned}y(t)&=s(t)*h(t)\\&=\frac{1}{\sqrt{12}}\sum_{k=-26}^{26}S_k\left(\sum_{i=0}^{L_p-1}\alpha_i\exp\left[j2\pi k\Delta_F\left(t-\tau_i\right)\right]\right)+n(t)\\&=\frac{1}{\sqrt{12}}\sum_{k=-26}^{26}S_k\left(\sum_{i=0}^{L_p-1}\alpha_i\exp\left(-j2\pi k\Delta_F\tau_i\right)+N_k\right)\cdot\exp\left(j2\pi k\Delta_F t\right)\end{aligned} \tag{4}$$

Comparing equation ***(2)*** and ***(4)***, received signal can be described as

$$y(t)=\frac{1}{\sqrt{12}}\sum_{k=-26}^{26}Y_k\exp\left(j2\pi k\Delta_F t\right) \tag{5}$$

where $$Y_k = S_k\left(\sum_{i=0}^{L_p-1}\alpha_i e^{-j2\pi k\Delta_F\tau_i} + N_k\right) = S_k\left(\sum_{i=0}^{L_p-1}|\alpha_i(t)|e^{j\theta_i}e^{-j2\pi k\Delta_F\tau_i} + N_k\right) \quad (6)$$

represents the received data on the *k-th* subcarrier, and $N_k \sim N(0,\sigma^2)$ is the complex additive white Gaussian noise(AWGN).

3 LDCC-PDF HIERARCHICAL TIME DELAY ESTIMATION

LDCC-PDF hierarchical TDE method divides TOA into two parts, coarse TOA (sample level) and fine TOA (sub-sample level), according to the sampling period. A three-step method is used to detect the leading-edge of the FAP's peak to get the coarse TOA estimate. The fine estimate can be obtained from the linear fit to the phase delay difference between the transmitted and received data subcarriers.

3.1 Three-step Detection for Coarse TOA Estimation

In the ideal single-path channel, accurate coarse TOA estimate can be obtained by finding the maximum correlation peak. However, LOS path is almost non-existent in the real indoor world. Power attenuation and phase offset are inevitable and the received signal appears to be the superposition of the received multipath components. Thus shifted peaks will appear in the cross-correlation result. What's more, the component of FAP is not always the most powerful one under the NLOS propagation situation, so the powerful NLOS path components and weak FAP one and will lead to great estimation error. Now we improve the conventional cross-correlation method by using a three-step detection to get a more accurate coarse TOA estimate in the multipath environment.

Step 1: Maximum peak detection

It is assumed that radio frequency symbol is received at the antenna and directly down converted to baseband signal which is sampled at f_s Hz to give the received signal $y(n)$. Then the cross-correlation of $y(n)$ and a locally synthesized copy of the transmitted signal $s(n)$ is defined as

$$r(n) = \left|\sum_{l=0}^{L-1} y(n+l)s^*(l)\right| \quad (7)$$

where *L* is the correlation length. Define p_m and d_1 as the maximum peak value and corresponding sample index

$$\begin{cases} p_m = \max(r(n)) \\ d_1 = \arg\left\{\max_n(r(n))\right\} \end{cases} \quad (8)$$

Step 2: FAP peak detection

The purpose of cross-correlation is actually to find the first peak which contains the signal component, which is called FAP peak detection. In most cases, the FAP peak is generally the first adjacent peak before the maximum peak, and the further earlier ones seem to be caused by noise. So we can determine a FAP peak searching scope by selecting an optimal threshold, $D = \{p | p > \xi_{opt} \cdot p_m\}$. Then the first peak appeared in the scope can be considered as the FAP peak with a peak value p_ξ and corresponding sample delay index d_2.

Wherein the optimal threshold is obtained from a priori environmental experiments. Experiments in [17] show that for a particular indoor NLOS environment, there is always an optimal threshold that can minimize the TOA estimation error if the bandwidth of the system is a constant. Therefore, for the target environment to be localized, ξ_{opt} can be set in advance by experimental tests.

Step 3: Leading-edge detection of the FAP

Since the signal received by multipath has an additional delay compared with the one by LOS path, the leading-edge of the FAP is less affected and the TOA estimate based on it is much more reliable. Set a new threshold η in the FAP correlation peak and the smaller time delay that corresponds to the threshold can be considered to be the final TOA coarse estimate. Record the sample index d_3.

Finally, determine if the result of these sample values obtained from each step meets the chronological order ($d_3 \le d_2 \le d_1$). Do corresponding adjustment if necessary to make sure that d_3 is the desired sample index, then the final coarse TOA estimate is calculated as

$$\tau_{coarse} = \frac{d_3}{f_s} \tag{9}$$

3.2 Data Sub-carrier Phase Difference Fit for Fine TOA Estimation

Cross-correlation TDE can only get the sample level TOA value and its accuracy is related with the system bandwidth. The sampling period of 802.11n is 0.05μs, which means a deviation of one sample will lead to a range error of 15 m. Thus TOA fine estimation of sub-sample level is necessary.

According to the received data (see Equation ***(6)***, phase noise ignored) and the phase change periodicity, phase shift caused by the sub-sample delay from the transmitted signal to the received signal of the k *-th* subcarrier in the *i-th* path can be expressed as

$$\varphi_k = \theta_i - 2\pi k \Delta_F \tau_{i,fine} \tag{10}$$

where $\tau_{i,fine}$ is the fine TOA estiamte of the *i-th* path. Likewise, the phase shift of the $k+\delta$ *-th* subcarrier is

$$\varphi_{k+\delta} = \theta_i - 2\pi(k+\delta)\Delta_F \tau_{i,fine} \tag{11}$$

and the phase shift difference between the δ consecutive sub-carriers is

$$\Delta\varphi_\delta = |\varphi_{k+\delta} - \varphi_k| = 2\pi\delta\Delta_F \tau_{i,fine} \tag{12}$$

Then the fine delay of the *i-th* path can be estimated by

$$\tau_{i,fine} = \frac{\Delta\varphi_\delta}{2\pi\delta\Delta_F} \tag{13}$$

We can see from ***(12)*** that the phase difference between different sub-carriers is linear with the sub-carrier spacing when the delay is fixed. Thus, a linear fit to the phase difference between the transmitted and received adjacent sub-carriers ($\delta=1$) can be utilized. The fine delay can then be obtained based on the gradient of the fit m. It is known that time delay of the first path ($i=1$) refers to the TOA estimate desired, then the final sub-sample level fine TOA estimate is

$$\tau_{fine} = \frac{m}{2\pi\Delta_F} \tag{14}$$

On the basis of foregoing analysis, the process of fine TOA estimation is as follows.

Step1: Calculate the frequency domain received signal by conducting fast Fourier transform (FFT) of the time aligned received data according to the coarse TOA estimate

$$Y_k = \sum_{n=0}^{L-1} y(d+n)e^{-j2\pi kn/L} \tag{15}$$

Step2: Calculate the phase shift of each sub-carrier

$$\varphi_k = \angle Y_k - \angle S_k \tag{16}$$

Step3: Get the linear fit to the phase shift variation with different sub-carriers in the minimum mean square error (MMSE) sense. Then obtain the fine TOA estimate according to ***(14)***.Thus, the final TOA estimate τ is

$$\tau = \tau_{coarse} + \tau_{fine} \tag{17}$$

4 SIMULATION TESTS AND ANALYSES

MATLAB simulations are given to verify the performance of the proposed method. Parameters used in each simulation test are set based on IEEE 802.11n physical layer protocol [18], as shown in Table 1.

Table 1: Simulation Parameters Setting

System Bandwidth/B	20MHz
Sampling Interval /Ts	0.05μs
IFFT Length/point	64
CP Length/point	32
Data Sub-carrier Number	52
Monte Carlo Simulation Number	1000
Single Path Channel(AWGN model)	τ_0=10.5*Ts
Multipath Channel(6 paths) (AWGN and Rayleigh fading model)	Path delays: τ =[10.5, 10.8, 11, 11.5, 12.2, 12.6]*Ts Average path gains: pdb=[0, -1.6, -4.7, -10.1, -17.1, -21.7] dB Maximum Doppler shift: fd=5Hz

Test 1: Coarse TOA estimation

The proposed method is applied in the multipath channel condition (a Rayleigh fading model with 6 paths is used to simulate the real indoor office environment). The optimal threshold ξ_{opt} for FAP detection is obtained by preliminary tests, the value of ξ_{opt} is 0.4(see Fig 1). Another threshold η for leading-edge detection is set to be 0.8 by experience. Conduct Monte Carlo simulations according to the steps mentioned in section 2, then the correct coarse TOA estimation probabilities of maximum peak detection, FAP detection and leading-edge detection are shown in Fig 2. We can see that the maximum peak detection can hardly give the correct coarse TOA value. After the processing of FAP detection and leading edge detection, the correct probability increased dramatically up to 85%.

Noticing that when SNR is less than 0 dB, the performance is seriously affected for noise is the main component of received signal. But when SNR is greater than 0dB, the probability tends to be steady, proving that the anti-noise performance of the method is good when desired signal is dominant.

Test 2: Fine TOA estimation

The received signal after time aligned according to the coarse estimation is transformed into frequency domain by FFT. Phase difference between transmitted and received data of each sub-carrier can be seen in Fig 3(a), and the phase shift is basically linear with sub-carrier index. A deviation of 2π occurs for some shift values, which is caused by phase rotation and will be considered as phase noise. Thus, proper adjustment is needed to eliminate the deviation before the MMSE linear fit. Phase difference after adjustment and its linear fit is shown in Fig 3(b). Fine TOA estimate can then be obtained according to the gradient of the fit. The method is tested in both single path and multipath environment, and the root mean square error (RMSE) of the fine estimates are shown in Fig 4. Performance in single path is superior to that in multipath, and barely influenced by SNR. In the multipath, the RMSE is less than 0.1ns with high SNRs (>20dB) and reaches ns level with low SNRs (<20dB), which can satisfy the needs of high accuracy indoor localization.

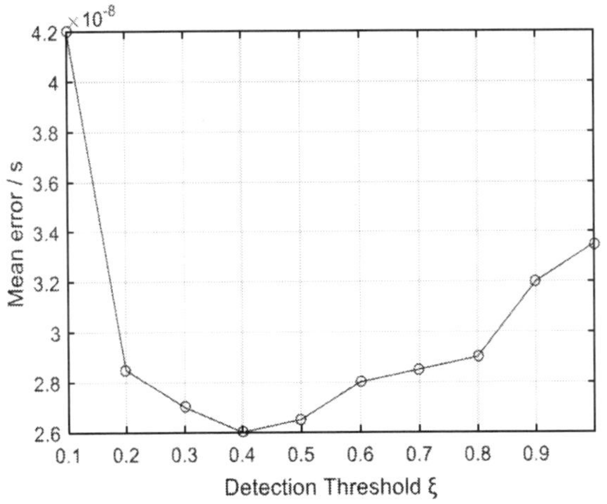

Fig 1: Optimal Threshold Test

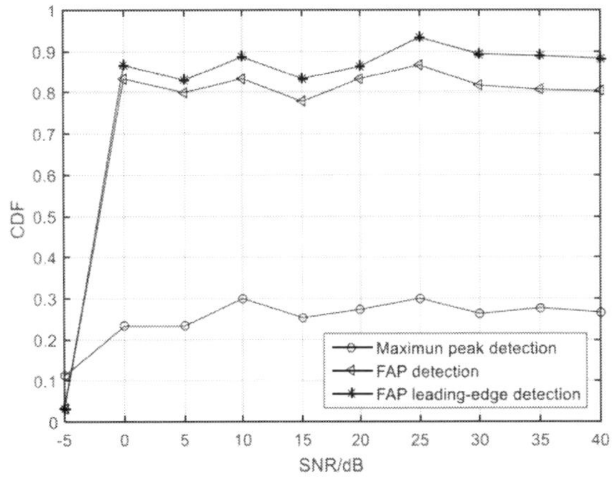

Fig 2: Correct Probabilities of Coarse TOA Estimate for 3 Detection Steps

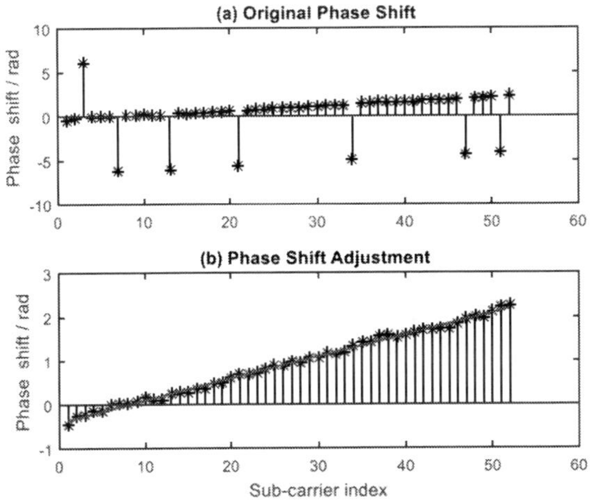

Fig 3: Phase Difference with Data Sub-carriers and Linear Fit (red)

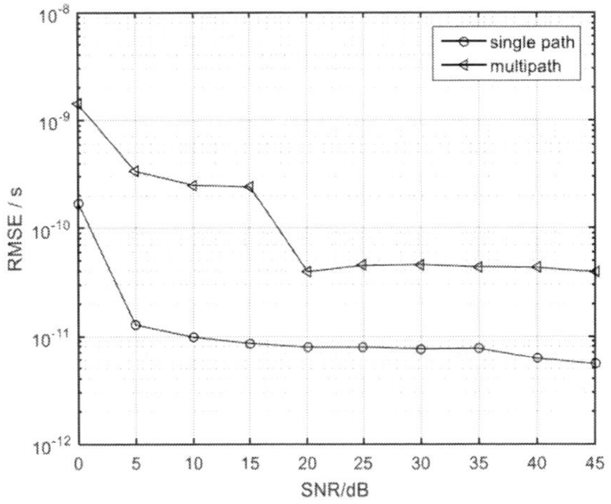

Fig 4: Fine Estimation Error

Test 3: LDCC-PDF Hierarchical Time Delay Estimation

In order to verify the practical performance of the proposed LDCC-PDF method, the classical super-resolution algorithm MUSIC and the method proposed in [11] are used for comparison, simulations are performed in both LOS single path and NLOS multipath conditions to test their performances.

Fig 5 shows that the TOA estimation RMSE of the three methods can all reach ns level under single path condition. The LDCC-PDF algorithm and method in [11] perform better than MUSIC with an average error of less than 1m when SNR is low. But when SNR tends to high, they all show good performance with a range error of less than 0.3m.

Fig 6 shows their performances in multipath environment. MUSIC almost failed at low SNR since the estimated prior information for the algorithm is greatly influenced by noise. Method in [11] can't meet the location need either for the failure in coarse estimation. The proposed algorithm can still keep good and stable accuracy in spite of a little worse than that in single path. It can reach an estimation RMSE of 3ns, i.e., 1m in range.

Taking computational complexity into consideration, MUSIC requires most and mainly includes covariance matrix estimation, eigenvalue decomposition and spectral peak searching. The more peak searching grids, the higher estimation accuracy, but the higher cost. Complexity of method in [11] is mainly caused by once cross-correlation and twice FFT (IFFT) calculations which is much smaller than the super-resolution algorithm needs. Our proposed algorithm needs twice more correlation value searching process compared to method in [11], however, the complexity does not increase too much thanks to the small searching scope based on the threshold.

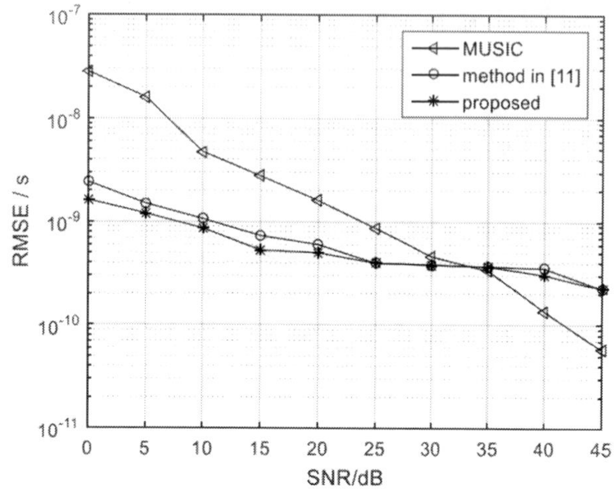

Fig 5: Methods Comparison in LOS Single Path

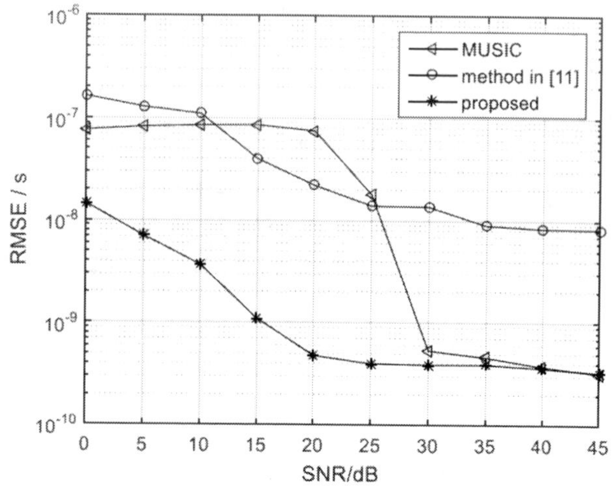

Fig 6: Methods Comparison in NLOS Multipath

5 CONCLUSION

In order to address the problem that existing TDE algorithms for indoor positioning perform poorly in accuracy under the real world and need high complexity as well, a novel hierarchical TOA estimation method called LDCC-PDF was proposed in this paper. Simulation tests proved that its performance could approach to the super-resolution algorithm but with lower computational complexity in the ideal single path propagation environment. It also performed well under the multipath environment even at low SNR while the other methods failed. The algorithm finally reached a range accuracy of sub-meter under the NLOS Rayleigh fading multipath channel. In addition, using single LTS of 802.11n as the positioning signal ensured the real time performance due to the low complexity. The accuracy and reliability of the algorithm could be enhanced by repeated measurements. What's more, it is easy to implement in both software and hardware, without need to change the existing WLAN facilities. Thus, the proposed LLDC-PDF TOA estimation method has a practical meaning to the high accuracy indoor positioning application service.

In the future work, we plan to make some improvements to the proposed method. Investigating data processing and channel compensation methods in both time and frequency domain to enhance the robustness in low SNR is necessary. And MIMO techniques of 802.11n can be utilized to further enhance the overall performance of the algorithm.

6 REFERENCES

[1] **Farid Z, Nordin R, and Ismail M. 2013.** Recent Advances in Wireless Indoor Localization Techniques and System[J]. Netw. Commun., vol.2013, Aug., Art. ID 185138.

[2] **Zhang L.W, Yang G. 2013.** Ultra-Wide-Band Based Indoor Positioning Technologies [J]. Journal of Data Acquisition and Processing, 28(6),pp 706-713.

[3] **Liu P, Lu T.C, Gao X. 2014.** A review of Radio Frequency Identification based indoor localization technology [J]. Journal of Terahertz Science and Electronic Information Technology, 12(02),pp 345-348.

[4] **Bian H.S. 2015.** Low-power indoor positioning research based on Bluetooth 4.0 [D]. Beijing: Beijing University of Posts and Telecommunications.

[5] **Ahn H. S. and Yu W. 2009.** Environmental-adaptive RSSI-based Indoor Localization [J]. IEEE Trans. Autom. Sci. Eng., 6(4), pp 626-633.

[6] **Su J, Jin Y.X. 2007.** Indoors Location Technology Research Based on WLAN[C]. Proc. 6th WSEAS Int. Conf. Instrum., Meas., Circuits Syst., pp 109-114.

[7] **Chen L.N.2014.** Research on major technology of using WLAN and fingerprint for indoor positioning [D]. Doctoral dissertation, Shanghai: East China Normal University.

[8] **Lymberopoulos D, Liu J, Yang X, et al.2015.** A Realistic Evaluation and Comparison of Indoor Location Technologies: Experiences and Lessons Learned [C]. Proc. 14th Int. Conf. Inf. Process. Sensor Netw., pp 178-189.

[9] **He M, Zu L, Sun H, et al. 2010.** Generalized Cross Correlation Time Delay Estimation Based on LMS [J]. Audio Engineering, 34(9),pp 46-48.

[10] **Sun C.Q. 2009.** Research on OFDM ranging techniques [D]. Master's dissertation, Xi'an: Xidian University.

[11] **Makki A, Siddig A, Saad M.M., et al. 2015.** High-resolution Time of Arrival Estimation for OFDM-based Transceivers [J]. Electronics Letters,51(3), pp 294-296.

[12] **Nur K, Feng S, Ling C, et al. 2012.** Application of the Improved FOCUSS for Arrival Time Estimation (IFATE) Algorithm to WLAN High Accuracy Positioning Service [J]. Proc. Ubiquitous Positioning, Indoor Navigation Location Based Service, pp 1-8.

[13] **Schmidt R.O. 1986.** Multiple Emitter Location and Signal Parameter Estimation [J]. IEEE Trans. Antennas Propag., 34(3), pp 276-280.

[14] **Roy R, KailathT. 1989.** ESPRIT-estimation of Signal Parameters via Rotational Invariance Techniques [J]. IEEE Trans. Acoust., Speech, Signal Process., 37(7), pp 984-995.

[15] **Wang F.Q, Zhang X.F, Wang F. 2014.** Root-MUSIC-based joint TOA and DOA estimation in IR-UWB[J]. Journal on Communications, 35(2).pp 137-145.

[16] **Zhao F, Yao W, Logothetis C.C, et al. 2006.** Comparison of Super-resolution Algorithms for TOA Estimation in Indoor IEEE 802.11 Wireless LANs [C]. Proc. Int. Conf. Wireless Commun., Netw. Mobile Comput., Sep, pp 1-5.

[17] **Chen D.W. 2015.** Analysis and optimization of pseudolite location accuracy under indoor multipath environment [D]. Dissertation for the Master degree,Harbin: Harbin Institute of Technology.

[18] **Yang F, He L.F, Pan C.Y.** 2013. OFDM Principles and Standands: Evolution of Communication Modulation Tenchniques [M]. Beijing: Publishing House Of Electronics Industry.

MULTIOBJECTIVE FLEXIBLE JOB SHOP SCHEDULING WITH TOTAL ENERGY CONSUMPTION

Deming Lei* and Shenpei Zhou
School of Automation, Wuhan University of Technology,
Wuhan, China
deminglei11@163.com

ABSTRACT

In this study fJSP with the minimization of total energy consumption and maximum workload is considered and a novel shuffled frog-leaping algorithm (SFLA) is proposed based on a three-string coding method. Memeplexes are constructed by using population and non-dominated set according to tournament selection. The search process of each memeplex is done on its non-dominated member. Extensive experiments are conducted to test the search performance of SFLA and the results show the conflicting between two objectives of fJSP and the promising advantages of SFLA on the considered fJSP.

Keywords: Flexible job shop scheduling, Total energy consumption, Shuffled frog-leaping algorithm, Maximum Workload

1 INTRODUCTION

The reductions of energy consumption or carbon footprint have attracted some attention in recent years [1]. Mouzon et al. [2] collected the operational statistics of four machines in a workshop, found that the non-bottleneck machines consume a considerable amount of energy when left idle and proposed a turn-on and turn-off scheduling framework. Mouzon and Yildirim [3] proposed a multiobjective genetic algorithm (GA) with hybridization with dominance rules and a heuristic to minimize energy consumption and total completion time on a single machine. Shrouf et al. [4] used the same framework by Mouzon et al. [2] to minimize total energy consumption cost in single machine.

Regarding flow shop scheduling problem (FSP) with energy related objective, Fang et al. [5] presented a mathematical programming model for flow shop scheduling problem with peak power load and carbon footprint and demonstrated the model by using a simple case study. Dai et al. [6] reported a genetic-simulated annealing algorithm to make a significant trade-off between makespan and total energy consumption in flexible flow shop scheduling. Liu et al. [7] developed a branch-and-bound method for minimizing total wasted energy consumption in permutation flow shop scheduling. Tang et al. [8] proposed an improved particle swarm optimization for energy-efficient dynamic scheduling in flexible flow shop. Ding et al. [9] considered a permutation flow shop scheduling with carbon emission and designed a modified multiobjective iterated greedy algorithm. Mansouri et al. [10] provided a mixed integer linear multi-objective optimization model, lower bounds and a heuristic for two-machine flow shop scheduling with makespan and total energy consumption. Few papers are about reducing energy consumption in job shop scheduling. Lei and

* Corresponding Author: Deming Lei deminglei11@163.com

Guo [11] presented a dynamical neighborhood search for a dual-resource constrained job shop with interval processing time and two objectives including carbon footprint. Zhang and Chiong [12] developed a multi-objective GA to minimize total weighted tardiness and total energy consumption of job shop scheduling.

Flexible job shop scheduling problem (fJSP) is an extension of job shop scheduling by allowing an operation to be processed on more than one machine. fJSP has attracted much attention in the past decade and a number of results have been obtained [13-16]; however, energy consumption or carbon footprint are seldom considered and the frequently optimized objectives are always related to completion time or workload in the literature on fJSP. With the intensive concern on environmental problem and the advent of green manufacturing, it is vital to focus on energy-efficient scheduling in flexible job shop.

Shuffled frog-leaping algorithm (SFLA) is a meta-heuristic originally developed by Ensuff and Lansey [17] and has been applied to various production scheduling problems [18-20]; however, SFLA is seldom applied to solve fJSP with multiple objectives [18], let alone solve fJSP with energy-related objective.

In this study, fJSP with total energy consumption and maximum workload is considered and a novel SFLA is proposed. A three-string coding method is applied and memeplexes are constructed by using solutions in population and non-dominated set. The search process within each memeplex is done on its non-dominated member and consists of two kinds of searches: global search and multiple neighborhood search. Many experiments are executed to test the performance of SFLA. The computational results demonstrate the effectiveness of SFLA on the considered fJSP.

2 PROBLEM DESCRIPTION

fJSP is composed of n jobs $J_i\ (i=1,2,\cdots,n)$ and m machines $M_k\ (k=1,2,\cdots,m)$. Each job J_i has h_i operations. Operation o_{ij} indicates the jth operation of job J_i and can be processed on a set S_{ij} of compatible machines. There is a finite and discrete set of d different processing speeds for a machine, $V=\{v_1,v_2,\cdots,v_d\}$. The processing speed of a machine cannot be changed during its execution of a job. It is assumed that operation o_{ij} on machine M_k has a basic processing requirement measured by p_{ijk}. When operation o_{ij} is processed on a machine $M_k \in S_{ij}$ at speed v_l, the processing time p_{ijkl} is equal to p_{ijk}/v_l and the energy consumption per unite time is denoted by E_{kl}.

With respect to relation between p_{ijkl} and E_{kl}, there is an assumption [9]. It is assumed that if an operation o_{ij} is processed at a higher speed on a machine M_k, the energy consumption increases while its processing time decreases. This implies that

$$\text{For } \forall v_l > v_g, l, g \in \{1,2,\cdots,d\},\quad p_{ijkl} < p_{ijkg},\ E_{kl} \times p_{ijkl} > E_{kg} \times p_{ijkg} \tag{1}$$

A machine cannot be turned off completely until all jobs are finished. Each machine M_k will be on a stand-by mode with instantaneous energy consumption of SE_k per unit time during the idle periods.

There are several constraints on jobs and machines, such as: each machine can process at most one operation at a time, no jobs may be processed on more than one machine at a time,operations cannot be interrupted, setup times and remove times are included in the processing times et al.

The considered fJSP is made up of three sub-problems: machine assignment sub-problem for choosing an appropriate machine for each operation; scheduling one for sequencing the operations on machines and speed selection for deciding processing speed of machine used to process operation.

The following two objectives are minimized simultaneously.

$$Min\ TEC = \int_0^{C_{\max}} \left(\sum_{i=1}^{n} \sum_{j=1}^{h_i} \sum_{k=1}^{m} \sum_{l=1}^{d} E_{kl} y_{ijkl}(t) + \sum_{k=1}^{m} SE_k z_k(t) \right) dt \quad (2)$$

$$Min\ MB = \max \{W_k\}$$

(3)

where TEC indicates total energy consumption. $y_{ijkl}(t)$ is a binary variable. If machine $M_k \in S_{ij}$ is in processing mode at speed v_l at time t, then $y_{ijkl}(t)$ is equal to 1; otherwise $y_{ijkl}(t)$ is 0. $z_k(t)$ is 1 if machine M_k is in stand-by mode at time t and 0 otherwise. MB denotes maximum workload and W_k is workload of machine M_k.

3 SFLA FOR MULTIOBJECTIVE FJSP WITH TOTAL ENERGY CONSUMPTION

Lei and Guo [19,20] presented SFLA by using some new principles, which are as follows: the search process is done on x_b of each memeplex and all solutions in the memeplex can be combined together with x_b in the same probability; tournament selection is used to decompose the population P. In this study, we present some new paths to design SFLA for multi-objective fJSP.

3.1 Initial solution

A three-string coding method is used to represent three sub-problems independently. For fJSP with n jobs, m machines and d speeds, its solution is represented by using a doublet string $((\theta_1,r_1),(\theta_2,r_2),\cdots,(\theta_i,r_i),\cdots,(\theta_h,r_h))$, a machine assignment string $(q_{11},q_{12},\cdots,q_{1h_1},\cdots,q_{nh_n})$ and a speed selection string $(z_{11},z_{12},\cdots,z_{1h_1},\cdots,z_{nh_n})$, $h=\sum_{i=1}^{n} h_i$.

The first string is for scheduling sub-problem, the doublet (θ_i,r_i) indicates operation $o_{\theta_i r_i}$ and the whole string is an ordered operation list, $\theta_i \in \{1,2,\cdots,n\}$, $1 \le r_i \le h_{\theta_i}$. In other two strings, gene $q_{ij} \in S_{ij}$ indicates a compatible machine for operation o_{ij} and z_{ij} is the speed of the machine $q_{ij} \in S_{ij}$ when it processes operation o_{ij}.

An initial population is randomly produced. To obtain a feasible schedule, the following decoding procedure is adopted for each solution: (1) Translate the first string into a list of the ordered operations, and assign a machine for each operation and decide the speed of the assigned machine for each operation according to other two strings; (2) The first operation of the operation list is arranged first, and the second one is then done and so on; each operation is allocated in the best available time on its assigned machine. The procedure is repeated until a schedule is obtained.

3.2 Memeplex construction

In this study, a new method is applied to construct memeplexes by using population P and a non-dominated set Ω. The set Ω is used to store the non-dominated solutions generated by SFLA. To construct s memeplexes, solutions x_1, x_2 are randomly chosen from $P \cup \Omega$; if $x_1 \succ x_2$ or $x_2 \succ x_1$, the dominating solution is directly added into memeplex $\mathcal{M}_1$; if x_1 and x_2 are non-dominated each other, then they can be added into $\mathcal{M}_1$ according to the same probability and one of them is randomly chosen, where $x_1 \succ x_2$ means that x_1 dominates x_2; the unselected solution goes back to $P \cup \Omega$, the above steps are repeated until the size of $\mathcal{M}_1$ reaches. The same procedure is done to construct $\mathcal{M}_2$, $\mathcal{M}_3$, $\cdots \mathcal{M}_s$ sequentially.

3.3 Search in each memeplex

1. Global search

There are three global searches. The first global search is about doublet string and described as follows [21]: for the doublet string of x_b and x, start with the first position, stochastically generate a random number α, if $\alpha < \delta$, the first doublet of x_b is selected; else the first doublet of x is chosen; suppose that the doublet (θ_i, r_i) is selected, it is appended to the offspring and then delete (θ_i, r_i) from x_b and x, if the doublet (θ_i, r_i) is the gth doublet of x, then shift the doublets of x between position $g+1$ and last position left once. The above step is repeated until the doublet string of x_b and x is empty; as a result, a new doublet string is obtained, where δ is a probability.

The second one is done between machine assignments of two solutions, which is shown as follows: for x_b and x, two positions $g_1, g_2 \in [1, h]$ are stochastically decided and then the machines of x_b between two positions are replaced with those of x in the same position.

The third global search is similar with the second one and described below: for x_b and x, two positions $g_1, g_2 \in [1, h]$ are stochastically decided and then the speed of x_b between two positions are replaced with those of x in the same position. Unlike the works of Lei and Guo [19,20], the execution of three global searches is decided by two probabilities.

Global search is summarized as follows: if $\alpha < \beta$, the first global search is applied; if $\alpha \in [\beta, \eta]$, then the second one is done; if $\alpha > \eta$, the third global search is executed.

2. Multiple neighborhood search

Neighborhood structure $insert$ is designed for generating neighborhood solutions of doublet string. For doublet string $((\theta_1, r_1), (\theta_2, r_2), \cdots, (\theta_i, r_i), \cdots, (\theta_h, r_h))$, a string $(\theta_1, \theta_2, \cdots, \theta_i, \cdots, \theta_h)$ is first obtained then a element θ_j and a position $k \neq j$ are decided and the element θ_j is inserted into the position, a new doublet string is constructed.

Neighborhood structure $change$ is conducted on the string $(q_{11}, q_{12}, \cdots q_{1h_1}, \cdots, q_{n1}, \cdots, q_{nh_n})$, which is shown as follows: a set $\Theta = \{q_{ij} \| S_{ij}| > 1, i = 1, 2 \cdots, n, 1 \le j \le h_i\}$ is first determined, some elements are stochastically selected from Θ, suppose that q_{ij} is selected, then q_{ij} is replaced with a randomly chosen machine from S_{ij}.

Neighborhood structure $speed$ is used to select a new speed for some chosen machines. It obtains new solution by randomly choosing α elements from the third string and assigning a new speed to these elements in a random way.

Let $\mathcal{N}_1, \mathcal{N}_2, \mathcal{N}_3$ be *insert*, *change* and *speed*. $\mathcal{N}_i(x)$ represents the neighborhood of x obtained by using $\mathcal{N}_i$. After global search and multiple neighborhood search are executed on x_b, a new solution z is produced, a replacement principle is used to decide if x_b can be replaced with z : if z is dominated by x_b, then x_b is not changed; otherwise, z substitutes for x_b.

The non-dominated set Ω is updated in the following way: after x_b is replaced with a new solution, it is included into the set Ω and all members of Ω are compared according to Pareto dominance. After the dominated members are deleted from the set Ω, if the actual size of Ω exceeds its maximum size H, crowding distance is calculated for each member in terms of the methods proposed by Deb et al. [22] and some members with the smallest crowding distance are removed from Ω to make the actual size be equal to the maximum size.

3.4 Algorithm description

The detailed procedure of SFLA is described as follows.

(1)Randomly produce the initial population P, construct the initial set Ω, $\sigma_i \leftarrow 1$, $i = 1,2,\cdots,s$.
(2)Repeat the following steps (3-5) until the termination condition is met.
(3)Apply tournament selection to construct s memeplexes using solutions in $P \cup \Omega$.
(4)For each memeplex $\mathcal{M}_i, i = 1,2,\cdots,s$, $w \leftarrow 1$, if $w \le \mu$, repeat the following steps:
4.1) Randomly select a non-dominated solution x_b and a solution from $\mathcal{M}_i$, $k \leftarrow 1$.
4.2) Global search between x_b and x is done, a solution z is obtained and compared with x_b; if the replacement condition is met, then x_b is replaced with z, the set Ω is updated, $w \leftarrow w+1$, go to step (4.1);
4.3) A new solution $z \in \mathcal{N}_{\sigma_i}(x)$ is obtained and compared with x_b, if the replacement condition is met, then x_b is replaced with z and the set Ω is updated, $w \leftarrow w+1$ and go to (4.1); else $\sigma_i \leftarrow \sigma_i + 1$, if $\sigma_i \ge 4$, then let $\sigma_i \leftarrow 1$.
4.4) A new solution $z \in \mathcal{N}_{\sigma_i}(x)$ is obtained and compared with x_b, if the replacement condition is met, then x_b is replaced, the set Ω is updated, $w \leftarrow w+1$ and go to (4.1);
(5) Shuffle the evolved memeplexes.
(6) Output the set Ω.
where μ is an integer. The termination condition is the predetermined number of objective function evaluations. In step (5), after the search in each memeplex is done, all evolved memeplexes are combined together and a new population P is constructed.

4 COMPUTATIONAL EXPERIMENTS

To test the performance of SFLA for the considered fJSP, extensive experiments are conducted on a set of problems. All experiments are implemented by using Microsoft Visual C++ 6.0 and run on 4.0G RAM 1.70GHz CPU PC.

We use 13 instances MK1-13. These instances are designed for testing flexible job shop scheduling algorithms and extended them by adding machine speed information: $E_{kl} = 4 \times v_l^2$, $SE_k = 1$ $V = \{1.00, 1.30, 1.55, 1.80, 2.00\}$.

Distance metric DI_R is applied to measure the performance of the non-dominated solution set Ω_l relative to a reference set Ω^*.

$$DI_R(\Omega_l) = \frac{1}{|\Omega^*|}\sum_{y\in\Omega^*} \min\{d_{xy} | x \in \Omega_l\} \tag{4}$$

where d_{xy} is the distance between a solution x and a reference solution y in the normalized objective space, $d_{xy} = \sqrt{(f_1^*(x) - f_1^*(y))^2 + \cdots + (f_D^*(x) - f_D^*(y))^2}$, D is the number of objectives, f_i^* is the ith normalized objective, the details of normalization is shown in Ishibuchi et al. [23]. The reference set Ω^* consists of the non-dominated solutions of $\bigcup_{l=1}^{A} \Omega_l$, A is the total number of algorithms.

Metric ρ_l indicates the ratio of number of the elements in the set $\{x \in \Omega_l | x \in \Omega^*\}$ to $|\Omega^*|$.

As stated above, the conflicting between two objectives of the considered fJSP is not studied, so there are no existing methods for our fJSP. In this study, non-dominated sorting genetic algorithm-II (NSGA-II, [22]) and VNS are chosen.

Ahmadi et al. [16] applied NSGA-II to fJSP with machine breakdown. We extended this version of NSGA-II to our fJSP in the following way: the representation for machine assignment and scheduling and crossovers and mutation related to this representation are adopted directly, the handling part on machine breakdown is deleted, a speed selection string is added for each individual, two-point crossover and mutation using $speed$ are also added for speed selection sub-problem.

VNS is directly constructed by using $\mathcal{N}_1, \mathcal{N}_2, \mathcal{N}_3$ and described as follows.

(1)Produce an initial solution x and an initial set Ω. $k \leftarrow 1$

(2)Repeat the following steps until the termination criterion is met.

(3)Randomly generate a solution $z \in \mathcal{N}_k(x)$ and compare with x, if the replacement condition is met, the current solution x is replaced with z, the set Ω is updated using the strategy shown in Section 4.3 and $k \leftarrow 1$; otherwise $k \leftarrow k+1$ and if $k = 4$ then $k \leftarrow 1$

We choose the following settings for SFLA: $N = 40$, $\mu = 100$, $H = 20$, $s = 5$ and $max\_it = 10^5$. The parameters of VNS and NSGA-II are listed below. For NSGA-II, population scale of 100, crossover probability of 0.8, mutation probability of 0.1 and maximum generation of $max\_it/100$; For VNS, max_it is its unique parameter. When max_it solutions are generated, the search of each algorithm is terminated. All parameters of three algorithms are set based on many experiments.

Table 1 The Computational Results and Times of Three Algorithms

instance	$DI_R(\Omega)$			ρ			Time(s)		
	SFLA	VNS	NSGA-II	SFLA	VNS	NSGA-II	SFLA	VNS	NSGA-II
MK1	0.16	14.4	5.61	0.88	0.00	0.12	3.23	3.00	4.58
MK2	0.26	17.1	5.68	0.89	0.00	0.11	3.25	2.98	4.69
MK3	0.00	25.6	11.6	1.00	0.00	0.00	8.58	7.78	9.31
MK4	0.85	15.3	4.32	0.78	0.00	0.22	4.55	4.06	5.84
MK5	3.09	20.5	0.45	0.05	0.00	0.95	7.13	7.62	8.40

MK6	0.00	22.8	14.9	1.00	0.00	0.00	7.54	6.39	8.67
MK7	1.17	21.9	6.73	0.50	0.00	0.50	5.79	5.77	6.99
MK8	2.19	19.8	9.78	0.83	0.00	0.17	16.5	15.2	17.5
MK9	3.35	1.70	17.8	0.31	0.69	0.00	13.8	15.9	17.3
MK10	0.68	17.3	18.1	0.84	0.16	0.00	15.8	14.1	16.1
MK11	4.53	18.9	1.52	0.21	0.00	0.79	13.4	13.9	14.9
MK12	1.19	17.2	11.3	0.66	0.34	0.00	15.8	14.9	16.7
MK13	0.35	8.67	24.6	0.71	0.29	0.00	15.9	15.9	16.5

The computational results and times of three algorithms are reported in Table 2, in which Ω^* is a set of the non-dominated solutions in the set $\bigcup_{l=1}^{3}\Omega_l$. $\Omega_1,\Omega_2,\Omega_3$ are the non-dominated set of SFLA, VNS and NSGA-II. $DI_R(\Omega_l)$ and ρ_l are the corresponding metric values of Ω_l .

As shown in Table 1, SFLA generates better results than NSGA-II and VNS on most of instances. VNS can converge to members of the set Ω^* only for four instances and its solutions are always far away from the solutions in Ω^* . Most of members of Ω^* are produced by SFLA on 9 of 13 instances, NSGA-II can approximate most of members of Ω^* for only 3 instances; moreover, $DI_R(\Omega_3)$ is always bigger than $DI_R(\Omega_1)$ by at least 3 percentage on 11 instances. Based on the above discussions, SFLA is an effective method for the considered fJSP.

5 CONCLUSIONS

In this study fJSP with total energy consumption and maximum workload is investigated and a novel SFLA is proposed based on a three-string coding method. Population and the non-dominated set are used to construct memeplexes according to tournament selection. The search process within each memeplex is done on its non-dominated member. Extensive experiments are conducted and the results show the promising advantages of SFLA on the considered fJSP.

Acknowledgement: This work is supported by the National Natural Science of Foundation of China (61573264,71471151)

6 REFERENCES

[1] **Gahm C. Denz F. Dirr M. Tuma A.** Energy-efficient scheduling in manufacturing companies: a review and research framework. European Journal of Operational Research 248(2016), 744-757

[2] **Mouzon G. Yildrim M.B. Twomey J.** Operational methods for minimizing of energy consumption of manufacturing equipment. International Journal of Production Research 45(2007), 4247-4271

[3] **Mouzon G. Yildirim M.B.** A framework to minimize total energy consumption and total tardiness on a single machine. International Journal of Sustainable Engineering 1(2008), 105-116

[4] **Shrouf F. Ordieres-Meré J., García-Sánchez A. Ortega-Mier M.** Optimizing the production scheduling of a single machine to minimize total energy consumption costs. Journal of Cleaner Production 67(2014), 197-207

[5] **Fang K. Uhan N., Zhao F.Sutherland J.W.** A new approach to scheduling in manufacturing for power consumption and carbon footprint reduction Journal of Manufacturing Systems 30(2011), 234-240

[6] **Dai M. Tang D.B. Giret A. Salido M.A. Li, W.D.** Energy-efficient scheduling for a flexible flow shop using an improved genetic-simulated annealing algorithm Robotics and Computer-Integrated Manufacturing 29(2013), 418-429

[7] **Liu G.S. Zhang B.X. Yang D.D. Chen X. Huang G.Q.** A branch-and-bound algorithm for minimizing the energy consumption in the PFS problem. Mathematical Problems in Engineering 2013

[8] **Tang D.B. Dai M. Salido M.A. Giret A.** Energy-efficient dynamic scheduling for a flexible flow shop using an improved particle swarm optimization. Computers in Industry 30(2015), 223-232

[9] **Ding J.Y. Song S.J. Wu C.** Carbon-efficient scheduling of flow shops by multi-objective optimization European Journal of Operational Research 248(2016), 758-771

[10] **Mansouri S.A. Aktas E. Besikci U.** Green scheduling of a two-machine flowshop: trade-off between makespan and energy consumption. European Journal of Operational Research 248(2016), 772-788

[11] **Lei D.M. Guo X.P.** An effective neighborhood search for scheduling in dual resource constrained interval job shop with environmental objective. International Journal of Production Economics 159(2015), 296-303

[12] **Zhang R. Chiong R.** Solving the energy-efficient job shop scheduling problem: a multi-objective genetic algorithm with enhanced local search for minimizing the total weighted tardiness and total energy consumption Journal of Cleaner Production 112(2016), 3361-3375

[13] **Li J.Q. Pan Q.K.Tasgetiren M.F.** A discrete artificial bee colony algorithm for the multi-objective flexible job-shop scheduling problem with maintenance. Applied Mathematical Modelling 38(2014), 1111-1132

[14] **Wang L. Wang S.Y. Liu, M.** A Pareto-based estimation of distribution algorithm for the multi-objective flexible job-shop scheduling problem International Journal of Production Research 51(2013), 3574-3592

[15] **Yuan Y. Xu H.** Multiobjective flexible job shop scheduling using memtic algorithms IEEE Transactions on Automation Science and Engineering 12(2015), 336-353

[16] **Ahmadi E. Zandieh M. Farrokh M. Emami S.M.** A multi objective optimization approach for flexible job shop scheduling problem under random machine breakdown by evolutionary algorithm Computers and Operations Research 73(2016), 56-66

[17] **Eusuff M.M. Lansey K.E. Pasha F.** Shuffled frog-leaping algorithm: a memetic meta-heuristics for discrete optimization Engineering Optimization 38(2006), 129-154

[18] **Li J.Q. Pan Q.K.Xie S.X.** An effective shuffled frog-leaping algorithm for multi-objective flexible job shop scheduling problems. Applied Mathematics and Computation 218(2012),9395-9371

[19] **Lei D.M. Guo X.P.** A shuffled frog leaping algorithm for hybrid flow shop scheduling with two agents Expert Systems with Applications 42(2015), 9333-9339

[20] **Lei D.M. Guo X.P.** A shuffled frog leaping algorithm for job shop scheduling with outsourcing options International Journal of Production Research 2016 in press

[21] **Mattfeld D.C. Bierwirth C.** An efficient genetic algorithm for job shop scheduling with tardiness objectives. European Journal of Operational Research 155(2004), 616-630

[22] **Deb K. Pratap A. Agarwal S. Meyarivan T.** A fast and elitist multiobjective genetic algorithm: NSGA II . IEEE Transactions on Evolutionary Computation 6(2002), 182-197

[23] **Ishibuchi H. Yoshida T. Murata, T.** Balance between genetic search and local search in memetic algorithms for multi-objective permutation flowshop scheduling. IEEE Transactions on Evolutionary Computation 7(2003), 204-223

A CUSTOMIZED SHOE LAST SELECTION FOR PATIENTS WITH DIABETIC FOOT

C.C. Wang[1], C.H. Yang[2], T.R. Chang[3], C.S. Wang[4*], P.S. Huang[5]

[1]Department of Multimedia and Game Science
Chung-Chou University of Science and Technology, Chung-Hwa, Taiwan
ccwang@dragon.ccut.edu.tw

[2]Department of Industrial Engineering and Enterprise Information
Tung-Hai University, Taichung, Taiwan
gemlake@hinet.net.tw

[3]Department of Industrial Management
Nan-Kai University of Technology, Nan-Tou, Taiwan
t237@nkut.edu.tw

[4*]Department of Industrial Design
Tung-Hai University, Taichung, Taiwan
cswang@thu.edu.tw

[5]Department of Industrial Design
Tung-Hai University, Taichung, Taiwan
pshuang@thu.edu.tw

ABSTRACT

Foot ulceration is the one of complication for most people with diabetes. To design a bespoke footwear has become the basic requirement for the patients with diabetic foot. This research addressed in the selection of shoe lasts for the footwear design to aid people relief the pain from the diabetic neuropathy and foot ulcer. The reverse engineering (RE) technique was utilized to convert the point cloud of the scanned shoe lasts and the diabetic foot data into STL (STereo Lithograph) meshes. A slicing algorithm was developed to find the important feature girths that addressed both the diabetic foot and the shoe lasts. Seven characteristic girths are the key features of shoe lasts and individual foot shapes were calculated. An artificial neural self-organizing map (SOM) network was automated to classify the 60 samples of shoe lasts in order to categorize the range of similar groups. A test of 3 patient's foot shape was then entered into the SOM feature categories to match it with the suitable shoe lasts. Through conducting the expert questionnaire analysis to the characteristic girths feature data with the analytic hierarchy process (AHP), the weight of the girths can be obtained. Then, the grey relational analysis (GRA) is applied to calculate the related correlation for the patient's foot girths with the closest range of the shoe lasts. Finally by calculating the relative fitness function of the patient, the most suitable shoe last to the patient can be determined.

Keywords: Diabetic foot, Shoe last, Slicing algorithm, Self-organizing map, Analytic hierarchy process, Grey relational analysis, Fitness function.

---

* Corresponding Author

1 INTRODUCTION

The foot infections often occur in patients with diabetes, and can lead to foot ulcers, infections, gangrene, amputation, etc. Over the lifespan of patients with diabetes, at least 15% will encounter a foot ulcer once, and this accounts for 50% of all cases of limb amputations without any injury[1]. Based on investigations in the literature, diabetic foot ulcers result from nerve infection and/or ischemia in blood vessels, wherein around 65% of foot ulcers are triggered by nerve involvement. Amputation cases because of foot ulcers account for a percentage as high as 85%, which is fifteen times that of non-diabetic patients[2]. Foot ulcers affect patient quality of life, and the process often leads to a very poor outcome. Therefore, it is important for researchers to investigate possible technologies that can help to prevent foot ulcers.

The most important factor in the manufacture of a pair of shoes is the shoe last. The best clinical result and fit is obtained when the shoe last is appropriate to a person's foot shape and size[3]. For the production of diabetic bespoke shoes, a customized 3D shoe last is needed. Instead, the most appropriate shoe last design does not necessarily just conform to the foot shape of a patient, but rather should satisfy particular biomechanical conditions, and reduce the higher plantar and dorsal pressure to the extent possible the development of foot ulcer[4].

2 LITERATURE REVIEW

2.1 Diabetes and diabetic foot

In order to develop a common standard for diagnosis of patients with diabetes, the Wagner grading system was developed and is at the moment the best accepted and most widely used classification system for diabetic foot affections. The degree of diabetic foot ulcers can be correctly identified thru the wound classifications. This system includes Grade 0 to 5 for an ulcer scales. Patients in the therapeutic process may receive varying medical treatments suitable for the various degrees of severity. The emphasis of this study is based on the Wagner classification for classifying diabetic, neuropathic foot lesions, as shown in Table 1[2]. In the Grade 0, where patients are at a state of a high risk but as yet without any ulcers. Proper treatment can reduce the incidence of ulcers along with lower extremity amputation.

Table 1 Wagner Classification for Grading Foot Ulcers

Grades	Characteristics
Grade 0	No ulcer at the moment, but the risk for foot ulcers.
Grade 1	Superficial ulcer of cutis or hypoderm.
Grade 2	Deeper ulcer, possibly extended to bones, tendons, ligaments, or articulus.
Grade 3	Deep ulcers that cause affections of bone organizations or abscess.
Grade 4	Partial gangrene (toe or sole).
Grade 5	Extensive gangrene.

2.2 Shoe Last Design for Diabetic Shoes

The shoes should fit the feet, where not the feet to fit the shoes. To make a comfortable pair of shoes, the shoe last should be of its good design[5, 6]. Bernabéu, et al., teamed up with doctors, technicians, and researchers in the shoe-manufacturing industry to define a broad set of measurements for shoe lasts of diabetic shoes based on typical methods of measuring feet[7]. In their experiments, there were 150-foot shapes along with 130 measurement parameters. During the measurement process, they conducted analyses and standardization. Their study discovered that, in terms of functional patterns of diabetic foot, many measurement parameters are surplus and similar to others. They also defined the most typical measurement characteristics in diabetic shoes as the following: last length, last width, ball girth, last girth, medium instep girth, high instep girth, toe girth, as shown in Figure 1.

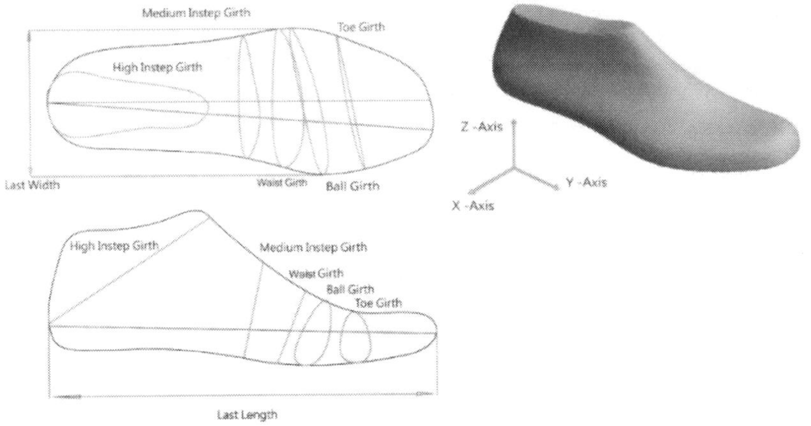

Figure 1: Characteristic girths for shoe last design of diabetic shoes.

2.3 Self-Organizing Feature Map

Self-organizing map (SOM) is a kind of unsupervised learning in the artificial neural network. The training of the network is entirely data-driven and no target results for the input data vectors are provided. SOM provided a topology preserving mapping from a high dimensional space onto a two-dimensional plane to serve as a cluster analyzing tool for reducing the dimensions[8]. Shieh and Yeh [9] performed running shoes as the primary research object. Kansei questionnaires, principle component analysis (PCA) and partial least squares (PLS) and neural network were used to develop a design support system for the exterior form of running shoes. In this research, SOM is proposed to achieve the reduction of dimensionality and to build a topological feature map of inputted shoe lasts. All the 60 shoe lasts are categorized into a 6x6 matrix.

2.4 Grey Relational Analysis

Grey relational analysis (GRA) is suitable for small amounts of samples and uncertain situations[10]. The GRA is an analysis tool for system dynamic development process based on the similarities or differences among factors. The GRA aims to find the correlation between a serial and other serials or the degree of correlation between changes in various serial factors and certain serial factors from a group of dispersed sequence data. In this research, a comfort fitness function is calculated by using the GRA for the related and correspondent shoe lasts with the diabetic foot. The most suitable shoe last can be found in a diabetic foot.

3 METHODOLOGIES

3.1 Measurements by reverse engineering

Nonintrusive laser scanners were utilized in this study in the reverse engineering measurements, and acquired data for point cloud precisely and rapidly to derive the 3D space data of shoe lasts and patient foot profiles as shown in Figure 2. The large set of data for point cloud from shoe lasts and foot profiles were merged in a point cloud with noise cancellation, and point cluster reduction was obtained as shown in Figure 3. Subsequently, the data was transformed into triangular meshes and saved as STL files in ASCII format for follow-up reading and handling.

(a)Reverse scanning for a shoe last (b) Scanning for a diabetic foot

Figure 2: Establishing data of shoe lasts and foot profiles

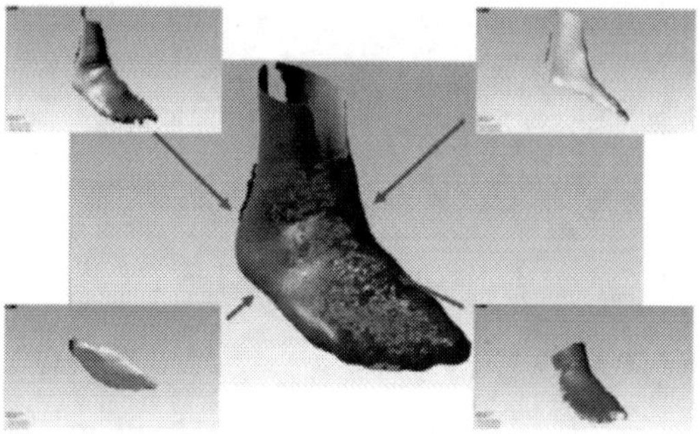

Figure 3: Point cloud merged

3.2 Capturing foot shape and shoe last characteristics

Before carrying out an analysis of foot characteristics and shoe last samples, the STL files must first be placed in a space to be rotated, and moved to the correct position and direction, which allows for the determination of correct characteristic data as shown in Figure 4.

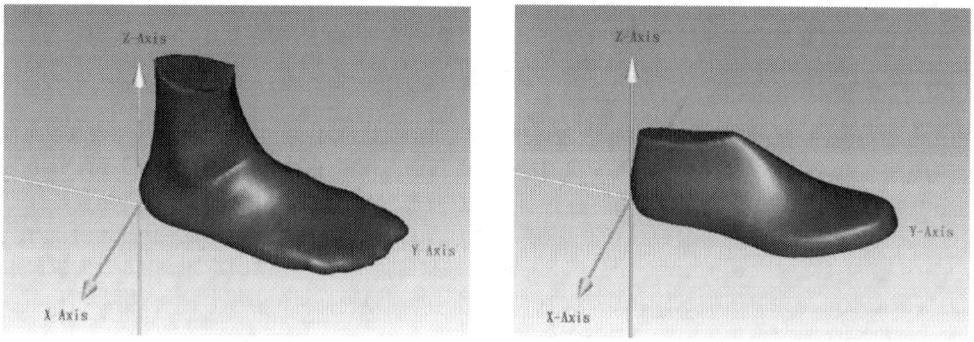

(a)A diabetic foot (b) A shoe last

Figure 4. Reference coordinate positions of shoe last and foot

A profile system was developed that automatically captured foot and shoe last characteristic features. The important characteristic girths, including length, width, ball girth, waist girth, medium instep girth, and high instep girth and toe girth can be calculated by using the slicing algorithm. Subsequently, a slicing algorithm was employed. Utilizing MATLAB software, a graphic user interface (GUI) is established to set of automatic functions was. The STL files for the foot shape, along with a shoe last profile, were utilized to determine each dimension, with calculations and capture of the characteristic feature girths as shown in Figure 5[11].

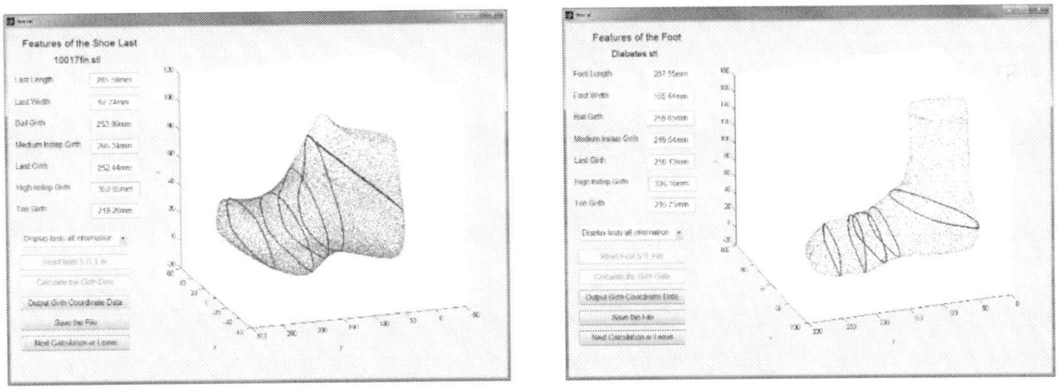

(a)Feature girths of the shoe last (b) Feature girths of the diabetic foot

Figure 5. Girths data of the last and a diabetic foot shape

3.3 SOM neural network for Diabetic Foot

Self-organizing feature map provided the clustering of 60 shoe lasts into related categories. Then, the analysis of the characteristic diabetic foot data can be corresponded into the most suitable shoe last group in the topological feature map.

4 IMPLEMENTATION

4.1 Extraction of Girths Dimensions

We collected 60 shoe lasts, which included from No. 8, 8.5, 9, 9.5, 10, 10.5 in US size and each No. with 10 shoe lasts. Three mild patients in diabetic foot are as the examinees. The reverse engineering process is implemented to measure the lasts and foot information by scanning. The 3D point cloud data in STL format are collected.

Numerical data of all characteristics are available in an Excel file format (*.xls). This file was used in the subsequent SOM neural classification process. The feet of three persons were used as data, along with a collection of 60 shoe lasts. This served as a data for experimental validation, with each feet undergoing coordinate transformation and matching with shoe last profile information from the STL file. The results are shown in Table 2 and Table 3.

Table 2. Foot girth data of subjects

	Length (mm)	Width (mm)	Ball girth (mm)	Waist girth	Medium instep girth	High instep girth	Toe girth (mm)
1	257.56	105.45	259.85	250.13	249.55	336.16	235.75
2	259.44	99.53	242.44	244.89	235.96	343.33	216.48
3	260.66	102.55	251.82	254.03	243.67	345.44	232.43

Table 3. Girth data of shoe lasts.

	Length (mm)	Width (mm)	Ball girth	Instep girth (mm)	Medium instep girth	High instep girth (mm)	Toe girth (mm)
1	285.58	97.74	245.99	257.34	244.45	353.06	218.26
2	282.12	98.66	252.94	262.54	253.69	342.00	215.14
3	287.28	97.04	254.46	264.46	251.82	351.78	213.24
:	:	:	:	:	:	:	:
60	295.71	101.95	261.27	272.99	260.94	366.67	222.06

4.2 SOM Reconstruction

SOM provides a topology preserving mapping from a high dimensional space onto a two-dimensional map. It is very suitable for clustering the selected 60 shoe lasts. These results can group all input lasts into related categories. In the study, the related self-organizing feature maps parameters can be set and the structure includes input layer, output layer and the weighting values for transferring between input and output layers. Figure 6 shows the SOM training process. The input layer is set to the 7 characteristic girths. The competitive learning rule finds the winner-takes-all neuron by calculating the short Euclidean distance between the input vector and the weights of each node. After the SOM learning process, every sample can find a best-winning neuron in the output topology map. All 60 lasts with seven inputs can get a mapping point in the 6x6 for the clustering algorithms. The weights of the winning neurons will close to the input vector after each iteration process. After repetitive iterations, SOM converges so that the similar neurons are grouped together. Normally, the output neurons for SOM can be set to 60% to 70% for the samples. Using the SOM, the 60 shoe lasts can be output onto 6x6 neurons in the output layers.

Finally, MATLAB was used to code a network framework for SOM. This system used a large shoe last dataset to run a learning network model. Three diabetic foot-type characteristics were input to obtain a relationship to the shoe lasts feature map. The results indicated that examinee 1 and examinee 3 matched with shoe lasts No. 27 and 36. Examinee 2 matched with shoe last No. 7 and No. 16, as shown in Figure 7.

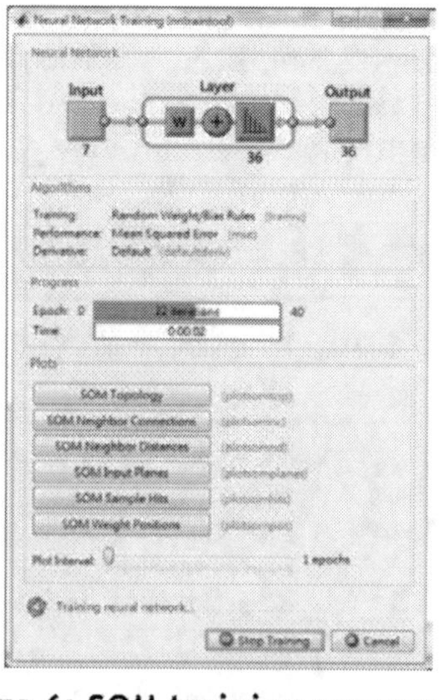

Figure 6: SOM training process

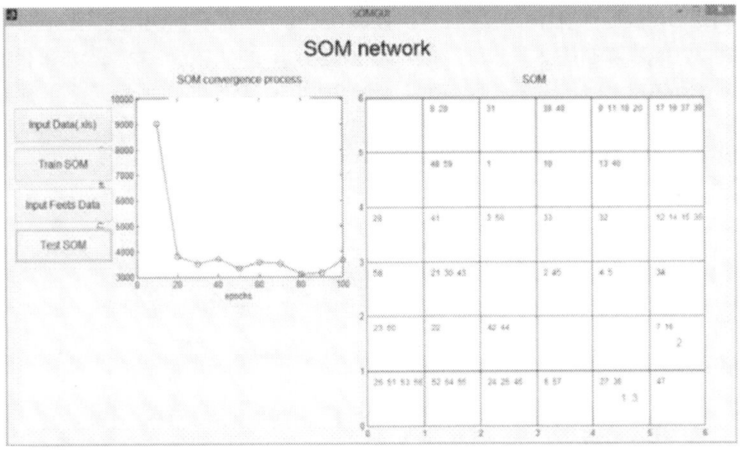

Figure 7: Clustering for shoe lasts and foot styles

4.3 Fitness Function

The analytic hierarchical process (AHP) is used to deal with the weights of the important girths[12]. The grey relational analysis (GRA) is to stress the relation in diabetic foot with the suitable shoe lasts in the corresponding girths. Finally, the fitness function examined the optimum shoe last for the patient with diabetes.

$$R = w_B F_B + w_W F_W + w_M F_M + w_H F_H + w_T F_T \tag{1}$$

Where R: the fitness function; w_B: the weight of ball girth; w_W: the weight of waist girth; w_M: the weight of medium instep girth; w_H: the weight of high instep girth; w_T: the weight of toe girth; F_B = grey relational correlation for ball girth; F_W = grey relational correlation for waist girth; F_M = grey relational correlation for medium instep girth; F_H = grey relational correlation for high instep girth; F_T = grey relational correlation for toe girth.

4.4 Results and Discussion

The ranking of the shoe lasts can be obtained by the estimation obtained from the proposed method using self-organizing feature map, and, using AHP, we can determine the weights for all girths. The best fit of the shoe last for each examinee's foot can be determined from the fitness function calculation. For the case of the first patient, according to the data recorded, the most suitable one can be estimated using the accurate fitness function, $R = w_B F_B + w_W F_W + w_M F_M + w_H F_H + w_T F_T$, As a result, $R_{27} > R_{36} > R_{47} > R_{16}$, which means that, among the 60 shoe lasts, No. 27 would be the most suitable for the first diabetic foot.

5 CONCLUSIONS

Shoe customization for individuals still has its limitation owing to higher cost both in sales and manufacturing. However, it is should be considered in foot pathologies, especially for diabetic foot. This study proposed a mass customized shoe last selection algorithm which is very suitable and practical for mild diabetes.

By utilized measurements obtained by reverse engineering to rapidly acquire accurate profile data for shoe lasts and patient feet. Characteristic points were then used to establish characteristic planes utilizing a slicing algorithm with triangular meshes. The developed system automatically acquires such characteristic data based on the algorithm and reduces errors possible from manual measurements. The system then generates related data and image data that can assist in establishing a database of the shoe last design information. Clustering calculations by neural networks were also done on the shoe last data, and the system provides results that can rapidly determine the most suitable shoe last for any particular patient. This approach was successfully verified experimentally. This study achieved the following key aims:

1. To establish 3D point cloud space data of shoe lasts and diabetic foot.
2. To grab the important characteristic girths in shoe lasts and diabetic foot.
3. To create the automated classified groups in shoe lasts.
4. To find out the most suitable shoe last for the diabetic foot.

ACKNOWLEDGEMENTS

This research was sponsored by the Ministry of Science and Technology, Taiwan under the Projects No. MOST 104-2221-E-029-028 for supporting this research.

REFERENCES

[1] King, H., Aubert, R.E., & Herman, W.H. 1998. Global burden of diabetes, 1995-2025, prevalence, numerical estimates and projections. *Diabetes Care*, 21(9), 1414-1431.

[2] Jeffcoate, W.J., & Harding, K.G. 2003. Diabetic foot ulcers, The Lancet, 361, 1545-1551.

[3] Germani, M., Mandolini, M., Mengoni, M., Raffaeli, R., Montiel, E., & Davia, M. 2011. Methods and tools dedicated to shoe customization for people with diabetes. *International Conference on Innovative Methods in Product Design*. 698-706.

[4] Rupérez, M. J., Monserrat, C., Alemany, S., Juan, M. C., & Alcañiz, M. 2010. Contact model, fit process and, foot animation for the virtual simulator of the footwear comfort. *Computer-Aided Design*, 42(5), 425-431.

[5] Davia, M., Jimeno-Morenilla, A., & Salas, F. 2013. Footwear bio-modelling: an industrial approach, *Computer-Aided Design*, 45, 1575-1590.

[6] Germani, M., Bernabeu, J.A., Mandolini, M., Mengoni, M., Raffaeli, R. 2012. A Knowledge-based design process for diabetic shoe lasts. *International Design Conference - Design 2012*. 889-900.

[7] Bernabéu, J.A., Germani, M., Mandolini, M., Mengoni, M., Nester, C., Preece, S., & Raffaeli, R. 2013. CAD tools for designing shoe lasts for people with diabetes. *Computer-Aided Design*, 45(6), 977-990.

[8] Kohonen, T. 1989. *Self*-Organization *and Associative Memory*, 3rd Ed. New York: Springer-Verlag.

[9] Shieh, M.D. & Yeh, Y.E. 2013. Developing a design support system for the exterior form of running shoes using partial least squares and neural networks. *Computers & Industrial Engineering*, 65, 704-718.

[10] Deng, J.L. 1989. Introduction to grey system theory. *Journal of Grey System*, 1(1), 1-24.

[11] Wang, C.S., 2010. An analysis and evaluation of fitness for shoe lasts and human feet, *Computers in Industry*. 61(6), 532-540.

[12] Saaty, T.L. 2008. Decision-making with the analytic hierarchy process. *International Journal of Services Sciences*, 1(1), 83-98.

Applying sentiment analysis for the hotel recommendation based on uncertainty theory

Lihua Sun[1], Junpeng Guo [2*] and Yanlin Zhu [3]
[1] College of Management and Economics, Tianjin University, P.R. China,
sunlh68@tju.edu.cn
[2] College of Management and Economics, Tianjin University, P.R. China,
guojp@tju.edu.cn
[3] College of Management and Economics, Tianjin University, P.R. China,
mouzhiben@163.com

ABSTRACT

In this study, we propose to employ the users' reviews into the hotel recommender systems that have a more comprehensive grasp of users' opinion. Because of the uncertainty of users' opinions that are usually written in the unnatural language we utilize uncertain set to calculate users' opinion polarities, and uncertain variable to acquire the attention degree based on sentiment analysis. Then, user-based collaborative filtering method is used to generate recommendation results with the proposed methods mentioned above. Furthermore, the proposed approach recommender is evaluated with real online Chinese reviews of hotels. The experimental results show that the proposed method can capture riche information than ratings. It also demonstrates that uncertainty theory is a useful tool to characterize the indeterminate problem of uncertain expression with unnatural language in Chinese.

Keywords: Recommender system, Sentiment analysis, Product reviews, Uncertain set, Uncertain variable

1 INTRODUCTION

Recommender systems have been developed for a wide range of applications to address the information overload. There are two common approaches to generate recommendations: content-based filtering and collaborative filtering[1]. Collaborative filtering technique performs well when there is sufficient rating information[2]. It usually contributes to poor coverage of the recommendation space. To solve the problem, some types of information are used to alleviate the data sparsity problem and increase the accuracy of recommenders[3, 4].

Recently, many e-commerce websites encourage users to write reviews describing their shopping experiences. But it is not easy to directly acquire consumers' opinions from the large collection of users' reviews. Thus, it is a key issue that users' opinion-features need to be automatically detected. A wide variety of sentiment analysis techniques are utilized to analyze users' opinions. Therefore, through combining ratings with reviews, opinion mining techniques are utilized in recommender systems to capture internal information[3]. With respect to the characteristics of online Chinese reviews in this study, owing to the ambiguous expressions of human language[6], the meanings of users' expressions with unnatural language, especially Chinese, are vague. Therefore, uncertain set and uncertain variable[5] are used to characterize users' opinions by sentiment analysis.

* Corresponding Author: guojp@tju.edu.cn

We aim to detect the effectiveness of modelling opinion-feature by uncertain sets and uncertainty variable using our proposed recommendation system. The main contributions of this paper are as follows. (1) A new method is proposed to characterize users' sentiment polarities. (2) The uncertain variable is used to describe the attention degree which users have on some opinion-feature.

The remainder of this paper is organized as follows. In section 2, we briefly introduce the related work. In section 3, some definitions and models are presented which will be used to detect users' opinions. To test the performance of the proposed approach, an experiment is set up to evaluate the outperformance of the recommendation results in section 4. A brief summary of the study is given in section 5, along with future research directions.

2 RELATED WORK

In this section, we describe the related work on recommender systems as well as some relevant contributions in sentiment analysis and opinion mining. Some basic concepts and properties of uncertainty theory are also introduced.

2.1 Recommender systems

Collaborative filtering is one of the most well-known recommendation approaches that explores relevant information to satisfy users' requirements[2]. It works well when there is sufficient rating information. However, in the real world, collaborative filtering method usually faces extremely sparse data problem. Therefore, the performance of the recommender systems may be seriously limited by this drawback. A literature survey on this topic reveals that an increasing number of studies have mainly combined product ratings with reviews to infer users' opinions[7].

In some e-commerce sites, consumers are usually encouraged to express their experiences with their own words. Highly personalized reviews are expressed to show consumption experiences, such as ratings, short reviews, and figures. However, it is difficult for users to acquire useful information when there is a great amount of reviews, or to make a better decision in real time. Nowadays, sentiment analysis and opinion mining techniques are applied to detect users' opinions in text mining systems. Ratings and reviews are utilized to identify user preferences in recommender systems. The recommender results demonstrate that sentiment analysis can make further to explore users' opinions[8]. The relationship between user reviews and products was inferred using the technique of information retrieval[9]. Furthermore, the accuracy of recommender results can be improved on the basis of sentiment analysis. To date, a few studies have examined Chinese reviews in recommender systems to improve the accuracy of recommended results.

2.2 Uncertain set and uncertainty variable

Uncertainty theory[5] is a new branch of mathematics introduced in 2007. In recent years, theory and practice have demonstrated that uncertainty theory is an efficient tool for dealing with non-deterministic information, especially subjective estimation. The concepts of uncertain measures, uncertain variables, and uncertain set theory have been theoretically established. From a practical viewpoint, uncertainty statistics uncertain processes, uncertain differential equations etc. have also developed quickly[10].

Users' reviews that are feedbacks of shopping experiences are explored to reflect users' opinions about relevant products. To the best of our knowledge, little work has been done to introduce uncertainty theory into recommender systems based on sentiment analysis. A word-to-review approach is proposed to detect the sentiment polarities and attention degree for each user. The proposed method is applied into a hotel recommender system to test the effectiveness.

3 DEFINITION AND MODEL

The proposed hotel recommender system in this paper generates recommender results based on sentiment analysis. The details are as the following.

3.1 Definition

Definition 1 (*Opinion-feature*) A hotel is called an item h_i. Item h_i is associated with a set of opinion-feature words $F = \{f_1, f_2, \cdots, f_L\}$, $F = \{f_1, f_2, \cdots, f_L\}$ is the set of L features users comment on hotels. Each of them $f_L \in F$ is a component or subcomponent element of the item h_i. $H = \{h_1, h_2, \cdots, h_i\}$ is a set of items which are commented by users.

Definition 2 (*Polarity*) The polarities of users' opinions show their sentiment orientations, which may be positive and negative. We use two uncertain sets to characterize positive ξ_{pos} and negative ξ_{neg}, respectively. To estimate the polarity values, positive and negative can be characterized by two membership functions, respectively.

Definition 3 (*Attention degree*) Given a set of reviews D, and a opinion-feature word $f_L \in F$ is given by user u_N, the concern degree of the feature is defined as $concern(u_N, h_i, f_L), f_L \in h_i$. $U = \{u_1, u_2, \cdots, u_N\}$ is the set of users N who wrote these reviews D.

3.2 Opinion-feature extraction

A review is made up of a set of term $T = \{t_1, t_2, \cdots, t_j\}$. The term t_j is a component or subcomponent element of a review. In this study, we consider items H, and apply term frequency-inverse document frequency (TF-IDF)[11] method to acquire opinion-feature words which is shown as

$$\mathrm{TF-IDF}(h_i, t_j, H) = \frac{\mathrm{tf}(t_j, h_i)}{\sum_{t_j \in h_i} \mathrm{tf}(t_j, h_i)} \times \mathrm{idf}(t_j, H) \quad (1)$$

where the term frequency $\mathrm{tf}(t_j, h_i)$ and the inverse document frequency $\mathrm{idf}(t_j, H)$ are be written as: $\mathrm{tf}(t_j, h_i) = n_{(t_j, h_i)}$, $\mathrm{idf}(t_j, H) = log\left(\frac{|H|}{|\{h_i \in H : t_j \in h_i\}|}\right)$. Briefly, the weight of a term t_j in an item h_i, with respect to a set of items H is proportional to the frequency of occurrence of t_j in H overall. Therefore, considering opinions of users', these opinions are beneficial to discriminate h_i from other items in the collection. At last, a set of opinion-feature words will be collected which is denoted as F.

Before extracting features, we use Chinese word segmentation repository (IKAnalyzer) proposed by the Community of China (COC) to process each review in document. In the experiments, we obtain the feature words by the method of term frequency-inverse document frequency (TF-IDF).

3.3 Model

According to definition 2, the membership function of the uncertain set ξ_{pos} for a review is defined as $\mu\left(x_{pos}\right)$. Let:

$$\xi_{pos} = \begin{cases} [\mathrm{p}_{1,1}, \mathrm{p}_{1,2}] \text{ with uncertain measure } \alpha_{11} \\ [\mathrm{p}_{2,1}, \mathrm{p}_{2,2}] \text{ with uncertain measure } \alpha_{12} \\ \quad \vdots \qquad\qquad \vdots \\ [\mathrm{p}_{n,1}, \mathrm{p}_{n,2}] \text{ with uncertain measure } \alpha_{1n}, \end{cases} \quad (2)$$

then, we immediately have $\mu\left(x_{pos}\right) = \mathcal{M}\{x_{pos} \in \xi\}$ by (2). According to this definition[12], we have the expected value of the positive polarity ξ_{pos}:

$$E\left[\xi_{pos}\right] = \frac{1}{2}\int_0^1 \left(\inf \mu_{pos}^{-1}(\alpha) + \sup \mu_{pos}^{-1}(\alpha)\right) d\alpha, \quad (3)$$

where $\inf\mu_{pos}^{-1}(\alpha)$ and $\sup\mu_{pos}^{-1}(\alpha)$ are the infimum and supremum of the set $\mu_{pos}^{-1}(\alpha)$ for each α, respectively. In the same way, we can also acquire the negative polarity $E(\xi_{neg})$. Let $\mathcal{M}(f_1) = \alpha_1, \mathcal{M}(f_2) = \alpha_2, \cdots, \mathcal{M}(f_L) = \alpha_L$. If a review contains the opinion-feature word $f_L \in h_i$ which is written by user u_N. Then the user u_N may pay more attention to the opinion-feature word f_L. A subset F_{sub} of F is given by user u_N about the item h_i. We can obtain uncertainty distribution Φ of the subset F_{sub}, and calculate the attention degree $concern(u_N, (f_L, \alpha_L), h_i)$ for user u_N based on uncertain statistics [10].

Consequently, the expected value of user u_N for the item h_i is estimated according to the polarities and attention degree. The expected value is

$$E_{u_N,h_i} = \left(E\left[\xi_{pos,u_N,h_i} - \xi_{neg,u_N,h_i}\right] \times E\left[concern\left(u_N, h_i, (f_L, \alpha_L)\right)\right]\right) \quad (4)$$

4 EXPERIMENT

The evaluation of the effectiveness of the proposed algorithm of recommending hotels has been carried out. The experimental designs and results are presented in the remainder of this section.

4.1 Experimental data and evaluation metrics

The experiment dataset about hotel was crawled from a well-known E-commerce website in China (http://t.dianping.com/hotel/tianjin). We collected a total of 69774 reviews in which users described their consumption experiences, with each review including the user's ID (UserId), hotel's ID (ShopId), user's reviews, total rating (Rating). As mentioned above, the scores range from 1 to 5 which were expressed by users after their consumption experiences. The dataset consists of users' reviews, total rating. It contains 608 shops and 47376 users, the density of the records being only 0.24%.

Since most users usually expressed their overall consumer experiences, users' opinion-feature could not be directly acquired. Therefore, some data are manually selected to label the opinion-feature words. And then semi-supervised learning technique is applied to acquire the opinion-feature words.

The evaluation metrics applied in our study are Precision (the fraction of extracted features or opinions are correct), Recall (the fraction of correct features are extracted), and F1 ($\mathrm{F1} = (2 \times precision \times recall)/(precision + recall)$).

In order to evaluate the effectiveness of the proposed method, we conduct a five-fold cross-validation and compute the three metrics for each fold. To do the five-fold cross-validation, the users who commented on less than two hotels are removed. Consequently, 31961 reviews are written by 9563 users who express their shopping experiences about 608 hotels. The dataset is split into five partitions, where four partitions identify opinions and one part tests the opinions.

4.2 Recommendation results

The expected values are inferred for each user that will give the score to the hotel. We then compare the proposed method with user-based collaborative filtering.

It can be clearly seen from the figures Fig.1, Fig.2, Fig.3 that the recommender results of the proposed algorithm are better than CF (user-based). In the process of searching KNN neighbours, the similarities between users are computed by Cosine similarity to search the K nearest neighbours for the aim user. The length of recommendation lists are varied as 5, 10, 15, 20, 25, 30, respectively. The results of the proposed method outperforms CF(user-based) which combines only ratings. With the increasing length of the recommender lists, the experiment results are much better than the user-based CF. Therefore, the gap between the two methods becomes more significantly. The ratings from 1 to 5 typically do not show obvious fluctuations, since users often give high ratings without much thought.

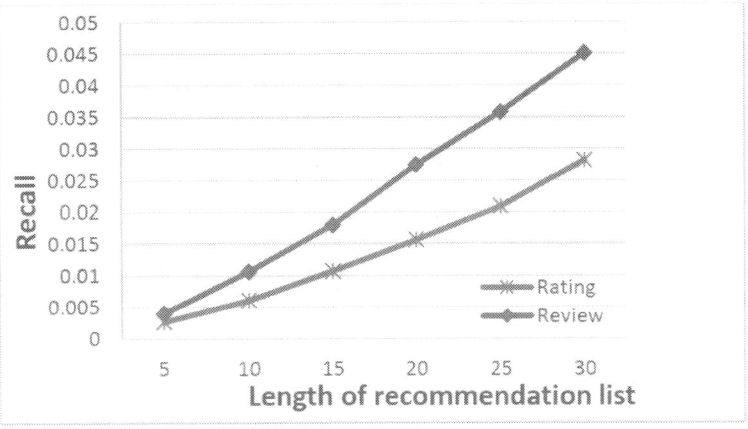

Figure1: Recall

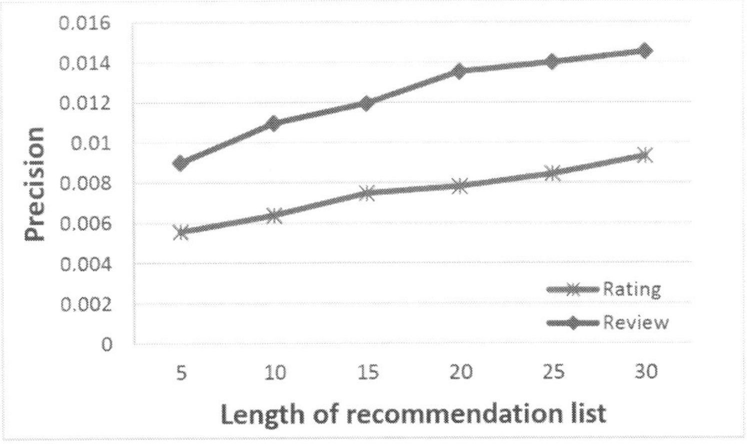

Figure 2: Precision

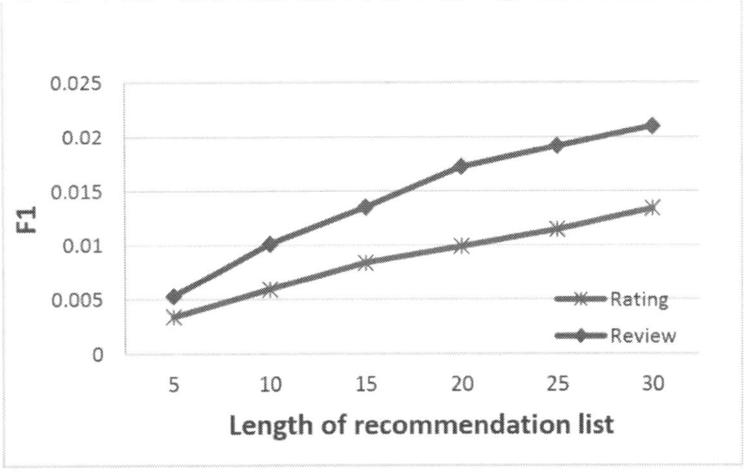

Figure 3: F1

5 CONCLUSION AND FUTURE WORKS

A novel method is introduced into recommender system based on uncertainty theory. The uncertain set and variable are applied into sentiment analysis to characterize users' opinions polarities and attention degree on opinion-feature, respectively. To estimate the opinion polarities, the uncertain membership functions of uncertain sets are presented to acquire the expected values of the polarities. The attention degree is determined by uncertain distribution of the uncertain variable. The value of attention degree can be calculated by the uncertainty distribution. According to the values of polarities and attention degree, the expected value will be given by users' that can be estimated. The experiment results are also shown that reviews include more rich information than ratings. Therefore, utilizing reviews into recommender system can alleviate the problem of data sparseness.

Some future works are planned to follow this study. First, uncertainty theory applies into opinion mining to further detect users' opinion-feature. For example, user's opinions, sentiment polarities, time and location may ~~be~~ affect a person to make decisions. Second, we will study how to apply the proposed method into recommender systems to deal with the data sparsity problem.

REFERENCES

[1] **Adomavicius, G., & Tuzhilin, A. 2005**. Toward the next generation of recommender systems:A survey of the state-of-the-art and possible extensions. Knowledge and Data Engineering, IEEE Transactions on, 17(6), pp 734-749.

[2] **Su, X., & Khoshgoftaar, T. M. 2009**. A survey of collaborative filtering techniques. Advances in artificial intelligence, 2009, 4.

[3] **Chen, L., & Wang, F. 2013**. Preference-based clustering reviews for augmenting e-commerce recommendation. Knowledge-Based Systems, 50, pp 44-59.

[4] **Gao, Y., Yu, W., Chao, P., Zhang, R., Zhou, A., & Yang, X. 2015**, April. A Restaurant Recommendation System by Analyzing Ratings and Aspects in Reviews. In Database Systems for Advanced Applications. Springer International Publishing. pp 526-530.

[5] **Liu, B.** 2007. Uncertainty theory. Springer Berlin Heidelberg. pp 205-234.

[6] **Liu, B. 2010**. Uncertain set theory and uncertain inference rule with application to uncertain control. Journal of Uncertain Systems, 4(2), pp 83-98.

[7] **Chen, L., Chen, G., & Wang, F. 2015**. Recommender systems based on user reviews: the state of the art. User Modeling and User-Adapted Interaction, 25(2), pp 99-154.

[8] **Liu, H., He, J., Wang, T., Song, W., & Du, X. 2013**. Combining user preferences and user opinions for accurate recommendation. Electronic Commerce Research and Applications, 12(1), pp 14-23.

[9] **Esparza, S. G., O'Mahony, M. P., & Smyth, B. 2012**. Mining the real-time web: a novel approach to product recommendation. Knowledge-Based Systems, 29, pp 3-11.

[10] **Lin, B. 2010**. Uncertainty theory: A branch of mathematics for modeling human uncertainty.

[11] **Salton, G., & McGill, M. J.** 1986. Introduction to modern information retrieval.

[12] **Liu, B. 2012**. Membership functions and operational law of uncertain sets. Fuzzy Optimization and Decision Making, 11(4), pp 387-410.

An energy-efficient two-stage hybrid flow shop scheduling problem with batch machining

Shijin Wang[1*] and Wenhui Zhang[2]
[1]School of Economics and Management
Tongji University, Shanghai, China
shijinwang@tongji.edu.cn

[2]School of Economics and Management
Tongji University, Shanghai, China
1122alicezhang@tongji.edu.cn

ABSTRACT

Nowadays, with the growing energy demand, increasing energy price and higher requirement of environment protection, energy-efficient production and manufacturing has become an increasingly urgent task for academic researches and manufacturing companies, especially for energy-intensive industries. In this paper, we study a two-stage hybrid flow shop problem to minimize the total energy consumption cost, which consists of several identical parallel machines at stage 1 and a batching machine at stage 2. A set of jobs with nonidentical sizes can be processed on one of machines at stage 1 and then have to be grouped to form batches, in order to be processed on the batch processing machine at stage 2. An mixed integer optimization model is developed for the problem. Finally, a numerical example is carried out to illustrate the application of the model.

Keywords: Energy-efficient, Hybrid flow shop, Batch machining

1 INTRODUCTION

In recent decades, energy demand and environmental protection regulation in industry have been growing worldwide, therefore, energy-efficient scheduling and energy consumption reduction has attracted more and more attentions both in academia and industry (Gahm et al. 2016; Giret et al. 2015; Apostolos et al. 2013). Since total energy consumption costs occupies a large part of total production costs, it is essential to develop an energy-efficient scheduling to reduce energy consumption costs.

In the traditional production scheduling, decision-makers in industry and researchers in academy focus on objectives of the productivity, such as the makespan, total completion time and total tardiness (Kaplan, Rabadi. 2012; Joo, Kim. 2015). However, objective of energy should be also taken into consideration in some cases. In the modern production scheduling, the fluctuation of energy prices during a production period has a significant impact on energy consumption costs. This paper develops a mathematical programming model to minimize total energy consumption for a two-stage hybrid flow shop scheduling problem with non-identical job sizes.

This problem is originated from a glass-ceramic manufacturing shop floor. The process of glass decoration and ceramization contains two stages in which stage 1 has several identical parallel printing machines and stage 2 has a batch machine which is used to process the glasses in batch.

* Corresponding Author

The remainder of the paper is organized as follows: Section 2 provides a literature review. Section 3 gives the problem description and presents the model formulation. A numerical example is given to make our results clear in section 4. Finally, in Section 5, we end the paper with some concluding remarks and point out some further research directions.

2 LITERATURE REVIEW

Energy efficiency and energy consumption reduction have been widely considered as an important factor in recent literatures on production scheduling, such as single machine scheduling problems (Mouzon et al. 2007; Yildirim, Mouzon. 2011; Shrouf et al. 2014), flow shop scheduling problems (Fang et al, 2011; Zhang et al, 2014; Mansouri et al, 2015), and job shop scheduling problems (Liu et al, 2014; Zhang and Chiong, 2015). In the following, we mainly review the studies in the flexible flow shop environment and the flexible job shop environment, since the problem in this paper is a hybrid flow shop scheduling which is highly related to flexible flow shop and flexible job shop.

Dai, Tang, Giret, Salido, Li(2013) proposed an energy efficient model for flexible flow shop scheduling. The mathematical model is described to make a significant trade-off between the makespan and the total energy consumption to implement a feasible scheduling and they also adopted an improved, genetic-simulated annealing algorithm. Dai, Tang, Zhang, Yang(2014) presented a bi-objective scheduling model for a flexible flow shop problem and they investigated a new algorithm that combines genetic algorithm and simulated annealing for a flexible flow shop problem. The model discovered a production sequence with energy savings for jobs on machines to optimize objectives of makespan and total energy consumption.

He, Li, Wu, Sutherland(2015) formulated an energy-saving model that consideres an energy consumption objective combined with an objective of makespan for a flexible job shop. They adapted a weighted sum method to combine the two objectives into a single objective and utilized Nested Partitions algorithm to solve the problem. Liu, Tiwari(2015) developed a model for a bi-objective flexible job shop problem that minimise total processing energy consumption and makespan with lot sizing circumstance. They proposed an optimisation approach based on Non-dominant Sorting Genetic Algorithm (NSGA-II) to solve this problem. They considered energy reduction together with makespan minimization for recycling carbon fiber reinforced polymer (CFRP).

According to the literature review mentioned above, many previous studies have investigated energy consumption reduction in production scheduling problems. Meanwhile, various optimization models and algorithms have been developed to solve related scheduling problems. However, problems that consider energy consumption in a hybrid flow shop with a batch machine have not been researched thoroughly yet. Furthermore, in existing literature, energy consumption is usually used as a limit and is embedded as a constraint rather than a goal. In this context, we investigate a two-stage hybrid flow shop scheduling problem with a batch machine to minimize the total energy consumption cost. In this problem, the different energy consumption rate of machines and the fluctuating electricity prices, i.e, time-of-use (TOU) prices are integrated.

3 PROBLEM DESCRIPTION AND FORMULATIONS

3.1 Problem description

The problem can be described as follows. Several parallel machines are used to process a set of jobs with nonidentical size on the first stage and the next stage is a batch processing machine in which several jobs that come from the previous stage can be processed simultaneously in a batch. We have to make sure that every job is finished at stage 1 before it can be delivered to stage 2.

The parallel machines at stage 1 have a production availability constraint to each job which means not all machines are available for each job. The single batch machine at stage 2 can process a batch as long as a batch's total size is less than or equal to the capacity of the batch machine. It is not permitted to add or remove a job from a batch once the batch is formed at stage 2. Since energy consumption of a machine at working and processing status and idle status are different, therefore, two energy consumption rates are introduced.

The work flow of the manufacturing process of the two-stage hybrid flow shop problem is depicted in Figure1.

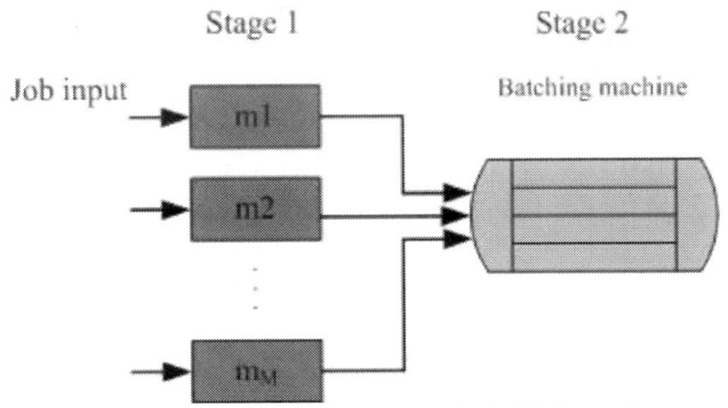

Figure 1. Work-flow of the two-stage hybrid flow shop problem

3.2 Mathematical program formulation

An MIP model can be formulated for the problem as follows:

Indices:

i: index of job, $i = 1, \ldots, n$.
k: index of machine, $k = 1, \ldots, m$.
t: index of period, $t = 1, \ldots, |T|$.
b: index of batch, $b = 1, \ldots, |B|$.

Parameters:

p_i^1: the processing time of job i in stage 1;
p_i^2: the processing time of job i in stage 2;
n: number of jobs;
m: total number of machines at stage 1;
m_i: number of machines that can be used to process job i in stage 1;
α_{ik}: equal to 1 if job i is available for processing at machine k, 0 otherwise;
q_i: size of job i, $i = 1, \ldots, n$;
c_t: unit energy price for time period $t \in T$;
Q: the capacity of the batch machine;
N: the set of jobs, i.e., $N = \{1, \ldots, n\}$;
M: the set of machines , i.e., $M = \{1, \ldots, m\}$;
M_i: set of machines that can be used to process job i, $M_i = \{1, \ldots, m_i\}$;
B: the set of batches, i.e., $B = \{1, \ldots, |B|\}$;
T: the set of time periods, i.e., $T = \{1, \ldots, |T|\}$;
F_W^b: the energy consumption rate of the batch machine in working;
F_I^b: the energy consumption rate of the batch machine in idle;
E_W^k: the energy consumption rate of the machine k in working, $k = \{1, \ldots, m\}$;

E_I^k: the energy consumption rate of the machine k in idle, $k = \{1, ..., m\}$;
L: a large enough number.

Decision variables:

x_{ik}^t: equal to 1 if job i is processed on machine k during time period t, 0 otherwise;binary variable.
y_k^t: equal to 1 if machine k is working during time period t, 0 otherwise;binary variable.
z_{ik}: equal to 1 if job i is processed on machine k, 0 otherwise;binary variable.
s_i: the start time of job i in stage 1;non-negative integer.
u_{ib}: equal to 1 if job i is assigned to batch b, 0 otherwise;binary variable.
v_b: equal to 1 if batch b is formed with at least one job, 0 otherwise;binary variable.
w_b^t: equal to 1 if batch b is processing at period t, 0 otherwise;binary variable.
R_b: the start time of batch b in stage 2;non-negative integer.
P_b: the processing time of batch b;non-negative integer.

Then the **MIP** model for the problem is as follows:

$$\min \quad f = \sum_{t\in T}\sum_{k\in M} c_t\left[y_k^t E_W^k + \left(1 - y_k^t\right) E_I^k\right] + \sum_{t\in T}\sum_{b\in B} c_t\left[w_b^t F_W^b + \left(1 - w_b^t\right) F_I^b\right] \tag{1}$$

$$\sum_{k\in M} z_{ik} = 1, \quad \forall i \in N \tag{2}$$

$$z_{ik} \le \alpha_{ik}, \quad \forall i \in N, k \in M \tag{3}$$

$$y_k^t = \sum_{i\in N} x_{ik}^t, \quad \forall k \in M, t \in T \tag{4}$$

$$z_{ik} \ge x_{ik}^t, \quad \forall i \in N, k \in M, t \in T \tag{5}$$

$$z_{ik} \le \sum_{t\in T} x_{ik}^t, \quad \forall i \in N, k \in M \tag{6}$$

$$\sum_{k\in M}\sum_{t\in T} x_{ik}^t = p_i^1, \quad \forall i \in N \tag{7}$$

$$(1 - \sum_{k\in M} x_{ik}^t)L + t \ge s_i, \quad \forall i \in N, t \in T \tag{8}$$

$$(1 - \sum_{k\in M} x_{ik}^t)L + s_i + p_i^1 \ge t + 1, \quad \forall i \in N, t \in T \tag{9}$$

$$\sum_{b \in B} u_{ib} = 1, \quad \forall i \in N \tag{10}$$

$$\sum_{i \in N} q_i u_{ib} \leq Q v_b, \quad \forall b \in B \tag{11}$$

$$P_b \geq p_i^2 u_{ib}, \quad \forall i \in N, b \in B \tag{12}$$

$$\sum_{t \in T} w_{bt} = P_b, \quad \forall b \in B \tag{13}$$

$$v_b L \geq P_b, \quad \forall b \in B \tag{14}$$

$$\sum_{b \in B} w_{bt} \leq 1, \quad \forall t \in T \tag{15}$$

$$v_b \geq v_{b+1}, \quad b = 1, \ldots, |B| - 1 \tag{16}$$

$$v_b \geq w_b^t, \quad b \in B, \quad \forall t \in T \tag{17}$$

$$v_b \leq \sum_{t \in T} w_b^t, \quad \forall b \in B \tag{18}$$

$$(1 - u_{ib})L + R_b \geq s_i + p_i^1, \quad \forall i \in N, b \in B \tag{19}$$

$$R_b + P_b \leq R_{b+1}, \quad b = 1, \ldots, |B| - 1 \tag{20}$$

$$R_b + P_b \leq T, \quad \forall b \in B \tag{21}$$

$$(1 - w_b^t)L + t \geq R_b, \quad \forall b \in B, t \in T \tag{22}$$

$$(1 - w_b^t)L + R_b + P_b \geq t + 1, \quad \forall b \in B, t \in T \qquad (23)$$

The objective function (1) aims at minimizing the total energy consumption costs, which depends on the working or idle status of the parallel machines and the batching machine, and the energy price in every period. Constraint (2) ensures that a job can be only processed on one machine. Constraint (3) makes sure that machine k is available when job i assigned to it. Constraint (4) ensures that at time period t , machine k can either process or be idle. Constraints (5) and (6) coordinate the values of x_{ik}^t and z_{ik}. The constraint (7) defines the processing time of job i. Constraints (8) and (9) guarantee that the processing of job i is not interrupted. Constraint (10) ensures that each job is assigned to exactly one batch. Constraint (11) limits the total size of jobs assigned to one batch not exceeding the capacity of that batch. Constraints (12)-(14) determine the batch processing time. The processing time of a batch is the longest processing time of all the jobs in the batch. Constraint (15) ensures that each period t has exactly one condition, working or idle. Constraint (16) excludes the circumstance that batch $b + 1$ is formed while batch b is not. Constraints (17) and (18) coordinate the values of v_b and w_b^t. Constraint (19) ensures that the start time of a batch is at least larger than the sum of start time and processing time (i.e the completion time) of any job in that batch. Constraint (20) ensures that the start time of batch $b + 1$ is larger than the sum of start time and processing time (i.e the completion time) of batch b. Constraint (21) makes sure that the completion time of each batch is no more than the time boundary. Constraints (22) and (23) make sure that the processing of batch b on the furnace is not interrupted.

4 COMPUTATIONAL EXPERIMENTS

To illustrate the application of the model, we provide a numerical example in this section. The example has 10 jobs and 5 parallel machines. The processing time of each job are {1,2,4,1,5,1,1,2,1,2} and {4,2,5,2,2,3,3,4,3,4} at stage 1 and stage 2, respectively. Sizes are {1,3,1,2,3,1,3,2,2,1} for each job and {(1,4);(3,4);(1,2,5);(1,2,3,5);(1,2,3,4,5);(4);(2,4,5);(2,5);(5);(2,3,4)} is the subset of available machines that can process job i. The batch machine has a capacity 10 and the number of periods 20 with energy price for each period are {1,4,4,3,3,2,2,1,3,4,3,2,1,3,4,1,4,1,2,3}. Energy consumption rate of machine k are {2,2,3,2,2} in working and {1,1,1,1,1} in idle. The batch machine has energy consumption rates of 7 in working and 2 in idle.

CPLEX 12.6 is used to solve the instance and it's computation time is 17.16 seconds. The results are shown in Figure 2. The total energy consumption cost is 1390.

5 CONCLUSION

This paper presents a mathematical formulation for a two-stage hybrid flow shop in which there are several parallel machines at stage 1 and a batching machine at stage 2. A numerical example is presented to demonstrate the application of the model.

There are several improvements that deserve future investigations. First, techniques such as heuristic algorithm and approximation algorithm may be devised to solve large instances since solving large instances of the problem using a commercial solve would be a challenge. Considering a case of 15 jobs takes 1859.77 seconds to find the optimal solution. Second, it is necessary to consider machine setup and limited intermediate storage between stages since they should not be ignored in practical production systems. Finally, objectives such as makespan, total tardiness and total completion time are encouraged to be integrated with the objective of TEC minimization in the problem, i.e. to consider the multiple objective optimization.

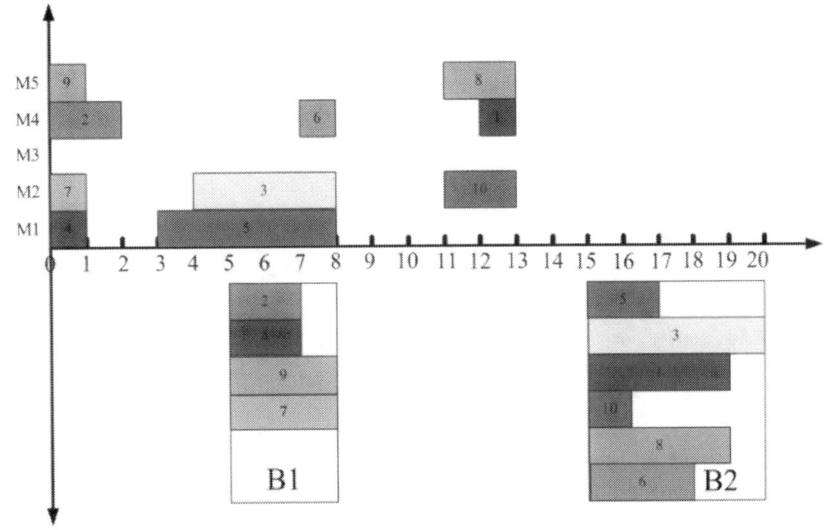

Figure 2. The schedule results for the instance

6 REFERENCES

[1] **Gahm, C., Denz, F., Tuma, A. 2016.** Energy-efficient scheduling in manufacturing companies: A review and research framework, *European Journal of Operational Research*, 248(3), pp 744-757.

[2] **Giret, A., Prabhu, V. 2015.** Sustainability in manufacturing operations scheduling: a state of the art review, *Journal of Manufacturing System* , 37(1), pp 126-140.

[3] **Apostolos, F., Alexios, P. 2013.** Energy Efficiency of Manufacturing Processes: A Critical Review, 46th CIRP Conference on Manufacturing Systems, Sesimbra, Portugal, May 29-31, pp 628-633.

[4] **Kaplan, S., Rabadi, G. 2012.** Exact and heuristic algorithms for the aerial refueling parallel machine scheduling problem with due date-to-deadline window and ready times, *Computers and Industrial Engineering*, 62(1), pp 276-285.

[5] **Joo, C.M., Kim, B.S. 2015.** Hybrid genetic algorithms with dispatching rules for unrelated parallel machine scheduling with setup time and production availability, *Computers and Industrial Engineering*, 85, pp 102-109.

[6] **Mouzon, G., Yildirim, M.B., Twomey, J. 2007.** Operational methods for minimization of energy consumption of manufacturing equipment, *International Journal of Production Research*, 45(18-19), pp 4247-4271.

[7] **Yildirim, M.B., Mouzon, G. 2012.** Single-machine sustainable production planning to minimize total energy consumption and total completion time using a multiple objective genetic algorithm, *IEEE Transactions on Engineering Management*, 59(4), pp 585-597.

[8] **Shrouf, F., Ordieres-Mere, J. 2014.** Optimizing the production scheduling of a single machine to minimize total energy consumption costs, *Journal of Cleaner Production*, 67(15), pp 197-207.

[9] **Fang, K., Uhan, N., Zhao, F., Sutherland, J.W. 2011.** A new approach to scheduling in manufacturing for power consumption and carbon footprint reduction, *Journal of Manufacturing Systems*, 30(4), pp 234-240.

[10] **Margaux, N., Christian, A., Pierre, L., Rosa, M., Victor, P., Lorena, P. 2015.** A batch sizing and scheduling problem on parallel machines with different speeds, maintenance operations, setup times and energy costs, 6th IESM Conference, Universite de Valenciennes, Hainaut, Belgium, Oct 30, pp 889-891.

[11] **Liu, Y., Tiwaria, A. 2015.** An investigation into minimising total energy consumption and total completion time in a flexible job shop for recycling carbon fiber reinforced polymer, 22nd CIRP Conference on Life Cycle Engineering, Sydney, Australia, Apr 7-9, pp 722-727.

[12] **Zhang, H., Zhao, F., Fang, K., Sutherland, J.W. 2014.** Energy-conscious flow shop scheduling under time-of-use electricity tariffs, *CIRP Annals-Manufacturing Technology*, 63(1), pp 37-40.

[13] **Mansouri, S.A., Aktas, E., Besikci, U. 2015.** Green scheduling of a two-machine flowshop: Trade-off between makespan and energy consumption, *European Journal of Operational Research*, 248(3), pp 772-788.

[14] **Liu, Y., Dong, H., Lohse, N., Petrovic, S. 2014.** An investigation into minimising total energy consumption and total weighted tardiness in job shops, *Journal of Cleaner Production*, 65, pp 87-96.

[15] **Zhang, R., Chiong, R. 2015.** Solving the energy-efficient job shop scheduling problem: a multi-objective genetic algorithm with enhanced local search for minimizing the total weighted tardiness and total energy consumption, *Journal of Cleaner Production*, 112(4), pp 3361-3375.

[16] **Damodaran, P., Srihari, D. 2004.** Mixed Integer Formulation to Minimize Makespan in a Flow Shop with Batch Processing Machines, *Mathematical and Computer Modelling*, 40(13), pp 1465-1472.

[17] **Dai, M., Tang, D., Zhang, H., Yang, J. 2014.** Energy-aware Scheduling Model and Optimization for a Flexible Flow Shop Problem, 26th Chinese Control and Decision Conference, Changsha, China, 31 May - 2 June, pp 323-326.

[18] **He, Y., Li, Y. Wu, T., Sutherland, W. 2015.** An energy-responsive optimization method for machine tool selection and operation sequence in flexible machining job shops, *Journal of Cleaner Production*, 87, pp 245-254.

DESIGNING A BI-OBJECTIVE MODEL FOR A BLOOD BANK LOCATION-ALLOCATION PROBLEM IN EMERGENCY LOGISTICS UNDER UNCERTAINTY

R. Tavakkoli-Moghaddam[1*], S. Mortazavi[1], S.H. Madani-Saatchi[2] and M. Fallah[3]
[1]School of Industrial Engineering, College of Engineering
University of Tehran, Tehran, Iran
{tavakoli, s.m.mortazavi}@ut.ac.ir

[2]Department of Industrial Engineering
Kharazmi University, Karaj, Iran
std_h.madani@khu.ac.ir

[3]Department of Industrial Engineering, Central Tehran Branch
Islamic Azad University, Tehran, Iran
mohammad.fallah43@yahoo.com

ABSTRACT

Blood management is a problem of specific concern for the human race. Up to now, unlike other substitutable resources, the need for blood donation and its derived products have been always existed. Blood supply chains play a key role in healthcare systems. Any improvement in management of these chains will have a direct impact on the supply of blood as a life-saving product. In this research, we present a novel multi-period location-allocation model for blood banking in humanitarian relief logistics. The components considered in this study are donor groups, fixed blood collection facilities, temporary blood collection centers, research laboratories and demand points that can be hospitals and clinics. The model consists of a bi-objective mathematical programming model that minimizes the total cost and time. The ε-constraint method is used to solve small and medium-sized problems to optimality. For larger-sized problems, we propose the non-dominated sorting genetic algorithm (NSGA-II) as a well-known meta-heuristic algorithm. Several numerical examples with managerial insights are evaluated. Finally, the obtained results are discussed.

Keywords: Blood supply chain, Location-allocation, ε-constraint, NSGA-II, Uncertainty.

1 INTRODUCTION

A blood supply chain is one of the most important categories in health and treatment, which needs precise planning and management. Thus, balancing and managing the blood supply chain are important critical need in order to ensure that the demand satisfaction degree can appropriately be met. Blood is a perishable goods and contains different components, which are medically important. It is usually collected as whole blood (WB), and after some processes its components are red and white cells, platelets and plasma. The blood supply chain network consists of main and temporary /mobile facilities and lab centers. The main centers are in charge of blood collections with great

---

* Corresponding Author.

capacity of storage and technology. According to the above-mentioned descriptions, the efficient management of the blood supply chain in strategic level necessitates to take several decisions simultaneously. These decisions include: (1) locating/relocating of temporary/mobile facilities among a given set of potential sites, (2) establishing main collective centers (MCC) among a set of candidate locations, (3) selecting lab centers among established MCCs, (4) assigning different donors to collection centers, (5) assigning different products to a lab centers, and (6) allocating the derived components to demand zones while considering the corresponding perish time.

In this paper, a five-echelon supply chain network is designed. These echelons are donators, main collective centers, temporary collective centers, laboratories and demand zones that can be hospitals, clinics or research labs. Due to establishment constraints, some specific numbers of main collective centers are equipped with labs. In real world, sometimes some of centers and labs may be inaccessible because of some problems. Hence, in this paper, facilities work under different scenarios that are applied as binary matrices in the model. The reminder of this paper is organized as follows. Section 2 represents a review on the relevant literature. Problem definition and mathematical model formulation are given in Section 3. Section 4 is dedicated to some numerical and sensitivity analyses. Finally, Section 5 represents the conclusion and some future research directions.

2 LITERATURE REVIEW

Widespread studies have been done in a case of a blood supply chain. The relevant literature can be classified into two main categories, namely location-allocation of healthcare facilities and blood supply chain network design. Applications of operational research in healthcare are fully discussed by Papageorgiuo [1] and Rais and Viana [2]. Syam and Cote [3] significantly focused on location-allocation of healthcare facilities and presented a model for location-allocation of specialized healthcare systems. Bidhandi et al [4] presented a mixed-integer linear supply chain model considering multiproduct assumption under uncertainty. Rajagopalan et al [5] presented a multi-period set covering model for determining the location of ambulances. Melkote and Daskin [6] simultaneously optimized uncapacitated transportation network design facilities and locations. Melkote and Daskin [7] on the other work assumed capacity limitation for a location-allocation problem. Also, Daskin and Snyder [8] prepared a review for facility location problems.

In the field of the blood supply chain, Belien and Force [9] presented a comprehensive review on the supply chain (SC) of blood products to support different decision making issues and determine required future studies. Piereskalla and Brailer [10] and Pierskalla [11] addressed both strategic and tactical issues related to blood supply chains. Sahin et al [12] addressed several location-allocation models for the blood SC network. Cetin and Sarul [13] presented a hybrid set covering model to address facility location decisions of blood banks. Zahiri et al [14] developed a mixed-robust integer supply chain model considering strategic and tactical decisions. Arvan et al [15] designed a two-objective blood supply chain under the amount of blood waste and cost minimization assumptions.

As the literature reveals, a small proportion of the literature is dedicated to mathematical programming models and none of them address the multi-product property of the blood SCs. Inefficiency of some blood collector facilities in the real world is one of the important factors leads to unmet demands in different time periods and this factor is not mentioned in the literature. In this paper, this problem is taken into account. Also, diverse transportation facilities are considered in order to drop the duration of holding blood products in the network.

3 PROBLEM DEFINITION AND FORMULATION

In this paper, we propose a new multi-product, multi-period mathematical model for a blood collection and distribution network design problem considering consisting of both temporary and fixed collection sites and lab centers. The addressed model aims to minimize the total cost and total transportation time in the network. Different scenarios are considered for utilizing fixed collection centers in different periods. Blood donations are commonly taking place via main or temporary facilities. In case of donating via temporary facilities, an additional shipment will be occurred to main centers at the end of each period. All the received blood packs will be sent to lab centers, which are located within the main centers, to derive useful components from blood packs. Therefore, the proposed network is a single product network before the lab centers, and thereafter it transfers into a multi-product supply chain. Finally, all the derived components are sent to demand zones. Figure 1 shows a schematic view of the proposed network.

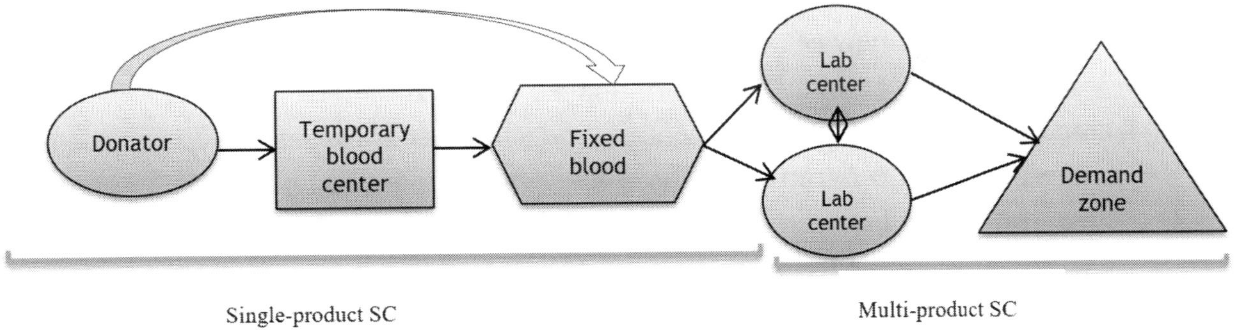

Figure 1: Schematic View of the Proposed Network

3.1 Model Assumptions

In this section, a novel mathematical programming model is proposed for the blood supply chain network design. The considered assumptions and simplifications used in the presented model are as follows:

- Location of recipient zones are fixed and known, while only candidate locations for temporary and main centers are predetermined.
- Single product is shipped through the network before derivation and thereafter multiple products are transported within the supply chain.
- Dynamic nature of supplies and demands needs to consider planning decisions in multiple periods.
- Due to the uncertain nature of parameters, some input parameters are considered fuzzy.
- Different transportation modes for shipments between fixed collection centers are taken into account.

3.2 Indices, Parameters and Variables

The following indices, parameters and variables are used to formulate the generalized multi-product, multi-period location-allocation model for blood supply chain system. Noteworthy, the uncertain parameters are differentiated from the deterministic ones with the tilde on them.

3.2.1 Indices

i Index of donors, $i \in I$

j Index of candidate locations for temporary facilities, $j \in J$

k Index of candidate locations for main facilities, $k, k' \in K$

p Index of derived products, $p \in P$

t	Index of time periods, $t \in T$
h	Index of demand zones, $h \in H$
r	Index of transportation modes, $r \in R$
s	Index of scenarios for main facilities

3.2.2 Parameters

$\widetilde{c_{j_1 \cdot j_2}}$	Cost of moving a temporary facility from j_1 to j_2
$\widetilde{c'_k}$	Cost of locating a main center in the candidate location k
$\widetilde{c''_k}$	Cost of establishing a lab center in the candidate location k
$\widetilde{c_{ij}}$	Transportation cost between donator i and temporary center j
$\widetilde{c'_{ik}}$	Transportation cost between donator i and main center k
$\widetilde{c''_{jk}}$	Transportation cost between temporary center j and main center k
$\widetilde{b_{kk'r}}$	Transportation cost between main center k and lab center k' by transportation mode r
$\widetilde{b'_{kk'r}}$	Transportation cost between lab center k and k' by transportation mode r
$\widetilde{b''_{khr}}$	Transportation cost between lab center k and demand zone h by transportation mode r
$\widetilde{De^t_{ph}}$	Total demand of zone h for product p in period t
$\widetilde{H_{kk'r}}$	Travel time between main center k and lab center k' by transportation mode r
$\widetilde{H'_{kk'r}}$	Travel time between lab center k and k' by transportation mode r
$\widetilde{H''_{khr}}$	Travel time between lab center k and demand zone h by transportation mode
$\widetilde{r_{ij}}$	Travel time between the donator i and temporary facility j
r'	Maximum coverage radius of temporary facilities, if $r_{ij} \leq r'$, i is covered by j
$\widetilde{w_{ik}}$	Travel time between the donator i and main center k
w'	Maximum coverage radius of main centers, if $w_{ik} \leq w'$, i is covered by k
$\widetilde{q_{jk}}$	Travel time between the donator i and main center k
q'	Maximum coverage radius of main centers, if $q_{jk} \leq q'$, j is covered by k
VJ	Maximum capacity of each temporary facility
VK_k	Maximum capacity of main center k
$\widetilde{VI^t_i}$	Maximum blood volume that can be donated by donator i at time period t
M	Reasonably large number
ϑ	Fraction of valid blood packs for derivation process
F^t_{ks}	If main center k

3.2.3 Variables

$$X^t_{ij} \quad \begin{cases} 1, \text{ if donator } i \text{ is assigned to temporary facility } j \text{ at period } t; \\ 0, \text{ otherwise} \end{cases}$$

$$X'^t_{ik} \quad \begin{cases} 1, \text{ if donator } i \text{ is assigned to main center } k \text{ at period } t; \\ 0, \text{ otherwise} \end{cases}$$

$$X''^t_{jk} \quad \begin{cases} 1, \text{ if temporary facility } j \text{ is assigned to main center } k \text{ at period } t; \\ 0, \text{ otherwise} \end{cases}$$

$y^t_{j_1,j_2}$ $\begin{cases} 1, \text{if a temporary facility moves from } j_1 \text{ to } j_2 \text{ in two consecutive periods;} \\ 0, \text{otherwise} \end{cases}$

z_k $\begin{cases} 1, \text{if a main facility is located in location } k; \\ 0, \text{otherwise} \end{cases}$

L_{kts} $\begin{cases} 1, \text{if a lab center is established within main facility } k \text{ at period t and scenario } s; \\ 0, \text{otherwise} \end{cases}$

s^t_{ij} Blood volume, in which donator i donates to temporary facility j in period t

s'^t_{ik} Blood volume, in which donator i donates to main center k in period t

s''^t_{jk} Blood volume shipped from temporary facility j to the main center k in period t

$La^t_{kk'r}$ Blood volume shipped from main center k to the lab center within main center k' in period t

$v^t_{pkk'r}$ Volume of derived product p shipped from lab center within main center k to the lab center k' in period t

u^t_{pkhr} Volume of derived product p shipped from lab center k to the demand zone h in period t

3.2.4 Proposed Model

By considering the above-mentioned notations, the proposed mathematical model can be formulated by:

$$\begin{aligned} \text{Min } Z_1 = & \sum_{k} z_k \widetilde{c'_k} F^t_{ks} + \sum_{k} L_k \widetilde{c''_k} F^t_{ks} + \sum_{j_1,j_2,t} y^t_{j_1,j_2} \widetilde{c_{j_1,j_2}} + \sum_{i,j,} \widetilde{c_{ij}} s^t_{ij} \\ & + \sum_{i,k,} \widetilde{c'_{ik}} s'^t_{ik} F^t_{ks} + \sum_{j,k,t} \widetilde{c''_{jk}} s''^t_{jk} F^t_{ks} + \sum_{k,k',r,t} \widetilde{b_{kk'r}} La^t_{kk'r} F^t_{ks} \\ & + \sum_{k,k',r,p,t} \widetilde{b'_{kk'r}} v^t_{pkk'r} F^t_{ks} + \sum_{k,h,p,r,t} \widetilde{b''_{khr}} u^t_{pkhr} F^t_{ks} \end{aligned} \quad (1)$$

$$\text{Min } Z_2 = \sum_{k,k',r,t} \widetilde{H_{kk'r}} La^t_{kk'r} + \sum_{k,k',r,p,t} \widetilde{H'_{kk'r}} v^t_{pkk'r} + \sum_{k,h,r,p,t} \widetilde{H''_{khr}} u^t_{pkhr} \quad (2)$$

s.t.

$$\sum_{j_1 \in J} y^t_{j_1,j_2} \leq 1 \qquad \forall j_2 \in J, t \quad (3)$$

$$\sum_{j_2 \in J} y^t_{j_1,j_2} \leq 1 \qquad \forall j_1 \in J, t \quad (4)$$

$$\sum_{j_2 \in J} y^t_{j_1,j_2} \leq \sum_{j} y^{t-1}_{j,j_1} \qquad \forall j_1 \in J, t \geq 2 \quad (5)$$

$$\sum_j X_{ij}^t + \sum_k X_{ik}'^t \leq 1 \qquad \forall i,t \qquad (6)$$

$$X_{ij}^t \widetilde{r_{ij}} \leq r' \sum_{j_1 \in J} y_{j_1,j}^t \qquad \forall i,j,t \qquad (7)$$

$$X_{ik}'^t \widetilde{w_{ik}} \leq w' z_k F_{ks}^t \qquad \forall i,k,t,s \in S_t \qquad (8)$$

$$X_{jk}''^t \widetilde{q_{jk}} \leq q' \sum_{j_1} y_{j_1,j}^t \, z_k F_{ks}^t \qquad \forall j,k,t,s \in S_t \qquad (9)$$

$$s_{ij}^t \leq M X_{ij}^t \qquad \forall i,j,t \qquad (10)$$

$$s_{ik}'^t \leq M X_{ik}'^t F_{ks}^t \qquad \forall i,k,t,s \in S_t \qquad (11)$$

$$s_{jk}''^t \leq M X_{jk}''^t F_{ks}^t \qquad \forall j,k,t,s \in S_t \qquad (12)$$

$$\sum_j s_{jk}''^t + \sum_i s_{ik}'^t \leq VK F_{ks}^t \qquad \forall k,t,s \in S_t \qquad (13)$$

$$\sum_j s_{ij}^t + \sum_k s_{ik}'^t \leq \widetilde{VI_i^t} \qquad \forall i,t \qquad (14)$$

$$\sum_i s_{ij}^t \leq VJ \qquad \forall j,t \qquad (15)$$

$$\sum_i s_{ij}^t = \sum_k s_{jk}''^t \qquad \forall j,t \qquad (16)$$

$$\sum_j s_{jk}''^t + \sum_i s_{ik}'^t = \sum_{k' \in K} La_{kk'r}^t \qquad \forall k,t,r \qquad (17)$$

$$\sum_h u_{pk'hr}^t \leq \left(\vartheta \sum_k La_{kk'r}^t + \sum_k v_{pkk'r}^t - \sum_k v_{pk'kr}^t \right) L_{k'} F_{k's}^t \qquad \forall p,k',t,r,s \in S_t \qquad (18)$$

$$L_k \leq Z_k F_{ks}^t \qquad \forall k,s \in S_t \qquad (19)$$

$$\sum_r \sum_{k'} u_{pk'hr}^t \geq \widetilde{De_{ph}^t} \qquad \forall p,h,t \qquad (20)$$

$$X_{ij}^t, X_{ik}'^t, X_{jk}''^t, y_{j_1,j_2}^t, Z_k, L_k \in \{0,1\} \qquad \forall i,j,k,t \qquad (21)$$

$$s_{ij}^t, s_{ik}^t, s_{jk}''^t, u_{pk'hr}^t, La_{kk'r}^t \geq 0, integer \qquad \forall i,j,k,k',t,r \qquad (22)$$

Objective function (1) minimizes the total cost including establishment cost of main and lab centers (strategic-level costs) and locating and movement costs of temporary centers in each period as well as material flow costs (tactical-level costs). Objective function (2) tends to minimize total amount of time that products travel by different transportation modes. Constraint sets (3) to (5) refer to movement of temporary facilities in each period. Constraint (6) restricts the donors to only donate either to temporary or main facilities. Constraints (7) to (9) are the coverage constraints. Constraints (10) to (12) force the maximum allowable flow of blood. Constraints (13) to (15) assure the capacity constraints. Constraints (16) and (17) ensure the flow balance at temporary and main centers, respectively. Constraint (18) addresses the volume of derived product p in each lab center and assures that they should be sent to demand zones at the end of each time period. Constraint (19) guarantees the establishment of a main center when the associated lab center is opened. Constraint (20) ensures the derived products from lab centers to be transported to demand zones. Finally, Constraint (21) and (22) enforce the binary, integer and non-negativity restrictions on decision variables.

4 SOLUTION METHODOLOGY AND EVALUATION

In the previous section, a hybrid linear integer model with two objectives location-allocation problem for blood centers is demonstrated. According to its objective function, the ε-constraint method is selected as exact solving method. Hence, the model is coded in GAMS software with a core i3 computer with 2 GHz RAM. Also, as it is obvious in the formulation, this model is NP-hard and optimal results are unreachable while expanding the problem size. Therefore, a well-known multi-objective evolutionary algorithm, namely NSGA-II, is chosen to solve this problem at large sizes. Tables 1 and 2 shows the performance of these approaches for this problem. Regarding these tables, the time for solving with the exact method surges when the size of problem increases. So, this approach is inefficient at medium and large sizes.

Table 1: Results of the ε-constraint Method for Small-sized Problems

Problem size $\lvert i\rvert\times\lvert j\rvert\times\lvert k\rvert\times\lvert p\rvert\times\lvert t\rvert\times\lvert h\rvert\times\lvert r\rvert\times\lvert s\rvert$	OF1	OF2	Running time (s)
5×2×2×2×1×3×2×1	107379	221263	20
5×2×3×3×2×5×3×4	536188	370212	103
8×5×5×3×3×8×3×16	1271922	1046783	983
10×5×6×3×3×10×4×22	4822961	1672497	10613

Table 2: Results of NSGA-II for Small-sized Problems

Problem size	Objective function value		Running time (s)
$\lvert i\rvert\times\lvert j\rvert\times\lvert k\rvert\times\lvert p\rvert\times\lvert t\rvert\times\lvert h\rvert\times\lvert r\rvert\times\lvert s\rvert$	OF1	OF2	
5×2×2×2×1×3×2×1	282949	121338	16.18
5×2×3×3×2×5×3×4	461098	546912	23.86
8×5×5×3×3×8×3×16	1134537	1869725	30.29
10×5×6×3×3×10×4×22	1784689	6848605	54.47

5 CONCLUSION

In this paper, a bi-objective model for location-allocation blood supply chain elements was presented and solved by the ε-constraint method. The performance of this method decreases while considering large-scale problems, a multi-objective evolutionary algorithm, namely NSGA-II, was proposed to tackle such limitations. For future studies, considering different blood groups can be suggested. Also, using different meta-heuristic algorithms and making comparisons and sensitivity analysis on critical parameters leads to cope with such a complex problem in large-sized instances.

6 REFERENCES

[1] **Papageorgiou, J.C.** 1978. Some operations research applications to problems of health care systems (a survey), *International Journal of Bio-Medical Computing*, 9, pp 101-114.

[2] **Rais, A., Viana, A.** 2011. Operations Research in Healthcare: A survey, *International Transactions in Operational Research*, 18, pp 1-31.

[3] **Syam, S.S., Côté, M.J.** 2010. A location-allocation model for service providers with application to not-for-profit health care organizations, *Omega*, 38, pp 157-166.

[4] **Mohammadi Bidhandi, H., Mohd Yusuff, R., Megat Ahmad, M.M.H., Abu Bakar, M.R.** 2001. Development of a new approach for deterministic supply chain network design, *European Journal of Operational Research*, 118, pp 121-128.

[5] **Rajagopalan, H.K., Saydam, C., Xiao, J.** 2008. A multi period set covering location model for dynamic redeployment of ambulances, *Computers & Operations Research*, 35, 814-826.

[6] **Melkote, S., Daskin, M.S.** 2001. An integrated model of facility location and transportation network design, *Transportation Research - Part A*, 35, pp 515-538.

[7] **Melkote, S., Daskin, M.S.** 2001. Capacitated facility location/network design problems, *European Journal of Operational Research*, 121, pp 481-415.

[8] **Daskin, M.S., Snyder, L.V., Berger, R.T.** 2005. Facility location in supply chain design, In: *Logistics systems: Design and Optimization*, Springer, pp 31-65.

[9] **Beliën, J., Forcé, H.** 2012. Supply chain management of blood products: A literature review. *European Journal of Operational Research*, 217, pp 1-16.

[10] **Pierskalla, W.P., Brailer D.J.** 1994. Applications of operations research in health care delivery. *Handbooks in OR & MS*, 6, pp 4-7.

[11] **Pierskalla, W.P.** 2005. Supply chain management of blood banks, In: *Operations Research and Health Care*, Springer, pp 103-145.

[12] **Şahin, G., Süral, H., Meral, S.** 2007. Locational analysis for regionalization of Turkish Red Crescent blood services, *Computers & Operations Research*, 34, pp 692-704.

[13] **Cetin, E., Sarul L.S.** (2009). A blood bank location model: A multi objective approach. European Journal of Pure and Applied Mathematics, 2:112-124.

[14] **Zahiri, B., Torabi, S.A., Mousazadeh, M., Mansouri, S.A.** 2015. Blood collection management: Methodology and application, *Applied Mathematical Modelling*, 39, pp. 7680-7696.

[15] **Arvan, M., Tavakkoli-Moghaddam, R., Abdollahi, M.** 2015. Designing a bi-objective and multi-product supply chain network for the supply of blood, *Uncertain Supply Chain Management*, *3*(1), pp. 57-68.

MYTHS OF THE THEORY OF CONSTRAINTS AND THE CLARIFICATION

Kan Wu[1*]
[1]School of Mechanical and Aerospace Engineering
Nanyang Technological University, Singapore
kan626@gmail.com

ABSTRACT

The theory of constraints was proposed in the mid-1980s and has significant impact on the productivity improvement in manufacturing systems. While it is intuitive and easy to understand, its conclusions are mainly derived in deterministic settings or based on the first moment results. Since production systems are stochastic in general, some of its conclusions are not rigorous and have to be modified. In this study, we show that the process of ongoing improvement may lead to unfavorable outcomes and the throughput bottleneck should be planned on certain types of machines. Specifically, if a system has no bottleneck, every station will be a bottleneck. Furthermore, through capturing the dependence among stations, we show that improving the variability of a frontend machine in a production line can be more effective than improving the variability of a throughput bottleneck.

Keywords: Theory of constraint, Queueing theory

1 INTRODUCTION

Simulating by the invention of computers and automation, new theories for production control, such as material requirement planning and just-in-time, were developed in the 1960s and 1970s. The manufacture industry entered the era of the third industrial revolution. Along the same line, theory of constraints (TOC) has been proposed in the book, The Goal, by Goldratt and Cox [1] for companies to achieve competitive advantages. According to Rahman [2], the concept of the TOC can be summarized as (1) Every system must have at least one constraint, and (2) The existence of constraints represents opportunities for improvement.

The working principle of TOC consists of five focusing steps.

1. Identify the system's constraint(s).
2. Decide how to exploit the system's constraint(s).
3. Subordinate everything else to the above decision.
4. Elevate the system's constraint(s).
5. Warning! If in the previous steps a constraint has been broken, go back to step 1, but do not allow inertia to cause a system's constraint.

In the TOC literature, the above five steps are referred to as the process of ongoing improvement (POOGI). The process ensures that ongoing improvement efforts are centered on the system's constraints, where a constraint is anything that prevents the system from achieving its goal. As a

---

* Corresponding Author

result, a bottleneck in a manufacturing system is defined as “any resource whose capacity is equal to or less than the demand placed upon it” [1].

The above definition specifically limits the scope of a bottleneck to a throughput bottleneck, and is not rigorous from the viewpoint of queueing theory, since the inventory of a stochastic system will diverge to infinity if its utilization is equal or greater than one as specified by Kingman's *G/G/1* heavy traffic approximation [3]:

$$QT(G/G/1) \cong \frac{c_a^2+c_s^2}{2}\frac{\rho}{1-\rho}\frac{1}{\mu}, \quad (1)$$

where ρ is utilization ($=\lambda/\mu$), λ is arrival rate, μ is service rate, c_a^2 is the squared coefficient of variation (SCV) of arrival intervals, c_s^2 is the SCV of service time, and QT is mean queue time. Cycle time is the sum of queue time and service time.

To overcome this shortcoming, Wu [4] and Wu and Zhao [5] extended the definition of bottlenecks from a throughput bottleneck (TPBN) to a cycle time bottleneck (CTBN), where a cycle time bottleneck is the workstation which prevents a production system from achieving its mean cycle time target. Since system cycle time is contributed by all workstations, all workstations are cycle time bottlenecks with different levels of contribution. With the same mean cycle time target, reducing the mean queue time of any workstation would allow the queue time increases of the others, and potentially lead to a higher throughput rate of the system under the same system capacity. By defining bottlenecks from the viewpoint of cycle time, the concept of TOC has been extended from a deterministic system to a stochastic one. Although all workstations can be cycle time bottlenecks, there are still major and minor ones, where a major one has a higher impact on system cycle time. The question becomes which workstation is the major cycle time bottleneck and we should improve first?

Since demand is fluctuating, a station which has demand more than its capacity in the hot season can have excess capacity in the cold season. Hence, it is more general and reasonable to define a throughput bottleneck as the station with the highest utilization [6, 7], so that productivity improvement can be continued in the cold season to prepare for the high demand in the hot season. Based on POOGI, the throughput bottlenecks in a production line should be removed one by one. However, machine costs have large variations. For example, while the cost of a photolithography machine in a modern semiconductor production line can be up to one hundred million US dollars, the costs of other machines are much lower. An important question to ask is if it is reasonable to make the most expensive machine less busy and allocate the bottleneck at the other cheaper machines? Should the throughput bottleneck be planned in advance on specific machines, or be allocated later based on the result of POOGI? Should we only focus on the throughput bottleneck or improve the productivity of the other machines at the same time?

The above questions and myths caused by TOC will be clarified in the following sections. Section 1 gives the motivation of this study. Section 2 introduces the concept of intrinsic ratios and contribution factors. Variability of production lines will be discussed in Section 3. The second moment result of the TOC is given in Section 4 and Section 5 concludes the study.

2 VARIABILITY OF PRODUCTION LINES

The concept of variability is often used by practitioners and researchers to represent the stochastic effect in a manufacturing system. The common sources of variability in manufacturing systems are machine breakdowns, setups, reworks, product mixes, operator availability, batching and fluctuation in process time and arrival intervals [7-9]. Reducing variability decreases job queue time, improves system service level [10], and is essential in well-known production control techniques, such as just-in-time (JIT) production [11], theory of constraints [1] and six sigma [12].

Hence, quantifying variability and understanding its basic properties play a key role in achieving effective control of manufacturing systems.

When a manufacturing system consists of a single machine with general service times and arrival intervals, the mean queue time of the system can be approximated using Kingman's heavy traffic approximation.

$$QT \leq \frac{\sigma_a^2\mu^2+c_s^2}{2}\frac{\rho}{1-\rho}t \cong \frac{c_a^2+c_s^2}{2}\frac{\rho}{1-\rho}t = \frac{c_a^2+c_s^2}{2}QT_{M/M/1}, \quad (2)$$

where QT is the mean queue time of the system, and $QT_{M/M/1}$ is the mean queue time of an $M/M/1$ queue with the same mean arrival rate and service rate as the single machine, c_a^2 is the squared coefficient of variation of arrival intervals, c_s^2 is SCV of service times, μ is service rate, σ_a is the standard deviation of arrival intervals, ρ is utilization ($= \lambda/\mu$), λ is arrival rate and t is mean service time (or $1/\mu$).

In terms of a specific random variable, variability can be simply regarded as the squared coefficient of variation (SCV) (or sometimes the coefficient of variation). Inspired by Kingman's approximation [13] and Friedman's reduction method [14], Wu et al. [15] define system variability of a single-server production line as

$$\alpha_f = \frac{QT_f}{QT_{M/M_B/1}}, \quad (3)$$

where QT_f is mean queue time of a production line, $QT_{M/M_B/1}$ is the mean queue time of an M/M/1 queue with the same arrival rate and service rate as the system bottleneck of the production line. The system bottleneck of a production line is the station with the highest utilization. Throughout the paper, the subscript B is used to denote the system bottleneck.

Let $\alpha_B = \left(c_a^2 + c_{s_B}^2\right)/2$, $\mu_B = \min_{1\leq i\leq N}\mu_i$, and $\rho_B = \lambda/\mu_B$. Based on the concept of intrinsic ratios [16] and contribution factors [17], α_f can be written as

$$\alpha_f = \frac{QT_f}{QT_{M/M_B/1}} = \alpha_B + \frac{1-\rho_B}{\rho_B}\mu_B\sum_{i\neq B}f_i\,\alpha_i\frac{\rho_i}{1-\rho_i}\frac{1}{\mu_i} = \alpha_B + \left(\frac{\mu_B^2}{\lambda} - \mu_B\right)\sum_{i\neq B}f_i\,\alpha_i\frac{\lambda}{\mu_i-\lambda}\frac{1}{\mu_i}, \quad (4)$$

where f_i is contribution factor.

Variability of a production line can be decomposed into two parts: the bottleneck variability and the gross effect from the non-bottlenecks, where the contribution from the non-bottlenecks converges to zero in heavy traffic. That is, system variability is dominated by the bottleneck in heavy traffic, which is consistent with the heavy-traffic bottleneck phenomenon by Iglehart and Whitt [18].

2.1 Properties of System Variability

According to Wu et al. [15], when a production line consists of multiple single-server stations in series, it has the following properties.

(I) Lower variability for higher arrival rate

From Eq. (4), we have

$$\frac{\partial\alpha_f}{\partial\lambda} = \mu_B\sum_{i\neq B}\frac{f_i\alpha_i}{(\mu_i-\lambda)^2}\frac{\mu_B-\mu_i}{\mu_i} < 0. \quad (5)$$

Property 1: Variability of production lines with single-server stations is monotonically decreasing in the external arrival rate.

(II) Lower variability for a small increment of μ_i $(i \neq B)$

Because intrinsic ratios change with μ_i, f_i also depends on μ_i. However, when the value of μ_i does not change much, we can assume f_i in Eq. (4) is (nearly) a constant. In the following, we would like to investigate how system variability changes when the service rate of a non-bottleneck has a small increment (i.e., where the value of f_i does not change much). Since $\frac{\mu_B}{\lambda} - 1 > 0$, we have

$$\frac{\partial \alpha_f}{\partial \mu_i} = -\mu_B\left(\frac{\mu_B}{\lambda} - 1\right) f_i \alpha_i \left(\frac{\lambda}{(\mu_i-\lambda)^2\mu_i} + \frac{\lambda}{(\mu_i-\lambda)\mu_i{}^2}\right) < 0. \qquad (6)$$

Property 2: When the non-bottleneck service rate has a small increment, variability of a production line with single-server stations decreases.

(III) Higher variability for a small increment of μ_B

Following the previous assumption, since $\frac{2\mu_B}{\lambda} - 1 > 0$, we have

$$\frac{\partial \alpha_f}{\partial \mu_B} = \left(\frac{2\mu_B}{\lambda} - 1\right) \sum_{i \neq B} f_i \, \alpha_i \frac{\lambda}{\mu_i - \lambda} \frac{1}{\mu_i} > 0. \qquad (7)$$

Property 3: When the system bottleneck service rate has a small increment, variability of a production line with single-server stations increases.

(IV) Gap effect

Assume F1 and F2 are two almost identical production lines, except that there are small differences in their system bottleneck service rates and non-bottleneck service rates, (so that the value of f_i does not change much). The service rates of system bottleneck and non-bottleneck are μ_{B1} and μ_{L1} in F1, and μ_{B2} and μ_{L2} in F2. The service rate gap is defined as $\mu_{Li} - \mu_{Bi}, i = 1,2$.

According to properties 2 and 3, when μ_{B2} and μ_{L2} satisfy one of the following three conditions, F1 has a lower variability:

i. $\mu_{L2} < \mu_{L1}$ and $\mu_{B2} = \mu_{B1}$;
ii. $\mu_{L2} = \mu_{L1}$ and $\mu_{B2} > \mu_{B1}$;
iii. $\mu_{L2} < \mu_{L1}$ and $\mu_{B2} > \mu_{B1}$.

Property 4: For two almost identical production lines with single-server stations, when one's service rate gap inwardly bounds the other, the one with larger service rate gap has the lower variability.

We often hear from production managers that a production line with multiple bottlenecks is much harder to control than a production line with a distinct bottleneck. Property 4 indeed confirms this intuition and gives a theoretical foundation for this claim. While a throughput bottleneck is defined based on the average utilization in the long run and is commonly unique, due to stochastic effects, those non-bottlenecks with higher utilizations may also sometimes face congestions and create a transient bottleneck in the short period of time. Hence, when the utilizations of non-bottlenecks are closer to the bottleneck utilization, variability of a production line becomes higher with longer queue time.

According to the POOGI, one has to remove the throughput bottlenecks in a production line continuously. Based on Property 4, the POOGI will indeed cause higher variability of a production line. No matter what their costs are, all stations will have the same utilization and all stations will be short of capacity in the hot seasons. Hence, if a system has no bottleneck, every station is a bottleneck (in the hot seasons).

Based on the G/G/m approximation of Sakasegawa [19], it can be shown that the above properties are also applicable to a production line with multi-server stations.

3 THE SECOND MOMENT RESULTS ON THE THEORY OF CONSTRAINTS

In a stochastic system, the price of higher throughput rate is longer queue time. When the throughput rate approaches capacity, the queue time goes to infinity. Since no customer would accept infinite cycle time, a bottleneck defined by throughput rate cannot occur in a stochastic production line. Hence, the bottleneck in manufacturing is typically defined as the workstation with the highest level of utilization [6, 7]. However, due to the dependence among workstations, the station with the highest utilization may not have the most impact on system cycle time.

In practical manufacturing systems, cycle time reduction is often achieved through improving job scheduling or reducing the mean or variance of service times. Delp et al. [20] suggested that reducing service time variability is a cost-effective way to achieve shorter cycle time. Assuming the cost of reducing service time variability (in percentage) is the same for all stations, an important question to ask is how to identify the second moment CTBN, which is the most effective station to improve system cycle time if its variability is reduced.

Through identifying the blocking and diffusion effects, Wu and Zhao [5] presented a way to quantify the dependence among single server queues in series. When the variability of station i is reduced, the mean queue times at station i and its downstream stations are shortened. Denote the mean queue time of the original system by $QT_f(0,0)$. If the service time SCV of station i is reduced by p (in percentage), the new system mean queue time is denoted by $QT_f(i,p)$, $1 \leq i \leq N, 0 \leq p \leq 1$. If the mean service times of all stations are unchanged, the second moment CTBN is the station resulting in the shortest system mean queue time, i.e., $min_{1 \leq i \leq N}\{QT_f(i,p)\}$.

Rather than improving the throughput bottleneck (with the highest utilization), we show that it can be more effective to start with some other stations when one would like to reduce the system mean cycle time. An example is presented in the following. All the cases are examined at 10 different utilizations (ρ ranges from 0.1 to 0.95). Thirty replications are conducted at each utilization. Each replication consists of 200,000 jobs after discarding the first 400,000 jobs for warm-up. The sample size is sufficiently large so that the half width of 95% confidence intervals of the mean simulation queue time is less than 2%.

Example: A tandem queue consists of five single-server stations in series. The SCV of the external arrival interval at the first station is 0.3. The mean service times from the first station to the last are 28, 26, 24, 22 and 30 min. The service time SCVs from the 1st station to the last is SCV(0) = [0.8,0.8,0.8,0.8,0.8]. Let system utilization $\rho = \lambda / \mu_5$ and $p = 50\%$. The second moment CTBN is identified by reducing the service time SCV of each station. Let SCV(1) = [0.4,0.8,0.8,0.8,0.8], SCV(2) = [0.8,0.4,0.8,0.8,0.8], SCV(3) = [0.8,0.8,0.4,0.8,0.8], SCV(4) = [0.8,0.8,0.8,0.4,0.8] and SCV(5) = [0.8,0.8,0.8,0.8,0.4]. The system mean queue times of the six tandem queues with SCV(i), i = 0, 1, ..., 5 are compared.

The system mean queue times from simulation are presented in Table 1. Among the six different configurations, the shortest queue time at each utilization is underlined. Because the SCV of the external arrival intervals is less than the SCVs of service times, diffusion effects occur. The main system bottleneck is station 5 as μ_5 is the least among the five stations. Simulation results show that when $0.1 \leq \rho \leq 0.9$, the system mean queue time becomes the shortest if the variability of the first station (rather than the main system bottleneck) is reduced by 50%. The second moment CTBN is station 1.

Table 1: Identification of CTBN for the tandem queue

ρ	$QT_f(0,0)$	$QT_f(1,0.5)$	$QT_f(2,0.5)$	$QT_f(3,0.5)$	$QT_f(4,0.5)$	$QT_f(5,0.5)$
0.1	1.63	1.33	1.38	1.43	1.48	1.42
0.2	7.90	6.62	6.88	7.07	7.24	7.05
0.3	18.36	15.57	16.17	16.64	17.04	16.59
0.4	33.23	28.48	29.54	30.35	30.97	30.20
0.5	54.12	46.39	48.27	49.52	50.58	49.15
0.6	84.08	72.13	75.15	77.27	78.79	76.28
0.7	129.37	111.49	116.70	119.78	122.08	116.80
0.8	208.70	179.53	188.95	194.80	197.70	186.96
0.9	388.79	336.96	357.86	368.96	374.53	340.62
0.95	652.69	573.47	610.73	625.14	635.92	548.87

To explore the reason, dependence among the tandem queues with SCV(1) and SCV(5) at $\rho = 0.8$ is analyzed. For the system with SCV(1), due to the diffusion effects, the reduction of variability at the first station not only reduces the queue time at station 1, but also reduces queue times at the downstream stations. However, in the system with SCV(5), the reduction of variability at the fifth station only reduces the queue time at station 5. As shown in Figures 3 and 4, the diffusion effect is less in the system with SCV(1), but is larger in the system with SCV(5). For example, the diffused queue time from station 1 to 2 is 2.39 under SCV(1), but it is 9.63 under SCV(5).

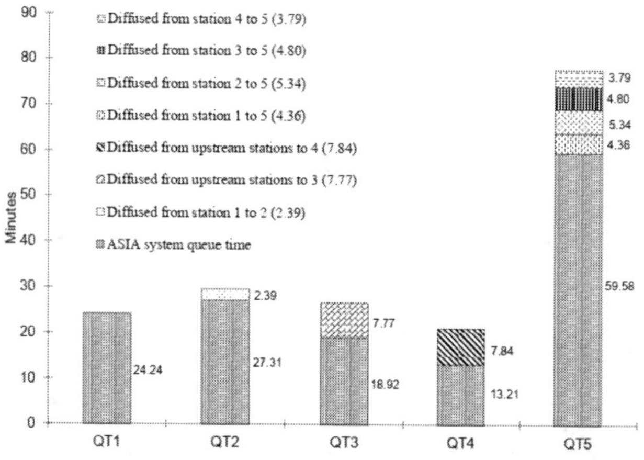

Figure 1: Queue time analysis for the tandem queue with SCV(1) and $\rho = 0.8$

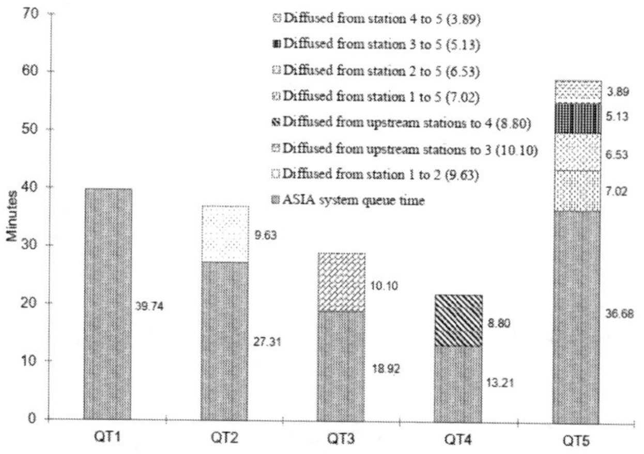

Figure 2: Queue time analysis for the tandem queue with SCV(5) and $\rho = 0.8$

However, at $\rho = 0.95$, the system mean queue time is the shortest if the variability of station 5 is reduced by 50%. This is because the queue time at station 5 is much longer than the others, and system queue time is dominated by the main system bottleneck in heavy traffic. Shortening system queue time by reducing the service time variability of station 5 becomes more effective in this situation.

Maintaining short cycle time plays a key role to achieve fast response to customer demand. To avoid long cycle time, machines are not favored to be operated in heavy traffic. In this situation, the most effective place to reduce system cycle time may not be the throughput bottleneck. Different from the claim of TOC, due to the dependence among stations, improving service time variability of a frontend machine in a production line can be more effective.

4 CONCLUSION

The theory of constraints has been advocated for decades and has significant contribution to the productivity improvement in practical manufacturing systems. While it is intuitive and easy to understand, its conclusions are mainly derived in deterministic settings or based on the first moment results. Since production systems are stochastic in general, some of its conclusions are not rigorous and have to be modified.

Due to the deviation of machine costs, the capacity of different machines has different value. The capacity of an expensive machine is more valuable and its lost should be minimized. In a stochastic production line, in order to ensure the expensive machine always have enough jobs to work on, the cheaper machines should prepare enough excess capacity to serve those expansive ones. Otherwise, those expensive machines may suffer from the unscheduled breakdowns of the cheaper machines. There should be a considerable capacity gap between the expansive machines and the cheaper ones. Hence, the throughput bottleneck should be planned based on machine costs rather than the POOGI result. As pointed out by Property 4, a production line has higher variability if the capacity gap is smaller. The key insight is if a system has no bottleneck, every station is a bottleneck.

Although the existence of dependence among stations is well recognized, it is difficult to be modeled exactly in general and independence is commonly assumed directly or indirectly in queueing models. In order to improve the performance evaluation of queueing networks, accurately capturing the dependence into the mathematical model is of critical importance. Through contribution factors, the dependence among single server queues in series is analyzed. Due to the dependence among stations, improving service time variability of a frontend machine in a production line can be more effective.

Through contribution factors and simulation results, we find the dependence among workstations can be either blocking or diffusion. However, the conditions for blocking and diffusion effects have not been fully understood. To understand the dependence better, rather than relying on simulation, computing intrinsic ratios analytically is necessary and left for future research

5 REFERENCES

[1] Goldratt, E. M. and Cox, J. 1984. *The goal: a process of ongoing improvement*, North River Press.

[2] Rahman, S. 1998. Theory of constraints: a review of the philosophy and its applications, *International Journal of Operations & Production Management*, vol. 18, pp. 336-355.

[3] Kingman, J. F. C. 1965. The heavy traffic approximation in the theory of queues, Proceeding of the Symposium on Congestion Theory, University of North Carolina at Chapel Hill, pp. 137-159.

[4] Wu, K. 2005. An examination of variability and its basic properties for a factory, *IEEE Transactions on Semiconductor Manufacturing*, vol. 18, pp. 214-221.

[5] Wu, K. and Zhao, N. 2015. Dependence among single stations in series and its applications in productivity improvement, *European Journal of Operational Research*, vol. 247, pp. 245-258.

[6] Lozinski, C. and Glassey, C. R. 1988. Bottleneck starvation indicators for shop floor control, *IEEE Transactions on Semiconductor Manufacturing*, vol. 1, pp. 147-153.

[7] Hopp, W. J. and Spearman, M. L. 1995. Factory physics, Irwin/McGraw-Hill Burr Ridge, IL.

[8] Wu, K. 2014. Classification of queueing models for a workstation with interruptions: a review, *International Journal of Production Research*, vol. 52, pp. 902-917.

[9] Wu, K. 2014. Taxonomy of batch queueing models in manufacturing systems, *European Journal of Operational Research*, vol. 237, pp. 129-135.

[10] Jacobs, J., Etman, L., Van Campen, E. and Rooda, J. 2003. Characterization of operational time variability using effective process times, *IEEE Transactions on Semiconductor Manufacturing*, vol. 16, pp. 511-520.

[11] Ohno, T. 1982. How the Toyota production system was created, *Japanese Economy*, vol. 10, pp. 83-101.

[12] Barney, M. 2002. Motorola's second generation, *Six Sigma Forum Magazine*, vol. 1, pp. 13-16.

[13] Kingman, J. F. C. 1962. Some inequalities for the queue GI/G/1, *Biometrika*, vol. 49, pp. 315-324.

[14] Friedman, H. D. 1965. Reduction methods for tandem queuing systems, *Operations Research*, vol. 13, pp. 121-131,.

[15] Wu, K., Zhou, Y., and Zhao, N. 2016. Variability and the fundamental properties of production lines, *Computers & Industrial Engineering*, to appear.

[16] Wu, K. and McGinnis, L. 2013. Interpolation approximations for queues in series, *IIE Transactions*, vol. 45, pp. 273-290,.

[17] Wu, K. and McGinnis, L. 2012. Performance evaluation for general queueing networks in manufacturing systems: Characterizing the trade-off between queue time and utilization, *European Journal of Operational Research*, vol. 221, pp. 328-339.

[18] Iglehart, D. L. and Whitt, W. 1970. Multiple channel queues in heavy traffic. II: Sequences, networks, and batches, *Advances in Applied Probability*, vol. 2, pp. 355-369.

[19] Sakasegawa, H. 1977. An approximation formula $L_q = \alpha \cdot \rho^{\beta} / (1-\rho)$, *Annals of the Institute of Statistical Mathematics*, vol. 29, pp. 67-75.

[20] Delp, D., Si, J., and Fowler, J. W. 2006. The development of the complete X-factor contribution measurement for improving cycle time and cycle time variability, *IEEE Transactions on Semiconductor Manufacturing*, vol. 19, pp. 352-362.

INTEGRATED SCHEDULING OF PRODUCTION AND RAIL TRANSPORTATION WITH COMPRESSIBLE PROCESS TIMES

M. Hajiaghaei-Keshteli[1*], M. Aminnayeri[2], R. Tavakkoli-Moghaddam[3]
[1]Department of Industrial Engineering, University of Science and Technology of Mazandaran, Behshahr, Iran
mostafahaji@mazust.ac.ir

[2]Department of Industrial Engineering and Management Systems, Amirkabir University of Technology, Tehran, Iran
mjnayeri@aut.ac.ir

[3]School of Industrial Engineering, College of Engineering, University of Tehran, Tehran, Iran
tavakoli@ut.ac.ir

ABSTRACT

This paper proposes a new mathematical model for a simultaneous production scheduling and rail transportation allocation in a supply chain. The aim is to determine both production schedule and rail transportation allocation of orders, along with deciding on processing speed mode, to optimize customer service at a minimum total cost. To the best of our knowledge, there is no research reported to date, developing and addressing the problem of integrated of production and rail transportation with compressible process times in production site. Among others, one of our realistic assumptions is dealing with flexible machines enabling the production manager to change and choose their working speeds. Besides, the developed model is linearized.

Keywords: Integrated Production-Distribution, Rail Transportation, Linearization

1 INTRODUCTION

Coordination of production and delivery schedules has been one of the most interesting issues in supply chain literature [6,7,25,26]. Traditionally, the relevant decisions are often made in a sequential manner with little or no integration where the production schedule is set first by the operations department, and then, the distribution schedule is set for finished products by the logistics department based on the production schedule [6]. Obviously, such a sequential approach is suboptimal. Chen and Vairaktarakis [7] and Pundoor and Chen [25] showed a significant benefit by using the optimal integrated production-distribution schedule in comparison to sequential approach in their models. Over 70% of the worldwide companies now rely on third party logistics (3PL) providers for their daily distribution and other logistics needs [16]. According to the Chen [6], as a recent survey in this research area, the rail transportation in the production and outbound distribution scheduling literature is categorized as the problems with fixed delivery departure dates (See also References [29,19,20, 26,13,14]).

Generally, most of the decisions in these areas are made with the hope of cost reduction and usually updated via a rolling horizon framework. Having different rates/modes in production site is ordinarily

* Corresponding Author.

observed in lots of real world shop floors and provides a degree of flexibility for a production manager, but it may also complicate determining the optimal policy [15]. However, this may lead to mixed integer programming (MIP) formulations that are hard to solve in reasonable time for sizes close to real world [24]. This paper tries to fill in the gap by providing a problem formulation along with linearization. In a nutshell, the main contributions of the current research may be developing a well-structured and linear model, contrary to previous related complicated models, that can be solved by commercial software in a short time.

2 LITERATURE REVIEW

Supply chain models focusing on integrating production and distribution operations are known in the recent literature under the name of integrated production-distribution problems [7,25,27].The authors refer the readers to review the related papers in integrated analysis of production-distribution systems [1,2,5,6,7,9,10,12,23,28]. Research on integrated scheduling models of production and outbound distribution is almost recent but is growing very rapidly [6]. There have been considerable researches conducted in production-transportation integration with emphasis on the road transportation and vehicle routing problem [4,5,17,30,30].

There are a few researches on production scheduling considering vehicles with fixed departure dates. Li *et al.* [18,19,20] studied the synchronization of single machine scheduling and air transportation with single destination. Li *et al.* [21] extended their work by considering parallel machines in production and using simulated annealing (SA) based heuristic algorithm to solve the problem. Besides, Zandieh and Molla-Alizadeh-Zavardehi [30, 31] and Rostamian Delavar *et al.* [26] proposed some mathematical models for scheduling of production and air transportation and utilized metaheuristics to solve their problem. Recently, Hajiaghaei-Keshteli *et al.*, [13] firstly developed an integrated production and transportation model, which considers rail transportation and delivers the orders from a facility to the customers. So, from the literature, it can be concluded that this research area is an important topic both in academia and industry.

Three sections follow this Introduction. The next section describes the mathematical model. In the section 4, the proposed model is linearized. Finally, conclusions are provided in section 5.

3 MATHEMATICAL MODEL AND DESCRIPTIONS

Here, initially we develop a new mathematical model for the integrated production and rail transportation scheduling problem considering compressible process times. The notations that will be used to describe the problem and in the rest of this paper, are as follows:

Indices

i, i'	order / job index, $i, i' = 1,2,...,N$;
t, k	ordinary train index, $t = 1,2,...,T$; order destination index, $k = 1,2,...,K$;
L, z	pass (train destination) index, $l = 1,2,...,L_i$; machine's speed $Z=1,...,\mu$.
j, j'	position or sequence of order i $j, j' = 1,...,N$;

Parameters

D_t	departure time of ordinary train **t** at the local rail station; without loss of generality, it is assumed that $D_1 \le D_2 \le ... \le D_T$
A_{it}	arrival time of ordinary train **t** at order *i*'s destination;
α_{1i} / α_{2i}	departure time earliness/delivery earliness penalty cost (/unit/h) of order i;
β_i	delivery tardiness penalty cost (/unit/h) of order i;
π_{it}	earliness/ tardiness penalty cost (/unit/h) of order i transported by train **t**;

Q_i, d_i	quantity of order i; due date of order i;
Cap_t / Tc_i	capacity of train t; transportation cost of order i;
Tc_{it}	transportation cost of order i when allocated to train t;
Des_{it}	0 if order i can be transferred by train t, $\lambda \neq 0$ otherwise;
des_t	destination of train t: the train's final destination (number);
des_{il}	the final destination of lth pass which an order i
L_i	maximum number of rail passes which an order can be transported
c_{iz} / p_{iz}	processing cost/time of order i when the machine works at its zth mode (speed level)
$c_i^{\max}$, $c_i^{\min}$	Maximum/ minimum processing cost of order i on the machine;
μ / ξ	number of modes/speed of the machine/the per hour plant costs

Constants

LN / θ	a large positive number; a real digit between 0 and 1;

Variables

x_{it}	quantity of portion of order i allocated ordinary train t;
p_i / c_i	processing time of order i (/unit) ; completion time of order i;
I_i	the idle time before order i in the production schedule
$c_{\max}$ / C_j	maximum completion time (CT) of orders; CT of order allocated to position j;
pc_{iz}	processing cost of the machine working at its mode z on order i;
u_{ij} / x'_{it}	1 if order i is in position j, 0 otherwise; 1 if x_{it}>0, 0 otherwise;
v_{iz}	1 if order i is processed on speed position z, 0 otherwise;
P_j	processing time of order allocated to position j;

With respect to the problem defined above and mentioned assumptions, the mathematical model is formulated as follows:

$$\text{Min} \sum_{i=1}^{N}\sum_{t=1}^{T}\Big((Tc_i \times x_{it}) + (\alpha_{1i} \times (D_t - c_i) \times x_{it}) + (\alpha_{2i} \times \max(0, d_i - A_{it}) \times x_{it}) + (\beta_i \times \max(0, A_{it} - d_i) \times x_{it})\Big) + \sum_{i=1}^{N}\sum_{z=1}^{\mu} pc_{iz} + \xi \times c_{\max} \quad (1)$$

s.t.

$$x_{it} \times \prod_{l=1}^{L_i}(des_{il} - des_t) = 0 \qquad i = 1,\dots,N; t = 1,\dots,T \quad (2)$$

$$\sum_{i=1}^{N} x_{it} \le Cap_t \qquad t = 1,\dots,T \quad (3)$$

$$\sum_{t=1}^{T} x_{it} = Q_i \qquad i = 1,\dots,N \quad (4)$$

$$\sum_{j=1}^{N} u_{ij} = 1 \qquad i = 1,\dots,N \quad (5)$$

$$\sum_{i=1}^{N} u_{ij} = 1 \qquad j = 1,\dots,N \tag{6}$$

$$\sum_{z=1}^{\mu} v_{iz} = 1 \qquad i = 1,\dots,N \tag{7}$$

$$p_i = \sum_{z=1}^{\mu} \upsilon_{iz} p_{iz} Q_i \qquad i = 1,2,\dots,N \tag{8}$$

$$\sum_{j=1}^{N} \left(u_{ij} \left(p_i Q_i + I_i + \sum_{j'=1}^{j-1} \sum_{i'=1}^{N} u_{i'j'} (p_{i'} Q_{i'} + I_{i'}) \right) \right) = c_i \qquad i = 1,\dots,N \tag{9}$$

$$c_{\max} = \sum_{i=1}^{N} \sum_{j=1}^{N} u_{ij} (p_i Q_i + I_i) \tag{10}$$

$$pc_{iz} \geq Q_i c_{iz} - \left(c_i^{mac} Q_i\right)\left(1 - v_{iz}\right) \quad i = 1,\dots,N; z = 1,\dots,\mu \tag{11}$$

$$c_i \leq \left(\min_t \left(\frac{D_t}{\left(\frac{\max_t\left(0, x_{it} - \theta\right)}{x_{it} - \theta} \right) + \frac{1}{LN}} \right) \left(1 + \frac{1}{LN}\right) \right) \qquad i = 1,2,\dots,N \tag{12}$$

$$u_{ij} \in \{0,1\} \qquad i = 1,\dots,N; j = 1,\dots,N \tag{13}$$

$$v_{iz} \in \{0,1\} \qquad i = 1,\dots,N; z = 1,\dots,\mu \tag{14}$$

$$x_{it} = \textit{Non–negative integer variable} \tag{15}$$

Constraint (2) ensures that order i can be allocated to trains which travel to or pass from the order's destination. Constraint (3) states that the capacity of train t is not exceeded. Constraint (4) ensures that order i is completely allocated. Constraints (5) and (6) state that each job has to be assigned to a position and each position has to be covered by a job. Constraint (7) state that each order is processed on the machine in one speed mode. Constraint (8) determine the processing time of each order. Constraint (9) calculates completion time of the jobs in the facility. Constraint (10) determines the C_{max} and also Constraint (11) evaluates the processing cost of order i at z^{th} speed level. Constraint (12) states that the completion time of each order i, should be less than or equal to the departure time of all their allocated trains.

4 MODEL LINEARIZATION

Obviously, the proposed model is a nonlinear mathematical model. An attempt is made in this part to linearize the objective function and nonlinear constraints via introducing some auxiliary variables and consequently, adding some additional constraints.

4.1 Focusing on the Objective Function

Here, we focus on each part of the objective function:

Proposition 1. The expression $\sum_{i=I}^{N} \sum_{t=1}^{T} (\alpha_{1i} \times (D_t - c_i) \times x_{it})$ in the objective function is linear.

Proof. the obtained expression is linear:

$$\sum_{i=1}^{N}\sum_{t=1}^{T}(\alpha_{1i}\times(D_t-c_i)\times x_{it})=\sum_{i=1}^{N}\sum_{t=1}^{T}(\alpha_{1i}\times D_t\times x_{it})-\sum_{i=1}^{N}\sum_{t=1}^{T}(\alpha_{1i}\times c_i\times x_{it})=\sum_{i=1}^{N}\sum_{t=1}^{T}(\alpha_{1i}\times D_t\times x_{it})-\sum_{i=1}^{N}(\alpha_{1i}\times c_i\times Q_i).$$

Proposition 2. The expression $(\alpha_{2i}\times \max(0,d_i-A_{it})\times x_{it})+(\beta_i\times \max(0,A_{it}-d_i)\times x_{it})$ in the objective function is altered to $\pi_{it}\times x_{it}$ and consequently, it will be linear.

Proof. The claim is on hand by introducing the following parameters; E_{it} and T_{it}:

$$(\alpha_{2i}\times E_{it}\times x_{it})+(\beta_i\times T_{it}\times x_{it}) \tag{16}$$

Due to the following relation between delivery earliness and tardiness of orders:

$$E_{it}\times T_{it}=0 \tag{17}$$

We can simplify the expression (16) into the following:

$$\pi_{it}\times x_{it} \tag{18}$$

where $\pi_{it}=(\alpha_{2i}\times E_{it})+(\beta_i\times T_{it})$.□

Corollary 1. The following obtained objective function is linear.

$$\min \sum_{i=1}^{N}\sum_{t=1}^{T}((\alpha_{1i}\times D_t+Tc_i+\pi_{it})\times(x_{it}))-\sum_{i=1}^{N}(\alpha_{1i}\times c_i\times Q_i)+\sum_{i=1}^{N}\sum_{z=1}^{\mu}pc_{iz}+\xi\times c_{\max}. \tag{19}$$

4.2 Focusing on the Constraints

To detail more on constraint (2), here, we show this constraint by another notation and concept. The parameter Des_{it} is defined to show that the orders can be allocated to the trains which transfer them to their corresponding destination. If an order can be transported by a train, Des_{it} gains zero value and otherwise it gains a nonzero value (λ) in order to ensure that the order cannot be allocated to a train which does not transfer the order to its corresponding destination. So, the following linear constraint (20) can be used instead of constraint (2):

$$x_{it}\times Des_{it}=0 \qquad i=1,\dots,N; t=1,\dots,T \tag{20}$$

Proposition 3. The constraint (2) or (20), can be removed by replacing Tc_{it} instead of Tc_i.

Proof. By introducing Tc_{it}, the transportation cost is allocated to the trains which can transfer the orders to their destinations, and also, by introducing large number, *LN*, transferring the orders are forbidden for the trains which do not go to those orders' destinations.

Proposition 4. Linear form of constraint (9) is the simultaneous use of the following constraints:

$$P_j\geq p_i-LN(1-u_{ij}) \qquad i=1,2,\dots,N, j=1,2,\dots,N \tag{21}$$

$$C_j=\sum_{j'=1}^{j}P_{j'} \qquad j=1,2,\dots,N. \tag{22}$$

$$c_i\geq C_j-LN(1-u_{ij}) \qquad i=1,2,\dots,N, j=1,2,\dots,N. \tag{23}$$

Proof. The claim is on hand by using C_j and P_j.

Proposition 5. The constraint (10) can be removed by using c_N instead of c_{max}.

Proof. Since $c_N = c_{\max}$, the constraint (10) can be removed and replaced with c_N.

The constraint (12) also can be rewritten to the following constraint (24):

$$c_i \le \left(\frac{D_t}{\left(\frac{\max(0, x_{it} - \theta)}{x_{it} - \theta} \right) + \frac{1}{LN}} \right) \left(1 + \frac{1}{LN} \right) \qquad i = 1,2,\dots,N, t = 1,2,\dots,T. \tag{24}$$

The constraint (24) checks all the trains for all orders.

Besides, we can use constraint (12) or (24) instead of constraint sets (25) or (26).

$$x_{it} \times c_i \le x_{it} \times D_t \qquad i = 1,2,\dots,N, t = 1,2,\dots,T. \tag{25}$$

$$x_{it} \times (c_i - D_t) \le 0 \qquad i = 1,2,\dots,N, t = 1,2,\dots,T. \tag{26}$$

The constraints 12, 24, 25 and 26 act the same and all of them are nonlinear.

Proposition 6. Linear form of constraints 12, 24, 25 or 26 is the simultaneous use of the following constraints:

$$D_t - c_i \ge LN(x'_{it} - 1) \qquad i = 1,2,\dots,N, t = 1,2,\dots,T. \tag{27}$$

$$x_{it} \le x'_{it} Q_i \qquad i = 1,2,\dots,N, t = 1,2,\dots,T. \tag{28}$$

Proof. The claim is on hand by using a new auxiliary and binary variable x'_{it}.

Corollary 2. The whole model is altered to the linear

5 CONCLUSION

This paper formulated a new mathematical model for a simultaneous production scheduling and transportation allocation in a supply chain, by considering several processing speed mode, to optimize customer service at a minimum total cost. The model is linearized by adding a new set of constraints to the previous basic models about rail transportation and compressible processing times. There are potentially unlimited research opportunities for in this area. For future researches, it is possible to add new assumptions to the model and also develop solution approaches such as decomposition, Lagrangean relaxation, and/or propose heuristics or metaheuristics.

REFERENCES

[1] **Armentano, V.A., Shiguemoto, A.L., Løkketangen, A., 2011,** Tabu search with path relinking for an integrated production-distribution problem, Computers and Operations Research, 38, 1199-1209.

[2] **Ashoka Varthanan, P., Murugan, N., and Mohan Kumar, G., 2012,** A simulation based heuristic discrete particle swarm algorithm for generating integrated production-distribution plan, Applied Soft Computing, 12, 3034-3050.

[3] **Calvete, H., I., Gale, C., Oliveros, M., 2011,** Bilevel model for production-distribution planning solved by using ant colony optimization, Computers and Operations Research, 38, 320-327.

[4] **Chang, Y.C., and Lee, C.Y., 2004,** Machine scheduling with job delivery coordination. European Journal of Operational Research, 158, 470-487.

[5] **Chen, B., Lee, L., 2008**, Logistics scheduling with batching and transportation, European Journal of Operational Research, 189, 871-876.

[6] **Chen, Z., 2010,** Integrated Production and Outbound Distribution Scheduling: Review and Extensions, Operations Research, 58, 1, 130-148.

[7] **Chen, Z. -L., and Vairaktarakis, G.L., 2005,** Integrated scheduling of production and distribution operations, Management Science, 51(4), 614-628.

[8] **Condotta, A., Knust, S., Meier, D., Shakhlevich, N. V., 2013,** Tabu search and lower bounds for a combined production-transportation problem, Computers and Operations Research, 40, 886-900.

[9] **De Matta, R., Miller, T., 2004,** Production and inter-facility transportation scheduling for a process industry, European Journal of Operational Research, 158, 72-88.

[10] **Erenguc, S.S., Simpson, N.C. and Vakharia, A.J., 1999,** Integrated production / distribution planning in supply chain: An invited review, European Journal of Operational Research, 115, 219-236.

[11] **Fahimnia, B., Zanjirani Farahani, R., Marian, R., Luong, L., 2013,** A review and critique on integrated production distribution planning models and techniques, Journal of Manufacturing Systems, 32, 1- 19.

[12] **Goetschalckx, M., Vidal, C.J. and Dogan, K., 2002,** Modeling and design of global logistics systems: A review of integrated strategic and tactical models and design algorithms. European Journal of Operational Research, 143, 1-18.

[13] **Hajiaghaei-Keshteli, M., Aminnayeri, M., Fatemi Ghomi, S. M. T., 2014,** Integrated Scheduling of Production and Rail Transportation, Computers and Industrial Engineering, 74, 240-256.

[14] **Hajiaghaei-Keshteli, M., Aminnayeri, M., 2014,** Solving the Integrated Scheduling of Production and Rail Transportation Problem by Keshtel Algorithm, Applied Soft Computing, 25, 184-203.

[15] **Karimi-Nasab, M., and Seyedhoseini, S.M., 2013,** Multi-level lot sizing and job shop scheduling with compressible process times: A cutting plane approach, European Journal of Operational Research, 231, 598-616.

[16] **Langley, C. J., E. van Dort, and S. R. Sykes., 2006,** 2005 Third-Party Logistics: Results and Findings of the 10th Annual Study.

[17] **Lee, C.Y., and Chen, Z.L., 2001,** Machine scheduling with transportation considerations. Journal of Scheduling, 4, 3-24.

[18] **Li, K.P., Sivakumar, A.I., Mathirajan, M. and Ganesan, V.K., 2004,** Solution methodology for synchronizing assembly manufacturing and air transportation of consumer electronics supply chain, International Journal of Business, 9, 361-380.

[19] **Li, K.P., Ganesan, V.K. and Sivakumar, A.I., 2005,** Synchronized scheduling of assembly and multi-destination air transportation in a consumer electronics supply chain. International Journal of Production Research, 43, 2671-2685.

[20] **Li, K.P., Ganesan, V.K. and Sivakumar, A.I., 2006,** Methodologies for synchronized scheduling of assembly and air transportation in a consumer electronic supply chain. International Journal of Logistic System and Management, 2, 52-67.

[21] **Li, K.P., Sivakumar, A.I. and Ganesan, V.K., 2008,** Complexities and algorithms for synchronized scheduling of parallel machine assembly and air transportation in consumer electronic supply chain. European Journal of Operational Research, 187, 442-455.

[22] **Li, C., Vairaktarakis, G., Lee, C., 2005,** Machine scheduling with deliveries to multiple customer locations, European Journal of Operational Research, 164, 39-51.

[23] **Melo, R. A., and Wolsey, L. A., 2010,** Optimizing production and transportation in a commit-to-delivery business mode, European Journal of Operational Research, 203, 614-618.

[24] **Pochet, Y., and Wolsey, L. A., 2006,** Production planning by mixed integer programming, New York: Springer.

[25] **Pundoor, G., and Chen, Z.-L. 2005,** Scheduling a production-distribution system to optimize the tradeoff between delivery tardiness and total distribution cost, Naval Research Logistics, 52, 571-589.

[26] **Rostamian Delavar, M., Hajiaghaei-Keshteli, M., Molla-Alizadeh-Zavardehi, S., 2010,** Genetic algorithms for coordinated scheduling of production and air transportation, Expert Systems with Applications, 37, 8255-8266.

[27] **Russell, M., Chiang, W., and Zepeda, D., 2008,** Integrating multi-product production and distribution in newspaper logistics, Computers and Operations Research, 35, 1576-1588.

[28] **Sarmiento, A.M. and Nagi, R., 1999,** A review of integrated analysis of production-distribution systems. IIE Transaction, 31, 1061-1074.

[29] **Stecke, K. E., Zhao., X., 2007,** Production and transportation integration for a make-to-order manufacturing company with a commit-to delivery business mode, Manufacturing Service Operation Management, 9, 206-224.

[30] **Zandieh, M., and Molla-Alizadeh-Zavardehi, S., 2008,** Synchronized production and distribution scheduling with due window, Journal of Applied Sciences, 8, 2752-2757.

[31] **Zandieh, M., and Molla-Alizadeh-Zavardehi, S., 2009,** Synchronizing production and air transportation scheduling using mathematical programming models, Journal of Computational and Applied Mathematics, 230, 546-558.

A SIMULATION METHOD WITH VIRTUAL 4D REALITY FOR A HIGH-TECH INDUSTRIAL FAB CONSTRUCTION

Teng-Sheng Su[1*] and Luh-Maan Chang[2]
[1]Department of Marketing and Logistics Management
Chaoyang University of Technology, Taichung City 41349, Taiwan
tengshengsu@gmail.com

[2]High-Tech Facility Research Center, Department of Civil Engineering
National Taiwan University, Taipei City, 10668, Taiwan
luhchang@ntu.edu.tw

ABSTRACT

The paper provides a simulation method that makes high-tech industrial fab construction plan more efficient and precise in the supervision of cost, resources, and construction duration to minimize the risk of space interface conflict. The technique of an entity-driven, a basic concept in a manufacturing production line to assign each entity its attribute, is proposed to regard every objective in which the construction activities develop as an entity throughout the construction process. Accompanying a process simulation computing software, each time slot of the construction activity can be recorded in accordance with the construction process flow. After the computing is completed, the scheduling information of each construction activity can be exported explicitly. Combined into a three-dimensional (3D) model with building information modeling (BIM), a four-dimensional (4D) plant construction model with virtual reality (VR) is developed to project the resource usage efficiency over the construction duration accurately. The performance of diverse range of construction projects in each level can be compare to each other by the 4D model to choose a satisfactory one. Finally, it is our hope that a high-tech industrial fab construction plan can be achieved and profitable according to the technique proposed in this study.

Keywords: Entity-driven simulation, BIM, Virtual 4D reality

1 INTRODUCTION

In the face of stiff competition from the high-tech industries, production cycle time plays one of the key factors to succeed. A fab construction completed earlier can lead the products that into manufacturing process as quickly as possible to remain the advantage in markets. How to reduce the construction duration of a new fab has been one of imperative topics. The construction duration in a traditional construction plan is projected by each fixed activities to estimate the construction duration of a sub-structure, setup time of microelectromechanical systems (MEMS), move-in time of the machinery and equipment. However, it is difficult to perceive the potential space conflict in in a real construction project by such method. Apart from a persuasive scheduling model that evaluates the performance of the personnel, machines, and materials, one may suffer from a problem to justify which construction project is more satisfactory in the early planning phase.

* Corresponding Author

In order to solve this problem, a construction simulation model which takes all resources of the personnel, machines, and materials into consideration is proposed in this paper. The simulation method makes high-tech industrial fab construction plan more efficient and precise in the supervision of cost, resources, and construction duration to minimize the risk of space interface conflict. The technique of an entity-driven, a basic concept in a manufacturing production line to assign each entity its attribute, is proposed to regard every objective in which the construction activities develop as an entity throughout the construction process.

Accompanying a process simulation computing software, each time slot of the construction activity can be recorded in accordance with the construction process flow. After the computing is completed, the scheduling information of each construction activity can be exported explicitly. Combined into a three-dimensional (3D) model with building information modeling (BIM), a four-dimensional (4D) plant construction model with virtual reality (VR) is developed to project the resource usage efficiency, examine the space interface conflict, and even correct the mistakes of the scheduling logic over the construction duration accurately. The performance of diverse range of construction projects in each level can be compare to each other by the 4D model to choose a satisfactory one. Finally, it is our hope that a high-tech industrial fab construction plan can be achieved and profitable according to the technique proposed in this study.

The remainder of this paper is organized as follows. In section 2, a review of the literature pertinent to this design problem is conducted. In section 3, the design methodology we proposed is explained. An example is given in section 4 to illustrate the proposed design method. Finally, section 5 concludes and summarizes the results of this study.

2 LITERATURE REVIEW

Lu and Chan [1] introduced an enhanced version of the Simplified Discrete Event Simulation Approach (SDESA) by incorporating a concurrent interruptions model. The authors intended to make construction simulations more realistic without compromising the simplicity of the original SDESA. A construction model was validated by the well-established CYCLONE method to check the check the results given by SDESA. Chan and Lu [2] presented a university-industry joint endeavor for improving the effectiveness of the materials handling system on a precast viaduct construction project in Hong Kong by implementing the simplified discrete-event simulation approach (SDESA). The results showed that knowledge derived from simulation can add to experiences of site managers in materials handling system design. With the aid of simulation, even junior engineers would be capable to draw up an actionable construction plan that leads to enhancement of cost effectiveness and productivity. Zhang *et al.* [3] proposed a simulation-based methodology to handle the time constraints including the cyclical break, preemption, and overtime use. An algorithm to determine the execution of the time-constrained activities and the concepts of time cycle and time window were also introduced by them. Some examples illustrated and validate the algorithm, and highlighted the effectiveness of the developed construction simulation.

Song and Eldin [4] proposed an adaptive real-time tracking and simulation of heavy construction operations for look-ahead scheduling during construction field operations. They used tacking sensors to capture construction operation data and the data was then fed into a simulation model for automatic model updating. Compared to the traditional offline simulation, the capability of the proposed real-time simulation can improve the accuracy of project look-ahead scheduling. Akhavian and Behzadan [5] described a methodology for collecting and mining of spatio-temporal data corresponding to the interactions of queue entities in a construction simulation. They argued that there are many situations in which formation of waiting lines or queues is inevitable, the queuing systems simulation for analysis of queuing systems is employed to identify the best operational strategies to reduce the time wasted in queues. Zhang *et al.* [6] presentd a combination of the fuzzy set theory and a fuzzy ranking measure with discrete-event simulation to model uncertain activity duration in simulating a real-world system. The uses of the fuzzy activity duration and the probability distribution-modeled duration were compared through a series of simulation experiments in authors' study. The results showed that the fuzzy simulation outputs are arrived at through only one cycle of fuzzy discrete-event simulation. Sadeghi *et al.* [7] provided a methodology to consider subjective uncertainty in analyzing the fuzzy queues in construction fuzzy discrete event simulation (FDES) models. They argued that FDES provides a framework to consider subjective uncertainty in construction simulation models, and the project complete time is only considered. To enhance the applicability of FDES in construction projects, they further incorporated the fuzzy queuing theory with FDES to illustrate practical aspects of proposed methodology. Cheng *et al.* [8] presented the information constraint net (ICN) to represent the complex information constraint relations among design activities involved in the building design process. They developed an algorithm to transform the information constraints throughout the ICN which simulates and optimizes resource allocation into a Petri net model. The experiment demonstrated that approach can shorten the development cycles and optimal allocation of resources. Koo and Fischer [9] proposed 4D models that are a useful alternative to project scheduling tools like CPM, bar chart, and arrow diagram. The authors argued that a 4D model can detect the incompleteness of the original schedule, find inconsistencies in the level of detail among the schedule activities, discover an impossible schedule sequence, and anticipate potential time-space conflicts and accessibility problems. A case study showed that 4D models are effective in evaluating the executability of a construction schedule.

3 THE PROPOSED METHODOLOGY

The aim of this study is to propose a methodology that the scheduling simulation and analysis of the precast and cast-in-place construction of a sub-structure for a high-tech industrial fab construction was implemented under various routes and criteria. Due to the various aspects of activities between the precast and cast-in-place construction, lots of space conflict will highly take place during the real construction period. Given the reason to avoid the efficiency reduction from the construction conflict, a well-prepared plan becomes more substantial.

Figure 1 gives an overview of the proposed construction method. As shown in Figure 1, the proposed construction method has two steps - construction simulation model and virtual 4D reality integration. The purpose of the construction simulation is to provide a model that can simulate appropriate construction logic of the precast and cast-in-place activities to reflect conditions in a real construction project. To achieve this purpose, the project engineer is required to identify in-&out transportation routes and attributes for vehicles, materials and equipment in precast and cast-in-place construction. Then a construction model can be developed and programmed by a process simulation computing software, Arena® [10][11]. After conducting the simulation experiment, each time slot of the construction activity can be recorded as the schedule information that will be an input to virtual 4D reality integration. The purpose of the virtual 4D reality integration is to create a model that can inspect the visible space interface conflict. To achieve this purpose, a 3D model is built up by a kind of 3D virtualization software to find the space and location requirements in the sub-structure and temporary work. The objects of the 3D model are brought about, piled up, and assigned with attribute according to the building information that the construction project defined. Combined into the 3D model with scheduling information obtained by simulation in Step 1 of Figure 1, a 4D plant construction model is developed by virtual reality technique to project the resource usage efficiency, examine the space interface conflict, and even correct the mistakes of the scheduling logic over the construction duration accurately.

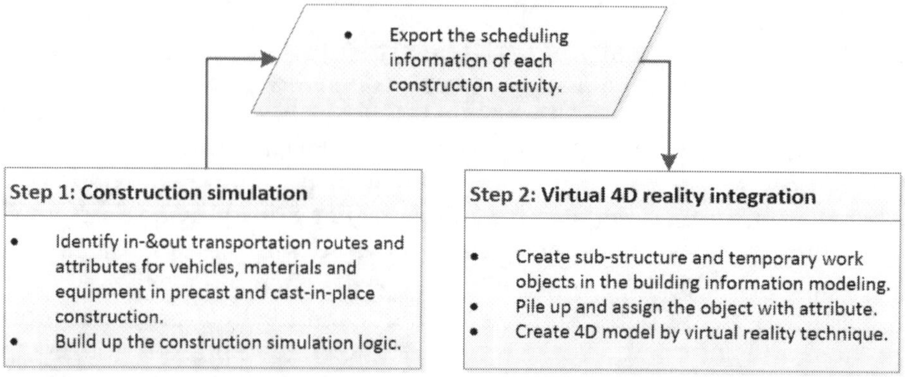

Figure 1: The flow chart of the proposed construction method

3.1 The study scope of the fab construction

A semiconductor fab construction of the high-tech industry can be categorized into three projects - sub-structure, super-structure, and cleanroom. The study scope in this paper will focus on the sub-structure that accounts for the majority of the construction duration. Figure 2 shows the layout of a semiconductor fab which contains a precast and cast-in-place construction area.

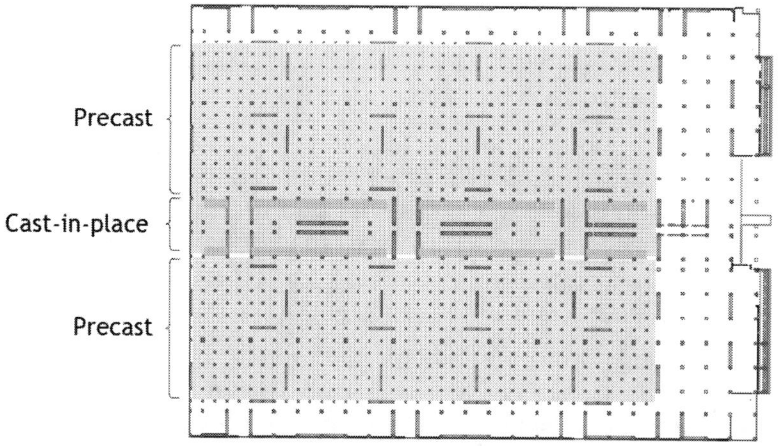

Figure 2: The layout of a precast and cast-in-place construction for a semiconductor fab

3.2 The construction simulation

Given various parameters in terms of construction items, resources, sequences, and priorities, two models of the construction process simulation are proposed to stand for the precast and cast-in-place activities respectively. If the construction activities at the operations specifically present the visualization of objects, attributes, and the logic over the construction duration, the potential problems are easily identified in the processes simulated by the process simulation method. Therefore, we convert some construction elements into entity units used for driving the construction process, assign each entity its attribute that presents meaningful criteria for construction, and implement the simulation to record the experiment result as the schedule information.

3.2.1 The model in the precast construction area

The object in the precast activities is comprised of columns, beams and slabs. The beams can be divided into big and small ones. The hoisting logic for the object is formulated in accordance with the of construction convenience which takes into account the space factor in the hoisting activity. The hoisting activities consider some factors involving established sequence, quantity limitation, and hoisting behaviour to give priority to columns, beams and slabs and. For example, the priority in order will be column - big beam with x axis - big beam with y axis - small beam - slab based on the established sequence, one beam is supported by two columns based on the quantity limitation, and objects will be hoisted far away from the hoisting crane rather than close to it based on the hoisting behaviour to avoid the interruption of the crane boom movement.

Figure 3 demonstrates an entity unit in the precast construction field. An entity unit grouped together by dotted line is formed from amount of columns, beams and slabs. The process sequence by hoisting crane for the entity is labelled from one to nine. The process completion time of each entity is calculated on the total number of the columns, beams and slabs the hoisting crane operated expressed by Equation (1). Figure 4 shows the simulation model of a precast construction process in which an entity stands for precast units to be created and disposed throughout the construction duration.

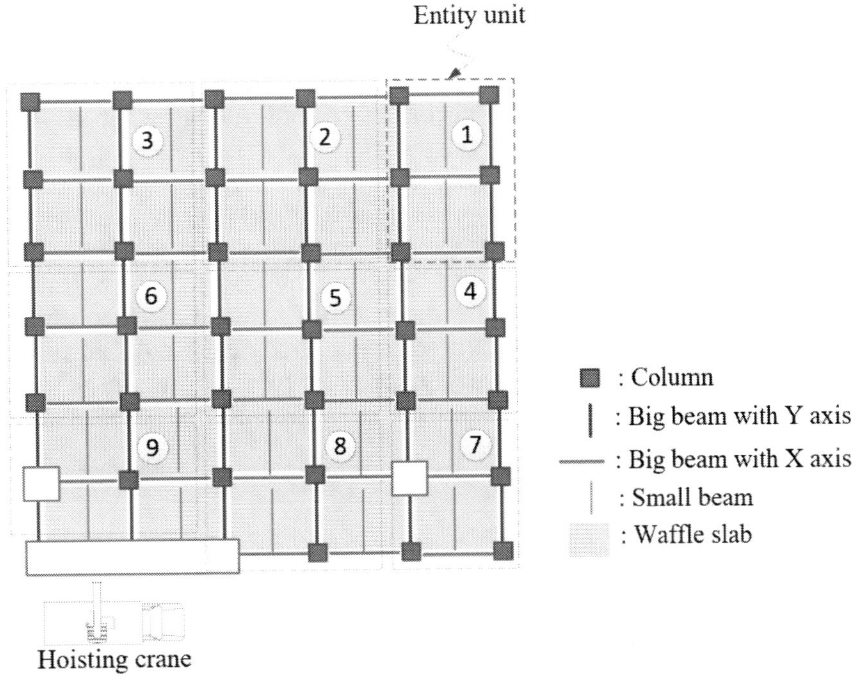

Figure 3: The demonstration of an entity unit in the precast construction field

$$PT = \sum_{i=1}^{m} \left(P_i + W_{ij}\right) \quad \forall j \neq i \tag{1}$$

where

PT = the total completion time of an entity in the precast,

P_i = the process time of the ith object,

W_{ij} = the waiting time of the ith object for the jth object the hoisting crane been operating, and

m = the number of the sets of objects

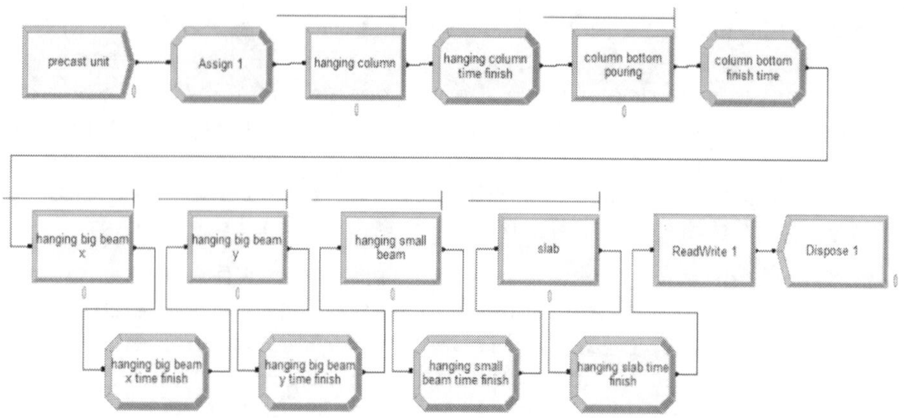

Figure 4: The simulation model of a precast construction process

3.2.2 The model in the cast-in-place construction area

Unlike the construction logic in the precast activities, the object in the cast-in-place only contains columns and shear walls. The logic for the object is formulated associated with the construction size of area divided into distinct types. The construction activities relied on an enormous and skilled labour supply at the operations abide by a process in a fixed sequence by loft, prolong, scaffolding, stirrups, template, pouring, remove template, remove scaffold, heavy bracket, slab template, slab stirrups, and slab pouring. Figure 5 demonstrates an entity unit for dissimilar types in the cast-in-place construction field. An entity unit grouped together by dotted line is formed from amount of columns and shear walls and characterized by three types in the both top and bottom half of the cast-in-place area. The process completion time of each entity is calculated on the total number of the columns and shear walls the labour operated expressed by Equation (2).

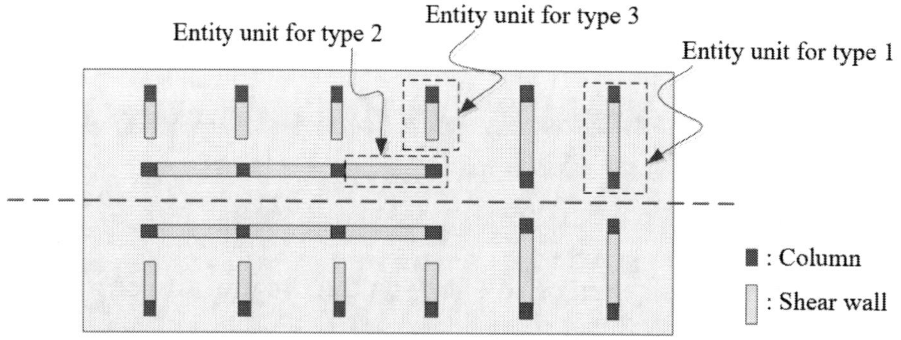

Figure 5: The demonstration of an entity unit in the cast-in-place construction field

$$CT = \sum_{i=1}^{m} \left(P_i + W_{ij}\right) \ \forall j \neq i \qquad (2)$$

where

CT= the total completion time of an entity in the cast-in-place,
P_i = the process time of the ith object,
W_{ij} = the waiting time of the ith object for the jth object the labour been operating, and
m = the number of the sets of objects

3.3 The virtual 4D reality integration

After conducting the simulation experiment, each time slot of the construction activity of precast and cast-in-place is recorded as the schedule information used for integrating into a 3D model. Therefore, the 3D model are brought about, piled up, and assigned with attribute by a kind of 3D virtualization software according to the building information that the construction project defined. Figure 6 shows a 3D model of hoisting crane and objects in precast and the virtual 4D reality integration result in the sub-structure of a semiconductor fab. While the 4D plant construction model is developed, one can project the resource usage efficiency, examine the space interface conflict, and even correct the mistakes of the scheduling logic over the construction duration accurately.

Figure 6: The demonstration of a 3D model of hoisting crane and objects in precast and the virtual 4D reality integration result in the sub-structure of a semiconductor fab

4 EXPERIMENTAL EXAMPLE

According to the construction activities of the cast-in-place in fixed sequence, the completion time for each of them can be obtained on a daily basis. Table 1 lists the results of the construction simulation experiment for the twelve activities in the both top and bottom half of the cast-in-place area. Except for the completion time, there are only five activities which correspond to the 3D model built up so that the level of detail in the virtual 4D reality integration is restricted.

Table 1: Results of the construction simulation experiment

Activity	Completion time of bottom half	Completion time of top half	3D model mapping
Loft	15th, July	15th, July	
Prolong	15th, July	16th, July	
Scaffolding	17th, July	19th, July	Scaffold
Stirrups	20th, July	23th, July	Steel reinforcing bar
Template	24th, July	28th, July	
Pouring	25th, July	29th, July	
Remove template	27th, July	31th, July	Column , wall
Remove scaffold	31th, July	4th, August	
Heavy bracket	4th, August	8th, August	Heavy bracket
Slab template	7th, August	11th, August	
Slab stirrups	8th, August	12th, August	Beam, slab, and Steel reinforcing bar
Slab pouring	10th, August	14th, August	

5 CONCLUSION

In this paper, a simulation method with virtual 4D reality that takes all resources of the personnel, machines, and materials into consideration to make high-tech industrial fab construction plan more efficient and precise in the supervision of cost, resources, and construction duration to minimize the risk of space interface conflict is proposed. The proposed method is capable of justifying more objectively which construction project is satisfactory in the early planning phase.

6 REFERENCES

[1] **Lu, M., and Chan, W.H. 2004.** Modeling concurrent operational interruptions in construction activities with simplified discrete event simulation approach (SDESA), Proceedings of the 2004 Winter Simulation Conference, Dec. 5-8, pp 1260-1267.

[2] **Chan, W.H., and Lu, M. 2008.** Materials Handling System Simulation in Precast Viaduct Construction: Modeling, Analysis, and Implementation, *Journal of Construction Engineering and Management*, 134(4), pp 300-310.

[3] **Zhang, H., Li, H., and Lu, M. 2008.** Modeling Time-Constraints in Construction Operations through Simulation, *Journal of Construction Engineering and Management*, 134(7), pp 545-554.

[4] **Song, L., and Eldin, N.N. 2012.** Adaptive real-time tracking and simulation of heavy construction operations for look-ahead scheduling, *Automation in Construction*, 27, pp 32-39.

[5] **Akhavian, R., and Behzadan, A.H. 2014.** Evaluation of queuing systems for knowledge-based simulation of construction processes. *Automation in Construction*, 47, pp 37-49

[6] **Zhang, H., Tam, C.M., and Li, H. 2005.** Modeling uncertain activity duration by fuzzy number and discrete-event simulation, *European Journal of Operational Research*, 164, pp 715-729.

[7] **Sadeghi, N., Fayek, A.R., and Seresht, N.G. 2015.** Queue performance measures in construction simulation models containing subjective uncertainty. *Automation in Construction*, 60, pp 1-11.

[8] **Cheng, F.F., Li, H., Wang, Y.W., Skitrnore, M., and Forsythe, P. 2013.** Modeling resource management in the building design process by information constraint Petri nets, Automation in Construction, 29, pp 92-99.

[9] **Koo, B., and Fischer, M. 2000.** Feasibility Study of 4D CAD in Commercial Construction, *Journal of Construction Engineering and Management*, 126(4), pp 251-260.

[10] **Rockwell Automation.** 2004. *Arena user's guide*, Milwaukee: Rockwell Softwarem.

[11] **Kelton, W. D., Sadowski, R. P., and Sturrock, D. T.** 2003, *Simulation with arena*, Boston: McGraw-Hill.

DIGITAL SYSTEM CYBER SECURITY IN NUCLEAR INDUSTRIES

Li Jing[1], Bian Yufang[1], Yang Anyi[1*]
[1] Nuclear and Radiation Safety Center
Ministry of Environmental Protection of the People's Republic of China, Beijing, China
lijing@chinansc.cn
yanganyi@chinansc.cn

ABSTRACT

The increasing reliance on computer systems and networks within control systems at nuclear power plants has presented new potential cyber security threats such as hacking, scanning, viruses, worms, and unauthorized access. This paper analyzes the security risks of digital instrumentation and control systems in nuclear power plants. Subsequently, the key enhancements to cyber security required on a global level are summarized and proposals for improvements are suggested.

Keywords: Cyber Security, Nuclear Industry, Digital Instrumentation and Control Systems

1 INTRODUCTION

For nuclear power plants, digital computers, communication systems, and networks are strategic assets. Nuclear facilities use digital instrumentation and control systems to monitor and operate equipment. In addition, a digital system is used to obtain and store vital information. Therefore, the digital system is regarded as a plant's "control center" and "nerve center." In addition, many plant computer systems are connected to digital networks that extend across the plant and performing safety, security, and emergency preparedness functions. The increasing reliance on computer systems and networks within control systems has presented new potential cyber security threats such as hacking, scanning, denial-of-service (DoS) attacks, new exploitations in applications, viruses, worms, and unauthorized access at nuclear power plants. These threats are listed in Table 1.

Studies have shown that the development of open standards, such as Ethernet, TCP/IP, and web technologies, provides opportunities for hackers to take advantage of the control industry's ignorance[1]. Hacking communities are becoming increasingly aware of the industrial control systems and process systems and are beginning to focus their attention on them. In particular, the increasing trend of cyber attacks on nuclear power plants is disquieting[2]. Hacking, in general, and attacks on protected computer systems are becoming increasingly common and sophisticated. The process of defending against this class of failures has been made more challenging by this rapidly evolving industry that continues to develop new attack methods.

---

* Corresponding Author

The potential effects that such cyber security threats might pose to real-time visibility and control, along with real-time responses and deployment of necessary resources in a plant on a daily basis, are not acceptable. Nuclear power plants are vulnerable to cyber attacks, which might, in extreme cases, lead to substantial releases of radioactive material and therefore cause loss of lives, radiation sickness, psychotrauma, extensive property destruction, and economic upheaval. Therefore, it is a realistic issue affecting people's livelihood and is related to national security and social stability.

Given the potential disastrous effects, digital instrumentation and control systems in nuclear power plants have been the focus of increased scrutiny by global regulatory and legislative bodies. Protecting these critical digital assets and the information they contain from sabotage or malicious use is called cyber security. The purpose of cyber security is to detect and eliminate or mitigate vulnerabilities that could be exploited either from outside or inside the digital system's protected area.

This paper analyzes the security risk of digital systems in the nuclear industry. Then, brief reviews of the key enhancements to cyber security required by the International Atomic Energy Agency (IAEA) and the U.S. Nuclear Regulatory Commission (NRC) are presented. Finally, this paper presents some suggestions to improve cyber security.

2 CYBER SECURITY RISK ANALYSIS OF THE NUCLEAR INDUSTRY

2.1 Digital instrumentation and control systems

To provide higher reliability and better performance with additional diagnostic capabilities in operating NPPs, digital instrumentation and control systems have been introduced to replace analogue systems, including Supervisory Control and Data Acquisition (SCADA), the Distributed Control System (DCS), Programmable Logic Controller (PLC), and the Fieldbus Control System (FCS). Digital instrumentation and control systems used in nuclear power plants are based on the industry-wide implementation of distributed real-time network-centric systems. This type of system adopts decentralized control, centralized management of the hierarchical distributed control structure, including the operation and control center system, power plant control systems, protection and security detection systems, diversification drive system, the data display and processing system, and the core instrumentation system, as well as other subsystems. The entire system is divided into three layers of networks: Scene Control, Monitoring, and Management. The Management network uses TCP/IP Ethernet; the Monitoring network uses industrial Ethernet between the engineering station, central processing server, and different systems and has a strong networking capacity; and the Scene Control network uses a high-speed fieldbus control layer. As can be seen, current digital instrumentation and control systems use open standards such as Ethernet, TCP/IP, and web technologies, which allow hackers to take advantage of the control industry's ignorance. The TianWan Nuclear power station instrument control digital system in JiangSu was the first case of digital instrumentation and control application in the China Nuclear Power field. Currently, nuclear power plants under construction in China all use digital systems. The usage statistics are shown in Table 2.

2.2 Threats to the digital control systems

With the development of computer and network technologies, digital systems are increasingly using a common protocol and common hardware and software to connect to the Internet or other public networks to improve system integration and the informationized management level. The adoption of general Internet technologies gradually breaks the seal of the digital control system, realizes the integration of management and control, and improves the efficiency of production and management; however, it also increases the threat and run risks. To analyze the safety risk, we examined the characteristics of the instrumentation and control system and compared it with a traditional information system, as shown in Table 3.

In fact, many cyber security incidents have not been made public due to associated state secrets. According to USA ICS-CERT[5], industrial control system security incidents have displayed a rapidly increasing trend with time, as shown in Figure 1.

2.2.1 *Existing defects in the Communication protocol*

Digital systems used in nuclear power plants concentrate on functionality and reliability, not security. The communication protocol of instrumentation and control systems, such as Mod bus and DNP3, are bus protocols based on serial links, which are used in industrial control networks isolated from other computers. For these systems, security is not an important functional requirement. Therefore, these communication protocols have birth defects in terms of cyber security protection, lacking authentication mechanisms, authorization mechanisms, and information encryption processing.

2.2.2 *System vulnerabilities are updated slowly*

With the development of technology, the communication protocol loopholes of industrial control systems have increasingly attracted attention. There are patches for vulnerabilities, which is our standard practice to manage system vulnerabilities. However, in the industrial system, particularly in nuclear power plants, digital instrumentation and control system vulnerability patch management is very cautious. First, before patch installation, full and complete tests and experiments, even an expert technical review of the assessments, need to be performed to determine that the patch will not affect other functions of the system. Second, some equipment vulnerabilities are difficult for maintenance and require specialist technical support. Third, system patching usually needs to be performed in a downtime during an overhaul with a long interval. In special cases, some older version systems have no available patches.

2.2.3 *Difficult to protect from viruses*

For digital control systems in nuclear power plants, the use of anti-virus software is also very conservative. First, common anti-virus software are not suitable for use in a nuclear power plant's digital control system and professional protection software support is required. Professional protection software requires adequate testing and evaluations, which are difficult to coordinate. Second, the use of protection software may lead to unexpected interruptions of the control system; therefore, installations are not permitted. Third, some systems are in 7 × 24 × 365 continuous operation; as a result, there is no good time for a comprehensive scan and installations cannot be performed. Finally, virus databases update slowly and are only able to protect against viruses with known characteristics; therefore, the system faces an endless stream of new viruses and their variants without protection. In particular, for APT types of attacks, it is nearly impossible for traditional anti-virus software to play a role.

2.2.4 *Security monitoring auditing abilities are weak*

Currently, there are few types of specialized surveillance equipment that can monitor the status of a network in real time to ensure availability. There is no way to promptly detect network anomalies and the abnormal flow of information and to determine whether the system is suffering from a safety hazard. No unified security management center exists and all security is single point. No in-depth analysis and positioning functions exist for relevant potential security threats that have been detected. To connect between the digital control system and external systems, the relevant requirements have been developed from a management point of view and the relevant security devices are used; however, there is no effective monitoring and filtering for the process and the content of information transmissions and cyber security problems cannot be found in time.

A Log Server is deployed in digital control systems to record system operation log information. However, for completeness, timeliness, and effectiveness of the log information, there are no clear requirements and operating procedures to audit; therefore, it is not possible to discover system abnormalities from the log information and the log does not play a preventative role. The log is usually only checked after a problem occurs.

2.2.5 Cyber security awareness is weak

Currently, security protection uses the concept of passive security protection, using physical protection as the primary security measure to strengthen the physical security and personnel operating procedures, with a lack of awareness concerning in depth defense and professional cyber security knowledge. Digital systems in nuclear power plants are generally managed by the production department, who are experts in the control system but not experts in IT or networks and lack cyber security capabilities and awareness. There are still some security protection awareness misconceptions in the system, e.g., the digital system is isolated from the outside world, others cannot attack me, no one will attack a digital system, one-way communication can guarantee a security of 100%, and security is uncontrollable for most digital control systems produced by foreign manufacturers. After security incidents, there is no way to locate and process the incidents in an efficient and timely manner. The general cyber security responsibilities are in the information document department, and the related technical personnel are usually not familiar with industrial control systems; therefore, it is difficult to give guidance recommendations for industrial control system cyber security. Enterprise cyber security management systems are not suited to the professional needs of industrial control systems; therefore, industrial control systems are divorced from the enterprise security management systems, and may be a blind spot or weakness in the security management system.

3 INTERNATIONAL COMMUNITY MEASURES

3.1 International Atomic Energy Agency

The International Atomic Energy Agency (IAEA) pays a great deal of attention to cyber security threats in the nuclear power system. In April 2011, at the Fifth Review Conference of the “Convention on Nuclear Safety,” there were ten issues of common concern to parties proposed by the IAEA; one item was that member states should strengthen their focus on the nuclear power industry’s cyber security and network security, should enact and implement relevant policies, laws, and regulations, and should assess and determine measures for the threats their own cyber security had suffered while increasing the number of cyber security professionals.

In early 2013, the IAEA published a document called the “IAEA Coordinated Research Activities in 2013”[6], which made it clear that nuclear power plant network security research would be listed as one of the key research projects in 2013. Currently, the IAEA nuclear power systems cyber security regulation and guideline documents are still in the process of being prepared, as shown in Table 5[7].

3.2 The United States

The United States has always considered nuclear power plants to be a key critical infrastructure requiring cyber security safeguards. The federal government issued several executive orders emphasizing nuclear power plants as requiring the protection of cyber security[8]. On January 2, 2008, the United States established a comprehensive national cyber security plan, in which nuclear power was a key target. In 2009, the United States promulgated the “protection strategy of industrial control systems,” and the same year, the industrial control systems Network Emergency Response Team established on-site support of incident responses and forensics analyses. The US Department of Homeland Security initiated a control system security program (Control System Security Program, CSSP) relying on the industrial control system simulation platform to analyze and validate the vulnerability of industrial control system products. The US National Institute of

Standards and Technology (NIST) and the Federal Energy Regulatory Commission issued the "Guide to Industrial Control Systems Security" and "21 measures to improve the SCADA cyber security," respectively, and other related control systems security standards guidelines or best practices documents. Traditional domestic industrial manufacturers in the USA, such as Rockwell, GE, and cyber security vendors, such as Symantec, MCAFEE, and Cisco, also conducted in-depth research, practice, and industrialization studies in industrial control system security products and services.

The US Nuclear Regulatory Commission (NRC) is responsible for the cyber security supervision of nuclear power units[9]. Since the events of September 11, 2001, NRC joined multiple agencies to conduct research on the regulations and standards of nuclear power system cyber security and published a series of relevant documents[10], as shown in Table 6.

4 SUGGESTIONS

Nuclear safety is an important component of national security. There is no way to immediately solve the cyber security problems of nuclear power plants. The solution will need to mobilize a wide range of state and industry regulators, system development manufacturers, industrial control systems cyber security services, nuclear power plants, and other users, establish multi-disciplinary and cross sectorial cooperation and collaborative mechanisms, and gradually establish a comprehensive legal system, technical standards, technical protection systems, personnel development system, and the supervision and inspection of the system.

State and industry regulators need to further strengthen national policy research and development, regularize guidance and requirements of cyber security at the system level, and further establish and improve the regulations and standards system for nuclear cyber security, which will provide the cyber security of nuclear power systems with regulations and constraints. In addition, state and industry regulators need to strengthen their coordination with domestic and foreign government departments, research institutions, manufacturers, and other organizations to address the cyber security threat, which is in a constant process of change and development, organize safety supervision and safety inspections, build mechanisms to strengthen cyber security notifications and cyber security incident emergency responses, and build an information sharing platform for Trojans, vulnerabilities, viruses, and other threats[11].

Systems research and development firms need to integrate relevant technical requirements for security design, security coding, and security testing into all stages of the development life cycle, such as requirement analysis, architecture design, development, implementation, testing, and validation, and should strengthen the security requirements in the building process to reduce systemic risk from the source[12]. In addition, they need to consider the relationship between integrity, availability, and confidentiality, maximize the use of security communication protocols and norms with authentication, encryption, and authorization mechanisms if possible, and establish credible security systems and equipment to provide timely and reliable vulnerability solutions that provide users with better technical support services.

Cyber security services for Industrial control systems, via cooperation with relevant manufacturers, research organizations, or institutions, need to strengthen their research concerning vulnerability analyses of the digital control system, system patch management, and vulnerability analyses. They need to study and propose cyber security protection solutions for industrial control systems, research and develop all types of security equipment that meet the needs of industrial control systems, improve cyber security protection, detection, and processing capabilities, and ensure normal operation of the industrial control system.

Nuclear power plant users need to further strengthen the nuclear safety culture, raise awareness of cyber security, and constantly improve the management system and practices[13]. Cyber system security needs to be a key feature that is considered when building and integrating a system. Cooperation with cyber security services needs to be strengthened, the appropriate cyber security program needs to be selected, and a defense system needs to be established for security risk detection, system management, and emergency responses. In addition, the risk assessment and performance monitoring for digital control system in operation and the network anomaly detection and inspection process needs to be strengthened, and threat perception needs to be improved. Finally, the nuclear power plant users need to strengthen log audit management, create protocol session processes for recording and auditing, and effectively supervise illegal operations, unauthorized access, and other acts to provide tracing abilities after security incidents and to ensure system security.

Table 1: The main cyber security events in the field of nuclear industry

Date	Incident
Jan 2001	Hackers attacked a British nuclear power plant.[1]
Jan 2003	The Slammer worm infected computer systems at the USA Davis-Besse nuclear power plant. For four hours and fifty minutes, plant personnel could not access the Safety Parameter Display System (SPDS).[3]
Aug 2006	Hackers attacked the USA Browns Ferry nuclear power plant. Unit 3 was manually shutdown after the failure of both the reactor recirculation pumps and the condensate demineralizer controller.[2]
May 2007	Hackers attacked a Russian nuclear power plant site. Hackers blocked all access to the automatic radiation environment control system.[3]
March 2008	Unit 2 of the Hatch nuclear power plant near Baxley, Georgia, automatically shut down after an engineer applied a software update to a single computer on the plant's business network.[4]
June 2010	Stuxnet attacked Iran's nuclear plants; 30% of the Natanz computer facilities were controlled.[3]
Feb 2012	Hackers attacked a British nuclear power plant resulting in confidential data leakage.[4]
Dec 2014	Hackers attacked a Korean nuclear power plant resulting in information leakage.[4]

Table 2: Nuclear power plant digital control system list for China

Nuclear power plant	DCS (security level)	DCS (non-security level)
Tianwan Unit 1, Unit 2 Lingao Unit 3, Unit 4 Taishan	German AREVA (formerly Siemens) Teleperm XS	Teleperm XP, Siemens AG
Sanmen, Haiyang	Westinghouse	SNPAS

Fuqing, Fangjiashan	Tricon, Invensys	I/A, Invensys
Hongyanhe,Ningde Yangjiang Unit 3,Unit 4 Fangchenggang	MELTAC, Mitsubishi	HOLLiAS-N, CETC

Table 3: Comparison of digital control systems to general information systems

Comparative item	Digital control system	General information system
system structure	composed of sensors, valves, cont rollers, PLC, and RTU via communic ation protocol	composed of hardware and sof tware via communication protoco l
operating system	widely use embedded operating s ystems such as VxWorks and WinCE	widely use general-purpose op erating system such as windows, linux, and unix
communication protocol	special communication protocols, such as OPC, Modbus and DNP3 TCP /IP	TCP/IP protocol stack, HTTP, a nd FTP
reliability	continuous operation, does not no rmally allow stops Failure should be dealt with as an emergency response	allows for the occasional failur e or restart; fault response level according to system requirement s; allows for Beta testing
security	strict physical separation and syst em stability	physical security is not strict; n etwork interconnections
System upgrade	proprietary platform, weak comp atibility, difficult to upgrade	general system, good compatib ility, easy to upgrade
attack method	organized and purposeful attacks, multi-patterns, difficult to detect	common attacks
Security defense ability	difficulty in system software upgr ades, patch management, leak prot ection, and inadequate monitoring and auditing means; security protec tion system is not sound, leading to insufficient protection	complete mature security prot ection system, system security g uarantee, management strategy, operation methods, and a huge s ecurity service industry chain

Table 4: IAEA regulations and guidance documents

Name	Year
NSS 17 - Computer Security at Nuclear Facilities	Published 2011
Protection and Confidentiality of Nuclear Security Sensitive Information	120 day review
Conducting Computer Security Assessments for Nuclear Facilities	Draft
Applying Computer Security Controls to Instrumentation and Control Systems at Nuclear Facilities	Draft
Incident Response Planning for Computer Security Events at Nuclear/Radiological Facilities	Draft
Cyber Security Regulation	New start 2013
Computer Security Incident Response Exercises for Nuclear/Radiological Facilities	New start 2013
Conducting Computer Security Incident Investigation and Forensics at Nuclear/Radiological Facilities	New start 2014

Table 5: NRC regulations and standards

Name	Year
Interim Safeguards and Security Compensatory Measures for Nuclear Power Plants (NRC Order EA-02-026)	2001
Design Basis Threat for Radiological Sabotage (NRC Order EA-03-086)	2003
Design Basis Threat for Radiological Sabotage (NUREG/CR-6847)	2004
Cyber Security Program for Power Reactors	2005
Criteria For Use of Computers in Safety Systems of Nuclear Power Plants (RG1.152)	2006
Cyber Security Guidance for Nuclear Power Plants	2007
Cyber Security DI&C-ISG	2007
Protection of Digital Computer and Communication Systems and Networks 10 CFR 73.54	2009

Regulatory Guide 5.71, "Cyber Security Programs for Nuclear Facilities"	2010

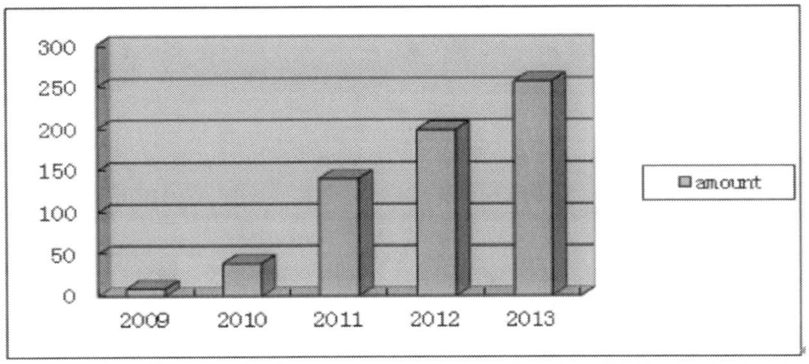

Figure 1: Security events according to ICS-CERT statistics

5 REFERENCES

[1] **Byres E., Lowe J. 2003.** The Myths and Facts behind Cyber Security Risks for Industrial Control Systems, ISA Process Control Conference, Columbia, British, pp 1-6.

[2] **Kesler B. 2010.** The Vulnerability of Nuclear Facilities to Cyber Attack, *Strategic Insights*, 10 (1), spring 2010, pp 15-25.

[3] Nuclear power plant cyber-security Incidents. 2011. [Online].Available: http://www.safetyinengineering.com/FileUploads/Nuclear%20cyber%20security%20incidents_1349551766_2.pdf. Last visited: 2013.

[4] **Krebs B.** Cyber Incident Blamed for Nuclear Power Plant Shutdown. 2008. [Online]. Available: http://seclists.org/isn/2008/Jun/25. Last visited: 2013.

[5] ICS-CERT. Monthly Monitor Newsletters,ICS-MM201404,http://ics-cert.us-cert.gov/monitors/ICS-MM201404

[6] IAEA.Coordinated Research Activities in 2013. http://www.varans.vn/Traodoi/13-03591E2_Attachment.pdf.

[7] **Dudenhoeffer D.**Office Of Nuclear Security Cyber Security Programme. http://www.iaea.org/NuclearPower/Downloadable/Meetings/2013/2013-05-22-05-24-TWG-NPE/day-2/4.cyber_security_introduction.pdf.

[8] **Erlanger C., Simonds G.2012.**Cyber Security for Nuclear Facilities: An Overview of the NRC Regulatory Framework. http://pbadupws.nrc.gov/docs/ML1227/ML12276A251.pdf

[9] NRC 10 CFR 73.54: Protection of digital computer and communication systems and networks. http://www.nrc.gov/reading-rm/doc-collections/cfr/part073/part073-0054.html.

[10] **Spaulding-Yeoman W.** Overview of Digital System Safety and Cyber Security. http://www.mwftr.com/cneF11/WK09_NRC_Cybersecurity.pdf.

[11] **Shea T., Gaycken S., Martellini M. 2013.** Cyber Security for Nuclear Power Plants, *Springer Briefs in Computer Science*, pp 25-35.

[12] **Kim DY.2014.** Cyber security issues imposed on nuclear power plants, *Annals of Nuclear Energy*, 65(65), pp 141-143.

[13] **Song, J. G., Lee, J. W., Park, G. Y., Kwon, K. C., Lee, D. Y., & Lee, C. K. 2013.** An analysis of technical security control requirements for digital i&c systems in nuclear power plants. *Nuclear Engineering & Technology*, 45(5), pp 637-652.

KNOWLEDGE MANAGEMENT PRACTICE IN THE NATIONAL NUCLEAR SAFETY ADMINISTRATION OF CHINA

Li Jing[1] and Chen Feng[2*]

[1]Nuclear and Radiation Safety Centre

Ministry of Environmental Protection of the People's Republic of China, Beijing, China
lijing@chinansc.cn

[2]Department of Information Resources Management, Business School,

Nankai University, Tianjin, China
chenfengwork@163.com

ABSTRACT

The National Nuclear Safety Administration of China (NNSA) implemented a knowledge management (KM) program in 2015. Following the implementation, the administration encountered two problems: the loss of tacit knowledge and finding the right method to organize the sharing and coordination of documents. Thus, we designed a KM system that is suitable for an administration institute. The objectives of KM are: to acquire, use, share, disseminate, and retrieve knowledge in the NNSA using a systematic approach and develop a strategic approach with action plans to collect valuable data and information. This paper introduces the details of the program, including the background of the KM initiative, the KM system, policy, human resource, and challenges. The challenges and any proposal for future work are discussed at the end of the paper.

Keywords: Knowledge Management, Nuclear Industry, Data Mining

1 INTRODUCTION

As the nuclear regulatory authority in China, the National Nuclear Safety Administration (NNSA) has accumulated a large number of technical documents, such as research reports and project supervision reports. The accumulated knowledge plays an important role in the supervision of nuclear and radiation safety. One of the greatest challenges is with tacit knowledge. NNSA is hiring and training a lot of skilled inspectors to keep pace with reactor construction. As many nuclear experts are on the verge of retirement, their knowledge and experience have not been transferred to the younger employees. Tacit knowledge represents the fruitful result of the interaction between an individual and the external environment. Tacit knowledge sharing can enhance employees' innovative capabilities and have a positive impact on their work performance. However, tacit knowledge sharing may have many organizational and technical barriers. A research done on

* Corresponding Author

Romanian companies demonstrated that the organizational culture is very important when it involves tacit knowledge sharing [1].

In 2014, the NNSA of China had planned to organize the records and archives of 30 years. Their staff had encountered many problems during the data collection process: most of the important data were stored in various places, making them difficult to use and facilitate the knowledge sharing and transfer of information.

Chinese cultural influences on knowledge management (KM) practice is also complex[2]. Although KM research has significantly matured, interdisciplinary research involving psychology and human resources are still inadequate [3]. For a Chinese adminstration office, it may be a big challenge to develop and put to use a KM system.

Knowledge needs to be preserved for future use. It is necessary to increase the awareness of the need to develop a strategic approach and action plans to address the potential loss of critical knowledge and skills. A documented and centralized knowledge database can prevent the loss of valuable information when an employee leaves.

Knowledge is going to disappear unless it is managed using a systematic KM approach. Access to existing nuclear knowledge can be improved in many cases, and the sharing and pooling of knowledge can contribute towards ideas for new development and innovation as well as facilitate decision making process.

2 OBJECTIVES OF THE KNOWLEDGE MANAGEMENT INITIATIVE

There are several perspectives and objectives of KM when it involves for example, human resources [4], strategy management [5] and information systems [6]. Since making decisions and promoting efficiency is threatened by the risk of knowledge loss, we have the responsibility to not only establish and maintain adequate technical knowledge and experience in our nuclear organizations but also to make good use of it.

The objectives of a KM mission are to acquire, store, use, share, disseminate and retrieve knowledge in the NNSA using a systematic approach and to develop a strategic approach and action plans to address the potential loss of critical knowledge and skills. The NNSA of China is composed of three administrations departments, six nuclear and radiation safety supervision station and a technical center (Nuclear and Radiation Safety Centre). So, all the members of the NNSA of China must be part of the KM initiative.

3 DESCRIPTION OF THE KNOWLEDGE MANAGEMENT INITIATIVE

The schedule of KM project is showed in table1.

Table 1: Time schedule of KM project

Date	Activity
2015 January-May	prophase investigation symposium on special topic, questionnaire, former study
2015 May-June	the technical schemes of KM system was compared and optimized in aspects of function, security, compatibility, price and so on policy for KM domain knowledge architecture

2015 July-August	system installation, deployment, debugging, and personalized interface design
2015 Sep-Oct	system on line
2015 OCT	user training and administration training
2016 March	updated KM system and improved database
2016 March	user training for the second time

3.1 Regulations

In order to maintain the persistent effect mechanism of KM, we have made regulations in implementing our strategy in the KM area. The Social Identity, Organizational Identity, and Corporate Identity have a great impact on an individual's compliance with the set rules [7]. In the process of making the regulations, we have designed various methods to rationalize the rules in place so that they can be easily accepted by employees. The rules have been communicated to all the staff in the organization to seek further advice and ideas before a formal administrative document is produced. Those who are responsible have been clearly identified in advance.

3.2 Human resource

The human resource factors play an important role in management [8]. We set up a KM working group responsible for managing the formulation and implementation of the KM strategy, which will consist of the leadership team and the implementing team. The leadership team is made up of the chief of NNSA and its members are the leaders from the eight member units who are in charge of KM. The leading team is responsible for strategic planning, decision making and supervision. The implementing team is responsible for building and maintaining the KM platform.

3.3 KM system

We built a KM system to effectively retrieve, share, and utilize knowledge. The system provides a range of advanced KM features as follows:

3.3.1 Data security

The knowledge database is self-owned with three copies stored, and the hardware and software are installed on our own server.

3.3.2 Personal network backup

The system provides storage that is user-accessible from any computer using either a web-based application or a client-installed application.

3.3.3 Acquire, store, and share knowledge easily

The system provides the means for individuals, groups or departments to innovate, build new knowledge and share knowledge in a group. This makes it easy for them to share, search, rate and print documents, procedures and more. Now, everyone will have access to the same single version document from the same location. The web-based interface makes it easy for customers to find answers to their problems instead of submitting emails or calling the concerned support department.

Staffs and partners can access information locally or over the Internet. The Knowledge Manager's powerful group-based permission architecture makes it easy to share knowledge with only the people or groups he/she chooses.

3.3.4 *Gain knowledge easily*

Staff will be able to gain the job knowledge faster, hence reducing their training time significantly and producing better job performance. By providing new staff with a list of company-wide policies, procedures and how-to guides to read in your knowledge base, they can be up and running quicker. It will give new staff a "hands on" approach to learning.

3.3.5 *Flexible Security & User Role Management*

Although the objective of KM is opening and sharing knowledge, extra attention should be given to Information security. There are four basic paradigms about information security management: functionalist paradigm, interpretive paradigm, radical humanist paradigm and radical structural paradigm [9]. Based on previous research, we have designed a flexible security system which includes the following measures:

- Access Levels: Ability to define where articles can be viewed
- User Roles: allows customizable user roles to define what your users may and may not do
- Multi Department/Group Knowledge Management: allows knowledge base articles to be managed at the department or group level. This allows each department to create and manage their own information
- Portal Views: The public view can be set so that authenticated contacts can only view their designated information

3.3.6 *Searchable File Attachments*

Upload company procedures and documents (Microsoft Office or PDF) which are instantly indexed and searchable. Categories can be password protected and restricted.

- Multiple Formats: Ability to upload any file format as an attachment
- Full Text Search Capability: Searches ALL text

3.3.7 *Content Quality Features*

Version Control: Revisions are created and saved automatically when an item is updated

3.3.8 *Others*

- Group-based permissions and access
- Both web-based and client-installed applications are available.
- Easily Installed and Accessible

3.4 Training and Performance improvement

We have organized several training courses utilizing computer and multi-media simulations. We also provide coaching and mentoring to members.

We have updated KM system and improved database usability twice.

4 DISCUSSION

We have benefited by implementing KM strategies and techniques. The KM system helps us acquire, store and utilize knowledge conveniently. The practice of KM strategies and techniques is already providing benefits. It allows the staff to locate, capture and share information with each other. It reduces the time in searching for information. Our domain knowledge architecture is designed just for nuclear engineering field. Except it, most of the strategies and techniques are also suitable for common KM systems, especially more suitable for administration institute.

During the KM initiative, we encountered many difficulties:

Managing knowledge depends upon people's willingness to share and reuse knowledge. It was discovered that the Chinese prefer to keep their knowledge implicit and are only willing to share it informally. A series of contributing factors derived from the Chinese culture includes the fear of losing face, a sense of modesty, competitiveness and a preference for face-to-face communication; these can act as barriers to the KM initiative [2].

Although knowledge sharing is evident from the KM activities in different departments, limitations such as resistance to change due to the implementation of the KM systems are impeding the effectiveness of the KM process. The success of the KM initiative is likely to be critically dependent on having suitably motivated people playing an active role in the process.

Enterprises have begun to research and deploy the intelligent KM system, lifting the value of knowledge through data mining [10]. Zhang et al. proposed the foundations and analysed the process of intelligent knowledge management [11]; however, previous researches mostly focused on business intelligence, which seems to be unsuitable for the government or the research institute. In our KM system, the majority of knowledge is stored in the form of text; text mining is also an important and challenging research field, wherein consumption patterns are gathered through postings on Twitter [12]. How to connect or link the knowledge in the text is a big challenge for the Intelligence KM system.

Next, we will adopt the user feedback mechanisms. More training and incentives are needed to increase knowledge creation and sharing. Therefore, we should acquire the following information:

- Most Popular Articles
- Most Helpful Articles
- Recently Added Articles or Recently Updated Articles
- Article Hit Tracking: Ability to track number of times each knowledge base article is viewed
- Article Rating System: Ability for web users to rate and comment on an article's helpfulness
- Article Discussion/Comments

Moreover, we are going to build feedback loops to help staff willing to share and improve knowledge. Search engine can be performed on more content types such as picture and video, providing the most relevant search results. Text mining built in the KM helps to link the knowledge and a Recommendation algorithm could be developed to push useful information to staff.

5 REFERENCES

[1] **Bratianu, C., Orzea, I. 2010.** Tacit Knowledge Sharing in Organizational Knowledge Dynamics, The Proceedings of the European Conference on Intellectual Capital, Lisboa, Portugal, Mar 29-3 0, Vol.24, pp 107-114.

[2] **Jin, T., Mitra, A. 2009.** Chinese Cultural Influences on Knowledge Management Practice. *Jou rnal of Knowledge Management*, 13(2), pp 49-62.

[3] **Akhavan, P. , Ebrahim, N. A. , Fetrati, M. A. , Pezeshkan, A. 2016.** Major Trends in Knowledge Management Research: A Bibliometric Study. *Scientometrics*, (107)4, pp 1-16.

[4] **Swan, J., Scarbrough, H., Preston, J. 1999.** Knowledge Management: the Next Fad to Forget People. European Conference on Information Systems, Copenhagen, Denmark, June 23-25, pp 668-678.

[5] **Newell, S Robertson, Scarbrough, M H Swan J 2009.** *Managing Knowledge Work & Innovation*, 1st edition, Palgrave Macmillan, Basingstoke, Hampshire.

[6] **Mertins, K., Heisig, P. Vorbeck, J. 2000.** Knowledge Management: Best Practices in Europe, *Springer-Verlag*, New York.

[7] **Cornelissen, J. P., Haslam, S. A., Balmer, J. M. T. 2007.** Social identity, Organizational Identity and Corporate Identity: Towards an Integrated Understanding of Processes, Patternings and Products. *British Journal of Management*, 18(Supplement s1), pp 1-16.

[8] **Sfeir, E. 2009.** Human Resources Management Strategies to Support Organizational Changes. *Information Processing & Management*, *32*(1), pp 164-165.

[9] **Lin, R., Xie, Z., Liu, Q. 2014.** Information Security Management Research Review, Carding and Future Prospects (In Chinese). *China Journal of information systems* 15(2), pp 81-84.

[10] **Hou, J. L., Sun, M. T., Chuo, H. C. 2005.** An Intelligent Knowledge Management Model for Construction and Reuse of Automobile Manufacturing Intellectual Properties. *International Journal of Advanced Manufacturing Technology*, 26(1-2), pp 169-182.

[11] **Zhang, L., Li, J., Li, A., Zhang, P., Nie, G., Shi, Y. 2009.** A New Research Field: Intelligent Knowledge Management. International Conference on Business Intelligence and Financial Engineering, Beijing, China, July24-26, pp 450-454.

[12] **Carlos Alberto Ochoa-Zezatti.2016.** Identifying consumption patterns in Twitter using text mining to classify trends in shopping. 17th International Conference on Intelligent Text Processing and Computational Linguistics, Konya, Turkey, April 3-9.

DYNAMIC JOINT CONSTRUCTION AND OPTIMAL STRATEGY OF MULTI-OBJECTIVE MULTI-PERIOD MULTI-STAGE GOVERNMENT-ENTERPRISE REVERSE LOGISTICS NETWORK: A CASE STUDY OF LEAD BATTERY IN SHANGHAI

Guo Jianquan[1*], Cao Feng[2], Mitsuo Gen[3]
[1] Shanghai-Hamburg College
University of Shanghai for Science and Technology, Shanghai, China.
jasonguousst@126.com
[2] Business school
University of Shanghai for Science and Technology, Shanghai, China.
caofeng0126@163.com
[3] Fuzzy Logic Systems Institute
Tokyo University of Science, Tokyo, Japan.
gen@flsi.or.jp; gen@rs.tus.ac.jp

ABSTRACT

An integrated multi-objective multi-period multi-stage government-enterprise joint model is constructed to recycle used lead battery effectively. The feasibility of the model is validated with a numerical example by adopting a two-phase heuristic algorithm nesting GA/PSO. The results show that the negative effect of the operation of the system to the surroundings is minimized, and the facility resources are fully utilized because of the multi-period arrangement of recycling activities in the system. In addition, the arrangement of the locations and quantity of its nodes in different stages not only reduces the uncertain risk of the system, but optimizes the input of money, manpower and material during the whole operation process. The study provides a reference for the design of joint construction of government-enterprise reverse logistics network and its optimal operation, especially for the recovery of this kind of used products with high negative externality.

Keywords: Government-enterprise reverse logistics network, Dynamical joint construction, Optimal strategy, Two-phase heuristic algorithm nesting GA/PSO, Used lead battery

1. INTRODUCTION

Reverse logistics network planning includes the sites selection of the recycling centers and the recycling plants, the design of transportation routes, and the arrangement of traffic volume [1]. The purpose of reverse logistics network planning is to reduce the total cost or increase the total revenue of the system in the process of regaining product value or handling products appropriately [2].

To study the planning of reverse logistics network, numerous studies were proposed by both researchers and practitioners at home and abroad. Reverse logistics network referring to waste pickers as a Brazilian legal requirement for municipal solid waste management was proposed to determine the recycling facility location, the recycling facility quantity, and the transportation arrangement by Ferri et al. (2015)[3]. Considering simultaneously production, distribution and reverse logistics activities under stochastic demands, supply chains with reverse flows has been

* Corresponding Author

designed and planned within a mixed integer linear programming Equation, which makes the maximization of the expected net present value by Cardoso et al. (2013) [4]. Robust and reliable design of forward-reverse logistics network and facility disruptions in which the objective function of the proposed model minimized the nominal cost under uncertain demand was rendered by Hatefi et al. (2014) [5]. These studies only considered the economy object of the system, however, its negative effect, which is called social cost in this circumstance, to the surroundings should also be considered and minimized [6].

Garg et al. (2015) explained a multi-criteria optimization approach to manage environmental issues through the setup of an efficient recovery system in extending closed loop supply chain (CLSC) network design [7]. A case-oriented approach to the design of a lead battery closed-loop supply chain network with the risk and uncertainty has been studied by Subulan, et al. (2015) [8]. However, these studies only deal with single period instead of multi-period during the recovery process. In fact, multi-period recovery is more meaningful for it can make full use of the facilities, equipment and resources in the system, moreover, the dynamic multi-period network is more adaptive to reality [9]. Therefore, the redesign of recovery period is necessary for an effective reverse logistics network.

Gao et al. (2015) studied the closed-loop supply chain by considering the balance of total cost and carbon emission [10]. Weng et al. (2014) analyzed the reverse logistics network through B2C apparel e-business enterprises [11]. The reverse logistics network design of above studies is based on the single stage model. However, considering many uncertainties and risks in the system, the types, locations and quantity of the notes in different stages should be arranged respectively for avoidance of too much input initially.

At present, an effective recycling system of used lead battery is still scarce in China. There are approximate 3.3 million tons of used lead batteries per annum, and the recovery rate is less than 30 percent [12]. Moreover, there exists plenty of so called "disassembling randomly", "lying randomly" and "discharging randomly" in recycling process, which results in a series of environmental problems after long-term accumulation [13]. Therefore, strict laws and regulations are necessary during the recycling of used lead batteries, and government should also play a positive role.

Based on the analysis above, an integrated multi-objective multi-period multi-stage government-enterprise joint model is constructed with the consideration of both economy cost and social cost. And the feasibility of the model is to be verified by adopting a two-phase heuristic algorithm nesting GA/PSO with a case study of the recovery of used lead battery in Shanghai.

2. MULTI-OBJECTIVE MULTI-PERIOD MULTI-STAGE GOVERNMENT-ENTERPRISE JOINT MODEL

Figure 1 shows the seven elements in the government-enterprise joint model. The solid line represents the material flow, and the dotted line represents the information flow.

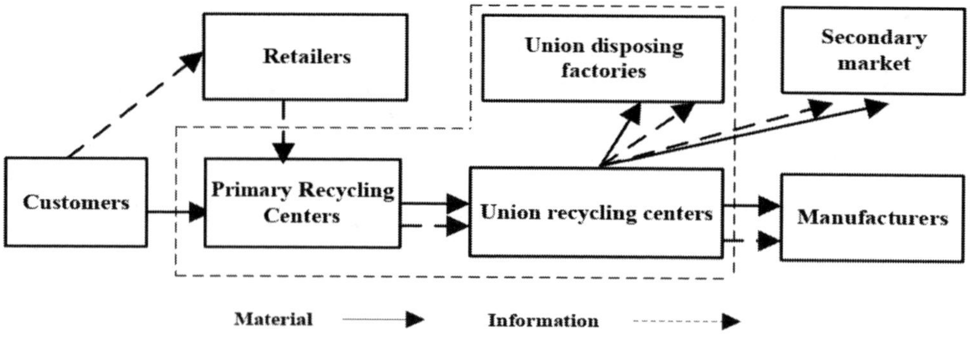

Figure 1: Structure chart of government-enterprise joint model

2.1 Assumptions

The following assumptions are rendered for multi-objective multi-period multi-stage government-enterprise joint model:

(1) The location and quantity of enterprises are determined. (2) The capacity of potential nodes is all limited within their assigned value. (3) The returned lead batteries can only be transported from primary recycling centers, and the recycling information transmission starts from retailers. (4) The location and quantity of potential primary recycling centers, potential union recycling centers and potential union disposing factories are determined. (5) The delivery costs are linearly proportional to the quantity and distance, and the information processing costs between nodes are the same. (6) The single period time length is one year, while the time length of one period in multi-period system is 6 months (According to the actual situation, the exact time length of each period may vary.)

2.2 Indices

$g \in G$, customer index; $i \in I$, primary recycling center index; $j \in J$, union recycling center index; $k \in K$, union disposing factory index; $p \in P$, manufacturer index; $s \in S$, secondary market index; $e \in E$, retailer index; $t \in T$, period index.

2.3 Parameters

f_i^t, f_j^t, f_k^t: represent the fixed cost for primary recycling center, union recycling center and union disposing factory at the period of t;

ma_i, ma_j, ma_k: represent the maintenance expense for each corresponding element node after one period;

ca_i, ca_j, ca_k, ca_s, ca_p: represent the capacity of each corresponding element node;

me_i, me_j, me_k, me_e, me_s, me_p: represent the unit cost of information processing for each corresponding element node;

u_i, u_j, u_k: represent the unit cost of handling goods for each corresponding element node;

α, β, γ: represent the ratio of abandon, second hand and raw material in the union recycling center;

d_{gi}, d_{gj}, d_{gk}, d_{ij}, d_{jk}, d_{js}, d_{jp}: represent the distance between two points;

tr_{gi}, tr_{ij}, tr_{jk}, tr_{js}, tr_{jp}: represent the unit cost of transportation between two points;

nc_{gi}^t, nc_{ij}^t, nc_{jk}^t, nc_{js}^t, nc_{jp}^t: represent the quantity of good from one point to another at the period of t;

ns_i^t, ns_j^t, ns_k^t, ns_p^t, ns_e^t, ns_s^t: represent the quantity of information processing from one point to another at the period of t.

2.4 Decision variables

x_i^t, x_j^t, x_k^t: If primary recycling *i*, union recycling *j*, disposing factory *k* is open at the period of *t*, the value of x_i^t, x_j^t, x_k^t is 1, otherwise the value is 0.

2.5 The mathematical model

The first objective function of the model: the economic cost of reverse network

$$EC = u_1 + u_2 + u_3 + u_4 \tag{1}$$

The fixed cost and maintenance expense

$$u_1 = \sum_i^I \mathrm{f}_i^t x_i^t + \sum_i^I \mathrm{ma}_i\, x_i^t + \sum_j^J \mathrm{f}_j^t x_j^t + \sum_j^J \mathrm{ma}_j\, x_j^t + \sum_k^K \mathrm{f}_k^t x_k^t + \sum_k^K \mathrm{ma}_k\, x_k^t \tag{2}$$

The cost of information processing

$$u_2 = \sum_i^I \mathrm{me}_i\, ns_i^t + \sum_j^J \mathrm{me}_j\, ns_j^t + \sum_k^K \mathrm{me}_k\, ns_k^t + \sum_e^E \mathrm{me}_e\, ns_e^t + \sum_s^S \mathrm{me}_s\, ns_s^t + \sum_p^P \mathrm{me}_p\, ns_p^t \tag{3}$$

The cost of disposing returned goods

$$u_3 = \sum_g^G \sum_i^I \mathrm{u}_i\, nc_{gi}^t + \sum_i^I \sum_j^J \mathrm{u}_j\, nc_{ij}^t + \sum_j^J \sum_k^K \mathrm{u}_k\, nc_{jk}^t \tag{4}$$

The cost of transportation

$$u_4 = \sum_g^G \sum_i^I tr_{gi} nc_{gi}^t d_{gi} + \sum_i^I \sum_j^J tr_{ij} nc_{ij} d_{ij} + \sum_j^J \sum_k^K tr_{jk} nc_{jk}^t d_{jk} + \sum_j^J \sum_s^S tr_{js} nc_{js}^t d_{js} + \sum_j^J \sum_p^P tr_{jp} nc_{jp}^t d_{jp} \tag{5}$$

The second objective function of the model: social cost

$$SC = \sum_g^G \sum_i^I \left(nc_{gi}^t\right)^{\phi} \left(d_{ij}\right)^{\eta} x_i^t + \sum_g^G \sum_i^I \sum_j^J \left(nc_{ij}^t\right)^{\phi} \left(d_{gj}\right)^{\eta} x_j^t + \sum_g^G \sum_j^J \sum_k^K \frac{\left(nc_k^t\right)^{\phi} x_k^t}{\left(d_{gk}\right)^{\gamma}} \tag{6}$$

Analytic hierarchy process (AHP) is used to index the system for weight assignment, and the economic effectiveness coefficient is measured to the second objective function. Then the two objective functions are transformed into a single objective function with the help of turning the social cost into the economic cost by the coefficient (Equation 7).The value of setting parameter ϕ, η is 1 without loss of generality [14].

The transformation is as follows:

$$\lambda_1(u_1+u_2+u_3+u_4)+\lambda_2\xi\left\{\begin{array}{l}\sum_{g}^{G}\sum_{i}^{I}\left(nc_{gi}^{t}\right)^{\phi}\left(d_{gi}\right)^{\eta}x_i^t+\\ \sum_{g}^{G}\sum_{i}^{I}\sum_{j}^{J}\left(nc_{ij}^{t}\right)^{\phi}\left(d_{gj}\right)^{\eta}x_j^t+\sum_{g}^{G}\sum_{j}^{J}\sum_{k}^{K}\frac{\left(nc_k^t\right)^{\phi}x_k^t}{\left(d_{gk}\right)^{\eta}}\end{array}\right\} \tag{7}$$

Subject to:

$$\alpha\sum_{i}^{I}\sum_{j}^{J}nc_{ij}^{t}=\sum_{j}^{J}\sum_{k}^{K}nc_{jk}^{t},\ \beta\sum_{i}^{I}\sum_{j}^{J}nc_{ij}^{t}=\sum_{j}^{J}\sum_{s}^{S}nc_{js}^{t},\ \gamma\sum_{i}^{I}\sum_{j}^{J}nc_{ij}^{t}=\sum_{j}^{J}\sum_{p}^{P}nc_{jp}^{t}$$

$$\sum_{g}^{G}\sum_{i}^{I}nc_{gi}^{t}=\sum_{i}^{I}\sum_{j}^{J}nc_{ij}^{t}=\sum_{j}^{J}\sum_{k}^{K}nc_{jk}^{t}+\sum_{j}^{J}\sum_{s}^{S}nc_{js}^{t}+\sum_{j}^{J}\sum_{p}^{P}nc_{jp}^{t},\ \sum_{j}^{J}\sum_{s}^{S}nc_{js}^{t}=\sum_{s}^{S}ns_{s}^{t},\sum_{j}^{J}\sum_{p}^{P}nc_{jp}^{t}=\sum_{p}^{P}ns_{p}^{t}$$

$$\sum_{g}^{G}\sum_{i}^{I}nc_{gi}^{t}=\sum_{e}^{E}ns_{e}^{t},\sum_{g}^{G}\sum_{i}^{I}nc_{gi}^{t}=\sum_{i}^{I}ns_{i}^{t},\ \sum_{i}^{I}\sum_{j}^{J}nc_{ij}^{t}=\sum_{j}^{J}ns_{j}^{t},\sum_{j}^{J}\sum_{k}^{K}nc_{jk}^{t}=\sum_{k}^{K}ns_{k}^{t} \tag{8}$$

$$\sum_{i}^{I}x_i^t\geq 1,\sum_{i}^{I}x_j^t\geq 1,\sum_{k}^{K}x_k^t\geq 1 \tag{9}$$

$$\sum_{j}^{J}\sum_{k}^{K}nc_{jk}^{t}\leq\sum_{k}^{K}ca_k,\sum_{j}^{J}\sum_{s}^{S}nc_{js}^{t}\leq\sum_{s}^{S}ca_s,\sum_{j}^{J}\sum_{p}^{P}nc_{jp}^{t}\leq\sum_{p}^{P}ca_p \tag{10}$$

Equation (1) shows the economic cost of reverse network of the model. Equation (2), (3), (4), (5) shows the fixed cost and the maintenance expense, the cost of information processing, the cost of disposing goods, the cost of transportation respectively. Equation (6) shows the social cost: The first and second item of the formula are proportional to the distance. The third of that is proportional to the quantity and is inversely proportional to distance. Equation (7) shows that the two objective functions are transformed into a single objective function. Constraint (8) shows the balance of goods flow and the balance of message flow. Constraint (9) shows that the quantity of potential nodes is at least more than one. Constraint (10) ensures that the quantity of the products transferred from one point to other points do not exceed their capacities.

3. ALGORITHM INTRODUCTION

3.1 GA and PSO

Genetic algorithm (GA) was proposed in the 1970s to simulate the process of natural biological evolution [15,16]. It is a global probabilistic intelligent algorithm which has characteristics of high parallelism, robustness, expansibility and universality based on genetic variation and natural selection [17]. Particle swarm optimization algorithm (PSO) was proposed to mimic the foraging process of a bird swarm by Eberhart and Kennedy of the United States [18].

3.2 Two-phase heuristic algorithm nesting GA/PSO

The planning of Two-phase heuristic algorithm nesting GA/PSO is as follows: the location and quantity of the primary recycling centers and union recycling centers are determined at the first stage, namely calling the PSO or GA to search for the optimal solution which makes the value of target function (Equation 11) to be minimum during the process of iteration. Then it turns into the second stage to determine the location and the number of the union disposing factories, namely calling the PSO or GA to search for the optimal solution which makes the value of target function

(Equation 12) to be minimum during the process of iteration. For the following arrangement, it returns to the first stage and re-determine the position and the quantity of the primary recycling centers and union recycling centers depending on the planning of the second stage. The two stages affect each other. The algorithm will not be finished until the position and quantity of the primary recycling centers, union recycling centers and union disposing factories remain unchanged.

The target function at the first stage:

$$\lambda_1 (u_1 + u_2 + u_3 + u_4) + \lambda_2 \xi \left\{ \sum_g^G \sum_i^I (nc_{gi}^t)^{\phi} \left(d_{gi}\right)^{\eta} x_i^t + \sum_g^G \sum_i^I \sum_j^J (nc_{ij}^t)^{\phi} \left(d_{gj}\right)^{\eta} x_j^t \right\} \quad (11)$$

The target function at the second stage:

$$\lambda_1 (u_1 + u_2 + u_3 + u_4) + \lambda_2 \xi \left\{ \sum_g^G \sum_j^J \sum_k^K \frac{(nc_k^t)^{\phi} x_k^t}{(d_{gk})^{\eta}} \right\} \quad (12)$$

4. INPUT DATA FOR THE MODEL

Take the used lead batteries recycled in Shanghai as an example. 7 points are preset as initial points (g_1, g_2 ... g_7), 3 potential primary recycling centers (i_1, i_2, i_3) are selected; 2 potential union recycling centers (j_1, j_2) are selected; 2 potential disposing factories (k_1, k_2) are selected; 1 manufacturer (p_1) is determined [19].

In the model, the simulated values of parameters [14] are as follows: the capacity of primary recycling center, union recycling center, union disposing factory , manufacturer and primary market is 20000, 30000, 30000, 20000 and 105000; The unit handling cost for primary recycling center, union recycling center, union processing factory and manufacturer is 5,4,6,7 respectively; The unit cost of transportation between two points is 1; The fixed cost for primary recycling center, union recycling center, union processing factory is 400000,600000,500000; The unit cost of information processing for each node is 4; The value of parameters α, β, γ is 0.2, 0.2, 0.55; The maintenance expense for primary recycling center, union recycling center, union disposing factory is 30000,45000,35000 respectively. Considering the limitation of length of paper, the initial value of location and recycling quantity are omitted.

5. COMPUTATIONAL RESULTS BY TWO-PHASE HEURISTIC ALGORITHM NESTING GA/PSO

The results of the two-phase heuristic algorithm nesting GA:

Table 1: Results of the Single period/ the multi-period

Optimal Solution	Single period	The first period	The second period	The multi-period
	4677288	2241316	2313224	4554540

Table 2: Results of the single stage/ the multi-stage

Optimal solution	Single stage	The first stage	The second stage	The multi-stage
	4677289	3244523	1432765	4677288

The results of the two-phase heuristic algorithm nesting PSO:

Table 3: Results of the Single period/ the multi-period

Optimal solution	Single period	The first period	The second period	The multi-period
	4676275	2241205	2313213	4554418

Table 4: Results of the single stage/ the multi-stage

Optimal solution	Single stage	The first stage	The second stage	The multi-stage
	4676275	3243512	1432763	4676275

The figures of the network location distribution

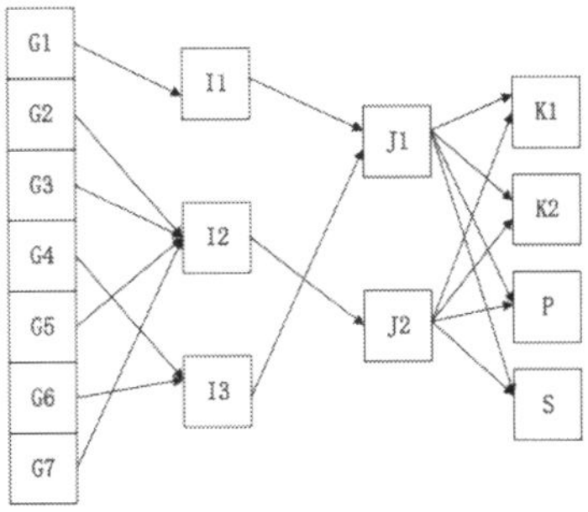

Figure 2: Single period location distribution.

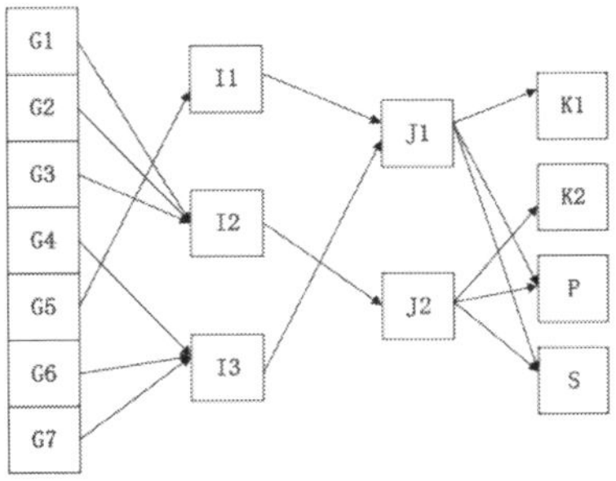

Figure 3: multi-period location distribution(GA).

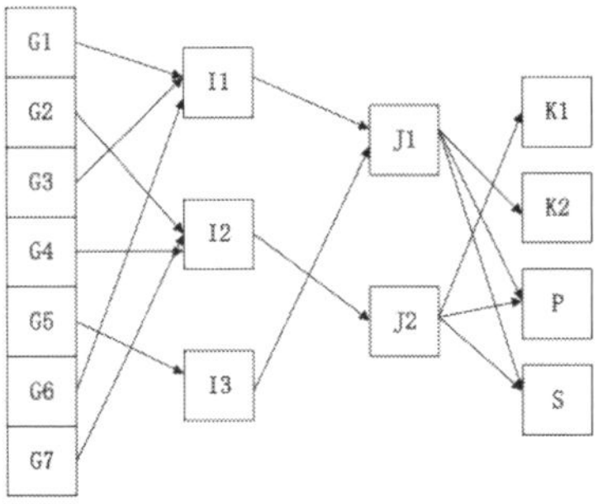

Figure 4: multi-period location distribution(PSO)

The computational result verifies the following:

(1) Because of the variety of recycling quantity of used lead battery, the location of the nodes in multi-period system is dynamic while the location in single period system is stationary, which is more accordant with the reality.

(2) The total system cost (Table1, Table3) and the quantity (Fig.2, Fig.3, Fig.4) of the chosen nodes in the multi-period system are less than those in the single period system.

(3) This kind of arrangement of the locations and quantity of its nodes in different stages can not only reduces the uncertain risk of the system, but optimizes the input of money, manpower and material during the whole operation process.

6. CONCLUSION

In this paper a multi-objective multi-period multi-stage government-enterprise reverse logistics network joint model is constructed, and a two-phase heuristic algorithm nesting GA/PSO is designed. Then the feasibility of the model is verified with a case study of used lead batteries in Shanghai. The study provides a reference for the design of joint construction of government-enterprise reverse logistics network and its optimal operation, especially for the recovery of this kind of used products with high negative externality.

7. ACKNOWLEDGEMENT

This work was partly sponsored by the National Natural Science Foundation, China (No. 71071093, 71471110), the Social Science Foundation of Shannxi Province, China (No. 2015D060) and by JSPS: Grant-in-Aid for Scientific Research (C; No. 15K00357).

REFERENCES

[1] **Zuo, Z., Tan, G., Xu, B. 2012.** Review and outlook on adverse logistics network, *Science and Technology Management Research*, 6, pp.75-78.

[2] **Zhang, Z., Liu, X. 2013.** Several issues about the basic concepts of logistics, *China Business and Market*, 2, pp.39-45.

[3] **Giovane Lopes Ferri, Gisele de Lorena Diniz Chaves, Glaydston Mattos Ribeiro. 2015.** Reverse logistics network for municipal solid waste management: The inclusion of waste pickers as a Brazilian legal requirement, *Waste Management*, 40, pp.173-191.

[4] **Sónia R. Cardoso, Ana Paula F.D. Barbosa-Póvoa, Susana Relvas. 2013.**Design and planning of supply chains with integration of reverse logistics activities under demand uncertainty, *European Journal of Operational Research*, 226 pp.436-451.

[5] **S.M. Hatefi, F. Jolai. 2014.** Robust and reliable forward-reverse logistics network design under demand uncertainty and facility disruptions, *Applied Mathematical Modelling*, 38, pp.2630-2647.

[6] **Ferri G L., Chaves GL D, Ribeiro G M. 2015.**Reverse logistics network for municipal solid waste management: The inclusion of waste pickers as a Brazilian legal requirement, *Waste Management*, 40, pp.173-191.

[7] **Kiran Garg, Devika Kannan, Ali Diabat, P.C. Jha. 2015.** A multi-criteria optimization approach to manage environmental issues in closed loop supply chain network design, *Journal of Cleaner Production*, 100, pp.297-314.

[8] **Kemal Subulan, Adil Baykasoğlu, Fehmi B. Özsoydan, A. Serdar Taşan, Hasan Selim. 2015.** A case-oriented approach to a lead/acid battery closed-loop supply chain network design under risk and uncertainty, *Journal of Manufacturing System*, 37, pp.340-361.

[9] **Maria Elbek, Sanne Wøhlk. 2016.** A variable neighborhood search for the multi-period collection of recyclable materials, *European Journal of Operational Research*, 249, pp.540-550.

[10] **Gao J., Wang R., Wang H. 2015.** Closed-loop supply chain network design under carbon subsidies, *Computer Integrated Manufacturing Systems*, 11, pp.3033-3040.

[11] **Weng C., Guo J. 2014.**Research on Reverse Logistics Network Joint Construction in B2C Apparel E-business Enterprises, *Resource Development & Market*, 30(9), pp.1046-1050.

[12] **China online. 2016.** Specification lead battery recovery system cannot be delayed, *http://www.ciaps.org.cn/news/show-htm-itemid-18148.html.*

[13] **Li J., Nie Y., Bai Q., Niu D. 2000.** Current Status and Future Strategic Focuses on Recycling of Waste Storage Batteries in China, *Environmental Protection*, 04, pp.40-42.

[14] **Li Changing, Zhnag F. 2014.** Reverse logistics network optimization of integrated location-routing-inventory problem, *Computer Integrated Manufacturing Systems*, 7, pp.1793-1798.

[15] **Yu X., Gen M. 2010.** *Introduction to evolutionary algorithms*, 430, pp, Springer, London.

[16] **Gen M., Cheng R. 2000.** *Genetic algorithms and engineering optimization*, 512, pp, John Wiley & Sons, New York.

[17] **Gen M., Cheng R., Lin L. 2008.** *Network models and optimization: Multi-objective Genetic algorithms approach*, 710, pp, Springer, London.

[18] **Koyuncu E, Erol R. 2015.** PSO based approach for scheduling NPD projects including overlapping process, *Computers & Industrial Engineering*, 85, pp.316-327.

[19] **Guo J, Liu X, Jo J. 2015.** Dynamic joint construction and optimal operation strategy of multi-period reverse logistics network: a case study of Shanghai apparel E-commerce enterprises, *Journal of Intelligent Manufacturing*, DOI: 10.1007/s10845-015-1034-8.

CLASSIFYING AND SCREENING THE ACTIVE FACTORS WITH SEQUENTIAL BIFURCATION IN SIMULATION EXPERIMENTS

Lijun Liu[1], Yizhong Ma[2*] and Yanjing Zhang[3]
[1]School of Economics and Management
Nanjing University of Science and Technology, Nanjing, China
lijun_liu06@163.com

[2]School of Economics and Management
Nanjing University of Science and Technology, Nanjing, China
yzma-2004@163.com

[3]School of Economics and Management
Nanjing University of Science and Technology, Nanjing, China
zhangyanjing_1211@163.com

ABSTRACT

Sequential bifurcation (SB) is a kind of group screening method that has been considered for identifying the active factors which are influential to the location effects of the response, using as few runs as possible in simulation experiments. However, in some industrial or simulation experiments, dispersion effects should also be considered. In this article, a new sequential bifurcation procedure considering both location and dispersion effects simultaneously is firstly proposed to identify the significant control factors and classify them into three types. Then a stopping rule is presented to control two types of error during the screening procedure. Finally, the efficiency and robustness of the proposed method is verified through a Monte Carlo experiment.

Keywords: Sequential bifurcation, Factor screening, Location effects, Dispersion effects, Robust parameter design

1 INTRODUCTION

Factor screening defined as searching for important input factors, is the first step of design of experiments[1]. According to the sparsity principle, there are only a few input factors among many input factors that are significant to the response during the process of simulation model or real system. Factor screening is an efficient way of identifying active factors with as few runs as possible. Factor screening methods include Resolution-III designs, supersaturated designs, group-screening designs and frequency domain experiments[2]. Sequential bifurcation (SB) proposed by Bettonvil [3] is used in the simulation experiments with many factors, which is the most efficient and effective method if the following assumptions are satisfied: (i) second-order polynomials are adequate approximations of the input/output functions implied by the simulation model; (ii) the signs of all first-order effects are known. These assumptions indicate the need of priori information about the system. Moreover, it is

* Corresponding Author

shown that efficiency would be further improved if factors are in the ascending order of their effects (Bettonvil and Kleijnen [4]).

Recently, SB draws much attention from many researchers. The original SB method assumes deterministic simulation. Next, Cheng [5] extended SB to the situation of random discrete-event simulation, which is more common in the real-world problem (Shi et al. [6]). And on this basis, controlled sequential bifurcation (CSB) was proposed by Wan et al. [7], which can restrict two types of error during the screening procedure. Then a generalized procedure of CSB is addressed to improve screening efficiency and efficacy, and the hypothesis testing procedure is also extended to the model with first-order interaction[8].

The naive SB above only considers location effects. While inspired by the robust parameter design by Taguchi, the dispersion effects often play a more important role than location effects. In the screening procedure, ignoring the dispersion effects may lead to the loss of some significant factors and lead to estimating a wrong metamodel. Ankenman et al. [9] firstly presented a screening procedure for dispersion effects by SB, which provides a difficult statistical proof of stopping rules and importance test procedure to control two types of error. However, they did not consider the location and dispersion effects simultaneously. According to the classification of control factors by Taguchi, different control factors implement in different steps of robust parameter design. For instance, adjustment factors are used to make the response close to nominal value but maintain the variation fixed; while robust factors play a role in reducing variation but keeping the location constant.

The rest of the article is organized as follows. Section 2 addresses a new screening procedure of SB for considering both location and dispersion effects simultaneously, which includes stopping rules and importance test for the compound problem based on a result in Ankenman et al. [9]. Section 3 gives a Mote Carlo experiments which shows the efficiency and robustness of the proposed method and concluding remarks are given in Section 4.

2 SB FOR LOCATION AND DISPERSION EFFECTS

2.1 Modeling and screening procedure

Based on the assumption of naive SB, we assume that the adequate metamodel for both location and dispersion effect is a first-order polynomial.

The metamodel is denoted by

$$y_i = \beta_0 + \sum_{j=1}^{k} \beta_j x_{ij} + \varepsilon_i = \beta_0 + \boldsymbol{\beta}^T \boldsymbol{x}_i + \varepsilon_i(\mathbf{x}_i) \quad (1)$$

$$log\ \sigma(\mathbf{x}_i) = \gamma_0 + \sum_{j=1}^{k} \gamma_j x_{ij} = \gamma_0 + \boldsymbol{\gamma}^T \mathbf{x}_i \quad (2)$$

$i = 1,2,\cdots,n$ and $j = 1,2,\cdots,k$, where n denotes the total simulation runs, which have been made. Here, y_i is the observed response in the ith run; there are k factors, and x_{ij} is the code level of factor j in the ith run which is 0 or 1; β_0 is a general mean; and β_j is the jth unknown factor location effects. In the vector form, we have written $\boldsymbol{\beta} = (\beta_1, \beta_2, \cdots, \beta_k)^T$ and $\mathbf{x}_i = (x_{i1}, x_{i2}, \cdots, x_{ik})^T$.

$\varepsilon_i(\mathbf{x}_i)$ is the observation error normally distributed $N(0, \sigma^2(\mathbf{x}_i))$. The standard deviation takes the form of formula (2) where γ_j denotes the jth unknown factor dispersion effect. And its vector form is $\boldsymbol{\gamma} = (\gamma_1, \gamma_2, \cdots, \gamma_k)^T$.

According to the second assumption of naive SB, operators can adjust the two levels of the control factor to make the location and dispersion effects non-negative: $\beta_j \geq 0$, $\gamma_j \geq 0$. So we present the following factor screening and classification algorithm focused on both location and dispersion effects.

Factor Screening and Classification Algorithm

Step 0 Initialization: Putting all the factors in one single group.

Step 1 Error controlling: Generating initial number of experiments n_0, and adding one experiment point at a time until the stopping rule is satisfied.

Step 2 Importance tests: Calculating the confidence interval with the experiment data. There will be three cases.

(a) If the group is not significant to both location effects and dispersion effects, then remove it. And all the factors in this group are regarded as unimportant.

(b)If it is significant to either location or dispersion effects, and there are more than one significant factor in this group, this group is divided into two subgroups, and return to Step1.

(c) If it is significant to either location or dispersion effects, but there is only one factor in this group, then turn to next step.

Step 3 Factor classifications: According to the importance test, classifying the significant factors into three kinds, important factors, adjustment factors and robust factors.

As illustrated above, the main process is based on the SB procedure. Furthermore the screening and classification are interlaced in this algorithm, therefore factors can be screened and classified simultaneously. The stopping rules and importance tests are the two critical issuess which we will investigate more.

2.2 Error controlling stage

For estimating an arbitrary group location effect, we firstly take observations in pairs, one at $\mathbf{x}(k_1)$, and the other at $\mathbf{x}(k_0 - 1)$.

$$y(k_1) = \beta_0 + \sum_{j=1}^{k} \beta_j + \varepsilon[\boldsymbol{x}(k_1)] \tag{3}$$

$$y(k_0 - 1) = \beta_0 + \sum_{j=1}^{k_0-1} \beta_j + \varepsilon[\boldsymbol{x}(k_0 - 1)] \tag{4}$$

Their difference is

$$d(k_0, k_1) = y(k_1) - y(k_0 - 1) = \sum_{j=k_0}^{k} \beta_j + \varepsilon(k_0, k_1) = \mu(k_0, k_1) + \varepsilon(k_0, k_1), \tag{5}$$

which is clearly a normal distributed random variable. Taking n pairs of observations, and $d(k_0, k_1)$, $i = 1,2, \cdots, n$ can be calculated, and their average difference is obtained as

$$D_n(k_0, k_1) = n^{-1} \sum_{i=1}^{n} d_i(k_0, k_1) = \mu(k_0, k_1) + n^{-1} \sum_{i=1}^{n} \varepsilon_i(k_0, k_1) \tag{6}$$

which can be used to estimate $\mu(k_0, k_1)$ with improved accuracy as n increases. Consequently, the problem is the determining of n, which depends on the two types of specified error.

In this paper, the fixed-width confidence interval method is used in the stopping rules. Let the confidence interval (CI) be of the form $(D_n - w_1, D_n + w_2)$, $w_1, w_2 > 0$. We can obtain such a CI from the probability statement

$$P\left(z_\beta \le \sqrt{n}(D_n - \mu)/\sigma \le z_{1-\alpha}\right) = 1 - \alpha - \beta \tag{7}$$

By setting $w_1 = \sigma z_{1-\alpha}/\sqrt{n}$ and $w_2 = -\sigma z_\beta/\sqrt{n}$, the probability statement in (7) can be rewritten as

$$P(D_n - w_1 \le \mu \le D_n + w_2) = 1 - \alpha - \beta \tag{8}$$

If the fixed width confidence interval is $w = w_1 + w_2$, we have

$$w_1 = \frac{z_{1-\alpha}}{z_{1-\alpha} - z_\beta} w \text{ and } w_2 = -\frac{z_\beta}{z_{1-\alpha} - z_\beta} w \tag{9}$$

Then we can obtain the formula which is also called *Known Sigma Stopping Rule*

$$n^{-1}\sigma^2 = \frac{w^2}{\left(z_{1-\alpha}-z_\beta\right)^2} \quad \textbf{(10)}$$

Consider now the situation where σ^2 is not known, then a substitute of equation (10) is

$$n^{-1}S_n{}^2 = \frac{w^2}{\left(z_{1-\alpha}-z_\beta\right)^2} \quad \textbf{(11)}$$

where $S_n{}^2 = \frac{1}{n-1}\sum_{i=1}^{n}(d_i - D_n)^2$.

As we know, the choice of n in (11) has first-order accuracy, Anscombe [10] further discussed the second-order accurate version of (11) in the case of $\alpha = \beta$, which was extended to where α and β are not necessarily equal [9], and the corresponding stopping rule of

$$\{n[n - 2.676 - (\tau_0/2)]\}^{-1}\sum_{i=1}^{n-1} U_i \le \frac{w^2}{\left(z_{1-\alpha}-z_\beta\right)^2} \quad \textbf{(12)}$$

where

$$U_i = \frac{1}{i(i+1)}\left\{id_{i+1} - \sum_{j=1}^{i} d_j\right\}^2 \; i = 1,2,\cdots,n-1, \quad \textbf{(13)}$$

and

$$\tau_0 = \left[z_{1-\alpha}\varphi(z_{1-\alpha}) - z_\beta\varphi(z_\beta)\right]^{-1}\left[z_{1-\alpha}^3\varphi(z_{1-\alpha}) - z_\beta^3\varphi(z_\beta)\right] \quad \textbf{(14)}$$

For the dispersion problem, we can use *Known Sigma Stopping Rules* to determine the value of simulation numbers based on (10), and $\sigma^2 = \pi^2/4$. So the number of observations needs to be

$$n = \frac{\pi^2\left(z_{1-\alpha}-z_\beta\right)^2}{4w^2} + 1 \quad \textbf{(15)}$$

It should be pointed out that the purpose here is to handle the stopping rules for location and dispersion problems simultaneously. For the proof details for the single dispersion problem, it can refer to Ankenman et al. [9].

As illustrated above, (12) and (15) are the two separate stopping rules for location and dispersion problems. Here we specify n in (12) as $n^{(\mu)}$ and n in (15) as $n^{(\sigma)}$, which denote the observations needed for location and dispersion problems, respectively. So the overall stopping rule is

$$n = max\{n^{(\mu)}, n^{(\sigma)}\} \quad \textbf{(16)}$$

which shows that the error of either problem can be controlled.

In the stage of error controlling, we mainly attempt to design the stopping rules to control the two types of error. And in the next stage, the objective is to verify the group of factors significant or not, which is called importance test.

2.3 Importance test and factors classification

For the location problem, we can calculate the upper and lower limits of CI according to (12)

$$C_U = D_n - w\left(z_{1-\alpha} - z_\beta\right)^{-1} z_\beta \quad \textbf{(17)}$$

and

$$C_L = D_n - w\left(z_{1-\alpha} - z_\beta\right)^{-1} z_{1-\alpha}. \quad \textbf{(18)}$$

It is obvious that $C_U - C_L = w = \Delta_1 - \Delta_0$, which satisfies the precondition. Δ_1 and $\Delta_0$denote the upper and lower specification limits, respectively.

For handling the location and dispersion problems simultaneously, we use superscript to distinguish the two problems. Then we can rewrite equations (17) and (18) as

$$C_U^{(\mu)} = D_n - w^{(\mu)}\left(z_{1-\alpha} - z_\beta\right)^{-1} z_\beta \tag{19}$$

and

$$C_L^{(\mu)} = D_n - w^{(\mu)}\left(z_{1-\alpha} - z_\beta\right)^{-1} z_{1-\alpha} \tag{20}$$

where $w^{(\mu)} = \Delta_1^{(\mu)} - \Delta_0^{(\mu)}$.

For the dispersion problem, the upper and lower limits of the CI are

$$C_U^{(\sigma)} = H_n - w^{(\sigma)}\left(z_{1-\alpha} - z_\beta\right)^{-1} z_\beta \tag{21}$$

and

$$C_L^{(\sigma)} = H_n - w^{(\sigma)}\left(z_{1-\alpha} - z_\beta\right)^{-1} z_{1-\alpha} \tag{22}$$

where $w^{(\sigma)} = \Delta_1^{(\sigma)} - \Delta_0^{(\sigma)}$. H_n here is a statistic analogy to D_n in the location problem. And the calculation of H_n is as follow.

Using the n pairs of observations at points $\mathbf{x}(k_0 - 1)$ and $\mathbf{x}(k_1)$, we can get the statistic sequences

$$V_i[\boldsymbol{x}(k)] = \frac{1}{i(i+1)}\left\{iy_{i+1}[\boldsymbol{x}(k)] - \sum_{j=1}^{i} y_j[\boldsymbol{x}(k)]\right\}^2 \quad i = 1,2,\cdots,n-1, k = k_0 - 1, k_1 \tag{23}$$

From the equations of (4) and (5), it is obvious that

$$0.5logV_i[\boldsymbol{x}(k)] = \gamma_0 + \sum_{j=1}^{k} \gamma_j + \xi_{ik} \quad i = 1,2,\cdots,n-1 \quad k = k_0 - 1, k_1 \tag{24}$$

where ξ_{ik} is distributes as a $\frac{1}{2}\log(\chi_1^2)$ variable.

The sequence

$$h_i(k_0,k_1) = 0.5\{logV_i[\boldsymbol{x}(k_1)] - logV_i[\boldsymbol{x}(k_0 - 1)]\} = \sum_{j=k_0}^{k_1} \gamma_j + \delta_i, \; i = 1,2,\cdots,n-1 \tag{25}$$

can be used to get the analog to D_n, that is,

$$H_n = \frac{1}{n-1}\sum_{i=1}^{n-1} h_i(k_0,k_1) = \lambda(k_0,k_1) + \frac{1}{n-1}\sum_{i=1}^{n-1} \delta_i, \tag{26}$$

which can be used to test the importance of $\lambda(k_0,k_1)$ in the factor dispersion effects problem.

Using the analysis and calculation results of the CI for the location and dispersion effects problem in the importance test stage, the subgroup are classified as follows:

(a) The subgroup is unimportant, if $C_U^{(\mu)} < \Delta_1^{(\mu)}$, $C_U^{(\sigma)} < \Delta_1^{(\sigma)}$.

(b) The subgroup is important for both the location and dispersion effects, if $C_L^{(\mu)} \geq \Delta_0^{(\mu)}$, $C_L^{(\sigma)} \geq \Delta_0^{(\sigma)}$.

(c) The subgroup is important only for the location effect, if $C_L^{(\mu)} \geq \Delta_0^{(\mu)}$, $C_U^{(\sigma)} < \Delta_1^{(\sigma)}$.

(d) The subgroup is important only for the dispersion effect, if $C_U^{(\mu)} < \Delta_1^{(\mu)}$, $C_L^{(\sigma)} \geq \Delta_0^{(\sigma)}$.

As illustrated in the factor screening and classification algorithm, a subgroup is considered significant if either the location or the dispersion effect is significant, while it is insignificant if both the two effects are insignificant. So the subgroup will be removed if the result of importance test belongs to category (a). While the subgroup matches the category (b), (c), or (d), the bifurcation of this subgroup begin. As it shows in section 2.1, it is an iteration procedure and stops when there is only one factor in one subgroup. Then we can classify the factors as Figure 1.

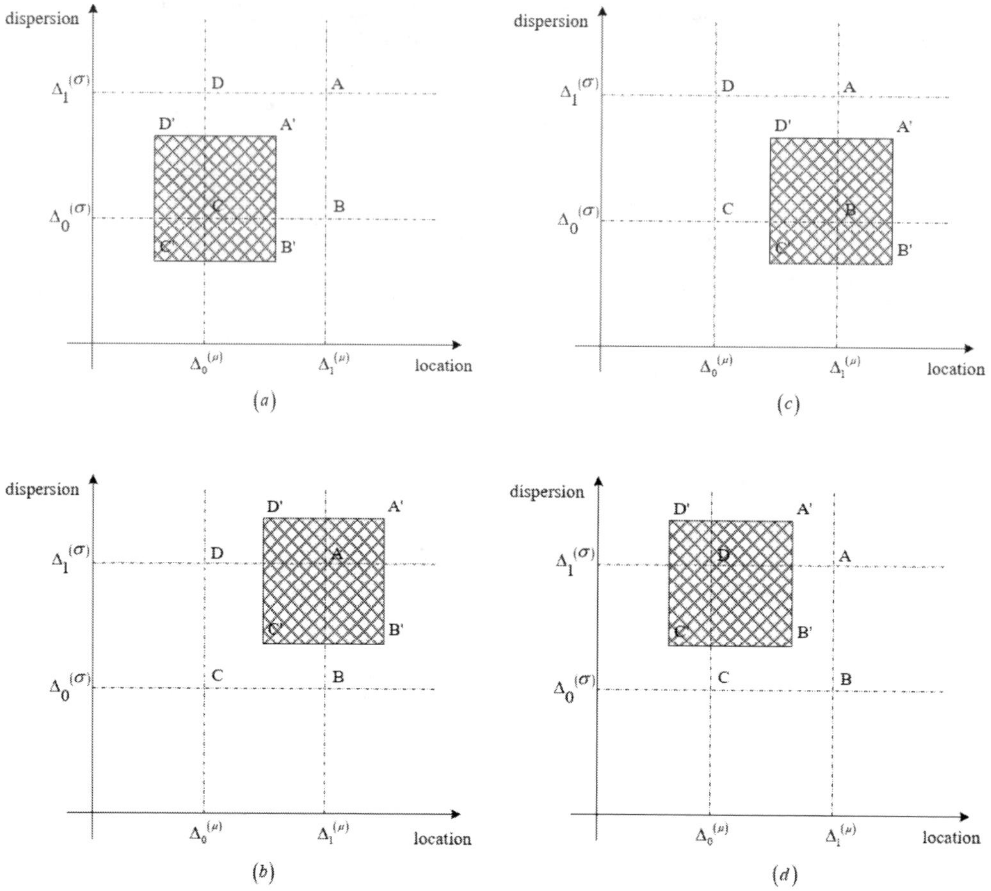

Figure1: Four Categories of Control Factors

In Figure 1, (a), (b), (c), and (d) are the four types of factors; unimportant, important, adjustment, and robust factors. The rectangular $ABCD$ is the specified region, and x axis denotes the location effect and y axis the dispersion effects. According to the importance test, a confidence region which consists of location and dispersion effects is shown as $A'B'C'D'$. In Figure 1(a) the confidence region $A'B'C'D'$ is in the lower-left from the specified region $ABCD$, which shows that the two kinds of effects are insignificant. And hence Figure 1(a) indicates the category of unimportant factors. Note that most of the unimportant factors are identified in the first situation of Step 2, which can interpret the high efficiency of SB Algorithm. The other three categories can be interpreted similarly.

3 NUMERICAL EXAMPLE

In order to illustrate the efficiency and robustness of the proposed approach, a Monte Carlo experiment is employed in this section. Like Ankenman et al. [9], here we choose the probabilities of two types of errors $\alpha = \beta = 0.1$, and the lower and upper limits for location effects are $\Delta_0^{(\mu)} = 2$, $\Delta_1^{(\mu)} = 4$. Note that, for the dispersion problem, the limits are $\Delta_0^{(\sigma)} = \log(1.1) \approx 0.0953$, $\Delta_1^{(\sigma)} = \log(1.25) \approx 0.2231$. And this pair of limits indicates that there has to be a 25% increase in the standard deviation for the change to be significant, while an increase of up to 10% would to be considered insignificant. In this section, we consider two kinds of order for the factors, effect order and random order, both scenarios with the same 32 factors. Eight values for the location effects are $\beta = (2, 2.44, 2.88, 3.32, 3.76, 4.2\ 4.64, 5.08)$, and four values for the dispersion effects are $\gamma = (0.05, 0.0953, 0.2231, 0.4)$. As we know there are 16 factors which have the distinct classes, the results of these factor classifications in two experiment scenarios are in Table 1 and Table 2, each with 1000 independent experiments.

Table 1: The Results of Factor Classification in Effect Order

	0.05	0.0953	0.2231	0.4
2	Unimportant[0.984]	Unimportant[0.958]	Robust [0.959]	Robust [0.998]
4.2	Adjustment [0.992]	Adjustment [0.951]	Important [0.947]	Important [0.997]
4.64	Adjustment [0.978]	Adjustment [0.943]	Important [0.962]	Important [0.997]
5.08	Adjustment [0.983]	Adjustment [0.941]	Important [0.951]	Important [0.996]

Table 2: The Results of Factor Classification in Random Order

	0.05	0.0953	0.2231	0.4
2	Unimportant [0.987]	Unimportant [0.959]	Robust [0.945]	Robust [0.997]
4.2	Adjustment [0.987]	Adjustment [0.956]	Important [0.957]	Important [0.996]
4.64	Adjustment [0.984]	Adjustment [0.955]	Important [0.957]	Important [0.999]
5.08	Adjustment [0.981]	Adjustment [0.960]	Important [0.946]	Important [0.999]

In the two Tables, the first row and the first column denotes the location and dispersion effect of the factor, respectively. So by considering the four location effects and four dispersion effects, there are the 16 factors with distinct categories. In every cell, the corresponding factor category is written, and the number in the bracket is the ratio of right classification in the 1000 replication. In Table 1, we can see that four types of factors can be classified correctly, and the errors of mis-classification are all controlled within the specified thresholds. The same conclusion can be obtained in the Table 2, which shows the efficiency and robustness of the proposed algorithm.

In addition, if the naive SB procedure is employed, the adjustment factors and important factors can also be screened out, whereas the robust factors may perform insignificance. However, all the factors can be screened out by the proposed screening method, which shows its advantage over the traditional SB.

4 CONCLUSION

Inspired by the theory of robust parameter design, we consider both location and dispersion effects simultaneously during the SB procedure and classify all the significant factors into three categories according to the classification of control factors by Taguchi. All the factors can be rightly screened and classified using the proposed screening procedure. And the simulation results show the efficiency and robustness of the proposed method. Further study may include the screening of robust factors under the circumstance of multiple responses, which is more practical in the simulation experiments.

5 REFERENCES

[1] **Dean, A., Lewis, S.** 2006. Screening: methods for experimentation in industry, drug discovery, and genetics, Springer Science & Business Media.

[2] **Kleijnen, J.P.C.** 2015. Design and Analysis of Simulation Experiments, 2rd Edition, Springer.

[3] **Bettonvil, B. 1990.** Detection of Important Factors by Sequential Bifurcation, Ph.D. dissertation, Tilburg University Press, Tilburg.

[4] **Bettonvil, B., Kleijnen, J.P.C. 1997.** Searching for important factors in simulation models with many factors: Sequential bifurcation, *European Journal of Operational Research*, 96(1), pp 180-194.

[5] **Cheng, R.C.H. 1997.** Searching for important factors: Sequential bifurcation under uncertainty, Proceedings of the 29th conference on Winter simulation, pp 275-280.

[6] **Shi, W., Shang, J., Liu, Z., Zuo, X. 2014.** Optimal design of the auto parts supply chain for JIT operations: sequential bifurcation factor screening and multi-response surface methodology, *European Journal of Operational Research*, 236(2), pp 664-676.

[7] **Wan, H., Ankenman, B.E., Nelson, B.L. 2006.** Controlled sequential bifurcation: A new factor-screening method for discrete-event simulation, *Operations Research*, 54(4), pp 743-755.

[8] **Wan, H., Ankenman, B.E., Nelson, B.L. 2010.** Improving the efficiency and efficacy of controlled sequential bifurcation for simulation factor screening, *INFORMS Journal on Computing*, 22(3), pp 482-492

[9] **Ankenman, B.E., Cheng, R.C.H, Lewis, S.M. 2014.** Screening for Dispersion Effects by Sequential Bifurcation, *ACM Transactions on Modeling and Computer Simulation*, 25(1), pp 1-27.

[10] **Anscombe, F.J. 1953.** Sequential estimation, *Journal of the Royal Statistical Society, Series B*, 15(1), pp 1-29.

Acknowledgements

This work is partly supported by the Nation Natural Sciences Foundation of China under Grants NSFC 71471088 and Innovation project of Jiangsu Province KYLX16_0406 in China.

PRODUCT DESIGN PROBLEM SOLVING BASED ON CONTRADICTION MATRIX ABOUT DESIGN RATIONALE MODULE

A. ZHOU Jianhui[1] B. YANG Haicheng[2] and C. LIU Jihong[3*]
[1]School of Mechanical Engineering and Automation
BeiHang University, Beijing, China
zhjh19851101@163.com

[2]China Aerospace Science and Technology Corporation
Beijing, China
13911253139@139.com

[3]School of Mechanical Engineering and Automation
BeiHang University, Beijing, China
ryukeiko@buaa.edu.cn

ABSTRACT

Product design is a particular process, in which designers use their knowledge, experience, inspiration, etc. to create the product. While the design style is freedom, the design consideration is constrained by the performance of product, especially industrial product. There are countless impact factors in product design, these factors is not changeless. Different products have different factors, and different designers focus different factors. Due to these differences, although in the same product, the problems needed to be solved is different. TRIZ can provide the abstract solution for these problems. Designers should make efforts to transform the abstract solution to the concrete solution. If we can make use of the design case knowledge, which exist in the knowledge base, designers could get the concrete solution by other designers directly. In this paper, we make module to represent the design case knowledge, we name it design rationale (DR), and then, structure the contradiction matrix about DR, this matrix replaces the inventive principles with the solution methods, finally use this matrix to provide the concrete solution for designers. Meanwhile, designers can focus the factors they assign, i.e. they can get the appropriate solution, which correspond the appropriate factors. Therefore, the method presented in this paper help designer making decisions more quickly, more accurately, and more reasonable.

Keywords: Problem solving, Contradiction matrix, Design rationale, TRIZ

1 INTRODUCTION

The first aim of product design is to satisfy the functional requirements and performance requirements of the product. The functional requirements decide the function modules of the

* LIU Jihong, born in 1966, is currently a professor at Beihang University, China. His research interests include complex product engineering, digital design and manufacturing, knowledge management and design theory and method.

Acknowledge: The work supported by National Key Technology R&D Program through approval No. 2015BAF18B01 and National Science Foundation of China through approval No.51475027.

product in the process of design. However, whether the performance index of each module reaches the desired requirements or not involves many factors, and it is also the main basis of trade-offs and judgment decision for the design personnel in designing. There is not absolute good or bad for a product design scheme. One which can meet the expected requirements of all aspects is thought to be a successful design. Therefore, in the process of product design, a deciding problem involving multiple indexes is the crucial question to decide whether the design scheme is successful or not.

Design rationale represents all the knowledge existing in the process of design and it is the process knowledge of design[1]. The reuse of design rationale can be seen as the reuse of knowledge, especially as the reuse of process knowledge. The methods of knowledge reuse mainly rely on the knowledge retrieval and knowledge mining[2].

Because of the difference of knowledge, retrieval can be classified into content retrieval and case retrieval. Content retrieval can be seen as the retrieval of text in this paper, absolutely it also has the retrieval of picture, media, etc. The technologies of the retrieval of text include participle, natural language processing (NLP), ontology, etc. They are mainly applied in the retrieval of the web information, documents, and so on. By these methods, designers can only get the results, which contain the keywords or the related keywords[3], not the principles or rules existing in the results. This confinement cause that the knowledge reuse only by retrieval have not been very effective. The other mode of retrieval is to retrieval the cases, which have been existed in the knowledge database[4]. Design rationale can be seen as a kind of case. Recently, there are some methods to retrieval the design rationale. They analyze the semantic relationships of the classes in the design rationale, and then extract the information or sub-module according to the requirements of designers[5]. The results reflect why do it and how to do it in every case – every design rationale. However, when designers do work, they often want to get the principles or rules which have been reflected by all or a part of design rationales, not only one. Therefore, neither the text retrieval nor the cases retrieval have been very effective.

TRIZ (Teoriya Resheniya Izobreatatelskikh Zadatch) is a knowledge-based systematic methodology of inventive problem solving[6]. Fey and Rivin described TRIZ as a methodology for the effective development of new systems, in addition to it being a set of principles that describe how technologies and systems evolve. Also, it has been described by Gadd as a toolkit consisting of methods which cover all aspects of problem understanding and solving. This toolkit is regarded by some as the most comprehensive, systematically organized for invention and creative thinking methodology known to man[7].

Problem solving theory and contradiction matrix theory in TRIZ is a kind of abstract results formed after a comprehensive analysis, summarizing and reasoning of a large number of existing cases[8]. The abstract results have a very distinctive advantage: the applicability is very good. Aiming at the problem of input, instead of giving solutions to the existing cases, problem solving gives an abstract solving method[9,10]. Contradictions are indicative of inventive problems arising from the apparent incompatibility of desired features within a system. Resolving the contradictions is solving the problems. There are two major types of contradictions: technical contradictions and physical contradictions[11]. The contradiction matrix correspond the inventive principles. These principles can direct designer to solve the problem they faced.

This paper refers to the thoughts in TRIZ and forms the abstract contradiction matrix after a conclusion of knowledge for trade-offs in many indexes in the process of design thinking. TRIZ more focuses on product innovation design question, and as a result 40 items of invention principle are summarized. But in problems of trade-offs in many indexes, it does not involve product innovation problem, as long as giving a new method for trade-offs, namely the solving method, it can guide the design personnel to solve the problem very well. The contradiction matrix takes impact factor (namely performance index) as parameters, and it can more accurately represent the problem and solve the problem. The solving method is a kind of summary and abstract for large amount of case solution methods, and the applicability is also very good. We can find that the reuse efficiency of this method is higher than other commonly used retrieval and mining methods for design knowledge.

2 DESIGN RATIONALE MODULE

2.1 Design Rationale

Design rationale module provides an efficient approach to the representation of design knowledge. It is consists of concepts and the sematic relations of concepts. In this module, the "evolve-to" segment represents a kind of design knowledge, which includes the decisions made by designers for meeting the performance requirements. Design rationale module represents almost all knowledge and the relation of this in the design thinking process. It is structured by four child modules: Design Intention, Design Decision, Decision Justification, and Design Operation.

In design rationale module, "evolve-to" segment represents the relation of design intention and decision justification. It is that designer make the transformation in the process of product design in order to meeting the requirements.

2.2 Ternary Module

Design rationale consists of design intention, design decision and decision justification, etc. No matter design intention, or design decision, they both have their reason – decision justification at the decomposition to the next segment. The reason of decision justification can be treated as impact factors. This paper structures a ternary module to represent the intention driving decision, by splitting the "evolve-to" segment of design rationale module.

Ternary module include different decisions, because it need to meet different reasons – different impact factors. Therefore, ternary module is a module which is constrained by impact factors. We use the formula to represent this ternary module in the paper.

$$M = \{M_1, M_2, \dots, M_n\} \tag{1}$$

$$I = \{i_1, i_2, \dots, i_n\} \tag{2}$$

$$D = D_F \cup D_B = \{d_{f1}, d_{f2}, \dots, d_{fn}\} \cup \{d_{b1}, d_{b2}, \dots, d_{bn}\} \tag{3}$$

$$M_k = M_{kf} \cup M_{kb} = \{E = k|(i_1, d_{f1}), (i_1, d_{f2}), \dots, (i_1, d_{fs})\} \cup \{E = k|(i_2, d_{b1}), (i_2, d_{b2}), \dots, (i_2, d_{bt})\} \tag{4}$$

$$E = g(2^J) = \{x, y \in N|(f_x, f_y)\} \tag{5}$$

Where: M is *set of all ternary modules*, I is *set of all intentions*, J is *set of all impact factors*, f_k is *impact factor*, D is *set of all justifications*, D_f is *set of all parent justifications*, D_b is *set of all child justifications*, $g(A)$ is *function of the impact factor contradiction*, E is *set of all impact factor contradictions*.

In ternary module, "intention" and "justification" is not equal to design intention and decision justification, and they are two indexes of every segment in design rationale module, one is the active index, the other is the passive index. For example, Figure 1 shows the segment of turbine blades design rationale. By abstracting this instance, a ternary module about turbine blades is structured, and we can use the formula to represent it.

$$E = g(f_1, f_2, f_3) = \{(f_1, f_2), (f_1, f_3)\} \tag{6}$$

$$M_{1f} = \{E = l|(i_1, d_1)\} \tag{7}$$

$$M_{1b} = \{E = l|(i_2, d_2)\} \qquad (8)$$

Where: i_1=i_2="*quantity*", d_1="*ten blades*", d_2="*fifteen blades*", f_1="*stability*", f_2="*cost*", f_3="*energy*".

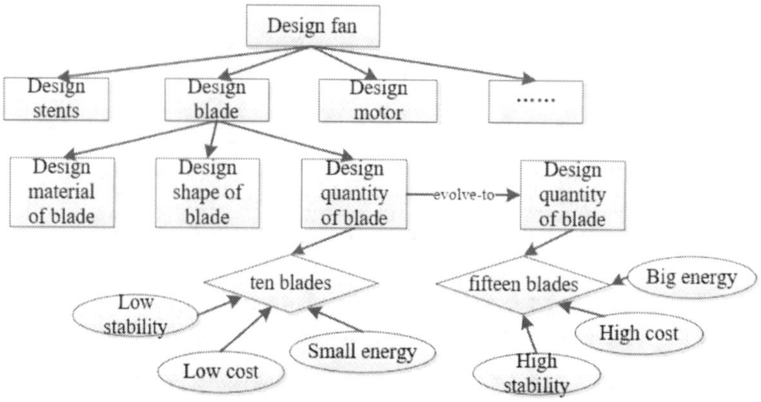

Figure 1: Segment of fan design rationale

3 CONTRADICTION MATRIX ABOUT DESIGN RATIONALE

TRIZ has good universal applicability, contradiction matrix, inventive principle and solving methods are summarized and abstracted to become standard solutions according to it, and it is suitable for all kinds of invention. Therefore, with the aid of TRIZ, this article abstracts the design problem converting part in design rationale and builds design rationale contradiction matrix after summarizing, so that designers can effectively reuse it later. As shown in Figure 2 is the overall framework of the structure method of design rationale contradiction matrix.

Figure 2 shows that in the contradiction matrix about design rationale, it corresponds to the contradiction between the impact factors are the solution abstractly obtained by the ternary module evolution, rather than the inventive principles. This method focuses on the impact factor rather than the inventive principle, the former is more applicable to the product innovative design, and the analysis of the impact factor is more suitable for the reuse of existing cases in the product conceptual design.

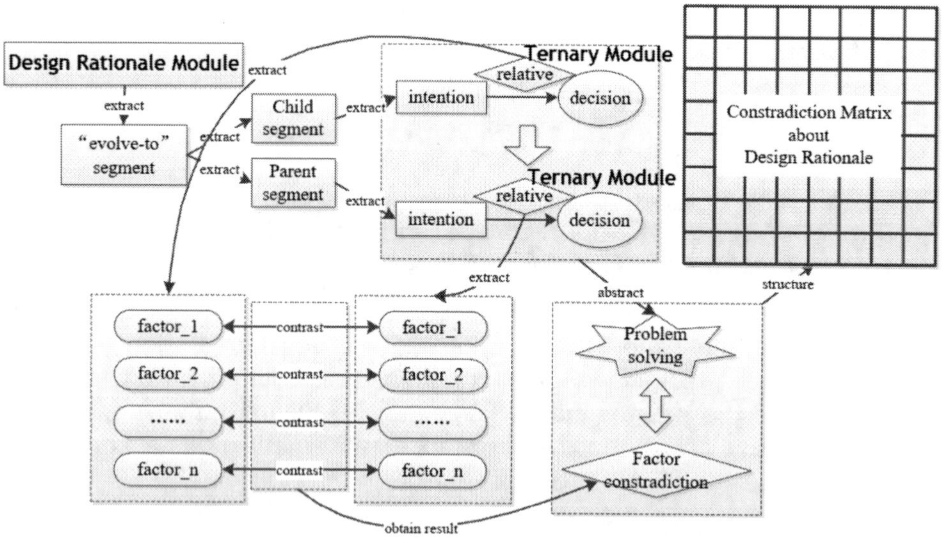

Figure 2: Method of structure contradiction matrix about design rationale

4 SOLUTION APPROACHES

4.1 Abstract Solution Approaches

Every "evolve-to" segment represents an evolution of intention, and it can be seen as an evolution of the ternary module. By referencing the problem solving theory, this paper summarizes five abstract approaches:

First Solution: "intention transformation", transform the intention to another, in order to meeting design requirements.

$$M' = G(M) = M_f \cup M_b = \{E|(i,d)\} \cup \{E|(i',d')\} \qquad (9)$$

Second Solution: "intention decomposition", decompose the intention to one or more child intention, in order to meeting design requirements.

$$M' = G(M) = M_f \cup M_b = \{E|(i,d)\} \cup \{E|(i_1,d_1),\dots,(i_n,d_n)\} \qquad (10)$$

Third Solution: "justification transformation", transform the justification to another, in order to meeting design requirements.

$$M' = G(M) = M_f \cup M_b = \{E|(i,d)\} \cup \{E|(i,d')\} \qquad (11)$$

Fourth Solution: "justification increase", increase one or more justifications, in order to meeting design requirements.

$$M' = G(M) = M_f \cup M_b = \{E|(i,d)\} \cup \{E|(i,d+d_1+\dots+d_n)\} \qquad (12)$$

Fifth Solution: "justification removal", remove unnecessary justifications, in order to meeting design requirements.

$$M' = G(M) = M_f \cup M_b = \{E|(i,d)\} \cup \{E|(i,d-d_1-\dots-d_n)\} \qquad (13)$$

Figure 3 illustrates that how the approaches are abstracted by analysis of the relation in design rationale module. And then it shows the rules of this method.

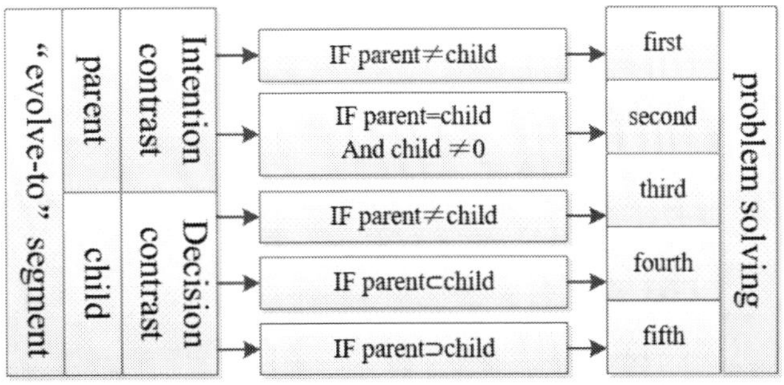

Figure 3: Process of abstracting solution approaches

4.2 Mining based on Frequency

Every contradiction of the factors has one or more problem solutions in one design rationale, so with abstracting amount of design rationales, varieties of problem solutions can be corresponded to one contradiction of the factors. In order to get the only one standard solution, this paper uses the method named mining based on frequency. This method statistics the accounts of cases which corresponding it, and then we can calculate the frequency of each problem solution. Table 1 is an example of the contradiction of factors, which are ***stability*** (optimized parameter) and ***mass*** (deteriorated parameter). **Third solution** is the finally solution, because of the highest frequency.

Table 1: Frequency of problem solution of the contradiction of stability and mass

Problem Solution	Corresponded Case	Account	Frequency
First Solution	*Gas turbine blades design, Turbine engine design*	2	12.5%
Second Solution	*NULL*	0	0
Third Solution	*Turbine blades design, High turbo design, Gas turbine blades design, ...*	9	56.25%
Fourth Solution	*High turbo design, Turbine engine design, Turbine design, ...*	4	25%
Fifth Solution	*Wind turbine design*	1	6.25%

This example illustrate that the frequency of the problem solution is depending on the account of the cases, which have the corresponded solution. Therefore, we can obtain the only standard problem solution by the mining based on frequency.

5 CONTRADICTION MATRIX STRUCTURE

The contradictory factors of contradiction matrix in TRIZ are 39 engineering parameters obtained after summarizing and induction, which is invariance. And the contradictory factors of contradiction matrix in design rationale are impact factors. The number of the impact factors is dynamic, and it equals to the total number of impact factors of all samples in design rationale.

In the contradiction matrix of design rationale, the transverse axis direction is on behalf of the optimized impact factors, and the longitudinal axis represents the deteriorative impact factors. The judgment of optimized and deteriorative is based on fuzzy expression of impact factors. Table 2 shows the fuzzy expression of impact factors.

Based on the fuzzy expression, we can contrast the parent design rationale with the child design rationale. Assuming that there are n numbers impact factors: $f_1, f_2, \dots, f_n$. If:

$$f_{1parent} - f_{1child} < 0(optimization) \quad \textbf{(14)}$$

$$f_{2parent} - f_{2child} < 0(deterioration) \quad \textbf{(15)}$$

$$f_{3parent} - f_{3child} < 0(deterioration) \quad \textbf{(16)}$$

Then we receive two contradictions: (f_1, f_2) and (f_1, f_3). Therefore, with iterative arithmetic, the contradiction matrix about design rationale can be structured. Figure 4 shows the segments of contradiction matrix about turbine blades design rationale.

This part reflects that when the designers meet the contradiction of ***energy*** and ***cost*** in the design of blade, they can consider adopting the **Third Solution**. It means to change the decision again to achieve goals and requirements without changing the intention. If there is further need, you can refer to them, because each problem solution corresponds to its practical cases.

Table 2: Fuzzy expression of impact factors

impact factor	best	better	common	worse	worst
fuzzy expression	1	0.75	0.5	0.25	0

optimized parameters / deteriorated parameters	load	cost	energy	mass	machin-ability	thermost-ability	stability
load				the third solution			the first solution
cost			the third solution			the fourth solution	the fourth solution
energy							
mass	the third solution				the second solution	the second solution	the third solution
machin-ability				the second solution			
thermost-ability		the fourth solution		the second solution			the fifth solution
stability	the first solution	the fourth solution		the third solution		the fifth solution	

Figure 4: Contradiction matrix about turbine blades design rationale

6 EXAMPLES

Blade in a lot of mechanical equipment and product is in an important position, especially in the turbine, the performance of the turbine mainly depend on the blade, including material, shape, quantity, etc. We assume that we want to optimize the machinability, the thermostability, and the stability of turbine blade; the deteriorated parameters we can tolerant are the cost and the mass of it. Six contradictions are obtained by Optimized Parameters and Deteriorated Parameters. They are: (o_1, d_1), (o_1, d_2), (o_2, d_1), (o_2, d_2), (o_3, d_1), (o_3, d_2).

Where: o_1 is "*machinability*", o_2 is "*thermostability*", o_3 is "*stability*", d_1 is "*cost*", d_2 is "*mass*".

Designers can transform these solutions to concrete solution, or they can also get concrete solution by the cases, examples are shown in Figure 5 and Figure 6. Solution of contradiction (o_3, d_2) is the **Third Solution**, transform *the number of blades*. Solution of contradiction (o_2, d_1) is the **Fourth Solution**, increase one justification: *drilling on the blades*.

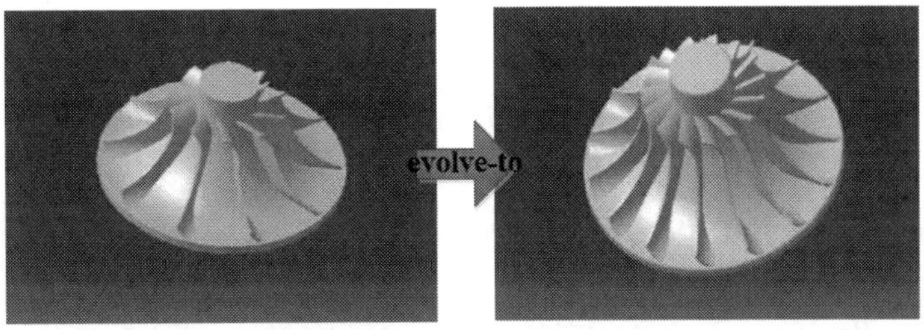

Figure 5: Solution of contradiction (o_3, d_2)

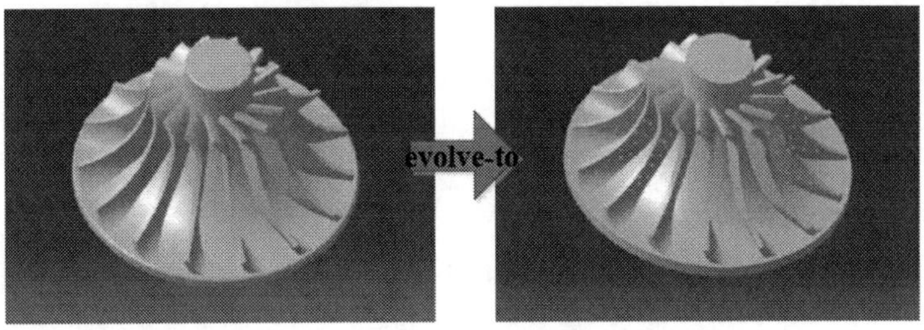

Figure 6: Solution of contradiction (o_2, d_1)

7 CONCLUSION

Aiming at solving the problem of reusing design rationale that has the same or the same kind of design intention, this paper set up a ternary module for the evolution part in design rationale, summarized five standard solving method of design rationale, and finally abstract the design rationale to the contradiction matrix dynamically through mining based on frequency, referring to contradiction matrix and problem solving theory in TRIZ.

Contradiction matrix about design rationale is mainly used to solve some problems that designers encounter in the process of design thinking when indicators cannot meet the requirements, and can act as a guide for providing a solution to designers through the existing design rationale samples.

This method can not only be used in product design, it can be used in similar process as process design. We structure the contradiction matrix about the cases, dig the standard solution about it, and give the concrete solution finally. The possible constrain of this method is linguistic code. It is important to the data mining and the semantic computing. In the future, we should research on the linguistic code method and the mining algorithm, attempt to improve the precision of this method.

8 REFERENCES

[1] **Liu, J., Hu, X. 2013.** A reuse oriented representation model for capturing and formalizing the evolving design rationale, *Artificial Intelligence for Engineering Design, Analysis and Manufacturing*, 27(4), pp 401-413.

[2] **Ruiz, P.A.P., Kamsu-Foguem, B., Noyes, D. 2013.** Knowledge reuse integrating the collaboration from experts in industrial maintenance management, *Knowledge-Based Systems*, 50, pp 171-186.

[3] **Ma, S., Tian, L. 2015.** Ontology-based semantic retrieval for mechanical design knowledge, *International Journal of Computer Integrated Manufacturing*, 28(2), pp 226-238.

[4] **Behbahani, M., Saghaee, A., Noorossana, R. 2012.** A case-based reasoning system development for statistical process control: Case representation and retrieval, *Computers & Industrial Engineering*, 63(4), pp, 1107-1117.

[5] **Li, L.F., Qin, S., Gao, Liu, Y. 2014.** An Approach for Design Rationale Retrieval Using Ontology-Aided Indexing, *Journal of Engineering Design*, 25(7-9), pp, 259-279.

[6] **Savransky, S.D.** 2000. *Engineering of creativity: Introduction to TRIZ methodology of inventive problem solving*, CRC Press.

[7] **PAUL, P., IVAN, A. 2012.** The development of a modified TRIZ Technical System ontology, *Computers in Industry*, 63(3), pp 252-264.

[8] **Roberto D, NOEL L, HUMBERTO A, DAVID S. 2011.** Inventive problem solving based on dialectical negation, using evolutionary algorithms and TRIZ heuristics, *Computers in Industry*, 62(4), pp 437-445.

[9] **Ko, Y.T. 2016.** Modeling a hybrid-compact design matrix for new product innovation, *Computers & Industrial Engineering*, in Press.

[10] **Pokhrel, C., Cruz, C., Ramirez, Y. 2015.** Adaptation of TRIZ contradiction matrix for solving problems in process engineering, *Chemical Engineering Research and Design*, 103, pp 3-10.

[11] **Choi, S. 2015.** Review and Application of Creative Problem-Solving Processes for Technical and Physical Contradictions Using Cause-And-Effect Contradiction Tree and Integrated Principles of TRIZ, *Journal of the Korea Safety Management and Science*, 17(2), pp 215-228.

A METHOD OF SEARCH INTENTION RECOGNITION IN SPECIFIC DOMAIN

A. ZHOU Jianhui[1] B. YANG Haicheng[2] and C. LIU Jihong[3*]
[1]School of Mechanical Engineering and Automation
BeiHang University, Beijing, China
zhjh19851101@163.com

[2]China Aerospace Science and Technology Corporation
Beijing, China
13911253139@139.com

[3]School of Mechanical Engineering and Automation
BeiHang University, Beijing, China
ryukeiko@buaa.edu.cn

ABSTRACT

The researches on new generation search engine more and more focus on the search intention, which answers the question that why people search, instead of how they search or what they search for. Most existed investigations on search intention analysis considered only the information embedded in the search context; however, the information contained in the search query is more important for analyzing the search intention, especially in specific domain. This paper presents a novel search intention recognition method, which makes full use of the information extracted from both the search context and the search query. Firstly, the characteristics of the two datasets are satisfied independently by two algorithms; and then, the search intention is obtained from those characteristics and datasets through an improved algorithm. In addition, by classifying the keywords of the search query on the basis of the artificial vocabularies, the number of learning sample datum has significantly increased, which makes the algorithm more precisely. Finally, a case is studied to verify the validation and precision of the new method.

Keywords: Search Intention, Search Context, Search Query

1 INTRODUCTION

Knowledge search has become an important part in our daily work. Users can find many types of knowledge to meet different needs through search engine. There is an assumption: when users search knowledge in search engine, they must need help, especially in specific domain. Every search behavior has its search intention. Automatically recognizing search intention is a part of Search Engine Optimization (SEO) [1-3]. It can be used to rank the search results, make the search engine more intelligent.

---

* LIU Jihong, born in 1966, is currently a professor at Beihang University, China. His research interests include complex product engineering, digital design and manufacturing, knowledge management and design theory and method.

Acknowledge: The work supported by National Key Technology R&D Program through approval No. 2015BAF18B01.

Search intention recognition is making search engine to know why users are searching [4,5]. Without it, search engine determines the search results by matching the query that users input with the index about the knowledge [6]. This process is mostly based on matching model and ranking algorithm, which reflects the intelligence by semantic technology. These models and algorithms operate by using frequency information in the indices and queries [7-10]. But, users input query according their own intention, not frequency information. They cannot know a keyword is most frequent in which knowledge or which type of knowledge. For example, the keyword of "Process Optimization" corresponds to two types of knowledge, they are papers and cases. Researchers probably want the papers, but processors probably want the cases. No matter keyword in which type of knowledge is more frequent, it cannot satisfy the users both. The "researchers" and "processors" mentioned here are the instances of search context in specific domain. How to choose context and use context are problems which this paper should solve.

Search intention is not only in context, but also in query. In other words, when users are searching, they input query with their own intention. The query that users input are mostly very short, only two or three words, even one word [11,12]. These words themselves or their categories contain the search intention of users. For example, considering two types of knowledge about "winded missile", one is wikis, the other is news. The query of "winded missile" probably wants the wikis, but the query of "winded missile 2014" probably wants the news. Although both index of them contain "winded missile" and "2014". Two keywords mentioned here have the different categories, this paper focus on the words categories, not words themselves. This process has two advantages: the logic of words categories in specific domain is complex, and there are many rules about them; on the other hand, in specific domain, the learning samples of one word are less, categorizing the words can enlarge the learning samples of one category, this can make the learning results more precisely.

In this paper, we present a method of search intention recognition, this method consider both context and query. We categorize the keywords of specific domain by artificial vocabulary, and classify the knowledge, i.e. classify the search intention of specific domain. At last, we design an experiment by the datum of a military company.

2 SEARCH INTENTION LEARNING MODEL

This section presents the process of search intention learning; it is composed of context search intention learning and query search intention learning, as shown in Figure 1. The former requires the learning samples of users' characteristics; and the latter requires the learning samples of keyword categories.

Which search intention the context or query correspond is maximum probability of it. Therefore, the search intention learning samples is consist of context or query as the input and probability as the output, which are defined as follows:

Definition 1: Context search intention learning sample, denoted as *CSI*, is defined as a tuple:

$$CSI = < V, P(SI) > \quad (1)$$

Where: *V* is the set of users' characteristics; *P*(*SI*) is the set of probabilities of search intention.

Definition 2: Query search intention learning sample, denoted as *QSI*, is defined as a tuple:

$$QSI = < K, P(SI) > \quad (2)$$

Where: *K* is the set of keyword categories; *P*(*SI*) is the set of probabilities of search intention. In order to composite the two probabilities of search intention; we use the General Linear Model to simulate both of them. The inputs of this model are two sets of probability of search intention, and the output is the final probability of search intention. We define this model as follows:

Definition 3: Search intention learning model, is defined as a General Linear Model:

$$F = w_0 p_0 + w_1 p_1 + w_2 p_2 \qquad (3)$$

Where: F is the final probability of search intention, as output; w_0 is a parameter of model, need to be learnt; w_1 is the weight of context search intention, need to be learnt; w_2 is the weight of query search intention, need to be learnt; p_0 is a constant, equals 1; p_1 is the probability of context search intention, as input; p_2 is the probability of query search intention, as input.

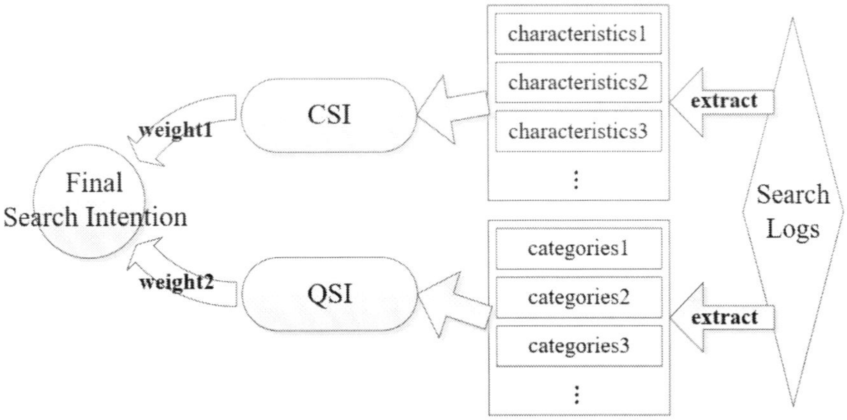

Figure 1: Process of Search Intention Learning

3 PRELIMINARY PREPARATION OF LEARNING

3.1 Categories of Keywords

In most cases, the keywords of query in specific domains are terminology along with a small amount of natural language. Based on the characteristics of specific domain, the natural language can be divided into the following categories: person, institution, time, etc. The main purpose of categorizing keywords in specific domain is to recognize the search intension which is corresponded with different types of keywords combination. This paper takes the example of the knowledge in a military company, the keywords that the users probably enter could be divided into the following categories: term, refers to the appellation sets that are used to indicate conception, such as "Winded Missile", "Process Optimization", etc.; person, refers to the people names, they also can be abbreviations or nickname, such as "Qian Xuesen", "Liang Shoupan", etc.; institution, refers to the names of the companies or groups in military field, they can be abbreviations, such as "CASIC", "BJSASC", "NASA", etc.; time, refers to the year, month, day, etc. and others.

The categories of keywords could be achieved through building artificial vocabularies. Since there are only five categories, five artificial vocabularies are needed. In each vocabulary, the method of building the vocabulary is facility and accurate because the keywords in this field are limited and relatively.

3.2 Classification of Search Intention

Since the structure of knowledge in specific domain is stronger than that in the web search field, through collecting 200 thousands knowledge of a military company, this paper generalize 18 properties of knowledge and cluster the knowledge. As shown in Table 1. If the property appears in knowledge, then we mark it as "1", otherwise, mark it as "0".

We can abstract the knowledge as multidimensional vector. Since the difference between classes is obvious, using the k-means method can make a fast convergence. This paper classifies the knowledge of a military company into six classes: URL, News, Wiki, Standard, Technology Document and Paper, they are classified based on the classes of Navigational, Informational and Resource. The names are used to distinguish different classes, it does not matter whether the names are accurate or not. In order to write conveniently, we mark the classes mentioned above as "URL.", "New.", "Wik.", "Sta.", "Tec.", "Pap.".

Table 1: Classes of Search Intention

Search Intention		Description	Examples
Navigational	URL.	I want go to a known website I already have in mind.	Aerospace college
Informational	New.	I want know some news.	Winded missile 2014 Wingmanship today
	Wik.	I want learn something about my topic.	Passion ratio Electrolysis
Resource	Sta.	I want get the standard document about something.	Flight speed of standard Choose gear
	Tec.	I want get the technology document about something.	Navigation technology Missile model
	Pap.	I want get some papers about my keywords.	Process optimization method

4 SEARCH INTENTION LEARNING

4.1 Context Search Intention Learning

As context in specific domain is more explicit than it in web search, it is more facility to choose context, which should satisfy search intention. In this paper, we implement this method in a military company; main job of the company is R&D of guided missile model. Users' search intention is related to instances of context, such as "*users' identity*", "*users' duty*" and "*users' work*". The data type of their values is string, not integer. For processing this data type reasonably, we adopt the Decision Trees.

ID3 algorithm is deployed in this paper, and transformation of ID3 is output, which is not one decision, but the vector of probabilities of decision, i.e. probabilities of search intention, in order to be composited with probabilities of query search intention.

The inputs of ID3 are *Examples* (the set of context), *Target_attribute* (the set of vectors of probabilities of search intention), *Attributes* (the learning samples extracted from search logs). We define them as follows:

$$Attributes = \{v_i | \forall i = 1,2,3, \dots, n\} \quad (4)$$

$$Target\_attribute = \{(p_i(Url), p_i(New), p_i(Wik), p_i(Sta), p_i(Tec), p_i(Pap)) | \forall i = 1,2,3, \dots, n\} \quad (5)$$

$$Examples = \{((v_{1k}, v_{2k}, \dots, v_{nk}), SI) | \forall k = 1,2,3, \dots, m\} \quad (6)$$

Where: *SI* is the set of search intention, as ***(7)***.

$$SI = \{Url, New, Wik, Sta, Tec, Pap\} \quad (7)$$

4.2 Query Search Intention Learning

Keywords of queries in web search are numerous, and the correlations of them each other are numerous also, contrary to the specific domain. Therefore, we categorize the keywords in specific domain into five categories, such as "*term*", "*person*", "*institution*", "*time*" and "*others*". This is according to the military company, which we implement the method in. Most of keywords extracted from the search logs are in the first four categories, the remains are in the "*others*", such as some auxiliary words, symbols, classifier, and so on.

Different categories input by users representative different search intention. We use the numbers of every category to learning query search intention. The data type of their values is integer, not string. Moreover, in specific domain, there are some rules about search intention; for example, if the category is "person", its equal 1, then the search intention corresponded is "wiki". We can add rules according to our own needs. Based on these characteristics, Maximum Entropy Method is the most reasonable learning method.

The inputs of Maximum Entropy Method are *x* and *y*. We define them as follows:

$$T = \{(x_i, y_i) | \forall i = 1,2,3, \dots, n\} \quad (8)$$

$$x = \{\#TermType_1, \#TermType_2, \#TermType_3, \dots, \#TermType_n\} \quad (9)$$

$$y = \{y \in SI | SI = \{Url, New, Wik, Sta, Tec, Pap\}\} \quad (10)$$

T is the learning samples, *x* is the vector of numbers of categories, *y* is one instance of search intention. The restriction of specific domain is defined as follows:

$$p_i(Url) + p_i(New) + p_i(Wik) + p_i(Sta) + p_i(Tec) + p_i(Pap) = 1 \quad (11)$$

$$f_i(x, y) = 1(if\ x = x_i\ then\ y = y_i) \quad (12)$$

$$f_i(x, y) = 0(otherwise) \quad (13)$$

Every $p(y_i | x_i)$ is obtained by Maximum Entropy Method, and then we can obtain the vector of probabilities of query search intention by compositing $p(y_i | x_i)$. The results can predict users' query search intention.

4.3 Final Search Intention Learning

The learning results of context search intention and query search intention are vectors of probabilities. Each of them has a corresponded weight in final search intention. This is a General Linear Model, in order to do this, we define T as the learning samples, V as the set of context:

$$T = \{(V_i, \boldsymbol{TermType}_i, \boldsymbol{P}_i) | \forall i = 1,2,3, \dots, M\} \quad (14)$$

$$V = \{v_i | \forall i = 1,2,3, \dots, n\} \quad (15)$$

The learning results of context search intention and query search intention are vectors of probabilities. Each of them has a corresponded weight in final search intention. This is a General Linear Model, in order to do this, we define T as the learning samples, V as the set of context:

Where ***TermType*** is the vector of numbers of categories, ***P*** is the vector of probabilities of final search intention:

$$\boldsymbol{TermType} = \{\#TermType_1, \#TermType_2, \#TermType_3, \dots, \#TermType_n\} \quad (16)$$

$$\boldsymbol{P} = (p_i(Url), p_i(New), p_i(Wik), p_i(Sta), p_i(Tec), p_i(Pap)) \quad (17)$$

The General Linear Model of final search intention is defined as follow:

$$\boldsymbol{P} = w_0\boldsymbol{x}_0 + w_1\boldsymbol{x}_1 + w_2\boldsymbol{x}_2 \quad (18)$$

Where: $x_0=E$; x_1 is the vector of context search intention; x_2 is the vector of query search intention.

We deploy the ***ordinary least squares*** algorithm, and obtain the parameters of w_0, w_1, w_2, and then the users' search intention can be recognized by this model.

5 EXPERIMENTS

The method we present is implemented in a military company, main job of the company is R&D of guided missile model. Knowledge in databases of the company has subjects of Guided Missile, Winded Missile, Process Optimization, Process Design, Structure Design, Reliability Analysis, Feasibility Analysis, etc. We classify the types of knowledge into six classes, as URL, News, Wiki, Standard, Technology Document and Paper. They are the search intentions in its way. We categorize the keywords into five categories, as term, person, institution, time and others. These works is according to the knowledge and search logs itself. We list the instances of context as follows: users' identity, users' duty and users' work. We take 200 thousands records to learning context search intention, query search intention and final search intention. The datum of results of experiments is shown in the next sections.

Table 2: Learning Results of Context Search Intention

Identity	Duty	Work	(p(URL.),p(New.),p(Wik.),p(Sta.),p(Tec.),p(Pap.))
Processor	Staff	Process Optimization	(0.13, 0.02, 0.18, 0.21, 0.38, 0.08)
Processor	Staff	Process Design	(0.17, 0.02, 0.12, 0.31, 0.29, 0.09)
Processor	Staff	Project Check	(0.46, 0.13, 0.03, 0.22, 0.16, 0.00)
Processor	Engineer	Project Check	(0.24, 0.13, 0.02, 0.03, 0.57, 0.01)
Processor	Engineer	Product Optimization	(0.13, 0.02, 0.18, 0.21, 0.38, 0.08)
Researcher	Staff	Product Optimization	(0.05, 0.03, 0.18, 0.13, 0.19, 0.42)
Researcher	Staff	Model R&D	(0.09, 0.02, 0.13, 0.07, 0.07, 0.62)
Designer	Staff	Model R&D	(0.16, 0.23, 0.08, 0.06, 0.26, 0.21)
Designer	Staff	Project Check	(0.09, 0.24, 0.02, 0.21, 0.39, 0.05)
Designer	Staff	Product Design	(0.07, 0.16, 0.22, 0.25, 0.27, 0.03)
Designer	Director	Project Check	(0.34, 0.51, 0.01, 0.10, 0.02, 0.02)
Designer	Director	Product Design	(0.12, 0.21, 0.06, 0.26, 0.33, 0.02)

Table 3: Learning Results of Query Search Intention

(#term,#person,#institution,#time,#others)	(p(URL.),p(New.),p(Wik.),p(Sta.),p(Tec.),p(Pap.))
(1, 1, 0, 0, 0)	(0.06, 0.12, 0.20, 0.02, 0.02, 0.58)
(0, 1, 1, 0, 0)	(0.25, 0.33, 0.15, 0.01, 0.02, 0.24)
(2, 0, 0, 0, 0)	(0.03, 0.01, 0.38, 0.11, 0.14, 0.33)
(0, 1, 0, 1, 0)	(0.08, 0.36, 0.13, 0.07, 0.06, 0.30)
(0, 0, 1, 1, 0)	(0.18, 0.12, 0.03, 0.31, 0.24, 0.12)
......	......

Table 2 illustrates that one composition of context corresponds one vector of probabilities of search intention. We will predict the context search intention of the next new composition of context, by the way of similarity technology, which is not emphasis in this paper.

Table 3 illustrates that one composition of category corresponds one vector of probabilities of search intention. We will predict the query search intention of the next new composition of category.

Final search intention is simulated by General Linear Model, which inputs are the vectors of probabilities of context search intention and query search intention. Iteration results of Least Square Method are the final search intention. The datum of them is not listed here, they can be calculated by the model as follow:

$$\boldsymbol{P} = -0.03571 + 0.34453\boldsymbol{x}_1 + 0.65547\boldsymbol{x}_2 \qquad \textbf{\textit{(18)}}$$

This model is learnt by the samples we designed, it is not absolutely correct in any field, it is only used to verify the effectiveness of the method this paper present.

In specific domain, there are two key factors that affect the learning results; they are the instances of context and the method of learning. We compare it with web search; take the accuracy of recognition results as the evaluation criterion. Results of experiments are shown in Figure 2. They illustrate that the method this paper present is more effective, actually, it is the truth that choosing the proper instances of context and adopting the proper method or algorithm is better.

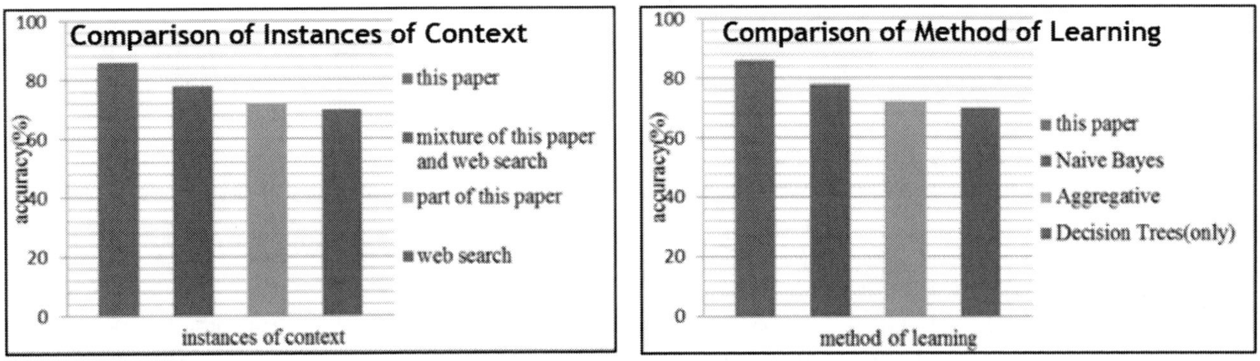

Figure 2: Results of Experiments

6 CONCLUSIONS

Search intention is a problem in SEO, application of it is that re-ranking or filtering the search results to make users more satisfied. It is the type of knowledge in some way. Contrary to the contents of keywords users input and knowledge, search intention hid in the context and query, learning search intention is the problem of data mining or machine learning, there are many methods, models, or algorithms to solve the problem, but in specific domain, proper methods, models, or algorithms should be adopted with the special conditions, such as the types of knowledge, characteristics of users, categories of keywords, etc. Verification of search intention is another problem, intention is users subjective factors, it is not absolute correct or error. Whether effective or not, the satisfaction of users is the only target. This paper selects some characteristics of users, actually, in specific domain, consecutiveness of search is another characteristic of users, technology of process mining will be used to consider it in future, it will make search results more satisfaction, and it will make search engine more intelligent.

7 REFERENCES

[1] **Amin, G.R. Emrouznejad, A. 2011.** Optimizing search engines results using linear programming, *Expert Systems with Applications*, 38(9), pp 11534-11537.

[2] **Gokhan Egri, Coskun Bayrak. 2014.** The Role of Search Engine Optimization on Keeping the User on the Site, *Procedia Computer Science*, 36, pp 335-342.

[3] **Nursel Y., Utku K. 2010.** What is search engine optimization: SEO?, *Procedia-Social and Behavioral Sciences*, 9, pp 487-493.

[4] **Xiang Ren, Yujing Wang, Xiao Yu, Jun Yan, Zheng Chen, Jiawei Han.** Heterogeneous graph-based intent learning with queries, web pages and Wikipedia concepts, Proceedings of the 7th ACM international conference on Web search and data mining, New York, the United States, February 24-28, pp 23-32.

[5] **Kinam Park, Hyesung Jee, Taemin Lee. 2012.** Automatic Extraction of User's Search Intention from Web Search Logs, *Multimed Tools Appl.*, 61, 145-162.

[6] **Sebastian Schmidt, Steffen Schnitzer, Christoph Rensing. 2016.** Text classification based filters for a domain-specific search engine, *Computers in Industry*, 78, pp 70-79.

[7] **Luis Ramos. 2015.** Semantic Web for manufacturing, trends and open issues Toward a state of the art, *Computers & Industrial Engineering*, 90, pp 444-460.

[8] **Zhang Xutang, Hou Xin, Chen Xiaofeng, Zhuang Ting. 2013.** Ontology-based semantic retrieval for engineering domain knowledge, *Neurocomputing*, 116, pp 382-391.

[9] **Tibor Kocsis, Stéphane Negny, Pascal Floquet, Xuân Meyer, Endre Rév. 2014.** Case-Based Reasoning system for mathematical modelling options and resolution methods for production scheduling problems: Case representation, acquisition and retrieval, *Computers & Industrial Engineering*, 77, pp 46-64.

[10] **Sharifullah Khan, Jibran Mustafa. 2014.** Effective semantic search using thematic similarity, *Journal of King Saud University-Computer and Information Sciences*, 26(2), pp 161-169.

[11] **David A.G. Frieder, O.** 2004. *Information Retrieval: Algorithms and Heuristics*, 1st Edition, Springer.

[12] **Luh, C. Yang, S. Huang, T.D. 2016.** Estimating Google's search engine ranking function from a search engine optimization perspective, *Online Information Review*, 40(2), PP 239-255.

APPLICATION OF RBF NEURAL NETWORK IN GAS LOAD FORECASTING

Xing BI[1*], Zhao-qiang DING[1] and Hai-tian BI[1]
[1] College of Management and Economics, Tianjin University, Tianjin, China
bistar@126.com

ABSTRACT

This paper introduces gas load forecast methods at present, such as time series method, regression analysis method, neural network method, grey system method, etc. The paper focuses research on the neural network method and proposes a gas load forecast model based on RBF neural network. Then the paper discusses the influence of data normalization parameters on the accuracy of load forecast. Using the RBF neural network method, the paper forecasts the gas load of the example company, and the results show that the method is effective and reliable.

Keywords: RBF neural network, Data normalization, Gas load forecast

1 INTRODUCTION

Gas load forecast has great influence on economy and safety of the gas system operation. There are many methods for gas load forecast at present, such as time series method [1], regression analysis method [2], neural network method [3], grey system method, etc. In recent years, some researchers use the support vector machine model [4] and EMD-PSR-LSSVM combined model [5] to predict the gas load, and achieve a good result. Traditional load forecast model is described by explicit expressions, and this determines the limitations of traditional forecast models. In fact, the nature of load changes is difficult to state by an explicit mathematical formula. Neural network method is a major breakthrough in this area. Using the traditional explicit functions of the variable and dependent variable as input and output of the network, the method translates the traditional function into higher dimensional nonlinear mapping. Neural network is a highly nonlinear large scale continuous time dynamical systems. It can map any complex nonlinear relationship and identify samples with noise or distortion. Through learning, it distributes implicit sample characteristics in the neural network connection weights. So that in recent years, it has been widely used in the field of load forecast [6][7]. Literature [8], based on the characteristics of short-term load of city gas, taking into account the influence of factors such as weather, temperature and holiday, used artificial neural network back propagation (BP) model to predict the short term load of city gas pipe network. Literature [9], proposed a gas load forecast model based on BP neural network, the model using Levenberg-Marquardt optimization method, taking the day average temperature of forecast day, and gas load of k (k is less than 15) days before and one day before as input parameter, the output parameter being load forecast of predicting day. Literature [10] proposed to mix the SOFM network and MLP network, establishing short-term city gas load forecast method. Literature [11] introduced the momentum to accelerate convergence through the weights in neural network prediction model, and introduced noise in data input to improve network generalization.

---

* Corresponding Author

This paper proposes a gas load forecast model based on RBF neural network, and predict the gas load of example company in the next 12 months.

2 RBF NETWORK STRUCTURE AND CHARACTERISTICS

RBF neural network, also known as the radial basis function (RBF) neural network, is a kind of feed forward neural network with multiple inputs and single output. Its schematic diagram is shown in figure 1. RBF neural network composes of input layer, hidden layer and output layer. The hidden layer uses radial basis function as activation function, performing a fixed nonlinear transformation to map input space into a new space. The output layer achieves linear combination in the new space, and the adjustable parameter is the weights of the linear combination. The advantage of this method is that the linear least squares method can be used to determine these parameters. The network, having been proved in theory, may provides the approximation capability of general three layers' network, and may approximates any continuous function with any degree of accuracy.

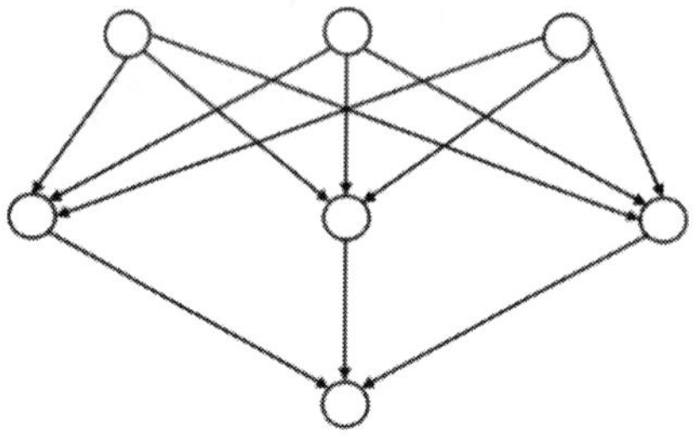

Fig1: RBF neural network structure

RBF network has the following features and benefits:

(1) Activation function in the hidden layer units usually is the function of local receptive field, which means that only when the input falls in a small designated area in the input space, hidden layer unit make meaningful nonzero response. This avoids the arbitrary partition caused by BP network hyperplane segmentation.

(2) Neural network error back propagation algorithm (Back Propagation, hereinafter referred to as "BP) is a kind of multilayer feedforward neural network, which the weights adjustment rules use BP learning algorithm. RBF network is a kind of radial basis function neural network, and network hidden layer using radial basis function (usually a Gaussian function) as activation function. The former has slow convergence rate, long training time due to fixed learning rate and can't guarantee weight convergence to the global minimum of error plane. The latter avoids these weaknesses in principle.

(3) About the RBF network, due to the linear characteristic of the output unit, its parameters adjustment is very simple and there is no partial minima problem.

(4) The form of nonlinear activation functions used by general RBF networks impacts on network performance is not critical. The key factor is the selection of the centre of the basis function.

(5) RBF network output is independent of initial weight, and has strong mapping functions of input and output. It has been proved theoretically that RBF network is optimal to complete the network mapping function in the feedforward neural network.

3 DATA NORMALIZATION PROCESS

In order to improve the learning speed of network and the prediction accuracy, a normalization processing for the loading data participated in the training is necessary. It eliminates the characteristics of data itself by mathematical method to make all data in the model obtain the same competitiveness, and make the input load data locate in [0,1]. Usually the following formula is used to perform load data normalization at time *t*.

$$Y = A + B\frac{X - X_{min}}{X_{max} - X_{min}} \quad (1)$$

Following formula is used to calculate back the actual load value in the output layer.

$$X = X_{min} + (Y - A)\frac{X_{max} - X_{min}}{B} \quad (2)$$

In the formula:

Y——the load data after normalization processing, located in the [0, 1]

A——normalization coefficient

B——normalization coefficient

X——loading sample raw data at time *t*

X_{min}——the minimum value of the load in the training sample set

X_{max}——the maximum value of the load in the training sample set

Normalization coefficient interpretation: Because $(X-X_{min})/(X_{max}-X_{min})$ values within the range from 0 to 1, and *A* and *B* take different values between 0 and 1, so the original data value at different discrete levels after normalization. Normalization coefficient *A* and *B* are coefficients adjusting the discrete degree of normalized data. In principle, the sum of *A* and *B* is the best when it is near to 1, and keeps one decimal. *A* values from low to high.

4 NETWORK DESIGN AND IMPLEMENTATION

4.1 Network Design

In this paper, the neurons of the network input layer and output layer are 12, to represent respectively input historical data over the past 12 months and output forecasting data in the next 12 months. With the MATLAB7 platform, an accurate RBF Neural network is set up, which automatically selects the number of hidden layers, allowing for error being 0.
The gas load data from January 2008 to December 2014 of an example gas company, are shown in table 1, in which all data has been normalized using formula *(1)* (take *A* = 0.1, *B* = 0.8).

Table 1: The sample data after the processing of a company's gas load *(Y)*

Year	Jan	Feb	Mar	Apr	May	Jun	Jul	Aug	Sep	Oct	Nov	Dec
2008	0.1530	0.1502	0.1158	0.1339	0.1091	0.1074	0.1000	0.1072	0.1140	0.1228	0.1573	0.1788
2009	0.1751	0.1987	0.1795	0.1824	0.1550	0.1584	0.1415	0.1438	0.1566	0.1603	0.2029	0.2199
2010	0.2820	0.2462	0.2095	0.2331	0.2094	0.2050	0.1756	0.1900	0.2027	0.2247	0.2749	0.3273
2011	0.3768	0.3610	0.3478	0.3210	0.2772	0.2641	0.2435	0.2630	0.2843	0.2852	0.3738	0.3954
2012	0.4606	0.4631	0.3946	0.4178	0.3528	0.3478	0.3570	0.3483	0.3558	0.3809	0.4741	0.5365
2013	0.6519	0.6180	0.4918	0.5382	0.4585	0.4437	0.4113	0.4599	0.4717	0.5206	0.6389	0.6728
2014	0.8602	0.7265	0.6528	0.6524	0.5275	0.5511	0.4932	0.5157	0.5473	0.5783	0.7426	0.9000

Using the data of six years from 2008 to 2013 as the training sample of network, taking monthly data of annual gas load as input vectors, the next year's gas load as target vectors, five group samples are obtained. Monthly load in 2014 is taken as the test sample of the network to compare with the forecast data.

4.2 Network Implementation

The comparison between the gas load forecasting value using the enterprise's samples and actual value in 2014 is presented in table 2 by using the above model.

Table 2: The comparison between the gas load forecasting value and actual value in 2014 (A=0.1, B=0.8)

Time	The forecasting value of Y	The actual value of Y	Relative error/%
Jan	0.7900	0.8602	-8.15
Feb	0.7406	0.7265	1.93
Mar	0.6168	0.6528	-5.51
Apr	0.6172	0.6524	-5.40
May	0.5257	0.5275	-0.34
Jun	0.4930	0.5511	-10.53
Jul	0.4141	0.4932	-16.04
Aug	0.5314	0.5157	3.02
Sep	0.5610	0.5473	2.52
Oct	0.5983	0.5783	3.48
Nov	0.7553	0.7426	1.71
Dec	0.7092	0.9000	-21.20
Total	7.3526	7.7476	-5.10

4.3 Predicted Results Analysis

According to the table 2, the analysis of forecasting relative error is shown in table 3. Δ is the absolute value of relative error.

Table 3: The analysis of forecasting relative error

The relative error with the maximum absolute value/%	-21.20
The mean relative error/%	-5.10
The frequency of $\Delta \in [0,5\%]$	6
The frequency of $\Delta \in (5\%,10\%]$	3
The frequency of $\Delta \in (10\%,15\%]$	1
The frequency of $\Delta \in (15\%,20\%]$	2

The relative error with the maximum absolute value is -21.20%, and the mean of relative error is -5.10%. There are 6 forecasting values in the range of [-5%, 5%], accounting for 50% of the total. The forecasting result meet the company's actual production requirement.

5 MODEL OPTIMIZATION

In order to further improve the model predictive accuracy and wide application, this paper discusses the sample data normalization. Table 4 shows the comparison of the gas load forecast results using normalization formula *(1)* with the coefficient of A and B are 0.2 and 0.7 and actual value in 2014. More comparison when the coefficient of A and B are 0.3 and 0.6, 0.4 and 0.5, 0.5 and 0.4 are omitted for lack of space. The table 5 shows the relative error comparison of the five predictions.

From table 5, as the normalized coefficient changes regularly, the predictive value comparative indicators also change regularly. Considering the comparative indicators, the final forecast model may select the RBF network model which the normalized coefficient A and B is 0.4 and 0.5. The relative error with the maximum absolute value is -13.79%, the mean of relative error was -4.39%. There are 7 forecasting values in the range of [-5%, 5%], accounting for 58.33% of the total.

Table 4: The comparison between the gas load forecasting value and actual value in 2014 (A=0.2, B=0.7)

Time	The forecasting value of Y	The actual sample value of Y	Relative error/%
Jan	0.7962	0.8651	-7.96
Feb	0.7543	0.7482	0.82
Mar	0.6452	0.6837	-5.63
Apr	0.6492	0.6833	-4.99
May	0.5689	0.5741	-0.91
Jun	0.5413	0.5947	-8.98
Jul	0.4747	0.5441	-12.76
Aug	0.5738	0.5638	1.77
Sep	0.5987	0.5914	1.23
Oct	0.6321	0.6185	2.20
Nov	0.7673	0.7623	0.66
Dec	0.7309	0.9000	-18.79
Total	7.7326	8.1292	-4.88

Table 5: Analysis of prediction error

The normalized coefficient A	0.1	0.2	0.3	0.4	0.5
The normalized coefficient B	0.8	0.7	0.6	0.5	0.4
The relative error with the maximum absolute value/%	-21.20	-18.97	-14.10	-13.79	-13.39
The mean relative error/%	-5.10	-4.88	-4.42	-4.39	-4.47
The frequency of $\Delta \in [0,5\%]$	6	7	7	7	7
The frequency of $\Delta \in (5\%,10\%]$	3	3	4	4	4
The frequency of $\Delta \in (10\%,15\%]$	1	1	1	1	1
The frequency of $\Delta \in (15\%,20\%]$	2	1	0	0	0

6 CONCLUSION

A RBF network has the unique characteristics of the best approximation, and there is no partial minimum problem. RBF neural network has a strong function of input and output mapping, and it is proved theoretically that the RBF network is the optimal network to complete the mapping function. The example company's gas load forecast using RBF neural network method shows that the forecast accuracy is satisfied.

7 REFERENCES

[1] **Miao Yanshu, Duan Changgui, Zhang Shuhong. 2005.** The application of data mining technology in gas load forecast, *Gas & Heat*, 25(11), pp1-4.

[2] **Yuan Shuming, Lai Jianbo, Zhang Dekun, etc. 2009.** Prediction of city gas pipeline network long term planning of gas, *Gas & Heat*, 29(5), pp24—27.

[3] **Li Chijia, Jiao Wenling, Zhu Jianhao, etc. 2004.** Gas load forecast during the Spring Festival based on artificial neural network, *Gas & Heat*, 24(9), pp477—480.

[4] **Zhang Chao, Liu Yi, Zhang Hui, etc. 2014.** Study on urban short-term gas load forecasting based on support vector machine model, *J Tsinghua Univ(Sci & Technol)*, 54(3), pp320-325,333.

[5] **Gong Cheng-zhu, Li Lan-lan, Yang Juan, etc. 2014.** An integrated short-term load forecasting approach for urban gas pipeline network based on EMD, PSR, and LSSVM, *Systems Engineering-Theory & Practice*, 34(11), pp3001—3008.

[6] **K.Metaxiotis, A.Kagiannas, D.Askounis.** 2003. Artificial Intelligence in Short Term Electric Load Forecasting: A State-of-art Survey for Researcher. *Energy Conversion and Management*, 44, pp1525-1534.

[7] **T.Yalcinoz, U.Eminoglu.** 2005. Short Term and Medium Term Power Distribution Load Forecasting by Neural Networks. *Energy Conversion and Management*, 46, pp1393-1405.

[8] **Tan Yu-fei, Chen Jia-xin, Jiao Wen-ling, etc. 2001.** Short-term load forecast of city gas based on artificial neural network. *Gas & Heat*, 21(3), pp199-202.

[9] **Yang Zhao, Liu yan, Miao Zhi-bin, etc. 2003.** Application of artificial neural network in prediction of gas load. *Gas & Heat*, 23(6), pp331-332.

[10] **Liu Han, Liu Ding, Zheng Gang, etc. 2005.** Short-term load prediction of urban natural gas. *The natural gas industry*, 25(7), pp105~107.

[11] **Dou Lian-wang, Feng Liang. 2005.** Short-term city gas load forecasting based on neural network. *Gas & Heat*, 25(12), pp10-14.

MAINTENANCE OPTIMIZATION POLICIES OF MULTI-STATE SYSTEMS: A REVIEW

H.W. Liu[1*] and Y. Wang[2]
[1]College of Management and Economics
Tianjin University, Tianjin, China
hw_liu999@tju.edu.cn

[2]College of Management and Economics
Tianjin University, Tianjin, China
15022657161@163.com

ABSTRACT

With the development of industrialization, industrial equipments become more and more complex, and the user demand becomes more and more important. Traditional binary-state system cannot accurately describe the equipment status. A multi-state system (MSS) oriented to customers' demands can describe the operation status of equipments more accurately. Maintenance optimization policies of MSSs have become a practical and imperative research topic. However, no relevant review has been carried out in this field and the development of this field is not enough. To address those issues, a review of current research on MSSs was carried out in this paper. The review result shows that maintenance optimization policies of MSSs can be categorised in four classes according to the difference of maintenance actions: perfect maintenance, imperfect maintenance, selective maintenance and integrated maintenance. According to different maintenance scenarios, perfect maintenance can be distinguished by perfect continuous monitoring and inspection. Then imperfect maintenance models on MSSs—from binary-state elements to multi-state elements—have been briefly reviewed. In addition, more complex maintenance optimization problems including selective maintenance and integrated maintenance are touched, which usually consider more maintenance actions or more complex maintenance scenarios. After induction and comparison for the existing related literature, several potential research directions are proposed.

Keywords: Industrialization, Multi-state system, Maintenance optimization policies, Maintenance actions, Maintenance scenarios

1 INTRODUCTION

With the increasing complexity of industrial equipment and system function structure, a great number of research contributions have been carry out on the multi-state reliability theory like the representation, modelling and quantification techniques of MSSs. A multi-state system (MSS) is regarded as a system with a range of discrete performance levels (from perfectly functioning to complete failure) [1, 2]. For example, the overall performance of manufacturing, production and power generation can settle on different levels (e.g. 100%, 80%, 50% of the nominal capacity), depending on the operative conditions of elements [3].

Compared to traditional binary-state system which only considers perfectly functioning and complete failure, the reliability of an MSS is expressed by the state distribution. An MSS needs to be repaired once its performance level, determined by the multiple elements and the configuration, falls below the user demand level. Instead, it is considered as in a normal operating state. In general, systems that conform to the following characteristics can be referred to as an MSS [1-3]:

* H.W. Liu

1. The system is composed of a number of discrete units, and every unit has a cumulative effect on the overall performance of the system.
2. The current states of system components are changing with the deterioration and maintenance.

Maintenance is defined as a set of actions or tasks used to restore an item to a state in which it can perform its designated functions [4]. It's one of the main methods to improve the system reliability. Maintenance actions used in reliability optimization can be mainly divided into three categories [5]: perfect maintenance, minimal maintenance and imperfect maintenance. Some reviews have been carried out on the maintenance optimization problems, however, the above them are only based on the binary-state systems. The literature on binary-state systems is mostly classified according to different maintenance scenarios, including: both single-unit and multi-unit systems [6], different maintenance objectives [7] and the dependence (stochastic, structural or economic) between components [8, 9]. In addition, there are several reviews classifying the related literature by different maintenance actions [5, 10]. So far no research has synthetically considered maintenance actions and maintenance scenarios and no research has been specifically devoted to reviewing the maintenance optimization problem of MSSs.

The maintenance optimization policies based on MSSs are a natural extension of the maintenance studies of binary-state systems. In the last few years, there have been many studies on the maintenance policies of MSSs, where the correlation, the uncertainty and the decision variables are more complicated. Yet there has been no research to review the maintenance policies of MSSs. According to the difference of maintenance actions and maintenance scenarios, maintenance optimization problems of MSSs are divided into four types in this paper: perfect maintenance, imperfect maintenance, selective maintenance and integrated maintenance. The perfect maintenance is usually regarded as the simplest maintenance action. The imperfect maintenance models are more complex and realistic. The selective maintenance policies generally need to consider multiple maintenance actions and scenarios. The integrated maintenance represents more complex maintenance scenarios. The remaining part of the paper is organized as follows. Section 2-5 respectively review four different maintenance optimization problems of MSSs, followed by a brief conclusion in section 6.

2 PERFECT MAINTENANCE

Perfect maintenance means a maintenance action restoring the system operating condition to "as good as new" like the component replacement policy and the major repair policy. It is the most common and simplest maintenance policy used to achieve the maintenance time optimization. Perfect maintenance policies are distinguished by different maintenance scenarios: perfect monitoring and inspection.

2.1 Perfect maintenance and perfect monitoring

Perfect maintenance policies are mostly based on replacement policy, which means to replace the failed system by a brand new one. Some replacement models are based on the assumption of perfect and continuous monitoring. It means that the performance rate of the system can be obtained accurately at any given time. Previous approaches include the age replacement policy (ARP) [11], the fuzzy reliability optimal replacement policy [12], the global optimal multi-state element (MSE) replacement policy [13], the replacement policy under economic dependence[9, 14], the replacement schedule optimization [15], and the reliability block diagram method with minor failures and repairs [16]. The above approaches are under infinite horizon assumptions. In practice, it is important to consider maintenance policies under a finite horizon because the life cycles of most systems are finite. A finite life-cycle MSS that is subject to both degradation and Poisson failures has been considered [17]. The maintenance optimization problems under a finite horizon should attract more attention in the future.

2.2 Perfect maintenance and inspection

Above maintenance models are mostly based on the assumption of continuous state monitoring. However, continuous monitoring is impossible and system's real degradation level can only be known through inspection [18]. So, researchers gradually recognized the importance of inspection [5]. Yet previous studies usually only considered replacement policy and inspection based on the assumption that components are independent. [19-22]. In recent years, the dependence between elements are considered including the maintenance model under economic dependence and state-dependent inspection intervals [23] or competing failure models [24]. Nowadays, considering the combination between inspection and continuous monitoring is necessary in the real maintenance scenarios and more realistic inspection models should be studied.

3 IMPERFECT MAINTENANCE

Imperfect maintenance means a maintenance action which can't make a system like as good as new, but younger. In the field of MSSs, imperfect maintenance can restore the system to any intermediate state, which seems to be more realistic. So it is of great practical significance to consider the imperfect maintenance of MSSs. A multi-state system is more complex than a binary-state system, yet binary approaches can be extended to MSSs. Basically, there are four classes of imperfect maintenance models for binary-state systems [5, 25]: a probabilistic approach, the improvement factor method, the virtual age model and the method based on the system degradation. However, only virtual age model has been extended to MSSs. At present, there have been several studies on the imperfect maintenance of MSSs.

3.1 Imperfect maintenance considering binary-state elements

From the view of component states, MSSs can be composed of binary-state elements or multi-state elements. The virtual age model [26] is extended to describe the imperfect maintenance of MSSs, which uses an age reduction factor to describe the maintenance effect. However, so far it has only been used in MSSs consisting of binary-state elements [27]. Previous studies include: a preventive maintenance optimization problem under imperfect maintenance [27, 28] and an optimal preventive maintenance (PM) policy considering perfect, minimal, and imperfect repair actions [29]. Besides, some general monotone process models have been developed based on geometric process considering imperfect maintenance, which arrange maintenance plans through different maintenance policies including: the policy by time and number of failures (T, N) [30], the policy by time only (T) [31] and the policy by failure number only (N) [32-34]. On the whole, a multi-state system composed of binary-state elements is not the general form of MSSs, elements containing multiple discrete states should be considered.

3.2 Imperfect maintenance considering multi-state elements

In addition to the virtual age model extended, there are some new attempts to describe the imperfect maintenance of MSSs, in which multi-state elements are considered. An imperfect maintenance model [35] assumes that each individual element would be restored to its best functional state, yet not in "as good as new" condition according to a quasi-renewal process [36]. In fact, the state transition intensities increase proportionally after repair. Based on the "system perspective" concept, a random restoration factor (RF) concept is used to describe the imperfect maintenance of MSSs [25]. The concept of RF has led to a discussion on the imperfect maintenance of MSSs [37, 38]. Besides, several papers consider the combination between imperfect maintenance and inspection [18, 39]. It seems that both of maintenance and monitoring are imperfect. The combination optimization problem between them should be a promising research direction.

Obviously, only several research has focused on the imperfect maintenance model of MSSs, which rarely considers the dependences between elements. New models that can accurately characterize complex stochastic behaviors of repairable MSSs should be studied and more complex maintenance scenarios should be considered.

4 SELECTIVE MAINTENANCE STRATEGY

Most maintenance models of MSSs assume that maintenance resources are infinite and unnecessary [40]. In fact, the maintenance resources are finite and they are impossible to perform the whole desirable maintenance actions. Selective maintenance policies mean that the decision maker needs to identify the subset of feasible maintenance actions to be carried out [41], including the selectivity of components and maintenance actions[41, 42]. Previous studies tend to consider some complex maintenance scenarios including selective maintenance models considering economic dependence [43, 44] and stochastic dependence between elements [40, 45] or the integrated model considering maintenance actions and optimal load distribution [46]. Most selective maintenance models consider most maintenance actions including perfect and imperfect maintenance. Perhaps in a certain way, selective maintenance strategies are more realistic.

5 INTEGRATED MAINTENANCE

Maintenance and redundancy are regarded as two main methods used to provide a required level of system reliability, which are usually considered separately [47]. Redundancy allocation optimization of MSSs can be defined as optimal combinations of functional components and redundant components under certain constraints. The versions and the corresponding redundancy are selected in every subsystem in order to achieve the system cost minimum (or maximum reliability) [48]. There are some studies for the integrated models between redundancy and maintenance [47, 49]. However, most of them only consider minimal repair and replacement actions [47, 49-51], or the limited maintenance resources [52, 53]. The maintenance actions and scenarios considered are both simple.

In addition, there are several research contributions considering more complex maintenance scenarios including the maintenance model considering multiple states and failure interactions [54], the optimal allocation and maintenance model with common bus performance sharing [55] and the integrated optimal model considering two interconnected MSSs [56]. And some integrated models even consider more complex maintenance actions like imperfect maintenance [57, 58]. Obviously, the combination of multiple optimization problems can comprehensively consider more aspects related to equipment reliability and are more close to the reality.

6 CONCLUSION

There have been many studies on the reliability optimization of MSSs in recent years. The maintenance policy is one of the main methods to improve system reliability [30]. In this paper, the relevant literature on four types of maintenance policies of MSSs is summarized. After induction and comparison, several future research directions can be put forward:

1. There are few literature focusing on the imperfect maintenance models of MSSs and new imperfect maintenance models should be developed. In addition, imperfect maintenance should be taken into account together with more complex maintenance scenarios like economic dependence and stochastic dependence.
2. Most maintenance models of MSSs are under the assumption of continuous monitoring, which is not in conformity with real scenarios. Actually, inspection action is still necessary. The

combination between inspection and maintenance action, especially imperfect maintenance, should be a promising research direction.

3. Most maintenance models of MSSs usually focus on the consideration of maintenance cost, yet repair time and replacement time are ignored. The time distribution of maintenance actions should be studied and average cost per unit time should be contained in the future.
4. Most maintenance models of MSSs only consider identical components. Non-identical components are rarely considered.
5. The combination of multiple optimization problems like maintenance and redundancy can make the maintenance scenario more complex and more realistic. Multiple optimization problems should become the focus of research in the future.
6. Most maintenance models are based on the assumption of infinite life cycle. Actually, maintenance policies under a finite horizon are more realistic and should be studied more in the future.

7 REFERENCES

[1] **Levitin, G., Lisnianski, A. 2003.** *Multi-state system reliability: assessment, optimization and applications*. World Scientific Publishing Co. Pvt. Ltd.

[2] **Levitin, G. 2004.** A universal generating function approach for the analysis of multi-state systems with dependent elements, *Reliability Engineering & System Safety*, 84(3), pp 285-292.

[3] **Zio, E. 2009.** Reliability engineering: Old problems and new challenges, *Reliability Engineering & System Safety*, 94(2), pp 125-141.

[4] **Ahmad, R., Kamaruddin, S. 2012.** An overview of time-based and condition-based maintenance in industrial application, *Computers & Industrial Engineering*, 63(1), pp 135-149.

[5] **Pham, H., Wang, H. 1996.** Imperfect maintenance, *European journal of operational research*, 94(3), pp 425-438.

[6] **Wang, H. 2002.** A survey of maintenance policies of deteriorating systems, *European journal of operational research*, 139(3), pp 469-489.

[7] **De Almeida, A.T., Ferreira, R.J.P., Cavalcante, C.A.V. 2015.** A review of the use of multicriteria and multi-objective models in maintenance and reliability, *IMA Journal of Management Mathematics*, pp 1-23.

[8] **Dekker, R., Wildeman, R.E., Van der Duyn Schouten, F.A. 1997.** A review of multi-component maintenance models with economic dependence, *Mathematical Methods of Operations Research*, 45(3), pp 411-435.

[9] **Nicolai, R.P., Dekker, R. 2008.** *Optimal maintenance of multi-component systems: a review.* Springer London.

[10] **Jardine, A.K., Lin, D., Banjevic, D. 2006.** A review on machinery diagnostics and prognostics implementing condition-based maintenance, *Mechanical systems and signal processing*, 20(7), pp 1483-1510.

[11] **Sheu, S.-H., Zhang, Z.G. 2013.** An optimal age replacement policy for multi-state systems, *Reliability, IEEE Transactions on*, 62(3), pp 722-735.

[12] **Liu, Y., Huang, H.-Z. 2011.** Optimal Replacement Policy for Fuzzy Multi-State Element, *Multiple-Valued Logic and Soft Computing*, 17(1), pp 69-92.

[13] **Liu, Y., Huang, H.-Z. 2010.** Optimization of multi-state elements replacement policy for multi-state systems, Reliability and Maintainability Symposium (RAMS), California, USA, January 25-28, pp 1-7.

[14] **Zhou, Y., Zhang, Z., Ma, L. 2012.** Maintenance optimisation of a series-parallel system with multi-state components considering economic dependence, Quality, Reliability, Risk, Maintenance, and Safety Engineering (ICQR2MSE), Chengdu, China, June 15-18, pp 427-431.

[15] **Levitin, G., Lisnianski, A. 2000.** Short communication optimal replacement scheduling in multi - state series - parallel systems, *Quality and Reliability Engineering International*, 16(2), pp 157-162.

[16] **Lisnianski, A. 2007.** Extended block diagram method for a multi-state system reliability assessment, *Reliability Engineering & System Safety*, 92(12), pp 1601-1607.

[17] **Wu, J., Ng, T.S.A., Xie, M., et al. 2010.** Analysis of maintenance policies for finite life-cycle multi-state systems, *Computers & Industrial Engineering*, 59(4), pp 638-646.

[18] **Le, M.D., Tan, C.M. 2013.** Optimal maintenance strategy of deteriorating system under imperfect maintenance and inspection using mixed inspectionscheduling, *Reliability Engineering & System Safety*, 113, pp 21-29.

[19] **Lam, C.T., Yeh, R. 1994.** Comparison of sequential and continuous inspection strategies for deteriorating systems, *Advances in Applied Probability*, pp 423-435.

[20] **Machihara, F. 1990.** A new approach to the fundamental period of a queue with phase-type Markov renewal arrivals, *Stochastic Models*, 6(3), pp 551-560.

[21] **Yeh, R.H. 1997.** Optimal inspection and replacement policies for multi-state deteriorating systems, *European Journal of Operational Research*, 96(2), pp 248-259.

[22] **Chiang, J.-H., Yuan, J. 2001.** Optimal maintenance policy for a Markovian system under periodic inspection, *Reliability Engineering & System Safety*, 71(2), pp 165-172.

[23] **Zhou, Y., Zhang, Z., Lin, T.R., et al. 2013.** Maintenance optimisation of a multi-state series-parallel system considering economic dependence and state-dependent inspection intervals, *Reliability Engineering & System Safety*, 111, pp 248-259.

[24] **Zhang, J., Huang, X., Fang, Y., et al. 2016.** Optimal inspection-based preventive maintenance policy for three-state mechanical components under competing failure modes, *Reliability Engineering & System Safety*, 152, pp 95-103.

[25] **Tan, C.M., Raghavan, N. 2008.** A framework to practical predictive maintenance modeling for multi-state systems, *Reliability Engineering & System Safety*, 93(8), pp 1138-1150.

[26] **Kijima, M. 1989.** Some results for repairable systems with general repair, *Journal of Applied probability*, pp 89-102.

[27] **Levitin, G., Lisnianski, A. 2000.** Optimization of imperfect preventive maintenance for multi-state systems, *Reliability Engineering & System Safety*, 67(2), pp 193-203.

[28] **Nahas, N., Khatab, A., Ait-Kadi, D., et al. 2008.** Extended great deluge algorithm for the imperfect preventive maintenance optimization of multi-state systems, *Reliability Engineering & System Safety*, 93(11), pp 1658-1672.

[29] **Sheu, S.-H., Chang, C.-C., Chen, Y.-L., et al. 2015.** Optimal preventive maintenance and repair policies for multi-state systems, *Reliability Engineering & System Safety*, 140, pp 78-87.

[30] **Zhang, Y.L., Yam, R.C., Zuo, M.J. 2007.** A bivariate optimal replacement policy for a multistate repairable system, *Reliability engineering & system safety*, 92(4), pp 535-542.

[31] **Zhang, Y.L., Wang, G.J. 2010.** An optimal replacement policy for a multistate degenerative simple system, *Applied Mathematical Modelling*, 34(12), pp 4138-4150.

[32] **Zhang, Y., Yam, R., Zuo, M. 2002.** Optimal replacement policy for a multistate repairable system, *Journal of the Operational Research Society*, 53(3), pp 336-341.

[33] **Zhang, Y.L. 2004.** An optimal replacement policy for a three-state repairable system with a monotone process model, *Reliability, IEEE Transactions on*, 53(4), pp 452-457.

[34] **Yeh, L. 2005.** A monotone process maintenance model for a multistate system, *Journal of Applied Probability*, 42(1), pp 1-14.

[35] **Liu, Y., Huang, H.-Z. 2010.** Optimal replacement policy for multi-state system under imperfect maintenance, *Reliability, IEEE Transactions on*, 59(3), pp 483-495.

[36] **Wang, H., Pham, H. 1996.** A quasi renewal process and its applications in imperfect maintenance, *International journal of systems science*, 27(10), pp 1055-1062.

[37] **Liu, Y., Huang, H.-Z. 2009.** Comment on "A framework to practical predictive maintenance modeling for multi-state systems" by Tan CM and Raghavan N. [Reliab Eng Syst Saf 2008; 93 (8): 1138-50], *Reliability Engineering & System Safety*, 94(3), pp 776-780.

[38] **Tan, C.M., Raghavan, N. 2009.** Reply to comments on "A framework to practical predictive maintenance modeling for multi-state systems", *Reliability Engineering & System Safety*, 94(3), pp 781-782.

[39] **Lu, Z., Chen, M., Zhou, D. 2015.** Periodic inspection maintenance policy with a general repair for multi-state systems, Chinese Automation Congress (CAC), Wuhan, China, November 27-29, pp 2116-2121.

[40] **Dao, C.D., Zuo, M.J. 2015.** Selective Maintenance for Multistate Series Systems With S-Dependent Components, *Reliability, IEEE Transactions on*, pp 1-15.

[41] **Liu, Y., Huang, H.-Z. 2010.** Optimal selective maintenance strategy for multi-state systems under imperfect maintenance, *Reliability, IEEE Transactions on*, 59(2), pp 356-367.

[42] **Cassady, C.R., Pohl, E.A., Paul Murdock, W. 2001.** Selective maintenance modeling for industrial systems, *Journal of Quality in Maintenance Engineering*, 7(2), pp 104-117.

[43] **Dao, C.D., Zuo, M.J., Pandey, M. 2014.** Selective maintenance for multi-state series-parallel systems under economic dependence, *Reliability Engineering & System Safety*, 121, pp 240-249.

[44] **Dao, C.D., Zuo, M.J. 2015.** *Selective Maintenance for Multi-state Systems Considering the Benefits of Repairing Multiple Components Simultaneously. In Engineering Asset Management-Systems, Professional Practices and Certification (pp. 413-425).* Springer International Publishing.

[45] **Dao, C.D., Zuo, M.J. 2015.** Reliability analysis of multi-state systems with s-dependent components, Reliability and Maintainability Symposium (RAMS), Florida, USA, January 26-29, pp 1-6.

[46] **Chen, C., Liu, Y., Huang, H.-Z. 2012.** Optimal load distribution for multi-state systems under selective maintenance strategy, Quality, Reliability, Risk, Maintenance, and Safety Engineering (ICQR2MSE), Chengdu, China, June 15-18, pp 436-442.

[47] **Levitin, G., Lisnianski, A. 1999.** Joint redundancy and maintenance optimization for multistate series-parallel systems, *Reliability Engineering & System Safety*, 64(1), pp 33-42.

[48] **Chaaban, W., Schwarz, M., Börcsök, J. 2013.** Cost and redundancy optimization of homogeneous series-parallel multi-state systems subject to availability constraints using a Matlab implemented genetic algorithm, *Recent Advances in Circuits, Systems and Automatic Control, WSEAS*, pp 53-61.

[49] **Li, Y., Peng, R. 2014.** Availability modeling and optimization of dynamic multi-state series-parallel systems with random reconfiguration, *Reliability Engineering & System Safety*, 127, pp 47-57.

[50] **Nourelfath, M., Dutuit, Y. 2004.** A combined approach to solve the redundancy optimization problem for multi-state systems under repair policies, *Reliability Engineering & System Safety*, 86(3), pp 205-213.

[51] **Tian, Z., Levitin, G., Zuo, M.J. 2009.** A joint reliability-redundancy optimization approach for multi-state series-parallel systems, *Reliability Engineering & System Safety*, 94(10), pp 1568-1576.

[52] **HUANG, H.-z. 2012.** JOINT OPTIMIZATION OF REDUNDANCY AND MAINTENANCE STAFF ALLOCATION FOR MULTI-STATE SERIES-PARALLEL SYSTEMS OPTYMALIZACJA ŁĄCZONA ALOKACJI NADMIAROWOŚCI ORAZ ALOKACJI PRACOWNIKÓW SŁUŻB UTRZYMANIA RUCHU W WIELOSTANOWYCH SYSTEMACH SZEREGOWO-RÓWNOLEGŁYCH, *Terms and Conditions of Publication*, 14(4), pp 312.

[53] **Nourelfath, M., Ait-Kadi, D. 2007.** Optimization of series-parallel multi-state systems under maintenance policies, *Reliability Engineering & System Safety*, 92(12), pp 1620-1626.

[54] **Wang, J., Li, M. 2015.** Redundancy Allocation Optimization for Multistate Systems With Failure Interactions Using Semi-Markov Process, *Journal of Mechanical Design*, 137(10), pp 101403.

[55] **Xiao, H., Peng, R. 2014.** Optimal allocation and maintenance of multi-state elements in series-parallel systems with common bus performance sharing, *Computers & Industrial Engineering*, 72, pp 143-151.

[56] **Lisnianski, A., Ding, Y. 2009.** Redundancy analysis for repairable multi-state system by using combined stochastic processes methods and universal generating function technique, *Reliability Engineering & System Safety*, 94(11), pp 1788-1795.

[57] **Nourelfath, M., Châtelet, E., Nahas, N. 2012.** Joint redundancy and imperfect preventive maintenance optimization for series-parallel multi-state degraded systems, *Reliability Engineering & System Safety*, 103, pp 51-60.

[58] **Liu, Y., Huang, H.-Z., Wang, Z., et al. 2013.** A joint redundancy and imperfect maintenance strategy optimization for multi-state systems, *Reliability, IEEE Transactions on*, 62(2), pp 368-378.

SERVICE-ORIENTED TRACEABILITY PLATFORM FOR PACKAGED FOODS SUPPLY CHAIN

Guo Liu[1*]. Zhi Li[2]. Xinjun Lai[3]. Yi Jing[4]
Guangdong Provencal Key Lab of Computer Integrated Manufacturing
Guangdong University of Technology
13724076781@163.com

ABSTRACT

Quality and safety of packaged foods is involved in every participant of supply chain, during which problems concerning quality and safety may result in serious consequences. Good traceability system not only can provide manufacturers with the product's end-of-life activities and processes, but also provide consumers with inquiry service about the product's traceability information. Aiming at realizing the packaged foods life-cycle visibility and traceability from its production, circulation and consumption, this paper proposes a systematic architecture and an implementation method. Firstly, service-oriented traceability platform(SOTP)used in packaged foods supply chain has been proposed. The proposed SOTP provides four visualization services, supply chain management service, life-cycle traceability service, evaluation and warning service, and big data statistics service, respectively. Secondly, based on the combination of internet of things and cloud computing technologies,the SOTP achieves real-time dynamic data acquisition and processing of packaged foods information, creating a ubiquitous environment in packaged foods supply chain. Finally, this study develops a prototype of SOTP to validate the proposed framework and approach. In this paper,the case study proves that the proposed SOTP has huge advantage over the traditional traceability method in the services of supply chain participants.

Key words: Service-oriented platform, Traceability , Food supply chain, Internet of things

1 INTRODUCTION:

Packaged foods supply chain is a set of entities (i.e. manufactures, distributors, retailers and consumers) working together to maximize the benefits of the whole chain. The food industry should have customer-oriented concept and a faster response dealing with food scandals and incidents[1].Information sharing and information technology enable and support logistics and quality management in global food supply chains through sophisticated information technologies and virtualization of product flows[2].Traceability's application is considered to be a vital issue for all stakeholders in food supply chains. It can optimize business processes along the supply chain in order to improve efficiency and reduce lead times and food waste[3].

Many scholars have focused on the research about the food supply chain based on the internet of things(IOT), as the technology of IOT applied to the supply chain of fresh vegetables can provide an intuitive perspective for users(including the terminal consumers) to understand the information after collecting the data. When customers buy foods and supervise the whole supply chain, these data will help customers make decision more easily[4]. RFID technology based on IOT in the field of manufacturing industry has been applied widely. The application of RFID for industrial products traceability can help consumers learn the history of the product and the path of flow and realize the products being controlled and managed[5]. When virtualization technology of the food supply chain is realized, it is very convenient for operators in each node of supply chain to monitor,control,plan and manage the entire food business process[6].Establishing a simulation platform of supply chain based on a IOT technology to provide customers with a query service can realize the visual management in the process of circulation and trace the whole process to achieve the effective management of the supply chain[7].

According to the literature research above, food safety and food supply chain management have become a hot topic at home and abroad[8]. The application of IOT in the supply chain has gradually become wider and wider, which improves the management of food supply chain and food safety, but there are many problems remained as follow:

1. How food management services are achieved to control the food safety? Food management services are to achieve the sharing of information in the whole supply chain to strengthen the supervision of food quality.
2. How traceability services are achieved to trace the person responsible for packaging food from production to consumption in each participant of supply chain? Traceability service for packaged food security is to trace the source where the problem of food quality and food safety appears.
3. How the evaluation and warning services is realized to be able to quickly respond to consumer? Evaluation and warning services is to facilitate the collection of consumer information on the feedback of packaging food and solve the problem of consumers encountering with the quality problems.
4. How large data statistics is achieved to facilitate the correct decision and maximize the benefits of the supply chain? The collection of information in each node of the supply chain(including producers and consumers) and the mining of related information about supply chain based on statistics is helpful to achieve the coordination schedule of supply chain.

In order to solve the above problems, this paper proposes a service-oriented traceability platform to solve the problems of four visual services and realize the supervision of the quality and safety of the food supply chain. The rest of the paper is organized as follow: Section 2 is about the overview of the service-oriented traceability platform. Section 3 describes the four visualization services to facilitate operations and decision making of packaged foods participants. In Section 4, we describe the implementation methods through a case study. Lastly, conclusion and future works are given in Section 5 to summarize this research.

2 OVERVIEW OF SERVICE-ORIENTED TRACEABILITY PLATFORM

Fig. 1 describes the system architecture of the proposed SOTP. There are four layers: Perception layer, Data layer , Service layer and Application layer. From the bottom-up, the perception layer provides the physical assets and resources virtualization. The quick response two-dimensional barcode (QR code) and radio frequency identification (RFID)applied in packaged foods supply chain are to establish the precondition and foundation for traceability management.Attaching Qr code to the item not only can provide producers with the packaged foods' end-of-life activities and processes to improve the product traceability management capacity, but can provide consumers with inquiry service about product source information. RFID is a non-contact automatic identification technology, which mainly consists of reading/writing device and radio tags.Attaching RFID tags in packages, cases, transportation units, or even delivery trucks not only can detect objects and obtain the related data via radio frequency without human intervention, greatly reducing the potential delay and improving the operation efficiency, but can detect multi objects per time, reducing work load and labor cost, showing its convenience of collecting and sharing data in supply chain effectively and efficiently[9]. In addition to the classic cloud resources (storage, server and network), smart gateway enable the application of IoT technologies. Smart gateway mainly contains a gateway server deployed in the cloud and a set of smart assets, which not only processes caches and exchanges real-time data and events locally and temporally, but provides facilities for service definition, configuration and execution locally at workstations[10].

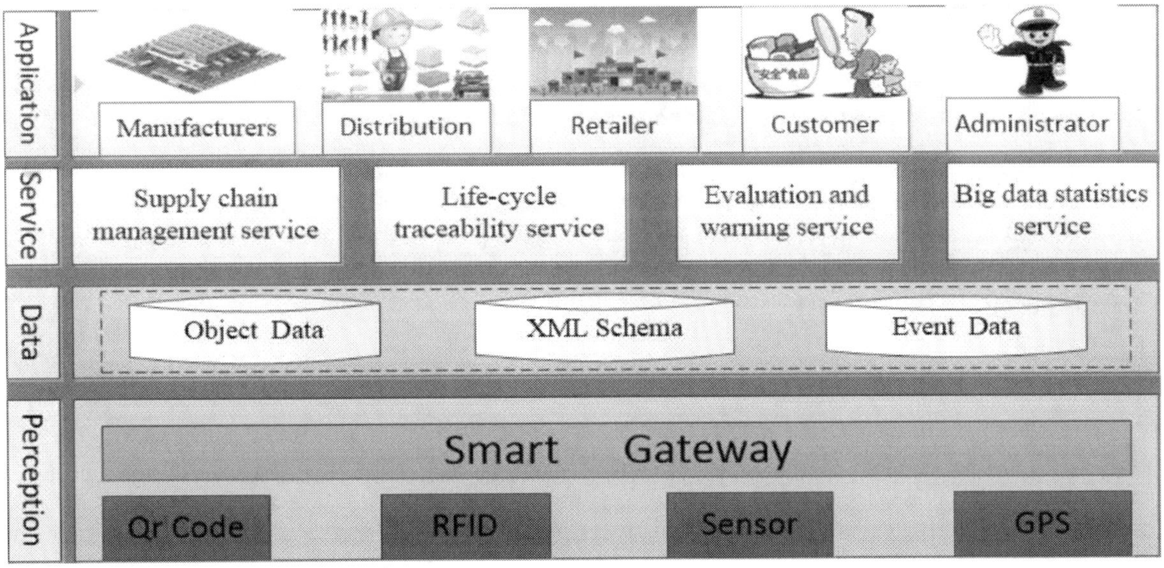

Figure 1: Architecture of SOTP

The captured life-cycle traceability information needs to be stored appropriately in data layer so that the further regulation and query can be executed precisely. Packaged foods data repository stores object and event data. Object data is the information that generally can not be modified or changed through supply chains including but not limited to company registering information, packaged foods information, and packaged foods regulatory codes information. Event data is dynamically generated or distinguished by packaged foods supply chain. For example, a particular pallet holds with different boxes at different times. Therefore, the boxes on the pallet are dynamic and have time stamps.It is also in line with international standards of XML(Xtensible Markup Language) schema to facilitate a variety of data exchange between complex system architectures. It this way, the historical information can be obtained easily, stored properly, and exchanged compatibly.

At the top of this architecture, five types of services are developed and deployed in application layer. These services fulfill the management and execution requirements from the four visualization services in SOTP including supply chain management service, life-cycle traceability service, evaluation and warning service, and big data statistics service. The following innovative components have been implemented in the proposed SOTP infrastructure and will be discussed in detail below.

3 DEVELOPMENT OF CLOUD-ENABLED APPLICATION SERVICES

The innovative components, which are discussed in this section, have been implemented in the proposed SOTP infrastructure:

- Cloud service for supply chain management;

- Cloud service for life-cycle traceability;
- Cloud service for evaluation and warning;
- Cloud service for big data statistics;

3.1 Cloud service for supply chain management

3.1.1Assembling parent-child relation management

Assembling parent-child relation management is mainly responsible for assembling the relationship of packaged foods and its each-level packages from small package to large package in Fig.2. Generally, unique regulatory codes of QR code and RFID are downloaded from SOTP. The information of packaged foods is decoded and stored in QR code which will then be labelled on the smallest-level packaged foods . If middle-level package such as box is necessary, several smallest packages will be placed together into the middle level packages ,meanwhile the RFID tag are attached in the middle-level package . This assembling relation will be captured by smart asset like PDA. Afterwards, all associated data created in the above-mentioned procedure will be stored for further usage.

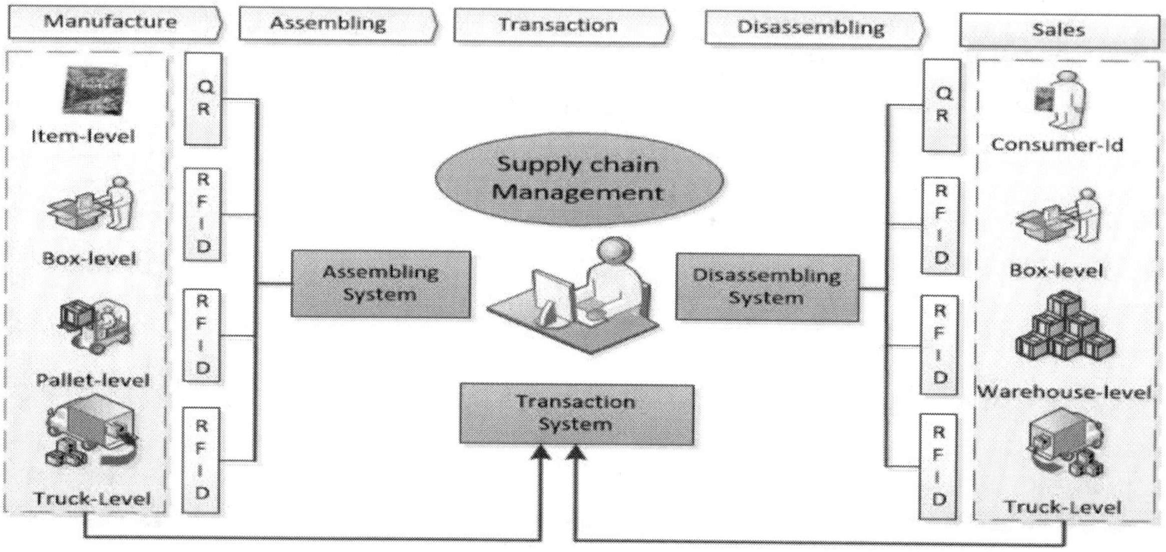

Figure 2: Supply Chain Management

3.1.2 Disassembling parent-child relation management

Disassembling parent-child relation management is primarily in charge of disassembling ultimate package into lower level small packaged foods in sequence in Fig.2. It is actually a reverse operation of assembling. The disassembling process may also happen in retail shop where the retailer disassembles the larger packages into final goods that can be sold one by one. Similarly, the disassembling parent-child relation will be captured by smart asset and stored in database.

3.1.3 Transactions parent-child relation management

Transactions parent-child relation management associated with a particular transaction at some time to identify the origination and destination so that the life-cycle traceability can be initiated from manufacturer to final consumer in Fig.2. In this way, the traded product can be tracked backward and traced forward throughout the packaged foods supply chain. Transaction relation management may well cope with the movement of packaged foods and clearly depict the routine of them, providing the key evidence for life-cycle traceability.

3.2 Cloud service for life-cycle traceability

Implementation of life-cycle traceability service is that manufacturers attach a unique QR code to packaged foods item and attach RFID tags to packages, cases, transportation units. Traceability system can facilitate products in the distribution network flow orderly and achieve strict supervision and control to improve the level of enterprise management and avoid the risks.

Through life-cycle traceability service, manufacturers can capture the product's end-of-life activities and processes. If manufacturers receive the notification about the suspected packaged foods, they will adopt measures to minimize nominal and economic loss as much as possible. As for distributors, they can identify the suspected packaged foods simultaneously. Without safety promise, distributors will not put the suspected packaged foods into circulation until the all-clear signal. Consumers can scan the QR code by the smart phone, and capture from the packaged foods a series of genuine security information(production date, production place). Administrative authorities can send notification about suspected packaged foods before they pose threat to the public health, and issue recall commands when defective packaged foods are confirmed or safety pledge to claim all things back to normal.

3.3 Cloud service for evaluation and warning

In order to obtain a good evaluation of packaged foods for enterprises, the most important is to continuously improve the quality of products and services.Through this platform, consumers can evaluate the purchased packaged foods , which enables the enterprise to capture the product quality problems and shortcomings, and timely treatment.The main content of the consumer review is the authenticity, reliability, and other related information about the packaged foods. Through this evaluation system, the content of the review can bring more useful information about the product, which not only can help other consumers with purchase decisions, but can help enterprises with production decisions.

The platform sets up a packaged foods warning services for citizens, professionals and regulatory agencies to share the warning services for the poor packaged foods, which achieve the life-cycle supervision of packaged foods to the closed feedback.Through analyzing and mining the traceability data of packaged foods, regulators can timely send early warning information to prevent consumers from eating bad foods, and at the same time inform the relevant enterprises of reaction in a timely manner.

3.4 Cloud service for big data statistics

Big data statistics service is a data view of the way to show the historical data of packaged foods supply chain to the manufacturer.These data are captured by the consumer or operator's scanning the QR code attached to the item and RFID attached to the package.Through SOTP, the enterprise can obtain the following aspects of packaged foods traceability data statistics service.

- Enterprises can clearly grasp the flow of products, such as where and what people are using, according to the two-dimensional code and RFID.

- Through the data view, enterprises can view the sales progress of packaged foods in different regions.If there is a group of packaged food which will soon expire in the A area, the enterprise will transfer them to B area or conduct a promotion.
- According to consumer evaluation data, enterprises can also analyze the quality of the packaged foods, and timely process or recall to avoid the impact of consumer health.

4 Case study

This section demonstrates how the proposed SOTP infrastructure and services are deployed in a case company for packaged foods supply chain.

4.1 Motivating scenario description

This case study is based on a typical packaged foods of Guiling Gao enterprise which is located in China.The enterprise leading the Guiling Gao packaged foods have amounted to more than 100, which include the collection, production, packaging, logistics, sales in the integration of food enterprises. Quality and safety of Guiling Gao is involved in every participant of supply chain, during which problems concerning quality and safety may result in serious consequences. In order to continuously improve product quality and management level,the enterprise want to develop a traceability system to realize the Guiling Gao life-cycle visibility and traceability from its production, circulation and consumption.On the one hand this traceability system can provide enterprise with the product's end-of-life activities and processes; on the other hand, it can provide consumers with inquiry service about the product's traceability information.Therefore, this section demonstrates how the proposed SOTP infrastructure and services are deployed in a case company for the packaged foods of Guiling Gao life-cycle visibility and traceability.

4.2 Real-time Visibility and traceability service

1) **Product management:** the enterprise need to edit the product information into the QR code.This information includes producing area, production batch, production date and the city to be sent to.Then the operator scans the QR code to store the information in the system through the smart reading device.

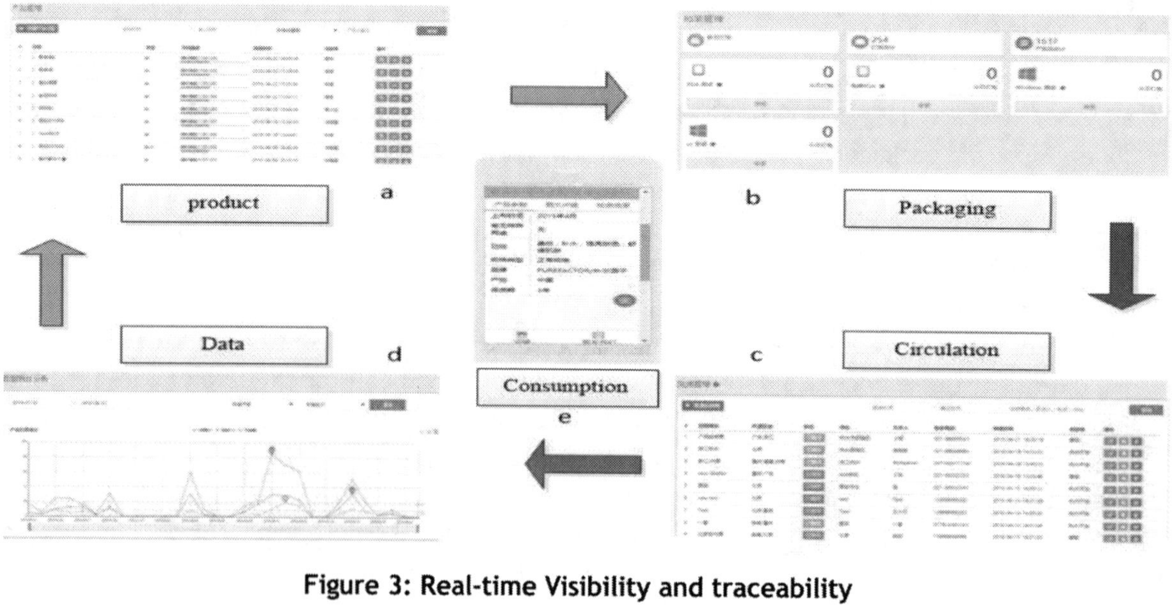

Figure 3: Real-time Visibility and traceability

2) **Packaging management:** Unique regulatory codes of QR code and RFID are downloaded from SOTP. These information of packaged foods are decrypted and stored in QR code which will then be labelled on the smallest-level packaged foods . If middle-level package such as box is necessary, several smallest packages will be placed together into the middle level packages; meanwhile the RFID tag are given in the middle-level package . This parent-child relation will be captured by smart asset like PDA as shown in Fig. 3 (b).

3) **Circulation management:** Circulation records those packaged foods that are associated with a particular transaction at some time, or records a specific transaction that is associated with an unexpected packaged foods at some time as shown in Fig. 3 (c). It is important to determine the origin and destination, for the purpose of realizing the life-cycle traceability from manufacturing to final consumers. Circulation management may well cope with the movement of packaged foods and clearly depicts the routine of them, providing the key evidence for life-cycle traceability.

4) **Consumption management:** After purchase, consumers can scan QR code attached to packaged foods to verify the authenticity via smart phones, and consumers can also inquire the product 's life-cycle activities and processes as shown in Fig. 3 (e). If there are quality and safety problems about the purchased products, consumers can depend on the evaluation and warning service to avoid further purchase of these products by other consumers. According to the evaluation,the enterprise can analyze what the problems are and timely process and recall to avoid greater losses.

5) **Data management:** The enterprise can not only show the historical data of packaged foods supply chain in the way of data view from SOTP, but can clearly grasp the flow of products, such as where and what people are using according to the QR code and RFID as shown in Fig. 3 (d).Through the data view, the enterprise can also view the sales of packaged food in different regions.If there is a group of packaged food to expire in the A area, the enterprise will transfer them to B area or conduct a sales promotion.According to consumer evaluation data, enterprises can also analyze the quality of the packaged foods and timely process or recall to avoid the impact of consumer health.

5 CONCLUSION

The service-oriented traceability platform is packaged foods supply chain management system which is based on IOT real-time visibility and traceability. Four contributions are important in this research and development effort.The first contribution is the QR code and RFID used in the perception layer of packaged foods supply chain, which will greatly reduce the potential delay and improve the operation efficiency. The second contribution is packaged foods data repository stores object data and event data to protect information privacy and business interests.It also complies with the international standard of XML schema to facilitate the data exchange.The third contribution is that the proposed SOTP provides four visualization services, supply chain management service, life-cycle traceability service, evaluation and warning service, and big data statistics service, respectively.The last contribution is this study develops a prototype of SOTP to validate the proposed framework and approach.

However ,the current work provides a basic traceability platform to support packaged foods supply chain. The research will be further extended in several aspects. The proposed SOTP has only developed a generic system architecture and addressed the most urgent and basic challenges based on investigations at a case company. A great effort should be needed to improve the performance and enrich the functions.

6 REFERENCES

[1] **Myo Min Aung, Yoon Seok Chang.2014.** Traceability in a food supply chain: Safety and quality perspectives,39(1),Food Control, pp172-184

[2] **J.H. Trienekens, J.G.A.J. van der Vorst.2014.** Global Food Supply Chains, Encyclopedia of Agriculture and Food Systems, pp499-517

[3] **H. Scholten, C.N. Verdouw,2016.** Defining and Analyzing Traceability Systems in Food Supply Chains, Advances in Food Traceability Techniques and Technologies,pp 9-33

[4] **Yi Liu, Weili Han,2016.** An Internet-of-Things Solution for Food Safety and Quality Control: A Pilot Project in China, Journal of Industrial Information Integration

[5] **Diana M. Segura Velandia,2016.** Towards industrial internet of things: Crankshaft Monitoring,traceability and tracking using RFID,Robotics and Computer-Integrated Manufacturing,41,pp 66-77

[6] **C.N.Verdouw,J.Wolfert,2016.**Virtualization of food supply chains with the internet of things,Journal of Food Engineering,176(1), pp128-136

[7] **Qingqi Long, Wenyu Zhang,2014.** An integrated framework for agent based inventory-production-transportation modeling and distributed simulation of supply chains, Information Sciences, 277(2), pp 567-581

[8] **WANG Dianhua,HUANG Douxuan,2010.**Food Supply Chain Management Under Conditions Of Food Safety, Management and Service Science (MASS), 2010 International Conference, pp1-4

[9] **Qin Zhifeng,2010.**RFID-enabled life-cycle traceability in pharmaceutical supply Chain

[10] **Xiang T.R. Kong, Ji Fang, Hao Luo, George Q. Huang,2014.**Cloud-enabled real-time platform for adaptive planning and control in auction logistics center, Computers& Industrial Engineering84(C), pp79-90

A SURVEY OF HUB LOCATION PROBLEM FOR THE NEW NETWORK STRUCTURE RESEARCH

Kai Zhang[1,2], Zhibin Ma[1],Canrong Zhang[1*] and Lixin Miao[1,2]
[1]Division of Logistics & Transportation
Graduate School at Shenzhen
Tsinghua University, Shenzhen, China
zhangkai@sz.tsinghua.edu.cn

[2] Environment Science and New Energy Technology Center
Tsinghua-Berkeley Shenzhen Institute, Shenzhen, China
lxmiao@tsinghua.edu.cn

ABSTRACT

Hub location problem is a typical quadratic assignment problem (QAP). QAP has been investigated over the years, and the hub location theories are used in various areas. Hub location problem has many research branches, while this paper focuses on the network structure design. The objective of this paper is to provide a literature review of network structure design of hub location problem, especially the advanced researches in the field since 2013, and to present the current research status and several applications of new network structure. The paper also discusses some shortcomings of current designs, and refers to advanced researches to recommend ideas of feasible directions or demands for future efforts.

Keywords: Hub location, Network structure design, Survey

1 INTRODUCTION OF HUB LOCATION PROBLEM

Hub location problem is a hot research part of location science. It was first raised in transportation area and then spread to many other areas like telecommunications, geography and economics. The problem was studied by many experts and has many branch research directions until now. Some review papers introduced the researches, such as Alumur and Kara (2008)[1] provided a review paper on hub location models until 2007. While Farahani et al. (2013)[2] focused on reviewing the researches from 2007 to 2013. This paper will review some typical researches of hub location problem and focus on the part of new network design. The basic assumptions and research conditions will be introduced in section 1and the different hub location network structures will mainly be shown in section 2. Section 3 will analyze some recent papers and discuss the future trends and the last section is a summary of the paper.

1.1 Basic models of hub location problem

In classic hub location problem, there is a set of nodes including the origins, destinations and potential hubs. Demands are between origin and destination nodes which can be treated as OD pairs. That's an obvious difference from many other location problems which demands are from individual

* Corresponding Author Tel.: +86075526036021.

E-mail address: crzhang@sz.tsinghua.edu.cn (Canrong Zhang).

node. Demand flows are transported from the origin nodes to the hub facilities and then transported to the destination nodes. Hub facilities can redirect the flows from the origin nodes as well as combine or split flows from different sources. The flows can be transported through one or two hubs in the paths, but direct connections of OD pairs are not allowed. Figure 1 shows that demands from origin 1 to destination 2 can be transported through path 1-h1-2 which passes one hub and the other path 1-h1-h3-3 passes two hubs. But the path directly connects origin 1 and destination 6 is forbidden. The structure using hubs needs fewer links than the one directly connect any two nodes. Hubs can also perform better in dealing with the uncertain demands so the network is more stable. The main decisions of hub location problem conclude two aspects: the location of hub facilities and the allocating of the non-hub nodes. The target is to find the optimal design to make the cost of the system least. As the location and allocation problems are not independent, they must be considered together to get the optimal network design.

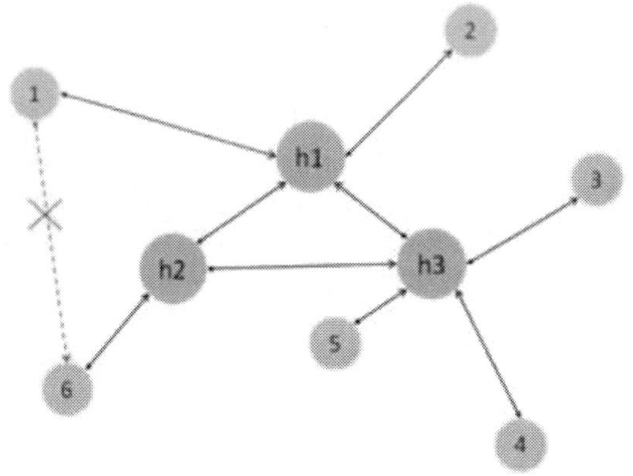

Figure 1: Hub Location Network Example

Hub location problem is studied as a distinct research area since the research of O'Kelly in 1986[3]. Based on the above assumptions, O'Kelly (1987)[4] presented the first mathematical formulation which is a discrete integer quadratic formulation as follows.

$$Min\ z = \sum_i \sum_j W_{ij}\ \left(\sum_k X_{ik} C_{ik} + \sum_m X_{jm} C_{jm} + \alpha \sum_k \sum_m X_{ik} X_{jm} C_{km}\right) \quad (1)$$

$$s.t.\ \ (n-p+1)X_{jj} - \sum_i X_{ij} \geq 0 \ \ for\ all\ j, \quad (2)$$

$$\sum_j X_{ij} = 1 \quad for\ all\ i, \quad (3)$$

$$\sum_j X_{jj} = p \quad for\ all\ j, \quad (4)$$

$$X_{ij} \epsilon \{0,1\} \quad for\ all\ i,j, \quad (5)$$

$$0 \leq \alpha \leq 1 \quad (6)$$

In the model Wij means unit flow and Cij means unit transportation cost from node i to j. α is the economies of scale factor. p represents the hubs and n is the total number of demand points. The model was used in an airline passenger transportation case between 25 US cities and the objective function minimized the total transportation costs incurred over all links. Constraint (2) ensures that a node can only be allocated to a selected hub as well as the number of nodes assigned to the same hub cannot exceed the amount of non-hub nodes. Constraint (3) ensures that a non-hub can and only can be assigned to one hub while constraint (4) limits the number of hubs.

1.2 Research directions of hub location problem

The following research fields of hub location are mainly in new models, new algorithms and new application cases. Various new models are raised through different ways. A main idea is to consider new factors that influence the cost like the fix cost of establish hubs[5], the hub capacity [6], service time limits of hubs, and etc. New network design is also a hot research field to form new models because the traditional network cannot be applied to all industries. This paper also focuses on these researches and the problem will be discussed in detail in section 2.

From the objective function of O'Kelly's model we can see it is a quadratic assignment problem (QAP) which is a typical NP-hard problem. It is hard to solve with exact algorithm when the number of nodes rises. The solving methods of the models also attracted many attentions and new heuristic algorithms were raised like tabu search[7], simulated annealing[8], genetic algorithm[9] and etc. Linearization of the quadratic model raised by Campbell (1992)[10] is a direct idea to transform the model to a linear one which is not NP-hard any more. It is also a useful way to slack or tighten the constraints so that we can get the upper or lower bound of the solution.

The traditional applications of hub location are mainly in transportation and telecommunication systems. But the theories are also used in different industrials and their demands stimulate the development of new models. Air transportation network is always an attractive area and many papers introduced the application examples in this area. There are two classic application cases in hub location area: the Civil Aeronautics Board (CAB) problem and Australia Post (AP) problem. Many following researches use these data sets to test their new theories or algorithm. Express delivery networks are also studied in the early development stage of hub location and Federal Express is a classic application. Bostel et al. (2015)[9] also studied a postal service systems of freight transport providers which used a Less Than Truckload logistic network. They used a meta-heuristic method based on a genetic algorithm to solve the problem and performed better than the standard solver. Another important application is electronic area. Klincewicz (1998)[11] provided a summary of the relative prior researches in electronic communications network design and suggested to solve problems with greater complexity for future research. In some new application areas hub location theories can provide a different thought to solve the problem and may achieve an unexpected better result than the existing methods.

2 RESEARCH STATUS OF HUB LOCATION NETWORK STRUCTURE

As we can see from the development process of hub location researches, there are many branches of this area until now. To satisfy the demands of different industries, new networks were designed. Campbell (1994)[12] concluded the fundamental features of hub location problems and defined four fundamental facility location problems as p-hub median, uncapacitated hub location, p-hub center and hub covering. There are also other classifications of the researches, such as discrete models and continuous models. The discrete p-hub median problem has the closest relationship with the network structure so most examples are in this field. Researches can be allocated into the following classifications according to the different connection parts or factors in the network.

2.1 Research on hub-to-non-hub node connections

Hub location networks can be divided into two basic types - single allocation and multiple allocation problems according to how many hubs that the non-hub nodes can connect to. For some express system, a non-hub node is severed by the nearest regional center which is a single allocation application. While the airline system usually employs the multiple allocation network structure because the destinations of the passengers are various, so the non-hub cities need to be severed by several or all the hubs. Another case was shown by Yaman (2011)[13] who limited the number of node's access connections in a multiple allocation model. The author called it r-allocation p-hub median problem in which each node can be connected to at most r hubs. This can be seen as a simplified multiple allocation problem. Some researchers studied the detail supply routes of hubs. For example Çetiner et al. (2010)[14] used TSP method to design the route to supply the connected nodes of each hub in a Turkish postal delivery system. This is an extension of traditional network design for it takes the actual path distance in consideration.

2.2 Research on inter-hubs connections

In most hub location problems hubs are all connected with each other, this is called fully interconnected network or complete hub network. But there are many shortcomings of the complete network in some areas. For example in large trucking networks a hub cannot connect to all the other hubs because of insufficiency of infrastructure or policy constraints. Many researchers study incomplete hub network which relaxes this assumption. Campbell et al. (2005)[15] interpreted the motivation and benefit to design less than fully interconnected network and different hub arc location models are presented. Alumur et al. (2009)[16] presented the single allocation incomplete p-hub median problems with fixed cost. Each hub has to connect to at least one other hub but does not need to connect to the whole remaining ones. The result showed that the service provided by a complete hub network can also be provided an incomplete hub network with the same efficiently and less cost. Campbell (2010)[17] studied another models for time definite hub location and network design with isolated and connected hubs. In some other researches the hubs are not only less than fully interconnected but also need to be arranged in a specific shape to satisfy the demand. For example in hierarchical hub location problem there is a centre hub and all the other hubs need to connect to this hub. Yaman (2008)[18] studied this hub structure called a star-star network. A heuristic algorithm based on Lagrangian relaxation and local search was used to solve the 81 cities network problem in Turkey. In 2012[19], they studied an extended problem which bounded path lengths.

Another hot incomplete hub structure studied by many experts is the tree of hubs location problem. Contreras et al. (2010)[20] presented the tree of hubs location problem which aimed at minimizing the transportation costs over the network. The hubs are connected by means of a tree and the model was tested with AP and CAB data sets. Then Martins et al. (2013)[21] used a Benders decomposition approach to solve this type of problem.

2.3 Research allowing direct connections between non-hub nodes

It is acknowledged that the flows between hubs are usually larger than that between other paths. Because the economies of scale, the inter-hub discount factor is used to calculate the cost between hubs. However, in real cases some closely related non-hub node pairs may have big flows which even exceed those between hubs. So some studies allow the connections between non-hub nodes if the demands are large enough. Groothedde et al. (2005)[22] studied a multi-allocation collaborative incomplete hub network which relax the restriction of non-hub nodes connections. In their model direct connections between non-hub nodes are allowed to achieve economies of scale. Recently a real hub location problem was studied by Aygun et al. (2015)[23] for a public institution which also allows direct flows between nodes.

2.4 Research about network with connection restrictions

In some studies connection restrictions are considered in designing the network structure. In section 2.1 the r-allocation p-hub median problem is also a restricted situation which limits the number of node's access connections to hubs in multiple allocation models. Another common factor that will influence the structure of network is the service time. The overnight restrictions and time zones institution was researched by Hall (1989)[24] in an air freight network. They considered the time limits in the transportation process and added the constraints in the design.

On the other aspect the traditional hub location network can be seen as a 2-stop network which allows an Origin-Destination path passes through one or two hubs. In a study by Sasaki et al. (1999)[25] a 1-stop network was presented which allows at most one hub among each route. The network is more practical than the 2-stop one in small size network such as domestic airline network in small countries. The simplified route will also improve the service efficiency and reduce cost. They also added the arc capacity constraints and hub capacity constraints in the 1-stop model to make it more practical in 2003[26].

3 FUTURE TRENDS OF NEW NETWORK STRUCTURE DESIGN

Nowadays many new industries appeared and got booming development. A considerable proportion of the industries need more detail and efficient designs of network in the more intense competition environment. This section analyzes the actual situation of some industrials combined with new research papers to provide several feasible and valuable directions for future efforts of new network structure design.

3.1 Hub line location network

The concept of hub line location was raised by Contreras [27] in 2015. The idea was raised in the design of public transportation and rapid transit systems. As we know the cost of establish the infrastructure in these systems is expensive and complete hub network is not a feasible scheme. So the purpose of the design is to select p hubs and connect them in a line using p-1 hub arcs which is also a common structure of real-life subway hub structure. Figure 2 is extracted from the author's paper and shows the structure of a hub line network with six hub nodes and five hub arcs. The model is a multiple allocation type for each node can be assigned to more than one hub. Direct connections between origin and destination are also allowed in the network. Although hub lines have the advantages of high transportation speed, passengers also have to bear the time wastes of access, exit and waiting. The goal is to minimization the total weighted travel time which is different from the traditional cost minimization objective. The author proposed an exact algorithm based on a Benders decomposition of a strong path-based formulation and confirm the algorithm could solve instances with up to 100 nodes with considerable efficiency. Later Contreras (2015)[28] extended the research to multiple lines case which is more closer to actual transportation network. The research is a realistic modification of hub location network in actual demand. Future studies can apply and modify the model in real life applications, or develop new heuristic algorithm to achieve better performance.

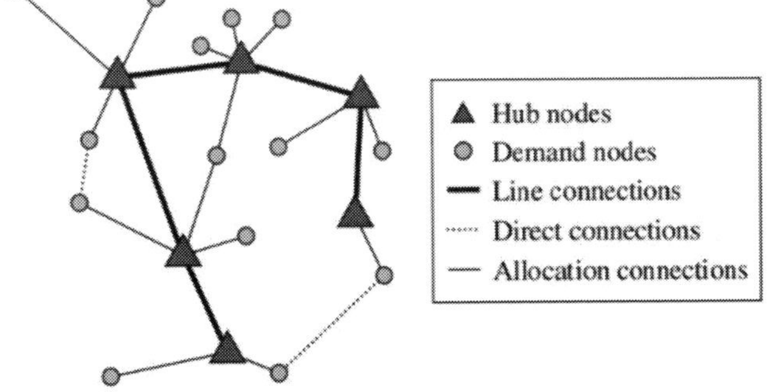

Figure 2: A Hub Line Network Example

3.2 New network with multi-routes

In recent years more studied focus on the integrated problem of hub location network design and route planning. However in the real logistic network there is always more than one feasible path between two nodes. For example Fontes (2015)[29] considered alternative paths of the stable rates for long periods to optimize the network design global cost. Another research by Niakan et al (2015)[30] thought that there are several feasible paths connecting two nodes which have different capacity, transportation mode, travel time, transportation and construction costs. They used a multi-objective model combined the objectives of time, cost and service reliability. New network with multi-routes is a future research direction for more and more industries need detail design in supply routes. More routes and transportation modes are available in logistic system and rational supply routes can help save a rather proportion of total transportation cost.

3.3 New network considering uncertainty factors

In real world, when designing a network for a new project or some secretive projects, the data is usually not easy to get or cannot be guaranteed to be accurate. Under this uncertain environment some fuzzy variables are needed to simulate the uncertain factors. Lium et al. (2009)[31] studied a case with uncertain demand and compared the solutions with those under deterministic demand. Their research showed the design considering demand stochasticity is more robust and had more operational flexibility. Similar researches can be seen in Davari et al (2013)[32].

Another uncertainty factors are the disruptions of hubs which may be caused by natural disasters, labor dispute, severe weather and etc. Parvaresh (2014)[33] took the disruptions of hubs into consideration when designing the hub network and thought that the occasional failure may cause excessive costs to current system. For future researches, new uncertain factors and combinations of different factors can be studied like Mohammadi (2014)[34] who considered the hub location under mixed uncertainty. The design of uncertain network is closer to real world condition than deterministic case and more efforts are needed in this area. We look forward to new discoveries in uncertain model researches and more applications.

3.4 Sustainable development network

Previous hub location researches usually deal with a current design mission or make a scheme for a short period. But the flow traffic increases a lot at short intervals, other factors may also change. Although the result may work well under current situation, the scheme cannot fit well when the factors change in the future. Nowadays, some experts start to research the sustainable development network. Wang (2015)[35] studied the possible change of hubs in the network such as opening new hubs, adjusting their capacity level or closing some obsolete hubs. The costs of changing hubs were also added to the objective to achieve a long term minimized total cost and they used the China Deppon Logistics case to test their model. The design of sustainable development network is a trend for future network design and new attention can be paid to the changes of other parts of the system.

4 CONCLUSION AND REMARKS

This paper surveys the development of hub location research and mainly focuses on the part of network design. The prior researches are classified according to the changed parts compared with the classic hub location network. It introduced some frequently used data sets as well as new application cases of different areas in recent years. By analyzing the condition of modern industries and new ideas from recent published papers, the paper also provides several feasible directions for future hub location network design.

5 ACKNOWLEDGMENT

This work is supported by the National Nature Science Foundation of China under Grant nos. 71472108 and 71102011.

6 REFERENCES

[1] **Alumur, S. and Kara, B.Y. 2008.** Network Hub Location Problems: The state of the art, European Journal of Operational Research, 190 (1) , pp 1-21.

[2] **Farahani, R.Z. Hekmatfar, M. Arabani, A.B. and Nikbakhsh, E. 2013.** Hub Location Problems: A Review of Models, Classification, Solution Techniques, and Applications, Computers & Industrial Engineering, 64 (4), pp 1096-1109.

[3] **O'Kelly, M.E. 1986.** The Location of Interacting Hub Facilities, Transportation Science, 20, pp 92-106.

[4] **O'Kelly, M.E. 1987.** A Quadratic Integer Program for the Location of Interacting Hub Facilities, European Journal of Operational Research, 32 (3), pp 393-404.

[5] **O'Kelly, M.E. 1992.** Hub Facility Location with Fixed Costs, Papers in Regional Science, 71 (3) , pp 293-306.

[6] **Labbé, M. Yaman, H. and Gourdin, E. 2005.** A Branch and Cut Algorithm for Hub Location Problems with Single Assignment. Math, Programming, 102 (2) , pp 371-405.

[7] **Skorin-Kapov, D. and Skorin-Kapov, J. 1994.** On Tabu Search for the Location of Interacting Hub Facilities, European Journal of Operational Research, 73 (3). pp 501-508.

[8] **Ernst, A.T. and Krishnamoorthy, M. 1996.** Efficient Algorithms for the Uncapacitated Single Allocation p-hub Median Problem, Location Science, 4 (3) , pp 139-154.

[9] **Bostel, N. Dejax, P. and Mi Zhang. 2015.** A Model and a Metaheuristic Method for the Hub Location Routing Problem and Application to Postal Services, 2015 International Conference on Industrial Engineering and Systems Management (IESM), Seville, Spain,October 21-23, pp 1383-1392.

[10] **Campbell, J.F. 1992.** Location and Allocation for Distribution Systems with Transshipments and Transportation Economies of Scale, Annals of Operations Research, 40 (1), pp 77-99.

[11] **Klincewicz, J.1998.** Hub Location in Backbone/Tributary Networks: A Review, Location Science, 6 (1-4),307-335.

[12] **Campbell, J.F. 1994.** Integer Programming Formulations of Discrete Hub Location Problems, European Journal of Operational Research,72, pp 387-405.

[13] **Yaman, H. 2011.** Allocation Strategies in Hub Networks, European Journal of Operational Research,211 (3),442-451.

[14] **Çetiner, S. C. Sepil, and Süral, H. 2010.** Hubbing and Routing in Postal Delivery Systems, Annals of Operations Research, 181 (1), pp 109-124.

[15] **Campbell, J.F. Ernst, A.T. and Krishnamoorthy, M. 2005.** Hub Arc Location Problems: Part I—Introduction and results, Management Science, 51 (10) , pp 1540-1555.

[16] **Alumur, S. Kara, B.Y. and Karasan, O.E. 2009.** The Design of Single Allocation Incomplete Hub Networks, Transportation Research Part B Methodological, 43 (10) , pp 936-951.

[17] **Campbell, J.F. 2010.** Designing Hub Networks with Connected and Isolated Hubs, 43rd Hawaii International Conference on Systems Sciences (HICSS 2010), Honolulu, HI, January 05-08, pp 1105-1114.

[18] **Yaman, H. 2008.** Star p-hub Median Problem with Modular Arc Capacities, Computers and Operations Research, 35 (9) , pp 3009-3019.

[19] **Yaman, H. and Elloumi, S. 2012.** Star p-hub Center Problem and Star p-hub Median Problem with Bounded Path Lengths, Computers and Operations Research, 39 (11),pp 2725-2732.

[20] **Contreras, I. Fernández, E. and Marín, A. 2010.** The Tree of Hubs Location Problem, European Journal of Operational Research, 202 (2) , pp 390-400.

[21] **Sá, E. M. D. Camargo, R. S. D. and Miranda, G. D. 2013.** An Improved Benders Decomposition Algorithm for the Tree of Hubs Location Problem, European Journal of Operational Research, 226 (2), pp 185-202.

[22] **Groothedde, B. Ruijgrok, C. and Tavasszy, L. 2005.** Towards Collaborative, Intermodal Hub Networks—A Case Study in the Fast Moving Consumer Goods Market, Transportation Research Part E Logistics & Transportation Review,41 (6) , pp 567-583.

[23] **Aygun, S. Kose, E. and Aplak, H.S. 2015.** Capacitated Multiple Allocation Hub Location Problem Which Allows Direct Flow between Nodes, Journal of the Faculty of Engineering and Architecture of Gazi University, 30 (4), pp 649-658.

[24] **Hall, R.W. 1989.** Configuration of an Overnight Package Air Network, Transportation Research Part A General, 23 (2), pp 139-149.

[25] **Sasaki, M. Suzuki, A. and Drezner, Z. 1999.** On the Selection of Hub Airports for the Airline Hub-and-spoke System, Computers and Operations Research, 26, pp 1411-1422.

[26] **Sasaki, M. and Fukushima, M. 2003.** On the Hub-and-spoke Model with Arc Capacity Constraints, Journal of the Operations Research Society of Japan, 46 (4), pp 409-428.

[27] **Sá, E. M. D. Contreras, I. Cordeau, J.F. Camargo, R. S. D. and Miranda, G. D. 2015.** The Hub Line Location Problem, Transportation Science, 49 (3), pp 500-518.

[28] **Sá, E. M. D. Contreras, I. and Cordeau, J.F. 2015.** Exact and Heuristic Algorithms for the Design of Hub Networks with Multiple Lines, European Journal of Operational Research,246 (1), pp 186-198.

[29] **Costa Fontes, F.F. D. and Goncalves, G. 2015.** Hub Location and Routing Problem with Alternative Paths, 2015 4th International Conference on Advanced Logistics and Transport (ICALT), Valenciennes, France ,May 20-22,pp 317-339.

[30] **Niakan, F. Vahdani, B. and Mohammadi, M. 2015.** A multi-objective optimization model for hub network design under uncertainty: An inexact rough-interval fuzzy approach, Engineering Optimization, 47 (12) , pp 1670-1688.

[31] **Lium, A-G. Crainic, T.G. and Wallace, S.W. 2009.** A Study of Demand Stochasticity in Service Network Design, Transportation Science. 43 (2) , pp 144-157.

[32] **Davari, S. and Zarandi, M.H.F. 2013.** The Single-Allocation Hierarchical Hub-Median Problem with Fuzzy Flows, Soft Computing Applications. Springer Berlin Heidelberg,195, pp 165-181.

[33] **Parvaresh, F. Golpayegany, S.A.H. Husseini, S.M.M. and Karimi, B. 2013.** Solving the p-hub Median Problem Under Intentional Disruptions Using Simulated Annealing, Network and Spatial Economics, 13 (4), pp 445-470.

[34] **Mohammadi, M. and Jula, P. 2014.** Multi-objective Design of an Organ Transplant Network under Uncertainty, Transportation Research Part E Logistics & Transportation Review, 72, pp 101-124.

[35] **Wang, Y.F. and Huang, Y.F. 2015.** Incremental Optimization of Hub and Spoke Network for the Spokes' Numbers and Flow, Mathematical Problems in Engineering , pp 1-12

THE RESULTS OF RESEARCH ON NEURO-INDUSTRY ENGINEERING: THE NEW STAGE OF IE

Xiaojie Xu[1*] and Juan Shi[1]

[1]The School of Management

Tianjin University of Technology, Tianjin, China

947964973@qq.com

studentcont@163.com

ABSTRACT

With the improvement of science technology and productivity, IE had made a rapid development greatly. Its application area and research content have been expanded and enriched constantly. Neuro-Industry Engineering (Neuro-IE) is established upon Cognitive Neuroscience and traditional Industry Engineering (Traditional IE). And it becomes known to people after it was an emerging interdisciplinary science. Thus, a great deal of researches are done in many areas. This paper reviewed the three development stages of Industry Engineering briefly and gave a detailed account of the conception of Neuro-IE. Meanwhile, this paper detailedly introduced application and results of research on Neuro-IE in some areas.

Keywords: Neuro-Industry Engineering, Industry Engineering, Cognitive Neuroscience

1 INTRODUCTION

Industry Engineering(IE) is an engineering technology of integrated design, improvement and implementation of a whole system containing human, material, equipment, energy and information. With the development of science and technology, IE has been increasingly expanded and integrated in research fields and methodologies. Neuro-Industry Engineering (Neuro-IE) was firstly proposed by Qingguo Ma and Xiaoyi Wang (2006)[1], it was an emerging research area formed by the intersection of Cognitive Neuroscience and traditional IE. And it becomes known to people after the exposure of relevant research results. As an emerging interdisciplinary science, Neuro-IE has to do with research on product, environment and human, and discovery of hidden neural mechanism. Then it'll find the secrets of decision made by human brain and create a safe and friendly environment for people's work and life. And there has been a great number of related research results.

2 DEVELOPMENT STAGES OF INDUSTRY ENGINEERING

First, IE emphasized on improving the productivity, reducing the costs, ensuring the quality and safety of the production, as well as tried its best to make the whole production system under the optimal running state and get the highest overall benefits. Looking back on its development, we divide it into three main stages.

The first stage, when IE germinated and laid the foundation for its development with taking division of labor based on specialization, time and motion study, standardization approach, etc. as its main contents, was "Scientific Management Era" or "traditional Industry engineering". Representatives of this stage are Taylor and the Gilbreth couples, who respectively put forward principle of scientific

* Corresponding Author; E-mail address: 947964973@qq.com

management and motion study. Herry L. Gantt is also an important pioneer of this period, during which he created "Gantt Chart" which has been widely used in Production Planning & Control up to now.

The second stage was "the Comprehensive Improvement" or "Modern Industry Engineering ", and some disciplines like Operation Research (OR), Systems Engineering (SE), Computer Science were widely used in IE. Furthermore, a great deal of new ideas and approaches such as Enterprise Resource Planning (ERP), Total Quality Management (TQM), Toyota Production System (TPS), Computer Integrated Manufacturing (CMIS) emerged pushing forward the development of IE. During this period, the theoretical basis and scientific methods came into being in IE due to its continuous absorption of modern scientific and technological achievements.

The third stage was "Human Factors Period"[2] or "Neuro-Industry Engineering", which is characterized by adding physiological sensing technology and comprehensive processing status information of production and human physiological information, so as to achieve the human-oriented scientific production. Compared with traditional IE, Its essential progress includes the human physiological information (EEG, EDA, EMG, ECG, TEMP, HR, etc) as the objective data in the scope of scientific analysis, and regards this objective data as an important basis for the implementation of management. Neuro-IE not only covers the ingredients of traditional IE (materials, energy, equipment, information and the relationship of them), but also considers human as the main part of the system and is committed to the unity and coordination of people, technology, environment.

3 CONCEPTION OF NEURO-IE

Neuro-IE, put forward by Qing-guo Ma and Xiao-yi Wang(2006), is developed on the basis of traditional IE. It studies how human brain neural activities in traditional IE react to various visual and auditory signals and then integrates traditional IE with cognitive neuroscience. It is related with three disciplines of life science, engineering and management science[4]. Neuro-IE optimizes the whole process of manufacture, operation and production according to objects' neural activities and physiological information including brain wave, galvanic skin response(GSR), myoelectricity and heart rate, etc. Compared with traditional IE, Neuro-IE pays more attention to workers' physiological and psychological changes in the process of production so that people can adjust the production effectively.

4 STUDY ON FATIGUE OF WORKERS IN PRODUCTION LINE

Traditional IE is always trying to reduce people's leisure time and wasted movements to reduce the cycle time and increase the outputs. But it is easy to cause accidents in production activities or reduce product quality. Neuro-IE has really reached to the unity and coordination of people-technology-environment and radically improved the productivity from the perspective of human factors.

We can dig out human psychological information from human physiological information, and then clarify the relevance of them to provide a reference for setting rest mechanism and making standard operation plan. According to the characteristics of different work station and workers' physiological and psychological information, managers can arrange the appropriate work period and rest period to ensure high working efficiency and low error rate, so as to improve the work efficiency and implement humanized production management.

Jianhui Fu, Jun Bian (2012)[7] conducted a research on "input program" and "maintenance", two workstations in a company's communication power production workshop, and extracted "thera mental fatigue indicator" and "SMR attention fatigue indicator" from measured data. EEG curve is used to reflect the changes of the two indicators. Research results show that the maintenance station is more precise and occupied more mental resources than the input program station. So the EEG curve values (about "thera mental fatigue indicator" and "SMR attention fatigue indicator") of maintenance station are higher than that of the input program station. That indicates that the workers in the maintenance station have a greater degree of fatigue. While the EEG curve values of input program station increased first and then decreased, and finally increased at the time of getting close to getting off work. That indicates that the workers' efficiency is close to zero after the fatigue degree reaches a certain value. These results shows that it is necessary to add a short break and establish corresponding dynamic SOP for the complex work station. While In traditional management patterns, time point for rest is usually set at the latter fatigue point, neglecting the increase of the probability of errors caused by no rest at the first fatigue point for workers at stations that require more brain resource. Thus this research facilitates the design of rest system for the management[8].

5 DESIGN AND APPLICATION OF WARNING SIGNS

Recent years, Neuro-IE has been combining traditional IE and neuroscience, studying the neural response characteristics on various types of signal (visual or auditory signals) and the appropriate decision-making process of the neural activity. It also can open up brain's black box behind warning signs identification.

The design and application of warning signs are mostly based on questionnaires and behavioral experiments, which are lack of theoretical research on brain cognition and electrophysiology. Warning sign design should conform to people's cognitive and subjective experience, in order to more effectively play a warning role——this is the focus of theoretical research of Neuro-IE in the field of safety production. Xianwei Tang(2010), Jun Bian(2014)[5] compare different ERPs component indicators (such as P2, N2, LPP) in different experiments by using Event-Related Potentials (ERPs), a method of neuroscience method, as well as explore what kind of cognitive activities people response to warning sign with different characteristics, thus providing a theoretical basis for the design and effectiveness of warning signs.

Xianwei Tang's research(2010) gives positive and negative stimulus to the test subjects, and shows that mean amplitude of P200 induced by negative stimulus is remarkably higher than that induced by positive stimulus. The result means that compared with positive stimulus, negative stimulus in the same level of intensity can arouse stronger emotional reactions. From the perspective of cognitive neuroscience, it is confirmed that the effect of the safety sign make people be more alert to the risk information when the safety sign is a negative stimuli. What's more, the amplitude of P200 can be used to measure the effectiveness of the warning sign and help us improve the design of warning signs.

Jun Bian(2014) used warning signs with different color background as the experimental materials. According to the experimental results, different background color of warning sign can affect the automatic processing of human brain in warning sign. The results shows that incubation period of P1, N170, P2 induced by warning sign of yellow background is remarkably shorter than that induced by warning sign of blue background, and amplitude of N170, P2 induced by warning sign of blue background is significantly lower than that induced by warning sign of yellow background. The incubation period of P1, N170, P2 induced by warning sign of blue background is longer. It indicates that subjects need to put more attention resources to identify them, so warning sign of yellow background have its own advantage. During the midanaphase of classification of color perception processing, the P2 component in frontal area mainly reflects the rapid processing of the warning signs with yellow background that people are familiar with. So it is concluded that people's perceptive performance of blue color is significantly worse than that of yellow color in the automatic processing of brain cognition.

Qingguo Ma, Jing Jin, Lei Wang(2010)[3] used event-related potentials (ERPs) to explore neural temporal features during the processing of warning signal words in human brain, and found that there were two stages involved in processing warning signal words, providing an electrophysiological evidence for a previous warning information processing model, the Communication-Human Information Processing Model (C-HIP).

Combined with the research results mentioned above, it can be seen that the Neuro-IE provides a theoretical and practical reference for the effective safety production management of enterprise.

6 THE DEVELOPMENT OF EMERGING IE SOFWARES

The modern production in the 21st century is scientific, rational and humanistic. So when the enterprises made intensive and heavy work, managers must take care of the employees' emotional, psychological and physiological factors, so as to enhance the working efficiency and job satisfaction and avoid accidents arisen from fatigue issues and mental depression. Therefore managers should consider not only the removal of the invalid motion but also the physical and mental work intensity when they make Standard Operation Plan(SOP). Most of industrial engineering management software didn't take the mental expenditure into account .

Therefore, Ma(2012) suggest integrating the Biofeedback Technology with the IE software frame, and then we could measure the physiological indicators (e.g. EEG, GSR, HR, EMG, EDA etc.) of workers in work stations, analyze the relevance between the physiological functions and station demand, and decide whether to remove the invalid motion. In other words, it remains corresponding "invalid motion" as necessary short break in operation to satisfy physiological needs. Obviously, this new kind of software introduces the computer video analysis technology, neuroscience and biofeedback technology to IE management. It could leave IE engineers more time to care about the workers' working conditions and emotional changes in the production line and actually help them to arrange production plans more rationally and humanistically, and finally establish a good environment (tend to reach the optimal combination of human, machine and environment) for workers. This will increase the productivity and promote the sustainable development of enterprises. Finally it establishes a good environment for sustainable production and long-term development of the enterprise, as well as promotes its social image and company value.

7 OTHER EXISTING STUDIES ON NEURO-IE

ERPs is one of the most commonly used research methods in Neuro-IE. Mingliang Chen(2009)[9] used event-related potentials (ERPs) to investigate the neural and psychological bases of consumer herding decision in purchasing books online. The behavioral data revealed that the higher the consistency of the customer reviews, the higher the herd rate. It is possible that customer reviews with higher consistency let participants make herding decisions more resolutely. The present results suggest that the LPP, a component of ERP sensitive to categorization processes, may be regarded as an endogenous neural signal of the herding mechanism in a sense and that the LPP amplitude is potentially a measure of consumers' herd tendency in purchase decisions.

Neural Industrial Design is one of research applications of Neuro-IE. Customers' perception of the product function or quality of the product can be truly demonstrated through the EEG signals. According to the EEG, we can obtain the customers' true evaluation of design, so as to objectively screen out the designs of the products that are welcomed by customers.

Safety production is a seriously stubborn problem in modern industry engineering. Qingguo Ma, Jun Bian(2011)[10] make use of neuroscience methods to study Warnings from the basically cognitive levels and have acquired preliminary achievements, which provide new evidence, discussion and introductions for former researches. Therefore, with the new perspective and methods of Neuro-IE, we could try to solve some basic scientific issues from the brain-cognitive level, and then promote the new design and use of Warnings for better caution and instructions in modern Industrial Production, especially for Mine Exploitation.

In the field of management science, Neuromarketing is an important interdisciplinary subject with neuroscience. Neuromarketing has fundamentally changed the research methods of the traditional marketing, such as questionnaire, interview, behavior experiment and so on. These methods are used to obtain the subjective judgment data of the consumers. However Neuromarketing can get objective data of consumers by the means of neuroscience, which makes the marketing science more scientific[11]. Neuroscience also intersect with Logistics Engineering and supply chain management, and conduct study from the new angle[6].

8 CONCLUSION

The proposition of Neuro-IE sets up a multi-way bridge among structures of human brain functions and production decisions, physiological and psychological indicators, working state and safe production, which can not only add fresh blood for the original theoretical system, but also have profound influence on enterprise for health and rapid development in the long term. The application of Neuro-IE will continue to be expanded to other frontiers. For example, user's satisfaction and demands for a product can be examined by physiological measurement techniques, and through which the designing of products can get improved. Most of the researches in the same domain are limited in questionnaires and interviews, which have certain limitations. Whereas, this defect can be improved by Neuro-IE.

The rise of Neuro-IE has been more than ten years. Its precise and objective study on human behavior will be popular among various research fields, and the content of Its research will be further enriched.

ACKNOWLEDGMENT

This thesis is supported by Ministry of Education Philosophy and Social Science Research Project (15JZD021): sub topic 3rd of "Several major issues in the implementation of the collaborative development strategy of Beijing, Tianjin and Hebei ".

9 REFRERENCES

[1] **Q.G.Ma, X.Y.Wang. 2006.** Cognitive Neuroscience Neuroeconomics and Neuromanagement, Management Word,no.10, pp 139-149.

[2] **Jun Bian, Huijian Fu, Qian Shang, Xiangyang Hou, Qingguo Ma.2012.** The Study on Neuro-IE Management Software in Manufacturing Enterprises, Physics Procedia, no. 33, pp. 1608-1603.

[3] **Qingguo Ma, Jing Jin, Lei Wang. 2010.** The neural process of hazard perception and evaluation for warning words: Evidence form event-related potentials, Neuroscience Letters, 483(3), pp. 206-210.

[4] **Zhou Xiao-Hong, Wang Liang, Hu Yi-Cheng. 2013.** Neuro-industrial engineering: Research

and practice,2nd International Conference on Applied Mechanics, Materials and Manufacturing, ICAMMM, Changsha, China, November 17-18, pp 171-179.

[5] **Jun Bian. 2014.** Research on the Perception Mechanism of Safety Signs in Safety Management: A Neuro-Industrial Engineering Perspective, Zhejiang University.

[6] **Juan Shi, Qian Wang , 2015.** Research on the risk analysis of Supply Chain Finance from the perspective of encoding function forecast, Metallurgical and Mining Industry, 7(9), pp. 525-530.

[7] **Qingguo Ma, Jianhui Fu, Jun Bian. 2012.** Neuro-Industry Engineering: The new development stage of Industry Engineering, Management Word, no. 6, pp 163-179.

[8] **Qingguo Ma, Qian Shang, Jun Bian, Huijian Fu. 2011.**A Research on the Application of Physiological Status Information to Productivity Enhancement, Electrical Engineering and Control, vol. 98, pp 801-808.

[9] **Mingliang Chen, 2009.**The Neural and Psychological Basis of Herding in Purchasing Books Online: An Event-Related Potential Study, Cyberpsychology Behavior & Social Networking, 13(3), pp 321-328.

[10] **Qingguo Ma, Jun Bian, Wenjing Ji, Qinrui Tang, Qing Xu. 2011.** Research on Warnings with New Thought of Neuro-IE, Procedia Engineering, vol. 26, pp. 1633-1638.

[11] **Qingguo Ma, Liangchao Shu, Xiaoyi Wang. 2007.** Innovative marketing thinking Neuromarketing found “Buy Button”, Enterprise Management, no. 4, pp. 10-13.

OPTIMAL STRATEGY OF THE SUPPLY CHAIN WITH SINGLE-SUPPLIER MULTI-RETAILER UNDER RETAILER- OR VENDOR-MANAGED INVENTORY

Feng LIN[1*], Yu WANG[2] and Tao JIA[3]

[1]School of Management, Xi'an Jiaotong University, Xi'an 710049, China
lfxddz@foxmail.com

[2]School of Management, Xi'an Jiaotong University, Xi'an 710049, China
wy_sail@stu.xjtu.edu.cn

[3]School of Management, Xi'an Jiaotong University, Xi'an 710049, China
jiatao@mail.xjtu.edu.cn

ABSTRACT

The coordination of operational strategies among members of the supply chain is essential to the smooth running of the multi-echelon distribution system, whose common mechanisms include retailer-managed inventory (RMI) and vendor-managed inventory (VMI) policies. Meantime, the wide implementation of distribution center (DC) adds to the difficulty of maintaining the flexibility for the integrated inventory system. Therefore, this paper intentionally takes advantage of RMI and VMI policy to reconsider the decision process of the supply chain with single-supplier multi-retailer incorporating the location of a single DC. The retailers are assumed to adopt the common cycle to synthesize their ordering policies, and then the average cost functions including the supplier, the retailers and the supply chain are developed, respectively. Under the cooperative setting, the optimal operational strategies of RMI and VMI policies can be obtained while the solution of the integrated production-inventory policy is served as the comparison. In view of the application, numerical examples and sensitivity analyses are provided. And it can be found that the optimal location of the DC relies on which member plays the dominant role in the supply chain.

Keywords: Retailer-managed inventory, Vendor-managed inventory, Distribution center, Single-supplier multi-retailer supply chain

1 INTRODUCTION

Due to the increasingly fierce market competition, the decentralized decision process may not guarantee the maximum profit of the whole supply chain. Thus, upstream suppliers and downstream retailers have devoted to coordinating their strategies to gain more earnings. Actually, as one of common forms of the inventory system, the supply chain with single-supplier multi-retailer has continuously attracted considerable attention. To deal with this type of supply chain, one of the prominent synchronization mechanisms is to adopt VMI policy. From a vertically view, VMI policy is an approach for the supplier to monitor and control inventory, which is recognized as the effective collaborative initiatives to save considerable operational costs (Yao et al. [1]). The previous

* Corresponding Author

extensive researches have built solid theoretical foundations for VMI policy under various dimensions, such as the inventory-level-dependent demand (Chen et al. [2]), unequal-sized batches (Hoque [3]) and multi-level (Ben-Daya et al. [4]). As another form of coordination method, RMI policy, which denotes that the decision of the supply chain's stock level and the location of DC are arranged by the downstream retailers (Ru and Wang [5]), has also attracted considerable attention.

In a single-supplier multi-retailer supply chain, the DC is applied to receive items from the supplier and transfer them to retailers. The popularity of the DC can be ascribed to three significant factors: (1) the DC can lower total safety stock by risk-pooling benefits and line-haul shipping benefits (Shen et al. [6]); (2) the DC is effective to shorten delivery time to balance the dilemma between inventory cost reduction and customer responsiveness motivation (Nozick and Turnquist [7]); and (3) especially for deteriorating item, "the first mile" problem requires that the pre-processing should be conducted to meet the requirements of cold chain transportation, such as pick, classification, precooling and package. Hence, the location of the DC would be critical for the design and the operation of the supply chain. Sheu [8] proposed an integrated supply chain based spatial interaction model to determine facility locations of the manufacturing centers and regional product distribution centers. Haq and Kannan [9] presented an integrated approach for supplier selection and the multi-echelon distribution inventory model using a fuzzy analytical hierarchy process. Gendron and Semet [10] implemented two models (an arc-based model and a path-based model) to consider a multi-echelon location-distribution problem in fast delivery service.

For the joint supplier-retailers supply chain, transportation cost plays a major role of the total operational cost (Ertogral et al. [11]). Especially when introducing the DC, the transportation stage would fall into two phases with different speeds, which consequently influences the transportation time and transportation cost of whole supply chain. Thus, in this paper, we take account of the transportation stage in the context of the single-supplier multi-retailer supply chain by considering the location of the single DC. The distinguished features of this paper consist of the following aspects: (1) the optimal operational strategies of RMI and VMI policies are developed to compare with the solution of the integrated production-inventory policy, and (2) the optimal location of the single DC relies on which member plays the dominant role in the supply chain.

The remainder of the paper is organized as follows. The notations and assumptions are established in Section 2. The average cost functions including the supplier, the retailers and the supply chain are developed in Section 3. Section 4 characterizes the theoretical results of the average cost functions under RMI, VMI and the integrated production-inventory policy. Section 5 provides several numerical examples, while conclusions are presented in Section 6.

2 NOTATIONS AND ASSUMPTIONS

The following notations and assumptions are used to develop the single-supplier multi-retailer supply chain throughout the paper.

A_s	the major setup cost of the supplier per production cycle
a_i	the minor setup cost of the supplier for the retailer i per production cycle
Ar_i	the fixed setup cost of the retailer i per ordering cycle
h_s	the unit holding cost for the supplier per item per unit time during the supplier's production and transportation stages
hr_i	the unit holding cost of the retailer i per item per unit time during the retail stage
h_t	the unit holding cost of the retailers per item per unit time during the retailers' transportation stage
D_i	the demand rate of the retailer i per unit time
D	the total demand rate of the retailers
P	the supplier's production rate per unit time, where $P \geq D$

F_j the fixed transportation cost per shipment, where $j = s$ for the supplier and $j = r$ for the retailers
u_j the transportation cost per unit time, where $j = s$ for the supplier and $j = r$ for the retailers
v_j the transportation speed per shipment, where $j = s$ for the supplier and $j = r$ for the retailers
T the common ordering cycle of the retailers (decision variable)
Q_i the ordering quantity of the retailer i
R_i the distance between the retailer i and the supplier
L_i the transportation time from the DC to the retailer i, where $L_i = \frac{R_i - R_{dc}}{v_r}$
R_{dc} the distance between the DC and the supplier (decision variable), where $R_{dc} \in [0, R_1]$
L_{dc} the transportation time from the supplier to the DC, where $L_{dc} = \frac{R_{dc}}{v_s}$
$TC_s(T, R_{dc})$ the average cost of the supplier per unit time
$TC_r(T, R_{dc})$ the average cost of the retailers per unit time
$JTC(T, R_{dc})$ the joint average cost of the supply chain per unit time

1. The inventory system involves single supplier, multiple retailers and single item. Referring to Viswanathan and Piplani [12], multiple retailers adopt common cycle T to coordinate their ordering policies.
2. The transportation stage can be divided into two stages: the part concerning the supplier and the DC is denotes as TS-1, while the part concerning the DC and the retailers is denoted as TS-2.
3. The DC is located between the supplier and the retailer 1. And during the TS-1 stage, the supplier would spend L_{dc} to deliver the items to the DC.
4. During the TS-2 stage, the shipment sequence of the retailers is on the basis of the value of R_i; thus, R_i is used to depict the relative locations between the supplier and the retailer i. For convenience, we assume that $R_i < R_{i+1}$, i.e., the retailer i would receive the item earlier than the retailer $i + 1$. An example for the supply chain with single-supplier and two-retailer is illustrated in Figure 1.
5. To reflect the scale-effect of the transportation, the transportation cost is assumed to be the sum of the fixed ordering cost per shipment and the variable cost which is square with the transportation time per shipment, i.e., $F_j + u_j L_j^2$. And the transportation equipment has enough capacity to transport the items.
6. Time period is infinite and shortages are not allowed.

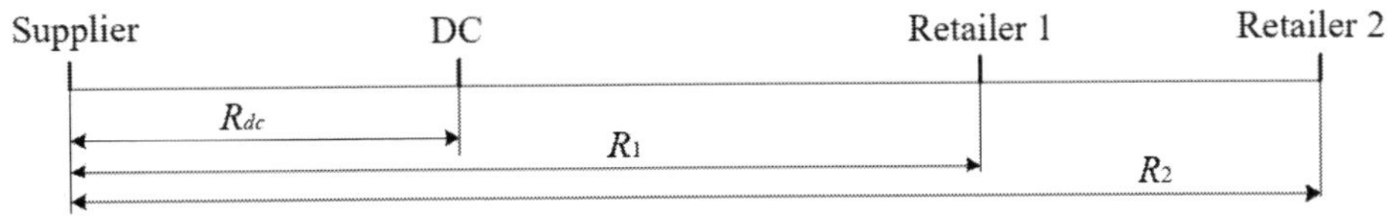

Figure 1: Graphical Representation of the Supply Chain (Single-supplier Two-retailer)

3 MODEL FORMULATION

In this section, we depict the decision process of the supply chain with single-supplier multi-retailer. Since retailers adopt common cycle T to place a joint order, the supplier should spend $\frac{\sum_{i=1}^{n} D_i T}{P}$ to complete the production process. And due to the introduction of the DC, it would take L_{dc} for the supplier to deliver the item to the DC. And then the retailer i would receive their individual shipment after $\sum_{j=1}^{i} L_j + L_{dc}$. The whole operational process of the supply chain can be illustrated by Fig. 2.

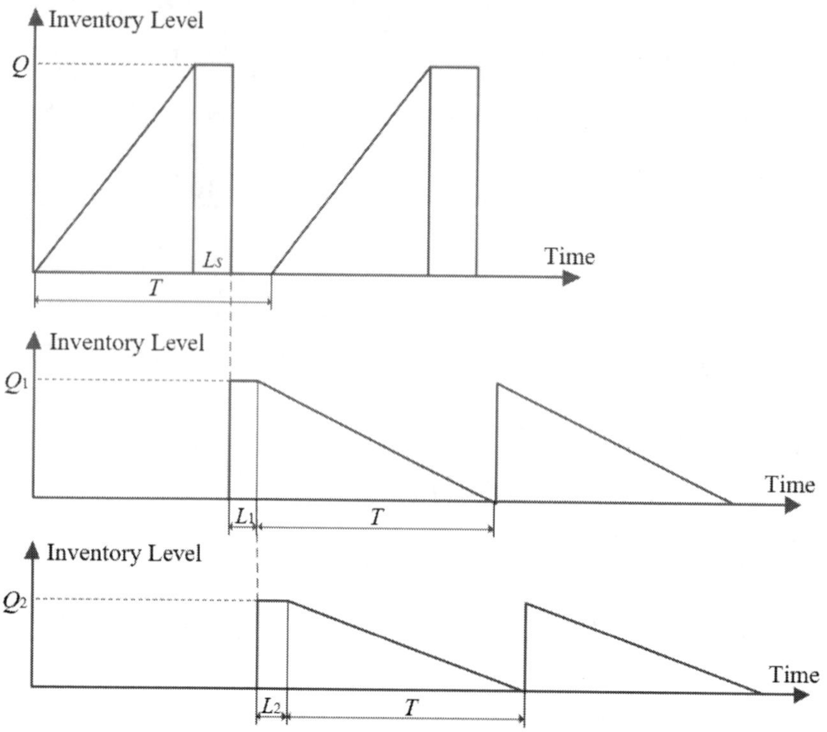

Figure 2: The Variation of the Supply Chain's Inventory Level (Single-supplier Two-retailer)

The supplier's average cost consists of the following elements: (1) the set up cost per unit time = $\frac{A_s+\sum_{i=1}^{n}a_i}{T}$; (2) the holding cost per unit time during the production stage and TS-1 stage = $\frac{h_s\sum_{i=1}^{n}D_iT}{T}\left(L_{dc}+\frac{\sum_{i=1}^{n}D_iT}{2P}\right)=h_sD\left(\frac{R_{dc}}{v_s}+\frac{DT}{2P}\right)$; and (3) the transportation cost per unit time during TS-1 stage = $\frac{F_s+u_sL_{dc}^2}{T}=\frac{1}{T}\left[F_s+u_s\left(\frac{R_{dc}}{v_s}\right)^2\right]$. To conclude, the supplier's average cost per unit time

$$TC_s(T,R_{dc})=\frac{A_s+\sum_{i=1}^{n}a_i}{T}+h_sD\left(\frac{R_{dc}}{v_s}+\frac{DT}{2P}\right)+\frac{1}{T}\left[F_s+u_s\left(\frac{R_{dc}}{v_s}\right)^2\right]. \tag{1}$$

The retailers' average cost function includes: (1) the set up cost per unit time = $\frac{\sum_{i=1}^{n}Ar_i}{T}$; (2) the holding cost per unit time during the TS-2 stage and retail stage= $\frac{h_t\left[\sum_{i=2}^{n}\sum_{j=i}^{n}Q_j(L_i-L_{i-1})+\sum_{j=1}^{n}Q_jL_1\right]}{T}+\frac{\sum_{i=1}^{n}hr_iD_iT}{2}=\frac{h_tD(R_1-R_{dc})}{v_r}+\frac{h_t\sum_{i=2}^{n}\sum_{j=i}^{n}D_j(R_i-R_{i-1})}{v_r}+\frac{\sum_{i=1}^{n}hr_iD_iT}{2}$; and (3) the transportation cost per unit time during TS-2 stage = $\frac{F_r+u_r(L_n)^2}{T}=\frac{1}{T}\left[F_r+u_r\left(\frac{R_n-R_{dc}}{v_r}\right)^2\right]$. To sum up, the retailers' average cost per unit time

$$TC_r(T,R_{dc})=\frac{\sum_{i=1}^{n}Ar_i}{T}+\frac{\sum_{i=1}^{n}hr_iD_iT}{2}+h_t\left[\frac{D(R_1-R_{dc})}{v_r}+\frac{\sum_{i=2}^{n}\sum_{j=i}^{n}D_j(R_i-R_{i-1})}{v_r}\right]+\frac{1}{T}\left[F_r+u_r\left(\frac{R_n-R_{dc}}{v_r}\right)^2\right]. \tag{2}$$

To conclude, the average cost of the supply chain per unit time

$$\begin{aligned}JTC(T,R_{dc})=&\frac{A_s+\sum_{i=1}^{n}a_i}{T}+h_sD\left(\frac{R_{dc}}{v_s}+\frac{DT}{2P}\right)+\frac{1}{T}\left[F_s+u_s\left(\frac{R_{dc}}{v_s}\right)^2\right]+\frac{\sum_{i=1}^{n}Ar_i}{T}+\frac{\sum_{i=1}^{n}hr_iD_iT}{2}\\&+h_t\left[\frac{D(R_1-R_{dc})}{v_r}+\frac{\sum_{i=2}^{n}\sum_{j=i}^{n}D_j(R_i-R_{i-1})}{v_r}\right]+\frac{1}{T}\left[F_r+u_r\left(\frac{R_n-R_{dc}}{v_r}\right)^2\right].\end{aligned} \tag{3}$$

4 THEORETICAL RESULTS

Under the cooperative setting, RMI and VMI policies are applied to determine the properties of $TC_s(T,R_{dc})$ and $TC_r(T,R_{dc})$, respectively. Simultaneously, we regard the integrated production-inventory policy as the benchmark to present their comparison results explicitly.

4.1 Retailer managed inventory policy

Under RMI policy, the supplier's decision should be conformed to the retailers' policy. Thus, via analyzing the property of $TC_r(T,R_{dc})$, the following lemma can be obtained.

Lemma 1

$TC_r(T,R_{dc})$ is jointly convex in T and R_s.

Proof. See Appendix A for details. ■

By taking the first derivative of $TC_r(T,R_{dc})$ with respect of R_{dc} and setting it equals zero, we can obtain

$$R_{dc}^1 = \frac{h_t v_r DT}{2u_r} + R_n. \qquad (4)$$

Since $R_{dc}^1 > R_n$, the retailers would let $R_{dc}^1 = R_1$ to minimize the $TC_r(T,R_{dc})$ and satisfy the constraint $R_{dc} \in [0,R_1]$ based on Lemma 1. And then the optimal ordering cycle can be obtained by taking the first derivative of $TC_r(T,R_{dc})$ with respect to T:

$$T_1 = \sqrt{\frac{2\left[F_r v_r^2 + \sum_{i=1}^{n} Ar_i v_r^2 + u_r (R_n - R_1)^2\right]}{\sum_{i=1}^{n} hr_i D_i v_r^2}}. \qquad (5)$$

Thus, under RMI policy, we can regard T_1 and R_{dc}^1 as the optimal solution, and denote $TC_{s1}(T_1,R_{dc}^1)$ and $TC_{r1}(T_1,R_{dc}^1)$ as the minimum cost of the supplier and the retailers, respectively. Additionally, we define it as DR case when the DC is located at the retailer 1, i.e., $R_{dc}^1 = R_1$.

4.2 Vendor-managed inventory policy

In VMI policy, the supplier rather than the retailers is assumed to determine the supply chain's stock level and the location of the DC. And the retailers would adjust their policies to achieve the coordination of the whole supply chain. Through analyzing the property of $TC_s(T,R_{dc})$, the following lemma can be obtained.

Lemma 2

$TC_v(T,R_{dc})$ is jointly convex in T and R_{dc}.

Proof: The proof is similar to that of Lemma 1. Hence, we omit it. ■

By taking the first derivative of $TC_s(T,R_{dc})$ with respect of R_{dc} and setting it equals zero, we can obtain

$$R_{dc}^2 = -\frac{v_s DT h_s}{2u_s} < 0. \qquad (6)$$

Since $R_{dc}^2 \in [0,R_1]$, the supplier may take its dominance to let $R_{dc}^2 = 0$. Based on Lemma 2, the optimal ordering cycle can be obtained by taking the first derivative of $TC_s(T)$ with respect to T,

$$T_2 = \sqrt{\frac{2P\left(A_s + F_s + \sum_{i=1}^{n} a_i\right)}{h_s D^2}}. \qquad (7)$$

Thus, under VMI policy, T_2 and R_{dc}^2 can be regarded as the optimal solution of the supply chain. $TC_{s2}(T_2, R_{dc}^2)$ and $TC_{r2}(T_2, R_{dc}^2)$ are the minimum cost of the supplier and the retailer, respectively. Additionally, we define it as DS case when the DC is located at the supplier, i.e., $R_{dc}^2 = 0$.

4.3 Integrated production-inventory policy

From the view of the integrated production-inventory policy, the supplier and the retailers are assumed to form a centralized inventory system to jointly formulate the optimal solution. Via analyzing the property of $JTC(T, R_{dc})$, the following lemma can be obtained.

Lemma 3

$JTC(T, R_{dc})$ is jointly convex in T and R_{dc}.

Proof. The proof is similar to that of Lemma 1. Hence, we omit it. ■

By taking the first derivative of $JTC(T, R_{dc})$ with respect of R_{dc} and setting it equals zero, we can obtain

$$R_{dc}^3 = f_1 + f_2 T. \tag{8}$$

where $f_1 = \frac{R_n u_r v_s^2}{u_r v_s^2 + u_s v_r^2}$ and $f_2 = \frac{-D v_r v_s (h_s v_r - h_t v_s)}{2(u_r v_s^2 + u_s v_r^2)}$. However, due to the constraint $R_s \in [0, R_1]$, there are three cases to occur based on the value of R_{dc}^3. By substituting R_{dc}^3 into $JTC(T, R_{dc})$ and taking the first-order derivative of T_3, the following corollary can be obtained.

Corollary 1

(1) DS case: $R_{dc}^3 = 0$ and $T_3 = \sqrt{\frac{2P[v_r^2(A_s + \sum_{i=1}^n a_i + F_s + F_r + \sum_{i=1}^n Ar_i) + u_r R_n^2]}{(h_s D^2 + \sum_{i=1}^n hr_i D_i P) v_r^2}}$.

(2) DR case: $R_{dc}^3 = R_1$ and $T_3 = \sqrt{\frac{2P[v_r^2 v_s^2(A_s + \sum_{i=1}^n a_i + F_s + F_r + \sum_{i=1}^n Ar_i) + u_r v_s^2 (R_1 - R_n)^2 + u_s v_r^2 R_1^2]}{(h_s D^2 + \sum_{i=1}^n hr_i D_i P) v_r^2 v_s^2}}$.

(3) DB case: DC is located at between the supplier and the retailer 1, i.e., $R_{dc}^3 \in (0, R_1)$ and $T_3 = \sqrt{\frac{2P[v_r^2 v_s^2(A_s + \sum_{i=1}^n a_i + F_s + F_r + \sum_{i=1}^n Ar_i) + u_r v_s^2 (f_1 - R_n)^2 + u_s v_r^2 f_1^2]}{2 f_2 P D v_r v_s (h_s v_r - h_t v_s) + (h_s D^2 + \sum_{i=1}^n hr_i D_i P) v_r^2 v_s^2 + 2 f_2^2 P (u_r v_s^2 + u_s v_r^2)}}$.

5 NUMERICAL EXAMPLES

In this section, the default values of parameters are to illustrate theoretical results.

Example 1. Given $A_s = 10$, $h_s = 1$, $h_t = 2$, $P = 100$, $F_s = 20$, $F_r = 30$, $u_s = 1.5$, $u_r = 2$, $v_s = 3$, $v_r = 5$, $a_1 = 4$, $a_2 = 3.5$, $a_3 = 4.5$, $a_4 = 2.5$, $Ar_1 = 12$, $Ar_2 = 8.5$, $Ar_3 = 10$, $Ar_4 = 9$, $hr_1 = 2.5$, $hr_2 = 3$, $hr_3 = 4$, $hr_4 = 3$, $D_1 = 11$, $D_2 = 20$, $D_3 = 8$, $D_4 = 24$, $R_1 = 15$, $R_2 = 30$, $R_3 = 40$, $R_4 = 60$.

Thus, we have $T_1 = 1.55$, $JTC_1 = 1328.36$, $T_2 = 1.50$, $JTC_2 = 1451.55$, $T_3 = 1.65$, $R_{s3} = 15$ and $JTC_3 = 1327.73$. To further characterize the optimal solutions of the supply chain, we remain other parameters the same, while vary the values of R_1, R_4 and P, respectively. The results are presented in Table 1.

Table 1: The Impact of Changes of R_1, R_4 and P on the Optimal Solutions

Parameters		RMI		VMI		Integrated decision		
		T_1	JTC_1	T_2	JTC_2	T_3	R_{s3}	JTC_3
R_1	10	1.68	1337.03	1.50	1429.55	1.69	10	1337.02
	15	1.55	1328.36	1.50	1451.55	1.65	15	1327.73
	20	1.44	1328.94	1.50	1473.55	1.63	20	1325.78
	25	1.32	1340.93	1.50	1495.55	1.65	25	1331.35

Parameters		RMI		VMI		Integrated decision		
		T_1	JTC_1	T_2	JTC_2	T_3	R_{s3}	JTC_3
R_4	45	1.22	1127.38	1.50	1223.41	1.39	15	1124.47
	50	1.32	1192.53	1.50	1296.79	1.47	15	1190.65
	55	1.44	1259.63	1.50	1372.84	1.56	15	1258.49
	60	1.55	1328.36	1.50	1451.55	1.65	15	1327.73
P	80	1.55	1336.07	1.33	1471.61	1.61	15	1335.82
	90	1.55	1331.79	1.42	1460.32	1.63	15	1331.35
	100	1.55	1328.36	1.50	1451.55	1.65	15	1327.73
	110	1.55	1325.55	1.57	1444.68	1.66	15	1324.75

1. The location of retailers exerts influences on the operational strategies of the supply chain: (i) the increase of R_1 would make retailers adopt a shorter ordering cycle; (ii) High values of R_4 would result in longer length of the retailers' ordering cycle, while JTC_1, JTC_2 and JTC_3 present growing trends; and (iii) the retailers' optimal ordering cycle remains the same under the VMI.
2. A higher value of the supplier's production rate P leads to a longer length of the ordering cycle and a lower value of the average cost of the supply chain. Since the increase of P makes it possible for the supplier to provide retailers with items as soon as possible, the supplier's holding cost could be reduced significant. And it should be noted that the retailers' optimal ordering cycle remains the same under the RMI.

6 CONCLUSIONS

Upstream suppliers and downstream retailers have long recognized the importance to synthesize their strategies to achieve the win-win situation. To maintain the flexibility of the supply chain, the DC, which receive items from the supplier and transfer them to retailers, have been widely implemented in the multi-echelon distribution system. In this paper, we focus on the optimal operational strategies of the supply chain with single-supplier multi-retailer under RMI or VMI contract. And the retailers are assumed to adopt common cycle to coordinate their ordering cycle. Hence, the location of the DC would be critical for the total cost system design and the smooth operation of the supply chain. And via analyzing the average cost functions including the supplier, the retailers and the supply chain, these following characteristics distinguishing this paper from previous researches: (1) the optimal operational strategies of RMI and VMI policies are developed to compare with the solution of the integrated production-inventory policy, and (2) the optimal location of the single DC relies on which member plays the dominant role in the supply chain, which includes three cases: DR case, DS case and DB case.

Acknowledgements

The research presented in this paper is supported by the National Natural Science Foundation Project of China (No: 71271168).

Appendix A. Proof of Lemma 1

By taking the first derivative of $TC_r(T, R_{dc})$ with respect of T and R_{dc}, we can find that

$$\frac{\partial^2 TC_r(T,R_{dc})}{\partial T^2} = \frac{\sum_{i\in R} 2Ar_i}{T^3} + \frac{2[F_r v_r^2 + u_r(R_n - R_{dc})^2]}{v_r^2 T^3}, \quad \text{(A1)}$$

$$\frac{\partial^2 TC_r(T,R_{dc})}{\partial R_{dc}^2} = \frac{2u_r}{v_r^2 T}, \quad \text{(A2)}$$

and

$$\frac{\partial^2 TC_r(T,R_{dc})}{\partial T \partial R_{dc}} = \frac{2u_r(R_n - R_{dc})}{v_r^2 T^2}. \quad \text{(A3)}$$

Thus, we can obtain Hessian matrix $H = \frac{2u_r}{v_r^2 T}\left\{\frac{\sum_{i\in R} 2Ar_i}{T^3} + \frac{2\left[F_r v_r^2 + u_r(R_n - R_{dc})^2\right]}{v_r^2 T^3}\right\} - \frac{4u_r^2(R_n - R_{dc})^2}{v_r^4 T^4} > \frac{2u_r}{v_r^2 T}\frac{2u_r(R_n - R_{dc})^2}{v_r^2 T^3} - \frac{4u_r^2(R_n - R_{dc})^2}{v_r^4 T^4} = 0$. That is, $H > 0$ and $TC_r(T, R_{dc})$ is jointly convex in T and R_{dc}.

7 REFERENCES

[1] **Yao Y. L., Evers P. T., Dresner M. E. 2007**. Supply chain integration in vendor-managed inventory. *Decision Support Systems*, 43(2), pp 663-674.

[2] **Chen J. M., Lin I. C., Cheng H. L. 2010**. Channel coordination under consignment and vendor-managed inventory in a distribution system. *Transportation Research Part E-Logistics and Transportation Review*, 46(6), pp 831-843.

[3] **Hoque M. 2011**. Generalized single-vendor multi-buyer integrated inventory supply chain models with a better synchronization. *International Journal of Production Economics*, 131(2), pp 463-472.

[4] **Ben-Daya M., Hassini E., Hariga M., Aldurgam M. M. 2013**. Consignment and vendor managed inventory in single-vendor multiple buyers supply chains. *International Journal of Production Research*, 51(5), pp 1347-1365.

[5] **Ru J., Wang Y. Z. 2010**. Consignment contracting: Who should control inventory in the supply chain? *European Journal of Operational Research*, 201(3), pp 760-769.

[6] **Shen Z. J. M., Coullard C., Daskin M. S. 2003**. A joint location-inventory model. *Transportation Science*, 37(1), pp 40-55.

[7] **Nozick L. K., Turnquist M. A. 2001**. Inventory, transportation, service quality and the location of distribution centers. *European Journal of Operational Research*, 129(2), pp 362-371.

[8] **Sheu J. B. 2003**. Locating manufacturing and distribution centers: An integrated supply chain-based spatial interaction approach. *Transportation Research Part E-Logistics and Transportation Review*, 39(5), pp 381-397.

[9] **Haq A. N., Kannan G. 2006**. Design of an integrated supplier selection and multi-echelon distribution inventory model in a built-to-order supply chain environment. *International Journal of Production Research*, 44(10), pp 1963-1985.

[10] **Gendron B., Semet F. 2009**. Formulations and relaxations for a multi-echelon capacitated location-distribution problem. *Computers & Operations Research*, 36(5), pp 1335-1355.

[11] **Ertogral K., Darwish M., Ben-Daya M. 2007**. Production and shipment lot sizing in a vendor-buyer supply chain with transportation cost. *European Journal of Operational Research*, 176(3), pp 1592-1606.

[12] **Viswanathan S., Piplani R. 2001**. Coordinating supply chain inventories through common replenishment epochs. *European Journal of Operational Research*, 129(2), pp 277-286.

THE FORECASTING OF CENTRAL TECHNOLOGY PROBLEM AND DEVELOPMENT DIRECTION ——— A STUDY BASED ON THE TEXT DATA FROM SCIENTIFIC RESEARCH FOUNDATION

Zhi Li. Tao Zhou. Xinjun Lai.
Guangdong Provencal Key Lab of Computer Integrated Manufacturing
Guangdong University of Technology
zhoumiek123@163.com

ABSTRACT

The scientific research foundation is a vital channel to support fundamental research in many countries, which is open to meet the needs of high-level talents, especially to subsidize and support the valuable research projects. Every government-funded project involves high-level talents' consideration about the directions of future development and their explorations about the heated and difficult technology. In the process of analyzing national fund, we can grasp the information of the research conducted by high level talents. By extracting some useful data and combining them with technological management, we are able to predict the future development objectively and reasonably. Based on the practical significance and demand, we propose an analytical model to explore the central problem and development direction, based on the data from the national research foundation documents. In this study, we employ the technology of text mining, which mainly captures the text data from the titles of the national fund projects. Next, this research combines clustering analysis, statistical inference to construct our analysis model to forecast the development direction and central problem in one field. The research result facilitates a better awareness and understanding of the status and problems in the field, meanwhile the result can provide researchers with a scientific research direction and build an R&D policy effectively. We perform a case study to verify how our research can be practically applied, using the data about the National Natural Science Foundation of China to construct the analytical model. The rationality of the model is validated through the analysis of National Natural Science Foundation of China from 1999 to 2014(total of 15 years) in the field of mechanical engineering.

Keywords: Scientific research foundation, Central forecasting, Text mining, Web crawler

1 INTRODUCTION

Management of technology(MOT) reveals an important role in the field of R&D management, and the fruitfulness of its integration into strategic management has become clear[1]. Through MOT, the department of Research & Development can monitor the development more reasonably and promote their competitive force.[2] Application of the management of technology, makes the future trend will be forecast able, and better services the technology development. Research foundation is a vital channel to support fundamental research in many countries. It has been designed to promote research activity.[3] National foundation has revealed the high-level talents ' thinking of the field what he is proficient. The purposes of research foundations are leading the future technology and solve the difficult technology problem. In recent years, the funding scale and funding budget have increased dramatically. Benefited from this, all kind of research has also made considerable progression.[4] With the increase of the number of the research foundations, a problems appears in front of the researchers and scholars ,that is, which technology is the most dominant in a field, and which research is the central in this field. Central technology has significant effects on the development of other technologies. Central technology model is a technology analysis approach to proceed technology forecasting.[5] Researchers can better seize the central technology, and better seize the hot research spot and future development trend; thus the proposed topic and research direction have more practical meaning and value, and also they can better promote the development of their research field. Hence obtaining an objective technology forecasting through technology analysis has a practical significance. In general, there are three approaches to technology forecasting and analysis: qualitative, quantitative, and combine of the two. The qualitative analysis relies on the expert's knowledge, and the quantitative analysis relies on the statistics or Mathematical model. [6] Based on the obtained data, we applied the quantitative analysis to construct a forecasting model. In this paper, the model is composed of two indexes which are concern-degree and correlation-degree, applied the linear combination of the two indexes to evaluate the central-degree. [7,8] In order to verify how our research can be practically applied, the National Natural Science Foundation of China from 1999 to 2014(total of 15 years) in the field of mechanical engineering has been analyzed as the target technology field.

The rest of the paper is organized as follow:
Section 2 presents the model and methodology
Section 3 takes a case to implement the methodology, and the case analysis illustrate the China's central technology and future development in the field of machine engineering.
Section 4 take a conclusion and illustrate the work what will be did in the next.
Section 5 shows the references of this paper.

2 PROPOSED MODEL

This paper proposes a model for central technology forecasting based on the data of China National Natural Science Foundation in the field of mechanical engineering, and analyses the future development trend by the central technology. This proposed model is divided into four steps:

2.1 Step 1: Obtaining of text data

Web-crawler is an effective method of data mining [9]. In this paper, we apply the method of web-crawler to obtain text data from the foundation information web-site. Every foundation title is composed of several key words, which is the concentration of the whole foundation. Therefore, the foundation title can reflect related technologies that the foundation authors expect to research. So we only collect the data of foundation title. In this part, we apply Java program to capture text data in the internet and store the text data in the table.

2.2 Step 2: The calculation of concern degree

In order to apply for the foundation convenient, the foundation competent authorities have divided a big research field into several smaller areas, for example the field of machine engineering in China national natural Science Foundation having been divided into 13 research direction, and each foundation belongs to a research direction among 13 categories. Fig 1 shows the research direction classification of machine engineering. Everyone's research direction is a gather of the same technology category in a big research field. In this paper, we analyses the central technology from the technology categories. In general, the foundation application system had been divided into categories. Based on the statistic text data, that's capture in the Internet, and we defined:

$$Q(C_i) = 1 - \frac{|S_{\max} - S_i|}{S_{\max} - S_{\min}} \quad (1)$$

Where C_i represents the ith category, and S_i represents the foundation number of ith category, and $S_{\max}$ represents the maximum number among n categories, and $S_{\min}$ represents the minimum number among n categories.

2.3 Step 2: The calculation of concern degree

2.3.1 Automatic text segmentation

Every foundation title is composed of a number of key words and conjunctions. The key word will be used to parse foundation, so we should apply automation tool to capture key words. For each foundation title use lucene segmentation system and bidirectional matching algorithm to complete automated segmentation. Considering the foundation title contain a lot of professional vocabulary and the segmentation system dictionary don't have those vocabulary, so a manual modification process is necessary. Next removing the connection and commonly-used words, that only preserve the 3-5 keywords which is the concentration of the whole foundation. The keywords will be used in the step 4 to analyse the foundation.

2.3.2 The semantic analysis of foundation title

Most traditional text mining techniques are designed to handle long text documents. However, Individual foundation's text data is too little to use the traditional statistical method to take the similarity computation between the two foundations. To understand short messages, we observe that it is often possible to find topically related long texts, which can be utilized as the auxiliary data when mining the target short texts data. In this paper, we adopt the keywords which are captured in the foundation title to search in an internet text library [10], thus obtain enough text data to fit the foundation meaning. The internet text library must have enough related text data to couple the key words. We choose different library according to the different data type. For example, the case study is China national foundation data, so the internet text library choose China-hownet data system (http://www.cnki.net/). Each foundation key word searches in the China-hownet and uses web-crawler to capture the first ten pages of text data, the first ten pages data is the most similarity data which is . Therefore, we can use the China-hownet matched data to do analyze. The enough data can apply the traditional statistical method and implement similarity calculation.

2.3.3 Similarity degree calculation of two foundations

After finishing the interpretation of foundation, we used auto-programming to calculate similarity [11]. In this paper, the Google open source Doc2Vec was chosen. Fig 1 shows the progress of the two foundations' similarity, where f_i represents a foundation title, and A represents the matched text data that acquired in the step of 2.3.2. And A was mapped to a multidimensional vector ($Vector_A$) via the Google open tool Doc2vec(f_j has the same process). The similarity of two foundations defined as:

$$sim(f_i, f_j) = \frac{Vector_A \times Vector_B}{\sqrt{Vector_A^{\ 2}} \times \sqrt{Vector_B^{\ 2}}} \tag{2}$$

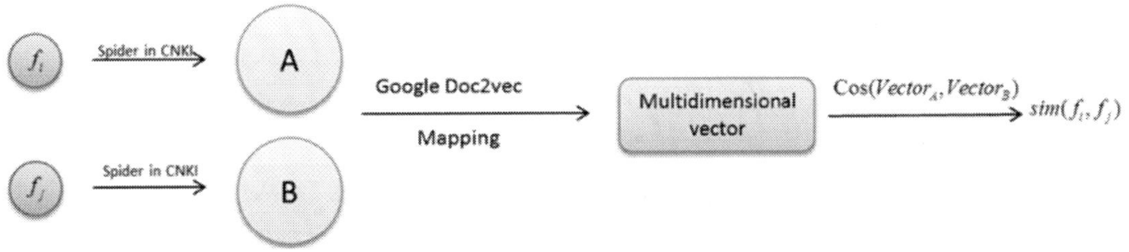

Figure 1: Similarity degree calculation of two foundations

2.3.4 Similarity degree calculation of two foundation categories

Each foundation belongs to a certain category which is classified during the application process. After calculating the similarity of two foundation, we will use the $sim(f_i, f_j)$ to calculate the similarity of the two foundation categories. The similarity of the two foundation categories was defined as:

$$Sim(C_i, C_j) = Sim(C_j, C_i) = \frac{\sum_{i=1}^{m}\sum_{j=1}^{n} sim(f_i, f_j)}{m \times n} \tag{3}$$

Where C_i represent the i th foundation category, and total m foundations in this category.

C_j represent the j th foundation category, and total n foundations in this category.

when $i = j$ the $sim(C_i, C_j) = 0$

$i = j$ the $Sim(C_i, C_j) = \frac{\sum_{i=1}^{m}\sum_{j=1}^{n} sim(f_i, f_j)}{m \times n}$

2.3.5 Correlation degree calculation of each foundation category

In this paper, we construct an (i*i) matrix, Fig.2 show the categories similarity matrix.

$$\begin{array}{c|cccc} & C_1 & C_2 & \cdots & C_n \\ \hline C_1 & sim(C_1, C_1) & sim(C_1, C_2) & \cdots & sim(C_1, C_n) \\ C_2 & sim(C_2, C_1) & sim(C_2, C_2) & \cdots & sim(C_2, C_n) \\ \vdots & \vdots & \vdots & \ddots & \vdots \\ C_n & sim(C_i, C_1) & sim(C_i, C_2) & \cdots & sim(C_n, C_n) \end{array}$$

Fig.2 The categories similarity matrix

The rows and columns of this matrix are the categories in a technology field, the $sim(C_i, C_j)$ is calculated from the step 6. The i th categories correlation defined as:

$$G(C_i) = 1 - \frac{\left| \sum_{j=1}^{n} sim(C_i, C_j)_{\max} - \sum_{j=1}^{n} sim(C_i, C_j) \right|}{\sum_{j=1}^{n} sim(C_i, C_j)_{\max} - \sum_{j=1}^{n} sim(C_i, C_j)_{\min}} \tag{4}$$

Where $\sum_{j=1}^{n} sim(C_i, C_j)_{\max}$ and $\sum_{j=1}^{n} sim(C_i, C_j)_{\min}$ respectively represents the Maximum and minimum values among form $\sum_{j=1}^{n} sim(C_0, C_j)$ to $\sum_{j=1}^{n} sim(C_n, C_j)$.

2.4 Step 4: Central degree calculation and develop forecasting

We apply the $Q(C_i)$ calculated in step 2 and the $G(C_i)$ calculated in step 3 to calculate the central degree, and the central degree Key_i defined as:

$$Key_i = Q(C_i) + G(C_i) \tag{5}$$

The Key_i is the comprehensive evaluation of central degree.

Fig.3 shows our proposed model.

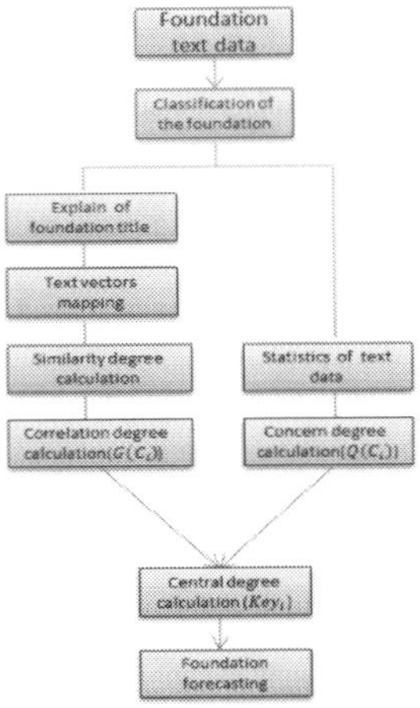

Fig.3 Proposed model

3 EXPERIMENTAL RESULTS

To verify the performance of our approach, we conducted a case study using National Natural Science Foundation of China from 1999 to 2014(total of 15 years) in the field of mechanical engineering. In this paper, the total 7000 foundations title had been collected in the publically website (resource from the web of LetPub: http://www.letpub.com.cn/).

3.1 The calculation of concern degree

The National Natural Science Foundation of China in the field of machine had been divided into 13 categories. In this paper, we take out the category of C_{13} , because the category is not a state category and can't represent a technology field. Through the statistics data, the concern degrees is calculated. Table.1 shows concern degrees results（Keep three valid Numbers）of each category, From the table x the largest category is $C_4(S_{max} = S_4 = 975)$, the smallest category is $C_8(S_{min} = S_8 = 217)$.

Table 1: The result of concern degree

Category code	Chinese name of category	Number	Concern degree
C_1	机械结构强度学	269	0.0686
C_2	机械摩擦学与表面技术	525	0.406
C_3	机械测试理论与技术	508	0.384

C_4	零件成型制造	975	1
C_5	传动机械学	469	0.332
C_6	机械动力学	910	0.914
C_7	机械设计学	664	0.590
C_8	机械仿生学	217	0
C_9	零件加工制造	854	0.840
C_{10}	制造系统与自动化	487	0.356
C_{11}	微/纳机械系统	410	0.255
C_{12}	机构学与机器人	437	0.290

The greater the number of foundations is proportional to the value of concern degree. The larger concern-degree indicates the more thorny problem in this technology field. Concern degree is an index of central degree.

3.2 The calculation of correlation degree

First, the similarity degree of two foundations is calculated, and then the similarity degree of two categories is calculated. The Fig.4 shows the similarity degree matrix of two categories. In the (12*12) matrix, the value between 0 and 1, the lager value indicates the stronger link between the two categories.

	C_1	C_2	C_3	C_4	C_5	C_6	C_7	C_8	C_9	C_{10}	C_{11}	C_{12}
C_1	0	0.7354	0.6522	0.5682	0.6202	0.5767	0.6089	0.7651	0.5913	0.1478	0.6191	0.5962
C_2	0.7354	0	0.7731	0.5913	0.6202	0.5827	0.6079	0.7639	0.6093	0.1602	0.6304	0.5976
C_3	0.6522	0.7731	0	0.0452	0.5579	0.3060	0.5534	0.5687	0.0871	0.0608	0.5677	0.0322
C_4	0.5682	0.5913	0.0452	0	0.6112	0.5630	0.5949	0.5977	0.8252	0.5447	0.6242	0.5821
C_5	0.6202	0.6202	0.5579	0.6112	0	0.6564	0.7414	0.5378	0.5469	0.1173	0.7384	0.5758
C_6	0.5767	0.5827	0.3060	0.5630	0.6564	0	0.6478	0.7541	0.6183	0.2666	0.6541	0.8411
C_7	0.6089	0.6079	0.5534	0.5949	0.7414	0.6478	0	0.5429	0.5510	0.6467	0.7243	0.5768
C_8	0.7651	0.7639	0.5687	0.5977	0.5378	0.7541	0.5429	0	0.6025	0.1807	0.7617	0.6691
C_9	0.5913	0.6093	0.0871	0.8252	0.5469	0.6183	0.5510	0.6025	0	0.7177	0.6000	0.5569
C_{10}	0.1478	0.1602	0.0608	0.5447	0.1173	0.2666	0.6467	0.1807	0.7177	0	0.1288	0.1488
C_{11}	0.6191	0.6304	0.5677	0.6242	0.7384	0.6541	0.7243	0.7617	0.6000	0.1288	0	0.5802
C_{12}	0.5962	0.5976	0.0322	0.5821	0.5758	0.8411	0.5768	0.6691	0.5569	0.1488	0.5802	0

Figure 4 Similarity degree matrix of two categories

In this paper , we applied the formula (4) to calculate the correlation degree . The table lists the correlation and ranking of the 12 categories according to their correlation value. The value is proportional to central degree. Table.2 shows the results of correlation degree.

Table 2: The results of correlation degree

Category code	Correlation-degree	Ranking
C_1	0.914	5
C_2	0.966	3
C_3	0.295	11
C_4	0.824	9
C_5	0.871	7

C_6	0.910	6
C_7	1	1
C_8	0.986	2
C_9	0.867	8
C_{10}	0	12
C_{11}	0.955	4
C_{12}	0.717	10

Under the of correlation degree, the C_7 C_8 C_2 C_{11} has the biggest correlation degrees, and the bigger possibility states the central position.

3.3 The calculation of central degree

The central degree (Key_i) is composed of concern-degree and correlation-degree, and the calculation results have been showed in table 3.

Table 3: The results of central degree

Category code	Central-degree	Total ranking
C_1	0.9826	10
C_2	1.372	5
C_3	0.679	11
C_4	1.824	1
C_5	1.203	7
C_6	1.824	1
C_7	1.590	4
C_8	0.986	9
C_9	1.707	3
C_{10}	0.356	12
C_{11}	1.210	6
C_{12}	1.007	8

In this paper, we divided the 12 categories into 3 grades, as is shown in table 4

Table 4: The results of central forecasting

Central grades	Key_i intervalle	Categories
High	$Key_i \geq 1.500$	C_4、C_6、C_7、C_9
media	$1.000 \leq Key_i < 1.500$	C_2、C_5、C_{11}

low	$Key_i < 1.000$	C_1、C_3、C_8、C_{10}、C_{12}

4 CONCLUSION

In this paper, we proposed a model for central technology forecasting based on the concern-degree and the correlation-degree, and the two indexes are liner adding. We computed the central degree and ranked the results to extract the central technology field. From the experimental, we concluded that the C_4,C_6,C_7,C_9 situate the high central grade, the C_2,C_5,C_{11} situate the media central grade, and the $C_1,C_3,C_8,C_{10},C_{12}$ situate the low central grade. The high central grade represents the basic areas, those areas have tricky problem of research and enough number of researcher take into the field.

Our model also have many problems to be solved, there are mainly two aspects: (1) the accuracy of text similarity calculation; (2) the accuracy of central-degree calculation accuracy. These problems will be discussed in subsequent work and study.

5 REFERENCES

[1] K Brockhoff. 1998. Technology management as part of strategic planning - some empirical results, R&D Management, 28(3), pp 129-138.

[2] CV Trappey, HY Wu, F Taghaboni-Dutta, AJC Trappey. 2011. Using patent data for technology forecasting: China RFID patent analysis, Advanced Engineering Informatics, 25(1), pp53-64.

[3] A Pouris. 2007. The National Research Foundation's rating system: Why scientists let their ratings lapse, South African Journal of Science, 103(11-12), pp 697-697.

[4] Q Wan, Y Xu, E Dong. 2015. Diabetic nephropathy research in China: data analysis and review from National Natural Science Foundation of China, Journal of Diabetes, 7(3)

[5] S Jun, 2012. Central technology forecasting using social network analysis, Communications in Computer and Information Science, pp 1-8.

[6] S Park, J Kim, H Lee, D Jang, S Jun. 2016. Methodology of technological evolution for three-dimensional printing, Industrial Management & Data Systems, 116(1), pp 122-146.

[7] B Balasundaram, S Butenko, IV Hicks. 2011. Clique Relaxations in Social Network Analysis: The Maximum k-Plex Problem, Operation Research, 59(1), pp 133-142.

[8] AC Adamuthe, AV Surve, GT Thampi. 2016. Technology Forecasting: A Case Study of Software Technology Product Families, I.J. Information Engineering and Electronic Business,1, pp 11-20.

[9] Thelwall, 2001. A Web crawler design for data mining, Journal of Information Science, 27(5), pp 319-325.

[10] Jin, O., Liu, N. N., Zhao, K., Yu, Y., & Yang, Q. 2011. Transferring topical knowledge from auxiliary long texts for short text clustering, InProceedings of the 20th ACM international conference on Information and knowledge management, pp. 775-784.

[11] Xu, L., Jiang, C., Ren, Y., & Chen, H. H. 2016. Microblog Dimensionality Reduction—A Deep Learning Approach, IEEE Transactions on Knowledge and Data Engineering, 28(7), pp 1779-1789.

RETAILER'S STRATEGY CHOICE FOR DEFECTIVE ITEMS UNDER TEMPORARY PRICE REDUCTION: INCURRING DEFECT SALES RETURN OR INCREASING THE SCREENING RATE

Guiyang ZHU[1*], Tao JIA[2] and Yu WANG[3]
[1]School of Management
Xi'an Jiaotong University, Xi'an, China
zhuguiyang@stu.xjtu.edu.cn

[2]School of Management
Xi'an Jiaotong University, Xi'an, China
jiatao@mail.xjtu.edu.cn

[3]School of Management
Xi'an Jiaotong University, Xi'an, China
wy_sail@stu.xjtu.edu.cn

ABSTRACT

Suppliers offer retailers temporary price reduction for various reasons, such as a need to reduce excess inventory or the marketing strategy of getting retailers to pass some of the price discounts to the final consumers to stimulate demand. This paper investigates two inventory models for defective items under temporary price reduction in a situation where the screened good items cannot satisfy the increased consumer demand during the screening period. The retailer has two strategies in this situation. (1) The retailer incurs defect sales return because he satisfies the consumer demand with both screened good items and unscreened items which include a proportion of defective items; (2) The retailer increases the screening rate through investment to quality checking process. Two mathematical models are developed corresponding to the two strategies. Then the properties of each model are analysed. Finally, numerical examples are provided to compare the characteristics of these two models.

Keywords: temporary price reduction, defective items, defect sales return, quality checking

1 INTRODUCTION

Sales promotion is widely used by practitioners as an effective tool to increase short-run sales. As one of the most commonly used trade promotion tools, temporary price reduction is offered by manufacturers or suppliers to retailers in an aim to encourage them to order a larger than usual quantity and reduce their retail price. Online retailers are also using temporary price reduction as a marketing tool to boost sales and drive profits. For example, during the double eleven shopping festival of China's largest online retailer Taobao, the trading volume achieved within only one day of promotion accounts for three percent of the whole year trading volume in 2015.

Due to the practical significance, a stream of research investigates inventory models under temporary price reduction. Some studies assume a constant consumer demand rate and optimize the special lot size for the retailer. Sarker and Al Kindi[1] derived the optimal special lot size in five

* Corresponding Author

scenarios. Taleizadeh et al.[2] determined the optimal special order and shortage quantities of a perishable item when a special sale is offered. Other studies assume a price-dependent demand rate and jointly determine the pricing and ordering policy for the retailer. Abad[3] analysed retailer's optimal promotion price and ordering policies by incorporating retailer's forward buying strategies. Ardalan[4] developed an inventory model under a general price-demand relationship. Arcelus et al.[5] further extended this work by incorporating forward-buying using a linear price-demand relationship. For more details, the readers can refer to a review provided by Ramasesh[6] .

Besides determining promotion strategies in the retailing context, the retailer faces the problem that items received in a lot usually contain a proportion of defective ones. Another stream of research investigates inventory models with defective items. Salameh and Jaber[7] assumed that the defective items are sold as a single batch at a lower price by the end of the 100% screening process and derived the economic lot size quantity. Some researchers proposed the concept of **defect sales return**. Khan et al.[8] assumed there are inspection errors during the screening process and classified the errors into two types, accepting a defective item (type I error) which leads the inventory system to incur defect sales return and rejecting a non-defective item (type II error). Then they determined the optimal order size. On the other hand, Hauck and Vörös[9] assumed that the percentage of defective items is a random variable and stated that it is possible that the screened good items cannot satisfy the consumer demand. They addressed the approach of speeding up the quality checking process and then proposed two inventory models.

The problem of defective items becomes even more severe during a promotional period because the amount of the defective items becomes larger when the retailer orders a larger quantity than usual. Hsu and Yu[10] developed an EOQ model for defective items under one-time-only discount. They assumed that the consumer demand rate is constant and the retailer can always satisfy the consumer demand with the screened good items. While our paper considers a price-sensitive demand rate, so the retailer faces a situation in which the screened good items can not satisfy the increased consumer demand during a promotional period. The retailer has two strategies to deal with this situation. The first strategy is to meet the consumer demand with both screened good items and unscreened items which contain a proportion of defective items. As a result, the retailer incurs defect sales return. While the second strategy is to increase the screening rate through investment to insure the screened good items can satisfy the increased consumer demand. Consequently, the retailer incurs extra cost of increasing the screening rate for the second strategy.

The remainder of this paper is organized as follows. Section 2 lists the notation and assumptions used in this paper and briefly reviews Salameh and Jaber's model. Section 3 develops two inventory models corresponding to retailer's two strategies. Section 4 analyses the properties for the proposed models. Section 5 provides numerical examples. Finally, section 6 concludes the paper with managerial insights.

2 NOTATION, ASSUMPTIONS AND A BRIEF REVIEW OF SALAMEH AND JABER'S MODEL

2.1 Assumptions

1. Under the normal ordering policy, the screened good items can satisfy the consumer demand during the screening period, namely, $(1-p)x_0 > D_0$. While during the special sale time, the screened good items cannot satisfy the increased consumer demand, $(1-p)x_0 < D_1$.
2. The consumer demand rate is price-sensitive.
3. The percentage of defective items p is a constant for a lot size.

2.2 Notation

A	the ordering cost per order
r	the holding cost per unit/per unit time
c	the purchasing cost per unit
k	supplier's unit discount to the retailer
p	percentage of defective items for each order
d	unit screening cost
x_0	the screening rate per day
x	the screening rate per day after the retailer's investment in a sales promotion
N	the number of working days in a year
v	unit selling price of defective items, $v < c$
s_0	the regular unit selling price of items of good quality
s_1	the special unit selling price of items of good quality
D_i	the retailer's demand per day at unit selling price s_i, $i = 0,1$
T_i	the replenishment cycle length for alternatives $i = 0,1$ (days)
Q_i	the retailer's order quantity for alternatives $i = 0,1$
β	$0 \le \beta \le 1$, the probability a defective item is returned by the consumer
t_b	the time when the consumers begin to return the defective items they receive
w	the retailer's unit handling cost for the defective items returned
z	the ratio of demand to the screening rate, i.e. $z = D_1/x$. let $z_{max} = D_1/x_0$
$g(z)$	the daily cost of increasing the screening rate from the current x_0 to x

2.3 A brief review of Salameh and Jaber's model

We briefly review Salameh and Jaber's model as the basis for our model. The inventory level is illustrated in Figure 1, where $T_0 = (1-p)Q_0/D_0$. We denote the total revenue per cycle as $TP(Q_0)$ which consists of sale revenue, ordering cost, procurement cost, screening cost, and holding cost. $TP(Q_0)$ is given by

$$TP(Q_0) = s_0(1-p)Q_0 + vpQ_0 - \left\{A + cQ_0 + dQ_0 + rc\left[\frac{Q_0^2(1-p)^2}{2D_0} + \frac{Q_0^2 p}{x_0}\right]\right\} \quad (1)$$

We assume that the number of working days in a year is N, so the ordering frequency is $\frac{N}{T_0} = \frac{ND_0}{(1-p)Q_0}$. The average annual profit is given by

$$TPU(Q_0) = ND_0\left(s_0 + \frac{vp}{1-p}\right) - \frac{ND_0}{(1-p)}\left\{\frac{A}{Q_0} + c + d + rc\left[\frac{Q_0(1-p)^2}{2D_0} + \frac{Q_0 p}{x_0}\right]\right\} \quad (2)$$

The optimal order quantity Q_0^* can be obtained as

$$Q_0^* = \sqrt{\frac{2AD_0x_0}{rc[(1-p)^2x_0 + 2pD_0]}} \quad (3)$$

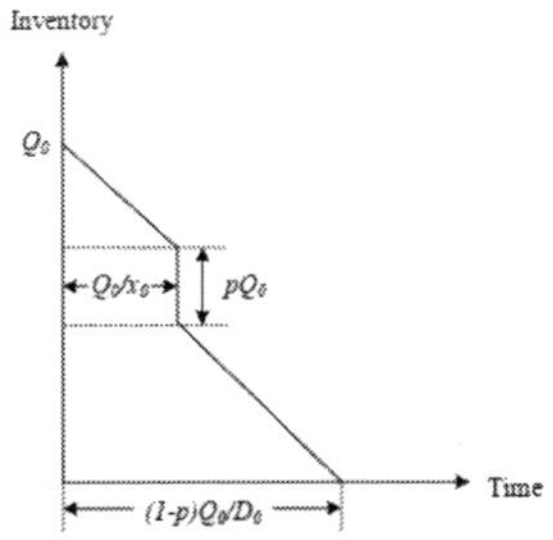

Figure 1: Behaviour Of Inventory Level Over Time For Salameh And Jaber's Model

Based on their model, we come up with two models to deal with a situation where the retailer's screened good items can not satisfy the increased consumer demand during a promotional period.

3 MATHEMATICAL MODEL

3.1 Model I: The retailer incurs defect sales return during the screening period

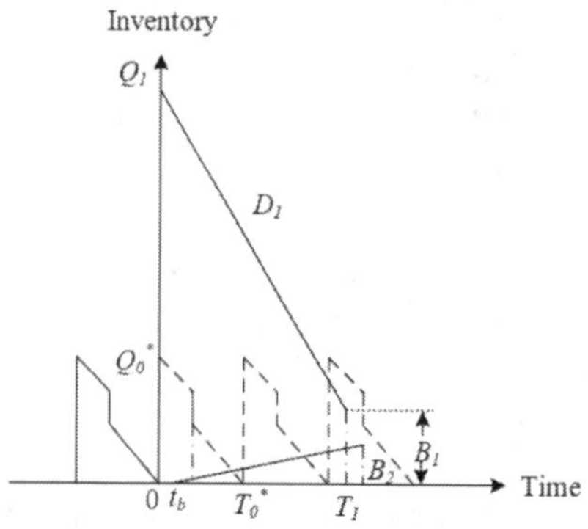

Figure 2: Inventory Level Over Time When The Retailer Incurs Defect Sales Return

When the supplier offers temporary price reduction, the retailer has two policies: (1) placing a large, special order Q_1 at the reduced wholesale price and passing some of the supplier's discounts on to the consumers (called policy S) and (2) ordering the EOQ and offering no discount (called policy N). During the special sale time T_1, we denote the profit earned under policy S as $TP_s^{(1)}(Q_1)$ and the profit under N as $TP_n^{(1)}(Q_1)$. The optimal Q_1 is obtained by maximizing the difference between policy S and N, which is written as $D^{(1)}(Q_1) = TP_s^{(1)}(Q_1) - TP_n^{(1)}(Q_1)$.

Under policy S, the behaviour of the inventory level is illustrated in Figure 2. The supplier offers temporary price reduction at time 0. The retailer determines the special order quantity Q_1 at time 0 at a discounted wholesale price $(c-k)$. Since we assume that $(1-p)x_0 < D_1$, a proportion of consumer demand is satisfied with the screened good items at the rate $(1-p)x_0$, while the remaining proportion is satisfied with unscreened items at the rate $D_1-(1-p)x_0$. Moreover, the unscreened items incorporate a proportion of defective items $p[D_1-(1-p)x_0]$. Since the probability that a defective item is returned by the consumer is β, the return rate during the special sale time is calculated as $\beta p[D_1-(1-p)x_0]$.

The defective items include two batches. The number of defective items screened out by the retailer is B_1 and they are sold as a single batch at a lower price at time T_1. The number of the defectives returned from the market is B_2 and they are sold as a single batch at time T_1+t_b. From Figure 2, we have $B_1 = px_0T_1$ and $Q_1 = D_1T_1 + B_1$. So that the special sale time can be written as

$$T_1 = \frac{Q_1}{D_1+px_0} \tag{4}$$

Also, B_1 and B_2 are given by

$$B_1 = \frac{px_0Q_1}{D_1+px_0} \tag{5}$$

$$B_2 = \frac{\beta p[D_1-(1-p)x_0]Q_1}{D_1+px_0} \tag{6}$$

The relevant revenue consists of the following elements under policy S:

1. The sale revenue is

$$\begin{aligned} s_1(Q_1-B_1-B_2)+v(B_1+B_2) &= s_1\{D_1-\beta p[D_1-(1-p)x_0]\}\frac{Q_1}{D_1+px_0} \\ &+v\{px_0+\beta p[D_1-(1-p)x_0]\}\frac{Q_1}{D_1+px_0} \end{aligned} \tag{7}$$

2. The handling cost of the defective items returned is $wB_2 = w\frac{\beta p[D_1-(1-p)x_0]Q_1}{D_1+px_0}$.
3. The procurement cost is $A + (c-k)Q_1$.
4. The screening cost is dQ_1.
5. The inventory holding cost is

$$\frac{r(c-k)}{2}[(Q_1+B_1)T_1 + B_2T_1] = \frac{r(c-k)}{2}\left(Q_1 + \frac{px_0Q_1}{D_1+px_0} + \frac{\beta p[D_1-(1-p)x_0]Q_1}{D_1+px_0}\right)\frac{Q_1}{D_1+px_0}.$$

According to the above revenue and cost, we can obtain

$$\begin{aligned} TP_s^{(1)}(Q_1) &= s_1\{D_1 - \beta p[D_1-(1-p)x_0]\}\frac{Q_1}{D_1+px_0} + v\{px_0 + \beta p[D_1-(1-p)x_0]\}\frac{Q_1}{D_1+px_0} \\ &\quad - \left[w\frac{\beta p[D_1-(1-p)x_0]Q_1}{D_1+px_0} + A + (c-k)Q_1 + dQ_1 \right. \\ &\quad \left. + \frac{r(c-k)}{2}\left(Q_1 + \frac{px_0Q_1}{D_1+px_0} + \frac{\beta p[D_1-(1-p)x_0]Q_1}{D_1+px_0}\right)\frac{Q_1}{D_1+px_0}\right] \end{aligned} \quad (8)$$

Under policy N, the retailer's total profit during the special sale time is

$$TP_n^{(1)}(Q_1) = TP_1(Q_0^*) + \left[\frac{Q_1}{D_1+px_0} - \frac{(1-p)Q_0^*}{D_0}\right]\frac{TPU(Q_0^*)}{N} \quad (9)$$

Where $TP_1(Q_0^*) = s_0(1-p)Q_0^* + vpQ_0^* - \left\{A + (c-k)Q_0^* + dQ_0^* + r(c-k)\left[\frac{(Q_0^*)^2(1-p)^2}{2D_0} + \frac{(Q_0^*)^2p}{x_0}\right]\right\}$.

Hence,

$$\begin{aligned} D^{(1)}(Q_1) &= TP_s^{(1)}(Q_1) - TP_n^{(1)}(Q_1) \\ &= s_1\{D_1 - \beta p[D_1-(1-p)x_0]\}\frac{Q_1}{D_1+px_0} + v\{px_0 + \beta p[D_1-(1-p)x_0]\}\frac{Q_1}{D_1+px_0} \\ &\quad - \left[w\frac{\beta p[D_1-(1-p)x_0]Q_1}{D_1+px_0} + A + (c-k)Q_1 + dQ_1 + \frac{r(c-k)}{2}\left(Q_1 + \frac{px_0Q_1}{D_1+px_0}\right.\right. \\ &\quad \left.\left. + \frac{\beta p[D_1-(1-p)x_0]Q_1}{D_1+px_0}\right)\frac{Q_1}{D_1+px_0}\right] - TP_1(Q_0^*) - \left[\frac{Q_1}{D_1+px_0} - \frac{(1-p)Q_0^*}{D_0}\right]\frac{TPU(Q_0^*)}{N} \end{aligned} \quad (10)$$

3.2 Model II: The retailer increases the screening rate through investment

The second strategy for the retailer is to increase the daily screening rate to screen out enough good items to meet the increased demand rate D_1. It can be achieved by investing new facilities or hiring more employees. Consistent with Hauck and Vörös[9], we let $z = D_1/x$ and let $g(z)$ denote the daily cost of increasing the screening rate from the current x_0 to x. $g(z)$ is convex and decreasing in z, moreover $g(z_{max}) = 0$.

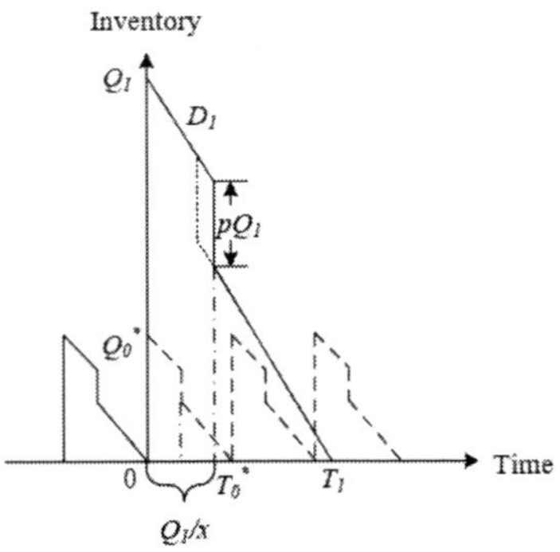

Figure 3: Inventory Level Over Time When The Retailer Increases The Daily Screening Rate

The behavior of the retailer's inventory level is shown in Figure 3. The retailer chooses to increase the daily screening rate through investment at least to meet the demand during the screening period. His action of further increasing the screening rate can lead the inventory system to decrease the screening time Q_1/x and decrease the inventory holding cost (it is shown with the dashed lines). While at the same time, the retailer has to incur more investment cost.

The special sale time can be written as $T_1 = (1-p)Q_1/D_1$.

The cost of increasing the screening rate from x_0 to x is $\frac{Q_1}{x}g(z) = \frac{Q_1}{D_1}zg(z)$.

Under policy S, the retailer's profit during the special sale time can be expressed as

$$TP_s^{(2)}(Q_1,z) = s_1(1-p)Q_1 + vpQ_1 - \Big\{\frac{Q_1}{D_1}zg(z) + A + (c-k)Q_1 + dQ_1 \\ +r(c-k)\left[\frac{{Q_1}^2(1-p)^2}{2D_1} + \frac{{Q_1}^2pz}{D_1}\right]\Big\} \tag{11}$$

$TP_s^{(2)}(Q_1,z)$ consists of five elements: the sale revenue, the cost of increasing the screening rate from x_0 to x, the procurement cost, the screening cost and the inventory holding cost.

Under policy N, the retailer's total profit during the special sale time is

$$TP_n^{(2)}(Q_1) = TP_1(Q_0^*) + \left[\frac{(1-p)Q_1}{D_1} - \frac{(1-p)Q_0^*}{D_0}\right]\frac{TPU(Q_0^*)}{N} \tag{12}$$

Hence,

$$\begin{aligned} D^{(2)}(Q_1,z) &= TP_s^{(2)}(Q_1,z) - TP_n^{(2)}(Q_1) \\ &= s_1(1-p)Q_1 + vpQ_1 - \Big\{\frac{Q_1}{D_1}zg(z) + A + (c-k)Q_1 + dQ_1 \\ &\quad +r(c-k)\left[\frac{{Q_1}^2(1-p)^2}{2D_1} + \frac{{Q_1}^2pz}{D_1}\right]\Big\} - TP_1(Q_0^*) - \left[\frac{(1-p)Q_1}{D_1} - \frac{(1-p)Q_0^*}{D_0}\right]\frac{TPU(Q_0^*)}{N} \end{aligned} \tag{13}$$

Subject to

$$z = D_1/x \leq 1-p$$

4 ANALYSIS

4.1 Analysis of Model I

Taking the first derivative of $D^{(1)}(Q_1)$ with respect to Q_1 and set the result to zero. By solving the equation, the optimal solution Q_1^* can be obtained as follows:

$$Q_1^* = \frac{s_1D_1 + vpx_0 - (s_1+w-v)\beta p[D_1-(1-p)x_0] - (c-k+d)(D_1+px_0) - TPU(Q_0^*)/N}{r(c-k)\left(1+\frac{px_0}{D_1+px_0}+\frac{\beta p[D_1-(1-p)x_0]}{D_1+px_0}\right)} \tag{14}$$

Taking the second derivative of $D^{(1)}(Q_1)$ with respect to Q_1, we obtain $\frac{\mathrm{d}^2D^{(1)}(Q_1)}{\mathrm{d}(Q_1)^2} < 0$. As $D^{(1)}(Q_1)$ is concave with respect to Q_1, Q_1^* is the unique value that maximizes $D^{(1)}(Q_1)$.

4.2 Analysis of Model II

Consistent with Hauck and Vörös[9], we let $g(z) = \frac{M}{z^2}$ for $z < 1-p < z_{max}$ and $g(z_{max}) = 0$. Substituting $g(z) = \frac{M}{z^2}$ into Eq. (13) yields

$$D^{(2)}(Q_1, z) = s_1(1-p)Q_1 + vpQ_1 - \left\{\frac{Q_1}{D_1}\frac{M}{z} + A + (c-k)Q_1 + dQ_1 + r(c-k)\left[\frac{{Q_1}^2(1-p)^2}{2D_1} + \frac{{Q_1}^2pz}{D_1}\right]\right\} - TP_1(Q_0^*) - \left[\frac{(1-p)Q_1}{D_1} - \frac{(1-p)Q_0^*}{D_0}\right]\frac{TPU(Q_0^*)}{N} \quad (15)$$

Proposition 1. When $g(z) = \frac{M}{z^2}$, $D^{(2)}(Q_1, z)$ has at most two stationary points which simultaneously satisfy $\frac{\partial D^{(2)}(Q_1,z)}{\partial Q_1} = 0$ and $\frac{\partial D^{(2)}(Q_1,z)}{\partial z} = 0$.

5 NUMERICAL RESULTS

Parameter values: $A = 100$, $r = 0.2$, $c = 10$, $s_0 = 20$, $x_0 = 15$, $d = 1$, $v = 9$, $N = 365$, $k = 2$, $s_1 = 18$, $D_0 = 10$, $D_1 = 50$ ($D_i = 410 - 20s_i$).

Based on the parameter values, we implement the sensitivity analysis as follows.

Table 1: Effects of an increase in β on the optimal solution for Model I ($w = 0.5$, $p = 0.05$)

β	Q_0^*	T_0^*	Q_1^*	T_1^*	$D^{(1)}(Q_1^*)$
0.2	32.1219	3.0516	258.8489	5.1005	895.0063
0.4	32.1219	3.0516	255.0135	5.0249	870.4822
0.6	32.1219	3.0516	251.2303	4.9504	846.4287

Table 2: Effects of an increase in M on the optimal solution for Model II ($p = 0.05$)

M	Q_0^*	T_0^*	Q_1^*	z^*	T_1^*	$D^{(2)}(Q_1^*, z^*)$
2	32.1219	3.0516	267.0721	0.3060	5.0744	880.6419
4	32.1219	3.0516	261.6429	0.4372	4.9712	852.1571
6	32.1219	3.0516	257.5532	0.5396	4.8935	830.8871

Table 3: Effects of an increase in p on the optimal solutions for Model I and Model II ($\beta = 0.5$, $M = 5$)

p	Q_0^*	Model I		Model II		
		Q_1^*	$D^{(1)}(Q_1^*)$	Q_1^*	z^*	$D^{(2)}(Q_1^*, z^*)$
0.04	32.0268	255.3291	873.3904	258.7546	0.5495	850.3139
0.06	32.2145	250.9358	843.6969	260.4005	0.4472	832.4449
0.08	32.3920	246.6770	815.1627	262.6198	0.3857	817.3182
0.10	32.5587	242.5511	787.7654	265.2226	0.3433	803.7945

Table 1 shows that when β increases, for model 1, the special order quantity decreases and the extra profit also decreases.

Table 2 shows that when M increases, for model 2, the special order quantity decreases and the extra profit also decreases.

From Table 3, for model 1, as p increases, the special order quantity decreases and the extra profit decreases. While for model 2, as p increases, the special order quantity increases and the screening rate during the promotion period increases (z^* decreases), and the extra profit also decreases. So when p is large, it is more beneficial for the retailer to adopt strategy 2 under a promotion.

6 CONCLUSIONS

This paper investigates the inventory models for defective items under temporary price reduction for a situation in which the screening good items cannot satisfy the increased consumer demand rate. Two strategies are given for the retailer to deal with this situation, namely, incurring defect sales return and increasing the screening rate. Then two models are developed to determine the retailer's special order quantity and screening rate. Numerical examples indicate that when the percentage of defective items is large, it is more beneficial for the retailer to choose to increase the screening rate. Furthermore, a larger p motivates the retailer to invest more to increase the screening rate under a promotion. For future research work, it is meaningful to consider the production-delivery policy for a two-stage supply chain with defective items reworked.

7 REFERENCES

[1] **Sarker B. R., Al Kindi M. 2006.** Optimal ordering policies in response to a discount offer. *International Journal of Production Economics*, 100(2), pp 195-211.

[2] **Taleizadeh A. A., Mohammadi B., Cárdenas-Barrón L. E., Samimi H. 2013.** An EOQ model for perishable product with special sale and shortage. *International Journal of Production Economics*, 145(1), pp 318-338.

[3] **Abad P. L. 2003.** Optimal price and lot size when the supplier offers a temporary price reduction over an interval. *Computers & Operations Research*, 30(1), pp 63-74.

[4] **Ardalan A. 1995.** A comparative analysis of approaches for determining optimal price and order quantity when a sale increases demand. *European Journal of Operational Research*, 84(2), pp 416-430.

[5] **Arcelus F., Shah N. H., Srinivasan G. 2001.** Retailer's response to special sales: price discount vs. trade credit. *Omega*, 29(5), pp 417-428.

[6] **Ramasesh R. V. 2010.** Lot-sizing decisions under limited-time price incentives: A review. *Omega*, 38(3), pp 118-135.

[7] **Salameh M., Jaber M. 2000.** Economic production quantity model for items with imperfect quality. *International journal of production economics*, 64(1), pp 59-64.

[8] **Khan M., Jaber M. Y., Bonney M. 2011.** An economic order quantity (EOQ) for items with imperfect quality and inspection errors. *International Journal of Production Economics*, 133(1), pp 113-118.

[9] **Hauck Z., Vörös J. 2015.** Lot sizing in case of defective items with investments to increase the speed of quality control. *Omega*, 52, pp 180-189.

[10] **Hsu W.K. K., Yu H.F. 2009.** EOQ model for imperfective items under a one-time-only discount. *Omega*, 37(5), pp 1018-1026.

A DISCRETE-EVENT SIMULATION APPROACH TO PREDICT TRANSPORTATION COST IN A MULTI-OBJECTIVE SUPPLY CHAIN NETWORK DESIGN CONSIDERING DISRUPTION AND PRICING

Seyed Mohamad Mahdi Kazemi[1*] and Peyman Taki[2] and Sayed Abbas Kazemi[3]
[1,2,3]Department of Industrial Engineering
Damavand Branch, Islamic Azad University, Tehran, Iran
kazemi@damavandiau.ac.ir, peymantaki@gmail.com,s.abbas.kazemi@gmail.com

ABSTRACT

A well-designed network can has enormous effects on the performance, efficiency and effectiveness of the supply chains to achieve the expectable profit and meet the customer needs. In this paper, a multi-objective multi-echelon supply chain network design with regard to pricing and disruption is considered. Transportation cost for each vehicle is defined as a dynamic stochastic function not a parameter. Therefore, discrete event simulation is used to evaluate the transportation cost. Due to the important role of the concept of risk and disruption in supply chain network design, minimization of risks along with maximization of the total profit is defined as objective function. Demand of the supply chain is considered as a linear function which depends on the product price. At last, the risk-pricing-simulation-supply chain network design is solved with NSGAII algorithm. Results are validated with MOSA.

Keywords: supply chain network design; discrete event simulation; disruption; pricing

1 INTRODUCTION

1.1. Supply chain network design

Nowadays, many cases such as disruption, disasters, Perishable materials problem and some other topics are important for supply chain and pricing (Kazemi, Taki, & Kazemi, 2012). Today's competitive business world creates huge impacts on markets, the customers' expectations and replaces the observations from individual firms to supply chains (Farahani, Rezapour, Drezner, & Fallah, 2014). The aims of the network design of the supply chains are often defined as maximization goals such as benefits or service level or minimization ones as cost, risks or environmental and social damages. Thus, to increase the efficiency more than before, companies have to choose the appropriate and practical models for their networks and promote their cooperating. Supply chain network design is totally defined as to form facilities such as suppliers, manufacturers, distribution centers, retailers and their markets connected together considering their systems, sub-systems, operations and relations(Shapiro, 2006).

1.2. Discrete event simulation (DES)

DES is a powerful tool conceptualized around events. In a DES, the model moves forward through the time to further points when the next events are incident. Therefore, it can provide more flexibility to the supply chain network design with respect to analyzing input data and model the effects of capacity constraints and queues through representing the characteristics which are not generally observed in the mathematical modeling(Karnon & Afzali, 2014). Due to new needs of supply systems,

* Corresponding Author

new solutions must be developed and evaluated. One of the important tools is discrete simulation1 model(Taki & Kazemi, 2013). Simulation as a modeling tool is widely used in operational research, where computer models are employed to know and experiment with a system (Pidd, 1998). Companies and manufacturers can use DES to perform their analyses to estimate the impact of their decisions performance on the overall before they made any real system changes (Kazemi & Taki, 2012). They investigated the impact of applying integrated inventory-distribution strategy.

1.5. Motivation

In this survey, the concepts of disruption and coordinating in the field of pricing in supply chains along with DES are integrated to develop the field of designing supply chain networks. However a wide range of researches have considered the concept of supply chain network design and modeling coordination on pricing and disruption, but according to the best of our knowledge and the related literature, there are no related researches in the mentioned area. Yu, Sun, & Chen (2007) has used simulation to analyze supply disruption whether coordination is formed in the supply chain or not.

Using pricing concept is an economic effort to increase the profit generated from a limited supply of supply chain assets. To increase the total margin earned from these assets, managers must use all available levers, including price. Therefore, the adjustment between the pricing and available supply of assets can maximize the total profit of the supply chains.

To overcome the derived gap from the literature, the proposed paper presents an applicable multi-echelon multi-objective supply chain network design which is able to 1) consider the concept of pricing to the proposed area, 2) evaluate the transportation costs by discrete event simulation and 3) bring the disruption as one of the common concepts of risk as objective function along with the total revenue of the supply chain.

2 PROBLEM FORMULATION

2.1. Problem definition

A multi-echelon supply chain including plants and retailers is considered. There are some predefined points for plants and retailers. Plants produce the product internally under consideration of their capacity limits. Plants are single product and all vehicles using for transportation activities are as the same. The average cost of the transportation activities depends on some variables such as travel time, mean of waiting time, and the services of loading and unloading. Therefore, to estimate the transportation activities cost, DES is applied. Four events are defined which are as follows. 1) The event of traveling from plant i to retailer j, 2) the event of traveling from retailer j to plant i, 3) end of unloading event for retailer j and 4) end of loading event for plant i.

Notations M and N are defined as the number of predefined locations for plants and retailers, respectively. Therefore, it can be concluded easily that $M \times N$ different cases are created between plants and retailers. Other assumptions of the proposed problem are as follows. Each retailer can be supplied by only a plant.

A cooperative pricing policy is considered for the proposed supply chain network design. Therefore, products are sold by a unique price.

Inventory policies are implemented in the proposed supply chain. Therefore, each retailer orders products to plants based on the assigned inventory policies for the proposed supply chain.

2.2. Model formulation

Notations using in the model formulation is as the following.

Indices	
i	Candidate locations for plants $i = 1,\ldots,N$
j	Candidate locations for retailers $j = 1,\ldots,M$
Parameters	
C_i	Per unit production cost at plant i
B_i	Fixed cost of opening plant i
CAP_i	Maximum capacity of production at plant i
D_j	Demand of retailer j
h_j	Per unit holding cost for each product at retailer j
A_j	Ordering fixed cost for retailer j
E_j	Fixed cost of opening retailer j
S	Maximum capacity of transportation activities
Variables	
U_i	1 if a plant is open and located in candidate location i ; 0 otherwise
r_i	Disruption probability for plant i
V_j	1 if an open retailer is located in candidate location j ; 0 otherwise
ρ_j	Disruption probability for retailer j
Q_j	Order quantity of retailer j
M_{ij}	Unit transportation cost from plant i to retailer j
Ψ_{ij}	Disruption probability of transportation from plant i to retailer j
X_{ij}	1 if plant i is connected to retailer j ; 0 otherwise
P	Per unit selling price of each product

$$Max\, f_1 = \sum_{j=1}^{N} PD_jV_j - \sum_{i=1}^{M}\sum_{j=1}^{N} C_iD_jX_{ij}U_iV_j - \sum_{j=1}^{N} h_j \frac{Q_j}{2} V_j - \sum_{j=1}^{N} A_j \frac{D_j}{Q_j} V_j - \sum_{j=1}^{N} E_jV_j - \sum_{i=1}^{M} B_iU_i - \sum_{i=1}^{M}\sum_{j=1}^{N} M_{ij} \left\lceil \frac{Q_j}{Q_r} \right\rceil \frac{D_j}{Q_j} X_{ij}U_iV_j \quad (1)$$

$$Min\, f_2 = \sum_{i=1}^{M} r_iU_i + \sum_{j=1}^{N} \rho_jV_j + \sum_{i=1}^{M}\sum_{j=1}^{N} \Psi_{ij}X_{ij}U_iV_j \quad (2)$$

$$\sum_{i=1}^{M} X_{ij} \le V_j \qquad \forall j \qquad (3)$$

$$X_{ij} \le U_i \qquad \forall i,j \qquad (4)$$

$$\sum_{j=1}^{N} Q_jX_{ij}U_iV_j \le CAP_i \qquad \forall i \qquad (5)$$

$$V_j, U_i, X_{ij} \in \{0,1\} \qquad \forall i,j \qquad (6)$$

$$P, Q_j \ge 0 \qquad \forall j \qquad (7)$$

Due to above notations, mathematical formulation of the proposed model is presented as the above.

Equation (1) as the first objective function of the mathematical model maximizes the benefit of the supply chain according to predefined strategies and assumptions. The first term of the first objective function is the revenue of selling products in supply chain. The second term illustrates the production costs of open manufacturer centers. The third one shows inventory costs in open retailers. The fourth one is considered to evaluate the ordering cost for any open retailer. The fifth term shows the total transportation cost based on the number of used vehicles and the output results of DES. The last two terms show the setting up costs for open suppliers and retailers. Equation (2) as the second objective function minimizes disruption risk of the supply chain. The first term is equal to the total probability of disruption for open manufacturers and the second term is equal to the total probability of disruption for open retailers and the third one is defined for the total probability of disruption of relations between open manufacturers and retailers. Equation (3) shows the relationship between plant i and retailer j if they are open. Therefore, it means each retailer can only be supplied by a plant but each plant can supplies multiple retailers. Equation (4) shows that each plant without any connection should not be open. Equation (5) demonstrates the capacity constraint for plants. Equation (6) presents binary variables and Equation (7) shows the non-negative ones.

Demand of the supply chain is considered as a deterministic function which depends on the product price. Equation (8) shows the demand-price function of the supply chain.

$$D_j = \alpha - \beta P \qquad \forall j \qquad (8)$$

Where, notation β presents the price elasticity for customers. The market scale of products is depicted by notation α depicts.

2.3. DES modeling

To evaluate the transportation costs between the facilities of the supply chain, DES is implemented. Therefore, to reach the amounts of unit transportation cost from plant i to retailer j (parameter M_{ij}), four events are defined. These events are as follows; 1) End of loading event in plants, 2) end of traveling event from plant i to retailer j, 3) end of unloading event in retailers, and 4) end of traveling event from retailer j to plant i. Flowcharts of the defined events are demonstrated by Fig.s 1 to 4.

3 SOLUTION APPROACH

The presented mathematical model is a mixed integer non-linear programming model that cannot be solved by the exact solving methods or the time of solving is too huge. Thus, Non-dominated Sorting Genetic Algorithm-II (NSGA-II) is applied to derive a Pareto front of near optimal solutions for the proposed mathematical model which is coded by Visual Basic 2012. An operating system with Core i5-2.5 GHz and 8 GB Ram is used to run the code. To validate and verify the derived Pareto front, Multi-Objective Simulated Annealing (MOSA) is implemented. Fig. 5 shows the flowchart of NSGA-II.

To model the proposed problem with NSGA-II, 100 chromosomes are considered as initial population. Also, the probability for crossover is assumed equal to 0.9. The probability of mutation is considered equal to 0.1.

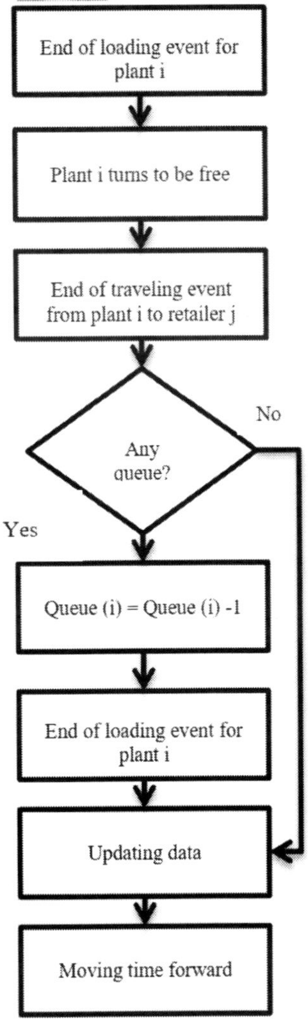

Fig.1. Flowchart of DES model for end of loading event

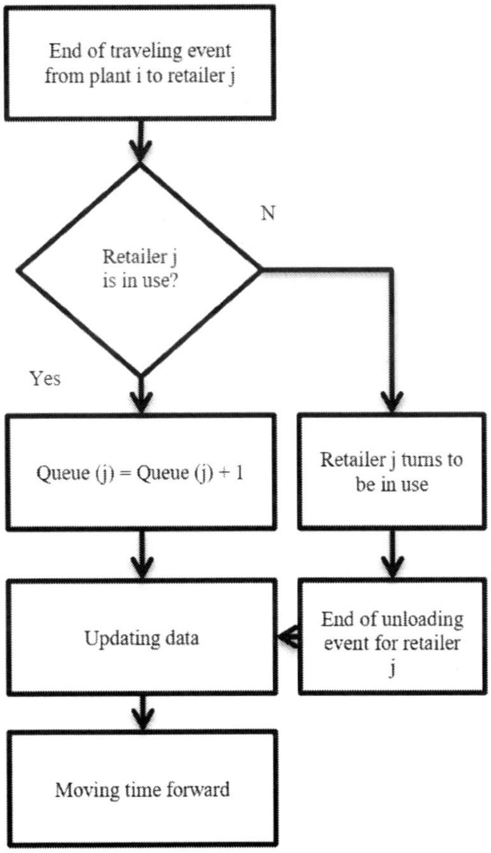

Fig.2. Flowchart of DES model for end of traveling event from plant i to retailer j

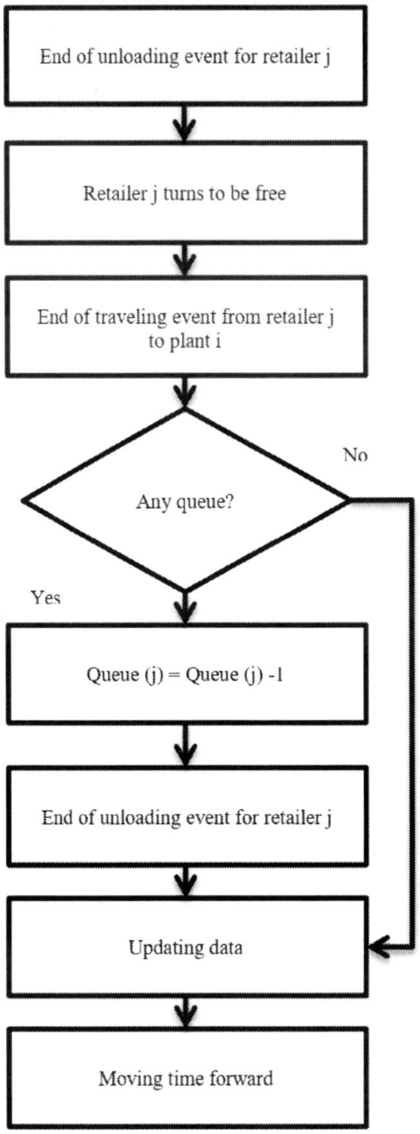

Fig.3. Flowchart of DES model for end of unloading event

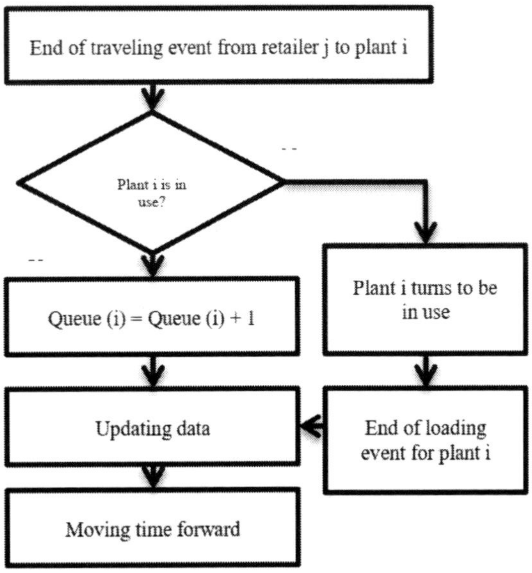

Fig.4. Flowchart of DES model for end of traveling event from retailer j to plant i

NUMERICAL RESULTS

At the first step, to determine the amount of unit transportation cost from plant i to retailer j (M_{ij}) a DES is used. In this example, we assumed that loading time in all manufacturers is the same and has normal distribution function with the mean of 20 minutes and the variance of 5 minutes. In a similar way, unloading time in all retailers is the same and has normal distribution function with the mean of 15 minutes and the variance of 3 minutes. Also, traveling time is considered as five different variables by uniform distribution function. These five groups are defined as follows. ~ U (120 min, 150 min), ~ U (100 min, 130 min), ~ U (70 min, 90 min), ~ U (30 min, 50 min) and ~ U (90 min, 110 min).

Unit loading cost per minute is assumed equal to 20, unit unloading cost per minute is considered equal to 10. Unit traveling cost per minute and unit waiting cost per minute are determined equal to 5 and 2, respectively.

The fixed cost of transportation between the facilities is defined in the range of 720 to 1370. Also, a set of 20 pre-defined points are considered for manufacturers and retailers. Maximum capacity of each vehicle is 121 products.

As the second step, the mathematical model is coded by NSGA-II and MOSA. The derived solutions are presented in Table

To model the proposed problem with NSGA-II, 100 chromosomes are considered as initial population. Also, the probability for crossover is assumed equal to 0.9. The probability of mutation is considered equal to 0.1.

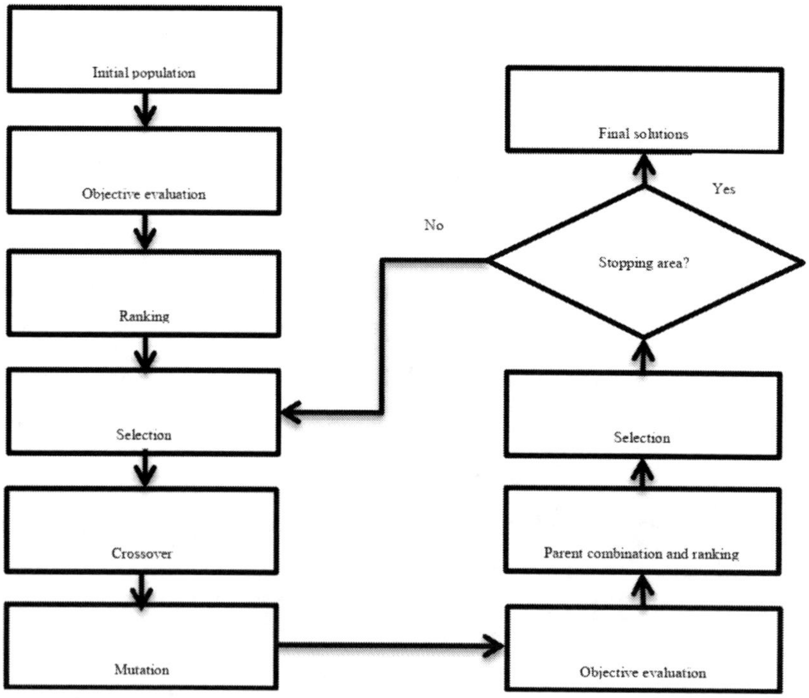

Fig.5. Flowchart of NSGA-II

4 NUMERICAL RESULTS

At the first step, to determine the amount of unit transportation cost from plant i to retailer j (M_{ij}) a DES is used. In this example, we assumed that loading time in all manufacturers is the same and has normal distribution function with the mean of 20 minutes and the variance of 5 minutes. In a similar way, unloading time in all retailers is the same and has normal distribution function with the mean of 15 minutes and the variance of 3 minutes. Also, traveling time is considered as five different variables by uniform distribution function. These five groups are defined as follows. ~ U (120 min, 150 min), ~ U (100 min, 130 min), ~ U (70 min, 90 min), ~ U (30 min, 50 min) and ~ U (90 min, 110 min).

Unit loading cost per minute is assumed equal to 20, unit unloading cost per minute is considered equal to 10. Unit traveling cost per minute and unit waiting cost per minute are determined equal to 5 and 2, respectively.

The fixed cost of transportation between the facilities is defined in the range of 720 to 1370. Also, a set of 20 pre-defined points are considered for manufacturers and retailers. Maximum capacity of each vehicle is 121 products.

As the second step, the mathematical model is coded by NSGA-II and MOSA. The derived solutions are presented in Table1.

P_2	NSGA-II	f_1	601887.4	603093.4	634887.2	637101.7	663248	702851.3	----	----	25
		f_2	15.8	15.9	16.4	16.7	16.8	17.3	----	----	
		P	106.1	104.5	119.2	96.0	154.6	170.8	----	----	
	MOSA	f_1	565312.2	578135.2	584304.3	609630.7	611738.1	687925.9	----	----	33
		f_2	7.2	8.3	10.8	10.9	11	12.2	----	----	
		P	31.5	69.3	54.6	26.1	91.8	15.5	----	----	
P_3	NSGA-II	f_1	396756.3	399223.1	405512.2	450273.8	482768.2	521023.6	584710.7	594003.3	29
		f_2	3.2	4.4	4.5	5.4	7.3	7.5	7.7	9.4	
		P	49.6	37.8	36.6	61.1	97.4	36.3	176.7	59.9	
	MOSA	f_1	397707.5	406608.4	412876.6	430778.3	479145.7	532261.0	550660.7	599646.8	41
		f_2	6.7	7.7	8.1	8.2	9.2	9.4	9.6	10.2	
		P	88.0	54.1	86.4	87.5	31.9	39.7	135.4	71.7	
P_4	NSGA-II	f_1	296912.2	315581.5	331286.0	351287.0	450840.6	----	----	----	32
		f_2	3.4	4.3	4.7	5.7	5.8	----	----	----	
		P	58.1	77.0	88.7	18.6	70.8	----	----	----	
	MOSA	f_1	236547.8	259125.7	266585.1	328585.7	375409.2	479889.3	----	----	33
		f_2	4.3	4.8	6.2	6.7	9.3	9.4	----	----	
		P	87.1	96.8	22.9	35.9	58.3	21.1	----	----	
P_5	NSGA-II	f_1	308037.4	336992.8	363493.7	438118.2	----	----	----	----	26
		f_2	4.3	4.4	5.3	5.5	----	----	----	----	
		P	86.0	63.2	54.9	32.8	----	----	----	----	
	MOSA	f_1	328080.3	374994.0	391289.0	429305.1	----	----	----	----	28
		f_2	4.7	5.2	5.4	5.6	----	----	----	----	
		P	71.0	90.6	36.6	49.9	----	----	----	----	

* Non-dominated solutions Prob.: Problem

5 CONCLUSIONS

In his paper, a multi-objective multi-echelon supply chain network with regard to pricing and disruption has been design. It has been considered that transportation cost for each vehicle is a dynamic stochastic function not a parameter. Therefore, discrete event simulation is used to evaluate the transportation cost with regard to four events including loading, unloading, traveling from plant to retailer and vice versa. Maximization of the total profit of the supply chain and minimization of the risk are defined as objective functions. Due to the important role of pricing in supply chains and networks design, the concept of pricing is used to determine the optimal price of products to maximize the total profit of the supply chain. For future research, we suggest to use mathematical optimization models, use DES to simulate the other cost parameters of the model and apply the presented mathematical formulation for real cases as applicable case studies such as supply chains for perishable products or materials.

References

[1] Farahani, R.Z.; Rezapour, S.; Drezner, T.; Fallah, S. (2014): Competitive supply chain network design: An overview of classifications, models, solution techniques and applications. Omega 45(0), 92-118. doi:http://dx.doi.org/10.1016/j.omega.2013.08.006

[2] Shapiro, J.: Modeling the supply chain. Cengage Learning, (2006)

[3] Karnon, J., Haji Ali Afzali, H.: When to Use Discrete Event Simulation (DES) for the Economic Evaluation of Health Technologies? A Review and Critique of the Costs and Benefits of DES. PharmacoEconomics 32(6), 547-558. doi:10.1007/s40273-014-0147-9

[4] Pidd, M. (1998). Computer simulation in management science, John Wiley & Sons, Inc.

[5]Kazemi, S. M. M. and P. Taki (2012). Discrete event simulation of Packed groceries logistics supply system. 2012 4th International Conference on Computer Modeling and Simulation (ICCMS 2012). HongKong, IPCSIT. 22.

[6] Taki, P. and S. M. M. Kazemi (2012). A THREE ECHELON SUPPLY SYSTEM: A DISCRETE EVENT SIMULATION. 2012 4th International Conference on Computer Modeling and Simulation (ICCMS 2012). HongKong, IPCSIT. 22.

[7] Li, X. and Q. Wang (2007). "Coordination mechanisms of supply chain systems." European Journal of Operational Research 179(1): 1-16.

[8] Fugate, B., F. Sahin, et al. (2006). "Supply chain management coordination mechanisms." Journal of Business Logistics 27(2): 129-161.

[9] Xu, L. and B. M. Beamon (2006). "Supply Chain Coordination and Cooperation Mechanisms: An Attribute-Based Approach." Journal of Supply Chain Management 42(1): 4-12.

[10]Sunil, C. (2010). Supply Chain Management, 4/E, Pearson Education India.

[11]Chopra, S., Meindl, P.: Supply chain management. Strategy, planning & operation. In: Das Summa Summarum des Management. pp. 265-275. Springer, (2007)

[12]Xu, N., & Nozick, L. (2009). Modeling supplier selection and the use of option contracts for global supply chain design. Computers & Operations Research,36(10),2786-2800, doi:http://dx.doi.org/10.1016/j.cor.2008.12.013.

[13] Azad, N., Saharidis, G.D., Davoudpour, H., Malekly, H., Yektamaram, S.: Strategies for protecting supply chain networks against facility and transportation disruptions: an improved Benders decomposition approach. Ann Oper Res 210(1), 125-163 (2013). doi:10.1007/s10479-012-1146-x

[14] Azad, N., Davoudpour, H., Saharidis, G.D., Shiripour, M.: A new model to mitigating random disruption risks of facility and transportation in supply chain network design. Int J Adv Manuf Technol 70(9-12), 1757-1774 (2014). doi:10.1007/s00170-013-5404-01

Singh, A., Mishra, P.K., Jain, R., Khurana, M.K.: Design of global supply chain network with operational risks. Int J Adv Manuf Technol 60(1-4), 273-290 (2012). doi:10.1007/s00170-011-3615-9

[15] Fang, H. and R. Xiao (2014). "Cycle quality chain early warning network with e-channel lead time disruption." International Journal of Systems Science: Operations & Logistics(ahead-of-print): 1-21.

[16] Giri, B. and S. Bardhan (2014). "Coordinating a supply chain with backup supplier through buyback contract under supply disruption and uncertain demand." International Journal of Systems Science: Operations & Logistics(ahead-of-print): 1-12.

[17] Xu, M., X. Wang and L. Zhao (2014). "Predicted supply chain resilience based on structural evolution against random supply disruptions." International Journal of Systems Science: Operations & Logistics 1(2): 105-117.

[18] Yu, H., Sun, C., Chen, J.: Simulating the supply disruption for the coordinated supply chain. J. Syst. Sci. Syst. Eng. 16(3), 323-335 (2007). doi:10.1007/s11518-007-5045-7

PRICING STRATEGY OF GENERATING COMPANY UNDER ABSOLUTE AND RELATIVE EMISSION ALLOWANCE ALLOCATION POLICIES

Xiao Ke[1], Zhigui Guan[1*] and Yanchun Pan[2]
[1]China Center for Special Economic Zone Research
Shenzhen University, P.R. CHINA
zhiguiguan@126.com

[2]College of Management
Shenzhen University, P.R. CHINA
panyc@szu.edu.cn

ABSTRACT

Cap-and-trade (C&T) is regarded as one of the most effective ways to control carbon emissions and implemented all over the world. In this paper, two pricing models are proposed for monopoly-type generating company based on different allocation methods of carbon emission allowance, including absolute and relative means. Our research indicates that different allocation policies will observably affect the company's choice of optimum pricing strategy. Comparing with traditional basic pricing model where no emission constraint is introduced, under the conditions of allocating absolute allowance, the optimal product price will increase, and the production quantity will decrease as well as the carbon emission. However, the target emission intensity allocated by the government has significant impact on the optimal pricing strategy, production quantity and carbon emission when relative allowance policy is employed. If it is not set properly, more carbon emission may be produced.

Keywords: Cap-and-Trade, Emission Allowance Allocation, Pricing Strategy

1 INTRODUCTION

Global warming has been recognized worldwide as a major cause for many negative and significant environmental changes [1]. It is also clear that, based on scientific evidence collected so far, the global warming is caused primarily by unrestricted greenhouse gas emissions often due to over-developed human activities [2]. To effectively reduce the greenhouse gas emissions (i.e. carbon dioxide equivalent or CO2-e emission), many mechanisms have been developed. Among which Cap-and-Trade (C&T) is regarded as one of the widely implemented programs based on eco-economic theory [3]. Known as emissions trading, C&T is a market-based policy for controlling CO2-e emission. Under a C&T program, manufacturers or "emission generating companies" (GCs) in a region are allocated emission allowances to offset their pollutant emission. Initial allowance can be grandfathered or auctioned through a regulatory agency, and then be traded later among the generating companies through market transactions [4]. The total of emission allowances issued is restricted by a pre-determined cap, which is gradually tightened over time. Well established C&T systems around the world include European Union Emissions Trading System (EU-ETS), CRC Energy Efficiency Scheme UK (CRC), Western Climate Initiative (WCI), and Tokyo Cap and Trade (TMG). In

* Corresponding Author

China, the government has committed to cut its CO2-e emission per unit of gross domestic product (GDP) by 40% to 45% of 2005 level by 2020 [5], and is aggressively pushing for a compulsory carbon emission reduction program based primarily on administrative penalty and market trade of emission allowances [6]. Shenzhen was chosen as one of the earliest seven experimental cities in China where a C&T program is being established and some empirical studies have been done [7]. Different from the other C&T systems, instead of allocating an absolute emission allowance, Shenzhen allocates a relative emission allowance, i.e. target emission intensity mainly based on production value, to the GCs in the beginning of the year, and determines the actual/absolute emission allowance according to the total production value in the end of the year. By doing this, the government thinks that the C&T scheme will not restrict the development of GCs too much when emission reduction policies are implemented since more emission allowance can be obtained if the GC produces more values. An interesting question is then raised as follows. How do these different emission allowance allocation methods affect the pricing strategy of GCs as well as the total carbon emission?

Some research attention has been paid to the area of emission allowance allocation under C&T mechanism. For example, Maeda [8] estimated the distortion degree of emission allowance on market by constructing multi-agent decision model based on the initial emission allowance allocation method. He [9] analyzed China's initial allocation of carbon emission allowance from the perspective of financial engineering. Carbon option is introduced in his allocation model of initial emission allowance. The author put forward that generating companies can get a certain proportion of emission allowance for free and pay for carbon option. Li and Qu [10] used partial equilibrium analysis to study the economic effects of three different allocation policies of initial emission allowance. They held the opinion that in the initial trading stage of China's carbon emission allowance, the combination of upgradable free allocation and auction should be applied. Luo et al. [11] studied the carbon emission allocation in China's petrochemical industry and pointed out that among the three allocation policies including economic development priority, historical emission priority, and economic and environment balanced policy, the historical emission priority exerted the smallest influence on GDP's total amount. By constructing a dynamic game model in the electric power industry. Lu and Fang [12] concluded that emission allowance allocation policies based on historical emission is beneficial for companies with high carbon emission, while emission allowance allocation policies based on output is beneficial for companies with low carbon emission. Wu et al. [13] analysed the influence of carbon market under different emission allowance allocation policies on multi-regional macro economy as well as key emission-reduction companies through building the CGE model of China. They believed that the combination of free allocation of emission allowance in energy industry and auction of emission allowance in high energy-consuming industry serves as an optimum allocation policies. Fan [14] conducted research on the company's cost and revenue structure under three different allocation methods of initial emission allowance, including grandfathered, auctioned and output based standards. Fan thinks that the major reasons for companies to participate in carbon-emission reduction system include product's price elasticity of demand, price supply elasticity of carbon emission allowance and coal's price elasticity of demand.

It can be seen from the literature review that previous studies mainly focus on selection of different emission allowance allocation policies. However, studies on products' pricing strategy of generating companies under absolute and relative emission allowance allocation policies as well as the factors that influence the pricing strategy of generating companies are quite insufficient. In this paper, two pricing models for monopoly-type generating company based on absolute and relative emission allowance allocation policies are proposed to compare with the basic pricing model where no C&T mechanism is considered. The paper is organized as follows. In section 2, we describe the notations and propose the basic model (M0) and the absolute allowance model (M1). Relative allowance model

(M2) is shown in section 3. Section 4 illustrates the model analysis and finally the conclusions are drawn in section 5.

2 BASIC PRICING MODEL (M0) AND ABSOLUTE ALLOWNACE MODEL (M1)

Suppose a monopoly-type generating company produces a kind of product to satisfy the market. The decision maker has to determine the product price to maximize the revenue. The market demands follow a basic law, i.e. the higher the price is, the lower the demands will be. To simplify our problem, the linear demand function widely used in pricing models is introduced in this paper. The following notations are defined for problem modelling.

a: Market capacity of the product.

b: Demand price elasticity of the product.

p: Price of the product.

P: Price of carbon emission allowance.

A: Absolute emission allowance allocated.

e: Relative emission allowance (target emission intensity) allocated.

c: Production cost for producing per unit of product.

u: Carbon emission for producing per unit of product.

R: Total revenue.

In the case of no carbon market (model M0), the pricing decision problem can be modelled as follows.

$$\begin{aligned} &\max R_0 = (p-c)\cdot(a-bp) \\ &s.t.\ \ p \geq 0 \end{aligned} \tag{M0}$$

Take the first and second derivative of R_0 with respect to *p*, we have:

$$\frac{dR_0}{dp} = a + bc - 2bp\,,\ \ \frac{d^2R_0}{dp^2} = -2b$$

Since the second derivative is negative, there must exist a unique optimum price where the first derivative is equal to zero and the revenue reaches its maximum value. Consequently, we have the following optimum for M0.

$$p_0^* = \frac{a+bc}{2b}\,,\ \ R_0^* = \frac{(a-bc)^2}{4b} \tag{1}$$

Under the C&T mechanism, the generating company has to take the emission cost into consideration. If the total carbon emission is larger than the total emission allowance allocated, the generating company will purchase allowance from the carbon market for emission offsetting. Otherwise, the company can earn more revenue by selling surplus emission allowance to the carbon market. In the model of absolute allowance (M1), the decision problem can be described as follows.

$$\begin{aligned} &\max R_1 = (p-c)\cdot(a-bp) - P\cdot[u\cdot(a-bp) - A] \\ &s.t.\ \ p \geq 0 \end{aligned} \tag{M1}$$

Similarly, we can obtain the following optimum solution for model M1.

$$p_1^* = \frac{a+bc+buP}{2b},\ R_1^* = \frac{(a-bc)^2 + bP\cdot[4A - u\cdot(2a-2bc-buP)]}{4b} \tag{2}$$

3 RELATIVE ALLOWNACE MODEL (M2)

Under the condition of allocating relative emission allowance (target emission intensity), the pricing model is described as follows.

$$\begin{aligned} &\max R_2 = (p-c)\cdot(a-bp) - P\cdot[u\cdot(a-bp) - ep(a-bp)] \\ &s.t.\ p \geq 0 \end{aligned} \tag{M2}$$

The following optimum solution for model M2 can then be obtained.

$$p_2^* = \frac{a+bc+P(ae+bu)}{2b(1+eP)},\ R_2^* = \frac{[(a-bc)+P(ae-bu)]^2}{4b(1+eP)} \tag{3}$$

4 MODEL ANALYSIS

According to the above optimum solutions of model M0, M1 and M2 shown in equations (1), (2) and (3), we have the following propositions.

Proposition 1. $p_1^* > p_0^*$.

Proof: Based on (1) and (2), it is easy to know that $p_1^* - p_0^* = uP/2 > 0$. Proposition 1 is proved.□

From proposition 1 we can see that the oligopoly tends to increase the product price to compensate for the emission cost (*uP*/2) under C&T scheme where absolute emission allowance is allocated. As a result, both the total carbon emission and the production quantity decrease due to the increase of the product price, which indicates that absolute emission allowance allocation policy stimulates the generating company to reduce carbon emission. However, at the same time, emission cost is passed on to the consumers by pricing decision.

Proposition 2. If $u < ce$, $p_2^* < p_0^*$; Otherwise, $p_2^* \geq p_0^*$.

Proof: Based on (1) and (3), we have:

$$p_2^* - p_0^* = \frac{P(u-ce)}{2(1+eP)} \tag{4}$$

If $u < ce$, then $p_2^* - p_0^* < 0$; Otherwise, $p_2^* - p_0^* \geq 0$. Proposition 2 is proved.□

In the point of view of economics, proposition 2 can be explained as follows. Firstly, the government allocates emission allowance according to enterprise's output in model M2. In addition, to gain profit, the product price (*p*) should be larger than the production cost (*c*). Therefore, the minimum emission allowance allocated for producing per unit of product is *ce*. When the carbon emission for producing per unit of product (*u*) is less than the allocated emission allowance (*ce*), the generating company shall obtain benefits by decreasing product price and increasing production quantity, and selling the surplus emission allowance to the carbon market. Otherwise, the best way for the generating company is to increase product price and decrease production quantity so as to make up the emission cost. Furthermore, it can be proved that the absolute value of product price difference ($\Delta = | p_2^* - p_0^* |$) is increasing in carbon price (*P*) since $d\Delta / dP = | u - ce | / [2(1+eP)^2] \geq 0$. Namely, the higher the carbon price is, the larger the product price difference will be. Secondly, the allocation of targeted emission intensity (e) is of great significance. Only $e<u/c$ can stimulate emission reduction behaviour. If $e>u/c$, it shows that the carbon emission allowance allocated by government is more than the carbon emission generated by the company, i.e. the government has given subsidies to companies for producing more and releasing emissions more. Thirdly, for given e, when $u<ce$, it means that the company is a low carbon emission company, then it can benefit from relative emission allowance allocation policies. Otherwise, the company needs to purchase emission allowance from carbon market, which is the original intention of C&T scheme and it makes sense.

Proposition 3. $p_1^* > p_2^*$.

Proof: Based on (2) and (3), we have:

$$p_1^* - p_2^* = \frac{eP(c+uP)}{2(1+eP)} > 0 \tag{5}$$

Proposition 3 is proved.□

Due to the linear demand function, the higher the product price is, the lower the production quantity and carbon emission will be. Consequently, according to the above propositions, we can see that for monopoly-type generating company, absolute emission allowance allocation policy is in favour of emission reduction and stricter than relative emission allowance allocation policy which may lead to higher carbon emission if target emission intensity is not properly set up.

As for the relationships among revenues of model M0, M1 and M2, it is vague because not only enterprise related parameters (u and c) and government related parameters (A and e) but also market related parameters (a, b and P) have significant impacts on the revenues.

5 CONCLUSION

How to make pricing strategy under different emission allowance allocation policies in C&T scheme is very important for monopoly-type generating companies. In this paper, two pricing models based on absolute and relative allowance allocation are proposed and compared with traditional basic pricing model. The optimum solutions under different conditions are obtained. By analysing the pricing strategies, it can be concluded that absolute emission allowance allocation leads to higher product price, lower production quantity and lower carbon emission. However, it is not so clear about the impacts of relative emission allowance allocation on product price and carbon emission. Relative emission allowance can be regarded as a subsidy assigned to generating companies by government. If the target emission intensity is not set properly, it may results in higher carbon emission, i.e. the monopoly-type decision maker may decrease product price and increase production quantity to achieve more subsidy of emission allowance and earn more revenue by selling surplus allowance through the carbon market.

We only take the oligopoly generating company into consideration in this paper. For future research, more attention can be paid to the multiple oligopolies competition problems and perfect competition problems.

6 ACKNOWLEDGEMENT

This research is partially supported by National Natural Science Foundation of China (Grant No. 71472126) and a foundation of Guangdong Province, China (Grant No. YQ2014152).

7 REFERENCES

[1] IPCC, Climate Change 2007: the Physical Science Basis. www.ipcc.ch/publications_and_data/publications_ipcc_fourth_assessment_report_wg1_report_the_physical_science_basis.htm (last accessed 20.9.14.), 2007.

[2] Q. Zhang, X. Min, Y. Chen, "Reflection on the development of low-carbon cities from the perspective of systems engineering", Systems Engineering, 2011, vol. 29, pp. 1-7.

[3] J. H. Dales, Pollution. Toronto: University of Toronto Press, Canada, 1968.

[4] Environmental Protection Agency (EPA), www.epa.gov/capandtrade (last accessed 20.9.14.), 2012.

[5] W. J. Yi, L. L. Zou, J. Guo, K. Wang, Y. M. Wei,. "How can China reach its CO2 intensity reduction targets by 2020? A regional allocation based on equity and development", Energy Policy, 2011, vol. 39, pp. 2407-2415.

[6] EN (Enigin News), China set to Use Iron Hand to Force Energy Efficiency. news appeared on Thursday 6th May 2010, www.enigin.com/about/enigin-plc/news/-/enigin-news-china-set-to-use-ironhand-to-force-energy-efficiency. (last accessed 20.9.14.), 2010.

[7] Shenzhen Emission Exchange, China Shenzhen Emission Exchange (last accessed 20.9.14.). www.szets.com, 2012.

[8] Akira maeda, "The emergence of market power in emission rights markets: the role of initial permit distribution", Journal of Regulatory Economics, 2003, vol. 24, pp. 293-314.

[9] M. S. He, "Research about initial distribution of China carbon emission allowance: from the perspective of finance engineering", Management World, 2011, vol. 11, pp. 172-173.

[10] K. J. Li, R. X. Qu, "Economic impacts and lessons of initial distribution in carbon emission allowance", IEC, 2012, vol.3, pp.21-24.

[11] R. L. Luo, T. J. Fan, S. X. Li, "The study of carbon emission permits allocation in China's petrochemical industry", China Soft Science, 2014, vol. 2, pp. 171-178.

[12] M. Lu, X. N. Fang, "Game analysis on carbon market from allocation of allowance", Chinese Journal of Management Science, 2015, vol. 23, pp. 807-811.

[13] J. Wu, Y. Fan, Y. Xia, "Impacts of initial quota allocation on regional macro-economy and industry competitiveness", Management Review, vol. 27, pp. 18-26.

[14] D. S. Fan, "Analysis on firm's cost and earning with initial distribution methods of tradable carbon permits", Technoeconomics & Management Research, 2014, vol. 4, pp. 59-69.

OPTIMAL 3D COMPACT AS/RS STORAGE RACK DESIGN WITH FULL TURNOVER-BASED STORAGE POLICY CONSIDERING THE OPERATING CHARACTERISTICS OF S/R MACHINE

Peng Yang[1,2*], KaidongYang[1,2], Shilu Wang[1,2], LixinMiao[1,2], and Bin Ye[1,2*]
[1] Research Center for Modern Logistics
Graduate School at Shenzhen, Tsinghua University, Shenzhen, China

[2]Shenzhen Logistics Engineering &Simulation Laboratory
Graduate School at Shenzhen, Tsinghua University, Shenzhen, China
yangpeng.lotus@gmail.com

ABSTRACT

Because three-dimensional (3D) compact (multi-deep) automated storage and retrieval systems (AS/RSs) are widely used in newly designed warehouses, optimal AS/RS designs requiring the minimal operation time are becoming increasingly critical. This paper describes the design of an optimal rack for a 3D compact AS/RS under the full turnover-based storage policy. The acceleration and deceleration characteristics of storage and retrieval (S/R) machine are considered along the horizontal and vertical directions to more accurately represent a real-world 3D compact AS/RS operation scenario. We derive the expected retrieval time for single-command cycles adopting more accurate speed model and develop optimally designing storage rack model for a 3D compact AS/RS. From the model, we can derive the optimal rack dimensions for various scenarios of any given capacity, S/R machine characteristics of the system and the ABC curve skewness characteristics of the stock keeping units (SKUs). Through applying this model, we also show that the full turnover-based storage policy is more efficient than the random policy and use a practical example to demonstrate this finding. Finally, we discuss the shape factors and future research directions for the speed model.

Keywords: compact storage rack design, AS/RS, travel time model, acceleration/deceleration, turnover-based storage

* Corresponding author.

1 INTRODUCTION

In recent decades, newly designed automated storage and retrieval systems (AS/RS) have been increasingly employed by warehouse managers to enhance warehouse operation efficiency. In traditional AS/RSs, multiple aisles are required between pairs of adjacent racks, which is an inefficient use of warehouse space. The three-dimensional (3D) multi-deep compact AS/RS that has been presented by De Koster et al. (2008)[3], Yu and De Koster (2009a)[10] utilizes depth space with the assistance of a depth movement mechanism (DMM) in Figure 1. This warehouse system can reduce the number of aisles required for operation and can increase the storage density. Because of the merits of high-density storage and lower land occupation, the 3D multi-deep compact AS/RS became especially popular in crowded cities.

The design of the 3D compact AS/RS storage rack dimensions is the first decision-making issue for construction. The optimality of the rack significantly affects warehouse efficiency. Under the random, full turnover-based, two-class-based storage policy, De Koster et al. (2008)[3], Yu and De Koster (2009a) [10], and Yu and De Koster (2009b)[11] have explored the optimal storage-rack design problem for 3D compact AS/RSs. However, extant research has always assumed that storage/retrieval (S/R) machines move horizontally and vertically at a constant speed, overlooking the influence of its acceleration and deceleration. Yang et al. (2014)[9] extended the work of De Koster et al. (2008)[3] in investigating the optimal design of a storage rack in a 3D compact AS/RS that uses the random storage policy by considering the acceleration and deceleration of the S/R machine. According to our research, scenarios considering the operational characteristics of the S/R machine under the other two storage policies have not been investigated.

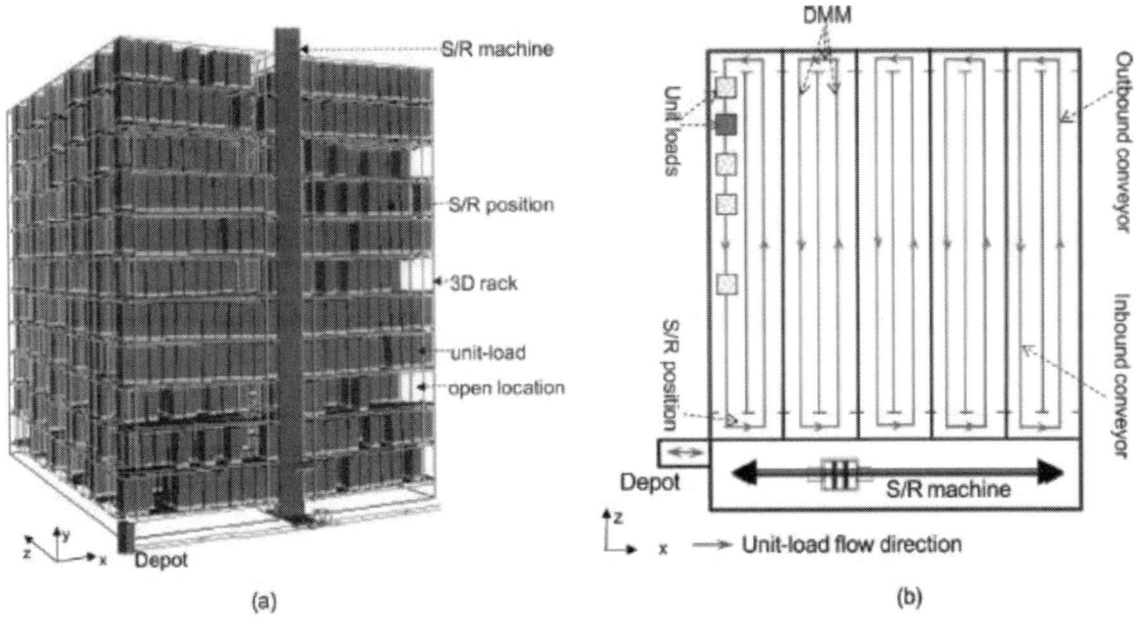

Figure 1: A multi-deep compact AS/RS: (a) overall view and (b) top view (Yu and De Koster (2011)[12])

Previous studies have shown that a full turnover-based storage policy is the most efficient policy. Under this policy, unit loads are assigned to locations in accordance with the unit-load turnover frequency. Unit loads with a high turnover frequency are stored in locations closer to the input/output (I/O) point, whereas those with a lower turnover frequency are stored in farther locations. However, the inaccuracy resulting from ignoring the influence of the acceleration and deceleration of an S/R machine increases under the full turnover-based storage policy when the storage rack is optimally designed. Chang et al. (1995)[2] presented an exponential speed curve approximating the acceleration and deceleration speed characteristics. However, it is inaccurate, particularly when applied under the full turnover-based storage policy. When an item is stored on the basis of its turnover rate, higher weightings is given to items with higher turnovers when determining which locations are closer to I/O point. In the aforementioned approximate speed model, particularly when considering the interval closest to the I/O point, the full turnover-based storage policy causes considerable deviation from the real-life speed characteristics. This policy causes considerable inaccuracy when optimizing a 3D compact AS/RS rack. Following is a detailed example to demonstrate this phenomenon. Given a rack length of 30 m, acceleration and deceleration of an S/R machine of 2 m/s^2(in the horizontal direction), and maximum speed of the S/R machine of 2 m/s, it can be determined that the items with a top 5% turnover rate require1.732s to be stored or retrieved under the full turnover-based storage policy. Under the approximate speed model presented by Chang et al. (1995)[2], storage or retrieval takes 1.040 s, and the relative deviation reaches 40.0%.

Based on this observation, our paper applies a non-approximation speed model to describe the acceleration and deceleration characteristics of S/R machines. Similar to Yu and De Koster (2009a)[10], we consider the expected cycle time of the typical retrieval travel, and the calculated travel time for the storage operation is similar to the retrieval time. With the objective of minimizing the expected retrieval time, we designed a 3D compact AS/RS with optimal functionality when using the full turnover-based storage policy.

The remainder of the paper is organized as follows. We derive the single-command retrieval travel-time model for a 3D compact AS/RS as a function of the rack dimensions considering the speed characteristics of an S/R machine in Section 2. In Section 3, to demonstrate applying the travel time model, we analyze and optimize the rack dimensions to minimize the travel time with a given rack storage capacity by using a non-approximation speed model to describe the real-life S/R operational characteristics. We illustrate the findings with a practical example in Section 4. Finally, in Sections 5 and 6, we draw conclusions, provide a discussion, and identify directions for future research.

2 ASSUMPTIONS AND THE GENERAL MODEL

2.1 Assumptions

The proposed system, which is illustrated in Figure 1, is identical to that of Yu and De Koster (2011)[12]. We adopted most of the major assumptions mentioned by Yu and De Koster (2009a)[10]. However, one assumption that differed concerns the speed characteristics of the S/R machine: the operating characteristics of the S/R machine, which represent real-life operation scenarios, are provided in Figure 2. The S/R machine can simultaneously move in vertical and horizontal directions at a uniform acceleration and deceleration. The other assumptions are identical to those of Yu and De Koster (2009a)[10].

2.2 Problem description and the general model

Assume that a quantity of goods has a total volume of V and that their turnover rate is known. The problem is how to then arrange the length (L), high (H), and depth (P) ($L * H * P = V$) of the storage rack of a 3D compact AS/RS optimally to minimize the average travel time when using a full-turnover storage policy by considering the operational characteristics of the S/R machine.

The goods are stored according to their turnover rate (index coefficientδ). The higher the rate is, the closer the goods are stored to the I/O point (namely a lowerESC). Therefore, the storage location of the same turnover goods can form an isochronous surface.

Suppose the conveyor is rotated at the constant speed s_c in the depth direction. In the horizontal or vertical direction, the stacker accelerates uniformly (the accelerations are a_h, a_v) until reaching the maximum speed(maximum speeds are s_h, s_v), at which it moves only some of the time. Finally, it decelerates uniformly (the absolute value of the accelerations are also a_h, a_v) to a standstill, covering only the length or height required. Alternatively, in the horizontal or vertical direction, if there is no requirement to accelerate to the maximum speed, it accelerates to a certain speed and then decelerates to a standstill. Hwang and Lee (1990)[5] outlined a similar model, which is illustrated in Figure 2.

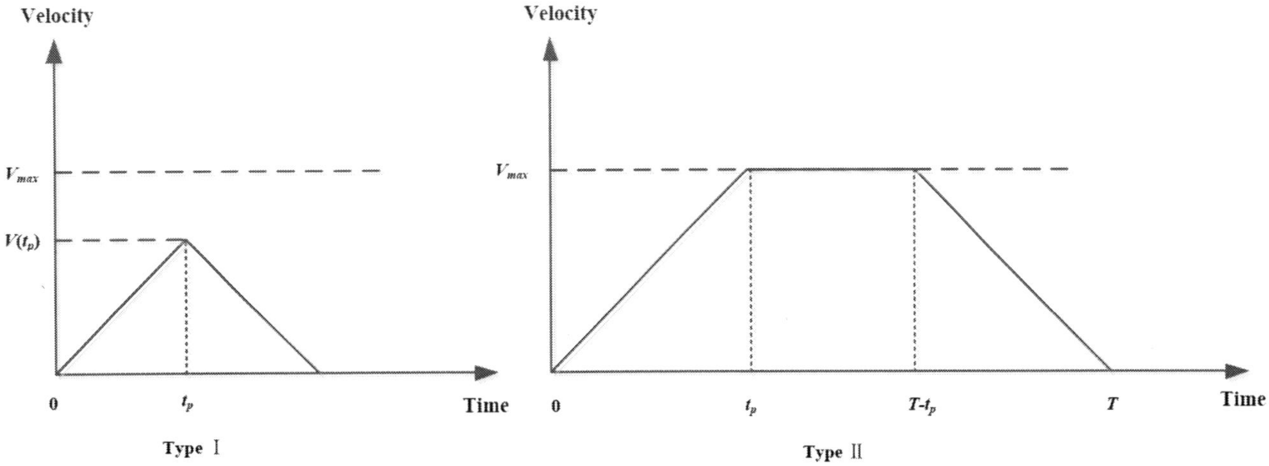

Figure 2: Speed model of the S/R machine travelling in the horizontal or vertical direction(Hwang and Lee (1990)[5])

In the storage space, a certain percentage of goods (such as j, sorted by the turnover rate in descending order), must be stored on the same isochronous surface. Hausman et al. (1976) derived the turnover rate of goods in the jth percentile of the turnover distribution: $\lambda(j) = \sqrt{\frac{2\delta}{K}} j^{\frac{\delta-1}{\delta+1}}$, $(0 < j \leq 1)$, where K is the ratio of the order cost to the holding cost.

The expected single command travel time under a full-turnover storage policy is expressed as follows: $T_T' = \frac{\int_{j=0}^{1} \lambda(j) y(j) dj}{\int_{j=0}^{1} \lambda(j) dj}$, where $y(j)$ represents the distance from I/O point to a location in the jth percentile of the distance distribution.

3 OPTIMAL DIMENSIONS FOR THE COMPACT RACK

When time is employed to measure the rack dimensions, the length, height, and depth of rack in time is denoted as t_h, t_v, and t_c, respectively. Without loss of generality, it can be assumed that the maximum time used for traversing the length is no less than that of the height; that is $t_h \geq t_v$. When the maximum time used in traversing the depth is considered, there are three conditions that must be considered.

Case (1): Depth is the minimum of the three dimensions (as measured in the maximum time used), namely $t_c \leq t_v \leq t_h$;

It can be determined that

$$y(j) = \begin{cases} \sqrt{\frac{a_h}{L}} \sqrt[5]{\frac{jLHP}{a_h a_v s_c}}, & 0 < j < \frac{a_h a_v P^4}{LHs_c{}^4} \\ \sqrt[4]{\frac{jHa_h}{La_v}}, & \frac{a_h a_v P^4}{LHs_c{}^4} < j < \frac{Ha_h}{La_v} \\ \sqrt{j}, & \frac{Ha_h}{La_v} \leq j \leq 1 \end{cases} \tag{1}$$

We adopted the method of Yu and De Koster (2009a) and assumed $s = \frac{2\delta}{\delta+1} \in (0,1]$. This enabled obtaining $\lambda(j) = \sqrt{\frac{2\delta}{K}} \cdot j^{s-1}$, $(0 < j \leq 1)$, then

$$T_T' = \frac{\int_{j=0}^{1} \lambda(j) y(j) dj}{\int_{j=0}^{1} \lambda(j) dj} = s\left[\frac{1}{(4s+1)(5s+1)}\left(\frac{a_h}{L}\right)^{s+\frac{1}{2}}\left(\frac{a_v}{H}\right)^{s}\left(\frac{P}{s_c}\right)^{4s+1} + \frac{2}{(2s+1)(4s+1)}\left(\frac{Ha_h}{La_v}\right)^{s+\frac{1}{2}} + \frac{2}{2s+1}\right] \tag{2}$$

Therefore, the average time used from I/O point to the retrieval location can be calculated as

$$E(W) = T_T' \cdot T_{max} = s\left[\frac{1}{(4s+1)(5s+1)}\left(\frac{32V}{a_h a_v s_c}\right)^{4s+1} \frac{1}{t_h{}^{10s+2} t_v{}^{10s+2}} + \frac{2}{(2s+1)(4s+1)} \frac{t_v{}^{2s+1}}{t_h{}^{2s}} + \frac{2}{2s+1} t_h\right] \tag{3}$$

The average time used from the retrieval location to I/O point can be calculated as $E(U) = T_T'' \cdot T'_{max} = s\left[\frac{2}{(2s+1)(4s+1)} \frac{t_v{}^{2s+1}}{t_h{}^{2s}} + \frac{2}{2s+1} t_h\right]$, where T_T'' and T'_{max} are similar to T_T' and T_{max} in the returning process.

The expected single-command time can be calculated as:

$$ESC = E(W) + E(U) = f_1(t_h, t_v) = s\left[\frac{1}{(4s+1)(5s+1)} T_1{}^{20s+5} \frac{1}{t_h{}^{10s+2} t_v{}^{10s+2}} + \frac{4}{(2s+1)(4s+1)} \frac{t_v{}^{2s+1}}{t_h{}^{2s}} + \frac{4}{2s+1} t_h\right] \tag{4}$$

Because$t_c = \frac{2P}{s_c} = \frac{32V}{a_h a_v s_c t_h^2 t_v^2}$, the value of$T_1 = \sqrt[5]{\frac{32V}{a_h a_v s_c}} > 0$can be obtained. Because$t_c \le t_v \le t_h$, the specific conditions can be obtained: $t_h \ge t_v > 0$ and $t_h^2 t_v^3 \ge T_1^5$, which contains an invisible constraint, $t_h \ge T_1$.

The expected function can then be obtained:

$$f_1(t_h, t_v) = s[\frac{1}{(4s+1)(5s+1)} T_1^{20s+5} \frac{1}{t_h^{10s+2} t_v^{10s+2}} + \frac{4}{(2s+1)(4s+1)} \frac{t_v^{2s+1}}{t_h^{2s}} + \frac{4}{2s+1} t_h] \quad (5)$$

in which $s \in (0,1]$, and $T_1 \in (0, +\infty)$.

Because the binary function $f_1(t_h, t_v)$ is continuously differentiable, the global minimum can only exist where the local minimaare or within the boundary lines.

The boundary lines are the half-line $t_h = t_v (t_h = t_v \ge T_1)$, curve $t_h^2 t_v^3 = T_1^5 (t_h \ge T_1 \text{and } 0 < t_v \le T_1)$, and virtual boundary of the half-line without an extreme point $t_h = +\infty (t_v > 0)$. Therefore, we can determine the feasible region encircled by the three mentioned boundary lines, which is denoted as ***D1***, which is shown in Figure 3.

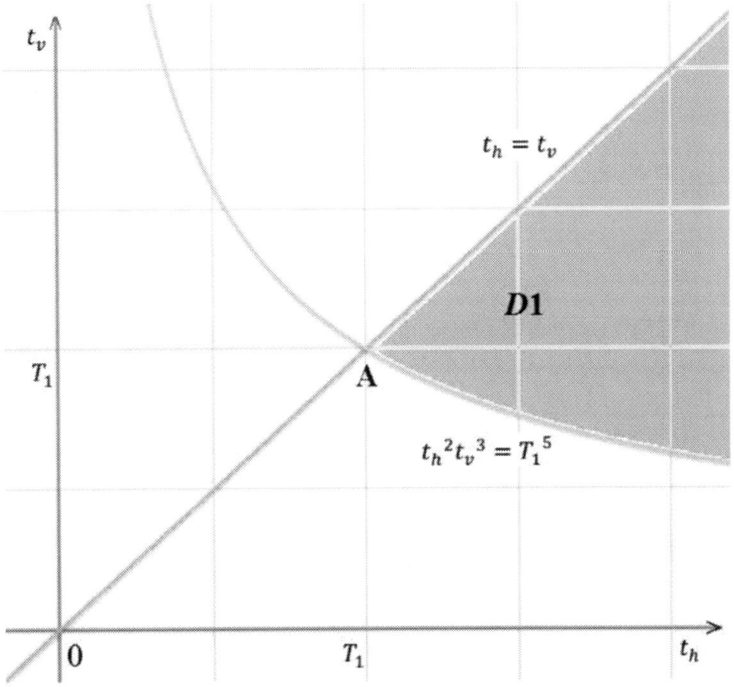

Figure 3: Feasible region under the constraints expressed in Case (1)

First, the global minimum may equal the local minimum. Accordingly, let $f_1{'}_{t_v}(t_h, t_v) = 0$ and $f_2{'}_{t_h}(t_h, t_v) = 0$, and we can get $t_h = [\frac{s(11s+4)2^{\frac{10s+2}{12s+3}}}{(4s+1)(5s+1)}]^{\frac{12s+3}{44s^2+28s+3}} T_1^{\frac{20s+5}{22s+3}}$, $t_v = 2^{-\frac{12s+1}{44s^2+28s+3}}[\frac{s(11s+4)}{(4s+1)(5s+1)}]^{-\frac{10s}{44s^2+28s+3}} T_1^{\frac{20s+5}{22s+3}}$. The stationary point$(t_{h_1}, t_{v_1})$ may be the only possible local minimum point. However, we can obtain a contradiction as follows: if $s \in (0,1)$, then $\frac{t_h}{t_v} = 2^{\frac{1}{2s+1}}[\frac{s(11s+4)}{(4s+1)(5s+1)}]^{\frac{1}{2s+1}} = [\frac{2s(11s+4)}{(4s+1)(5s+1)}]^{\frac{1}{2s+1}} < 1$, (because $\frac{2s(11s+4)}{(4s+1)(5s+1)} - 1 = \frac{(2s+1)(s-1)}{(4s+1)(5s+1)} < 0$) .Furthermore, if $s = 1$, then$t_h = \frac{1}{2^{\frac{1}{25}}} T_1 < T_1$. So regardless of the value of s is in (0,1], the stationary pointcannot exist in region **D1**, thus ensuring that the global minimum of $f_1(t_h, t_v)$ is not the local minimum. Therefore, the global minimum of $f_1(t_h, t_v)$ must exist in the boundarylines.

Second, consider the curve ${t_h}^2 {t_v}^3 = {T_1}^5 (t_h \geq T_1 \text{ and } 0 < t_v \leq T_1)$. It can be determined that $t_v = \frac{T_1^{\frac{5}{3}}}{t_h^{\frac{2}{3}}}$, $t_h \in [T_1, +\infty)$, and $ESC = \frac{s}{(2s+1)(4s+1)(5s+1)}[(22s+5)\frac{T_1^{\frac{5}{3}(2s+1)}}{t_h^{\frac{2}{3}(5s+1)}} + 4(4s+1)(5s+1)t_h]$. Let $m(t_h) = (22s+5)\frac{T_1^{\frac{5}{3}(2s+1)}}{t_h^{\frac{2}{3}(5s+1)}} + 4(4s+1)(5s+1)t_h$, providing $m'(t_h) = -\frac{2}{3}(5s+1)(22s+5)\frac{T_1^{\frac{5}{3}(2s+1)}}{t_h^{\frac{5}{3}(2s+1)}} + 4(4s+1)(5s+1) \geq -\frac{2}{3}(5s+1)(22s+5) + 4(4s+1)(5s+1) > 0$. The minimum of $m_1(t_h)$ in the condition of $t_h = T_1$ can then be obtained. Thus, we can obtain$f_1(t_h, t_v) = f_1(T_1, T_1) = \frac{s(40s+9)}{(4s+1)(5s+1)} T_1$, with $t_h = t_v = t_c = T_1$.

Third, consider the half-line $t_h = t_v (t_h = t_v \geq T_1)$. Similarly, we can get the minimum of $n_1(t_h)$ as $f_1(T_1, T_1)$, under the condition of $t_h = T_1$, namely $t_h = t_v = t_c = T_1$, which is shared with curve ${t_h}^2 {t_v}^3 = {T_1}^5 (t_h \geq T_1 \text{ and } 0 < t_v \leq T_1)$.

Finally, the virtual boundary with the half-line without an extreme point is $t_h = +\infty (t_v > 0)$.There are always $f_1(t_h, t_v) = +\infty$ on this half-line.

In summary, the intersection of the first two boundary lines, namely(T_1, T_1), is the global minimum point, which we denote as Point A. Therefore, the minimum value is obtained when the depth is the minimum of the three dimensions: $f_1(t_h, t_v) = f_1(T_1, T_1) = \frac{s(40s+9)}{(4s+1)(5s+1)} T_1$. However, this critical condition $t_h = t_v = t_c = T_1$ can be shared with the two other conditions.The term $f_1(t_h, t_v)_{min}$ is no less than the minimum values of the other two. Therefore this condition can be disregarded, with only the two other conditions requiring consideration.

Case (2): Depth is the medium of the three dimensions (measured in the maximum time used), namely $t_v \le t_c \le t_h$;

Referring to the previous case, it can be determined that

$$y(j) = \begin{cases} \sqrt{\frac{a_h}{L}} \sqrt[5]{\frac{jLHP}{a_h a_v s_c}}, & 0 < j < \frac{Ha_h}{La_v}\frac{s_c}{P}\sqrt{\frac{H}{a_v}} \\ \sqrt[6]{\frac{j^2P^2a_h}{Ls_c^{\,2}}}, & \frac{Ha_h}{La_v}\frac{s_c}{P}\sqrt{\frac{H}{a_v}} < j < \frac{P^2a_h}{Ls_c^{\,2}} \\ \sqrt{j}, & \frac{P^2a_h}{Ls_c^{\,2}} \le j \le 1 \end{cases} \quad (6)$$

$$ESC = s\left[\frac{2}{(3s+1)(5s+1)}\frac{t_v^{\,5s+1}}{T_1^{\,5s}} + \frac{1}{(2s+1)(3s+1)}\frac{T_1^{\,10s+5}}{t_h^{\,6s+2}t_v^{\,4s+2}} + \frac{2}{(2s+1)(4s+1)}\frac{t_v^{\,2s+1}}{t_h^{\,2s}} + \frac{4}{2s+1}t_h\right] = f_2(t_h,t_v) \quad (7)$$

Similar to Case (1), the global minimum can exist only where the local minima are or within the boundary lines. The boundary lines are the curves $t_h^{\,3}t_v^{\,2} = T_1^{\,5}(t_h \ge T_1\ and\ 0 < t_v \le T_1)$ and $t_h^{\,2}t_v^{\,3} = T_1^{\,5}(t_h \ge T_1\ and\ 0 < t_v \le T_1)$,as well as the virtual boundary of the line segment without an extreme point, $t_h = +\infty(0 < t_v < T_1)$.Therefore, the feasible region ***D2*** encircled by the three aforementioned boundary lines is obtained, which is shown in Figure 4.

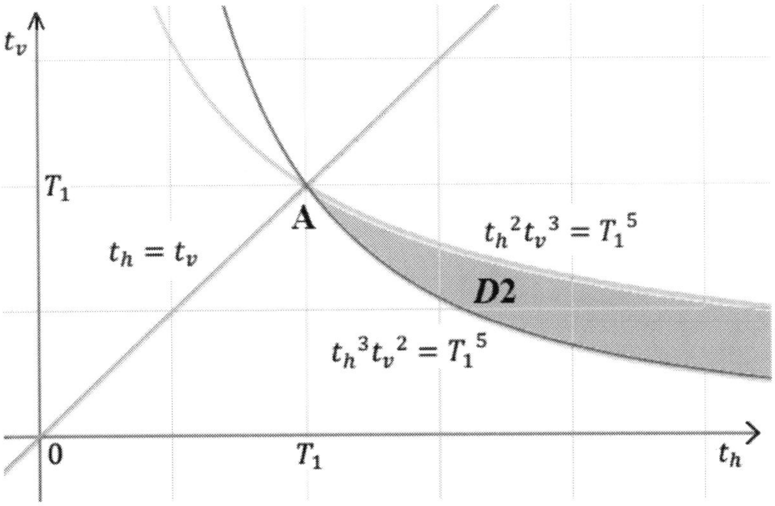

Figure 4: Feasible region under the constraints expressed in Case (2)

First, $f_2{}'_{t_h}(t_h,t_v) > 0$ can always be, which proves that there are no local minima of $f_2(t_h,t_v)$ in region ***D2***.

Second, if the global minimum of ESCcan be obtained on the boundary lines, it must be no less than the global minimum of ESC of Case (3).

Therefore, when this is combined with the conclusion of Case (1), it can be concluded that the minimum value of ESC in the three parts must be obtained in Case (3).

Case (3): Depth is the maximum of the three dimensions (measured in the maximum time used), namely $t_v \le t_h \le t_c$;

It can be determined that

$$y(j) = \begin{cases} \frac{s_c}{P}\sqrt[5]{\frac{jLHP}{a_h a_v s_c}}, & 0 < j < \frac{Ha_h}{La_v}\frac{s_c}{P}\sqrt{\frac{H}{a_v}} \\ \sqrt[3]{\frac{jLs_c^2}{P^2 a_h}}, & \frac{Ha_h}{La_v}\frac{s_c}{P}\sqrt{\frac{H}{a_v}} \le j \le \frac{s_c}{P}\sqrt{\frac{L}{a_h}} \\ j, & \frac{s_c}{P}\sqrt{\frac{L}{a_h}} \le j \le 1 \end{cases} \quad (8)$$

The expected single-command time can be obtained as:

$$ESC = s[\frac{2}{(3s+1)(5s+1)}\frac{t_v^{5s+1}}{T_1^{5s}} + \frac{2}{(s+1)(3s+1)}\frac{t_h^{3s+1}t_v^{2s}}{T_1^{5s}} + \frac{1}{s+1}\frac{T_1^5}{t_h^2 t_v^2} + \frac{2}{(2s+1)(4s+1)}\frac{t_v^{2s+1}}{t_h^{2s}} + \frac{2}{2s+1}t_h] = f_3(t_h, t_v) \quad (9)$$

Because the binary function $f_3(t_h, t_v)$ is continuously differentiable, the global minimum can exist only where the local minimaare or within the boundary lines. The boundary lines are the line segments without an extreme point,$t_h = t_v (0 < t_h = t_v \le T_1)$, the curve $t_h^3 t_v^2 = T_1^5 (t_h \ge T_1 \text{ and } 0 < t_v \le T_1)$, the virtual boundary of the half-line without an extreme point $t_v = 0 (t_h > 0)$, andthe virtual boundary of the line segment without an extreme point $t_h = +\infty (0 < t_v < T_1)$.Therefore, the feasible region ***D3*** encircled by the four aforementioned boundary lines is obtained ,which is shown in Figure 5.

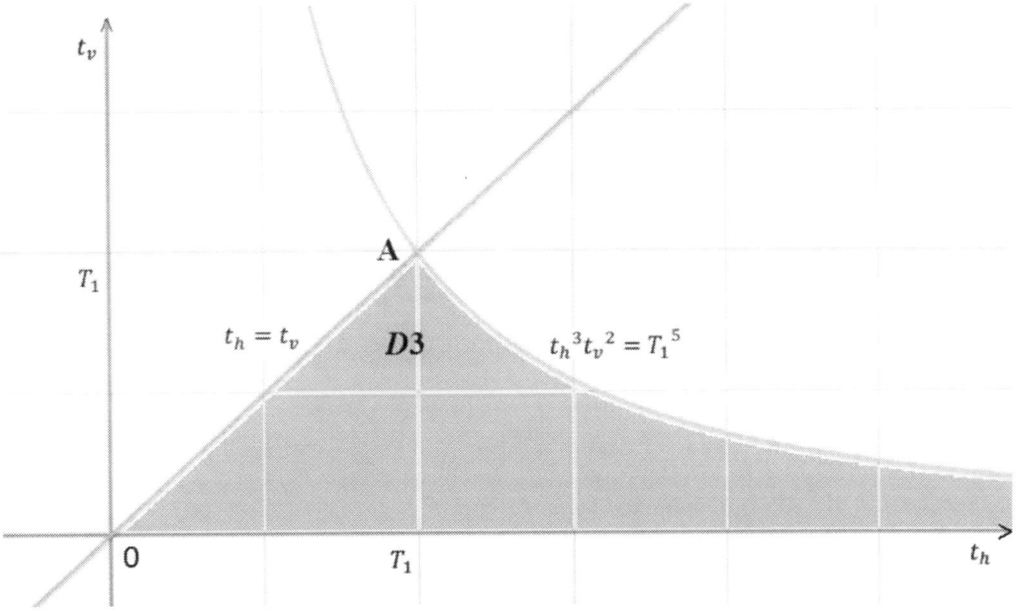

Figure 5: Feasible region under the constraints expressed in Case (3)

First, the global minimum may equal the local minimum. But we calculate that $f_3{}'_{t_h}(t_h, t_v) - f_3{}'_{t_v}(t_h, t_v) = 0$ can be obtained only if $t_h = t_v$. Therefore, the local minimum points must be on the line segment without an extreme point $t_h = t_v (0 < t_h = t_v \le T_1)$if local minimum points exist. Thus, these local minima are no less than the minimum value obtained on the line segment.

Second, consider the line segment without an extreme point $t_h = t_v (0 < t_h = t_v \le T_1)$. As $ESC = s[\frac{4}{(s+1)(5s+1)} \frac{t_h{}^{5s+1}}{T_1{}^{5s}} + \frac{1}{s+1} \frac{T_1{}^5}{t_h{}^4} + \frac{4}{4s+1} t_h]$, $ESC'(t_h) = \frac{4s}{s+1}\left[(\frac{t_h{}^5}{T_1{}^5})^s - \frac{T_1{}^5}{t_h{}^5} + \frac{s+1}{4s+1}\right]$, Let $x_1 = \frac{t_h{}^5}{T_1{}^5} \in (0,1)$, and we can get $ESC_{min} = \frac{5s}{5s+1} T_1 x_1{}^{-\frac{4}{5}} (1 + \frac{4s}{4s+1} x_1) \le ESC(T_1, T_1) = \frac{s(40s+9)}{(4s+1)(5s+1)} T_1$ ($t_h = t_v = x_1{}^{\frac{1}{5}} T_1$, and $t_c = x_1{}^{-\frac{4}{5}} T_1$).

Third, consider the curve $t_h{}^3 t_v{}^2 = T_1{}^5 (t_h \ge T_1 \; and \; 0 < t_v \le T_1)$, in which$t_v = \frac{T_1{}^{\frac{5}{2}}}{t_h{}^{\frac{3}{2}}}$ and $t_h \in [T_1, +\infty)$.

Similarly, let $x_2 = \frac{t_h{}^{\frac{5}{2}}}{T_1{}^{\frac{5}{2}}} \in [1, +\infty)$, and we can get $ESC_{min} = 5sT_1 x_2{}^{\frac{2}{5}} (\frac{1}{5s+1} \frac{1}{x_2{}^{3s+1}} + \frac{1}{4s+1} \frac{1}{x_2{}^{2s+1}}) \le ESC(T_1, T_1) = \frac{s(40s+9)}{(4s+1)(5s+1)} T_1$, when $t_h = t_c = x_2{}^{\frac{2}{5}} T_1$, $t_v = x_2{}^{-\frac{3}{5}} T_1$.

Fourth, the virtual boundary of the half-line without an extreme point is$t_v = 0 (t_h > 0)$, in which $f_3(t_h, t_v) = +\infty$ is constant.

Finally, the virtual boundary of the line segment without an extreme point is $t_h = +\infty (0 < t_v < T_1)$ in which $f_3(t_h, t_v) = +\infty$ is constant.

In summary, it can be concluded that

$$ESC_{min} = f_3(t_h, t_v)_{min} = \min\left\{f_3\left(x_1^{\frac{1}{5}}T_1, x_1^{\frac{1}{5}}T_1\right), f_3\left(x_2^{\frac{2}{5}}T_1, x_2^{-\frac{3}{5}}T_1\right)\right\} \quad (10)$$

However, this can also be proven with a non-strict numerical method. We list many cases of this, including both commonly used and extreme ones in a table. Thus, $f_3\left(x_2^{\frac{2}{5}}T_1, x_2^{-\frac{3}{5}}T_1\right)$ is always higher than $f_3\left(x_1^{\frac{1}{5}}T_1, x_1^{\frac{1}{5}}T_1\right)$. Furthermore, we have listed many other cases with a step length of 1% of the ABC curve, the result of which are always the same.

Therefore, the resulting equation is as follows:

$$ESC_{min} = f_3\left(x_1^{\frac{1}{5}}T_1, x_1^{\frac{1}{5}}T_1\right) = \frac{5s}{5s+1}T_1 x_1^{-\frac{4}{5}}(1 + \frac{4s}{4s+1}x_1) = \frac{5s}{5s+1}x_1^{-\frac{4}{5}}(1 + \frac{4s}{4s+1}x_1)\sqrt[5]{\frac{32V}{a_h a_v s_c}} \quad (\text{where } x_1^{s} - \frac{1}{x_1} + \frac{s+1}{4s+1} = 0) \quad (11)$$

4 EXAMPLE

In this section we explain the design of a real-world 3D turnover-based compact system similar to the example provided by Yu and De Koster (2009a)[10]. The data are presented in Table 1.

Table 1: System parameters

Total system capacity in pallets		1000 pallets
Storage policy		Full turnover-based storage
Pallet size(width× length× height)	Net	1.0m×1.0m×1.5m
	Gross	1.2m×1.2m×2.0m
S/R machine	operating policy	Single-command cycle
	Horizontal acceleration	0.7m/s2
	Vertical acceleration	0.3 m/s2
Conveyor speed		1.6m/s
Cases of ABC curve considered		20%/80%

The required capacity is $V = 1000 * (1.2m * 1.2m * 2m) = 2880\ m^3$. Thus, the results obtained are as follows: $T_1 = \sqrt[5]{\frac{32V}{a_h a_v s_c}} = 12.24s, t_h = t_v = 11.24s,\ t_c = 17.21s,\ t_h\!:t_v\!:t_c = 0.653 : 0.653 : 1, L = 22.09m$, $H = 9.47m, P = 13.77m$ and the optimal expected single-command time is $ESC_{min} = 12.49s$.

However, the real-world rack dimensions must be multiples of the pallet dimensions. Accordingly, the practical approximate optimal dimensions are designed as follows: $L = 21.6m$, $H = 10.0m$, and $P = 14.4m$, which allow the compact storage system to contain 1080 total pallets (18×5×12 pallets).

In this practical example, the maximum horizontal, vertical, and conveyor speeds are 3.89, 1.73, and 1.60 m/s respectively, which are easily reached by the automated system.

5 DISCUSSIONS ON SHAPE FACTORS AND OPERATING CHARACTERISTICS OF S/R MACHINES

The shape factors of an optimal rack system could reveal crucial information during calculation and proofing. Such information could facilitate designing a more complex rack system for which the optimal design may be difficult to calculate. Before we started to solve the problem in this article, we assumed that we may obtain the final conclusion of $t_h = t_v$, which implies that the pick surface is always a square when measured according to the maximum time used in traversing the horizontal and vertical dimensions. Finally, our calculations prove that these assumptions were accurate for a 3D compact AS/RS when using a turnover-based storage policy with an S/R machine that accelerates and decelerates uniformly.

Then on strict conclusion obtained from experience and guesswork could contribute toward solving more difficult cases in the future. For example, the assumption that $t_h = t_v$would yield the optimal rack design for a 3D compact AS/RS when using a turnover-based storage policy and both uniform acceleration and deceleration as well as uniform speed (i.e., the maximum time used to accelerate and decelerate is also the same, namely $\frac{2s_h}{a_h} = \frac{2s_v}{a_v}$).

Moreover, acceleration (denoted as a_{ha}) and deceleration (denoted as a_{hd}) in the horizontal direction need not be equal. This is also true for the vertical direction. Therefore, $a_{va} = a_{vd}$is not required either. The result can be extended, replacing a_h with $\frac{a_{ha}+a_{hd}}{2}$, and a_v with $\frac{a_{va}+a_{vd}}{2}$, to achieve the same conclusion (equation 11). This makes the results more flexible for optimizing the design of compact AS/RSs.

6 CONCLUSION

In this paper, we discuss a 3D compact AS/RS with a full turnover-based storage policy by considering the operating characteristics of the S/R machine. This discussion is a crucial extension for optimizing the design of compact AS/RSs. We conclude with some relevant and practicable findings:

- The optimal rack dimensions can be obtained when $t_h = t_v = \sqrt[5]{\frac{32V}{a_h a_v s_c}} x_1^{\frac{1}{5}}$, and $t_c = \sqrt[5]{\frac{32V}{a_h a_v s_c}} x_1^{-\frac{4}{5}}$, in which x_1 is the only solution of equation $x_1^{s} - \frac{1}{x_1} + \frac{s+1}{4s+1} = 0$ $(x_1 \in (0,1])$,and the optimal travel time is $ESC_{min} = \frac{5s}{5s+1} x_1^{-\frac{4}{5}} \left(1 + \frac{4s}{4s+1} x_1\right) \sqrt[5]{\frac{32V}{a_h a_v s_c}}$.
- When the turnover case becomes 80%/20%, $x_1 = 0.653$ can be obtained, as well as$t_h = t_v = 0.918 \sqrt[5]{\frac{32V}{a_h a_v s_c}}$, $t_c = 1.406 \sqrt[5]{\frac{32V}{a_h a_v s_c}}$, and $ESC_{min} = 1.021 \sqrt[5]{\frac{32V}{a_h a_v s_c}}$.
- However, when the turnover case becomes 20%/20%, $x_1 = 0.820$ is obtained, as well as$t_h = t_v = 0.961 \sqrt[5]{\frac{32V}{a_h a_v s_c}}$, $t_c = 1.172 \sqrt[5]{\frac{32V}{a_h a_v s_c}}$, and$ESC_{min} = 1.617 \sqrt[5]{\frac{32V}{a_h a_v s_c}}$. This case can be considered as a random policy. Compared with the previous conclusion, this one shows that the full turnover-based storage policy is more efficient compared with the random policy.
- The optimal ratio of the three dimensions is $t_h : t_v : t_c = x_1 : x_1 : 1$, which can bedetermined from the turnover rate alone.
- Acceleration and deceleration do not to be identical in the horizontal or vertical direction. Therefore, $a_h$can be replaced with $\frac{a_{ha}+a_{hd}}{2}$, and a_v with $\frac{a_{va}+a_{vd}}{2}$ in the conclusions.
- The assumption can be made that $t_h = t_v$ is applicable in many optimal designs of AS/RSs, if the two dimensions involve the same type of motion and travel time in the whole cycle.

When the maximum horizontal and vertical speeds of the S/R machine are limited and the rack dimensions are sufficiently large, the S/R machine must move at a constant speed for a moment after accelerating to the maximum speed. Under this scenario, our presented model is limited in application. The most complex scenario, which would involve a 3D system when using a full turnover-based storage policy and motion with both uniform acceleration and deceleration and uniform speed, is a topic to be explored in future research.

7 ACKNOWLEDGMENTS

This work was supported by the National Natural Science Foundation of China (Grant No. 71202022), the National Natural Science Foundation of China (Grant No. 71572090), Guangdong Nature Science Foundation (Grant No. 2015A030313824), and the Shenzhen City Special Strategic Emerging Industry Development (Grant No. JCYJ20150331151358130).

8 REFERENCES

[1] **Cardin, O. and P. Castagna, et al. 2012.** Performance evaluation of In-Deep Class Storage for Flow-Rack AS/RS, *International Journal of Production Research*, 50 (23), pp 1-17.

[2] **Chang, D. and U. Wen, et al. 1995.** The impact of acceleration/deceleration on travel-time models for automated storage/retrieval systems, *IIE Transactions*, 27 (1), pp 108-111.

[3] **De Koster, R. B. M. and T. Le-Duc, et al. 2008.** Optimal storage rack design for a 3-dimensional compact AS/RS, *International Journal of Production Research*, 46 (6), pp 1495-1514.

[4] **Hausman, W. H. and L. B. Schwarz, et al. 1976.** Optimal Storage Assignment in Automatic Warehousing Systems, *Management Science*, 22 (6), pp 629-638.

[5] **Hwang, H. and S. B. Lee 1990.** Travel-time models considering the operating characteristics of the storage and retrieval machine, *International Journal of Production Research*, 28 (10), pp 1779-1789.

[6] **Hwang, H. and Y. Song, et al. 2004.** The impacts of acceleration/deceleration on travel time models for carousel systems, *Computers & Industrial Engineering*, 46 (2), pp 253-265.

[7] **Sari, Z. and C. Saygin, et al. 2005.** Travel-time models for flow-rack automated storage and retrieval systems, *The International Journal of Advanced Manufacturing Technology*, 25 (9), pp 979-987.

[8] **Wen, U. P. and D. T. Chang, et al. 2001.** The impact of acceleration/deceleration on travel-time models in class-based automated S/R systems, *IIE Transactions*, 33 (7), pp 599-608.

[9] **Yang, P. and L. Miao, et al. 2014.** Optimal storage rack design for a multi-deep compact AS/RS considering the acceleration/deceleration of the storage and retrieval machine, *International Journal of Production Research*, 53 (3), pp 929-943.

[10] **Yu, Y. and M. B. M. De Koster 2009a.** Designing an optimal turnover-based storage rack for a 3D compact automated storage and retrieval system, *International Journal of Production Research*, 47 (6), pp 1551-1571.

[11] **Yu, Y. and M. B. M. De Koster 2009b.** Optimal zone boundaries for two-class-based compact three-dimensional automated storage and retrieval systems, *IIE Transactions*, 41 (3), pp 194-208.

[12] **Yu, Y. and R. B. M. De Koster 2011.** Sequencing heuristics for storing and retrieving unit loads in 3D compact automated warehousing systems, *IIE Transactions*, 44 (2), pp 69-87.

STUDY ON MODEL AND METHOD OF PACKAGING COST ESTIMATION BASED ON KNOWLEDGE

Yunqi GONG*
College of Light Industry
Harbin University of Commerce, Harbin, China
gongyunqi1@sohu.com

ABSTRACT

Quick and efficient estimation of product packaging cost could be used in a timely response to customer inquiry, price quotation and other aspects of demand. A semantic model of packaging cost knowledge and its semantic representation were proposed. In the support of the model, cost estimation model and methods were presented for the whole process of product packaging design, which includes scheme design stage, detailed design stage and construction stage, corresponding to case-based reasoning, activity-based costing and simulation respectively. Finally, the packaging cost estimation of mechanical and electronic product was used as example to verify the application.

Keywords: Packaging cost estimation, Semantic model, Case-based reasoning, Neural network, Simulation

1 INTRODUCTION

When the packaging enterprises participate in the market competition, a reasonable product packaging quotation should be given in time for the customer's inquiry or order. Packaging cost estimates could be used for technical and economic evaluation in the packaging design process of product, for support of the reduction of packaging cost also [1].

Cost estimation is one of the most basic content of product cost research [2], there are some studies on the combination of knowledge and cost [3,4]. Product packaging contains relatively few factors, the product packaging cost model [5] and forecasting methods which were used to realize the cost estimation system is relatively easy. Ontology is an explicit specification of a conceptualization. Ontology-based knowledge representation approach was used for cost estimation in [6]. There are a number of research results of cost estimation based on case-based reasoning [7,8] which could be used for reference.

With the development of artificial intelligence and computer technology, it is possible to solve the problem that the packaging cost result could not be calculated accurately by mathematical method, so model and method of packaging cost estimation based on knowledge was studied in this paper, and the architecture of research was shown in figure 1.

* Corresponding Author: Yunqi GONG

Sponsor: Project supported by the Heilongjiang Provincial Department of Education, China (No.12541203)

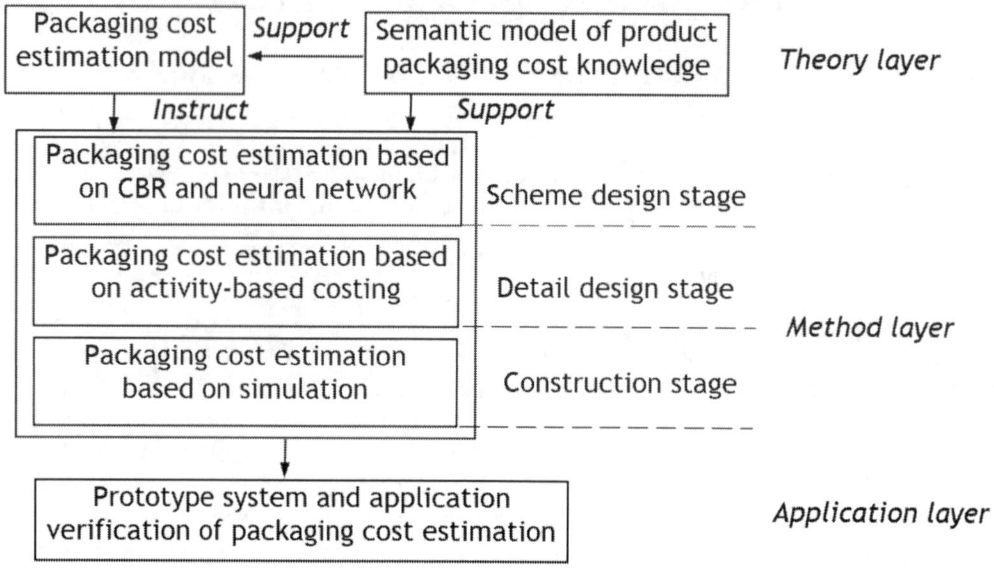

Figure 1: Research Architecture of Packaging Cost Estimation

2 KNOWLEDGE OF PRODUCT PACKAGING COST

2.1 Classification and Organization

Knowledge of packaging cost could be divided into definitional knowledge, reasoning knowledge and the corresponding packed product knowledge. A large number of empirical knowledge used in product packaging cost estimating is in a state of disorder, lack of systematic arrangement. Knowledge acquisition, expression method, the relationship and potential conflicts between the knowledge have not been digging, and effective packaging costs knowledge expression and application system should be constructed.

2.2 Semantic Model of the Packaging Cost Knowledge

Different people have different view at cost, for example, from the design process, product structure, cost structure and so on, this also reflects the multidimensional nature of the cost. On the basis of the classification and organization of product packaging costs knowledge, in order to support the packaging cost estimation based on knowledge, there are two problems to be solved: one is sharing different areas, different aspects of knowledge, to achieve the goal of knowledge and its semantic representation "unified understanding" [9]; Another is to reveal the relationship of variable parameters in the knowledge level used by different estimation method.

Considering the packaging cost knowledge from the perspective of ontology, and by reference of grouping and coding of Packing Character [10], packaging cost knowledge ontology was divided into three levels as shown in figure 2.

1) Top-level ontology is an explicit specification of general concepts such as thing, object, event, behaviour and task etc. These concepts are independent on specific domain so as to be shared in a large scale.

2) Meta knowledge in the second layer means knowledge about itself, for example, rule, framework and semantic network etc. which was used in packaging cost knowledge representation.

3) The application ontology in the third layer is a specification of specific packaging cost knowledge. It could cite concept of domain ontology and of meta knowledge ontology also.

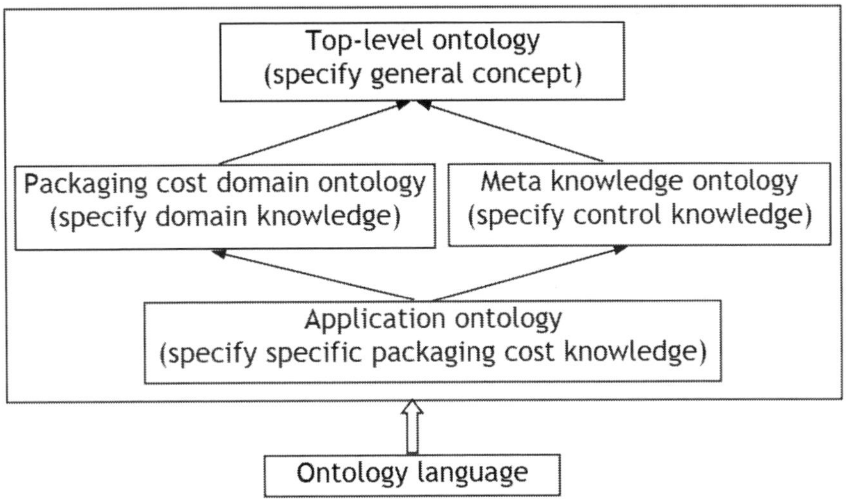

Figure 2: Research Layer of Packaging Cost Knowledge Ontology

Ontology language provides tool to represent and develop the ontology above. The layer of packaging cost knowledge ontology was shown in table 1.

Table 1: The Layer of Packaging Cost Knowledge Ontology

level	instance
concept of top-level ontology	thing object type ...
concept of packaging cost domain ontology	packed goods packaging material ...
term of meta knowledge ontology	framework rule ...
term of specific packaging cost knowledge	packaging case packaging decision ...

Packaging cost knowledge comes from information collected, which is associated with packaging case and packaging characteristics variables. By explicating the concepts in packaging cost estimation and their relationship, the semantic model of packaging cost estimation knowledge was presented. After comparing several semantic representation languages, OWL is used to describe entities, properties and relationships of concepts which are related with packaging cost estimation. Axioms and constraints for system of packaging cost estimation concepts are represented in First Order Logic. The model laid a foundation for comprehensive analysis of product packaging costs.

3 PACKAGING COST ESTIMATION MODEL

The dynamic model of cost and related non cost information can be fully expressed in the process of product packaging design. This cost model is the basis of the cost estimation. The product packaging cost model which supports the whole process of design is progressive, dynamic, open and hierarchical. Progressive product packaging cost model means refinement with the continuous deepening of the design, dynamic and openness ensure that the product packaging cost model can reflect the intention of designer's design changes and their impact on the packaging cost of the product. The combination representation of numerical value and knowledge can be realized in the process of cost modelling.

Based on the mature modelling method, a product packaging cost estimation model was put forward, the three tier structure of the model includes process layer, logic layer and application layer, and the model can provide support for product packaging cost estimation.

The process layer of the first layer is the definition of product packaging. It records the process of product packaging definition and the data structure of information, which provides data for the subsequent level of the model. The process layer is the main information of the packaging cost estimation model, and the other levels are based on the process level.

The logical layer of the second layer is the description of the product packaging cost information from different aspects, so it forms sub-models, such as the description of the product package cost information, packaging management cost information and use cost information of packaging, product package cost estimation is the focus of this paper and its sub-models include package goods characteristics model, packaging material model, cost structure model and so on. These sub-models contain product packaging cost estimation information, and these models have complex coupling relations, building process of cost model must clarify the relationship which is reflected in the cost estimation model, logical layer is the characterization of the relationship between these models and the relationship between the information and these models will be passed to the application layer.

The organization and expression of packaging cost knowledge and data was made in the application layer to meet the needs of packaging cost estimation. Application layer's task is to organize products' packaging definition information systematically which is in the process layer, to organize the relationship of sub-models in the logic layer systematically. Application model can be directly applied to the packaging cost estimation procedure.

4 PACKAGING COST ESTIMATION METHODS

It is found that not enough has been done on the packaging cost estimation research. According to the characteristics of the product packaging, how to apply mature theory of cost estimation methods is a problem that needs to be solved.

By the combination of information with packaging case, parameters and characteristics of variables, packaging cost estimation knowledge was formed. To solve the cost estimation problem in whole process of product packaging design, namely, the scheme design, detailed design and construction design phase, with the support of semantic model for packaging cost estimation, the theory and method of artificial intelligence and knowledge engineering was used.

Research achievements of activity based costing[11,12] could be used for reference in this paper, according to the characteristics of the product packaging, a set of cost estimation methods oriented packaging design process was put forward, including case-based reasoning and neural network technology used in estimation in conceptual design stage, activity based costing in detailed design stage based on the characteristics and cost drivers of product packaging, and application system simulation analysis method for cost estimation in construction stage.

4.1 Packaging Cost Estimation in Scheme Design Stage

In view of the characteristics of product packaging, the case retrieval process was presented. The idea of hierarchical Case-based Reasoning and the method of combination of knowledge instruction strategy and nearest neighbourhood strategy were adopted. The method of product packaging cost estimation based on similar cases was proposed [13].

From the view of packaging, type of product is a main factor to determine the packing container, such as solid state products to use boxes packaging, and liquid products to use barrels, cans and other packaging. Moreover, it also affects sealing fixation methods and protection method and grade of product packaging. The structure of the product is closely related to the fixed way in the process of transportation. The classification of the types of products for mechanical products, electronic products, instrumentation, chemical products, medicine, gas items and so on, the first three types could be divided into two types of complex structure and general structure. For the selected set of products of the same type and structure, neural network could be used to estimate the cost of packaging.

The selection of input variable and output variable and its normalization processing method should be made firstly in packaging cost estimation method based on neural network, take mechanical products for example, product precision level which is expressed as a crisp value, weight and size of product could be selected as input variable and packaging cost as output variable. The selection of number of hidden layer nodes and the transfer function was made secondly, the packing cost variable combination and the corresponding cost according to the features of history data as neural network training sample set, nonlinear relationship between packaging cost characteristic variable combination scheme of input and packaging cost result of output was established to forecast result of the new packaging cost variable combination.

4.2 Packaging Cost Estimation Based on Cost Drivers in Detailed Design Stage

Packaging cost results should be detailed estimated in this stage. The estimate starting point is the processing information, the activity based costing method focuses on the decomposition of the packing "activity", "cost drivers" mining which caused the packaging cost, and the number of "activity" to calculate the cost of cause, the product packaging cost was collected afterward.

Take mechanical products for example, closed packing box or steel edge packing box for general use, packaging process mainly includes wood cutting activity, drilling activity, spraying activity and assembly work. Therefore, it can be determined that 4 operations centres, namely, 4 cost centres, and thus the cost drivers. Combining the cost driver of each activity, and determining the cost driver rate of two kinds of packing boxes, the cost estimation function of unit parts work was established, and the value can be brought into the function.

4.3 Packaging Cost Estimation Based on Simulation in Construction Stage

At present, simulation is used for cost estimation including simulation model of product life cycle cost from the perspective of manufacturer [14], simulation based outsourcing procurement cost optimization [15] and so on. Application of production system simulation software “Plant Simulation” of “Siemens digital plant” was made. Cost simulation model of production line was set up, and the simulation model can be changed by some adjustments in the design of construction, it can be used to estimate the cost not yet occurred.

In construction phase, product packaging costs in addition to the cost of packaging materials, which was decided by selection form of packaging and product specifications, packaging process cost also includes man hours of labour and equipment costs, work time of each activity is not stable, some fluctuations occur, it is random variable obeying a certain random distribution. Therefore, it is necessary to carry out the system simulation to estimate the cost of operation. In application of the simulation model, according to different products, different packaging process, the system parameters and data input should be adjusted.

5 THE APPLICATION VALIDATION OF PRODUCT PACKAGING COST ESTIMATION

Product packaging cost estimation prototype system could be developed based on the study in this paper. According to the system design idea of independent function and modular, the structure was divided into three levels of the development of the prototype system, which is data acquisition layer, data management layer, user service layer.

With the support of packaging cost knowledge semantic model, the packaging cost estimation of a mechanical product which is steam turbine bypass device made in a power equipment plant verified the feasibility and validity of the approach. One of simulation interface was shown in figure 3.

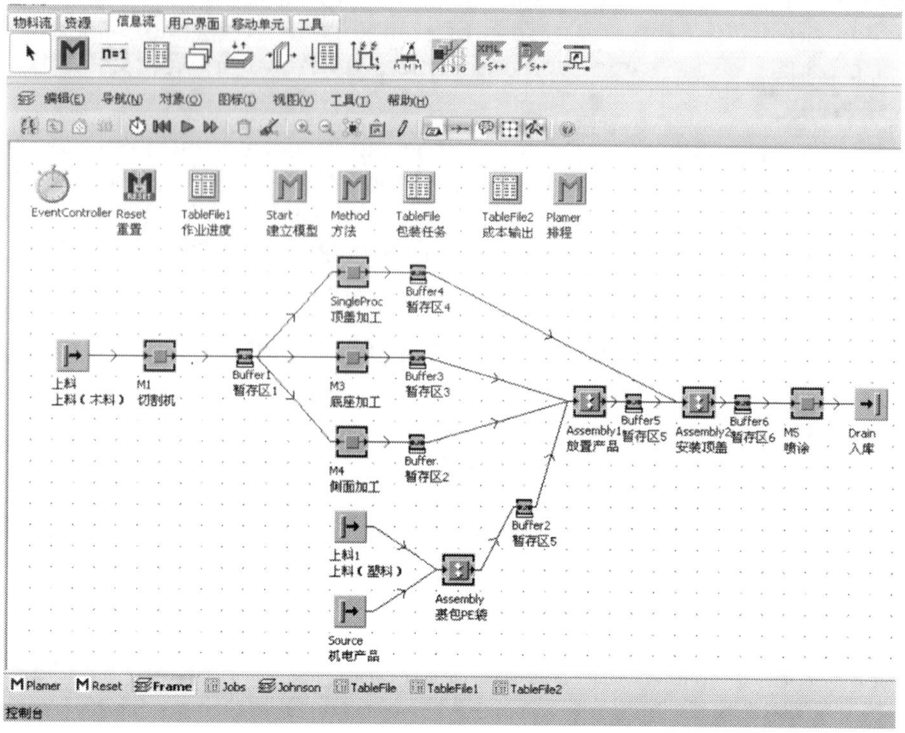

Figure 3: Simulation Model Interface of Cost Estimation

6 CONCLUSION

In view of the product packaging cost estimation, for the whole process of product packaging design, packaging cost knowledge semantic model platform was set up to provide support for cost estimation. For parameters and cost information in detail and accurate degree is not consistent with different product packaging design stages, a set of estimation method based on knowledge of the product packaging cost was put forward, which includes that case-based reasoning was used in scheme design stage, activity-based costing was used in detailed design stage, and simulation was used in construction stage. The research makes up for the deficiency of mature cost estimation model and method used in packaging cost estimation.

7 REFERENCES

[1] **Gong Gui-fen, Lan Ming, Yin Ke. 2013.** Research on comprehensive evaluation AHP model of total packaging cost, (in Chinese), *Packaging Engineering*, 34(3), pp 100-103.

[2] **Caroline Bramklev, Patrik Ström. 2011.** A conceptualization of the product/service interface: case of the packaging industry in Japan, *Journal of Service Science Research*, 3, pp 21-48.

[3] **Nathan Gnanasambandam, Anthony Primavera, K. Srihari. 2005.** The reliability prediction of electronic packages- an expert systems approach, *Int J Adv Manuf Technol*, 27, pp 381-391.

[4] **George J. Besseris. 2013.** Improving quality cost performance with qualimetrical methods: a case from a high-speed packaging process, *Int J Adv Manuf Technol*, 69, pp 1775-1789.

[5] **Jin Da-xin, Chen Xiao-hui. 2015.** Combined cost model of industrial package recycling in service life, (in Chinese), *Packaging Engineering*, 36(1), pp 129-133.

[6] **Tobias Walter, Fernando Silva Parreiras, Steffen Staab. 2014.** An ontology-based framework for domain-specific modeling, *Software System Model*, 13, pp 83-108.

[7] **Liu Xin, Jiang Shao-hua, Li Zhong-fu. 2015.** Ontology-based knowledge representation approach for building cost estimation, *Journal of Engineering Management*, 29(3), pp 19-24.

[8] **Gong Jian, Zhang Hong, Zhou Min. 2010.** Application of CBR in product packaging design system, (in Chinese), *Packaging Engineering*, 31(11), pp 77-79.

[9] **S. Lee, K. Kim, J. Yu. 2014.** BIM and ontology-based approach for building cost estimation, *Automation in Construction*, 41(5), pp 96-105.

[10] **Ma Shuang, Yan Ru-ming, Xu Wen-cai. 2001.** Probing into the Grouping and Coding of Packing Character, (in Chinese), *Packaging Engineering*, 22(4), pp 13-16.

[11] **Waghmode L, Sahasrabudhe A, Kulkarnip. 2010.** Life cycle cost modeling of pumps using an Activity Based Costing methodology, *Journal of Mechanical Design*, 132, pp 121-128.

[12] **Zou Xin-pin, Li Dong-sheng, Zhang Xin-chang. 2008.** Research and application of the recycle packaging system based on ABC, (in Chinese), *Packaging Engineering*, 29(1), pp 108-110,129.

[13] **Gong Yunqi. 2012.** Research on packaging cost estimation of product based on CBR, (in Chinese), *Packaging Engineering*, 33(15), pp 64-67,87.

[14] **Han Qinglan, Zhang Xiaoyan. 2015.** Simulation model of product life cycle cost from the perspective of manufacturer, (in Chinese), *Science and Technology Management Research*, 8, pp 210-215.

[15] **Jia Jiang-ming, Pan Xiao-hong, Wang Zheng-xiao. 2008.** Simulation based outsourcing procurement cost optimization, (in Chinese), *Journal of Zhejiang University (Engineering Science)*, 42(5), pp 869-874.

EARTHQUAKE MEDICAL RELIEF CONCEPTUAL MODEL DESIGN AND EFFICIENCY EVALUATION

Bo PENG[1*]

[1]Institute of Health Service and Medical Information,

Academy of Military Medical Sciences China

plapengbo@yeah.net

ABSTRACT

Purpose: This study aims to address the lack of standardization and evaluation system of the medical response to an earthquake and to improve emergency medical services. **Methods:** Consulting domestic and overseas literature and taking into account conditions in China and experience learned in real emergency medical rescue services, this paper adopts methods of the analytic hierarchy process and the queueing theory in designing a conceptual model of the earthquake medical response and presenting an efficiency evaluation model for disaster relief. **Results:** Four model systems have been established: echelons of medical care, early treatment, specialized care, patient evacuation and efficiency evaluation. Simulations are run to assess 4 different medical care solutions. **Conclusion:** The model standardizes and parameterizes the complex emergency medical services and evaluation process afterwards. It helps assisted decision making on disaster medical relief.

Keywords: Earthquake, Medical relief, Procedural modeling, Evaluation simulation

China is prone to natural disasters. Major disasters occurred in recent years have caused immeasurable damage to people's lives and properties. We propose a conceptual modeling that standardizes echelons and process of earthquake medical relief. It simulates and evaluates the medical relief process. The model helps to improve the efficiency of medical relief and lay the theoretical groundwork for standardization of relevant disaster relief processes.

1 DATA SOURCE AND MODELING BASIS

This model is based on relevant literature and data collected by medical teams served in three major earthquakes that attacked Wenchuan in 2008, Yushu in 2010 and Lushan in 2013.[1] The indexes in the evaluation model are results of consultation with rescuers in the three earthquakes and experts in relevant fields. Historical data have been integrated into the model as well.

2 EARTHQUAKE MEDICAL RELIEF CONCEPTUAL MODEL DESIGN

2.1 Echelon model

Earthquake medical relief can be arranged in three echelons: field first aid, early treatment and specialized care. Echelon I, the field first aid is performed by medical squads attached by field clinics. Echelon II, the early treatment can be administered by field clinics set by hospitals. Echelon

* Corresponding Author

III, the specialized care is provided by hospitals. In order to simulate conditions in reality, it is necessary to standardize the treatment capacity of each echelon: base hospital treatment capacity = 2 * field clinic treatment capacity = 4 * field medical squad treatment capacity. A patient can be transferred by air, rail or road or these means combined. The model is established as in Figure 1.

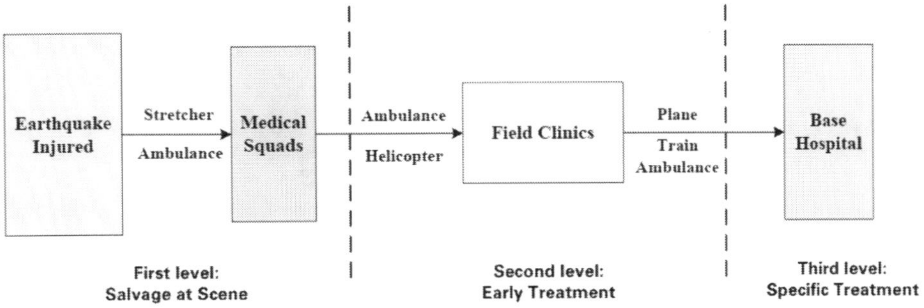

Figure 1:Earthquake medical relief echelon model

2.2 Field first aid model

Field first aid is performed as follows: Firstly, medical squads participate in searching and rescuing patients in the earthquake-stricken areas, including hemostasis, bandaging and fracture fixation, etc. Secondly, patients are moved to a designated place where basic wound care is provided, including anti-shock and anti-infection treatment. Thirdly, run quick examinations on patients; provide emergency care to the badly injured; offer support and protection to those who have excessive hunger and thirst or photophobia; provide psychological counseling to panic patients. Fourthly, fill in medical charts and help initiate evacuation. Based on basic process, we suggest a field first aid model as in Figure 2. [2]

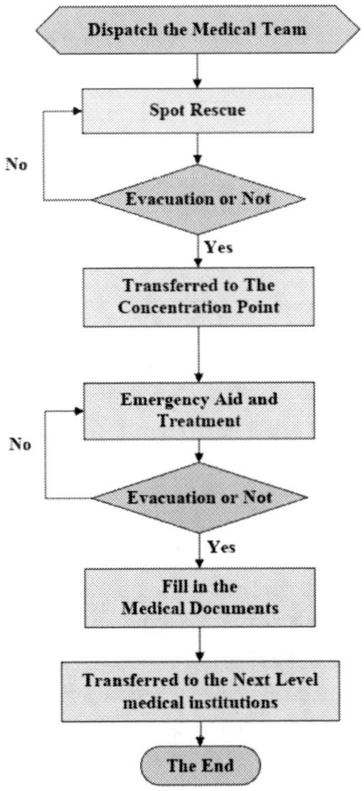

Figure 2:Field first aid conceptual model

2.3 Early treatment model

The basic process of early treatment and evacuation is as follows: Firstly, receive the injured or those who have been given emergency care; receive the wounded and divide them into groups. Secondly, inquire about the patients' conditions and check their medical tags to learn about their symptoms. Thirdly, divide patients into different groups to help them to be attended by different teams. Lastly, it is necessary to choose medical agencies for future treatment, vehicles to evacuate patients and evacuation sequences and put patients into different evacuation categories. (see Figure 3).

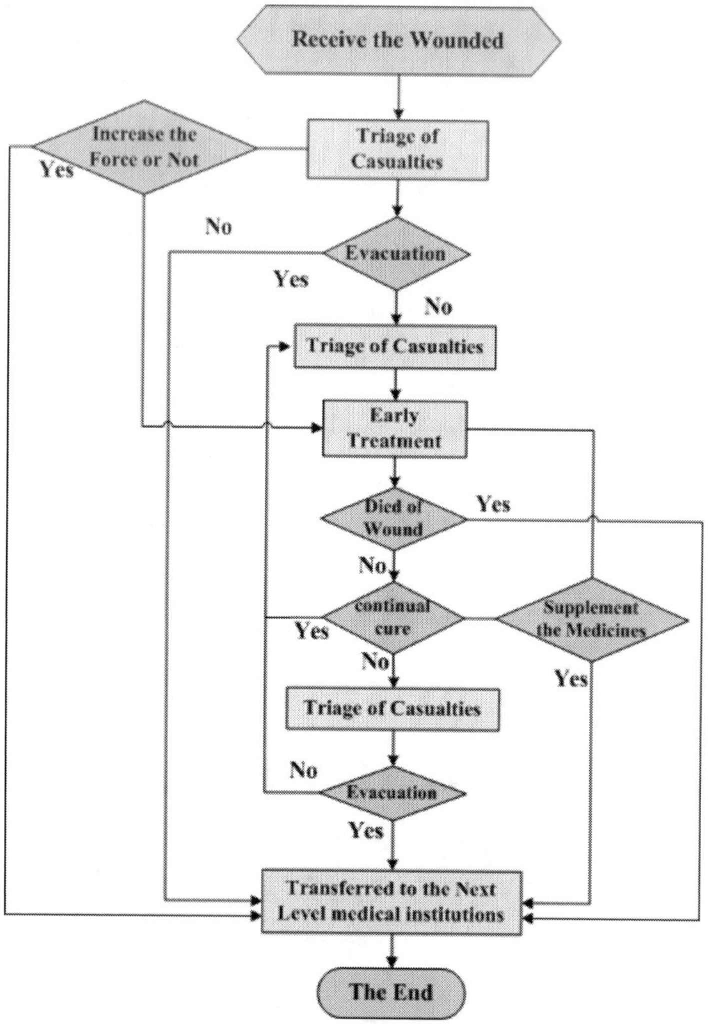

Figure 3:Early treatment conceptual model

2.4 Specialized care model

The basic procedure of the specialized care is as follows: Firstly, pick up a patient from an airport or a station and transfer the patient to a hospital. Secondly, set up categorizing receptions at the entrance of the hospital where the medical record of the patient is checked and the patient is grouped under a more specific category. Thirdly, move the patient to a relevant department or an operating room for treatment including all kinds of thorough specialized treatment and definitive surgery, comprehensive treatment for complications, comprehensive anti-shock and general anti-infection treatment, and certain major and medium function recovery surgeries and orthopedic operations (as in Figure 4). [3] [4]

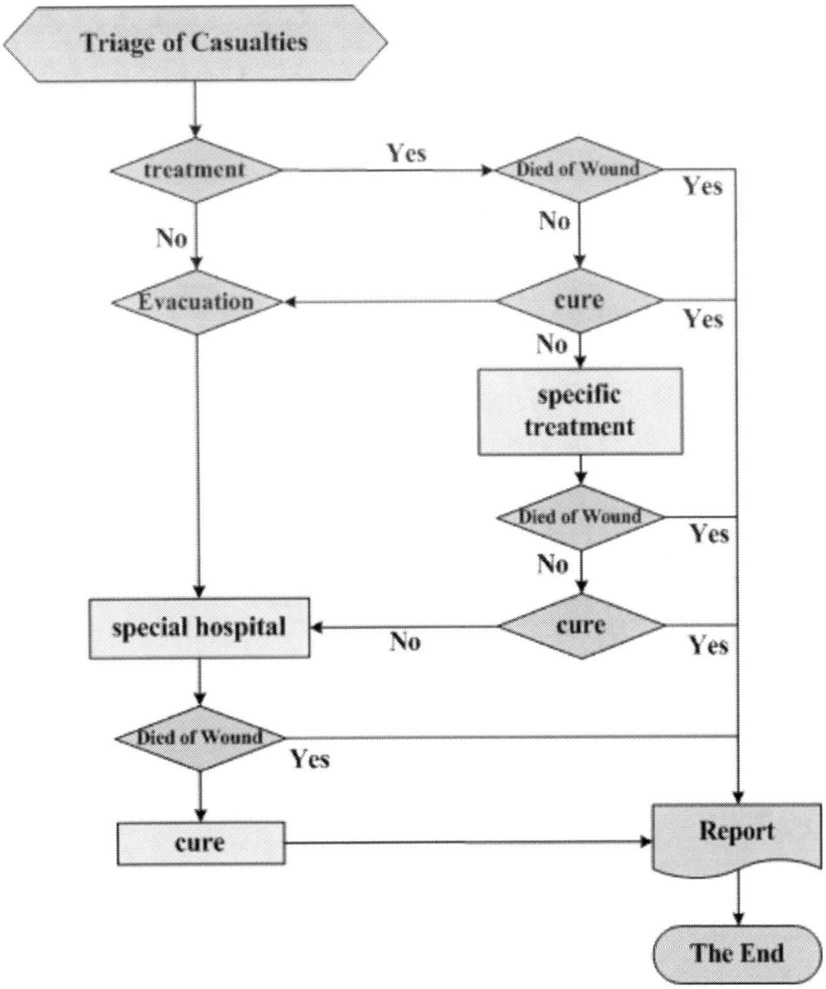

Figure 4:Specialized care conceptual model

3 EARTHQUAKE MEDICAL RELIEF EVALUATION MODEL DESIGN

3.1 Evaluation index system establishment

A comprehensive evaluation method is adopted that aggregates and processes multiple efficiency-revealing indexes. As a result, the advantages and disadvantages of the objects can be exposed on the overall level. Specifically, the method includes establishing an evaluation index system, determining the weight coefficient, designing an evaluation formula, obtaining simulation data and calculating the evaluation results. Medical relief data are generated by simulation modelling. When some parts of the relief cannot be represented by simulated parameters, relevant questions and data need to be employed to reveal the efficiency of these parts in question. [5]

After the evaluation index system is established, the real situations of earthquake medical relief are graded on the following scale: excellent (0.9), good (0.7), average (0.5), pass (0.3) or fail (0.1). All indexes are quantitative which are assessed by default parameters quantitatively. [6] Every grade of each index has a distinctive definition which is not elaborated on due to space restraints.

Table 1:Evaluation index system

First-grade indexes	Second-grade indexes	Rating scale				
		0.9	0.7	0.5	0.3	0.1
Index of the timely cure (B1)	Arriving time to the first medical agency (C1)	<20min	20-40min	40-60min	60-120min	>120min
	Time to get the definitive treatment (C2)	<25min	25-45min	45-65min	65-125min	>125min
	Time to finish the emergent operations (C3)	<30min	30-60min	60-80min	80-150min	>150min
	Time to finish the definitive surgery (C4)	<2h	2-4h	4-8h	8-24h	>24h
Index of the effective cure (B2)	Dead rate after the medical cure (C5)	<1%	1%-3%	3%-5%	5%-8%	>8%
	Disabled rate after the medical cure (C6)	<3%	3%-6%	6%-10%	10%-15%	>15%
	Health rate after the medical cure (C7)	>95%	90%-95%	85%-90%	80%-85%	<80%
	Operating rate of the heavy injured (C8)	>85%	75%-85%	65%-75%	50%-60%	<50%
Index of the economic cure (B3)	Utilization of the medical personnel (C9)	>90%	80%-90%	70-80%	60%-70%	<60%
	Utilization of the operating table (C10)	>90%	80%-90%	70-80%	60%-70%	<60%
	Utilization of the nursing bed (C11)	>90%	80%-90%	70-80%	60%-70%	<60%
	Utilization of the evacuation vehicle (C12)	>90%	80%-90%	70-80%	60%-70%	<60%

3.2 Index weight calculation

Judgement matrices are constructed based on the analytic hierarchy process. According to the evaluation criteria, calculate the relative importance between items in a matrix. Normalize the result and calculate the relative importance calculations and consistency tests. [7]

The proportion quotiety of the relative importance

To calculate the index weight, we have:

$$w_i = \overline{w_i} \Big/ \sum_{i=1}^{n} \overline{w_i}$$

To calculate the maximum eigenvalue of a judgment matrix, we have:

$$\lambda_{max} = \frac{1}{n} \sum_{i=1}^{n} \frac{(AW)_i}{w_i}$$

To calculate the consistency index (CI), we have:

$$\text{C. I.} = \frac{\lambda_{max} - n}{n - 1}$$

Table 2: Index weights

Index (B/C)	B1	B2	B3	Synthetic Weight of C
	0.558	0.320	0.122	
C1	0.558*0.4723			0.264
C2	0.558*0.2854			0.160
C3	0.558*0.1697			0.095
C4	0.558*0.0725			0.040
C5		0.320*0.5789		0.185
C6		0.320*0.1592		0.050
C7		0.320*0.2055		0.065
C8		0.320*0.0565		0.018
C9			0.122*0.4626	0.056
C10			0.122*0.3044	0.037
C11			0.122*0.1123	0.014
C12			0.1228*0.1207	0.015

4 EARTHQUAKE MEDICAL RELIEF EFFICIENCY EVALUATION

4.1 Simulated plans

Generate 200 standard patients using the patient generating model. Simulate the structure of injury in real earthquakes. Presume that medical agencies of the same echelon share the same rescue capacity. Set locations of all medical teams and distances in between according to those in real earthquake-stricken areas. The three echelons employ different numbers of medical squads, field clinics and base hospitals. All plans share the same general rescue capacity.

The 200 patients are given medical care by medical agencies of each echelon accordingly. Based on the different services provided by each echelon, 4 simulation plans are proposed (as in Table 5). The whole process of a medical response to an earthquake is simulated based on the system introduced in this paper. [8] [9]The results of different simulated plans are evaluated using the efficiency evaluation model.

Table 3: Simulated plans

Cure level or Scheme	First level The number of the medical squads	Second level The number of the field clinics	Third level The number of the base hospitals	Total of the treatment capacity (based in the medical squads)
A1	10	5	1	24
A2	8	4	2	24
A3	6	3	3	24
A4	4	2	4	24

4.2 Evaluation result

Because the model is evaluated quantitatively, data collected from the simulations can be used to directly grade the efficiency of the four rescue plans. The results are as follows (Table 6):

Table 4: Evaluation results

Evaluation elements		C1	C2	C3	C4	C5	C6	C7	C8	C9	C10	C11	C12	Score
Synthetic Weight (W)		0.264	0.16	0.095	0.04	0.185	0.05	0.065	0.018	0.056	0.037	0.014	0.015	
The Scheme	A1	0.7	0.7	0.3	0.3	0.7	0.5	0.9	0.3	0.7	0.9	0.7	0.5	0.64
	A2	0.5	0.3	0.5	0.3	0.7	0.5	0.7	0.3	0.5	0.9	0.7	0.5	0.52
	A3	0.5	0.3	0.5	0.5	0.5	0.3	0.7	0.5	0.3	0.5	0.5	0.7	0.46
	A4	0.3	0.1	0.7	0.9	0.5	0.3	0.7	0.7	0.1	0.3	0.5	0.9	0.40

To conclude, the four solutions can be ranked as A1→A2→A3→A4. This is generally in line with a real medical relief situation that life-saving is the top priority. Therefore, on-site first aid should be put first in a medical response to a disaster. The key to a successful medical response that efficiently saves lives is to improve the success rate of on-site emergency treatment.

5 CONCLUSION AND THINKING

The earthquake medical relief and efficiency evaluation model design standardizes and parameterizes complex medical responses and evaluation afterwards. The model provides a procedure and reference for simulation and rescue in the future. It ensures the efficiency of a rescue plan by simulating the combination of different teams and suggesting the optimal solution. Medical resources are maximized using this model.[10] However, attention needs to be paid to a few questions during the designing of the model. Firstly, further expert inquiry and reference to real situations are needed for all the parameters in the models. Constant development and upgrading helps optimize the medical relief process. Secondly, the consumption of medical resources and specific treatment technique models are not involved in the designing, which should be taken into consideration in the next step.

6 REFERENCES

[1] **TONG Hu.** 2007.Research on Key Technology of Emergency Research Manage System for Civil Airport, Nanjing University of Aeronautics and Astronautics,pp20-25.

[2] **HUANG Jianfa.** 2003.The United Nations held in Algeria earthquake rescue experience sum-up meeting, Recent Developments in World Seismology, 38(12),pp17-21.

[3] **SyiSu, Chung-Lang Shih.** 2003. Modeling an Emergency Medical Services System Using Computer Simulation, International Journal of Medical Informatic, 12(1),pp57-72.

[4] **Ray Mitehell, Kathleen Onofrio, Paula Konoske, Johnny Broek.** 2005.Ship to Objective Maneuver (STOM): Medical Analysis Using the NHRCT actical Medical LoGIStics(TML+) Planning Tool in Support of the Marine Corps War-fighting Laboratory(MCWL), Technical Report, pp05-33.

[5] **Park YT.** 2010. Impact of task, structure, and environment on electronic health record adoption, use, and interoperability in hospitals. University of Minnesota, pp 56-60.

[6] **WU Sijing, GUO Qing.**2011. Status and barriers of application of electronic health records at home and abroad, CGP Chinese General Practice, 14(1), pp 545-547.

[7] **WU Feng, XU Lei, LI Li.** 2012. the idea of establishing Army's electronic health records, People's Military Surgeon, 55(1), pp 32-34.

[8] **PENG Bo, ZHANG Peng, LI Wenhao.** 2013.Disaster relief medical evacuation simulation and evalution of requirements analysis, Military Medical Sciences,37(6)pp401-403.

[9] **PENG Bo, HE Zhen,** GE Yi. 2012.Problems existing in the emergency medical rescue information collection, People's Military Surgeon,55(3), pp202-203.

[10] **Ministry of health of China P.R.** 2009.The guide of information platform construction based on the record(trial implementation).

ON LOCATION DESIGN PROBLEM OF DISTRIBUTION CENTRES WITH AREA CONSTRAINTS

Tsubasa Suzuki[1*], Hiroshi Katayama[2], Nobuyuki Hayashi[3], Keisuke Harigaya[3] and Norio Ishiwatari[3]

[1]Department of Industrial and Management Systems Engineering,
Graduate School of Creative Science and Engineering,
Waseda University, Tokyo, Japan
tbs-m.66mj@akane.waseda.jp

[2]Faculty of Science and Engineering,
Waseda University, Tokyo, Japan
kata@waseda.jp

[3]Sankyu Inc., Tokyo, Japan
nhayashi@sankyu.co.jp

ABSTRACT

Settlement of distribution centre (DC) problem in order to support large-scale logistic operation was offered by the joint research company, in which derivation of optimal placement assuming the shuttle transportation between multiple DCs and customers is desired. In this study, area constrained DC placement problem is often discussed in logistics industry is picked up and, aiming to the effectiveness of the proposed solution, the solving model based on genetic algorithm (GA) is developed followed by its application to the case problems. Various cases are supposed such as the case that part of the position of DC is specified, the case that the area should be covered by the DC and the case that there is more than one DC. Logistics industry, in general, desires to developing the design model corresponding to these conditions.

Keywords: Supply chain, Distribution center, Area constraint, Performance evaluation, Genetic algorithm, Chromosome design

1 INTRODUCTION

Recently, distribution centre (DC) location problem is coming up as a critical issue among logistics companies which stand for large-scale operations. It is also a strategically critical matter for the long-term joint research company with our affiliated academic laboratory. Especially in actual logistics business scene, area constrained DC placement problem is often discussed from the view point of the effective operations under the constraints of their divisional asset such as customers with long-term contract, number of vehicles and drivers for their operations.

Further, in this problem solving, derivation of optimal DC placement under the multiple-to-multiple shuttle transportation between plural DCs and customers is desired because of large numbers of

Corresponding Author: Tsubasa Suzuki (e-mail: tbs-m.66mj@akane.waseda.jp)

parcel transportation. In this study, therefore, area constrained DC placement with multiple-to-multiple transportation routing problem is tackled to design and develop relevant models based on genetic algorithm (GA), then applied to various problems such as part of the DC position is specified, the problem that the specified area should be covered by a DC and the problem to settle more than one DC to cover entire territory.

2 RELATED STUDIES

Many studies on the DC settlement problem have been carried out for many years, *e.g.* Takizawa *et al.* [3]. Especially, Katayama *et al.* [1] devised six models focusing on chromosome design in delivery planning problem. Suzuki *et al.* [2] proposed two design models based on GA for DC location problem. Based on these studies, this paper proposes how to derive solution for DC location problem with area constraints by expanding the model developed by Suzuki *et al.* [2], which is described below.

Figure 1 illustrates the chromosome structure of Model 1 (Two-step hierarchical planning model) and it consists of the following parameters.

- Number of genes: Number of customers
- Numerical value in each gene: Cluster No. (=DC code number)
- Gene position code: Customer number
- Parameters decided by fitness function: Customer number nominated as a DC

Customer No.	1	2	3	4	5	6	7	8
Number of DC=1	1	1	1	1	1	1	1	1
Number of DC=2	1	1	1	1	2	2	2	2
Number of DC=3	1	2	1	3	2	1	1	3

Figure 1: Example of chromosome structure of Model 1.

Model 1 takes the following two-step hierarchical procedure.

Phase 1: Splitting the total customers (I) into N clusters.

Phase 2: determining the DC position by a fitness function of the DC candidates in each cluster. Finally, the Z value defined the aggregated criterion value, is minimized to determine the specified number of DC.

Suzuki *et al.* proposed Model 1 first, but this model has problem on its efficiency. Therefore, they proposed another model, for efficiency improvement, simplicity and general versatility.

Figure 2 illustrates the chromosome structure of Model 2 and it consists of following parameters.

- Number of genes: Number of DC
- Numerical value in each gene: Customer No., which becomes candidate DC
- Gene position code: DC No.
- Parameters are decided by the value of fitness function: DC No., which each customer being delivered from

This model derives the aggregated criterion value Z (sum of ton-kilometer values) by setting N, namely number of DCs, onto chromosome structure as described in the figure and determines the set of DC candidate C that minimizes Z. It is obvious that Model 2 is far simpler than Model 1 for GA procedure and therefore, Model 2 can derive the solution more efficient than Model 1.

DC No.	1
Number of DC=1	3

DC No.	1	2
Number of DC=2	3	6

DC No.	1	2	3
Number of DC=3	2	4	6

Figure 2: Example of chromosomal structure of Model 2

3 RESEARCH PROCEDURE

STEP 1: Description of objective logistics system

Structure of objective logistics problem is re-investigated in collaboration with the joint research company.

STEP 2: Defining area constraint

Area constraints are defined and described.

STEP 3: Modeling

Models, which derive the optimal location of DCs with area constraints, are proposed. These models are developed through extension of the model presented in Suzuki *et al.* [2]

STEP 4: Simulation and discussion

Simulation experiments and comparison analysis in area constraints problems is taken place for discussion.

4 OBJECTIVE LOGISTICS SYSTEM (STEP 1)

Conventional DC location problem proposed by the joint research company is worked on with the following inputs.

- Number of customers (I), Customer number (i=1, ..., I)
- Distance between two customers (M_{ij})
- Demand of each customers (D_i)
- Number of DCs (N), DC number (n=1,..., N)

The parameter "D_i" could has weight, units or volume dimensions, and unit is supposed in this paper. The parameter "M_{ij}" can be distance (kilometres) or time (hours) and kilometre measure is supposed. It is premised that transportation is performed by one to many direct transport between DC and corresponding customers as each customer's demand supposed to be mass volume. Further, based on the practical business situation, possible DC locations are assumed to be the geometric locations adjacent to appropriate customers.

5 DEFINING AREA CONSTRAINT (STEP 2)

In this chapter, two types of area constraint problems are defined.

5.1 Case 1: Some DC locations are settled as designated DCs

This problem is, under given designated DCs, optimal locations of non-designated DCs must be determined. This problem is similar problem examined in Suzuki *et al.* [2], but some DC locations are given as the constraint. Figure 3 shows the example of Case 1.

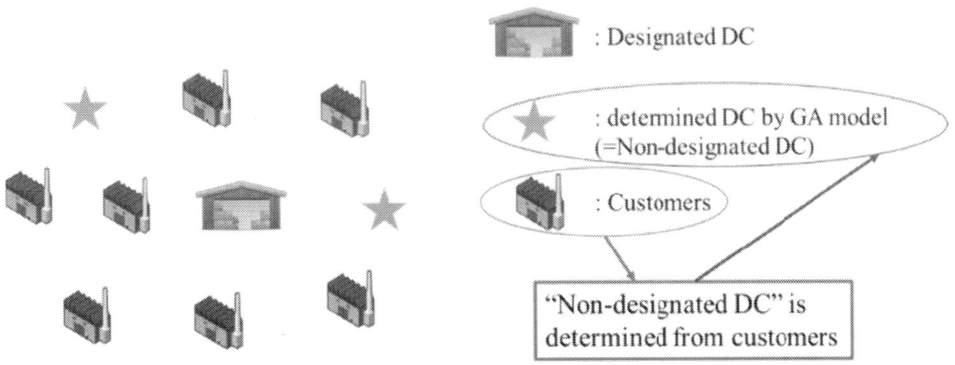

Figure 3: The example of Case 1 (Number of customers: 10)

5.2 Case 2: There are some specified delivery areas

This problem is, under given specified delivery area, optimal locations of DCs in such area and other DCs out of this area must be determined. This problem consists of 2 problems examined in Suzuki *et al.* [2], *i.e.* DC location problem for inside of the specified delivery area and that for outside of the specified delivery area. Figure 4 shows the example of Case 2.

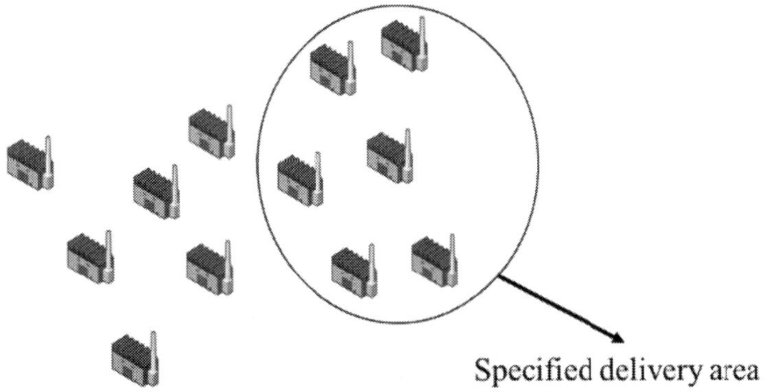

Figure 4: The example of Case 2 (number of customers: 12)

6 MODEL DESCRIPTION OF THE PROBLEMS (STEP 2)

In this section, the corresponding two types of area constraint model descriptions are proposed. The evaluation measure for adequacy of the solution, the optimal DC locations, is sum of "demand-distance of each delivery pass", which is known as "ton-kilometre value" between DC and customers for delivery.

6.1 Case 1: Some DC locations are settled as designated DCs

Case 1 can be classified into following two patterns.

- Case 1-1: All of the DCs are designated.
- Case 1-2: DCs are partially designated.

In the Case 1-1, corresponding model calculates total ton-kilometre value and derives its minimum value to determine the delivery area of each DC. Therefore, GA algorithm is not necessary for this problem from calculation time point of view. On the other hand, in the Case 1-2, GA based model is useful and is extended version of Model 2 discussed in Suzuki *et al.* [2]. Figure 5 shows the chromosome structure of this model for solving the problem of Case 1-2 and it has the following parameters.

- Number of genes: Number of DC
- Numerical value in each gene: Customer number, which can be candidate DC
- Gene position code: DC number
- Parameters decided by the value of fitness function: DC number which each customer is delivered from

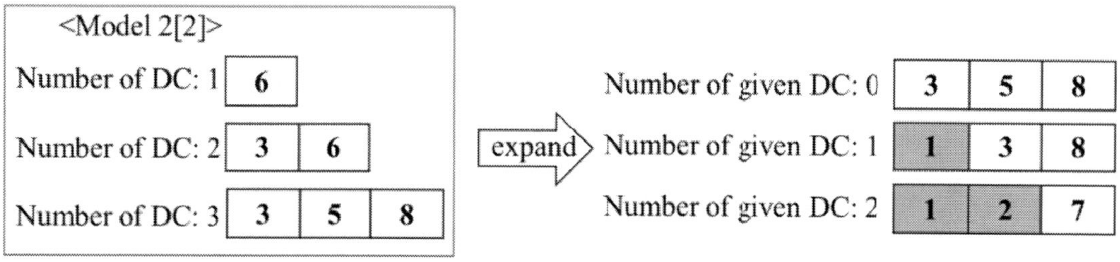

Figure 5: Example of chromosome structure of Model for Case 1-2

Model for Case 1-2 derives the aggregated criterion value Z with chromosome structure of Figure 5 and minimizing Z to determine DC locations.

Grey genes in chromosome represent designated DC and numerical values in grey genes are not changed even if chromosomes go through for mutation and crossover operations. On the other hand, white genes in chromosome represent non-designated DCs and numerical value in these cells can be changed by mutation and crossover. This model determines non-designated DC locations through minimizing value of aggregated criterion Z that includes designated DC locations.

Formulating this model, it is premised that chromosome has l locations of designated DCs and p locations of non-designated DCs which are chosen from the set of customers I. Equations (1)-(5) respectively express how to obtain Z under numerical value in each gene of each chromosome with constraints.

$$Z = \sum_{i \in I} \min_{c \in C} M_{ic} D_i \quad (1)$$

$$C = L \cup P \quad (2)$$

$$c \in C \quad (3)$$

$$l \in L \quad (4)$$

$$p \in P \quad (5)$$

<Symbols>

C	: Set of candidate DC
c	: Candidate DC
L	: Set of designated DC
l	: Designated DC
P	: Set of non-designated DC
p	: Non-designated DC
I	: Set of customers
i	: Customer code number

This model determines solutions below that minimize *Z* in (6).

- Non-designated DCs $p \in P$
- Delivery area of each DC

$$\min_{p \in P} Z \qquad (6)$$

6.2 Case 2: There are some specified delivery areas

Figure 6 shows the chromosome structure of the model for solving the Case 2 problem and it consists of the same parameters as the Case 1-2 model. This model is also developed by expanding Model 2 of Suzuki *et al.* [2].

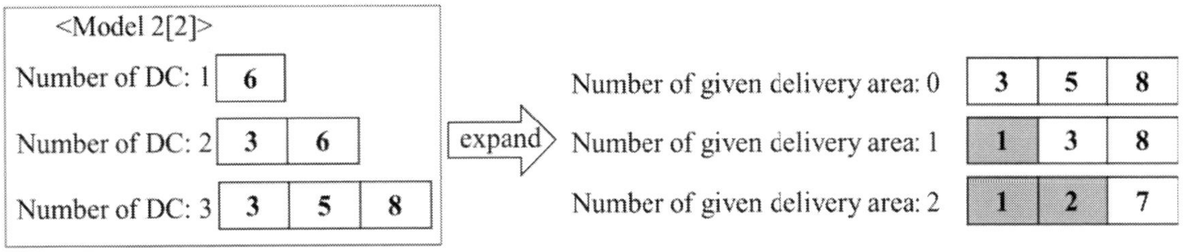

Figure 6: Example of chromosome structure of the Model for Case 2

Grey genes in chromosome represent candidate DCs in each specified delivery area and white genes in chromosome represent candidate DCs out of all specified delivery area.

The solution of the model for case 2 is described here. First, customers must be classified according to the fact whether they belong each specified delivery area, or not. Second, DCs can be determined in each area. Remaining locations of DCs are determined in the outside area of each specified delivery area.

Formulation of this model is described here. As a prerequisite, specified delivery area must be defined. Moreover, chromosome has *K* locations for DC number in designated delivery area and the remaining locations for DC number outside of designated delivery area.

Equation (7) is the function which derives DC location in each designated delivery area. On the other hand, Equation (8) is the function which derives DC location outside of designated delivery area. Equation (9) is the constraint for Equations (7) and (8).

$$Z_1 = \sum_{R_k \in S} \sum_{q \in Q} M_{ih_k} D_i \qquad (7)$$

$$Z_2 = \sum_{i \in I-S} \min_{q \in Q} M_{iq} D_i \qquad (8)$$

$$C = \{\cup_k h_k\} \cup Q \qquad (9)$$

<Symbols>

R_k	: kth specified delivery area (k=1, ..., K)
S	: Set of specified delivery area
h_k	: Candidate DC in kth specified delivery area
q	: Candidate DC outside of S
Q	: Set of Candidate DC outside of S
Z_1	: Aggregated criterion value in S
Z_2	: Aggregated criterion value outside of S

This model derives solutions by minimizing sum of Z_1 and Z_2. This process can be drawn by Equation (10) through following 2 operations

- Creating combination of h_k and q
- Clustering the customers outside of S

$$\min_{h_k, q \in C} Z = Z_1 + Z_2 \qquad (10)$$

7 SIMULATION AND DISCUSSION (STEP 3)

In this section, one case problem of determining DC location with area constraints is described. The detail of the simulation condition is noted below.

- Number of customer: 64 (Considered area is country of Japan except Okinawa Prefecture)
- Number of DCs: 1-3
- Average demand per customer (units): 7,276.5
- Average distance between two customers (km): 607.4

Parameters used by GA model are following

- Population size (PopSize): 500
- Maximum generation (Maxgen): 8,000
- Crossover probability: 70%
- Mutation probability: 1%

Described issues below are the case problems which have various specific area constraints

- Case 0: No area constraint
- Case 1 (1-1): Osaka is the designated DC.
- Case 1 (1-2): Osaka and Miyagi are the designated DCs.
- Case 2 (2-1): One of 3 areas Kanto, Tohoku and Hokkaido area, which are northern area, is set as the specified delivery area.

- Case 2 (2-2): Besides the area defined in (2-1), one of the south areas including Kyushu and Shikoku area is another designated delivery area.

Table 1 shows the aggregated criterion value of each case.

Table 1: Comprehensive evaluation value (unit-km)

Case		Number of DC		
		1	2	3
Case 0		165,412,261	90,944,616	59,858,812
Case 1	(1-1)	217,439,329	92,333,002	59,951,106
	(1-2)	-	150,178,485	79,211,223
Case 2	(2-1)	-	103,039,164	63,712,631
	(2-2)	-	-	66,953,061

It is noticed from this table that the more the number of area constraints (in other words, the lower flexibility is), the worse aggregated criterion value is. The larger the gap between designated DC/area and the result of Case 0 (which has no area constraints), the worse solution is. Moreover, it is noticed that gap of each solution from Case 0 could be an index of validity to settle some DCs or delivery areas.

8 CONCLUDING REMARKS

In this study, GA based models for determining DC locations with area constraints were proposed, which is often discussed by logistics companies. There are two types of area constraints. One is the case that the DCs are partially designated and the other is the case that there are designated delivery areas. This study focuses on devising particular chromosome strucuture to solve these problems and it is significant for obtaining the effective solutions.

REFERENCES

[1] Katayama, H. and Suganuma, S., "On Application of Genetic Algorithm to Distribution Planning Problems -Classification of Encoding/Decoding Methods and their Comparison-", Proceeding of the 14th Annual Meeting of Japan Society of Logistics Systems, pp. 59-64 LEC Tokyo Regal Mind University, Tokyo, 2011.

[2] Suzuki, T., Katayama, H., Takizawa, G., Takezawa, K., Ishiwatari, N. and Hayashi, N., "On Location Design Problem of Distribution Centres for Large-scale Logistics Operation", Proceeding of the 2016 International Symposium on Semiconductor Manufacturing Intelligence (ISMI2016) (forthcoming), National Tsing Hua University, Hsinchu, Taiwan, 2016.

[3] Takizawa, G., Morito, S. and Katayama, H., "A Study on Design of Distribution Systems - Hierarchical Planning of Locating Distribution Centre and Routing-", Proceeding of the 17th Annual Meeting of Japan Society of Logistics Systems, pp. 107-114, Faculty of Science and Engineering, Waseda University, Tokyo, 2014.

SUPPLY CHAIN ARCHITECTURE FRAMEWORK (ITHPBGA*) PRODUCTS/SERVICES BASED ON IT INFRASTRUCTURE

Kaveh M. Cyrus (PhD.)[1] and Nader Nematpour Kojal (MSc.)[2†]
1Department of Industrial Engineering and Management Systems
University of Amirkabir (Tehran Polytechnic), Iran
cyrusk@aut.ac.ir

2Department of Industrial Engineering and Management Systems
University of Amirkabir (Tehran Polytechnic), Iran

nnkojil@aut.ac.ir

ABSTRACT

Nowadays competitive business environment information obviously plays a specific role on both macro (strategical) and micro (operational) company's insights. Banking industry is not exception in this environment and its complexity increases by time goes by. According to rapid changing environment, IT companies providing services to banks are always under pressure, and should respond quickly and creatively. However, this has led IT to be looked at not only as a facilitator, but as an enabler tool for organizational goals achiever. Although IT hardware, networks and methodology have had good progress in country's banking field, but comprehensive, integrated programs and tailored applications for IT infrastructure did not well developed. Covering those deficiencies, Data Processing (DP) Bank Group Corp tried to rely on enterprise architecture issues, which provides a suitable approach to face. IT based infrastructure (ITPBDPGA) provided. Concepts and tools presented in previous CIEs articles showed parts of ITPBDPGA architect. However, the aim of this paper is to review all those parts emphasizing on information architecture. Offers the frame work for three tier supply chain architecture with the help of BI-art technologies. A case study of local active in online banking field corporation was employed to show how this presented methodology helped bank administration to manage better and more sufficient on their organizations. This paper describes those processes are used to build Enterprise Architect including Business, Information, Applications (CPM and BI systems) and IT Infrastructure.

Keywords: Enterprise, Architecture

1 INTRODUCTION:

Features modern world organizations, have recognized that information is the lifeblood of the digital economy and the key to success in the information age, make decisions that consistently, better and faster than the competition does. Bad business decisions and decisions based on inadequate or incomplete information at least, the loss of millions of dollars to bring. As a result, their market share outside the main and majority or even bankrupt the company.

---

* Information Technology Parsian Bank Data Processing Group Architecture

† Corresponding Author

Iranian banks are looking after proper tools for developing their own technological solutions. So, first they traditionally centered approach to solve business problems using applications that after a while, a large number of applications and application of IT was established Each with several sub-sets of business operations cover the overlapping and sometimes they Failure to integrate new applications with existing applications were complicated conditions As a result, banks are now faced with disparate information technology systems that are often not flexible and are inefficient and not aligned with business needs.

At the same time, banks continued to keep pace with the changing needs of business, process optimization and infrastructure were old To reduce complexity, improve agility, standardization and alignment of business and IT and business and promoting effective competitive advantage It is why a growing infrastructure-based IT and use of tools in the field of electronic banking products are always taken into consideration and the IT industry of electronic banking Iran with regard to planning and investments made appreciable progress in the is Constantly looking to create business transformation, migration to modern banking change of attitude in architecture, frameworks and models of international banking standard.

Lack of an IT architectural comprehensive map used in the banking industry is one reason to delays in service delivery. The results of 94 banks that have to define or redefine its organizational architecture project, Shows only three banks failed to implement enterprise architecture and governance of their IT Department. This means. [1]

Introducing ITHPBGA framework (including plans that are in line with banking industry development and use of effective IT infrastructure running), the following basic questions were answered:

DP Group Corp Despite the development and implementation of business strategies and investment programs for multiple applications on CPM & B system seek the effectiveness ensure of these systems. To ensure strategic alignment and operational plans, identification and selection of key indicators and strengthen the rule of IT, acting three layers architecture with a focus on business architecture, information architecture as well as architectural applications in accordance with the expectations of its stakeholders, DP Group Corp acted as following overview as figure 1.

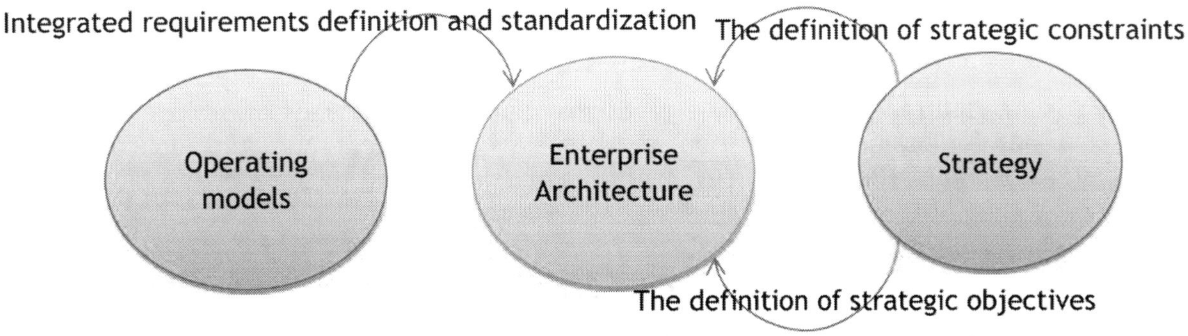

Figure 1: Enterprise Architecture as Strategy

The above figure is enterprise architecture as a bridge between the organization's strategic and operational modules. In other words, this architectural is required addition to the strategic objectives and stakeholders needs. And the role of IT governance in order to achieve the organization's objectives and thereby ensure the necessary information is available to decision-makers in the organization. [2].

Looking back, to align and integrate strategies, enterprise architecture and operational model, special BI systems utilized management information system is required. Such systems will provide, integrate and multi- dimensional analysis intelligent data gathered from different sources. BI systems integrate interior and exterior gathered data. [6]

BI process helps organization's decision making to improve its performance in three strategically, tactical and operation's levels and bring it to a better market competitive advantage place. Proper decision is impossible with no related data in hand and ability to analyse those data. [6]

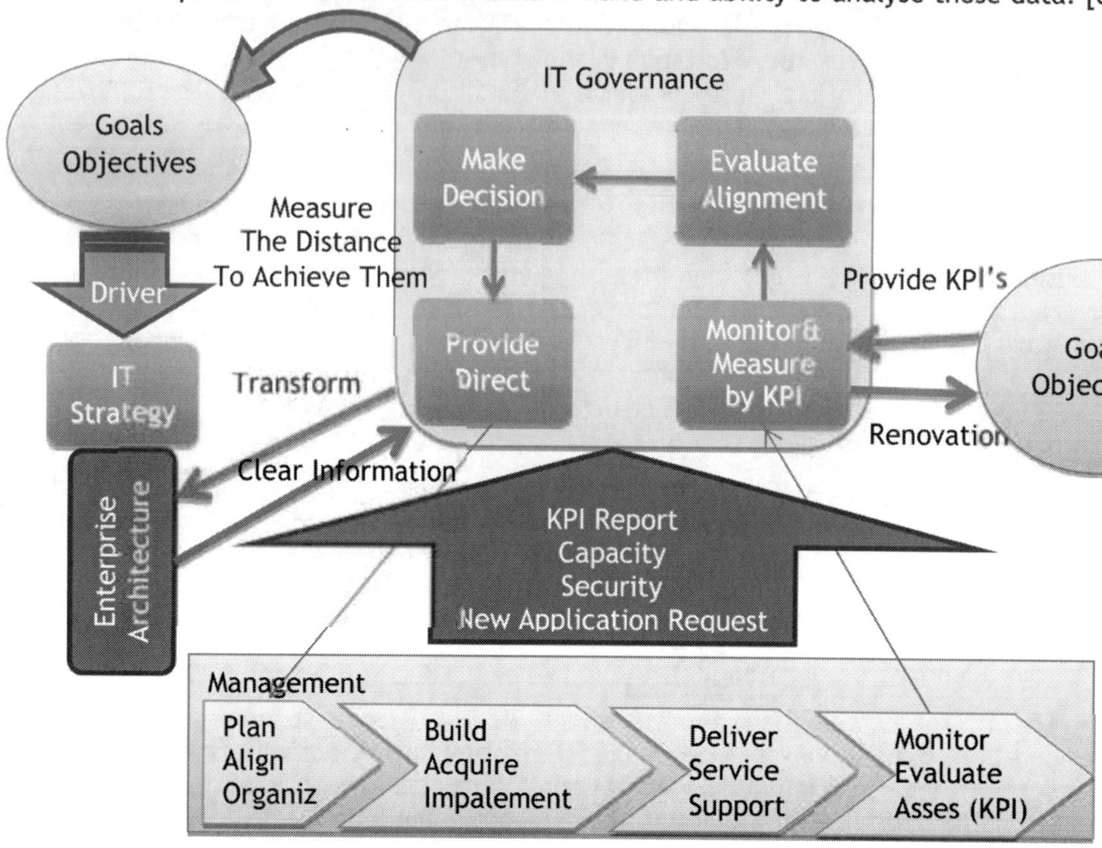

Figure 2: Comprehensive Map

The business and information technology BI Logic subjects has been big issue for long time and brought enough attention. By general definition, BI Logic is broad terms that refer to a set of software systems which analyse data, convert those into usable information. [6]

The main goal of enterprise architecture problems supplying the needs of stakeholders Offering suitable solutions for business integration and process management And run applications in the following areas of information technology, which has been referred to some of the most important remaining problems [1].

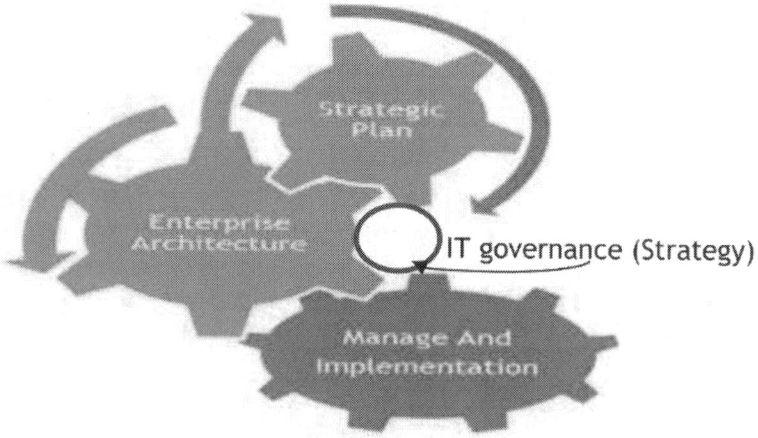

Figure 3: Missing Ring

But for continuity and stability of system of CPM & BI within the organization as well as the calibration of the alignment of the device with the desired objectives, strategic priorities, creating system architecture systematically need to be found removing; implementation and improvement of processes and infrastructure objectives will be provided. [1]

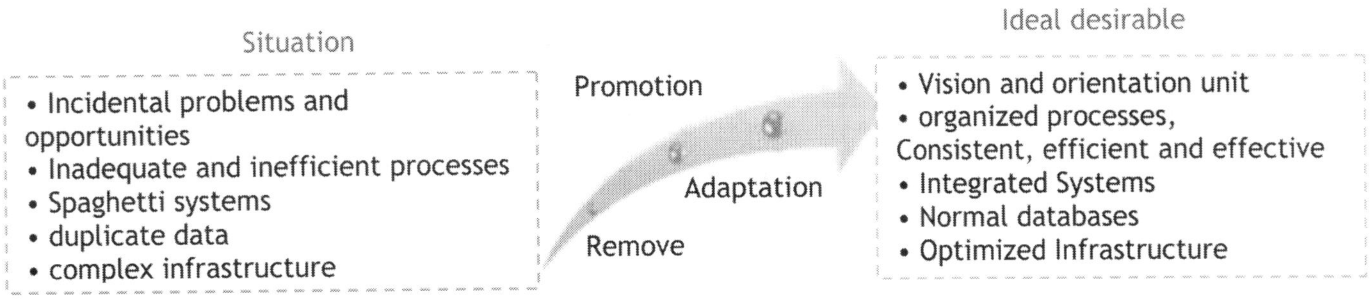

Figure 4: Enterprise Architect Process

2 THE ROLE OF ENTERPRISE ARCHITECTURE

As shown below, beginning the process of measuring and monitoring centers section defines IT governance, To do with evaluating and monitoring indicators, preparations for upgrade or remove unnecessary processes and provide improved control system with the requirements of the government team is data processing enterprise information architecture provides a clear and transparent then pivot capability for in line and well-integrated. [7]

Following describes the constituent elements of enterprise architecture will be discussed Pyramid. ITHPBGA framework describes how the different layers of technology planning and deployment DP Information Group Corp shows the enterprise architecture group DP as pyramid is named Parsian Bank (Information Technology Parsian Bank Data Processing Group Architecture).

This framework is appropriate method for integrating different viewpoints and providing organization's needs aspects of. As showed in figure below, based on to information technology layered approach enterprise architecture has been applied bottom up. Substrate and its upper layer carrier is planning to implement top down. [1]

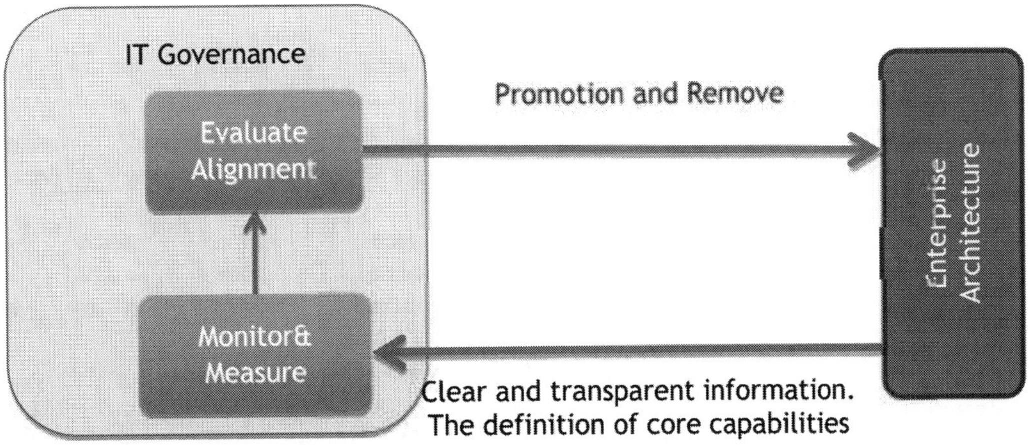

Figure 5: Enterprise Architect IT Governance [3]

Based on IT enterprise architecture, the pyramid show the participation of all strategic and operational levels of the organization is in the process architecture is required. [3]

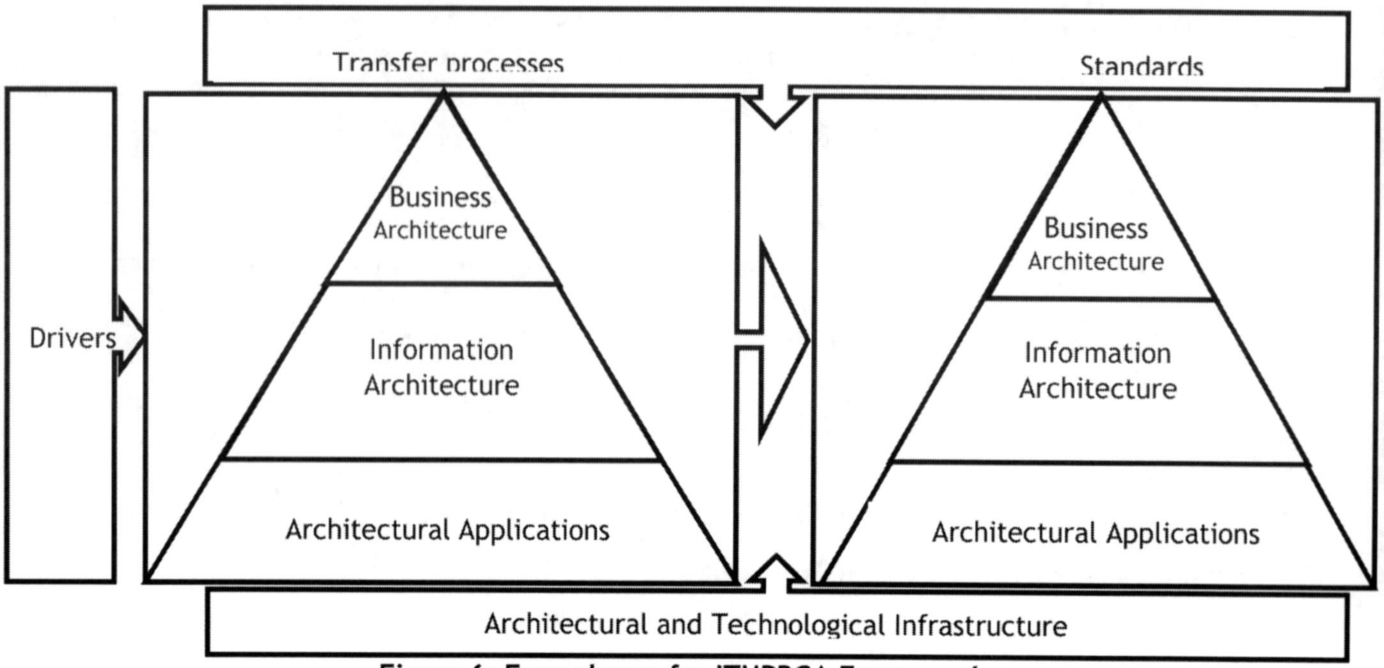

Figure 6: Four phases for ITHPBGA Framework

3 INTRODUCTION OF THE MAIN CAUSES OF THE ITHPBDPGA

The ITHPBDPGA, a framework that aims to meet the needs of the banking, especially electronic banking supply chain using IT capabilities, is offered with three different layers. The framework describes how IT can e-banking in line with the objectives, needs and problems of supply chain product (goods and services) Group Data Processing Parsian Bank was used and how it is structured. In this section each of the layers and their components is described. To reach our objective in IT Dept., we had to take some steps. As you see in Figure 6, 3 Steps were introduced. These steps include: [8]

3.1 The first layer

The first layer includes 5 essential architecture components. ITHPBDPGA, offers:

3.1.1 Drivers

Several drivers were considered for the architecture, but the major ones are the needs and problems of electronic banking industry and lack of suitable infrastructure technology enablers are appropriate. Budgetary constraints have forced finances the investment on information technology act carefully and save more and solid economic reasons and justifications provided for the costs.

3.1.2 Architectural Current Situation

Although the field of IT is considered by senior managers of banks, but look at it is less a process view. Therefore, systematic efforts were made to elicit processes and improve them. The framework ITHPBDPGA with a different perspective to understanding the process, the relationship between them and the flow of information is dealt with that can be used in the form of business architecture output. Organization of special architectural interest are not architectural past the status quo Therefore, selection criteria and application deployment needs of senior managers was cross-sectional and a maximum of two of the pyramid architecture business architecture and information architecture constituted. As a result, for the status quo, merely to study the processes related to the main business architecture and information flow architecture is investigated.

3.1.3 Ideally Architecture

Business Architecture desired status is determined based on the output of the current situation

And core processes and support this level due to the mentioned problems, the organization's goals, expectations of various stakeholders DP Group Corp and learning from APQC model was revised. Then, new information requirements and the needs of the organization, specific and desired status information architecture was completed. At the end of the third level enterprise architecture required applications as business requirements and architecture information system architecture level of cover offered.

3.1.4 IT Infrastructure Architecture

The remarkable thing about architecture is an infrastructure that cannot be seen as in most architectural framework; only at the lowest level of the pyramid architecture and architectural components were affected by high levels. Because this architecture based on what banks can be seen in the field of IT, architecture is predetermined. This level of architecture in the present situation has a series of network infrastructure and the ideal situation According to what is prescribed in the design of electronic banking should include a series of new infrastructure facilities that they have access to the Internet and telecommunications networks (mobile) services. So have the high-level architecture design and equipped in such a way that due to the architecture of the IT infrastructure are implemented. So the overall architecture of the IT infrastructure is affected more than the upper levels, these levels are impressive. [7]

3.1.5 Standards and Transportation Processes

This component of the framework ITHPBDPGA, including extensive discussions and further research is needed therefore not discussed in this article. Transfer processes when they are faced with the implementation of the framework ITHPBDPGA, is on the agenda. The process of how to move from the status quo to desired status set. The transition should be a series of principles and standards architecture was explained at different levels in the design, implementation and maintenance of architecture over time to be considered. These principles and standards based architecture interact with each other and higher levels of configuration management architecture is impressive. These principles and standards, information security, technical, and technological.

3.2 The second layer

Architecture Framework ITHPBDPGA, the second layer models that have business architecture, IT infrastructure and applications develop, it becomes clear. In this layer as a set of products and methods to localize their output architecture, develop. This layer has an output matrix which version of Zachamnis endemic. Zachamn-as the best tool to extract a general framework-the architecture was selected output levels. [7]

3.2.1 Describe the components of Pyramid Architecture

Business architecture layer: tip of the pyramid business IT architecture and business aspects related to the organization explains. At this level over such issues as business strategies and technology organization, the decision on commercial paradigms field of information technology, electronic banking business is done.

- Business architecture, information architecture, diverse base
- What information is needed to support the company's processes?

Information architecture Layer: the information needed for effective application of IT in the organization's processes optimization and modelling. Then, based on the clustering of information and communications, intelligence groups with the working process of the system required is determined. The main information needed to perform organizational tasks identified in the information architecture layer. This layer handles data models, logical data repositories and their relationship with the organization and functions of the system are identified and defined applications and issues such as knowledge management mechanisms and procedures set forth in this layer.

- Architecture information, basic stop for architectural applications
- What systems are necessary to manage the information?

Application architecture layer: This layer consists of systems applications, such as performance management system (CPM), system management dashboards (BI), Enterprise Resource Planning (ERP), customer relationship management (CRM), management information systems (MIS), supply chain management systems (SCM) is. Identifies and describes applications and modules and their relationship with the organization's processes and other applications in architecture, application layer is done.

- Architecture applications, based architecture arrest for IT infrastructure
- What infrastructure is needed to support these systems?

This technological aspects it is Excessive fascination to this layer is a mirage that attracted the eyes of many of our organizations and prevent them from understanding the philosophy and principles (upper layers) have been unaware information technology applications. How many organizations are equipped with the latest model of computing technologies but the lowest percentage of benefits they absorb it. [3]

3.3 The third layer

Architecture ITHPBDPGA, in the third layer, the planning process will determine the mining enterprise architecture. Details are shown in below figure. [8]

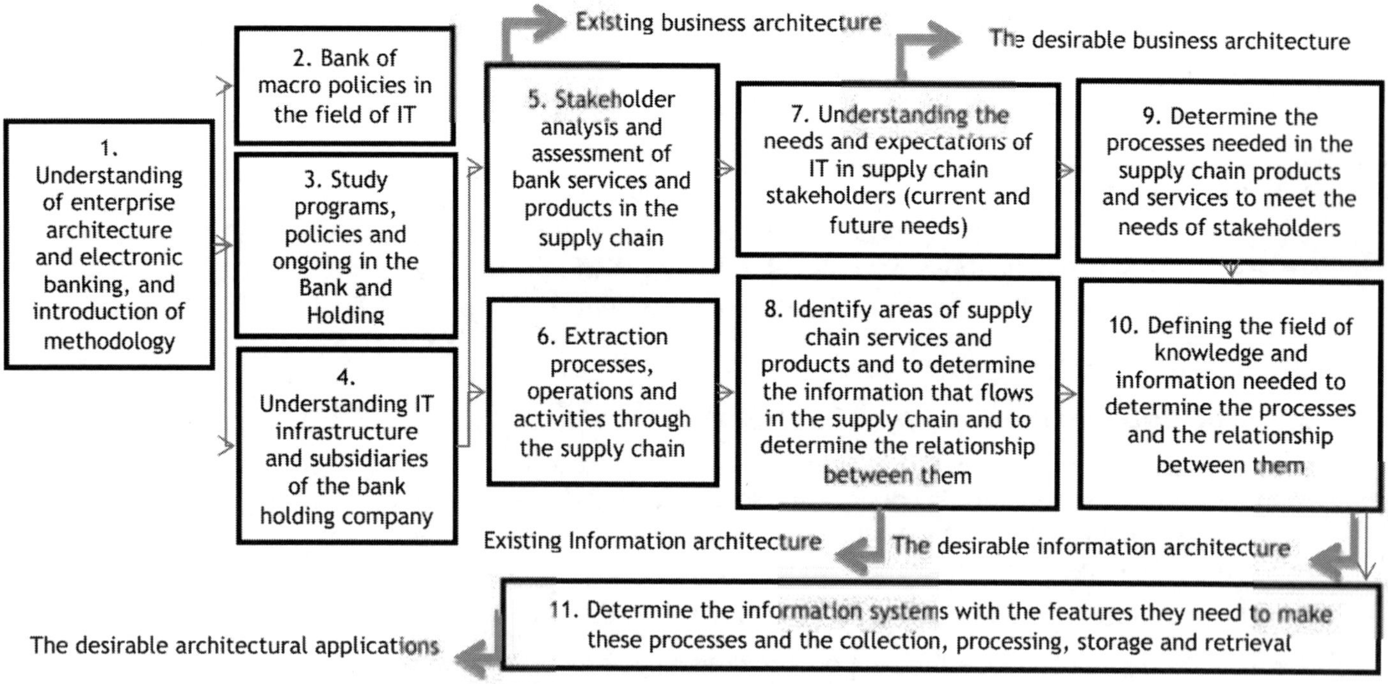

Figure 7: Framework ITHPBDPGA

4 CONCLUSION

There are certain requirements more than any other factor in the formation of electronic banking IT projects Parsian Bank Group data processing of the stimuli applied in the field of enterprise architecture bank. Therefore, the main objective of the framework ITHPBDPGA, trying to use IT infrastructure to solve problems and address the needs of the banking sphere with an emphasis on supply chain products / services.

This framework is primarily based on the drivers of the existing architecture, including business architecture and information architecture is the operations and processes in the supply chain of goods / services were extracted. The information contained in this operation and process flow status quo was extracted in the form of information architecture. According to the bank's IT infrastructure capabilities Corp, Ed optimal business architecture. Moving from the status quo to desired status and implementation of the architecture ideally require some transfer mechanisms and standards. In general, we can provide a framework for all banking systems cited as the most important feature of this framework this makes the framework generalize to other similar areas and have the ability to implement. The following drivers and influencing factors that are shaping this context and the relationship between its components most banks are in the same country.

5 REFERENCES

[1] **Weill, P. & Ross, J. W.,** "IT Governance: How Top Performers Manage IT Decision Rights for Superior Results", Harvard Business School Press, Boston.

[2] Exclusive Interview with John Zachman **by Rogers Session**, 2007.

[3] **Klaus D. Niemann,** From Enterprise Architecture to IT Governance, ISBN 3-528-05856-0

[4] **Kaveh M. Cyrus2 (PhD.) & Nader Nematpour Kojal1 (MSc.).**2013.Introducing a Road map to Deploy strategic process in local organizations successfully; Case study: Iranian Large Car Manufacturing Co., CIE43 Proceedings, 16-18 October 2013, University of Hong Kong, pp 183.

[5] **Kaveh M. Cyrus2 (PhD.) & Nader Nematpour Kojal1 (MSc.).**2014. Creating Corporate Performance Management Road Map For Filling Gaps Between Our Plans (Where We Want To Be) And Our Executions (Where We Are)Case Study: Iranian Car Manufacturing Large Company, CIE44 & IMSS'14 Proceedings, 14-16 October 2014, Istanbul / Turkey, PP 241.

[6] **Kaveh M. Cyrus2 (PhD.) & Nader Nematpour Kojal1 (MSc.).**2015. B I's Performance Management System in Strategy base Corporations (BI Logic) Case Study: Iranian Bank Data processing group Company, CIE45 & IMSS'14 Proceedings, 28-30October 2015, Metz/France, pp 80.

[7] **Weill, P., Ross, J. W., Robertson, D.** "Enterprise Architecture as Strategy: Creating a Foundation for Business Execution"

[8] Chief Information Officers Council; Federal enterprise architecture framework version 1.1, September 1999.

ROBUST MODELING AND PLANNING OF RFID NETWORK FOR PORT LOGISTICS UNDER UNCERTAIN CONDITIONS

XU Bo-wei[1] and LI Jun-jun[2*] and YANG Yong-sheng[1]
[1]Logistics Research Center
Shanghai Maritime University, Shanghai, China
xubowei138@126.com

[2]Merchant Marine College
Shanghai Maritime University, Shanghai, China
jsliljj@163.com

ABSTRACT

To realize higher coverage rate, lower reading interference, and cost efficiency of RFID network when the tags' position is uncertain in port logistics, a novel robust RFID network planning (RNP) model is built, and a new robust particle swarm optimization (RPSO) is proposed. In RNP model, coverage model is established by referring the probability sensing model of sensor with uncertain sensing range; reading interference is calculated by Concentric Map based Monte Carlo method; cost efficiency is described with the quantity of readers. In RPSO, a sampling method, whose sample size varies with the number of iterations, is put forward to improve the robustness of RPSO within limited sample size. In specific, the exploitation speed in the prophase of RPSO is quickened by small expected sample size; the exploitation precision in the anaphase of RPSO is ensured by large expected sample size. Simulation results show that compared to other three methods, the planning solution obtained by this work is more conducive to enhance coverage rate, reduce interference and cost.

Keywords: Radio Frequency Identification (RFID) network planning, Port logistics, Uncertain environment, Robust, Particle swarm optimization (PSO)

1 INTRODUCTION

With the development of information technology, there is an increasing usage of Radio Frequency Identification (RFID) network in port logistics. In RFID network system for port logistics, tags are usually put on one side of items. As the tagged items may be placed randomly, there is great uncertainty in tags' positions. Moreover, the RFID coverage performance is largely influenced by the number and positions of the RFID devices. Thus, the balance among coverage rate, interference and cost in RFID network construction need to be comprehensively considered [1,2]. Therefore, how to adopt the intensive way of deploying the readers to obtain higher coverage rate and less interference within the budge becomes a key issue for the popularization and application of RFID network in port logistics.

* Corresponding Author

How to deploy the minimum number of readers for covering all tags in the entire space is known as the RFID network planning (RNP) problem [3]. RNP problem is one of fundamental problems in large-scale RFID networks [4]. In the past decades, extensive efforts have been made to model and plan RFID network under certain conditions [1,2,5-7], and coverage problem is one of fundamental issues in RNP.Gong, Shen, Zhang, Kaynak, Chen, & Zhan [1] formulated a model of RNP, in which tag coverage, the number of readers, interference and the sum of transmitted power were considered. Liu & Ji [2] proposed an optimization model of RFID network system to solve the problem of how to place readers, taking the coverage rate and load balance into consideration. Giampaolo, Forni, & Marrocco [6] developed a simple and effective model, in which the performance indicators consisted of coverage efficiency, overall overlapping, total network power and cost.

The RFID network planning problem has been proved to be NP-hard for its nonlinearity and complexity. In recent years, evolutionary computation (EC) and swarm intelligence (SI) have become an effective tool for solving the problem of RFID network planning, such as genetic algorithm (GA)[3,5], plant growth simulation algorithm (PGSA) [4], particle swarm optimization (PSO) [6,7], artificial colony algorithm (ACA) [8], etc. Yang, Wu, Xia, & Qin [5] proposed a genetic algorithm based RNP method, which included mapping RFID network, planning issues into genetic algorithms, presenting problem states by using gene and chromosome, and implementing the mechanisms of individual selection and genetic operation. Simple and effective models of electromagnetic elements involved in RNP were developed and included in the frame of a particle swarm optimization algorithm [6]. An improved particle swarm optimization based on genetic algorithm was proposed to solve the problem of how to place readers so that the readers can effectively get the informations of multiple tags [7]. A k-coverage model, which was formulated as a multi-dimensional optimization problem with constraint conditions, was developed to evaluate the network performance. And the plant growth simulation algorithm (PGSA) is used to optimize RFID networks by determining the optimal adjustable parameters in the model[4]. Ma, Hu, Zhu, & Chen [8] proposed a cooperative multi-objective artificial colony algorithm to find all the pareto optimal solutions and to achieve the optimal planning solutions by simultaneously optimizing four conflicting objectives (tag coverage, reader interference, economic efficiency, load balance) in multi-objective RNP. Lin & Tsai [3] proposed a micro genetic algorithm (mGA) with novel spatial crossover and correction schemes to cope with this constrained three-dimensional reader network planning problem. The proposed algorithm was computationally efficient, which allowed a frequent replacement of the RFID readers in the network to account for the fast turnaround time of the stored objects in the warehouse, and guaranteed 100% tag coverage to avoid missing the records of objects.

RFID is one of the recent wireless communication technologies that is increasingly being used in various fields, such as logistics, supply chain management, etc [9,10]. It is very important to consider uncertain conditions in wireless sensor network system for the performance improvement. Due to some uncertainties, the sensor deployment problem of wireless sensor networks is a hot issue. Li, Xu, Pan, & Pan [11]; Ozturk, Karaboga, & Gorkemli [12] proposed probability sensing model when the detection region of sensor was uncertain. Vu & Zheng [13] presented a systematic study of location uncertainty's impacts on the coverage properties of wireless sensor networks, and devised an efficient polynomial algorithm. Vu & Zheng [14] carried out a rigorous study of location uncertainty's impacts on the accuracy of target localization and tracking, and proposed an effective algorithm based on order-k max and min Voronoi Diagrams. However, there are several different properties which make these approaches used in WSNs inapplicable to RFID networks [4]. What is more, the researches on RNP under uncertain conditions are rare up to now.

To deal with complex uncertain problem, it is simple and effective to combine intelligent optimization algorithms and robust optimization [15-18]. However, the robust optimization is generally based on Monte Carlo Integral, while cyclic iteration is usually used in intelligent optimization algorithms. This straight-forward method needs many times of fitness evaluation, which brings large calculation and long computing time. Therefore, how to keep search performance while reducing the computational complexity, in other words, how to improve the search performance in the case of limited sample size, is a problem deserving study.

Although the modelling and planning of RFID network has drawn extensive attention, many scholars only focus on RNP under certain condition. Few studies have considered several objectives under uncertain conditions about the decision-making for RNP problem. To enhance the reliability and stability of RFID network system for port logistics, the uncertain position of tags is considered, and the modelling and planning of RFID network under uncertain conditions is studied in this work. Firstly, a robust planning model of RFID network for port logistics is built, in which coverage rate is analysed on the basis of the probability sensing model of wireless sensor network under uncertainty; Secondly, Concentric Map based Monte Carlo method is applied to calculate the interference; Thirdly, to enhance the searching capability, a robust particle swarm optimization (RPSO), which can trade off exploitation speed and precision, is put forward to solve the robust planning problem of RFID network. Lastly, simulation results indicate that the proposed method possesses a better robust optimization capability.

The rest of this work is organized as follows: Section 2 formulates robust modelling of RFID network planning under uncertain conditions. Section 3 designs a robust particle swarm optimization, and gives the pseudo-code of RPSO for RFID network optimization problem. Simulation experimental tests are carried out to verify proposed model and algorithm in Section 4, with results thoroughly analysed. In Section 5, conclusions are drawn.

2 RFID NETWORK PLANNING PROBLEM UNDER UNCERTAINTIES

The key to the RFID network planning problem is the deployment of readers to satisfy multi-objective requirements due to the limited range of reader-tag communication. First, it is hoped that the tags can be identified as many as possible within the budge. Second, reading interference is closely related to the reader collision problem [19], which may occur when tags are located in the overlapping area of any two readers' interrogation zones, and both readers read tags simultaneously [3]. Consequently, the number of tags located in the overlapping area of interrogation zones should be as small as possible. Third, the smaller the number of placed readers is, the better the cost. Therefore, the proposed planning model of RFID network aims to optimize a set of objectives, such as optimal tag coverage, minimize reading interference, economic efficiency simultaneously by adjusting the system control variables (the coordinates of readers, the number of readers, etc.). In view of the great uncertainty in tags' positions, coverage and interference analysis in our work is different from prior studies.

The deployment region of RFID network system is supposed as a two-dimension square domain which consists of several tags. Assuming that the uncertain area of tag's position is a circle with radius R_b, a conceptual view of tags' uncertain positions is shown in Fig.1, in which coarse dots indicate tags and small circles are ranges of tags' uncertain positions.

2.1 Nomenclature

N_b	Number of tabs
N_{max}	Max number of readers deployed
R_R	Sensing radius of reader
N_R	Number of readers deployed
(x_S^i, y_S^i)	Coordinate of the ith reader
(x_O^j, y_O^j)	Coordinate of the jth tag

$d(s_i, o_j)$	Distance between the ith reader and the jth tag
R_b	Radius of uncertain position of tag
$reader_{ij}$	Capability of the ith reader to identify the jth tag
c_O^j	Coverage of tag which is overlapped by several readers
$o_k^j(x'^j_k, y'^j_k)$	The kth sampling site within the uncertain range of the jth tag
$reader_{ik}^j$	Capability of the ith reader to identify the sampling site o_k^j
$Over_k^j$	Reading overlap of the kth sampling site within the uncertain area of the jth tag

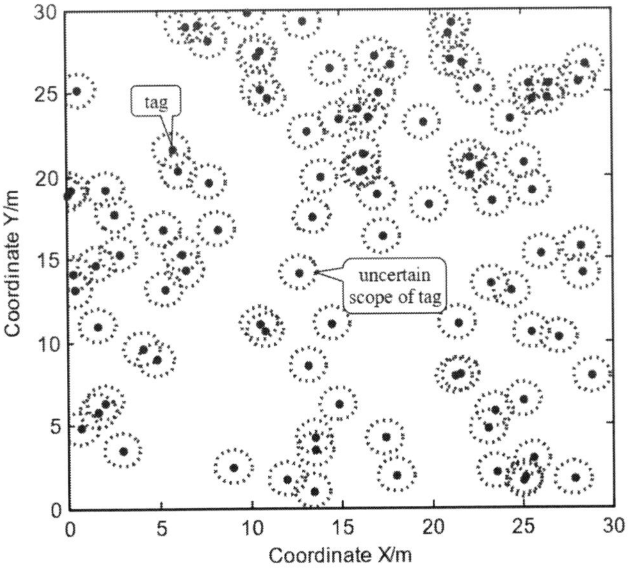

Figure 1: Tags and ranges of their uncertain positions

2.2 Coverage rate

Coverage rate is the most important for RFID network. By referring the probability sensing model for wireless sensor networks in references [11,14], $reader_{ij}$ is described as Eq. (1):

$$reader_{ij} = \begin{cases} 0, & R_R + R_b \le d(s_i, o_j) \\ e^{(-\lambda_1 \alpha_1^{\beta_1} \alpha_2^{\beta_2} + \lambda_2)} & R_R - R_b < d(s_i, o_j) \le R_R + R_b \\ 1 & d(s_i, o_j) \le R_R - R_b \end{cases} \quad (1)$$

Where $\alpha_1 = R_b - R_R + d(s_i, o_j)$, $\alpha_2 = R_b + R_R - d(s_i, o_j)$, λ_2 is disturbing effect, $\beta_1, \beta_2, \lambda_1$ are measuring parameters of detection probability, $d(s_i, o_j) = \sqrt{(x_S^i - x_O^j)^2 + (y_S^i - y_O^j)^2}$. Then, c_O^j can be described as Eq. (2):

$$c_O^j = 1 - \prod_{i=1}^{N_R} (1 - reader_{ij}) \quad (2)$$

In order to make the coverage rate comparable in cases of different scales, f_1 which denotes the coverage of RFID network can be obtained through Eq. (3) by referring to references [1,2,6].

$$f_1 = \frac{100}{N_b}\sum_{j=1}^{N_b} c_o^j \qquad (3)$$

2.3 Interference

Interference mainly occurs in a dense reader environment, where several readers try to interrogate tags at the same time in the same area. This results in an unacceptable level of misreads [20]. Due to the uncertain positions of tags, it is complicated to use geometric analysis method to deal with interference problem. And it is hard to build approximate mathematical model like the coverage model. Therefore Monte Carlo sampling method is adopted here.

If the ith reader can identify sampling site o_k^j, $reader_{ik}^j = 1$; otherwise, $reader_{ik}^j = 0$. Where $k = 1,2,\cdots,K$, K is the number of sampling sites. When $R_R + R_b < d(s_i,o_j)$ or $R_R - R_b \geq d(s_i,o_j)$, $reader_{ik}^j$ can be attained straightly by Eq. (4):

$$reader_{ik}^j = \begin{cases} 0, & R_R + R_b < d(s_i,o_j) \\ 1, & R_R - R_b \geq d(s_i,o_j) \end{cases} \qquad (4)$$

When $R_R - R_b < d(s_i,o_j) \leq R_R + R_b$, Concentric Map method [21] is applied. For a unit circle, whose center is in origin of coordinate, polar coordinates of sampling site are:

$$r = \varepsilon_1, \theta = \frac{\pi}{4}\cdot\frac{\varepsilon_2}{\varepsilon_1}, if \quad \varepsilon_1 > |\varepsilon_2| \qquad (5)$$

$$r = \varepsilon_2, \theta = \frac{\pi}{4}\cdot(2 - \frac{\varepsilon_1}{\varepsilon_2}), if \quad -\varepsilon_2 < \varepsilon_1 \leq \varepsilon_2 \qquad (6)$$

$$r = -\varepsilon_1, \theta = \frac{\pi}{4}\cdot(4 + \frac{\varepsilon_2}{\varepsilon_1}), if\ \varepsilon_1 \leq -\varepsilon_2, \varepsilon_1 < \varepsilon_2 \qquad (7)$$

$$r = -\varepsilon_2, \theta = \frac{\pi}{4}(6 - \frac{\varepsilon_1}{\varepsilon_2}), if\ \varepsilon_2 \leq \varepsilon_1 \leq -\varepsilon_2, \varepsilon_2 \neq 0 \qquad (8)$$

$$r = 0, \theta = 0, if \quad \varepsilon_1 = \varepsilon_2 = 0 \qquad (9)$$

Where, ε_1 and ε_2 are random real numbers in the range of $[-1,1]$. Cartesian coordinate of sampling site is $(x,y) = (r\cos\theta, r\sin\theta)$. In this work, Cartesian coordinate of the jth tag is $(x,y) = (x_O^j + R_b r\cos\theta, y_O^j + R_b r\sin\theta)$, because the jth tag is located in (x_O^j, y_O^j) and the uncertain area of tag's position is a circle with radius R_b.

The precision of random sampling is low, for $\varepsilon_1, \varepsilon_2$ are random real numbers in the range [-1,1]. A low discrepancy sampling, Korobov Lattic [22] is adopted to enhance coverage calculation precision. Korobov Lattic is defined as: $P_K = \{\frac{m}{K}(1, a \bmod K, \cdots, a^{s-1} \bmod K) \bmod 1 : m = 0,\cdots,K-1\}$. Here, it is set that $a = \lceil\sqrt{K}\rceil$.

If $\sqrt{(x_S^i - x'^j_k)^2 + (y_S^i - y'^j_k)^2} \leq R$, $reader_{ik}^j = 1$; otherwise, $reader_{ik}^j = 0$. $Over_k^j$ can be described in Eq. (10). Referring to reference [6], f_2 which denotes the average interference, is described in Eq. (11).

$$Over_k^j = \begin{cases} 0, & \sum_{i=1}^{N_R} reader_{ik}^j \leq 1 \\ -1 + \sum_{i=1}^{N_R} reader_{ik}^j, & \sum_{i=1}^{N_R} reader_{ik}^j \geq 2 \end{cases} \qquad (10)$$

$$f_2 = \prod_{j=1}^{N_b} \frac{1}{1 + \frac{1}{K}\sum_{k=1}^{K} Over_k^j} \tag{11}$$

2.4 Cost

The quantity of reader is a key impact factor of RFID network cost. f_3 denotes RFID network cost.

$$f_3 = (N_{\max} - N_R) / N_{\max} \tag{12}$$

2.5 RFID network planning model under uncertainties

Robust planning model of RFID network is built, in which coverage rate, interference and cost are considered. Eq. (13) is taken as objective function of robust RFID network planning model. Where, $\gamma_1, \gamma_2, \gamma_3$ are weight coefficients, $\gamma_1 + \gamma_2 + \gamma_3 = 1$.

$$max f = \gamma_1 f_1 + \gamma_2 f_2 + \gamma_3 f_3 \tag{13}$$

Any deployed reader must be in the region of RFID network. Let A be the region of RFID network. Then feasible region is shown in formula (14).

$$(x_S^i, y_S^i) \in A, \ \forall i \in \{1, 2, \cdots, N_R\} \tag{14}$$

Constraints of robust RFID network planning model are formulas (1)~(3), (10)~(12) and (14).

3 ROBUST PARTICLE SWARM OPTIMIZATION

When radius R_b is large in the RFID network, there is a trade-off between interference calculation and sampling sites. In order to solve this dilemma, a novel robust particle swarm optimization is proposed.

3.1 Particle swarm optimization (PSO)

PSO is a population-based heuristic search technique, in which a group of particles find the best solution by iterations. The iteration formulations [23] are as follows:

$$V_{ld}^{t+1} = \omega V_{ld}^t + c_1 R_1 (X_{ld}^{pb} - X_{ld}^t) + c_2 R_2 (X_d^{gb} - X_{ld}^t) \tag{15}$$

$$X_{ld}^{t+1} = X_{ld}^t + V_{ld}^{t+1} \tag{16}$$

where $d = 1, 2, \cdots, D$, D is the number of particles' dimensions. t is the iteration number, $t = 1, 2, \cdots, T$, T is the maximum iteration. ω is the inertia weight, c_1 and c_2 are acceleration constants, R_1 and R_2 are random numbers between 0 and 1. X_l^{pb} is the best solution of the lth particle, and X_{ld}^{pb} is the dth dimention of X_l^{pb}. X^{gb} is the best solution of particle swarm, and X_d^{gb} is the dth dimention of X^{gb}. Besides, the velocity V_{ld} is limited by the maximum velocity $V_{max,d}$.

3.2 Basic idea of RPSO

In the prophase of iterations, PSO should explore the search space to achieve the optimum region as fast as possible. The smaller the sampling size is, the faster the exploring velocity is. In the anaphase of iterations, PSO is supposed to develop the better region to find the optimal or near optimal solution as soon as possible. The bigger the sampling size is, the higher the exploitation precision is.

According to this line of thinking, a robust particle swarm optimization is proposed. In specific, two sampling sizes N_1 and N_2 (N_1 is significantly less than N_2) are employed. In the prophase of iterations, the probability of N_1 is larger than that of N_2; and vice versa in the anaphase of iterations. Then, the exploring velocity in the prophase of iterations, and the exploitation precision in the anaphase of iterations can be ensured. Besides, N_2 is kept in low probability in the prophase of iterations to get some high reliable solutions, which can reduce blind exploration; and N_1 is kept in low probability in the anaphase of iterations to get some unreliable solutions, which can help RPSO to escape local extremum.

3.3 Sampling size selection

$\varphi(N_1,t)$, which is described in Eq. (17), denotes the probability of sampling size N_1 in the tth iteration; $\varphi(N_2,t)$ denotes the probability of sampling size N_2 in the tth iteration, $\varphi(N_2,t)=1-\varphi(N_1,t)$. Where, $1\le t_1 \le t_2 \le T$, $0\le \varphi_3 < \varphi_1 \le 1$, and $\varphi_1 + k(t_2-t_1)=\varphi_3$.

$$\varphi(N_1,t)=\begin{cases}\varphi_1, & 1\le t\le t_1\\ \varphi_1+k(t-t_1), & t_1+1\le t\le t_2\\ \varphi_3 & t_2+1\le t\le T\end{cases} \tag{17}$$

It can be seen that this setting is consistent with the idea talked in Sect. 3.2. Sampling size is dominated by N_1 in the prophase of iterations. On the other hand, sampling size is dominated by N_2 in the anaphase of iterations.

3.4 Expected sampling size

Expected sampling size is relevant to computational complexity of RPSO. $E(N,t)$ denotes expected sampling size of the tth iteration. $E_1(N,t)$, $E_2(N,t)$ and $E_3(N,t)$ denote expected sampling sizes when $1\le t\le t_1$, $t_1+1\le t\le t_2$ and $t_2\le t\le T$, respectively. $\bar{E}(N,t)$ denotes average expected sampling size in T iterations. By referring to Sect. 3.3, $E(N,t)$, $E_1(N,t)$, $E_2(N,t)$, $E_3(N,t)$ and $\bar{E}(N,t)$ are derived in Eqs. (18), (19), (20), (21) and (22), respectively. Obviously, $E_1(N,t)$ and $E_3(N,t)$ are constant when $E_2(N,t)$ linearly increases with t.

$$E(N,t)=N_1\cdot\varphi(N_1,t)+N_2[1-\varphi(N_1,t)]=N_2-(N_2-N_1)\varphi(N_1,t) \tag{18}$$

$$E_1(N,t)=N_2-(N_2-N_1)\varphi_1 \tag{19}$$

$$\begin{aligned}E_2(N,t)&=N_2-(N_2-N_1)[\varphi_1+\frac{\varphi_3-\varphi_1}{t_2-t_1}(t-t_1)]\\ &=N_2+(N_1-N_2)(\varphi_1-\frac{\varphi_3-\varphi_1}{t_2-t_1}t_1)+(N_1-N_2)\frac{\varphi_3-\varphi_1}{t_2-t_1}t\end{aligned} \tag{20}$$

$$E_3(N,t)=N_2-(N_2-N_1)\varphi_3 \tag{21}$$

$$\overline{E}(N,t)=\frac{1}{T}[t_1\cdot E_1(N,t)+(T-t_2)E_3(N,t)+\sum_{t=t_1+1}^{t_2}E_2(N,t)] \tag{22}$$

From Eq. (20), the sum of expected sampling sizes of the $(t_1+1+t_x)^{th}$ and $(t_2-t_x)^{th}$ iterations can be obtained by Eq. (23), where $t_x=0,1,2,\cdots,t_2-t_1-1$. Then the third part of Eq. (22) can be calculated detailedly in Eq. (24). $\bar{E}(N,t)$ can be described concretely in Eq. (25) by substituting Eqs. (19), (21) and (24) into Eq. (22).

$$\begin{aligned} & E_2(N,t_1+1+t_x)+E_2(N,t_2-t_x) \\ & = 2N_2+2(N_1-N_2)(\varphi_1-\frac{\varphi_3-\varphi_1}{t_2-t_1}t_1)+(N_1-N_2)\frac{\varphi_3-\varphi_1}{t_2-t_1}(t_1+1+t_x+t_2-t_x) \\ & = 2N_2+2(N_1-N_2)(\varphi_1-\frac{\varphi_3-\varphi_1}{t_2-t_1}t_1)+(N_1-N_2)\frac{\varphi_3-\varphi_1}{(t_2-t_1)}(t_1+1+t_2) \end{aligned} \tag{23}$$

$$\begin{aligned} & \sum_{t=t_1+1}^{t_2} E_2(N,t) = \frac{t_2-t_1}{2}[E_2(N,t_1+1+t_x)+E_2(N,t_2-t_x)] \\ & = (t_2-t_1)[N_2+(N_1-N_2)(\varphi_1-\frac{\varphi_3-\varphi_1}{t_2-t_1}t_1)]+(N_1-N_2)\frac{\varphi_3-\varphi_1}{2}(t_1+1+t_2) \end{aligned} \tag{24}$$

$$\overline{E}(N,t) = N_2-(N_2-N_1)\varphi_3+\frac{1}{T}[(N_2-N_1)(\varphi_3-\varphi_1)(\frac{t_1+t_2-1}{2})] \tag{25}$$

3.5 Variance of sampling size

Different from the expected sampling size, sampling size variance helps to understand whether RPSO focuses on one side of searching speed and precision, or both of them. If the sampling size variance is bigger, RPSO gives consideration to both searching speed and precision; if the sample size variance is smaller, RPSO only focuses on the speed or precision of searching.

The sampling size variance of the tth iteration is:

$$\begin{aligned} D(N,t) & = E\{[N-E(N,t)]^2\} \\ & = [N_1-E(N,t)]^2\varphi(N_1,t)+[N_2-E(N,t)]^2[1-\varphi(N_1,t)] \end{aligned} \tag{26}$$

$D(N,t)$ can be described concretely in Eq. (27) by substituting Eq. (18) into Eq. (26).

$$\begin{aligned} D(N,t) & = [N_1-N_2+(N_2-N_1)\varphi(N_1,t)]^2\varphi(N_1,t) \\ & \quad +[N_2-N_2+(N_2-N_1)\varphi(N_1,t)]^2[1-\varphi(N_1,t)] \\ & = (N_2-N_1)^2[1-\varphi(N_1,t)]\varphi(N_1,t) \\ & = (N_2-N_1)^2\{0.25-[\varphi(N_1,t)-0.5]^2\} \end{aligned} \tag{27}$$

It can be seen that $D(N,t)$ is mainly determined by $|N_2-N_1|$ and $|\varphi(N_1,t)-0.5|$. $D(N,t)$ will be larger if the difference between N_1 and N_2 is bigger, and $\varphi(N_1,t)$ is closer to 0.5.

$D(N,t)$ of $1\le t\le t_1$, $t_1+1\le t\le t_2$ and $t_2\le t\le T$ can be obtained in Eqs. (28)-(30) after substituting Eq. (17) into Eq. (27).

$$D_1(N,t) = (N_2-N_1)^2[0.25-(\varphi_1-0.5)^2] \tag{28}$$

$$D_2(N,t) = (N_2-N_1)^2\{0.25-[\varphi_1+k(t-t_1)-0.5]^2\} \tag{29}$$

$$D_3(N,t) = (N_2-N_1)^2[0.25-(\varphi_3-0.5)^2] \tag{30}$$

3.6 Parameter setting

Here, the selection method of sampling size mentioned in Sect. 3.3 is illustrated. Fig. 2 discloses the selection probability change of sample size, where $N_1=8$, $N_2=40$, $T=100$, $t_1=50$, $t_2=80$, $\varphi_1=0.7$ and $\varphi_3=0.1$. From Fig.2, it is clearly seen that the smaller sampling size is more probable than the bigger one in the prophase of iterations. And the bigger sampling size is more probable than the smaller one in the anaphase of iterations.

Figure 3 discloses the change of $E(N,t)$. It can be seen from Fig.3 that the expected sampling size in the prophase of iterations is smaller, or vice versa. $E(N,t)$ linearly increases with t when $t\epsilon(t_1,t_2)$.

Figure 4 discloses the change of $D(N,t)$. The sample size variance is smaller in the prophase of iterations, then increases, and decreases finally. It means that RPSO firstly focuses on searching speed, then both searching speed and precision, and searching precision finally.

All these are consistent with the descriptions in Sects. 3.2, 3.4 and 3.5.

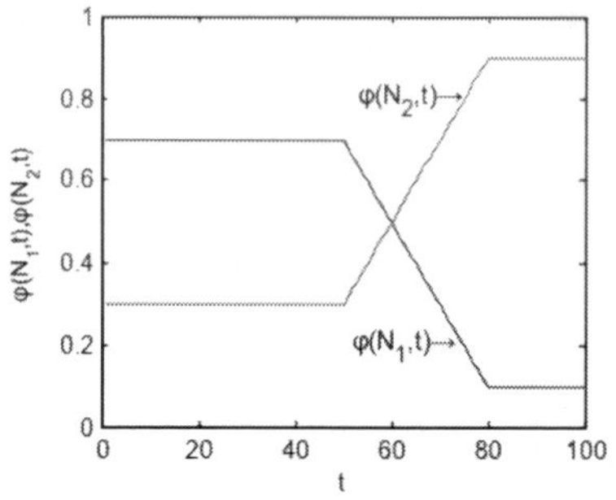

Figure 2: Selection probability change of sample size

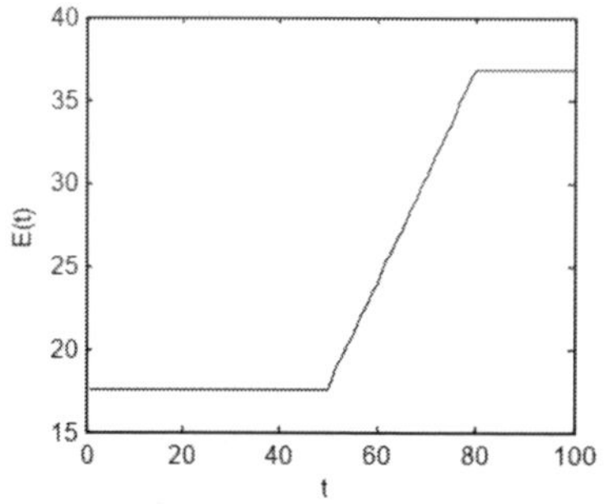

Figure 3: Change of $E(N,t)$

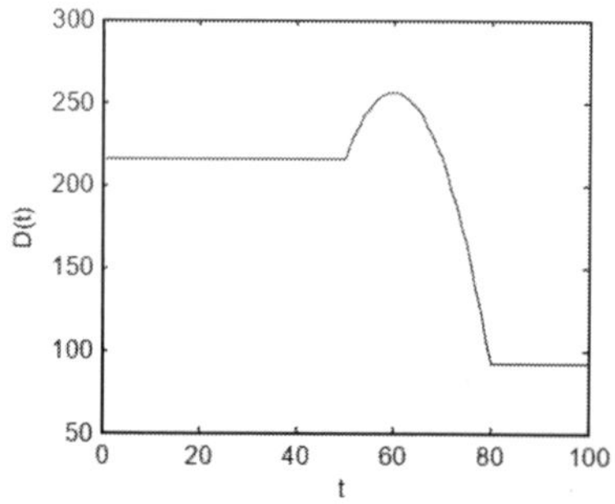

Figure.4: Change of $D(t)$

3.7 RPSO for RFID network planning problem

Here, the proposed RPSO is applied in RFID network optimization problem. The objective function, Eq. (13), is set as fitness function. Coverage of RFID network is calculated by referring to Sect. 2.2; interference is determined by Monte Carlo sampling method introduced in Sect. 2.3; cost is counted by Eq. (12) in Sect. 2.4. The pseudo-code of RPSO is as following:

```
Initialize population
For gen=1: maxgen
    For i=1: population size
        Evaluate coverage rate according to Sect. 2.2
        Evaluate interference according to Sect. 2.3, in which the sampling size is selected based
        on Sect. 3.3
        Evaluate cost according to Sect. 2.4
        Evaluate fitness according to Sect. 2.5
        If f(X_l^t)>f(X_l^pb) then X_l^pb = X_l^t
        X^gb = max{X_l^pb | l=1,2,···,}
        for d=1:Dimension
            update V_ld^(t+1) and X_ld^(t+1) according to Sect. 3.1
        end
    end
end
```

4 CASE STUDY

In a RFID network area of $30\text{m} \times 30\text{m}$, 100 RFID tags are random distributed. In Eq. (1), $R_R = 5\text{m}$ [1], $R_b = 1\text{m}$, $\lambda_1 = 1$, $\lambda_2 = 0$, $\beta_1 = 1$, $\beta_2 = 0.5$ [12]. In Eq. (12), $N_{max} = 20$. In Eq. (13), $\gamma_1 = 0.08$, $\gamma_2 = 0.91$ and $\gamma_3 = 0.01$.

To verify the performance of the proposed method, four algorithms are implemented and compared, among which RGA_MC is real-coded genetic algorithm (GA) based on traditional Monte Carlo method, RPSO_MC is PSO based on traditional Monte Carlo method, RGA_SC is real-coded GA based on the sampling method introduced in this work, RPSO_SC is RPSO described in Sect. 4. In all these four algorithms, coverage of RFID network is calculated by referring to Sect. 2.2; cost is counted by Eq. (12); interference is determined by traditional Monte Carlo method in RGA_MC and RPSO_MC, while calculated by Monte Carlo method based on the sampling method of this work in RGA_SC and RPSO_SC. The computer configuration is Pentium Dual 2.00GHZ, 2GHz memory. Programming tool is Matlab7.1. The population size M is 20, the number of iterations T is 100. In RGA_MC and RGA_SC, the uniform crossover and uniform mutation are implemented, where crossover and mutation rates are 0.7 and 0.01 respectively. Inertia weight ω is 0.729. Acceleration coefficients c_1 and $c_2$are both 1.49445 in RPSO_MC and RPSO_SC. Every chromosome or particle is coded as a $3 \times N_{max}$ matrix. Each reader has 3 codes: abscissa, ordinate and whether to be deployed. The third code is a real number in the range of $(0,1)$. If the third code is greater than 0.5, the reader is deployed, or vice versa. Parameters on sampling size selection in RGA_SC and RPSO_SC are set according to Sect. 3.6. From Eq. (25), $\bar{E}(N,t) \approx 24.416$.

To compare these four algorithms, let the calculation quantity of the sampling method introduced in this work be the same as traditional Monte Carlo method. In view of additional selection probability calculation for sampling size in RGA_SC and RPSO_SC, the sampling sizes of RGA_MC and RPSO_MC are both set as 26, which are slightly higher than $\bar{E}(N,t)$ in RGA_SC and RPSO_SC.

These four algorithms continuously compute 50 times. The best results and average results are shown in Tabs. 1 and 2 respectively, where f^{exp} is obtained by traditional Monte Carlo method in which the sampling size is 1000. Tab. 3 discloses $Error(f^{exp})$ of interference and fitness ($Error(f^{\exp}) = |f^{\exp} - \hat{f}^{\exp}|$, where $\hat{f}^{exp}$ is obtained by the corresponding algorithm.). The best results acquired by RPSO_AS are shown in Fig.5, where * represents reader, coarse dots indicate tags, large circles are identification range of readers, small circles are uncertain scopes of tags.

From Tabs. 1 and 2, RGA_SC and RPSO_SC display better than RGA_MC and RPSO_MC in the best and average results. RPSO_SC encloses more excellent interference and fitness than RGA_SC, although this trait is not found in every subgoal.

Table 1: The best results' f^{exp} of four methods

Performance index	RGA_MC	RPSO_MC	RGA_SC	RPSO_SC
Coverage	93.61903	83.71875	97.99482	90.61990
Interference	0.00174	1.00000	0.00002	1.00000
Network cost	0.40000	0.55000	0.30000	0.45000
Fitness	7.49511	7.61300	7.84260	8.16409

Table 2: Average results' f^{exp} of four methods

Performance index	RGA_MC	RPSO_MC	RGA_SC	RPSO_SC
Coverage	82.248890	83.824990	85.180260	86.262060
Interference	0.014683	0.031458	0.034402	0.357385
Network cost	0.200000	0.200000	0.260000	0.320000
Fitness	6.595273	6.736626	6.848327	7.229385

Table 3: $Error(f^{exp})$ of interference degree and fitness

	Performance index	RGA_MC	RPSO_MC	RGA_SC	RPSO_SC
Beat results	Interference ($\times10^{-3}$)	5.60	3.52	2.05	1.51
	Fitness ($\times10^{-3}$)	5.10	3.20	1.87	1.37
Average results	Interference ($\times10^{-3}$)	4.04	2.49	2.03	1.33
	Fitness ($\times10^{-3}$)	3.68	2.27	1.85	1.21

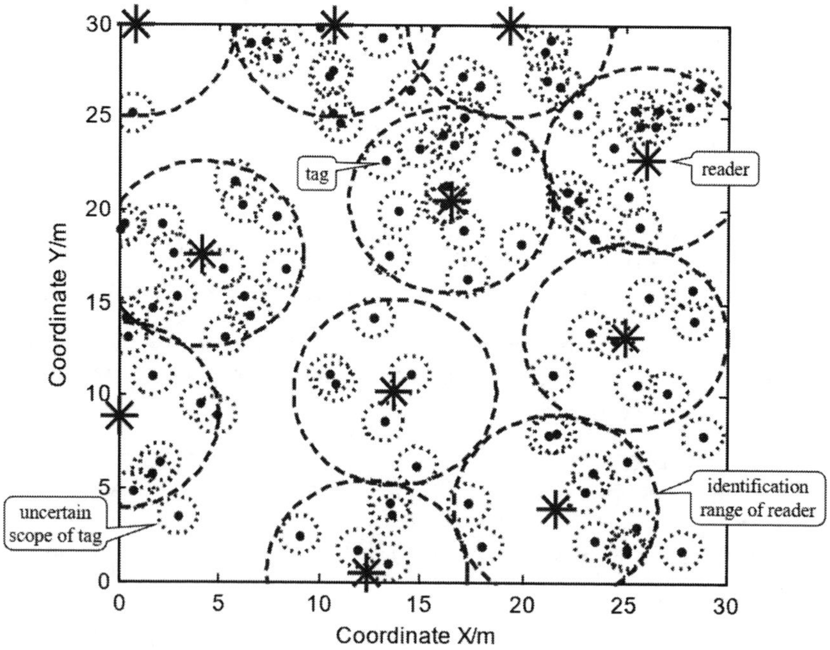

Figure 5: Diagram of best RFID network obtained by RPSO_SC

From Tab. 3, individual evaluation errors in the best and average results of RGA_SC and RPSO_SC are significantly smaller than RGA_MC's and RPSO_MC's. It can be reflected from the expected sampling sizes from the 81th to 100th iterations, in which the expected sampling size of RGA_SC and

RPSO_SC, 36.8, is drastically larger than the expected sampling size of RGA_MC and RPSO_MC, 26. In addition, it can be seen that the individual evaluation error in RPSO_SC is smaller than that of GA_SC.

In order to verify the model and method proposed in this work comprehensively, RFID network planning is done in different tag sizes. In areas of 50m×50m, 50m×100m and 100m×100m, N_{max} is 40, 75 and 140 respectively. And there are 250, 500 and 1000 tags respectively, which are random distributed. Tab. 4 shows average optimization results of these three examples with four algorithms. In different tag sizes, RPSO_SC is more excellent than other three algorithms in interference and fitness.

Table 4: Average planning results of three examples

Size of area Network scale	Performance index	RGA_MC	RPSO_MC	RGA_SC	RPSO_SC
50m×50m 250 tags	Interference	0.0123	0.0407	0.0807	0.3191
	Fitness	6.1660	6.2140	6.3906	6.4971
50m×100m 500 tags	Interference	0.0117	0.0376	0.0642	0.2254
	Fitness	6.0389	6.0999	6.2371	6.3468
100m×100m 1000 tags	Interference	0.0108	0.0261	0.0513	0.1595
	Fitness	5.9520	6.0194	6.1985	6.3121

These results and analysis suggest that RPSO_SC has better optimization performance in solving the RFID network planning problem when the tags' position is uncertain.

5 CONCLUSIONS

To sum up, this work builds robust planning model for the RFID network for port logistics under uncertain condition. Monte Carlo method is applied to calculate the interference. In order to reduce computational complexity and improve search performance, a robust particle swarm optimization with a new sample size selecting method is proposed. Expected sampling size and sample size variance are studied. Several case studies in different network scales show that the approach proposed in this work can get better RFID network planning scheme.

Acknowledgment

This work is supported by the National Natural Science Foundation of China (No. 51579143, 61540045), Action Plan for the Innovation of Science and Technology supported by Shanghai Science and Technology Committee (No. 14170501500), The Ministry of education of Humanities and Social Science project(No. 15YJC630145, 15YJC630059), Natural Science Foundation supported by Shanghai Science and Technology Committee (No. 15ZR1420200). Here we would like to express our gratitude to them.

6 REFERENCES

[1] **Gong, Y. J., Shen, M., Zhang, J., Kaynak, O., Chen, W. N.,** & **Zhan, Z. H.** 2012. Optimizing RFID network planning by using a particle swarm optimization algorithm with redundant reader elimination, *IEEE Transactions on Industrial Informatics*, 8(4), pp 900-912.

[2] **Liu, K.**, & **Ji, Z. C.** 2012. RFID network deployment based on hybrid particle swarm optimization, *Application Research of Computers*, 29(4), pp 1326-1328.

[3] **Lin, S.Y.**, & **Tsai, H.F.** 2016. Micro genetic algorithm with spatial crossover and correction schemes for constrained three-dimensional reader network planning, *Expert Systems With Applications*, 44, pp 344-353.

[4] **Lu, S. L.**, & **Yu, S. Z.** 2014. A fuzzy k-coverage approach for RFID network planning using plant growth simulation algorithm, *Journal of Network and Computer Applications*, 39, pp 280-291.

[5] **Yang, Y. H.**, **Wu, Y. J.**, **Xia, M.**, & **Qin, Z. J.** 2009. A RFID Network Planning Method Based on Genetic Algorithm, International Conference on Networks Security, Wireless Communications & Trusted Computing, Wuhan, China, April 25-26, pp 534-537.

[6] **Giampaolo, E. D.**, **Forni, F.**, & **Marrocco, G.** 2010. RFID-Network Planning by Particle Swarm Optimization, Proceedings of the 4th European Conference on Antennas and Propagation, Barcelona, Spain, April 12-16, pp 1-5.

[7] **Liu, K.**, **Shen, Y. X.**, & **Ji, Z. C.** 2011. RFID network optimization based on improved particle swarm optimization algorithm, *Journal of Central South University (Science and Technology)*, 42(s1), pp 900-904.

[8] **Ma, L.B.**, **Hu, K.Y.**, **Zhu, Y.L.**, & **Chen, H.N.** 2014. Cooperative artificial bee colony algorithm for multi-objective RFID network planning, *Journal of Network and Computer Applications*, 42, pp 143-162.

[9] **Ngai, E. W. T.**, **Cheng, T. C. E.**, **Au, S.**, & **Lai, K. H.** 2007. Mobile commerce integrated with RFID technology in a container depot, *Decision Support Systems*, 43, pp 62-76.

[10] **Kuo, R.J.**, **Chen, C.M.**, **Warren, L. T.**, & **Tien, F.C.** 2013. Hybrid of artificial immune system and particle swarm optimization-based support vector machine for Radio Frequency Identification-based positioning system, *Computers & Industrial Engineering*, 64, pp 333-341.

[11] **Li, S. J.**, **Xu, C. F.**, **Pan, W. K.**, & **Pan, Y. H.** 2005. Sensor Deployment Optimization for Detecting Maneuvering Targets, Proceedings of 7th International Conference on Information Fusion, Piscataway, NJ, USA , July 25-28, pp 1629-1635.

[12] **Ozturk, C.**, **Karaboga, D.**, & **Gorkemli, B.** 2011. Probabilistic Dynamic Deployment of Wireless Sensor Networks by Artificial Bee Colony Algorithm, *Sensors*, 11(6), pp 6056-6065.

[13] **Vu, K.**, & **Zheng, R.** 2011. Robust coverage under uncertainty in wireless sensor networks, International Conference on Computer Communications, Shanghai, China, April 10-15, pp 2015-2023.

[14] **Vu, K.**, & **Zheng, R.** 2012. Geometric Algorithms for Target Localization and Tracking Under Location Uncertainties in Wireless Sensor Networks, International Conference on Computer Communications, Orlando, Florida, USA, March 25-30, pp 1835-1843.

[15] **Jin, Y. C.**, & **Branke, J.** 2005. Evolutionary Optimization in Uncertain Environments-A Survey, *IEEE transactions on evolutionary computation*, 9 (3), pp 303-317.

[16] **Yu, X.**, **Jin, Y. C.**, **Tang, K.**, & **Yao, X.** 2010. Robust Optimization over Time-A New Perspective on Dynamic Optimization Problems, IEEE Congress on Evolutionary Computation, Barcelona, Spain, July 18-23, pp 3998-4003.

[17] **Zhu, Y. F., Luo, B., Zheng, J. H.,** & **Cai, Z. X.** 2011. Research on Increasing the Performance of Evolutionary Algorithm in Searching Robust Optimal Solutions Based on Quasi-Monte Carlo Method, *Pattern recognition and artificial intelligence*, 24(2), pp 201-209.

[18] **Jin, Y. C., Tang, K., Yu, X., Sendhoff, B.,** & **Yao, X.** 2013. A framework for finding robust optimal solutions over time, *Memetic Computing*, 5(1), pp 3-18.

[19] **Li, Z., Li, J.,** & **He, C.** 2014. Artificial immune network-based anti-collision algorithm for dense RFID readers, *Expert Systems with Applications*, 41(5), pp 4798-4810.

[20] **Chen, H. N., Zhu, Y. L., Hu, K. Y.,** & **Ku, T.** 2011. RFID network planning using a multi-swarm optimizer, *Journal of Network and Computer Applications*, 34, pp 888-901.

[21] **Shirley, P.,** & **Chiu, K.** 1997. A low distortion map between disk and square, *Journal of Graphics Tools*, 2(3), pp 45-52.

[22] **Gill, H. S.,** & **Lemieux, C.** 2007. Searching for extensible Korobov rules, *Journal of Complexity*, 23(4-6), pp 1-13.

[23] **Shi, Y. H.,** & **Eberhart, R.** 1998. A modified particle swarm optimizer, IEEE International Conference on Evolutionary Computation, Anchorage, AK, USA, May 4-9, pp 69-73.

CLOSED-LOOP SUPPLY CHAIN NETWORK MODEL WITH MODULE MANUFACTURE AND RECOVERY PROCESS[+]

Xing Chen[1], Chuluunsukh Anudari[1], YoungSu Yun[1*] and Mitsuo Gen[2]
[1]Department of Business Administration
Chosun University, Korea
chenxing1982@naver.com, anudr11@naver.com, ysyun@chosun.ac.kr

[2] Department of Research, Fuzzy Logic Systems Institute, Japan
[2]Research Institute of Science & Technology, Tokyo University of Science, Japan
gen@flsi.or.jp

ABSTRACT

In this paper, we consider the forward logistics (FL) with module manufacture and the reverse logistics (RL) with recovery process for constructing closed-loop supply chain (CLSC) network model. For the FL with module manufacture, some suppliers in each area send several types of part to module manufacturer and then some modules using the parts are assemble in module manufacturer. Products are produced at product manufacturer using some modules from module manufacturer. Finally the products are sent to retailer through distribution center. For the RL with recovery process, the products returned from customers are collected at collection center. They are tested and checked according to their quality conditions and then disassembled into recoverable products, recoverable modules, recoverable parts and unrecoverable parts. Each of them is sent to recovery center, module manufacturer, suppliers, and waste disposal centers. The proposed CLSC network model is designed using a nonlinear mixed integer programming and implemented using a hybrid genetic algorithm (HGA). For proving the performance of the HGA, some conventional algorithms are presented and their performances are compared with each other in numerical experiment. Finally, the experiment result shows that the HGA outperforms conventional algorithms.

Keywords: Forward logistics (FL), Reverse logistics (RL), Closed-loop supply chain (CLSC) network model, Module manufacturer, Recovery process, Hybrid genetic algorithm (HGA)

1 INTRODUCTION

How to design a closed-loop supply chain (CLSC) network model considering forward logistics (FL) and reverse logistics (RL) can provide an additional revenue opportunity for enterprise [1]. The reason is related to four aspects which are governmental legislation, customer awareness of recycling, social responsibilities, and economic benefits [2]. By the three aspects of governmental legislation, customer awareness of recycling, and social responsibilities, enterprises are required for a right

[+] This work was supported by the National Research Foundation of Korea Grant funded by the Korean Government (NRF-2014S1A5A2A01010951) and the Grant-in-Aid for Scientific Research (C) of Japan Society of Promotion of Science (JSPS: No. 15K00357).

[*] Corresponding Author

treatment of their end of life (EOL) products before starting their production activities. By the last aspect of economical benefits, enterprises can enjoy cost saving and profit increasing effects.

According to Chung et al. [3], module design has a positive effect on interaction of product life cycle (PLC) and the CLSC. In Newcomb et al. [4], many researchers used a new perspective to focus on module production instead traditional thought of production activities in order to extend their PLCs. They suggested that module production is a very effective way to extend the PLC. This means that enterprise not only can catch this point by using module production system in producing product, but also can take care of the recycling activities. Therefore, designing the CLSC network model both with module manufacture and recovery process is a very critical issue for most of enterprise.

There are many literatures about CLSC network model [5-8]. Georgiadis and Besiou [5] proposed the rules on the recycling products with the Waste Electrical and Electronic Equipment (WEEE) in EU. They also introduced many other environment regulations adapted in USA, Japan, and Hong Kong. In the conclusion, they insisted that environmental problem has become an unavoidable problem in many manufacturing industry. Hamed et al. [6] suggested the designing and planning problem of a CLSC network model. In their study, the returned product is classified into three categories. The returned product of the first category is resold at secondary market. That of the second category is recycled to be remanufactured, and that of the last category is disposed at waste disposal center. They considered the reselling activities at secondary market, but they did not consider a module production system. Fujita et al. [7] proposed a module production system to achieve flexibility and productivity simultaneously, but they did not considered them at CLSC network model. Chen et al. [8] proposed the CLSC network model considering the quality of returned product and classified it into three categories as follows. The returned products with good-quality are recycled to be remanufactured. Those of poor-quality are either wasted at waste disposal center or disassembled as raw materials. However, in this study, they did not consider the reselling activities of returned products at secondary market for economic benefits.

To find optimal solution in the CLSC network model, many researchers adopted metaheuristics [9-12]. Min et al. [9] proposed a CLSC network model which reinforces the collective function of initial collection center. They implemented the CLSC network model using a genetic algorithm (GA) and proved that the performance of the GA is superior to those of conventional competing approaches. Chen et al. [8] introduced many applications of GA approach to the CLSC network model. They also proposed the CLSC network model with various recovery activities at collection center and implemented the model using GA and Lingo [18]. In numerical experiment, they proved that the GA approach outperforms Lingo.

However, in spite of successful applications of GA approach to many CLSL network models, GA approach has some weakness such as the premature convergence of solution and the absence of local search during its search process. Therefore, many researchers have developed various hybrid GA (HGA) approaches using GA and other local search techniques and proved the HGA approaches are more efficient in locating optimal solutions than GA approach [14-17].

Therefore, in this paper, we propose a new type of the CLSC network model using a HGA approach. The FL with module manufacture and the RL with various recovery processes such as reuse, reselling, waste disposal activities are simultaneously considered in the proposed CLSC network model. The proposed CLSC network model is differ from the conventional CLSC network models mentioned above, since the latter does not use the concept of module manufacture and various recovery processes. For constructing the HGA, we improve the conventional HGA approach by Kanagaraj et al. [17].

The structure of this paper is as follows: Section 2 shows a conceptual structure of the proposed CLSC network model. In Section 3, the proposed CLSC network model is designed using a nonlinear mixed integer programming (NMIP). The NMIP is implemented using an improved HGA (iHGA) in Section 4. Various scales of the proposed CLSC network model are used for comparing the performances of the iHGA and other competing approaches in Section 5. Finally, some conclusions and future study are followed in Section 6.

2 CLSC NETWORK MODEL WITH MODELE MANUFACTURE AND RECOVERY PROCESS

In this section, the CLSC network model with module manufacture and recovery process is proposed. The conceptual structure of the proposed CLSC network model is described in Figure 1.

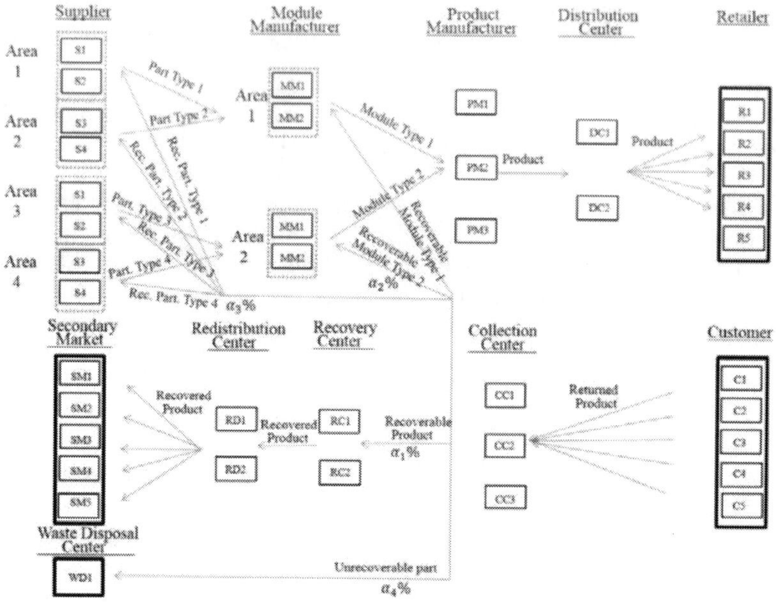

Figure 1: Conceptual Structure of the Proposed CLSC Network Model

In Figure 1, FL network consists of five stages including suppliers, modules manufacturers, product manufacturers, distribution centers, and retailers. Firstly, four types of part are prepared from one supplier in each area and then send them to module manufacturers in each area. Module manufacturers assemble each part and then make two types of module. The two types of module are sent to a product manufacturer and one product is made in it. The product is sent to retailers through a distribution center. RL network consists of six stages including customers, collection centers, recovery centers, redistribution centers, secondary markets and waste disposal centers. Firstly, all returned products from customers are collected at a collection center. In the collection center, the returned products are checked and tested and then classified into the recoverable and unrecoverable products. The recoverable products with $\alpha_1\%$ are sent to a recovery center to be recovered and then finally sent to secondary markets to be resold. The unrecoverable products are disassembled into the recoverable and unrecoverable modules. The recoverable module with $\alpha_2\%$ are sent to a module manufacturer to be recovered. The unrecoverable modules are disassembled into the recoverable and unrecoverable parts. The recoverable parts with $\alpha_3\%$ are sent to suppliers in each area to be recovered and the unrecoverable parts with $\alpha_4\%$ are sent to waste disposal centers to be disposed.

3 MATHEMATICAL FORMULATION

Before designing the proposed CLSC network model using mathematical formulation, some assumptions are presented as follows:

- Only single product is taken into consideration.
- The numbers of all facilities considered at each stage are already known.
- Only one facility at suppliers, module manufacturers, product manufacturers, distribution centers, collection centers, recovery centers and redistribution centers is opened, but all facilities at retailers/customers, secondary markets and waste disposal centers are always opened.
- Fixed costs of all facilities considered at each stage are different and already known.
- Unit handling costs of all facilities considered at same stage are identical and already known.
- Unit transportation costs between each stage are different and already known.
- The handling capacities of all facilities considered at a stage are the same or greater than those of all facilities considered at the previous stage.
- The return rate of returned products from all customers is 100%.
- The qualities of recovered modules and parts are identical with new ones.

The mathematical formulation is designed using the following index set, parameters and decision variables.

- **Index Set**

  a : index of area at suppliers; $a \in A$
  b: index of area at module manufacturers; $b \in B$
  h : index of supplier; $h \in H$
  i : index of module manufacturer; $i \in I$
  j : index of product manufacturer; $j \in J$
  k : index of distribution center; $k \in K$
  l : index of retailer/customer; $l \in L$
  m : index of collection center; $m \in M$
  n : index of recovery center; $n \in N$
  o : index of redistribution center; $o \in O$
  p : index of secondary market; $p \in P$
  q : index of waste disposal center; $q \in Q$

- **Parameters**

  FI_S_{ha}: fixed cost at supplier *h* of area *a*
  FI_MM_{ib} : fixed cost at module manufacturer *i* of area *b*
  FI_M_j : fixed cost at product manufacturing center *j*
  FI_DC_k : fixed cost at distribution center *k*
  FI_CC_m : fixed cost at collection center *m*
  FI_RC_n : fixed cost at recovery center *n*
  FI_RD_o : fixed cost at redistribution center *o*
  UH_S_{ha} : unit handling cost at supplier *h* of area *a*
  UH_MM_{ib} : unit handling cost at module manufacture center of area *b*
  UH_PM_j : unit handling cost at product manufacture center
  UH_DC_k : unit handling cost at distribution center
  UH_CC_m : unit handling cost at collection center
  UH_RC_n : unit handling cost at recovery center
  UH_RD_o : unit handling cost at redistribution center
  UT_SMM_{hai} : unit transportation cost from supplier *h* of area *a* to module manufacture center *i*
  UT_MMD_{ibj} : unit transportation cost from module manufacturer center *i* of area *b* to product manufacturer center *j*
  UT_PMD_{jk} : unit transportation cost from product manufacture center *j* to distribution center *k*
  UT_DR_{kl} : unit transportation cost from distribution center *k* to retailer/customer *l*
  UT_RCC_{lm} : unit transportation cost from retailer *l* to collection center *m*
  UT_CCMM_{mib} : unit transportation cost from collection *m* center to module manufacturer *i* of area *b*
  UT_CCS_{mha} : unit transportation cost from collection *m* center to module manufacturer *h* of area *a*

UT_CCRC_{mn} : unit transportation cost from collection center m to recovery center n
UT_CCWD_{mq} : unit transportation cost from collection center m to waste disposal center q
UT_RCRD_{no} : unit transportation cost from recovery center n to redistribution center o
UT_RDSM_{op} : unit transportation cost from redistribution center o to secondary market p

- **Decision Variables**

HCs_{ha} : handling capacity at supplier h of area a
$HCmm_{ib}$: handling capacity at module manufacture center i of area b
$HCpm_j$: handling capacity at product manufacture center j
$HCdc_k$: handling capacity at distribution center k
HCr_l : handling capacity at retailer/customer l
$HCcc_m$: handling capacity at collection center m
$HCrc_n$: handling capacity at recovery center n
$HCrd_o$: handling capacity at redistribution center o
$HCsm_p$: handling capacity at secondary market p
$HCwd_q$: handling capacity at waste disposal center q
x_s_{ha} takes the value of 1 if suppler h at area a is opened and 0 otherwise
x_mm_{ib} takes the value of 1 if module manufacturer i is opened and 0 otherwise
x_pm_j takes the value of 1 if product manufacturer j is opened and 0 otherwise
x_dc_k takes the value of 1 if distribution center k is opened and 0 otherwise
x_cc_m takes the value of 1 if collection center m is opened and 0 otherwise
x_rc_n takes the value of 1 if recovery center n is opened and 0 otherwise
x_rd_o takes the value of 1 if redistribution center o is opened and 0 otherwise

Minimize Total Cost (TC) =

$$\sum_a \sum_h (FI\_S_{ha} \cdot x\_s_{ha}) + \sum_b \sum_i (FI\_MM_{ib} \cdot x\_mm_{ib}) + \sum_j (FI\_PM_j \cdot x\_pm_j) + \sum_k (FI\_DC_k \cdot x\_d_k) + \sum_k \sum_m (FI\_CC_m \cdot x\_cc_m) + \sum_n (FI\_RC_n \cdot x\_rc_n) + \sum_o (FI\_RD_o \cdot x\_rd_o) + \sum_a \sum_h (UH\_S_{ha} \cdot HCs_{ha} \cdot x\_s_{ha}) + \sum_b \sum_i (UH\_MM_{ib} \cdot HCmm_{ib} \cdot x\_mm_{ib}) + UH\_PM_j \sum_j (HCm_j \cdot m_j x) + UH\_DC_k \sum_k (HCdc_k \cdot x\_dc_k) + UH\_CC_m \sum_m (HCcc_m \cdot x\_cc_m) + UH\_RC_n \sum_n (HCrc_n \cdot x\_rc_m) + UH\_RD_o \sum_n (HCrd_o \cdot x\_rd_o)$$
$$\sum_a \sum_h \sum_i (UT_{\_SMM_{hai}} \cdot HCs_{ha} \cdot x\_s_{ha} \cdot HCmm_{ib}) + \sum_b \sum_i \sum_j (UT_{MM_{ibj}} \cdot HCmm_{ib} \cdot x\_mm_{ib} \cdot x\_pm_j) + \sum_i \sum_k (UT\_MD_{ik} \cdot HCmm_i \cdot x\_mm_i \cdot x\_dc_k) + \sum_k \sum_l (UT\_DR_{kl} \cdot HCdc_k \cdot x\_dc_k) + \sum_l \sum_m (UT\_RCC_{lm} \cdot HCr_l \cdot x\_cc_m) + \sum_m \sum_a \sum_h (UT\_CCS_{mha} \cdot HCs_{ha} \cdot x\_s_{ha} \cdot x\_cc_m) + \sum_m \sum_i \sum_b (UT\_CCMM_{mib} \cdot HCmm_{ib} \cdot x\_mm_{ib} \cdot x\_cc_m) + \sum_m \sum_n (UT\_CCRC_{mn} \cdot HCrc_n \cdot x\_cc_m) + \sum_m \sum_n (UT\_RCRD_{mo} \cdot HCrd_o \cdot x\_rc_n) + \sum_m \sum_n (UT\_RDSM_{np} \cdot HCsm_p \cdot x\_rc_n) + \sum_m \sum_n (UT\_CCWD_{mq} \cdot HCwd_q \cdot x\_cc_m) \quad (1)$$

Subject to

$$\sum_h x\_s_{ha} = 1, \quad \forall a \in A \quad (2)$$
$$\sum_i x\_mm_{ib} = 1, \quad \forall b \in B \quad (3)$$
$$\sum_j x\_pm_j = 1 \quad (4)$$
$$\sum_k x\_dc_k = 1 \quad (5)$$
$$\sum_m x\_cc_m = 1 \quad (6)$$
$$\sum_n x\_rc_n = 1 \quad (7)$$
$$\sum_o x\_rd_o = 1 \quad (8)$$
$$\sum_h (HCs_{ha} \cdot x\_s_{ha}) - \sum_i (HCmm_i \cdot x\_mm_{ib}) = 0, \ \forall a \in A \quad (9)$$
$$\sum_i (HCmm_{ib} \cdot x\_mm_{ib}) - \sum_j (HCpm_j \cdot x\_pm_j) = 0, \ \forall b \in B \quad (10)$$
$$\sum_i (HCpm_j \cdot x\_pm_j) - \sum_k (HCdc_k \cdot x\_dc_k) = 0 \quad (11)$$
$$\sum_k (HCdc_k \cdot x\_dc_k) - \sum_l HCr_l = 0 \quad (12)$$
$$\sum_l HCr_l - \sum_m (HCcc_m \cdot x\_cc_m) = 0 \quad (13)$$
$$\sum_i (HCmm_{ib} \cdot x\_mm_{ib}) - a2 \sum_m (HCcc_m \cdot x\_cc_m) \geq 0 \quad (14)$$
$$\sum_h (HCs_{ha} \cdot x\_s_{ha}) - a3 \sum_m (HCcc_m \cdot x\_cc_m) \geq 0 \quad (15)$$
$$\sum_n (HCrc_n \cdot x\_rc_n) - \alpha 1 \sum_m (HCcc_m \cdot x\_cc_m) \geq 0 \quad (16)$$
$$\sum_a (HCwd_a - \alpha 4 \sum_m (HCcc_m \cdot x\_cc_m) \geq 0 \quad (17)$$
$$\sum_n (HCrc_n \cdot x\_rc_n) - \sum_o (HCrd_o \cdot x\_rd_o) = 0 \quad (18)$$
$$\sum_n (HCrd_o \cdot x\_rd_o) - \sum_p HCsm_p = 0 \quad (19)$$
$$x\_s_{ha} = \{0,1\}, \ \forall h \in H, \ a \in A \quad (20)$$

$$x\_mm_{ib} = \{0,1\}, \ \forall i \in I, \ b \in B \tag{21}$$
$$x\_pm_j = \{0,1\}, \ \forall j \in J \tag{22}$$
$$x\_dc_k = \{0,1\}, \ \forall k \in K \tag{23}$$
$$x\_cc_m = \{0,1\}, \ \forall m \in M \tag{24}$$
$$x\_rc_n = \{0,1\}, \ \forall n \in N \tag{25}$$
$$x\_rd_o = \{0,1\}, \ \forall o \in O \tag{26}$$
$$HCs_{ha}, HCmm_{ib}, HCpm_j, HCdc_k, HCcc_m, HCrc_n, HCrd_o, HCsm_p, HCwd_q \geq 0,$$
$$\forall h \in H, \ a \in A, \ \forall i \in I, \ b \in B, \ j \in J, \ \forall k \in K, \ \forall l \in L, \ \forall m \in M, \ \forall n \in N, \ \forall o \in O, \ \ \forall p \in P, \ \forall q \in Q \tag{27}$$

The objective function shown in equation (1) is to minimize the sum of total cost which is consisted of fixed costs, handling costs and transportation costs. Equations (2) to (8) mean that only one facility at each stage should be opened. Equations (9) to (19) indicates that the capacity of the facility opened at a stage is the same or greater than that of facility opened at the previous stage. Equations (20) to (26) restrict that the decision variables should have integers value of 0 or 1. Equation (27) means non-negativity.

4 IHGA APPROACH

As we already know, GA approach was widely used to solving many optimization problems [18-21]. However, GA has some weaknesses such as the premature convergence of the solution and the absence of local search. Therefore, many researchers have developed various types of HGA approach. Recently, Kanagaraj ***et al.*** [17] proposed a new type of HGA which combines GA and cuckcoo search (CS). They used Levy flight concept to improve the quality of the solutions resulting from GA search process. However, the Levy flight is adapted to only one individual among the offspring reslting from GA, and the others do not have any chance of the adaptation of Levy flight. This may decrease the occurance probability of respective (or good) individuals.

Therefore, in this study, we proposed the iHGA approach to improve the HGA approach by Kanagaraj ***et al.*** [17]. In the iHGA approach, all individuals of offspring resulting from GA are adapted by Levy flight. The detailed implementation procedure is as follows:

Step 1: generate initial population randomly.
Step 2: the following GA operators are used.
  Selection: elitist selection operator in enlarged sampling space [20]
  Crossover: two-point crossover operator [20]
  Mutation: random mutation operator [20]
Step 3: produce offspring after GA operators.
Step 4: apply Levy flight scheme to an individual (*Levy_Ind*) randomly selected from the offspring. Randomly select another individual (*Ano_Ind*) from the offspring.
Step 5: Compare the fitness value of *Levy_Ind* with that of *Ano_Ind*, if the former outperforms the latter, then the *Levy_Ind* is inserted into the offspring and the *Ano_Ind* is deleted from the offspring.
Step 6: The fraction (p_a) of the individuals with lower fitness values is abandoned and randomly regenerate new individuals so that the lost individuals are to be replaced.
Step 7: If pre-defined total number of generations is satisfied, then produce best solution and exit all steps, otherwise, go to step 2.

5 NUMERICAL EXPERIMENT

In numerical experiment, three scales of the proposed CLSC network model as shown in Table 1 is presented. In Table 1, various numbers of suppliers (S), module manufacturers (MM), product manufacturers (PM), distribution centers (DC), retailers (S), collection centers (CC) recovery centers (RC), redistribution centers (RD), secondary markets (SM) and waste disposal centers (WD) are shown.

For various comparisons, two conventional approaches (GA, HGA) and one benchmark program, Lingo [18] are used. For the GA, the steps 1, 2, 3 and 7 suggested in Section 4 are adapted. The HGA by Kanagaraj *et al.* [17] is used. The GA, HGA and iHGA approaches were programmed using Matlab R2015. All approaches including Lingo were ran under IBM compatible PC 3.40GHZ processor (Inter Core i7-3770 CPU), 8GB RAM and Window 7. The parameters used for GA, HGA and iHGA approach are as follows: Population size is 20, crossover rate 0.5, mutation rate 0.3, generation number 10,000. Total 30 trials were carried out to reduce the randomness during the search process of each approach. 1,000 products are produced in FL and treated in RL.

Table 1: Three Scales of the Proposed CLSC Network Model

	No. of S Area 1,2,3,4	No. of MM Area 1,2	No. of PM	No. of DC	No. of R	No. of CC	No. of RC	No. of RD	No. of SM	No. of WD
Scale 1	4	4	6	4	10	6	4	4	10	2
Scale 2	12	10	14	10	25	14	10	10	25	5
Scale 3	20	20	28	20	50	20	20	20	50	10

The performances of each approach are compared using some measures of performance as follows: i) best solution is the best value of the objective functions under satisfying all constraints, ii) average solution is the averaged value of the objective function under satisfying all constraints, iii) average iteration is the average number of iteration after all trials and iv) run time is the average running time (unit: sec.) after all trials. The computation results of each approaches for scales 1 2, and 3 are summarized in Table 2.

Table 2: Computation Results for Each Approach

	Scale 1				Scale 2				Scale 3			
	GA	HGA	iHGA	Lingo	GA	HGA	iHGA	Lingo	GA	HGA	iHGA	Lingo
Best Solution	154664	154776	154664	144562	154201	154142	153477	149830	151485	151498	151232	138713
Average solution	155246	155224	155172	-	154636	154736	154461	-	152404	152304	152188	-
Average iteration	9973	9848	9400	-	5320	5543	4970	-	5137	5566	5377	-
Run Time	24.66	26.61	67.90	-	26.06	30.40	76.10	-	20.83	26.03	72.30	-

For the scale 1, in terms of the best solution, although the result of Lingo is significantly better than those of the others, the results of GA and iHGA approaches are the same and they are slightly better than that of the HGA approach. However, in terms of the average solution and average iteration, the iHGA approach outperforms the GA and HGA approaches. The GA approach is the quickest and the iHGA approach is the slowest in terms of the run time.

For the scale 2, the iHGA approach is more efficient than the GA and HGA approaches in terms of best solution, average solution, and average iteration. However, the search time of the former is about two times slower than those of the others. Similar results are also shown in the scale 3. The performance of the iHGA approach is superior to those of the GA and HGA approaches in terms of the best solution and average solution. However, in terms of the run time, the performance of the iHGA approach is worse than those of the GA and HGA approaches.

For more detailed comparison, Figure 2 shows the convergence behaviors of the GA, HGA and iHGA approaches until the generation number is researched to 4,500. In Figure 2, the GA and HGA approaches show rapid and various convergence behaviors until the generation number is reached to about 1,500, but after the generation number, they do not show any convergence behaviors. The search ability of the iHGA approach is slightly worse than those of the GA and HGA until the generation number is reached to about 3,500, but after the generation number, the iHGA approach shows a better convergence behavior rather than the others.

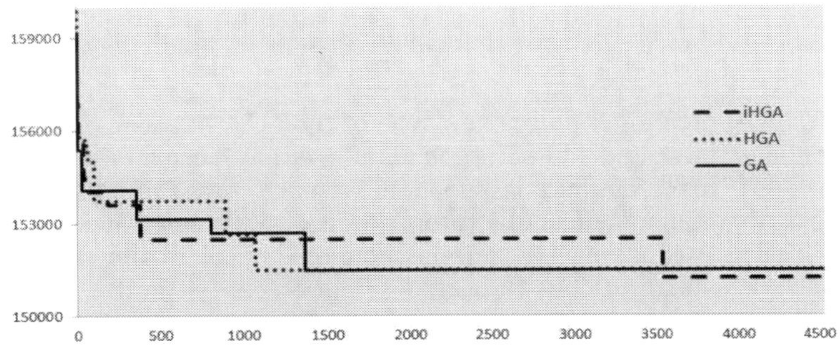

Figure 2: Convergence Behaviours of Each Approach

The computation results in Table 2 and Figure 2 shows the following. Firstly, the iHGA approach proposed in this paper outperforms the GA and HGA approaches in most measures of performance, which implies that the search scheme used in the iHGA approach is more efficient than those of the others. Secondly, in terms of the run time, the iHGA approach does not shows any benefit compared with the GA and HGA approaches, this means that the iHGA approach consumes more times in search process than the GA and HGA approaches, although the former is able to locate better solution than the latter. Therefore, an effort to reduce the search time of the iHGA approach will be required.

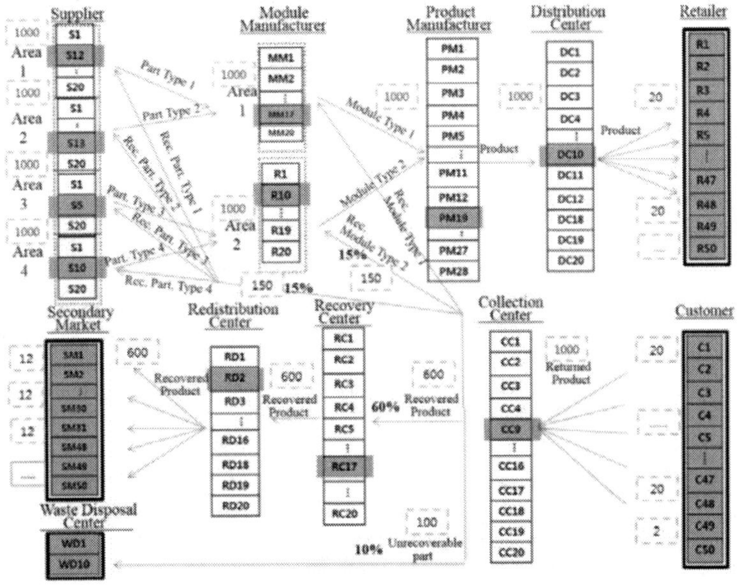

Figure 3: Location and Allocation Decision of iHGA for Scale 3

Figure 3 shows the location and allocation decisions of the iHGA approach for scale 3. For the location decision, the suppliers (12, 13, 5, and 10) at each area, module manufacturers (17, 10) at each area, product manufacturer (19), distribution center (10), collection center (9), recovery center (17), and redistribution center (2) are opened at each stage. For the allocation decision, 1,000 parts at each area are produced and sent to module manufacturer. 1,000 modules are manufactured at module manufacturer using 1000 parts from suppliers. Finally, 1,000 product are assembled at product manufacturer and they are all sent to retailers through distribution center. In RL, 1,000 returned products are collected at collection centers. 600 (60% of the 1,000) recoverable products are sent to recovery center and then sold at secondary markets through redistribution center. 150(15% of the 1,000) recoverable modules are sent to module manufacturer. 150(15% of the 1,000) recoverable parts are sent to suppliers and 100(10% of the 1,000) unrecoverable parts are sent to the waste disposal centers.

6 CONCLUSION

First objective of this paper is to propose a new type of the CLSC network model. The proposed CLSC network model considers module manufacture in FL and various recovery processes in RL. In the FL, four types of parts are manufactured in suppliers at each area, and they are sent to module manufacturers. Using them, module manufacturers produce two types of modules and then sent them to product manufacturer. Product manufacturer assembles a product using the two types of modules. Finally, all products are sent to retailers through distribution center. In the RL, all returned products are collected at collection center. Collection center checks and disassembles them into recoverable products, recoverable modules, recoverable parts, and unrecoverable parts. The recoverable products are sent to recovery center and then their functions are recovered. Finally the recovered products are resold at secondary market through redistribution center. The recoverable parts and modules are respectively sent to suppliers and module manufacturers to be used for assembling a product. The unrecoverable parts are sent to waste disposal center.

Second objective of this paper is to suggest the iHGA approach for implementing the proposed CLSC network model. Firstly, the proposed CLSC network model is designed using mathematical formulation. Secondly, the mathematical formulation is implemented in the iHGA approach. The iHGA approach is a revised version of the conventional HGA approach by Kanagaraj *et al.* [17]. In numerical experiement, three scales of the proposed CLSC network model is presented to compare the performances of the iHGA approach and the other competing approaches. the experimental results show that the iHGA approach outperforms the others. Howerver, the search time of the iHGA approach is very slower than the those of the others. Therefore, an effort to reduce the search time will be required.

7 REFERENCES

[1] **Guide, V.D.R., Van Wassenhove, L.N.** 2009. The evolution of closed-loop supply chain research, *Journal of Operations Research*, 57(1), pp 10-18.

[2] **Beamon, B.M.** 1999. Designing the greeen supply chain. *Journal of Logistics Information Management*, 12(4), pp 332-342.

[3] **Chung, W.H. Okudan, G. and Richard, A.W.** 2011. Modular Design to Optimize Product Life Cycle Metrics in a Closed-loop Supply Chain, Proceedings of the 2011 Industrial Engineering Research Conference.

[4] **Newcomb, P.J., Bras, B. Rosen, D.W. 1998.** Implication of modularity on product design for the life cycle. *Journal of Mechanical Design*, 120(3), pp 483-490.

[5] **Chen, T.T., Chan, F.T.S. and Chung, S.H.** 2015. An integrated closed-loop supply chain model with location allocation problem and product recycling decisions, *International Journal of Production Research*, 53(10), pp 3120-3140.

[6] **Hamed, S., Mirmehdi, S.E. and Mohsen, A.S.** 2013. Designing and planning a multi-echelon multi-period multi-product closed-loop supply chain utilizing genetic algorithm, *International Journal Advanced Manufacturing Technology*, 68, pp 917-931.

[7] **Georgiadis, P. and Besiou, M.** 2010. The impact of two-product joint lifecycles on capacity planning of remanufacturing networks, *European Journal of Operational Research*, 202(2), pp 420-433.

[8] **Fujita. K., Amaya, H. and Akai, R.** 2013. Mathematical model for simultaneous design of module commonalization and supply chain configuration toward global product family, *Journal of Intelligent Manufacturing*, 24, pp 991-1004.

[9] **Yang, B. C., Ren, M.M.** 2007. A fuzzy Multi-objective Programming for Optimization of Reverse Logistics for Solid Waste Through Genetic Algorithm, Proceedings of the 4th International Conference on Fuzzy Systems and Knowledge Discovery, pp 416-420.

[10] **Wang, H. F., Hsu, H. W.** 2010. A closed-loop logistics model with a spanning tree based genetic algorithm, *Computers and Operations Research*, 37(2), pp 376-389.

[11] **Kanagaraj, G., Phonnambalam, S. G. and Jawahar, N.** 2013. A hybrid cuckoo search and genetic algorithm for reliability-redundancy allocation problems, *Computers and Industrial Engineering*, 66, pp 1115-1124.

[12] **Min, H., Ko, C.S. and Ko, H.J.** 2006. The spatial and temporal consolidation of returned products in a closed-loop supply chain network, *Computers and Industrial Engineering*, 51(2), pp. 309-320.

[13] **Lee, J. E. and Lee, K.D.** 2012. Integrated forward and reverse logistics model: a case study in distilling and sale company in Korea, *International Journal of Innovative Computing Information and Control*, 8(7A), pp. 4483-4495.

[14] **Zhang, J., Liu, X. and Tu, Y.L.** 2011. A capacitated production planning problem for closed-loop supply chain with remanufacturing, *International Journal of Advanced Manufacturing Technology*, 54(5-8), pp. 757-766.

[15] **Demirel, N., Ozceylan, E. Paksoy, T. and Gokcen, H.** 2014. A genetic algorithm approach for optimising a closed-loop supply chain network with crisp and fuzzy objectives, *International Journal of Production Research*, 52 (12), pp. 3637-3664.

[16] **Yun Y.S. Chung H. S. and Moon C.U.** 2013. Hybrid genetic algorithm approach for precedence-constrained sequencing problem, *Computers and Industrial Engineering*, 65, pp.137-147.

[17] **Yun Y.S. Moon C.U. and Kim D.H.** 2009, Hybrid genetic algorithm with local search scheme for solving multistage-based, supply chain problems, *Computers and Industrial Engineering*, 56, pp. 821-838.

[18] **Lingo**, Lindo Systems, 2016, www.lindo.com.

[19] **Gen, M. and Cheng, R.** 1997. *Genetic Algorithms and Engineering Design*, John Wiley and Son, New York.

[20] **Gen, M. Cheng, R. and Lin, L.** 2008. *Network Models and Optimization: Multiobjective Genetic Algorithm Approach*, Springer, London.

[21] **Syarif A., Yun, Y.S. and Gen. M.** 2002. Study on multi-stage logistic chain network: a spanning tree based genetic algorithm approach, *Computers and Industrial Engineering*, 43(1-2), pp 299-314.

[22] **Wang, H.-F. and Hsu, H.-W.** 2010. A closed-loop logistic model with a spanning-tree based genetic algorithm, *Computers and Operations Research*, 37, pp 376-389.

Yard Crane Scheduling with Reshuffles in An Import Container Block

Feifeng Zheng[1], Xiaoyi Man[1*], Ming Liu[2], and Feng Chu[3]
[1] Glorious Sun School of Business & Management Donghua University, Shanghai, China
ffzheng@dhu.edu.cn, manxiaoyi8996@163.com

[2]School of Economics & Management Tongji University, Shanghai, China
mingliu@tongji.edu.cn

[3]Laboratoire IBISC, Université d'Evry Val d'Essonne, 40 rue du Pelvoux, 91020 Evry Cedex, France
feng.chu@ufrst.univ-evry.fr

ABSTRACT

This paper studies the yard crane scheduling problem to minimize the maximum tardiness of tasks in an import container block, in which two yard cranes process both storage and retrieval tasks simultaneously. Since reshuffle is common in the process of retrieval tasks, we focus on the scenario with reshuffle operations for retrieval tasks, and the constraint of inter-crane interference is involved. An integer linear programming model is established, and a genetic algorithm is developed to solve the problem.

Keywords: Yard Crane Scheduling, Tardiness, Reshuffle, Genetic Algorithm

1 INTRODUCTION

In recent decades, there is a rapid growth of research on the planning and scheduling of major resources in container terminals, such as berth allocation, quay crane (QC) assignment and scheduling, and integrated scheduling of the resources. Besides berths and QCs, in some cases yard cranes (YCs), behave as a critical resource and become the bottleneck during the processing of loading/unloading containers. YCs handle the containers that are temporarily stored in the blocks of the yard on the land side of a terminal.

In the container yard there are typically three kinds of container blocks, i.e., *import block*, *export block* and *transshipment block*. Each block is dedicated to the temporary storage and transfer of the corresponding type of containers. In an import block, there are exclusively import containers that carry cargoes from oversea and are transported via yard trucks from vessels to the yard. Later on they are delivered to domestic customers by customer trucks. In an export block, export containers are sent to the yard by domestic customer trucks, and later on they are loaded onto vessels for oversea destinations. In a transshipment block, the containers from some original foreign countries are unloaded from vessels, and they will be loaded again onto other vessels for their oversea destinations.

In an import block, the container loading/unloading operations between the block and either yard trucks or customer trucks are mainly handled via YCs. It will cause *inter-crane interference* between

* Corresponding Author

YCs if they are very close to each other, resulting in a negative impact on their processing efficiency. So there is a safety distance requirement between YCs. Moreover, due to the space limitation within a block, there are usually no more than two YCs simultaneously serving container tasks in the block at a time. Therefore, in this work we study the scheduling of two YCs in an import block with the objective of minimizing the maximum tardiness of tasks. Note that the container processing in an export block can be similarly discussed.

It is quite often that during the processing of import containers, each YC unloads containers from yard trucks and places them on the top of some container stacks in the block. Meanwhile it retrieves some other containers from the stacks and loads them onto customer trucks. That is, there may exist two kinds of container tasks concurrently. We refer to the tasks for moving containers from yard trucks to the container stacks in the block as *storage tasks*, and refer to those for moving containers from the container stacks to customer trucks as *retrieval tasks*. The corresponding containers are respectively called *storage containers* and *retrieval containers*. Especially in dealing with a retrieval task, *reshuffles* occur provided that the target container is beneath some other containers. A reshuffle operation is to move away an involved container that is above the target container, or to move it back to its original stack after the retrieval of the target container. A retrieval task includes all the container moves for the target and involved containers. Hence, the processing time of a retrieval task may be much longer than that of a storage task which needs a single container move from the yard truck to its destination stack.

In this work we consider the scenario where there requires a safety distance between the two YCs at any time and there occur both storage tasks and retrieval tasks concurrently. That is, both reshuffle operations and interference constraint are well considered, and they make it nontrivial to produce an optimal schedule of the considered problem.

2 LITERATURE REVIEW

In the last decade, besides a large amount of literature considering berth allocation, quay crane assignment and scheduling on the seaside (Carlo et al. [1]), there is also a lot of work focusing on YC scheduling in container yards (Ng et al. [2]; Zhen et al. [3]). Below we give a brief literature review of YC scheduling in three parts: inter-crane interference, simultaneous storage and retrieval and reshuffles.

2.1 Inter-Crane Interference

Many authors consider the YC scheduling problem with inter-crane interference, in which two or more YCs may process tasks simultaneously within a single block or neighbour blocks in a line (Bazzazi et al., [4]; Angeloudis et al. [5]).

Ng [2] studied the multiple YCs scheduling problem with inter-crane interference in two container blocks. The objective is to minimize the total waiting time of trucks. It is assumed that the time for one YC to process a task is constant. An integer programming model together with a two-phase scheduling heuristic was proposed to solve the problem. Guo and Huang [6] extended the problem of Ng [2] to the case with several container blocks in a line. They gave a dynamic time and space partitioning algorithm which outpesrforms that of Ng [2] by numerical experiments.

Jin et al. [7] studied the storage yard manage problem where traffic congestion is considered. They established an integer linear programming model with the objective of minimizing both operating cost and inter-block movement cost of YCs, and proposed a divide-and-conquer approach. Wu et al. [8] addressed the YC scheduling problem to produce an optimal sequence of storage and retrieval operations. The objective is to minimize the deviation between the actual handling time and the ideal handling time, such as retrieval lateness, retrieval earliness and storage lateness.

2.2 Simultaneous Storage and Retrieval

Many authors considered the scenario where both storage and retrieval containers are requested during YC processing, and they investigated various models and approaches (Wu et al. [8]; Chen et al. [9]).

Li et al. [10] addressed the YC scheduling problem where the YCs keep fixed distances between them during processing storage and retrieval containers. Assuming that any container move consumes three minutes, they develop a mixed-integer linear programming model to minimize a linear combination of the retrieval earliness and retrieval delays. Gharehgozli et al. [11] presented the makespan minimization problem of scheduling twin YCs which carry out a set of storage and retrieval requests in a block. The block is divided into a seaside zone and a landside zone. Besides the requirement of a safety distance between the two YCs, they assumed that stacking and retrieving containers are of different priorities. An adaptive neighborhood search heuristic algorithm is developed to solve the problem.

2.3 Reshuffle

During container moving, reshuffle is an inevitable operation in most situations and it is time consuming and increases a vessel's turnaround. In many works the reshuffle operation is well involved (Huang et al. [12]; Gharehgozli et al. [13]).

Huang et al. [12] investigated the problem of YC sequencing where the due dates of tasks and the arrival times of relevant vehicles are known a prior. The objective is to minimize the total tardiness of tasks. They considered the case with a single YC in each exclusive operation zone, and thus there is no inter-crane interference. Gharehgozli et al. [13] proposed a decision-tree heuristic to minimize the expected number of reshuffles. Through the decision trees, they concluded that the shared-stacking policy appears to outperform dedicated-stacking policy in reducing the number of reshuffles.

Summing up, we conclude from literature that: (i) the types of blocks in the container yard are often ignored, (ii) the processing time of each task or container is usually assumed to be identical, and (iii) the specific reshuffles of containers to process a task is also omitted. Hence, to improve the practical solutions for yard crane operations, we consider the following scenario in this work: (i) both storage and retrieval are involved in an import block, (ii) the processing time of each task depends on the tier location of the target container and thus it is not a constant, and (iii) the number of containers to be reshuffled and the processing time of each task are calculated accordingly.

3 PROBLEM STATEMENT

We are given two yard cranes YC1 and YC2 to process a set of storage and retrieval tasks in an import block. All the task information is known beforehand. The import block consists of $l_{\max}$ container slots, and each slot contains $a_{\max}$ rows of containers. Given any index combination of slot and row, there stow at most $b_{\max}$ containers in a vertical stack. Each container occupies one tier of the stack. The number of tiers in a stack varies during processing provided that there is one or more tasks in the stack.

Each task i corresponds to a target container with three dimension location (l_i, a_i, b_i), representing the indices of slot, row, and tier respectively. $b_i = 1$ if the container is in the lowest tier of the stack, and the container in a higher tier is of a larger value of b_i. As each task corresponds to a unique target container, we sometimes use container i and task i alternatively.

The processing time p_i of a storage task i is the time for a YC to unload the import container from a yard truck beside the slot l_i and place it on the top of the (l_i, a_i) stack in the block. That is, $p_i = h$ where h is the time of a single container move. For a retrieval task i, its processing is much more time consuming for the case where container i is not on the top of a stack and reshuffle occurs. We refer to those containers stowed above container i as *interfering containers*. The YC needs to temporarily move away all the interfering containers, load the target container i onto the customer truck, and then move back all the interfering containers to the stack. In this case, p_i is the total time of moving the container i and its interfering containers, i.e., $p_i = h(1 + 2\eta_i)$ where η_i the number of interfering containers. Notice that the value of η_i increases over time if some storage tasks located in the same stack as task i are processed before i.

The tardiness of task i is denoted by $T_i = c_i - d_i$ where c_i and d_i are the completion time and respectively the due date of task i. The objective of the considered problem is to minimize the maximum tardiness, i.e., $\min\max\{T_i\}$. Due to the inter-crane interference between YC1 and YC2, we assume that there requires a safety distance $f_{\min}$ between the cranes at any time. The remainder of this paper is under the following fundamental assumptions.

3.1 Assumptions:

1) The due date d_i and location (l_i, a_i, b_i) of any task i are known at the beginning.
2) The two cranes YC1 and YC2 are originally located at the two ends of the import block, and YC1 is always at a location with a smaller slot index than YC2.
3) Each YC occupies one slot at any time, and the minimum required safety distance between YC1 and YC2 is eight slots, i.e., $f_{\min} = 8$.
4) Each YC moves along slots in a constant speed, and thus the time of moving a YC between any two consecutive tasks is determined by their slot distance.

3.2 Mathematical Formulation

In this section we present an integer linear model for the considered problem. We give preliminary notations and then formulate the objective function and the constraints as well.

3.2.1 Indices:

- i, j indices of storage or retrieval tasks
- k index of yard crane, and $k \in \{1, 2\}$
- t index of time period

3.2.2 Input parameters:

- B the set of slots, and $B = \{1, 2, \ldots, 40\}$
- N the set of tasks, and $N = \{1, 2, \ldots\}$
- Γ_j the set of tasks which are of slot distance less than $f_{\min}$ to task j ($\in N$), i.e., for any $i \in \Gamma_j$, $|l_i - l_j| < f_{\min}$
- N^{out} the set of retrieval tasks, $N^{out} \subset N$
- τ_j^{in} the set of storage tasks in the same stack of a retrieval task j ($\in N^{out}$), i.e., for any $i \in \tau_j^{in}$, $\beta_i^{in} = 1, l_i = l_j$, and $a_i = a_j$
- τ_j^{out} the set of retrieval tasks in the same stack of a retrieval task j ($\in N^{out}$), i.e., for any

$i \in \tau_j^{out}$, $\beta_i^{out} = 1, l_i = l_j$, and $a_i = a_j$

- $\mathcal{T}$ the set of time periods, and $\mathcal{T} = \{1, 2, ...\}$
- $f_{\min}$ the minimum number of slots required between YC1 and YC2, and $f_{\min} = 8$ in this paper
- g an upper bound of the difference between the number of tasks processed by YC1 and that of YC2
- h the handling time for a YC to make a single move of one container
- q_i the initial number of interfering containers stowed above the target container i
- l_i the slot index of task i, and $1 \le l_i \le l_{\max}$=40 in this paper
- a_i the row index of task i, and $1 \le a_i \le a_{\max}$=6 in this paper
- b_i the tier index of task i, and $1 \le b_i \le b_{\max}$=5 in this paper
- r_i the release time of task i, and a related yard truck or customer truck arrives slot l_i at the time
- d_i the due date of task i
- $m_{i,j}$ the time for a YC to move from the slot of task i to that of task j, i.e., from l_i to l_j

3.2.3 *Decision variables:*

- α the maximum tardiness of tasks
- s_i the start time of processing task i
- p_i the process time of task i that includes the handling times of both target and reshuffling containers
- c_i the completion time of processing task i, and $c_i = s_i + p_i - 1$
- η_i the number of interfering containers stowed above the target container i at time s_i
- β_i^{in} equal to 1 if i is a storage task, 0 otherwise
- β_i^{out} equal to 1 if i is a retrieval task, 0 otherwise
- $X_{i,j,k}$ equal to 1 if YC k processes task i before task j, 0 otherwise
- $Y_{i,k}$ equal to 1 if task i is assigned to YC k, 0 otherwise
- $Z_{i,k,t}$ equal to 1 if YC k starts to process task i at period t, 0 otherwise
- $U_{k,l,t}$ equal to 1 if YC k is located at slot l at period t, 0 otherwise
- $W_{i,t}$ equal to 1 if task i starts processing at period t or $s_i = t$, 0 otherwise
- $V_{i,t}$ equal to 1 if task i is being processed at period t, 0 otherwise
- $O_{i,j}$ equal to 1 if task $i \in \Gamma_j$ is processed before task j, 0 otherwise

3.3 Mathematical model

We define two dummy tasks $v1, v2$ with locations $l_{v1} = 0, l_{v2} = l_{\max}$, and $s_j = c_j = 0$ for $j \in \{v1, v2\}$. Tasks $v1$ and $v2$ are processed by YC1 and YC2 respectively. For any two tasks i, j with $s_i < s_j$ that are processed consecutively by the same YC, the start time of task j is equal to $s_j = \max\{c_i + m_{i,j}, r_j\}$. The two cranes YC1 and YC2 are initially located at slot 1 and slot $l_{\max}$ respectively, i.e., $U_{1,1,1} = 1$ and $U_{2,l_{\max},1} = 1$.Denote by α the maximum tardiness, i.e., $\alpha = \max\{T_i\}$. We formulate the ILP model as follows.

(1) The Objective Formulation

By the definition of α, the objective of the problem is formulated as follows.

$$\min \alpha$$

(2) Objective Function Constraints

Together with $T_i = c_i - d_i$ and $\alpha \ge T_i$, we have

$$\alpha \ge c_i - d_i \quad i \in N \tag{1}$$

(3) Inter-crane Interference Constraints

$$\sum_{l' \in B} U_{2,l',t} \cdot l' - \sum_{l \in B} U_{1,l,t} \cdot l \geq f_{\min} \qquad t \in \mathcal{T} \qquad (2)$$

Constraints(2) ensure that the slot difference between YC1 and YC2 is at least $f_{\min}$.

(4) Simultaneous Task Processing Constraints

$$X_{i,j,k} + X_{j,i,k} \leq 1 \qquad i,j \in N, i \neq j, k \in \{1,2\} \qquad (3)$$
$$X_{i,j,k} + X_{j,i,k} \geq 1 - M(2 - Y_{i,k} - Y_{j,k}) \qquad i,j \in N, i \neq j, k \in \{1,2\} \qquad (4)$$
$$Y_{i,k} \geq X_{i,j,k} \qquad i,j \in N, i \neq j, k \in \{1,2\} \qquad (5)$$
$$Y_{j,k} \geq X_{i,j,k} \qquad i,j \in N, i \neq j, k \in \{1,2\} \qquad (6)$$
$$Y_{i,1} + Y_{i,2} = 1 \qquad i \in N \qquad (7)$$
$$\sum_{i \in N} Y_{i,1} - \sum_{j \in N} Y_{j,2} \leq g \qquad i \neq j \qquad (8)$$
$$\sum_{j \in N} Y_{j,2} - \sum_{i \in N} Y_{i,1} \leq g \qquad i \neq j \qquad (9)$$
$$\sum_{i \in N} Z_{i,k,t} \leq 1 \qquad k \in \{1,2\}, t \in \mathcal{T} \qquad (10)$$
$$\sum_{i \in N} W_{i,t} \leq 2 \qquad t \in \mathcal{T} \qquad (11)$$
$$Z_{i,k,t} \leq Y_{i,k} \qquad i \in N, k \in \{1,2\}, t \in \mathcal{T} \qquad (12)$$
$$Z_{i,k,t} \geq Y_{i,k} + W_{i,t} - 1 \qquad i \in N, k \in \{1,2\}, t \in \mathcal{T} \qquad (13)$$
$$Z_{i,1,t} + Z_{i,2,t} = V_{i,t} \qquad i \in N, t \in \mathcal{T} \qquad (14)$$
$$Z_{i,k,t} - Z_{i,k,t+1} \geq -M(2 - V_{i,t} - V_{i,t+1}) \qquad i \in N, k \in \{1,2\}, t \in \mathcal{T} \qquad (15)$$
$$Z_{i,k,t} - Z_{i,k,t+1} \leq M(2 - V_{i,t} - V_{i,t+1}) \qquad i \in N, k \in \{1,2\}, t \in \mathcal{T} \qquad (16)$$
$$\sum_{l \in B} U_{k,l,t} = 1 \qquad k \in \{1,2\}, t \in \mathcal{T} \qquad (17)$$
$$U_{k,l_i,t} \leq 1 + M(1 - Z_{i,k,t}) \qquad k \in \{1,2\}, i \in N, t \in \mathcal{T}, l_i \in B \qquad (18)$$
$$U_{k,l_i,t} \geq 1 - M(1 - Z_{i,k,t}) \qquad k \in \{1,2\}, i \in N, t \in \mathcal{T}, l_i \in B \qquad (19)$$

Constraints(3)-(19) state the constraints when there are simultaneous storages and retrievals.

(5) Reshuffle-related Constraints

$$s_i \geq r_i \qquad i \in N \qquad (20)$$

$$s_i = \sum_{t \in \mathcal{T}} (t \times W_{i,t}) \qquad i \in N \qquad (21)$$

$$c_i = s_i + p_i - 1 \qquad i \in N \qquad (22)$$

$$s_j \geq c_i + m_{i,j} + 1 - M(1 - X_{i,j,k}) \qquad i,j \in N, i \neq j, k \in \{1,2\} \qquad (23)$$

$$s_j \geq l_j - 1 - M(1 - Y_{j,1}) \qquad j \in N \qquad (24)$$

$$s_j \geq l_{\max} - l_j - M(1 - Y_{j,2}) \qquad j \in N \qquad (25)$$

$$\sum_{t \in \mathcal{T}} W_{i,t} = 1 \qquad i \in N \qquad (26)$$

$$O_{i,j} + O_{j,i} \leq 1 \qquad i \in \Gamma_j, j \in N \qquad (27)$$

$$O_{i,j} \geq X_{i,j,k} \qquad i \in \Gamma_j, j \in N, k \in \{1,2\} \qquad (28)$$

$$\eta_j = q_i + \sum_{i \in \tau_j^{in}} O_{i,j} - \sum_{i \in \tau_j^{out}} O_{i,j} \qquad j \in N^{out} \qquad (29)$$

$$p_i = (1 + 2\eta_i) \times h \qquad i \in N \qquad (30)$$

$$p_i = \sum_{t \in \mathcal{T}} V_{i,t}, \qquad i \in N \qquad (31)$$

$$V_{i,t} \geq W_{i,t} \qquad i \in N, t \in \mathcal{T} \qquad (32)$$

$$\sum_{t'=1}^{t-1} V_{i,t'} \leq M(1 - W_{i,t}) \qquad i \in N, t \in \mathcal{T} \qquad (33)$$

$$V_{i,t+1} + V_{i,t-1} - V_{i,t} \leq 1 \qquad i \in N, t \in [2, \mathcal{T} - 1] \qquad (34)$$

Constraints(20)-(34) give the relationship of s_i, p_i and c_i.

(5) Variable Ranges

$$s_i, p_i, c_i, \eta_i \in Z^+ \qquad i \in N \qquad (35)$$

$$\beta_i^{in}, \beta_i^{out}, X_{i,j,k}, Y_{i,k}, Z_{i,k,t}, U_{k,l,t}, W_{i,t}, V_{i,t}, O_{i,j}, \delta_{i,j} \in \{0,1\}$$
$$i,j \in N, i \neq j, k \in \{1,2\}, t \in \mathcal{T}, l \in B \qquad (36)$$

Constraints(35)-(36) are variable ranges.

4 NUMERICAL EXPERIMENTS

We solve the above model with solver CPLEX for small instances. It is assumed that there are the same number of storage and retrieval tasks or containers in the import block during the time horizon. YC1 and YC2 are initially located on the leftmost and rightmost ends of the block, respectively. The values of parameters l_i, a_i, b_i, r_i and d_i for each task i are generated randomly. For an instances with 30 tasks, the solver consumes 2036 seconds to output an optimal solution, and it spends much more time to solve larger instances. Therefore, we propose a genetic algorithm (GA) for larger instances. For the instance with 30 tasks, GA produces a solution in 3 seconds with a difference less than 3.9%. A series of numerical experiments are conducted to validate the efficiency of the algorithm. Due to the page limitation, we omit the detailed experimental results here.

5 CONCLUSION

In this paper, we present a scheduling problem of twin YCs in an import block to minimize the maximum tardiness of tasks. We consider the inter-crane interference and there exist both storage and retrieval tasks simultaneously. Our main contribution is an integer linear programming model of the problem, and a genetic algorithm for large instances is also devised to solve the problem.

6 ACKNOWLEDGMENT

This work was partially supported by the National Science Foundation of China (71571134 and 71428002), and the Cai Yuanpei Program between the French Ministries of Foreign and European Affairs and the Higher Education and Research and the Chinese Ministry of Education (27927VE). The work was also supported by DHU Distinguished Young Professor Program (A201305) and the Fundamental Research Funds for the Central Universities.

7 REFERENCES

[1] **Carlo, H. J., Vis, I. F. A., Roodbergen, K. J. 2014.** Storage yard operations in container terminals: literature overview, trends, and research directions. *Flexible Services & Manufacturing Journal*, 235(2), 1-13.

[2] **Ng, W. C. 2005.** Crane scheduling in container yards with inter-crane interference. *European Journal of Operational Research*, 164(1), 64-78.

[3] **Zhen, L., Jiang, X., Lee, L. H., Chew, E. P. 2013.** A review on yard management in container terminals. *Industrial Engineeering & Management Systems*, 12(4), 289-304.

[4] **Bazzazi, M., Safaei, N., Javadian, N. 2009.** A genetic algorithm to solve the storage space allocation problem in a container terminal. *Computers & Industrial Engineering*, 56(1), 44-52.

[5] **Angeloudis, P., Bell, M. G. H. 2010.** An uncertainty-aware AGV assignment algorithm for automated container terminals. *Transportation Research Part E*, 46(3), 354-366.

[6] **Guo, X., Huang, S. Y. 2012.** Dynamic Space and Time Partitioning for Yard Crane Workload Management in Container Terminals. *Transportation Science*, 46(1), 134-148.

[7] **Jin, J. G., Lee, D. H., Cao, J. X. 2014.** Storage yard management in maritime container terminals. *Transportation Science*, Articles in Advance, pp.1-14.

[8] **Wu, Y., Li, W., Petering, M. E. H., Goh, M., Souza, R. D. 2015.** Scheduling multiple yard cranes with crane interference and safety distance requirement. *Transportation Science*, 49(4), 1-16.

[9] **Chen, L., Langevin, A., Lu, Z. 2013.** Integrated scheduling of crane handling and truck transportation in a maritime container terminal. *European Journal of Operational Research*, 225(1), 142-152.

[10] **Li, W., Wu, Y., Goh, M. 2009.** Discrete time model and algorithms for container yard crane scheduling.*European Journal of Operational Research*, 198(1), 165-172.

[11] **Gharehgozli, A. H., Laporte, G., Yu, Y., de Koster, R. 2014.** Scheduling twin yard cranes in a container block. *Transportation Science*, 49(3), 686-705.

[12] **Huang, S. Y., Guo, X., Wen, J. H., Wei, L. L. 2012.** Yard Crane Dispatching to Minimize Job Tardiness in Container Terminals. *International Conference on Logistics and Maritime Systems*, 2, 1-11.

[13] **Gharehgozli, A. H., Yu, Y., de Koster, R., Udding, J. T. 2014.** A decision-tree stacking heuristic minimising the expected number of reshuffles at a container terminal. *International Journal of Production Research*, 52(9), 2592-2611.

DISPATCHING RULE-BASED SCHEDULING ALGORITHM FOR A DYNAMIC FLEXIBLE FLOW SHOP WITH TIME-DEPENDENT PROCESS DEFECT RATE

B.J. Joo*, T.J. Chua and T.X. Cai
Planning and Operations Management Group,
Singapore Institute of Manufacturing Technology (SIMTech), Singapore
joobj@simtech.a-star.edu.sg, tjchua@simtech.a-star.edu.sg, txcai@simtech.a-star.edu.sg

ABSTRACT

This paper focuses on a dynamic scheduling problem in a multi-product three-stage flexible flow shop where new jobs continuously arrive over time. Each of the parallel machines in the second stage requires a setup to changeover job types. As soon as a setup is done at a machine, its defect rate becomes stable. However, after unknown time periods, the machine's defect rate becomes unstable due to some technical reasons. Jobs are inspected at the final stage to send quality feedback to the previous stage when the average quality rate becomes lower than a predetermined tolerance. A new dispatching rule-based scheduling algorithm (DRSA) is developed to achieve better quality and tardiness-related performance by extending the previous research of Joo *et al.* [1]. Results of computational experiments show that the new DRSA outperforms previous algorithms.

Keywords: Scheduling, Flexible Flow Shop, Time-dependent Process Defect, Setup, Dispatching Rule

1 INTRODUCTION

The real life scheduling environment is extremely complicated due to the dynamic nature of manufacturing systems such as dynamic arrival of customer orders and random process defect rate. To remain competitive, generating effective production schedules in real-time is essential. The dynamic flexible flow shop (DFFS) under consideration comprises three sequential process and inspection stages. The most interesting point is the time-dependent process defect rates of the parallel machines in stage-2. Each of the parallel machines requires a setup whenever there is a job type change, and the setup for a new job type resets the machine's process defect rate to a stable phase which lasts for unknown periods. However, after the unknown periods, its process defect rate turns to an unstable phase which lasts until a new setup is done for any job type. This kind of quality problem usually occurs in manufacturing processes with process parameters that are sensitively fluctuating by the uncontrollable or hard-to-control factors, such as mechanical stress, vibrations, variations in thermal conditions and impurities accumulated.

This research is an extension of Joo *et al.* [1] developed two DRSAs for the same DFFS scheduling problem. However, there is a limitation in the research, which is any of the two suggested DRSAs cannot outperform the previous algorithms in terms of both quality and tardiness performance. To overcome the limitation, this paper aims to develop a new DRSA that can outperform the previous algorithms by extending the outputs of the previous research.

As reviewed in Joo *et al.* [1], there are many research papers on a few relevant research topics: flexible flow shop scheduling problems [2-5], dispatching rules for scheduling problems with setups

* Corresponding Author

[1, 6-11], scheduling with the process quality or yield [1, 11-20], and the integrated scheduling and preventive maintenance problems [21-24]. See Joo *et al.* [1] for more details about the papers listed above. According to the authors' extensive literature survey, there has been no advance in the research problem under consideration until now.

This paper is organized as follows. The definition of the DFFS scheduling problem is introduced in detail in section 2, and a new DRSA is proposed in section 3. The results of the computational experiments are given in section 5, and the final section is devoted to conclusions with the future research directions.

2 PROBLEM DEFINITION

The DFFS is composed of three sequential process stages: stage-1 with a single machine, stage-2 with M identical parallel machines and stage-3 with an inspection machine as shown in Figure 1.

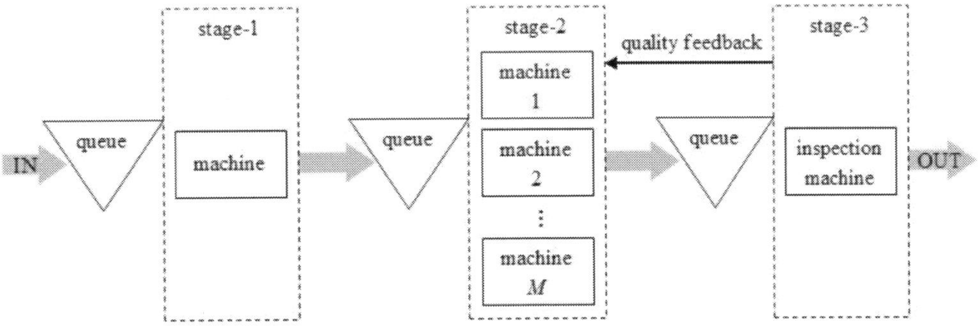

Figure 1: Three-stage flexible flow shop

The processing time of type-f jobs at stage-i is denoted by $p_{f,i}$ while the setup time of type-f jobs at stage-2 is denoted by s_f, where f and i are job type and stage index. Each job should be processed first at the single machine in stage-1, next at one of M identical parallel machines in stage-2 and finally at the inspection machine in stage-3. The completion time of job j at stage-i is denoted by $C_{j,i}$, and the inspection time at stage-3, $p_{f,3}$, is identical for all job types and set to a relatively shorter than the processing times at the upstream stages.

Jobs arrive dynamically over time at the queue of stage-1. The inter-arrival times of jobs are exponentially distributed with mean $\mu = (\alpha\bar{s} + \bar{p}_2)/M$, where α is the inter-arrival time ranging parameter and $\bar{s}$ and $\bar{p}_2$ are the average setup and processing times for all job types at stage-2. When a job arrives, its job type is given randomly and its due date is randomly determined from a uniform distribution with a range $[a_j + \bar{p}_f, a_j + \beta p_f]$, where a_j, $\bar{p}_f$ and β denote the arrival time of job j, the total processing and inspection time of type-f job and the due date ranging parameter. When the single machine becomes available, jobs waiting in the queue of stage-1 are prioritized first by the slack per remaining work with setup ratio (S/RW-SR) rule introduced in Joo *et al.* [1]. Then, the job with the highest priority is selected and started at the machine.

At stage-2, when one of the M parallel machines becomes available, jobs waiting in the common queue of stage-2 are prioritized and the job with the highest priority is selected and started at the machine. At any machine in stage-2, a sequence-independent setup time s_f should be inserted before processing a type-f job immediately after a job of different type. Once a setup for job type-f is completed at machine k at time $t_{f,k}$, machine k's defect rate for type-f jobs becomes phase-I with a low and stable defect rate, and it is maintained until $t_{f,k} + \Delta$, where $\Delta = \gamma(\bar{s} + \bar{p}_2)$ and γ is the parameter for the duration of phase-I. The duration of phase-I, Δ, is unknown in advance and is not dependent on the job types since the M machines are identical. From $t_{f,k} + \Delta$, the machine's defect rate becomes phase-II with a relatively high and unstable defect rate, and it is maintained until a new setup is done for any job type at the machine. The new DRSA for stage-2 machines is described in the next section.

At stage-3, jobs are inspected at the inspection machine by first-come, first-served rule. Since the defect rates of jobs are dependent on the latest completion times of setups at the machines in stage-2, the defect rate of job j, r_j, processed at machine k is determined as the following way. If the duration between the latest setup completion time for type-f at machine k and the completion time of job j at the same machine is no longer than Δ, i.e., if $C_{j,2} - t_{f,k} \leq \Delta$, r_j is obtained from a normal distribution for phase-I with the mean μ_f^{P1} and the standard deviation σ_f^{P1}. Otherwise, if $C_{j,2} - t_{f,k} > \Delta$, it is obtained from a normal distribution for phase-II with μ_f^{P2} and σ_f^{P2}. For generality, it is assumed that $\mu_f^{P1} < \mu_f^{P2}$ and $\sigma_f^{P1} < \sigma_f^{P2}$. After the inspection, job j is scrapped if $r_j > \tau$, where τ denotes the predetermined tolerance level for jobs. Once a job is scrapped, a new job which is identical to the scrapped job will be input to the queue of stage-1 at once in order to satisfy the customer orders.

In order to send quality feedback from stage-3 to stage-2, the average defect rate (ADR), $R_{f,k}$, for each combination of f (job type) and k (machine in stage-2) is computed by accumulating r_j's of jobs of type-f processed at machine k. If $R_{f,k} > \tau^{ADR}$ where τ^{ADR} denotes the predetermined tolerance level for ADR, a quality feedback is sent to machine k. After machine k gets a quality feedback, a setup should be done for the job type to be processed next (after completing the job in progress as pre-emption is not allowed) if machine k is still operating with the same setup. This is because the status of any machine with a quality feedback is regarded as unstable. Note that $R_{f,k}$ is reset to 0 once a setup is done for any job type at machine k and there is no time delay for sending the quality feedback.

As performance measures, the average quality ratio, the ratio of the number of quality jobs to the number of jobs inspected, and the mean tardiness of all jobs are considered. The two measures are practically very important in terms of the quality-related production cost, the customer satisfaction and the operational efficiency. Obviously, there exists a trade-off between the two measures, that is, more setups can improve the quality ratio but delay jobs' completion times due to the setups.

3 NEW DISPATCHING RULE-BASED SCHEDULING ALGORITHM

The apparent tardiness cost with quality (ATCSQ) rule proposed by Ko *et al.* [11] for a non-identical parallel machines scheduling problem with sequence-dependent setups and quality restrictions is a good benchmark rule if it is slightly modified. Nevertheless, a fundamental limitation is expected since it computes jobs' process quality-related priority values by using the real-time quality measurement data. Hence, in order to obtain a good schedule, the data needs to be reliable. However, in case that the time duration of phase-I defect rate is not long enough to obtain the reliable quality measurement data, any dispatching rule using the data may not be work effectively. Also, it should be considered that the real-time quality feedback system is not usually being used in practice. Moreover, it may not be possible to obtain the quality information as soon as each process is completed since each process does not have an inspection function in many practical situations. Therefore, there exists a time gap between the completion time of each process step and the completion time of the inspection step, and the quality information may not be up-to-date although it could be transferred in real-time.

Two previous DRSA algorithms developed by Joo *et al.* [1] cannot outperform the other benchmark algorithms. They show a trade-off between the two performance measures depending on the parameter settings, i.e., the quality performance is better while the tardiness performance is worse than the other benchmark algorithms in one parameter setting and vice versa.

To overcome the limitations in the previous research, a new algorithm, named new algorithm for time-dependent quality combined with ATCS (ATCS-TQN), is developed by extending the ATCS-TQ2 rule of Joo *et al.* [1] for scheduling the parallel machines in stage-2 dynamically. In the ATCS-TQN, two priority rules, the TQN rule developed in this research and the ATCS rule modified from the research of Lee & Pinedo [8], are used to select a job to process at the available machine. More specifically, when a machine in stage-2 becomes available, the TQN rule prioritize job types, each of which has at least one job in the queue, in order to select one job type with the highest priority. After a job type is selected, the ATCS rule prioritizes the jobs of the selected type in order to select one job with the highest priority among the jobs of the selected type. Finally, the selected job is scheduled at the available machine and processed with or without setup depending on the previously processed job type. In the two priority rules, the following information is taken into account: (a) real-time information on the elapsed time after the latest setup of each machine instead of the real-time quality measurement data, (b) real-time information on the queue, and (c) information on jobs, i.e., due dates, processing times and setup times of jobs. The priority of job-j of type-f at time t, $ATCS_{j,f,t}$, is computed by the following equation:

$$ATCS_{j,f,t} = \frac{1}{p_{[j],2}+p_{[j],3}} \cdot \exp\left(-\frac{\max\{d_j-t-p_{[j],2}-p_{[j],3},0\}}{k_1\bar{p}^R}\right) \cdot \exp\left(-\frac{s_f}{k_2\bar{s}}\right) \qquad (1)$$

Equation (1) is a slightly modified version from the ATCS rule in Lee & Pinedo [8] in order to take the remaining processing time at both stage-2 and 3 into account for more accurate calculation of the slack times since the remaining processing time at stage-2 is only considered in Joo *et al.* [1]. In the equation, $p_{[j],2} + p_{[j],3}$ is the total remaining processing time of job-j from stage-2 to stage-3, where $[j]$ is the index of job-j's type, and $\bar{p}^R$ denote and the average remaining processing time of all job types from stage-2 to stage-3, computed as $\sum_{f=1}^{F}(p_{f,2} + p_{f,3})/F$. Then, the priority of job type-f at time t, $TQN_{f,t}$, is computed by the following equation:

$$TQN_{f,t} = \frac{|J_{f,t}|}{|J_t|} \cdot \frac{|G_{f,t}|}{M} \cdot \sum_{j\in J_{f,t}} \frac{ATCS_{j,f,t}}{|J_{f,t}|} \qquad (2)$$

In equation (2), $J_{f,t}$ and J_t denote the set of type-f jobs and the set of all jobs waiting in the queue at time t, respectively, and $|x|$ denotes the cardinality of set x. $G_{f,t}$ denotes the set of stable machines in stage-2 that are setup for job type-f at time t, where machine-k is defined to be stable if the elapsed time after its latest setup is no longer than π at time t, i.e., $t - t_{f,k} \leq \pi$ holds. Here, π is the phase-I duration estimating parameter, by which it can be assumed that the phase-I of each stable machine will last for a while after t. Based on the definition of π, if the value of π becomes too small, it will be difficult to differentiate the TQN values of the job types because most of the machines will be included in the set of stable machines. On the other hand, if its value becomes too large, the average quality ratio of jobs will be greatly increased because the machines are regarded as stable although they are about to shift to the phase-II.

The first two terms in the equation of the TQN, the ratio of the number of type-f jobs to the number of all jobs waiting in the queue and the ratio of the number of stable machines that are setup for job type-f to the number of all machines in stage-2, are devised to take the quality ratio measure into account. The last term, the average ATCS priority value of the type-f jobs waiting in the queue-2, is devised to take the tardiness measure into account. Unlike the ATCS-TQ2 rule of Joo *et al.* [1], to enhance the overall performance, the three terms are multiplied for emphasizing the effect of the quality factor in the equation of the TQN. If the value of the first, second, or third term for a job type is relatively higher than the other job types, it means that there are more jobs of that type in the queue, there are more stable machines that are setup for that job type, and/or the average ATCS value of the waiting jobs of that type (i.e., the urgency of the waiting jobs of that job type) is relatively higher than the others. Note that since each of the three terms has a real value in the range of 0 to 1, the TQN value has also a real value in the range of 0 to 1. The step by step procedure of the ATCS-TQN is presented in Procedure 1.

Procedure 1. *New algorithm for time dependent quality combined with ATCS (ATCS-TQN)*

Step 1. When a machine becomes available at t, prioritize jobs in the queue with the ATCS rule.

Step 2. Prioritize job types (with at least one job of that type) in the queue with the TQN rule.

Step 3. Select the job type-f^* with the maximum TQN value among all prioritized job types.

Step 4. Select the job-j^* with the maximum ATCS value among all type-f^* jobs waiting in the queue.

Step 5. Schedule the job-j^* at the machine in order to be started at t.

4 COMPUTATIONAL EXPERIMENTS

In the experiments, four benchmark algorithms and the ATCS-TQN are included for scheduling the parallel machines in stage-2. The first benchmark algorithm is the well-known slack per remaining work (SLACK/RW) with equation $-\{d_j - (t + s_f + r_j)\}/(s_f + r_j)$, and the modified ATCSQ (MATCSQ) with equation $\sqrt{\left[\exp\left(-\min\left\{\frac{USL-\mu_{k,f,t}}{3\sigma_{k,f,t}}, \frac{\mu_{k,f,t}-LSL}{3\sigma_{k,f,t}}\right\}\right)-1\right]^2} \cdot ATCS_{j,f,t}$ is included as the second benchmark algorithm. The MATCSQ is modified from the ATCSQ of Ko *et al.* [11] to be able to prioritize jobs. In the equation, $\mu_{k,f,t}$ and $\sigma_{k,f,t}$ denote the mean and standard deviation of the average defect rate for job type-f at machine k at time t, and the lower and upper specification limits, LSL and USL, are set to 0 and τ, respectively. For the other two benchmark algorithms, the ATCS-TQ1 and ATCS-TQ2 proposed in Joo *et al.* [1] are included. Note that Equation (1) is commonly used to calculate the ATCS values in the last three benchmark algorithms for more accuracy in the slack times of jobs.

The phase-I duration estimating parameter π for the ATCS-TQN is set as 150 (when M = 3) and 200 (when M = 5) by a preliminary test. Figure 2 shows the performance changes in terms of the two performance measures according to the different levels of π, and the ATCS-TQN with the selected parameter values shows a good and balanced performance for both measures in both cases with M = 3 and M = 5. Note that the selected values of π are also used in the ATCS-TQ2 rule. Also, from another preliminary test, the scaling parameter ω of the ATCS-TQ2 rule is commonly set as 0.1 regardless of the number of machines in stage-2 under the selected values of π, and the scaling parameter ρ for the ATCS-TQ1 rule is set as 2.0 and 2.5 when M = 3 and M = 5, respectively. For the two scaling parameters, k_1 and k_2, of the ATCS rule, their values are computed as suggested in Lee & Pinedo [8] and commonly used in all the algorithms using the ATCS rule.

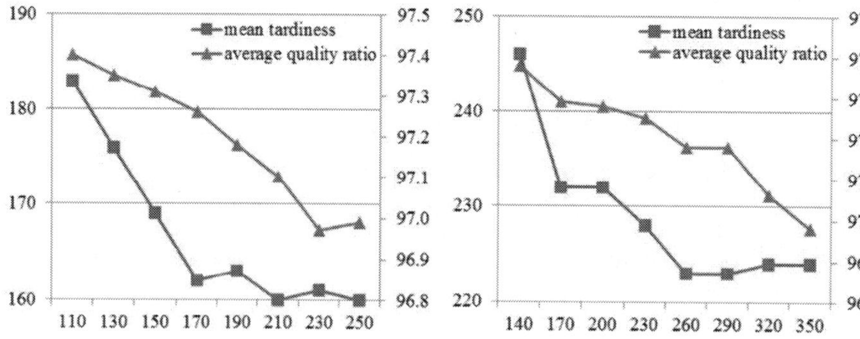

(a) when m = 3 (b) when m = 5

Figure 2: Performance of ATCS-TQN for different π values

In the following experiment, there are 3 job types, the inspection time is set to 5 for all job types, the processing time at stage-1 is randomly generated from a uniform distribution [10, 20], and the setup and processing time at stage-2 is randomly generated from a uniform distribution [10, 20] and either [35, 55] (when M = 3) or [65, 85] (when M = 5), respectively. The duration of phase-I defect rate is set by randomly choosing γ (parameter for setting the phase-I duration) among 5, 7 and 9, and the mean and standard deviations of phase-I and phase-II defect rates are set to (3, 1) and (5, 2), respectively. The tolerance of defect rate for each job and ADR are set to 5 and 3.5. In total, 8 test scenarios are generated according to the variations in three factors, the number of machines (M) in stage-2 (3 and 5), the inter-arrival time ranging parameter (0.5 and 0.7), and the due date ranging parameter (2 and 4). For each scenario, 20 problem instances are randomly generated, and for each problem instance, jobs are generated until the arrival time of the last job is no later than 10,000 time periods.

Table 1 shows the summary of the computational experiments. Each cell shows the mean (upper row) and standard deviation (lower row, in parenthesis) of two performance measures, the average quality ratio (Q) and the mean tardiness of jobs (T), for each algorithm and scenario.

Table 1: Performance of scheduling algorithms

Scenario			Scheduling Algorithm									
			SLACK/RW		MATCSQ		ATCS-TQ1		ATCS-TQ2		ATCS-TQN	
M	α	β	Q	T	Q	T	Q	T	Q	T	Q	T
3	0.5	2	95.79	333	92.96	303	92.91	309	93.24	292	**97.14**	**255**
			(2.06)	(133)	(2.95)	(165)	(2.90)	(167)	(3.61)	(169)	**(1.13)**	**(112)**
		4	95.69	311	92.39	257	92.82	243	94.34	204	**97.15**	**197**
			(2.05)	(132)	(3.42)	(166)	(3.04)	(161)	(3.08)	(149)	**(1.11)**	**(106)**
	0.7	2	97.11	195	93.96	175	94.29	156	94.55	150	**97.46**	**135**
			(0.94)	(82)	(3.12)	(126)	(2.84)	(103)	(3.22)	(103)	**(0.62)**	**(71)**
		4	96.88	170	94.33	101	94.59	97	94.94	94	**97.50**	**89**
			(1.29)	(85)	(2.95)	(86)	(2.74)	(85)	(2.65)	(81)	**(0.60)**	**(65)**
5	0.5	2	96.08	392	92.20	427	91.57	441	97.02	321	**97.16**	**321**
			(1.85)	(149)	(3.32)	(208)	(3.18)	(177)	(1.19)	(148)	**(1.18)**	**(139)**
		4	96.12	331	91.40	385	91.56	367	97.06	**228**	**97.04**	255
			(1.72)	(134)	(4.06)	(219)	(3.97)	(211)	(0.93)	**(127)**	**(1.29)**	(124)
	0.7	2	96.65	269	93.08	290	93.35	281	97.19	**193**	**97.45**	208
			(1.60)	(112)	(3.33)	(175)	(3.51)	(168)	(1.13)	**(97)**	**(0.78)**	(98)
		4	97.11	207	92.47	245	91.99	244	97.31	**120**	**97.48**	143
			(1.10)	(104)	(4.12)	(182)	(3.95)	(180)	(0.88)	**(80)**	**(0.79)**	(82)
Average			96.43	276	92.85	273	92.89	267	95.71	200	**97.30**	**200**
			(1.68)	(138)	(3.48)	(194)	(3.26)	(157)	(2.09)	(119)	**(0.94)**	**(99)**

As expected, when the due date ranging parameter β becomes smaller, the mean tardiness of jobs becomes larger because the average due date of jobs becomes tighter. Also, when the inter-arrival time ranging parameter α becomes smaller, the mean tardiness becomes larger because the expected average waiting time of jobs becomes longer and the average slack time of jobs becomes shorter.

In overall, the ATCS-TQN shows the best performance in terms of the mean and standard deviation of both performance measures, Q and T. By using the ATCS-TQN, the average quality ratio increases from 0.87% to 4.45% than the other four benchmark algorithms. In practice, the improvement in the quality ratio is quite significant in terms of the manufacturing cost, i.e., the quality cost. Also, the mean tardiness decreases from 33.5% to 38% than the first three benchmark algorithms. Comparing with the ATCS-TQ2, the ATCS-TQN shows decreased standard deviation of the mean tardiness by 20.2% although the average of the mean tardiness shows no difference. The improvement in the average and standard deviation of the mean tardiness is practically significant in terms of the customer satisfaction, i.e., the tardiness cost. Note that the ATCS-TQ2 shows worse quality performance than the ATCS-TQN while it shows a slightly better or similar tardiness performance when $M = 5$.

One interesting point is that the SLACK/RW shows better quality performance than the other three benchmark algorithms. It is because it generates a schedule incurring more setups (since it only considers the slack times of jobs, without considering the defect rate) which resets defect rates of the machines to phase-I more often. Another point is that the MATCSQ showed the worst quality and tardiness performance. Due to the use of real-time quality measurement data to prioritize jobs, it inevitably results in some quality loss during the time gap between the completion at stage-2 and inspection at stage-3. That is, some jobs must be processed or in process at a certain machine in the phase-II defect rate because it can detect the machines' shift to phase-II only after the inspection stage sends a quality feedback.

5 CONCLUSION

In this research, a dynamic flexible flow shop scheduling problem with setup timing-dependent process defect rate and a quality feedback mechanism is considered. A new DRSA, ATCS-TQN, focusing on the latest setup timings of the machines as well as the real-time shop information is developed by extending a previous DRSA. From a series of computational experiments, the ATCS-TQN rule showed the best and balanced performance in terms of both the average quality ratio and the mean tardiness of jobs. For the shop floor scheduling problems with time-dependent process defect rate, the suggested ATCS-TQN is expected to be easily applied in practice. As future research issues, the proposed DRSA can be extended to the dynamic flexible job shop or flow shop scheduling problems with more complicated process flows, such as rework or re-entrant flows. Also, the proposed DRSA can be extended for the scheduling problems with more realistic probabilistic models of time-dependent defect rates.

6 REFERENCES

[1] **Joo, B.J., Choi, Y.C., Xirouchakis, P., 2013**, Dispatching rule-based algorithms for a dynamic flexible flow shop scheduling problem with time-dependent process defect rate and quality feedback, *Procedia CIRP*, 7, pp 163-168.

[2] **Kianfar, K., Fatemi Ghomi, S.M.T., Karimi, B., 2009**, New dispatching rules to minimize rejection and tardiness costs in a dynamic flexible flow shop, *International Journal of Advanced Manufacturing Technology*, 45, pp 759-771.

[3] **Kia, H.R., Davoudpour, H., Zandieh, M., 2010**, Scheduling a dynamic flexible flow line with sequence-dependent setup times: a simulation analysis, *International Journal of Production Research*, 48, pp 4019-4042.

[4] **Choi, H-S., Kim, J-S., Lee, D-H., 2011**, Real-time scheduling for reentrant hybrid flow shops: a decision tree based mechanism and its application to a TFT-LCD line, *Expert Systems with Applications*, 38, pp 3514-3521.

[5] **Kianfar, K., Fatemi Ghomi, S.M.T., Oroojlooy Jadid, A., 2012**, Study of stochastic sequence-dependent flexible flow shop via developing a dispatching rule and a hybrid GA, *Engineering Applications of Artificial Intelligence*, 25, pp 494-506.

[6] **Vepsalainen, A.P.J., Morton, T.E., 1987**, Priority rules for job shops with weighted tardiness costs, *Management Science*, 33(8), pp 1035-1047.

[7] **Lee, Y.H., Bhaskaran, K., Pinedo, M., 1997**, A heuristic to minimize the total weighted tardiness with sequence-dependent Setups, *IIE Transactions*, 29, pp 45-52.

[8] **Lee, Y.H., Pinedo, M., 1997**, Scheduling jobs on parallel machines with sequence-dependent setup times, *European Journal of Operational Research*, 100, pp 464-474.

[9] **Yang, Y., Kreipl, S., Pinedo, M., 2000**, Heuristics for minimizing total weighted tardiness in flexible flow shops, *Journal of Scheduling*, 3, pp 89-108.

[10] **Pfund, M., Fowler, J.W., Gadkari, A., Chen, Y., 2008**, Scheduling jobs on parallel machines with setup times and ready times, *Computers & Industrial Engineering*, 54, pp 764-782.

[11] **Ko, H-H., Kim, J., Kim, S-S., Baek, J-G., 2010**, Dispatching rule for non-identical parallel machines with sequence-dependent setups and quality restrictions, *Computers & Industrial Engineering*, 59, pp 448-457.

[12] **Lee, H.L., Yano, C.A., 1988**, Production control in multistage systems with variable yield losses, *Operations Research*, 36(2), pp 269-278.

[13] **Akella, R., Rajagopalan, S., Singh, M.R., 1992**, Part dispatch in random yield multistage flexible test systems for printed circuit boards, *Operations Research*, 40(4), pp 776-789.

[14] **Sloan, T.W., Shanthikumar, J.G., 2002**, Using in-line equipment condition and yield information for maintenance scheduling and dispatching in semiconductor wafer fabs, *IIE Transactions*, 34, pp 191-209.

[15] **Liao, G-L., 2007**, Optimal production control and maintenance policy for imperfect process, *European Journal of Operational Research*, 182, pp 1140-1149.

[16] **Kazaz, B., Sloan, T.W., 2008**, Production policies under deteriorating process conditions, *IIE Transactions*, 40, pp 187-205.

[17] **Trikel, I., Reshef, N., Rabinowitz, G., 2009**, In-line inspection impact on cycle time and yield, *IEEE Transactions on Semiconductor Manufacturing*, 22(4), pp 491-498.

[18] **Dhouib, K., Gharbi, A., Ben Aziza, M.N., 2012**, Joint optimal production control/preventive maintenance policy for imperfect process manufacturing cell, *International Journal of Production Economics*, 137, pp 126-136.

[19] **Raviv, T., 2013**, An efficient algorithm for maximizing the expected profit from a serial production line with inspection stations and rework, *OR Spectrum*, 35(3), pp 609-638.

[20] **Tsao, Y-C., 2013**, Combined production-maintenance decisions in situations with process deterioration, *International Journal of Systems Science*, 44(9), pp 1692-1700.

[21] **Cassady, C.R., Kutanoglu, E., 2005**, Integrating preventive maintenance planning and production scheduling for a single machine, *IEEE Transactions on Reliability*, 54(2), pp 304-309.

[22] **Sortrakul, N., Nachtmann, H.L., Cassady, C.R., 2005**, Genetic algorithms for integrated preventive maintenance planning and production scheduling for a single machine, *Computers in Industry*, 56, pp 161-168.

[23] **Chung, S.H., Lau, H.C.W., Ho, G.T.S., Ip, W.H., 2009**, Optimization of system reliability in multi-factory production networks by maintenance approach, *Expert Systems with Applications*, 36, pp 10188-10196.

[24] **Wong, C.S., Chan, F.T.S., Chung, S.H., 2013**, A joint production scheduling approach considering multiple resources and preventive maintenance tasks, *International Journal of Production Research*, 51(3), pp 883-896.

ACCURACY EVALUATION METHOD FOR INTERMITTENT DEMAND FORECASTS

Alexandra Canda and Xue-Ming Yuan[1]
[1] Singapore Institute of Manufacturing Technology
2 Fusionopolis Way, #08-04 Innovis, Singapore 138634
xmyuan@SIMTech.a-star.edu.sg

ABSTRACT

Having accurate forecasts of demand requirements is essential for production planning and practising good inventory management. Demand data often manifest themselves with intermittences, finding the right forecasting method for such data and especially a proper way to assess the accuracy of the results is not an easy task. In this paper, we propose a new two dimensional accuracy evaluation method which we show to be a better fit for intermittent demand than other conventional accuracy evaluation methods, such as Mean (Absolute) (Scaled) Error.

Keywords: Forecasting Algorithms, Accuracy Evaluation Methods,, Intermittent Demand ME, MAE, MASE, EMID

1 INTRODUCTION

Forecasting is encountered in many fields and holding good forecasts is of great importance. Whether we are talking about forecasting the number of tourists expected to arrive at a certain moment in a specific area, security prices, the evolution of currency exchange rates, real estate prices or demand expected to be requested at some point in time, not having accurate forecasts will bring various negative implications. Our area of interest is demand forecasting, this is what we are going to focus further on in this paper.

Having excess of stock involves having capital tied up as well as various inventory costs, such as costs with storing, renting, utilities, insurance. Having insufficient stock will lead to backlog costs, self-image impairment or even loss of customers. Ideally, the smallest amount of inventory sufficient to cover the demand should be preserved. Not holding accurate forecasts about potential demand requests will not only involve several financial losses but will affect the entire flow of the supply chain.

Demand data are quite often having an intermittent pattern. Intermittent demand is encountered in several areas like the aerospace industry (aircraft components), as well as for various types of products such as luxury goods or very specialized ones. Such type of data is characterized by demand requests placed rarely in time, existing periods with no demand at all. Demand occurrence is not the only element that varies greatly in case of intermittent demand data; the size of the placed demands is also highly variable. These characteristics lead to difficulties in forecasting such trends of data.

Our aim is not to develop a new method of forecasting intermittent patterns, but to find a proper way of checking the accuracy of forecasts results for this kind of data. Having accurate forecasts of expected demand requirements will lead to a high level of fulfilment and implicitly to a high degree of customer satisfaction. Other benefits include: sales increasing (and thus revenue increment), lowering inventory costs and, therefore, enhancement of profits, which increases management's confidence towards investment.

There is no accuracy measure generally accepted for intermittent demand and opinions on which would be most reasonable highly differ. As already stated, our focus is to figure out which error measures are most reasonable for the case of intermittent data patterns.

The following section of the paper presents some literature review. The third one exposes the selected accuracy measures chosen to be applied to our own data as well as the new error measure that we are proposing. The fourth section presents the analysed results, while the last one concludes our work.

2 LITERATURE REVIEW

Several authors consider, and there is clear evidence at least for the case of intermittent type of data, that the performance of the forecasting methods is influenced by the accuracy measures used [1, 4, 6].

Syntetos and Boylan turned out to be very passionate about this area. Their work has resulted in a considerable number of journal articles. One of their papers [2] states that Croston's procedure is a robust one but still has quite poor forecasting performance. In an attempt to identify the reason for this, a mistake has been mentioned, which is assumed to contribute to the modest accuracy of the procedure. Subsequently, an adjustment has been brought, which was giving approximately unbiased demand estimates. The achieved improvement was tested through a simulation experiment and further confirmed by means of the statistical t-test.

Further, Syntetos and Boylan supported the idea of also considering the variance of intermittent demand estimation methods aside from their bias [3]. These two elements were incorporated in the calculus of MSE (one step ahead as well as over a fixed lead time). The estimation procedures used for the exposure of the proposed idea are Single Exponential Smoothing (SES), Croston and two methods proposed by Syntetos (SY) and Syntetos and Boylan (SBA), which are advocated to be (approximately) unbiased contrary to Croston's original procedure.

The same authors have proposed a Modified Croston method [4] and have been testing its accuracy through comparison to single exponential smoothing and the original Croston procedure by applying the mean square error to a set of simulated stationary demand data. The tested method has proved to perform worst in most of the cases (exception making the data being not highly intermittent). The original Croston's method got the lowest measurement errors and it was suggested to be further considered as a reference method.

Another estimation method was proposed as a deviation of Croston's procedure [5]. The original Croston's method, SES and Simple Moving Average (13 periods) are the other procedures considered for the testing and comparison of the newly proposed method's accuracy. The measures of accuracy considered were mean signed (ME), relative geometric root-mean-square error (RGRMSE), Percentage Better and Percentage Best (based on the former two error measures per series). Two scenarios were examined: all data points being considered to originate forecasts or concentrating only on the points in time soon after a demand occurred. The new method brought in was proven to perform better than the others for both absolute accuracy measures used and most smoothing constant values applied. SMA (13) was also competitive, performing best in several cases, while SES was mostly more accurate than Croston.

Prestwich et al. [6] chose a few forecasting methods, namely: Croston's, Single Exponential Smoothing, random walk (which considers as forecast the demand of the former period) and ZF (all zero forecasts) and considered that Croston's method beats SES, which outperforms ZF. The paper exposes a generous number of error measures clustered into several groups: measures which are scale-dependent, percentage ones, scaled, relative and relative error-based. For each existing error measure, a new one was proposed, which was achieved by assessing predictions towards the mean of the demand process rather than demands themselves. Several error measures from different classes were chosen and applied. The newly-proposed ones seemed to be more reliable, while among their original counterparts PB (Percent Better) gave the expected ranking most of the times.

Wallstrom and Segerstedt [7] have selected Single Exponential Smoothing, Croston's method and two modified versions of it and have used various error measures to test the accuracy of the applied forecast methods. Besides the traditional measures, mean absolute deviation (MAD) and mean square error (MSE), it was proposed to take into account only the time points with existent demand. These measures were called MAD_n and MSE_n. All these are scale dependent measures. A non-scale dependent measure used is the symmetric mean absolute percentage error (sMAPE). Other measures were proposed with the purpose of testing the potential existing bias, namely cumulated forecast error (CFE), number of shortage (NOS) and periods in stock (POS). One of the modified versions of Croston's method performed worst, having the most biased error and highly overestimating the demand.

Having the belief that improved accuracy of forecasts does not necessarily imply a high degree of customer satisfaction or a diminished inventory cost, Syntetos and Boylan were keen on seeing which of few selected forecasting procedures can better perform a control of the stocks [8]. They selected four estimation methods, SMA, SES, Croston and a modified version which they had formerly developed, and considered three constraints: a stipulated customer service level and two cost policies. The measures used to check the estimators' accuracy are percentage best (PBt) and average percentage regret (APR). It was shown that the method proposed by the authors of this paper has an overall superior performance of inventory control as well as the lowest total inventory cost. Croston's method and SMA had a similar performance while EWMA was found to perform worst.

3 PROBLEM DESCRIPTION AND ACCURACY EVALUATION METHOD

Intermittent demand occurs in many industries such as remanufacturing, MRO, and service industries. Items with intermittent demand include service (spare) parts, aircraft engines, heavy machinery, etc. Such items are often described as slow moving items. The data we used in our paper are from a retail store. The data consist of over 100 products with daily intermittent demands. Through forecasting these product demands, this paper is to conduct a critical assessment of various methods for forecasting intermittent demand. The comparative study shows that our accuracy evaluation method is able to well reflect the forecasting errors for intermittent demands. Keeping optimal inventories for slow-moving items is a very challenging issue. Producers/distributors of such type of items continuously struggle between minimizing costs (incurred by excess of stock), while at the same time trying to fulfill to a large extent customer requirements.

The incipient point for efficient inventory management and optimization is represented by having accurate information on future customer demands. Thus keeping accurate forecasts of demand is essential, but since customer demand often manifests sporadically, it makes predictions of this type of demand quite tough, as well as accuracy assessment difficult and challenging.

Many authors have emphasized the importance of using the forecasting algorithm that suits best for the case of each data pattern. No matter which method is used, one has to admit that any forecast results have a certain level of error. Knowing the error level of the forecast results that we are holding is also very important.

As one has to be able to recognize the pattern of the data he is dealing with and to apply a suitable forecasting algorithm in accordance (with that pattern), the accuracy measure that best suits each forecasting method has to be identified and used. Thus, it is essential that the error measure considers the specific characteristics of the data which are explored, just like the forecasting method applied should also do.

Some quite frequently used error measure indicators, which we have also computed and considered for comparison purpose, are Mean Error, Mean Absolute Error and Mean Absolute Scaled Error.

For a fixed moment of time t, the forecasting error is the difference between the observed value (Y_t) and the estimated one $(\hat{Y}_t)$:

$$e_t = Y_t - \hat{Y}_t \quad (1)$$

$$ME = \frac{1}{n}\sum_{t=1}^{n} e_t \, , \quad (2)$$

where *n* is the number of forecast data points (same as the number of data points existent in the estimation set).

$$MAE = \frac{1}{n}\sum_{t=1}^{n}|e_t| \quad (3)$$

$$MASE = \frac{1}{n}\sum_{t=1}^{n} \frac{|e_t|}{\frac{1}{n-1}\sum_{i=2}^{n}|Y_i - Y_{i-1}|} = \frac{\sum_{t=1}^{n}|e_t|}{\frac{n}{n-1}\sum_{i=2}^{n}|Y_i - Y_{i-1}|} \quad (4)$$

As long as intermittent data have several special features, the way of evaluating the accuracy of forecast results for this data pattern should be given a particular treatment. Thus, we propose a new index of assessing the error of forecasting results for this type of data.

$$EMID = \frac{\sigma}{\alpha} + \beta \cdot \varepsilon \quad (5)$$

$$\alpha = \frac{a}{N} \quad (6)$$

$$\beta = \frac{b}{N} \quad (7)$$

$$\sigma = \frac{1}{a}\sum_{t=1}^{a} e_t \quad (8)$$

$$\varepsilon = \frac{1}{b}\sum_{t=1}^{b} e_t \quad (9)$$

where *N* represents the number of data points in the evaluation set; *a* is the number of matching points in the sets of actual and forecasted data; *b* is the number of non-matching points in the sets of actual and forecasted data; σ is the mean error for the matching points; ε is the average error of the non-matching points.

Compared to the conventional ways of quantifying the error measures, the approach we propose is two-dimensional, considering the number (ratio) of matching and non-matching points besides errors. By matching point we mean that for the same moment in time actual and forecast data points are both either zero or non-zero.

4 RESULTS AND ANALYSIS

The dataset used consists of series representing customer demand which have an intermittent pattern (for each series more than half of the data points are represented by zeros). The data are on a daily basis (one week consisting of 6 working days) with an available horizon of time of 1 year and 1 month comprising around 330 data points and the length of the inter-demand interval being of maximum 217 days. From the whole dataset available an amount of around 80% was used to get the forecasts, while the remaining was utilized for testing purpose.

The forecasting methods which we applied are Single Exponential Smoothing (SES), Simple Moving Average (SMA, 10 periods), Croston's method and Croston's approach with SMA used instead of SES for demand size and demand interval, as well as the naïve approach ZF.

Below we have exposed the plots comparing the actual data and each set of forecast results for one SKU as well as the error measure values and their ranking.

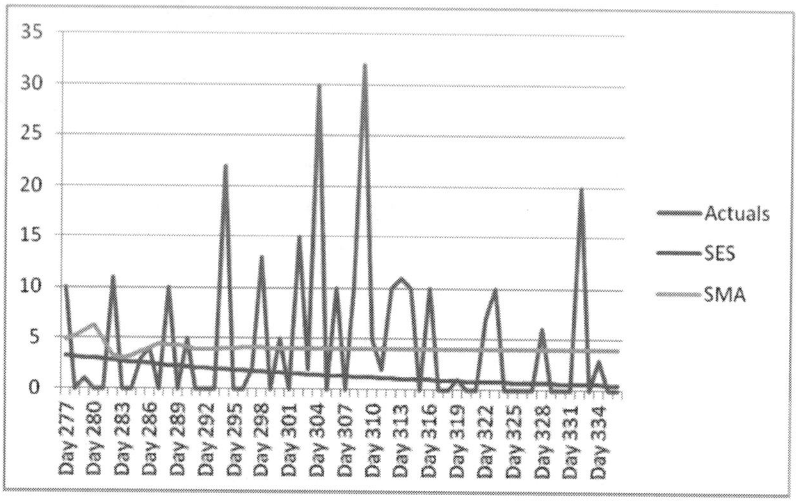

Figure 1: Actual data versus forecasts obtained using Single Exponential Smoothing and Single Moving Average Methods

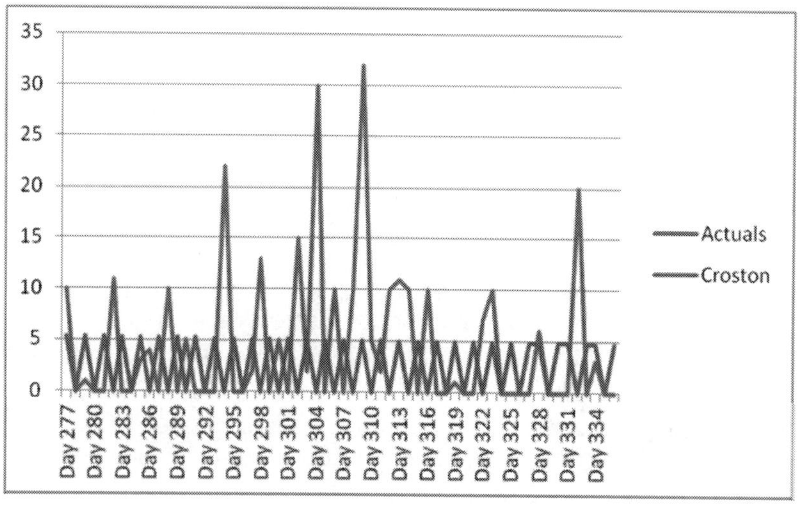

Figure 2: Actual data compared to Croston's forecast method results

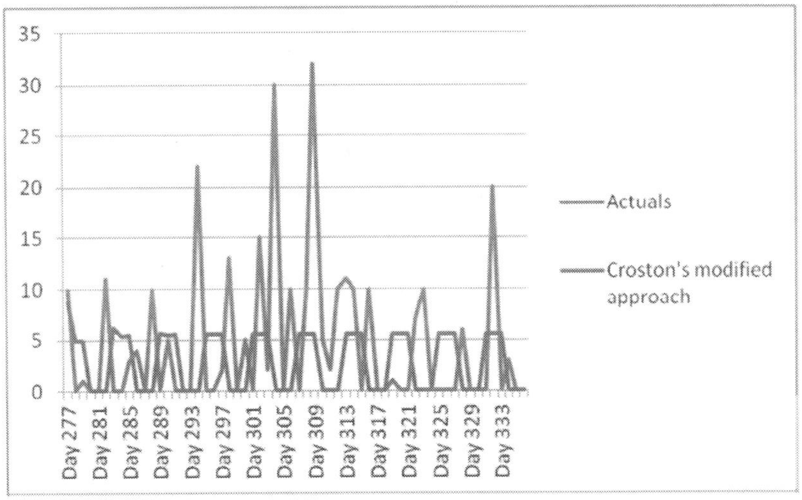

Figure 3: Actuals versus Croston's modified approach forecast results

From a visual perspective the forecasts given by Croston's method are able to best capture the actual data. Furthermore, we will focus on the values of the accuracy measures and their interpretation. In the table below the first row corresponding to each error measure contains the values of these for each forecasting algorithm, while the second row consists of the ranking of errors corresponding to each set of results among all forecasting methods applied. We want to specify that we did not consider whether the forecasts have over or underestimated the actual demands and only looked upon the absolute values of Mean Error and new accuracy evaluation method which we have proposed.

Table 1: Values and ranking of error measures among forecast methods used

	SES	SMA	Croston	Sma_Sma	ZF
MAE	4.76	5.34	6.05	5.6	4.67
	2	**3**	**5**	**4**	**1**
ME	3.13	0.56	1.99	1.88	4.67
	4	**1**	**3**	**2**	**5**
MASE	0.64	0.72	0.81	0.75	0.62
	2	**3**	**5**	**4**	**1**
EMID	15.95	9.37	4.17	5.03	4.67
	5	**4**	**1**	**3**	**2**

Looking at the conventional error measures one can notice that none of these three accuracy measures rates Croston's method as best. Furthermore the absolute error measures assess it as worst. According to the first three error measures exposed, the best results are given by the naïve ZF approach, the Simple Moving Average and Single Exponential Smoothing. This could be easily put under doubt.

One can notice that the new index proposed by us evaluates Croston's method as best and SES and SMA the worst ones. One can blame the rating of ZF approach as second best. This can be associated with the highest number of matching points existent in the case of this algorithm.

Table 2: Number of matching points between actual data and forecast results

SES	SMA	Croston	Sma_sma	ZF
29	29	23	27	31

To further convince the reader of the importance and relevancy of considering the two dimensions (error and ratio of matching and mismatching points) in the calculation of our accuracy evaluation method (EMID), we are going to further provide one more example.

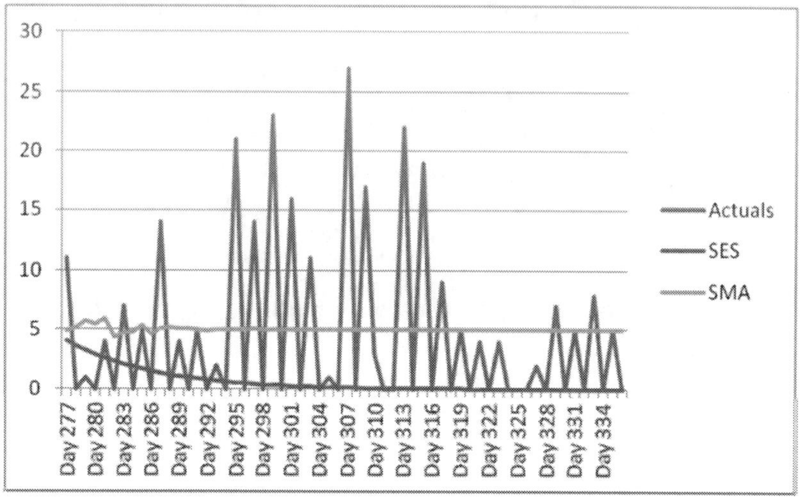

Figure 4: Actual values compared to forecast results obtained using Single Exponential Smoothing and Single Moving Average methods

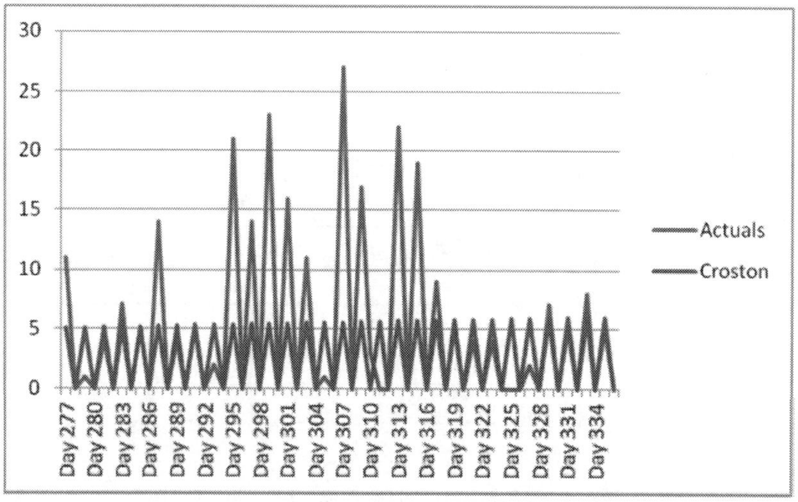

Figure 5: Actual data versus Croston's forecast results

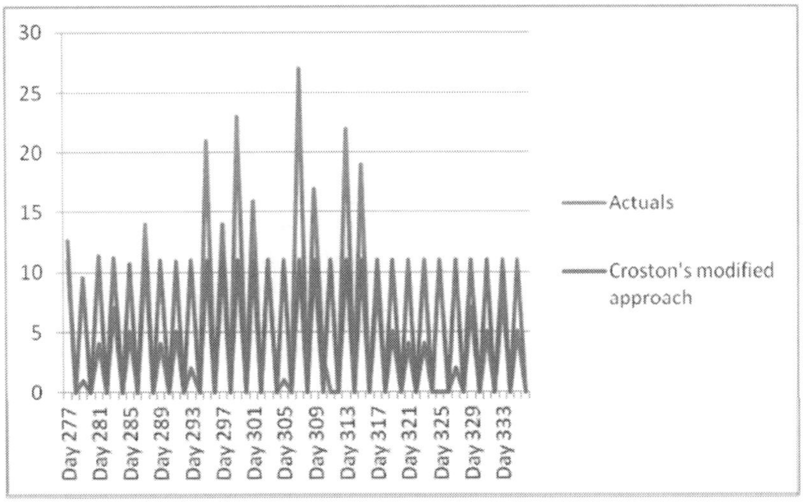

Figure 6: Actuals versus Croston's modified approach forecasts results

For this example the results given both by Croston's method and its modified approach follow the pattern of the actual demand. Visually, on the whole, Croston's modification looks like having given the best forecast results for this SKU.

MAE and MASE give identical rankings. It is easy to note that for these two error measures and the new proposed method, the battle for the top two positions is given between the two prior mentioned algorithms. Only the method which we propose gives the expected ranking. MAE and MASE give a reversed ranking for the two forecasting methods under discussion, while much worse, Mean Error rates Simple Moving Average as the best.

Although the best two forecasting methods have same number of matching points, we are able to prove that taking into account both number of matching/mismatching points as well as errors gives better results compared to the conventional measures which only consider error amounts.

Table 3: Error measure values and their ranking

	SES	SMA	Croston	Sma_Sma	ZF
MAE	4.64	5.46	3.05	3.47	4.60
	4	**5**	**1**	**2**	**3**
ME	3.95	-0.42	1.85	-0.91	4.6
	4	**1**	**3**	**2**	**5**
MASE	0.51	0.60	0.34	0.38	0.50
	4	**5**	**1**	**2**	**3**
EMID	17.92	6.65	2.06	-0.98	4.6
	5	**4**	**2**	**1**	**3**

Table 4: Number of matching points among forecast results

SES	SMA	Croston	Sma_sma	ZF
29	29	57	57	31

5 CONCLUSIONS

Accuracy measures play a very important role in quantifying the size of the errors existent for the forecast results. Led by this thought, we have decided to create a new accuracy evaluation method of assessing the forecast accuracy for intermittent demand because many popular accuracy evaluation methods are no longer suitable to evaluate the forecasting accuracy of intermittent demand.

The accuracy evaluation method that we have proposed takes into consideration the ratio of matching /mismatching points, besides the amount of errors. For comparison purpose, the Mean Error, Mean Absolute Error and Mean Absolute Scaled Error have been computed. The forecasting algorithms applied are Single Exponential Smoothing, Simple Moving Average, the naïve ZF approach, Croston's method and its variation. We have shown that the new accuracy evaluation method that we have proposed gives more reasonable results, in comparison with the other conventional error measures used.

6 REFERENCES

[1] R. J. Hyndman, G. Athanasopoulos, 2013, *Forecasting: principles and practice*, OTexts

[2] A.A. Syntetos, J.E. Boylan, 2001, "On the bias of intermittent demand estimates", *International Journal Production Economics*, 71, pp. 457-466

[3] A.A. Syntetos, J.E. Boylan, 2010, "On the variance of intermittent demand estimates", *International Journal Production Economics*, 128, pp. 546-555

[4] J.E. Boylan, A.A. Syntetos, 2007, "The accuracy of a Modified Croston procedure", *International Journal Production Economics*, 107, pp. 511-517

[5] A.A. Syntetos, J.E. Boylan, 2005, "The accuracy of intermittent demand estimates", *International Journal of Forecasting*, 21, pp. 303-314

[6] S. D. Prestwich, R. Rossi, S. Armagan Tarim, B. Hnich, 2014, "Mean-Based Error Measures for Intermittent Demand Forecasting", *International Journal of Production Research*, 52, pp.6782-6791

[7] P. Wallstrom, A. Segerstedt, 2010, "Evaluation of forecasting error measurements and techniques for intermittent demand ", *International Journal Production Economics*, 128, pp. 625-363

[8] A.A. Syntetos, J.E. Boylan, 2006, "On the stock control performance of intermittent demand estimators", *International Journal Production Economics*, 103, pp. 36-47

THE IMPACT OF PRODUCT RETURNS ON DELIVERY TIME DECISION*

Si-si Zhao[1,†], Feng Wu[2], Lei Shu[3] and Tao Jia[4]
[1]School of Management
Xi'an Jiaotong University, Xi'an, China
zhaosisi@stu.xjtu.edu.cn

[2] School of Management
Xi'an Jiaotong University, Xi'an, China
The Key Lab of the Ministry Of Education for Process Control & Efficiency Engineering
fengwu@mail.xjtu.edu.cn

[3]School of Management
Xi'an Jiaotong University, Xi'an, China
shulei@stu.xjtu.edu.cn

[4]School of Management
Xi'an Jiaotong University, Xi'an, China
The Key Lab of the Ministry Of Education for Process Control & Efficiency Engineering
jiatao@mail.xjtu.edu.cn

ABSTRACT

The increasing quantity of product returns has been one of the major problems faced by many businesses in online retailing. Besides the quality of products and other factors such as price, the inconsistency between promised delivery lead time (PDL) and actual realized lead time could increase product returns to some extent. Thus, this paper studies the impact of product returns on the business's joint decisions of PDL and service level when both demand and returns are leadtime-sensitive. Through revealing the tradeoff among the shorter PDL to capture more demand, the higher service level or the longer PDL to decrease the probability of returns due to delay and the cost in online retailing, we show that the PDL is not affected by returns when the returns sensitivity to delay is relatively low and increases with returns while the delay sensitivity increases to a threshold. In addition, the threshold decreases in the return rate to demand, which means the impact of product returns would be more serious on the business with a high return rate with respect to demand.

Keywords: Promised delivery lead time, Service level, Product returns, Online retailing

1 INTRODUCTION

Due to the fierce competition in online retailing, competing on price is no longer enough. Since consumers have to wait after ordering online, many e-commerce businesses are offering promised delivery lead time (PDL) as a marketing weapon to attract customers [1-3]. Although making a shorter

* This work was supported by the National Natural Science Foundation under Grant No. 71471144, 71428003, 71271168.

† Corresponding Author

PDL decision could increase the sales, it also increases the possibility of inconsistency, namely delivery-delay effect. Recently, Rao [4] indicates that the occurrence of product returns depends on the delay in the delivery of orders to consumers relative to the PDL made by the retailers. That is, a shorter PDL leads to not only a better performance of sales, but also an increasing quantity of product returns once delayed. In fact, the problem of increasing product returns has become one of the major problems faced by businesses in online retailing [5]. Therefore, businesses should carefully decide the PDL while taking the delay effect on product returns into consideration to maximize their profits.

Although it is inevitable of delivery delay due to uncertainties in real delivery process, businesses could take some measures to ensure the timeliness of delivery to some extent. For instance, for businesses who have their own distribution systems, they could decrease the possibility of delay through investing to enhance the efficiency of the systems; for businesses using third-party logistics (3PL), they could choose different 3PL firms that differentiate in delivery performance, etc. In a word, it is the service level that decided by businesses to ensure timeliness of delivery and generate cost at the same time.

Hence, in this paper, we investigate the trade-off among the shorter PDL to capture more demand, the higher service level or the longer PDL to decrease the probability of returns due to delay and the cost in online retailing. The previous research about PDL overlook the delay impact on product returns, and take it as the general penalty cost instead. This will lead to non-optimal decisions and underrate the effect of product returns on PDL decisions and profits in online retailing.

There are two streams of research that are relevant to this paper. The first is about PDL. Earlier research focus on individual businesses, especially in their production and assembling systems. For instance, Yano [6] investigates to optimize planned leadtimes in serial production systems where the actual procurement and processing times is stochastic. Based on that, some research extends to study PDL in the context of competition. So [1] analyses the equilibrium price and PDL decisions in a multiple-firm competition. Shang and Liu [7] study firms' competition both in PDL and capacity. Meanwhile, there are also research discussing PDL from the perspective of supply chain. Liu et al. [8] investigate the price and PDL decisions in decentralized supply chains by constructing a Stackelberg game. Xiao and Qi [9] examine how to coordinate a two-stage supply chain, where the demand rate depends on selling price, PDL and delivery reliability.

Overall, the literature study PDL from the perspective of individual business, multi-firm competition, supply chain or competitive supply chain, but most of them do not take the delay effect on returns into consideration. Once the order is delayed from PDL, the common way for previous research is using a general penalty cost to evaluate. In this case, there could not estimate the PDL on profit accurately. Besides, Xiao et al. [3] assume that shortening PDL will accompany with higher extra cost than before. However, this paper is to examine whether it is always true when businesses make joint decision of PDL and service level in online retailing.

The second is research on product returns. With the increasing number of product returns, especially in online retailing, the costly impact on businesses' profit is widely noticed by researchers. Many studies investigate the optimal return policies, which refer to the money back policy offered by retailers to their consumer. Su [10] develops a model of consumer returns policies and investigate the impact of full/partial returns policies on supply chain performance. Similarly, Chen and Bell [11] examine how a firm that faces customer returns can enhance profit through offering different returns policies. Li et al. [12] derive the retailer's optimal pricing and refund policies accompanied by different advanced selling strategies. Li et al. [13] and Xu et al. [14] also provide optimal return policies when considering product returns. However, although a full refund policy may lead to unnecessary returns [10], the retailer has to adopt this with no choice due to the competition, especially in online retailing. In this case, Chen and Bell [15] study the impact of returns on pricing and inventory replenishment when the firm use full-credit return policy.

In this paper, we follow the idea that businesses offer a 100% money-back guarantee when facing fierce competition in online retailing with price exogenous. Shang and Liu [7] also study PDL while assuming the price is exogenous in order to focus on the impact of time dimensions and make model tractable. In this case, our research aims to study the impact of product returns on joint decision of PDL and service level when both the demand and product returns are PDL-sensitive.

2 BASIC MODEL AND ANALYSIS

2.1 Notations

Decision variables	
T	promised delivery lead time (PDL)
λ	service level indicator
Parameters	
D	demand
R	product returns
t	actual realized lead time with p.d.f. $f(\cdot)$ and c.d.f. $F(\cdot)$
d_0	market scale
s_1	the demand sensitivity to PDL
α	the return sensitivity to demand (return rate)
β	the return sensitivity to delay
p	the exogenous price per unit
w	unit wholesale price
c	cost sensitivity to service level
s	unit residual value of product returns

2.2 Problem description

In this paper, we consider a business selling products online for one period. The market is under perfect competition so that the price is exogenous. The demand is deterministic and depends only on PDL. Further, we assume demand linearly depends on PDL, which could be expressed by $D(T) = d_0 - s_1 T$, where d_0 represents the market scale, namely the maximum demand when PDL is the shortest, and s_1 denotes the demand sensitivity to PDL. The consumers would be expected a full credit return policy when returning products with a residual value of s. There are two assumptions about the returns function R: (1) product returns increase with demand; (2) product returns also depend on the delay between PDL and the actual realized lead time. The greater the extent of delay is, the more the possibility of products to be returned would be. The first assumption is used by most of the previous research related to returns from both theoretical and empirical perspectives, while the second assumption follows the latest empirical findings aforementioned by Rao [4]. Then, we could further assume the general form of returns function is:

$$R(T) = \alpha D(T) + \beta(t - T)^+ \tag{1}$$

, where α, β denote to the return sensitivity to demand and delay, respectively. Note that $0 < \alpha < 1, \beta > 0$. The actual realized lead time is denoted by t, which is stochastic distribution within the interval $[\underline{t}, \overline{t}]$. $(t - T)^+ = \max\{(t - T, 0)\}$ represents the extent of delay between PDL and the actual realized lead time. Further, we assume that t follows a uniform distribution. Without loss of generality, we have $t \sim U(0, \lambda)$, where λ denotes the latest lead time when the order is fulfilled. As aforementioned, the businesses could ensure the timeliness through increasing the service level in fulfilment process. In this paper, the effort is made through deciding a proper λ. When choosing a relative low λ, it means businesses intend to increase the service level to ensure timeliness and increase the cost as well. If $C(\lambda)$ denotes the cost or investment when the upper bound of t is λ, then we could know that the higher of λ, the less of cost. From this property, we could assume the cost function is $C(\lambda)=\frac{c}{\lambda}$, where c denotes the cost index for λ.

2.3 Model analysis

Assume that the risk attitude of decision maker is neutral and the objective is to maximize the expected profit. To make the PDL and λ decisions simultaneously, the business faces the trade-off among the shorter PDL to capture more demand, the longer PDL or the higher λ to guarantee the promise in order to decrease the expected product returns, and the cost. Thus, the objective function with a full credit return policy can be written as follows:

$$\max_{T,\lambda} \Pi_R = \big(p-w-C(\lambda)\big)D(T)-(p-s)E[R(T)] \tag{2}$$

$$\text{s.t. } 0\le T\le \frac{d_0}{s_1}, 0\le\lambda\le\frac{d_0}{s_1}$$

Note that: (1) to ensure demand is always non-negative, we have the constraints of $0\le T\le\frac{d_0}{s_1}, 0\le \lambda\le\frac{d_0}{s_1}$. (2) when $\alpha>\frac{d_0(p-w)-cs_1}{d_0(p-s)}$, the expected profit would be always nonpositive regardless of any T or λ *giving* $p>w>s>0$. Thus, in this paper we only discuss the case where the returns sensitivity to demand is not too high, namely $\alpha<\frac{pd_0-cs_1}{d_0(p-s)}$.

Proposition 1: when $\alpha<\frac{d_0(p-w)-cs_1}{d_0(p-s)}$, the optimal joint decisions of T and λ *is* as follows:

(1) when $0<\alpha\le\frac{d_0(p-w)-3cs_1}{d_0(p-s)}$,

i. when $0<\beta<\frac{2cs_1^2}{d_0(p-s)}$, $(T^*,\lambda^*)=\left(0,\frac{d_0}{s_1}\right)$;

ii. when $\frac{2cs_1^2}{d_0(p-s)}\le\beta<\beta_1$, $(T^*,\lambda^*)=(0,\lambda_1^*)$;

iii. when $\beta\ge\beta_1$, $(T^*,\lambda^*)=(T_1^*,\Phi)$;

(2) when $\frac{d_0(p-w)-3cs_1}{d_0(p-s)}<\alpha<\frac{d_0(p-w)-cs_1}{d_0(p-s)}$,

i. when $0<\beta<\frac{s_1d_0(p-w-p\alpha+s\alpha)-cs_1^2}{d_0(p-s)}$, $(T^*,\lambda^*)=\left(0,\frac{d_0}{s_1}\right)$;

ii. when $\frac{s_1d_0(p-w-p\alpha+s\alpha)-cs_1^2}{d_0(p-s)}<\beta<\frac{s_1d_0(p-w-p\alpha+s\alpha)+cs_1^2}{2d_0(p-s)}$, $(T^*,\lambda^*)=\left(\frac{d_0A+cs_1^2}{s_1\beta(p-s)},\frac{d_0}{s_1}\right)$;

iii. when $\beta>\frac{s_1d_0(p-w-p\alpha+s\alpha)+cs_1^2}{2d_0(p-s)}$, $(T^*,\lambda^*)=(T_1^*,\Phi)$,

where $T_1^*=\frac{cs_1+A\Phi}{\beta(p-s)}$, $A=\beta(p-s)-s_1(p-w-\alpha(p-s))$, $\Phi=\sqrt{\frac{c(cs_1^2-2\beta d_0(p-s))}{A^2-\beta^2(p-s)^2}}$, $\lambda_1^*=\sqrt{\frac{2cd_0}{\beta(p-s)}}$, $\beta_1=\frac{cs_1^2+4d_0s_1(p-w-\alpha(p-s))-s_1\sqrt{c^2s_1^2+8d_0s_1c(p-w-\alpha(p-s))}}{d_0(p-s)}$.

Proof: It is easy to verify that Π_R is a concave function in T and λ. Since this is an optimization problem with inequality constraints, we use Karush-Kuhn-Tucker Conditions (KKT) to solve it. Let $L(T,\lambda,\lambda_1,\lambda_2,\lambda_3)=\Pi_R+\lambda_1\left(\frac{d_0}{s_1}-T\right)+\lambda_2\left(\frac{d_0}{s_1}-\lambda\right)+\lambda_3T$. According the KKT Conditions, the optimal solutions should satisfy the following conditions:

$$\frac{\partial L}{\partial T}=0,\frac{\partial L}{\partial \lambda}=0 \tag{3}$$

$$\lambda_1\left(\frac{d_0}{s_1}-T\right)=0,\lambda_2\left(\frac{d_0}{s_1}-\lambda\right)=0,\lambda_3T=0 \tag{4}$$

$$0 \leq T, \lambda \leq \frac{d_0}{s_1} \quad (5)$$

$$\lambda_1, \lambda_2, \lambda_3 \geq 0 \quad (6)$$

Note that in this paper we assume $p>w>s>0$ and $\alpha < \frac{d_0(p-w)-cs_1}{d_0(p-s)} < \frac{p-w}{p-s}$. After discussing in different cases, we obtain the optimal T and λ as mentioned in Proposition 1. □

From proposition 1, we could deduce that the optimal joint decisions of PDL and λ depend on the returns sensitivities to both demand and delay. More specifically, when β is relatively low, which means delay is not susceptible to cause product returns, all the businesses would make the shortest PDL as well as the longest λ regardless of α. As β increases, the businesses with a relatively low α would remain PDL unchanged and begin to decrease λ via investing. Other businesses whose products are more easily to be returned with respect to demand, would lengthen PDL first when β exceeds a threshold value $\frac{s_1 d_0(p-w-p\alpha+s\alpha)-cs_1^2}{d_0(p-s)}$, rather than decreasing λ. Furthermore, in the case where β is relatively large, all the businesses would increase the PDL and decrease λ simultaneously. From Proposition 1, we can obtain the following properties.

Corollary 1: when $\alpha < \frac{d_0(p-w)-cs_1}{d_0(p-s)}$, T^* has the following relationships with returns parameters.

(1) When $0 < \alpha \leq \frac{d_0(p-w)-3cs_1}{d_0(p-s)}$: T^* is not affected by α and β when $0 < \beta < \beta_1$; otherwise, T^* monotonically increases in both α and β.

(2) When $\frac{d_0(p-w)-3cs_1}{d_0(p-s)} < \alpha < \frac{d_0(p-w)-cs_1}{d_0(p-s)}$: T^* is not affected by α and β when $0 < \beta < \frac{s_1 d_0(p-w-p\alpha+s\alpha)-cs_1^2}{d_0(p-s)}$, otherwise T^* monotonically increases in both α and β.

The corollary above is derived using the first-order conditions of results in Proposition 1.

Corollary 1 shows the impact of both return rate and the returns sensitivity to delay on the optimal PDL. Since the business would take the shortest PDL when β is relatively low, T^* has no relationships with both α and β. When β is beyond the threshold, the business would increase PDL as β increases to reduce the probability of delay. In addition, the threshold is different in the two cases where businesses have high or low return rates, and the thresholds always decrease in α in both cases.

Corollary 2: when $\alpha < \frac{d_0(p-w)-cs_1}{d_0(p-s)}$, λ^* has the following relationships with returns parameters.

(1) When $0 < \alpha \leq \frac{d_0(p-w)-3cs_1}{d_0(p-s)}$n: λ^* is not affected by β when $0 < \beta < \frac{2cs_1^2}{d_0(p-s)}$; otherwise, monotonically decreases in β; λ^* is not affected by α when $0 < \beta < \beta_1$; otherwise, λ^* monotonically increases in α.

(2) when $\frac{d_0(p-w)-3cs_1}{d_0(p-s)} < \alpha < \frac{d_0(p-w)-cs_1}{d_0(p-s)}$: λ^* is not affected by α and β when $0 < \beta < \frac{s_1 d_0(p-w-p\alpha+s\alpha)+cs_1^2}{d_0(p-s)}$, otherwise λ^* increases in α and decreases in β, respectively.

As can be seen from Corollary 2, λ^*, which is the upper bound of the actual realized delivery time to represent the service level, have different properties as α and β increase. The business would decide the service level as low as possible when delay does not easily to cause returns, thus λ^* is not affected by both α and β.In the case $0 < \alpha \leq \frac{d_0(p-w)-3cs_1}{d_0(p-s)}$, as β increases to the threshold of $\frac{2cs_1^2}{d_0(p-s)}$, the business should begin to take the delay effect into consideration and would enhance the service level to guarantee the timeliness to avoid returns as products are easier to be returned once delayed. While β exceeds the threshold of β_1, the return rate starts to affect the service level decision. Rather than increasing in α, the service level would be lowered. The reason in understandable, the saved cost by lowering the service level contributes more to profit than other factors in the trade-off in this case. We omit the discuss about the case when $\frac{d_0(p-w)-3cs_1}{d_0(p-s)} < \alpha < \frac{d_0(p-w)-cs_1}{d_0(p-s)}$, since the only difference between high and low return rates businesses is that the threshold is the same for α and β when $\frac{d_0(p-w)-3cs_1}{d_0(p-s)} < \alpha < \frac{d_0(p-w)-cs_1}{d_0(p-s)}$.

3 NUMERICAL EXAMPLES

In this section, numerical examples were carried out to better understand the impact of product returns on joint decisions and profit in online retailing. The impact of product returns (α and β) on PDL, λ , demand, total returns and profit is summarized in Tables 1 where $d_0 = 20, s_1 = 4, c = 10, p = 20, w = 10, s = 8$. Note that in this paper we assume the return rate is not too high, which is $\alpha < \frac{d_0(p-w)-cs_1}{d_0(p-s)} = \frac{2}{3}$ in this section, thus we let $\alpha = 0.1, 0.4$ and $\beta = 0.8, 1.2, 1.6, 2.0, 5.0$ to explain the returns' impact.

Table 1: The impact of customer returns on decisions and profit

	β	T	λ	D	R	Π_R
$\alpha = 0.1$	0.80	0.00	5.00	20.00	4.00	112.00
	1.20	0.00	5.00	20.00	5.00	100.00
	1.60	0.00	4.56	20.00	5.65	88.36
	2.00	0.00	4.08	20.00	6.08	78.02
	5.00	1.80	2.74	12.80	2.09	56.25
$\alpha = 0.4$	0.80	0.00	5.00	20.00	10.00	40.00
	1.20	0.56	5.00	17.78	9.48	28.44
	1.60	1.74	4.08	13.03	6.28	22.96
	2.00	2.17	3.76	11.33	5.21	20.69
	5.00	2.82	3.29	8.72	3.66	16.83

The results illustrate that PDL is not affected by returns when the returns sensitivity to delay is low ($\beta = 0.8$) regardless of the return rate α, but the businesses with high return rate are susceptible to be affected by PDL as β increases ($\alpha = 0.4$). Furthermore, the profit monotonically decreases in both α and β.

4 CONCLUSION

We investigated the impact of product returns on the business', which selling products online, joint promised delivery lead time and service level decisions making for a single period with deterministic leadtime-sensitive demand and consider the case where product returns depend not only on demand but also the delay between PDL and the actual realized lead time.

Our findings suggest that PDL is not affected by returns when the returns sensitivity to delay is relatively low, but as the sensitivity to delay increases to a threshold, the business would lengthen PDL as well as product returns increase. In addition, the threshold decreases in the return rate to demand, which means the businesses with high return rate are susceptible to be affected by PDL.

5 REFERENCES

[1] **So, K. C. 2000.** Price and time competition for service delivery, *Manufacturing & Service Operations Management*, 2(4), 392-409.

[2] **Li, Y., Lin, Q., & Ye, F. 2014.** Pricing and promised delivery lead time decisions with a risk-averse agent, *International Journal of Production Research*, 52(12), 3518-3537.

[3] **Xiao, T., Shi, J., & Chen, G. 2014.** Price and leadtime competition, and coordination for make-to-order supply chains, *Computers & Industrial Engineering*, 68, 23-34.

[4] **Rao, S., Rabinovich, E., & Raju, D. 2014.** The role of physical distribution services as determinants of product returns in Internet retailing, *Journal of Operations Management*, 32(6), 295-312.

[5] **Huang, X., Choi, S. M., Ching, W. K., Siu, T. K., & Huang, M. 2011.** On supply chain coordination for false failure returns: A quantity discount contract approach, *International Journal of Production Economics*, 133(2), 634-644.

[6] **Yano, C. A. 1987.** Setting planned leadtimes in serial production systems with tardiness costs, *Management science*, 33(1), 95-106.

[7] **Shang, W., & Liu, L. 2011.** Promised delivery time and capacity games in time-based competition, *Management Science*, 57(3), 599-610.

[8] **Liu, L., Parlar, M., & Zhu, S. X. 2007.** Pricing and lead time decisions in decentralized supply chains, *Management Science*, 53(5), 713-725.

[9] **Xiao, T., & Qi, X. 2012.** A two-stage supply chain with demand sensitive to price, delivery time, and reliability of delivery, *Annals of Operations Research*, 1-22.

[10] **Su, X. 2009.** Consumer returns policies and supply chain performance, *Manufacturing & Service Operations Management*, 11(4), 595-612.

[11] **Chen, J., & Bell, P. C. 2012.** Implementing market segmentation using full-refund and no-refund customer returns policies in a dual-channel supply chain structure, *International Journal of Production Economics*, 136(1), 56-66.

[12] **Li, Y., Xu, L., Choi, T. M., & Govindan, K. 2014.** Optimal advance-selling strategy for fashionable products with opportunistic consumers returns, *IEEE Transactions on Systems, Man, and Cybernetics: Systems*, 44(7), 938-952.

[13] **Li, Y., Xu, L., & Li, D. 2013.** Examining relationships between the return policy, product quality, and pricing strategy in online direct selling, *International Journal of Production Economics*, 144(2), 451-460.

[14] **Xu, L., Li, Y., Govindan, K., & Xu, X. 2015.** Consumer returns policies with endogenous deadline and supply chain coordination, *European Journal of Operational Research*, 242(1), 88-99.

[15] **Chen, J., & Bell, P. C. 2009.** The impact of customer returns on pricing and order decisions, *European Journal of Operational Research*, 195(1), 280-295.

PORT TRUCK SCHEDULING ON A DEDICATED TRANSPORTATION ROUTE AT A CONTAINER TERMINAL

Feifeng Zheng[1], Jidan Huang[1*], Ming Liu[2], and Feng Chu[3]
[1] Glorious Sun School of Business and Management
Donghua University, Shanghai, China
ffzheng@dhu.edu.cn

[1] Glorious Sun School of Business and Management
Donghua University, Shanghai, China
huangjd@dhu.edu.cn

[2]School of Economics and Management
Tongji University, Shanghai, China
mingliu@tongji.edu.cn

[3] Laboratoire IBISC, Université d'Evry Val d'Essonne, 40 rue du Pelvoux, 91020 Evry Cedex, France
feng.chu@ufrst.univ-evry.fr

ABSTRACT

This paper studies the problem of container truck scheduling on a dedicated transportation route, which connects the container yard and the outside of a container terminal. We consider two objectives to minimize the total waiting time of tasks and the total fuel consumption simultaneously. It is motivated by practical applications in some container terminals which are located at islands tens of miles away from continents and connect to the outsides via long distance bridges. Each container truck either sending an outbound container to the terminal yard or fetching an inbound container from the yard suffers halfway empty driving during the round trip.

With the assumption that all the tasks of container transportation on the bridge segment are served by port trucks, we propose the vehicle scheduling problem which balances the trade-off between service quality and energy consumption. Our main contributions include the formulation of a bi-objective integer linear programming (ILP) model and an evolutionary heuristic which applies NSGA-II algorithm in searching non-dominated solutions for either objective. Experimental results show that the proposed heuristic algorithm efficiently solves the problem.

Keywords: Pork Truck Scheduling, Integer Linear Programming, Container Terminals

1 INTRODUCTION

With the rapid development of container shipping industry and to meet the minimum annual throughput capacity of millions of TEUs, some container terminals locate at islands and make use of the

* Corresponding Author

advantages of island natural conditions. Shanghai Yangshan terminal and Zhoushan terminal in China are two typical examples. They are connected to continents or the outsides via tens of miles long bridges. Thousands of container trucks run back and forth on the bridges everyday, either sending outbound containers to the terminal yard or fetching inbound containers from the yard.

It is observed that almost all the trucks are in halfway empty run during the round trip between the continent and the terminal since the trucks, which serve for different customers, are not in the centralized management of the port. It inevitably occurs a large amount of fuel or energy waste during transportation. Moreover, each truck consumes the round trip time to satisfy one container transport request, excluding the waiting time at the terminal for loading or unloading the container.

In a terminal the trucks that move around mainly in the container yard are also called *port trucks* or *yard trucks*. A decision of port truck scheduling is dynamically made in accordance with the schedule of other resources such as quay cranes and yard cranes. Most yard truck involved scheduling problems for container terminals focus on improving the service efficiency of resources such as the terminal throughput or decreasing the turnaround time of a container vessel (Tang et al. [1]). In this paper, assuming that the container transportation on the bridge segment is under central control of the terminal port, we study the problem of port truck scheduling on a dedicated route at a container terminal. Regarding each container transport request as one task, we aim at minimizing total waiting time of tasks and total fuel consumption which is equivalently represented by the empty-run rate.

The considered problem in this work is related to the VRP (Vehicle Routing Problem) problem. The requests in the former problem locate at either end of a single route, which is a special case of the network in the latter problem. Similar to the OVRP (Open VRP), each container truck needs not go back to its original location after serving a task. Compared to previous VRP problems and port truck scheduling problems, the model in this literature is new in the sense:

(i) This study considers port trucks and customer trucks together in container transportation such that customer trucks request services dynamically and wait for response by port trucks, while port trucks need to serve the requests timely under the requirement of fuel consumption minimization. There is few literature addressing inland container transportation problems in which port trucks and customer trucks are both considered.

(ii) Most VRP models consider minimal total traveling costs (the total distance, the number of vehicles) or maximal customer satisfaction, while we focus on minimizing energy utilization (denoted by empty run times) as well as total waiting time of customer trucks in this study.

(iii) In this study one truck serves at most one task at a time, but in VRP each vehicle may complete several requests in one trip whose route depends on the combination of the requests as well as their service sequence.

The organization of the rest of this paper is as follows. Section 2 gives a brief review of related studies. We formulate the considered problem as a integer linear programming (ILP) in Section 3, and propose an evolutionary heuristic to solve it in Section 4. Section 5 presents some experimental results of the heuristic, and conclusions are included in the final section.

2 LITERATURE REVIEW

Etsuko et al. [2] modeled the yard truck scheduling problem as a trailer routing problem, in which they investigated both single-trailer and multi-trailer cases for the pickup and delivery of containers in the terminal. Cao et al. [3] proposed a mixed-integer programming model for the yard truck and yard crane scheduling problem of loading containers in the terminal yard. Kim and Bae [4] presented a look-ahead dispatching method for AGVs (Automatic Guided Vehicles) in an integrated vehicle and quay crane scheduling problem, which was translated into static and dynamic cases of the one-to-one assignment problem. Chen et al. [5] studied the interaction between yard truck transportation and quay crane handling where yard trucks are shared among ships to reduce empty-run rate in the terminal area. Some authors further investigated the integrated scheduling problem of yard cranes, quay cranes and yard trucks since the performances of these resources are much related in serving container vessels (Chen et al. [6], Kaveshgar and Huynh [7]).

Some authors combined the factor of fuel consumption reduction into vehicle scheduling and container port transportation problems (He et al. [8]). Especially in green logistics it aims at improving environmental sustainability by reducing fuel consumption in freight transportation systems, and it has been a subject undergoing intense study in operation research. Xiao and Konak [9] used total CO_2 emissions as the objective function to be minimized by various operational decisions including customer-vehicle assignment, route selection, and travel time scheduling. The reduction of emissions from trucks has become an important part of world wide efforts to reduce CO_2 emissions. Carbon emission optimization is also involved in the VRP problem (Ehmke et al. [10], Zhang et al. [11]).

The VRP problem is one of the most related studies that was first introduced by Dantzig and Ramser [12]. Kovacs et al. [13] presented the multi-objective generalized consistent VRP with three objective functions: arrival time consistency, driver consistency, and travel cost. Beheshti et al. [14] addressed multi-objective VRP with multiple prioritized time windows in which a distributer proposes a set of non-overlapping time windows with equal or different lengths and the customers prioritize these delivery time windows. Voratas et al. [15] proposed two solution representations for solving the generalized multi-depot vehicle routing problem with multiple pickup and delivery requests, and developed several real-valued PSO algorithms. Nossack and Pesch [16] addressed a truck scheduling problem that arises in intermodal container transportation where containers need to be transported between customers (shippers or receivers) and container terminals (rail or maritime) and vice versa. A fleet of homogeneous trucks must be routed and scheduled to minimize the total truck operating time under hard time window constraints imposed by the customers and terminals.

3 PROBLEM STATEMENT

There are a number of port trucks delivering containers and running back and forth on a single route, which connects the container terminal yard and the toll gate locating at the continent side. A port truck may load an outbound container, which is sent by a customer truck, at the outside of the toll gate and delivers it to the other end of the route, i.e., to the terminal yard. It may also load an inbound container at the terminal yard and delivers it to the toll gate place, at which the container is transferred to the customer truck. It takes τ units of time for a single ride on the dedicated route.

At the toll gate place, tasks are released at different time points, at which the corresponding customer trucks arrive at the outside of the toll gate. We call it *an outbound container task* if the involved customer truck brings/unloads an outbound container at the gate, and *an inbound container task* if the customer truck arrives and releases a request at the gate for fetching/loading an inbound container from the terminal yard. That is, the task set I consists of two types of tasks that correspond to loading inbound containers and unloading outbound containers, respectively.

On the arrival of an outbound container task, if there is no available port truck at the toll gate, then one of the available trucks at the terminal yard may take an empty-loaded ride to the gate, load the outbound container and then deliver it to the yard, or the port truck which firstly arrives at the toll gate satisfies the task. It can be similarly discussed for the case with no available port trucks at the terminal yard on the arrival of an inbound container task. The remainder analyses of the considered problem are based on the fundamental assumptions as follows.

3.1 Assumptions:

1) Each truck delivers at most one container at a time.
2) The running speed of either a loaded or an empty-loaded truck is constant, and it takes τ (> 0) time units for a single ride on the route.
3) The time for loading (or unloading) a container onto (or from) either customer truck or port truck is negligible.

3.2 Indices:

i, j: indices of tasks/containers;
k, h: indices of port trucks;

3.3 Input parameters:

I: the set of tasks, and it also denotes the size of the set (instead of notation $|I|$, for simplicity);
m: the initial number of port trucks at the toll gate;
n: the initial number of port trucks at the terminal yard;
K: The set of port trucks, and it also denotes the size of the set, i.e., $K = m + n$;
r_i: the release time of task i;
P_i: equals 1 if i is an inbound container task, and 0 if it is an outbound container task, $i \in I$;
τ: the single ride travel time on the dedicated route;
Q_k: equals 1 if truck k is located at the terminal yard origin, and 0 if it is located at toll gate, $k \in K$; where $Q_k = 0$, for $1 \le k \le m$ and $Q_k = 1$, for $m + 1 \le k \le K$
M: a sufficiently large positive integer.

3.4 Decision variables:

S_i: the start time of processing task i at which the corresponding container has been loaded onto the involved port truck;
C_i: the completion time of processing task i;
x_{ik}: equals 1 if truck k processes task i without empty run origin, and 0 otherwise, $k \in K, i \in I$;
β_{ik}: equals 1 if truck k processes task i immediately following an empty run origin, and 0 otherwise, $k \in K, i \in I$;
y_{ijk}: equals 1 if truck k performs consecutively tasks i and j which are at different locations, implying $C_i \le S_j$, and 0 otherwise, $k \in K; i, j \in I$;(location including toll gate and terminal yard).
z_{ijk}: equals 1 if truck k processes consecutively tasks i and j which are at the same location, implying an empty run of k during the service, and 0 otherwise, $k \in K; i, j \in I$;
u_{ik}: equals 1 if truck k processes task i, and 0 otherwise, $k \in K, i \in I$;

Without loss of generality we assume that all the parameters are integers. Below we establish the corresponding **ILP model** for the problem under consideration.

Objectives:

There are two objectives, one of which is to minimize the total waiting time of tasks, i.e.,

$$\min f_1 = \sum_{i \in I} (S_i - r_i) \tag{1}$$

The other objective is to minimize the times of empty-load single ride running by the port trucks, i.e.,

$$\min f_2 = \sum_{i \in I} \sum_{k \in K} \beta_{ik} + \sum_{i \in I} \sum_{j \in I} \sum_{k \in K} z_{ijk} \tag{2}$$

Constraints:

The complete time of task i is equal to its start time plus the single ride travel time τ.

$$S_i + \tau = C_i, \quad i \in I \tag{3}$$

Task i can only be processed after it's release time.

$$S_i \geq r_i, \quad i \in I \tag{4}$$

If truck k processes tasks i and j consecutively without empty run in between, i.e., $y_{ijk} = 1$, then j is started not earlier than the completion time of i.

$$S_j - C_i + M(1 - y_{ijk}) \geq 0 \quad i,j \in I, k \in K \tag{5}$$

If truck k processes tasks i and j consecutively with empty run in between, i.e., $z_{ijk} = 1$, then j is started not earlier than the completion time of i.

$$\tau \leq S_j - C_i + M[1 - z_{ijk}], \quad i,j \in I, k \in K \tag{6}$$

For each port truck k, it process at most one task initially.

$$\sum_{i \in I} x_{ik} \leq 1, \quad k \in K \tag{7}$$

The following inequalities guarantee that if a task is allocated to truck k, there must be one and only one succeeding task and respectively one preceding task or empty run before this task.

$$x_{ik} \leq u_{ik}, \quad i \in I, k \in K \tag{8}$$

$$\beta_{ik} \leq u_{ik}, \quad i \in I, k \in K \tag{9}$$

$$x_{ik} + \beta_{ik} \leq u_{ik}, \quad i \in I, k \in K \tag{10}$$

$$\sum_{j \in I} y_{ijk} \leq u_{ik}, \quad i,j \in I, k \in K \tag{11}$$

$$\sum_{i \in I} y_{ijk} \leq u_{jk}, \quad i,j \in I, k \in K \tag{12}$$

$$\sum_{j\in I} z_{ijk} \le u_{ik}, \quad i \in I, k \in K \tag{13}$$

$$\sum_{i\in I} z_{ijk} \le u_{jk}, \quad j \in I, k \in K \tag{14}$$

For any two tasks i and j, if $y_{ijk} = 1$, then there must be $u_{ik} = 1$ and $u_{jk} = 1$.

$$M(1 - y_{ijk}) + u_{ik} + u_{jk} \ge 2, \quad i \in I, k \in K, l \in L \tag{15}$$

For any two tasks i and j, if $z_{ijk} = 1$, then there must be $u_{ik} = 1$ and $u_{jk} = 1$.

$$M(1 - z_{ijk}) + u_{ik} + u_{jk} \ge 2, \quad i \in I, k \in K, l \in L \tag{16}$$

Each truck k can only be processed by a single task origin

$$\sum_{i\in I} (x_{ik} + \beta_{ik}) \le 1, \quad \forall k \in K \tag{17}$$

If truck k processes both tasks i and j, then i is either before or after j.

$$y_{jik} + y_{ijk} \le 1, \quad i,j \in I, k \in K \tag{18}$$

$$z_{jik} + z_{ijk} \le 1, \quad i,j \in I, k \in K \tag{19}$$

$$\sum_{k\in K}\sum_{j\in I} y_{ijk} \le 1, \quad i \in I \tag{20}$$

$$\sum_{k\in K}\sum_{i\in I} y_{ijk} \le 1, \quad j \in I \tag{21}$$

For any two tasks i and j, if $z_{ijk} = 1$, then the completion time of i is not later than the start time of task j.

$$M(1 - z_{ijk}) + S_j - C_i \ge \tau, \quad i \in I, k \in K, l \in L \tag{22}$$

If task i is loaded by truck k without empty run origin, the location of the truck k satisfies $Q_k = P_i$.

$$M(1 - x_{ik}) \ge 1 - (P_i + Q_k - 1)^2, \quad i \in I, k \in K \tag{23}$$

If task i is loaded by truck k with empty run back origin, then the location of the truck satisfies $Q_k \ne P_i$.

$$M(1 - \beta_{ik}) \ge (P_i + Q_k - 1)^2, \quad i \in I, k \in K \tag{24}$$

For any two tasks i and j, if $y_{ijk} = 1$, then the containers i and j are in different locations.

$$M(1 - y_{ijk}) \ge (P_i + P_j - 1)^2, \quad i,j \in I, k \in K \tag{25}$$

For any two tasks i and j, if $z_{ijk} = 1$, then the containers i and j are in the same location.

$$M(1 - z_{ijk}) \ge 1 - (P_i + P_j - 1)^2, \quad i,j \in I, k \in K \tag{26}$$

For the relation between tasks i, j, and parameters P_i and P_j, if $y_{ijk} = 1$, then $P_i \ne P_j$.

$$y_{ijk} \le 1 - (P_i + P_j - 1)^2, \quad i,j \in I, k \in K \tag{27}$$

If $z_{ijk} = 1$, then $P_i = Pj = 0$ or $P_i = Pj = 1$.

$$z_{ijk} \le (P_i + P_j - 1)^2, \quad i,j \in I, k \in K \tag{28}$$

If $y_{ijk} = 1$, then tasks i and j can not be at the same location, and vice versa.

$$1 + z_{ijk} \ge (P_i + P_j - 1)^2 + y_{ijk}, \quad i,j \in I, k \in K \tag{29}$$

4 NUMERICAL EXPERIMENTS

We propose an improved NSGA-II algorithm in this work as NSGA-II is an efficient and powerful multi-objective evolutionary algorithm for solving large instances. According to the practical data of Shanghai Yangshan deepwater port, there are about $1000 \sim 5000$ container trucks passing the 20 miles long Donghai bridge every day. Therefore, we set the number of tasks $I = 3000$, and set a single ride time $\tau = 30$ minutes(Due to the length of the Donghai bridge about 33km, the truck speed limit of 80km/hour in the Donghai bridge). For the number of port trucks initially located at the two ends of the bridge, we set $m = n = 50$, and thus $K = 100$. As we consider the tasks in a work day, i.e., in 1440 $(= 24hour \times 60)$ minutes, the release time r_i of each task i is randomly generated in $1 \sim 1440$. According to the actual situation, we randomly generated 10 sets of data of 3000 tasks.

We use the proposed NSGA-II algorithm to solve 10 randomly generated instances, and calculate minimum objective values as well as the average values over the instances. The results are shown in Table 1 where objective 1 denotes total waiting time and objective 2 is the times of empty-load single ride running. By the table data, the total waiting time is less than 17000 minutes on average, and there are less than 40 times of empty-load running in total for any input instance. Notice that in the corresponding solution by customer trucks, each truck spends 60 minutes to satisfy each task and the 3000 tasks consume 180,000 minutes together with 3000 times of empty-load ride. Hence, the proposed algorithm improves that the solution greatly reduces energy consumption by adopting port trucks than customer trucks to serve the tasks.

Table 1. Experimental results with 3000 tasks, 100 trucks, $\tau = 30$

		1	2	3	4	5	6	7	8	9	10
Objective 1	Min	1000	200	0	500	100	600	30	2000	1000	0
	Ave	12000	2350	2750	4500	2600	12300	6250	10500	17000	6000
Objective 2	Min	25	16	16	16	8	23	21	24	25	22
	Ave	40	28	22	24	30	33	34	33	39	36

5 CONCLUSION

In this paper, we study the problem of port truck scheduling on a dedicate route in a container terminal. We mainly establish an ILP model, and present an improved NSGA-II algorithm to solve the problem. Experimental results show that the proposed algorithm performs well in practice.

6 ACKNOWLEDGMENT

This work was partially supported by the National Science Foundation of China (71571134 and 71428002), and the Cai Yuanpei Program between the French Ministries of Foreign and European Affairs and the Higher Education and Research and the Chinese Ministry of Education (27927VE). The work was also supported by DHU Distinguished Young Professor Program (A201305) and the Fundamental Research Funds for the Central Universities.

7 REFERENCES

[1] **Tang, L., Zhao, J., Liu, J. 2014.** Modeling and solution of the joint quay crane and truck scheduling problem, *European Journal of Operational Research*, 236, pp 978-990.

[2] **Etsuko, N., Akio, I., Stratos, P. 2005.** Yard trailer routing at a maritime container terminal. *Transportation Research Part E*, 41, pp 53-76.

[3] **Cao, J., Lee, D., Chen, J., Shi, Q. 2010.** The integrated yard truck and yard crane scheduling problem: Benders' decomposition-based methods, *Transportation Research Part E*, 46, pp 344-353.

[4] **Kim, K.H., Bae, J.W. 2004.** A look-ahead dispatching method for automated guided vehicles in automated port container terminals. *Transportation Science*, 38(2), pp 224--234.

[5] **Chen, L., Langevin, A., Lu, Z. 2013.** Integrated scheduling of crane handling and truck transportation in a maritime container terminal. *European Journal of Operational Research*, 225, pp 142-152 Contents lists.

[6] **Chen, L., Bostel, N., Dejax, P., Cai, J., Xi,L.2007.** A tabu search algorithm for the integrated scheduling problem of container handling systems in a maritime terminal. *European Journal of Operational Research*, 181, pp 40-58.

[7] **Kaveshgar, N., Huynh, N. 2015.** Integrated quay crane and yard truck scheduling for unloading inbound containers,*International Journal of Production Economics*, 159, pp 168-177.

[8] **He, J., Huang, Y., Yan, W., Wang, S. 2015.** Integrated internal truck, yard crane and quay crane scheduling in a container terminal considering energy consumption, *Expert Systems with Applications*, 42, pp 2464-2487.

[9] **Xiao, Y., Konak, A. 2016.** The heterogeneous green vehicle routing and scheduling problem with time-varying traffic congestion. *Transportation Research Part E*, 88, pp 146-166.

[10] **Ehmke, J., Campbell, A., Thomas, B.W. 2016.** Vehicle routing to minimize time-dependent emissions in urban areas. *European Journal of Operational Research*, 251, p-p 478-494.

[11] **Zhang, J., Zhao, Y. 2015.** Vehicle routing problem with fuel consumption and carbon emission. *International Journal of Production Economics*, 170, pp 234-242.

[12] **Dantzig, G.B., Ramser, J.H. 1959.** The Truck Dispatching Problem. *Management Science*, 6(1), pp 80-91.

[13] **Kovacs, A.A., Parragh, S.N., Hartl, R.F. 2015.** The multi-objective generalized consistent vehicle routing problem. *European Journal of Operational Research*, 247(1), pp 441-458.

[14] **Beheshti, A., Hejazi, S., Alinaghian, M. 2015.** The vehicle routing problem with multiple prioritized time windows: A case study. *European Journal of Operational Research*, 90, pp 402-413.

[15] **Voratas, K., Pandhapon, S., Siwaporn, K. 2015.** Two solution representations for solving multi-depot vehicle routing problem with multiple pickup and delivery requests via PSO. *Computers And Industrial Engineering*, 89, pp 125-136.

[16] **Nossack, J., Pesch, E. 2013.** A truck scheduling problem arising in intermodal container transportation. *European Journal of Operational Research*, 230, pp 666-680.

BP NEURAL NETWORK BASED TAGUCHI METHOD AND ITS APPLICATION IN THE PREPARATION OF COMPOSITE PHOTOCATALYST

Chunling Zhu*and Yihai He
School of Reliability and Systems Engineering
Beihang University, Beijing, China
zcl1514@buaa.edu.cn

ABSTRACT

The traditional Taguchi design method has been widely accepted and recognized as an important tool for improving the quality of a product or a process. However, the optimal setting of a design parameter may sometimes be very impractical since it is determined only among the levels included in the parameter design experiment. In this paper, based on the capability of self-learning, self-training and output prediction of BP (Back Propagation) neural network, we combine the Taguchi method with BP neural network to optimize the preparation parameters of composite photocatalyst TiO2-MWCNTs.Firstly, we construct the BP neural network model using the Taguchi method to determine the number of neurons, learning rate and momentum values. Then, the BP neural network is trained using sample data and is applied to optimize the preparation parameters of the photocatalyst. Finally, we evaluate the fitting precision of the BP neural network and verify the validity of this method by comparing the results with the optimized results of classical Taguchi method.

Keywords: Taguchi method, BP network,TiO_2-MWCNTs,Preparation parameters

1 INTRODUCTION

Taguchi method is mainly used to achieve insensitivity of products or processes to sources of unwanted variation which has been widely accepted and recognized as an important tool for improving the quality of a product or a process[1]. However, the optimal setting of a Taguchi design parameter may sometimes be very impractical since it is determined only among the levels included in the parameter design experiment. Therefore, how to improve the Taguchi method by combining it with other heuristic methods is of great importance.

Many approaches are now available in the literature for improving the Taguchi method. Chiang, et al[2]. proposed an approach to improve the yield of Chrome (Cr) thin-film sputtering process and to find the robust parameters of the process by using the Taguchi method combined with the grey relational analysis.Shukla et al[3]stated a hybrid approach incorporating simulation, Taguchi method, robust multiple non-linear regression analysis and the Psychoclonal algorithm to determine the optimal operating conditionsof supply chain network.For optimization of processes parameters for SMT, Jou et al[4]. applied TRIZ and Taguchi's method to improve the product quality. Gu et al[5]. applied the Taguchi method and principal component analysis to improve the mechanical properties of recycled polypropylene (PP), with detailed assessments performed on each method. However, the operation of orthogonal experimental method is more complicated actually, and do not have the function of the output forecast to multiple experiments input.

---

*Corresponding Author: Zhu Chunling , School of Reliability and Systems Engineering, Beihang University, Beijing

Due to the strong learning ability and simple structure, back propagation (BP)neural network, composed of input, hidden and output-layers, is broadly utilized in engineering field and has wide application prospect[6].However, the setting of neural network structure parameters has a great influence on the prediction result and is usually set according to experience. Applying BP neural network to the experimental prediction system can resolve the shortcoming of the Taguchi method. In turn, Taguchi method can offer a scientifically disciplined methodology to explore the structure parameters of the BP neural network, which can improve the prediction accuracy. Accordingly, in this paper, the BP neural network based Taguchi method is proposed, and it is applied to optimize the preparation parameters of the photocatalyst.

2 BP NEURAL NETWORK BASED TAGUCHI METHOD

2.1 BP neural network

BP network can learn and store a large number of input-output model mapping relationships, without prior revealing of the mathematical equations of the mapping relationships. Its learning rule is the steepest descent method, and constantly adjust the network weights and threshold by back propagation to minimize the error sum of squares of the network. The BP neural network model for topological structures include input layer, the hidden layer (single layer or multilayer) and output layer. Figure 1 shows a BP neural network, and the information forward propagation embodies the basic features of feed forward network, the error back propagation is the core idea of BP algorithm.

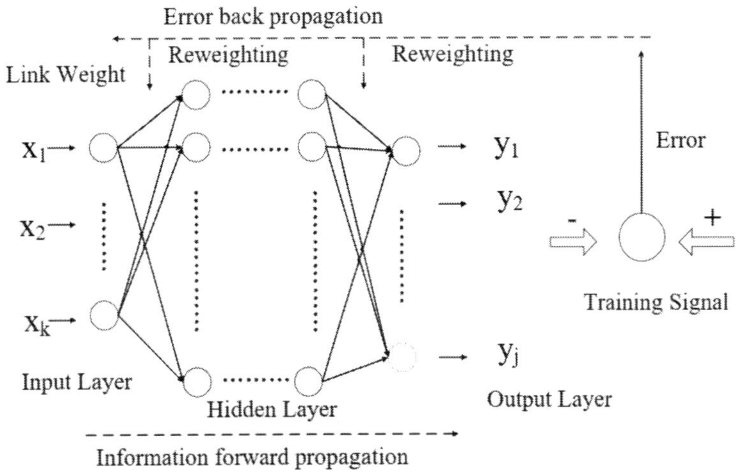

Figure 1: BP neural network structure

Use The basic characteristics of BP neural network can be summarized as follows:

1) BP network can contain multiple hidden layers with full connection between layers being achieved and no connections between neurons in the same layer. Multilayer network design can make the BP network model dig more information.

2) The mission of the activation function f(·)of BP network is a kind of transformation between the input and output values in essence. Assume that the input vector of BP net is= $(x_1, x_2,\cdots, x_k)^T$; $\omega = (\omega_1, \omega_2,\cdots, \omega_K)^T$is the weight vector; b is input deviation of neurons ,then the output of neurons in this layer y can be expressed as:$y_j = f(\omega \cdot x + b) = f\left(\sum_{i=1}^{k} \omega_i x_i + b\right)$; $\omega = (b, \omega_1, \omega_2,\cdots, \omega_K)^T$andy = f$(\omega \cdot x)$.The activation functionf(·)usually use Sigmoidal function, namely:

$$f(x) = \frac{1}{1+e^{-x}} \quad (1)$$

3) Using the error Back Propagation Algorithm for model training is a kind of supervised learning Algorithm and its basic idea is that the data information propagate backward from the input layer through the hidden layer, then revise the weights between neurons and the network layer along the direction of the error reduction through the middle from the output layer forward when training network weights. With the development of learning, the error is smaller and smaller.

4) The mean squared error (MSE) is chosen to be the performance function.

$$\mathrm{MSE} = \frac{\sum_{h=1}^{H}\left(y_h-\hat{y}_h\right)^2}{H-1} \quad (2)$$

where the H is the number of output predicted value.

2.2 Taguchi method

Taguchi method can offer a scientifically disciplined methodology to explore the structure parameters of the BP neural network, which can improve the prediction accuracy.

For measuring the response variability, Taguchi proposed the use of signal-to-noise (S/N) ratios. The S/N formula depends on the goal of the experiment; however, three formulas are considered standard and are generally applicable when the response can be classified as 'Target-is-Best', 'Larger-is-Better' or 'Smaller-is-Better'. These formulas are transformed so that the objective always is to maximize the S/N ratio.

To obtain optimal parameters, the smaller-the-better quality characteristics for MSE should be taken. The S/N ratio η for the smaller-the-better quality characteristic can be expressed as below:

$$S/N = -10[(1/n)\sum_{i=1}^{n} {y_i}^2] \quad (3)$$

Where ${y_i}^2$ is the value of the MSEs of the i_{th} experiment and n is the number of tests of data points in an experiment.

2.3 The optimized BP model

Based on the BP neural network and Taguchi method, the optimal BP model to optimize the preparation of Composite Photocatalyst can be expressed as Figure 2.

First, we set the basic parameters of BP neural network.

1) Network layer. The existing researches show that, in general, BP neural network can achieve nonlinear mapping by appropriately increased neuron nodes and in this paper we adopted the single hidden layer of BP network.

2) Activation function. We have discussed the selection of the activation function above, choosing s-shaped function for the hidden layer, linear function for the output layer.

3) Number of input layer nodes. The number of input layer nodes depends on the structure characteristics of the input data. In this article, each node of the input layer corresponded to an experimental design parameters. Therefore, in this paper, the input data can be depicted using the matrix$X_{3\times n}$.

$$X_{3\times n} = \begin{pmatrix} x_{11} & \cdots & x_{1n} \\ \vdots & \ddots & \vdots \\ x_{31} & \cdots & x_{3n} \end{pmatrix} \quad (4)$$

Where n is the number of the experimental group.

4) Number of output layer nodes. Generally, the number is determined according to the requirements of practical problems, here we set output node value to two.

Second, we constructed the BP neural network model using the Taguchi method to determine the number of neurons, learning rate and momentum values. And the mean squared error (MSE) is chosen to the response. Because the number of hidden layer nodes has a great influence on the effect of BP network. In general, the number of hidden layer nodes selection is more subjective, and many studies mostly determine it by the empirical formula. In this article, the optimal number of the hidden layer nodes was chosen through the Taguchi experiment design method. And we adopted the standard steepest descent method and introduced additional momentum. As for the setting of values of the learning rate and additional momentum, we also used the Taguchi method to determine.

Third, the BP neural network is trained using sample data and its fitting precision is evaluated by the evaluation function.

Then, we applied the optimal BP neural network to optimize the preparation parameters of the photocatalyst and verified the validity of this method.

Finally, we study the photocatalytic activity of the optimal composite photocatalyst.

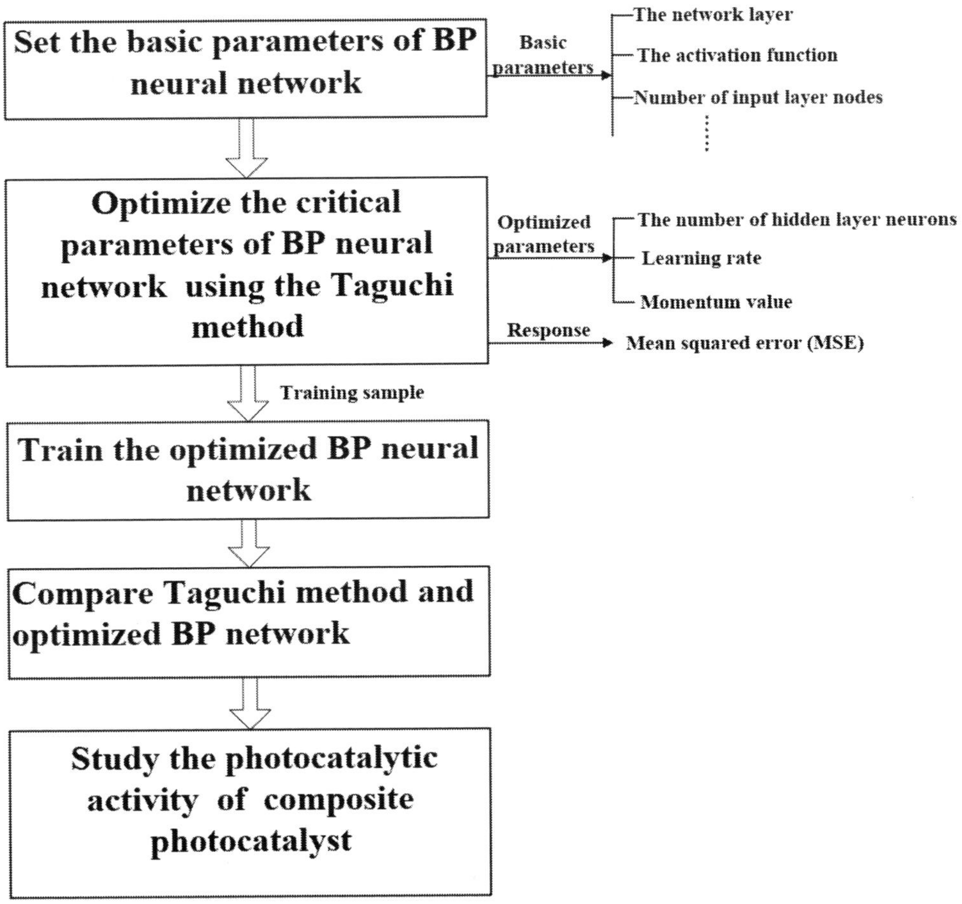

Figure 2: The optimized BP model

3 CASE STUDY

With the rapid development of the chemical industry and the resultant widespread use of chemicals, many types of pollutant are produced, which are potentially harmful to the environment[7].For the processing of air and water pollutants, synthesizing the composite photocatalyst is a kind of very good method, especially the synthesis of MWCNTs/TiO_2 composite photocatalyst by sol-gel methods [8][9].The process parameters such as added amount of carbon nanotubes, time of magnetic stirring and the temperature of heat treatment are high relevant with the preparation process and can affect the photocatalytic effect of the composite material directly. Conventional approaches of parameter optimization are variation of one parameter at a time keeping other parameters constant, orthogonal experiment method which tend to have some limitations[10]. In this paper part, we apply the BP neural network based on the Taguchi method to optimize the preparation parameters.

3.1 The Taguchi design of the BP network

After setting the basic parameters of BP neural network, we consider three parameters of the BP neural network: the number of hidden layer neurons A, learning rate B and momentum value C, we took three levels of each parameter, as shown in table 1.

Table1: BP network parameter selection level

Parameters	Level1	Level 2	Level3
A	1	3	6
B	0.1	0.2	0.3
C	0.1	0.3	0.5

Because the interaction between the number of hidden layer neurons A and learning rate B was considered , the interaction of two factors can be treated as a new factor. For four factors with three levels, the experimental layout of an $L_9(3^4)$ orthogonal array was selected. The sum of MSEs (the mean squared error) of the increasing rate of specific surface area and the rate of TiO_2 remaining mass is regarded as the response.

The larger the S/N ratio is, the smaller the variance of performance characteristics around the desired value is. The S/N ratio for each parameter at different levels is plotted in Figure3.From Figure3 , it can be easily find that the optimum factor level combination is A2B2C3.

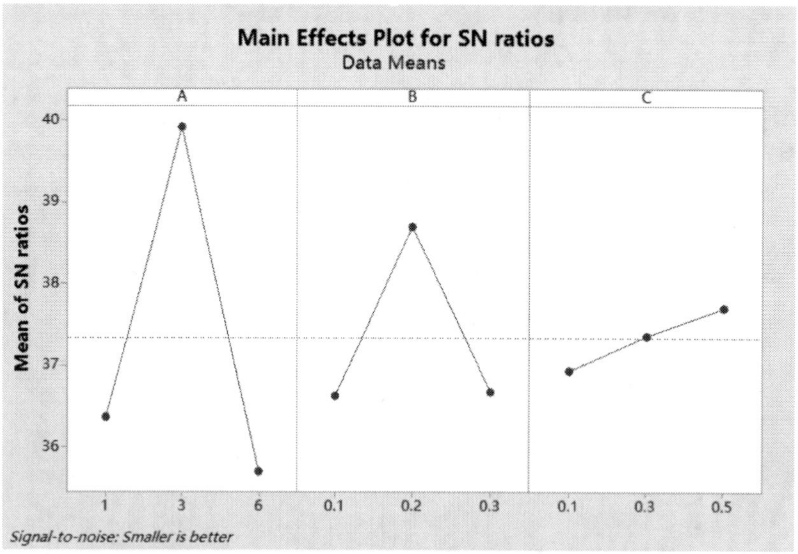

Figure 3: Average values of S/N rations

Moreover, the Table2 shows the result of the ANOVA analysis.

Table2: Results of ANOVA

Source	Sum of squares	DOF	Mean square	F-ratio	Contribution(%)
A	0.0000702	2	0.0000351	234	72.15
B	0.0000221	2	0.0000110	73.67	22.71
C	0.0000043	2	0.0000021	14.33	4.42
A×B	0.0000004	2	0.0000002	1.33	0.41
Error	0.0000003	2	0.00000015		0.31
Total	0.0000973	10			100

A larger F indicates a reduced interaction. This equation can be used to evaluate the importance of the effect of each factor, relative to the experimental error. Only when the importance reaches a specific level will the related factors be used to predict the S/N ratio - the factors above this level are called important factors. The effects of the remaining factors can only be regarded as occasional effects caused by experimental error; these factors are called light factors. After the ANOVA is made, Table 2 shows that, in the Taguchi design of the BP network, the number of hidden layer neurons most significantly affects the effect of BP network-its contribution is 72.15 per cent. The next most important factor is the learning rate, and momentum value affects the least.

So the optimal parameters are as follows: The number of hidden layer neurons is 3. the learning rate is 0.2. And t momentum value is 0.5.

3.2 Training of the constructed BP network

The established neural network is trained to optimize the preparation parameters of the composite photocatalyst based on the experimental results. Added amount of carbon nanotubes, time of magnetic stirring and the temperature of heat treatment were chosen to be the input layer which can be expressed in terms of x1, x2 and x3 in Figure 1.And the increasing rate of specific surface area and the rate of TiO_2 remaining mass were chosen to be the output layer which can be expressed in terms of y1, y2 in Figure 1. In neural network fitting process, we used the 25 groups of data in table3 as the training sample and loaded them into the constructed training BP neural network model. Through the independent learning and memory ability of the system to train the neural network, the training stopped when the system training satisfied the approximation precision or achieved the maximum number of iterations. After completion of training, we tested its ability of recognizing.

Table3:Test parameters design and the statistical resultof the experiment

Num.	Added amount of carbon nanotubes(g)	time of magnetic stirring(h)	the temperature of heat treatment(℃)	the increasing rate of specific surface area	the rate of TiO2 remaining mass
1	0.015	0.5	400	0.40346	0.6015

2	0.015	0.75	450	0.85076	0.6138
3	0.015	1	500	1.22803	0.6849
4	0.015	1.25	550	0.95075	0.6233
......	......	......	......	......	......
25	0.055	1.5	550	1.0679	0.4965

The tendency of the prediction results and the desired outputs is the same as shown in Figure 4. The comparison between the experimental data and the BP neural network prediction results depicts a good consistency. The BP neural network prediction model reflects the influences of key factors and is applicable to assess the photocatalytic effect of the composite.

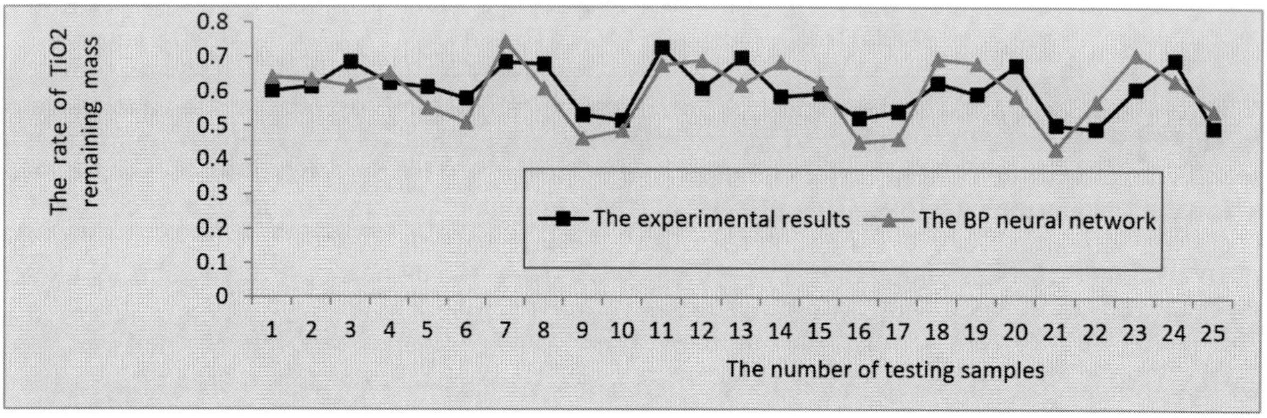

Figure 4: The increasing rate of specific surface area prediction results of BP neural network

3.3 The result comparison between Taguchi method and optimized BP network

We use the particle swarm optimization algorithm in the neural network prediction, input the variables that combine randomly in the range of all possible parameters, and predict the output using the neural network trained by the the experimental data. By continuously iterate calculation and reverse the global optimal value iteration, we found that thebest parameters input is as follows: The added amount of carbon nanotubes is 0.025g.The time of magnetic stirring is 1h.And the temperature of heat treatment is 500°C . The optimal parameters results obtained by using Taguchi method and the neural network method are shown in the table .The expected values obtained by the neural network method is better than the Taguchi methods. Moreover, its optimal parameters levels don't have to appear in the original experiment set levels.

Table4:The results comparison of BP neural network and Taguchi method

Method	D	E	F	the increasing rate of specific surface area	the rate of TiO2 remaining mass
BP	0.038	1.25	515	2.03681	0.8452
Taguchi	0.035	1	500	1.83697	0.7874

4 CONCLUSION

A BP Neural Network based Taguchi Method that requires fewer experimental runs is proposed which can achieve a high predictive precision. Firstly, we construct the BP neural network model using the Taguchi method to determine the number of neurons, learning rate and momentum values. And the optimal parameters are gotten as follows: The number of hidden layer neurons is 3. the learning rate is 0.2. And t momentum value is 0.1. Then, the BP neural network is trained using sample data. Next, we evaluate the fitting precision of the BP neural network and the comparison between the experimental data and the BP neural network prediction results depicts a good consistency. Then, by continuously iterate calculation and reverse the global optimal value iteration, we found that the best parameters input is as follows: The added amount of carbon nanotubes is 0.025g. Finally, we verify the validity of this method by comparing the results with the optimized results of Taguchi method. The final result shows that the BP Neural Network based Taguchi Method have practical effects and it's very promising to apply the BP neural network model based on the Taguchi method into the material preparation industrial. Although, of course, neural network can accurately grasp the experiment response variable and complex nonlinear relationship between the parameters and get the relatively scientific results, however, the complexity of this approach may

encounter certain difficulties in the promotion.

5 REFERENCES

[1] **Hasenkamp T, Arvidsson M, Gremyr I.** 2009.A review of practices for robust design methodology, Journal of Engineering Design, 20(6): 645-657.

[2] **Chiang Y M, Hsieh H H.**2009.The use of the Taguchi method with grey relational analysis to optimize the thin-film sputtering process with multiple quality characteristic in color filter manufacturing, Computers & Industrial Engineering, 2009, 56(2): 648-661.

[3] **Shukla S K, Tiwari M K, Wan H D, et al.** 2010. Optimization of the supply chain network: Simulation, Taguchi, and Psychoclonal algorithm embedded approach, Computers & Industrial Engineering, 58(1): 29-39.

[4] **Jou Y T, Lin W T, Lee W C, et al.** 2013. Integrating the TRIZ and Taguchi's Method in the Optimization of Processes Parameters for SMT, Advances in Materials Science and Engineering, 2013.

[5] **Gu F, Hall P, Miles N J, et al.** 2014.Improvement of mechanical properties of recycled plastic blends via optimizing processing parameters using the Taguchi method and principal component analysis,Materials & Design, 62: 189-198.

[6] **Kurd Z, Kelly T P.** 2007. Using fuzzy self-organising maps for safety critical systems,Reliability Engineering & System Safety, 92(11): 1563-1583.

[7] **Xing Z P, Sun D Z, Yu X J, et al.** 2014. Treatment of antibiotic fermentation - based pharmaceutical wastewater using anaerobic and aerobic moving bed biofilm reactors combined with ozone/hydrogen peroxide process,Environmental Progress & Sustainable Energy, 33(1): 170-177.

[8] **Cao W, Suzuki T, Elsayed-Ali H E, et al.** 2015. Synthesis of Cerium-Doped Titania Nanoparticles and Nanotubes,Journal of Nanoscience and Nanotechnology, 15(3): 2508-2513.

[9] **Kausar A, Iqbal A, Hussain S T.** 2014.Preparation and properties of polyamide/epoxy/multi-walled carbon nanotube nanocomposite,Journal of Plastic Film and Sheeting, 30(2): 205-224.

[10] **Hasenkamp T, Arvidsson M, Gremyr I.2009.**A review of practices for robust design methodology[J]. Journal of Engineering Design, 20(6): 645-657.

COMPARISON OF DATA MINING TECHNIQUES FOR ANALYSIS OF PARETO OPTIMAL SOLUTIONS IN LAYOUT PLANNING PROBLEMS IN MANUFACTURING SYSTEMS

D. Ishizuka[1] K. Izui[1], T, Yamada[1] and S. Nishiwaki[1]
[1]Department of Mechanical Engineering and Science
Kyoto University, Kyoto, Japan
ishizuka.daisuke.58z@st.kyoto-u.ac.jp

ABSTRACT

In this paper, we propose and compare two data mining techniques for layout planning problems of robotic cellular manufacturing systems, one of which considers the correlation of variables for the analysis of non-dominated solutions. First, we develop a new multiobjective optimization method that can provide a large number of non-dominated solutions, to facilitate later analysis of correlations among the design variables. Then we develop two clustering techniques, *k*-means clustering and EM clustering, and compare their effectiveness. The proposed technique is applied to a numerical example to examine its effectiveness.

Keywords: Data mining, Clustering, Multiobjective optimization, Robotic cellular manufacturing system, EM clustering

1 INTRODUCTION

The layout planning process is regarded as a particularly important process in the design of manufacturing systems since the layout design has a major impact on the ultimate productivity of such systems. To assist this design process, many facility layout optimization techniques that aim to provide efficient arrangement of manufacturing components or facilities under constraints imposed according to various design requirements have been proposed [1][2].

In realistic layout design processes, layout-planning problems often have multiple objective functions that deal with material handling and equipment flows. In such cases, decision-making becomes harder due to conflicts among the objective functions, and finding the best compromising solution is a crucial problem. One simple way to handle multiple objective functions is to convert a multiobjective optimization problem into a single objective function using aggregating techniques such as a weighted-sum method [3]. However, these methods provide only a single solution that highly depends on parameters settings chosen by decision-makers prior to optimization computation.

Another way to handle multiobjective optimization problems is to obtain a set of non-dominated solutions [4][5]. Decision-makers can then observe trade-off information among the objective functions and can select a preferable solution from among the obtained solutions. Furthermore, decision-makers can extract information about underlying design principals from the distribution of non-dominated solutions. However, selecting a single solution from the large number of non-dominated solutions distributed in the two- or multi-dimensional objective function space is not a trivial problem, and development of systematic ways to handle the big data of non-dominated solutions is required.

Information reduction techniques can be used to simplify the process of investigating the set of non-dominated solutions. Objective reduction approaches [6] that explore the smallest set of conflicting objective functions can reduce the number of dimensions of objective function space. Morse [7], and

Rosenman and Gero [8] utilized clustering techniques to reducing the size of the non-dominated solution set. The latter approach, which uses clustering techniques, is promising since it can significantly reduce the number of solutions that must be investigated. Sugimura et al. [9] and Ulrich et al. [10] demonstrated the effectiveness of using clustering techniques in design variable space. Furthermore, Ulrich applied a clustering technique to both objective and design variable space [11].

However, the clustering techniques described in the above literature have a limitation because the correlation among design variables in obtained non-dominated solutions is not considered, since these techniques use simple clustering techniques such as the *k*-means method. Real-world optimization problems such as layout-planning problems often have a large number of design variables and the design variables in non-dominated solutions are strongly correlated. Therefore, in this paper, we propose the use of clustering techniques that can consider the correlation of design variables in analyses of non-dominated solutions in layout planning problems for robotic cellular manufacturing systems (RCMS) [15], and we investigate the effectiveness of these techniques. We propose two techniques in this paper. First, a new multiobjective optimization technique that can provide a large number of non-dominated solutions is developed, since a large number of solutions is required to accurately analyse correlations among design variables. Next, we propose a technique to use clustering methods that consider correlations among design variables.

2 MUTIOBJECTIVE OPTIMIZATION TECHNIQUE FOR LAYOUT DESIGN OF ROBOTIC CELLULAR MANUFACTURING SYSTEMS

2.1 Problem formulation

Robotic cellular manufacturing systems (RCMS) are a new kind of manufacturing system where one or more industrial robots perform complex assembly operations. RCMS have similar advantages to those of human cellular manufacturing systems, such as high flexibility for product changes and low material handling costs. Furthermore, RCMS can take advantage of factory automation to achieve low manufacturing costs for high product volumes and a low rate of process errors, compared to human manufacturing systems. However, since industrial robots perform very complex motions, the design process for RCMS is highly complex, so the development of a design support system for RCMS is of great practical necessity. Therefore, in this paper, we develop a new RCMS layout planning support system.

A layout design problem for RCMS can be formulated as a three-objective optimization problem [12]. Design variables for this optimization problem are the positions of manufacturing components such as industrial robots, an assembly base and parts boxes. The objective functions are the cycle time, manipulability, and layout area, formulated as follows.

In multi-robot systems, cycle time T is the maximum value of the sum of each operational time for each robot.

$$T = \max_m \left[\sum_i T_i \right] \tag{1}$$

where m is the robot number. T_i is the operation time of the i-th robot and the i-th movement time is evaluated using the maximum joint motion time as follows:

$$T_i = \max_k \left[\frac{|\bar{q}_k^i|}{\omega_k} \right] \tag{2}$$

where ω_k indicates the maximum angular velocity of the k-th joint and $\bar{q}_k^i$ is the angle through which the k-th joint moves during the i-th motion. In this formulation, we assume that all joints can simultaneously change their angle and that the joint motion has a uniform velocity based on the maximum angular velocity, so that approximate robot movement times can be estimated in the layout planning stage.

A second objective function for the manipulability [13] is evaluated to ensure that the industrial robots can conduct the very complex required assembly tasks. In this paper, the manipulability measure W is calculated as

$$W = \sum_{j=1}^{h} a^j w^j, \tag{3}$$

where w^j is the manipulability at the j-th task position and a^j denotes a weighting coefficient. w^j can be calculated based on the robot posture at each task position.

Furthermore, layout area S is evaluated using the area of the smallest rectangle that can contain all allocated components.

Using the above three design criteria for the RCMS layout, a multiobjective optimization problem for its design can be formulated as

$$\begin{cases} \text{minimize} \quad T, S \\ \text{maximize} \quad W. \end{cases} \tag{4}$$

Additionally, in this problem, constraints for collision avoidance between multiple robots is considered.

2.2 Multiobjective optimization problem solution approach

To avoid local optima due to overlapping constraints among manufacturing components, and obtain a set of non-dominated solutions for the above three-objective optimization problem, we develop a two-step optimization technique.

Step 1: Multiobjective genetic algorithm

At the first step, approximate positions of manufacturing components are determined using a multiobjective genetic algorithm. In this step, the sequence-pair technique [14], used to represent the relative positions of manufacturing components, encodes the layout of the RCMS components as pairs of permutation variables, and any overlapping among the components is automatically avoided during the genetic algorithm optimization process. To obtain non-dominated solutions, Non-dominated Sorting Genetic Algorithm-II (NSGA-II) [15] is used as the selection operation, but, as a special feature of our technique, all individuals under the n-th rank, where n is a parameter value assigned by a decision-maker, are stored during the optimization process, so that a large number of solution candidates is generated.

Step 2: Multiobjective gradient-based search

Using the solutions under the n-th rank obtained by the multiobjective genetic algorithm as initial points, a multiobjective gradient-based search method is applied to improve the detailed positions of the RCMS components. The objective functions for this problem can be written as

$$\text{minimize} \quad T, S, -W \tag{5}$$

Note that the maximization of manipulability is converted to a minimization problem by introducing a negative sign in the above formulation. Since the aim of applying the multiobjective gradient-based search is to improve the solutions previously obtained by the genetic algorithm, all objective functions must be improved. To do so, the three objective optimization problem is converted into the following single objective optimization problem,

$$\text{minimize} \quad \max\{(T-T_0)/T_0, (S-S_0)/S_0, -(W-W_0)/W_0\} \tag{6}$$

where T_0, S_0 and W_0 are values for the operation time, area and manipulability, respectively, of an initial point obtained by the genetic algorithm. This problem can be further converted into the following optimization problem using a single slack variable, t, and weighting coefficients w_T, w_S, and w_W.

$$\begin{aligned} &\text{minimize} \quad t \\ &\text{subject to:} \begin{cases} (T-T_0)/T_0 - w_T t \le 0 \\ (S-S_0)/S_0 - w_S t \le 0 \\ -(W-W_0)/W_0 - w_W t \le 0 \end{cases} \end{aligned} \tag{7}$$

If the weighting coefficients are all equal, all objective functions are improved equally. Here, to obtain a large number of non-dominated solutions, the multiobjective gradient-based search is carried out using various values for the weighting coefficients. The fundamental concept of this two-step optimization technique was previously proposed [12]. However, since a large number of non-dominated solutions is required in order to analyse the correlations among design variables, we modified the steps of this approach.

3 DATA MINING TECHNIQUES FOR ANALYSIS OF PARETO OPTIMAL SOLUTIONS

3.1 Clustering non-dominated solutions

The proposed multiobjective optimization technique produces a large number of non-dominated solutions that are then analysed with a clustering technique, to reveal essential information of the optimization problem while reducing redundant information of the non-dominated solutions. This facilitates subsequent scrutiny of the obtained solutions by design engineers. In real-world optimization problems, design variables have correlations that cannot be known *a priori*. To handle these correlations, expectation-maximization (EM) clustering is applied to the non-dominated solutions. EM clustering is a technique that allows a dataset to be displayed as clusters of deviations and correlations. To compare the efficacy of two clustering techniques, *k*-means clustering and EM clustering are applied to obtained non-dominated solutions. The details of these two methods are described below.

3.1.1 K-means clustering

The *k*-means algorithm divides a given dataset into K clusters based on the similarity of data, so that some clusters contain similar data while other clusters contain dissimilar data. The similarities are calculated based on Euclidean distance.The details of the clustering algorithm are as follows:

1. Define parameter K as the number of clusters, and randomly select K data points. The selected points are treated as the initial centroids of K clusters.
2. Divide the dataset into the nearest clusters. Each datum is assigned to the cluster whose centroid is closest to it.

3. Calculate the mean value vector $\boldsymbol{\mu}_k$ where $k = 1, 2, \ldots, K$. The mean value vector $\boldsymbol{\mu}_k$ becomes the new centroid of the k-th cluster.
4. Re-divide the dataset into the nearest clusters based on the new centroids.
5. Iterate steps 3-4 until convergence.

3.1.2 *EM clustering*

The EM algorithm [16] derives the Gaussian mixture model with maximum likelihood and presents the dataset as a superposition of Gaussian distributions. In an EM algorithm-based clustering method, the contents of a dataset are assigned to the clusters based on the results of a maximizing likelihood function. The details of the algorithm are as follows:

1. Define parameter K as the number of clusters, and randomly select K data points.
2. Assume a Gaussian mixture distribution that has K components. The mean values are the selected K data points and the variance matrices contain diagonal elements calculated from the entire dataset while the other elements have a value of zero. In addition, the mixture weights of the assumed distributions are set as equal. The vector $\boldsymbol{\theta}$ contains all of these parameters.
3. Calculate the conditional expected value of the log likelihood function based on the estimated conditional distributions under the parameter vector $\boldsymbol{\theta}$.
4. Find the new parameter vector $\boldsymbol{\theta}^{new}$ that maximize the expected value, and update the parameter vector so that $\boldsymbol{\theta} = \boldsymbol{\theta}^{new}$
5. Iterate steps 3-4 until convergence

4 CASE STUDY

The proposed optimization technique and clustering techniques are applied to an example problem for a manufacturing system comprised of three industrial robots, whose outline is shown in Fig. 1. In this robotic system, a part from a collection of identical parts randomly piled in Part box 1 is picked up and transferred by three industrial robots in sequence, as its orientation is adjusted. The part is then accurately placed along a line in Part box 2. In this case study, the location of Part box 1 is confined to the left vertical edge of the layout area and location of Part box 2 is confined to the upper horizontal edge. Figure 2 shows the configuration of the industrial robots, RV-2SQ models manufactured by Mitsubishi Electric Corporation.

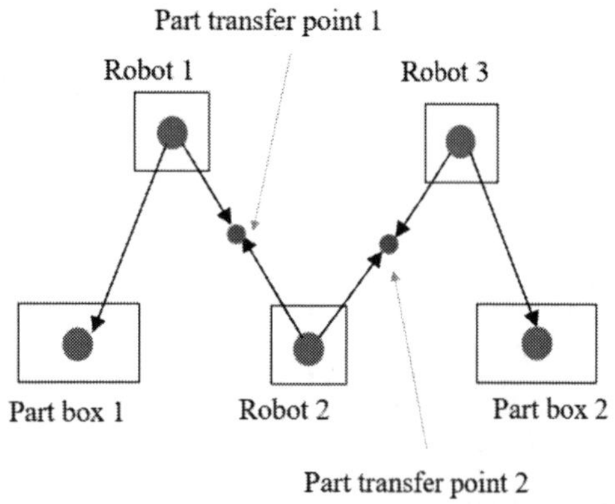

Figure 1: Three-robot manufacturing system

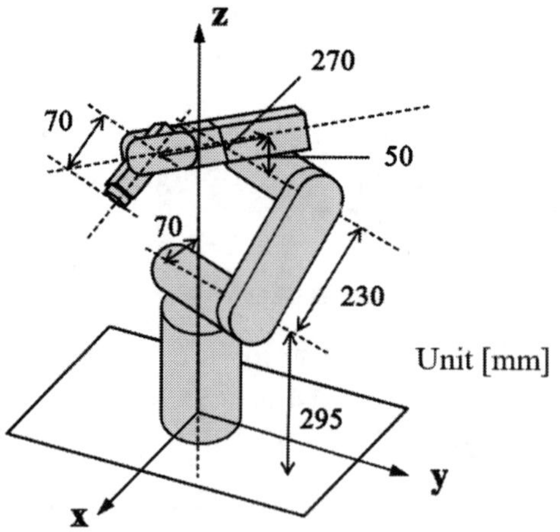

Figure 2: Configuration of RV2SQ-type industrial robot

The non-dominated solutions obtained with the multiobjective optimization technique are shown in Fig. 3. Since this optimization problem has three objective functions, the solutions are distributed in a three-dimensional objective function space. In this figure, blue points indicate the 40 solutions obtained by the previous multiobjective technique proposed in [12], and the 89 solutions obtained by the developed multiobjective optimization technique are indicated by red points. Clustering techniques were then applied to these non-dominated solutions and the obtained cluster distributions are shown in Fig. 4, with four clustering patterns shown to enable a comparison of the clustering techniques. Layout designs for seven characteristic solutions indicated by arrows in Fig. 4 are shown in Fig. 5.

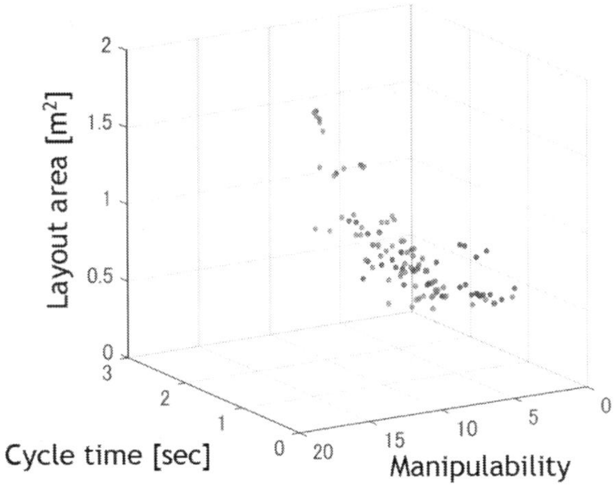

Figure 3: Obtained solutions (Blue points: previous technique in [12]; Red points: developed technique in this paper)

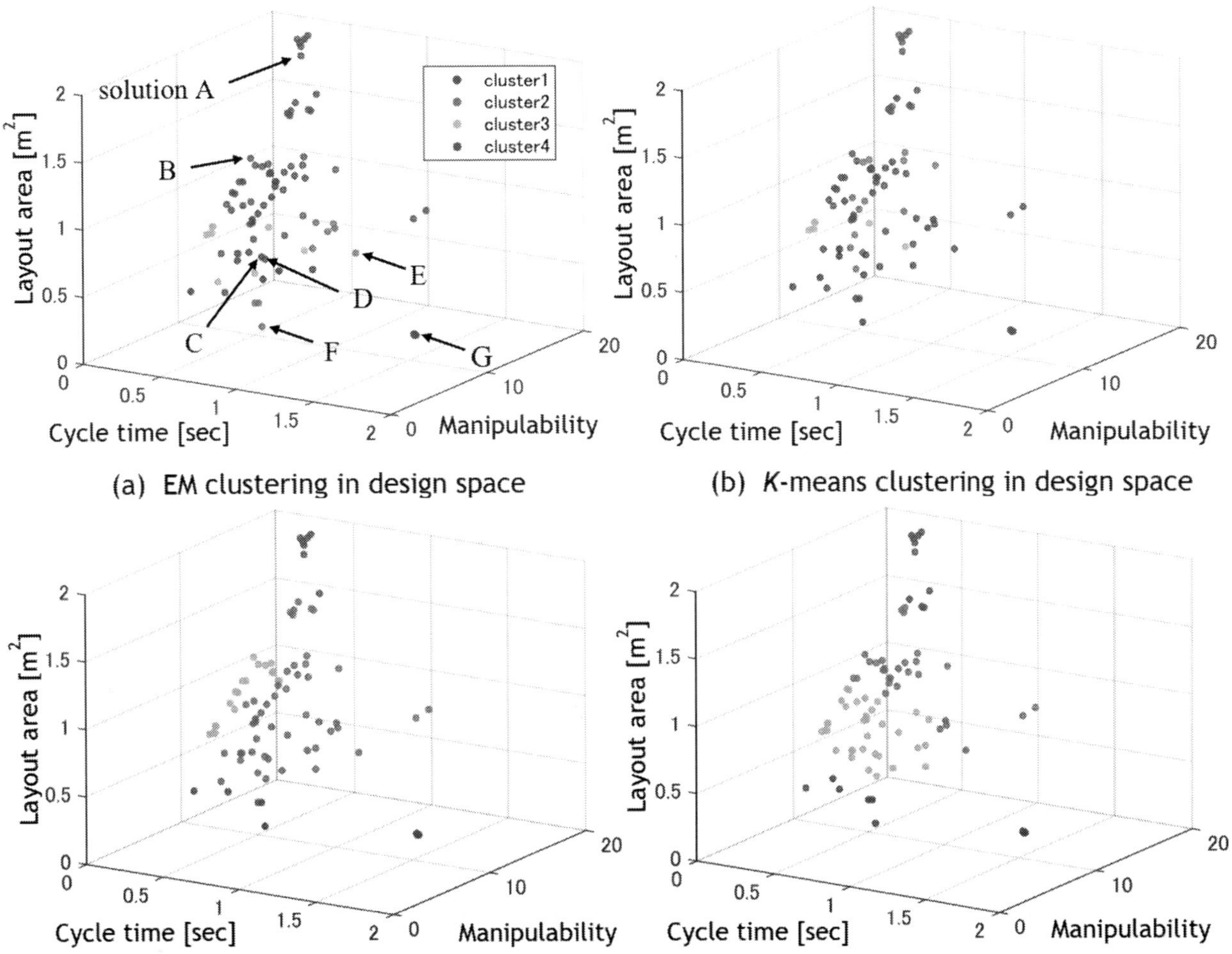

(a) EM clustering in design space

(b) *K*-means clustering in design space

(c) EM clustering in objective space

(d) *K*-means clustering in objective space

Figure 4: Cluster distribution mapped in objective function space

Firstly, we compare the results of clustering obtained by the two clustering techniques in design space, (a) and (b) in Fig. 4, and corresponding clustering in objective space, (c) and (d) in Fig. 4. In the objective space clustering, solutions have been appropriately placed in the various clusters, but dissimilar design solutions that have similar objective values, such as solutions C and D, cannot be distinguished. On the other hand, in the design space clustering, dissimilar design solutions are separated and similar designs that are distant in objective space, such as solutions E and F, can be identified.

Next, comparing the results (a) and (b) in Fig. 4, which are the results of EM clustering and *k*-means clustering in design space, the clusters derived by EM clustering clearly illustrate the deviation of the non-dominated solutions. Cluster 2 in Fig. 4 (a) contains solutions E and F that feature a linear arrangement of robot 1, Part box 1 and robot 2. On the other hand, in solution G of cluster 4, the layout is compact without a linear orientation. However, cluster 4 obtained through *k*-means clustering in Fig. 4 (b) ignores such features with correlated arrangements and gathers the solutions E, F and G together. In this layout-planning problem, differences of the layout area sizes have a large effect on the design variables, so in *k*-means clustering based on dissimilarities measured by Euclidian distances, the dissimilarities of the layout areas are stand out, and they bury other slight differences that indicate features with correlated arrangements. On the other hand, EM clustering handles the dissimilarities with correlations of the design variables, and extract features with correlated arrangement.

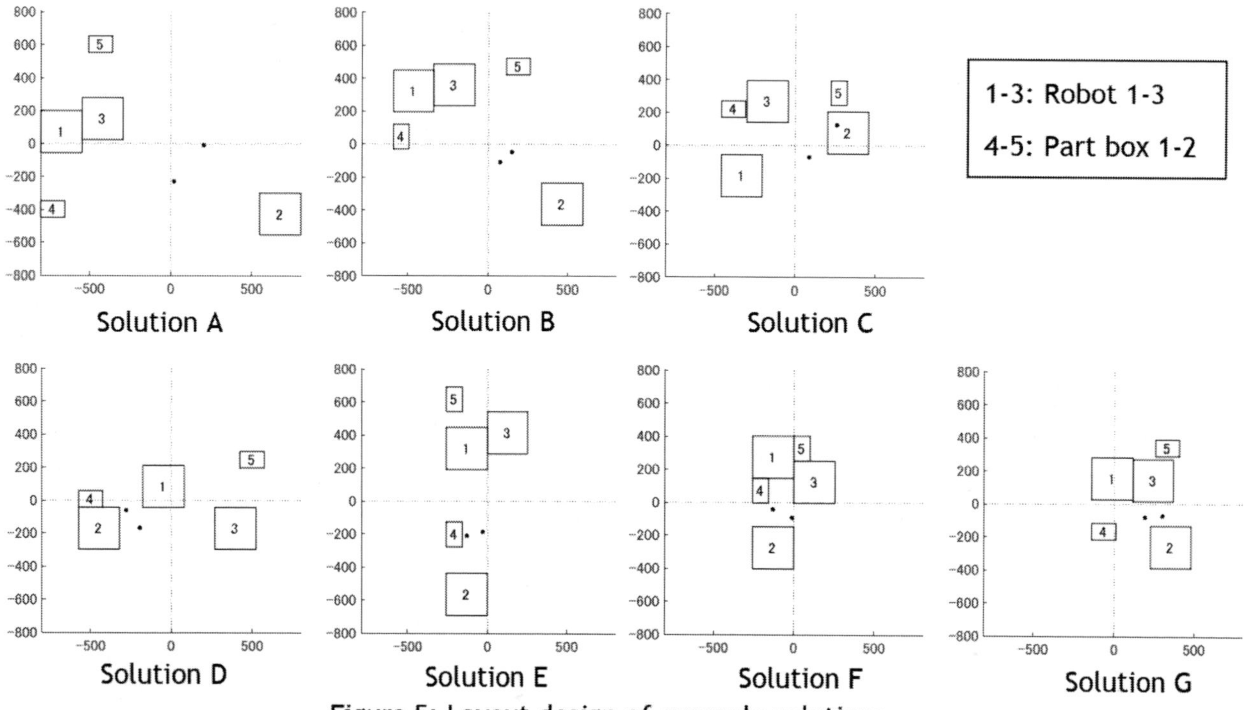

Figure 5: Layout design of example solutions

5 CONCLUSION

This paper proposed a new multiobjective optimization technique and demonstrated an approach to analyse the obtained solutions. The following were achieved.

(1) A new multiobjective optimization method was developed to obtain a large number of non-dominated solutions.

(2) Two clustering methods were demonstrated to analyze the obtained non-dominated solutions.

(3) The proposed method was applied to a RCMS layout multiobjective optimization problem, and the importance of the correlation consideration in data analysis of the obtained non-dominated solutions was illustrated.

6 REFERENCES

[1] **Meller, R. D., and Gau, K.-Y., 1996.** The facility Layout Problem: Recent and Emerging Trends and Perspectives, *Journal of Manufacturing Systems*, 15(5), pp. 351-366.

[2] **Drira, A, Pierreval, H., and Hajri-Gabouj, S., 2007.** Facility layout problems: A survey, *Annual Reviews in Control*, 31, pp.255-267.

[3] **Chen, C. W., Sha, D. Y. 2005.** Heuristic approach for solving the multiobjective facility layout problem. *International Journal of Production Research*, 43(21), 4493-4507.

[4] **Aiello, G., Enea, M., & Galante, G. (2006).** Multi-objective approach to facility layout problem by genetic search algorithm and Electre method. *Robotics and Computer-Integrated Manufacturing*, 22, pp. 447-455.

[5] **Izui, K., Murakumo, Y., Suemitsu, I., Nishiwaki, S., Noda, A., and Nagatani, T., 2013.** Multiobjective layout optimization of robotic cellular manufacturing systems, *Computers & Industrial Engineering*, 64(2), pp. 537-544.

[6] **Brockhoff, D. and Zitzler, E., 2009.** Objective reduction in evolutionary multiobjective optimization: *Theory and applications, Evolutionary Computation*, 17(2), pp. 135-166.

[7] **Morse J. N. 1980.** Reducing the size of the nondominated set: pruning by clustering. *Computers and Operations Research 7*, pp. 55-66.

[8] **Rosenman M. A, Gero J. S. 1985.** Reducing the Pareto optimal set in multicriteria optimization. *Engineering Optimization 8*, pp. 189-206.

[9] **Sugimura, K, Jeong, S, Obayashi, S, Kimura, T. 2009.** Kriging-model-based multi-objective robust optimization and trade-off-rule mining using association rule with aspiration vector, Proceedings of the 2009 IEEE Congress on Evolutionary Computation, pp. 522-519.

[10] **Ulrich T, Brockhoff D, Zitzler E. 2008.** Pattern identification in Pareto-set approximations. Proceedings of the 10th Annual Conference on Genetic and Evolutionary Computation, pp. 737-744.

[11] **Ulrich T., 2013.** Pareto-Set Analysis: Biobjective Clustering in Decision and Objective Spaces, *Journal of Multi-Criteria Decision Analysis*, 20, pp. 217-234.

[12] **Izui, K., Bai, X., Yamada, T., Nishiwaki, S., Noda, A., Nagatani, T., 2015.** A bi-level multiobjective optimization approach for layout design of robotic cellular manufacturing systems, CIE45 Proceedings, Paper No. 175, 28-30 October 2015, Metz, France.

[13] **Yoshikawa, T. 1985.** Manipulability of robotic mechanisms, *International Journal of Robotics Research*, 4(2), pp. 3-9.

[14] **Murata, H., Fujiyoshi, K., Nakatake, S. and Kajitani, Y. 1991.** VLSI module placement based on rectangle-packing by the sequence-pair, *IEEE Transactions on Computer-Aided Design of Integrated Circuits and Systems*, 15(12), pp. 1518-1524.

[15] **Deb, K., Pratap, A., Agarwal, S., and Meyarivan, T. 2002,** A fast and elitist multiobjective genetic algorithm: NSGA-II, *IEEE Transaction on Evolutionary Computation*, 6(2), pp. 182-197.

[16] **Dempster, A. P., Laird, N. M., Rubin, D. B. 1977,** Maximum Likelihood from Incomplete Data via the EM Algorithm, *Journal of the Royal Statistical Society,* Series B 39(1), pp. 1-38.

AN EFFICIENT MIXED INTEGER LINEAR PROGRAMMING MODEL FOR ENERGY-CONSCIOUS SINGLE MACHINE SCHEDULING UNDER TIME-OF-USE TARIFFS

Shibohua ZHANG, Xueqi WU, Ada CHE*
School of Management
Northwestern Polytechnical University, Xi'an, China
ache@nwpu.edu.cn

ABSTRACT

This paper investigates an energy-conscious single machine scheduling problem to minimize total electricity cost under the time-of-use (TOU) electricity pricing scheme. In such a scheme, the electricity consumed in different periods of a day requires different prices. Each job to be processed on the machine is characterized by their processing time and electricity consumption per hour. The objective is to arrange the processing of the jobs under TOU electricity tariffs so as to minimize the total electricity cost. An improved mix-integer linear programming model with less number of constraints is formulated for the problem. Computational results demonstrate that our model can solve the problem more effectively.

Keywords: single machine scheduling, time-of-use tariffs, mixed-integer linear programming model, total electricity cost

1 INTRODUCTION

With the mounting concern over the energy consumption in the manufacturing sector, energy efficiency has become another crucial criterion for production managers to consider when formulating production plans apart from production efficiency. Among a wide range of energy forms, the electricity is regarded as the main form utilized in production activities [1]. However, the process of generating electricity can take a heavy toll on environment protection. For example, generating 1 kwh of electricity means to release two pounds of carbon dioxide[2], which is the main contributor of the global warming process. On the other hand, with the increase in demand and price for fossil fuels and the decrease in their reserves [3], energy-saving has become increasingly urgent. Since the manufacturing sector is the main consumer of electricity in most countries around the world, for example, according to [4], manufacturing occupies about half of the total electricity generated each year in China, containing the usage of electricity in the manufacturing sector is of vital significance.

To achieve the improvement of electricity efficiency in the manufacturing sector, there are mainly three strategies on different levels, which are called structure energy-saving, technical energy saving and management energy-saving[5], respectively. The first two strategies focus on bringing in more energy-efficient production equipment, while the third one tries to achieve the purpose via

This work was in part supported by National Natural Science Foundation of China under grant no. 71471145.

* Corresponding Author.

scientific management, which is quite prevalent in many electricity companies nowadays. Since the demand of electricity may vary greatly in different periods during a day, extra pressure has been put on the electricity suppliers. To tackle the condition, several time-dependent electricity pricing schemes have been proposed and adopted, such as the time-of-use (TOU) pricing scheme.

The TOU pricing scheme works by setting different prices for electricity in different periods during a day. Specifically, each day is divided into different types of periods according to the demand for electricity, such as on-peak periods, mid-peak periods and off-peak periods. In on-peak periods, the electricity prices are set at the highest level because the demands for electricity in these periods are the highest. Accordingly, the electricity prices in the mid-peak and off-peak periods are moderate and the lowest, respectively, because of the moderate and the lowest demand in these periods. Obviously, effective TOU pricing scheme will help shift the workloads from on-peak periods to mid-peak periods and off-peak periods [6]. Hence the demand for electricity during a day may be effectively balanced, thus reducing the burden on the electricity suppliers.

To improve the efficiency of electricity as well as of production under the TOU tariffs, production managers may need the help of scheduling. According to [7], scheduling is to allocate resources to tasks with the goal of optimizing one or several objectives. A special case of scheduling is the single machine scheduling, which is the basis of more complicated machine environments and has been studied widely in the literature. For example, Hariri and Potts [8] developed a branch and bound algorithm to minimize the weighted number of tardy jobs. Boland, Clement and Waterer [9] presented a bucket indexed (BI) MILP model for non-preemptive single machine scheduling problems. Computational results demonstrate that the BI model outperforms significantly the classical time indexed model.

Besides, single machine scheduling problems considering energy consumption have been a hot topic lately. Basically, there are two typical types of energy-saving mechanisms, which are the power-down mechanism (e.g. [3, 10-12]) and the speed-scaling problems (e.g. [13-15]). For single machine scheduling under TOU tariffs, Shrouf, Ordieres-Meré, García-Sánchez and Ortega-Mier [16] formulated a discrete-time integer programming model based on power-down mechanism with the objective of minimizing the total electricity cost. In their work, the processing sequence of jobs is assumed to be pre-determined and a genetic algorithm is proposed to obtain the approximate optimal solutions. Fang, Uhan, Fu and Sutherland [15] studied several speed-scaling versions of the single machine scheduling problem to minimize the total electricity cost. Another work was conducted by Che et al.[17], who proposed a continuous-time mixed-integer linear programming (MILP) model and an efficient greedy insertion heuristic. In this paper, we propose a new MILP model for single machine scheduling problem to minimize the total electricity cost under the TOU tariffs. We will also compare our model with that presented by Che et al.[17].

The remander of the paper is organized as follows. In section II, we give a clear description of the problem addressed and formulate a new MILP model for it. The computational results and the comparison with a previous model are presented in section III. Finally, we draw a conclusion in section IV.

2 PROBLEM DESCRIPTION AND FORMULATION:

2.1 Problem description

This paper addresses a single machine scheduling problem with the objective of minimizing total electricity cost under the time-of-use tariffs. Its NP-hardness was proven by Fang, Uhan, Fu and Sutherland [15]. In this problem, a set of N jobs to be processed on the machine are specified by their own processing time t_i and electricity consumption rate p_i, $1 \le i \le N$. All the jobs are available at the beginning of the processing course and so is the machine. No preemption is allowed, that is, once a job is processed on the machine, no interruption is permitted until the processing of this job comes to an end. The machine can only process one job each time and breakdowns are not considered.

In terms of the TOU pricing scheme, the electricity prices vary in different periods during a day. Obviously, the same workload assigned to different periods leads to different amount of electricity cost. To better illustrate the problem addressed, without loss of generality, suppose that the whole processing horizon is divided into K periods. We use s_j to denote the start time of period j, $1 \le j \le K+1$, and note that s_{K+1} represents the end time of the last period. The electricity price in period j is denoted by c_j, $1 \le j \le K$. For simplicity, we assume that $s_1=0$. To guarantee the feasibility of the problem addressed, the value of s_{K+1} should be larger than the sum of the processing times of all the jobs.

To achieve the goal of minimizing the total electricity cost, the majority of efforts should be focused on the assignment of jobs among different periods of the machine. Hence we need to develop the decision variables to describe whether job *i* is handled in period *j* and further know how much processing time of job *i* is assigned to period *j*. Since the processing time of job *i* may be larger than the duration of period *j* or job *i* is assigned to be processed across two adjacent periods, we propose another decision variable to describe whether job *i* is processed across periods to better formulate the problem addressed.

2.2 Problem formulation

As is discussed above, the decision variables of the MILP model for the problem addressed are:

x_{ij}: The assigned processing time of job *i* in period *j*, for $1 \le i \le N, 1 \le j \le K$

y_{ij}: Binary variable. If job *i* is processed in period *j*, then $y_{ij}=1$; otherwise, $y_{ij}=0$, for $1 \le i \le N, 1 \le j \le K$

z_{ij}: Binary variable. If job *i* is processed across period *j* and period *j*+1, then $z_{ij}=1$; otherwise $z_{ij}=0$, for $1 \le i \le N, 1 \le j \le K-1$

With the notations presented above, the MILP model can be written as follows:

$$\min \sum_{i=1}^{N} \sum_{j=1}^{K} p_i c_j x_{ij}$$

$$\text{s.t.} \sum_{j=1}^{K} x_{ij} = t_i, 1 \le i \le N \quad (1)$$

$$\sum_{i=1}^{N} x_{ij} \le s_{j+1} - s_j, 1 \le j \le K \quad (2)$$

$$x_{ij} \le t_i y_{ij}, 1 \le i \le N, 1 \le j \le K \quad (3)$$

$$z_{ij} \ge y_{ij} + y_{i,j+1} - 1, 1 \le i \le N, 1 \le j \le K\text{-}1 \quad (4)$$

$$\sum_{i=1}^{N} z_{ij} \le 1, 1 \le j \le K\text{-}1 \quad (5)$$

$$\sum_{l=j+2}^{K} y_{il} \le K(1-y_{ij}+y_{i,j+1}),\ 1 \le j \le K\text{-}2 \quad (6)$$

$$x_{ij} \ge (y_{i,j-1}+y_{i,j+1}-1)(s_{j+1}-s_j),\ 1 \le i \le N,\ 2 \le j \le K\text{-}1 \quad (7)$$

The objective is to minimize the total electricity cost represented by $\sum_{i=1}^{N}\sum_{j=1}^{K} p_i c_j x_{ij}$. Constraint (1) ensures that the sum of the assigned processing time of job *i* to different periods is equal to its total processing time. Constraint (2) states that the total processing time of all jobs assigned to period *j* should be no larger than its own duration $s_{j+1}-s_j$. Constraint (3) shows that if $y_{ij}=0$, then the processing time of job *i* assigned to period *j* should be zero as well.

Constraint (4) tells that if job *i* is processed both in period *j* and period *j*+1 (i.e., $y_{ij}=1$ and $y_{i,j+1}=1$), then z_{ij} should take the value of 1. Constraint (5) ensures that at most one job can be processed across two adjacent periods. Constraint (6) is the processing continuity constraint ensuring that the processing of a job in different periods are continuous. To be more specific, if job *i* is processed in period *j* but not in period *j*+1 (i.e., $y_{ij}=1$ and $y_{i,j+1}=0$), then the end time of processing job *i* must be in period *j*, that is, job *i* cannot be processed in the periods afterwards anymore. We note that constraint (6) can also be equivalently written as:

$$\sum_{l=1}^{j-1} y_{il} \le K(1-y_{i,j+1}+y_{ij}),\ 2 \le j \le K\text{-}1.$$

Constraint (7) ensures that if job *i* is assigned to both period *j*–1 and period *j*+1 (i.e., $y_{i,j-1}=1$ and $y_{i,j+1}=1$), namely it is processed across at least three periods, then the assigned processing time of job *i* to period *j* should be no less than the duration of this period.

We note that Che et al.[17] only defined the first two groups of binary variables (i.e. x_{ij} and y_{ij}) to formulate the single machine scheduling problem under TOU tariffs. In this work, the number of constraints is greatly reduced due to the use of the third group of binary variables (i.e. z_{ij}). We also note that similar definitions of the three groups of binary variables given above and their corresponding constraints can also be found in Ding et al. [18], who proposed an interesting MILP model for unrelated parallel machine scheduling problem under TOU Tariffs. In fact, they defined five groups of binary variables to formulate their problem. In this work, we need to define three groups of binary variables due to improvements of the formulation.

It is worthwhile to note that we improve the processing continuity constraint given by Che et al. [17] and Ding et al. [18], which is a very key constraint for machine scheduling problems under TOU tariffs. In the work of Che et al. [17], the processing continuity constraints formulated as $O(NK^2)$ inequalities. However, in this work, it is formulated as $O(NK)$ inequalities without introducing any additional binary variables. In Ding et al [18], for a given unrelated parallel machine, the processing continuity constraint is also formulated as $O(NK)$ inequalities by introducing additional *NK* binary variables. From the above analysis, our formulation of the processing continuity constraint is more compact than Che et al.'s and Ding et al.'s, which speeds up the resolution of the model. It is worthwhile to point out that the formulation for processing continuity constraint developed in this work can also be applied to other machine scheduling problems under TOU tariffs.

3 COMPUTATIONAL RESULTS

In this section, we compare the performance of our MILP model with a previous one developed by Che et al.[17] through randomly generated instances. The model formulations are coded and implemented in C++ and solved by the MILP solver of CPLEX (Version 12.6). All the experiments are performed on a Lenovo PC with 1.7 GHz quad-core Intel i5 CPU and 8GB RAM.

The instances are generated in the following way. The processing time t_i for job i follows a uniform distribution between 10 and 100 minutes, while the power consumption per hour p_i between 30 and 100kW/h. The TOU pricing scheme adopted in this paper is the one implemented in Shaanxi Province, whose parameters are shown in [16]. To test the performances of the two models, we utilize 8 groups of random instances with different number of jobs from 30 to 100 with an increment of 10. Note that the chosen processing horizon length (i.e., the makespan) can affect the performances of the two models as well as the feasibility of the problem. Obviously, when the processing time of each job in an instance takes the value of its upper bound (i.e., 100 minutes), we can obtain the least makespan needed to keep each problem of that scale feasible. Therefore, to better test the performances of the two models, we use different lengths of processing horizon. To achieve that, we test each problem under several given K periods, and in each setting the feasibility of the problem is guaranteed by the above rule.

We test each setting of the problem with 10 instances, and the computational results are shown in Table I. To better demonstrate the performance of our model compared to the previous one under different settings, we propose the following measurement, which reflects how much our model is faster than the previous one under certain setting. Let $t_{O,N,K}$ denotes the average computational time of our model to solve the problem with N jobs and K periods and $t_{C,N,K}$ represents that of the previous model under the same setting. We define

$$F_{N,K} = \frac{t_{C,N,K}}{t_{O,N,K}}.$$

From Table I, we can see that our model outperforms the previous one in terms of computational time, especially for large-size problems. For example, when dealing with problems with 100 jobs and 48 periods, the previous model cannot find the optimal solutions for all the random instances within one hour, while our model find the optimal solution within about 230 seconds on average. Specifically, the previous model cannot solve two of the ten instances to optimality within one hour. But the gaps from the lower bounds when the two instances are solved for one hour are both below 0.03%. Here we regard the computational time for the two instances both as 3600 seconds and calculate their mean. Even under such a circumstance, our model still runs nearly 5 times faster than the previous one. This is because the improved model is more compact in terms of the number of constraints.

4 CONCLUSION

This paper considers an energy-conscious single machine scheduling problem to minimize total electricity cost under the TOU tariffs. To tackle the problem, we formulate an improved mixed-integer linear programming model and compare it with a previous one. The computational results on randomly generated instances demonstrate that our model is more efficient in dealing with the problems, especially for large-size problems. In the future, efficient heuristics shall be developed, which can increase the problem scale solved greatly.

Table 1. Comparison of the two models under different settings

Problem		CPU Time(s)		$F_{N,K}$
N	*K*	Our Model	Previous Model	
30	12	0.38	0.62	1.63
	14	0.25	0.38	1.52
	16	0.65	1.17	1.8
40	16	1.95	2.19	1.12
	20	2.47	3.11	1.26
	24	0.48	1.17	2.44
50	20	3.47	5.8	1.67
	24	0.62	1.54	2.48
	28	9.18	11.46	1.25
60	24	4.39	12.85	2.93
	28	7.21	20.78	2.88
	32	17.44	31.03	1.78
70	28	7.58	24.84	3.28
	32	22.94	44.55	1.94
	36	48	73.3	1.53
80	32	17.2	64.1	3.73
	36	75.64	112.52	1.49
	40	91.47	167.01	1.83
90	36	42.93	149.87	3.49
	40	79.18	242.05	3.06
	44	29.81	172.04	5.77
100	40	107.9	278.08	2.58
	44	238.09	401.18	1.68
	48	229.97	1138.44*	4.95

*: Two of ten instances cannot be solved to optimality within 3600 seconds.

5 REFERENCES

[1] **Park, C.W., Kwon, K.S., Kim, W.B., Min, B.K., Park, S.J., Sung, I.H., Yoon, Y.S., Lee, K.S., Lee, J.H., and Seok, J. 2010.** Energy consumption reduction technology in manufacturing — A selective review of policies, standards, and research, International Journal of Precision Engineering & Manufacturing, 10(5), pp 151-173.

[2] **Group, T.C.1998.** Regional Electricity Emission Factors Final Report.

[3] **Mouzon, G., and Yildirim, M.B. 2008.** A framework to minimise total energy consumption and total tardiness on a single machine, International Journal of Sustainable Engineering, 1(2), pp 105-116.

[4] **Tang Decai, L.L., Du Kai. 2006.** On the development path of Chinese manufacturing industry based on resource restraint, Jiangsu Social Sciences, 4, pp 51-58.

[5] **Cheng, J., Chu, F., Xia, W., and Ding, J. 2014.** Bi-objective optimization for single-machine batch scheduling considering energy cost, International Conference on Control, Decision and Information Technologies, Metz, France, November 3-5, pp 236 - 241.

[6] **Wang, Y., and Li, L. 2013.** Time-of-use based electricity demand response for sustainable manufacturing systems, Energy, 63(1), pp 233-244.

[7] **Pinedo, M.L. 2012.** Scheduling: Theory, Algorithms, and Systems, Springer-Verlag New York.

[8] **Hariri, A.M.A., and Potts, C.N. 1994.** Single Machine Scheduling with Deadlines to Minimize the Weighted Number of Tardy Jobs, Management Science, 40(12), pp 1712-1719.

[9] **Boland, N., Clement, R., and Waterer, H. 2016.** A Bucket Indexed Formulation for Nonpreemptive Single Machine Scheduling Problems, INFORMS Journal on Computing, 28(1), pp 14-30.

[10] **Liu, C., Yang, J., Lian, J., Li, W., Evans, S., and Yin, Y. 2014.** Sustainable performance oriented operational decision-making of single machine systems with deterministic product arrival time, Journal of Cleaner Production, 85, pp 318-330.

[11] **Yildirim, M.B., and Mouzon, G. 2012.** Single-Machine Sustainable Production Planning to Minimize Total Energy Consumption and Total Completion Time Using a Multiple Objective Genetic Algorithm, IEEE Transactions on Engineering Management, 59(4), pp 585-597.

[12] **Mouzon, G., Yildirim, M.B., and Twomey, J. 2007.** Operational methods for minimization of energy consumption of manufacturing equipment, International Journal of Production Research, volume, 45(18), pp 4247-4271.

[13] **Che, A., Lv, K., Levner, E., and Kats, V. 2015.** Energy consumption minimization for single machine scheduling with bounded maximum tardiness, 12th IEEE International Conference on Networking, Sensing and Control, Taipei, April 9-11, pp146-150.

[14] **Dürr, C., Jeż, Ł., and Vásquez, O.C. 2014.** Scheduling under dynamic speed-scaling for minimizing weighted completion time and energy consumption, Discrete Applied Mathematics, 196(C), pp 20-27

[15] **Fang, K., Uhan, N.A., Zhao, F., and Sutherland, J.W. 2016.** Scheduling on a single machine under time-of-use electricity tariffs, Annals of Operations Research, 238(1-2), pp 199-227

[16] **Shrouf, F., Ordieres-Meré, J., García-Sánchez, A., and Ortega-Mier, M. 2014.** Optimizing the production scheduling of a single machine to minimize total energy consumption costs, Journal of Cleaner Production, 67(6), pp 197-207

[17] **Che, A., Zeng, Y., and Lyu, K. 2016.** An efficient greedy insertion heuristic for energy-conscious single machine scheduling problem under time-of-use electricity tariffs, Journal of Cleaner Production, 129, pp 565-577.

[18] **Ding, J.Y., Song, S., Zhang, R., and Chiong, R. 2016.** Parallel Machine Scheduling Under Time-of-Use Electricity Prices: New Models and Optimization Approaches, IEEE Transactions on Automation Science & Engineering, 13 (2), pp 1138-1154.

MULTIOBJECTIVE HYBRID EVOLUTIONARY ALGORITHM BASED ON MULTIOBJECTIVE LOCAL SEARCH FOR FLOW SHOP SCHEDULING PROBLEM

Wenqiang Zhang[1*], Yu Wang[1], Jiaming Lu[1] Chunxiao Wang[1] and Mitsuo Gen[2]
[1] College of Information Science and Eng., Henan University of Technology, China
zhangwq@haut.edu.cn

[2]Fuzzy Logic Systems Institute & Tokyo University of Science, Japan
gen@flsi.or.jp

ABSTRACT

In this paper, we proposed a hybrid evolutionary algorithm to solve bi-criteria flow shop scheduling problem (FSP) with the makespan minimization and max-tardiness minimization. Multi-objective local search (Mo-LS) is a multi-objective local search strategy, which combines two neighborhood structure, multi-objective insert (Mo-INS) and multi-objective exchange (Mo-EXC). The original idea of Mo-LS is systematic change neighborhoods in the local search procedure. It can achieve two goals while reducing makespan and maximum tardiness. Moreover, Mo-LS does not need to set parameters. Mo-LS utilize reasonable iterative of the two neighborhood structure to control search direction which improve searching space. The Mo-LS combined with improved the multi-objective evolutionary algorithm—multi-objective hybrid evolutionary algorithm (MOHEA) is called as "MOHEA-MOLS". To illustrate the efficacy of proposed approach, some different scaled FSP are used to test the performance of MOHEA-MoLS. Simulation results show that the improved algorithm MOHEA-MOLS have been strengthened in the performance of convergence and distribution.

Keywords: Flow Shop Scheduling, Multiobjective Hybrid Evolutionary Algorithm, Multiobjective Local Search, Maximum Tardiness, Makespan

1 INTRODUCTION

Shop scheduling problem is known as NP-hard problem. It is a decision-making process, which aims to optimize one or more objectives[1]. FSP as a typical optimization model in shop scheduling problem has a broad background[2][3].

In recent years, minimizing makespan and maximum tardiness are the most commonly criteria studied and discussed[4]. Makespan is a measure of system utilization while maximum tardiness can be considered as measure of performance in customer's due dates. Makespan minimization can benefit many industrial fields, such as saving electricity, machinery manufacturing and intelligent automation etc. Eksioglu[5] and Grabowski[6] have addressed FSP on the single criterion of minimizing makespan. In addition, many researchers consider that the tardiness cost is great importance for manufacturing and production managers. Similarly, Grabowski had published some typical papers[7]. Gu et al. improved adaptive particle swarm optimization algorithm for solving the minimum makespan problem of FSP[8]. However, most of above studies prefer single objective rather than multi-objective. Actually, the multi-objective FSP meet the demand in real life. A.Nagar et at.[9] had summarily surveyed that the most widely criteria were related to mimimizing makespan and job

* Corresponding Author

tardiness in FSP. Although above studies focused on the multi-objective, they didn't consider the algorithm performance of convergence and distribution.

The multi-objective evolutionary algorithm—multi-objective hybrid evolutionary algorithm (MOHEA) [10][11] is improved on the basis of classical algorithm VEGA. It not only has VEGA advantages of fast convergence and low computational complexity, but also make up for the lack of VEGA distribution. Next, improving the MOHEA can consider adding the local search mechanism to increase the algorithm to search the solution space. MOLS[12] is a local search strategy which is basis of multi-objective exchange and multi-objective insert. It is used to deal with the FSP of makespan and the maximum tardiness. MOLS (different from other local search strategy) do not need to set parameters, .i.e. it is not controlled by the people, and its search direction is determined by the multi-objective neighborhood structure. To aim at solving the maximum completion time and the maximum tardiness of FSP, this paper improved the MOHEA by joining MOLS to achieve the improvement of the algorithm.

The structure of this paper is as follows: the second part describes and constructs the mathematical model of makespan minimization and maximum tardiness minimizing bi-objective FSP; the MOHEA-MoLS is described in detail in the third part; the fourth part gives the numerical simulation experiments and the results are analyzed and discussed; the conclusion and the prospect of the future work are given in the fifth part.

2 PROBLEM DESCRIPTION

Flow shop scheduling can be described as follows [12]: each job from the set $J = \{J_1, J_2, \ldots, J_n\}$ should be processed on the m machines from set $\{M_1, M_2, \ldots, M_m\}$. Assuming that the jobs are processed as the indexed sequential of the machines, O_{ji} is defined as the i_{th} operation of the job J_i and the processing time of O_{ji} on the machine M_i is p_{ji}. The objective is to find a set of non-dominated solutions set, minimizing the makespan and maximum tardiness simultaneously.

2.1 Minimizing makespan and Minimizing Maximum Tardiness

Let $\pi = (\pi(1), \pi(2), \ldots, \pi(n))$ be a set of processing permutations for the job, where π_a is the a_{th} job of the permutation π,n is number of jobs. U represents all possible process combinations, and $\pi \in U$. $C^i_{\pi(j)}$ is the completion time of job π_j on the machine M_i and t_{ji} is the completion time of operation O_{ji} on the machine M_i. If operation O_{ji} is processed on the machine M_i, $x_{ji} = 1$, otherwise, $x_{ji} = 0$; if operation O_{ji} is processed before O_{kh}, $y_{jikh} = 1$, otherwise, $y_{jikh} = 0$. The mathematical formulation of FSP can be described as below:

$$
\begin{aligned}
&C^1_{\pi(1)} = p_{\pi(1)1} \\
&C^i_{\pi(1)} = C^{i-1}_{\pi(1)} + p_{\pi(1)i} \\
&C^1_{\pi(j)} = C^1_{\pi(j-1)} + p_{\pi(j)1} \\
&C^i_{\pi(j)} = \max\left\{C^i_{\pi(j-1)}, C^{i-1}_{\pi(j)}\right\} + p_{\pi(j)i} \\
&j = 2,3,\ldots n,\ i = 2,3,\ldots m.
\end{aligned}
\tag{1}
$$

The purpose of FSP with makespan criterion is to find a permutation π^* from U.

$$
\begin{aligned}
&\pi^* = \arg\min_{\pi\in U} C_{\max}(\pi) \\
&C_{\max}(\pi) = C^m_{\pi(n)}
\end{aligned}
\tag{2}
$$

All the constraint conditions are as follows:

$$\text{s. t.} \quad \left(t_{kh} - p_{kh} - t_{ji}\right) x_{ji} x_{kh} y_{jikh} > 0, \forall (j,i),(k,h)$$
$$y_{jiji} = 0, \forall (j,i)$$
$$\sum_{i=1}^{m} x_{ji} = 1, \forall (j,i)$$
$$\sum_{j=1}^{n} x_{ji} = 1, \forall (j,i) \tag{3}$$
$$y_{jikh} \in \{0,1\}, \forall (j,i),(k,h)$$
$$x_{ji} \in \{0,1\}, \forall (j,i)$$
$$t_{ji} \geq p_{ji} \geq 0, \forall (j,i)$$

The first four formulas provide the jobs machining of all machines on order, ensure the feasibility of the machining permutation, the uniqueness of the jobs and the uniqueness of the machines respectively. The remaining equations define non-negative conditions.

The minimizing of maximum tardiness is the most common criteria in FSP. Suppose that $T_{max}(\pi)$ is maximum tardiness of permutation π, $T_{\pi(j)}$ and $D_{\pi(j)}$ mean the tardiness and project time of job $\pi(j)$, respectively. The project time is determined by makers. Similarly, the mathematical description of FSP with maximum tardiness criterion as follows:

$$T_{\max}(\pi) = \max_{j \in [1,n]} \{T_{\pi(j)}\}$$
$$T_{\pi(j)} = \max \{C^{m}_{\pi(j)} - D_{\pi(j)}, 0\} \tag{4}$$

The bi-criteria optimization objectives can be formalized as:

$$\min\{C_{\max}(\pi), T_{\max}(\pi)\}, \pi \in U \tag{5}$$

2.2 Key Job and Activity Block

For a set of machining permutation π, makespan $C_{\max(\pi)}$ value is calculated by the following equation:

$$C_{\max}(\pi) = \max_{1 \leq t_1 \leq \cdots \leq t_{m-1} \leq n} \left[\sum_{j=1}^{t_1} p_{\pi(j)1} + \sum_{j=t_1}^{t_2} p_{\pi(j)2} + \cdots + \sum_{j=t_{m-1}}^{n} p_{\pi(j)m} \right] \tag{6}$$

Integer permutations $t = (t_1, t_2, \ldots, t_{m-1})$ defines that a path $P(\pi,t)$ that the first operation to the last operation. Where t_i ($t_i \in t$) is the machine index number of processing permutation π. Suppose $L(P(\pi,,t))$ is the path length of $P(\pi,,t)$, then:

$$L\left(P(\pi,t)\right) = \sum_{j=1}^{t_1} p_{\pi(j)1} + \sum_{j=t_1}^{t_2} p_{\pi(j)2} + \cdots + \sum_{j=t_{m-1}}^{n} p_{\pi(j)m} \tag{7}$$

And all processing permutation has the following relationship:

$$C_{\max}(\pi) \geq L\left(P(\pi,t)\right) \tag{8}$$

$P(\pi,t)$ is the arbitrary path of π. Among that the length of the longest path $P(\pi,h)$ is the makespan for π.Where $h=(h_1,h_2,\ldots,h_{m-1})$:

$$C_{max}(\pi) = L\left(P(\pi,\mathrm{h})\right) \tag{9}$$

Path $P(\pi,h)$ is called the critical path i.e., the job on the critical path is called the key artifacts π[13]. A set of key artifacts are shown formula(10) and Activities block can be defined as formula (11):

$$A=\left\{\pi(\eta_i)\middle|\pi(\eta_i)\neq\pi(\eta_j)\quad \eta_i,\eta_j\in\eta\right\} \tag{10}$$

$$B_\kappa=\left\{\pi(h_{k-1}),\pi(h_{k-1}+1),\cdots,\pi(h_k)\right\} \tag{11}$$

Where $\pi(h_k)$ is known as k^{th} block Activity of processing permutation π. $k\in\{1,2,...|A|\}$. When $h_0=1$, $\pi(h_k\text{-}1)$ and $\pi(h_k)$ are the adjacent jobs in the processing permutation π.

3 HYBRID MULTIOBJECTIVE EVOLUTIONARY ALGORITHM BASED ON LOCAL SEARCH

3.1 Two theorems about FSP

Suppose (u, v) is one pair of position index in a processing permutation π, where $u < v$. The neighbourhood structure, INS and EXC, can be described:

INS: Suppose that permutation π is $\pi = (\pi(1), \pi(2),..., \pi(t),..., \pi(n))$.The job $\pi(t)$ has maximum tardiness and there are k active blocks in π, such as $\{B_1, B_2 ,...B_{h-1} ,B_h ,B_{h+1} ,...,B_k\}$. The job $\pi(t)$ in the h_{th} *active block* $B_h(\pi(t)\in B_h)$).The value of u and v are list as Table 1.First, randomly generate an integer u from [1,t). According to the Table 1, identify the job $\pi(u)$ and determine the parameter v based on categorization of $\pi(u)$.Then to insert job $\pi(u)$ to the position v. According to category 1, specific steps are as follows in Figure 1.

Table 1: Movement of Job Positions For MO-INS

Category	Job π(u)	Possible positions of v
1	$\pi(u)\in B_i\ \ i=1,2,...,h-1$	$v\in[t+1,n]$
2	$u\in\{x\mid(\pi(x)\in B_h)\cap(x<t)\}$	$\pi(v)\in B_j,\ \ i=h+1,h+2,...,k$

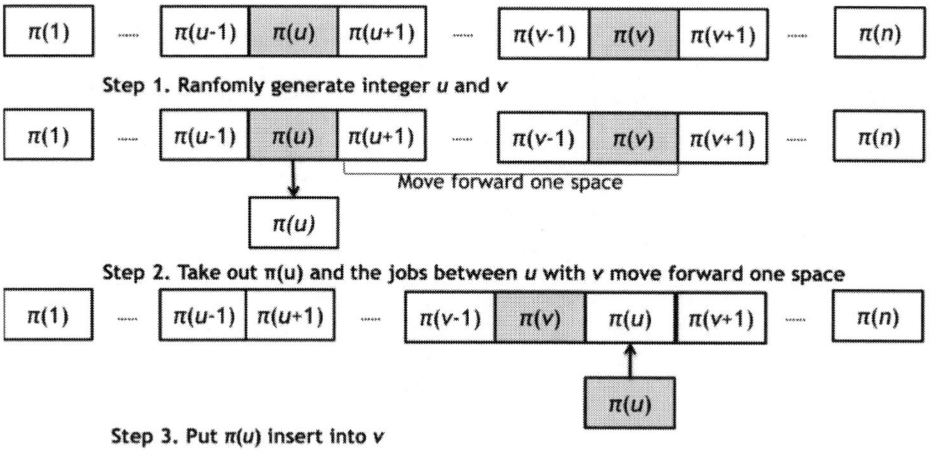

Figure 1: MOINS Neighborhood Structure

EXC: Suppose π is a current permutation. The job π(u) has maximum tardiness in π.Randomly select a job π(v) from B_i, where I = 1,2,3,...,h-1.Then to interchange the positions of π(t) and π(v).The new permutation π'' = (π(1), π(2), π(t), π(4),..., π(t-1), π(v), π(t+1),..., π(n)).

Theorem 1: Assuming processing permutation π'' is obtained by moving the job. Moving the job is the exchange of $\pi(u)$ and $\pi(v)$ or insert $\pi(u)$ into the position v. If $\pi(u)$ and $p(v)$ in the same activity block, then:

$$C_{max}(\pi'') \geq C_{max}(\pi) \tag{12}$$

Theorem 2: For any processing permutation π, T_{max} (π) is the maximum tardiness of π and $\pi(t)$ is the longest of the job. The permutation π'' is obtained by the movement of $\pi(t)$ to v. If $v > t$, then:

$$T_{max}(\pi'') \geq T_{max}(\pi) \tag{13}$$

$\pi(u)$ and $\pi(v)$ represents the ith and n^{th} job in the permutation π respectively. All of proofs can be seen the reference [12].

3.2 The basic idea of multi-objective local search

Mo-LS integrate the Mo-EXC and Mo-INS. The basic idea is to use local search operation for a particular individual. The specific algorithm of Mo-LS is shown in Figure 2(a).

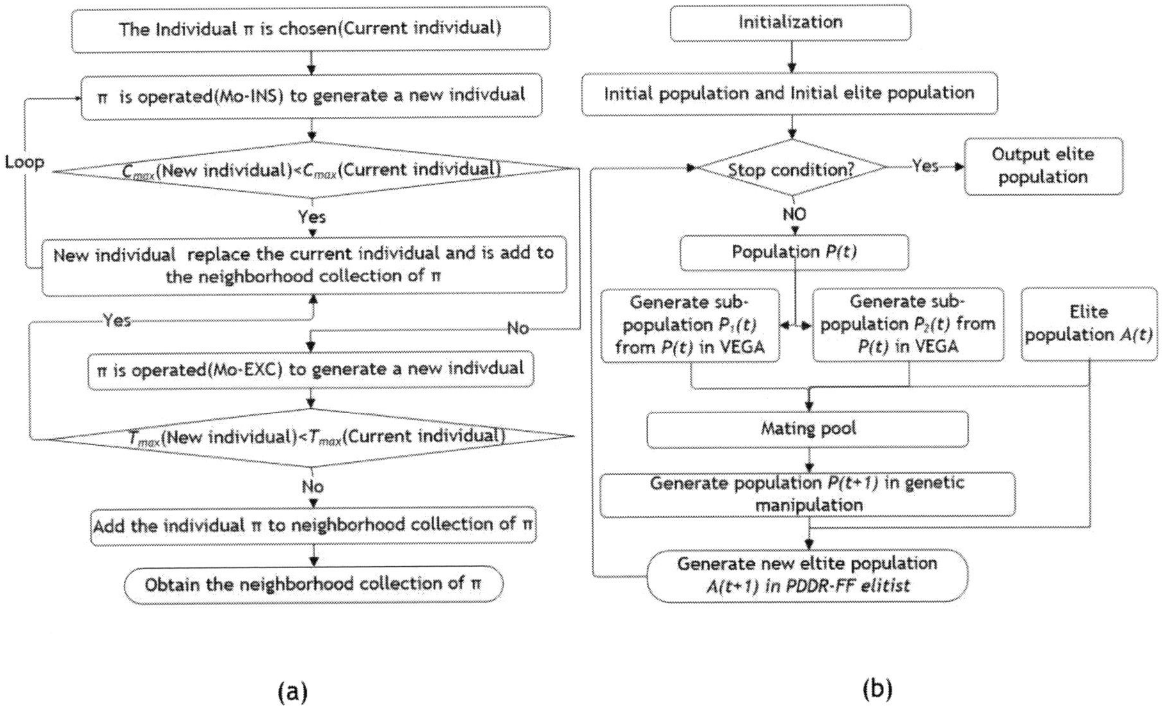

Figure 2: Mo-LS Flow chart and MOHEA Flow chart

When individual π implement local search operation, each operation will produce a valid individual π". To compare the π, if any target value of π" become bigger, π" is invalid individual and will be discarded. Then all effective individuals will be retained as the current individual of next operation. Frist, π is selected as the current individual .To reduce the maximum completion, π will be implemented by Mo-INS. If a new individual's makespan is smaller than the current individual, the new individual will replace the current individual and be added to the neighborhood collection π. When a new individual's makespan can no longer be reduced, the current individual will be implemented by Mo-EXC to reduce the maximum tardiness. Otherwise, add the individual π to neighborhood collection. The local search ended.

3.3 MOHEA and MOHEA-MOLS

Fitness value allocation policy has a very important role in the MOEAs. The MOHEA improves the VEGA by combing a new sampling strategy according to the Pareto dominating and dominated relationship-based fitness function. Mixed sampling mechanism not only has a convergence toward Pareto frontier central region, but also but also to reduce the time complexity as much as possible. Simulation experiments on multi-objective test problems show that, compared with NSGA-II and SPEA2, the fast multi-objective hybrid evolutionary algorithm is better in the two aspects of convergence and distribution, and has obvious advantages in the efficiency [10]. The specific algorithm of MOHEA is shown in Fig 2(b).

The fusion algorithm of Mo-LS and MOHEA is called MOHEA-MoLS. In the process of population evolution, only elite individuals implement Mo-LS. Each of individual that implement Mo-LS will generate an additional set of neighbourhood. In order to maintain the good distribution, all the individuals of neighbourhood collection and the population that complete the genetic operation combine to a new temporary population. Provisional population generates progeny population by screening fitness selection mechanism.

4 EXPERIMENTS AND DISCUSSION

In order to verify the effectiveness of the Mo-LS, the bi-objective FSP with minimizing makespan and maximum tardiness is implemented by MOHEA-MoLS and MOHEA under the same conditions. FSP scale problem as follows: FSP01: Job=20, Machine=5; FSP02: Job=50, Machine=10; FSP03: Job=50, Machine=20;

MOHEA-MoLS and MOHEA use the same encoding and genetic manipulation, specific parameter values as follows: popSize=100; Elite population=50; Crossover rate=0.4; Mutation rate=0.8; Maximum iterations=50000; Evolutional Generation=500; the only difference is that elite population of MOHEA-MoLS must be implemented by MoLS operation in each generation. The distribution and convergence indicators are used *SP* and *C*. The results of MOHEA-MoLS and MOHEA run 30 times are generated box diagram for comparison.

Table 2: *SP* and *C*

Problem	*SP*(MOHEA)	*SP*(MOHEA-MOLS)	*C*(MOHEA,MOHEA-MOLS)	*C*(MOHEA-MOLS,MOHEA)
FSP01	$7.54e\text{-}01_{5.1e\text{-}01}$	$6.85e\text{-}01_{4.2e\text{-}01}$	$5.00e\text{-}01_{5.0e\text{-}01}$	$7.00e\text{-}01_{4.6e\text{-}01}$
FSP02	$3.50e{+}00_{1.2e{+}00}$	$3.21e{+}00_{1.4e{+}00}$	$2.59e\text{-}01_{3.2e\text{-}01}$	$4.94e\text{-}01_{4.0e\text{-}01}$
FSP03	$1.26e{+}00_{8.2e\text{-}01}$	$8.76e\text{-}01_{7.3e\text{-}01}$	$4.00e\text{-}01_{4.9e\text{-}01}$	$4.94e\text{-}01_{4.0e\text{-}01}$

The mean and variance of SP indicators as shown Table 2 are obtained by MOHEA and MOHEA-MoLS running 30 times respectively. The C indicator is the same condition. From the Table 2, the distributed of MOHEA-MoLS is better than MOHEA which indicate Mo-LS step up the search capabilities for solution space. Compared with MOHEA, MOHEA-MoLS have an advantage on convergence. The solution quality obtained by MOHEA-MoLS is better than MOHEA description that Mo-LS can search for a valid solution. It speeds up the convergence to Pareto frontier. Figure 3 SP and C indicators of box-plot can also come to the same conclusion.

5 CONCLUSIONS AND FUTURE PERSPECTIVES

In this paper, a multi-objective hybrid evolutionary algorithm is proposed to solve multi-objective flow shop scheduling problem, considering minimization of makespan and max-tardiness minimization. MOHEA has good convergence and good distribution. By joining MoLS to enhance MOHEA local search ability and increase the number of outstanding individuals. The results of the experimental data show that MOHEA-MoLS is outstanding in convergence and distribution performance.

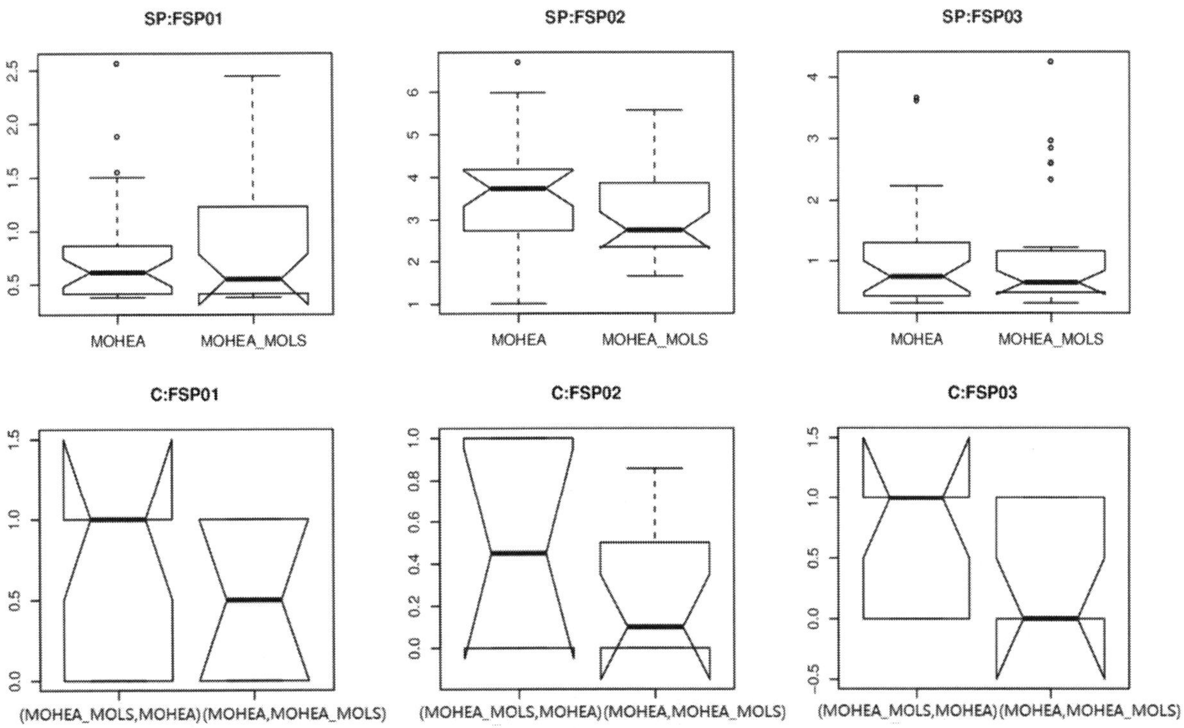

Figure 3: C and SP by MOHEA-MoLS and MOHEA

ACKNOWLEDGMENTS

This research work is supported by the National Natural Science Foundation of China (U1304609), Foundation for Science & Technology Research Project of Henan Province (162102210044, 152102110076), Program for Innovative Research Team(in Science and Technology) in University of Henan Province (your 17IRTSTHN011), Fundamental Research Funds for the Henan Provincial Colleges and Universities(2014YWQQ12, 2015XTCX03, 2015XTCX04), Research Funds for Key Laboratory of Grain Information Processing and Control (Henan University of Technology)(KFJJ-2015-

106), Ministry of Education and the Grant-in-Aid for Scientific Research (C) of Japan Society of Promotion of Science (JSPS): No. 15K00357.

REFERENCES

[1] **Pinedo, M.L. 2008.** *Scheduling: theory, algorithms, and systems*, 1rd Edition, *Potts C.n.van Wassenhove L.n.*

[2] **Gen, M., Cheng, R. & Lin, L., 2008.** *Network Models and Optimization: Multiobjective Genetic Algorithm Approach*, Springer, London.

[3] **Yu, X. & Gen, M., 2010.** *Introduction to Evolutionary Algorithms*, Springer, London.

[4] **Linn, R, W. Zhang. 1999.** Hybrid flow shop scheduling: a survey.*Computers & Industrial Engineering, 37*(1-2), pp. 57-61.

[5] **Ekşioğlu, B., S.D. Ekşioğlu, P. Jain. 2008.** A tabu search algorithm for the flowshop scheduling problem with changing neighborhoods. *Computers & Industrial Engineering, 54*(1), pp 1-11.

[6] **Grabowski, J., Wodecki, M. 2004.** A very fast tabu search algorithm for the permutation flow shop problem with makespan criterion.*Computers & Operations Research, 31*(11), pp. 1891-1909.

[7] **Grabowski, J., Skubalska, E. and Smutnicki, C., 1983.** On flow shop scheduling with release and due dates to minimize maximum lateness. *Journal of the Operational Research Society, 34*(7), pp.615-620.

[8] **Gu, W., D. Tang, K. Zheng. 2014.** Solving job-shop scheduling problem based on improved adaptive particle swarm optimization algorithm. *Transactions of Nanjing University of Aeronautics & Astronautics, 31*(5), pp. 559-567.

[9] **Nagar, A., J. Haddock, S. Heragu. 1995.** Multiple and bicriteria scheduling: a literature survey. *European Journal of Operational Research, 81*(1), pp 88-104.

[10] **Zhang, W., Gen, M. and Jo, J., 2014.** Hybrid sampling strategy-based multiobjective evolutionary algorithm for process planning and scheduling problem. *Journal of Intelligent Manufacturing*, 25(5), pp. 881-897.

[11] **Zhang, W., Lu, J., Zhang, H., 2016.** Fast multi-objective hybrid evolutionary algorithm for flow shop scheduling problem. *Journal of Computer Applications*, 36(4), pp 1015-1021.

[12] **Wei, X., Zhang, W., Weng, W., Fujimura, S., 2012.** Multi-objective local search combined with NSGA-II for bi-criteria permutation flow shop scheduling problem. *IEEJ Transactions on Electronics, Information & Systems, 132*(1), pp 32-41.

[13] **Taillard, E. 1993.** Benchmarks for basic scheduling problems. *European Journal of Operational Research, 64*(2), pp 278-285.

AN IMPROVED MULTIOBJECTIVE EVOLUTIONARY ALGORITHM COMBINING WITH DIFFERENTIAL EVOLUTION FOR PROCESS PLANNING AND SCHEDULING PROBLEM

Wenqiang Zhang[1*], Hao Li[1], Yu Wang[1] Chunxiao Wang[1] and Mitsuo Gen[2]
[1] College of Information Science and Eng., Henan University of Technology, China
zhangwq@haut.edu.cn
[2]Fuzzy Logic Systems Institute & Tokyo University of Science, Japan
gen@flsi.or.jp

ABSTRACT

Process planning and scheduling (PPS) problem is a most complicated and practical scheduling problem in intelligent manufacturing systems. PPS processes a set of prismatic parts into completed products effectively and economically by determining the optimal process plans and the most appropriate moment to execute each operation with competitive resources. Many research works use multiobjective evolutionary algorithm (MOEA) to solve such problems. This paper proposes an improved hybrid sampling strategy-based multiobjective evolutionary algorithm (HSS-MOEA) combining with differential evolution (HSS-MOEA-DE) to solve the PPS problem. HSS-MOEA-DE divides population and elitism into two parts: center and edge individuals closing to Pareto frontier by special selection strategies according to their objectives and special designed fitness function. Moreover, differential evolution tries to improve the convergence and distribution performances of individuals by converging to the different directions of Pareto frontier through sampling randomly from different parts, in which can be easily distinguished the relationships between dominated and dominating individuals. Numerical comparisons show that the HSS-MOEA-DE outperforms the traditional HSS-MOEA without DE in convergence and distribution performances.

Keywords: Process Planning and Scheduling, Multiobjective Evolutionary Algorithm, Differential evolution

1 INTRODUCTION

Process planning and scheduling (PPS) processes a set of prismatic parts into products effectively in a manufacturing system. Each operation of feature of a prismatic part includes the selection of machines, tools and tool access directions (TADs). The precedence relationship constraints must be satisfied with the geometrical and technological considerations. Process planning and scheduling (PPS) is the determination of the optimal process plan and schedule simultaneously within the precedence relationship constraints and manufacturing resources. It is very important but difficult to allocate the resources in the manufacturing system for PPS problem since seeking an optimal solution rapidly and effectively from all of the permutations, combinations of all of the tasks, manufacturing resources according to specified criteria

Many research studies have addressed the PPS problem by different approaches in the past decade. These approaches include simulated annealing (SA) algorithm, tabu search (TS) algorithm, agent-based approach, particle swarm optimization (PSO) algorithm and genetic algorithm (GA).

* Corresponding Author

Multiobjective evolutionary algorithms (MOEAs) have been recognized to be well-suited for solving multiobjective optimizations (MOP) [1][2]. Vector evaluated genetic algorithm (VEGA) divides the population into sub-populations, each of which evolves toward a single objective [5]. As two classical MOEAs, NSGA-II [3] and SPEA2 [4] have been proven to get better quality in solving multiobjective optimization problems. NSGA-II obtains better quality owing to its Pareto ranking and crowding distance mechanism. SPEA2 depends on raw fitness assignment mechanism and density mechanism.

Hybrid sampling strategy-based multiobjective evolutionary algorithm (HSS-MOEA) combines the sampling strategy of VEGA and the sampling strategy according to a new Pareto dominating and dominated relationship-based fitness function (PDDR-FF) [6]. The sampling strategy of VEGA prefers the edge area of the Pareto front and PDDR-FF-based sampling strategy has the tendency converging toward the central area of the Pareto front. The hybrid sampling strategies preserve both the convergence rate and the distribution performance.

Storn and Price proposed the Differential Evolution (DE) to solve real-parameter optimization problems in 1995 [7]. DE is a population-based approach, mutation and crossover are the variation operators used to generate new solutions. DE uses a simple mutation operator based on differences of individuals as main operator to find a search direction according to the distribution of solutions in the current population. The replacement mechanism of DE provides capabilities to maintain the population. The advantages in simplicity and effectiveness of DE has attracted many researchers interest that its use has been extended to single objective optimization and multiobjective optimization in the past years [8]-[16]. Especially, Iorio and Li propose three DE variants incorporating the directional information to improve the convergence and spread performance. For directional convergence, non-dominated sorting differential evolution with directional convergence (NSDE-DC) calculates the differential vectors that converges towards regions where better ranked individuals are located. For directional spread, non-dominated sorting differential evolution with directional spread (NSDE-DS) prefers along the Pareto frontier where individuals come from same rank. The non-dominated sorting differential evolution with directional convergence and spread (NSDE-DCS) combines both of the features of the NSDE-DC and NSDE-DS. Experiments results indicate that all variants outperform the NSGA-II. It is very interesting topic that combines the DE with ranking information to improve the efficacy in MOP.

This paper proposes hybrid sampling strategy-based multiobjective evolutionary algorithm combining with differential evolution (HSS-MOEA-DE) to solve the complicated PPS problem that combines the HSS-MOEA and DE. The HSS-MOEA could improve the convergence and distribution performance and the DE use to enhance the efficacy after applying the HSS-MOEA.

The paper is organized as follows: Section 2 describes the PPS problem description and formulates the mathematical model; Section 3 presents the detailed HSS-MOEA-DE approach; Section 4 gives a discussion and analysis of numerical experiments results; finally, the conclusion and future work are given in Section 5.

2 PROCESS PLANNING AND SCHEDULING PROBLEM

As an example of PPS problem in a manufacturing system, there are many parts to be processed by machines with a number of tools and different TADs. Each part has several operations, and each operation can be performed on more than one available machine with different processing times.

The PPS problem herein is to determine a proper process plan and schedule, which provides solution to decision maker how, when and in which sequence to allocate these operations of parts to suitable manufacturing resources effectively. When determining the process plan, the operation sequence should be decided firstly. Then, manufacturing resources such as machine, tools and tools access directions should be assigned to each operation in operation sequence. All the manufacturing resources are assumed as available in this phase. The determination of schedule is to determine the most appropriate moment to execute each operation with competitive resources. Precedence and resource constraints should be satisfied while deciding process plan and schedule. Moreover, the process plan and schedule should also satisfy the multiple objectives concurrently while maintaining the feasibility.

Two important and frequently-used objectives are minimizing makespan and variation of workload for each machine. The mathematical model of the problem is expressed in the following notations.

Indices

i,k : indices of part, $(i,k=1,2,\ldots,I)$.
j,h: indices of operation for part i, $(j,h=1,2,\ldots,J_i)$.
m : index of machine, $(m=1,2,\ldots,M)$;
l: index of tool, $(l=1,2,\ldots,L)$.
d : index of TAD, $(d=1,2,\ldots,D)$.

Parameters

I : number of parts.
J_i : number of operations for part i.
M : number of machines.
L: number of tools.
D : number of TADs.
O_i :set of operations for part i, O_i={o_{ij} | j=1,2,...,J_i}.
o_{ij} : the j-th operation of part i.
m_m : the m-th machine.
t_l : the l-th tool.
a_d : the d-th TAD.
M_{ij} : set of machines that can process o_{ij}.
A_m : set of operations that can be processed on machine m.
S : operation sequence. $S=(s_1,s_2,\ldots,s_w,\ldots,s_W), s_w=o_{ij}, W=\sum_{i=1}^{I} J_i$.
r_{ijh} : precedence constraints. if o_{ij} is predecessor of o_{ih}, $r_{ijh}=1$; otherwise, 0.
t_{mij}^{M} : machining time of o_{ij} by machine m.
t^{MCI} : machine change time index. It is the same for each machine change.
t_{mij}^{MC} : machine change time of o_{ij} by machine m. A machine change is described in article by Zhang et al. [6]. Once machine change is occurred, $t_{mij}^{\mathrm{MC}}=t^{\mathrm{MCI}}$; otherwise, $t_{mij}^{\mathrm{MC}}=0$.
t^{TCI} : tool change time index. It is the same for each tool change.
t_{mij}^{TC} : tool change time of o_{ij} by machine m. A tool change is described in article by Zhang et al. [6] If tool change is occurred, $t_{mij}^{\mathrm{TC}}=t^{\mathrm{TCI}}$; otherwise, $t_{mij}^{\mathrm{TC}}=0$.
t^{SCI} : set-up change time index. It is the same for each set-up change.

t^{SC}_{mij} : set-up change time of o_{ij} by machine m. A set-up change is described in article by Zhang et al. [6] If set-up change is occurred, $t^{SC}_{mij} = t^{SCI}$; otherwise, $t^{SC}_{mij} = 0$.

t^{PRE}_{mij} : preparation time of operation o_{ij} by machine m. The preparation time for an operation consists of machine change time, tool change time and set-up time for the operation.

$$t^{PRE}_{mij} = t^{MC}_{mij} + t^{TC}_{mij} + t^{SC}_{mij} \tag{1}$$

t^{P}_{mij} : processing time of operation o_{ij} by machine m. The processing time for an operation consists of the preparation time and the machining time for the operation.

$$t^{P}_{mij} = t^{PRE}_{mij} + t^{M}_{mij} \tag{2}$$

u_m : workload of machine m.

$$u_m = \sum_{i=1}^{I} \sum_{j=1}^{J_i} t^{M}_{mij} x^{M}_{mij} \tag{3}$$

$\bar{u}$: average workload of machine.

$$\bar{u} = \frac{1}{M} \sum_{m=1}^{M} u_m \tag{4}$$

t^{C}_{mij} : completion time of o_{ij} by machine m, it should satisfy the inequality $t^{C}_{m'i(j-1)} + t^{P}_{mij} \le t^{C}_{mij}$ that means for every operation, its direct predecessor's completion time plus its processing time might be smaller that its completion time.

Decision variables

$$x^{M}_{mij} = \begin{cases} 1, & \text{if } o_{ij} \text{ is performed by machine } m, \\ 0, & \text{otherwise} \end{cases} \qquad x^{T}_{lij} = \begin{cases} 1, & \text{if } o_{ij} \text{ is performed by tool } l, \\ 0, & \text{otherwise} \end{cases}$$

$$x^{D}_{dij} = \begin{cases} 1, & \text{if } o_{ij} \text{ is performed by TAD } d, \\ 0, & \text{otherwise} \end{cases} \qquad y_{ijkh} = \begin{cases} 1, & \text{if } o_{ij} \text{ is performed directly before } o_{kh}, \\ 0, & \text{otherwise} \end{cases}$$

$$\Omega(X,Y) = \begin{cases} 1, & \text{if } X \ne Y, \\ 0, & \text{otherwise} \end{cases}$$

The mathematical model can be formulated as the following bicriteria nonlinear mixed integer programming (NMIP) model:

$$\min t_M = \max_{m,i,j} \{t^{C}_{mij}\} \tag{5}$$

$$\min w_P = \sqrt{\frac{1}{M} \sum_{m=1}^{M} (u_m - \bar{u})^2} \tag{6}$$

$$\text{s.t.} \ (t^{C}_{mkh} - t^{P}_{mkh} - t^{C}_{mij}) x^{M}_{mij} x^{M}_{mkh} y_{ijkh} \ge 0, \ \forall (i,j),(k,h),m \tag{7}$$

$$r_{ijh} y_{ihij} = 0, \ \forall (i,j), h \tag{8}$$

$$y_{ijij} = 0, \ \forall (i,j) \tag{9}$$

$$\sum_{m=1}^{M} x^{M}_{mij} = 1, \ \forall (i,j) \tag{10}$$

$$x^{M}_{mij} = 0, \ \forall (i,j) \notin A_m, \forall m \tag{11}$$

$$y_{ijkh} \in \{0,1\}, \ \forall (i,j),(k,h) \tag{12}$$

$$x^{M}_{mij} \in \{0,1\}, \ \forall m,(i,j) \tag{13}$$

$$t_{mij}^{C} \geq 0, \ \forall m,(i,j) \tag{14}$$

Equation (5) illustrates the minimization of makespan t_M. Makespan t_M is the last operation's finishing time, i.e., the maximization of completion time among all the operations. Minimization of workload variance w_p is defined as Equation (6). Workload variance w_p is defined as standard deviation of workload of all the machines. The left equatoin (7) to (14) impose the precedence relationship constraints, resource constraints and nonnegative condition.

3 HYBRID SAMPLING STRATEGY-BASED MULTIOBJECTIVE EVOLUTIONARY ALGORITHM COMBINING WITH DIFFERENTIAL EVOLUTION

The evolving process of one generation of HSS-MOEA-DE is shown in Figure 1. *A*(*t*) represents the archive as an external population at generation *t* and *P*(*t*) represents the population at generation *t*. The solution procedure of one generation includes 5 phases.

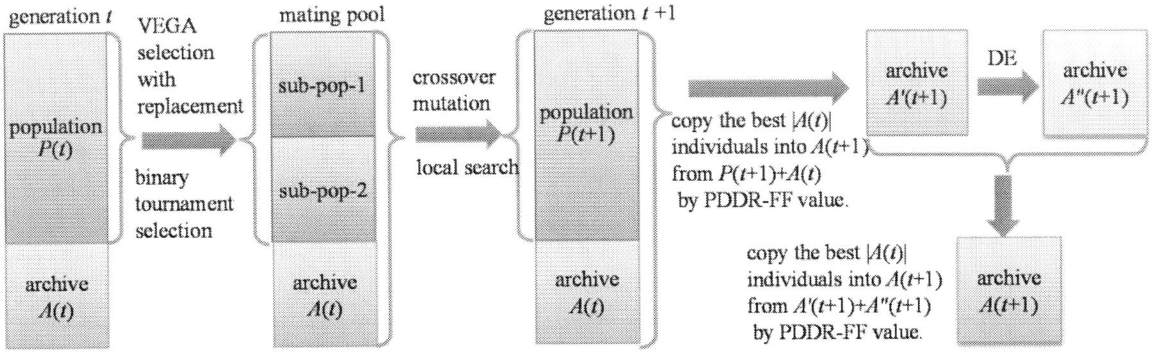

Figure 1: The evolving process in one generation

Phase 1: Selection (Sampling Strategy of VEGA)

In this phase, the sampling strategy of VEGA is used to select the better individuals into part of mating pool.

Phase 2: Generation of Mating Pool (Hybrid Sampling)

In this study, the sub populations and *A*(*t*) are combined to form a mating pool. In the mating pool, sub-pop-1 stores the good individuals for one objective, and sub-population 2 holds the good individuals for the other objective. The archive saves the individuals with good PDDR-FF values. Therefore, in the mating pool, one-third of the individuals serve one objective, one-third the other objective, and the left one-third both the two objectives.

Phase 3: Reproduction and Local Search

Some problem-dependent crossover and mutation operators are used to reproduce new chromosomes. Moreover, a local search mechanism is proposed to improve the quality of individuals after the reproduction process. The details of reproduction and local search is described in article by Zhang et al. [6].

Phase 4: Archive Maintenance (PDDR-FF based Elitist Sampling Strategy)

The individuals of *A*(*t*) and *P*(*t*) are combined to form a temporary archive *A'*(*t*). Thereafter, the PDDR-FF values of all individuals in *A'*(*t*) are calculated and sorted in an ascending order. In PDDR-FF, the smaller value is better. Moreover, the individual belongs to nondominated one, its fitness value will not exceed one. The fitness value of individual which is dominated by others will exceed one. Furthermore, even for the all nondominated individuals they have also different PDDR-FF values. The nondominated individuals will have smaller values (near to 0) than the edge points (near to 1).

After calculating the PDDR-FF value, the smallest $|A(t)|$ individuals in $A'(t)$ are copied to form $A'(t+1)$. This archive updating mechanism is similar to an elitist sampling strategy to keep the better individuals with better PDDR-FF values.

Phase 5: Archinve Enhancement by Combinating DE

After updating the archive to obtain the temporary archive $A'(t+1)$, the DE mutation and crossover operators are applied on it to create the temporary archive $A''(t+1)$. Then, the individuals in $A'(t+1)$ and $A''(t+1)$ are combined to form the last archive $A(t+1)$ according to the PDDR-FF values. Different from the traditional DE, the archive maintenance mechanism same as HSS-MOEA is applied to update the archive based on the PDDR-FF values. Moreover, the crossover operator is same as traditional DE and the crossover propobility is shared with HSS-MOEA in framework of HSS-MOEA-DE.

The DE/rand/1 is used to generate new mutants that the each individual s_i in $A'(t+1)$ and a randomly individual s_1 different from s_i are picked up to do mutation operator according to the following schemes ($eval(s_i)$ means the fitness function value of s_i).

Scheme 1 : the current individual s_i is nondominated solution that means the $eval(s_i)<= 1$.

If the $eval(s_1) <= 1$, the individual s_1 is also nondominated solution, then we compare the fitness function values of them to decide which rule will be used. If $eval(s_i)<eval(s_1)$, then $v_i = s_i+K(s_1-s_i)$, otherwise, $v_i = s_i+K(s_i-s_1)$. This mutation means the search direction starts from the nondominated solution and follows the direction from edge region to center region(or from center region to edge region) along the Pareto frontier to improve the distribution performance.

If the $eval(s_1) > 1$, the individual s_1 is dominated solution, then then $v_i = s_i+K(s_i-s_1)$. This mutation means the search direction starts from the nondominated solution and follows the direction from dominated region to nondominated region towards the Pareto frontier to improve the convergence performance.

Scheme 2 : the current individual s_i is dominated solution that means the $eval(s_i)> 1$.

If the $eval(s_1) > 1$, the individual s_1 is nondominated solution, then we also compare the fitness function values of them. If $eval(s_i)<eval(s_1)$, then $v_i = s_i+K(s_i-s_1)$, otherwise, $v_i = s_i+K(s_i-s_1)$. This mutation means the search direction starts from the current dominated solution and follows the direction from edge region to center region (or from center region to edge region) along the Pareto frontier to improve the distribution performance.

If the $eval(s_1) <= 1$, the individual s_1 is nondominated solution, then then $v_i = s_i+K(s_1-s_i)$. This mutation means the search direction starts from the dominated solution and follows the direction from dominated region to nondominated region to improve the convergence performance.

Even though the HSS-MOEA could improve the convergence and distribution performance, the combination of HSS-MOEA and DE could much improve the performance more.

4 EXPERIMENTS AND DISCUSSION

For experimental data, 4 parts (each part having 20, 16, 14 and 7 operations, respectively) problem are used. All the simulations are performed on Core i5 processor (2.6 GHz clock), and the program is written in C# language. The adopted parameters are listed as follows: population size =100; maximum generation=500; archive size=50; crossover probability=0.70; mutation probability=0.30 and the scale factor K=0.9. HSS-MOEA-DE and HSS-MOEA are run 30 times to compare the results.

Let S_j be a solution set for each method (j=1, 2). PF^* is a known reference Pareto set which comes from combining all of the obtained Pareto set by 2 methods with 30 runs. In this study, Coverage $C(S_1, S_2)$, Average distance $AD(S_j)$, Hypervolume $HV(S_j)$ and Spacing $SP(S_j)$ are considered to compare the performances of algorithms.

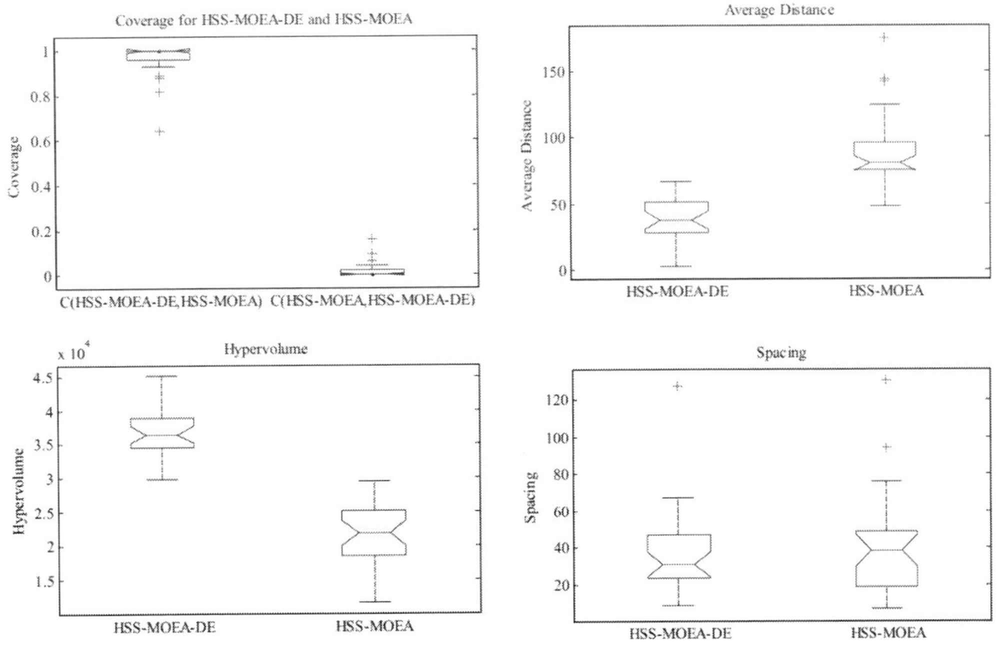

Figure 2: *C*, *AD*, *HV*, and *SP* by HSS-MOEA-DE, and HSS-MOEA

Figure 2 shows the numerical comparison of the box-and-whisker plots for *C*, *AD*, *HV*, *SP* and the CPU times by HSS-MOEA-DE, and HSS-MOEA. From Figure 2 it is easy to see that the HSS-MOEA-DE is better than HSS-MOEA in terms of *C* measure. The *AD* measure as shown in Figure 2 also indicates that HSS-MOEA-DE can get smaller value than HSS-MOEA. For *HV* measure as shown in Figure 2, HSS-MOEA-DE is better than HSS-MOEA. The distribution performances, *SP*, as shown in Figure 2 indicates that HSS-MOEA-DE is better than HSS-MOEA. The PDDR-FF could give different values among nondominated and dominated solution in elitism, and the combination of DE for individuals in elitism could significantly improve the convergence and distribution performances.

5 CONCLUSIONS

In this study, a HSS-MOEA-DE approach was proposed to solve the multiobjective PPS problem while considering minimization of makespan and minimization of variation of machine workload. The sampling strategy of VEGA has a preference for the edge region of the Pareto front and the PDDR-FF-based sampling strategy has the tendency converging toward the center area of the Pareto front. Moreover, the DE operators applied for the elitism has the capabilities of converging toward the true Pareto frontier much more and even distributing individuals along the Pareto frontier further. Numerical comparisons indicated that HSS-MOEA-DE outperform the HSS-MOEA on convergence and distribution performance.

ACKNOWLEDGEMENT

This research work is supported by the National Natural Science Foundation of China (U1304609), Foundation for Science & Technology Research Project of Henan Province (162102210044,

152102110076), Program for Innovative Research Team(in Science and Technology) in University of Henan Province (17IRTSTHN011), Fundamental Research Funds for the Henan Provincial Colleges and Universities(2014YWQQ12, 2015XTCX03, 2015XTCX04), Research Funds for Key Laboratory of Grain Information Processing and Control (Henan University of Technology)(KFJJ-2015-106), Ministry of Education and the Grant-in-Aid for Scientific Research (C) of Japan Society of Promotion of Science (JSPS): No. 15K00357.

REFERENCES

[1] **Gen, M., Cheng, R. & Lin, L., 2008.** *Network Models and Optimization: Multiobjective Genetic Algorithm Approach*, Springer, London.

[2] **Yu, X. & Gen, M., 2010.** *Introduction to Evolutionary Algorithms*, Springer, London.

[3] **Deb, K., Pratap, A., Agarwal, S., & Meyarivan, T., 2002.** A fast and elitist multiobjective genetic algorithm: NSGA-II, *IEEE Transactions on Evolutionary Computation*, 6(2), pp. 182-197.

[4] **Zitzler, E., Laumanns, M. and Thiele, L., 2001.** SPEA2: Improving the strength Pareto evolutionary algorithm. In Proceedings of *Eurogen*, vol. 3242, no. 103, pp. 95-100.

[5] **Schaffer, J., 1985.** Multiple objective optimization with vector evaluated genetic algorithms, *Proceedings of the 1st International Conference on Genetic Algorithms*, pp. 93-100, L. Erlbaum Associates Inc. Hillsdale, NJ, USA.

[6] **Zhang, W., Gen, M. and Jo, J., 2014.** Hybrid sampling strategy-based multiobjective evolutionary algorithm for process planning and scheduling problem. *Journal of Intelligent Manufacturing*, 25(5), pp. 881-897.

[7] **Storn, R. and Price, K., 1997.** Differential evolution-a simple and efficient heuristic for global optimization over continuous spaces. *Journal of global optimization*, 11(4), pp. 341-359.

[8] **Babu, B.V. and Jehan, M.M.L., 2003.** Differential evolution for multi-objective optimization. In Proceedings of IEEE *Congress on Evolutionary Computation* (CEC 2003), vol. 4, pp. 2696-2703.

[9] **Li, H. and Zhang, Q., 2006.** A multiobjective differential evolution based on decomposition for multiobjective optimization with variable linkages. In Proceedings of Parallel Problem Solving from Nature (PPSN IX), pp. 583-592, Springer Berlin Heidelberg.

[10] **Abbass, H.A. and Sarker, R., 2002.** The Pareto differential evolution algorithm. *International Journal on Artificial Intelligence Tools*, 11(04), pp. 531-552.

[11] **Santana-Quintero, L.V. and Coello, C.A.C., 2005.** An algorithm based on differential evolution for multi-objective problems. *International Journal of Computational Intelligence Research*, 1(1), pp. 151-169.

[12] **Ali, M., Siarry, P. & Pant, M., 2012.** An efficient Differential Evolution based algorithm for solving multi-objective optimization problems. *European Journal of Operational Research*, 217(2), pp. 404-416.

[13] **Guo, W. & Yu, X., 2014.** Non-dominated sorting differential evolution with improved directional convergence and spread for multiobjective optimization. In *Proceedings of the 2014 Annual Conference on Genetic and Evolutionary Computation* (GECCO 2014), ACM, pp. 87-88.

[14] **Iorio, A.W. & Li, X., 2006.** Incorporating Directional Information within a Differential Evolution Algorithm for Multi-objective Optimization. In *Proceedings of the 8th Annual Conference on Genetic and Evolutionary Computation* (GECCO 2006). New York, ACM, pp. 691-698.

[15] **Qu, B.Y., Suganthan, P.N. & Liang, J.J., 2012.** Differential Evolution with Neighborhood Mutation for Multimodal Optimization. *IEEE Transactions on Evolutionary Computation*, 16(5), pp. 601-614.

[16] **Sindhya, K. et al., 2011.** A new hybrid mutation operator for multiobjective optimization with differential evolution. *Soft Computing*, 15(10), pp. 2041-2055.

REUSE OF TERTIARY PACKAGING IN PRODUCTION DISTRIBUTION NETWORK

S.H. Chung[1*], H.L. Ma[2], and H.K. Chan[3]
[1]Department of Industrial and Systems Engineering
The Hong Kong Polytechnic University, Hong Kong, China
mfnick@polyu.edu.hk

[2]Department of Industrial and Systems Engineering
The Hong Kong Polytechnic University, Hong Kong, China
hllamma@polyu.edu.hk

[3]Nottingham University Business School China
University of Nottingham Ningbo China, China
Hingkai.Chan@nottingham.edu.cn

ABSTRACT

Tertiary packaging is necessary for transportation in production distribution network because of the benefits in enhancing the logistic efficiency. However, in the meantime, it produces a lot of packaging waste every day. In fact, some tertiary packaging after transportation may still be in good condition and can be collected back for reuse. However, this has not been widely studied in the existing literature. Accordingly, this paper proposed an optimization methodology to maximize the collection of used tertiary packaging for reuse, meanwhile minimizing the total operating cost by taking the advantages of simultaneous optimization of a multi-day planning.

Keywords: Reuse, Tertiary Packaging, Production Distribution, Genetic Algorithm

1 INTRODUCTION

Recycling of packaging waste can directly reduce the consumption of raw materials. Meanwhile it can minimize the demand on landfill and land pollutions. Earlier before the issue of the Directive 94/62/EC on Packaging and Packaging Waste (PPW) in 1994, many research studies have already working on the minimization of packaging usage and more importantly, is the maximization of recycling of packaging. By that time, the focus was mainly on taxation approach [1]. Nowadays, the research focus is shifting from the legislation point of view onto the economic and financial benefits that can be obtained [2].

In general, a packaging system consists of three main parts knowns as primary, secondary, and tertiary packaging [3]. Primary packaging is regarded as the first envelop to protect directly the product. Secondary packaging is used to protect the primary packaging. Lastly, tertiary packaging is used for bulk handling in warehousing, and transportation. Tertiary packaging will affect the logistic efficiency in supply chains and induce different requirements on the handling equipment, vehicles, etc. [4]. In a typical production and distribution network, products will usually be sent from the source points with large batch size into different transit points, known as Distribution Centers (DC),

* S.H. Chung

for further dispatching in smaller sizes to the demand points. Accordingly, products will usually be packed with different tertiary packaging, such as plastic films, polystyrene foam, carton box, net, rope, etc. in order to reduce damages and tight different products together for the ease of handling.

This paper is divided into the following sections. Section 2.0 gives a literature review of the recent works in the field. Section 3.0 describes the problem to be studied. Section 4.0 presents the proposed optimization approach by using Genetic Algorithm. Section 5.0 discusses the resulting and findings, and lastly will be concluded by a conclusion.

2 LITERATURE REVIEW

In general, optimization of the recycling and reuse of items in production and distribution network will usually involve the simultaneous planning of both the forward and reverse flow of the items. In which, many of them focus on designing an optimal transportation route, facility location and allocation, etc. [5].

Practical applications of production and distribution network in recycling and reuse of products can be easily found in many literatures. For example, Krikke et al. [6] studied the recyclability and the design of the logistics network structure for electronic home appliances products (refrigerators). Jayaraman [7] studied a remanufacturing model of mobile phone. They considered the recycling and remanufacturing of several core components, and proposed a Remanufacturing Aggregate Production Planning approach to minimize various costs including inventory, dissembles, and remanufacturing. Olugu and Wong [8] proposed a fuzzy rule based system to measure the green supply chain management performance of automotive industry.

Other than recycling and reuse of products or components, there are many papers specially focusing on the recycling of plastic related items. For example, Kartalis et al. [9] studied the recycling of post-used polyethylene packaging film and their possibility of being reuse. Gomes et al. [10] also studied the recycling of plastic waste initiated by the huge consumption of plastic in Brazil with an estimation of 1150 thousand tons per year. Lee and Lee [11] studied to determine the optimal delivery route for the reuse of plastic bottle for distilling water. They proposed a multi-criteria decision support system to support recycling decisions. However, there are not many papers studying in the recycling and reuse of territory packaging.

3 MODEL DESCRIPTION

Set

S	set of source points
D	set of demand points
Ø	set of DCs
I	set of Items
φ	set of Days

Index

s	no. of source points, s = 1, 2, ..., S
i	no. of Item types, i = 1, 2, ..., I
j	no. of demand points, j = 1, 2, ..., D
k	no. of DCs, k = 1, 2, ..., Ø
d	no of days, d = 1, 2, ..., φ

Parameter

q_{ijd} demand quantity of item *i* at demand point *j* on day *d*.

c_k maximum handling capacity of DC *k*.

p_s maximum production capacity of source point *s*.

r_{ik} recycle rate of item type *i* at DC *k*.

$c_{sk}^{S\emptyset}$ travelling cost between source point *s* to DC *k*.

c_{ik}^{H} handling cost of item type *i* at DC *k*.

c_s^{I} storage unit cost of collected packaging at source point *s*.

$c_{kj}^{\emptyset D}$ travelling cost between DC *k* to demand point *j*.

Variable

σ_{isd} quantity of collected packaging of item *i* to be sent back to source point *s* on day *d*.

y_{isd} quantity of unused packaging for item *i* to be stored at source point *s* on day *d*.

Decision Variable

x_{iskjd} =1, if the item type *i* is supplied by source point *s* transited through DC *k* to demand point *j* on day *d*, otherwise = 0.

The distribution network studied is shown as in Figure 1. It consists of source points (**S**), distribution centers (**∅**), and demand points (**D**). The demand of each item type *i* at each demand point *j* is supplied by only one source point through only one distribution center. The demand quantity (q_{ijd}) for different types of item *i* at each demand point *j* is different along the planning horizon *d*. For the reason of batch delivery, items will usually be further packaged by using various simply packaging, such as carton box, plastic wrap, foam box, etc. for the ease of transport and protection in practice. Each source point *s* and each DC *k* has its maximum production (p_s) and handling capacity (c_k). Moreover, the skills of labor in each DC *k* on handling different types of items are varied. Therefore, the handling cost (c_{ik}^{H}) and the rate of collecting the packaging back for reuse (r_{ik}) is varied. In this model, the collected packaging for the items will be sent back to the source point during the return of the trip. Therefore, no extra transportation will be considered. However, if the collected packaging will not be reused on the next day, storage cost (c_s^{I}) will be induced. In addition, demand splitting is not being considered.

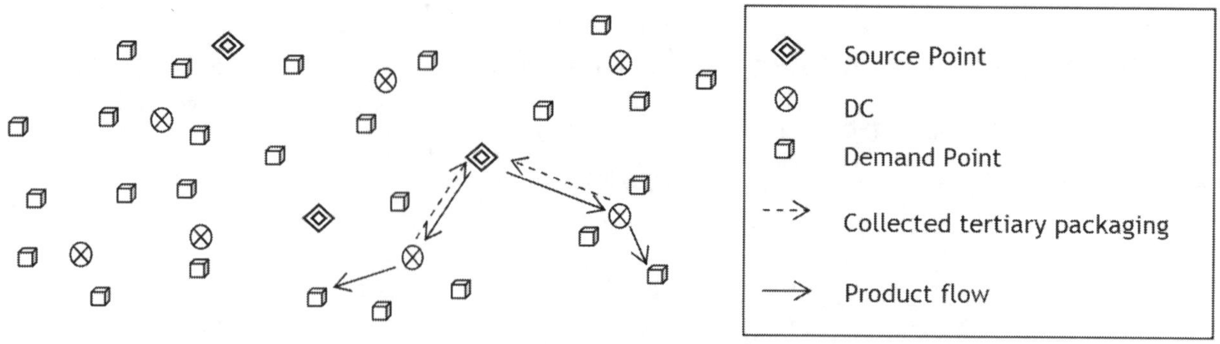

Figure 1: Outline of the distribution network model

The objective of the system is to minimize the total operating cost and maximize the total quantity of packaging being collected at DCs for reuse as shown in Equation (1). Among which, Z_1 is the total operating cost consisting of the total traveling cost from source points to DCs, total handling cost at DCs, total traveling cost from DCs to demand points, and total storage cost for collected packaging as shown in Equation (2). Z_2 is total quantity of packaging being collected at DCs for reuse as shown in Equation (3). Z_1' and Z_2' are used to normalize the total operating cost and total quantity of reused items. They can be the best known (or optimal) solution obtained by the optimization of only Z_1 and Z_2 as objective function respectively.

$$\textbf{Min } Z = \alpha(Z_1/Z_1') + (1-\alpha)(Z_2'/Z_2) \tag{1}$$

$$Z_1 = \sum_{i\in I}\sum_{s\in S}\sum_{k\in\emptyset}\sum_{j\in J}\sum_{d\in\varphi} x_{iskjd}\left(c_{sk}^{S\emptyset} + c_{ik}^{H} + c_{kj}^{\emptyset D}\right) + \sum_{i\in I}\sum_{s\in S}\sum_{d\in\varphi} y_{isd}\left(c_s^I\right) \tag{2}$$

$$Z_2 = \sum_{i\in I}\sum_{s\in S}\sum_{d\in\varphi} \sigma_{isd} \tag{3}$$

, where $y_{isd} = \max\{[\sum_{k\in\emptyset}\sum_{j\in J}(x_{iskjd}q_{ijd}) - \sigma_{is(d-1)} - y_{is(d-1)}],0\}$

, where $\sigma_{isd} = \sum_{k\in\emptyset}\sum_{j\in J} x_{iskjd}q_{ijd}r_{ik}$.

s.t.

Network flow constraints:

$$\sum_{s\in S}\sum_{k\in\emptyset}\sum_{j\in J} x_{iskjd} = 1 \qquad \forall i\in I,\ \forall d\in\varphi \tag{4}$$

Constraints (4) ensure each demand will be supplied by exactly one source point and transit through only one single DC.

Capacity constraints:

$$\sum_{i\in I}\sum_{k\in \emptyset}\sum_{j\in J}(x_{iskjd}q_{ijd}) - p_s \le 0 \qquad \forall s \in S,\ \forall d \in \varphi \qquad (5)$$

$$\sum_{i\in I}\sum_{s\in S}\sum_{j\in J}(x_{iskjd}q_{ijd}) - c_k \le 0 \qquad \forall k \in \emptyset,\ \forall d \in \varphi \qquad (6)$$

Constraints (5) ensure the items supplied by source point *s* will not exceed the maximum production capacity of the source point. Constraints (6) ensure the items passing through the DC will not over the handling capacity of the DC.

4 OPTIMIZATION METHOD: GENETIC ALGORITHM

The decision in this problem is to determine the flow of the supply of items to demand points by passing through which DC. In addition, this paper aims to maximize the collection of packaging for reuse meanwhile minimizing the storage cost of the reused packaging induced due to not being used. As the demands of items at demand points along the planning days are different. Therefore, this characteristic is modeled in the encoding of the chromosome. Accordingly, a new encoding of chromosome is designed as follows.

Encoding and Decoding of Chromosome

The chromosome consists of $D \times J \times I$ number of genes as shown in Figure 1. Each gene consists of two values, representing i) the source point with value (1 to S), and ii) the DC (1 to $\emptyset$). Accordingly, the first gene with value of (4,2) in the chromosome value row represents that the Item 1 at Demand Point 1 will be supplied by Source Point 4 via DC 2. If there is no demand for a particular item at the demand point, a "-" will be given, such as Item 1 at Demand Point 2 on Day 1.

Crossover Operation

Two types of crossover mechanism are applied. The first type is traditional uniform crossover, in which a predefined number of chromosomes are randomly selected for crossover from the same Day *d*, which is also randomly selected. The reason of not selecting genes from different days is to avoid too random search. Figure 3 shows an example of crossover mechanism, in which two values are being selected from Parents A and B at the same position for crossover. After crossover, validation will be carried out to avoid invalid chromosome, which will be explained in later part.

Chromosome *n*

Day	1												
Demand Points	1				2				...	J			
Item	1	2	...	I	1	2	...	I	...	1	2	...	I
Chromosome value	4,2	3,1	...	3,3	-	-	...	2,3	...	4,4	1,2	...	-
...	...												
...	...												
...	...												
...	...												
Day	D												
Demand Points	1				2				...	J			
Item	1	2	...	I	1	2	...	I	...	1	2	...	I
Chromosome value	2,2	-	...	3,4	5,2	2,1	...	1,3	...	3,5	4,2	...	-

Figure 2: A sample of chromosome encoding

Mutation Operation

Two types of mutation will be applied. The first type is traditional uniform mutation operation. Similarly reason as in tradition uniform crossover, all the selected genes for mutation will be selected among the same day. Accordingly, a day will be randomly selected first. Then a predefined number of genes will be selected according to the mutation rate. The purpose is to prevent solution prematurity and increase solution diversity. In here, any selected gene will undergo mutation either in the value of source point or DC. For example as in Figure 5, the gene with value (3,3) and (2,3) being selected are undergo mutation into (4,3) and (2,2) in the value of source point and DC respectively.

5 NUMERICAL EXPERIMENT

This experiment aims to study the changes in terms of cost and the corresponding number of packaging(s) collected by reduction of the α value by using the same multi-day planning approach. The results are summarized in Table 1, in which, one can see that the total costs generally increased along with the reduction of the α value in small, medium, and large scale problems. This can be expected because the cost weighting is being reduced. Accordingly, on the other hand, the number of return packaging(s) collected is then increased. This can be seen in the all problem scales.

Increased Cost for each α

$$= \frac{\text{Total cost obtained by using } \alpha \ - \text{Total cost obtained by using } \alpha{=}1}{\text{Total cost obtained by using } \alpha{=}1}$$

Increased Returns for each α

$$= \frac{\text{Total packaging collected by using } \alpha \ - \text{Total packaging collected by using } \alpha{=}1}{\text{Total packaging collected by using } \alpha{=}1}$$

Table 1: Summary of the results for adjusting cost and reuse weighting

	Small		Medium		Large	
α	Increased Cost	Increased Returns	Increased Cost	Increased Returns	Increased Cost	Increased Returns
1	0.0	0.0	0.0	0.0	0.0	0.0
0.9	0.7	1.1	1.3	2.2	1.6	3.1
0.8	0.8	1.3	2.1	3.1	2.6	4.7
0.7	0.8	1.3	2.7	3.6	3.4	6.3
0.6	1.1	2.3	3.1	4.9	4.6	8.2
0.5	1.4	2.5	3.4	5.2	5.8	8.6
0.4	1.5	2.8	3.9	6.1	6.5	9.7
0.3	1.8	2.9	4.3	6.8	7.1	10.5
0.2	2.1	3.0	4.4	7.0	7.2	10.6
0.1	2.2	3.1	4.6	7.1	7.4	10.8

6 CONCLUSION

Although many papers have been studied on recycling and reuse problem, recycling and reuse of tertiary packaging is not being studied thoroughly. In logistic operation, tertiary packaging helps in bulk transportation. However, this packaging will induce packaging waste and increase the demand on landfill and produce land pollution. By taking into the consideration of the variety in handling ability and recyclability of DCs, this paper proposed a multi-day planning approach to determine the product flow for every day so that the number of packaging can be collected for reuse can be maximize meanwhile the total operating cost can be minimized. The results also demonstrate that by the consideration of multi-day planning, the total operating cost can be reduced.

7 REFERENCES

[1] **Pearce, D., Turner, R.K.** 1992. Packaging Waste and the polluter pays principle: a taxation solution. Journal of Environmental Planning and Management, 35(1), pp 5-15.

[2] **Cruz, N.F., Simões, P., Marques, R.C.** 2012. Economic cost recovery in the recycling of packaging waste: the case of Portugal. Journal of Cleaner Production, 37, pp 8-18.

[3] **Palsson, H., Hellström, D.** 2016. Packaging logistics in supply chain practice - current state, trade-offs and improvement potential. International Journal of Logistics: Research and Applications http://dx.doi.org/10.1080/13675567.2015.1115472.

[4] **Jahre, M., C.J. Hatteland.** 2004. Packages and Physical Distribution: Implications for Integration and Standardisation. International Journal of Physical Distribution and Logistics Management, 34(2), pp 123-139.

[5] **Ozceylan, E., Paksoy, T.** 2013. A mixed integer programming model for a closed-loop supply-chain network. International Journal of Production Research, 51(3), pp 718-734.

[6] **Krikke, H., Bloemhof-Ruwaard, J., Van Wassenhove, L. N.** 2003. Concurrent product and closed-loop supply chain design with an application to refrigerators. International Journal of Production Research, 41(16), pp 3689-3719.

[7] **Jayaraman, V.** 2006. Production planning for closed-loop supply chains with product recovery and reuse: an analytical approach. International Journal of Production Research, 44(5), pp 981-998.

[8] **Olugu, E. U., Wong, K. Y.** 2012. An expert fuzzy rule-based system for closed-loop supply chain performance assessment in the automotive industry. Expert Systems with Applications, 39(1), pp 375-384.

[9] **Kartalis, C.N., Papaspyrides, C.D., Pfaendner, R.** 2000. Recycling of post-used PE packaging film using the restabilization technique. Polymer Degradation and Stability, 70, pp 189-197.

[10] **Gomes, C. F. S., Nunes, K. R. A.** Helena Xavier, L., Cardoso, R., Valle, R., 2008. Multicriteria decision making applied to waste recycling in Brazil. Omega, 36(3), pp 395-404.

[11] **Lee, J. E., Lee, K. D.** 2012. Integrated Forward and Reverse Logistics Model: A Case Study in Distilling and Sale Company in Korea. International Journal of Innovative Computing Information and Control, *8*(7A), pp 4483-4495.

OPTIMIZED TOOL PATH PLANNING FOR FIVE-AXIS FLANK MILLING OF RULED SURFACE USING GEOMETRIC DECOMPOSITION STRATEGY AND MULTI-POPULATION HARMONY SEARCH ALGORITHM

Jin Yi[1], Chih-Hsing Chu[2*], Xinyu Li[1] and Liang Gao[1]
[1] The State Key Laboratory of Digital Manufacturing Equipment and Technology,
School of Mechanical Science and Technology,
Huazhong University of Science and Technology, Wuhan 430074, PR China
lixinyu@mail.hust.edu.cn

[2] Department of Industrial Engineering and Engineering Management
National Tsing-Hua University, Hsinchu,Taiwan
chchu@ie.nthu.edu.tw

ABSTRACT

The optimization of tool path planning plays an important role in 5-axis flank milling. One of the aims is to reduce the machining error, which is a challenging task because it involves large scale of decision variables. This paper proposes a novel approach for generating optimized tool path in 5-axis flank milling of ruled surfaces based on geometric decomposition strategy and multi-population harmony search algorithm. In the proposed method, the ruled surfaces are geometrically divided into several segments first. And then, the tool paths on these sub-surfaces are independently optimized by the multi-population harmony search algorithm. At last, the tool paths are combined together to form the complete tool path for the machined surface. The test results of several representative surfaces show that the proposed approach obtains higher machining accuracy within much less computational time than the previous methods in comparison.

Keywords: Geometric decomposition, Harmony search algorithm, Tool path planning, Flank milling

1 INTRODUCTION

Nowadays, 5-axis CNC machining is widely used in the modern manufacturing industry. It provides better shaping capability for complex mechanical parts with free form surfaces, such as impeller of aircraft engine, screw propeller for ship, impeller of steam turbine. There are two kinds of operations in 5-axis CNC machining, end milling and flank milling, while this paper will focus on the flank milling. The flank milling operation is more efficient than end milling for ruled surface machining in terms of material removal rate and machining efficiency. The flank milling operation can also provide higher surface quality by generating an appropriate tool path. In practice, the machined surface quality is considered acceptable as long as the amount of the machining error can be limited within a given tolerance [1]. The machining error in 5-axis flank milling can be reduced by adjusting all cutter locations of tool path simultaneously, but it's not an easy task. Many different approaches have been proposed in previous studies to achieve this goal, such as ant colony systems (ACS) [2], particle swarm

* Corresponding Author

optimization (PSO) [3], Electromagnetism-like (EM) algorithm [4]. However, the solution qualities are not satisfied and these methods also involve high computation cost.

The harmony search algorithm (HS) is one of the recent meta-heuristics that was first proposed by Geem [6]. HS imitates the music improvisation process, in which musicians continuously adjust the pitch of their instruments to achieve better harmony. The optimization process is similar to the music improvisation process, each decision variable continuously changes its value during the search process to converge to the global best. HS has the characteristics of few mathematical requirements, easy implementation, and fast convergence speed. Hence, HS has caught researchers' attention and has been applied in many areas, such as renewable energy systems [7], experiment designs [8], wireless sensor networks [9] and robotics [10].

To improve the quality of the final solutions and decrease the computation cost as well, this paper proposes a novel approach based on geometric decomposition strategy and the multi-population harmony search algorithm, which is abbreviated as GD-MPHS. The core idea of GD-MPHS is inspired from the "divide and conquer" strategy. The GD-MPHS consists of three steps. In the first step, the machined surface is geometrically divided into several segments. And in the second step, a multi-population harmony search algorithm is employed to obtain the optimal CLs independently for each segment. In the last step, the CLs for each segment are combined together to form the final tool path for the whole machined surface. The performance of GD-MPHS is tested on several representative surfaces and compared with previous methods, the results show that the GD-MPHS not only can achieve higher machining accuracy but also save much computational time.

2 TOOL PATH PLANNING IN 5-AXIS FLANK MILLING

A tool path in the flank milling is defined by a set of CLs, and the cutter follows a linearly interpolation tool motion between the consecutive CLs. The machining error in 5-axis flank milling can be reduced by adjusting all cutter locations to appropriate positions simultaneously.

Figure 1 illustrates the determination of a cutter location. Points p_A and p_B are the two contact points between the cutter and the surface boundary curves. They can be determined by parameters u_A and u_B, respectively. And offsetting points p_A and p_B along the surface normal direction with a distance of the cutter radius produces two cutter center points. In this study, the cutter center is also allowed to deviate from the surface along the normal, tangent and the bi-normal directions at points p_A and p_B. So the cutter location can be expressed as:

$$c_A = p_A(u_A) + N_A n_A + t_A t_A + B_A b_A \qquad \textit{(1a)}$$

$$c_A = p_B(u_B) + N_B n_B + t_B t_B + B_B b_B \qquad \textit{(1b)}$$

where $c_A$and c_B are the cutter center points; p_A and p_B are the two contact points corresponding to the curve parameters u_A and u_B, respectively; n_A, t_A and b_A are the unit vector at p_A from normal, tangent and bi-normal directions, respectively; N_A, T_A and B_A are the magnitudes of the above three unit vectors; n_B, t_B and b_B are the unit vector at p_B from normal, tangent and bi-normal directions, respectively; N_B, T_B and B_B are the magnitudes of the above three unit vectors.

The objective function is to minimize the accumulated machining error on the machined surface. However, an exact estimation of the geometrical machining deviations is actually very difficult and might be unnecessary. Here, an approximate approach called the stock height method is adopt in this study [5], and the problem is formulated as:

$$Minimize : \sum_{i=0}^{N-1} \delta^*(CL_i \overset{interpolation}{\longrightarrow} CL_{i+1}) \qquad \textit{(2a)}$$

$$subject\ to: \begin{cases} 0 \le u_{A_i} \le 1 \\ 0 \le u_{B_i} \le 1 \\ u_{A_i} \le u_{A_{i+1}} \\ u_{B_i} \le u_{B_{i+1}} \\ lb \le N_{A_i}, N_{B_i}, T_{A_i}, T_{B_i}, B_{A_i}, B_{B_i}, \le ub \end{cases} \quad \textbf{(2b)}$$

where δ^* is the approximation of machining error induced by the cutter motion M from CL_i to CL_{i+1}. lb and ub are the lower and upper boundary for N_A, N_B, T_A, T_B, B_A and, B_B. A complete tool path consists of N CLs. An optimal tool path with minimized machining error can be obtained by adjusting a set of parameters (u_A, u_B, N_A, N_B, T_A, T_B, B_A, B_B) for each cutter location simultaneously. So the total number of decision variables of this problem is $8N$, which means the complexity of this problem depends on the number of CLs. Normally, a larger number of CLs will generate smaller machining error, but the search space is also become larger. However, the selection of an appropriate number of CLs is not the topic of this study, so it is kept as a constant value.

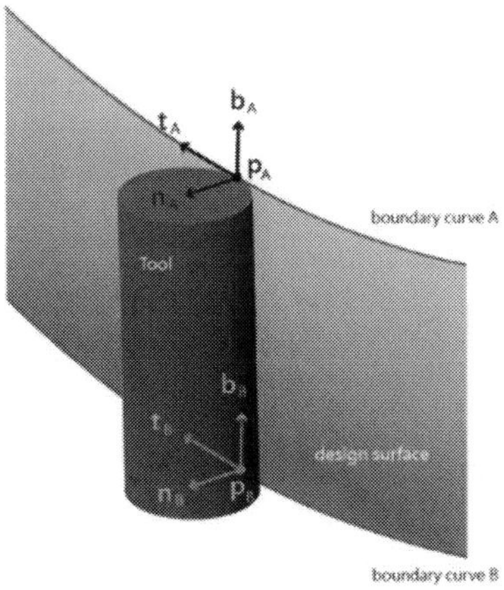

Figure 1: Determine a cutter location respect to the ruled surface [4]

3 PROPOSED APPROACH

3.1 Geometric decomposition strategy

As mentioned above, the tool path planning problem has very large scale of decision variables. For example, the number of CLs N in this study is set to 40, so the total number of decision variables reaches 320. Searching for optimal solutions in such a high-dimensional solution space is actually a very difficult task.

One of the efficient solutions to solve large scale problems is to break down the large scale problems into many sub-problems based on the "divide and conquer" strategy. In this study, a geometric decomposition strategy is proposed. The main idea of the geometric decomposition strategy is to divide the machined surface into several segments and the tool path of each segment is adjusted independently. In this study, the surface is evenly divided by curve parameters u_A and u_B. The proposed geometric decomposition strategy is also based on the founding that the interaction only appears between two consecutive CLs. As mentioned in section 2, cutter motion between consecutive cutter locations is approximated by linearly interpolations. So the machined error between consecutive CLs is only determined by the two CLs, but not by other CLs. Thus, tool paths distributed on different segments can be optimized independently. It should be noted that the interaction between two consecutive CLs which distributed on different segments should be also taken into consideration. For example, there are two consecutive segments k and k+1, the end cutter location on segment k is CL_i, and the start cutter location on segment k+1 is CL_{i+1}. So CL_{i+1} should be also taken into account when calculating the machined error on segment k. When the machined surface is divided into several segments, the next work is to allocate all the CLs to different segments. Here, the number of CLs on each segment is kept as uniform as possible.

3.2 A multi-population HS optimization framework

Since the machined surface is divided into several segments, a multi-population HS optimization framework is proposed to adjust the CLs on all segments simultaneously. There will be NS HS populations, if the machined surface is divided into NS segments, and each HS population is responsible for only one segment. In our previous study, a modified variant of HS was proposed [12]. The modified HS variant outperformed other HS variants in comparison. So it is adopted as the basic optimizer in this study. Besides, a simple local search method is proposed in this study. The local search method is applied on the current best harmony in the HM at every iteration, which helps enhance the exploitation capability of the modified HS algorithm. The proposed local search method also adopts a self-adaption mechanism to adjust the step length. If a random search successfully updated the current best harmony, the step length of the next random search will be shrink. However, if a random search failed in updating the current best harmony, the step length of the next random search will be enlarged. The pseudo code of the local search is showed in Figure 2.

Algorithm 1 Pseudo code of the local search

```
Begin
Set the max iteration number of local search, Iter
Set the step length, S
x_temp = x_best
for i = 1 : Iter do
   k = randi(D) /*Randomly choose a dimension*/
   x^k_temp = x^k_best + rand(-1,1) * S
   if fitness(x_temp) ≤ fitness(x_best) then
      x^k_best = x^k_temp
      fitness(x_best) = fitness(x_temp)
      S = S * 0.9
   else
      S = S/0.7
   end if
end for
End
```

Figure 2: Pseudo code of the local search

To show the proposed GD-MPHS approach from a comprehensive perspective, the flowchart of the whole procedure is summarized in Figure 3. As mentioned above, the interaction between two consecutive CLs which distributed on different segments should be also taken into consideration when calculating the machining error. Here, we introduced a greedy method to complete this work. For example, as is shown in Figure 3, the first cutter location CL_{i+1} of the current best CLs in the second segment is used when estimating the machined error in the first segment, while the last cutter location CL_i of the current best CLs in the first segment is used when estimating the machined error in the second segment.

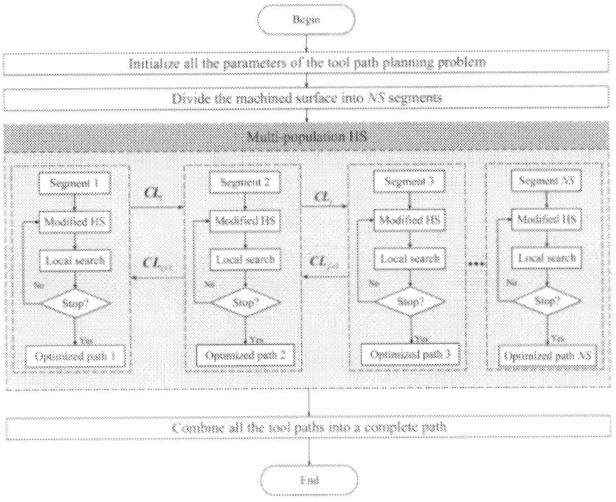

Figure 3: The overall flowchart of the proposed GD-MPHS approach

4 COMPUTATION RESULTS AND ANALYSIS

4.1 Experiment setup

To validate the effectiveness of the proposed GD-MPHS approach, a set of representative ruled surfaces constructed in previous study are tested [3]. These surfaces vary on properties of curvature (C), twist (T) and length difference (L), and alternatively change each property from a lower to an upper bound. The cutting parameters are set as the same as previous study [6]. The values of these parameters are set as the same as the settings in previous study [4] in order to perform a fair comparison with the results. The setting of parameters in the proposed GD-MPHS approach areshown in Table 1. The parameters related to HS algorithm are set as recommended in previous study [12]. Our GD-MPHS approach is programmed in C/C++ and compiled using Microsoft VS2013 on a PC with 3.40 GHz CPU and 4 G RAM. Since the problem is too time consuming, our GD-MPHS method is implemented only five independent runs for each test surface.

4.2 Computation results

Table 2 lists the best and average values obtained by PSO, EM, simplified EM (SEM), EM with move solution screening (EM-MSS), EM with sampling techniques and the proposed GD-MPHS method. The results of the compared methods are taken from the previous literature [4]. Since the best values obtained by EM with sampling techniques are not given in the literature, only the average values are used in the comparison. It can be seen that GD-MPHS outperforms other methods for all test surfaces except surface 6. More specific, the results obtained by GD-MPHS show in average 77.87%, 69.77%, 69.39%, 43.29% and 11.54% improvement to PSO, EM, SEM, EMEM-MSS and EM with sampling techniques on the average performance, respectively. Although, the results on surf-6 obtained by GD-MPHS are worse than EMEM-MSS and EM with sampling techniques, they are better than those obtained by PSO, EM and SEM.

Table 1: Parameter settings in the GD-MPHS approach

Parameter name	Value
Number of divided segments NS	10
Maximum number of iterations NI	100
Harmony memory size HMS	100
Harmony consideration rate $HMCR$	0.99
Pitch adjustment rate PAR	0.9
Proportion of harmonies in the better group M	0.4
Mutation parameter F	0.5
Maximum number of local search iterations $iter$	100
Upper bound of variable setting ub	1.0
Lower bound of variable setting lb	-1.0
The Initial step length of local search S	$0.1 * (ub - lb)$

The computational time spent by GD-MPHS is also compared with above methods except EM with sampling techniques (it was not given in the literature). Table 3 shows the statistics on the computational time spent by different methods. Again, GD-MPHS shows its superiority to other methods, spending much less computational time. In detail, GD-MPHS show in average 81.14%, 94.61%, 93.94% and 95.39% savings to PSO, EM, SEM and EMEM-MSS on the average computational time, respectively.

Table 2: Comparison between different methods

Surface	PSO		EM		SEM		EM-MSS		Sampling+EM		GD-MPHS	
	Best	*Mean*	*Best*	*Mean*	*Best*	*Mean*	*Best*	*Mean*	*Best*	*Mean*	*Best*	*Mean*
Surf-1	8.534	12.119	6.334	7.845	6.486	7.737	4.642	6.323	–	3.11	1.060	1.328
Surf-2	15.812	20.562	9.456	11.170	9.196	11.155	9.275	10.234	–	7.72	6.444	6.513
Surf-3	7.315	11.691	9.456	11.922	9.325	11.876	4.678	7.244	–	3.31	1.639	1.686
Surf-4	7.601	12.321	11.879	15.046	11.235	14.757	6.539	8.431	–	4.45	2.148	2.278
Surf-5	25.753	32.449	10.101	12.652	11.437	12.121	7.919	11.257	–	4.05	2.006	2.391
Surf-6	27.969	30.953	20.049	22.124	20.135	22.121	4.831	5.267	–	4.09	11.394	11.856
Surf-7	21.364	23.451	13.424	14.231	12.676	14.206	7.602	12.367	–	4.39	2.052	2.375
Surf-8	25.352	28.087	24.435	26.657	22.365	25.829	22.288	24.125	–	16.99	12.486	12.830

Table 3: The computational time of different methods (in seconds)

Surface	PSO		EM		SEM		EM-MSS		GD-MPHS	
	Best	*Mean*	*Best*	*Mean*	*Best*	*Mean*	*Best*	*Mean*	*Best*	*Mean*
Surf-1	9634	10,123	33,231	34,592	29,078	29806	36,257	37,369	1892	2018
Surf-2	9782	10,243	32,796	35,693	31,044	32,018	42,396	45,056	1935	2067
Surf-3	9843	10,356	34,288	37,119	33,462	35,961	44,199	47,264	1896	2064
Surf-4	9756	10,342	34,531	37,654	32,344	32,845	41,056	42,965	1867	1997
Surf-5	10,156	10,643	33,998	38,852	31,137	31,732	38,921	42,337	1767	1976
Surf-6	10,537	10,789	35,413	38,752	30,560	31,440	38,200	41,463	1839	2010
Surf-7	10,342	10,974	32,168	35,575	31,727	34,005	42,695	44,223	1709	2008
Surf-8	10,432	10,622	34,468	35,857	33,038	34,677	41,095	44,747	1672	1700

The test surfaces were divided into 10 segments in the above computations using the proposed GD-MPHS method. It is an interesting work to explore how the number of segments NS affects the performance of GD-MPHS. So in the rest of this section, the computation results obtained by GD-MPHS with different number of segments NS are compared. Table 4 lists the results with different number of segments. “Without dec.” represents applying the modified HS algorithm and local search without the geometric decomposition strategy. It can be seen that by introducing the geometric decomposition strategy, the machined errors are reduced. To show the trend more clearly, Figure 4 plots the average errors vary with the number of segments NS. It can be seen that for most of the test surfaces, the machined errors keep decreasing along with the increasing of NS. However, for some surfaces with very large length difference, e.g. surf-2,surf-6 and surf-8, the machined errors keep decreasing at the beginning, but they begin to increase when the number of segments NS is too large. Hence, we can conclude that a mediate number of segments NS is more appropriate.

Table 4: Results with different number of segments

Surface	Without dec		2 segments		4 segments		6 segments		8 segments		10 segments		20 segments	
	Best	*Mean*	*Best*	*Mean*	*Best*	*Mean*	*Best*	*Mean*	*Best*	*Mean*	*Best*	*Mean*	*Best*	*Mean*
Surf-1	1.595	2.328	1.653	1.907	1.285	1.773	1.525	1.698	1.342	1.488	1.060	1.328	0.737	0.792
Surf-2	7.210	8.157	6.451	7.148	6.597	6.688	6.335	6.680	6.269	6.549	6.444	6.513	6.805	6.916
Surf-3	2.093	2.492	1.765	2.211	1.812	2.341	1.703	1.831	1.659	1.743	1.639	1.686	1.490	1.577
Surf-4	3.079	3.375	2.458	3.770	2.367	2.681	2.488	2.604	2.401	2.580	2.148	2.278	2.062	2.147
Surf-5	2.901	3.962	2.278	3.733	2.034	3.198	2.142	2.826	1.995	2.487	2.006	2.391	1.417	1.551
Surf-6	11.694	12.015	11.622	12.077	11.527	11.696	11.724	12.011	11.489	11.693	11.394	11.856	12.290	12.345
Surf-7	3.661	4.446	2.385	2.818	2.772	3.318	2.387	2.934	2.302	2.533	2.052	2.375	1.587	1.640
Surf-8	13.46	14.65	12.38	13.07	12.13	12.57	12.13	12.42	12.27	12.51	12.49	12.83	13.15	13.25

5 CONCLUSION

This study developed a systematic approach for generating optimized tool path for 5-axis flank milling based on the geometric decomposition strategy and multi-population harmony search algorithm. The main idea of the proposed approach is to divide the machined surface into several segments and employ the multi-population harmony search algorithm to obtain the optimal CLs independently for each segment. The geometric decomposition makes the complex problem divided into several sub-problems, which are much easier to solve. Compare to previous methods in literatures, the proposed approach can improve 11.54% to 77.87% in terms of the sum of the machined errors. Moreover, the proposed approach also save a lot of computation cost compared with previous methods. One of the potential future work is to further accelerate the optimization procedure by GPU parallel computing techniques.

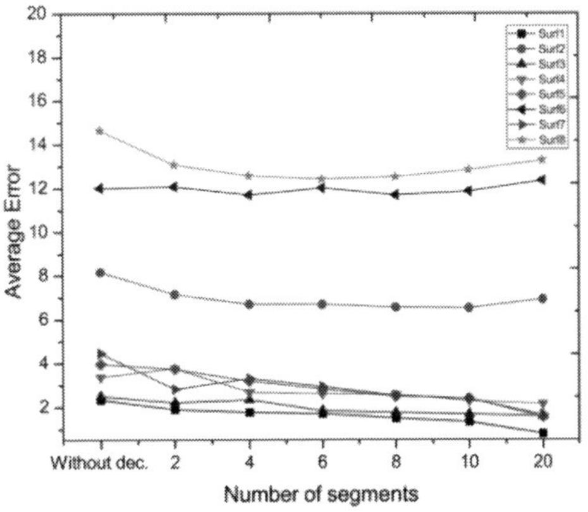

Figure 4: Average error vary with the number of segments NS

6 REFERENCES

[1] **E. Budak, E. Ozturk, L. 2009.** Modeling and simulation of 5-axis milling processes, CIRP Annals-Manufacturing Technology, 58(1), pp 347-350.

[2] **C.-H. Chu, C.-T. Lee, K.-W. Tien, C.-J. Ting.** 2011. Efficient tool path planning for 5-axis flank milling of ruled surfaces using ant colony system algorithms, International Journal of Production Research, 49(6), pp 1557-1574.

[3] **H.-T. Hsieh, C.-H. Chu. 2013.** Improving optimization of tool path planning in 5-axis flank milling using advanced pso algorithms, Robotics and Computer-Integrated Manufacturing,29(3), pp 3-11.

[4] **C.-L. Kuo, C.-H. Chu, Y. Li, X. Li, L. Gao. 2015.** Electromagnetism-like algorithms for optimized tool path planning in 5-axis flank machining, Computers & Industrial Engineering, 84, pp 70-78.

[5] **P.-H. Wu, Y.-W. Li, C.-H. Chu. 2008.** Optimized tool path generation based on dynamic programming for five-axis flank milling of rule surface, International Journal of Machine Tools and Manufacture, 48(11), pp 1224-1233.

[6] **Z. W. Geem, J. H. Kim, G. Loganathan. 2001.** A new heuristic optimization algorithm: harmony search, Simulation,76(2),pp 60-68.

[7] **A. Chauhan, R. Saini. 2016.** Discrete harmony search based size optimization of integrated renewable energy system for remote rural areas of uttarakhand state in india, Renewable Energy,94,pp 587-604.

[8] **J. Yi, X. Li, M. Xiao, J. Xu, L. Zhang. 2016.** Construction of nested maximin designs based on successive local enumeration and modified novel global harmony search algorithm, Engineering Optimization,pp 1-20.

[9] **B. Zeng, Y. Dong. 2016.** An improved harmony search based energy-efficient routing algorithm for wireless sensor networks, Applied Soft Computing,41,pp 135-147.

[10] **S. Kundu, D. R. Parhi. 2016.** Navigation of underwater robot based on dynamically adaptive harmony search algorithm, Memetic Computing,pp 1-22.

[11] **J. Yi, L. Gao, X. Li, J. Gao. 2015.** An efficient modified harmony search algorithm with intersect mutation operator and cellular local search for continuous function optimization problems, Applied Intelligence,pp 1-29.

CLOSED-LOOP SUPPLY CHAIN NETWORK MODEL WITH PRODUCT RECOVERY, RESELLING AND WASTE DISPOSAL+

YoungSu Yun[1*], Chuluunsukh Anudari[1], Xing Chen[1] and RaeKook Hwang[2]
[1] Division of Business Administration, Chosun University, Korea
ysyun@chosun.ac.kr, anuka5658@naver.com, chenxing1982@naver.com

[2] Industry & Strategy Department I, Samsung Economics Research Institute, Korea
rkhwang02@gmail.com

ABSTRACT

In this paper, a closed-loop supply chain (CLSC) network model with product recovery, reselling and waste disposal is proposed. The proposed CLSC network model considers both forward logistics and reverses logistics. A mathematical formulation using nonlinear mixed integer programming is designed in order to effectively represent the proposed CLSC network model. For implementing the mathematical formulation, a hybrid genetic algorithm (pro-HGA) approach is proposed. The pro-HGA combines conventional genetic algorithm with revised cuckoo search. In numerical experiment, three scales of the proposed CLSC network model are presented to compare the performances of the pro-HGA and conventional competing approaches under various measures of performance. Finally, we can confirm that i) the proposed CLSC network model is more efficient in consideration of real-world situation than conventional CLSC network models, and ii) the pro-HGA outperforms the conventional competing approaches.

Keywords: Closed-loop supply chain network model, Product recovery, Reselling, Waste disposal, Nonlinear mixed integer programming, Hybrid genetic algorithm, Revised cuckoo search.

1 INTRODUCTION

The necessity of the closed-loop supply chain (CLSC) network model is mentioned in some aspects of product life cycle (PLC), environmental regulation, and customer requirement. In PLC, most of products have their own PLCs. They are initiated from forward logistics (FL) network for product production and then finished in reverse logistics (RL) network for treatment of the returned product. Therefore, considering both FL and RL in PLC is more reasonable than considering FL and RL respectively. In environmental regulation, many countries and governments enforce the companies involved in CLSC to produce product without environmental problem and to treat the returned product effectively. Also, as customer conception level to environmental problem is higher, customer is willing to buy more environment-friendly products than environment-unfriendly products.

+ This work was supported by the National Research Foundation of Korea Grant funded by the Korean Government (NRF- 2014S1A5A2A01010951).
* Corresponding Author

According to the increased necessity of the CLSC network model, many researchers have studied various CLSC network models [1-3, 6, 12]. Fleischmann *et al.* [3] proposed a conceptual CLSC network model which combines FL with RL. In FL, supply, production and distribution are considered for product production. In RL, collection, selection, reprocessing, and redistribution are taken into consideration for treating the returned product. Wang and Hsu [12] proposed a CLSC network model including both FL and RL. In FL, suppliers, manufactures, distribution centers and customers are considered. In RL, recycler and landfill area are considered. The returned product from customer is collected at distribution center and then sent to recycler. The recycler checks and classifies it into reusable and unusable materials. The reusable material is sent to manufacturer to produce product, and the unusable material is buried in landfill area. The CLSC network model proposed by Amin and Zhang [1] considers various facilities such as part supplier, manufacture, distribution center and retailer in FL and disassembly center, refurbishing center and disposal center in RL. Especially, for product production, they used two types of part. First type is to use a new part from part supplier in FL and second type is to use the reusable part from refurbishing center in RL. Similar to Amin and Zhang [1], Chen, et al. [2] also considered various facilities such as supplier, manufacturer, warehouse and retailer in FL and customer, collection center, recycling center and waste disposal plant in RL. They used the function of recycling centers which can classify the returned product into reusable and unusable products. The reusable product is sent to manufacturer in FL and the unusable product is disassembled into reusable and unusable raw materials. The reusable raw material is sent to supplier in FL and the unusable raw material is treated in waste disposal center in RL.

In the above mentioned studies for the CLSC network model, Fleischmann et al. [3] considered the reselling in reuse market and the disposal in disposer market, but they did not considered the reuse of the returned product in FL. On the other hand, Wang and Hsu [12], Amin and Zhang [1], and Chen, et al. [2] considered the reuse situation of returned product in FL, while the reselling situation of the recovered product did not taken into consideration in RL. Analysis of the conventional studies shows that both the reuse of the returned product in FL and the reselling of the recovered product in RL should be considered simultaneously.

Therefore, the CLSC network model proposed in this paper considers the reuse situation of the returned product and the reselling situation of the recovered product in FL and RL, respectively. The proposed CLSC network model is consist of part supplier, module manufacturer, product manufacturer, distribution center and retailer in FL and customer, collection center, recovery center, redistribution center, secondary market and waste disposal center in RL. Especially, opening/closing decisions is considered in part supplier, module manufacturer, product manufacturer, distribution center, collection center, redistribution center and recovery center. A mathematical formulation is designed for effectively representing the proposed CLSC network model. For implementing the mathematical formulation, a hybrid genetic algorithm (pro-HGA) approach is proposed.

The rest of this paper is organized as follows. Section 2 describes the proposed CLSC network model. A mathematical formulation for the proposed CLSC network model is designed in Section 3. The mathematical formulation is implemented in the pro-HGA approach in section 4. The numerical experiments using three scales of the proposed CLSC network model are done for comparing the performances of the pro-HGA approach and other competing approaches in Section 5. Finally, some conclusions are drawn in Section 6.

2 PROPOSED CLSC NETWORK MODEL

A conceptual structure of the proposed CLSC network model is shown in Figure 1. In Figure 1, each part supplier in areas 1 and 2 sends part types 1 and 2 to module manufacturer, whereas the part suppliers in areas 3 and 4 sends part types 3 and 4 to product manufacturer, respectively. Module manufacturer assembles part types 1 and 2 to make modules and then send the modules to product manufacturer. In product manufacturer, a product using the modules from module manufacturer and the part types 3 and 4 from the part suppliers in areas 3 and 4 is produced and then sends it to retailer through distribution center. The returned products from customer are collected at collection center and then they are tested and classified into recoverable and unrecoverable products. The recoverable products with α_1% are sent to recovery center and after recovering their all functions, they are sent to secondary market to be resold. The unrecoverable products are disassembled into recoverable and unrecoverable modules. The recoverable modules with α_2% are sent to module manufacturer for recovering their functions. The unrecoverable modules with α_3% are sent to waste disposal center to be varied or incinerated.

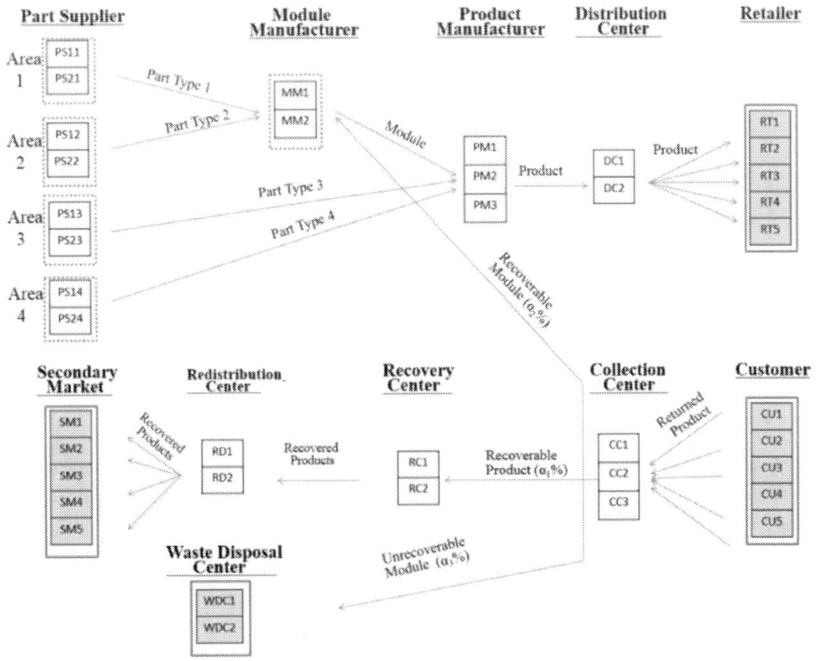

Figure 1: Conceptual CLSC Network Model

3 MATHEMATICAL FORMULATION

For the design of the proposed CLSC network model, the following assumptions are considered.

- The single item production is considered in FL and RL.
- The numbers of the part suppliers in areas 1, 2 3 and 4, module manufacturers, product manufacturers, distribution centers, retailers(=customers), collection centers, recovery centers, redistribution centers, secondary market, and waste disposal centers are known in advance. Of the numbers, all the retailers, customers, secondary markets, and wasted disposal centers are always opened, while among all the part suppliers in areas 1, 2, 3 and 4, module manufacturers, product manufacturers, distribution centers, collection centers, recovery centers, and redistribution center, only one facility at each stage should be opened and the remainders should be closed.
- Fixed costs for operating each facility are different and are already known.
- Unit treating costs of all facilities at same stage are identical with each other even though which facility is opened or closed. All the unit handling costs are already known.
- Unit transportation costs between all facilities are different and are already known

- The treating capacity of the facility opened in a stage is the same or greater than that of the facility opened in the previous stage.
- The returned products from all customers are collected with the rate of 100% at collection center.
- The quality of the recovered module which is made by the recoverable modules from collection center is identical with that of the module from the suppliers at areas 1 and 2.

Based on the above assumptions, a mathematical formulation for designing the proposed CLSC network model is developed. Minimization of the cost which is consisted of total sum of fixed costs, treating costs and transportation costs is considered as objective function under various constraints. First, Indies, parameters and decision variables are defined.

Indices:
a : index of area; $a \in A$
i : index of part supplier; $i \in I$
j : index of module manufacturer; $j \in J$
k : index of product manufacturer $k \in K$
l : index of distribution center $l \in L$
m: index of retailer(customer); $m \in M$
n : index of collection center; $n \in N$
o : index of recovery center; $o \in O$
p : index of secondary market; $p \in P$
q: *index of* waste disposal center; $q \in Q$
e : index of redistribution center; $e \in E$
Parameters
FS_{ia}: fixed cost at part supplier i of area a
FM_j : fixed cost at module manufacturer j
FP_k : fixed cost at product manufacturer k
FD_l : fixed cost at distribution center l
FC_n : fixed cost at collection center n
FR_o : fixed cost at recovery o
FE_e : fixed cost at redistribution center e
TS_a: unit treating cost at part supplier of area a
TM : unit treating cost at module manufacturer
TP : unit treating cost at product manufacturer
TD : unit treating cost at distribution center
TC : unit treating cost at collection center
TR : unit treating cost at recovery center
TE : unit treating cost at redistribution center
TSM_{iaj}: unit transportation cost from part supplier i of area a to module manufacturer j
TMP_{kj} : unit transportation cost from module manufacturer j to product manufacturer k
TPD_{kl} : unit transportation cost from product manufacturer k to distribution center l
TDR_{lm}: unit transportation cost from distribution center l to retailer(customer) m
TCC_{mn} : unit transportation cost from customer m to collection center n
TCM_{nj} : unit transportation cost from collection center n to module manufacturer j
TCR_{no} : unit transportation cost from collection center n to recovery center o
TRE_{oe} : unit transportation cost from recovery center o to redistribution center e
TCW_{nq} : unit transportation cost from collection center n to waste disposal center q
TES_{ep} : unit transportation cost from redistribution center e secondary market p
Decision variables:
cs_{ia} treating capacity at part supplier i :of area a

cm_j : treating capacity at module manufacturer j
cp_k : treating capacity at product manufacturer k
cd_l : treating capacity at distribution center l
ct_m : treating capacity at retailer/customer m
cc_n : treating capacity at collection center n
cr_o : treating capacity at recovery center o
ce_e : treating capacity at redistribution center e
cs_p : treating capacity at secondary market p
cw_q : treating capacity at waste disposal center q
xs_{ia} takes the value of 1 if part supplier i at area a is opened and 0 otherwise
xm_j takes the value of 1 if module manufacturer j is opened and 0 otherwise
xp_k takes the value of 1 if product manufacturer k is opened and 0 otherwise
xd_l takes the value of 1 if distribution center l is opened and 0 otherwise
xc_n takes the value of 1 if collection center n is opened and 0 otherwise
xr_o takes the value of 1 if recovery center o is opened and 0 otherwise
xe_e takes the value of 1 if redistribution center e is opened and 0 otherwise

Using the above mentioned parameters and decision variables, a mathematical formulation is designed for effectively representing the proposed CLSC network model as follows:

$$Minimize\ Total\ cost\ (TC) = total\ fixed\ cost\ (TF) + total\ handling\ cost\ (TH) + total\ transportation\ cost\ (TP) \quad (1)$$

$$TF = \sum_a \sum_i (FS_{is} \cdot xs_{ia}) + \sum_j (FM_j \cdot xm_j) + \sum_k (FP_k \cdot xp_k) + \sum_l (FD_l \cdot xd_l) + \sum_n (FC_n \cdot xc_n) + \sum_o (FR_o \cdot xr_o) + \sum_e (FE_e \cdot xe_e) \quad (2)$$

$$TH = \sum_a \sum_i (TS_a \cdot cs_{ia}\, xs_{ia}) + TM \sum_j (cm_j \cdot xm_j) + TP \sum_k (cp_k \cdot xp_k) + TD \sum_l (cd_l \cdot xd_l) + TC \sum_n (cc_n \cdot xc_n) + TR \sum_o (cr_o \cdot xr_o) + TE \sum_e (ce_e \cdot xe_e) \quad (3)$$

$$TP = \sum_a \sum_i \sum_j \left(TSM_{iaj} \cdot cs_{ia}\, xs_{ia} \cdot xm_j\right) + \sum_k \sum_j (TMP_{jk} \cdot cm_j \cdot xm_j \cdot xp_k) + \sum_l \sum_k (TPD_{kl} \cdot cp_k \cdot xp_k \cdot xd_l) + \sum_m \sum_l (TDR_{lm} \cdot cd_l \cdot xd_l) + \sum_m \sum_n (TCC_{mn} \cdot cc_n \cdot xc_n) + \sum_j \sum_n TCM_{nj} \cdot cc_n \cdot xc_n \cdot xm_j) + + \sum_o \sum_n TSR_{no} \cdot cc_n \cdot xc_n \cdot xr_o) + \sum_e \sum_o TRE_{oe} \cdot cr_o \cdot xr_o \cdot xe_e) + \sum_q \sum_n TCW_{nq} \cdot cc_n \cdot xc_n) + \sum_p \sum_e TES_{ep} \cdot ce_e \cdot xe_e) \quad (4)$$

subject to

$$\sum_i xs_{ia} = 1, \quad \forall a \epsilon A \quad (5)$$
$$\sum_j xm_j = 1 \quad (6)$$
$$\sum_k xp_k = 1 \quad (7)$$
$$\sum_l xd_l = 1 \quad (8)$$
$$\sum_n xc_n = 1 \quad (9)$$
$$\sum_r xr_o = 1 \quad (10)$$
$$\sum_e xe_e = 1 \quad (11)$$
$$\sum_i (cs_{ia} \cdot xs_{ia}) - \sum_j (cm_j \cdot xm_j) = 0, \quad \forall a \epsilon A \quad (12)$$
$$\sum_i (cm_j \cdot xm_j) - \sum_k (cp_k \cdot xp_k) = 0 \quad (13)$$
$$\sum_k (cp_k \cdot xp_k) - \sum_l (cd_l \cdot xd_l) = 0 \quad (14)$$
$$\sum_l (cd_l \cdot xd_l) - \sum_m cr_m = 0 \quad (15)$$
$$\sum_m cr_m - \sum_n (cc_n \cdot xc_n) = 0 \quad (16)$$
$$\sum_o (cr_o \cdot xr_o) - \sum_p cs_p = 0 \quad (17)$$
$$\sum_o (cr_o \cdot xr_o) - \sum_n (ce_e \cdot xe_e) = 0 \quad (18)$$
$$\sum_j (cr_o \cdot xr_o) - \alpha_1 \sum_n (cc_c \cdot xc_n) \geq 0, \quad \forall a \epsilon A \quad (19)$$

$$\sum_l (cm_l \cdot xm_l) - \alpha_2 \sum_n (cc_c \cdot xc_n) \geq 0, \quad \forall a \epsilon A \tag{20}$$
$$\sum_q cwd_q - \alpha_3 \sum_n (cc_c \cdot xc_n) \geq 0 \tag{21}$$
$$xs_{ia} = \{0,1\} \quad \forall\, i \in I, a \epsilon A \tag{22}$$
$$xm_m = \{0\,,1\} \, \forall j \epsilon J \tag{23}$$
$$xp_k = \{0\,,1\} \, \forall k \epsilon K \tag{24}$$
$$xd_l = \{0\,,1\} \, \forall l \epsilon L \tag{25}$$
$$xc_n = \{0\,,1\} \, \forall n \epsilon N \tag{26}$$
$$xr_o = \{0\,,1\} \, \forall o \epsilon O \tag{27}$$
$$xe_e = \{0\,,1\} \, \forall e \epsilon E \tag{28}$$
$$cs_{ia}, cm_j, cp_k, cd_l, cc_n, cr_o, ce_e, cs_p, cw_q \geq 0$$
$$\forall i \epsilon I, \forall a \epsilon A, \forall j \epsilon j, \forall k \epsilon K, \forall l \epsilon L, \forall m \epsilon M, \forall n \epsilon N, \forall o \epsilon O, \forall e \epsilon E, \forall p \epsilon P, \forall q \epsilon Q \tag{29}$$

The objective function of equation (1) is composed of the equations (2) to (4) which minimize the total sum of fixed costs, treating costs and transportation costs. Equation (5) means that the each one part supplier at areas 1 and 2 should be opened, and the same meaning is considered in equations (6) to (11). Equation (12) implies that the sum of the treating capacity at each part supplier in areas 1 and 2 is the same as that of the module manufacturer. The same meaning is considered in equations (13) to (18). Equation (19) restricts that the sum of the treating capacity at all recovery centers is the same or greater than that of the recoverable products with $\alpha_1\%$ at all collection centers. Same concepts are also shown in equations (20) to (21). Equations (22) to (28) restrict the variables to integers 0 and 1. Equation (29) means non-negativity.

4 PRO-HGA APPROACH

As already known, GA has been proved to be an effective approach for solving many optimization problems [4-5, 11, 12]. However, GA has also some weaknesses in premature convergence of the solution and the absence of local search ability. Therefore, many researchers have developed various hybrid approaches using GA and other local search approaches to overcome the weaknesses [5, 8, 13].

Recently, Kanagaraj ***et al.*** [7] developed an HGA using GA and cuckoo serarch(CS). In the HGA, they used the CS as a local search to overcome the premature convergence of GA so that the HGA is converged to global optimal solution quickly. Main scheme of the HGA is as follows. In first stage, GA approach is adapted and produces new population (offspring). The second stage is to apply CS to GA. That is, Levy flight is adapted to an individual randomly selected from the offspring in order to produce a new individual (*New_Ind*). The *New_Ind* competes with the individual (*Old_Ind*) randomly selected from offspring. If the *New_Ind* is more efficient in fitness value than the *Old_Ind*, the former is inserted into the offspring and *Old_Ind* is deleted from the offspring. Using these two stages, the improvement of solution is continuously proceeded during all generations. However, this approach has some weakness. That is, we cannot guarantee the improvement of solution using Levy flight in all trials since the fitness value of the *New_Ind* is always superior to that of the *Old_Ind*. Therefore, we improve this weakness in the pro-HGA.

In the pro-HGA, we also use two stage approaches, similar to the HGA by Kanagaraj ***et al.*** [7]. In first stage, a offspring is generated using GA operator. In secdond stage, all the individuals of the offspring are adapted in CS. That is, an individual randomly selected from offspring is suffered from Levy flight and then a *New_Ind* is produced and it competes with the *Old_Ind*. This process is repeated until all individuals from offspring are adapted in Levy flight. Therefore, the possible occurrence of the *New_Ind* with higher fitness value is more increased than that of the *New_Ind* with high fitness in the HGA approach by Kanagaraj ***et al.*** [7].

Detailed implementation procedure is as follows:

Step 1: GA approach
  Step 1-1: initilization
    Randomly generate initial population.
  Step 1-2: GA operator
    Elitist selection strategy in an enlarged sampling space [4], two-point crossover operator [10], and random mutation operator [4] are used for selection, crossover and mutation, respectively.
Step 2: CS approach
  Step 2-1: Comparison by Levy flight
    Compare the fitness value of *New_Ind* with that of *Old_Ind*. If the former is more efficient than the latter, the *New_Ind* is inserted into Offpsring and the *Old_Ind* is deleted in it.
    This procedure is repeated until all individuals are sufferred from Levy flight.
  Step 2-2: generating new population
    After sorting all individuals using their fitness values, a fraction (p_a) of the individuals with lower fitness values is abandoned and regenerate new individuals randomly to replace the lost individuals.
Step 3: Termination condition
    If pre-defined termination condition is satisfied, then stop all steps , otherwise, goto step 1-2.

5 NUMERICAL EXPERIMENT

Three scales of the proposed CLSC network model are presented in numerical experiment. The detailed sizes of each scale are summarized in Table 1. For each scale, we assume that 3,000 products are produced in FL and treated in RL. Handling rates at collection center are divided into $\alpha_1 = 60\%$, $\alpha_2 = 30\%$, and $\alpha_3 = 10\%$ for the recoverable products, recoverable modules and the unrecoverable modules.

Table 2: Three Scales of Proposed CLSC Network Model

Scale	Suppliers				Material Manufacturers	Product Manufacturers	Distribution Center	Retailers	Collection Center	Recovery Center	Redistribution Center	Secondary Markets	Waste Disposal Centers
	1	2	3	4									
1	10	10	10	10	8	10	8	15	10	8	8	15	1
2	15	15	15	15	10	15	10	20	15	10	10	20	2
3	20	20	20	20	15	20	15	25	20	15	15	25	3

For various comparisons, two conventional approaches (GA, HGA) and a commercial version of Lingo [9] are used for comparing their performances with that of the pro-HGA. In the GA approach, the step 1 of the detailed implementation procedure used in Section 4 is adapted. The HGA approach developed by Kanagaraj et al. [7] is used. The GA, HGA and pro-HGA was programmed under MATLAB version 2014b and ran under a same computation environment (Mac OS version 10.11 with a 1.3 GHz processor (Intel core i5) and 4 GB RAM). The used parameters are as follows. population size 50, crossover rate 0.5, mutation rate 0.2, the fraction (p_a) 50% for all scales and total generation number is 1,000 for scales 1 and 2 and 5,000 for scale 3. Total 30 independent trials were done to eliminate the randomness of each approach. The performance of each approach is compared using various measures as follows. i) best solution is the best one among all solutions, ii) average solution is the averaged value of all solutions, iii) average time is the average CPU time (unit: sec.) and iv) average iteration is the average number of generations. The computation results of each approach using three scales of the proposed CLSC network model is described in Table 2.

Table 2: Computation Results of Each Approach

	Scale 1				Scale 2				Scale 3			
	GA	HGA	pro-HGA	Lingo	GA	HGA	pro-HGA	Lingo	GA	HGA	pro-HGA	Lingo
Best solution	319007	318434	317666	317228	320322	318534	318517	321736	322234	321950	321172	322373
Average iteration	22	31	38	-	25	32	32	-	88	137	229	-
Average solution	321724	321857	320674	-	322448	323147	322316	-	326565	325991	325242	-
Average time	14.0	15.0	33.0	-	16.0	17.0	46.0	-	41.0	45.0	88.0	-

For Table 2, in terms of the best solution of the scale 1, the performance of the Lingo is the best, and that of the GA is the worst. Among the GA, HGA and pro-HGA, the pro-HGA is more efficient in terms of the best solution than the others. However, the pro-HGA does not show any benefit in terms of the average solution, average iteration, and average time, when compared with the others. Especially, the search speed of the pro-HGA is about two times slower than those of the others.

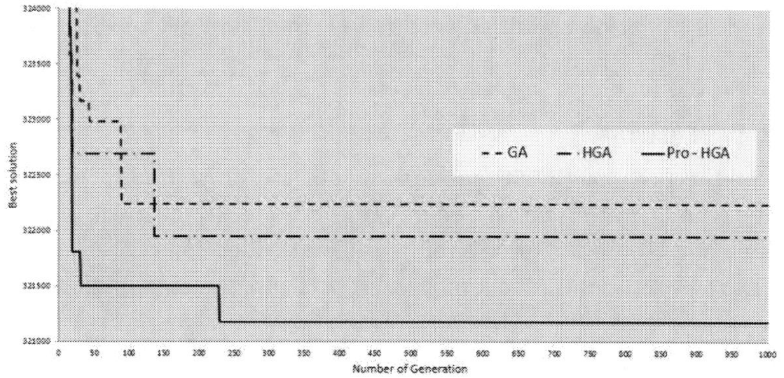

Figure 2: Convergence Behaviors of GA, HGA, and pro-HGA

In scale 2, the GA, HGA and pro-HGA outperforms the Lingo in terms of the best solution. Among the GA, HGA and pro-HGA, all the approaches show similar results in terms of the average solution and average iteration. However, similar to the scale 1, the pro-HGA is the slowest in terms of the average time. In scale 3, the performance of the Lingo is not superior to those of the others in terms of the best solution. The pro-HGA shows the best performance in terms of the best solution, average iteration. However, similar to the scales 1 and 2, the search speed of the pro-HGA is very slower than the others.

For additional comparison, the convergence behaviors of the GA, HGA, and pro-HGA are described in Figure 2. In Figure 2, all the approaches are converged very quickly in initial generations and also show various convergence behaviors until about 150 generations. After about 150 generations, the GA and HGA do not show any convergence behaviors, but the pro-HGA shows an additional convergence behaviour and its performance is significantly better than those of the GA and HGA. Figure 3 shows optimal location and allocation of the pro-HGA for Scale 3. In Figure 3, the facilities in grey colour are opened at each stage. Total 3,000 products are produced and delivered in FL and they are collected at collection center. At collection center, 1,800 recoverable products (=3,000 × 60%) are sent to the recovery center, 900 recoverable modules (=3,000 × 30%) to the module manufacturer, and 300 unrecoverable modules (=3,000 × 10%) to the waste disposal center.

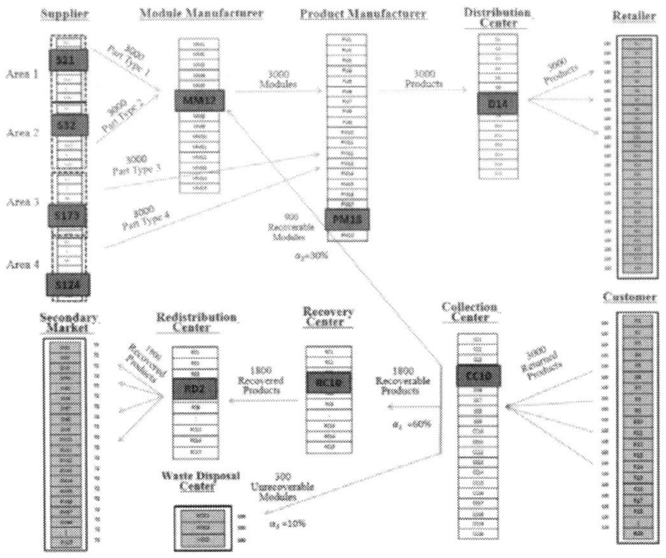

Figure 2: Optimal Location and Allocation of pro-HGA for Scale 3

6 CONCLUSION

In this paper, we have proposed a new type of CLSC network model. The proposed CLSC network model considers various facilities such as part suppliers in various areas, module manufacturers, product manufacturers, distribution centers and retailers in FL and customers, collection centers, recover centers, redistribution center, secondary markets and waste disposal centers in RL. Especially, in the proposed CLSC network model, a product production using module and part in FL and various treatments using recoverable products, recoverable modules and unrecoverable modules in RL have been considered.

The proposed CLSC network model has been designed using a mathematical formulation and implemented in the pro-HGA approach. For the pro-HGA approach, conventional GA approach and a revised CS approach have been adapted. In numerical experiment, three scales of the CLSC network model have been presented and the performance of the pro-HGA approach has been compared with those of some conventional approaches (GA and HGA) using various measures of performance. Experimental results have shown that the pro-HGA approach outperforms the GA and HGA in most of measures of performance.

For our future study, larger-scaled CLSC network models will be presented and practical data such as real transportation cost will be used for comparing the performance of the pro-HGA. Also various hybrid approaches using evolutional computation approaches such as particle swarm optimization and biogeography-based optimization will be developed and applied for optimizing the CLSC network model.

7 REFERENCES

[1] **Amin, S. H. and Zhang, G.** 2012. An integrated model for closed-loop supply chain configuration and supplier selection: multi-objective approach, *Journal of Expert Systems with Applications*, 39, pp 6782-6791.

[2] **Chen, T. T., Chan, F. T. S. and Chung, S. H.** 2015. An integrated closed-loop supply chain model with location allocation problem and product recycling decisions, *International Journal of Production Research*, 53(10), pp 3120-3140.

[3] **Fleischmann, M., Krikke, H. R., Dekker, R. and Flapper, S. D. P.** 2000. A characterization of logistics networks for product recovery, *Omega*, 28(6), pp 653-666.

[4] **Gen, M. and Cheng, R. 1997**. *Genetic Algorithms and Engineering Design*, John Wiley and Son, New York.

[5] **Gen, M. Cheng, R. and Lin, L.**, 2008. *Network Models and Optimization: Multiobjective Genetic Algorithm Approach*, Springer, London.

[6] **Georgiadis P. and Besiou, M.**, 2008. Sustainability in electrical and electronic equipment closed-loop supply chains: a system dynamics approach, *Journal of Cleaner Production*, 16, pp 1665-1678.

[7] **Kanagaraj G., Phonnambalam S. G. and Jawahar N.** 2013. A hybrid cuckoo search and genetic algorithm for reliability-redundancy allocation problems, *Computers and Industrial Engineering*. 66, pp 1115-1124.

[8] **Lee, J. E., Gen, M. and Rhee, K. G.**, 2009. Network model and optimization of reverse logistics by hybrid genetic algorithm, *Computers and Industrial Engineering*, 56, pp 951-964.

[9] **Lingo**, 2015. Lindo Systems, www.lindo.com.

[10] **Michalewicz, Z.**, 1994. *Genetic Algorithms + Data Structures = Evolution Program*, Spring-Verlag.

[11] **Syarif A., Yun, Y. S. and Gen. M.** 2002. Study on multi-stage logistic chain network: a spanning tree based genetic algorithm approach, *Computers and Industrial Engineering*, 43(1-2), pp 299-314.

[12] **Wang, H.-F. and Hsu, H.-W.**, 2010. A closed-loop logistic model with a spanning-tree based genetic algorithm, *Computers and Operations Research*, 37, pp 376-389.

[13] **Yun, Y. S., Chung, H. S. and Moon, C. U.** 2013. Hybrid genetic algorithm approach for precedence-constrained sequencing problem, *Computers and Industrial Engineering*, 65, pp 137-147.

A TWO-STAGED ALGORITHM TO THE STOWAGE PLANNING FOR CONTAINER SHIPS

Canrong Zhang[1*], Guodong Zhu[2] and Shaojun Liu[3]
Research Center for Modern Logistics, Graduate School at Shenzhen, Tsinghua University
[1]crzhang@sz.tinghua.edu.cn
[2]zhugd15@mails.tsinghua.edu.cn
[3]liu-shaojun@foxmail.com

ABSTRACT

The quality of the stowage planning for the container ship has a significant impact on the cost and efficiency of the port and the ship. How to decrease the re-shift number of containers and how to reduce the imbalance of the ship during sailing are two important influential factors among others. Although great efforts have been made to this field, these two influential factors are rarely simultaneously considered in the previous literature. This paper proposes a two-staged algorithm to solve the problem which integrates these two influential factors. In the first stage, an integer programming model with the objective of minimizing the number of re-shift movements is constructed, and a heuristic algorithm is developed to solve large-scale instances. In the second stage, a mixed integer programming model with the objective of minimizing the imbalance is constructed. Numerical experiments are conducted to test the performance of the proposed two-staged algorithm.

Key Words: stowage planning; re-shift; imbalance; heuristic algorithm; integer linear programming

1 INTRODUCTION

The stowage planning for container ship is a difficult decision especially when the container ship is large. Normally, a container ship will have to visit several ports during its voyage route. The container ship is empty at the beginning. It will load containers at the starting port and leaves for the next one with loaded containers. When the ship arrives at the following port, the containers bound for this port will be unloaded and the containers starting from this port will be loaded. While arriving at the final port, all containers will be unloaded and the ship will be empty again for next voyage route.

Early researches are aimed to build a stowage system relying on rules or experience, such as Shields (1984) and Perakis and Dillingha (1987). Botter and Brinati (1992) built up a 0-1 linear programing model for the first time aimed to reduce shift number, but their model did not take deck constrain into consideration. Avriel et al. (2000) defined this problem as k-shift problem and proved it to be NP-hard. Kevin et al. (2014) also proved the stowage problem considering the deck constraint NP-hard. As the stowage problem is NP-hard, the main research turns to using heuristic algorithm to attain a feasible solution. Avriel et al. (1998) provided an algorithm named suspensory heuristic procedure to solve the k-shift problem and attained a solution with a low shift number. Opher et al. (2002) used a genetic algorithm, which can both take shift and stability into consideration. Wilson and Roach (2000) also proposed a decompose algorithm to reduce the complexity of the problem.

This work is supported by the National Nature Science Foundation of China under Grant No. 71472108 and Shenzhen Municipal Science and Technology Innovation Committee under Grant No. JCYJ20160531195231085.

*Corresponding Author

Ding and Mobel (2015) provided a heuristic algorithm which can get a better result than Avriel et al. (1998). Those researches solved stowage problems but most of them did not take into consideration both deck constraint and ship stability.

Container ship can often hold thousands of containers. The containers will be placed in different cargos. A cargo is separated into two parts by the deck: the part above deck and the part below deck. Each slot inside the cargo is measured by a three-dimensional coordinates: row, column and layer. To simplify this problem, all stacks in the same cargo can be reallocated along the same row, and then the slot coordinate will be reduced to a two-dimensional one.

At each port, the containers bound for this port will be unloaded and the containers starting from this port will be loaded. As containers are operated one by one, a container with a farther destination may be placed upon a container that should be unloaded at the current port. Then, this farther container has to be unloaded temporarily and be loaded again later. This temporary unloading operation is called re-shift. As we all know, the re-shift operation is costly. Our main purpose of the stowage planning is to reduce this kind of costly re-shift operation.

The other important purpose of the stowage planning is to reduce the imbalance of the whole ship for the voyage safety. After arranging containers into the cargos, each cargo has a weight associated with the loading containers. If the stowage plan generates a bad placement for the cargos, then the ship may suffer an imbalance risk. At this moment, we have to add ballast water to balance the ship, which only brings in operation time and sailing cost. Therefore, it is of significance to properly consider the imbalance to make a high-quality stowage plan.

In order to solve the stowage problem considering the deck constraint and the ship balance, we design a two-stage algorithm. In the first stage, it is aimed at the placement of the containers at the cargos to decrease the number of re-shift movement. In the second stage, a linear algorithm is constructed to solve the cabin placement problem to reduce the imbalance of the ship during the voyage sailing.

2 PROBLEM FORMULATION

As presented previously, each cargo of the container ship corresponds to a storage space separated into two parts by the deck: above-deck part and below-deck part. As pointed previously, each cargo is a two-dimensional structure and containers can be placed at each slot. Under this simplification, we can get a bird-looking view of the ship (Figure1) and the cargo view (Figure2).

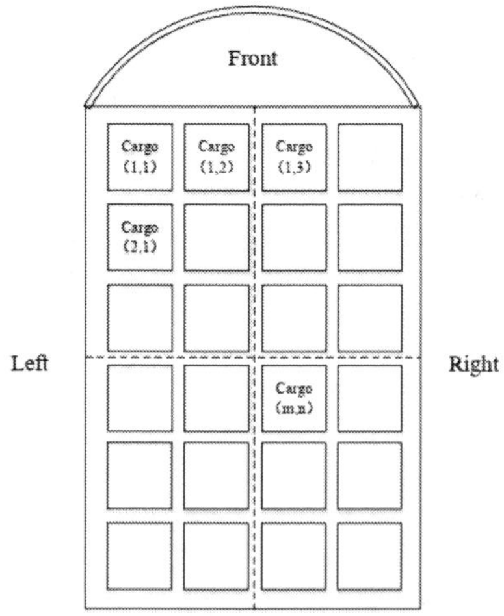

Figure 1: Ship view

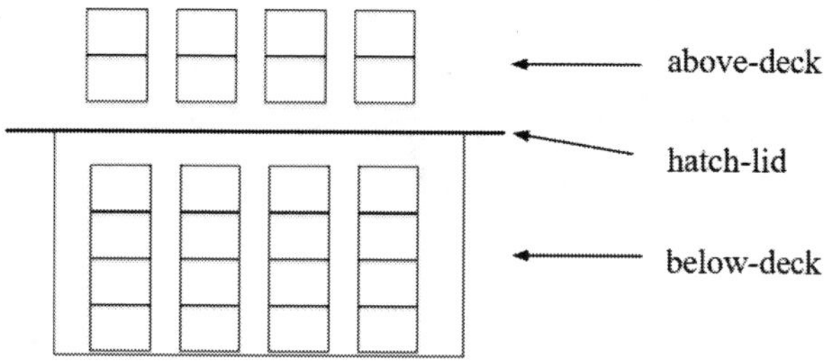

Figure 2: Cargo view

In the first stage we aim at the placement of the containers into the cargos. It is noted that these cargos are abstract because we do not designate a ship-position for it at this stage. Its ship position will be determined in the second-stage decision. It is assumed that there are T cargos, corresponding to T decks on the ship, and for each cargo there are C columns (after relocating other stacks to the same row) both below and above the separated deck. There are R_0 and R_1 layers below and above the deck, respectively.

Until now, we can use $slot(t,h,r,c)$ to define a container position. t means the cargo, h means above or below deck, c means column and r means layer. In the whole voyage, the ship will visit N ports and the transport information is stored in a two dimensional matrix: the number of containers transported from each port to the other ports.

We define the binary variable $x_{ijv}(t,h,r,c)$ as follows: $x_{ijv}(t,h,r,c) = 1$ if there is a container at $slot(t,h,r,c)$ which is loaded at port i whose destination is port j but should be unloaded at port v. Binary variable $y_i(t,h,r,c) = 1$ means there is a container placed at $slot(t,h,r,c)$ when the ship leaves port i. Hence we get the following binary linear program:

$$\text{Min}\sum_{i=1}^{N-1}\sum_{j=i+1}^{N}\sum_{v=i+1}^{j}\sum_{t=1}^{T}\sum_{h=0}^{1}\sum_{r=1}^{R_h}\sum_{c=1}^{C}x_{ijv}(t,h,r,c) \tag{1}$$

$s.t.$

$$\sum_{v=i+1}^{j}\sum_{t=1}^{T}\sum_{h=0}^{1}\sum_{r=1}^{R_h}\sum_{c=1}^{C}x_{ijv}(t,h,r,c) - \sum_{k=1}^{i-1}\sum_{t=1}^{T}\sum_{h=0}^{1}\sum_{r=1}^{R_h}\sum_{c=1}^{C}x_{kji}(t,h,r,c) = T_{ij} \qquad i=1,\dots,N-1,\ j=i+1,\dots,N,\ t=1,\dots,T \tag{2}$$

$$\sum_{k=1}^{i}\sum_{j=i+1}^{N}\sum_{v=i+1}^{j}x_{kjv}(t,h,r,c) = y_i(t,h,r,c) \qquad i=1,\dots,N-1, t=1,\dots,T, h=0,1, r=1,\dots,R_h, c=1,\dots,C \tag{3}$$

$$y_i(t,h,r+1,c) - y_i(t,h,r,c) \le 0 \qquad i=1..N, t=1,\dots,T, h=0,1, r=1,\dots,R_h-1,, c=1,\dots,C \tag{4}$$

$$\sum_{i=1}^{j-1}\sum_{p=j}^{N}x_{ipj}(t,h,r,c) + \sum_{i=1}^{j-1}\sum_{p=j+1}^{N}\sum_{v=j+1}^{p}x_{ipv}(t,h,r+1,c) \le 1 \qquad j=2,\dots,N, t=1,\dots,T, h=0,1, r=1,\dots,R_h-1, c=1,\dots,C \tag{5}$$

$$\sum_{i=1}^{j-1}\sum_{p=j}^{N}x_{ipj}(t,0,r,c) + \sum_{i=1}^{j-1}\sum_{p=j+1}^{N}\sum_{v=j+1}^{p}x_{ipv}(t,1,rs,cs) \le 1 \qquad j=2,\dots,N, t=1,\dots,T, r=1,\dots,R_h-1, c=1,\dots,C, rs=1,\dots,R_h-1, cs=1,\dots,C \tag{6}$$

The object function is the total number of shifting operations for all ports including the must-do loading and unloading operations. The re-shift number is the object function subtracting a fixed number which corresponds to the must-do operations. Constraint (2) is the flow conservation constraint. Constraint (3) guarantees that each slot can only hold at most one container. Constraint

(4) guarantees if one slot has a container, the slots below it must have a container, which means all containers should be supported. Constraint (5) is a movement constraint, which guarantees when one container is loaded, the containers above it should be moved aside. Constraint (6) is also a movement constraint, which guarantees the containers above deck should all be moved aside first before the container below is loaded.

In the second stage, we aim at the placement of the cargos onto the ship. Normally, the shape of the ship is rectangular. Here it is assumed that a ship has M cargos in a line and N cargos in a row. The number of total cargos is set to T, which is equal to M*N. The number of ports is set to P. We can get the weight of each abstract cargo from the first stage, denoted by $weight[i][t]$. Each cargo may encounter both a horizontal unbalance and a longitudinal unbalance. The method we calculate unbalance is using the weight of the cargo to multiply the sum of its distance to the horizontally and longitudinally symmetric lines. At each port, the unbalance of the ship is the absolute value of longitudinal unbalance and horizontal unbalance. The total unbalance is the sum of the unbalances of all the ports.

We define integer variable $x(i,j)$ as the unbalance value of the ship when it leaves port i. When j=0, it means the horizontal unbalance and if j=1 it means longitudinal unbalance. Binary variable $y(m,n,t) = 1$ means cargo t is put at slot (m,n) on the ship. Nonnegative integer variable $z(i,m,n)$ means the imbalance caused by slot (m,n) when the ship leaves port i. Nonnegative integer variable $u(i,j)$ and $v(i,j)$ are two auxiliary variables to make sure there are no absolute-value operation in the object function. Hence the mixed integer linear program is shown below:

$$Min \sum_{i=0}^{P-1}\sum_{j=0}^{1}\left(u(i,j)+v(i,j)\right) \tag{7}$$

$s.t.$

$$x(i,j)+u(i,j)-v(i,j)=0 \qquad i=0,...,P-1, j=0,1 \tag{8}$$

$$\sum_{t=1}^{T} y(m,n,t)=1 \qquad m=1,...,M, n=1,...,N \tag{9}$$

$$\sum_{m=1}^{M}\sum_{n=1}^{N} y(m,n,t)=1 \qquad t=1,...,T \tag{10}$$

$$z(i,m,n)= \sum_{t=0}^{T-1} weight(i,t)\times y(m,n,t) \qquad i=0,...,P-1, m=1,...,M, n=1,...,N \tag{11}$$

$$x(i,0)=\sum_{m=0}^{M-1}\sum_{n=0}^{N/2-1} z(i,m,n)\times(N/2-n) - \sum_{m=0}^{M-1}\sum_{n=N-N/2}^{N-1} z(i,m,n)\times(n+N/2-N+1) \qquad i=0,...,P-1 \tag{12}$$

$$x(i,1)=\sum_{n=0}^{N-1}\sum_{m=0}^{M/2-1} z(i,m,n)\times(M/2-m) - \sum_{n=0}^{N-1}\sum_{m=M-M/2}^{M-1} z(i,m,n)\times(m+M/2-M+1) \qquad i=0,...,P-1 \tag{13}$$

3 SOLUTION ALGORITHM

As the stowage problem is NP-hard, it is not easy to obtain a good feasible solution or even a feasible solution in limited time. We analyze the running time of the linear program by CPLEX, a commercial solving package, for stage one and stage two, and find the that the stage one can't get a feasible solution in hours using CPLEX, prompting to design a heuristic algorithm for this stage. But the second stage can easily attain a feasible solution with a gap smaller than 20% in minutes. Therefore, CPLEX is viable to solve the second-stage problem. Here we concentrate our focus on the design of the heuristic algorithm for the first-stage problem.

The cargos are classified, according to the stowage state, into six states. In the algorithm, the containers will be assigned to different kinds of cargos by their priorities. The six states of the cargos are (Figure 3):

State one: The total cargo is empty.

State two: The below-deck part has some containers but not full, and the above-deck part is empty.

State three: The below-deck is full or nearly full, and the above-deck part is empty.

State four: The below-deck is full, and the above-deck has some containers.

State five: The below-deck is full, and the above-deck is nearly full.

State six: The cargo is full.

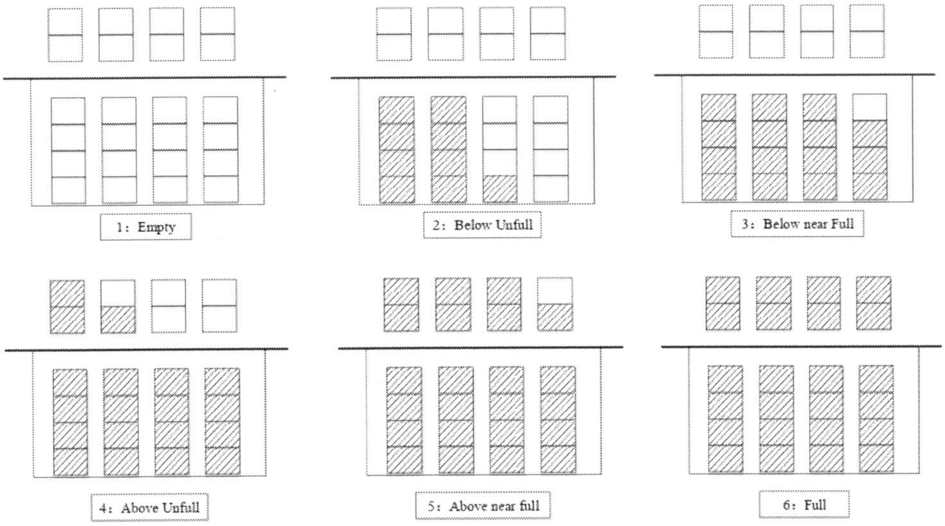

Figure 3: Cargo states

The ideas behind our algorithm are based on the principles below:

1. When a container on below-deck should be unloaded, we have to unload the containers on above-deck first.
2. We will not put a container on above-deck when the below-deck is not full, because this arrangement will not lead to a better result. This principle makes us classify the states of the cargos into six types.
3. If a container has a further destination, it tends to be put at a bottom position.
4. If some containers have the same destination and they can share the same cargo, we tend to make them placed in the same cargo, because containers of different destinations may cause re-shift when they are mixed together.

The flow chart of the algorithm is shown as Figure 4.

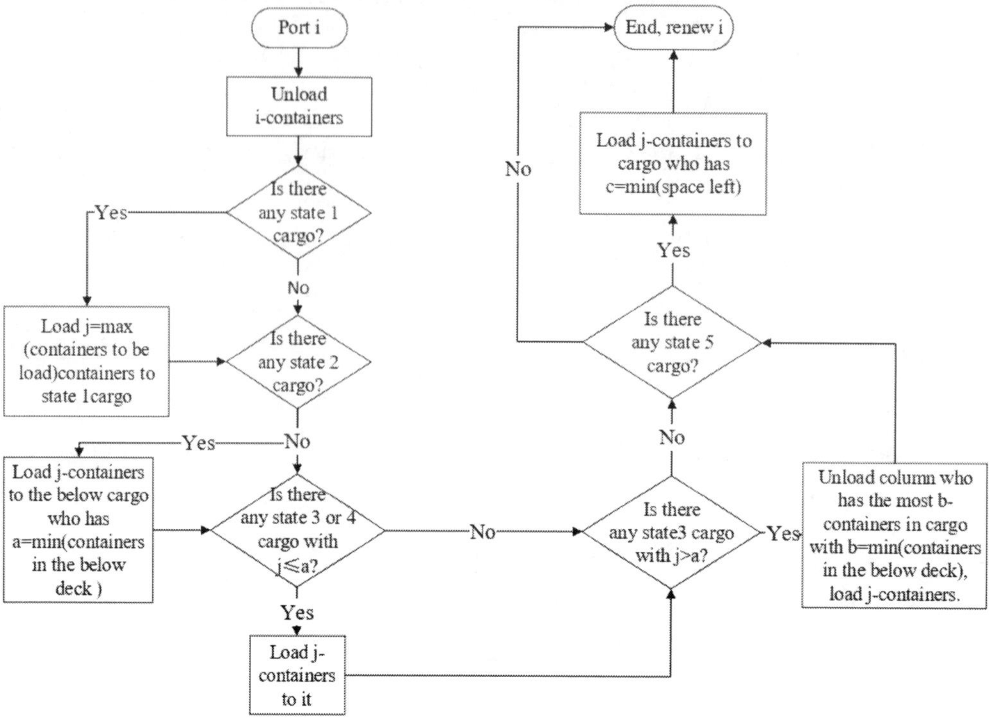

Figure 4: Algorithm flow chart

4 NUMERICAL EXPERIMENTS

4.1 Parameters setting

Since a container ship may visit several ports during its voyage route, we set the number of ports ranging from 3 to 10. The container ship varies as well, so that we can test the impact of ship scale on algorithm performance. Here eight kinds of ships are tested whose capacities ranges from hundreds to more than ten thousands. Here are the parameters of the ship as shown in Table 1:

Table 1 : Parameter setting

Ship Type	Cargo Num	Col Num	Below Layer	Above Layer	Capacity
Type1	4	6	4	3	168
Type2	8	8	5	4	576
Type3	12	10	6	5	1320
Type4	16	12	7	6	2496
Type5	20	14	8	7	4200
Type6	24	16	9	8	6528
Type7	28	18	10	9	9576
Type8	32	20	11	10	13440

4.2 Stage one experiment

In the first stage, we assign the containers to the abstract cargos. To know how well our algorithm can optimize the solution, we contrast it to a basic stowage algorithm, which searches for available slots and loads containers following a sequence from the farthest destination to the nearest one while arriving at each port. We run an instance for 15 times and obtain the average running time and re-shift rate. The running time is limited to one hour.

The running time ratio of heuristic to basic stowage algorithm is shown in Figure 5 and the re-shift rate ratio of heuristic to basic stowage algorithm (The difference of re-shift rate between the basic stowage to heuristic algorithm) is shown in Figure 6.

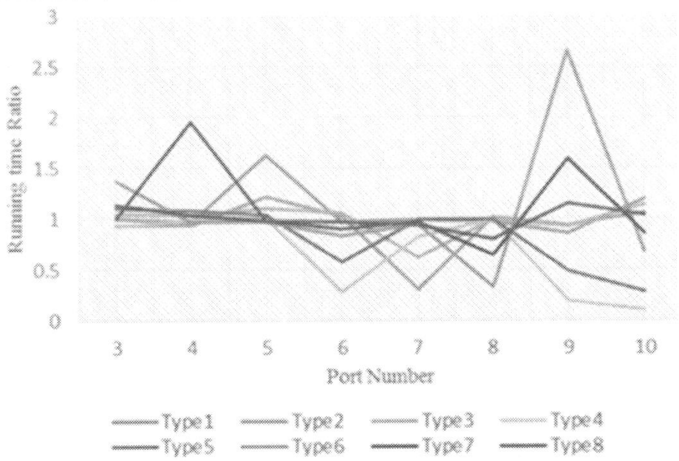

Figure 5: Running time Ratio

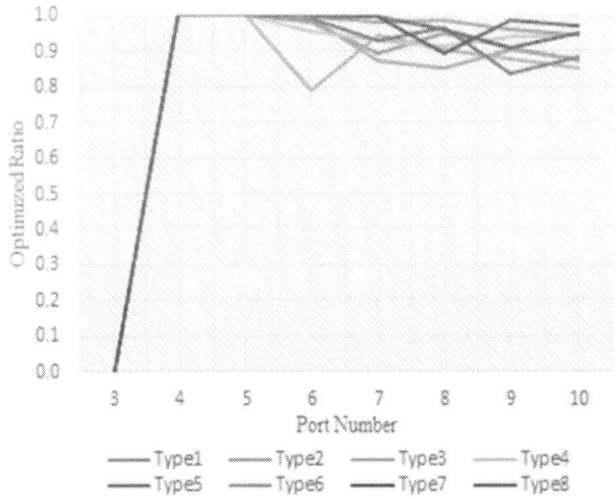

Figure 6: Optimized Ratio

From the result we can know that in terms of running time, the two algorithms have a similar efficiency. The running time ratio is mostly in the vicinity of 1.

In terms of the re-shift ratio, our heuristic algorithm can normally have an optimization more than 80% except when the port number is 3. When the port number is 3, the two algorithms can both get the optimal solutions. The comparison of our heuristic algorithm with the basic algorithm shows that our algorithm can get a large optimization on re-shift ratio within a short time.

4.3 Stage two experiment

In the first stage, we have got the assignment of the containers to the cargos aimed at lowering the re-shift movements. In the second stage, we assign these cargos to the ship with the objective of avoiding imbalance. The time limit is also set as one hour and each parameter is run for 15 times to obtain the average value.

The running time of the integer programming for stage two is shown in Figure 7 and the imbalance value of the integer programming for the second stage is shown in Figure 8.

From the result we know that when the ship and the port number are of middle size, we can get the optimal solution in one hour. But when the ship becomes larger and the port number enlarges, we can only get a feasible solution. It is shown that the imbalance value increases in a linear style when the problem becomes larger, which means even when it's just a feasible solution for a large-scale problem, the value does not turn too bad. So we believe that the feasible solution is not far from the optimal one. To verify this, we now conduct a more detailed experiment to look at the convergence speed of the integer programming. We test it in a middle-scale situation and a large-scale situation, respectively. The middle scale situation has 10 ports and uses the type 4 ship and the large situation has ten ports and uses the type 8 ship. The information of running time, best integer and best bound are recorded. We track the results with the running time and find a typical convergence.(Figure 9 and Figure 10)

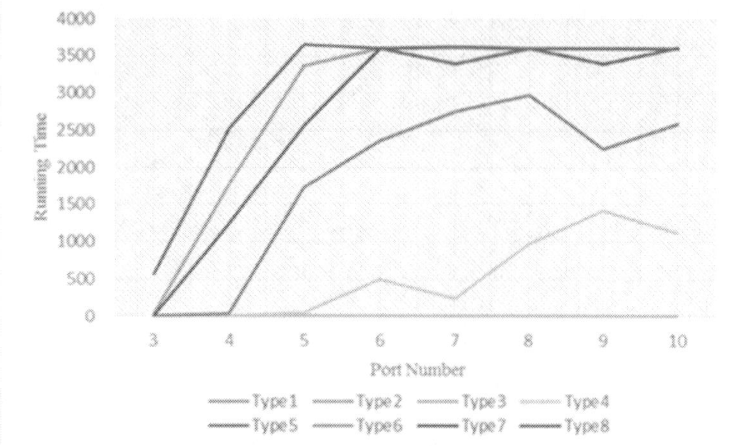

Figure 7: Running time of Stage 2

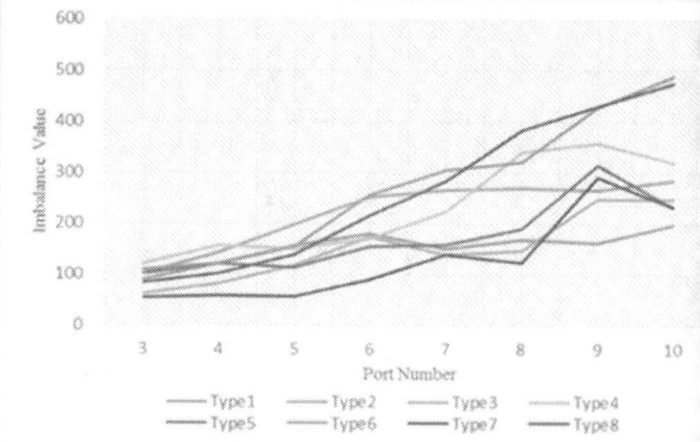

Figure 8: Imbalance value of Stage 2

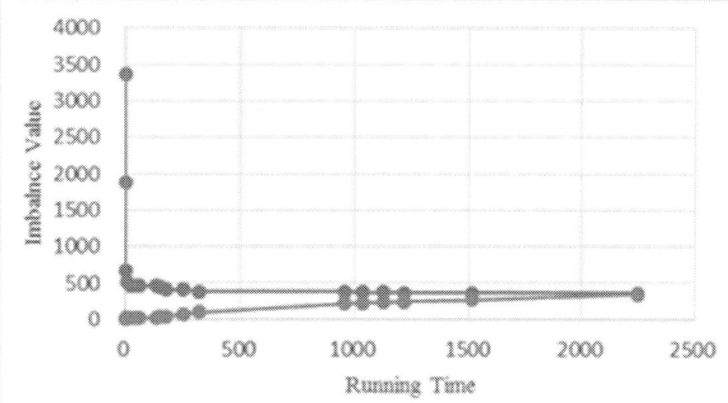

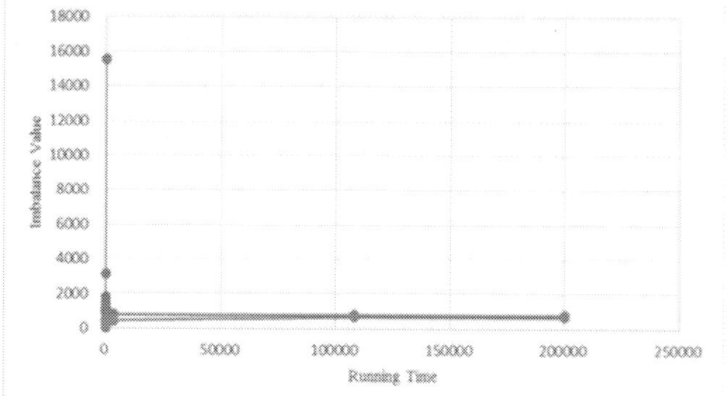

Figure 9: Convergence of Middle scale problem

Figure 10: Convergence of Large scale problem

In the medium-scale situation, we can always get the optimal solution in less than one hour and find that the optimal solution can often be found in five minutes.

In the large-scale situation, it's more difficult to get the optimal solution. In fact we have run the program for 200000 seconds and still cannot get the optimal solution. But fortunately, the program provides a best bound and a feasible solution with a GAP less than 20% can be obtained in five minutes

5 CONCLUSION

In this paper, we consider the stowage planning problem for container ships, paying special attention to deck constraint and the balance of the ship. A two-stage algorithm is developed to solve this problem. In the first stage, a heuristic algorithm is proposed to solve the assignment of the containers into the abstract cargos. Numerical experiments show that the heuristic algorithm can attain a good solution in a very short time and significantly reduce the re-shift rate when compared to the basic stowage method. In the second stage, we develop an integer linear programming to solve the assignment of abstract cargos to the ship with the objective of decreasing the imbalance as the scale of this problem is limited. Numerical experiments show that the integer linear programming can usually generate a feasible solution which has a GAP less than 20%. Totally, the proposed two-stage algorithm is viable to solve this problem.

6 REFERENCES

[1] **Shields J. J, 1984.** Container-ship stowage: a computer-aided preplanning system. Marine Technol, 21, 370-383.

[2] **Perakis A. N., Dillingham J. T., 1987.** The application of artificial intelligence techniques in marine operations. Ship Operations, Management and Economics International Symposium USMMA, Kings Point, New York.

[3] **Botter R. C., Brinati M. A., 1992.** Stowage container planning: a model for getting an optimal solution. IFIP Trans B (Applications in Technology), B-5, 217-229.

[4] **Avriel M., Penn M., Shpirer N., 2000.** Container ship stowage problem: complexity and connection to the coloring of circle graphs. Discrete Applied Mathematics, 103(1), 271-279.

[5] **Tierney K., Pacino D., Jensen R. M., 2014.** On the complexity of container stowage planning problems. Discrete Applied Mathematics, 169, 225-230.

[6] **Avriel M., Penn M., Shpirer N., Witteboon S., 1998.** Stowage Planning for Container Ships to Reduce the Number of Shifts. Annals of Operations Research, 76, 55-71.

[7] **Dubrovsky O., Levitin G., Penn M., 2002.** A genetic Algorithm with a Compact Solution Encoding for the Container Ship Stowage Problem. Journal of Heuristics, 8(6), 585-599.

[8] **Wilson I. D., Roach P. A., 2000.** Container stowage planning: a methodology for generating computerized solutions. Journal of the Operational Research Society, 51(11), 1248-1255

[9] **Wilson I. D., Roach P. A., Ware J. A., 2001.** Container stowage pre-planning: using search to generate solutions, a case study. Knowledge-Based Systems, 14(3), 137-145.

[10] **Steenken D., Vob S., Stahlbock R., 2004.** Container terminal operation and operations research - a classification and literature review. OR Spectrum, 26(1), 3-49.

[11] **Avriel M., Penn M., 1993.** Exact and Approximate Solutions of the Container Ship Stowage Problem. Computers and Industrial Engineering, 25(1), 271-274.

[12] **Aslidis A., 1990.** Minimizing of Overstowage in Container Ship Operations. Operational Research, 90, 457-471.

METHODOLOGY FOR REMANUFACTURABILITY EVALUATION OF WASTED MACHINE TOOLS

X.Y. Jiang[1], B.X. Song[1*], L. Li[1], M.M. Dai[1], H.Y. Zhang[1], Y.F. Li[1]
[1]School of mechanical engineering
Shenyang University of Technology, Shenyang, China
nightwaltz.harold@foxmail.com

ABSTRACT

The remanufacturability evaluation will be difficult for the uncertainties of the original condition or the remaining life of the components. The main factors affecting the remanufacturability are identified by analyzing the remanufacturing process. Then, the remanufacturability evaluation system is established including technical indexes, economic indexes, sustainability indexes and comprehensive performance improvement indexes. Finally, the remanufacturing evaluation method based on G1 and fuzzy comprehensive evaluation is proposed. The remanufacturability evaluation of the TPX6113 in a machine tool plant is taken as an example and the result shows the proposed method can effectively reduce the subjective and the TPX6113 has the enough value to be remanufactured according to the evaluation result.

Keywords: Remanufacturability, Evaluation, Wasted machine tools

1 INTRODUCTION:

As a typical wasted electromechanical product, wasted machine tools have high value to be remanufactured, which can realize the resource cycle utilization to 80%, the performance improvements to 20%, the noise decrease to 10%, the emission decrease of oil mist, oil pollution and dust to 90% [1]. China has the largest number of machine tools up to 8.5 million and most of them are under service beyond 10 years. Thus, large number of machine tools will be eliminated in the next 10 years for out-of-date performance and information level.

The wasted machine tools are the blank for remanufacturing with uncertainties for the quality itself, the working condition and the residual life of the parts. Thus, the remanufacturability of wasted machine tools should be evaluated to decide the remanufacturing value..

Aiming to the remanufacturing evaluation problems, Bras[2] et al. proposed a quantitative method by analyzing the remanufacturing processes; ZHANG Zongxiang[3] et al. established the remanufacturing evaluation system by analyzing the production characteristics effect to the remanufacturability; ZHOU Li[4] et al. proposed a forecasting method using graphical evaluation and review technique; MAO Guoping[5] et al. proposed a model including resources utilization, environment effect, economic feasibility and technical feasibility; LU Chao[6] et al. proposed a ANP method to establish an evaluation indicator system for heavy machine tool remanufacturing scheme. Although there much researches about the remanufacturability evaluation, the evaluation systems in the articles above are mainly based on the performance restore, rather than the comprehensive performance improvements. Also, most subjective evaluation methods such as AHP [7] are wildly used to verify the weight of evaluation indexes, which exists high subjectivity and low efficiency.

In this paper, the main remanufacturability factors are verified concerning the remanufacturing processes and the remanufacturability index system are established including economic attributes, technical attributes, sustainability attributes and comprehensive performance improvements attributes. Then, a remanufacturability evaluation method based on G1 and fuzzy evaluation is proposed, which can effectively reduce the subjectivity effect to the evaluation results.

2 ANALYSIS OF REMANUFACTURING PROCESSES

The remanufacturability means whether a wasted machine tool has enough value to be remanufactured. A remanufactured machine tool should be with high resource utilization, high sustainability and low remanufacturing cost. There are many factors that affect the remanufacturability. Thus, to precisely evaluate the remanufacturability, the remanufacturing processes should be analyzed.

The remanufacturing processes mainly include recycling, disassembly, cleaning, detection, redesign, reassembly and re-sale [8]. The processes are illustrated in figure 1. The wasted machine tools are recycled from customers and disassembled for cleaning and detecting. Then the remanufacturing plan will be established according to the customer requirements and the structural characteristics. Remanufacturing techniques such as laser cladding, surface sticking plastic and brush plating are implemented in the remanufacturing process. Finally, the remanufactured parts and the alternative parts are assembled.

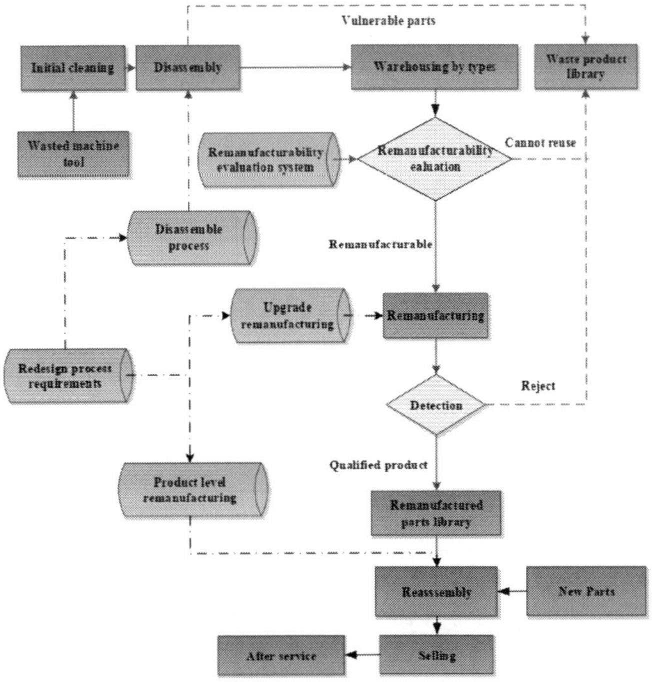

Figure 1: The remanufacturing process

As can be seen, besides the original condition of the wasted machine tools, the disassembly process and the rationality of evaluation, the economic attributes and the sustainability attributes are also the main factors affecting the remanufacturability. The comprehensive performance improvements should also be considered that these factors will decide whether the remanufactured machine tools have the enough competitiveness. Thus, the evaluation index system should be established including the factors above.

3 REMANUFACTURABLITY EVALUATION INDEX SYSTEM

According to the remanufacturing processes analysis, the remanufacturability evaluation system for wasted machine tools is established as shown in figure 2.

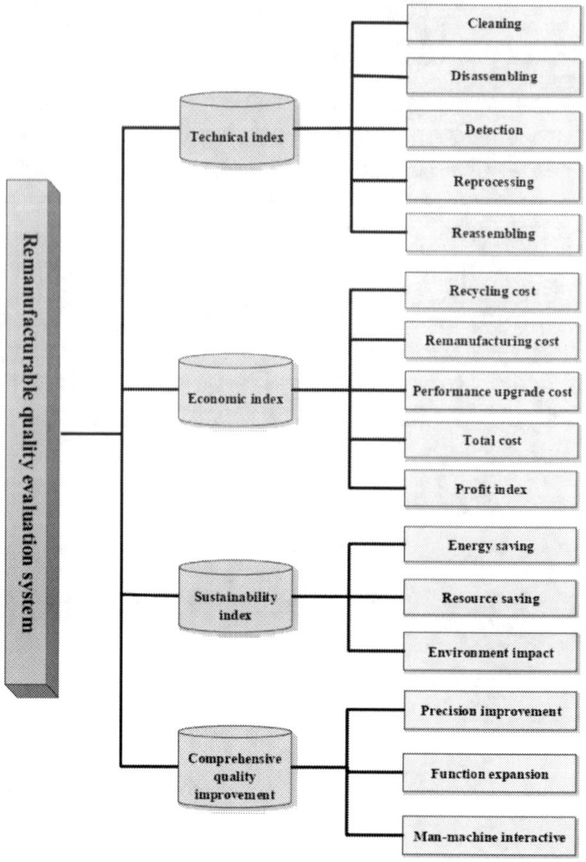

Figure 2: The remanufacturability evaluation system

3.1 Technical index

3.1.1 Cleaning index

The surface attachments will be eliminated in cleaning process. The surface attachments include oil, rust, besmirch and dust, and the cleaning methods include blowing, scrape, washing and baking according to the different residue characteristics. The importance of every cleaning method can be quantitatively defined by the relative importance matrix. For example, the relative importance of every cleaning method are displayed in table 1. The score can be obtained by summarizing the relative importance in a row and standardized as the weight. If the minimum weight (0.0065) is set as 1 cleaning score, the others can be obtained by the same way as shown in table 1. The cleaning method with minimum cleaning score is the ideal cleaning method.

Table 1: The evaluation matrix of the cleaning methods

Cleaning method	Blowing(A)	Scrape(B)	Washing(C)	Baking(D)	Score	Weight	Cleaning Score
Blowing(A)	1	1/3	1/5	1/5	1.73	0.065	1
Scrape(B)	3	1	1/3	1/3	4.67	0.177	3
Washing(C)	5	3	1	1	10	0.379	6

Baking(D)	5	3	1	1	10	0.379	6
Total					26.4	1	

Thus, the cleaning index can be defined as:

$$\mu_c = \frac{n_{ideal} \times 1}{s_{cleaning}} \tag{1}$$

Where $n_{ideal} \times 1$ is the ideal cleaning score, n_{ideal} is the number of the parts needed to be cleaned and $s_{cleaning}$ is the actual cleaning score.

3.1.2 Disassembly index

To avoid the variety of part types, connecting types and processes, the disassembly time is applied to measure the disassembly index.

$$\mu_d = \frac{n_d \times t_d}{t_{dideal}} \tag{2}$$

Where n_d is the parts that need to be disassembled, t_d is the disassemble time and t_{dideal} is the ideal disassemble time.

3.1.3 Detection index

Similarly, the disassembly index, the detection index can be measured by the detection time to avoid the variety of parts and devices.

$$\mu_{de} = \frac{n_{de} \times t_{de}}{t_{deideal}} \tag{3}$$

Where n_{de} is the parts that need to be detected, t_{de} is the detection time and $t_{deideal}$ is the ideal detection time.

3.1.4 Remanufacturing index

The remanufacturing index is measured by the proportion of the number of the parts that need to be remanufactured to the number of all parts.

$$\mu_r = 1 - \frac{n_r}{n} \tag{4}$$

Where n_r is the number of the parts that need to be remanufactured and n is the number of all parts.

3.2 Economical index

3.2.1 Recycling cost

The remanufacturing plant needs to recycle wasted machine tools, which will generate the recycling cost. The recycling cost consists of the purchase cost C_{11} and the transporting cost C_{12}.

$$C_1 = C_{11} + C_{12} \tag{5}$$

3.2.2 Remanufacturing cost

The Remanufacturing cost should include the personnel cost, the replacement cost for new parts, the consumables cost and the treatment cost for abandoned parts.

$$C_2 = C_{21} + C_{22} + C_{23} + C_{24} \quad (6)$$

3.2.3 *Performance improvement cost*

The performance improvement cost will be generated to meet the customer requirements for performance improvements and mainly includes the CNC upgrade cost, the human-machine interaction transformation cost, the high precision parts cost and the safeguard cost.

$$C_3 = C_{31} + C_{32} + C_{33} + C_{34} \quad (7)$$

3.2.4 *Comprehensive cost*

The comprehensive cost maintains the normal operation of the remanufacturing plant and includes the tax cost, the plant management cost, and the design-sale cost.

$$C_4 = C_{41} + C_{42} + C_{43} \quad (8)$$

3.2.5 *Profit indext*

The profit index includes the recycling and utilization profit, the drop-step application cost and the sales profit.

$$R_1 = R_{11} + R_{12} + R_{13} \quad (9)$$

The net profit of the remanufactured machine tools is

$$P_1 = R_1 - C_1 - C_2 - C_3 - C_4 \quad (10)$$

The net profit of the original machine tool is

$$P_2 = R_2 - C' \quad (11)$$

Where R_2 is the sales profit of the original machine tools and C' is the manufacturing cost. Thus, the economic index can be measured as:

$$\mu_e = \frac{P_1}{\gamma P_2} \quad (12)$$

Where γ is a correct coefficient obtained by the remanufacturing statistics, and it's always set as 2.5.

3.3 Sustainability index

3.3.1 *Resource recycling rate*

The resource recycling rate can be measured as:

$$\mu_{rr} = \frac{W_r}{W_M} \quad (13)$$

Where W_r is the weight of the recycled parts and W_M is the original weight of the machine tool.

3.3.2 *Energy consumption*

The parts in the remanufacturing process can be divided into the directly reused parts, the remanufactured parts and the new parts. Where, the directly reused parts have no energy consumption; the energy consumption of the new parts is the same as they are in the original manufacturing processes [9]; the remanufactured parts will consume the energy. Different energy consumption will cause different influences to the environment. Thus, different energy consumption can be integrated into the standard energy consumption index [10].

$$\mu_{ec} = 1 - \frac{\sum_{i=1}^{n} E_i \times N_i}{\sum_{j=1}^{n} E_j \times N_j} \qquad (14)$$

Where N_i is the i th energy consumption in remanufacturing process, N_j is the j th energy consumption in manufacturing process, n is the number of the parts that need to be remanufactured, m is the number of the total parts and E is the standard energy consumption index.

3.3.3 *Environment impact*

The environment impact can be measured as:

$$\mu_{ei} = 1 - \frac{Q_r}{Q_m} \qquad (15)$$

Where Q_r is the emission of the remanufacturing process and Q_m is the emission of manufacturing process.

3.4 Comprehensive performance improvement index

3.4.1 *Precision upgrade*

The precision upgrade means the remanufactured machine tools should own higher machining precision than the original ones. The precision vector can be defined as A= (spindle precision, guide precision, transmission precision, position precision, process precision).

3.4.2 *CNC transformation*

Some machine tools are difficult to implement CNC transformation for the constrains of their fixed structure and size. Therefore, the expert meeting method is applied to evaluate CNC transformation index. The vector of CNC transformation index can be defined as B= (the selection of CNC systems, improvement of transmission structure, control precision improvement, the selection of assist device).

3.4.3 *Human-machine interaction*

The human-machine interaction mainly includes the color design, the control panel design and the effectiveness of machine tool layout. The vector of man-machine coordination can be defined as C= (bed color, structure layout, the height of operating floor, the selection of control method, control panel layout, protection).

4 REMANUFACTURABILITY EVALUATION METHOD BASED ON G1 AND FUZZY COMPREHENSIVE EVALUATION

At present, AHP or fuzzy comprehensive evaluation methods are most applied. However, these methods contain much subjectivity that there may be error with the actual situation. Besides, complex matrixes need to be established and consistency analysis should be implemented that the evaluation efficiency and precision will be effected. Thus, aiming to simplify these complexities and the uncertainties in remanufacturing, an evaluation method based on G1 and fuzzy comprehensive is proposed. The main steps are as follows:

Step 1: Verifying the evaluation index set

If there are m first level indexes, the corresponding first level index set are established as $\mathrm{U} = (U_1, U_2, \dots, U_m)$. Thus, $U_1, U_2, \dots, U_m$ can be regarded as the second level indexes and displayed as $U_i = (u_{ij}), i = 1 \sim m$, the range of j depends on the number of the second level indexes contained in the corresponding first level index.

Step 2: Verifying the comment set

A comment set is the probable evaluation of the valuators and can be displayed as $\mathrm{V} = (V_1, V_2, \dots, V_k)$. The corresponding grade is given to each comment and the grade vector can be obtained as $\mathrm{C} = (C_1, C_2, \dots, C_k)$. The content of the comment set depends on the evaluation object such as $\mathrm{V} = (\text{strong}, \text{medium}, \text{weak})$ or $\mathrm{V} = (\text{better}, \text{good}, \text{medium}, \text{bad}, \text{worse})$.

Step 3: Verifying sequence relationships among the evaluation indexes such as $U_1 > U_2 > \cdots > U_m$.

Step 4: Verifying the rate of relative importance between adjacent evaluation indexes.

$$r_k = \frac{w_{k-1}}{w_k} \tag{16}$$

$\mathrm{k} = \mathrm{m}, \mathrm{m} - 1, \dots, 2, 1$

Step 5: Calculating the weight of the evaluation indexes

$$w_m = (1 + \sum_{k=2}^{m} \prod_{i=k}^{m} r_i)^{-1} \tag{17}$$

$\mathrm{k} = \mathrm{m}, \mathrm{m} - 1, \dots, 2, 1$

Weight calculating includes weight distribution calculation for the first level evaluation indexes, also includes weight distribution calculation for the second level evaluation indexes and the third evaluation level indexes.

Step 6: Verifying the membership degree and establishing the evaluation matrix

Firstly, the corresponding membership degree for each evaluation index should be verified, and then the evaluation matrix ***R*** can be established for every level evaluation indexes. The number of rows is the number of evaluation indexes and the number of columns is the number of evaluation levels.

Step 7: Establishing comprehensive evaluation results

The fuzzy calculation can be operated according to the weight of a first evaluation index $w_i$and the evaluation matrix ***R***. Then the results can be standardized and the result B_i can be obtained.

$$\boldsymbol{B}_i = w_i \boldsymbol{R}_i \tag{18}$$

Step 8: Calculating comprehensive results and quantitative processing

The final evaluation results can be obtained by the evaluation matrix ***B*** and equation 21. Then it can be processed with the grade vector ***C*** to get the final results ***F***.

$$\mathrm{F} = \boldsymbol{B}\boldsymbol{C}^t \tag{19}$$

5 CASE STUDY

Aiming to popularize the remanufactured machine tools, plenty of wasted machine tools are recycled by Shenyang Machine Tool Co., LTD. However, the original condition of the wasted machine tools is much different, which will bring difficulties to remanufacturing processes. The remanufacturing cost will he high for a highly damaged machine tool. Thus, the remanufacturability evaluation should be implemented to determine whether the wasted machine tools have the enough value to be remanufactured.

The remanufacturability evaluation data of the wasted TPX6113 is shown in table 2.

Table 2: The remanufacturability evaluation data

Type	TPX6113	Number of parts	2321
Number of disassembly parts	2321	Disassembly time	4.4 hours
Number of cleaning parts	1957	Number of A type cleaning parts	261
Number of B type cleaning parts	322	Number of C type cleaning parts	1157
Number of D type cleaning parts	235	Number of detected parts	1957
Total detection time	20.3 hours	Number of fixed parts	29
Number of assembly parts	2372	Assembly time	11.3 hours
Recycling cost	77500 yuan	Transporting cost	1230 yuan
Re-processing cost	5235 yuan	Change cost of purchased parts	2200 yuan
Consumables cost	570 yuan	Process cost of wasted parts	70 yuan
Protection cost	1520 yuan	CNC upgrade cost	4500 yuan
Man-machine transformation cost	2600 yuan	High precision parts cost	3760 yuan

5.1 Calculating evaluation index

The number of every evaluation index can be obtained according to the index equations from 1 to 15: disassembly index is 0.2213, cleaning index is 0.2043, detection index is 0.2678, re-processing index is 0.9906, re-assembly index is 0.1749, economic index is 0.702, resource saving index is 0.8, energy saving index is 0.9361, and environment impact index is 0.95.

5.2 Verifying index set and comment set

The index set can be verified according to the evaluation system as: the first level evaluation index set U= (technology index, economic index, green index, comprehensive improvement index). The second level evaluation index set of technology is U_1= (disassembly index, cleaning index, detection index, re-processing index, re-assembly index). The second level evaluation index set of economy is U_2= (recovery cost, remanufacturing cost, performance improvement cost, comprehensive cost, profit index). The second level evaluation index set of sustainability is U_3= (resource saving, energy saving, environment impact). The second level evaluation index set of comprehensive performance improvement is U_4= (precision improvement, function expand, design for man-machine coordination). The third level evaluation index set of precision improvement is U_{31}= (spindle precision, guide precision, transmission precision, positioning precision, process precision). The third level evaluation index set of function expand is U_{32}= (the selection of CNC systems, improvement of transmission structure, improvement of control precision, the selection of assist device). The third level evaluation index set of design for man-machine coordination is U_{33}= (bed color, structure layout, the height of operating floor, the selection of control method, control panel layout, protection).

The comment set is defined as V= (best, better, good, medium, bad). The corresponding scores are given and the grade vector can be obtained as C= (95, 85, 75, 60, 50).

5.3 Verifying the weight of evaluation indexes

There are 10 experts to separately evaluate the weight and the average value were obtained. The weight of evaluation indexes can be verified by equations from 16 and 17. For example, the first level evaluation index (technology index, economic index, green index, comprehensive improvement index) can be verified by the following steps:

Step 1: Verifying sequence relationships: technology index > economic index > comprehensive improvement index > green index.

Step 2: Verifying the importance of adjacent indexes

$$r_2 = \frac{u_1}{u_2} = 1.4, r_3 = \frac{u_2}{u_3} = 1.6, r_4 = \frac{u_3}{u_4} = 1.2$$

Step 3: Calculating the weight of evaluation indexes

$$w_4 = (1 + r_2r_3r_4 + r_3r_4 + r_4)^{-1} = 0.147$$

$$w_3 = w_4r_4 = 0.176, w_2 = w_3r_3 = 0.282, w_1 = w_2r_2 = 0.395$$

Therefore, the weight of technology index is 0.395; the weight of economic index is 0.282; the weight of sustainability index is 0.147; the weight of comprehensive improvement index is 0.176.

In the same way, the weights of the second level evaluation index of technology W_1 are 0.124,0.254,0.133,0.278,0.229. The weights of the second level evaluation index set of green W_3 are 0.333,0.332,0.345. The weights of the second level evaluation index set of comprehensive performance improvement W_4 are 0.379,0.296,0.326. The weights of the third level evaluation index set of precision improvement are 0.190,0.173,0.175,0.201,0.258. The weights of The third level evaluation index set of CNC transformation are 0.262,0.265,0.294,0.178. The weights of the third level evaluation index set of design for man-machine coordination are 0.165,0.180,0.154,0.154,0.167,0.177.

5.4 Verifying evaluation matrix

(1) Qualitative indexes are voted by experts and the results are summarized. For example, the number of V_1, V_2, V_3, V_4, V_5 for u_{ij} in the index set U_i is relatively $v_{1ij}, v_{2ij}, v_{3ij}, v_{4ij}, v_{5ij}$. Therefore, the corresponding membership degree is $r_{1ij}, r_{2ij}, r_{3ij}, r_{4ij}, r_{5ij}$, and $r_{tij} = {}^{v_{tij}}\!/_{\sum_{t=1}^{5} vtij}$. Thus, the fuzzy evaluation matrix of quantitative index set U_i is

$$R_i = \begin{bmatrix} r_{1i1} & r_{2i1} & r_{3i1} & r_{4i1} & r_{5i1} \\ r_{1i2} & r_{2i2} & r_{3i2} & r_{4i2} & r_{5i2} \\ \cdots & \cdots & \cdots & \cdots & \cdots \\ r_{1ij} & r_{2ij} & r_{3ij} & r_{4ij} & r_{5ij} \\ \cdots & \cdots & \cdots & \cdots & \cdots \\ r_{1in} & r_{2in} & r_{3in} & r_{4in} & r_{5in} \end{bmatrix}$$

Where $\sum_{t=1}^{5} r_{ijt} = 1$.

Similarly, fuzzy evaluation matrixes of precision improvement R_{41}, CNC transformation R_{42} and man-machine interactive R_{43} are:

$$R_{41}=\begin{bmatrix}0.4&0.3&0.2&0.1&0\\0.6&0.3&0.1&0&0\\0.5&0.4&0&0.1&0\\0.7&0.3&0&0&0\\0.5&0.2&0.3&0&0\end{bmatrix},R_{42}=\begin{bmatrix}0.4&0.1&0.3&0.2&0\\0.3&0.2&0.5&0&0\\0.5&0.3&0.1&0&0.1\\0.4&0.4&0.2&0&0\end{bmatrix}R_{43}=\begin{bmatrix}0.6&0.6&0.2&0&0\\0.7&0.3&0&0&0\\0.5&0.3&0.1&0.1&0\\0.4&0.5&0.1&0&0\\0.6&0.2&0.2&0&0\\0.7&0.1&0.2&0&0\end{bmatrix}$$

As for quantitative indexes, the membership degree function can be structured by reviewers and the membership degree of comment set can be verified as V = $(V_1, V_2, V_3, V_4, V_5)$. The comment set is defined as V= (best, better, good, medium, bad) = (≥0.75, 0.55 - 0.75, 0.3 - 0.55, 0.15 - 0.35, ≤0.15). The maximum value is 0.95 and the minimum value is 0. The membership degree of every level in the comment set is $y_1 \cdot y_2 \cdot y_3 \cdot y_4$. The membership degree function is defined as figure 3.

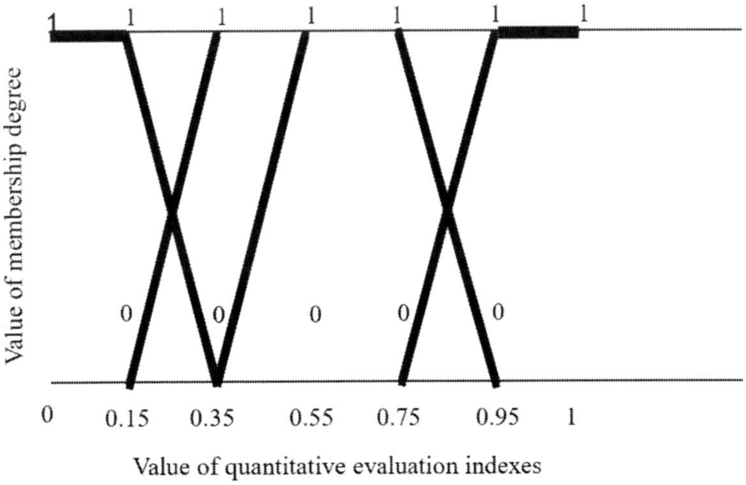

Figure 3: The membership degree function

The vector of membership degree for disassembly index is $R_{11} = (0,0,0,0.3565,0.6435)$, The vector of membership degree for cleaning index is $R_{12} = (0,0,0,0.2715,0.7285)$, The vector of membership degree for detection index is $R_{13} = (0,0,0,0.589,0.411)$, The vector of membership degree for re-processing index is $R_{14} = (1,0,0,0,0)$, The vector of membership degree for re-assembly index is $R_{15} = (0,0,0,0.1245,0.8755)$, The vector of membership degree for economic index is $R_2 = (0,0.762,0.238,0,0)$, The vector of membership degree for energy saving index is $R_{31} = (0.25,0.75,0,0,0)$, The vector of membership degree for resource saving index is $R_{32} = (0.931,0.069,0,0,0)$, The vector of membership degree for environment impact index is $R_{33} = (1,0,0,0,0)$.

5.5 Fuzzy composition

In the quality evaluation system, single factor evaluation should be firstly implemented, and then the fuzzy comprehensive evaluation should be implemented. The fuzzy composition is according to the equation 18.

$$\boldsymbol{B_{41}} = w_{41}\boldsymbol{R_{41}} = (0.5371,0.2909,0.0366,0)$$

$$\boldsymbol{B_{42}} = w_{42}\boldsymbol{R_{42}} = (0.4029,0.2387,0.2764,0.0526,0.0294)$$

$$\boldsymbol{B_{43}} = w_{43}\boldsymbol{R_{43}} = (0.5877,0.2614,0.1327,0.0155,0)$$

$$R_4 = \begin{bmatrix} 0.5371 & 0.2909 & 0.1327 & 0.0366 & 0 \\ 0.4029 & 0.2387 & 0.2764 & 0.0526 & 0.0294 \\ 0.5777 & 0.2614 & 0.1327 & 0.0155 & 0 \end{bmatrix}$$

$$\boldsymbol{B}_4 = w_4\boldsymbol{R}_4 = (0.5142,0.2659,0.1751,0.0344,0.0009)$$

5.6 Comprehensive evaluation

Similarly, the following vectors can be obtained:

$$\boldsymbol{B}_1 = (0.278,0,0,0.174,0.548)$$

$$\boldsymbol{B}_2 = (0,0.76,0.24,0,0)$$

$$\boldsymbol{B}_3 = (0.718,0.282,0,0,0)$$

According to the index weight of technology index, economic index, green index and comprehensive performance improvement, and the evaluation matrix of the first evaluation index $\mathbf{B} = [\boldsymbol{B}_1, \boldsymbol{B}_2, \boldsymbol{B}_3, \boldsymbol{B}_4]^T$:

$$\mathrm{G} = \mathrm{w}\begin{bmatrix} B_1 \\ B_2 \\ B_3 \\ B_4 \end{bmatrix} = \begin{bmatrix} 0.315 \\ 0.325 \\ 0.104 \\ 0.068 \\ 0.188 \end{bmatrix}^T$$

The final result is $\mathrm{F} = GC^T = 79.2$, showing that the remanufacturability of TPX6113 is good according to the comment set and the corresponding grade vector. This result means the TPX6113 have the value to be remanufactured. The original machine tool and the remanufactured machine tool are shown in figure 4. As can be seen, the wasted TPX6113 is well remanufactured and the methods proposed are verified with effectiveness and feasibility by the result.

Figure 4: The comparison of the original machine tool and the remanufactured machine tool

6 CONCLUSION

(1) The main factors affecting the remanufacturability of wasted machine tools are identified as the original condition, the disassembly process, the rationality of evaluation, the economic attributes and the sustainability attributes. And the comprehensive performance improvements should also be considered to rise the competitiveness.

(2) The remanufacturability evaluation system is established including technical indexes, economic indexes, sustainability indexes and comprehensive performance improvement indexes. Every index are defined by qualitative or quantitative methods.

(3) The remanufacturability evaluation method based on G1 and fuzzy comprehensive evaluation is proposed and applied to the evaluation process of the TPX6113 in a machine tool plant. The results show the TPX6113 has the enough value to be remanufactured.

Besides, the evaluation method proposed belongs to the theoretical research and needs to be improved by the actual situations. The evaluation software can be also developed based on the proposed method to improve the evaluation effectiveness.

7 ACKNOWLEDGEMENT

We gratefully acknowledge the financial support of National Natural Science Foundation of China, No. 51305279.

8 REFERENCES

[1] **Cao H.J.** 2010. The key technologies and industrial application of waste machine tool remanufacturing, China Plant Engineering, pp7-9.

[2] **Bras B., Hammond R.** 1996. Towards design for remanufacturing metrics for assessing remanufacturability, Proceedings of the 1st International Workshop on Reuse, Eindhoven, The Netherlands.

[3] **Zhang Z.X, Xiao S.M. et al.**, Shi Y.Q. 2010. Assessment of remanufacture ability based on property of product, Machine building and automation, 39 (1), pp74-76.

[4] **Zhou L., Xie J., et al.** 2015. Forecasting return of used products for remanufacturing using Graphical Evaluation and Review Technique (GERT), *International Journal of Production Economics*, article in press.

[5] **Mao G.Q., Zhu Y.W., Wu C.** 2009. A remanufacturability assessing model for waste electromechanical products, Modern Manufacturing Engineering, 06, pp114-118.

[6] **Lu C., Pan S.F., et al.** 2014, Evaluation model for heavy machine tool remanufacturing scheme based on ANP, *Applied Mechanics and Materials*, 635-637, pp1836-1840.

[7] **Satty T.L., Tran L.T.** 2007. On the invalidity of fuzzy numerical judgments in the Analytic hierarchy process, 48(7-8), pp962-975.

[8] **Wang R.B.** 2013. Study on the green manufacturing model and support technology of machine tool industry, Guide of Sci-tech Magazine, 17, pp313.

[9] **Zhu S.H., Xu B.S.** 2006. Resources and environmental assessment of waste machinery remanufacturing, China Surface Engineering, 19(02), pp6-12.

[10] **Zhang Z.X.** 2010. Research of the assessment of remanufacturability based on green manufacture, Sichuan: Southwest University.

ON APPLICATION OF ANDON TOOL AS INFORMATION-SHARING SYSTEM USED IN CONSTRUCTION INDUSTRY

K. Murata*
Department of Industrial Engineering and Management,
College of Industrial Technology, Nihon University, Chiba, JAPAN
murata.kouichi30@nihon-u.ac.jp

ABSTRACT

This paper analyses the ANDON tool, one representative tools for visual management (VM), within a construction company. The analysis is based on the Toyota Production System (TPS), a type of lean management developed in the manufacturing industry. The analysed tool manages the progress of executed construction projects among work-site, local office, and head office locations. The purpose of this paper is to understand the application of TPS to the construction industry through the analysis. An analysis framework of VM systems is used to understand the mechanisms of the analysed object. The results of the analysis show that an application of the concept of TPS to the case company is certainly effective for systematically promoting sharing of varied information within construction industry. The systems within the company are also based on a relationship of trust among individuals to improve the capabilities related to safety and management of their jobs. This paper can serve as a good guide to researchers who are interested in a transfer of lean management practices from the manufacturing industry to the construction industry. This research also contributes to other construction companies that hope to improve information sharing within a single company and among a group of companies.

Keywords: Technology Transfer, Manufacturing Industry, Construction Industry, Visual Management, Information Sharing System, Toyota Production System (TPS)

1 INTRODUCTION

Visual management (VM) is now an important tool for realizing smooth management processes. Toyota Motor Corporation is said to be the founder of VM (Endo 2005), with the original purpose being a classification of normal or abnormal (Ono 1988). Moreover, lean management, which was formed as a concept from Toyota Production System (TPS), made a wider interpretation as follows;

'the placement in plain view of all tools, parts, production activities, and indicators of production system performance, so the status of the system can be understood at a glance by every involved. Used synonymously with transparency' (Womack and Jones 2003).

In addition, many cases of VM are found in other industries beyond manufacturing (Nakajima and Murata 2015).

Based on the above situation, the purpose of this study is to understand the novel possibilities of VM. In particular, this study focuses on ANDON tool used in the management system of the construction

* Corresponding Author

company. Finally, through comparing with ANDON tool of a production system, new possibilities of VM are considered.

2 RESEARCH BACKGROUND

2.1 Visual Management of Lean Construction

Lean construction was born from the TPS, which was renamed as lean management (Womack et al. 2003). This concept has been introduced to six countries including the United Kingdom, Russia, Israel, Denmark, Germany, Norway, Australia and Finland by the Lean Construction Institute (LCI) founded in 1997 (LCI 2016). Sacks et al. (2010) illustrate the concept as follows: 'As in the TPS, the focus in lean construction is on reduction in waste, increase in value to the customer, and continuous improvement'. The TPS has developed many useful methods and Visual Management (VM) is one of them. Andon and Kanban are representative examples and VM is also used on construction sites. Tezel et al. (2010) reported 187 VM cases in the construction sites of high-rise buildings in Brazil. These cases were classified into the following 18 groups:1) Site layout and fencing, 2) Standardisation of the workplace elements, 3) In the warehouse, 4) The 5s, 5) In the elevators, 6) Pull production through the Kanban, 7) Production levelling through the heijunka box, 8) In-station quality (jidoka) through the Andon, 9) Prototyping and sampling, 10) Visual signs, 11) Visual work facilitators, 12) Improvisational visual management, 13) Explaining the work schedule, 14) Performance management through visual management, 15) Distributing system wide information, 16) Human resources management, 17) Safety management and 18) Poka-yoke (mistake proofing) and prefabrication. Murata et al. (2015b) also focused on VM for operation management and conducted a comparative analysis of VM cases from the construction industry and those from manufacturing to explore the transfer of this technology between these industries. This paper focuses on ANDON, one representative of VM tools as a method of sharing information related to project management.

2.2 Kaizen activity of case company

The company studied in this case analysis was founded in 1921 and presently employs about 400 people. It is a construction company with two business units: a civil engineering unit and a construction unit. In 2008, they started Kabuki Construction System (KCS) using a TPS system (Kabuki 2016b). KSC has two purposes and three systems based on the company's approach to VM, as shown in Figure 1 (Kabuki 2016a). The first purpose is information sharing within the value chain and there are two systems used to realize the first purpose. The first system (a) is designed to share projects among departments within the company, such as the business department, design department, operations department, and management department. The second system (b) is designed to share the progress of the project among the work-site, the local office to take charge of plural work-sites, and the head office to manage all local offices. The second purpose of the system is to promote information sharing on the work-site. In order to realize the second purpose, system (c) contributes to visualizing executed projects from three viewpoints: the safety environment, high quality construction, and smooth operation.

Moreover, a forum (d) has been created for exchanging practical experiences of construction, including activities related to KCS. Participants include the staff of the company and members of companies that have a business relationship with the company.

In this paper, (b) system are focused on and analysed by the methods described in Section 3.

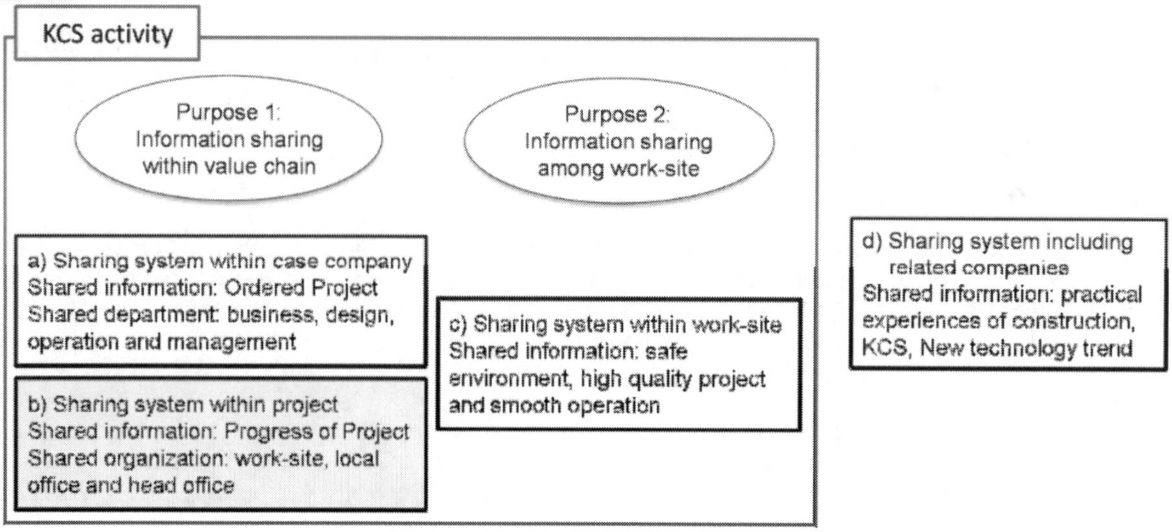

*Grey hutching: Focused part of this paper

Figure 1: Framework of KCS activity

3 METHODOLOGY

Figure 2 demonstrates the analytical method used in this paper. It shows the basic mechanism of VM systems. Within the framework, VM is thought of as a tool that supports communication between humans and the production system that is managed by humans. This framework has three components: a) humans, b) the production system managed by humans, and c) a tool to realize VM. This framework hopes to design two items: 1) a request action (RA) from the production system to a human, and 2) a function for the tool.

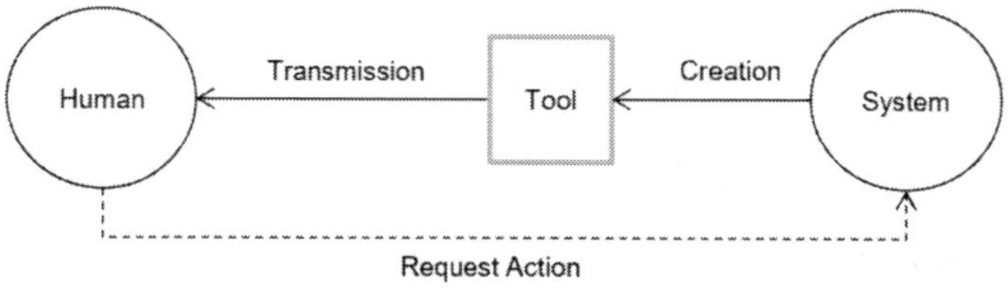

Figure 2: Basic mechanism for VM system (Murata et al. 2015a)

RA is a human action requested by the production system when it fails. When designing RA, its comprehensibility needs to be considered. In this design, it is necessary to determine the range of information offered from a tool for solving problems that occur in the production system. If the operator who manages the production system is an expert, all information offered from the tool is sufficient during emergency situations. However, if the operator is a novice, the tool needs to offer a procedure for handling emergencies. When multiple information items need to be visualized, the tool's design is more deliberate.

For the second item, two tool functions, namely creation of information (F-1) and transmission of information (F-2), need to be designed.

In the first function, a visualized item created by the production system is designed and sent from the system to a tool. When designing this function, the 'non-ordinariness' and 'continuity' of the created information need to be considered. 'Non-ordinariness' needs to be designed to notice when a visualized item needs to be created. When the production system is in the ordinary condition, the tool is not required to work. This tool's important role is to handle a special situation within a production system. With regard to 'continuity', the tool must have a structure for monitoring the typical conditions of the production system in order to grasp special situations at any time as they arise.

In the second function, the visualized and created item is provided from the tool to a manager. When designing this function, the 'inducibility' and 'immediacy' of the delivered visualized item need to be considered. With regard to the former, the tool needs to have appeal for operators because their physical and mental conditions are not always stable due to stress, exhaustion, or fatigue, etc. In order to improve the capability of 'inducibility' in the tool, it is necessary to build a device that attracts the operator's attention to the tool. With regard to the latter, the delivery timing of a created visualized item has to be when immediate response is required to the prevailing condition of the production system.

This paper focuses on the following system used in the one activity described in Section 2.2.

Analysis: (b) A management system for construction projects

For the analysis, the company studied uses a system that is similar to Andon. Therefore, in this analysis, this system is compared with Andon as utilized in manufacturing industry. The analysis procedure is as follows: Step 1) describing the two systems, Step 2) analysing the two systems with the framework outlined in Figure 2, and Step 3) comparing the two sets of analysis results. Resources for the analysis are collected by a document analysis and an interview related to the system.

4 ANALAYSIS

4.1 Description of analysed objects (Step 1)

4.1.1 Andon in production line

'Just-in-time' and 'Automation with a human element (Jidoka)' are two pillars of TPS (Spear and Bowen 1999). A stack light is classified for realizing the latter pillar (Ono 1988). TPS thinks of this as a VM tool and calls it 'Andon'. In a production line, the stack light has rich colour variation, such as two, three, and four colours. In this study, a stack light with three colours is considered. The stack light operation follows the rule of colours, which contributes to the maintenance and improvement of the utilization rate for the production line. The light's role is to monitor signals related to mechanical problems in the production line. The meaning of each colour in the stack light of a production line is summarized in Table 1. If the production line operates normally, the green light stays on. However, if any problems occur in the production line, the red light immediately turns on, and a line manager and/or operator inspects the present condition of the production line. Once the manager and/or operator resolve the issue, the yellow light turns on, and when the production line restarts, the green light turns on again.

Table 1: Colour meaning for Andon in production line

Colour	Status of production line per colour
Green	Production line operating normally.
Yellow	Manager and/or operator have resolved the issue.
Red	Problems occurring in production line.

4.1.2 Andon in case company

Construction projects are managed among three players at a work-site, a local office, and a head office. The actors at the work-site actually execute the construction projects. They must report the progress of their projects to the local office to which each of them belongs every month. In this report, they describe the present condition of their projects from six viewpoints such as plan progress, cost control, workmanship of the project, safe maintenance, environmental friendliness, and others. They also evaluate each item with three colours: green, yellow, and red. Green means that the project is on schedule without problems. Yellow means that the project may not work without problems. Red means that the project has problems. The meaning of each colour in the system is summarized in Table 2. The local office evaluates the present condition of the project in the report and sends it to the head office. The head office performs a final evaluation and decides on a support level, which consists of three levels. The first level (green) designates keeping watch on a project. The second level (yellow) requires support by an expert on a problem needed to solve. The third level (red) requires support by a team of experts on a problem needed to solve.

Table 2: Colour meaning for Andon in case company

Colour	Status of production line per colour
Green	Project is on schedule without problems.
Yellow	Project may not work without problems.
Red	Project has problems.

4.2 Analysis of described objects (Step 2)

4.2.1 Andon in production lines

The results analysed by this method are illustrated in Figure 3. The type of system is a production line. The human attribute is a line manager or operator. The tool is a stack light with three colours.

RA from the production line to human is designed to immediately understand the occurrence of any problems in the production line. Naturally, in order to recover the performance of the production line, a line manager or operator should understand the nature of the problem. However, many stack lights do not have functions that can identify the cause of the problem or deliver countermeasures because the problem can result from a wide range of causes.

For 'how to create information' (F-1), the stack light monitors a signal related to mechanical problems in the production line, detects changes in condition, and automatically switches to a suitable colour. For 'how to transmit information' (F-2), the moment the production line stops, the stack light immediately transmits this information to a manager or operator by switching the colour instantaneously.

4.2.2 Andon in the company

The results analysed by this method are illustrated in Figure 4 and the type of system is a work-site. The human attribute is both a local office and a head office. The tool is a report of the present condition of the project.

RA from the work-site to a local office and a head office is designed to deliver quick and adequate support for the project.

For 'how to create information' (F-1), the work site describes the present condition of the project from the aforementioned six perspectives every month and evaluates each item using three colours. For 'how to transmit information' (F-2), a local office and a head office evaluate the judgment of the present condition of the project by work-site and decide support levels based on this evaluation.

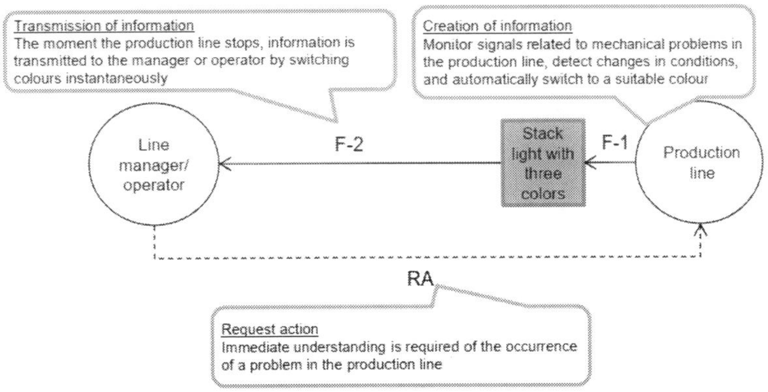

Figure 3: Analysis results from Andon in the production line

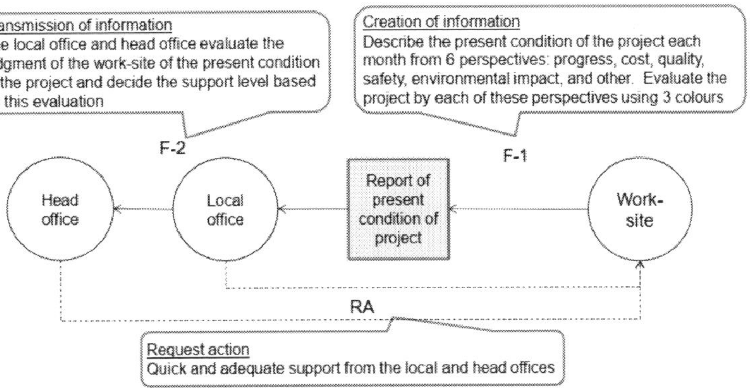

Figure 4: Analysis results from Andon in the company

4.3 Comparison of the two analysed systems (Step 3)

Table 3 shows the comparison result between two Andons of the production line and the company.

Within the production line, abnormal situations can occur. The detection mechanism for an abnormal situation is a mechanical stack light. The solution is human-made by a line manager or operator. Visual information is created by electric signal from one viewpoint related to machine operation and is mechanically transmitted directly to the line manager or operator signified by colours.

Within the company, abnormal situations can occur on a work-site. The detection mechanism for abnormal work-site situations is a report by a manager. The solution is a systematic procedure involving work-sites, local offices, and the head office of the company. Visual information is created by each work-site using an evaluation based on the six viewpoints related to the executed project. Information is systematically transmitted to the local office and the head office in multiple steps and signified by colours.

The Andon system of the company detects an abnormal situation within a part of the project managed by a human being. This is an unstable mechanism compared to mechanical detection such as the Andon of the production line. In order to strengthen this function, the company must have multiple steps for detecting an abnormal situation by three separate organizations, the work-site, local office, and head office. A plurality of viewpoints is necessary to judge an abnormal situation of a managed part of the project. Based on interviews of staff in the company, these systems are a support system for a continuously operating Andon system and are based on trust among employees. In the old days, the work-site corresponded to an abnormal situation alone without any reporting to superior organizations. As a result, on many occasions serious damage occurred to the project. Owing to the Andon system, abnormal situations are now resolved by staff of the company while respecting the judgement of the work-site.

Table 3: Quantified data of two Andons

Compared item	Production line	Case company
Manager of abnormal situations	Production line	Work-site
Detection mechanism	Mechanical, three colours	Human-made, three colours
of abnormal situations	(Static light)	(Report)
Solution mechanism	Human-made	Systematic
for abnormal situations	(Line manager/operator)	(Work-site, Local office, and Head office)
Creation of information	Electric signal	Judgement by work-site
	One viewpoint	Six viewpoints
	(Line stops)	(Plan progressing, Cost control, Workmanship of the project, Safety maintenance, Environmentally friendly and Others)
Transmission of information	Mechanical	Systematic
	Colour	Colour
	Single step	Multiple steps

5 CONCLUSION

In this paper, one information-sharing systems of a construction company are analysed. The system is based on TPS, a lean management system developed by the manufacturing industry. It is a management system for construction projects quouted by ANDON tool, a key tool of TPS. As the result of the analysis, the result shows that an application of a lean manufacturing scheme, focusing on VM, within the company is certainly effective for systematically sharing various types of information. Lean activities within the company are based on a relationship of trust and individuals can work to improve capabilities related to safety and efficient management of their projects.

This research has practical application for other construction companies that hope to improve information sharing within a single company or among a group of companies. This paper can also serve as a guide to researchers who are interested in a transfer of lean management from the manufacturing industry to the construction industry.

6 ACKNOWLEDGMENT

This research was undertaken with the support of Kabuki construction company Ltd. in Japan. We sincerely appreciate the excellent research opportunity that the company provided and wish to thank the staff members for their kind support during interviews and in providing valuable materials. We would also like to recognize Mr. Riku Sasaki who supported our data collection efforts and prepared interviews as a member of our research laboratory. This paper was also supported by JSPS KAKENHI Grant Number JP 16K03897.

7 REFERENCES

[1] **Endo, E.** 2005. *Visualization - Mieruka [In Japanese]*, Toyo Keizai Inc.

[2] **Kabuki construction company Ltd.** 2016a. *A booklet of Kabuki Construction System (in Japanese).*

[3] **Kabuki construction company Ltd.** 2016b. Homepages of Kabuki construction company Ltd. (in Japanese), URL: http://www.kabuki.co.jp/company/ (Access day: 3rd March, 2016b).

[4] **Lean Construction Institute.** 2016. Homepages of Lean Construction Institute, URL: http://www.leanconstruction.org/ (Access days 3rd March, 2016).

[5] **Murata, K. and Katayama, H.** 2015. A study on improving corporate lean problem in manufacturing industry through learning visual management cases in other industries, Symposium Proceedings of 19th Cambridge International Manufacturing Symposium, Moeller Centre, Churchill College, University of Cambridge, Cambridge, United Kingdom, 24th-25th September, 11 pages in USB.

[6] **Murata, K., Nakajima, K., Kinoshita, K., Tezel, A., Koskela, L., Tzortzopoulos, P. and Katayama, H.** 2015. Investigation of Visual Management Cases in Construction by An Analytical Framework from Manufacturing, Proceedings of 45th International Conference on Computers & Industrial Engineering (CIE45), Université de Lorraine, Metz, France, 28th -30th October, 8 pages in USB.

[7] **Nakajima, K. and Murata, K.** 2015. A Study on Multi-purpose Stack Light with Three Colors: Through Analysis of the Management System of a Barber and Beauty Store of a Case Company, Proceedings of the International Conference on Innovation, Entrepreneurship and Technology (ICONIET2015), Hotel Mercure Serpong Alam Sutera, Tangerang, Indonesia, 25th-26th, pp. 301-306.

[8] **Ohno, T.** 1988. *Toyota Production System: Beyond Large-scale Production*, Portland: Productivity Press.

[9] **Sacks, R., Koskela, L., Dave A. B., Owen, R.**, 2010. Interaction of Lean and Building Information Modeling in Construction, *Journal of Construction Engineering and Management*, Vol. 136, No. 9, pp 968-980.

[10] **Spear, S. and Bowen, H. K.** 1999. Decoding the DNA of the Toyota Production System, *Harvard Business Review*, September/October, pp. 96-106.

[11] **Tezel, A., Koskela, L. and Tzortzopoulos, P.** 2010. Visual Management in Construction: Study Report on Brazilian Cases, University of Salford.

[12] **Womack, J.P. and Jones, D.T.** 2003. *Lean thinking: Banish waste and create wealth in your corporation*, Free Press.

A CASE STUDY FOR KNOWLEDGE SHARING OF KAIZEN ACTIVITY IN CONSTRUCTION INDUSTRY

K. Murata*
Department of Industrial Engineering and Management,
College of Industrial Technology, Nihon University, Chiba, JAPAN
murata.kouichi30@nihon-u.ac.jp

ABSTRACT

This paper describes the knowledge-sharing system of the kaizen activity in case construction company. The kaizen activity of the company is based on Toyota Production System (TPS), a type of lean management developed in the manufacturing industry. The type of analysed system is a forum for exchanging practical experiences of construction among all company staff, including related companies' members. The purpose of this paper is to understand the kaizen activity of the construction company which has applied to TPS. This paper considers the aforementioned knowledge sharing system as visual management (VM) system. Subsequently, an analysis framework of VM systems is used to understand the mechanisms of the analysed object. To collect data from the analysed object, and participation in the system, an analysis of relevant documents and interviews with key personnel is performed. The results of the analysis show that an application of the concept of TPS to the case company is certainly effective for systematically promoting sharing of varied information/knowledge.

Keywords: Knowledge Sharing, Construction Industry, Kaizen, Toyota Production System (TPS)

1 INTRODUCTION

In contemporary Japan, buildings and social infrastructure need to be rebuilt and repaired due to lapses in maintenance for years after their construction. The Ministry of Land, Infrastructure, Transport and Tourism (MLIT) announced that 2013 was the 'First Year of Social Capital Maintenance Activity' in Japan (MLIT 2014). Furthermore, the need for new construction is rising because the next summer Olympic Games are to be held in Tokyo, in 2020. In order to meet both requirements, construction companies require experienced workers, the development of young engineers, and a realization of more effective and safety-related management technologies.

Based on these recent trends within the construction industry, this study analyses the knowledge sharing system of the kaizen activity within a company utilizing a scheme of Toyota Production System (TPS) (Ono 1988, Womack and Jones), which is based on a lean management scheme developed in the manufacturing industry. The object of analysis is the forum for exchanging practical experiences of construction. Moreover this paper describes the movement of applying a management scheme developed in the manufacturing industry to the construction industry by describing and analysing this object.

2 KAIZEN ACTIVITY OF CASE COMPANY

* Corresponding Author

The company studied in this case analysis was founded in 1921 and presently employs about 400 people. It is a construction company with two business units: a civil engineering unit and a construction unit. In 2008, they started Kabuki Construction System (KCS) using a TPS system (Kabuki 2016b). KSC has two purposes and three systems based on the company's approach to VM, as shown in Figure 1 (Kabuki 2016a). The first purpose is information sharing within the value chain and there are two systems used to realize the first purpose. The first system (a) is designed to share projects among departments within the company, such as the business department, design department, operations department, and management department. The second system (b) is designed to share the progress of the project among the work-site, the local office to take charge of plural work-sites, and the head office to manage all local offices. The second purpose of the system is to promote information sharing on the work-site. In order to realize the second purpose, system (c) contributes to visualizing executed projects from three viewpoints: the safety environment, high quality construction, and smooth operation.

Moreover, a forum (d) has been created for exchanging practical experiences of construction, including mentioned above three activities related to KCS. Participants include the staff of the company and members of companies that have a business relationship with the company.

In this paper, (d) systems are focused on and analysed by the methods described in Section 3.

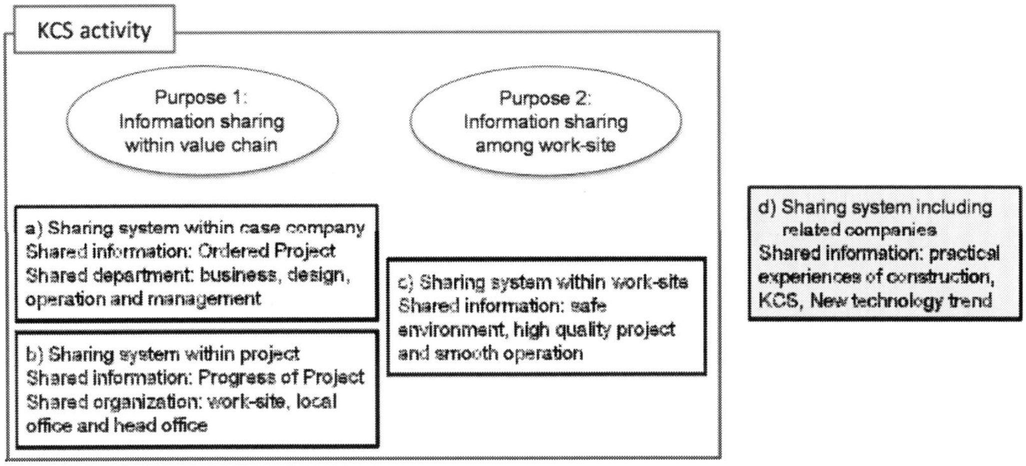

*Grey hutching: Focused part of this paper

Figure 1: Framework of KCS activity

3 METHODOLOGY

Figure 2 demonstrates the analytical method used in this paper. It shows the basic mechanism of VM systems. Within the framework, VM is thought of as a tool that supports communication between humans and the production system that is managed by humans. This framework has three components: a) humans, b) the production system managed by humans, and c) a tool to realize VM. This framework hopes to design two items: 1) a request action (RA) from the production system to a human, and 2) a function for the tool.

RA is a human action requested by the production system when it fails. When designing RA, its comprehensibility needs to be considered. In this design, it is necessary to determine the range of information offered from a tool for solving problems that occur in the production system. If the operator who manages the production system is an expert, all information offered from the tool is sufficient during emergency situations. However, if the operator is a novice, the tool needs to offer a procedure for handling emergencies. When multiple information items need to be visualized, the tool's design is more deliberate.

For the second item, two tool functions, namely creation of information (F-1) and transmission of information (F-2), need to be designed.

In the first function, a visualized item created by the production system is designed and sent from the system to a tool. When designing this function, the 'non-ordinariness' and 'continuity' of the created information need to be considered. 'Non-ordinariness' needs to be designed to notice when a visualized item needs to be created. When the production system is in the ordinary condition, the tool is not required to work. This tool's important role is to handle a special situation within a production system. With regard to 'continuity', the tool must have a structure for monitoring the typical conditions of the production system in order to grasp special situations at any time as they arise.

In the second function, the visualized and created item is provided from the tool to a manager. When designing this function, the 'inducibility' and 'immediacy' of the delivered visualized item need to be considered. With regard to the former, the tool needs to have appeal for operators because their physical and mental conditions are not always stable due to stress, exhaustion, or fatigue, etc. In order to improve the capability of 'inducibility' in the tool, it is necessary to build a device that attracts the operator's attention to the tool. With regard to the latter, the delivery timing of a created visualized item has to be when immediate response is required to the prevailing condition of the production system.

This paper focuses on the following activity of the four activities described in Section 2.

Analysis: (d) A forum for exchanging practical experiences of construction projects

For the analysis, the analysed object considered consists of the three elements of the VM framework described in Figure 3: the system of the company studied in this case, the forum tool, and the human participants. Based on this framework, the analysis of this paper clarifies a) RA by the forum and b) two of the forum's functions, which are a creation function and a transmission function from the analysed forum to its participants. In order to collect forum data, participation within the forum, an analysis of relevant documents, and interviews with key personnel were performed. Based on the data collection, the following analysis is performed.

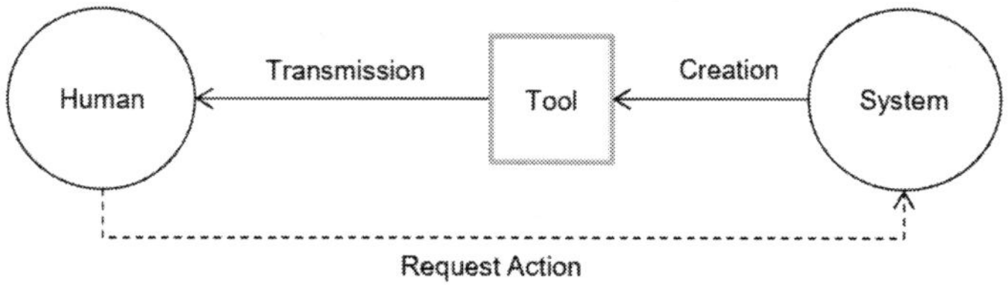

Figure 2: Basic mechanism for VM system (Murata et al. 2015a)

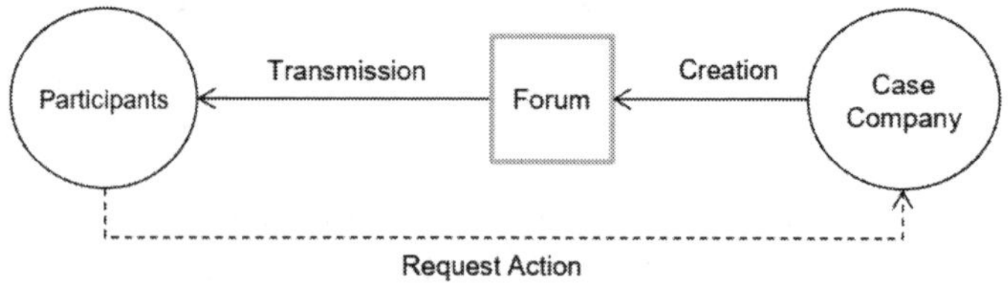

Figure 3: Basic mechanism for analysing the forum of the company

4 ANALYSIS

4.1 Outline of the forum

A forum has been planned once every two years, and the seventh forum was held on 21 November 2016. The place of the forum is a venue outside of the company. The participants are board members, the president, current employees, incoming employees, and associated corporate members. The main purpose of the forum is to improve the presentation skills of company employees because a competition to obtain new projects is won though a presentation of its plan. Preparation of the presentation began six months prior to the scheduled date of the forum. The forum's preparation committee checked the quality of the presentation several times until the day of the event. At the forum, eight presentations of different construction projects were performed and the presentation time was 10 minutes. Afterwards, all presentations were ranked and one presentation was chosen to win a 'best' prize, two presentations were chosen to receive an 'excellent' prize, and the others were chosen for an 'effort' prize. After finishing all presentations, contemporary topics, new technologies related to construction industry, and a history of Kabuki Construction System (KCS) are shared among participants of the forum.

4.2 Analysed result 1: The relationship between the forum and the request action (RA)

Figure 4 analyses result of the relationship between the forum and the request action. The forum consists of four main parts: 1) a greeting by president, 2) a presentation, 3) an introduction of a recent topic and new technology, and 4) a report by KCS. Each part contains an RA. The first part is designed to penetrate the philosophy of the company. The second part is designed to share information of executed projects. The third part is designed to promote study to improve the capabilities of all forum participants. The fourth part is designed to discuss Kaizen.

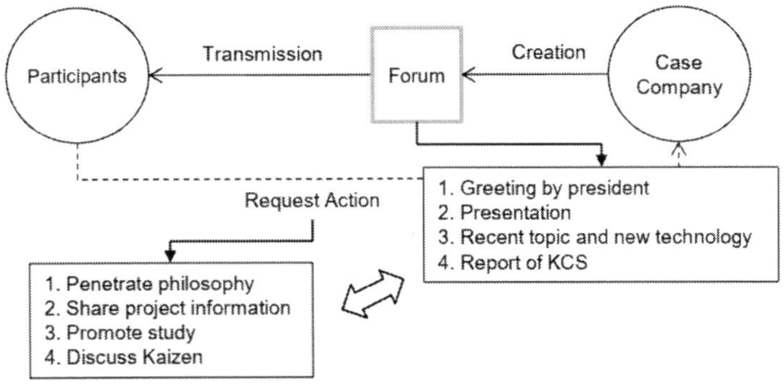

Figure 4: The Forum's tool and request action

These four parts are arranged on a matrix as shown in Figure 5 on two axes. One axis is related to the business itself, which has two components: a foundation for daily jobs and daily jobs. The other axis is related to business innovation, which has two components: succession/continuation and growth/effort. Consequently, the penetration of the philosophy of the company lies in the second quadrant, sharing of project information lies in the third quadrant, the promotion of study to improve capabilities lies in the first quadrant, and the enlightenment of Kaizen lies in the fourth quadrant, as a sequence of forum's program.

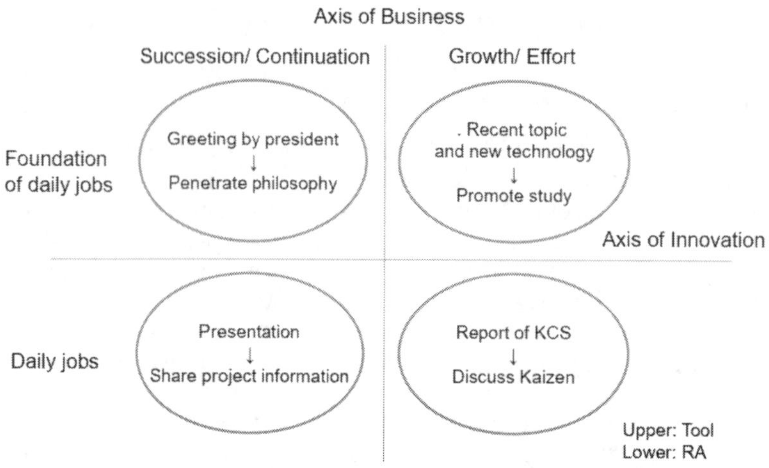

Figure 5: The relationship between the forum and the request action (RA)

4.3 Analysis 2: Two functions of the forum

Two functions of the four portions of the forum, the creation of information and the transmission of information are summarized in Tables 1-5.

The first portion was a program designed to penetrate the company's philosophy, which consists of an opening ceremony and closing ceremony. In the two ceremonies, a board member of the company and the president talk about recent changes in the external environment of the company, the meaning of the forum, an evaluation of the presentations during the forum, and so on. Board members and the president devise thoughtful phases in order to convey the purpose of the forum to staff. For example, the president used the phrase 'presentation full of the human touch' in order to transmit purpose. Two awards were also given during the forum. One was for the presentations in the forum and the other was for the daily project itself. Two different awards at the same time contributed to improving presentation skills and motivation for daily jobs.

The second portion was a program designed to share project information about projects. This was the main program within the forum. Eight presentations were performed as shown in Table 5. There were three presentations by the civil engineering department, four presentations by the construction department, and one presentation that represented a joint effort of the two departments. In each presentation, the object of the project, project needs, problems that occurred during the project, and methods used to solve the problems were introduced. The style of the majority of the presentations was oral using a slideshow and notes on the stage.

The third portion was a program was designed to promote study of project competencies. In this part of the program, recent topics of concern and a new technology for improving performance of daily projects were discussed. The recent topic discussed was environmental concerns and the new technology was a virtual simulation.

The fourth portion of the forum was a program designed to discuss Kaizen. It is conventionally difficult to continue Kaizen. Since one countermeasure is to prevent against routine, this portion illustrated a background of KCS, told a story of KCS, and discussed the recent situation of KCS.

Table 1: Summary of the functions of the program to penetrate company philosophy

Agenda of analysed forum		Creation of information (What does the forum inform or visualize?)	Transmission of information (How does the forum transmit created information?)
Opening ceremony	Opening address	Declaration of opening	By general manager for civil engineering (orally, on the stage)
	Speech from president	Recent topics in the construction industry (Need for sincere communication etc.) Meaning of the forum (Improving job skills, etc.) Need for continuation of the forum Quality of presentation ('Presentation full of the human touch') KCS activity	By president (orally, on the stage)
	Introduction of judging committee	Judging committee How to judge a presentation	By moderator (orally)
Closing ceremony	Award ceremony for projects	Award of general manager for civil engineering Award of general manager for construction	By each general manager (orally, on the stage)
	Announcement of the results	Judgement result	By moderator (orally)
	Award ceremony for presentations	Best prize, Excellent prize, Effort prize Awarding of each prize	By president (orally, on the stage)
	Speech for review	Comment on the presentations (Evaluation of the presentation at this time, Suggestions for improving skills and materials of presentation, Impact to relevant skill)	By head of engineer (orally, on the stage)
	Closing address	Expression of appreciation for presenters Comments on the forum (Evaluation of presentations at this time, Status of daily projects)	By vice president (orally, on the stage)

Table 2: Summary of the function of the program designed to share project information 1

Agenda of analysed forum	Creation of information (What does the forum inform?)	Transmission of information (How does the forum transmit created information?)
Presentation 1	Object: Improvement of the interchange of expressway system Needs: Reduction of the construction period Problem: Consolidation settlement by soft ground with banking Method: Precast method	By Civil engineering department, Tokyo head office (slide, orally with a notes, on the stage)
Presentation 2	Object: Construction of nursing home Needs: Reduction of construction period Problem: Steep slope and many rolling stones Method: Improvement of both erosion barrier method and Soil improvement method	By Construction department, Osaka branch (slide, orally with a notes, on the stage)
Presentation 3	Object: Maintenance of deteriorated pier Needs: Strengthening of pier function Problem: Shortage of bearing power of reused steel tube stake Method: Improving of reused steel tube stake	By Civil engineering department, Tokyo head office (slide, orally with a notes, on the stage)

Table 3: Summary of the function of the program designed to share project information 2

Agenda of analysed forum	Creation of information (What does the forum inform?)	Transmission of information (How does the forum transmit created information?)
Presentation 4	Object: Construction of gymnasium Needs: Control of large roof under wind and snow Problem: Construction within narrow space Method: High precision of construction	By Construction department, Ibaraki head office (slide, orally with a notes, on the stage)
Presentation 5	Object: Seismic retrofitting of filtration plant Needs: Exchange of earthquake-proof water pipe Problem: No detailed drawing Method: By basic theory and expert knowledge	By Construction department, Osaka branch (slide, orally without a notes, on the stage)
Presentation 6	Object: Construction of apartment complex Needs: Safety management of construction vehicle and Retention of existing breast wall Problem: Built-up area and construction site with different elevations Method: Announcement of traffic procedure of construction vehicle and modification of erosion barrier method etc.	By Construction department, Tokyo head office (slide, orally with a notes, on the stage)
Presentation 7	Object: Repair of deteriorated bridge Needs: Identification of injury points and suggested repair Problem: Indefiniteness of injury points Method: Visualization of injury points based on detailed survey, etc.	By Civil engineering department, Ibaraki head office (slide, orally with a notes, on the stage)
Presentation 8	Object: Construction of public housing after Great East Japan earthquake Needs: Construction in mid-slope of steep terrain with rias coast Problem: Shortage of bearing capacity of the soil and Existence of water vein under bearing ground Method: Suitable soil improvement method and draining method	By Construction department and Civil engineering department, Tokyo head office (slide, orally with a notes, on the stage)

Table 4: Summary of the function of the program designed to promote a study of competencies

Agenda of analysed forum	Creation of information (What does the forum inform?)	Transmission of information (How does the forum transmit created information?)
Special lecture 1	Case of CO2 reduction with biomass Recycling system of garbage in a town	By business department and associated corporation (slide, orally, on the stage)
Special lecture 2	4-dimensional simulation model for improving project planning	By engineering department (slide, orally, on the stage)

Table 5: Summary of the function of the program designed to enlighten Kaizen

Agenda of analysed forum	Creation of information (What does the forum inform?)	Transmission of information (How does the forum transmit created information?)
Special lecture 3	Background of KCS (Diminished by the collapse of bubble the economy, Trouble in work-site, etc.) Story of KCS (Conventional management by foreman, Antipathy to cycle-time management, etc.) Recent situation of KCS (In total of the company, On each work-site, etc.)	By management department (slide, on the stage)

4.4 Discussion

The overall effects of the forum are a sharing of daily job duties with all staff, including associated corporate members. As shown in Figure 5s, this forum has four components. The first is the presentation of daily projects as the main event of the forum. The second is a greeting by the board members and the president to establish a foundation for daily job activity. The third is a report of kaizen activity for the continuous strengthening of operations. The fourth is an introduction of recent topics in order to recognize future directions for daily job activity. These components are systematic in order for members to share their experiences and knowledge of their daily jobs.

5 CONCLUSION

In this paper, a knowladge-sharing system of a construction company are analysed. The case comanpy have performed kaizen activites based on TPS, a lean management system developed by the manufacturing industry. The system is a forum for exchanging practical construction experiences. As the result of the two analyses, the result shows that the analysed system contributes to be certainly effective for systematically sharing various types of knowlagde/information. Lean activities within the company are based on a relationship of trust and individuals can work to improve capabilities related to safety and efficient management of their projects. This research has practical application for other construction companies that hope to improve knowlage sharing within a single company or among a group of companies. This paper can also serve as a guide to researchers who are interested in a transfer of lean management from the manufacturing industry to the construction industry.

6 ACKNOWLEDGMENT

This research was undertaken with the support of Kabuki construction company Ltd. in Japan. We sincerely appreciate the excellent research opportunity that the company provided and wish to thank the staff members for their kind support during interviews and in providing valuable materials. We would also like to recognize Mr. Riku Sasaki who supported our data collection efforts and prepared interviews as a member of our research laboratory. This paper was also supported by JSPS KAKENHI Grant Number JP 16K03897.

7 REFERENCES

[1] **Kabuki construction company Ltd.** 2016a. *A booklet of Kabuki Construction System (in Japanese).*

[2] **Kabuki construction company Ltd.** 2016b. Homepages of Kabuki construction company Ltd. (in Japanese), URL: http://www.kabuki.co.jp/company/ (Access day: 3rd March, 2016b).

[3] **Ministry of Land, Infrastructure, Transport and Tourism (MLIT)**, 2014. *WHITE PAPER ON LAND, INFRASTRUCTURE, TRANSPORT AND TOURISM IN JAPAN, 2014.*

[4] **Ohno, T.** 1988. *Toyota Production System: Beyond Large-scale Production*, Portland: Productivity Press.

[5] **Womack, J.P. and Jones, D.T.** 2003. *Lean thinking: Banish waste and create wealth in your corporation*, Free Press.

AN α ALGORITHM-BASED SHARED DESIGN CONTEXT EXTRACTION METHOD

Jihong Liu[1*] and Kejian Wang[1]
[1]School of Mechanical Engineering and Automation
BeiHang University, Beijing, China
ryukeiko@buaa.edu.cn

[1]School of Mechanical Engineering and Automation
BeiHang University, Beijing, China
wkj920218@.163.com

ABSTRACT

The current study which focuses on the collaborative design lacks the method to rapidly form and express effective design consensus, which makes it difficult to find out cognitive rules and mechanisms of consensus formation. This paper presents an α algorithm-based shared design context extraction method to gain the context of shared design to support knowledge-fused collaborative design. We firstly define the corresponding relationship between DR model and petri net, then remove the influence of noise in the context to the mining process, next scan each design context to extract shared design context through α algorithm, finally transform it into the well-understood shared design rationale model. An application example is demonstrated to verify the effectiveness of the shared design context extraction method.

Keywords: Collaborative design, Shared design context, Petri net, α Algorithm

1 INTRODUCTION

With the increasingly complex product design, in order to improve product quality, response to market rapidly, reduce development costs and improve user satisfaction, collaborative design has become a mainstream way of product development under the impetus of concurrent engineering and computer information technology [1]. Co-design has the characters of treating human-body as the main part in design team collaboration which is embodied as a set of "collecting intellectual behavior": participants generate shared comprehension of the team design by sharing their design intent, design justification, design criteria and other knowledge. Then form a relative stable design consensus through continuous mutual inspiration and gradual iterations, finally solve design problems [2].

In the process of solving the same problems, different designers tend to make different design schemes. However there lack of collaborative design methods for iterations and fusion of different design schemes to obtain an accurate and complete expression of design thinking process. To solve this problem, this paper characterizes the design process through design rationale model, and achieves collaborative design of design thinking integration through process mining method.

Design rationale (DR) knowledge is treated as an important design process knowledge which plays a guiding role in the process of making decisions. Extracting the design context effectively, reusing DR knowledge plays an important auxiliary role in helping design modifications and guiding other

---

* Jihong Liu, born in 1966, is currently a professor at Beihang University, China. His research interests include complex product engineering, digital design and manufacturing, knowledge management and design theory and method.
Acknowledge: The work is supported by National Science Foundation of China through approval No.51475027.

designers to comprehend the design process [3]. The current study on DR knowledge mainly focused on improving DR knowledge expression [4, 5], capturing DR knowledge [6], as well as improving the efficiency and accuracy of retrieval DR model [7], but lacking of effective support for the collaborative design. The design context of DR model represents design scheme, however, design context of the same design intent may be different due to different designers' idea. Thus the accuracy and fitness for the DR model of a single instance cannot be guaranteed, it is difficult to provide a reliable design reference to the designers.

Process mining means extracting knowledge from event logs to discover, monitor and improve the actual process [8]. Process mining is established on the process model-driven approach and data mining. One of the important applications is discovery technology which does not use any prior information of the event log to generate models. DR model records the details of design activity [9], but the majority of information is stored in unstructured form, which can be converted to structured events through rational algorithm, finally the model will be transformed to achieve the ultimate goal [10].

Therefore, we propose an α algorithm-based shared design context extraction method to excavate shared design context from a plurality of DR models which describe identical design tasks. We firstly define the shared design context then extract design context from single DR model and convert it into petri net according to the rules between DR model and petri net. Finally we remove the influence of noise in the context to the mining process and extract the shared design context to realize the knowledge-fused collaborative design.

2 SHARED DESIGN CONTEXT

The thinking ideas of designers when solving design tasks are embodied as design activities context in the design process, which including the design modification and the addition of design activities [11]. DR models are often too large and complex for designers to understand the overall design concept. The extraction of design context can help designers understand the design process and grasp the overall design ideas. In DR model, design context is a kind of design scheme which logically combined all the design elements to realize the design intent. The design context is shown in Figure 1 with dotted lines, the rejected design scheme nodes which can only provide information and support to the designer's decisions during the design process do not belong to the design context.

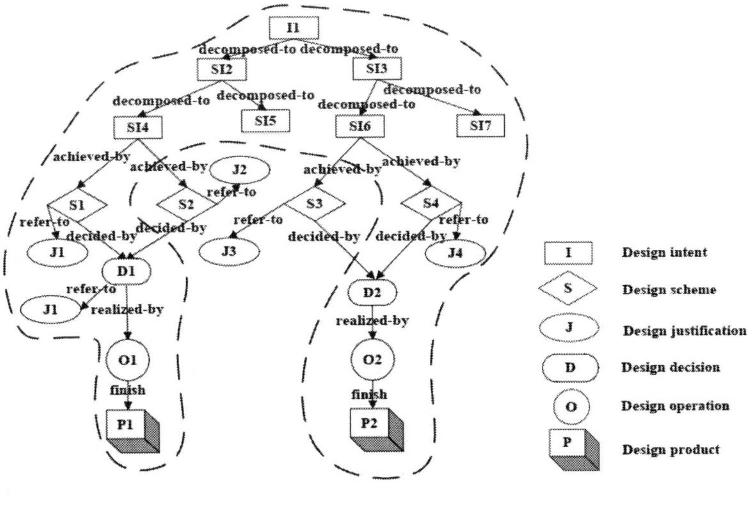

Figure 1: Design Context Model

Design context provides a design scheme to meet the design intent. It is difficult to ensure the suitability of the design scheme for the limit of designers' level of knowledge. Therefore we propose the concept of shared design context which means the integration of different design contexts that reflect the designer's shared design thinking process when solving the same problems. The organization of these models performs as a set of collecting intellectual behaviour which includes: 1) dynamic cognitive processes such as design problems definition, design concepts production, design reasoning and decision-making. 2) individual cognitive activities such as thinking, reasoning, weighing and decision-making. 3) the communication process between multiple designers in the demonstration, negotiation, compromise and other typical team cognitive processes.

3 DESIGN CONTEXT PETRI NET CONSTRUCTION

Before applying α algorithm for the mining of the shared design context, we need to transform the single design context to the petri net.

3.1 Petri Net

Petri net is a workflow modelling tool which represents the executable workflow through a simple and intuitive graphical way [8]. During the operation of complex systems, there are four typical relationships: sequential, parallel, selection and iteration among different business activities. Petri net express parallelism and selection relationship among business activities by four different route structures such as AND-Split, AND-Join, XOR-Split and XOR-Join, and express sequential route and iteration route through directional arcs.

Currently the DR can be represented in informal, semi-formal and formal ways. The informal representation is difficult to identify by a computer, such as design documents, etc. The semi-formal is difficult to excavate, such as IBIS model. The formal representation such as DMML language which is based on the OWL ontology language uses logic rules and the reasoning methods to represent design thinking process. However, the formal representation of DR requires designers gain some understanding about the ontology language which limits the designers' ability of representation. Petri net can represent DR model in a simple and intuitive graphical way which can be identified by the computer, so it is more suitable for the needs of design context representation.

3.2 Mapping Relationship between Petri Net and DR Model

The design context model which is extracted from DR model contains elements and their inter-relationship. Elements are embodied in the form of nodes while the relationship between elements in the model is expressed in the form of cable. In this paper, we transform the design context model into petri net by building the mapping between two models' elements. The nodes of DR model are mapped into transitions and the relationship between nodes is mapped into the route. The mapping relationship is shown in Figure 2.

Design intents, accepted design options, design decisions, design operation and design artefacts of the design context model are mapped into transitions of the petri net. The relationship "decomposed-into" is mapped into AND-Split means output to multiple places when a task is completed simultaneously, allowing follow-up concurrent tasks being executed. The relationship "merged-into" is mapped into AND-Join which is used for mission combination that means the follow-up tasks can be performed only when all inputs are true. The relationship "achieved-by" is mapped into XOR-Split which represents only one mission is allowed to occur when there is plurality of choices. The relationship "decided-by" is mapped into XOR-Join which represents that the subsequent mission occurs if one of the input missions has occurred. The remaining relationship, "realized-by", "initiate", "remove-to", "after-of", "finish" are mapped into sequential route which reflects a causal relationship between the main tasks in order to execute in a fixed order.

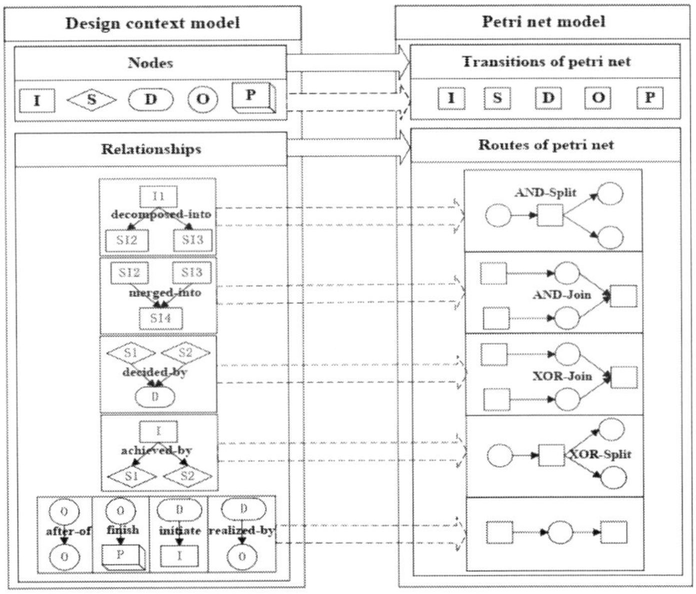

Figure 2: Mapping Between Petri Net and Design Context Model

4 α ALGORITHM-BASED SHARED DESIGN CONTEXT MINING

Based the constructed design context petri net, the extraction of shared context is mainly based on identifying specific patterns through α algorithm among the design contexts, then convert the specific pattern petri net into a corresponding DR model.

4.1 Noise Processing

In the α algorithm, the noise may cause errors between task relationships, and output error workflow model. There are ways to deal with noise specifically, such as heuristic mining, genetic mining and fuzzy mining. In the DR model, the noise refers to infrequent design behaviours which do not represent the typical behaviour of the design process. In the process of DR modelling, designers' individual difference will also bring the noise to DR model.

This paper adopts association rules mining method to identify the infrequent design mode in the design context. Association rules define the standard of support and confidence. α algorithm mining process is based on sequential relationship *a>b* The support of the rule *a>b* represents its applicability that shows proportion of the premise and the conclusion realize at the same time. The confidence of the rule *a>b* represents its reliability; the high degree of confidence indicates that the rule is very reliable. If the support or confidence is below the threshold which was set before mining process then it will be considered as noise rules. The definition of the support and confidence are as follows:

$$support(a > b) = N_{a>b}/N \quad (1)$$

$$confidence(a > b) = N_{a>b}/N_a \quad (2)$$

$N_{a>b}$ represents the number of sequential relationship *a>b* occurrence. N_a represents the number of sequential relationship which was caused by *a*, *N* means the number of all sequential relationship.

4.2 Shared Design Context Mining Process

This paper extracts shared design context from multiple design context by α algorithm. The sequential relationship in the design context is mainly based on elements relationship, which can infer causal relationship, parallel relationship and non-dependence relationship. The definitions of task order relationship among shared design context are as follows:

(1) Sequential relationship $I_1 > I_2$: two design elements nodes exist direct semantic link such as "decomposed-into", "achieved-by", "decided-by", "realized-by", are determined as sequential relationship.

(2) Causal relationship $I_1 \rightarrow I_2$: if $I_1 > I_2$ and $I_2 \not> I_1$, task I_2 will occur immediately after task I_1.

(3) Parallel relationship $I_1 // I_2$: if $I_1 > I_2$ and $I_2 > I_1$, task I_1 and I_2 will occur at the same time.

(4) Non-dependence relationship $I_1 \# I_2$: if $I_1 \not> I_2$ and $I_2 \not> I_1$, only one of the tasks will occur.

The route model can be found by four kinds of relations which are based on the log.

(i) XOR-split: $I_1 \rightarrow I_2$, $I_1 \rightarrow I_3$ and $I_2 \# I_3$; (ii) XOR-join: $I_1 \rightarrow I_3$, $I_2 \rightarrow I_3$ and $I_1 \# I_2$; (iii) AND-split: $I_1 \rightarrow I_2$, $I_1 \rightarrow I_3$ and $I_2 // I_3$; (iv) AND-join: $I_1 \rightarrow I_3$, $I_2 \rightarrow I_3$ and $I_1 // I_2$;

Sequential relationship is the foundation of the mining process. α algorithm will get a better starting point by removing the noise of sequential rules and representative tracks. The steps of the shared design context mining process are as follows:

Step 1: Construct tasks set Rw;

Step 2: Construct start tasks set Ri;

Step 3: Construct end tasks set Ro;

Step 4: Construct sequential relationship set *A*, causal relationship set *B*, parallel relationship set *C* and non-dependence relationship set *D* according to design context in the form of petri net;

Step 5: Calculate the support and confidence of all the sequential relationship. Sequential relationship will be redefined as the non-dependence relationship if the support or confidence does not reach the threshold, and the node will be deleted if it only has non-dependence relationship with other nodes;

Step 6: Select the largest causal relationship set and update causal relationship set *B* and tasks set Rw;

Step 7: Construct places set PT according to the tasks set Rw;

Step 8: Construct flow relation set FT according to tasks Rw and places set PT;

Step 9: Construct petri net according to Rw, PT and FT.

The result shown in the petri net form cannot take advantage of the unambiguous structure and rich semantics of DR model. Therefore, after the completion of mining algorithms we restore the petri net to the structural DR context model to facilitate designers' understanding. The transition is based on the mapping relationship between petri net and DR model.

5 EXAMPLE VERIFICATION

The effectiveness of α algorithm to excavate shared design context is verified by the design process of automatic marking machine. Two different design contexts for automatic marking machine are

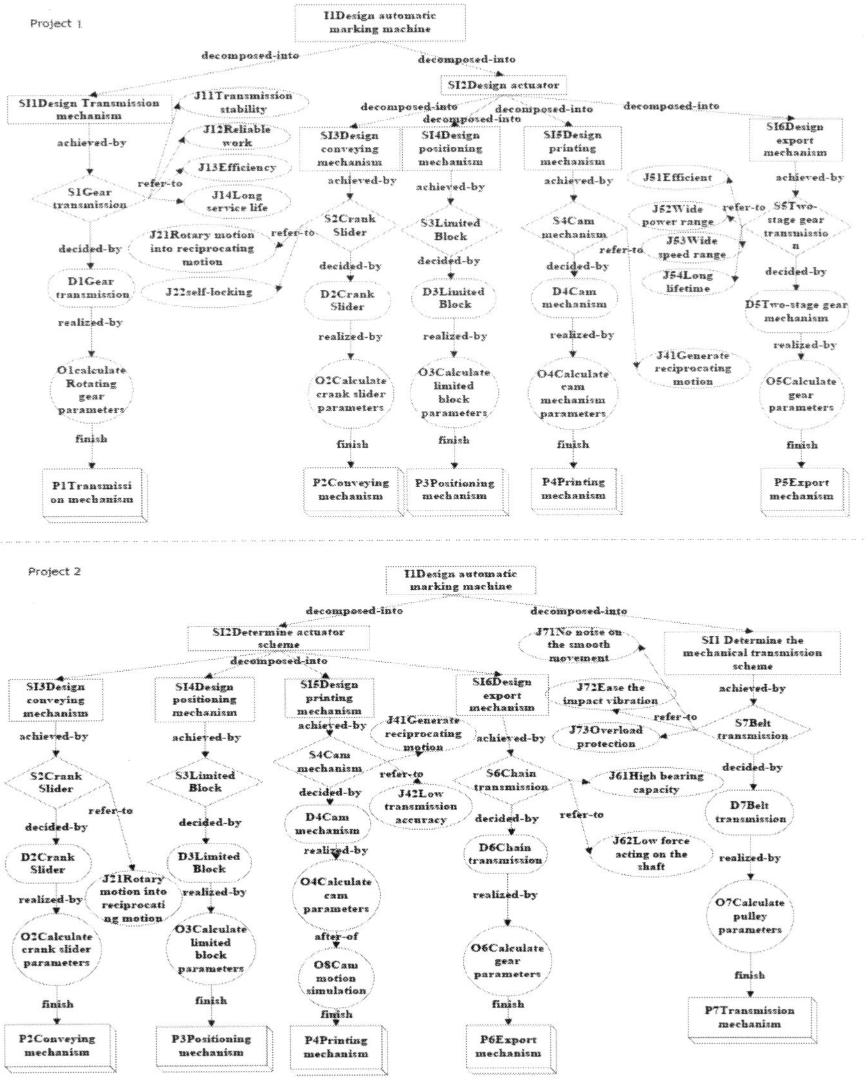

Figure 3: The Design Context of Two Automatic Marking Machines

shown in Figure 3. Set the support and confidence threshold 1/80 and 0.5 respectively. Two noises S2>J22 and S4>J42 can be found in the model, remove noise nodes J22 and J42, merge the repeated design processes and establish new relationships between the design nodes by reasoning two design contexts. The mining process is shown in Figure 4.

Restore the petri net to the shared DR context to facilitate designers' understanding by transforming transitions to the corresponding nodes and the route to the corresponding relationship in DR model. The shared design context model is shown in Figure 5. The result shows:

(1) The mining process of shared design context is an incremental design process which enriches the design process to provide designers with a more comprehensive solution to a design task;

(2) The mining process eliminates the influence of noises, which ensures the accuracy of the shared design context;

(3) The mining process reserves different operations and merges the same operations without influencing the relationship between the nodes.

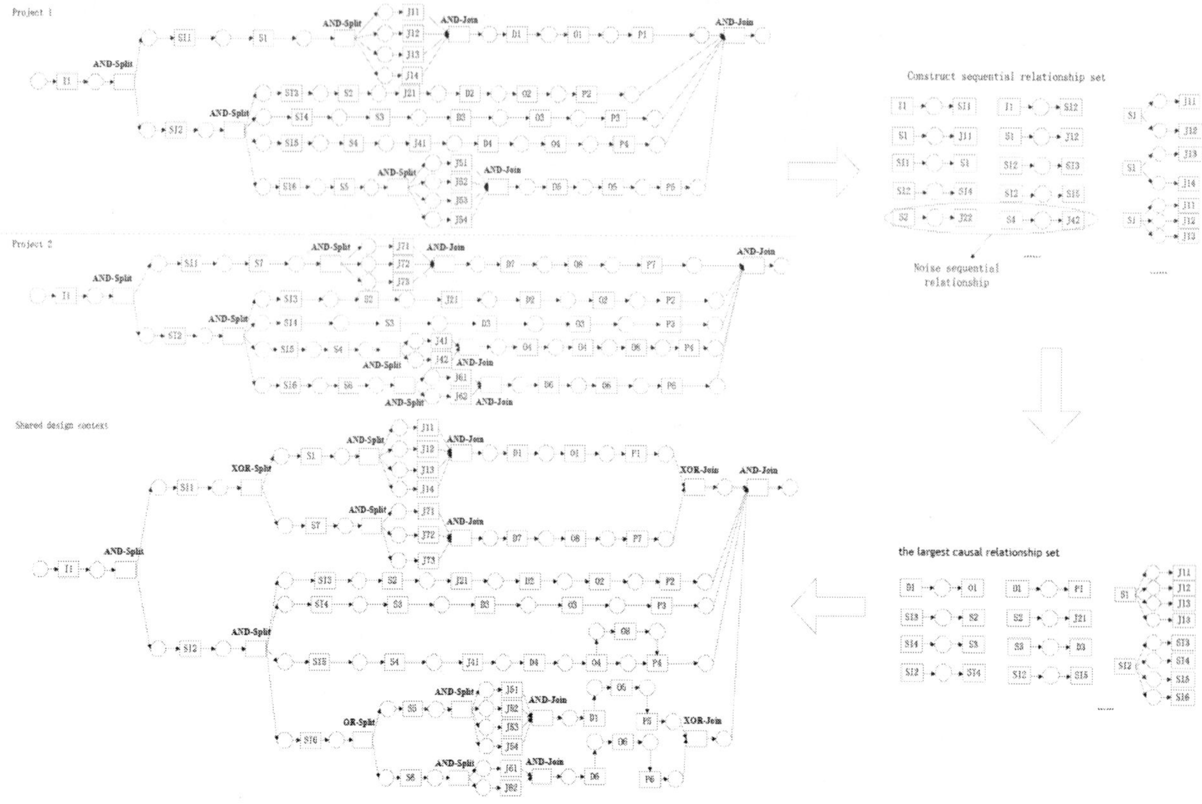

Figure 4: The Mining Process in the Form of Petri Net

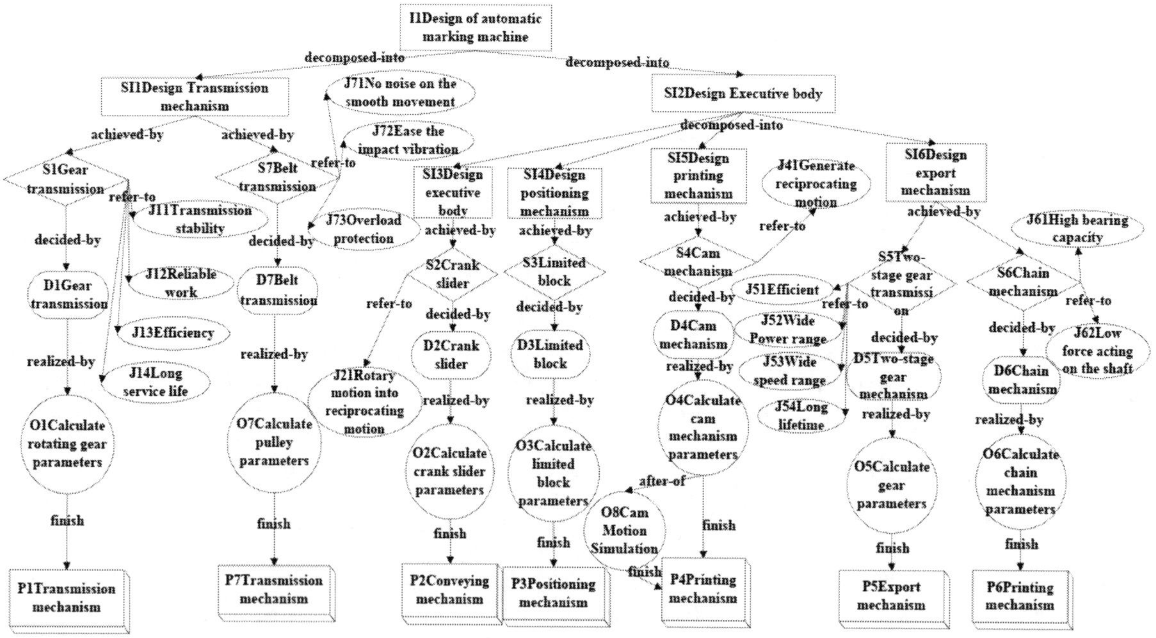

Figure 5: The Shared Design Context of Automatic Marking Machine

6 CONCLUSION

The design context contracted from a single DR model instance can only describe the design process of specific design which is difficult to provide the universal design context for a class of design problems to the designers. This article presents the concept of shared design context and the finding method according to process mining. Firstly, transform design context model into petri net. Secondly, excavate the shared design context in the form of petri net using the improved α algorithm. Thirdly, translate the petri net model into shared design context model in order to facilitate understanding and reusing for designers. At the end of this article, we demonstrate the effectiveness of this method by excavating the shared design context of two automatic marking machines' design context model.

The extracting of design context helps designers to simplify the design rationale model while the shared design context contributes to the preliminary implement of collaborative design. The future work will focus on the method to effectively express the design context of designers in collaborative design environment, and the method to evaluate DR context model.

7 REFERENCES

[1] **W Shen, J P Barthès, J Luo. 2015.** Computer Supported Collaborative Design: Technologies, Systems, and Applications. *Contemporary Issues in Systems Science and Engineering*, pp 537-573.

[2] **S. McNeil. 2015.** Visualizing mental models: understanding cognitive change to support teaching and learning of multimedia design and development. *Educational Technology Research and Development*, 63(1), pp 73-96.

[3] **Bracewell R, Wallace K, Moss M. 2009.** Capturing design rationale. *Computer-Aided Design*, 41(3), pp 173-186.

[4] **Y Liu, Y Liang, Kwong C K. 2010.** A new design rationale representation model for rationale mining. *Journal of Computing and Information Science in Engineering*, 10(3), pp 1-10.

[5] **Y Zhang, X Luo, Q Chen. 2012.** Semantic Representation for Design Rationale of Product. *Journal of Mechanical Engineering*, 48(17), pp 135-143.

[6] **Rohde D, Štorga M, Marjanović D. 2015.** DESIGN RATIONALE CAPTURING MODEL FOR USE DURING THE EMBODIMENT PHASE OF THE PRODUCT DESIGN. *Transactions of FAMENA*, 39(1), pp 27-42.

[7] **L Li, F Qin, S Gao. 2013.** An Extended Design Rationale Representation for Retrieval of Design Knowledge. *Journal of Computer-Aided Design& Computer Graphics*, 25(10), pp 1514-1522.

[8] **J Wang, L Wen. 2014.** *Process Mining: Discovery, Conformance and Enhancement of Business Process*, 1rd Edition, Tsinghua University press.

[9] **J Liu, X Hu. 2013.** A reuse oriented representation model for capturing and formalizing the evolving design rationale. *Artificial Intelligence for Engineering Design, Analysis and Manufacturing*, 27(04), pp 401-413.

[10] **H. Jiang, J Liu, K Wang. 2015.** Design process knowledge mining based on design rationale models. *Computer Integrated Manufacturing Systems*, 21(2), pp 289-197.

[11] **J Liu, H Zhan. 2013.** A reconstruction method of the design rationale model based on design context. Fuzzy Systems and Knowledge Discovery (FSKD), 2013 10th International Conference on. IEEE, shenyang, China, July23-25, pp 492-497.

DESIGN OF VISUAL MANAGEMENT TOOLS FOR TWO ADJACENT SIMILAR SWITCHES IN CHEMICAL PLANT AND THEIR EVALUATION

H. Kikukawa[1*], I. Saito[2], M. Mori[2], H. Takahashi[2], Y. Honda[3], K. Murata[1] and H. Katayama[4]

[1]Department of Industrial Engineering and Management,
College of Industrial Technology, Nihon University, Chiba, JAPAN

[2]Nippon Nyukazai Co.,Ltd., Kanagawa, JAPAN

[3]JMA Consultants Inc., Tokyo, JAPAN

[4]Department of Industrial and Management Systems Engineering,
School of Creative Science and Engineering, Waseda University, Tokyo, JAPAN

Corresponding e-mail: haru8141121@yahoo.co.jp

ABSTRACT

In this paper, concepts of visual management are designed to prevent the operational error of two adjacent similar switches, one typical problem of the process industry, in the chemical plant of the case company. The procedure of this study consists of three steps: Step 1) description of developed concepts, Step 2) design of standard for evaluating developed concepts and Step 3) evaluation of developed concepts. As the result of the evaluation of 7 developed concept, it is found that there are tread-off relationship between the usefulness of them and the easiness to install them. And then, concepts just before the operation are more important than concepts before the start of the operation. These findings are considered as one step to realize systematic development of a visual management tool.

Keywords: Process industry, Visual Management Design, Prevention System

1 INTRODUCTION

In the case that a chemical plant is constructed in a process industry, the switches which control some equipment are located at one location from the perspective as the centralization of pipe laying and/or the efficiency of equipment operation. However, the similarity of the switches loses workers' safety and a quality of product is lowered by operating wrong switch. This paper discusses the countermeasure to solve the problem with visual management tool.

Visualization for operation management, which is called visual control (VC)/visual management (VM), is the communication tool between human and a system which is not able to speak. In order to deliver the voice of a production system to human, there are used a lot of cases of visualization in factories around the world. Andon and Kanban are representative examples of VC/VM in Toyota Production System (TPS). In addition, a visual factory (Greif 1991) is an innovative concept that extends the application of VM from the production sector to other corporate sectors, including

* Corresponding Author

administration, engineering, and sales. Parry et al. Turner (2006) reported that VM tools are powerful tools that can be used beyond manufacturing. For example, lean construction is proposed as an extended theory of lean management in the production sector (Sacks 2010). Further, VM is considered as one of the support methods for lean construction (Tezel et al. 2010). In their study, Tezel et al. (2010) classify the relevant cases in the construction sector in Brazil. The reason for this extended application of VM is that it contributes to the maintenance and improvement of the relevant KPIs, i.e., reliability, productivity and safety.

However it is seems that a study of how to design a useful VM is not enough in an academic side. It is a basic problem in VM. As the above mentioned, VM is confused with a mere notice board. Light-hearted design and implementation of VM will invite simplistic result and unessential management. A systematic procedure is needed for designing effective technology. Murata et al. (2015) discusses a basic mechanism of a visual management system. However this paper is a concept level of how to design VM and it is necessary to accumulate cases to design VM. The challenge of this paper belongs to the problem.

2 RESEARCH PROCEDURE

The procedure of this research procedure consists of the following three steps.

Step 1: Description of developed concepts (Section 3)

This paper evaluates 7 concepts for the visual management to solve the problem of two adjacent similar switches. After illustrating the problem to be solved, this step illustrates how to create them and the background of the design of each concept.

Step 2: Design of standard for evaluating developed concepts (Section 4)

This step proposes the evaluation standard of the capability of the developed concept from the two perspectives such as the usefulness of it and the easiness to install it.

Step 3: Evaluation of developed concepts (Section 5)

This step evaluates 7 developed concepts by the evaluation standard proposed in Step 2. And then the discussions performs based on the evaluation results.

3 STEP 1: DESCRIPTION OF DEVELOPED CONCEPTS

3.1 The problem to be solved

Figure 1 expresses the situation of the problem to be solved. The two switches are used to operate the different equipment. However the forms of two switches are similar and the installation location of each switch is very close. Thereby, there is a high probability that operate a wrong switch. Generally, high level of control operations is made automatically by distributed control system (DCS). However, a lot of operations has been done by hand because of the cost-effectiveness of investment in plant and equipment and the necessity of visual confirmation. It has become a typical problem in such situations. This operation is usually performed to reduce a mistake by two operators. However, in this paper, this operation is performed by one operator because of the simplification of the problem to be solved.

Figure 1: The situation of the problem to be solved

3.2 Development of Concepts

There are the following three conditions in the development of the concepts; 1) one switch is not able to be separated from the other switch. 2) Large investment and a big conversion are not be able performed. 3) A visual management is applied for resolving the focused problem in kaizen technologies. A basic policy to develop the concepts is the realization of the prevention of the operation miss by a worker. Thereby the development uses simple motions study to disassemble some motions of the operation for designing more suitable concepts. As the result of the motions study, four sub-operations are clarified as shown in Figure 2 (Kikukawa 2015).

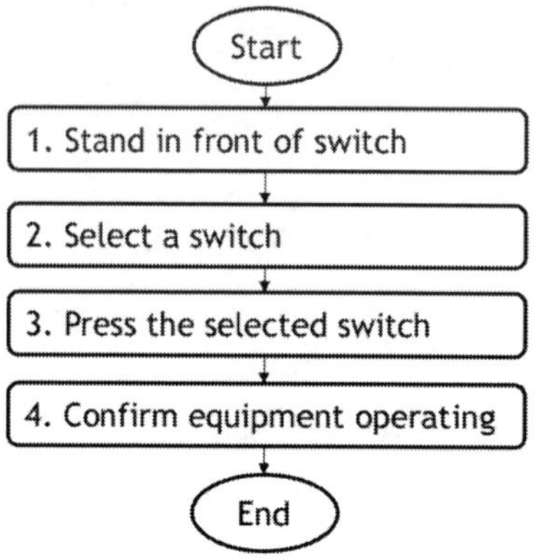

Figure 2: The result of the motions study

Concepts are developed in every four sub-operations. a) Reflection and b) Prediction are two concepts for the first sub-operation. c) Differentiation and d) Unification are two concepts for the second sub-operation. e) Procedure and f) Double-check are two concepts for the third sub-operation. g) Post-confirmation is one concept for the fourth sub-operation. A relationship between 4 sub-operations and 7 concepts is illustrated from Table 1.

Table 1: Concept of visual management system in each action

Sub-operations	Concepts
1. Stand in front of switch	a. Reflection
	b. Prediction
2. Select a switch	c. Differentiation
	d. Unification
3. Press the selected switch	e. Procedure
	f. Double-check
4. Confirm equipment operating	g. Post-confirmation

Also, illustrations of 7 developed concepts are as follows and their images are Figure 3:

a. Reflection

In the concept, a worker can recognize the possibility of erroneous operation to recall the past cases by displaying the defective and the situation occurred in a previous accident near the switches.

b. Prediction

In the concept, a worker can recognize the operation in need of a caution by displaying bulletins against the background of heterogeneous state like beautiful landscape and magnificent view, and phases and tables to awaken a caution.

c. Differentiation

The point of the similarity between two switches is resolved by the visual difference between them switches. The purpose of this concept is to make quite different two switches. If the similarity between them remains after installing the concept, it is possible to happen an error.

d. Unification

In the concept, the mechanism which operations for two switches can be realized by one simple operation is developed. It help a worker to select operated switch without the confusion by the similarity between two switches.

e. Procedure

In this concept, two switches are regarded as one system. The system provides the order of a series of the operation to a worker.

f. Double-Check

In this concept, two switches are regarded as one system. The system provides the information to confirm right orders a worker does next.

g. Post-Confirmation

In this concept, a worker can check equipment states by developed display after a worker operates.

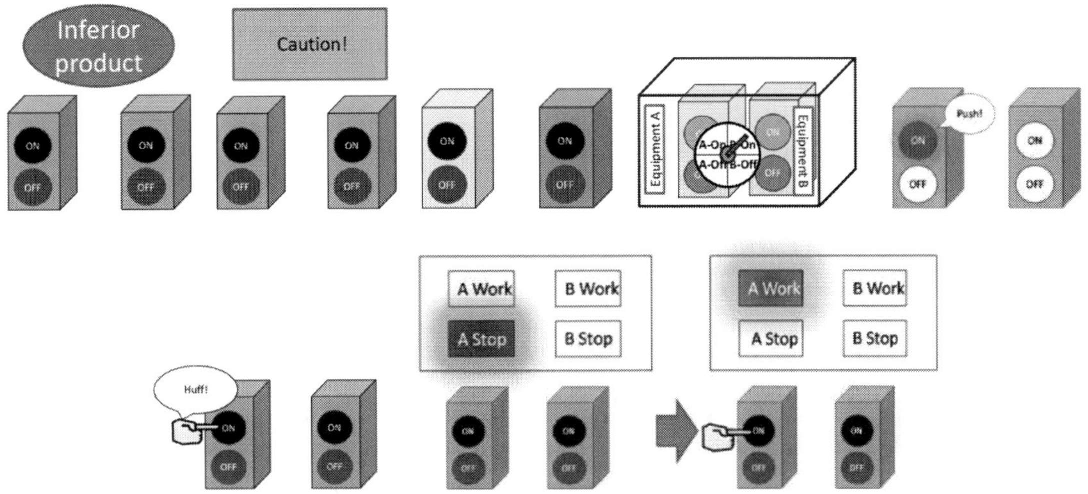

Figure 3: Images of 7 developed concepts

(Upper row: concept a to e, Lower row: concept f and g)

4 STEP 2: DESIGN OF STANDARD FOR EVALUATING DEVELOPED CONCEPTS

4.1 Evaluation standard of developed concept

In the step, the evaluation standard of the capability of the developed concept is proposed from the two viewpoints such as the usefulness of it and the easiness to install it. It is summarized in Table 2. Two sub-standard of the first viewpoint are the expectation to prevent accidental operation by developed concept and the durability of the effect of developed concept. Two sub-standard of the second viewpoint are the Economy of the development of developed concept and the reconstruction degree of the installation place of developed concept.

Table 2: Evaluation standard of developed concept

Viewpoint	Evaluation standard	Explanation
Usefulness	The expectation to prevent accidental operation by developed concept	Do you have an expectation to prevent from operation mistake by developed concept?
	The durability of the effect of developed concept	How long is the effect of developed concept?
Easiness	The Economy of the development of developed concept	Cost and time concerned installing developed concept
	The reconstruction degree of the installation place of developed concept	The degree of reconstruction from the current device to the device after remodelling

4.2 Formulation for evaluating developed concept

Two formulas are as follows. They are established from the viewpoints mentioned above. s_i, t_j, c_{hi} and c_{hj} are on the basis of 5 points.

$$e1_h = \sum_{i=1}^{m} s_i c_{hi} (h = 1,...,l) \qquad (1)$$

(Evaluation formula of the usefulness of each concept)

$$e2_h = \sum_{j=1}^{n} t_j c_{hj} (h = 1,...,l) \qquad (2)$$

(Evaluation formula of the easiness for installing each concept)

s_i : Weight of evaluation standard *i* of usefulness of developed concepts

t_j : Weight of evaluation standard *j* of easiness of developed concepts

c_{hi} : Evaluation results of evaluation standard *i* of developed concept *h*

c_{hj} : Evaluation results of evaluation standard *j* of developed concept *h*

l : Number of concepts

m : Number of evaluation standard of usefulness of developed concepts

n : Number of evaluation standard of easiness of developed concepts

h : Suffix of a concept

I : Suffix of an evaluation standard of usefulness of developed concepts

j : Suffix of an evaluation standard of easiness of developed concepts

5 STEP 3: EVALUATION OF DEVELOPED CONCEPTS

5.1 Expectation of evaluation of developed concepts

After the evaluation of developed concepts, its expectation are performed as shown in Table 3. In the expectation, the evaluation of the usefulness of the concepts for early sub-operations such as standing in front of two switches or selecting a switch is high because the original purpose of a visual management is to early detect the abnormal situation. The evaluation of the easiness of two concepts, b. Prediction and c. Differentiation, is high because the cost to install them is low.

Table 3: Expectation of evaluation of developed concepts

Concepts	Usefulness		Easiness	
a. Reflection	A worker can remember the state after the operation by this concept.	High	It cost a few to prepare inferior product.	Middle
b. Prediction	A worker can remember the state after the operation by this concept.	High	It does not take an effort.	High
c. Differentiation	This concept has the possibility not to entirely remove the problem of the similarity.	Middle	It does not take an effort.	High
d. Unification	It is possible of a worker to become familiar with the concept.	Middle	It is necessary to improve the relevant system.	Low
e. Procedure	It is possible of a worker to become familiar with the concept.	Middle	It is necessary to improve the relevant system.	Low
f. Double-check	It is possible of a worker to become familiar with the concept.	Middle	It is necessary to improve the relevant system.	Low
g. Post-confirmation	The concept is a countermeasure after the fact.	Middle	It is necessary to improve the relevant system.	Low

High: 4.0 and over, Middle: 4.0 through 2.5, Low: 2.5 less than

5.2 Evaluation Results

5 evaluators are the plant director of the case company, the head of the production division, 3 secretariats of total productive maintenance and management (TPM). Evaluation results shows from Table 4 to Table 6.

Table 4: Evaluation results of usefulness of developed concepts

Concepts	The expectation to prevent accidental operation by developed concept					The durability of the effect of developed concept				
	A	B	C	D	E	A	B	C	D	E
a. Reflection	3	3	2	1	3	2	2	1	1	2
b. Prediction	3	3	4	1	3	2	2	2	1	2
c. Differentiation	4	3	3	3	4	2	3	3	3	3
d. Unification	4	5	5	5	4	4	3	4	5	3
e. Procedure	5	5	4	4	5	5	5	5	4	5
f. Double-check	4	4	4	2	5	4	4	4	2	5
g. Post-confirmation	5	1	4	5	3	5	1	5	5	3

Table 5: Evaluation results of easiness of developed concepts

Concepts	The economy of the development of developed concept					The reconstruction degree of the installation place of developed concept				
	A	B	C	D	E	A	B	C	D	E
a. Reflection	5	5	5	5	5	2	5	5	5	2
b. Prediction	4	5	4	5	5	3	5	4	3	3
c. Differentiation	4	5	4	5	5	3	4	5	5	4
d. Unification	3	2	2	3	3	4	2	3	3	3
e. Procedure	2	1	2	2	2	4	1	2	2	5
f. Double-check	2	2	1	4	2	5	2	1	3	5
g. Post-confirmation	1	1	1	1	2	5	1	1	1	3

Table 6: Evaluation results of weight of usefulness and easiness of developed concepts

Evaluation standard	A	B	C	D	E
The expectation to prevent accidental operation by developed concept	3	5	5	4	5
The durability of the effect of developed concept	4	5	4	3	3
The Economy of the development of developed concept	5	3	3	3	4
The reconstruction degree of the installation place of developed concept	4	5	3	4	5

5.3 Discussion

Figure 4 shows usefulness and easiness of 7 developed concepts. The relationship between two viewpoints is considered as a trade-off relationship. Based on the discussion of the evaluators, their company often use visual managements followed by the concept of double-check. The concept is well balanced from two viewpoint. In spite of only one case, it is found from the analysed result that practitioners have developed visual management by the reasonable concept.

In addition, Table 7 shows the comparison between the expectation of evaluation of developed concepts and its result. As the remarkable point, the evaluation results of the usefulness of a. Reflection and b. Prediction are lower than their expectations. And the evaluation results of the usefulness of d. Unification, e. Procedure and f. Double-check are higher than their expectations. From the analysed results, it is found that concepts just before the operation is important compared with that of concept before the operation. A conventional idea of a visual management is to detect early abnormal situation. However the visual management for a countermeasure is needed according to circumstances.

The future study are actual installation of selected concepts to the plant of the case company based on the evaluation results and so on.

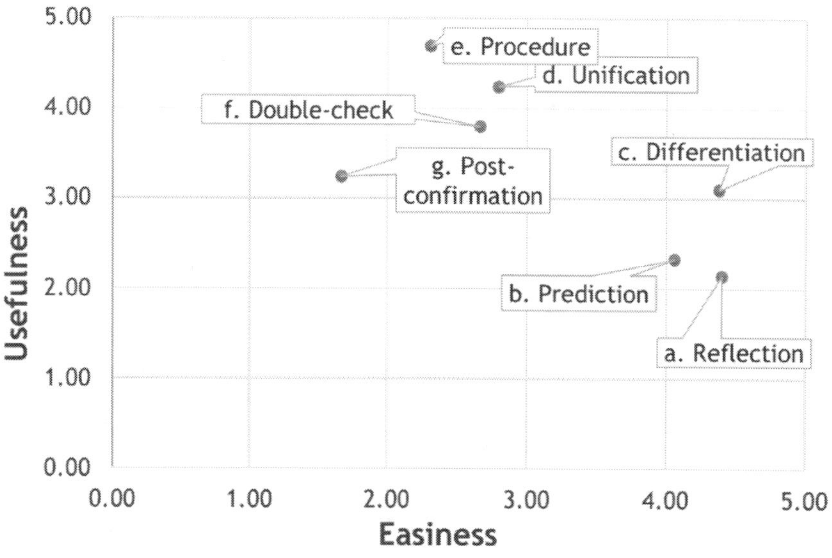

Figure 4: Usefulness and easiness of 7 developed concepts

Table 7: Comparison between expectation and result of evaluation of developed concepts

Concepts	Usefulness		Easiness	
	Expectation	Result	Expectation	Result
a. Reflection	High	Low	Middle	High
b. Prediction	High	Low	High	High
c. Differentiation	Middle	Middle	High	High
d. Unification	Middle	High	Low	Middle
e. Procedure	Middle	High	Low	Low
f. Double-check	Middle	High	Low	Middle
g. Post-confirmation	Middle	Middle	Low	Low

6 CONCLUSIONS

This paper tackles with the design of concepts of visual management to prevent the operational error of two adjacent similar switches in the chemical plant of the case company. In the study, 7 concepts are developed. They are evaluated from the two proposed viewpoints, the usefulness of them and the easiness to install them. It is found from the result of the evaluation that there are tread-off relationship between two viewpoints. And then, before evaluating developed concepts, concepts before the start of the operation will be evaluated well. However concepts just before the operation is important from the evaluation result. These findings are obtained from mere one case. However they are considered as one step to realize systematic development of a visual management tool.

7 ACKNOWLEDGMENT

This paper was also supported by JSPS KAKENHI Grant Number JP 16K03897.

8 REFERENCES

[1] **Greif, M.** 1991. *The Visual Factory: Building Participation through Shared Information*, Boca Raton, FL: CRC Press.

[2] **Kikukawa, H. and Murata, K.** 2014. A Study on Visual Management for Preventing Miss-operation of Adjoining Similar Switches: Quoting Operation Analysis, Proceedings of 48th academic symposium of college of industrial technology of Nihon University, college of industrial technology, Nihon University, Chiba Japan, December 5th, pp. 1007-1010.

[3] **Murata, K. and Katayama, H.** 2015. A study on improving corporate lean problem in manufacturing industry through learning visual management cases in other industries. Symposium Proceedings of 19th Cambridge International Manufacturing Symposium: International Manufacturing - Revisited: Embracing New Technologies, Capabilities and Markets, Moeller Centre, Churchill College, University of Cambridge, Cambridge, United Kingdom, 24th-25th September, 11 pages in USB.

[4] **Parry, C. E., and G. C. Turner.** 2006. Application of Lean Visual Process Management Tools, *Production Planning and Control*, 17(1), pp. 77-86.

[5] **Sacks, R., Koskela, L., Dave A. B., Owen, R., 2010.** Interaction of Lean and Building Information Modeling in Construction, *Journal of Construction Engineering and Management*, 136(9), pp 968-980.

[6] **Tezel, A., Koskela, L. and Tzortzopoulos, P., 2010.** *Visual Management in Construction: Study Report on Brazilian Cases*, University of Salford.

CLOUD SERVICE QOS MULTI-PARAMETERS QUANTITATIVE EVALUATION BASED ON COMPREHENSIVE WEIGHT

A. Bai Yaxin* and B. Zhang Hong
Science and Technology on Reliability and Environmental Engineering Laboratory
School of Reliability and Systems Engineering, Beihang University, China
baiyaxin2010@163.com

ABSTRACT

Cloud computing is a new technology, which combines grid computing, utility computing, SOA and other technologies. Cloud computing effectively integrates computing, storage, data and application resources. So it is widely used, however, Amazon, Google, Microsoft and other service providers have occurred service interruption, service delays, data loss and other failures when they provide services. In this situation, the QOS (quality of service) of cloud services is questioned by users. Providing reliable cloud services which meet users' requirements becomes a key constraint of cloud service development. A comprehensive QOS evaluation method of cloud service is proposed in this paper. Not only objective factors but also users' need are taken into account in this method. Firstly, an extended QOS parameters system is established and the calculation method of these parameters is given. Secondly, analyses users' preference vector, then according to the similarity of preference vector and objective vector determine the weight of these parameters. Finally establish the QOS evaluation model, and evaluate the QOS of cloud services.

Keywords: Cloud service, QOS parameters, Evaluation, User preference, Variable weight.

1 INTRODUCTION

With the development of cloud computing, more and more applications are available to users in the form of cloud services. Cloud services can provide powerful function, make full use of shared resources, speed up system development, reduce costs, and fast meet the needs of users. So cloud computing and cloud service get widespread concern in IT, manufacturing and education fields. At the beginning of cloud computing launching, the expansion of cloud computing service capabilities and application of collaborative computing resources are more concerned, but the cloud computing continues to provide reliable application services ability is ignored. However, with the adoption of cloud services, Amazon, Google, Microsoft and other major vendors occur service interruption, service delays and other failures when providing services. Due to the complexity and dynamic nature of cloud systems, the QOS evaluation of cloud services is difficult. Cloud computing QOS problem has become one of the major constraints in the promotion and application of cloud service.

Some scholars have carried out lots of researches on the QOS evaluation of cloud service. However, there also exist many problems. Some study [1-4] evaluated the QOs of cloud service proposed some methods but they only using one parameter they cared. The study [5] proposed some metrics for the QOS of cloud service, but it did not a certain evaluation method. The study [6-7] [9] proposed the evaluation structure and method, but did not consider the need of users. The study [8] [10-11] consider the users' influence on the QOS evaluation, but it only use one parameter. The study [12-

* [1]Bai Yaxin, baiyaxin2010@163.com

13] show the service is recommended according to QOS. So in this paper, the comprehensive evaluation method is proposed, and this method uses many parameters users care.

The paper is organized as follows: introduction is described in Sections I. it describes present research on QOS evaluation of cloud service and the existing problems. Cloud service QOS parameters system is described in Section II. QOS quantitative evaluation process is described in Section III. Application and analysis of the QOS evaluation model is described in Section IV. Finally, we present the conclusion and acknowledgment of this article in Section V.

2 CLOUD SERVICE QOS PARAMETERS SYSTEM

As is known, QOS is affected by several factors. And it is difficult to use one parameter to assess the QOS. Before establishing the cloud service QOS evaluation model, it is need to define QOS of cloud service. According to the QOS parameters in software, the QOS parameters of cloud service can be defined. Different users in different fields have different understanding of QOS definition, so they use different parameters to evaluation QOS. Therefore, the QOS attributes system should be extended, so as to ensure the scalability of QOS model.

2.1 Service failure rate $\lambda(t)$.

The failure rate of cloud services can be defined as the conditional probability of failure occurring within unit time of service in the time interval from t to $t+\Delta t$ (Δt is very small). Assume that the service does not fail at the time of t and the service fails at $t+\Delta t$. $\lambda(t)$ represents the failure rate. The failure rate is calculated in the mathematical formula 1:

$$\lambda(t) = \lim_{\Delta t \to 0} \frac{P_r\{t+\Delta t \geq T > t\}}{\Delta t} \quad (1)$$

2.2 Service average response time MTTA (Means Time to Answer).

The average response time of service means: the average time delay from receiving a request to output execution result. For MTTA, its result depends not only on the processing rate of service itself but also on the number of request for the same service at the same time and a service processing rules for the implementation of all requests. Assuming that the service call request obeys Poisson distribution and parameter is λ. The processing time obeys exponential distribution and parameter is μ. The processing rules of the service request obey queuing rules. And it can handle up to k service requests at the same time. MTTA is calculated in the mathematical formula 2.

$$E(MTTA) = \frac{\rho\pi_k}{(1-\rho)^2} + \frac{1}{\mu} \text{ Where } \rho = \frac{\lambda}{k\mu},\ \pi_0 = [\sum_{i=0}^{k-1}\frac{(k\rho)^i}{i!} + \frac{(k\rho)^k}{k!(1-\rho)}]^{-1},\ \pi_k = \frac{\pi_0(k\rho)^k}{k!}. \quad (2)$$

2.3 Reliability $R(t)$

The $R(t)$ of cloud services reflects the reliability level of the service. Assuming that a_i represents the time interval between the i th service failure and the $i+1$ th service failure. $a_1, a_2, \ldots, a_n$ can be obtained through the cloud service reliability test manager. $R(t)$ is calculated in formula 3.

$$R(t) = \sum_{i=1}^{n} a_i \Big/ n \quad (3)$$

2.4 Mean Time to Failure Recovery (MTTFR).

MTTFR of service represents the average time required for service recovery when monitoring the service failure. The shorter average recovery time indicates that the service failure time is shorter. Assuming that b_i represents the time interval between the i th service failure and the i th failure recovery. $b_1, b_2, ..., b_n$ can be obtained through the cloud service QOS test manager. MTTFR is calculated in the mathematical formula 4.

$$E(MTTFR) = \sum_{i=1}^{n} b_i \Big/ n \tag{4}$$

2.5 Service successful visiting rate $R(f)$.

$R(f)$ is the ratio of the number of successful service visiting times and the total number of visiting times. The rate of service successful visiting describes the ability to access the service immediately when it is needed. The cloud detection manager records total number of visiting times N_r and the service successful visiting times N_S within time t. $R(f)$ is calculated in the mathematical formula 5.

$$R(f) = N_s / N_r \tag{5}$$

2.6 Service successful executing rate $R(d)$.

$R(d)$ is the ratio of the number of service successful executing times and the total number of executing times. The rate of service successful executing describes the ability to execute the service successfully. The cloud detection manager records total number of executing times N_S and the successful service executing times N_d within time t. $R(d)$ is calculated in the mathematical formula 6.

$$R(d) = N_e / N_s \tag{6}$$

3 QOS QUANTITATIVE EVALUATION OF CLOUD SERVICE

3.1 QOS parameters model

The users request for a service. The service is not a specific service but an abstract service. In the cloud service environment there are many services which provide the same function which can meet users' functional requirement but the QOS of different optional services is different. How to evaluate the QOS of these services is important. First, choose some parameters from the QOS parameters system according to users' preference. Then the QOS model can be established. For example we can choose m parameters from the QOS parameters system, and the QOS model can be expressed by the following matrix 8. Every row of the matrix represents all the QOS parameters of a service which choose by the users. And each column represents a QOS parameter of all the services. The matrix support expansion. When increase or decrease an optional service that is to add or delete a row in the matrix. And when increase or decrease a parameter that is to add or delete a column in the matrix. QOS parameters normalized

$$Q(cs) = \begin{pmatrix} C_{1,1} & \cdots & C_{1,m} \\ \vdots & \ddots & \vdots \\ C_{n,1} & \cdots & C_{n,m} \end{pmatrix} \tag{7}$$

After establishing QOS model, you can evaluate the QOS and compare the result. However, the representation of the QOS parameters is different and the quantization unit is different. So before evaluation, it is needed to harmonize these QOS parameters. The service QOS parameters include reduction type (average service response time, service failure recovery time, etc.) and incremental type (QOS, service access success rate, etc.). In order to find the best service match customers QOS demand, these two types of QOS parameters approach is not the same. For the reduction type parameter, the small index is better. For the incremental type parameter, the big index is better. So these two types need to be normalized.

Services normalization refers to remove QOS parameters dimension by converting the number to eliminate dimensions impact on evaluation of the QOS. Convert the parameter value into a normalized parameter values. For incremental parameter normalized quantified by the formula 8 and 9. For reduction type parameter normalized quantified by the formula 10and 11.

When $\max_{j=1}^{n} C_j - \min_{j=1}^{n} C_j \neq 0$, $c(i,j) = \dfrac{C_{i,j} - \min_{j=1}^{n} Cj}{\max_{j=1}^{n} C_j - \min_{j=1}^{n} Cj}$ *(8)*

When $\max_{j=1}^{n} C_j - \min_{j=1}^{n} C_j = 0$, $c(i,j) = 1$ *(9)*

When $\max_{j=1}^{n} C_j - \min_{j=1}^{n} C_j \neq 0$ $c(i,j) = \dfrac{\max_{j=1}^{n} Cj - C_{i,j}}{\max_{j=1}^{n} C_j - \min_{j=1}^{n} Cj}$ *(10)*

When $\max_{j=1}^{n} C_j - \min_{j=1}^{n} C_j = 0$ $c(i,j) = 1$ *(11)*

For basic services, $c(i,j)$ in the above formula is normalized j parameter value of the i th service in the service pool. The quantized parameter values range in the interval $[0,1]$. By quantifying the QOS parameters achieve the normalization, then the normalized QOS parameters can be used for QOS evaluation.

3.2 Parameter weight determination

After establishing service QOS model, in order to select the service which satisfies users' QOS requirement better, first it is need to evaluate the QOS according to the selected parameter value. Different users prefer different QOS parameters. So it is important to get users' preference vector to determine every parameter's weight. Namely, it is important to get the weight vector $W_1, W_2, ..., W_m (\sum_{i=1}^{m} W_i = 1)$. There are two service QOS preference weight calculation methods: subjective weight method and objective weight method. Subjective weight methods have more human involvement, such as expert scoring method, AHP, experience judgment method. These methods require experts in the field of QOS to determine the weight. However these methods reply on users too much, it cannot reflect the fact if the users don't have much knowledge on the field. The objective weight method calculated QOS parameters weight according to the objective data. These methods are relative object compared with subjective method. These objective methods calculate the weight in the absence of subjective experience interference. However they can't reflect the users' preference. Combine the above two cases, the service QOS evaluation need comprehensive weight. The weight determination process is not only according to the objective information but also according to the users' preference vector.

3.2.1 QOS parameters objective weight determination

This paper use positive and negative ideal solution least weighted squares to calculate the weight of cloud service parameters. Based on the above method, the QOS parameters values of all the services can be represented as a matrix. And the matrix is formula 13.

$$Q(cs)=\begin{pmatrix} c_{1,1} & \cdots & c_{1,m} \\ \vdots & \ddots & \vdots \\ c_{n,1} & \cdots & c_{n,m} \end{pmatrix} \quad (12)$$

The positive and negative ideal solution refers to service whose value of QOS parameter is best or worst. In the above, the QOS parameters are normalized. So the QOS parameter value of positive ideal solution is 1. And the QOS parameter value of negative ideal solution is 0.Assume the weight of the QOS parameters can be noted as $W_1,W_2,...,W_m(\sum_{i=1}^{m} W_i=1)$. So the weight vector is $W=<W_1,W_2,...,W_m>$, then the weighted square of the positive and negative ideal solution can be expressed as formula 13.

$$f_i(W_1,W_2,...,W_m)=\sum_{j=1}^{m} W_j^2(1-c_{ij})^2+\sum_{j=1}^{m} W_j^2 c_{ij}^{\ 2} \quad (13)$$

Generally choose a suitable weight vector to make the value of $f_i(W_1,W_2,...,W_m)$ smallest. Then we can establish multi objective programming formula 14. Then use the Lagrange multiplier method to solve the optimization problem with constraints. The Lagrange function is formula 15. Do differential operation respectively for W_j and λ. Then differential equations are as formula 16.

$$\begin{cases} Min\ f(W)=\sum_{i=1}^{n} f_i(W_1,W_2,...,W_m) \\ s.t. \sum_{j=1}^{m} W_j=1, W_j\geq 0\ 1\leq \mathrm{j}\leq \mathrm{m} \end{cases} \quad (14)$$

$$F(W,\lambda)=\sum_{i=1}^{n}\sum_{j=1}^{m} W_j^2[(1-c_{ij})^2+c_{ij}^2]+\lambda(1-\sum_{j=1}^{m} W_j) \quad (15).$$

$$\begin{cases} \frac{\partial F}{\partial W_j}=2\sum_{i=1}^{n} W_j[(1-c_{ij})^2+c_{ij}^2]-\lambda=0, 1\leq j\leq m \\ \frac{\partial F}{\partial \lambda}=1-\sum_{j=1}^{m} W_j=0 \end{cases} \quad (16)$$

Solv these two equations in formula 17, then the result of weight can be calculated. The result is expressed as formula 17.

$$W_j=\frac{\rho_j}{\sum_{j=1}^{m}\rho_j}, \quad \rho_j=\frac{1}{\sum_{i=1}^{n}[(1-c_{ij})^2+c_{ij}^2]} \quad (17)$$

Through the above steps, the objective weights of the i th service QOS parameters are calculate d. The weights can be expressed as W_j. Order the weights from large to small and record the p arameters according to the weights sequence. Then the ordered QOS parameters vector is noted as $Q_c=<V_{i1},V_{i2},...,V_{im}>$. And c_{ij}' is the QOS parameters matrix corresponding to Q_c.

3.2.2 *User preferences similarity weight*

According to users' subjective preference, the QOS parameters vector based on users' preference is easy to get. The vector $Q_s=<U_{i1},U_{i2},...,U_{im}>$ records the positions of QOS parameters, according to the importance users thinks from large to small order. Each service user preference vector weight vector and objective comparison, by the objective similarity weight vector and user preference vector metric for each basic service, to determine the relative importance of each QOS index can be employed based on the offset the degree of similarity measure. Compare the QOS parameters vector based on the users' preference $Q_S = <U_{i1}, U_{i2}, ..., U_{im}>$ with the QOS parameters vector $Q_C = <V_{i1}, V_{i2}, ..., V_{im}>$ based on the objective factors. Calculate similarity of these two vectors to determine the relative importance of the QOS parameters. This paper adopts similarity measurement method based on deviation. For a service, record the position of a QOS parameter in the two vectors. Assume the QOS parameter is k, and the position of the parameter in the two vectors respectively noted i and j. Then the similarity of parameter k in the two vectors is noted as formula 18. $|i-j|$ represents the deviation in the two vectors of one parameter. m represents the total number of QOS parameters. Dis_k represents the similarity in the two vectors of one parameter. Then the user preferences similarity weight is noted as formula 21.

$$Dis_k = m - |i-j| \quad , \quad U_k = \frac{Dis_k}{\sum_{k=1}^{m} Dis_k} \tag{18}$$

3.2.3 *Comprehensive weight determination*

The QOS parameters vector is noted as $Q_C = <V_{i1}, V_{i2}, ..., V_{im}>$ according to the objective weight order from large to small. And the objective weight is expressed as $V = <V_1, V_2, ..., V_m>$. The weight based on users' preference is noted as $U = <U_1, U_2, ..., U_m>$ Integrate these two weight vector, then the comprehensive weight is expressed as formula 19.

$$W_i^{'} = \frac{U_i * V_i}{\sum_{i=1}^{m} U_i * V_i} \tag{19}$$

3.2.4 *Multi-parameters QOS evaluation*

Using QOS parameters model matrix and weight vector to calculate the QOS of cloud service. Users provide QOS parameters preference vector. If the users don't provide it, the specified vector is used to calculate the comprehensive weight.

$$QOS = \sum_{i=1}^{m} c_{ij}^{'} * W_i^{'} \tag{20}$$

3.3 QOS evaluation process

The cloud service QOS evaluation is carried out by the following step: first, select out QOS parameters from parameters system; second, normalized these parameters; third, calculate the objective weight and objective vector using normalized parameters; fourth, using objective vector and users' preference vector to calculate the subjective weight; fifth, combine the objective weight and subjective weight to calculate the comprehensive weight; finally, calculate QOS. The process of QOS evaluation of cloud service is shown in Figure 1.

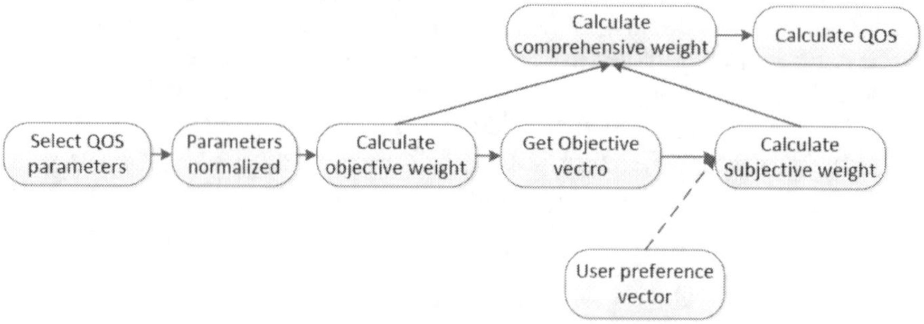

Figure 1: QOS evaluation process

4 APPLICATION AND ANALYSIS OF QOS EVALUATION MODEL

A travel service system is shown to illustrate cloud service QOS evaluation model proposed in this paper is feasible and effective. First, choose QOS parameters and get the original value of these parameters. The table 1 shows querying tour path service provided by different providers. The following call them provider 1, provider 2, provider 3, provider 4.

Table 1: Service QOS Parameters Original Value

	MTTA	λ(t)	R(t)	R(f)	MTTFR
1	30	0.05	0.95	0.85	120
2	40	0.01	0.96	0.89	180
3	25	0.03	0.90	0.92	60
4	35	0.02	0.92	0.95	200

Using the method proposed above, when the preference vector is $Q_{s1} = [R(t), R(f), MTTA, \lambda(t), MTTFR]$, the calculated comprehensive weight is $W_{i1}' = [0.389, 0.139, 0.138, 0.171, 0.163]$. Using $QOS = \sum_{i=1}^{m} c_{ij}' * W_i'$ to calculate the QOS of these four services, the result is $QOS_1 = [0.508, 0.638, 0.484, 0.440]$. When the preference vector is $Q_{s2} = [MTTA, R(t), MTTFR, R(f), \lambda(t)]$, the calculated comprehensive weight is $W_{i2}' = [0.394, 0.176, 0.132, 0.173, 0.125]$. Using to calculate. Using $QOS = \sum_{i=1}^{m} c_{ij}' * W_i'$ to calculate the QOS of these four services, the result is $QOS_2 = [0.514, 0.637, 0.479, 0.450]$.

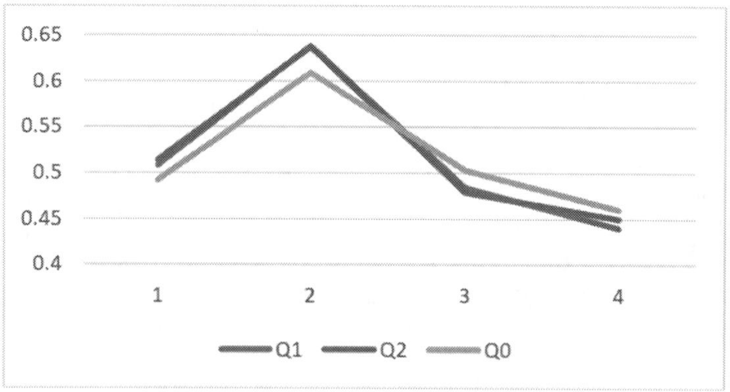

Figure 2: QOS result comparison

The Figure 2 shows the QOS comparison of these four services. The line 0 is the QOS value evaluated based on the objective weight. Line 1 and line 2 are the QOS values evaluated based on the different comprehensive weight.

5 CONCLUSION

The QOS parameters model is proposed in this paper. The model is reasonable and scalable. The parameters quantitative method is also given in this paper, so the parameters' values can be got. A QOS quantitative evaluation method of cloud service is proposed in this paper. This method not only considers the objective values of these parameters, but also considers users' preference to these parameters. By using this method we can get quantitative QOS value, and the evaluation result is closer to users' need of QOS.

6 REFERENCES

[1] **Liu, Zhi-Zhong. Chu, Dian-Hui. 2016.** Social learning optimization (SLO) algorithm paradigm and its application in QoS-aware cloud service composition, pp315-334.

[2] Singh, **Sukhpal. 2015.** Q-aware: Quality of service based cloud resource provisioning, pp138-161.

[3] **Tang, Mingdong. Dai, Xiaoling. 2016.** Towards a trust evaluation middleware for cloud service selection, Future Generation Computer Systems.

[4] **Mao, Chengying. Chen, Jifu. 2015.** Search-based QoS ranking prediction for web services in cloud environments, PP111-127.

[5] Amid **Khatibi Bardsiri;Seyyed Mohsen Hashemi.** QoS Metrics for Cloud Computing Services Evaluation[J].International Journal of Intelligent Systems and Applications,2014,No.12

[6] **Hong Jin. Xi Fan Yao *. Jie Zhang. 2014.** An Execution Framework for Domain-Oriented QoS Evaluation of Manufacturing Cloud Services [J].Applied Mechanics and Materials.

[7] **Lianyong Qi. Wanchun Dou. Jiancheng Ni. Xiaona Xia. Chunmei Ma. Jinfeng Liu. 2014.** A Trust Evaluation Method for Cloud Service with Fluctuant QoS and Flexible SLA [A].

[8] **Shuai Ding. Shanlin Yang. Youtao Zhang. Changyong Liang. Chenyi Xia. 2014.** Combining QoS prediction and customer satisfaction estimation to solve cloud service trustworthiness evaluation problems [J].Knowledge-Based Systems.

[9] **Nadanam, Padmapriya. 2012.** QoS evaluation for web services in cloud computing[A].

[10] **Ding S, Yang S, Zhang Y. 2013.** Combining QoS prediction and customer satisfaction estimation to solve cloud service trustworthiness evaluation problems[J]. Knowledge-Based Systems, 56(3):216-225.

[11] **Zhang X, Dou W. 2010.** Preference-aware QoS evaluation for cloud web service composition based on artificial neural networks[C]// Web Information Systems and Mining - International Conference, WISM 2010, Sanya, China. Proceedings.410-417.

[12] **Pan Y, Ding S, Fan W, et al. 2015.** Trust-Enhanced Cloud Service Selection Model Based on QoS Analysis[J]. Plos One, 10(11).

[13] **Jayapriya K, Mary N A B, Rajesh R S. 2015.** Cloud Service Recommendation Based on a Correlated QoS Ranking Prediction[J]. Journal of Network Systems Management, 1-28.

RESEARCH ON COUPLING SCHEDULING OF QUAY CRANE DISPATCH AND CONFIGURATION IN THE CONTAINER TERMINAL

Chengji Liang[1], Lubin Fan[1,2*], Dehong Xu[3], Yi Ding[4], Mitsuo Gen[5]
[1]Logistics Research Center, Shanghai Maritime University, Shanghai, China.
liangcj@shmtu.edu.cn,
[2]Shanghai International Port Group Co., Ltd, Shanghai, China.
helloflb@yeah.net
[3]Logistics Institute, Xi'an International University, Xi'an, Shaanxi,China.
delhxu75@163.com
[4]Logistics Research Center, Shanghai Maritime University, Shanghai, China.
dingyi1018@gmail.com
[5]Fuzzy Logic Systems Institute, Tokyo University of Science, Tokyo, Japan.
gen@flsi.or.jp

ABSTRACT

This paper studied the QCs scheduling problem from two aspects: QCs task dispatch and QCs quantity configuration. The QCs task dispatch problem main research the QCs handling which container tasks and the tasks handling sequence. QCs quantity configuration problem mainly studied how many numbers of QCs should be configured for the tasks. Through the analysis of the relationship between the two problems, the coupling model of two problems is established. Then, a large scale calculation example of multiple ships and mixed loading and unloading is designed, and the model is solved by using CPLEX. And the results are compared with the genetic algorithm for solving the quay crane scheduling.

Keywords: Quay crane, Coupling scheduling, Dispatch, Configuration

1 INTRODUCTION

Quay crane (QC) is an important equipment in container terminal which provides loading and unloading operations for containers ships. Because of the scarcity of berth resources, it is required to arrange the loading and unloading operations of QCs reasonably, and to improve the utilization efficiency of berth by improving loading and unloading efficiency and reducing the berthing time of ships. At the same time, QC is a kind of expensive equipment, and its use cost and maintenance cost are high. Therefore how to finish the loading and unloading operations with less time and less number of QCs is the key to the problem of the QCs scheduling problem.

The basic quay crane scheduling problem has been intensively studied and many solving methods have been proposed. Al-Dhaheri et al.(2015)[1] focused on the QCs scheduling problem by minimizing the differences between the container loads stacked over a number of bays. Fu et al. (2015)[2] proposed a mathematical formulation for the integrated QCs assignment and scheduling problem. Lagrangian relaxation was used for near optimal solutions. Kenana et al.(2015)[3] developed a branch-and-price algorithm to effectively solve the QCs scheduling problem with very large instances. Ali Diabat et al.(2014)[4] presented a formulation for the QCs assignment and scheduling Problem, which accounts for crane positioning conditions. J. Chen et al.(2014)[5]

*Corresponding Author

focused on the study of a special strategy for the cluster-based QCs scheduling problem that forces QCs to move unidirectionally during the scheduling. M. Ji et al.(2015)[6] established a mathematical model to integrate the loading sequence and the rehandling strategy under the parallel operation of multi-QCs. In a container terminal, coordination of QC and other equipment affects total operation time. L. Tang et al.(2014)[7] addressed the joint quay crane and truck scheduling problem considering the coordination of the two types of equipment. Türkogullari et al.(2016)[8] researched the integration of berth allocation, QCs assignment and QCs scheduling problems. J. He et al.(2015)[9] developed a MIP model for the integrated scheduling of cranes and trucks which the energy-saving objective is considered in the container terminal operations. Dynamics and uncertainty make the problem more complicated. Al-Dhaheri et al.(2016)[10] considered the uncertainty inherent to container handling process and proposed a stochastic mixed integer programming model for the QCs scheduling problem.

Most of the existing researches solve equipment scheduling problem by using one-way coordination. That is to say, they determine the equipment quantity first, and then calculate the optimal equipment dispatch scheme. Few researches determine the equipment dispatch scheme first, and then study the optimal configuration quantity of equipment. This paper studied the QCs scheduling problem from two aspects: QCs task dispatch and QCs quantity configuration. According to this, the QCs scheduling problem is transformed into a two-objective optimization problem with optimal QCs task dispatch scheme and optimal QCs quantity.

2 PROBLEM DESCRIPTION

The container terminal assigns QCs according to the number of container loading/unloading tasks. Thus the total operation time depends on the completion time of the final QC which is fundamental to the total stay time of ships. This paper proposed a dispatch model [M1] which set minimizing completion time of the final QC as its objective and a configuration model [M2] which set minimizing the number of QCs as its objective. Four assumptions were considered, including:

1. Berth allocation plan, the number of ships, the number of tasks are all known;
2. The types of tasks at a certain period including loading and unloading;
3. All containers to be operated are 20 TEU and operation time of QC is set as a fixed value;
4. The QCs are not allowed to cross one another but safety distance between QCs isn't considered. It means two QCs can operate on two adjacent bays at the same time.

In two sub models, time windows and the number of QCs are set as public variables. The framework of coupling is shown as figure 1.

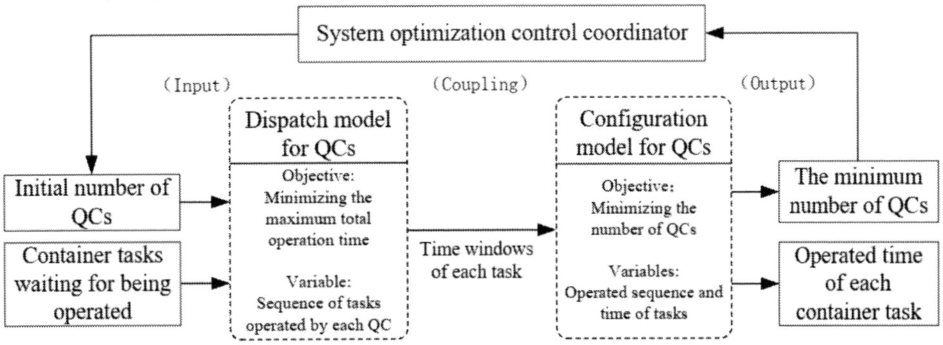

Figure 1: Framework of dispatch and configuration coupling model for QCs

3 DISPATCH MODEL FOR QCS

The set of unloading tasks is denoted by U ; the set of loading tasks is denoted by L ; the loading and unloading tasks is denoted by $J = U \cup L$, indexed by i and j; the set of all container operation tasks is denoted by $\overline{J} = \{I, F\} \cup J$, $I \in \{0\}$ represents a virtual initial task and $F \in \{J+1\}$ represents a virtual final task; the set of QCs is denoted by K, indexed by k.The following notation is considered:

- n the number of containers that need to be operated, $i, j \in J,\ i, j = 1, \ldots, n$;
- m the number of QCs, $k \in K, k = 1, \ldots, m$;
- b_i bay number, a QC picks up the container task i from bay b_i or stacks it onto bay b_i;
- qt_{ij}^k the time interval between task i and task j operated by the QC k ,from the moment that QC spreader arrive at the position of task i for handling task i to the moment that spreader arrive at the position of task j for handling task j;
- d_{ij}^k the idle time of spreader between task i and task j operated by the QC k ,if task i and task j are loading container and unloading container respectively, $d_{ij}^k = 0$;
- c_{ij}^k the move time of QC k from the bay where task i stacks to the bay where task j stacks, if task i and task j stack in the same bay, $c_{ij}^k = 0$;
- hq_i^k the time of QC k handling task i;
- qr_i^k the moment of QC k complete the task i;
- Q_J a constant that all QCs must complete all tasks before the moment of Q_J;
- M a constant that represents a large enough number.

In this model, a binary variable x_{ijk} is considered. It is equal to 1 if task j is operated directly after task i by the QC k and 0 otherwise. Another binary variable y_{ik} is considered. It is equal to 1 if task i is operated by the QC k and 0 otherwise.

The time interval qt_{ij}^k between tasks is determined according to the types of tasks, there are four kinds of container tasks relationships: (a)Unloading→Unloading; (b)Loading→Loading; (c)Unloading →Loading; (d) Loading→Unloading.

For case (a) and (b), the method of calculating qt_{ij} is shown in equation (1), while for case (c) and (d), the method of calculating qt_{ij} is shown in equation (2).

$$qt_{ij} = hq_i + c_{ij} + d_{ij} \qquad (1)$$

$$qt_{ij} = hq_i + c_{ij} \qquad (2)$$

The QCs dispatch problem can be formulated as follows [M1].

$$\min \max \sum_{j=1}^{n} \sum_{i=1}^{n} qt_{ij}^k \cdot x_{ijk} \qquad (3)$$

$$\text{s.t.} \quad \sum_{j=1}^{n} \sum_{i=1}^{n} qt_{ij}^k \cdot x_{ijk} \le Q_J \quad \forall k \in K \qquad (4)$$

$$\sum_{k=1}^{m} y_{ik} = 1 \quad \forall i \in J \qquad (5)$$

$$\sum_{j=1}^{n} x_{ijk} = \sum_{j=1}^{n} x_{jik} \quad \forall i \in J; k \in K \qquad (6)$$

$$\sum_{j=1}^{n} x_{ijk} = y_{ik} \quad \forall i \in J; k \in K \tag{7}$$

$$\sum_{i=1}^{n} x_{ijk} = y_{jk} \quad \forall j \in J; k \in K \tag{8}$$

$$\sum_{k=1}^{m} y_{Ik} = m \tag{9}$$

$$\sum_{k=1}^{m} y_{Fk} = m \tag{10}$$

$$qr_i + qt_{ij} + hq_j - qr_j \le M(1 - x_{ijk}) \quad \forall i, j \in J; \forall k \in K \tag{11}$$

$$k \cdot \sum_{k=1}^{m} y_{ik} \le l \cdot \sum_{l=1}^{m} y_{jl} \quad if\ b_i \le b_j \quad \forall i, j \in J \tag{12}$$

$$\mathrm{qt}_{Ij}^{k} + hq_j - qr_j \le \mathrm{M}(1 - x_{Ijk}) \quad \forall j \in J; \forall k \in K \tag{13}$$

$$qr_j + \mathrm{qt}_{jF}^{k} - \sum_{j=1}^{n}\sum_{i=1}^{n} qt_{ij}^{k} \cdot x_{ijk} \le \mathrm{M}(1 - x_{jFk}) \quad \forall j \in J; \forall k \in K \tag{14}$$

$$x_{ijk} \in \{0,1\} \quad \forall i, j \in J; k \in K \tag{15}$$

$$y_{ij} \in \{0,1\} \quad \forall i, j \in J \tag{16}$$

4 CONFIGURATION MODEL FOR QCS

The configuration model aimed at calculating a smaller number of QCs with meeting all the time windows of shipment tasks. The QCs can be scheduled for the shortest operation time by dispatch model [M1], then get the time window $[Et_i,\ Lt_i]$ of task i. As shown in figure 2, the new operation sequence of tasks of each QC and the new number of QCs resulted from reselecting operation time for each task, g and h are examples of operation time for two tasks. In order to improve the solving speed, this paper transformed the continuous problem to the discrete problem by dividing each time window into equal parts. g and h are both equidistant points, the function (17) represents corresponding time.

$$st_{ig} = Et_i + (Lt_i - Et_i)(g-1)/m \tag{17}$$

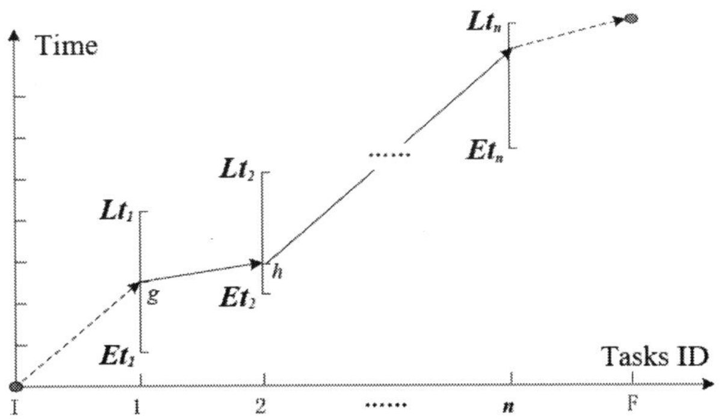

Figure 2: Time windows for tasks of QCs

m	The number of equidistant parts of each time window;
u	The number of QCs for all operation tasks;
$u_{\max}$	The maximum number of available QCs;
$[Et_i, Lt_i]$	The time window of task i;

st_{ig} The corresponding time of equidistant point g in the time window of task i.

In this model, a binary variable Z_{igjh} is considered. It is equal to 1 if task i is finished in equidistant point g and the next task j is finished in equidistant point h, and 0 otherwise. The model is formulated as [M2].

$$\min u \tag{18}$$

$$\text{s.t.} \quad \sum_{h=1}^{m}\sum_{j=1}^{n+1}\sum_{g=1}^{m} Z_{Igjh} = u \tag{19}$$

$$\sum_{h=1}^{m}\sum_{g=1}^{m}\sum_{i=0}^{n} Z_{igFh} = u \tag{20}$$

$$\sum_{g=1}^{m}\sum_{i=0}^{n} Z_{igab} = \sum_{h=1}^{m}\sum_{j=1}^{n+1} Z_{abjh} \quad \forall a \in J; b \in \{1,2,\cdots,m\} \tag{21}$$

$$\sum_{h=1}^{m}\sum_{g=1}^{m}\sum_{i=0}^{n} Z_{igjh} = 1 \quad \forall j \in J \tag{22}$$

$$\sum_{h=1}^{m}\sum_{j=1}^{n+1}\sum_{g=1}^{m} Z_{igjh} = 1 \quad \forall i \in J \tag{23}$$

$$Z_{igjh}\left|b_i - b_j\right| \le 1 \quad \forall i,j \in J; \forall g,h \in \{1,2,\cdots,m\} \tag{24}$$

$$\left|st_{jh} - st_{ig}\right| \ge Z_{igjh} \cdot \mathrm{qt}_{ij} \quad \forall i,j \in \bar{J}; \forall g,h \in \{1,2,\cdots,m\} \tag{25}$$

$$st_{jh} \ge st_{ig} \cdot Z_{igjh} \quad \forall i,j \in \bar{J}; \forall g,h \in \{1,2,\cdots,m\} \tag{26}$$

$$Z_{igjh} \in \{0,1\} \quad \forall i,j \in \bar{J}; \forall g,h \in \{1,2,\cdots,m\} \tag{27}$$

5 COORDINATION ALGORITHM FOR THE TWO MODELS

5.1 Calculation of time window

QCs dispatch model [M1] can finish the operation of all tasks at the fastest speed. Therefore, qr_i can be regarded as the earliest available time of the task i, as shown in the equation (28). In figure 3, we assume that a QC has 4 container task to operate. Through the [M1], it is concluded that the QC completed the 4 tasks at the fastest speed at the moment Q_k .If the QC must complete these 4 tasks before the moment of Q_J ,the QC at least to start the 1st task at the moment $Q_J - Q_k$ .The computing method of Lt_i is shown as equation (29).

$$Et_i = qr_i \tag{28}$$

$$Lt_i = Q_J - (Q_k - qr_i) \tag{29}$$

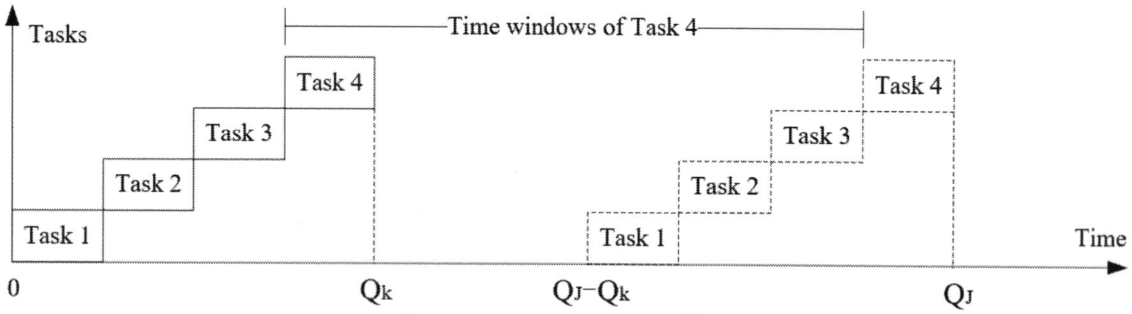

Figure 3: Computation of time windows for tasks

5.2 Setting of coordinator

The number of QCs input to [M1] is denoted by u_1 ,the minimum number of QCs solved by [M2] is denoted by u_2. Constraint (30) defines the configuration number of the QCs.

$$u_2 \le u_{\max} \tag{30}$$

If the u_1 and u_2 are equal, as shown in the equation(31),it imply that the optimal number of QCs for QCs scheduling problem is obtained. [M2] is also able to get a dispatch scheme and the processing moment st_{ig} of task i .For all tasks, if the equation(32) is true, it indicates that the QCs to complete the tasks with the fastest allocation scheme. If the calculation results of [M1] and [M2] meet the equation(31) and equation(32), stop the iteration calculation. That is to say, at this time, we not only get the best QCs dispatch scheme, but also get the optimal QCs configuration quantity.

$$u_1 = u_2 \tag{31}$$

$$st_{ig} = Et_i, \forall i \in J \tag{32}$$

The specific iterative calculation process is shown in figure 4.

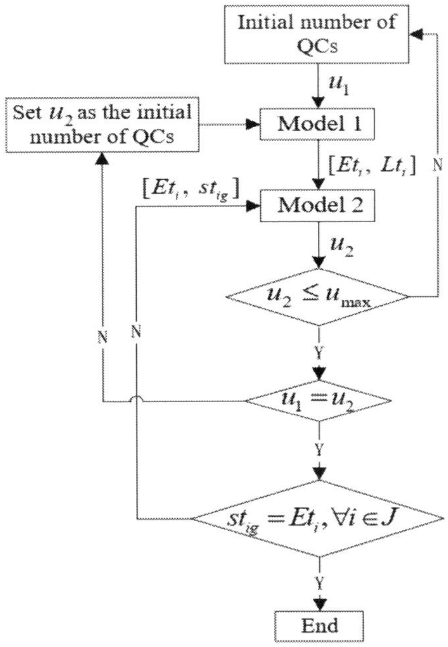

Figure 4: Flow chart of loop iteration

6 SOLUTION METHOD AND MODEL IMPROVEMENT

6.1 Parameters

The number of container tasks at a certain period is set as 60. These containers are allocated on different bays and their types are shown as table 1: U represents unloading and L represents loading. The number of available QCs is 9; h_{qi} is 1.2 min/TEU; d_{ij} is 0.8 min; c_{ij} is 1.5 min; Q_J is 30. The time window of each task is divided into 4 parts.

Table 1: Types (load/unload) and allocation of container tasks

ID	Type	Bay	ID	Type	Bay	ID	Type	Bay	ID	Type	Bay
1	U	1	16	U	5	31	L	9	46	U	13
2	L	1	17	L	6	32	U	9	47	U	13
3	L	1	18	L	6	33	U	10	48	L	14
4	U	1	19	L	6	34	L	10	49	L	14
5	L	2	20	U	6	35	L	10	50	U	14
6	U	2	21	U	6	36	U	11	51	U	15
7	U	2	22	U	7	37	U	11	52	L	15
8	L	3	23	U	7	38	L	11	53	L	16
9	L	3	24	U	7	39	L	11	54	L	16
10	U	3	25	L	7	40	U	11	55	U	16
11	U	3	26	L	8	41	L	12	56	U	17
12	U	4	27	L	8	42	L	12	57	U	17
13	L	4	28	U	8	43	U	12	58	L	17
14	L	5	29	U	9	44	U	12	59	U	18
15	L	5	30	L	9	45	L	13	60	L	18

6.2 Numerical Results

The models are formulated and solved in a CPLEX 12.2. When calculating the [M2], the result isn't ideal. The key reason is there are two problems with [M2]:(1)After solving the minimum number of QCs, it happens that tasks on one bay are operated by two QCs. (2)Some QCs begin to operating tasks on other bays without finishing tasks on certain bay.

It's too complex to consider the relationship between bays and QCs in existing models. An improvement was made on [M2] to solve above problems which set all tasks on a bay as one task group. If the number and type of tasks on a bay are known, the operation time of this bay can be solved by [M1]. Then to study configuration of QCs using this operation time by setting operating for one bay as one task. It avoids confusion caused by setting unit container as one task and avoids above two problems.

After iteration, the number of QCs stabilizes in 6 so it can say 6 is the optimal number of QCs. The optimal dispatch plan is shown as table 2, the number of QCs is 6 and the maximum total operation time is 17 min.

Table 2: The optimal scheduling plan of QCs

QC ID	Sequence of tasks	Makespan(min)
1	1→3→4→2→7→5→6→9→10→8→11	16.2
2	13→12→14→16→15→19→20→17→21→18	15.8
3	24→22→25→23→26→28→27→29→30→32→31	17
4	35→33→34→40→39→37→38→36	11.1
5	43→41→44→42→47→45→46→49→50→48	15.0
6	52→51→53→55→54→57→58→56→60→59	16.5

7 COMPUTATIONAL RESULTS AND ALGORITHM COMPARISON

The table 3 shows comparison between optimal solution of CPLEX and GA.

Table 3: Comparison between results solved by CPLEX and GA

The number of QCs	CPLEX results(min)	CPLEX time(s)	GA results (min)	GA time(s)
3	32	23.760	32.6	129.216
4	24	33.446	25.2	110.240
5	18.9	36.449	22.9	104.463
6	17.0	69.622	17.4	75.490
7	15	52.905	15.8	100.063
8	12.6	36.338	13.8	98.561
9	12.3	43.355	12.3	95.761

8 CONCLUSION

Through the establishment of coupling model of QCs dispatch and QCs configuration, this paper puts forward new ideas for equipment coordination scheduling in container ports and has enlightening significance for future research of equipment coordination scheduling problem in container terminals. Moreover, the view of according to the container task time window to study the configuration of the equipment, providing a theoretical basis for the configuration of the port equipment, it has certain application value.

9 ACKNOWLEDGEMENT

This research is supported by National Natural Science Foundation of China(71471110,71301101, 61540045), Science and Technology Commission Foundation of Shanghai (14170501500, 16DZ1201 402), Shanghai Education Commission Leading Academic Discipline Project (J50604) and Social Sc ience Foundation of Shaanxi Province(2015D060).

REFERENCES

[1] **N. Al-Dhaheri and A. Diabat. 2015.** The Quay Crane Scheduling Problem, *Journal of Manufacturing Systems*, 36,pp 87-94.

[2] **Y.-M. Fu and A. Diabat. 2015.** A Lagrangian relaxation approach for solving the integrated quay crane assignment and scheduling problem, *Applied Mathematical Modelling*, 39, pp 1194-1201.

[3] **N. Kenana and A. Diabatb. 2015.** A Branch-and-Price Algorithm to Solve a Quay Crane Scheduling Problem, *Procedia Computer Science*, 61, pp 527-532.

[4] **A. Diabat and E. Theodorou. 2014.** An Integrated Quay Crane Assignment and Scheduling Problem, *Computers & Industrial Engineering*, 73, pp 115-123.

[5] **J. H. Chen, D.-H. Lee, and M. Goh. 2014.** An effective mathematical formulation for the unidirectional cluster-based quay crane scheduling problem, *European Journal of Operational Research*, 232, pp 198-208.

[6] **M. Ji, W. Guo, H. Zhu, and Y. Yang. 2015.** Optimization of loading sequence and rehandling strategy for multi-quay crane operations in container terminals, *Transportation Research Part E*, 80, pp 1-19.

[7] **L. Tang, J. Zhao, and J. Liu. 2014.** Modeling and solution of the joint quay crane and truck scheduling problem, *European Journal of Operational Research*, 236, pp 978-990.

[8] **Y. B. Türkogulları, Z. C. Taskın, N. Aras, and I. K. Altınel. 2016.** Optimal berth allocation, time-variant quay crane assignment and scheduling with crane setups in container terminals, *European Journal of Operational Research*, 1, pp 1-17.

[9] **J. He, Y. Huang, W. Yan, and S. Wang. 2015.** Integrated internal truck, yard crane and quay crane scheduling in a container terminal considering energy consumption, *Expert Systems with Applications*, 42, pp 2464-2487.

[10] **N. Al-Dhaheri, A. Jebali, and A. Diabat. 2016.** A simulation-based Genetic Algorithm approach for the quay crane scheduling under uncertainty, *Simulation Modelling Practice and Theory*, 66, pp 122-138.

TOWARDS AN OPERATOR 4.0 TYPOLOGY: A HUMAN-CENTRIC PERSPECTIVE ON THE FOURTH INDUSTRIAL REVOLUTION TECHNOLOGIES

David Romero[1,4*], Johan Stahre[2], Thorsten Wuest[3], Ovidiu Noran[4], Peter Bernus[4], Åsa Fast-Berglund[2], Dominic Gorecky[5]

[1]Tecnológico de Monterrey, Mexico
david.romero.diaz@gmail.com

[2]Chalmer University, Sweden
johan.stahre@chalmers.se , asa.fasth@chalmers.se

[3]West Virginia University, USA
thwuest@mail.wvu.edu

[4]Griffith University, Australia
O.Noran@griffith.edu.au , P.Bernus@griffith.edu.au

[5]German Center for Artificial Intelligence, Germany
dominic.gorecky@dfki.de

ABSTRACT

This paper presents early concepts and future projections of the so-called 'Operator 4.0', understood as a smart, skilled operator who performs not only cooperative work with robots but also aided work by machines as and if needed by means of human cyber-physical systems, advanced human-machine interaction technologies and adaptive automation towards achieving human-automation symbiosis work systems. This research introduces an Operator 4.0 typology as well as exploring a set of key enabling technologies that can support the development of human-automation symbiosis work systems for the Operator 4.0 in the Factory of the Future defined within the Industry 4.0 framework.

Keywords: Industry 4.0, Operator 4.0, Human Cyber-Physical Systems, Advanced Human-Machine Interaction Technologies, Adaptive Automation, Human-Automation Symbiosis, Work Systems, Socially Sustainable Manufacturing, Exoskeletons, Augmented Reality, Virtual Reality, Wearable's, Intelligent Personal Assistants, Collaborative Robots, Social Networks, Big Data Analytics.

1 INTRODUCTION

The German program Industrie 4.0 (or Industry 4.0) and corresponding international initiatives (e.g. Smart Manufacturing, USA & Smart Factory, South Korea) will continue to transform the industrial workforce and their work environment through 2025 [1]. This will have significant implications on the nature of work in industry as Industry 4.0 will transform design, manufacture, operation, and service of products and production systems [2]. At the same time, the demography is changing, especially in Europe and Japan, which brings forth additional challenges for manufacturing companies. Increasing immigration may relieve some of the effects of demographic change.

* Corresponding Author

However, integration of new migrant workers with a high variety of technical skill and educational levels and different culture is considered a great challenge. As a consequence, near-future manufacturing enterprises, and in particular 'smart factories' as socio-technical systems, will need to form and adapt a *social perspective* to be proficient in assisting ageing, disabled and apprentice operators by using advanced digital and industrial enabling technologies to help people to remain in, return to or incorporate into the modern manufacturing workforce. Meanwhile, considering the developments from a *technical perspective*, new connectivity and interaction technologies among parts (*cf.* smart products), machines (*cf.* smart machines) and humans (*cf.* smart operators) will make production systems more lean, agile, traceable, and adaptable [2] [3] [4].

To successfully embrace the Industry 4.0 paradigm in a *socially sustainable way*, manufacturing enterprises will need to accompany its technological transformations with training and development programs for their workforce, in new tools and technologies that skilled labor uses and by which the operators are directly and indirectly affected. Furthermore, new working environments such as the 'cyber-physical factory' will directly affect the operator and the nature of work, creating new interactions not only between humans and machines, but also between digital and physical worlds. Therefore, socio-technical transformation towards the factory of the future (*cf.* factory 4.0 / smart factory) will need new design and engineering philosophies for twofold 'human-centric' and 'cyber-physical' production systems where automation, robotics, and other advanced manufacturing technologies are seen as possibilities for the further enhancement and augmentation of the human's physical, sensorial and cognitive capabilities rather than for unmanned, autonomous factories [5]. This new approach requires the rethinking of simple work design, moving towards design methods similar to those used to design the working environment for aircraft pilots, process industry operators and military personnel. These would typically include human supervisory control and human situation awareness.

This paper explores early manifestations and future projections of the *Operator 4.0* [5] as a smart and skilled operator who performs not only - 'cooperative work' with robots - but also - 'aided work' by machines as and if needed - by means of *human cyber-physical systems, advanced human-machine interaction technologies* and *adaptive automation* towards "human-automation symbiosis work systems". This research work introduces an *Operator 4.0* typology as well as explores a set of key enabling technologies for supporting the development of 'human-automation symbiosis work systems' for the Operator 4.0 in the Factory of the Future.

The vision of the *Operator 4.0* aims to build trusting relationships (interaction-based) between humans and machines, making possible for those smart factories to capitalize not only on smart machines' strengths and capabilities, but also empower their 'smart operators' with new skills and gadgets to fully capitalize on the opportunities being created by Industry 4.0 technologies. Hence, a *socially sustainable factory* within the Industry 4.0 framework is a workplace where work systems design and engineering uses collaborative robotics, kinematics, human-in-the-loop control systems, sensors, manipulation, navigation and adaptive automation to improve the knowledge and capabilities of operators. In this sense, a *human-centric production system* is characterized by allowing a unification of planning and implementation, expecting the operator to be in control of the work process and the technology and fostering the utilization of human competencies [6]. Furthermore, a *human cyber-physical production system (H-CPPS)* is defined as a work system that improves operators' abilities thanks to a dynamic interaction between humans and machines in the cyber- and physical-worlds by means of 'intelligent' human-machine interfaces. H-CPPS will be using human-computer interaction techniques designed to fit the operators' cognitive and physical needs, and improve human physical-, sensing- and cognitive-capabilities, by means of various enriched and enhanced technologies [5]. Moreover, these cyber-physical and human-machine interactions are supervised and controlled by adaptive control systems [7] [8] [9] within the human-in-the-loop [10] paradigm and based on sharing and trading of control strategies [11] (e.g. critical-event, measure-based and/or modeling-based) in order to allow a dynamic and seamless transition of functions (tasks) allocation between humans and machines, always aiming for the operator inclusiveness without compromising the production objectives [5].

2 TOWARDS AN OPERATOR 4.0 TYPOLOGY

This section presents an *Operator 4.0* typology that depicts how the Industry 4.0 technologies can assist operators to become 'smarter operators' in their future factory workplaces (see Figure 1), from a social manufacturing perspective. Furthermore, it is important to mention that these types of Operators 4.0 may exist on the shop-floor as either single- or hybrid- types. A selection of various augmentations of the original human capabilities are presented below; note however, that there might be multiple other aspects that are part of the Operator 4.0. Those augmentations do not only come in a variety of levels but also can be combined. It is also very likely that the future Operator 4.0 may only be augmented in one specific area whereas the other aspects are neglected. In some cases that will not even be possible (e.g. augmented reality functionality necessarily needs a 'connected operator' to perform).

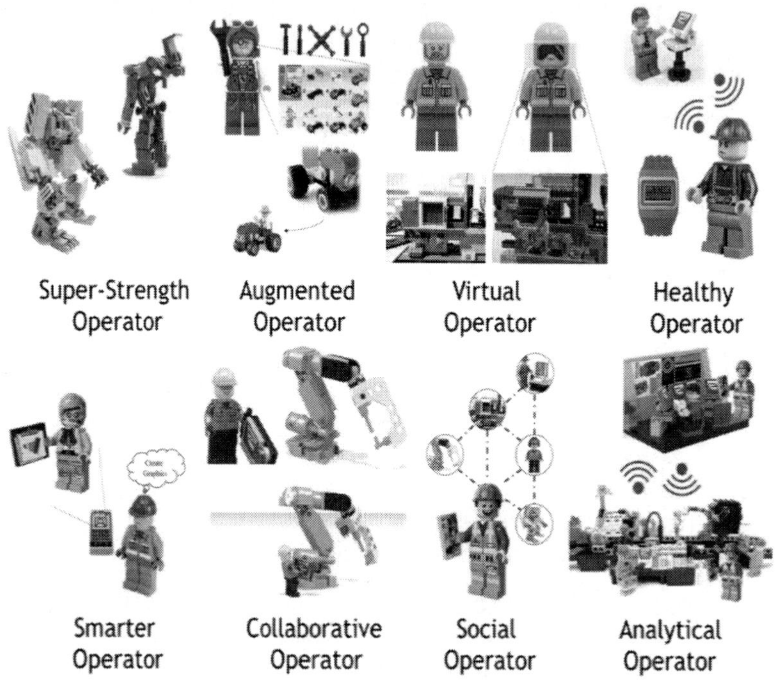

Figure 1: Operator 4.0 Typology

2.1 Operator + Exoskeleton = Super-Strength Operator [physical interaction]

Powered (Industrial) Exoskeletons are wearable lightweight, flexible and mobile, representing a type of biomechanical system where the human-robotic exoskeleton powered by a system of motors, pneumatics, levers or hydraulics works cooperatively with the operator to allow for limb movement, increased strength and endurance. The idea of wearable exoskeletons has been around for decades in industry, aiming to use powered mechanics to increase the strength of a human operator for effort-less manual functions (tasks) (e.g. [12]).

Powered exoskeletons can help to reduce the trade-offs between manual and automated operations in production systems - in other words, between flexibility and efficiency in balanced automation systems as well as to increase the social sustainability of factories in the long-term, especially with the outlook of a larger proportion of elderly workers due to changing demography.

Powered exoskeletons may allow (e.g. in an assembly area, where most manual work takes place) to have humans and technology cooperate in order to simplify the job and reduce the physical stress, thus making the overall system more efficient and productive by auxiliary processes such as safe lifting and moving of heavy items, enhancement of physical capabilities (endurance) to stay longer in a demanding position or additional strength for handling weighty tools. In general, powered exoskeletons can offer additional protection, support and strength to the operators and contribute to the social sustainability of the workforce by improving the ergonomics of manual operations (occupational health), helping to reduce injuries and accidents because of heavy work (safety), boosting productivity and quality of work by improving the operator workload-handling capabilities (e.g. decreasing the physical workload and relocating the energy to the sensorial and cognitive capabilities), and assisting to keep elder/experienced operators longer in critical positions by compensating for their loss of strength due to aging, while still capitalizing on their knowledge and experience. An additional advantage is the possibility to equip the exoskeleton with tools and (heavy) supporting equipment, which in the exoskeleton absence would reduce the humans' productivity.

An example of this type of Operator 4.0 is the *Robo-Mate* system, defined as: "a user-friendly intelligent cooperative light weight wearable human-robotic exoskeleton for manual handling work" [13].

2.2 Operator + Augmented Reality = Augmented Operator [cognitive interaction]

Augmented Reality (AR) is a technology enriching the real-world factory environment of the smart operator with digital information and media (sound, video, graphics, GPS data, etc.) that is overlaid in real-time in his/her field of view (e.g. head-gear, smart-phones, tablets or spatial AR projectors). Hence, AR can be considered a key enabling technology for improving the transfer of information from the digital to the physical world of the smart operator in a non-intrusive way.

AR technology may offer significant advantages (e.g. faster cycle times, reliability, reduced failure rate and traceability) to support the smart operator in real-time during manual operations by becoming a digital assistance system for reducing human errors and at the same time reducing the dependence on printed work instructions, computer screens and operator memory, which need to be interpreted first by a skilled worker. AR for example can enable 'digital poka-yokes systems' for work-intensive functions (tasks) in order to reduce defects, rework and redundant inspection by offering intuitive information and combining operator intelligence and flexibility with error-proofing systems to increase the efficiency of manual work steps, while improving the quality of work [14]. Moreover, AR technology can incorporate a new human-machine interface to manufacturing IT applications and assets, displaying real-time feedback about smart manufacturing processes and machines to the smart operator in order to improve decision-making [15]. This can be implemented at machine level using traditional Programmable Logic Controllers (PLCs) and Supervisory Control & Data Acquisition (SCADA) systems but also emerging Internet of Things (IoT) technologies for assets condition monitoring. AR can be implemented also at mid-level operations like Manufacturing Execution Systems (MES), novel production line simulations and big data-driven quality controls and at higher levels such as the Enterprise Resource Planning (ERP) systems.

At machine level, AR can redefine the maintenance and repair of equipment, by means of 'diagnostic intelligence' derived from real-time sensor data about a machine or part performance; at operations and enterprise levels, AR can allow production managers to view production KPIs and have an intra-factory overview of workstations and production lines in real-time for monitoring, identifying, analyzing, diagnosing and resolving problems and flaws (e.g. alerting on deviations) to keep manufacturing processes moving towards operational efficiency. Furthermore, AR technology acting as 'tag reader' may also create new human-product interactions enabled by QR codes, GPS, OCR, barcodes, RFID and NFC technologies, allowing the smart operator to retrieve current and historical information about a product and monitor and configure data and settings about it.

Presently, an early-stage example of this type of Operator 4.0 is the *Satisfactory system*, defined as "an augmented-enabled ecosystem for increasing satisfaction and working experience in smart factory environments" [16].

2.3 Operator + Virtual Reality = Virtual Operator [cognitive interaction]

Virtual Reality (VR) is an immersive interactive multimedia and computer-simulated reality that can digitally replicate a design, assembly or manufacturing environment and allow the smart operator to interact with any presence within (e.g. a blueprint, a hand-tool, a product, a machine tool, a robot, a production line, a factory), with reduced risk and real-time feedback.

VR technology can provide a combination of interactive virtual reality and advanced simulations of realistic scenarios for optimized decision-making and training for the smart operator. For example, at product design and engineering stage, VR will transform blueprints into 3D virtual models where all types of design rules, guidelines and methodologies (*cf.* Design for eXcellence) can be digitally threaded to upstream design and engineering decisions and check their impact along the product lifecycle (e.g. design for manufacturability, design for assembly, design for serviceability, design for maintainability, design for disassembly, design for repair-reuse-recyclability, etc.); at product assembly stage, CAD models of parts, hand-tools and assemblies can be transformed into interactive virtual simulations (assembly sequences) for training operators in complex assembly tasks; and at product manufacturing stage, VR brings to life the 'virtual factory' as an integrated simulation model of the major sub-systems of a factory in order to evaluate different factory layouts (arrangements of machinery, equipment and inventories for smooth flow of work, material and finished products), production line configurations (manufacturing processes sequences), production balance (automation vs. mechanization) and production schedules (work and workloads scheduling) in order to optimize the production master plan by means of what-if analyses, decision support systems and estimation methods.

Several commercial software tools for Virtual Product Design (VPD), Computer-Aided Design (CAD), and Computer-Aided Engineering (CAE) are available on the market. VR systems for dynamic representation of humans are also available. Some examples of software supporting the Operator 4.0 are the *VISTRA system:* "a virtual simulation and training system, which is used to train operators and test manual assembly processes" [17], and the *VFF system:* "a holistic, extensible, scalable and standard virtual factory framework and integrated simulation environment that considers the factory as a whole and provides advanced planning, decision support and validation capability features to production managers" [18].

2.4 Operator + Wearable Tracker = Healthy Operator [physical and cognitive interaction]

Wearable Trackers are devices designed to measure exercise activity, stress, heart rate and other health-related metrics as well as GPS location and other personal data (e.g. biometrics). With the dawn of commercially available solutions like the Apple Watch, Fit-bit and Android wear, many people all over the world are already using aspects of this envisioned system. Military applications are going a step further and employ data analytics on bio-data to predict potentially problematic situations before they emerge, e.g., for Special Forces combat divers during missions. Currently there are already first steps being made to take this to the next level, with the potential to track the complex nature of the human brain during different activities. While this might take decades to be applicable in industry, it gives an indication of the potential this might offer.

Without entering into data privacy issues, rules and employment as well as law implications, wearable trackers (bio-data sensors) can drive positive change via improved productivity, well-being and proactive safety measures for the workforce. Different application levels are possible, e.g. both at the individual and collective level, with the system boundaries for the collective level being flexible.

A smart operator for example can use 'personal analytics' to plan and schedule his/her work-shifts, rest-breaks and overtime based on health-related metrics, can monitor his/her physical workload (exercise activity) and cognitive workload (mental effort) during the work-shift and set alerts and warnings to manage proper levels of occupational effort and stress. Using advanced data analytics on bio-data might allow for utilizing subconscious cognitive states and thus allow for a warning mechanism for imminent danger and/or potential harm to the worker him-/her-self and/or others.

As an aggregation of operators' personal analytics, 'workforce analytics' can prevent urgent threats to operators' safety and also production quality by monitoring health-related metrics and workloads and alert decision-makers (e.g. production line supervisors and factory managers) for example if the stress levels are or workload is too high, which may lead to human errors (e.g. accidents or poor quality of labor) or on the contrary if energy levels of operators are high, controlled aggressive production targets can be set and pushed to the workforce. Moreover, operators' location inside a big factory or warehouse can improve internal logistics (e.g. response time) and help to locate and allocate the closest operator to an urgent function (task) in-hand.

A present example of this type of Operator 4.0 includes the use of smart-watches to leverage awareness of biometrics, so a smart operator can make better decisions in regards to his/her occupational health care self-management (e.g. fitness, wellness, medical). In this basic example, however, the required technology is available today and the advanced data analytics capability needed is progressing fast.

2.5 Operator + Intelligent Personal Assistant = Smarter Operator [cognitive interaction]

An *Intelligent Personal Assistant (IPA)* is a software agent or artificial intelligence that has been developed to help a smart operator in interfacing with machines, computers, databases and other information systems as well as managing time commitments and performing tasks or services in a human-like interaction [19].

One of the main features of IPAs is their capability to offer a voice-interaction technology (a natural language interface) to the smart operator, which induces productivity and operational efficiency by allowing the operator to go hands-free to complete certain tasks. Some scenarios where IPAs can create advantages for the smart operator and offer personal assistance are: in searching and retrieving from a digital library, based on a voice request, the repair or maintenance manual of a machine tool or part and reading the instructions to the operator while he/she performs the task; in scheduling and setting reminders for actions or critical events in operations, inventory or assets management (e.g. re-certifications, check stocks, preventive maintenance); planning activities, where human creativity can be applied to solve problems of routing or staffing while utilizing the IPA to store and visualize the underlying planning data; in providing mobility and location assistance for logistics (e.g. GPS-based geographical navigation) and warehouse stocks whereabouts (e.g. IPS-indoor positing system); interfacing with connected devices through voice commands (e.g. voice user interfaces); detecting and diagnosing errors and problems and suggesting troubleshooting tools and strategies in smart, connected machines and systems; and building predictive models by tracking operator or machine behavior and alerting for proactive actions.

A present example of this type of Operator 4.0 are Apple's Siri, Android's 'Hey Google' and especially Amazon's Alexa. Especially Amazon's Alexa allows external developers to access the API and build additional apps and services using the existing functionalities and infrastructure. Alexa can already perform tasks like 'finding suitable recipes' and writing a shopping list - tasks that translate easily in a shop-floor environment.

2.6 Operator + Collaborative Robot = Collaborative Operator [physical interaction]

Collaborative Robots (CoBots) are industrial robots (*cf.* humanoid, robotic arm or SCARA configurations) capable of performing a variety of repetitive and non-ergonomic tasks and that have been specially designed to work in direct cooperation with the smart operator by means of safety (e.g. force sensing and collision) and intuitive interaction technologies, including easy shop-floor programming. Popular examples are Rethink-Robotics' Baxter & Sawyer, which promise low-cost and easy to use collaborative robots.

CoBots will allow co-working spaces and interaction with their human counterparts without the need for traditional safety barriers. These possibilities will create benefits such as recovering shop-floor space normally lost due to safety barriers (*cf.* safety cage) and savings for the costs associated to their implementation; increasing the smart operator productivity and job satisfaction by augmenting him/her to accomplish a task more effectively and relieving him/her from tedious, non-ergonomic and vulnerable tasks (e.g. difficult placement of parts, "third-hand" functionality for assembly, heavy and repetitive lifting, loading and handling of hazardous materials). Another societal positive effect emerging from close collaboration of humans and robots in the workplace may be the increasing acceptance of robotic help in the healthcare domain, again, being estimated to being highly relevant due to demographic change in some areas.

A present example of this type of Operator 4.0 is the LIAA CoBot system: "a hybrid assembly systems combining manual and automatic workstations in symbiotic human-robot collaborations to achieve a balance between investment costs, batch size and flexibility in the assembly line" [20]. The INSA project is another example of applied research leveraging advanced image recognition to create safe collaborative workspaces for close human robot interaction [21].

2.7 Operator + Social Networks = Social Operator [cognitive interaction]

Enterprise Social Networking Services (E-SNS) focuses on the use of mobile and social collaborative methods to connect the smart operators at the shop-floor with the smart factory resources. Such connections include 'social relations' among the workforce (*cf.* social network services) and between operators and smart things (*cf.* social Internet of Industrial Things [22]) to interact, share and create information for decision-making support.

Social networking between smart operators, enabled by real-time mobile communication capabilities, can empower the workforce to contribute their expertise across the production line and to the shop-floor, can accelerate ideas generation for product and processes innovation and can facilitate problems-solving by bringing together the right people with the right information and especially knowledge management and knowledge creation within the enterprise. Knowledge creation and management is (and always has been) challenging as there is still no 'ideal way' to proceed. Research and industrial practice (e.g. Airbus) suggests that a personal approach to sharing, communicating and managing knowledge within the enterprise (e.g. deriving knowledge from a future retiree) is more successful than a technical and highly structured approach. Social networks embedded in the companies' knowledge system might present a chance to utilize the 'social' component and still allow for storing and accessing collective knowledge. Meanwhile, the social IoIT can connect, through 'interactive machine learning', smart operators with smart things (*cf.* intelligent assets) in social networks for sharing information and exchanging messages about their location, condition, operation status and availability for improving (for example) at machine level the asset reliability (e.g. intelligent maintenance) and at production line level the material flows and resources productivity (e.g. spotting bottle-necks) towards social problems-solving and optimization of the production system.

In both cases, the final aim of enterprise social networking is to communicate and enable cooperation between smart operators and smart machines via social relations to accomplish production goals.

An example of this type of Operator 4.0 can be considered the internal (intranet) forums and wikis (such as e.g. used at VW). However, this only utilizes parts of what is ultimately envisioned for the Operator 4.0. Some companies and their employees make use of existing Social Network Services (e.g. Facebook or LinkedIn) to communicate and connect. This is risky however, as the data is then in the hand of private entities outside of the company and thus should be accompanied by an in-depth risk analysis.

2.8 Operator + Big Data Analytics = Analytical Operator [cognitive interaction]

Big Data Analytics is the process of collecting, organizing and analyzing large sets of data (big data) to discover useful information and predict relevant events. Its application to the smart factory has given birth to manufacturing real-time analytics at the shop-floor, also known as 'smart manufacturing'.

Big Data analytics may help smart operators (e.g. production managers) to achieve better forecasts, understand the smart factory performance (shop-floor control), fuel continuous improvement (Six Sigma), provide greater visibility of KPIs (data visualization and interactive dashboard) and real-time alerts based on predictive analytics (fault detection and quality improvement) in order to leverage real-time information for driving the right response to prevent mistakes, quickly identify problems and call for the right decisions to improve operational efficiency.

Data analytics and machine learning have several applications in manufacturing and are already fairly widely employed [23]. However, the increase in available data through cheap sensors and the Industrial IoT (connected devices) and also fast progress in unsupervised methods like deep learning will bring forth even more powerful and applicable solutions in the near future.

The 'analytical operator' is somewhat connected to several other applications as many of them rely on advanced data analytics. So does the 'collaborative operator', who often utilizes image recognition to allow working in close proximity to CoBots, the 'healthy operator' who relies on the analytics of the bio-data collected and the 'smarter operator' using Artificial Intelligence embedded in a virtual assistant.

This type of Operator 4.0 is illustrated by the variety of monitoring and control tools currently using machine learning and data mining algorithms to improve quality, lead time, etc. [24].

3 RELATED SYSTEMS, STRATEGIES AND ENABLING TECHNOLOGIES TO/FOR THE OPERATOR 4.0

This section briefly introduces related systems, strategies and enabling technologies to/for the *Operator 4.0*. Thus, Figure 2 presents a generic *human cyber-physical system control loop* depicting the interactions between humans and machines and cyber and physical worlds.

3.1 Towards Human Cyber-Physical Systems

Human Cyber-Physical Systems (H-CPS) are the new frontier of human-machine interactions and physical-digital worlds interfacing for the augmentation or enhancement of human performance.

For the case of the Operator 4.0, H-CPS aim to become safety (fault tolerance) engineered systems of systems with the human-in-the-loop, using context-sensitive, advanced communication and adaptive control technologies to support inter-agent systems of humans, machines and software to interface in the virtual and physical worlds towards a sustainable and human-centric production system (see Figure 2).

H-CPS will be deployed in the near future at the shop-floor to optimise the outcome of a production system considering the social sustainability of the manufacturing.

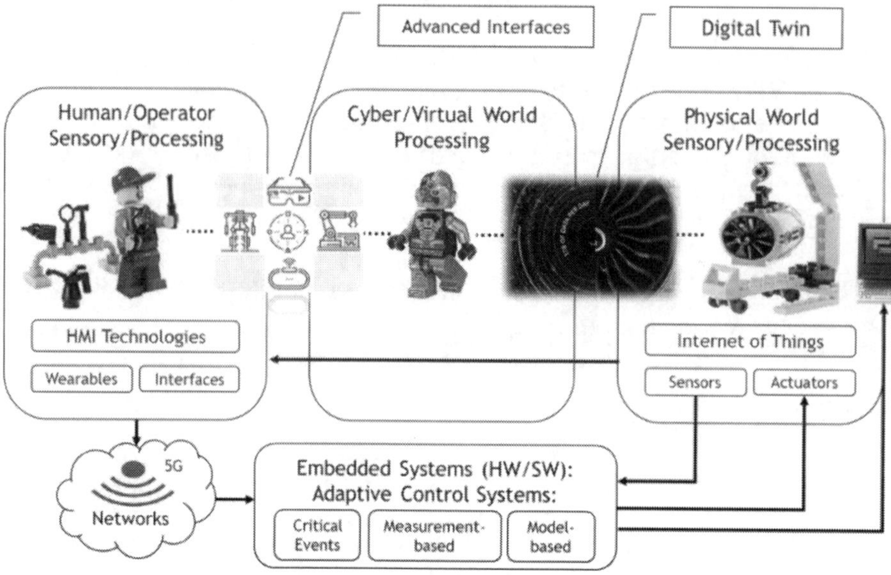

Figure 2: Generic Human Cyber-Physical System Control Loop

3.2 Adaptive Automation Control Systems Strategies

Adaptive Automation Control Systems rely on three main strategies for adaptive function and task allocation in human-machine interactions towards supporting the Operator 4.0 [11]: (1) based on 'critical event(s)' that will trigger the sharing or trading of control between human-machine agents in order to guarantee the completion of a critical to task by one or the other or by a hybrid or emergent agent, or when the functioning of the overall system is endangered by the action(s) of an agent; (2) 'measurement-based' as human and machine agents are continuously monitored in order to detect deviations from their tasks or ideal performance and trigger the proper countermeasures; and (3) 'model-based' on cognitive human and machine models for handling effective and efficient distribution of workload.

3.3 Human-Machine Interaction Technologies

Some relevant HMI technologies that will support the Operator 4.0 are: *Dialogue Systems* - handling the dialogue between humans and machines (e.g. using natural language interface); *Control Devices* - physical (e.g. keyboards, mouse, joystick, trackball, steering wheel, pedals, knobs and switches) and digital ones (e.g. buttons, sliders and menu buttons) enhanced with haptics technology; *Multimedia-Multimodal Displays* - providing the smart operator multiple modes of interacting with a system; and *Adaptive Interfaces* - which adapt their layout and elements to the needs of the smart operator and/or context.

4 CONCLUSIONS AND FUTURE WORK

Human-centric manufacturing has been a core topic for most previous manufacturing paradigms; that same is true for Industry 4.0. Computer-based manufacturing control requires the human to handle complexity emerging beyond the imagination of manufacturing system designers. Therefore, the full potential of Industry 4.0 and the achievement of a socially sustainable manufacturing industry will only be realized if the Operator 4.0 is at its heart (*cf.* human-centric) and interacting with the machines through physical and cognitive means. As mentioned by [25], *human-automation symbiosis* [3] [5], manifested as the Operator 4.0 vision, is necessary for achieving sustainable development in human society. However, it can only be secured through the use of intelligent (smart) automation systems and human-machine interfacing technologies. There, the assumed 'intelligence' allows inclusion of the explicit representation of human goals and plans (e.g. productivity, occupational health, safety, job inclusion and job satisfaction, etc.), thus constituting the basis of human-machine interaction and 'social' interaction with technology. The Operator 4.0 typology is useful in order to increase the understanding of the future roles of humans and machines in the factories of Human Cyber-Physical Systems. By creating a typology and a transcript of available assets and skills, traditional manufacturing companies can easily adopt the future contributions of humans in Industry 4.0. Future work will identify and address the specific challenges of the Operator 4.0 typology types.

5 REFERENCES

[1] **BCG Group.** 2015. *Report on Man and Machine in Industry 4.0: How Will Technology Transform the Industrial Workforce Through 2025?*

[2] **BCG Group.** 2015. *Report on Industry 4.0: The Future of Productivity and Growth in Manufacturing Industries.*

[3] **Romero, D., Noran, O., Stahre, J., Bernus, P., Fast-Berglund, Å.** 2015. *Towards a Human-Centred Reference Architecture for Next Generation Balanced Automation Systems: Human-Automation Symbiosis.* Innovative Production Management towards Sustainable Growth, Service, Manufacturing, and Resilient Value Chain, S. Umeda et al. (Eds.), IFIP, Part II, AICT 460, Springer, pp. 556-566.

[4] **De Weck, O.L., Ross, A.M., Rhodes, D.H.** 2012. *Investigating Relationships and Semantic Sets amongst System Lifecycle Properties (Ilities).* 3rd Int'l Engineering Systems Symposium, CESUN 2012, Delft University of Technology, 18-20 June 2012, pp. 18-20.

[5] **Romero, D., Bernus, P., Noran, O., Stahre, J., Fast-Berglund, Å.** 2016. *The Operator 4.0: Human Cyber-Physical Systems & Adaptive Automation towards Human-Automation Symbiosis Work Systems.* Production Management Initiatives for a Sustainable World. Springer. TO APPEAR.

[6] **Charles, T., Charles, R., Roulstone, A.** 1990. *Prospects and Conditions for Anthropocentric Production Systems in Britain,* APS-Research Papers Series, Vol. 8, Commission of the European Communities, Bruxelles.

[7] **Hancock, P.A., Chignell, M.H. 1987.** *Adaptive Control in Human-Machine Systems.* Human Factors Psychology, pp. 305-345, North Holland: Elsevier Science Publishers.

[8] **Hancock, P.A., Jagacinski, R.J., Parasuraman, R., Christopher D. Wickens, C.D., Wilson, G.F., Kaber, D.B.** 2013. *Human-Automation Interaction Research: Past, Present and Future.* Ergonomics in Design, 21(2):9-14.

[9] **Sheridan, T., Parasuraman, R. 2006.** *Human-Automation Interaction.* Human Factors and Ergonomics, 1(1):89-129.

[10] **Munir, S., Stankovic, J.A., Liang, C-J.M., Lin, S.** 2013. *Cyber-Physical System Challenges for Human-in-the-Loop Control.* Feedback Computing.

[11] **Inagaki, T.** 2003. *Adaptive Automation: Sharing and Trading of Control.* Chapter 8 - Handbook of Cognitive Task Design, pp. 147-169.

[12] **Sylla, N., Bonnet, V., Colledani, F., Fraisse, P.** (2014). *Ergonomic Contribution of ABLE Exoskeleton in Automotive Industry.* Int'l J. of Industrial Ergonomics, 44(4): 475-481.

[13] **Robo-Mate** (Intelligent exoskeleton based on human-robot interaction for manipulation of heavy goods in Europe's factories of the future) http://www.robo-mate.eu/

[14] **Gorecky, D.; Campos, R.; Chakravarthy, H.; Dabelow, R.; Schlick, J.; Zühlke, D.** 2013. *Mastering Mass Customization - A Concept for Advanced, Human-Centered Assembly.* Assembly Manufacturing Engineering Journal, 11(2):62-67.

[15] **Gorecky, D.; Schmitt, M.; Loskyll, M.; Zühlke, D.** 2014. *Human-Machine-Interaction in the Industry 4.0 Era.* 12th IEEE International Conference on Industrial Informatics, pp. 289-294.

[16] **Satisfactory** (A collaborative and augmented-enabled ecosystem for increasing SATISfaction and working experience in smart FACTORY environments) http://satisfactory-project.eu/

[17] **VISTRA** (Virtual Simulation and Training of Assembly and Service Processes in Digital Factories) http://vistra-project.eu/

[18] **VFF** (Holistic, extensible, scalable and standard Virtual Factory Framework) http://www.vff-project.eu/

[19] **Myers, K., Berry, P., Blythe, J., Conley, K., Gervasio, M., McGuinness, D.L., et al.** 2007. *An Intelligent Personal Assistant for Task and Time Management* AI Magazine, 28(2):47-61.

[20] **LIAA** (Lean Intelligent Assembly Automation) http://www.project-leanautomation.eu/

[21] **INSA** (Protection and Integral Security Concepts in Cyber-Physical Systems) http://www.insa-projekt.de - In German.

[22] **Wuest, T., Hribernik, K., Thoben, K.-D.** 2012. *Can a Product Have a Facebook? A New Perspective on Product Avatars in Product Lifecycle Management.* Rivest, L., Bouraz, A. & Louhichi, B. (Eds.), PLM 2012, IFIP AICT 388, Heidelberg Berlin: Springer, pp. 400-410.

[23] **Wuest, T., Weimer, D., Irgens, C., Thoben, K.-D.** 2016. *Machine Learning in Manufacturing: Advantages, Challenges and Applications.* Production & Manufacturing Research, DOI: 10.1080/21693277.2016.1192517

[24] **Wuest, T., Irgens, C., Thoben, K.-D.** 2014. *An Approach to Quality Monitoring in Manufacturing using Supervised Machine Learning on Product State Data.* Journal of Intelligent Manufacturing, 25(5):167-1180.

[25] **Tzafestas, S.** 2006. *Concerning Human-Automation Symbiosis in the Society and the Nature.* Int'l J. of Factory Automation, Robotics and Soft Computing, 1(3):16-24.

A GA based parameter meta-optimization of ACO algorithm for solving assembly sequence optimization

Atul Mishra[1] and Sankha Deb[2]
FMS and Computer Integrated Manufacturing Laboratory,
Department of Mechanical Engineering,
Indian Institute of Technology Kharagpur, Kharagpur -721302, India.
[1] atulmishra_chdi@yahoo.in
[2] sankha.deb@mech.iitkgp.ernet.in

ABSTRACT

Assembly sequence planning deals with finding the sequence of operations to assemble the components and sub-assemblies into the final product. With advent of Artificial Intelligence, several soft-computing based evolutionary optimisation algorithms had been used by researchers to solve the problem of finding the best feasible assembly sequence. In the proposed paper, an Ant Colony Optimisation (ACO) algorithm based approach has been used for assembly sequence optimisation based on minimizing the number of direction changes, while conforming to the precedence constraints between components. Because of the graph-based nature of ACO, it is relatively less computationally expensive than many of the other soft computing based approaches reported in the literature, and moreover it is also well suited to the nature of the problem on hand. However, the ACO algorithm has a lot of parameters such as rate of pheromone evaporation, pheromone decay parameter, number of iterations, etc. each of which needs to be varied in order to obtain the best convergence rate of the algorithm, Furthermore, the combination of the parameters also needs to be optimised. In the present paper, a binary coded Genetic Algorithm (GA) has been proposed to optimise the parameters of the ACO algorithm. Some of the reasons for choosing GA include the fact that the number of parameters in GA is less than ACO and thus it takes less time and efforts to optimise the GA than ACO. Moreover the modeling of the solutions (i.e. chromosomes) in the form of binary strings is easier. The optimized ACO parameters have been used for solving a problem of assembly sequence optimization for a sixteen component assembly to demonstrate the effectiveness of our proposed meta-optimization procedure.

Keywords: Computer-Aided Process Planning, Assembly Sequence Optimization, Ant Colony Optimisation, Meta-optimisation by Genetic Algorithm.

1 INTRODUCTION

Assembly sequence planning deals with finding the sequence of operations to assemble the components into final product. Since assembly costs contribute to a significant amount of total manufacturing cost, proper assembly sequence planning plays a crucial role in reducing cost. Significant expertise and knowledge are necessary to determine the optimal assembly sequence. Moreover, with increase in number of components, the number of feasible assembly sequences possible also increases, making manual process planning laborious and time consuming. To overcome these drawbacks, various Computer-Aided Process Planning approaches have been developed. With advent of Artificial Intelligence, several soft-computing based evolutionary optimization algorithms had been used for assembly sequence optimization, based on criteria like changes in orientation, assembly tools, and stability. They possess number of parameters and their values can have significant impact on the effectiveness of an algorithm. Moreover, the parameter values need to be tuned for each new problem instance. Naturally a key challenge is determining the optimum combination of parameters (known as meta-optimization), which if done manually, may involve lot of time and effort. In the present paper, an ACO algorithm has been used for assembly sequence optimization based on number of orientation changes, while conforming to the part precedence constraints. Because of graph-based nature, it is less computationally expensive than other soft computing approaches, and moreover it is also well suited to the nature of the problem on hand. However, ACO has a lot of parameters such as Q value to adjust the increment of pheromone, rate of pheromone evaporation, pheromone decay parameter, number of iterations required for convergence. Each of these needs to be varied to obtain the best algorithm performance. Furthermore, the parameter combination also needs to be optimized. In the present paper, a binary coded GA has been proposed to optimise the ACO parameters. Reasons for choosing GA include the fact that the number of GA parameters is less than ACO and modeling of the solutions in form of binary strings is easier.

A brief review of different approaches for assembly sequence optimization is given below. Hong and Cho [1] developed an approach for robotic assembly sequence optimization by Hopfield neural network and an expert system to infer the assembly constraints from the liaison data of the product. Bonneville et al. [2] developed a GA based approach. The authors reported that the proposed GA could generate all valid and good assembly plans but its performance was slow, and does not necessarily guarantee optimum plans. Chen et al. [3] proposed a three-stage integrated approach with heuristic working rules to assist the planner to develop a better assembly plan. Above Graph and transforming rules were used to create a correct Explosion Graph of the assembly models, followed by a three-level relational model to create a complete relational model graph and an incidence matrix, a mathematical model based on a penalty index was formulated, and a revised minimum spanning table method was used to generate and evaluate a feasible assembly sequence. Wang et al. [4] developed an ACO approach for optimisation of assembly sequences. The optimal solution was with respect to the least reorientations during assembly. The concept of assembly by disassembly was adopted in that paper and disassembly matrix was used to guarantee the validity and feasibility of sequences. The parameters of this ACO were chosen after trial computations. Cao and Xiao [5] explored Immune Optimisation Algorithm (IOA). It was based on the bionic principles of Artificial Immune Systems (AIS). Assembly sequences were evaluated on the basis of total number of components, number of changes in assembly directions and tools, base component location, and feasibility degree. Chang et al. [6] showed the application of AIS for assembly sequence planning exploration using connector concepts. Combination property, tool property, direction property, precedence relationships were considered. Wang and Liu [7] developed a PSO based approach. Assembly cost was subjected to geometrical constraints and five assembly process constraints namely, local assembly precedence, number of the unstable parts, assembly direction and tool changes, connector changes. Lv and Lu [8] presented the application of discrete PSO. They considered total number of tool changes, orientation changes, and operation type changes and interference times in the product assembly. Gao et al. [9] developed an approach based on Memetic Algorithm where a chromosome represented an assembly sequence consisting of genes containing the part number and the direction variable. They considered times of the assembly direction changes and assembly feasibility. Zhou et al. [10] combined the Bacterial Chemotaxis with GA. Assembly sequences were encoded as chromosomes, where gene in the chromosome is treated as a bacterium. Fitness function comprised of length of longest sub-sequence, number of orientation changes, number of gripper changes. Xing and Wang [11] presented a hybrid PSO and GA based optimisation for compliant assemblies based on graph theory. Liaison graph and adjacency matrix were used to describe the geometry of the compliant assemblies. Assembly sequences were evaluated on the basis of assembly variation due to dimensional tolerance. Recently, Li et al. [12] developed an algorithm based on Improved Harmony Search

(IHS). They have proposed new aspects like an initial harmony memory (HM) established using the opposition based learning (OBL) strategy, a way to improvise a new harmony and a local search strategy. They considered changes of assembly direction and assembly tool and stability criteria in the fitness function.

In the following sections, first the approach of assembly sequence optimisation by using ACO is described, followed by the proposed approach for parameter meta-optimization of ACO by GA. An illustrative example is given to demonstrate the application and the results and discussions are presented. Finally important conclusions and scope for future work are given.

2 APPROACH FOR ASSEMBLY SEQUENCE OPTIMIZATION BY USING ACO

The ACO approach developed by Wang et al. [4] has been applied to determine the best feasible assembly sequence with least number of direction changes. The concept of assembly by disassembly has been adopted, where a disassembly sequence has been represented as an ordered list of disassembly operations (DO). Each DO is denoted as a duple DO = (N, D), where N is the component number and D is its disassembly direction i.e. either of ±X/±Y/±Z. The feasibility of an assembly sequence is ensured using a Disassembly Matrix (DM). The number of rows of DM is equal to number of parts n. The number of columns of DM is equal to 3n, where each column of corresponding component refers to one of the disassembly directions, +X/+Y/+Z. If the component c_i does not pose interference with the component c_j in the direction of +X, then DM (i, j) value of +X column will be equal to 0, otherwise 1. Conversely, if the component c_i does not pose interference with the component c_j in the direction of -X, then DM (j, i) value of +X column will be equal to 0, otherwise 1. A typical Disassembly Matrix (DM) has been presented in Table 1 for a four component assembly, shown in Figure 1 (Wang et al. [4]).

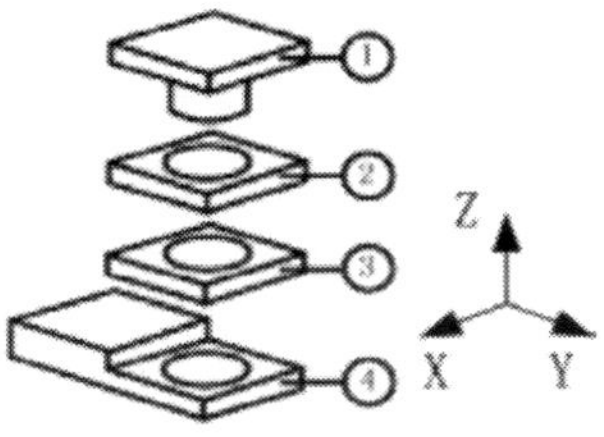

Figure 1: Four component assembly

Table 1: Disassembly Matrix for Four component assembly

Directions➧	X	Y	Z	X	Y	Z	X	Y	Z	X	Y	Z
1	0	0	0	1	1	0	1	1	0	1	1	0
2	1	1	1	0	0	0	0	0	0	0	0	0
3	1	1	1	0	0	1	0	0	0	0	0	0
4	1	1	1	0	0	1	0	0	1	0	0	0

The problem of assembly sequence optimization has been modeled using a graph, with each node representing a component number and its disassembly direction. In ACO, first of all, using the DM it is necessary to identify those components which can be used to start the disassembly operation. The number of these components decides the number of ants for the ACO algorithm. These ants are placed on the starting nodes and from the START node, the ants start moving towards the END node, gradually connecting the intermediate nodes. The next node of the graph is selected by the ants using a probability function given in eq. 1. The probability with which ant k in DO i chooses DO j is calculated as follows.

$$p_k(i,j) = \begin{cases} \frac{\tau(i,j)[\eta(i,j)^{\beta}]}{\sum_{u \in C_k} \tau(i,j)[\eta(i,j)^{\beta}]}, if\ j\ \in C_k(i) \\ 0,\ Otherwise \end{cases} \quad (1)$$

The ants release some pheromone while traversing the path; this information is passed onto subsequent generations using a matrix, namely Pheromone Matrix (PhM). During the course of algorithm, this PhM is updated using global and local updating rule. Gradually, the pheromone on the paths that are less traversed by ants is evaporated because of more distance, and the pheromone on the more traversed paths concentrates since they are traversed by more number of ants due to shorter distance. This is how the shortest path is selected, which refers to the feasible and optimal solution. The ACO algorithm has number of parameters such as Q, ρ (rate of pheromone evaporation), and gamma (pheromone decay parameter), and the number of iterations required for convergence. The values of the above parameters need to be varied in order to obtain the best performance of the algorithm, and furthermore, the combination of the parameters also needs to be optimized, which if done manually, may involve lot of time and efforts. Keeping the above in mind, a binary coded GA has been proposed here to optimize ACO parameters, details of which are given in section 2.1.

3 PROPOSED APPROACH FOR PARAMETER META-OPTIMIZATION OF ACO BY GA

The GA consists of three steps: 1) Selection, 2) Crossover, and 3) Mutation. The first is to encode the solutions in the form of chromosomes. In the present paper, the binary coded GA has been proposed. A typical chromosome of GA is [0 1 0 1 0 0 0 1]. The length of each chromosome is the summation of length of each parameter when converted into binary string. The initial population for GA is generated randomly. A typical population of chromosomes of population size "3" is as follows: [0 1 0 1 0 0 1 1 0; 1 0 1 0 0 1 0 1 0; 0 1 1 1 0 0 0 1 0].
The selection operator is used to choose the good quality chromosomes. In the present work, Tournament Selection has been applied to generate the mating pool. It provides selection pressure by holding a tournament among "s" competitors, i.e. tournament size. The winner of the tournament is the individual with the highest fitness. The winner is then inserted into the mating pool. The mating pool, being comprised of tournament winners, has a higher average fitness than the average population fitness. Increased selection pressure can be provided by simply increasing the tournament size "s", as the winner from a larger tournament will, on average, have a higher fitness than the winner of a smaller tournament (Miller and Goldberg [13]). After making the mating pool, the population individual undergoes the crossover operation, based on a probability, known as crossover probability. A simple 2-point crossover has been used to generate the offsprings possessing the mix quality of their parents. Then, based on the mutation probability, the population of GA undergoes the mutation. A simple swap mutation operator has been used. After this process, the offsprings are evaluated again by converting the binary chromosomes into the real parameter values, followed by running the ACO algorithm ten times. The fitnesses of offsprings are compared with their parents, which if improved, are retained and carried over to next generation, otherwise they are discarded. The GA cycle is repeated number of times, until the prespecified number of iterations of GA is completed. For the conversion of binary values of GA to real values of ACO parameters, the entire binary string of GA chromosome is divided into substrings, where each substring represents different ACO parameters namely, Q (a constant number used to adjust the increment of pheromone), ρ (rate of pheromone evaporation), gamma (pheromone decay parameter) and the number of iterations required for convergence. Wang et al. [4] suggests that the Q should be a discrete number and the value of ρ and gamma should lie between 0 and 1. The binary substring is then converted by using the formulae given in eq. 2 to its decimal equivalent which represents the actual parameter value of the ACO algorithm.

$$\text{actual number of iteratons for ACO} = \text{round}\left(\text{minimum number of iteratons} + \frac{(\text{maximum number of iteratons} - \text{minimum number of iteratons})}{(2^{\text{length of substring for number of iteratons}} - 1)}\right) \quad \mathbf{(2)}$$

Likewise, the actual value of Q, ρ, and gamma can also be determined. These values are then used as parameters of the ACO algorithm. The following fitness function of GA has been taken into account considering the average fitness provided by all the ants in the last iteration. The fitness function, FF can be thus given as follows:

$$FF = mean\ fitness\ of\ all\ simulation\ runs\ of\ ACO$$

The flow chart of the GA based parameter meta-optimization of ACO algorithm is given in Figure 2.

4 RESULTS AND DISCUSSIONS

An example of a punching machine assembly consisting of 16 components as shown in Figure 3, has been considered here to demonstrate the application of the proposed approach for parameter meta-optimization of ACO algorithm. Figure 4 presents an extract of the disassembly matrix, which is a 16 x 48 matrix for the given 16 component assembly. Here, the GA is used to determine the best parameter combination for ACO algorithm for solving the given assembly sequence optimization problem. The main parameters which affect the performance of GA are population size, crossover and mutation probability. A sensitivity analysis of the three GA parameters was performed to determine the influence of these parameters on the optimal solution given by GA. The following ranges of variation of different ACO algorithm parameters have been assumed: Q value is varied between 1 and 8, ρ value is varied between 0.1 and 0.8, gamma value is varied between 0.1 and 0.8 and the number of iterations is varied between 50 and 190. To study the influence of GA crossover probability, it was varied from 0.5 to 0.95 in steps of 0.05 keeping the mutation probability fixed at 0.1 and population size fixed at 5. Each binary chromosome of GA yields a particular combination of real parameters of the ACO algorithm. Using this parameter combination, the ACO algorithm has been run ten times to evaluate the fitness each GA chromosome, and then their average fitness value over these ten runs has been calculated. This average indicates the actual performance of the ACO algorithm for a certain parameter combination. Each GA simulation is carried out for 50 iterations. Figure 5 shows the variation of fitness value of GA with the crossover probability. It is observed that a crossover probability of 0.7 gives the optimal fitness value. Next, to study the influence of mutation probability, it was varied from 0.05 to 0.15 in steps of 0.025 keeping the crossover probability fixed at 0.7 and population size fixed at 5. Figure 6 shows the variation of fitness value with the mutation probability. It is observed that an optimal fitness value is obtained for mutation probability of 0.1. We further increased the population size of GA from 5 to 10. However, there was no improvement of the optimal solution given by GA. Therefore from the above results, it may be concluded that the optimum GA parameters are crossover probability of 0.7 and mutation probability of 0.1, when we use a GA population size of 5 and the corresponding optimum parameters of the ACO algorithm are as follows: Q = 7, ρ = 0.1, gamma = 0.2, number of iterations = 170. Figure 7 shows the convergence plot of the GA when it was run with the above set of optimal parameters. Figure 8 shows the convergence plot of the ACO algorithm for the given assembly sequence optimization problem when the ACO algorithm was run with the above set of optimal parameters recommended by the proposed GA based meta-optimization approach. The optimal disassembly sequences along with their directions given by the meta-optimized ACO algorithm are as follows. [(12, -Y) (11, -Y) (14, Y) (13, Y) (2, -X) (9, Z) (15, Z) (8, Z) (7, Z) (5, Z) (6, Z) (10, Z) (16, Z) (4, Z) (3, Z) (1, Z)] and [(13, Y) (14, Y) (12, -Y) (11, -Y) (2, -X) (9, Z) (6, Z) (7, Z) (5, Z) (8, Z) (10, Z) (16, Z) (15, Z) (3, Z) (4, Z) (1, Z)]. Each of these sequences requires three direction changes.

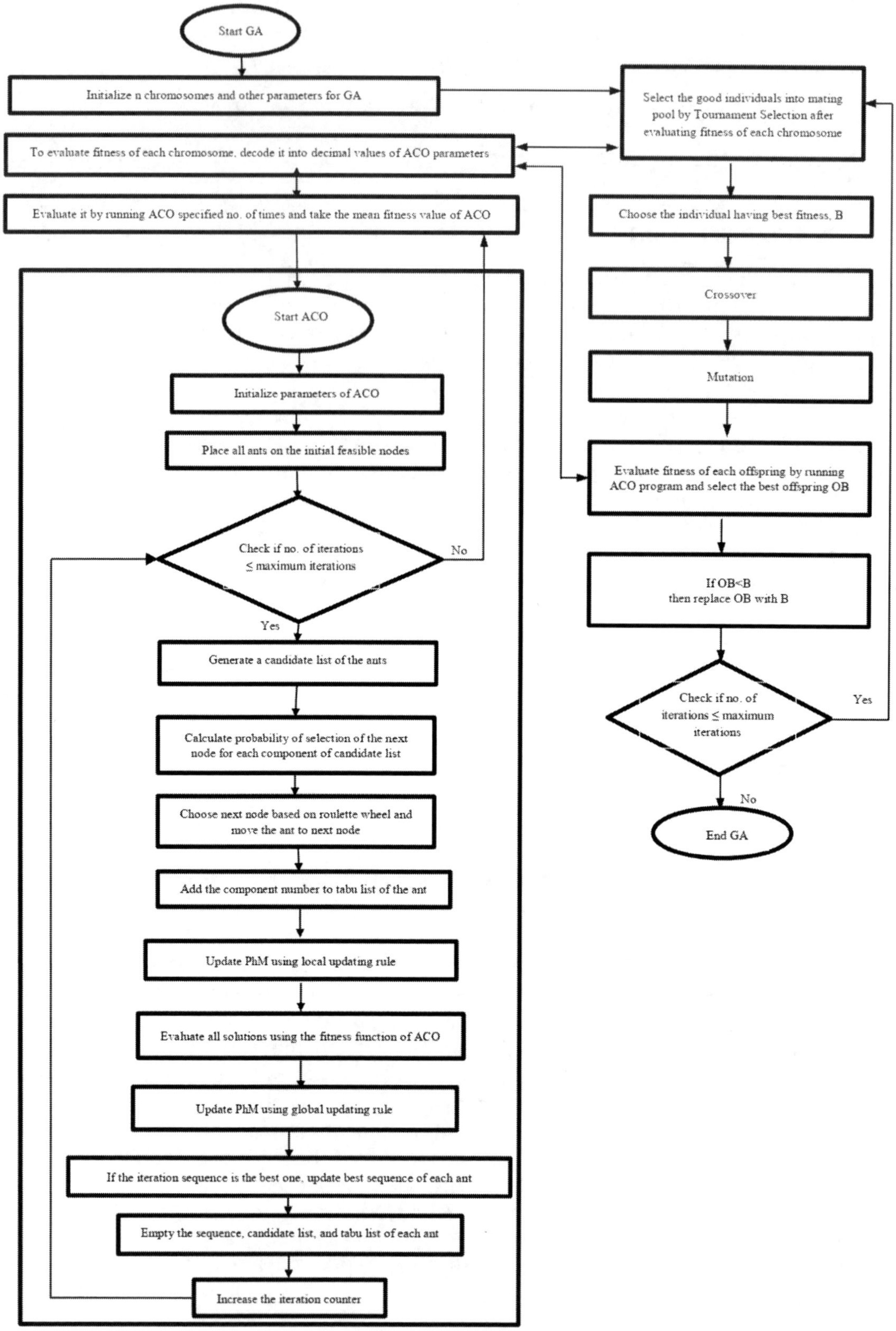

Figure 2: Flowchart of the proposed Meta-optimization algorithm

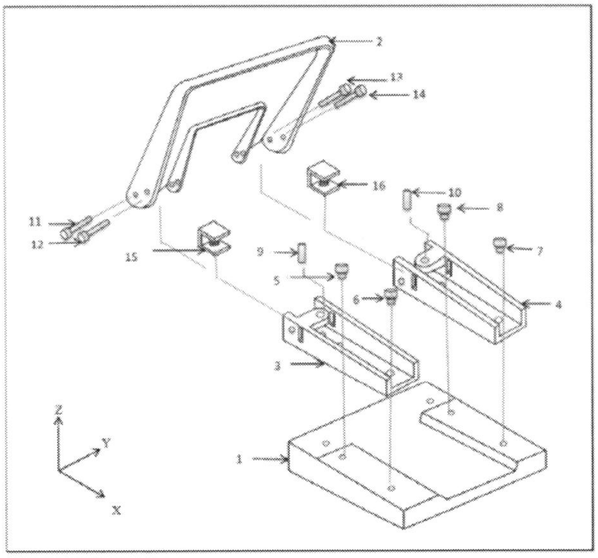

Figure 3: Punching Machine assembly

Dir. \ Part no.	1 X	1 Y	1 Z	2 X	2 Y	2 Z	3 X	3 Y	3 Z
1	[0,	0,	0],	[0,	0,	1],	[0,	0,	1],
2	[0,	0,	0],	[0,	0,	0],	[1,	1,	1],
3	[0,	0,	0],	[0,	0,	1],	[0,	0,	0],
4	[0,	0,	0],	[0,	0,	1],	[0,	0,	0],
5	[1,	1,	0],	[0,	0,	0],	[1,	1,	0],
6	[1,	1,	0],	[0,	0,	0],	[1,	1,	0],
7	[1,	1,	0],	[0,	0,	0],	[0,	0,	0],
8	[1,	1,	0],	[0,	0,	0],	[0,	0,	0],
9	[0,	0,	0],	[0,	1,	1],	[1,	1,	0],
10	[0,	0,	0],	[0,	1,	1],	[0,	0,	0],
11	[0,	0,	0],	[1,	1,	1],	[1,	1,	1],
12	[0,	0,	0],	[1,	1,	1],	[1,	1,	1],
13	[0,	0,	0],	[1,	0,	1],	[0,	0,	0],
14	[0,	0,	0],	[1,	0,	1],	[0,	0,	0],
15	[0,	0,	0],	[0,	1,	1],	[1,	1,	0],
16	[0,	0,	0],	[0,	1,	1],	[0,	0,	0],

Figure 4: An extract of the Disassembly Matrix for the Punching Machine assembly

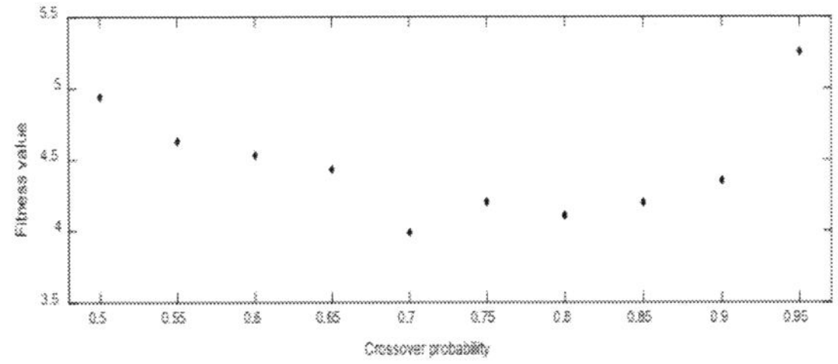

Figure 5: Variation of fitness value of GA with the crossover probability

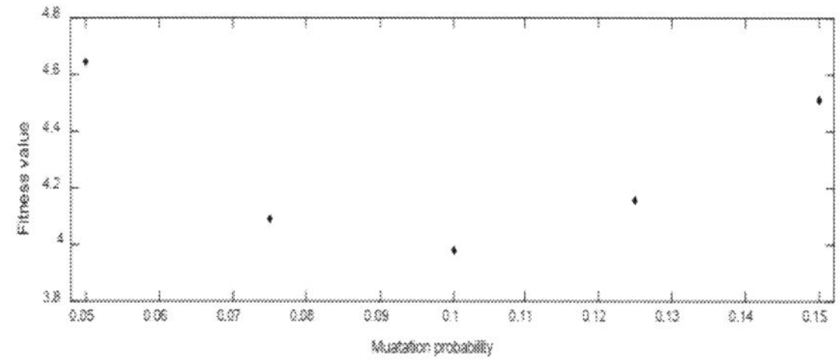

Figure 6: Variation of fitness value of GA with the mutation probability

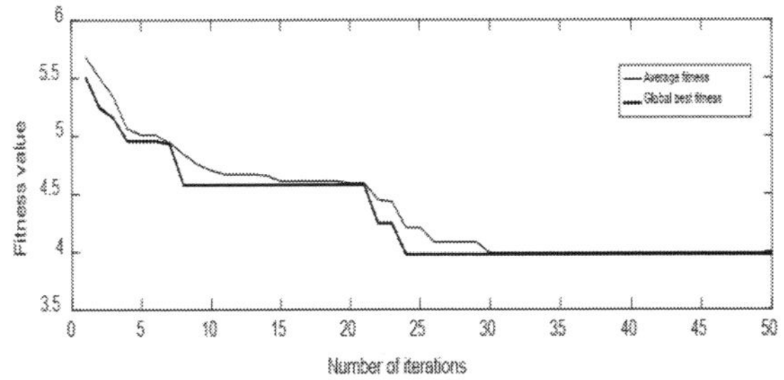

Figure 7: Convergence plot of GA used for meta-optimization of ACO

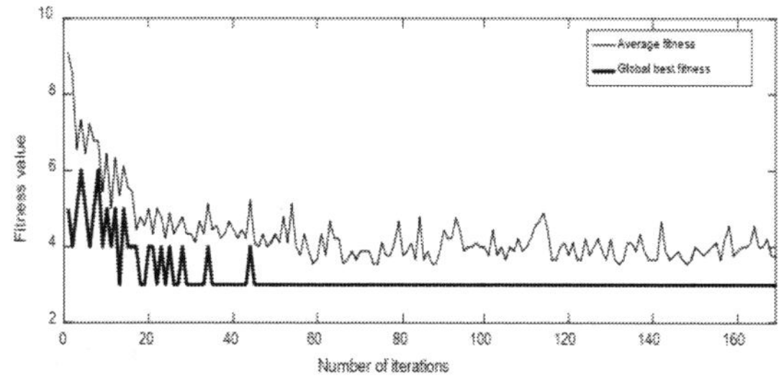

Figure 8: Convergence plot of ACO for assembly sequence optimization

The corresponding assembly sequences can be obtained by reversing the disassembly sequences, which are as follows. [(1,-Z) (3, -Z) (4, -Z) (16, -Z) (10, -Z) (6, -Z) (5, -Z) (7, -Z) (8, -Z) (15, -Z) (9, -Z) (2, X) (13, -Y) (14, -Y) (11, Y) (12, Y)] and [(1, -Z) (4, -Z) (3, -Z) (15, -Z) (16, -Z) (10, -Z) (8, -Z) (5, -Z) (7, -Z) (6, -Z) (9, -Z) (2, X) (11, Y) (12, Y) (14, -Y) (13, -Y)]. It should be noted that the directions of the components would have to be reversed. The above assembly sequences are found to be feasible as they are in accordance with the precedence constraints. The global best fitness found at the end of all iterations of ACO is 3 i.e. the minimum number of direction changes required by optimal assembly sequence is 3. The average fitness value of all ants at the end of all iterations of ACO was found to be 3.9911.

5 CONCLUSIONS

In the present paper, an ACO algorithm based approach has been used for assembly sequence optimisation based on minimizing the number of direction changes, while conforming to the precedence constraints between components. Further, a binary coded GA has been proposed for meta-optimization of ACO in order to optimize the different ACO parameters such as rate of pheromone evaporation, pheromone decay parameter, number of iterations, etc. Some of the reasons for choosing GA include the fact that the number of parameters in GA is less than ACO and thus it takes less time and efforts to optimise the GA than the ACO. Moreover the modeling of the solutions (i.e. chromosomes) in the form of binary strings is easier. The optimized ACO parameters have been used for solving a problem of assembly sequence optimization for a sixteen component assembly to demonstrate the effectiveness of our proposed meta-optimization procedure. The future scope of work will include considering some other criteria for optimization such as stability, minimizing assembly tool changes to reduce handling time, etc.

REFERENCES

[1] Hong, D S. Cho, H S. 1993. Optimization of robotic assembly sequences using neural network, IEEE/RSJ Int Conf on Intelligent Robots and Systems, Japan, July 26-30, pp. 232-239.
[2] Bonneville, F. Perrard, C. Henrioud, J M. 1995. A genetic algorithm to generate and evaluate assembly plans, INRIA/IEEE Symposium on Emerging Technologies & Factory Automation, France, October 10-13, pp 231-239.
[3] Chen, R S. Lu, K Y. Tai, P H. 2004. Optimizing assembly planning through a three-stage integrated approach, *International Journal of Production Economics*, 88, pp 243-256.
[4] Wang, J F. Liu, J H. Zhong, Y F. 2005. A novel ant colony algorithm for assembly sequence planning, *International Journal of Advanced Manufacturing Technology*, 25, pp 1137-1143.
[5] Cao, P B. Xiao, R B. 2007. Assembly planning using a novel immune approach. *Int J Adv Manuf Technol*, 31(7), pp 770-782.
[6] Chang, CC. Tseng, HE. Meng, LP. 2009. Artificial immune systems for assembly sequence planning exploration, *Engineering Applications of Artificial Intelligence*, 22, pp 1218-1232.
[7] Wang, Y. Liu, JH. 2010. Chaotic particle swarm optimization for assembly sequence planning. *Robot Computer Integrated Manufacturing*, 26(2), pp 212-222.
[8] Lv, HG. Lu, C. 2010. An assembly sequence planning approach with a discrete particle swarm optimization algorithm, *Int J Adv Manuf Technol* 50(5-8), pp 761-770.
[9] Gao, L. Qian, W R. Li, X Y. Wang, J F. 2010. Application of memetic algorithm in assembly sequence planning, *Int J Adv Manuf Technol* 49(9-12), pp 1175-1184.
[10] Zhou, W. Zheng, J R. Yan, J J. Wang, J F. 2011. A novel hybrid algorithm for assembly sequence planning combining bacterial chemotaxis with genetic algorithm, *Int J Adv Manuf Technol*, 52(5-8), pp 715-724.
[11] Xing, Y F. Wang, Y S. 2012. Assembly sequence planning based on a hybrid particle swarm optimisation and genetic algorithm, *Int Journal of Prod Research*, 50(24), pp 7303-7312.
[12] Li, X. Qin, K. Zeng, B. Gao, L. Su, J. 2016. Assembly sequence planning based on an improved harmony search algorithm, *Int J Adv Manuf Technol*, 84, pp 2367-2380.
[13] Miller, B L. Goldberg, D E. 1995. Genetic Algorithms, Tournament Selection, and the Effects of Noise, *Complex Systems*, 9, pp 193- 212.

NETWORK-FLOWS OPTIMIZATION IN THE SEMICONDUCTOR MANUFACTURING SUPPLY CHAIN

Jun-Der Leu [1,*] ,Hans Ehm[2], Larry Jung-Hsing Lee[3] and Yi-Wei Huang[4]
[1,3,4] Department of Business Administration
National Central University, Taoyuan City, Taiwan
leujunder@mgt.ncu.edu.tw;hans.ehm@infineon.com; larrylee@mgt.ncu.edu.tw; hiw@ey.gov.tw

[*,1] Department of Business Administration
Munich University of Applied Sciences, Munich, Germany
leu@hm.edu

[2] Infineon Technologies AG
Am Campeon 1-12, D-81726 Munich, Germany

ABSTRACT

The semiconductor industry focuses on the integration and coordination of resources either inside or outside an enterprise, and in the same or different countries. The integration of material-, cash-, and information-flows is complex. The semiconductor industry is one of the most complex manufacturing and logistic environments to plan and control. Hence, the development of a decision model incorporating electronic information exchange is required. Based on the Advanced Planning Systems (APS) framework a hybrid solution model is proposed, which implements the concepts of the Enterprise Resource Planning (ERP) system. Mathematical models, a heuristic method, and simulation models, are integrated, and the model considers the logistics of the semiconductor industry primarily through demand planning, master planning, and production planning. Master planning is central to the process. The inbound and final assembly logistics of the semiconductor industry can then be organized and integrated. The proposed hybrid solution model supports the planning tasks, which decide the optimal product allocation ratio and the allocation ratio from wafer fabrication to final assembly and testing, and is validated by a set of numerical data.

Keywords: Semiconductor Manufacturing; Supply Chain; Advanced Planning Systems (APS)

1 INTRODUCTION

The semiconductor industry is capital-intensive, requiring a tremendous amount of capital investment in production equipment and in the construction of wafer fabrication. Characteristics of the industry include high market demand variation, a complex product mix, frequent price variation, complex production processes, a short product life cycle, rapidly changing technology, long value-added process, etc. The concept of global logistics management (GLM) deals with issues concerning global integration and coordination of the material flows, information flows, and money flows, particularly in the semiconductor industry. It also involves the concept of resource, process, and strategy coordination; factors that may be within the enterprise or related to different companies, and in the same or in different countries. Companies can develop their business value by properly coordinating a network of geographically dispersed enterprises and using global resources and markets. Wafer fabrication is the highest value production process in the semiconductor industry due to its complexity. The value-added process of the industry is long and the business activities involve different companies in different countries, so a global semiconductor manufacturing company faces problems of resources acquisition, wafer fabrication, and distribution to markets. An integrated approach to capacity allocation, material requirement planning, and demand fulfillment

[*] Corresponding Author

is therefore required. The business operations of a global semiconductor manufacturing company, particularly the collaboration of inbound logistics and final assembly plans, require an efficient decision support solution. The Advanced Planning Systems (APS) model of Stadtler et al. [1] provides a clear decision framework and decision processes for a semiconductor manufacturing company. From the general concept, Vidoni and Vecchietti [2] develop a characterization of the APS from a systemic point of view. Based on the APS decision framework and middle term business operation planning, this study will develop a collaborative decision model for inbound wafer fabrication and the final assembly plan of a global semiconductor enterprise. The paper is organized as follows: in Section 2, a literature review of previous research is presented. The mathematical model, solution heuristic, and simulation model of the underlying problem are given in Section 3. A numerical example and a comparison with the hybrid solution method are given in Section 4. Finally, the conclusion and further research are presented in Section 5.

2 LITERATURE REVIEW

A semiconductor company faces two major issues; the production planning of single wafer fabrication (known as single fab) and of multiple fabrication (known as multi-fabs). Research into single fab production planning has been undertaken, but multi-fab production planning has seldom been considered. Kim and Lee [3] suggest an iterative synchronization approach to coordinate the input and output quantity of the production plan. Computational experiments show that a production planning model can be generated with practical parameters, and can be implemented at the scheduling and the dispatching levels. Bunn and Oliveira [4] formulate a computational learning model, based upon the heuristic of strategic slack valuation on capacity planning, for growth in market share. Iwata and Wood [5] provide an equation-based model to estimate the relationship of capacity and production cost, divided into capacity-independent and capacity-dependent costs. Higher capacity-independent costs motivate larger fabs, and are significant for high product-mix fabs. Capacity planning is a challenge in the semiconductor industry. The characteristics of high market demand variation and limited capacity mean that accurate forecasts of the actual demand and how to best use the capacity are required in a semiconductor company. Yang and Yang [6] focus on the estimation and variable selection of varying coefficient, partially nonlinear models to develop an efficient variable selection procedure based on smooth-threshold estimating equations. Mallik and Harker [7] propose a game theory model of the incentive problems arising from demand forecasting and capacity allocation. Wang et al. [8] use a heuristic preemptive dispatching method to tackle the problem of semiconductor manufacturing traffic-jams. The experimental results with simulation model show that the proposed method outperforms original method in terms of average delivery time, and appears to better solve the traffic-jam problem and reduce transportation time. Previous research into semiconductor production planning has focused on single fab production without considering the multi-fabs situation and the collaborations of the front-end, back-end, and raw materials. All global enterprises, including semiconductor companies, face the problem of the integration of material-flows, cash-flows, and information-flows. Leng and Parlar [9] analyze both simultaneous-move and leader-follower games to determine the Nash and Stackelberg equilibria, and find a globally-optimal solution that maximizes system-wide profit. None of the above-mentioned studies reviewed consider the characteristics of the semiconductor industry. The value chain of the industry focuses on the production of front-end wafer fabrication and back-end final assembly and testing. A global semiconductor company also faces problems of resource acquisition, transportation, and distribution to customers, so integrating capacity allocation, material requirement planning, and demand fulfillment should be considered. A genetic algorithm-based simulation optimization methodology has been used for semiconductor manufacturing, and is found to significantly improve the performance of wafer fabrication [10].

3 DEVELOPMENT OF A DECISION MODEL

Information technology can help enterprises process corporate related information more rapidly, and to deal with a complex business environment. The electronic exchange of information leads to a reduction of errors and increased efficiency in the enterprise operation. The Enterprise Resource Planning (ERP) information system has thus been widely applied. The concept of ERP is to instantly integrate the information of an enterprise operation, but the ERP system is not designed for planning, and hence the APS has been developed. A global semiconductor company's orders may come from different countries, and local or foreign manufacturing and assembly plants will then fulfill the demand. By applying the APS, the planning of semiconductor manufacturing can be carried out hierarchically, ranging from long-term demands to detailed short-term planning, which can overcome the complexity of the tasks. The optimizing of an entire supply chain to perform all planning tasks is impossible, but the planning tasks in the supply chain can be compatible with the information flows of the ERP system.

We propose a hybrid solution model that includes a mathematical model, a simulation model, and a heuristic method based on the APS planning framework, mid-terms, and master operations planning. Required information is derived from the ERP system, and implemented by the hybrid solution model. Figure 1 represents the information structure of the hybrid solution model. In the APS planning framework, the demand module collects the order information and periodically aggregates the demand of customers in terms of product groups, technology requirements, manufacturing processes, due date constraints, quantity, and manufacturing cost. The master planning module integrates the inbound and outbound logistics of the semiconductor industry.

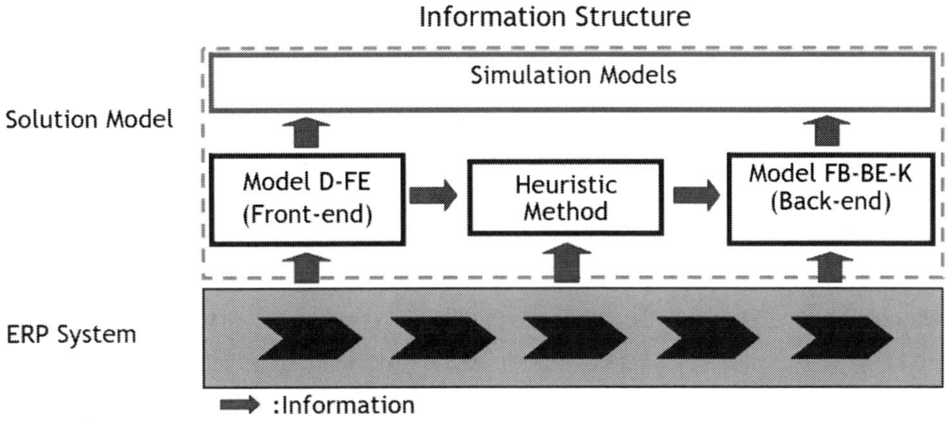

Figure 1: Information Structure of the Hybrid Solution Model

3.1 Development of the Solution

The semiconductor industry is one of the most complex manufacturing and logistic environments to plan and control. The industry has the characteristics of a long value added process, a complex product mix and production process, a short product life cycle, rapidly changing technology, and limited production capacity and capability. The two main value added processes of the semiconductor industry are front-end wafer fabrication and back-end assembly and final testing. Each stage has its own manufacturing capability and limited service boundary. Collaboration between the front- and back-ends is challenging, and capacity is an important factor. A global semiconductor company may attempt to best utilize its available capacity to fulfill customer demand, the business operation goals, and to maximize profit. Correct decisions are therefore required. Thus, a hybrid solution is proposed, made up of a mathematical model, a heuristic method, and a simulation model, for the logistics of a global semiconductor company. In this research, we divide the mathematical model into two stages, model D-FE and model FE-BE-K, to simplify the variables and the time complexity of the solution. To avoid operational obstacles resulting from the lack of final assembly and test capacity, and the capacity lost because of the complex product mix, a heuristic method is proposed, to adjust the wafer fabrication plan by considering both front- and back-end capacities. If the demand exceeds the capacity constraint, there will be a forward planning mechanism until all demand meets the capacity constraint. After models D-FE and FE-BE-K generate the logistic plan, we will proceed to the simulation model and evaluate the feasibility of the logistic plan in the real world system. The main purpose of this research is to decide the optimal logistic strategy of a global semiconductor company based on different performance indices, such as the best allocation ratio and capacity utilization.

3.2 Mathematical model

For a global semiconductor manufacturing company, the most important objectives of operation management are minimizing the total value of the inventory, which includes resource, work-in-process, and end product inventories. We propose a two-stage integrated mathematical model, covering collaboration in demand and production planning, and specifically focusing on the capacity problem, ranging from front-end wafer fabrication, though back-end assembly and testing and distribution, to customers. A global semiconductor manufacturing company will allow outsourcing either when demands cannot be met after fully utilizing the available capacity, or because of other factors, such as transportation cost, geographical reasons, production cost, etc.

3.2.1 Model D-FE

Index:

i: Index of product, divided into three groups; FLASH, LOGIC, and DRAM
j: Index of manufacturing fab
k: Index of technology; k = k1, k2, k3
t: Index of planning period

Data:

C_{ij}: The capacity requirement of producing one unit of product i in fab j

CA_j The capacity of fab j

CT_{ij} The production cycle time of product i in fab j

y_{ij} :The yield of product i in fab j

$u_{ik} = 1$:If product i uses manufacturing technology k, $i \in \{\text{FLASH, LOGIC, DRAM}\}$
$k \in \{k1, k2, k3\}$

$v_{jk} = 1$: If fab j has the manufacturing technology k,

CN_j :The capacity of the inventory in manufacturing fab j

DJ_{it} :The quantity of demand for product i in period t

PO_i :The value of one unit of product i

Decision variable:

X_{ijt} :The quantity of finished wafers of product type i in fab j in period t

N_{ijt} The inventory of product i in fab j in period t

D_{ijt}: The level of fulfilled demand of product i in fab j in period t

Objective function:

$$\text{Min} \sum_i \sum_j \sum_t N_{ijt} \square PO_t \quad (1)$$

Constraints:

$$\sum_i \left[\frac{X_{ijt} \square C_{ij}}{y_{ij}} \right] \leq CA_j \qquad \forall j, t \quad (2)$$

$$N_{ijt} = X_{ij,(t-CT_{ij})} + N_{ij,(t-1)} - D_{ijt} \qquad \forall i, j, t \quad (3)$$

$$u_{ij} \geq (1 - v_{jk}) \square X_{ijt} \qquad \forall i, j, k, t \quad (4)$$

$$\sum_i N_{ijt} \leq CN_j \qquad \forall j, t \quad (5)$$

$$\sum_j D_{ijt} = DJ_{it} \qquad \forall i, t \quad (6)$$

$$X_{ijt} \geq 0 \quad (7)$$

$$D_{ijt} \geq 0 \quad (8)$$

$$N_{ijt} \geq 0 \quad (9)$$

The objective equation (1) minimizes the total value of the inventory. Equation (2) considers the capacity constraint. Equation (3) represents the inventory balance equation for the products. Equation (A-4) considers the product and manufacturing technology. Equation (5) considers the inventory capacity constraint. Equation (6) is concerned with demand fulfillment. Equations (7)-(9) are non-negative constraints.

3.2.2 Model FE-BE-K

Index:

i: Index of product
j: Index of manufacturing fab
m: Index of assembly and test plant
l: Index of customers
n: Index of final assembly and test technology, n = n1, n2, n3
t: Index of planning period

Data:

X_{ijt} : The quantity of product i produced in fab j in period t

CAA_m :The capacity of assembly plant m

$un_{in} = 1$:If product i uses assembly technology n, $n \in \{n1, n2, n3\}$

$vn_{mn} = 1$: If assembly plant m has the final assembly and test technology n, $n \in \{n1, n2, n3\}$

CTA_{im} :The production cycle time of product i in assembly plant m

DF_{ilt} : The demand quantity of product i of customer l in period t

PO_{im} The value of one unit of product i stock in assembly plant m

CNA_m :The capacity of the inventory in assembly plant m

CR_{im}: The capacity requirement of product i assembled in assembly plant m

Decision variable:

Y_{imt} : The quantity of product i produced in assembly fab in period t

NF_{imt} The inventory of wafer i in assembly fab m in period t

YO_{imlt} :The level of fulfilled demand of product i in assembly plant m for customer l in period t

DA_{ijmt}: The amount of product i allocated from fab j to assembly m in period t

NA_{imt} The inventory of product i in assembly plant m in period t

Objective function:

$$\text{Min} \sum_i \sum_m \sum_t NA_{imt} \cdot PO_{im} \quad (10)$$

Constraints:

$$un_{in} \geq (1 - vn_{mn}) \cdot Y_{imt} \qquad \forall\, i,\, m,\, n,\, t \quad (11)$$

$$\sum_i Y_{imt} \cdot CR_{im} \leq CAA_m \qquad \forall\, m,\, t \quad (12)$$

$$X_{ij,(t-CT_{ij})} = \sum_m DA_{ijmt} \qquad \forall\, i,\, j,\, t \quad (13)$$

$$NF_{imt} = NF_{im,t-1} + \sum_j DA_{ijmt} - Y_{imt} \qquad \forall\, i,\, m,\, t \quad (14)$$

$$NA_{imt} = Y_{im,(t-CTA_{im})} + NA_{im,(t-1)} - \sum_l YO_{imlt} \qquad \forall\, i,\, m,\, t \quad (15)$$

$$\sum_i NA_{imt} \leq CNA_m \qquad \forall\, m,\, t \quad (16)$$

$$\sum_m YO_{iml,(t-LT_{ml})} = DF_{ilt} \qquad \forall\, i,\, l,\, t \quad (17)$$

$$Y_{imt} \geq 0 \quad (18)$$

$$N_{imt} \geq 0 \quad (19)$$

$$DA_{ijmt} \geq 0 \quad (20)$$

$$DF_{ilt} \geq 0 \quad (21)$$

$$NW_{ijt} \geq 0 \quad (22)$$

Objective equation (10) minimizes the total value of the inventory. Equation (11) considers the product and assembly technology. Equation (12) considers the manufacturing capacity constraint, which considers the yield of product in different fabs. Equations (13) and (14) represent the product allocation from fab to assembly plant and the wafer inventory of the assembly fab, and equation (15) also represents the wafer inventory. Equation (16) represents the inventory balance equation for the end product. Equation (17) considers the inventory capacity constraint. Equation (18) is concerned with the demand fulfillment. Equations (19)-(22) are non-negative constraints.

3.3 Solution Heuristic

We propose a heuristic method to solve the capacity collaboration problem for a global semiconductor manufacturing company. As the final assembly and test capacities are scarce, the purpose of this heuristic solution is to coordinate the front-end and back-end capacity. This heuristic is based on the idea of balancing the demand quantity by both front-end and back-end capacity to fulfill the customer demand. The front-end wafer fabrication and back-end final assembly and test capacity plan are integrated.

3.4 Simulation Model

The simulation model describes how all parts of the supply chain will operate over time as a function of the parameters, to accommodate the gap between the real-world system and the optimization model. Though the optimization method uses the optimal result of any given problem, it neglects the real-world dynamic factors of a system, such as unexpected delays, queuing, breakdowns, and operation time. The characteristics of the semiconductor industry are long manufacturing processing times, a complex product mix, and high production restrictions. Here, we provide three semiconductor manufacturing control mechanisms; an individual model (hot-lot model), a one-week model (normal model), and a batch model, to determine if the results from the optimization model apply to a real-world system. The hot-lot model implies that customers' orders will be handled immediately and the batch model implies that fabs will be manufactured until the target quantity of customers' orders has been reached. Generally, fabs manufacture wafers under a weekly plan, known as the normal model. These simulation models are used to determine the operational decisions under different demand situations.

4 RESEARCH RESULTS

A numerical example is constructed using the LP software LINGO, Visual C++, and the simulation software eM-Plant to validate the hybrid solution model. The output of model D-FE is used as input data to model FE-BE-K. However, there will be no feasible solution in model FE-BE-K, as there is a gap between the demand and the current final assembly and test capacities. To fill this gap, we use the heuristic to adjust the production plan of model D-FE. The average capacity utilization using the heuristic method is flatter than the original. The exceeding-capacity-wafers will be manufactured in the previous planning period. The average capacity utilization is indicated to be near 90%, which is reasonable in a real-world global semiconductor company, as the capacity for bad wafers is typically retained.

From the above, we find that the LINGO program will generate the optimal solution of model D-FE. If the demand quantities in any planning period exceed the back-end final assembly and test capacities, there would be no feasible solution due to the lack of assembly capacity. Hence, a heuristic method is applied to solve the gap between the front-end and back-end capacities.

The total inventory value of the planning period of model FE-BE-K is $671,960, which initially may appear too high, but it is reasonable when considering the shortage of back-end assembly capacity. The above solution uses a deterministic model to solve the logistic problem in a global semiconductor company, but it ignores real-world factors. Depending on the solution results, other final assembly and test capacities may be sought, or the company may decide to do nothing. This decision is too rough, so we use the simulation model to determine whether the logistic plan is applicable and then the decision can be made. We find from the simulation results that the hot lot model has better production efficiency and less production cycle-time than the other models. In contrast, the normal model is inefficient and requires more production cycle-time. All models are very close to the target demand quantities, with around +5% ~ +2% difference, which is in a reasonable range.

From the above, the logistic plan is proven to be effective and acceptable and the gap between the mathematical and the simulation models is very small. A global semiconductor company can apply this hybrid method to solve the capacity collaboration of front-end manufacturing fabs and back-end final assembly and test fabs, under the condition of a demand boom, and when the demand exceeds the back-end final assembly and test capacity.

5 CONCLUSIONS AND FURTHER RESEARCH

An ERP-based business planning for the global IC industry has been proposed. Based on the APS framework, a hybrid solution model including mathematical models, an heuristic method, and simulation models is provided. The framework considers the logistics of the semiconductor industry primarily in demand planning, master planning, and production planning. The inbound and final assembly logistics of the industry can then be integrated. Master planning is central to global operations management, where the collaboration of demand, production, and final assembly is considered. A hybrid solution model is developed to support the planning tasks, which decides the optimal logistics plan, such as the optimal product allocation ratio and the allocation ratio from wafer fabrication to final assembly and testing. The performance of the hybrid solution model has also been validated by a set of numerical data. The results of the simulation models also prove the practicality of the logistics plan by the hybrid solution model. Many global semiconductor companies have already implemented ERP systems as their information infrastructure. The provided hybrid solution model can be integrated with these systems to enhance logistics planning. Finally, this study does not consider the collaboration of resource suppliers and inbound logistics, or the collaboration of the final assembly and the resource suppliers. Future research should consider all these in an integrated manner and in more detail, to obtain practical solutions to logistical problems in the semiconductor industry. The machine setup time is also a critical factor affecting the operational performance of a semiconductor company when the product type changes. By considering this setup time, the simulation model can provide more accurate evaluations for the logistics of a global semiconductor company.

6 REFERENCES

[1] **Stadtler, H. 2005.** Supply chain management and advanced planning-basics, overview and challenges. *European Journal of Operational Research*, 163 (3), pp 575-588.

[2] **Vidoni, M.C., & Vecchietti, A.R. 2016.** A systemic approach to define and characterize Advanced Planning Systems (APS). *Computers & Industrial Engineering*, 90, pp 326-338.

[3] **Kim, S. H., & Lee, Y.H. 2016.** Synchronized production planning and scheduling in semiconductor fabrication. *Computers & Industrial Engineering*, 96, pp 72-85.

[4] **Bunn, D.W., & Oliveira, F.S. 2016.** Dynamic capacity planning using strategic slack valuation, *European Journal of Operational Research*, 253(1), pp 40-50.

[5] **Iwata, Y., & Wood, S. C. 2002.** Simple cost models of high-process-mix wafer fabs at different capacities. *IEEE Transactions on Semiconductor Manufacturing*, 15 (2), pp 267-273.

[6] **Yang, J., & Yang, H. 2016.** Smooth-threshold estimating equations for varying coefficient partially nonlinear models based on orthogonality-projection method, *Journal of Computational and Applied Mathematics*, 302, pp 24-37.

[7] **Mallik, S., & Harker, P. T. 2004.** Coordinating supply chains with competition: capacity allocation in semiconductor manufacturing. *European Journal of Operational Research*, 159 (2), pp 330-347.

[8] **Wang, C.N., Lee, Y.H., Hsu, H.P., & Nguyen, D.H. 2016.** The heuristic preemptive dispatching method for convey-based automated material handling system of 450 mm wafer fabrication. *Computers & Industrial Engineering*, 96, pp 52-60.

[9] **Leng, M., & Parlar, M. 2010.** Game-theoretic analyses of decentralized assembly supply chains: Non-cooperative equilibria vs. coordination with cost-sharing contracts, *European Journal of Operational Research*, 204(1), pp 96-104.

[10] **Chang, X., Dong, M., & Yang, D. 2013.** Multi-objective real-time dispatching for integrated delivery in a Fab using GA based simulation optimization, *Journal of Manufacturing Systems*, 32(4),pp 741-751.

A STUDY ON STABLIZATION OF ELECTRIC POWER SUPPLY UNDER SMART GRID ENVIRONMENT: UTILIZING STORAGE FUNCTION OF ELECTRIC VEHICLES

Kazuki Takeyama[1*], Koichi Murata[2] and Hiroshi Katayama[3]

[1]Department of Industrial and Management Systems Engineering,
Graduate School of Creative Science and Engineering,
Waseda University, Tokyo, Japan
ktakeyama@fuji.waseda.jp

[2]Department of Industrial Engineering and Management,
College of Industrial Technology, Nihon University, Chiba, Japan
murata.kouichi30@nihon-u.ac.jp

[3]Faculty of Science and Engineering,
Waseda University, Tokyo, Japan
kata@waseda.jp

ABSTRACT

Drastic changes in global warming and natural disasters all around the world have yielded an interest in usage of renewable and sustainable energy. Out of all the types of renewable energy, the usage of solar power generation has been rapidly growing in recent years. However, since solar power is acutely affected by weather conditions, smart grid systems that can stabilize each house's electricity supply and demand balance are accumulating interest. In smart grid systems, storage batteries have an important role in stabilization, and its costs and management problems have emerged. From the points above, this study will take a look at utilizing electric vehicle's storage batteries, which its technology has been rapidly advancing recently, and construct a model of such smart grids. Upon further research, this study will point out the relationship between the necessary solar panels and storage capacity when acknowledging about stabilized smart grids. Its effects and problems on the application of such systems in real life will be also discussed.

Keywords: Smart Grid, Sustainable Energy, Electric Power Supply, Electric Vehicle, Management System

1 INTRODUCTION

In recent years, there has been an unprecedented amount of natural disasters all around the world, such as global warming and earthquakes. Because of this, the interest in renewable energy, such as solar power, has been growing rapidly. However, solar panel has its downsides. Since solar energy is deeply affected by weather conditions, the production amount of solar energy varies when solar panels are implemented. Thus, electricity cannot be stably provided every day. To cope with this problem, smart grids can be implemented to stabilize electricity demand and supply characteristics of each house. Thus, smart grids have been gaining an increasing popularity as a solution to such

* Corresponding Author: Kazuki Takeyama, e-mail: ktakeyama@fuji.waseda.jp

problems. In order to construct a stable smart grid, on the other hand, storage battery is compulsory equipment. Maintaining storage battery, both its condition and costs, has been an uprising problem within smart grids.

This study focuses in on utilizing storage batteries in electric vehicles (EV). Technology of electric vehicle has been growing rapidly in the recent years [4]. There are multiple applications already and tests about such smart grids around the world. For example, there is a project in Hawaii that consists of more than 30 EVs interlinked together to construct a smart grid [5]. This study constructs a model about smart grid system that utilizes EV equipped batteries, and find out the parameters, including necessary number of panels and battery capacity, that are needed to maintain a stable smart grid. Additionally, the effects and problems of implementing such smart grids will be discussed and analyzed.

2 RELATED STUDIES

There are number of studies about smart grids that utilize solar energy. Murata et al. [2] figured out the interactions between long-distance electricity transfer (e.g. power plants) and facilities that use solar power generation. Furthermore, Ka et al. [1] analyzed the stable conditions for smart grids consist of number of households with solar panels. These models have problems such as low energy conversion rate of storage battery and unstable electricity supply. Therefore, in this study, EV equipped batteries are involved in the proposed model, which has higher energy conversion rate than that for the household.

3 STUDY PROCEDURE

STEP 1: Constructing a solar energy panel model

The amount of solar radiation is gathered and calculated into the amount of solar energy produced by each house.

STEP 2: Generating electricity demand

A model representing the amount of electricity demand is constructed.

STEP 3: Modeling the storage battery

After analyzing the charge/discharge mechanism, a model representing such mechanism is constructed.

STEP 4: Simulation and discussion

A simulation linking the above models is undergone.

4 MODEL

4.1 Solar panel model

During time period *t*, House *i*'s energy production can be calculated by the following equation (1) which is referred from Ka et al. [1] model.

$$sp_{ih} = max\left\{0, min\left(\varepsilon_h^C \int_h^{h+\Delta} \gamma \frac{p_t}{3.6} dt * s_i\,,\quad p_{max} * s_i\right)\right\} \qquad (1)$$

<Symbols>

---

sp_{ih}: House *i*'s energy production (time *t*)

ε_h^c : Solar energy panel power coefficient according to temperature C (time t)

γ: Solar energy panel's energy conversion ratio

p_t: Solar radiation intensity (time t)

s_i : Number of panels on house i

p_{max}: Panel power by area

The amount of solar radiation can be referred from NEDO's homepage [6]. The amount was then multiplied by the conversion ratio. Conversion ratio of each solar panel, which is highly affected by temperature, was referred from the model by company T.

4.2 Generating electricity demand

Each household's electricity demand depends on the type and pattern of daily electricity usage within each house. The data of average electricity usage of houses in Japan was referred from the statistics of Ministry of Economy, Trade, and Industry of Japan [3]. Then, electricity demand of each house was generated using random number generator to show variance among houses.

Additionally, each EV's electricity usage, when running the EV, was calculated. This was calculated by researching the average driving distance for cars used by standard household, and multiplying the amount by electricity used for running such distance.

4.3 Creating a storage battery model

The sum of electricity within storage batteries in a smart grid system can be calculated by equations (2) and (3) which are modified version of Ka et al. [1] model. This study particularly emphasizes on the standby percentage α of EV, which represents the number of cars that are parked at home and ready to use their storage batteries.

$$ep_h = \left(\sum_{i=1}^{n} sp_{ih} - \sum_{i=1}^{n} cp_{ih}\right)\mu_p - (1-\alpha)\mu_{out} * \sum_{i=1}^{n} vp_{ih} \quad (2)$$

$$\mu_p = \begin{cases} \mu_{in} & (\text{charge} \quad \sum_{i=1}^{n} sp_{ih} - \sum_{i=1}^{n} cp_{ih} \geq 0) \\ \mu_{out} & (\text{discharge} \quad \sum_{i=1}^{n} sp_{ih} - \sum_{i=1}^{n} cp_{ih} < 0) \end{cases} \quad (3)$$

<Symbols>

---

ep_h : Amount of electricity in storage battery (time t)

sp_{ih}: House i's energy production (time t)

cp_{ih}: House i's electricity demand (time t) Note: Does not include EV's electricity demand

vp_{ih}: House i's car's electricity demand

μ_{in}, μ_{out} : Storage battery's energy conversion ratio (charge and discharge)

α: EV standby percentage in smart grid

---

5 EXPERIMENT CONDITIONS

In the experimental examination, following conditions are supposed.

1. Solar panel's energy conversion rate: 0.21.

2. Solar panel's direction: Azimuth 0 degree, Radiation Angle 30 degree
3. Battery specification: 244KWh by nickel ion battery (energy conversion ratio 0.9 both charge and discharge) by Company N.
4. Number of EV per household : 1

6 SIMULATION RESULTS AND ANALYSIS

6.1 Energy production data

Figures 1, 2 and 3 represent different amounts of energy produced by solar panels on each house throughout one year. Each figure has corresponding number of solar panels. According to Figures 1, 2 and 3, the peak day of energy production is around spring and autumn, which is just around day 90 and day 240 (note: January 1st is counted as day 1).

One might think, the peak comes during summer time, but this is not the case. This is because the solar panel energy conversion coefficient related to temperature in equation (1) becomes too high. In other words, the temperature of panel surface becomes too high, which weakens the power of panels. Also, since the amount of solar radiation is low during winter, the energy production becomes low. Note that these measures are taken in Japan.

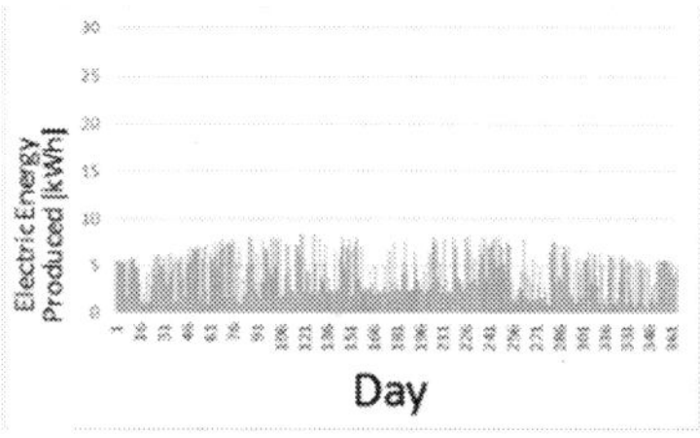

Figure 1: Energy produced (7 panels)

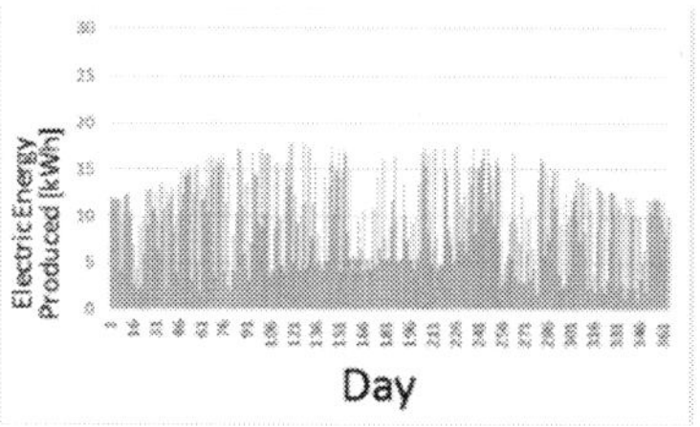

Figure 2: Energy produced (13 panels)

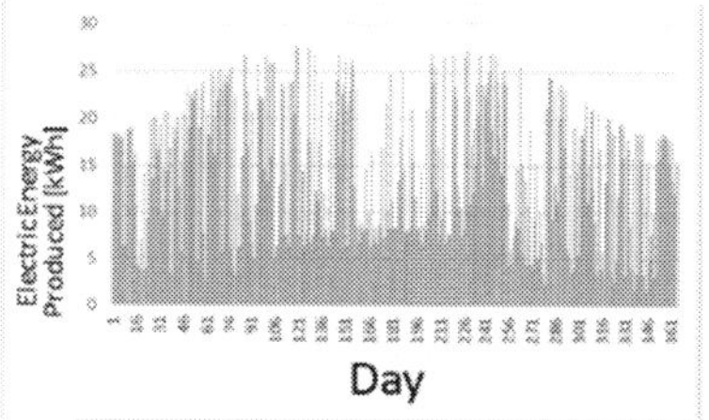

Figure 3: Energy produced (20 panels)

6.2 Sample pass of electricity demand

Figure 4 represents an example of electricity demand within a house throughout one year. It can be taken from this figure that there is some difference among seasons, and also these patterns are unlike the patterns that can be seen in the production data. Namely, the peak days do not coincide, difference between produced and demand electricity levels must be overcome. Thus, this study attempts to utilize smart grids to solve this problem.

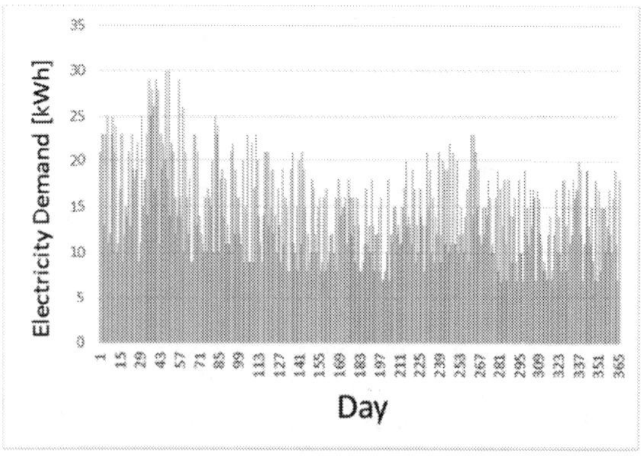

Figure 4: Electricity Demand

6.3 Figuring out the characteristics of electric shortage and availability

In this section, let's take a look at how the number of panels and the number of standby EVs will affect the performance of smart grid system. Remember that each EV is equipped a storage battery that a smart grid can use, thus number of standby EVs determines overall battery capacity in the grid.

Figure 5 represents the relationship between the number of panels and the EV standby ratio that affects electricity shortage rate being occurred in the grid. Shortage rate is defined as the ratio of days which occurs in a year. For instance, 0.1 means that shortage occurs 10% of days in a year (around 36 days). In this particular simulation, it is assumed that the allowance range of electricity shortage is supposed around 0~10%, and thus marked where allowance area is. According to this figure, it is noticed that as the number of panels increase, the less the EV standby ratio needs to be.

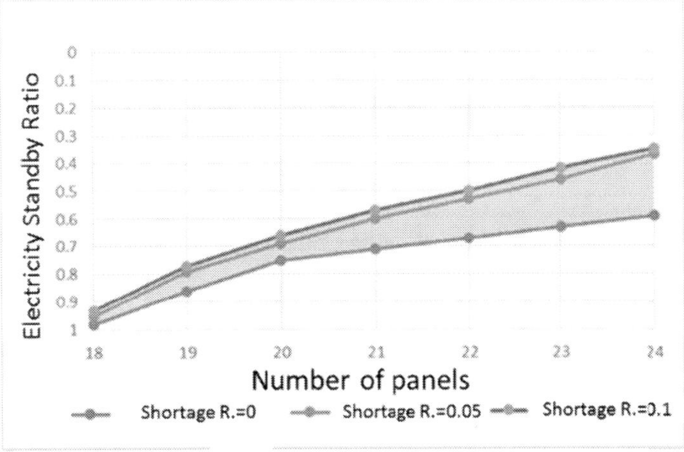

Figure 5: Relationship between the number of panels and the EV standby ratio

After confirming the amount of panels and the EVs needed, we then looked closer at the exact number of panels needed.

Table 1 summarizes characteristics of shortage days under the delicate parameter conditions, i.e. for 17, 18, and 19 panels with 1, 0.9 and 0.8 EV standby ratios. According to this table, the difference between 18 and 19 panels is so significant that the number of shortage days greatly differs when the EV standby ratio decreases. It is said that the EV standby ratio is around 0.9, which means that 90% of the time, cars are parked at their houses. Note that this includes cars parked overnight. As a result of this table, the case of 19 panels is enough to generate electricity every day without shortage.

Table 1: Number of shortage days in relation to parameters

Number of Panels	17			18			19		
EV Standby Ratio	1	0.9	0.8	1	0.9	0.8	1	0.9	0.8
Battery Capacity Sum[Kwh]	7320	6588	5856	7320	6588	5856	7320	6588	5856
Shortage Days (w/EV)	276	317	321	4	242	315	0	0	15
Shortage Days (w/o EV)	321			284			21		

From the above points, let's take further look into this issue by dividing panels into ten pieces. Figure 6 illustrates the difference between 18 and 19 panels in this case, and the corresponding number of shortage days if EV storage batteries were not used.

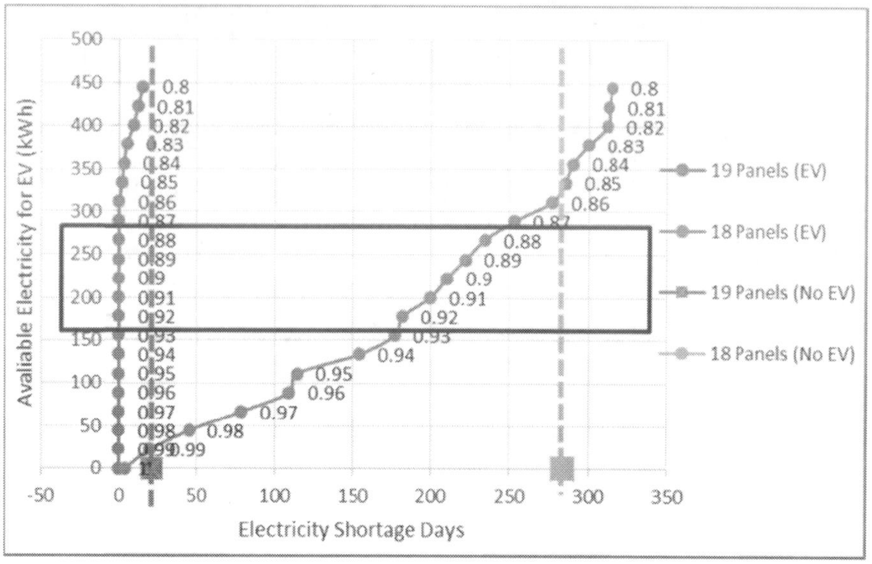

Figure 6: Number of shortage days in relation to parameters

Figure 7 is the zoomed in version of Figure 6 and panels 18 to 19 are split up into 10 parts. From this figure, it can be said that the points inside the circle guarantee acceptable shortage days. As a reasonable case, panel size 18.7 is picked up, because the electricity shortage days is under 10 days when the EV standby ratio is 0.9

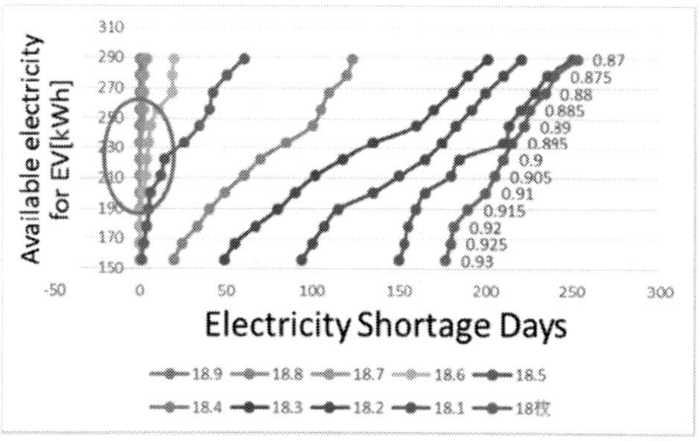

Figure 7: Split panels

6.4 Confirming electric power interchangeability

After pinning down the number of panels and the EV standby ratio that is required, a simulation was underwent to confirm electric power interchangeability. The whole point of using a smart grid system is to interchange electricity power from houses with extra electricity to houses that are in shortage of electricity. Figure 8 illustrates the number of electricity shortage days in relation to the number of houses in a smart grid.

According to this figure, the larger the grid scale is, the more it becomes stabilized. This is because interchange of electricity among houses becomes frequent along with increase of houses in the smart grid. From this figure, it can be assumed that around 20 houses is the appropriate number of houses to design a stable grid.

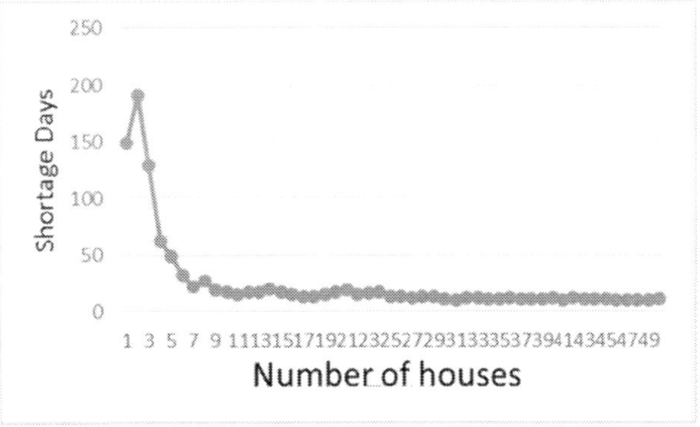

Figure 8: Characteristics of electricity shortage under the grid scale

6.5 Confirming the amount of electricity in storage battery for a year

After finding out the number of panels and houses required, Figure 9 was drawn to see how the different number of panels affect the amount of electricity in storage batteries.

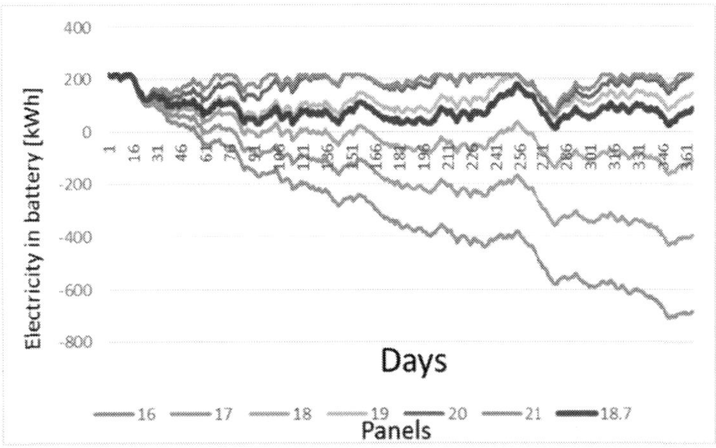

Figure 9: Confirming storage amount in relation to panels

Looking at this figure, it is noticed that 18 panels is just the right amount of panels to not have any or too many shortage days when the storage reaches its capacity. When the amount of electricity reaches its capacity, the excessive amount is considered to be discharged and wasted.

When there are fewer than 17 panels, the amount of electricity in storage batteries continues to decrease. Thus, this case leads to energy shortage on almost every day. Even one panel (comparing 17 and 18) can make huge difference because smart grid systems are unable to recover after a shortage occurs in such case. In other words, after there is a shortage, recovery is quite difficult as it is not possible to procure more energy from the sun.

7 CONCLUDING REMARKS

This study confirmed the combinations of necessary parameters for constructing a stable smart grid. Additionally, this study found out how the number of panels and EV standby ratio affects the quality of electricity service provided to households in a smart grid.

For future studies, there needs to be a more in-depth research with shorter time frames. Due to privacy issues and limited data available, this study was only able to undergo simulations day by day. Additionally, there needs to be further research on the costs for constructing a smart grid with the calculated parameters.

REFERENCES

[1] Ka, H., Murata, K., Takeyama, K. and Katayama, H., "Study on stabilization of electric supply within solar production systems", Public Interest Incorporated Group Corporation Japan Management Engineering Association 2015 Autumn Contest, pp. 282-283, Kanazawa Industrial University, Kanazawa, 2015/11/29.

[2] Murata, K., Katayama, H., "Study on efficiently utilizing smart grid systems: Decreasing lowering electric loads by focusing on electric supply chain", General Incorporated Association Japan Logistics System, Vol. 12, No. 1, pp. 79-86, 2012.

[3] Ministry of Economy, Trade, and Industry HP, URL: http://www.meti.go.jp/ (Accessed on 2015/12/5).

[4] Nissan LEAF HP, URL: http://ev.nissan.co.jp/LEAF/ (Accessed on 2016/7/1).

[5] JumpSmartMaui HP (Hitachi Smart Grid Project), URL: http://www.jumpsmartmaui.com/ (Accessed on 2016/7/1).

[6] NEDO Solar Energy Database System HP, URL: http://app0.infoc.nedo.go.jp/ (Accessed on 2016/7/1).

TEMPORAL PARTITIONING OPTIMIZATION FOR DYNAMICALLY RECONFIGURABLE ARCHITECTURE

Fadel Abdallah, Camel Tanougast* , Imed Kacem, Camille Diou and Daniel Singer
Université de Lorraine, LCOMS, EA 7306, 7 rue Marconi, Metz, F-57070, France
{Fadel.Abdallah, Camel.Tanougast}@univ-lorraine.fr

ABSTRACT

Reconfigurable computing systems are evolving rapidly, due to their flexibility and high performance. To solve the temporal partitioning problems, we focus on scheduling length (Makespan) of applications in order to implement an algorithm based on genetic algorithms. The main goal is cut to a complete application described in the data flow graph form (DFG) into sub-applications successively executed by the logic area referred through dynamic reconfiguration, taking into account the surface constraint in order to minimize the makespan criteria (C_{max}). The proposed algorithm was tested on several examples and the results show that it is robust, efficient, and gives important solutions in reasonable execution times.

Keywords: Temporal partitioning, Genetic algorithm, Data flow graph, Scheduling, Optimization, Field-Programmable Gate Arrays (FPGA), Real time, Makespan, Dynamic reconfiguration (DR).

1 INTRODUCTION

Reconfigurable computing is a new paradigm for the development of new strategies which allow getting a trade-off between the adaptability, area optimization and high performance. Reconfigurable hardware devices, such as field-programmable gate arrays (FPGAs), are increasingly used and have become ubiquitous, especially in embedded systems that operate under real time constraints. As designs become large and complex, it is necessary to perform parallel design by temporal partitioning or architectural synthesis [14-16]. The temporal partitioning divides the target application into a number of specification segments that are destined to be executed one by one in the target reconfigurable hardware device [3-4]. This allows minimizing the logical area which needs to implement the application while increasing the silicon effectiveness [4].

In this paper, we focus on the dynamic reconfiguration (DR) that consists of the division and successive execution of an application or algorithm with successive reconfiguration of configurable material part. The advantages of DR have been shown several times in [1, 2]. This approach minimizes the logic area required for the implementation of an algorithm or application while increasing the efficiency of silicon processing [3, 4].This is very important for the development of embedded applications as it allows to avoid oversizing of the required resources for an application and thus optimizes its overall cost. More precisely, the implementation of dynamic reconfiguration consists to identify the different parts sequentially executed (temporal partitions) according to a scheduling and dependency data.

Combinatorial optimization has become increasingly used in many research fields such as operational research, computer science and discrete mathematics. Indeed, optimization problems [6] are

* {Fadel.Abdallah, Camel.Tanougast}@univ-lorraine.fr

difficult and many practical applications can be formulated as a combinatorial optimization problem [7] such as hardware partitioning. Generally, combinatorial optimization problems are easy to identify but difficult to solve. In fact, most of these problems are considered as NP-hard problems and there is no effective algorithmic solution for all data [8]. In this context, many approaches have been developed in operational research and artificial intelligence [5]. These methods can be grouped into two categories: exact methods that ensure the completeness of the resolution and approximate methods which lose completeness to gain efficiency.

In the literature, many approaches and algorithms have been proposed and developed by different authors to solve the temporal partitioning problem [4, 9-11]. Tanougast et al. [4] proposed a temporal partitioning method of a FPGA by using the dynamic reconfiguration feature. This method estimates the number of possible partitions by considering a characterized FPGA technology (speed and area) in order to increase the silicon efficiency. On the one hand, Ouni et al. [9] have presented a typical temporal partitioning methodology that temporally cuts a data flow graph onto a reconfigurable system by minimizing the data transfer required between design partitions, and the routing cost between FPGA modules. This algorithm aims to find temporal partitioning of the graph and to select the optimal solution in terms of communication cost. On the other hand, they proposed an algorithm to solve the same problem which aims to optimize the latency of the graph and the number of partitions [10]. In [11] the authors used ILP algorithm which is a mathematical model for determining a way to achieve the best outcome, such as lowest latency. The main problem of the ILP approach is its high execution time. Therefore, the algorithm can only be applied to small examples.

In this paper, we focus on meta-heuristic methods to solve the temporal partitioning by applying adapted and approximated method which provides optimal and good solutions in a short execution time while ensuring the specific performance criteria such as makespan. The main goal is to propose new optimization method based on genetic algorithms (GAs, [12][13]) called Genetic Temporal Partitioning Algorithm (GTPA). The method aims to solve the temporal partitioning problems while ensuring a high performance and responding to most application requirements (real time, etc.). Therefore, we apply GTP algorithm to partition the input application described in a data flow graph form into various partitions of small sizes that fit inside FPGA area in order to ensure the minimization of the makespan.

An overview of the paper is as follows. Section 2 illustrates and explains our partitioning problem by an example. The Genetic Temporal Partitioning Algorithm (GTPA) is deeply explained in Section 3, where the pseudo-code and the overall performed steps are fully elucidated through an example. Experimental results are shown and discussed in Section 4. Finally, the conclusion is presented and further work is envisaged in Section 5.

2 POSITION AND DESCRIPTION OF THE PROBLEM

The main goal of this work is to be able to execute into a dynamically reconfigurable FPGA, an application described in the data flow graph form that is too large to be completely mapped onto FPGA. Therefore, the application has to be partitioned in such a way that allows mapping into the target hardware using the dynamically reconfigurable capability (Run Time Reconfiguration - RTR).This section is intended to give an overview about the main problem. The notations of the graph task and the dynamic reconfiguration architecture are presented by considering one specific example for our problem.

An illustrative example is presented in the Figure 1 giving an overview of the problem. The first part of the Figure 1 presents an application that consists of eleven tasks (Tasks graph) to be executed onto the hardware FPGA where the processing time and area required for each task are given in a table (Processing time & Areas).The link from tasks T7 to T10 indicates that the task T10 needs a data size of 3 from task T7, before starting the execution. The time cost to exchange this amount of data depends on the communication rate (suppose the communication access cost is negligible). To explain this example, let 800 CLB (configurable logic Block) be the FPGA area and 2 is the communication rate of the device FPGA with reconfiguration time RTR equal to 1.

The second part of the Figure 1 shows two solutions of the considered partitioning problem thanks to (a) and (b). We have two types of communication cost: intra-partitions and inter-partitions communication. In the first side, we note that in (a) the delay to send three data units from T7 to T10 in the same partition P_5 is 0 + 3/2 = 1.5 time units (access + communication costs). On the other side, each partition can execute when it receives all the data from all predecessor tasks of their tasks. For example, in the partition P_3 (Figure 1 (a)), the communication cost from all previous tasks of P_3 (T1, T2, T4, T5) to the tasks in P_3 (T5, T6) is the maximum between 3/2, 4/2, 3/2 and 5/2 is 5/2 = 2.5 time units. The makespan and number of partitions in (a) are 192 time units and 6, respectively. However, the makespan and number of partitions for the solution 2 of the same partitioning problem are 160 time units and 5, respectively. Hence, the difference between solutions (a) and (b) depends on the mapping of tasks graph on the partitions. Indeed, to find the optimal solution, the mapping should be very effective.

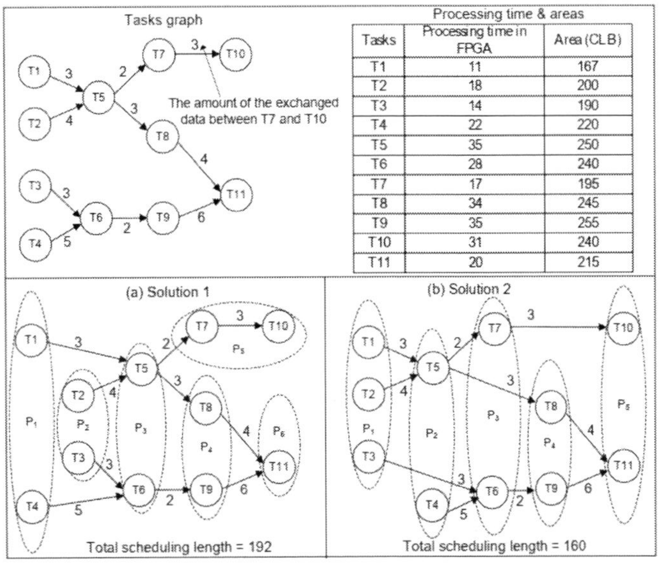

Processing time & areas

Tasks	Processing time in FPGA	Area (CLB)
T1	11	167
T2	18	200
T3	14	190
T4	22	220
T5	35	250
T6	28	240
T7	17	195
T8	34	245
T9	35	255
T10	31	240
T11	20	215

Figure 1. General overview of the problem.

3 GENETIC TEMPORAL PARTITIONNING ALGORITHM

We model and solve the temporal partitioning problem through a Genetic Algorithm (GA) [12, 13]. Our goal is to achieve the cutting of a complete application described in the data flow graph form (GFD) into sub-applications successively executed by the logic area referred through dynamic reconfiguration. The main objective is to take into account the logic area constraint in order to minimize the makespan criterion (C_{max}). In this section, we show the proposed GTPA algorithm which inherits almost the same steps from the genetic algorithms. Thereby, the elements constituting this algorithm and the principle of their operations will be addressed in the below subsections in order to create the GTP algorithm. In the rest of this section, we consider the following notations in common:

- *V*: Set of *n* tasks (T_0, T_1, T_2, .., T_{n-1});

- $G = (V, E)$: A given directed acyclic graph, where V is the set of nodes $|V| = n$ and E is a set of edges. A directed edge $e_{ij} \in E$ represents the dependence between two nodes (T_i, T_j);
- *NP*: Number of chromosomes (solutions) in the population;
- a_{FPGA}: Total surface of FPGA;
- a_i: The area of task T_i;
- $Area(P_i)$: The area of partition P_i;
- t_i: Processing time of task i on FPGA;
- c_{ij}: Cost of direct communication between task i and task j;
- *Index(i)*: the index of partition where the task i is assigned;
- *Pred_same(i)*: Set of tasks that precede task i in the same partition;
- *Pred_before(i)*: Set of tasks that precede task i in the previous partitions of the partition *Index(i)*;
- *AFT(i)*: Actual Finish Time of task i;
- T_{rec}: The reconfiguration time of FPGA (total reconfiguration);
- T_p: Execution time of partition p.

3.1 Encoding

In the considered problem (Temporal Partitioning Problem), the chromosome represents a solution to the problem where the encoding of a chromosome has a combination of tasks and partitioning as shown in Figure 2. More precisely, each task in the first part of chromosome (Task List) is linked to the corresponding partition in the second part, for example the task T_i belongs to, and is processed on the partition 2 (index 3).

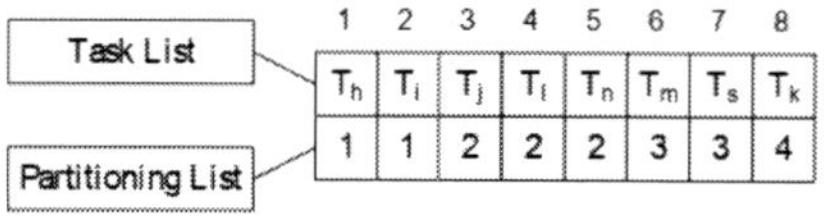

Figure 2. Chromosome that encodes task and partitioning lists.

In addition, a sequence (Task List) should be realizable in order to respect the dependencies between tasks. Consequently, a sequence is considered to be realizable if and only if:

- The order of tasks in a given sequence must respect the data dependencies between them, such as: if T_i precedes $T_j \Rightarrow T_i$ exists before T_j in a sequence.

According to this realizable sequence, the second part of the chromosome is calculated by considering the proposed algorithm 1 (Partitioning). Accurately, each partition must respect the surface constraint of logic circuit (FPGA logic area) as follows:

$$\forall k \in [1,n], \quad Area(P_k) \leq a_{FPGA} \tag{1}$$

Algorithm 1: Partitioning.

```
Input  : Task List TL                        /* TL is a realizable sequence with n tasks */
Output: Chromosome        /* Chromosome = Task List (TL) + Partitioning List (PL) */
Val = 0;
Index = 1;                                               /* the index of partitions */
for p = 0; p < n; p++ do
    Val = Val + a_TL[p];
    if Val ≤ a_FPGA then
        PL[p] = Index;
    else
        Val = a_TL[p];
        Index++;
        PL[p] = Index;
    end
end
```

A key component of the genetic algorithm is the evaluation function. At any given iteration of the GTP algorithm, the evaluation function executes how to find the makespan of the chromosomes in the current population. The pseudo code of the evaluation function is shown in the proposed Algorithm 2. Finally, a chromosome containing a realizable sequence is called realizable chromosome.

Algorithm 2: Evaluation of chromosome.

```
Input  : Chromosome                                  /* Chromosome = Task List (TL) + Partitioning List (PL) */
Output: the makespan C_max
C_max = 0;                                                                /* initialize the makespan to zero */
S1 = S2 = S3 = 0;                                       /* this variables used to calculating the makespan */
I = 1;                                                                          /* the index of partitions */
Val =0;
Step 1:                                              /* calculate the sum of execution time for all partitions */
for i = 0; i < n; i++ do
    if I ≠PL[i] then
        S1 = S1 + Val;
        T_I = Val;                                                       /* execution time of the partition I */
        Index = PL[i];
        Val = 0;
    end
    AFT(i) = max_{j∈Pred_same(i)}{AFT(j) + c_ji} + t_i
    if Val < AFT(i) then
        Val = AFT(i);
    end
end
S1 = S1 + Val;                                             /* the sum of execution time for all partitions */
T_I = Val;                                                               /* execution time of the partition I */
Step 2:                                             /* calculate the communication time between partitions */
Val = 0;
P = 2;
for i = 0; i < n; i++ do
    if P == PL[i] then
        Val = max{Val, max_{j∈Pred_before(i)}{T_Index(j) + c_ji}};
    else
        S2 = S2 + Val;
        Val = 0; P = PL[i];
    end
end
S2 = S2 + Val;
Step 3:                                   /* calculate the sum of reconfiguration time for all partitions */
S3 = I * T_rec;                                            /* I contains the total number of partitions */
End:
C_max = S1 + S2 + S3;                                                                        /* Makespan */
```

3.2 Initial population

Several methods exist to define the mechanisms of the generation of the initial population which represents the starting point for the constitution of future generations such as random section, heuristic, etc. In our study, we have created randomly the initial population containing the number of realizable chromosomes (*NP*).

3.3 Crossover and Mutation on the task list

We have implemented our approach using the crossover 1.X, where two realizable chromosomes (parents) give birth to two new realizable chromosomes (children). Let E1 and E2 be the two realizable children obtained by crossing two realizable parents P1 and P2. p is a randomly selected crossover point. The crossover is performed as follows:

- The child E1 receives the same genes as Task List of P1 between the position 1 and the position p; the rest will be completed according to the order of the missing genes in Task List of P2. The Partitioning List of E1 is determined by Algorithm 1.
- The child E2 receives the same genes as Task List of P2 between the position 1 and the position p; the rest will be completed according to the order of the missing genes in Task List of P1. The Partitioning List of E2 is determined by Algorithm 1.

There are many ways to apply the mutation on a given chromosome. In our case, due to the previous constraints, we propose a mutation operator that allows creating realizable chromosomes. The required steps for this operator are:

- Determination of E, the set of pairs of tasks that can be exchanged between them in the realizable sequence of a given parent chromosome without losing the realizability.
- Among the couples found in the set E, choose a random element e= (a, b). The child chromosome receives the same tasks as Task List of parent chromosome with permutation between tasks 'a' and 'b'. The Partitioning List of child chromosomes is also determined by Algorithm 1.

3.4 Reproduction and Evaluation

The application of the two previous operators (NP/2 crossover and NP mutation) on the current population P(t) in the preceding step causes birth of a population P'(t). This step constitutes the population of the next generation P(t+1) from the parents P(t) and the children P'(t) of the current generation. A fraction of the population is replaced by its offspring in each generation. The gap between generations indicates the size of the population of Parents replaced by children. More precisely, an evaluation phase (Evaluation of chromosome) is integrated to calculate the Makespan and number of partitions for all chromosomes of the population P(t) and P'(t). Then, we add the best NP chromosomes in terms of Makespan among them on the population P(t+1) of the next generation, while always ensuring to select the smaller number of partitions.

3.5 Termination criteria

Several stopping criteria of the algorithm are possible. Therefore, the number of generations can be either fixed a priori (time constant) or the algorithm can be stopped when the population is changing rapidly enough. In our study, the process is executed for a fixed number of iterations and where the best obtained chromosome is taken as the optimal one for our problem. However, it is not trivial to fix the number of generations required for the execution of a GA.

4 COMPUTATIONAL RESULTS AND DISCUSSION

The proposed GTPA in this paper has been tested for several DFGs in order to evaluate its performance. Each DFG consists of n tasks with data dependencies, and where the number of tasks n varies from 5 to 100 with different graph topologies, as mentioned in the Table 1. We have developed the proposed GTP algorithm using the C language. This algorithm has been executed on a machine equipped with 8 Intel processor i7-4700MQ cores and 8GB RAM memory. The operating system is a 64-bits Windows 8 Pro.

Table 1. Set of data used in simulation.

Data Flow Graph	Nodes (*n*)	Edges (|*A (G)*|)
DFG1	5	4
DFG2	10	22
DFG3	20	29
DFG4	30	67
DFG5	50	65
DFG6	100	145

4.1 General results

In this section, an analysis of the proposed approach is carried out to demonstrate its performance. We present the results obtained when applying the GTP algorithm on the data set defined in Table 1. Table 2 shows the results of our proposed GTP algorithm in terms of C_{max} makespan, Number of partitions, improvement rate expressed in percentage and representing founded improvement of C_{max} between the best solutions in the first population (initial population) and in the population of the last generation. We notice, in almost all cases, that the makespan and the number of partitions are reduced in the end of GTP algorithm, except in the case of DFG1 where the optimal solution is found by the initial population (makespan = 61 time units and number of partitions = 4) which shows that the creation of the initial population is effective. In the case of DGF3, the obtained makespan and number of partitions in the initial population are 345.50 time units and 17, respectively. These values decreased in the end of algorithm (last generation) to 307.00 time units and 14, respectively. Thereby, the improvement rate of the makespan value and number of partitions are 11.14 % and 17.65 %, respectively. In addition, the reduced makespan value causes a decrease of the number of partitions showing that there is a direct relationship between these two criteria. Based on the obtained results, we can conclude that the GTP algorithm works effectively and provides good solutions, in particular when we consider large DFGs.

Table 2. The results obtained to study the performance of proposed algorithm.

	Best solution in the initial population		Best solution in the last population		
DFG	C_{max}	Number of partitions	C_{max}	Number of partitions	Improvement (%)
DFG1	61.00	4	61.00	4	0.00
DFG2	166.00	7	159.00	7	4.22
DFG3	345.50	17	307.00	14	11.14
DFG4	440.50	21	402.00	19	8.74
DFG5	763.00	39	692.50	36	9.24
DFG6	2081.50	76	1864.00	70	10.45

Figure 3 plots the computational time in milliseconds with respect to the number of tasks. We notice that we were able to solve all instances, and the solving time is about few seconds in almost all cases. For example, in the case of DFG5 (50 tasks), the execution time is 15572.03 milliseconds. In summarize, the GTP algorithm provides an approximate solution for our temporal partitioning problem in a very reasonable execution time. This proves that our algorithm performs efficiently as expected.

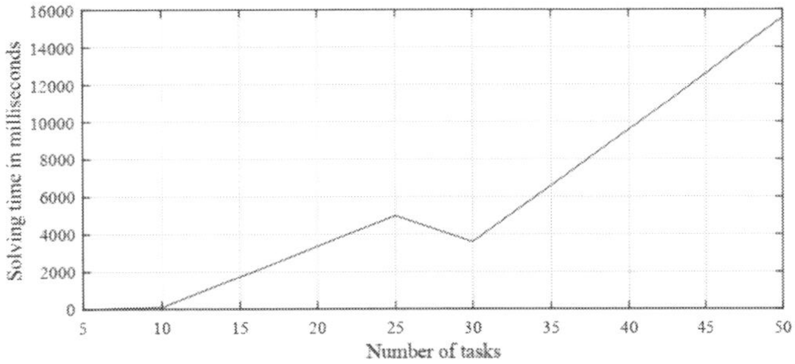

Figure 3. Solving time with number of tasks from 5 to 50.

5 CONCLUSION

In this paper, a Genetic Temporal Partitioning algorithm has been developed to solve the temporal partitioning problem existing in FPGA with logic area constraints. This algorithm aims to reduce the makepan in order to get the number of partitions. Implementation results of the GTP algorithm have demonstrated that this algorithm works efficiently and provides good solutions in short execution times. In conclusion, the proposed GTP algorithm is a very promising method that allows obtaining approximate solutions in reasonable execution times.

As future works, we can compare our proposed algorithm with other state-of-the art methods to prove its efficiency and performance. In addition, this algorithm can be easily and effectively adapted to other targets to further increase their performance (power consumption, load balancing, etc.).

6 REFERENCES

[1] **Wirthlin, M.J., Hutchings, B.L. 1998.** Improving functional density using run-time circuit reconfiguration [FPGAs], *IEEE Transactions on Very Large Scale Integration (VLSI) Systems*, 6(2), pp 247-256.

[2] **Tanougast, C., Berviller, Y., Weber, S., Brunet, P. 2003.** A partitioning methodology that optimises the area on reconfigurable real-time embedded systems, *EURASIP Journal On Applied Signal Processing (J.A.S.P.), Special Issue on Rapid Prototyping of DSP Systems*, 2003(6), pp 494-501.

[3] **Brunet, P., Tanougast, C., Berviller, Y., Weber, S. 2003.** Hardware partitioning software for dynamically reconfiureable SoC design, *in Proc. the 3rd IEEE International Workshop on System-on-Chip for Real-Time Applications, IEEE circuits and Systems Society's Techincal Committee on VLSI and on Communication*, pp 106-111.

[4] **Tanougast, C., Berviller, Y., Brunet, P., Weber, S., Rabah, H. 2003.** Temporal partitioning methodology optimizing FPGA resources for dynamically reconfigurable embedded real-time system, *Microprocessors and Microsystems*, Elsevier 27(3), pp 115-130.

[5] **Mitchell, T.** 1997. *Machine Learning*, McGraw-Hill Science/Engineering/Math.

[6] **PAPADIMITRIOU, C.H., STEIGLITZ, K.** 1982. *Combinatorial optimization: algorithms and complexity*, Prentice Hall.

[7] **RIBEIRO, C.C., MACULAN, N.** 1994. *Applications of combinatorial optimization*, Annals of Operations Research 50.

[8] **GAREY, M.R., JOHNSON, D.S.** 1979. *Computers and intractability: A guide to the theory of NP-completeness*, A series of books in the mathematical sciences, W.H. Freeman and Company, New York.

[9] **Ouni, B., Ayadi, R., Mtibaa, A. 2011.** Combining temporal partitioning and temporal placement techniques for communication cost improvement, *Advances in Engineering Software*, 42(7), pp 444-451.

[10] **Ouni, B., Ayadi, R., Mtibaa, A. 2011.** Temporal partitioning of data flow graph for dynamically reconfigurable architecture, *Journal of Systems Architecture*, 57(8), pp 790-798.

[11] **Wu, G-M., Lin, J-M., Chang,Y-W. 2001.** Generic ILP-based approaches for time-multiplexed FPGA partitioning, IEEE Transactions on Computer-Aided Design of Integrated Circuits and Systems, *IEEE Transactions on Computer-Aided Design of Integrated Circuits and Systems*, 20(10), pp 1266-1274.

[12] **Mitchell, M.** 1996. *An Introduction to Genetic Algorithms*, Cambridge, MA: MIT Press.

[13] **Holland, J. H.** 1975. *Adaptation in Natural and Artificial Systems*, University of Michigan Press, Ann Arbor, MI.

[14] **Kaul, M.,Vemuri, R.** 1998. Optimal temporal partitioning and synthesis for reconfigurable architectures, *Design, Automation, and Test in Europe*, p 389-396.

[15] **Cardoso, J.M.P.** 2003. On combining temporal partitioning and sharing of functional units in compilation for reconfigurable architectures, *IEEE Transactions on Computers*, 52(10), p 1362-1375.

[16] **Vemuri, R., Govindarajan, S., Ouaiss, L., & el.** 2001. Automated Design Synthesis and Partitioning for Adaptive Reconfigurable Hardware, *Hardware implementation of intelligent systems*, 74, p 3-52.

ON HEIJUNKA MANUFACTURING SYSTEM FOR AGRICULTURAL PRODUCTS: APPLICATION OF LEANIZED PRODUCTION MANAGEMENT MODEL FOR AGRI-BUSINESSES

Hiroshi Katayama[1*] and Ritsuko Aoki[2]

[1]Faculty of Science and Engineering,
Waseda University, Tokyo, Japan
kata@waseda.jp

[2]Department of Industrial and Management Systems Engineering,
Graduate School of Creative Science and Engineering,
Waseda University, Tokyo, Japan
may-mar.1103@ruri.waseda.jp

ABSTRACT

Recognizing recent drastic change of agriculture business environment in Japan happened after the revision of the Japanese Agricultural Land Act enforced in 2009, challenging farmers are trying to learn and implement the way of management developed in Japanese manufacturing industries such as Lean Management Scheme. This study, through investigation of the case agricultural process of a collaborating company, proposes some models of loss visualization with a way of level production (Heijunka) which is a critical concept for coping with customer satisfaction issues and profitable business making simultaneously. In addition, performance of the proposed system is evaluated for its validation through simulation experiments.

Keywords: Lean management, Operations and production management, Supply chain, Agri-businesses, Heijunka production, Parameter estimation and design, Performance evaluation

1 INTRODUCTION

The environment of agriculture businesses in Japan is changing drastically because of the shrinking number of farmers with rising average age, lowering the self-sufficiency rate of the domestic products, fiercer competition with imported goods by the agreement of TPP (Trans-Pacific Strategic Economic Partnership Agreement) and increase of exported agricultural products due to the boom of Japanese food [4].

On the other hand, the number of activities which try to adopt the management approach developed in manufacturing industries such as TPS (Toyota Production System), TPM (Total Productive Maintenance and Management), TQM (Total Quality Management), other members of Lean Management Scheme [3][5], is increasing after the revision of the Japanese Agricultural Land Act enforced in 2009 [4]. However, scientific approach provided by these schemes, namely, analysing and improving performance by countermeasure implementation against hidden problems

Corresponding Author: Hiroshi Katayama (e-mail: kata@waseda.jp)

on the PDCA platform was not satisfactory performed up to now due to insufficient managers' leadership and employees' understanding, intuitions and skills.

Based on this understanding, this study proposes some models of loss visualization with level production called Heijunka operations [3] through collaboration with a company producing agricultural goods which aggressively introduces above mentioned management approach.

2 RELATED STUDIES

Ito [2] contributes scientific approach as an early stage research work for agri-business management by proposing mathematical programming model. Chiyoma *et al.* [1] proposed the production planning model for agricultural supply chain by the following mathematical Formulas (1)-(3) which is based on the standard periodic ordering system developed by Vassian [6].

$$P_t = \sum_{i=1}^{L} \hat{D}_{t,t+i} - \sum_{i=1}^{L-1} P_{t-i} - I_t + S \quad (1)$$

$$I_t = I_{t-1} + Q_t - D_t \quad (2)$$

$$Q_t = P_{t-L} + \varepsilon_t \quad (3)$$

<Symbols>

P_t	: The production quantity planned at period t, completed at period $t+L$
$\hat{D}_{t,t+i}$	: The demand quantity of period $t+i$ predicted at the end of period t
I_t	: The inventory quantity at the end of period t
S	: The safety stock level
L	: The production lead time +1
D_t	: The demand quantity at period t
Q_t	: The yield quantity at period t, where it is not always coincides with the planned quantity.
ε_t	: The difference between Q_t and P_{t-L}

3 BACKGROUND OF THE CASE COMPANY AND THE PRODUCTION PROCESS OF THE OBJECTIVE PRODUCTS

This study focuses on the baby leaf growing and supply process at the collaborating company T, which are now booming foods among people who will be careful about their health. In fact, this business is under rapid growth in Japan, and many SMEs have been newly established including this company. As a result, competitiveness in this field becomes serious. On the other hand, production and delivery activities have burden on lack of reliable human resources and useful tools/weapons relevant for farms' management.

Trying to mitigate such obstacles, some companies start making effort to learn from businesses in other industrial sectors and implement their way of management as well as technologies, *e.g.* TPS, one representative lean management scheme born in Japan. Background of the collaborating company with our laboratory is remarkable. This company launched TPS learning project in 2006 with Toyota Motor headquarters under supervision of former Chairperson Mr. Cho and Senior Technical Executive Mr. Hayashi, who made suggestions on how to implement TPS during 2006-2008. Based on their historical efforts such as 5S, visualization, yield improvement, loss elimination, this study focuses on reaching the next stage of lean, *i.e.* realizing level production and supply (Heijunka) operations with fair concept of lean.

The process consisting of 4 major portions, *i.e.* seeding, germination, harvesting and shipping, is illustrated in Figure 1. Note that, in each sub-process, there are some unknown mechanisms creating losses. Major KPIs on these losses are called germination yield rate, harvest yield rate, weight gain rate and sorting yield rate. Management burden here is Heijunka operation is not easy to realize due to these losses, which harass quality, quantity and time on production activity even if it is well planned.

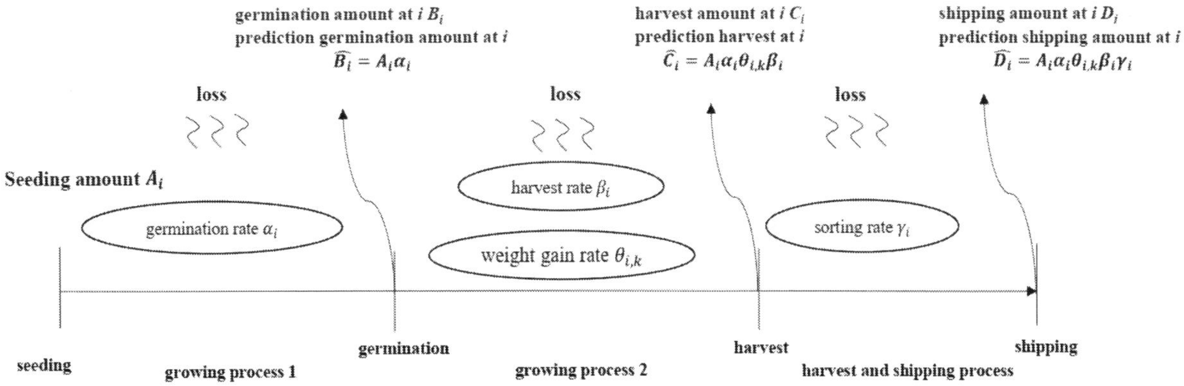

Figure 1: Outline of the case production process

4 APPROACH

To cope with this tough problem, the following 3-step approach is adopted in this study.

Step 1: Data collection

The data about "Beat" and "Pinot Green" leafs is collected and arranged, which are major 2 products of this farm in terms of shipping quantity.

Period to investigate: January 1st , 2013-December 31st , 2015 (3 year terms)

Data items: quantity of seeding, harvest and shipment

Step 2: Model of parameter estimation

A model presuming harvest quantity is formulated that utilizes estimated germination yield rate and actual seeding data.

Then the estimated values δ, θ_k and γ, which are the mean of population parameter δ_i ,$\theta_{k,i}$, γ_i (See definition table of symbols below), are defined as $\hat{\delta}, \hat{\theta}_k, \hat{\gamma}$ respectively, and these statistics are evaluated by the equations described in the following steps. In addition, because of that δ and γ fluctuate in every seeding operation, standard deviations of these parameters are estimated for effective production management.

- Step 2-1-1: calculation formulas of $\hat{\delta}$, $\hat{\theta}_k$

Estimates $\hat{\delta}$ and $\hat{\theta}_k$ can be evaluated by least square method through minimizing the total difference between actual data of harvest and its estimated quantity as shown in Formulas (4)-(7). Where, based on concerned manager's long-term experience, population parameters δ and θ_k, which denote the product of germination yield rate and harvest yield rate, and weight gain rate, are supposed to satisfy conditions (6) and (7) respectively.

$$\min \sum_{i \in N} (\hat{C}_i - C_i)^2 \qquad (4)$$

$$\hat{C}_i = A_i \times \hat{\delta} \times \hat{\theta}_k \ (k = 1,2) \qquad (5)$$

$$s.t.\ 0.64 \le \delta \le 1 \qquad (6)$$

$$100 \le \theta_{1,k} \le 300 \qquad (7)$$

- Step 2-1-2: calculation formulas of standard derivation

Estimating standard derivation of $\hat{\delta}$ is also important as this value represents the degree of internal turbulence that harms stable production.

$$\sigma_{\hat{\delta}} = S(\hat{\delta}) \qquad (8)$$

Where, sample data of $\hat{\delta}$ is obtained by the formula $\hat{\delta}_i = \frac{C_i}{A_i \hat{\theta}_k}$, which is deduced from formula (5) and S is the operator for standard deviation.

- Step 2-2-1: calculation formula of $\hat{\gamma}$

Point estimate of sorting rate $\hat{\gamma}$ can be obtained by the quotient of actual data of shipment and harvest quantities as shown in formula (9). Where, *E* is the operator for mean value.

$$E(\hat{\gamma}) = E(\frac{D_i}{C_i}) \qquad (9)$$

- Step 2-2-2: calculation formula of standard derivation

As the same meaning described in Step 2-1-2, standard derivation of $\hat{\gamma}$ is also important to estimate.

$$\sigma_{\hat{\gamma}} = S(\hat{\gamma}) = S(\frac{D_i}{C_i}) \qquad (10)$$

<Symbols>

A_i	:Seeding quantity at term *i* [kg/a]	α_i	: Germination yield rate at term *i*
C_i	: Harvest quantity at term *i* [kg/a]	β_i	: Harvest yield rate at term *i*
$\hat{C}_i$	: Estimated harvest at term *i* [kg/a]	γ_i	: Sorting yield rate at term *i*
D_i	: Shipping quantity at term *i* [kg/a]	$\theta_{i,k}$	: Weight gain rate in *k* (*k*=1:summer,*k*=2:winter)
$\hat{D}_i$	: Estimated shipping quantity at term *i* [kg/a]	δ_i	: $\alpha_i \times \beta_i$
N	: period [day]		

Step 3: Model of level control of harvest quantity

Firstly, let's clarify the relationship between total seeding quantity (input) and harvest quantity (output) as it is not possible to design proper level-production plan without such information.

The formula representing the relation between seeding interval and harvest quantity is given in Formula (11). Where, seeding interval is also fluctuated due to lack of workers, unstable schedule of using farm field *etc*. By using Formula (5) given in Step 2-1-1, interaction of expected seeding quantity and harvest quantity is obtained in (12)

$$R_k = \frac{Q}{E(x_i)} \times E(\hat{C}_i) \quad (11)$$

$$\hat{C}_i = A_i \times \hat{\delta} \times \hat{\theta}_k \ (k=1,2) \quad (5)$$

$$E(A_i) = \frac{R_k}{Q} \times \frac{1}{\hat{\delta}\hat{\theta}_k} E(x_i) \quad (12)$$

<Symbols>

R_k	: Total harvest quantity in *k* [kg/a] (*k*=1:summer,*k*=2:winter)	A_i	: Seeding quantity at term *i* [kg/a]
Q	: Period (180days)	x_i	: Seeding interval at term *i* [days]
		C_i	: Harvest quantity at term *i*[kg/a]

Secondly, utilizing this formulas and knowledge, standard derivation and average of variables must be controlled by this order to realize continuous and stable supply through level production principle. This procedure is illustrated in Figure 2. As variables and parameters in this system is very volatile, stochastic simulation is a possible way to analyse the characteristics. Then, in the first stage, variance control must be focused, which consists of 2 experiments, *i.e.* causal analysis between growing period + seeding interval and harvest interval (experiment 1) and that of loss parameters $\delta(=\alpha \times \beta)$, γ, θ and harvest quantity (experiment 2). In the second stage, average control can be followed and level production and supply operations become much easier than before, because internal turbulence caused by fluctuated variables and parameters were controlled in the first stage.

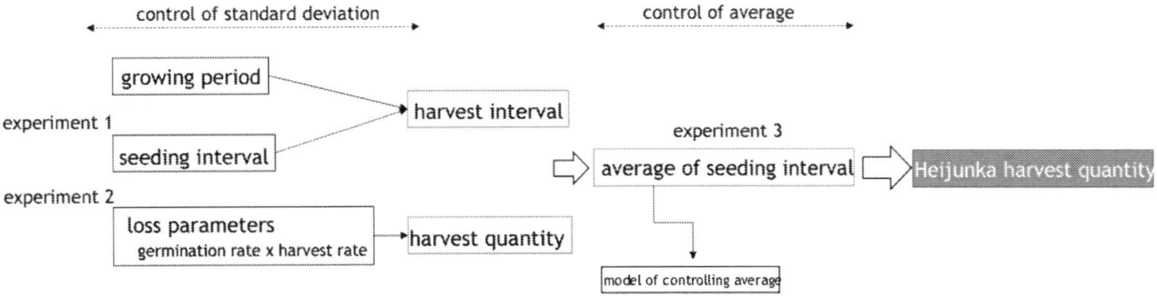

Figure 2: Model of level production (Heijunka operation)

5 RESULT

In this chapter, 3 results derived from the model/procedure described in the previous chapter are summarized.

5.1 Result of parameter estimation (Step 2 in Chapter 4)

The result of estimated parameters of "Beat" is shown in Table 1.

Table 1: Estimated parameters of "Beat" production
(Data: summer: 96 days/ winter: 61 days)

	$\hat{\delta}$	$\hat{\theta}_1$	$\hat{\theta}_2$	$\hat{\gamma}$
average	0.74	100.0	146.0	0.57
standard deviation	0.24	-	-	0.21

From this result, it is noticed that average of $\hat{\gamma}$ (Sorting yield rate) is smaller than $\hat{\delta}$ (Germination yield rate x Harvest yield rate) and both of the standard derivation are roughly equal. Therefore, first priority for raising productivity must be improvement of $\hat{\gamma}$ (Sorting yield rate), which may be tackled by industrial engineering methods.

5.2 Result of levelling control model (Step 3 in Chapter 4)

Outline of each investigation is summarized below and detail characteristics are given in Table 2-4.

<Experiment 1>

Purpose: Stability evaluation of harvest performance by improving fluctuation of variables

Variable Conditions: Seeding quantity ⇨ average 1.5 (fixed), SD 0.0 (fixed)
Seeding interval ⇨ average 7 (fixed), SD 3.11- 0.0 (reduced)
Growing period ⇨ average 64 (fixed), SD 18-0.0 (reduced)

Parameter Conditions: all parameters ⇨ average (See Table 1) SD 0.14 (fixed)

Result: Harvest interval ⇨ average 7 (unchanged), SD 7.0-0.0 (decreased)
Harvest quantity ⇨ average 120 (unchanged), SD 30-25 (fluctuated)

Remarks: Reduction of variance of "seeding interval" and "growing period" contributes to reduction of harvest interval fluctuation, but not to harvest quantity. This means variable control is effective for controlling harvest interval, but not for controlling harvest quantity.

Table 2: Result of experiment 1(300 times)

Case 1	Seeding quantity	Seeding interval	Growing period	Harvest interval	Harvest quantity [kg/a]
average	1.53	7.16	63.25	7.24	125.53
standard deviation	0.00	3.11	18.13	6.98	29.43

Case 2	Seeding quantity	Seeding interval	Growing period	Harvest interval	Harvest quantity [kg/a]
average	1.49	6.98	63.25	6.99	119.91
standard deviation	0.00	2.17	12.38	5.97	25.25

Case 3	Seeding quantity	Seeding interval	Growing period	Harvest interval	Harvest quantity [kg/a]
average	1.51	7.05	64.17	7.05	123.88
standard deviation	0.00	1.26	6.08	5.24	29.05

Case 4	Seeding quantity	Seeding interval	Growing period	Harvest interval	Harvest quantity [kg/a]
average	1.50	7.00	64.00	7.00	120.57
standard deviation	0.00	0.00	0.00	0.00	26.41

<Experiment 2>

Purpose: Stability evaluation of harvest performance by improving fluctuation of parameters

Variable Conditions: same as Experiment 1

Parameter Conditions: all parameters ⇨ average (See Table 1), SD 0.05 (reduced)

Result: Harvest interval ⇨ average 7 (unchanged), SD 7.0-0.0 (decreased)
Harvest quantity ⇨ average 120 (unchanged), SD 10-9.9 (stabilized)

Remarks: Reduction of parameter variance of "$\hat{\delta}$ (Germination yield rate x Harvest yield rate)" and "$\hat{\gamma}$ (Sorting yield rate)" contributes to reduction of harvest quantity fluctuation. Therefore, parameter control is effective for controlling harvest quantity. It is recognized from this investigation both variance of variables and parameters must be reduced for high grade level production.

Table3: Result of experiment 2

Case 1	Seeding quantity	Seeding interval	Growing period	Harvest interval	Harvest quantity [kg/a]
average	1.53	7.16	63.25	7.24	123.98
standard deviation	0.00	3.11	18.13	6.98	10.14

Case 2	Seeding quantity	Seeding interval	Growing period	Harvest interval	Harvest quantity [kg/a]
average	1.49	6.98	63.25	6.99	121.66
standard deviation	0.00	2.17	12.38	5.97	9.88

Case 3	Seeding quantity	Seeding interval	Growing period	Harvest interval	Harvest quantity [kg/a]
average	1.51	7.05	64.17	7.05	122.06
standard deviation	0.00	1.26	6.08	5.24	9.56

Case 4	Seeding quantity	Seeding interval	Growing period	Harvest interval	Harvest quantity [kg/a]
average	1.50	7.00	64.00	7.00	121.35
standard deviation	0.00	0.00	0.00	0.00	9.90

<Experiment 3>

Purpose: Realization of frequent small-lot production (Level production)

Variable Conditions: Seeding quantity ⇨ average 0.9 (reduced), SD 0.0 (fixed)
Seeding interval ⇨ average 4 (reduced), SD 2.23- 0.0 (reduced)
Growing period ⇨ average 64 (fixed), SD 19-0.0 (reduced)

Parameter Conditions: all parameters ⇨ average (See Table 1) SD 0.05 (fixed)

Result: Harvest interval ⇨ average 4 (decreased), SD 4.0-0.0 (decreased)
Harvest quantity ⇨ average 70 (decreased), SD 6.46-5.17 (decreased)

Remarks: Reduction of average with frequency increase is right way for level production which is composed by reducing average seeding quantity and its interval together. As a result, harvest interval and quantity are both decreased. In addition, turbulent level of the entire system is not notable as proper countermeasure must be performed beforehand.

Table4: Result of experiment 3

Case 1	Seeding quantity	Seeding interval	Growing period	Harvest interval	Harvest quantity [kg/a]
average	0.88	4.09	63.07	4.14	71.12
standard deviation	0.00	2.23	18.98	4.03	6.46

Case 2	Seeding quantity	Seeding interval	Growing period	Harvest interval	Harvest quantity [kg/a]
average	0.87	4.06	63.73	4.11	71.80
standard deviation	0.00	1.50	11.46	3.69	5.41

Case 3	Seeding quantity	Seeding interval	Growing period	Harvest interval	Harvest quantity [kg/a]
average	0.86	4.04	64.06	4.02	70.07
standard deviation	0.00	0.74	6.07	3.42	5.30

Case 4	Seeding quantity	Seeding interval	Growing period	Harvest interval	Harvest quantity [kg/a]
average	0.86	4.00	64.00	4.00	69.96
standard deviation	0.00	0.00	0.00	0.00	5.17

6 CONCLUDING REMARKS

In this paper, a procedure of lean management for achieving Heijunka operation is reviewed and applied to a production management division of collaborating agri-company. Discussed procedure consists of ①Investigating performance of current system, detecting unstable behaviour of variables and extracting root causes of these phenomena,② Eliminating causes of unstable factors such as internal turbulence and then, ③ Controlling input (average) for stable and reliable operations. Obtained result can be a fine example of lean management transfer to agriculture industry.

REFERENCES

[1] Chiyoma, H., Murata, K. and Katayama, H. 2014. On Performance Analysis of Sustainable closed Logistics System in Agricultural Industry (Nousakubutsu Sapurai Chein ni okeru Junkangata Seisan Sisutemu ni kansuru Kenkyu: in Japanese), Proceedings of the 17th Annual Conference of Japan Society of Logistics Systems, Waseda University, Japan, May 10-11, pp.101-106.

[2] Ito, K. 2006. Mathematical Programming Problems in Agricultural Management (Mougyou Keiei ni okeru Suri Keikaku Mondai: in Japanese), *Operations Research: Management Science (Opereisyonzu risachi: Keiei no Kagaku: in Japanese)* 51(5), pp. 264-267.

[3] Katayama, H. 2006. Lean Thecnology and its Extensions -A case of Target Chasing Method-, Proceedings of The 11th Cambridge Symposium on International Manufacturing (CSIM2006), Ed. Gregory, M., Institute for Manufacturing, University of Cambridge, UK, September 27-28, 15 pages in CD-ROM.

[4] Ministry of Agriculture, Forestry and Fisheries HP, URL: http://www.maff.go.jp/, January 25, 2015.

[5] TKF (Tsukuba Kimura Farm) Co. Ltd. HP, URL: http://www.t-k-f.jp/company.html, , October 8, 2015.

[6] Vassian, H. J. 1955. Application of Discrete Variable Servo Theory to Inventory Control, *Operations Research*, 3(3), pp. 272-282.

ANALYSIS OF THE RISK AND TIME-WINDOW FOR CHINESE PRIVATE ENTERPRISES LISTED IN UNITED STATES

Zhang Caixia[1-],Zhang Yu[2]
[1]World Economy,School of Business
China University of Political Science and Law, Beijing, China
108575528@qq.com

Department of Fresh Projects
[2]Beijing Renwo-Online Electronic Commerce co., LTD
zcx830322@163.com

ABSTRACT

Capital markets in United States are the largest, most liquid, highest efficiency of markets all over the world, they also have the most perfect legal system and the best protection for investors. Chinese enterprises began to list in United States in 1992, and so far has experienced 24 years. There are more risks for Chinese private enterprises in the process of listing in United States, which can be divided into macro risk, medium risk and micro risk according to the sources and their influence. In this paper, an innovative theory of time window is put forward, and is mainly to describe the macro risks faced by private enterprises in the process of listed in United States. Also, we analyzed mainly from the capital market, the macro economic situation and the capital market investment confidence and so on. The capital market can be assumed as a function and mainly measured by NASDAQ composite index, the New York Stock Exchange general index and American stock exchange index, and supplemented by the NASDAQ China Index. The macroeconomic situation as a function can be predicted mainly by GDP growth rate and inflation rate and the confidence of capital market investment can be measured by Sentix confidence index. So Time Window for private enterprises listed in United States is a comprehensive function which should take capital market expectations, macroeconomic and investor confidence index into consideration.

KEYWORDS: Private enterprises, Listed in United States, Risks, Time-Window

1 THE BASIC SITUATION OF THE PRIVATE ENTERPRISES LISTED IN U.S.

Capital markets in United States are the largest, most liquid, highest efficiency of markets all over the world, they also have the most perfect legal system and the best protection for investors. So the capital markets in United States have great attraction for financing companies all over the world. Chinese enterprises began to list in United States in 1992, and so far has experienced 24 years, and the primary listing mode are two ways, namely IPO and reverse acquisition. IPO, that is to say

[1-] Zhang Caixia, a doctor in China University of Political Science and Law.Email:zcx830322@163.com

initial public offerings, is refers to the mode of raising funds through securities exchange firstly open to investors; while reverse acquisition is refers to the mode in the form of a private placement of shell companies or directly convertible shares for listing. On the analysis of listed companies in United States from 1999 to 2015, we can mainly reflects the following characteristics:

The absolute number of listed companies in United States gradually increases, but the trend experiences a wave development, which is shown in Fig.1. In 1999, there was just one company listed in United States by form of IPO, and the growth rate in 2000-2004 was not much; from 2004 to 2007, the number of listed enterprises has increased year by year, because of the influence of red chips "10th notice", at the end of 2007, the number of listed enterprises in our country ushered in the peak, and real estate and Internet is the hot industry at the time. In 2008, due to the outbreak of the sub-prime mortgage crisis in the United States, the number suffered a low rate, throughout the year only two listed successfully. From 2010 to 2015, as a result of the recovery of global economy, especially American economic recovery and China's strong economic growth, more and more enterprises listed in United States in form of IPO successfully, including McCaw, Dangdang, Youku, Alibaba, Thunder, Zhaopin, Jingdong, Sina, Tianhua sunlight and so on.

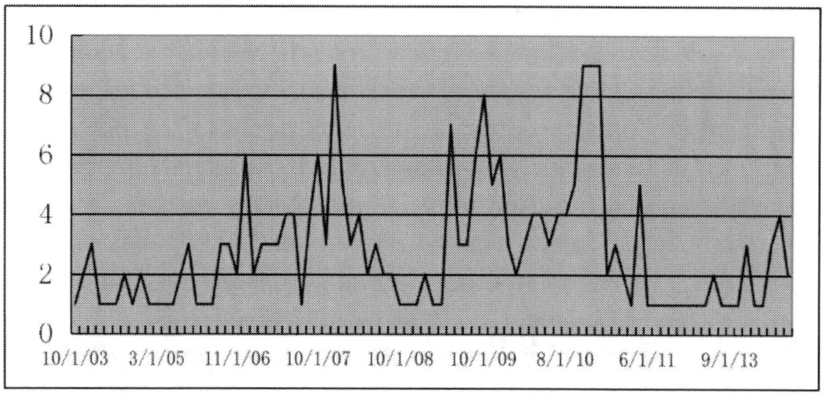

Figure 1: The number of private enterprises listed in U.S. in 1999-2014

Reverse acquisition has become an important mode for private enterprises listed in American capital market due to its features of short period, high guaranteed, low cost and so on. According to the statistics of the Dealflow, since 2001, a total of 267 Chinese enterprises indirectly listed in United States by this form, while there just only 63 Chinese enterprises listed in United States in the form of IPO.

Many companies in U.S. suffered serene Securities group litigation, which is the legal supervision power for illegal acts in securities, which applied to all listed companies and managers. And also securities group litigation is the most severe indictment which most of Chinese listed companies are facing in U.S. According to statistics, from 2001 to 2006, six Chinese companies suffered group litigation; in 2007, 10 Chinese companies were accused of group litigation; in 2008, four Chinese companies were sued. In 2009, three Chinese companies were sued; and in 2010, 13

Chinese companies were sued, while in 2011, this number increased to 39, which accounting for about 18% of the total number of Securities group litigation in 2011.

2 THE RISK OF PRIVATE ENTERPRISES LISTED IN U.S.

In general speaking, we can understand risks as uncertainties in future which will impact on the company's stated goals, or the possibility of losses. The definition of risks can be concentrated into four aspects. The first one is future, that is to say, risks must influence the behavior and income in the future. The second one is uncertainty, that is the source of risks, which can be result from external environment or the management of related enterprises. The third one is the established goal, that is to say, risks influence the income or utility of related enterprises, and degree of influence depends on the enterprise's objective function. Fourthly, the key factor is influence. There must be a deviation between the real goal and the established goal because of risks, which could be a positive effect or a negative effect.

The risks of Chinese Private enterprise listed in United States refer to the possibility of unexpected failure or losses in the process of listed in United States. Firstly, the listing process should include pre-listing, listing and the success of sustainable management after-listing. So the so-called successful listed is refers to the completion of equity financing, system specification and the continued operating after listing. This goal is comprehensive, just listed in United States Exchange but cannot be sustained operating is not successful in sense. Secondly the objectives of listing in United States is not only to widen the channels of financing and lift the continuous operation and financing ability, but also to set up the standard system of business management and to make itself more well-known. Due to various factors influenced, the objectives of listed companies are often uncertain, which may form certain risks. Finally, the so-called losses include directly or indirectly failure in listing on the market, non-

contiguous operating or losing the ability of sustainable financing in the capital market. In the process of listed in United States for private enterprises, there exists more and more possibilities not up to the expected goal, due to some uncertainty of the economic environment, capital market, intermediary selection and corporate structure, which will bring losses to the enterprise. Therefore, in order to prevent the losses or reduce the degree of losses, we should diagnose and identify these risks, find some reasons and take corresponding preventive measures. In summary, the risks of private enterprises listed in United States is universal, long-term, hidden, dynamic, conductivity, multilevel and so on.

In this paper, we can divided risks into macro risk, medium risk and micro risk according to the sources and their influence degree. Among them: Macro risk refers to systemic risk, such as the financing environment, the overall economic situation in United States, the access conditions and preferences for different Exchange Market , legal risks, political risks, social culture and so on, all of which may caused system risks. Secondly, medium risks are mainly derived from a certain industry or a region, such as business prospects for the development of the industry, not only the development in the domestic situation, but also investors' awareness of the industry in American. Thirdly, micro risk is refers to the enterprise's own risk, including the operating risk (governance structure, financial system, development planning and expected performance, the ability of independent innovation, etc.), the risks of selecting intermediary, risks of listed scheme (time to market, location, costs, reorganization and restructuring), information disclosure and so on. In this paper, the analysis of time-window are mainly concentrated on the macro risks.

3 THE ANALYSIS OF TIME WINDOW FOR PRIVATE ENTERPRISES LISTED IN U.S.

3.1 The theory of Time Window

In stock market , we often hear the word Time Window. In fact, Time Window is not unique to stock market, in the foreign exchange market and other financial market, and even every things are all related to cycle time. Time Window in physics is occurring in the four-dimensional space (i.e. a time coordinate). Time Window in stock market is a series of theoretical analysis and empirical study around the time element, involving Wave Theory, Cycle Theory and Gann Time Window Theory. Wave theory is invented by Ralph. Nelson. Eliot (R.N.Elliott), who is a technical analysis master, and it is one of the most commonly known tools in the trend analysis. The theory is analyzed by Eliot based on the index "Dow Jones Industrial Average", and then Eliot found the price structure and used it to explain the market behavior. Cycle theory is mainly analyzed the special factor-time, its practical effect is to speculate dates of highs and lows in the cycle and to formulate sale strategy from the past cycle length. Gann time window theory issued that time factor and time cycle are most important in the trend of stock market. And when the time reached a certain point, the volume can be amplified and forced the price reversed in the future.

3.2 Time Window for Listed Company

There are two meanings for Time window. The first one refers to the time period to a certain stage, when will be the turning point. The second one refers to a critical period, if some specific things occurred or taking some actions, the probability of success will increase. In this paper, time window is refers to the key period of time in the process of Chinese private enterprises listed in United States. In this time, if more effective measures or actions were taken, the probability of successful listed in United States will increase. In this paper, we analyzed mainly from the capital market, the macro economic situation and the capital market investment confidence and so on. The capital market can be assumed as a function and mainly measured by NASDAQ composite index, the New York Stock Exchange general index and American stock exchange index, and supplemented by the NASDAQ China Index. And we can represent the basic situation of the capital market mainly by the cycle of rising or falling, which can be represented by Y_i. The macroeconomic situation is a function can be predicted by GDP growth rate and inflation rate, which can be represented by Y_{GDP} and Y_{if}; and the confidence of capital market investment can be measured by Sentix confidence index, which can be represented by Y_{ic}. Thus the function of Time Window for private enterprises can be expressed as:

$$T = F(Y_i, Y_{GDP}, Y_{if}, Y_{ic})$$

Among them:

① Y_i represents the function of the expected capital market, which mainly used to determine whether the capital market is expected to rise or fall in future. As follows

Assuming that any one month's capital market index is I_0 (expected capital market can be measured by a monthly line chart)

$$\text{If } I_1 > I_0, I_2 > I_1, I_3 > I_2, \cdots I_n > I_{n-1} (n = 1,2,3 \cdots 6)$$

In the rising period, it is suitable for listing;

$$\text{If } I_n < I_{n+1}, I_{n+1} < I_{n+2}, \cdots I_{n+m} < I_{n+m-1} (m = 1,2,3, \cdots 6),$$

In the declining period, it is not suitable for listing;

Where: I_n is the peak of the cycle, while I_{n+m} is the trough of the cycle.

② Y_{gdp} and Y_{if} represent the comprehensive function of the macroeconomic expectation, which are specifically measured by two indicators such as annualized rate of GDP and inflation rate (CPI price index) ;

③ Y_{ic} is a function of investor confidence, measured by Sentix confidence index.

As follows we will make empirical analysis based on some data from January in 2001 to June in 2014 which can be described the situation for listing.

Firstly, we can analyze the cycle of stock market based on NYSE composite index, NASDAQ Composite and American Stock Exchange Index, which are shown in Fig.2.

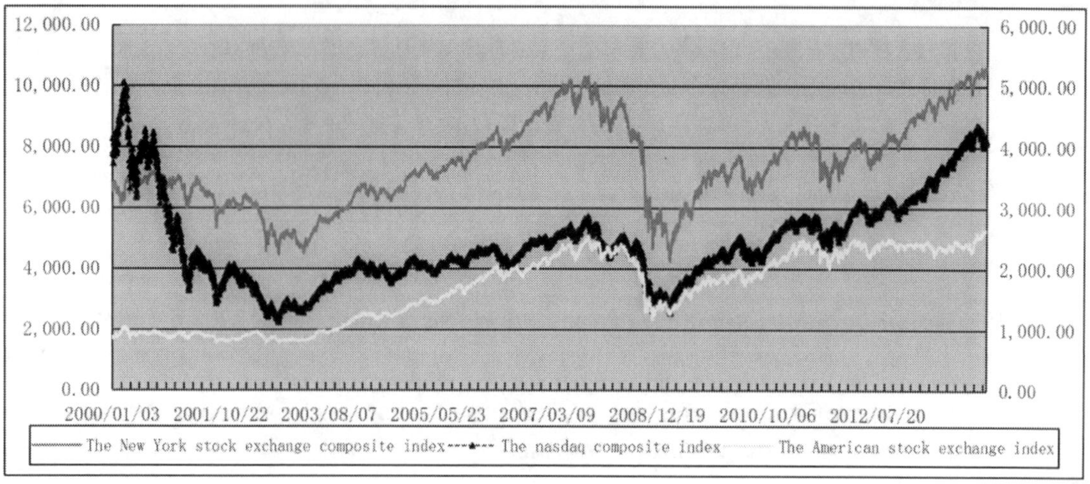

Figure 2:Three major stock index chart in United States in 2000-2014

2001.01-2001.05, capital market is expected to rise;

2001.06-2002.10, capital market is expected to decline;

2002.11-2007.10, capital market is expected to rise;

2007.11-2009-03, capital market is expected to decline;

2009.04- so far, the capital market is expected to rise.

Secondly, we can analyze the overall economic situation based on annualized rate of GDP and inflation rate, which are shown in Fig.3 and Fig.4.

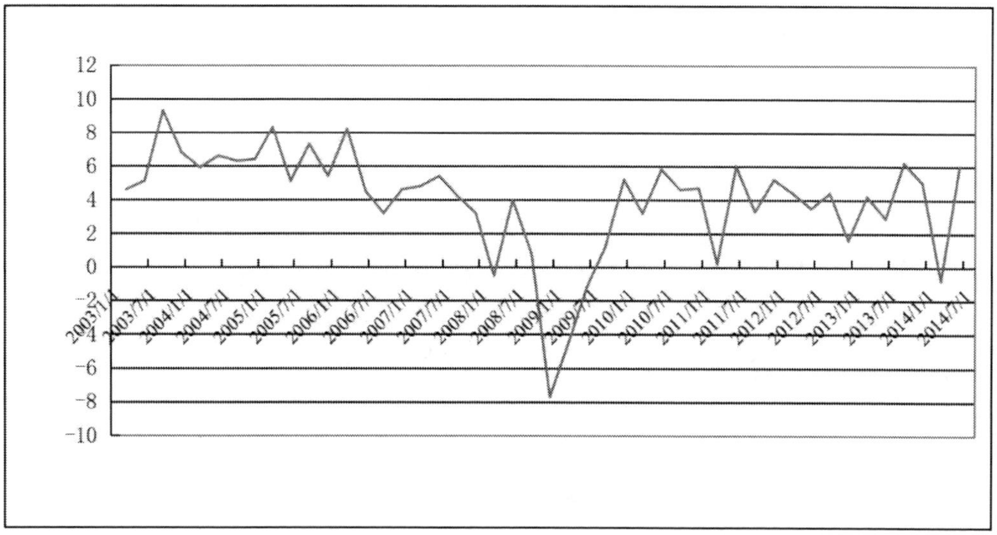

Figure 3: The annualized rate of GDP in United States in 2000-2014

2001.01-2003.12, basically in the rising period and reached to the highest point at the end of 2003;

2004.01-2007.09, basically in shock period, the late falling in decline

2007.09-2008.12, in a rapid decline due to the impact of the economic crisis;

2009.01-2010.09, basically to be restored, and a sustained shock.

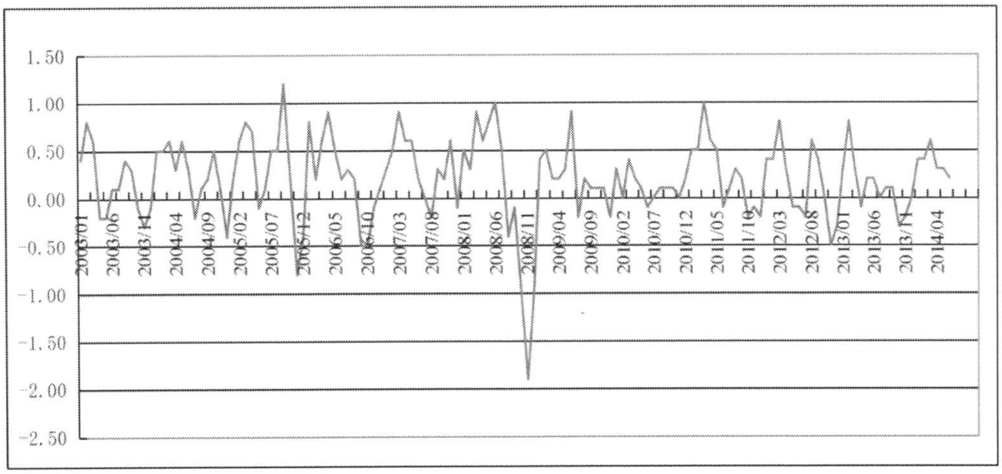

Figure 4:The trend chart of inflation rate in United States in 2000-2013

From the CPI price index, in addition to the rapid decline in 2007.06-2008.12, the other is in a normal shock.

Thirdly, we should analyze investor confidence index based on Sentix investor confidence index, which is shown in Fig.5.

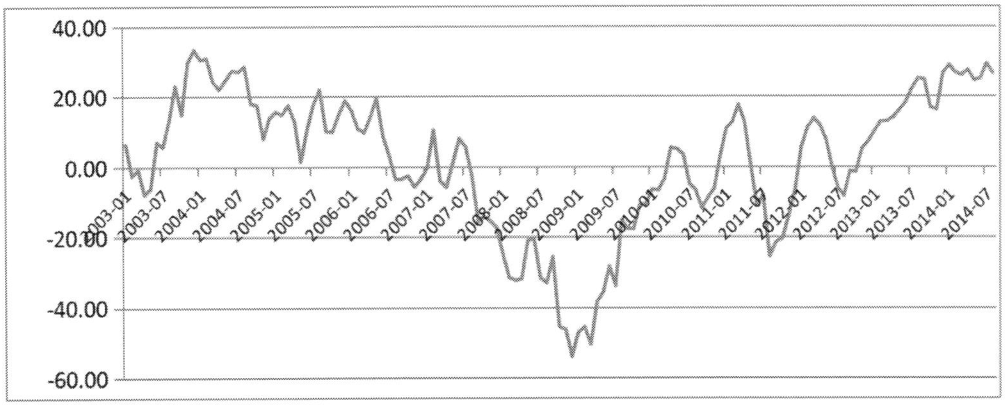

Figure 5: The trend chart of Sentix investor confidence index in U.S. in 2000-2013

Here choose Sentix investor confidence index, specifically:

2003.01-2004.01, investor confidence index was on the rise and more than 1;

2004.02-2009.01, investor confidence index showed a downward trend: 2004.02-2007.08, investor confidence index is more than 0; 2007.09-2009.01, investor confidence index is less than 0;

2009.02-2014.06, investor confidence index was in a rising trend, which 2009.01-2010.05, investor confidence index is less than zero; 2010.06-2012.09, overall in the state of the cycle; 2012.10-2014.07, investor confidence in the rising period and index greater than 0.

Generally speaking, from the trend of various factors above, we can be drawn that there are two best time-window from the fourth quarter of 2002 to the fourth quarter of 2003 and the fourth quarter of 2012 to the first quarter of 2014. From the third quarter of 2004 to the third quarter of 2007, although the development of microscopic economy and the investor information index were declined slightly, with the healthy development of the securities market operation cycle, those are also better time-windows which suitable for listing. Similarly, from the second quarter of 2009 to the fourth quarter of 2012, although the investor confidence was in the state of shock, securities market expectations and the overall economic were recovered already, so it also can be chosen as the second time-window for listing in United States. From the fourth quarter of 2007 to the second quarter of 2009, due to the serious impact of the economic crisis, market expectations, the overall economic and the confidence of investors were declined rapidly, therefore it is not suitable for listing. This is consistent with the time choice of Chinese enterprises listed in United States in practice. For example, most of Chinese enterprises were listed in United States in 2004 and 2007. After the financial crisis of 2008, with the gradual recovery of the overall economic, the securities market expectations and the investor confidence in 2011, there formed another peak in financial market. Because of generally suppression and the phenomenon of financial fraud, private enterprises in our country experienced another trough from 2012 to 2013. After 2014, more and more enterprises has successfully listed in United States, such as Locke holdings, Ikang, Sina, Micro-blog, cheetah mobile, Tuniu, Jingdong, Zhaopin and etc.. Maybe another time-window for listing in United States has been coming.

4 CONCLUSION

Therefore Time Window for private enterprises listed in United States is a comprehensive function which should take capital market expectations, macroeconomic and investor confidence index into consideration. if the capital market expectations is good, macroeconomic situation and investor confidence index experience a rising trend, that must be a best time-window for private enterprises for listing in United States. If the capital market expectations is in a downward trend, or the macroeconomic situation is not ideal or investor confidence index is in the decline, maybe it is not the better time-window for listing which should be analyzed in-depth.

5 REFERENCES

[1] **Wang Lisong.2011.**Risk Analysis and Countermeasure Research of private enterprises listed overseas, University of Electronic Science and technology

[2] **Cheng Yunchuan.2012.**Research on the risk control of China's listed companies,Wuhan University of Technology, pp 29-31.

[3] **Ding Yuhua.2013.**Research on China's small and medium enterprises listedin the United States, Suzhou University,pp 5-10.

[4] **Qiu Yonghong.2012.**Research on the legal risks and Countermeasures of Chinese enterprises listed in the United States,Law Forum,27(2),pp 40-47.

[5] **Zhang Qian.2010.**Research on overseas financing of small and medium enterprises, Southwestern University of Finance and Economics.

[6] **Jiang Wei, Zhou Yu. 2002.**An empirical analysis of the "size effect" and "time effect" of China stock market -- a case study of Shanghai stock market,Economic Research Journal,10.

[7] **Yang Yubo, Wei Hua.2005.**An empirical analysis of The"time effect" of China securities market, Financial Aspect, 06,pp 61-64.

[8] **Zhang Sujia.2006.**The Research on Reverse Merger in the U.S.by Chinese private enterprises and Case-study,Tongji University,pp 25-34.

[9] **Cheng Yunchuan.2012.**Study on the Risk Control of Chinese listed Enterprises, Wuhan University of Technology,pp 29-31

[10] **Liu Dongyang.2010.**Research on the Core Thoughts of Gann's Investment Theory and Applications-Based on the Price, Time and Trading Volume,pp15-20

[11] **Wang Wei,Zhou Yu.2010.**An Empirical Study on the Size and Time Effect of the Stock Market in China-Based on Shanghai Stock Market,Economic research,10,pp 16-21

CENSORED DATA ORIENTED ESTIMATION OF THE CHANGE POINT IN THE INFANT PERIOD OF A BATHTUB CURVE BY MAXIMUM LIKELIHOOD ESTIMATION

Yi Wei*, Yihai He, Chunling Zhu
School of Reliability and Systems Engineering
Beihang University, Beijing, China
weiyi@buaa.edu.cn

ABSTRACT

The change point evaluation and monitoring of failure rate is an inevitable task for product reliability assurance. Since the fact that the evolution mechanism of product infant failure is fuzzy due to its complexity, it is infeasible to determine the accurate burn-in time in advance when applying burn-in testing for electronic product, the bottleneck lies in the accurate identification of the change point for infant failure. Accordingly, in this paper, the connotation of the change point of infant failure in product early lifetime is discussed, and the processing method of the censored life testing data from burn-in testing is presented. Then, based on the principle of maximum likelihood estimation(MLE) for censored life testing data, the convergent change point identification technique for Weibull distribution is proposed to be more applicative for the analysis of censored characteristics. Finally, the validity of the proposed method is verified by a change point identification example for a control board reliability analysis.

Keywords: Infant failure rate, Weibull model, change-point estimation, burn-in process

1 INTRODUCTION

Burn-in testing, as a common measure to exclude the products that may break down in early lifetime, can always lead to huge cost. To determine the necessary time for burn-in testing, the mechanism of infant failure should be thoroughly considered by designers. Unfortunately, the mechanism of infant failure is fuzzy now [1][2], it is not applicable to determine the accurate burn-in time in advance. Therefore, in order to improve the effectiveness of the burn-in testing, the change point of the infant failure monitoring is needed, which is a beneficial supplement for product production reliability assurance as well [3].

Weibull models are the most widely applicable models for describing the whole lifetime of a product, especially for electronic goods [4]. The well-known bathtub curve can be viewed as a combination of three parts from different specific graphs of Weibull functions [5]. Considering the fact that the shape of a graph from Weibull function can be easily changed just by modifying the shape parameter, it's a practicable way to monitor the composition of a bathtub curve by concentrating on the changing trend of the parameters and some related numerical derivations derived from Weibull functions.

*Corresponding Author

Regarding to change point evaluation and monitoring of failure rate, there are substantial papers. Hadidoust et al. [6] used a kind of control chart to recognize the out-of-control point as the change point. Gijbels and Gurler [7] considered the problem of estimating a single change point in a piecewise constant hazard function when the observed variables are subject to random censoring. Zamba and Hawkins [8] proposed a method to monitor change point without knowing abundant exact historical data which are used to estimate the distributions' parameters. Jing [9] proposed an approach for detecting certain change-points, via a two-stage failure model and stochastic time-lagged regression functions. Azevedo et al. [10] discussed shape and change point analyses for the hazard rate of the BS-t distribution. Niaki and Hekmati [11] proposed a new multi-attribute control chart based on two transformation methods to monitor the parameter vector of multi-attribute Poisson processes. Bae et al. [12] proposed an informational change-point approach for the pattern of recurrent failures in repairable artillery systems, which provides additional information about the locations of change-points over rates of occurrence of failures as well as failure trends. Niaki and Khedmati [13] considered the multivariate Poisson process and studied the MLE of change point for monotonic changing process.

Although the preceding literature review shows that the change point evaluation and monitoring of failure rate is an inevitable task for product reliability assurance, few studies have examined the infant failure [2]. Accordingly, in this paper, the connotation of the change point of infant failure in product early lifetime is discussed, and the evaluation and detection method of infant failure change point from the censored life testing data is presented.

2 CHANGE POINT OF INFANT FAILURE IN PRODUCT EARLY LIFETIME

2.1 Basics of Change Point

The bathtub curve is considered to be a typical model of the whole lifetime for a product. Three significant periods are regarded as the compositions of a completed bathtub curve, and the first two periods are supposed to denote the infant high-failure-rate period and the stable period. Apparently, customers would not want to receive a product which will break at the initial usage time. The burn-in tests in factories is specialized for this problem [14]. However, as shown in figure 1, the longer the tests last, the more costs will be paid. Therefore, producers hope this process can be finished as soon as possible.

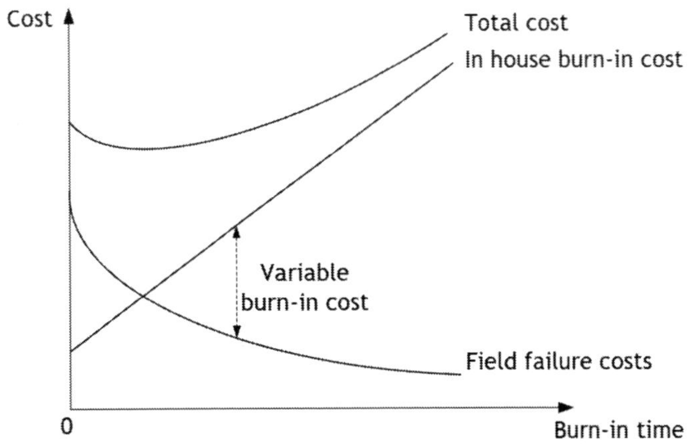

Figure 1: The relevance between costs and burn-in time

One of the most significant characteristic of the data collected in burn-in tests is that they are all censored. Unfortunately, the study on treatment of censored data has got slack development in the past decades. Consequently, few efficient rules have been made to help experimenters to control the time of burn-in testing. To solve this problem, the accurate estimation of the change point (shown in figure 2) that the failure rate turns to be stable from gradually decreasing is urgently needed. [15]

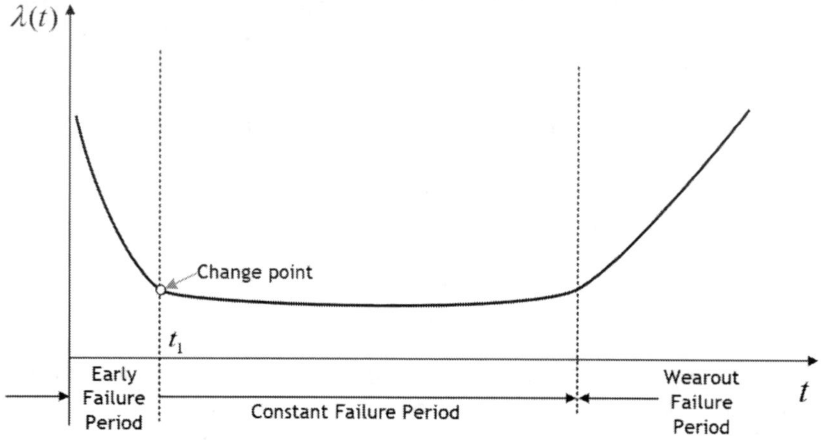

Figure 2: the change point in infant period of a classic bathtub curve

As to the problem of modelling the failure rate of bathtub curve, Duan Q. and Liu J. [16] proposed a method based on a specific distribution to study it. Opposite to the situation that one goes into a changeable status from a steady process, the change point from the infant quickly decreasing status to a stable one is still a remaining problem.

2.2 Estimations of Weibull parameters from Censored lifetime data

To detect the change of failure rate in infant failure, a reasonable monitoring index is needed. In this paper, we are going to study the method of estimating the continuing length of infant failure period with the data collected in burn-in tests. Due to the practicability of tests, Type I censored data will be used in it this paper.

2.2.1 The distribution assumption for population

Weibull distribution involves forms of two-parameter and three-parameter, of which the two-parameter one is mainly used for describing lifetimes of rolling bearings and material fatigue suffering high stress; and the three-parameter one is mainly used for characterizing lifetimes of materials or some special parts under low stress. Here, this paper primarily focuses on two-parameter Weibull distribution.

The cumulative distribution function(CDF) of two-parameter Weibull distribution, denoted by $W(\alpha,\beta)$, is expressed as

$$W(\alpha,\beta)=\begin{cases}\frac{\beta}{\alpha}\left(\frac{x}{\alpha}\right)^{\beta-1}e^{-\left(\frac{x}{\alpha}\right)^{\beta}}, & x\geq 0\\ 0, & x<0\end{cases} \quad (1)$$

2.2.2 MLE Calculation for censored data

Maximum-likelihood estimation is a classic method to assume the statistical indicator from a sample set from a large population. For Type I censored data, the likelihood function is given by

$$L(\theta) = \prod_{i=1}^{n} [f(t_i)]^{\delta_i} [1 - F(v)]^{1-\delta_i} \quad (2)$$

Since the basic Weibull model has two parameters, estimates of the model parameters can be obtained using the sample mean $\bar{t}$and sample variance s^2 . In thecase of complete data these are given by

$$\bar{t} = \sum_{i=1}^{n} \frac{t_i}{n} \quad (3)$$

$$s^2 = \sum_{i=1}^{n} \frac{1}{n-1} (t_i - \bar{t})^2 \quad (4)$$

However, it is not easy to clearly observe a change of the process by these two statistics. The idea of transforming into monitoring the parameter of Weibull function will be used in our study.

For complete data, the MLE of two parameters can be calculated from the likelihood function as following.

$$L(\alpha, \beta) = \prod_{i=1}^{n} \left(\frac{\beta t_i^{(\beta-1)}}{\alpha^\beta} \right) exp\left\{ -\left(\frac{t_i}{\alpha} \right)^\beta \right\} \quad (5)$$

For the situation in burn-in tests, data collected are censored and the calculation will be modified correspondingly. For a reordered sample set in which the first K are uncensored data and the others are censored, the likelihood function should be [2]

$$L(\alpha, \beta) = \frac{\beta^K}{\alpha^{\beta K}} \prod_{i=1}^{K} t_i^{\beta-1} exp\left\{ -\frac{1}{\alpha^\beta} \left[\sum_{i=1}^{K} t_i^\beta + (n-K) v^\beta \right] \right\} \quad (6)$$

The estimator of β is the solution which fits the equation

$$\frac{\sum_{i=1}^{n} t_i^{\hat{\beta}} lnt_i}{\sum_{i=1}^{n} t_i^{\hat{\beta}}} - \frac{1}{\hat{\beta}} - \frac{1}{k} \sum_{i=1}^{k} lnt_i = 0 \quad (7)$$

And alpha is given by

$$\hat{\alpha} = \left\{ \frac{1}{n} \left[\sum_{i=1}^{k} t_i^{\hat{\beta}} + (n-k) v^{\hat{\beta}} \right] \right\}^{\frac{1}{\hat{\beta}}} \quad (8)$$

3 DETECTION OF CHANGE POINT

3.1 Monitoring index

As it is an important index to save the costs of producers, the detection of change point is the research emphasis of this paper. After the estimator of parameters recalculated (7) and (8), the reasonable use of them will help experimenters find out the accurate change point.

As shown in figure 1, the constant failure rate of a product is supposed to be stable. From a mathematical perspective, the shape parameter, alpha, in the formula (1) turns to be equal 1, and the function of failure rate for this period could be expressed by an exponential distribution.

Let d denote the difference between 1 and the estimator of β.

$$d = |\hat{\beta} - 1| \quad (9)$$

When the infant failure period tends to end, m would asymptotically converge to zero. Thus, it would be an easy way to detect the trends in burn-in tests.

3.2 Determination for the end of burn-in process

With the going of the burn-in tests, the failure rate tends to be stable. As shown in figure 3, it can be expressed by the later part of the graph of a Weibull function. Actually, the longer the tests last, the closer a function with a shape parameter less than 1 will be to one with beta equal 1.

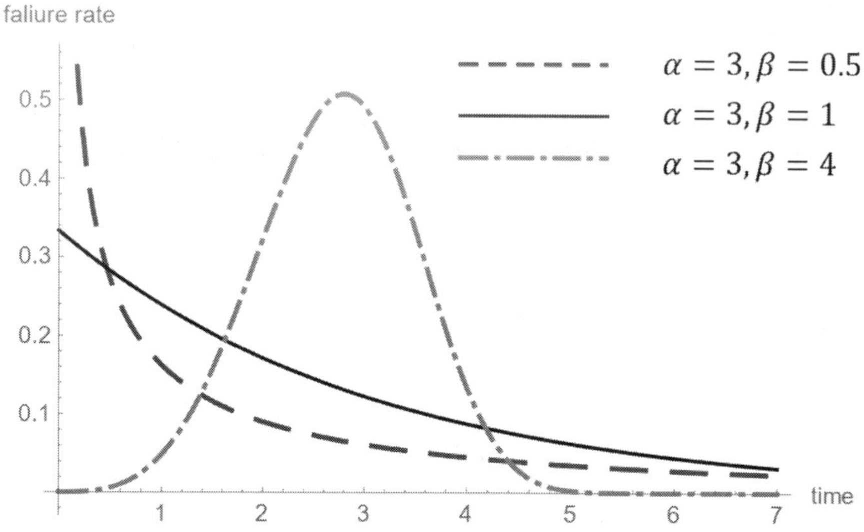

Figure 3: function graphs with a different shape parameter β while other parameters are constant

Since the fact that people always consider a product which is in the healthy useful status has a stable failure rate, the basis of determination for the end of burn-in process could be that the estimated parameter from collected data become similar with one in a stable process. The specific implementation steps are as follows:

Step1. Collecting failure data from burn-in testing

Conduct the burn-in test and keep notes of the time when products broke down. Then the failure rate in a certain time period can be calculated. Let N denote the number of all the items used in a test. Let t_i denote the time when the i-th failure happens. Until the time of t_i the mean failure rate is as formula (10)

$$\lambda_i = \frac{i}{t_i} \qquad \textit{(10)}$$

Step2. Estimating the shape parameter from existed data

Since the lifetime of remaining items are all unknown, the data used in calculation for these remaining items are all censored. Then the estimator of β is the solution which fits the equation

$$\frac{\sum_{i=1}^{n} t_i^{\beta} \ln t_i}{\sum_{i=1}^{n} t_i^{\beta}} - \frac{1}{\beta} - \frac{1}{K}\sum_{i=1}^{K} \ln t_i = 0 \qquad \textit{(11)}$$

Step3. Comparing the estimator with ideal value

We can set an accuracy standard as the basis to determine whether the tests should be last for a longer time. As a tentative in this paper, we set the acceptable difference between estimator and ideal value to be 10^{-2}. If the estimator got in step2 fulfils (12), then the infant failure period can be viewed to be near the end. Otherwise, go to the next step.

$$d < 10^{-2} \tag{12}$$

Step4. Repeat the Step 1 to 3.

Refreshing the estimator with added data until the formula (12) is proved to be true.

As a supplement for step 3, the decreasing speed of failure calculated by (10) can also prove the fact whether the lifetime has got into the stable period.

Step5. Finish the burn-in testing.

As the shape parameter is already very close to the one in a stable useful process, the tests could be stopped immediately. And the data can be noted as a reference for later tests conducted on products in similar type.

4 CASE STUDY

4.1 Background

There is a set of sample data, with the size of 100, collected from the reliability testing on a type of control board for a domestic washing machine. To verify the practicability of the derived method in this paper, the analysis on completed data will help us to know the real change point of lifetime for the product. Strictly speaking, this data is also censored because there are still a few samples were working properly when the test was stopped. But it's considered to be a credible approximation to the truth because this test has been lasted for much longer than usual burn-in tests. To do analysis of the censored data, it can easily be achieved by replacing the latter data with the stopping time.

4.2 Numerical Example

Here is the data from the test mentioned above. The left two columns are originally noted from the tests, and the other columns are the related calculated results by (10)-(12) . Except the initial data in the first 100 hours are insufficient to calculate the mean value of failure rate, other outputs are separately computed.

Table 2: Sample data and related calculations

Order	Time /hour	Estimator of beta	Difference	Mean failure rate in recent 100hrs	Order	Time /hour	Estimator of beta	Difference	Mean failure rate in recent 100hrs
1	5	null	null	null	26	816	0.94	0.06	0.02
2	27	0.41	0.59	null	27	875	0.96	0.04	0.02
3	27	0.44	0.56	null	28	885	0.96	0.04	0.03
4	37	0.45	0.55	null	29	1049	0.98	0.02	0.01
5	48	0.45	0.55	null	30	1113	0.99	0.01	0.01
6	64	0.49	0.51	null	31	1118	0.99	0.01	0.01
7	111	0.51	0.49	0.07	32	1208	0.99	0.01	0.01

8	122	0.53	0.47	0.07	33	1267	0.99	0.01	0.01
9	159	0.55	0.45	0.04	34	1299	0.99	0.01	0.01
10	201	0.55	0.45	0.04	35	1325	0.99	0.01	0.01
11	207	0.54	0.46	0.05	36	1399	0.99	0.01	0.01
12	249	0.59	0.41	0.04	37	1452	0.99	0.01	0.01
13	260	0.60	0.40	0.04	38	1701	0.99	0.01	0.01
14	297	0.59	0.41	0.05	39	1765	1.00	<0.01	0.01
15	323	0.62	0.38	0.04	40	1866	1.00	<0.01	0.01
16	350	0.66	0.34	0.04	41	2025	1.00	<0.01	0.01
17	413	0.70	0.30	0.03	42	2215	1.00	<0.01	0.01
18	440	0.75	0.25	0.03	43	2374	1.00	<0.01	0.01
19	477	0.77	0.23	0.03	44	2671	1.00	<0.01	0.01
20	488	0.80	0.20	0.04	45	2809	1.00	<0.01	0.01
21	525	0.80	0.20	0.04	46	3053	1.00	<0.01	0.01
22	551	0.82	0.18	0.04	47	3095	1.00	<0.01	0.01
23	594	0.88	0.12	0.03	48	3270	1.00	<0.01	0.01
24	631	0.90	0.10	0.03	49	3604	1.00	<0.01	0.01
25	747	0.94	0.06	0.01	50	5406	1.00	<0.01	0.01

By observing the calculated difference corresponding to the 39th data, it could be noticed that the failure rate has tend to be stable, which proved that the change point to be find out is near 1765hr. Moreover, the comparison between the estimating output and the factual could be achieved by realizing the asymptotic behavior of the difference value.

5 CONCLUSION

In this paper, a practicable method based on parameter estimation was introduced to monitor the change point in the infant period of a bathtub curve. First, the censored characteristic of data collected in burn-in testing was studied. Then, the MLE theory was used to estimate the two parameter of Weibull functions. Considering the factor of censoring, some improvements was studied to optimize the estimator. Next, the estimated shape parameter was chosen as the index for monitoring the developing trend of a curve. Through the detection of shape parameter, the change point of Weibull function can be observed, thus saving the time of burn-in tests as well as extending the useful life for customers.

The real case showed that it is accessible to find the change point. There are some remaining problems to be discussed for further study. In common issues, a bathtub curve was treated as a combination of three independent parts form three separate graph of Weibull functions. However, in real cases, one part of the bathtub such as the infant period, may not be a strictly graph of a Weibull function which has constant parameters. Moreover, Although MLEs have nice asymptotic properties, they are biased for small and censored samples. Corrections to compensate for the bias have received some attention in the literature. The algorithm of parameter estimation can be replaced to other methods and the performance comparison among them can be studied.

6 REFERENCES

[1] **He, Y. H., Wang, L. B., He, Z. Z., & Xie, M.2016.**A fuzzy TOPSIS and rough set based approach for mechanism analysis of product infant failure. *Engineering Applications of Artificial Intelligence*,47, pp 25-37.

[2] **He, Y.H., Wang, L.B., He, Z.Z., & Xiao, X. 2016.**Modelling infant failure rate of electromechanical products with multilayered quality variations from manufacturing process.*International Journal of Production Research*.pp 1-19.

[3] **He, Y.H., Wang, L.B., He, Z.Z., & Wei, Y..2015.**Product reliability oriented design scheme of control chart based on the convergent CEV for censored characteristics.*Mathematical Problems in Engineering*. Volume 2015, Article ID 128491, 11 pages.

[4] **Murthy, D. P., Xie, M., & Jiang, R.** 2004. *Weibull models*, John Wiley and Sons.

[5] **Jiang R. and Murthy.2011.**A study of Weibull shape parameter: Properties and significance." *Reliability Engineering and System Safety*, 96(12), pp 1619-1626.

[6] **Hadidoust, Z., Samimi, Y., & Shahriari, H. 2015.**Monitoring and change-point estimation for spline-modeled non-linear profiles in phase II. *Journal of Applied Statistics*, 42(12), pp 2520-2530.

[7] **Gijbels, I. and U. Gurler.2003.**Estimation of a change point in a hazard function based on censored data. *Lifetime Data Anal*, 9(4), pp 395-411.

[8] **ZambaK.D. and D.M. Hawkins.2006.**A Multivariate Change-Point Model for Statistical Process Control. *Technometrics*, 48(4), pp 539-549.

[9] **Jing, L.2008.** A Two-Stage Failure Model for Bayesian Change Point Analysis. *IEEE Transactions on Reliability*, 57(2), pp 388-393.

[10] **Azevedo, C., Leiva, V., Athayde, E., & Balakrishnan, N.2012.** Shape and change point analyses of the Birnbaum-Saunders-t hazard rate and associated estimation. *Computational Statistics and Data Analysis*, 56(12),pp 3887-3897.

[11] **NiakiS.T.A. and M. Khedmati.2012.** Estimating the change point of the parameter vector of multivariate Poisson processes monitored by a multi-attribute T^2 control chart. *The International Journal of Advanced Manufacturing Technology*, 64(9-12), pp 1625-1642.

[12] **Bae, S.J., B.M. Mun, and K.Y. Kim.2013.** Change-point detection in failure intensity: A case study with repairable artillery systems. *Computers and Industrial Engineering*, 64(1), pp 11-18.

[13] **Niaki, S.T.A. and M. Khedmati.2013.** Monotonic change-point estimation of multivariate Poisson processes using a multi-attribute control chart and MLE. *International Journal of Production Research*, 52(10), pp 2954-2982.

[14] **Kececioglu D. and Fengbin Sun ,1997.** *Burn-in testing: its quantification and optimization*. Prentice Hall.

[15] **Olteanu, D.A., 2010,** *Cumulative sum control charts for censored reliability data*. Virginia Polytechnic Institute and State University.

[16] **Duan Q. and Liu J.2016.** Modelling a bathtub-shaped failure rate by a Coxian distribution. *IEEE Transactions On Reliability*, 65(2), pp 878-885.

THE EMPIRICAL STUDY: INVENTORY FINANCING BUSINESS PLEDGE RATE UNDER FLUCTUATION

Xiaojun Liu[1,2*], Yingshan Tan[1], and Xiangyu Jin[1,2]

[1]Department of Logistics Management and Engineering
Zhuhai College of Jilin University, China
Lxj02041820@163.com

[2]School of Business
Macau University of Science and Technology, Macau, China

ABSTRACT

The paper Applied the risk assessment strategies "subject + debt" and analyzing with the fluctuation of the market to carry out the pledge of the stock. At the same time, the final price of pledge of the stock depend the general distribution and several specific distribution of the pledge rate. And the conditions show that: as long as we know final price of the pledge stock that it allows what kind of distribution. It can be obtained by the pledge rate. And in the paper, the empirical case is showed.

Keywords: Pledge rate, Inventory pledge financing, Supply chain finance

1 INTRODUCTION

Small and medium enterprises at the present stage "financing" the issue remains high, with a focus on financing high risk, high risk comes from ability to pay is not determined. Banks for lending to small and medium enterprises, there are some concerns: lower cost of default and tax, a relatively small, weak anti-risk capability and operating stability, different management levels, and so on. Therefore, based on this problem, this paper focuses on inventory financing, that is common in the financial supply chain warehouse receipt. And inventory financing operations as long as the pledge rate is obtained, we can provide the scientific basis for the decision of the parties involved.

The Static goods value financing often compared simple, so compared difficult of is in dynamic goods value financing in the of analysis, here has a index is pledge rate, can full of reflect inventory pledge financing business of risk profile, the index is loan amount and pledge inventory value of ratio, it and pledge inventory characteristics, and business mode, and regulatory way, and Enterprise default rate, and loan interest rate, index is closely related to related. However, in practice, the Bank has mostly relied on experience in valuations, which can't make the Bank accurately calculate that mortgage rates, which cannot be effective control of business risk. This

---

* Corresponding Author: Xiaojun Liu

Email:lxj02041820@163.com

Foundation: The key scientific projects of university in Guangdong—Young innovative project(2015WQNCX174) paper analyses the pledge rate of quantitative models, aimed at providing scientific basis for banking and logistics-related decisions, so great and practical value of the research.

2 THEORETICAL BACKGROUND

The theoretical basis of this paper is that the enterprise obtains loans. By using collateral material to transfer its credit risk into the pledge of price changes randomly, so it need to maintain sufficient collateral to reduce bank's losses caused by default of enterprise.

In recent years, scholars have done some studies: Bai Shizhen and other scholars analyzed the pledge of the stock collateral financing business under the fluctuation of the price at 2013[1]; Zhang Yuanyuan and other scholars analyzed the key risk control index of inventory financing under delegation and supervision at 2011[2]; Li Yi-xue, Wang Shouyang and other scholars studied decision-making of seasonal inventory financing pledge rate in the logistics financing field at 2011[3]; Buzacott and Zhang analyzed how to determine the credit limits and interest rates of bank under inventory financing at 2004[4]; What's more, some scholars were analyzed and discussed by different ways in the early years, one of which is in accordance with structured way of Merton[5]. For instance, Stulz and Johnson analyzed the impact of the pledge of collateral security debt pricing at 1985[6]; Jokivuolle and Peura focused on the pledge rate of hypothecated loan[7] and this year Cossin and Hickok adopted the structured method to determine the discount rates in the pricing of mortage-backed collateral risk tool[8]. Maihami and Karimi explained integrating option pricing and decision analysis appro].aches[9]

3 EMPIRICAL ANALYSIS

3.1 RESEARCH METHOD

3.1.1 BACKGROUND ASSUMPTION

Assumed that in stock at a fixed loan after the final sales, loans will return to the designated account loan, borrower should take the probability of (1-Q) to make up the balance. To be specific, Borrowing enterprise with q_0 for the price of stock B_0 as the pledge to the specified logistics warehousing enterprises, in order to apply for loans from bank, and we can assume that the bank uses the price "B_0" of opening as a reference price which is in the hypothecated loan. Setting pledge rate as "ω", and the amount of loan at the beginning is V_0 that equals $\omega q_0 B_0$, if:

R: loan rate
T: loan time

So we can get that when the time is T, the relationship between the loan principal and interest:

$$V_T=V_0e^{RT}=\omega q_0 B_0$$

The pledge of inventory that was entrusted logistics company by bank will produce storage costs during the period of loan, and the bank can incorporate storage costs into loan rates. Finally, changed to the borrower companies together, and assumes the bank's funding costs as deposit rates "r". Assumes the default probability of the enterprise in the loan period is "Q", and obey the uniform distribution, the value of the specific number "Q" will be made by bank after investigation and research for the borrowing enterprise to rate. The following is the model by the use of analysis.

3.1.2 GENERAL DISTRIBUTION OF THE BASIC MODEL

The price of the ending inventory was assumed as a random variable named "B_T", and obtained the price by deriving which obeys general distribution "$F(x)=P(B_T<x)(x \in (0,+\infty))$", so bank profit is:

$$\Pi\ (\omega)=\begin{cases}\omega q_0 B_0\left(e^{RT}-e^{rT}\right), B_T\geq\omega B_0 e^{RT} & \text{The probability of not default to Q}\\ q_0 B_T-\omega q_0 B_0 e^{rT}, 0<B_T<\omega B_0 e^{RT} & \\ \omega q_0 B_0\left(e^{RT}-e^{rT}\right), 0<B_T<\omega B_0 e^{RT} & \text{The probability of default to Q}\end{cases} \quad (1)$$

So there:

$E_{(\Pi\omega)}$

$$=\omega q_0 B_0\left(e^{RT}-e^{rT}\right)\left(1-F\left(\omega B_0 e^{RT}\right)\right)+Q\int_0^{\omega B_0 e^{RT}} q_0 x f(x)dx-Q\omega q_0 B_0 e^{rT}F\left(\omega B_0 e^{RT}\right)+(1-Q)\omega q_0 B_0\left(e^{RT}-e^{rT}\right)F(\omega B_0 e^{RT})$$

$$=\omega q_0 B_0\left(e^{RT}-e^{rT}\right)-Q\int_0^{\omega B_0 e^{RT}} q_0 F(x)dx \quad (2)$$

The biggest loss for the bank to bear is assumed as "L", which is defined as the bank's determined loan loss and consider the bank's risk limits. There is the relationship"P(loss>L)=QP(loss>L)≤β". The requirement of the bank for the probability is less than its risk tolerance level as "β", the final stock price volatility loan losses for the enterprise funds back when not in default bank and final sales inventory is defined as the difference between the proceeds. There: $loss=\omega q_0 B_0 e^{RT}-q_0 B_T$, so we can get the decision model of pledge rate for the bank:

$$\max E_{(\Pi(\omega))}=\omega q_0 B_0\left(e^{RT}-e^{rT}\right)-Q\int_0^{\omega B_0 e^{RT}} q_0 F(x)dx \quad (3)$$

s. t. $P(loss>L)\leq\beta$

If the loan enterprises use inventory of "q" as the pledge to loan, "B_T" is the end of the stock price of the loan that follows general distribution $F(x)$, and the bank has given the company's pledge rate is:

$$\omega^*=\begin{cases}\bar{\omega},\ \text{if}\ QF(\bar{\omega}\ B_0(e^{RT}-l))\leq\beta\\ \frac{F^{-1}}{B_0(e^{RT}-l)}\left(\beta/Q\right),\ \text{if}\ QF(\bar{\omega}\ B_0(e^{RT}-l))>\beta\end{cases} \quad (4)$$

Among them:

$$F\left(\bar{\omega}\ B_0 e^{RT}\right)=\frac{e^{RT}-e^{rT}}{Qe^{RT}} \quad (5)$$

3.2 CASE ANALYSIS

3.2.1 ENTERPRISE BACKGROUND

On December 2003, approved by the Administration for Industry and Commerce, Shanghai, A Company became China's first batch of company specialized in third party steel electronic trading market. In 2006, A Company innovatively introduced a new model of "Electronic Supermarket." On May 29th, A Company Steel Electronic Supermarket began to operate officially. On November 24th, the electronic supermarket turnover reached 3433 tons. On April 3, 2007, electronic supermarket turnover reached 5335 tons. On July 27, the total transaction volume of electronic supermarket reached the highest value, reached 8411 tons. Meanwhile, A Company began to get involved in the supply chain of financial services. On December 27, 2006, A Company signed the "Bank Enterprise Cooperation Agreements" with Hua Xia Bank. On January 8, 2011, A Company launched a cooperative of online supply chain financing project with Shenzhen Development Bank based on the electronic commerce platform. On November 2012, AMLS limited liability company recombined A Company and became the controlling shareholder of A Company. A Company began to focus on the mass production of electronic cash transactions. The predecessor of AMLS was the AMLS big market founded in 2006. AMLS big market was set up to promote development of social logistics services.

Jiangsu Shen Te steel Co., Ltd. was an enterprise mainly engaged in iron and steel. Production projects included coking, sintering, pelletizing, lime, iron and steel, steel rolling, oxygen production, power generation and coal washing, in which the Main products were carbon structural steel and alloy steel. AMLS was a special agent of Jiangsu Shen Te steel Co., Ltd. Special Steel Co., Ltd. Shanghai Shen was an distributor of Jiangsu Shen Te steel Co., Ltd. in Shanghai. The steel of Special Steel Co., Ltd. Shanghai Shen was purchased through AMLS.

In 2013, the steel market was in downturn. There was a backlog of steel inventory in AMLS. The steel of Special Steel Co., Ltd. Shanghai Shen did not have relative money to purchase the batch of steel. Shang hai Shen Te steel Co., Ltd needed money to run the company. At this moment, AMLS consulted with Minsheng Bank. Credited by the AMLS, Special Steel Co., Ltd. Shanghai Shen pledged 30 thousand tons of stock of the goods to Minsheng Bank. Minsheng Bank designated the third party storage management company responsible for the supervision of the rights.

3.2.2 The Problems in Enterprises

According to this case, the problem is a typical warehouse receipt pledge. Then in the warehouse receipt pledge, corporate finance was deeply affected by cash liquidity and price volatility of goods. In pledge loan, it is very important to guard against the price risk of this kind of pledge. Even so, the pledge rate can provide scientific basis for the problems encountered by the enterprise in this paper.

3.2.3 Solution of Pledge Rate

The price data of rebar steel is calculated which was selected from July 2011 to July 2015 from A Company official website. From July 2011 to July 2015, rebar price had changed as shown below:

Figure 1: The price trend of rebar steel from July 2011--July 2015

On the official website, steel was a sharp point as the calculating unit. It was converted into a kilogram unit for calculation. We can know that the steel price distribution function is:

$F(x)=-0.057083\,x+4.82$

From the perspective of quantitative risk management, risk tolerance should be quantified. Quantitative risk management requires banks to have a relatively high risk quantitative ability and to quantify the risk tolerance. According to the requirements of the new capital accord, 99.9% risk tolerance was established by the Basel Committee which was equal to A level in SP. China's banks can also refer to the risk tolerance coefficient. According to YM Pan in the study about the risk tolerance of commercial banks in China, we can get $\beta=0.01$[10].

According to the actual situation in the case and the relevant data: **Q=0.1,R=0.15,I=0.2, T=0.3**, we can get a pledge rate of 0.57395651 with the help of formula(4).This data is in line with the standard requirements.

4 CONCLUSION

Although inventory pledge financing business can solve the shortage of funds in the supply chain, yet there is still a lot of risk. Therefore, in the stock pledge financing business, risk control is very important. Therefore, the characteristic of this paper lies in that the price of pledge will fluctuate with the market fluctuation, which applied the risk assessment method of "subject + debt." The pledge rate is analyzed according to market fluctuations. The pledge rate can provide scientific basis for decision making of banks and other parties, which can well control the risk. Because the market volatility is uncertain, this paper does not limit the pledge rate of definition. Therefore, in the future we will pledge rates on the upper and lower limits of this study in order to better prevent the risk.

5 REFERENCES

[1] **Bai Shizhen,Xu Na. 2011.** Study of loan-to-value ratios decisions on inventory financing, *Journal of Systems Engineering*, 31(4),pp587-598.

[2] **Li Yixue,Feng Gengzhong,Zhang Yuanyuan. 2011.** Key risk control indicator of inventory pledge financing under consigning supervision, *System Engineering Theory and Practice | System*, 31(4),pp587-598.

[3] **Li Yixue,Wang Shouyang,Feng Gengzhong. 2011.** Decision of loon-to-calue katios of seasonal inventon/pledge financing based on logistics finance, *Journal of Management Scince*,14(11).

[4] **Buzacott J A, Zhang R Q. 2004.** Inventory management with asset-based financing, *Management Scince*, 24,pp1274-1292.

[5] **Merton R. 1974.** On the pricing of corporate debt:The risk structure of interest rates, *The Journal of Finance*, 49,pp449-470.

[6] **StulzR,JohnsonH. 1985.** An analysis of secured debt, *Journal of Financial Economics*, 14,pp501-521.

[7] **JokivuolleE, PeuraS. 2003.** Incorporating collateral value uncertainty in loss given default estimates and loan-to-value ratios, *European Financial Management*, 9(3),pp299-314.

[8] **CossinD,HrickoT. 2003.** Astructural analysis of credit riskwith risky collateral:Amethodology.forhaircutdetermination, *Ecomomic Notes*, 32(2),pp243-282.

[9] **Maihami, Karimi, B. 2016.** Effect of two-echelon trade credit on pricing-inventory policy of non-instantaneous deteriorating products with probabilistic demand and deterioration functions, *Annals of Operations Research*, pp1-37.

[10] **YM Pan, LI Zi-Wei. 2015.** Research on the Risk of Inventory Financing of Logistics Enterprise under the Unified Credit Mode,*Value Engineering*, 61(1).

A SWITCHING PROBLEM FOR SMART FACTORY CONSIDERED PRODUCTION, DUE DATE AND QUALITY

Jing Sun[1*], Hisashi Yamamoto[2] and Masayuki Matsui[3]
[1]Department of Civil Engineering and Systems Management, Nagoya Institute of Technology, Gokiso-cho, Showa-ku, Nagoya, 466-8555, Japan
sun.jing@nitech.ac.jp

[2]Department of System Design, Management Systems Engineering, Tokyo Metropolitan University, 6-6 Asahigaoka, Hino Tokyo, 191-0065, Japan
yamamoto@tmu.ac.jp

[3]Faculty of Engineering, Kanagawa University, Yokohama, Japan
matsui@kanagawa-u.ac.jp

ABSTRACT

- This paper aims to derive an optimal switch model considered production, due date and quality for parallel production system with multiple periods in smart factory environment. In the achievement of Smart Supply Chain, optimal operation management for horizontal integration of production network has been paid to attention recently. Due to the customer needs of reducing cost and delivery date shorting, prompt change in the production plan became more important. In the multi period system (For instance, production line.) where target processing time exists, production, idle and delay risks occur repeatedly for multiple periods. In such situations, delay of one process may influence the delivery date of an entire process. This kind of problem is called "a limited-cycle problem with multiple periods", and is seen in production lines, time-bucket balancing, and production seat systems and so on. In this paper, we discuss minimum expected cost including production, due date and quality in a parallel production process, where the risk depends on the previous situation and occurs repeatedly throughout multiple periods. Also, the policy of optimal switching for parallel production system will be analyzed.

Keywords: limited-cycle problem, optimal switching problem, production networks, manufacturing line, smart factory.

1 INTRODUCTION

This paper aims to derive an optimal switch model considered production, due date and quality for parallel production system in smart factory environment. To achieve the smart factory at which IoT (Internet of Things) aim, the technical improvement in various fields is requested [1]. The horizontal integration of cooperation by information systems of enterprises or factories that are related to make a product is an important problem in Supply Chain Management.

* Corresponding Author

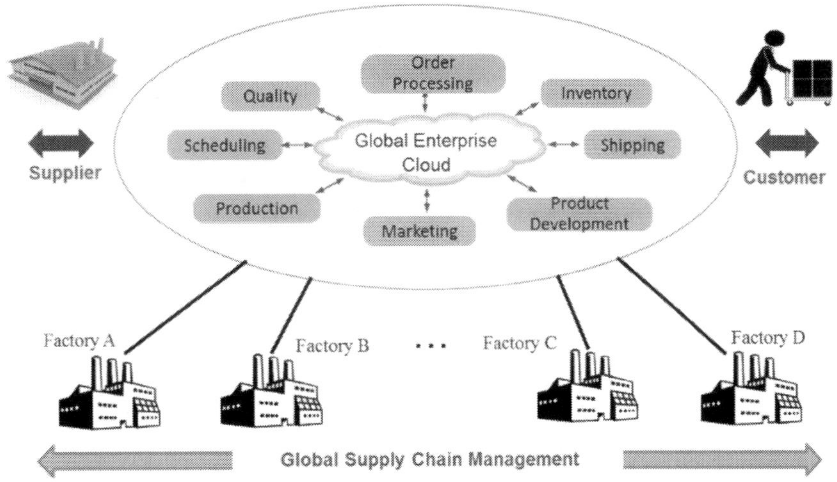

Figure 1: Horizontal Integration Example for Smart Factory

On the other hand, variable production networks is not only one of the most important issues in manufacture management and operation research but also a significant factor affecting supply chain (SC). The supply collaborate concept requires formation and optimization of production or supply networks, characterized by intensive communication between the distributed entities. The goal is to allocate among the collaborating partners the production demand. This capability provides the entire network with the required flexibility to respond quickly to demand disruption for the products and services [2]. This paper analyzes an optimal switching model for parallel production system with multiple periods in SCS.

On the other hand, variable production networks is not only one of the most important issues in manufacture management and operation research but also a significant factor affecting supply chain (SC). The supply collaborate concept requires formation and optimization of production or supply networks, characterized by intensive communication between the distributed entities. The goal is to allocate among the collaborating partners the production demand. This capability provides the entire network with the required flexibility to respond quickly to demand disruption for the products and services [2]. This paper analyzes an optimal switching model for parallel production system with multiple periods in SCS.

In any social system or production process with multi periods and predetermined target time, system idleness and process delay risks exists throughout the multiple periods. In such situations, delay of one process sometimes affect the delivery date of the entire process. This kind of problem is called "a limited-cycle problem with multiple periods", and is seen in production lines, time-bucket balancing, production seat systems and so on [3] [4] In this paper, we discuss minimum expected risk (cost) in a parallel production process, where the risk depends on the previous situation and occurs repeatedly throughout multiple periods in smart supply chain environment.

As limited-cycle problems of production line [5], analyzed the element and construction of the production system. [6] presented a framework for the analysis of delays within the production system. [7] gave a review for the origin and solution of period batch control system. [8] presented an easily implemented hybrid algorithm for the multi-objective flexible job-shop problem. Recently [9], concerned with the problem in scheduling a set of jobs associated with random lead-time on a single machine so as to minimize the expected maximum lateness in stochastic environment.

As limited-cycle problems of operation management, [10] proposed an integrated mathematical model of the multi-period cell formation and production planning in a dynamic cellular manufacturing system to minimize costs through a mixed integer programming technique. [11] proposed a mixed integer linear programming based capacitated lot sizing models that included carryovers incorporating set-up times with associated costs. [12] employed an optimal multi-period production plan for a single product over a finite planning horizon to minimize the total inventory and backlog costs by solving a nonlinear programming problem. [13] described an optimal solution structure by the dynamic programming approach for a joint manufacturing and remanufacturing system in a multi-period horizon.

Under uncertain conditions, the result and efficiency of a certain production cycle period and a certain process are influenced not only by the risks that exist in the current period but also by the risks that existed in the foregoing periods. Therefore, we discuss the minimum expected risk of the case mentioned above, in which the risk depends on the foregoing situation and occurs repeatedly for multiple periods. Whether the process (period or site) satisfies the time limit (restriction) usually depends on the state of the foregoing process, as seen in [14], [15], [16], [17].

In this paper, first, the optimal switching problem is systematically classified in smart factory and an optimal switching model for parallel production system is presented. Next, the mathematic formulation of the total expectation considered production, due date and quality for the parallel production system is proposed. Finally, the optimal switching point is investigated by numerical experiments.

2 OPTIMAL SWITCHING PROBLEM FOR TANDEMLY PRODUCTION SYSTEM CONSITERED PRODUCTION, DUE DATE AND QUALITY

This paper considers cases in which the above two risks not only occur in the single period, but also in multiple periods repeatedly. The problem of minimizing the expected risk in such a situation is a limited-cycle problem with multiple periods. The multi period problem could be classified according to whether the periods are independent or not.

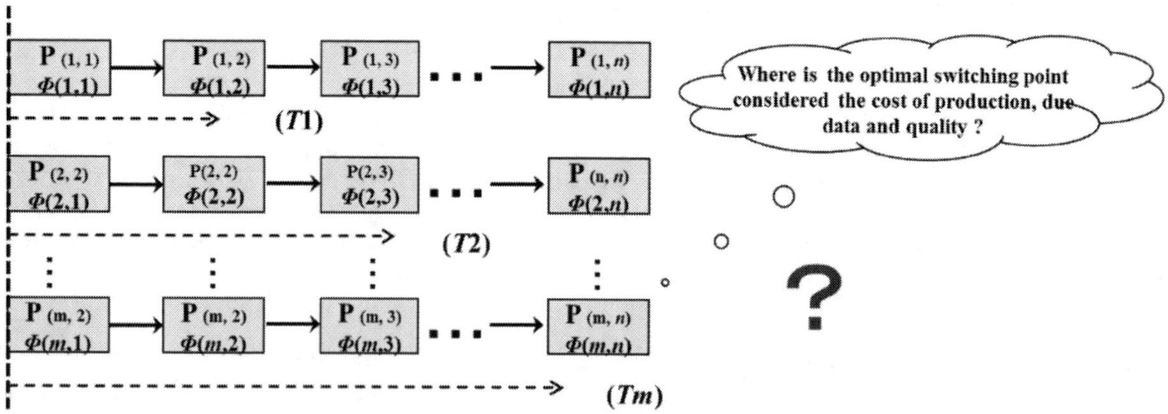

Figure 2: Optimal Switching Problem for Parallel Production System with Multiple Periods

For this problem, one result is the general form of production rate and waiting time by a station-centered approach as discussed in [4]. The explicit form is obvious and consists of the product form in the period-independent case, such as a single line, but it is untraceable in the period-dependent case such as a mixed or tandem line. The mixed line has a absorbing barrier, but the tandem line has a reflective barrier at the end. This paper presents a cost approach for the latter.

The purpose of this paper is discussing the switching problem for parallel production system shown in Figure 2.

3 OPTIMAL SWITCHING MODEL FORPARALLEL PRODUCTION SYSTEM

3.1 The assumption and notation

The Optimal switching model for the parallel production system with multiple periods is considered based on the following assumptions:

(1) One product is made by a process with *n* processes.
(2) For $i=1, 2, \ldots, n; j=1, 2, \ldots, m$, the production time of process i of line j is denoted by T_{ji} which is assumed to be statistically independent, respectively. The usual processing rate is μ_{j1}, and the emergency processing rate is μ_{j2}.
(3) For $i=1, 2, \ldots, n$, the target production time of process i is denoted by iT, and the due time of the entire process (n periods) is nT.
(4) k is switching point.
(5) The cost per unit time ($C_s^{(h)}$) occurs when a process is executed before the target production time of the process (h=1 means before switching and h=2 means after switching).
(6) The cost per unit time ($C_p^{(h)}$) occurs when a process is executed after the target production time of the process (h=1 means before switching and h=2 means after switching).
(7) When $X_i = \sum_{l=1}^{i} T_l$ > due time (U_n) of n periods, the delay cost C_p occurs.
(8) When $X_i = \sum_{l=1}^{i} T_l$ < due time (U_n) of n periods, the idle cost C_s occurs.

Some notations are also defined.

For $i=1, 2, \ldots, n$,

$C(T_1, T_2, \ldots, T_n)$: the total cost of the production process.

$C(i)$: the production cost of period i.

T_i: the production time of period i.

X_i: the production time of i periods ($X_i = \sum_{l=1}^{i} T_l$).

$\Pr(X_n > U_n)$: the probability of delay.

$\Pr(X_n \le U_n)$: the probability of idle.

β: Probability of quality problem occurring in production process ($\Pr(\sum_{i=1}^{n} T_i > I)$)

3.2 The assumption and notation

The objective of proposed model is shown as following:

$$f(k^*;Z_1,Z_2,...Z_m,)=\min_k(k;\sum_{j=1}^{m}E[Z_m]) \quad (1)$$

Where,

$$E[Z_j]=\min_k(E[C_1(k;T_1,T_2,...T_n)+C_2(k;T_1,T_2,...T_n)+C_3(k;T_1,T_2,...T_n)]) \quad (2)$$

Z_j is the total expected cost of line *j*, $C_1(k;T_1,T_2,...T_n)$ is the production cost, $C_2(k;T_1,T_2,...T_n)$ is the due date penalty cost and $C_3(k;T_1,T_2,...T_n)$ *is* the quality penalty cost.

(i) Production Cost

From assumptions (1)-(7) mentioned in Section3.1, we can easily see,

$$E[C_1\ (k;T_1,T_2,...,T_n)]=\sum_{i=1}^{n}E[C(i)] \quad (3)$$

Where, $E[C(i)]$ is the expected cost of period *i*.

In this research, the production time *Ti* is assumed to be exponential distributed and statistically independent, respectively.

Then, for $i=1,2,...,k$,

$$\begin{aligned}E[C(i)]&=E[(C_p^{(1)}-C_s^{(1)})\cdot S_i+C_s^{(1)}\cdot T_i]\\&=(C_p^{(1)}-C_s^{(1)})E[S_i]+C_s^{(1)}E[T_i]\\&=(C_p^{(1)}-C_s^{(1)})E[S_i]+\frac{C_s^{(1)}}{\mu_1}\end{aligned} \quad (4)$$

and for $i=k+1,k+2,...,n$,

$$\begin{aligned}E[C(i)]&=(C_p^{(1)}-C_s^{(1)})E[S_i\mid\sum_{i=1}^{k}T_i\le U_k]\cdot\Pr\{\sum_{i=1}^{k}T_i\le U_k\}\\&+(C_p^{(2)}-C_s^{(2)})E[S_i\mid\sum_{i=1}^{k}T_i>U_k]\cdot\Pr\{\sum_{i=1}^{k}T_i>U_k\}\\&+\frac{C_s^{(1)}}{\mu_1}\cdot\Pr\{\sum_{i=1}^{k}T_i\le U_k\}+\frac{C_s^{(2)}}{\mu_2}\cdot\Pr\{\sum_{i=1}^{k}T_i>U_k\}\end{aligned} \quad (5)$$

(ii) Due Date Penalty Cost

$$E[C_2\ (k;T_1,T_2,...,T_n)]=C_p\Pr(X_n>U_n)+C_s\Pr(X_n\le U_n) \quad (6)$$

where, $C_p\Pr(X_n>U_n)$ is the delayed expected cost, $C_s\Pr(X_n\le U_n)$ is the idle expected cost.

$$\Pr\{\sum_{i=1}^{k}T_i>U_k\}=\sum_{l=0}^{k-1}\frac{(\mu_1U_k)^l}{l!}e^{-\mu_1U_k} \quad (7)$$

$$\begin{aligned}\Pr\{\sum_{i=1}^{k}T_i\le U_k\}&=1-\Pr\{\sum_{i=1}^{k}T_i>U_k\}=1-\sum_{l=0}^{k-1}\frac{(\mu_1U_k)^l}{l!}e^{-\mu_1U_k}\\&=\sum_{l=k}^{\infty}\frac{(\mu_1U_k)^l}{l!}e^{-\mu_1U_k}\end{aligned} \quad (8)$$

(ii) Quality Penalty Cost

For $i = 1,2,\dots,k$,

$$E[C_3\ (k;T_1,T_2,\dots,T_n)] = C_{01}\ (E[\max(U_k - I, 0)])\Pr\{\sum_{i=1}^{n} T_i > U_k\} \tag{9}$$

and for $i = k+1, k+2, \dots, n$,

$$E[C_3\ (k;T_1,T_2,\dots,T_n)] = C_{02}\ (E[\max(\sum_{i=1}^{n} T_i - I, 0)])\Pr\{\sum_{i=1}^{n} T_i \le U_k\} \tag{10}$$

where, C_{01} and C_{02} are the quality cost per unit time.

In this research, the in-control time of a production process (I) is assumed to be exponential distributed which the mean value is $1/\lambda$.

4 EXPERIMENTAL CONSIDERATION

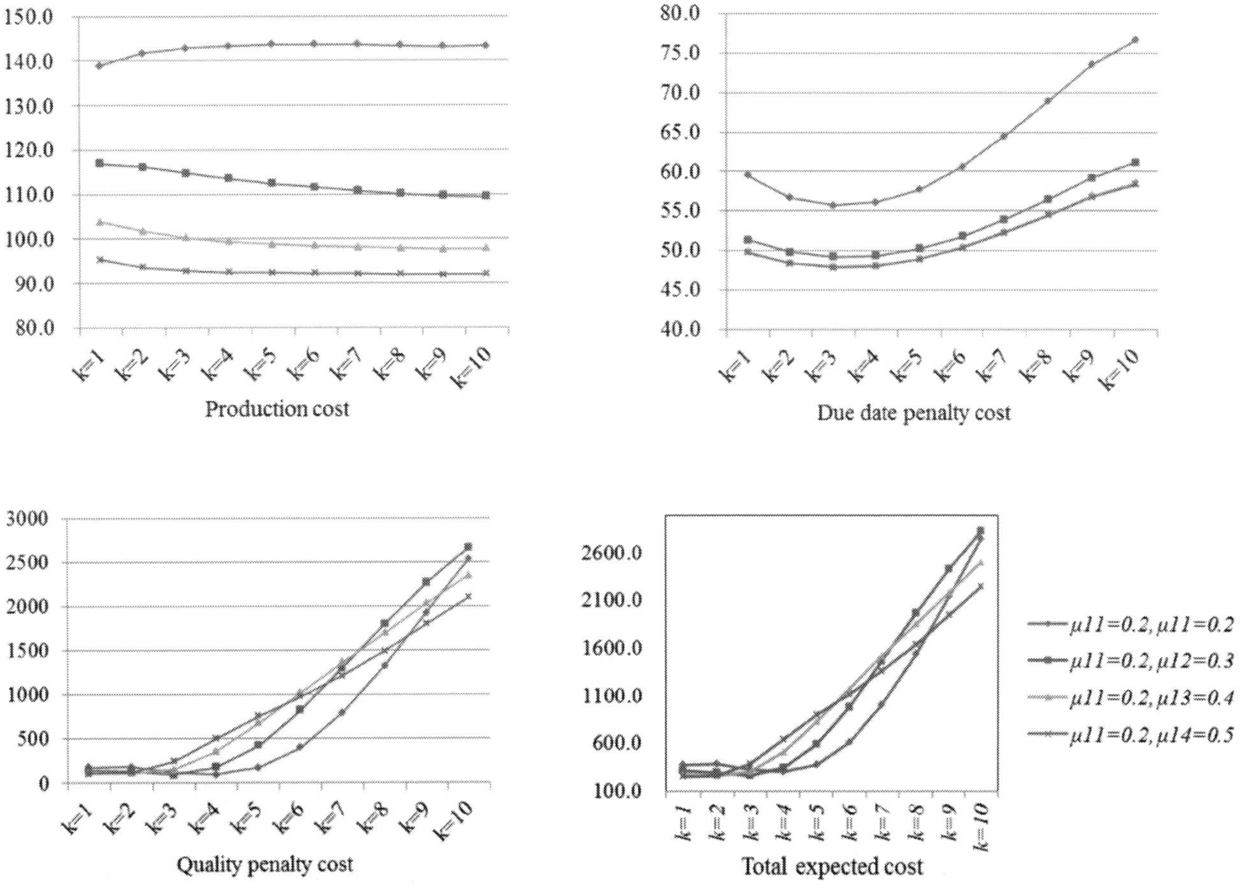

Figure 4: Behavior of the optimal switching time by change of the usual processing rate

In this section, we consider the optimal switching time to minimize the total expected cost by numerical experiments, where $C_s^{(1)}=1$, $C_p^{(1)}=2$, $C_s^{(2)}=3$, $C_p^{(2)}=6$, $C_p=60$, $C_s=20$, $C_{01}=C_{02}=100$, $\beta=0.5$, $\lambda=0.3$, $T=5$, $m=2$ and $n=10$.

Figure 4 show the behavior of the optimal switching time by change of the usual processing rate when emergency processing rates of system is 0.6. From Figure 3, it can be noted that when usual processing rates of line 1 and line 2 are (0.2, 0.2), (0.2, 0.3), (0.2, 0.4) and (0.2, 0.5), the optimal switching times are 4T, 3T, 2T and 1T, respectively.

From Figure 4, it also can be noted that when usual processing rates of line 1 is constant, the optimal switching time parameter *k* decreases with the increase of the usual processing speed of line 2. This is because the quality penalty cost is large in this case, so the The behavior of total expected cost follows the behavior of quality penalty cost.

5 CONCLUSIONS

In this paper, we considered the optimal switching point (time) to minimize the total expected cost considered production, due date and quality in parallel production system with multiple periods. First, we systematically explained the multi period problem and optimal switching problem for smart factory. Next, we proposed an optimal switching time model and showed the corresponding mathematical formulations. Finally, by investigating behaviours of the optimal switching time, the optimal switching point could be found.

6 REFERANCES

[1] **Leonard, H., Eduardo, L. and Stefan, V. 2016.** A cloud brokerage approach for solving the resource management problem in multi-cloud environments. *Computers & Industrial Engineering*, 95, pp 16-26.

[2] **Shimon, Y. Nof·Jose, Ceroni Wootae Jeong·Mohsen Moghaddam 2015.** Revolutionizing Collaboration through e-Work, e-Business, and e-Service. Springer.

[3] **Yamamoto, H., Matsui, M. and Liu, J. 2006.** A basic study on a limited-cycle problem with multi periods and the optimal assignment problem, *Journal of Japan Industrial Management Association*, 57, pp 23-31(in Japanese).

[4] **Matsui, M. 2008.** Manufacturing and Service Enterprise with Risk: A Stochastic Management Approach. Springer.

[5] **Verzijl, J. J. 1976.** Production Planning and Information Systems, Macmillan, London, 16-17.

[6] **S.T.Enns, 2001.** MRP performance effects due to lot size and planned lead time settings, *International Journal of Production Research*, 39, pp 461-480.

[7] **Benders, J. 2002.** The origin of period batch control (PBC), *Int. J. Prod. Res*, 40, pp 1-6.

[8] **W. J. Xia and Z. M. Wu. 2005.** Hybrid particle swarm optimization approach for multiobjective flexible job-shop scheduling problems, *Control and Decision*, 20(2), pp. 137-141.

[9] **X. Wu and X. Zhou. 2008.** Stochastic scheduling to minimize the expected maximum lateness, *European Journal of Operational Research*, 190, pp. 103-115.

[10] **Safaei, N., Tavakkoli-Moghaddam, R., 2009.** Integrated Multi-Period Cell Formation and Subcontracting Production Planning in Dynamic Cellular manufacturing Systems, *International Journal of Production Economics*, 120(2), pp 301-314.

[11] **Porkka, P.; Vepsalainen, A.P.J., Kuula, M., 2003.** Multi-period Production Planning Carrying Over Set-up Time, *International Journal of Production Research*, 41(6), pp 1133-1148.

[12] **El Hafsi, M.; Bai, S.X., 1998.** Multi-period Production Planning with Demand and Cost Fluctuations, *Journal of Mathematical and Computer Modelling*, 28(3), pp 103-119.

[13] **Li, Y., Zhang, J., Chen, J. and Cai, X. 2010.** Optimal Solution Structure for Multi-period Production Planning with Returned Products Remanufacturing, *Asia-Pacific Journal of Operational Research (APJOR)*, 27(5), pp 629-648.

[14] **Verzijl, J. J. 1976.** Production Planning and Information Systems. Macmillan, London, 16-17.

[15] **Benders, J. 2002.** The origin of period batch control (PBC), *Int. J. Prod. Res*, 40, pp 1-6.

[16] **Sun, J., Yamamoto, H. and Matsui, M. 2010.** A study of optimal switching problem in limited-cycle with multiple periods, *Journal of Zhejiang University-SCIENCE A (Applied Physics & Engineering)*, 11, pp 933-937.

[17] **Sun, J., Kong, X., Yamamoto, H., and Matsui, M. 2014.** Optimal switching polity on Limited-Cycle Problem with Multiple Periods, *J.SOPE Japan*, 95(1), pp 1-6.

A MIXED INTEGER LINEAR PROGRAMMING APPROACH FOR CLOSED-LOOP FOOD SUPPLY CHAIN WITH RETURNABLE TRANSPORT ITEMS

Yipei ZHANG[12*], Feng CHU[2], Ada CHE[1]
[1]School of management
Northwestern Polytechnical University, Xi'an, China
peipei19890114@gmail.com
ache@nwpu.edu.cn

[2]Laboratory IBISC
University of Evry-Val d'Essonne, Evry, France
feng.chu@ibisc.univ-evry.fr

ABSTRACT

This paper studies a multi-period, single-manufacturer and single-retailer closed-loop food supply chain with returnable transport items (RTIs) which are used for shipping perishable products. The quality of the products is distinguished according to its age over time. After commodities delivered to the retailer, the emptied returnable transport items are returned to the manufacturer after certain disposes and will be reused in the subsequent shipments. During this process, the manufacturer has to decide when and how many new RTIs to be purchased so as to avoid their shortage and deterioration of products. The objective is to maximize the total expected profit of the supply chain. To cope with the problem, a mixed integer linear programming model is developed and commercial software CPLEX12.6 is applied to solve it. Computational results based on randomly generated instances illustrate the validity and effectiveness of the proposed model.

Keywords: Closed-loop supply chain (CLSC), Perishable products, Returnable transport items, Mixed integer linear programming (MILP)

1 INTRODUCTION

Returnable transport items (RTIs), such as pallets, containers, crates or boxes, refer to a special kind of reusable packaging materials that can be re-used for shipping materials or finished products [1][2]. Using high quality RTIs in supply chain can not only lower the negative impact on environment incurred by one-way packing materials, but also improve the protection of products during the transport, storage, handling process and reduce the cost of products in the long term [1]. A recent study by Goellner and Sparrow has shown that compared to the one-way containers, reusable containers generate 75% lower CO_2 emissions over their lifecycle [3]. Besides, many companies have found that using RTIs can significantly reduce costs [4]. In recent years, the RTIs have been frequently studied and introduced in various industries for these advantages. However, in comparison with various researches on returnable finished products in CLSC domain, e.g., [5][6], the studies related to the optimization of RTIs in a closed loop supply chain, especially the coordination of finished products and RTIs, are relatively limited [7].

* Corresponding Author. Tel.:33 0647697352.

E-mail addresses: peipei19890114@gmail.com (Yipei ZHANG)

The existing RTI researches focus on two categories. One is about the management of RTIs. Hellström and Johansson [1] studied the influence of different control strategies, such as using a switch pool, on the management of RTIs in CLSCs. The authors developed a simulation model based on an empirical case. The simulation analysis indicated that the choice of control strategy can significantly influence the investments for RTIs and their operating costs. Bottani et al. [7] investigated a real CLSC making up of a pallet supplier, a manufacture and 7 retailers to optimize the management process of pallets. They developed a mathematical model and used Microsoft Excel simulator to help the company with palette reorder decision. Another one is about RTI inventory policy in CLSC. Cobb [8] proposed an inventory model for RTIs aiming at optimizing inspection, repair, and purchase closed loop cycle in uncertain RTI return environment. Glock and Kim [9] studied a single vendor and multiple retailers CLSC with RTIs. They considered two production-shippement strategies with different lot sizes and the corresponding lot size models are developed. Glock and Kim [10] studied CLSC under stochastic RTI return times. They proposed a joint economic lot size model to show that safety return times and RTI safety stock policies can avoid RTI Stockout. Hariga et al. [11] investigated an integration of product and container inventory problem for a single-vendor single-buyer supply chain with both owned and rented RTIs.The problem was formulated into a mixed-integer non-linear programming model to minimize the costs of the supply chain. Above researches address a general CLSC, the developped models and methods have not considered perishable food characteristic and cannot be used directly to the studied problem in the paper.

Dairy products such as fruits and vegetables are perishable, whose quality will deteriorate over time [12]. It's recognized that the quality issue of food product should be preserved in transportation and inventory process and RTIs return can significantly impact the food CLSC performance. Farahani et al. [13] studied an integrated production and distribution system for perishable food aiming to improve its quality by cutting down the delay between production and distribution. Wang and Li [14] developed a pricing model based on dynaminc quality evaluation by applying tracking and tracing technologies in the perishable food supply chain. Unfortunately, they did not take into consideration the RTIs that is often used in food supply chain and can affect the supply chain performance. To our best knowledge, the only work reltating to the food CLSC with RTIs in litterature is Kim et al.'work [12]. They investigated the influence of the RTI's size in a single supplier and single buyer CLSC for deteriorating products under stochastic container return times. A joint economic lot size model was developed for this problem and a simulation was conducted. However, this paper assumed the demands of the buyer are constant and the finished products will deteriorate when the shortage of the RTIs occurs. The last assumption means only good quality and deteriorating state of products were distinguished in the paper. But in reality, a food product can have several qualities during its lifetime and can be sold with different prices.

We investigate an intergarated production-distribution-transport CLSC for perishable food with RTIs return. The final objective is to improved the food CLSC perfomance by efficiently decision aid. This paper is our first work in which we consider a multi-period, single-manufacturer, single-retailer's perishable food closed loop supply chain with returnable containers. The problem is to determine the poduction and shippment quanitity in each period, and the number of new containers need to be purchased to avoid disrupting the supply chain activities to maximize the total profit of the supply chain. Then the complexity of the model is provided. Finally, randomly generated instances are used to simulate the system.

The rest of this paper is organized as follows: we describe the problem in detail and propose the mixed integer linear programming model in Section 2. Computational experiments are conducted and analysed in Section 3. Finally, Section 4 concludes the paper and sheds light on the future avenue.

2 PROBLEM DESCRIPTION AND THE MILP MODEL

2.1 Problem description

This article studies a multi-period, single-manufacturer, single-retailer closed loop supply chain for returnable containers which are used for shipping deteriorating food products. The manufacture produces products whose qualities (called age hereinafter) belong to a discrete set and deteriorates by one unit in each period. After a shipment arrives at the retailer, the containers are emptied and returned after being cleaned and repaired (if necessary) to the manufacturer for reusing. Before the issued containers return, a shortage may occur in the container inventory at the manufacturer side, which leads to be unable to fulfill the following delivery. In this case, the manufacturer has to decide when and how many new containers to be purchased to satisfy the demand at the retailer. The objective is to maximize the total expected profit of the holistic CLSC instead of the individual members.

2.2 The MILP Model

2.2.1 Assumptions

To formulate the problem, we have the following assumptions:

- The horizon we consider spans T periods.
- The supplier has a limited and deterministic production capacity and set up cost in each period, whereas an unlimited inventory capacity.
- The retailer has a limited inventory capacity and the demand at the retailer in each period is deterministic.
- The containers are initially owned by the manufacturer at the beginning of the study.
- The initial inventory of the products at both the manufacturer and the retailer is 0 without loss of generality.
- Only one type of RTI is considered.
- No backorders are permitted.
- All entities of the closed loop supply chain are considered as an integral system having a cooperated interest.
- The demand of the customer can be satisfied by various aged products with corresponding prices.
- New containers are purchased from nearby container provider and the lead time is neglected.

2.2.2 Notations

To formulate the model, we have the following notations:

- **Parameters:**

T: Set of periods, indexed by t, t=1, 2,...,T

G: Set of product quality, indexed by g, g=0, 1, 2,...,G, where 0 represents products newly produced and after G periods they become spoiled.

C^p_t: Production capacity in period t

C^r: Inventory capacity at the retailer

C^{Truck}: Truck capacity in unit of containers

C^c: Container capacity in unit of products

I^c_0: Initial container inventory at the manufacturer, we set I^c_0= 0 to better explain the purchase decision for new containers

sc_t: Production set up cost at the manufacture in period t

CC^m: Unit storage cost for container at the manufacturer

CP^m: Unit storage cost for product at the manufacturer

CP^r: Unit storage cost for product at the retailer

C^{pur}: Unit purchasing cost for new container

PC_t: Unit production cost in period t

TC^f: Fixed transportation cost per truck

TC^{lc}: Transportation cost per loaded container

TC^{ec}: Transportation cost per empty container

L_t: Percentage of containers never returned (lost or irreparable) in period t with $0 \leq L_t \leq 1$

s_g: Selling revenue of product with quality $g \in G$ at the retailer

d_t: Demand at the retailer in period t.

k: Return time for empty containers to the manufacture

- **Decision variables:**

p_t: Production quantity in period t

q_{gt}: Quantity of products of quality g delivered to the retailer in period t

d_{gt}: Quantity of demand for products of quality g at the retailer in period t

y_t: Number of new containers purchased in period t

CI^m_t: Container inventory level at the manufacturer at the end of period t with $CI^m_0 = I^c_0$

CI^r_t: Container number at the retailer at the end of period t

PI^m_{gt}: Product inventory level of quality g at the manufacturer at the end of period t with PI^m_{g0}=0

PI^r_{gt}: Product inventory level of quality g at the retailer at the end of period t with PI^r_{g0}=0

n^c_t: Number of containers used for shipping products to the retailer in period t, note that $n^c_{t-k} = 0, \ \forall t \leq k \in T$

n^{Truck}_t: Number of trucks used in period t

$$w_t = \begin{cases} 1, & \text{if production is needed in period } t \\ 0, & \text{otherwise} \end{cases}$$

2.2.3 *Model formulation*

As already mentioned, the objective of this study is to maximize the total profit of the holistic CLSC, which equals to the selling revenue minus the cost incurred by the production, inventory, purchasing, and transportation activities. With the above assumptions and notations, the MILP model can be formulated as follows:

$$Max\ z = \sum_{g\in G}\sum_{t\in T} d_{gt}\cdot s_g - \left[\begin{array}{l} \sum_{t\in T}(p_t\cdot PC_t + sc_t\cdot w_t) + \sum_{t\in T} y_t\cdot C_{pur} \\ +\sum_{t\in T}[2\cdot TC^f\cdot n^{Truck}_t + TC^{lc}\cdot n^c_t + TC^{ec}\cdot(1-L_t)\cdot n^c_{t-k}] \\ +\sum_{g\in G}\sum_{t\in T}(CP^m\cdot PI^m_{gt} + CP^r\cdot PI^r_{gt}) + \sum_{t\in T} CC^m\cdot CI^m_t \end{array}\right] \quad (1)$$

Subject to:

$$p_t \le C^P{}_t \cdot w_t, \forall t \in T \quad (2)$$

$$CI^m{}_t = CI^m{}_{t-1} + (1-L_t)\cdot n^c{}_{t-k} - n^c{}_t + y_t, \forall t \in T \quad (3)$$

$$PI^m{}_{0t} = p_t - q_{0t}, \ \forall t \in T \quad (4)$$

$$PI^m{}_{gt} = PI^m{}_{g-1,t-1} - q_{gt}, \forall t \in T, q \in G \setminus \{0\} \quad (5)$$

$$CI^r{}_t = CI^r{}_{t-1} + n^c{}_t - (1-L_t)\cdot n^c{}_{t-k}, \forall t \in T \quad (6)$$

$$PI^r{}_{0t} = q_{0t} - d_{0t}, \forall t \in T \quad (7)$$

$$PI^r{}_{gt} = PI^r{}_{g-1,t-1} + q_{gt} - d_{gt}, \forall t \in T, q \in G \setminus \{0\} \quad (8)$$

$$\sum_{g \in G} PI^r{}_{gt} \le C^r, \forall t \in T \quad (9)$$

$$\sum_{g \in G} d_{gt} = d_t, \forall t \in T \quad (10)$$

$$\sum_{g \in G} q_{gt} \le C^c \cdot n^c{}_t, t \in T \quad (11)$$

$$n^c{}_t \le C^{Truck} \cdot n^{Truck}{}_t, t \in T \quad (12)$$

$$w_t \in \{0,1\}, \forall t \in T \quad (13)$$

$$p_t, q_{gt}, d_{gt}, y_t \ge 0, \forall t \in T, g \in G \quad (14)$$

$$n^c{}_t, n^{Truck}{}_t \ge 0 \text{ and integer}, \forall t \in T \quad (15)$$

The objective function (1) aims to maximize the total profit of the CLSC. Constraint (2) is used to restrict the production capacity. Constraint (3) represents the container flow balance at the manufacture side, which ensures that the storage of container in period ***t*** is equal to the storage in period ***t***-1 plus the containers returned from period ***t-k*** and containers purchased in period ***t*** and then minus the container used in period ***t***. Constraints (4) and (5) are product storage balance constraints at the manufacture side. Constraint (6) is nearly the same as (3) but for the retailer side. Constraints (7) and (8) correspond to the product storage balance constraints at the retailer side. Constraint (9) guarantees that the summation of products with different ages cannot exceed the storage capacity at the retailer side in each period. Constraint (10) declares that in each period the demand at the retailer side can be fulfilled by products with different ages. Constraints (11) and (12) restrain the number of containers and trucks used in each period, respectively. Constraints (13)-(15) are variable constraints.

Note that, if we assume that the container inventory at the manufacture side is large enough, the proposed model will be transferred to the capacitated single item lot sizing problem, which has already proven to be NP-hard problem. Correspondingly, the problem we study is also NP-hard.

3 COMPUTATIONAL EXPERIMENTS

3.1 Parameter settings

We code the model in C++ language and run it using IBM Concert Technology CPLEX (version 12.6.0) with randomly generated instances. All the experiments are tested on a PC with CPU 2.40 GHz and 12.0 GB RAM. The parameters are generated as follows:

Note that U (a, b) and V (c, d) are uniform distribution functions used for generating integer numbers between parameter a and b, and real number between parameter c and d, respectively. The demand is randomly generated by d_t= U (300,500), the production capacity in each period is C^P_t= α_t*max { d_t }, where α_t is randomly generated by α_t = V (2,3). The unit production cost in each period is randomly generated by PC_t =V (10, 12). The set up cost in each period is randomly generated by sc_t= U (1000, 3000). The lost and damaged percentage in each period is randomly generated by L_t=V (0.05, 0.2).The Selling revenue of product with quality $g \in G$ at the retailer s_g= β-(β-λ)*g/G, where β and λ are randomly generated by β= U(20, 30) and λ=U(8, 12), respectively. The inventory capacity at the retailer C^r=γ* max { d_t }, where γ is randomly generated by γ= V(2, 3). Note that the storage capacity at the retailer is only related to finished products for the reason that in this study we assume the containers and the finished products are stored separately. When the containers loaded with finished products arrive at the retailer side, they are emptied and cleaned and repaired (if necessary), which we only consider the time but not the storage at the retailer side. The capacity of containers is randomly generated by C^c= U (20, 40). The capacity of each truck is randomly generated by C^{Truck}= U (20, 50). The unit storage cost for containers at the manufacture side is randomly generated by CC^m= U (3, 7). The unit storage cost for finished product at the manufacture and retailer side are randomly generated by CP^m= V (0.5,1.5) and CP^r= V (1.5,2.5), respectively. The purchase cost for new container is randomly generated by C^{pur}= U (20, 30). The fixed transportation cost for a truck is randomly generated by TC^f = U (10, 30). The transportation costs for an empty container is randomly generated by TC^{ec}= U (4, 8). The transportation cost for a single container loaded with finished products are randomly generated by TC^{lc}= U (TC^{lc}, TC^{lc}+10). We consider two kinds of scheme to generate the return times to investigate the impact of return times on the CLSC. The first one namely S1 is randomly generated by k= U (1, T/2) and the second one namely S2 is randomly generated by k= U (1, G). In addition, we set the number of periods to be 10, 20, 30, 40, 50, 60 and the corresponding product age 5, 5, 10, 10, 15, 15, respectively.

3.2 Computational results and analysis

For each combination of parameters T, G and two schemes of k generation, five instances are generated and partial computational results are present in what follows.

From Table 1, we can see that, CPLEX is of high efficiency in solving the proposed model. When the problem scale increases up to 60 periods and 15 ages, it also can be solved within 1 second. Moreover, it is clear that when the return time of empty container is shorter using the second generation scheme, the average times of production set up are unaffected because that no backlogging is allowed and once the containers are not enough, the manufacturer will purchase new ones to guarantee the transportation of the finished products. However, the average times of new container purchasing decline because of the quicker return of the containers. In this case, the total profit of the supply chain increases because that although the transportation cost and inventory cost goes up, the container purchasing cost significantly reduced while the selling revenue is not changed.

In addition, we present the line chart of three instances to show the container purchasing quantity in each period in Figure 1.

From Figure 1, we can learn that in each instance, the purchasing quantities of the first few periods are very high for the reason that the initial container inventory is assumed to be zero and no containers return. After that, because of the containers returning, the quantity of purchasing decreases, where very small or even zero quantities occur. When the purchasing quantity is very small, making purchase decision may not be economical. In this case, the manufacturer can choose to wait for RTIs return with the permission of backorder, which will be studied in our following work.

Table 1: Computational Results of Random Instances

Instance size (*T*-*G*)	*k* generate scheme	Average Return time (period)	Average times of production set up	Average times of container purchasing	Average CPU time (s)
10-5	S1	3.40	8.20	6.60	0.21
	S2	3.40	8.20	6.60	0.21
20-5	S1	5.20	15.80	13.20	0.29
	S2	2.20	15.80	12.40	0.30
30-10	S1	7.40	20.40	20.20	0.50
	S2	6.40	20.40	19.80	0.51
40-10	S1	12.40	27.80	30.60	0.72
	S2	4.40	27.80	25.00	0.73
50-15	S1	15.20	29.00	36.80	0.76
	S2	7.20	29.00	32.60	0.74
60-15	S1	19.80	32.40	44.60	0.87
	S2	7.80	32.40	36.00	0.84

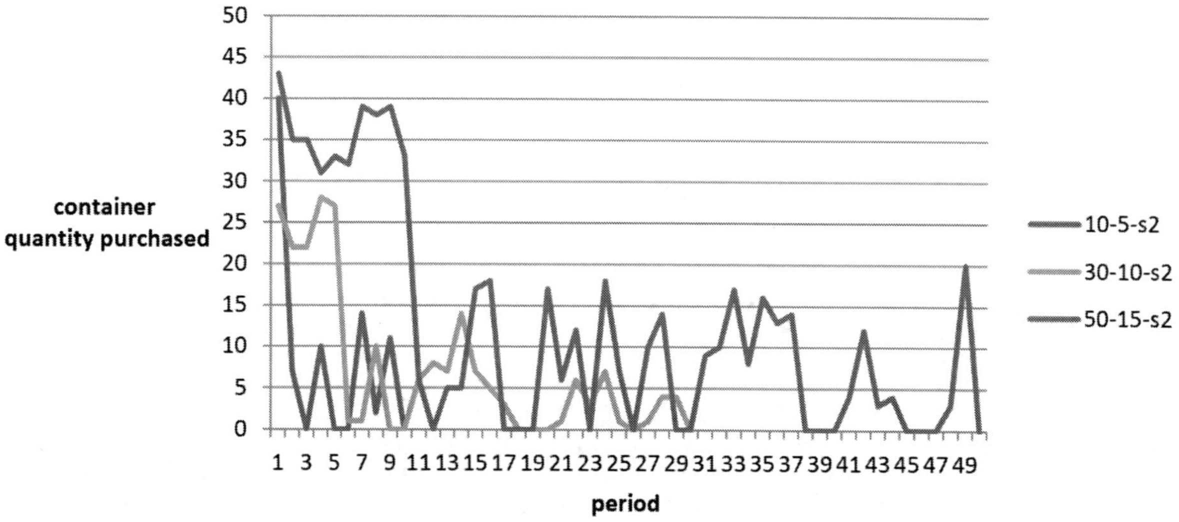

Figure 1: Line Chart of Container Purchasing Quantity in Each Period

4 CONCLUSION

In this paper, we studied an integrated closed-loop food supply chain with returnable containers and proposed a new mixed integer linear programming model. The results of the randomly generated instances have demonstrated the efficiency of the model. In the future, we will devote ourselves to extending this first work step by step in the following four directions. Firstly, backorder will be permitted so that the manufacturer can choose to wait for containers return instead of purchasing new containers instantly since the customers can be satisfied with products of different ages. Secondly, a multi-manufacture multi-retailer CLSP model which is more common in practice as well as the corresponding efficient solving algorithm should be developed. Thirdly, the vehicle routing problem will be considered simultaneously in the problem. Finally, we will add other objectives, such as minimizing the environment impact, to the model and address multi-objective problem.

ACKNOWLEDGMENTS

This work was partially supported by the Cai Yuanpei Program between the French Ministries of Foreign and European Affairs and Higher Education and Research and the Chinese Ministry of Education, under Grant 34644SB.

5 REFERENCES

[1] **Daniel Hellström, Ola Johansson. 2010.** The impact of control strategies on the management of returnable transport items, *Transportation Research Part E*, 46(6), pp 1128-1139.

[2] **Twede, D., Clarke, R. 2004.** Supply chain issues in reusable packaging, *Journal of Marketing Channels*, 12 (1), pp 7-26.

[3] **Goellner, K. N., and E. Sparrow. 2014.** An Environmental Impact Comparison of Single-use and Reusable Thermally Controlled Shipping Containers, *The International Journal of Life Cycle Assessment*, 19 (3), pp 611-619.

[4] **Diane Mollenkopf, David Closs, Diana Twede, Sangjin Lee, Gary Burgess. 2005.** Assessing the viability of reusable packaging: A relative cost approach, Journal of Business Logistics, 26(1), pp 169-197.

[5] **Özkır, V.,Baslıgil, H.2013.** Multi-objective optimization of closed-loop supply chains in uncertain environment. *Journal of Cleaner Production*, 41, pp 114-125.

[6] **Zhang, S., Zhao, X., Zhang, J. 2014.** Dynamic model and fuzzy robust control of uncertain closed-loop supply chain with time-varying delay in remanufacturing. *Industrial and Engineering Chemistry Research*, 53(23), pp 9805-9811.

[7] **Eleonora Bottani, Roberto Montanari, Marta Rinaldi and Giuseppe Vignali. 2015.** Modeling and multi-objective optimization of closed loop supply chains: A case study. *Computers and Industrial Engineering*, 87, pp 328-342

[8] **Barry R. Cobb.2016.** Inventory control for returnable transport items in a closed-loop supply chain. *Transportation Research Part E*, 86, pp 53-68

[9] **Christoph H. Glock, Taebok Kim. 2014.** Container management in a single-vendor-multiple-buyer supply chain. *Logistics Research*. 7, pp 112

[10] **Christoph H. Glock and Taebok Kim. 2015.** Safety measures in the joint economic lot size model with returnable transportant items, *International Journal of Production Economics*.

[11] **Moncer Hariga, Christoph H. Glock, Taebok Kim. 2016.** Integrated product and container inventory model for a single-vendor single-buyer supply chain with owned and rented returnable transport items. *International Journal of Production Research*, 54(7), pp 1964-1979

[12] **Taebok Kim, Christoph H. Glock, Yongjang Kwon. 2014.** A closed-loop supply chain for deteriorating products under stochastic container return times, *Omega*, 43, pp 30-40

[13] **Farahani P, Grunow M, Günther H-O. 2012.** Integrated production and distribution planning for perishable food products. *Flexible Services and Manufacturing Journal*, 24(1), pp 28-51.

[14] **Xiaojun Wang, Dong Li. 2012.** A dynamic product quality evaluation based pricing model for perishable food supply chains. *Omega*, 40(6), pp 906-917

EFFECT OF PHASE I ESTIMATION ERROR ON THE PHASE II MONITORING OF SIMPLE LINEAR PROFILE WITH WITHIN-PROFILE AUTOCORRELATION

Yang Zhang[1,2], Nini Gao[1] and Qing Wang[1,2*]
[1]School of Business, Tianjin University of Commerce, Tianjin, 300134, China
[2]Management Innovation and Evaluation Research Centre
Tianjin University of Commerce, Tianjin, 300134, China
wangqingtjcu@163.com

ABSTRACT

Profile monitoring is the utilization of the control charts to monitor the quality of a product or process which is characterized by a functional relationship. This functional relationship is referred as a 'profile'. In most cases of profile monitoring, the values of in-control parameters are assumed to be known. However, this is usually invalid in practice. Therefore, the in-control parameters should be estimated in phase I before starting online monitoring in phase II. This paper illustrates how phase I estimation error affect the performance of phase II monitoring schemes where the observed data in each simple linear profile are correlated. The standard deviation of the average run length is adopted as a performance metric, which can take account of the practitioner-to-practitioner variation. Finally, some conclusions are given on the recommendation of phase I sample size and the evaluation and selection of control charts.

Keywords: Estimation error; Phase II analysis; Profile monitoring; Standard deviation of average run length; Statistical Process Control

1 INTRODUCTION

In some practical applications, the quality of a process or product is characterized by a relationship between a response variable and one or more variables (in time or space) which is called profile. Profile monitoring is to use statistical process control methods to check whether or not this relationship has changed over time. Overviews on profile monitoring can be found in Woodall et al. [1], Woodall [2] and Noorossana et al. [3].

In profile monitoring, a profile can be described reasonably well by a simple linear regression model in some cases. Kang and Albin [4] firstly studied monitoring simple linear profiles in Phase II with two different control charts, a multivariate T^2 chart and a combined exponentially weighted moving average (EWMA) and range (R) chart based on residuals. Another method consisting of three separate univariate EWMA control charts was proposed by Kim et al. [5] to monitor the Y-intercept, the slope and the error variance based on transformed linear profile model. Other control charts have also been proposed for monitoring linear or multivariate multiple linear profiles, including Mahmoud and Woodall [6], Zou et al. [7], and Noorossana et al. [8]. However, in some other cases, the relationship cannot be adequately represented by a linear profile, but can be by a nonlinear regression model. Nonlinear profile monitoring approaches include parameteric methods [9-11] and deviation-metric-based methoeds [12-13].

In the previous references, it is assumed that the random errors associated with the measurements within a profile are independent of each other. However, this is usually invalid in practice for many

This research was supported in part by National Natural Science Foundation of China (grant nos. 71401123 and 71672122).

* Corresponding Author

types of data. Successive measurements may exhibit obvious serial or spatial correlation over time. To monitoring the profiles with the within-profile correlation (WPC), the linear mixed models (LMM) and nonlinear mixed models (NLMM) are suggested by Jensen et al. [14] and Jensen and Birch [15] to model the linear and nonlinear profiles with WPC. Amiri et al. [16] also suggested using LMM to model the autocorrelation within polynomial profiles and proposed a T^2 control chart to monitor the second-order polynomial relationships. Zhang et al. [17] suggested using Gaussian process models to model the WPC, and developed two control chart to monitor the linear trend and the WPC, respectively. While a first order autoregressive (AR(1)) model was used by Soleimani et al. [18] to describe the correlation among observations within each simple linear profile.

In their paper, Soleimani et al. [18] applied a transformation of the observations to eliminate within-profile autocorrelation (WPA) for the case of simple linear profiles, and then developed four Phase II methods. The transformation is based on the autocorrelation coefficient which is assumed to be known. However, this is usually invalid in practice, and the autocorrelation coefficient should be estimated first from Phase I data obtained by practitioner when the process is in-control (IC). When the estimates are used in place of the known parameters, however, the variability of the estimators can result in a significant impact on the chart performance in Phase II. Due to using different Phase I datasets obtained by different practitioners, there would always be a practitioner-to-practitioner variation in the values of the ARL [19]. Consequently, the ARL becomes a random variable when parameters are estimated. Thus, it is necessary to take the average ARL (AARL), i.e., the expected value of the ARL, and the standard deviation of the ARL (SDARL) as the important performance metrics.

In this paper, we study the performance of the control charts developed by Soleimani et al. [18] when the autocorrelation coefficient is estimated using the AARL and SDARL metrics. The process model is presented in Section 2, and the control charts with known and estimated autocorrelation coefficient is introduced in Section 3. Section 4 presents the marginal performance of the control charts with estimated parameters. Finally, some conclusions are concluded in Section 5.

2 PROCESS MODEL FOR LINEAR PROFILES WITH WPA

Assume that for the jth sample collected over time the observations on each profile are $(x_i, y_{ij}), i = 1, 2, \dots, n$. It is further assumed that when the process is in statistical control, the jth profile can be modelled as:

$$y_{ij} = A_0 + A_1 x_i + \epsilon_{ij}, \epsilon_{ij} = \rho\epsilon_{(i-1)j} + a_{ij}, \qquad (1)$$

where ϵ_{ij}'s are the correlated error terms and a_{ij}'s are independently and identically distributed normal random variables with mean zero and variance σ^2. ρ is the autocorrelation coefficient. It is assumed that the x-values are fixed and constant among profiles. Here, the parameters A_0, A_1, and σ^2 are assumed to be known and have known values.

To eliminate the effect of autocorrelation within linear profiles, the transformation applied by Soleimani et al. [18] is as follows:

$$y'_{ij} = y_{ij} - \rho y_{(i-1)j}. \qquad (2)$$

Based on this transformation, the process model in Equation (1) can be transformed into a simple linear profile model with independent error terms shown as follows:

$$y'_{ij} = A'_0 + A'_1 x'_i + a_{ij}, \qquad (3)$$

where $y'_{ij} = y_{ij} - \rho y_{(i-1)j}$, $x'_i = x_i - \rho x_{i-1}$, $A'_0 = A_0(1-\rho)$, $A'_1 = A_1$, and a_{ij}'s are independent random variables with mean zero and variance σ^2.

3 CONTROL CHARTS TO MONITOR LINEAR PROFILES WITH WPA

3.1 Control charts with known autocorrelation coefficient

Recall the process model in Equation (3), the transformation of the observations should be based on the autocorrelation coefficient ρ. While ρ is assumed to be a known parameter, Soleimani et al. [18] proposed four control charts based on the transformed model to monitor linear profile with WPA. Only two of them are considered in our paper, the T^2 method and the EWMA-3 method, which are introduced in the following subsections. See more details about these two approaches in Soleimani et al. [18].

3.1.1 T^2 method

The T^2 method is similar to that in Kang and Albin [4], using the intercept and the slope in the transformed models instead in order to reduce the effect of autocorrelation that exists between error terms within a profile. The T_j^2 statistic used in the control chart is as follows:

$$T_j^2 = \left([\hat{A}'_{0j}\ \hat{A}'_{1j}] - [A'_{0j}\ A'_{1j}]\right)' S^{-1}\left([\hat{A}'_{0j}\ \hat{A}'_{1j}] - [A'_{0j}\ A'_{1j}]\right), \quad (4)$$

where $S = \begin{pmatrix} \sigma^2\left(\frac{1}{n-1} + \frac{\bar{x}'^2}{S_{x'x'}}\right) & -\frac{\sigma^2\bar{x}'}{S_{x'x'}} \\ -\frac{\sigma^2\bar{x}'}{S_{x'x'}} & \frac{\sigma^2}{S_{x'x'}} \end{pmatrix}$.

When the process is IC, T_j^2 follows a central chi-square distribution with two degrees of freedom. Hence, the upper control limit (UCL) for the T^2 chart is $UCL = \chi^2_{2,\alpha}$, where $\chi^2_{2,\alpha}$ is the $100(1-\alpha)$ percentile of the chi-square distribution with two degrees of freedom.

3.1.2 EWMA-3 method

The EWMA-3 method is similar to the approach proposed by Kim et al. [5]. Another transformation, which scales the x-values so that the average x-value is zero, is adopted to make the estimators of the intercept and slope independent. The transformed model is given by

$$y'_{ij} = B_0 + B_1 x''_i + a_{ij}, \quad (5)$$

where $x''_i = x'_i - \bar{x}'$, $B_0 = A'_0 + A'_1\bar{x}'$ and $B_1 = A'_1$. Therefore, three separate EWMA charts can be used simultaneously to detect shifts in the intercept, the slope or the error variance, respectively. The statistics used in the EWMA-3 method are

$$EWMA_I(j) = \theta b_{0j} + (1-\theta)EWMA_I(j-1), \quad (6)$$

$$EWMA_S(j) = \theta b_{1j} + (1-\theta)EWMA_S(j-1), \quad (7)$$

$$EWMA_E(j) = max\{\theta(MSE_j - 1) + (1-\theta)EWMA_E(j-1), 0\}, \quad (8)$$

where θ $(0 < \theta \le 1)$ is the smoothing constant and set at 0.2 in the simulations, and $MSE_j = \frac{1}{n-1}\sum_2^n \left(y'_{ij} - B_0 - B_1 x''_i\right)^2$. Furthermore, $EWMA_I(0) = B_0$, $EWMA_S(0) = B_1$, $EWMA_E(0) = 0$, and the control limits for these three EWMA charts are as follows:

$$LCL_I = B_0 - L_I\sigma\sqrt{\theta/[(2-\theta)(n-1)]}, UCL_I = B_0 + L_I\sigma\sqrt{\theta/[(2-\theta)(n-1)]},$$

$$LCL_S = B_1 - L_S\sigma\sqrt{\theta/[(2-\theta)\sum_2^n x''^2_i]}, UCL_S = B_1 + L_S\sigma\sqrt{\theta/[(2-\theta)\sum_2^n x''^2_i]},$$

$$LCL_E = 0, UCL_E = L_E\sqrt{\theta Var(MSE_j)/(2-\theta)},$$

where $Var(MSE_j) = 2\sigma^4/(n-1)$, and L_I, L_S and L_E are chosen to obtain a specified IC average run length (IC ARL, denoted by ARL_0).

3.2 Control charts with estimated autocorrelation coefficient

The two control charts above are developed based on the transformation of observations shown in Equation (2). However, as pointed out by Soleimani et al. [18], Equation (2) can not be used directly since the autocorrelation coefficient ρ is usually unknown and should be estimated first from Phase I data in practice. The Cochrane-Orcutt approach is suggested by Soleimani et al. [18] to estimate ρ, in which the error term $\epsilon_{ij} = \rho\epsilon_{(i-1)j} + a_{ij}$ is considered as a regression model. The steps for carrying out the Cochrane-Orcutt procedure are as follows:

Step 1 Estimate Equation (1) by ordinary least square (OLS) and compute its residuals e_{ij}.

Step 2 Fit a simple linear regression model with e_{ij} and $e_{(i-1)j}$ and no intercept. The estimated slope from this regression is the estimate of the autocorrelation coefficient, ρ.

Step 3 Fit the transformed model in Equation (3) by OLS, and obtain the estimates for A_0' and A_1'. Then we have $A_0 = A_0'/(1-\rho)$ and $A_1 = A_1'$.

Step 4 Use these estimates in Step 3 for A_0 and A_1 in Equation (1) and obtain a new set of estimates of the residuals.

Step 5 Go back to Step 2 to get a new estimate of ρ, and repeat Step 2-4 until the difference between two successive estimates of ρ is no more than some predetermined value, such as 0.001. Then the final estimate of ρ is the estimated autocorrelation coefficient.

To reduce the estimation error, we suggest using m in-control Phase I samples to estimate the autocorrelation coefficient. Suppose for the jth sample the estimate of autocorrelation coefficient is $\hat{\rho}_j$ using the Cochrane-Orcutt method, then the estimator of ρ is

$$\bar{\hat{\rho}} = \frac{1}{m}\sum_{j=1}^{m}\hat{\rho}_j. \qquad (9)$$

Then, when the autocorrelation coefficient is unknown, the estimate $\bar{\hat{\rho}}$ is used in the transformed process model in Equation (3) instead. Then the charting statistics in the T^2 method and the EWMA-3 method are developed based on the Equation (3) by replacing ρ by $\bar{\hat{\rho}}$. In our paper, we aim to investigate the effect of the estimate error and the phase I sample size on the two control charts developed by Soleimani et al. [18].

4 SIMULATION STUDIES

In the simulation study, without loss of generality, it is assumed that $A_0 = 3$, $A_1 = 2$, and $\sigma^2 = 1$. The x-values used in our simulation is $x_i = 2 + 6(i-1)/(n-1)$ with $n = 1201$. Moreover, it is assumed that the sample size $m = (30, 50, 100, 200, 500, 1000, 5000)$. Furthermore, three different autocorrelation coefficients, $\rho = (0.1, 0.5, 0.9)$, are considered here. To investigate the effect of estimation error on the IC performance of the control charts, the Monte Carlo simulation method is used and includes the following steps.

Step 1 The control limits with known ρ should be determined first to achieve a specified $ARL_0 = 200$. According to Soleimani et al. [18], the UCL for the T^2 chart is set to 10.597, and the values of L_I, L_S and L_E equal to 3.014, 3.012 and 2.87, respectively, with $\theta = 0.2$ for the EWMA-3 method.

Step 2 An estimate, $\bar{\hat{\rho}}$, is obtained from m IC Phase I data by using the Cochrane-Orcutt method.

Step 3 Generate profiles with n observations using the transformed model in Equation (3) using $\bar{\hat{\rho}}$ instead of ρ.

Step 4 Calculate the charting statistics in the T^2 chart or the EWMA-3 method, and compare them with their corresponding control limits.

Step 5 Record the number of random profiles sampled until the control chart triggers a signal, which is the run length.

Step 6 Repeat Steps 3-5 *N* iterations and the in-control ARL can be obtained by averaging the run length. In the simulation, *N* is set at 1000.

Step 7 Repeat Steps 2-6 *M* times, then the in-control AARL and SDARL values can be get by using the obtained *M* ARL values. Here, *M* equals to 100.

The in-control AARL and SDARL values under various sample sizes and different values of ρ for the T^2 chart and the EWMA-3 method are given in Tables 1 and 2, respectively. Several results can be obtained from Tables 1 and 2. Firstly, as the value of m increases, the AARL values for specified ρ are close to each other, which is not as expected to that the AARL values increase. This is mainly because that the mean of the sampling distribution of the estimator keeps almost the same for different m, i.e., is not seriously affected by the Phase I sample size. Moreover, the SDARL values for specified ρ decrease when increasing the value of m. The reason behind this is that the variability in the sampling distribution of the estimator decreases as m increases.

Secondly, as the autocorrelation coefficient ρ increases, the IC AARL values of the T^2 chart decrease and the SDARL values increase for particular m. Furthermore, the IC AARL values are smaller than the designed $ARL_0 = 200$, and become much smaller as ρ increases. This is mainly resulted by that the estimator in Equation (9) using the Cochrane-Orcutt procedure underestimates ρ, especially when the errors are positively autocorrelated. The larger ρ is, the larger the estimation error is, and the larger the variability in the sampling distribution of the estimator is. Similar results can also be found for the EWMA-3 method.

Thirdly, comparing the in-control performance of the T^2 method and the EWMA-3 method shown in Tables 1 and 2, the EWMA-3 method performs uniformly better than the T^2 method, using either the AARL metric or the SDARL metric.

Table 1: In-control Performance of the T^2 method for various m and different ρ

m	$\rho = 0.1$		$\rho = 0.5$		$\rho = 0.9$	
	AARL	SDARL	AARL	SDARL	AARL	SDARL
30	194.1	13.3	185.7	18.8	125.5	29.3
50	197.0	9.3	186.7	14.9	126.5	22.0
100	195.8	8.5	184.6	10.4	126.4	17.5
200	194.7	7.9	185.4	8.8	127.5	10.9
500	196.2	7.3	187.1	7.1	126.1	7.8
1000	195.8	7.0	186.7	7.0	125.0	6.6
5000	195.2	5.7	186.1	5.4	125.9	4.4

Table 2: In-control Performance of the EWMA-3 method for various m and different ρ

m	$\rho = 0.1$		$\rho = 0.5$		$\rho = 0.9$	
	AARL	SDARL	AARL	SDARL	AARL	SDARL
30	196.3	10.2	191.1	13.4	150.1	22.4
50	198.3	8.1	193.0	9.9	151.2	17.8
100	199.1	8.3	192.9	7.8	152.7	12.0
200	197.5	5.2	191.7	6.2	151.4	9.2
500	199.7	5.7	194.3	6.4	152.5	6.9
1000	198.2	6.8	193.0	7.0	151.3	6.0
5000	198.1	5.8	192.4	5.2	152.9	5.7

Finally, based on the SDARL value, the Phase I sample size could be determined and the control chart could be selected. To seek a fairly small SDARL, Zhang et al. [19] suggested that the SDARL should be within 5%-10% of the desired IC ARL value. From Tables 1 and 2, to achieve an SDARL value close to 10, the T^2 method requires $m = 46, 100, 200$ for $\rho = 0.1, 0.5, 0.9$, respectively, whereas the EWMA-3 method requires $m = 30, 50, 170$ for $\rho = 0.1, 0.5, 0.9$, respectively. Moreover, it is can be seen that larger m is needed for larger ρ. In the case of small number of phase I samples, the EWMA-3 method is highly recommended.

5 CONCLUSION

In the paper, we have studied the effects of estimation error of the autocorrelation coefficient ρ on the in-control performance of the T^2 method and the EWMA-3 method that used to monitor the linear profiles with within-profile autocorrelation. Our results show that the AARL values of these two control charts are not obviously affected by the Phase I sample size, and the variability of the sampling distribution of the estimator significantly influences the SDARL. Moreover, because the estimator underestimates ρ and the estimation error gets larger as ρ becomes large, the AARL values are smaller than the designed ARL_0 and become much smaller as ρ gets larger. Furthermore, in case of either the AARL metric or the SDARL metric, the EWMA-3 method is highly recommended, and smaller size of Phase I samples is needed for the EWMA-3 method.

6 REFERENCES

[1] **Woodall, W.H., Spitzner, D.J., Montgomery, D.C. and Gupta, S. 2004.** Using control charts to monitor process and product quality profiles, *Journal of Quality Technology*, 36(3), pp 309-320.

[2] **Woodall, W.H. 2007.** Current research on profile monitoring, *Produção*, 17(3), pp 420-425.

[3] **Noorossana, R., Saghaei, A., and Amiri, A. 2011.** *Statistical analysis of profile monitoring*, Hoboken, NJ: John Wiley & Sons.

[4] **Kang, L. and Albin, S.L. 2000.** On-line monitoring when the process yields a linear profile, *Journal of Quality Technology*, 32(4), pp 418-426.

[5] **Kim, K., Mahmoud, M.A. and Woodall, W.H. 2003.** On the monitoring of linear profiles, *Journal of Quality Technology*, 35(3), pp 317-328.

[6] **Mahmoud, M.A. and Woodall, W.H. 2004.** Phase I analysis of linear profiles with calibration applications, *Technometrics*, 46(4), pp 380-391.

[7] **Zou, C., Tsung, F. and Wang, Z. 2007.** Monitoring general linear profiles using multivariate exponentially weighted moving average schemes, *Technometrics*, 49(4), pp 395-408.

[8] **Noorossana, R., Eyvazian, M., and Vaghefi, A. 2010.** Phase II monitoring of multivariate simple linear profiles, *Computer & Industrial Engineering*, 58(4), pp 563-570.

[9] **Williams, J.D., Woodall, W.H. and Birch, J.B. 2007.** Statistical monitoring of nonlinear product and process quality profiles, *Quality and Reliability Engineering International*, 23(8), pp 925-941.

[10] **Zou, C., Tsung, F. and Wang, Z. 2008.** Monitoring profiles based on nonparametric regression methods. *Technometrics*, 50(4), pp 512-526.

[11] **Chuang, S.C., Hung, Y.C., Tsai, W.C., and Yang, S.F. 2013.** A framework for nonparametric profile monitoring, *Computer & Industrial Engineering*, 64(1), pp 482-491.

[12] **Vaghefi, A., Tajbakhsh, S.D. and Noorossana, R. 2009.** Phase II monitoring of nonlinear profiles, *Communications in Statistics: Theory and Methods*, 38(11), pp 1834-1851.

[13] **Zhang, Y., He, Z., Fang, J. and Zhang, M. 2012.** Nonparametric control scheme for monitoring phase II nonlinear profiles with varied argument values, *Chinese Journal of Mechanical Engineering*, 25(3), pp 587-597.

[14] **Jensen, W.A., Birch, J.B., and Woodall, W.H. 2008.** Monitoring correlation within linear profiles using mixed models, *Journal of Quality Technology*, 40(2), pp 167-183.

[15] **Jensen, W.A. and Birch, J.B. 2009.** Profile monitoring via nonlinear mixed models, *Journal of Quality Technology*, 41(1), pp 18-34.

[16] Amiri, A., **Jensen, W.A., and Kazemzadeh, R.B. 2010.** A case study on monitoring polynomial profiles in the automotive industry, *Quality and Reliability Engineering International*, 26(5), pp 509-520.

[17] **Zhang, Y., He, Z., Zhang, C., and Woodall, W.H. 2014.** Control charts for monitoring linear profiles with within-profile correlation using Gaussian process models, *Quality and Reliability Engineering International*, 30(4), pp 487-501.

[18] **Soleimani, P., Noorossana, R., and Amiri, A. 2009.** Simple linear profiles monitoring in the presence of within profile autocorrelation, *Computers and Industrial Engineering*, 57, pp 1015-1021.

[19] **Zhang, M., Megahed, F.M., and Woodall, W.H. 2014.** Exponential CUSUM charts with estimated control limits, *Quality and Reliability Engineering International*, 30(2), pp 275-286.

INTEGRATED MACHINE SCHEDULING AND DISTRIBUTION WITH FIXED DELIVERY DEPARTURE TIME AND CARBON EMISSION CONSIDERATION

Peng Chen, Haitao Liu, Qu Wei, and Zhaoxia Guo*
Business School, Sichuan University
Chengdu 610065, Sichuan, China
zxguo@scu.edu.cn

ABSTRACT

This paper addressed an integrated machine scheduling and green distribution problem with some important realistic characteristics. Solving the problem effectively is crucial to minimizing the total supply chain cost. A memetic algorithm-based optimization approach is presented to handle this problem, in which a memetic algorithm is combined with heuristic procedures. The effectiveness of the proposed approach is evaluated by using numerical experiments. The experimental results show that (1) the proposed approach has a good optimum-seeking capacity, and (2) the approach is efficient in solving the investigated problem.

Keywords: Single Machine; Green Transportation; Hybrid Algorithm; Integrated Scheduling

1 INTRODUCTION

The integrated production and transportation scheduling (IPTS) problem investigates how to schedule production orders and how to deliver the finished products to customers in a joint and integrated manner, in which production and transportation operations are highly integrated to enhance the supply chain performance and therefore achieve the higher supply chain sustainability. In real world, it is common that transportation is performed by a third-party logistics company such as UPS or DHL, which provides multiple transportation modes such as air, sea and land transportation. Reducing energy consumption and carbon emission in transportation is very critical. This paper addresses an IPTS problem with the consideration of multiple transportation modes and carbon emissions, called as integrated machine scheduling and green distribution problem.

Many researchers have studied the IPTS problems from different perspectives. Moon et al. [9] study the IPTS problem with the objective of minimizing the maximum completion time, and establish a mixed integer linear programming model. Considering the particularity of concrete transportation, Garcia et al. [5] deal with an IPTS problem in a scenario of no-wait, immediate delivery to the customer site. In 2010, Chen [4] made a comprehensive survey on IPTS problems. Lee et al. [6] construct an IPTS model in a make-to-order producer-buyer supply chain with the objective of minimizing the total cost including transportation cost. Most of previous studies on IPTS problems focused on simple realistic characteristic. Previous studies seldom considered such more complicated realistic features as fixed vehicle departure times, multiple transportation modes and green distributions.

* Corresponding Author

These various complicated realistic features exist in practice. Most companies worldwide now rely on third-party logistics providers for their daily distribution and transportation needs. Many third-party service providers have daily fixed package pickup times. The IPTS problems thus have to consider fixed vehicle departure times. In addition, multiple transportation modes are usually exist in practice, each of which corresponds to a certain combination of transportation speed and capacity. However, IPTS problems, which consider fixed delivery departure times, multiple transportation modes and green distribution simultaneously, have not been investigated so far although these problems are widespread in some real-world supply chains such as apparel and footwear. This paper thus investigates an integrated machine scheduling and green distribution (IMSGD) problem with the consideration of these realistic features.

Various optimization techniques have been used to solve these IPTS problems, which involve mathematical programming, heuristics and traditional intelligent algorithms, etc. Memetic algorithms, which are also referred to as hybrid genetic algorithm or genetic local search algorithm, can obtain good performance with faster computation time and has excellent performance in solving various complex optimization problems [8]. Memetic algorithms have also been used to solve complex IPTS problems effectively [2, 10]. This paper thus proposes a memetic algorithm-based optimization (MAO) approach to solve the IMSGD problem investigated.

2 PROBLEM FORMULATION

2.1 Problem description

At the beginning of a scheduling horizon, the plant receives a set of orders from customers all around the world and commits a delivery date for each order. The plant needs to produces these orders on a dedicated machine of this plant and delivers the finished products to customers by a third-party logistics company. Each order contains a set of jobs. Different orders could have different order sizes and the complexities of different jobs could be different. The plant could produce multiple orders at the same time. If so, the production capacity for each order in parallel is the same. The jobs within an order must be processed continuously in turn. To respond quickly to customer orders, the plant integrates the machine scheduling and distribution operations together, which needs to determine firstly the production beginning time of each order in the plant. After the production of an order is completed, finished products need to be delivered to customer destinations. In each order, the jobs with the same destination are defined as a product batch. Some product batches may arrive in advance or late, which lead to earliness or tardiness penalty respectively. Third-party logistics companies are responsible for transporting finished products to customer-specified destinations, which provide multiple transportation modes including sea, land and air transportation. Different transportation modes correspond to different unit transportation costs, time and carbon emissions. The plant needs to select a suitable transportation mode dynamically with the objective of minimizing the total cost of supply chain.

With different transportation tools, fuel consumptions are apparently different to transport products from one place to another. For sea transportation, carbon emissions are mainly from shipping fuel consumption. The amount of fuel consumed by shipping depends on its load factor, frequency of sailing, speed, distance involved, and fuel efficiency [7]. For land transportation, the definition of energy consumption is given by Bektas [1]. For air transportation, carbon emissions are generated in two main parts: landing/take-off cycle (LTO) and cruise of aircrafts. The analysis about these 2 parts has been given by Chao [3].

2.2 Mathematical model

Let i (or i') index order ($i \in (1,2,...,I)$), k index delivery destination ($k \in (1,2,...,K)$), m index transportation mode ($m \in (1,2,...,M)$), and (i,k) index the kth product batch of order i. Let d_i denote the expected delivery date of order i, DC the standard daily production capacity (number of standard products that can be completed per day), DE_{ik} (DT_{ik}) the daily earliness (tardiness) penalty of product batch (i,k), E_i the efficiency percentage of processing order i, HR_i the average daily holding rate of the single product in order i, Q_{ik} the products quantity of product batch (i,k), TC_{ikm} the transportation cost of product batch (i,k) via transportation mode m, and TT_{ikm} the transportation time of product batch (i,k) via transportation mode m. Let c_l denote the unit labor cost, and λ the unit carbon emission penalty. Let AD_{ik} denote the actual delivery date of product batch (i,k), B_i the production beginning date of order i, CE_{ik} the carbon emission penalty of product batch (i,k), C_i the production completion date of order i, C_{ik} the production completion date of product batch (i,k), CP_i the carbon emission penalty of order i, EP_{ik} the earliness penalty of product batch (i,k), HC_{ik} the holding cost of product batch (i,k), PC_i the production cost of order i, T_i the processing time of order i, TC_{ik} the transportation cost of product batch (i,k), TP_{ik} the tardiness penalty of product batch (i,k), TT_{ik} the transportation time of product batch (i,k), W_i the workloads of order i (equal to the number of standard products.

Three decision include $R_{ii'}$, S_{ik} and M_{ikm}. $R_{ii'}$ is 1 if order i' is the immediate succeeding of order i, otherwise it is 0. S_{ik} is the departure time of product batch (i,k). M_{ikm} is 1 if the product batch (i,k) is transported via transportation mode m, otherwise it is 0.

The mathematical model of the investigated problem can be formulated as follows.

$$\min \; F\left(R_{ii'}, S_{ik}, M_{ikm}\right) = \sum_{i=1}^{I} (PC_i + HC_i + TC_i + EP_i + TP_i + CP_i) \tag{1}$$

with

$$PC_i = c_l \cdot Q_i \tag{2}$$

$$HC_i = \sum_{k=1}^{K} \left((S_{ik} - C_{ik}) \cdot HR_i \cdot Q_{ik} \right) \tag{3}$$

$$TC_i = \sum_{k=1}^{K} \sum_{m=1}^{M} \left(M_{ikm} \cdot TC_{ikm} \right) \tag{4}$$

$$EP_i = \sum_{k=1}^{K} \left(\max\left(0, (d_i - AD_{ik}) \cdot DE_{ik}\right) \right) \tag{5}$$

$$TP_i = \sum_{k=1}^{K} \left(\max\left(0, (AD_{ik} - d_i) \cdot DT_{ik}\right) \right) \tag{6}$$

$$CP_i = \lambda \cdot \sum_{i=1}^{I} \sum_{k=1}^{K} CE_{ik} \tag{7}$$

subject to

$$C_i = B_i + T_i, \;\; \forall i \tag{8}$$

$$T_i = \frac{\sum_{k=1}^{K} Q_{ik}}{DC \cdot E_i}, \quad \forall i,k \tag{9}$$

$$B_{i'} = C_i, \quad \forall R_{ii'} = 1 \tag{10}$$

$$C_i \le S_{ik}, \quad \forall i,k \tag{11}$$

$$\sum_{m=1}^{M} M_{ikm} = 1, \quad \forall i,k,m \tag{12}$$

$$AD_{ik} = S_{ik} + TT_{ik}, \quad \forall i,k \tag{13}$$

$$TT_{ik} = \sum_{m=1}^{M} \left(M_{ikm} \cdot TT_{ikm} \right), \quad \forall i,k,m \tag{14}$$

$$CE_{ik} = M_{ik1} \times \phi d_{ik} \times Q_{ik} + M_{ik2} \times (r_1 \cdot (\rho Q_{ik} + f) + r_2) \cdot \theta \cdot d_{ik} + M_{ik3} \times cf \times \rho d_{ik} Q_{ik}, \quad \forall i,k \tag{15}$$

$$R_{ii'}, M_{ikm} \in \{0,1\}, \quad \forall i,i',k,m \tag{16}$$

$$AD_{ik}, W_i, S_{ik}, TT_{ik} \in N, \quad \forall i,k \tag{17}$$

$$B_i, C_i, T_i, r_i \in R^+, \quad \forall i,k,m \tag{18}$$

The objective function (1) minimizes the total cost of the supply chain. Formulae (2)-(7) complement the objective function. Constraints (8)-(9) calculate the production completion date and the production time. The orders are proceed in a given production sequence and requires a certain time for production. Constraint (10) indicates that the production is not allowed to interrupt. Constraints (11)-(14) are about transportation related constraints. Constraint (11) indicates that the departure time S_{ik} of product batch (i,k) must be greater than its completion date C_{ik}; Constraint (12) states that product batch (i,k) can select only one transportation mode; Constraint (13) calculates the actual delivery date of each product batch; Constraint (14) states that TT_{ikm} depends on the transportation mode m for product batch (i,k). Constraint (15) calculates the total carbon emission of each product batch via transportation mode m. Constraints (16)-(17) indicate the constraints on decision variables and intermediate variables.

3 PROBLEM SOLVING METHODOLOGY

The memetic algorithm-based optimization (MAO) approach combines memetic algorithm with some heuristic procedures, which aims to generate the best solutions to the investigated IMSGD problem, including the values of $R_{ii'}$, S_{ik} and M_{ikm}.

3.1 Overview of memetic algorithm-based optimization approach

The investigated IMSGD problem needs to determine the values of 3 decision variables, $R_{ii'}$, S_{ik} and M_{ikm}. These values are interdependent. For instance, S_{ik} depends on the production completion date of product batch (i,k), which is determined by the production sequence $R_{ii'}$ of orders. To effectively solve this problem, an MAO approach is developed by integrating a memetic optimization process and some heuristic procedures. The memetic optimization process is used to find the best order sequence solutions $R_{ii'}$ to the investigated IMSGD problem. The heuristic procedures are proposed to calculate the supply chain cost (including carbon emission penalty) and values of other variables based on the candidate production sequence solution. Figure 1 shows the flow chart of the MAO approach. The steps involved are described as follows in detail.

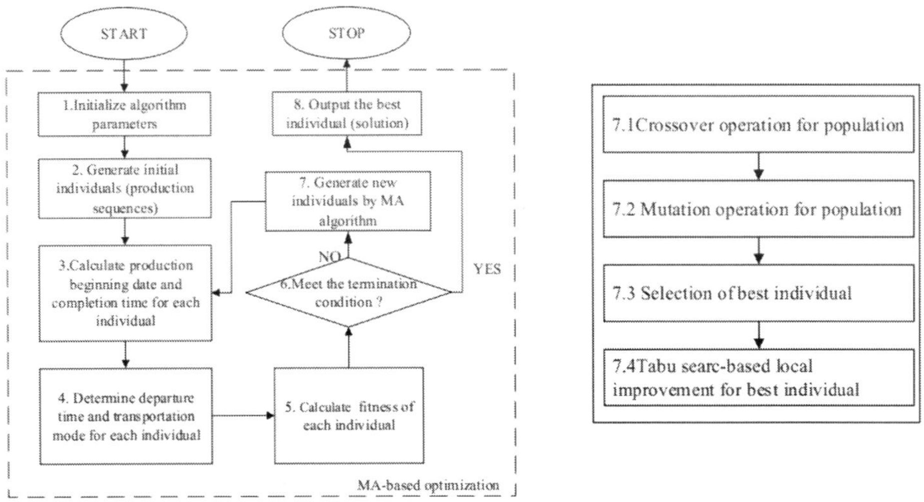

(a) main process **(b) procedure of step 7**

Figure 1: Flow chart of the MAO approach

The main process of the MAO approach is shown in Figure 1a. First, the values of algorithm parameters are initialized, including population size, mutation rate, crossover rate, tabu size, and so on. In step 2, we randomly generate the population of initial individuals, each of which indicates the production precedence of orders in the plant. Next, steps 3-7 constitute the iterative process for memetic optimization, which iteratively find the best values of 3 decision variables. Each iteration represents a generation of the evolutionary process of MAO approach. In steps 3-5, the values of decision variables $R_{ii'}$, S_{ik} and M_{ikm} and the corresponding objective value are calculated based on the solution individual, the detail of which will be described in section 3.3. If the termination condition is satisfied in step 6, the optimization process is terminated and the best solution obtained is returned as the best integrated scheduling solution in step 8. Otherwise, the process returns to step 7 for generating the population of next generation. As shown in Figure 1b, step 7 consists of 4 sub-steps. Step 7.1-7.2 indicates that a new population is generated based on a crossover operation and a mutation operation. Step 7.3 states that the best individual of the population is selected out and preserved. Step 7.4 states that a tabu search-based local improvement is performed based on the best individual.

Let N denote the population size. Let x_n denote the nth ($n \in [1, N]$) individual in the population, which represents a production sequence solution of orders in the plant. Set $x_n = (S^n, F(S^n))$, where S^n denotes the candidate production sequence solution and $F(S^n)$ denotes the fitness of S^n. Set $S^n = (a_1^n, a_2^n, ..., a_I^n)$ and the value of a_i^n denotes the order number of the ith order to be produced. For example, x_n =((3,2,1,4,5),0.001) states that orders 3, 2, 1, 4 and 5 are produced in turn and the corresponding fitness is 0.001.

The initial population is generated randomly for the first generation. The individuals in the population are evaluated using a fitness value. The selection operator used is the tournament selection. This research utilizes the uniform crossover operator and the inversion mutation operator to generate the individuals in the offspring population.

3.2 Calculation of values of S_{ik} , M_{ikm} and fitness function

The individual in the evolutionary process of the MAO approach determines the production sequence $R_{ii'}$ of orders, the completion date C_i can then be determined by formula (8). According to the values of expected delivery date d_i and production completion date C_i, the values of variables S_{ik} and M_{ikm} can be determined by the following 3 rules.

Rule 1: If no transportation mode can transport products to the destination by the expected delivery date d_i, then M_{ikm} is set as the transportation mode with the shortest transportation time, S_{ik} is equal to the production completion date C_i of order i.

Rule 2: If multiple transportation modes can complete the transportation task by the expected delivery date d_i, then M_{ikm} is set as the transportation mode with the longest transportation time (shortest transportation cost), S_{ik} is determined by Rule 3.

Rule 3: If the holding cost of product batch (i,k) is greater than earliness penalty, set $S_{ik} = \lceil C_i \rceil$; otherwise, set $S_{ik} = d_i - \lceil C_i \rceil - TT_{ikm}$.

After the values of these variables are determined, the value of the objective function (the total cost of the supply chain) can be calculated according to formula (9). The value of its fitness function can then be set as the reciprocal of the objective value.

4 NUMERICAL EXPERIMENTS

This section presents the numerical experiments to validate the performance of our approach.

4.1 Experimental data and algorithm parameters

A series of numerical experiments have been conducted to evaluate the effectiveness of the proposed MAO approach. This section presents one representative experiment in practice. Similar tasks are widespread in the global labor-intensive enterprises. Experimental data were collected from a global make-to-order manufacturing enterprise in China.

The plant-related parameters used are given as follows: the plant production capacity (pieces/day) is 6000, the standard daily production cost ($/pieces) is 10, the daily holding cost ($/piece·day) is 0.12, and the plant production efficiencies for different order complexities are 120%, 100% and 95% respectively. This experiment handles 8 orders with 38 jobs. Three transportation modes are included in total. Due to the page limit, the detail of these orders and transportation modes are not presented.

The MAO approach is implemented in MATLAB version R2013a.The experiments were carried out on a laptop with Intel Core i5-5200U CPU @2.2 GHz and 4 GB of RAM, running on Windows 7 Professional.

4.2 Numerical results

The best production sequence of orders for processing in experiment 1 is (1, 2, 3, 8, 5, 4, 6, 7). The optimal total supply chain cost, generated by the MAO approach is 3221091. The proposed MAO approach can easily eliminate the tardiness penalty with objective of minimizing tardiness penalty, but the total cost of the supply chain is not optimal. The reason why earliness penalty occurs is that the unit earliness penalty of the product batch is lower than average daily holding cost, so more product batches are transported in advance to reduce total cost, and vice versa.

To validate the optimum-seeking performance of the proposed MAO approach, we compare this approach with the enumeration method, which is able to find the optimal solution. The solution obtained by the MAO approach is the same as the optimal solutions obtained by enumeration method. In addition, the average computation times of the MAO approach is 5.39, which is much less than the computation times of the enumeration method. The comparison results show that, in terms of computation time, the performance of MAO approach is far better than enumeration method without losing solution quality.

5 CONCLUSIONS

This paper presented an IMSGD problem considering a variety of realistic features. These features include mainly fixed delivery departure times, multiple transportation modes and green distribution. This paper established the mathematical model of the problem.

A MAO approach was proposed to solve this problem, in which the optimal production sequence of orders is obtained by a memetic algorithm, and the values of other variables are then determined by some heuristic rules. In order to verify the effectiveness of the MAO approach, various numerical experiments have been carried out and validated the effectiveness of the proposed approach.

Future study may consider IPTS problems with multiple plants under uncertain environments and study more powerful algorithms for IMSGD problems.

6 REFERENCES

[1] **Tolga Bektaş, and Gilbert Laporte. 2011.** 'The Pollution-Routing Problem', *Transportation Research Part B: Methodological*, 45, 1232-50.

[2] **M. Boudia, and C. Prins. 2009.** 'A Memetic Algorithm with Dynamic Population Management for an Integrated Production-Distribution Problem', *European Journal of Operational Research*, 195, 703-15.

[3] **Ching-Cheng Chao. 2014.** 'Assessment of Carbon Emission Costs for Air Cargo Transportation', *Transportation Research Part D: Transport and Environment*, 33, 186-95.

[4] **Zhi-Long Chen. 2010.** 'Integrated Production and Outbound Distribution Scheduling: Review and Extensions', *Operations Research*, 58, 130-48.

[5] **J. M. Garcia, S. Lozano, and D. Canca. 2004.** 'Coordinated Scheduling of Production and Delivery from Multiple Plants', *Robotics and Computer-Integrated Manufacturing*, 20, 191-98.

[6] **Shine-Der Lee, and Yen-Chen Fu. 2014.** 'Joint Production and Delivery Lot Sizing for a Make-to-Order Producer-Buyer Supply Chain with Transportation Cost', *Transportation Research Part E: Logistics and Transportation Review*, 66, 23-35.

[7] **Chun-Hsiung Liao, Po-Hsing Tseng, and Chin-Shan Lu. 2009.** 'Comparing Carbon Dioxide Emissions of Trucking and Intermodal Container Transport in Taiwan', *Transportation Research Part D: Transport and Environment*, 14, 493-96.

[8] **Raúl Mencía, María R. Sierra, Carlos Mencía, and Ramiro Varela. 2015.** 'Memetic Algorithms for the Job Shop Scheduling Problem with Operators', *Applied Soft Computing*, 34, 94-105.

[9] **Chiung Moon, Young Hae Lee, Chan Seok Jeong, and YoungSu Yun. 2008.** 'Integrated Process Planning and Scheduling in a Supply Chain', *Computers & Industrial Engineering*, 54, 1048-61.

[10] **Martin Dario Arango Serna, Conrado Augusto Serna Uran, Julian Andres Zapata Cortes, and Andres Felipe Alvarez Benitez. 2014.** 'Vehicle Routing to Multiple Warehouses Using a Memetic Algorithm', *Procedia - Social and Behavioral Sciences*, 160, 587-96.

ECONOMIC DESIGN OF ARMA CHARTS WITH VARIABLE SAMPLING INTERVALS

Zhang Li[1] and Dong Xiaoyang[2*]

School of Management Science and Engineering

Zhengzhou university of Aeronautics, Zhengzhou, China

[1]lizhang89@hotmail.com

[2]13027618837@163.com

ABSTRACT

To improve the sensitivity of statistical process control while reducing costs for process monitoring, the economic design of autoregressive moving average process with Variable Sampling Intervals (VSI ARMA) was discussed. Markov chain was used to calculate the average time to signal (ATS). Using genetic algorithm to search for the optimal combination of seven parameters of VSI ARMA model made the cost of economic model at unit time to be minimum at the same time. The results of a numerical example adopted in an actual case indicate that the loss cost of VSI ARMA control charts is consistently lower than that of the FSI scheme. In addition, the sensitivity analysis of the cost parameters of the model was carried out, which provide theoretical basis for enterprise to reduce cost while enhance productivity effect.

Keywords: ARMA control chart, Markov chain, Variable sampling interval, Economic design

1 INTRODUCTION

Statistical process control (SPC) is a quality management technical which is an assessment and monitoring of the process, so as to ensure the products and services in accordance with the provisions. The principal tool of SPC is control chart. Traditional control charts only consider the statistical performance, whose evaluation index is the average run length (ARL). However, such a design always causes higher cost. Therefore, the economic design of control charts is first proposed by Duncan in 1956 to ensure that such a design actually lowers the cost[1]. Afterwards, Lorenzen and Vance presented a general method of economic design from another perspective[2]. Lupo proposed a muti-objective design approach for the c chart based on Lorenzen and Vance's model[3]. Shams et al used a cost function to determine the parameters of the hotelling's T2 monitoring chart and the proposed approach was used to monitoring the products in the Tennessee Eastman process (TEP) [4]. Hu et al monitored the mean value and variance of process quality feature with a supplementary sample X (SSX) control chart, and then researched the economic performance of SSX control chart. The pattern search algorithm was used to optimize the economic model[5]. The traditional control charts were done in the assumption that the quality feature is independent identically distributed. Since the 1980s, with the plethora of mixed industrial process, the output data of the process were generally characterized by autocorrelation, which run counter to the assumption above and shorten the average run

*Corresponding Author

length (ARL) in control. Under the influence of short ARL, the frequency of false alarms of control chart would increase, thus the cost of control chart would grow up. The study of auto-correlated process monitoring system is divided into two categories: the one category is to use the residual to make the control charts, the other is to adjust the control lines to monitor the auto-correlated process directly. But both methods produce significant change with the different degree of autocorrelation of the process. Hence, Jiang et al proposed the autoregressive moving average (ARMA) control chart for the first time[6]. The economic design model of ARMA control chart was constructed by Lin et al, and the appropriate design parameters were chosen to minimize the expected unit time cost, and the sensitivity analysis was carried out[7]. Numerous studies have indicated that the ARMA control chart is an effective tool for monitoring the auto correlation process.

Most economic design of control charts focused on a fixed sampling interval (FSI), while some others were done for variable sampling interval (VSI) control charts. The average control chart with variable sampling intervals (VSI) was first proposed by Renolds et al[8] . Yu et al proposed the economic design of MA and EWMA control charts with variable sampling interval. The Researches showed that MA and EWMA control charts with variable sampling interval has lower costs than fixed sampling interval control charts[9,10]. Chen et al considered an economic design model of the variable sampling interval Cumulative Conformance Count Charts [11]. Numerous studies have indicated that the variable sampling interval control chart can detect the process faster than the traditional control chart with lower costs[12-15]. So this study builds the economic design of VSI ARMA control chart. First we proposed the economic design model of ARMA control chart with variable sampling interval, and then a specific example is given to solve the model, and Markov chain method was used when calculating the ATS of control chart. Finally, the sensitivity analysis of the cost parameters of the model was carried out to find the cost parameters associated with unit time cost.

2 THE ECONOMIC DESIGN OF VSI ARMA CONTROL CHARTS

2.1 The ARMA control charts

The statistic of ARMA control chart is $Z_t=\theta_0X_t+\phi\theta_0X_{t-1}-\theta X_{t-1}$, where $\theta_0=1+\theta-\phi$, ϕ is the regression coefficient, and θ is the average migration parameter. When the control chart is used to monitor a process obeys ARMA(1,1): $X_t=uX_{t-1}+a_t-va_{t-1}$, we can deduce that: $\sigma_X^2=\frac{1-2uv+v^2}{1-u^2}\sigma_a^2$. Based on the general design of control chart to construct of ARMA control chart, the first is to calculate the variance statistics σ_Z. Then we can deduce an equation based on the statistic of ARMA control chart that:

$$\sigma_Z^2=\left\{\theta_0^2+\frac{\alpha^2}{1-\phi^2}+2\left(\theta_0\alpha+\frac{\phi\alpha^2}{1-\phi^2}\right)\frac{\rho(1)}{1-\phi u}\right\}\sigma_X^2$$

Hence, set the mean value as μ_0, control coefficient of the control chart as k, sample size as n, the control lines of the ARMA(1,1) control chart can be expressed as:

$$UCL,LCL=\mu_0\pm k\left(\frac{\sqrt{\theta_0^2+\frac{\alpha^2}{1-\phi^2}+2\left(\theta_0\alpha+\frac{\phi\alpha^2}{1-\phi^2}\right)\frac{\rho(1)}{1-\phi u}}}{\sqrt{n}}\right)\sigma_X$$

2.2 The VSI ARMA control charts

The rational of VSI control charts is that the next sampling interval will be short if the sample point falls far away from the centre line but not outside the control lines, then you may wonder if the next sampling point will fall outside the control lines. The next sampling interval will be long if there is no such indication. Hence VSI control chart added an upper warning line and a lower warning line: $UWL=\mu_0+\varpi\sigma_Z$, $LWL=\mu_0-\varpi\sigma_Z$, where ϖ is warning coefficient of the control chart, and $0<\varpi<k$. In this study, we choose two sampling interval h_1 and h_2 , and $h_1>h_2>0$. When the sample point falls in the safe zone, the corresponding sampling interval values h_1 , otherwise, if the sample point falls in the warning zone, the corresponding sampling interval values h_2. If the sample point falls out the control lines, the control chart alarms. The warning zone and the safe zone are shown in the figure 1

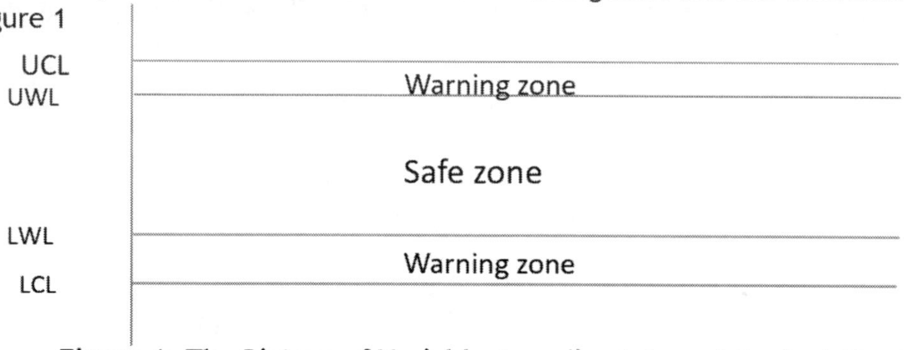

Figure 1: The Picture of Variable Sampling Interval Control Charts

2.3 Economic model of the VSI ARMA control chart

We assumed that: (1) the time that assignable cause's occurrence is to be exponential distribution with a mean time of $1/\lambda$; (2) the mean of quality characteristic value is μ, the variance is σ. When the process is in control, $\mu=\mu_0$, $\sigma=\sigma_0$, where the μ_0 and σ_0 is the target value of mean and variance. When the process is out of control, the variance is fixed, while the mean turns into $\mu=\mu_0+\delta\sigma$; (3) the production process continues when it is searching the assignable reason and recovering the process.
The expected average cost of a cycle equals to the total cost divided by the total time. The total time includes the time to alarm, the time to sampling, and the time to identify and recover the assignable cause; the total cost contains the cost of the quality in the production, the cost for false alarm, the cost for sampling and examining, and the cost for identifying and recovering the assignable cause. Based on the Lorenzen and Vance's economic model, we established VSI economic model as follows:

$$C=\frac{c_1/\lambda+c_2[-\tau+nd+ATS_1+T_1]+sg/ANSS_0+m}{1/\lambda-\tau+nd+ATS_1+T_1}+\frac{a+bn}{h_0}$$

Where, c_1 means the quality cost per unit in-control time
c_2 means the quality cost per unit out of control time
T_1 means the time to identify and recover the assignable cause
s means the expected value of sampling in control
g means the cost for false alarming
$ANSS_0$ means the sampling size the control chart needed in control for alarming
m means the cost for identifying and recovering the assignable cause
d means the time to sample and plot every time
ATS_1 means the time during the process out of control until the control alarming
a means the fixed cost every time to sample
b means the variable cost per item for sampling
h_0 means the average sampling interval

τ means the time between the sampling point just prior to occurrence of the assignable cause and the occurrence itself, and the expected value of τ is presented as $\tau=\frac{1-(1+\lambda h)e^{-\lambda h}}{\lambda(1-e^{-\lambda h})}$

3. EXAMPLE

We set the model established in the 2.2 as objective function, limiting the control chart parameters $[k_1, k_2, n, d_1, d_2, \theta, \phi]$ in a scope. And then the genetic algorithm is used to search the optimal parameter combination and the minimum target function.

3.1 Solution steps

1) Setting the parameters of the genetic algorithm for initial solution number as 40, mating rate as 0.8, and mutation rate as 0.1;
2) Setting parameter values are as follows: $\lambda=0.01$, $c_1=20$, $c_2=100$, $d=0.05$, $t_1=2$, $t_2=2$, $g=50$, $m=20$, $a=0.5$, $b=0.1$, $\delta=3$; With MATLAB programming using Markov chain method to calculate the $ATS_0$and ATS_1 of the control chart and cost functions;
3) Setting the scope of the control chart parameters: the minimum values of $[k_1, k_2, n, d_1, d_2, \theta, \phi]$ as [1 0.01 1 0.01 1 0.01 0.01], the maximum values as [4 3 20 1 3.5 1 1]. There are constrains that the sampling size of n is an integer, and that $k_1 \geq k_2$;
4) Using MATLAB genetic algorithm Toolbox to search the optimal parameter combination.
5) After the process is completed, outputting the results.

3.2 Computing ATS of ARMA control charts with Markov chain

The way to compute ATS with Markov chain is first putting forward by Brook and Evans in 1972, whose main idea is approximation the test statistics into a finite state Markov chain, while corresponding the range of each statistic into each state space of Markov chain, afterwards, writing a further transition probability matrix. But the statistic of ARMA control chart doesn't have the property of Markov chain. Hence, Jiang et al calculated the transition probability matrix by constructing a two dimensional random vector, in which the in control zone is discretized into (2m+1)×(2m+1) rectangles. Every rectangle stands for a state[12]. Huang et al viewed control charting methods as the charting of the output a linear filter applied to the process data, and presented that the in control zone can be discretized into a series of parallelogram to reduce the control line[13]. According to the theory of Jiang, this paper constructed a three dimensional random vector $W_t=(Z_t,X_t,a_t)'$, which is

$$\begin{pmatrix} Z_t \\ X_t \\ a_t \end{pmatrix} = \begin{pmatrix} \phi & \theta_0 u_1-\theta & -\theta_0 v_1 \\ 0 & u_1 & -v_1 \\ 0 & 0 & 0 \end{pmatrix} \begin{pmatrix} Z_{t-1} \\ X_{t-1} \\ a_{t-1} \end{pmatrix} + \begin{pmatrix} \theta_0 \\ 1 \\ 1 \end{pmatrix} a_t$$

To calculate the ATS, we need to set the control lines of X_t and a_t as L_X and L_a. A three dimensional space is constructed in the controlled area, $[-L_Z, L_Z]\times[-L_X,L_X]\times[-L_a,L_a]$, which is divided into $M=(2m+1)\times(2m+1)\times(2m+1)$ copies. And the length of every copy is $(2\delta_1)\times(2\delta_2)\times(2\delta_3)$. The random variable $W_i=(Z_i,X_i,a_i)'$ stands for the state in the time i, and will switch into the state j. If $SZ_j-\delta_1<Z_i\leq SZ_j+\delta_1$, $SX_j-\delta_2<X_i\leq SX_j+\delta_2$, $Sa_j-\delta_3<a_i\leq Sa_j+\delta_3$, then the transition probability matrix P can be shown as:

$$p_{ij}=\begin{Bmatrix} SZ_j-\delta_1<Z_t\leq SZ_j+\delta_1, SX_j-\delta_2<X_t\leq SX_j+\delta_2, Sa_j-\delta_3<a_t\leq Sa_j+\delta_3 | Z_{t-1}=SZ_i, \\ X_{t-1}=SX_i, a_{t-1}=Sa_i \end{Bmatrix}$$

Supposed $Q=(I-P)^{-1}=c_{ij}$, then the ATS in control and out of control can be shown as: $ATS=\sum_{j=1}^{2m+1} c_{m+1,j} \cdot b_j$. Where b_j means the sampling interval. When a state Z_j's centre lines in the safe zone, the next sampling interval $b_j=d_2$, if the centre lines in the warning zone, the next sampling interval $b_j=d_1$

3.3 Results

After five times to solve, the optimal combination of parameters and the lowest total cost are obtained as shown in table 1:

Table 1: Example Results Table

Number	k_1	k_2	n	d_1	d_2	θ	ϕ	Cost
1	1.525	1.525	1	0.708	3.337	0.902	1	34.062
2	2.403	0.162	1	1	3.412	0.094	0.874	34.854
3	4	0.028	1	0.953	3.472	0.495	0.43	29.409
4	3.184	0.013	1	1	3.305	0.145	0.671	34.766
5	4	0.01	1	0.982	3.412	0.042	0.04	29.383

Through a comparative analysis of the results, we obtained the minimum total cost of the economic design models and the corresponding combination of parameters:

$$cost(k_1,k_2,n,d_1,d_2,\theta,\phi)=cost(4,0.01,1,0.982,3.412,0.042,0.04)=29.383$$

3.3 The optimal test

This section is to compare economic design of the VSI control charts and FSI control charts when the cost parameters are in the same situation. As shown in table 2

Table 2: Comparison Between Economic Design of VSI and FSI ARMA Control Charts

Number	k_1	k_2	n	d_1	d_2	θ	ϕ	Cost	Percentage
1	4	0.01	1	0.982	3.412	0.042	0.04	29.383	43.18%
2	0.931		1	0.931		0.021	0.75	68.049	100%

It can be drawn from table 2, compared with the FSI ARMA control charts, the economic design of VSI ARMA control charts' average costs greatly reduced.

4 SENSITIVITY ANALYZE

The economic model of variable sampling interval ARMA control charts have a lot of cost parameters, which affect the output of the cost function, so we used orthogonal experiment design and statistical regression to sensitivity analysis for the model. First of all, we selected 12 parameters, and three levels for each parameter, as shown in table 3. Secondly, there are three levels of each factor, so we selected $L_{27}(3^{13})$ orthogonal table.

Finally, utilizing the numerical calculation of the MATLAB program and test each set of parameter value, and the results as shown in table 4.

Table 3: Sensitivity Analysis of Parameters For Horizontal Table

Factor	Parameter	Meaning of parameter	Level 1	Level 2	Level 3
A	u	Auto-regressive parameters	0.1	0.5	0.9
B	v	Moving average parameter	0.2	0.6	0.8
C	$1/\lambda$	The time that assignable cause's occurrence	0.01	0.03	0.05
D	d	The time to sample and plot every time	0. 05	0.25	0.5
E	T_1	The time to identify and recover the assignable cause	4	20	40
F	m	The cost for identifying and recovered the assignable cause	20	25	30
G	g	The cost for false alarming every time	100	150	200
H	c_1	The quality cost per unit in-control time	5	10	15
I	c_2	The quality cost per unit out of control time	80	100	120
J	a	The fixed cost every time to sample	0.5	2.5	5
K	b	The variable cost per item for sampling	0.1	0.5	1
L	δ	The deviation ratio of the process	0.5	1	1.5

Table 4: The Results of The Sensitivity Analysis Experiment

Number	k_1	k_2	n	d_1	d_2	θ	ϕ	Cost
1	3.997	0.247	1	1	3.094	0.853	0.92	12.029
2	3.876	0.027	1	0.9	3.367	0.553	0.736	32.786
3	1.013	0.566	1	0.208	3.48	0.444	1	62.122
4	1.891	0.681	1	0.313	3.409	1	1	44.976
5	1.327	0.478	1	0.753	3.463	0.574	1	76.4191
6	1.53	0.725	2	0.104	3.463	0.375	1	63.25
7	3.768	0.056	1	0.973	3.342	0.237	0.098	69.186
8	3.965	0.224	1	0.996	2.805	0.554	0.664	73.357
9	3.994	0.096	2	0.997	3.382	0.986	0.034	97.021
10	3.407	0.308	1	0.968	3.449	0.168	0.511	78.22
11	3.823	0.275	1	0.992	3.449	0.57	0.964	71.257
12	3.811	0.166	1	0.992	3.463	0.57	0.934	71.888
13	3.734	0.044	1	0.485	3.298	0.282	0.954	49.721
14	3.734	0.275	1	0.996	3.298	0.282	0.934	54.123
15	3.824	0.044	1	0.996	3.472	0.01	0.082	75.128
16	3.823	0.044	1	0.996	3.472	0.698	0.683	57.944

Number	k_1	k_2	n	d_1	d_2	θ	ϕ	Cost
17	3.826	0.222	1	0.968	3.486	0.698	0.683	41.634
18	3.823	0.044	1	0.996	3.472	0.698	0.683	63.899
19	1.864	0.291	19	0.996	2.421	0.176	0.684	146.475
20	1.864	0.044	1	0.996	3.472	0.01	0.939	146.271
21	1.864	0.297	1	0.996	1.757	0.01	0.939	105.59
22	3.545	0.275	1	0.968	3.472	0.57	0.939	96.226
23	3.545	0.275	1	0.968	3.23	0.57	0.939	128.064
24	3.617	0.044	2	0.968	3.291	0.282	0.954	82.744
25	3.407	0.044	1	0.913	3.291	0.35	0.964	49.708
26	3.259	0.044	1	0.909	3.291	0.282	0.964	69.15
27	3.407	0.044	1	0.913	3.291	0.35	0.964	56.568

In order to study the effect of parameters on the optimal solutions of the economic model, regression analysis was carried out for each parameter. Taking significance level as 0.1, conclusions are as follows:

1) k_1 and v are positively related. k_2 and u are negatively related. d_1 and u are positively related, and negatively related to b. d_2 is negatively related to λ, and is positively related to a. θ is negatively related to u, and is positively related to λ; ϕ is negatively related to λ.

2) The total cost is positively related to u, λ, d, g, n, and d_1, and is negatively related to k_1.

5 CONCLUSIONS

To effectively monitor the autocorrelation processes, the economic design model of ARMA control chart with variable sampling interval was established, and the model is solved in a practical case with genetic algorithm. ARL of the model was calculated with Markov chain method. Finally, sensitivity analysis for 12 model parameters were done to find out the effect of optimal combination of parameters $[k_1,k_2,n,d_1,d_2,\theta,\phi]$ on the control chart and the impact of the total cost per unit time, providing theoretical basis for enterprises to reduce process cost and improve production efficiency. Moreover, the economic statistical design is another question worth further researching.

ACKNOWLEDGMENTS

This research was supported by the National Natural Science Foundation of China Grants U1204702 and the Aeronautical Science Foundation 2013ZG55024 and the Graduate Education Innovation Fund of Zhengzhou University of Aeronautics 2016CX014.

REFERENCE

[1] **Ducan A.J.1956.** The economic design of X-bar charts used to maintain current control of a process, Journal of American Statistical Association, 51(274), pp 228-242.

[2] **Lorenzen, Thomas J., Vance C. L. 1986.** The economic design of control charts: a unified approach, Technometrics, 28(1), pp 3-10.

[3] **Lupo, Toni. 2014.** A Multi-Objective Design Approach for the c Chart Considering Taguchi Loss Function, Quality and Reliability Engineering International, 30(8), pp 1179-1190.

[4] **Shams M.B., Elkamel A., Moorthy K., et al, 2015.** Optimal design of T2 monitoring chart for chemical processes, International Journal of Process Systems Engineering, 3(4), pp 232-247.

[5] **Hu X., Sun J. 2015.** Economic design of X control chart with supplementary samples, Computer Integrated Manufacturing Systems, 1, pp 019.

[6] **Jiang W., Tsui K.L., Woodall, et al. 2000.** A new SPC monitoring method: The ARMA chart, Technometrics. 42(04), pp 399-410.

[7] **Lin S.N., Chou C.Y., S.L. Wang, et al. 2012.** Economic design of autoregressive moving average control chart using genetic algorithms, Expert systems with Applications, 39(2), pp 1793-1798.

[8] **Renolds M.R., Amin R.W., Arnold J.W., et al. 1988.** X charts with variable sampling intervals, Technometrics, 30, pp 181-192.

[9] **Yu F.J., Wu H.H.. 2004.** An economic design for variable sampling interval MA control charts, The International Journal of Advanced Manufacturing Technology, 24(1-2), pp 41-47.

[10] **Yu F.J., Chen Y.S.. 2005.** An Economic design for a variable sampling interval $\overline{X}$ control chart for a continuous flow process, The International Journal of Advanced Manufacturing Technology, 25(3-4), pp 370-376.

[11] **Chen Y.K., Chen C.Y.. 2012.** Economic Design of Cumulative Conformance Count Charts with Variable Sampling Intervals. Arabian Journal for Science and Engineering, 37(7), pp 2079-2088.

[12] **Noorossana R., Toosheghanian M., Gazaneh F.M. 2013.** Using genetic algorithm and response surface methodology for statistically constrained optimization of VSI X-bar control charts under multiple assignable causes and non-normality, The International Journal of Advanced Manufacturing Technology, 67(9-12), pp 2325-2342.

[13] **Lee M.H., Khoo M.B.C.. 2014.** Design of a multivariate exponentially weighted moving average control chart with variable sampling intervals, Computational Statistics, 29(1-2), pp 189-214.

[14] **Faraz A., Kazemzadeh R.B., Parsian A., et al. 2012.** On the advantages of economically designed the Hotelling's T2 control chart with variable sample sizes and sampling intervals, Quality & Quantity, 46(1): 39-53.

[15] **Faraz A, Kazemzadeh R B, Moghadam M B, et al. 2012.** On the advantages of economically designed variable sample sizes and sampling intervals T2 control chart: double warning lines scheme, Quality and Quantity, 46(4),pp 1323-1336.

[16] **Jiang W.. 2001.** Average run length computation of ARMA charts for stationary processes, Communications in Statistics-Simulation and Computation, 30(3), pp 699-716.

[17] **Huang X., Xu N., Bisgaard S.. 2013.** A class of Markov chain models for average run length computations for auto-correlated processes, Communications in Statistics-Simulation and Computation, 42(7), pp 1495-1513.

ISSUES IN HUMAN PERFORMANCE-BASED EVALUATION TEST FOR IMPROVING SYSTEM SAFETY

Xiaolu Dong[1*], Peng Liu[2], Le Liu[1†]
[1]Nuclear and Radiation Safety Center
Ministry of Environmental Protection, Beijing, China
dongxlchinansc@163.com
liule@chinansc.cn

[2]College of Management and Economics,
Tianjin University, Tianjin, China
lp527965873@163.com

ABSTRACT

Human plays an important role in social-technical systems. The reliability of human performance and the ability of human handling of abnormality are crucial to system safety. By conducting performance-based evaluation tests, evaluators will be able to identify potential problems and validate integrated system design. This paper first summarizes basic elements for the human performance-based evaluation tests, including task sampling, test team, participants, testbeds, as well as techniques for human performance measurements, data collection, and data processing. Examples in nuclear industry are provided on corresponding issues with recommendations for the evaluation tests. Based on the review, this paper then discusses the relationship between cognitive process in task performing and human performance measurements. Finally, further discussions are made on how the performance measurements are connected with evaluation objectives. Though the examples reviewed and recommendations suggested are from nuclear industry, they can be applied to the design and evaluation of human-system interface in other safety-critical systems.

Keywords: human performance, performance measurement, evaluation test, human-system interface design

1 INTRODUCTION:

The effect of human activities on safety is essential in safety-critical systems such as nuclear power plants, petrochemical plants, and aviation industry, etc. In socio-technical systems where people interact with technology, the reliability of human performance and the ability of human handling of abnormality are of vital importance [1]. For instance, modern control rooms in nuclear power plants are equipped with digital instrumentation and control systems, which results in the appearance of many task support systems such as computerized procedure system, digital alarm system, and safety function surveillance system [2]. Major activities for operators working in such control rooms include surveillance, monitoring, and diagnosis. In order to accomplish tasks in this sophisticated system, operators should coordinate between each other and cooperate as a team. Because most commands, especially safety-related ones, are coming from the control room, careful design and

* Corresponding Author

† Corresponding Author

thorough testing of human system interface in the control room are indispensible for the overall safety of the plant.

In order to improve system safety, designers generally conduct evaluation tests that involve system prototypes, tasks, and human operators during the design process. Similarly, performance-based validation tests are required to provide evidence on safe operation of the system, but with simulators and more realistic tasks. Considering the importance of human involvement and the characteristics of advanced technology in human-system interaction, the cognitive tasks should be scrutinized in the tests for evaluation and validation of interfaces and integrated systems [3][4].

This paper first summarizes major elements for a human performance-based test, which covers the stages from preparation, execution, to post-test analysis. Techniques for human performance measurements, data collection, and data processing are reviewed and provided with recommendations. It then discusses the relationship between cognitive process in task performing and human performance measurements. Finally, further discussions are made on how the performance measurements are connected with evaluation objectives.

2 MAJOR ELEMENTS FOR A HUMAN PERFORMANCE-BASED TEST

To conduct a performance-based test requires sufficient planning, careful execution, and detailed analysis. The following sections provide reviews on major elements.

2.1 Sampling of tasks

One of the keys to a successful performance-based test is the appropriate sampling of corresponding tasks. Task sampling should consider the purpose of the evaluation and the maturity of the system design. In nuclear industry domain, NUREG-0711 provides guidelines on the sampling of operational conditions from the aspects of plant conditions, personnel tasks and error-forcing contexts in the verification and validation process [3]. In order to comprehensively evaluate and validate the integrated system, tasks should cover as wide range as possible under the restriction of resources such as time and money. In engineering practices, the influences of such restrictions are so strong that the designers often have to make serious trade-offs.

Nevertheless, certain rules can be followed to obtain a reasonable sample of tasks. The following aspects can be considered: 1) functions specified in the requirement and design documents; 2) risk-important conditions not directly described in documents but have possible occurrence which require human handling; 3) error-forcing contexts and malfunctioning of supportive systems. Tasks satisfied the first criteria are easy to identify since they are clearly specified during the design process. However, because many new designs are based on the previous models, including only the design-based tasks in the performance-based tests will benefit little for the overall system improvement. As for the second aspect, analysts can apply methods of probabilistic safety analysis, operating experience review, and subject matter experts' (SMEs) judgment to select risk-important tasks that might be missed by the first rule. Finally, error-forcing contexts such as high workload, low situation awareness, and complicated diagnostic situations can be selected or embedded in the task scenarios developed under the first two rules. With the considerations of system malfunctioning, abnormalities can be added to the as-developed scenarios. Therefore, the evaluation team can observe the effect of system disturbances on human performance.

2.2 Test team, participants, and testbeds

It has been readily acknowledged that the test team should better be independent from the design team. Although the designers' opinions are necessary for the modification, independent evaluators could be more objective and thus could unveil more potential problems. Since the focus of the test is on human performance, experts specialized in system operation, procedure, human factors engineering, human-system interface, and training should be included in the team.

Generally, participants in the tests need to be representative of the operators in real situations, with the consideration of variability. However, tests conducted in preliminary design stage are sometimes suitable for domain experts to participate. Because when the final design has not been stabilized and training for the target users has not been completed, expert participants can provide useful and efficient input for the design process.

As for the testbed, early tests can be conducted on the engineering simulator or other simplified testbeds, but the final validation tests should be conducted on a simulator close to the real system, for example, full scale simulators in nuclear industry.

2.3 Human performance measurement

In nuclear industry, a comprehensive, hierarchical set of performance measures is recommended which includes the plant performance measures, personnel task measurement, cognitive factors measurement, and anthropometric & physiological factors [4]. Other researchers also suggested that teamwork performance could be included [5]. While the system performance measures are maintained within the acceptable range, latent problems could be revealed by unnecessary demands placed on operators or non-optimal decision making process. In general, the performance measures selected should acceptably exhibit the characteristics such as construct validity, reliability, resolution, sensitivity, diagnosticity, simplicity, and objectivity, etc. [6]

Human performance can be measured by task results, which are system conditions and task features such as time, accuracy, error, and frequency, etc. [3][4][6] From the cognitive perspective, major measures include cognitive workload and situation awareness. They can be measured based on performance, subjective rating, and physiological measurement [7][8][9]. In proving a proper system design, operators should be able to successfully accomplish the task with appropriate levels of workload and situation awareness.

2.3.1 Personnel task measurement

The personnel task measurement can be conceptualized along many dimensions. The measures used to quantify tasks should be chosen to reflect the important aspects of the task with respect to system performance, such as time, accuracy, frequency, errors, amount achieved or accomplished, consumption or quantity used, subjective reports of participant, behavior categorization by observers, etc. [3][4][6]. For some simple scenarios, measuring time to complete a task may be sufficient. For some more complicated tasks, especially those knowledge-based ones, a more fine-grained analysis may be appropriate. SMEs are indispensable in the evaluation.

2.3.2 Cognitive workload

Cognitive workload could influence human performance and human reliability in that overload may cause errors and performance deterioration. The measure techniques of workload can be categorized into predictive and empirical methods. Predictive techniques are based on mathematical modelling, task analysis, simulation, and expert's opinions. Whereas empirical techniques are based on experiments, and can be categorized into three categories: 1) performance, e.g. primary and secondary tasks; 2) subjective rating, e.g. NASA-TLX (NASA Task Load Index) [10][11] and SWAT (Subjective Workload Assessment Technique)[12][13], etc. ; 3) physiological measures, e.g. heart rate, heart rate variability, ongoing electroencephalographic (EEG), and eye movement [8][14][15][16].

2.3.3 Situation awareness

Situation awareness (SA) is the perception of the elements in the environment within a volume of time and space, the comprehension of their meaning, and the projection of their status in the near future. The measurement techniques can be categorized into four groups: performance-based measurement, subjective rating measurement, physiological measurement and simulation. SAGAT (Situation Awareness Global Assessment Technique) is an objective performance-based measurement that requires the participants to answer questions on three levels of SA during short stops in the test. The developers have conducted reliability and validity test for this technique [17][18]. SART (Situation Awareness Rating Technique) is a subjective rating technique that consists of ten dimensions and three dimensions for the simplified version of 3-D SART [9]. These two techniques are relatively more popular in measuring SA.

2.3.4 Team performance

There are many ways to describe important characteristics of a successful team, such as communication, cooperation, coordination, adaptability, leadership, assertiveness, situation assessment, and decision making, etc.[19] Measurements of team performance usually include behavior checklists and SMEs' judgment, for example, BARS (Behaviorally anchored rating scales)[5][20].

In the control room of NPPs, though each individual operator has certain task assignment and responsibility, the crew still must cooperate closely together to achieve the safe operation of the plant. Facing the dynamic system conditions during abnormalities, simple coordination may not be sufficient. It is better to have a mutual situational awareness among team members, so that the team can handle unexpected situations under high pressure. In order to do so, appropriate trainings are necessary in cultivating accurate understanding of team requirements and in developing a shared mental model of the team [21].

2.4 Data collection and processing

2.4.1 Data collection

Further down the line of data collection, recording techniques such as video and audio recording, screen recording, behavior observation, operation log, observer checklist, subjective rating, questionnaire, and interview, etc. can be considered [22]. Physiological indicators such as heart rate, EEG (electroencephalographic), and eye movement (blink rate and duration, fixation frequency and duration, gaze area, saccades, and pupil diameter etc.) can be adopted to facilitate the analysis on task performance, cognitive workload, and interface design[8][14][15][16][22]. Physiological records can be obtained continuously and simultaneously with the task execution.

Observers are usually included in the test for efficient evaluation and expert opinion. The observers' notes can contain information on the operators' behaviors, task performance, and errors made during the test. There can be check lists or pre-printed tables with necessary items to avoid omission. It is better to make the notes comparable with other recording data. Also, especially in validation tests, the observer should not interfere with the task operation. Walking around or talking to the participants would not only be distractive but also send a strong message to the participants that they are being observed, which is against the objectives of evaluation. Using a single sided glass or video surveillance cameras can be good alternatives [22].

Interviews after the test are important in understanding operators' cognitive process. Errors identified, difficulties encountered, and confusions appeared during the test need clarification through operators' recall of the process and explanation of the decision making. While the cognitive process cannot be directly measured, a carefully prepared interview can reveal much information.

However, it is not easy for the interviewer to motivate the participants to speak freely and completely. Although topics and questions can be prepared beforehand, the effectiveness of interview largely relies on the experience of the interviewer.

2.4.2 Data processing

Data processing requires much effort, because it contains steps from data exporting, conversion, and formatting, to data transcription and translation. In order to extract information from large amount of raw data, amendment or multiple transformations of the data may be required because data from different sources are not compatible with each other or convenient for further analysis. In data processing, the information available can be connected with each other. The data can be reconstructed according to the task timeline and cognitive models. Usability testing software can be helpful in the process because many of the standard modules can be directly applied and hierarchical data can be easily exported [22].

Data processing needs participation of SMEs. In many cases, for instance data transcription of interviews, video, and audio recordings, etc. requires experts' involvement, because inexperienced analysts are unfamiliar with the task operations and terminology. Furthermore, deep domain knowledge is indispensible for the content analysis and data coding, for example, in activities such as defining error type, categorizing operation behavior, and decomposing communication contents. Even when handling more hierarchical data, experts' judgments are still required in identifying important system conditions and critical human actions.

In the control room of NPPs, complicated tasks have high demands on coordination. The operators are required to work closely together under the surveillance of senior supervisor. Also, they are encouraged to read aloud the contents of the procedures and raise questions whenever necessary. The heavy dependency on verbal communication provides benefits in using protocol analysis technique [23][24]. Through this method, evaluators will be able to get qualitative and quantitative data on the process of situation assessment, diagnosis, planning, and decision making along with task operations. The analysis of communication can be a good way to link data collected from other resources.

3 DISCUSSION

By measuring human performance, one can assess the appropriateness of function assignment, interface design, training, and organization management. The results can facilitate corresponding design of the interface and training course, which can better support operators' task performance and improve system safety in return. Synthesizing related information is necessary for thorough analysis of important tasks and processes.

There are four stages in task performing from cognitive perspective: 1) information collection and interpretation; 2) situation assessment and diagnosis; 3) planning and decision making; and 4) task execution. The results of execution will become feedbacks to the "information collection and interpretation" stage and initiate a new iterative process. This cognitive process can be affected by many factors, including external factors such as environment, automation level, interface design, and organization management, as well as internal factors such as stress, physical conditions, and level of knowledge, etc. The task results, human behavior, subjective opinions on workload and situation awareness, and changes in physiological indicators are reflections of these factors on operators' cognitive process. Therefore in engineering practices, corresponding measurements can be taken to evaluate human performance as well as system design, as illustrated by figure 1.

In data analysis, the situational conditions under which human activities take place are important, in that the measurements can only be accurately interpreted with the considerations of the situational factors. In addition, the measures are sometimes influenced by same factors and thus are usually correlated. In order to have a comprehensive assessment of the system, all measurements should be considered in an integrated manner.

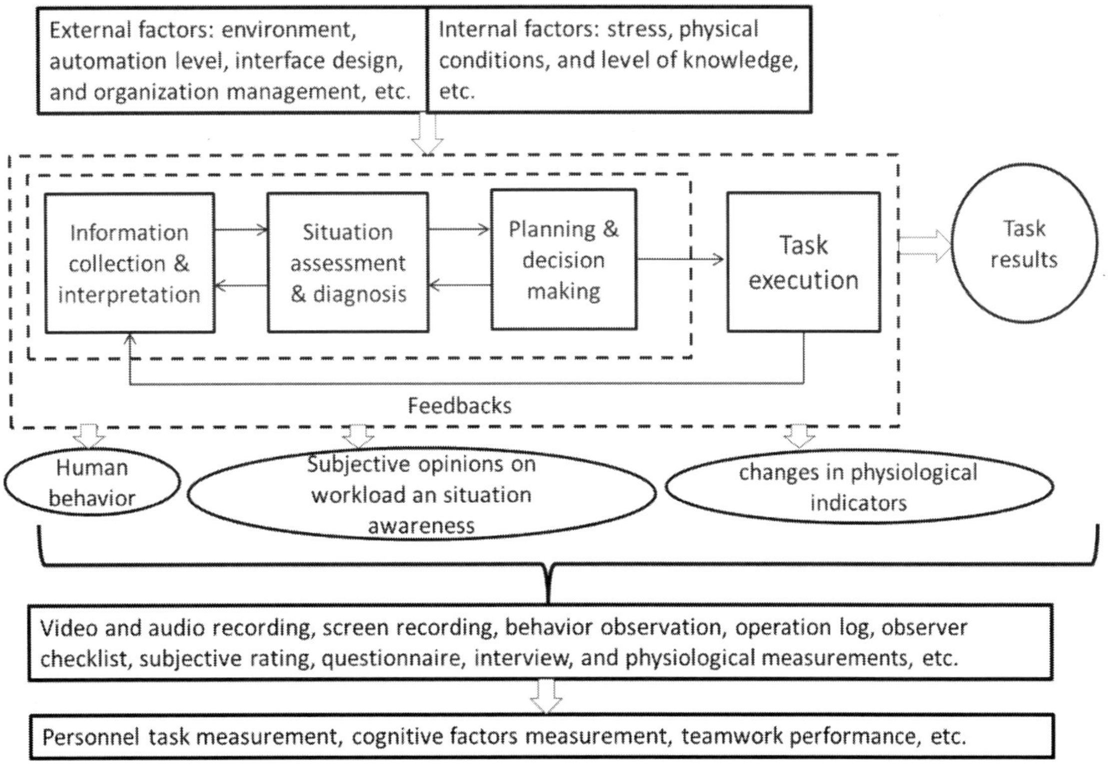

Figure 1: Cognitive process and human performance measurements

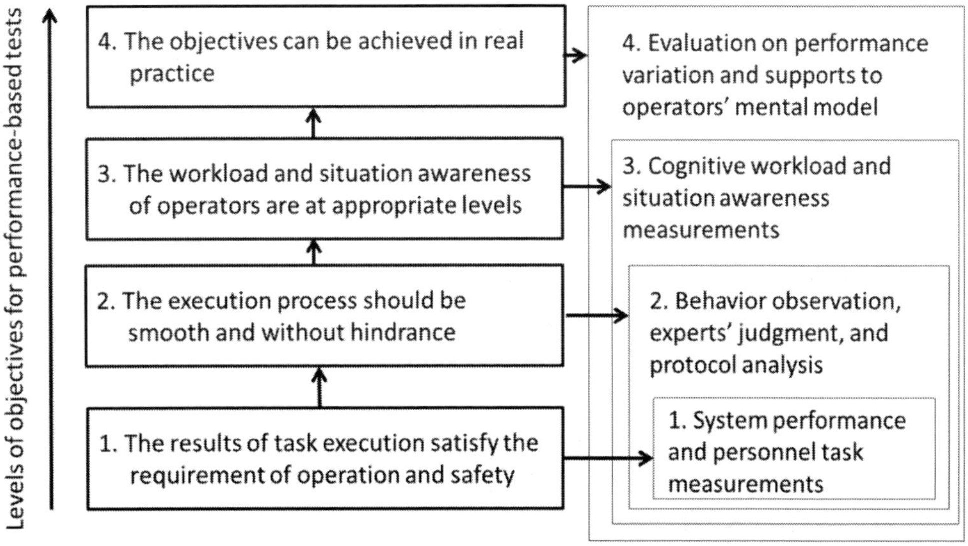

Figure 2: Levels of objectives and performance measurements

There are four levels of objectives in evaluation: 1) the results of task execution satisfy the requirement of operation and safety; 2) the execution process should be smooth and without hindrance; 3) the workload and situation awareness of operators are at appropriate levels; 4) the above objectives can be achieved in real practice, that is, can be generalized to conditions of greater scale. These four levels of objectives are gradually enhanced, which considers not only the task requirement, experts' evaluation, and conventions, but also the variation of human performance and uncertainties in real conditions. For the first level of objective, measurements of system performance and personnel tasks can be taken. For the second level, additional techniques of behavior observation, experts' judgment, and protocol analysis could be applicable. The measurements of cognitive workload and situation awareness belong to the third level. Whether these three levels of objectives can be achieved in real practice depends on the variation of performance and supports to operators' mental model. Apart from synthesizing results from measurements taken, further research on the cognitive process could provide supportive evidence. Figure 2 demonstrates the relationship between levels of objectives and various measurements.

4 CONCLUSION

Conducting human performance-based tests is critical for improving safety level of social-technical systems. In doing so, important actions can be validated and potential problems can be unveiled. This paper summarizes basic elements for the performance-based evaluation tests. Elements include tasks sampling, test team, participants, and testbeds in preparation stage, as well as human performance measurement and data analysis afterwards. Examples in nuclear industry are provided on each issue with recommendations for the evaluation tests.

Human performance measurements could be used to assess influencing factors like the interface design. Raw data like video and audio recording, screen recording, behavior observation, questionnaire, and physiological measurements, etc. describe changes in human activities, subjective opinions, and physiological status. Because these changes contain information on how human cognitive process is influenced by the external and internal factors, inference therefore can be made on the influencing factors. Measurements of tasks, cognitive factors, and teamwork could be taken using the above methods. They can be organized according to the four levels of evaluation objectives. In this way, a hierarchical set of performance measures can be constructed and evaluators would have a clearer idea in analysing and interpreting related data.

In general, the assessment of human system interface in social-technical systems has common objectives in spite of domain differences. Also, the techniques for evaluating human performance and interface design in modern systems are similar in nature. Though the examples reviewed and recommendations suggested in this paper are from nuclear industry, they can be applied to the design and evaluation of human-system interface in other safety-critical systems.

5 REFERENCES

[1] **Hollnagel, E. 1998.** *Cognitive Reliability and Error Analysis Method (CREAM)*, Elsevier.

[2] **National Research Council, U. S. 1997.** *Digital Instrumentation and Control Systems in Nuclear Power Plants: Safety and Reliability Issues*, National Academy Press.

[3] **O'Hara, J. M., Higgins, J. C., Persensky, J. J., Lewis, P. M., and Bongarra, J. P. 2012.** *Human Factors Engineering Program Review Model*, NUREG-0711.

[4] **O'Hara, J. M., Stubler, W., C.Higgins, J., and Brown, W. 1997.** *Integrated System Validation: Methodology and Review Criteria*, NUREG/CR-6393.

[5] **Ha, J. S., Seong, P. H., Lee, M. S., and Hong, J. H. 2007.** Development of Human Performance Measures for Human Factors Validation in the Advanced MCR of APR-1400, *IEEE transactions on nuclear science*, 54(6), 2687-2700.

[6] **Meister, D. 1985.** *Behavioral Analysis and Measurement Methods*, Wiley: New York.

[7] **Gawron, V. J. 2008.** *Human performance, workload, and situational awareness measures handbook*, Second Edition, CRC press: Boca Raton.

[8] **Tran, T. Q., Boring, R. L., Dudenhoeffer, D. D., Hallbert, B. P., Keller, M. D., and Anderson, T. M. 2007.** Advantages and Disadvantages of physiological assessment for next generation control room design, IEEE 8th Human Factors and Power Plants and HPRCT 13th Annual Meeting, Vol. 1, pp. 259-263: IEEE.

[9] **Endsley, M. R., and Garland, D. J. 2000.** *Situation Awareness: Analysis and Measurement*, CRC Press.

[10] **Hart, S. G., & Staveland, L. E. 1988.** *Development of NASA-TLX (Task Load Index):results of empirical and theoretical research*, North Holland Press: Amsterdam.

[11] **Hart, S. G. 2006.** NASA-Task Load Index (NASA-TLX): 20 years later, Proceedings of the Human Factors and Ergonomics Society 50th Annual Meeting, Santa Monica.

[12] **Reid, G. B., and Nygren, T. E. 1988.** *The subjective workload assessment technique: A scaling procedure for measuring mental workload*, Elsevier: Amsterdam.

[13] **Potter, S. S., and Bressler, J. R. 1989.** *Subjective workload assessment technique (SWAT): A user's guide*, Wright-Patterson Air Force Base.

[14] **Kramer, A. F. 1991.** *Physiological metrics of mental workload: A review of recent progress*, in Multi-task environments, Taylor and Francis: London.

[15] **Wilson, G. F., and Eggemeier, F. T. 1991.** *Psychophysiological Assessment of Workload*, in Multi-task environments, Taylor and Francis: London.

[16] **Lin, Y., Zhang, W.J., Watson, L.G., 2003.** Using eye movement parameters for evaluating human-machine interface frameworks under normal control operation and fault detection situations, *International Journal of Human-Computer Studies*, 59 (6),837-873.

[17] **Endsley, M. R. 1995.** Measurement of Situation Awareness in Dynamic Systems, *Human Factors: The Journal of the Human Factors and Ergonomics Society*, 37(1), 65-84.

[18] **Endsley, M. R. 1990.** Predictive Utility of an objective measure of situation awareness, Proceedings of the Human Factors Society 34th Annual Meeting, Santa Monica, CA.

[19] **Brannick, M. T., Salas, E., and Prince, C. W. 2009.** *Team Performance Assessment and Measurement: Theory, Methods, and Applications*, Psychology Press, Taylor & Francis group: New York, London.

[20] **Roth, E. M., Mumaw, R. J., and Lewis, P. M. 1994.** *An empirical investigation of operator performance in cognitively demanding simulated emergencies*, Washington, D.C.

[21] **Cannon-Bowers, J., Salas, E., and Converse, S. A. 1993.** *Shared Mental Models*, in Expert Team Decision Making, Lawrence Erlbaum Associates: Hillsdale, NJ, England.

[22] **Dong, X., Song, F., Li, Z., and Zhang, S. 2013.** Data extraction and analysis for integrated system validation of a nuclear power plant, *Nuclear Engineering and Design*, 265, 826-832.

[23] **Ericsson, K. A., and Simon, H. A. 1993.** *Protocol analysis: Verbal reports as data*, MIT Press: Cambridge, MA.

[24] **Carvalho, P.V.R., Vidal, M.C.R., Carvalho, E.F.D., 2007.** Nuclear power plant communications in normative and actual practice: a field study of control room operators' communications, *Human factors and ergonomics in manufacturing*, 17(1), 43-78.

MINING EMPIRICAL AND TEMPORAL MEDICAL DATA FOR MULTISCALE DECISION-MAKING

Haiyan Yu[1*], Man Xu[2], Jiang Shen[3] and Ya Xu[1]
[1]School of Economics and Management, Chongqing University of Posts and Telecommunications, Chongqing, 400065, China
yhy188@tju.edu.cn
[2] College of Management and Economics, Tianjin University, Tianjin, 300072, China
[3] Business School, Nankai University, Tianjin, 300071, China

ABSTRACT

In healthcare recommender systems, making decisions is a complex task for physicians at all levels of the empirical hierarchy. To achieve an optimal course of action, a physician has to qualify the new additive information from the temporal data and the historical decisions of other physicians, engaging in a stochastic, belief-theoretical reasoning process. Furthermore, lower-level physicians seek to align the empirical information of their seniors by social network theory. Mining competent patient profiles and their temporal monitoring data need to include the labeling efforts of higher-level physicians on their expertise in the query class label predictions. In this paper, we present a multiscale decision-making model that accounts for the new additive information from the temporal data and empirical information in the historical case base. Multiscale decision-making combines the base rate probability from unipartite network and the true positive rate of the temporal attribute to model and solve empirical- temporal -scale problems. This is the first model that unifies the temporal and empirical scales and can solve an belief updating problem. We implement the model to a critical care decision challenge that illustrates the applicability and efficiency of the model.

Keywords: Multiscale decision-making, Patient profile, Social network, Recommender system, Temporal data mining

1 INTRODUCTION

In the context of big data, the externality of data sources extends the traditional industrial engineering boundaries, leading to a new perspective of decision making through cross-boundary association and internal-external data fusion. With increased volume and acquisition speed of data from both wearable and implantable sources, new automated algorithms are needed to reduce false alarms such that they are sufficiently robust to support large-scale deployment, particularly or free-living environments. Automatic classifications are necessary since the dataset sizes are beyond the capability of manual interpretation within a reasonable time period. Most of this work focused on the analysis of very large multidimensional longitudinal patient data collected over many years. However, most clinical databases provide low temporal resolution information due to the difficulty in collecting rich long-term time-series data. In healthcare, data heterogeneity and variety arise as a result of linking the diverse range of biomedical data sources available. Sources can be either

* Corresponding Author. yhy188@tju.edu.cn

This work was supported in part by National Natural Science Foundation of China (71601026, 71571105) and Science and Technology Research Project of Chongqing Municipal Education Commission(KJ1600401).

quantitative (e.g., sensor data, images, gene arrays, laboratory tests) or qualitative (e.g., free text, demographics). The objectives underlying this data challenge are to support the basis for observational evidence to answer clinical questions, which would not otherwise been solved via studies based on randomized trials alone. In addition, the issue of generalizing results based on a narrow spectrum of participants may be solved by taking advantage of the potential of big data for deploying longitudinal studies.

Wemz (2010) first proposed multiscale decision (MSD), which built upon the theory of distributed decision making, hierarchical, multi-level systems and dependency graphs. MSD has been applied to service operations by unifying temporal and organizational scales. Yu (2015) has subsequently extended MSD in clinical decision-making by unifying medical temporal and empirical scales. In this study, diagnostic decisions have been achieved by fusing the knowledge from temporal data and the empirical knowledge from medical experts. Generally, MSD implements modeling and solving techniques from Markov decision processes, through which transition probabilities are deduced for the decision states. Moreover, differences lie in those two kinds of research, Wemz (2010) applies stochastic game theory to model the objectives of the process from the subjective agent, while Yu (2015) applies belief aggregation theory to model its objectives from the partially observed data.

We present a multiscale decision-making model that accounts for the new additive information from the temporal data and empirical information in the historical case base. Multiscale decision-making combines stochastic belief updating with partially observed Markov decision processes to model and solve multi-time-scale and multi-empirical-scale problems. We also implement the model to a critical care decision challenge that illustrates the applicability and efficiency of the model. This paper attributes to the foundation of multiscale decision theory with data mining and represents a key step towards solving the belief updating problem with empirical and temporal data.

2 MINING EMPIRICAL AND TEMPORAL MEDICAL DATA

Personalized recommendation systems have emerged as essential elements of medical decision support practice and business analytics initiatives. However, empirical and temporal data with various structures may prevent these systems from achieving their full potential. To formally describe the business implications of this predicament, we proposed the multiscale decision making framework.

2.1 MULTISCALE DECISION-MAKING FRAMEWORK

In process of MSD, three types of decision period include short term, middle term and long term. The data collected in those decision periods has different data granularities, which can be split and reorganized by time scale.

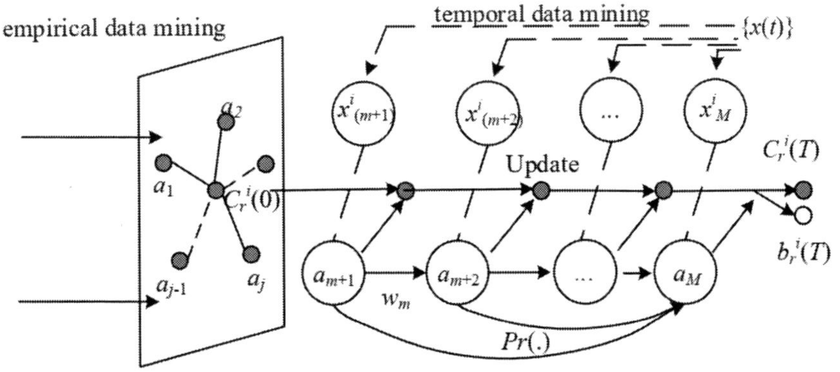

Figure 1: Multiscale decision-making framework with empirical and temporal data

To illustrate the data acquisition problem in internal-external data fusion, we follow Zheng and Padmanabhan's notation, such that the local data D_L (or $S_0^{(q)}$) contains m attributes, $x^{(i)}$ ={ $a_1^{(i)}$, $a_2^{(i)}$, ... , $a_m^{(i)}$ }, where $c^{(i)}$ is a specific target variable ($c^{(i)}$ can be defined as whether or not a patient has heart disease within a certain period) and i is the index of the entities. After the cross-boundary data association process, more temporal attributes including inspection items in EMRs and other data outside of EMRs (i.e., medicare data and environmental data) are integrated with the local data. Let D_G be the global data set that contains attributes { $a_1^{(i)}$, $a_2^{(i)}$,..., $a_m^{(i)}$, $a_{m+1}^{(i)}$,..., $a_N^{(i)}$, $c^{(i)}$ }. Here assume that the values for attributes { $a_{m+1}^{(i)}$,..., $a_N^{(i)}$ } are sparse and with different granularities, but can be represented with time window. Let $A(t)$ denote the observing value matrix of temporal attributes at moment t , $A(t)=\left[a_j^{(i)}(t)\right]_{i=1,j=m+1}^{i=n,j=N}, 0 \le t \le T$, which is the values of the j th attribute of the i th instance in the period $[0,T]$. $a_j^{(i)}(t)$ is the observing values at the moment t . $C^{(i)}(t)$ denotes the decision label of $x^{(i)}$ at t , $C_i(t)$ $\in$ { C_{k_1} , C_{k_2} } for the binary classification in $[0,T]$.

Accuracy (Alizamir et al. 2013) is defined as the probability that the patient label is correctly classified. To achieve high accuracy, two stages of inference are designed to construct the reasoning machines. First, with the empirical data, we assume that all patients start with a base rate probability $\beta^{(q)}$ of being label C_{k_1} . Subsequently, with the sensitivity and specificity of each new attribute's discrimination ability, this temporal attribute data updates the subjective probability of patient label according to Dempster' rule (Fu, L. et al, 2015).

2.2 Mining Empirical Data with Unipartite Network

In bipartite network (Zhou T et al, 2007, Landon B.E., et al. 2012) , the nodes can be partitioned into two sets with all their ties(also called links, edges or connections) linking nodes from one set to the other and no ties within one set. In medical decision system, the node sets can refer to physicians and patients. Basic elements of network diagrams are demonstrated in Figure 2 and 3.

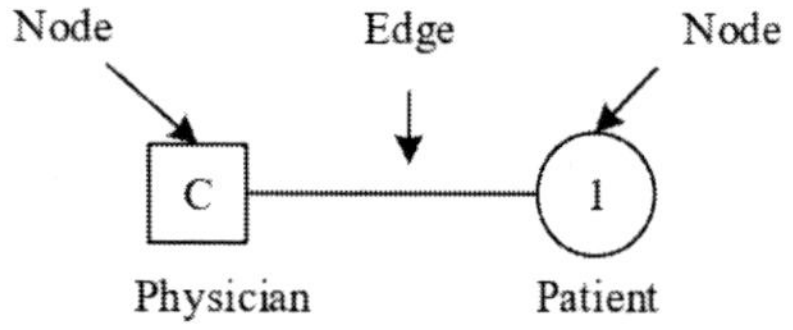

Figure 2: A network between 2 types of nodes (physicians and patients)

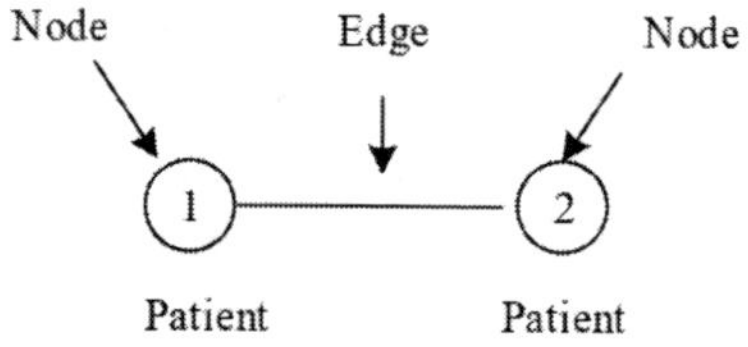

Figure 3: A network between a single type of node (patients)

In more familiar networks, unipartite representation depicts the people (i.e., physicians or patients) as nodes, with ties between those on the same board, forming cliques. Although this approach may discard important information and conflate various structures, we converted the physician-patient bipartite network to a unipartite network of linked patients, shown as Figure 4. Based on Electronic Medical Records (EMRs), we can construct patients' social network which profiles the patient-centric data base. Figure 5 illustrates Patient 6's social network, which consists of five patient neighbour nodes and edges designating communication weight between any two.

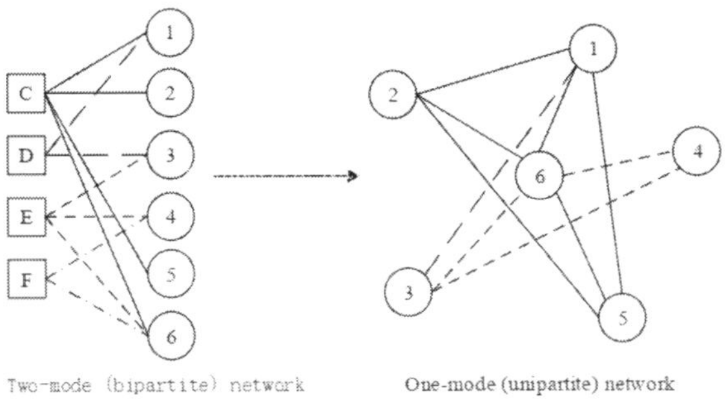

Figure 4: Projection of a unipartite network from a bipartite network

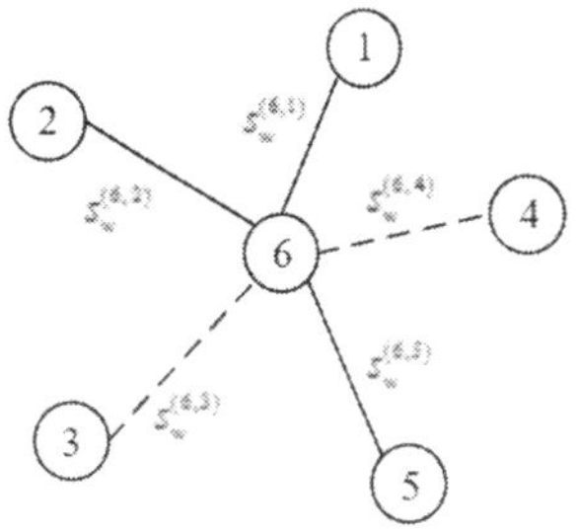

Figure 5 : EMRs for Patient 6's social network

A patient node's probability of class membership is determined by the membership of the unipartite network's neighbours (Park et al., 2012). In Patient-centric network, the weights are determined by the inspection items which those two patients are sharing during the hospital duration. The time is the period that the patient stayed during this time.

Due to uncertainty of the unipartite network, similarity is adopted to describe its distribution. We rely upon the framework suggested by Yu, H. et al.(2016), the base rate probability $\beta^{(q)}$ of the query being label C_{k_1} :

$$\beta^{(q)}=\frac{\sum_{k=1}^{K} s_w^{(q,i_k)}\cdot\beta^{(i_k)}}{\sum_{k=1}^{K} s_w^{(q,i_k)}} \tag{3}$$

where $s_w^{(q,i_k)}$ is the similarity between the node i_k and q, $\beta^{(i_k)}$ is the belief of the node i_k being label C_{k_1} and K is the size of the query's neighbours. The function $s_w^{(q,i)}$ is a non-negative real value similarity function, satisfying the properties of symmetry, transitivity and multiplicity. The form of similarity can be chosen due to its applications (Man XU et al, 2012; Yu H. et al, 2015). Here, we can adapted the Euclidean distance to demonstrate the dissimilarity of these neighbours, and adapted the rate of the number of the common inspection items between any two nodes by the maximum.

2.3 Classifying Temporal Data with Its Discrimination Ability

In coarse-grained process, multiple entries of the raw temporal data are mapped onto an entry of the new coarse-grained temporal data, by increasing the length of the non-overlapping time windows. Given the multi-variate temporal data $A(t)\,|\,t=1,\ldots,T$, where T is denoted as the number of samples in each dimension of the temporal data, ε as the scale factor, the process is a type of multi-coarse (e.g., ε :1). Assuming $f(\cdot)$ the data mapping or aggregation function, the multi-coarse data is reconstructed as $B(\tau)=\{b_{ij}^{\varepsilon}(\tau)\}$, $b_{ij}^{\varepsilon}(\tau)=f(a_j^{(i)}(t)),\ t=t=(\tau-1)\varepsilon+1,\ldots,\tau\varepsilon$. If the mapping rule is the average of the data, then the coarse-grained data of the jth attribute of the ith case in the raw temporal data: $b_{ij}^{\varepsilon}(\tau)=\frac{1}{\varepsilon}\sum_{t=(\tau-1)\varepsilon+1}^{\tau\varepsilon} a_j^{(i)}(t)$, $1\le\tau\le[T/\varepsilon]$, abbreviated as. $[T/\varepsilon]$ is an integer not greater than T/ε, $1\le\varepsilon\le T/2$.

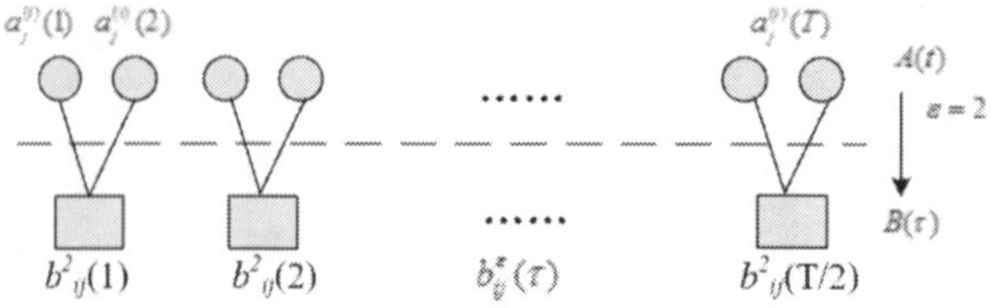

Figure 6: The coarse-grained temporal data (take $\varepsilon = 2$ for example)

Given that $B(\tau)$ is the multi--grained attribute set, the discrimination ability $h(a_j^k, r_i^j)$ for the temporal data is such that discrimination ability increases (decreases) with data aggregation efforts (data sparsity). To formally describe the distribution of temporal data, true positive rate of the query patient is

$$P_d(b_{ij}^{\varepsilon}(\tau)) = x^a P_f^a(b_{ij}^{\varepsilon}(\tau)) + (1 - x^a) P_f(b_{ij}^{\varepsilon}(\tau)), 0 \le a \le 0.5 \qquad (1)$$

where a is a domain parameter that illustrates the difficulty of diagnosis. A domain with a higher value of a demonstrates where diagnosis is more difficult. More properties of this parameter has been detailed in Bensoussan et al. (2009).

2.4 *Hybrid Mining Model of Empirical and Temporal* Data

With the base rate probability $\beta^{(q)}$ and the true positive rate, the posterior probability of the query can be achieved with belief updating mechanism. As Dempster' rule has the ability of reducing the inconsistency between heterogeneous data, we adapted this rule to update the belief of the query patient. If the forthcoming state of the class label is positive, the posterior probability becomes

$$\tilde{\beta}^{(q)} = \frac{\beta^{(q)} \cdot P_d(b_{ij}^{\varepsilon}(\tau))}{1 - \Gamma} \qquad (13)$$

where $\Gamma = \beta^{(q)}(1 - P_d(b_{ij}^{\varepsilon}(\tau))) + (1 - \beta^{(q)}) P_d(b_{ij}^{\varepsilon}(\tau))$. This hybrid mining model (HMET) leverages the similarity knowledge of the social network based on EMR empirical data and the discrimination ability of the temporal data with multiple scales.

3 NERMERICAL EXPERIMENT

The sharing of patient information between hospitals and many other health care related organizations has been gaining hot attentions recently (Mikhno et al., 2015). The role of medical data fusion process can attribute to reduce its misdiagnosis and improve the system discrimination ability. Electronic medical records (EMRs) systematically collect patients' data electronically, sharing across different health care settings. EHRs may include a range of data, including empirical data (patient profiles, i.e., demographics, medical history) and temporal data (inspection items, i.e., CT images, vital signs). Due to the characteristics of temporal, fragmentation and uneven distribution of EMRs, various types of decision period could occur during the patient stay duration in the hospital. The data collected in those decision periods has different data granularities, which can be split and reorganized by time scale.

Our experimental data is obtained from Framingham Heart Study (Fowler, J. H. et al, 2008), in which we acquired an experimental data set with 4240 entities and 13 attributes. We have validated the proposed model HMET and compared the results with two related models, TkNN(Yu H, 2016) with temporal case and Bayesian network(BN). The receiver operating characteristic (ROC) curves, representing system prediction performances, are demonstrated in Figure 7.

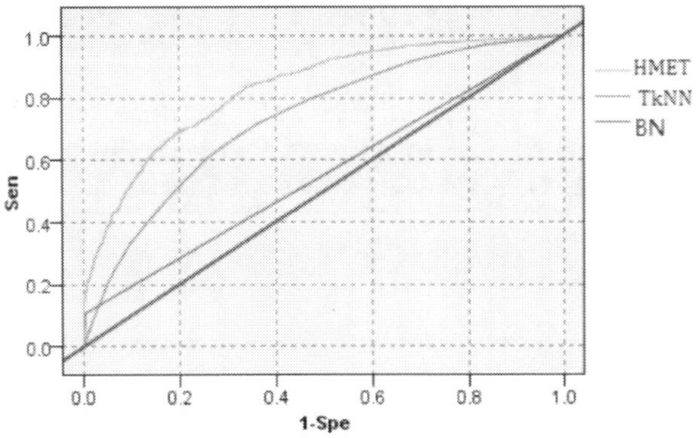

Figure 7: Comparison Results of Related Models

The AUC of the ROC curve of HMET is larger than the AUC of the ROC curves of the other two models. Although we cannot simply conclude that training HMET model is the best which can help the system achieve the best prediction performance, we used P value to compare the AUC of the ROC curves and measure the significance level of the system prediction performance improvement after each training round. The results of $P<0.05$ demonstrated that the HMET model is more effective in predicting the patients' labels and reducing the misdiagnosis rate.

4 CONCLUSION

Due to the characteristics of temporal, fragmentation and uneven distribution of the clinical data in evidential reasoning, the medical data fusion process has the interesting and challenging issue of reducing its misdiagnosis and improving the system discrimination ability. However, the previous data fusion approaches on multi-source heterogeneous data cannot deal with the issues for providing resilient decision support. To solve these issues, we first introduce the concept of multiscale on three dimensions of organizational structure, decision time and its locations. The multi-granularity multi-attribute data space was constructed with the social network theory in probability graph model, then the temporal medical data were analyzed with pattern recognition, segmentation and associative rules. The model of belief fusion was constructed based on Dempster' rule, achieving the confidence fusion of fragmented evidence. Through the parameter inference mechanism of multiscale decision-making model, the recommendation of identifying the labels of medical instances were achieved with the belief trajectory of the reasoning model. Finally, experiments on clinical test data verified the effectiveness and resilience of the evidence-based reasoning models and their multi-scale decision-making mechanism, improving the efficiency and quality of medical services.

5 REFERENCES

[1] Alizamir, S., de V, Ricourt, F., & Sun, P. (2013). Diagnostic accuracy under congestion. *Management Science*, 59(1), 157-171.

[2] Bordley, R. F. (2011). Using bayes' rule to update an event's probabilities based on the outcomes of partially similar events. *Decision Analysis*,*8*(2), 117-127.

[3] Bensoussan, A., Mookerjee, R., Mookerjee, V., & Yue, W. T. (2009). Maintaining diagnostic knowledge-based systems: a control-theoretic approach. Management Science, 55(2), 294-310.

[4] CHEN Guoqing (2016). Big Data Driven Management and Decision Sciences - Overall Framework and Highlights of the NSFC Grand Research Plan, NSFC-CUHK Symposium on Big Data Driven Management & Decision Research,HongKong

[5] Fowler, J. H., & Christakis, N. A. (2008). Dynamic spread of happiness in a large social network: longitudinal analysis *over* 20 years in the framingham heart study. *Bmj*, 337(49), a2338-a2338.

[6] Fu, L., Shen, M., Hai, Y., & Xu, M. (2015). Random disturbance reasoning model of decision-making system and its anti-interference capability. *International Journal of Industrial Engineering*,22(5), 645-660

[7] XU, Man, YU, Haiyan, & Shen, Jiang. (2012). New algorithm for CBR-RBR fusion with robust thresholds. Chinese *Journal of* Mechanical *Engineering*, *25*(6), 1255-1263.

[8] XU, Man, YU, Haiyan, & Shen, Jiang. (2013). New approach to eliminate structural redundancy in case resource pools using α mutual information. Journal of System Engineering and Electronics, 24(4), 625-633.

[9] Mikhno, A., Zanderigo, F., Ogden, R. T., & Mann, J. J. (2015). Toward non-invasive quantification of brain radioligand binding by combining electronic health records and dynamic pet imaging data. *IEEE Journal of Biomedical & Health Informatics*, *19*, 1271-1282.

[10] Landon, B. E., Keating, N. L., Barnett, M. L., Onnela, J. P., Paul, S., & O'Malley, A. J., et al. (2012). Variation in patient-sharing networks of physicians across the united states. *Jama the Journal of the American Medical Association*, *308*(3), 265-73.[12]

[11] Park, S. H., Huh, S. Y., Oh, W., & Han, S. P. (2012). A social network-based inference model for validating customer profile data. *Mis Quarterly*, *36*(4), 1217-1237.

[12] Shen, J., Yu, H. Y., & Xu, M. (2015). Heterogeneous evidence chains based fusion reasoning for multi-attribute group decision making.*Acta Automatica Sinica*, *41*(4), 832-842.

[13] Wernz, C., & Deshmukh, A. (2010). Multiscale decision-making: bridging organizational scales in systems with distributed decision-makers.*European Journal of Operational Research*, 202(3), 828-840.

[14] YU, Haiyan, Shen, Jiang & XU, Man. (2015). Ecs-based reasoning for group decision analysis in the mislabeled classification context. System Engineering and Electronics, 37(11), 2546-2553

[15] Yu, H., Shen, J., & Xu, M. (2016). Temporal case matching with information value maximization for predicting physiological states.*Information Sciences*, *367-368*, 766-782.

[16] Zheng, Z., & Padmanabhan, B. (2006). Selectively acquiring customer information: a new data acquisition problem and an active learning-based solution. *Management* Science, *52*(5), 697-712.

[17] Zhou, T., Ren, J., Medo, M., & Zhang, Y. C. (2007). Bipartite network projection and personal recommendation. *Physical Review E Statistical Nonlinear & Soft Matter Physics*, *76*(2), 70-80.

THE SIMULATION AND OPTIMIZATION OF STEERING KNUCKLE PRODUCTION LINE BASED ON PLANT SIMULATION

Danting.Li[*]

Department of Industrial Engineering
Harbin University of Commerce, CHINA
ldtjing@sina.com

ABSTRACT

Layout problem is the key problem in the production planning of enterprise. The optimization of production logistics system is an important way to improve enterprise's productivity, cut down cost and improve equipment's utilization. For this demand of enterprises, this paper used genetic algorithm and the computer simulation technology to seek out a most optimal layout design in an actual plant.The Plant Simulation is applied to simulate the steering knuckle production line to achieve the simulation and optimization of the production line balance.The simulation and verification can offer great decision making support for the production line's production logistics optimization, which ensure the scientific management decisions.

Keywords: Production logistics, Plant simulation, Simulation, Optimization

1 INTRODUCTION

Simulation technology is increasingly becoming the main support tool for many fields to solve and analyze the problems. It plays a huge role in many aspects,such as the advance planning of project, investment balance analysis and operation and control system of the production logistics, supply-chain management and inventory control,scheduling, resource allocation, process analysis and improvement[1].

Many enterprises has production logistics problem including low productivity, long production cycle and waste of resources.Production logistics system is highly randomized, complicated and discrete system, which can hardly be modeled and analyzed in the traditional way, so computer simulation technology and plant simulation simulation platform were introduced in the paper to model and analyze production logistics system,more than modeling plant simulation also can analyze production logistics system's equipment's utilization, working efficiency, production capacity statistics effectively, which can offer useful guideline for production logistics system optimization and scientific decision-making.The computer simulation technology is a effective way to find and solve these problems, it based on the production line balancing theory and method ,it can shorten the bottleneck process of assignments and preparation time, reduce the time wasted.It can improve the production situation of production line, increase the output and reduce the labor cost,In the case of enterprises with no investment or less investment and no increase in labor intensity or even lower intensity of labor [2, 3] .

2 ANALYSIS OF THE STEERING KNUCKLE PRODUCTION PROCESS ROUTE

The steering knuckle is one of the main parts of the automobile steering axle. Its main function is to bear the load of the front of the vehicle, support and drive the front wheel to rotate around the

[*] Corresponding Author

main pin.The whole steering knuckle production process includes 32 procedures,which is listed in Fig.1.

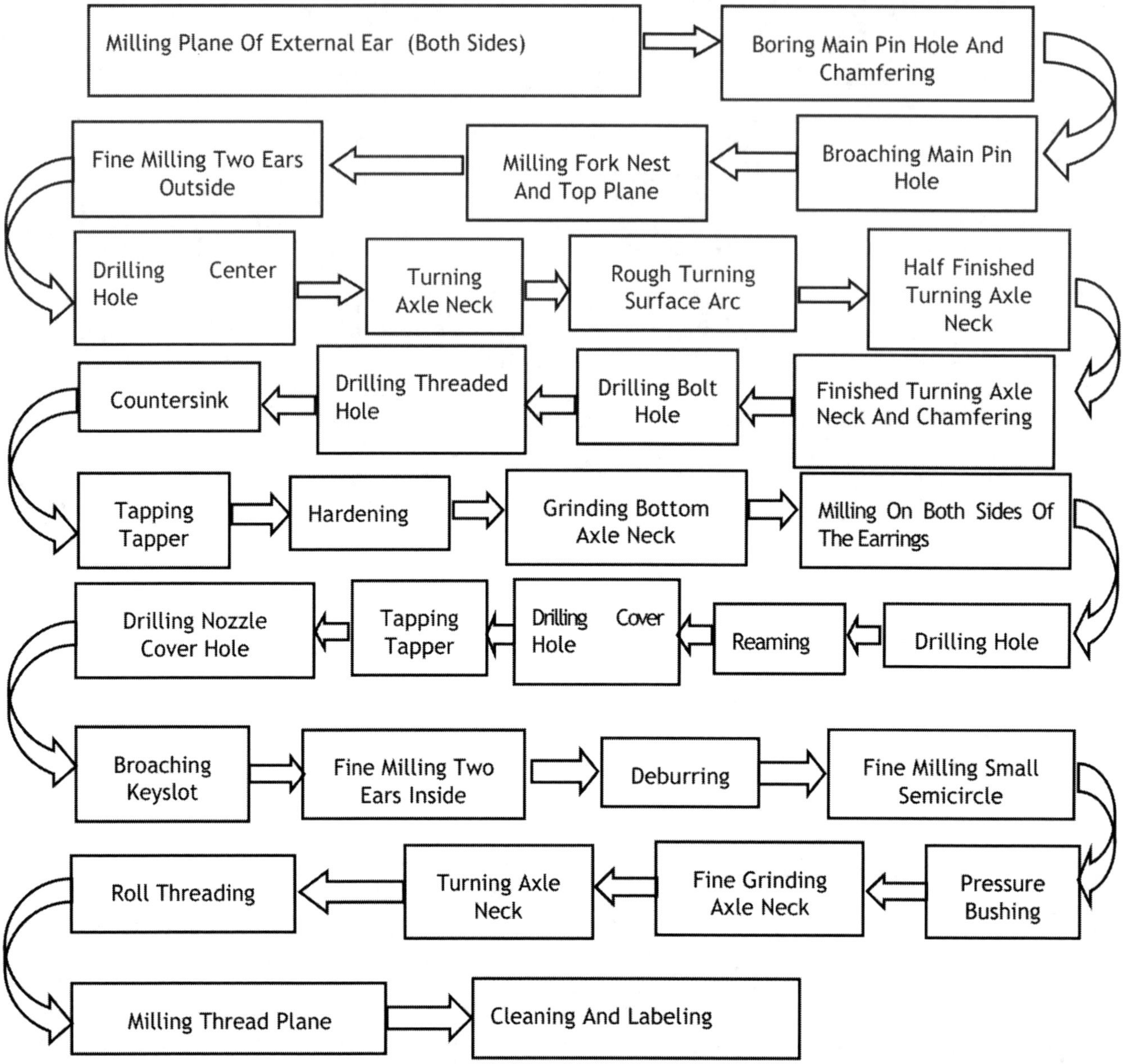

Figure 1: Steering Knuckle Production Route

3 PRODUCTION LOGISTICS SIMULATION MODEL

3.1 Software System Modeling Principles

The Plant Simulation software uses genetic algorithm to assign the work place, and begins to seek the balance of the production line. Model first starts "Start" to carry on the initial setting, starts "GAOptimization" to produce an initial code. Exchange and adjustment of coded positions by "SeqChg". And then a reasonable processing order will return to the "GAOptimization"as the father, to the "GASequence" tools through crossover, mutation and other genetic operations and produce offspring. All of these codes must be checked by "Evaluate". Reasonable coding by "Function" to carry out the distribution of the work station, fitness evaluation, a generation of excellent coding by the "GAOptimization" to retain. When the simulation is finished, "GAOptimization" will call the method "Termination", the optimal set of coding records in the "BestSolution" [4-7],the process can be expressed in Fig. 2.

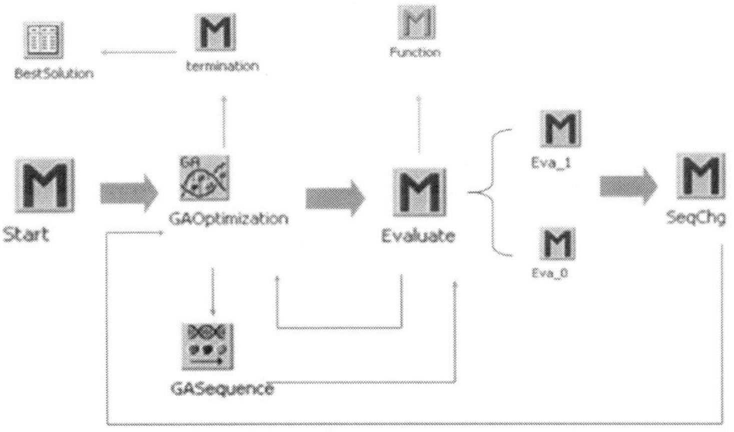

Figure 2: Software Operation Process

3.2 Production line Simulation Based on Plant Simulation

According to the steering knuckle production process route, edit the definition in the software. Then right-click the start icon. After the operation, the balance of the production line is optimized, and the simulation model of the production line is generated automatically .the simulation model of the production line build in plant simulation is showed in Fig.3[8,9].

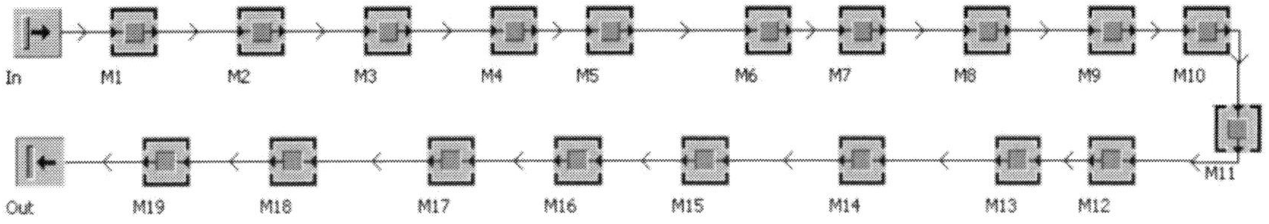

Figure 3: The Plant Simulation Model Of The Steering Knuckle Production Line

3.3 The Simulation Results by Plant Simulation

Once the production line's simulation model is completed, run the simulation model,Plant simulation can offer many kinds of statistics table and chart for us to analyze and optimize. Here we take the workstation's working balance histogram to analyze the workstation working load, which is show in Fig.4. tells us that there is huge balance loss in the production line, which results in waste of resources and time, and the production line must be optimized[10].

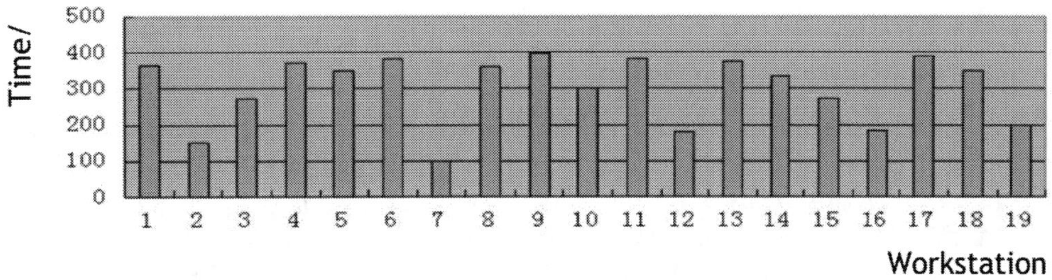

Figure 4: Production Line Balance Histogram

4 THE OPTIMIZATION AND SIMULATION OF PRODUCTION LOGISTICS

4.1 Production Line Process Route Optimization

By the above evaluation, you can see that in Plant Simulation modeling software Simulation of steering knuckle line balancing result and improve space, mainly to improve the work station for workstation 1 and 2, 6, 7, 11, 12, 16 and 19, among them the key to improve workstation 2 and 7. The operation of the process time is standard time for work, excluding all kinds of the external factors influencing the operating time, assuming change personnel arrangement does not affect the completion time for process. With the basic principle of line balancing, that is establishing the least amount of work station and the average working time of each workstation.With the rest of the workstation status arrangements remain unchanged, you can get the improvement of the steering knuckle production line work station arrangements, such as Table 1.

Table 1: Steering Knuckle Production Workstation After Optimization

NO	Workstation	NO	Workstation
1	Milling Plane Of External Ear (Both Sides)	11	Drilling And Reaming Hole
2	Boring Main Pin Hole And Chamfering, Broaching Main Pin Hole	12	Drilling Cover Hole
3	Milling Fork Nest And Top Plane	13	Tapping Tapper
4	Fine Milling Two Ears Outside And Drilling Center Hole And Turning Axle Neck	14	Drilling Nozzle Cover Hole
5	Rough Turning Surface Arc And Half Finished Turning Axle Neck And Finished Turning Axle Neck And Chamfering	15	Broaching Keyslot And Fine Milling Two Ears Inside
6	Drilling Bolt Hole And Drilling Threaded Hole	16	Deburring And Fine Milling Small Semicircle
7	Countersink And Tapping Tapper	17	Pressure Bushing And Fine Grinding Axle Neck
8	Hardening	18	Turning Axle Neck And Roll Threading
9	Grinding Bottom Axle Neck	19	Milling Thread Plane ,Cleaning And Labeling
10	Milling On Both Sides Of The Earrings		

4.2 Simulation Results After Optinuzation

Once the optimization simulation model is completed, run the simulation model, we can get the interesting statistics after the simulation, hear we take the workstation's working balance as an example.The number of the entire steering knuckle workstation is unchanged, reduce a machine for milling on both sides of the earrings and tapping machine, add a pressure bushing machine. After the optimization, the workstation's working balance histogram is shown in Fig.5.

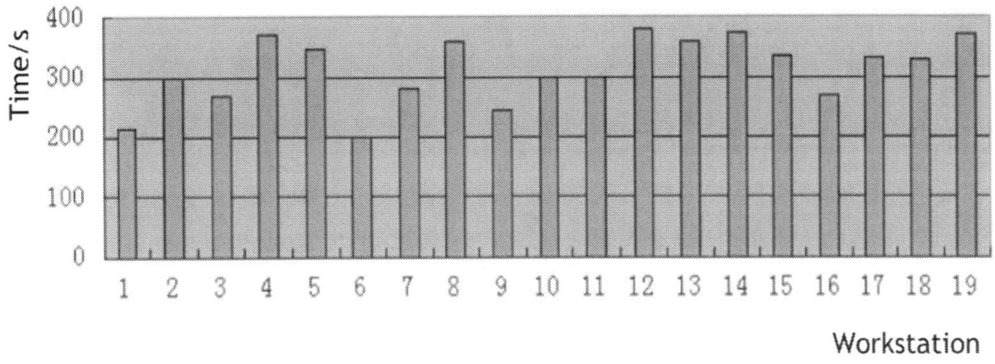

Figure 5: Production Line Balance Histogram After Optimization

After optimization, the rate of the steering knuckle production line balance has greatly improved from 75.99% to 82.29%.And also,the steering knuckle line of line balance loss rate has greatly reduced from 24.01% to 17.71%.On the average, which speeds up the enterprise's logistics working efficiency greatly.

5 CONCLUSIONS

With Plant simulation software that is set up the steering knuckle production line model . By analyzing the output results of the simulation model, find out the bottleneck of the production line location. And use the Plant simulation software to optimize the model simulation analysis, contrast and balance production line balancing rate before and after the improved loss rate are improved.The results show that the steering knuckle of production line production status is well.The ultimate objective of this system is to find through simulation the optimal set of operation parameters and optimized scheduling results for guiding the field production, while making full use of just the existing manufacturing resources.Manufacturing enterprise's production line is a large, complex discrete dynamic system. Using the method of simulation modeling, auxiliary production logistics process.It has the obvious effect for improving the process, remove bottlenecks, improve the level of logistics .

6 REFERENCES

[1] **HAN Xiao-dong,LIU Dong,LONG Ming,YUAN Fu-gui,HE Feng.2015.**Simulation and Analysis of Engine Testing Line Based on Plant Simulation,Modular Machine Tool&Automatic Manufacturing Technique,(11),pp 58-64.

[2] **GAN Weihua, WANG Ruhong, CHE Pengfei， YOU Jinning.2015.**Production Throughput Improvement for Plant Streamline N.Journal of Nanchang Hangkong University,17(4),pp 62-71.

[3] **Zhang Qing.2011.**Research on Discrete Manufacturing System Based on Plant Simulation,Design &Manufacture of Diesel Engine,17(2),pp 43-46.

[4] **Chen Xiaofeng, Xiao Tianyuan, Hao Xin, Qiao Guixiu.2001.**Assembly Line Balancing Using Genetic Algorithms, Computer Engineering and Application,(23),pp81-83.

[5] **FAN Weibo, ZHOU Jun, XU Zhengliang.2010.**Appication of Genetic Algorithm to Assembly Line Balancing,Computer Technology And Development,20(2),pp194-201.

[6] **Jinpin Zhou.2011.**Production system simulation—Plant Simulation Application of tutorial,1st Edition, Publishing house of electronics industry.

[7] **Xiaolan Yuan, Ziqi Xu, Fan Shu.2015.**Based on the Genetic Algorithm Simulation and Optimization of Engine Assembly Line Balancing Problem,Logistics Engineering and Management, 37(7),pp213-220.

[8] **HUANG Jian, LIU Liang, QI Er-shi.2014.** Multiple Objective Sequencing Problems in Mixed Model Assembly Line with Consideration of Line Balancing,Modular Machine Tool & Automatic Manufacturing Technique, 8(8),pp156-160.

[9] **Jinbo Zhou,Liuqun Fan. 2012.** Optimization and Simulation of Assembly Line Material Distribution Based on eM-Plant,Mechatronics,18(6),pp17-20.

[10] **Haifeng YAO,Xunsheng FENG,GUO Ding-jun.2010.**Simulation Study on Balancing Technology of Automobile Spare Parts Production Line Using eM-plant,Logistics Technology,29(12),pp114-117.

Mutual Effect Between Online and Offline Interaction: From a Perspective of Chinese College Students

Peihan Wen*, Dongqun Zhou and Shuping Yi
College of Mechanical Engineering
Chongqing University, Chongqing, 400044, China
wenph@cqu.edu.cn

ABSTRACT

Online interaction has become an important communication form with the emergence of social networking sites. Mutual effect between online and offline interaction was explored by introducing concepts of I-Intention and We-Intention. A research model was developed based on the uses and gratification paradigm together with theories of social influence and social presence. Relative impact of I-Intention and We-Intention, personal and group acts, online and offline interaction was examined. An empirical study of Chinese college students revealed that two kinds of We-Intention driven interaction form a positive cycle.

Keywords: Social networking sites, Intentional social action, Social interaction, I-intention, We-Intention

1 INTRODUCTION

The number of Chinese netizens has reached 618 millions as of 2013, of which 86.2% use social networking sites (SNSs), such as QQ, Douban, Renren, Weibo, and WeChat, and over 50% are at the age of 20 to 29 years old [1-2], which is just the age group of college students.

Although college students tend to maintain and enhance existing offline social ties in contrast with formation of new connections, online interaction through SNSs could help them constructing and enhancing social credits incorporating bridging bonding and maintained social capital [3]. Furthermore, a comprehensive research model synthesizing We-Intention, social influence, uses and gratifications (U&G) paradigm, and social presence was developed to explain why college students use SNSs like facebook, and emphasized the importance of social presence and social related factors (maintaining interpersonal connectivity and social enhancement) of U&G paradigm[4].

SNSs related research used to mostly focus on Western sites like facebook and twitter. Recently, Chinese researchers have begun to pay attention to Chinese SNSs. Extant research of SNSs mainly concerns about contents' popularity and users' behavior. Renren, similar to Facebook, focuses mainly on users' relationship and posts' sharing. Social influence of users and posts can be predicted by introducing user ranking denoting who shares a post will make its diffusion range maximized in a social network, and post ranking denoting what post a user share will maximize his influence among friends [5]. Popularity of content on Weibo, similar to Twitter, depends not only on its quality, interface to highlight, and collective behavior of users, but also upon structural characteristics of networks spanned by adopters who view or forward contents at the early stage [6].

* Corresponding Author

However, There is little research about how *online (virtual) interaction* through SNSs impacts traditional *offline (face-to-face) interaction*, especially for college students, who are not mature enough and at the stage of building personality and values. Thereby, we took a perspective of Chinese college students and conducted an empirical study to explore the mutual effect of online and offline interaction in terms of intentional social action.

The rest is organized as follows. Section 2 gives a brief review of related theories, then a research schema is developed in Section 3. Investigation with online and offline questionnaire is briefed, with data analyzed and discussed in Section 4. Finally, some conclusions are given in Section 5.

2 THEORETICAL BACKGROUND

Concepts of I-Intention and We-Intention were introduced to explore differences of motivation and forms between online and offline interaction.

I-Intention is defined as "a personal intention to perform an individual act by oneself" [7], with two types of impetus: individual-level reasons, and perceived interpersonal pressure. Individual-level reasons assumes that users are aware of their needs, and their behaviors are goal-directed. Purposive value (PV), self-discovery (SD), entertainment value (EV), social enhancement (SE), and maintaining interpersonal connectivity (IC) are widely adopted to determine the use of virtual communities [8]. Perceived interpersonal pressure can be analyzed with social influence theory, which elucidates different processes of social influence and determines different levels of changes in attitudes and actions, including three widely adopted processes of subjective norm (SN), group norms (GN), and social identity (SI) [4].

We-Intention, which is defined as a "commitment of an individual to engage in joint action and involves an implicit or explicit agreement between the participants to engage in that joint action", dwells when a person regards the self as part of a social representation in performing a group act [9]. Besides U&G paradigm and social influence theory, We-Intention can also be explained in terms of social presence theory defined as "the degree of salience of the other person in the interaction and the consequent salience of interpersonal relationships" [10]. Individuals participate and keep using a SNS only if they perceive presence of online friends. Therefore, SNSs with more presence of users will be more powerful to attract new comers and retain current ones. We-Intention is an appropriate approach to study online social activities, and affected positively by social presence strongly, as well as SN, GN, SI, and levels of PV, SD, EV, SE, and IC [4].

3 RESEARCH SCHEMA

A research schema depicted in Fig. 1 was developed to explore mutual effect between online and offline interaction, with ten hypotheses put forth.

3.1 I-Intention and We-Intention vs. personal act and group act

For offline interaction, people get used to face-to-face communicating, which is helpful for individuals with strong I-Intention to persuade others to accept their intention and set up a common objective. Thereby, We-Intention may be enhanced by some strong I-Intention (H1 as listed in Table 1). As to online interaction, it is not easy to persuade others to change their minds or accept a specific objective with text or verbal message, image sharing, voice or video chatting. For acquaintance, perceived interpersonal pressure plays an important role. Social influence such as individuals' SN and SI will promote the evolution from I-Intention to We-Intention (H2 as listed in Table 1). Relationship resulting from common interest or close geographic location between strangers is so crisp that can not be counted on (H3 as listed in Table 1).

It only depends on I-Intention whether or not to perform a personal act, whereas it depends on both I-Intention and We-Intention whether or not to perform a group act. With regard to the evolution process from I-Intention to We-Intention, the possibility to perform a group act can be depicted by We-Intention (H4 and H5 as listed in Table 1).

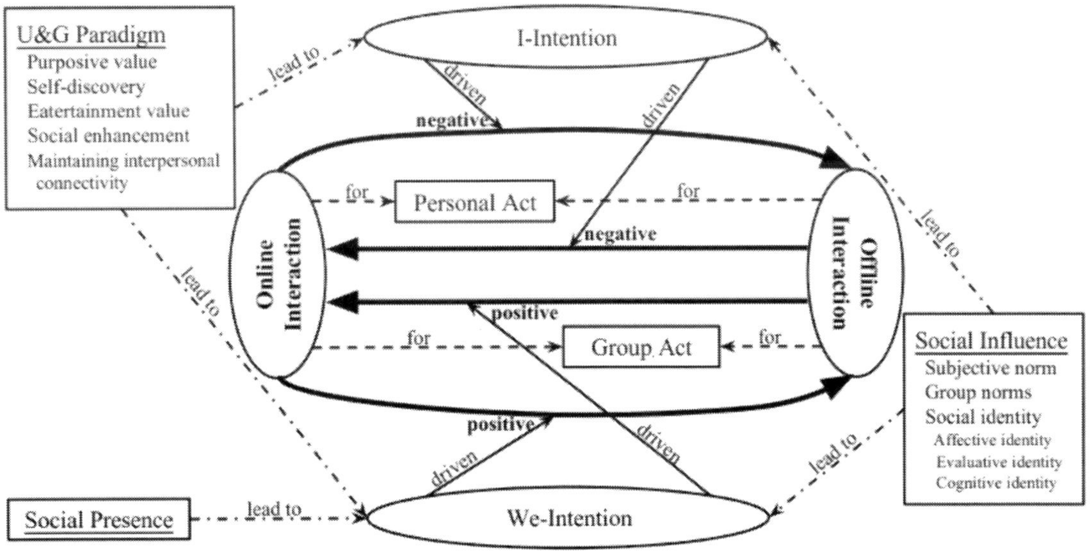

Figure 1. Research Schema of Mutual Effect Between Online and Offline Interaction

3.2 Mutual effect between online and offline interaction

In contrast with offline interaction, online interaction eliminates restraints of time, space, and cost. People living in either part of the earth can communicate for free, and usually don't need to spare specific time period for casual conversation. However, people may not concentrate to one contact or topic, thinking carefully, deeply or thoroughly. Moreover, kinds of fancy functions divert users' attention and become an important reason to select SNSs, especially for young people like college students (H6 as listed in Table 1).

An interaction driven singly by I-Intention cares more about me, and focuses only on "my" objective. Once the objective is accomplished by either online or offline means, no impetus will be stimulated to convert to the other. Only if the objective can not be realized by one form of interaction (online or offline), the other will be put into action (H7 and H8 as listed in Table 1). Whereas a We-Intention driven interaction cares more about us, focusing on "our" goal. Participants with similar interests will agree on a collective opinion or objective by a process of acceptance and argument, which is indeed an evolution from "my" goal to "our" goal. With the accomplishment of "our" goal, participants will perceive and get used to interpersonal influence, encouraging them to keep in touch. If cooperating online, they will be eager to meet face-to-face to know better; otherwise, they will exchange SNS information for further communication (H9 and H10 as listed in Table 1).

4 INVESTIGATIVE ANALYSIS AND DISCUSSION

The most popular SNSs in China, in chronological order of launching, are QQ, Douban, Renren, Weibo, and WeChat. With Chinese college students, the mainstream to uses SNSs, chosen as target subjects, a controlled investigation was conducted to test the ten hypotheses proposed in Section 3, restricting SNSs to the above 5 ones.

Table 1. Proposed Hypotheses

	Hypotheses
H1	A stronger I-Intention leads to a stronger We-Intention during offline interaction.
H2	A stronger I-Intention leads to a stronger We-Intention during online interaction between acquaintance.
H3	A stronger I-Intention can not lead to a stronger We-Intention during online interaction between strangers.
H4	A stronger I-Intention results in more possibility to perform a personal act.
H5	A stronger We-Intention results in more possibility to perform a group act.
H6	SNSs with more functions attract college students to participate more online interaction.
H7	An I-Intention singly-driven successful online interaction negatively affects a related offline interaction.
H8	An I-Intention singly-driven successful offline interaction negatively affects a related online interaction.
H9	A We-Intention driven successful online interaction positively affects a related offline interaction.
H10	A We-Intention driven successful offline interaction positively affects a related online interaction.

4.1 Design of investigation

A questionnaire with 12 multiple-choice questions and 16 Likert scale questions was designed and published on SoJump (`http://www.sojump.com`). Multiple choice questions were mainly used to collect respondents' basic information and trends on SNSs, with 3 questions about demographic information, 5 about using experience of SNSs, 2 with regard to online-made friends, and 2 related to SNSs' functions. Likert scale questions included 13 ones related to hypotheses and 3 about respondents' comprehensive attitudes towards online and offline interaction.

The URL of the online questionnaire was disseminated on the 5 objective SNSs, explicitly notifying expectants were restricted to college students. 154 online responses was collected in one month. Meantime, 100 printed questionnaires were dispensed on campus, and 86 ones were taken back. Finally, a total of 240 online and offline responses from 38 universities all over China was acquired.

4.2 Data analysis

In a total of 240 responses, 65% (156) were male and 35% (84) were female; where, 55% (132) were from countrysides or towns, 36% (86) were from cities, and 9% (22) were from metropolis. In terms of behaviors on SNSs, 3% (8) just read posts, 66% (157) comment or repost besides reading, and 31% (75) also publish new posts. A statistical analysis on respondents' behaviors is shown in Fig. 2. More than 86% have been using SNSs for over one year. More than 83% use SNSs every week, and nearly a half uses every day. No less than 76% post, comment or repost every day, with almost a half more than once a day. Over 84% cannot live without SNSs for a week.

Fig. 3 illustrates a statistical analysis in regard to respondents' trends on SNSs. Nearly one third never make new friends online, and over 90% have more offline-made friends in their online contact lists. The result shows that people are very cautious about developing friendship through SNSs. It can be inferred that it is more difficult for strangers meeting online to trust each other. 38% choose to use a SNS because of their friends, and 28% think it is the best online fulfilled requirement to maintain interpersonal connectivity. Both kinds of data validate that people are mainly prone to keep in touch with acquaintance through SNSs. The second important factor affecting people to choose SNSs is about functions, in accordance with "H6".

Statistics of Likert scale questions are shown in Table 2. Column "Approval" denotes indeed percentages of respondents at least not disagreeing, and Column "Approval+" indicates percentages of respondents agreeing. "Approval" and "Approval+" of "QH1" and "QH2" increase significantly relative to that of "BL1-BL3", strongly supporting "H1" and "H2". Compared with "BL1-BL3",

"Approval" and "Approval+" of "QH3" decrease significantly, supporting "H3". "Approval" and "Approval+" of "QH4", as well as "Approval" of "QH5" and "QH6", are all over 60%, meaning "H4" to "H6" should be accepted. "Approval+" of "QH5" is less than that of "QH4", which may imply people are not so sure to differ "our" from "my" behavior. "Approval" and "Approval+" of "QH7" and "QH8" slightly decrease relative to related baselines, supporting "H7" and "H8" in a sense, but not so explicitly, perhaps further research is needed. In contrast, "Approval" and "Approval+" of "QH9" and "QH10" increase explicitly compared with related baselines, meaning 'H9" and "H10" can be accepted confidently. "CA1" to "CA3" investigate respondents' comprehensive attitudes to online and offline interaction. Most people (over 60%) are prone to interact online with friends, although they are aware of that it may be not good for real-world connection. However, few people (only 14%) feel closer with online-made friends than the offline-made.

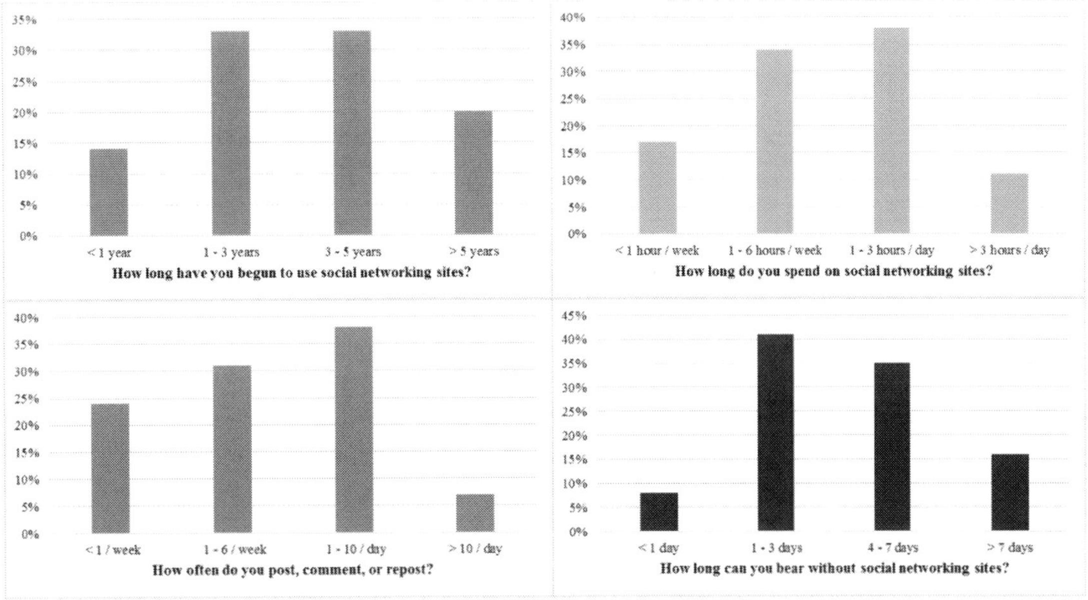

Figure 2. Statistics of Respondents' Behaviors on SNSs

4.3 Theoretical and practical implications

Data analysis shows that all 10 hypotheses are supported in some degree. "H1" to "H3" are about the evolution from I-Intention to We-Intention, i.e., how "my" objective can be accepted as "our" collective objective. Results imply it is easier to build and lead the collective spirit and behaviors with face-to-face contact than virtual connection, between acquaintance than strangers.

"H4" and "H5" are in relation to the causality between intention and action. Individually, people are confident about that they will take action if they intend to. However, individual intention will not work directly on group acts. Only if people in a group find common points of their individual goals and form a common objective, i.e., promote I-Intention to We-Intention, a group action could be taken. Due to the difficulty of driving I-Intention to We-Intention among strangers, especially with online interaction, people should be first classified by some similar features such as interest, vocation, education, to increase the possibility to perform a group act. Therefore, almost all SNSs provide a function of "Groups", "Circles", or something like that. Accordingly,fancy functions are important characteristics of SNSs, not only to fulfill users' 5 key values, but also to attract them to keep logging and spend more time on the site, since active users per day and pageviews per user are two important indexes to solicit commercial advertisement.

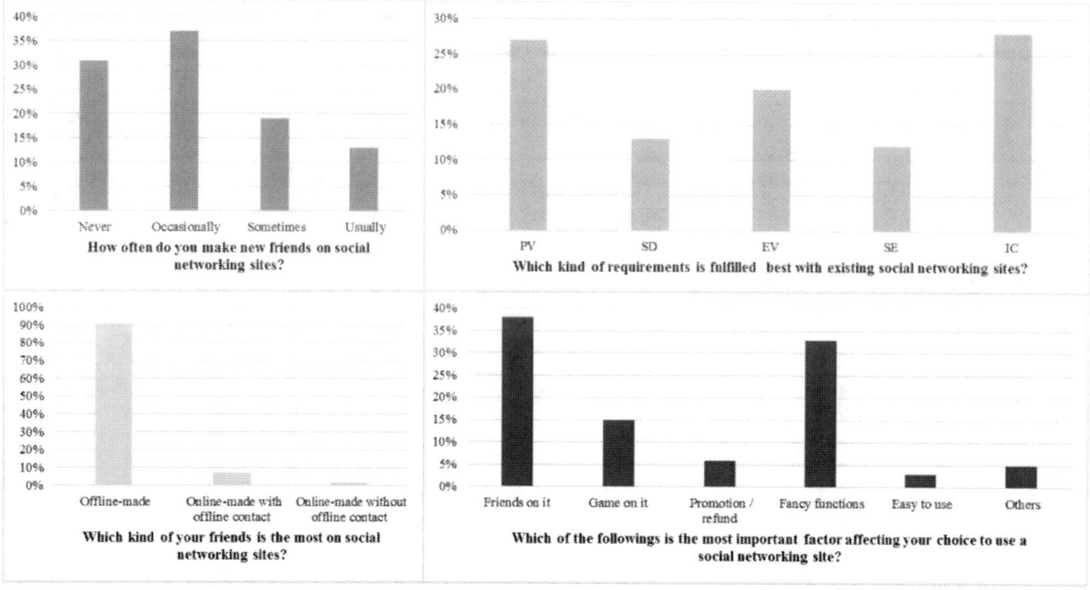

Figure 3. Statistics of Respondents' Trends on SNSs

Although there is only a little tendency that I-Intention driven online (offline) interaction will negatively affect related offline (online) interaction, it is evidently supported that We-Intention driven online (offline) interaction will positively affect related offline (online) interaction. This finding validates the importance of collectivity and team spirit, which helps to form a virtuous interactive cycle to avoid the inhibitive effect of online interaction on offline interaction by means of encouraging group acts based on "our" collective goal. It is especially necessary for college students who are forming personality and values, because the increasing of virtual connection and corresponding decreasing of face-to-face contact will have some negative effect. There are kinds of differences between online and offline communications such as words, gestures, tongue, accents, or even some salutation. And there are lots of feelings that people cannot acquire with virtual connection. Lacking of face-to-face contact may bring some psychological problems to college students, causing disability to communicate in real world, or even resulting in tragedies like suicide or murder. It is extremely important for college students to balance between online and offline interaction.

5 CONCLUSION

Online interaction through SNSs have been becoming an important approach of intentional social action. The objective of this study is to explore mutual effect between online virtual interaction and traditional offline face-to-face interaction from a perspective of Chinese college students. Concepts of I-Intention and We-Intention were employed to build a research schema based on the U&G paradigm, social influence and social presence theories. By means of questionnaire investigation, an empirical study was conducted and four main findings are presented. Firstly, a stronger I-Intention will lead to a stronger We-Intention with either online or offline interaction between acquaintance. Secondly, personal behaviors depend on I-Intention, whereas We-Intention is more appropriate to speculate group acts. Thirdly, functions are important for SNSs to attract users and

Table 2. Statistics of Likert Scale Questions†

Likert scale questions‡	1	2	3	4	5	Approval	Approval+
BL1-BL3: *To what extent do you agree that you are an easily-persuaded person?*							
	21%	25%	38%	11%	5%	54%	16%
QH1: *To what extent do you agree that you will be persuaded by an offline friend to do a specific thing together?*							
	8%	17%	26%	44%	5%	75% (21%)	49% (33%)
QH2: *To what extent do you agree that you will be persuaded by an online friend to do a specific thing together if you are also offline friends?*							
	13%	16%	31%	37%	3%	71% (17%)	40% (24%)
QH3: *To what extent do you agree that you will be persuaded by an online friend to do a specific thing together if you are not offline friends?*							
	27%	29%	38%	5%	1%	44% (-10%)	6% (-10%)
QH4: *To what extent do you agree that you, yourself, will set out to do something if you personally intend to?*							
	5%	13%	22%	47%	13%	82%	60%
QH5: *To what extent do you agree that you, together with couples of friends, will set out to do something if all of you intend to?*							
	11%	20%	37%	22%	10%	69%	32%
QH6: *To what extent do you agree that SNSs with more functions will attract you more?*							
	9%	21%	24%	34%	12%	70%	46%
BL7 & BL9: *To what extent do you agree that you will have a face-to-face contact with an online-made new friend?*							
	24%	33%	32%	10%	1%	43%	11%
BL8 & BL10: *To what extent do you agree that you will have a virtual contact with an offline-made new friend?*							
	10%	11%	40%	31%	8%	79%	39%
QH7: *To what extent do you agree that you will have a face-to-face contact with a friend you make online to help you to solve a problem?*							
	27%	35%	31%	5%	2%	38% (-5%)	7% (-4%)
QH8: *To what extent do you agree that you will have a virtual contact with a friend you make offline to help you to solve a problem?*							
	12%	12%	41%	29%	6%	76% (-3%)	35% (-4%)
QH9: *To what extent do you agree that you will have a face-to-face contact with an online-made new friend if you participate a common interest group and solve a problem together?*							
	17%	28%	40%	13%	2%	55% (12%)	15% (4%)
QH10: *To what extent do you agree that you will have a virtual contact with an offline-made new friend if you have common interest and solve a problem together?*							
	4%	8%	34%	34%	20%	88% (9%)	54% (15%)
CA1: *To what extent do you agree that it hinders your face-to-face communication with friend to participate SNSs?*							
	18%	20%	40%	16%	6%	62%	22%
CA2: *To what extent do you agree that you feel more comfortable to communicate with friends online virtually than offline face-to-face?*							
	14%	26%	39%	19%	2%	60%	21%
CA3: *To what extent do you agree that online-made friends are closer than offline-made friends?*							
	35%	51%	7%	4%	3%	14%	7%

† Columns of "1" to "5" are percentages of Likert scales; columns of "Approval" and "Approval+" are a sum of percentages scaled "3" to "5", and scaled "4" and "5", respectively, with percentages in parentheses indicating increments relative to related baseline questions.

‡ Questions beginning with "QH*" are about related hypotheses, with "BL*" are baseline questions, and with "CA*" are about comprehensive attitudes. Also, each question occupies a whole row, with related percentages in the next row.

realize commercial profit. Finally, We-Intention driven online and offline interaction positively affect each other, which helps leveraging between them to eliminate negative effect.

The above findings can be a basis for further theoretical research, and practitioners can also benefit. However, there are some issues one must be careful of. Personality, which is an important factor influencing social action, was not taken into account in this study, and will be incorporated to perform a cross analysis in further research. Also, the study subjects are Chinese college students. Different groups (e.g., nations, vocations, ages) of users may have different trends. More effect about both presence of online and offline interaction should be studied, such as the extent to which online interaction influences people's real-world behavior and everyday life, study, work, and judgement, how we can promote education/management in use of online and offline mutual interaction.

6 ACKNOWLEDGEMENTS

The research work was granted by the National Natural Science Foundation of China (No.71501020).

7 REFERENCES

[1] **China Internet Network Information Center. 2014.** The 33rd statistical reports on the Internet development in China, 5, p 24. `http://www.cnnic.net.cn/hlwfzyj/hlwxzbg/hlwtjbg/201403/P020140305346585959798.pdf`

[2] **China Internet Network Information Center. 2013.** Research report on users' behavior of Chinese social networking service applications of 2013, 10, p 18. `http://www.cnnic.net.cn/hlwfzyj/hlwxzbg/sqbg/201312/P020131225358386313453.pdf`

[3] **N. B. Ellison, C. Steinfield, and C. Lampe. 2007.** The benefits of Facebook 'friends': Social capital and college students' use of online social network sites, *J. Comput. Commun.*, 12, pp 1143-1168.

[4] **C. M. K. Cheung, P.-Y. Chiu, and M. K. O. Lee. 2011.** Online social networks: Why do students use facebook?, *Comput. Human Behav*, 27, pp 1337-1343.

[5] **P. Cui, F. Wang, S. Liu, M. Ou, S. Yang, and L. Sun. 2011.** Who should share what? Item-level social influence prediction for users and posts ranking, Proc. 34th Intl. ACM SIGIR Conf. on Research and Development in Information Retrieval, pp 185-194.

[6] **P. Bao, H. Shen, J. Huang, and X. Cheng. 2013.** Popularity prediction in microblogging network : a case study on Sina Weibo, Proc. 22nd Intl. Conf. on World Wide Web Companion, pp 177-178.

[7] **R. P. Bagozzi and K.-H. Lee. 2002.** Multiple routes for social influence: The role of compliance, internalization and social identity, *Soc. Psychol. Q*, 65, pp 226-247.

[8] **U. M. Dholakia, R. P. Bagozzi, and L. K. Pearo. 2004.** A social influence model of consumer participation in network- and small-group-based virtual communities, *Int. J. Res. Mark*, 21, pp 241-263.

[9] **R. Tuomela.** 1995. *The importance of us: A philosophy study of basic social notions*, Stanford, CA: Stanford University Press.

[10] **J. Short, E. Williams, and B. Christie. 1976.** Theoretical approaches to differences between media, The social psychology of Telecommunication, London, New York, pp 61-76.

STRATEGIC, TACTICAL AND OPERATIONAL FLEXIBILITY IN DESIGN AND EVALUATION OF A DECENTRALIZED ON-SHORE LNG PRODUCTION SYSTEM

M. Ranjbar-Bourani[1*] and M.-A. Cardin[2]
[1]National Elites Foundation and School of Industrial Engineering, College of Engineering
University of Tehran, Tehran, Iran
ranjbar.mehdi@ut.ac.ir

[2]Department of Industrial and Systems Engineering, National University of Singapore, Singapore
macardin@nus.edu.sg

ABSTRACT

The conceptual design phase of capital-intensive and long-lasting engineering systems is very important as crucial decisions need to be made at this stage. These systems generally are subject to various sources of uncertainty throughout the system lifetime. Flexibility is a proactive way to deal with such uncertainty. This paper presents a three-step methodology to quantify flexibility in engineering systems design: 1) developing a deterministic quantitative performance model; 2) developing a quantitative performance model under uncertainty and 3) developing a quantitative performance model for flexibility. The proposed methodology is then applied to a case study in designing and evaluation of a decentralized on-shore Liquefied Natural Gas (LNG) production system. Different types of flexibility are embedded in strategic, tactical and operational level decision-making of the case study. Results show that the value of flexibility increases for strategies that combine different types of flexibility. The analysis provides insight for decision makers to quantify the value of flexibility when there are different types of flexibility and motivate the use of flexibility in engineering design as a paradigm to deal with uncertainty affecting the lifecycle performance of engineering systems.

Keywords: Flexibility, Design and Evaluation, Strategic, Tactical, Operational, LNG

1 INTRODUCTION

In the design and management of complex engineering systems, literature shows that the notion of flexibility is at the center of attention for improving economic performance under uncertainty. Due to the importance of evaluating flexibility in engineering designs, various evaluation methods have been developed. Of these methods, a quantitative performance model based on the simulation approach, first developed by de Neufville, Scholtes [1], for the evaluation of flexibility under uncertainty has gained wide attention from academia and industry.

Flexibility in engineering design is an interdisciplinary field for research and practice [2]. It adapts the concept of financial options to real engineering systems, with the goal of increasing the expected economic value by providing the "right, but not the obligation to change a system" to respond to uncertainties most profitably [3]. Flexibility in engineering design has been widely used in different domains such as aerospace, airport design, the automotive industry, defense, energy, healthcare, mining, public infrastructure and management. Several examples of flexibility in

* Corresponding Author

engineering design, including urban infrastructures [1], satellite systems [4], water resource systems [5], automotive manufacturing systems [6] and petroleum exploration and production systems [7], have been summarized by de Neufville and Scholtes [2] in their book. The emerging literature on flexibility in engineering design has shown that flexibility can improve the economic performance of a project from 10% to 30% as compared to standard methods [2, 8]. In this paper, two main axes of research opportunities are explored: 1) flexibility analysis that accounts for both design variables and decision rules; 2) different types of flexibility, considering strategic, tactical, and operational level decisions. The proposed methodology has been applied to design and evaluation of a decentralized on-shore LNG production system for the first time.

The reminder of the paper is structured as follows: in section 2 the proposed three-step methodology for uncertainty and flexibility is presented. Subsequently in section 3, a decentralized LNG production system is investigated. Results and discussions are provided in section 4. Section 5 summarizes the major findings, provides conclusions, and gives insights into further research.

2 RESEARCH METHODOLOGY

The proposed methodology is a simulation-based flexibility and uncertainty analysis that consists of the three following steps:

2.1 Step 1: Develop deterministic quantitative performance model

This step builds a baseline quantitative performance model to evaluate the design alternatives. To do so, the scope of the problem and underlying assumptions about problem modeling and parameters need to be determined, such as market parameters, design variables, key costs and revenue drivers of the system. Following this, a deterministic quantitative performance-based model is developed to represent the relations among components of the system and to measure the lifecycle performance of the design alternatives using different metrics, such as Net Present Value. At the end of this step, the model can generate single-point outputs in terms of different values of design variables. This model, however, provides unrealistic solutions as it does not recognize uncertainty.

2.2 Step 2: Develop quantitative performance model under uncertainty

In this step, the deterministic quantitative performance model is extended into the model under uncertainty. Once the main sources of uncertainty have been identified, the corresponding historical trends need to be analyzed by understanding the data, developing the overall pattern and assessing the uncertainty in their trends. Stochastic functions, such as Geometric Brownian Motion (GBM), s-curve function and Mean Reverting Process, can be used to model the uncertainty behavior over the evaluation period. By incorporating these stochastic behaviors into the deterministic model using the Monte Carlo simulation, a large number of possible scenarios can be generated. In this step, hence, one deals with distributions of outcomes in terms of different input variables.

2.3 Step 3: Develop quantitative performance model for flexibility

The process here mainly focuses on flexibility valuation, but it is augmented by engineering design tools in systems design concept generation that help identify the valuable flexibility strategies. For instance, flexible strategies such as capacity expansion/reduction, switching inputs/outputs and deferring investment can be considered. In this step, for the "generating the flexibility" that requires expert domain knowledge, systematic processes such prompting, as suggested by Cardin, Kolfschoten [9], or the procedure explained in Hu and Cardin [10] are all ways to generate the flexible alternatives as part of this step. When the effective flexible strategies have been identified, corresponding decision rules need to be explicitly embedded in the quantitative performance model under uncertainty. Similar to Step 2, there are different performance metrics to evaluate design alternatives in this step.

3 CASE STUDY

The case study that will be explained in this section focuses on the development of a decentralized on-shore LNG production system to provide fuel for on-road transportation in southeast Australia. Figure 1 shows the schematic representation of the LNG production system from a fixed design towards a more flexible design. All five sites have access to the existing on-shore natural gas pipeline. Figure 1(a) depicts a fixed design where optimum capacity is deployed at each demand point. Figure 1(b) shows a flexible modular design where the initial capacity is built at each demand point depending on the market situation. Using this design, extra capacity will be deployed based on the managerial capacity expansion policy. Figure 1(c) shows a flexible modular design similar to the design in Figure 1(b), but with extra operational flexibility. This flexibility enables the production system to rebalance its capacity among the demand points using a fuel truck fleet transportation system.

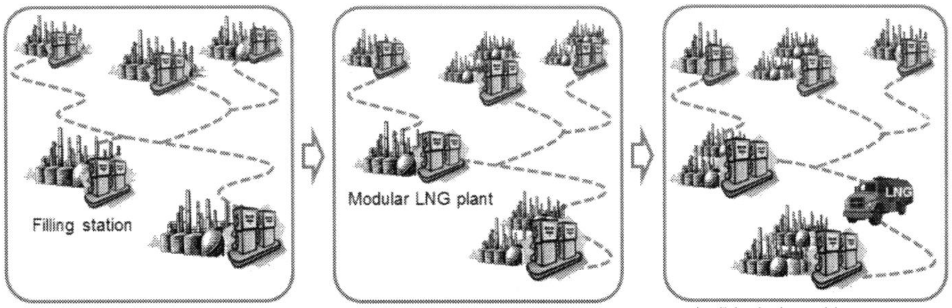

a: Fixed design b: Flexible design without transport c: Flexible design with transport

Figure 1: A fixed LNG system design towards a more flexible decentralized LNG system design

3.1 Modeling assumptions

The following assumptions are made for problem modeling. Demand is assumed to be unevenly distributed in the region over five distinct demand sites. Essentially the distributions of the demand at different sites are independent and identical where the same distribution parameters are assumed. All sites have access to an on-shore natural gas pipeline in the region, but this access has a cost due to gas tie-in operations and acquiring extra land. For all demand sites, the time needed to build the first plant is two years and the capacity expansion takes only one year. In other words, if one decides to expand the current capacity in year t, the extra capacity will be available for production in year t+1. Regarding the financial parameters, the project lifetime is assumed to be 20 years and each year is considered to be 350 working days. A ten-year straight-line depreciation method is used for all LNG production facilities with a zero salvage value. The discount rate as an after-tax MARR is assumed to be 10% and the corporate tax rate is 30%. Essentially, the quantitative performance of the design is evaluated based on an After Tax Cash Flow (ATCF) analysis. The details of the modelling assumptions can be found in [11].

3.2 Step 1: Develop deterministic quantitative performance model

The problem modeling starts with the deterministic analysis. The aim is to understand the key components of the system that influence its lifecycle performance. The performance metric used in this problem is NPV, calculated as the sum of discounted cash flows throughout the project lifecycle T = 20 years - see equation 1. Variables TR_t and TC_t are the total revenues and costs incurred in years t = 1, 2, ..., T, and r is the discount rate, Tax is the effective income tax on ordinary income and d_t is the sum of all noncash, or book, costs during year t, such as depreciation.

$$NPV = \sum_{t=1}^{T} \frac{(1 - Tax)(TR_t - TC_t) + Tax\ d_t}{(1+r)^t} \quad (1)$$

LNG demand is identified as the key driver of system performance. A deterministic s-curve function is assumed to simulate LNG demand over the study period, as shown in equation 2. The rationale is that LNG demand initially grows slowly; it then increases exponentially, and finally tapers as it approaches a saturation limit. Variable M_T^D is the maximum expected demand for LNG, b^D is the sharpness parameter that determines how fast demand grows over time to reach the upper bound for demand. The parameter a^Dtranslates the curve horizontally.

$$D_t^D = \frac{M_T^D}{1 + a^D e^{-b^D t}} \quad (2)$$

where a^D is calculated using equation 3.

$$a^D = \frac{M_T^D}{D_0^D} - 1 \quad (3)$$

In general, the conventional DCF model is built to assess the performance of a system under deterministic conditions. This step captures the standard industry practice in terms of design and project evaluation.

3.3 Step 2: Develop a quantitative performance model under uncertainty

This step enables the analyst to explicitly recognize, characterize and model the major uncertainty drivers affecting the future lifecycle performance. The analysis under uncertainty considers a distribution of outcomes instead of a single performance output, which can be modeled using different techniques (e.g., Monte Carlo simulation, decision trees or binomial lattice). Here, NPV_s, which refers to NPV under demand scenario s, is calculated in terms of different realized and uncertain demand scenarios via simulation. A stochastic s-curve function is used to simulate LNG demand over the system's lifecycle using additional uncertainty factors, as shown in equation 4.

$$\mathrm{D_t^U} = \frac{\mathrm{M_T^U}}{1 + \mathrm{a^U e^{-(b^U)t}}} \quad (4)$$

Equation 5 shows how a^U is calculated.

$$a^U = \frac{M_T^U}{D_0^U} - 1 \quad (5)$$

Realized demand at time t + 1 equals realized demand at time t plus annual volatility multiplied by the growth rate at time t, as equation 6 shows.

$$D_{t+1}^U = D_t^U + (Av \times G_t) \quad (6)$$

In this equation, G_t is the annual growth rate assuming adherence to a standard normal distribution, (G_t ~ Normal (0, 1)), and Av is assumed as a fixed parameter throughout the project lifetime calibrated using historical data. Monte Carlo simulation is used to simulate a wide range of LNG demand scenarios. This analysis recognizing uncertainty provides designers with a more realistic overview of system performance as compared to the deterministic analysis in Step 1. In order to extend the deterministic model into the model under uncertainty, the uncertainty drivers that significantly affect the economic performance of the project must first be identified. Through a sensitivity analysis, LNG demand is again treated as the main source of uncertainty. Next, the Monte Carlo simulation technique can be used to simulate a wide range of LNG demand scenarios.

3.4 Step 3: Develop *quantitative performance model for flexibility*

To deal with uncertain LNG demand, three different flexible strategies are proposed: 1) strategic; 2) tactical and 3) operational flexibility. In the strategic level, flexible decision rules for determining initial capacities are applied. In the tactical level, capacity expansion flexibility is identified as the most relevant strategy to cope with demand uncertainty. To embed these flexible strategies, a set of managerial decision rules is embedded in the programming of the Excel® spreadsheet DCF model under uncertainty. These three flexible strategies are described in greater detail in the following subsections.

3.4.1 Strategic level flexibility

In this case study, determining the initial plant capacity is considered as a strategic level decision. This decision is made by relying on a short-term, forward-looking forecast.

Given the practical modular capacities available to the company, the three decision rules used in the simulation model are as follows: 1) IF "realized demand in year of forecast t <= bound 1" THEN "Initial capacity = 0"; 2) IF "realized demand in year of forecast t > bound 1" AND "realized demand in year of forecast t <= bound 2" THEN "Initial capacity = 25"; and 3) IF "realized demand in year of forecast t > bound 2" THEN "Initial capacity = 50", where bound 1, bound 2 and year of forecast t are parameters of the decision rules. Band 1, Band 2 and Band 3 are evenly distributed between the lower and upper bounds of LNG demand and the short-term, forward-looking year of forecast is set to year 6.

3.4.2 Tactical level flexibility

In this case study, tactical level capacity expansion policy was considered the most appropriate flexibility for responding to demand uncertainty. For capacity expansion, it is important to know when capacity should be expanded given the available modular designs (i.e., 25 tpd and 50 tpd). Hence, an appropriate managerial decision rule was embedded at each production site.

The decision rule is defined in a logical form as: IF *"observed demand is higher than current capacity"* AND *"the difference between the observed demand and current capacity reaches a threshold value that is a certain percentage of the modular design used at each production site in year t"* THEN *"build extra modular plant that will be available for production in year t+1"* ELSE *"do nothing"*. The threshold value determines when the extra capacity should be built. For example, decision-makers may decide to add another modular plant as soon as the difference between the realized and current capacity reaches 60% of the modular plant capacity for the site.

3.4.3 Operational level flexibility

To further improve the lifecycle performance of the flexible LNG production system, an operational flexible strategy can be applied to better meet unmet demand with unused capacity in the system. The LNG production at each site is planned to be absorbed by the demand at the same location. However, if the capacity at any operational period (i.e., one year, in this case) is not matched with the realized demand, this will lead to either an unused capacity or unmet demand situation at any given location. The proposed operational flexibility (also referred to as the rebalancing problem) is concerned with the situation in which LNG is transported from sites with unused capacity to locations with unmet demand. The rebalancing transportation problem can be formulated as a linear programming, transportation problem. By embedding this flexibility for each set of demand scenarios for each operational time period, the optimum amount of LNG that should be transported from the supply sites (i.e., those with unused LNG capacity) to the demand sites (i.e., those with unmet LNG demand) can be determined.

The value added by enabling operational flexibility in year t under demand scenario s is calculated using equation 7. It is assumed that operational flexibility is enabled in year t under scenario s when such flexibility is worthwhile.

$$\mathrm{VAO}_{t,s} = \max\left(0, TRO_{t,s} - TCO_{t,s}\right) \qquad \forall\, t,s \tag{7}$$

The NPV of the design under demand scenario s is calculated using equation 8. To generate the distribution of NPV outcomes, Monte Carlo simulation, which allows one to consider a large enough number of scenarios (e.g. 2000), is used.

$$NPV_s = \sum_{t=1}^{T} \frac{(1 - Tax)\left(VAO_{t,s} + TR_{t,s} - TC_{t,s}\right) + Tax\ d_{t,s}}{(1+r)^t} \qquad \forall\, s \tag{8}$$

4 RESULTS AND DISCUSSION

Table 1 demonstrates a full factorial experimental design matrix of the problem (i.e., Bound 1 and threshold value). The three levels of flexibility correspond to three factors of the experimental design. Each factor has two potential values: Y - with flexibility, or N - no flexibility. Thus, there are 2^3 flexible strategies in total. Last column of the table shows the design vectors. Each design vector has three elements: 1) initial capacity for the fixed design or the value of Bound 1 in case of flexibility shown as InCap/ B1%; 2) no capacity expansion or threshold value of modular design capacity shown as 0%(0) /TV%(tpd); 3) state of operational flexibility shown as No/Full.

Table 1: Design of experiment for strategies 1 to 8 under 2000 demand scenarios

Flexible strategy	Strategic flexibility	Tactical flexibility	Operational flexibility	Design vector
	Initial capacity (Y/N)	Capacity expansion (Y/N)	transportation flexibility (Y/N)	[InCap/B1% - 0%(0) /TV%(tpd) - No/Full]
1	N	N	N	Initial Cap. 5×30=150 - 0%(0) - No
2	N	N	Y	Initial Cap. 5×30=150 - 0%(0) - Full
3	N	Y	N	Initial Cap. 5×30=150 - 65%(25) - No
4	N	Y	Y	Initial Cap. 5×30=150 - 65%(25) - Full
5	Y	N	N	B1=20% - 0%(0) - No
6	Y	N	Y	B1=20% - 0%(0) - Full
7	Y	Y	N	B1=45% - 65%(25) - No
8	Y	Y	Y	B1=45% - 65%(25) - Full

Each strategy is simulated using 2000 sets of demand scenarios for different locations. The first strategy represents the optimum fixed design with initial capacity 5×30=150 tons per day. An exhaustive enumeration was conducted to find the optimum design variables and decision rule parameters of all flexible strategies. To analyze the contribution of each type of flexibility to the value of flexibility, the NPVs of all strategies are compared to the ENPV of strategy 1. Strategy 1, which represents the optimum fixed design, is considered the baseline design.

Figure 2 shows the cumulative density function of strategies one to eight. A pairwise t-test is applied to see whether the ENPVs of any two strategies are statistically different from each other. Given a large number of scenarios, pairwise t-tests show that the differences among strategies are statistically significant. As can be seen, strategy 8 with operational, tactical, and strategic flexibility is the best of the eight flexible strategies in terms of ENPV that is shown in bold figure.

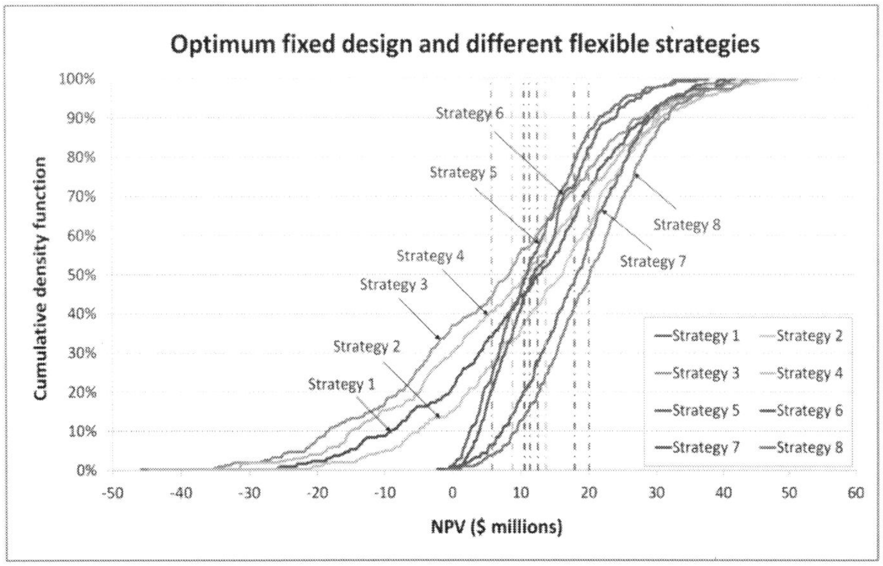

Figure 2: Cumulative density function curves for strategy 1 to 8

4.1 Multi-criteria decision-making

Table 2 shows a multi-criteria decision-making table to compare and contrasts the different solutions above. Best flexible strategies in terms of different criteria are shown in bold figures in the table. It compares results for the optimum fixed design (i.e., strategy number one) and flexible designs (i.e., strategies two to eight). Strategy number 8 outperforms among the others in terms of ENPV ($20M) and Value at Risk 10%, by protecting the downside risk by more than $8M. In terms of value at gain 90%, flexible strategy 3 gains the most value, more than $33M, from the upside opportunities among the rest. Strategy number 5 provides the least standard deviation, with less than $7.7M.

Table 2: Multi-criteria decision-making table, numbers in million dollar

Criteria	Flexible strategy								Value of flexibility	Best strategy
	1	2	3	4	5	6	7	8		
ENPV	10.56	13.67	5.83	8.82	11.32	12.48	17.90	**20.05**	9.49	8
$VaR_{10\%}$	-8.87	-4.99	3.71	-14.79	2.41	3.55	6.66	**8.53**	17.40	8
$VaG_{90\%}$	28.89	30.49	**33.53**	30.98	21.47	22.56	30.09	32.28	4.64	3
STD	14.69	13.94	11.01	17.72	**7.65**	7.73	9.02	9.20	0	5

5 CONCLUSION

In the proposed methodology, attempts were made to demonstrate the economic value of flexibility in the long-term design and deployment of production facilities subject to demand growth uncertainty. Results demonstrated promising improvement on economic lifecycle performances where ideas of flexibility are exploited in the different levels of strategic, tactical and operational decision-making. This approach proves to be superior to a baseline design concept developed from standard design and evaluation approaches. The significance of the approach used is that it motivates the use of flexibility in engineering design as a paradigm to deal with uncertainty affecting the lifecycle performance of engineering systems. In this respect, the study represents an argument for a shift in the design paradigm, away from the frequent focus on economies of scale and on to the development and deployment of unitary large facilities that embody this advantage. The concepts introduced in the problem modeling phase are general and can be applied to other distributed engineering systems sharing similar characteristics. However, consideration of flexibility adds another layer of complexity to the analytical problem making the simulation model computationally intensive.

6 ACKNOWLEDGEMENT

The authors would like to thanks colleagues in KOMtech for their professional contribution and appreciate the financial support from the Singapore Agency for Science, Technology and Research (A*STAR) contributed a Singapore International Graduate Award (SINGA). The first author also thanks the National Elites Foundation and the Science and Technology Vice President of Iran.

7 REFERENCES

1. **de Neufville, R., S. Scholtes, and T. Wang. 2006.** *Real options by spreadsheet: parking garage case example*. Journal of infrastructure systems. **12**(2): p. 107-111.

2. **de Neufville, R. and S. Scholtes. 2011.** *Flexibility In Engineering Design*. Cambridge, MA, United States: MIT Press.

3. **Trigeorgis, L. 1996.** *Real Options: Managerial Flexibility and Strategy in resource Allocation*. Cambridge, MA, United States: MIT Press.

4. **de Weck, O.L., R. de Neufville, and M. Chaize. 2004.** *Staged Deployment of Communications Satellite Constellations in Low Earth Orbit*. Journal of Aerospace Computing, Information, and Communication. **1**: p. 119-136.

5. **Wang, T. 2005.** *Real options in projects and systems design - identification of options and solutions for path dependency*. Massachusetts Institute of Technology: Cambridge, MA, United States.

6. **Yang, Y. 2009.** *A screening model to explore planning decisions in automotive manufacturing systems under demand uncertainty*, in *Engineering Systems Division*. Massachusetts Institute of Technology: Cambridge, MA, United States.

7. **Lin, J. 2009.** *Exploring flexible strategies in engineering systems using screening models - applications to offshore petroleum projects*. Massachusetts Institute of Technology: Cambridge, MA, United States.

8. **Cardin, M.-A. 2014.** *Enabling flexibility in engineering systems: a taxonomy of procedures and a design framework*. ASME Journal of Mechanical Design. **136**(1).

9. **Cardin, M.-A., et al. 2013.** *Empirical evaluation of procedures to generate flexibility in engineering systems and improve lifecycle performance*. Research in Engineering Design. **24**(3): p. 277-295.

10. **Hu, J. and M.-A. Cardin. 2015.** *Generating flexibility in the design of engineering systems to enable better sustainability and lifecycle performance*. Research in Engineering Design. **26**(2): p. 121-143.

11. **Cardin, M.-A., M. Ranjbar-Bourani, and R. de Neufville. 2015.** *Improving the lifecycle performance of engineering projects with flexible strategies: example of on-shore LNG production design*. Systems Engineering. **18**(3): p. 253-268.

RESEARCH ON PRICING STRATEGY CONSIDERING PRODUCT DESIGN UNDER DIFFERENT REMANUFACTURING MODES

Meng-meng LI[1] Ju-hong GAO[1*] Bing HAN[1]
[1]College of Management and Economics
Tianjin University, Tianjin, China
gaojuhong@tju.edu.cn

ABSTRACT

Differential pricing strategies were respectively studied under original manufacturer (OM), independent operator (IO) and OM/IO remanufacturing modes by Stackelberg theory with the consideration of the level of interchangeability in the process of product design and different consumers' willingness to pay. The impacts of remanufacturing cost weights and the level of interchangeability on prices and market demands are comparatively analyzed under different remanufacturing modes when OM or IO takes high or low pricing strategy. In order to provide guidance for enterprises, numerical example was calculated. And the results show that: Under OM or IO remanufacturing mode, it is better to take high pricing strategy. However, under OM/IO remanufacturing mode, it is better to take low pricing strategy.

Keywords: Closed-loop supply chain, Product design, Pricing, Remanufacturing mode

1. INTRODUCTION

As an effective way of end-of-use products dispose methods, remanufacturing can save production cost, reduce greenhouse gas emissions and improve resources utilization. More and more firms have pay attention to remanufacture. There are different remanufacturing modes in the widen remanufacture market. Original manufacturer (OM) remanufacturing mode is the mainstream mode. For example, YUCAI, WEICAI, Volkswagen and Xerox as OMs are positive in remanufacturing products. Independent operator (IO) remanufacturing mode also exists because of the economies of scale, the distinct advantage of price, low remanufacture costs and other advantages. There is no doubt that IO competes with OM for products market and profits. In order to deter IO and protect new products market, OM often determine the level of interchangeability carefully in new product design process. Because the interchangeability is considered as a degree to which the product can be disassembled without force. Decreasing the level of interchangeability would increase an IO's cost in cannibalizing used items, but it would also increase the production costs of the OM. The decisions about the level of interchangeability would be different in different remanufacturing modes. Thus, How to make optimal pricing decisions when considering the level of interchangeability in product design under OM, IO and OM/IO remanufacturing modes is a practical problem that confronts OMs and IOs now.

Hatcher et al. [1, 2, 3] made a literature review of the design for remanufacture, summarized the theoretical system, research methods and research trends. Hua et al. [4] established a two period model and analysed the impacts of product design for remanufacture on decision-making and profits between upstream and downstream enterprises. Wu [5] established a two period model composed of an OM and a remanufacturer, investigated the OM's product-design strategy and the remanufacturer's pricing strategy. Wu [6] made a further study in the product design decision of the OM and both chain members' competitive pricing strategies, and analysed the effects of production cost and consumers' willingness to pay (WTP) on the equilibrium decisions and profits.

Ferguson and Toktay [7] studied the competition between OM and IO in a two period CLSC. Ferrer and Swaminathan [8] assumed that consumers' willingness to pay (WTP) for remanufactured product and new product were same, then investigated the remanufacturing strategies of OM considering the competition between OM and IO. They [9] made a further study on the production and pricing decision-making in multi-period OM model considering the remanufactured product is differentiated from the new product. Lu et al. [10] made analysis and comparisons of the pricing strategies under no remanufacturing model, OM remanufacturing model and IO remanufacturing model. Wu [11] established a CLSC structure in which OM produce new product and IO produce remanufactured product. Chen and Chang [12] analysed the effects on OM profit in OM remanufacturing model and IO remanufacturing model considering the competition between new product and remanufactured product. Bulmus et al. [13] investigated the pricing problems considering a two-period model with OM and IO both participated into remanufacturing.

To the best of the author's knowledge, no work has been reported in the literature considering differential pricing strategies and different remanufactured mode in making pricing decisions with the consideration of product design for remanufacture. With the consideration of the consumer's WTP, OM remanufacturing model and IO remanufacturing model are established to study differential pricing strategies and analyse the impacts of remanufacturing cost savings and the level of interchangeability in product design on the products market and the OM profit.

2. MODEL

The structures of CLSC under OM, IO and OM/IR remanufacturing modes are shown in Fig.1. In OM mode, OM produces new product, recycles used products from consumers and remanufactures. In IO mode, OM only produces new product. IO recycles used products and remanufactures. Considering OM acts as a Stackelberg game leader. We assume that remanufactured product remanufactured by OM and IO are homogeneous. Consumers are classified into ordinary consumers who prefer to purchase new product and green consumers. Green consumers who have environmental awareness and consider remanufactured product green typically equally value the remanufactured and new products.

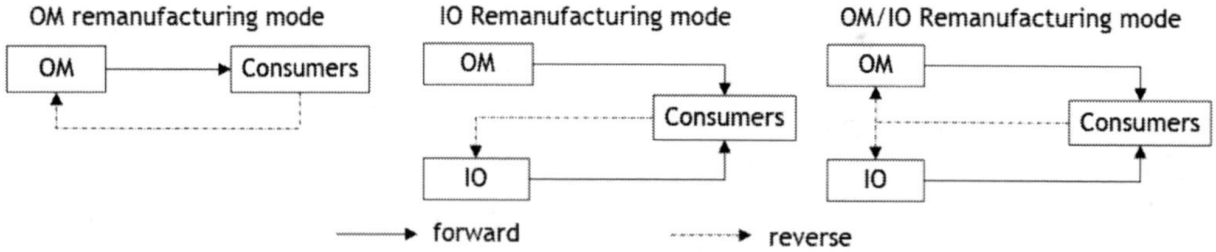

Fig.1: CLSC structures under different remanufacturing modes

Sets and parameters are defined as follows:

i:n,o, r denote new product, OM-remanufactured product and remanufactured product

p_i^t, D_i^t: the price, the demand under different remanufacturing mode (t=1,2,3 represents OM remanufacturing mode, IO remanufacturing mode and OM/IO remanufacturing mode respectively)

η: the sensitive coefficient of a fixed cost of the interchangeability adjustment

l: the level of interchangeability, $0<l<1$ means the increase of the level of interchangeability,

$-1<l<0$ means the decrease and $l=0$ means no adjustment. We assume that the cost of adjusting the interchangeability is $0.5\eta l^2$[11]

c: the initial production cost of new product

λ: the initial production cost weights of remanufactured product, it means the ratio of initial production cost between remanufactured product and new product is λ

δ_n, δ_r: the impact factors of the degree of interchangeability on the production costs of new and OM-remanufactured or remanufactured products. With a l-level of interchangeability in the product design, the unit production cost of the new product is $c-\delta_n l$ and the unit production cost of the OM-remanufactured or remanufactured is $\lambda c-\delta_r l$.

ρ: ordinary consumers' value-discount factor for OM-remanufactured product compared with new product.

σ: ordinary consumers' value-discount factor for remanufactured product compared with new product. Assuming $\sigma<\rho$ by considering OM-remanufactured product has the brand advantage over remanufactured product.

A: the number of consumers in total market

B: the percentage of the green consumers

2.1 Demand functions

Considering a high-priced remanufactured product is usually accepted by green consumers. Therefore, the remanufacturer can make a strategic choice with regard to price: a high-pricing strategy that focuses only on the green market segment or a low-pricing strategy that penetrates the whole market. According to it, the hypotheses of low pricing strategy under different modes is ($p_o^1\le\rho p_n^1$, $p_r^2\le\sigma p_n^2$, $p_o^3\le\rho p_n^3$ and $p_r^3\le\sigma p_n^3$) and high pricing strategy is ($p_o^1\ge\rho p_n^1$, $p_r^2\ge\sigma p_n^2$, $p_o^3\ge\rho p_n^3$ and $p_r^3\ge\sigma p_n^3$). And $\sigma p_o^3>\rho p_r^3$ is assumed to ensure the remanufactured product can be purchased under OM/IO remanufacturing mode. According to utility function of ordinary and green consumers, the total demand for new and remanufactured products can be obtained are shown in table 1.

Table 1: Total demand under different remanufacturing mode and different pricing strategies

demand	l pricing strategy	h pricing strategy
D_n^1	$(1-B)A(1-\frac{p_n^1-p_o^1}{1-\rho})$	$(1-B)A(1-p_n^1)$
D_o^1	$\frac{(1-B)A(\rho p_n^1-p_o^1)}{\rho(1-\rho)}+BA(1-p_o^1)$	$BA(1-p_o^1)$
D_n^2	$(1-B)A(1-\frac{p_n^2-p_r^2}{1-\sigma})$	$(1-B)A(1-p_n^2)$
D_r^2	$\frac{(1-B)A(\sigma p_n^2-p_r^2)}{\sigma(1-\sigma)}+BA(1-p_r^2)$	$BA(1-p_r^2)$
D_n^3	$(1-B)A(1-\frac{p_n^3-p_o^3}{1-\rho})$	$(1-B)A(1-p_n^3)$
D_o^3	$\frac{A(1-B)(p_n^3-p_o^3)}{1-\rho}-\frac{A(1-B-\rho B)(p_o^3-p_r^3)}{\rho-\sigma}+BA$	$BA(1-\frac{\rho(p_o^3-p_r^3)}{\rho-\sigma})$
D_r^3	$\frac{A(1-B+\rho B)(p_o^3-p_r^3)}{\rho-\sigma}+\frac{A(1-B-\rho B)p_r^3}{\sigma}$	$BA(\frac{\rho(p_o^3-p_r^3)}{\rho-\sigma}-\frac{\rho p_r^3}{\sigma})$

2.2 OM remanufacturing model

OM make the equilibrium price decisions based on purchase intention and profit maximization principle.
The objective function:

$$\max \pi_m^{o}=[p_n^{1}-(c-\delta_n l)]\,D_n^{1}+[p_o^{1}-(\lambda c-\delta_r l)]\,D_o^{1}-\frac{\eta l^2}{2} \tag{1}$$

2.3 IO remanufacturing model

The objective function:

$$\max \pi_m^{R}=[p_n^{2}-(c-\delta_n l)]\,D_n^{2}-\frac{\eta l^2}{2}$$

$$\text{s.t } \max \pi_r^{R}=[p_r^{2}-(\lambda c-\delta_r l)]\,D_r^{2} \tag{2}$$

2.4 OM/IO remanufacturing model

The objective function:

$$\max \pi_m= [p_n^{3}-(c-\delta_n l)]\,D_n^{3}+ [p_o^{3}-(\lambda c-\delta_r l)]\,D_o^{3}-\frac{\eta l^2}{2}$$

$$\text{s.t } \max \pi_r= [p_r^{3}-(\lambda c-\delta_r l)]\,D_r^{3} \tag{3}$$

Put the total demand functions in table 1 into (1) and (3), the equilibrium pricing decisions can be obtained through stackelberg game theory. The results are shown in table 2.

Table 2: Equilibrium pricing decisions under different remanufacturing modes

Mode	price	*l* pricing strategy	*h* pricing strategy
OM	p_n	$\frac{c-\delta_n l+1-\rho}{2}+\frac{\rho}{2[1-\beta(1-\rho)]}$	$\frac{c-\delta_n l+1}{2}$
	P_o	$\frac{\lambda c-\delta_r l}{2}+\frac{\rho}{2[1-\beta(1-\rho)]}$	$\frac{\lambda c-\delta_r l+1}{2}$
IO	p_n	$\frac{c-\delta_n l+1}{2}$	$\frac{c-\delta_n l+1}{2}$
	p_r	$\frac{\lambda c-\delta_r l}{2}+\frac{2\beta\sigma(1-\sigma)+\sigma(1-\beta)(c-\delta_n l+1)}{4\beta\sigma(1-\sigma)+4(1-\beta)}$	$\frac{\lambda c-\delta_r l+1}{2}$
OM/IO	p_n	$\frac{(2Z-3Y)X(\lambda c-\delta_r l)-Y}{8XZ-6XY}+\frac{1}{2X}+\frac{1-\rho+c-\delta_n l}{2}$	$\frac{c-\delta_n l+1}{2}$
	P_o	$\frac{(2Z-3Y)X(\lambda c-\delta_r l)-Y}{8XZ-6XY}+\frac{1}{2X}+\frac{\lambda c-\delta_r l}{2}$	$\frac{2(\rho-\sigma)+3\rho(\lambda c-\delta_r l)}{4\rho-\sigma}$
	p_r	$\frac{(2Z-3Y)X(\lambda c-\delta_r l)-Y}{4XZ-3XY}$	$\frac{\sigma(\rho-\sigma)+\rho(2\rho+\sigma)(\lambda c-\delta_r l)}{\rho(4\rho-\sigma)}$

Where $X=\frac{1-\beta-\rho\beta}{\rho-\sigma}$, $Y=\frac{1-\beta+\rho\beta}{\rho-\sigma}$, $Z=\frac{1-\beta-\rho\beta}{\sigma}$

3. ANALYSIS OF PRICING DECISIONS AND DEMANDS

3.1 Effects of remanufactured product cost weights (λ) on price and demand

From the functions in table 1-2, we can obtain that:

1) When OM or IO takes high pricing strategy under three different remanufacturing modes, the new product price and the demand are independent with λ. The price of OM-remanufactured product or remanufactured product increases as λ increase and the demand is negative with λ.
2) When low pricing strategy is taken under OM or IO remanufacturing mode, the new product price and the demand for new product is independent with λ and the price of OM-remanufactured or remanufactured product increases as λ increases. The demand for new product increases with λ and the demand for OM-remanufactured or remanufactured product decreases with λ. When low pricing strategy is taken under OM/IO remanufacturing mode, the prices of three products increase as λ increases. The demand for new product increases with λ. The main reasonable explanation is that the remanufacturing cost saving contributes not significantly towards improves the competitiveness of remanufactured product in price.

3.2 Effects of the level of interchangeability (l) on price and demand

Form table 1-4, we can obtain:

1) When OM or IO takes high or low pricing strategy under three different remanufacturing modes, the prices of the three products decrease as l increases. Because improving the level of interchangeability can reduce the production costs and thus the prices increase.
2) When OM or IO takes high pricing strategy, the demand for the three products increase. The reasonable explanation is that the demand for new product and the demand for OM-remanufactured or remanufactured product come from different markets. Thus, the influences of l on demands only affected by their own impact factor of l on production cost (δ_n, δ_r).
3) When OM or IO takes low pricing strategy, the relationship between the demand and l depends on the magnitude relation of δ_n and δ_r under different remanufacturing modes. The details are as follows ($k=\delta_r/\delta_n$):
   Under OM remanufacturing mode, when $k<1$ the demand for new product increases and when $k>\frac{(1-\beta)\rho}{1-\beta+\beta\rho(1-\rho)}$ the demand for OM-remanufactured product increases as l increases.Under IO remanufacturing mode, when $k<1-\frac{(1-\beta)\sigma}{2\beta\sigma(1-\sigma)+2(1-\beta)}$ the demand for new product increases and when $k>\frac{\beta\sigma}{\beta+(1-\beta)\sigma(1-\sigma)}-\frac{(1-\beta)\sigma}{2[1-\beta+\beta\sigma(1-\sigma)]}$ the demand for remanufactured product increases as l increases. Under IO/OM remanufacturing mode, when $k<1$ the demand for new product increases, when $k>\frac{(4Z-3Y)(1-\beta)(\rho-\sigma)}{(4Z-3Y)(1-\beta)(\rho-\sigma)+2(1-\rho)(1-\beta-\rho\beta)}$ the demand for OM-remanufactured product increases.

4. NUMERICAL ANALYSIS

An example about an air-conditioning enterprise is analysed in this section. Specific data are as follows: A=50, c=0.5, η=0.4, β=0.3, ρ=0.6, λ=0.5.

The profit of OM under OM remanufacturing mode is shown in Fig.2. Considering the value of k

would affect the relationship between new and OM-remanufactured products' demand and l, we take the various values of k (0.2, 0.8 and 1.2). The profit of OM and IO under IO remanufacturing mode are respectively shown in Fig.3 and Fig.4. We take the various values of k (0.8 and 1.2).

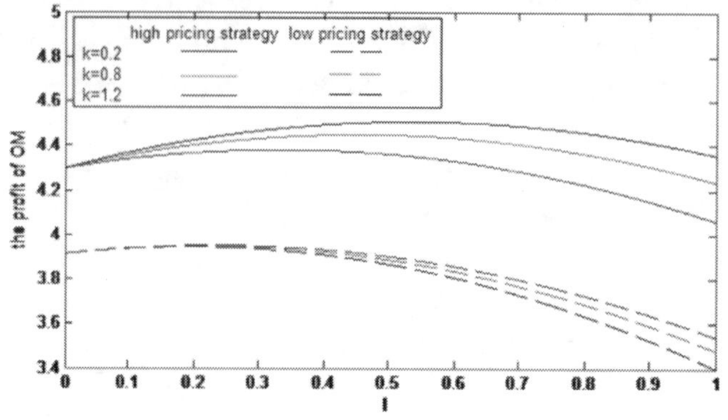

Fig.2 the impacts of l on the profit of OM under OM remanufacturing mode

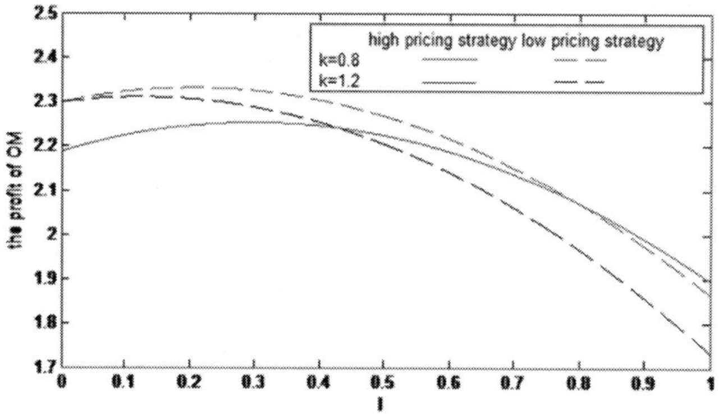

Fig.3 the impacts of l on the profit of OM under IO remanufacturing mode

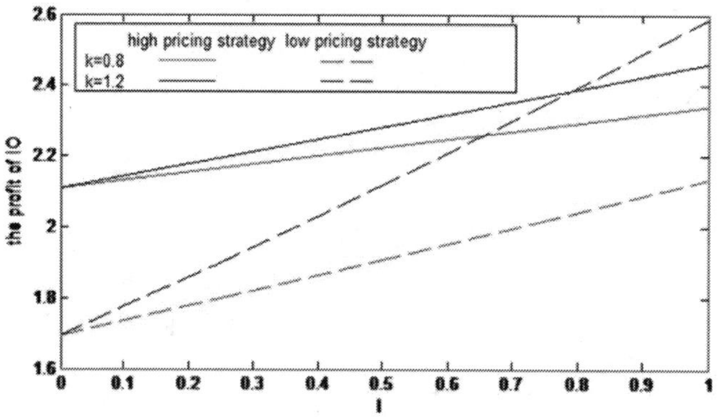

Fig.4 the impacts of l on the profit of IO under IO remanufacturing mode

From Fig.2-4, we can see that, when the profit of OM increases first and then decreases as l increases under OM or IO mode. However, the IO profit keeps on increasing. It is better for OM to take high pricing strategy and adjust the level of interchangeability to the optimal value.

The profit of OM and IO under OM/IO remanufacturing mode are respectively shown in Fig.5 and Fig.6. We take the various values of k (0.8, 1.2 and 1.6).

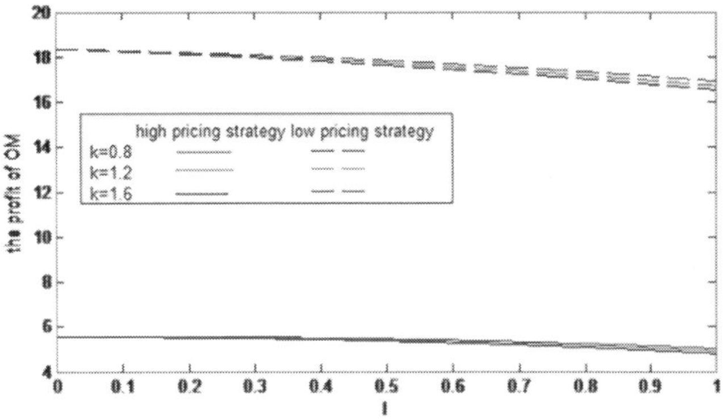

Fig.5 the impacts of *l* on the profit of OM under OM/IO remanufacturing mode

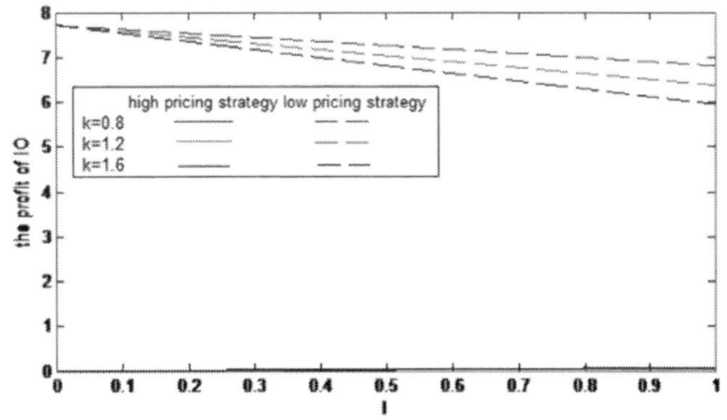

Fig.6 the impacts of *l* on the profit of IO under OM/IO remanufacturing mode

Form Fig.5 and Fig.6, we can obviously find that, the low pricing strategy is good for members' profits. And the profit of OM or IO decreases as l increases. The value of k does not affect the influence rules of l.

5. CONCLUSION

Three decision-making models with the consideration of the product's interchangeability and different consumers' WTP study the optimal pricing strategies. The impacts of production cost weight of remanufactured product and the level of interchangeability on different pricing decisions and demands are analysed through analysing the equilibrium solutions. And the impacts of the level of interchangeability on members' profits are investigated through numerical examples. We find that: under OM or IO remanufacturing strategy, there exists an optimal value of the level of interchangeability for the profit of OM. Under OM/IO remanufacturing mode, it is better to decrease the level of interchangeability as much as possible. And the relationship of between the impact factors of the degree of interchangeability on the production costs of new and OM-remanufactured or remanufactured products does not affect the influence rules of the level of interchangeability on members profits under three different remanufacturing modes and different pricing strategies. Under OM or IO remanufacturing mode, it is better to take high pricing strategy. However, under OM/IO remanufacturing mode, it is better to take low pricing strategy.

6. ACKNOWLEDGMENT

This research is supported by MOE (Ministry of Education in China) Project of Humanities and Social Sciences (no.16YJAZH010). The suggestions of the reviewers are also gratefully acknowledged.

7. REFERENCES

[1] **Hatcher, G.D., Ijomah, W.L., Windmill, J.F.C. 2011.** Design for remanufacture: a literature review and future research needs, *Journal of Cleaner Production*, 19(17), pp 2004-2014.

[2] **Hatcher, G.D., Ijomah, W.L., Windmill, J.F.C. 2013.** Integrating design for remanufacture into the design process: the operational factors, *Journal of Cleaner Production*, 39(1), pp 200-208.

[3] **Hatcher, G.D., Ijomah, W.L., Windmill, J.F.C. 2014.**A network model to assist 'design for remanufacture' integration into the design process, *Journal of Cleaner Production*, 64, pp 244-253.

[4] **Hua, Z., Zhang, X., Xu, X. 2011.** Product design strategies in a manufacturer-retailer distribution channel, *Omega*, 39(1), pp 23-32.

[5] **Wu, C.H. 2013.** OEM product design in a price competition with remanufactured product, *Omega*, 41(2), pp 287-298.

[6] **Wu, C.H. 2012.** Product-design and pricing strategies with remanufacturing, *European Journal of Operational Research*, 222(2), pp 204-215.

[7] **Ferguson, M.E., Toktay, L.B. 2006.** The effect of competition on recovery strategies, *Production and Operations Management*, 15 (3), pp 351-368.

[8] **Ferrer, G., Swaminathan, J.M. 2006.** Managing new and remanufactured products, *Management Science*, 52(1), pp 15-26.

[9] **Ferrer, G. Swaminathan, J.M. 2009.** Managing new and differentiated remanufactured products, *European Journal of Operational Research*, 203 (2010), pp 370-379.

[10] **Hongjuan, L.U., GUO. 2011.** Research on Remanufacturing Production Mode Based on Game Theory, *Journal of Mechanical Engineering*, 47(20), pp 18-22.

[11] **Wu, C.H. 2012.** Price and service competition between new and remanufactured products in a two-echelon supply chain, *Int. J. Production Economics*, 140 (1), pp 496-507.

[12] **Chen, J.M., Chang C.I. 2012.** The co-opetitive strategy of a closed-loop supply chain with remanufacturing, *Transportation Research Part E*, 48 (2), pp 387-400.

[13] **Bulmus, S.C., Zhu, S.X., Teunter, R. 2014.** Competition for cores in remanufacturing, *European Journal of Operational Research*, 233(1), pp 105-113.

STABILITY ANALYSIS AND OPTIMIZATION ON MANUFACTURING PROCESS OF COMPLEX EQUIPMENT BASED ON SMALL-WORLD NETWORK

T. Yang[1*], H. Li[1] and G.H. Ma[2]
[1]School of Economics and Management
Xidian University, Xi'an, China
longyang719@163.com

[2]Xi'an Aero-engine (Group) LTD, Xi'an, China
shuaihaitun@126.com

ABSTRACT

The complex equipment is usually manufactured in a mixed production organized mode, accordingly various products and manufacturing resource will interact with each other. So in this paper small-world network is introduced to describe these complicate relationships within different elements. Moreover manufacturing resources are regarded as vertices, while relationship between vertices is defined as directed edges. And different products pass these vertices to accomplish the manufacturing task. Then the mathematical model is built to characterize basic elements and network architecture of manufacturing process. And some topological characteristics are defined to describe the whole status of network model. Then, through analysis on the global characteristics, it can be seen that the manufacturing process of complex equipment shows small-world property. In a small-world network a stochastic invalidation of vertex will lead to the catastrophic failure of the whole manufacturing process. Thereby stability is analyzed through comparing with ER random network in some topological characteristics. And the optimization method on selecting optional vertices in some manufacturing steps is studied to raise the stability of manufacturing process. Finally, an actual example is introduced as an example. It shows that the robustness and reliability of manufacturing process are improved through selecting an optimal manufacturing process route.

Keywords: Stability, Small-world network, Manufacturing process, Analysis, Optimization

1 INTRODUCTION

Complex equipment is a kind of special products with complex customer needs, complex configuration, complex technology, complex manufacturing process, complex maintenance and complex working environment [1], such as large aircraft, manned spacecraft, numerical controlling machinery, aero-engines, and so on. Complex equipment is generally composed by a large number of parts with complicate characteristics. For example, as shown in Figure 1, an aero-engine comprises fan disc, shaft, spool, turbine casing and so on [2].

It could be seen all of those parts are difficult to manufacture owing to their complicate geometrical shape and configuration. Usually these parts need pass many steps and their manufacturing processes of these parts are every complicate. And manufacturing processes of various parts of complex equipment are mixed in a same production. So there are complicate relationships within different

* Corresponding Author, The paper is supported by China National Soft Science Research Project (2014GXS2D031).

parts in their manufacturing processes. Doubtlessly the manufacturing process of complex equipment is a complex system with all kinds of factors affecting each other.

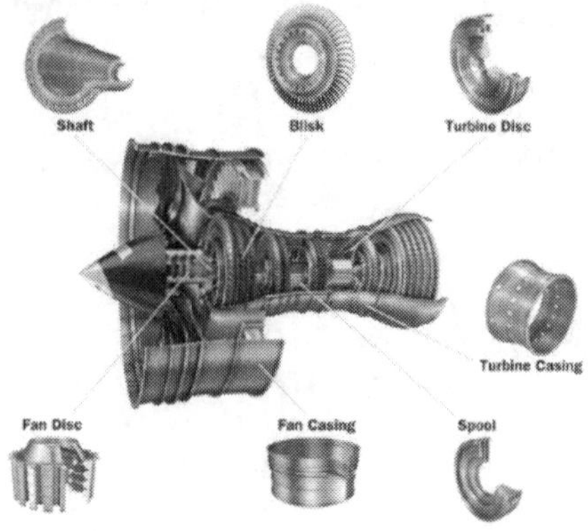

Figure 1: Configuration and shape of aero engine

Parts, manufacturing processes and manufacturing resources are the most important elements of production line. Among them manufacturing process is the controlling element to drive the parts to pass various manufacturing resources to accomplish manufacturing task. In production management area manufacturing process planning is to make choice from the flexible manufacturing process routes to meet the given optimization goal, which is the foundation for detailed process planning and production scheduling [3]. As the intermediate procedure between design and manufacturing, manufacturing process planning is to select the resource and technique involved in the manufacturing process, which is the essential step [4]. Thereby a great deal of research work has been completed on manufacturing process planning. Among them, most research focuses on the optimization of machining process schedule [5-7]. These papers usually take the manufacturing capability and production cost as constraint conditions, and then intelligent searching algorithms such as genetic algorithm and particle swarm optimization, are used to adjust manufacturing schedules to balance the duty load and shorten the production cycle time. However manufacturing process not only is to arrange manufacturing activity but also is to define the relationships with different manufacturing resources. As we know, for complex equipment these relationships have the important effect on the stability of the production capacity and quality because the complicate relationships within different parts. So the stability of manufacturing process must be emphasized in planning phase.

And the research method and theory on complex system should be introduced to analysis and optimization on manufacturing process of complex equipment. The small-world network is an efficient tool to study the relationships with various units [8][9]. The small network with short average path length and high cluster coefficient was proposed by Watts and Strogatz in 1998. And the disastrous collapse caused by expiration of the key nodes was released in the article of Nature named Achilles' heel of the Internet [10]. Small-world network is influenced by different factors, which even the minimum dynamic change may lead to fierce collapse [11]. Moreover small-world network could analyze effects on stability in general by means of network architecture and topological properties. Consequently, in this paper small-world is introduced to describe the complicate relationships with manufacturing resources in mixed production line. Furthermore the small-world network is introduced to analyze and optimize the stability of manufacturing process of complex equipment.

The paper is organized as follows. The small-world network model is introduced and constructed in section 2, while the mathematical definition and topological properties of network model are given. Moreover in section 3 the stability optimization index is given to construction the optimization model. Then the stability of manufacturing process is optimized through selecting manufacturing route in manufacturing process planning. In section 4 an actual production line is studied to validate the method and algorithm presented in this paper. Finally the paper is concluded.

2 SMALL-WORLD NETWORK MODEL OF MANUFACTURING PROCESS

2.1 Small-world network model

In the production line manufacturing resources such as machinery, tools and human resources are arranged to undertake the manufacturing tasks. In resource configuration and capacity planning some gathered resources which could accomplish one certain duty of manufacturing process can be regarded as a basic unit as in Figure 2. Consequently in network model these units are defined as manufacturing vertices. Therefore a part accomplishes its manufacturing process pass through various manufacturing vertices as shown in Figure 2, and then a manufacturing process route come into being which is UX235 → SAJO → ... → examination vertex.

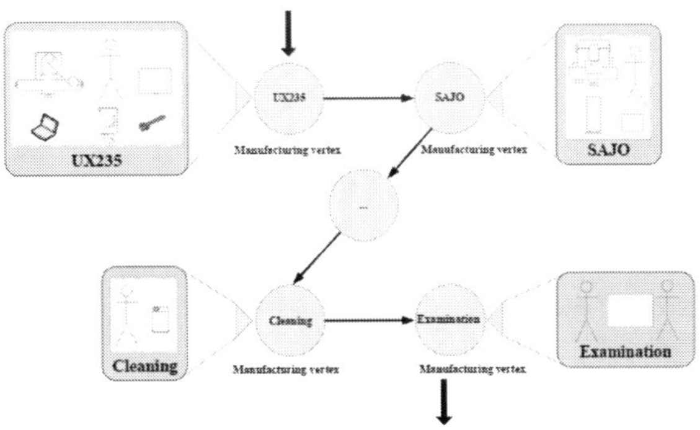

Figure 2: Manufacturing process route of one part

Moreover the relationships with various vertices are defined as directed edges. Then manufacturing process of different parts will cross at some manufacturing vertices. Therefore in mixed production line the network architecture of manufacturing processes would appear as shown in Figure 3.

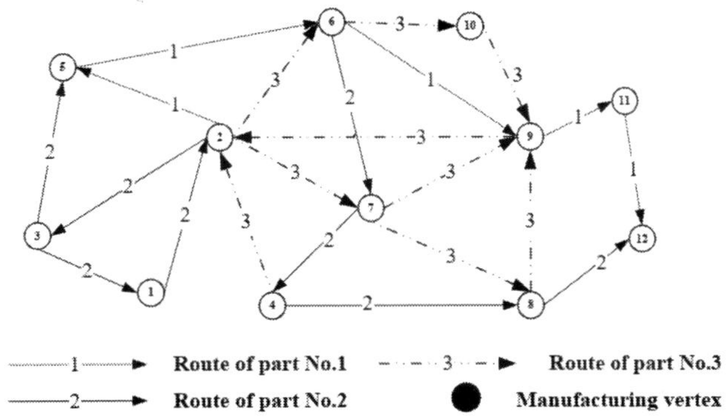

Figure 3: Network of manufacturing processes

Consequently a mathematical model might be built to describe the complicate relationships within various parts. Now the small-world network model $G(V, E)$ is defined in detail as follows.

- Manufacturing vertices, $V\{V_i\}$ (i=1, ..., n), is defined as the n manufacturing vertices.
- Directed edges, E $\{E_{ij}\}$, (i,j=1,...,n), is regarded as manufacturing process route from manufacturing vertex V_i to manufacturing vertex V_j. E_{ij} equals to 1 or 0.When two vertices are connected E_{ij} =1, otherwise E_{ij} =0. It's noted that $E_{ij} \neq E_{ji}$ because of difference in their direction.

Moreover the other elements in mixed production line are defined as:

- Parts, $P\{P_i\}$ (i=1, ..., m), is used to describe m parts maufactured in mixed production line.

- Manufacturing process, $Pr\{ Pr_{ij} \}$, ($Pr_{ij} \in V$), Manufacturing process drives the part to flow through manufacturing vertices. In this paper a matrix Pr is defined to describe all the manufacturing processes in the mixed production line. Meanwhile the element Pr_{kj} is equal to the code of the manufacturing vertex on j-step of part encoded k which is P_k.

2.2 Topological properties

Then the topological properties are defined as follows to describe the global characteristics of network.

- Weight, $w\{w_{ij}\}$, (i,j=1,...,n). Some parts usually pass the same vertexs several times, and then the weight is used to represent this repeatability. The element w_{kl} is calculated as:

$$\begin{cases} \text{If } Pr_{ij} = k, Pr_{ij+1} = l, \text{ then } w_{kl} = 1 \\ \text{If } Pr_{lj} = k, Pr_{lj+1} = l, \text{ then } w_{kl} = w_{kl} + 1 \ (l>i) \\ \text{If there is no relationship from vertex } k \text{ to vertex } l, \text{ then } w_{kl} = 0 \\ w_{kk} = 0 \\ w_{kl} \neq w_{lk} \end{cases} \tag{1}$$

- Degree of vertices, $k\{k_i\}$, (i=1,...,n). k_i is defined as the number of other vertices linked to vertex V_i. In directed network, out-degree k_{out} means the number of edges from the vertex to other vertexes, otherwise in-degree k_{in}. They are defined as:

$$k_{out}:\ k_i = \sum_{j=1}^{n} E_{ij} \qquad k_{in}:\ k_j = \sum_{i=1}^{n} E_{ij} \tag{2}$$

- Path length, $d\{d_{ij}\}$ (i,j =1,...,n). Path length d_{ij} is the distance from the manufacturing vertex V_i to vertex V_j. It is defined as total amount of directed edges in the shortest path which could be obtained by Dijkstra algorithm [11]. The average path length D of the whole network is defined as:

$$D = \frac{1}{\frac{1}{2} n(n-1)} \sum_{i,j} d_{ij} \tag{3}$$

- Clustering coefficient of vertices, $c\{c_i\}$(i=1,...,n). It is supposed that there are l manufacturing vertices is connected to the given manufacturing vertex V_i. And the number of actual edges within these l manufacturing vertices is defined as NE_i, while the total number of possible edges within these l manufacturing vertices is $l_i(l_i$-1)/2. Then clustering coefficient of V_i is defined as:

$$c_i = \frac{NE_i}{\frac{1}{2} l_i (l_i - 1)} \tag{4}$$

The equal (4) is the ratio between the quantities of triangle and that of triples connected to manufacturing vertex V_i. It could be computed through validity of all connected triples [12], which is

$$c_i = \frac{1}{k_i (k_i - 1)} \sum_{j,h} a_{ij} a_{ih} a_{jh} \tag{5}$$

The clustering coefficient of the whole network is equal to the average of all vertices, which is

$$C = \frac{1}{n}\sum_{i=1}^{n} c_i \tag{6}$$

3 STABILITY ANALYSIS AND OPTIMIZATION ON MANUFACTURING PROCESS

3.1 Stability analysis on manufacturing process based on small-world network

According to the preceding research, the specific steps on stability analysis on the manufacturing process are given as follows:1) Define the set of manufacturing vertices *V* and the set of pars *P* according to the organization of the mixed production line. 2) Translate the manufacturing process of all parts to manufacturing process matrix *Pr*, while the manufacturing processes are expressed as manufacturing vertices code according to the code in *V* and *P*. 3) Compute the weight of directed edges by equation (1) through relationships between the vertices described in *Pr*. Consequently the availability of directed edges could be validated through Boolean computation on weight of directed edges. And the set of directed edges *E* would be obtained. 4) Establish the small-world network model *G*(*V*,*E*). Then the network topology of manufacturing process could be generated.

3.2 Stability optimization on manufacturing process based on small-world network

3.2.1 Stability index

Through analysis mentioned above specific steps, if the network model of manufacturing process shows small-world characteristics, this phenomenon must be eliminated to increase anti-destroying ability through optimization on manufacturing process planning. Within many the complex network model, the ER random network model has with short average path length, but not high clustering features. Therefore, the ratio between the clustering coefficient C_{ER} of ER random network and the clustering coefficient *C* of manufacturing process could be used as the stability index to measure the deviation extent between the network model and the ER random network.

$$\chi = \frac{C_{ER}}{C} = \frac{K/n}{C} \tag{7}$$

When the stability index χ is smaller, it means *C* is larger than C_{ER}, network model of manufacturing process has obvious small-world characteristics. And the key vertices have large impact on relevant manufacturing vertices, while the manufacturing process is low stable.

3.2.2 Optimization method

In the manufacturing process planning of complex equipment, manufacturing process of complex equipment is usually configured some equipment with similar processing capacity, which provides a great choice space for equipment selection. And the manufacturing vertices formed by such equipment might constitute the optional manufacturing vertices to meet the manufacturing duty. As shown in Figure 4, in a step of manufacturing process the manufacturing vertex could be replaced by the optional vertices which are 2, 15, 16, 19, 20, 36 and 42.

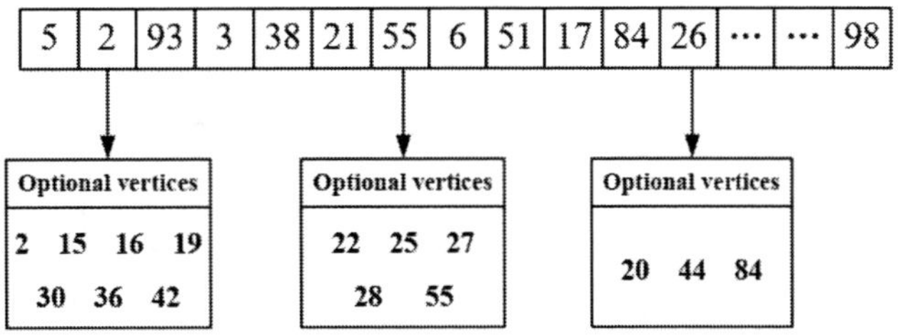

Figure 4: Optional vertices of manufacturing process

Because of the optionality of manufacturing vertices in certain manufacturing step, the manufacturing process matrix *Pr* will change consequently. And the topological architecture and properties of the network model will change, which could eliminate small-world characteristics and improve the stability of manufacturing vertices. Therefore the stability index could be regard as the function of manufacturing process *Pr*. Then the stability optimization model could be built as:

$$\chi_{\max} = \mathrm{Max}[c(Pr)] \tag{8}$$

Usually the feasible solution space of the manufacturing process is abnormal huge, and intelligent searching algorithm must be introduced, and in this paper the genetic algorithm is adopted.

4 NUMERICAL EXAMPLE

In this paper an aero-engine casing is introduced as a numerical example. It has 105 manufacturing vertices and yields 81 kinds of casing parts. According to the specific steps to build the network model as described in paragraph 3.1, the topology graph could be got in Figure 5.

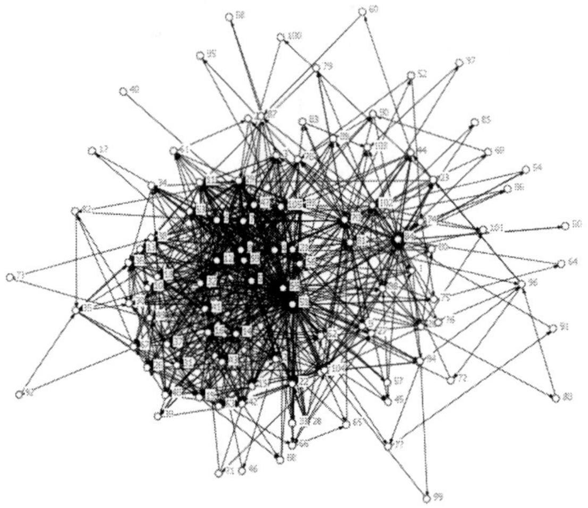

Figure 5: The network topology of manufacturing process

Then the average path length *D* and clustering coefficient *C* are calculated according to (3), (5) and (6).Thus, for aero-engine casing production line, the average path length *D* is:

$$D = 2.3523 \approx D_{ER} \approx \frac{\ln n}{\ln K} = 2.3850 \tag{9}$$

The clustering coefficient *C* is:

$$C = 0.3493 \quad >> \quad C_{ER} = \frac{K}{n} = 0.0670 \tag{10}$$

As shown in (9) and (10), through comparing with ER random network with the same vertices number and average degree, the clustering coefficient of casing production line is much larger than ER random network. And the short average path length is approximate to ER random network.. Therefore the stability index χ of manufacturing process of casing is:

$$\chi = \frac{C_{ER}}{C} = \frac{K/n}{C} = 0.1919 \tag{11}$$

According to the given method, within the optional vertices the manufacturing process is optimized through genetic algorithm. And the genetic evolution is executed with crossover probability of 70%

and mutation probability of 20%. As shown in Figure 6, passing through 643 iterations, the algorithm has converged, and the optimal targets continuously increased.

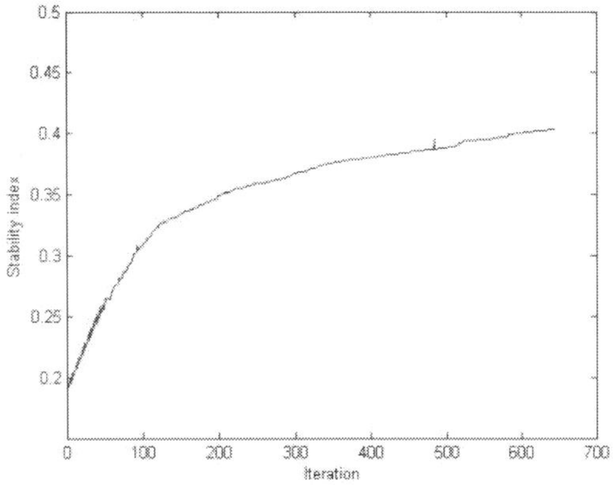

Figure 6: Evolution process of Stability index of manufacturing process

At the moment, the clustering coefficient C is 0.1666, the stability index is:

$$\chi = \frac{C_{ER}}{C} = \frac{K/n}{C} = \frac{0.1666}{0.0675} = 0.4051 \quad (12)$$

The average path length D is:

$$D = 2.4716 \approx D_{ER} \approx \frac{\ln n}{\ln K} = 2.3768 \quad (13)$$

The stability index of manufacturing process significantly increases from 0.1919 to 0.4051 through optimization. The evolutionary process of the clustering coefficient of manufacturing process and ER random network with the same number of vertices and average degree is shown in Figure 7.

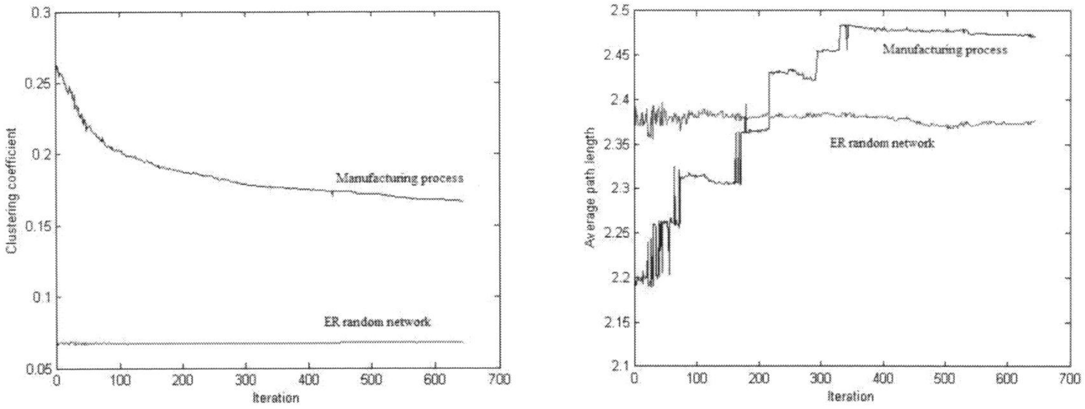

(a) Clustering coefficient (b) The average path length

Figure 7: Evolution process of the clustering coefficient and the average path length

It could be seen that the clustering coefficient of manufacturing process continuously decreases in Figure 7(a) through cutting off partial connections with the key vertices. Consequently the key vertices' effects on their adjacent vertices are weakened, so the stability is improved.

In the optimization process, during the reduction of clustering coefficient, the connection distance becomes longer. As shown in Figure 7(b), the average path length of manufacturing process increases slightly, but it still has maintained at a short level, which is similar with the ER random network.

5 CONCLUSION

Facing the complicate characteristics of manufacturing process a small-world network is introduced to describe the complicate relationships with various manufacturing resources and parts. Then the network model is built to analyse the stability of manufacturing process according to its topological properties. Meanwhile the specific analytical steps are given in the paper. Then a stability index is proposed to reflect the small-world phenomenon of manufacturing process with high risk. And an optimization model is construction to eliminate small-world characteristics and improve the stability of manufacturing process through planning the manufacturing process. Finally, using an aero-engine casing production line as an example, the reliability of methods proposed in this paper is proved.

6 REFERENCES

[1] **B. H. Li. 2006.** Important technology of manufacturing information of complex equipment, *Manufacture Information Engineering of China*, **(6), pp19-23.**

[2] **H. Sean.** Manufacturing method and knowledge of advanced aero engine production, [EB/OL] (2016-2-16) [2013-11-15]. *http://www.mmindiaonline.com/DetailNews/428a-Manufacturing-Methods-for-Advanced-Aerospace-Engine-Production.aspx.*

[3] **Z. Y. Zhang, C. T. Tang, J.R. Zhang, et al. 2009.** Genetic Algorithm for Integration of CAPP and Production Scheduling, *Computer Integrated Manufacturing Systems*, 15(7), pp 621-624.

[4] **X. Y. Li, L. Gao, X.Y. Shao, et al. 2010.** Mathematical modeling and evolutionary algorithm-based approach for integrated process planning and scheduling, *Computers & Operations Research*, 37, pp 656-667,

[5] **H. F. Rahman, R. Sarker, D. Essam. 2015.** A genetic algorithm for permutation flow shop scheduling under make to stock production system, *Computers & Industrial Engineering*. 90(11), pp 12-24.

[6] **B.Y. Cheng, J.Y. Leung, K. Li. 2015.** Integrated scheduling of production and distribution to minimize total cost using an improved ant colony optimization method, *Computers & Industrial Engineering*. 83(4), pp 217-225.

[7] **C. Low, C.-M. Chang, R.-K. Lib, et al. 2014.** Coordination of production scheduling and delivery problems with heterogeneous fleet, *International Journal of Production Economics*. 153(5), pp 139-148.

[8] **D. J. Watts, S. H. Strogatz. 1998.** Collective dynamics of 'small-world' networks, *Nature*, pp 393440-442.

[9] **A. Barrat, M. Barthélemy, R. Pastor-Satorras, el at. 2004.** The architecture of complex weighted networks, *Applied physical sciences*, 101, pp 3747-3752.

[10] **Y. H. Tu. 2000.** How robust is the Internet? *Nature*, pp 406353-354,

[11] **S. K. Peer, D.K. Sharma. 2007.** Finding the shortest path in stochastic networks, *Computers and Mathematics with Applications*, 53, pp 729-740.

[12] **A. Barrat, M. Barthélemy, A. Vespignani. 2004.** Modeling the evolution of weighted networks, *Physical Review E*, 70, pp 066149.

[13] **X. F. Wang, X. Li, G.R. Chen. 2006.** The theory and application of complex network, Tsinghua University Press.

A STUDY OF CAPABILITY EVALUATION FOR SERVICE-ORIENTED MANUFACTURING NETWORK NODE

Wei ping Zeng[1] and Liang qing Feng[1,2*]
[1]Department of Industrial Engineering
Nanchang Hang kong University, Nanchang, China
928601147@qq.com

[2]Department of Industrial Engineering
Tianjin University, Tianjin, China
flq9848@163.com

ABSTRACT

The evaluation and selection of node capability is the key part for service-oriented manufacturing network to optimize and improve its capability, which plays an important role on the improvement of operational effect and efficiency. Under the basis of analyzing that the service-oriented manufacturing network is a modular service network, constructing the node capability index system and evaluation model that their features of integration, value added and innovation are considered. According to experts' opinions, the precedence relation judgment matrix is constructed. And the weight of evaluation index is determined by using the eigenvalue method. The node capability is evaluated comprehensively with the TOPSIS method. At last, the feasibility of the method is verified by a certain case.

Keywords: SMN, Node capability, AHP, Fuzzy TOPSIS

1 INTRODUCTION

The rapid development of information, networking, and digital technology, provides convenience for the manufacturing enterprises to share manufacturing resources and to achieve business information in global scope, which greatly promote the cooperation between manufacturing enterprises and integration of manufacturing and service[1-3]. But the difference of resource endowment, operation mode, innovation level of node enterprise result in reduction of cooperative operation capability. Perishable characteristics such as customer demand make the whole service manufacturing network face more uncertainty[4], which increasing the node capability selection difficulty. So far there are many methods on the node evaluation, such as AHP (analytic hierarchy process), IE (information entropy method). PCA (the principal component analysis), TOPSIS(technique for order preference by similarity to ideal solution) , grey evaluation method, DEA (data envelopment analysis) and so on[5-9]. Above all kinds of methods have their own advantages, for example, AHP can effectively measure decision maker's judgment and comparison; IE to determine the weights according to the degree of variation of the relationship between each index and the index value, which can avoid the deviation caused by human factors; Excluding a lot of subjective factors, DEA method has strong objectivity. But due to the complexity and uncertainty features of service-oriented manufacturing network, making a variety of single evaluation method is difficult to meet the requirements of the service oriented manufacturing network node capability assessment.

This paper is based on the completely perspective of the service oriented manufacturing network, combining AHP and fuzzy theory and TOPSIS method to evaluate node capability comprehensively. We constructed the node capability assessment model. It can solve the problem which node capability

* Corresponding author: Tel.: (86) 791-83953338; E-mail: flq9848@163.com

evaluation is not easy to quantify. Through the combination evaluation method, combining the expert's experience and fully consider the dynamic and non quantitative nature of the node capability indexes, at last, using TOPSIS method to sort the program. It is simple, reliable, and also can ensure the objectivity of the results. To sum up, the method is practical for leading enterprises in service-oriented manufacturing network to evaluate the cooperative node enterprises capability.

The remainder of this paper is organized as follows. The second part explains node capability index selection principle and connotation of index system. In the third part the node capability evaluation methods are described, AHP and fuzzy TOPSIS method included. The fourth part gives an application case to verify the effectiveness of the evaluation method. The fifth part gives the corresponding conclusions and points out the application value of the method.

2 NODE CAPABILITY INDEX SYSTEM

The node capability evaluation is not only based on the perspective of a single enterprise, but also based on the perspective of value module and the cooperation level of node enterprises should be addressed. The effective cooperation of the node enterprises can achieve the maximum value of the module node, and then realize the optimization of the whole service manufacturing network operation. Node capability evaluation index selection should fully reflect the capability which need to be assessed, and relatively independent and objective [10] .Especially in service-oriented manufacturing network with the node capability has the complexity feature. The evaluation of the node capability not only consider the capability structure of the node itself, but also the relationship between the node capability should be considered. Comparing with the traditional supplier capability, service oriented manufacturing network node capability has new features such as integration, value added, innovation[11-13]. Customization of customer and demand uncertainty, put forward higher requirements on the node capability. Considering the new background, new features, new requirements, which service oriented manufacturing network node capability faced, this paper proposed new node capability evaluation elements to ensure the comprehensive and accuracy of the result.

It is based on the collaborative perspective to examine the capability of node enterprises to realize their functions in the service network which taking leading enterprises as the core. Node capability elements are divided into the four elements including Service flexibility, integration capability, cooperation capability, innovation capability, which can be subdivided into 12 indexes[14],and three sub-index are included for every first level index.Moreover, the evaluation index system of node capability are presented on the table 1.

Table 1: Evaluation index system of node capability

Evaluation goal	First level index	Sub-index	Index Description
Node capability Q	service flexibility Q_1	batch flexibility Q_{11}	capability to adapt batch changes
		time flexibility Q_{12}	response speed to unexpected situation
		market flexibility Q_{13}	response to changes in market demand
	Integration capabilities Q_2	hardware resource integration Q_{21}	production or service facilities integration
		human resources integration Q_{22}	enterprise personnel cooperation efficiency
		information resources integration Q_{23}	customer demand information integrate
	collaboration capabilities Q_3	production standardization Q_{31}	the same level of production standards as the leading enterprises
		organizational culture matching Q_{32}	matching degree with leading enterprise organizational culture
		strategic goal consistency Q_{33}	the strategic objectives consistency degree with the leading enterprises
	innovation capability Q_4	institution innovation Q_{41}	innovation of production or service institution
		process innovation Q_{42}	innovation of production or service process
		collaborative innovation Q_{43}	the overall synergy effect is realized through coupling the innovation elements

3 NODE CAPABILITY EVALUATION METHOD

Non quantitative is considered when the node capability evaluation index constructed in this research. General evaluation method can not objectively and accurately measure the weight of the index and make a judgment. TOPSIS method has the characteristic of intuitive analysis principle, the small sample size requirement and little affected by subjective factors. In view of this, we construct judgment matrix by the experts' opinion, and use the AHP to calculate the weight of each evaluation index, fuzzy theory and TOPSIS combined.The relative closeness degree is the evaluation standard to judge the capability of the node enterprise.

3.1 AHP method to determine the weight of index

There are experts subjective judgment in the evaluation of each index. In order to reduce the error, semantic judgment is used in this paper. It enable the expert fully express its willingness evaluation value[15]. selecting the Delphi method and 1 to 9 scale method to construct the judgment matrix Q. and meet: $q>0$, $q_{ij}=1/q_{ji}$, $q_{ii}=1$

After the judgment matrix Q obtained, the eigenvalue method is used to solve the weight of each index vector.

The first level index weight set w and second level index weight set w_i can be obtained respectively by the eigenvalue method. That is:

$$w=(w_1, w_2,\cdots, w_n)^T \tag{1}$$

$$w_i=(w_{i1}, w_{i2},\cdots, w_{in})^T \tag{2}$$

Calculate the index weight of each level, the weight of the second level indexes is further obtained with respect to the upper level index relative weight.

$$w^*=w\times w_i \tag{3}$$

3.2 Fuzzy TOPSIS method evaluation index

TOPSIS evaluation method first determines the positive and negative ideal solution .After that,it also need to determine a measure method to indicate the degree that the target value of each scheme which close to the ideal solution and far away from the negative ideal solution. This measure is relative closeness. Therefore the relative closeness is the sequencing criterion of the ideal solution. In the evaluation of the node capability, taking the subjectivity and uncertainty of the index into account, semantic judgment is used to express it. Specifically divided into high, higher, common, lower and low 5 levels of standard[16]. Let the 5 semantic variable expressed by interval fuzzy numbers, as shown in Table 2.

Table 2: The semantic variables and interval fuzzy numbers of evaluation index

Judgment standard	Interval fuzzy number
High	(0.8, 1.0)
Higher	(0.6, 0.8)
Common	(0.4, 0.6)

Lower	(0.2, 0.4)
Low	(0.0, 0.2)

According to the evaluation set, the initial fuzzy matrix is constructed.

$$X = \begin{bmatrix} x_{11} & x_{12} & \cdots & x_{1q} \\ x_{21} & x_{22} & \cdots & x_{2q} \\ \vdots & \vdots & \vdots & \vdots \\ x_{p1} & x_{p2} & \cdots & x_{pq} \end{bmatrix}$$

Where $i = 1, 2, \cdots, p \quad j = 1, 2, \cdots, q$.

Among them, x_{pq} is the semantic value of enterprise p which in the Two level evaluation index q.

The matrix X is standardized as a matrix R, according to the weight of the evaluation index W^* and the standardization fuzzy matrix R, establish the weighted fuzzy matrix V.

$$V = [v_{ij}] = r_{ij} \times w_j^* \tag{4}$$

Determine the positive ideal solution and the negative ideal solution

$$\begin{aligned} v^+ &= \{(max\, v_{ij} \mid j \in J^+), (min\, v_{ij} \mid j \in J^-)\} = \{v_1^+, v_2^+, \cdots, v_n^+,\} \\ v^- &= \{(min\, v_{ij} \mid j \in J^+), (max\, v_{ij} \mid j \in J^-)\} = \{v_1^-, v_2^-, \cdots, v_n^-,\} \end{aligned} \tag{5}$$

Among them J^+ ={ Benefit index set }, J^- ={ Cost index set }

The distance to the positive ideal solution and to the negative ideal solution is calculated.

The distance to the positive ideal solution is:

$$S_i^+ = \sum_{j=1}^{m} d(v_{ij}, v_j^+) = \sqrt{\sum_{j=1}^{m} \left[(v_{ij}^l - v_j^{+l})^2 + (v_{ij}^u - v_j^{+u})^2 \right]} \tag{6}$$

The distance to the negative ideal solution is:

$$S_i^- = \sum_{j=1}^{m} d(v_{ij}, v_j^-) = \sqrt{\sum_{j=1}^{m} \left[(v_{ij}^l - v_j^{-l})^2 + (v_{ij}^u - v_j^{-u})^2 \right]} \tag{7}$$

The relative closeness of each solution is calculated.

$$C_i^* = \frac{s_i^-}{s_i^- + s_i^+} \tag{8}$$

According to the size of the relative closeness, the corresponding schemes are sorted. The relative closeness degree big is good, and the relative closeness small is bad.

4 APPLICATION CASE

With the increasingly fierce competition in the domestic automobile industry, a brand of car manufacturing enterprises in order to enhance the competitiveness of products to expand market share, need to take their own advantages in resources to transfer to core competition capability, and the non core business to achieve professional outsourcing. It is necessary to evaluate the capability of each node enterprise so as to select cooperative node enterprise in 5 parts manufacturing enterprises. Employing 6 industry experts, the use of the extraction of the node capability evaluation index and evaluation model to give an evaluation.According to the experts' original evaluation, using statistical methods to manipulate it appropriately, the relative importance of the first level index and the relative importance of the second index at the corresponding level index are obtained.

4.1 AHP method to determine the weight of index

According to the experts' opinion, the node capability of first and second index of the judgment matrix Q, Q_i established.

$$Q=\begin{bmatrix} 1 & 1/1.7 & 1/1.9 & 1.8 \\ 1.7 & 1 & 1/1.6 & 2.2 \\ 1.9 & 1.6 & 1 & 2.4 \\ 1/1.8 & 1/2.2 & 1/2.4 & 1 \end{bmatrix}$$

By the eigenvalue method obtained $W=(0.200, 0.287, 0.381, 0.132)^T$

$$Q_1=\begin{bmatrix} 1 & 1/1.7 & 1.8 \\ 1.7 & 1 & 2.1 \\ 1/1.8 & 1/2.1 & 1 \end{bmatrix} \qquad Q_2=\begin{bmatrix} 1 & 1/1.6 & 1/2.1 \\ 1.6 & 1 & 1/1.8 \\ 2.1 & 1.8 & 1 \end{bmatrix}$$

$$Q_3=\begin{bmatrix} 1 & 2.3 & 2.5 \\ 1/2.3 & 1 & 1.7 \\ 1/2.5 & 1/1.7 & 1 \end{bmatrix} \qquad Q_4=\begin{bmatrix} 1 & 1.9 & 1/2.6 \\ 1/1.9 & 2 & 1/3.2 \\ 2.6 & 3.2 & 1 \end{bmatrix}$$

In the same way, the weight of the two level index can be obtained $W_1=(0.320, 0.479, 0.201)^T$ $W_2=(0.211, 0.302, 0.487)^T$ $W_3=(0.539, 0.274, 0.188)^T$ $W_4=(0.260, 0.159, 0.580)^T$.

And by the consistency check, consistency ratio of each judgment matrix $CR \le 0.1$, indicates that the judgment matrix meets the satisfactory consistency. According to the weight of the index at all levels, reference formula (3) can be further obtained second index relative to the weight of the first level index.

$W^*=(0.064, 0.099, 0.040, 0.061, 0.087, 0.140, 0.205, 0.104, 0.072, 0.034, 0.021, 0.077)^T$

4.2 Using TOPSIS method to calculate the evaluation index

After obtaining the weight of each level index by the AHP method, based on the evaluation criteria of table 2, the second level indexes of the 5 enterprise's capability are evaluated. Then it can get the fuzzy evaluation matrix X.The standardization of the matrix X is R.

According to the formula (4), the weighted normalized fuzzy matrix V is constructed.

$$V=\begin{bmatrix} (0.032,0.048) & (0.048,0.064) & (0.016,0.032) & (0.032,0.048) & (0.048,0.064) \\ (0.074,0.099) & (0.025,0.050) & (0.025,0.050) & (0.050,0.074) & (0.050,0.074) \\ (0.010,0.020) & (0.020,0.030) & (0.020,0.030) & (0.030,0.040) & (0.010,0.020) \\ (0.046,0.061) & (0.031,0.046) & (0.046,0.061) & (0.015,0.031) & (0.031,0.046) \\ (0.022,0.044) & (0.065,0.087) & (0.065,0.087) & (0.044,0.065) & (0.044,0.065) \\ (0.105,0.140) & (0.035,0.070) & (0.070,0.105) & (0.105,0.140) & (0.105,0.140) \\ (0.103,0.154) & (0.103,0.154) & (0.051,0.103) & (0.154,0.205) & (0.051,0.103) \\ (0.052,0.078) & (0.078,0.104) & (0.052,0.078) & (0.026,0.052) & (0.052,0.078) \\ (0.018,0.036) & (0.036,0.054) & (0.054,0.072) & (0.018,0.036) & (0.018,0.036) \\ (0.026,0.034) & (0.009,0.017) & (0.026,0.034) & (0.017,0.026) & (0.026,0.034) \\ (0.011,0.016) & (0.011,0.016) & (0.016,0.021) & (0.005,0.011) & (0.016,0.021) \\ (0.058,0.077) & (0.039,0.058) & (0.039,0.058) & (0.019,0.039) & (0.019,0.039) \end{bmatrix}^T$$

Because the selected node capability evaluation index are benefit index, the bigger the numerical optimization, according to the formula (5), the positive and negative ideal solutions of second level evaluation index for the 5 enterprise node capabilities are obtained and shown in table 3.

Table 3: positive and negative ideal solution of the second level evaluation index

Second level index	Positive ideal solution	Negative ideal solution	Second level index	Positive ideal solution	Negative ideal solution
Q_{11}	(0.048, 0.064)	(0.016, 0.032)	Q_{31}	(0.154, 0.205)	(0.051, 0.103)
Q_{12}	(0.074, 0.099)	(0.025, 0.050)	Q_{32}	(0.078, 0.104)	(0.026, 0.052)
Q_{13}	(0.030, 0.040)	(0.010, 0.020)	Q_{33}	(0.054, 0.072)	(0.018, 0.036)
Q_{21}	(0.046, 0.061)	(0.015, 0.031)	Q_{41}	(0.026, 0.034)	(0.009, 0.017)
Q_{22}	(0.065, 0.087)	(0.022, 0.044)	Q_{42}	(0.016, 0.021)	(0.005, 0.011)
Q_{23}	(0.105, 0.140)	(0.035, 0.070)	Q_{43}	(0.058, 0.077)	(0.019, 0.039)

From the table 3 and formula (6) to (7)the distance between each target with the positive and negative ideal solution can be calculated: $S_1^+=0.119$, $S_1^-=0.166$; $S_2^+=0.150$, $S_2^-=0.136$; $S_3^+=0.181$, $S_3^-=0.118$; $S_4^+=0.126$, $S_4^-=0.186$; $S_5^+=0.177$, $s_5^-=0.130$.

According to the formula (8), the relative closeness of each target is: $C_1^*=0.582$; $C_2^*=0.476$; $C_3^*=0.395$; $C_4^*=0.596$; $C_5^*=0.423$. So it can get the results of the program sort: $C_4^*>C_1^*>C_2^*>C_5^*>C_3^*$. By the sorting results show that the capability of the fourth node enterprise is strongest, the capability of the third node enterprise is weakest. Leading enterprises in the choice of cooperative node can be based on the results of the sorting, on the one hand, give priority to the node enterprises which comprehensive evaluation showed the higher capability to cooperate. On the other hand, node enterprises can also improve their capability structure and quality based on the comprehensive evaluation results.

5 SUMMARY

On the basis of thorough analysis the connotation of service oriented manufacturing network node capability, we construct the node capability evaluation model and evaluation index system. Combining AHP, interval fuzzy theory and TOPSIS evaluation method, a comprehensive evaluation of the service manufacturing network node capability is carried out. The weight of evaluation index is determined by AHP method, and the evaluation set is determined by interval fuzzy theory. The positive and negative ideal solution and the relative closeness of each scheme are calculated. The the objectivity and accuracy of the evaluation results is improved. Through the implementation of the evaluation of node capability, the rationality and practicability of the evaluation index are verified. And we solve the problem which node capability not easy to be quantified.

Through the construction of the evaluation index system, it not only makes the service manufacturing network node capability choosing more effective, but also provides a comprehensive reference for the node enterprise capability improvement. In the future, the choice of node capability, only according to the situation of the specific node enterprises then using the index system for comprehensive evaluation that results can be obtained,it has a certain effectiveness.

6 ACKNOWLEDGMENTS

This work is funded by the National Natural Science Foundation of China (Grant No. 71362019), Jiangxi Science Fund for Distinguished Young Scholars and Jiang Xi Provincial Education Department of Science and Technology Project (Grant No. GJJ13513).

7 REFERENCES

[1] **Wang, Q, Zhao, X, Voss, C. 2015.** Customer orientation and innovation: A comparative study of manufacturing and service firms, International Journal of Production Economics, 171,pp 221-230.

[2] **Valilai, O F, Houshmand M. 2013.** A collaborative and integrated platform to support distributed manufacturing system using a service-oriented approach based on cloud computing paradigm, Robotics and Computer-Integrated Manufacturing, 29(1),pp 110-127.

[3] **Mourtzis, D, Doukas, M. 2014.** Design and Planning of Manufacturing Networks for Mass Customisation and Personalisation: Challenges and Outlook, Procedia Cirp, 19,pp 1-13.

[4] **Zheng, Z, Ren,J. 2011.** Optimal pricing policy research for perishable hi-tech products with uncertain demand, Journal of Systems Engineering, 26(1),pp 113-120.

[5] **Hu, Z, Li, C, Xiong, Y. 2011.** Research on Selection Model of Strategic Emerging Industries Based on Combination Weights: AHP-IE-PCA, Science Of Science And Management Of S & T, 32(7),pp 104-110.

[6] **Sun, X, Jiao, Y, Hu ,J. 2006.** Grey Correlation Based on Combinational Weight and Its Application, Industrial Engineering and Management, 11(1), pp 62-66.

[7] **Zhou, F. 2013.** Research on the Evaluation System of Regional Ecosystem for Entrepreneurial Talent of Science and Technology Based on AHP-DEA, Journal of Industrial Engineering/Engineering Management, 27(1), pp 8-14.

[8] **Wang, D, Wang, X. 2010**. Research on the Green Vendor Selection Index Weight of Iron and Steel Enterprises Base on AHP and Entropy Method, Soft Science, 24(8), pp 117-122.

[9] **Lin, Z. 2012**. City insurance social responsibility based on an entropy and AHP combination model, science research management, 33(3), pp 142-147.

[10] **Xia, S. 2012**. The capability criteria system and assessment module of contractors and its application, chinese journal of management science,20(S), pp 74-79.

[11] **Jin, Q, Ding, Z, Zhang, Z. 2013**. The capability knowledge system architecture of Service-oriented manufacture mode, Science & Technology Progress and Policy, 30(22), pp 57-64.

[12] **Li, G, Sun, L, Gao, J. 2010**. The research for system architecture of Service-oriented manufacture mode and its implementation model, Science & Technology Progress and Policy, 27(7), pp 45-50.

[13] **Hao, X, Yun, S, Tong, S. 2015**. The research for Value creating network with Service-oriented manufacture mode, Science & Technology Progress and Policy, 32(9), pp 60-66.

[14] **Xie, X. 2013**. Operation mechanism of synergic innovative effect: a perspective from metropolitan region, Studies in Science of Science, 31(12).

[15] **Guo, L, Yu, B. 2013**. Evaluation of enterprise technology integration capability from dynamic perspective—Based on AHP and Fuzzy TOPSIS Method, Science Research Management, 34(12), pp 75-84.

[16] **Zhang, S, Sun, S. 2011**. Evaluation of High-Tech Research Project Based on Internal Fuzzy TOPSIS and AHP, journal of shang hai jiao tong university, 45(1),pp 134-138.

RESEARCH ON CONSUMERS' ATTITUDES AND INFLUENCING FACTORS TO GENETICALLY MODIFIED FOOD IN BEIJING

Jia Li Wenjun Song
Economics and Management School
Beijing University of Agriculture, Beijing, China
mzjb522@163.com

ABSTRACT

Based on the three areas of Beijing 300 consumers on-site questionnaire survey, this paper reveals that 27.67% of consumers do not know about genetically modified foods, 39.67% of them don't know too much, and 32.67% of them are understood. Overall, In terms of attitude,33% of the respondents were neutral, the attitude of the general, 36% of respondents said they liked, 31% of respondents said they did not like.In the influencing factors of consumer cognition and attitude, the relationship between the nature of the work unit and the monthly income of the family and the overall attitude is higher.In terms of consumer behavior tendency, low-income people are the most affected by the price.

Key words:Genetically Modified food, Consumer, Cognition, Attitude, Behavior tendency

1 INTRODUCTION

Genetically modified food (GMF) is the use of modern molecular biology technology to some biological gene transfer to other species, in order to transform the biological genetic material, its shape and the nutrition quality of people need to change goal.Genetically modified organisms into commercial production since1995 in china, transgenic organisms and their products used as food or food additives have been listed for 18 years old. However, the influence of genetically modified organisms on the environment and the genetically modified food on human health risk are in controversy, due to lack of adequate scientific basis, it is inconclusive now.Genetically modified foods may bring huge economic benefits and social benefits for the human, may also pose potential risks and threats to the health of mankind and the ecological environment, therefore, consumer attitudes to GM food is a commonly concerned problem, consumers of genetically modified food acceptance will directly affect the transgenic organisms and their products in the market, and ultimately affect the development of transgenic technology.This study selected Beijing three typical areas, trying to through the questionnaire survey to understand the Beijing market consumer cognition of genetically modified food, attitude, purchase intention, and analysis of the development of genetically modified food, in order to provide the basis for the safety management of genetically modified food.

2 THE RESEARCH DESIGN

Carries on the investigation and study to the investigation to the citizens of Beijing for the overall sample, a total sample of 300 valid samples selected from Beijing is representative of Chaoyang District, Huilongguan, Changping village.In the sample, the respondents selected from the gender, age, the nature of the job and position of the respondents considered representative.Of the students and school units to access, excluding food professionals.This research mainly uses the Excel, Spss software to carry on the reorganization and the analysis to the data, obtains the related survey result, thus proposes the effective suggestion.

2.1 Sex, age and cultural degree

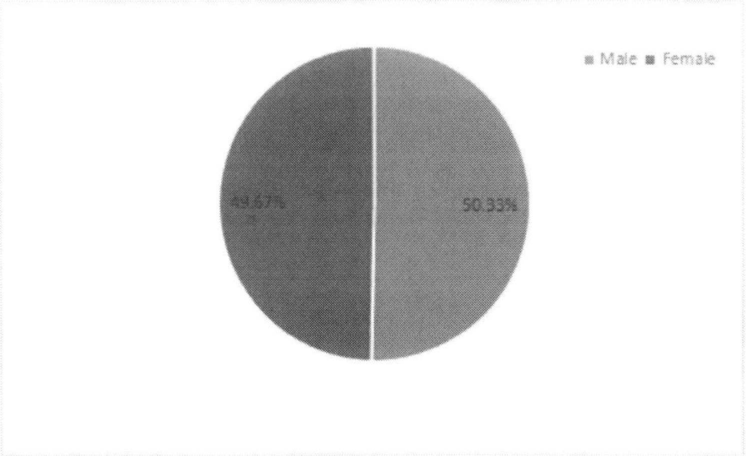

Figure 1:sample gender distribution

Source: Questionnaire

From the gender perspective, the proportion of men and women were interviewed basically in accordance with the 1:1 (Figure 1). From the age, 25 years of age accounted for 21%, 25-30 years of age accounted for 19.77%, 21-40 years of age accounted for 17%, 41-50 years of age accounted for 23.33%, 50 years of age or older accounted for 19%. From the point of view of the culture of the respondents, high school and below accounted for 45.33%, University 45.33%, master degree or above accounted for 6.33% .

2.2 The nature of the unit and the region

Look from the work nature of the respondents, state-owned enterprises accounted for 11.0%, accounted for 10.33 percent of institutions, schools accounted for 17.33%, foreign-funded enterprises accounted for 11.00%, private enterprises accounted for 20.00%, accounted for 10.33 percent of self-employed and other properties was 20.00%. From the year of the family income, less than 5 million yuan accounted for 40.67%, 5-15 million yuan, accounting for 48.00%, 15 million yuan or more accounted for 11.33% .From the regional perspective, the proportion of samples in each region is 33.33%, the specific situation as shown below:

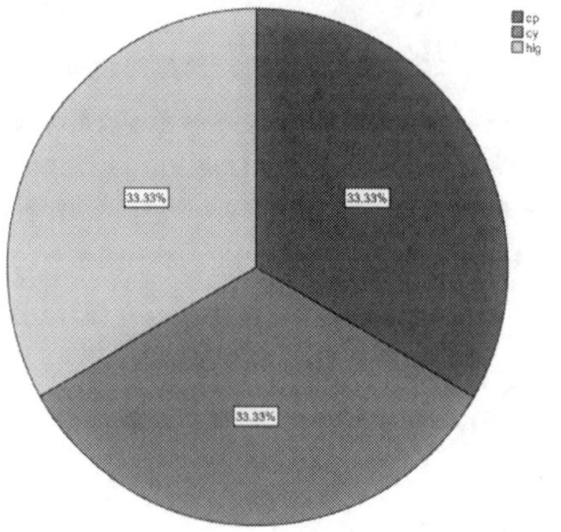

Figure 2: regional distribution

Source: Questionnaire

3 THE RESEARCH RESULTS

This research adopts the measure of homemade structured questionnaire, the design of the problem from three aspects of attitude, cognition, attitude and behavior orientation for consumers to ask specific questions.

3.1 Analysis of the cognition of consumers on genetically modified foods

Overall, China's consumers are not very understanding of genetically modified foods. Genetically modified products have a more thorough understanding of consumers accounted for only 2.67% of the surveyed consumers. It can be seen: of genetically modified foods have a smattering of knowledge of consumers accounted for 70%, nearly one-third of consumers (5.67% of people do not understand completely genetically modified food, do not understand the 22.00%) do not understand the genetically modified food.

3.2 The attitudes and influencing factors of consumers on genetically modified foods

3.2.1 Consumer attitudes towards genetically modified foods

The overall attitude of consumers on genetically modified foods is broadly divided into three types: do not like, general, like. These three types of almost all of the total number of 1/3, from the pie chart is difficult to see the general trend of consumer attitudes toward genetically modified foods.

Attitude to the variable assignment: don't like = 1, general= 2, like = 3, statistical analysis of consumer attitudes towards genetically modified food (as shown in chart 4) : from the 300 respondents overall attitude of genetically modified foods can see the proportion of respondents of genetically modified food positive attitude is slightly higher than that of the proportion of negative attitudes and overall attitude mean 2.05, more than 2, overall, a positive attitude began to occupy the mainstream. The table fully shows that more and more consumers are beginning to slowly accept genetically modified foods.

Table 1: consumer attitudes toward genetically modified foods

Value	N	Minimum value	Maximum value	Mean value	Standard deviation
	300	1	3	2.05	.818

Source: Questionnaire

Overall, the respondents were clear, the support and opposition to different attitudes are very clear, consistent with the trend of the overall attitude.

3.2.2 The impact of consumer attitudes to genetically modified food factors

From the analysis of table 2, we can see that, from the perspective of relevance, the influence factors of attitude, education, family annual income is higher. There was a significant correlation between the nature of the work unit and the annual income of the family and the overall attitude (P<0.01). And age, sex, education and regional, and the overall attitude of the respondents were not significantly correlated.

Table 2: Influence Factors of consumers' attitudes towards genetically modified foods

Variable name	Correlation coefficient
Demographic variables	
Gender	-0.082
Age	0.053
Education	-0.063
Unit property	0.135**
Region	-0.017
Material variable	
Annual household income	-1.77**

Note: * * in confidence (double test) is 0.01, the correlation is significant

The correlation is significant when the confidence level (double test) is 0.05.

Source: Questionnaire

Through the correlation analysis, we will consumers annual household income (x1) and work units (x2) properties as independent variables, to consumers of genetically modified food general attitude (y) as the dependent variable and multivariable logistic regression analysis. The results are as follows:

Table 3: multivariate regression analysis

Variable	Weight	N	Marginal percentage
Overall attitude	Dislike	93	31.0%
	Commonly	99	33.0%
	Like	108	36.0%
Annual household income	Fifty thousand below	122	40.7%

		N	Percentage
	5--150000	144	48.0%
	More than 150000	34	11.3%
Nature of work unit	State-owned enterprise	33	11.0%
	Institution and institution	31	10.3%
	School	52	17.3%
	Foreign enterprise	33	11.0%
	Private enterprise	60	20.0%
	Individual merchant	31	10.3%
	Other	60	20.0%
Effective		300	100.0%
Defect		0	
Total		300	
Sub population		21[a]	
The A. dependent variable has only one value observed in the 3 (14.3%) sub population.			

Source: Questionnaire

Table 4: regression model fitting information

Model	Model fitting standard	Likelihood ratio test		
	-2 times log likelihood	Chi square	DF	Significant level
Only the intercept	62.754			
Final	25.850	36.904	4	.000

Data source: SPSS computing

Table 5: likelihood ratio test of regression model

Effect	Model fitting standard	Likelihood ratio test		
	Simplified log likelihood value of -2	Chi square	DF	Significant level
intercept	107.382[a]	.000	0	.
Annual household income	140.844	33.461	4	.000
Nature of work unit	133.382	25.999	12	.011

Source: Questionnaire

As can be seen from the above results, the model fit well, the likelihood ratio test is also very high, and the model is reasonable. From the parameter estimates can be seen on genetically modified food choice between negative and positive attitude, low-income families and middle-income families in high-income families are more likely to choose a positive attitude; in the state-owned enterprises, consumers than consumers in other various work unit are more likely to choose a negative attitude; between neutral and positive attitude, middle-income consumers high-income consumers are more likely to choose neutral attitude.Based on the above parameters estimation, the following functions can be obtained:

Ln (P negative attitude /p positive attitude) =0.443-1.163* low income consumers -1.377* middle income consumers +1.688* state owned enterprises work consumers

Ln (P neutral attitude /p positive attitude) +2.244* =2.523 middle-income consumers

Among them: your family annual income is =3 and the nature of your work unit is =7 as a reference, so its parameter is 0

Effects of comprehensive annual household income factors of genetically modified food overall attitude, you can see that low - and moderate income of genetically modified food attitude than in high-income consumer attitudes are more negative, but in buying behavior, it can be seen that low-income consumers and moderate income consumers but are more likely to choose to follow the crowd, from the point of view of the data, attitude and behavior of low - and moderate income consumers exist certain contradiction.

4 CONCLUSION AND SUGGESTION

4.1 The consumer lack of understanding of genetically modified foods, media plays an important role in the spread of genetically modified foods

In the present circumstances, consumer awareness of genetically modified food in general is still relatively superficial and attitudes to GM food in a certain extent by the awareness of the impact. Therefore, the propaganda of knowledge of genetically modified food is to promote one of the tasks of the genetically modified food.In this regard, the media to increase the knowledge of the promotion of genetically modified food, not only to increase the breadth, more important is to increase the depth, so that consumers really understand genetically modified foods.And in the case of understanding is not deep, consumer attitudes are more easily affected by media attitude, so in the uncertain about the benefits and disadvantages of genetically modified food under the premise should reduce extreme reported, because the media to the public understanding of science plays a pivotal role.In October 2000 in the United States, the media disclosed "Star group" of genetically modified corn in food (" Star "corn events Star Link corn Episode came) a few months later, 44% of consumers get a lot of genetically modified food knowledge. Notable is that some consumers to "star" event's memory is long, when referring to the incident, consumers know that this is about the safety of genetically modified crops. It can be seen that the impact of the media on the incident to the consumer is far-reaching.

4.2 The development of genetically modified foods should consider the needs of the end consumer

The development of genetically modified food not only decided in transgenic crop growers, genetically modified food processors acceptance, but decided to end consumer acceptance. Therefore, the development of genetically modified food is more important to take into account the wishes of the end consumer. Consumers' willingness to purchase mainly focus on the quality of genetically modified foods, more concerned about the quality and safety, resistance to insect resistance and other types of demand is not strong. The current development of genetically modified crops is mainly focused on reducing production costs and other aspects of the crop growers (such as resistance to herbicides, pest resistance, etc.), but not very good to meet the needs of the end consumer. The development of transgenic crops should be more consideration to the consumer's interests, so that they can be keenly aware of the advantages of transgenic food. Although consumers have a larger view of the current advantages and disadvantages of genetically modified foods have a greater difference, but in general on the continued development of genetically modified foods to support the attitude of the future of genetically modified foods optimistic attitude and high hopes.

4.3 Genetically modified food labeling system needs to be strengthened

The results of the survey show that the public is not satisfied with the current implementation of the identification system, although many manufacturers have been marked on the product, but mostly take a fuzzy logo, logo is not eye-catching, consumers are difficult to identify. The survey found that 54% of respondents do not know the country has to implement the identification of genetically modified products, while 36.7% of respondents know that there is a logo, but can not identify the signs of genetically modified products market. This shows that the system needs to be standardized and strengthened. "The protocol on Biosafety" provisions came into effect on September 11, 2003, transgenic organisms and their products the transboundary marker, at present more than 30 countries around the world of genetically modified organisms and product implementation of the identity management system, China's Ministry of agriculture in 2002 promulgated the agriculture genetic identity management measures ", the requirements of 5 types of genetically modified organisms 17 products implementation of identity. But the survey shows that consumers in the Beijing market is still not satisfied with the current identification of genetically modified organisms.

4.4 The tendency of consumers' consumption will affect the development of transgenic technology

The United Nations Food and Agriculture Organization (FAO) pointed out that "GM can not is only in isolation as a technical achievement", as a result of the promotion of transgenic technology, "the international community is necessary to ensure that GM crops in food safety, food safety, food quality and sustainability aspects can make positive contribution", we need to establish adequate international consensus on security ". But in any case, genetically modified technology has been used in agricultural production in China and many other countries, genetically modified products have already appeared in our lives. Scientists see the technology as a new challenge, and entrepreneurs see it as another source of profit, and consumers are likely to support it, and may be rejected. However, the trend of consumer spending will greatly affect the development of transgenic technology. In Europe, due to consumer rejection of genetically modified food, genetically modified organisms and commercial production of no market, even running of gene technology research is also limited, consumers tend to dominated the market demand and the research direction. The survey found that only 10% of the respondents said to hold positive attitudes for the genetically modified food. This attitude with Europeans have similarities, the GM industry sector should be a lot of listening to consumers voice, and consumer attitudes as its decision-making reference.

5 ACKNOWLEDEGMENT

I am thankful and acknowledge the support by Beijing Social Science Fund Project“empirical research on the influential factors of purchase intention of GMF in the suburban residents of Beijing（15JGB063）”.

6 REFERENCES

[1]**Lone Bredahl.2001.**Determinants of consumer attitudes and purchase intentions with regard to genetically modified foods - results of a cross-nation survey . Journal of Consumer Policy , 24,pp 23-61.

[2]**Huangjikun, Chou Huanguang, et al. 2006.**Chinese urban consumers' cognition, acceptance and purchase intention of genetically modified food. Chinese soft science, 2, pp 61-67.

[3]**Feng Liangxuan, Qi Zhenhong, Zhou Hui et al.2012.** An Empirical Study on the willingness of consumers to purchase genetically modified foods -- Taking Chongqing city as an example [J]. Journal of Huazhong Agricultural University (SOCIAL SCIENCE EDITION), 2,pp 1 - 6.

[4]**Wang Ruidong, Wang Jingxu, Zhou Zhu et al. 2010.**Study on the influence factors of consumers" cognition and acceptance of genetically modified foods -- Taking Wuhan city as an example [J]. Tianjin agricultural science, 2,pp 44 - 48.

[5]**Wang Rui, Qi Zhenhong.2010.** A review of the research on the cognition and attitude of Chinese and foreign species of genetically modified food [J]. international trade issue, 12,pp 115-119.

[6]**Chen Ying, Yang Xinglong, Wang Lin et al. 2011.**Analysis on the cognition degree and purchase intention of genetically modified foods by consumers in Yanbian [N]. Journal of Yanbian University, 4,pp 257-261.

[7]**Li Q . Curtis K R , McClusky J J ,etal .2003.** Comsumer Attitudes toward Genrtically Modified Food in Beijing China [J] .AgBioForum , 5(4),pp 145-152.

[8]**Zhang X Y.2002** .CHINESE Consumer over Food Safety [D] . Working Paper , Agricultural Economics Institute , Hague , the Netherlands.

[9] **Yu Ting.2011.** investigation and analysis of the awareness of genetically modified food. BBS, science and technology of China, 7,pp 1-6.

[10]**YanZhiyang, Lin Deming, Shi Mengyao, Li Na.2011.** Abroad on consumer attitude toward genetically modified (gm) food research review and the latest progress. 5 ,pp 1-3.

METADATA SCHEME FOR OPEN DATA IN CHINESE LOCAL GOVERNMENT

Yu Mengyue* Zhai Jun and Lin Yan
Department of Management Science and Engineering
Dalian Maritime University, Liaoning, China

1985176785@qq.com

ABSTRACT

Open government data can encourage innovation and promote economic growth, but users finding, understanding and using open data can't depart from the support of high quality metadata. Metadata scheme for open data in Chinese local government still exists insufficient in such aspects as normalization and standardization, affecting the interoperability of data catalogs. This paper first introduces the W3C formal recommendation DCAT, then on the basis of investigating and analyzing Metadata scheme for open data in Chinese local government, design metadata scheme for dataset based on DCAT. This work is a beneficial attempt to refer to international metadata standards in the construction of the metadata in the open government data.

Keywords: Open government data, Metadata, Data portals, DCAT, XML/RDF dataset

1 INTRODUCTION

All levels of government hold 80% of the whole social data resources [1]. The opening of government data, in order to have both social and economic value generated from them, has attracted the attention and interest of both researchers and practitioners from various disciplines [2]. Worldwide rise of the "open government data movement" began in the "Open Government Action Plan" in 2009 which the US government promoted [3], and open the craze of open government agencies to use open data for sharing data and information [4]. Today, more than 250 governments at national, sub national and city levels; almost 50 developed and developing countries; and entities such as the World Bank and United Nations have launched Open Data initiatives—and more are launched every year [5].

A typical implementation to opening government data is to collect relevant datasets and their respective metadata and publish them on an Open Government Data Portal [6]. Open Government Data Portal is an official platform to manage, open, query, and then reuse the Government data. From the beginning of 2012, Beijing, Shanghai, Zhejiang, Wuhan, Qingdao and other local governments have launched an open data portal in succession [7]. On September 5, 2015, the State Council issued "Promote the development of Big Data Platform for Action" which clearly states to build a national unity Open Government Data Portal before the end of 2018.

Open data is the development and deepening of disclosed government information in the era of big data [8]. As the "metadata has a very important role for the description, discovery and management of

---

* Yu Mengyue

This work was supported in part by the National Natural Science Foundation of China under Grant 71571025 and 51404052.

government public information"[9], core function of data catalog platform is metadata management for open datasets [7, 10-12].With the deepening of the open data movement, metadata from improving the availability of open datasets [13] to helping users understand and select the dataset [14], is developing to standard, semantic and machine-readable direction.

In the field of government practice and academic research, China has begun to accelerate to catch up the pace of the global wave of open data [15]. As for metadata, although now the local governments can provide basic metadata information for open data [16], the metadata schema varies, a description of the dataset is not yet comprehensive and metadata mainly presents in HTML web pages format and text files. There is still not a small gap reaching standardization, interoperability and machine readable, and without a doubt it will affect data public effect [16]. So in this paper, we draw on advanced international experience, analyze the lack of metadata for local government open data platform, design metadata schema based on W3C DCAT standard, and provide reference for the construction and upgrading of all levels of government open data platform.

2 W3C METADATA SPECIFICATION FOR OPEN GOVERNMENT DATA

W3C leaded by the Tim Berners-Lee is an active promoter of open government data movement [17]. On January 16, 2014, the member of W3C the Government Linked Data Working Group finally released a metadata recommended standard for open data-- DCAT(Data Catalog Vocabulary)which was initiated at the Digital Enterprise Research Institute (DERI) and modified by e-Government Interest Group[18].

DCAT is an RDF vocabulary designed to facilitate interoperability between data catalogues published on the Web. DCAT is mainly used to describe the datasets, improve the discoverability of datasets, and enable applications easily to read metadata from multiple catalogs. DCAT defines seven classes and seventeen properties, and the main classes and properties are shown in Figure 1[18]. In describing the datasets, data distributions, data catalogs and other objects, multiplexing related vocabulary such as Dublin Core, FOAF ontology and SKOS ontology where necessary.

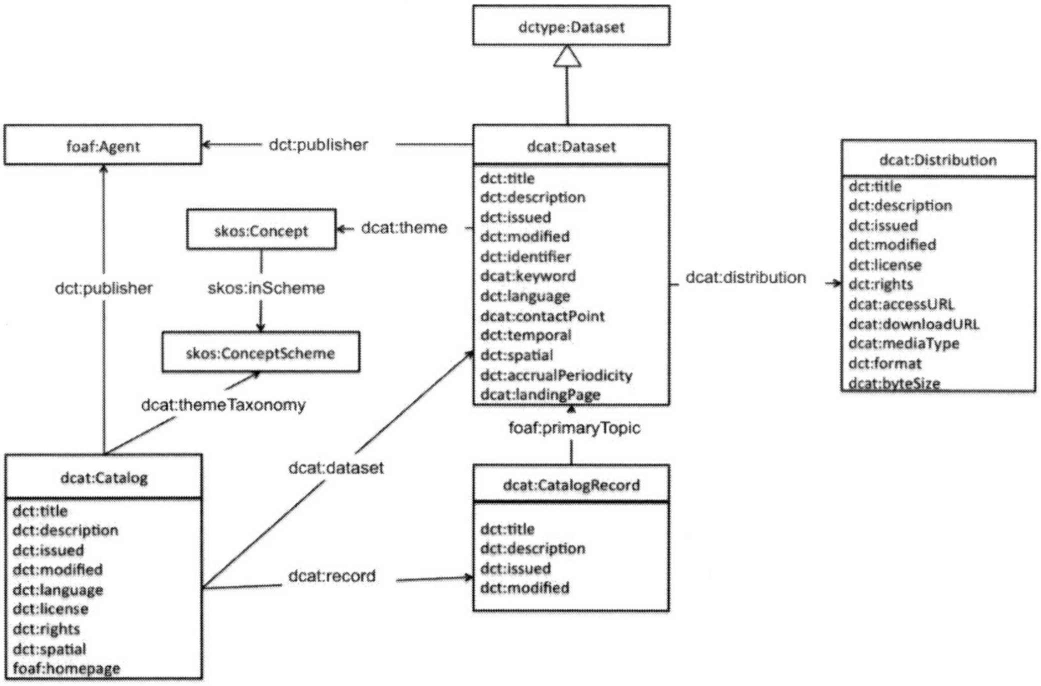

Figure 1: The Main Classes and their Attributes of DCAT Vocabulary

According to the W3C statistics, DCAT and its Application Profile has been widely used by the European Union, the United States, Britain, Spain, Ireland and Italy and other countries[19].DCAT Application Profile for data portals in Europe has been applied to the 28-nation harmonized data portals for EU. Metadata for 87% of the datasets conformed to DCAT-AP specification on British national Data Portal (Data.Gov.UK), the proportion of Italy (Dati.Gov.IT) was 53%, and the proportion of Ireland (Data.Gov.IE) was 36% [20]. The Swedish national data portal（www.opengov.se）will support DCAT-AP as their metadata standard [21].

3 ANALYSIS OF METADATA FOR OPEN LOCAL GOVERNMENT DATA

3.1 Metadata Schema

Unlike the development path of United States, Britain and other countries from the central government to local government, Chinese open data portal mainly originates and practices in local governments. After a thorough investigation of metadata for each data portal, as the limited space of this article, the analysis results of Beijing Province are given.

"Beijing Government Data Resources Network"(http://www.bjdata.gov.cn/) launched the beta in October 2012 and officially opened by the end of 2013. So far it has been open more than 300 datasets from the City Board of Education, Municipal Transportation Commission, the Municipal Tourism Commission nearly 40 departments, and provides metadata file in txt text format for each dataset. Table 1 summarizes the metadata case where the instance data is taken from the description of "public welfare library" dataset.

Table 1: Metadata for Beijing Government Data Resources

element	label	instance data	Explanation
Resource Name		public welfare library	
Resource Classification	Subject Classification	Sports and entertainment / cultural venues / Cultural Collections	multiple choice
Resource Description	Data Introduction	The data is about welfare library information registered by the Beijing Municipal Cultural Bureau, including name, address, telephone, bus routes and other property field.	Text Type
Keyword Description	Keyword	Cultural collections; library; cultural sites	multiple choice
Resource ownership units	Data Sources	Beijing Municipal Cultural Bureau	Enumerated types: Beijing Municipal Commission of Tourism Development
Resource update cycle		annually	Text type, such as a year, one year
Resource Type		table	Enumerated types: text, table, etc.

element	label	instance data	Explanation
Resource Language		ZH	Chinese
Data security restrictions rating		publicly	Text Type: publicly, publicly class
Online Resources link address		WWW.BJDATA.GOV.CN	Links to homepage of the Beijing Government Data Resources Network (only value)

Data Sources	file name
	downloads
	Upload Date
	Download raw data
	Download spatial data
	Resource Records
	Size of data files
		file name: public welfare library; number of downloads:166;
		upload date:2015-12-25;
		original data download(links);
		spatial data download(links);
		the number of resource records:25;
		the size of data file:3kb
			Seven elements qualifier: file name, number of downloads, upload date, original data download, spatial data download, the number of resource records, the size of data file

Historical Data	Historical Data	20151225154819697369.zip (links)	URL links

3.2 Insufficient of Metadata

Although the current metadata may meet the basic needs of local government open data, there are some following insufficient compared with the international trends.

(1) The difference between metadata schemas of the local governments is large, expressed as the inconsistence of the number, name and format of elements. It brings difficulty for the application across multiple data directories reading the metadata of datasets, increasing the complexity and costs.

(2) The description of the dataset is not comprehensive enough, there is no standard, long-lasting, stable unique identifier, there is no temporal coverage of data, there is no problem feedback channels, no open licensing and the like.

(3) No to distinguish between datasets and data distributions (downloadable files or access the endpoint, etc.), confusing metadata between the two. For example, "resource type", "resource records", " download " should be described as a data file rather than the dataset itself.

(4) Generally not using machine readable metadata format is not conducive to program automatic processing.

(5) In the same portal, the value of metadata is not standardized. For example, the size of data file in Zhejiang Province's site has both values with unit (such as 172KB) and without unit (such as 10). Resource update frequency in Beijing municipal site has both values of a year and every year. Online resources link address value is "WWW.BJDATA.GOV.CN", causing users cannot find resources data directly, losing the value of metadata.

4 METADATA DESIGN

Starting from the existing metadata, towards to W3C DCAT standardized direction, to meet the core functions (i.e. dataset discoverability, accessibility and interoperability between the data catalogs) as the target, design metadata schema for our local government data portal. This article makes "dataset" as described object using metadata. DCAT defines dataset as a collection of data, published or curated by a single agent, and available for access or download in one or more formats [18], corresponding to dcat: Dataset class of Fig 1.

Table 2 shows the metadata of the dataset. The first column is element name corresponding DCAT, and the last two columns corresponding to metadata actually used by Beijing and Zhejiang.19 elements are all from DCAT vocabulary and DC metadata, not only covering the vast majority of metadata currently used by Beijing, Zhejiang and the others, but also providing space for future extensions.

Table 2: Metadata for datasets

element	explanation	Beijing	Zhejiang
title	A name given to the dataset	Resource Name	Resource Name
description	free-text account of the dataset	Resource Description	Resource Description
identifier	A unique identifier of the dataset		
theme	The main category of the dataset. A dataset can have multiple themes	Resource Classification	Data fields
keyword	A keyword or tag describing the dataset	Keyword Description	Keyword
contributor	An entity responsible for making contributions to the resource	Resource ownership units	Information resource provider
accuralPeriodicity	The frequency at which dataset is published	Resource update cycle	Update frequency
issued	Date of formal issuance (e.g., publication) of the dataset		Information Resources issued Date
modified	Most recent date on which the dataset was changed, updated or modified		
source	A related resource from which the described resource is derived		Data Sources
temporal	The temporal period that the dataset covers		
spatial	Spatial coverage of the dataset		

element	explanation	Beijing	Zhejiang
contactPoint	Link a dataset to relevant contact information which is provided using VCard		
license	A legal document giving official permission to do something with the resource		
distribution	Connects a dataset to its available distributions.		
landingPage	A Web page that can be navigated to in a Web browser to gain access to the dataset, its distributions and/or additional information	Online Resources link address	Online Resources link address
provenance	A statement of any changes in ownership and custody of the resource since its creation that are significant for its authenticity, integrity, and interpretation	Historical Data	
accessRights	Information about who can access the resource or an indication of its security status	Data security restrictions rating	
language	The language of the dataset	Resource Language	

5 CONCLUSION

Open and sharing data can't depart from the support of metadata. Additional high quality metadata description for open datasets is the basis to achieve interoperability between the data catalogs, is more conducive to finding, understanding and using data. The normalization and standardization of the metadata schema for open data in our local government is not high, making reference to international common metadata standard will be an important direction for development. This paper has done a preliminary attempt to build a metadata schema for open dataset based on W3C DCAT. Obviously, there will be many problems carrying out the metadata schema in the practical, for example, the need to supplement and improve the encoding schemes qualifier, specify the type for the value of metadata, design specification of dataset identifier, increase new objects and attributes according to the demand and the like.

6 REFERENCES

[1] **Xu Zhang, Jianman Wang. 2008.** Exploration into Metadata Scheme for Government Information Openness, *Library Development*, 12, pp10-14.

[2] **Yannis Charalabidis, Charalampos Alexopoulos, Euripidis Loukis. 2016.** A taxonomy of open government data research areas and topics, *Journal of Organizational Computing and Electronic Commerce*, 26(1-2), pp 41-63.

[3] **Han Zhang, Zhong Wang. 2015.** Comparison Research on Open Data of International Government, *Journal of Intelligence*, 34(8), pp 142-146.

[4] **Ruixian Yang, Chunlei Mao, Ze Zuo. 2016.** Comparative Study on Chinese and Foreign Government's Data Opening Situation, *Journal of Intelligence*, 35（5）, pp167-172.

[5] World Bank. Open Data in 60 Seconds[EB/OL].

http://opendatatoolkit.worldbank.org/en/open-data-in-60-seconds.html.

[6] **Attard J, Orlandi F, Scerri S, et al. 2015.** A systematic review of open government data initiatives, *Government Information Quarterly*, 32(4), pp 399-418.

[7] **Lei Zheng, Feng Gao. 2015.** Chinese Open Government Data Platform: Framework Research, Current Situation and Suggestions, *E-GOVERNMENT*, 7, pp 8-16.

[8] **Jun Zhai.** 2016. *Principle and Application of Linked Government Data-Technology and Practice of Open Data in Big Data Era,* 1st Edition, Publishing House Of Electronics Industry.

[9] **Rongrong Zhao, Huiwei Liang, Lei Sa. 2013.** Public Library Government publicly Metadata Research- A Case of Chinese government public information service platform for the integration, *Researches In Library Science*, 15, pp 63-67.

[10] **Priebe S, Kannicht M, Jacob M, et al. 2013.** The complementarity of open data infrastructures: An analysis of functionalities, Proceedings of the 14th Annual International Conference on Digital Government Research, Quebec City, QC, Canada, June 17 - 20, pp166-171.

[11] **Xiaohong Qian, Mangguo Hu. 2014.** Building an Open Government Data Platform and Analysing Its Technical Characteristics, *Document, Information & Knowledge*, 3, pp124-129.

[12] **Zhifeng Zhou, Ruhua Huang. 2013.** An Analysis on Service Function of Foreign Government Open Data Portals, *Journal of Intelligence*, 3, pp144-147.

[13] **Erickson J S, Viswanathan A, Shinavier J, et al. 2013.** Open Government Data: A Data Analytics Approach, *IEEE Intelligent Systems*, 28(5), pp19-23.

[14] **Anneke Zuiderwijk, Marijn Janssen &IrynaSusha. 2016.** Improving the speed and ease of open data use through metadata, interaction mechanisms, and quality indicators, *Journal of Organizational Computing and Electronic Commerce*, 26(1-2), pp116-146.

[15] **Haiqun Ma, Pan Pu. 2015.** Status Analysis and Comparison of Research Achievements on Open Data Policy and an Estimation of the Research Trends in China, *Journal of Library Science In China*, 41(5), pp76-86.

[16] Chinese e-government information network. What is the status of open government data [EB/OL].

http://www.cegov.cn/tabid/77/InfoID/3610/frtid/38/Default.aspx,2015-10-19.

[17] Tim Berners-Lee. Putting Government Data online[EB/OL]. http://www.w3.org/DesignIssues/GovData.html, 2009-06.

[18] W3C. Data Catalog Vocabulary (DCAT)[EB/OL].

http://www.w3.org/TR/vocab-dcat/,2014-01-16.

[19] W3C. DCAT Implementations[EB/OL].

https://www.w3.org/2011/gld/wiki/DCAT_Implementations

[20] European Data Portal. Metadata Quality Dashboard[EB/OL].

http://www.europeandataportal.eu/mqa-service/en/dashboard.html

[21] DCAT application profile for data portals in Europe[EB/OL].

https://joinup.ec.europa.eu/asset/dcat_application_profile/description

AN ACCURACY EVALUATION METHOD FOR INTERMITTENT DEMAND FORECASTING: DATA ANALYTICS APPROACH

Prerna Mishra[1] and XUE-Ming Yuan[2*]

[1]Charles River Development,
700 District Avenue, Burlington MA, USA, 01803
prernam@bu.edu

[2]Singapore Institute of Manufacturing Technology,
Fusionopolis Way Innovis #08-04 Singapore 138634
xmyuan@SIMTech.a-star.edu.sg

ABSTRACT

Intermittent demand is a demand pattern where there are periods of no demand interspersed with some periods of demand. It is difficult to forecast it, the most popular method of forecasting being the Croston method. Calculation of error of the forecasts is also as difficult a challenge. The usual methods of error calculation like Mean Average Percentage Error and Root Mean Squared Error are not suitable. In this paper a new method is proposed to calculate the error of intermittent demand forecasted using Croston method.

Keywords: Intermittent Demand, Error Calculation, Croston Method, EMID, MAPE, RMSE

1. INTRODUCTION

Demand forecasting is an essential requirement in supply chain management. It plays an important role in synchronized planning and bullwhip effect [1]. It is true that organizations function in an environment of uncertainty, the most important uncertainty being that of demand. Hence forecasting is used to make educated guesses about demand. A typically difficult case of demand pattern is intermittent demand. Intermittent demand is characterized by periods of zero demand followed by a sudden spike in demand. This nature of the demand makes forecasting a very challenging task. This type of demand is typically found in industries like aircraft manufacturing and heavy machinery [2]. An equally important, but often overlooked aspect of the process of intermittent demand forecasting is the measurement of error. It is critical because of the presence of a large number of zero values, which makes measures like Mean Absolute Percentage Error (MAPE) not a very useful measure, since the calculation requires zero to be in denominator.

This raises the need for a more accurate measure of error in intermittent demand forecasting. This paper tries to develop a method to determine the same. Then, a comparative assessment is made between the proposed method and the commonly used method like mean absolute error and mean squared error. The dataset used to evaluate the error consists of 700 products with intermittent demand pattern.

The paper is organized as follows. The following section is a review of some of the previous work done on this topic. Section III lays out the framework for a new method of accuracy

*Corresponding Author

calculation. Section IV describes our experiment. Section V presents some of the observations that we draw from our experiment. Section VI concludes our observations and lays foundation for some possible future work.

2. LITERATURE REVIEW

The methodology most commonly used for intermittent demand forecasting was proposed by Croston [3]. The main motivation behind this modified method was that Croston observed that the most commonly used method of exponential smoothing placed maximum weight on the most recent demand, hence the forecast must be the highest just after an actual high demand occurrence, which mostly is not accurate in case of intermittent demand. Croston method uses separate estimates for the size and time periods between demands by applying exponential smoothing on these two individually. This addresses the bias problem [4].

Most accuracy measurement methods are sensitive to extremes. It has been suggested that scale free methods be used to compare different forecasting methods [5]. The most commonly used scale free method is MAPE, which is of the form $\frac{|E|_t}{D_t}$and therefore could have zero in the denominator. It was proposed to use an alternative of MAPE, which is a ratio estimate:

$$\frac{\sum_{t=1}^{n}|E|_t}{\sum_{t=1}^{n}D_t} X\ 100 \quad (1)$$

This avoids non zero denominator. Syntetos and Boylan [6] used two methods to compare the accuracy of intermittent demand forecast: Relative Geometric Root Mean Square Error (RGRMSE) and Percentage Best (PB). RGMSE does not incorporate error due to time period, whereas PB is scale free. There has been use of symmetric MAPE; however it is known to suffer for asymmetry problems [7].

Most of these measures are either relative-to-a base or relative-to-the series. To our best knowledge, no method has been proposed till now which treats the matching points between the actual and the forecast demand and the non-matching point separately, which is what this paper intends to do.

3. NEW ACCURACY EVALUATION METHOD

We propose the following accuracy evaluation method to quantify the forecasting accuracy of intermittent demand:

$$EMID = \frac{\sigma}{\alpha} + \beta \cdot \varepsilon \quad (2)$$

where we define $\alpha = \frac{a}{N}$; $\beta = \frac{b}{N}$; $\sigma = \frac{1}{a}\sum_{t=1}^{a} e_t$; $\varepsilon = \frac{1}{b}\sum_{t=1}^{b} e_t$;

and N represents the number of periods of the evaluation set;

$e_t = |Y_t - \hat{Y}_t|$ is the forecasting error is the difference between the observed value (Y_t) and the estimated one ($\hat{Y}_t$);

a is the number of matching points (among the sets of actual and forecast data);
b is the number of non-matching points (among the sets of actual and forecast data);
σ is the mean error for the matching points;
ε is the mean error of the non-matching points.

The formula of *EMID* includes the information on the number of matching points, the number of non-matching points, the errors from the matching points and non-matching points. When the number of matching points is bigger, the value of *EMID* will become smaller. When the number of non-matching points is bigger, the value of *EMID* will become bigger. When the errors from either matching points or non-matching points go bigger, the value *EMID* will become bigger.

4. EXPERIMENTAL DESIGN AND RESULTS

We consider a set of 700 products. The demand of these products is forecasted using Croston method. The forecasting error is calculated using our proposed method. We also calculate Root Mean Square Error (RMSE). RMSE and EMID are compared for 700 products some observations are made.

We present our results in four different subsets.

4.1 Case I:

The first subset that we look at is where *RMSE* very high and *EMID* is very low. The criterion for this subset is:

$$\frac{RMSE}{EMID} > 2 \qquad (3)$$

This implies that *RMSE* is at least 200% greater than *EMID*. Out of the set of 700 products that we considered, there were 89 cases where this criterion was met. A sample of the demand and their forecasted values of this case are presented in table 1. 24 time periods were considered for calculation, only 12 have been depicted here and in also tables 3,5 and 7.

Table 1: Products in which *RMSE* high and *EMID* is low

Product 8		Product 28		Product 155		Product 437		Product 510	
Actual	Forecast	Actual	Forecast	Actual	Forecast	Actual	Forecast	Actual	Forecast
0	0	1	0.371	0	2.2	0	0.29	0	0
0	0	0	0.4339	0	2.2	0	0.29	0	0
0	0	0	0.4339	3	2.2	0	0.29	0	0
0	0	0	0.4339	0	2.28	0	0.29	0	0
0	0	2	0.4339	3	2.28	0	0.29	0	0
0	0	0	0.59051	0	2.352	0	0.29	0	0
0	0	0	0.59051	2	2.352	0	0.29	0	0
0	0	1	0.59051	0	2.3168	0	0.29	10	0
0	0	0	0.631459	10	2.3168	0	0.29	5	1
0	0	0	0.631459	0	3.08512	20	0.29	0	1.4
1000	0	1000	0.631459	9	3.08512	0	2.261	0	1.4
0	100	0	100.5683	0	3.676608	0	2.261	0	1.4

RMSE and *EMID* of these products are in table 2:

Table 2: RMSE and *EMID* of products where RMSE/EMID > 2

	RMSE	*EMID*	*RMSE/EMID*
Product 8	214.26407	4.54545	47.138095
Product 28	214.14332	6.213862	34.46219345
Product 155	5.36382	2.515787	2.132065213

Product 437	4.28579	0.467335	9.170689915
Product 510	2.35372	0.25881	9.094383908

It can be seen that in these products, the demand pattern is such that there is an extremely high demand in one or two periods followed by zero. *RMSE* is high because the few the forecast of the high demand doesn't immediately reflect the actual demand. Hence, the difference between demand and forecast is high and hence *RMSE* is quite high. Since *EMID* method considers the average error of matching/not matching points separately, the high error value present in the period of high actual demand is averaged out.

4.2 Case II:

We consider the case where *RMSE* is very low whereas *EMID* is very high. The criterion for this is:

$$\frac{RMSE}{EMID} < 0.5 \qquad (4)$$

This implies that *EMID* is at least 200% greater than *RMSE*. Out of the 700 products that we considered, 309 met this criterion. Table 3 represents a sample of this case:

Table 3: Products in which *RMSE* and *MAE* is low and *EMID* high

Product 11		Product 42		Product 96		Product 160		Product 278	
Actual	Forecast	Actual	Forecast	Actual	Forecast	Actual	Forecast	Actual	Forecast
0	3.00	0	5.5	0	1	0	1	0	1.9
1	3.00	0	5.5	0	1	0	1	0	1.9
0	3.00	0	5.5	0	1	0	1	0	1.9
0	3.00	0	5.5	0	1	0	1	0	1.9
0	3.00	0	5.5	2	1	0	1	0	1.9
0	3.00	0	5.5	0	1.1	0	1	0	1.9
0	3.00	1	5.5	0	1.1	0	1	0	1.9
1	3.00	0	5.05	0	1.1	0	1	0	1.9
0	3.00	2	5.05	2	1.1	1	1	0	1.9
0	3.00	0	4.745	4	1.19	0	1	0	1.9
1	3.00	0	4.745	0	1.471	0	1	2	1.9
0	3.00	0	4.745	2	1.471	0	1	0	1.91

RMSE and *EMID* of these products are in table 4:

Table 4: *RMSE* and *EMID* of products where *RMSE*/*EMID*< 0.5

	RMSE	*EMID*	*RMSE*/*EMID*
Product 11	2.68696	18.05241	0.14884236
Product 42	5.40034	45.24124	0.11936758
Product 96	1.16345	7.399602	0.157232037
Product 160	0.95346	11.90909	0.080061744
Product 278	1.85360	23.205	0.079879217

It can be noted that in these cases, the demands spikes in some periods, but is not very high. The forecast reflects the demand in consecutive periods, however since neither the demand nor the forecast is very high, *RMSE* is low. On the contrary, *EMID* behaves exactly the same way as it did in case I.

4.3 Case III:

In this case we consider the products, where *RMSE* and *EMID* is approximately the same. The criterion for this is:

$$0.5 < \frac{RMSE}{EMID} < 1.5 \qquad (5)$$

Out of 700 products, 249 satisfied this condition. Table 5 represents a sample of this case:

Table 5: Products in which *RMSE* and *EMID* are around the same

Product 92		Product 98		Product 301		Product 524		Product 615	
Actual	Forecast	Actual	Forecast	Actual	Forecast	Actual	Forecast	Actual	Forecast
0	0.3439	2	0.19	0	0.3439	0	0.1	0	0.3439
0	0.3439	0	0.371	0	0.3439	1	0.1	0	0.3439
0	0.3439	0	0.371	1	0.3439	1	0.19	0	0.3439
0	0.3439	0	0.371	0	0.40951	0	0.271	2	0.3439
0	0.3439	0	0.371	0	0.40951	0	0.271	2	0.50951
0	0.3439	0	0.371	0	0.40951	0	0.271	0	0.658559
0	0.3439	0	0.371	0	0.40951	0	0.271	0	0.658559
0	0.3439	0	0.371	0	0.40951	0	0.271	0	0.658559
1	0.3439	0	0.371	0	0.40951	0	0.271	0	0.658559
0	0.40951	0	0.371	0	0.40951	0	0.271	0	0.658559
0	0.40951	3	0.371	0	0.40951	0	0.271	1	0.658559
0	0.40951	0	0.6339	0	0.40951	0	0.271	1	0.658559

RMSE and *EMID* of these products are in table 6:

Table 6: *RMSE* and *EMID* of products where *RMSE/EMID*>0.5 and *RMSE/EMID*<1.5

	RMSE	*EMID*	*RMSE/EMID*
Product 92	0.48130	0.7758	0.620393399
Product 98	0.79358	0.814085	0.974813335
Product 301	0.49178	0.534486	0.920104145
Product 524	0.38002	0.263795	1.440573557
Product 615	0.71962	1.212225	0.593634879

It should be noted that these products have an actual and forecasted demand of 0 and 1 for most of the time periods. This causes absolute error to be between 0 and 1. Hence *RMSE*, is also mostly between 0 and 1. Since *EMID* averages the error between matching and non-matching cases, the values are also around 1. Hence the ratio between the two types of errors is around 1.

4.4 Case IV:

In this case we consider the products, where *RMSE* and *EMID* vary slightly. The criterion for this is:

$$1.5 < \frac{RMSE}{EMID} < 2.0 \qquad (6)$$

This means that *RMSE* is approximately 150% to 200% of the value of *EMID*. There are 52 such cases in our dataset. Table 7 represents a sample of this case.

Table 7: Products in which *RMSE* and *EMID* vary slightly

Product 51		Product 65		Product 301		Product 344		Product 478	
Actual	Forecast	Actual	Forecast	Actual	Forecast	Actual	Forecast	Actual	Forecast
1	0.19	0	0.2	1	0.1	0	0.19	1	0.00
0	0.271	0	0.2	1	0.19	7	0.19	0	0.00
1	0.271	0	0.2	0	0.271	0	0.871	0	0.00
0	0.3439	0	0.2	0	0.271	0	0.871	1	0.00
0	0.3439	1	0.2	0	0.271	0	0.871	0	0.00
0	0.3439	0	0.28	4	0.271	0	0.871	0	0.00
0	0.3439	0	0.28	0	0.6439	0	0.871	0	0.00
0	0.3439	0	0.28	0	0.6439	0	0.871	0	0.00
0	0.3439	0	0.28	0	0.6439	0	0.871	0	0.00
0	0.3439	1	0.28	0	0.6439	0	0.871	0	0.00
2	0.3439	0	0.352	0	0.6439	0	0.871	0	0.00
0	0.50951	3	0.352	0	0.6439	0	0.871	0	0.00

RMSE and *EMID* of these products are in table 8:

Table 8: *RMSE* and *EMID* of products where *RMSE/EMID*>1.5 and *RMSE/EMID*<2.0

	RMSE	*EMID*	*RMSE/EMID*
Product 51	0.56089	0.353069	1.588608267
Product 65	0.76657	0.425636	1.800994482
Product 81	0.93273	0.58267	1.600779831
Product 344	1.59506	0.846932	1.883343518
Product 478	0.94167	0.498671	1.888360384

It can be seen that in this case the demand usually follows a pattern of 0 and 1, but there is a slightly high demand in one or two periods. However, as the demand is not very high the error and therefore *RMSE* is not very high. The *EMID* method follows the same pattern as before, averaging out the errors which are between 0 and 1 mostly, higher in case of high demand.

It is an intuitive deduction that the higher the number of matching points in a given product, the lower should be the value of *EMID*. While we did not assume this while deducing the formula, the results in different cases that we considered corroborate this observation as can be seen in figure 1.

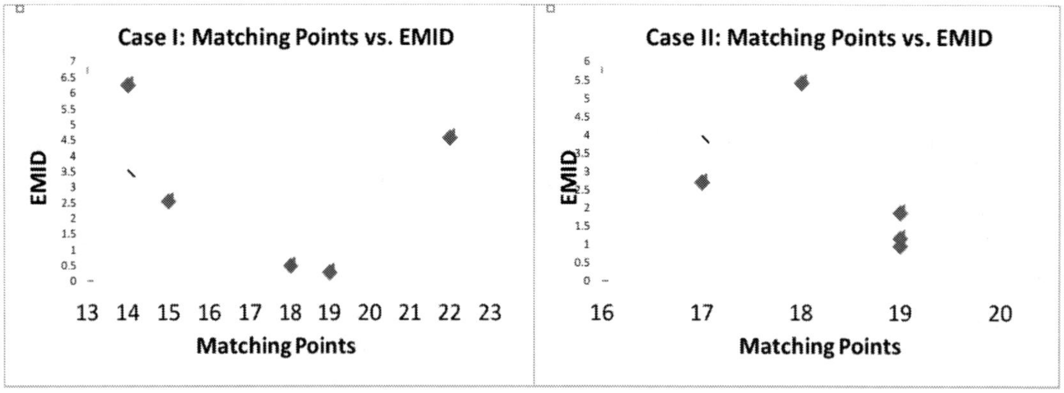

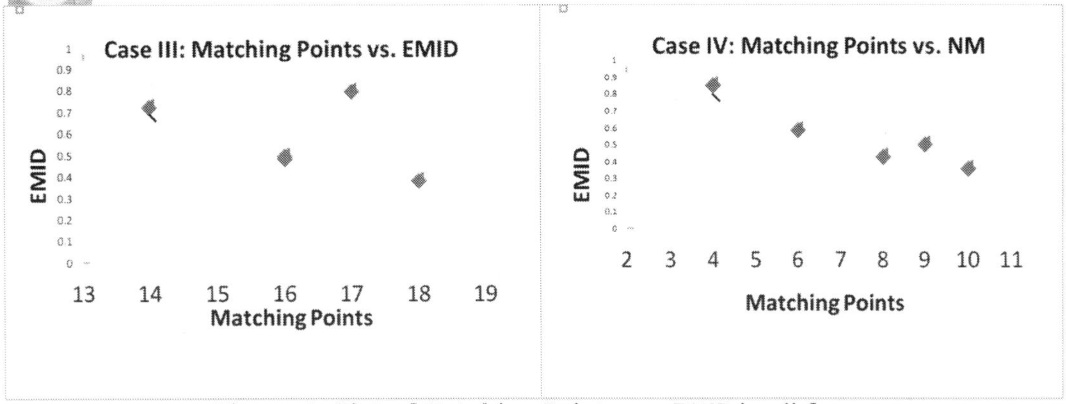

Figure 1: Plot of Matching Points vs. *EMID* in all four cases

As can be seen in the figure, the trend line slopes downwards, implying that the more the number of matching points, the lesser the value of *EMID*. However, there are some exceptional points such the fifth point in Case I, the second point in Case II, the third point in Case III and the fourth point in Case IV. In Case I, the demand is very high, i.e. 1000 in the second last period, which leads to higher error as compared to other four points in which case the products do not have such demands. Similar is the issue in Case II and Case IV, the demands, 6 and 7 respectively, not being that high, but higher compared to other products considered in the case. There is no plausible explanation for the anomaly in Case III, it could just be an outlier.

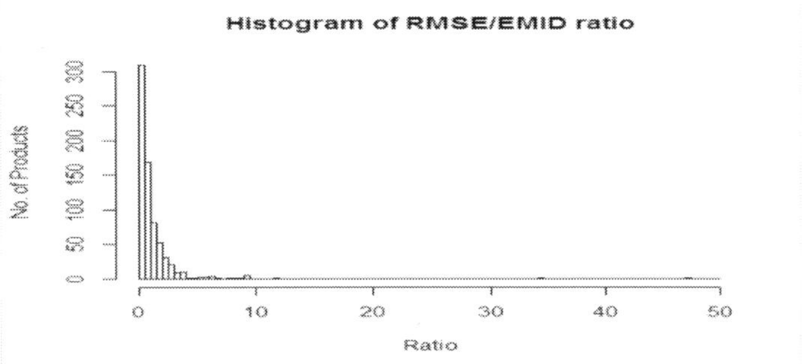

Figure 2: Histogram of RMSE/EMID ratio

Figure 2 is a histogram which shows the number of products across against the ratio of *RMSE/EMID*. As can be seen in the chart, most concentration is below 2.

5. CONCLUSION

Intermittent demand presents a tremendous challenge in many industries, and the calculation of errors between actual values and forecasted values constitutes a very significant portion of this challenge. The presence of a large number of zero values deems the usage of some of the most commonly used error measures vis. *MAPE* unsuitable. In this paper we used data reflecting the demand of 700 products in the manufacturing in the manufacturing industry. We proposed a novel method of calculating error, called *EMID*. It treats matching points and non-matching points differently and averages the error. We compare it against one of the most commonly used method, *RMSE* and explore different scenarios. The results show that *EMID* is able to well reflect the forecasting errors for intermittent demand, the more the matching points the lesser the error. Most of the outliers can be explained too. We also show that unlike *RMSE*, *EMID* values do not fluctuate severely when the demand suddenly spikes to a very high number in a particular time period. This provides consistency in error calculation and there are no extreme numbers to be explained. It also treats matching and non-matching points differently, which is necessary give the intermittent demand pattern.

There is certainly a lot of scope for future work. We have just explored the results with Croston method. A lot of work has been done in enhancing this method and quite a few modifications are now used in the industry. We need to explore if these observations hold consistent calculated while using some other forecasting method.

6. REFERENCES

[1] **Mishra, P., Yuan, X. M., Huang, G., & Duc, T. T. H. (2014).** Intermittent Demand Forecast: Robustness Assessment for Group Method of Data Handling. In *2014 International Conference on Industrial Engineering and Operations Management Bali, Indonesia, January 7 - 9, 2014* (pp. 338-347)

[2] **Mishra, P., Yuan, X. M., Huang, G., & Xu, X. X. (2013, December).** Remanufacturing intermittent demand forecast: A critical assessment. In *2013 IEEE International Conference on Industrial Engineering and Engineering Management Dec 10, 2013 (pp. 1012-1016).*

[3] **Croston, J. D. (1972).** Forecasting and stock control for intermittent demands. *Journal of the Operational Research Society*, *23*(3), 289-303.

[4] **Mukhopadhyay, S., Solis, A. O., & Gutierrez, R. S. (2012).** The Accuracy of Non-traditional versus Traditional Methods of Forecasting Lumpy Demand. *Journal of Forecasting*, *31*(8), 721-735.

[5] **Carbone, R., & Armstrong, J. S. (1982).** Carbone, Robert, and J. Scott Armstrong. "Note. Evaluation of extrapolative forecasting methods: Results of a survey of academicians and practitioners." *Journal of Forecasting* 1.2 (1982): 215-217

[6] **Syntetos, A. A., & Boylan, J. E. (2005).** The accuracy of intermittent demand estimates. *International Journal of Forecasting*, *21*(2), 303-314.

[7] **Goodwin, P., & Lawton, R. (1999).** On the asymmetry of the symmetric MAPE. *International Journal of Forecasting*, *15*(4), 405-408

A NEW AGVS TASK ALLOCATION MODEL BASED ON BOUNDED RATIONALITY AND GAME THEORY

Guomin Li and Liang Gao*
State Key Lab of Digital Manufacturing Equipment & Technology
School of Mechanical Science and Engineering
Huazhong University of Science and Technology, Wuhan, China
Kevin_YF_lee@hust.edu.cn, gaoliang@hust.edu.cn

ABSTRACT

Automatic Guided Vehicle System (AGVS) with limited sensing range demands an adaptable dynamic dispatching rule considering its discrete and dynamic properties. Each task cost is formulated as a utility function taking into account the distance and local scope task number. A new model with two steps is proposed to optimize task allocation in AGVS. Firstly, the model is initialized by a heuristic scheduling algorithm to generate the initial scheduling strategy. And then, from the bounded rationality perspective, the task allocation problem of AGVS with limited sensing range is investigated, which means each AGV can't sense all other AGV in the system. A task allocation algorithm based on game theory optimizes the initial scheduling strategy. In other word, each AGV updates its behavior strategies by maximizing the utility function in its limited dynamic neighborhood. The utility function takes into account the distance and local scope task number, which raises the efficiency of the system. Finally, the feasibility and effectiveness of the model is verified by a simulation example.

Keywords: AGVS, Game Theory, Bounded Rationality, Dynamic Neighborhood

1 INTRODUCTION

With the development of production and automation, the traditional manufacturing methods have changed deeply; meanwhile, the material conveyance's system should meet the demands of automation, flexibility and just in time. Automatic Guided Vehicle has played an important role in the material conveyance's system. With the increase of the degree of the manufacturing automation and the scale of production, single AGV system can't meet the requirements[1]. Thus people turn to the research of AGVS, which means larger scale[2][3]. AGV will become a more highly self-programming and adaptive capacity robot, which is suitable for working in the complex unstructured environment. AGVS is developing towards the networked material conveyance's system, which is dispersed, different, relax coupling. AGVS includes intricate software and hardware scheme, and every portion harmonizes complicatedly. The traditional, concentrated supervising system is not satisfied, so we must take a distributed management system.

AGVS is a distributed multiple robot system, the methods of task allocation includes market, capability classification, fitness and swarm intelligence[4][5][6][7]. Some analysis shows that larger scale task allocation is a major research area and swarm intelligence will be a good choice to solve this kind of task allocation[8]. Now, more and more people aim to solve the problem by introducing

* Corresponding Author

E-mail address: gaoliang@mail.hust.edu.cn (L Gao)

game theory and fine results have been obtained[9][10]. In the Mass Production, AGV can't sense all other AGV in the system. Bounded rationality is the idea that when individuals make decisions, their rationality is limited by the available information, the tractability of the decision problem, the cognitive limitations of their minds, and the time available to make the decision. In this view, decision-makers act as satisfiers, seek a satisfactory solution rather than an optimal one. So the method of bounded rationality is put forward to solve the problem of task allocation. In order to find a reasonable and effective task allocation algorithm, this paper builds the AGV based on agent technology, game theory, bounded rationality and then studies the task allocation for AGVS, which can alleviating the calculation pressure with scale extending. Simulation results demonstrate that the algorithm has good real-time performance, control quality and robustness.

2 MODELING AGVS TASK ALLOCATION

AGVS is a fully automated system that provides logistic services in an industrial environment such as a warehouse or a factory. A fundamental problem of AGVS control is to determine which vehicles should convey and when[11]. The AGVS scheduling and the establishment of conflict-free routing are issues that concern and drive the system real-time operation. Now, this paper only considers the AGVS scheduling. Two main approaches have been used to task allocation. They are bounded rationality and game theory. Although they have been applied multi-agent system, to the best of our knowledge, the application reported in this paper is the first account of the use of them for real world AGV control.

2.1 Bounded Rationality

An increasing number of applications are being implemented as multi-agent systems in which individual agents interact with one another in pursuit of both individual and system goals. To build successful applications, designers need their agents to be rational and to do the right thing. However, the agents always do not have unlimited time or computational resources[12]. Hence, agents should be able to modify how they make decisions based on their current context. Account for limited resource, people introduces the concept of bounded rationality. Bounded rationality refers to the concept that human problem-solvers are rarely able to identify all possible solutions to the problems at hand, therefore, settle for choices that seem to satisfy the required solution properties of a problem.

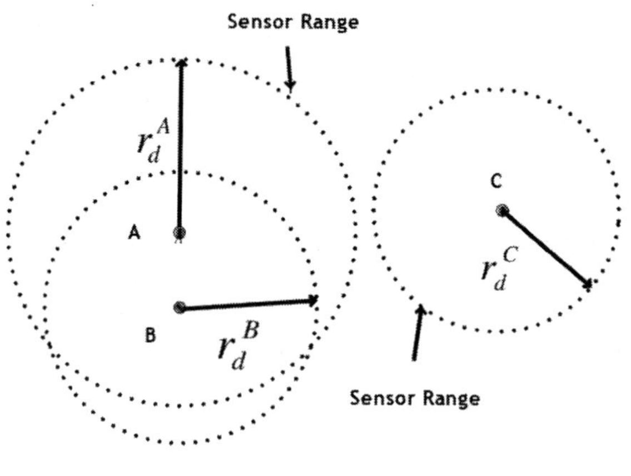

Figure 1: The interaction of AGVS

Figure.1 shows that the AGV B in the sensor range of the AGV A and can use the information of its. The AGV A and the AGV C have different neighborhood sizes. The AGV A can exchange information like speed, destination with other AGV, which is in the sensor range of its, and decides its direction of movement at each interation. The sensor range will be a function of the radial sensor range.The collaboration among the AGV is implemented by a common database. If each AGV exchange task and travels to the nearest destinations and drops off loads, AGVS may be able to move without conflict and which will increase the performance of system.

In practice, we can control the sensing range of AGV, called local decision domain, by adding dynamic constraint. Let r_0 be the initial neighbourhood range of each AGV (that is,$r_0^i(0) = r_{0,}\forall i$). To adaptively update the neighbourhood range of each AGV, the following rule is applied[13]:

$$r_d^i(t+1) = min\{r_0, max\{0, r_d^i(t) + \beta(n_t - |N_i(t)|)\}\} \quad (1)$$

Where β is a constant parameter and n_t is a parameter used to control the number of neighbours. $N_i(t)$ is the set of neighbours of AGV i at time t.

This paper introduces the theory of bounded rationality which will help to build the dynamic dispatch method for AGVS.

2.2 Game Theory

Game theory is a branch of economics that studies interactions between self-interested agents. In the last few years, there have been increasing interests in the use of game theory for analysing and implementing of multi-robot system[14][15]. An agent plays a game through actions. An action is a choice that a player makes, according to his own strategy. In game theory, the Nash equilibrium is a solution concept of a non-cooperative game involving two or more players. Nash equilibrium refers to a set of such strategies from which no individual players can increase their payoffs by changing their strategies unilaterally. There are n players participating in a game. A game $G = \{S_1, S_2, \cdots, S_n : \mu_1, \mu_2, \cdots, \mu_n\}$. Each player can take an individual action represented by a vector S_i. The μ_i is the payoff function of the player. All players, when acting together, can take a collective action, which is a vector S_i.

For $i = 1,2,\cdots,n$, here

$$\mu_i(s_i^*,\ s_{-i}^*) \geq \mu_i(s_i, s_{-i}^*)\ \ \forall s_i \in \textstyle\sum i \quad (2)$$

The point $s^* = \{s_1^*, s_2^*,\ \cdots,\ s_n^*\}$ is called the Nash equilibrium point.

Most models using game theory suppose that all agents have perfect information and perfect rationality for computing the Nash equilibrium. In reality, AGV has some interactions through the information flow that exits only among that the neighbouring AGVS. The sensor range of AGV is the information exchange range. The system may change tasks of AGVS to get better scheduling. So, the above concepts of game theory are applied to the task allocation. Two revenue functions are defined for each AGV as following.

Distance suitability: This function will make AGV picks up loads and travels to the nearest destinations.

$$W_{ik}^d = 1/\sqrt{D_{ik}} \quad (3)$$

D_{ik} is the shortest distance between AGV i and destination k.

Task density suitability: f_k is the total task of units which destination k belongs to. f_h is the total task of units which destination h belongs to.

$$W_{hk}^f = (f_h - f_k) \quad (4)$$

The bigger the W_{hk}^{f}, AGV i and AGV j may more possibility change task.

In this paper, the AGVS try to maximize their total benefits. That is to say, each AGV faces the following optimization problem.

$$Max\ Profit = \alpha * \frac{1}{\sqrt{D_{ik}}} + \beta * (f_k - f_h) \quad (5)$$

Where $\alpha \geq 0,\ \beta \geq 0$.

Sometimes, AGV can work alone on a task and the total task of units which destination belongs to, but sometimes not. Following (5) and the ability to complete tasks independently, the competitive strategy can be described as.

Strategy (1): AGV i can work alone and AGV j can work alone. If $W_{ik}^{d}+W_{jh}^{d} \geq W_{ih}^{d}+W_{jk}^{d}$, AGV i and AGV j need not change task.

Strategy (2): AGV i can't work alone and AGV j can work alone. If $W_{ik}^{d}+W_{jh}^{d} \geq W_{ih}^{d}+W_{jk}^{d}+y*W_{kh}^{f}$, AGV i and AGV j need not change task.

Strategy (3): AGV i and AGV j both can't work alone. If $W_{ik}^{d}+W_{jh}^{d} \geq W_{ih}^{d}+W_{jk}^{d}+y*W_{hk}^{f}$, and $W_{ik}^{d}+W_{jh}^{d} \geq W_{ih}^{d}+W_{jk}^{d}+y*W_{kh}^{f}$, AGV i and AGV j need not change task.

2.3 Task Allocation Algorithm

The task allocation algorithm (Figure.2) is based on the bounded rationality and game theory. The bounded rationality will divide the allocation problem into smaller and simpler problems. And then, based on game theory, the algorithm is applied to optimizing the initial scheduling strategy.

Step (1): The algorithm starts by generating the new tasks.

Step (2): The system receives a list of requests as input. Each request is a task defined by destination node, tasks of the neighbouring destination node, etc. The machine will assign task to the nearest AGV. If the number of the nearest AGV is more than one, we will select one by roulette algorithm[16].

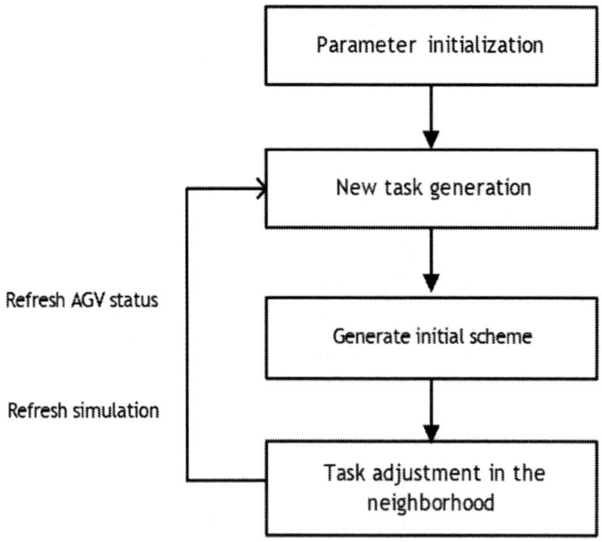

Figure 2: Task Allocation Algorithm

Step (3): In the sensor range of the AGV, the strategy 1-3 will be used to optimizing the initial scheduling strategy.

Step (4): Refresh AGV status and simulation clock. Go to step 1 for another cycle.

The scheduling objective is to meet all material requirements in a minimum time.

3 EXPERIMENTATION AND ANALYSIS

To evaluate the performance of the proposed algorithm, a computational experiment is conducted in this section. The task allocation algorithm is coded in C# and run on Intel Pentium(R) on a personal computer with 2.806 GHz CPU and 1.69 GB RAM in Windows XP.

In order to describe the marrow of our task allocation algorithm, this paper makes a simple but general model in which one yard block has only one station in the middle of adjacent intersections of pathways (refer to Figure3). In this mesh layout, there are in total 20×20 blocks, namely 20 blocks in each column and 20 blocks in each row. Each block has the same size. There are two paths with different directions between two adjacent blocks. Each Block has one Pick up-Drop off station (or P/D station for short), located at the bottom centre of the block. The 20 ×20 blocks grouped in sets of 5×5, and every set has a vehicle park for one AGV station initially. We do not care the details for AGVS routing, the size of the junction, the length of the AGV, etc. It is reasonable and realistic for us to simplify the mesh model for convenience of analysis and discussion.

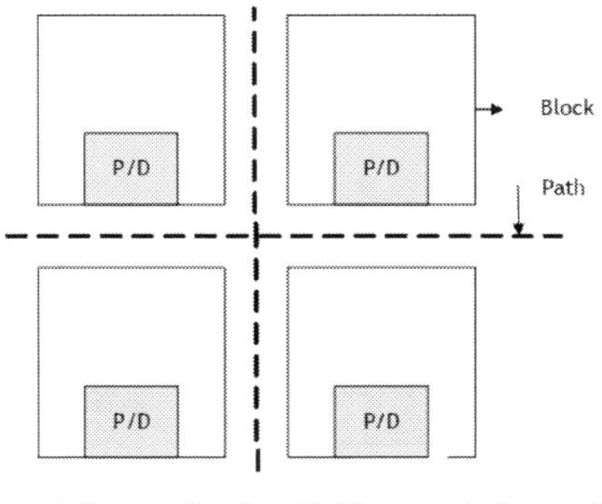

Figure 3: Realistic mesh layout

In the simplified mesh layout model, shown as Figure.4, there are 20×20 junctions. The P/D station is regard as a node. Each node is to assign it with the coordinates (x, y) as its address, where x and y represent respectively the row and column address. Task are generated obey the [100,120] uniform distribution. r_0 =60. The load-carrying capacity of each AGV is 50 and the simulation time is 5000. The average speed of AGV is 5.

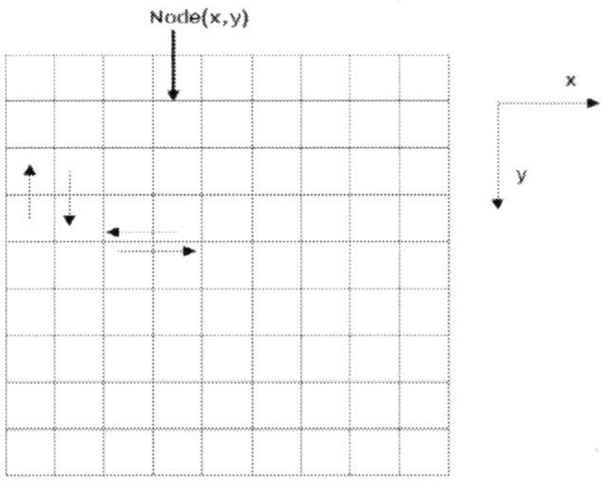

Figure 4: Simplified mesh routing model

According to the discussions in the previous sections, the assumptions made in this work are as follows:

- An AGV will not take part in the execution of another task, until it has finished the present task assigned.
- No manufacturing or transportation operation can be pre-empted.
- The materiel on AGV at least can meet the need of one task.
- AGV are all bidirectional AGV.
- Do not consider the conflict problem.

To verify the proposed algorithm, the "First in First out (FIFO)" algorithm was used as the compared algorithm. There are many ways to carry out the comparison of algorithm performance, some obvious approaches are : to compare the average task time, to compare their task quantity, to compare their distance, or to compare equipment output. Here, we have used all of them. In our simulations, we have runed each algorithm for 20 times so that we can do meaningful statistical analysis. The main comparison results are listed in table 1. The average task time of the proposed algorithm is higher 2.7% than the compared algorithm. The task quantity grow 2.5%, which is relatively high. The performance of this algorithm is based on a very simple modification of the task allocation and reduces the transport time. AGV will exchange task and travel to the nearest destinations and drop off loads. From the results, the proposed algorithm presents a better performance of task allocation.

Table 1: COMPARISON OF SIMULATION RESULTS

Layout	**Proposed Algorithm**	**FIFO**
Average task time	11.51	11.83
Task quantity	396	386
distance	20820	20949
Equipment output	16.42	16.06

4 CONCLUSION

AGVS is considered as one of the most appropriate modes for material handing support of contemporary flexibly automated production environments. In this paper, the task allocation of AGVS has been analyzed based on bounded rationality and game theory. The bounded rationality is a simple way to describe the sensor range of the AGV, which also is the decision region and is sufficient to avoid the problem of calculated quantities. The game theory is better suited to the task allocation of making some sense out of the AGV's rationality and its bounded manifestations. As shown in this paper, extensive research has been conducted to solve the problem of task allocation; further research is needed on the issues of conflict-free routing, urgent task and communication fault.

5 ACKNOWLEDGMENTS

This research work was supported by Fund Committee innovation group projects under Grant no.51421062.

6 REFERENCES

[1]**Ho, Y. C., Liu, H. C., & Yih, Y. 2012**. A multiple-attribute method for concurrently solving the pickup-dispatching problem and the load-selection problem of multiple-load AGVs, *Journal of Manufacturing Systems*, 31(3), pp 288-300.

[2]**Xu, L., Jia, W., & Ming, H. 2008**. New coding method for massive production scheduling problem, *Computer Integrated Manufacturing Systems*, 14(10), pp 1974-1977.

[3]**Feng, J., & Cheng, W. U. 2006**. Research status and prospects for massive production scheduling, *Computer Integrated Manufacturing Systems*, 12(2), pp 161-167.

[4]**Xin-Yue, Q. I., Tian, Y. T., Yang, M., & Yang, Y. M. 2009**. Market-based multi-robot task allocation for fire-disaster response, *Journal of Jilin University(Information Science Edition)*, 27(5), pp 506-513.

[5]**Liu, S. H., Zhang, Y., Fu, S., & Wu, H. Y. 2010**. Multi-robot task allocation based on swarm intelligence, *Journal of Jilin University (Engineering and Technology Edition)*, 40(1), pp 123-129.

[6]**Dong, Y. B., Jiang, J. P., & Yan, H. E.. 2007**. Fitness-based task allocation strategy for multi-robot system, *Journal of Zhejiang University (Engineering Science)*, 41(2), pp 272-277.

[7]**Gerkey, B. P., & Mataric, M. J. 2004**. A formal analysis and taxonomy of task allocation in multi-robot systems, *The International Journal of Robotics Research*, 23(9), pp 939-954.

[8]**Zhang, Y., & Liu, S. H. 2008**. Survey of multi-robot task allocation, *CAAI Transactions on Intelligent Systems*, 3(2), pp 115-120.

[9]**Han, Q. Y., Wang, S., & Dang, J. W. 2007**. Research of task allocation under competitive environment, *Computer Engineering and Design*, 28(3), pp 517-579.

[10]**Liu, X. M., Tian, Y. T., & Yang, M. 2010**. Game theory based multi-Robot task allocation algorithm, *Journal of Jilin University (Information Science Edition)*,28(3), pp 256-263.

[11]**Bartholdi, J., Platzman, L.1989**. Decentralized control of affixed route automatic guided vehicle system, *IIE Transactions*, 21(1), pp 76-81.

[12]**Wooldridge, M. 2001**. The control of reasoning in resource-bounded agents, *Knowledge Engineering Review*, 16(3), pp 215-240.

[13]**Zhou, Y., Wang, Y., He, S., & Wu, J. 2014**. A novel double glow-worm swarm co-evolution optimization algorithm based levy flights, *Applied Mathematics & Information Sciences*, 8(1), pp 355-361.

[14]**Bayram, H., & Bozma, H. I.** 2015. Assistance networks for dynamic multi-robot tasks, *Autonomous Robots*, 40(4), pp 615-630.

[15]**Hernández, E., Barrientos, A., & Cerro, J. D. 2012**. Selective Smooth Fictitious Play: An approach based on game theory for patrolling infrastructures with a multi-robot system, *Expert Systems with Applications*, 41(6), pp 2897-2913.

[16]**Wang, Y. 2012**. Glow-worm Swarm Optimization algorithm based on fluorescein diffusion, *Computer Engineering and Applications*, 48(10), pp 34-38.

RELEASE OF METADATA OF OPEN DATA OF CHINESE LOCAL GOVERNMEN BASE ON DRUPAL

WangDi[1*] and ZhaiJun[2]
Dalian Maritime University, Da Lian, China
dlmu_dean@163.com

ABSTRACT

Since 2009, "The Open Government Initiative" has been launched by US government, a number of open data movements sprung up around the world, with transparency and data reuse as purpose. Thanks for that, until 2015 the number of data sets released by governments worldwide has reached nearly 2 million. In order to manage the huge number of open data and for convenience to users to search data, one of the key initiatives is release of open data catalogs portals. For the local government in China it is an big issue that the open data format has been low stars rating which base on "5 star rating scheme" proposed by the president of W3C Tim Berners- Lee. This paper taking DCAT which is the standard of data catalogs approved by W3C as standard, build a open government data catalogs platform based on Drupal, and support open data formats of 5 stars rating. It is provides a part of the solution for the construction and upgrade of our local government open data platform.

Keywords: Drupal; Government Open Data; Open Data Platform; Metadata; DCAT

1 INTRODUCTION OF OPEN DATA PLATFORM

The platform which be used to release open data by the governments of the world has be divided into two categories: Open-Source and commercial purpose [12]. Junar (junar.com) and Socrata that are open data platform based on cloud-service SaaS (Software as a Service), which belongs to the commercial category; CKAN (Comprehensive Knowledge Archive Network), DKAN , ODPL are Open Source Software to establish an open data platform.

Currently, the most widely used platform is CKAN and DKAN [13,14]. CKAN applies to large data constructs, such as the United States (Data.Gov), United Kingdom (Data.Gov.UK) National Open Data portal [10]. However, the deployment and implementation of CKAN are more complex, require high professional for the staffs. DKAN (Drupal + CKAN) is method based on Drupal , to achieve data catalog service, it can be compatible CKAN and more functionality, such as support for linked open data. The main technical DKAN and CKAN comparison in Table 1.

---

[*] Corresponding Author

Table 1: Comparing DKAN and CKAN

Features	**DKAN**	**CKAN**
Primary Language	PHP	Python
Platform	Drupal	Python
Database	MySQL,CartoDB	PostgreSQL
Data Preview	Recline.js	Recline.js
No. of modules	33728	129

As an open source content management system, Drupal is easy to develop and maintain, all functions of Drupal can be increased or decreased by start up or shut down the modules to achieve. Until April 2016, modules has reached 33,728 on Drupal site, and attracted 100,654 subscribers to contribute the modules. Drupal has been widely used in the construction of all types of library portal [15, 16] as well as teaching resources management [17]. More and more governments data portal are choosing Drupal, for example, nearly one-third of the African Government Open Data Platform utilizes DKAN [14]. The focus of this paper is Drupal Application of Local Government open data portal.

2 ANALYSIS OF THE OPEN DATA PLATFORM

Open data is data which can be used by anyone free-free access, use and re-released, and not subject to copyright, patent restrictions, or other mechanisms [11]. The fundamental of open data is to promote the maximum reuse of the data, the best way to achieve this purpose is that all countries publish a variety of "machine-readable"(machine-readable, that is, data can be programmed automatic processing) high-value data, based on the Web Open Data platform [6]. In order to make open data to machine readable, W3C • President Tim Berners- Lee who is an active promoter of open government data, in 2010 in Washington, DC Gov 2.0 Expo was first proposed openness (Openness) of "5 Star rating model "(5 star rating scheme) [2,15]:

(1) ★ Data is available on the Web, in whatever format (for example, a scanned image).

(2)★★ Data is available as machine-readable structured data (for example, an Excel spreadsheet).

(3) ★★★ Data is available in a non-proprietary format (for example, CSV).

(4)★★★★ Data is published using open data standards from the World Wide Web Consortium.

(5)★★★★★ All of the above apply, plus links to other people's data.

The 5-Star system is cumulative. Each additional star presumes that your data meets the criteria for the previous steps. The model focus on two main aspects of open data: Law and technology, in term of law open government data initiatives is needed, in term of technology data should be published as machine readable. The model points out the upgrading route of data format: from the proprietary format to the non proprietary format, from the open standard to the related data (i.e. data World Wide Web).

In the world, the number of 5-stars government open data has grown[2].The open data which released by the government of New York City on website (data.cityofnewyork.us) has reached the level of 5-stars [8]. Until2014the UK as an champion of "Open Data index", has already released 150 data sets which reached 5-stars on national data website (Data.Gov.UK). In contrast, the local governments in China's has no data released as 5-stars, Shanghai (www.datashanghai.gov.cn) and Zhejiang (data.zjzwfw.gov.cn) which are the best provided still in the 2-stars, 3-stars rating [7,8].

In the Open Data Platform, a set of data files gather to a dataset [9], a list of the dataset is data catalog, and the data catalog manage the metadata of datasets [6]. For this, the core function of the data platform is metadata management of a various formats of data sets and the API service of data access, see Figure 1. Other functions of data platform, such as data search, data storage, data visualization, user interaction, etc. [9], are serve the core function or built based on it.

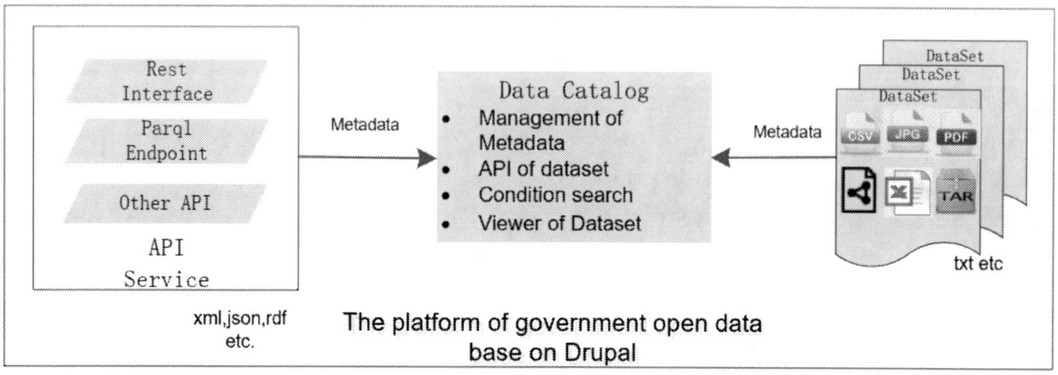

Figure 1: The Core Functions Of Open Data Platform

3 METADATA MANAGEMENT

3.1 Metadata pattern

With the rapid growth of open data portals and thedatasets, for the users it is an challenge to find the appropriate data set, especially search across frommultiple data directories [2]. For this, an appropriate description of the data set is necessary, the same as to provide Metadata. Metadata is Data about Dataor message about information. It is the most basic function of government data catalog metadata to open the management of Metadata of datacatalog. used to being structured, standardized, machine-readable semantic and direction [6,10,16,17].DCAT (Data Catalog Vocabulary) is common metadata standard which be supported and released by the W3C ,can support the data catalog interoperability . DCAT has been adopted by the UK , the United States and the European Union's national data portals and datahub.io etc. In China, the results of survey about the local government open data show that [7,8]: each local government data portals use different metadata schema, the metadata formats always be HTML , not machine-readable, because of this, the availability and reuse of the open data has been impeded. [6].

It is the greatest feature of Drupal to organize and manage structured data through content type" . Content type can be understood as a data pattern having a specific structure, same as the concept of class or table structure of the database, a specific data that accords the demand of content type defined as node in Drupal. DKAN treat dataset as a kind of "content type", and to establish the correspondence between DKAN field / property and DCAT vocabulary / elements by mapping (see Figure 2), thereby forming a set of metadata which describ the data model, a part of them shown in Table 3.

Table 3: Metadata schema of data sets in data directory based on DCAT

Element	DKAN	Request
dc:title	Title	Yes
dcat:keyword	Tags	Yes
dc:modified	Changed	Yes
dct:publisher	Pulisher	Yes
dcat:theme	Groups	Recommend
dc:identifier	Identifier	Yes
dct:spatial	Spatial/ Geographical Coverage Area	Recommend

The interface of metadata management shown in Figure 2 is located Drupal platform Configuration-> Web services-> RDF publishing settings below. It is worth noting that the above fields are DKAN provided by default. If you need the data set described in more detail, DKAN data directory platform allows users to extend the existing vocabulary or define the ontology according to users' needs.

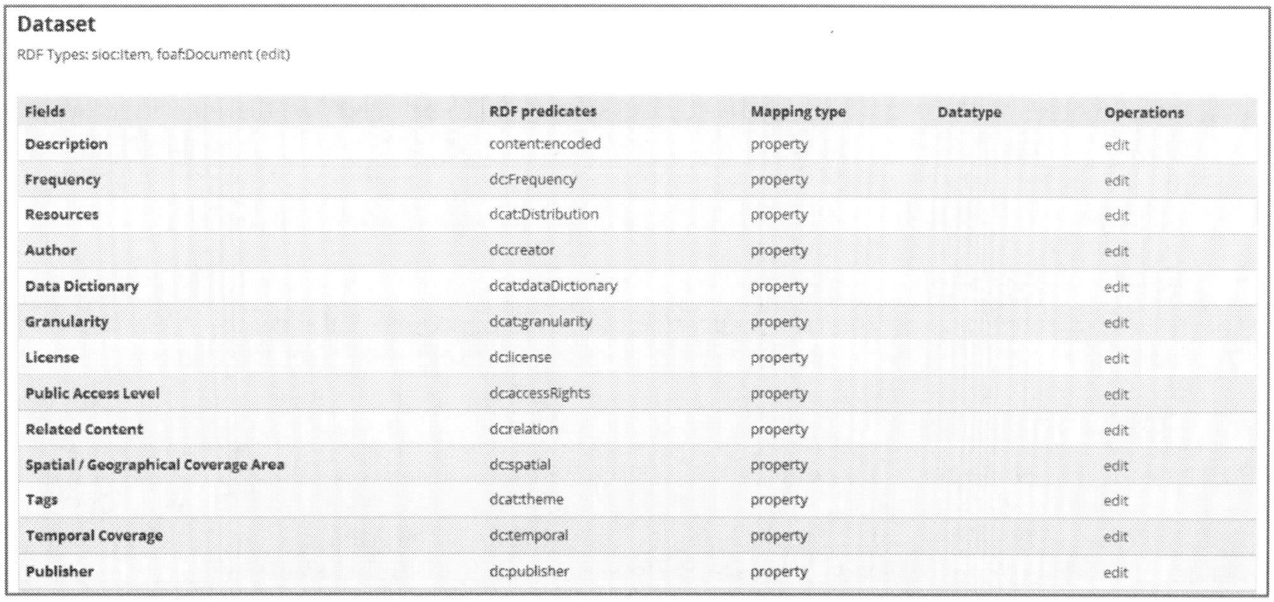

Dataset

RDF Types: sioc:Item, foaf:Document (edit)

Fields	RDF predicates	Mapping type	Datatype	Operations
Description	content:encoded	property		edit
Frequency	dc:Frequency	property		edit
Resources	dcat:Distribution	property		edit
Author	dc:creator	property		edit
Data Dictionary	dcat:dataDictionary	property		edit
Granularity	dcat:granularity	property		edit
License	dc:license	property		edit
Public Access Level	dc:accessRights	property		edit
Related Content	dc:relation	property		edit
Spatial / Geographical Coverage Area	dc:spatial	property		edit
Tags	dcat:theme	property		edit
Temporal Coverage	dc:temporal	property		edit
Publisher	dc:publisher	property		edit

Figure 2: The Core Functions Of Open Data Platform

3.2 The instance of Metadata

When add a data set or create an instance of the node in DKAN platform, the system will automatically generate metadata for the data set in two formats, json and rdf now available, both of them are machine-readable. It is shown by Figures 3 that the metadata instance which is "Teachers Library data set" on Zhejiang Open Data website (data.zjzwfw.gov.cn) are automatically generated by the Druple. Therefore the format of original data set on the website is HTML, It is hardly be used by application automatically. How to publish detailed process data sets on the platform DKAN see below "Application Examples" section.

Figure 3: The Metadata Of RDF Format

On the data catalog platform, the access address of these two metadata files were "/node/nodeid.rdf" and "/ api / 3 / action / package_show? Id = cd1be547-7fae-4d91-bbae-ef37661a1bc4" respectively . Platform provides json format metadata by program processing, the metadata will be encapsulated as a result set which be provided a unique identifier. Users can access the metadata which is corresponded by the specific data sets through the identifier. Whether rdf format, or json format, user applications are able to read and parse the metadata for more information about the data set, that will improves the interoperability between the open data platform and users' applications [6] .

4 APPLICATION EXAMPLE

The following categories of Zhejiang Educational Technology Open Data on the Web site of the "Teachers Library data set" for example, the use of DKAN platform release of the data set.

4.1 The instance of Metadata

As shown in Figure 4 select "Content Types" Content-> Add content for Dataset, and then add the description of the data set as metadata. Dataset "field" values are taken from Zhejiang Open Data website (data.zjzwfw.gov.cn), but there is no "open licenses" information on the site, shown in Figure 5. DKAN provides 17 kinds of open licenses, such as OGL, ODbl, PDDL, etc., and each open licensing can link to the appropriate help page.

Figure 4: Add Dataset On Platform Of DKAN

Figure 5: The Dataset Released On Open Data Portal Of Zhejiang

(http://data.zjzwfw.gov.cn/catedetail.action?resid=210562/20150619105946000432&catecode=SJLY006)

4.2 The instance of Metadata

After adding data sets, it is needed to add specific data resources. Data resources also be treat as text format on DKAN , because of the loosely coupled structured data can be managed efficiency on the platform. A data set may correspond to a plurality of data resource file, and the data set and the data resource file will be connected by identifying field. DKAN provides three ways to registrant data resources: API, data file uploads and links to the URL data source file. This example is using API to release data resource file. DKAN can manage data set data files effectively in multiple formats, users can access or download them as needed.

After the completion of the operation, showing page platform of the data set shown in Figure 7. On the left side of the page contains the dataset Open License (License), and machine-readable formats (RDF and JSON) metadata file download icon, details of which are shown in Figures 3 and 4; the right page side shows the data file information data set, dataset info section on the right below the HTML table lists all metadata.

Figure 6: The Dataset Released On DKAN Platform

Comparison of Figures 5 and Figures 6 shows that DKAN platform has obvious advantages on management of open dataset metadata: not only provides people read the metadata in HTML format, the program also provides machine-readable metadata file for the application software process automatically .

5 CONCLUSION

In recent years, open government data has become a hot topic, in information management, e-government, and other areas of Library and Information Science field[1]. It has important practical significance for our government open data which is in the development stage at all levels that how to design and build a high-quality open platform to make the data of data set be discovered more facilitate and be used more effectively.

Drupal because of its open source, modular features and DKAN modules, can provides the technical foundation to build an open data platform. This essay shows that the platform based on Drupal supports common metadata standard and machine-readable metadata format, and provides an API interface to a variety of formats of data sets (particularly 5-star-related data sets), it overcomes some of the lack of domestic open data site. The next step will focuse on increasing the data search and navigation, data storage, data visualization, user interaction, statistical analysis and other functions.

for books:

[1] **Zipei Xu**.Big Data: The coming data revolution, and how does it change the government, business and our lives (Third Edition) [M].Guilin: Guangxi Normal University press,2015: 345-351.

[2] **Jun Zhai**. The principle and application of related data of government -- the technology and practice of open data in the era of big data [M].BeiJin: PHEI,2016:4-12.

[3] **Han Zhang,Zhong Wan**. Comparison reserch on open data of international government[J]. Journal of Intelligence, 2015,34(8):142-146.

[4] World Bank. Open Data in 60 Seconds[EB/OL].

[5] World Bank.Benefits of Open Data[EB/OL].

[6] **Attard J, Orlandi F, Scerri S**, et al. A systematic review of open government data initiatives[J]. Government Information Quarterly, 2015, 32(4): 399-418.

[7] **Lei Zheng, Feng Gao**. Research of the platform of open data of China's government: framework, current situation and suggestions [J].E-Goverment, 2015, (7):8-16.

[8] **Huina Xu,Lei Zheng**. The Open Government Data Platform for oriented use: a comparative of New York and Shanghai [J]. E-Goverment, 2015, (7):37-45.

[9] **Xiaohong Qian,Manggu Hu**. Building an Open Government Data Platform and Analysing Its Technical Characteristics [J]. Documentation, Information and Knowledge, 2014, (3):124-129.

[10] **Zhifeng Zhou, Ruhua Huang**. An Analysis on service function of foreign government open data portals[J]. Journal of Intelligence, 2013,32(3):144-147.

[11] **Haiqun Ma, Hongshuai Wang**. SLEPT analysis and strategic deployment of Chinese government's open data strategy [J]. Information Science, 2016,34(3):3-8.

[12] **Priebe S, Kannicht M, Jacob M**, et al. The complementarity of open data infrastructures: An analysis of functionalities[C].Proceedings of the 14th Annual International Conference on Digital Government Research, ACM, 2013:166-171.

[13] World Bank. Technical Assessment of Open Data Platforms for National Statistical Organisations[EB/OL].http://documents.worldbank.org/curated/en/2014/10/20451797/technical-assessment-open-data-platforms-national-statistical-organisations,2014-10.

[14] **Bello O, Jolayemi O, Akinwande V**, et al. Open Data Portals in Africa An Empirical Analysis of Open Government Data Initiatives[C]. Conference: 4th iSTEAMS Research Nexus,2015:595-606.

[15] **Tim Berners-Lee**. Linked Data[EB/OL].http://www.w3.org/DesignIssues/LinkedData.html, 2009-06-18.

[16] **Zuiderwijk, A., Helbig, N., Gil-García, J.R.A., & Janssen, M.** Special issue on innovation through open data: Guest editors' introduction[J]. Journal of Theoretical and Applied Electronic Commerce Research, 2014,9(2).

[17] W3C. Data Catalog Vocabulary (DCAT)[EB/OL].http://www.w3.org/TR/vocab-dcat/,2014-01-16.

A HYBRID HEURISTIC FOR HUB LOCATION AND ROUTING PROBLEM IN INTERMODAL LOGISTIC NETWORKS

Hai Ling Bi[1,2], Kai Kang[2] and Xu Tao Zhang[1*]
[1] Department of Equipment Support
Military Transportation University, Tianjin, China
lonewind1981@126.com

[2] College of Economics and Management
Hebei University of Technology, Tianjin, China
bhlryf@126.com

ABSTRACT

Hub location problem in intermodal logistic networks is much more complex than it is for single mode logistics. In this paper, the design of intermodal logistic networks is decomposed into location/allocation problems. Some simplifications are proposed when we derive the integer programming model. A hybrid heuristic is proposed to solve the complex model. The tabu search is used to determine the location problem, and the Floyd shortest path algorithm is employed to tackle the routing problem. An illustrated example evaluates the performance of the algorithm.

Keywords: Hub networks, Intermodal logistics, Integer programming, Heuristic

1 INTRODUCTION

Intermodal transport refers to combined use of two or more modes of transport in a single transport chain (Macharis and Bontekoning [1]). It is a major transportation form for modern logistics in its own right. A hub-and-spoke structure is generally adopted in the intermodal transport network. The network is composed of a few central locations which act as hubs. The hubs play key roles in the network because the economies-of-scale is realized due to the concentration of flows between the hubs. The hub location is a key problem in design of an intermodal logistic network. However, the problem in intermodal logistic networks is much more complex than it is for single mode logistics.

Most researchers have addressed hub location problems from a single transport network. Alumur, Kara and Karasan [2] studied a single allocation incomplete hub networks for LTL trucking in Turkey. Yaman [3] discussed single allocation and multiple allocation strategies and their effects on total routing costs in hub networks. Single transport network is a simple network, that is, there is only one link between nodes in the same direction. Since different transport modes can be used in a pair of nodes, there are at least two links between nodes in the same direction, which makes the hub location problem becoming more complicated. Ishfaq and Sox [4] doubled the nodes of a road-rail intermodal network by using the method of the extended nodes, which can transfer a complex network containing multiple edges into a simple network. However, it is unreasonable the research hypothesized that a node could only be built a highway hub or a railway hub in reality. Other studies (Ishfaq and Sox [5], Alumur and Kara [6]) usually directly derive some complicated integer

* Corresponding Author: Xu Tao Zhang. Military Transportation University. lonewind1981@126.com.

programming formulations and employ some heuristics for model solution. Unfortunately, since these approaches need to solve the sub problem repeatedly in the search process, the performance of these algorithms is not well and the efficiency is low.

In fact, if we know the weights for a given hub set of intermodal network edges, we can simplify the network through taking the minimum weight between the nodes. In this paper, we derives an uncapacitated multiple allocation integer programming formulations for hub location problem in intermodal logistics networks, and decomposes it to location/allocation problems. Then a heuristic algorithm based on tabu search and the Floyd shortest path algorithm is proposed. Finally, a simulation example is presented.

2 MODELING DESCRIPTION AND HYPOTHESIS

The intermodal transportation network can be viewed as hub-spoke network. There are two modes of transportation of roadway and railway in the network. Nodes have pairwise traffic demands. Some assumptions are made as follows.

1. Hubs are connected by a complete network.
2. A non-hub node can be allocated to more than one hub nodes, and no direct links exist between non-hub nodes.
3. The inter-hub transportation has a cost discount factor.
4. Railway is only for hub transportation.
5. Drayage charges and the operation cost are consolidated into the cost of inter-hub transportation.
6. Hubs and links are uncapacitated.

3 MATHEMATICAL MODEL

3.1 Model parameters

Notations employed in the model are introduced as follows.

$N = \{1, 2, \cdots n\}$ Set of all nodes

p Number of hubs

$S = \{1, 2\}$ Set of transportation modes between hub pairs, where $s = 1$ for roadway transportation, $s = 2$ for railway transportation

w_{ij} Traffic demand from origin node $i \in N$ to destination node $j \in N$, where $w_{ii} = 0$

F_k Fixed cost of opening and operating a hub at node *k*

c_{ij}^s Unit transportation cost from node *i* to node *j* using transportation mode *s*, where $c_{ii}^s = 0$

$\hat{c}_{km}^s$ Drayage and operating cost of one unit transportation via hubs *j* and *m*, where $\hat{c}_{kk}^s = 0$

$C_{ijkm}^s = c_{ik}^1 + \hat{c}_{km}^s + \alpha^s c_{km}^s + c_{mj}^1$ Unit transportation cost from origin node *i* to destination *j* via hubs *k* and *m* using mode *s*

α^s Transportation cost discount factor between hubs using mode *s*

3.2 Decision variables

There are two set of decision variables in the model.

$$y_k = \begin{cases} 1, if\ node\ k\ is\ a\ hub; \\ 0, otherwise. \end{cases}$$

$$X_{ijkm}^{s} = \begin{cases} 1, if\ flow\ from\ i\ to\ j\ via\ hub\ pair(k,m)\ using\ mode\ s; \\ 0, otherwise. \end{cases}$$

For X_{ijkm}^{s}, when $k=m$, it means that there is only one hub from i to j. Freight for origin-destination pair will be transported by roadway only and does not need drayage. When $i=k$, it means that the original nodes is a hub. When $i=k, j=m$, it means that the O-D pairs are both hubs.

3.3 Objective function and constraints

Using the variables above, the hub location problem in intermodal transport networks from a minimum cost perspective is modeled as follows.

$$\text{Minimize} \sum_{i\in N}\sum_{j\in N}\sum_{k\in N}\sum_{m\in N}\sum_{s\in S} w_{ij} X_{ijkm}^{s} C_{ijkm}^{s} + \sum_{k\in N} F_k y_k \qquad (1)$$

$$subject\ to \quad \sum_{k\in N} y_k = p \qquad (2)$$

$$\sum_{k\in N}\sum_{m\in N}\sum_{s\in S} X_{ijkm}^{s} = 1, \quad \forall i,j\in N \qquad (3)$$

$$X_{ijkm}^{s} \le y_k, \quad \forall i,j,k,m\in N,\ s\in S \qquad (4)$$

$$X_{ijkm}^{s} \le y_m, \quad \forall i,j,k,m\in N,\ s\in S \qquad (5)$$

$$y_k, X_{ijkm}^{s} \in \{0,1\} \quad \forall i,j,k,m\in N,\ s\in S \qquad (6)$$

Function (1) is the objective function of the mathematical model which seeks to minimize the total network cost, the first term in right hand is the network transportation cost, and the second term is the fixed costs of hubs. Constraint (2) requires that there are p hubs to be located. Constraints (3) refer to the choice of a specific arc for inter-hub travel, which ensure that each O-D pair (i, j) is assigned to exactly one using one transportation mode between hubs. Constraints (4) and (5) guarantee that each O-D pairs can only be assigned to an existing hub pair. Constraints (6) are binary requirements.

4 A HYBRID HEURISTIC ALGORITHM

The main work to solve the model is as follows. First we select a set of reasonable hubs. Then we determine the connections among non-hubs and hubs and the transport routes and modes from O-D pairs. Since the model above is a combinatorial and optimized problem, it is very difficult to solve using quantitative method such as Branch and bound method (Cánovas, Garcia and Marín [7]), decomposition of Benders (Camargo, Miranda and Luna [8]). And LINGO, CPLEX and other commonly used planning software can only deal with Small scale problem. General practices are using the heuristic algorithms (Martins, Contreras and Cordeau [9]). In this paper, we develop a heuristic approach based on Floyd shortest path algorithm and tabu search (Calık et al. [10]) to solve the model.

4.1 Floyd shortest path algorithm

Let H be the set of hubs. The weight matrix of the network is denoted as $D=(d_{ij})_{n\times n}$. There is

$$d_{ij}=\begin{cases}\min\left\{\hat{c}_{ij}^{1}+\alpha^{1}c_{ij}^{1},\hat{c}_{ij}^{2}+\alpha^{2}c_{ij}^{2}\right\} & \forall i,j\in H, i\neq j\\ c_{ij}^{1} & \forall i\in H, j\notin H\ \vee\ i\notin H, j\in H\\ \infty & \forall i,j\notin H\end{cases} \tag{7}$$

Denote $R=(r_{ij}^{(k)})_{n\times n}$ as the route matrix of the network using vertices only from the set $\{1,2,\cdots k\}$ as intermediate points along the way, where $r_{ij}^{(0)}=j$ for $k=0$.

Then, basic steps for Floyd shortest path algorithm are as follows.

(1) Let $D^{(0)}=D$.

(2) Calculate $D^{(k)}=(d_{ij}^{(k)})_{n\times n}\quad(k=1,2,\cdots n)$, where $d_{ij}^{(k)}=\min\left[d_{ij}^{(k-1)},d_{ik}^{(k-1)}+d_{kj}^{(k-1)}\right]$;

Calculate $R^{(k)}=(r_{ij}^{(k)})_{n\times n}\quad(k=1,2,\cdots n)$, where $r_{ij}^{(k)}=\begin{cases}r_{ij}^{(k-1)}, & \text{if } d_{ij}^{(k-1)}\leq d_{ik}^{(k-1)}+d_{kj}^{(k-1)}\\ r_{ik}^{(k-1)}, & \text{if } d_{ij}^{(k-1)}> d_{ik}^{(k-1)}+d_{kj}^{(k-1)}\end{cases}$.

(3) The elements in $D^{(n)}=(d_{ij}^{(n)})_{n\times n}$ are the unit transportation cost of the shortest path between all pairs of vertices. The allocation relationship between non-hub nodes and hub nodes can be determined by the route matrix $R^{(n)}=(r_{ij}^{(n)})_{n\times n}$. The shortest path from i to j is $i\rightarrow r_{ij}^{(n)}\rightarrow r_{ji}^{(n)}\rightarrow j$.

Using the unit transportation cost matrix of the shortest path in the specific network, we can easily calculate the transportation cost terms in the objective function (1).

4.2 Tabu search

We employ tabu search to determine a set of hub nodes to minimize the total network cost. The search procedure starts with a randomly set of hubs as the initial solution, denoted by H. The neighborhood $N(H)$ is constituted by a swap of a hub node with a non-hub node. The objective function (1) is used to be the fitness functions for search procedure. The move is made according to the fitness and the aspiration criterion. The aspiration criteria used in this research are: (a) a tabu move can be accepted if it generates a best solution so far in the search and (b) the best move can be made if all the moves are tabu. The search restarts from a new initial solution for several times for diversification to get the global optimal solution.

4.3 Steps of the hybrid heuristic

The steps of the algorithm are described below with some notions.

- Step 1: Initialize the tabu list as empty, set the maximum run times of the search, and the maximum iteration numbers in each run. Generate an initial solution as the current solution.
- Step 2: Generate the neighborhood set for current hubs. Solve the shortest path problem of all the hub sets in the neighborhood. Solutions in the neighborhood are sorted according to their fitness, then selected as the candidate solutions.

- Step 3: Choose the non-tabu (or if it is tabu, but it satisfy the aspiration criteria) candidate solution with the minimum fitness value as the best candidate solution.
- Step 4: Update the current solution, tabu list and tabu frequency. The iteration has to be terminated if the maximum iteration numbers without improving the solution.
- Step 5: The search restarts from a new initial solution for diversification
- Step 6: If the run times reach the maximum limit, terminate the search and output the best solutions.

5 ILLUSTRATED EXAMPLE

5.1 Computational data

Suppose that there are 15 city nodes, the horizontal and vertical coordinates are sampled randomly from [0, 1000], as shown in Table 1.

Table 1: Serial number and coordinates of the city nodes

serial number / coordinates	1	2	3	4	5	6	7	8	9	10	11	12	13	14	15
X	50	100	250	300	350	450	550	650	670	730	750	780	800	850	920
Y	900	300	600	200	400	500	50	530	360	200	790	650	460	800	920

The distance matrix is obtained using the Euclidean distance between nodes, that is $d_{ij} = \sqrt{(X_i - X_j)^2 + (Y_i - Y_j)^2}$. Suppose that the road transportation cost is 0.6 ¥/t·km. The railway transportation cost is set as 0.2 ¥/t·km. The traffic demands between O-D pairs are generated from the uniform distribution [200, 600], the unit is t. The fix cost of opening and operating a hub is generated by sampling from uniform distribution [100, 120], the unit is ten thousand ¥. The unit transit and operating cost for highway transportation between nodes are generated by sampling from a uniform distribution according to 5%-10% of the road transportation cost. The unit transit and operating cost for railway transportation is set as two times corresponding to the cost of highway for the same nodes pair. The inter transportation cost discount factor using highway is set as $\alpha^1 = 0.6$, and the factor using railway is set as $\alpha^2 = 0.5$.

5.2 Computational results and analysis

The proposed hybrid heuristic is used to solve the problem. The optimal hub set is [1, 5, 8, 10, 14], and the corresponding objective function value is 2.2581×10^7 ¥. The routing matrix for the optimal hub sets are shown in Table 2. From Table 2, we can conclude not only the distribution of the non-hub nodes to hub nodes, but also routes for each O-D pairs. For example, non-hub node 2 is all allocated to hub node 5; non-hub node 9 is mainly allocated to hub node 8 and some to hub node 10. $3 \to 5 \to 8 \to 9$ is the route from node 3 to node 9,and $1 \to 8 \to 10 \to 7$ is the route from node 1 to node 7. The transport mode of the hubs can be obtained through the routing matrix of the optimal hub sets, which can be backward inferred from the formula (7).

The location of the hubs and the distribution with the non-hubs are shown in Figure 1. Some interesting things can be found in Figure 1, that is, although node 1 is a relatively isolated point in the network, it still is selected as a hub node. The cause of setting a hub in the node 1 can be summed up to three points: (1) There are flows between node 1 to any other nodes; (2) The distance from node 1 to any other nodes is long; (3) If the flows transit from other hubs, the total cost of the network will increase because of the triangle inequality.

Table 2: Routing matrix for the optimal hub sets

nodes	1	2	3	4	5	6	7	8	9	10	11	12	13	14	15
1	1	5	3	5	5	5	8	8	8	8	14	8	8	14	14
2	5	2	5	5	5	5	5	5	5	5	5	5	5	5	5
3	1	5	3	5	5	5	5	5	5	5	5	5	5	5	5
4	5	5	5	4	5	5	5	5	5	5	5	5	5	5	5
5	1	2	3	4	5	6	10	8	8	10	8	8	8	8	8
6	5	5	5	5	5	6	5	8	8	5	8	8	8	8	8
7	10	10	10	10	10	10	7	10	10	10	10	10	10	10	10
8	1	5	5	5	5	6	10	8	9	10	14	12	13	14	14
9	8	8	8	8	8	8	10	8	9	10	8	8	8	8	8
10	8	5	5	5	5	5	7	8	9	10	14	8	13	14	14
11	14	14	14	14	14	14	14	14	14	14	11	14	14	14	14
12	8	8	8	8	8	8	8	8	8	8	14	12	8	14	14
13	8	8	8	8	8	8	10	8	8	10	8	8	13	8	8
14	1	8	8	8	8	8	10	8	8	10	11	12	8	14	15
15	14	14	14	14	14	14	14	14	14	14	14	14	14	14	15

6 CONCLUSION

The uncapacitated multiple allocation integer programming formulations are derived for hub location problem in intermodal logistics networks. The problem is decomposed into two sub problems: the hub location problems and traffic flow assignment problems. A heuristic algorithm based on tabu search is proposed to hub location, and the Floyd shortest path algorithm is employed to solve assignment sub problem efficiently. Through simplifying the specific intermodal network, the computation of the algorithm is greatly reduced. At last, a simulation example is designed. Simulation results show that the proposed heuristic algorithm can solve the model quickly, and the results can provide some guidance for the design of intermodal network. In further research, some assumptions in the model can be changed or be relaxed, such as each node can send and receive traffic through exactly one hub node, incomplete network and the scenarios with capacity constraint.

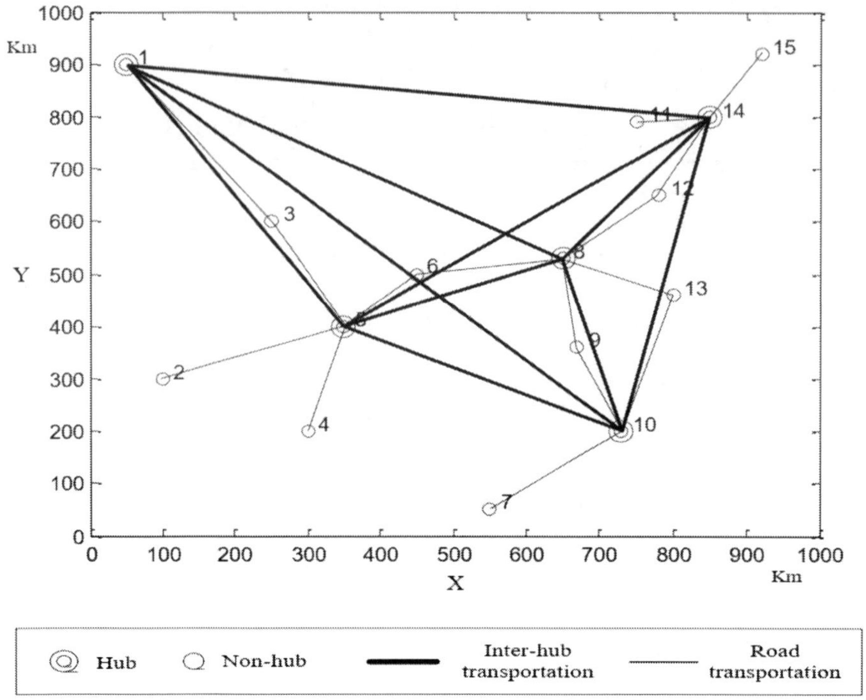

Figure 1: The location of the hub and the distribution with the non-hub

7 REFERENCES

[1] **Macharis, C., Bontekoning, Y. M. 2004.**Opportunities for OR in intermodal freight transport research: A review, *European Journal of operational research*, 153(2), pp 400-416.

[2] **Alumur S A, Kara B Y, Karasan O E. 2009.** The design of single allocation incomplete hub networks, *Transportation Research Part B: Methodological*, 43(10), pp 936-951.

[3] **Yaman H. 2011.** Allocation strategies in hub networks, *European Journal of Operational Research*, 211(3), pp 442-451.

[4] **Ishfaq R, Sox C R. 2011.** Hub location-allocation in intermodal logistic networks, *European Journal of Operational Research*, 210(2), pp 213-230.

[5] **Ishfaq R, Sox C R. 2012.** Design of intermodal logistics networks with hub delays, *European Journal of Operational Research*, 220(3), pp 629-641.

[6] **Alumur S A, Kara B Y, Karasan O E. 2012.** Multimodal hub location and hub network design, *Omega*, 40(6), pp 927-939.

[7] **Cánovas L, Garcia S, Marín A. 2007.** Solving the uncapacitated multiple allocation hub location problem by means of a dual-ascent technique, *European Journal of Operational Research*, 179(3), pp 990-1007.

[8] **de Camargo R S, Miranda G, Luna H P. 2008.** Benders decomposition for the uncapacitated multiple allocation hub location problem, *Computers & Operations Research*, 35(4), pp 1047-1064.

[9] **Martins de Sá E, Contreras I, Cordeau J F. 2015.** Exact and heuristic algorithms for the design of hub networks with multiple lines, *European Journal of Operational Research*, 246(1), pp 186-198.

[10] **Calık H,** Alumur **S A, Kara B Y, et al. 2009.** A tabu-search based heuristic for the hub covering problem over incomplete hub networks, *Computers & Operations Research*, 36(12), pp 3088-3096.

A HIERARCHICAL ROBUST OPTIMIZATION APPROACH FOR MUTIPLE RESPONSE PROBLEMS BASED ON DESIRABILITY FUNCTION AND BAYESIAN METHOD

Xu Tao Zhang and Hai Ling Bi*
Department of Equipment Support
Military Transportation University, Tianjin, China
bhlryf@126.com

ABSTRACT

The robustness issue in multiple response optimization problems is studied. The variation and uncertainty in the process are considered by taking the model parameters as random variables which follow some distributions. The robustness of a solution is accessed by the Bayesian probability that the predicted responses satisfy the quality specifications. A two-stage optimization strategy based on desirability function and Bayesian approach is proposed to solve the model. It can simplify the optimization procedure and make tradeoffs between optimality and robustness. Illustrated example taken from the literature shows that the proposed approach can achieve more reliable solutions which have practical benefits.

Keywords: Design of experiment, quality, Robustness, Multiple response, Parameter design

1 INTRODUCTION

Response surface method (RSM) is widely used in the optimization for product with multiple quality characteristics (Myers and Anderson-Cook [1]). It is a collection of tools of design of experiment (DOE), fitting surface models and numerical optimization. There have been many classical methods discussed in the literatures for dealing with multi-response optimization (MRO) problems, such as desirability function (Derringer, [2]) and quadratic loss function [3]. Traditional optimization methods usually focus on the optimality issue. For example, the solution with the highest satisfaction, or the solution with the minimum loss function. However, input variables in practice are hard to be fixed without fluctuations, and noises in the process cannot be ignored. Furthermore, when we obtain the model fitting from limited experiment data, the uncertainty of parameter estimation should be taken into account. If the variations and uncertainty were ignored, the optimal solution given by the MRO approach would be not robust. It is clear that both robustness and optimality are of most important in industry. The robust optimal solutions which make the responses better meet the quality specifications as well as insensitive to variations have more practical benefits.

The major focus of existing studies on robust parameter design is on the impact of the noise factors (Vining and Meyers [4], Miró-Quesada et al. [5], Goethals and Cho [6], Del Castillo et al. [7], de Paiva et al. [8], etc.). Some methods also consider the model uncertainty (Vining [9], Stinstra and Hertog [10], He et al. [11], etc.) for robust parameter optimization. Considering the fluctuation of the controllable factors in process, He et al. [12] used the changes in responses caused by the

The research is supported by National Natural Science Foundation of China (grant number 71225006 and 71171180)

* Corresponding Author

fluctuations of input variables to measure the robustness of a solution. Peterson [13] gave a posterior predictive approach considered about the model uncertainty. The advantage of this approach is that it takes into account the uncertainty in the parameter estimates. However, it doesn't make balance between robustness and optimality. There is no criterion for tradeoffs between two solutions that have the same probability [14].

In this paper, a more intuitive approach for multi-response robust optimization based on desirability function and Bayesian approach is proposed. The balance between robustness and optimization of the solutions is achieved by a two-stage optimization strategy. An example taken from the literature shows the advantages of the proposed approach.

2 DESIRABILITY FUNCTION APPROACH FOR MULTI-RESPONSE OPTIMIZATION

Suppose $y_1, y_2 \cdots y_m$ are quality characteristics of the process or product. The regression model for the ith response is given by

$$\mathbf{y}_i = \mathbf{X}_i \boldsymbol{\beta}_i + \boldsymbol{\varepsilon}_{i,} \quad i = 1, 2, \ldots, r \tag{1}$$

where $\mathbf{y}_i$ is the $n \times 1$ ector of observations on the ith response, $\mathbf{X}_i$ is the $n \times p_i$ data matrix, $\boldsymbol{\beta}_i$ is the $p_i \times 1$ coefficient vector, $\boldsymbol{\varepsilon}_i$ is a random error vector, $\varepsilon_i \sim N(0, \sigma_i^2)$.

For a response with a target value level (nominal-is-best), the individual desirability function is given by

$$d_i(\hat{y}_i) = \begin{cases} (\dfrac{\hat{y}_i - L_i}{T_i - L_i})^{s_i} & L_i \le \hat{y}_i \le T_i \\ (\dfrac{U_i - \hat{y}_i}{U_i - T_i})^{t_i} & T_i < \hat{y}_i \le U_i \\ 0 & \hat{y}_i < L_i \text{ or } \hat{y}_i > U_i \end{cases} \tag{2}$$

where L_i is the lower bound of the response, U_i is the upper bound, T_i is the target value, s_i and t_i are positive value for weight, which determine the shape of the desirability function. The closer the value of the response is near the target value T_i, the larger d_i *is*.

Similarly, for the cases of larger-the-better and smaller-the-better, the individual desirability functions are defined as Eq. (3) and Eq. (4), respectively.

$$d_i(\hat{y}_i) = \begin{cases} 0 & \hat{y}_i < L_i \\ (\dfrac{\hat{y}_i - L_i}{T_i - L_i})^{s_i} & L_i \le \hat{y}_i < T_i \\ 1 & \hat{y}_i \ge T_i \end{cases} \tag{3}$$

$$d_i(\hat{y}_i) = \begin{cases} 1 & \hat{y}_i < T_i \\ (\dfrac{U_i - \hat{y}_i}{U_i - T_i})^{t_i} & T_i \le \hat{y}_i < U_i \\ 0 & \hat{y}_i \ge U_i \end{cases} \tag{4}$$

Considered the different importance of each response, the overall desirability function can be calculated using the weighted geometric mean

$$D = (d_1^{w_1} d_2^{w_2} \cdots d_m^{w_m})^{1/\sum w_i} \quad (5)$$

where w_i is the weight of the i_{th} response, $0<w_i<1$, and $\sum w_i = 1$.

3 ROBUSTNESS EVALUATION OF FEASIBLE SOLUTION

The basic purpose of DOE is to estimate the unknown parameters $\boldsymbol{\beta}$ and σ^2 in Eq. (1) through a limited number of trials. Traditional methods generally treated $\boldsymbol{\beta}$ and σ^2 as some fixed values. However, considering the variance of the process and the uncertainty of parameter estimation in model fitting, it is more sensible to take the model parameters as random variables which follow some distributions [15]. Based on a diffuse prior distribution, the predictive density of y at the regressor $\mathbf{x}$ is derived by Bayes' theorem (Press, [16])

$$p(y|\mathbf{x},\mathbf{y}) = \int_{\sigma^2}\int_{\boldsymbol{\beta}} L(y|\mathbf{x},\mathbf{y},\boldsymbol{\beta},\sigma^2)h(\boldsymbol{\beta},\sigma^2|\mathbf{y})\mathrm{d}\boldsymbol{\beta}\mathrm{d}\sigma^2 \quad (6)$$

where $L(y|\mathbf{x},\mathbf{y},\boldsymbol{\beta},\sigma^2)$ is the likelihood function.

Given the experiment data **y**, if the error term ε in Eq. (1) is assumed normal distributed, the predictive distribution of y at the given regressor $\mathbf{x}$ follows the equation bellow (Press, [19])

$$\frac{(y|\mathbf{x},\mathbf{y}) - \mathbf{x}\hat{\boldsymbol{\beta}}}{\hat{\sigma}\sqrt{1+\mathbf{x}'(\mathbf{X}'\mathbf{X})^{-1}\mathbf{x}}} \sim t_{n-p} \quad (7)$$

where t_{n-p} is a *t*-distribution with $\upsilon = n-p$ degree of freedoms, *n* is the number of the observations of the response and *p* is the number of estimators. $\hat{\boldsymbol{\beta}}$ is the least squares estimator, and

$$\hat{\sigma}^2 = \frac{(\mathbf{y}-\mathbf{X}\boldsymbol{\beta})'(\mathbf{y}-\mathbf{X}\boldsymbol{\beta})}{n-p} \quad (8)$$

That is, the predictive distribution of y given $\mathbf{x}$ is associated with a *t*-distribution.

Then the posterior mean of the response at $\mathbf{x}$ is

$$E(y|\mathbf{x},\mathbf{y}) = \mathbf{x}'\hat{\boldsymbol{\beta}} \quad (9)$$

and the posterior variance is

$$Var(y|\mathbf{x},\mathbf{y}) = \frac{n-p}{n-p-2}\hat{\sigma}^2(1+\mathbf{x}'(\mathbf{X}'\mathbf{X})^{-1}\mathbf{x}) \quad (10)$$

Eq. (10) reflects the accuracy of the predicted response. The larger the variance is, the more dispersed the distribution of the response is.

Suppose d_i^* be the target desirability of the response y_i. For a given **x** value, the predicted desirability of the response is d_i. From Eqs. (2)-(4), we can deduce that $d_i \ge d_i^*$ corresponds to the value interval $y \in [l,u]$. Thus, the probability of $d_i \ge d_i^*$ equals to that of $y \in [l,u]$, which is the cumulative integration of the predictive density of **y**.

One numerical way of performing the integration is Monte Carlo simulation from the *t*-distribution in Eq. (7). Generate *N* values of y_i by sampling from the distribution in Eq. (7). If there are *n* times that y_i falls into the interval $[l,u]$, according to the weak law of large numbers, there is

$$p(d_i \geq d_i^* | \mathbf{x}, \mathbf{y}) = p(y_i \in [l,u] | \mathbf{x}, \mathbf{y}) \approx \frac{n}{N} \quad (11)$$

The posterior probability $p(d_i \geq d_i^* | \mathbf{x}, \mathbf{y})$ gives a measure of the robustness of the response given the solution **x**.

Suppose D^* is the target overall desirability for the multi-response $\mathbf{y}$. Similarly, we can obtain the posterior probability of $D \geq D^*$ given **x**. Sampling from the posterior distributions of $y_1, y_2 \cdots y_m$ respectively, if $D \geq D^*$, set $I(D \geq D^*) = 1$, otherwise $I(D \geq D^*) = 0$. Suppose that the simulations are done *N* times, there is

$$p(x) \approx \frac{1}{N} \sum_{s=1}^{N} I_s(D \geq D^*) \quad (12)$$

Then the objective function for the robust optimization for multi-response is given by

$$\max p(\mathbf{x}) = P(D \geq D^* | \mathbf{x}, \mathbf{y}) \quad (\mathbf{x} \in \mathbf{R}) \quad (13)$$

where **R** is the feasible operating region.

4 TWO-STAGE OPTIMIZATION STRATEGY

Usually, the experimenter is interested in finding the robust optimal solutions. They hope that the responses not only achieve the target well but also are not sensitive to the variation of the process and the uncertainty of parameter estimation. In this case, it is feasible to employ a two-stage optimization strategy. In the first stage, considering the optimality of the solutions, one can find the optimal solutions and local optimal solutions by desirability function method. In the second stage, the robustness of the solution is taken into consideration. The robust optimization process can be carried out by maximizing the probability given by Eq. (13). Details are as follows.

Stage 1:

1) Analyze the experiment data, and fit the regression model for each response.

2) Give the weights of the responses according to the experimenter's preference information and engineering experience, and substitute the values and the regression model into Eq. (2) - Eq. (5), then we can obtain the overall desirability function associated with the input factors.

3) Optimize the overall desirability function in Eq. (5) to find the overall optimal solution and local optimal solutions, denoted by D_{opt} and D_i, respectively.

Stage 2:

4) Estimate the variance of the random error term for each response by Eq. (8), then the posterior distribution of each response at a given regressor **x** can be given by Eq. (7).

5) Set the objective overall desirability $D^* = \alpha D_{opt}$, where 0< α<1. Sampling from the posterior distributions of each response respectively, the posterior probability of $D \geq D^*$ given **x** can be calculated by Eq. (12).

6) Optimize the posterior probability function given by Eq. (13) in the feasible regions. The robust optimization process can be carried out by the pattern search algorithm with taking the optimal solution and the local optimal solutions as the initial search points.

5 COMPUTATIONAL EXAMPLE

The experiment data are given by Montgomery [17] in a chemical process. There are three responses and two design factors. The chemical engineer is interested in determining the operating conditions of reaction time (x_1) and reaction temperature (x_2), and simultaneously optimizing the yield (y_1), viscosity (y_2), and molecular weight (y_3) of a chemical product. Response type of y_1 is larger-the-better, with the specification limit (78.5, 80, -). Response type of y_2 is nominal-is-best, with the specification limit (62, 65, 68). Response type of y_3 is smaller-the-better, with the specification limit (-, 3100, 3450).The experiment design and observed data are shown in Table 1.

Table 1 Designed Experiment and Observed Data

standard order	Natural variables		Coded variables		responses		
	ξ_1	ξ_2 (°F)	x_1	x_2	y_1	y_2	y_3
1	80	170	-1	-1	76.5	62	2940
2	90	170	1	-1	78.0	66	3680
3	80	180	-1	1	77.0	60	3470
4	90	180	1	1	79.5	59	3890
5	77.928	175	-1.41421	0	75.6	71	3020
6	92.071	175	1.414214	0	78.4	68	3360
7	85	167.929	0	-1.41421	77.0	57	3150
8	85	182.071	0	1.414214	78.5	58	3630
9	85	175	0	0	79.9	72	3480
10	85	175	0	0	80.3	69	3200
11	85	175	0	0	80.0	68	3410
12	85	175	0	0	79.7	70	3290
13	85	175	0	0	79.8	71	3500

Suppose that each desirability function of the three responses is linear-shaped, corresponding that all the s_i and t_i equal to 1. The weight parameters are assumed to be w_1=0.7, w_2=0.2, w_3=0.1. The overlapping contour plots of the responses are shown in Figure 1, and the white areas represent the feasible regions for optimization. As shown in Figure 1, there are two feasible regions, which are denoted by Region A and Region B, respectively.

Select several random points from the two feasible regions as initial search points, and use the simplex search algorithm for optimizing the overall desirability function in Stage one. Finally we obtain a global optimal solution in Region A and a locally optimal solution in Region B, denoted by Solution 1 and Solution 2, respectively.

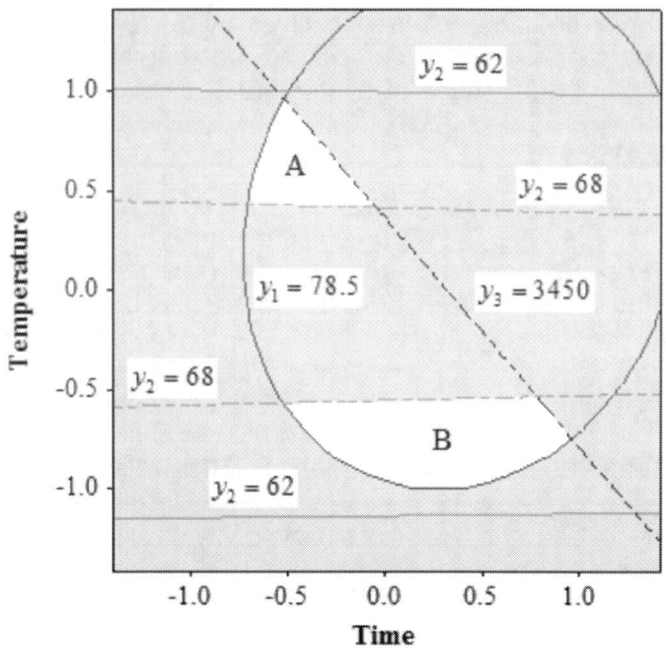

Figure 1 Overlaid Contour Plots of the Reponses (in Coded Variables)

Suppose that $D^* = \frac{1}{2} D_{opt} = \frac{1}{2} \times 0.460 = 0.230$ is the target overall desirability. Taking Solution 1 in Region A and Solution 2 in Region B as the initial search points, a pattern search algorithm is used for the robust optimization process in Stage two. Through searching the solutions that maximize the posterior probability $P(D \geq D^* | x)$ in Eq. (14), two robust solutions are obtained, denoted by Solution 3 and Solution 4, respectively. The four solutions before and after are lumped together in Table 2 for an intuitive quantitative analysis.

Table 2 Select Robust and Optimal Solution by Tradeoff

Stages	Solutions	Feasible Regions	x_1	x_2	$\hat{y}_1$	$\hat{y}_2$	$\hat{y}_3$	$D(\hat{y})$	$P(D \geq D^* \| x)$
I	1	A	-0.243	0.557	79.56	66.98	3435.15	0.460	31.50%
	2	B	0.243	-0.671	79.26	67.15	3316.89	0.440	43.03%
II	3	A	-0.4613	0.6195	79.05	66.47	3401.40	0.362	36.78%
	4	B	0.0926	-0.7940	78.96	66.10	3264.30	0.382	51.54%

From Table 2 we get some useful information for tradeoffs between robustness and optimality of the solutions. Solution 1 found in Stage one has the highest desirability but the lowest probability. Solution 4 found in Stage two has the highest probability as well as a sensible desirability. Therefore, considering the balance between robustness and optimality of the solutions, Solution 4 can be considered to be the robust optimal solution for this multi-response problem. Note that Solution 1 and Solution 3 are all in region A which has a smaller range of feasible region. In addition, the predicted values of response y_3 at Solution 1 and Solution 3 are all close to its upper bound, which may easily exceed the specification limit when the varieties and predicted errors exist. Compared with Solution 1 and Solution 3, Solution 2 and Solution 4 are in region B which has a larger range of feasible region, and the predicted values of each response are all close to their target values. Even if the process fluctuations or predicted errors exist, we can still get acceptable responses at a high probability, indicating that the solutions have a better robustness.

6 CONCLUSIONS AND DISCUSSIONS

Traditional optimization methods for multi-response have not taken into account the variation of the process and the uncertainty of parameter estimation in model fitting. The so-called optimal conditions given by the analysis of experiment do not always perform well in practice. In this paper, the variation and the uncertainty are considered through taking the model parameters as random variables which follow some distributions. The robustness of the solution is evaluated by a Bayesian probability that the responses meet the quality specifications. A two-stage optimization strategy is proposed to make balance between the robustness and optimality. The proposed method is based on desirability function, which is well known to quality engineers. The illustrated example shows that the proposed method can achieve more reliable solutions which have practical benefits. However, note that the posterior probability of the best solution is 51.54%, which is not sufficiently large. There may be two reasons for the poor reliability in this example. On one hand, the number of the experiment runs is too small compared to the number of the regression parameters to be estimated. The degree of freedoms of the t-distribution in Eq. (7) is too few, which makes of the shape of the distribution too decentralized. On the other hand, the variety of the process may be too large, which can cause large model uncertainty and posterior variance in Eq. (10). Therefore, the increase of reliability can be obtained by reducing the variation of the process or increasing the sample size, e.g. doing replication experiments.

7 REFERENCES

[1] **Myers, R. H., and Anderson-Cook, C. M.** 2009. *Response Surface Methodology: Process and Product Optimization Using Designed Experiments*. Wiley, New York.

[2] **Derringer G.** 1994. A balancing act: optimization a product's properties, *Quality Progress*, 27(6), pp 51-58.

[3] **Ko Y H, Kim K J, Jun C H.** 2005. A new loss function-based method for multiresponse optimization, *Journal of Quality Technology*, 37(1), pp 50-59.

[4] **Vining, G. G., and Myers, R. H.** 1990. Combining Taguchi and response surface philosophies: a dual response approach, *Journal of Quality Technology*, 22(1), pp 38-45.

[5] **Miró-Quesada, G., Del Castillo, E., and Peterson, J. J.** 2004. A Bayesian approach for multiple response surface optimization in the presence of noise variables, *Journal of Applied Statistics*, 31(3), pp 251-270.

[6] **Goethals, P. L., and Cho, B. R.** 2012. Extending the desirability function to account for variability measures in univariate and multivariate response experiments, *Computers & Industrial Engineering*, 62(2), pp 457-468.

[7] **Del Castillo, E., Colosimo, B. M. and Alshraideh, H.** 2012. Bayesian modeling and optimization of functional responses affected by noise factors, *Journal of Quality Technology*, 44(2), pp 117-135.

[8] **de Paiva, A. P., Gomes, J. H. F., et al.** 2014. A multivariate robust parameter optimization approach Based on principal component analysis with combined arrays, *Computers &*

Industrial Engineering, 74, pp 186-198.

[9] **Vining G G.** 1998. A compromise approach to multiresponse optimization, *Journal of Quality Technology*, 30 (4), pp 309-313.

[10] **Stinstra, E, and Den Hertog, D.** 2008. Robust optimization using computer experiments, *European Journal of Operational Research*, 191(3), pp 816-837.

[11] **He Z, Wang J, et al.** 2010. Robust optimization for multi-response using response surface methodology, *Applied Stochastic Models in Business and Industry*, 26(2), pp157-171.

[12] **He Z, Zhu P F, Park S H.** 2012. A robust desirability function method for multi-response surface optimization considering model uncertainty, *European Journal of Operational Research*, 221(1), pp241-247.

[13] **Peterson J J.** 2004. A posterior predictive approach to multiple response surface optimization, *Journal of Quality Technology*, 36(2), pp139-153.

[14] **Peterson J J, Miro-Quesada G, Del Castillo E.** 2009. A Bayesian reliability approach to multiple response optimization with seemingly unrelated regression models, *Quality Technology and Quantitative Management*, 6(4), pp353-369.

[15] **Rajagopal R, Del Castillo E.** 2007. A Bayesian approach for multiple criteria decision making with applications in Design for Six Sigma, *Journal of the Operational Research Society*, 58(6), pp779-790.

[16] **Press S J.** 2003. *Subjective and objective Bayesian statistics: principles, models, and applications*, 2rd Edition, New York- John Wiley.

[17] **Montgomery D C.** 2005. *Design and Analysis of Experiments*, 6rd Edition, New York- John Wiley.

RELATIONSHIPS AMONG MANUFACTURING QUALITY, TECHNOLOGY INNOVATION, AND POLICY: A SYSTEM DYNAMICS MODEL

Jian Wang[1], Peipei Wang[1], Chen Lu[2], and Zhixin Chen[1]
[1]School of Management
Shanghai University, Shanghai, China
jwang@t.shu.edu.cn, 503533365@qq.com, 2472988869@qq.com

[2]China Mobile
Zhejiang, China
luchen_flyer@sina.com

ABSTRACT

In order to analyse the status of Chinese manufacturing quality, an early research established an index system for manufacturing industry which includes five criteria and the corresponding sub-criteria. In order to clarify the development of manufacturing quality, the relationships among the key criteria and related factors were analysed based on System Dynamics Models. Referring to the index system, three factors were designed as the internal motivation for the development of manufacturing quality, i.e., technological, political, and demographic factors. Data were collected from Chinese manufacturing for the system dynamic models. The results suggest that a higher investment on R&D brings a higher improvement in manufacturing quality, a lower technical conversion rate brings a less due return of investment, and the demographic factor has little effect on the improvement.

Keywords: Manufacturing, Quality, System dynamics method

1 INTRODUCTION

In order to measure the macro-quality of manufacturing industry from the national level, a macro-quality index was constructed, and applied to predict the trend of the macro-quality in five countries, which included China, Japan, South Korea, USA and Germany [1]. This article is a follow-up study which analysed internal factors that have the influence on China's manufacturing industry with System Dynamics method.

The paper is organized as follows. Section 2 presents a brief literature review. Section 3 illuminates the basic research methodology and simulation process. Section 4 presents our conclusions.

2 LITERATURE REVIEW

In this section, a brief description of the manufacturing industry development and macro-quality index are presented.

2.1 Manufacturing industry development in different countries

The manufacturing industry has been valued by different countries. The German government announced Industry 4.0 to establish smart factories that are the ultimate realization of Smart Manufacturing based on CPS (Cyber-Physical System), IoT (Internet of Things) and IoS (Internet of Services) [2]. In U.S., General Electric's Industrial Internet is a main driving force for the digitalization of manufacturing industry [3] and additive manufacturing, colloquially known as 3D printing, is currently promoted as the spark of a new industrial revolution [4]. Japan proclaimed that it will lead the world with the robots and realize a robot revolution in Japan's Robot Strategy in the manufacturing industry [5]. The State Council released 'Made in China 2025' plan on May 19, 2015, laying out a 10-year blueprint to increase China's manufacturing competitiveness and combat the slowdown in economic growth. In recent years, several indexes have been developed to assess the competitiveness countries' manufacturing sector. Three of the most notable ones are as follows [6]:

- Competitive Industrial Performance Index (CIP) (A UNIDO index). It is composed of 8 indicators assessing industrial performance based on an economy's ability to competitively produce and export manufactured goods. The index takes both domestic and international factors into account, but lacks sustainable ability evaluation.
- Global Manufacturing Competitiveness Index (GMCI). Based on a global survey of 550 CEOs, GMCI ranks the competitiveness of 40 individual components agglomerated in 10 categories. It is a subjective index based on managerial perception, and used to justify current practices. To some degree it lacks sustainable ability evaluation.
- Global Competitiveness Index (A World Economic Forum index). It is a weighted average of components grouped into "12 pillars of competitiveness" and represents an extensive mix of objective and subjective criteria. It underlines that competitiveness is a proxy to economic, social, and technological development.

2.2 Improvement of manufacturing quality

In addition, since 1999, the Chinese government has developed 'Quality Competence Index' (QCI) to evaluate China's manufacturing quality. It applies 12 indexes from different industries and areas, and calculates QCI value in two perspectives. One is the quality level that mainly reflects the current situation of quality. The other is the development capacity to describe the suitability of country or firms. It is a reflection of China's manufacturing quality, but with a lack of comparison between China and other countries.

The following five items have been used for measuring the manufacturing quality [1]:

1) Gross manufacturing output value which can be measured by Value-added of Manufacturing industries.

2) Export quality level which can be measured by Export Unit Value (EUV). This measurement scale has been suggested by several studies [7] [8].

3) Production efficiency which can be measured by Overall Labour Productivity (OLP).

4) International competitiveness which can be measured by Trade Competitive Index (TCI).

5) Sustainable development capacity which can be measured by Return on Sales (ROS).

Among these measurement scales, how to improve the product quality level is a key question. Chinese manufacturing industry faces new challenges from many aspects such as the increasing-cost of labor force, the decrease of foreign investment, structural changes and safety issues. Several proposals can be found on this topic, i.e., whether government investment poses positive effect on technology which can help improve product quality, whether technological conversion rate has an enormous effect on manufacturing's macro-quality and how the quality practice would influence the quality level of China. A system dynamics model is built to show various internal motivations of Chinese manufacturing industry under different scenarios.

3 THE SYSTEM DYNAMICS PROCESS

3.1 Method

To the best of our knowledge, we analysed internal factors that have an influence on China's manufacturing industry with System Dynamics method. The following are the steps of the method.

Step 1: Identification of goals.

In this paper, the system dynamics method is used to analyse internal factors which influence China's manufacturing industry's macro-quality since it can explain the underlying reason why China can surpass the other four advanced countries in manufacturing macro-quality in the near future.

Step 2: Identification of variables and constructing the flow loop and influence diagram.

Taking our national conditions into account, the overall model considers four main factors: political, technical, market and demographic factors. The next step in the modelling process is to construct the causal loop diagram as figure 1 shows.

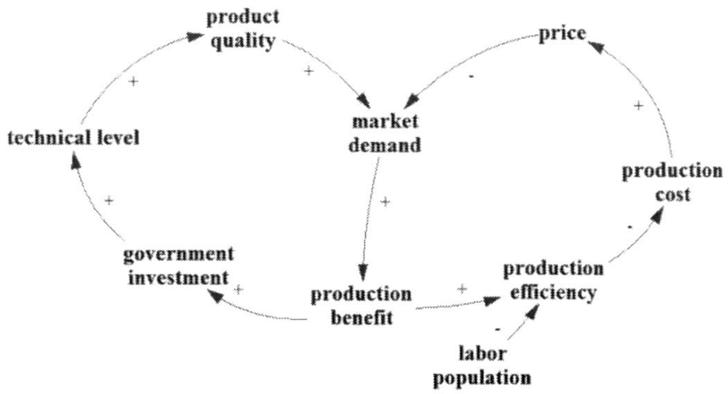

Figure 1: Technology and politics causal loop diagram

There are two distinct loops in figure 1. The first loop is the technical feedback loop. It shows that the government investment has a positive effect on technology which could help improve product quality. When product quality increases, the market demand would also increase. Finally, the production benefit improvement would be raised and then promote government investment. The second one shows the relationship between production benefit and demographic factor. Labor population has a counteractive effect on the production efficiency. It requires higher production efficiency. Higher production efficiency could reduce production cost, so does the product price. When price decreases, market demand would be higher and competitiveness would be enhanced.

Based on the influence diagram, auxiliary variables are added to simplify and clarify the relationships between elements. Let export quality value, gross manufacturing output value and R&D investment ratio be *level* variables. The factors included in UV value, value-added of manufacturing industries, R&D investment growth ratio are identified as *rate* when others are *auxiliary* variables. (As shown in figure 2)

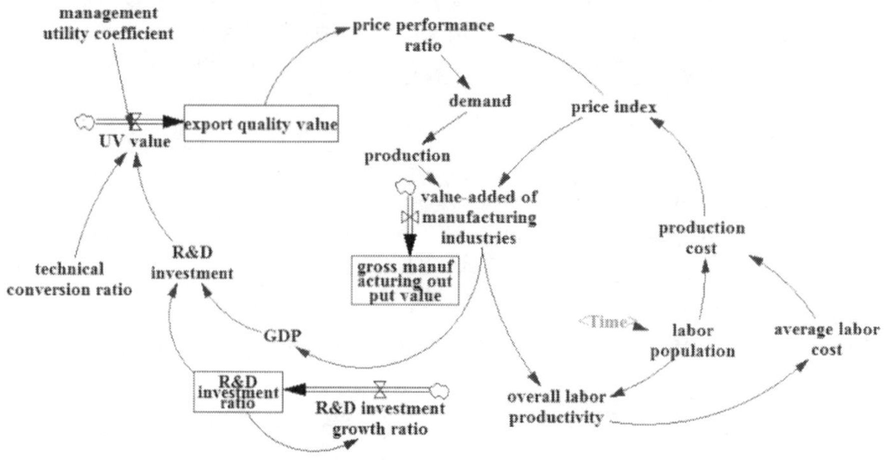

Figure 2: Flow diagram

Step 3: Data sources and accuracy test

To construct the main variables in our paper, we use data drawn from official databases for the period 2004-2013. Table 2 shows where data we used in this paper come from with system dynamics. The simulation step is 1 year and especially, production cost was replaced by main business cost of industrial enterprises, where export quality level was with UV value of main export products.

Table 2: Data sources

Variable	Source	Notes
Labouring population	International Labor Organization	/
GDP	World Bank	Current dollars
Manufacturing added	World Bank	Current dollars
R&D ratio	World Bank	/
Average labour cost	Reference	/
Price index	National Bureau of Statistics	/
Production cost	National Bureau of Statistics	Main business cost of industrial enterprises
Export quality value	National Bureau of Statistics	Main export products UV-weighted

For space limitation, the complete list of equation and relations can be access by contacting the authors.

The error between simulated and the real value of four main parameters is examined. Since all the errors are within 10%, the system dynamics was considered to be practically significant.

3.2 Discussion

Sensitivity analysis is the study of how the uncertainty of the output of a system can be apportioned to different sources of uncertainty in its inputs. The results of sensitivity analysis may allow the model builder to determine which of the model parameters are more important for the simulation output [9]. This paper mainly focuses on sensitivity analysis of parameters, including political, technical and demographic factors.

3.2.1 Political effect

The ratio of R&D investment to GDP is applied to measure government effort from a political perspective. In 2013, the ratio was 1.98%, remaining a relatively low level compared with 3% of the advanced countries. At first, we set a 0.084% growth based on the average growth speed from 2004 to 2013. If the growth changes to 0.07% and 0.1%, the curves are obtained respectively.

As can be seen from the figure, change for R&D investment makes a small difference. As time goes by, gaps get bigger. When R&D ratio is 0.1%, MQMI value makes a 19.6% improvement compared with 0.084%, 31.2% improvement than 0.07%. The sensitivity of R&D growth ratio to the model parameters indicates the parameter that is important for MQMI level. Such information is useful while developing policies for manufacturing.

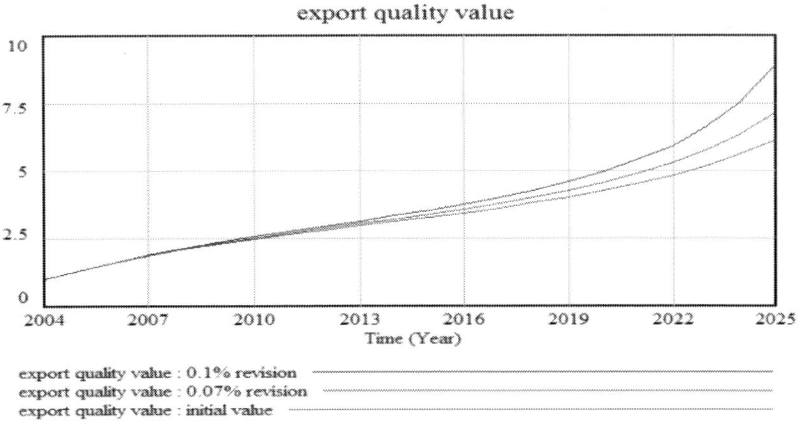

Figure 4: Sensitivity graph for R&D growth ratio value

3.2.2 Technological effect

A report from Ministry of Science and Technology revealed that technology accounted 39% of economic development, among which high-tech makes up 20 per cent. It contributed well below developed countries. Technical conversion ratio was employed to measure technical effects. Based on existing data, we consider a 20% ratio. When the ratio changes to 18% and 22%, the simulations results are as follows.

When conversion rate is raised up to 22%, MQMI value gets an 88.8% growth relative to 20%, 161% growth relative to 18%. So it's apparent that technological conversion rate has an significant effect on MQMI level. The country should lay great emphasis to improve the conversion ratio.

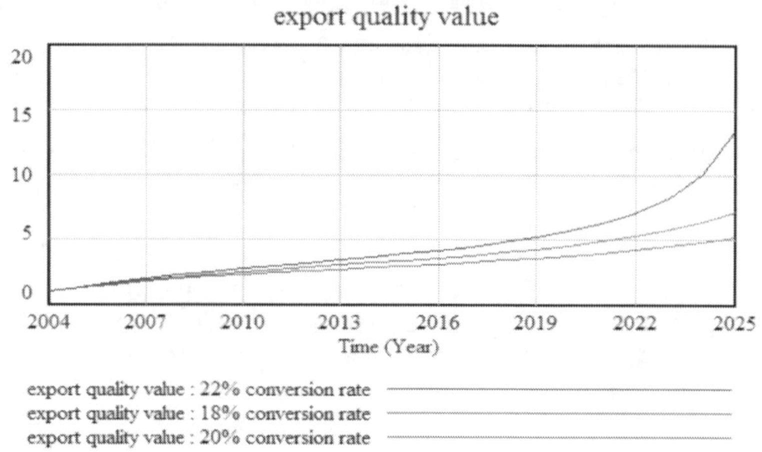

Figure 5: Sensitivity graph for technological ratio value

3.2.3 Demographic effect

Demographic dividend has contributed a lot to Chinese economic growth during the past 30 years [10]. As the population ages, the absolute quantity of the working age population will begin to shrink after 2015 [11]. In contrast, the result of the fifth census (2011) showed it would last until 2030. So not only manufacturing industry but also the whole Chinese economy will face a distress without cheap labour.

In order to examine how the workforce influences macro quality level, two cases are considered when workforce maintains a present level or shows slow decrease. Two functions are listed respectively:

*IF THEN ELSE (Time>2013, RANDOM UNIFORM (1.193, 1.253, 2), 0.093*LN (Time-2003) +0.994),* ***(1)***

*IF THEN ELSE (Time>2013, 1.223+RAMP (-0.02, 2013, 2025), 0.093*LN (Time-2003) +0.994).* ***(2)***

The following graph shows how labour forces change and how it influences macro quality level. The three curves basically coincide and therefore, the demographic factor is considered insensitive.

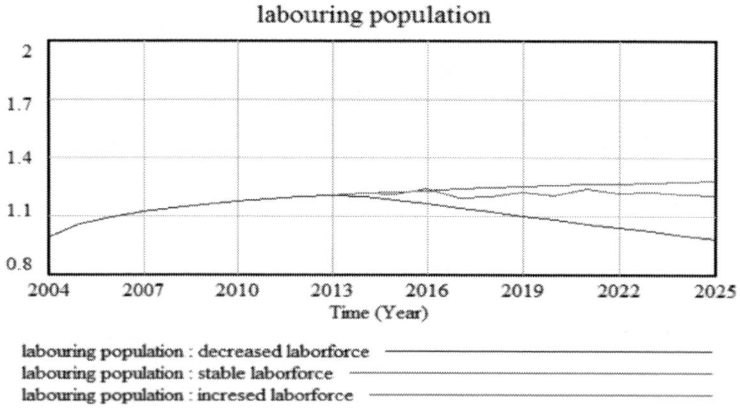

Figure 6: Labor forces

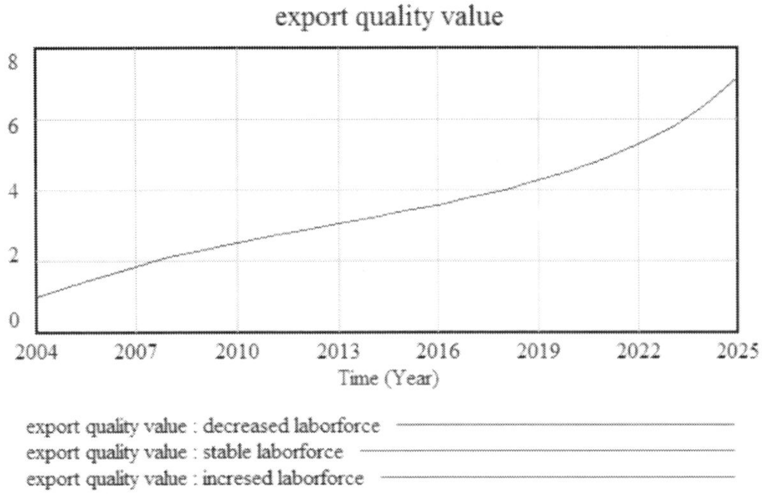

Figure 7: Sensitivity graph for labor forces

4 CONCLUSION

The SD model has been built to study the three internal motivations of Chinese manufacturing industry. According to the results, we can come to the following conclusions:

(1) The results suggest that a higher investment on R&D brings a higher improvement in manufacturing quality, a lower conversion rate brings a less due return of investment, and the demographic factor has little effect on the improvement.

Based on the above conclusions, we present the following suggestions:

(1) China is stepping into a fast developing stage while the big advanced countries in mature stage. To maintain high speed development more efforts from government are required. The government should rise up R&D investment to promote the manufacturing macro quality. Like the developing process of Japan, keeping better quality-price relation is a good way to win favour of consumers and further open up overseas market and then to strength the brand via improving design and quality.

(2) The present level of technical conversion rate is low, which indicates the low level of technology into productivity. Not only the government but the research institute should put emphasis on ways to improve conversion rate, skilled employed are encouraged to apply technologies effectively for examples.

(3) During the research, labor productivity of manufacturing industry has made little progress for years. Though it's not obvious that labour quantity would make quite a difference on the quality level, it's also a long-term task to enhance the quality of workers, improve production technology and upgrade manufacturing equipment to improve manufacturing efficiency so that we can transform China from "a factory for the world" into a world manufacturing power.

5 ACKNOWLEDGMENT

Thanks for the support of NSFC (71572104).

6 REFERENCES

[1] **J. Wang et al. 2015.** Development trend forecasting on Chinese manufacturing industry with focus on manufacturing quality: An empirical study, *Science and Technology Progress and Policy*, 32(18), pp 43-50.

[2] **H. S. Kang et al. 2016.** Smart manufacturing: past research, present findings, and future directions, *International Journal of Precision Engineering and Manufacturing-Green Technology*, 3(1), pp 111-128.

[3] **N. Mikel. 2015.** Business Understanding, Challenges and Issues of Big Data Analytics for the Servitization of a Capital Equipment Manufacturer, Proceedings of 2015 IEEE International Conference on Big Data, Santa Clare, CA, Oct 9-Nov 01, pp 1368-1377.

[4] **Christian Weller, Robin Kleer, Frank T. Piller. 2015.** Economic implications of 3D printing: Market structure models in light of additive manufacturing revisited, *International Journal of Production Economics*, 164, pp 43-56.

[5] **Ministry of Economy, Trade and industry, Ministry of Health, Labour and Welfare, Ministry of Education, Culture, Sports, Science and Technology. 2015.** *Summary of the White Paper on Manufacturing Industries*, pp 31.

[6] **United Nations Industrial Development Organization. 2015.** *Competitive Industrial Performance Report 2014*, pp 12-14.

[7] **J. C. Hallak. 2003.** The Effect of Cross-Country Differences in Product Quality on the Direction of International Trade, *RSIE Working Paper*, pp 493.

[8] **P. K. Schott. 2004.** Across-Product versus Within-Product Specialization in International Trade, *Quarterly Journal of Economics*, 119, pp 647-678.

[9] **D. J. Power. 2014.** Competence and Capability in Quality in the High-Tech Sector: an international comparison, *International Journal of Operations and Production Management*, 34(9), pp 1184-1209.

[10] **S. B. Song, Q. Song, T. Zhang. 2011.** Demographic Dividend and Economic Growth in China Mainland, Proceedings of International Conference on Applied Social Science, Changsha, China, Mar 19-20, pp 305-310.

[11] **F. Cai, M. Wang. 2006.** Challenge facing China's economic growth in its aging but not affluent era, *China & World Economy*, 14(5), pp 20-31.

REDUNDANCY ALLOCATION PROBLEMS OF MULTI-STATE SYSTEMS: A REVIEW

H.W. Liu[1*] and Y. Wang[2]
[1]College of Management and Economics
Tianjin University, Tianjin, China
hw_liu999@tju.edu.cn

[2]College of Management and Economics
Tianjin University, Tianjin, China
15022657161@163.com

ABSTRACT

With the development of industrialization, industrial equipments become more and more complex, and the user demand becomes more and more important. Traditional binary-state system cannot accurately describe the equipment status. A multi-state system (MSS) oriented to customers' demands can describe the operation status of equipments more accurately. The reliability of MSSs has been repeatedly researched and related reliability optimization is mainly designed based on redundancy strategy. However, no relevant review has been carried out in this field. To address those issues, a review of current research on MSSs was carried out in this paper. The review result shows that redundancy allocation problems of MSSs can be categorised in three classes according to different complexity: active redundancy allocation, standby redundancy allocation and redundancy allocation considering maintenance. Then active redundancy allocation on MSSs—from single objective optimization to multi-objective optimization—have been briefly reviewed.In this paper, the existing literature on redundancy allocation problems of MSSs can be summarized and the future research direction is proposed.

Keywords: multi-state system, redundancy allocation problem, active redundancy allocation, standby redundancy allocation

1 INTRODUCTION

A multi-state system(MSS) is regarded as a system with a range of discrete performance levels (from perfectly functioning to complete failure) [1, 2]. For example, the overall performance of manufacturing, production and power generation can settle on different levels (e.g. 100%, 80%, 50% of the nominal capacity), depending on the operative conditions of elements [3]. Different from the traditional binary-state system (only consider perfectly functioning and complete failure), the reliability of an MSS is expressed by the state distribution. And the MSS needs to be repaired once its performance level, determined by the multiple elements and the configuration falls below the user demand level, instead, it's considered as in the normal operating state. In general, systems that conform to the following characteristics can be referred to as an MSS [1-3]:

1. The system is composed of a number of discrete units, and every unit has a cumulative effect on the overall performance of the system.
2. The current states of system components are changing with the deterioration and maintenance.

* H.W. Liu

Multi-state system reliability optimization design has been studied widely in current research. Two methods can be used to achieve the optimal design of MSSs: redundancy allocation optimization and the change of components' state distribution. Redundancy allocation problem has been regarded as the main method to improve and ensure the reliability of MSSs. It can be defined as optimal combinations of functional components and redundant components under certain constraints [4]. The versions and the corresponding redundancy are selected in every subsystem in order to achieve system cost minimum (or maximum reliability). A great number of research contributions have been devoted to the redundancy allocation problems of MSSs. In recent years, the development of this field has been gradually mature. However, no relevant review has been carried out.

Redundancy allocation problem usually refers to the active redundancy problem. Yet standby redundancy considering different active states of components is more in line with the actual situation [5, 6]. In addition, some redundancy allocation problems are considered together with maintenance optimization activities. Therefore, depending on the different redundancy complexity, redundancy allocation problems can be divided into three types: active redundancy, standby redundancy and redundancy considering maintenance. In this paper, three types of redundancy allocation problems are summarized and several potential research directions are proposed. The remaining part of the paper is organized as follows. Section 2-4 respectively review three different redundancy allocation problems of MSSs, followed by a brief conclusion in section 5.

2 ACTIVE REDUNDANCY ALLOCATION

Active redundancy allocation assumes that redundant components operate the same way as working components [5]. In other words, redundant and work components have the same failure rate, if they are identical. According to the different optimization methods, redundancy allocation problem can be divided into two classes: single objective optimization and multi-objective optimization methods. Depending on whether different versions of redundant components are considered in a certain subsystem, redundancy allocation can be categorised in two classes: homogenous redundancy allocation problem and heterogeneous redundancy allocation problem. In the existing literature, most research on redundancy allocation problems of MSSs focused on algorithm optimization and innovation.

2.1 Single objective optimization

Most redundancy allocation problems are based on minimizing the cost or maximizing the reliability under the assumption of independence between components. This kind of single objective optimization problem is usually realized by the optimization algorithm. Genetic algorithm has become the most common algorithm in the homogenous redundancy allocation problem of MSSs. The combination of universal generating function(UGF) and genetic algorithm(GA) has been widely used [4, 7-12].

However, compared to the homogenous redundancy allocation, heterogeneous redundancy allocation problem is more complex and the feasible region is larger. The heterogeneous redundancy allocation problem represents that non-identical components can be used in parallel to improve the system availability by providing redundancy in subsystems [13]. In such a situation, the effectiveness of genetic algorithm is usually limited so that genetic algorithm needs to be used in combination with other methods like the combination of genetic algorithms (GA) and tabu search (TS) [13], MARG algorithm based on GA [14], a genetic algorithm based on the penalty function [15] and a biologically inspired genetic algorithms [16]. In fact, the disadvantages of genetic algorithm have been revealed facing with more complex problem.

In addition, more effective intelligent algorithms and heuristic methods are provided to solve this single objective optimization problem like ant colony algorithm [17, 18], tabu search algorithm [19, 20] and other hybrid heuristic algorithms [21-24]. Even some new hybrid heuristic algorithms like parallel metaheuristic algorithms (SA-parallel and VDO-Parallel) [25] have been used to the heterogeneous redundancy allocation problem . So it can be seen that research on active redundancy allocation problem of MSSs concentrate on algorithm optimization. The dependence between components is rarely considered.

2.2 Multi-objective optimization

Compared to single objective optimization problems, multi-objective optimization problems have been carried out based on the homogenous redundancy allocation problem, where two objectives of minimizing cost and maximizing reliability are considered synthetically [26, 27]. Even several papers have been carried out on the multi-objective optimization of heterogeneous redundancy allocation problem based on a multi-state genetic multi-objective algorithm [13, 28] or a new MOMS-GA algorithm [29]. Yet these research contributions are still not involved in the independence between components.

3 STANDBY REDUNDANCY ALLOCATION

Standby redundancy allocation problems can be divided into three categories: namely cold, warm and hot [5, 6]. In the cold standby systems, components do not fail before they are used, which is usually considered together with the maintenance policy [6, 30] or aging and random shocks [31]. The failure probability of warm standby redundant components is more than zero, but less than the failure probability of the operating components, which has been considered together with the multi-objective optimization [5]. In the hot standby redundancy systems, the failure pattern of components does not depend on whether it is working or in the standby mode, which is quite similar to the active redundancy strategy in terms of mathematical formulation [32]. Obviously, although standby redundancy strategy is closer to the real industrial situation, there are few papers on standby redundancy strategy of MSS. Standby redundancy allocation problem of MSSs should be considered more in future research.

4 REDUNDANCY ALLOCATION CONSIDERIND MAINTENANCE ACTIVITIES

Redundancy allocation and maintenance optimization are two main methods used to provide a required level of system reliability. Engineers typically try to achieve this level with minimal cost by solving the problems of redundancy optimization or maintenance optimization separately [33]. Maintenance is defined as a set of actions or tasks used to restore an item to a state in which it can perform its designated functions [34]. It's one of the main methods to improve the system reliability. There are some studies for the integrated models between redundancy and maintenance [33, 35]. However, most of them only consider minimal repair and replacement actions [33, 35-37], or the limited maintenance resources [38, 39]. The maintenance actions and scenarios considered are both simple.

In addition, there are several research contributions considering more complex maintenance scenarios including the maintenance model considering multiple states and failure interactions [40], the optimal allocation and maintenance model with common bus performance sharing [41] and the integrated optimal model considering two interconnected MSSs [42]. And some integrated models even consider more complex maintenance actions like imperfect maintenance [43, 44]. Obviously, the combination of multiple optimization problems can comprehensively consider more aspects related to equipment reliability and are more close to the reality.

5 CONCLUSION

Many studies on the reliability optimization of MSSs have been carried out in recent years. The redundancy allocation is a main method on reliability optimization design of MSSs. In this paper, the relevant literature on three types of redundancy allocation problems on MSSs is summarized. After induction and comparison, several future research directions can be put forward:

1. The research on active redundancy allocation focuses on the application algorithm optimization. More complex scenarios are rarely considered like the dependence between components, especially stochastic dependence and structural dependence.
2. The existing literature only considers the dependence between components in a same subsystem. The dependence between different subsystems is rarely considered.
3. The combination of multiple optimization problems like redundancy and maintenance can make the redundancy scenario more complex and more realistic. Multiple optimization problems should become the focus of research in the future
4. Standby redundancy optimization problem especially warm standby redundancy optimization problem should be considered more, which is more realistic.

6 REFERENCES

[1] **Levitin, G., Lisnianski, A. 2003.** *Multi-state system reliability: assessment, optimization and applications.* World Scientific Publishing Co. Pvt. Ltd.

[2] **Levitin, G. 2004.** A universal generating function approach for the analysis of multi-state systems with dependent elements, *Reliability Engineering & System Safety*, 84(3), pp 285-292.

[3] **Zio, E. 2009.** Reliability engineering: Old problems and new challenges, *Reliability Engineering & System Safety*, 94(2), pp 125-141.

[4] **Chaaban, W., Schwarz, M., Börcsök, J. 2013.** Cost and redundancy optimization of homogeneous series-parallel multi-state systems subject to availability constraints using a Matlab implemented genetic algorithm, *Recent Advances in Circuits, Systems and Automatic Control, WSEAS*, pp.

[5] **Azadeh, A., Shoja, B.M., Ghanei, S., et al. 2015.** A multi-objective optimization problem for multi-state series-parallel systems: a two-stage flow-shop manufacturing system, *Reliability Engineering & System Safety*, 136, pp 62-74.

[6] **Moghaddass, R., Zuo, M.J., Pandey, M. 2012.** Optimal design and maintenance of a repairable multi-state system with standby components, *Journal of Statistical Planning and Inference*, 142(8), pp 2409-2420.

[7] **Tian, Z., Zuo, M.J., Huang, H.-Z. 2007.** *Optimal redundancy allocation of multi-state systems with genetic algorithms*, edition.

[8] **Lisnianski, A., Levitin, G., Ben-Haim, H. 2000.** Structure optimization of multi-state system with time redundancy, *Reliability Engineering & System Safety*, 67(2), pp 103-112.

[9] **Levitin, G., Lisnianski, A. 2001.** A new approach to solving problems of multi-state system reliability optimization, *Quality and reliability engineering international*, 17(2), pp 93-104.

[10] **Levitin, G., Lisnianski, A., Ben-Haim, H., et al. 1998.** Redundancy optimization for series-parallel multi-state systems, *IEEE Transactions on Reliability*, 47(2), pp 165-172.

[11] **Levitin, G. 2001.** Redundancy optimization for multi-state system with fixed resource-requirements and unreliable sources, *IEEE transactions on reliability*, 50(1), pp 52-59.

[12] **Levitin, G., Lisnianski, A. 2001.** Structure optimization of multi-state system with two failure modes, *Reliability Engineering & System Safety*, 72(1), pp 75-89.

[13] **Ouzineb, M., Nourelfath, M., Gendreau, M. 2011.** A heuristic method for non-homogeneous redundancy optimization of series-parallel multi-state systems, *Journal of Heuristics*, 17(1), pp 1-22.

[14] **Gupta, R., Agarwal, M. 2007.** Heuristic Approach for Heterogeneous Redundancy Optimization in Multi-State Series-Parallel System, *International Journal of Reliability, Quality and Safety Engineering*, 14(04), pp 327-359.

[15] **Gupta, R., Agarwal, M. 2006.** Penalty guided genetic search for redundancy optimization in multi-state series-parallel power system, *Journal of combinatorial optimization*, 12(3), pp 257-277.

[16] **Chaaban, W., Schwarz, M., Börcsök, J.** Cost optimization and redundancy allocation of availability constrained heterogeneous series-parallel systems using genetic computing, pp.

[17] **Zeblah, A. 2008.** Reliability maximization of power system using ant colony approach, *International Journal of Electrical and Power Engineering*, 2(3), pp 192-201.

[18] **Nourelfath, M., Nahas, N., Zeblah, A. 2003.** An ant colony approach to redundancy optimization for multi-state system, International conference on industrial engineering and production management (IEPM'2003), Porto-Portugal.

[19] **Ouzineb, M., Nourelfath, M., Gendreau, M. 2008.** Tabu search for the redundancy allocation problem of homogenous series-parallel multi-state systems, *Reliability Engineering & System Safety*, 93(8), pp 1257-1272.

[20] **Ouzineb, M., Nourelfath, M., Gendreau, M. 2006.** Availability Optimization of Series-Parallel Multi-State Systems Using a Tabu Search Meta-heuristic, 2006 International Conference on Service Systems and Service Management.

[21] **Agarwal, M., Sharma, V.K. 2009.** Ant colony optimization algorithm for heterogeneous redundancy allocation in multi-state series-parallel systems, *International Journal of Reliability, Quality and Safety Engineering*, 16(02), pp 163-186.

[22] **Zuo, M., Choy, L., Yam, R. 1999.** A model for optimal design of multi-state parallel-series systems, Electrical and Computer Engineering, 1999 IEEE Canadian Conference on, pp 1751-1754.

[23] **Ramirez-Marquez, J.E., Coit, D.W. 2004.** A heuristic for solving the redundancy allocation problem for multi-state series-parallel systems, *Reliability Engineering & System Safety*, 83(3), pp 341-349.

[24] **Agarwal, M., Gupta, R. 2007.** Homogeneous redundancy optimization in multi-state series-parallel systems: A heuristic approach, *IIE Transactions*, 39(3), pp 277-289.

[25] **Sharifi, M., Mousa Khani, M., Zaretalab, A. 2014.** Comparing Parallel Simulated Annealing, Parallel Vibrating Damp Optimization and Genetic Algorithm for Joint Redundancy-Availability Problems in a Series-Parallel System with Multi-State Components, *Journal of Optimization in Industrial Engineering*, 7(14), pp 13-26.

[26] **Tian, Z., Zuo, M.J. 2006.** Redundancy allocation for multi-state systems using physical programming and genetic algorithms, *Reliability Engineering & System Safety*, 91(9), pp 1049-1056.

[27] **Tian, Z., Zuo, M.J., Huang, H. 2008.** Reliability-redundancy allocation for multi-state series-parallel systems, *IEEE Transactions on Reliability*, 57(2), pp 303-310.

[28] **Taboada, H., Coit, D. 2007.** Multiple objective design allocation problems: development of new evolutionary algorithms, Proceedings of the European Safety & Reliability Conference (ESREL).

[29] **Taboada, H., Espiritu, J., Coit, D. 2008.** Design allocation of multistate series-parallel systems for power systems planning: a multiple objective evolutionary approach, *Proceedings of the Institution of Mechanical Engineers, Part O: Journal of Risk and Reliability*, 222(3), pp 381-391.

[30] **Hsieh, C.-C., Chiu, K.-C. 2002.** Optimal maintenance policy in a multistate deteriorating standby system, *European Journal of Operational Research*, 141(3), pp 689-698.

[31] **Hsieh, C.-C. 2005.** Replacement and standby redundancy policies in a deteriorating system with aging and random shocks, *Computers & operations research*, 32(9), pp 2297-2308.

[32] **Tavakkoli-Moghaddam, R., Safari, J., Sassani, F. 2008.** Reliability optimization of series-parallel systems with a choice of redundancy strategies using a genetic algorithm, *Reliability Engineering & System Safety*, 93(4), pp 550-556.

[33] **Levitin, G., Lisnianski, A. 1999.** Joint redundancy and maintenance optimization for multistate series-parallel systems, *Reliability Engineering & System Safety*, 64(1), pp 33-42.

[34] **Ahmad, R., Kamaruddin, S. 2012.** An overview of time-based and condition-based

maintenance in industrial application, *Computers & Industrial Engineering*, 63(1), pp 135-149.

[35] **Li, Y., Peng, R. 2014.** Availability modeling and optimization of dynamic multi-state series-parallel systems with random reconfiguration, *Reliability Engineering & System Safety*, 127, pp 47-57.

[36] **Nourelfath, M., Dutuit, Y. 2004.** A combined approach to solve the redundancy optimization problem for multi-state systems under repair policies, *Reliability Engineering & System Safety*, 86(3), pp 205-213.

[37] **Tian, Z., Levitin, G., Zuo, M.J. 2009.** A joint reliability-redundancy optimization approach for multi-state series-parallel systems, *Reliability Engineering & System Safety*, 94(10), pp 1568-1576.

[38] **HUANG, H.-z. 2012.** JOINT OPTIMIZATION OF REDUNDANCY AND MAINTENANCE STAFF ALLOCATION FOR MULTI-STATE SERIES-PARALLEL SYSTEMS OPTYMALIZACJA ŁĄCZONA ALOKACJI NADMIAROWOŚCI ORAZ ALOKACJI PRACOWNIKÓW SŁUŻB UTRZYMANIA RUCHU W WIELOSTANOWYCH SYSTEMACH SZEREGOWO-RÓWNOLEGŁYCH, *Terms and Conditions of Publication*, 14(4), pp 312.

[39] **Nourelfath, M., Ait-Kadi, D. 2007.** Optimization of series-parallel multi-state systems under maintenance policies, *Reliability Engineering & System Safety*, 92(12), pp 1620-1626.

[40] **Wang, J., Li, M. 2015.** Redundancy Allocation Optimization for Multistate Systems With Failure Interactions Using Semi-Markov Process, *Journal of Mechanical Design*, 137(10), pp 101403.

[41] **Xiao, H., Peng, R. 2014.** Optimal allocation and maintenance of multi-state elements in series-parallel systems with common bus performance sharing, *Computers & Industrial Engineering*, 72, pp 143-151.

[42] **Lisnianski, A., Ding, Y. 2009.** Redundancy analysis for repairable multi-state system by using combined stochastic processes methods and universal generating function technique, *Reliability Engineering & System Safety*, 94(11), pp 1788-1795.

[43] **Nourelfath, M., Châtelet, E., Nahas, N. 2012.** Joint redundancy and imperfect preventive maintenance optimization for series-parallel multi-state degraded systems, *Reliability Engineering & System Safety*, 103, pp 51-60.

[44] **Liu, Y., Huang, H.-Z., Wang, Z., et al. 2013.** A joint redundancy and imperfect maintenance strategy optimization for multi-state systems, *Reliability, IEEE Transactions on*, 62(2), pp 368-378.

Application of Integrated Product Development to the Organizational Structure of Aero Engine Development Project

Tao Du, Yanhong Ma, Qiuhong Zhao*
School of Economics and Management, Beihang University, Beijing, China
dutao@skyrizon.com.cn; yhmabeihang@outlook.com; qhzhao@buaa.edu.cn

ABSTRACT

Aero engine development is a critical link in the aero engine industry, and the organizational structure of aero engine development project impacts the efficiency and quality of the aero engine development. As integrated product development (IPD) has becoming a widely used product development concept, this paper analyzes the applicability of IPD to the organizational structure of aero engine development project based on the analysis of the characteristics of aero engine development project. A case of the application of IPD at Pratt & Whitney, which is an American aero engine manufacturer is illustrated. Moreover, the new requirements of the application of IPD to the organizational structure of aero engine development project are proposed, such as strengthening the international cooperation, combining integrated risk management and optimizing the project management methods.

Keywords: Integrated Product Development (IPD); Aero Engine Development; Project Management; Organizational Structure

1 INTRODUCTION

Aero engine is a typical high-tech product which is technology intensive and high value-added. It is an industry that matters homeland security, economic development and technology development of a country. And it can reflect a country's comprehensive national strength, industrial base and technology level. Also, the aero engines business is a highly competitive market and the related gas turbines are at the leading edge of the technology development (Karl et al., 2004). Therefore, many countries invest lots of money in aero engine industry. With the rapid development of aeronautical science and technology and the growing requirement of the airplane on aero engine performance, the aero engine is becoming more and more complex.

Aero engine development is one of the most important parts in the whole industry. Engineering and manufacturing firms face increasing pressure to continuously improve product development performance in terms of time to market, development cost, and customer satisfaction (Mascoli, 1999), this is because product development can be a crucial competitive lever in the global marketplace (Browning and Ramasesh, 2007). A well project management has a significant effect on the success of the product development. As the organizational structure plays an important role as an antecedent to the development of radical and incremental product innovation capabilities (Menguc and Auh, 2010), the product development organization continues to evolve and a lot of concepts come into being used in product development organization. For example, matrix management, integrated product development (IPD) and some other advanced concepts have been used widely in building the project management organization. In these advanced concepts, the application of IPD to building organizational structure of product development has improved the efficiency of product development significantly. And it is also applied to the organizational structure of aero engine development project.

Many scholars have done research on the concept of IPD and its application. IPD is a managerial approach proposed with the goal of improving product development performance through managing the overlapping, parallel execution and concurrent workflow of activities (Gerwin and Barrowman, 2002; Naveh, 2005; Sommer, 2014). The concept of IPD puts its emphasis on the concurrent activities, which means the overlap and interaction of certain activities (Gerwin and Barrowman,

2002; Duhovnik et al., 2009). Some researchers consider the impact of some performance-affecting elements on the implementation of IPD, such as project management itself (Danilovic and Browning, 2007), organizational structure (Menguc and Auh, 2010). Rauniar and Rawski (2012) analyze the effect of organizational structuring on IPD performance, which is indicated by indices such as development time and development cost. In return, the adoption of the concept of IPD also affects the formation of the organization structure in the company, especially the organization structure of the product development project.

Against this background, in this paper, we seek to analyze the application and the impact of the application of IPD to the organization structure of the aero engine development project. Overall, this paper is organized as follows. In Section 2, the characteristics of aero engine project are analyzed. Section 3 gives a brief introduction of IPD. Section 4 illustrates the application of IPD in aero engine development project in the company Pratt & Whitney. Finally, in Section 5, we draw some conclusions.

2 CHARACTERISTICS OF AERO ENGINE DEVELOPMENT PROJECT

The main characteristics of the aero engine development project are long development cycle, complicated and difficult technology, huge amount of investment, high risk and wide range of coordination, which are described in detail as follows.

2.1 Long development cycle

The development cycle of an aero engine is from the time when product demand is determined to the time when product model is finalized. It usually takes about 10 years. With the constant improvement of the aero engine technology and the increasing international competition, aero engine companies have to speed up the process of aero engine development. Also, aero engine companies have to adjust organization structure in order to shorten the development cycle.

2.2 Complicated and difficult technology

Aero engines are repeatedly used over a long period of time under harsh conditions, such as high temperature, high pressure, high speed, alternating load, and big range of stress. Therefore, they should meet a lot of conflicting and demanding requirements. For example, big power, low fuel consumption, light weight, long service life, low noise, less pollution, strong reliability, low research and follow-up maintenance cost and so on. The development of aero engines involves a wide range of disciplines and departments. And the aero engine is becoming more and more complex, which makes the complexity of project management becomes much higher than before.

2.3 Huge amount of investment

As the development cycle of the aero engine is long, and technology is complex, and the requirements on performance, quality, safety, environmental protection are high, the amount of money which needs to be invested in the aero engine development should be huge. And the amount of money needed is increasing recently. Normally, the expenditure on aero engine development accounts for about 15% of sales. For example, the F135 engine developed by Pratt & Whitney costs more than seven billion dollars on research and development.

2.4 High risk

Aero engine development is the embodiment of sophisticated technology and the exploration of its research is extremely strong. Therefore, some new technology or new technique will be used and lots of manpower, material resources, money and time will be used, which means there are many uncertain factors in the process of the development of aero engines. These uncertain factors may bring risk to the development of aero engines, which is usually very high. The high risk of aero engine development indicates the project management organization should have strong consciousness of risk management and transformative capability.

2.5 Wide range of coordination

Different from ordinary products, an aero engine is composed of dozens of systems and thousands of parts, which means the development of an aero engine involves a number of departments in the company and it also involves a number of industries and companies. Within one company, the aero engine development organization should coordinate different departments. And it should also coordinate a number of companies during the process of aero engine development.

These characteristics of the process of aero engine development put forward some requirements on the project management of aero engine development. One of the most important requirements is that concurrent product development should be applied during the process of aero engine development. Consequently, the project management organization of aero engine development should meet the requirement of concurrent product development.

On the one hand, for the development of a single product, representatives from departments such as development department, process design department, production department, test department, purchasing department, financial department, customers, subcontractors, and suppliers should be integrated to form a development center and an authorized leader should be assigned to the center. These representatives participate in the processes from product design, trial production until the product is finalized according to their respective division of labor.

On the other hand, the organization of aero engine development should also achieve concurrent development of multi products. According to the needs of multi project management, the company should reduce the intermediate link and build a flat organization structure. For example, the company can form a project centered matrix flexible organization structure, in which a project management office is built. This project management office is responsible for the planning, organizing, monitoring of each project.

3 ANALYSIS OF THE APPLICABILITY OF IPD TO THE ORGANIZATION OF AERO ENGINE DEVELOPMENT PROJECT

As discussed in Section 2, the characteristics of the aero engine development project put forward some requirement on the organization of aero engine development project and one of the most important requirements is that the organization should achieve concurrent development, not only for a single product development, but also for multi products development. Concurrent development is one of the core concepts of IPD. In this section, a brief introduction of IPD is given and from which the applicability of IPD to the organization of aero engine development project can be analyzed.

3.1 Definition of concurrent engineering and IPD

A widely accepted definition of concurrent engineering can be described as follows: Concurrent engineering is a systematic approach, by using which the manufacturing process and support process are implemented integrated and concurrently. The goal of adopting concurrent engineering is to enable product developer to consider all the possible factors during the whole product life cycle at the beginning of the product design, including quality, cost, project scheduling and customer's demand (Davis, 1991; Cleetus, 1992; Chen et al., 1994).

As a guiding ideology of product development, the essence of concurrent engineering is enabling people from different disciplines to develop the product concurrently and cooperatively under a certain coordination and control mechanism. It is a process that can reflect the integration and cooperation during the process of product development. In the computer integrated environment, the integration of the related processes of the product is based on concurrent development. In short, integrated product development is based on the concept of concurrent engineering.

IPD is an extension of concurrent engineering. By using the principles and methods of concurrent engineering, IPD can achieve the timely cooperation among product developers during the whole product life cycle. IPD is also a management strategy, it can improve the process of product development through customer consultation, cross-functional team and integrated technology.

3.2 Framework of IPD

The process of IPD is not only a process of product development. It is a cross-functional departments business process. IPD integrates all the main activities required for product management to ensure the success of project plan and achieve the company's business goal. And project management is the core in IPD beneath project governance in the project organization hierarchy (Sommer et al., 2014), where project governance is broadly defined as a set of management systems, rules, protocols, relationships, and structures that provide the framework within which decisions are made for project development and implementation to achieve the intended business or strategic motivation (Bekker and Steyn, 2007).There are influencing elements external to the IPD organization, which are the basis organization, the involved human resources, and the external stakeholders. The core of IPD includes an interactive, concurrent project model and the project team. And there are seven obvious characteristics of IPD: parallelness of activities, standardized IPD process, process integration, built-in instability, self-organizing project teams, subtle control, and multi-learning. Based on the above analysis, the framework of IPD can be shown by Figure 1.

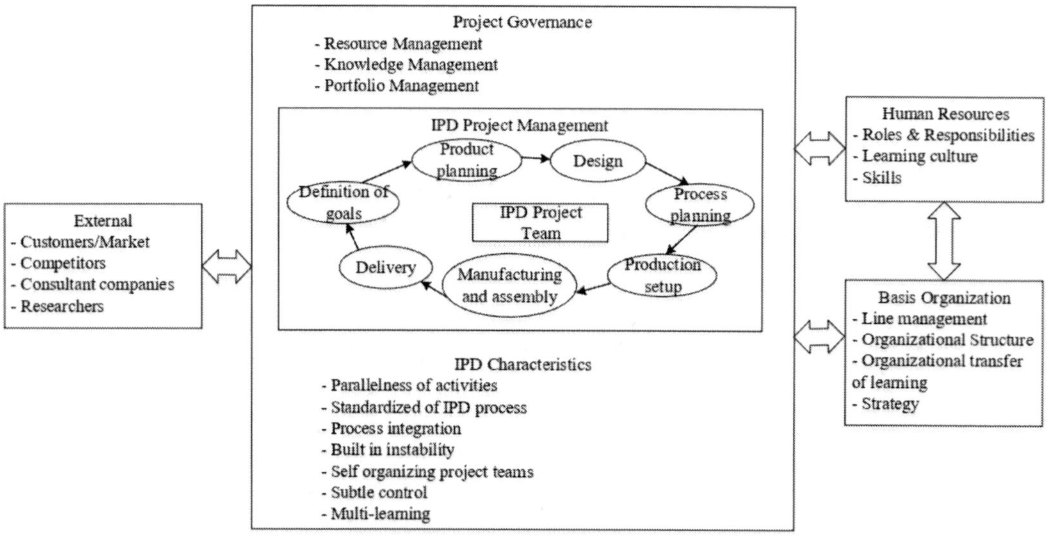

Figure 1: Framework of Integrated Product Development (IPD)

3.3 Integrated Product Team (IPT)

As pointed above, project management is the core of IPD. And there is an IPD project team which is responsible for project management of integrated product development. This team is named as integrated product team (IPT) in IPD. In the organization structure, IPD emphasizes the role of IPT in product development. The integration of product is reflected by the integration of project management organization. The emphasis of IPT is improving team spirit and cooperation method, and also the management of product information (Department of Defense, DoS). The integrated development of a product involves many departments in a company, it is cross-functional. Consequently, matrix organization structure should be used when building IPT.

4 APPLICATION OF IPD TO THE ORGANIZATION STRUCTURE OF AERO ENGINE DEVELOPMENT PROJECT: A CASE OF PRATT & WHITNEY

Pratt & Whitney is an American aerospace manufacturer with global service operations. It is a subsidiary of United Technologies Corporation (UTC). Pratt & Whitney's aircraft engines are widely used in both civil aviation (especially airlines) and military aviation. Its headquarters are in East Hartford, Connecticut. As one of the 'big three' aero engine manufacturers, it competes with General Electric and Rolls-Royce, although it has also formed joint ventures with both of these companies. In addition to aircraft engines, Pratt & Whitney manufactures gas turbines for industrial and power generation, and marine turbines.

Pratt & Whitney engine development is organized into module centers (Mullooly, 2002). Before this, Pratt & Whitney was organized as a classic vertically organization since its beginnings in 1925 until approximately 1990. As complexity in engine design is increasing all the time, the organization of engine development gradually involved into module centers. Pratt & Whitney has five module centers which are segmented by engine section. The five module centers and their location are shown in Table 1.

Table 1. Five module centers and their location of Pratt & Whitney

Module Center	Location
Engine Center, Assembly and Test	Middletown, CT
Electronic & Mechanical Systems Module Center	East Hartford, CT
Compression System Module Center	Middletown, CT
Combustor, Augmentor & Nozzle Module Center	East Hartford, CT
Turbine Module Center	North Haven, CT

The module center is responsible for the product related activities from raw material to customer delivery. Therefore, in each module center, there are work force including all disciplines. In Pratt & Whitney, the organizations contained in each module center include: Human resources, Environment Health and Safety, Facilities, Machine Services, Information Services, Finance, Quality, Continuous Improvement, Mechanical Design, Manufacturing Engineering, Procurement, Commodity Management, Project Engineering, Secondary Flow Analytical Engineering, Heat Transfer Engineering, Aerodynamics and Structures Engineering (Mullooly, 2002).

In each module center, the development of aero engine adopts IPD in Pratt & Whitney. IPD was first adopted in Pratt & Whitney in 1991. Different from the organization structure used before, the IPD used in the module center can make sure that the communication of product development is within a module center, which makes the product development more efficient. The specific organization structure for an aero engine development by using IPD in Pratt & Whitney is illustrated by an example of RL60 aero engine.

At Pratt & Whitney, the concept of system engineering is applied within the framework of the IPD process. By adopting IPD, customer and internal business requirements are integrated across the business functions through the entire program life cycle, which is from conceptual design of the product to the end of the product service. This process makes it possible that program management, engineering, operations, product support, finance, and quality are all unified to meet the customer requirements. An Integrated Program Management Team (IPMT) is built, and the members in the IPMT are management-level representatives from each of the above areas. A program manager will be assigned for the IPMT, and led by the program manager, the IPMT has the authority to make decisions relative to direction, scope, allocation of budget, establishment of goals, and approval of configuration changes (Arbogast et al., 2003).

Under the IPMT, programmatic and technical responsibilities are allocated through Work Breakdown Structure (WBS) at the sub-system or component levels. Component Integrated Product Team (CIPT) is organized, which is responsible for the management of an individual sub-system or a single component. Within each CIPT, the sub-system or component is further subdivided, and Integrated Product Team (IPT) is organized to be responsible for an individual element or a single part. The organization structure in a module center at Pratt & Whitney can be shown in Figure 1.

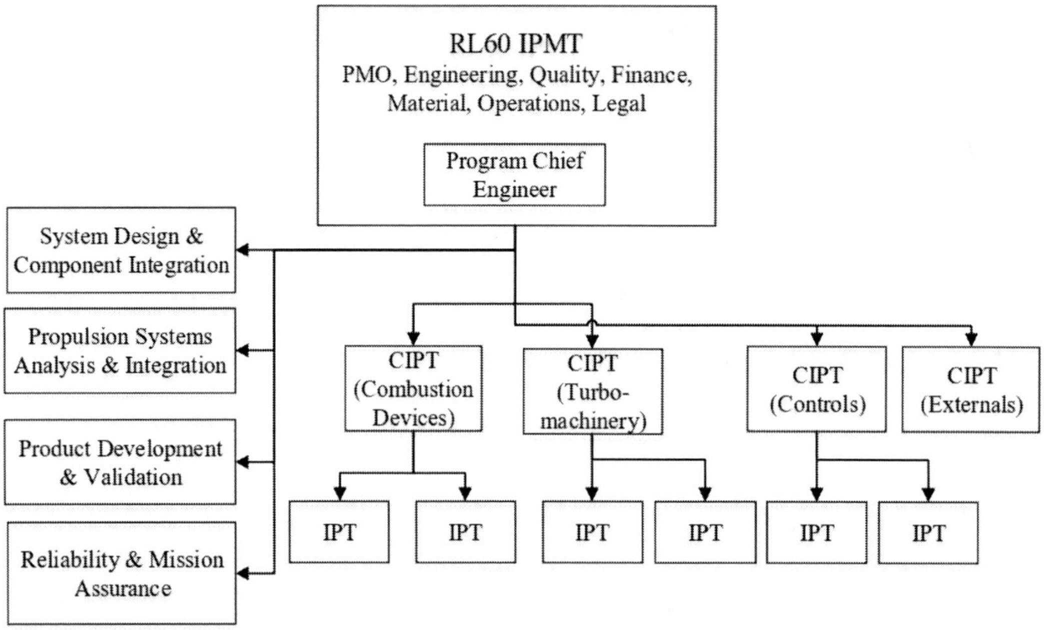

Figure 2: IPD organization structure of RL60 in Pratt & Whitney

In the organization structure, all the engineering efforts related to the product are led by a program chief engineer. The work of program chief engineer is supported by engineers from four systems engineering disciplines, which are system design and component integration, propulsion systems analysis and integration, product development and validation, reliability and mission assurance (Arbogast et al., 2003). These engineers work together with CIPTs and IPTs under the goal of finishing the product development.

The organization structure of aero engine development project built through adopting IPD at Pratt & Whitney helped improve the efficiency and reliability of aero engine development. It helped Pratt & Whitney maintain its leading position in the aero engine industry.

5 NEW REQUIREMENTS ON THE APPLICATION OF IPD TO THE ORGANIZATIONAL STRUCTURE OF AERO ENGINE DEVELOPMENT PROJECT

The product development faces challenges that stem from organizational structure (Lakemond et al., 2015). As a product which needs multi-party cooperation, this challenge in the aero engine development project is even larger. Because of the high development cost, high development risk and fierce market competition, extensive international cooperation is the focus of aero engine companies at present (Liu et al., 2015). How to control the development risk is also a focus. In addition, the project management methods, for example, the technology acquisition mode, is also a challenge that affecting aircraft industry, especially for the latecomers in the industry (Lee and Yoon, 2015). Because of these challenges, there are new requirements on the application of IPD to the organizational structure of aero engine development project, which can be summarized as follows.

5.1 Strengthening international cooperation

The internationalization of the aviation industry is becoming more and more obvious, which is reflected in the cooperation of development, production, and sale of aviation system and parts. The development of aero engine also involves development of many parts, and international cooperation has a great help to improve the quality of the aero engine. In the application of IPD, aero engine company should pay more attention to international cooperation and should adjust their organization structure according to the relationships of international cooperation.

5.2 Combining integrated risk management

Compared with the development of a common product, the risk of aero engine development is much higher. Controlling the high risk during the whole development process is now a focus for the companies. Integrated risk management is now considered to be an effective method to control risk. Consequently, when building the organizational structure of the aero engine development project, the concept of integrated risk management should be combined into IPD.

5.3 Optimizing project management methods

As discussed in Section 3, project management is the core of the concept of IPD. And the efficiency of product development depends on project management methods significantly. Technology is becoming more and more complex, also it is developing faster and faster, especially in aero engine area. Consequently, more emphasis should be put on the application of new technology of project management methods. Furthermore, the competition between the aero engine companies is increasing. Shortening the product development cycle can help the company to gain competitive advantage during the competition. Therefore, project management methods should be optimized and the organization structure of product development should be changed according to the new project management methods.

6 CONCLUSIONS

The organizational structure of development project plays and important role during the process of product development. As integrated product development (IPD) has become a widely used concept in product development area, its application affects the formation of organizational structure in the company. This paper analyses the application of IPD to the organizational structure of the aero engine development project. First, the characteristics of aero engine development project are analysed, such as long development cycle, complicated and difficult technology, huge amount of investment, high risk and wide range of coordination. Based on the characteristics, one of the most important requirements on the organization of aero engine development project are proposed, which is concurrent engineering. As concurrent engineering is the core concept of IPD, the applicability of IPD in aero engine development project is given in Section 3. And subsequently it is illustrated by a case study of Pratt & Whitney, who changes its organizational structure of aero engine development project after adopting IPD. With the development of technology and the economic environment, product development faces challenges that stem from organizational structure, especially for aero engine development, which put forward some new requirements on the application of IPD in aero engine development project. This paper proposes some new requirements that the company should focus on when adopting the IPD into aero engine development, for example, strengthening international cooperation, combining integrated risk management and optimizing project management methods. The conclusions of this paper can guide the establishment of organization structure of aero engine development project for the company to some extent.

7 ACKNOWLEDGEMENTS

This work is supported by National Natural Science Foundation of China (NSFC) under grant Nos. 91224007 and 71471006.

8 REFERENCES

[1] **Arborgast, T., Doughty, D., Hoffman, K., et al. 2003.** Enhanced Systems Engineering Practices as Applied to the Pratt & Whitney RL60 Demonstrator Engine. Aiaa/asme/sae/asee Joint Propulsion Conference and Exhibit.

[2] **Bekker, M.C., Steyn, H. 2007.** Defining 'project governance' for large capital projects. South African Journal of Industrial Engineering, 20(2), pp 1-13.

[3] **Browning, T.R., Ramasesh, R.V. 2007.** A Survey of activity network-based process models for managing product development projects. Production & Operations Management, 16(2), pp 217-240.

[4] **Chen, Y.M., Miller, T., Dik, L.L. 1994.** Object-Oriented part model for geometric reasoning. Journal of Computer-Aided Engineering, 1(5), pp 375-395.

[5] **Cleetus, J. 1992.** Definition of concurrent engineering, Morgantown, WV: Concurrent Engineering Research Center.

[6] **Danilovic, M., Browning, T.R. 2007.** Managing complex product development projects with design structure matrices and domain mapping matrices. International Journal of Project Management, 25(3), pp 300-314.

[7] **Davis, T.A., Trapp, G. 1991.** Advancing concurrent engineering using STEP. CERC Technical Report. CERC-TR-RN-91-012. West Virginia University, Concurrent Engineering Research Center.

[8] **Duhovnik, J., Zargi, U., Kusar, J., et al. 2009.** Project-driven Concurrent Product Development. Concurrent Engineering, 17(3), pp 225-236.

[9] **Gerwin, D., Barrowman, N.J. 2002.** An Evaluation of Research on Integrated Product Development. Management Science, 48(7), pp 938-953.

[10] **Karl. A., Lisiewicz. S., et al. 2004.** Optimization and robust design in aero engine development. ASME 2004 International Design Engineering Technical Conferences and Computers and Information in Engineering Conference. Salt Lake City, Ulah, USA, September 28-October 2.

[11] **Lakemond, N., Magnusson, T., Johansson, G., et al. 2015.** Assessing Interface Challenges in Product Development Projects[J]. Research Technology Management, 56(1), pp 40-48.

[12] **Lee, J.J., Yoon, H. 2015.** A comparative study of technological learning and organizational capability development in complex products systems: Distinctive paths of three latecomers in military aircraft industry. Research Policy, 44(7), pp 1296-1313.

[13] **Liu, H., Zhang, S., Liang, C., et al. 2015.** Development Strategy Study for Commercial Aero Engine Companies in European Countries and the United States. Procedia Engineering, 99, pp 804-811.

[14] **Mascoli, G.J. 1999.** A systems engineering approach to aero engine development in a highly distributed engineering and manufacturing environment. Massachusetts Institute of Technology.

[15] **Menguc, B., Auh, S. 2010.** Development and return on execution of product innovation capabilities: The role of organizational structure. Industrial Marketing Management, 39(5), pp 820-831.

[16] **Mullooly, J.F. 2002.** A system engineering approach for implementation of a corporate growth strategy. Massachusetts Institute of Technology.

[17] **Naveh, E. 2007.** The effect of integrated product development on efficiency and innovation. International Journal of Production Research, 43(13), pp 2789-2808.

[18] **Rauniar, R., Rawski, G. 2012.** Organizational structuring and project team structuring in integrated product development project[J]. International Journal of Production Economics, 135(135), pp 939-952.

[19] **Sommer, A.F., Dukovska-Popovska, I., Steger-Jensen, K. 2014.** Barriers towards integrated product development — Challenges from a holistic project management perspective. International Journal of Project Management, 32(6), pp 970-982.

[20] **US Department of Defense.** Guide to integrated product and process development.

MEDICAL RELIEF SHELTER LOCATION CONSIDERING THE SEVERITY OF PATIENTS UNDER LIMITED RELIEF BUDGET

Jeongmin Gu,Yanjie Zhou and Gyu M. Lee*
Department of Industrial Engineering
Pusan National University, Pusan, Republic of Korea
glee@pnu.edu

ABSTRACT

It is an essential but difficult problem to provide the effective and efficient medical or relief services to the people in the impact areas by various anthropogenic and natural disasters. This study proposes a mathematical programming model to determine the locations of temporary medical relief centers and provide the required medical supplies from medical deployment centers effectively and efficiently under the limited relief budget, considering the severities and geographical locations of patients. To demonstrate the validity of the proposed model for various situations, computational results under different scenarios are provided as well.

Keywords: disaster management, medical relief shelters, medical deployment centers, location problem, injury severity

1 INTRODUCTION

Anthropogenic and natural disasters which may cause the losses of life, injuries, property damages, social and economic disruptions, or environmental degradation have become regular occurrences. The Haiti earthquake has caused casualties about 300,000 people on January 12, 2010. The earthquake in Sichuan on May 12, 2008 killed and injured approximately 69,000 and 374,000 people, respectively. It also damaged the assets of more than 40 million people. Hurricane 'Patricia' in 2015 was responsible for an estimate of 7300 deaths. Cyclone 'PAM' in March, 2015 took place in the island of the South Pacific, Vanuatu and attacked the Port Vila, Vanuatu's capital where over the 80 % of houses were damaged and 3,300 victims had to been evacuated to 37 evacuation centers at that time. The tsunami and earthquake which has occurred in Japan in March, 2011 have attacked the nuclear power plants in Fukushima and their damages leaked the dangerous radiative materials. It caused the numerous victims and prolonged damages to Japan as well as serious concerns of neighboring countries in environment, health and safety.

Without considering the fincinal damages caused by various disasters, they affect a large number of people, frequently leading to deaths or injuries. In respone to the disaster impact, it is essential to recognize the severity and distribution of the patients, establish the medical relief shelters and transport them for appropriate medical treatments, considering the location medical deployment centers, which provides the medical supplies (medical staffs, medical equipments, medicines, etc.), in order to decrease the number of deaths. Note that medical staffs are also considered as one of medical supplies in this study without loss of generality. In any circumstance of disasters, it is critical

* Corresponding Author

to deliver the right supplies to the right people at the right place in the right quantities. Balcik and Beamon [1] considered facility location decisions for a humanitarian relief chain responding to quick-onset disasters. They developed a model which determines the number and locations of distribution centers in a relief network and the amount of relief supplies to be stocked at each distribution centers to meet the needs of people affected by the disasters. Afshar and Haghani [2] proposed a mathematical model that controls the flow of several relief commodities from sources through the supply chain and until they are delivered to recipients. A humanitarian and disaster relief supply chains within the broad field of supply chain management were studied by Day et al. [3]. Sheu and Pan [4] proposed a novel relief supply collaboration approach to address the issue of post-disaster relief supply-demand imbalance in emergency logistics operations.

In this study, the locations of medical relief shelters are determined in consideration of the severity and distribution of the patients and the locations of medical deployment centers. Facility location problem which concerns about the optimal placement of facilities to minimize transportation costs while considering various factors is an important one in terms of strategic planning and has been studied extensively by many researchers. Medical facility location problem to determine the medical facilities' locations and an optimal allocation of the impacted patients is a practical and challenging problem. Jia et al. [5] proposed a maximal covering facility location model which efficiently finds locations for tremendous demands of medical services in emergency situations. Jia et al. [6] studied about characteristics of large-scale disasters and proposed the large-scale emergency situations model and shortly reviewed the traditional facility models which are covering models, p-median models and p-center models for emergency services. Farahani and Hekmatfar [7] introduced a variety of facility location problems such as P-median, P-center, covering problem and explained the basic concept of each facility location problems. Mete and Zabinsky [8] proposed a stochastic optimization approach for emergency and non-emergency cases to select the storage locations of medical supplies and required inventory levels for each type of medical supply under a wide variety of possible disaster types and magnitudes, in preparation for disasters. Farahani et al. [9] presented an extensive review on covering problem in facility location. They introduced a variety type of maximal covering location problem (MCLP) model such as planer maximal covering and capacitated MCLP.

In this study, it is assumed that candidate locations of medical relief shelters are determined in advance before the disaster impact. It is also assumed that the severity and distribution of the patients in the impacted areas via emergency call centers or estimation. There are two types of patients; emergency and non-emergency patients. The severity of a patient determines whether he or she is an emergency one or not by using a threshold on the severity. Normally, the seriously injured people need to be taken care of with high priority. Emergency patients requires an immediate medical service and needs to be transported to nearest medical shelters quickly. It is assumed that more medical supplies are required for emergency patients. The proposed mathematical programming model in this paper can be easily extended to accommodate different numbers of patient types and different ways to determine them.

The objective of this study is to maximize the number of patients who gets medical services at multiple medical relief centers under the limited relief budget, while considering the severity of patients and distances to medical relief shelters. The costs to establish temporary medical relief shelters and to procure and tranport the medical supplies including medical staff are subject to the maximum relief budget. To construct the medical relief shelters, there are variable construction costs, which is proportional to the capapcity of shelters, in addition to the fixed construction cost. The quantify of required medical supplies at a medical relief shelter is determined by the number and type of patients who are assigned to it. Medical deployment centers provide the medical staffs and medical supplies to medical relief centers and have their own maximum capaicy for each supplies. Transportation cost includes the vehicle costs and variable costs. Nicholl et al. [10] surveyed the relationship between the distance to hospital and the patient death in emergencies. It is natural to see that the fast evacuation is crucial to save the human lives. Najafi, et al. [11] proposed a multi-objective, multi-mode, Multi-commodity, and multi-period stochastic model to manage the logistics of both of commoditise and injured people in the earthquake response. They minimize the total unserved patients and the unsatisfied demands during the planning horizon. In our study, the optimal locations of temporary medical relief shelters are determined as well from the candidate locations which can be pre-selected at the peace time. In addition, the objective function in the proposed mathematical programming model assigns as many patients as possible considering the severity, proximity and the relief budget. Hence, the problem under consideration in this study is a variant of the MCLP in disaster situation.

The paper is organized as follows. In Section 2, a variant of MCLP is proposed which considers injury degree of patients and distance from individual patient to temporary medical shelter, as well as allocation of the medical supplies. A short introduction of computational results is provided in Section 3. Finally, the conclusions are given in Section 4.

2 MATHEMATICAL FORMULATIONS

2.1 Mathematical formulation

In this paper, a variant of MCLP is proposed to maximize the number of patients who gets medical services at newly-established medical relief centers under the limited relief budget, while considering the severity of patients and distances to medical relief shelters. This problem has been explained in the previous section in detail. In this section, a list of assumptions for this problem under consideration are provided as well as a mathematical programming model, which includes the parameters and decision variables.

Problem assumptions

- The degrees of injury (i.e., severity) and the locations for patients have been known.
- The severity classified patients into two types (emergency and non-emergency) by the pre-set threshold.
- The emergency and non-emergency patients require different quantities of medical supplies.
- The medical deployment centers provide the medical supplies to medical relief shelters at the maximum of pre-set quantities.
- Each type of medical supplies has a fixed volume and a vehicle transporting medical supplies has the maximum capacity in volume.
- The maximum budget which can be used for medical relief operations are limited, including the construction and operating costs of medical relief shelters as well as the procurement and transportation costs of medical supplies from medical deployment centers.

Parameters

I	Set of individual patients, $i \in I$
K	Set of types of medical supplies, $k \in K$
L_R	Set of candidate locations for medical relief shelters, $j \in L_R$
L_s	Set of medical deployment centers, $l \in L_s$
A_k^e	Quantity of medical supply k, required for an emergency patient
A_k^n	Quantity of medical supply k, required for non-emergency patient
MK_{lk}	Maximum quantity of medical supply k at medical deployment center l
V_k	Volume of medical supply k

MV	Maximum capacity of a vehicle in volume
S_i	Severity of patient i
D_{ij}	Distance from the location of patient i to medical relief shelter j
CV	Vehicle cost per vehicle
CT	Transportation cost per vehicle per unit distance
CCF_j	Fixed construction cost of a medical relief shelter at candidate location j
CCV_j	Variable construction cost per additional capacity of a medical relief shelter at candidate location j
CO_j	Operating cost of a medical relief shelter at candidate location j
CP_k	Procurement cost of medical supply k
B	Total relief budget
M	large number
E_i	$\begin{cases} 1 & \text{if patient } i \text{ is in emergency condition} \\ 0 & \text{otherwise} \end{cases}$

Decision variable

y_j	$\begin{cases} 1 & \text{if a relief center is constructed at candidate location } j, \text{for } j \in L_R \\ 0 & \text{otherwise, for } j \in L_R \end{cases}$
cy_j	Capacity of medical relief shelter at candidate location j
z_{ljk}	Quantity of medical supply k, required to be transported from medical deployment center l to medical relief shelter j, $l \in L_s$
x_{ij}	$\begin{cases} 1 & \text{if a patient is assigned to medical relief shelter } j \\ 0 & \text{otherwise} \end{cases}$

The proposed mixed integer programming formulation for the problem under consideration in this study is presented in the following.

Formulations

$$\text{Maximize } \sum \sum \frac{S_i}{D_{ij}} x_{ij} \qquad (1)$$

$$\sum_i x_{ij} \leq cy_j \qquad \forall j \in L_R \qquad (2)$$

$$A_k^e \sum_i x_{ij} E_i + A_k^n \sum_i x_{ij} (1 - e_i) \leq \sum_l z_{ljk} \qquad \forall k \in K,\ \forall j \in L_R \qquad (3)$$

$$\sum_j z_{ljk} \leq CK_{lk} \qquad \forall k \in K,\ \forall l \in L_S \qquad (4)$$

$$\sum_j CCV_j\, cy_j + \sum_j CCF_j\, y_j + \sum_j CO\, cy_j + \sum_l \sum_j \sum_k \mathrm{CP}_k\, z_{ljk} + \frac{1}{\mathrm{MV}} \sum_l \sum_j \sum_k z_{ljk}\, \mathrm{V}_k (CV + CT * D_{lj}) \leq B \qquad (5)$$

$$cy_j \leq My_j \qquad \forall j \in L_R \qquad (6)$$

$$x_{ij} \leq y_j \qquad \forall j \in L_R, \forall i = 1, \ldots, n \qquad (7)$$

$$\sum_i x_{ij} = 1 \qquad \forall j \in L_R \qquad (8)$$

$$y_j = \{0,1\} \qquad \forall j \in L_R \qquad (9)$$

$$x_{ij} = \{0,1\} \qquad \forall j \in L_R, \forall i = 1, \ldots, n \qquad (10)$$

$$cy_j \geq 0 \text{ , } z_{ljk} \geq 0 \qquad \forall k \in K, \ \forall j \in L_R, \forall i = 1, \ldots, n \qquad (11)$$

The objective function (1) maximizes the number of patients who gets medical services at multiple medical relief centers under the limited relief budget, while considering the severity of patients and distances to medical relief shelters. Constraint (2) indicates that total number of assigned patients in medical relief shelter j cannot exceed its capacity. Constraint (3) constrains the quantity of medical supplies k, transported from medical deployment center l must be more than required for emergency and non-emergency patients at medical relief center j. Constraint (4) defines the maximum amount of medical supplies which can be transported from medical deployment center l. The maximum relief budget including fixed construction, variable construction, operating, procurement, vehicle and transportation costs is limited in constraint (5). Constraint (6) reflects that the capacity of medical relief shelter j can be allowed only when its candidate location is chosen for the construction. Similarly, constraint (7) indicates that the patient can be assigned to medical relief shelter j only when the medical relief shelter is constructed at candidate location j. Constraint (8) indicates a patient is assigned to only one medical relief shelter. Finally, constraints (9), (10) and (11) define the nature of decision variables.

3 COMPUTATIONAL RESULTS

An extensive effort to test the proposed model for problems of various sizes under different scenarios has been conducted. The proposed formulation is modelled and implemented in Lingo 10 on a PC with 3.50 GHz Intel core i5-4690 CPU and 8GB RAM. A problem generator has also been programmed in Visual C++ 2013 to generate the locations of patients (x, y), where $0 \leq x, y \leq 100$ and the severity ranging 0 to 100 randomly. The program generator can accept the customized values for number of patients under impact, distribution ranges, severity ranges, threshold for emergency patient, candidate locations for medical relief shelters and locations medical deployment centers, etc.

For simplicity, Euclidian distance is used. The fixed construction cost of medical relief shelter and the variable construction cost per additional capacity are $20,000 and $50, respectively. The operating cost of medical relief shelter per patient is $100. The vehicle cost is $1,000 and the transportation cost per unit distance is $10. The procurement costs of medical supplies for three types are $30, $50 and $30, respectively. The volume for three types of medical supplies is 2, 3 and 5, respectively. The capacity of vehicle in volume is 20.

An example in Figure 1 shows 100 randomly distributed patients with their own severities, 5 candidate locations for medical relief shelters and 2 medical deployment centers. There are 3 types of medical supplies. It is assumed that the quantities of 3, 2 and 4 for three types are necessary for emergency patients, and 1, 1 and 1 for non-emergency patients. Black empty dots indicate the locations of patients, circles with cross in the middle candidate locations for medical relief shelters, and circles with triangle in the middle medical deployment centers. The number of shoulder of candidate locations indicates the index.

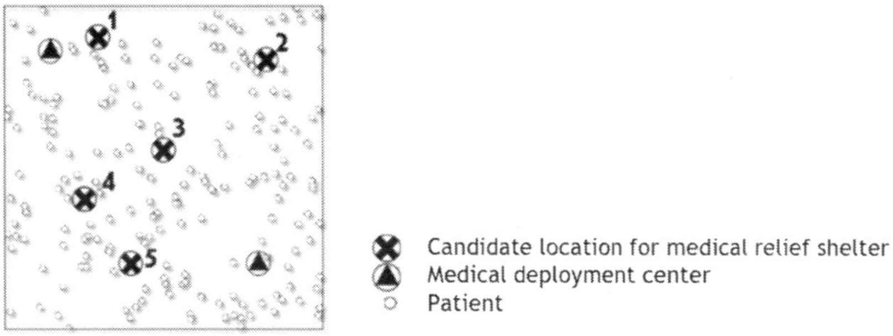

Figure 1: An example of test problem

Given the configuration of example in Figure 1, three scenarios under different relief budgets of $300,000, $250,000 and $200,000 are considered to validate the proposed model. In this test, the number of patients assigned (or relieved) and the locations of established medical relief shelters are studied under different amount of relief budget.

Table 1: Computational results for scenarios under different relief budget

	Relief Budgets		
	$300,000	$250,000	$200,000
Number of established medical relief shelter	5	5	4
Candidate location 1	12	12	11
Candidate location 2	23	11	0
Candidate location 3	18	16	17
Candidate location 4	22	19	18
Candidate location 5	25	15	13
Total number of relieved patients	100	73	59

The computational results for scenarios under different relief budgets are summarized in Table 1. Note that the relieved patients decrease as the relief budgets go down. Under the relief budget of $300,000, all patients have been assigned to the medical relief shelters.

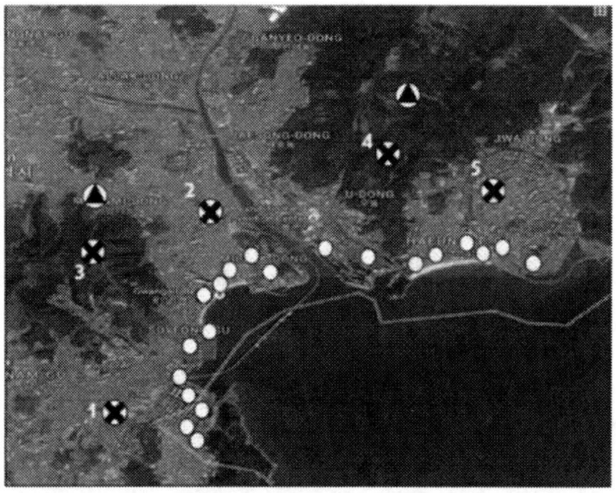

Figure 2: Coastal areas of Busan, Korea

Another example shown in Figure 2 is given for coastal areas of Busan, Korea. It includes 20 patients, 5 candidate locations for medical relief shelters, and 2 medical deployment centers. The patients are distributed in the waterfront since it is assumed that the disaster like tsunami has attacked the areas. The relief budget in this example is $50,000. The fixed construction cost of medical relief shelter and the variable construction cost per additional capacity is $8,000. Other parameters are kept as example in Figure 1.

In this example, three scenarios under the different degrees of impacts by disasters are considered to see how the situations are changed. It is natural to assume that a stronger impact by disasters will lead to more aggravating situations, where the severities of patients get higher. In three scenarios, the ranges of severity vary over 20~60, 30~80, and 40~100, respectively. Since emergency patients require more medical supplies as assumed, the urgency of medical treatments and the required costs including vehicle, transportation, and procurement costs get higher.

As shown in Table 2, the numbers of relieved patients under three different scenarios are 20, 19 and 15, respectively. Note that these numbers have decreased as the severity of patients get higher under the fixed relief budget. Depending on the cost information, it is advisable to decide whether the budget would be spent on the transportation or the construction of additional medical relief shelter wisely. The additional experiments have been performed to see the effects of medical supplies' quantities, which are required for emergency and non-emergency patients. If the required quantities of medical supplies for emergency and non-emergency patients are identical, the objective function in the proposed model of Section 2 puts the higher priority on emergency patients and the proposed model has more flexibility to relieve more patients.

Table 2: Computational results for scenarios under different degree of impacts

	Relief Budgets		
	20~60	30~80	40~100
Number of established medical relief shelter	3	2	2
Candidate location 1	13	14	15
Candidate location 2	3	0	0
Candidate location 3	0	0	0
Candidate location 4	0	0	0
Candidate location 5	4	5	0
Total number of relieved patients	20	19	15

4 CONCLUTION

The natural disasters such as hurricanes, typhoons, earthquake, and tsunami as well as the anthropogenic disasters such as chemical explosions, nuclear accidents, and terrors become more frequent in the recent years. In such emergency situations, the authorities need to augment its medical response capability by using the scarce medical resources to provide emergency medical service. Therefore, in this study, we proposed a variant of MCLP model to manage the logistics of both medical supplies and patients.

In this study, the locations of medical relief shelters are determined in consideration of the severity and distribution of the patients and the locations of medical deployment centers. Through a rigorous experimental study, the cost information and other input parameters, which are introduced in Section 2 had a great influence on the decisions on where to establish medical relief centers, who to be relieved by medicfal relief centers, How much capapcities each medical relief center should have, how much resources emergency and non-emergency patients need, etc. Therefore, it is important to have a concrete decision-making model to support the authorities in the emergency situations.

In emergency management system, the planners want to serve as many injured people and dispatch the scarce medical resources as quickly as possible with the lowest cost. The objective of this study is to maximize the number of patients who gets medical services at multiple medical relief centers under the limited relief budget, while considering the severity of patients and distances to medical relief shelters. The aim is to find optimal locations of temporary medical relief shelters to offer medical service to as many injured people as possible and minimize the total costs under limited relief budget simultaneously.

The computational experiences indicate the proposed model produces the optimal solutions quickly and the results were significantly varied in accordance to the input parameters. It is important to have a concrete model for the authorities to make a quick decision in the emergency situations. The timely response to disasters that call for medical service is very important. In future research, multi-time period will be incorporated, which may lead to the computational intractability. In such cases, an effective and efficient heuristics algorithm for disaster emergency can be proposed.

5 REFERENCES

[1] Balcik, B., Beamon, B. M. 2008. Facility location in humanitarian relief, International Journal of Logistics Research and Applications, 11(2), pp 101-121.

[2] Afshar, A., Haghani, A. 2011. Modeling integrated supply chain logistics in real-time large-scale disaster relif operations, Sosio-Economic planning Sciences, 46(2), pp 327-338.

[3] Day, J.M., Steven A. Melnyk., Paul D. Larson, Edward W. Davis, D. Clay Whybark 2012. Humanitarian and disaster relief supply chains: a matter of life and death, Journal of Supply Chain Management, 48(2), pp 21-36.

[4] Sheu, J.B., Pan, C. 2015. Relief supply collaboration for emergency logistics responses to large-scale disasters, Transportmetrica A: Transport Science, 11(3), pp 210-242.

[5] Jia, H., Fernando Ordonez, Maged M. Dessouky. 2007. Solution approaches for facility location of medical suplies for large-scale emergencies, Computers & industrial Engineering, 52(2), pp 257-276.

[6] Jia, H., Fernando Ordonez, Maged M. Dessouky. 2007. A modelling framework for facility location of medical services for large-scale emergencies, IIE Transactions, 39(1), pp 41-55.

[7] Farahani, R.Z., Hakmatfar, M. 2009. Facility location: Concept, Model, Algorithm and Case studies, 2009th Edition, Physica.

[8] Mete, H.O., Zabinsky, Z.B. 2010. Stochastic optimization of medical supply location and distribution in disaster management, International Journal of Production Economics, 126(1), pp 76-84.

[9] Farahani, R.Z., Asgari, N., Heidari, N., Hosseininia, N., Goh, M. 2012. Covering problems in facility location: A review, Computers & Industrial Engineering, 62(1), pp 368-407.

[10] Nicholl, J., West, J., Goodacre, S., Turner, J. 2007. The relationship between distance to hosptital and patient mortality in emergencies: an observational study, Emergency Medichine Journal, 24(9), pp 665-668.

[11] Najafi, M., Eshghi, K., Dullaert, W. 2013. A multi-objective robust optimization model for logistics planning in the earthquake response phase, Transportation Research Part E, 49(1), pp 217-249.

RESEARCH ON CAPABILITY EVALUATION METHOD FOR MULTI-STAGE MEASUREMENT SYSTEM

XU Yishu[1],MA Lin[2], Shi Liangxing[3*]
[1,2]School of Reliability and Systems Engineering
Beihang University, Beijing, China
lvdianzhang@sina.com; malin@buaa.edu.cn

[3]College of Management and Economics
Tianjin University, Tianjin, China
shi@tju.edu.cn

ABSTRACT

In the process of measurement system capability evaluation, the multi-stage measurement system that applies for a continuous stage of measurement of the same products with different quality characteristics is of great importance. The multi-stage multivariate measurement system model based on one-factor measurement system analysis (MSA) theory and the state space model for the multistage manufacturing process is built in this paper. The new model identifies the variance of local stage that is affected by upstream stage in the measurement system. After using partial least squares regression (PLSR) algorithm to eliminate correlation between the measurement system stages, one factor variance analysis and P/T %, R&R % indexes are used to measure the capability of system analysis. This paper proves its advantages with a case study in which four stages of the single variable measurement system capability analysis is carried out. Further simulations are also carried out to validate the model under four different conditions.

Keywords: MSA, state and space model, multi-stage measurement system

1 INTRODUCTION

In the process of measurement system analysis (MSA), there is one kind of multi-stage measurement system that does not measures quality characteristics simultaneously. The multi-stage measurement system measures different quality characteristics continuously based on the design characteristics of measurement equipment.

In the research of MSA, there are few related researches. Only Greg [1] focused on two kinds of errors for multi-stage measurement system, conducted simulation research and summed up the formula. However, there is still a blank in basic measurement system capability evaluation. Research starts from multi-stage manufacturing system. This paper made an effort to treat each measurement stage as special manufacturing stage without output for related MSA.

In the study of multi-stage manufacturing system, one of the key challenges is the complex relationship between quality characteristics during different stages. Product quality characteristics variation depends not only on the local stage, but also on the upstream stages that include correlative quality characteristics. The conventional statistical monitoring method distinguishes no difference of variations between local stage and upstream stages. As a result, a lot of false alarms are generated. Monitoring may mistakenly treat variation transmitted through the upstream stage as the one that local stage produces, generating an improved demand actually caused by other stages variations, and affecting decision-making. Zantek etc. [2] validated the model and procedure that measured changes and influence of quality characteristics of each stage. It clearly quantified the influence of the quality characteristics variation of each stage for the first time for evaluating stage performance. Ding, Dariusz and Shi [3] integrated the state-space model and the perturbation theory to build a new model. In the new model, quality characteristics are transmitted between stages through linear relationship, three-dimensional space model is transferred into two-dimensional vector set accompanied with system errors and principal component analysis (PCA) is used for linear regression in multi-stage manufacturing process model. Wang [4] proposed to use partial least squares regression (PLSR) algorithm to solve the multicollinearity between the quality characteristics. This kind of regression modelling method can solve more problems comparing with the ordinary multiple regressions, achieving an integration of a variety of data analysis method.

In this paper, the main contents of each part are as follows. The second part introduces basic concept and statistical properties involved in MSA. The third presents MSA algorithm based on the multi-stage manufacturing process (MMP). Through extension of multi-stage manufacturing process model, a new multi-stage measurement system evaluation model is provided with the details of the model structure and specific algorithm. In the fourth part, both the actual multi-stage measurement data and simulation data are used to verify the feasibility and advantages of the model. The summary and limitations are presented in the last part.

2 MEASUREMENT SYSTEM ANALYSIS

Measurement system includes one measurement equipment at least, in addition there are operators, environment, measurement method and calibration activities and so on. MSA aims to find out variation of measurement system in the total variance of whole observations, analyse total variation, process variation and measurement variation from the perspective of statistical analysis, and separate them from each other [5].

For one factor measurement system, the main method is analysis of variance (ANOVA)[6]. Then the measurement system evaluation indexes are calculated to judge the measurement system.

The focus of MSA is whether the measurement values truly reflect the fluctuation of tested product quality characteristics, so the evaluation of measurement system capability is the core of MSA. Common evaluation indexes are precise-to-tolerance ratio (P/T%) and percentage of repeatability and reproducibility(R&R%). P/T% is designed to measure whether a product conforms to the design specification, and R&R% is the ratio between the measurement precision estimate and the process width.

3 A MSA ALGORITHM BASED ON MULTI-STAGE MANUFACTURING PROCESS ALGORITHM

3.1 The Model

Suppose there are total q quality characteristics in N stages, whole multi-stage process model is as shown in figure 1.

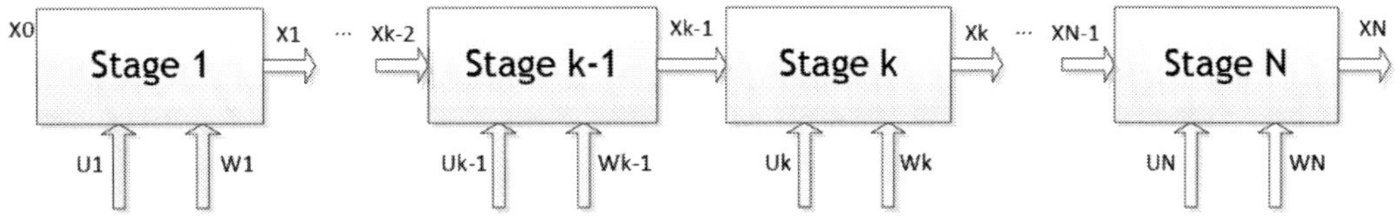

Figure 1: Multi-stage Process Model

In figure 1, X_k, quality characteristic of stage k, is composed of X_{k-1} and U_k together, where X_{k-1} represents for quality characteristic of stage k-1 and U_k represents for effect on quality characteristic of stage k. X_{k-1} is both the output of quality characteristic state of stage k-1, and the input of quality characteristic state of stage k, where U_k of each stage are independent of each other. Random vector W_k represents for random error with the mean of zero, and it is independent of U_k respectively.

In other words, the linear model of whole stages in word form is as follows: X = effect of product manufacturing of upstream stage + effect of product manufacturing of local stage + random error of local error. Based on that, this paper extends the state space model to the multi-stage measurement system. Regarding the measurement step as a manufacturing process without processing, consider measurement system impact of upstream stage on measurement system of local stage while taking measurement error into account an extended linear model in multi-stage measurement system is proposed: X_j = (product manufacturing quality effects + product measurement error effects) of (j-1)th stage + (product manufacturing quality effects + product measurement error effects) of jth stage + the random error of jth stage = (product manufacturing quality effects + product measurement error effects) of (j-1)th stage + **(product manufacturing quality effects + product measurement error effects + the random error)**of jth stage = measurement system observation effect of (j-1)th stage+ measurement system observation of jth stage. The bold words in the second part is the three components of MSA, namely measurement system observation of jth stage, which is set to M. At the stage j, regression algorithm is used to process the observations for the response coefficient calculation, and independent measurement data which are not affected by upstream stages are extracted and named as residuals. The processed measurement system is equal to k independent single factor measurement system for MSA and evaluation index calculation.

Define X_{ji} as the ith quality characteristic at the stage j. In the whole k stages, the output quality characteristics of each stage is i_j, and total number of output quality characteristics is q_j. With X_0 as initial states, the jth state space model in multi-stage measurement system is as follows:

$$X_{ji_j} = A_{(j-1)1ji_j} X_{(j-1)1} + A_{(j-1)2ji_j} X_{(j-1)2} + \cdots + A_{(j-1)i_{j-1}ji_j} X_{(j-1)i_{j-1}} + \cdots + A_{(j-1)i_t ji_{j-1}} X_{(j-1)i_t} + M_{ji_j}$$

(t=qi) *(1)*

Where $A_{(j-1)itjij-1}$ represents the effect coefficient that (j-1)i_t^{th} quality characteristic in upstream stage impact on ji_{j-1}^{th} quality characteristic in local stage, X_{ij} represents for ith quality characteristic of stage j.

Due to multiple quality characteristics during output of each stage, the final output has a certain correlation with quality characteristics in different stages. Therefore linear model of whole stages is as follows:

$$X_{ji_j} = \sum_{i=1}^{t} A_{(j-1)i_{j-1}ji_j} X_{(j-1)i_{j-1}} + M_{ji_j} \quad \textbf{(2)}$$

3.2 The Algorithm

For the model calculation, Ding, Dariusz and Shi proposed PCA. Though PCA has an advantage of finding out relevance information hidden in the independent variables, and when relevance between useful variables is tiny, the selected principal components might be ignored, which leads to decreases of the reliability of prediction model. PLSR, otherwise, is the integration and development of multiple linear regression, canonical correlation analysis and PCA. It can reduce dimension under the condition of minimum information loss. For PLSR, first of all, extract independent components from the independent variable set, and then build the regression equation based on components and the independent variables. Effective component extraction is the key of PLSR. Different from PCA, extracted components well summarize the information in the independent variable system, better explain the dependent variable, and eliminate the noise. Thus it effectively solves the regression modelling problem under consideration of multiple correlation between independent variables. In PLSR algorithm, firstly standardize the observations, then extract the first , second until h[th] component which is identified by the principle of cross-validation.Finally, the PLSR model is as (10).

$$\widehat{y^*} = \alpha_1 x_1^* + \alpha_2 x_2^* + \cdots + \alpha_m x_m^* \tag{3}$$

ANOVA and P/T %, R&R% obtain M, measurement residuals, as revised data analysis and calculation in (11).

$$M = y^* - \widehat{y^*} = y^* - \alpha_1 x_1^* - \alpha_2 x_2^* - \cdots - \alpha_m x_m^* \tag{4}$$

4 EMPIRICAL APPLICATION AND SIMULATION

4.1 Process Description

In a multi-stage measurement system, when there is a certain relationship between stages, the operations of measurement system in upstream stage may affect the ones in downstream stage. Using QTM instrument to measure is a multi-stage measurement process. In the system, single cigarette goes through 4 stages and respectively outputs 4 data: weight (g), circumference (mm), the suction resistance (Kpa) and hardness (%).Historical studies have shown that these indicators has certain relevance. The hardness is mainly influenced by weight of single cigarette and filling value of cut tobacco, with positively related to the weight of single cigarette. The suction resistance is mainly influenced by tobacco packing density, filling value of cut tobacco, the dust content of cut tobacco, and hardness, with positively related to the hardness [7].The four quality characteristics shall be regarded as a multi-stage response, thus do multi-stage MSA, study system ability, so as to get a pair of overall indexes (P/T% and R&R%) to evaluate the measurement system.

4.2 Model Foundation

Measurement with QTM instrument is a four stage single variable measurement system. Set the observation of weight (g), circumference (mm), the suction resistance (Kpa) and hardness (%) to X_1, X_2, X_3, X_4 respectively. The residual of four stages are U_1, U_2, U_3, U_4 respectively. The relationships between quality characteristic in 4 stages are as follows:

$$X_1 = U_1 \tag{5}$$

$$X_2 = \alpha_1 X_1 + U_2 \tag{6}$$

$$X_3 = \alpha_2 X_2 + U_3 \tag{7}$$

$$X_4 = \alpha_3 X_3 + U_4 \tag{8}$$

4.3 MSA

4.3.1 The Experiment Design and Data

For equipment, there is a calibrated QTM instrument where measure quality characteristics in different dimensions. Operators are only responsible for instrument operation and reading results, who has been well trained. The enterprise provides cigarette standard parts as measurement samples, due to bad reproducibility [8].The procedures are as follows: (1) Randomly select 10 cigarette standard parts, label number 1 to 10.(2) Repeat in measurement 3 times for every samples.(3) Collect measurement and standardize Data.(4) Carry out SIMCA-P software for PLSR.(5) Output the measurement residuals of each stage using correlation coefficient of fitting linear equation and the model.(6) Do MSA with measurement residuals of each stage.(7) Calculate P/T %,R&R % indexes and evaluate every stage of the measurement system combined with ANOVA.

At the same time, the measurement environment is based on national tobacco measurement standard [9]. Considering PLSR is adapted to for fitting measurement system model, experiment data acquisition can be divided into two parts. The first part acquires data to establish a regression equation model, and the second part acquires data for MSA.

4.3.2 Analysis and Evaluation

In measurement system capability analysis, the standardization of data makes the variables in the same direction, and no dimensional impact[10].The data subtract the mean and then are divided by the standard deviation, making four different dimensional data subjected to $N(0, 1)$. It can be proved by the definition that the standardization has no impact on ANONA and the result of the P/T % and R&R % indexes.

Using SIMCA-P software on the processed data modelling analysis. The software can carry out data fitting and explain the results. In accordance with the model set up before, set X_2, X_3, X_4 as the dependent variables and calculate the three independently. The relationships between stages that get from software results are as follows:

$$X_2=0.911711X_1-0.291755 \quad \textit{(9)}$$

$$X_3=0.881502X_2-0.600780 \quad \textit{(10)}$$

$$X_4=0.860815X_2-0.225035 \quad \textit{(11)}$$

Using the fitting equation to calculate the measurement residuals as the source of data for ANOVA. Measurement residual computation formulas of each stage are as follows:

$$X_1= \text{Measurement residual1} \quad \textit{(11)}$$

$$X_2-0.911711X_1= \text{Measurement residual2} \quad \textit{(12)}$$

$$X_3-0.881502X_2= \text{Measurement residual3} \quad \textit{(13)}$$

$$X_4-0.860815X_2= \text{Measurement residual4} \quad \textit{(14)}$$

The measurement system capability indexes are calculated after multivariate ANOVA by MINITAB. Use the formula to calculate the P/T % while the USL and LSL also need to be standardized. In the actual measurement system operations, to simplify the calculation, a fixed total standard deviation can be determined in the way of repeated measurement for better standardization. So the results of the two indexes are shown in table 1.

By table 1, the measurement capacity of circumference, suction resistance and hardness is sufficient, but the weight factor is affected by the measuring system, with a large value of P/T %. The enterprise should check and adjust the weight measurement part.

4.4 Simulation

In order to further verify the advantages of multi-stage multivariate measurement system model, this paper also adopts the method of simulation to analyse the measurement system under different conditions, and then achieves the comparison results. In a word, stage n influences stage n+2 through passing the effect on stage n+1.So according to the cases before, the simulation can be divided into four stages where the second stage is associated with the first stage, the third stage is only associated with the second stage, the fourth stage is only related to the third stage. The simulation can be seen as 1st-2nd, 2nd-3rd, 3rd-4th three independent stages. In order to simplify the simulation, only study 1st-2nd stage. To compare the usage of the multi-stage multiple measurement system model and four single measurement system model conveniently, the simulation assumes for our kinds of circumstances of 1st-2nd stage (when in out-of-control condition, the distribution of measurement system error meets *N* (0, σ_G'), σ_G' is from historical data from measuring system). See in table 2.

Table 1: Results of measurement system capability indexes

Measurement factors	P/T%	R&R%
Weight	37.81%	1.80%
Circumference	16.91%	5.27%
Suction resistance	13.52%	6.63%
Hardness	2.05%	7.94%

Table 2: Simulation state table

Test serial number	1st stage state	2nd stage condition
1	In control	In control
2	In control	Out of control
3	Out of control	Out of control
4	Out of control	In control

Using data characteristics of empirical analysis to simulate randomly the process data and measurement error, get 3 times repeated in 100 samples of the simulated experimental results, and by using the empirical analysis method to the simulation process analysis, thus draws the conclusion of measurement system capacity evaluation. In simulation, measurement data are composed by true value and measurement error. Taking advantage of the enterprise manufacturing standards and the data characteristics in the empirical analysis for parameter setting. By setting whether the P/T % is in control to determine to backstep the standard deviation of measurement system error. There is no need to consider the operators and interaction between the operators and parts because of the unobvious effect results in enterprise analysis. Add the two generated random numbers together, and each process true value is corresponding to three measurement errors, in order to simulate the three times repeatedly. Data will be standardized in the same way before. Still use SIMCA-P for linear fitting of standardized data, obtain the fitting equation and get measured residual values. Use residuals and standardized observations for the second stage of ANOVA. The first stage has no need for residuals calculation because of the lack of the influence factors. The final results of the two indexes are shown in table 3.

Though the simulation indexes results, when the second stage is out of control (state 2, 3), using the model is significantly better than do not use the model. By comparing σ_G' and σ_G, the former is closer than the latter between the true value of standard deviation. In test 2, the true value of the P/T % is 32.19%. According to the judgment rules of AIAG, it means the measurement system is out of control. However, when the influence of the upstream stage does not remove, the P/T % values of not using the model are less than 30%, just in an acceptable state of the system, and the P/T % values is relatively close to the value of true, reflecting that the measurement system is out of control. Test 3 is in the same way. In test 1 and 4, the P/T % results are similar, and using the model is a little better than not to use. In the comparison of R&R %, the value of using the model have the consistency with the P/T % value (both in a state of greater than or less than 30%), and when the P/T % don't detect the problem, its corresponding R&R % all well show out-of-control state of the measurement system.

Table 3: Simulation index analysis table

Test serial number	Stages		σ_G'	P/T% true value	σ_G	P/T% calculated value	R&R% calculated value
1	1st		0.005	21.45%	0.0049	20.96%	28.34%
	2nd	Out of model	0.01	12.88%	0.011263	14.04%	29.69%
		In model	0.01	12.88%	0.010125	13.50%	26.58%
2	1st		0.005	21.45%	0.0047	20.14%	27.25%
	2nd	Out of model	0.025	32.19%	0.02388	29.74%	43.88%
		In model	0.025	32.19%	0.02485	33.48%	52.40%
3	1st		0.01	42.92%	0.01012	43.44%	58.72%
	2nd	Out of model	0.025	32.19%	0.02266	28.04%	42.48%
		In model	0.025	32.19%	0.02455	34.48%	50.44%
4	1st		0.01	42.92%	0.01012	43.44%	58.72%
	2nd	Out of model	0.01	12.88%	0.010623	13.44%	28.63%
		In model	0.01	12.88%	0.010075	12.94%	26.18%

5 SUMMARY AND CONCLUSIONS

Based on MSA and measurement system capability evaluation method, this paper puts forward a new model to identify and eliminate measurement error transferred from upstream stages, combined with the state space model in the multistage manufacturing process. The measurement errors that do not belong to local stage are eliminated while the residuals are treated as calculation observations. The "residuals" are used for ANOVA calculation. In the multi-stage multivariable measuring system model, it uses PLSR for linear fitting, which makes each stage characteristics independent of each other, and then calculates the P/T % and R&R % indexes, completing measurement system capability evaluation. The model makes up a blank for multi-stage multivariable measurement system to evaluate the system capability more accurate and complete.

There are still some limitations in this paper. First of all, this paper still cannot do better fitting for destructive measurement stage. Measurement residuals of local stage are found based on the observation of local stage; therefore it cannot conclude that the destructive effect upstream stage on the local measurement when there are no more precise measurement tools. Secondly, empirical cases are lacked. It has not yet found a multi-stage multivariable measurement system in reality for empirical analysis and simulation test.

6 REFERENCES

[1] **Larsen,G.A. 2003**. Measurement system analysis in a production environment with multiple test parameters, *Quality Engineering*, 16(2), pp 297-306.

[2] **Zantek,P.F. Wright,G.P. Plante,R.D. 2002**.Process and product improvement in manufacturing systems with correlated stages, *Management Science*, 48(5), pp 591-606.

[3] **Yu,D. Dariusz,C. Shi,J. 2002**.Fault diagnosis of multistage manufacturing processes by using state space approach. *Journal of Manufacturing Science and Engineering*, 124(2), pp 313-322.

[4] **Wang,N. Xu,J. Yang,J. 2013**. Method of Identifying Key Quality Characteristics in Multistage Manufacturing Process Based on PLSR, *Operations Research and Management Science*, 22(5), pp 226-232.

[5] **Montgommery,D.C. 2008**. *Statistical Quality Control: a Modern Introduction*. John Wiley&Sons.

[6] **Zhou,S.Qiang,H.Shi,J. 2003**. State space mopropagation in multistage machining process using differential motion vectors, *Robotics and Automation*, 19(2), pp 296-309.

[7] **Ni,K. Fan,T. 2002**. Correlation Analysis of Hardness, Draw Resistance and Unit Weight of Cigarettes, *Tobacco Science & Technology*, (3), pp 9-13.

[8] **Fan,S. Yu,X. Fan,X. 2014**. Analysis of Cigarette Measurement System, *Tobacco Science & Technology*, 4, pp 10-14.

[9] **GB/T 16447-2004**. Tobacco and Tobacco Products-Atmosphere for Conditioning and Testing.

[10] **Guo,Y. 1998**. Sensibility and Practice Analysis of Comprehensive Evaluation Result, *Journal of Management Sciences in China*, 1(3), pp 28-35.

,

AUTHOR INDEX

AUTHOR INDEX

AUTHOR INDEX

AUTHOR INDEX

46th International Conference on Computers & Industrial Engineering 2016 (CIE46)

Tianjin, China
29 – 31 October 2016

Volume 2 of 2

ISBN: 978-1-5108-3478-1

Printed from e-media with permission by:

Curran Associates, Inc.
57 Morehouse Lane
Red Hook, NY 12571

proceedings.com

Some format issues inherent in the e-media version may also appear in this print version.

Printed by Curran Associates, Inc. (2017)

Computers & Industrial Engineering
c/o Mohamed Dessouky
3715 McClintock Avenue, GER 240
Los Angeles CA 90089-0193

Phone: (949) 425-1686
Fax: (949) 425-1696

dessouky@usc.edu

Additional copies of this publication are available from:

Curran Associates, Inc.
57 Morehouse Lane
Red Hook, NY 12571 USA
Phone: 845-758-0400
Fax: 845-758-2633
Email: curran@proceedings.com
Web: www.proceedings.com

TABLE OF CONTENTS

VOLUME 1

VOLUME 2

A COMPARATIVE ANALYSIS OF CANCER HOSPITALS THROUGH DATAMINING OF PATIENTS SATISFACTION

Irfan Anjum Manarvi*
Department of Engineering Management,
Prince Sultan University, Riyadh, Kingdom of Saudi Arabia
irfanmanarvi@psu.edu.sa

ABSTRACT

Cancer is one of the deadliest diseases which require immediate attention by the sufferer as well as his doctors and hospital support staff. Its cure may not be completely possible for a number of reasons. Patients and their attendants undergo series of decision making processes to carry out its treatment. Experiences of patients are therefore considered very important for the best handling of this disease. In third world countries like Pakistan; choice of hospitals, expenses, amount of travel to the hospital by patients and treatment given by the doctors and nurses at hospital plays a vital role in early recovery. Data of patients is recorded in medical logs as well as reports of their handlers. This information is of considerable importance for new patients aspiring to have sound treatment in any hospital. Present research was focused on 30 cancer patients undergoing cancer treatment at two premier hospitals of Islamabad Pakistan for over an year. Their experiences were recorded through a questionnaire answered. It lead to inferences about various factors such as waiting time for treatment, expenses on tests, travel to hospitals, attention given by the hospital staff and doctors and administrative problems faced. This research provides information on various factors for selecting a hospital for patients and hospital management staff to understand the expectations to help them provide better services.

Keywords: Cancer treatment , patients satisfaction, Data mining

1 INTRODUCTION:

Cancer has always been considered as a very serious disease all over the world. A lot of research has been carried out in establishing its causes at early stages as well as medication and treatment that could help cure a wide variety of its patients. A study was carried out [1] to compare the costs of two regimens of chemotherapy for lung cancer. Weighted costs concepts were applied to observe the influence of two different types of treatments on a number of patients. Direct costs were applied to nursing time and the dosage of medicines and also the side effects costs were calculated. Economic analysis showed that such studies could help in rational therapeutic decisions. In another research the affects of hospital type on surgical outcomes of ovarian cancer patients were investigated [2]. Effect of hospital type on chemotherapy efficacy was investigated. It was found that hospital type did not influence the outcomes of first-line chemotherapy in ovarian cancer patients. Overall survival was better in hospitals with two or more medical oncologists. A hospital

* Corresponding Author

wide survey of pain in a tertiary cancer center in Hanoi, Vietnam was carried out [3]. Moderate to severe pain was reported in 50% respondents with 21% reporting severe ratings of pain and 33% reported severe impairment in their ability to work due to pain. Liver cancer disparities were studied to establish measurement tools such as surveys assessing knowledge, attitudes and behaviors of patients [4]. Same questionnaire were conducted in three different languages with same constructs including Vietnamese, Hmong and Korean languages. It was observed that different cultures and languages did not hinder in the processes of questionnaire based surveys and other researchers could also use this methodology across different populations. Treatment of cancer in childhood was investigated as a costly affair [5] because of factors such as periods of hospitalization, use of physician, professional services, laboratory, diagnostic tests, administration of chemotherapy, and pharmaceuticals. It was also found that treatment required frequent outpatient visits which imposed serious financial and administrative burden on the families. Distress in patients after curative surgery for colon cancer was investigated to establish how this corresponds with the need for supportive care [7]. A survey recorded symptoms, experienced problems, and (un)expressed needs. Satisfaction with supportive care was recorded. It was observed that patients with physical problems seek supportive care in one in three cases, while patients with psychosocial problems only seek help in one in eight cases.A study was conducted to to investigate whether clinical or patient reported outcomes are most important for perceived quality of care by breast cancer patients [8]. Avoiding severe breast symptoms rather than 2 year longer disease free survival is considered most important in our population of breast cancer patients for evaluation of quality of care. Research was conducted to empirically test a model of association linking hospitalized cancer patients' health status, nursing care quality, perceived individuality in care and trust in nurses. A cross-sectional, exploratory and correlational study design was used [9]. A model of trust in nurses was developed. Health status, individualized care, and nursing care quality were found to be associated with trust. The model highlighted the complexity of caring for cancer patients. Trust in nurses was found to be s influenced by the provision of individualized care. Patient, Family-Centred Care (PFCC) is internationally advocated as a way to improve patient care. An integrative review was conducted to extend the knowledge and understanding by synthesising empirical evidence of PFCC interventions within the adult intensive care unit (ICU) setting [10]

A lot research has been carried out for specific types of cancer itself, limited information exists that could assist in decision making to individuals requiring treatment for selection of a hospital among available alternatives. Present research was therefore focused on survey of two hospitals, located at Islamabad and Rawalpindi area Pakistan. The objective was to establish factors which patients considered critical for their decision making process to choose a hospital. This data could be termed as tacit knowledge gained through patient's and attendants individual experiences rarely shared or published for analysis in a structured manner.

2 METHODOLOGY

A survey of two hospitals was carried out in which the patients and their attendants were interviewed to seek information on their problems for treatment of cancer. About 300 persons out of which 30 patients undergoing treatment for over an year at the two hospitals were interviewed. Following research questions were framed for collection of data and its analysis:

1. Which hospital had more number of days for diagnostics based on its facilities.
2. Which hospital had higher expense for tests.
3. Which hospital had higher waiting time for the patients .
4. Which hospital provided better consultation time with doctors.

5. Which hospital provided better quality services in terms of nurses and doctors response and visits.
6. How did travel cost influence the selection of a particular hospital.
7. Which hospital had more convenience and satisfaction level for the patients as well as attendants.

Primary data included variables such as number of diagnostic days, expenses on tests, waiting time for check up, consultation time for doctors, travel and medical expenses, nurses rounds, follow up visits by doctors, patients convenience and satisfaction level in each of the hospital by respective respondents which was statistically analysed. The two hospitals are indicated as H1 and H2 respectively.

3 PERIOD FOR DIAGNOSIS:

Pakistan has a population of well over 160 million out of which 80% are living in towns and villages with limited health care facilities. However; cancer of any form is considered as one of the deadliest diseases among the populace. Factors such as lack of awareness, self medication and nominally qualified medical staff at rural health centers result into very late identification of disease. Exact diagnosis starts when such patient reports to a suitable hospital. The two hospitals under study have elaborated facilities for diagnosis and are funded by the government. Both also have the possibilities of sharing each other's laboratory facilities. The doctors normally advise a series of tests to the patients on their arrival as a part of their routine medical advice. The patients or their attendants are required to seek these test results as soon as possible with the help of hospital staff. Considering the complexity of disease the number of days for diagnostics could range from 10 to 50 days. Data collected for diagnostic days in both hospitals for the patients understudy is shown in Figure 1.0 below:

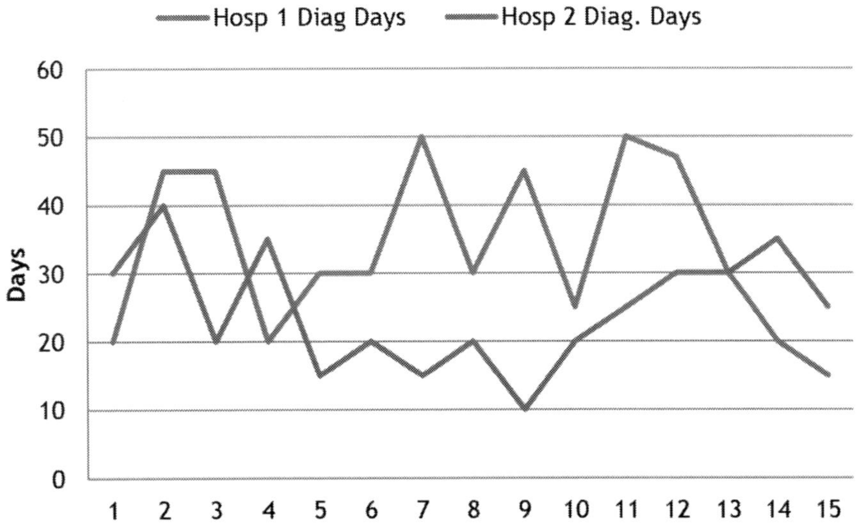

Figure 1.0 Comparison of diagnostic time period prior to start of treatment

Following inferences could be drawn from above figure:

(a) Hospital H1 was taking more number of diagnostic days among 67% of the patients for similar type of cancer. Quality and type of tests and equipment was similar in both hospitals.

(b) Exact number of days taken by H1 was twice that of H2 among selected respondents.

(c) Only four patients got their diagnosis carried out in H1 earlier than similar patients in H2.

Performance of H2 appeared better as compared to H1 with respect to no. of days for diagnosis

4 TEST EXPENSES:

Both hospitals were funded by the government for government employees and the members of armed forces. However. Present study was focused on self funded patients. A comparison of test expenses for the two hospitals is shown in figure 2.0 as follows:

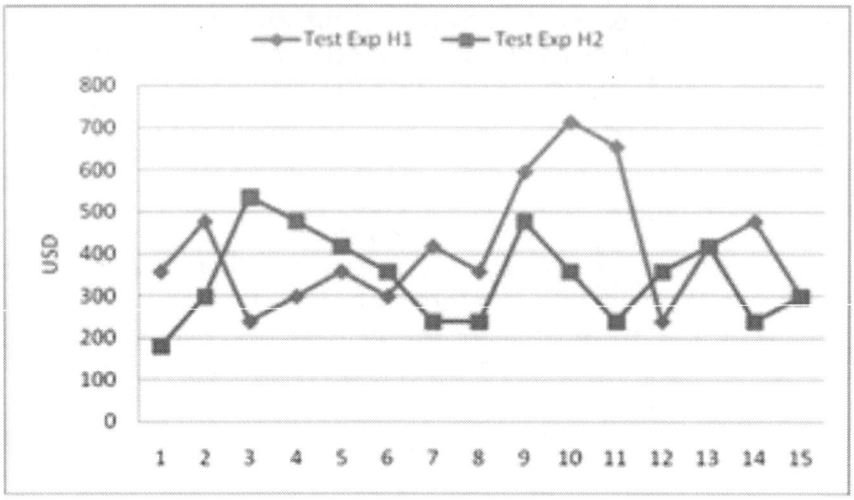

Figure 2.0 Cost of tests at the two hospitals

Above figure lead to following inferences:

(a) Test expenses showed varying trends in both hospitals depending upon the type of cancer and the stage a patient reported it at the hospitals.

(b) Expenses in H1 were substantially higher for patients 9,10 and 11 due to late reporting by the patient and having entered various complexities prior to diagnosis.

(c) Mean expenses for tests at H1 were 15 to 20% higher as compared to H2 regardless of the type of cancer.

(d) Overhead charges of H2 were paid by different financial heads of the government whereas H1 was provided lump sump funds by government and hence did not get extra funding for overheads.

Possibility of getting tests from either hospital by the patients could reduce their costs by using testing facilities from H2. However they were forced to take tests from same hospital if it was inconvenient to move the patient from one to another

5 WAITING TIME AND CONSULTATION:

A huge trend of late reporting to a specialist cancer hospital was observed among the patients. This factor directly attributed to the waiting time by a patient and consultants time allocated for advice on treatment. Both the hospitals had different trends for these parameters based on the number of doctors and time taken for diagnostic tests. Waiting time was calculated on the basis of arrival of a patient with or without an appointment. Consultation time was based on actual time spent by the doctor studying the case of an individual patient and arriving at some conclusion for further advice. A comparison of the two hospitals is shown in figure 3.0(a) and 3.0(b) below:

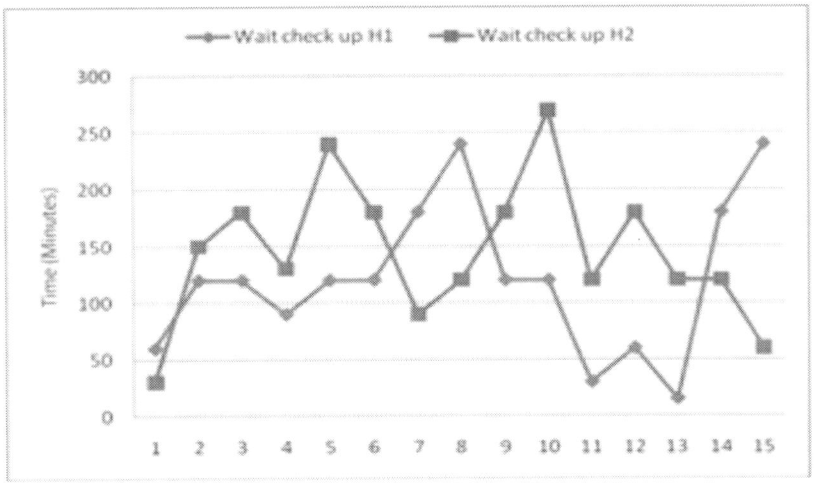

Fig 3.0 (a) Patients waiting time

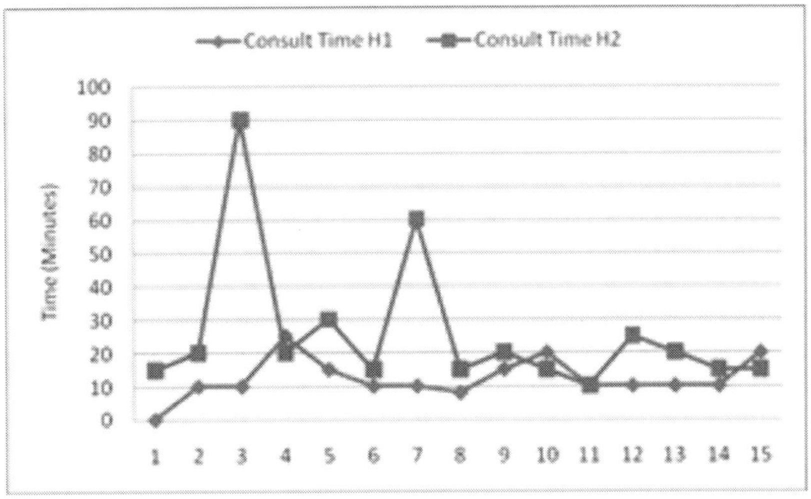

Figure 3.0 (b) Doctors consultation time for specific patient

Above figures revealed following observations:

(a) Patients had to wait much lesser time for consultation with the specialist doctors in H1

(b) Specialist doctors had other engagements in H2 which resulted in more waiting time for patients on a usual business day.

(c) Average waiting time for H2 patients was approximately 20% higher than patients at H1

(d) Consultation time given by specialist doctors was similar in both except few exceptions.

(e) Average consultation time in H1 was 10 to 15 minutes and from 20 to 25 minutes for H2

Patients however did not appreciate long waiting times to have consultation from a specialist doctor and also wished to be explained more about their condition.

6 MEDICAL AND TRAVEL EXPENSES:

Cancer treatment is considered as very expensive and when coupled with travel expenses; it is a burden on the family. Patients and attendants travel from far off places to arrive at these hospitals.

H2 was well linked with transportation systems whereas H1 required private transportation which was costly. Data collected for transportation expenses for two hospitals was compared as shown in Figures 4.0below:

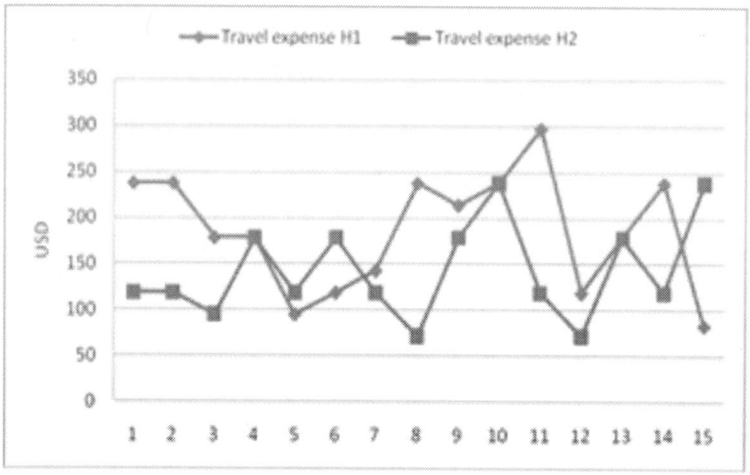

Fig 4.0 (a) Travel Expenses comparison

Reviewing the trends one could arrive at the following inferences:

(a) Medical expenses in both hospitals were the same.

(b) There were isolated cases observed where expenses of H2 patient were grossly higher as compared to H1 attributed to late diagnosis of a disease requiring specialized treatment.

(c) Travel expenses were higher for H1 patients because location was in accessible through public or mass transit systems.

Patients and their attendants did not give much importance to travel expenses vs. quality of treatment.

7 SERVICE QUALITY:

Hospital service quality was considered as of vital importance by the patients. They considered behavior of nursing staff as well as doctors as a primary factor for evaluating service quality of the hospitals. A comparative analysis of nurses rounds/day and follow up visits by specialist doctors is shown in Fig 5.0 (a) and (b) respectively.

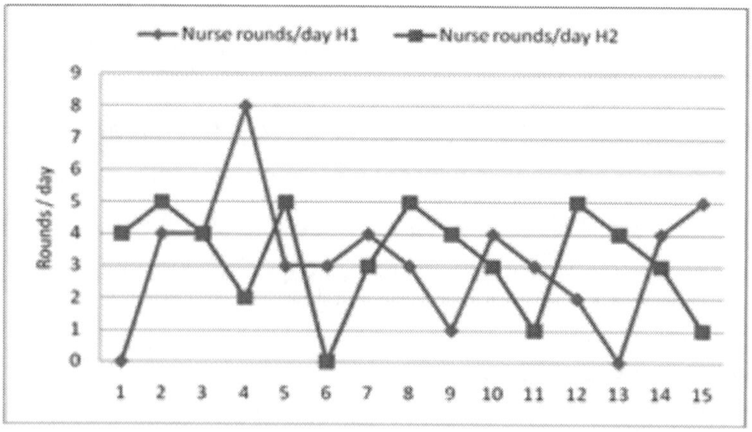

Figure 5.0 (a) Services quality assessment by patients

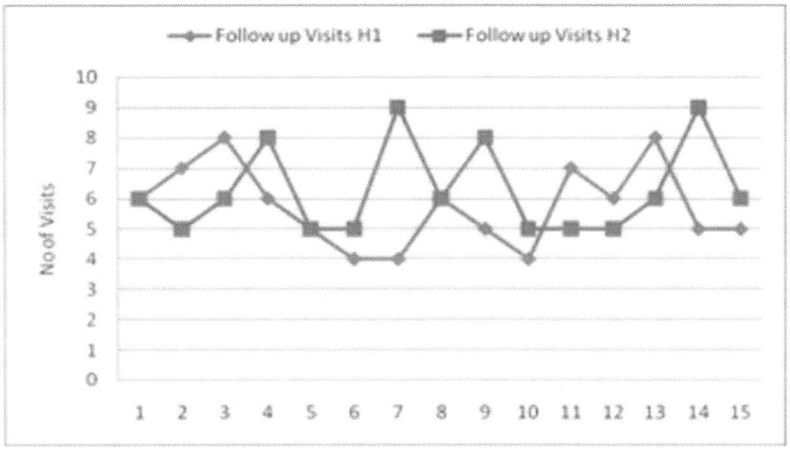

Figure 5.0 (b) Services quality assessment by patients

Above figures lead to following inferences:-

(a) Nurses rounds per day were 4-6 on the average for H1 and 4-5 for H2.

(b) Severity of a patients disease contributed to the nurses rounds/day.

(c) Follow up visits were higher in H2 as compared to H1.

Patients considered the services provided by H2 better than that of H1.

8 FINDINGS:

Present research lead to the following findings:

(a) Hospital H2 carried out quicker diagnostics as compared to H1 with lower test expenses.

(b) Waiting time for patients at H2 were higher than H1 which provided specialists consultancy to the patients much quicker. Treatment started earlier in H1 as compared to H2.

(c) Self financed patients had higher expenses in H2 as compared to H1 due to its location.

(d) Patients at H1 had higher number of nurses rounds and follow up visits than H2.

(e) For satisfaction level of patients, H1 scored higher among the patients as compared to H2.

Present research collected data from patients and their attendants for key parameters for selection of a hospital. Analysis of two hospitals helped to establish these parameters for patients for decisions based on quality of service.

9 CONCLUSIONS:

Patients suffering cancer are likely to undergo various cycles of depression even if the disease is diagnosed at an early stage. Their frustration increases due to administrative delays in start of actual treatment especially in third world countries. This delay causes their disease to worsen if not promptly handled. Patients primarily need an assurance of cure by the nurses and doctors on regular basis. When they do not see timely response, they like to move from one hospital to another regardless of the expenses that may incur. Hospital management is thus required to give reasonable importance to the convenience and satisfaction levels of a patient through efficient management systems. Further research in this area may be carried out for cancer as well as other chronic diseases to help hospital managers to improve upon this vital sector of services industry.

10 ACKNOWLEDGEMENT:

The author greatly acknowledges the work done by Ms Kiran Fatma for this research project.

11 REFERENCES:

[1] **A Vergnenegre, M. Perol, E.Pham. 1996.** Cost analysis of Hospital treatment-two chemotherapic regimes for non surgical non small cell lung cancer'. Lung Cancer, Elsevier, Vol 14, 31-44.

[2] **Flora Vernooij, Petronella O Witteveen, Eva Verweij, Yolanda van der Gaaf, A. Peter M Heintz. 2009**, Impact of hospital type on the efficacy of chemotherapy treatment in ovarian cancer patients.` Gynecologic Oncology, Vol 115 , pp 343-348.

[3] **Cielito C, Reyes-Gibby, DrPH, Nguyen Ba Duc, Nguyen Phi Yen, Nguyen Hoai Nga, Thuan Van Tran, Hong Guo, Samrat Bhat, Charles Clleland. 2006,** Status of cancer Pain in Hanoi, Vietnam: A hospital wide survey in a tertiary Cancer treatment center', Journal of Pain and Symptom Management, Vol 31 No 5 pp 431- 439.

[4] **Annette E. Maxwell, Roshan Bastani, Moon S Chen Jr, Tung T. Nguyen, Susan L Stewart, Vicky M Taylor, 2010,** Constructing a theoretically based set of measures for liver cancer control research studies', Preventive Medicine, Elsevier, Vol 50, pp 68-73.

[5] **Ronald D Barr, David Feey, William Furlong, 2004,** Economic evaluation of treatments for cancer in childhood' , European Journal of Cancer Vol 40, pp 1335-1344.

[6] **J.L. Kuijpens, B. Hansen, J.F. Hamming, J.G Ribbot, H.R. Haak and J.-W.W. Coebergh, 1998**, Trends in treatment and long term survival of Thyroid cancer in Southeastern Netherlands, 1960-1992', European Journal of Cancer Vol 304, No 8, pp 1235-1241.

[7] **T. Wieldraaijer, L.A.M. Duineveld, K.M. van Asselt, A.A.W. van Geloven, W.A. Bemelman, H.C.P.M. van Weert, J. Wind, 2016**, Follow-up of colon cancer patients; causes of distress and need for supportive care: Results from the ICARE Cohort Study, European Journal of Surgical Oncology (EJSO).

[8] **Melissa Kool, Joost R.M. van der Sijp, Judith R. Kroep, Gerrit-Jan Liefers, Ilse Jannink, Onno R. Guicherit, Robbert Vree, Esther Bastiaannet, Cornelis J.H. van de Velde, Perla J. Marang - van de Mheen,** Importance of patient reported outcome measures versus clinical outcomes for breast cancer patients evaluation on quality of care, The Breast, Volume 27, June 2016, Pages 62-68

[9] **Andreas Charalambous, Laurel Radwin, Agneta Berg, Katarina Sjovall, Elisabeth Patiraki, Chryssoula Lemonidou, Jouko Katajisto, Riitta Suhonen,** An international study of hospitalized cancer patients' health status, nursing care quality, perceived individuality in care and trust in nurses: A path analysis, International Journal of Nursing Studies, Volume 61, September 2016, Pages 176-186

[10] **Marion L. Mitchell, Fiona Coyer, Susanne Kean, Renee Stone, Jenny Murfield, Toni Dwa,** Patient, family-centred care interventions within the adult ICU setting: An integrative review, Australian Critical Care, 1 September 2016

ETCHING PROCESS DIAGNOSE BASED ON DATA-SPLIT K-MEANS CLUSTERING

Jungang Yang[1] and Jie Zhang [1*]
[1]School of Mechanical Engineering
Shanghai Jiao Tong University, Shanghai, 200240, P. R. China
yjg@sjtu.edu.cn, zhangjie@sjtu.edu.cn

ABSTRACT

With the development of tool monitoring technology, more and more data records are collected as time series from etching process in semiconductor manufacturing. Traditional statistic theory based process control methods are challenged by large-scale data and limited response time. In order to analyse so much data while keeping effective and rapid feedback, a parallel clustering based process diagnose method was proposed. A k-Means clustering algorithm was used to group process data into normal or fault status without manual supervision. This clustering algorithm was enhanced by parallel center calculation aiming to deal with large-scale time series. After cleaning and normalizing raw streaming data at first, a data split procedure was introduced to slice data into several segments. Cluster centers initialization of the proposed clustering algorithm was improved by arranging each segment to an independent working thread and calculating its center separately. These parallel center calculations made it able to process mass data. After all of segment centers were obtained, following clustering steps were executed. Experiments show that this improved clustering method could do with large-scale data set and distinguish process status, which may help keep high performance of etching tools.

Keywords: Process Diagnose, Fault Classification, Data Analysis, k-Means Clustering, Semiconductor Manufacturing

1 INTRODUCTION

To produce micro structure of circuit, an etching process is used to selectively remove unmasked portions of a layer, which presents circuit features transferred from lithography process [1]. Plasma etching is one of the etching craft in which a solid thin film is removed by a physical and chemical reaction with ground-state or excited-state neutral species using plasma [2]. This process is carried out in an etching tool named plasma etching tool. It is the beginning of "real fabricating" because once the etching starts, the processed wafer cannot return to its original shape as materials have been removed. This unique characteristic makes it a critical step in entire semiconductor manufacturing procedure, and also impacts on the final yield greatly.

Because of its significant influence to the semiconductor manufacturing, Fault Detection and Classification (FDC) is the most common technology used to monitor etching tools and diagnose etching process status [3]. Nowadays a plasma etching tool is combined with hundreds of sensors, which report its status through a large number of parameters. With data collected from these sensors, traditional statistical process control (SPC), such as control charts [4], Hotelling T^2 [5] and principal component analysis (PCA) [6] are the most common used FDC methods. As univariate SPC

* Corresponding Author

has its own weakness to tell relationships between several parameters, improved SPC methods like multiway PCA [7] and partial least squares discriminant analysis (PLS-DA) [8], classification methods like k-nearest neighbor (kNN) [9] and support vector machine (SVM) [10], and artificial intelligence like artificial neural network (ANN) [11], have been introduced into FDC system.

The "big data" characteristics [12] of etching process data bring challenges to traditional process diagnose and fault classification methods. A significant problem is how to analysis and find a group of failure data in such a huge data set with a limited time. SPC, kNN and ANN spend much more time and require more resources while the scale of data set increases. Hence a data-split based process diagnose and fault classification approach is proposed in this paper. Raw data is first pre-processed and cleaned by selectively picked out fragmentary data. Then moving average is used for data normalization. Finally, an improved k-Means algorithm working with data split approach is developed to cluster the pre-processing data into normal and fault part. Numerical and industrial example are used to demonstrate practicability and effectiveness of the proposed process diagnose and fault classification method.

2 DATA PRETREATMENT AND CLUSTERING METHODS

As the special data characteristics mentioned above, previous single node data process achievement encounters problems like low effectiveness and limited capacity. The core idea of big data analysis is to slice a huge data package into small segments and process by large amount of servers, instead of a powerful minicomputer. This idea helps extend existed solutions by parallel computation. Hence a data-split pretreatment and multi-thread computation is used to improve k-Means clustering algorithm. The major steps are as follows:

2.1 Data normalization

Since etching data shows a great variety in data shape, dimensions and scale, data normalization is an important data pre-processing step. Because the etching data are sent second by second, former methods of calculating average and standard deviation using all data is infeasible. Here a moving average is used to calculate $\overline{x}$ and σ.

From time point t_0 to time point t_n, there are several samples gathered, for example, $X_n = [x_1, x_2, ..., x_n]$. The mean and variance of these samples are as follows.

$$\overline{X} = \frac{1}{n}\sum_{i=1}^{n} x_i, \sigma = \sqrt{\frac{\sum_{i=1}^{n}(x_i - \overline{X})^2}{n-1}} \quad (1)$$

Data record x_i can be normalized as $x'_i = \frac{x_i - \overline{X}}{\sigma}, i = 1, 2, ..., n$, and normalized sample data set is $X'_n = [x'_1, x'_2, ..., x'_n]$.

On time point t_{n+1}, a new data record x_{n+1} is added into the data set, and x_1 is removed from the data set. Now the data set is $X_{n+1} = [x_2, x_3, ..., x_{n+1}]$. The mean and variance of these samples are updated as follows.

$$\overline{X'} = \frac{1}{n}\sum_{i=2}^{n+1} x_i, \sigma' = \sqrt{\frac{\sum_{i=2}^{n+1}(x_i - \overline{X'})^2}{n-1}} \quad (2)$$

Data record x_i can be normalized as $x'_i = \frac{x_i - \overline{X'}}{\sigma'}, i = 2,3,...,n+1$ Normalized sample data set comes into $X'_{n+1} = [x'_2, x'_3, ..., x'_{n+1}]$.

In this way, all of the data can be normalized into a normal distribution follows $N[0,1]$ with similar scale.

2.2 Data cleaning

All of the etching process data is collected from sensors installed in an etching tool via monitoring network in a fab. Interference from workshop, overload of data writing, failures of sensors and monitoring network may cause data loss or singular point. For a partial blank data record, it can be simply removed from the entire data set because a single sample has tiny impact to the entirety in the situation of big data. For a singular point in data sets which follows a normal distribution, a 6σ rule is used to find it out.

The 6σ range is defined bellow:

$$x \in [\bar{x} - 6\sigma, \bar{x} + 6\sigma] \quad (3)$$

Where $\bar{x}$ is the mean of data, and σ is variance.

Data records are eliminated if they exceed range of 6σ according to the theory of normal distribution. As the 6σ range covers 99.9999998% of the entire data space, the rest data has a minor contribution to the final result and can be treated as singular points and then excluded.

2.3 Process diagnose

As etching tools run under normal status or fault status, etching processing data is divided into two parts. Thus a process diagnose and fault classification issue can be treated as a cluster issue. K-Means clustering algorithm is a popular and effective method for cluster analysis in data mining [13]. The original k-Means clustering algorithm first select k objects in total n objects randomly, each object is initially representing a clustering center. The remaining objects are assigned to the nearest cluster according to its distance from the center of each cluster. Then center of each cluster is recalculated. This process is repeated until all of the unclustered objects have been labeled.

Because the original k-Means clustering algorithm calculates all of the distances from each unlabeled object to each cluster center, and then update this center later, the problem is computationally difficult (NP-hard). In big data situation, it could be a curse of dimensionality. Hence a data-split k-Means clustering algorithm is proposed to reduce number of unlabeled objects in a single classification.

The main idea of the proposed method is to cluster a data set by computing on several parallel threads. Data set is firstly split to several segments according to certain rules. Then each segment is dispatched to a working thread and parallel calculate its center. Once all the centers are obtained and sent back, a final cluster job carries out and outputs classification results.

The proposed data-split k-Means clustering algorithm presents under 4 steps.

2.3.1 Data splitting

The number of data records contained in a split data segment can be calculated based on Guha S, Rastogi R and Shim K's sampling method [14, 15].

$$s = f \times n + \frac{n}{n_i} \times \log(\frac{1}{\delta}) + \frac{n}{n_i}\sqrt{\log(\frac{1}{\delta})^2 + 2 \times f \times n_i \times \log(\frac{1}{\delta})} \quad (4)$$

Where f is the sampling rate, $0 \le f \le 1$;

n is the number of data records;

n_i is the number of data records in cluster C_i;

Eq. (4) means the number of data records contained in a split data segment s with the probability $1-\delta(0 \le \delta \le 1)$ to drawing $f \times n_i$ pieces of data.

2.3.2 *Initializing clustering centers*

Based on the data splitting of step 1, the center of samples is similar to the raw data centers. Forgy's method [16] is used here to randomly choose k observations from the split data segment and use these observations as initial centers. Then these clustered segments are combined together as segment-clustered data set with an average of every segment cluster. At last a second clustering procedure is carried out to find the final center of the original data set.

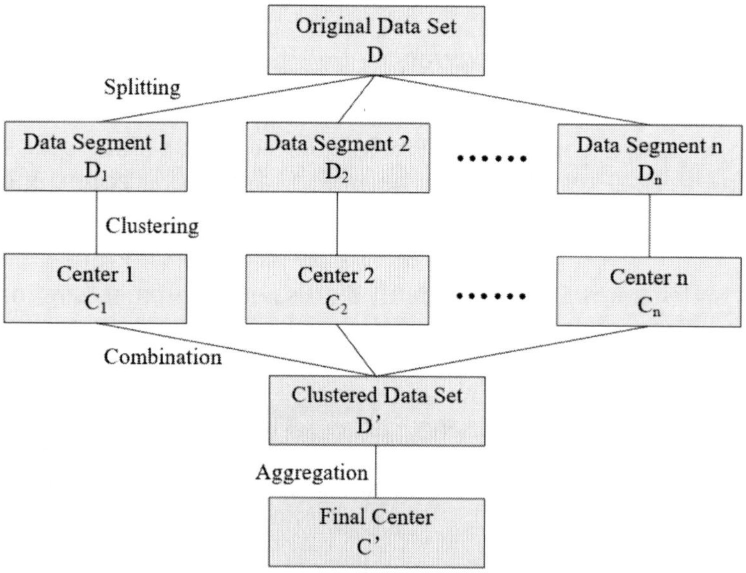

Figure 1: Process of initializing clustering centers

2.3.3 *Updating center*

The remaining data after data splitting is used here to update clustering centers. As soon as an unlabeled data record is allocated into a cluster C_i, the center of this cluster is updated as follows:

$$\begin{cases} x_i \leftarrow \dfrac{x_i \times n_i + x_\Delta}{n_i + 1} \\ n_i \leftarrow n_i + 1 \end{cases} \quad (8)$$

Where x_Δ is the unlabeled data record, and n_i is the data number of this cluster.

2.3.4 Clustering data

The k-Means clustering algorithm can be trained under an unsupervised learning mechanism. Newly collected data can be input into this classifying model and divided into several patterns. In a simple process diagnose situation, there are only two patterns. The etching process status can be monitored as normal or failure pattern.

3 EXPERIMENT AND CASE STUDY

This section presents numerical and actual experiment studies to demonstrate the practicability and effectiveness of the proposed approach. Prototype system is developed by C# combine with .Net Framework 4.0 under Microsoft® Visual Studio® 2010 IDE. All experiments were carried out on a server with configurations of an Intel® Xeon® E7-4807 CPU (1.87GHz @ 6 cores, 12 hyper-threadings), 32GB memory and 6TB hard disk.

3.1 Illustrative experiment

During the illustrative experiment, all of the test data records are generated randomly. Each data record contains 6 parameters like $d = [p1, p2, p3, p4, p5, p6]$. A total of 1000 data records, in which 700 records belongs to class A and the rest belongs to class B, are conducted in this test. This data set are used in 4 different runs, with 1, 2, 4, 12 data splitting and computing threads respectively. Test results can be found in Table 1 and Figure 2.

Table 1: Test results of data splitting

Exp. No.	No. of working threads	Computing time (ms)	Performance (kilo steps/s)	Avg. computing time (ms)	Avg. performance (kilo steps/s)	Avg. performance rate	Avg. performance rate vs single thread
1		120704	0.8285				
2	1	113775	0.8789	117290.7	0.8531	=1	=100%
3		117393	0.8518				
4		65679	1.5226				
5	2	65327	1.5308	65716.0	1.5217	1.7838	89.19%
6		66142	1.5119				
7		33957	2.9449				
8	4	33071	3.0238	33897.3	2.9512	3.4594	86.49%
9		34664	2.8848				
10		12746	7.8456				
11	12	12576	7.9517	12456.0	8.0329	9.4164	78.47%
12		12046	8.3015				

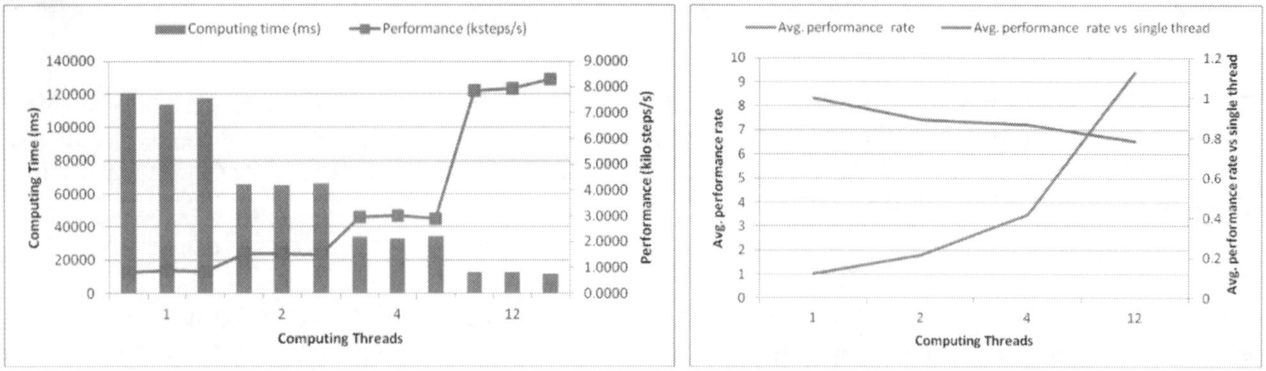

Figure 2: Data processing comparison with different number of threads

The proposed method is applied to the same data set with different number of working threads. Results show that there is an approximate positive linear correlation between computing threads and performance. However, calling more computing threads leads to more communication overhead among these threads. Thus effective workload of a single thread decreases. In this case, using 4 concurrent threads is optimal both on overall and single thread performance.

3.2 Industrial case

Here an industrial example is used to demonstrate the performance of the proposed data-split k-Means clustering based process diagnose and fault classification method. The original data set is collected from an Aluminum stack etching process performed on a commercial scale LAM 9600 plasma etching tool at Texas Instrument, Inc. [7, 17]. The data set is stored by the machine status sensor system at 1-s intervals during etching. And it consists of 108 normal wafers and 21 fault wafers. Each wafer's data set contents about 100 groups of records, and each records contains 20 parameters. The total number of all the records is more than 75,000. A more detailed description on the data can be found in Eigenvector Res. Inc. [17].

The process diagnose result is shown in Figure 3.

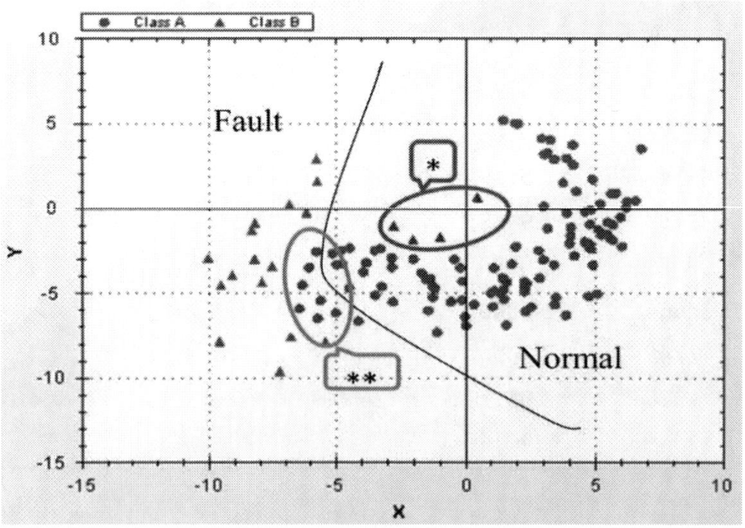

Figure 3: Process diagnose results

*: Type I Error **: Type II Error

From the total 20 records of fault data, 16 faults are successfully classified. Besides, 8 records of normal data are wrongly selected as fault data. A detailed statistic result is shown in Table 2.

Table 2: Statistics of process diagnose results

Item	Count	Total	Percentage(%)
Fault Diagnosed	24	20	120.00%
Type Ⅰ Error	4	20	20.00%
Type Ⅱ Error	8	107	7.48%
Correct Fault Diagnosed	16	20	80.00%

From the result it can be found that the machine data is self-divided into two parts after running data-split k-Means clustering algorithm. The relationship of different data records is measured by the distance of sample to center of cluster, instead of computing and finding possibility or probability of different root causes. Also, if a data point is located in the boundary of two clusters, it is rather difficult to tell which cluster it shall belong to. That is why some type Ⅰ & Ⅱ errors occurred.

4 CONCLUSION

In this paper, a process diagnoses and fault classification method of plasma etching tools using data-split k-Means clustering is developed to monitor process status in mass amount of data collected from sensors. The data pre-processing procedure consists of two stages: data cleaning and data normalization. The purpose of first stage is to prevent adverse impact of missing point and singular point contained in raw data, and the second stage is used to provide a normalized data with unified units and dimension. A k-Means clustering algorithm is used to treat this clean data. Considering with characteristics of big data, a data-split mechanism is introduced to reduce processing scale by allocating more working threads. Practicability of the proposed method is proved by numerical tests, and an industry case shows that the proposed method is effective. As similarity in pattern recognition of process status between different semiconductor manufacturing tools, this fault classification method not only can be used for different types of etching tools, but also suitable for different manufacturing process after a minor improvement. This can great help to improve Overall Equipment Efficiency (OEE), reduce unexpected machine break down, and increase daily move of workshop. In this paper we highlight the obvious advantage of the proposed method in mass data processing, which leads to a new data analysis field of process diagnose and fault classification. With further research and improving of the proposed method, it may extend to more tools and processes. Also there are still some type Ⅰ & Ⅱ errors in the final process diagnose result, studies of reducing percentages of miss classification are under way.

5 ACKNOWLEDGEMENTS

This work was supported by the State Key Program of the National Natural Science Foundation of China under Grant No. 51435009. The authors would like to thank for their support.

6 REFERENCES

[1] **Chang, C. Y. and Sze, S. M.** 1996. *ULSI technology*. McGraw-Hill.

[2] **May, G. S. and Spanos, C. J.** 2007. *Fundamentals of Semiconductor Manufacturing and Process Control*. Wiley.

[3] **Barna, G. G. 1996.** APC in the semiconductor industry, history and near term prognosis, Proceedings of Advanced Semiconductor Manufacturing Conference and Workshop, Cambridge, MA, United States, November 12-14, pp 364-369.

[4] **Schirru, A., Pampuri S. and Nicolao, G D, 2010.** Multilevel statistical process control of asynchronous multi-stream processes in semiconductor manufacturing, 66(262):57-62.

[5] **Chen, M. S., Yen, T. and Coonan, F. B. 2004.** Real-time fault detection and classification for manufacturing etch tools, Semiconductor Manufacturing Technology Workshop Proceedings, Hsinchu, Taiwan, September 9-10, pp 103-106.

[6] **Fan, S. K. S., Lin, S. C., Tsai, P. F. 2016.** Wafer fault detection and key step identification for semiconductor manufacturing using principal component analysis, AdaBoost and decision tree, *Journal of Health Science*, 31(2), pp 1-18.

[7] **Wise, B. M. Gallagher, N. B., Butler, S. W. *et al.* 1999.** A comparison of principal component analysis, multiway principal component analysis, trilinear decomposition and parallel factor analysis for fault detection in a semiconductor etch process, *Journal of Chemometrics*, 13, pp 379-396.

[8] **Barker, M. and Rayens, W. 2003.** Partial least squares for discrimination, *Journal of Chemometrics*, 17(3), pp 166-173.

[9] **Verdier, G. and Ferreira, A. 2011.** Adaptive Mahalanobis distance and k-nearest neighbour rule for fault detection in semiconductor manufacturing, *IEEE Transactions on Semiconductor Manufacturing*, 24(1), 59-68.

[10] **Tafazzoli, E. and Saif, M. 2009.** Application of combined support vector machines in process fault diagnosis, Proceedings of American Control Conference, St. Louis, MO, United States, June 9-12, pp 3429-3433.

[11] **Kim, J. W., Han, Y. S., and Lee, J. S., 2016.** Euclidean distance based feature selection for fault detection prediction model in semiconductor manufacturing process. *Advanced Science and Technology Letters*, 133, pp 85-89.

[12] **Laney, D. 2012.** *The importance of "big data": A definition*, Gartner.

[13] **Wu, J. J. 2012.** *Advances in k-means clustering: a data mining thinking*, Springer.

[14] **Guha, S. Rastogi, R. and Shim, K. 2001.** CURE: an efficient clustering algorithm for large databases, *Information Systems*, 26(1), pp 35-58.

[15] **Lu, Z. M., Feng, J. M., Fan, D. M., et al. 2014.** Novel partitional clustering algorithm for large data processing, *Systems Engineering and Electronics*, 36(5), pp 1010-1015.

[16] **Forgy, E. W. 1965.** Cluster analysis of multivariate data: efficient versus interpretability of classifications, *Biometrics*, 21(3), pp 768.

[17] **Eigenvector Res. Inc. 1999.** Metal etch data for fault detection evaluation, Information on http://software.eigenvector.com/Data/ Etch/index.html.

THE INDUSTRY TECHNOLOGY ROADMAP PUBLISHING PLATFORM FOR THE KNOWLEDGE INTERACTION OF UNIVERSITY-INDUSTRY COLLABORATION

Tang Yongli [1] Lu Zhuobin[1*] Hu Xinyue [1] and Zhang Haiwen[1]

[1]School of Management

Jinan University, Guangzhou, China

zbluzhuobin@163.com

ABSTRACT

The development activities of the industry technology roadmap in Guangdong are characterized by the knowledge interaction of University-industry collaboration. Through participating and observing some industry technology roadmap projects, we find that the products of these activities do not been diffused widely, also not been economized. This situation impedes the expansion of the knowledge interaction of University-industry collaboration. However, the literatures on technology roadmap are mostly focus on the aspect of method research. It is still lack of the research on the popularization and application. Through the integration of the theories from multi-filed, this paper proposes to construct an industry technology roadmap publishing platform for Guangdong to enhance the popularization efficiency. This paper introduces the conceptual framework of this platform.

Keywords: Technology roadmap, Knowledge interaction, University-industry collaboration

* Corresponding Author

1 INTRODUCTION

The technology roadmap (TRM) is a technology management and planning method which is derived from enterprise practice [1]. It is a system framework base on time and a process to acquire and integrate the interdisciplinary knowledge and information from different sources through knowledge interaction [1]. In this process, the participants can exploit, express, communicate and formulate the objective and the strategic plan, also can explore the dynamic relation of the technological resource, the organizational objective and the environment evolution [1]. With the popularization, TRM has been adopted in different levels and different contexts [2, 3]. It includes product level, technology level, enterprise level, industry level, department level, national level and international level [2, 3].

The TRM development activities initiated by government in Guangdong are mainly at the industry level. The purpose of these activities is to facilitate the knowledge interaction of University-industry collaboration and the development of the industry. However, the authors find that the publishing of the TRM is defective through participating series development activities of the industry technology roadmap (ITRM). The products of a large portion of ITRMs do not been diffused into the University-industry alliance. Consequently, the efficiency of the knowledge interaction of University-industry collaboration is poor. Thus, the products of ITRMs can't be economized. How to efficiently publishing ITRMs to facilitate the knowledge interaction of University-industry collaboration, that is the research question discussed in this paper.

The literatures on technology roadmap are mostly focus on the aspect of method research. It is still lack of the research on the popularization and application of technology roadmap. To fill this gap, this paper tries to combine the practical situation and the relevant theory, and proposes a platform for publishing, so as to use network to publish the product of ITRM. We think that is useful to improve the work of ITRM publishing.

This paper is organized as follows: In section 2, the relevant theory is cited to prove the availability of platform. In section 3, the conceptual framework of the platform is presented. The section 4 is conclusions.

2 LITERATURE REVIEW

2.1 The value of TRM

Bray indicates that TRM provides a tool for organization to choose responding technology depend on the time frame [4]. Probert and Radnor indicate that a TRM represents the opinion integrated from the stockholders about how to reach organizational objective [5]. It can help an organization ensure appropriate capacity can be use in appropriate place and appropriate time to reach the objective [5]. Gerdsri and Kocaoglu think that TRM is a kind of integrated program for strategic planning to integrate the development of science/technology to products or business [6]. The framework of TRM is divided into 16 kinds by Probert *et al.* (2003), as shown in table 1 [7]. The framework of TRM provides different perspectives to capture the knowledge, include the 'Know-when' perspective, the 'Know-why' perspective, the 'Know-what' perspective, the 'Know-how' perspective, the 'Know-who' perspective and the 'Know-where' perspective [7].

Table1: The kinds of the framework of TRM

According to the purpose	According to the format
Product planning	Multiple layers
Service\capacity planning	Bars

According to the purpose	According to the format
Strategic planning	Tables
Long-term planning	Graphs
Knowledge asset planning	Pictorial
Project planning	Flow charts
Process planning	Single layer
Integration planning	Text

There is a series of workshops in the development process of a TRM [8]. In these workshops, the stockholders and the domain experts will get to gather to capture, share and structure the knowledge about the organizational strategic problems [8]. Through these workshops, the organizational strategic context can be visualized by TRM [8]. TRM provides a mechanism for participants to think, express and communicate to make a consensus, and the tacit knowledge can be communicated sufficiently [9].

Therefore, we can deduce that the kinds of the framework of TRM and the perspectives it contains decide the knowledge store in the TRM is abundant but complex. Also, we can know the TRM activities will generate abundant explicit knowledge and tacit knowledge. Only this knowledge can be spread and economized, the utility of TRM can be maximization.

2.2 The software support of TRM

Probert *et al.* (2003) indicate that the software support can help to develop the TRM efficiently [7]. They also said that the software must can support the standard process and customized process of the TRM, comprise the multiple layers framework and be convenient to both 'novice' and advanced user [7]. Similarly, Choomon and Leeprechanon proposed that the software is an important role to support the development of TRM [10]. They think TRM software can help the different roles in the organization to clarify his/her responsibility and improve the communication between the individuals [10]. Lee and Park think the previous TRMs are finished in the papers by hands, which results in the difficulty of the update and the maintenance [11]. Though the later TRMs are finished by PowerPoint or Visio drawing, it is also hard to maintain them [11]. Thus, they propose that software can simplify the work of TRM and the network technique is an important role [11]. Therefore, we can know clearly that software can help the TRM activities become more efficient, and the application of network has positive influence.

2.3 The knowledge interaction of University-industry collaboration

Nonaka and Konno indicate the concept of 'ba' and the SECI model to clarify the transfer mechanism between explicit knowledge and tacit knowledge, include Socialization, Externalization, Combination and Internalization [12]. They think the knowledge creation is based on the sustaining knowledge interaction [12]. The knowledge creation contain three factors, (1) the SECI process, (2) 'ba'- a shared context for knowledge creation, and (3) knowledge asset, the inputs, outputs and moderators of the knowledge-creating process [13]. The knowledge-creating process is a spiral process- organization uses existing knowledge assets to create new knowledge through the SECI process that takes place in ba and new knowledge becomes in turn the basis for a new spiral [13]. The knowledge interaction of University-industry collaboration is a complex mechanism. Schartinger *et al.* (2002) find there are lots of enterprises and science fields take places knowledge interaction which is not restricted to a few enterprises or disciplines [14]. In addition, they find there are multiple channels for knowledge interaction between industry and university, which can ensure the efficiency of R&D and satisfy various requirements of knowledge [14]. Yusuf proposes it requires different kinds of intermediaries to work in concert to realize efficient knowledge interaction in the industry-university-research institute community, if it wants to realize the commercialization of a new technology [15].

The ITRMs of Guangdong are the activities have knowledge interaction nature. The participants come from enterprises, universities, research institutions and government. However, the number of participants is limited, which results in the knowledge interaction takes place in a small-scale. Thus, to realize a large-scale knowledge interaction of University-industry collaboration, the products and relevant documents of ITRMs need to be spread in the industry-university-research institute community.

2.4 The knowledge transfer

The knowledge transfer, translation and transformation need methods or tools as support [16]. Lang *et al.* (2014) claim, for the manufacture, the knowledge transfer will reduce the cost, but this process needs fee [17]. Thus, they indicate that to reduce the fee of knowledge transfer will increase the performance of knowledge transfer [17]. Information technology (IT) has become a powerful, cheap tool to transfer the structural or non-structural knowledge in the organization. Iyengar et al. indicate that there is a positive effect on the efficiency of knowledge transfer and organizational absorbing ability if IT is used in the organizational learning, and these will improve the financial performance of the organization [18]. Szulanski *et al.* (2016) indicate that if the knowledge needs to be transfer is hard to understand, to take a convenient channel to transfer it from the source to the receiver will reduce the difficulty of knowledge transfer [19].

From these we can know, IT is a efficient and low-cost tool for knowledge transfer. To adopt IT can reduce the difficulty of knowledge transfer and enhance the performance of knowledge transfer. To combine above-mentioned, we think, to adopt IT to transfer the products and documents of ITRMs in the industry community will improve the situation of ITRMs publishing and enhance the performance of ITRMs.

3 THE PUBLISHING PLATFORM OF ITRM

3.1 The knowledge interaction of the follow-up phase

Combining the follow-up phase of ITRM of Guangdong and the SECI model, we will analyse the knowledge interaction in different activities of follow-up phase, as show in figure 1. Base on this analysis, we present the deficiency of the follow-up phase and the tools needed.

The publishing activity is to publish the products and relevant documents of ITRM to the industry community. Then the organizations in the industry community can acquire and utilize these documents. The publishing activity contains the process of knowledge externalization. The publisher integrates the tacit knowledge embedded in the products of ITRM and the documents of workshops. Then, they publish in the form of explicit knowledge. This activity makes the organizations accept the knowledge embedded in ITRM more easily.

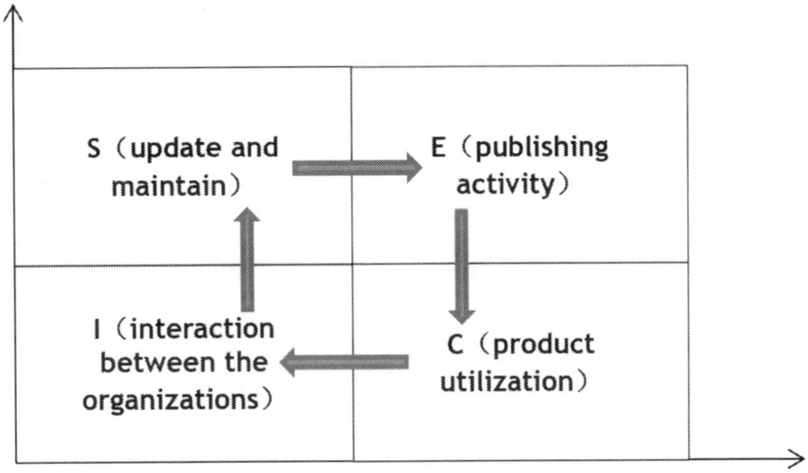

Figure 1: The process of knowledge creation in the follow-up phase

The products of ITRM indicate the factors of market requirement, industry development goals and technology. The industry community can carry on the R&D of the generic technology efficiently base on the factors of the products of ITRM. Different kinds of organizations in the community can utilize the products of ITRM according to their self-demand. The product utilization transfers the explicit knowledge of the published documents to a systematic form through the process of combination. This knowledge is showed in the structured form, such as industry generic technology.

Through consulting the government and experts or cooperating with other organizations, each organization in the community can deepen their comprehension of the products of ITRM. Those knowledge interactions are the process of internalization. Through these activities, the organizations transfer the explicit knowledge of the products of ITRM to the tacit knowledge belongs to them.

In the update and maintain activity, some organizations of the industry community and the government get together to discuss. These organizations share the tacit knowledge with each other in this activity. Through this process of socialization, the tacit knowledge of each organization is increased. The knowledge spiral takes place in the follow-up phase continually to create new knowledge.

However, there are some shortcomings in the activities of the follow-up phase of ITRM in Guangdong. It results in that the knowledge creation process is impeded. Currently, the products of ITRM are published in the form of book, which is not convenient for the organizations to acquire and utilize the products of ITRM. This situation weakens the knowledge interaction between the organizations. In addition, to show the products of ITRM in the form of book would increase the difficultly to update and maintain. Thus, current activities of the follow-up phase affect the process of knowledge creation from several aspects, which results in the inefficient utilization of the products of ITRM.

Therefore, we present that it needs tools to assist the activities of the follow-up phase to promote the extension of the knowledge spiral. For the publishing activity, it needs to construct an appropriative platform base on the network technology for the organizations acquire the products of ITRM conveniently. Meanwhile, the platform should have the function to inform the organizations about the new products of ITRM through instant messaging tools, such as e-mail or wechat. In addition, it needs a knowledge base to store the products of ITRM and relevant knowledge. This platform also needs to have the function for the interaction between the organizations. For the update and maintain activities, the manager that is the government can upload the new version of the products of ITRM to take place the old version.

3.2 The conceptual framework of the platform

Base on the process of knowledge creation in the follow-up phase, we present a conceptual framework of the ITRM publishing platform, as figure 2.

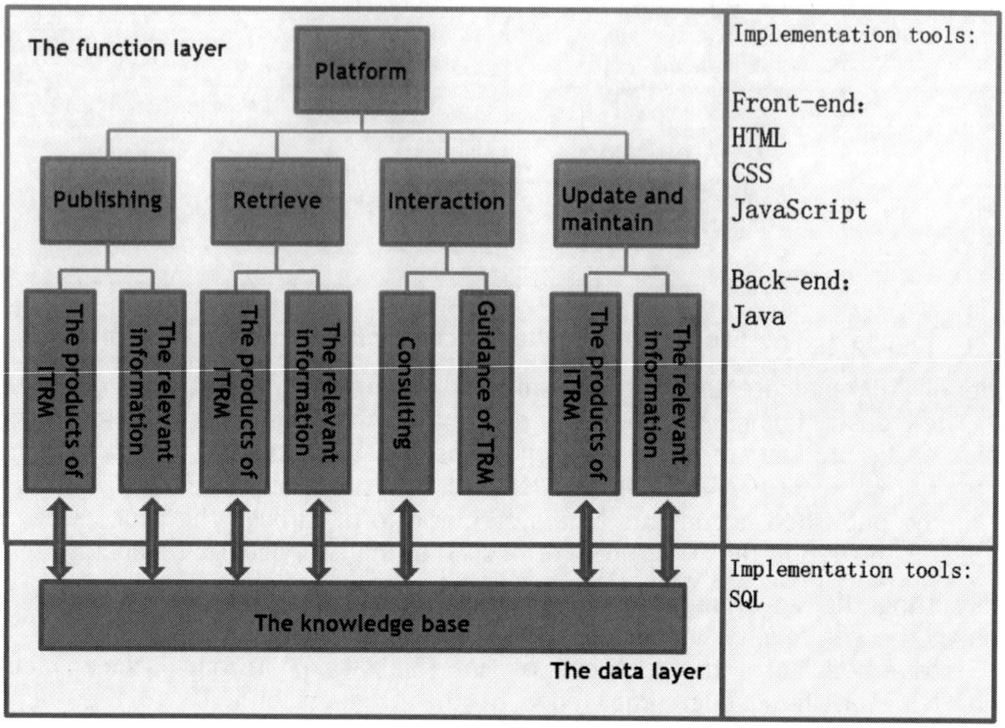

Figure 2: The conceptual framework of the ITRM publishing platform

The conceptual framework of this platform can be divided into function layer and data layer. The function layer is constructed by all functions needed in the follow-up phase of the ITRM. The data layer is used to store the data of the platform, include the products of ITRM and relevant information. The component of the platform can be divided into front-end, back-end and data base. The information is showed to the users through the front-end and managed through the back-end by the manager.

To implement the platform, we adopt Java, HTML, CSS, JavaScript and SQL. HTML, CSS and JavaScript are adopted to implement the front-end. HTML is adopted to define the elements of the front-end. CSS is adopted to define the display pattern of these elements. JavaScript is adopted to determine the operation logic. The back-end is implemented by Java. The knowledge base is implemented by SQL.

The platform mainly contains four functions, include publishing, retrieve, interaction, update and maintain.

(1) Publishing. After the development phase of the ITRM, the managers need to integrate the products and the file of the workshops and then publish these documents through the platform. In addition, the managers can search and then publish the relevant information in the field of TRM, such as development method, the research trends at home and abroad, the cases at home and abroad. Meanwhile, all the documents mentioned above will be showed in the front-end and stored

in the knowledge base. According to the requirement of the users, this platform can send the latest information about ITRM to them through instant messaging tools, such as e-mail or wechat.

(2) Retrieve. When the managers and the users want to retrieve the published documents about TRM, they can use this function. The retrieval mode is the combination of the classification retrieval and the keyword retrieval. According to the keywords and the selected category of the documents, such as the products of ITRM or the research trends at home and abroad, the platform will screen the corresponding documents and then show them in the front-end.

(3) Interaction. When different organizations acquire and utilize the documents of ITRM, it will create new questions or knowledge. Through this function, the organizations can raise their questions by the way of leaving messages. Then, the managers will ask the experts to answer these questions. Meanwhile, each organization can answer other organizations' questions. All the answers of the questions will be showed in the front-end and stored in the knowledge base. These activities realize the knowledge interaction between different organizations and government. In addition, users can fill the form to apply for the guidance about developing TRM from the Science and Technology Agency.

(4) Update and maintain. Some ITRMs need to be updated according to the industry development situation. Thus, after the development activities of ITRM, the old version of products will be replace by the latest version. The manager can upload the new version of product through the back-end and show it in the front-end. In addition, if mistakes are found in the products of ITRM after the publishing, managers can modify the products through the platform. Similarly, the same operation is applicable to other documents which are published by the platform.

4 CONCLUSION

This paper proposes the deficiencies of the follow-up phase of ITRM in Guangdong. First, the scope of the publishing of products of ITRM is small and the number of audiences is few. Second, the way to manage the products of ITRM is laggard. Third, it lacks a specialized platform for the work of the follow-up phase. To handle these deficiencies, this paper presents an ITRM publishing platform base on relevant theories. This platform can promote the efficiency of the follow-up phase of ITRM in Guangdong through the software technology and network technology.

The literatures on technology roadmap are mostly focus on the aspect of method research. However, this paper is focus on the follow-up phase. This paper presents and proves that the software technology and network technology can promote the efficiency of the follow-up phase of ITRM and facilitate the knowledge interaction in the industry community. It provides the theoretical basis for the later research and policy-making.

In addition, the platform presented by this paper is significant for the practical application. So far, the first version of the platform has been developed successfully and is being test phase. This platform integrates the functions which are required for the work of follow-up phase, include publishing, retrieve, interaction, update and maintain. This platform makes the management of the products of ITRM become specialization and electronization for first time.

However, there are some deficiencies of this paper. First, this paper proposes that to construct a publishing platform can promote the efficiency of the follow-up phase, but it doesn't discuss how to optimize the final implement effect of the follow-up phase from the aspects of process, rules and policies. Because the platform is just a tool, it can only help to maximum the effect through combining the process, rules and policies. Second, the current publishing platform just integrates the required functions simply. It needs to be improved according to the design philosophy of software and webpage.

5 ACKNOWLEDGEMENT

The research is supported by National Science Foundation of China (71401063; 71571088), Soft Science Research Program of Guangdong Province (2014A070703020), S&T Programs of Guangdong Province (2014A080802004; 2015A080803003).

6 REFERENCES

[1] **PHAAL R, FARRUKH C J P, PROBERT D R. 2004.** Technology roadmapping - A planning framework for evolution and revolution. Technol Forecast Soc Chang, 71(1-2): 5-26.

[2] **PHAAL R, MULLER G. 2009.** An architectural framework for roadmapping: Towards visual strategy. Technol Forecast Soc Chang, 76(1): 39-49.

[3] **AMER M, DAIM T U. 2010.** Application of technology roadmaps for renewable energy sector. Technol Forecast Soc Chang, 77(8): 1355-70.

[4] **BRAY O H. 1997.** Technology roadmapping: The integration of strategic and technology planning for competitiveness. New York: I E E E

[5] **PROBERT D, RADNOR M. 2003.** Frontier experiences from industry-academia consortia. Res-Technol Manage, 46(2): 27-30.

[6] **GERDSRI N, KOCAOGLU D F. 2007.** Applying the Analytic Hierarchy Process (AHP) to build a strategic framework for technology roadmapping. Math Comput Model, 46(7-8): 1071-80.

[7] **PROBERT D R, FARRUKH C J P, PHAAL R. 2003.** Technology roadmapping - developing a practical approach for linking resources to strategic goals. Proc Inst Mech Eng Part B-J Eng Manuf, 217(9): 1183-95.

[8] **PHAAL R, FARRUKH C J P, PROBERT D R. 2007.** Strategic roadmapping: A workshop-based approach for identifying and exploring strategic issues and opportunities. Emj-Engineering Management Journal, 19(1): 3-12.

[9] **KERR C, PHAAL R, PROBERT D. 2012.** Cogitate, articulate, communicate: the psychosocial reality of technology roadmapping and roadmaps. R D Manage, 42(1): 1-13.

[10] **CHOOMON K, LEEPRECHANON N. 2011.** A literature review on technology road-mapping: A case of power-line communication [J]. Afr J Bus Manag, 5(14): 5477-88.

[11] **LEE S, PARK Y. 2005.** Customization of technology roadmaps according to roadmapping purposes: Overall process and detailed modules. Technol Forecast Soc Chang, 72(5): 567-83.

[12] **NONAKA I, KONNO N. 1998.** The concept of "ba": Building a foundation for knowledge creation. California Management Review, 40(3): 40-+.

[13] **NONAKA I, TOYAMA R, KONNO N. 2000.** SECI, ba and leadership: a unified model of dynamic knowledge creation. Long Range Planning, 33(1): 5-34.

[14] **SCHARTINGER D, RAMMER C, FISCHER M M, et al. 2002.** Knowledge interactions between universities and industry in Austria: sectoral patterns and determinants. Res Policy, 31(3): 303-28.

[15] **YUSUF S. 2008.** Intermediating knowledge exchange between universities and businesses. Res Policy, 37(8): 1167-74.

[16] FIGUEIREDO M C, DE SOUZA C R B, PEREIRA M Z, et al. Knowledge transfer, translation and transformation in the work of information technology architects. Information And Software Technology, 56(10): 1233-52.

[17] **LANG M, DEFLORIN P, DIETL H, et al. 2014.** The Impact of Complexity on Knowledge Transfer in Manufacturing Networks [J]. Production And Operations Management, 2014, 23(11): 1886-98.

[18] **IYENGAR K, SWEENEY J R, MONTEALEGRE R. 2015.** INFORMATION TECHNOLOGY USE AS A LEARNING MECHANISM: THE IMPACT OF IT USE ON KNOWLEDGE TRANSFER EFFECTIVENESS, ABSORPTIVE CAPACITY, AND FRANCHISEE PERFORMANCE. Mis Quarterly, 39(3): 615-+.

[19] **SZULANSKI G, RINGOV D, JENSEN R J. 2016.** Overcoming Stickiness: How the Timing of Knowledge Transfer Methods Affects Transfer Difficulty. Organization Science, 27(2): 304-22.

COMPLEX HARNESS PRODUCT MODELING AND PROCESS KNOWLEDGE INTEGRATION BASED ON ONTOLOGY

Jingqian Wen[1*], Chunyang Wang[1], Pei Gao[1], Peng Wu[2],and Liping Zou[2]
[1]The Institute of Industrial and System Engineering
Beijing Institute of Technology, Beijing, China
wenjq@bit.edu.cn

[2]Jiangxi Hongdu Aviation Industry Group, Jiangxi,China
l15_ping@163.com

ABSTRACT

To adapt the rapid market changes, product development process is required to be shortened as much as possible and carried on parallel design. The production mode of complex harness is multi-varieties and small-batch. And the process design is difficult due to frequent engineering changes and knowledge isolation between different platforms. In this paper, a product modelling method was put forward based on ontology to support knowledge integration and reuse. The enterprise semantic information layer, which can be understood both by human and computer, was built with ontology model and knowledge rules. And an inter-operation mechanism between heterogeneous platforms was designed to face the challenge of agile manufacturing. The construction process of the domain-specific ontology model and semantic expression of process knowledge was described in detail. Finally, taking some processing constraints as example, the semantic rules were established and intelligent reasoning was performed.

Keywords: Ontology, Product modeling, Knowledge integration, Semantic reasoning

1 INTRODUCTION

To improve the manufacturability of complex harness, designers should consider processing constraints concurrently in design phase to make sure that problems in different fields can be found as early as possible. Besides, the differences between engineering software in data structure definition and expression of the wiring logic of complex harness [1], have brought puzzles for the processing of design information. In addition, the transmission of harness design information is mainly in the form of engineering drawing, which is supplemented by three-dimensional entity model. Relevant information is distributed in different data sources. Technologists have to carry on complex computation and verification because the systems lack reasoning and decision support, leading low design efficiency.

Model based definition (in short, MBD) has become a mainstream in aviation field, but model based process is lack of application, especially for wire harness. A feasible method is, constructing product model of complex harness with necessary process knowledge, to provide a unified description for different platforms, and to verify design information and meet the requirement of parallel design by performing logic reasoning [2]. Many scholars have researched on modelling of wire harness for simulation to solve the routing problem. Conru [3] established a structure model of wire harness in the form of an undirected graph. Hergenröther [4] proposed dividing wire harness into simple segments to express the whole structure based on the kinematics model of fixed length wire harness. On the basis of the previous work, mass-spring model added torsional spring was put forward by Loock [5]. Ng [6] used multi-point line method to construct wire harness structure model and perform human-computer interactive route design. The topology information or geometric information can be described by the models above to some extent, however, it's not easy to transfer model data between systems. Product data exchange standard (in short, STEP) is a set of standards defined by ISO, providing a system-independent data format. The Extensible Markup Language (in short, XML) can describe and exchange structured data between applications with a uniform method to adapt network transmission. Whereas, these two kinds of method still remain on the syntax level and do not support processing logic description.

The development of ontology model and reasoning rule provide an effective way to brake the communication barrier between platforms and realize knowledge sharing [7]. Product data model based on ontology uniformly explains domain terminologies without ambiguity, and semantic reasoning is a helpful to intelligent decision, process design and knowledge reuse [8]. This paper aims at constructing a complete and inferable product ontology model [9], combined with the semantic web rule language (in short, SWRL) and logical reasoning to integrate process verification with electrical logic design for complex wire harness.

2 COMPLEX HARNESS BACKGROUND

2.1 Characteristic analysis of complex harness

Avionics devices are connected by Electrical Wiring Interconnection System (in short, EWIS) , which mainly composed by wire harness. The design of EWIS is a systematic project, not only because it involves large amounts of electrical connection data, but also concerns both electronic and mechanical fields. Moreover, wire harness is mainly manufactured by hand for its deformation feature. And processing equipment is lack of intelligent control and data exchange interface. With the rapid development of avionics devices, the manufacturing of harness is facing new challenges:

1. Compressed development cycle: Harness is at the end of the electrical system in design phase, but at the beginning of assembly stage, production preparation always is inadequate.
2. Frequent design changes: Continuous updating of avionics makes frequent changes of harness, which caused inconsistent data for lack of implementation criteria.
3. Knowledge sharing difficult: Harness electrical logic designers and process designers use different tools, the gap between electrical design tool (such as Capital Harness System) and CAPP system is not conducive to knowledge sharing.

In order to realize concurrent design, the EWIS designers should sharing experience and reuse knowledge, that need a computer aided system which realize data integration and automatic verification.

2.2 Application framework of ontology model

A harness manufacturing enterprise has been taken as the research target, considering the actual demands of data sharing and knowledge reuse, the existing software, hardware and the process are analyzed. An application framework, which based on the construction of ontology model and support integration of process knowledge, is put forward for concurrent design and intelligent reasoning, as shown in Figure 1. The logic structure of this framework consists of four layers, which are data layer, semantic information layer, interface layer and application layer from bottom to top.

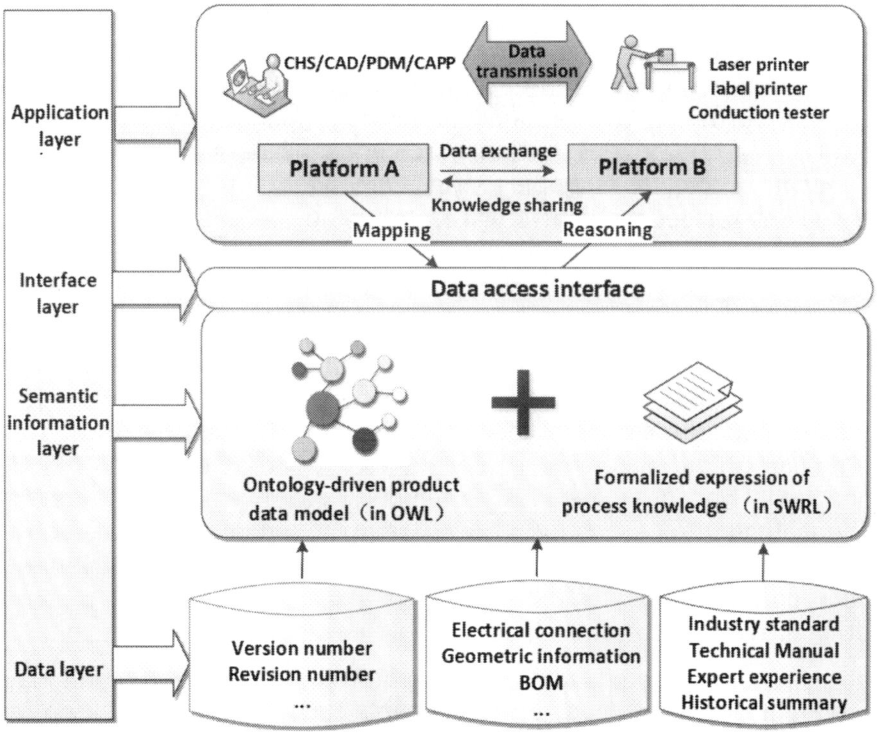

Figure 1: Ontology-based product data model application framework

1. Data layer: This layer is designed to store data and relevant technical standards of complex harness, providing design example for the ontology file generation and data transmission. As foundation for design, product, materials, equipment and environment data are contained in this layer.
2. Semantic information layer [10]: This layer can give consist definitions for domain concepts by ontology, and reasoning rule can be expressed by SWRL. Thus can eliminate ambiguity between different platforms.
3. Interface layer: This layer provides interfaces for each platform to access the ontology files.
4. Application layer: This layer includes various software and systems for electrical design and process planning of wire harness, such as CHS, CAD, PDM, CAPP etc. and equipment for manufacturing include laser printer, label printer, conduction tester etc.

3 PRODUCT MODELING BASED ON ONTOLOGY

Domain-specific ontology for complex harness is a description of related concepts and relationship between EWIS parts. According to the object-oriented method, the ontology model can be defined as a six-tuple made up of concept "C", attribute set of concept "D_C", relationship "P", attribute set of relationship "D_P", concept hierarchy "H" and constraint assertions "R". Namely, the ontology model can be expressed as$\{C, D_C, P, D_P, H, R\}$. In general, the construction process can be divided into three stages: ontology planning, is to determine the ontology scope and acquire domain knowledge through demand analysis; conceptualization, is to extract key concepts and clarify the attributes and the relations, forming a concept structure chart; realization, is to encode the conceptual model with formal language to be understood and applied by computers.

3.1 Ontology planning

Design data of complex harness mainly describe the electrical connection, processing requirements and component information. On one hand, harness product data is large amount and hard be processed by hand. On the other hand, the constraint rules in design, process planning, manufacturing and inspection are same but data sizes and contents are different, that provides a basis for acquisition and application of rule-based reasoning. Through demand analysis, the ontology model was built up to form the semantic information layer of enterprise, make intelligent decision and correct deviation in data transfer. The harness drawings and data files, which transferred from design system to processing system, focus on the description of component parameters. Therefore, the complex harness domain-specific ontologies embodies wires, components, topological structure, and electrical connection. The sources for getting knowledge from include industry standards, technical manuals, expert experience and the historical summary.

3.2 Conceptualization

Complex harness is a product with wires, conductors and metal terminals pressed or welded together to transfer power and signal between the avionic devices. Conductors will be inserted to a certain pinhole of terminal realizing physical connection. Electric wire can be divided into various types: single-core, multi-core, shield according to the number of cored wired or function. Terminal is a component connecting two active devices, commonly known as "plug", consisting of electrical connector and the matching tail accessory. There are many pinholes on the electrical connector and the different arrangement of pinhole forming the "type spectrum" attribute of the electrical connector. In addition, there are some components attached on harness or terminals which play marking, protection or fixation roles, such as marker sleeves, soldering tab, splice, etc.

Through the analysis of harness components and the relationship between them, main concept class"C" of the ontology model can be obtained, other concepts such as terminal, electrical connector, tail accessory, conductor, electric wire, multi-core wire, single-core wire and so on. The concept itself has certain properties. Taking the core concept "conductor", "electrical connector" and "component" as examples, attribute set of the concepts in ontology model can be defined as following Definition 1 to 3. Additionally, the relationship between concepts "P" is embodied as connected relation in the ontology model, being defined as Definition 4.

- Definition 1: Attribute set of conductor "O_C" can be described by a 3-tuple $\{I_C, Awg, Clr\}$, in which I_C is the conductor number, Awg is the conductor specification and Clr represents the conductor color.
- Definition 2: Attribute set of electric wire "O_E" can be described by a 2-tuple $\{G_E, N_E\}$ in which G_E is the grade of electric wire and N_E represents the number of cored wires contained.
- Definition 3: Attribute set of component "O_{Com}" can be described by a 2-tuple $\{G_{Com}, Spec\}$, in which G_{Com} is the grade of component and $Spec$ represents the component's specification.
- Definition 4: Attribute set of wiring relationship "O_p" can be described by 5-tuples $\{F_T, F_{Pin}, I_C, T_T, T_{Pin}\}$, in which F_T, F_{Pin} , T_T , T_{Pin} represent the from-terminal, from-pinhole, to-terminal, to-pinhole respectively, and I_C is the conductor number.

The hierarchy of the concept "H" is designed by Top-Down methodology, that is, starting from the common concept in the domain of complex harness, then add subclasses to refine these concepts. Combining the above summary of concepts, concept properties and relationship between concepts, the structure diagram of the class-hierarchy of complex harness ontology was drawn as Figure 2:

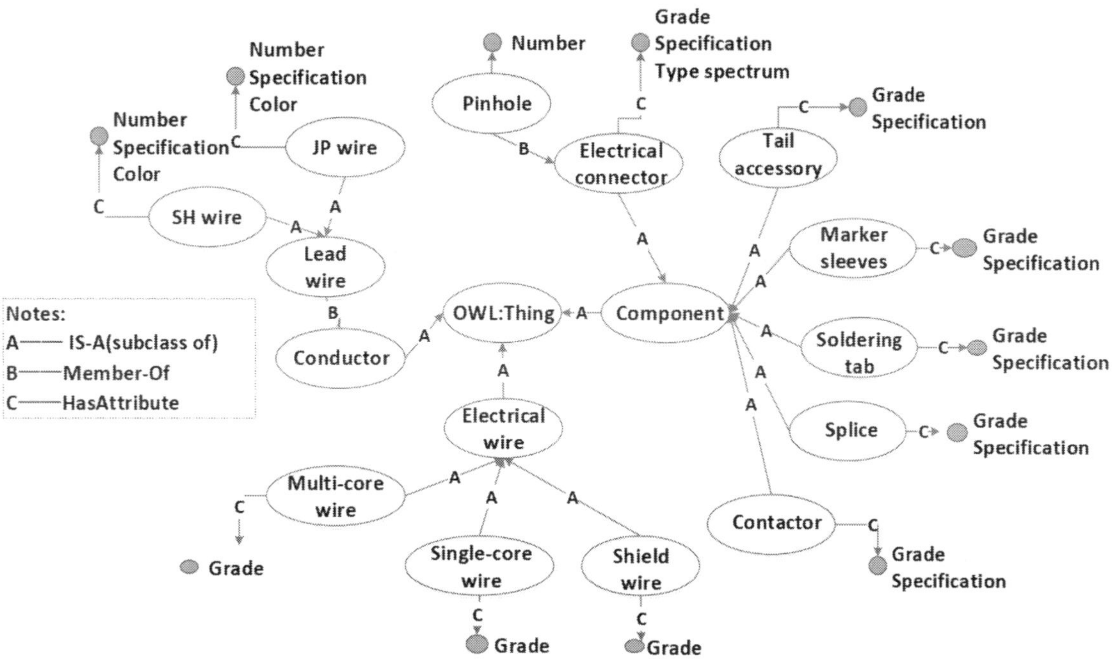

Figure 2: Structure of the class-hierarchy of complex harness ontology

3.3 Ontology Realization

The harness conceptual model should be descripted properly by certain language in order to let it be understood by computer. OWL (Web Ontology Language), which recommended by W3C, can express facts, axioms, rules, and judgments based on semantics. OWL has been widely used due to its strong ability of description and the general applicability. However, describing ontology model directly with OWL is a complex project, and hard to integrate with other functional modules as well. Protégé is one of popular ontology editing tools. It is a free, open source tool which was developed by Stanford University, and can directly construct object-oriented model [11]. Therefore, the complex harness ontology model developed by protégé in the paper. The hierarchy of classes and properties are represented as tree structure so that it's very convenient for engineers to check or edit. Moreover, the function of knowledge integration and reasoning can be executed by protégé plug-ins.

4 PROCESS KNOWLEDGE INTEGRATION BASDE ON SEMANTIC WEB RULE LANGUAGE

The process knowledge of complex harness is the information and facts related to harness process design. Process knowledge integration is to collect and classify constraints, and describe them in the form of semantic rules. For each harness design instances, process knowledge shared and reused by executing reasoning engine to verify model automatically.

However, the OWL expression of rules is limited to description logic, for "if...then...". The Semantic Web Rule Language (SWRL), which was put forward in 2003 by DARPA groups, can support complex rule reasoning. SWRL, which evolved from OWL DL and OWL-Lite(sub-languages of OWL), can use existing classes, properties and instances of the ontology to form a rule. In addition, SWRL regards XML as entity syntax, with characteristics of higher cross-platform.

A rule of SWRL is made up of reasoning premise "body" and reasoning result "hand", like:

Body (premise) - > hand (results)

Each premise is composed of several "Atom", containing lines of expression, which can be class "C" or property "P" of the ontology. Therefore, harness ontology built in chapter 3 can be used to form SWRL rules. An example shows as follows:

> *"Contactor (?x) ∧ Connect (?x, ?a) ∧Conductor (?a) ∧Contactor (?y) ∧ Connect (?y, ?b) ∧Conductor (?b) ∧Pinhole (?x, ?m) ∧Pinhole (?y, ?n) ∧ From-Terminal (?a, ?p) ∧ From-Terminal (?b, ?q) ∧ swrlb:equal (?m, ?n) ∧ swrlb:equal (?p, ?q)*
>
> *→ Details (?a, "more than one conductors in a pinhole ") ∧ Details (?a, "more than one conductors in a pinhole ") "*

It will be regarded as wrong that more than one conductor is inserted to the same pinhole in process design phrase, which is likely to occur in the electrical logic design phrase. In this case, the reasoning premise is two wires (?a/?b) enter the same pinhole (?m/?n) of the same terminal (?p/?q) , then come to the conclusion that one pinhole has more than one conductor. "Details" of conductor properties will be rewrite, helping process designers to find the errors in harness product data. Engineers can edit the rules easily without any programming. In this way, repetitive verification of complex harness can be executed by computers instead of process designers.

5 A CASE STUDY

To realize computer aided verification when product data transferred from design platform to manufacturing system, ontology model integrated with process knowledge are used to detect the inconsistency of data and check manufacturing correctness. Helpful messages will be prompted by rule based reasoning and attached on original data. This chapter take harness process design as an example to show the semantic rule reasoning for manufacturing constraint verification.

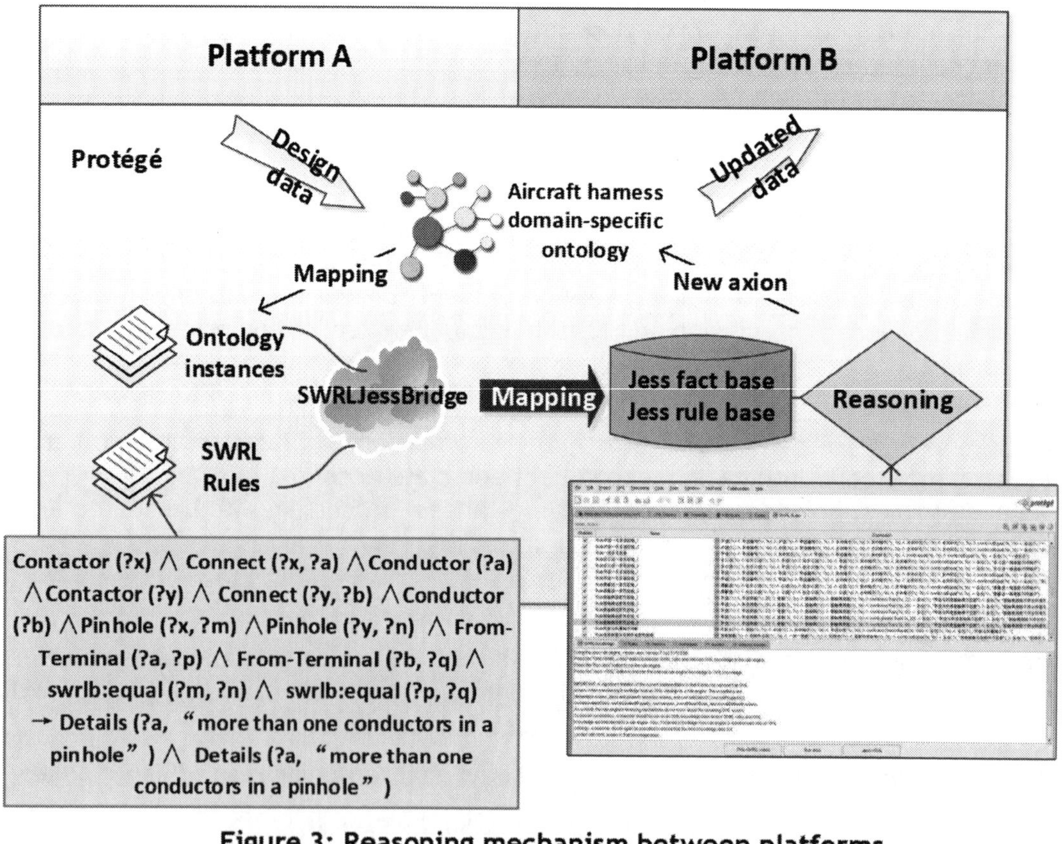

Figure 3: Reasoning mechanism between platforms

The paper adopts Jess as reasoning engine, use SWRL to descript the verification rules, and take complex harness domain-specific ontology model as seasoning base. The reasoning mechanism is shown in Figure 3. Firstly, load design data as instances of ontology model. Secondly, respectively map the instance and SWRL rules to the Jess's fact base and rule base through SWRL Jess Bridge, namely OWL + SWRL→ Jess. Thirdly, execute reasoning, produce new axiom, and rewrite it back into the ontology, namely Jess → OWL. Finally, export updated ontology instances, and return the results to process designers. Thus designers can make quick response to incorrect product data, and report Engineering change Request as early as possible.

Taking the rule mentioned above as an example, the harness instance, which includes all relevant contactors, conductors and terminals, was automatically loaded by Jess. The rule Engine screened out all records which meet reasoning premise, and added message “more than one conductors in a pinhole” in the attributes of these records. As showed in Table1, two conductors with number “242L0-0072” and “242L0-0076” both entered the No.3 pinhole of the connector “PX16-T”, which is not in conformity with the rule. Reasoning result indicates that the ontology model along with integrated process knowledge can play an effective role in product data checking, and realize common understanding for exchanged information.

Table 1: CHANGES IN INSTANCE'S PROPERTY AFTER REASONING

Instance					Passed?	Details
...	...	...	...	...	...	...
241L0-0209	PX16-T	2	NX63	20		
242L0-0072	PX16-T	3	NX63	21	No	“more than one conductors in a Pinhole”
242L0-0076	PX16-T	3	NX63	22	No	“more than one conductors in a Pinhole”
245L0-0158	PX16-T	4	NX51-T	X		
245L0-0159	PX16-T	5	NX51-T	X		
...	...	...	...	...	...	...

6 CONCLUSION

Ontology has gotten more and more attention for its advantages in knowledge representation and semantic modelling. A domain ontology model can provide an effective solution to eliminate information communication barriers between different platforms. Taking complex harness manufacturing industry as the background, the key techniques of product modelling was researched from knowledge collection, classification to representation. And process knowledge was described in SWRL. With ontology and semantic rules, knowledge reasoning was performed to promote knowledge sharing between heterogeneous systems. At last, a case study showed the feasibility of such method.

However, there are still many limitations. First, the completeness and representativeness of the model should be further verified in practice, that whether the model has covered enough concepts, relationships and constraints. Second, this paper only collected part of the process knowledge. In order to make the semantic information framework can be widely used, it needs to find out more explicit and implicit process knowledge to build enterprise's knowledge base. Finally, the efficiency of the semantic reasoning should be improved to meet application requirement.

7 ACKNOWLEDGMENT

The authors would like to thank The National Defense Basic Scientific Research- Project A0920132002.

8 REFERENCES

[1] **Praba, K. 2001.** Multiple viewpoint design models for automotive body-in-white design, International Journal of Production Research, 39(8), pp 1689-1705.

[2] **Jianxin, S. 2003.** Research on PLM system framework and key technologies, *Journal of Nanjing University of Aeronautics & Astronautics*, 16(5), pp 353-374.

[3] **Conru, A.B. 1994.** A genetic approach to the cable harness routing problem, IEEE World Congress on Computational Intelligence, Proceedings of the First IEEE Conference on, Orlando, USA, June 27-29, pp 200— 205.

[4] **Hergenröther, E. 2000.** Real-Time Virtual Cables Based on Kinematic Simulation, Conference Proceedings of the WSCG, Pilzen, Czech Republic, February 7-10, pp 402—409.

[5] **Loock, A. 2001.** A virtual environment for interactive assembly simulation: From rigid bodies to deformable cables, 5th World Multiconference on Systematics, Cybernetics and Informatics, Orlando, USA, July 22—25, pp 325—332.

[6] **Ng, FM. 2000.** Designing cable harness assemblies in virtual environments, *Journal of Materials Processing Technology*, 107(1-3), pp 37-43.

[7] **Usman, Z. 2011.** A Manufacturing Core Concepts Ontology for Product Lifecycle Interoperability, *Lecture Notes in Business Information Processing*, 76, pp 5-18.

[8] **Chungoora, N. 2013.** A model-driven ontology approach for manufacturing system interoperability and knowledge sharing, *Computers in Industry*, 64(4), pp 392-401.

[9] **Qiaoxiang, G.2007.** Modeling techniques for product data ontology based on meta-data, *Journal of Zhejiang University (Engineering Science)*, 41(5), pp 736-741.

[10] **Song, F. 2013.** An ontology-driven framework towards building enterprise semantic information layer, *Advanced Engineering Informatics*, 27(1), pp 38-50.

[11] **Matsokis, A.** 2010. An ontology-based approach for Product Lifecycle Management, Computers in Industry, 61(8), pp 787-797.

A TWO-LEVEL ALGORITHM FOR CAPACITATED VEHICLE ROUTING PROBLEM WITH THREE-DIMENSIONAL LOADING CONSTRAINTS

Ming-yao QI*, Zu-jian WANG, Zi-li WU and Qi LUO
Division of Logistics and Transportation, Graduate School at Shenzhen,
Tsinghua University, Shenzhen 518055, Guangdong, China

ABSTRACT

This paper addresses a variation of the capacitated vehicle routing problem, known as the Capacitated Vehicle Routing Problem with Three-Dimensional Loading Constraints (3L-CVRP), which is a highly complex combinatorial optimization problem. We solve this problem with a two-level algorithm, consisting of an outer stage and an inner stage. A tabu search algorithm is developed for the outer stage which handles the vehicle routing problem, whereas a loading check algorithm for searching the residual capacity of a vehicle is developed for the inner stage. We consider a new space structure based on the support area for the design of the loading check algorithm. The space structure, coupled with three loading strategies, provides more options for handling different loading scenarios. The proposed algorithm is tested by solving some benchmark instances and yields results that are better on average than the existing best solutions obtained from other approaches.

Keywords: Capacitated Vehicle Routing, Three-Dimensional Bin Packing, Tabu search, Two-Level Algorithm, Neighborhood Search

1 INTRODUCTION

The Capacitated Vehicle Routing Problem with Three-Dimensional Loading Constraints (3L-CVRP) is the combination of two well-known optimization problems: The Capacitated Vehicle Routing Problem (CVRP) and the Bin Packing Problem (BPP). First proposed by Gendreau et al. [1], this problem assumes that a fleet of identical vehicles deliver goods to scattered customers with deterministic demands. The package of goods is assumed to be cubic shaped. Each route assigned to a vehicle is determined under the constraint that goods packed should not exceed the vehicle space. The problem is a NP-hard one since both the CVRP and the BPP are NP-hard problems.

In practice, this problem has a broad range of applications in logistics and supply chain management, especially for shipping companies to make decisions on delivering inhomogeneous items to non-identical customers. For example, a household appliance distribution center needs to determine the optimal order combination and vehicle routing to deliver goods to individual buyers at minimum costs. The solution of 3L-CVRP can potentially lead to savings for the shipping company, reduction of times for loading and unloading, and improve the overall efficiency of the delivery process.

2 LITERATURE REVIEW

The three-Dimensional Bin Packing Problem (3D-BPP) is to put 3D rectangular items of different size into one or multiple cubic bins in order to minimize the number of bins used, and maximize the loading ratio of each bin. Subsequently, 3D-BPP can be considered as an extension of the knapsack

* Corresponding Author. Tel.:+86 755 26036130; fax:+86 755 26036005; E-mail address: qimy@sz.tsinghua.edu.cn

problem, or knapsack problem can be viewed as 1D-BPP. However, 2D-BPP and 3D-BPP have more applications in real life. Martello and Toth [3] summarized the algorithms and computer implementations of knapsack problem in detail. A problem which is similar to 3D-BPP is the Container Loading Problem (CLP). CLP calls for loading a subset of rectangular boxes into a rectangular container of fixed dimensions such that the volume of the packed boxes is maximized. A heuristic algorithm by Pisinger et al. [4] can be referred by readers.

Martello et al. [5] probably are the first who proposed an exact method for solving 3D-BPP. They developed a lower bound for the number of boxes to be used and designed an exact algorithm to solve the problem. A concept named corner point was introduced to design the loading process for items, in which the direction of items was assumed to be fixed. For items that can be turned 90 degrees, Boschetti [6] further separated 3D-BPP into 3BP|O|F and 3BP|M|F.

In addition to exact algorithms, a few heuristic algorithms were proposed to solve 3D-BPP. Faroe et al. [7] proposed a guided local search heuristics to solve 3D-BPP. Crainic et al. [8] introduced the Extreme Point concept and present a new Extreme Point-based rule for packing items inside a three dimensional container.

In 2006, Gendreau et al. [1] proposed a group of benchmark problems for 3L-CVRP research. The author also proposed a two-level algorithm in which the outer and inner algorithms were both based on tabu search algorithm. Tanrantilis et al. [9] proposed a hybrid metaheuristic algorithm for this problem. The main contribution of that paper was that it improved the update method of item placement (named basic points) in loading check. The outer routing planning algorithm used in this paper was a guided tabu search algorithm and the inner loading check algorithm was composed of six feasibility checking strategies. Fuellerer et al. [10] designed a two level heuristics in which ant colony algorithm and neighborhood search algorithm were used separately in the outer and inner level. In 2010, Lei et al. [11] reported an algorithm for 3L-CVRP which mainly consists of two heuristic loading methods: Deepest-Bottom-Left-Fill and Maximum Touching Area. Bortfeldt [13] also adopted a hybrid algorithm including a tabu search algorithm for routing and a tree search algorithm for loading.

In this paper, the structure of the two-level algorithm is adopted for algorithm development. We consider a new space structure which can provide more options to place items. The proposed space structure is embedded into the algorithm to enhance the performance. An extensive test has been conducted to examine the performance of the proposed algorithm.

3 PROBLEM DESCRIPTION

The 3L-CVRP model considered in this paper was first proposed by Gendreau et al.[1] in 2006. The problem can be described as follows. Suppose there is a depot which provides delivery services. Given a fleet of v identical vehicles, each of which has a weight capacity Q and a three-dimensional box-shaped loading space with width W, length L and height H. The set of vehicles is denoted by K. There are n customers who need delivery services and the requirements of each customer are already known. The 3L-CVRP can be defined on a complete directed graph $G=\{V,E\}$, where $V=\{0,1,2,...n\}$ is the set of the depot (vertex 0) and n customers (vertex 1,2... n), and $E=\{(i,j)|i\in V, j\in V\}$ is the set of links connecting vertex pairs. Each arc $(i,j)\in E$ is associated with a cost (or distance) c_{ij}. Each customer i has a demand of m_i items. The demand set $I_{ir}, r=1,2,...m_i$, contains three-dimensional items, with attributes length l_{ir}, width w_{ir}, height h_{ir} and weight q_{ir} for all of the items. Denote $d_i=\sum_{r=1}^{m_i} q_{ir}$ as the total weight of goods required by customer i. Most of the notations are adopted from [1]. Items can be fragile or non-fragile: fragile items cannot be overlapped by other items. f_{ir} is equal to 1 if item r for customer i is fragile, and 0 otherwise.

The 3L-CVRP consists of finding at most v routes with minimum cost, defined as the sum of distances of all edges included in the routes. Two kinds of constraints should be satisfied, one for vehicle routing, and the other for item loading. The vehicle routing constraints are as follows:

1) Each vehicle starts from and returns to the depot after servicing all customers in its route.

2) All goods required by a customer must be delivered by the same vehicle.

3) No vehicle carries a total weight or volume that exceeds its capacity.

4) The number of routes generated must not exceed the number of vehicles available.

The items which are loaded in the same vehicle should meet the following loading constraints:

1) Non-overlap constraint: items cannot overlap and are completely contained inside the bin.

2) Orientation constraint: items have a fixed orientation with respect to the height. Namely, when items are placed in the bins, their height directions are always parallel to the height directions of bins. But the items can be rotated by 90° on horizontal plane.

3) Last-in-first-out (LIFO) constraint: when a client is visited, it must be possible to unload the items through a set of straight movements, without moving the items yet to be delivered.

4) Supporting-area constraint: In order to guarantee the item's stability, the ratio of supporting area to the item's base area must overpass a given threshold percentage α.

5) Fragility constraint: no non-fragile items are placed on top of fragile items.

4 ALGORITHM DESIGN

This paper proposes a two-level meta-heuristic algorithm to solve the aforementioned 3L-CVRP, which consists of an outer stage and an inner stage. We design a tabu search algorithm for the outer stage to iteratively construct feasible routes until the stop criterion is reached. We also design a local search algorithm for the inner stage that checks whether a solution is feasible for loading. The loading check algorithm will be called repeatedly during the iteration of the outer stage algorithm. The core of this two-level algorithm is the proposed residual space which defines all possible locations an item may be put; therefore, we denote this algorithm as RSA.

4.1 Loading check Algorithm

4.1.1 The residual space structure

Although the x, y and z coordinates of the available space of a bin are continuous, people often select only a few finite points to place the items in mathematical formulation of the problem, see Gendreau et al. [1] and Tarantilis et al. [8]. However, some feasible loading scenarios cannot be realized by these methods. We propose a support area based space structure (hereinafter referred as the residual space) to define the available space fractions that provides more options for placing items into the bin. The residual space is the available space in the bin, in other words, the available space of a bin is a union of a set of residual spaces in the form of cubes, which are supported by either the top of a loaded non-fragile item or the bottom of the bin. Clearly, the residual space of an empty bin is the entire bin itself. Once an item is put into the bin, the residual spaces which overlap with this item will be cut and then divided into more residual spaces. The maximum number of cuts is six, by left, right, back, front, bottom and top of the item respectively. Notice that residual spaces may overlap.

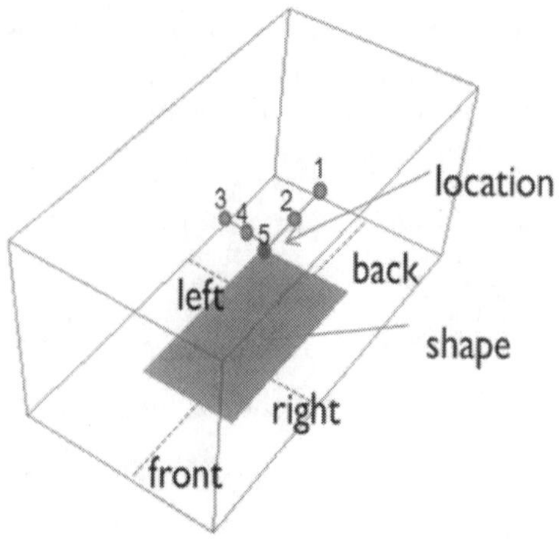

Figure 1: Support area based space structure

The residual space is defined by seven attributes: location(x,y,z), shape(w,l,h), back, left, front, right and fragility, as shown in figure 1. location(x,y,z) represents the coordinates of the left-back-bottom corner of the residual space shown as the point 5 in Figure 1; shape(w,l,h) is the size of the residual space with w for the width, l the length and h the height. The gray part in the figure represents the base of a residual space, or the support area, which can be either the top of a loaded non-fragile item or the bottom of the bin. Attributes back, front, left and right of a residual space are used to define the space an item can be extended out of the residual space. Attribute fragility is an indicator for the fragility of the item below that provides a support area. It is set to 0 if the support area is the bottom of the bin. When an item is placed into a bin, the original space structure of the bin will be changed. The remaining space of the bin can always be represented by a set of new residual spaces with those seven attributes. If an item cannot be placed into a single or multiple residual spaces, the item may not be added into bin.

4.1.2 Loading check algorithm

Suppose a set of items is given. The loading check algorithm is designed to determine whether these items can be packed into an empty bin in some sequence, satisfying all the loading constraints defined earlier. Since this sub-process is to be called by the outer main algorithm every iteration, its computational efficiency needs to be considered. A neighbourhood search algorithm is proposed here to find a feasible load sequence. Table 1 illustrates the steps in the loading check algorithm.

Table 1: Process of the loading check algorithm

Begin

1. Sort the items by *item_sorting_criteria*, and store the initial loading sequence by an item set *List<I>*; create an empty *residual space set List<RS>*, and add be bin itself into *List<RS>* as the first residual space; set *Iter* = 0.
2. If *Iter* > *InerMax (=1000)*, go to step 5;
3. For each *item* in *List<I>* ,
   3.1. Sort the residual spaces by *residual_space_sorting_criteria*, and get the residual space set *List<RS>*;
   3.2. For each residual space in *List<RS>*, try to locate the *item* at *five basic points* consecutively; if the loading satisfies all the loading constraints, then add the item to the bin, and record its position;

3.3. Update the residual space set *List<RS>*, go to step 3.1;

4. If failed to load the item at any ***residual space*** and any ***basic point***, exchange the sequence of two items in *List<I>*, set *Iter = Iter+1*, go to step 3;
5. If all items are loaded successfully, output the loading sequence and position of each item, otherwise, return false.

End

Item_sorting_criteria: the initial load sequence of a vehicle can be determined as follows. First, sort customers by their service time in a descending order so that items for customers to be served earlier will be loaded later, following the last-in-first-out principle. Then, sort the items by the fragility constraint so that the fragile items will be placed on the top. Finally, for those items for customers with the same service time and the equal fragility level, sort them by their volume, with smaller items being loaded later.

Residual_space_sorting_criteria: for each item, a residual space with enough space (including the neighborhood spaces that it can extend) will be selected to place it. In order to increase the chance for packing all items into the bin, we designed a set of strategies so that alternative strategies can be tried in case a particular one fails. Each strategy defines the criteria to sort residual spaces. It is a heuristic process and is hard to say whether one strategy is superior to another. The following three sorting strategies are utilized:

1) ZXY strategy. This strategy selects the most appropriate residual space according to their location(x,y,z) value. The residual space with minimum z will be selected first. In the situation that there are multiple residual spaces with minimum z, choose the one with minimum x. Similarly, for residual spaces with minimum z and minimum x, choose the one with minimum y.

2) Maximum touch area strategy. When a particular residual space is selected, the item to be placed into that space may touch with the items already loaded on the side or beneath it, or the surface area of the bin. With this strategy, the residual space that can provide the maximum touch area will be selected.

3) Best matching strategy. The residual space with maximum fitness will be selected. fitness describes the level of similarity between the item and the available residual space in a three dimensional sense. Suppose that the length, width and height of an item are w, l and h, respectively. The shape parameters of the residual space are W, L, H, respectively. fitness can be calculated by $\min(w/W, W/w) * \min(l/L, L/l) * \min(h/H, H/h)$. The smaller the value is, the higher the fitness level.

4.2 Vehicle routing algorithm

The vehicle routing algorithm designed in this paper is essentially a tabu search algorithm. An initial solution is generated by extending the Clarke-Wright saving algorithm [2]. The initial solution will be improved with the proposed tabu search algorithm iteration by iteration until the stopping criterion is met. In the process, the loading check algorithm will be called in each iteration ensuring that the resulting loading constraints are met. To make the paper concise, the details of the tabu search is skipped.

4.3 Acceleration Strategies

Since 3L-CVRP is the combination of two complex optimization problems, the computation is very time consuming. Tarantilis et al. [9] proposed an acceleration strategy that is able to reduce the computation time by three times. This strategy employs a memory structure that records the loading feasibility information of all previously encountered routes, which can eliminate duplicate executions of loading check. In this paper, we also store all the checked customer sequences, whether feasible or not, in a list, named SEQ. Furthermore, we use a mapping method that coverts this sequence into a single positive integer, serving as the index to retrieve the sequence on the list with a classical binary search method. Our test shows that the proposed acceleration strategy can further improve the computation time for retrieving a sequence by three to eight times.

The indexing strategy can be described as follows. Suppose there is a route R1 with its customer service sequence being 2-9-12-17-8. Different mapping schemes can be considered as long as the resulting mapping is one-to-one so that the original sequence can be recovered from the index. For example, given a 3L-CVRP instance of 50 customers, the index of the service sequence for route R1 is $f(R1) = 2\times 50^4 + 9\times 50^3 + 12\times 50^2 + 17\times 50^1 + 8\times 50^0$. The index generated will be used to store the information from loading check. We generated two lists to store the results of loading check of a service sequence. If the result suggests that the service sequence is feasible, then add the associated index to $ListA$, otherwise to $ListB$. When checking a new route R_{NEW}, we first search $ListA$ or $ListB$ to determine if the sequence has already been evaluated. If $f(R_{NEW})$ is not in $ListA$ or $ListB$, the loading check algorithm will be called, and then the result will be added either to $ListA$ or to $ListB$.

5 COMPUTATIONAL RESULTS FOR THE BENCHMARK PROBLEMS

The performance of the proposed algorithm RSA is evaluated by the 27 benchmark problems proposed by Gendreau [1] (available in http://www.or.deis.unibo.it/research.html). Our program is implemented in c# and run on a computer with 2.4GHz Intel Core 2 Duo CPU, 2GB of RAM and 32 bit Windows XP.

We compare the result with those in recent literatures in Table 1. Denote C_{RSA}, C_{TS}, C_{GTS}, C_{ACO}, C_{DIMS}, C_{VRLH1} as the solution cost that derived from this paper, TS[1], GTS[9], ACO[10] , DMTS[12] and VRLH1[13], respectively. The improvement of our algorithm is calculated by:

$improvement = \frac{\min(C_{TS}, C_{GTS}, C_{ACO}, C_{DTMS}, C_{VRLH}) - C_{RSA}}{\min(C_{TS}, C_{GTS}, C_{ACO}, C_{DTMS}, C_{VRLH})} \times 100\%$. Positive improvement indicates that the cost derived by our algorithm is smaller than the minimal cost calculated by existing algorithms.

Table 1: Compare with existing result of 3L-CVRP

instance	TS[1]	GTS[9]	ACO[10]	DMTS[11]	VRLH1[12]	RSA	*Improvement*
1	316.32	321.47	305.35	301.77	302.02	304.13	-0.78%
2	350.58	334.96	334.96	334.96	334.96	**334.96**	**0.00%**
3	447.73	430.95	409.79	387.91	401.44	387.97	-0.02%
4	448.48	458.04	440.68	438.59	437.54	**437.19**	**0.08%**
5	464.24	465.79	453.19	440.23	451.03	**436.4**	**0.87%**
6	504.46	507.96	501.47	499.48	498.38	499.08	-0.14%
7	831.66	796.61	797.47	771.09	772.49	773.47	-0.31%
8	871.77	880.93	820.67	805.95	821.35	807.75	-0.22%
9	666.1	642.22	635.5	630.9	645.81	**630.13**	**0.12%**
10	911.16	884.74	841.12	832.46	827.29	832.55	-0.64%
11	819.36	873.43	821.04	781.85	815.62	**781.52**	**0.04%**
12	651.58	624.24	629.07	614.78	630.46	**612.25**	**0.41%**
13	2928.34	2799.74	2739.8	2715.82	2694.81	2728.12	-1.24%
14	1559.64	1504.44	1472.26	1456.13	1413.59	**1391.76**	**1.54%**
15	1452.34	1415.42	1405.48	1371.26	1355.5	**1344.06**	**0.84%**
16	707.85	698.61	698.92	699.54	705.05	**698.61**	**0.00%**
17	920.87	872.79	870.33	875.19	917.96	871.63	-0.15%

18	1400.52	1296.59	1261.07	1248.28	1228.98	1234.82	-0.48%
19	871.29	818.68	781.29	776.35	753.87	**753.35**	**0.07%**
20	732.12	641.57	611.26	593.17	596.42	**590.38**	**0.47%**
21	1275.2	1159.72	1124.55	1121.6	1107	**1106.6**	**0.04%**
22	1277.94	1245.35	1197.43	1176.76	1171.49	**1165.5**	**0.51%**
23	1258.16	1231.92	1171.77	1148.02	1135.46	**1125.65**	**0.86%**
24	1307.09	1201.96	1148.7	1144.56	1128.82	1145.75	-1.50%
25	1570.72	1457.46	1436.32	1457.09	1428.8	**1387.83**	**2.87%**
26	1847.95	1711.93	1616.99	1616.61	1625.31	1644.06	-1.70%
27	1747.52	1646.44	1573.5	1574.23	1550.85	**1548.05**	**0.18%**
Ave. Cost	1042.26	997.18	966.67	956.1	953.79	**947.17**	**0.69%**
Ave. time	2058.86	1471.64	1746.55	820.02	228.6	462.25	**-102.21%**

Compared with TS[1], GTS[9] and ACO[10], which are reported in relatively earlier papers, our algorithm updates or equals to the solution cost of almost all benchmark problems when solving 3L-CVRP problems, except in problems 17 and 26, which the cost is slightly higher than the best result in these three papers. Even compared with the best of DMTS[11] and VRLH1[12], proposed in the more recent two papers, our proposed algorithm reduces the solution cost by an average rate of 0.69% and maximum rate of 2.87%. Among the 27 test instances, the solution cost of 16 instances is updated or remains the same, as shown in bold text. The computation time decreases significantly from 820.02s to 462.25s when compared with DMTS[11], however, it is almost doubled compared with VRLH1[12]. It may be attributed to different computing hardware to some extent, since according to VRLH1[12], the computer CPU used is Intel 3.17G, while in this paper we use Intel 2.4G.

Although we cannot conclude that the proposed RSA algorithm outperforms all published algorithms, it makes sense that RSA is comparable to state of the art algorithms for solving 3L-CVRP. The contribution of this paper lies that we found a new method to express the loading space. As shown earlier, the residual space structure can provide more options for placing an item compared with the 3 or 7 corner point methods in the literature since it includes not only the corner points but also other points within a residual space.

6 CONCLUSION

3L-CVRP is a highly complex combinatorial optimization problem that combines vehicle routing and 3D loading decisions. This paper proposed a two-level algorithm to solve this problem, a loading check algorithm and a tabu search based vehicle routing algorithm. Instead of using corner points, a new space structure is proposed in which the available space is partitioned into residual spaces. Items are loaded into the residual spaces inside the bin. The residual spaces ensure that the loading check algorithm can examine more possible positions to accommodate different loading scenarios, including those that cannot be realized with corner point methods. In order to speed up the loading process, acceleration strategies are considered to map a customer sequence to an integer number that can be retrieved and determine whether a customer sequence to be evaluated is checked or not. This strategy can save much memory, and furthermore, can improve computation time by three to eight times. The proposed algorithm was tested by solving some benchmark instances and yielded results that outperform the existing best solutions obtained from other approaches.

The proposed support area based space structure can be adopted for a broad range of applications dealing with 3D loading issues. In the real world, more variations can be considered that would lead to different loading constraints (e.g. pallet-packing vehicle routing problems). Further improvement may be achieved by using inter-route exchange to decrease the number of vehicles used, so that lower transportation cost may be deduced.

7 REFERENCES

[1] **Gendreau, M., Iori, M., Laporte, G., 2006.** A Tabu Search Algorithm for a Routing and Container Loading Problem. *Transportation Science* 40(3): 342-350.

[2] **Clarke, G., Wright, J. V., 1964.** Scheduling of vehicles from a central depot to a number of delivery points. *Operations Research* 12: 568-581.

[3] **Martello, S., Toth, P., 1990.** Knapsack Problems: *Algorithms and Computer Implementations. John Wiley & Sons,* New York.

[4] **Pisinger D.** Heuristics for the container loading problem. *European Journal of Operational Research,* 141(2): 382-392

[5] **Martello, S., Pisinger, D., Vigo, D., 2000.** The three-dimensional bin packing problem. *Operations Research* 48(2): 256-267.

[6] **Boschetti, M. A., 2004.** New lower bounds for the three-dimensional finite bin packing problem. *Discrete applied Mathematics* 140: 241-258.

[7] **Faroe, O., Pisinger, D., Zachariasen, M., 2003.** Guided local search for the three-dimensional bin packing problem. *Informs Journal On Computing* 15(3): 267-283.

[8] **Crainic, T. G., Perboli, G., Tadei, R., 2008.** Extreme point-based heuristics for three-dimensional bin packing. *INFORM Journal on Computing* 20(3): 368-384.

[9] **Tarantilis, C. D., Zachariadis, E. E., Kiranoudis, C. T., 2009.** A Hybrid Metaheuristic Algorithm for the Integrated Vehicle Routing and Three-Dimensional Container-Loading Problem. *IEEE Transactions on intelligent transportation systems* 10(2): 255-271.

[10] **Fuellerer, G., Doerner, K. F., Hartl, R. F., Iori, M., 2010.** Metaheuristics for vehicle routing problems with three-dimensional loading constraints. *European Journal Of Operational Research* 201(3): 751-759.

[11] **Lei, W., Guo, S., Shi, C., Zhu, W., Lim, A., 2010.** Two Natural Heuristics for 3D Packing with Practical Loading Constraints. in: *Proceedings of the Pacific Rim International Conference on Artificial Intelligence,* pp.256-267.

[12] **Bortfeldt, A., 2012.** A hybrid algorithm for the capacitated vehicle routing problem with three-dimensional loading constraints. *Computers & Operations Research,* 39(9): 2248-2257.

THE STUDY OF REVENUE SHARING ABOUT TWO TOW-LEVEL SUPPLY CHAINS BASED ON THE SHAPLEY VALUE

Bo Huang[1,2] and Huajun Tang[2*]
[1]Department of Logistics Management and Engineering
Zhuhai College of Jilin University, Zhuhai, China
bo6636836@126.com
[2]School of Business, Macau University of Science and Technology, Taipa, Macau
hjtang@must.edu.mo

ABSTRACT

This paper focuses on the two tow-level supply chains composed of a manufacturer and a retailer. Firstly, it analyzes four modes about Stackelberg game between two different supply chains. Then, the study uses the backward induction to solve four modes and present the analytic solutions for Stackelberg game. Finally, some concepts and models of the Shapley Value are proposed and applied to the profit allocation among two tow-level supply chains. The results shows that each enterprise can get the optimal profit, and the profit allocation can achieve "Pareto optimal" status.

Keywords: Tow-level supply chain, Stackelberg game, Backward induction, Revenue sharing, Shapley value algorithm

1 INTRODUCTION

Supply chain coordination is a very important research area in supply chain management. Revenue sharing is the key to the success of relationship coordination. It means that the participants make their decisions so as to achieve the optimal profit of the total supply chain. A revenue sharing contract is beneficial to the manufacturer and the retailer within a supply chain (see [1] and[2]). Then many scholars conducted diverse in-depth studies. The most important ones in the method of revenue sharing are Shapley value, profit allocation models based on resources and contribution rate, and risk pooling problems. For example, Pang et al. [3] focused on the distributor's effort, and improved well to revenue-sharing in a three-level supply chain. Qiet al. [4] gave an income distribution model of the modified Shapely value, considering the income distribution principles of the collaborative product innovation. He and Zhang [5] studied sharing in a two-level supply chain with random yield risk. Xia et al.[6], Sammi and Panos [7] presented a sharing model with demand and supply risk in a supply chain.

Shapley value algorithm is a distribution way, which is a quantitative study of profit allocation. Van den Brink [8] showed that the difference between the Shapley value and the equal division. Hence the Shapley value neither distributes equally nor proportional about investment cost, but the importance in the production of economic benefits. Shapley [9][10] proposed a mathematical method to solve the profit allocation problem of "*n*" personal. This approach presumes that "*n*" person pursues economic activity, and can receive benefits. When the interests of members are not in conflict with each other, the new member will not lead to a decline in profits. In this way, all "*n*"

* Corresponding Author

person cooperation will bring the maximum benefits. Shapley value is a method to allocate the benefits.

The defined about Shapley value as follows:

Let I be a set $I = \{1, 2, \cdots, n\}$, if corresponds to a function $v(s)$ for any subset S of I .It is satisfies:

$$v(\phi) = 0$$

$$v(s_1 \cup s_2) \geq v(s_1) + v(s_2),\ s_1 \cap s_2 = \phi \quad (s_1 \subseteq I, s_2 \subseteq I)$$

$[I, v]$ is cooperative game of n ; v is a characteristic function for game.

$$w(|s|) = \frac{(n-|s|)!\,(|s|-1)!}{n!}$$

$$\varphi_i(v) = \sum_{s \in s} w(|s|)\,[v(s) - v(s \setminus i)],\ i = 1,2,\cdots,n$$

$$\Phi(v) = (\varphi_1(v), \varphi_2(v), \cdots,\ \varphi_n(v))$$

in which, s_i is all the subsets of I;

$|s|$ is the number of elements for S;

n is the number of I ;

$w(|s|)$ is the weighting factors;

$v(s)$ is the benefit of S;

$v(s \setminus i)$ is the benefit excluding enterprise i .

Previous researches about Shapley value have focused on a single supply chain. For example, Hu et al. [11] discussed a cooperative game problem of four-level supply chain. Ma and Wang [12] dealt with the Shapley value with a three-level supply chain. Now competition has been transferred from traditional competition among enterprises to competition among supply chains. In this paper, we use the Shapley value to solve the problem of profit allocation about two tow-level supply chains.

2 MATHEMATICAL MODEL

In real life, some brands are to start give pricing. Then, the competitive brands make the best response according to the previous. This paper develops two two-level supply chains, SC_1 and SC_2. Both supply chains contained a manufacturer $(M_1 \setminus M_2)$ and a retailer $(R_1 \setminus R_2)$, and manage alternative products. The figure as shown below.

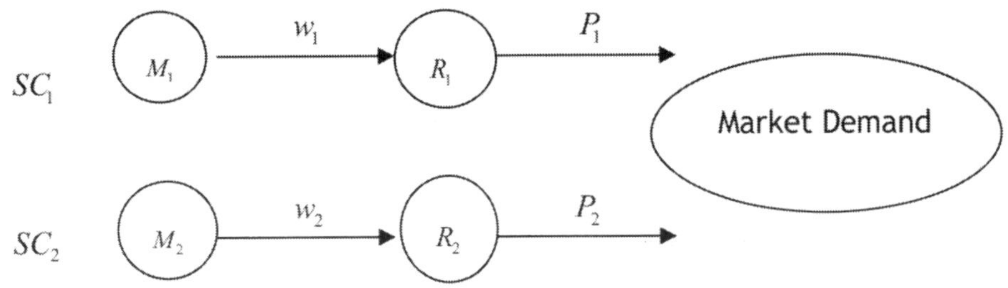

Figure 1: The Structure Chart of Tow Supply Chains

2.1 Related Description

The possible alliance has the following four most common alliance forms.

$[M_1, R_1, M_2, R_2]$, $[(M_1, R_1), M_2, R_2]$, $[M_1, R_1, (M_2, R_2)]$, $[(M_1, R_1), (M_2, R_2)]$,

2.1.1 Variable Definition

Related variables are defined as follows.

- W_1: The wholesale price of SC_1
- W_2: The whole sale price of SC_2
- P_1: The retail price of SC_1
- P_2: The retail price of SC_2
- C_1: The unit production cost of M_1
- C_2: The unit production cost of M_2
- θ: The product alternative.
- a_1: The market size of SC_1
- a_2 :The market size of SC_2
- D_1 :The market demand of SC_1
- D_2 :The market demand of SC_2
- Π : The optimal total profit

2.1.2 Model Assumptions

To build the model, some relation assumptions are described below.

a. People are always economically rational in the supply chain, and share information completely each other.

b. The retailer decides the retail price according to the manufacturer's wholesale price. SC_1 is the leader, and SC_2 is the follower.

c. $D_1 = a_1 - P_1 + \theta P_2$, $D_2 = a_2 - P_2 + \theta P_1$

2.2 Model Description

According to the above, we can get:

- $\Pi_{M_1} = (W_1 - C_1) \times D_1 = (W_1 - C_1) \times (a_1 - P_1 + \theta P_2)$ **(1)**
- $\Pi_{R_1} = (P_1 - W_1) \times D_1 = (P_1 - W_1) \times (a_1 - P_1 + \theta P_2)$ **(2)**
- $\Pi_{M_2} = (W_2 - C_2) \times D_2 = (W_2 - C_2) \times (a_2 - P_2 + \theta P_1)$ **(3)**
- $\Pi_{R_2} = (P_2 - W_2) \times D_2 = (P_2 - W_2) \times (a_2 - P_2 + \theta P_1)$ **(4)**
- $\Pi_{SC_1} = (P_1 - C_1) \times (D_1) = (P_1 - C_1)(a_1 - P_1 + \theta P_2)$ **(5)**

- $\prod_{SC_2} = (P_2 - C_2) \times (D_2) = (P_2 - C_2)(a_2 - P_2 + \theta P_1)$ ***(6)***

We solve the Stackelberg game by backward induction.

2.2.1 $[M_1, R_1, M_2, R_2]$

Take the derivative of P_2 about $\prod_{R_2} = (P_2 - W_2) \times (a_2 - P_2 + \theta P_1)$, Obviously, $\left(\frac{\partial^2 \prod_{R_2}}{\partial^2 P_2} < 0\right)$ The function exhibits a maximum.

Take $\frac{\partial \prod_{M_2}}{\partial P_2} = 0$, we can get: $P_2 = \frac{a_2 + \theta P_1 + W_2}{2}$ ***(7)***

Take eq. (7) into the eq. (3) , Take the derivative of W_2 and we get:

$$\frac{\partial \prod_{M_2}}{\partial W_2} = (W_2 - C_2) \times \{a_2 - [\frac{a_2 + \theta P_1 + W_2}{2}] + \theta P_1\} = -W_2 + \frac{a_2 + \theta P_1 + C_2}{2}$$

Take $\frac{\partial \prod_{M_2}}{\partial W_2} = 0$, we can get: $W_2 = \frac{a_2 + \theta P_1 + C_2}{2}$ ***(8)***

Take eq. (8) into the eq. (7)we can get: $P_2(P_1) = \frac{3a_2 + 3\theta P_1 + C_2}{4}$ ***(9)***

Take eq. (9) into the eq. (2), Take the derivative of P_1 and we get:

$$\frac{\partial \prod_{R_1}}{\partial P_1} = (P_1 - W_1) \times (a_1 - P_1 + \theta P_2) = \frac{6\theta^2 - 8}{4} P_1 + a_1 + \frac{3a_2\theta}{4} + \frac{C_2}{4} + \frac{4W_1 - 3W_1\theta^2}{4}$$

Take $\frac{\partial \prod_{R_1}}{\partial P_1} = 0$, we get: $\frac{-6\theta^2 + 8}{4} P_1 = a_1 + \frac{3a_2\theta}{4} + \frac{C_2}{4} + \frac{4W_1 - 3W_1\theta^2}{4}$

$$P_1 = \frac{W_1}{2} + \frac{4a_1 + 3\theta a_2 + \theta C_2}{8 - 6\theta^2} \quad \textbf{(10)}$$

Take eqs. (9)(10)into the eq. (1)we can get:

$$\prod_{M_1} = (W_1 - C_1) \times \{a_1 - [\frac{W_1}{2} + \frac{4a_1 + 3\theta a_2 + \theta C_2}{8 - 6\theta^2}] + \theta[\frac{3a_2 + 3\theta\left(\frac{W_1}{2} + \frac{4a_1 + 3\theta a_2 + \theta C_2}{8 - 6\theta^2}\right) + C_2}{4}]\}$$

Take its partial respect to P_1, we can get:

$$\frac{\partial \Pi_{M_1}}{\partial W_1}=\left\{a_1-\left[\frac{W_1}{2}+\frac{4a_1+3\theta a_2+\theta C_2}{2\left(4-3\theta^2\right)}\right]+\frac{3\theta a_2}{4}+\frac{3\theta^2 W_1}{8}+\frac{3\theta^2}{4}\times\frac{4a_1+3\theta a_2+\theta C_2}{2\left(4-3\theta^2\right)}+\frac{C_2\theta}{4}\right\}+\left(W_1-C_1\right)\left(-\frac{1}{2}+\frac{3\theta^2}{8}\right)$$

$$=-W_1+\frac{6\theta^2}{8}W_1+a_1+\frac{3\theta-4}{4}\times\frac{4a_1+3\theta a_2\theta C_2}{2\left(4-3\theta^2\right)}+\frac{3\theta a_2+C_2\theta}{4}+C_1\left(\frac{1}{2}-\frac{3\theta^2}{8}\right)$$

Take $\frac{\partial \Pi_{M_1}}{\partial W_1}=0$, we can get:

$$W_1\left(\frac{8-6\theta^2}{8}\right)=\frac{4a_1+3\theta a_2+C_2\theta}{4}+\frac{3\theta^2-4}{4}\times\frac{4a_1+3\theta a_2+\theta C_2}{2(4-3\theta^2)}+\frac{C_1}{2}\left(\frac{8-6\theta^2}{8}\right)$$

$$W_1=\frac{8\left(4a_1+3\theta a_2+C_2\theta\right)}{8(4-3\theta^2)}-\frac{4a_1+3\theta a_2+\theta C_2}{2(4-3\theta^2)}+\frac{C_1}{2}$$

$$W_1=\frac{C_1}{2}+\frac{4a_1+3\theta a_2+\theta C_2}{8-6\theta^2} \quad \textbf{(11)}$$

Take eq.(11) into the eq.(10)we can get the optimal price of SC_1

$$P_1^*=\frac{C_1(4-3\theta^2)+3\left(4a_1+3\theta a_2+\theta C_2\right)}{4(4-3\theta^2)} \quad \textbf{(12)}$$

Take eqs. (12)(11)(9)into the eq. (1)-(4)we can get:

$$\Pi_{M_1}{}^*=(W_1-C_1)\times(a_1-P_1+\theta P_2)$$

$$=\frac{1}{8}C_1{}^2+\frac{1}{2}\frac{a_1B}{A}+\frac{3}{8}C_1a_1\theta+\frac{3}{8}\frac{B\theta a_2}{A}-\frac{3}{32}(C_1\theta)^2-\frac{1}{8}C_1C_2\theta+\frac{9}{32}\left(\frac{B\theta}{A}\right)^2-\frac{1}{2}C_1a_1-\frac{17}{32}\frac{BC_1}{A}-\frac{15}{32}\frac{BC_1\theta^2}{A}+\frac{3}{8}BC_2\theta-\frac{3}{4}C_1\theta a_2-\frac{3}{8}\left(\frac{B}{A}\right)^2$$

$$\Pi_{R_1}{}^*=(P_1-W_1)\times(a_1-P_1+\theta P_2)$$

$$=\frac{1}{8}C_1{}^2+\frac{1}{4}\frac{a_1B}{A}-\frac{6}{16}C_1a_2\theta+\frac{3}{16}\frac{B\theta a_2}{A}-\frac{3}{64}(C_1\theta)^2-\frac{1}{16}C_1C_2\theta-\frac{9}{32}\left(\frac{B\theta}{A}\right)^2-\frac{1}{4}C_1a_1+\frac{5}{4}\frac{BC_1}{A}-\frac{3}{32}\frac{BC_1\theta^2}{A}+\frac{51}{64}\left(\frac{B}{A}\right)^2\theta-\frac{1}{16}C_1-\frac{3}{8}ABC_1-\frac{9}{16}(AB)^2+\frac{1}{16}\frac{B\theta C_2}{A}$$

$$\Pi_{M_2}{}^*=(W_2-C_2)\times(a_2-P_2+\theta P_1)$$

$$=\frac{1}{8}C_2{}^2+\frac{3}{16}C_1a_2\theta++\frac{3}{16}\frac{B\theta a_2}{A}+\frac{1}{128}(C_1\theta)^2-\frac{1}{16}C_1C_2\theta-\frac{1}{4}C_2a_2+\frac{3}{64}\frac{BC_1\theta^2}{A}-\frac{1}{8}C_1C_2\theta+\frac{9}{128}\left(\frac{B\theta}{A}\right)^2-\frac{3}{16}\frac{B\theta C_2}{A}$$

$$\Pi_{R_2}{}^*=(P_2-W_2)\times(a_2-P_2+\theta P_1)$$

$$=\frac{5}{16}C_2{}^2+\frac{1}{32}C_1a_2\theta+\frac{3}{32}\frac{B\theta a_2}{A}+\frac{31}{256}(C_1\theta)^2-\frac{1}{32}C_1C_2\theta+\frac{81}{256}\left(\frac{B\theta}{A}\right)^2-\frac{3}{8}C_2a_2+\frac{3}{128}\frac{BC_1\theta^2}{A}-\frac{9}{32}\left(\frac{B}{A}\right)^2-\frac{3}{32}\frac{BC_2\theta}{A}+\frac{1}{16}a_2{}^2$$

Among them, $A=4-3\theta^2$, $B=4a_1+3\theta a_2+\theta C_2$

2.2.2 $[(M_1, R_1), M_2, R_2]$

In this alliance form, the leader SC_1 will get the main to maximize the profit of the whole supply chain. Can infer the same token,

$$\Pi_{SC_1}^* = (P_1 - C_1)(a_1 - P_1 + \theta P_2)$$

$$= \frac{1}{4}C_1^2 + \frac{1}{2}\frac{Ba_1}{A} + \frac{3}{8}\frac{BC_1}{A} + \frac{3}{8}\frac{C_1\theta^2 B}{A} + \frac{3}{8}\frac{B\theta a_2}{A} + \frac{1}{8}\frac{B\theta C_2}{A} + \frac{3}{16}\left(\frac{B\theta}{A}\right)^2 - \frac{1}{2}C_1 a_1 - \frac{1}{4}\left(\frac{B}{A}\right)^2 - \frac{3}{16}(C_1\theta)^2 - \frac{3}{8}C_1\theta a_2 - \frac{1}{8}C_1 C_2\theta - \frac{3}{8}\frac{\theta^2 BC_1}{A}$$

Among them, $A = 4 - 3\theta^2$， $B = 4a_1 + 3\theta a_2 + \theta C_2$

2.2.3 $[M_1, R_1, (M_2, R_2)]$

In this alliance form, both SC_1 SC_2 are get the main to maximize the profit of the whole supply chain. Can infer the same token,

$$\Pi_{SC_2}^* = (P_2 - C_2)\times(D_2) = (P_2 - C_2)(a_2 - P_2 + \theta P_1) = \left[\frac{\theta(2C_1 - C_1\theta^2 + 6a_1 - \theta a_2 + 7\theta C_2) + 8a_2 - 8C_2}{8(2-\theta^2)}\right]^2$$

Among them, $A = 4 - 3\theta^2$， $B = 4a_1 + 3\theta a_2 + \theta C_2$

2.2.4 $[(M_1, R_1), (M_2, R_2)]$

In this alliance form, the follower SC_2 will get the main to maximize the profit of the whole supply chain. Can infer the same token,

$$\Pi_{SC_1}^* = (P_1 - C_1)\times(D_1) = (P_1 - C_1)(a_1 - P_1 + \theta P_2) = \left[\frac{\theta(3a_2 + 3C_1\theta + C_2) - 4(C_1 - a_1)}{4(4-3\theta^2)}\right]^2$$

$$\Pi_{SC_2}^* = (P_2 - C_2)\times(D_2) = (P_2 - C_2)(a_2 - P_2 + \theta P_1) = \left[\frac{(C_1\theta + 2a_2 - 2C_2)(2-\theta) + \theta(2a_1 + \theta a_2 + \theta C_2)}{4(2-\theta^2)}\right]^2$$

3 SHAPLEY VALUE

For two supply chains are composed of composed of manufacturer and retailer, there have four most common alliance forms. According to Shapley Value, we can get the profit of each enterprise.

Table 1: The calculation of M_1 distribution $\varphi(v)[M_2, R_2]$

S	M_1	$M_1 \cup R_1$

$V(S)$	$\frac{1}{8}C_1^2+\frac{1}{2}\frac{a_1B}{A}+\frac{3}{8}C_1a_1\theta+\frac{3}{8}\frac{B\theta a_2}{A}-\frac{3}{32}(C_1\theta)^2-\frac{1}{8}C_1C_2\theta$ $+\frac{9}{32}\left(\frac{B\theta}{A}\right)^2-\frac{1}{2}C_1a_1-\frac{17}{32}\frac{BC_1}{A}-\frac{15}{32}\frac{BC_1\theta^2}{A}+\frac{3}{8}BC_2\theta$ $-\frac{3}{4}C_1\theta a_2-\frac{3}{8}\left(\frac{B}{A}\right)^2$	$\frac{1}{4}C_1^2+\frac{1}{2}\frac{Ba_1}{A}+\frac{3}{8}\frac{BC_1}{A}+\frac{3}{8}\frac{C_1\theta^2B}{A}+\frac{3}{8}\frac{B\theta a_2}{A}+\frac{1}{8}\frac{B\theta C_2}{A}$ $+\frac{3}{16}\left(\frac{B\theta}{A}\right)^2-\frac{1}{2}C_1a_1-\frac{1}{4}\left(\frac{B}{A}\right)^2-\frac{3}{16}(C_1\theta)^2-\frac{3}{8}C_1\theta a_2$ $-\frac{1}{8}C_1C_2\theta-\frac{3}{8}\frac{\theta^2BC_1}{A}$
$V(S\setminus M_1)$	0	$\frac{1}{8}C_1^2+\frac{1}{4}\frac{a_1B}{A}-\frac{6}{16}C_1a_2\theta+\frac{3}{16}\frac{B\theta a_2}{A}-\frac{3}{64}(C_1\theta)^2-\frac{1}{16}C_1C_2\theta-\frac{1}{4}C_1a_1$ $-\frac{9}{32}\left(\frac{B\theta}{A}\right)^2+\frac{5}{4}\frac{BC_1}{A}-\frac{3}{32}\frac{BC_1\theta^2}{A}+\frac{51}{64}\left(\frac{B}{A}\right)^2\theta-\frac{1}{16}C_1-\frac{3}{8}ABC_1$ $-\frac{9}{16}(AB)^2+\frac{1}{16}\frac{B\theta C_2}{A}$
$V(S)-V(S\setminus M_1)$	$\frac{1}{8}C_1^2+\frac{1}{2}\frac{a_1B}{A}+\frac{3}{8}C_1a_1\theta+\frac{3}{8}\frac{B\theta a_2}{A}-\frac{3}{32}(C_1\theta)^2-\frac{1}{8}C_1C_2\theta$ $+\frac{9}{32}\left(\frac{B\theta}{A}\right)^2-\frac{1}{2}C_1a_1-\frac{17}{32}\frac{BC_1}{A}-\frac{15}{32}\frac{BC_1\theta^2}{A}+\frac{3}{8}BC_2\theta$ $-\frac{3}{4}C_1\theta a_2-\frac{3}{8}\left(\frac{B}{A}\right)^2$	$\frac{1}{8}C_1^2+\frac{1}{4}\frac{Ba_1}{A}+\frac{15}{32}\frac{C_1B\theta^2}{A}+\frac{3}{16}\frac{B\theta a_2}{A}+\frac{1}{16}\frac{B\theta C_2}{A}+\frac{3}{16}\left(\frac{B\theta}{A}\right)^2$ $+\frac{9}{32}\left(\frac{B\theta}{A}\right)^2+\frac{1}{16}C_1+\frac{3}{8}ABC_1+\frac{9}{16}(AB)^2-\frac{7}{8}\frac{BC_1}{A}-\frac{1}{4}C_1a_1$ $-\frac{9}{16}(C_1\theta)^2-\frac{1}{16}C_1C_2\theta-\frac{3}{8}\frac{\theta^2BC_1}{A}-\frac{51}{64}\left(\frac{B}{A}\right)^2$
$\lvert S\rvert$	1	2
$W(\lvert S\rvert)$	$\frac{1}{2}$	$\frac{1}{2}$
$W(\lvert S\rvert)\times$ $V(S)-V(S\setminus M_1)$	$\frac{1}{16}C_1^2+\frac{1}{4}\frac{a_1B}{A}+\frac{3}{16}C_1a_1\theta+\frac{3}{16}\frac{B\theta a_2}{A}-\frac{3}{64}(C_1\theta)^2-\frac{1}{16}C_1C_2\theta$ $+\frac{9}{64}\left(\frac{B\theta}{A}\right)^2-\frac{1}{4}C_1a_1-\frac{17}{64}\frac{BC_1}{A}-\frac{15}{64}\frac{BC_1\theta^2}{A}+\frac{3}{16}BC_2\theta$ $-\frac{3}{8}C_1\theta a_2-\frac{3}{16}\left(\frac{B}{A}\right)^2$	$\frac{1}{16}C_1^2+\frac{1}{8}\frac{Ba_1}{A}+\frac{15}{64}\frac{C_1B\theta^2}{A}+\frac{3}{32}\frac{B\theta a_2}{A}+\frac{1}{32}\frac{B\theta C_2}{A}+\frac{3}{32}\left(\frac{B\theta}{A}\right)^2$ $+\frac{9}{64}\left(\frac{B\theta}{A}\right)^2+\frac{1}{32}C_1+\frac{3}{16}ABC_1+\frac{9}{32}(AB)^2-\frac{7}{16}\frac{BC_1}{A}-\frac{1}{8}C_1a_1$ $-\frac{9}{32}(C_1\theta)^2-\frac{1}{32}C_1C_2\theta-\frac{3}{16}\frac{\theta^2BC_1}{A}-\frac{51}{128}\left(\frac{B}{A}\right)^2$

Thus we can get the drives profit of M_1

$$\varphi_{M_1}(\mathrm{v})=$$

$$=\frac{3}{8}\left(\frac{B\theta}{A}\right)^2+\frac{9}{32}\left(\frac{B\theta a_2}{A}\right)+\frac{3}{8}\frac{a_1B}{A}+\frac{1}{8}C_1^2-\frac{75}{128}\left(\frac{B}{A}\right)^2-\frac{5}{64}(C_1\theta)^2-\frac{3}{32}C_1C_2\theta+\frac{1}{32}C_1-\frac{45}{64}\frac{BC_1}{A}-\frac{3}{8}C_1a_1+\frac{3}{16}C_1a_1\theta+\frac{3}{16}BC_2\theta-\frac{3}{8}C_1\theta a_2+\frac{1}{32}\frac{B\theta C_2}{A}+\frac{3}{16}ABC_1+\frac{9}{32}(AB)^2-\frac{3}{16}\frac{\theta^2BC_1}{A}$$

$$\varphi_{M_1}(\mathrm{v})-\Pi_{M_1}$$

$$=\frac{1}{64}(C_1\theta)^2+\frac{1}{32}C_1C_2\theta+\frac{3}{32}\left(\frac{B\theta}{A}\right)^2+\frac{1}{8}C_1a_1+\frac{9}{32}\frac{\theta^2BC_1}{A}+\frac{3}{8}C_1\theta a_2+\frac{3}{8}\frac{a_1B}{A}+\frac{1}{32}C_1+\frac{1}{32}\frac{B\theta C_2}{A}+\frac{3}{16}ABC_1+\frac{9}{32}(AB)^2$$

$$-\frac{1}{2}\frac{a_1B}{A}-\frac{3}{16}C_1a_1\theta-\frac{3}{32}\frac{B\theta a_2}{A}-\frac{11}{64}\frac{BC_1}{A}-\frac{3}{16}BC_2\theta-\frac{27}{128}\left(\frac{B}{A}\right)^2>0$$

$$\varphi_{M_1}(\mathrm{v})>\Pi_{M_1}$$

Can infer the same token,

$$\varphi_{R_1}(\mathrm{v})>\Pi_{R_1}\ ;\quad \varphi_{M_2}(\mathrm{v})>\Pi_{M_2}\ ;\quad \varphi_{R_2}(\mathrm{v})>\Pi_{R_{21}}$$

$$\varphi_{M_1}(\mathrm{v})+\varphi_{R_1}(\mathrm{v})>\Pi_{M_1}+\Pi_{R_1}([M_2,R_2]);\quad \varphi_{M_1}(\mathrm{v})+\varphi_{R_1}(\mathrm{v})>\Pi_{M_1}+\Pi_{R_1}([(M_2,R_2)])$$

$$\varphi_{M_2}(\mathrm{v})+\varphi_{R_2}(\mathrm{v})>\Pi_{M_2}+\Pi_{R_2}([M_1,R_1]);\quad \varphi_{M_2}(\mathrm{v})+\varphi_{R_2}(\mathrm{v})>\Pi_{M_2}+\Pi_{R_2}([(M_1,R_1)])$$

The profit allocation can achieve "Pareto optimal" state, and each firm can get the optimal profit.

4 CONCLUSION

This paper focus on the two two-level supply chains composed of a manufacturer and a retailer. The results show that each enterprise can get the optimal profit, when they are in a union. In addition, the profit allocation can achieve "Pareto optimal" status, when using the Shapley Value to distribute profits and the results can be used for reference. In the practice, it should be noted that there is a difference like undertaken risks, innovative level, investment and so on for every enterprise. Therefore, we need to modify the Shapley Value in the future.

5 ACKNOLWEDGEMENT

This research is supported by Macau Foundation under Grant No. 0485n.

6 REFERENCES

[1] Cachon, G.P. Lariviere, M.A. 2005. Supply Chain Coordination with Revenue-Sharing Contracts: Strengths and Limitations, Management Science, 51, pp 30-44.

[2] Li, S.J. Zhu, Z.B. Huang,.H. 2009. Supply Chain Coordination and Decision Making under Consignment Contract with Revenue Sharing, International Journal of Production Economics, 120, pp 88-99.

[3] Pang, Q.H. Wu, X.Y. Tan, M.L. Cao, X.Y. 2015. Supply chain Coordination using revenue sharing contract with distributor's effort dependent demand, International Journal of Simulation Modelling ,14(2), pp 335-348.

[4] Qi, X.G. Lv, B. Fu, Q. 2008. Profit Allocation in Collaborative Product Innovation Supply Chain Enterprises Based on Improved Shapely Value,Industrial Engineering Journal,18(1), pp 102-109.

[5] He, Y. Zhang, J. 2008. Random yield risk sharing in a two-level supply chain, International Journal of Production Economics, 112(2), pp 769-781.

[6] Xia, Y. Ramachandran, K. Gurnani, H. 2011. Sharing demand and supply risk in a supply chain. IIE Transactions, 43(6), pp 451-469.

[7] Sammi, Y.T. Panos, K. 2016. Pay-Back-Revenue-Sharing Contract in Coordinating Supply Chains with Random Yield, Production and Operations Management Society, 23(12), pp 2089-2103.

[8] Van den Brink, R. 2007. Null or nullifying players: the difference between the Shapley value and equal division solutions, Journal of economic theory, 136(1), pp 767-775.

[9] Shapley, L. 1953. A value four-person games.In: Roth AE(ed) The Shapley value. Cambridge University Press, Cambridge, pp 41-48.

[10] Maschler, M. Peleg, B. Shapley, L.S. 1979. Geometric *Properties* of the Kernal,Nucleolus,and Related Solution Concepts, Mathematics of Operational Research,4, pp303-338.

[11] Shiqiang, H. Bixi, Z. Yingying, G. 2009. Profit Allocation about the Four-level Supply chain Based on the Shapley Value, Systems Engineering, 27(9), pp 49-54.

[12] Shihua, M. Peng, W. 2006. The Study of Profit Allocation among Partners in Supply Chain Based on the Shaley Value, Industrial Engineering and Management, 4, pp 43-49.

BLOOD SUPPLY CHAIN DESIGN FOR URGENT RELIEF DEMANDUNDER LIMITED FUNDS AND TIME

Xiangyu Jin[12],Huajun Tang[1*], Xiaojun Liu[2]
[1]School of Business
Macau University of Science and Technology, Macau, China
hjtang@must.edu.mo

[2]Departmentof Logistics and Information Management
Zhuhai College of Jilin University, Zhuhai, China
jinxiangyu_2006@126.com

ABSTRACT

Natural and man-made disasters have caused unexpected losses in recent years. Blood plays an essential part in disaster rescue. This paper designs a blood supply model to fulfil the urgent demand of blood in an efficient and effective way, so as to minimize the overall cost under certain delivery time requirement. To solve the most critical problem and balance the uncertain need of blood, minimum demand is introduced into the model. Finally,one numerical case is illustrated to test the feasibility of the model.

Keywords: Blood supply chain, Cost minimization, Time requirement, Minimum demand

1 INTRODUCTION

Owing to the increasing of population and environmental pollution, more unconventional disasters and incidents have occurred. Both natural and man-made disasters bring unexpected injuries and deaths.

Because of the challenge of obtaining resources and service in accidents, people in mounting numbers put great emphasis on disaster relief management.Most of the researches focused on the demand forecasting and model construction. Modelling efforts are mainly in optimization and simulation areas[1][2][3]. A multi-objectivemodel for radioactive materials'transportation was established by Badriet et al.[4]. The model involved the determination of the destination and the choice of delivery routes.Hong et al. [5] used a simulation method to study on the emergency logistics in supply chain. BarbarosoğluandArda[6] designed a two-stage dynamic programming framework model, which was mainly used in the overall planning and design of rescue materials distribution.Ukkusuriand Yushimito[7]studied the choice and optimization of distribution route in the humanitarian relief.

Transfusionof bloodand blood products savepeople worldwide every year. Most of studies focused on the inventory management of blood. However, researches on emergency blood supply in disasters were scarce.The earliest research on health care facilities location started by Rahman and Smith [8]. Then several papers were proposed in the following years. Sha and

* Corresponding Author: Huajun Tang

Email: hjtang@must.edu.mo

Huang[9] presented a location-allocation and scheduling model for supply of emergency blood in Beijing. Sahin et al.[10] developed a hierarchical location-routing model of the Turkish Red Crescent Society. Most of them presenteda single-objectivemodel, but a modelling effort for blood supply chain should consider tradeoffs between multiple goals such as time and cost. A stochastic bi-objective supply chain design model wasproposed and a hybrid solution approach was developed by Behnam et al.[11]. However, it has a great challenge to solve models with conflicting objectives.

In this paper, wedesign ahierarchical blood supply model to fulfil urgent demand of blood in an efficient and effective way. "Urgent demand" means that the minimum of demand must be met in disasters. The "efficient" concerns time requirement and the "effective" implies cost saving.We aim to determine the number and location of mobile blood facilities and delivery the collected blood to hospitals timely and costly.

2 MODEL DEVELOPMENT

2.1 Problem statement

The hierarchical blood supply chain in this studyincludesfour parts: blood donors, mobile blood donation facilities, local blood centres, and hospitals.

Mobile units and local centres both have functions of blood collecting and testing, and local centres have more equipment with processing units.Blood donors can donate at either a mobile facility or a local blood centre.Blood collected at mobile blood donation facilitiesshouldbesent to local blood centres for further processing. Safe blood will then satisfy the demand of hospitals.

2.2 Parameters and decision variables

In this section, we establish a multi-period hierarchical blood supply model to improve performance of blood supply. This model explores possible locations and blood delivery decisions for the purpose of cost minimization in certain time requirements.

Before establishmentofthe location-distribution model, related notations are defined as follows.

Indices:

- $\boldsymbol{I}$ set of blood donators,indexed by i
- $\boldsymbol{J}$ set of candidate locations for mobile blood donation facilities,indexed by j
- $\boldsymbol{K}$ set of blood centers,indexed by k
- $\boldsymbol{H}$ set of possible hospitals,indexed by$h$
- $\boldsymbol{T}$ set of time periods,indexed by t

Parameters:

- $\boldsymbol{e}$ fixed cost of building a mobile blood donation unit
- $\boldsymbol{m_{jlt}}$ moving cost of amobile blood donation unit from location l to location j during period t
- $\boldsymbol{OJ}$ unit operational cost of mobile blood donation units
- $\boldsymbol{OL}$ unit operational cost of local blood centers
- $\boldsymbol{AJ}$ unit transportation cost from mobile blood facilities to local blood centers
- $\boldsymbol{AL}$ unit transportation cost from local blood centers to hospitals
- $\boldsymbol{d_{ht}}$ minimum demand of hospital h during period t
- $\boldsymbol{b}$ capacity of a mobile blood donation facility
- $\boldsymbol{S_i}$ maximum blood supply quantity of donators i

- r_{jl} distance between each candidate location of mobile blood donation unit
- r_{ik} distance between donatorsi and local blood center k
- r_{jk} distance between mobile facility j and local blood center k
- r_{kh} distance from local blood center k to hospital h
- rc maximum service distance for donator groups
- LT time requirement of blood delivery in certain situations
- V speed of blood delivery
- N *a very large number*

The following decision variables are defined:

- X number of mobile blood donation(or collecting) facilities
- M_{jlt} a binary variable,which equals to 1 if a mobile blood donation unit$j$ is located at site l during period $t-1$,and ships to $j$during period t;0,otherwise.
- Y_{ijt} a binary variable, which equals to 1 if blood donor group i chooses mobile blood collectingunit$j$ to donate during period t;0,otherwise.
- U_{ikt} a binary variable, which equals to 1 if blood donor group i chooses blood collecting center$k$to donate during period t;0,otherwise.
- G_{jkt}a binary variable, equals to 1 if the blood collected in mobile blood donation unit$j$ is delivered to blood center$k$during period t;0,otherwise.
- L_{kht} a binary variable, equals to 1 if the blood in blood center$k$ is assigned to hospital $h$during period t;0,otherwise.
- Q_{ijkt} quantity of blood collected at mobile donationunit$j$ from donor i ,and shipped to local center k in period t.
- Q_{ikt} quantity of blood collected at local center $k$from donation group$i$ during period t
- Q_{kht} quantity of blood shipped from local center k to hospital $h$during period t

2.3 Objective functions

Equation (1) presents the objective function of this paper, which minimizesoverall cost of the blood supply chain.It is composed of four parts: the cost of building mobile blood units, the cost of moving mobile facility from one site to another, operational cost of blood donation facilities(including mobile units and local blood centres) and transportation cost.

$$minF = BC + MC + OC + TC \tag{1}$$

$$BC = eX \tag{2}$$

$$MC = \sum_{j=1}^{J}\sum_{l=1}^{L}\sum_{t=1}^{T} m_{jlt} M_{jlt} \tag{3}$$

$$OC = OJ\sum_{i=1}^{I}\sum_{j=1}^{J}\sum_{k=1}^{K}\sum_{t=1}^{T} Q_{ijkt} + OL\sum_{k=1}^{K}\sum_{t=1}^{T}(\sum_{i=1}^{I} Q_{ikt} + \sum_{h=1}^{H} Q_{kht}) \tag{4}$$

$$TC = AJ\sum_{j=1}^{J}\sum_{k=1}^{K} r_{jk}(\sum_{i=1}^{I}\sum_{t=1}^{T} Q_{ijkt}) + AL\sum_{k=1}^{K}\sum_{h=1}^{H} r_{kh}(\sum_{t=1}^{T} Q_{kht}) \tag{5}$$

Equation (2) shows the establishing cost of mobile blood donation units, which is related to the number of mobile facilities. Equation (3) calculates the moving cost of mobile blood donation units in each period. Equation (4) specifies total operational costs of mobile blood units and local blood centers. OL in this equation contains unit inventory cost of local blood centers.Equation (5) formulates overall cost of transportation, including cost of shipping blood from mobile facilities to local centers, and from local blood centers to hospitals in demand.

2.4 Model Constrains

$$\sum_{j=1}^{J}\sum_{l=1}^{L} M_{jlt} \le X \qquad \forall t \in T \tag{6}$$

$$\sum_{l=1}^{L} M_{jlt} \le 1 \qquad \forall j \in J \quad \forall t \in T \tag{7}$$

$$\sum_{l=1}^{L} M_{ljt} \le \sum_{l=1}^{L} M_{jlt-1} \qquad \forall j \in J \quad \forall t \in T \tag{8}$$

$$Y_{ijt} \le \sum_{l=1}^{L} M_{jlt} \qquad \forall i \in I \quad \forall j \in J \quad \forall t \in T \tag{9}$$

$$\sum_{j=1}^{J}\sum_{k=1}^{K}\sum_{t=1}^{T} Q_{ijkt} + \sum_{k=1}^{K}\sum_{t=1}^{T} Q_{ikt} \le S_i \qquad \forall i \in I \tag{10}$$

$$\sum_{i=1}^{I}\sum_{k=1}^{K} Q_{ijkt} \le b \qquad \forall j \in J \quad \forall t \in T \tag{11}$$

$$Q_{ijkt} \le NY_{ijt} \qquad \forall i \in I \quad \forall j \in J \quad \forall k \in K \quad \forall t \in T \tag{12}$$

$$Q_{ikt} \le NU_{ikt} \qquad \forall i \in I \quad \forall k \in K \quad \forall t \in T \tag{13}$$

$$Q_{kht} \le NL_{kht} \qquad \forall k \in K \quad \forall h \in H \quad \forall t \in T \tag{14}$$

$$\sum_{k=1}^{K} Q_{kht} \ge d_{ht} \qquad \forall h \in H \quad \forall t \in T \tag{15}$$

$$r_{ij} Y_{ijt} \le rc \qquad \forall i \in I \quad \forall j \in J \quad \forall t \in T \tag{16}$$

$$r_{ik} U_{ikt} \le rc \qquad \forall i \in I \quad \forall j \in J \quad \forall t \in T \tag{17}$$

$$\sum_{i=1}^{I}\sum_{j=1}^{J} Q_{ijkt} \le \sum_{i=1}^{I} Q_{ikt} \qquad \forall k \in K \quad \forall t \in T \tag{18}$$

$$\frac{r_{kh}L_{kht} + r_{jk}G_{jkt}}{V} \le LT \qquad \forall k \in K \quad \forall h \in H \quad \forall t \in T \tag{19}$$

Equations (6)-(19)present constraints of model in this paper. Constraint (6) guarantees the number of facilities which are moved should be less than the number of mobile units in each period. Constraint (7) ensures at most one mobile facility can be built in each area. Constraint (8) means that mobile blood donationunits will not be shipped to the site where a mobile donation facility has been builtpreviously. Constraint (9) shows the fact that donors cannot be allocated to facilities which are not built. Constraint (10) ensures the blood collected from donor group i do not exceed their maximum blood supply volume. Constraint (11) represents capacity limits of blood collected at mobile units. Constraints (12)-(13) mean that blood can only be collected if donators are assigned to certain blood facilities. Constraint (14) expresses blood cannot be sent out if a local center is not assigned to a certain hospital.Constraint (15) ensures the minimum demand of each hospital should be fulfilled. Constraints (16)-(17) make sure all bloodcollecting units only accept donators withinthe scope of service area. Constraint (18) reflects that donors prefer to donate at mobile blood facilities than local blood centers. Constraint (19) is the time requirement of blood delivery in certain situations.

3 CASE STUDY AND COMPUTATIONAL RESULTS

Data in this section is based on the current structure of blood service in a city of China. As is shown in Figure 1, we consider fifteen donor groups, eight candidate locations for mobile facilities and three local blood centres to support need of blood from seven hospitals.

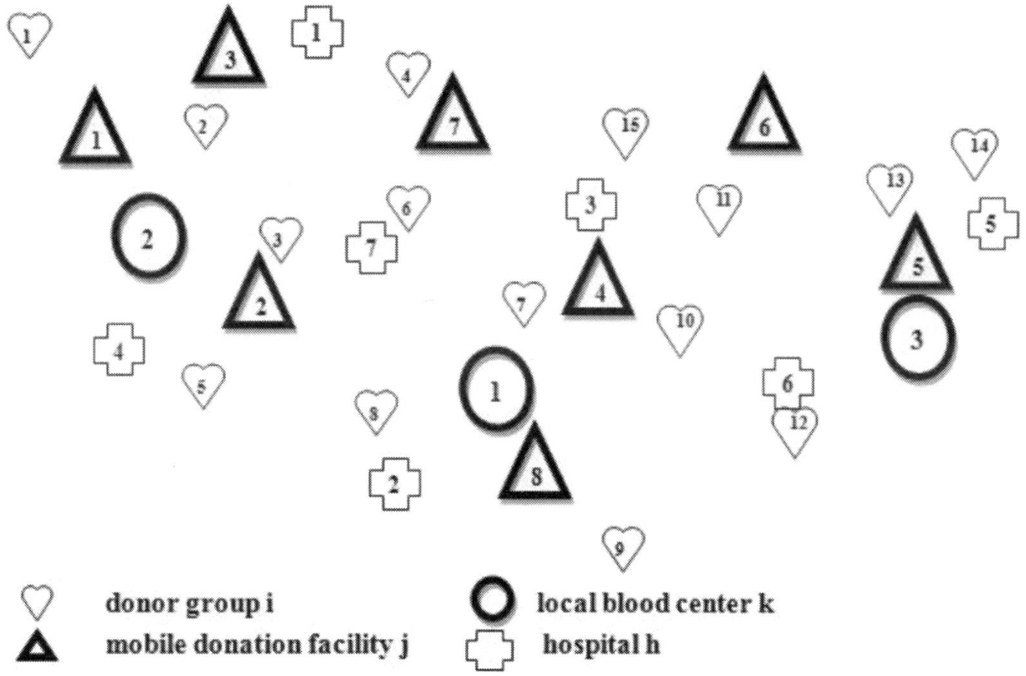

Figure 1:Positions of Nodes in Blood Supply Network

Demand for blood at different periods appears to be uncertain. We ensure that the minimum demand of blood in certain situationshould be fulfilled. Based on the number and damage degree of people wounded, minimum blood supply volumes for different hospitals in each period are assumed in Table 1.

Table 1: Minimum Blood demand in Hospitals during Different Period

Period t	Hospitals in demand						
	h1	h2	h3	h4	h5	h6	h7
t=1	10	8	9	7	4	9	12
t=2	12	13	10	9	7	12	20
t=3	25	29	28	12	10	15	35
t=4	80	70	60	23	15	21	90

A set of computational tests are designed to evaluate the solutions, results shows as follows. To meet the urgent need of blood, four mobile blood donation facilities are needed, and the optimal total cost is 58582 dollars.

Overall cost of blood supply will change when other parameters are varied. We investigate the changes in overall cost in blood supply chain as the delivery time and service distances of blood facilities are altered.

3.1 Changingthe delivery time requirement

Time requirement of blood delivery fordifferent type of disasters varies. Accidents which are serious may need more strict time requirement. Table 2 shows the impacts of changed delivery time requirement on overall cost of blood supply chain. It presents that time requirement of blood delivery increases from 2 hours may gain more delivery speed improvements at only minor changes in overall cost. When the time decreases from 2 hours, the overall cost increases sharply.

Table 2: Impact of Changing Delivery Time on Overall Cost

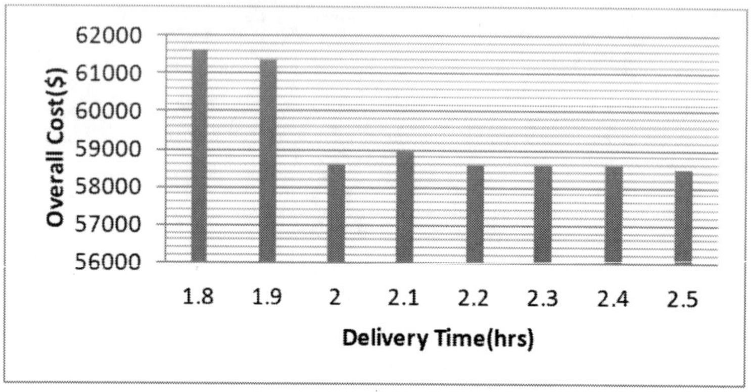

3.2 Changingthe capacity of mobile facilities

Table 3 implies that altering the capacity of mobile blood donation facilities will cause a significant change in overall cost under certain delivery time. However, we should expand the capacity properly, as the excessive increase will lead to more establishment cost of mobile facilities.

Table 3: Impact of Changing Capacity of Mobile Facilities on Overall Cost

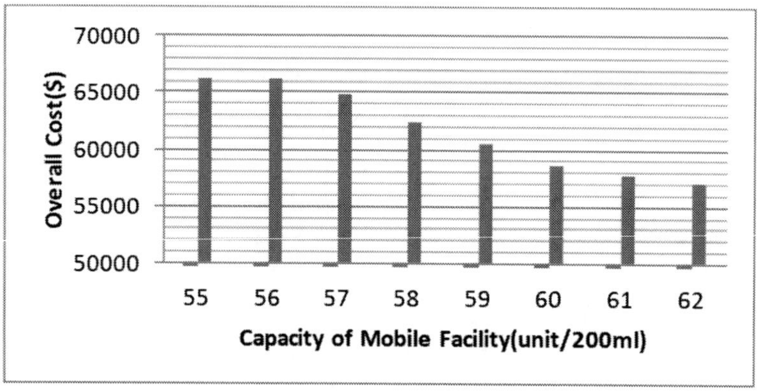

4 CONCLUSIONS

A stochastic supply chain network design model is presented in this paper. Our goal is to make the collected blood fulfilminimum demand of hospitals in an effective and efficient way. The objective minimizes the overall costs of blood supply chain, and the constraints limit time requirement under certain situations. As the demand for blood during different period appears to be uncertain in actual situations, we investigate the minimum need of blood in certain situation and try best to fulfil the most urgent need. In order to solve the model, a series of data is designed. Numerical results are explored in different insights. What we provided is that (1) Serious accidentsneed more strict time requirement. Otherwise, we should consider more about the limited funds. Changing the time requirement of delivery properly can improve the performance of overall cost in blood supply chain.(2)Appropriate expanding the capacity of mobile facilities decreases the overall cost of blood supply chain under certain situations.

In the future research, shortage cost could be taken into account. The demand of safe blood is increasing while the quantity of supply cannot meet it. Shortage of blood causes serious consequences. Furthermore, we could explore some new optimizationalgorithms to improve the running quality and speed.

5 ACKNOLWEDGEMENT

This research is supported by Macau Foundation under Grant No. 0485n.

6 REFERENCES

[1] **Beliën J, Forcé H. 2012.**Supply chain management of blood products: A literature review. *European Journal of Operational Research*, 217(1),pp1-16.

[2] **Özdamar, L., &Demir, O. 2012.**A hierarchical clustering and routing procedure for large scale disaster relief logistics planning. *Transportation Research Part E Logistics & Transportation Review*, 48(3), pp 591-602.

[3] **Sheu, J. B. 2010.**Dynamic relief-demand management for emergency logistics operations under large-scale disasters. *Transportation Research Part E Logistics & Transportation Review*, 46(1), pp1-17.

[4] **Badri, M. A., Mortagy, A. K., & Alsayed, C. A. 1998.** A multi-objective model for locating fire stations, *European Journal of Operational Research*, 110(2), 243-260.

[5] **Hong-Minh, S. M., Disney, S. M., &Naim, M. M. 2013.** The dynamics of emergency transhipment supply chains, *International Journal of Physical Distribution & Logistics Management*, 30(9), pp 788-816.

[6] **G. Barbarosoğlu, Y. Arda.2004.** A two-stage stochastic programming framework for transportation planning in disaster response, *Journal of the Operational Research Society*, 55(1),pp43-53.

[7] **Ukkusuri, S. V., & Yushimito, W. F. 2008.** Location routing approach for the humanitarian prepositioning problem, *Transportation Research Record Journal of the Transportation Research Board*, 2089(6), pp18-25.

[8] **Rahman S, Smith DK. 2000.**Use of location-allocation models in health service development planning in developing nations, *European Journal of Operational Research*,123,pp437-52.

[9] **Sha, Y., Huang, J. 2012.**The multi-period location-allocation problem of engineering emergencyblood supply systems,*Systems Engineering Procedia*, 5, pp21-28.

[10] **Sahin G, Süral H, Meral S.2007.** Location analysis for regionalization of Turkish Red Crescent Blood services, *Computers & Operations Research*,34,pp692-704.

[11] **Fahimnia, B., Jabbarzadeh, A., Ghavamifar, A., & Bell, M.2015.** Supply chain design for efficient and effective blood supply in disasters, *International Journal of Production Economics*, 11.007,pp1-10.

A STUDY OF MODULAR QUALITY MATCHING SELECTION IN AVIATION EQUIPMENT DEVELOPMENT PROJECT

Liangqing Feng[1, 2*], Dali Huang[2*], Zhen He[2] and Wenchuan Li[1]
[1]School of Economics and Management
Nanchang Hangkong University, Nanchang, China
flq9848@163.com

[2]College of management and Economics
Tianjin University, Tianjin, China
zziahll@126.com

ABSTRACT

Under the analysis of modular quality characteristics for aviation equipment development project, to decrease the subjectivity of evaluation value, this paper extends the traditional two-dimension House of Quality to multi dimensions and proposes a model, Multi-dimension House of Quality (MHOQ). The MHOQ includes three dimensions, they are, product quality requirements dimension, cooperation requirements dimension and service requirements dimension. With the MHOQ method, a mathematical model, termed as supplier evaluation function, is applied to measure the ability of suppliers. To reduce subjective while evaluating requirements, the similarity calculation based on Hausdorff distance is used to determine the influence weight of each dimension. At last, according to the numerical analysis and contrastive analysis based on a certain model development project case, this MHOQ approach is demonstrated to be a more effective way than the traditional HOQ on solving problems of modular quality matching selection in aviation equipment development project. It would be possibly to help the project managers and suppliers' selectors to make a more comprehensive and wiser decision.

Keywords: Aviation equipment development project, modular quality matching, multi-dimension House of Quality, supplier selection

1 INTRODUCTION

In general, aviation equipment development product refers to a large product of complex structure, highly technical content and highly integrated system modules; it owns the characteristic of modularization. It is characterized by a complex system of engineering so that many companies have taken part in these development activities around the aviation equipment development product. At present, the aviation equipment development project mainly follows "module integrator—module supplier" model, which makes the module integrators (the mainly development enterprise) as the core strategy enterprise to realize the collaborative innovation development, the quality of each module depends on the quality level of module suppliers, hence the problem about how to find a way to realize modular quality matching selection plays an important role in improving quality of development project.

The project quality management research on aviation equipment development is a great significance. With the continued improvements in various kinds of certification system of international organization, project quality management is striding forward to this direction. The

* Corresponding Author: Email: flq9848@163.com ; zziahll@126.com.

project quality management can be described as a series of quality control and quality assurance activities cover every stage of project to ensure the quality of product and meet the quality level of customer expectations [1]. Aviation equipment development project management research aims to control each subsystem and strengthen coordination between them in order to achieve benefit maximization [2]. Based on complex characteristics of aviation equipment enterprises, it is more important on how to realize quality enhancement of corporate management systems [3].

As multiple-level characteristics of module suppliers in aviation equipment supplier development project, in an effort to better solve this problem, many evaluation methods about supplier selection were put forward. After a lot of research, Dickson [4] designed 23 rules for identity suppliers' behaviours. J. Chen studied the suppliers' evaluation and selection methods [5]. From supplier selection perspective, Li and Liu [6] proposed a multi-level hierarchical planning model to control the quality of complex equipment development project. According to the supply chain of complex product, a multi-level model was development [7]. A point was made about supplier selection problem by Mou and Cao [8] that the product quality issues should be considered from the whole production stages not after design stage. Vijay and Ravindran [9] model the vendor selection problem as a multi-objective optimization problem in outsourcing. However, due to the incomplete information characteristics of supplier selection, the multi assessment theory should be used to solve these issues [10-13]. The Quality Function Deployment (QFD) successfully achieved more attention and become a popular tool to help decision makers to connect the customers' requirements with product decision. AHP and ANP are introduced to determine the relative weight of requirements and decision [14, 15] under uncertain or in complete information.

While those ways still ignore that the requirements should be divided into different dimensions, as the different importance of each indicator in different dimensions should be considered while in complex system. To let the project managers make a more comprehensive decision, after confirming indictors of different modules, from the perspective of modular quality, with the House of Quality (HOQ) [16], this paper designs Multi-dimension House of Quality (MHOQ) replaces the traditional two-dimension HOQ.

This paper contributes to the literature in the following ways. First, it provides a MHOQ model with three dimensions to solve the different importance of each indicator in different dimensions. Second, this paper provides an approach that combines Hausdorff distance and similarity function and interval estimation method to obtain the optimal matching scheme. Finally, numerical analysis and contrastive analysis demonstrate the effectiveness of MHOQ in solve thus problems.

The remainder of the paper is organized as follows. Section 2 designs the MHOQ model for modular quality matching selection. Section 3 describes the matching steps based on MHOQ. Section 4 takes a certain model aircraft development project as a case and designs a MHOQ for it. The modular quality matching process is shown and the optimal matching schemes are obtained in section 5. The contrastive analysis demonstrates that the MHOQ method is useful than the traditional methods. Section 6 concludes the paper and provides future research directions.

2 THE MHOQ DESIGN FOR MODULAR QUALITY MATCHING SELECTION

2.1 MHOQ design for modular quality

The modular quality is a dynamic form of the product or service module decomposition, also the inherent characteristics' form of product or service model. The modular quality in this paper refers to a degree that product or service module decomposed via a standard interface to meet the requirement of inherent characteristics [17].

HOQ is the mainly analysis tool of Quality Function Deployment (QFD). QFD is come up by two Japanese quality experts in 1970s [18] and is widely used in many areas [19-21]. which is a strategic tool to help companies in product development that satisfy the desires of customers by relevant planning matrix whose function is translate the requirements of customers or market demand to the technical attributes of products and production requirements. While it still cannot solve different important rate questions in different dimensions. Therefore, this paper models a MHOQ structure. Fundamentally, the MHOQ refers to classify the quality requirements to different categories according to their attributions, and design their own HOQ for different quality attributions by considering their different effects on the final modular quality matching selection. In order to avoid the subjectivity of cross—dimensional decisions, according to the up analysis, the MHOQ is designed as shown in Figure 1.

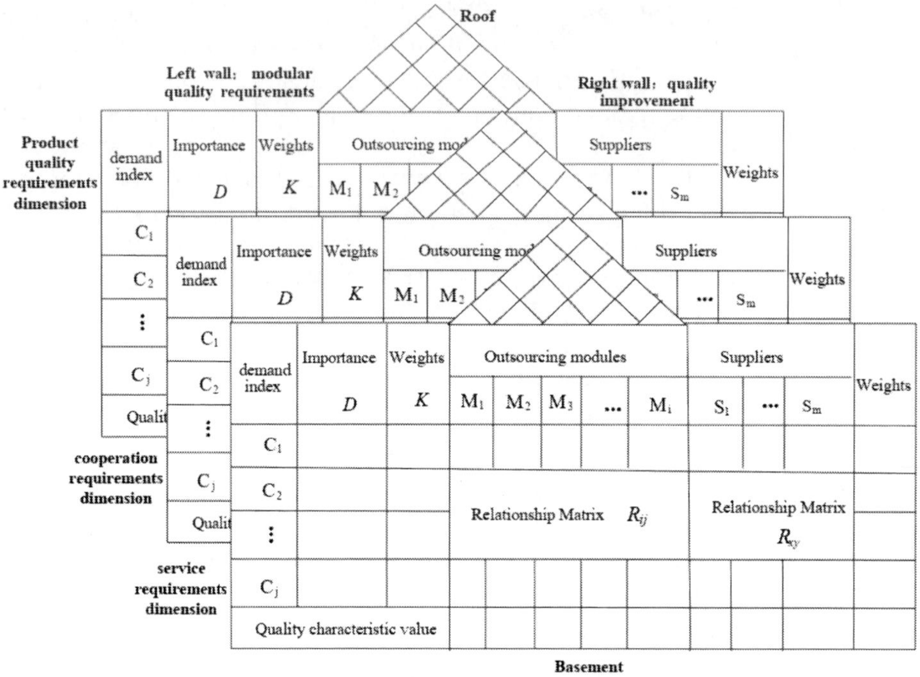

Figure 1: MHOQ model of modular quality.

From the Figure 1, it could be briefly described as following: (1) left wall, the quality requirement indexes of modules. It is used to describe the mainly quality performance of outsourcing modules and suppliers. (2) The ceiling, it mainly includes names of outsourcing modules. (3) The roof, it is used to describe the connection degree of different modules quality. (4) Right wall, mainly consists the compete information of suppliers. (5) Room, it is used to describe relevance between outsourcing modules and quality requirements. (6) Basement, it refers to important degree of module characteristics.

2.2 MHOQ characteristics of aviation equipment development product

The industry chain of aviation equipment development product usually all over the word, the manufacturers typically set up the development cooperation policy with global suppliers that own product development ability to share the product development task to benefit from low development cost and low development risk. Its outsourcings generally could be decomposed into lots modules and sub-modules. So in addition to following quality management standard, the key process quality requirements of product to determine optional suppliers, it still needs to make a comprehensive evaluation about development and design product capability, ability of improving new technology to match the suppliers with modules.

From the product quality view, there is a certain risk for aviation equipment development product to develop new product. Therefore, it is necessary for manufacturer to not only control product quality strictly, but also evaluate the development design ability and new production technology adaptability comprehensively. From the cooperation perspective with suppliers, it needs to consider problems of communication because of globalization production causes and relationship management costs, in addition to on time delivery ability of suppliers, corporate reputation and so on. From the service requirements of product, in addition to the demand of basic service level and service cycle need to be considered, the service flexibility while facing the high uncertainty service and transportation costs because of the dispersed geographical position also need to be considered to satisfy the customer requirement.

2.3 The modular quality matching steps based on MHOQ

Step 1, constructing a set of supplier selection.

Let $S=\{S_i \mid i=1,\ 2,\ \ldots,\ m\}$ be a set of suppliers.

Step 2, quality characteristics calculation of sub-modules.

Define an outsourcing system M consists of v sub-modules and whose indicators evaluation set U consists of u element. Where $M=\{M_x \mid x=1,2,\ \ldots,\ v\}$, $U=\{U_y | y=1,\ 2,\ \ldots,\ u\}$ and $v \leq m$ (it means there are plenty of suppliers could be chosen). Denote the HOQ correlation matrix between M_x and U_y as R_{xy}, so the weight W_{U_y} of indicator U_y could be shown as following:

$$W_{U_y} = \sum_{x=1}^{v} R_{xy} \Big/ \sum R_{xy} \qquad (1)$$

Let $K_M = \{K_{M_x} \mid x=1,\ 2,\ \ldots,\ v\}$ be a quality characteristics value set of sub-module, hence the quality characteristics value K_{M_x} of sub-module M_x could be obtained according to the Eq. (2).

$$K_{M_x} = \sum_{y=1}^{u} R_{xy} W_{U_y} \qquad (2)$$

Step 3, determining the index weight of MHOQ.

Assume quality requirements C could be divided into e dimensions, where $C=\{C_t \mid t=1,\ 2,\ \ldots,\ e\}$, C_{t_j} is defined to be the second grade assessment indicators of C_t, where $C_t=\{C_{t_j} | t=1,\ 2,\ \ldots,\ e,\ \ j=1,\ 2,\ \ldots,\ n\}$, n refers to the number of indicators. Define correlation matrix between supplier S_i and requirement indictor C_{t_j} as R_{ij}, the each index weight of the t dimension as D_t, $D_t=\{D_{t_j} \mid t=1,\ 2,\ \ldots,\ e,\ \ j=1,\ 2,\ \ldots,\ n\}$. Let $K_t=\{K_{t_j} | t=1,\ 2,\ \ldots,\ e,\ \ j=1,\ 2,\ \ldots,\ n\}$ be the normalized vector of t dimension, which can be expressed as the Eq.(3).

$$K_{t_j} = D_{t_j} \Big/ \sum_{j=1}^{j=n} D_{t_j} \qquad (3)$$

If the suppliers of t dimension need to consider a coefficient of improving situation ability, define $H_t=\{H_{t_i} \mid t=1,\ 2,\ \ldots,\ e,\ \ i=1,\ 2,\ \ldots,\ m\}$ as evaluation values set of each supplier on this ability, and $SH_t=\{SH_{t_i} \mid t=1,\ 2,\ \ldots,\ 3,\ \ i=1,\ 2,\ \ldots,\ m\}$ as the evaluation coefficient, while the equation could be described as Eq. (4).

$$SH_{t_i} = H_{t_i} / \sum_{i=1}^{i=m} H_{t_i} \quad (4)$$

Where, if there is no evaluation value about thus coefficients of improving situation ability in the dimension t, then $SH_{t_i} = 1$, $(i = 1, 2, ..., m)$.

Step 4, the weights determination of each dimension.

(1) Expressions of uncertain linguistic variable and numerical evaluation information.

Determining remark set P consists of q orderly assessment set, where $P = \{P_a | a = 1, 2, ..., q\}$. Assuming P is linguistic variable, A refers to a interval number set, then mapping function φ of P to A could be represented as $\varphi: P \to A$, where $\varphi(P_a) = [P_a^L, P_a^U] \in P$. Defining there has d experts are involved in the evaluation indexed by z $(z = 1, 2, 3..., d)$, the evaluation matrix $A^{(z)}$ of expert z could be expressed by interval numbers according to the mapping function φ.

(2)Similarity calculation based on Hausdorff distance.

Define symmetrical interval number $A = [a^L, a^U]$ and $B = [b^L, b^U]$, so the classic Hausdorff distance between interval numbers A and B could be obtained by Eq.(5) [22].

$$H(A, B) = |c(A) - c(B)| + |r(A) - r(B)| \quad (5)$$

In which, X is A or B. $c(X)$ refers to the middle of interval number X and $c(X) = (X^U + X^L)/2$; $r(X)$ refers to the radius of X, and $r(X) = (X^U - X^L)/2$.

Weights of experts are determined based on similar degree of experts' evaluation, which follows the gathering principle that the minority should submit to the majority. The average Hausdorff distance about evaluation value of expert z to other experts' evaluation value is taken as an evaluation measure. The longer average Hausdorff distance, the lower similarity, therefore, the strength of support is lower. The average similarity degree of expert z could be derived from the following equation [23].

$$\eta_t(z) = \frac{1}{1 + \frac{1}{d-1} \sum_{\substack{u=1 \\ u \neq z}}^{d} H(A_t(z), A_t(u))} \quad (6)$$

Where, $A_t(z)$ is evaluation value of dimension t given by expert z; $A_t(u)$ is evaluation value of dimension t given by expert u $(u = 1, 2, ..., d)$; $\eta_t(z)$ refers to the similarity degree of the evaluation value of dimension t given by expert z compared with other experts. The normalized weight of dimension t given by expert z could be calculated as the following Eq. (7) [24]:

$$w_t(z) = \frac{\eta_t(z)}{\sum_{u=1}^{d} \eta_t(z)} \quad (7)$$

(3) Construction of an interval value evaluation method.

$A = [a^L, a^U]$ is an interval value. In order to get more precise result, the deviation coefficient of interval should be considered. As the deviation coefficient shown in Table (1), the interval number A of linguistic variable could be turned into interval number denoted as $\overline{A}$ following the Eq. (8) [25].

Table 1:Linguistic variables and interval deviation coefficient

	Left deviation coefficient α					Right deviation coefficient β				
Linguistic variables	Extremely close to left	Close to left	Mid -dle	Far from left	Extremely far from left	Extremely Close to right	Close to right	Middle	Far from right	Extremely far from right
Evaluation factors	1	0.75	0.5	0.25	0	1	0.75	0.5	0.25	0

$$\overline{A}=[a^L,\ a^L+\alpha\times(a^U-a^L),\ a^U+\beta\times(a^L-a^U),\ a^U] \quad (8)$$

In which, $\overline{A}$ refers to the interval number, a^L and a^U are the upper limit and inferior limit. α and β are deviation coefficients, and $\alpha\geq 0,\ \beta\geq 0,\ \alpha+\beta\leq 1$.

According to Eq. (8), interval evaluation of all experts could be represented as $\overline{A}_t(z)=[a_t(z),\ b_t(z),\ c_t(z),\ d_t(z)]$. The comprehensive evaluation value could be computed by weighted arithmetical average method:

$$W_t=[a_t,\ b_t,\ c_t,d_t]=[\sum_{u=1}^{d}w_t(z)a_t(z),\ \sum_{u=1}^{d}w_t(z)b_t(z),\ \sum_{u=1}^{d}w_t(z)c_t(z),\ \sum_{u=1}^{d}w_t(z)d_z(t)] \quad (9)$$

Weight of dimension t denoted as $\overline{W_t}$, normalized value $\overline{W_t}$ and denoted it as W_t' following Eq. (10) and Eq.(11) .

$$\overline{W_t}=\frac{1}{4}(a_t+b_t+c_t+d_t) \quad (10)$$

$$W_t'=\overline{W_t}\Big/\sum_{t=1}^{e}\overline{W_t} \quad (11)$$

Where $W'=\{W_t'\mid t=1,\ 2,\ ...,\ e\}$.

Step 5, constructing a decision matrix for a single dimension.

Constructing a decision matrix G_t for the single dimension t, which is shown in (12).

$$G_t=(G_{t_i})\ (i=1,\ 2,\ ...,\ m) \quad (12)$$

$$G_{t_i}=SH_{t_i}\times\sum_{j=1}^{n}K_{t_j}R_{ij} \quad (13)$$

Step 6, construction of a comprehensive decision matrix.

Let $G_t'=(G_{t_i}')$ be the normalized value of decision matrixes from t dimension, then the comprehensive decision matrix G can be expressed according to ordinary weighting method in the Eq. (14), here $G=(G_{S_i})$, $(i=1,\ 2,\ ...,\ m)$.

$$G=\sum_{t=1}^{e}w_t\ G_t' \quad (14)$$

Step 7, the quality characteristics matching selection of sub-modules.

According to the Eq. (8), it needs to sort the suppliers and select the top v suppliers by following the principle that the larger dates should be considered first. Then we can match suppliers with sub-modules according the decision matrix G_t of t dimension that has same quality requirements both suppliers and sub-modules.

3 CASE STUDY

3.1 Module division of a certain model aircraft development project

Based on the production pattern of a large aviation equipment manufacturing enterprise, we take a certain model development project as the example. According to the structure of aircraft parts and following the function - structure analysis method, this paper divides the aircraft configuration into 5 modules after lot of data collection [19, 26-27], they are airframe module, power system module, flight control module, aviation electronics module and environmental control module. Each module of those 5 modules consists of many sub-modules (as shown in Figure 2).

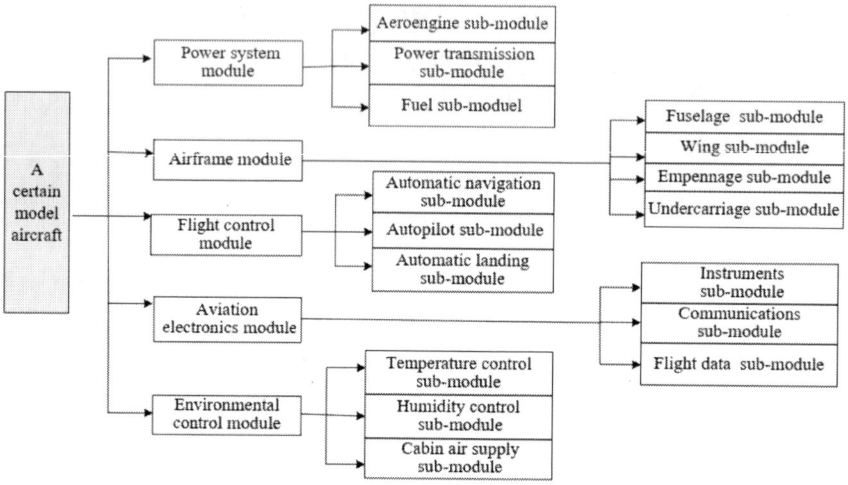

Figure 2: Structural schematic diagram of a certain model aircraft

3.2 MHOQ design of a certain model aircraft development project

The project integrator needs to analyze the main factors of influencing modular quality, gather views of aviation development departments and characteristics of cooperation while cooperating with exiting module suppliers, then transfer them factors into operational indicators. In addition, the abilities for suppliers to provide qualified products, the adaptability to new technology, the ability of risk treatment and ability to meet the service requirements, they all should be shown in index system to comprehensively evaluate module suppliers.

Concerns with the principles of comprehensiveness, objectivity, relevance and systematises, it needs to determine indicators through expert interview and the collection of references [4, 10] and according to characteristics and function of each module [28]. For a large aviation equipment development project, due to the reason that module suppliers usually participate in the new product development, it needs to evaluate the production adaptability of suppliers to meet the new production technology while choosing the modular quality matching ways, in addition to considering the existing basic levels. Based on this, this paper comes up with the idea that it needs to consider the production adaptability coefficient of each module supplies in product quality requirements dimension while selecting suppliers. The higher of adaptability coefficient, the lower risk for development project integrator to select the supplier. Though there are different characteristics in different modules, the modular quality matching process and methods are similar. Therefore, this paper takes the airframe module and its sub-modules as the study object to analysis the modular quality matching selection method.

3.2.1 *The HOQ design for the quality requirements of airframe module*

Based on the quality requirements of airframe module and characteristics of each sub-module, the following important factors are determined to evaluate importance of each sub-module. Using interval evaluation method and following the Five Mark Scoring to give the most important index 5 score and the very not important index 1 score, based on the reference [28] fully illustrates functions of each sub-module, quality standards of aircraft parts AS9100 [29] developed by International Aerospace Quality Group （ISQG） and quality requirement to parts that defined as two dimensions matrix code in AS9132 [30] quality system are regarded as the evaluation standard to determine the relationship matrix between index and sub-module. Then matrix between quality requirements and airframe module is designed by a module integrator from aviation equipment development as shown in Table (2).

Table 2: Quality planning HOQ of airframe module

Quality requirements		**Airframe module**				**Absolute weight**	**Relative Weight**
Criteria layer	Indicators	Fuselage sub-module M_1	Wing Sub-module M_2	Empennage sub-module M_3	Undercarriage sub-module M_4		
Quality requirem-ents of product	Corrosion resistance U_1	5	5	5	4	19	0.10
	Stiffness U_2	4	5	5	3	17	0.09
	Fatigability U_3	3	5	5	3	16	0.09
	Weight U_4	5	4	4	2	15	0.08
	Aero elastic U_5	4	3	3	2	12	0.07
	Flight stability U_6	3	4	4	1	12	0.07
	Service Life U_7	4	5	5	3	17	0.09
	Pollution U_8	3	3	3	3	12	0.07
	Flight speed U_9	4	4	5	1	14	0.08
	Lifts steadily U_{10}	1	1	1	5	8	0.04
	Heat stability U_{11}	5	5	5	2	17	0.09
	Operability U_{12}	2	4	5	3	14	0.07
	Nice appearance U_{13}	3	3	3	2	11	0.06
Quality characteristic value		3.72	4.14	4.29	2.58		

3.2.2 *The MHOQ design*

To help designers make a comprehensive assessment of suppliers about how to outsource modules to suppliers, it needs to divide the influence factors to three classes, the product quality requirements, coordination requirements and service requirements[31,32]. Therefore, this paper extends the traditional QFD to three dimensions to evaluate the abilities of suppliers, and matches suppliers with modules according to the quality characteristics of each module. There are 15 suppliers(S_1, S_2, ..., S_{15}), while only 8 suppliers(S_1, S_2, ..., S_8) fill the manufacturing requirements to produce this model airframe by checking their core technical ability, enterprise credibility and quality certifications. Five experts are invited to evaluate suppliers' abilities from production management department, sales management department and service management department. The evaluation matrixes are shown in Table (3-5).

Table 3: HOQ of product quality requirements dimension

Quality requirements		Weight	Name of supplier							
Criteria layer	Indicators		S_1	S_2	S_3	S_4	S_5	S_6	S_7	S_8
Quality requirements of product C_1	Corrosion resistance C_{11}	0.10	4	4	4	5	4	5	4	5
	Stiffness C_{12}	0.09	4	3	4	5	5	4	5	4
	Fatigability C_{13}	0.09	4	5	4	4	4	4	5	4
	Weight C_{14}	0.08	4	4	4	4	5	4	4	4
	Aero elastic C_{15}	0.07	3	4	2	4	2	4	3	3
	Flight stability C_{16}	0.07	3	3	4	3	4	4	4	4
	Service Life C_{17}	0.09	4	3	3	3	3	3	4	3
	Pollution C_{18}	0.07	2	2	3	2	4	2	4	3
	Flight speed C_{19}	0.08	3	4	3	4	2	3	3	3
	Lifts steadily C_{110}	0.04	4	2	3	4	3	4	4	2
	Heat stability C_{111}	0.09	3	3	4	3	2	4	2	4
	Operability C_{112}	0.07	4	5	4	4	4	4	5	4
	Nice appearance C_{113}	0.06	3	4	3	4	4	4	3	3
Quality characteristic value			3.49	3.6	3.52	3.8	3.56	3.79	3.86	3.65
production adaptability			6	7	5	7	5	6	8	4
production adaptability coefficient			0.13	0.15	0.1	0.15	0.1	0.13	0.16	0.08

Table 4: HOQ of cooperation requirements dimension

Quality requirements		Import-ance	Wei-ght	Name of supplier							
Criteria layer	Indicators			S_1	S_2	S_3	S_4	S_5	S_6	S_7	S_8
Cooperat-ion require-ments C_2	Deliver on time C_{21}	9	0.31	3	4	3	4	4	4	3	4
	Communicational ability C_{22}	4	0.14	4	4	5	3	2	4	3	2
	Capital financing capability C_{23}	6	0.21	2	4	3	5	4	3	4	3
	Enterprise credibility C_{24}	3	0.10	3	3	4	4	3	4	3	4
	Relation management cost C_{25}	7	0.24	3	4	4	4	2	3	4	2
Quality characteristic value				2.93	3.9	3.62	4.07	3.14	3.55	3.45	3.03

Table 5: HOQ of service requirements dimension

Quality requirements		Import-ance	Weight	Name of supplier							
Criteria layer	Indicators			S_1	S_2	S_3	S_4	S_5	S_6	S_7	S_8
Service requirem-ents C_3	Maintenance cost C_{31}	9	0.31	4	4	3	4	2	4	3	4
	Service interval C_{32}	7	0.24	3	3	4	3	3	4	2	3
	Service flexibility C_{33}	7	0.24	4	2	4	2	3	3	4	3
	Transport advantage C_{34}	6	0.21	3	2	3	4	2	4	3	4
Quality characteristic value				3.55	2.86	3.48	3.28	2.48	3.76	3	3.52

4 NUMERICAL ANALYSIS

4.1 The modular quality matching process based on MHOQ

4.1.1 Decision-making process of a single dimension

According to the Eq. (4) and Eq. (13), the decision matrix of each dimension is normalized as following:

$$G_1 = \begin{bmatrix} 0.44 & 0.53 & 0.37 & 0.55 & 0.37 & 0.47 & 0.64 & 0.30 \end{bmatrix};$$

$$G_2 = \begin{bmatrix} 2.93 & 3.9 & 3.62 & 4.07 & 3.14 & 3.55 & 3.45 & 3.03 \end{bmatrix};$$

$$G_3 = \begin{bmatrix} 3.55 & 2.86 & 3.48 & 3.28 & 2.48 & 3.76 & 3.00 & 3.28 \end{bmatrix};$$

The decision matrix of each dimension is normalized as following:

$$G_1' = \begin{bmatrix} 0.119 & 0.143 & 0.100 & 0.151 & 0.100 & 0.129 & 0.175 & 0.083 \end{bmatrix};$$

$$G_2' = \begin{bmatrix} 0.106 & 0.141 & 0.131 & 0.147 & 0.113 & 0.128 & 0.125 & 0.109 \end{bmatrix};$$

$$G_3' = \begin{bmatrix} 0.137 & 0.110 & 0.134 & 0.126 & 0.096 & 0.145 & 0.116 & 0.136 \end{bmatrix}$$

After calculation, there has no abnormal result tested by the evaluation deviation.

4.1.2 The weight determination of each dimension

Evaluation dimensions including product requirements C_1, cooperation requirements C_2 and service requirements C_3. Here 3 experts are invited to make an importance evaluation of each dimension following the Five Mark Scoring, and determine interval evaluation value based on deviation coefficient. According to the statistical, all the 3 experts used triangular evaluation value to express their idea, no one adopt interval trapezoidal evaluation method. The results are as following in Table (6):

Table 6 : Evaluation values of experts

	Opinions of experts		
Dimensions	EXP_1	EXP_2	EXP_3
Product requirements C_1	[4, 5]; right; 1	[1, 3]; 0.5	[2, 4]; left; 0.75
Cooperation requirements C_2	[3, 5]; 0.5	[2, 3]; 0.5	[2, 4]; 0.5
Service requirements C_3	[4, 5]; right; 0.75	[2, 4]; left; 1	[3, 4]; right; 0.25

The weight of different dimension from each experts are given from the Eq.(6):

$$\eta = \begin{bmatrix} 0.667 & 0.5 & 0.667 \\ 0.5 & 0.5 & 0.667 \\ 0.667 & 0.5 & 0.5 \end{bmatrix}.$$

With Table (6), calculating the triangular interval evaluation value according to Eq. (8) and denote it as $\overline{A}$:

$$\overline{A} = \begin{bmatrix} [4, & 5, & 5] & [1, & 2, & 3] & [2, & 3.5, & 4] \\ [3, & 4.5, & 5] & [2, & 2.5, & 3] & [2, & 3, & 4] \\ [4, & 4.5, & 5] & [2, & 2, & 4] & [3, & 3.75, & 4] \end{bmatrix}.$$

Hence weight of each dimension W' could be calculated by Eq. (9-11):

$$W' = [0.445 \quad 0.236 \quad 0.319].$$

4.1.3 *Decision-making process of multi-dimensional comprehensively*

Following the Eq. (13, 14), the finally decision making vector is obtained; here we magnify the data 100 times in order to make it more convenience to observer results.

$$G = [G_{S1},\ G_{S2},\ G_{S3},\ G_{S4},\ G_{S5},\ G_{S6},\ G_{S7},\ G_{S8}];$$
$$G = [12.17,\ 13.20,\ 11.82,\ 14.21,\ 10.18,\ 13.38,\ 14.44,\ 10.60].$$

So the matching scheme as shown in Table 7:

Table 7: The matching scheme

Outsourcing modules	Name of supplier	Outsourcing modules	Name of supplier
Fuselage sub-module M_1	S_2	Empennage sub-module M_3	S_7
Wing sub-module M_2	S_4	Undercarriage sub-module M_4	S_6

4.2 Contrastive analysis

If only judging thus suppliers matching selection problem by the original data obtained from experts, without any methods, the results of matching selection will less accurate. From the original data and with the simple analysis way, normalizing the selection matrix magnify the data 100 times, denoted it as G_E'.

$$G_E' = [G_{S1},\ G_{S2},\ G_{S3},\ G_{S4}\ G_{S5},\ G_{S6},\ G_{S7},\ G_{S8}]$$
$$G_E' = [11.99,\ 12.32,\ 12.65,\ 13.29\ 11.51,\ 13.29,\ 12.80,\ 12.15]$$

So the top suppliers owning high modular quality ability are S_3, S_4, S_6, S_7, and $G_{S4} = G_{S6} > G_{S7} > G_{S3}$.

Traditional HOQ has no division dimensions of requirements indicators. Without considering the production adaptability of suppliers and following the traditional HOQ, the decision matrix of suppliers G_T is shown obtained, normalizing G_T and magnify the data 100 times, the matrix of supplier evaluation G_T' could be represented as following:

$$G_T' = [G_{S1},\ G_{S2},\ G_{S3},\ G_{S4}\ G_{S5},\ G_{S6},\ G_{S7},\ G_{S8}]$$
$$G_T' = [11.95,\ 12.29,\ 12.22,\ 13.31\ 11.92,\ 12.91,\ 13.01,\ 12.39]$$

To improve the proposed method is effective, Figure 3 and Figure 4 are shown to make contrasts.

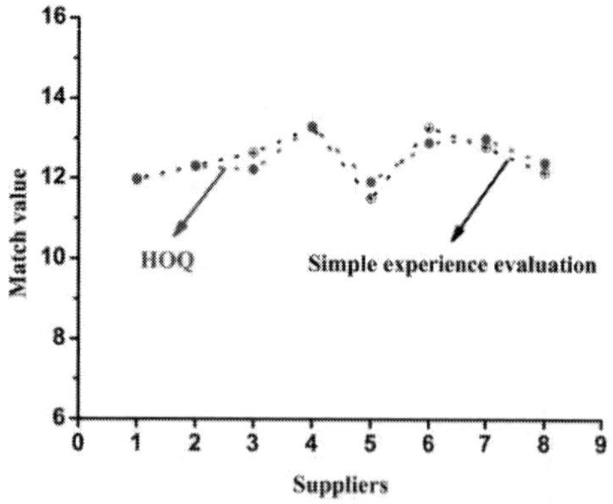

Figure 3: Supplier selection scheme both HOQ and simple experience evaluation

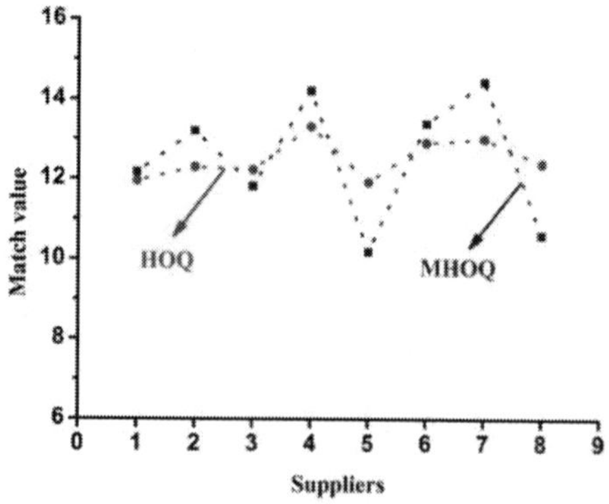

Figure 4: Supplier selection scheme both MHOQ and HOQ

As shown in Figure 3, there has same evaluation value between G_{S_4} and G_{S_6} based on experience evaluation, it is difficult for design makes to determine how to match modular with suppliers. Following the principle of supplier matching selection, for example, the supplier S_7 should match the sub-module M_1. However, based on the G_1 matrix, the production quality requirements of S_7 is the most high, it should match with sub-module M_3 but not M_1. From those, the selection scheme is more convincing based on the HOQ method than experience evaluation method. As we can see from the matrix $G_T{}'$, however the similar decision value makes it difficult to distinguish the virtues or defect degree from each suppliers.

From the Fig. 4, it could be easily distinguished that the supplier selection scheme is different based on MHOQ from the selection by HOQ, the distance between supplier weights is farther by MHOQ than the result calculated by HOQ, which makes it easier for decision makers to evaluate suppliers' abilities. The main difference of decision result is the selection of S_2 and S_8 between the traditional way and this way proposed by this paper. It can be seen from Table 3 that the product quality requirement weight of supplier S_2 is a little lower than the weight of supplier S_8, however, there are higher cooperation requirements ability and stronger production adaptability in supplier S_2 compared with supplier S_8. Therefore, from the perspective of comprehensiveness, it is wiser choose for integrators of projects to make a decision to choose S_2 rather than S_8 .The decision-making method by MHOQ and considering production adaptability is more objective than traditional way in evaluating suppliers' ability.

5 CONCLUSION

The characteristics of high technical, high complexity and modular structure in aviation equipment development project make it difficult to control its modular quality. By considering the quality characteristic requirements of modules,this paper studies the problem of the modular quality matching selection by the way of constructing a MHOQ and combing trapezoidal interval value evaluation. This paper proposes to study the module suppliers' selection in aviation equipment development project from the perspective of modular quality. considering that different category of properties owns different influence weight on modular quality, MHOQ is constructed to replace traditional HOQ, and the evaluation method based on Hausdorff distance and triangular comprehensive evaluation are used to determine each dimension weights.

The following conclusions from that case study could to be summarized easily: (1) since the different importance in each evaluation dimension should be considered, MHOQ could help the solve thus problem, it is a more effective way to solve problems module quality matching selection in aviation equipment development project by the MHOQ method rather than traditional QFD method. (2) As the expressions of uncertain linguistic variable and numerical evaluation information by similarity calculation based on Hausdorff distance, the subjectivity of evaluation value could be decreased in some way by combining comprehensive evaluation method. (3) The risk due to a new product development could be reduced if the influence of production adaptability in suppliers on modular quality were considered.

6 ACKNOWLEDGEMENTS

We would like to extend our gratitude to those experts who did contribute to the data acquisition. This work is partially sponsored by Aeronautical Science Foundation of China (Grant No. 2014ZG56028); National Natural Science Foundation of China (Grant No. 71362019); China Scholarship Foundation (Grant No. 201408360080) and Jiangxi Provance Science Fund for Distinguished Young Scholars.

7 REFERENCES

[1] **Wang, H.J. 2007.** The study of the quality management of Bora HB project, *Master Dissertation*, Jilin University.

[2] **Yang, W.P. 2002.** One plan and management of aeronautical models research, Master Dissertation, Northwestern Polytechnical University.

[3] **Giron, K., & Liang, Y. 2014.** Quality enhancement of corporate management systems: an overview of best management practices, *Journal of Service Science and Management*, 7, pp 302-312.

[4] **Dickson, G.W. 1996.** An analysis of vendor selection systems and decisions, *Journal of Purchasing*, 2(1), pp 5-17.

[5] **Chen, J. 2008.** Research on comprehensive evaluation and selection of Aero-engine MRO supplies, *Master Dissertation*, Nanjing University of Aeronautics and Astronautics.

[6] **Li, Y.P., & Liu, S.F. 2014.** Hierarchical planning-based quality coordination model for complex equipment development, *Journal of Systems & Management*, 23(1), pp 36-40.

[7] **Liu, Y., Hiple, K.W., & Fang, Z.G. 2011.** Networked model for managing quality in supply chain of complex product, *Control and Decision*, 11, pp 1685-1693.

[8] **Mou, L.F., & Cao, Y. 2015.** A multi-objective method for supplier pre-selection of platform-based products, *Operations Research Transaction*, 19(01), pp 31-44.

[9] **Vijay, W., & Ravindran, A.R. 2007.** Vendor selection in outsourcing, *Computers & Operations Research*, 34, pp 3725-3737.

[10] **Zhao, X.H., & Zhao, X.M. 2002.** Vendor-selection method based on fuzzy decision making, *Industrial Engineering and management*, 4, pp 27-29.

[11] **Kahraman, C., Cebeci, U. & Ulukan, Z. 2003.** Multi-criteria supplier selection using fuzzy AHP, *Logistics Information Management*, 16, pp 382-394.

[12] **Semih, Ö., & Selin, S. K. 2009.** Long term supplier selection using a combined fuzzy MCDM approach: A case study for a telecommunication company, *Expert Systems with Applications*, 36, pp 3887-3895.

[13] **Hu, H.Y. 2011.** Supplier evaluation method based on incomplete information and its application, *Master Dissertation*, Nanjing University of Aeronautics and Astronautics.

[14] **Alptekin, E., & Isiklar, G. 2005.** An application of quality function deployment: Evaluation of e-learning products in Turkey, In Proceedings of the 35th international conference on computers and industrial engineering, pp 99-104.

[15] **Zaim, S., Sevkli, M., & Hatice, C. A. 2014.** Use of ANP weighted crisp and fuzzy QFD for product development, *Expert Systems with Applications*, 41, pp 4464-4474.

[16] **Hauser, J.R., & Don, C. 1988.** The House of Quality, *Harvard Business Review*, 3, pp 63-73.

[17] **Feng, L.Q.** 2012. *Study on quality behavior of the service-oriented manufacturing network node*, Economic Science Press, Beijing.

[18] **Yoji, A.** 1990. *Quality Function Deployment: integrating customer requirements into product design*, cam-bridge.

[19] **Jia, G.Z., & Bai, M. 2010.** An approach for manufacturing strategy development based on fuzzy-QFD, *Journal of Computers & Industrial Engineering*, 60(3), pp 445-450.

[20] **Wang, Y.M., & Chin, K.S. 2011.** Technical importance ratings in fuzzy QFD by integrating fuzzy normalization and fuzzy weighted average, *Computers and Mathematics with Applications*, 62(11), pp 4207-4221.

[21] **Bevilacqua, M. & Ciarapica, F. E. 2012.** Development and test of a new fuzzy-QFD approach for characterizing customers rating of extra virgin olive oil, *Food Quality and Preference*, 24, pp 75-84.

[22] **De Carvalho, F. A.T & Lechevallier Y. 2009.** Partitional clustering algorithms for symbolic interval data based on single adaptive distances, *Pattern Recognition*, 42(7), pp 1223-1236.

[23] **Nie, W.B., Liu W.D & Hu, K. 2015.** Process failure risk assessment based on group decision-making and the generalized hausdorff distance, *Computer Integrated Manufacturing Systems*, 21(9).pp. 2484-2493.

[24] **Lee, J. W., & Kim, S. H. 2000.** Using analytic network process and goal programming for interdependent information system project selection, *Computers and Operations Research*, 27, pp 367-382.

[25] **Chu, T.C. 2002.** A fuzzy number interval arithmetic based fuzzy MCDM algorithm, *Int of Fuzzy systems*, 4(4), pp 867-872.

[26] **Tang, H.C.** 2013. Commercial aircraft projects, Aviation industry Press.

[27] China civil aviation maintenance network, 2008. Main Component and Function of Plane, *http://www.china-cam.cn/oblog/u/214/archives/2008/254.html.*

[28] **Wang, Z.J., & Yao, W.X. 2009.** Aircraft structure design, National Defence Industry.

[29] **Society of Automotive engineering**, 2009. AS9000 Quality management systems-requirements for aviation, space and defense organization. Aerospace Industry.

[30] **Society of Automotive engineering**, 2005. AS9132 Quality management systems-data matrix quality requirements for parts markin. Aerospace Industry.

[31] **Feng, H. 1996.** Assessment of aviation enterprise supplier quality assurance ability and management approach.

[32] **De Brito, M. P., & Van Der Laan, E. A.** 2008. Supply chain management and sustainability: Procrastinating a holistic integration, *IEMC - Europe : International engineering management conference proceedings*, pp 389-392.

WAVELET BASED DIABETIC RETINOPATHY IMAGE ENHANCEMENT IN BLOOD VESSELS

S. Deva Kumar[1*] and Dr N.Gnaneswara Rao[2]
[1,2]Department of Computer Science and Engineering
VFSTR University, Andhra Pradesh, India
[1]deva.248@gmail.com,
[2]gnani.nitta@gmail.com

ABSTRACT

Diabetic Retinopathy (DR) is the problem for long-standing diabetes and the major cause for visual loss is because of variations in blood vessels of the retina. Vision loss is due to DR is preventable with regular checkups at the earlier stages. The presence of exudates is the basic signs of DR. Hence, exudates detection becomes a major problem-solving task, in which retinal images plays a vital role. If detected accurately, eye doctor can start treating the patients to avoid vision loss. Image processing techniques reduces the effort of eye doctor plus tools are utilized in order to find the exudates. Automated examination methods on retinal images became a main zone in research to develop evaluating programmers. MATLAB is used for implementing the paper. In this paper, a methodology is proposed based on an edge detection operator and wavelet transforms which are in image processing is used to detect blood vessels from retinal image. At the starting phase, the retinal image must be preprocessed, where at first the image which is in RGB color format is altered to the image of grayscale color format and then in one case histogram equalization is performed to normalize the image. In subsequent phase, blood vessels are detected using Sobel edge detection operation and then wavelet transform is applied. Finally coming into the results, this proposed technique evaluates edge detection of blood vessels on retinal images mainly based on the Peak to Signal Noise Ratio when only Sobel operator is performed on retinal image with histogram equalized image and discrete wavelet transform. So, the values of PSNR for the same retinal image are 253.31 and 246.76.

Keywords: Segmentation, Diabetic Retinopathy (DR), Retinal Image, Histogram Equalization, Edge detection operator (Sobel Operator), Discrete Wavelet Transforms, Peak to Signal Noise Ratio.

1 INTRODUCTION

Image Segmentation technique divides an image into its constituent objects or the parts or even the parts of the object. A segmentation technique takes the method a long way towards the successful result of imaging complications that need objects to be identified separately. In general, the recognition of an object is succeeded, when the segmentation is done more accurately. The result of the segmentation phase is the representation of the objects and description of those objects suitable for computer processing, and recognition of individual objects from the overall image, which is generally a raw pixel data, containing either the edge of an object (i.e., the set of pixels separating one image object from another) or all the points in the object itself.

[*]Department of Computer Science and Engineering, VFSTR University, deva.248@gmail.com

The common effect of vision loss in people is because of diabetes which is the Diabetic retinopathy and mainly leads to blindness among elders who are working. Human with diabetes are at danger for causing DR. Disease increases when someone has diabetes for longer time. Among all the Americans diagnosed with diabetes have certain phase of diabetic retinopathy which is about 40-45 percent, out of them only about partial are aware of DR. Women who are having diabetes in the course of pregnancy may have quick effect of diabetic retinopathy. When someone has high blood sugar levels, the blood vessels in the retina can become damaged, so that it can be prevented through frequent checkups and treatment, and proper follow-up on diabetic eye disease can check against vision damage. Therefore, by early recognition reduces the severity of the disease [1]. For disease analysis, the greatest method is to take victim's image of retina between definite intervals of time and match these images to realize the variations of the retinal image [2, 3].

In processing of a retinal image, fundus images are used for primary analysis of diseases. The segmentation of retinal image structures [4] is the major part of this paper, which is used for analysis by ophthalmologist.

The fundus image in RGB model is a 24-bit color image which uses 8-bits for each component, where R=Red, G=Green, B=Blue are 3 components and are also called as Hues (Colors). An RGB color model is called additive color model space as shown in Fig 1.

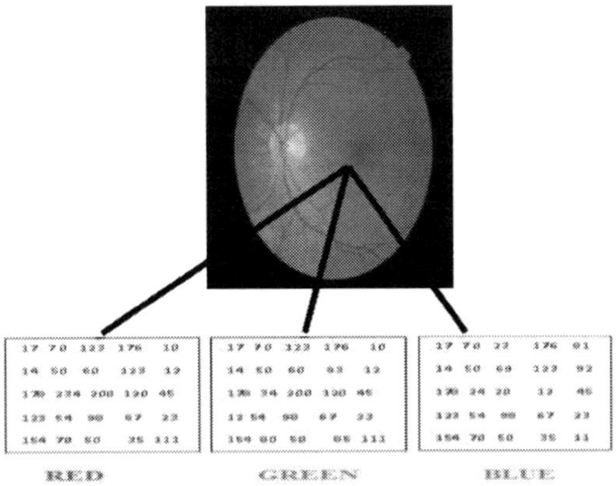

Fig1: Representing red, green, blue color bands of RGB image

A Grayscale Image is even called as gray level image or intensity as shown in Fig 2. The intensity value range of values between [0, 255] is for uint8. The intensity value range of values between [0, 65535] is for uint16.

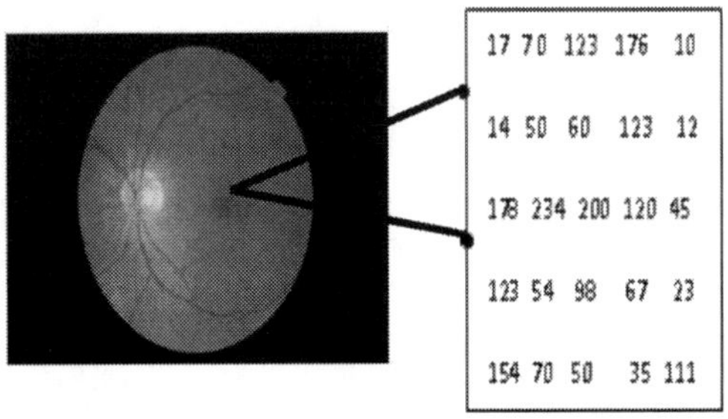

Fig2: Sub image of the image

Histogram Equalization is an preprocess technique to improve the quality (contrast of an image). It's not compulsory that contrast will ever be rise in this. There may be certain circumstances were histogram equalization can be of low quality. In such circumstances the contrast is diminished. On performing the histogram equalization the entire shape of the histogram alters, while in histogram stretching the entire shape of histogram is same. Usage of histogram equalization is to get the pictorial representation of an image in a better manner that is suitable for inspection by means of human or through an automated machine.

Edges are the variations of discontinuities in an image. Major changeovers in an image are called as edges. At first identify the edges in an image because the information about shape is surrounded in edges. Sobel operator [5] is a derivate mask for discovering edges in both directions i.e. horizontally plus vertically. Thus this is a great method for extending the features also to categorize the pixels either as a blood vessel or as a non-blood vessel.

The wavelet transform is known as a mathematical method in which a specific signal is examined (or synthesized) in the domain of time through various versions of a dilated (or contracted) also translated (or shifted) basis function named the wavelet prototype or even the mother wavelet. A wavelet function $\psi(t)$ is a small wave that is oscillatory in certain way to discriminate a various frequencies. The Discrete Wavelet Transform (DWT) [6] is an employment of the wavelet transform that's makes use of a discrete set of the wavelet scales plus translations ensuing few defined rules. The symlet wavelet is expanded as "symmetrical wavelets". They are built in such a way that the least asymmetry and extreme amount of disappearing moments for the specified compact support. Quality of compressed images starts degrading slowly while rapid degrading occurs on higher order when sym4 is used as shown in Fig 3. The sym4's performance on compression is superior and the pictorial quality of recreated image is superior even at the equal threshold.

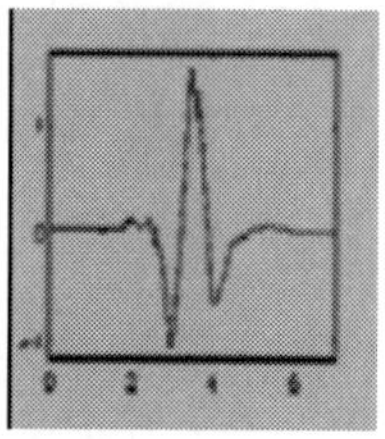

Fig3: sym4

The Peak signal to Noise Ratio (PSNR) [7] is called as the ratio of the power of an original signal and the power of distorting noise signal that results to the quality of its image. PSNR is defined by:

$$PSNR = 20\log_{10}\left(\frac{MAX_f}{\sqrt{MSE}}\right)$$

Where, $MAX_f$is the maximum variation in the image.

TheMean Square Error(MSE) [8] is the cumulative squared error difference among the original and the encoded image. When error in the image is lower it indicates that MSE shows the lower value. MSE is defined by:

$$MSE = \frac{1}{mn}\sum_{0}^{m-1}\sum_{0}^{n-1}\|f(i,j) - g(i,j)\|^2$$

Where, f is the original image plus g is the encoded image and m x n is the dimension of the image i.e. m is the rows of the original image and n is the columns of the original image. So, for the better performance of compression MSE must be as lower as possible.

A. MOTIVATION

As the medical images are the vital for human survivance and proper screening is must in order to diagnose a disease. So, in this paper, segmented the blood vessels from the retinal image using Sobel edge detection operator and discrete wavelet transforms by this diabetic retinopathy disease can be reduced in a right way.

B.STRUCTURE

The remaining part of the paper is organized as follows- Section II shows the entire methodology of Detecting Blood Vessels. Subsequently, Section III represents the result after execution of methodology and Section IV presents the final conclusion. Section V includes the references, which are used for survey to accomplish this whole paper.

2 METHODOLOGY

The proposed method detects blood vessels in retinal image using the concepts of Sobel edge detection operator. In the proposed technique, the entire process can be separated into two sub-stages, namely, preprocessing and detecting blood vessels are shown in Block Diagram 1. The entire proposed methodology is shown as a Block diagram 2.

2.1. Preprocessing

2.1.1. Image Capture

The initial step in the image examination is to retrieve retinal image. Image capturing is done by a camera which includes an embedded digital camera at the back. The digital camera works like the normal camera, only difference is that it is embedded with an image sensor.

2.1.2. Retina Image

Initially, the retinal image which is in RGB color format i.e. which is taken from camera is transformed to Gray scale image [9].

2.1.3. Image Enhancement

Histogram equalization is a process which uses histogram of an image for contrast correction. This technique raises overall contrast of an image; mainly the data which is important in an image is shown in low contrast values. The intensity values of the pixels can be better scattered on the histogram through this histogram equalization [10, 11]. This method is useful when both background and foreground of an image are bright or dark.

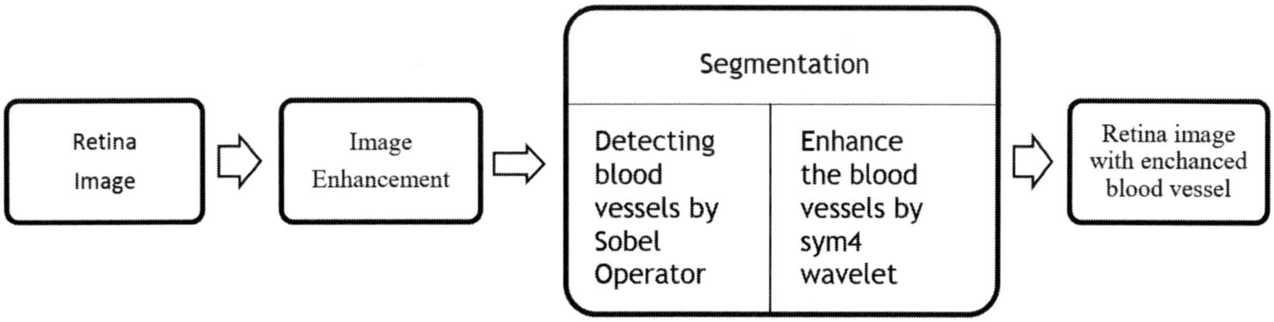

Block diagram 2: Entire Process of the Proposed Technique

2.2. Detection of Blood Vessels

2.2.1. Sobel Edge Detection Method

The Sobel method is for detecting of blood vessels, exudates, blobs, also optic disk. The method of finding pel intensity changeovers is the Edge detection. The Sobel operator is a procedure that determines the borders among regions and moreover separates objects from background part of an image. Discovering features plus objects in an image is a vital part for medical field for analyzing the disease. The Sobel operator identifies edges in an image by the Sobel operator approximation up to the derivative. When the gradient of I is extreme it yields edges at that points. In generally edge detection operations, a dimension of 3 × 3 gradient matrices isused for the horizontal and vertical directions. Consider A as the original image, also Gx and Gyare even the two images where at every point holds the horizontal plus the vertical derivative approximations, the calculations are as follows-

$$G_x = \begin{bmatrix} -1 & 0 & +1 \\ -2 & 0 & +2 \\ -1 & 0 & +1 \end{bmatrix} * A \quad \text{and} \quad G_y = \begin{bmatrix} -1 & -2 & -1 \\ 0 & 0 & 0 \\ +1 & +2 & +1 \end{bmatrix} * A$$

$$G = \sqrt{Gx + Gy}\,; \quad \text{Where, } |G| = |Gx| + |Gy| \ \&$$

$$G_x = (W_7 + 2W_8 + W_9) - (W_1 + 2W_2 + W_3)$$

$$G_y = (W_3 + 2W_6 + W_9) - (W_1 + 2W_4 + W_7)$$

This evaluates the gradients in both the horizontal (x) direction as well as vertical (y) direction and the magnitude is the gradient is just the sum of G_x and G_y gradients.

2.2.2. Discrete Wavelet Transform (sym4)

Wavelets are for representing the signals also the functions. Mean of infinite sequences of wavelets in the wavelet space is symbolized as a function. Symmetric filters are chosen since these are utmost valuable for diminishing the edge special effects in the wavelet demonstration of discrete wavelet transform (DWT) of a function; false edges results to large coefficients so that periodization can be evaded. Symlets are "symmetrical wavelets". Here in this sym4 is used.

2.2.3. Mean Square Error (MSE) and Peak to Signal Noise Ratio (PSNR)

In order to measure the performance of the resultant images, original image and encoded/uncompressed image are compared, following equations are used to compute:-

In the image of edge detection, the MSE includes statistical features of noise as well as the degradation function. It evaluates the cumulative squared difference among the parameter plus the estimator. MSE insist on the average difference of the pels all over the real ground truth image thru the image of edge detected. The higher the MSE value the greater is the difference among the original and the processed images [12,13]. When handling the image for restoration, rebuilding and compression the value of MSE must be a smaller amount. But when coming to edge detection on an image, the Mean Squared Error might be greater for finding more number of edge points in the image plus it can also discover weak edge points.

The PSNR is conveyed in terms of the decibel (dB) scale. PSNR is an assessment on human visual quality on rebuilt [14,15]. Even though a greater PSNR value normally specifies that the rebuilt is of greater quality in image compression. However in certain circumstances, PSNR must be less to attain correct results like in edge detection. The PSNR calculated based on the MSE.

3 RESULTS

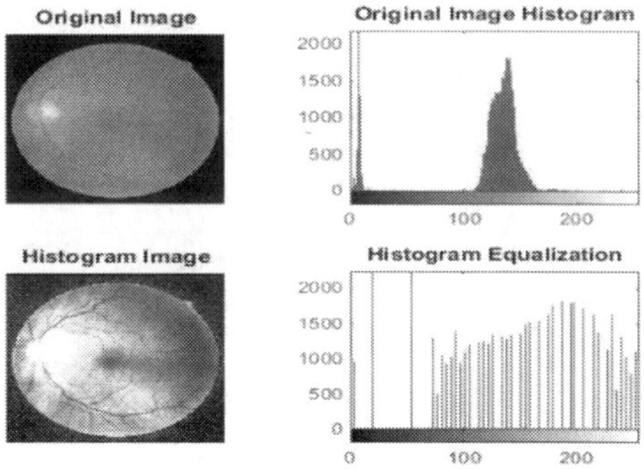

Fig 1:Original Retinal Image normalized to histogram equalized image

In the above shown Fig 1, on the original retinal image and histogram image of original image histogram equalization is applied to enhance the image.

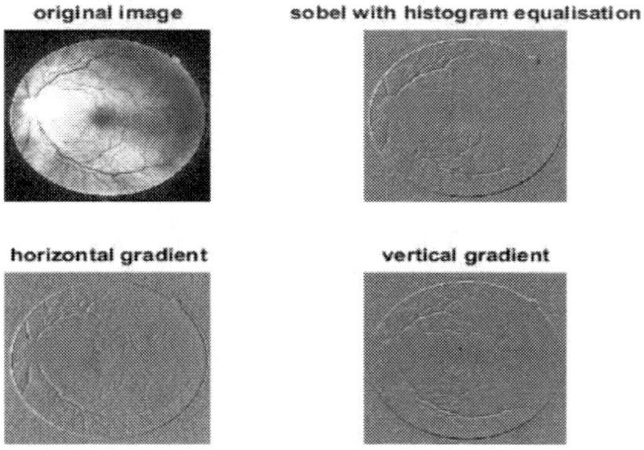

Fig 2: Sobel edge detection with histogram equalization

In the above shown Fig 2, on the original retinal image Sobel edge detection operation is performed with the histogram equalization.

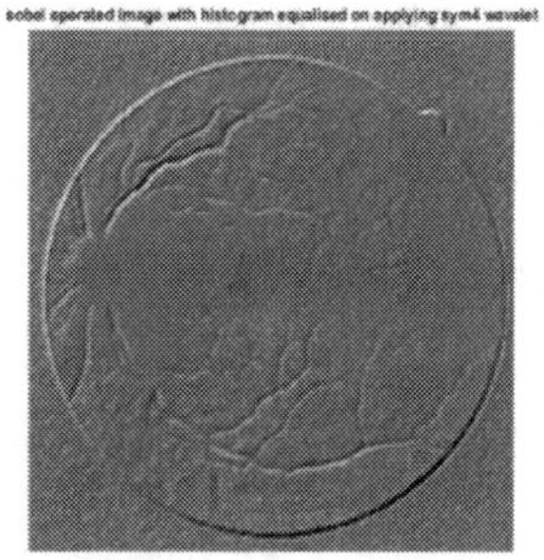

Fig 3: Sobel operated image with histogram equalization on applying sym4 wavelet

In the above shown Fig 3, sym4 wavelet is applied on Sobel edge detection operation with the histogram equalization.

TABLE 1: RESULTS OF DETECTING BLOOD VESSELS USING SOBEL EDGE DETECTION OPERATION BASED ON PSNR AND DISCERE WAVELET TRANSFORM (SYM4) PSNR VALUES

Image Number	Sobel Edge Detection	Discrete Wavelet Transform (sym4)
	PSNR	PSNR
Retinal Image 1	245.40	251.79
Retinal Image 2	246.76	253.31
Retinal Image 3	247.27	252.65
Retinal Image 4	246.69	252.42

In the above Table 1 Results are shown as the PSNR values between the Sobel edge operator and Discrete wavelet transform for Detecting Blood Vessels.

4 CONCLUSION

The proposed method is an innovative procedure to detect blood vessels by means of Sobel edge detection operation, discrete wavelet transform and measuring the performance by MSE plus PSNR values. In this whole procedure, the RGB color image is altered to a grayscale image and then for enhancing used histogram equalization to equalize all the regions in an image for showing irregular contrast. Subsequently, on this Sobel edge detection operation is performed to discover blood vessels and then discrete wavelet transform of type sym4 is applied. Finally, this paper evaluates edge detection on images based on the Mean Squared Error and mainly based on the Peak to Signal Noise Ratio where when Sobel operator is performed on retinal image with two different ways one without histogram equalized image and another with histogram equalized image i.e. on comparing Sobel edge detection operator with same image but may or may not utilize the histogram equalization, then the operator which provides resultant image with lesser PSNR value and higher MSE value will have the higher edge detection ability. This procedure can be applied to every retinal image. The MSE and the PSNR values for same image on detecting the blood vessels using the proposed methodology gives the result on utilizing the histogram equalization as 1.7827e-20 and 245.65 plus on not utilizing the histogram equalization as 3.0454e-21.

5 REFERENCES

[1] S. Sekhar, W. Al-Nuaimy and A. K. Nandi, 2008. *Automated Localisation of Optic Disk and Fovea in Retinal Fundus Images*,16th European Signal.

[2] AlauddinBhuiyan, BaikunthNath, KotagiriRamamohanara and Tien Y. Wong,*Retinal Image Matching Using Hierarchical Vascular Features*,The University of Melbourne, Australia, VIC 3002.

[3] M. Zubair ; Dept. of Comput. Eng., NUST, Rawalpindi, Pakistan; A. Yamin ; S. A. Khan; 2013. Automated detection of Optic Disc for the analysis of retina using color fundus image, Published in: Imaging Systems and Techniques (IST), IEEE International Conference.

[4] Jayakumari, C., Santhanam, T., 2007. Retinal blood vessel segmentation for assessment of diabetic retinopathy using a two-dimensional model, Asian J. Inf. Technol., 6, (12), pp. 1205-1211.

[5] D. Poobathy, Dr. R. ManickaChezian, 2014. Edge Detection Operators: Peak Signal to Noise Ratio Based Comparison, *I.J. Image, Graphics and Signal Processing*, 10, 55-61 Published Online September 2014 in MECS.

[6] Meenakshi Chaudhary1, AnupmaDhamija, 2013. A BRIEF STUDY OF VARIOUS WAVELET FAMILIES AND COMPRESSION TECHNIQUES, *Journal of Global Research in Computer Science*, April, Volume 4, No. 4.

[7] Huynh-Thu, Q.; Ghanbari, M., 2008. Scope of validity of PSNR in image/video quality assessment.*ElectronicsLetters* 44 (13): 800. doi:10.1049/el:20080522.

[8] Lehmann, E. L.; Casella, George, 1998. *Theory of Point Estimation*,2nd ed. New York: Springer. ISBN 0-387-98502-6.MR 1639875.

[9] Sopharak, A., Uyyanonvara, B., Barman, S., Williamson, T.H, 2008. "Automatic detection of diabetic retinopathy exudates from non-dilated retinal images using mathematical morphology methods, Comput. Med. Imaging Graph, 32, (8), pp. 720-727.

[10] Acharya and Ray, 2005. *Image Processing: Principles and Applications*, Wiley-Interscience, ISBN 0-471- 71998-6.

[11] Russ, 2002.*The Image Processing, Handbook: Fourth Edition*, CRC, ISBN 0-8493-2532-3.

[12] Wackerly, Dennis; Mendenhall, William; Scheaffer, Richard L. 2008. *Mathematical Statistics with Applications,* 7th ed., Belmont, CA, USA: Thomson Higher Education. ISBN 0-495-38508-5.

[13] Steel, R.G.D, and Torrie, J. H., 1960. *Principles and Procedures of Statistics with Special Reference to the Biological Sciences*, McGraw Hill, page 288.

[14] Thomos, N., Boulgouris, N. V., & Strintzis, M. G, 2006. Optimized Transmission of JPEG2000 Streams Over Wireless Channels, IEEE Transactions on Image Processing ,15 (1), January.

[15] Salomon, David (2007).*Data Compression: The Complete Reference*,4thed., Springer.p.281.ISBN 978 1846286025.Retrieved 26 July 2012.

MULTI-OBjECTIVE DEMAND FULFILLMENT PROBLEM FOR SOLAR CELL INDUSTRY

J.-N. Zheng[1], C.-F. Chien[2*], and J.-Z. Wu[3]

[1] Department of Business Administration
Asia University, Taiwan
jianian@asia.edu.tw

[2] Department of Industrial Engineering and Engineering Management
National Tsing Hua University, Taiwan
cfchien@mx.nthu.edu.tw

[3] Department of Business Administration
Soochow University, Taiwan
jzwu@scu.edu.tw

ABSTRACT

There are many companies with vertical integration and outsourcing process. Their different levels of capacity in different stages create complex coopetition relationships. Different business models differently influence on the corresponding business objectives including revenue, profit, and gross margin. Because of the expensive capacity and scarce resources, the Available-to-promise (ATP) becomes a crucial strategy. This study proposes a multi-objective demand fulfillment genetic algorithm (modfGA) to optimize the ATP. The nonlinear objective functions could be computed by simply process after decoding. The Pareto approach tradeoffs objectives without sacrificing other benefits. Results of empirical study show that the proposed modfGA can improve gross margin while maintaining the desired revenue and profit. With effective solutions and reasonable computation effort, the modfGA has become a regular decision support system in the case company.

Keywords: Demand fulfillment, Solar cell, Multi-objective genetic algorithm

1 INTRODUCTION

With petroleum supply uncertainty increasing and environmental consciousness rising, energy conservation and green energy become popular issues. Market demand and supply of green energy are dramatically increasing under energy policies. However, the possible changes of government policy and severe competition with low entry barrier have forced a company to improve operating excellence. The instability makes capacity installation in high-tech industry highly risky and thus most of companies seek strategy combination of making by oneself, strategic outsourcing, transformational outsourcing, insourcing, strategic alliance, and vertical integration. The industry structure evolution yields unbalanced supply chain integration where companies install unbalanced capacities in different supply chain stages where companies may seek opportunities of coopetition [1]. In solar supply chain, players constitute different levels of vertical integration that supply unbalanced capacity in different stages.

Only some product bins of cells can be allocated for customers' order, and the remaining products are saved as the inventory. The solar cells must be classified into different bins (product level) because of variation from different manufacturing process and materials. In the solar industry, the term bin refers to classifying solar products according to various criteria, such as efficiency and surface (grade). Indeed, the downstream companies (module) require similar cells to improve benefits during the assembly. The modules with mixed efficiency provide medium wattage which profits are lower than ones with classified efficiency for individual price. The grade not only affects the appearance, but also affects the product life and attrition rates of efficiency. However, little study has been done in this regard.

The planning process that determines how the actual customer demand is fulfilled is called demand fulfillment. The demand fulfillment process strongly influences the order lead-time and the on time delivery. In today's competitive markets, it is important to generate fast and reliable order promises in order to retain customers and increase market share [2]. The most important issues of manufacturing strategy in high-tech industry are to sustain profitability and growth while fulfilling demand with the invested capacity.[3]. However, business objective is not unique. [4] used multi-criteria decision models to solve an order selecting problem. The criteria of case study are (1) financial risk, (2) customer satisfaction, (3) benefiting, (4) profitability.

This study aims to propose a multi-objective demand fulfillment genetic algorithm (modfGA) to optimize the ATP in solar cell industry. The model determines ATP to tradeoffs revenue, profit and margin rate by Pareto approach without sacrificing other benefits. The model considers the resource constraints, including capacity and material. The nonlinear functions could be computed by simply process after decoding. The modfGA applies multi-population strategy and fuzzy logic controller to improve efficiency and effectiveness. The empirical study was conducted in a leading green energy company in Taiwan. We compared some method in different scenarios. A decision support system with the graphic user interface is further developed to provide comparative analysis and decision making assistance for the executives.

The remainder of this paper is organized as follows. In the section 2, we review related studies contributing to the fundamental of this approach. In the section 3, this study develops programming model to provide the optimal solution as a benchmark. In the section 4, we describe the proposed genetic algorithm for demand fulfillment problem. Subsequently, Section 5 presents the experiments for validation. Finally, Section 6 concludes with discussions and future research directions.

2 LITERATURE REVIEW

2.1 Demand fulfillment

The traditional approach of order promising is to search for inventory and to quote orders against it. If there is no inventory available, orders are quoted against the production lead-time. This procedure may result in non-feasible quotes, because a quote against the supply lead-time may violate other constraints, e. g. available capacity or material supply. [2]

Many studies have discussed demand forecast, demand planning, and order allocation in high-tech industry. However, little of them addressed multi-market demand planning in a vertically integrated company who supplies unbalanced capacity. In the existing studies, objectives of capacity planning are to minimize costs, to deliver on time, to fulfil demand, or to consider the maximum total profit assuming the cost or price function is linear with limited capacity to find the optimal goal [5]. When the structure of problem is non-linear, the optimal or near optimal solution could be found by a specific algorithm. Weighting procedure is a common idea to deal with multi-objective problems by accessing the weights for different objectives and then transforming the multi-objective problem into the single objective problem. In addition, to extract the preference structure of different objectives, some studies use goal programming to deal with multi-objective problems in a hierarchical order. In such a solution process, the optimal value of the previous stage becomes the restrictions of the current stage [6][7]. The gross margin objective (the ratio of profit and revenue) belongs to linear fractional programming problem. Fraction is a common type of objects that exist in the real life, such as productivity, return on investment (ROI), efficiency and other indicators.

2.2 Multi-ob ective genetic algorithm

Each computation iteration of the genetic algorithm is called a generation, denoted by t, and contains a set of solutions that collectively corresponds to a population, P(t). Each solution is called a chromosome that consists of genes, the encoded genotypes for evolution. Each chromosome is evaluated during each generation based on specific measures of fitness. New offspring C(t) are generated in one of three ways, i.e., copy, crossover, and mutation. A new generation of chromosomes is created by selecting some of the parents and offspring, according to the fitness values selection and elitist strategy while maintaining a constant population size. Following several generations of evolution, the algorithms improve solution quality and, in some cases, converge to optimal solutions of the problem. Increasing the population size, generation size, crossover rate, or mutation rate explores more of the solution space while requiring greater computational effort.

[8] used genetic algorithm to solve multi-objective problems, developed non-dominated sorting, and further brought up the idea of using the elitist strategy to yield efficient convergences. This study develops a modfGA and applies the Pareto front, so that revenue, profit, and gross margin can be simultaneously achieved subject to capacity constraints.

3 DEMAND FULFILLMENT PROBLEM IN SOLAR CELL INDUSTRY

Index

i : order number

s : workstation number

b : business model type

m : material type

Parameter

C_{ib} : unit cost of order i in business model b

P_{ib} : unit price of order i in business model b

UC_{ibs} : unit capacity of workstation s for order i in business model b processes at

TC_s : the available capacity at workstation s

UM_{ibm} unit usage amount of material m for order i in business model b

TM_m the available quantity of material m

L_i : the minimal committed amount of order i

D_i : demand of order i

Decision Variable

q_{ib} : available-to-promise (ATP) of order i in business model b

Like the semiconductor industry, we can divide the solar supply chain into five main functions, including silicon, wafer, cell, module, and system companies. First, the silicon and wafer companies provide the rectangular raw wafer as material supplier. The cell and module companies are main component manufacturers. They usually play OEM roles, such as the foundry in semiconductor industry. At final stage, system companies provide end product to customers, such as water heater, power plant.

A component (module) manufacturer can generate demand from different markets each of which has distinct impact on business operating goals. Since capacity is limited, it is very important to determine the amount of fulfilled orders and allocate them to specific capacity to achieve multiple objectives harmonically. Table 1 depicts three distinct markets contributing to revenue, total profit, and gross margin in different levels. In free market, the green energy component manufacturer purchases raw materials. The component is sold by the market price. In tolling market, the green energy component manufacturer plays an OEM role, such as the foundry in semiconductor industry. Since raw materials are provided by the customer, the manufacturer only charges the process cost. In the last one, called exchange market, the green energy component manufacturer also plays an OEM role. Raw materials are sold to the manufacturer by customers at a lower price. At the same time, the customer promises to repurchase at a reasonable price. Nevertheless, the transaction model hides the potential risk that is breach of undertaking.

Table 1: The contributed level between market and ob ective

Market	Revenue	Total profit	Gross margin
Free	Highest	Highest	Moderate
Tolling	Low	Moderate	Highest
Exchange	High	High	Moderate

The revenue is a critical index for evaluating a company's market share and bargaining power in the industry; the profit is used to evaluate whether the company is making money or not and its ability to return earnings to investors; the gross margin relates to earning per share (EPS), the finance efficiency of running a business and the application of funds. These three objectives are key performance indices from both business operation and stock market perspectives. However, they involve nonlinear and complex trade-offs.

In practice, the company sales staff and customers have recorded order transactions and potential deals according to customer characteristics of market segment. Customers have their own product demands, minimal promising and maximal requirement. The unit price, profit and gross margin are different among different orders. The decision is to allocate appropriate ATP to orders on the premise of the upper limit of capacity utilization.

Different order combinations and product mixes have different levels of impact on revenue, total profit, and gross margin. According to assigned ATPs, we can derive total revenue, net profit, and gross margin rate. Therefore, orders mapped by customers have been already signed up to the corresponding market which affects operation goals with different levels. The order belonging to the market has specific unit price, profit, and gross margin. The most important issue for the manufacturer is to decide the ATPs to customers.

The multi-objective demand fulfilment problem can be modelled as a multi-objective fractional programming model. Thus, for this problem, the three objectives of revenue, total profit, and gross margin are presented as in Eq. (1), Eq. (2), and Eq. (3), respectively. The capacity constraints are shown in Eq. (6). The material constraints are shown in Eq. (7).

$$\text{Max } O_1 = \sum_i \sum_b (P_{ib} q_{ib}) \quad (1)$$

$$\text{Max } O_2 = \sum_i \sum_b (P_{ib} q_{ib} - C_{ib} q_{ib}) \quad (2)$$

$$\text{Max } O_3 = {}^{O_1}/_{O_2} \quad (3)$$

s.t.

$$\sum_b q_{ib} \leq D_i \ \ \forall \text{i} \quad (4)$$

$$\sum_b q_{ib} \geq L_i \ \ \forall \text{i} \quad (5)$$

$$\sum_s (q_{ib} \times UC_{ibs}) \leq TC_s \ \ \forall \text{s} \quad (6)$$

$$\sum_m (q_{ib} \times UM_{ibm}) \leq TM_m \ \ \forall \text{m} \quad (7)$$

4 MULTI-OB ECTIVE DEMAND FULFILLMENT GA

This study proposes a Multi-objective demand fulfillment genetic algorithm to deal with the demand fulfillment problem. In particular, the modfGA contains specific parameters including maximum generation (*max_gen*), the number of the population (*pop_num*), population size (*pop_size*), elitist rate (r_e), crossover rate (r_c), and mutation rate (r_m). The modfGA is composed of four sections, namely initial, crossover, mutation, and selection. We also apply some advance mechanisms to improve it. The following subsections present the details of modfGA.

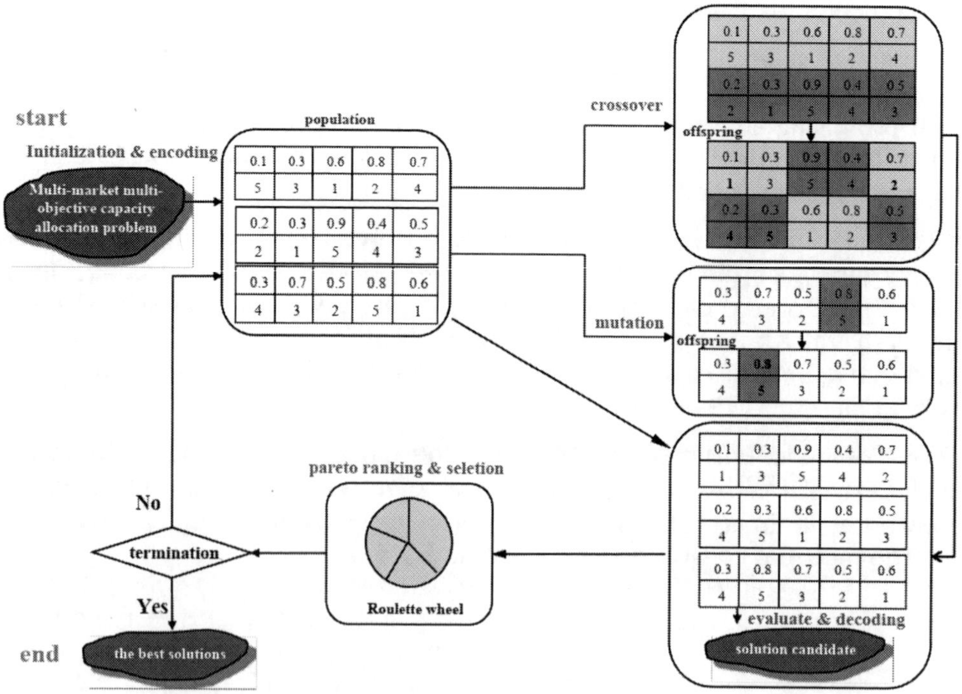

Figure 1: Conceptual framework of the proposed moGA

Encoding

Each chromosome contains two parts; one is the remaining proportion of orders, the other is the priority of orders. They are presented by float and integer, respectively.

z_i	0.25	0.69	0.36	0.95	0.72
priority	1	3	2	5	4

Figure 2: The encoding of modfGA

Since the configuration of resources, there are capacity and material constraints. The ATPs are restricted by the specific bottleneck workstation. According to the presentation of chromosomes, they might emerge infeasible condition. The corresponding repairing mechanism is thus designed to ensure full utilization of bottleneck workstations which are dynamically determined.

Crossover and Mutation

The second and third sections, crossover and mutation, are very important GA operators. They can help GA find more diversified chromosomes from searching space by modified some genes during the evolution. The proposed crossover mechanism is reformed by partial-matched crossover (PMX) and the mutation mechanism employed the insertion mutation.

Evaluation

We applied an algorithm for efficient bi-objective Pareto solutions generation [9]. The Pareto ranking process iteratively finds, ranks, and removes the non-dominated solutions until it ranks the entire population. The resulting rank numbers enable different rank individuals with different levels of chance to be selected in the next generation.

Selection

In genetic algorithm, chromosomes with higher fitness value have higher survival rate. Additionally, we applied the selection with elitist strategy. The top *pop_size**r_e chromosomes were selected from offspring$_t$ to parent$_{t+1}$ directly. Its advantage traditional probabilistic reproduction is that the best solution is monotonically improving from one generation to the next.

Besides, to prevent too many duplicated chromosomes from overweighting in each generation, we remove duplicated chromosomes with totally the same genes before the selection process.

Fuzzy Logic Controller

The behaviors of GA are characterized by the balance between exploitation and exploration in the search space. The balance is strongly affected by the strategy parameters such as: population size, maximum generation, crossover probability, mutation probability, and other parameters.

Fuzzy logic controller (FLC) updated them based on average fitness Instead of determine these values at the beginning. FLC for the crossover rate (r_c), and mutation rate (r_m) is to consider the different of average fitness in the continuous generations. It increases the rc, and rm if operators consistently produces a better offspring; otherwise, it reduces these rates if it produces a poorer offspring.

Multi-population Strategy

In the multi-population strategy used in this paper, several populations are evolved independently in parallel. After a pre-determined number of generations, all the chromosomes in each population are ranked by fitness. Then, the overall top chromosomes are added into each population. Some studies observed that exchanging too many chromosomes, or exchanging them too frequently, often leads to the disruption of the evolutionary process [10].

5 EXPERIMENT

To estimate the validity of the proposed modfGA, an empirical study was constructed for a manufacturer of green energy components in Taiwan. Without losing the generality, the data of one month collected in real setting is transformed to follow the non-disclosure agreement that protects the proprietary information of the case company. The data included 75 customers, 183 orders, 7 work stations, 5 main material types. Numerical analysis was performed on a personal computer with a 3.30 GHz Intel(R) Xeon(R) CPU E3-1230 V2 and 8 GB RAM.

Numerical analysis

We generated the 9 scenarios by problem scale and utilization. For the problem scale, we use three levels: small (half of original), normal, large (twice of original). The setting of material, capacity and order number were changed by problem scale level. For the utilization, we also use three levels: low (80 of original), normal, high (120 of original). The setting of order number was changed by utilization level.

We compared proposed model with (1) simple GA with the same parameter setting and (2) package software in limited time. The commercial software LINGO 8.0 was used to generate a reference set of non-dominated solutions, by utilizing embedded integer nonlinear programming (INLP) packages. For each scenario, LINGO solved the weighted-sums problem with two objectives which was recognized as INLP' and terminated at local optimal solutions. For each problem instance, a local optimal solution was collected after 1 hours of computation time. Hence, each scenario had a set of non-dominated solutions comprising 11 points with weight increasing from 0 to 1 by 0.1.

In summary, A and B are two sets of non-dominated solutions in the criterion space to be compared, and R is reference optimal Pareto front. A good non-dominated solution can provide solutions that are on, or close to, the optimal Pareto front. So the decision maker can make a final judgment based on his or her preference structure [11]. However, we did not know the real Pareto front. We use the union of sets from all compared models as reference set.

Table 2: The result of numerical analysis

Scenarios (scale, Util.)			modfGA		simple GA		Lingo in 1 hr	
1	small	low	15/15	100.0	11/12	91.7	11/11	100.0
2	small	normal	22/22	100.0	14/14	100.0	11/11	100.0
3	small	high	27/27	100.0	18/18	100.0	11/11	100.0
4	normal	low	34/34	100.0	11/13	84.6	11/11	100.0
5	normal	normal	47/47	100.0	27/29	93.1	11/11	100.0
6	normal	high	62/62	100.0	31/31	100.0	10/11	90.9
7	large	low	37/38	97.4	14/19	73.7	8/11	72.7
8	large	normal	55/55	100.0	30/37	81.1	8/11	72.7
9	large	high	83/83	100.0	35/44	79.5	9/11	81.8
Summary			382/383	99.7	191/217	88.0	90/99	90.9

6 CONCLUSION

This paper developed multi-objective demand fulfilment genetic algorithm (modfGA) to optimize the ATP. The model tradeoffs objectives with by Pareto approach without sacrificing other benefits. Results of empirical study show that the proposed modfGA can improve gross margin while maintaining the desired revenue and profit. Since there are several non-dominated solutions, the decision maker can pick the best one obtained by his/ her preference. We also implement modfGA into the decision support system. It will be a more flexible reference for making a decision. With effective solutions and reasonable computation effort, the modfGA has become a regular decision support system in the case company.

Nevertheless, the effort of selecting the ideal solution will increase by Pareto frontier size. If the candidate size of solutions excesses bounded rationality of the users, it is better to develop a systematic method.

REFERENCES

[1] **Chien, C.-F. and Kuo, R.-T. 2013.** Beyond make-or-buy: Cross-company short-term capacity backup in semiconductor industry ecosystem, *Flexible Services Manufacturing Journal*, 25(3), pp. 310-342.

[2] **Kilger, C., and Meyr, H. 2008.** Supply Chain Management and Advanced Planning, 4th Edition, Springer.

[3] **Shih, W., Shih, C., and Chien, C.-F. 2008.** Horizontal specialization and modularity in the semiconductor industry, *Harvard Business School Case Study*, 9-609-001.

[4] **Parsaei, M., Shahraki, A, and Shahgholian, K. 2014.** Selecting the best responsive option to unexpected orders at the time of capacity-shortage using multi criteria decision models. *Decision Science Letters* , 3(3), pp. 375-390.

[5] **Zhang, Z., Zhang, M.T., Niu S., and Zheng, L. 2006.** Capacity planning with reconfigurable kits in semiconductor test manufacturing, *International Journal of Production Research*, 44(13), pp.2625-2644.

[6] **Zhang, M.T., Niu, S., Deng, S., Zhang, Z., Li, ., and Zheng, L. 200 .** Hierarchical Capacity Planning With Reconfigurable Kits in Global Semiconductor Assembly and Test Manufacturing, *IEEE Transactions on Automation Science and Engineering*, 4(4), pp.543-552.

[7] **Wu, .-Z. and Chien, C.-F. 2008.** Modeling strategic semiconductor assembly outsourcing decisions based on empirical settings, *OR Spectrum*, 30(3), pp.401-430.

[8] **Deb, K., Pratap, A., Agarwal, S., and Meyarivan, T. 2002.** A Fast and Elitist Multiobjective Genetic Algorithm: NSGA-II, *IEEE Transactions on Evolutionary Computation*, 6(2), pp.182-197.

[9] **Gen, M.,Cheng, R., and Lin, L. 2008.** Network Models and Optimization: Multiobjective Genetic Algorithm Approach, Springer.

[10] **Gon alves, .F., amd Resende, M.G.C. 2012.** A parallel multi-population biased random-key genetic algorithm for a container loading problem. *Computers & Operations Research*, 39(2), pp. 179 190.

[11] **Li, B.B., Wang, L., and Liu, B. 2008.** An effective PSO-based hybrid algorithm for multi-objective permutation flow shop scheduling, *IEEE Transactions on Systems, Man and Cybernetics - Part A: Systems and Humans*, 38(4), pp.818-831.

UNCERTAIN DECISION MODEL FOR STRATEGIC CAPACITY PLANNING UNDER UNCERTAIN ENVIRONMENT IN SEMICONDUCTOR INDUSTRY

Wenhan Fu[1] and Chen-Fu Chien[1*]
[1]Department of Industrial Engineering and Engineering Management
National Tsing Hua University, Taiwan
cfchien@mx.nthu.edu.tw

ABSTRACT

Semiconductor manufacturing is a capital intensive industry that capacity utilization significantly affect the capacity effectiveness and final profit. Capacity portfolio planning strategy must coordinate capacity expansion and migration decisions in balance for profit maximization. Manufacturers have to determine capital investments based on various demand forecasts of products in advance for long-lead time in manufacturing process. The rapid rate of change in semiconductor technology makes it difficult for companies to estimate future tool need especially in new-generation products planning decisions.

Considering human prediction uncertainty that makes predictive belief degrees far away from probability frequency when lack-of-data, this study aims to develop an uncertain regret decision (URD) strategy framework based on uncertainty theory for capacity expansion and migration problem. The objective of proposed model is to minimize the potential regret of capacity surplus/shortage under demand uncertainties. Besides, an intelligent algorithm is designed for solving the uncertain regret model. A numerical experiment was conducted to estimate the effectiveness and robustness in uncertain environment.

Keywords: Capacity planning, Decision analysis, Uncertainty theory, Uncertain Programming, Semiconductor manufacturing, Demand uncertainty

1 INTRODUCTION

In semiconductor industry, production technology updates rapidly. Following Moore's law [1], the number of transistors on an integrated circuit doubled every 18 months during the past forty years. The process of semiconductor manufacturing comprises four primary phases: wafer fabrication, wafer probe, assembly and packing, and final test [2]. Wafer fabrication is the most technologically complex and capital intensive among the four phases. Wafer fabrication normally requires 100-500 operation steps over a total cycle time of many weeks. A modern 300mm wafer fabrication facility (fab) costs in excess of US$3 billion with a variety of tools that value between US$1-13 million. Equipment cost accounts for more than 60% of the total in common.

* Corresponding Author

Surplus or shortage of products in wafer fabrication process will lead to loss of capacity utilization or sales order.

Capacity planning problem is the configuration decisions to manage tool capacity and technology demand effectively [3]. Semiconductor capacity planning is basically consists of capacity expansion and migration. Capacity expansion is the process to add new facilities over time into manufacturing process to meet the rising demand [4]. Capacity migration is the transition between technologies to balance capacity output temporarily. It is especially difficult for semiconductor companies to make capacity expansion and migration decision due to several reasons: rapid technology change, complex fabrication process, long lead time of increment, high cost in equipment, and demand uncertainties [5]. A single semiconductor manufacturing equipment may charges tens of millions of dollars. Wafer fabrications need to equip multiple generational technologies to maintain stable manufacturing plans.

Strategic capacity planning focuses on the investment of productive resources (such as tools, equipment, facilities, etc.) for expansion. Corporations strategically focus on core competences and outsource other activities to enhance capital effectiveness in semiconductor industry [6]. Manufacturers in high-tech industry with high level of horizontal specialization must invest in the latest technology to keep market competitiveness. Normally it requires 6-9 months for semiconductor manufacturers from placing a tool procurement order to tool operation. The long-term planning decisions normally are based on demand forecast due to long lead time of tool procurement.

Geng and Jiang [7] give a comprehensive literature review of strategic capacity planning in semiconductor industry and classify the methods into three categories: static capacity method, neighborhood search technique and mathematical programming models. In semiconductor industry, main sources of uncertainty in long-term capacity planning problems consist of demand forecast for technologies and capacity estimations for fabrication facilities [3].

Demand uncertainty is usually modeled by a set of demand scenarios with associated occurrence probabilities. Stochastic programming model recently is a widely used method in capacity problems for its strength in handling such problems [8]. Stochastic methods solve capacity planning problem under demand uncertainty by setting different goals and performance index to get several finite scenarios [9]. A variety of ways is useful to generate scenarios such as enumeration, scenario trees, and Monte Carlo simulation [10]. Chou, Cheng, Yang and Liang [11] modeled the uncertainty of demand that follows a geometric Brownian motion process.

In practice, the predicted demand figures made by domain experts are based on practical experience rather than enough observed data. Kahneman and Tversky [12] showed that human beings usually estimate much wider than real events. Without enough samples, probability distributions (ex. normal distribution) will be far away from true frequency. In such situation, a premise of probability theory is invalid and all of methods based on this mathematical system do not work. An uncertainty theory

[13] without frequency assumption was found by to model uncertain events. It can handle the variations of human's belief degrees well.

This research aims to propose an uncertain regret decision strategy framework based on uncertainty theory for capacity expansion and migration to minimize the potential regret of capacity surplus/shortage under uncertain environments. The model objective called capacity regret is the capacity loss between proposed decision and the ideal one.

The structure of the paper is organized as follows. Section 1 introduces the background, significance and motivation of the research. In section 2, the fundamental concepts of uncertainty theory and related theorems are introduced. Section 3 expounds the framework of proposed model and decision process. Section 4 presents a numerical example to illustrate the algorithm designed in section 3. Section 5 concludes with discussions.

2 PRELIMINARIES

Uncertainty Theory was founded by [13] to measure and model uncertainty in human's belief degree. It has been applied in some optimization problems such as facility location-allocation problem [14]. However, no prior work has been done in practical manufacturing problems like capacity planning problem before.

Let Γ be a nonempty set, and $\mathcal{L}$ be a σ-algebra over Γ. Each element $\Lambda \in \mathcal{L}$ is called an event.

Definition 1 A set function $\mathcal{M}$ from $\mathcal{L}$ to [0, 1] is called an uncertain measure if it satisfies the following axioms:

Axiom 1 (Normality Axiom) $\mathcal{M}\{\Gamma\} = 1$ for the universal set Γ;

Axiom 2 (Duality Axiom) $\mathcal{M}\{\Lambda\} + \mathcal{M}\{\Lambda^C\} = 1$ for any event Λ;

Axiom 3 (Subadditivity Axiom) For every sequence of events $\{\Lambda_i \in \mathcal{L}\}$, we have

$$\mathcal{M}\left\{\bigcup_{i=1}^{\infty} \Lambda_i\right\} \leq \sum_{i=1}^{\infty} \mathcal{M}\{\Lambda_i\}$$

Axiom 4 (Product Axiom) Let $(\Gamma_k, \mathcal{L}_k, \mathcal{M}_k)$ be uncertain space for $k = 1,2,\cdots$, Then the product uncertain measure $\mathcal{M}$ is an uncertain measure satisfying

$$\mathcal{M}\left\{\prod_{k=1}^{\infty} \Lambda_k\right\} = \bigwedge_{k=1}^{\infty} \mathcal{M}_k\{\Lambda_k\}$$

where Λ_k are arbitrarily chosen events from $\mathcal{L}_k$ for $k = 1,2,\cdots$, respectively.

Definition 2 The uncertainty distribution Φ of an uncertain variable ξ is defined by

$\Phi(x) = \mathcal{M}\{\xi \leq x\}$ for any real number x.

Definition 3 Let ξ be an uncertain variable. Then the expected value of ξ is defined by

$$E[\xi] = \int_0^{+\infty} \mathcal{M}\{\xi \geq x\}dx - \int_{-\infty}^{0} \mathcal{M}\{\xi \leq x\}\,dx$$

Definition 4 The uncertain variables$\xi_1, \xi_2, \cdots, \xi_n$ are said to be independent if

$$\mathcal{M}\left\{\bigcap_{i=1}^{n} \xi_i \in B_i\right\} = \bigwedge_{i=1}^{n} \mathcal{M}\{\ \xi_i \in B_i\}$$

for any Borel sets $B_1, B_2, \cdots, B_n$.

Assume that x is a decision vector, and ξ is an uncertain vector. Liu [13] proposed the following uncertain programming model with uncertain objective function $f(\boldsymbol{x}, \xi)$, uncertain constraints $g_j(\boldsymbol{x}, \xi) \leq 0$, and confidence levels α_j,

$$\begin{cases} \min\limits_{x} E[f(\boldsymbol{x}, \xi)] \\ subject\ to: \\ \qquad \mathcal{M}\{g_j(\boldsymbol{x}, \xi) \leq 0\} \geq \alpha_j,\ \ j = 1,2,\cdots,p \end{cases} \tag{1}$$

Definition 5 An uncertainty distribution Φ of an uncertain variable ξ is said to be regular if its inverse function $\Phi^{-1}(\alpha)$ exists and is unique for each $\alpha \in (0,1)$. The inverse function$\Phi^{-1}(\alpha)$ is called the inverse uncertainty distribution of ξ.

Theorem 1 Let $\xi_1, \xi_2, \cdots, \xi_n$ be independent uncertain variables with regular uncertainty distributions $\Phi_1, \Phi_2 \cdots \Phi_n$, respectively. If f is a strictly increasing function, then

$$\xi = f(\xi_1, \xi_2, \cdots, \xi_n)$$

is an uncertain variable with an inverse uncertainty distribution

$$\Phi^{-1}(\alpha) = f\left(\Phi_1^{-1}(\alpha), \Phi_2^{-1}(\alpha), \cdots, \Phi_n^{-1}(\alpha)\right) \tag{2}$$

Theorem 2 Assume that $\xi = (\xi_1, \xi_2, \cdots, \xi_n)$ where $\xi_1, \xi_2, \cdots, \xi_n$ are independent uncertain variables with uncertainty distributions $\Phi_1, \Phi_2 \cdots \Phi_n$, respectively. Without loss of generality, we also assume that f is monotone increasing with respect to $\xi_1, \xi_2, \cdots, \xi_k$, and strictly decreasing with respect to $\xi_{k+1}, \xi_{k+2}, \cdots, \xi_n$, and $g_j$is strictly increasing with respect to $\xi_1, \xi_2, \cdots, \xi_{k_j}$, and strictly decreasing with respect to $\xi_{k_j+1}, \xi_{k_j+2}, \cdots, \xi_n$ for $j = 1,2,\cdots,p$. Then the uncertain programming model (1) is equivalent to a crisp model as follows,

$$\begin{cases} \min\limits_{x} \int_0^1 f\left(\boldsymbol{x}, \Phi_1^{-1}(\alpha), \cdots, \Phi_k^{-1}(\alpha), \Phi_{k+1}^{-1}(\alpha), \cdots, \Phi_n^{-1}(1-\alpha)\right) d\alpha \\ \qquad subject\ to: \\ \qquad g_j\left(\boldsymbol{x}, \Phi_1^{-1}(\alpha), \cdots, \Phi_k^{-1}(\alpha), \Phi_{k+1}^{-1}(\alpha), \cdots, \Phi_k^{-1}(1-\alpha)\right) \leq 0,\ \ j = 1,2,\cdots,p \end{cases} \tag{3}$$

3 UNCERTAIN REGRET DECISION FRAMEWORK

This section illustrates the proposed methodology to deal with the capacity expansion and migration problem under demand uncertainty in high-tech industry like semiconductor fabrication.

3.1 Problem Definition

The strategic capacity planning problem is to find the expansion and migration plan of capacity resources for different products. Corporations adopt a rolling mechanism to update the prediction of demand referring to the latest business information in semiconductor industry. Suppose the lead time of tool procurement is 2 quarters, the capacity decision of Quarter 3 should be confirmed in Quarter 1.

In practice, semiconductor companies usually have the demand and capacity planning forecast made by multiple sources for different technologies from the beginning of technologies' lifetime to the decision epoch. The actual demand will be get 2 quarters later from the decision making.

3.2 Data preparation and Parameters Setting

The parameters and decision variables used in the proposed model are listed as follows:

e_{it}	Capacity expansion for product i in time t
c_{it}	Accumulate capacity for product i in time t
m_{abt}	Capacity migration from product a to product b in time t.
$u_{itt'}$	Upper bound of capacity level with maximum capacity expansion speed for product i in period t' refer to time t.
$l_{itt'}$	Lower bound of capacity level without capacity expansion for product i in period t' refer to time t.
$b_{ijtt'}$	Condition binary variable which is 1 if demand in period t' is greater than capacity in period t for product i, and vice versa.
ϕ_{it}	Uncertain distribution of demand for product i in time t.
Φ_{it}^{-1}	Inverse uncertainty distribution of ϕ_{it}
D_{it}	Actual demand for product i in time t.
W^o	Loss weight of capacity oversupply.
W^s	Loss weight of capacity shortage.
W_i	Weight of product i.
ψ_i	Uncertain distribution of maximal expansion for product i .
Ψ_i^{-1}	Inverse uncertainty distribution of ψ_i
ϑ_t	Uncertain distribution of maximal expansion at time t.
Θ_t^{-1}	Inverse uncertainty distribution of ϑ_t
ω_i	Uncertain distribution of maximal total migrations for product i.
Ω_i^{-1}	Inverse uncertainty distribution of ω_i

3.3 Uncertain Regret Decision Model

The process of uncertain regret decision model is illustrated as followed:

Step 1. Input the parameters W^o, W^s, E_i, ψ_i, F_t, ϑ_t, M_i, η_i.

Step 2. Compute stochastic demand d_{it} and its uncertain distribution Φ_{it}

Step 4. Compute capacity parameters

$$c_{it} = c_{i(t-1)} + e_{it} + \sum_{a \neq i} m_{ait} - \sum_{b \neq i} m_{ibt}$$

$$\sum_i W_i^P = 1 \quad W^S + W^o = 1$$

Step 5. Solve the uncertain regret model

Step 6. Input new data and do back to Step 1

Step 7.No new data and done.

In practice, the expansion and migration limits of product provided by manufacturers may not be certain. Machines in semiconductor fabs usually maintain round-the-clock work, which is based on stable continuously electricity supply. Once there is a break in municipal power pipes, companies normally use the diesel generator engine unit as the backup power to ensure an uninterrupted operation of facilities.

$$\begin{cases} \min_x E\left[\sum_i W_i^P \sum_t \sum_{t'} b_{itt'} W^S(\phi_{it'} - u_{itt'}) + (1 - b_{itt'j}) W^o(l_{itt'} - \phi_{it'})\right] \\ \text{Subject to} \\ \qquad u_{itt'} = c_{it} + (t - t')(\psi_i + \omega_i) \\ \qquad l_{itt'} = c_{it} - (t - t')\omega_i \\ \qquad \mathcal{M}\{e_{it} \le \psi_i\} \ge \tau \\ \qquad \mathcal{M}\left\{\sum_i e_{it} \le \vartheta_t\right\} \ge \beta \\ \qquad \mathcal{M}\left\{\sum_{a \neq i} m_{ait} + \sum_{b \neq i} m_{ibt} \le \omega_i\right\} \ge \gamma \end{cases} \tag{4}$$

According to (4), we can translate the by uncertain measure to inverse uncertain distribution. As shown below:

$$
\begin{cases}
\min_x \int_0^1 \left[\sum_i W_i^P \sum_t \sum_{t'} b_{itt'} W^S (\Phi_{it}^{-1}(\alpha) - r_{itt'}) + (1 - b_{itt'j}) W^o \left(l_{itt'} - \Phi_{it}^{-1}(\alpha) \right) \right] d\alpha \\
Subject\ to \\
\qquad u_{itt'} = c_{it} + (t - t') \left(\Psi_i^{-1}(\tau) + \Omega_i^{-1}(\gamma) \right) \\
\qquad l_{itt'} = c_{it} - (t - t') \Omega_i^{-1}(\gamma) \\
\qquad e_{it} \le \Psi_i^{-1}(\tau) \\
\qquad \sum_i e_{it} \le \Theta_t^{-1}(\beta) \\
\qquad \sum_{a \ne i} m_{ait} + \sum_{b \ne i} m_{ibt} \le \Omega_i^{-1}(\gamma)
\end{cases}
\quad (5)
$$

3.4 Intelligent algorithm

Genetic algorithm [15] is a powerful tool in solving programming and optimization problems. An intelligent algorithm is designed based on GA to solve the proposed model (5).

Step 1: Representation. From the potential capacity decision choice $\boldsymbol{dec} = (e_{it}, m_{ilt}), l = 1,2,\cdots i-1, i+1 \cdots, n$, initialize pop_size chromosomes

$$V^k = (\boldsymbol{e}^k, \boldsymbol{m}_i^k) = (e_1^k, e_2^k, \cdots, e_n^k, m_{12}^k, m_{13}^k, \cdots, m_{1n}^k, m_{21}^k, \cdots, m_{n,n-1}^k),$$

$k = 1,2,\cdots, pop\_size$ which denote the types of products.

Step 2: Initialization. Calculate the objective values R^k for all chromosomes $V^k, k = 1,2,\cdots, pop\_size$, to get the optimal capacity regret.

Step 3: Evaluation. Calculate fitness function of all chromosomes $V^k, k = 1,2,\cdots, pop\_size$. Evaluation function is $\mathrm{Eva}(V^k) = \beta(1-\beta)^{k-1}, k = 1,2,\cdots, pop\_size$ where the chromosomes $V^1, V^2, \cdots, V^{pop\_size}$ to be rearranged from good to bad due to objective values U^k and $\beta \epsilon (0,1)$ is a parameter in the genetic system.

Step 4: Selection. The process is based on spinning the roulette wheel characterized to select the chromosomes for a new population. By calculating the fitness of all chromosomes for pop_size times, a single chromosome is selected each time in the roulette wheel. Then the pop_size chromosomes $V^k, k = 1,2,\cdots, pop\_size$ updates.

Step 5: Crossover. Update the chromosomes $V^k, k = 1,2,\cdots, pop\_size$ by crossover operation. The probability of crossover P_c calculates the expected number $P_c \cdot pop\_size$ chromosomes undergoing the crossover operation.

Step 6: Mutation. Update the chromosomes $V^k, k = 1,2,\cdots, pop\_size$ by mutation operation. The probability of mutation P_m gives the expected number of $P_m \cdot pop\_size$ chromosomes undergoing the mutation operations.

Step 7: Repeat. Repeat the above steps for default cycles.

Step 8: Solution. Make the best chromosome $V^* = (x^*, y^*)$ as optimal solution for capacity decision suggestion.

4 NUMERICAL EXAMPLE

Suppose there is a semiconductor manufacturing company in the capacity situation as in Table 1. Only two products involved in this process and the length of forecasting provided is 3. The loss cost weight of oversupply and shortage are W^o=0.8, W^s = 0.2. Consider product 1 is a mature tool with trend of while product 2 is an advanced technology. Forecast demand and capacity at beginning time are showed in Table 2.

Table 1: Manufacturing situation setting

Product i	Weight W_i	Maximal expansion ψ_i	Maximal migration η_i
1	0.6	$\mathcal{N}(30,1)$	$\mathcal{N}(15,1)$
2	0.4	$\mathcal{N}(50,1)$	$\mathcal{N}(20,1)$

Table 2: Demand and Capacity Prediction

Decision variables	Time distance		
	1	2	3
$x_{1,t}$	170	200	180
$x_{2,t}$	100	140	200
c_{1t}	175		
c_{2t}	120		

Without loss of generality, we set initial belief degree α_0 =0.5 and the uncertain distribution of demand is $\Phi_1(x)\sim\mathcal{Z}(0,170,500)$ and $\Phi_1(x)\sim\mathcal{Z}(0,100,500)$.

Running the algorithm for 400 cycles yields that the optimal schedule for capacity planning decisions are showed in table 3.

Table 3: Results of the numerical example

Decision variables	Time distance	
	1	2
$e_{1,t}$	30	0
$e_{2,t}$	26	0
m_{1t}	0	0
m_{2t}	13	20

The experiment result means that at decision time point 1, product 1 needs 30 units of capacity expansion and the other product needs 26 unit for future development. In the meantime,

5 CONCLUSION

This study proposed a framework based on uncertainty theory in capacity regret optimization to solve the capacity planning problem with uncertain demand and capacity. By transformed into a crisp equation form, the proposed model was solved through an intelligent algorithm. A numerical experiment was given to illustrate the efficiency of the model and algorithm.

Reference

[1] **Moore, G. 1965.** Cramming More Components Onto Integrated Circuits, Electronics,(38) 8.

[2] **Uzsoy, R., Lee, C. Y., & Martin-Vega, L. A. 1992.** A review of production planning and scheduling models in the semiconductor industry part I: system characteristics, performance evaluation and production planning. *IIE transactions*, *24*(4), 47-60.

[3] **Christie, R. M., & Wu, S. D. 2002.** Semiconductor capacity planning: stochastic modeling and computational studies. *IIE Transactions*, *34*(2), 131-143..

[4] **Davis, M. H. A., Dempster, M. A. H., Sethi, S. P., & Vermes, D. 1987.** Optimal capacity expansion under uncertainty. *Advances in Applied Probability*, 156-176.

[5] **Swaminathan, J. M. 2000.** Tool capacity planning for semiconductor fabrication facilities under demand uncertainty. *European Journal of Operational Research*, *120*(3), 545-558.

[6] **Wu, J. Z., & Chien, C. F. 2008.** Modeling strategic semiconductor assembly outsourcing decisions based on empirical settings. *OR Spectrum*,*30*(3), 401-430.

[7] **Geng, N., & Jiang, Z. 2009.** A review on strategic capacity planning for the semiconductor manufacturing industry. *International Journal of Production Research*, *47*(13), 3639-3655.

[8] **Geng, N., Jiang, Z., & Chen, F. 2009.** Stochastic programming based capacity planning for semiconductor wafer fab with uncertain demand and capacity. *European Journal of Operational Research*, *198*(3), 899-908.

[9] **Laguna, M. 1998.** Applying robust optimization to capacity expansion of one location in telecommunications with demand uncertainty. *Management Science*, *44*(11-part-2), S101-S110.

[10] **Hood, S. J., Bermon, S., & Barahona, F. 2003.** Capacity planning under demand uncertainty for semiconductor manufacturing. *IEEE Transactions on Semiconductor Manufacturing*, *16*(2), 273-280.

[11] **Chou, Y. C., Cheng, C. T., Yang, F. C., & Liang, Y. Y. 2007.** Evaluating alternative capacity strategies in semiconductor manufacturing under uncertain demand and price scenarios. *International Journal of Production Economics*, *105*(2), 591-606.

[12] **Kahneman, D., & Tversky, A. 1979**. Prospect theory: An analysis of decision under risk. *Econometrica: Journal of the econometric society*, 263-291.

[13] **Liu, B. 2007**. *Uncertainty Theory*, Springer Berlin Heidelberg.

[14] **Wen, M., Qin, Z., & Kang, R. 2014**. The alpha-cost minimization model for capacitated facility location-allocation problem with uncertain demands. *Fuzzy Optimization and Decision Making*, *13*(3), 345-356.

[15] **Gen, M., Cheng, R. 2000**. Genetic algorithm and engineering optimization. *John Wily and Sons, New York*.

RESEARCH ON PRE-GRINDING HOB PRODUCTION PROCESS QUALITY CONTROL BASED ON PROCESS DESIGN

Lixin Ma

Department of Industrial Engineering , Harbin University of Commerce, Harbin 150028, China

13352502179@163.com

ABSTRACT

Making the pre-grinding hob for example, through the analysis of the process, according to the processing of the density of each processing machine tool can be obtained the key process of hob in the production process, countering the key procedures in the precision of the machine tool, the machining method, cutting the amount of use, material application and so on, the integrated use of process flow diagram, counting statistics, causality diagram, AutoCAD and other tools, finding out the key factors affecting the quality of gear hob, taking corresponding measures to control problem, to achieve the quality problems of gear hob, in order to improve the precision of gear hob.

Keywords: pre-grinding hob, technological process, key process, quality control tools

1 INTRODUCTION: THE BACKGROUND OF RESEARCH ON GEAR HOB

Widely used at home and abroad, the gear hob with high speed steel carbide materials, etc..These tools not only have high cutting speed and long service life but also can be put into mass production. In addition, coating technology in China is improving day by day, widely used titanium nitride coatings, etc., and constantly develop new coating material. However, foreign hob much more advanced than ours decades ago, they have long since mass production of high-speed steel hob, and the application of titanium nitride coating appeared two decades even three decades earlier than our country, many foreign countries have begun to make the gear hob structural change, and developed a variety of indexable hobs, this hob can reduce the number of re-grinding, after using blunt transposition[1], and greatly extend the durability of gear hob.

Pre-grinding hob, as one of the varieties of hob, which are mainly used for roughing hardened gears, which is characterized as being processed gear aside a certain amount of grinding allowance. Although the price of pre-grinding hob is very expensive, but pre-grinding hob are recommended at home and abroad.

2 THE KEY STEP IN THE PROCESS OF PRE-GRINDING HOB PROCESSING

2.1 Process flow of pre-grinding hob

Pre-grinding hob production line process is shown in figure 1.

Corresponding Author: Lixin Ma

Sponsor: Project supported by the Heilongjiang Provincial Department of Education, China (No.12541203)

No.	Process Description	Step Series: Process	Check	Transport	Wait	Store
1	Turning bore	●	□	→	D	▽
2	Move to the next step	○	□	➡	D	▽
3	Turning face and cylindrical	●	□	→	D	▽
4	Pull keyway	●	□	→	D	▽
5	Check	○	■	→	D	▽
6	Move to the next step	○	□	➡	D	▽
7	Milling supporting table on both sides	●	□	→	D	▽
8	Clamp processing	●	□	→	D	▽
9	Move to the next step	○	□	➡	D	▽
10	Milling tooth	●	□	→	D	▽
11	Check	○	■	→	D	▽
12	Milling edge groove (chip flutes)	●	□	→	D	▽
13	Move to the next step	○	□	➡	D	▽
14	Toothed shovel	●	□	→	D	▽
15	Check	○	■	→	D	▽
16	Move to the next step	○	□	➡	D	▽
17	Incomplete milling teeth	●	□	→	D	▽
18	Incomplete repair teeth	●	□	→	D	▽
19	Move to the next step	○	□	➡	D	▽
20	Mark	●	□	→	D	▽
21	Temporarily placed	○	□	→	◗	▽
22	Heat Treatment	●	□	→	D	▽
23	Move to the next step	○	□	➡	D	▽
24	Grinding bore	●	□	→	D	▽
25	Fine grinding bore	●	□	→	D	▽
26	Move to the next step	○	□	➡	D	▽
27	Grinding supporting table on both sides	●	□	→	D	▽
28	Move to the next step	○	□	➡	D	▽
29	Brothers front tooth	●	□	→	D	▽
30	Move to the next step	○	□	➡	D	▽
31	Grinding teeth	●	□	→	D	▽
32	Check	○	■	→	D	▽
33	Scratch	○	□	→	D	▼

Figure 1: Process Flow Of Pre-grinding Hob

2.2 The key working procedure and quality problem analysis of pre-grinding hob

2.2.1 The tooth profile of pre-grinding hob

The theory of gear hob and worm is Involute worm. Because of the involute worm axial cross-sectional shape is curve, as shown in figure 2 (a), machining and measuring are very difficult,so seldom use this worm hob processing gear. And combination with the characteristics of Archimedes worm grinding gear hob, as shown in figure 2 (b)[2], at the same time to ensure that they are in the gear machining to allow certain grinding allowance, so axial cross-sectional shape of pre-grinding hob as shown in figure 2 (c). This will produce the error to some extent--the model error. Modelling error needs to control in a reasonable range, but it will become smaller in diameter after repeated, the tooth top and tooth side joint rounded corners will be more and more big, modelling error will increase.

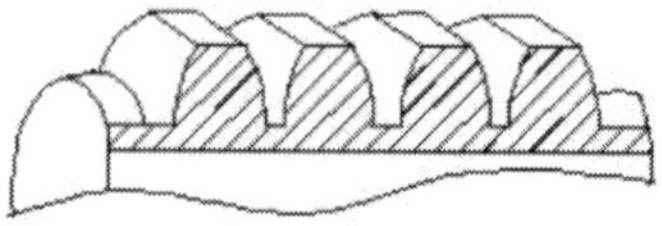

(a) Axial Cross-sectional Shape Of The Involute Gear Hob

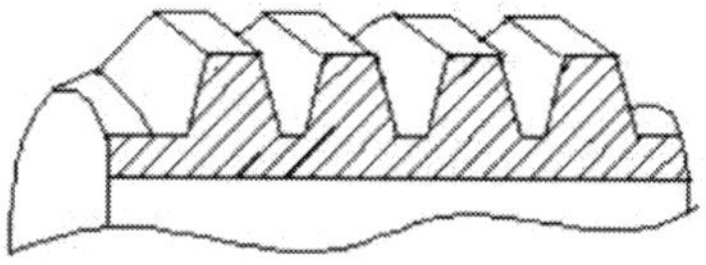

(b) Axial Cross-sectional Shape Of Archimedes Hob

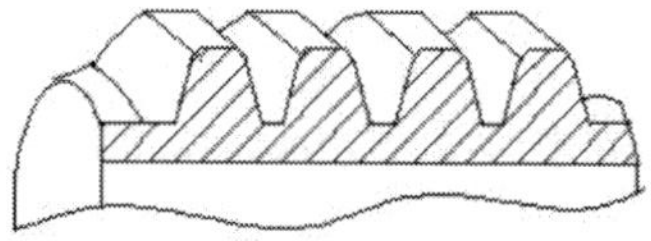

(c) Axial Cross-sectional Shape Of Before Grinding Gear Hob

Figure 2: Axial Cross-sectional Shape Hob

2.2.2 Inner hole of pre-grinding hob

As the benchmark of hob design, processing, measurement, using and the grinding , precision of inner hole directly affects the precision of the other processes. And there are many factors which are affected the accuracy of hob inner hole: such as the precision of the inner hole process equipment internal hole machining precision, included the tangible position tolerances, contact offset error, etc.; It can easily lead to inner hole happened swelling deformation ,because internal organization of gear hob has been changed before, during, and after quenching[3], as shown in figure 3. Processing personnel for clamping fixture, workpiece is not strong, not timely to lubrication of machine tools, measuring when measuring tool tilting, error operation is also one of the cause of inner hole problem.

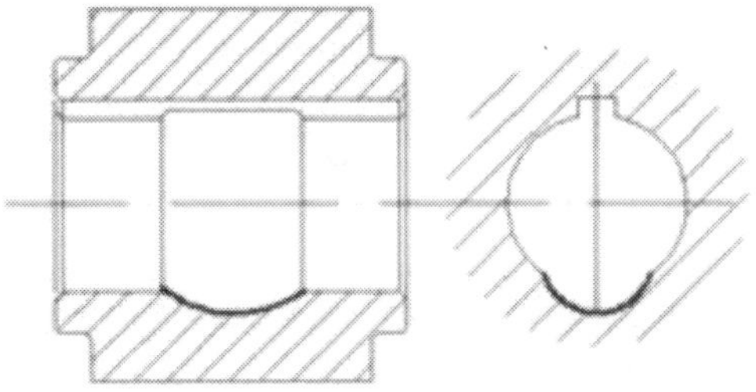

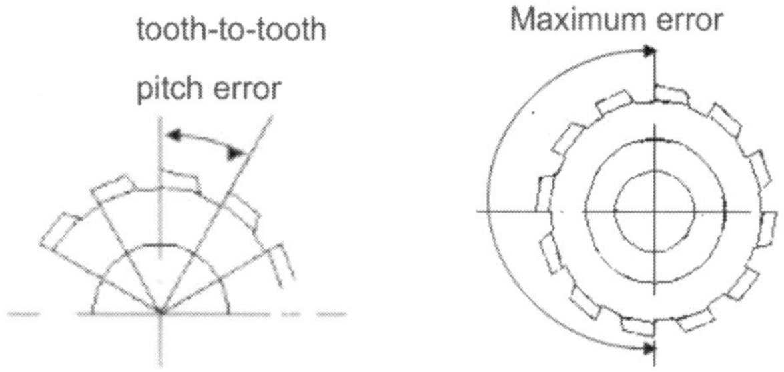

Figure 3: Inner Hole Swelling

Figure 4: Accumulated Error Of Weeks Section

2.2.3 Should the grooving of pre-grinding hob

To make gear hob cutting edge hob need slot milling capacity. Gear hob let crumbs slot is led by former knife surface and side of the shovel. Allow crumbs slot is divided into two kinds, one kind is straight flute, the other is a spiral groove. At present our country is with the more straight flute hob, for straight flute can improve the grinding accuracy and is easy to check the [4].

Problems should be paid attention to when chip the gear hob as follow:

(1) Cumulative error (as shown in figure 4) :Due to let crumbs slot consists of former knife surface and side of the shovel. Two of the intersecting line is the cutting edge, processing cutting edge usually use edge grinding machine. If blank after many grinding gear hob and adjustment, grinding machine precision indexing plate will reduce gradually, and it will lead to make hob pitch error accumulation become big[5].

(2) Let chip groove depth: The chip groove is used to accommodate chip. If let the crumbs channel shallow, chip accumulation will lead to cutter blade alert. If crumbs slot is deep, after repeated grinding , gear hob cutter durability will decrease.

(3) The crumbs slot number: If the crumbs slot number is too little, each cutter tooth on the cutting force will be too big.So it will lead to deformation of cutting tools. If the slot number is too much, cutting load will get more on the blade.So that each blade on the force should be reduced.

2.2.4 Relieving of pre-grinding hob

When the hob tooth flank of the axial curve for sectional shape, for every point on tooth slope is different, needing the axial relieving method to assure the tooth profile is changeless. The processing is shown in figure 5(a); Hob tooth profile of axial cross-sectional shape of straight line, with a radial relieving method, as shown in figure 5(b) [6]. In addition to relieving the side, but also relieving, and can only use the radial relieving method.

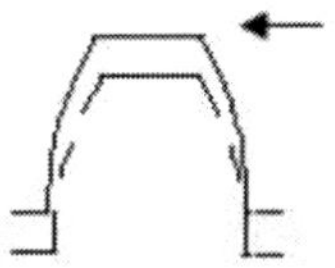

(a) Axial Direction

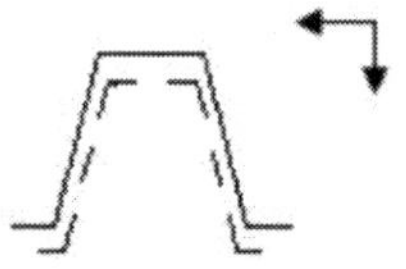

(b) Radial Direction

Figure 5 : Two Methods Of Relieving

2.2.5 Relief grinding of pre-grinding hob

Machining is a process with the end of the gear hob relief grinding. When hob relieve grinding, it do spiral motion relative to the grinding wheel. Gear hob is one revolution around the axis of itself. It not only must carry on the radial relief grinding of grinding wheel movement at the same time, also the distance along axis hob on a lead.

Problems in the Relief grinding process as follows:

(1)Distortion: Our country at present basically adopts the radial shovel shovel mill grinding. And in the process of relief grinding, tooth cutting blade oblong upward cannot guarantee relief grinding are basic worm of Archimedes spiral surface.This will cause the blade profile produce distortion, the gear hob qualified length is very short, so after a few times after grinding tooth thickness and pitch will become smaller, lower the hob durability[7].

(2)Radial error: When relief grinding, grinding wheel relative to the installation position of hob wrong words, grinding out of hob cutting face may deviate from the direction of the cutter radius, makes the hob generated after anterior horn or Angle, then make the hob produces the radial error.

As mentioned above the major problems affecting the quality of gear hob is shown in figure 6

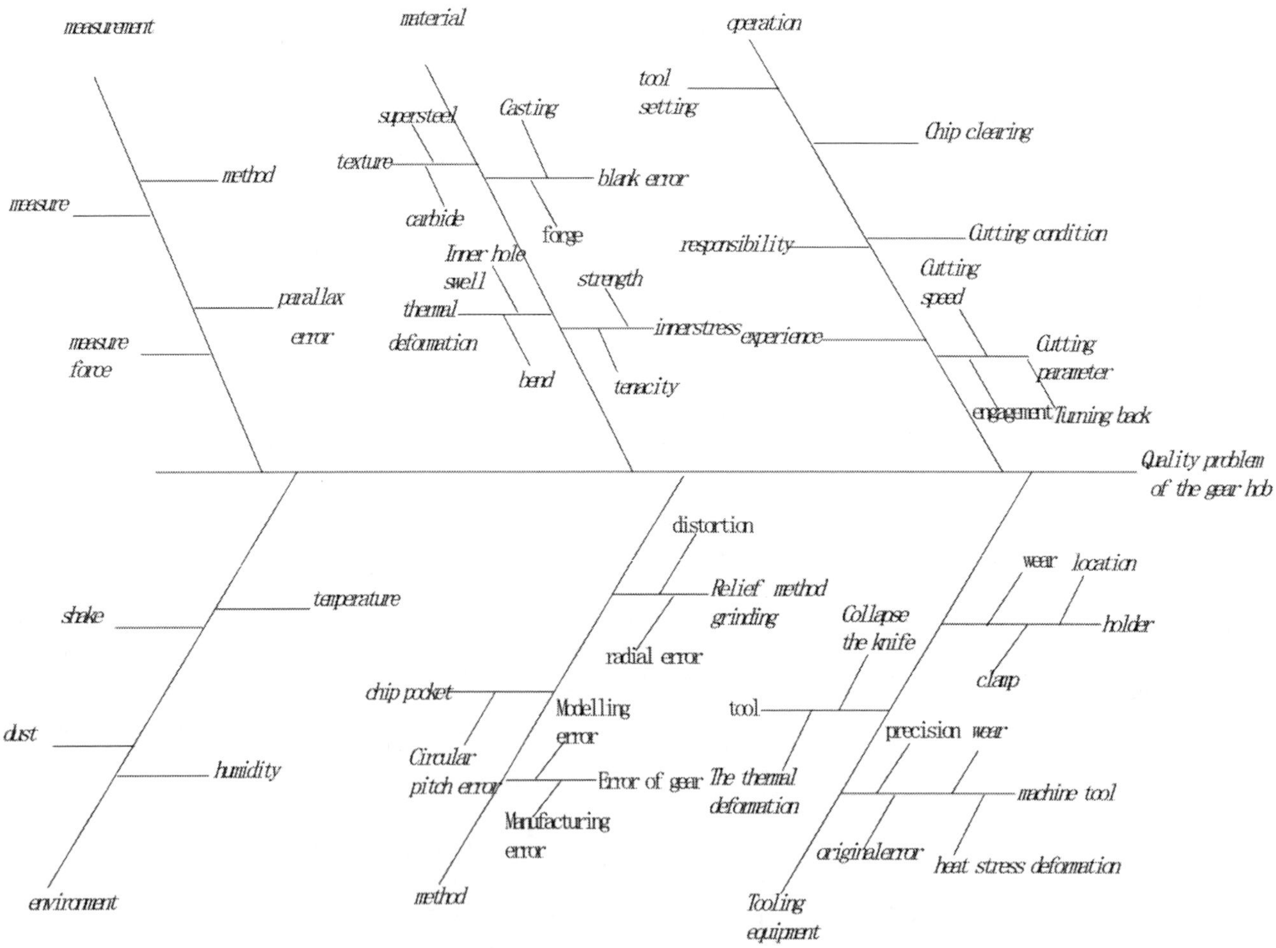

Figure 6: Analysis Diagram

3 IMPROVED METHOD OF PRE-GRINDING HOB'S PROBLEM

To sum up, the problem that impact gear hob's quality make gear hob's precision decline, life expectancy reduce, and also affect the quality of the gear to be machined. Specifically improved method as shown in Table 1.

Table 1: Improved Process Before Grinding Hob Problem

No.	Problems of gear hob		improvement method
1	Modelling error		As far as possible and processing involute hob tooth detection technology to measure and to continue with the grinding wheel tooth.
2	problem of Inner hole	dimension error, form-position errors	Improve the accuracy of machine tools and timely measurements recorded during processing, and constantly improve their skills
		swell easily occur during heat treatment	Try to use when selecting the blank uniformly distributed within the organization.
3	problem of chip pocket	circular pitch error	After working for some time stop adjustment of the grinding wheel and grinding machine precision indexing plate adjustment.
		depth of chip pocket, quantity chip pocket	Levels of development of different standards according to different gear.
4	problem of relief grinding	distortion	The use of CNC machining simulation, the cutting edge always in the face of the helical worm hob basic.
		radial error	You can install a knife angle template or spinner at the wheel at the grinding wheel and gear hob always maintain the correct relative positions.

4 CONCLUSION

Through the analysis to the key working procedure of pre-grinding hob: grinding gear hob tooth profile of the processing, the inner hole machining, crumbs slot and shovel and to five aspects respectively from the man-machine material method, we found out the problems affecting the quality of gear hob: modelling error and shape error, circular pitch error, radial error and so on. It can make the hob qualified rate increase through the use of grinding wheel dresser, reasonable choice of blank, dial, timely adjust and reasonable measures.

5 REFERENCES

[1] **Hanyibin,Zhangguofu. 2012.** *Several typical foreign indexable hob is introduced* [J],A mechanical engineer,pp 142 ~ 143.

[2] **Sheng-jun liu, Yang ling. 2012.** *Processing gear hob tooth profile in the process of how to improve the precision of tooth shape* [J], J tools technology, pp 87-90.

[3] **Li Dongbin, li bin. 2013.** *Gear hob bore and keyway heat treatment deformation research* [J], Metal processing: hot working,pp 23 ~ 25.

[4] **Yuan Zhejun, Liu Huaming, Tang Yisheng. 1983.** *Gear cutter design* (I) [M],Beijing: the new era of publishing house,pp 229 ~ 287.

[5] **Ma Lixin, liu xiang, da-yu zheng. 2002.** *After heavy grinding and grinding gear hob measurement study* [J], Journal of measurement test, pp 35 ~ 37.

[6] **Sun Menghui . 2012.** *After gear hob and Angle of relief grinding method research* [D],Chongqing: chongqing university.

[7] **LI Guolong, SUN Menghui, LI Xianguang, LIU Fei, and HUANG Chao. 2001.** *Improved Relief Grinding Method of Gear Hob with Equal Relief Angle*[J],Chinese Journal of Mechanical Engineering,pp 842-849.

Rebalancing of mixed-model two-sided assembly lines with incompatible task groups: An industrial case study

David Z. ZHANG[1*], Ibrahim KUCUKKOC[2], Aslan Deniz KARAOGLAN[2]

[1] College of Engineering, Mathematics and Physical Sciences,
University of Exeter, Exeter EX4 4QF, England, United Kingdom
d.z.zhang@exeter.ac.uk
[2] Department of Industrial Engineering, Faculty of Engineering,
Balikesir University, Cagis Campus, Balikesir, Turkey
i.kucukkoc@exeter.ac.uk, deniz@balikesir.edu.tr

ABSTRACT

Assembly lines, which are mostly used flow-oriented production techniques in mass production, are classified in different ways, e.g. considering their configurations and/or product variety produced on the line. Mixed-model two-sided assembly lines are usually constructed for the production of large-sized items, such as automobiles or trucks, in an inter-mixed sequence. While the literature on two-sided lines is extensive, there is no research concerning minimizing the cycle time in mixed-model two-sided lines, to the best of the authors' knowledge. Therefore, this paper introduces the type-II mixed-model two-sided assembly line balancing problem benefiting from the real data gathered through an industrial case study. This is the major contribution of this paper. This paper also contributes to knowledge by incorporating incompatible task groups, different from negative zoning constraints. Any two tasks existing in the same incompatible task group cannot be accomplished at the same time in the same mated-station.

Keywords: Assembly line balancing, type-II, mixed-model lines, two-sided lines, incompatible task groups.

1 INTRODUCTION

An assembly line consists of a sequence of workstations linked to each other via a conveyor or moving belt. Each workstation completes its operations on the product (or product model) produced on the line and the product moves to the downstream workstation [1]. The tasks are performed in workstations considering the precedence relationships between tasks and the capacity constraints. The capacity constraint is determined by the cycle time of the line, which is the maximum time each workstation is allowed to complete its tasks. The assembly line balancing (ALB) problem is to assign tasks to workstations considering the capacity and precedence relationship constraints such that a performance measure is optimized. The performance measure can be optimized by minimizing the number of workstations given the cycle time (referred to as type-I ALB problem), or minimizing the cycle time given the number of workstations (referred to as type-II ALB problem). In some cases, both performance measures (i.e. the number of workstations and the cycle time) are aimed to be minimized, which is called a type-E ALB problem [2].

The assembly lines are classified as one-sided lines and two-sided lines based on the formation of the workstations across the corresponding line. In a one-sided assembly line,

* Corresponding Author

the workstations are located in only one side of the line whereas the workstations are located in both (left and right) sides of the line in a two-sided assembly line [3]. Assembly lines are also divided into three groups based on the variety of the products assembled on the line: (i) single-model lines, (ii) mixed-model lines and (iii) multi-model lines. Single model lines are utilized to produce only one model of a product in mass quantities while two or more product models are assembled on mixed-model lines with no need for set-up between model changes [4]. Multi-model lines are also utilized to produce two or more product models. However, different from mixed-model lines, multi model-lines need set-up operations between model changes. Due to this drawback, mixed-model lines are favored in industry.

While there are numerous studies on single-model two-sided lines (e.g., Kim *et al.* [5], Lee *et al.* [6], Baykasoglu and Dereli [7], Ozcan and Toklu [8], Ozbakir and Tapkan [9], Tapkan *et al.* [10], Khorasanian *et al.* [11], Purnomo *et al.* [12]), very little attention has been paid to mixed-model two-sided lines, which consider the production of two or more product models on a two-sided assembly line. Simaria and Vilarinho [13] developed an ant colony optimization algorithm for solving the mixed-model two-sided assembly line balancing (MTALB) problem considering parallel workstations. The aim was minimizing the number of workstations considering some additional constraints, such as synchronous tasks and zoning constraints. Ozcan and Toklu [14] modelled the MTALB problem mathematically and developed a simulated annealing algorithm with the aim of minimizing the number of mated-stations and the number of workstations considering synchronous tasks, positional constraints and zoning constraints. Chutima and Chimklai [15] proposed a particle swarm optimization algorithm with negative knowledge when solving the MTALB problem considering work-relatedness and workload smoothness. Rabbani *et al.* [16] addressed the MTALB problem with multiple U-shaped layout and proposed a GA-based heuristic and a mixed integer program for minimizing the cycle time and the number of stations.

Kucukkoc and Zhang [17] developed a flexible agent-based ant colony optimization algorithm for the mixed-model parallel two-sided assembly line balancing problem (without model-sequencing mechanism) to minimize the line length and the number of workstations. Kucukkoc and Zhang [18] combined the model-sequencing and line balancing problems and introduced the mixed-model parallel two-sided assembly line balancing and sequencing problem, and Kucukkoc and Zhang [19] proposed a mathematical model and an agent-based solution approach for solving it. In their latter research [20], a GA-based model sequencing mechanism was integrated into the agent-based ant colony optimization algorithm to solve the problem more efficiently. This paper addresses to the mixed-model two-sided assembly line system and contributes to knowledge as follows:

i) Different from the common tendency to minimize the number of workstations in mixed-model two-sided lines, the cycle time is minimized in this research. Therefore, *type-II mixed-model two-sided assembly line balancing problem* is introduced.
ii) A new constraint between tasks, called *incompatible task groups*, is introduced. This concept will be explained in details in the following section.
iii) Furthermore, this research contributes to knowledge by an industrial case study conducted in a tractor cabin manufacturing plant. As stated by Baykasoglu and Dereli [7], "*real-life case studies can provide very useful research contributions*" for two-sided lines. In their research on mixed-model two-sided lines, Simaria and Vilarinho [13] also highlighted the need for additional efforts in "*matching theoretical procedures and practical applications*".

The remainder of this paper is organized as follows. The type-II mixed-model two-sided assembly line balancing problem with incompatible task groups is defined in Section 2. The solution method is described in Section 3 and an industrial case study is provided in Section 4 together with a discussion. Finally, the paper is concluded with some future research directions and practical implications of the study in Section 5.

2 PROBLEM DEFINITION

The configuration of the classical mixed-model two-sided assembly line is represented in Figure 1. As seen from Figure 1, two product models (model-I and model-II) are produced on the same two-sided line, simultaneously. Note that, in practice, there may be more than two product models on the line. The workstations are located on both left and right sides of the line, and one operator performs in each workstation. Opposite workstations are called the *companion* of each other, such as WS-1 is the companion of WS-2, or vice versa. The two companion workstations are called a *mated-station*; e.g. WS-1 and WS-2 constitute a mated-station. Thus, there are four mated-stations in Figure 1, which equivalents to eight workstations in total.

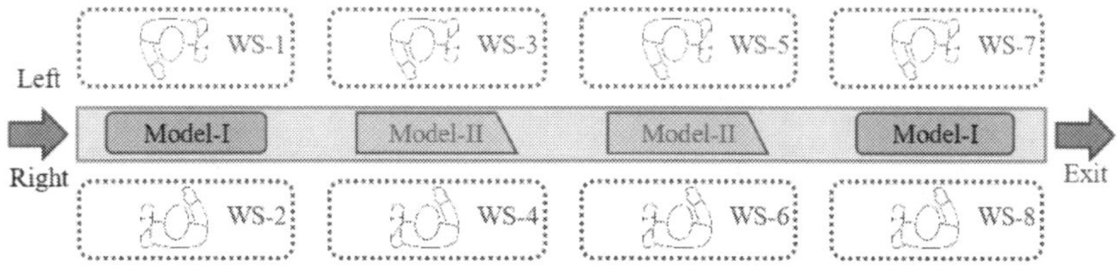

Figure 1. Representation of a mixed-model two-sided assembly line.

In two-sided lines, there are certain constraints; i.e. capacity constraints, precedence relationships constraints and operation side constraints to be satisfied to have a feasible solution. The capacity constraints are determined by the cycle time, which is the maximum time each workstation is allowed to complete tasks assigned to it. The precedence relationship constraints occur due to the technological requirements or organizational restrictions of the manufacturing environment. Considering the precedence relationships between tasks when building a line balancing solution is particularly important, or the line balance to be obtained eventually can be infeasible. Thus, a task can start after all of its predecessors have been completed. In two-sided lines, some tasks may require to be performed at a specific side, i.e. left or right side, while some others can be performed at either side. A task that needs to be performed on the left (right) side of the line is called a left (right) side task.

Table 1 presents the data of an example problem about the classical mixed-model two-sided assembly line balancing problem taken from Ozcan and Toklu [14]. The second column provides the information about which side the corresponding task must be assigned to. The letters L, R and E denote the left side, right side and either side, respectively. The third and fourth columns provide the processing time of the related task for each model, model-A and model-B, respectively. Processing times of tasks may vary from one model to another. For example, task 2 requires 3 time-units for product model A while 1 time-unit for product model B. The precedence relationships data are provided in the last column. For example, tasks 2 and 3 must be completed to initialize task 6.

The balancing solution of the example problem is given in Figure 2. The length of the bars represents the processing time of the task given in the corresponding bar. The gray shaded areas denote the unavoidable idle times. As seen from the figure, three workstations are needed when the cycle time is considered 5 time-units/item. Tasks 1, 3, 6 and 9 have been assigned to the left side of the first mated-station while tasks 2 and 5 have been assigned to the right. Tasks 4, 8 and 7 have been assigned to the left of the second mated-workstation. No tasks have been assigned to its opposite side as all tasks have already been assigned. As seen from the figure, the workload time of any workstation does not exceed the cycle time for any model.

The problem introduced in this paper is the type-II mixed-model two-sided assembly line balancing problem with incompatible task groups. The aim is to minimize the cycle time of a paced (or synchronous) mixed-model two-sided assembly line given the number of workstations.

Table 1. The data of the example problem [14].

Task No	Side	Processing Time (time-units)		Immediate Predecessor(s)
		Model A	Model B	
1	L	2	0	-
2	R	3	1	-
3	E	0	1	-
4	L	3	0	1
5	R	1	3	2
6	E	1	1	2,3
7	E	2	2	4,5
8	L	0	3	5
9	E	1	1	6

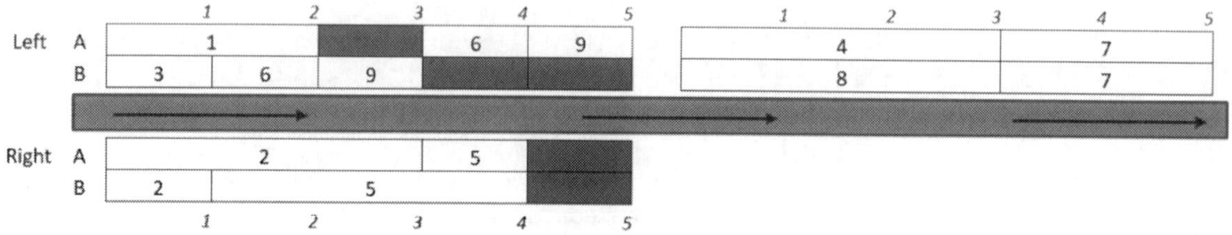

Figure 2. The balancing solution of the example problem [14].

In addition to the constraints explained above, this paper considers a new concept to represent the real-world manufacturing environment more efficiently: incompatible task groups (ITGs). In the literature, tasks have positive or negative zoning constraints. If there is a positive zoning constraint between two tasks, these two tasks must be assigned to the same workstation. On the contrary, the negative zoning constraint determines the two tasks that must be performed in different workstations. However, on certain occasions, it is not possible to perform two tasks at the same time on both sides of the line. For example, if the left-side operator needs to get in the product model being assembled on the line, the right-side operator cannot get in the product model at the same time as there is no enough space for two operators. Therefore, these two incompatible tasks must be performed at different times no matter they are assigned to the same or different workstations. Therefore, tasks do not have binary relationships between each other. Instead, a task existing in an ITG cannot be performed at the same time with another task existing in the same ITG. The ITGs can be categorized if there is more than one ITG for the same system.

3 SOLUTION METHOD

The pseudocode of the solution algorithm used in this research to conduct the case study (to be given in the next section) is presented in Figure 3. As seen from the figure, the cycle time value is initialized ($C \leftarrow C_{min}$) and a number of solutions ($maximumNumberOfSubIterations$ $= 200$ iterations) are produced for each C value using 10 different line balancing rules, most of which are used commonly in the line balancing domain: Computer Method of Sequencing Operations for Assembly Lines - Comsoal [21], Ranked Positional Weight Method - RPWM [22], Reverse Ranked Positional Weight Method - RRPWM (produced from RPWM), Longest Processing Time - LPT [23], Shortest Processing Time - SPT [24], Smallest Task Number - STN [25], Maximum Number of Predecessors - MNP (produced from Baykasoglu [24]), Least Number of Predecessors - LNP, (produced from MNP), Maximum Number of Successors - MNS (produced from Tonge [26]), and Least Number of Successors - LNS (produced from MNS). Each of these 200 balancing solutions is built using a randomly picked rule and the solution which gives the minimum number of stations is retrieved. If the minimum number of stations found by the algorithm is equal to the number of stations aimed, the algorithm is terminated and the solution which gives the minimum number of stations is designated as the best solution. If not, C is increased by $cycle\_time\_increment$

value and 200 solutions are built again using randomly picked heuristics. This cycle continues until $numberOfStationsAimed$ is achieved and the current cycle time is determined as the solution of the problem ($C^* \leftarrow C$). When there is more than one solution with the same number of stations, the solution which has smoother workload distribution is favored. Note that C_{min} is the theoretical minimum cycle time calculated as $C_{min} = \max_{j \in J}\left\{\left\lceil\frac{\sum_{i=1}^{N} t_{ij}}{K}\right\rceil^{+}\right\}$ where N is the total number of tasks, t_{ij} is the processing time of task i for model j ($j \in J$), K is the total number of workstations. $\lceil X\rceil^{+}$ denotes the least integer equals to or larger than X.

Start (a heuristic rule is selected randomly)
Initialize all parameters, $C \leftarrow C_{min}$
While ($numberOfStations > numberOfStationsAimed$) {
  While ($sub\_iteration\_number < maximumNumberOfSubIterations$) {
    Pick a heuristic rule randomly and build a balancing solution using the procedure given in Figure 4.
    Evaluate the solution obtained (calculate the $numberOfStations$)
  } **End while**
  $sub\_iteration\_number += 1$
} **End while**
$C += cycle\_time\_increment$
Terminate and retrieve C ($C^* \leftarrow C$)

Figure 3. The pseudocode of the heuristic algorithm coded in Java.

Start (a heuristic rule is selected randomly)
Initialize all sets (i.e., available tasks and unassigned tasks) and parameters (i.e., $st(k)j \leftarrow 0$, *where* $st(k)j$ *represents the station time of workstation k for model j*)
While (there is one or more unassigned task) {
  Select an operation side randomly and determine all available tasks for the current position
  If (there is at least one available task) {
    Select a task (i.e., task *i*) using the heuristic rule
    Assign task *i* to the current available station and increase the station time: $st(k)j \leftarrow st(k)j + (t_{ij})$
    Increase the earliest starting times of all successors of task *i*
    Remove task *i* from the unassigned tasks list of the relevant line
  } **else if** (there is no available task due to interference) {
    Increase the station time of the current workstation: $st(k)j \leftarrow st(\underline{k})j$, where $\underline{k}$ is the companion of k
  } **else if** (there is no available task due to insufficient capacity) {
      If (both sides of the line reached full capacity) {
        Increase the station number (*k++*)
      } **else if** (at least one side of the current line has not reached full capacity) {
        Alternate the operation side
      } **End if**
  } **End if**
} **End while**
Terminate

Figure 4. The procedure for building a balancing solution.

4 CASE STUDY

The case study has been conducted at an automotive company which produces safety cabins for tractors, excavators, and forklifts. The production operations of the company have been illustrated in Figure 5. As seen from the figure, the formed sheet metal and formed profile are welded in the welding line. The cabin frame is painted by a powder coating operation and moved to the assembly line, in which all functional units and accessories are sequentially mounted on to the cabin. More than five different cabin models are assembled on the same line which is constituted by 13 sequentially located mated-stations (which equivalents to a total of 26 workstations). Due to the page limit, the assembly process of the two mostly produced cabin models in the first eight workstations were selected for this case study. Thus, the aim is to minimize C given $K = 8$. The data on the processing times, operation sides and precedence relationships of the 49 tasks are provided in Table 2. The tasks existing in the same incompatible task group are marked with '1' in the 'ITG' column. The assignment configuration of these tasks in its current status is also presented in Table 3.

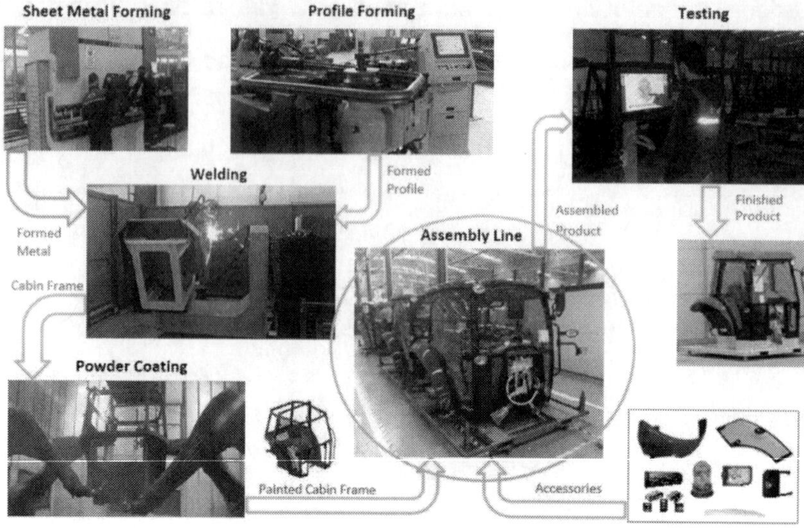

Figure 5. The production system of the company.

Table 2. Data for the case study.

Task	Processing Time (time-unit)		Side	ITG	Immediate Predecessor(s)	Task	Processing Time (time-unit)		Side	ITG	Immediate Predecessor(s)
	A	B					A	B			
1	29	29	E	1	-	26	17	17	E	1	24
2	16	16	R	-	7	27	4	4	L	-	12
3	10	10	R	-	7	28	28	28	E	-	27
4	4	4	E	-	-	29	7	10	E	-	28
5	18	18	E	-	4	30	9	9	E	-	28
6	15	15	E	-	-	31	17	17	E	-	29
7	9	9	E	-	-	32	16	16	E	-	30
8	11	11	E	-	7	33	15	15	E	-	10
9	7	7	E	1	6	34	16	10	R	-	12
10	4	4	E	-	-	35	17	17	R	-	12
11	13	13	E	1	2,9,10	36	10	20	R	-	35
12	6	6	E	-	5	37	10	30	R	-	36
13	13	13	E	-	12	38	11	11	R	-	37
14	8	8	E	-	13	39	8	8	L	-	10
15	65	69	E	1	14	40	10	10	E	-	12
16	6	6	E	-	5	41	11	17	E	-	12
17	5	5	E	-	12	42	10	10	E	1	40,41
18	11	11	L	-	12,17	43	8	8	E	1	40,41
19	7	7	E	1	12,15	44	9	9	E	1	43
20	31	31	E	1	11,19	45	5	10	E	1	44
21	7	7	E	1	19	46	11	11	E	1	2,3,11,15,43
22	22	22	E	-	12	47	4	4	E	1	42,45,46
23	24	24	E	1	22,27	48	6	6	L	-	12
24	14	14	E	1	23	49	18	27	E	-	47
25	15	15	E	1	22,24,27						

Figure 6 exhibits the change in the number of stations as the cycle time increases. As seen, the best solution obtained when the cycle time was 88 time-units requires 10 workstations. As the cycle time increases by one time-unit, the number of workstations decreases and the algorithm is terminated as soon as it reaches to eight workstations eventually.

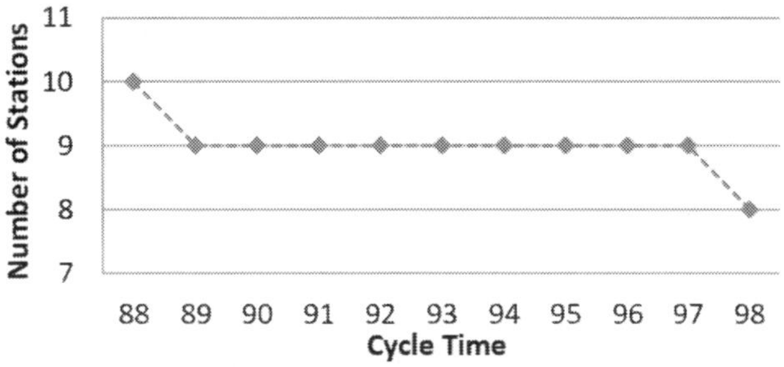

Figure 6. The number of stations obtained for the corresponding cycle time value.

The solution obtained when the algorithm was terminated is presented in Table 3 (see "Proposed Line Balance" column). As seen from the table, the maximum of the station workloads in the new solution is 98 time-units, which is much smaller than the one in current line balance (120 time-units). That means the cycle time of the line was reduced from 120 time-units to 98 time-units, which corresponds to a 22.4% improvement. Therefore, the weighted line efficiency is increased from 70% to 85.8%, with an enhancement of 22.4%. Moreover, in the current situation, the line requires 5 mated-stations as the left side of the second workstation (2-L) and the right side of the third workstation (3-R) are not utilized. However, the proposed solution requires four mated-stations since both of the operation sides have been utilized. Thus, the line length was also reduced in the proposed balance.

Table 3. The assignment configuration of the current and the proposed line balances.

Current Line Balance				Proposed Line Balance			
Station	Assigned Tasks	Station Workload		Station	Assigned Tasks	Station Workload	
		A	B			A	B
1-L	6,7,8,9,10,11,12	65	65	1-L	7,10,6,9,33,8,27,22,17,16	**98***	**98***
1-R	1,2,3,4,5	77	77	1-R	4,5,12,1,13,14,40	88	88
2-L	-	-	-	2-L	15,11,19	85	89
2-R	13,14,15	86	90	2-R	2,41,28,35,43	80	86
3-L	16,17,18,19,20,21	67	67	3-L	24,44,46,42,45,47,29,48	66	74
3-R	-	-	-	3-R	23,3,36,37,30	63	93
4-L	27,28,29,30,31,32	81	84	4-L	20,26,32,18,39	83	83
4-R	22,23,24,25,26	92	92	4-R	49,31,25,34,38,21	84	87
5-L	39,40,41,42,43,44, 45,46,47,48,49	100	120*				
5-R	33,34,35,36,37,38	79	103				

* The maximum station workload time for the corresponding situation

5 CONCLUSION

The mixed-model two-sided assembly line balancing problem was addressed considering ITGs. Different from the studies on mixed-model two-sided assembly lines in the literature, the main aim was to minimize the cycle time while maximizing the workload smoothness among the workstations. A case study was conducted using data gathered from an automotive company. The line was rebalanced using heuristic method coded in Java and the cycle time was reduced by 22.4%. The newly introduced ITG concept limits performing the tasks existing in the same ITG group on either side of the line. While there may be more than one ITG in some cases, only one ITG was considered in this research. This concept can also be implemented in other line concepts or industries where it is inconvenient to perform some group of tasks on the same item, simultaneously. As a limitation of the work, the number of tasks considered in the case study was kept to a minimum due to page limit.

ACKNOWLEDGMENT

The authors acknowledge engineers Mr. Korunay Ucok and Ms. Sevil Duran for their valuable collaboration in gathering the industrial data.

REFERENCES

[1] Kucukkoc I, Zhang DZ. Balancing of parallel U-shaped assembly lines. Computers and Operations Research. 2015;64:233-44, doi: http://dx.doi.org/10.1016/j.cor.2015.05.014.

[2] Kucukkoc I, Zhang DZ. Type-E parallel two-sided assembly line balancing problem: Mathematical model and ant colony optimisation based approach with optimised parameters. Computers and Industrial Engineering. 2015;84:56-69, doi: http://dx.doi.org/10.1016/j.cie.2014.12.037.

[3] Bartholdi JJ. Balancing 2-Sided Assembly Lines - a Case-Study. International Journal of Production Research. 1993;31(10):2447-61.

[4] Thomopoulos NT. Line Balancing-Sequencing for Mixed-Model Assembly. Management Science. 1967;14(2):B-59-B-75.

[5] Kim YK, Kim YH, Kim YJ. Two-sided assembly line balancing: a genetic algorithm approach. Production Planning & Control. 2000;11(1):44-53.

[6] Lee TO, Kim Y, Kim YK. Two-sided assembly line balancing to maximize work relatedness and slackness. Computers & Industrial Engineering. 2001;40(3):273-92.

[7] Baykasoglu A, Dereli T. Two-sided assembly line balancing using an ant-colony-based heuristic. International Journal of Advanced Manufacturing Technology. 2008;36(5-6):582-8.

[8] Ozcan U, Toklu B. A tabu search algorithm for two-sided assembly line balancing. International Journal of Advanced Manufacturing Technology. 2009;43(7-8):822-9.

[9] Ozbakir L, Tapkan P. Bee colony intelligence in zone constrained two-sided assembly line balancing problem. Expert Systems with Applications. 2011;38(9):11947-57.

[10] Tapkan P, Ozbakir L, Baykasoglu A. Modeling and solving constrained two-sided assembly line balancing problem via bee algorithms. Applied Soft Computing. 2012;12(11):3343-55.

[11] Khorasanian D, Hejazi SR, Moslehi G. Two-sided assembly line balancing considering the relationships between tasks. Computers & Industrial Engineering. 2013;66(4):1096-105.

[12] Purnomo HD, Wee HM, Rau H. Two-sided assembly lines balancing with assignment restrictions. Mathematical and Computer Modelling. 2013;57(1-2):189-99.

[13] Simaria AS, Vilarinho PM. 2-ANTBAL: An ant colony optimisation algorithm for balancing two-sided assembly lines. Computers & Industrial Engineering. 2009;56(2):489-506.

[14] Ozcan U, Toklu B. Balancing of mixed-model two-sided assembly lines. Computers & Industrial Engineering. 2009;57(1):217-27.

[15] Chutima P, Chimklai P. Multi-objective two-sided mixed-model assembly line balancing using particle swarm optimisation with negative knowledge. Computers & Industrial Engineering. 2012;62(1):39-55.

[16] Rabbani M, Moghaddam M, Manavizadeh N. Balancing of mixed-model two-sided assembly lines with multiple U-shaped layout. International Journal of Advanced Manufacturing Technology. 2012;59(9-12):1191-210.

[17] Kucukkoc I, Zhang DZ. Mixed-model parallel two-sided assembly line balancing problem: A flexible agent-based ant colony optimization approach. Computers & Industrial Engineering. 2016;97:58-72.

[18] Kucukkoc I, Zhang DZ. Simultaneous balancing and sequencing of mixed-model parallel two-sided assembly lines. International Journal of Production Research. 2014;52(12):3665-87.

[19] Kucukkoc I, Zhang DZ. Mathematical Model and Agent Based Solution Approach for the Simultaneous Balancing and Sequencing of Mixed-Model Parallel Two-Sided Assembly Lines. International Journal of Production Economics. 2014;158, :314-33.

[20] Kucukkoc I, Zhang DZ. Integrating ant colony and genetic algorithms in the balancing and scheduling of complex assembly lines. The International Journal of Advanced Manufacturing Technology. 2015: doi: http://dx.doi.org/10.1007/s00170-015-7320-y.

[21] Arcus A. COMSOAL, A Computer Method of Sequencing Operations for Assembly Lines. The International Journal of Production Research. 1966;4(4):259-77.

[22] Helgeson WB, Birnie DP. Assembly Line Balancing Using the Ranked Positional Weight Technique. The Journal of Industrial Engineering. 1961;12(6):394-8.

[23] Talbot FB, Patterson JH. An Integer Programming Algorithm with Network Cuts for Solving the Assembly Line Balancing Problem. Management Science. 1984;30(1):85-99.

[24] Baykasoglu A. Multi-rule multi-objective simulated annealing algorithm for straight and U type assembly line balancing problems. Journal of Intelligent Manufacturing. 2006;17(2):217-32.

[25] Arcus AL. An analysis of a computer method of sequencing assembly line operations: Ph.D. Dissertation, University of California, Berkeley, USA; 1963.

[26] Tonge FM. Summary of a Heuristic Line Balancing Procedure. Management Science. 1960;7(1):21-42.

RANKING COMMUNICATION ASSISTIVE TECHNOLOGY BY MCDM METHODS

O. Habert *, M. Gam, I. Kacem
Laboratoire de Conception, Optimisation et Modélisation des Systèmes
University of Lorraine, Metz, France
olivier.habert@univ-lorraine.fr

ABSTRACT

In this paper, we propose a solution to rank some chosen communication and information assistive technologies by using the well-known Multiple Criteria Decision-Making tool: fuzzy TOPSIS. There are many ways to restore or to assist disabled people in their daily communication and quest of information. Thanks to New Information and Communication Technologies (NICTs), communication assistive technology has started the path of a deep evolution. Indeed, numerous assistive devices using web technology and connected objects can be found. The main goal of this work is to propose assistive technology designer a tool to make prospective studies to guide them in their strategy and choice of new products' development. One of the advantages of MCDM method is to process with data coming from the disabled users, healthcare and social staff, researchers

Keywords: Assistive Technology, Fuzzy TOPSIS, AHP, Ranking

1 INTRODUCTION

In France and many other countries, children and adults with physical or cognitive disabilities are daily admitted in specialized establishment so-called medico-social centers. They offer various types of services in the context of short-term stays or permanent accommodation. Home services are also possible. Healthcare and social staff missions are to first maintain the well-being of the user by taking care of everyday life and then to provide better-being by adjusting and improving their daily environment. This can be achieved by adapted animations and activities alloying to merge leisure and physical and cognitive rehabilitation. French CAPs [1] institute is a typical association of medico-social centers, each dedicated to particular support missions. It is within this framework that numerous assistive technology (AT) are used to assist healthcare and social staff in mobility, cares, reeducation of communication, to reduce or offset the disabilities of the users. All AT are registered according ISO-9999-2001 standards [2] that provide a classification of assistive tools and devices, manufactured and especially available for people with disabilities. Section 22 of this classification covers all the communication and information AT in order to improve hearing or the visual field or to provide access to computer and phone facilities. Numerous input-output devices like adapted mice and keyboards, specific HMI, adapted readers can be found in this section. The section concerning the face to face communication grows rapidly with new software and devices (speech synthesis, therapist robots   ) that more and more use NICTs.

Assistive tools designers need reliable benchmarks to know the users expectations in NICTs products to communicate.

* Corresponding Author

The key question is the following: *Is providing the access to computer and phone is enough or do they have to go further in the integration of the latest technologies in future assistive tools?*

Only the disabled users, the medico-social staff and researchers working in this field can help the designer to answer this question. Users and personnel around them are not frozen face to ever-changing technology and it seems clear that assistive technology should change with them. It is therefore an important matter to propose the designer a method based on the judgment of involved persons. In this study, respondents have being restricted to two groups, the researchers on the AT field and the medico-social staff as the facilitators, occupational therapist, psychologists or Speech-Language Pathologists.

Decision-making consists in finding the best alternative among a set of feasible alternatives. Multi-Criteria Decision Making (MCDM) methods are used when the problem to solve is complex because it has to consider several criteria, often inconsistent. MCDM problems presented in Chen [3], Liang [4], Wang [5] often require the decision-makers to provide qualitative/quantitative evaluation for determining the interest of each criterion with respect to each alternative. MCDM methods seem to meet our needs, particularly fuzzy TOPSIS method.

This paper is organized in 5 sections. Chosen NICTs assistive technologies to evaluate are presented in Section 2. Section 3 explains the different steps of our problem in fuzzy TOPSIS approach. Section 4 analyses the first results and finally, Section 5 concludes and explains the future works.

2 ASSISTIVE TECHNOLOGIES

2.1 Which AT to evaluate

Assistive technology (AT) is any kind of technology (low-tech or high-tech) that can be used to enhance the functional independence and quality of life for a person with disabilities. AT for communication and information are daily used in socio-medical institute in most rehabilitation activities. These tools are very helpful for all the involved staff. Cardboard support, communication binder using alternative languages as pictograms are gradually replaced by systems using high technology, with screen, HIM and multimedia functions. First assistive electrical devices used to be autonomous boxes which act as a substitute in cases of severe speech impairment, throughout alternative languages using pictures and symbols communication board. Currently, electronic communication aids allow the user to use picture symbols, letters, and/or words and phrases to create messages which can be heard thanks to speech synthesis. These systems are usually stand-alone, proprietary and thus incompatible with other products.

Later, the evolution of computer sciences and the emergence of personal computers have enabled the development of numerous software applications without any physical constraints, easily increasing the amount of functionalities. The great challenge was then to develop hardware and software AT technology to help with routine computer tasks such as typing or using a mouse or an adapted HMI connected to all or nothing sensors.

It is now tactile age, with smartphone and touchpad. Such a mobile devices bring much more mobility and possibilities to develop light and low-cost applications with additional connected and multimedia capacities. Tactile HMI create new opportunities to disabled people communication. Researchers Hourcade [6] and De Leo [7] have shown the benefits of new generation HMI for children with autism disorder. These very new technological objects and new interaction mode are in line with everyone daily practices, including medico-social staff and some disabled. The latter can be recently physically or cognitively impaired because of an accident or a degenerative disease. In their previous life, these adults have used new interaction mode as social networks. In the near future, users of rehabilitation centers will more belong to the famous Z generation who was born with nowadays technology. Inevitably these new generations will feel comfortable with current technology and it will be obvious to use it.

Following the same reasoning, these generations will be familiar with techniques of distance learning like famous Massive Open Online Course (MOOC) and will be ready to use such a tools not for teaching but for online adapted rehabilitation activities. Some research works has shown that rehabilitation activities are better follow-up in the context of connected environment or video-games, see Lohse [8].

Therapist robots or interactive companions are emergent assistive tools. They have proved their efficiency in autist children rehabilitation and Alzheimer elderly institute as in Kawaguchi [9] and Robins [10].These multimodal interactive systems can bring help to solve social, cognitive or physical problems. Multimedia, color changing, gesture are used to interact with the user.

Table 1 summarizes the six selected AT after discussions with researchers and medico-social staff. The first two (1 and 2) are classical and well-tried. It is important to submit them for evaluation to know if evolutions are required. The second two (3 and 4) are now emerging as a result of mobile devices and artificial intelligence development. Clearly innovating, AT 5 and 6 are based on currently new interaction modes to communicate and to learn.

Table 1: Selected assistive technologies to rank

	Selected technical solutions for communication assistance	
1	**Proprietary systems**	Stand-alone specialized Device
2	**Computer softwares**	Software installed on desktop or laptop computer
3	**Mobile devices applications**	Applications installed on mobile devices
4	**Therapeutic devices**	Smart objects which interact with users
5	**Social network based applications**	User interaction with a group having the same profile to perform adapted activities
6	**Distance learning-based applications**	Online adapted activities for cognitive or physical rehabilitation.

2.2 Relevant criteria

It is necessary to identify a set of criteria that characterize as well as possible all the chosen alternatives. TOPSIS method needs to distinguish positive and negative criteria. Usually the cost represents a negative criterion. We have chosen ten criteria:

- Compatibility or Interoperability: It is the ability to work with other products, current or future with as little adaptation as possible. This ability needs the use of standard as communication protocols because it often concerns data exchanges. The lack of interoperability of a system can lead to its definitive disappearance.
- Connectivity: It is the ability of several systems to communicate with each other, locally or throughout internet. Wireless protocol like Wi-Fi, Bluetooth, Z-wave are more commonly used.
- Cost: This is the cost of the AT after taking into account different aids from social organisms. Cost is a very negative criterion which can discourage from buying an AT.

- Versatility: It is the ability of an AT to be used by several users who do not have the same profile. Indeed, it is important for a socio-medical center to standardize practices for efficiency and cost-effectiveness purposes.
- Mobility: It is the ability of the AT to be moved inside or outside the center.
- Data storage: Ability to store data concerning the profile and the usage data of the system is very important. Storing the profile improve the usability of the AT, important feature in a context of disabled people. Moreover, storing the usage data may allow some advanced analyses about temporal assessment of the capacities of the users.
- Batteries autonomy: This criterion concerns the period the AT is used without recharging the batteries. This feature is particularly appreciated by the users who can daily uses AT without taking into consideration the energy aspects.
- Interactivity: It concerns the exchanges between the user and the system which must be done in a natural and conversational way. A good interactivity makes the system more emphatic and increases its acceptability.
- Ergonomy: It concerns the adaptation of the system to the user to provide comfort, security, efficiency. This is all the more important that users are people with physical or cognitive disabilities.
- Personal data security: Using NICTs leads generation, processing and dissemination of a vast amount of personal data. The major risk is a diversion of these data for purposes other than those originally set out. Designing an AT must ensure compliance with ethics suggested by organizations addressing the issues of Information Technology and Civil Liberties.

3 HOW TO EVALUATE AND PRIORIZE

3.1 MCDM method

MCDM methods are very popular because there are applied on several fields as management, marketing, logistics when decisions must be taken in complex problems. Numerous MCDM variants can be found in the literature, when the uncertain and imprecise aspects of human data are taken into account or not, when pairwise comparison of criteria is used to compute a score for each alternative or when a weighted distance from each alternative to the best and the worst solution is computed. Most popular methods are Analytical Hierarchy Process (AHP), see Saaty [11] and Technique for Order Preference by Similarity to Ideal Solution (TOPSIS), see Hwang [12]. When considering uncertainties coming from human judgment with the use of fuzzy logic techniques, the two methods are called fuzzy AHP and fuzzy TOPSIS. A comparison between fuzzy AHP and fuzzy TOPSIS can be found in Lima [13] and in Ertuğrul [14]. Each method has its limitations and advantages and produces the same results when decision-makers are consistent with themselves. One of the arguments which have led us to choose TOPSIS Method is the ranking stability when adding or removing alternatives and criteria. AHP evaluates pairwise criteria interest and TOPSIS weighting each criterion with respect of each alternative. We believe that in the context of our study, with our decision-makers, the TOPSIS assessments are more natural and intuitive, and which will produce fewer inconsistencies. These are the reasons why we propose to use fuzzy TOPSIS method to solve our evaluation and ranking problem. TOPSIS needs to weight criteria. We propose to use AHP pairwise comparisons matrix to obtain these weights. The formalization of our problem follows TOPSIS approach in six steps:

1. Identify the technical solution (or alternative) to evaluate
2. Identify the relevant criteria

3. Assign a relative weight to each criterion
4. Identify the interest and performance of each criterion in relation to each alternative
5. Compute a based-distance score for each alternative
6. Sort or choose the best alternative

3.2 TOPSIS step by step

Step 1: This step has been discussed in Section 2.1 and is outline in Table 1. Six alternatives have been selected. Two are well-tried, two other are emerging and the two last are innovative.

Step 2: This step has been discussed in Section 2.2. Ten relevant criteria have been chosen.

Step 3: TOPSIS method requires assigning relative weight to each criterion. As in Farrokhi [15], it is possible to use AHP pairwise comparison of the criteria. First the comparisons are made with linguistic variables which are then changed into numerical values from one to five. A decision matrix is built and is then normalized. The vector of criteria weights can be computed from the normalized comparison matrix and then by a averaging of the entries on each row. For practical reasons due to computer access in the centers, we have chosen questionnaires in printed edition. Figure 1 shows the grid for pairwise criteria comparisons. Answer is written to the left or right according the more important criterion. Figure 2 shows a result of criteria relative weights. We have used the AHP excel template freely given by Goepel [16].

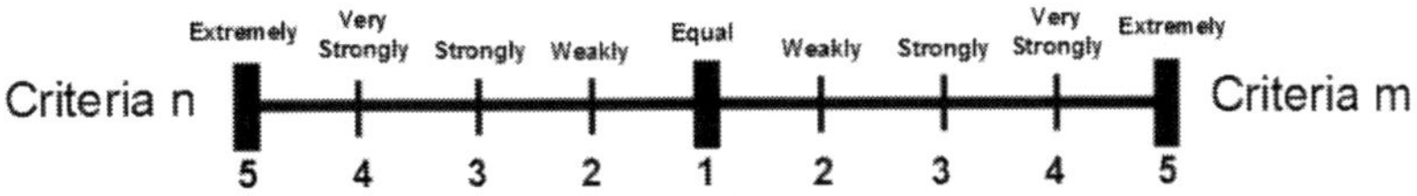

Figure 1: Grid for pairwise criteria comparison

	Criterion	Weights	Rk
1	Interoperability	3,8%	7
2	Connectivity	10,8%	4
3	Cost	1,5%	10
4	Versatility	9,7%	6
5	Mobility	2,9%	9
6	Data storage	18,9%	2
7	Batteries autonomy	3,7%	8
8	Interactivity	11,5%	3
9	Ergonomy	9,7%	5
#	Personal data protection	27,5%	1

Figure 2: Grid for pairwise criteria comparison

Step 4: TOPSIS method starts at this step. The questionnaire is now focusing on the interest of each criterion with respect to all the alternatives. Figure 3 shows the linguistic variables grid for evaluating a criterion interest.

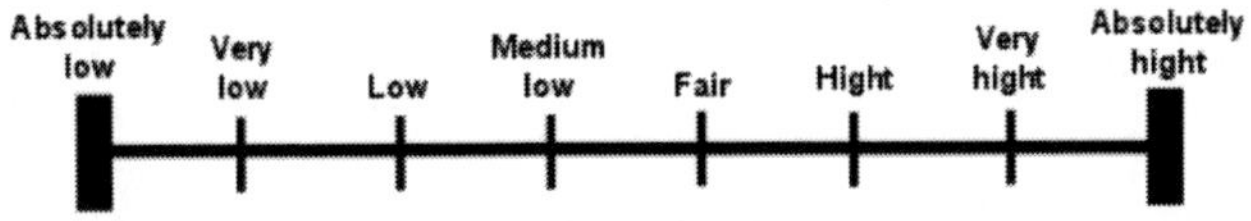

Figure 3: TOPSIS grid

In fuzzy TOPSIS method, linguistic variables are converted into triangular fuzzy numbers. Each fuzzy triangular number can be defined by a triplet (a, b, c) where b is the central value, a and c respectively the left and the right value. Table 2 gives the triplet of each linguistic variable.

Table 2: Fuzzy ratings for linguistic variables

Linguistics Variables	Triangular Fuzzy Numbers
Absolutely Low (AL)	$(1, 1, 2)$
Very Low (VL)	$(1, 2, 3)$
Low (L)	$(2, 3, 4)$
Medium low (ML)	$(3, 4, 5)$
Fair (F)	$(4, 5, 6)$
Hight (H)	$(5, 6, 7)$
Very Hight (VH)	$(6, 7, 8)$
Absolutely Hight (AH)	$(7, 8, 8)$

The fuzzy decision matrix $\widetilde{D}$, where each linguistic judgment has been converted as a triangular fuzzy number $\tilde{x}_{ij}$, $(i = 1, \dots, 6)$ *and* $(j = 1, \dots, 10)$ can be expressed as in (1).

$$\begin{array}{c} \\ \widetilde{D} = \end{array} \begin{array}{c} \\ A_1 \\ A_2 \\ A_3 \\ A_4 \\ A_5 \\ A_6 \end{array} \begin{array}{c} \begin{matrix} C_1 & C_2 & C_3 & \dots & C_{10} \end{matrix} \\ \begin{bmatrix} \tilde{x}_{11} & \tilde{x}_{12} & \tilde{x}_{13} & \dots & \tilde{x}_{1\,10} \\ \tilde{x}_{21} & \tilde{x}_{22} & \tilde{x}_{23} & \dots & \tilde{x}_{2\,10} \\ \tilde{x}_{31} & \tilde{x}_{32} & \tilde{x}_{33} & \dots & \tilde{x}_{3\,10} \\ \tilde{x}_{41} & \tilde{x}_{42} & \tilde{x}_{43} & \dots & \tilde{x}_{4\,10} \\ \tilde{x}_{51} & \tilde{x}_{52} & \tilde{x}_{53} & \dots & \tilde{x}_{5\,10} \\ \tilde{x}_{61} & \tilde{x}_{62} & \tilde{x}_{63} & \dots & \tilde{x}_{6\,10} \end{bmatrix} \end{array} \tag{1}$$

Each judgment $\tilde{x}_{ij}$is in fact the arithmetic average of the p judgments given by the p evaluators. Next, the normalized matrix $\widetilde{R}$ is computed from matrix $\widetilde{D}$. Each element $\widetilde{r_{ij}}$ of $\widetilde{R}$ is computed according (2) and (3).

$$\widetilde{r_{ij}} = \left(\frac{a_{ij}}{c_j^+}, \frac{b_{ij}}{c_j^+}, \frac{c_{ij}}{c_j^+}\right) \; if\ C_j\ is\ a\ positive\ criteria\ with\ c_j^+ = max_i\, c_{ij} \tag{2}$$

$$\widetilde{r_{ij}} = \left(\frac{a_j^-}{c_{ij}}, \frac{a_j^-}{b_{ij}}, \frac{a_j^-}{a_{ij}}\right) \; if\ C_j\ is\ a\ negative\ criteria\ with\ a_j^- = min_i\, a_{ij} \tag{3}$$

In our study, only the cost is a negative criterion.

We have to take into account the weights $W = [w_1, w_2, \dots, w_{10}]$ of each criterion, previously given by AHP questionnaire. Multiplying each column with respectively w_j of each criterion will give the weighted fuzzy normalized decision matrix $\tilde{V}$. The new step consists in computing the Fuzzy Positive Ideal Solution (FPIS) and the fuzzy Negative Ideal solution (FNIS) from matrix $\tilde{V}$. Equation (4) and (5) show how to determine FPIS and FNIS.

$$A^* = (\tilde{v}_1^+, \tilde{v}_2^+, \dots, \tilde{v}_{10}^+)\ with\ \tilde{v}_j^+ = \left\{\left(max_i\, \tilde{v}_{ij}\, | i = 1,2, \dots, 6\right), j = 1,2, \dots, 10\right\} \tag{4}$$

$$A^- = (\tilde{v}_1^-, \tilde{v}_2^-, \dots, \tilde{v}_{10}^-)\ with\ \tilde{v}_j^- = \left\{\left(min_i\, \tilde{v}_{ij}\, | i = 1,2, \dots, 6\right), j = 1,2, \dots, 10\right\} \tag{5}$$

Step 5: Several methods could be used to compute the distances between each alternative and the FPIS and FNIS. It is possible to compute distances directly with fuzzy numbers of $\tilde{V}$ and $A^*$and A^- or to previously make a defuzzification and then perform classical Euclidean distance computing. We have chosen the latter solution by the use of the graded mean integration representation (GMIR) as Ding [17] to solve the defuzzification problem of a triangular fuzzy number (6). GMIR is applied on all the terms of matrix $\tilde{V}$ and on A^* and A^-.

$$Let\ A_i = (a_i, b_i, c_i)\ be\ a\ triangular\ fuzzy\ number\ then\ the\ GMIR\ gives\ R(A_i) = \frac{a_i+4b_i+c_i}{6} \quad (6)$$

Equations (7) and (8) compute the distance between each alternative and FPIS and FNIS.

$$E^+ = [d_1^+, d_2^+, d_3^+, d_4^+, d_5^+, d_6^+]\ with\ d_i^+ = \sqrt{\sum_{j=1}^{10}(v_{ij} - v_j^+)^2}\ , i = 1,2,..6; j = 1,2,..,10 \quad (7)$$

$$E^- = [d_1^-, d_2^-, d_3^-, d_4^-, d_5^-, d_6^-]\ with\ d_i^- = \sqrt{\sum_{j=1}^{10}(v_{ij} - v_j^-)^2}\ , i = 1,2,..6; j = 1,2,..,10 \quad (8)$$

Step 6: Finally, computation of the different CC_i (9) allows us to rank the alternative or to choose the best AT.

$$CC_i = \frac{d_i^-}{d_i^+ + d_i^-} \quad (9)$$

4 FIRST RESULTS

Figure 4 shows the first results with a limited number of questionnaires. Thus, there do not have statistical representativeness. Nevertheless, it may be noted that Therapeutic devices and Mobile devices applications seem to dominate the other alternatives. It may be also noted that the differences between the rates is not very important and no conclusion can be drawn before being capable of collecting a sufficient amount of questionnaires.

	Cci	RANK	%
Computer softwares	0,55	4	16,04
Therapeutic devices	0,76	1	22,38
Social network–based applications	0,54	5	15,90
Proprietary systems	0,39	6	11,62
Distance learning-based applications	0,55	3	16,12
Mobile devices applications	0,61	2	17,95

Figure 4: First results with a limited number of respondents

5 CONCLUSION AND PERSPECTIVE

Contribution of this paper is to mainly show that it is possible to rank assistive technology with a classical MCDM method. We need to analyze further the detection of some inconsistencies in the questionnaires and some reflection has to be carried out about the fuzzy representation (shape and with) of each linguistic variable. The interest of this approach is to involve the main stakeholders in the research and medico-social sector. However, this interest also represents a major disincentive due to the difficulties to mobilize all these persons, particularly staff of medico-social centers whose time is precious. This is the reason why at this time, there is a lack of various questionnaires, making the results little bit representative. In coordination with the directors of structures grouping tens of medico-social centers, it is planned to carry out actions aimed at identifying technical aids used in the different centers. We will take advantage of these actions to mobilize and submit our questionnaires. When we will have a sufficient amount of questionnaires, coming from all the involved persons, it will be possible to observe if all the respondent groups share the same point of view.

6 REFERENCES

[1] http://www.caps-54-51.com/presentation.htm

[2] http://www.handicat.com/iso.php

[3] **Chen, C.T. 2000.** Extensions to the TOPSIS for group decision-making under fuzzy environment, Fuzzy Sets and Systems 114, pp 1 9.

[4] **Liang, G.S. 1 .** Fuzzy MCDM based on ideal and anti-ideal concepts, European Journal of Operational Research 112, pp 682 691.

[5] **Wang Y. . 200 .** Lee H.S. (2007). Generalizing TOPSIS for fuzzy multiple-criteria group decisionmaking,Computers and Mathematics with Applications 53, pp 1762 1772.

[6] **Hourcade, .P. 2013.** Williams, S.R. Miller, E.A. Huebner, K.E. Liang L.J. Evaluation of Tablet Apps to Encourage Social Interaction in Children with Autism Spectrum Disorders. Proceedings of the SIGCHI Conf. on Human Factors in Computing Systems, pp 3197-3206.

[7] **De Leo, G. 2008.** Leroy,G. Smartphones to Facilitate Communication and Improve Social Skills of Children with Severe Autism Spectrum Disorder, Proceedings of the 7th International Conference on Interaction Design and Children (IDC '08); Chicago, IL, USA, pp 45-48.

[8] **Lohse, K.R. 2013.** Lambert-Shirzad, N. Verster, A. Machiel Van der Loos, H.F. Video Games and Rehabilitation: Using Design Principles to Enhance Engagement in Physical Therapy, Journal of neurologic physical therapy: JNPT 37(4), pp 166-175.

[9] **Kawaguchi, Y. 2010.** Shibata,T. Wada, K. The effects of robot therapy in the elderly facilities, Alzheimer Dementia, Volume 6, Issue 4, Supplement, p 133.

[10] **Robins, B. 2005.** Dautenhahn, K. Te Boekhorst, R. Billard, A. Robotic assistants in therapy and education of children with autism: can a small humanoid robot help encourage social interaction skills, Universal Access in the Information Society, Volume 4, Issue 2, pp 105-120.

[11] **Saaty, T.L. 1 80.** The Analytic Hierarchy Process, McGraw-Hill, New ork.

[12] **Hwang, C.L. 1 81.** oon, K. Multiple Attribute Decision Making: Methods and Applications. New ork: Springer-Verlag.

[13] **Lima unior, F.R. 2014.** Osiro, L. Carpinetti, L.C.R. A comparison between Fuzzy AHP and Fuzzy TOPSIS methods to supplier selection, Applied Soft Computing Journal, pp 194-209.

[14] **Ertuğrul, İ. 2008.** Karakaşoğlu, N. Comparison of fuzzy AHP and fuzzy TOPSIS methods for facility location selection, Int J Adv Manuf Technol 39, pp 783 795.

[15] **Farrokhi,V. 2014.** Kacem, I. Pokoradi, L. Ranking the solution techniques for reactive scheduling problem in operating room, Proceedings - International Conference on Control, Decision and Information Technologies, pp 1-6.

[16] **Goepel, K. 2013.** Implementing the analytic hierarchy process as a standard method for multi-criteria decision making in corporate enterprises A new AHP excel template with multiple inputs, Conference: International symposium on the analytic hierarchy process, At Kuala Lumpur, Malaysia.

[17] **Ding, .-F. 2011.** An integrated fuzzy topsis method for ranking alternatives and its applications, Journal of Marine Science and Technology, Vol. 19, No. 4, pp. 341-352.

A PTAS FOR THE MAXIMUM LATENESS OPEN SHOP PROBLEM ON A FIXED NUMBER OF MACHINES

Imed Kacem[1*] and Christophe Rapine[2]
[1]LCOMS EA 7306
Université de Lorraine, Metz, France
imed.kacem@univ-lorraine.fr

[2]LGIPM
Université de Lorraine, Metz, France
christophe.rapine@univ-lorraine.fr

ABSTRACT

In this paper we consider the open shop scheduling problem where the jobs have delivery times. The minimization criterion is the maximum lateness of the jobs. This problem is knwon to be NP-hard, even restricted to only 2 machines. We establish that any list scheduling algorithm has a performance ratio of 2. For a fixed number of machines, we design a polynomial time approximation scheme (PTAS) which represents the best possible result due to the strong NP-hardness of the problem.

Keywords: Scheduling, Open Shop, Maximum Lateness, Approximation, PTAS

1 INTRODUCTION

1.1 Problem description

We consider the open shop problem with delivery times. We have a set $\mathcal{J} = \{1, 2, ..., n\}$ of n jobs to be performed on a set of m machines $\mathcal{M}_1, \mathcal{M}_2, \mathcal{M}_3, ..., \mathcal{M}_m$. Each job j consists of exactly m operations $O_{i,j}$ $(i \in \{1, 2, ..., m\})$ and has a delivery time q_j, that we assume non negative. For every job j and every index i, operation $O_{i,j}$ should be performed on machine $\mathcal{M}_i$. The processing time of each operation $O_{i,j}$ is denoted by $p_{i,j}$. At any time, a job can be processed by at most one machine. Moreover, any machine can process only one job at a time. Preemption of operations is not allowed. We denote by $C_{i,j}$ the completion time of operation $O_{i,j}$. For every job j, its completion time C_j is defined as the completion time of its last operation. The lateness L_j of job j is

* Corresponding Author

equal to $C_j + q_j$. The objective is to find a feasible schedule that minimizes the maximum lateness $L_{\max}$, where

$$L_{\max} = \max_{1 \le j \le n} \{L_j\} \qquad (1)$$

For any feasible schedule π, we denote the resulting maximum lateness by $L_{\max}(\pi)$. Moreover, $L_{\max}^*$ denotes the maximum lateness of an optimal solution π*, that is, $L_{\max}^* = L_{\max}(\pi^*)$. According to the tertiary notation, the problem is denoted as $O||L_{\max}$.

Recall that a constant approximation algorithm of performance ratio $\rho \ge 1$ (or a ρ-approximation) is a polynomial time algorithm that provides a schedule with maximum lateness no greater than $\rho L_{\max}^*$ for every instance. A polynomial time approximation scheme (PTAS) is a family of $(1+\varepsilon)$-approximation algorithms of a polynomial time complexity for any fixed $\varepsilon > 0$. If this time complexity is polynomial in $1/\varepsilon$ and in the input size, then we have a fully polynomial time approximation scheme (FPTAS).

1.2 Related approximation results

According to the best of our knowledge, the design of approximation algorithms has not yet been addressed for problem $O||L_{\max}$. However, some inapproximability results have been established in the literature. For a fxed number of machines, unless P=NP, problem $Om||L_{\max}$ cannot admit an FPTAS since it is NP-hard in the strong sense on two machines [7], [8]. The existence of a PTAS for a fixed m is an open question, that we answer positively in this paper. If the number m of machimes is part of the inputs, Williamson et al [10] proved that no polynomial time approximation algorithm with a performance guarantee lower than 5/4 can exist, unless P=NP, which excludes the existence of a PTAS. Several interesting results exist for some related problems, mainly to minimize the makespan:

- Lawler et al [7]-[8] presented a polynomial algorithm for problem $O2|pmtn|L_{\max}$. In contrast, when preemption is not allowed, they proved that problem $O2||L_{\max}$ is strongly NP-hard, as mentioned above.
- Gonzales and Sahni [4] proved that problem $Om||C_{\max}$ is polynomial for $m = 2$ and becomes NP-hard when $m \ge 3$.
- Sevastianov and Woeginger [9] established the existence of a PTAS for problem $Om||C_{\max}$ when m is fixed.
- Kononov and Sviridenko [6] proposed a PTAS for problem $Oq(Pm)|r_{ij}|C_{\max}$ when q and m are fixed.
- Approximation algorithms have been recently proposed for other variants such as the two-machine routing open shop problem. A sample of them includes Chernykh el al [2] and Averbakh et al [1].

1.3 Contribution

Unless P=NP, problem $Om||L_{\max}$ cannot admit an FPTAS since it is NP-hard in the strong sense on two machines. Hence, the best possible approximation algorithm is a PTAS. In this paper, we prove the existence of such an algorithm for a fixed number of machines, and thus gives a positive answer to this open problem. Moreover, we provide the analysis of some simple constant approximation algorithms when the number of machines is a part of the inputs.

2 APPROXIMATION RATIO OF LIST SCHEDULING ALGORITHMS

List scheduling algorithms are popular methods in scheduling theory. Recall that a list scheduling algorithm relies on a greedy allocation of the operations to the resources that prevents any machine to be inactive while an operation is available to be performed. If several operations are concurrently available, ties are broken using a priority list. We call a list schedule the solution produced by a list scheduling algorithm. We establish that any list scheduling algorithm has a performance guarantee of 2, whatever its priority rule.

Proposition 1 Any list scheduling algorithm is a 2-approximation algorithm for problem $O||L_{\max}$, and this bound is tight.

3 PTAS

In this section, we present the first PTAS for problem $Om||L_{\max}$, that is, when the number of machines is fixed. Our algorithm considers three classes of jobs as introduced by Sevastianov and Woeginger [9] and used by several authors for a variety of makespan minimization in shops (see for instance the extension by Jansen et al. for the job shop [5]). Notice that our approximation algorithm does not require to solve any linear program. It is a folklore that minimizing the maximum lateness on one resource can be achieved by sequencing the tasks in non-increasing order of their delivery times. This sequence is known as Jackson's order. Although Jackson's list does not perform better that any other list in the worst case, see Proposition 1, we use it extensively in our PTAS.

3.1 Description of the Algorithm

Let ε be a fixed positive number. We describe how to design an algorithm, polynomial in the size of the inputs, with a performance ratio of $(1 + \varepsilon)$ for problem $Om||L_{\max}$. As a shorthand, let $\varepsilon' = \varepsilon / 2m(m+1)$. Recall that $P = max_{i=1}^{m}\{\sum_{j=1}^{n} p_{ij}\}$ is the maximal workload of a machine. For a given integer k, we introduce the following subsets of jobs $\mathcal{B}$, $\mathcal{S}$ and $\mathcal{T}$:

$$\mathcal{B} = \{j \in J | max_{i=1}^{m} p_{ij} \geq \varepsilon'^{k} P\} \quad (2)$$

$$\mathcal{S} = \{ j \in J | \varepsilon'^{k} P > max_{i=1}^{m} p_{ij} \geq \varepsilon'^{k+1} P \} \qquad (3)$$

$$\mathcal{T} = \{ j \in J | \varepsilon'^{k+1} P > max_{i=1}^{m} p_{ij} \} \qquad (4)$$

By construction, for any integer k, sets $\mathcal{B}$, $\mathcal{S}$ and $\mathcal{T}$ define a partition of the jobs. For the ease of understanding, the jobs of $\mathcal{B}$ will be often called the big jobs, the jobs of $\mathcal{S}$ the small jobs, and the jobs of $\mathcal{T}$ the tiny jobs. Notice that the duration of any operation of a small job is less than $\varepsilon'^{k} P$, and less than $\varepsilon'^{k+1} P$ for a tiny job. The choice of k relies on the following proposition:

Proposition 2 There exists an integer $k \leq \lceil m/\varepsilon \rceil$ such that

$$p(\mathcal{S}) \leq \varepsilon P \qquad (5)$$

where $p(\mathcal{S}) = \sum_{j \in S} \sum_{i=1}^{m} p_{ij}$ is the total amount of work to perform the jobs of $\mathcal{S}$. Moreover, for the big jobs, we have:

$$|\mathcal{B}| \leq m / \varepsilon'^{k} P \qquad (6)$$

Notice that, for a fixed value m of machines, only a constant number $\lceil m/\varepsilon \rceil$ of values must be considered for k. Hence, an integer k verifying the conditions of Proposition 2 can be found in linear time. Assume from now that k has been chosen according to Proposition 2. In order to present our approach, let us explain how the different sets $\mathcal{B}$, $\mathcal{S}$ and $\mathcal{T}$ of jobs are scheduled in our PTAS. Since set $\mathcal{S}$ represents a very small work, we can schedule it first. Clearly, its last operation cannot complete after time $t(\mathcal{S}) \leq \varepsilon P$ in a list schedule. Since set $\mathcal{B}$ has a fixed number of jobs, we can afford to consider all the ways to sequence them. For that, we discretize the time, considering a time step $\delta = \varepsilon'^{k+1} P$. Finally, for each assignment of the big jobs, we schedule the tiny jobs using simply Jackson's list scheduling algorithm. One originality of our approach is the possibility for a tiny job to push a big job in order to fit before it. More precisely, if the tiny job the list scheduling algorithm is considering cannot complete before the start of the next big job on its machine, say b, then we force its schedule by shifting right the operation of job b as much as necessary. This shifting is special in two-folds: first, we also shift right of the same amount of time all the operations of the big jobs starting after job b. Second, the operation of job b is then *frozen*, that is, it cannot be pushed again by a tiny job. Hence, an operation of a big job can be pushed at most once by a tiny job, but can be shifted right a lot of times, due to the push of other operations of some big jobs. A more formal description of our algorithm can be given as follows:

ALGORITHM PTAS

1. Schedule first jobs of $\mathcal{S}$ using any list scheduling algorithm between time 0 to time $p(\mathcal{S})$ (the cost factor of this simplification will not be more than $1 + \varepsilon$).
2. Let $\delta = \varepsilon'^{k+1}P$. Consider all the time intervals between $p(\mathcal{S})$ and mP of length δ (the number of these intervals is a constant for a fixed ε).
3. Enumerate all the schedules of jobs in $\mathcal{B}$ between $p(\mathcal{S})$ and mP. Here, a schedule is reduced to an assignment of the operations to starting times of the time intervals defined in the previous step (the cost factor of this simplification will not be more than $1 + \varepsilon$).
4. Complete every partial schedule generated in the last step by adding the jobs of $\mathcal{T}$. The operations of $\mathcal{T}$ are added by applying a list scheduling algorithm using Jackson's order (i.e., when several operations are available to be performed we start by the one of the largest delivery time). Note that if an operation cannot fit in front of a big job b, then we translate b and all the next big jobs by the same necessary duration to make the schedule feasible. The operation of job b is then *frozen*, and cannot be shifted any more.
5. Return the best feasible schedule found by the algorithm.

Theorem 3 The algorithm is a PTAS for $Om||L_{\max}$.

To establish the guarantee $(1 + \varepsilon)$, we show that among all the possible assignments enumerated in Phase (3) for the big jobs, there exists one feasible assignment such that, for every job x, there exists a job y with a higher priority in the Jackson's order $(q_y \geq q_x)$ such that x does not complete later than the completion time of y in an optimal schedule, plus $2(|\mathcal{B}|+1)m\delta$.

4 CONCLUSION

In this paper we considered an open question related to the existence of PTAS to the m-machine open shop problem where m is fixed and the jobs have different delivery times. We answered successfully to this important question. This represents the best possible result we can expect due to the strong NP-hardness of the studied problem.
Our perspectives will be focused on the study of other extensions. Especially, the problem with release dates seems to be very challenging.

5 REFERENCES

[1] **Averbakh I., Berman O., Chernykh I.** 2005. A 6/5-approximation algorithm for the two-machine routing open shop problem on a 2-node network. *European Journal of Operational Research*, 166(1), pp 3-24.

[2] **Chernykh I., Kononov A., Sevastianov S.** 2013. Efficient approximation algorithms for the routing open shop problem. *Computers & Operations Research*, 40(3), pp 841-847.

[3] **Hall L.A., Shmoys D.B.** 1992. Jackson's rule for single machine scheduling: making a good heuristic better. *Mathematics of Operations Research,* 17, pp 22-35.

[4] **Gonzales T., Sahni S.** 1976. Open shop scheduling to minimize finish time. *Journal of Association Computing Machinery*, 23, pp 665-679.

[5] **Jansen K., Solis-Oba R., Sviridenko M.** 2003. Makespan minimization in job-shops: A linear time approximation scheme. *SIAM Journal on Discrete Mathematics*, 16(2), pp 288-300.

[6] **Kononov A., Sviridenko M.** 2002. A linear time approximation scheme for makespan minimization in an open shop with release dates. *Operations Research Letters*, 30, pp 276-280.

[7] **Lawler E.L., Lenstra J.K., Rinnooy Kan A.H.G.** 1981. Minimizing Maximum Lateness in a Two-Machine Open Shop. *Mathematics of Operations Research* 6, pp153-158.

[8] **Lawler E.L., Lenstra J.K., Rinnooy Kan A.H.G. 1982.** Erratum. *Mathematics of Operations Research* 7(4), pp 635-635. http://dx.doi.org/10.1287/moor.7.4.635

[9] **Sevastianov S.V., Woeginger G.J.** 1998. Makespan minimization in open shops: A polynomial time approximation scheme. Mathematical Programming 82, pp 191-198.

[10] **Williamson D.P. , Hall L.A., Hoogeveen J.A., Hurkens C.A.J., Lenstra J.K., Sevastianov S.V., Shmoys D.B.** 1997. Short shop schedules. *Operations Research*, 45, pp 288-294

A 2-STEPS PROCEDURE FOR SOLVING THE INVENTORY BALANCING PROBLEM WITH TIME CONSTRAINTS

I. Kacem[1*], A. Kadri[2*] and P. Laroche[3*]
(1, 2, 3) LCOMS, Laboratoire de Conception, Optimisation et Modélisation des Systèmes
Université de Lorraine, Metz, France
(1)imed.kacem@univ-lorraine.fr
(2)ahmed-abdelmoumene.kadri@univ-lorraine.fr
(3)pierre.laroche@univ-lorraine.fr

ABSTRACT

This study addresses a 2-steps procedure applied to the inventory regulation problem in bicycle sharing-systems. The objective is to balance a network consisting of set of stations by using a single vehicle, with the aim of minimizing the waiting times during which the stations remain imbalanced. The balancing operations can be complex, especially when the network size is large. To solve the problem, we propose an approach based on different steps. First, the network is split into multiple zones according to fixed heights and widths. Secondly, the balancing problem is solved for each zone. Finally, the order in which the zones must be visited is defined. To solve these problems, we propose two heuristic algorithms based on the genetic algorithm and the nearest neighbour search method. The proposed procedure is simple and efficient, and presents some advantages with respect to the balancing considering a single area, in particular for large networks. Indeed, the numerical results show that the proposed regulation methodology has a lower computational cost and provides improved results compared to those when balancing a single area.

Keywords: Bicycle sharing-systems, Combinatorial optimization, Vehicles routing problem, Genetic algorithms, Nearest neighbour search, Clustering.

1 INTRODUCTION:

Public Bike Sharing Systems (BSS) have been recently implemented in many big cities around the world. They are one of the solutions to face some public transport problems, including traffic congestion, air pollution, increasing of oil prices and global warming. Despite the apparent success of these systems, their exploitation and management imply crucial operational challenges while few scientific works are available to support such complex dynamical systems. The balancing of stations is the most crucial question for their operational efficiency and economic viability. Thus, the system requires constant control to balance the network. The monitoring system dispatches motorized redistribution vehicles to balance bicycles between stations which are emptying and those which are filling up.

Several works in literature address the balancing problem and particularly in the static mode. Static means that the balancing operations are done only when the usage rate' of the system is very low. This problem is the one faced by operators during the night when users' demands are negligible and that the network has to be prepared to satisfy the morning's demand. This problem has great similarities to some other routing problems which have been largely studied in the literature as the pickup and delivery problem (Forma et al. [2]).

Other works are based on combined methods as presented in (Kadri et al.[4], [5]), where an integrated stochastic Petri net and genetic algorithm approach is proposed. The approach is suitable for modeling and optimization of BSS in the static and the dynamic modes. In (Kadri el al. [6]), we proposed a branch-and-bound method for solving the vehicle routing problem in BSS considering the static case. Different lower and upper bounds are proposed and used in the developed branch-and-bound algorithm.

In this work, we extend our previous work (Kadri el al. [6]) aiming the minimization of the total waiting times of stations in an unbalanced state by considering multiple zones. In the remainder of the paper we use the abbreviation SZ-VRP to denote our problem. This paper is organized as follows: in Section 2, we describe the problem and we provide the associated mathematical formulation. In Section 3, we describe the proposed solving methods including the genetic algorithm, and the nearest neighbour search algorithm. In Section 4, we present and discuss the experimental results. Finally, we conclude the paper with a summary and perspectives.

2 PROBLEM STATEMENT AND MODEL FORMULATION

The balancing problem in a BSS in the static case consists in calculating a sequence that the balancing vehicle must follow in order to minimize one or multiple objectives under some constraints. A vehicle starts a tour at a central depot and visits the unbalanced stations in a specific order. The aim is to minimize the sum of the waiting times where stations remain unbalanced until the arrival of the balancing vehicle. As shown by Figure 1, an imbalance is measured as the gap between the current state (E_i) of the station and its fixed thresholds (R_i^-, R_i^+).

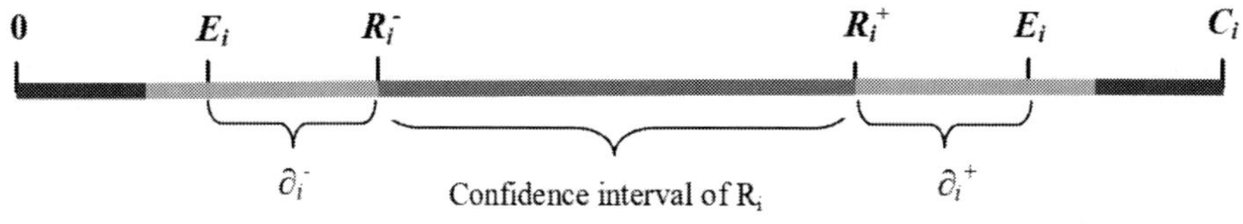

Figure 1: Balancing levels

This problem is strongly NP-hard (Kadri el al. [6]), and its complexity increases relatively to the network size. In real situations, the balancing operations for large networks are performed by decomposing a network into multiple zones (clusters). Many works in literature deal with the travelling salesman problem (TSP) by using clustering techniques. Ding et al. [1] addressed the large scale TSP, and proposed a two-level genetic algorithm for the clustered TSP.
With this motivation, we propose an approach based on the decomposition of the SZ-VRP, using the following process (see Figure 2):

i. **Zoning:** determine a set of zones with a fixed width and length, and calculate the coordinates of the local depot in the zone. The coordinates of the local depot represent the centre of gravity calculated from the coordinates of nodes in a zone.

ii. **Routing:** determine the sequence of the balancing vehicle in each zone starting from the zone depot (i.e., calculate the intra-zone paths), then calculate the sequence linking the zones (i.e., calculate the inter-zones path).

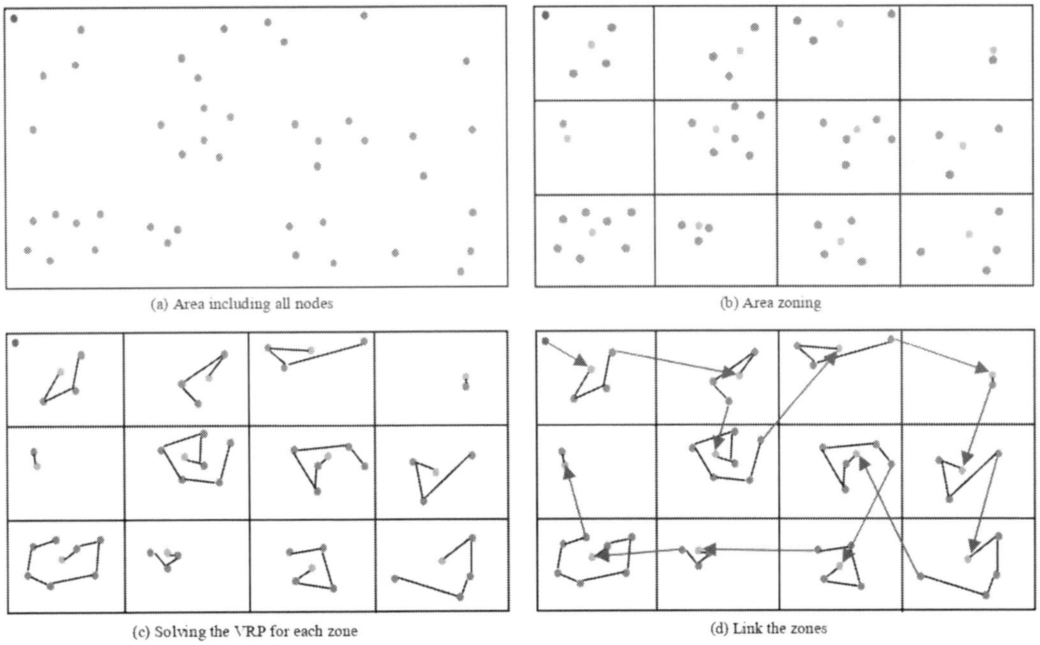

Figure 2: Solution process

2.1 Model formulation

Let us consider a bicycle sharing system consisting of a set of bicycle stations $N = \{S_1, S_2, .. , S_n\}$. A current state E_i indicates the number of available bicycles at the station S_i before the balancing operation, and C_i denotes its capacity, where $0 \leq E_i \leq C_i$. The target value R_i also known as the regulation threshold indicates how many bicycles each station should ideally hold to satisfy the provided users' demands. As shown by Figure 1, we consider a confidence target interval by defining the minimal and maximal levels denoted respectively R_i^- and R_i^+, where $0 \leq R_i^- < C_i$, $0 < R_i^+ \leq C_i$ and $R_i^- < R_i^+$.

Table 1: Notation

N	A set of nodes including the network stations, indexed by $i = 1,.., n$.
E_i	Current number of bicycles at the station i before the repositioning operation.
R_i^-	Minimal level of the required number of bicycles at a station i.
R_i^+	Maximal level of the required number of bicycles at a station i.
$d_{i,j}$	Vehicles traveling distance from a station i to a station j.
∂_i^-	Represents the gap between Ei and R_i^- ($\partial_i^- = R_i^- - E_i$ **if** $E_i < R_i^-$).
∂_i^+	Represents the gap between Ei and R_i^+ ($\partial_i^+ = E_i - R_i^+$ **if** $E_i > R_i^+$).

Decision variables

$x_{i,j}$ Binary variable which equals 1 if a station i is visited before j, and 0 otherwise. Where $i = 0,..,n$; $j=1,..,n$; and $i \neq j$.

t_i Arrival time of the vehicle to station i ($i = 1,.., n$).

Based on the notations, the following mathematical model can be formulated.

$$f = \text{Minimize } \sum_{i=1}^{n} t_i \left(\partial_i^- + \partial_i^+ \right) \quad (1)$$

$$\sum_{j=1}^{n} x_{0,j} = 1 \quad (2)$$

$$t_0 = 0 \quad (3)$$

$$t_j \geq t_i + x_{i,j} d_{i,j} + (x_{i,j} - 1) \times M \qquad \forall i \neq j,\ i,j \in N \quad (4)$$

$$\sum_{i=0}^{n} x_{i,j} = 1 \qquad \forall j \in N \quad (5)$$

$$\sum_{i=0}^{n} x_{j,i} = 1 \qquad \forall j \in N \quad (6)$$

$$x_{i,j} \in \{0,1\} \quad (7)$$

The objective function in Equation 1 minimizes the weighted sum of the arrival times of the vehicle, and the imbalance of stations are considered here as weights. Equation 2 forces the vehicle to start a tour at time 0 (Constraint 3). Constraints 4 insure the minimal time required to reach a station *j* from a previous station *i*. Constraints 5 and 6 are the vehicle flow constraints i.e. a station can be visited only one time by a vehicle in a tour. Finally, constraints 7 are the binary constraints.

2.2 Solving the problem

2.2.1 Genetic Algorithm

Due to the NP-hardness of the studied problem, we propose a genetic algorithm (GA) to solve our problem. GAs have been widely used and are well-known for giving good approximations to complex problems (Kacem [3]). A solution for the SZ-VRP must identify the order in which the zones and the stations of each of these zones must be visited. Hence, a suitable presentation is done by using two chromosomes. The first includes the solution linking zones, where each of these zones is represented by another chromosome containing the order in which the stations must be visited in a given zone. Figure 3 illustrates the representation of a feasible solution.

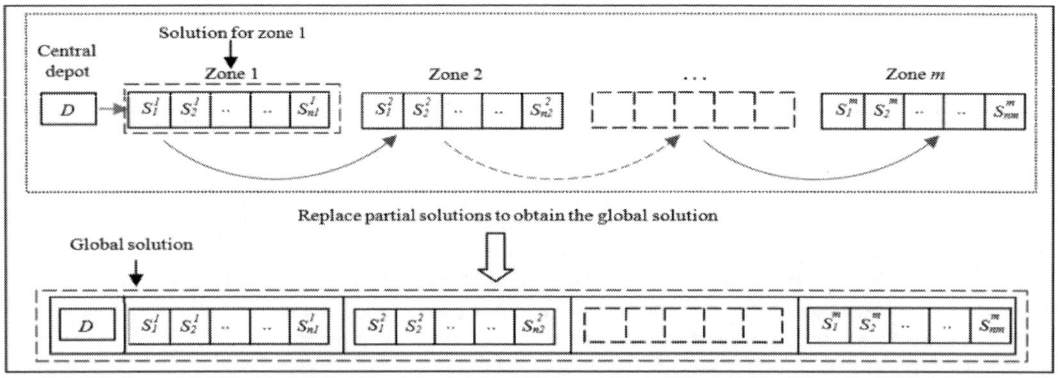

Figure 3 : Chromosome representation

We perform the 1-point modified crossover in order to obtain feasible solutions as shown by Figure 4. The mutation operator consists in swapping randomly two positions in a chromosome.

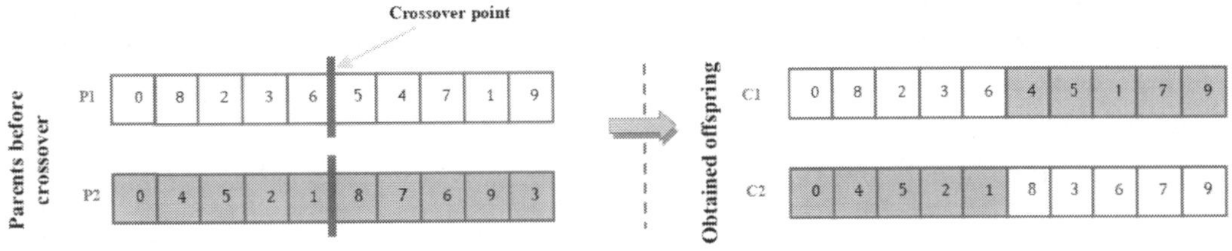

Figure 4 : Crossover operator

Table 2 : Pseudo-code of the proposed genetic algorithm (GA)

```
Begin
  /* Step 1: Calculate the solution linking the zones*/
  INITIALIZE population with random candidate solutions (POP_0)
        EVALUATE (POP_0)
  while Stop condition is not reached do
        Selection of individuals for reproduction
        if (P_cross) then Apply crossover operator
        if (P_mut) then Apply mutation operator
        for each obtained child chromosome do
                Evaluate the fitness of the chromosomes
                if (OBJ(Child) ≤ Best_Know_Solution_Zone) then
                |   Best_Know_Solution_Zone ← OBJ(Child)
                end if
        end for
        Constitute a new generation
  end while

  /* Step 2: For each zone, calculate the solution linking the stations*/
  For (1≤ k ≤m) do
      INITIALIZE population with random candidate solutions (POP_0-k)
           EVALUATE (POP_0-k)
           while (Stop_condition = false)
                Selection of individuals for reproduction
                if (P_cross) then Apply crossover operator
                if (P_mut) then Apply mutation operator
                for each obtained child chromosome do
                        Evaluate the fitness of the chromosomes
                        if (OBJ(Child) ≤ Best_Know_Solution_Stations) then
                        |   Best_Know_Solution_Stations ← OBJ(Child)
                        end if
                end for
                Constitute a new generation
           end while
  End for

  /* Step 3: Calculate the global solution */
  For (1≤ i ≤m) do
  |   Replace individual (i) by the best chromosome found for zone (i)
  |   Evaluate the fitness of the global solution
  End for
End
```

2.2.2 *The nearest neighbour algorithm*

The Nearest Neighbour Search (NNS) is an optimization algorithm where the objective is to find at each step the point that minimizes the total weighted travel time. However, such an algorithm can sometimes miss shorter routes because of its "greedy" nature. In the worst case, the algorithm gives a route that is much longer than the optimal one. For our problem, at each stage we must visit an unvisited station such that the partial $\sum t_i \left(\partial_i^- + \partial_i^+\right)$ is minimized. In the experiments, we use the NNS algorithm to calculate the local and the global solution.

3 EXPERIMENTS AND RESULTS ANALYSIS

In this section, we outline the numerical experiments performed in order to evaluate our algorithms. The developed algorithms are implemented in the C++ language and tested on an Intel Core i7 2.7 GHz processor and 8 GB of RAM, under Windows 8 environment. Due to the unavailability of real BSS data, we randomly generated our instances for every combination of parameters as follows. The network size n was chosen in {40, 60, 80, 100}, the grid coordinates were fixed to (0, 200, 0, 200), where the coordinates (x,y) of stations and the central depot are generated in the grid limits. We denote by m the number of obtained zones after the decomposition. For each combination of parameters n and m, we consider 10 groups, each one measures the average result of 10 instances. For easy viewing of results, we provide in Table 3 the average result of 10 groups. The value of the objective functions of algorithms GA (Resp. NNS) are given by OBJ_{GA} (Resp. OBJ_{NNS}).

For the decomposition, we consider different widths and lengths of zones. The weights representing the imbalances of stations were chosen in the interval [1, 10], and the travelling times of the vehicle between stations were uniformly chosen in the interval [5, 45].

For all experiments, we set the configuration of GA as follows: we perform one point crossover and we fix the rates of crossovers and (resp. mutations) to 0,9 (resp. 0.2). The number of generations is limited to 300. No specific requirements for NNS, we start all sequences for zones (resp. stations) from the central depot (resp. local depot), and we choose the zone (resp. station) which minimizes the partial cost of the objective function.

Table 3: Test results

Average result of Groups	Grid cutting				n	m	OBJ_{GA}	CPU_{GA}	OBJ_{NNS}	CPU_{NNS}
	L_{Grid}	H_{Grid}	L_{Zone}	H_{Zone}						
n=40, m=1	200	200	200	200	40	1	145864	0,7	91491	0
n=40, m=2	200	200	200	100	40	2	86894	0,2	67368	0
n=40, m=4	200	200	100	100	40	4	59224	0,1	63029	0
n=40, m=8	200	200	100	50	40	8	59103	0,1	62766	0
n=40, m=16	200	200	50	50	40	16	86676	0,1	73576	0
n=60, m=1	200	200	200	200	60	1	365520	1,7	168113	0
n=60, m=2	200	200	200	100	60	2	203075	0,4	114505	0
n=60, m=4	200	200	100	100	60	4	105646	0,2	101868	0
n=60, m=8	200	200	100	50	60	8	95564	0,1	99208	0
n=60, m=16	200	200	50	50	60	16	143506	0,2	119589	0

n=*80*, *m*=1	200	200	200	200	80	1	654724	3,2	258079	0
n=*80*, *m*=2	200	200	200	100	80	2	377008	0,7	169000	0
n=*80*, *m*=4	200	200	100	100	80	4	166762	0,3	143896	0
n=*80*, *m*=8	200	200	100	50	80	8	132213	0,2	133137	0
n=*80*, *m*=16	200	200	50	50	80	16	199155	0,2	161153	0
n=*100*, *m*=1	200	200	200	200	100	1	906561	5,3	353796	0
n=*100*, *m*=2	200	200	200	100	100	2	621653	1,1	227964	0
n=*100*, *m*=4	200	200	100	100	100	4	246219	0,5	186525	0
n=*100*, *m*=8	200	200	100	50	100	8	174474	0,3	171104	0
n=*100*, *m*=16	200	200	50	50	100	16	255623	0,3	204779	0

In our previous work (Kadri et al. [6], we solved the Lagrangian relaxation of the studied problem in order to obtain a valid lower bound. In that case, we considered a single zone and we relaxed the vehicle flow constraints. This problem is easier than the investigated one. Despite that, CPLEX (version 12.6) was not able to solve the problem in a time limited to 1 hour for problems with size of only 20 nodes.

From Figure 5, we observe that GA provides better results for small size problems/sub-problems (for an average of 10 nodes), and this is due to the small size of the search space and to the efficiency of the genetic operations (crossovers, mutations). However, for large values of *n*, NNS outperforms the GA and provide better results in negligible computation times. We observe that the gap between GA and NNS increases by the increasing of the size of problems/sub-problems, and this is due to the strong NP-hardness of the investigated problem.

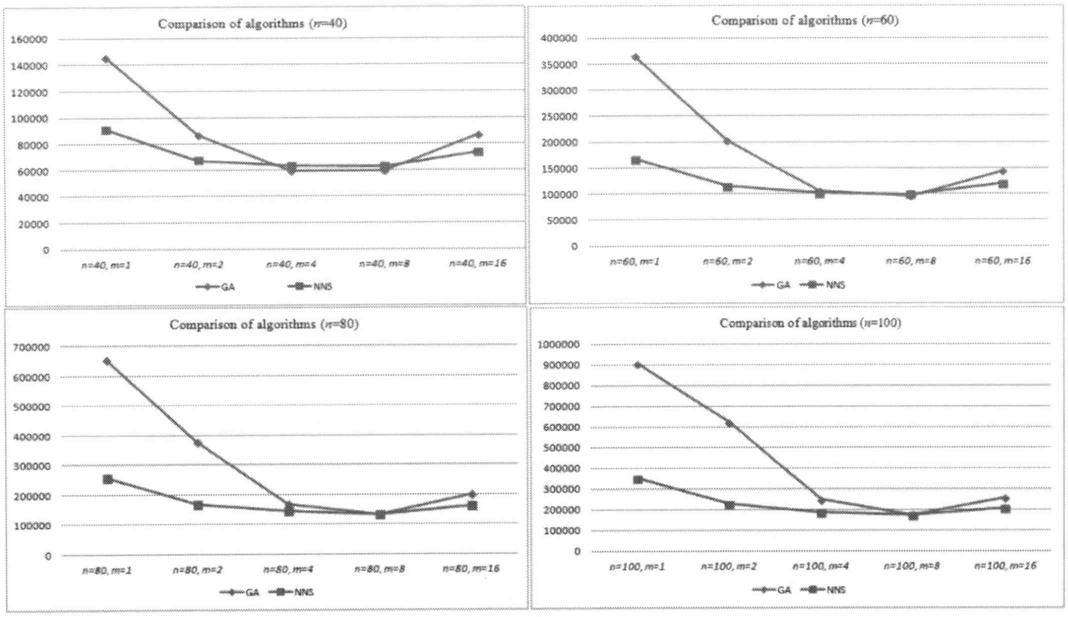

Figure 5: Comparison of algorithms for *n*=40, 60, 80 and 100

Beside this, we observe that the number of zones impacts the quality of solutions. Indeed, the decomposition of the network to multiple zones can be interesting in some cases, in particular when the number of stations per zone is large. However, for some cases, we observe that solutions are deteriorating after a number of decomposition, in particular when the number of stations per zone becomes small. Thus, the decomposition becomes useless.

As a conclusion, the proposed decomposition method and its associated algorithms seem to be an interesting method, providing solutions of good quality in short computation times despite the strong NP-hardness of the investigated problem.

5. Conclusion

This paper studied the problem of balancing of bicycle sharing systems considering the static case, which is similar to other known problems in the literature (e.g., the pick-up and delivery problem). The investigated problem is strongly NP-hard and requires the development of suitable methods able to provide solutions of good quality in short computation times.

In this work, we proposed a 2-steps approach for solving the balancing problem. First, the network is split into multiple zones according to fixed heights and widths. Secondly, the balancing problem is solved for each zone. Finally, the order in which the zones must be visited is defined.

We proposed two solving methods based on the genetic algorithms and the nearest neighbor search method. The experiments are conducted on a large set of instances in order to test the algorithms' performances as well as the impact of the decompositions on the quality of solutions. The obtained results showed that the decomposition of the network into multiple zones impacts the quality of solutions. Beside this, the proposed heuristic algorithms are able to solve different size of instances in short computation times.

In future work, we aim to develop other solving methods in order to improve the obtained results. We project to develop another decomposition methodology (clustering algorithm), which can probably improve the result obtained in this study, since the proposed method assigns nodes to zones according to their limits.

REFERENCES

[1] **Ding, C. Cheng, Ye., He, M.** 2007. Two-Level Genetic Algorithm for Clustered Travelling Salesman Problem with Application in Large-Scale TSPs, Tsinghua Science and Technology, Vol.12, No.4, 2007, PP459- 465.

[2] **Forma, I., Raviv, T., Tzur, M.** 2010. The Static Repositioning Problem in a Bicycle-Sharing System, Proc. of TRISTAN VII, 7th Triennial Symposium on Transportation Analysis, Tromsø, Norway, June 20-25, pp 279- 282.

[3] **Kacem, I.** Genetic Algorithms for Solving Flexible Job Shop Scheduling Problems. In book: Metaheuristics for Production Scheduling, Whiley, pp.19-44.

[4] **Kadri, A. A., Labadi, K., Kacem, I.** 2014. *An integrated Petri net and GA-based approach for performance optimization of bicycle sharing systems*. European Journal of Industrial Engineering, 9 (5), 638-663.

[5] **Kadri, A. A., Labadi, K., Kacem, I.** 2015. A bi-objective simulation-optimization based approach for solving the dynamic vehicles routing problem in bicycle sharing systems. The European Concurrent Engineering Conference, ECEC'2015, April 27-29, 2015, IST Campus, Lisbon, Portugal. ISBN 978-90-77381-88-5. Pages 34-41.

[6] **Kadri, A. A., Kacem, I., Labadi, K.** 2016. *A branch-and-bound algorithm for solving the static balancing in bicycle sharing systems*. Computers and Industrial Engineering 95: 41-52 (2016).

BIG DATA ANALYSIS FOR ROOT CAUSE DIAGNOSIS OF CYCLE TIME DELAY IN SEMICONDUCTOR WAFER FABRICATION SYSTEM

Junliang WANG[1] Jungang Yang[2] and Jie ZHANG[3*]

[1] School of Mechanical Engineering, Shanghai Jiao Tong University
800, Dongchuan, Minhang, Shanghai, China 200240
wjlwhut@163.com

[2] School of Mechanical Engineering, Shanghai Jiao Tong University
800, Dongchuan, Minhang, Shanghai, China 200240
yjg@sjtu.edu.cn

[3] School of Mechanical Engineering, Shanghai Jiao Tong University
800, Dongchuan, Minhang, Shanghai, China 200240
zhangjie@sjtu.edu.cn

ABSTRACT

Root cause detecting for delay in completion of wafer lot is crucial to keep high delivery reliability to enhance customer service and decrease the inventory cost. Since the cycle-time of wafer lot is affected by massive manufacturing factors, it is difficult to diagnose the accurate reason for product delay. This research aims to propose a novel big data approach for analyzing semiconductor manufacturing big data for high cycle-time diagnosis to detect process root causes for cycle-time reduction. Firstly, the candidate feature set is constructed for feature selection by analysing the material flow of wafer foundry. Secondly, a data pre-processing technique is designed to extract, transform and load (ETL) data from wafer lot transactions dataset. Additionally, a conditional mutual information based feature selection process is proposed to select key feature subset, which is highly correlated with wafer lots' CT. Finally, based on the selected key features, a real case has shown practical viability to efficiently detect possible root causes of cycle-time delay effectively.

Keywords: big data, semiconductor wafer manufacturing system, cycle-time, excursion

1 INTRODUCTION

Wafer fabrication is the most complex phase of semiconductor manufacturing, with such characteristics as the large production scale [2; 1], the diversity of machine types [3], the complex process route [4; 5] and highly dynamic environment of manufacturing system [6]. Cycle-time is a key performance indicator of the semiconductor wafer fabrication system (SWFS), which is mainly influenced by the real-time material flow characteristics in the SWFS. Affected by the dynamic workshop environment, the cycle-time of wafer lots changes greatly. The understanding of the cycle-time fluctuation is beneficial for the production control of SWFS to reduce the cycle-time of wafer lots.

Motivated by the realistic need in real settings, this study investigates the excursion diagnosis to detect the extraordinary process or equipment parameters that cause the cycle-time fluctuation. Full recognition of the material flow characteristics in a factory is the first step to estimation cycle-time [7]. However, the material flow characteristics emerge to be complicated and unpredictable,

* Corresponding Author

which is represented by the equipment and process properties, product mix and production control policies. Usually, there are more than one thousand factors correlated with the real-time material flow characteristics. Therefore, full accurate identification of the real-time material flow characteristics in the SWFS remains a major challenge. Indeed, many studies focus on developing simulate and analytical models to evaluate the system operation, and the root cause of cycle-time fluctuation can easily be captured in this model. However, as the complicated and unpredictable material flow characteristics, it is hard to construct a sophisticated model of the complex SWFS. Fortunately, deployed with a lot of sensors, a wealth of data is now available to the optimizing purposes in the SWFS. The emergence of massive data provides new opportunities for root cause diagnosis of cycle time delay. To make sense of this data, the challenge for manufacturing plants today is not only to integrate huge data into a single data repository, but also to extract useful information via data analysis to support the decisions [8].

Recently, a number of studies have developed different methods to exploit the value of manufacturing data to reduce cycle-time of wafer lots. For example, Kuo and his colleagues [9] proposed a neural networks model to exploit the value of the wealthy production data and tool data for predicting the WIP levels of the tool sets for cycle time reduction. Chen [10] fit the relationship between the controllable factors and the job cycle-time with a back propagation network. Based on this relationship, actions to shorten the job cycle time are planned. However, these method generally do not provide in-depth root cause analysis and leave the root cause detection to experts.

This study design a data driven approach to detect the root cause for cycle-time delay of wafer lots. First, a candidate feature set is constructed to collect all features may be correlative with CT. Second, an ETL approach is presented to pre-processing the raw data. Then, a conditional mutual information based feature selection is developed to select all key features. Subsequently, the root cause of cycle-time delay is detected in section 5. Next, numerical experiments are designed and the results are reported and discussed. Finally, conclusions and recommendations for future work are outlined in Section 7.

2 CHARACTERISTICS OF CANDIDATE FEATURES FOR ROOT CAUSE DIAGNOSIS

This article address the feature selection for root cause diagnosis of cycle time delay in a typical automatic semiconductor fab. Wafer lot is a job unit in wafer fab, which consists of 25 wafers with the same technical process. The configuration of the wafer fabrication analysed in this study is an 8-inch wafer fabrication in Shanghai, and the bottleneck workstation of this workshop is the lithography expose station, whose utilization is much higher than other stations. Batch processing stations are used in this fab, where a set of queued lots can be processed together. The scheduling rule used to sequence wafer lots is first come first service, which is stable enough to have little influence on the cycle time of a wafer order. The wafer producing line is an automated digital production line, which consists of 400 single-server or multi-server stations. The transportation system is an automatic material handling system, with Trams automatic feeding dollies to move the wafer lots. So the personnel factor can be ignored to the orders' cycle time. There are three categories of wafers are produced in this fab, which have totally different process routes. Every category has various kinds of wafer lots, which have somewhat difference from each other in technological parameter (etching depth, operating temperature, layer thickness, etc.). We study the wafer lot full of one type of wafers chips are produced in this fab, which involves 400 production procedures with re-enter processes.

Table 1: The candidate data used for root cause diagnosis

Type	Variables	Description

Order's Characteristics	$TP_1, \ldots, TP_{400}$	The processing times for process 1, ... , process n for a wafer lot.
	Pr	The Priority of a wafer lot.
Workshop Condition	$Q1, \ldots, Q_{400}$	The waiting queue length of machine 1, ... n.
	$U_1, \ldots, U_{400}$	The utilization of machine 1, ... n.
	N_{wip}	The total number of work in process in the fab.

The candidate features are featured by volume and variety. Each wafer lot has 1202 candidate features for selecting and there are ten million wafer lots work in process. Hence, the computing efficiency is one of the key performance indicators of feature selection methods. Moreover, the data are variety due to the different sampling frequency, multiple dimensionalities of data, which adds the complexity of correlation calculating. The highly complexity of wafer fabrication make it harder to find a model to descript the correlation with so many candidate variables.

3 DATA PRE-PROCESSING

In order to transform raw data into proper format or structure for analysis, an extract transform and load (ETL) technique is designed in this section. The raw dataset consists of 5 million wafer lot transactions extracted from the MES of an 8-inch wafer foundry during two months. The dataset is stored in the HDFS of Hadoop based big data platform and then uploaded to R studio for most of the analysis. Additionally, aiming to measure the correlation with all discrete and continuous variables uniformly, a discretization method is used to classify all data points of a continuous factor. There are many methods used in discretization such as equal width, equal frequency, and statistical analysis. To improve the homogeneity of each interval and computational efficiency of method, an equal frequency discretization (EFD) method is used to discretize those continuous candidates to ensure each interval has the same number of data points.

4 FEATURE SELECTION FOR ROOT CAUSE DIAGNOSIS

This section describes a data driven method, which selects key factors by estimating the correlation between massive candidate features and each wafer lots' CT with limited priori knowledge about the manufacturing. In correlation evaluation, regression-based method and mutual information (MI) method are mainly used in recent years. However, preselecting a regression function to fit the correlation between massive candidate factors and each wafer lots' CT is difficult, since the correlation always changes in the dynamic WMS. Therefore, a MI based method is used to estimate the probability that a candidate feature is related to the CT of a wafer lot. In this section, the feature selection can be divided into two parts: relationship measurement between CT and candidate features, and filtering based feature selection approach.

3.1 Relationship measurement between CT and candidate features

In this section, the correlation between CT and candidate features, the redundancy between different features and the complementary between features and CT-related feature set are considered in the feature selection of wafer Lot cycle time forecasting. In the feature selection of CTF, the relevance of candidate features and cycle time of wafer lots are measure by MI, which can be described as $I(f_i; CT)$. The redundancy of candidate features and feature subset is measure by $I(f_i; S)$, which means the correlation between the candidate feature and CT-related feature in the subset. The complementarity between candidate feature and feature subset is measure by conditional mutual information, which is detailed as $I(f_i;S|CT)$. Finally, the measurement of CT-related feature is design in Eq. 1. Where α and β are weight coefficient, $\alpha, \beta, (\alpha+\beta) \in [0,1]$.

$$Obj = max\{\alpha \times I(X;Y) - \beta \times I(f_i;S) + (1-\alpha-\beta)I(f_i;S|CT)\} \quad (1)$$

3.2 Filtering based feature selection approach

In this section, features with high score is selected to form a CT-related feature set. However, before the features are being filtering, the weight coefficients α and β should be assigned. In this study, the weight coefficients are determined by numerical experiment, where the assigned valuations for the α and β are drawn independently from a uniform distribution on the numbers of 0 and 1. Typically, feature selection algorithms iterate until a predefined maximum cardinality of subset. Each candidate subset is evaluated by a certain evaluation measure and compared with the previous best one with respect to this measure. The subset evaluation approach (SEA) can handle the minimum cardinality of subset. In this study, the SEA assess whether the selected subset can represent the whole candidate features. Consequently, the two terms $I(S;CT)$ and $I(F;CT)$ are directly compared in SEA and the stopping criterion is shown in Eq. 2.

$$\gamma < \frac{I(S;CT)}{I(F;CT)} \quad (2)$$

The procedure description of CMIFS is as follow

Step1: Initialize the coefficient $\alpha = 0$.

Step2: Initialize the coefficient $\beta = 0$

Step3: Calculate the value of *Obj* for all candidate features and sort features by *Obj* value from high to low.

Step 4: set the count variable i=0, clear the CT-related subset.

Step 5: select the ith feature into CT-related subset, calculate the value of $I(S;CT)$ and $I(F;CT)$. i=i+1.

Step 6: if $\gamma < \frac{I(S;CT)}{I(F;CT)}$ or $i \geq (\text{the number of all candidate features})$ then go to step 7, otherwise, go back to step 5. In this study $\gamma = 0.8$

Step 7: store the CT-related subset and the value of coefficient α and β. $\beta = \beta + 0.1$

Step 8: if $\alpha + \beta > 1$, then go back to step 3, otherwise, if $\alpha < 1$, then $\alpha = \alpha + 0.1$, go back to step 2, otherwise Optimize the coefficient α and β by select the highest $I(S;CT)$, The subset with highest $I(S;CT)$is selected to forecast CT of wafer lots.

5 EXCURSIONS AND CAUSE DETECTION

Each CT-related factor is monitored by a control chart to detect the abnormal situations. Data points would be regarded as anomalous when errors exceed the upper control limit and the lower control limit (e.g., fitting error within 5%). An analytic hierarchy process (AHP) is constructed according to the CT-related factors to identify the most possible factors that cause the cycle-time delay of wafer lot.

6 EXPERIMENT AND ANALYSIS

In this section, the proposed method is evaluated by using lot transaction dataset extracted from the manufacturing execution system of an 8-inch SWFS in Shanghai. The dataset is stored in HDFS of Hadoop based big data platform and then uploaded to R studio for further analysis. The dataset containing 800 wafer lots with the same recipe is collected in two month. The average cycle-time of these wafer lots is 72492 minutes. The wafer lot (ID: D84904) is selected out to be analysed, whose cycle-time exceeds more than 90,000 minutes.

The CT-related factors with 78 numbers were filtered from 1202 factors, which are illustrated in table 2. Among the 78 selected factors, 44 factors belong to the type of "the size of waiting queue for each machine", 29 factors are from the category: "the processing time of each operation", and the other 3 factors are enumerated as: the utilization of a machine, the priority of each wafer lot and the WIP level. In addition, two conclusions are obtained by analysing the makeup of CT-related factor subset, which is listed as follows.

Table 2: CT-related factors selected by the proposed method

	Number of Factors	Total Score	Average Score
Pr	1	4.392	4.392
CT	64	86.32486151	1.348826
Load	6	9.19575373	1.532626
Queueing	6	7.741863354	1.290311
WIP	1	1.474125517	1.474126

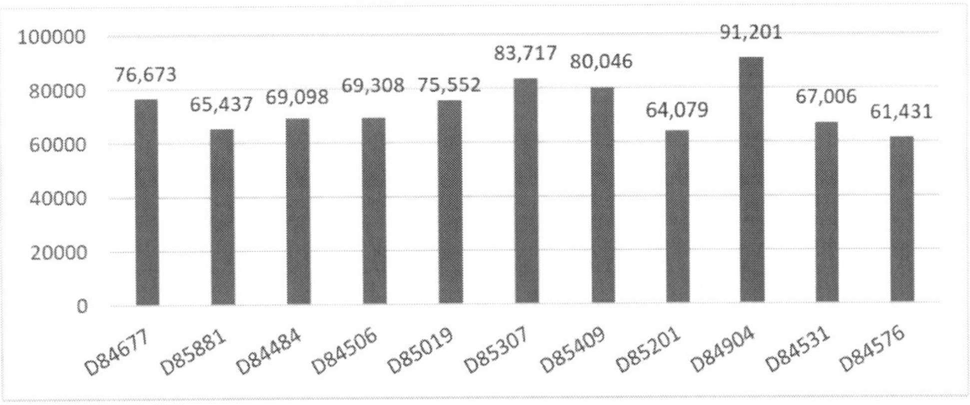

Figure 1: CT of 800 wafer lots collected in two months

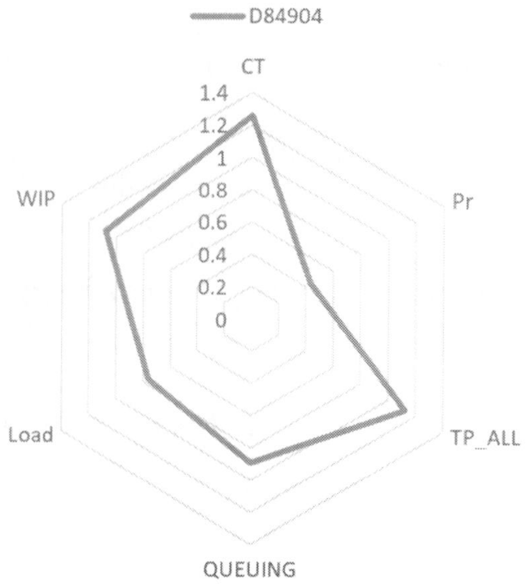

Figure 2: Factors deviation from the average level for D84904

Table 3: The processing time of the abnormal produces

	TP105	TP290	TP385	TP283	TP9	TP58
Average	44.81	46.01	44.44	39.71	161.17	59.85
D80904	68.37	74.23	63.92	105.95	519.68	399.17
	TP240	TP34	TP232	TP55	TP345	TP66
Average	44.13	42.13	39.06	37.88	53.92	30.69
D80904	66.08	555.12	73.37	147.27	85.37	68.82

The analysis of the wafer lot (ID: D84904) demonstrates that the processing time of some procedures is abnormal. Then, the processing time of some procedures contained in the CT-related factors are listed in the table 3. It is obvious that this wafer lot cost unusual processing time in the procedures, in order, TP283, TP9, TP58, TP34, and TP55. The processing time of "TP34" is even more than tenfold as the average level. The result suggests that this wafer may be reworked in several procedures, and the processing time should be controlled to reduce the cycle-time of this wafer lot.

7 CONCLUSION

This study presents and evaluates a BDA to detect root causes for cycle-time delay, which involves four parts: candidate feature set constructing, data pre-processing, MI based feature selection and excursions and cause detection. The contribution of this work is in designing a data-driven to detect root cause for cycle-time delay, which contains data gathering, data pre-processing, data analysis, and decision. Furthermore, the proposed feature selection method is unique in considering the relevance of candidate feature and CT, the redundancy and complementarity between candidate feature and CT-related subset to select the key features about the changes of cycle-time of wafer lots. In future research, we will address the cycle-time control models that consider relationship between the tuneable parameters and cycle-time of wafer lots.

8 ACKNOWLEDGEMENT

This work was supported by the State Key Program of the National Natural Science Foundation of China under Grant No. 51435009.

9 REFERENCES

[1] Zhang, J., W. Qin, L. H. Wu, AND W. B. Zhai. 2014. Fuzzy neural network-based rescheduling decision mechanism for semiconductor manufacturing. Computers in Industry 65, pp 1115-1125.
[2] Chang, P., J. Hieh, AND T. W. Liao. 2005. Evolving fuzzy rules for due-date assignment problem in semiconductor manufacturing factory. Journal of Intelligent Manufacturing 16, pp 549-557.
[3] Qin, W., J. Zhang, AND Y. Sun. 2013. Dynamic dispatching for interbay material handling by using modified Hungarian algorithm and fuzzy-logic-based control. The International Journal of Advanced Manufacturing Technology 67, pp 295-309.
[4] Lu, S. C., D. Ramaswamy, AND P. R. Kumar. 1994. Efficient scheduling policies to reduce mean and variance of cycle-time in semiconductor manufacturing plants. Semiconductor Manufacturing, IEEE Transactions on 7, pp 374-388.
[5] Chen, T. 2012. Intelligent scheduling approaches for a wafer fabrication factory. Journal of Intelligent Manufacturing 23, pp 897-911.
[6] Tai, Y. T., W. L. Pearn, AND J. H. Lee. 2012. Cycle time estimation for semiconductor final testing processes with Weibull-distributed waiting time. International Journal of Production Research 50, pp 581-592.
[7] Chung, S., AND H. Huang. 2002. Cycle time estimation for wafer fab with engineering lots. IIE Transactions 34, pp 105-118.
[8] Chien, C., W. Wang, AND J. Cheng. 2007. Data mining for yield enhancement in semiconductor manufacturing and an empirical study. Expert Systems with Applications 33, pp 192-198.
[9] Kuo, C., C. Chien, AND J. Chen. 2011. Manufacturing intelligence to exploit the value of production and tool data to reduce cycle time. IEEE Transactions on Automation Science and Engineering 8, pp 103-111.
[10]Chen, T. 2013. A systematic cycle time reduction procedure for enhancing the competitiveness and sustainability of a semiconductor manufacturer. Sustainability 5, pp 4637-4652.

OPTIMIZING THE SCHEDULING OF MAMMOGRAM PREVENTIVE SCREENING

Khaled S. Abdallah*
[1]Department of Logistics and Supply chains Management
College of International Transport and Logistics
Arab Academy for Science, Technology, and Maritime Transport, Cairo, Egypt
khaled.seif@aast.edu

ABSTRACT

Breast cancer is a major disease that affects 1 in every 8 females in the US. Mammogram screening is the standard method used for diagnosing breast cancer, and the earlier the discovery of the cancerous cells, the better the chances of healing. Despite that there are several risk factors that affect the chances of developing breast cancer, the American Cancer Society recommends a generally static breast cancer screening plan that does not vary per patient or per risk factor. In this research, we propose different dynamic screening plans that change the frequency of screening based on the risk of developing breast cancer. We use simulation to analyze the performance of the different plans. The proposed plans work well, and we recommend different approaches based on the insurance available fund, and the ability of the patient to afford extra screenings.

Keywords: Breast cancer screening scheduling, Simulation, ProModel

1 INTRODUCTION:

Breast cancer is a serious disease that affects 1 in every 8 females in the US with similar ratios for developed countries and more for developing countries. For women in the U.S., breast cancer death rates are higher than any other cancer, besides lung cancer, and is the most commonly diagnosed cancer beside skin cancer.

There are different risk factors that differ from one female to another including genetics, race, family history, and age. Due to these differences, each individual female has her own risk of developing breast cancer. For example, those females with a first-degree family member with breast cancer has double the risk of developing breast cancer themselves (McPherson [22]).

Despite the fact that there are numerous risk factors from one female to another as discussed in American Cancer Society (ACS [4]). ACS recommends a generally static breast cancer screening plan that does not vary per patient or per risk factor. The current one size fits all screening schedule may not work for patients who do not fall into the 'average' risk category and contributes more to the annual deaths due to false and misleading mammogram reports (Marron and Nerney [20])

Due to the large amount of data available in today's world, there is a need for personalized recommendations for breast cancer screening schedules. The purpose of this research is to evaluate the current screening schedule versus some proposed dynamic breast cancer screening schedules in terms of risks and rewards for each schedule to optimize the welfare of the female population.

* Corresponding Author

2 LITERATURE REVIEW

Breast cancer is a major disease that is spread worldwide affecting women more than men. Extensive interdisciplinary research is available in the literature with focus on diagnosis (Lim and Wiegmans [19]), treatment (Martin et al. [21]), devices (Puysseleyr [24]), statistical inference (Jemal et al. [15]) and planning & modeling (Danch-Wierzchowska et al. [10])

According to Anders et al. [5], 7% of the women with breast cancer in the US are diagnosed under the age of 40, about 1.4 million females, out of which about 1200 females die annually.

Certain qualities result in an increased risk of breast cancer such as white females over 45, African-American females under 45 (Kharazmi et al. [17]), some gene mutations (ACS [4]), family history (Pharoah et al., [23]), tobacco consumption, and alcohol consumption (Collaborative Group on Hormonal Factors in Breast Cancer [9]).

Operations research has long been used in healthcare in different fields of studies such as facility planning, resource allocation, and evaluation of treatment (Jun et al. [16] and Fone et al. [12]). One of the OR tools that is commonly used in healthcare modeling is simulation because of its power, flexibility, and ability to deal with variation (Davies and Davies [11]).

Some researchers note that the current standard screening schedule practice does not accommodate special arrangement for women at risk and left open to the practitioner. Ko et al. [18] recommends using risk assessment tools to introduce a personalized screening schedule to reduce the risk of breast cancer. Brailsford et al. [8] introduce a simulation model incorporating human behaviours such as psychological behaviours to simulate screening for breast cancer in the UK.

Other research related to the problem is scheduling the appointments for medical screening to reduce costs and improve utilization of medical doctors and machines as in Baker and Atherill [6] and Abo-Hamad and Arisha [3].

3 PROPOSED MODEL

3.1 Model Input parameters

There are several parameters needed for simulating the model including age, mortality rate, ethnicity, and breast cancer development rate for different ages and ethnicity. We focus our study on the US as it is has the highest breast cancer rate in the developed countries. Using information from the 2010 US census [25], the percentage of women in different age groups are given in table 1.

Table 1. 2010 Census, US population breakdown for females by age

Age Group	20-24	25-34	35-44	45-54	55-64	65-74	75-84	85+
Percent	9.1	17.6	17.7	19.7	16.2	10.0	6.5	3.2

Another factor to consider is the loss of population due to natural death. We use the mortality tables from Center of Disease Control and prevention, CDC [2], mortality table, shown in table 2, we assume that the percentages does not vary much between males and females.

Table 2: Death rates by 10 years age groups

Age	<30	30-40	40-50	50-60	60-70	70-80	80+
Rate	7.409	0.796	1.817	3.788	8.681	21.630	55.879

Table 3 shows the number of cases of females per 100,000 females that develop breast cancer as reported by the SEER Cancer Statistic Review 2010 per different age groups (Howlader et al. [14]).

There factors that increase the risk of breast cancer such as the presence of some genetic mutations such as BRCA1 or BRCA2 and the prevalence of family history. The presence of the BRCA1/2 gene is a recent risk factory discovery for breast cancer, about 1 in 500 women inherits one of these gene mutations from her parents. Per the CDC, half of the women with either gene mutation will develop breast cancer, many before 50 years old. When it comes to family history, if a woman has a first-degree female relative with breast cancer, the risk of breast cancer doubles but contrary to common belief, only about 13% of women have a first-degree female relative with breast cancer [1].

Table 3. Prevalence of breast cancer per 100,000 people per age group.

Age Group	20-30	30-40	40-45	45-50	50-55	55-65	65+
Rate per 100,000 females	10.2	87.4	121.7	188.6	224.3	608.8	423.5

This research evaluates the performance of the current breast cancer screening guidelines and contrasts it to some proposed dynamic schedules. For each case, the simulation model reports the number of cancer discoveries and the number screening occurred.

The standard static screening is dependent on the primary doctor for women under 45, one mammogram a year for women 45-55, and 1 mammogram every 2 years for women older than 55.Our proposed dynamic schedule is based on a priori of classifying each female as low and high risk groups. In our model, a woman is identified as high risk for developing breast cancer if 1) She bears BRCA 1/2, 2) she has a family history, or 3)she is older than 55. Table 4 shows five proposed dynamic schedules beside the standard plan, called plan 6. We use a constant equal to 1 if an annual screening is scheduled, and a uniform distribution when it is recommended for a women to perform one mammogram in a range of years (e.g., U(2,4) for 2 to 4 years). It implies no preference of a certain year, and fits the tendency that women might not be preferring to undergo screening. Moreover, we use a commitment indicator which we define as women chance of committing to a recommended schedule. So, with a commitment rate of 70%, a woman has a chance of 70% of adhering to the next schedule. There is no enough data available for this indicator, so we use an aggregated average for all the women.

Table 4. The standard plan and the proposed plans for breast cancer mammogram screening

Case No:	1		2		3		4		5		6. Standard	
Age Group / Risk	H	L	H	L	H	L	H	L	H	L	H	L
Young	1	U(2,4)	1	U(2,4)	1	U(2,4)	U(1,2)	U(2,4)	U(1,2)	U(2,4)	U(2,4)	U(2,4)
Middle	1	2	1	U(2,4)	1	U(2,4)	U(1,2)	2	U(1,2)	2	1	1
Old	U(1,3)		U(1,3)		1		U(1,2)		U(1,2)		U(1,3)	

The incidence of breast cancer per year for females have had a linear trend growth until 2002, then has evened out at an average incidence rate of 130 cases per year. The decline from the linear trend is thought to be a result of the decrease in using hormone therapy after menopause after the results of the Women's Health Initiative were published in 2002 that linked using hormone therapy to an increased risk of breast cancer and heart diseases (Heiss et al.[13]).

Table 5. Performance Measures for 1,838,372 Screening Mammography Examinations from 2004 to 2008 by Age (BCSC,2014).

	True Positive	True Negative	False Positive	False Negative
Total	84.40%	90.80%	9.200%	15.600%
Age 18-39	76.50%	83.80%	16.200%	23.500%
Age 40-44	73.40%	87.70%	12.300%	26.600%

Mammogram screening may result in true positive/negative when there is no error in the results of the screening, or false positive/negative when there is an error. False negative test results mean that the mammogram did not show any cancer present but the patient actually has cancer, and vice versa for the false positive. An example of the standard success rate for mammogram screening results published in the literature showing the results as either positive or negative along with whether it turned out to be true or false for some age groups is shown table 5 (BCSC [7]).

These false results have varying effects on a woman's mental stability and health with about 10% of those false positive cases will actually have cancer once a second mammogram is performed. Usually a second mammogram is requested when there is no physical appearance on evidence of breast cancer.

3.2 Model Logic

The model is applied to 100,000 females generated initially and simulated for their entire life. The model is replicated 10 times and average values are used. After they arrive, each female is assigned BRCA, family history, age normal death age, and cancer age according to the distributions defined above. Note that the distributions are truncated and adjusted if a partial distribution is required. When a female reaches her cancer age, she develops cancer. Likewise, if she reaches her normal death age, she exits the system. For the sake of our model, a woman is considered young, if she is younger than 45 years old, middle if she is between 45 and 55, or old otherwise.

The next step is assigning screening schedule which changes based on the age group and the screening plan used. If still alive and the patient is committed to the screening based on the average commitment rate used, the patients are then screened and assigned an inter-screening period based on their case and the plan under consideration (i.e., the original or proposed plan). If she is not committed, the same schedule is deferred for one loop and the same steps apply provided that the patient lives for the next year.

The results of the screening is either positive or negative then true or false based on the percentages discussed above. Depending on the results, the patient either exits the system to be treated for cancer if the result is true positive, exits the system due to death from undetected cancer if the result is false negative (after a second screening, amounts for 10% of the cases), exits the system due to natural death, or is scheduled for the next screening if the result is false positive or true negative. The simulation iterates once every year, thus, the patients go through the screening section at most once per year. Figure 1 shows a complete flowchart of the simulation logic.

4 NUMERICAL RESULTS

The simulation model is developed on a simulation program called Promodel. Several sample sizes and number of replications were tested without a significant difference from the 100,000 patients and 10 replications used in this research. With each iteration, the patient increases in age by one year and patients are switched from one group to another group with time. For the standard model, screening for young patients varies based on the patient's provider once a patient is over 40. As a result, it is assumed that a young patient under the age of 40 will not receive any mammograms. For the suggested plans, the screening schedule per year depends on each individual patient's risk factor. The risk factor is assumed to depend on the presence of a BRCA gene, family history, and age. BRCA is assumed to be generated with birth and not developed over time. Other risk factors are not included as the author was not able to find enough data for the simulation. Also, despite that breast cancer is depended on ethnicity, detailed info is again not available. Despite these assumptions, the model provides good insight to the operation and the results shouldn't differ much from what is found in this research as the above relaxed factors are 1) already included in the aggregated values obtained, and 2) small fractions. It is assumed that the cancer age is the breast cancer onset date.

As a verification for the model, several runs are performed using several input values and the results were not significantly different from the real life models values.

Figure 3 shows the number of mammograms conducted and the number of discovered cases for different planned at different commitment rate. The figure suggests that plans 3 and 4 result in significant larger number of mammograms than the other plans at any commitment level. The figure also suggests that the number of mammograms increases with increasing commitment, but the rate of increase is different from a plan to another. Obviously, plans 3 and 4 have larger rate of increase with the increase in commitment. Plans 1,2 and 5 seem to have close number to the standard plan (i.e., plan 6) but consistently require fewer mammograms. It should be noted that these numbers are for the life-span of women under study.

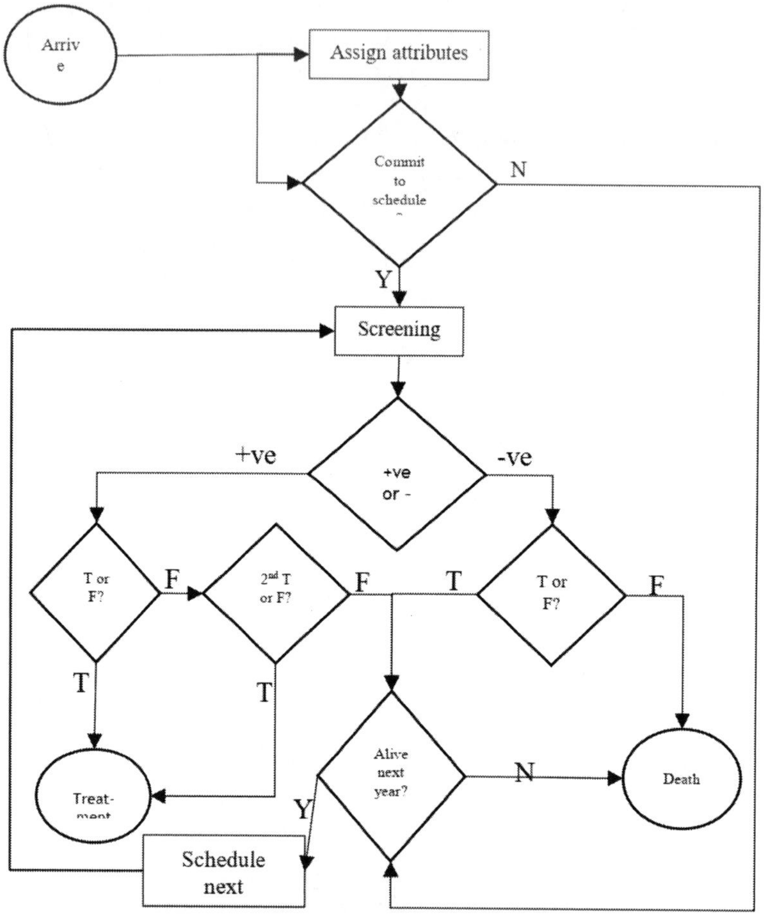

Figure 2. Logic Flow for the simulation

To contrast the number of mammograms against the number of cases discovered, figure 4 is constructed as a scatter diagram of the number of discoveries against the number of mammograms for different plans and different commitment rates. There isn't yet a way for conducting fewer mammograms and discover more cases. However, there is two distinct regions divided at around 70% commitment for plans 3 and 4 (i.e., at 130K mammograms). All the points to the right are from plans 3 and 4. However, the points to the left, despite having different outcomes, could be fine-tuned to select a plan that improves the screening process. For example, when the commitment is 50%, the number of cases discovered for plans 3 and 4 are similar to the number of cases discovered by the other plans at a commitment rate of 70%. Given that the fund insurance companies is eventually limited, it seems that adopting a dynamic schedule for women who are known to be committed much to the schedule would improve the discovery rate without ultimately affecting the number of mammograms required. This is especially valid for women at risk.

Other aspect insurance companies are interested in is the effectiveness of the plans, represented by the ratio of the number of discoveries to the number of mammograms as shown in figure 5. It represents the ratio of the returns, represented by cancer discovery, to the investment, represented by the number of mammograms. It shows that the effectiveness of plans 1,2,5, and 6 are slightly better than planes 3 and 4. Closer look at the upper part of the graph, zoomed in the right of the figure, shows plan 5 as the best plan for commitment rate between 50% and 70%, then the standard plan (plan 6) becomes better. When plane 3 is compared to 4 (those with significantly larger discoveries and larger number of mammograms), plan 4 is consistently better, specially at a commitment ratio of 70%. It is also noted that the plans 1, 2, 5, and 6 have their efficiency increase as women average commitment rate increases. However, it is the reverse for plans 3 and 4.

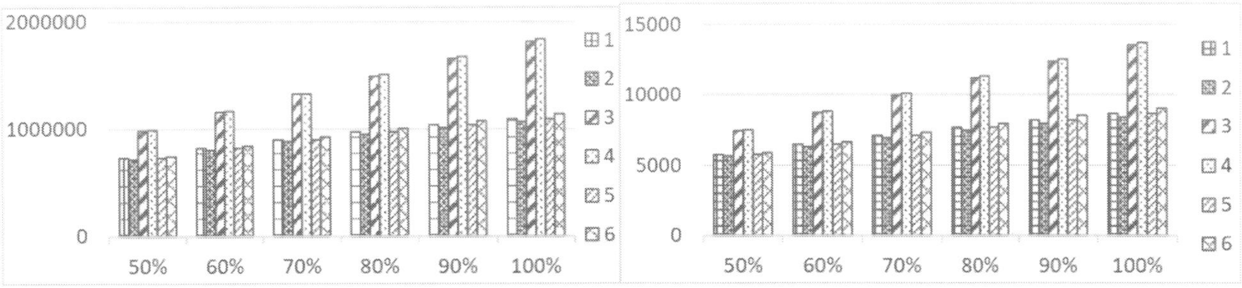

Figure 3. Number of mammograms (left) and discovered cases (right) for different plans under different commitment rates

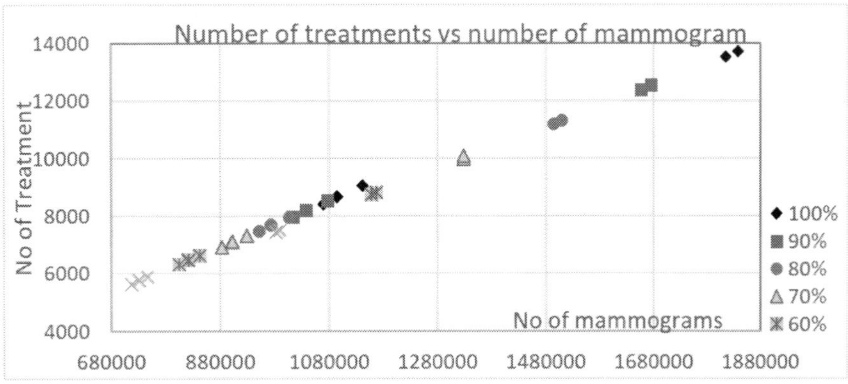

Figure 4. Number of treatments vs number of mammogram

5 DISCUSSION AND CONCLUSIONS

Whenever affordable, women with high risk should follow plan 3 or 4, but it is extremely expensive for the insurance companies to cover such extensive plans as yet the risk factors are not conclusive. So, women at risk who can afford it, should extend their screening beyond the insurance plan as recommended in plans 3 and 4.

Following plan 3 or 4 does not affect the number of mammograms required much when divided by the lifetime of women when the average commitment rate to screening is between 50-75%. But it provides the best discoveries within a limited budget insurance companies can afford, so a more viable alternative, than the current practice, that this research recommends is to follow a dynamic plan as follows: if the average patient commitment rate to mammograms screening is from 50 to 70-80%, then plan 3 or plan 4 should be recommended to them. Otherwise, the current practice is followed due to the exponential increase in the number of screenings required.

When it comes to efficiency represented as a percentage of discoveries to the number of mammogram screenings, plans 1, 2, and 5 along with the standard plan have better efficiency than plans 3 and 4, and the efficiency increases as women average commitment rate increases. Plans 5 and 6 provide the best efficiencies, with plan 5 being better when the commitment is less than 75%, then the standard plan becomes better. So for limited budgets that cannot afford plans 1 and 2, we recommend following plan 5 when the average commitment rate is below 75%; otherwise the standard plan is recommended.

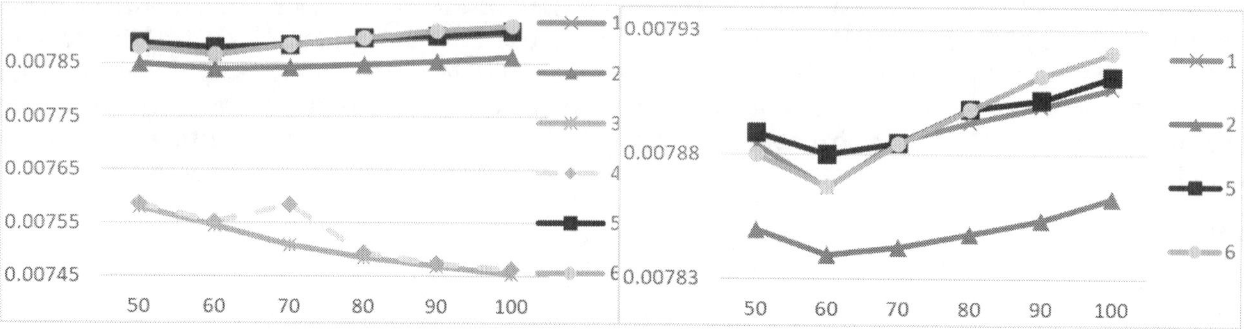

Figure 5. Effectiveness of different plans at different commitment rates with the upper part zoomed in the right.

The research assumed that the price of the mammogram screenings would not be affected by neither the number of screenings nor the advance in time. This opens a research question for future consideration as the decline in prices due to either case will render plan 3 and plan 4 feasible. Other future work includes the inclusion of human factors that affect risk and commitment into the simulation.

6 REFERENCES

[1] "BRCA1 And BRCA2 | Susan G. Komen®". Ww5.komen.org. N. p., 2016. Web. 17 July 2016.

[2] "NVSS - Mortality Tables". Cdc.gov. N. p., 2016. Web. 17 July 2016.

[3] **Abo-Hamad W., Arisha A. 2013.** Simulation-based framework to improve patient experience in an emergency department, *European Journal of Operational Research*, 224(1).

[4] **American Cancer Society. 2014.** What are the risk factors for breast cancer? <http://www.cancer.org/cancer/breastcancer/detailedguide/breast-cancer-risk-factors>

[5] **Anders CK, Johnson R, Litton J, Phillips M, Bleyer A. 2009.** Breast Cancer Before Age 40 Years, *Seminars in oncology*, 36(3).

[6] **Baker, R. D.; Atherill, P. L. 2002.** Improving appointment scheduling for medical screening, *IMA Journal of Management Mathematics*, v 13, n 4.

[7] **BCSC and National Cancer Institute. 2014.** Performance Measures for 1,838,372 Screening Mammography Examinations1 from 2004 to 2008 by Age -- based on BCSC data through 2009. <http://breastscreening.cancer.gov/statistics/performance/screening/2009/perf_age.html>

[8] **Brailsford S.C., Harper P.R., Sykes J. 2012.** Incorporating human behaviour in simulation models of screening for breast cancer, *European Journal of Operational Research*, Volume 219, Issue 3.

[9] **Collaborative Group on Hormonal Factors in Breast Cancer. 2002.** Alcohol, tobacco and breast cancer - collaborative reanalysis of individual data from 53 epidemiological studies, including 58 515 women with breast cancer and 95 067 women without the disease, *British Journal of Cancer*, 87(11).

[10] **Danch-Wierzchowska M., Borys D., Bobek-Bilewicz B., Jarzab M, Swierniak A. 2016.**

"Simplification of breast deformation modelling to support breast cancer treatment planning", *Biocybernetics and Biomedical Engineering*, Volume 36, Issue 4, Pages 531-536.

[11] **Davies R., Davies H. 1994.** Modelling patient flows and resource provision in health systems, *Omega*, 22 (1994).

[12] **Fone, D., Hollinghurst, S., Temple, M., Round, A., Lester, N., Weightman, A., Roberts, K., Coyle, E., Bevan, G., Palmer, S. 2003.** Systematic review of the use and value of computer simulation modelling in population health and health care delivery, *Journal of Public Health Medicine*, 25.

[13] **Heiss G, Wallace R, Anderson GL, et al. 2008.** Health Risks and Benefits 3 Years After Stopping Randomized Treatment With Estrogen and Progestin, *JAMA*, 299(9).

[14] **Howlader N, Noone AM, Krapcho M, Miller D, Bishop K, Altekruse SF, Kosary CL, Yu M, Ruhl J, Tatalovich Z, Mariotto A, Lewis DR, Chen HS, Feuer EJ, Cronin KA (eds).** SEER Cancer Statistics Review, 1975-2013, National Cancer Institute. Bethesda, MD, http://seer.cancer.gov/csr/1975_2013/.

[15] **Jemal, A., Bray, F., Center, M.M., Ferlay, J., Ward, E. Forman, D. 2011.** Global cancer statistics, *CA: A Cancer Journal for Clinicians*, vol. 61, no. 2.

[16] **Jun, J.B., Jacobson S.H., Swisher, J.R. 1999.** Application of discrete-event simulation in health care clinics: A survey, *Journal of the Operational Research Society*, 50.

[17] **Kharazmi E, Chen T, Narod S, Sundquist K, Hemminki K. 2014.** Effect of multiplicity, laterality, and age at onset of breast cancer on familial risk of breast cancer: a nationwide prospective cohort study, *Breast Cancer Res Treat*, 144(1).

[18] **Ko, M.G., Files, J.A., Pruthi, S. 2012.** Reducing the risk of breast cancer: A personalized approach, *Journal of Family Practice*, Vol. 61 Issue 6.

[19] **Lim, Y.C. Wiegmans A.P. 2016.** Tracking metastatic breast cancer: the future of biology in biosensors, *A.P. Med Oncol.*

[20] **Marron J., Nerney T. 2012.** 10,000 Deaths a Year Due to False and Misleading Mammogram Reports, *Institute for Health Quality and Ethics.*

[21] **Martin M, Lopez-Tarruella S, Gilarranz Y J. 2016.** Endocrine therapy for hormone treatment-naïve advanced breast cancer, *The Breast*, Volume 28.

[22] **McPherson K, Steel C M, Dixon J M. 2000.** Breast cancer—epidemiology, risk factors, and genetics, *BMJ2000.*

[23] **Pharoah P.D.P., Day N.E., Duggy S., Easton D.F., Ponder B.A.J.1997.** Family history and the risk of breast cancer: a systematic review and meta-analysis, *Int J Cancer*, 71(5).

[24] **Puysseleyr A. D., Neve W. D., Wagter C. D. 2016.** A patient immobilization device for prone breast radiotherapy: Dosimetric effects and inclusion in the treatment planning system, *Physica Medica*, Volume 32, Issue 6.

[25] **US Census Bureau. 2011.** "2010 Census Briefs:Age and Sex Composition 2010,and QuickFacts Tables".<https://www.census.gov/quickfacts/table/PST045215/00,http://www.census.gov/prod/cen2010/briefs/c2010br-03.pdf>

ACKNOWLEDMENT

I would like to thank Breanna Patrice Swan for suggesting the idea. To our mothers, wives, sisters, and daughters, we want you to stay forever healthy.

FACTORS AFFECTING THE BULLWHIP PHENOMENON WHEN REVERSE SUPPLY CHAIN FLOW EXISTS

Khaled S Abdallah[1*]
[1]Department of Logistics and Supply Chains Management
College of International Transport and Logistics
Arab Academy for Science, Technology and Maritime Transport, Cairo, Egypt
khaled.seif@aast.edu

ABSTRACT

Bullwhip effect is one of the most important factors that affects the performance of a supply chain. It results in a much larger variation going up in a supply chain resulting in consuming resources in anticipation of a demand that is not true, or a delay in production due to false sharp decrease in demand. In this research, we examine the factors affecting the bullwhip performance in a serial supply chain adopting a minimum/maximum inventory policy. We consider a reverse flow that includes returned products like what happens in peak seasons such as holidays. The bullwhip effect is large in these seasons. We build a simulation model and use design of experiment in performing a factorial experiment to examine the interactions among different factors.

Keywords: Bullwhip, Discrete event simulation, DOE, Factorial experiment.

1 INTRODUCTION:

The bullwhip effect is an amplification of the variability in demand going upward at upstream supply chain members. The bullwhip effect has a big impact in supply chain decisions. The bullwhip effect is sometimes referred to as 'demand amplification', 'variance amplification' or the 'Forrester effect'. This effect becomes significant when the cost from fluctuations in production/ordering outweighs the cost of holding inventory (Wang et al. [18]). The bullwhip effect often results in an over estimated demand at the manufacturer especially in the holidays seasons. Another characteristic of the holiday season is that there is a considerable amount of returns after the season that lasts for weeks.

Several factors affect the bullwhip, and most of the time these factors interact resulting in adverse results. This paper aims at analyzing the effect of some factors on the operation of supply chains with an inventory policy of (s,S). The factors considered are distributer and retailer up to quantities, demand rate, and return rate. The (s,S) inventory method is chosen as it is widely used in real life due to its simplicity, ease of use, and sort of good performance given its easiness of use.

In this research, we use the results of simulation using different factors affecting the operation of a supply chain in performing a factorial design analysis to examine the interaction between them and their effect on different bullwhip measurements when the (s,S) policy is used in the supply chain.

The remainder of the paper is divided as follows. In section 2, related literature review is introduced. Solution approach is discussed in section 3. We provide some observations and their analysis in section 4, then we conclude and provide insight for possible future work in section 5.

* Corresponding Author

2 LITERATURE REVIEW

Bullwhip effect was first introduced in Forrester [7], though not realized as bullwhip effect until it was first coined by Procter & Gamble (P&G) in the 1990s although similar phenomenon between P&G and its wholesalers has been documented during 1910s (Schisgall [16]). It is a rich topic in the literature with research to identify reasons and reducing it as in Geary et al.[14], Lee et al.[11] and [12]; and efforts in modeling, demand model, and measurement as categorized in Giard and Sali [9]. It is frequently studied with sort of focus on the effect of information sharing as in Sahin and Robinson [15] and sometimes with reverse logistics as in Govindan et al. [10].

Inventory strategies affect the bullwhip and the inventory costs. There is a trade-off between inventory and production costs (Baganha and Cohen [1] and Disney et al. [6]). Chen and Samroengraja [4] showed that when the cost function is concave, the replenishment policy that minimizes order fluctuations is not necessarily the one that minimizes total cost.

There are different measurements for the bullwhip effect. It can be measured by the coefficient of variation, variance, or standard deviation. Sometimes, empiricists use alternatives such as production quantity, sales and shipments as they are most of the time easier to observe (Blinder and Maccini [2]). When demand is non-stationary, bullwhip can be measured by the variance of order changes (West [20]), or the difference between order variances and demand variances (Gaalman and Disney [8]). Standard deviation is considered one of the major measures of bullwhip effect as it quantifies the true variability in order quantities at different levels of the supply chain (El-Beheiry et al. [21]).

Wang and Disney [17] categorizes the methodologies adopted in bullwhip research into: empirical, experimental, analytical and simulation-based. In empirical studies on bullwhip, historical data on demand, sales, shipment and production is collected and analysed which may lead to underlying causes; experimental research that uses laboratory experiments and management games to examine factors and mechanisms that affect bullwhip allowing theories to be tested; and mathematical modelling providing a good quantification of the bullwhip effect and its causes though often simplified when complex and, most of the times, simulation is used for more realistic bullwhip problems.

Although (s,S) ordering policy is still used in many industries due to its simplicity, it is little studied in the literature with regards to the bullwhip as in (Wang and Huang [19]) comparing the bullwhip effect under (s,S) versus vendor management inventory strategy using system dynamics. Likewise for reverse supply chain, it receives little attention in the literature. Cannella et al. [3] used a difference equation mathematical approach and design of experiment to perform a robust what-if analysis of a supply chain with reverse flow under different conditions. Noblesse et al. [13] characterize the order process of continuous review (s,S) and (r,nQ) inventory policies, and study the impact of the batching parameter on the variability in the order process. Derbel et al. [5] attempt to identify the critical parameters that affect the bullwhip when (s, S) is used using a simulation followed by factorial experiments, but they focused more on the desirability function without discussing the interaction within parameters. To the best of our knowledge, this work is among the first to attempt studying the interactions of several factors including reverse supply chains on the bullwhip performance.

3 PROBLEM DEFINITION

We consider a serial supply chain that consists of a factory, a distributer, and a retailer as shown in figure 1. The retailer sells to the consumer directly according to demand D that follows a Poisson distribution. There is no backlog and shortage results in lost sales. There is a reverse flow going back to the distributer from the consumer represented as a percentage r taking a number of days larger than the forward lead time between any chain members as consumers usually have a large window for returning products. We don't include any rework or remanufacture in the model, but we assume they, if needed, take place at the distributer, and their lead time is already included in the return lead time. So, items can be used for shipping at the distributer. When the inventory level reaches a re-order point of s_R, an order is placed for the distributer to deliver a quantity sufficient to return to maximum inventory level SR. It takes a relatively small time to deliver the products. If there is no enough products, the orders are backlogged until fulfilled. When the inventory level at the distributer reaches the re-order point of s_D, an order of SD is dispatched to the factory. The quantity is used to fill the inventory to the maximum level SD. The lead time is equal to the lead time between the retailer and the distributer. Factory has unlimited supply. The simulation logic is shown in figure 2.

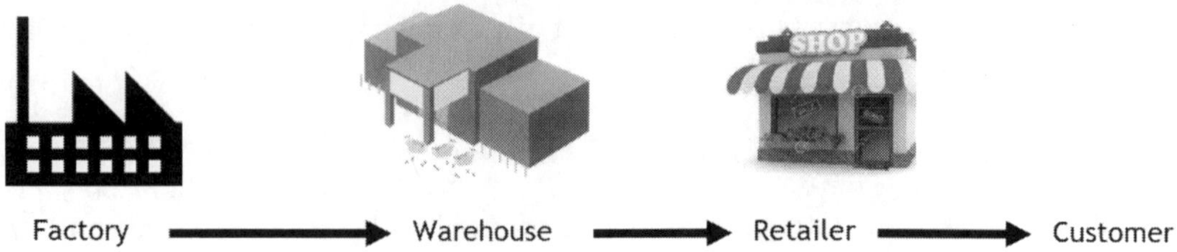

Figure 1. The supply chain under consideration

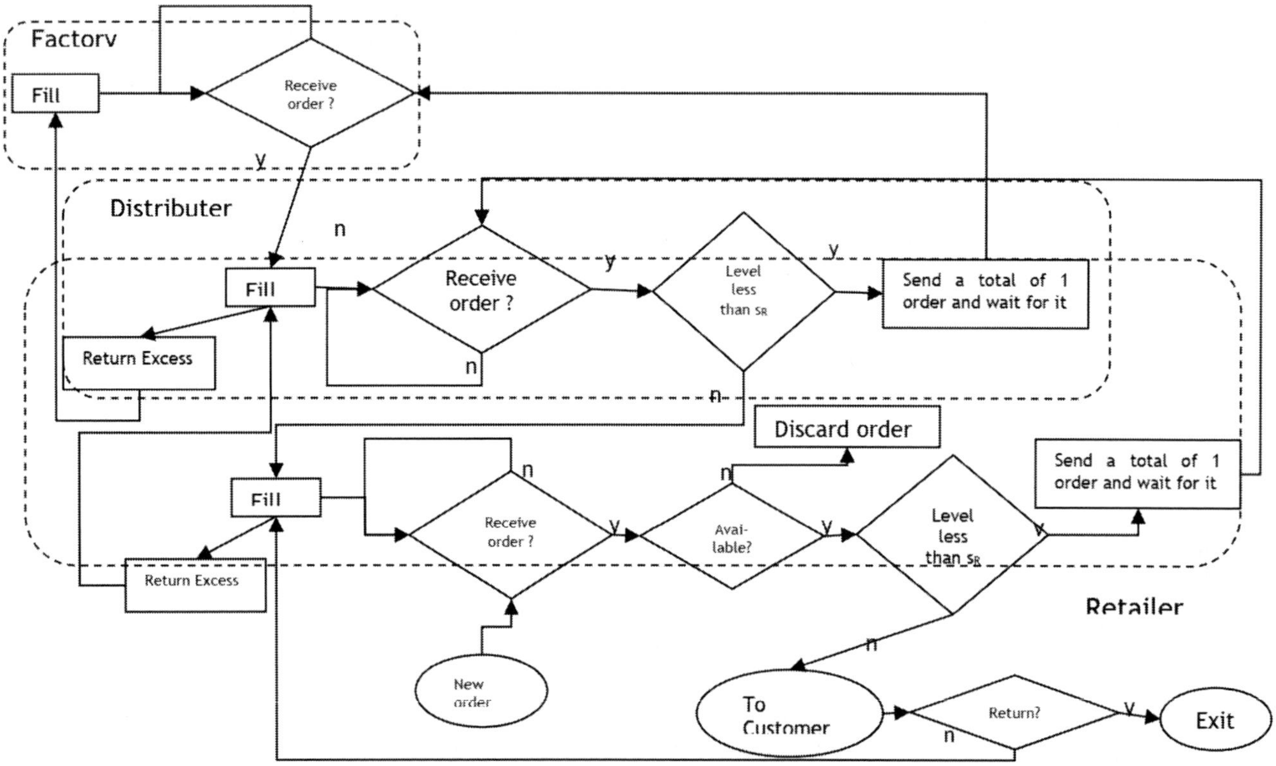

Figure 2. The flowchart of the model used.

We use ProModel 32bit on a notebook with Intel i3-4050u, 4GB RAM, and 64 bit Windows 10. We use the values from table 1. For each combination of the values in the table, 6 replications are run, each for 100 years with a similar warm up period. Same stream feed is used for distribution so that the comparison stays valid. After running the model, the standard deviation of delivery quantity and the standard of the inter order days for both of the distributer and the factory are computed(i.e., the quantity delivered to the retailer and the distributer, respectively).

Table 1. Factors used in the Model

Factor	SD	SR	r	D
Values	900, 1200, 1800, 2500, 4000	180,250, 400,600	0, 1, 2, 5, 10, 15%	2, 4, 8

The next step is to run a multi-factor factorial design experiment to test the significance of the interaction between different factors using different methods of measuring the bullwhip.

4 OBSERVATIONS

We divide this part into effect on standard deviation of the demand, and standard deviation of the inter-order duration (defined in some other researches as demand standard deviation and order standard deviation). For each case, we study the effect on factory, distributer, ratio between factory and distributer, then a ratio between the ratio between factory and distributer divided by the ratio between distributer and retailer, which, we define as bullwhip propagation factor. For all the cases, the interactions were actually significant. This means that the main effects should not be taken for granted and the interactions should be the ones that should be analysed. We are omitting the main effects plots. In the coming sections, the legend showed in Figure 3 is used in all the coming figures.

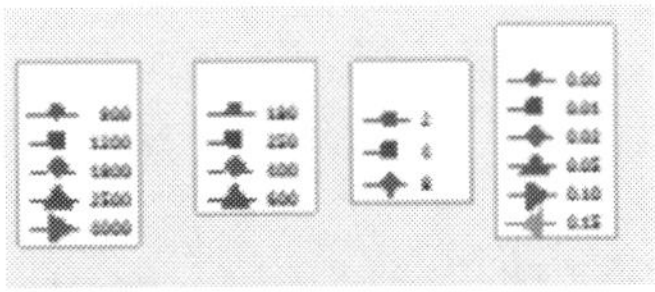

Figure 3. Legend used in all figures

4.1 Effect on the standard deviation of the demand

4.1.1 Effect on Factory

The increase of SD and SR decreases the factory standard deviation especially when SR is high. This is associated to the fact that the demand through the lead time would not result in higher standard deviation at the factory when the shipped quantity is very large (A*B Figure 4(L)).

The interaction between SD and D is significant when D is small. When D is small, increasing SD decreases the factory standard deviation. The reason is that the quantity consumed during the lead time is small, leading to a small variation to SR. This propagates leading to a smaller standard deviation at the factory specially when SD becomes large, in this case the factory standard deviation further decreases (A*C Figure 4(L)). This result is consistent with the interaction between D and SR which results in decreasing the factory standard deviation for a given demand when SR increases. (B*C Figure 4(L)). The interaction between r and D can be divided into two main effect regions with respect to r. When small, it decreases the factory standard deviation as it helps in smoothing the order quantity, then afterwards, it increases the standard deviation of the factory as too many items are returned. The exception here is when the D is small, an opposite effect takes places, this is because of small demand does not generate enough returns to even the demand.

4.1.2 Effect on Distributer

Interaction between SD and SR results in decreasing the distributer standard deviation in addition to the expected reduction from increasing SD. However, the slope decreases with the increase in SR as it flattens the variation of demand at the distributer (A*B Figure 4(R)). Increasing D results in increasing the distributer's standard deviation, but the rate is significantly higher for intermediate demand values then declines for higher demand (A*C Figure 4(R)). SD amplifies the standard deviation at intermediate demand values with the decrease of SD, and the rate of decrease declines for higher values (C*A Figure 4(R)). This could be interpreted as that when the demand rate is small, the return quantity does not contribute much in flattening the distributer's demand, then the intermediate demand is transitional, and finally the high demand results in a flattening to the demand. For any given D, increasing SR increases the distributer standard deviation (B*C Figure 4(R)), and for a given SR, increasing D increases the distributer standard deviation at a declining rate (C*B Figure 4(R)). The reason of the decline with the increase of the demand is that a stock out occurs with large demand which is considered lost sales. This extends to the factory and the order quantity stabilizes especially with the increase in SD. It seems that r is not significantly interacting with the remaining factors.

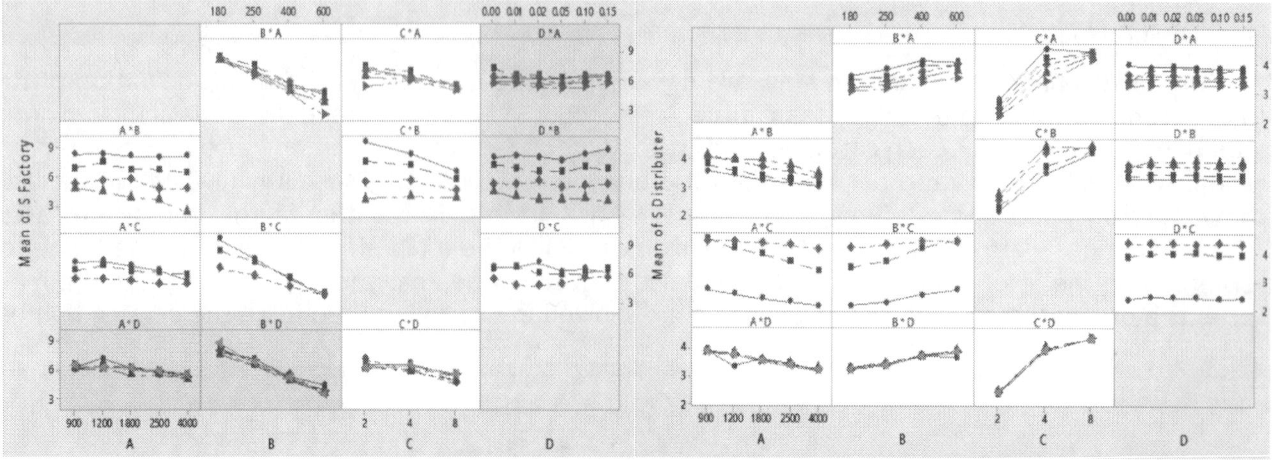

Figure 4. Factorial effects plots for S at the factory (left) and the distributer (right)

4.1.3 Effect on the ratio of the standard deviations of the factory to that of the distributer

Both of D and SR interact in decreasing the ratio at a declining rate in the ratio as the standard deviation at the factory increases at a larger rate than the increase of the standard deviation of the distributer (B*C Figure 5(L)). SR interacts with r in decreasing the ratio to certain point (different with different SR) then increasing it, except for small SR resulting in three region of increase, decrease then increase. This implies that a small reverse flow helps in reducing the factory standard deviation, then it becomes significant and affects the demand at the factory resulting in larger standard deviation again (D*B Figure 5(L)). It also interacts with D in the same manner, except that for small D, which results in a very small decrease followed by a relatively large increase.

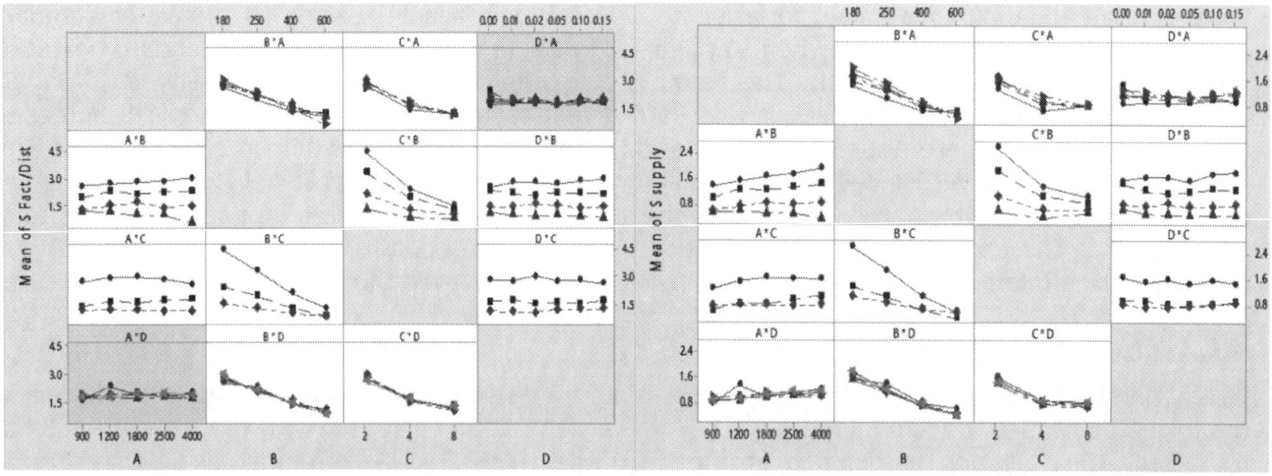

Figure 5. Factorial effects plots for the ratio of S (left) and propagation of S (right)

4.1.4 Effect on propagation of standard deviation

The increase of SD increases the ratio with the increase in SR. The rate of increase decreases as SR increases. It reaches a point when the ratio decreases when both quantities are large resulting in the variation in demand is small when compared to SR and SD (A*B Figure 5(R)). r seems to have the same pattern of decrease then increase as discussed before, but it moves the point of least ratio to the left with the increase of retail up-to quantity (D*B Figure 5(R)). The interaction between SD and D results in a decrease in the ratio by the increase of any of them. The rate of decrease decreases by the increase of D as the increase in the standard deviation at the distributer is large compared to that at the factory (B*C Figure 5(R)).

4.2 Effect on the standard deviation of inter-order time

4.2.1 Effect on Factory

The interaction between the SD and D increases the standard deviation of inter order time (SI). The increase of either parameter increases SI. It increases exponentially, but in a much slower rate when D is high. When the SD is large, small demand will take long time for a replenishment and will include large variation unlike large demand (F_S*C Figure 6(L)). SD and SR interact with r in the same way of decrease to a low point then increase. This is attributed to the effect of small returns which makes the inter-order duration more even (r*F_S, r*D_S Figure 6(L)).

4.2.2 Effect on distributer

There is only interactions between D and either of the SD or SR with the latter being more effective than the former. The increase of SR increases SI at the distributer. The rate of increase changes from exponential increases when D is small to a small increase when it is large (figure 6(R)).

4.2.3 Effect on ratio between SI at the factory to SI at the distributer

As SI at the distributer is not affected much except for interaction between the SR and D, the other factors stay as discussed for the factory except that interaction. That interaction results in a ratio that decreases with the increase of SR or the increase of D in a linear relationship(c*D_S Figure 7(L)). r still shows a similar decrease then increase trend as discussed above (R*D_S, R*F_S Figure 7(L)).

4.2.4 Effect on propagation of SI

One important observation is that the ratios have values that are often less than one. This means that the standard deviation of the inter order time at the distributer is quite often large compared to those at the factory and retailer. The interaction of the SD and SR increases the propagation with the increase of the former and decreases with the increase of the latter (F_S*D_S Figure 7(R)). Similar attitude results from the interaction of SD and D. While the interaction of SR and D results in a decrease in the ratio with an increase of any of them. r continues to show the pattern of decrease then increase.

5 CONCLUSIONS

This paper is among the first to examine the effect on bullwhip caused by the interactions of parameters including the returns of the reverse supply chains in a supply chain adopting the (s,S) inventory policy. We find that the effect on the performance is not limited to the variation of the demand quantity, but also affects the variation of the inter-order duration.

The returns, when not too large, reduces the variation of the demand and the inter-arrival time at the factory despite the difficulty of reducing both, but does not result in much change at the distributer. Afterwards, it shows a totally opposite effect.

Another interesting result is that the rate of increase in the variance of the demand at the distributer is sharp for intermediate end customer demand, possibly due to returns. But the increase in distributer up-to quantity or the decrease in the retailer up-to quantity helps in returning the increase into the regular rates. When it comes to the distributer, it is found that the distributer up-to quantity and the retailer up-to quantity show contradictory results for the variation of the demand at the distributer, so it is recommended that when one of them is large the other one should be small, especially when the customer demand is not large.

This work could be extended by considering other supply chains designs and structures. Also more factors could be added to the model.

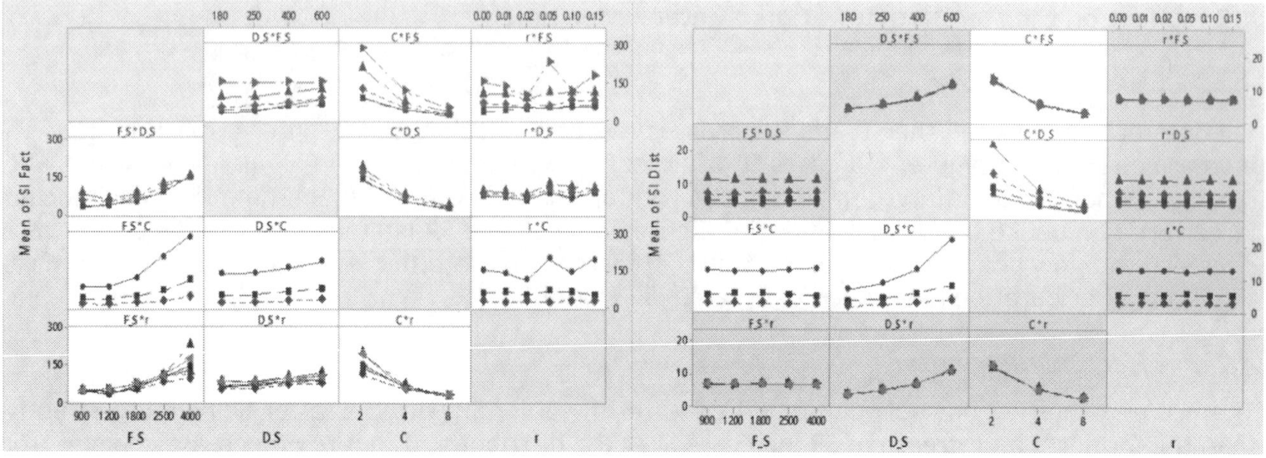

Figure 6. Factorial effects plots for SI at the factory (left) and at the distributer (right)

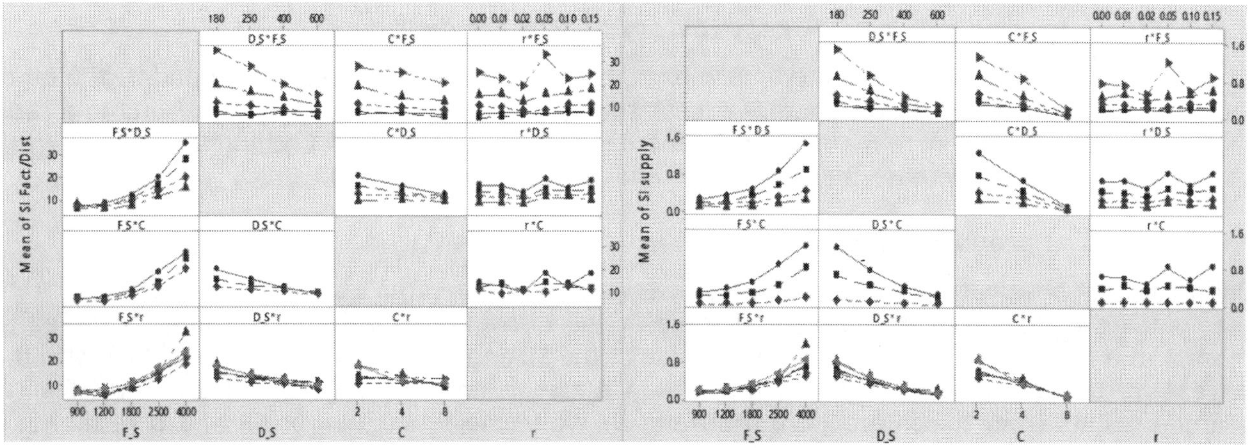

Figure 7. Factorial effects plots for ratio of SI at the factory to this at the distributer.

6 REFERENCES

[1] **Baganha M.P., Cohen M.A. 1998.** The stabilising effect of inventory in supply chains, *Operations Research*, 46 (3).

[2] **Blinder A.S., Maccini L.J. 1991.** Taking stock: a critical assessment of recent research on inventories, *Journal of Economic Perspectives*, 5 (1).

[3] **Cannella S., Ciancimino E. 2010.** On the bullwhip avoidance phase: supply chain collaboration and order smoothing, *International Journal of Production Research*, 48 (22).

[4] **Chen, F and Samroengraja, R. 2000.** The stationary beer game. *Production and Operations Management*, 9(1).

[5] **Derbel M., Chabchoub H., Hachicha W., Masmoudi F. 2013.** Measuring the impact of (s, S) ordering policy on the bullwhip effect by means of simulation optimization, *Advanced Logistics and Transport (ICALT)*, 2013 International Conference on, Sousse.

[6] **Disney S.M., Towill D.R., van de Velde W. 2004.** Variance amplification and the golden ratio in production and inventory control, *International Journal of Production Economics*, 90.

[7] **Forrester, J.W. 1961.** *Industrial Dynamics*, MIT Press, Cambridge, MA.

[8] **Gaalman G., Disney S.M. 2012.** Bullwhip behavior in the order-up-to policy with ARIMA demand, *4th World Conference on Production & Operations Management*, Amsterdam

[9] **Giard V., Sali M. 2013.** The bullwhip effect in supply chains: a study of contingent and incomplete literature, *International Journal of Production Research*, 51 (13).

[10] **Govindan K. Soleimani H. Kannan D.2014.**Reverse logistics and closed-loop supply chain: a comprehensive review to explore the future, *European Journal of Operational Research*, 240 (3).

[11] **Lee H, Padmanabhan, P and Whang, S. 1997.** Information distortion in a supply chain: the bullwhip effect. *Management Science*, 43(4).

[12] **Lee, H, Padmanabhan, P and Whang, S. 1997.** The bullwhip effect in supply chains. *Sloan Management Review*, 38(3).

[13] **Noblesse A.M., Boute R.N., Lambrecht M.R., van Houdt B. 2014.** Characterizing order processes of continuous review (s,S) and (r,nQ) policies, *European Journal of Operational Research*, 236.

[14] **Geary S., Disney S.M., Towill D.R. 2006.** On bullwhip in the supply chain - historical review, present practice and expected likely impact, *International Journal of Production Economics*, 101 (1).

[15] **Sahin R., Robinson E.P. 2002.** Flow coordination and information sharing in supply chains: review, implications, and directions for future research, *Decision Sciences*, 33 (4).

[16] **Schisgall O. 1981.** *Eyes on tomorrow: the evolution of Procter & Gamble*,J.G. Ferguson, NY.

[17] **Wang X., Disney S.M., 2016.** The bullwhip effect: Progress, trends and directions, *European Journal of Operational Research*, Volume 250, Issue 3.

[18] **Wang X., Disney S.M., Wang J. 2012.** Stability analysis of constrained inventory systems with transportation delay, European Journal of Operational Research, 223 (1).

[19] **Wang, X., Huang, X. (2011).** Supply chain bullwhip effect simulation under different, *Contemporary Logistics*, (2).

[20] **West K.D. 1986.** A variance bounds test of the linear quadratic inventory model, *Journal of Political Economy*, 94 (2).

[21] **Wong C.Y., El-Beheiry M.M., Johansen J., Hvolby H. 2007.** The implications of information sharing on bullwhip effects in a toy supply chain, *Int. J. of Risk Assessment and Management*, Vol 7 Issue 1.

AN EMPIRICAL STUDY ON THE CONTRIBUTION RATE AND ALLOCATIVE EFFICIENCY OF THE HUMAN CAPITAL OF LISTED COMPANIES

Huaming Wu[1] and Jin Wang[2*]
[1]Department of Business Management
Hefei University of Technology, Hefei, China
2221294666@126.com

[2]Department of Management Science and Engineering
Hefei University of Technology, Hefei, China
wangjinhfut@126.com

ABSTRACT

This paper utilizes the panel data of years 2011 to 2014 and the Lucas Model to analyze the influences of factors as material capital and human capital on the enterprise performance and to measure the contribution rates of the material capital and human capital to the enterprise performance, and analyzes the contribution rates of each industry to reveal the rules of the allocative efficiency of the human capital so as to provide references for decision-making for investment in the human capital.

Keywords: Enterprise performance, Contribution rate of the human capital, Risk management, Pareto Optimality, Marginal Rate of Technical Substitution

1 INTRODUCTION

The risk management is a decision-making process adopted by social organizations or individuals to reduce the negative results of risks. It utilizes risk identification, risk evaluation and risk assessment, and chooses and optimizes the combination of risk management technologies to implement effective control on risks and to properly handle losses caused by risks so as to maximize the security guarantee at lowest cost. In the decision-making for the investment in the human capital and the material capital, most enterprises currently in China tend to invest more in the material capital rather than human capital and consider the investment in the human capital as non-productive. But some far-sighted managers of enterprises have noted the great contribution of the investment in the human capital to the enterprise performance.

However, there are high risks in the investment in the human capital. The first is the characteristic risk. The development of the human capital investment has to be made according to the actual situations so as to ensure significant outcomes; otherwise, no significant outcomes may be gained. The second is the social "applicability risk". If the investment is not widely applicable, the rate of return on investment will not be high. The third is the market risk. If the investment in the human capital lasts long and cannot follow up the pace of the market, losses may occur in the investment in the human capital. The fourth is the "technological risk" caused by the rapid social development. If the investment in the human capital cannot keep pace with the increasingly rapid development of science and technology, the persons educated will not satisfy the requirements of the society. The

* Corresponding Author

fifth is the accidental risk, including two aspects. One is the health of the persons educated. They usually pay little attention to physical exercises, which leads to health-related issues and thus affects the effective use of the investment in the human capital. And the other is the environmental factor. Unexpected environmental factors may cause losses of the investment in the human capital, for example, illness and disability.

This paper selects 1419 listed companies as the sample and uses the panel data of years 2011 to 2014 and the Lucas Model to perform empirical study on the investment in the human capital. The paper analyzes the contribution rate and the allocative efficiency of the investment in the human capital to the enterprise performance in different industries from the angle of risk management so as to enable enterprises to know the importance of the investment in the human capital to the enterprise performance in terms of risk management and control.

2 LITERATURE REVIEW

Fleisher[1] studies the economic development of China and finds out that the human capital plays an important role in promoting sustainable and sound economic development of China, increasing economic development efficiency, reducing the regional economic differences and promoting balanced economic development. Klomp J[2] compares and analyzes the human capital factors of 123 countries and regions of years 2000 to 2008 and concludes multiple macro methods for measuring the human capital. Wang et al. [3] take the listed companies in the agriculture field in China of years 2007 to 2011 as samples to analyze the problems in the investment in the human capital and find out that the investment in the human capital can improve the performance of agricultural enterprises with the Student's t test and regression analysis. Wang[4], based on the resource-based view, thinks that, to adapt to the external environment, enterprises should utilize the strategic human resources to develop the strategic capabilities of employees. Chiang and Shih[5] think that the knowledge-oriented human resource management has positive influences on the innovation and performance. Enterprises that aim at differential strategies have high requirements for advancement and steadiness of techniques of employees in terms of skills. Santarelli et al.[6] discuss the influences of the human capital and social capital on the performance of new enterprises through survey on 1398 new enterprises in Viet Nam. Ganotakis et al.[7] perform study on the utilization of human resources by founders of new technological enterprises in UK.

Barnett[8] studies the stability of the Lucas Model and puts forward some guidelines for the economic stability. Gorostiaga et al.[9] study the optimal tax systems for external human resources with the Lucas Model. Bucci et al.[10] expand the growth modes of the Uzawa-Lucas Model with the randomization technology and analyze the strategy of increasing the human resources by increasing the investment in the human capital.

Currently, studies on the measurement of the contribution of the human capital and quantitative studies on the contribution value of the human capital are rarely found. Most studies focus on macro, theoretical and abstract aspects. Even studies on the measurement of the contribution rate of the human capital are empirical supplementary descriptions to the theories of the contributions of the human capital based on macro data samples. In general, studies on the contribution of the human capital of enterprises are insufficient and most of them are lack of empirical analysis. Currently, the most important thing is to measure the contribution of the human capital at the aspects of micro, application and concrete calculation so as to solve the problems in the contribution rate and the allocative efficiency of the human capital. This needs to measure and calculate the contribution rate of the human capital concretely. Quantitative analysis on the contribution rate of the human capital of enterprises is necessary.

3 STUDY DESIGN

Currently, researchers have found a positive correlation between the investment in the human capital and the enterprise performance through empirical studies. To further study the actual effect of the human capital on the enterprise performance, this paper utilizes the Lucas Model to calculate the contribution rate of the human capital of listed companies.

3.1 Model for measuring the contribution rate of the human capital

This paper uses the modified Lucas Model to determine the contribution rate of the human capital. The basic model is as follows:

$$Y = AK^{\alpha}H^{\beta}L^{\gamma}e^{\varepsilon} \quad (1)$$

In the model, Y stands for the total economic output and means the net profit for enterprises; K stands for the investment in the material capital and means the investment by enterprise owners in their enterprises, which is recorded as "owners' equity" in the balance sheet; H stands for the investment in the human capital and means the investment in the employees for enterprises, which is recorded as "cash paid to and on behalf of employees" in the cash flow statement; L stands for the level of the human capital and is marked as the total number of employees of enterprises; α is the flexibility of the total output of the material capital; β is the flexibility of the total output of the human capital; γ the flexibility of the number of employees to the total output; A stands for the technological level.

After taking the logarithm to the production model, model (2) is obtained:

$$LnY=lnA+\alpha lnK+\beta lnH+\gamma lnL+\epsilon \quad (2)$$

Whereas:

$$\text{Contribution rate of the human capital} = \beta lnH / lnY \quad (3)$$

$$\text{Contribution rate of the material capital} = \alpha lnH / lnY \quad (4)$$

3.2 Relevance and regression analysis of industries

To learn the influences of the investment in the human capital and the material capital on the financial performance in details of each industry, the regression analysis is performed respectively on the increase of the investment in the human capital, the increase of the investment in the material capital and the increase of the financial performance of each industry, and the following regression equations are obtained for major industries:

B for mining: LnY = -4.99 + 1.02lnK + 0.24lnH - 0.22lnL

(t=11.01) (t=1.25) (t=-1.28)

C for manufacturing: LnY = -2.48 + 0.71lnK + 0.43lnH - 0.20lnL

(t=24.36) (t=10.98) (t=-5.63)

E for building: LnY = -3.18 + 0.99lnK + 0.0045lnH - 0.08lnL

(t=9.33) (t=0.041) (t=-0.76)

F for wholesale and retail:

LnY = -2.82 + 0.76lnK + 0.23lnH + 0.02lnL

(t=11.75) (t=2.65) (t=0.33)

I for information transmission, software and IT service:

$LnY = -1.50 + 0.62lnK + 0.47lnH - 0.22lnL$

(t=6.28) (t=3.49) (t=-1.57)

K for real estate: $LnY = -1.88 + 0.36lnK + 0.95lnH - 0.30lnL$

(t=8.61) (t=9.16) (t=-3.61)

All industries: $LnY = -2.36 + 0.7\ 1lnK + 0.37lnH - 0.13lnL$

(t=45.04) (t=13.496) (t=-5.49)

According to the results of the regression analysis, in terms of all industries and major industries, the investment in the material capital is significantly positive-correlative to the financial performance. Meanwhile, the correlative coefficients of some industries are not significant, but the investment in the human capital is basically positive-correlative to the financial performance. In terms of the influence effects, the influence of the increase of the investment in the material capital on the financial performance is greater than that of the increase of the investment in the human capital on the financial performance; however, according to the results of the regression analysis of each industry, the influence of the investment in the human capital of the real estate industry is greater than that of the investment in the material capital, and the influence of the investment in the human capital on the enterprise performance of the IT industry is close to that of the investment in the material capital on the enterprise performance. Therefore, more and more enterprises have noted the influences of the investment in the human capital on the enterprise performance.

3.3 Calculation and analysis of the contribution rate of the human capital

To measure the contribution rate of the human capital, by leaving out extreme instances of the data of the selected listed companies in Shenzhen Stock Exchange and Shanghai Stock Exchange, 1419 listed companies are selected in total. Based on Formula (2), the flexibility coefficient β of the human capital and the flexibility coefficient α of the material capital of each listed company are calculated. Based on (3) and (4), the contribution rate H of the human capital and the contribution rate K of the material capital are obtained respectively. The number of employees is an absolute number and the number of employees has few influences on the net profit in the regression results Y, therefore, the number of employees is ignored in the residual allocation. The residual ϵ is directly allocated based on the weight ratio (r_1, r_2) of the human capital and the material capital and the error value h of the contribution rate of the human capital and the error value k of the contribution rate of the material capital are obtained; and finally the contribution rate H+h of the human capital and the contribution rate K+k of the material capital are obtained. The result is as shown in Table 1:

Table 1: Annual average contribution rate of the human capital

Year	2011	2012	2013	2014
Annual average contribution rate of the human capital	0.173057	0.270926	0.292539	0.387893

According to the above results, from 2011 to 2014, an annual increase is observed in the contribution rate of the human capital of listed companies and the contribution rate of the human capital in 2014 is higher than 35%. According to the data analysis, this, to some extent, indicates that more and more enterprises have noted the contribution of the investment in the human capital to the net profit and its importance to the operation and development of enterprises.

This paper further calculates the contribution rates of the listed companies as per industries. Some industries, such as agriculture, forestry, husbandry, fishery, electricity, heat, gas and water production and supply, irrigation, environment and public facility management, are vital to the national economy and the livelihood of the people and are usually state-owned; therefore, they do not meet the hypothesis of the Lucas Model and the contribution rates of the human capital of these industries are ignored. Meanwhile, the residual allocation exists in the regression process; therefore, the smaller the sample size, the higher the residual error. In this case, the calculation results may deviate from the actual situations and thus industries with small sample sizes are ignored. The calculation results of the contribution rates of the human capital of major industries are listed in Table 2:

Table 2: Average contribution rate of the human capital of major industries

Industry	Number of samples	2011	2012	2013	2014
B	41	0.12675	0.08977	-0.21322	0.403551
C	862	0.24907	0.309772	0.395136	0.464568
E	42	-0.08022	0.071529	0.073745	0.009489
F	111	0.090882	0.163636	0.112167	0.262928
I	53	0.31571	0.279871	0.517868	1.032993
K	101	0.302396	0.534559	0.366778	0.399679

Note: The classification of industries is based on the standard of China Securities Regulatory Commission. B: mining; C: manufacturing; E: building; F: wholesale and retail; I: information transmission, software and IT service; K: real estate.

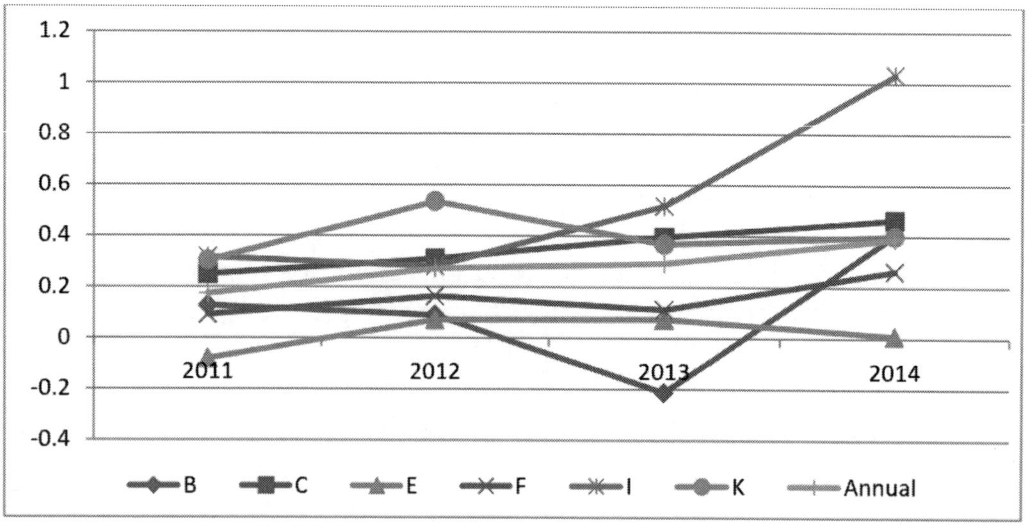

Note: The classification of industries is based on the standard of China Securities Regulatory Commission. B: mining; C: manufacturing; E: building; F: wholesale and retail; I: information transmission, software and IT service; K: real estate.

Figure 1: Average contribution rate of the human capital of major industries

According to Table 1 and Table 2, the average contribution rates of the human capital of major industries are positive values and close to the contribution rates calculated based on the total sample size except the contribution rates of the human capital of some industries in some years are negative. According to the calculation, we can conclude that the contribution rate of the human capital keeps stable or shows a tendency of increase, especially for manufacturing, wholesale and retail and real estate industries, though the contribution rates of the material capital of most industries are higher than that of the human capital. In addition, according to the chart, we can see that: for the IT service industry, the investment in the human capital shows a significant increase and the contribution rate of the human capital becomes more and more significant.

4 CRITERIA FOR VERIFYING THE OPTIMAL ALLOCATIVE EFFICIENT OF THE HUMAN CAPITAL

The condition to the Pareto Optimality of the allocation between the human capital and the material capital is that the Marginal Rate of Technical Substitution (MRTS) of the two factors must be equal, that is, $MRTS^{A}_{HK} = MRTS^{B}_{HK}$

whereas, $MRTS^{A}_{HK}$ is the RMTS of industry A when H substitutes K; $MRTS^{B}_{HK}$ is the RMTS of industry B when H substitutes K.

The Pareto Optimality between the human capital and the material capital is an ideal state. The condition to the Pareto Optimality between the human capital and the material capital mentioned above is the criteria for verifying the optimal allocation between the human capital and the material capital.

According to the analysis model of this paper: $Y = AK^{\alpha}H^{\beta}L^{\gamma}e^{\varepsilon}$, the marginal product ($MP_H$) of the human capital and the marginal product (MP_K) of the material capital can be obtained, that is:

$$MP_H = \beta AK^{\alpha}H^{\beta-1}L^{\gamma}e^{\varepsilon} \quad (5)$$

$$MP_K = \alpha AK^{\alpha-1}H^{\beta}L^{\gamma}e^{\varepsilon} \quad (6)$$

Thus, $$MRTS_{HK} = \frac{\beta AK^{\alpha}H^{\beta-1}L^{\gamma}e^{\varepsilon}}{\alpha AK^{\alpha-1}H^{\beta}L^{\gamma}e^{\varepsilon}} = \frac{\beta K}{\alpha H} \quad (7)$$

After substituting the inventory of the human capital, the inventory of the material capital and the flexibility of the material capital of the six industries into (7), the MRTS when H substitutes K of each industry can be obtained:

Table 3: MRTS of the human capital of major industries

Industry	MRTS
B for mining	0.2887
C for manufacturing	0.7840
E for building	0.0060
F for wholesale and retail	0.3938

I for information transmission, software and IT service	0.9419
K for real estate	3.8820
Sample size	0.6791

Based on the criteria to the optimal allocative efficiency of the human capital, it can be concluded that the allocation between the human capital and the material capital of the industries is not optimal and that the allocation of the human capital is imbalanced. According to the total samples selected, the MRTS of the human capital is 0.6791, indicating that the balance of the industries has not been reached. The MRTS of the building industry is significantly lower than the total MRTS, indicating that the allocative efficiency of the human capital is very low and large space exists for improvement of the allocation of the human capital. According to the equimarginal principle, the MRTS of the human capital indicates that the MRTS of the human capital of the IT industry is close to 1 and the MRTS of the human capital of the real estate industry is greater than 1, which means the investment in the human capital of the two industries needs to be limited and the allocative efficiency of the human capital needs to be considered.

5 CONCLUSION

According to the contribution rate model of the human capital investment built based on the Lucas Model, this paper mainly concludes that the contribution rate of the human capital of the IT industry is highest and that of the building industry is lowest. Although the contribution rates of the human capital of all industries are lower than that of the material capital, the contribution rates of the human capital show a tendency of increase. This explains that enterprises have paid more and more attention to the investment in the human capital, leading to an increase to the human capital. Currently, the allocation between the human capital and the material capital of the industries is not optimal and that the allocation of the human capital is imbalanced. The optimized allocation of the human capital can increase the allocative efficiency of the human capital and reduce the imbalance degree of the allocation of the human capital so as to promote a sound development of enterprises.

6 REFERENCES

[1] **Fleisher, B., Li, H.,Zhao, M. Q. 2010.** Human capital, economic growth, and regional inequality in China, *Journal of development economics*, *92*(2), pp 215-231.

[2] **Klomp, J. 2013.** The measurement of human capital: a multivariate macro-approach, *Quality & Quantity*, *47*(1), pp 121-136.

[3] **Wang, L. L., Wang, Y. W., Sun, W. M. 2013.** Study on the Relationship Between Human Capital Management and Agricultural Enterprise Performance—Based on Different Marketing Strategy, Proceedings of 20th International Conference on Industrial Engineering and Engineering Management, pp 827-835.

[4] **Wang, C. Y. P., Jaw, B. S., Tsai, C. H. C. 2012.** Building dynamic strategic capabilities: a human capital perspective, *The International Journal of Human Resource Management*, *23*(6), pp 1129-1157.

[5] **Chiang, Y. H., & Shih, H. A. 2011.** Knowledge-oriented human resource configurations, the new product development learning process, and perceived new product performance, *The International Journal of Human Resource Management*, *22*(15), pp 3202-3221.

[6] **Santarelli, E., & Tran, H. T. 2013.** The interplay of human and social capital in shaping entrepreneurial performance: the case of Vietnam, *Small Business Economics*, *40*(2), pp 435-458.

[7] **Ganotakis, P. (2012).** Founders' human capital and the performance of UK new technology based firms, *Small Business Economics*, 39(2), pp 495-515.

[8] **Barnett, W. A., Ghosh, T. 2014.** Stability analysis of Uzawa-Lucas endogenous growth model, *Economic Theory Bulletin*, 2(1), pp 33-44.

[9] **Gorostiaga, A., Hromcová, J., López-García, M. Á. 2013.** Optimal taxation in the Uzawa-Lucas model with externality in human capital, *Journal of Economics*, *108*(2), pp 111-129.

[10] **Bucci, A., Colapinto, C., Forster, M., La Torre, D. 2011.** Stochastic technology shocks in an extended Uzawa-Lucas model: closed-form solution and long-run dynamics, *Journal of Economics*, *103*(1), pp 83-99.

A BAYESIAN-BASED GROUPING MECHANISM FOR EVOLUTIONARY ALGORITHM IN SCHEDULING PROBLEM

Lu Sun[1], Lin Lin[1,2*], Mitsuo Gen[2,3] and Haojie Li[1]
[1] School of Software, Dalian University of Technology, Dalian, China
sunlu@evolab.cn , lin@dlut.edu.cn , hjli@dlut.edu.cn
[2] Fuzzy logic Systems Institute, Iizuka, 820-0067, Japan
gen@flsi.or.jp
[3] Tokyo University of Science, Tokyo, Japan
gen@rs.tus.ac.jp

ABSTRACT

This paper proposes a new grouping mechanism based on Bayesian optimization for particle swarm optimization algorithm (BOPSO). It is aimed to solve the discrete combinatorial optimization problems with high search space. Although the existing evolutionary algorithms can solve this kind of problems effectively through ingenious encoding and decoding mechanism, with the increase of the scale of data, the efficiency of EAs will be limited or even go down. So, we propose a co-evolutionary algorithm with a new grouping mechanism based on Bayesian optimization algorithm (BOA). Due to learning BN (Bayesian network) is also a NP hard problem, we use PSO to generate the BNs and sampling the training data sets to help learning network structures. We apply BOPSO to job shop scheduling problems (JSP) and the flexible JSP (fJSP) which belongs to a typical discrete combinatorial problem to verify the algorithm efficiency. Our experimental results and analysis suggest that BOPSO is a highly competitive optimization algorithm for the discrete combinatorial optimization problems, especially for JSP and fJSP.

Keywords: Bayesian optimization algorithm, Evolutionary algorithm, Flexible scheduling, Large search space

1 INTRODUCTION:

The process of optimization plays an important role in many real life engineering problem. There exists various features in the process of optimization, such as the search space is small or large (over 300 dimensions), the cost is cheap or expensive and so on. The large search space of problems lead to the exponential increase difficulty for both problem modeling and optimization. The references point out that to solve the problems with large search space contains three kinds of methods: the first one is to divide large search space into small search space; the second one is screening and the last one is mapping (Shan and Wang, 2001). However, most of the existing methods ignore learning the potential relationships among the random variables when they solve the problems with large search space. Job shop scheduling problem (JSP) and its extension flexible JSP are both the typical discrete combinatorial optimization problem. It is also a NP hard combinatorial problem under the constraint of priorities and resources (Gao et al, 2008). As the increasingly complex demand of production, to solve fJSP with large search space is particularly important (Clearwater et al, 1992).

* Lin Lin, Dalian University of Technology, lin@dlut.edu.cn

As we known, Bayesian network is one of typical probability graph model (PGM) which is a directed acyclic graph (DAG). It is not only suitable for expression and analysis of uncertainty and probabilistic events, but also can reflect the cause relationship between variables from the network structure (Larrañaga et.al,1996).

An estimation of distribution algorithm (EDA), a class of population-based optimization algorithms and proposed in 2002. A Bayesian optimization algorithm (BOA) belongs to EDAs and its estimation and sampling are based on a constructed Bayesian network (BN), which is designed to automatically estimate and construct the joint probability distribution of promising solutions in a search space. BOA is widely used in scheduling, such as, resource assignment, task scheduling and so on. In 2011, Yang et al present a novel scheduling algorithm based on the BOA for heterogeneous computing environments. Li et al (2014) proposed a Bayesian optimization algorithms to solve task assignment problems in heterogeneous computing systems. Hao et al (2011) proposed a novel cooperative Bayesian optimization algorithm (CoBOA) to overcome the challenges mentioned in the field of Multiple Resources Scheduling Problem (MRSP). However, they still have the weakness of solving the problems with large scaled search space. Inspired by Bayesian optimization, this paper proposed a new grouping mechanism, in which we apply the BOA to get the best BN structure for finding the potential relationships among variables and then for dividing the variables according to the optimal network structure.The remainder of this paper is organized as follows: Section 2 gives the formulating process of JSP and fJSP model and mathematical programming model; Section 3 describes the BOPSO combined BOA Particle swarm algorithm (PSO) with multiple subpopulations in detail; Section 4 gives kinds of detailed numerical experiments and computational results; finally, Section 5 comes to the conclusion of the paper.

2 PROPOSED ALGORITHM

We choose Particle swarm algorithm (PSO) as a basic algorithm for combining BOA because of its stronger searching ability (Bergh and Engelbrecht 2004; Yu and Gen 2010). BOPSO starts by generating a population solution, select a set of solutions from the initial population using a selection method, and the selected population is called “overall” pollution.

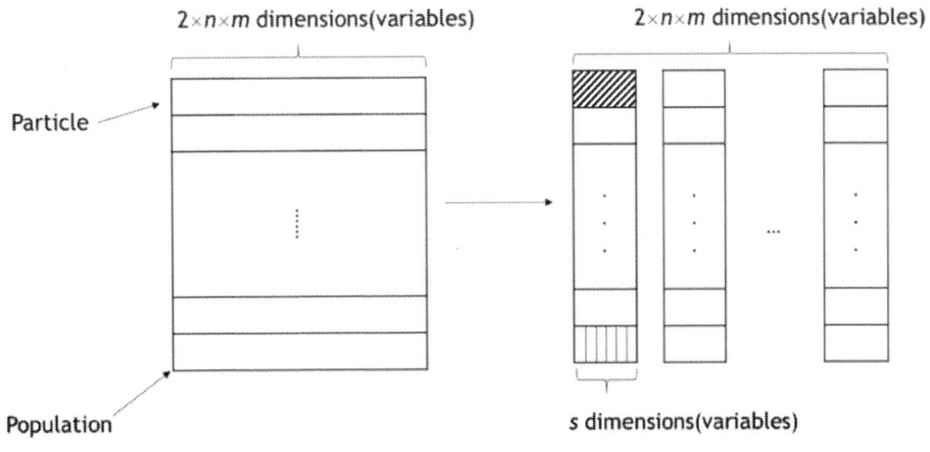

Figure 1: The network based for decoding the operations

Then, to group the “overall” population into several sub-populations which means it only contains part information of “overall” population randomly at initial stage. Meanwhile, to adjust group situation according to BN structure in the process of searching the optimal solutions. There are several key parts of BOPSO listed in the following subsections in detail. The grouping mechanism is shown in Figure 1.

2.1 Representation

2.1.1 *Encoding*

PSO is a typical metaheuristics algorithm in EAs, so the encoding and decoding methods will affect the performance partly (Yu and Gen 2010). We choose two-section real priority encoding method in order to explore larger solution space. The searching space is created in a $2 \times n \times m$ search space for n jobs on m machines. The position of a particle consists of $2 \times n \times m$ variables and is represented with real number. As mentioned above, the fJSP can be considered as two sub-vectors: operation sequences and machine allocation, in which the first part presents the operations; the second part presents the machines allocation corresponding to each operation, We use the network based decoding method to get fJSP solution from the particle. An example of representation and decoding network for one particle with 5 operations is shown Figure 2.

O_{11}	O_{12}	O_{21}	O_{31}	O_{32}	R_{11}	R_{12}	R_{21}	R_{31}	R_{32}
4.2	0.6	3.4	2.3	1.7	4.6	2.9	1.1	2.4	0.9
operation sequence					resource allocation				

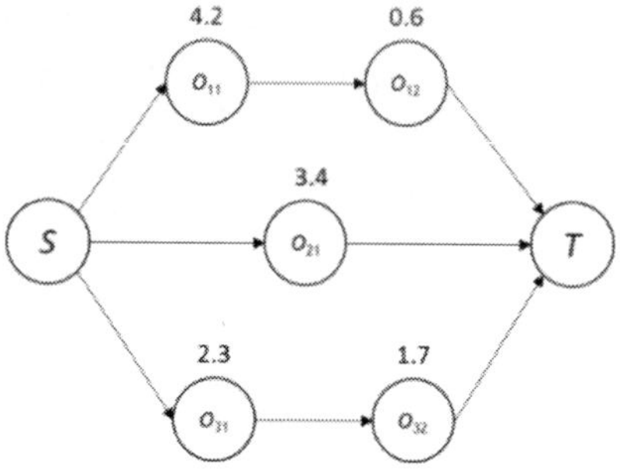

(a) an example of representation (b) network structure for decoding

Figure 2: An example of representation and decoding network for one particle

2.1.2 *Position and Velocity Update*

PSO uses the following formulas to update velocity and position in each generation t, respectively. The initial values set randomly. The ω presents inertia weight, $rand_1$ and $rand_2$ are random value within [0,1]. The bigger ω presents affecting by chromosome's own value; the smaller ω presents affecting by population's social factor. The adaptation of parameters used in PSO is adapted according to the methods studied in the paper (Yang, Tang & Yao, 2008).

$$v_i(t+1) = wv_i(t) + c_1 rand_1[p_{best}(t) - x_i(t)] + c_2 rand_2[l_{best}(t) - x_i(t)] \tag{8}$$

$$x_i(t+1) = x_i(t) + v_i(t+1) \tag{9}$$

In order to explore larger solution space, the vector and position update model with both Cauchy and Gaussian distributions for sampling is proposed in (Li and Yao 2012). The position updating formula is shown as follows:

$$v_i(t+1) = \begin{Bmatrix} p_{best}(t) + C(1)|p_{best}(t) - l_{best}(t)|, & if\ rand \le p \\ l_{best}(t) + N(0,1)|p_{best}(t) - l_{best}(t)|, & otherwise \end{Bmatrix} \tag{10}$$

p_{best} presents the local best of the population which has the definition that, the l_{best} is the best of three chromosomes containing the one before one, the one following and itself. *rand* and p are used for deciding which formula to be selected, *rand* is a random value from [0,1] and p is a given value at initial stage. $C(1)$ means the Cauchy distribution value with mean value 1 (Yu and Gen 2010); $N(0,1)$ means the normal distribution value with mean value 0 and the standard variance 1.When $rand \leqslant p$, the velocity will update follow Cauchy distribution update formula, otherwise, follow normal distribution update formula.

2.2 Bayesian-based grouping mechanism

2.2.1 Bayesian network structure learning

As mentioned above, BN structure learning is also a NP hard problem, so we use another PSO for construct the BNs. It is the same to the first PSO, each particle contains two parts, but they represent variable sequence and edge consideration respectively. We give an example with 4 variables one group in Figure 3. We can get variable sequence: x_4->x_2-> x_1-> x_3, then get the BN structure by transferring the real values in the second part particle into binary value for the first 4 (equals to the number of variables *s* in current group) values, then to randomly select from the got binary values for other 2 (equals to $s^*(s-1)/2-s$) binary values. In Figure 3, the binary value is 1->0->0->1 (rand and do modular arithmetic), plus randomly selected 2 binary values, finally, the binary sequence is 1->0->0->1->0->0

1	2	3	4	1	2	3	4
1.5	2.3	0.8	3.2	2.9	1.8	2.1	1.1
variable sequence				edge consideration			

Figure 3: Particle for learning BN structure

Then, we decode particle into a BN by a strictly upper triangular matrix which means to put the binary sequence in the shadow position successively from the bottom to the end in Figure 4. In which, 1 represents the variable in column is the parent of the variable in row.

	x_4	x_2	x_1	x_3
x_4		1	0	0
x_2			1	0
x_1				0
x_3				

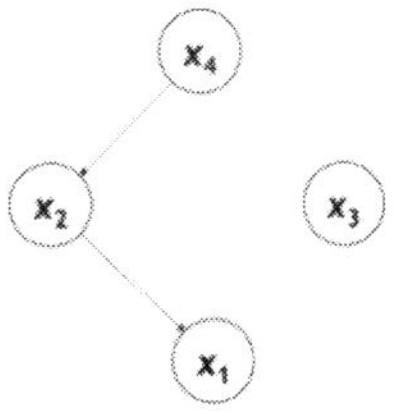

(a) a strictly upper triangular matrix (b) Bayesian network structure

Figure 4: A simple BN structure example get from a particle

Each generation can get one network structure, after *m* generation, we can *m* candidate networks. A simple BN is shown in Figure 4, it presents the network has 4 variables according to the particle. Then, we adjust the grouping situation depends on BN structure.

2.2.2 Sampling train data

We record down the change of each gene of solution chromosomes between the current generation and the last generation.

1	2	3	4
1.2	3.2	3.9	2.7
1.2	3.2	3.9	2.7
0.8	2.9	5.1	2.1
....			
1.0	2.7	3.0	2.3
1.2	2.5	2.9	2.7

1	2	3	4
0	0	0	0
0	0	1	0
...			
1	0	0	1
1	0	0	1

(a) recorded global best particle (b) training data

Figure 5: Sampling the training data

If the value of gene becomes bigger, then records 1, otherwise, becomes smaller or keeps invariant, records 0. Then we can get the network structure evaluation training data set, shown in Figure.5. After evaluating the network structure, we can find a best network with *BDe* score. Then, we readjust the grouping that put the discrete point into the other groups randomly. The process of BOPSO is in Procedure 1.

procedure 1: BOSPO

```
procedure: Co-Evolutional Particle swarm optimization with Bayesian Network (minim. problem)
input: f(x), v_i(0), hbest_i(0), gbest(0), b_1, b_2, r_1, r_2, BN parameters
begin
    t ← 0;
    while (not terminating condition) do
        initialize Pop(x_i(t)) by encoding routine;  //step 1 x_i(t): i^th particle is position x
        evaluate Pop(x_i(t)) by decoding routine and keep the best solution;    //step 2
        divide Pop(t) into subPop(t) by grouping randomly;          //step 3
        while (not cooperative generation) do
            for each particle x_i(t) in subPop(t) do
                v_i(t+1) = w(t)v_i(t) + b_1r_1(hbest_i(t) - x_i(t)) + b_2r_2(gbest(t) - x_i(t)); //step 4
                x_i(t+1) = x_i(t) + v_i(t+1);              //step 5 update position x_k(t+1)
                evaluate x_i(t+1) by decoding routine;     //step 6 evaluate x_k(t+1)
                if f(x_i(t+1)) < f(hbest_i(t)) ;           //hbest(t): historical best position
                    update hbest_i(t) = x_i(t+1);          // step 7
            end
            gbest(t+1) = argmin {f(hbest_i(t), f(gbest(t))} & update the best solution; //s8
            construct Bayesian network by estimating gbest(t+1); //s9 promising data
            create new offspring according to the joint prob. distribution by BN; //s9
            evaluate new offspring by decoding routine & update best solution; //s10
        end
        t ← t+1;
    end;
    output: the best solution gbest;
end;
```

Figure 6: The procedure in the pseudo code for BOPSO with multiple subpopulations

3 EXPERIMENTS AND RESULTS

In order to prove our proposed algorithm logically, we do four kinds of experiments for S-fJSP at the same time on 4 scales of data: 5×5 (50 dims), 10×10(200 dims), 15×15(450 dims) and 20×20 (800 dims). The dimensions of each scale of data are decided by encoding method mentioned in section 3. For data set, we use 5×5 as an example, it means 5 jobs, 5 machines and the same number of operations to the number of machines. These data instances are contributed to the OR-Library by Dirk et.al. In order to ensure the reliability of the experimental and avoid the shock of results, all experiments are repeated 30 times to get the mean value. The machine environment of running experiments is Intel(R) Core(TM) i7 -4770 CPU @ 2.2GHZ, 16GB.

3.1 Experiments Settings

The values for the parameters c_1, c_2, and w used in PSO were selected based on suggestions in other literature where these values are found, empirically, to provide good performance (Kennedy 2011; Li and Yao 2012). The crossover and mutation probability are set to 0.5. The initial parameter in DE is same as SaNSDE in Yan et al (2004). Besides the above parameters, there are still have the common parameters such as the population size, grouping size and the termination condition for all algorithms, to be seen clearly, they are shown in Table 1.

Table 1: The population parameters table

data	PopSize	Grouping Size	Termination condition
5×5	100	10	1000
10×10	1000	10	5000
15×15	1000	50	5000
20×20	5000	100	10000

3.2 Results and Discussion

3.2.1 *Experiment I*

1) In order to prove the effectiveness of Bayesian-based grouping mechanism, we did experiments focus on fJSP, we used PSO as the basic algorithm, experiment no grouping PSO, fixed grouping PSO, random grouping by set PSO and Bayesian-based grouping PSO respectively focusing the minimum value and the mean value of makespan for fJSP, the result are shown in Table 2.

Table 2: Results for experiment I focused on grouping method

scale		no	fixed	set	Bayes
5×5	min	245	282	212	157
	mean	273.43	340.47	279.866	217.2
10×10	min	742	631	694	569
	mean	748.966	686.47	766.86	696.1
15×15	min	1209	1080	1074	979
	mean	1277.4	1174.37	1232.7	1132.63
20×20	min	1784	1715	1677	1372
	mean	1794.83	418.9	1774.96	1555.06

We can get the conclusion that Bayesian-based grouping mechanism performs better than other grouping mechanisms, especially for the high dimension instances because as the scale increases, the dimensions of problems will increase too, among high dimensional variables to find the potential relationships can help the algorithms find the better solutions than the simple random grouping mechanism.

3.2.2 *Experiment II*

We compared BOPSO with a classic genetic algorithm (GA), the binary genetic algorithm (bGA; Pezzella, et al 2008), PSO (Kennedy 2011), the differential evolution algorithm (DE; Price et al 2006), the co-evolution particle swarm algorithm (CCPSO; Pezzella, et al 2008) and the self-adaptive grouping differential evolution (SaNSDE; Yang, et al 2008) on the minimum value, and the mean value, the results are shown in Table 3. Firstly we can get the conclusion that, for encoding methods, real number encoding method (GA) performances well than the binary encoding method (bGA). The reason is that real number can use only one gens to represent one variable meanwhile it has larger search space. Secondly, for the basic algorithms which contains: GA, bGA, DE, PSO, we can find that GA performs better than PSO in most time, however, due to the characters of GA, co-evolution GA performs bad unfortunately. Compared to improved algorithms: CCPSO and SaNSDE, we get the conclusion that BOPSO performs better than them on most evaluation indexes.

Table 3: Results for experiment 1 focused on JSP and fJSP

	scale	target	DE	bGA	GA	PSO	SaNSDE	CCPSO	BOPSO
JSP	6×6	min	57	60	55	55	61	55	55
		mean	59.1	60.6	57.93	57.6	70.37	56.23	56.23
	10×10	min	1331	1368	1310	1312	1424	1239	1210
		mean	1388.8	1418.2	1343.93	1391.13	1580.67	1314.8	1296.23
	15×15	min	1624	1678	1558	1584	1738	1518	1487
		mean	1661.33	1723.08	1621.13	1627	1887.8	1582.73	1556.23
	20×20	min	1064	1120	1030	1045	1159	1005	972
		mean	1088.6	1148.6	1072.8	1071.27	1219.43	1032.77	1026.57
fJSP	5×5	min	276	284	252	245	269	212	157
		mean	314.43	332.37	291.5	273.43	310.66	279.866	217.2
	10×10	min	755	817	689	742	746	694	569
		mean	809.9	856.3	778	748.966	805.7	766.86	696.1
	15×15	min	1284	1328	1238	1209	1231	1074	979
		mean	1316.06	1357.9	1301.5	1277.4	1238.17	1232.7	1132.63
	20×20	min	1730	1820	1711	1784	1706	1677	1372
		mean	1816.96	1856.88	1832.46	1794.83	1778.53	1774.96	1555.06

3.2.3 Experiment III

We record the fitness each 300 iterations for all algorithms especially for BOPSO to prove and compare the algorithm convergence on 4 scales of data set.

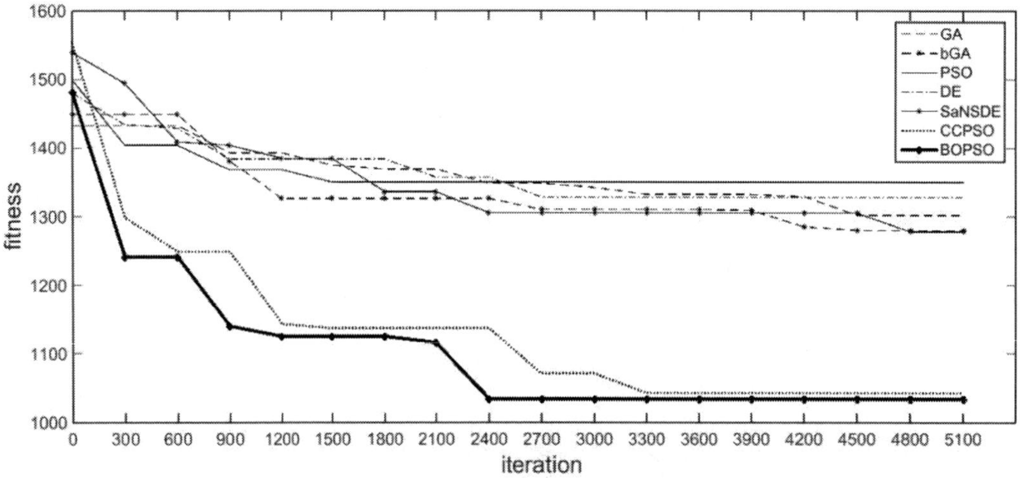

Figure 7: The algorithm convergence of proposed BOPSO

4 CONCLUSION

We proposed Bayian Optimization Algorithm combined Particle swarm optimization (BOPSO) with multiple subpopulations to solve job shop scheduling problems (JSP) and the flexible JSP (fJSP) in which it was based on with real-number encoding as the basic algorithm to increase the search space and avoid getting into local optimum solutions. Then, we used the optimal Bayseian Network (BN) structure to find out the relationship between the variables and according to the relationship to regroup, minimize the makespan of JSP and fJSP within a reasonable amount of calculation time.

Then, we got the results through a list of experiments that the Bayesian-based grouping mechanism was effective; BOPSO with multiple subpopulations can get better solutions and increase the robustness. In the future work, we are eager to process the sub-problems in distributed environment on the different and parallel processing machines in order to shorten the computational time more efficiency while optimize the optimal solution.

ACKNOWLEDGEMENT

This work is partly supported by the National Natural Science Foundation of China under Grant 61572100, the Fundamental Research Funds for the Central Universities No. DUT15QY10 and in part by the Grant-in-Aid for Scientific Research (C) of Japan Society of Promotion of Science (JSPS) No. 15K00357.

5 REFERENCE

[1] **Shan, S., Wang, G. G.2001.** Survey of modeling and optimization strategies to solve high-dimensional design problems with computationally-expensive black-box functions. *Structural and Multidisciplinary Optimization*, 41(2), pp 219-241.

[2] **Gao, J., Sun, L., Gen, M. 2008.** A hybrid genetic and variable neighbourhood descent algorithm for flexible job shop scheduling problems. Computers & Operations Research, 35(9), pp. 2892-2907.

[3] **Clearwater, S. H., Hogg, T.**, Huberman, B. A. 1992. Cooperative problem solving. Computation: The Micro and the Macro View, 275, pp. 33-70.

[4] **Larrañaga, P., Poza, M., Yurramendi, Y., Murga, R. H., Kuijpers, C. M. H. 1996.** Structure learning of Bayesian networks by genetic algorithms: A performance analysis of control parameters. *IEEE transactions on pattern analysis and machine intelligence*, *18*(9), pp 912-926.

[5] **Yang, J., Xu, H., Pan, L., Jia, P., Long, F., Jie, M. 2011.** Task scheduling using Bayesian optimization algorithm for heterogeneous computing environments. *Applied Soft Computing*, *11*(4), pp 3297-3310.

[6] **Li, J., Zhang, J., Kang, Q., Jiang, C. 2014.** Team of Bayesian optimization algorithms to solve task assignment problems in heterogeneous computing systems.2014 IEEE International Conference on Systems, Man, and Cybernetics (SMC), October, pp. 127-132.

[7] **Hao, X., Chen, X., Lin, H. W., Murata, T.2011.** Cooperative Bayesian optimization algorithm: a novel approach to simultaneous multiple resources scheduling problem. In Innovations in Bio-inspired Computing and Applications (IBICA), 2011 Second International Conference on IEEE, pp. 212-217.

[8] **Van den Bergh, F., Engelbrecht, A. P. 2004.** A cooperative approach to particle swarm optimization. IEEE Transactions on Evolutionary Computation, 8(3), pp. 225-239.

[9] **Li, X., Yao, X.** 2012. Cooperatively coevolving particle swarms for large scale optimization. IEEE Transactions on Evolutionary Computation, 16(2), pp. 210-224.

[10] **James Kennedy.** 2011. Particle swarm optimization, Encyclopaedia of Machine Learning, Springer, pp. 760-766.

[11] **Pezzella, F., Morganti, G., Ciaschetti, G.** 2008. A genetic algorithm for the flexible job-shop scheduling problem. Computers & Operations Research, 35(10), pp. 3202-3212.

[12] **Price, K., Storn, R. M., Lampinen, J.** A. 2006.Differential evolution: a practical approach to global optimization. Springer Science & Business Media.

[13] **Yu, X. & Gen, M.** 2010. Introduction to Evolutionary Algorithms, 430pp, Springer, London, July.

[14] **Yang, Z., Tang, K., & Yao, X. 2008.** Self-adaptive differential evolution with neighbourhood search. Proc. of IEEE Congress on Evolutionary Computation, June, pp. 1110-1116.

A CO-EVOLUTION BASED HYBRID EA FOR STOCHASTIC FLEXIBLE JOB SHOP SCHEDULING PROBLEM

Lu Sun[1], Lin Lin[1,2*], Mitsuo Gen[2,3] and Haojie Li[1]
[1] School of Software, Dalian University of Technology
Dalian, China
sunlu@evolab.cn , lin@dlut.edu.cn , hjli@dlut.edu.cn
[2] Fuzzy logic Systems Institute,
Iizuka, 820-0067, Japan
gen@flsi.or.jp
[3] Tokyo University of Science
Tokyo, Japan
gen@rs.tus.ac.jp

ABSTRACT

Flexible job shop scheduling, the core problem in manufacturing system aims to achieve flexibility and efficiency of production by efficient resource allocation under the constraint conditions. Flexible scheduling under uncertainty can not only increase flexibility but also reduce the impacts of uncertainty factors. So it can better meet production demands in practical applications, hence attracting more attentions. The existing algorithms aim to find the most compact structure on the left in the Gantt chart, however, this kind of methods have difficult in solving the stochastic scheduling problems. In this paper, we proposed coevolution hybrid evolution algorithm (ChEA), which chooses particle swarm optimization (PSO) as basic algorithm, then to use the set-based grouping methods to divide the "overall- population" into several sub-populations, evaluated by the co-evolution mechanism. Meanwhile, to self-adjust the parameters in the optimization process in order to satisfy the dynamic process. Through experiments, we get the conclusion that ChEA can get better solution of stochastic flexible job shop scheduling problems than original algorithms and improve the robustness of solution.

Keywords: Evolutionary algorithm, stochastic flexible job shop scheduling, co-evolution, parameter self-adaptive

1 INTRODUCTION

Stochastic flexible job shop scheduling (S-fJSP) attract more attention because it is closer to real application. It comes from the flexible job shop scheduling problem (fJSP) model which has been proved as a NP hard problem under the precedence and resource constraints [1]. fJSP has attracted researchers' attentions for tens of years, the references shows that EAs are well performed for fJSP [2]. However, fJSP aims to find the most compact structure on the left in the Gantt chart. S-fJSP aims to balance the operations to minimum the expected value of makespan. So the effective algorithms for fJSP couldn't be used for S-fJSP directly.

Recently, Gu et.al proposed a novel competitive co-evolutionary quantum genetic algorithm (CCQGA) a stochastic scheduling problem with the objective to minimize the expected value of makespan [3].Zhang et.al proposed a model for FJSP with transportation constraints and bounded processing times [4]. Horng et al proposed the evolution strategy (ES) into sequence evolution (SE) in order to minimize the sum of progress and expectations of stochastic scheduling with random processing time [5]. However, the methods mentions above still two weaknesses: with the increase of the search space of problems, the effectiveness of the evolutionary algorithm (EA) is limited even goes down, but, the search space of problem in real application is often large. So to overcome the weaknesses becomes very important. Recently, there exists series of EAs with grouping mechanism for solving large scale classical mathematical function optimization problems [6], especially, using co-evolution mechanism to increase the effectiveness of algorithms. Inspired by it, we want to use the co-evolution mechanism to help evolution algorithms (EAs) to optimize the solutions of the scheduling problems, especially the problems with large search space, and increase the robustness of optimal solutions. In this paper, we present the ChEA which is based the co-evolution framework to solve stochastic scheduling problem, using PSO as the basic evolutionary algorithm. Meanwhile, we also use the parameters self-adaptive mechanism to adjust parameters automatically in the optimization process in order to satisfy the dynamic process.

The remainder of this paper is organized as follows: Section 2 presents the model of stochastic flexible scheduling; Section 3 presents the proposed ChEA approach in the detail; Section 4 provides experimental setting and results analysis to prove the efficiency of ChEA for solving stochastic flexible job shop scheduling; and finally, Section 5 provides a conclusion.

2 FORMULATIONS OF S-FJSP

As mentioned above, we focus on the stochastic flexible job shop scheduling which means the processing time is not fixed but is accord with a certain distribution (such as the normal distribution, the uniform distribution and the exponential distribution), and the distribution is given in advance. There are two key parts in stochastic flexible job shop scheduling problem: tasks sequence and resources allocation. The notations which contain the indices, the parameters and the decision variables in stochastic flexible scheduling problem are expressed as follows:

Index:

i,h: indices of tasks, i,h=1,2.,...,n

j:indices of resources, index, j=1,2,...,m

k,g: indices of operations, k,g=1,2,..., n_i

Parameter:

n: number of total tasks

m: number of total resources

n_i: number of operations for task i

O_{ik}: kth operation of task i

ξt_{ikj}: the uncertain processing time of O_{ik} processed on jth resource

A_{ik}: the available resource set of O_{ik}

Decision variables:

ξc_{ik}: the uncertain completed time of O_{ik}

$$x_{ikj} = \begin{Bmatrix} 1, & \text{if } o_{ik} \text{ is performed on machine } j \\ 0, & \text{otherwise} \end{Bmatrix}$$

The objective function can be defined as: minimizing the expected value of makespan which is the maximum value of the completing time of all tasks. The objective function is shown as follows in (1):

$$min \quad E[C_M] = E[\max_i [\max_k [\max_j \xi \, c_{ikj}]]] \tag{1}$$

For the process of scheduling, the operations are not allowed to terminate until it is completed and for each resource, it could not be used for the next operation until the current operation is completed. The resource constraints are described as follows in (2):

$$\xi c_{ik} - \xi c_{i(k-1)} \geq \xi t_{ikj} \cdot x_{ikj}, k = 2, \dots, n_i, \forall i, j \tag{2}$$

For each task, it consists of several operations and each operation is allocated different machine, although each operation can be processed on different resources, the operation sequence within each task must be observed. The operation constraints are described as follows, note that the two constraints (3) and (4) only one need to be satisfied at the same time.

$$(\xi c_{hg} - \xi c_{i(k-1)} - \xi t_{ikj}) \cdot x_{hgj} \geq 0, \forall i, j, g, h \tag{3}$$

$$(\xi c_{ik} - \xi c_{hg} - \xi t_{ikj}) x_{ikj} \geq 0], \forall i, j, g, h \tag{4}$$

Besides the constraints mentioned above, there are some value constraints, such as for the processing time, in our case, we only consider the formal cases in which the processing time is positive real value, shown in (5).

$$\xi c_{ik} > 0, \forall i, k \tag{5}$$

For each resource, it contains only two cases: the first one is chosen to process one operation, the other is free, and the resource can only be chosen for only on operation at the same time, the constraints are shown as follows in (6) and (7).

$$x_{ikj} \in \{0,1\}, \ \forall i, k, j \tag{6}$$

$$\sum_{x_{ikj} \in A_{ik}} x_{ikj} = 1, \forall i, k, j \tag{7}$$

3 PROPOSED ALGORITHM

In this paper, we import the co-evolution mechanism to help PSO to overcome the problem of large search space. Due to the parameters also play an important role in EAs, and we aim to solve the stochastic flexible scheduling problem, so we use the parameter self-adaptive mechanism to adjust the parameters automatically. There are several key parts of ChEA listed in the following subsections in detail.

3.1 Particle Swarm Process

PSO is first proposed in 1995[8], which is a computational method that optimizes a problem by iteratively trying to improve a candidate solution with regard to a given measure of quality. In our algorithm, the position and velocity update formulas can be describe as follows:

$$v_i(k+1) = wv_i(k) + c_1 rand_1\left(p_{best}(k) - x_i(k)\right) + c_2 rand_2\left(g_{best}(k) - x_i(k)\right) \tag{8}$$

$$v_i(k+1) = wv_i(k) + c_1 rand_1\left(l_{lest}(k) - x_i(k)\right) + c_2 rand_2\left(g_{best}(k) - x_i(k)\right) \tag{9}$$

$$x_i(k+1) = v_i(k+1) + x_i(k) \tag{10}$$

In order to explore larger solution space, the velocity update formula is shown as follows:

$$v_i(k+1) = \begin{Bmatrix} Eq\ (8), \text{ if } rand(0,1) \geq p \\ Eq\ (9), \text{otherwise} \end{Bmatrix} \tag{11}$$

where $p_{best}(k)$ represents the personal best of the current particle of the kth generation. The definition of $g_{best}(k)$ is the particle with best fitness among the whole population in kth generation. *rand* is a random value from [0,1] and p is a given value in advance, the two value are used for deciding which equation is used to update particles for current population.

3.1.1 Encoding and decoding

Due S-fJSP can be considered as two sub-problems: operation sequences and resource allocation, so we choose two-stage encoding method: the position of each particle has two parts: the first part presents the operations; the second part presents the resources allocation corresponding to each operation, the representation of the position of one particle for 5 tasks on 5 resources is shown in Figure.1.

O_{11}	O_{12}	O_{21}	O_{31}	O_{32}	R_{11}	R_{12}	R_{21}	R_{31}	R_{32}
4.2	0.6	3.4	2.3	1.7	4.6	2.9	1.1	2.4	0.9
operation sequence					resource allocation				

Figure 1: Representation of the position of one particle

After evolution, we use real-priority based decoding method through the network in Figure 2.

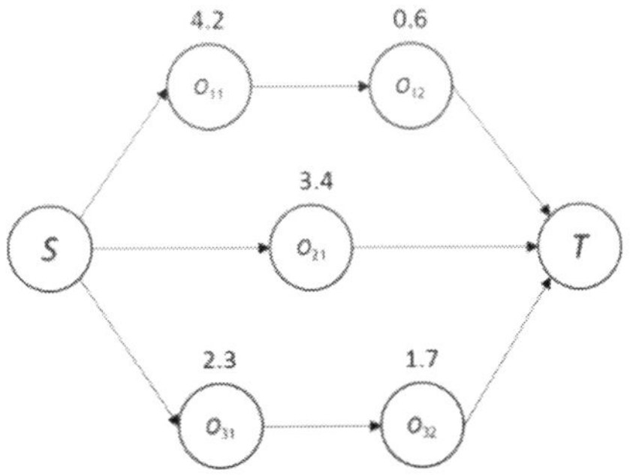

Figure 2: The network based for decoding the operations

3.1.2 *Co-evolution Framework*

The concept of co-evolution was first proposed in 1997 [7].The core idea is to decompose the overall problem (population in PSO) of dimension N into k sub-problems (each sub-problem has $s = N/k$ dimensions), and the divided sub-problems (shown in Figure.3) have the same essence and do not affect each other [6]. Then, get the solution of overall problem through the cooperation of k sub-problems. For any given variable x, it belongs to the same optimization process which means it has the equal probability to be put in the same sub-problems and the probability of two related variables put in the same sub-problem will increase with the increasing of the times of grouping. The process can be shown as in Figure 3.

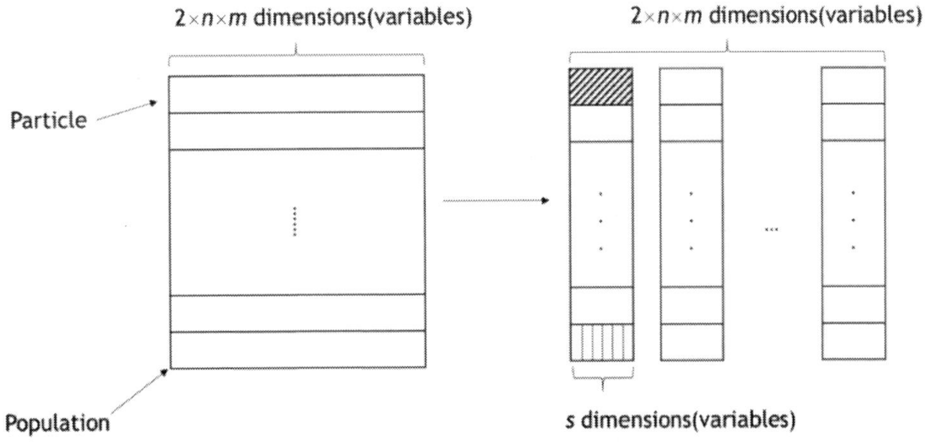

Figure 3. Representation of the position of one particle

3.1.3 Parameter self-adaptive mechanism

Due to the parameters have important influence on the results and decide the evolution operations, we assume that adjust dynamic according to the merits of the fitness value in the process of experiment, there will be a good influence on the performance of the algorithm. Adjust the parameters as follows: The random value $rand_1$ and $rand_2$ in equation (8) and (9) determines the influence of local best and global best. In the process of experiment, they are confirmed to the normal distribution with average value CR_{mean} and the standard deviation 0.1.

$$rand_{1or2} = N_i(CR_{mean}, 0.1) \tag{12}$$

CR_{mean} is firstly set as 0.5.These *CR* values for all individuals remain the same for 5 generations and then a new set of *CR* values is generated using the same equation. During every generation, the *CR* values associated with offspring successfully entering the next generation are recorded in an array CR_{sec}. After 25 generations CR_{mean} will be updated:

$$CR_{mean} = \frac{1}{|CR_{sec}|}\sum_{k=1}^{|CR_{sec}|} CR_{sec}(k) \tag{13}$$

Procedure 1: ChEA

```
begin
    t ← 0;
    initialize population p(t) by encoding routine randomly;
    get sub-p(t) randomly which means randomly select variables for each group;
    g(t) ← 0;
    while ( not meeting termination condition)
        if (meet the condition)
            adjust the grouping by selecting new dimension from given set S
        for each sub-p(t)
            for each chromosome
                if chromosome replace corresponding genes in sub-p(t) is better than the whole pbest(t)
                    replace corresponding genes in pbest(t) by genes in sub-p(t);
                if gbest(t) replace corresponding genes in sub-p(t) is better than the whole pbest(t)
                    replace corresponding genes in gbest(t) by genes in pbest(t);
            for each chromosome
                update its local best lbest(t) which means choose the best one of neibourhour particles;
        for each sub-p(t)
            for each chromosome
                update the position and velocity vector;
    adjust the parameters;
    g(t) ← g(t+1);
  output best solution of S-fJSP gbest(t)
end
```

The inertia weight ω is adapted by Equation (14), we first set the selection probability *fp* as 0.5, if the random value is smaller than *fp*, ω is confirmed to the normal distribution with average value 0.5 and the standard deviation 0.3, and otherwise, ω is confirmed to Cauchy distribution with parameter 1. The inertia weight ω is adapted each 15 generations.

$$w_i = \begin{Bmatrix} N_i(0.5,0.3), if\ U_i(0,1) < fp \\ \xi_i,\ otherwise \end{Bmatrix} \tag{14}$$

The selection probability *fp* is adapted by Equation (15), among them, After evaluation of all offspring, the number of offspring successfully entering the next generation while generated by Equation (8) and Equation (9) are recorded as ns_1 and ns_2, respectively, and the numbers of offspring discarded while generated by normal distribution and Cauchy distribution are recorded as nf_1 and nf_2. Those two pairs of numbers are accumulated within a specified 50 generations, called the “learning period”.

Then, the probability p is updated as: For easily understand, the pseudo-code for ChEA is presented in Procedure 1.

$$p = (ns_2 + nf_2)/(ns_2(ns_1 + nf_1) + ns_1(ns_2 + nf_2)) \tag{15}$$

4 EXPERIMENTS AND RESULTS

In order to prove our proposed algorithm logically, we do four kinds of experiments for S-fJSP at the same time on 4 scales of data: 5*5 (50 dims), 10*10(200 dims), 15*15(450 dims) and 20*20 (800 dims). The dimensions of each scale of data are decided by encoding method mentioned in section 3. For data set, we use 5*5 as an example, it means 5 jobs, 5 machines and the same number of operations to the number of machines. These data instances are contributed to the OR-Library by Dirk et.al. In order to ensure the reliability of the experimental and avoid the shock of results, all experiments are repeated 30 times to get the mean value. The machine environment of running experiments is Intel(R) Core(TM) i7 -4770 CPU @ 2.2GHZ, 16GB.

4.1 Experiments Settings

The values for the parameters c_1, c_2, and w used in PSO were selected based on suggestions in other literature where these values are found, empirically, to provide good performance [8]. The crossover and mutation probability are set to 0.5. The initial parameter in DE is same as SaNSDE in [9].Besides the above parameters, there are still have the common parameters such as the population size, grouping size and the termination condition for all algorithms, to be seen clearly, they are shown in Table 1.

Table 1. The population parameters table

data	PopSize	Grouping size	Termination condition
5×5	100	10	1000
10×10	1000	10	5000
15×15	1000	50	5000
20×20	5000	100	10000

4.2 Results and Discussion

4.2.1 Experiment I

We compared BOPSO with classic genetic algorithm (GA)[10], the binary genetic algorithm (bGA)[10], PSO[8], the differential evolution algorithm (DE)[9],the co-evolution particle swarm algorithm (CCPSO)[6] and the self- adaptive grouping differential evolution (SaNSDE)[13] on the mean value for both fJSP and S-fJSP. The results are shown in Table 2, and the best value is shadowed.

Table 2. Results for experiment 1 focused on JSP and fJSP

fJSP	target	GA	bGA	DE	PSO	SaNSDE	CCPSO	ChEA
5×5	mean	319.83	331.7	326.6	264.83	320.83	243.86	239.07
10×10	mean	800.4	847.3	818.13	759.47	813.0	622.8	619.8
15×15	mean	1330.4	1371.0	1329.5	1281.6	1318.1	1075.7	1064.1
20×20	mean	1848.5	1869.9	1821.3	1771.54	1819.5	1579.6	1520.5
S-fJSP	**target**	**GA**	**bGA**	**DE**	**PSO**	**SaNSDE**	**CCPSO**	**ChEA**
5×5	mean	403.78	401.78	432.4	426.8	408.52	348.91	337.56
10×10	mean	1054.36	3024.18	1115.08	1034.98	1061.18	763.1	749.19
15×15	mean	1801.31	6515.23	1761.39	1764.28	1670.83	1348.78	1276.08
20×20	mean	2273.17	9245.08	2219.34	2268.08	2232.11	1866.75	1777.91

Firstly we can get the conclusion that, for encoding methods, real number encoding method (GA) performances well than the binary encoding method (bGA). The reason is that real number can use only one gens to represent one variable meanwhile it has larger search space. Secondly, for the basic algorithms which contains: GA, bGA, DE, PSO, we can find that, GA performs better than PSO in most time, however, due to the characters of GA, co-evolution GA performs bad unfortunately [6]. Compared to improved algorithms: CCPSO and SaNSDE, we get the conclusion that, ChEA performs better than them on mean value for both fJSP and S-fJSP.

4.2.2 Experiment II

We compute the improvement percent to prove effectiveness and robustness of our algorithm, we respectively compute the improvement of mean fitness and variance fitness by the Equation (16) between CCPSO and PSO, ChEA and PSO, then compute the improvement (imp%)between ChEA (imp_Ch) and CCPSO (imp_CC), the table in shown in Table 3. We can see that with the increase of search space, ChEA performs better.

$$imp\% = \left| \frac{\Delta imp\_cc - \Delta mp\_Ch}{\Delta imp - cc} \right| * 100 \tag{16}$$

Table 3. Results for experiment 1 focused on JSP and fJSP

scale	Imp_CC	Imp_Ch	imp%
5×5	77.89	89.24	14.57
10×10	271.88	285.79	5.1
15×15	415.5	488.2	17.50
20×20	401.33	490.17	22.14

4.2.3 Experiment III

We record the fitness each 1000 iterations for all algorithms especially for ChEA to prove and compare the algorithm convergence on 4 scales of data set. The results are shown in Figure 4. We can find that although CCPSO convergences faster, ChEA still optimize after the first obvious convergence and get the better solutions than other algorithms.

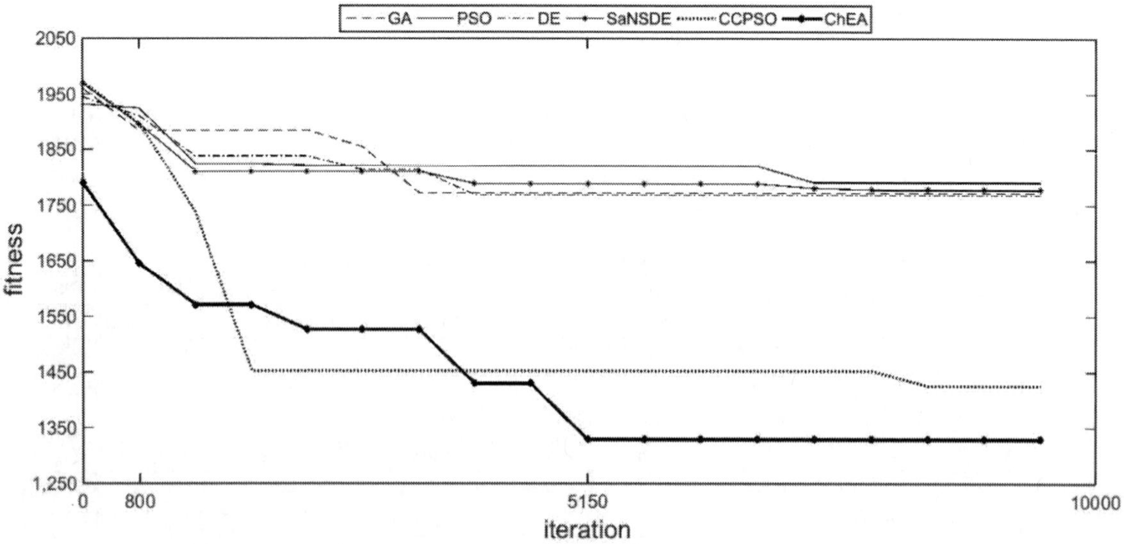

Figure 4 The algorithm convergence of all algorithms

5 CONCLUSTION

The proposed ChEA imports the co-evolution mechanism, using PSO as the basic algorithm with the parameter self-adaptive mechanism to solve the stochastic flexible scheduling problem effectively and increase the robustness of optimal solutions under uncertain. That proved that the features of the EAs and the encoded structures are suitable for the optimization problems under uncertain. The benchmarks used in the experiments are all full flexible problems which means they have larger search space than the partial flexible problems with the same scale. So, ChEA can solve the problems with both low search space and high search space, however, we due to the evolutionary mechanism, we still could not process the sub-problems in distributed environment on the different and parallel processing machines in order to shorten the time more efficiency while optimizing the optimal solution.

ACKNOWLEDGEMENT

This work is partly supported by the National Natural Science Foundation of China under Grant 61572100, the Fundamental Research Funds for the Central Universities No. DUT15QY10 and in part by the Grant-in-Aid for Scientific Research (C) of Japan Society of Promotion of Science (JSPS) No. 15K00357.

6 REFERENCE

[1] **Cheng, R., Gen, M., & Tsujimura, Y. 1996 .** A tutorial survey of job-shop scheduling problems using genetic algorithms—I. Representation. *Computers & industrial engineering*, 30(4), pp 983-997

[2] **Gao J, Gen M, Sun L. 2006.** Scheduling jobs and maintenances in flexible job shop with a hybrid genetic algorithm[J]. *Journal of Intelligent Manufacturing*, 2006, 17(4): 493-507.

[3] **Gu, J., Gu, M., Cao, C., & Gu, X. 2010.** A novel competitive co-evolutionary quantum genetic algorithm for stochastic job shop scheduling problem. *Computers & Operations Research*, 37(5), pp 927-937.

[4] **Zhang, Q., Manier, H., & Manier, M. A. 2012.** A genetic algorithm with tabu search procedure for flexible job shop scheduling with transportation constraints and bounded processing times. *Computers & Operations Research*, 39(7), pp 1713-1723.

[5] **Horng, S. C., Lin, S. S., & Yang, F. Y. 2012.** Evolutionary algorithm for stochastic job shop scheduling with random processing time. *Expert Systems with Applications*,*39*(3),pp 3603-3610.

[6] **Li, X., Yao, X. 2012.** Cooperatively coevolving particle swarms for large scale optimization. *IEEE Transactions on Evolutionary Computation*, *16*(2), pp 210-224.

[7] **Potter, M. A. 1997.**The design and analysis of a computational model of cooperative coevolution. Doctoral dissertation, George Mason University.

[8] **James Kennedy.2011.** Particle swarm optimization, Encyclopedia of machine learning, Springer, pp 760-766.

[9] **Price, K., Storn, R. M., Lampinen, J. A. 2006.***Differential evolution: a practical approach to global optimization*. Springer Science & Business Media.

[10] **Pezzella, F., Morganti, G., Ciaschetti, G. 2008.** A genetic algorithm for the flexible job-shop scheduling problem. *Computers & Operations Research*, *35*(10), pp 3202-3212.

[11] **Yang, Z., Tang, K., & Yao, X. 2008.** Self-adaptive differential evolution with neighbourhood search. 2008 IEEE Congress on Evolutionary Computation, June, pp. 1110-1116).

CLUSTER ANALYSIS STUDY ON QUALITY CREDIT EVALUATION INDEX OF ENTERPRISES

Zhou Li*, Liu Bisong and Zhao an
China national institute of standardization, Beijing, China
zhouli@cnis.gov.cn

ABSTRACT

To carry out the enterprise quality credit evaluation is important basis for the implementation of enterprise quality credit classified supervision. Through the analysis of elements and indicators which affect the enterprise quality credit level, the enterprise quality credit index classification and system framework have been constructed; index clustering method has been employed to classify evaluation index; typical index has been selected as evaluation basis, so as to reduce index dimension, simplify the evaluation process, quickly provide effective evaluation results. By the empirical analysis on quality credit evaluation of manufacturing industry enterprise, the effectiveness of index clustering method, which contributes to improve the evaluation efficiency, identify key elements, and promote the effectiveness of the quality level of credit, has been proved.

Keywords: uality credit, Evaluation index, Index cluster analysis

1 INTRODUCTION

Modern economics think that as a kind of contract relationship, the credit relationship's main characteristic is that the providers of credit transfer the property rights to the trustee [1]. Meanwhile, the trustee shall promise to give the providers of credit corresponding compensations in accordance with the prescribed limited time as regulated in the contract. In a certain sense, the essence of credit is the commitment during the process of property's cross-time transaction, and the transaction between the real property of creditor and future property of trustee, which enable the reasonable configuration of property between the reality and the future. The market economy is the credit economy in essence, and the development of the market economy is the process of the formation and the deepening of the credit relationship.

The quality competition is the most fundamental way of enterprise competition as well as healthy competition. The quality ethic and quality credit improve the core competitiveness of enterprises, strengthen the credit [2]. Good enterprise ethic and quality credit is the key to solve the quality issues. To reduce the quality dishonest behavior and create good quality credit environment, it is necessary to establish strict disciplinary system against quality dishonesty, improve the supervising ability and find the dishonest behavior in time. The development of the quality credit evaluation is beneficial to recognize and analyze the quality dishonest risk, to find the quality dishonest behavior, to realize the efficient classified regulation, to improve the efficiency of regulatory.

2 QUALITY CREDIT EVALUATION MODEL

According to the national standards of Credit-General Vocabulary, the credit in a narrow sense belongs to economics category, reflecting a kind of ability which is established on the basis of

* Corresponding Author

transaction. The credit in a broad sense refers to the wide application of good faith principle, including the moral category. The National uality Inspection [2006] No. 464 refers that the enterprise quality credit is the ability and degree of the enterprises during production and operation activities in compliance with quality laws and regulations, the implementation of standards, the honor the quality commitment. The definition of the quality credit in the General Rules of Grading Enterprise uality Credit is that the capacity of achieving and maintaining the quality credit. The capacity is realized by meeting the customers' requirements or expectations in the life cycle of providing products on the basis of abiding by the laws and regulations related with quality, executive standards and honoring the quality commitment. Thus, the core of quality credit is to abide by the laws, regulations, commitment and agreement in the aspect of product quality.

To evaluate the credit grading of enterprises scientifically, it is necessary to establish a scientific, reasonable and operational credit evaluation model. Namely, find the key elements of enterprises credit. In accordance with the general 3C evaluation model (namely the enterprises' character, capacity and capital, and 3C comes from the first letter of three words.) of enterprises credit, combined with the definition of enterprises quality credit, it is not difficult to see that the factors influencing the quality credit include quality trustworthiness willingness, quality contractual capacity, quality contractual performance [3]. Based on the three influencing factors, establish the enterprises quality credit evaluation basic model initially [4]. Among them, enterprises quality trustworthiness is the fundamental influencing factor, which determines the quality contractual performance and result. The quality contractual capacity is the basic influencing factor which ensures whether the quality contractual performance and result can be achieved. The quality contractual performance is the final influencing factor which is the behavior performance of honoring the quality commitment and the judgment criteria of occurrence of honoring the quality commitment [5].

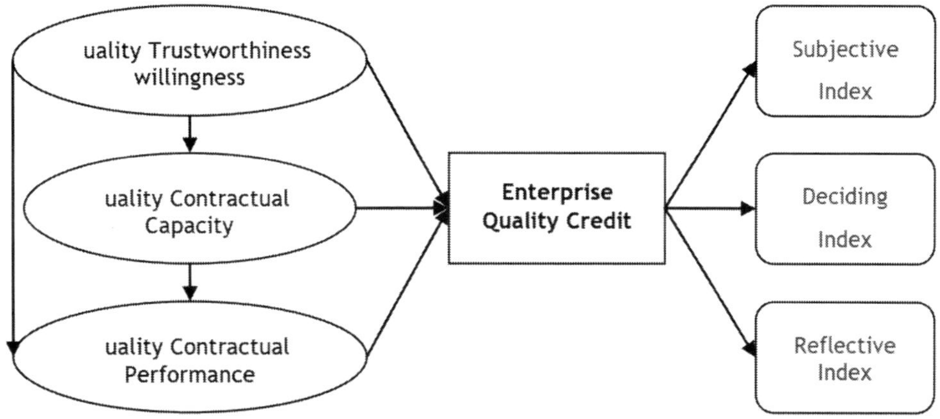

Figure 1: Enterprise Quality Credit Evaluation Model Diagram

In the evaluation model, the evaluation object is the status of enterprise quality credit. Therefore, the first credit factor is the status of enterprise quality trustworthiness willingness, including the necessary enterprise qualification, administrative licensing requirements, statutory and regulatory requirements and mandatory management requirements complied by enterprise and product. Whether the above requirements can be abided is the most fundamental requirement to ensure the product quality, which equals to the character of 3C model [6]. In addition, the corresponding quality capacity is required for enterprise, such as the sound quality management system, standard system and measurement system, which determines whether enterprise can make products meeting the customer requirements, equaling to the capacity of 3C model. Finally, the actual performance status of enterprise and product quality, such as product percent of pass, customer satisfaction, service, quality award status, quality punishment status, etc., which reflect the enterprise quality credit status directly [7]. All the above statuses can be collected via different ways, such as law enforcement supervision, supervision sampling, quality complaints, and quality files.

3 CLASSIFICATION OF QUALITY CREDIT EVALUATION INDEX

According to the enterprise quality credit evaluation model, determine the following indexes as the quality credit evaluation index from three aspects of quality trustworthiness status, quality contractual capacity and quality contractual actual performance, based on the product quality records.

3.1 Quality trustworthiness willingness index

The subjective willingness of enterprise quality credit, the decisive factor determining the quality credit level, refers the reasonable expectation of whether enterprises pay attention to the quality contract, respect consumers to abide the quality contractual to meet consumers [8]. The comparison of domestic and international economic and social environment and the development course of national market economy illustrate that the external environment of market condition, laws and regulations has an important influence on the high willingness of maintaining the enterprise quality credit. External environment, short-term and long-term goals are also the decisive factor of maintaining quality credit subjective willingness. The enterprise quality credit subjective willingness can be judged from whether the enterprise has ability to produce qualified products legally, namely whether the enterprise obtains relevant qualifications, reaches the national mandatory certification requirements and so on. If the enterprise continues to produce without reaching the above basic requirements, we can judge the enterprise has no subjective willingness of quality credit basically.

3.2 Quality contractual capacity index

Product is the medium of the quality contract and product quality is also an important way of whether the contract can comply with the quality contract and honor the reasonable expectations of consumers. The quality credit provides the capacity which includes the providing capacity of product quality and the quality credit guarantee.

The providing capacity of product quality refers to the capacity of providing the qualified product. It mainly includes the corresponding capacities of quality management system, standard system, measurement system, and so on. In addition, it also includes the advanced production condition, such as the advanced machinery equipment, production process, etc.

uality credit guarantee refers to the capacity of monitoring product quality and quality accident, as well as the control of accident and risk. For the products with quality risk, enterprises shall establish the relevant mechanism to monitor the quality status and quality risk, to control the potential quality accidents and to reduce the spread of happened accidents. For the quality accidents which have cause damage and loss to consumers, compensate in accordance with laws or enterprise standards higher than laws.

uality credit willingness and quality credit capacity both are the description and integration of internal characteristics of the enterprise quality credit and play a decisive role on the level of quality credit.

3.3 Quality contractual performance index

uality credit performance refers to the external performance of enterprise quality credit level, mainly including the quality contract status and the status of abiding by the contract. Namely, the indexes of qualified product status, the realization of product and service commitment, customer satisfaction index, customer complaints, quality supervision, and quality award and so on. If the quality credit level of an enterprise is high, it may perform as: the product quality meets requirements and maintains stable, consistency of external publicity and product quality, perfect after-sale service and other aspects. Only if the quality credit is high, it is possible to keep good external performance for a long time.

4 THE CLUSTER ANALYSIS ON QUALITY CREDIT EVALUATION INDEX OF ENTERPRISES

Enterprise quality credit evaluation index system is a multi-level complex system which consists of 3 first level indexes, some index items, wherein part of the index layer factor has fuzziness and the index layer factor will change according to the particularity of industry. To get effective and reasonable evaluation results and take into account the complexity and operability of the evaluation process, use index cluster method to classify the enterprise quality credit evaluation index for reducing the index dimensionality [9]. Select several typical indexes to represent the original numerous indexes for enterprise quality credit evaluation by calculating the average value of quantity of each classification index and other related indexes in the same classification, so as to simplify the evaluation dimensionality, innovate evaluation model, and strengthen evaluation quality [10].

Use the index cluster to classify the enterprise quality credit evaluation index, calculate the average value of all classification indexes' correlation coefficient and select the index of maximum average value of correlation coefficient as the typical index [11]. The clustering process is shown as below:

- Step1: View each index as the same classification, namely a classification includes an index. Calculate the similarity coefficient matrix wherein the element refers to the similarity coefficient of standard classification. The similarity coefficient matrix is a symmetric matrix.
- Step2: Merge the two classifications of maximum similarity coefficient (the minimum distance and maximum correlation coefficient) into a new classification and calculate the similarity coefficient of the new classification and other classifications.
- Step3: Repeat the step 2 till all indexes are merged as one classification.
- Step4: By analyzing the clustering process and the diagram and combining the professional knowledge and actual demands, determine the classification and index of each classification:
- Step5: Calculate the index correlation coefficient of each classification and select the maximum correlation coefficient index as the typical index.

5 THE EMPIRICAL STUDY ON QUALITY CREDIT EVALUATION INDEX CLUSTER

Take the production enterprise quality credit evaluation index as an example, for the convenience of calculation, only select 10 indexes of index layer and set them as X_1, X_2 X_{10} for index clustering analysis. According to the correlation data set by each index, calculate correlation coefficient between two indexes (take two decimal places) shown as table 1. The similarity coefficient of the index cluster is calculated by calculating the absolute value of the correlation coefficient between the two indexes, and the similarity coefficient of different classifications is calculated by the minimum similarity coefficient method. The correlation coefficient matrix obtained from calculating, namely the similarity coefficient matrix C_{ij} shown as Table 1.

Table 1: Similarity Coefficient C_{ij} Matrix of 10 Indexes

Index	X_1	X_2	X_3	X_4	X_5	X_6	X_8	X_9	X_{10}
X_2	0.89								

X_3	0.67	0.84							
X_4	0.40	0.35	0.26						
X_5	0.40	0.31	0.15	0.24					
X_6	0.44	0.37	0.21	0.21	0.87				
X_7	0.21	0.18	0.12	0.68	0.24	0.21			
X_8	0.62	0.54	0.37	0.28	0.73	0.82	0.22		
X_9	0.63	0.56	0.40	0.24	0.76	0.84	0.19	0.79	
X_{10}	0.34	0.31	0.22	0.89	0.23	0.20	0.80	0.26	0.23

First, 10 indexes will be 10 classifications separately. In the table 1, $C_{1,2}$=0.89 is maximum, merge the X_1 and X_2 to a new classification X_{11}, 9 classifications in total. Adopt the minimum similarity coefficient method to calculate the similarity coefficient of X_{11} and X_3, X_4, X_5, X_6, X_7, X_8, X_9 and X_{10}, $C_{3,11}$=min($C_{3,1}$, $C_{3,2}$)=min (0.67, 0.84)=0.67. Use the same method to calculate $C_{4,11}$=0.35, $C_{5,11}$=0.31, $C_{6,11}$=0.37, $C_{7,11}$=0.18, $C_{8,11}$=0.54, $C_{9,11}$=0.56, $C_{10,11}$=0.31. Compared with similarity coefficient of different classifications, $C_{4,10}$=0.89 is the minimum. Merge X_4 and X_{10} to a new classification X_{22}, 8 classifications in total. Continue to adopt the minimum similarity coefficient method to calculate the similarity coefficient of X_{11} and X_{22}, X_3, X_5, X_6, X_7, X_8, X_9, X_{22} and X_3, X_5, X_6, X_7, X_8, X_9. In this analogy, cluster the 10 indexes to a classification and the whole clustering process will be shown as Table 2.

Table 2: Clustering Process Table of 10 Indexes

Classification	Minimum Correlation Coefficient of Same Index	Clustering Status
10	1	X_1, X_2, X_3, X_4, X_5, X_6, X_7, X_8, X_9, X_{10}
9	0.89	(X_1, X_2), X_3, X_4, X_5, X_6, X_7, X_8, X_9, X_{10}
8	0.89	(X_1, X_2), X_3, (X_4, X_{10}), X_5, X_6, X_7, X_8, X_9
7	0.87	(X_1, X_2), X_3, (X_4, X_{10}), (X_5, X_6), X_7, X_8, X_9
6	0.79	(X_1, X_2), X_3, (X_4, X_{10}), (X_5, X_6), X_7, (X_8, X_9)
5	0.73	(X_1, X_2), X_3, (X_4, X_{10}), (X_5, X_6, X_8, X_9), X_7
4	0.68	(X_1, X_2), X_3, (X_4, X_7, X_{10}), (X_5, X_6, X_8, X_9)
3	0.67	(X_1, X_2, X_3), (X_4, X_7, X_{10}), (X_5, X_6, X_8, X_9)
2	0.19	(X_1, X_2, X_3), (X_4, X_5, X_6, X_7, X_8, X_9, X_{10})

1	0.12	(X_1, X_2, X_3, X_4, X_5, X_6, X_7, X_8, X_9, X_{10})

By analyzing the clustering process, the 10 indexes are clustered from three classifications to two classifications. The minimum similarity coefficient changes from 0.67 to 0.19, jumping from the large value to small value. Meanwhile, considering the three primary indexes of enterprise quality credit evaluation index system in the target layer, it is determined to be clustered to three classifications. The index including in each classification will be shown as table 3.

Table 3: Clustering Results of Three Classifications

Classification	Index	Minimum Correlation Coefficient
Classification 1	X_1, X_2, X_3	0.67
Classification 2	X_4, X_7, X_{10}	0.68
Classification 3	X_5, X_6, X_8, X_9	0.73

The minimum correlation coefficient of three classifications is 0.67, 0.68, and 0.73 individually without large differences which indicate the good clustering results. The three classification indexes, combining the specific meaning of index with the enterprise quality credit evaluation index system, individually reflect the evaluation contents of enterprise quality trustworthiness willingness, quality contractual capacity and quality contractual performance, which conform to the actual status. The clustering effect is satisfied.

Finally, select the typical index of each classification. The r_{ij} of indexes of three Classifications will be shown as table 4.

Table 4: r_{ij} of Indexes of Three Classifications

Classification 1			Classification 2			Classification 3			
	X_1	X_2		X_4	X_7		X_5	X_6	X_8
X_2	0.89		X_7	0.68		X_6	0.87		
X_3	0.67	0.84	X_{10}	0.89	0.80	X_8	0.73	0.82	
						X_9	0.76	0.84	0.79

Calculate the average $\overline{R_i^2}$ value of correlation coefficient of each index of each classification, the average value of the correlation coefficient of X_1 and X_2, X_3 of classification one is $\overline{R_i^2} = \frac{0.89^2 + 0.67^2}{3-1} = 0.62$, the average value of the correlation coefficient of X_2 and X_1, X_3 of classification one is $\overline{R_i^2} = \frac{0.89^2 + 0.84^2}{3-1} = 0.75$, the average value of the correlation coefficient of X_3

and X_1, X_2 of classification one is $\overline{R_i^2} = \frac{0.67^2 + 0.84^2}{3-1} = 0.58$. The $\overline{R_i^2}$ values of all indexes are shown as Table 5.

Table 5: $\overline{R_i^2}$ Value of Indexes of Each Classification

Classification	Index	$\overline{R_i^2}$ Value	Classification	Index	$\overline{R_i^2}$ Value	Classification	Index	$\overline{R_i^2}$ Value
1	X_1	0.62	2	X_4	0.63	3	X_5	0.62
1	X_2	0.75*	2	X_7	0.55	3	X_6	0.71*
1	X_3	0.58	2	X_{10}	0.72*	3	X_8	0.61
						3	X_9	0.64

Select the maximum value Xi of $\overline{R_i^2}$ from all classifications as the typical index. Therefore, select X_2 enterprise legal qualification index from classification 1, X_{10} quality credit management capacity index from classicist 2, and X_6 customer satisfaction index from classicist 3 as the typical index. The selected enterprise legal qualification index, quality credit management capacity index and customer satisfaction index can be used as the typical index of the production industry quality credit evaluation to grasp the key point of enterprise quality credit evaluation rapidly and effectively, and have important practical guiding significance to reduce the dimensionality of index, simplify the evaluation process and fast decision.

6 CONCLUSIONS

The development of enterprise quality credit evaluation is an important basis to prevent the quality dishonest risk and quality dishonest behavior, to grasp the enterprise quality credit status, to carry out the quality credit classification supervision effectively [12]. The enterprise quality credit evaluation, based on the index cluster, scientific and objective credit evaluation model and through the index clustering analysis, can simplify the evaluation content, strengthen the evaluation quality and innovate the evaluation model; find the key factor influencing the enterprise quality credit level rapidly and effectively; improve the efficiency of enterprise quality credit classification supervision, as well as promote recognize and analyze quality dishonest risk rapidly, strengthen the establishment of quality contractual capacity and progressively improve the enterprise quality credit level.

7 REFERENCES

[1] **Zhou Li, Li Bisong. 2014.** Research on construction and system Framework of China's social credit standardization, Standard science, 01,pp 11-14.

[2] **Zhou Li. 2014.** Understanding of Implement specifications for enterprise quality trustworthiness management', China quality supervision, 04,pp 18-21.

[3] **GB/T 31863-2015.** Evaluation index of enterprise quality credit.

[4] **M. Crouhy, D. Galai, R. Mark. 2000.** A Comparative Analysis of Current Credit Risk Models, Journal of Banking Finance, Vol. 24, No. 1-2, pp 59-117.

[5] **A. N. Berger, G. F. Udell. 2002.** Small Business Credit Availability and Relationship Lending: The Importance of Bank Organizational Structure, The Economic Journal, Vol. 112, No. 477, pp F32-F53.

[6] **A. N. Berger, W. S. Frame. 2007.** Small Business Credit Scoring and Credit Availability, Journal of Small Business Management, Vol. 45, No. 1, pp 5-22.

[7] **Zhou Li, Ye Ruyi, Xian kuitong, Wang Yanfeng. 2010.** Theory and practice of quality credit system construction, 1st Edition, China zhijian publishing house.

[8] **GB/T 31870-2015.** Guides for drafting enterprise quality credit report.

[9] **Ying Zhao,George Karypis. 2005.** Data clustering in life sciences, Molecular Biotechnology, 01, pp 55-80.

[10] **Tang xiaoqin, Dai ruyuan. 2003.** Technique of cluster analysis in Data mining, Microcomputer Information, 01, pp 3-4.

[11] **Zhou Li, Xian Kuitong, Li Bao. 2009.** Study on the Development Model of MDA-Based Enterprise uality Credit Information System, Information Engineering and Electronic Commerce, Ternopil, May 16-17, pp 423 427.

[12] **Yang Longguang, Wu Jingmei. 2014.** A research on the credit rating to SMEs in China based on the perspective of statistics and clustering, Journal of Sichuan University (Social Science Edition), 06, pp 89-97.

EVALUATION TECHNOLOGY STUDY OF ENTERPRISE QUALITY CREDIT BASED ON COMPREHENSIVE QUALITY RESPONSIBILITY

Zhou Li*, Liu Bisong, Jiang Zhou and Zhao an
China national institute of standardization, Beijing, China
zhouli@cnis.gov.cn

ABSTRACT

The main core of enterprise quality credit is to perform quality promise, as well as to fulfill quality responsibility. In this paper, enterprise quality credit evaluation based on total quality responsibility has been presented. Around the product life cycle production and according to the requirements of comprehensive quality responsibility, enterprise quality credit evaluation elements have been put forward, also, enterprise quality credit evaluation model has been constructed. According to the progressive layers of quality responsibility, enterprise quality credit basic evaluation index, development potential index and progressive evaluation index have been established respectively, aiming to improve enterprise quality credit level, as well as providing technical reference for implementing total enterprise quality credit evaluation continuously.

Keywords: uality responsibility, uality credit, Evaluation index

1 INTRODUCTION

Enterprise social responsibility has fundamental characteristics of voluntary, nobility, initiative, accountability, comprehensiveness, persistence and conventionality, which embody the relationship between the enterprise and interested party and social environment. It furthest creates comprehensive value of economy, society and environment, and takes laws and regulations as base line, to optimize the allocation of social resources. Enterprise quality credit is basically featured in spontaneity and binding character, which embody the relationship between the enterprise and consumers. It furthest creates products and services meeting consumers' demands, and takes laws and regulations as base line, to realize the enterprise's economic value. Therefore, as economic value of market subject, the enterprise emphasized in enterprise quality credit is one of enterprise's foundations performing social responsibilities [1]. To perform quality responsibilities is basic content for the enterprise to perform social responsibilities.

Enterprise's quality responsibility is a part of the enterprise's social responsibility, as well as one of the enterprise's basic and core social responsibilities [2]. The core of the enterprise's quality credit is to perform quality promise and the core of performing quality promise is to perform the enterprise's quality responsibilities. To systematically arrange enterprise's comprehensive quality responsibilities and clear enterprise's quality promise content is foundation and key of scientifically developing enterprise's quality credit evaluation.

2 QUALITY RESPONSIBILITY IS BASIC CONTENT OF ENTERPRISE'S SOCIAL RESPONSIBILITIES

* Corresponding Author

Basic feature of social responsibilities is a will that an organization considers the society and environment into its decision and undertakes responsibilities for influence of activities and activity decision on the society and environment [3]. The will embodies transparent and ethical behaviours, promotes sustainable development, considers interest of interested parties, conforms to applicable laws and brings into correspondence with international code of conduct, as well as embodies integration in the whole organization and implementation in various relations.

Based on recognition of connotation of enterprise's social responsibilities as well as consideration and judgment of domestic current economical situation development and current enterprise status, recognition of modern enterprise's social responsibilities shall take economy as foundation, take sociality as supplement, and take globalization as altitude. To take economy as foundation emphasizes that an enterprise must be an economic organization; this shall not be expected to finish non-economic purpose without financial stimulation. Therefore, to realize economy, an enterprise must realize its own basic responsibilities, that is to say, realize the enterprise' growth and interest, to provide products meeting customers' requirements and expectation [4]. To survive stably in the market is the most fundamental social responsibility.

To perform social responsibilities, an enterprise shall return origin, which is to meet consumers' demands, provide products and services conforming to consumers' demands, and performing enterprise's quality responsibilities [5]. Therefore, we can know that an enterprise's basic social responsibility is to provide products and services meeting consumers' demands, that is to say, to perform the enterprise's quality responsibility, which is basic content of the enterprise's social responsibilities.

3 QUALITY CREDIT IS TO PERFORM AN ENTERPRISE'S BASIC SOCIAL RESPONSIBILITIES

The nature of quality credit is to abide by quality contracts between an enterprise and consumers, and reflects the enterprise's ability to gain and keep consumers' trust in the enterprise's product quality before consumers' actual purchase (transfer of product value). Therefore, quality credit can be defined as an enterprise's will and ability to provide products meeting consumers' demands or expectation to gain and keep consumers' trust of product quality before purchase based on fulfil product quality contract (conforming to related laws and regulations of quality, executing standards and fulfilling quality promise and consumers' reasonable expectation).

uality credit mainly concerns consumers' satisfaction about products and services provided by an enterprise [6]. uality credit requests an enterprise to abide by laws and regulations, and operate in good faith, guarantee quality, meet consumers' demands and prevent risk, which is a part of an enterprise's social responsibilities. In the view of socialized mass production, only by meeting quality credit requirements and providing products and services meeting consumers' demands can an enterprise better perform other aspects of enterprise's social responsibilities.

Products are a link of quality contract and the unique medium, through which an enterprise conforms to related laws and regulations of quality, executes standards and fulfils quality promise and meets consumers' reasonable expectation, therefore, the key for an enterprise to perform social responsibilities is to be responsible for product quality. Consequently, the requirements that an enterprise meets quality credit are basic guarantee of realizing social responsibilities. An enterprise's quality credit is to perform the enterprise's quality responsibilities, which is the most fundamental social responsibility.

4 ENTERPRISE QUALITY CREDIT EVALUATION ELEMENTS BASED ON COMPREHENSIVE QUALITY RESPONSIBILITIES

For an enterprise, enterprise quality credit evaluation based on comprehensive quality responsibilities can review enterprise quality credit ability and level in the view of management of the whole life cycle of enterprise's products [7]. Whether the products and services provided by an enterprise are in good faith directly influences consumers' vital interests; what's more, in phases of product R D and design, purchase, production, sales and after-sales service, enterprise quality credit ability and level also directly influences multiple aspects such as enterprise organization governance, staffs' rights and benefits, environmental protection, resource consumption and social development, which are all problems concerned by enterprise's comprehensive quality responsibilities as well as influential factors of quality credit. The ring and process of enterprise's quality credit management are considered in the view of life cycle of products, including subject will of enterprise quality credit, product quality promotion ability and guarantee conditions in multiple phases of product R D and design, purchase, production, sales and after-sales service [8].

- Product R D and design: Market access, Administrative licensing, Compulsory certificate, Product design and development
- Purchase: Purchase system, Contract management process, Supplier management, uality management system, Standard system, Measuring system
- Production: uality monitoring, Production conditions, Production standard, uality risk monitoring, HR, Technical resources, Information resources, Communication mechanism, Monitoring and measuring, uality inspection
- Sales: Product quality, Product quality promise, Product quality conformance, uality award, Pre-sales advertisement propaganda, uality guarantee, Supervision and random checking
- After-sales service: After-sales service, Customer satisfaction, Self-assessment, Information collection, Supervision and inspection, Solve quality complaint, Product or service quality traceability, Recall or compulsory withdrawal of defective, Improved mechanism, Disposition of quality credit loss

5 ENTERPRISE QUALITY CREDIT EVALUATION TECHNOLOGY BASED ON COMPREHENSIVE QUALITY RESPONSIBILITY

5.1 Enterprise quality credit evaluation model based on comprehensive quality responsibility

In the view that an enterprise performs comprehensive quality responsibility, the quality credit evaluation model is established according to influence elements in phases of management of enterprise product life cycle, by combining main aspects considered in enterprise quality credit evaluation-contractual will, contractual capacity and contractual performance. Based on analysis of influence elements in the above phases, influence degree of elements is considered according to investigation aspect of elements in the model.

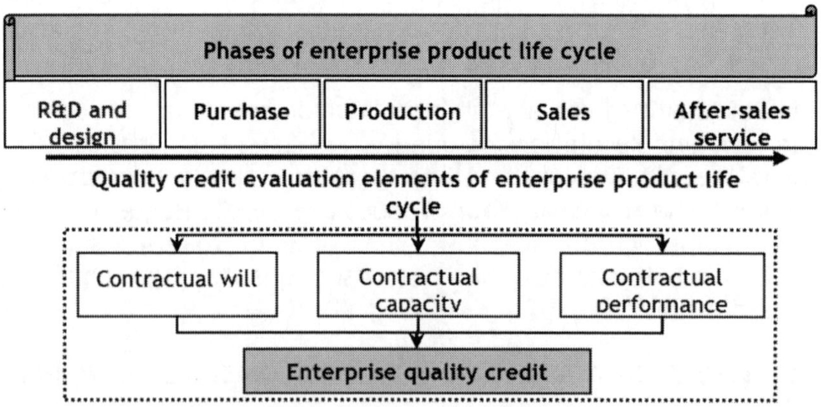

Figure 1: Enterprise Quality Credit Evaluation Model

5.2 Enterprise quality credit evaluation indicators based on comprehensive quality responsibilities

Based on establishment of enterprise quality credit evaluation model, establish the enterprise quality credit evaluation indicator system according to three main investigation aspects of enterprise quality credit evaluation. According to requirements of adaptation and development of evaluation indicators, the enterprise quality credit evaluation indicator system is divided into basic evaluation indicator and development potential indication, wherein the basic evaluation indicator is regarded as basic requirements for evaluation current enterprise quality credit and development potential evaluation indicator is mainly regarded as potential requirements for guiding continuous promotion of enterprise quality credit. Considering the range of comprehensive quality responsibilities and phased key point of enterprise quality credit evaluation, the evaluation indicator which may be increased in quality credit evaluation in the future is conducted anticipating exploratory research according to current national conditions as well as social and economic development trend [9]. Based on the above research on enterprise quality credit evaluation indicator system, progressive evaluation indicator is added as supplementary requirements of future social development trend for enterprise quality credit evaluation indicators.

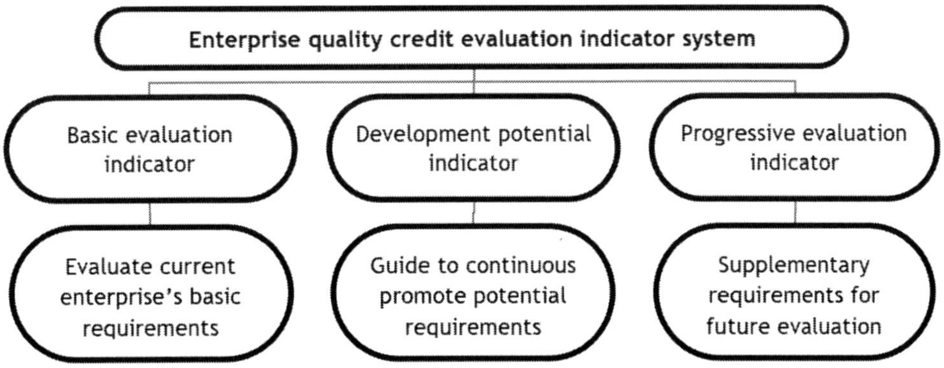

Figure 2: Enterprise Quality Credit Evaluation Indicator System

5.2.1 Basic evaluation indicators

According to requirements for comprehensive quality responsibility, analyze influence degree of quality information of enterprise product life cycle. Make comprehensive analysis of enterprise's comprehensive quality responsibilities in phases of product R D and design, purchase, production, sales and after-sales service. Implement the idea of quality integrity, perform quality promise capacity and fulfil quality promise, to establish basic evaluation indicators of enterprise quality credit respectively [10].

Table 1: Basic Evaluation Indicators of Enterprise Quality Credit

Level-1 indicator	Level-2 indicator
Implement the idea of quality integrity	Management policy of quality integrity
	Management target of quality integrity
	Cultural construction of quality integrity
Perform quality promise capacity	Production and operation qualification

	uality management
	uality integrity management
	Standard management
	Measuring management
	Certification management
	Inspection management
	Personnel qualification management
Fulfil quality promises	The quality supervision department's record of supervision and random checking
	Record of quality safety accidents
	Record of poor quality of exported products
	Spot check record of other related departments
	Market feedback and complaint record
	Award related to quality
	Other external evaluation
	Initiative after-sales management

5.2.2 *Indicators of development potential*

Based on current quality responsibilities, by expanding enterprise's concern on consumer problem and equal operation of enterprise, indicators of development potential of enterprise quality credit evaluation include 3 levels of indicator systems.

Table 2: Evaluation Indicators of Development Potential of Enterprise Quality Credit

Level-1 indicator	Level-2 indicator	Level-3 indicator
Consumers' problems	Fair trade	Disclosure of total price of product and tax
		Disclosure of using conditions and related expenses
		Disclosure of payment mode of credit consumption
		Words, audio or images are not used
		Consider optimal interest of disadvantaged groups in advertisement and marketing, not work on business activities damaging their interests
	Health and safety of	Design of product service information shall consider demands and capacity

	consumers	of different consumers
	Sustainable consumption	Provide products more attractive in price and quality
		Provide environment and social information related to production and product for consumers
	Protection of consumers' privacy	Collect information of consumers responsibly
		User consumer information responsibly
		Save consumer information responsibly
	Acquisition of basic service	Extend the time limit of basic life service for consumers who do not pay, and do not stop providing service for all consumers in the group for some consumers do not pay.
		Provide subsidy price list for consumers in need.
		Provide transparent information between price and charge
		Expand service coverage area and provide the same services for all consumers equally without discrimination
		Management reduction or interruption of service equally
		Maintain and update service system, to avoid damage to service supply
Fair operation	Fair competition	Conform to laws and regulations of fair competition
		Cooperate with related departments to maintain the environment of fair competition
		Prevent work on or assistance of anti-competition behaviours
		Improve staffs' awareness of conforming to laws and regulations of competition and conducting fair competition
		Establish perfect practice supporting anti-monopoly and anti-dumping and encourage public policies of fair competition
		Do not utilize special social conditions to conduct unfair competition

5.2.3 *Progressive evaluation indicators*

Progressive evaluation indicators of enterprise quality credit combine concrete requirements of related terms of international standards of ISO26000 Guidance of social responsibilities, to conduct detailed analysis of related terms of enterprise quality credit based on social responsibility in aspects such as organization governance, human right, labour practice, environment, fair operation practice, consumer problem, communication participation and development [11]. It selects 24 Level-3 evaluation indicators as progressive evaluation indicators of enterprise quality credit performing comprehensive quality responsibilities in the future.

Table 3: Progressive Evaluation Indicators of Enterprise Quality Credit

Level-3	Effectively use economic, natural and human resources (effective use of labour, finance and objects)

indicator	Balance organization decision maker's level of responsibilities, rights and capacity
	Analyze and control occupational health and safety risk of all activities of enterprise
	Provide necessary safety equipment
	Occupational health and safety training of all related personnel
	Analysis of/Evaluation on enterprise decision making and activity environment
	Whole control of pollution sources and waste of product and activity life cycle
	Measure, record and report major pollution resources and pollution, water consumption, waste generation and reduction of energy consumption
	Measures aiming to prevent pollution and waste
	Products have good property to reduce and minimize direct or indirect pollutants
	Effectively utilize energy, water and other resources, to save water, electricity and energy
	Take effective measures to reduce emission of greenhouse gas as possible
	Abide by laws and regulations of fair competition
	Prevent work on or assistance of anti-competition behaviours
	Improve staffs' awareness of conforming to laws and regulations of competition and conducting fair competition
	Establish perfect practice supporting anti-monopoly and anti-dumping and encourage public policies of fair competition
	Do not utilize special social conditions to conduct unfair competition
	Conform to laws and regulations of fair competition
	Promise to share related information in transparent way
	Disclose all price information related to product and service
	Consider optimal interest of disadvantaged groups in advertisement and marketing, not work on business activities damaging their interests
	Preferentially select materials contributing to sustainable development
	Provide sufficient and effective enterprise support and consulting system
	Maintain and upgrade service system, to avoid damage to service supply

6 CONCLUSIONS

The main purpose of quality credit is to promote an enterprise to provide products and services meeting consumers' demands, guarantee personal healthy safety of consumers, and maintain consumers' rights and interests [12]. These requirements can be understood to provide products and services through integrity business and to be sustainable development of enterprises contributing to meeting consumers' demands. Development and deepening of connotation and requirements of quality credit is deep embodiment and integration endowed to quality credit by the idea of comprehensive quality responsibility, therefore, the research on technology of enterprise quality credit evaluation based on comprehensive quality responsibility is to grasp the real body of enterprise quality promise, take enterprise product production life cycle as the centre, systematically arrange enterprise's comprehensive quality responsibility, to scientifically establish an indicator system of enterprise quality credit evaluation in the aspects of enterprise's will of keeping an agreement faithfully, construction of contractual capacity and expression of contractual behaviours. At the same time, considering development and containment of comprehensive quality responsibility, it proposes evaluation indicators of development potential and progressive evaluation indicators, to provide technical reference for evaluation on enterprise quality credit in the future.

7 REFERENCES

[1] **Zhou Li, Ye Ruyi, Xian kuitong, Wang Yanfeng. 2010.** Theory and Practice of uality Credit System Construction, 1st Edition, China zhijian publishing house.

[2] **Li Guoping, Wei Xiaoqian. 2014.** Definition, Measurements and Economic Consequences of Corporate Social Responsibility: A Survey on Theories of Corporate Social Responsibility, Accounting Research , 08, pp 33-40.

[3] **Li Weiyang, Xiao Hongjun. 2011.** Logics for Corporate Social Responsibility, China Industrial Economics, 10, pp 87-97.

[4] **Zhou Li. 2013.** Technology Research on uality Credit Evaluation of the Enterprise with Real-name System. Standard science. 05,pp 77-81.

[5] **Zhang Zhaoguo, Liang Zhigang, Yin Kaiguo. 2012.** Research on Corporate Social Responsibility from the Stakeholder Perspective, China Soft Science, 02, pp 139-146.

[6] **Zhou Li, Xian Kuitong, Li Bao. 2009.** Analysis of Developing Enterprise uality Credit Classified Supervision System Based on B/S Model. Information Engineering and Electronic Commerce, Ternopil, May 16-17, pp 500 505.

[7] **Yu Gaofeng, Luo Donghui, Zhang Jinmei, Feng Bingqiu, Qiu Jinming. 2016.** Credit Evaluation of Enterprise uality based on Factorial Variable Weight Analysis, Journal of Liaoning Technical University (Social Science Edition), 04, pp 503-507.

[8] **Ali Saeidi, Olivier Deck. 2013.** Thierry Verdel. Comparison of Building Damage Assessment Methods for Risk Analysis in Mining Subsidence Regions, Geotechnical and Geological Engineering, August, Volume 31, Issue 4, pp 1073 1088.

[9] **Yu Gao-feng, Liu Wenqi, Shi Mengting. 2015.** Credit Evaluation of Enterprise uality based on Local Variable Weight Model, Journal of Management Sciences in China, 02, pp 85-94.

[10] **GB/T 31863-2015,** Evaluation index of enterprise quality credit.

[11] **ISO 26000:2010.** Guidance on social responsibility.

[12] **Wu Chong,Wang Ying,Guo Yingjian. 2008.** The Model of Personal Credit Risk Assessment On Support Vector Machine, Operations Research and Management Science, 04, pp 96-99,105.

IMPROVEMENT OF MANPOWER UTILIZATION IN SUBLINE AND MAINLINE OF COMPANY X USING ASSEMBLY LINE BALANCING

Ryan Jeffrey P.Curbano* Keith Patrick S. Hernandez, Sairah Gail C. Palma, Ericka Joyce C. Panghulan,
Industrial Engineering Department, Lyceum of the Philippines University – Laguna
Corresponding email: *ryanjeffrey.curbano@lpulaguna.edu.ph

Abstract

The main objective of this study is to improve the manpower utilization in Subline of Company X using assembly line balancing. With the use of Stream Diagnostic Diagram, the main causes of the problem in the production line were identified such as high work-in-process, poor layout, low efficiency, and unbalanced processes. Time study and line balancing technique were used to determine the cycle time and number of operators in the assembly lines. Merging of lines and operations and adding of operators were the two alternatives proposed. These alternatives were validated using POM-QM software. Based on the findings of the study, merging of lines in Subline, the manpower reduces from 368 to 336 operators. The cycle time also reduces from 39.16 to 19.99 sec in Subline. The efficiency of the Subline has an increase of 17.39% .The savings of the company is increased by 10.53%.

Keywords: Assembly line balancing, manpower utilization, efficiency, cost-benefit, time and motion study, manpower improvement

1. Introduction

Productivity is considered as one of the most common measurements used in a production company to know the efficiency status of the production area. It is measured in terms of production output over time and/or total number of workers [1]. The line balancing technique is one of the tools to decrease the production cost, maximize the output, and minimizing the production cost of a product [2].

A balanced line enables the products to flow continuously without the presence of bottleneck. The task designation for each workstation is required to balance the production line under the sequential constraints, making the operation time at each station equally without the idle time. [3]

Productivity levels are also important for industries and companies. For companies, a higher productivity relative to their competitor gives them a competitive advantage over their competitors in the marketplace.

Every company aims to maximize the productivity and efficiencies of their labour forces. Effective utilization and allocation of manpower will result to cost reduction, clarity in performance tasks, and saves money, time, and wastes in resources. It can also lead to an easier achievement of the organizational goals [4].

Increased productivity and efficiency are the two major things to consider in a manufacturing company. The innovation team in Company X, Inc. is doing an intensive research on how to lessen the cost that will not sacrifice the quality of the product. The production area of the company is constantly hitting the company's target output but they are having a problem in the cost of production in terms of manpower and efficiency.

Company X has eight assembly lines. The production is operating eight hours with two shifts (Day and Night) with a 30 working days. This study focused in the two assembly lines of Project S: Sub line. The company's monthly average output is 5,381,808 units. In this case, the daily output per line should be 24,000 units. The assembly line has a takt time of 2.57 seconds which has an equivalent output of 1500 units per hour.

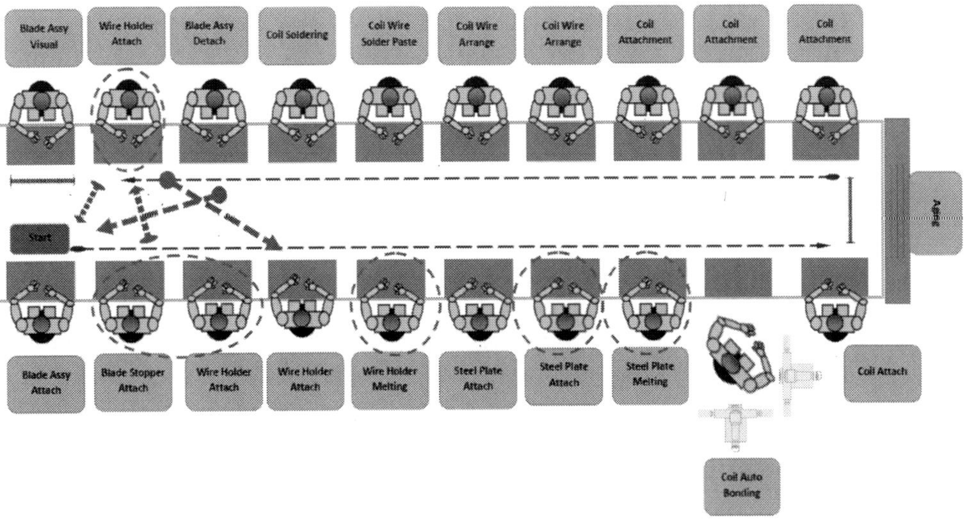

Figure 1. *Current Process in Subline*

Figure 1 shows the current condition of subline. There are 18 operations with a total of 23 operators. The Wire Holder Attachment, Wire Holder Melting, Steel Plate Meting Visual, Auto Coil Bonding, and Final Visual exceeds the required takt time per unit of 2.57 seconds.

2. Conceptual Framework

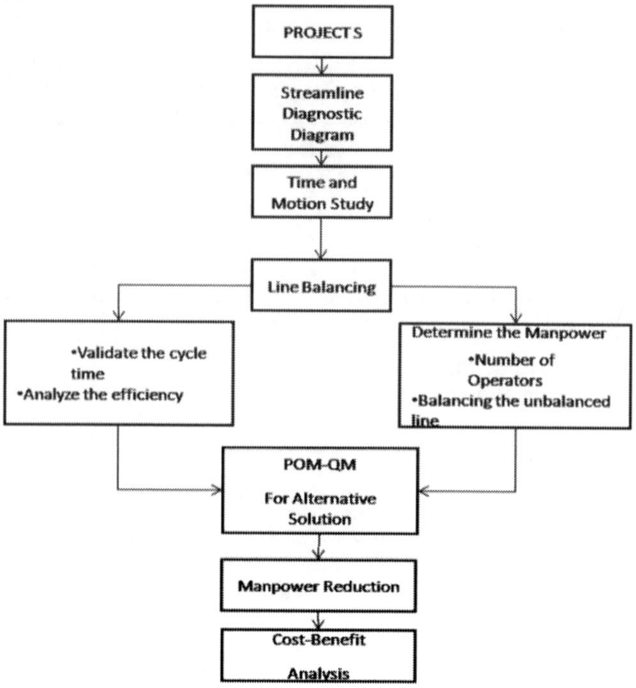

Figure 2. *Research Paradigm*

The researchers used Streamline Diagnostic Diagram to determine the main problem of project S. Time and motion study and line balancing are the effective tools essential to determine the optimal solution in the objectives. Through motion study they were able to validate the cycle time and efficiency and through line balancing the researchers were able to analyse the manpower and the assembly line. Cycle time is important in balancing the assembly line.

The number of operators and cycle time will be run in the software Production/Operations Management, Qualitative Methods (POM-QM) to validate if the solution is feasible and attainable. And with this, the researchers will be able to reduce the manpower, improve the productivity and solve the cost benefit analysis

3. Objectives of the Study

The main objective of this study is to improve the productivity in Subline of Project S in Company X. This study aims to attain the following objectives:

3.1 To validate the cycle time in Subline in Project S

3.2 To propose solution to improve the productivity in the assembly line in terms of manpower and efficiency.

3.3 To compute and analyze the efficiency of Subline in Project S

3.4 To assess the Cost-Benefit Analysis of the proposed alternatives.

4. Methodology

4.1 Research Design

The researchers used applied type of research. Applied research is a form of systematic inquiry involving the practical application of science.

4.2 Research Instrument

In calculating the standard time, cycle time, and efficiency of assembly line the proponents used time and motion study. It is needed to have the performance rating of each operator to solve the standard time which is calculated using the Performace Rating Table. The researchers gave 10% allowance to each operator based on International Labor Organization (ILO) Allowance Table. The 10% allowance is consists of 5% personal allowance, 4% basic allowance, and 1% monotonous activity.

5. Results and Discussion

5.1 Summary of cycle time

Table 1. Summary of cycle time

	Total No. of Operators	Total Standard Cycle Time (sec)	Total Standard Cycle Time/ Unit (sec)
Subline	23	94.1	53.4

Sub line can produce 1628 units per hour with a total of 23 operators working in the assembly line as shown in table 6. The company's target output is 1500 units, which means that the Sub Line exceeds the target hourly output by 128 units.

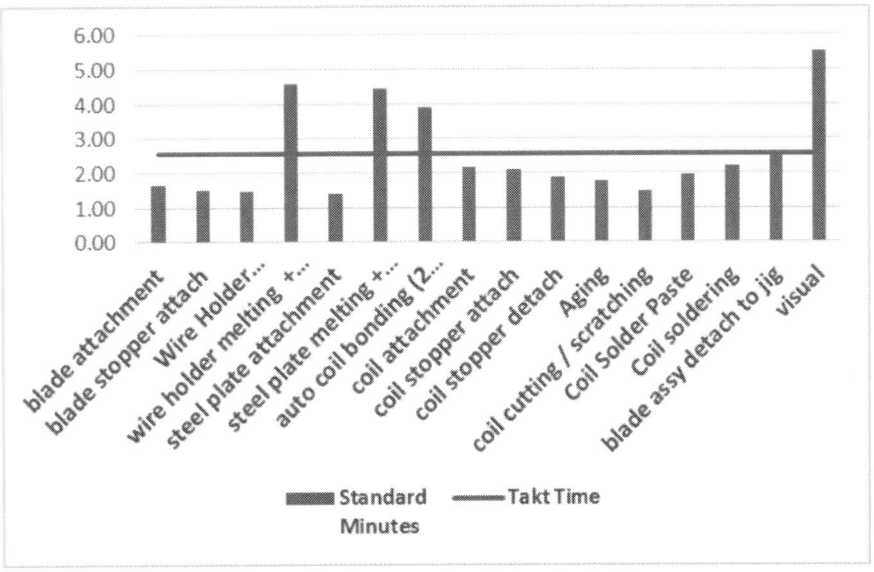

Figure 3. *Current Condition in Subline*

The figure above shows the graphical representation of the cycle time of Sub line along with the required takt time of 2.57 seconds per unit. Determining and comparing the cycle time and the takt time, improves the productivity of the workers and also one can establish the process that needs improvement [1]. Wire holder Melting + Visual, Steel plate attachment, Steel Plate Melting + Visual, Auto coil bonding and visual exceeds the required takt time. Operations having total cycle time more than the target takt time are the bottleneck stations that should be considered a priority for the improvement. Bottleneck elimination will help improve the target output of a company. But, if this problem will not be resolved, total productivity management will suffer. One of the concrete solutions to this problem is to come up with a strategic plan that will focus in line balancing. Having the right number of workers in an assembly line will result to greater efficiency and high productivity [5].

5.2. Problem Identification

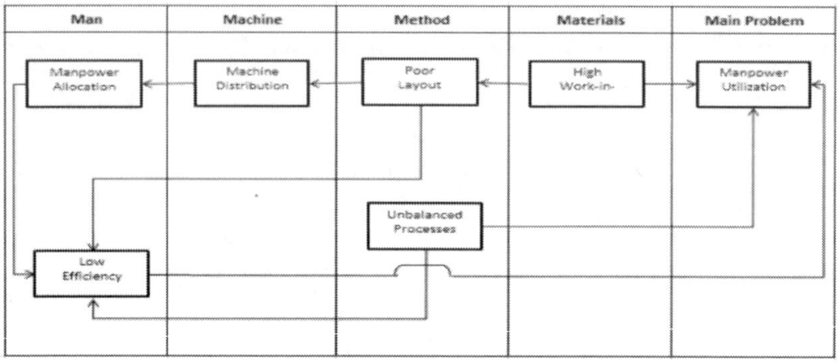

Figure 2. Streamline Diagnostic Diagram

The Stream diagnostic diagram determines the relationship of each problem. In the diagram above, manpower utilization is the main problem because of the high work-in-process, unbalanced line, and low efficiency. There is a low efficiency in the assembly line because of the unbalanced process, poor layout, unbalanced distribution of workload, and manpower allocation. Unbalanced Manpower allocation is the result of unbalanced machine distribution, poor layout and high work in process. Unbalanced process and high work-in-process (WIP) were the root causes of the problem since the down streaming lead to the Manpower utilization. Solving these two main reasons will solve the main problem.

5.3. Manpower Requirements

Table 2. Manpower Requirements in Subline before Line Balancing

Operations	Standard Time (Sec)	Number of Operators Theoretical		Actual
WireHolder Attachment	4.47	1.86	≈ 2	3
Wire Holder Melting + Visual	4.6	1.92	≈ 2	1
Steel Plate Meting + Visual	4.44	1.85	≈ 2	1
Auto Coil Bonding	4.88	2.03	≈ 2	1
Final Visual	5.53	2.3	≈ 2	1

Table 2 shows the theoretical and actual number of operators in Sub line. Standard time was used to compute for the theoretical number of operators. The total computed number of operators in theoretical is twenty-six (26) and the total actual operators is twenty-three (23). This means that the current assembly line is insufficient with three (3) operators.

5.4. Proposed Solution

Table 3. Manpower Requirements

Operation	Standard Time	Number of Operators
Blade Attachment	1.64	1.37 ≈ 2
Blade Stopper Attach	1.53	1.28 ≈ 2
Wire Holder Attachment	5.2	4.33 ≈ 5
Wire Holder Melting + Visual	4.6	3.83 ≈ 2
Steel Plate Attachment	2.78	2.32 ≈ 3
Steel Plate Melting + Visual	4.36	3.63 ≈ 2
Auto Coil Bonding (2 Machines + Manual Bonding + Visual)	4.79	3.99 ≈ 2
Coil Attachment	8.65	7.21 ≈ 7
Coil Stopper Attach	2.09	1.74 ≈ 2
Coil Stopper Detach	1.89	1.58 ≈ 2
Aging	1.83	1.53 ≈ 2
Coil Cutting / Scratching	3.01	2.51 ≈ 3
Coil Solder Paste	1.96	1.63 ≈ 2
Coil Soldering	2.53	2.11 ≈ 2
Blade Assy Detach To Jig	2.62	2.18 ≈ 2
Visual	2.53	2.11 ≈ 2
Total	**52.01**	**42**

Merging of assembly lines was proposed. One of the objectives of line balancing is to acquire tasks groupings that speak to around equivalent time prerequisites. This minimizes the idle time along the line and and results in high use of work [6]. Table 7 shows that the merged assembly lines will reduce the number of operators from 46 to 42. Subline has a total of eight (8) assembly lines with 23 operators. Four (4) operators will be reduced in merging of the assembly lines. In total, Subline will be composed of four (4) assembly lines with 336 operators (day and night shift).

Table 4. Output of Subline after Line Balancing

Operations	Remarks	Proposed Number of Operators	Standard Time (sec)	UPH Per operator	UPH per process	UPH of the line	Target UPH
coil stopper detach	Additional 1 operator	2	0.94	1918 1899	3818		
blade assy aging - walking / loading blade assy aging blade assy aging - unloading / walking	Additional 1 operator	2	0.92	2044 1895	3933		
coil cutting / scratching coil cutting / scratching	Additional 1 operator	3	1	1552 994 1161	3584		
Coil Solder Paste	Additional 1 operator	2	0.98	1847 1828	4381		
Coil soldering	Additional 1 operator	2	1.27	1429 1414	2843		
blade assy detach to jig	Additional 1 operator	2	1.31	1440 1311	2745		
Visual	Additional 1 operator	2	1.26	1446 1402	2847		
Total		**42**	**19.99**			**3236**	**3000**

The proposed solution in Subline is to merge the assembly lines. In the current process, it has 8 assembly lines but in the proposed solution, it now have 4 assembly lines. There are a total of 42 operators in each merged line. The target output is doubled from 1500 to 3000 units per hour since the lines are now merged. The proposed Subline can produce 3236 units per hour which exceeds the target output by 236 units.

5.5 POM-QM Software Validation Using Longest Operation Time

Table 5. Line Balancing Results

	Current	**Proposed**
Summary Statistics		
Cycle Time (sec)	88.48	34.5
Min(theoritical) # of stations	1	1
Actual # of workstations	1	1
Time Allocated (cycle time * # stations)	88.48	34.5
Task needed (sum of task time)	41.97	19.99
Idle time (allocated-neeeded)	46.51	14.51
Efficiency (needed/allocated)	47.43%	57.94%
Balance Delay (1-efficiency)	52.57%	42.06%

Table 5 shows the validated result of the alternative 1 in Subline using the POM-QM software. The proposed alternative 1 has a cycle time of 34.5 seconds. Compare to the current condition, the efficiency increased from 47.21% to 57.94% with the same number of workstation.

5.6 Efficiency

Table 6. Summary of Efficiency

	Current	Proposed Solution	Actual Efficiency
Subline	46%	54%	64.7%

The efficiency of the current Subline was 46% based on the time and motion study. The increase in efficiency in Subline is from 46% to 54% and. The 54% efficiency of the proposed solution in Subline is based on the simulated data. The proposed solution is already implemented and adapted by the company wherein the actual efficiency is 64.7%.

5.7 Cost-Benefit Analysis

Table 7. Cost- Benefit Analysis

	Current	Proposed
No. of Workers per Line	23	42
No. of Assembly Lines	8	4
No. of Working Days	30	30
Total Number of Workers(Day & Night Shift)	368	336
Cost per Month	₱127,666,560	₱116,565,120
Cost per Year	₱1,531,998,720	₱1,398,781,440

Table 7 shows the cost-benefit table of Subline. The current cost of the company in terms of the salary of the 368 operators in Subline (day and night shift) is ₱1,531,998,720 per year. In alternative 1 the company will cost ₱1,398,781,440 per year for paying the forty-two (42) operators per line in four (4) assembly lines in Sub line. If implemented the savings will be ₱133,217,280 per year. In alternative 2 with twenty-four (24) operators per line in eight (8) assembly lines the cost per year of the company is ₱1,665,216,000. The cost of the company in terms of labor cost will increase by ₱133,217,280per year if alternative 2 is implemented.

6. Conclusion

6.1 The researchers concluded that manpower utilization is the main problem of the two assembly lines using the stream diagnostic diagram. Some of the operations exceeds the required takt time of the process while some of the operators performs very fast. Based on theoretical computation, the manpower requirements for each Subline is 26 operators. Therefore, the current Subline is in lack of three (3) operators.

6.2 The recommended solution in Subline wherein merging of subline 1 and 2 reduced the cycle time from 39.16 to 19.99 seconds. Reducing the cycle time will lessen the work-in-process, which is a primary concern in utilizing the manpower. All the operations under Subline do not exceed the takt time of 2.57 seconds. The efficiency of the merged line has a 22.73% increase wherein from the current efficiency status of 47.43% it rises up to 57.94%. The idle time has a 68.8% decrease, meaning there is a proper allocation and utilization of manpower. In the balanced assembly line, there is a decrease in the number of manpower in the assembly line from 46 to 42 operators considering that there are 23 operators per line in Subline before merging. The total number of workers for Subline is 336 operators (day and night shift).

6.3 Through merging of lines, the efficiency increases from 46% to 54% in Sub line. After alternative 1 has been implemented in Sub line, the actual efficiency computed using the time study conducted by Company X rose up to 64.7% which has a difference of 18.7% compare to the current efficiency of the line.

6.4 Subline will have a 13.33% increase in the profitability rate of the company. The annual cost will decrease from Php 1, 531, 998, 270 to Php 1, 398, 781, 440 that results to a Php 133, 217, 280 increases in profit.

7. Recommendation

7.1 It is recommended to the future researchers to conduct further study regarding enhancement of manpower using assembly line balancing. It is also advisable to conduct a long term study to have a deeper analysis in the production assembly lines.

7.2 It is recommended to the future researchers to conduct several time and motion studies to have a deeper understanding in the processes and to gather more accurate data. It is also recommended to the future researchers to solve the problem regarding the efficiency of the other lines.

8. References

[1]. P. Jitchaiyaphum, S. Prombanpong, "A Productivity Improvement of a Packing Line", *Applied Mechanics and Materials, Vols. 789- 790, pp. 1240-1244*. Retrieved April 18, 2016 from *http://www.scientific.net/AMM.789-790.1240*.

[2].J. H. Gong, Q. Y. Zhu, L. J. Zhang, "Research on the Improvement of General Assembly Line Bottleneck for the Car", *Applied Mechanics and Materials,* Vol. 678, pp. 705-711, 2014
Retrieved April 18, 2016 from http://goo.gl/TdZpSY

[3] "Basic Research and Applied Research: Definitions and Differences".
Retrieved April 28, 2016 from http://goo.gl/dV0Bw4

[4].Y. Kurata ,A.Matias & L.Grepo, 2015. "Manpower Utilization in the Hotdog Meat Processing Production of a Meat Processing Company". European Journal of Operational Research Volume 3, 2015, Pages 340–345.
Retrieved April 13, 2016 from http://goo.gl/gIkw9U

[5]Jonard Brian Sancho, 2008. "Increasing Production Output through Line Balancing the Main Drive at Asian Transmission." Undergraduate Research, University of the Philippines-Los Banos

[6]"Toyota Production System (TPS) Terminology". Retrieved from
https://www.lean.uky.edu/reference/terminology/

RESEARCH ON FEM WITH PROCESSING DATA

Yuqing Zhang and Wei Dai
School of Reliability and Systems Engineering
Beihang University, BeiJing, China
zhangyq@buaa.edu.cn

School of Reliability and Systems Engineering
Beihang University, BeiJing, China
dw@buaa.edu.cn

ABSTRACT

In this paper, the connection between real processing and simulation is built. By creating a system which can acquire and transmit mechanical data, real data of stamp processing can be acquired and sent to computer. After sample management on primary data, stamp force-time figure is created and introduced into DEFORM where Finite Elements Method (FEM) analysis is launched. The result of FEM analysis should be more than traditional FEM ways corresponding to reality thus can be used in process improvement and manufacturing monitoring.

Keywords: processing data, FEM, DEFORM

1 INTRODUCTION

Precision machining plays a very important role in aviation industry for its wide effects on production. Processing of precision parts fundamentally influents production quality and it can be changed by many tiny turbulences in factors for accuracy. Monitoring processing factors can make contribution to help judging and predicting production quality, for instance, if parameter values deviate from normal values, it can be predicated that production could be scrap (Lee, D.E. 2006 [1]).

Besides, some hand-made parts have strict requirements on quality thus need accurate and stable processing which can't be achieved easily in hand-made manufacturing. To help stabilize factors in processing, parameters need to be monitored.

FEM analysis is widely used in many areas including manufacturing. However, FEM is always carried out with ideal parameters from which draws deviations into calculation at the first beginning and can reduce reliability of simulation. It can be predicted that FEM with real parameters will conform to reality better and will value more in guiding processing.

A conclusion can be drawn from paragraphs above that accepting real data and using data as conditions of FEM will be useful. Following works have been developed: building data acquirement system, obtaining data and simulating with collected data.

The plunger pump is widely used in plane and missile for its small volume and big flux (Huayong Yang 2008[4]). But, because of the high speed, the piston slipper is easily torn out and damaged. Some slippers are even snapped (LIANG Juan 2014 [2]). There are two ways closing the neck of slippers: push it through a half-mold and roll closing between wheels. The second way will leave less residual stress, but it can't be used for small size slippers (Changying Zhang 2013 [3]). Improving manufacturing accuracy, changing pressing force and reducing surface roughness can all extend the life of pump (Chunjiang Zhu, 2015 [1]). For half-mold closing, finish the closing processing in several steps instead of in one punch. This change will reduce residual stress (Hongpng Yang 2008 [5]). The closing processing is chosen to be the example of monitoring and analysis.

2 DATA ACQUIREMENT SYSTEM

To get the processing data, a data acquirement system is needed. An available system usually is composed of one or several sensers, an Analog-to-Digital Converter(ADC), a transmitter and a data processor. A force senser was chosen to monitor the pressing force in stamping process. An ATMega16A Microcontroller Unit(MCU) was chosen as a data acquirer which includes an ADC module, a Universal Synchronous/Asynchronous Receiver/Transmitter(USART) module and simple data processing ability. The collected data can be sent to PC then saved in Hard Disk Drive(HDD).

2.1 Codes on MCU

An ATMega16A chip produced by the Atmel company is a low-power16-bit MCU which is designed to meet many basic functions: ADC, USART, interrupt, and so on. To launch and invoke needed functions, programs must be designed properly and codes must be written correctly. To complete the work that includes fetching data from senser and sending it to a personal computer(PC), program must contain initiation of programs, invoking ADC and USART, execution of ADC and USART.

Design and write codes in ICCAVR using C program (E.Balagurusamy 2014[8]). After comparing with invoking ADC and USART by simple loop statements, invoking functional codes in interrupt statements is chosen because of its stability and convenience in debugging. Chose two interrupt source: Timer/Counter1(T/C1) and Timer/Counter0(T/C0), invoking reading ADC data and USART respectively because reading data from ADC registers must be invoked more frequently (ATMEL, 2006 [8]). Besides, T/C1 has a 16-bit counter register supporting a longer time interval which adapts to invoking USART. By assigning different value to counter register whose value pluses one to itself after each clock pulse, time interval of interrupt that caused by a filled counter register of T/C0 or T/C1 can be controlled. Another important thing is that despite the priority of each interrupt, one interrupt mustn't disturb another interrupt lest reading or sending wrong data. To make sure ADC and USART working correctly, they must be initiated by setting control bits.

Initiations of ADC and USART are necessary to ensure the MCU running properly. Set single-pin input mode, voltage on another pin as reference voltage, frequency of ADC and enable bit of ADC control bits to fetch proper data from senser. For USART, 9600 Baud rate, data pattern, and enable bit are needed to be set. After the initiation, a variable quantity is created to establish the connection between ADC results and USART and to convey data.

Since result of ADC is a 10-bit-binary data saved separately in ADCH(high 8-bit-binary of data) and ADCL(low 2-bit-binary of data), there has to be a variable quantity reading and storing data in ADCH and ADCL forming the true 10-bit data. For unknown errors, author's USART can only transmit 7-bit-binary data in an 8-bit-binary signal instead of 8-bit-binary data described in instruction, so one 10-bit-binary ADC data has to be sent in two signals. First signal is a 7-bit-binary data from highest 7-bit of the variable quantity, and second signal conveys the lowest 3-bit data of the variable quantity. Serial-port assistance in PC can recognize and transmit each signal to a 2-bit-hex data before save in HDD.

2.2 Senser circuit and fixture

Connecting senser and the MCU electrically can make sure that the MCU monitors senser. So a connection circuit is built to construct such connection.

Direct signal of senser is so weak that the MCU cannot recognize and transfer it directly, so one senser always has a transmitter that can amplify and convert senser's signal into current or voltage signal. Transferred signal can be suitable and strong enough to be made out by MCU. A transmitter need a DC-24V power supply and it needs to be onr common ground with MCU to reduce the disturbance and improve accuracy of measurement. After deploying and connecting senser, transmitter, power supply and the MCU correctly, a serial-port communication line is used to connect serial ports of the MCU and PC through which data will be transferred. After some tests and adjustment, program can run smoothly and the circuit works well.

The base of stamping mold does not fit with senser quite well, so a fixture that bear mold and convey the press to senser must be designed and produced. According to the mold sizes, the fixture was designed as in Figure 1.

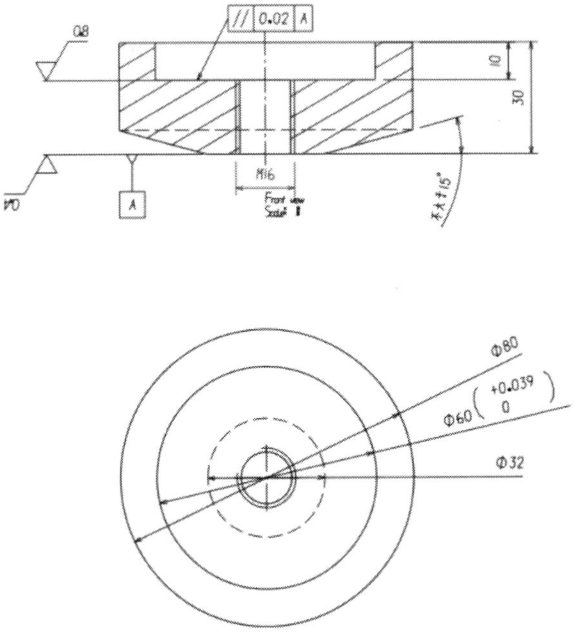

Figure 1 Drawing of Fixture

The pallet and the senser will be connected by a screw bolt. Assembled fixture and senser is as shown in Figure 2.

Figure 2 Real Assembled Senser

3 PROCESSING DATA OBTAINMENT

Measure stamping force with data acquirement system during manufacturing process in factory to obtain data of pressing force. First, assemble and adjust equipment. Then, set the senser under the stamping module.

Before the worker press the closing work piece, start the program in the MCU and the data is conveying continuously to PC. Worker press down the lever of manual press machine and raises the force constantly until stamps the piece through the closing module. The senser monitors the pressure exerting on one piece then the MCU fetches and transmits data before transfers it to PC. After one processing, collected data can be saved in PC in a TXT file.

4 FEM ANALYSIS WITH PROCESSING DATA ON DEFORM

The FEM analysis is based on some numerical methods such as least square to find the minimum remainders and to get the numerical solutions (Timothy Sauer 2014 [6]). To simplify the solving process, always develop potential energy functions and use the variational method to find the peak points (O. C. Zienkiewicz 2006 [7]).

Primitive data is composed of 2-bit-hex digits and each digit is transmitted from one 8-bit-binary signal conveyed through USART. Since FEM needs force data as calculative condition while primitive data is not press force but is linear with it, collected data has to be transmitted to force data.

First, convert all hex data to binary data which is as same as the data in USART. Second, shift binary data that represents the highest 7-bit of ADC result 3 bits to left—equals to timing 8—while do nothing to the data representing the lowest 3-bit of ADC result. Then, add two adjacent data together to form a 10-bit data that equals to ADC result. After that, convert binary data into decimal digits, minus base value which was acquired in the condition with 0 load. Then, all digit time a coefficient which can be calculated by comparing with a value of standard load to get the true force of press. Frequency of sampling data can be calculated according to baud rate. With the real force and frequency, the force-time chart can be drawn.

DEFORM is a FEM analysis software concentrating on metal forming, heat treatment, and so on. FEM analysis on DEFORM is similar with on other software and they both include importing model, meshing, set boundary conditions, set movement, calculating and post-treatment. Work piece is created in SolidWorks as Figure 3.

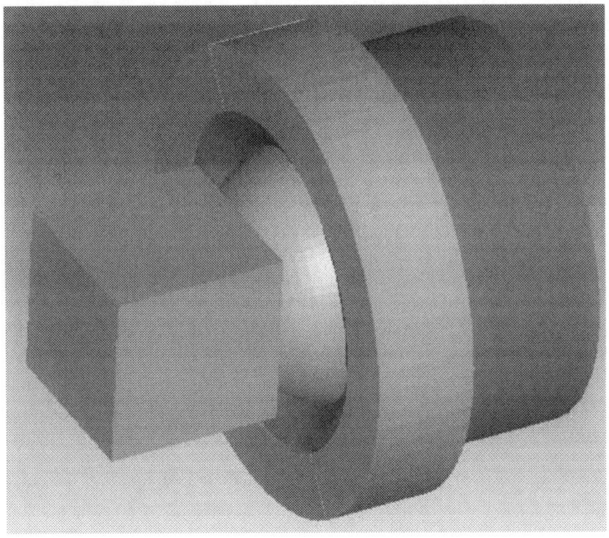

Figure 3 3-D Model of Work Piece

Import model of work piece and half-mold, then set upper die and lower die and every parts are rigid but slipper. The most important thing is importing collected data as movement control, then set stop condition to avoid infinite calculation. Imported data is shown in DEFORM as Figure 4. Then, calculate the processing.

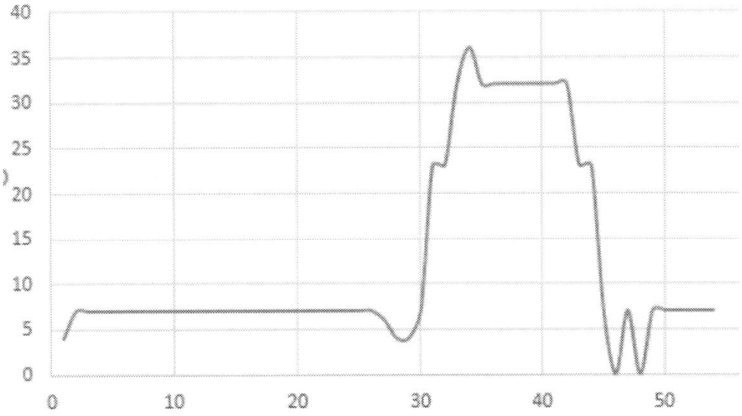

Figure 4 Stamping Force-Time Chart

After the calculation, obtained data can be used to analyse conditions of work piece. Closing force of piece is shown in Figure 5. Residual stress in closing part is shown in Figure 6.

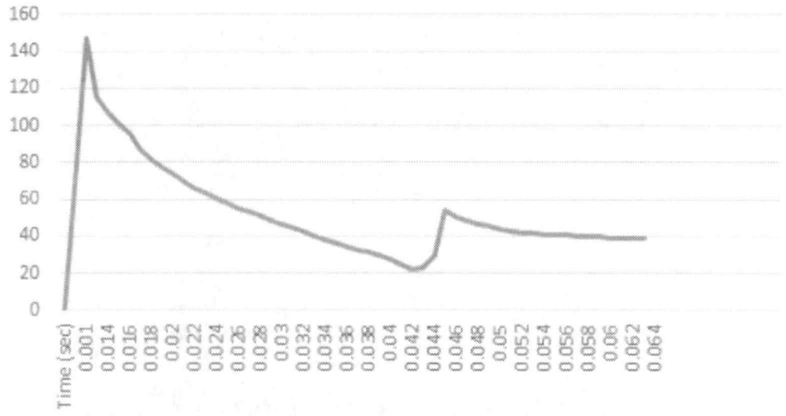

Figure 5 Closing Force Chart

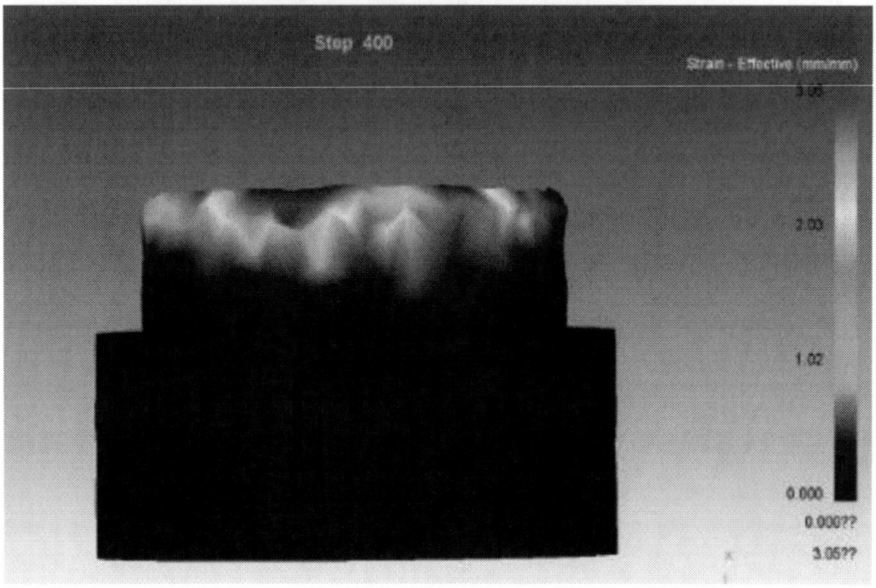

Figure 6 Chart of Residual Force

5 CONCLUSION

In this study, a senser system to monitor processing and acquire manufacturing data was designed and built. Then, real processing data has been monitored and acquired. Force-time chart was created and the real processing data was imported into FEM software. Use SolidWorks to create a 3-D CAD models and import them into DEFORM as calculative objects. Launch FEM analysis using measured force as calculative conditions. And the analysis should be a more accurate result that can be introduced in process improvement and help monitoring production.

By farther analysis of calculation, relation between processing conditions and product quality can be revealed. While by now the calculation and analysis are off-line, next phase work is to realize analysis in on-line condition, and try to use the result of analysis to introduce real processing.

6 REFERENCES

[1] **Lee, D.E. 2006.** Precision Manufacturing Process Monitoring With Acoustic Emission International *Journal of Machine Tools & Manufacture* P176-188

[2] **Chunjiang Zhu, 2015.** Technical Innovation of Shut Mode of Aeroengine Piston Slipper, *Aeronautial Manufacturing Technoligy*, PP129-131

[3] **LIANG Juan 2014.** Some Questions in Design of Piston and Shoe for Axial Hydraulic Piston Pumps, *Hydraulics Pneumatics & Seals*,

[4] **Changying Zhang 2013.** Analysis and Improvement of Plunger Components MACHINE TOOL & HYDRAULICS

[5] **Huayong Yang 2008.** Development of Plunger Pump/Motor *CHINESE JOURNAL oF MEC HANICAL ENGINEERIN G*

[6] **Hongpng Yang 2008.** Improvement on Plunger Closing Processing *AVIATION PRECISION MANUFACTURING TECHNOLOGY*

[7] **Timothy Sauer** 2014, *Numerical Analysis*, CHINA MECHINE PRESS

[8] **O. C. Zienkiewicz, 2006** The Finite Elements Method , 5th TSINGHUA UNIVERSITY PRESS

[9] **ATMEL,** 2006 *8-bit AVR Microcontroller with 16K Bytes In-System Programmable Flash ATmega16A Summary* ATMEL

[10] **E.Balagurusamy** 2014, *Programming in Ansi C*, TSINGHUA UNIVERSITY PRESS

[11] **Wei Dai** 2015, *Analysis and optimization of plunger necking-in process based on finite element simulation*

ASSESSING CHANGE INFLUENCES OF COMPONENTS ON COMPLEX PRODUCTS USING TOPSIS

Junchao Wei[1*] , Guoyuan Zhang[2], and Xiutian an[1,3]
[1] School of Mechanical Engineering
Northwestern Polytechnical University, Xi'an, China
weijunchao@mail.nwpu.edu.cn

[2] School of Electromechanical Engineering
Xidian University, Xi'an, China
gyzhang@xidian.edu.cn

[3]Design Manufacture Engineering Management
University of Strathclyde, Glasgow G1 1XJ, Scotland, UK.
x.yan@strath.ac.uk

ABSTRACT

Engineering changes are unavoidable in development process of complex products. Assessing the influences of the change propagation can provide designers with design support. The change propagation can be modeled using the network of the components. The existing centrality-based changeability indices can be used as criteria to assess and rank the influence of changes in a component on the product. However, different measures lead to different rankings. In order to combine measures into a single ranking, this paper proposes a ranking method with multiple criteria by using TOPSIS (the technique for order preference by similarity to ideal solution) algorithm. In the method, degree-changeability, reach-changeability and between-changeability are utilized as criteria. The proposed method is examined on a real-world example of helicopter. Results show that the suggested multi-criteria method is effective.

Keywords: Complex product, Change propagation, Network, Changeability indices, TOPSIS

1 INTRODUCTION

Engineering changes (ECs) are unavoidable in development process of products [1]. ECs can be triggered not only by many types of factors such as customer requirements and design mistakes. The occurrence of ECs covers the whole product lifecycle [2]. A change of one component of the product might have domino effects on other components leading to additional changes, which in turn can trigger new changes. This phenomenon is called change propagation [3], which is common to complex products due to the high interconnectivity between their parts. Furthermore, changes might propagate uncontrollably, resulting in change avalanches [4]. The change propagation often entails an increase in costs as well as a project deferment, leading to severe profit losses. Therefore, the prediction and management of ECs is very important for a company's competitiveness [2].

* Corresponding Author. Tel.: (+86) 153 535 95426

The assessment of change propagation can provide designers with design support. It can help to optimize product platform architecture [5], identify and prioritize product components for modularization [6], and support preventive design decision making in order to avoid changes to expensive subsystems [3]. Researchers have developed many frames and tools to analyze and predict the EC propagation, including the Change Prediction Method [3] and Bayesian network approach [7], etc. However, most of these frames and tools analyze the propagation paths from the change of a part to the target part; few are dedicated to analyzing the influence of a change on the whole product [8].

Since the change propagation in a complex product can be modelled with network, some centrality-based measures have been used to analyze the influence of a component on the product [8]. Centrality is an important notion in the social network analysis [9]. Centrality measures can identify the most important actors in a social network based on their social interactions [9, 10]. Freeman [10] firstly clarified the concept and general ways to measure it. According to Freeman, centrality measures can be classified into three categories: degree, closeness, and betweenness. The degree and closeness are based on the same idea that the centrality of a node in a network is related to how it is near to the other nodes; while the betweenness centrality is based on another idea that a central node is located at the centre of connection paths between other nodes. Cheng and Chu [8] proposed three centrality-based changeability indices to quantitatively assess the change influence of a component on the whole product: degree-changeability, reach-changeability and between-changeability. These indices are based on degree, reachability and betweenness respectively. The degree-changeability is used to capture the direct influence of changes in a part of the product. The reach-changeability can be used to assess indirect change influences. The reachability centrality is similar to the standard closeness centrality. Their difference lies in that in the definition of the reachability, the optimum path is defined using the maximum change propagation likelihood rather than geodesic distance. The between-changeability is used to predict which parts will change other parts and be changed by other parts dramatically. No single measure is enough for identifying an optimally efficient set of key nodes. For example, the degree-changeability cannot capture the indirect influences; reach-changeability cannot reflect the intermediary role in the change propagations; between-changeability cannot reflect how likely the change of a component can reach to others. So these indices should be utilized together [8]. However, different measures will lead to different rankings of the changeability of individual components, bringing about difficulties in assessment. It is necessary to effectively combine measures into a single ranking.

Confronted with the difficulty, in this paper, we apply a well-known multi-criteria decision making (MCDM) method called TOPSIS to the task of combing multiple criteria of the changeability of components. The remainder of this paper is organized as follows. Section 2 reviews product change network and the existing centrality-based component changeability indices in brief. In Section 3, we combine these changeability indices using the TOPSIS multi-criteria method. In Section 4, the proposed assessment approach is demonstrated in a real-world case of helicopter. Finally, the paper ends with conclusions.

2 RELATED WORK

2.1 Network model for change propagation in products

A product can be decomposed into n components. The component is just a generic expression; it can be a part, an assembly, or a sub-system. These components are represented by $C = \{c_1, c_2, \quad , c_n\}$. The weighted directed networks are used to model complex change propagation relationships between product components. A component corresponds to a node in the network. This network can be represented by the Design Structure Matrix (DSM) [11]. In this paper, the column headings of the DSM represent the *initiating* system components and the row headings represent the *affected* system components. The cell (i, j) stands for the weight w_{ij} of an arc in the network. It indicates how much the component c_i is sensitive to the change of another node c_j because of the direct connection between these two components. In other words, w_{ij} captures the direct likelihood between two. w_{ij} takes values in the interval [0,1]. The indirect EC propagation can be analyzed using these direct likelihood data.

The direct likelihood data can be gathered through staff interviews or by referring to the change documents created by the company [12]. These likelihood data can also be obtained using other types of couplings rather than the change data. These coupling types include the specification flows [5] and spatial/energy/material/information dependencies [13] between components. Martin and Ishii [5] proposed that the stronger coupling between two parts, the more likely the change of a part makes another part change. Based on this notion, coupling strengths can be normalized to the interval [0,1] as the likelihood data. Let S_{ij} stand for the direct coupling strength from j to i, then $w_{ij} = S_{ij} / \max\{S_{ij}\}$. Cheng and Chu [8] used this normalization method to get the likelihood data. If some standard centralities using the concept of distances are utilized, the weight of an arc should be mapped into the length. A simple mapping method is to use the reciprocal of the weight.

2.2 E isting changeability indices of components

Various node centrality indices can be used to measure how influential a component is in the change network. The higher the centrality index is, the greater the influence of the component on the change network is. As mentioned above, Cheng and Chu [8] use the changeability indices to measure the influence of the component on the product system. For a normalized centrality C_i, the corresponding changeability take a common form as follows:

$$C(C)_i = 1 - C_i \quad (1)$$

It can be seen that the lower the node changeability is, the greater the influence of the component on the product is.

2.2.1 *Degree-changeability*

The degree of a node is equal to the number of neighbors of the node. As for weighted networks, the concept of degree is extended to the strength [8,14]. The out node strength OD_i and the corresponding changeability index are defined as [8]:

$$OD_i = \sum_j \frac{w_{ji}}{n-1}$$
$$C(OD)_i = 1 - OD_i \quad (2)$$

2.2.2 *Between-changeability*

If n_{jk} is the number of the shortest paths linking the two nodes j and k, and $n_{jk}(i)$ is the number of the shortest paths linking the two nodes j and k through node i, the betweenness centrality and the corresponding changeability index of node i can be defined as [8,10] :

$$B_i = \frac{1}{(n-1)(n-2)} \sum_{j<k \in G} \frac{n_{jk}(i)}{n_{jk}}$$
$$C(B)_i = 1 - B_i \quad (3)$$

where j and k must be different from i, B_i takes on values between 0 and 1, and it reaches its maximum when node i falls on all optimal paths.

In [8], the shortest path is defined as the path with the maximum likelihood from component j to component i. Of course, the calculation of the betweenness can be based on the standard geodesic path definition; in this case, the connection weights should be mapped into connection lengths.

2.2.3 *Reach-changeability*

Propagation probability includes direct probability and indirect probability. The direct probability is the likelihood w_{ij} . If node i is affected by node j though a middle point k, the indirect probability P_{ij} are calculated by multiplying w_{ik} and w_{kj}. This reflects that the change propagation attenuates as the propagation step increases. Cheng and Chu [8] looked on the maximum likelihood $Max\{ P_{ij} \}$ as the criterion for the optimal change propagation path from node j to i.

The out-reachability OR_i of component i and the corresponding changeability are defined as [8]:

$$OR_i = \frac{1}{(n-1)}\sum_j Max\{P_{ji}\}$$
$$C(OR)_i = 1 - OR_i \qquad \textbf{(4)}$$

3 ASSESSING CHANGE INFLUENCES OF COMPONENTS ON COMPLEX PRODUCT BY USING TOPSIS

The ranking of the changeability of components can be employed using the TOPSIS method [15]. The basic principle of TOPSIS is that the chosen alternative should have the shortest distance from the positive ideal solution (S^+) and the farthest distance from the negative ideal solutions (S^-). As a well-known classical MCDM method, TOPSIS has been used in many areas such as design, engineering and manufacturing system [16].

3.1 Evaluation criteria of component changeability

In this study, we use the three existing changeability indices as criteria for assessment:

- Criterion 1: Degree-changeability
- Criterion 2: Between-changeability
- Criterion 3: Reach-changeability

3.2 The computational procedure of TOPSIS

Step 1: Given n choices, and m criteria, a data matrix A is obtained. This matrix has the form $(x_{ij})_{nxm}$. In present paper n is number of components and m (equal to 3) is the number of criteria. The element x_{ij} of A is the ratings of jth criterion of ith choice.

Step 2: Normalize the data matrix to get the normalized rating r_{ij} .

- For benefit criteria that larger is better, $r_{ij} = (x_{ij} - x_j^-)/(x_j^* - x_j^-)$, where $x_j^* = max_i x_{ij}$ and $x_j^- = min_i x_{ij}$.
- For cost criteria that smaller is better, $r_{ij} = (x_j^- - x_{ij})/(x_j^- - x_j^*)$.

Step 3: Calculate the weighted normalized data matrix $(v_{ij})_{nxm}$. Let $W=(w_1,...,w_m)$ be the relative weight vector about the criteria, satisfying $\sum_{j=1}^{m} w_j = 1$.

$$v_{ij} = w_j \times r_{ij},\ i = 1,2,...,n,\quad j = 1,...,m. \tag{5}$$

Step 4: Determine the positive ideal solution (PIS) and the negative ideal solution (NIS):

$$PIS = S^+ = \{v_1^+, v_2^+, ..., v_m^+\} = \{(\max_i v_{ij} \mid j \in J_1), (\min_i v_{ij} \mid j \in J_2) \mid i = 1,...n\} \tag{6}$$

$$NIS = S^- = \{v_1^-, v_2^-, ..., v_m^-\} = \{(\min_i v_{ij} \mid j \in J_1), (\max_i v_{ij} \mid j \in J_2) \mid i = 1,...n\} \tag{7}$$

where J_1 and J_2 are the benefit and the cost criteria, respectively. In this study, all criteria belong to J_1.

Step 5: Calculate the separation from the PIS and the NIS between alternatives. The separation values can be measured using the Euclidean distance, which is given as:

$$d_i^+ = \sqrt{\sum_{j=1}^{m} (v_{ij} - v_j^+)^2},\ i = 1,2,...,n. \tag{8}$$

$$d_i^- = \sqrt{\sum_{j=1}^{m} (v_{ij} - v_j^-)^2},\ i = 1,2,...,n. \tag{9}$$

Step 6: Calculate the relative closeness to the ideal solution, which is given as

$$C_i = \frac{d_i^-}{d_i^+ + d_i^-},\ i = 1,2,...,n. \tag{10}$$

Step 7: Ranking of the changeability of components based on C_i (i = 1,2, ,n).

4 CASE STUDY

This section presents the application of the approach to a more complex product. The product modeled here is EH101, a civil and military sea rescue and attack helicopter from Westland Helicopters Limited [12]. The change propagation model has been developed through 22 interviews of 21 senior engineers and one sales manager [12]. The direct likelihood matrix including key links between the 19 relevant sub-systems is represented in Figure 1.

The results of the three changeability indices are listed in Table 1. Here $C(B)_i$ is based on the standard betweenness definition. Results of TOPSIS are shown in Column 6 and the corresponding rankings of components' changeability are listed in the last column of Table 1. In this example, each of the single criteria is weighted equally. It is clear from the table that different measures lead to different rankings. For example, sub-system 14 (Cabling Piping) has the highest score of the degree-changeability; while sub-system 19 (Air conditioning) has the highest score of the between-changeability. TOPSIS combines the three measures into a single ranking. From the TOPSIS results, it is shown that the sub-systems 7, 6 and 13 (Tail rotor, Main rotor head, and Auxiliary electrics) have less influences on the whole product while the sub-systems 3, 1 and 10 (Engines, Bare fuselage and Engine auxiliaries) have strong influences on the whole product. As for sub-systems 7 (Tail rotor), its three changeability indices are all high, leading to the highest TOPSIS result. The sub-system 3 (Engines) has the strongest direct and indirect influences on others although its bridge role is not the strongest, leading to the lowest TOPSIS result. It is obvious that the results are effective and reasonable for this complex product. The results suggest that engineers should try to avoid changes to the sub-systems 3, 1 and 10.

		1	2	3	4	5	6	7	8	9	10	11	12	13	14	15	16	17	18	19
Bare fuselage	1		0.7	0.7	0.2	0.2		0.2	0.2	0.5		0.2	0.7	0.2	0.7	0.5	0.7	0.7	0.2	0.7
Fuselage additional items	2	0.2																		
Engines	3										0.7								0.2	
Transmission	4			0.5			0.2	0.2	0.2			0.2				0.2				
Main rotor blades	5						0.2												0.2	
Main rotor head	6				0.2	0.2						0.2				0.2			0.2	
Tail rotor	7				0.2	0.2						0.2							0.2	
Hydraulics	8	0.2	0.2		0.2			0.2				0.2				0.2				
Fuel	9	0.2	0.2	0.7							0.2									
Engine auxiliaries	10			0.7								0.2								
Flight Control system	11			0.2	0.2	0.2	0.2	0.2	0.2		0.2		0.2			0.2				
Avionics	12		0.5	0.7	0.2	0.5			0.2	0.2	0.5	0.5				0.5		0.2	0.2	0.2
Auxiliary electrics	13	0.2	0.5		0.2		0.2						0.2						0.2	0.2
Cabling & Piping	14	0.5	0.5	0.7	0.5				0.5	0.5	0.2	0.5	0.7	0.7		0.5		0.7	0.5	0.5
Weapons & Defence	15			0.2					0	0							0.2			
Equipment & Furnishings	16	0.2	0.2						0.2	0.2										0.2
Fire protection	17	0.2		0.2					0	0.2										
Ice and rain protection	18			0.5	0.2	0.2		0.2												
Air conditioning	19	0.2	0.2	0.2							0.2		0.2				0.2			

Figure 1: Direct Likelihood Matri for the Helicopter Product Model 12

Table 1: Changeability Ranking of Sub-systems in Helicopter Model

No.	Sub-system	$C(OD)_i$	$C(OR)_i$	$C(B)_i$	TOPSIS	Rank
1	Bare fuselage	0.849	0.880	0.636	0.518	18
2	Fuselage additional items	0.762	0.811	0.990	0.656	15
3	Engines	0.579	0.658	0.925	0.357	19
4	Transmission	0.833	0.848	0.932	0.741	12
5	Main rotor blades	0.881	0.851	0.993	0.823	7
6	Main rotor head	0.937	0.921	0.961	0.939	2
7	Tail rotor	0.921	0.911	0.998	0.958	1
8	Hydraulics	0.881	0.876	0.898	0.788	10
9	Fuel	0.873	0.881	0.997	0.861	4
10	Engine auxiliaries	0.841	0.741	0.940	0.602	17
11	Flight Control system	0.825	0.826	0.828	0.612	16
12	Avionics	0.841	0.837	0.870	0.680	14
13	Auxiliary electrics	0.929	0.894	0.982	0.931	3
14	Cabling Piping	0.944	0.897	0.861	0.791	9
15	Weapons Defence	0.818	0.828	0.958	0.716	13
16	Equipment Furnishings	0.913	0.866	0.950	0.849	5
17	Fire protection	0.873	0.857	1.000	0.827	6
18	Ice and rain protection	0.849	0.841	0.965	0.768	11
19	Air conditioning	0.857	0.861	1.000	0.817	8

5 CONCLUSION

The EC propagation in products can be modelled using network and centrality-based changeability indices can be used to assess the change influence of components on products. However, a single-criterion measure is not enough for an assessment. Different measures will lead to different rankings of the changeability of individual components. In order to combine measures into a single ranking, this paper proposes a ranking method with multiple criteria by using TOPSIS algorithm. These criteria include degree-changeability, reach-changeability and between-changeability.

To illustrate the utility of the approach, a real-world example of helicopter is used. The results demonstrate that TOPSIS method effectively captures the key components which are identified by single-criterion indices. In our example application, we have weighted each of the single criteria equally; this is a reasonable strategy if one is agnostic about which criterion is more important. But, in applications these weights can be tuned by decision makers to report different ranking as needed. For example, in a complex information system, bridge roles of components may be more important for decision makers; in this case, the weight for between-changeability can be set higher.

6 REFERENCES

[1] **Clark, K.B., Fu imoto, T.** 1991. *Product development performance: Strategy, organization and management in the world auto industry,* Harvard Business School Press.

[2] **Nichols, K. 1 0.** Getting engineering changes under control, *Journal of Engineering Design,* 1(1), pp 5 15.

[3] **Clarkson, P. ., Simons,C., Eckert, C. 2004.** Predicting change propagation in complex design. *Journal of Mechanical Design*, 126(9), pp 788 797.

[4] **Eckert, C.M., Clarkson, P. ., Zanker, W. 2004.** Change and customization in complex engineering domains. *Research in Engineering Design*, 15(1), pp 1 21.

[5] **Martin, M.V., Ishii, K. 2002.** Design for variety: Developing standardized and modularized product platform architectures. *Research in Engineering Design*, 13(4), pp 213 235.

[6] **Koh, Edwin C.Y., F rg, A., Kreimeyer, M., Lienkamp, M. 2015.** Using engineering change forecast to prioritise component modularization. *Research in Engineering Design*, 26(4), pp 337-353.

[7] **Hu, ., Cardin, M. A. 2015.** Generating flexibility in the design of engineering systems. *Research in Engineering Design*, 26(2), pp 121 143.

[8] **Cheng, H., Chu, X. 2012.** A network-based assessment approach for change impacts on complex product. *Journal of Intelligent Manufacturing, Journal of Intelligent Manufacturing*, 23(4), pp 1419-1431.

[9] **Wasserman, S., Faust, K.** 1994. *Social Network Analysis*, New ork: Cambridge University Press.

[10] **Freeman, L.C. 1   .** Centrality in social networks: Conceptual clarification. *Social Networks*, 1(3), 215 239.

[11] **Browning, T.R. 2001.** Applying the design structure matrix to system decomposition and integration problems: A review and new directions. *IEEE Transactions on Engineering Management*, 48(3), pp 292 306.

[12] **Hamraz, B., Caldwell, N. H.M., Clarkson P. . 2013.** A matrix-calculation-based algorithm for numerical change propagation analysis. *IEEE Transactions on Engineering Management*, 60(1), pp 186 198.

[13] **Pimmler, T.U. Eppinger, S.D. 1 4.** Integration analysis of product decompositions. Proceedings of the ASME design theory and methodology conference (DTM'94), Minneapolis, USA, September, pp 343 351.

[14] **Rubinov, M., Sporns, O. 200** . Complex network measures of brain connectivity: uses and interpretations. *NeuroImage*, 52, pp 1059-1069.

[15] **Hwang, C.L., Yoon, K.** 1981. *Multiple Attribute Decision Making: Methods and Applications*, Springer.

[16] **Behzadian, M., Otaghsara, S.K., Yazdani, M., Ignatius, . 2012.** A state-of the-art survey of topsis applications. *Expert Systems with Applications*, 39(17), pp 13051-13069.

STANDARDIZATION OF PRODUCTION OUTPUT OF THE 36-INCH CONCRETE PIPES AT COMPANY XYZ USING KAIZEN TECHNIQUE

Ryan Jeffrey P.Curbano* Althea C. Mendoza, Timothy John M. Flores,
Jovi Anne M. Necor, Jessa C. Onte,
Industrial Engineering Department, Lyceum of the Philippines University – Laguna
Corresponding email: *ryanjeffrey.curbano@lpulaguna.edu.ph

Abstract

The objective of this study is to standardize the production output per day at company XYZ using Kaizen Technique. Through Pareto Diagram, the main problem in the production was identified due to under/over production, raw materials, transportation, machine breakdown, output per day and manpower. Time and motion study were applied to determine the standard cycle time and number of output produce per day. Overall Equipment Effectiveness (OEE) was also applied to determine the efficiency of the machine. Standardizing output and reducing downtime is a plan to improve the production output, efficiency and performance of the existing machine. Standardized output which is 59 units per day per machine was used to prevent the underproduction/overproduction in the company. The OEE of each machine improves from 53.67% to 91.87% for machine A, 65.71% to 92.23% for machine B and 70% to 86.95% for machine C. The average OEE of all machines is from 63.34% to 90.35%. The production output of the company increase by 27.01%.

Keyword: Kaizen, Overall Equipment Effectiveness, Efficiency, Downtime, Standardization

1. Introduction

Kaizen is a lean manufacturing tool that energizes persistent change in quality, innovation, forms, profitability, security, and work environment society. It concentrates on applying little, day by day changes that outcome in real enhancements after some time [1]. The objective of Kaizen incorporate elimination of waste, maximum utilization and ideal assignment of assets or any proposal or thought that is implemented for the improvement of a product, process, framework or administration. An effective value chain perceives variation from the norm and finds a way to enhance by making a way to acknowledge recommendations [2].

Kaizen is a helpful tool in minimizing losses in the workplace that affect the company's efficiencies. Its objective is to attain and sustain zero losses with regards to minor stops measurement and adjustments, defects and inevitable downtimes [3]. It is a constant improvement procedure involving everyone, managers and workers alike. Through the use of Kaizen, it will improve also the problem solving capabilities of the workers.

Every company has their flaws, limitations and problems. Due to this fact, company needs to apply various and continuous study for improvements and solutions to problems. Company XYZ is a manufacturing company and has started out as a concrete hollow block producer and has expanded its products to reinforced concrete pipes, paving blocks and pre-cast concrete elements. The company's production head suggested to the researchers to concentrate in the production of 36 inches reinforced concrete pipes. Forming of concrete pipes has a group containing four (4) workers in one machine. The researchers asked the production head about the common problems that they encountered in the production and rate it from 0% up to 100%.

The problems encountered by the company are 1) raw materials; 2) transportation; 3) machine breakdown; 4) output per day; 5.) Manpower. The problems and ratings are analyzed with the use Pareto Chart which is a tool used to analyze and highlight the most important among a set of factors.

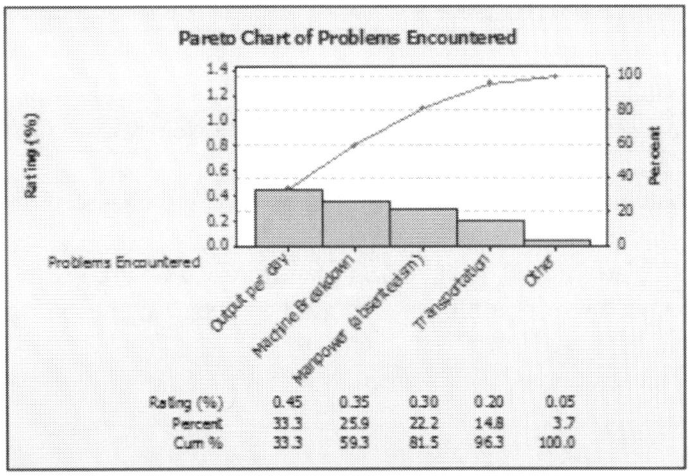

Figure 1. Pareto chart of problems encountered

Figure 1 shows that the most encountered problems in the company XYZ is the output per day. The researchers focus on output per day to prevent overproduction and underproduction that can lead to high inventory. Inventory is one of the seven (7) mudas which means that the cost of the product is on hold and manufacturing is not sure if the product will be sold at the end of the period [4].

2. Conceptual Framework

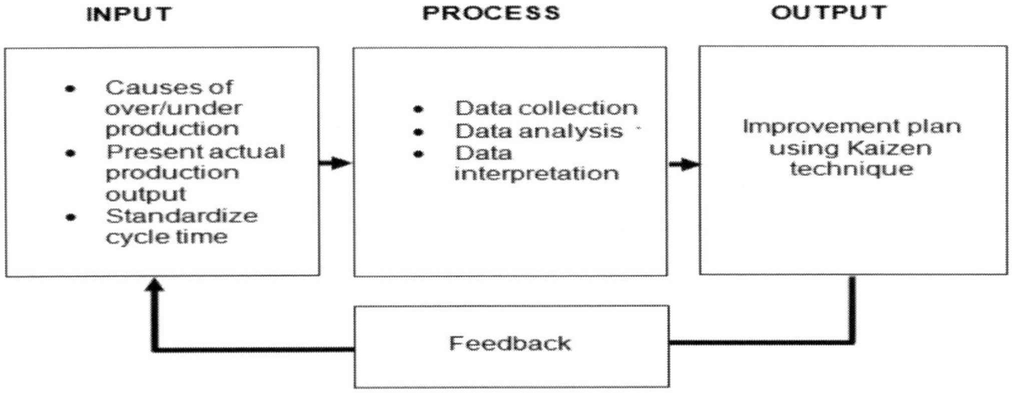

Figure 2. Research paradigm

Figure 2 shows the Application of Kaizen Technique to standardize the production output in 36" concrete pipe at company XYZ. The first box refers to the inputs. These are the information needed to be able to find the solution to the problem encounter in the production. It shows the causes of over/under production, present actual production output and standardizes cycle time. The second box which is the process, wherein, all the information needed is analyzed and interpreted by the researchers. The third box which is the output, the plan made by the researchers and aims that this study will be implemented and be successful.

3. Objectives of the study

The main objective of the study is to apply the Kaizen technique to standardize the production output in company XYZ.

Specifically the study will answer the following objectives:

3.1 To present the actual production output and compare the customer requirements for the past 12 months.
3.2 To conduct time study to determine the standard cycle time and standard output.
3.3 To analyze the root causes of manpower absenteeism and machine downtime and its effect on production output.
3.4 To calculate the Overall Equipment Effectiveness of the machine based on the machine downtime.
3.5 To calculate capacity plan using the improved Overall Equipment Effectiveness.
3.6 To test if there is a significant improvement in the output after improving Overall Equipment Effectiveness.
3.7 To propose action plan to sustain the improvement done.

4. Methodology

4.1 Research Design

The researchers used applied type of research. Applied research is a form of systematic inquiry involving the practical application of science.

4.2 Research Instrument

In calculating the standard time, normal time and standard output, the researchers used time study. It is needed to have the performance rating of each operator to solve the standard time which is calculated using the Performance Rating Table. The researchers gave 20% allowance to each worker based on International Labor Organization (ILO) Allowance Table. The 20% allowance is consists of 5% personal allowance, 4% basic allowance, 2% Standing Allowances, 2% Abnormal Position Allowances (bending), 7% Use of Force or Muscular Energy (Lifting, Pulling, or Pushing)

5. Results and Discussion

5.1 **Comparison of Production actual output and Customer requirements**

Table 1. Production actual output vs Customer requirements

Production Actual Output 2015-2016 (units)		**Customer Requirement for the whole Year 2015-2016**
May	3854	40000
June	3596	
July	3840	
August	2296	
September	1896	
October	1756	
November	2505	
December	308	

Table continues

February	4078	
March	5832	4000
April	5423	
Total	**37590**	**40000**

Table 1 show that the production actual output was not able to comply with the customer requirement which is 40,000 units of concrete pipes. The 40,000 was divided according to the customers' need every month. As the researchers observed in the table, there were months that have a low production and months that have a high production rate. According to this result, the researchers noticed that there were an inconsistency to their output and sometimes it has an underproduction and overproduction of output. The researchers conducted a time and motion study to compute the observed time, normal time and standard time. Those are factors to be considered in computing the cycle time or standard time that the researchers need to compute for the standard output per day of the production of 36 inches concrete pipes.

5.2 Summary of time study

In time and motion study, there were factors to be considered by the researchers to have a valid result of their study.

Table 2. Computed observed time, normal time and standard time

	Observed time	**Normal time**	**Standard time**
Total time	**7.16**	**6.80**	**8.16**

Table 2 shows that the average observed time is 7.16 minutes, the normal time is 6.80 minutes and the standard or cycle time is 8.16 minutes in one (1) operation.

Table 3. Results of time study

Results of Time Study	
Standard Time	8.16 minutes
Pieces/day per machine	58.82 ≈ 59 units/day
Pieces/day in 3 machines	176.47 ≈ 177units/day

In table 3, the standard or cycle time is 8.16 minutes. As stated above in the discussion of table 3.0, the standard pieces will be computed using 480 minutes divided by the standard cycle time which is 8.16 minutes. The computed standard units per day in one machine are 58.82 ≈ 59 units. Since there were three (3) machines in creating a 36 inches concrete pipe, their output per day will be a total of 176.47 ≈ 177 units of concrete pipes.

5.3 Cause and Effect Analysis of Manpower Absenteeism and Machine

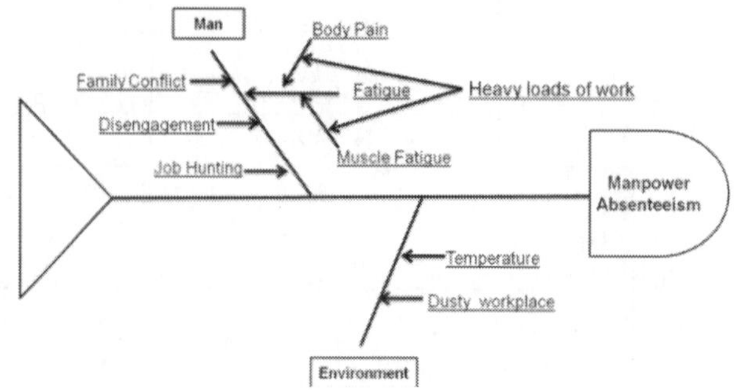

Figure 4. Causes of manpower absenteeism

In figure 4, creating a concrete pipe, the workers feed the machine with sand and gravel manually. The weight of one concrete pipe is 625 kg (kilograms) and the weight of the trolling cart is 950 kg according to the production head of company XYZ. The concrete pipes were transferred from one place to another after the molding process using a trolling cart. It is reported that workers with heavy physical workloads have a higher occurrence of muscle pain in most body parts [5]. It is a factor to be considered that doing heavy loads of work can cause absenteeism of the worker. As the researchers analyzed the fish bone diagram, the root cause of manpower absenteeism is heavy loads of work.

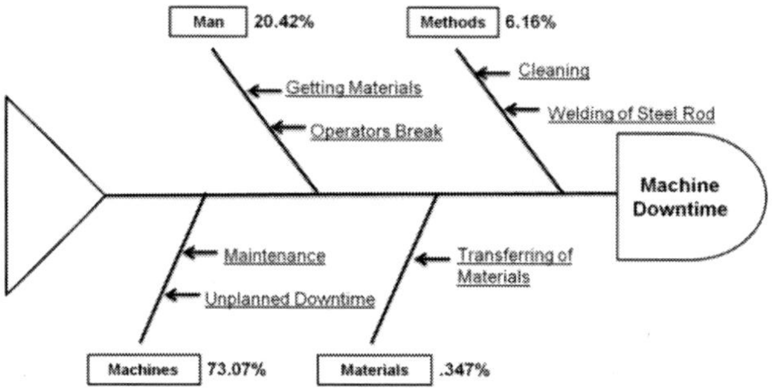

Figure 2. Causes of machine downtime

As the results being evaluated by the researchers, the highest percentage that causes machine downtime is the unplanned downtime and maintenance.

5.4 Calculations of Overall Equipment Effectiveness (OEE) of the Machines and their overall average OEE

In Table 4, the expected output is based on the data gathered by the researchers. The tables below show the actual OEE of the three (3) machines and their OEE.

Table 4. Actual Overall Equipment Effectiveness

	Machine A	Machine B	Machine C
Overall Equipment Effectiveness (OEE)	53.667%	65.71%	70%

Downtime is the most critical factor to improve the OEE because when the process is not running the other process will be affected and it might cause delayed on the production. Optimizing the performance of the existing machine may be more important than the major venture to buy new equipment. Reducing downtime can minimizing other activities such as setup up, cleaning and improving workers performance, a manufacturing company can unleash hidden capacity and benefit from monitoring OEE [6].

In improved Overall Equipment Effectiveness, the reduced planned downtime will be added to the operating time.

Table 5. Improved Overall Equipment Effectiveness

	Machine A	Machine B	Machine C
Overall Equipment Effectiveness (OEE)	91.87%	92.23%	86.95%

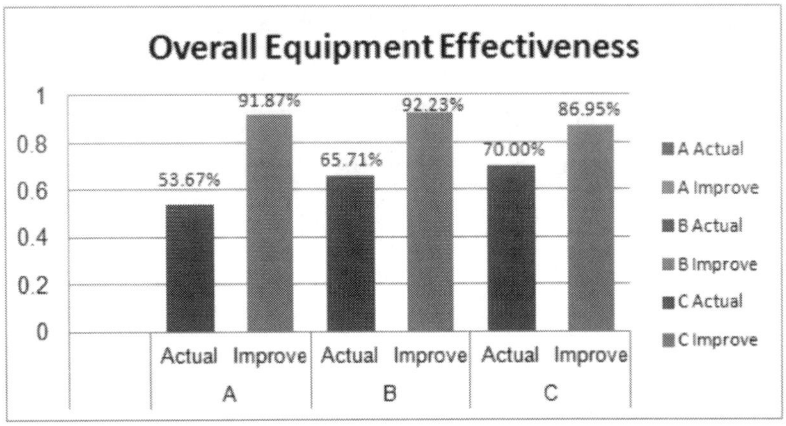

Figure 3.0 Actual OEE vs. Improved OEE

It shows the actual and improved ratings of all three machines and their overall equipment effectiveness. The OEE of machine A increases by 38.2%. For Machine B, the OEE increases by 26.51%. The OEE of machine C increases by 16.95%.

5.6 Capacity Planning

ALLIED CONCRETE PRODUCTS INC.
CAPACITY PLAN (2016-2017)

	May	Jun	Jul	Aug	Sep	Oct	Nov	Dec	Jan	Feb	Mar	Apr
A. Total Number of Days												
Total Days	31	30	31	31	30	31	30	31	31	28	31	30
Less:												
Sunday	5	4	5	4	4	5	4	4	5	4	4	5
Holiday	0	0	0	0	0	0	0	4	2	0	0	0
Working Days	26	26	26	27	26	26	26	23	24	24	27	25
Less:												
Regular Maintenance	0	0	0	0	0	0	0	1	0	0	0	1
Allowances	0	0	0	0	0	0	1	1	0	0	1	0
Ideal Operating Days	26	26	26	27	26	26	25	21	24	24	26	24
Less:												
Machine Breakdown	0	0	0	0	0	0	1	2	0	0	0	0
Material Shortage	0	0	0	0	0	0	0	1		0	0	0
Power Shortage	0	0	0	0	0	0	0	0	0	0	0	0
Normal Operation days	26	26	26	27	26	26	24	18	24	24	26	24
Less: Plant Adjustment Period	0	0	0	0	0	0	0	1	0	0	0	0
Possible Number of Days of Operation	26	26	26	27	26	26	24	17	24	24	26	24
B. Monthly Operation (Concrete pipes)												
Maximum Possible output	4602	4602	4602	4779	4602	4602	4248	3009	4248	4248	4602	4248
OEE percentage rate	90.35%	90.35%	90.35%	90.35%	90.35%	90.35%	90.35%	90.35%	90.35%	90.35%	90.35%	90.35%
Practical Output (unit measure)	4158	4158	4158	4318	4158	4158	3838	2719	3838	3838	4158	3838

Figure 4.0 Capacity plan for year 2016-2017

Figure 5.0 shows the possible flow of operation from May 2016 to April 2017. Comparing this improvement output to the actual output, there is a balance in every month's production output. There were no low production and high production.

5.6 Hypothesis testing

Statistical hypothesis testing is usually employed to determine if a result is statistically significant or not [7]. The researchers tested their improved output per day using a one-sample Z test if it is effective or not. It is used when the number of cycles is greater than 30.

Table 6.0 Result of Hypothesis Testing

	Null Hypothesis mean	Alternative Hypothesis Mean	N	Result
	Equal to 35	Not equal to 35	65 cycles	**Accept the alternative hypothesis**
Standard Deviation	21.05			
P-value	0.000			

Table 6.0 shows the result of the statistical testing. Since the P-value is 0.000, there is a significant improvement in the output. Based on the statistical decision rule, if the P-value is ≤ 0.05, it is considered to be statistically significant [7]. The computed standard output by the researchers was 59 units per day and is acceptable because the result of the hypothesis testing was accepting the alternative hypothesis. The standard output is based on the computed standard time by the researchers. Therefore, the effective increase in OEE can increase the output per day in company XYZ.

5.7 Action plan in sustaining the improvement done

Creating a sustainable improvement can help the life of the company [8]. Kaizen includes everyone in the company to create a better culture that can lead to success. It involves every employee and strongly encourages suggestions for enhancements, even if they are minor as long as enhancements continue [1].

Table7. Improvement Action Plan for sustaining the improvement done

Standardization of output	Action Plan	Responsible Persons	Time Frame
1. Introduce Kaizen and its Procedure to the worker	• The management should set a meeting/ seminar about Kaizen. • Give all the workers a guidelines/ handbook about the application of Kaizen. • Teach the workers on proper use of kaizen.	Top Management	Two (2) times a month
2. Create a Kaizen Group	• Set a meeting to all departments about what kaizen group is. • Talk about who will be the responsible or in charges in the kaizen group. • Prepare the things that are needed in creating the kaizen group.	Top Management	After introducing the Kaizen technique.

3. Train the worker	• Train the workers about Total Preventive Maintenance. • Require the workers to attend trainings for them to be more familiar about the process.	Production Head	Two (2) times a month
4. There should be a regular maintenance for each machine	• Establish a plan maintenance schedule to prevent unnecessary breakdown of the machines. • Establish equipment baseline exists with current and accurate process instructions.	Production Head	Every end of the month

Table 7 shows the action plan created by the researchers to sustain the improvement being done for the company being studied. Having trained workers is important. It means that the workers are learning new skills that can improve the production, reduce production costs, reduce mistakes and create a better working environment [9].

6. Conclusion

The following conclusions have been obtained through the gathered data and analysis of the researchers:

6.1. In the comparison of actual output and customer requirement, the production output does not meet the customer requirement due to some reasons and problems that affects the production area. By several visits and observation, the researchers noticed that there was a team that does not complete an 8-hour shift. Due to that, there were low productivity rates in the production.

6.2. The researchers determined that the cycle time for every operation is 8.16 minutes. The standard output per day that can be produced by the workers based on the calculation is 59 units of concrete pipes per day. Since there were three machines in creating a 36 inches concrete pipe, there is a total of 177 units of concrete pipes a day that they can produce.

6.3. In root cause analysis made by the researchers, one (1) reason for the inconsistency of their output is the absenteeism of the workers. Absenteeism takes place due to the heavy loads of works that can cause fatigue, disengagement of workers to their work and etc. Machine breakdown was also a reason for the inconsistency of output at company XYZ.

These problems could be resolved using the standardization of the production output per day to prevent the overworking of the workers that can cause for their absences and machine breakdowns.

6.4. The OEE of machine A increases 38.21%, machine B increases 26.51% and machine C increases 16.95%. The average OEE of all machines are improved and increased by 27.01%. It was done by minimizing the planned downtime because it has the controllable delays that can minimized by the management and maximizing the hours of working in one day which is 8 hours.

6.5. The researchers concluded that the workers should utilize their 8-hour shift so that they would come up to the maximum output. The practical output in the plan will be their target output every month. There will be no shortages of output and overproducing of output every day.

6. 6.Statistically, the effective increase in OEE made by the researchers can increase the output per day at company XYZ.

Overall, the study showed that it is effective to have a standard output per day and continuously implement it using Kaizen technique. Standardizing the work is an advantage that can get on implementing Kaizen technique in their entire organization.

7. Recommendation

The following recommendations are offered for company XYZ:

7.1. The Management in Allied Concrete Products Incorporated should have the continuous monitoring on this research study and should accept other researchers that will focus particularly on their inventory for further improvement. It can lead the production well and monitor the circumstances in the organization.

7.2. There should be a scheduling of rest days for the workers and attendance record for them to monitor of absences of the workers.

7.3. The researchers recommend to the management to follow their production plan. In terms of machines and equipment, the management should check and monitor the equipment if there are unnecessary problems especially on the trolling cart.

Bibliography

[1] B. J. Bradbury, "What Is Kaizen?," 20 may 2015. [Online]. Available: https://www.graphicproducts.com/articles/what-is-kaizen/. [Accessed 10 may 2016].

[2] "what is kaizen?," [Online]. Available: http://www.sohowebtools.com/index.php?option=com_content&view=article&id=56:what-is-kaizen-&catid=29:productivity-tools&Itemid=14. [Accessed 11 may 2016].

[3] G. e. a. Avenido, "Evaluation of 5s program in Lyceum of the," september 2013.

[4] M. P. Parikh Y., "Total Productive Maintenance: Need & Framework," *International Journal of Innovative Research in Advanced Engineering* , vol. 2, no. 2, 2015.

[5] B. Widanarko, S. Legg, M. Stevenson, J. Devereux, A. Eng, A. Mannetje, S. Cheng, J. Douwes, L. Ellison-Loschmann, D. McLean and N. Pearce, "Prevalence of musculoskeletal symptoms in relation to gender, age, and occupational/industrial group," International Journal of Industrial Ergonomics, Volume 41, Issue 5, September 2011. [Online]. Available: http://www.scopus.com/record/display.uri?eid=2-s2.0-80051471388&origin=inward&txGid=0. [Accessed 7 June 2016].

[6] E. /. D. Marquees, "The Complete Guide To Simple OEE," [Online]. Available: http://www.Exor-Rd.Com. [Accessed 11 May 2016].

[7] ""Statistically Significant Definition | Investopedia"," [Online]. Available: http://www.investopedia.com/terms/s/statistically_significant.asp#ixzz49vvQBLbc. [Accessed 28 may 2016].

[8] "Importance & formula of time and motion study.," 30 december 2010. [Online]. Available: http://thetimeandmotionstudy.blogspot.com/2010/12/time-and-motion-study.html. [Accessed 5 march 2016].

[9] B. Benton, " "Importance of Employee Training,"," 3 September 2014. [Online]. Available: https://lineshapespae.com/importance-of-employee-training/.. [Accessed 6 june 2016].

AN IMPROVING OFDM SIGNAL TIME OF ARRIVAL ESTIMATION METHOD

Y.L. Zong[1*], X.K. Ren[1], K. Li[1], and H.W. Li[1]
[1] Zhengzhou Information Science and Technology Institute, Zhengzhou 450001, China
754290573@qq.com rxkdlk@126.com happylike8023@163.com luckylhw66@163.com

ABSTRACT

In order to improve anti-noise performance of OFDM signal TOA estimation, TK-ESPRIT algorithm is proposed by introduce Teager-Kaiser energy operator based on ESPRIT (Estimation of Signal Parameters via Rotational Invariance Technique). TK operator can track the instantaneous change of the signal. With this character, TK-ESPRIT algorithm processes the signal correlation function initially and makes the correlation function peak more obvious. Then an eigenvalue matrix is built, and the ESPRIT algorithm is used to perform TOA estimation. The simulation results show that TK-ESPRIT algorithm can effectively improve estimation precision compared to currently algorithm.

Keywords: OFDM, Teager-Kaiser energy operator, TOA estimation, ESPRIT

1 INTRODUCTION

Orthogonal Frequency Division Multiplexing (OFDM) is the key technology of B3G and 4G wireless communication. The research of wireless location technology based on OFDM signal has attracted the attention of a large number of researchers in recent years. In related research literature, time of arrival (TOA) parameter estimation and direction of arrival (DOA) parameter estimation are the majority. For wireless signal TOA estimation, the location accuracy is interfered because of multipath, non-line-of-sight (NLOS), noise interference and so on. TOA estimation of OFDM wireless signal also faced with these problems[1]-[3].

To reduce the effect of multipath for TOA estimation error, researchers apply Modern Spectral Estimation to OFDM signal TOA estimation[4], such as Multiple Signal Classification(MUSIC), Estimation of Signal Parameters via Rotational Invariance Technique (ESPRIT). Moreover, [5] proposed to improve upon the classical ESPRIT algorithm by incorporating the iterative reduced rank Hankel approximation(RRHA) technique, but the algorithm complexity is too high to apply practically, and it's susceptible to noise interference. [6] proposed a two dimensional ESPRIT-like shift-invariant TOA estimation algorithm and have an outstanding performance in the indoor multipath channel, but the algorithm complexity is too high. In view of this situation, [7] proposed one-dimensional ESPRIT estimation by discrete Fourier transform with low complexity, could track signal instantaneous change to estimate TOA parameters, but the anti-noise capability is declined.

To improve algorithm's anti-noise capability and ensure accuracy, this paper introduced Teager-Kaiser energy operator to OFDM signal TOA estimation. TK energy operator was originally applied to measure energy in speech signal processing, using its feature that can track the instantaneous change of the signal, we improved our algorithm on the basis of ESPRIT and proposed TK-ESPRIT algorithm to estimate TOA parameters. The remainder of this paper is organized as follows: In Section II illustrates the TOA estimation system model of OFDM signal. In Section III, the proposed algorithm with TK operator is presented, together with the discussion about the TK operator and complexity issues. In

* Yulei Zong, M.S. of Zhengzhou Information Science and Technology Institute from 2014, E-Mail: 754290573@qq.com

Section IV, the performance of the improved algorithm is evaluated and compared to the RRHA-ESPRIT algorithm via simulations. Finally, Section V concludes the paper.

2 SYSTEM MODEL

TOA estimation system for OFDM signal is illustrated by following figure:

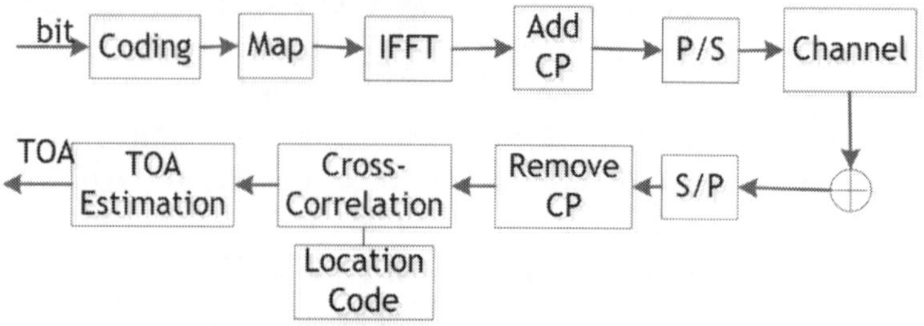

Figure 1: TOA estimation system for OFDM signal

A typical OFDM symbol transmitting over N subcarriers in time-domain can be expressed as

$$s(t)=\sum_{k=1}^{N} b_k e^{j2\pi f_k t} \tag{1}$$

where b_k is the modulated complex symbol of the kth subcarrier, $f_k = f_0 + k\Delta f$ is the frequency of the kth subcarrier, f_0 is the original frequency, $\Delta f = 1/T_s$ is frequency interval, T_s is OFDM symbol length, which is also the period of FFT/IFFT.

OFDM radio signal propagates in the form of electromagnetic wave, and the radio channel model of multipath can be expressed as

$$h(t)=\sum_{i=1}^{L_p} a_i(t)\delta(t-\tau_i) \tag{2}$$

where L_p is the number of multipath components, τ_i is ith path's signal propagation delay, $a_i(t)=|a_i(t)|e^{j\varphi_i(t)}$ is the ith multipath component's complex fading factor, $|a_i(t)|$ and $\varphi_i(t)$ is the amplitude and phase of complex fading factor, respectively.

OFDM received domain signal can be expressed as

$$\begin{aligned} y(t) &= \sum_{k=1}^{N} b_k[\sum_{i=1}^{L_p} a_i(t)e^{-j2\pi f_k \tau_i}]e^{j2\pi f_k t} + n(t) \\ &= \sum_{k=1}^{N} b_k[\sum_{i=1}^{L_p} a_i(t)e^{-j2\pi f_k \tau_i} + \mathrm{w}_k(\mathrm{t})]e^{j2\pi f_k t} \end{aligned} \tag{3}$$

where $w_k(t)$ is the additive white Gaussian noise of kth subcarrier. After FFT of received signal $y(t)$, the data of the kth subcarrier can be expressed as

$$Y(k)=b_k[\sum_{i=1}^{L_p} a_i(t)e^{-j2\pi f_k \tau_i} + w_k(t)] \tag{4}$$

and the channel frequency response estimation of the kth subcarrier can be expressed as

$$x_k(t) = H_k(t) + w_k(t) = \sum_{i=1}^{L_p} a_i(t) e^{-j2\pi f_k \tau_i} + w_k(t) \tag{5}$$

Now, we have ensured OFDM received signal form, next section, we will derive TK-ESPRIT algorithm.

3 TK-ESPRIT TOA ESTIMATION

At present, the main method of OFDM signal TOA estimation consist of super-resolution estimation algorithm[8] and TOA estimation algorithm using OFDM subcarrier signal characteristics[10]. Super-resolution algorithm can effectively deal with multipath problem, but couldn't guarantee estimation accuracy when noise interference is large. Using the cross-correlation of OFDM signal subcarriers' cyclic prefix to acquire TOA parameter has a low complexity, but this algorithm just focus on single path, the accuracy of multipath is not ideal. Considering of this, we introduced TK energy operator to increase the anti-noise capability of super-resolution algorithm.

Based on the Newton's law of motion, Teager and Kaiser firstly proposed a nonlinear quadratic operator, called Teager-Kaiser operator. It was originally applied to measure energy in speech signal processing, and the output includes the square of amplitude and frequency information. Because TK operator is simple, effective and can track the instantaneous change of the signal, it provides a good theoretical basis for the multi-path signal detection.

The continuous-time TK operator of a complex-valued signal $x(t)$ is

$$\psi_c[x(t)] = x'(t)x'^*(t) - \frac{1}{2}[x''(t)x^*(t) + x(t)x''^*(t)] \tag{6}$$

From (6), the discrete form of TK operator is given as

$$\psi_c[x(n)] = x(n-1)x^*(n-1) - \frac{1}{2}[x(n-2)x^*(n) + x(n)x^*(n-2)] \tag{7}$$

Add direct sequence spread spectrum (DSSS) before OFDM subcarrier FFT, such as CI code[9], could ensure OFDM signal transmission rate and at the same time to obtain frequency diversity gain and reduce signal's peak-to-average power rate (PAPR). Because it has a strong auto-correlation, the correlation result is similar to triangle, so after discrete sampling, the cross-correlation function of received subcarrier signal $x_k(t)$ and original signal $s(t-\tau)$ can be expressed as

$$R_{xs} = \sum_{k=0}^{N-1} s(k-\tau)\cdot x^*(k) = \sum_{i=1}^{L_p} \alpha_i \Delta(\tau - \tau_i, T_c) e^{j\theta_i} \tag{8}$$

where α_i is complex fading factor, τ_i is propagation channel delay, θ_i is phase offset, $\Delta(\tau - \tau_{i,} T_c)$ is a triangle equation.

$$\Delta(\tau - \tau_i, T_c) = \begin{cases} 1 - \dfrac{|\tau - \tau_i|}{T_c} & |\tau - \tau_i| \le T_c \\ 0 & other \end{cases} \tag{9}$$

Triangle equation processed by TK operator can be expressed as

$$\psi_c[\Delta(\tau - \tau_i)] = \frac{\prod(\tau - \tau_i, T_c)}{T_c^2} + \frac{\Delta(\tau - \tau_i, T_c)\delta(\tau - \tau_i)}{T_c} \tag{10}$$

After this, if (8) is processed by TK operator, it can conclude the following function

$$\psi_c[R_{xs}] = \frac{1}{T_c^2}\sum_{i=1}^{L_p}\alpha_i sign(\tau-\tau_i)\Pi(\tau-\tau_i,T_c) + \frac{1}{2T_c}[R_{xs}^*(\sum_{i=1}^{L_p}\alpha_i\delta(\tau-\tau_i)e^{j\theta_i}) + R_{xs}(\sum_{i=1}^{L_p}\alpha_i\delta(\tau-\tau_i)e^{-j\theta_i})] \quad (11)$$

Where sign(t) is a sign function. It can be concluded that when $\tau = \tau_i$, (11) will get a high peak value. So it is evident that TK operator can not only track the triangular peak well, but also reduce the coverage of the main lobe of correlation function. So this way provides an available way to detect objective function parameter and further improve the accuracy of the algorithm. Since the noise signal can be effectively suppressed, detection results after TK operator processed is much better than the original signal.

From (11) can derive cross-correlation function and construct correlation matrix to eigenvalue decomposition

$$\boldsymbol{R}_{TK} = E\{\mathrm{x_N x_N^H}\} = \boldsymbol{VAV}^H + \sigma^2 I \quad (12)$$

where, $\boldsymbol{A} = E\{aa^H\}$, $\alpha = \begin{bmatrix}\alpha_1 & \alpha_2 & \cdots & \alpha_{L_p}\end{bmatrix}^{\mathrm{T}}$ is complex fading factor, ***V*** is Vandermonde matrix about time delay and $v(\tau_i) = \begin{bmatrix}1 & e^{-j2\pi\tau_i/T_s} & \cdots & e^{-j2\pi(N-1)\tau_i/T_s}\end{bmatrix}^{\mathrm{T}}$. Assuming each τ_k different from others and furthermore they are literally with ascending order. Thus, matrix $\boldsymbol{R}_{TK}$ has principal eigenvalues larger than σ^2 and other $N-L_p$ eigenvalues equal to σ^2, the corresponding eigenvectors are called signal eigenvectors and noise eigenvectors, respectively. Therefore the subspace **U** spanned by L_p eigenvectors of $\boldsymbol{R}_{TK}$ can be classified into two subspaces, which defined as signal subspace $\boldsymbol{U}_{\mathbf{s}}$ and noise subspace $\boldsymbol{U}_{\mathbf{n}}$.

$$\boldsymbol{U} = [\boldsymbol{U}_{\mathbf{s}} | \boldsymbol{U}_{\mathbf{n}}] \quad (13)$$

Therefore (12) can be expressed as

$$\boldsymbol{R}_{TK} = [\boldsymbol{U}_{\mathbf{s}} \quad \boldsymbol{U}_{\mathbf{n}}]\begin{bmatrix}\varphi_0 & 0 & \cdots & 0 \\ 0 & \varphi_1 & \ddots & \vdots \\ \vdots & \ddots & \ddots & 0 \\ 0 & \cdots & 0 & \varphi_{N-1}\end{bmatrix}\begin{bmatrix}\boldsymbol{U}_{\mathbf{s}}{}^H \\ \boldsymbol{U}_{\mathbf{n}}{}^H\end{bmatrix} \quad (14)$$

signal subspace $\boldsymbol{U}_{\mathbf{s}} = \begin{bmatrix}u_0 & \cdots & u_{L_p-1}\end{bmatrix}$ concludes L_p eigenvectors, noise subspace $\boldsymbol{U}_{\mathbf{n}} = \begin{bmatrix}u_{L_p} & \cdots & u_{N-L_p}\end{bmatrix}$ concludes $N-L_p$ eigenvectors. $\boldsymbol{U}_{\mathbf{s}}$ can be expressed as

$$\boldsymbol{U}_{\mathbf{s}} = \begin{bmatrix}\boldsymbol{U}_1 \\ *\end{bmatrix} = \begin{bmatrix}* \\ \boldsymbol{U}_2\end{bmatrix} = \boldsymbol{AT} \quad (15)$$

* donates the last line and first line of matrix respectively and it's irrelevant value, *T* is mapping matrix and non-singular matrix,

$$A=\begin{bmatrix} 1 & 1 & \cdots & 1 \\ e^{-j2\pi\tau_0/T_s} & e^{-j2\pi\tau_1/T_s} & \cdots & e^{-j2\pi\tau_{Lp-1}/T_s} \\ \vdots & \vdots & \cdots & \vdots \\ e^{-j2\pi(N-1)\tau_1/T_s} & e^{-j2\pi(N-1)\tau_1/T_s} & \cdots & e^{-j2\pi(N-1)\tau_{Lp-1}/T_s} \end{bmatrix} \tag{16}$$

for (15), there are $U_1 = A_1T$, $U_2 = A_2T = A_1\phi T$, and $\phi = diag\begin{bmatrix}1 & e^{-j2\pi f_k\tau_1} & \cdots & e^{-j2\pi f_k\tau_{Lp-1}}\end{bmatrix}$, thus we define matrix ψ

$$\psi = T\phi T^{-1} \tag{17}$$

matrix ψ can be called as the similarity transformation of matrix ϕ, thus they have the same eigenvalue, which means τ_{k-TK} can be calculated by the following function

$$\tau_{k-TK} = \frac{\angle\phi_{k-TK}}{2\pi\Delta f} \tag{18}$$

ϕ_{k-TK} is the eigenvalue of matrix ϕ_{TK}, we can calculate the corresponding TOA parameters by calculate the phase.

Step 1: Construct cross-correlation function by received signal and local signal.

Step 2: Utilizing TK energy operator to process cross-correlation function.

Step 3: Construct TK operator processed cross-correlation function eigenvector matrix.

Step 4: Utilizing ESPRIT to calculate the eigenvalue of eigenvector matrix to ensure TOE parameters.

The costs of ESPRIT mainly consists of EVD, pseudo inverse and diagonalization, which can be derived is $O(4N+2RN+RN^2)+O(4L_p^2N+L_p^3)+O(RN^2)$. The costs of TK energy operator processed TK-ESPRIT is $O(4N+2R_{TK}N+R_{TK}N^2)+O(4L_p^2N+L_p^3)+O(R_{TK}N^2)+O(2L_p^2)$, and $R_{TK}<R$, thus the computation of the improved algorithm is lower.

4 SIMULATION

According to IEEE 802.11a to set the simulation parameters, wideband is 20MHz, the number of effective subcarriers is 52, in order to facilitate digital implementation, *N* is 64, subcarrier interval is 312.5KHz, SNR is from -4 to 16dB, time delay is 1.25 μs. Spreading code is CI Code, CI Code is a kind of spreading sequence with excellent orthogonality. For multi-user downlink, CI code could effectively decrease inter-cell interference since the spreading sequence cross-correlation. Experiment signal simulate as Figure 1, and is QPSK modulation.

Root Mean Square Error(RMSE) in single path can be expresses as

$$RMSE = \sqrt{\frac{1}{N_s}\sum_{k=1}^{N_s}(\tau^l - \hat{\tau}_k^l)^2} \tag{19}$$

where τ^l the mean value of total estimation results of ith path and $\hat{\tau}_k^l$ is the kth estimation value respectively. N_s is 1000 and utilizing Monte Carlo Simulation to calculate RMSE, simulation results as follows.

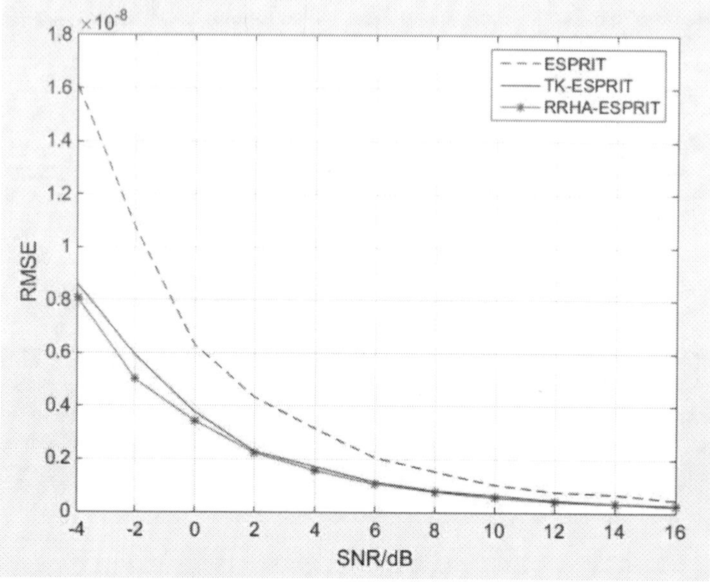

Figure 2: Comparison of time delay estimation algorithm RMSE

Figure 2 is the comparison of TK-ESPRIT, two-dimensional ESPRIT and RRHA-ESPRIT, the number of path is 1. It's obviously that in low SNR cases($SNR<12\text{dB}$), the performance of improved algorithm is better than original one, and in high SNR cases($SNR>12\text{dB}$), their performance is quite. The estimation accuracy of RRHA-ESPRIT in low SNR cases is better than improved algorithm, but with SNR increasing, the performance is quite with improved algorithm. Because of iteration, the complexity is much higher than improved algorithm.

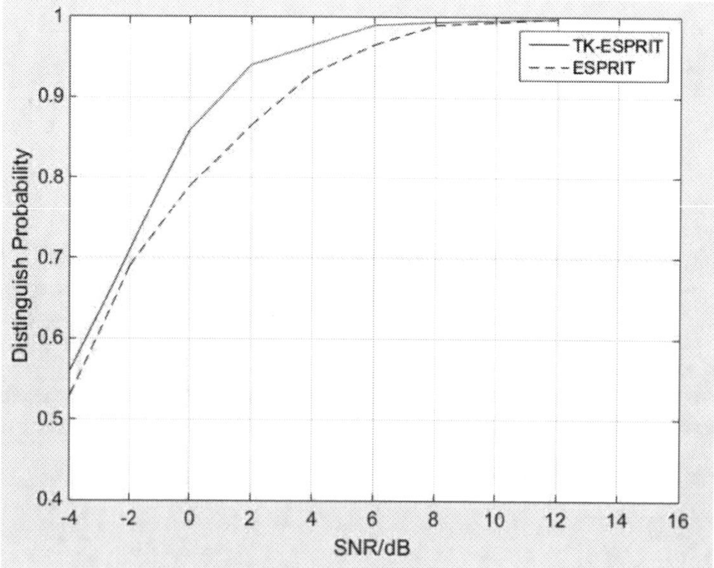

Figure 3: Comparison of time delay estimation algorithm distinguish probability

Figure 3 is the comparison of time delay estimation distinguish probability of improved algorithm and two-dimensional ESPRIT. Distinguish probability is the ratio of correct estimation results and total estimation number. Estimate 1000 times independently and compare both algorithms' performance in different SNR cases. It's obviously that in high SNR cases, both accuracy probability is nearly 100%, but in low SNR cases, improved algorithm is better than ESPRIT.

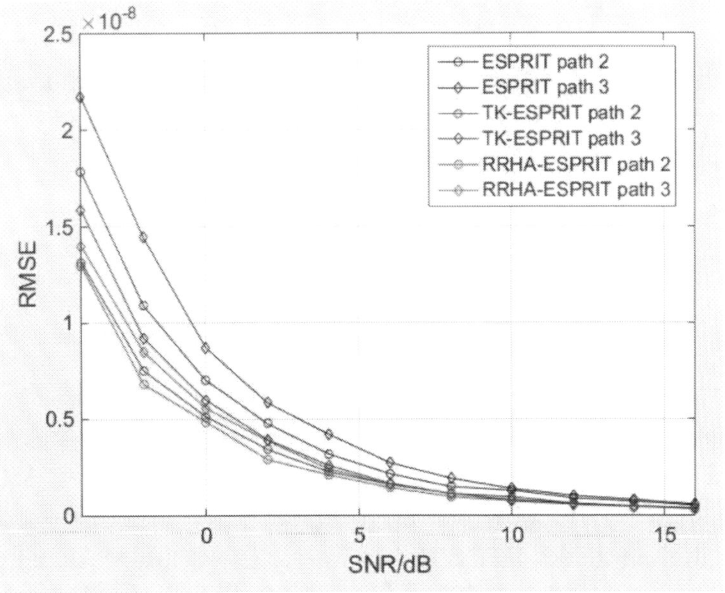

Figure 4: Algorithm comparison in multipath

Figure 4 is the comparison of improved algorithm, two-dimensional ESPRIT and RRHA-ESPRIT in multipath case. Set multipath number is 2 and 3, respectively. In low SNR cases($SNR<4\mathrm{dB}$), RRHA-ESPRIT is effected by multipath factor least, and accuracy is higher than improved TK-ESPRIT, but when SNR increased to 4dB, the two accuracy effected by multipath factors is nearly quite. Meanwhile, improved algorithm's variety is smaller and estimation accuracy is higher.

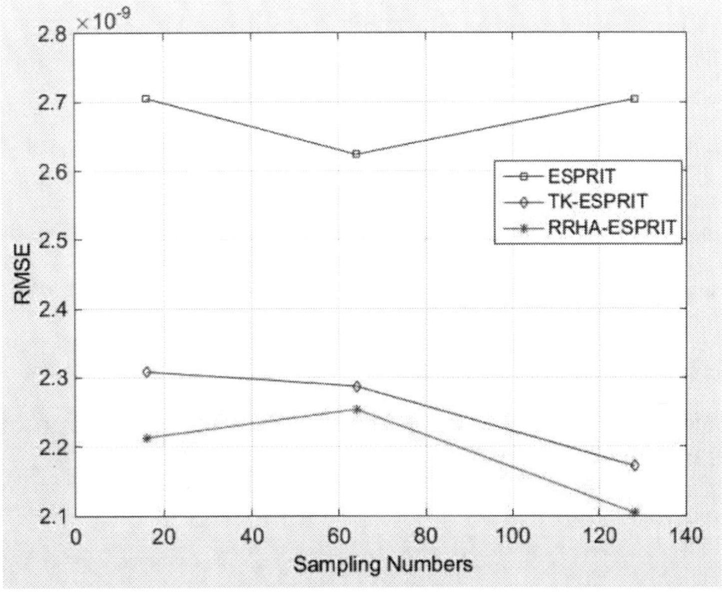

Figure 5: Algorithm comparison in different sampling numbers

Figure 5 is the comparison of improved algorithm, two-dimensional ESPRIT and RRHA-ESPRIT in different sampling numbers. Set sampling number is 16, 64, 128 respectively, SNR is 10dB. As shown in figure, with sampling number increasing, the estimation accuracy of TK-ESPRIT increasing, RRHA-ESPRIT remained relatively stable, but two-dimensional ESPRIT had obvious variety. Because the growth of sampling number leads the interference of AWGN is more serious. Improved algorithm could avoid variety by TK energy operator.

5 CONCLUSION

This paper focused on TOA estimation of OFDM signal, proposed an improved TK-ESPRIT based on ESPRIT. Improved the anti-noise capability through processing the cross-correlation of received signal and local signal by TK energy operator. At last, analysed its performance with stimulation. The results show that improved algorithm has the better resolution and RMSE, the improved algorithm can mitigate noise influence better.

6 REFERENCES

[1] **F. Zhou, L.K. Wang** and **X.Y. Fan. 2012.** An improving beam-space MUSIC time delay estimation algorithm for OFDM signal, 9th International Conference on Fuzzy Systems and Knowledge Discovery, Chongqing, China, May 29-31, pp 2115-2118.

[2] **A. Makki, A. Siddig, M Saad, et al. 2015.** Indoor Localization Using 802.11 Time Differences of Arrival, IEEE Transactions on Instrumentation and Measuremnet, 65(3), pp 1-10.

[3] **F. Zhao, W.B. Yao, C.C. Logothetis, et al. 2006.** Comparison of Super-Resolution Algorithms for TOA estimation in indoor IEEE 802.11 wireless LANs, International Conference on Wireless Communications, Networking & Mobile Computing, Wuhan, China, Sep 22-24, pp 1-5.

[4] **F. Zhao, W.B. Yao, C.C. Logothetis, et al. 2007.** Super-resolution TOA estimation in OFDM systems for indoor environment, IEEE International Conference on Networking, Sensing and Control, London, U.K., April 15-17, pp 723-727.

[5] **V.U. Prabhu, J. Jalihal. 2009.** An improved ESPRIT based time of arrival estimation algorithm for vehicular OFDM systems, IEEE Vehicular Technology Conference, Anchorage, USA, Sep 20-23, pp 1-4.

[6] **D.G. Oh, S.D. Kim, S.H. Yoon, et al. 2013.** Two-dimensional ESPRIT-like shift-invariant TOA estimation algorithm using multi-band chirp signals robust to carrier frequency offset, *IEEE Transactions on Wireless Communications*, 12(7), pp 3130-3139.

[7] **S.D. Kim, D.G. Oh, J.H. Lee. 2015.** Joint DFT-ESPRIT estimation for TOA and DOA in vehicle FMCW radars, IEEE Antennas and Wireless Propagation Letters, 14, pp 1710-1713.

[8] **H. Farrokhi. 2007.** TOA estimation using MUSIC super-resolution techniques for an indoor audible chirp ranging systems, IEEE International Conference on Signal Processing & Communications, Dubai, United Arab Emirates, Nov 24-27, pp 987-990.

[9] **X.L. Lu. 2012.** Research and implementation on OFDM spread spectrum communication technology algorithm, *University of Electronic Science and Technology.*

[10] **A. Makki, A. Siddig, M.M. Saad, et al. 2015.** High-resolution time of arrival estimation for OFDM-based transceivers, Electronic Letters, 51(3), pp 294-296.

TEXT CLASSIFICATION USING K-MEANS ALGORITHM

Hongqi Han* , Changqing Yao, Yiying Liu and Yongsheng Yu
Institute of Scientific and Technical Information of China, Beijing 100038, China
bithhq@163.com

ABSTRACT

Text clustering is an important and quick method to explore subject distribution in a large data set, however clustering results are usually difficult for human to understand. An improved k-means algorithm is presented to cluster text documents into pre-defined classes, or the algorithm is used to resolve text classification problems. In presented algorithm, class associated words are used to create initial clustering centroids, and supervise the generation of clusters during iteration process, where class associated words are words or phrases which represent the subject of classes in text corpus. Experiment results show that the technique can label documents into pre-defined classes with prior information provided by class associated words and achieves high classification accuracy.

Keywords: Class associated words; k-means; text clustering; text classification

1 INTRODUCTION

Along with the development of internet, the information contained within the web is increasing day by day. Acquiring a large quantity of unlabeled text document is becoming easy from web pages, web blogs, online databases, etc. [1, 2]. Simultaneously, organizing documents into a set of categories, and retrieving relevant text documents on a topic from a large document collection are becoming challenging tasks [1, 3], because there are simply too many texts, and rarely are people able to manually read all the texts in even moderately sized corpora, moreover, hiring coders to manually read all documents is still very expensive and impracticable [4, 5]. It is gratifying that some automated content analysis methods have been put forward to deal with these issues, such as text classification or text clustering.

Text classification and clustering are promising areas in data mining field. They have great practical value in application [6]. For a text document collection D, both classification and clustering algorithm can divide it into disjoint subsets, however they have some obvious distinctions. Class labels are known before the start of classification algorithm, but usually unknown for clustering. The classification method, known as supervised learning technique, which often needs a suitable scale of training set where each sample has a class label. A classifier is constructed after learning from the training set, and then is used to classify documents. Each document is assigned a class label after classification process. Whereas, known as an unsupervised learning technique, the clustering method does not need the training set and the class labels are usually not provided before learning. The goal of the unsupervised learning task is to find groups (clusters) in the documents according to similarities of document vector features. After clustering process, The data

* Corresponding Author

in same cluster have high similarity meanwhile the data in different cluster have high dissimilarity [7, 8]. Although the clustering method has the capability to divide documents into groups without labeled training data, it is often criticized to cluster blindly [6], i.e. it is very difficult to assign a category name for each cluster.

K-means is a well-known planar partitioning clustering algorithm. It is probably the most widely used algorithm in text clustering [9]. Given value k, k-means algorithm can be used to divide a document collection into k clusters [7, 8]. K-means, originally put forward by Mac- Queen [10], can be used to deal with big data set owing to its excellent performance [11]. Dhillon was the first researcher who applied k-means algorithm in text clustering. K-means used to be applied in clustering. In this research, it is employed to resolve text classification problem. The proposed method generates initial centroids by using prior information of class associated words (CAWs) contained in documents. CAWs are used to establish one-to-one correspondence between each centroid and each class label. During the clustering process of k-means algorithm, CAWs are used to supervise cluster generation and reduce the comparison numbers between document vectors and centroid vectors so that the clustering efficiency is improved. Finally each output cluster by algorithm is assigned to a predefined class label.

2 CLASS ASSOCIATED WORDS

Considering manual document classification process, it needs to use distinctive words to classify documents. For example, we have documents asked to be classified into two classes: sports, economy and management. Generally, if we find words or phrases like "basketball", "football", "table-tennis" and so on in a document, while we don't find other words correlated to economy or management, the document definitely belongs to the class of sports; similarly, if we find words or phrases like "finance", "marketing", "business management", "human resource management" in a document, and don't find other words correlated with sports simultaneously, then we can tag the document a label of economy and management class. Such words or phrases occurring in a document and closely related to a class are called class associated words (CAWs). Ideally, a document only contains CAWs corresponding to a class, however, the situation is not always the case. Perhaps there exist several CAWs belongs to two or more classes in a document. For example, possibly the term "finance" occurs in a document of sports. Under such circumstance, we may have to read it completely to decide its class label. CAWs were once used into a classification algorithm in one of our research [6]. Here we will introduce its usage into k-means clustering process.

3 ALGORITHMS

Traditionally, k-means is just a clustering algorithm used to divide data into k clusters (k is predefined). The presented method will employ k-means algorithm to implement classification. That is, Given a document collection, $D = \{d_1, d_2, \ldots, d_m\}$, and class label set, $C = \{c_1, c_2, \ldots, c_k\}$, we will use k-means algorithm and CAWs to cluster documents in D into k groups, and each document has a class label in C.

3.1 Obtaining Class Associated Words

It is a good way to select CAWs from subclasses, but it might be difficult to determine sub classes of a class. User's knowledge and experience are very helpful for us to select CAWs, however it may be inadequate, since user's recommendation may be incomplete or inconsistent with the real situation. Therefore, we should resort to more techniques or practices. According to our experiences, it is a better way to select candidates combining user's experience with machine recommendation. Usually, at least one CAW is selected for a class, but it doesn't mean you can select only one CAW for a class. Actually, you can select several CAWS for a class.

The basic principal of selecting a word or phrase as CAW is its discrimination ability. Practically, we should select those distinguishing and representative words with the help of evaluation measures, such as word frequency, information gain or mutual information. These evaluation metrics will rank all words for user to select. However, if some CAWs are very rare in all documents and no evaluation metric will give them high scores, the selection process needs user's provision. Each class should be provided one or more CAWs for the presented algorithms. Certainly, if a CAW provided by user can't be found in any document, it should not be included. Fig.1 demonstrates the correspondence relations between class labels and CAWs, where Z_i denotes a CAW.

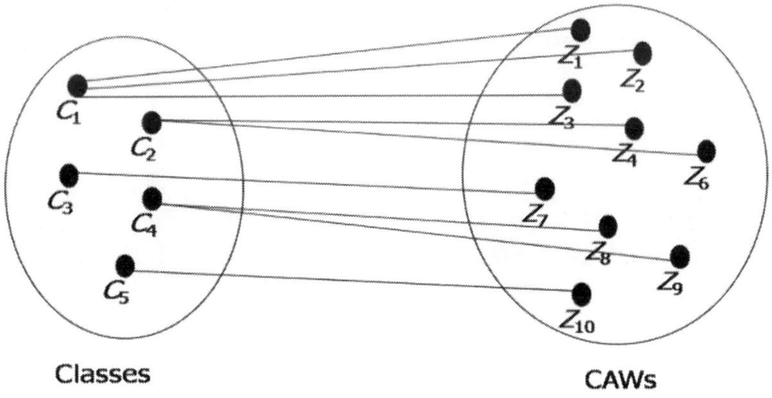

Figure 1: Mapping from class labels to class associated words

3.2 Generating Initial Centroids

Possibly, there have several CAWs belonging to one or more classes in a document. Also, a document may contain no CAWs. With CAWs, document collection D can be divided into three groups (Fig. 2): (1) all CAWs in a document belonging to a class, denote D_1; (2) CAWs in a document belonging to two or more classes, denote D_2; (3) no CAW in a document, denote D_3. If CAWs are selected prudentially, a document in D_1 must belong to a definite class; a document in D_2 is probably assigned to those classes decided by its CAWs, meantime, it is impossible to belong to any classes whose CAWs have never occurred.

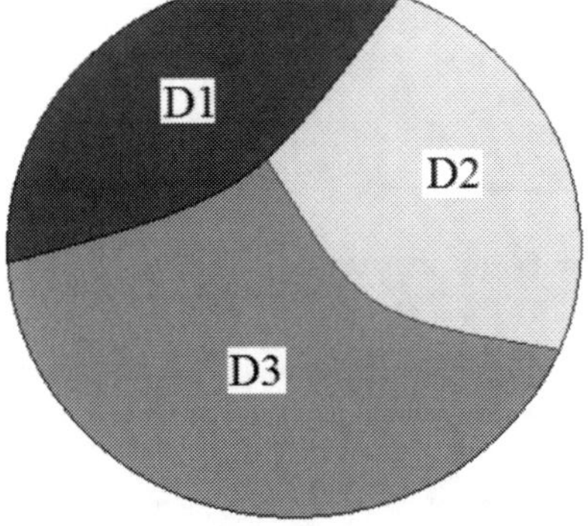

Figure 2: Document subsets divided by class associated words

The prior knowledge of CAWs can be used to generate initial centroids for k-means algorithm. For the documents in D_1, we can set a definite class label to each document because the CAWs in each just belonging to one class. For example, suppose $d \in D_1$, and the CAWs in d belong to class c_j, then document d can be assigned to class label c_j definitely. If there have documents in D_1 for a class c_j, the mean text vector value will be set to the initial centroid of class c_j. For every class which has not any corresponding documents in D_1, the initial centroid can be generated in two ways: the first is selecting the vector of any one document in D_2 which contains the CAWs of the class; the second is using the mean vector value of all documents in D_2 which contain the CAWs of the class.

3.3 Clustering

After generating initial centroids, k-means algorithm is used to implement clustering. On the premise of not causing misunderstanding, denote document vector set as $D=\{d_1, d_2, ..., d_n\}$, clustering centroid vector set as $C=\{c_1, c_2, ..., c_k\}$. Terms extracted from documents are set as vector characteristic items, values of which are the term frequency in one document, thus a document is represented as a vector using the vector space model (VSM). Then the clustering process using k-means algorithm is described as follows:

1. For each document vector d_i (i=1, 2, ..., n), its similarity with each clustering centroid in C is computed. Included angle cosine distance is used to measure the similarity between a document vector d_i and a centroid c_j. The similarity metric is given as equation 1.

$$sim(d_i, c_j) = \frac{d_i \cdot c_j}{\|d_i\| \cdot \|c_j\|} \quad (1)$$

2. For each document vector d_i in D, a centroid c_j is selected as its class label according to the maximum similarity, namely $\arg\max_{c_j \in C} sim(d_i, c_j)$.
3. Every centroid vector is recalculated (equation 2). The new centroid vector is the arithmetic mean value of all document vector having the class label.

$$c_j = \frac{1}{|c_j|} \sum_{d_i \in c_j} d_i, j = 1, 2, \cdots, k \quad (2)$$

Next, the objective function is computed (equation 3). The objective function is defined by the sum of square sum of distance between each document vector and its centroid vector.

$$E = \sum_{j=1}^{k} \sum_{d \in c_j} \|d - c_j\|^2 \quad (3)$$

Then judge the convergence of the function value. If the objective function value satisfies terminal condition, the iteration process will be stopped, otherwise it will go to step 1 and begin next iteration. The terminal condition is decided by the difference value between two successive objective function value. If the difference value is less than a threshold ε, the algorithm stops.

During each iteration, CAWs are used to supervise the clustering process and restrict the cluster generation. This will reduce the amount of calculation and accelerate algorithm convergence speed. For a document d in D_1, it is unnecessary to compute the similarity between document vector d and the centroid vector of corresponding class c because document d only contains CAWs of class c and can be assigned a definite cluster. For a document d in D_2, it is only necessary to compute the similarity between document vector d and those centroid vectors whose corresponding CAWs occur in document *d* in algorithm step 1, and it is unnecessary to compute similarity metrics with other centroid vectors. Therefore, the amount of computation is largely reduced, improving the efficiency of the algorithm. For the documents in D_3, they don't contain any CAWs, so we have to compute the similarity metrics with each centroid vector.

After several iterations, the convergence will be obtained and the algorithm will output *k* clusters when the termination condition is satisfied. Each cluster corresponds to a class label in class set *C*, and each document is assigned to a class label, thus implementing text classification.

4 EXPERIMENTAL BACKGROUND AND RESULTS

4.1 Experimental Background

The presented method was motivated by a patent analysis project. The task was to investigate innovation situation on research and development of fuel cell technology by patent analysis. We were given six domains of fuel cell technology: alkaline fuel cell (AFC), phosphate fuel cell (PFC), molten carbonate fuel cell (MCFC), solid oxide fuel cell (SOFC), solid polymer fuel cell (SPFC) and biological fuel cell (BFC). Our idea was to collect patent documents and assign each document to a specific domain. Then we could count patents on each domain, thereby we could estimate investment shares among different domains. Employing manpower was one way to finish the task, however financial expense and time consumption couldn't be accepted. Actually, this is a typical text classification problem. Machine learning algorithms have been proved to be capable of resolving it. Thus our task is transformed into classifying patent documents into 6 classes.

4.2 Selecting Class Association Words

Ordinary single word occurred in a document often provide few information for tagging the document, whereas technical terms usually give specific meanings of concepts and can provide efficient information to assign a definite label to the document. For example, if we give some words of a document to experts, like proton, membrane, exchange, fuel, and cell, it's difficult for them to decide the class label of the document. However, if we give term "proton membrane exchange fuel cell" to experts, they can easily decide that it is probable to be one document of class label SPFC. Considering these basic facts, we first used automatic term extraction algorithm to obtain technical terms in patent documents. In our experiment, totally 1123 technical terms were extracted from 2499 patent documents. Then we provided these terms to domain experts and asked them give probable relations between terms and class labels. Based on the suggestions of experts, we selected some terms as class associated words and built the mapping relations between terms and class labels shown in Table 1.

Table 1: Selected Class Associated Words and Corresponding Classes

No.	Classes	CAWs
1	AFC	alkaline fuel cell
2	PFC	phosphate fuel cell
3	MCFC	molten carbonate, carbonic acid
4	SOFC	solid oxide, solid fuel, oxidant, perovskite
5	SPFC	solid polymer, proton exchange membrane, polymer

		dielectric, electrolyte thin films
6	BFC	biology, microbiology, organic substance

4.3 Experimental Results

In order to testify the performance of presented methods, we collected Chinese patent documents of fuel cell from SIPO websites (http://www.sipo.gov.cn/). The collected patents are valid and granted so that the analysis can reflect actual investment shares. The presented algorithm is employed to implement the classification. In the experiment, the document collection contains 2499 patent documents; the quantities of class labels are 6 (k=6). First, selected CAWs were used to build mapping relations between documents and class labels. Next, initial centroids were generated according to the CAWs in Table 1. Then proposed algorithm was employed to start the iterative clustering process described in section 4.3. The iteration was stopped when the difference of successive objective function value was less than 0.5. Fig.3 illustrates variation of the objective function value along with the iteration number.

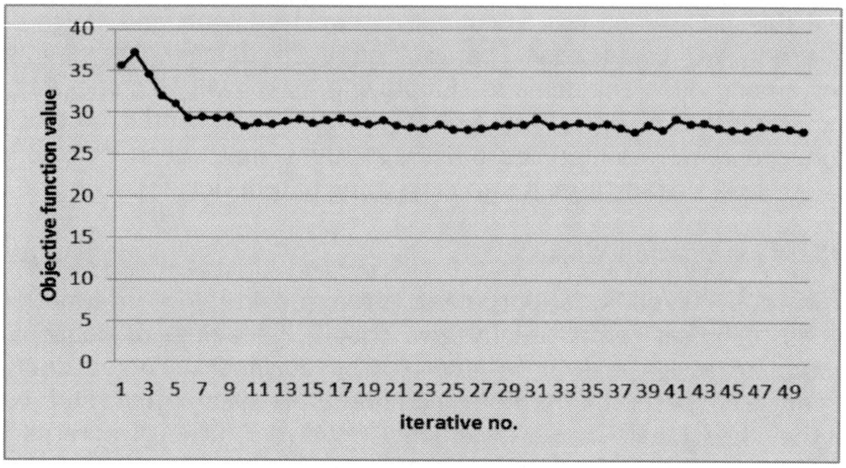

Figure 3: Variation of Objective function value with iterate number

It can be seen from Fig. 3, the value of objective function increases at the beginning of the iterative process. Under normal circumstance, the values of objective function decrease monotonously. Possibly the phenomenon owed to the influence of CAWs. Soon the values of objective function declines. After several iterations, the objective values become stable and fluctuate in a small range, and finally satisfy the terminal condition of algorithm. As a result, every document is assigned to a class label. Then, we counted patent documents in each class label. Fig. 4 illustrates the classification results.

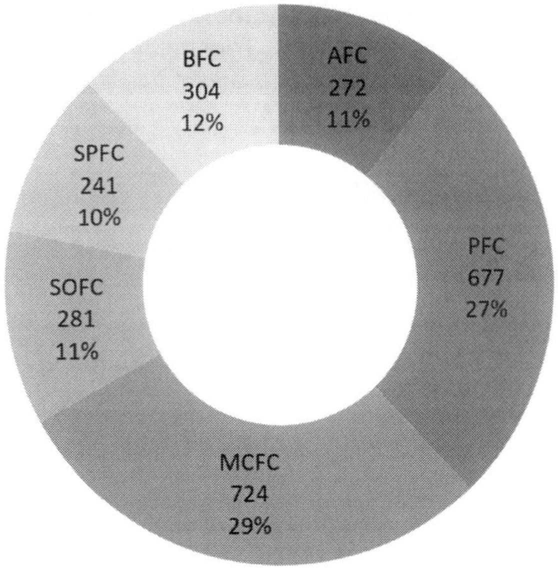

Figure 4: Classification result of fuel cell technology by electrolyte

Fig. 4 shows the proportion of patents in each fuel cell domain. There have 724 patents in the domain of the molten carbonate fuel cell, accounting for 29%, being the first and a dominating research technology in the fuel cell industry. The technology of the phosphate fuel cell is in the second place containing 677 patents, accounting for 27%. The patent quantities of other four technologies of fuel cell do not have significant differences; each proportion is ranged between 10% to 12%, showing that the four technologies have been paid the same attention.

In order to evaluate the accuracy of the presented algorithm, we randomly selected 200 patents from the patent documents and assigned class labels to each document manually. Then we compared the class label of each document tagged by the algorithm with manual labels. The comparison shows the accuracy of the presented algorithm is 89%.

5 CONCLUSION

K-means is a well-known clustering algorithm and is most widely used in text clustering, yet it is used to resolve text classification problem in the study. The presented algorithm can implement classification lies in the clues provided by CAWs in documents. The main advantage of the algorithm is that it can implement classification without training set. CAWs play important roles in the algorithm. CAWs should be decided according to the word importance evaluation rank and expert's prior knowledge. In practice, the selection of CAWs can combine machine recommendation and manual provision. CAWs are used to generate initial centroids and supervise cluster cohesiveness during k-means clustering process. The technique can improve clustering efficiency because it reduces computation amount in iterative process. Besides, the output clusters of algorithm are in one-to-one correspondence with related class labels, thus implementing the purpose of text classification. Especially, experimental results show the algorithm obtains a high precision and the performance is satisfactory.

6 ACKNOWLEDGEMENTS

This work is mainly supported by National Natural Science Foundation of China (Project 71473237), and partially supported by The National Key Technology R&D Program of Chinese 12th Five-Year Plan (2011-2015) (2015BAH25F01), and The Program of the China Knowledge Centre for Engineering Science and Technology (CKCEST-2016-2-10). Authors are grateful to National Natural Science Foundation of China, Ministry of Science and Technology of China and Chinese Academy of Engineering for financial support to carry out this work.

7 REFERENCES

[1] **P. Kaur. 2014.** Web content classification: A survey, *Eprint Arxiv*, 10(2).

[2] **S. C. Herring. 2010.** Web content analysis: Expanding the paradigm, *International handbook of Internet research*. Springer, pp 233-249.

[3] **N. Gupta. 2011.** Automatic generation of initial value k to apply k-means method for text documents clustering, *International Journal of Data Mining Modelling & Management*, 3(1), pp 18-41.

[4] **X. Zhu. 2006.** Semi-supervised learning literature survey, *Computer Science, University of Wisconsin-Madison*, 37(1), pp 63-67.

[5] **J. Grimmer and B. M. Stewart. 2013.** Text as data: The promise and pitfalls of automatic content analysis methods for political texts, *Political Analysis*, 21(3), pp 267-297.

[6] **H.-q. Han, D.-H. Zhu, and X.-f. Wang. 2009.** Semi-supervised text classification from unlabeled documents using class associated words, *Proceeding of the International Conference on Computers & Industrial Engineering*, Toyes, France, July 6-9, pp 1255-1260.

[7] **C. C. Aggarwal and C. Zhai. 2012.** A survey of text clustering algorithms, *Mining text data*. Berlin Heidelberg: Springer, pp 77-128.

[8] **A. K. Jain, M. N. Murty, and P. J. Flynn. 1999.** Data clustering: a review, *ACM computing surveys (CSUR)*, 31(3), pp 264-323.

[9] **H. Verma, E. Kandpal, B. Pandey, and J. Dhar. 2010.** A novel document clustering algorithm using squared distance optimization through genetic algorithms, *International Journal of Computational Science & Engineering*, 2(5), pp 1875-1879.

[10] **J. MacQueen.1967.** Some methods for classification and analysis of multivariate observations, *Proceedings of the fifth Berkeley symposium on mathematical statistics and probability*, California, USA: University of California Press, pp 281-297.

[11] **X. Cai, F. Nie, and H. Huang. 2013.** Multi-view k-means clustering on big data, *Proceeding of the International Joint Conference on Artificial Intelligence*, 2013.

USE OF A TWO-LEVEL DEA APPROACH FOR EVALUATING TWO-STAGE SYSTEMS WITH FUZZY DATA

F. Hosseinzadeh Lotfi[1*], A. Ghazi[2], R. Tavakkoli-Moghaddam[3], M. Fallah[4]
[1]Department of Mathematics, Science and Research Branch
Islamic Azad University, Tehran, Iran
farhad@hosseinzadeh.ir

[2]Department of Mathematics, Central Tehran Branch
Islamic Azad University, Tehran, Iran
emine.ghazi@gmail.com

[3]School of Industrial Engineering, College of Engineering
University of Tehran, Tehran, Iran
tavakoli@ut.ac.ir

[4]Department of Industrial Engineering, Central Tehran Branch
Islamic Azad University, Tehran, Iran
mohammad.fallah43@yahoo.com

ABSTRACT

A data envelopment analysis (DEA) is a method for measuring the relative efficiencies of a set of decision making units (DMUs) that use inputs to produce outputs. In series systems, outputs of the first stage are consumed as inputs of the second stage. The conventional DEA requires crisp input and output data; however, the observed values of the input and output data in real word applications are sometimes imprecise. This paper proposes a two-stage DEA model, in which input and output data are treated as fuzzy numbers and the two-level DEA approach is used for solving this model. A pair of two-level mathematical programs is formulated to calculate the upper and lower bounds of the fuzzy efficiency score. Then, this pair is converted to a pair of conventional mathematical programs to calculate the bounds of the fuzzy efficiency score.

Keywords: Data envelopment analysis, Two-stage approach, Two-level, Fuzzy Data

1 INTRODUCTION:

A data envelopment analysis (DEA) was suggested by Charnes et al. [2], based on the works of Farrell [4] and then developed by others [1]. DEA is a method for measuring the relative efficiency of decision making units (DMUs) with common inputs and outputs. DEA is a useful tool for measuring efficiency of DMUs in order to help execute management planning and control activities. Many production systems in the real world are composed of two processes connected in series, parallel, or both of them, simultaneously. Two-stage DEA models show the performance of individual processes and thus, they are more informative than the conventional one-stage DEA models for making decisions [7,9,10]. In real situation, the most of data are indefinite and are stated by natural language, so, some researchers have carried out studies in this field and used fuzzy set theory and comparison of fuzzy

* Corresponding Author

numbers [5]. Kao and Liu [8] transformed fuzzy input and output data into intervals data using $\alpha-$level sets. The $\alpha-$level set approach was extended by Saati et al. [11] who defined the fuzzy DEA model as a possibility-programming problem and transformed it into an interval programming. Hatami-Marbini et al. [6] proposed a four-phase fuzzy DEA framework based on the theory of displaced ideal. Moreover, Entani et al. [3] extended the $\alpha-$level set research by changing fuzzy input and output data into intervals. In this paper we develop a methodology based on the two-level approach is used. The new model is constructed based on the general form of the two-stage system. This research is organized as follows. Section 2 proposes a two-stage DEA model for the general form two-stage system. Then, the two-level approach for solving the proposed fuzzy two-stage DEA model is used. Also, the proposed method is applied for evaluating of 30 BANK branches in IRAN. Conclusions and future directions are summarized in the last section.

2 FUZZY TWO-STAGE DEA MODEL

2.1 Proposed Model

Let $\bar{x}_{ij}, \bar{z}_{tj}, \bar{y}^1_{rj}$ and $\bar{y}^2_{oj}$ $(i=1,...,m, t=1,...,g, r=1,...,s1, o=1,...,s2, j=1,...,n)$ be the fuzzy counterparts of input, intermediates and outputs, respectively. A two-stage system with series and paralell structure is portrayed in Figure 2.

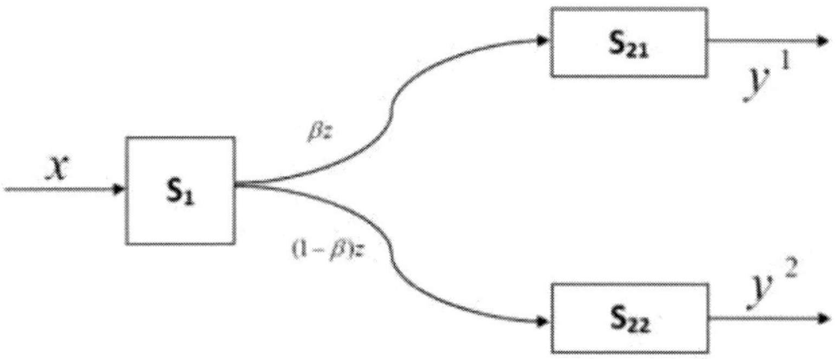

Figure 1: Two-stage System

A DEA model is proposed for the two-stage system in Figure 1, as follows:

$$
\begin{aligned}
\bar{E}_p = Max \quad & \sum_{r=1}^{s1} u^1_r \bar{y}^1_{rp} + \sum_{o=1}^{s2} u^2_o \bar{y}^2_{op} \\
s.t. \quad & \sum_{i=1}^{m} v_i \bar{x}_{ip} = 1, \\
& \sum_{t=1}^{g} w_t \bar{z}_{tj} - \sum_{i=1}^{m} v_i \bar{x}_{ij} \le 0, && j=1,...,n, \\
& \sum_{r=1}^{s1} u^1_r \bar{y}^1_{rj} - \sum_{t=1}^{g} w_t \beta \bar{z}_{tj} \le 0, && j=1,...,n, \qquad (1) \\
& \sum_{o=1}^{s2} u^2_o \bar{y}^2_{oj} - \sum_{t=1}^{g} w_t (1-\beta) \bar{z}_{tj} \le 0, && j=1,...,n, \\
& v_i, u^1_r, u^2_o, w_t \ge 0, && i=1,...,m, r=1,...,s1, o=1,...,s2, t=1,...,g.
\end{aligned}
$$

Where parameter $\beta \in [0,1]$. The resulting efficiency of $DMU_p\ (p \in \{1,\ldots,n\})$ is a fuzzy number because the observations are fuzzy number. The membership function $\mu_{\bar{E}_p}$ for $\bar{E}_p$ can be obtained as follows:

$$\mu_{\bar{E}_p(e)} = \sup \min \{\mu_{\bar{x}_{ij}}(x_{ij}), \mu_{\bar{z}_{tj}}(z_{tj}), \mu_{\bar{y}^1_{rj}}(y^1_{rj}), \mu_{\bar{y}^2_{oj}}(y^2_{oj})\}, \tag{2}$$

where $\mu_{\bar{x}_{ij}}, \mu_{\bar{z}_{tj}}, \mu_{\bar{y}^1_{rj}}$ and $\mu_{\bar{y}^2_{oj}}$ are membership function of $\bar{x}_{ij}, \bar{z}_{tj}, \bar{y}^1_{rj}$ and $\bar{y}^2_{oj}$, respectively. Denote $(x_{ij})_\alpha = [(x_{ij})^L_\alpha, (x_{ij})^U_\alpha], (z_{tj})_\alpha = [(z_{tj})^L_\alpha, (z_{tj})^U_\alpha], (y^1_{rj})_\alpha = [(y^1_{rj})^L_\alpha, (y^1_{rj})^U_\alpha]$ and $(y^2_{oj})_\alpha = [(y^2_{oj})^L_\alpha, (y^2_{oj})^U_\alpha]$ as the $\alpha-$cuts of $\bar{x}_{ij}, \bar{z}_{tj}, \bar{y}^1_{rj}$ and $\bar{y}^2_{oj}$, respectively. To find the membership function $\mu_{\bar{E}_p(e)}$, it is suffices to find the lower and upper bounds of the $\alpha-$cuts of $\bar{E}_p$, $(\bar{E}_p)_\alpha = [(\bar{E}_p)^L_\alpha, (\bar{E}_p)^U_\alpha]$, where

$$(\bar{E}_p)^L_\alpha = \min\{e \,|\, \mu_{\bar{E}_p}(e) \geq \alpha\} \tag{3}$$

$$(\bar{E}_p)^U_\alpha = \max\{e \,|\, \mu_{\bar{E}_p}(e) \geq \alpha\} \tag{4}$$

Above expression can be written as follows:

$$\begin{aligned} (\bar{E}_p)^L_\alpha = Min \quad & E_p(x, z, y^1, y^2) \\ s.t. \quad & (x_{ij})^L_\alpha \leq x_{ij} \leq (x_{ij})^U_\alpha \\ & (z_{tj})^L_\alpha \leq z_{tj} \leq (z_{tj})^U_\alpha \\ & (y^1_{rj})^L_\alpha \leq y^1_{rj} \leq (y^1_{rj})^U_\alpha \\ & (y^2_{oj})^L_\alpha \leq y^2_{oj} \leq (y^2_{oj})^U_\alpha \\ & \forall i, t, r, o, j, \end{aligned} \tag{5}$$

and,

$$\begin{aligned} (\bar{E}_p)^U_\alpha = Max \quad & E_p(x, z, y^1, y^2) \\ s.t. \quad & (x_{ij})^L_\alpha \leq x_{ij} \leq (x_{ij})^U_\alpha \\ & (z_{tj})^L_\alpha \leq z_{tj} \leq (z_{tj})^U_\alpha \\ & (y^1_{rj})^L_\alpha \leq y^1_{rj} \leq (y^1_{rj})^U_\alpha \\ & (y^2_{oj})^L_\alpha \leq y^2_{oj} \leq (y^2_{oj})^U_\alpha \\ & \forall i, t, r, o, j. \end{aligned} \tag{6}$$

Models (5) and (6) are called two-level programs. Two-level programs are used for modeling, and they must be converted to a one-level program to can be solved. Kao et al. [9] proposed a method to convert two-level program to one-level. Consequently, Models (5) and (6) can be expressed as:

$$
(E_p)_\alpha^U = \underset{\substack{(z_{tj})_\alpha^L \le z_{tj} \le (z_{tj})_\alpha^U \\ \forall j,t}}{Max} \begin{cases}
Max & E_p = \sum_{r=1}^{s1} u_r^1 (\bar{y}_{rp}^1)_\alpha^U + \sum_{o=1}^{s2} u_o^2 (\bar{y}_{op}^2)_\alpha^U \\
s.t. & \sum_{i=1}^{m} v_i (\bar{x}_{ip})_\alpha^L = 1, \\
& \sum_{t=1}^{g} w_t z_{tp} - \sum_{i=1}^{m} v_i (\bar{x}_{ip})_\alpha^L \le 0 \\
& \sum_{t=1}^{g} w_t z_{tj} - \sum_{i=1}^{m} v_i (\bar{x}_{ij})_\alpha^U \le 0, \qquad j = 1,...,n, j \ne p, \\
& \sum_{r=1}^{s1} u_r^1 (\bar{y}_{rj}^1)_\alpha^U - \sum_{t=1}^{g} w_t \beta z_{tj} \le 0, \\
& \sum_{r=1}^{s1} u_r^1 (\bar{y}_{rj}^1)_\alpha^L - \sum_{t=1}^{g} w_t \beta z_{tj} \le 0, \qquad j = 1,...,n, j \ne p \\
& \sum_{o=1}^{s2} u_o^2 (\bar{y}_{oj}^2)_\alpha^U - \sum_{t=1}^{g} w_t (1-\beta) z_{tj} \le 0, \\
& \sum_{o=1}^{s2} u_o^2 (\bar{y}_{oj}^2)_\alpha^L - \sum_{t=1}^{g} w_t (1-\beta) z_{tj} \le 0, \quad j = 1,...,n, j \ne p \\
& v_i, u_r^1, u_o^2, w_t \ge 0, \\
& i = 1,...,m,\ r = 1,...,s1,\ o = 1,...,s2,\ t = 1,...,g,
\end{cases} \tag{7}
$$

and,

$$
(E_p)_\alpha^L = \underset{\substack{(z_{tj})_\alpha^L \le z_{tj} \le (z_{tj})_\alpha^U \\ \forall j,t}}{Min} \begin{cases}
Max & E_p = \sum_{r=1}^{s1} u_r^1 (\bar{y}_{rp}^1)_\alpha^L + \sum_{o=1}^{s2} u_o^2 (\bar{y}_{op}^2)_\alpha^L \\
s.t. & \sum_{i=1}^{m} v_i (\bar{x}_{ip})_\alpha^U = 1, \\
& \sum_{t=1}^{g} w_t z_{tp} - \sum_{i=1}^{m} v_i (\bar{x}_{ip})_\alpha^U \le 0 \\
& \sum_{t=1}^{g} w_t z_{tj} - \sum_{i=1}^{m} v_i (\bar{x}_{ij})_\alpha^L \le 0, \qquad j = 1,...,n, j \ne p, \\
& \sum_{t=1}^{s1} u_r^1 (\bar{y}_{rj}^1)_\alpha^L - \sum_{r=1}^{g} w_t \beta z_{tj} \le 0, \\
& \sum_{r=1}^{s1} u_r^1 (\bar{y}_{rj}^1)_\alpha^U - \sum_{t=1}^{g} w_t \beta z_{tj} \le 0, \qquad j = 1,...,n, j \ne p \\
& \sum_{o=1}^{s2} u_o^2 (\bar{y}_{oj}^2)_\alpha^L - \sum_{t=1}^{g} w_t (1-\beta) z_{tj} \le 0, \\
& \sum_{o=1}^{s2} u_o^2 (\bar{y}_{oj}^2)_\alpha^U - \sum_{t=1}^{g} w_t (1-\beta) z_{tj} \le 0, \quad j = 1,...,n, j \ne p \\
& v_i, u_r^1, u_o^2, w_t \ge 0, \\
& i = 1,...,m,\ r = 1,...,s1,\ o = 1,...,s2,\ t = 1,...,g.
\end{cases} \tag{8}
$$

As mentioned before, the two-level programs must be converted to one-level ones. Since the inner and outer programs of Model (7) are maximization (same direction), therefore this model can be converted to one-level program, easily, as follow:

$$
\begin{aligned}
(E_p)^U_\alpha = Max \quad & E_p = \sum_{r=1}^{s1} u^1_r (\bar{y}^1_{rp})^U_\alpha + \sum_{o=1}^{s2} u^2_o (\bar{y}^2_{op})^U_\alpha \\
s.t. \quad & \sum_{i=1}^{m} v_i (\bar{x}_{ip})^L_\alpha = 1, \\
& \sum_{t=1}^{g} w_t z_{tp} - \sum_{i=1}^{m} v_i (\bar{x}_{ip})^L_\alpha \le 0 \\
& \sum_{t=1}^{g} w_t z_{tj} - \sum_{i=1}^{m} v_i (\bar{x}_{ij})^U_\alpha \le 0, \qquad j = 1,\dots,n, j \ne p, \\
& \sum_{r=1}^{s1} u^1_r (\bar{y}^1_{rj})^U_\alpha - \sum_{t=1}^{g} w_t \beta z_{tj} \le 0, \\
& \sum_{r=1}^{s1} u^1_r (\bar{y}^1_{rj})^L_\alpha - \sum_{t=1}^{g} w_t \beta z_{tj} \le 0, \qquad j = 1,\dots,n, j \ne p, \\
& \sum_{o=1}^{s2} u^2_o (\bar{y}^2_{oj})^U_\alpha - \sum_{t=1}^{g} w_t (1-\beta) z_{tj} \le 0, \\
& \sum_{o=1}^{s2} u^2_o (\bar{y}^2_{oj})^L_\alpha - \sum_{t=1}^{g} w_t (1-\beta) z_{tj} \le 0, \qquad j = 1,\dots,n, j \ne p, \\
& (z_{tj})^L_\alpha \le z_{tj} \le (z_{tj})^U_\alpha, \qquad t = 1,\dots,g, j = 1,\dots,n, \\
& v_i, u^1_r, u^2_o, w_t \ge 0, \\
& i = 1,\dots,m,\ r = 1,\dots,s1,\ o = 1,\dots,s2,\ t = 1,\dots,g.
\end{aligned}
\tag{9}
$$

Model (9) is nonlinear, and it can be easily transformed to linear programming by changing variables $\hat{z}_{tj} = w_t z_{tj}$ $(t = 1,\dots,g, j = 1,\dots,n)$ as follows:

$$
\begin{aligned}
(E_p)_\alpha^U = Max \quad & E_p = \sum_{r=1}^{s1} u_r^1 (\bar{y}_{rp}^1)_\alpha^U + \sum_{o=1}^{s2} u_o^2 (\bar{y}_{op}^2)_\alpha^U \\
s.t. \quad & \sum_{i=1}^{m} v_i (\bar{x}_{ip})_\alpha^L = 1, \\
& \sum_{t=1}^{g} \hat{z}_{tp} - \sum_{i=1}^{m} v_i (\bar{x}_{ip})_\alpha^L \le 0 \\
& \sum_{t=1}^{g} \hat{z}_{tj} - \sum_{i=1}^{m} v_i (\bar{x}_{ij})_\alpha^U \le 0, \qquad j = 1,\dots,n, j \ne p, \\
& \sum_{r=1}^{s1} u_r^1 (\bar{y}_{rp}^1)_\alpha^U - \sum_{t=1}^{g} \beta \hat{z}_{tp} \le 0, \\
& \sum_{r=1}^{s1} u_r^1 (\bar{y}_{rj}^1)_\alpha^L - \sum_{t=1}^{g} \beta \hat{z}_{tj} \le 0, \qquad j = 1,\dots,n, j \ne p, \\
& \sum_{o=1}^{s2} u_o^2 (\bar{y}_{op}^2)_\alpha^U - \sum_{t=1}^{g} (1-\beta) \hat{z}_{tp} \le 0, \\
& \sum_{o=1}^{s2} u_o^2 (\bar{y}_{oj}^2)_\alpha^L - \sum_{t=1}^{g} (1-\beta) \hat{z}_{tj} \le 0, \quad j = 1,\dots,n, j \ne p, \\
& w_t (z_{tj})_\alpha^L \le \hat{z}_{tj} \le w_t (z_{tj})_\alpha^U, \qquad t = 1,\dots,g, j = 1,\dots,n, \\
& v_i, u_r^1, u_o^2, w_t \ge 0, \\
& i = 1,\dots,m,\ r = 1,\dots,s1,\ o = 1,\dots,s2,\ t = 1,\dots,g.
\end{aligned} \tag{10}
$$

However, in Model (8) the directions for optimization for the inner and outer programs are different and thus, conversion of this model to a one-level program is not straightforward. According to duality theorem, the primal and dual programs have the same objective value at optimality. As a result, the inner program can be replaced by its dual to change the objective function from maximization to minimization. Dual of the inner program of Model (8) is as follow:

$$
\begin{aligned}
Min \quad & \theta \\
s.t. \quad & \theta (x_{ip})_\alpha^U - [\lambda_p^1 (x_{ip})_\alpha^U + \sum_{j=1, j\ne p}^{n} \lambda_j^1 (x_{ij})_\alpha^L] \ge 0, \qquad i = 1,\dots m, \\
& \sum_{j=1}^{n} \lambda_j^1 z_{tj} - \sum_{j=1}^{n} \lambda_j^2 \beta z_{tj} - \sum_{j=1}^{n} \lambda_j^2 (1-\beta) z_{tj} \ge 0, \qquad t = 1,\dots g, \\
& \lambda_p^2 (y_{rp}^2)_\alpha^L + \sum_{j=1, j\ne p}^{n} \lambda_j^2 (y_{rj}^1)_\alpha^U \ge (y_{rp}^1)_\alpha^L, \qquad r = 1,\dots,s1, \\
& \lambda_p^3 (y_{op}^2)_\alpha^L + \sum_{j=1, j\ne p}^{n} \lambda_j^3 (y_{oj}^2)_\alpha^U \ge (y_{op}^2)_\alpha^L, \qquad o = 1,\dots,s2, \\
& \lambda_j^1, \lambda_j^2, \lambda_j^3 \ge 0, \quad j = 1,\dots,n.
\end{aligned} \tag{11}
$$

Then, by replacing this problem to inner program of Model (8), the inner and outer of this program have same minimization and thus, this model can be converted to a one-level program as follows:

$$
\begin{aligned}
(E_p)^L_\alpha = Min \quad & \theta \\
s.t. \quad & \theta(x_{ip})^U_\alpha - [\lambda^1_p (x_{ip})^U_\alpha + \sum_{j=1, j\neq p}^{n} \lambda^1_j (x_{ij})^L_\alpha \geq 0, \quad i = 1,...m, \\
& \sum_{j=1}^{n} \lambda^1_j z_{tj} - \sum_{j=1}^{n} \lambda^2_j \beta z_{tj} - \sum_{j=1}^{n} \lambda^2_j (1-\beta) z_{tj} \geq 0, \quad t = 1,...g, \\
& \lambda^2_p (y^2_{rp})^L_\alpha + \sum_{j=1, j\neq p}^{n} \lambda^2_j (y^1_{rj})^U_\alpha \geq (y^2_{rp})^L_\alpha, \quad r = 1,...,s1, \qquad (12) \\
& \lambda^3_p (y^2_{op})^L_\alpha + \sum_{j=1, j\neq p}^{n} \lambda^3_j (y^2_{oj})^U_\alpha \geq (y^2_{op})^L_\alpha, \quad o = 1,...,s2, \\
& (z_{tj})^L_\alpha \leq z_{tj} \leq (z_{tj})^U_\alpha, \quad t = 1,...,g, j = 1,...,n, \\
& \lambda^1_j, \lambda^2_j, \lambda^3_j \geq 0, \quad j = 1,...,n.
\end{aligned}
$$

Model (12) is nonlinear and it cannot be easily converted to a linear programming.

2.2 Numerical Example

In this section, an example is presented to demonstrate how to use the proposed models in Subsection 2.2, to determine the fuzzy efficiency score for the 30 BANK branches in IRAN. The structure of these branches are exactly similar to Figure 1 and also they have 2 inputs, 2 intermediates and 4 outputs. The results of Models (12) and (10) are summarized in Table 1. With respect to results of Table 1, DMU21 and DMU23 have best and worst ranks, respectively.

Table 1: Results of the numerical example

No.	Fuzzy Efficiency Score	No.	Fuzzy Efficiency Score	No.	Fuzzy Efficiency Score
DMU1	(0.01846,1)	DMU11	(0.01993,1)	DMU21	(0.02677,1)
DMU2	(0.00054,1)	DMU12	(0.00092,1)	DMU22	(0.00237,1)
DMU3	(0.01162,1)	DMU13	(0.01761,1)	DMU23	(0.00020,1)
DMU4	(0.01298,1)	DMU14	(0.00271,1)	DMU24	(0.00130,1)
DMU5	(0.00856,1)	DMU15	(0.01550,1)	DMU25	(0.01409,1)
DMU6	(0.01957,1)	DMU16	(0.01321,1)	DMU26	(0.00449,1)
DMU7	(0.00244,1)	DMU17	(0.00152,1)	DMU27	(0.01687,1)
DMU8	(0.00150,1)	DMU18	(0.00885,1)	DMU28	(0.00220,1)
DMU9	(0.00976,1)	DMU9	(0.00210,1)	DMU29	(0.02146,1)
DMU10	(0.00074,1)	DMU20	(0.02342,1)	DMU30	(0.00213,1)

3 CONCLUSION

In this paper, a two-stage DEA model to evaluation of DMUs with inputs and outputs fuzzy data was proposed. For solving the proposed fuzzy DEA model the two-level DEA approach was used. Also, DMUs were considered as general form of a two-stage system which had both series and parallel structure. By conveniently changing the two-level DEA model converted to the one level model. Note that, one of the obtained models was nonlinear. The study of converting the proposed two-stage fuzzy model to the linear form without using the two-level DEA approach is recommended for future researchers.

4 REFERENCES

[1] **Banker, R.D., Charnes, A., Cooper, W.W.** 1984. Some models for estimating analysis, Management Science, 30, 1078-1092.

[2] **Charnes, A., Cooper W.W., Rhodes, E.** 1978. Measuring the efficiency of decision-making. European Journal of Operational Research, 2, 429-444.

[3] **Entani, T., Maeda, Y., Tanaka, H.** 2002. Dual models of interval DEA and its extension to interval data. European Journal of Operational Research, 136(1), 32-45.

[4] **Farrell, M.J.** 1957. The measurement of productive efficency, Journal of the Royal

a. Statistical Society, 120, 253-281.

[5] **Guo, P.,Tanaka, H.** 2001. Fuzzy DEA: A perceptual, evaluation method, Fuzzy Sets Systems, 119, 149-160.

[6] **Hatami-Marbini, A., Saati, S.** 2009. Stability of RTS of efficient DMUs in DEA with fuzzy under fuzzy data, Applied Mathematical Sciences, 3(44), 2157-2166.

[7] **Kao, C., Liu, S.T.** 2000a. Fuzzy efficiency measures in data envelopment analysis. Fuzzy Sets and Systems, 113, 427-437.

[8] **Kao, C., Liu, S.T.** 2000b. Data envelopment analysis with missing data: An application to university libraries in Taiwan, The Journal of the Operational Research Society, 51(8), 897-905.

[9] **Kao, C., Liu, S.T.** 2011. Efficiencies of two-stage systems with fuzzy data. Fuzzy Sets and Systems, 176, 20-35.

[10] **Kao, C., Lin, P.H.** 2012. Efficiency of parallel production systems with fuzzy data. Fuzzy Sets and Systems, 198, 83-98.

[11] **Saati, S., Memariani, A., Jahanshahloo, G.R.** 2002. Efficiency analysis and ranking of DMUs with fuzzy data. Fuzzy Optimization and Decision Making, 1(3), 255-267.

SIMULATION OF STRUCTURAL GFRP SECTIONS ENCASED IN CONCRETE COLUMNS SUBJECTED TO AXAIL LOAD

Marwa A. Yousef [1*], Mohamed I. El-Naggar [2], Sherine M. Swelem [3]

[1]Post graduate student
Marwaa_yousef@yahoo.com

[2]Professor in structural Engineering Department
Faculty of Engineering, Alexandria University, Egypt
elnaggarconsultants@yahoo.com

[3]Lecturer in structural Engineering Department
Faculty of Engineering, Alexandria University, Egypt
shswelem@yahoo.com

ABSTRACT

A new composite column composed of concrete, GFRP sections and reinforcement steel bar is presented in this paper. GFRP sections mimic steel sections and it is natural that construction follows what is seen in steelwork. Replacing GFRP section instead of steel section is a new trend in researches to get economic sections having special features such as light column weights which leads to relieve structure dead load and get rid of problems arising from steel corrosion. Standard structural GFRP shapes have been widely used for over half a century in aerospace, military, marine and automotive applications, but they have only seen limited application in the construction of buildings, bridges and other civil applications. Existence of design methods for GFRP profiles will encourage their use in civil building. The object from the research is to propose analytical method for calculating the strength of different GFRP profiles encased in concrete composite columns under concentric Load. Numerical model is developed to verify the analytical method. Initial overall (out-of-straightness) geometric imperfection is carefully incorporated in the model. Predicted values from numerical model are found to be in good agreement with the calculated values from analytical method. Finally, a comparison is conducted between the compressive load of concrete encased GFRP section column and the compressive load of concrete encased steel section column using the same dimensions and section properties and the results shows that the strength of concrete encased steel column decreases by 20% when replacing steel with GFRP.

Keywords: Composite column, Analytical method, GFRP pultruded profiles, Axial compressive strength, Numerical model, Initial geometric imperfection, Finite element method, Linear buckling analysis

1 INTRODUCTION:

Pultrusion is an economical process for manufacture of composite materials with prismatic profiles which have constant cross-section along the length. The benefits could be summarized as; weight

* Corresponding Author

saving (high strength to weight ratio), able to add to the old structures in the form of strengthening or repairing, low maintenance requirement, resistance to the environment effects, and ability to be formed into a complex shape. The necessity for design guidelines of GFRP sections is urgent because the use of these profiles is limited in civil building. The properties of GFRP sections at longitudinal direction are differed from those at transverse direction because of fiber direction which defines its mechanical characteristics. According to the fibers' direction, the strength and elastic constants are much higher than those in the orthogonal direction. The mechanical behavior is elastic up to failure. This study investigates replacement of steel section in concrete composite columns by GFRP section. ACI 318 stated that the strength of composite columns depends on the compatibility of the two materials, concrete and reinforcing material, to act together in resisting the external load. The reinforcing element has to undergo the same strain or deformation as the surrounding concrete in order to prevent the discontinuity or separation of the two materials under load. The ductility and the yield rupture strength of the reinforcement must also be considerably higher than those of concrete to raise the capacity of the reinforced concrete section to a meaningful level. Steel and fiber glass do possess the principal factors necessary: yield strength, ductility, and bond value.

2 OBJECTIVES

The objectives of this study can be summarized as follow:

- To represent a new composite section replacing steel with GFRP profiles which would be more economic and shed light on new type of material which expected to penetrate the markets due to its unique features which other materials couldn't meet.
- To propose a method for calculating the strength of various types of GFRP sections encased in concrete composite columns.
- To verify the analytical method by compare their results with ones obtained from a pin ended 3D finite element numerical model which developed to represent GFRP profiles encased in concrete columns.

3 PROPOSED ANALYTICAL METHOD FOR CALCULATING CRITICAL BUCKLING LOAD OF GFRP PROFILE ENCASED IN CONCRETE COLUMN SUBJECTED TO CONCENTRIC LOAD

The proposed design method follows the method mentioned in American institute of steel construction 2005 [7] for composite column design. The AISC use composite column design method based on steel column design. The idea behind this method is to turn the section properties of composite column into equivalent steel one. Also, the proposed method is developed based on pultruded GFRP column design method stated in Creative pultrusions design guide [2] which considers that Euler buckling is a general behavior of long column under axial compression load. The proposed method also depend on the same idea to convert the section properties of concrete encased GFRP section in to GFRP one. Figure (1) illustrates a comparison between FRP and mild-steel stress-strain curves. It is concluded from figure that the relation between stress and strain for GFRP material is linear until failure. That is mean the modulus of elasticity is constant until the stresses reach its ultimate value and on tangent modulus exist. Structural design of polymer composites Eurocomp design code [4] and Pultex [5] mentioned that the short GFRP columns generally fail in bearing deformation or local buckling mode and the long GFRP columns generally fail in the global buckling mode. The flexural buckling, known as Euler buckling, is the general behavior of long, slender GFRP columns under axial compression loads. GFRP is orthotropic material as the properties of GFRP sections at longitudinal direction are differed from those at transverse direction because of fiber direction which defines its mechanical characteristics. It is concluded that the flexural buckling is governed the axially loaded GFRP column design and this concept is largely similar to the design concept for axially loaded encased composite columns mentioned in ANSI/AISC 360-05 [7]. So, the critical axial load for GFRP section encased in concrete column can be given by the same equations using limit state of flexural buckling based on column slenderness as follow:

a) When $P_e \geq 0.44\, P_0$: $P_n = P_0\,(0.658)^{(P_0/P_e)}$ [1]

b) When $P_e < 0.44\, P_0$ $P_n = 0.877\, P_e$ [2]

The factor 0.877 is a reduction factor to consider initial crookedness (L/1500 per Galambos).

$$P_0 = A_{GFRP}\sigma_{comp,L} + A_{sr}f_{yr} + 0.85\ A_c f_c' \quad [3]$$

$$P_e = \frac{\pi^2 EI_{eff}}{(KL)^2} \quad [4]$$

EI_{eff}= effective rigity of composite section (N-mm2)

$$EI_{eff} = E_{GFRP,L} I_{GFRP} + 0.5\ E_{sr} I_{sr} + c_1\ E_c I_c \quad [5]$$

$$c_1 = 0.1 + 2\left(\frac{A_{GFRP}}{A_{GFRP} + A_c}\right) \leq 0.3$$

Where: A_{GFRP} =Area of GFRP section

$\sigma_{comp,L}$ = Compressive strength of GFRP in longitudinal direction

$E_{GFRP,L}$ = Elastic modulus of GFRP in longitudinal direction

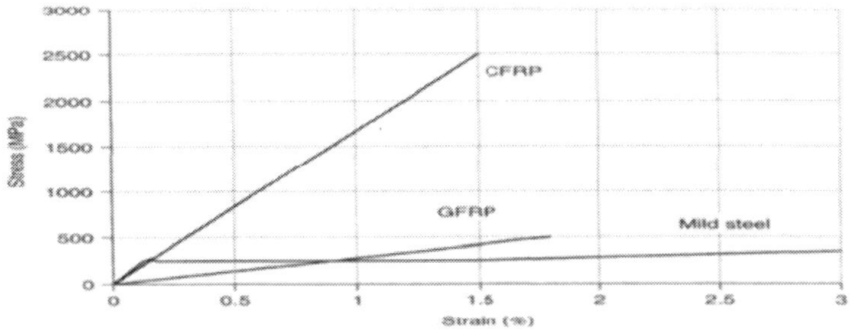

Figure 1: Typical FRP and mild-steel stress-strain curves [6]

4 FINITE ELEMENT MODEL

The study investigates the behavior of different GFRP pultruded sections encased in concrete column. An appropriate three-dimensional model is developed to accurately simulate GFRP profiles encased in concrete column under concentric Load. The finite element model for this new composite system is implemented using COMSOL Multiphysics 5.

4.1 Geometric Properties

4.1.1 Model Definition

The GFRP sections encased in concrete composite columns investigated in this study comprised of four components, GFRP profile, longitudinal reinforcement bars, transverse reinforcement bars and concrete. For GFRP profile, the sections considered are five different shapes. I- section, Wide flange section, square-section and tube section as shown in figure 2. 3D solid elements are selected to model the concrete block and GFRP section. The truss interface is used to model the steel reinforcements instead of 3D solid. This removed the necessity to model the bars with small mesh which saved computational time. Since steel reinforcements are relatively thin compared to the concrete structures, it is assumed that they are only capable of transmitting axial forces. The bending stiffness of each bar does not contribute much to the overall total bending stiffness of the section, therefore the reinforcement bars are modeled with truss elements instead of beam elements. The concrete was glued to the GFRP reinforcing bars, so bonding effects were not included.

4.1.2 Modelling of GFRP-Concrete Interaction

In this study, perfect bond between materials is assumed. Nodes of 3D GFRP solid elements are connected to those adjacent 3D solid concrete elements in order to satisfy the perfect bond assumption. The high strength of the epoxy use to glue the GFRP profile with concrete support the perfect bond assumption.

4.1.3 Modelling of End Boundary Condition

The column is assumed to be simply supported at its ends. For the loaded columns, translations at both ends are fixed, except for the vertical displacement at the top. Column is allowed to displace in the vertical direction only. Rotations are constrained around z-direction for both ends.

4.1.4 Modelling of initial overall (out-of-straightness) geometric imperfection

The critical buckling load of GFRP section encased in concrete composite column are obtained by perform linear buckling analysis include initial geometric nonlinearity which is taken as L/2000 where L is length of pin-ended. The value (L/2000) is an average measured value in the literature tests.

4.2 Material Properties

4.2.1 GFRP properties

The Glass fibre/polyester [solid, longitudinal (parallel to fiber axis)] was provided by COMSOL Multiphysics 5. Table 1 reported elastic constants from the pultruders' design manuals

Table 1: Elastic constants for GFRP profile

Source	E_LMPa	E_T MPa	G_{LT}MPa	υ_{LT}
Strongwell [1]	17200	5500	2900	0.33
EUROCOMP DC [3]	17200	5500	2900	0.33

Compressive strength of GFRP =210MPa

4.2.2 Concrete properties

Concrete strengths are specified in terms of the characteristic cube strengths measured at 28 days: F_{cu} = 25 MPa , Poisson ratio, υ =0.33 , Elastic modulus, E_c =200,00 MPa

4.2.3 Reinforced steel properties

Steel assumed to be elastic perfectly plastic material and identical in tension and compression

Yield stress: f_y = 240MPa , Poisson ratio: υ =0.3, Elastic modulus: E_s =200,000 MPa

4.3 Analysis Type

Linear Buckling study is carried out for estimating the critical load at which a structure becomes unstable. COMSOL Multiphysics 5 used the eigenvalue solver to compute the buckling modes and the associated critical load.

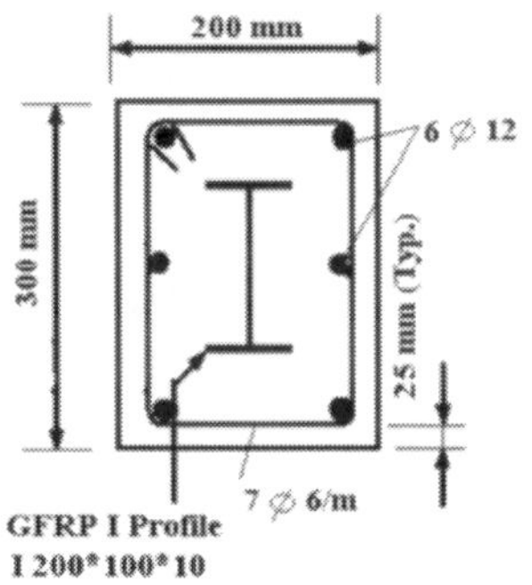

GFRP- I section encased in rectangular concrete column (RI-spec.)

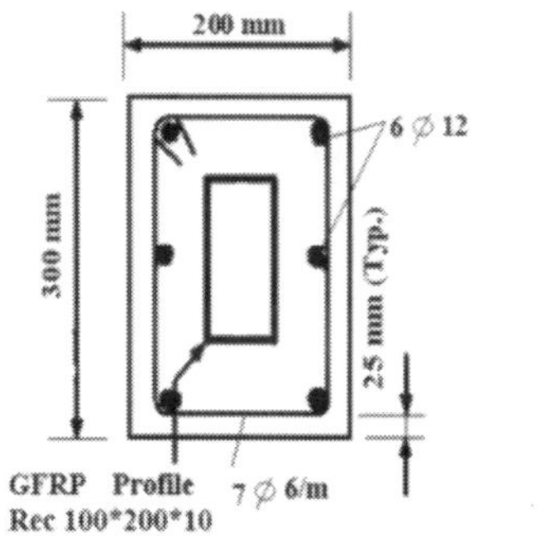

Rec. GFRP- section encased in rectangular concrete column (R-spec.)

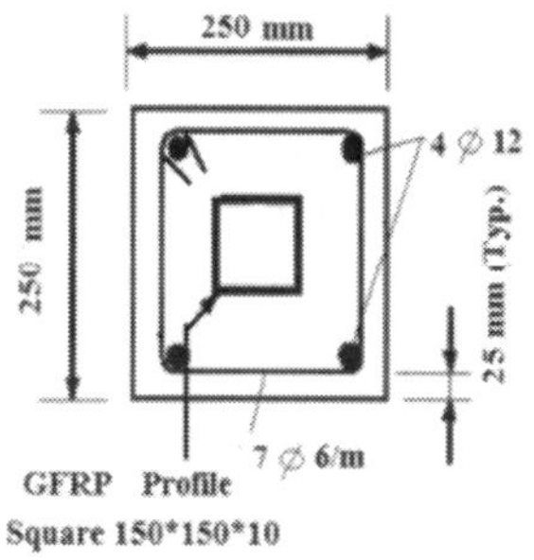

square GFRP- section encased in rectangular concrete column (S-spec.)

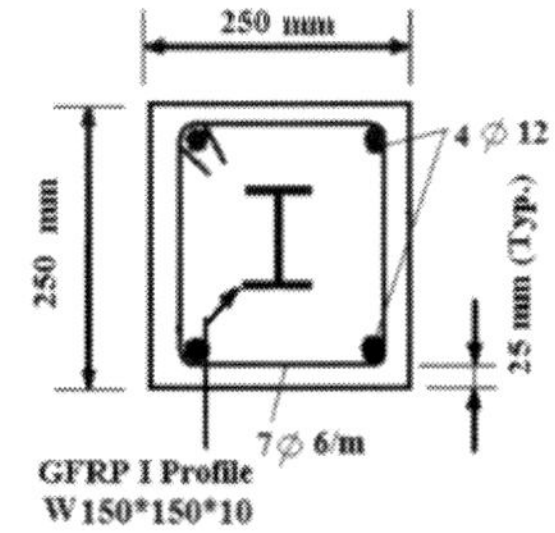

I GFRP- section encased in rectangular concrete column (S-spec.)

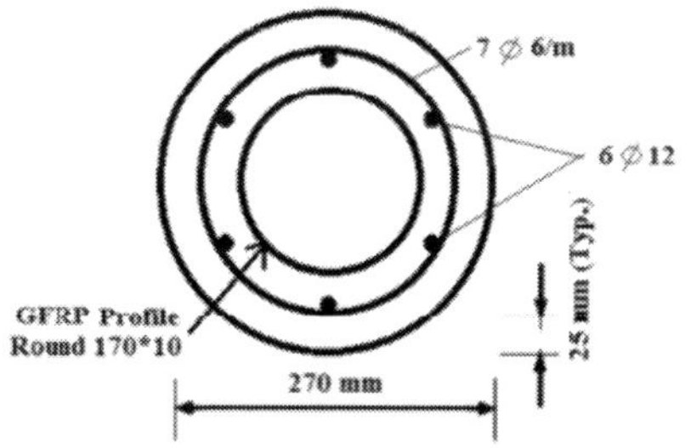

Round GFRP- section encased in Round concrete column (RS-spec.)

Figure 2: Different Cross Sections of Specimens

5 VERIFICATION OF FINITE ELEMENT MODEL

In this section, the finite element model of pin-ended GFRP encased in concrete composite columns developed in this study is verified against the analytical results. Table 2 presented a comparison between the axial ultimate loads calculated from proposed analytical method ($P_{analytical}$) and the critical buckling load obtained from finite element models (P_{FE}) with accompanied failure modes predicted from finite element model. A good agreement is shown between the calculated results from analytical model and obtained results from numerical model. The mean value of (P_{FE} $/P_{analytical}$) ratios equal 1.03 and corresponding coefficient of variance (COV) equal 0.008. It is noticed three modes of failure for GFRP section encased in concrete composite column predicted from numerical model. The failure modes are concrete crushing (CC), GFRP bearing failure (GB) and flexural buckling (B). To determine the failure mode from numerical model, the values of resulting stresses from concrete and GFRP elements in numerical model should be compared with ultimate strength values of concrete and GFRP materials. It is noticed that the failure of composite column due to axial load is concrete crushing failure CC and GFRP bearing failure GB which are arisen together for the column lengths 2000 mm and 2500 mm. The bearing deformation of end support appeared first and followed by concrete crushing. Whereas, the flexural buckling failure is observed for column having 3000mm length as shown in Figure (3) for all specimens which its length = 3500 mm. In flexural buckling mode, the stresses of concrete element at tension side reached its ultimate values before the GFRP element reached its compressive strength. It is observed from figure (4) that analytical load for GFRP rectangular and square section encased in concrete column is higher than the buckling load obtained from finite element model because the analytical load is affected by the second moment of inertia which appears too high for these sections.

Table 2: Comparison between Analytical and finite element model

Spec. No.	Cross Section	Length L (mm)	Analytical $P_{analytical}$ (KN)	Finite element P_{FE} (KN)	Finite element Failure mode	$\frac{P_{FE}}{P_{analytical}}$
RS-1	Round	2000	1941	2100	CC+GB	1.08
RS-2	Round	2500	1752	1923	CC+GB	1.09
RS-3	Round	3000	1546	1711	CC+GB	1.1
RS-4	Round	3500	1333	1273	B	0.95
RS-5	Round	4000	1124	1001	B	0.89
R-1	Rectangular	2000	2316	2413	CC+GB	1.04
R-2	Rectangular	2500	2192	2190	CC+GB	0.999
R-3	Rectangular	3000	2048	2100	CC+GB	1.025
R-4	Rectangular	3500	1891	1662	B	0.878
R-5	Rectangular	4000	1720	1319	B	0.766
S-1	Square	2000	2261	2300	CC+GB	1.017
S-2	Square	2500	2086	2150	CC+GB	1.03
S-3	Square	3000	1891	1960	CC+GB	1.03
S-4	Square	3500	1683	1740	B	1.03
S-5	Square	4000	1472	1380	B	0.93
RI-1	Rectangular	2000	1976	2201	CC+GB	1.11
RI-2	Rectangular	2500	1871	2001	CC+GB	1.069
RI-3	Rectangular	3000	1750	1950	CC+GB	1.11
RI-4	Rectangular	3500	1617	1468	B	0.907
RI-5	Rectangular	4000	1476	1383	B	0.936
SW-1	Square	2000	2017	2100	CC+GB	1.04
SW-2	Square	2500	1862	2050	CC+GB	1.1
SW-3	Square	3000	1688	1910	CC+GB	1.13
SW-4	Square	3500	1504	1444	B	0.96
SW-5	Square	4000	1316	1264	B	0.96
mean						1.03
COV						0.008

Note: CC denotes concrete crushing, GB denotes GFRP bearing failure, B denotes flexural buckling

6 COMPARISON BETWEEN CONCRETE ENCASED STEEL COLUMN AND CONCRETE ENCASED GFRP COLUMN USING ANALYTICAL METHOD

Table 3 notes the compressive capacity for various slenderness ratios of both concrete encased steel section and concrete encased GFRP section using analytical method. It is observed from the comparison conducted between the compressive load of concrete encased GFRP section column and the compressive load of concrete encased steel section column using the same dimensions and section properties that the decrease in composite column strength when replacing steel section with GFRP section depends on column length. It can be concluded that the strength of concrete encased steel column decreases by about 25% when replacing steel with GFRP and this percentage may reach to 50% for long column.

Table 3: Comparison between Concrete encased steel section column and concrete encased GFRP section column using analytical method

Spec. No.	Cross Section	L (m)	Concrete encased steel section column	Concrete encased GFRP section column	$\frac{P_{steel}}{P_{GFRP}}$
			$P_{analytical}$ (KN)	$P_{analytical}$ (KN)	
RS-1	Round	2	2290	1941	1.17
RS-2	Round	2.5	2190	1752	1.25
RS-3	Round	3	2073	1546	1.34
RS-4	Round	3.5	1943	1333	1.45
RS-5	Round	4	1803	1124	1.60
R-1	Rectangular	2	2625	2316	1.13
R-2	Rectangular	2.5	2565	2192	1.17
R-3	Rectangular	3	2493	2048	1.21
R-4	Rectangular	3.5	2411	1891	1.27
R-5	Rectangular	4	2319	1720	1.34
S-1	Square	2	2626	2261	1.16
S-2	Square	2.5	2539	2086	1.21
S-3	Square	3	2437	1891	1.28
S-4	Square	3.5	2321	1683	1.37
S-5	Square	4	2194	1472	1.49
RI-1	Rectangular	2	2204	1976	1.11
RI-2	Rectangular	2.5	2153	1871	1.15
RI-3	Rectangular	3	2092	1750	1.19
RI-4	Rectangular	3.5	2022	1617	1.25
RI-5	Rectangular	4	1945	1476	1.31
SW-1	Square	2	2319	2017	1.14
SW-2	Square	2.5	2242	1862	1.20
SW-3	Square	3	2152	1688	1.27
SW-4	Square	3.5	2051	1504	1.36
SW-5	Square	4	1940	1316	1.47

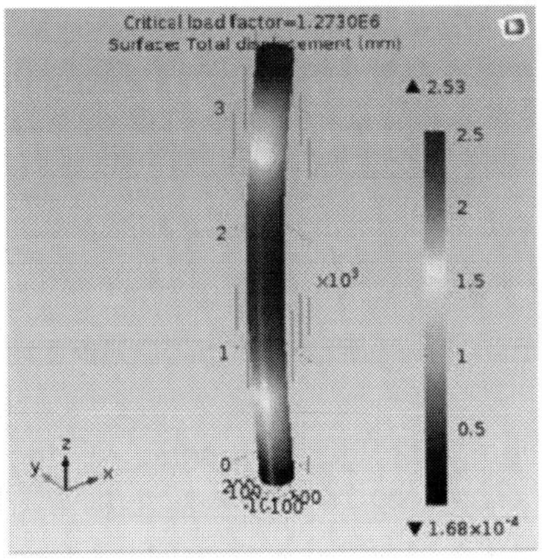

Critical buckling Load of Specimens RS-4

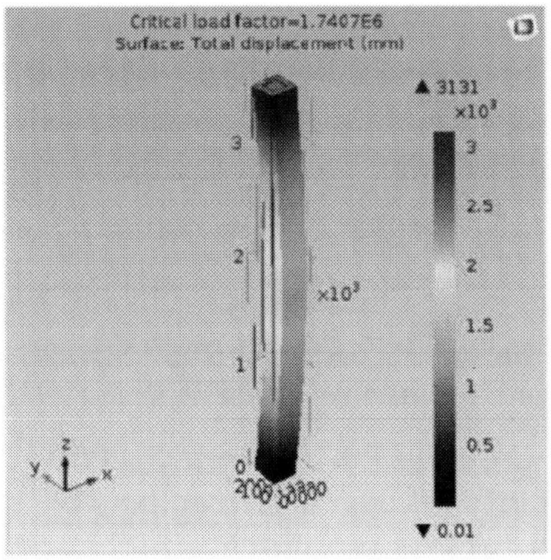

Critical buckling Load of Specimens S-4

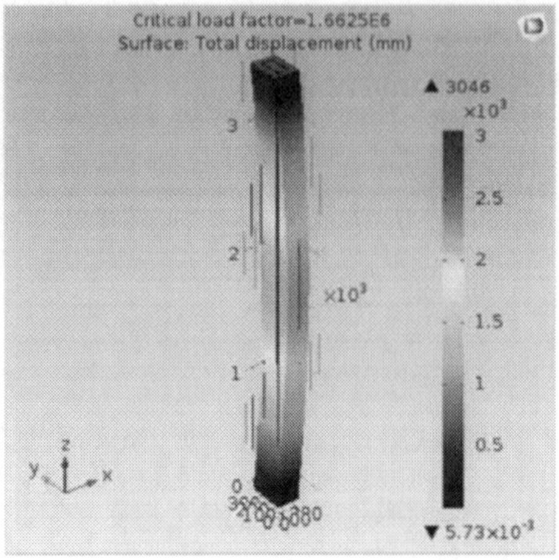

Critical buckling Load of Specimens R-4

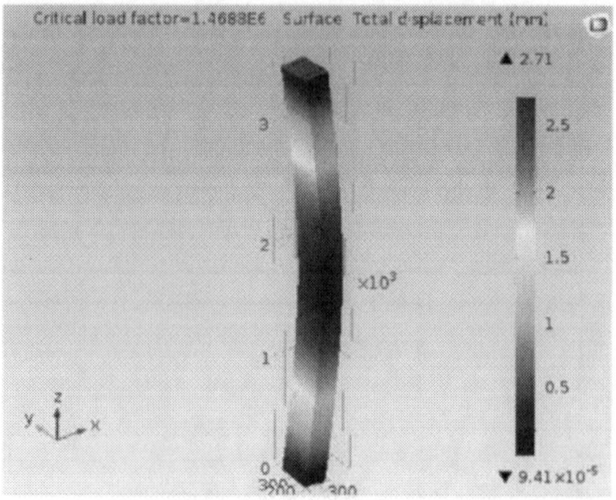

Critical buckling Load of Specimens RI-4

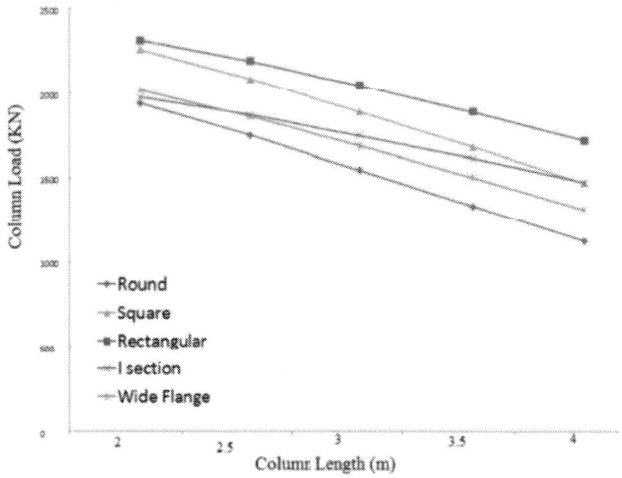

Figure 3: Buckling load for different GFRP section encased in concrete column

7 REFERENCES

[1] **Bristol *1998*.** *Strongwell design manual, VA:Strongwell*

[2] **Alum Bank, 1998** *Creative pultrusions design guide, PA: Creative Pultrusions*

[3] **Clarke JL, editor,1996.** *Structural design of polymer composites: EUROCOMP Design code and handbook. London: S. & F.N. Spon.*

[4] **Creative Pultrusion Inc. 2000.** *The Pultex pultrusion global design manual of standard and custom fiber reinforced polymer structural profiles. The Creative Pultrusion Inc.*

[5] **American Institute of Steel Construction (AISC), 1993.** *Load and resistance factor design specification for structural steel buildings. 2nd ed. Chicago (IL).*

[6] **J. G. Teng, J. F. Chen, S. T. Smith, L. Lam, 2002.** *FRP strengthened RC structures. John Wiley and Sons Ltd., Chichester, UK. 245p.*

[7] **American Institute of Steel Construction, 2010.** *Specification for Structural Steel Buildings, (AISC-LRFD), Chicago, Illinois.*

BIG OPERATIONAL DATA ORIENTED PREDICTIVE MAINTENANCE STRATEGY OF MANUFACTURING SYSTEM CONSIDERING MISSION RELIABILITY MODEL

HAN Xiao*, HE Yihai and GU Changchao
School of Reliability and Systems Engineering, Beihang University, Beijing, China.
buaahanxiao@buaa.edu.cn

ABSTRACT

According to the incomplete consideration of the impacts of the big operational data of state degradation and product quality of the multi-state manufacturing system in traditional maintenance studies, a new predictive maintenance technique based on the mission reliability state defined by the big operational data of manufacturing system is put forward. Firstly, the concept of big operational data of manufacturing system is defined, and its connotation is expounded focused on the operation state .Secondly, a mission reliability model of manufacturing system is defined to measure the operation performance state of manufacturing system, which is established from the big operational data. Thirdly, a predictive maintenance strategy is put forward considering the proposed mission reliability model for the manufacturing system. Finally, a case study of cylinder head manufacturing system has been presented to validate the proposed approach.

Keywords: big operational data, manufacturing system, mission reliability, predictive maintenance

1 INTRODUCTION

Manufacturing system, the material basis and foundation for the implementation of advanced manufacturing mode of intelligent manufacturing and etc., which directly determines the reliability of the product quality. However, traditional manufacturing system design is mainly driven by the productivity and production tasks, ignoring the requirements of the system's own reliability and the product reliability. The rare studies on the reliability of the manufacturing system only focus on the basic reliability of the single device of the manufacturing system, and the studies on the reliability of manufacturing systems running status is also rare, which constrains the real-time quality and effectiveness of quality control and maintenance decision of manufacturing process. Thus, the study on the mission reliability of manufacturing system [1-2] has becoming a hot and challenging topic in the field of reliability engineering and industrial engineering. Meanwhile, that the elements of various types of equipment of manufacturing systems are of great amount, and that there are interlinkages among them, which make the study on the dynamic reliability modeling of manufacturing systems more complex. Accordingly, the fuzzy reliability modeling hinders the accuracy of maintaining activities in the manufacturing enterprises. Once manufacturing systems shut down due to a sudden fault, the loss to the enterprises is huge. Therefore, how to carry out a predictive maintenance strategy to prevent the failures [3-5] has been a bottleneck problem to be solved for most of manufacturers.

* Corresponding Author

With the coming intelligent manufacturing era and the development of industrial big data [6], cloud computing and Cyber-physical system, the increasing automated manufacturing system is able to collect and deal with the big operational data dynamically in real-time, which makes calculating the mission reliability of the multi-state manufacturing system be possible. Li et al. [7] put forward that big data can be used for the study of manufacturing system, but how to apply the big operation data to the manufacturing process specific manufacturing task is the urgent problem to be solved. The traditional manufacturing system maintenance strategy is hard to consider the production process of the dynamic characteristics of time [8], or just simply to monitor the status of the whole manufacturing equipment [9], so that the accuracy of predictive maintenance model also needs to be improved. However, fast detection of the fault is the advantage of the advanced manufacturing mode. Thus, this study focuses on the predictive maintenance in terms of operational data of the manufacturing system based on the mission reliability model.

2 THE FOUNDATIONS OF BIG OPERATIONAL DATA AND MISSION RELIABILITY FOR MANUFACTURING SYSTEMS

2.1 Connotation of operational data

The operation data of manufacturing system is classified as follows.

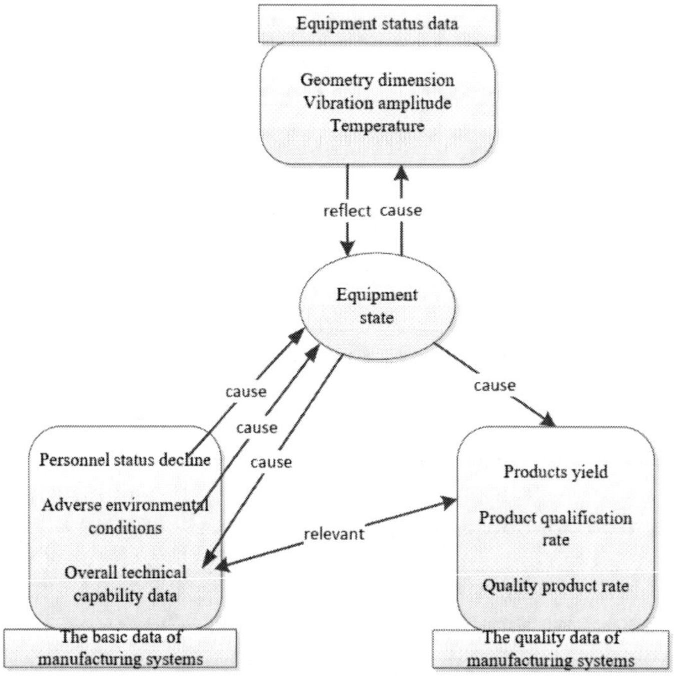

Fig. 1: Operation data relation

The mission reliability of composing equipment reliability directly determines the reliability of manufacturing system. The manufacturing system having a plurality of constituent units, the constituent unit performance state level, faults, maintenance history data and other coupling component units are important conditions for determining the operating state of the manufacturing system.

The basic data of manufacturing systems consists three kinds of data. The technical level of manufacturing system is the most important factor to affect the efficiency of the whole manufacturing system.In advanced manufacturing mode, the state of operator is also an important factor that can decide mission reliability of manufacturing system. In addition, environmental condition is also an important factor that can affect the normal operation of the system, temperature, dust, vibration and so on are likely to have an impact on the manufacturingequipment.

The quality data of manufacturing systems include two aspects of the data, respectively, the quantitative data and quality data of products. For the manufacturing systems, the vast majority of manufacturing tasks fail due to performance degradation and sporadic failure, which just can be bothreflected by the above two kinds of data.

2.2 Modelling of mission reliability of manufacturing systems

Mission reliability is the ability of the product to complete the specified function in the specified mission profile. It emphasizes more on the completion probability of each manufacturing system task in the "mission profile". Manufacturing system is a kind of typical multi-state system [10], the mission reliability of manufacturing system means that the products produced can meet the requirements and manufacturing system can continue product qualified products.

Based on the big operation data shown above, the mission reliability“ R_m ” can be modeled below:

$$R_M = \alpha_b \times R_m \times R_p \tag{1}$$

“ α_b ” is the influence coefficient of basic data, consisting the data of personnel, environment and the technical level of manufacturing system. “ R_p ” is to identify whether thequantitative data and quality data of productsmeet the requirement in the last manufacturing task. If the data meets the requirement, the will be 1, otherwise be 0, and the manufacturing system need to be repaired at once. “ R_m ” is the mission reliability of the whole manufacturing equipment. It can be calculated after the mission reliability of each Composite unit is calculated.

It is assumed that the index of unit i in the manufacturing equipment needs to be strictly monitored is Y ,whose range for working properly is from Y_{UL} to Y_{LL} .The mission reliability of this unit can be calculated as follows:

$$R_i = P\{Y_{LL} < Y < Y_{UL}\} \tag{2}$$

3 A PREDICTIVE MAINTENANCE STRATEGY BASED ON THE MISSION RELIABILITY MODEL

3.1 Constraints of maintenance decision for manufacturing systems

Maintenance decision constraintsshould to be considered when choosing the maintenance strategy. It is the precondition of the maintenance strategy selection to make the maintenance decision of the manufacturing system. For manufacturing system, the maintenance decision constraints mainly include the following aspects:

Safety. It is mainly to consider the damage caused by the failure of the manufacturing equipment to the operator and environment.

Mission reliability. Mission reliability is an important parameter to measure the performance status of manufacturing system and basis for the predictive maintenance of manufacturing system.

Maintenance cost. The fundamental purpose of manufacturing activities is to gain benefits, and maintenance is just to guarantee the production efficiency to be obtained.

Potential benefits. For the equipment unit of great importance in manufacturing system, the manufacturing equipment shutdown because of the long-time performance degradation could cause great loss of production and improve production costs, while timely predictive maintenance can solve such a problem, bring certain potential benefits.

3.2 System performance state prediction based on Markov model

The core technologies of predictive maintenance are performance state prediction and fault isolation. Based on the mission reliability model above, we can calculate mission reliability of manufacturing system in the next manufacturing performance task with the operational data when we know the states of composite units in advance. Take the equipment unit m as an example, whose performance state degradation is shown in Fig. 2

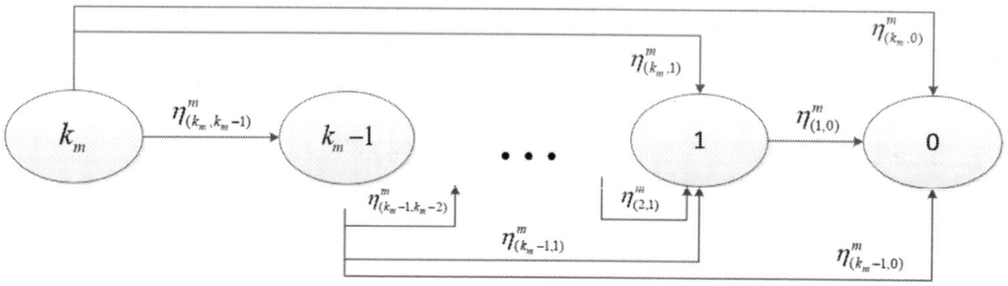

Fig. 2: Schematic diagram of performance state degradation

As shown in Fig. 2, k_m is the best performance status of equipment unit m, "0" is the unacceptable performance status. We denote the transfer rate of equipment unit m from performance status $i \in [1,\cdots,k_m]$ to performance status $j \in [0,\cdots,k_{m-1}]$ by $\eta^m_{(ij)}(t)$ $(i > j)$.

The probability of state k_m-2 of the equipment unit m is more complex to calculate than that of state k_m-1. The equipment unit has only one way to decline from the highest performance state to the state k_m-1:

$$k_m \to k_m - 1 \tag{3}$$

There are two ways for equipment unit to decline from the state k_m to the state k_m-2:

$$k_m \to k_m - 2 \tag{4}$$

And:

$$k_m \to k_m - 1 \to k_m - 2 \tag{5}$$

In the same way, the probability of equipment unit in any states in the next task can be expressed as follow:

$$\begin{aligned} p_{(m,i)} = & \int_0^t \cdots \int_0^{\tau_2} \exp[-\int_0^{\tau_1} \sum_{i=0}^{k_m-1} \eta^m_{(k_m,i)}(s)ds] \\ & \eta^m_{(k_m,k_m-2)}(\tau_2) \prod_r^{n-1} \left\{ \exp[-\int_{\tau_i}^{\tau_{r+1}} \sum_{i=0}^{x-1} \eta^m_{(k_x,i)}(s)ds] \eta^m_{(k_x,k_r+1)}(\tau_{r+1}) \right\} \\ & \exp[-\int_{\tau_n}^{\tau_i} \sum_{i=0}^{k_n-1} \eta^m_{(k_n,i)}(s)ds] \eta^m_{(k_n,i)} \exp[-\int_{\tau_i}^{t} \sum_{j=0}^{i=1} \eta^m_{(i,j)}(s)ds] d\tau_1 \cdots d\tau_{n+1} \end{aligned} \quad (6)$$

Therefore, if the condition of the initial performance of the equipment and the transition probability of each state are given, we can calculate the probability distribution of the performance status of the equipment in the next manufacturing task by the above formula.

3.3 Selection of maintenance unit group based on maintenance cost

Maintenance cost is the key factor in the research of maintenance strategy, calculating the maintenance costsmore comprehensive and accurate will make the maintenance activities more targeted.

For each equipment unit i, the cost of raw materials is C_{m_i} , the influence coefficient of the mission reliability is δ_i . The formula for calculation is as follows:

$$\delta_i = \frac{R_{m_i}{}'}{R_{m_i}} \quad (7)$$

" R_{m_i} "is the mission reliability of the equipment unit i before maintenance. " $R_{m_i}{}'$ " is the mission reliability after maintenance. The formula for $R_{m_i}{}'$ calculation is as follows:

$$R_{m_i}{}' = \Pr\{k_i > g'\} = \sum_{q=g'}^{k_m} p_{(m,q)} \quad (8)$$

" g' " is the performance status of the equipment unit after maintenance.

The cost of product quality loss is determined by the change of the qualified rate of the manufactured product, it can be calculated as follows:

$$C_p = \Delta p \times A \times C_0 \quad (9)$$

" Δp " is the change value of the qualified rate of the product before and after maintenance activities.

And the efficiency-cost ratiofor the maintenance of equipment unit can be calculated as follows:

$$ROI_i = I_i \times \frac{\delta_i}{C_m - C_p} \quad (10)$$

" I_i " is the importance of equipment units, which reflects the importance of each unit's reliability, avoiding the high risk of stopping manufacturing system operating normally.

The maintenance unit group $U\{u_1,u_2,u_3,u_4,\cdots\}$ is determined according to the descending order of efficiency-cost ratio for equipment unit until the mission reliability of the manufacturing system achieves the goal of value, just like the flow chart below.

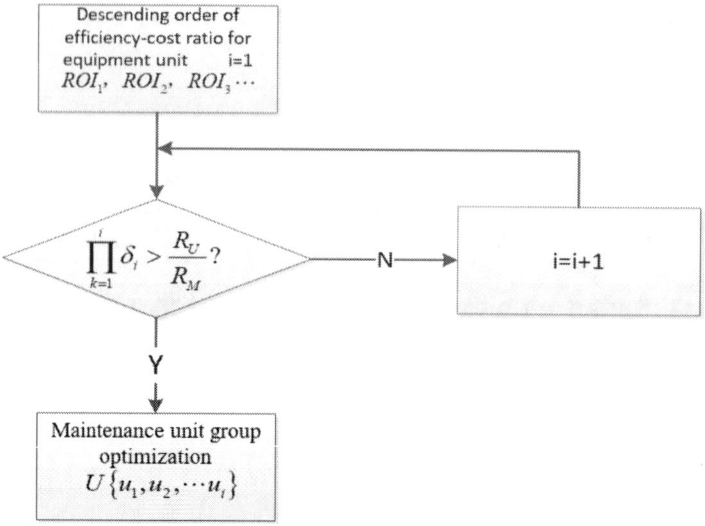

Fig. 3: The flow chart of maintenance unit group optimization

4 Case Study

In this part, operational data oriented predictive maintenance strategy manufacturing system is applied. Firstly, all the information of operational data that can be used and the performance state transition probability of the manufacturing equipment units should be summarized. In this case, the dust is classified as B, which means the state of environment is lower than the relatively high requirements, but can still meet the basic requirements and the influence factor of basic data is 0.9 because of dust.

Step1. Mission reliability calculation of equipment units

The state transition matrix of each equipment unit can be obtained from the given initial conditions. Take the unit m_1 for an example.

$$A^{m_1}=\begin{pmatrix}\eta_{(3,2)}^{m_1} & \eta_{(3,1)}^{m_1} & \eta_{(3,0)}^{m_1}\\ 0 & \eta_{(2,1)}^{m_1} & \eta_{(2,0)}^{m_1}\\ 0 & 0 & \eta_{(1,0)}^{m_1}\end{pmatrix}=\begin{pmatrix}0.002 & 0.001 & 0.002\\ 0 & 0.007 & 0.003\\ 0 & 0 & 0.001\end{pmatrix}$$

The steady-state probability of unit m_1 in each state is calculated as follows with the formula (6).

$$p_{(m_1,3)}=0.405; p_{(m_1,2)}=0.412; p_{(m_1,1)}=0.283; p_{(m_1,0)}=0.096$$

There the mission reliability of the equipment unit m_1 in the next manufacturing task is:

$$R_{m_1}'=\Pr\{k_1\geq 1\}=\sum_{q=1}^{4}p_{(m_1,q)}=1-0.096=0.904$$

Step2. Mission reliability calculation of manufacturing system

The mission reliability of manufacturing equipment can be calculated with the mission reliability of equipment units.

$$R_m = \prod_{i=0}^{5} R_i = 0.551$$

Then, we can calculate the influence coefficient of basic data "α_b" with the data in table 1.

$$\alpha_b = \alpha_t \times \alpha_h \times \alpha_e = 1 \times 1 \times 0.9 = 0.9$$

Finally, we can get the mission reliability of manufacturing system.

$$R_M = \alpha_b \times R_m \times R_p = 0.496$$

Step3. Predictive maintenance optimization process

The minimum acceptable mission reliability is set to 0.7, and the target of the mission reliabilityis 0.9.An effectiveness-cost ratio order of maintenance units needsto be identified by the formula (9) and (10).

We can learn useful information from the past operational data of the manufacturing system that when the equipment unit 2 in the best state, namely, the state 4, the qualified rate of manufactured products was 95%; when in state 3, the qualified rate was 90%.When equipment unit 5 was in the best state, the qualified rate was 96%, and 92% in state 3. We can calculate the efficiency-cost ratio more comprehensively with the data above.

Table 1: Effectiveness-cost ratio order of maintenance units.

Unit	Mode	Cost	Reliability effect	Importance	efficiency-cost ratio
Workshop	Dust treatment	0.6	1.111	1.3	2.407
m_1	Overall	0.3	1.106	1	3.687
m_2	Overall	0.5	1.200	1	2.400
m_2	Imperfect	0.4	1.111	1	2.778
m_3	Overall	0.4	1.200	1.2	3.600
m_4	Overall	0.5	1.067	1	2.134
m_5	Overall	0.4	1.067	1.2	3.201
m_5	Imperfect	0.3	1.043	1.2	4.172

With the data above, we learn that manufacturing system reliability will be more than 0.9 when $U\{\text{workshop}, m_1, m_{22}, m_3, m_4, m_{52}\}$ is selected as the maintenance unit group, the manufacturing system with the prerequisite of making theefficiency-cost ratio highest. Among them, the maintenance mode of equipment unit m_2 and m_5 is imperfect maintenance, which means that they only need to be repaired to state 3.

5 CONCLUSION

In this paper, the operational data of manufacturing system is expounded, and the mission reliability model of manufacturing system is built. Further, from the perspective of making the cost-effectiveness of maintenance activities maximum, a predictive maintenance strategy and optimization of maintenance unit group are proposed. In addition, it could be an effective tool to guide the personnel to carry out timely and effective maintenance decisions before the performance status of the manufacturing system is degraded to the failure.

However, there are some limitations in this study.In order to directly reflect the change of production capacity of manufacturing system with the product quality data (such as yield, qualified product rate, etc.), the study of relationship between the product quality data and mission reliability model manufacturing system is planned.

6 REFERENCES

[1] **Lin Li and Jun Ni. 2008.** Reliability Estimation BasedonOperationalData ofManufacturing Systems, Quality & Reliability Engineering, 24(7),pp 843-854.

[2] **Ji-Min Lun, Xiao-Yue Wu. 2014.** Reliability evaluation of generalized phased-mission systems with repairable components, Reliability Engineering and System Safety, 121,pp 136-145.

[3] **Cher Ming Tan, Nagarajan Raghavan. 2008.** A framework to practical predictive maintenance modeling for multi-state systems, Reliability Engineering & System Safety, 93, pp 1138-1150.

[4] **Xiaojun Zhou, Lifeng Xi, Jay Lee. 2007.** Reliability-centered predictive maintenance scheduling for a continuously monitored system subject to degradation.Reliability Engineering & System Safety, 92(4), pp 530-534.

[5] **Kim-Anh Nguyen, Phuc Do, Antoine Grall. 2015.** Multi-level predictive maintenance for multi-component systems.Reliability Engineering & System Safety, 144, pp 83-94.

[6] **Jay Lee, Hossein Davari Ardakani, Shanhu Yang, Behrad Bagheri. 2015.** Industrial big data analytics and cyber-physical systems for future maintenance & service innovation.Procedia Cirp, 38, pp 3-7

[7] **Jingran Li, Fei Tao, Ying Cheng, Liangjin Zhao.2015.** Big Data in product lifecycle management. International Journal of Advanced Manufacturing Technology, 81(1-4), pp 667-684.

[8] **Caetano Luis Filipe,Teixeira Paulo Fonsec. 2016.**Predictive Maintenance Model for Ballast Tamping. Journal of Transportation Engineering, 142(4).

[9] **Susto Gian Antonio,Schirru Andrea, Pampuri Simone, McLoone Sean,Beghi Alessandro. 2015.**Machine Learning for Predictive Maintenance: A Multiple Classifier Approach. IEEE Transactions on Industrial Informatics,11(3), pp 812-820.

[10] **Kim, Duck B, Denno, Peter O, Jones, Albert T. 2015.** A model-based approach to refine process parameters in smart manufacturing. Concurrent Engineering-Researchand Applications,23(4), pp 365-376.

A NEW MATHEMATICAL MODEL TO OPTIMIZE A GREEN GAS NETWORK USING A NOVEL HYBIRD GA/SA ALGORITHM: A CASE STUDY

M. Fallah*,A. Mohajeri,M.Barzegar-Mohammadi
Departmentof Industrial Engineering, Central Tehran Branch
Islamic Azad University,Tehran, Iran
mohammad.fallah43@yahoo.com, mohajeri.amir@gmail.com, barzegarmohamadi68@gmail.com

ABSTRACT

Global warming created by large scale emissions of Greenhouse Gases (GHG) are a worldwide concern.Due to this, the issue of green gas network has required more attention in the last decades. Here, we address the carbon footprint-based problem that arises in a gas network where gas flow is transferred from the Town Board Station (TBS) to consumers by pipeline systems. Given this environment,an optimization model for a gas network in which GHG emission is expressed in term of environmental constraints is developed. Here, we formulate a gas network considering profitability and ecological goals to achieve sustainable development. To solve the model accurately, in small and medium sizes, we use GAMS 23.2 software and compare their results with the result of a meta-heuristic algorithm (Hybrid GA/SA). The results show that the proposed algorithm is able to produce better answers in shorter time for large-scale problems. A case study in Mazandaran Gas Company in Iran is conducted to illustrate the validity and effectiveness of the proposed approach.

Keywords: Green gas network, Hybrid GA/SA algorithm, Greenhouse gas emission, Sustainable development.

1 INTRODUCTION

Regarding the development of gas industry in the recent decades, gas pipeline networks have evolved into huge and complex systems. Exploration, extraction, production, and transportation are stages which natural gas goes through to secure consumers. The network contains three phases. The transmission of gas from central production facility to City Gate Station (CGS) is called a transmission phase, while forwarding gas from CGS to Town Board Station (TBS) is called a feeding phase. Securement of gas to consumers is performed in the distribution phase. Pipelines in a natural gas network must be designed based on gas flow rate, length of pipe, gas maximum pressure drop allowance, and gas maximum velocity allowance. Integration of gas network with the issue of environment protection confirms sharp decline in pollution problem. Research on this approach has received considerable attention recently and led to the creation of new research agenda: green gas chain management. So, this concept is a new paradigm where the gas chain will have a direct relation to the environment. Nowadays, most research on green gas network has had a tendency to the transmission phase such as researches done by Azadeh et al [1], Pishvaee and Razmi [2], Jamshidi et al [3], Kashani and Molaei [4] and Subramanian et al [5]. These studies focused on transmission of gas and only formulated models corresponding to this field. Reviewing the literature on gas network design, it is concluded that only a few research papers have considered green gas network. None of these research papers focuses on distribution phase of gas network. In this study, we extend Mohajeri et al's [6 and 7] model to protect the environment. In addition to managing

*Corresponding Author.

properly gas distribution network to reduce negative impact of greenhouse gases emissions, we suggest a strategy for achieving an expected goal, simultaneously. Here, we focus on a different and important aspect of green gas network: we focus on pipelines and TBSs selection as a way to reduce emissions. For this, in addition to minimizing the total cost in the whole gas chain, we consider a regulation to reduce GHG emissions coming from gas flow. The proposed mechanism is a constraint on emissions. In this study, we pursue this scenario and develop problem formulation for it corresponding to the regulation. Overall, our main original contribution proposed in this paper is that we present the linear model based on Mohajeri et al's model to construct an optimal green gas distribution network and propose the Hybrid GA/SA algorithm to solve the large size problems. Here, we intend to conduct a case study of the natural gas network in Mazandaran Gas Company in Iran. To our knowledge, this study is the first paper which decides on the emissions of GHG. In other words, none of the above mentioned activity considers the green gas network in which the emission of GHG is a decision variable. Moreover, none of these research papers focuses on distribution of gas between TBSs and consumers. All of these aspects are covered in this research.

2 PROBLEM STATEMENT

2.1 Description

Here, we focus on the green gas distribution network comprising of the pipelines having various sizes, TBSs, and consumers. The pipeline and TBS belong to the two indispensable components. The pipeline is responsible to connect the two places and to transmit a gas between them, while the TBS provides the motive force to maintain the natural gas flow through the network systems. In this study, we design a green gas distribution network with the aim of both minimizing the total cost and reducing the environmental impact in the whole chain by choosing the optimal locations of the TBSs, the flows of gases, and the pipe diameter sizes along a distribution stage when the demands of customers are given. Figure 1 shows a schematic view of the proposed green gas network.

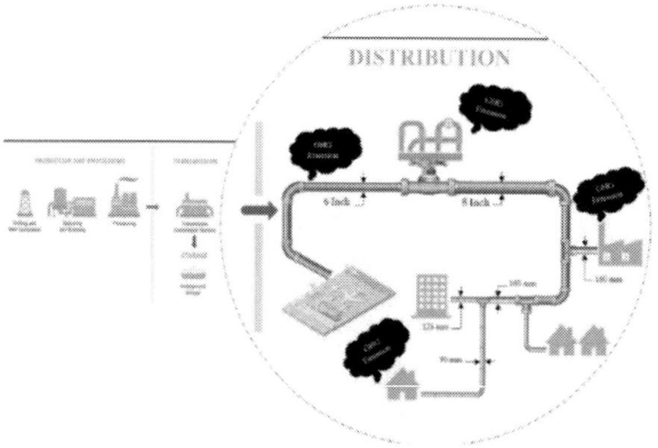

Figure 1: schematic view of the proposed green gas network

We use IPCC[8]equations to estimate fugitive emissions as shown below.

$$E_{gas,\ industry\ segment} = A_{industry\ segment} * EF_{gas,\ industry\ segment} \quad (1)$$

$$E_{gas} = \sum_{industry\ segment} E_{gas,\ industry\ segment} \quad (2)$$

where:

$E_{gas, industrysegement}$ = Annual emissions (Gg),

$EF_{gas,industrysegement}$ = emission factor (Gg/unit of activity),

$A_{industry\ segement}$ = activity value (units of activity).

The indicative factors presented in Table 1 are used to specific classification of emission factors.

Table 1: Classification of emission factors of natural gas

Facilities	Activity data	Yearly emission factors			
		Low	Medium	High	Units of Measure
Regulator Stations	Number of stations	1000	5000	50000	m3/station/yr
Distribution	Length of distribution network	100	1000	10000	m3/km/yr
Gas Use	Number of gas appliances	2	5	20	m3/appliance/yr

2.2 Formulation

Here, a mathematical programming model is proposed. To formulate this model mathematically, the following notations are necessary:

2.2.1 *Notations*

I = set of candidate TBSs
T = set of TBS types
Z = set of consumer/demand zones

2.2.2 *Parameters*

C = the cost of piping per distance unit with respect to gas flow rate
CT = the average cost of piping per distance unit among the TBS
S_t = establishing cost for TBS of type t
q_z = demand of consumer zone z
Q_{it} = capacity of TBS i of type t
d_{iz} = distance between TBS i and consumer zone z
$d'_{ii'}$ = the distance between TBS i and TBS i'
$d''_{zz'}$ = the distance between consumer zone z and consumer zone z'
M = a large number
EF_1 = emission factor for pipelines
EF_2 = emission factor for TBS
ME =Maximum allowable Annual Emissions

2.2.3 *Variables*

$$r_i = \begin{cases} 1, & \text{if TBS i is located} \\ 0, & \text{o.w} \end{cases}$$

$$h_{it} = \begin{cases} 1, & \text{if TBS i of type t is selected} \\ 0, & \text{o.w} \end{cases}$$

$$y_{itz} = \begin{cases} 1, & \text{if consumer zone z is connected t TBS i of type t} \\ 0, & \text{o.w} \end{cases}$$

$$w_{zz'} = \begin{cases} 1, & \text{if there is a direct link between consumer zone z to consumer zone z'} \\ 0, & \text{o.w} \end{cases}$$

$u_i = \begin{cases} 1, & \text{if TBS i is located} \\ 0, & \text{o.w} \end{cases}$

$x_{ii'} = \begin{cases} 1, & \text{if there is a direct link between TBS i to TBS i'} \\ 0, & \text{o.w} \end{cases}$

N_z = allocated number of consumers to consumer zone z

$f_{ii'}$ = amount of flow from TBS i to TBS i'

$f'_{zz'}$ = amount of flow from consumer zone z to consumer zone z'

$f''_{zz'}$ = amount of gas flow from consumer zone z to consumer zone z'

ew_z = amount of congested gas flow to be supplied to each consumer zone z

$g_{zz'}$ = amount of fugitive emissions from pipeline between consumer zone z to consumer zone z'

g'_{itz} = amount of fugitive emissions from pipeline between TBS i of type t to consumer zone z

g''_{it} = amount of fugitive emissions from TBS i of type t

ew'_{itz} = amount of gas flow from TBS i of type t to consumer zone z.

2.2.4 Objective function

$$\min f = f_1 + f_2 + f_3 + f_4 \quad (3)$$

where,

$$f_1 = \sum_{i \in I} \sum_{t \in T} h_{it} S_t, \quad (4)$$

$$f_2 = \sum_{i \in I} \sum_{t \in T} \sum_{z \in Z} y_{itz} d_{iz} C, \quad (5)$$

$$f_3 = \sum_{i \in I} \sum_{j \in J} x_{ii'} d'_{ii'} CT, \quad (6)$$

$$f_4 = \sum_{z \in Z} \sum_{z' \in Z} w_{zz'} d''_{zz'} C. \quad (7)$$

2.2.5 Constraints

$$u_i \le r_i, \qquad \forall i \in I, \quad (8)$$

$$\sum_{i \in I} u_i = 1, \quad (9)$$

$$r_i - M(r_{i'} - 1) \ge x_{ii'}, \qquad \forall i, i' \in I, \quad (10)$$

$$r_{i'} \ge x_{ii'}, \qquad \forall i, i' \in I, \quad (11)$$

$$\sum_{i \in I} x_{ii'} \ge (1 - u_{i'}) + M(r_{i'} - 1), \qquad \forall i' \in I, \quad (12)$$

$$\sum_{i \in I} x_{ii'} \le (1 - u_{i'}) - M(r_{i'} - 1), \qquad \forall i' \in I, \quad (13)$$

$$\sum_{i \in I} x_{ii'} \le r_{i'}, \qquad \forall i' \in I, \quad (14)$$

$$x_{ii'} \le f_{ii'}, \qquad \forall i, i' \in I, \quad (15)$$

$$f_{ii'} \le x_{ii'} M, \qquad \forall i, i' \in I, \quad (16)$$

$$\sum_{i \in I} f_{ii'} - \sum_{i'' \in I} f_{i'i''} \ge ((-u_{i'} M) + 1) + (r_{i'} - 1), \ \forall i' \in I, \quad (17)$$

$$\sum_{i \in I} f_{ii'} - \sum_{i'' \in I} f_{i'i''} \le ((u_{i'} M) + 1) + (r_{i'} - 1), \ \forall i' \in I, \quad (18)$$

$$\sum_{t\in T} h_{it} = r_i, \qquad \forall i \in I, \tag{19}$$

$$\sum_{z\in Z} y_{itz} \geq h_{it}, \qquad \forall i \in I, \forall t \in T, \tag{20}$$

$$\sum_{z\in Z} y_{itz} \leq h_{it} M, \qquad \forall i \in I, \forall t \in T, \tag{21}$$

$$\sum_{z\in Z} w_{zz'} + \sum_{i\in I}\sum_{t\in T} y_{itz'} = 1, \qquad \forall z' \in Z, \tag{22}$$

$$N_z - \sum_{z'\in Z} f'_{zz'} = 1, \qquad \forall z \in Z, \tag{23}$$

$$N_{z'} + M(w_{zz'} - 1) \leq f'_{zz'}, \qquad \forall z, z' \in Z, \tag{24}$$

$$N_{z'} - M(w_{zz'} - 1) \geq f'_{zz'}, \qquad \forall z, z' \in Z, \tag{25}$$

$$f'_{zz'} \leq w_{zz'} M, \qquad \forall z, z' \in Z, \tag{26}$$

$$f'_{zz'} \geq w_{zz'}, \qquad \forall z, z' \in Z, \tag{27}$$

$$(w_{zz'} - 1)M + (q_{z'} + ew_{z'}) \leq f''_{zz'}, \qquad \forall z, z' \in Z, \tag{28}$$

$$(w_{zz'} - 1)(-M) + (q_{z'} + ew_{z'}) \geq f''_{zz'}, \qquad \forall z, z' \in Z, \tag{29}$$

$$f''_{zz'} \leq w_{zz'} M, \qquad \forall z, z' \in Z, \tag{30}$$

$$f''_{zz'} \geq w_{zz'}, \qquad \forall z, z' \in Z, \tag{31}$$

$$\sum_{z'\in Z} f''_{zz'} = ew_z, \qquad \forall z \in Z, \tag{32}$$

$$(y_{itz} - 1)M + (q_z + ew_z) \leq ew'_{itz}, \qquad \forall i \in I, \forall t \in T, \forall z \in Z, \tag{33}$$

$$(y_{itz} - 1)(-M) + (q_z + ew_z) \geq ew'_{itz}, \qquad \forall i \in I, \forall t \in T, \forall z \in Z, \tag{34}$$

$$ew'_{itz} \leq y_{itz} M, \qquad \forall i \in I, \forall t \in T, \forall z \in Z, \tag{35}$$

$$ew'_{itz} \geq y_{itz}, \qquad \forall i \in I, \forall t \in T, \forall z \in Z, \tag{36}$$

$$\sum_{z\in Z} ew_{itz} \leq Q_{it}, \qquad \forall i \in I, \forall t \in T, \tag{37}$$

$$g_{zz'} = EF_1 w_{zz'} d''_{zz'} \forall z, z' \in Z \tag{38}$$

$$g'_{itz} = EF_1 y_{itz} d_{iz} \; \forall i \in I, \forall t \in T, \forall z \in Z \tag{39}$$

$$g''_{it} = EF_2 h_{it} \forall i \in I, \forall t \in T \tag{40}$$

$$\sum_z \sum_{z'} g_{zz'} + \sum_i \sum_t \sum_z g'_{itz} + \sum_i \sum_t g''_{it} \leq ME \tag{41}$$

$$r_i, h_{it}, y_{itz}, u_i, x_{ii'}, w_{zz'}, \in \{0,1\}, \qquad \forall i, i' \in I, \forall t \in T, \; \forall z, z' \in Z, \tag{42}$$

$$N_z, f_{ii'}, f'_{ii'}, f''_{zz'}, ew_z, ew'_{itz}, g_{zz'}, g'_{izt}, g''_{it} \geq 0 \qquad \forall i \in I, \forall t \in T, \forall z \in Z \tag{43}$$

Equations (4)-(7) are the cost functions corresponding to the location-allocation costs. Constraint (8) indicates that exactly one TBS must be defined as a root. Constraint (9) ensures that there is exactly one TBS as the root in the network. Constraints (10) and (11) show the link between two TBSs. Constraints (12)-(14) impose that each TBS receive exactly one link from other TBSs if it is not the root node. The amount of flow between each TBS i and TBS i' is represented by constraints (15) and (16). Constraints (17) and (18) guarantee that there is no closed loop in the network. Constraint (19) shows that each TBS can adopt only one type when it is selected to service consumers. Constraints (20) and (21) ensure that each TBS covers at least one consumer. Constraint (22) represents that each consumer receives service from one consumer or one TBS. Constraint (23) determines the allocated number of consumers to consumer zones. Constraints (24)-(27) express the flow between two consumers. Constraints (28)-(31) represent the amount of gas flow from consumer zone z to consumer zone z'. Constraint (32) indicates the amount of congested gas flow for supplying other consumers by each consumer. The amount of gas flow from TBS to consumer is shown by constraints (33)-(36). Capacity restriction is shown by constraint (37). Constraints (38)-(40) express the emission from pipelines and TBSs. Constraint (41) ensures that proposed network yearly emissions is less than Maximum allowable annual emissions. Constraint (42) imposes that the variables be binary. Non negativity of the variables is represented by constraint (43).

3 SOLUTION METHODOLOGY AND EVALUATION

Our proposed model is based on (MST) method which belongs to the NP-hard class of problem Garey and Johnson [9]. Because of the complexity of these problems, exact methods need excessive computation time. So, metaheuristic algorithms are essential tools for solving such problems in reasonable amounts of time. Here, the optimization model is solved by a proposed algorithm for large size problems. The hybrid GA/SA algorithm is proposed to attain near optimized solution, decrease computational complexity, and overcome the computer memory limitation.

3.1 THE PROPOSED ALGORITHM

Here, the proposed hybrid GA/SA algorithm is presented as fallows.

```
1.  Initialize GA parameters and SA parameters (nPop,MaxIt,Cper,Mper;T0,Alpha,MaxIt1,MaxIt2)
2.  Initialize SA_Database
3.  For Number of possible combinations of TBS
    3.1. Initialize GA population based on MST and calculate related costs and stored in GA_Database
    3.2. While Iteration< Maximum Generations
         3.2.1.  Apply crossover and calculate new child's costs and stored in Cross_Database
         3.2.2.  Apply Mutation and calculate mutate costs and stored in Mu_Database
         3.2.3.  Combines GA_Database, Cross_Database, Mu_Database and sort them according to FITNESS-FN
         3.2.4.  Truncated extra members

    END While 3.2

    3.3. Choose the best solution of GA
    3.4. Initialize T,T0,Alpha,MaxIt1,MaxIt2 for Simulate Annealing algorithm
         3.4.1.  BestSol= best solution of GA
         3.4.2.  While It1< MaxIt1
         3.4.3.  While It2< MaxIt2
         3.4.4.  X.Sol=Create neighbor solution of the BestSol and calculate related costs(X.Cost)
                 If (( X.Cost<BestSol.Cost)&(Total Emission<ME))
                     BestSol=X
                 Else
                     delta=X.Cost- BestSol.Cost;
                     p=exp(-delta/T)
                     a=rand (0,1)
                     if a<p
                             BestSol=X
                     END if

                 END if

           END While It1

         3.4.5.  T=T* Absorption rate
    END While It2
    3.5. Store BestSol in SA_Database
    END for Number of possible combinations of TBS (until terminating condition)
4.  Return the Best Solution that stored in SA_Database and show related results and graphs
```

4 CASE STUDY

A natural gas network case study of Mazandaran Gas Company in Iran is conducted to verify the proposed model. All the required information we used are as follows:

- Relationships among gas flow rates and pipe diameter sizes.
- The cost of piping per distance unit with respect to the gas flow rate.

- Consumers' demands.
- The establishing cost and capacity of different TBSs.
- Distance between the TBS and consumers, among the TBSs and among the consumers.

We applied GAMS 23.2 software package to facilitate computations in our Mixed Integer Programming (MIP) model and MATLAB R2015 software package for metaheuristic algorithm. The validity of proposed model is measured for gas distribution network case study of Mazandaran Gas Company as summarized in Table 2. We also make further experiments by changing the model to restrict the total number of pipes exiting a TBS by predefined values. For this, we need to change constraints (21) to:

$$\sum_{z} y_{itz} \leq np.h_{it} \qquad \forall i \in I, \forall t \in T \qquad \textbf{\textit{(44)}}$$

where, *np* is the maximum number of exiting pipes. We have used values of np equal to 2 and 3 and reported the results in Table 2 along with the optimal solution obtained before, for comparison purposes. Now, the validity and effectiveness of the proposed metaheuristic algorithm in comparison to the exact method is analyzed. We considered ten different problems. Also, each problem contained three types (restricted to np=2, restricted to np=3, and unrestricted). Consequently, we generated 30 test problems to test our proposed algorithm. The results are given in Table 2.The results show that our proposed algorithm is effective for solving the problems. The algorithm obtains good solutions in reasonable times. In comparison to the exact method, the algorithm obtains solutions closer to the optimal solutions with much less time than the time needed to be spent for obtaining exact optimal solution.

5 CONCLUSIONS

In this paper, an optimal mathematical model for a green gas network was established. The model considered all factors influencing the total cost of a gas network such as pipe diameter, length of pipe, etc. We also considered environmental impact of our proposed gas distribution network with constraint-based approach to achieve sustainable development. We used the actual data on Mazandaran Gas Company in Iran to conduct a case study. Optimal results were obtained applying GAMS 23.2 software package. Due to the inability of this software to provide solutions for large size problems in a reasonable time, the metaheuristic algorithm was proposed. As a result, with the proposed algorithm, we obtained excellent results. This will enable us to design a fast, effective and robust decision aid tool based on the suggested model.

Table2: Computational results

			Proposed Algorithm (Hybrid GA/SA)			Exact Model (GAMS)			GAP (%)		
NUM	No. of TBS	No. of Consumer	Without cons.	With cons<=3	With cons<=2	Without cons.	With cons<=3	With cons<=2	Without cons.	With cons<=3	With cons<=2
1	1	5	1.09E+08	1.09E+08	1.10E+08	108500385.5	108849094.7	109748998.4	4.71E-03	1.49E-03	1.48E-03
			0.71 sec	1.2 sec	1.03 sec	4.179 sec	3.556 sec	3.765 sec			
2	1	5	1.15E+08	1.15E+08	1.19E+08	114572467.6	114572467.6	119053166.5	4.70E-03	4.70E-03	4.86E-04
			0.95 sec	1.2 sec	1.03 sec	1.874 sec	3.962 sec	2.254 sec			
3	2	7	1.47E+08	1.48E+08	1.55E+08	145523793.2	146321990	153661789.3	6.88E-03	8.22E-03	5.62E-03
			1.13 sec	1.41 sec	1.76 sec	117.14 sec	89.42 sec	162.52 sec			

			Proposed Algorithm (Hybrid GA/SA)			Exact Model (GAMS)			GAP (%)		
NUM	No. of TBS	No. of Consumer	Without cons.	With cons<=3	With cons<=2	Without cons.	With cons<=3	With cons<=2	Without cons.	With cons<=3	With cons<=2
4	2	7	1.60E+08	1.60E+08	1.60E+08	158309288.5	158309289.1	159250897	7.68E-03	7.68E-03	1.72E-03
			1.68 sec	1.90 sec	1.87 sec	77.24 sec	76.21 sec	74.1 sec			
5	3	6	1.17E+08	1.18E+08	1.18E+08	116935096.6	116935096.6	117665355.6	9.04E-04	6.03E-03	5.74E-03
			2.01 sec	2.38 sec	2.02 sec	73.73 sec	59.325 sec	68.7 sec			
6	3	6	1.35E+08	1.35E+08	1.44E+08	134250942.2	134250942.2	143436671.7	5.88E-03	5.88E-03	4.21E-03
			1.93 sec	2.13 sec	2.40 sec	64.1 sec	49.87 sec	61.2 sec			
7	4	9	1.67E+08	1.67E+08	1.69E+08	165620613.6	166350994.3	168867128.7	8.76E-03	4.93E-03	2.99E-03
			2.22 sec	3.02 sec	3.71 sec	578.17 sec	646.94 sec	1278.3 sec			
8	4	9	1.85E+08	1.85E+08	1.85E+08	183824992.5	184450979.3	185111193.4	6.78E-03	3.91E-03	1.41E-03
			2.22 sec	3.04 sec	3.61 sec	1472 sec	1402 sec	1807 sec			
9	5	11	2.18E+08	2.21E+08	2.26E+08	217391385.8	220920376.4	225870999.9	2.91E-03	4.64E-04	6.73E-04
			3.65 sec	4.23 sec	4.80 sec	36808.4 sec	45409.6 sec	70759.6 sec			
10	5	11	2.20E+08	2.22E+08	2.27E+08	218733880.6	221101777.3	226593468.7	5.89E-03	4.17E-03	1.90E-03
			3.45 sec	4.03 sec	5.10 sec	37778.5 sec	73526.9 sec	97302.4 sec			

6 REFERENCES

[1] **Azadeh, A., Raoofi, Z., &Zarrin, M. 2015.** A multi-objective fuzzy linear programming model for optimization of natural gas supply chain through a greenhouse gas reduction approach. Journal of Natural Gas Science and Engineering, 26, 702-710.

[2] **Pishvaee, M.S., Razmi, J., 2012.** Environmental supply chain network design using multi-objective fuzzy mathematical programming. Appl. Math. Model. 36 (8), 3433e3446.

[3] **Jamshidi, R., Ghomi, S.F., Karimi, B., 2012.** Multi-objective green supply chain optimization with a new hybrid memetic algorithm using the Taguchi method. Sci. Iran. 19 (6), 1876e1886.

[4] **Kashani, A. H. A., &Molaei, R. 2014.** Techno-economical and environmental optimization of natural gas network operation. Chemical Engineering Research and Design, 92(11), 2106-2122.

[5] **Subramanian, R., Williams, L. L., Vaughn, T. L., Zimmerle, D., Roscioli, J. R., Herndon, S. C., ... & Sullivan, M. R. 2015.** Methane emissions from natural gas compressor stations in the transmission and storage sector: Measurements and comparisons with the EPA greenhouse gas reporting program protocol. Environmental science & technology, 49(5), 3252-3261.

[6] **Mohajeri, A., Mahdavi, I., Mahdavi-Amiri, N., &Tafazzoli, R. 2011.** Optimization of tree-structured gas distribution network using ant colony optimization: A case study. International Journal of Engineering-Transactions A: Basics,25(2), 141.

[7] **Mahdavi, I., Mohajeri. A, Arabmaghsudi, M., Yahyanejad, M. H.,Taghipourian, F.&Mahdavi-Amiri, N.** (2010, July). An optimal natural-gas network using minimum spanning tree. In 2010 40th International Conference on Computers and Industrial Engineering (CIE), (pp.1-5). IEEE.

[8] **IPCC** 2006, 2006 IPCC Guidelines for National Greenhouse Gas Inventories, Prepared by the National Greenhouse Gas Inventories Programme, Eggleston H.S., Buendia L., Miwa K., Ngara T., and Tanabe K. (Eds).

[9] **Garey, M. R., & Johnson, D. S.** 1979. Computers and intractability: a guide to the theory of NP-completeness. A Series of books in the mathematical sciences. WH Freeman and Company, New York, NY, 25(27), 141.

THE KNOWLEDGE NETWORK OF HUAWEI'S GLOBAL R&D ORGANIZATION: EVIDENCE FROM PATENT ANALYSIS

Hu Xinyue[1] Zhang Haiwen[1] Sun Fei[1*] Lu Zhuobin[1] and Tang Yongli[1]

[1]School of Management

Jinan University, Guangzhou, China

821713857@qq.com

ABSTRACT

R&D globalization has gradually become an important strategy for latecomer firms from developing countries to access the innovation resources worldwide in the process of technological catching up. This study explores how Huawei, one of the leading Chinese companies, organizes its global R&D units from the perspective of knowledge network. We identified Huawei's 37 R&D units and 37 external cooperative partners with its 18885 PCT patents applied during the period of 1999 to 2013. The knowledge network of Huawei's R&D units and external cooperative partners was constructed based on the joint patenting among them. Social network analysis was applied to investigate the network characteristics. We found that the headquarters of Huawei produced most PCT patents, and the joint patenting between R&D units was increasing over time. Domestic universities and foreign firms accounted for over 60% in Huawei's cooperative partners, while most of R&D units of Huawei located in developed countries and major domestic cities. The R&D unit in Huawei's headquarters and the one in US acted as the "cores" in the network since they broadly linked to others and intensively cooperated with them in patenting.

Keywords: R&D globalization, Huawei, Knowledge network, Social network analysis

1 INTRODUCTION

With its technological capabilities improved sustainably in past decades, the Chinese company Huawei Technologies Co., Ltd. (hereafter Huawei) has successfully transformed from a latecomer firm to one of the global leading firms in ICT industry. Firms' technological capabilities enable them to improve existing technologies as well as develop new ones [1]. And patents are usually regarded as one of the indicators reflecting firms' technological capabilities [2] or technological performance [3]. In terms of patent application, Huawei had not been included in the top 100 patent applicants in 1990s, but it progressed quickly to appear in the top 30 patent applicants worldwide in 2000s [4]. Huawei became the fourth-largest global patent applicant in 2007[5], and it filed 3898 PCT (Patent Cooperation Treaty) patents in 2015, ranking at the first among worldwide applicants [6].

Differing from the latecomer firms which relied on foreign technology transfer at the beginning of catching up, Huawei emphasized more on developing its in-house R&D capability and

---

[*] Corresponding Author

adopted R&D globalization as its important strategy to access the global innovation resources, thus enhance its technological capabilities [5]. Through setting up R&D units (e.g. research labs, subsidiaries and other R&D facilities etc.) in different domestic regions or foreign countries, latecomer firms in developing countries like Huawei can form the knowledge network among own R&D units and external organizations like universities, public research institutes, other firms etc., which drives the process of firms' technological learning[7]. The reasons or benefits for firms to set up research facilities abroad are in various aspects, e.g. adapting the product to local markets, tracing the development of emerging technologies and accessing the complementary resources from foreign countries [3; 5]. In this regard, R&D globalization also contributes to overcome two main disadvantages of latecomer firms, namely lacking of technological capabilities and links to the international markets [8; 9].

To our knowledge, although Huawei's catching up has attracted great research interest in previous studies (e.g. Fan [5]; Low [10]; Zhang and Vialle [11]), there are still insufficient empirical evidences to uncover the patterns of Huawei's R&D globalization, e.g. the geographic distribution of its R&D units, the collaboration among internal R&D units and with external organizations. Another research gap rests upon the studies on R&D internationalization or technology internationalization are basically at country level (e.g. Guellec and van Pottelsberghe [3]; Wang, Sutherland and Ning [7]; Duysters et al. [12]), in-depth studies on firm's global R&D organization are few. Thus, the primary contribution of our study is to fill these gaps by exploring how Huawei deploy R&D globalization from the perspective of knowledge network. More specifically, we identified Huawei's internal R&D units and external cooperative partners with the information extracted from PCT patents, and constructed a knowledge network among them based on their joint patenting activities. Social network analysis (SNA) was applied to explore the characteristics of this knowledge network.

The remaining of this paper are organized as follows: Section 2 illustrates the data source and methods. Section 3 gives the descriptive results of Huawei's R&D units and its external cooperative partners. Section 4 shows the results of networks analysis. Section 5 summarizes our findings and conclusions.

2 METHODS

2.1 Data collection

Patents have been usually used in investigating the technology internationalization of country level in previous studies (e.g. Guellec and van Pottelsberghe [3]; Wang, Sutherland and Ning [7]; De Prato and Nepelski [13]), mainly thanks to the rich information sourced from patents, such as the name and the address of the inventors as well as applicants (owners). In this study, the international PCT patents were used to investigate Huawei's global R&D activities. We used "Huawei" as the keyword for the patent applicant's name to search the PCT patent database of World Intellectual Property Organization (WIPO) in February 2016, and obtained more than 21000 patent records. Each PCT patent enclosed an international application status report, which provided the publication number, applicants and their address, inventors and their address, and other information about the patent. By going through the applicants in each PCT record, the irrelevant patents which were not owned (partly or solely) by Huawei were eliminated. In addition, to avoid the problem of data incompletion in our study, we selected the patents whose priority date was before 2014 since WIPO may not release the complete patent application information during 2014 to 2016 when we collected the data. As a result, 18885 patents of Huawei applied during 1999 to 2013 were retained.

The inventors' addresses of patent usually contained the information about working place, thus they were used to identify or distinguish different R&D units of Huawei. We extracted the inventors' addresses from 18885 PCT patents and merged similar addresses. As a result, 37 internal R&D units and 37 external cooperative partners of Huawei were discovered. Moreover, we classified the patents into two categories: joint patents applied by 2 or more Huawei's R&D units and/or external partners; and non-joint patents applied independently by one R&D unit. If the inventors of a patent were from different countries or different cities in China, the patent could be sorted to the joint patents. If a patent was invented by an individual researcher or more than one researchers from the same address, this patent was classified to the non-joint patents. Notably, Huawei's 33 R&D units had applied 766 PCT patents with 37 external organizations jointly, while 4 R&D units only independently applied PCT patents.

2.2 Social network analysis

Generally, a network can be defined by two basic elements, the actors and the linkages (relationships) among them [14]. The linkages based on specified social relationships can serve as the channels for knowledge or information flow between actors [15]. And the knowledge network allows the actors to search, access, transfer, and create knowledge or technology [16]. In this study, we define the knowledge network of Huawei as consisted of a set of internal R&D units and external organizations (actors) which are interconnected with the relationships based on joint patenting.

Social network analysis (SNA) was used to analyze the knowledge network of Huawei's global R&D organization. This method allows researchers to investigate the relationships among actors, and further explore the patterns and meanings of these relationships [17]. We constructed a 74*74 adjacent matrix of the Huawei's 37 R&D units and 37 external cooperative organizations, and the value of element a_{ij} in the adjacent matrix indicates the cooperation frequency between actor i and actor j. The cooperation frequency between any two actors (R&D units or external cooperative partners) is operatized as the number of times of two actors' address (inventors' address) appeared in the same patent among 766 joint patents. The adjacent matrix was imported into the SNA software Ucinet 6 for subsequent analysis.

3 AN OVERVIEW ON HUAWEI'S GLOBAL R&D ORGANIZATION

3.1 Trend of Huawei's joint patenting during 2002 to 2013

Collaboration among Huawei's R&D units and external partners are investigated via joint patenting. Figure 1 shows the trend of Huawei's joint patenting during 2002 to 2013. The first joint patent applied by Huawei's R&D units appeared in 2002, and the following 3 years (2003-2005) the joint patenting among Huawei's R&D units remained scarce. But since 2006, the number of joint patents of Huawei increased rapidly. In 2012 and 2013, Huawei applied nearly 200 joint patents, which were more than 6 times of those in 2006. We further distinguished the joint patents applied among Huawei's R&D units and external cooperative organizations. As shown in Figure 1, most of the joint patents of Huawei are created by its internal R&D units, while the cooperation of Huawei's R&D units and external organizations is relatively less. Thus, cooperation among internal R&D units plays the dominated role in joint patenting of Huawei.

3.2 The external cooperative partners of Huawei

Collaboration with different external organizations makes sense for Huawei's technological catching up since links to different external partners create opportunities for technological learning [18]. We distinguished 37 external partners by their organization types (firms, universities, or research institutes) and geographic location (domestic or foreign). Figure 2 shows that more than 60% of Huawei's R&D cooperative partners are universities, especially domestic universities account for the largest proportion (46%). Different from the scenes from the joint patenting with universities, Huawei links to more foreign research institutes than domestic ones. Similarly, the percentage of foreign firms in the portfolio of cooperative partners of Huawei is higher than domestic ones.

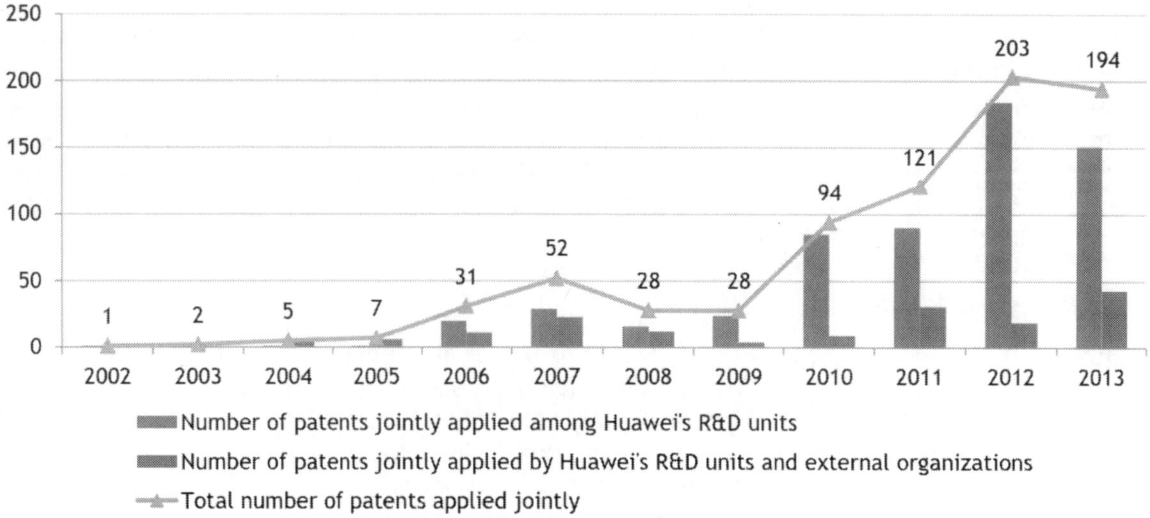

Figure 1: Trends of Huawei's joint patenting during 2002 to 2013

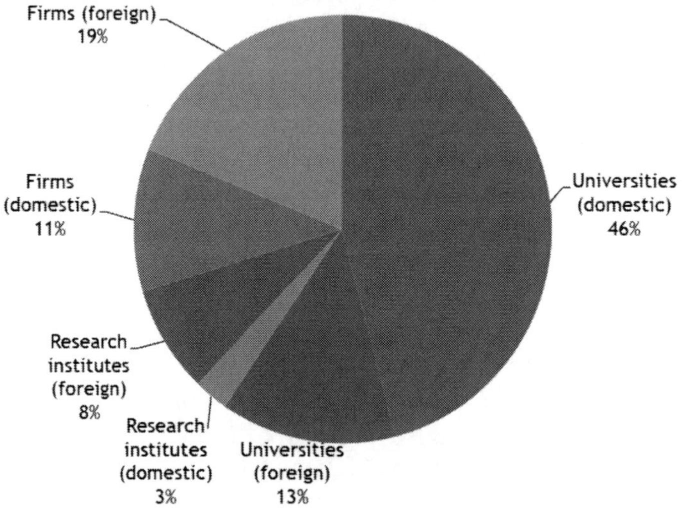

Figure 2: The composition of Huawei's external R&D cooperative partners

3.3 The global landscape of Huawei's R&D organization

Previous research has documented that Huawei's R&D units were strategically located not only in China but also around the world [5]. For firms in developing countries, setting up R&D organizations overseas represents an important channel to access the advanced technologies in developed countries [7]. During 2002 to 2013, nearly two thirds (25) of Huawei's R&D units were set up in foreign countries around the world, and the rests of them were in China (See Figure 3 below). More than half (16) of the overseas R&D units are located in Europe and North America, which cover the developed countries like Deutschland (DE), Sweden(SE), Great Britain(GB), France(FR), Canada(CA), United States(US) etc. Especially, the number of R&D units in Europe makes the largest share in overseas R&D units, indicating that Huawei tends to access the local talents, advanced technologies and sophisticated markets in European countries. Except the R&D unit in Huawei's headquarters in Shenzhen, other domestic R&D units distributes among Beijing(BJ), Shanghai(SH), Tianjin(TJ) and other relatively developed cities in China.

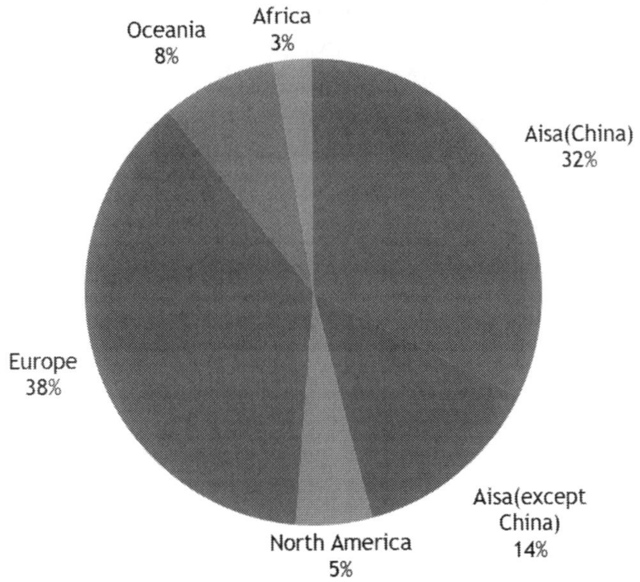

Figure 3: The geographic distribution of Huawei's R&D units

4 THE NETWORK OF HUAWEI'S R&D ORGANIZATIONS

4.1 Visualizing the knowledge network of Huawei's R&D organizations

We used the module NetDraw in Ucinet 6.0 to mapping the knowledge network (Figure 4). There are 74 nodes in the network, while 4 isolate nodes among them represent the R&D units had not joint patents during 2002 to 2013.

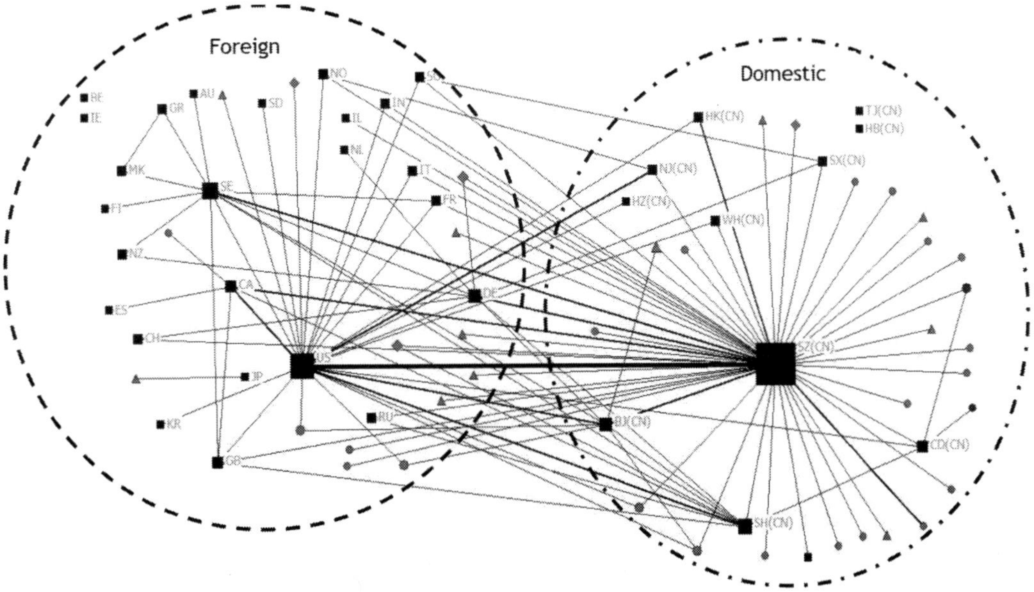

Notes: The legend of node in figure: ■ R&D units of Huawei; ● Universities; ▲ Firms; ◆ Research institutes. The label represents the location of R&D units (abbreviation of foreign countries or domestic cities in China). The size of a node in the network is weighted by the number of cooperative partners which had direct linkage with the focal node. The link between nodes is weighted by the cooperative frequency.

Figure 4: The knowledge network of Huawei's global R&D units

The cooperation among R&D units which locate in different regions or countries may give the clues about the inter-region or inter-nation knowledge or technology transfer within the firms group of Huawei. As seen in Figure 4, Huawei's R&D unit in headquarters (labeled as SZ(CN)) and the R&D unit in US (labeled as US) act as the hub in the network, since they have more links than other nodes in the network. Particularly, the R&D unit in its headquarters establishes the most number of links with other domestic as well as foreign R&D units and external cooperative partners. Thus, the knowledge transfer in the network is centered to the R&D unit in headquarters and the one in US.

Nearly 85% of linkages between nodes in the network have the cooperative frequency below 30 times. To manifest the main structure of knowledge network, we eliminated links which cooperative frequency is less than 30, and then remove the isolate nodes in the network. As a result, 11 nodes and the links among them have been kept in Figure 5. The R&D unit in headquarters of Huawei (SZ(CN)) seems to like the "backbone" in the network in Figure 5, and the other 9 R&D units of Huawei have direct links with it. And Tsinghua University is an important R&D cooperative partner for Huawei's headquarters.

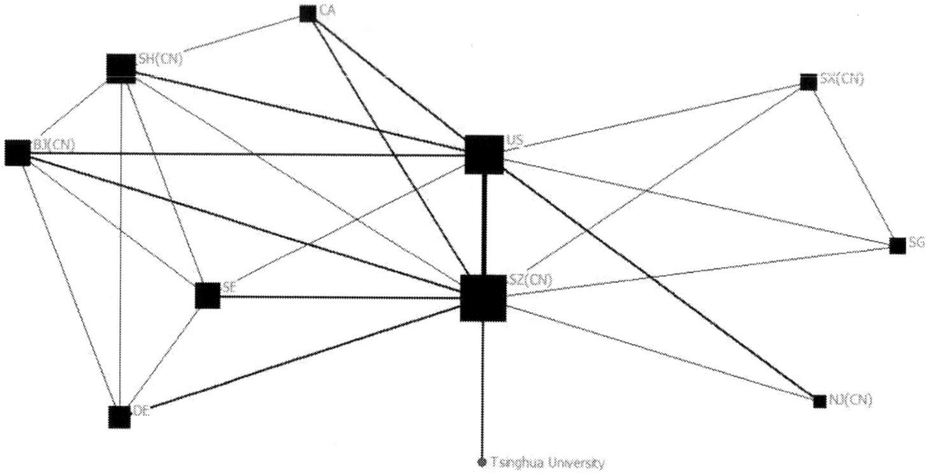

Notes: The legend of node in figure: ■ R&D units of Huawei; ● Universities. The label represents the location of R&D units (abbreviation of foreign countries or domestic cities in China). The size of a node in the network represents the number of cooperative partners which has direct linkage with the focal node. The link between nodes is weighted by the cooperative frequency.

Figure 5: The main structure of knowledge network of Huawei's R&D units

4.2 The characteristics of cooperation among Huawei's R&D units

R&D units may serve different functions or roles in Huawei's global R&D organization due to their locations or purposes [5], thus they may show different characteristics in R&D cooperation. Our study uses two indices, namely cooperation width and cooperation depth, to investigate the characteristics of cooperation among Huawei's R&D units. Only 33 R&D units of Huawei are included as the observations in this section since 4 R&D units had not applied any joint patents before 2014. The cooperation width of a R&D unit is defined as the number of cooperative partners that directly linked to it in the network. The cooperation depth of a R&D unit is the sum of cooperation times between the focal units and other units divided by the number of cooperative partners. The cooperative pattern of Huawei's 33 R&D units is depicts in Figure 6.

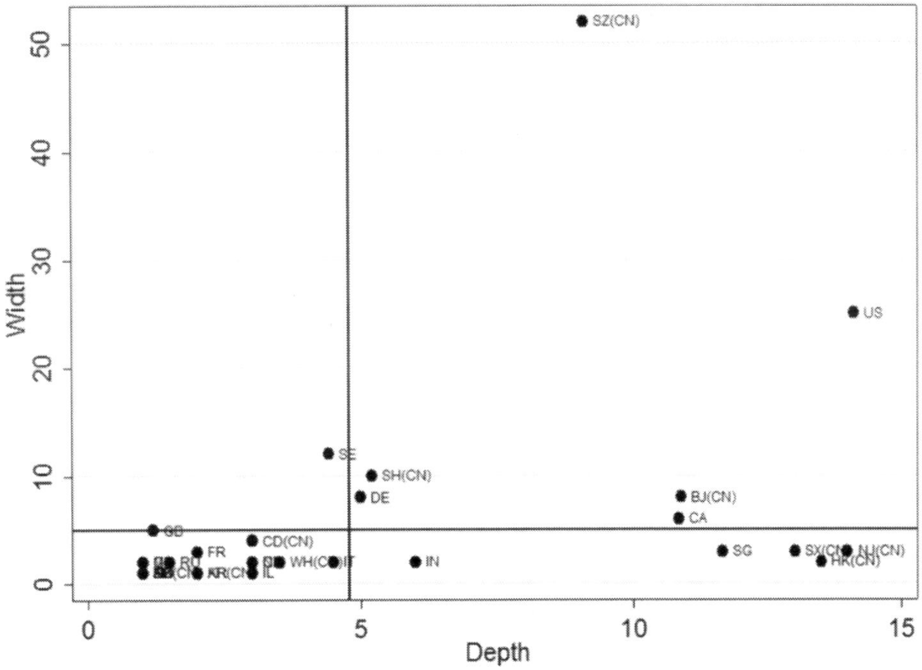

Figure 6: The cooperation width and depth of Huawei's 33 R&D units

In terms of cooperation width and cooperation depth, 6 R&D units of Huawei (labeled as SZ(CN), SH(CN), BJ(CN), DE, US, and CA) are at the relatively higher level than others. Especially, the R&D unit in Huawei's headquarters (SZ(CN)) owns the largest cooperation width, which means it links broadly to other R&D units and external cooperative partners. This finding also consists with the network shown in Figure 4. Besides, the R&D unit in United State (US) has the greatest cooperation depth among all R&D units, which implies it cooperates intensively with others. 5 R&D units (labeled as IN, SG, SX, HK, NJ) are characterized by "high in cooperation depth but low in cooperation width" since they intensively cooperate with a limited number of partners. The rest of R&D units have limited cooperative partners as well as low in cooperation frequency.

5 SUMMARY AND CONCLUSIONS

R&D globalization has been one of the important strategies for latecomer firms like Huawei in their technological catching up. The present study aims to investigate Huawei's R&D globalization in the view of knowledge network. We used the PCT patent data to identify Huawei's R&D units and external cooperative partners, and applied the social network analysis to explore the knowledge network among them. The findings of our study are concluded as follows:

In terms of PCT patent application, the R&D unit in headquarters solely produced most PCT patents of Huawei, and joint patenting among Huawei's R&D units dominated in joint patenting. This result confirms the similar findings appeared in previous work, e.g. in-house R&D played critical role in Huawei's technology innovation [5; 10], and the pattern of "independent development" in developing product and technology in Huawei's catching up[11].

In the case of external cooperative partners in patenting, domestic universities and foreign firms had been the main innovative partners for Huawei. Kang [19] also found that Huawei has concentrated in universities in patent co-application. Cooperation with other external organization also contributes to improve theirs technological capabilities, and universities and research institutes are valuable learning sources for Chinese firms [18]. Our study finds that Huawei had kept the close collaboration with domestic universities like Tsinghua University, Peking University in the joint patenting. And Huawei recruited a great number of graduates in telecommunications from domestic top universities in 2000[5]. In addition, cooperation with foreign firms creates the opportunity of "learning by doing" for Huawei.

Regarding to the R&D units of Huawei, a large proportion of them were established in developed countries and relatively developed cities in China. Setting up R&D units in technology advanced countries allows Huawei benefit from spillover effects and localized learning via taking advantage of geographic and relational proximities [5]. Huawei's R&D units locates in different countries could embed in the innovation system by interacting with local partners, attracting local talents, and access advanced technologies. And the knowledge network of R&D units can facilitate the knowledge exchange, transfer, and diffuse among Huawei's firm group.

Further, the "cores" in Huawei's global R&D structure have been identified. The R&D unit in Huawei's headquarters occupies the "central" position in the network, while the R&D unit in US connects broadly to other overseas R&D units. This may imply the centralized R&D structure within the firm's group of Huawei, i.e. the Central R&D Department of Huawei oversees the R&D activities within the firm group [9], and partly explain the enormous number of patent application comes from its headquarters. Moreover, the analysis on the characteristics of R&D cooperation among R&D units manifests that some R&D units played their unique roles in the network. Further study is needed for explaining the function of these R&D units in Huawei's technological learning.

There are some limitations in our study. Although our study tries to identify the R&D units of Huawei with patent data, some R&D units which had not applied any PCT patents may not be included in our analysis. To provide a more complete prospect of Huawei's R&D globalization, future works can combine various data sources, e.g. patents, press releases and reports etc. And the present study is descriptive in nature. An avenue for further studies could be exploring how the knowledge network evolves over time, and how the knowledge network would affect Huawei's technological innovation.

ACKNOWLEDGEMENT

The research is supported by National Natural Science Foundation of China (71401063; 71571088), Soft Science Research Program of Guangdong Province (2014A070703020), and S&T Programs of Guangdong Province (2014A080802004; 2015A080803003).

REFERENCES

[1] **KIM, L. 1997.** The Dynamics of Samsung's Technological Learning in Semiconductors, *California Management Review*, 39(3), pp 86-100.

[2] **FIGUEIREDO, P.N. 2014.** Beyond technological catch-up: An empirical investigation of further innovative capability accumulation outcomes in latecomer firms with evidence from Brazil, *Journal of Engineering and Technology Management*, 31(1), pp 73-102.

[3] **GUELLEC, D. and VAN POTTELSBERGHE, B. 2001.** The internationalisation of technology analysed with patent data, *Research Policy*, 30(8), pp 1253-1266.

[4] World Intellectual Property Organization. 2016. WIPO Economics & Statistics Series 2015. In *World Intellectual Property Indicators*. http://www.wipo.int/edocs/pubdocs/en/wipo_pub_941_2015.pdf.

[5] **FAN, P. 2011.** Innovation, globalization, and catch-up of latecomers: cases of Chinese telecom firms, *Environment and Planning A*, 43(4), pp 830-849.

[6] World Intellectual Property Organization. 2016. U.S. Extends Lead in International Patent and Trademark Filings. Retrieve from http://www.wipo.int/pressroom/en/articles/2016/article_0002.html.

[7] **WANG, Y.D., SUTHERLAND, D., and NING, L.T. 2014.** A Dynamic Comparative Analysis of International Innovation Networks in Emerging Market MNCs, *Industry and Innovation*, 21(6), pp 457-475.

[8] **HOBDAY, M. 1995.** East Asian latecomer firms: Learning the technology of electronics, *World Development*, 23(7), pp 1171-1193.

[9] **GAO, X. 2011.** Effective Strategies to Catch Up in the ERA of Globalization: Experiences of Local Chinese Telecom Equipment Firms, *Research-Technology Management*, 54(1), pp 42-49.

[10] **LOW, B. 2007.** Huawei Technologies Corporation: from local dominance to global challenge?, *Journal of Business & Industrial Marketing*, 22(2), pp 138-144.

[11] **ZHANG, J. and VIALLE, P. 2014.** Patterns of Innovation of a Successful Latecomer Firm: A Longitudinal Analysis of the Case of Huawei, Proceedings of the 2014 Ieee International Conference on Management of Innovation and Technology (ICMIT 2014), Singapore, pp 29-34.

[12] **DUYSTERS, G., JACOB, J., LEMMENS, C., and JINTIAN, Y. 2009.** Internationalization and technological catching up of emerging multinationals: a comparative case study of China's Haier group, *Industrial and Corporate Change*, 18(2), pp 325-349.

[13] **DE PRATO, G. and NEPELSKI, D. 2014.** Global technological collaboration network: network analysis of international co-inventions, *Journal of Technology Transfer*, 39(3), pp 358-375.

[14] **BORGATTI, S.P. and HALGIN, D.S. 2011.** On network theory, *Organization Science*, 22(5), pp 1168-1181.

[15] **OWEN-SMITH, J. and POWELL, W.W. 2004.** Knowledge networks as channels and conduits: The effects of spillovers in the Boston biotechnology community, *Organization Science*, 15(1), pp 5-21.

[16] **PHELPS, C., HEIDL, R., and WADHWA, A. 2012.** Knowledge, Networks, and Knowledge Networks: A Review and Research Agenda, *Journal of Management*, 38(4), pp 1115-1166.

[17] **WASSERMAN, S. and FAUST, K. 1994.** *Social network analysis: Methods and applications*. Cambridge university press.

[18] **CHEN, J. and QU, W.G. 2003.** A new technological learning in China, *Technovation*, 23(11), pp 861-867.

[19] **KANG, B. 2015.** The innovation process of Huawei and ZTE: Patent data analysis, *China Economic Review*, 36, pp 378-393.

WARRANTY COST PREDICTION BASED ON ACCELERATED DEGRADATION TESTING DATA

Xiujie Zhao* and Min Xie
Department of Systems Engineering and Engineering Management
City University of Hong Kong, Hong Kong, China
xiujizhao2-c@my.cityu.edu.hk
minxie@cityu.edu.hk

ABSTRACT

Warranty is an important marketing element for various customer products. Attractive warranty usually indicates higher reliability and an appropriate warranty policy provides a good balance in sales marketing and cost control, therefore results in better profit performance. For new products that have not been put on market, manufacturers usually want to predict the warranty cost to forecast the influence of warranty on future profit. In the test phase of products, reliability tests can be used to predict the reliability of highly reliable products. Among various reliability tests, accelerated degradation tests (ADT) have attracted the attention of researchers and engineers due to its advantages over accelerated life tests (ALT). In this paper, we present a method to predict the expected warranty cost under one-dimensional warranty with free replacement by analyzing ADT experimental data. By assuming the products are working under random stress conditions, the expected warrant cost and warranty risk are computed. Finally, numerical examples are given to illustrate the proposed method and compare with previous methods.

Keywords: Accelerated degradation test, warranty cost, maximum likelihood estimation, cost model

1 INTRODUCTION

A warranty is a contractual agreement between buyers and sellers in which the manufacturer or resellers usually commit to providing post-sale services concerning product failures for a discounted price or free of charge [1]. Warranty management is a key factor that influences the market performance and reputation of many manufacturers. Attractive warranty terms can usually promote product sales by indicating high quality and reliability [2], thus many companies highlight their product warranties as a marketing tool, for example, many computer hardware and electronic companies provide a free repair warranty for a period of time for their personal computers, cellphones etc. In addition, two dimensional warranty [3]-[6] and extended warranty [6]-[8] also receive wide applications for many products. However, servicing a warranty incurs additional costs to the manufacturer, thus it becomes a trade-off to balance the warranty cost and sales revenue when considering maximizing the profit. Manufacturers should consider the quality and reliability of products, customers use rates as well as sales simultaneously to design warranty policies.

Field failure data and previous warranty claims are important records to forecast warranty costs and understand products' reliability properties more precisely [4], [9]-[12]. Many previous studies focused on the analysis of warranty data, summarized in Wu [13][14]. Nevertheless, for new products that have not been put on market, it is very challenging to design warranty policies and predict warranty costs because there is no field failure data and historical warranty claim records. Murthy and Djamaludin [15] gave a literature review on new product warranty by considering

* Corresponding Author

marketing, logistics and warranty management. Huang *et al.* [16] proposed a joint optimization model that involves reliability warranty and price simultaneously for new products. However, as a key procedure to predict the number of failures during warranty period, the reliability prediction was not considered in these studies.

Accelerated life tests (ALT) could be conducted for new products to predict the reliability and then warranty cost. In Yang [17], optimal 3-level compromise ALT plans were discussed to minimize the asymptotic variance of maximum likelihood estimate of the warranty cost. Although ALT can accelerate failures, for some products, ALT cannot produce enough information of lifetime distribution because there may be no or too few failures occurring in the test. Alternatively, accelerated degradation testing (ADT) can predict the reliability by observing degradation measurements. In a constant stress accelerated degradation test (CSADT), test units are allocated under several stress levels. ADT can produce degradation information to estimate the lifetime distribution of test units. For an overview of accelerated reliability tests, one can be referred to Elsayed [18], Meeker *et al.* [19] and Meeker and Escobar [20].

By utilizing ADT data, field reliability prediction can be made. Liao and Elsayed [21] considered the stress variation in field reliability prediction by ADT data; Pan and Crispin [22] used a hierarchical method to analyze ADT data of LED lamps. In Wang *et al.* [23], a Bayesian framework was described to evaluate the reliability with both ADT data and field information.

In this paper, we propose a warranty cost analysis framework based on ADT experimental data. It is assumed that the use condition of the product is a random variable. Field reliability is predicted as a random variable as well and is used to calculate the warranty cost. The remainder of the paper is organized as follows: Section 2 presents the ADT data modeling framework and prediction of field reliability; Warranty cost model is made based on reliability prediction is described in Section 3; An illustrative example is shown in Section 4; Finally, Section 5 concludes the paper and discusses areas for future research.

2 MODELING OF ADT DATA AND PREDICTION OF FIELD RELIABILITY

For many products, the degradation process, $X(t)$, can be modeled by a continuous stochastic process [24]. Among various continuous stochastic process, Wiener process, gamma process and inverse Gaussian process are studied most intensively. In this paper, we use Wiener process as the degradation process for ADT as in Tseng and Wen [25] and Liao and Tseng [26].

2.1 Degradation and ADT Modeling

The basic Wiener process model is a drifted Brownian motion, which is given by

$$Y(t) = v\Lambda(t) + \sigma B(\Lambda(t)) \tag{1}$$

where v is the drift parameter, σ is the diffusion parameter, and $\Lambda(t)$ is a time scale function to describe the nonlinearity of the degradation path, $B(\cdot)$ is the standard Brownian motion. It is assumed that $\Lambda(t) = t$ in the following analysis. Denote L as the given failure threshold for the degradation process, then the first passage time of $X(t)$ to L, defined as T_L, follows an inverse Gaussian distribution, i.e., $T_L \sim IG(L/v, L^2/\sigma^2)$.

It is assumed that only one stress factor is considered in the ADT. S_0 is used to represent the use condition stress. Denote $S_1, \dots, S_m$ as the stress levels in the ADT, and number of test units allocated at $S_1, \dots, S_m$ are $n_1, \dots, n_m$ respectively. Let $n = \sum_{i=1}^{m} n_1$ denote the total number of test units. The following assumptions are made to model the ADT data.

1. $X(t|S_i)$ is the corresponding degradation process under stress S_i.
2. $X(t|S_i)$ follows a Wiener process with drift η_i and diffusion σ, that is

$$X(t|S_i) = \eta_i t + \sigma B(t) \tag{2}$$

3. The drift rate η_i is a function of stress level S_i based on the physical relationship. In this study, Arrhenius temperature relationship is used, thus under temperature stress S_i in Celsius scale,

$$\eta_i = exp\left(a + \frac{b}{273.15+S_i}\right) \tag{3}$$

4. The inspection interval τ_i and number of inspections o_i are pre-determined for each stress level S_i, where $i = 1, \ldots, m$.

In real engineering cases, testing time is usually limited, thus it is reasonable to pre-determine τ_i and m_i to put constraint on the total testing time. Under these assumptions, the increments of the degradation process for jth unit under stress S_i between time $(k-1)\tau_i$ and $k\tau_i$, denoted by X_{ijk} follows a normal distribution as follows,

$$X_{ijk} \sim N(\eta_i\tau_i, \sigma^2\tau_i) \tag{4}$$

Denote x_{ijk} as the observed increment data from the ADT, then the likelihood function of unknown parameters $\boldsymbol{\theta} = (a, b, \sigma^2)$ i

$$\mathcal{L}(\boldsymbol{\theta}) = \prod_{i=1}^{m}\prod_{j=1}^{n_i}\prod_{k=1}^{o_i}\frac{1}{\sqrt{2\pi\sigma^2\tau_i}}exp\left\{-\frac{[x_{ijk}-\eta_i\tau_i]^2}{2\sigma^2\tau_i}\right\} \tag{5}$$

By maximizing the likelihood (or log-likelihood) function, the maximum likelihood estimates (MLE) of $\boldsymbol{\theta}$ is obtained as $\boldsymbol{\theta} = \widehat{\boldsymbol{\theta}} = (\hat{a}, \hat{b}, \hat{\sigma}^2)$.

Recall that the lifetime under use condition T_L follows an inverse Gaussian distribution, i.e., $T_L \sim IG(L/\eta_0(S_0), L^2/\sigma^2,)$, where $\eta_0 = \exp\left(a + \frac{b}{273.15+S_0}\right)$. Given the estimator $\widehat{\boldsymbol{\theta}}$, the estimated MTTF under use condition is given by

$$\widehat{MTTF}(\hat{a}, \hat{b}, L, S_0) = \frac{L}{\widehat{\eta_0}} = \frac{L}{exp\left(\hat{a} + \frac{\hat{b}}{273.15+S_0}\right)} \tag{6}$$

2.2 Prediction of Field Reliability

In standard methods to predict product reliability based on ADT, the stress level under use condition is assumed to be fixed. However, many products are used under different stress levels due to the variety of customers and using environments. For example, LED lamps are widely used outdoors, where the temperature may vary significantly in different regions. The reliability prediction errors may be considerable by simply assuming the stress level is fixed. Therefore, in this study, S_0 is regarded as a continuous random variable with distribution function $F_{S_0}(s_0)$ and density function $f_{S_0}(s_0)$. Under this assumption the predicted MTTF is also a random variable with respect to S_0, for which the density function is given by

$$f_{\widehat{MTTF}(\hat{a},\hat{b},L,s)}(t) = f_{S_0}(g(t)) \tag{7}$$

Based on the density of predicted MTTF, the mean is given by

$$E_s\left[\widehat{MTTF}(\hat{a}, \hat{b}, L, s)\right] = \int_0^{\infty} f_{S_0}(g(t))t\,dt \tag{8}$$

Then the distribution function of lifetime T_L in terms of given $\boldsymbol{\theta}$ and use condition s is as follows,

$$F_{T_L}(t; \boldsymbol{\theta}, s_0) = \Phi\left[\sqrt{\frac{1}{\sigma^2 t}}(\eta_0(s_0)t - L)\right] + exp\left\{\frac{2\eta_0 L}{\sigma^2}\right\}\Phi\left[-\sqrt{\frac{1}{\sigma^2 t}}(\eta_0(s_0)t + L)\right] \tag{9}$$

with density function

$$f_{T_L}(t; \boldsymbol{\theta}, s_0) = \frac{L}{\sigma}\left[\frac{1}{2\pi t^3}\right]^{1/2} exp\left\{-\frac{(\eta_0(s)t-L)^2}{2\sigma^2 t}\right\} \tag{10}$$

3 A WARRANTY COST MODEL

A free-rectification warranty of period t_W is assumed to be attached to the products that we are interested, and the type rectification is replacement by a new product, i.e., a product with initial degradation level. Based on the lifetime distribution of which parameters are estimated by their MLE, the expected number of failures over $[0, t)$, $M(t)$, with given $\boldsymbol{\theta}$ and s_0 is

$$M(t;\boldsymbol{\theta},s_0) = F_{T_L}(t;\boldsymbol{\theta},s_0) + \int_0^t M(t-x;\boldsymbol{\theta},s_0) f_{T_L}(t;\boldsymbol{\theta},s_0)\,dx \quad (11)$$

Thus the expected number of failures over $[0, t)$ can be obtained by

$$M(t;\boldsymbol{\theta}) = E[M(t;\boldsymbol{\theta},s_0)] = \int_0^\infty M(t;\boldsymbol{\theta},s_0) f_{S_0}(s_0)\,ds_0 \quad (12)$$

Denote c_R as the expected cost of free replacement from the warranty for each claim, then the total expected warranty cost per unit

$$C_W = c_R M(t_W;\boldsymbol{\theta}) \quad (13)$$

We are also interested in the distribution of the expected warranty cost with respect to use conditions, thus we can derive the manufacturer's risk of warranty cost. For a given t_W, the density function of the C_W is given by

$$f_{C_W}(x) = f\left(M^{-1}(x/c_R)\right) \quad (14)$$

where $M^{-1}(x)$ is the inverse function of $M(t;\boldsymbol{\theta},s_0)$ with respect to s_0, i.e., $M^{-1}\left(M(t;\boldsymbol{\theta},x)\right) = x$. The corresponding cumulative density function is denoted as $F_M(x)$. Afterwards we can define the manufacturer's risk ρ as in Sahin and Polatoglu [27],

$$\rho = P[C_W > c_U] = \int_{C_U}^\infty f_{C_W}(x)dx = 1 - \int_0^{C_U} f_{C_W}(x)dx \quad (15)$$

4 ILLUSTRATIVE EXAMPLE

In this section we use a simulation example to illustrate our methods. We assume that the degradation process of a certain product is assumed well modeled by a Wiener process and the temperature is the only stress that influences the degradation process significantly. Other assumptions in the simulation work are given as follows,

1. The true parameter vector $\boldsymbol{\theta} = (a, b, \sigma^2) = (5.4, -2550, 0.00082)$, and they are unknown throughout the test and prediction, thus it can only be estimated by ADT data.
2. The failure threshold $L = 1$, and the initial degradation level for new product is assumed to be 0.
3. The test plan for ADT is $(S_1, S_2, S_3) = (45, 70, 105)$, $(n_1, n_2, n_3) = (40, 20, 10)$ and $o_i = 5, \tau_i = 0.5$ for $i = 1, 2, 3$. In other words, there are 70 test units in total, of which 40, 20, 10 of them are allocated to temperature level 40°C, 40°C and 40°C respectively; the measurement interval is 0.5 and for each test unit 5 measurements are taken, therefore the total test duration is 2.5 time units.
4. The stress under use condition S_0 across a large number of customers is normally distributed with mean 25°C and standard deviation 10°C, i.e., $S_0 \sim N(25, 100)$, and it is assumed that the distribution is known and true.
5. The warranty period is 40 time units, i.e., $t_W = 40$, and the cost for each claim is $c_W = 1$.

The ADT data is simulated from the true parameters and the based on the data, the MLE is obtained as $\hat{\boldsymbol{\theta}} = \left(\hat{a}, \hat{b}, \hat{\sigma}^2\right) = (5.3957, -2448.3, 0.0008182)$. By substituting $\boldsymbol{\theta}$ by $\hat{\boldsymbol{\theta}}$ in Eq. (11), the function of $M(t;\boldsymbol{\theta},s_0)$ with respect to s_0 can be obtained numerically [28], [29]. Figure 1 shows the relationship between the expected number of failures and stress level in the example.

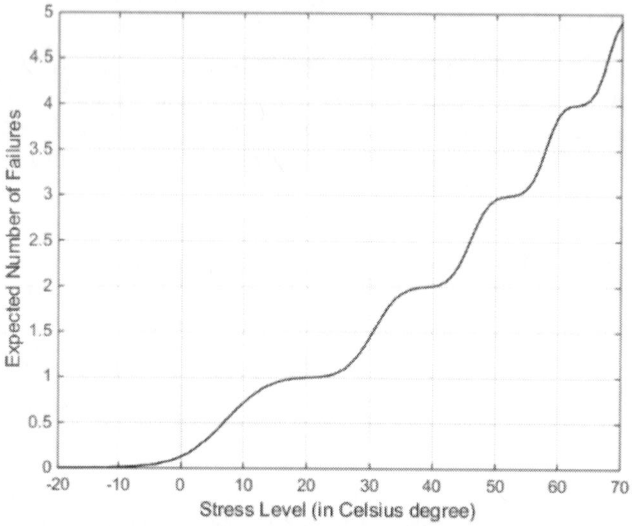

Figure 1: The expected number of failures vs. stress level

As is shown in the figure, the expected number of failures increases non-linearly when temperature stress becomes higher. There is an apparent "wavy" pattern in the plot and the distance between waves becomes smaller when temperature gets higher. By estimating C_W as in Eq. (13) by using the MLE $\widehat{\boldsymbol{\theta}}$, the distribution of C_W is obtained based on Eq. (14). The density plot of C_W is shown in Figure 2.

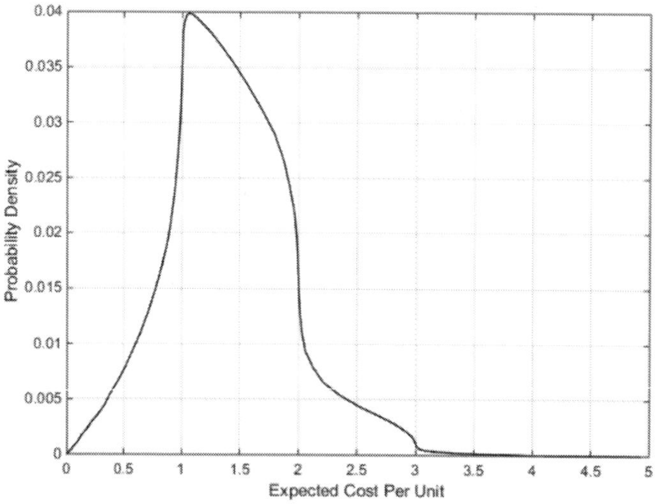

Figure 2: Density plot of C_W

The density plot shows a significant difference shape compared with normal distribution, and the plot is dissymmetric with a pointed mode instead of bell shaped. Based on the distribution of C_W, the expected value of C_W, $E(C_W)$ and the warranty risk can be obtained. In this example, we set the upper limit of the warranty cost $C_U = 3$ to compute the risk. Therefore, by numerically integrating, we can get $E(C_W) = 1.2778$ and $\rho = 0.0030$.

To compare the proposed methods with previous method, we simulate 10,000 Wiener processes based on the true parameter assumptions and use condition distribution, then calculate the mean value of warranty cost as well as the warranty risk. In addition, by assuming the use condition is fixed at 25°C, which is the mean of the use condition distribution, the prediction of warranty cost can be also computed, the results are summarized in Table 1.

Table 1: Comparison of warranty cost and risk prediction

	Proposed Method	Prediction based on fixed S_0	Simulated data
$E(C_W)$	1.2778	1.0597	1.2721
Error (compared with simulated data)	+4.5%	-16.7%	0
Warranty risk ρ	0.0030	Not applicable	0.0035

The result shows that the proposed method yields a warranty cost prediction based on ADT data with error rate 4.5% compared with the simulated results from true parameters. However, if the use condition is assumed to be fixed, the error rate becomes 16.7%, which is significantly larger. The comparison supports the advice to model the use condition as a distribution instead of a fixed value, because even when we set the use condition to the mean value, the warranty cost prediction has a significant error. In addition, the warranty risk with respect to the randomness of use condition can be also estimated from the proposed method.

5 CONCLUSION AND AREAS FOR FUTURE RESEARCH

This paper presents a method to predict warranty cost by utilizing data from an accelerated degradation test (ADT). The accelerated lifetime model parameters are estimated by MLE from ADT data, then a warranty cost distribution is derived due to the variability in use condition. A simulation study is carried out to illustrate the proposed framework and shows the advantages of predicting warranty cost by consider the randomness in use condition.

In future studies, more sources of variability can be considered, such as the precision of MLE estimates from ADT data, the diffusion properties in degradation processes that influence the randomness of number of failures during the warranty period, thus the model could be more general and practical, especially it is used predict the warranty risk due to several sources of variability. In addition, sensitivity analysis of such works and ADT optimal planning to predict warranty cost is also of interest to be explored.

6 REFERENCES

[1] **Blischke, W. 1995.** *Product warranty handbook*. CRC Press.

[2] **Murthy, D.N.P. and Blischke, W.R. 2006.** *Warranty management and product manufacture*. Springer Science & Business Media.

[3] **Jack, N., Iskandar, B.P., and Murthy, D.N.P. 2009.** A repair-replace strategy based on usage rate for items sold with a two-dimensional warranty. *Reliability Engineering & System Safety*, 94 (2), 611-617.

[4] **Gupta, S.K., De, S., and Chatterjee, A. 2014.** Warranty forecasting from incomplete two-dimensional warranty data. *Reliability Engineering & System Safety*, 126, 1-13.

[5] **Huang, Y.S., Gau, W.Y., and Ho, J.W. 2015.** Cost analysis of two-dimensional warranty for products with periodic preventive maintenance. *Reliability Engineering and System Safety*, 134, 51-58.

[6] **Huang, Y.-S., Huang, C.-D., and Ho, J.-W. 2016.** A Customized Two-Dimensional Extended Warranty with Preventive Maintenance. *European Journal of Operational Research*.

[7] **Bouguerra, S., Chelbi, A., and Rezg, N. 2012.** A decision model for adopting an extended warranty under different maintenance policies. *International Journal of Production Economics*, 135 (2), 840-849.

[8] **Ye, Z.-S. and Murthy, D.N.P. 2016.** Warranty Menu Design for a Two-Dimensional Warranty. *Reliability Engineering & System Safety*, 155, 21-29.

[9] **Hong, Y. and Meeker, W.Q. 2010.** Field-Failure and Warranty Prediction Based on Auxiliary Use-Rate Information. *Technometrics*, 52 (2), 148-159.

[10] **Majeske, K.D. 2007.** A non-homogeneous Poisson process predictive model for automobile warranty claims. *Reliability Engineering & System Safety*, 92 (2), 243-251.

[11] **Limon, S., Yadav, O., and Nepal, B. 2015.** Forecasting Warranty Claim by Using Past Warranty and Field Usage Data. In: Proceedings of the 2015 Industrial and Systems Engineering Research *Conference*. 1627-1636.

[12] **Yang, D., He, Z., and He, S. 2016.** Warranty claims forecasting based on a general imperfect repair model considering usage rate. *Reliability Engineering & System Safety*, 145, 147-154.

[13] **Wu, S., 2012.** Warranty data analysis: A review. *Quality and Reliability Engineering International*, 28 (8), 795-805.

[14] **Wu, S., 2013.** A review on coarse warranty data and analysis. *Reliability Engineering & System Safety*, 114 (1), 1-11.

[15] **Murthy, D.N.P. and Djamaludin, I. 2002.** New product warranty: A literature review. *International Journal of Production Economics*, 79 (3), 231-260.

[16] **Huang, H.-Z., Liu, Z.-J., and Murthy, D.N.P. 2007.** Optimal reliability, warranty and price for new products. *IIE Transactions*, 39 (8), 819-827.

[17] **Yang, G. 2010.** Accelerated Life Test Plans for Predicting Warranty Cost. *IEEE Transactions on Reliability*, 59 (4), 628-634.

[18] **Elsayed, E.A. 2012.** Overview of Reliability Testing. *IEEE Transactions on Reliability*, 61 (2), 282-291.

[19] **Meeker, W.Q., Sarakakis, G., Gerokostopoulos, A., Doganaksoy, N., Hall, D.B., Hahn, G.J., Kugler, D.L., Lawless, J., Pan, R., Trindade, D.C., and others 2013.** More pitfalls of accelerated tests. *Journal of Quality Technology*, 45 (3), 213.

[20] **Meeker, W.Q. and Escobar, L.A. 2015.** A review of recent research and current issues in AT. *International Statistical Review*, 61 (1), 147-168.

[21] **Liao, H. and Elsayed, E.A. 2006.** Reliability inference for field conditions from accelerated degradation testing. *Naval Research Logistics*, 53 (6), 576-587.

[22] **Pan, R. and Crispin, T. 2011.** A hierarchical modeling approach to accelerated degradation testing data analysis: A case study. *Quality and Reliability Engineering International*, 27 (2), 229-237.

[23] **Wang, L., Pan, R., Li, X., and Jiang, T. 2013.** A Bayesian reliability evaluation method with integrated accelerated degradation testing and field information. *Reliability Engineering & System Safety*, 112, 38-47.

[24] **Lu, C.J. and Meeker, W.Q. 1993.** Using Degradation Measures to Estimate a Time-to-Failure Distribution. *Technometrics*, 35 (2), 161.

[25] **Tseng, S. and Wen, Z., 2000.** Step-stress accelerated degradation analysis for highly reliable products. *Journal of Quality Technology*, 32 (3), 209-216.

[26] **Liao, C.M. and Tseng, S.T. 2006.** Optimal design for step-stress accelerated degradation tests. *IEEE Transactions on Reliability*, 55 (1), 59-66.

[27] **Sahin, I. and Polatoglu, H. 2012.** *Quality, warranty and preventive maintenance*. Springer Science & Business Media

[28] **Xie, M. 1989.** On the solution of renewal-type integral equations. *Communications in Statistics-Simulation and Computation*, 18 (1), 281-293.

[29] **Tortorella, M. 2005.** Numerical Solutions of Renewal-Type Integral Equations. *INFORMS Journal on Computing*, 17 (1), 66-74.

STUDY ON THE INFLUENCING FACTORS OF TECHNICAL EFFICIENCY OF MAIZE

D.M. Ni, R. H.Liu*[1]
School of Economics and Management
Beijing University of Agriculture,China
ndmty@163.com
ruihanliu@163.com

ABSTRACT

Based on the costs and benefits of maize production in Beijing and other business to make a research, we get input and output data from 2013 to 2014. Through the results obtained from Frontier4.1 software fitting stochastic frontier production function estimation, finding thatBeijing technical efficiency of maize production is generally higher and summer maizeproduction is slightly higher than the spring one. In production inputs, increasing investment in machinery makethe highest contribution rate to maize outputs. Finally, we provide a way to boost maize output efficiency to help producers improve the efficiency factors.

Key words: MaizeproductionCost-benefitScale operation

1.INTRODUCTION

Maize is the largest crop planting in Beijing, acreage, yields and total production rank first in crops and food production is a top priority.Maize plays an important role in urban modern agriculture in Beijing,and in addition to the production function,maize also has demonstration capabilities in life, ecology and technology fields.Beijing suburbs maize production hasexperiencedthree stages of development,thesteady growth of grain production,the yields fluctuate caused by crop structure adjustment and the modern agriculture, in the meantime through the integration of agricultural mechanization and agronomic technology,farming reforms , new varieties and cultivation techniques research and extension,improving land productivity,economic development and structural adjustment of the market after that Beijing embarked on a high-yield maize production, special, ecological and industrial management of the road, with a total production stabilized at a certain level, the qualityhas made a significant contribution in making farmers to get rich and improving the ecological environment .

Beijing maize acreage has experienced an increasing trend to decline from 2004 to 2014, after reaching a peak of 226 acres in 2009, it decreased year by year until133 acres in 2014, more than 22.7% decrease in 2013 of 172 acres.Maize acreage accounted 90.5% (counties) of the total area of Yanqing, MiyunTongzhou, Fangshan, Daxing, Shunyi District. Where 96.7% of the total area of ordinary maize, the sowing area proportion of Beijing spring maize and summer maizeisabout 2: 1, forage maize and fresh maize is less than 5 acres in total. Fresh maizeand waxy maize accounted for 3/4.

With the national emphasis on food security and the adjustment of implementation and consumption structure of food subsidiespolicy,Beijing suburbs

[1]* Corresponding author.

maize production has been into the high-quality, efficient, safe and eco targeted urban modern agriculture development stage.Final maize production problems is to increase maize production capacity, make an increase of the output level of maize.Many factors affect maize yields growth probably can be summarized into three categories, the first is inputs, the second is technological progress, the third is technical efficiency.In reality, the pressure in scarcity of economic resources and ecological environment determines the increase of maize production can not be depended on the factor inputs with unlimited expansion, in addition, general technical progress is difficult to achieve a breakthroughin a short time, therefore, the study ofmaize crop technology efficiency and influencing factors for improving the level of maize production in Beijing has important significance.

2.LITERATURE REVIEW

Maize production efficiency has been many scholars'research focus. By region, the country has to carry out research for the object, but also to a certain province as the object of analysis. In research methods, mainly using stochastic frontier production function approach and data envelopment analysis.

using data envelopment analysis method to research maize production has a lot of meaningful results.Ding Yanused DEA analysis and Malmquist Index Liaoning and Jilin provinces maize production was estimated TFP,we found a change in the provinces maize TFP increase or decrease depending on technical efficiency.Su Guihong using the two-step method DEA- Tobit measure the technical production efficiency inChinese main maize producing areas , the results show the scale maize production is mostly ineffective, and generally decreasing returns to scale will be in stages, the increasing of maize production TFPmain dues to technical advances faster than the rate of decline in technical efficiency.Zhang Honganalysis about maize production total factor input and output efficiency based on DEA and Malmquist Index Analysis in Jilin Province, found that the land productivity and labor productivity arehigher,the capital productivity islower,and inter-annual fluctuation is greater, but the overall shows downward trend, technological progress contribution rate isonly 52.6%, compared with developed countries (70% to 80%) , there is a large gap.

Stochastic frontier production function approach is another important way to analyze the technical efficiency of maize production.GuoZhichaoused ultra-logarithmic stochastic frontier production function, found that China's maize production average technical efficiency is 91.12%, there is no significant loss of efficiency.Zhao Guiyuuesd the stochastic frontier production function approach, found that the next steps should be taken to promote technological progress, and tap the potential of the main producing areas to promote the main producing areas of maize TFP growth.Wang et al. Using the stochastic frontier analysis, found that total factor productivity has become the most important source of power in the core areas of maize yield growth, contributing 69.3 percent of yield growth, technological progress and technical efficiency improve the contribution of 60.7%.In addition, Yang Guoqing based on 15 main maize producing provinces agricultural Costs and Returns Panel Data between 2001 and 2010,using stochastic frontier production function model and the DEA Malmquist index analysis method,from the view of time gradient and geographical differences,analyzed the efficiency of maize production in the main producing areas of maize (TFP) growth conditions.From 2001 to 2010,the main producing areas of maize TFP growth rate is -1.3%, and the growth trend showing a jagged profile, full TFP progress with the same trend of technological change, technological progress is a key factor in promoting China's main maize producing areas TFP growth.according to China's 15 main maize producing provinces of maize production costs panel data from 2006 to 2011, use of stochastic frontier production function technical efficiency of maize production, Goh Kai got theestimated results showed that: 15 main maize producing provinces of maize production the average technical efficiency of about 78.5%, labor costs and other factors into existence adversely affect technical efficiency of maize production, while maize prices, irrigation and other factors have a positive impact.Shi Junqing improved and achieved a balanced development of maize and wheat production technical efficiency were the main factors affecting technical efficiency of maize and wheat production from the overall situation of sub-regional and sub-annual perspective of the specific circumstances of the two to do a thorough analysis of its while on maize and wheat technical efficiency factors were analyzed and compared to find factors of technical efficiency of maize and wheat have a common direction of action and found that the direction of action of different factors,on this basis, has put forward in the future technology to improve the efficiency of maize and wheat production methods and measures.

The research for this study provide a useful reference, but there has little relativeresearchfor Beijing maize producing areas . in recent years, the Ministry of Agriculture are no longerpublish the statistical data compilationof Beijing food costs and benefits , the data related to the application of the Institute are from Beijing for the Beijing main maize producing areas of field research, using stochastic frontier analysis (Stochastic Frontier Analyze, SFA) and data envelopment analysis (Data EnvelopeAnalyzing, DEA) estimates maize production technology efficiency, and comparative analysis between the countries in order to provide policy reference for improving Beijing maize production capacity.

3.THE MATERIAL AND SERVICE FEE STRUCTURE OF BEIJING MAIZE PRODUCTION

3.1 SUMMER MAIZE

summer maize producersin Different districts have different sizes, machinery costs, fertilizer costs and seed costs account a dominant position in material elements inputs.

(1)More than 50 acres medium and large-scale producers of summer maize elements input costs sort as mechanical fee> fertilizer costs> seed fee. Other material costs remaining smaller proportion.

(2)Less than 50 acres small-scale producers of summer maize elements input costs sort as fertilizer costs> mechanical fee > seed fee. Other material costs remaining smaller proportion.

(3)Summer maize farmers of all sizes, the average cost of factor inputs sort, except higher fertilizer costs Fangshan machinery costs, the remaining regions are fertilizer costs> machinery costs> seed fee. Other material costs smaller proportion.

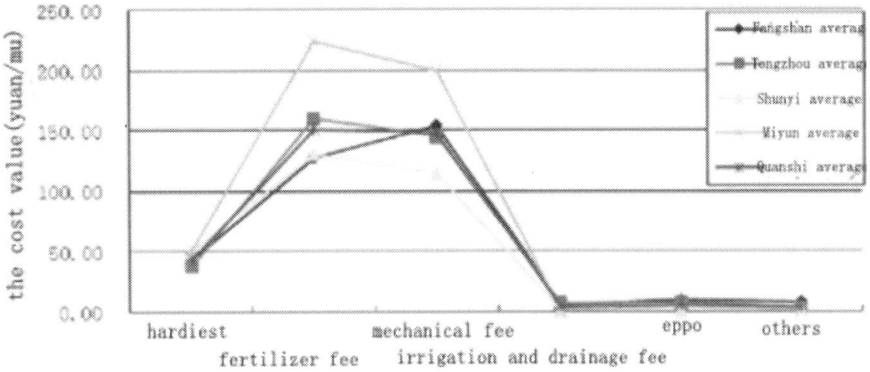

Fig.1the materialcost for summer maize producers inDifferent areas

Source: Based on the calculating results of field research project.

3.2 SPRING MAIZE

spring maize production in Different districts , different sizes, machinery costs, fertilizer costs and seed costs account a dominant position in material elements inputs and service costs.

(1)More than 50 acres medium and large-scale producers of spring maize elements input costs sort as mechanical fee> fertilizer costs> seed fee,and mechanical fee is higher than fertilizer costs apparently.Other material costs remaining smaller proportion (Figure 2).

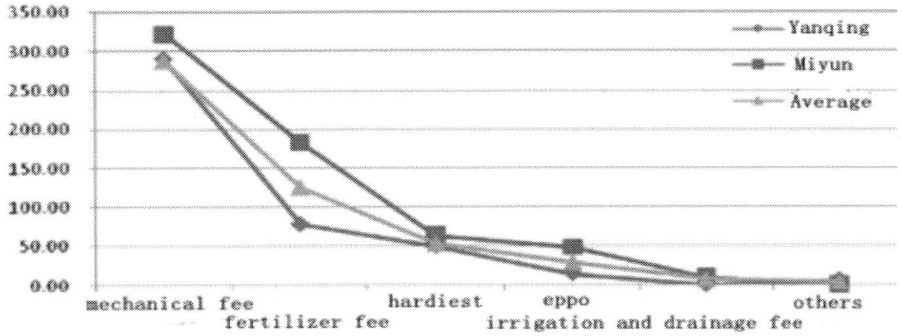

Fig. 2 2014-2015 >50 mu spring maize producers to pay the material and service

Source: Based on the calculating results of field research project.

(2)Less than 50 acres small-scale producers of spring maize elements input costs sort as mechanical fee > fertilizer costs > seed fee,and the using horizontal difference between machinery costs and fertilizer costs is not obvious. Other material costs remaining smaller proportion (Figure3).

Whether it is summer or spring maize production,in different districts and different production scale, the maize producers of machinery cost inputs, fertilizer and seed costs account for their fees and service charges substance into a dominant position.

Fully visible, in maize production, material and service charges invest in machinery, fertilizers and seeds accounted for the major proportion. Therefore, quantitative analysis below, material inputs usemachinery costs, fertilizer costs and seed costs, labor elements usemaize production labor costs (for comparison, the level of $ 100 man-days have been calculated)

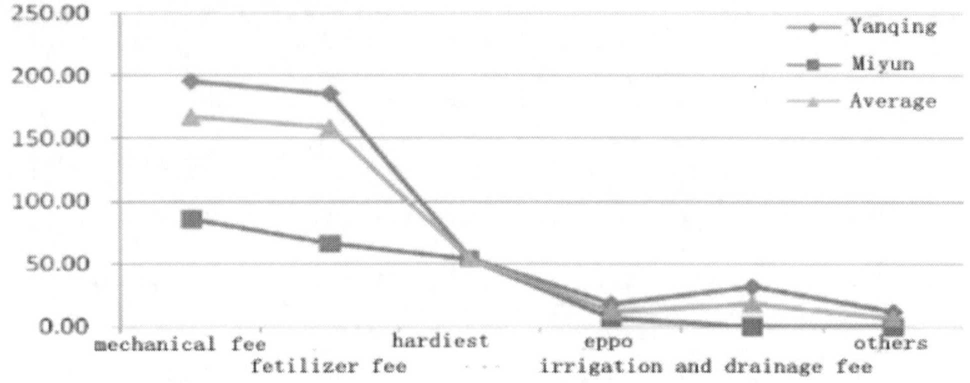

Fig.3 2014-2015 <50mu spring maize producers to pay the material and service
Source: Based on the calculating results of field research project.

4.MODEL

Technical efficiency frontier function was measured first by Farrell,the initial definition is based on its investment of resources,after the continuous improvement and development, there are two methods to determine the type ,parameters and non-parametric. Deterministic parameter method originally developed by Aigner and Chu propose,after a Timmer improved, raised the possibility of frontier production model, Aigner presented statistical production frontier deterministic parameter method.Thinking the disadvantage that type of determine parameters exist,Aigner, Lovell and Schmidt and Meeusen and Van Den Broeck also proposed to use a random-type parameter method to measure technical efficiency,stochastic parameter method is also known as random frontier analysis method (Stochastic Frontier Analysis, SFA).Thereafter,the stochastic frontier analysis method has been constantly improved,at present,the most commonstochastic frontier model been used was proposedby Battese and Coelli (1992,1995) ,its use panel data to estimate the production frontier.

Stochastic frontier modelproposed by Battese and Coelli used in this study , based on the CD production function select commonly expressed, in order to calculate the maize planting techniques efficiency of different producers.Specific expression of the model is:

$$Ln(Y_i) = \beta_0 + \beta_1 Ln(X_{1i}) + \beta_2 Ln(X_{2i}) + \beta_3 Ln(X_{3i}) + \beta_4 Ln(X_{4i}) + (V_i + U_i)$$

Y_i represents the i-th household acre acreage output, X_{1i} represents farmers invested seed costs per acre, X_{2i} represents fertilizer inputs per acre fee, X_{3i} represents labor costs per acre investment, X_{4i} represents machinery costs per acre investment,units are yuan / mu. $\beta_0 \sim \beta_{14}$ represent parameters to be estimated, V_i represents random variables, normal distribution $N(0,\sigma_v^2)$, for measure error and a variety of uncontrollable random factors,and independent non-efficiency variable U_i, U_i obeys the half-normal distribution,in estimating the two variables are error terms, the value can not be determined.

5.ANALYSIS OF FINDINGS

The data derived from field research in Beijing Miyun, Yanqing, Shunyi, Fangshan and Tongzhou and other suburban counties,due to time and energy constraints, data is only include 2014 to the first half of 2015. By Frontier4.1 software fitting the stochastic frontier production function to obtain the estimation results.

5.1SPRING MAIZE PRODUCTION TECHNICAL EFFICIENCY ANALYSIS

Spring maize (grain) production data mainly from Beijing Miyun and Yanqing suburban counties field research,using the 2013--2014 years of input and output data,the sample size is 154. By Frontier4.1 software fitting the stochastic frontier production function,the variables estimation results obtained are 6.39019, -0.06142, 0.00013, -0.06966, 0.19325, 0.03819, and 0.00011.Technical invalid interference term T-statistic is 0.00001 γ, through the significant test below the level of 1%, and the coefficient is close to zero,indicating the output gap between actual and theory is not obviously.Technical efficiency of 0.998552.

From the average elastic elements,the most one is the cost of the mechanical output elasticity for 0.19325,indicating growth of spring maize (grain) per unit of output increased need to rely on a large part of the cost of investment in machinery,for the every per unitthe cost increase, spring maize (grains) output will increase 0.19325 units.Fertilizers, seeds and elastic modulus from artificial inputs are small or negative:for every 1% increase in investment costs of fertilizer, spring maize (grain) output increased by only 0.00013%, indicating that the elastic contribution fertilizer on spring maize (grain) Income already little;seeds and artificial inputs to increase the contribution of spring maize output increases even in the negative direction.Visible output efficiency of fertilizer, seed and labor invested are close to saturation, continue to increase such investment will not contribute much to income.From the view of comparing two main maize producing areas Yanqing and Miyun,the technical efficiency of producers planted spring maize in Miyun (0.99899)is more than Yanqing (0.99898).Overall,the technical efficiency of Beijing spring maize planting is higher than average level.

5.2 SUMMER MAIZE(GRAIN) PRODUCTION TECHNICAL EFFICIENCY ANALYSIS

Summer maize (grain) production data mainly from Beijing Fangshan,Shunyi,Tongzhou and Daxing suburban counties field research, using the 2013--2014 years input and output data,the sample size is 104. By Frontier4.1 software fitting the stochastic frontier production function,the variables estimation results obtained are 5.92061, 0.04485, -0.13324, -0.04615, 0.24280, 0.02792, and 0.00009.Technical invalid interference term T-statistic is 0.00002 γ, through the significant test below the level of 1%, and the coefficient is close to zero,indicating the output gap between actual and theory is not obviously.Technical efficiency of 0.99914 is slightly higher than spring.

From the average elastic elements,the most one is the cost of the mechanical output elasticity for 0.24280,indicating growth of summer maize (grain) per unit of output increased need to rely on a large part of the cost of investment in machinery,for theevery per unit the cost increase, summer maize (grains) output will increase 0.24%.Fertilizers, seeds and elastic modulus from artificial inputs are small or negative,fertilizers and artificial inputs to increase the contribution of summer maize output increases even in the negative direction.Visible output efficiency of fertilizer and labor invested are close to saturation, continue to increase such investment will not contributed much to income.This is consistent with the DEA estimates of fertilizer and labor input redundancy of higher rate.

Technical efficiency comparison of different production scale, found more than 50 acres of cultivated area in mass production households, their technical efficiency (0.99945)is higher than the technical efficiency of 50 acres or less for small-scale retail (0.99892).Description scale cultivation of maize has a high technical efficiency.

6.CONCLUSIONS AND IMPLICATIONS

The paper based on the typical model household in 2014 and 2015 in Beijing Miyun, Yanqing, Shunyi, Fangshan and Tongzhou district survey and sampling survey of conventional maize farmers produce cost-benefit data,using stochastic frontier analysis of different types of maize products production technology efficiency calculations,analyze the contribution of technical efficiency of its products and the different elements of different inputs to output. The main conclusions are:

(1)Overall Beijing maize is higher in efficiency of the production technology, and summer maize is slightly higher than the spring. Constantly improve the reliability of the data is essential for accurate measurement of efficiency and output elements of the output elasticity.

(2)Production inputs, the increase in investment in machinery maize output increment highest contribution rate.Thus,to promote large-scale production and machinery and other elements of an alternative labor ratio optimization strategy implementation is an important path to maize income.

(3)Helping producers improve the efficiency factors using is an important way to promote scientific and technological workers maize output efficiency.

Output elasticity due to fertilizer, seed and labor and other factors of the smaller or even negative factors, indicating that their contribution to output incremental relative saturation are even counterproductive. Therefore, an appropriate reduction in seed and labor input, through science and technology experiments conducted fertilizer, reducing fertilizer use scientific or optimize the ratio of the amount of different fertilizers and optimization of fertilization time, these are important parts of improving the efficiency of maize output.

(4)Study limitations: The study was limited to microscopic internal analysis, not involving the macro external environment.

ACKNOWLEDGE

The authors wish to acknowledge financial support from Beijing Agriculture Innovation Consortium of Grain Industry(BAIC09-2016) and DaBeiNong young teachers' Scientific Research Fund Project(15SK002) .

REFERENCE

[1] Ding Y.,Zhai Y.L.,Zhou Y.B., Fan Q.A..Comparative study of two maize Liaoji TFP- a study based on the Morse inde[J],BusinessResearch.2008(12):181-182.

[2] Zhang H., Wang Z.H.,Jiang H.M..Analysis of the input and output efficiency of Maize Production -- An Empirical Study Based on the production of Maize in Jilin Province,Journal of Jilin Agricultural University.2011, 33(6):698- 702.

[3] Guo Z.C..Analysis of maize production function and technical efficiency in China[J].Journal of Economic Issues. 2009,(11):74- 78.

[4] Zhao G Y,Zhang Y J,Ge S P. Study on maize production efficiency in Heilongjiang province[J].Journal of Maize Sciences,2009,17(4):139- 143.

[5] WangJ.,Xu X.H.,Wang H.L.,Yao F.T..Growth and decomposition analysis of maize production efficiency in the core of China[J],Journal of Maize Sciences,2010,18(6):133-137, 142.

[6] Yang G.Q..China'smaize production efficiency evaluation since entering WTO: panel data analysis from 15 provinces in China[J].Guangdong Agricultural Sciences,2010,18(6):133-137, 142.

[7] Gao J.G..Research on technical efficiency of maize production in 15 main producing provinces of China[J].Western forum,2013, 6 (23) ,69-75.

[8] Shi J.Q..Analysis on technical efficiency of main grain crops in China.Chinese Academy of Agricultural Sciences.Doctoral dissertation,2010.

[9] Kang X.,Liu X. M.. Analysis of technical efficiency of grain production in China based on stochastic frontier analysis method[J],China Rural Survey2005(04):25-32.

[10] Ma J. L..The evolution, driving forces and efficiency of farm size of maizeplanting households in Hebei[D]. China Agriculture University,2015.

A HYBRID INTELLIGENT APPROACH FOR INTEGRATED PRODUCTION SCHEDULING AND VEHICLE ROUTING WITH TIME WINDOWS

Qu Wei, Qi Liu, Zhaoxia Guo*
Business School, Sichuan University, Chengdu 610065, PR China
(zx.guo@alumni.polyu.edu.hk)

ABSTRACT

This paper addresses the problem of integrated production scheduling and vehicle routing with time windows (IPS-VRTW). A novel hybrid intelligent approach is proposed to solve this problem, by which the investigated problem is decomposed into 3 sub-problems namely, order assignment problem, order scheduling problem and traveling salesman problem with time windows. This approach is implemented by integrating two genetic optimization process with the longest processing time heuristic to solve the 3 sub-problems integratedly. A series of experiments are conducted and validate the effectiveness of the proposed approach.

Keywords: Production scheduling; Vehicle routing with time windows; Genetic algorithm; Hybrid algorithm

1 INTRODUCTION

Research on integrated production and distribution scheduling (IPDS) problems have attracted more and more researchers' attention in recent years. This paper deals with an integrated production and distribution scheduling problem considering vehicle routing with time windows, called integrated production scheduling and vehicle routing with time windows (IPS-VRTW).

Research on IPDS can be traced back to 1980's [1]. Toptal et al. [2] consider an IPDS with heterogeneous vehicles. The manufacturer can use two types of vehicle to deliver goods. Seyedhosseini and Ghoreyshi [3] study an IPDS model for perishable products consisting of a production facility and multiple distribution centers. However, studies on IPS-VRTW are very limited so far. Ullrich [4] consider an IPS-VRTW problem with the objective of minimizing the total tardiness. This problem consists of two sub-problems, including a machine scheduling sub-problem and a distribution scheduling sub-problem. He constructed a genetic algorithm to solve the problem. However, their method is very time-consuming because of the high complexity of IPS-VRTW problems.

This research aims to propose a more efficient methodology to solve the IPS-VRTW problem. We present first the mathematical model of the investigated IPS-VRTW problem. To solve this problem a

* Corresponding Author

novel hybrid intelligent approach is then proposed, by which we decompose the problem into 3 sub-problems, including an assignment sub-problem of orders to vehicles, an order scheduling sub-problem with the objective of minimization makespan, and a sub-problem of traveling salesman problem with time windows (TSPTW).

2 PROBLEM STATEMENT

2.1 Problem description

In a make-to-order supply chain, the supplier has a set of identical parallel machines to produce customer orders. Orders cannot be splitted and each order must be continuously produced on one machine. After production, orders are delivered to customers located at different regions by homogeneous capacitated vehicles. Customers have their own time windows to accept orders. If an order is not delivered within the predefined time window, the supplier will incur a penalty.

A supplier has to find a joint production and distribution schedule, by deciding the optimal assignment of orders to machines, the optimal production sequence of orders, and the optimal vehicle routes to deliver the orders to customers. The objective is to minimize the total cost of production and distribution operations.

2.2 Notations

For easy reference, notations that will be used in this paper are listed as follows. Customer (order), machine and vehicle are indexed by i , m and k respectively. Let p_i denote the production processing time of order i, q_i the size of order i. Let e_i and l_i denote the starting and ending time of time window at customer i respectively. c_{ij} denotes the travel time between customers i and j, g the penalty parameter of violation time windows, α the relative weight placed upon the objective of total transportation cost, and β the relative weight placed upon the objective of violation penalty of time windows. The values of 3 decision variables need to be determined, which are o_{mi} , t_{ki} and x_{ijk}. t_{ki} is 1 if order i is delivered by vehicle k , otherwise it is 0. o_{mi} is 1 if order i is produced on machine m, otherwise it is 0. x_{ijk} is 1 if arc(i , j) is a part of vehicle k 's route, otherwise it is 0.

There are some intermediate variables as follows: a_i denotes the time arriving at customer i , et_{km} is the end time of producing the last order delivered by vehicle k on machine m , st_{km} is the start time of producing the first order delivered by vehicle k on machine m , and pt_{km} is the total production time of orders delivered by vehicle k on machine m. TD_k is the total delivery cost of vehicle k . TW_i is the violation penalty of time windows of order i and vst_k denotes the departure time of vehicle k

2.3 Mathematical model

$$Min\ Z = \alpha \sum_{k=1}^{K} TD_k + \beta \sum_{i=1}^{N} TW_i \tag{1}$$

Subject to:

$$\sum_j \sum_k x_{ijk} = 1, \qquad i = 1,...,n \tag{2}$$

$$\sum_i \sum_k x_{ijk} = 1, \qquad j = 1,...,n \tag{3}$$

$$\sum_i x_{ihk} - \sum_j x_{hjk} = 0 \quad h = 1,...,n \;\; k = 1,...,K \tag{4}$$

$$\sum_j x_{0jk} = 1 \qquad k = 1,...,K \tag{5}$$

$$st_{k+1,m} = et_{km}, \quad k = 1,...,K-1 \;\; m = 1,...,M \tag{6}$$

$$st_{1,m} = 0, \qquad m = 1,...,M \tag{7}$$

$$et_{km} = st_{km} + pt_{km} \quad k = 1,...,K \;\; m = 1,...,M \tag{8}$$

$$pt_{km} = \sum_i p_i \cdot t_{ki} \cdot o_{mi} \quad k = 1,...,K \;\; m = 1,...,M \tag{9}$$

$$vst_k = et_{km} \qquad k = 1,...,K \;\; m = 1,...,M \tag{10}$$

$$\sum_i q_i \cdot t_{ki} \le Q \qquad k = 1,...,K \tag{11}$$

$$a_j = a_i + c_{ij} \;\; if \; x_{ijk} = 1, \; i,j = 1,...,n \;\; i \ne j; k = 1,...,K \tag{12}$$

$$a_i = vst_k + c_{0j}, \;\; if \; x_{0ik} = 1, \; i = 1,...,n; \; k = 1,...,K \tag{13}$$

$$TD_k = \sum_i \sum_j c_{ij} \cdot x_{ijk} \qquad k = 1,...,K \tag{14}$$

$$TW_i = g \cdot (\max(e_i - a_i, 0) + \max(a_i - l_i, 0)) \quad i = 1,...,n \tag{15}$$

Objective (1) is to minimize the summation of the total delivery cost and violation penalty of time windows of all orders. Constraints (2)-(5) are vehicle routing problem constraints. Constraints (2) and (3) indicate that each customer need be serviced once and only be serviced by one vehicle at most. Constraint (4) shows that if a vehicle arrives one customer it must finally leave. Constraint (5) show that each vehicle can only leave from the supplier once. Constraints (6)-(10) are production scheduling constraints. Constraint (6) states that the end time et_{km} of producing the last order delivered by vehicle k on machine m must be earlier than the start time $st_{k+1,m}$ of producing the first order delivered by the next vehicle $k+1$ on machine m. Constraint (7) states that all machines are ready on time 0. Constraints (8) and (9) calculate et_{km} and the total production time pt_{km} of orders delivered by vehicle k on machine m. Constraint (10) states that vehicle k cannot departure before the production of all orders delivered by vehicle k are completed. Constraint (11) is a vehicle capacity constraint. Constraints (12) and (13) define the arriving time at customer j. Constraints (14) and (15) calculate the total delivered time TD_k of vehicle k and the violation penalty of time window of order i.

3.1 Problem decomposition and hybrid intelligent approach

In section 2, we propose a mathematical model of the IPS-VRTW problem. There are three decision variables $\{o_{mi}, t_{ki}, x_{ijk}\}$ in the model. In this part we decompose the IPS-VRTW into three sub-problem, an order assignment sub-problem of orders to vehicles, an order scheduling sub-problem and a TSPTW sub-problem, according to the three decision variables $\{o_{mi}, t_{ki}, x_{ijk}\}$.

To solve the investigated problem, we need to decide the values of decision variables $\{o_{mi}, t_{ki}, x_{ijk}\}$. Because every customer must be serviced and can only be serviced by one vehicle, we can consider decision variable t_{ki} first through assigning *N* orders to *K* vehicles to form an initial solution, which is called as the order assignment sub-problem of orders to vehicles. After the value of decision variable t_{ki} is decided, then we need to find the solutions to other variables o_{mi} and x_{ijk}. For the orders assigned to vehicle k, we thus need to solve a parallel machine scheduling problem with the makespan objective by determining the values of variable o_{mi}, which is so-called order scheduling sub-problem. After obtaining the values of t_{ki}, o_{mi}, we consider decision variable x_{ijk} at last. When orders delivered by vehicle k are completed, the vehicle can leave (i.e., vst_k is determined. For each vehicle, to deliver the orders to n customers with time windows is actually a traveling salesman problem with time windows (TSPTW), which aims to minimize the summation of the total delivery cost and the time windows violation penalty.

To solve the problem a novel hybrid intelligent approach is proposed as shown in Fig. 1, under which we decompose the problem into 3 simple sub-problems, including the order assignment sub-problem, the order scheduling sub-problem, and the traveling salesman problem with time windows (TSPTW). There are a variety of techniques in the literature can be used to solve the 3 sub-problems. For the first sub-problem, numerous variations of the generalized assignment problem have been studied, many methods can solve the problem such as branch and bound algorithm (Ross and Soland [5]), genetic algorithm (Chu and Beasley [6]), greedy algorithm (Romeijn [7]), tabu search (Dıaz [8]). For the order scheduling sub-problem with makespan minimization, there are also many methods to solve it. Pinedo [9] has made a detailed comparison and analysis with some different rules. For the TSPTW sub-problem, solution approaches range from exact mathematical-programming techniques to various heuristic approaches.

This research use a genetic algorithm to solve the vehicle assignment sub-problem and obtain the value of variable t_{ki}, use the LPT rule to solve the order scheduling sub-problem and obtain the value of variable o_{mi}, use another genetic algorithm to solve the traveling salesman problem with time windows sub-problem and obtain the value of variable x_{ijk}.

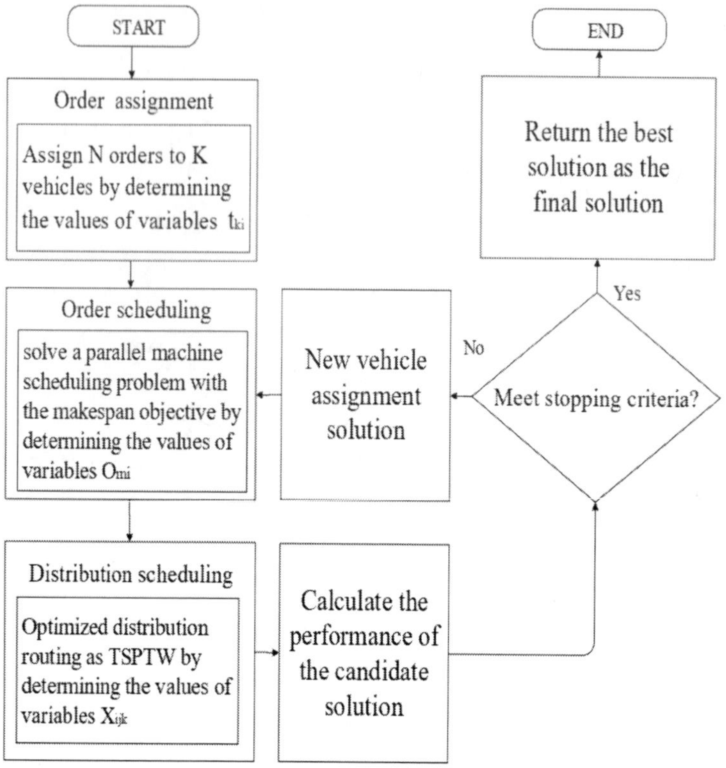

Fig.1. Hybrid intelligent approach for the IPS-VRTW problem

4 COMPUTATIONAL EXPERIMENTS

To evaluate the performance of the proposed novel hybrid framework to solve IPS-VRTW, a series of numerical experiments have been conducted between PGA and another two algorithms. Customers' information (including locations, order size, time windows) of the IPS-VRTW instances is created based on Solomon's [10] problem. We test different problem sizes selected from the 100 customers from RC201 (available at http://web.cba.neu.edu/~msolomon/problems.htm). In the numerical experiments the related parameter settings is $\alpha=1, \beta=0.5, g=0.5, p_i=q_i$. The results show that the PGA can solve the IPS-VRTW with 100 customers within 10 minutes. The experiments were carried out on a laptop with Intel Core i5-3210M CPU @2.5 GHz and 4 GB RAM using MATLAB version R2013a.

4.1 Optimum-seeking performance of the PGA

Due to the NP-hard and nonlinear nature of the investigated IPS-VRTW problem, commercial optimization software has difficulty in finding the optima. We therefore used the enumeration method (EM) to find the optimal solutions in small-size instances. The EM is a general problem-solving technique that systematically enumerates all possible solutions to find the global optimal solution to the problem. These problem instances consider problems with single (two) machine(s), 10 orders, and

1, 2 and 3 vehicles respectively. Ten trials of the PGA were performed for each of the test problems. Each trial terminated when the known optimal solution value was found or when 5000 non-duplicate solutions were generated without improving the best solution found. The PGA's population size is 100, the number of generations is 50, the crossover probability is 0.7 and the mutation probability is 0.3. Both the EM and the PGA (in each trial) found the optimal solutions to the 12 small-sized instances, respectively. For problem instances with 5 orders, the EM is faster than the PGA in terms of computation time. But with the increase enlarge of the orders, the computation time of EM become more and more long. For problem instances with 10 orders, the PGA is much faster than the EM in terms of computation time. Even for small-sized instances, solving the IPS-VRTW by the EM is very time-consuming. To handle a single machine, 10 orders and 3 vehicles instances, the EM used 1318.386 s, while the PGA used only 53.117s. When the instance considers two machines, 10 orders and 3 vehicles instances, the EM needs 6123.012 s, while the PGA needs only 143.335s.
On the basis of the experimental results on test problems, it is clear that the PGA and the proposed novel hybrid intelligent framework have the good ability in achieving the optimal solutions.

4.2 Performance comparison

To further evaluate the performance of the proposed novel hybrid framework, this research compared the optimization performance of the proposed method (called PGA) to the method proposed by Ullrich [4]. Ullrich [4] considered a similar problem with our problem. He proposed a genetic algorithm to solve the IPS-VRTW model as a whole with optimization objective is to minimize total tardiness. Because there are some differences between two models, we modified Ullrich's genetic algorithm (UGA) to adapt the investigated problem.
We solve the IPS-VRTW considering 3 machines, 3 vehicles and 25 orders first. Because the difference genetic code, the PGA has a much smaller search space than the UGA, in the formation of initial population the PGA has a smaller initial value than UGA. As shown in Fig.2 and Fig.3 , the PGA converged in 85 generations, while the UGA converged in 300 generations. The optimization result of the PGA is 861, while the NGA is 1143. Compared with the UGA , the PGA has a faster convergence rate and achieves a much better result.
We also solve the problem with 3 machines, 3 vehicles, 25 and 50 orders respectively. We test 10 times for the PGA and the UGA. Under the same iterative numbers, population size is 100 and iteration number is 100. Results are shown in table 1. The minimum generated by PGA is 861,2323 and 11160, while the minimum generated by UGA is 1143 and 3057. The PGA leads to the decrease of 24.6% and 24.0% respectively. The maximum generated by PGA are 973 and 2577, while the maximum generated by UGA are 1477 and 4510. The PGA leads to an increase of 34.1% and 42.8% respectively. It is clear that the proposed PGA is superior to the UGA. By converting the IPS-VRTW into three commonly seen sub-problems, it can reduce the search space of the IPS-VRTW largely, which makes the proposed novel hybrid framework perform well. It is also possible to use other techniques to solve the three sub-problems so that the overall performance can be improved further.

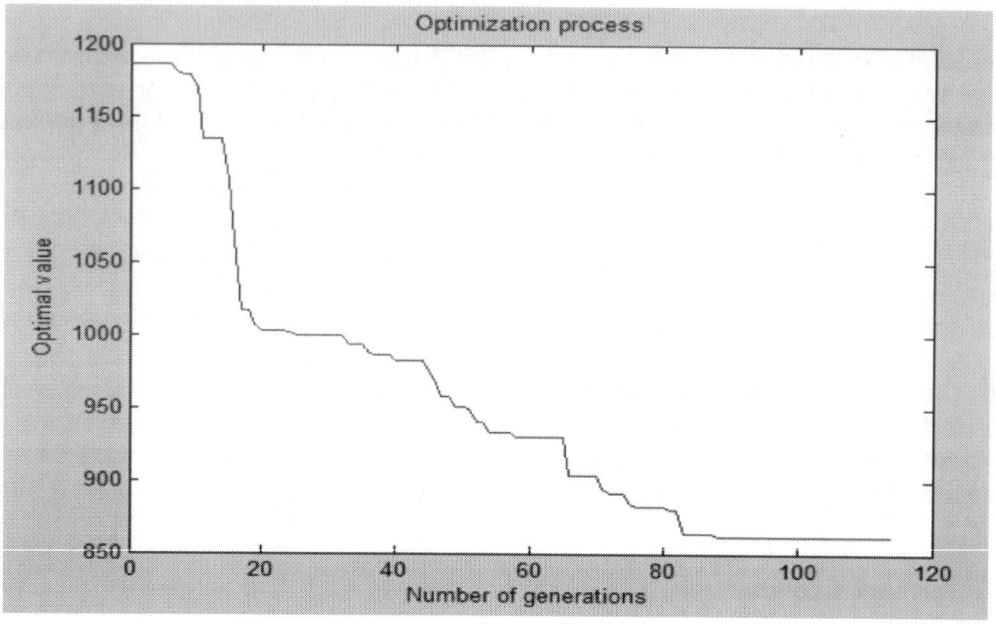

Fig.2. Iterative process of PGA

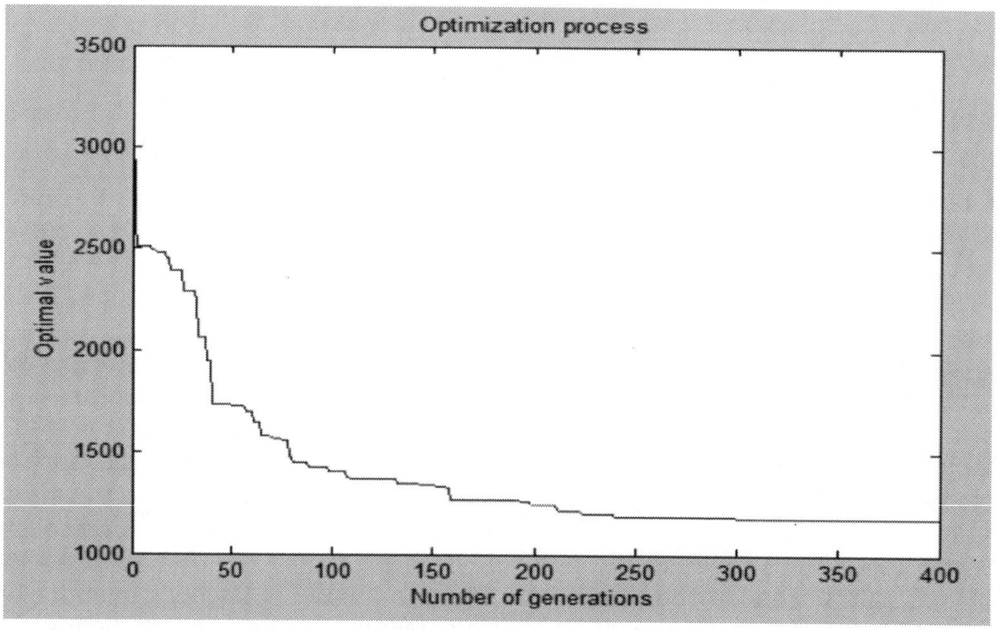

Fig.3. Iterative process of UGA

Table 1: COMPARISONS BETWEEN PGA AND UGA UNDER THE SAME ITERATIVE NUMBERS

Orders		UGA	PGA	GAP (%)
25	Minimum	1.143E+03	0.861E+03	24.6
	Maximum	1.477E+03	0.973E+03	34.1
50	Minimum	3.057E+03	2.323E+03	24.0
	Maximum	4.510E+03	2.577E+03	42.8

5 CONCLUTION

This paper addressed an IPS-VRTW problem with consideration of some important real-world features in make-to-order or time-sensitive products, for example demands of time windows, multiple order sizes, multiple machines and multiple distribution vehicles. The mathematical model of the IPS-VRTW problem was established. A novel hybrid intelligent approach was proposed to solve this problem, by which the investigated problem is decomposed into 3 relatively simple sub-problems including an assignment sub-problem of orders to vehicles, an order scheduling sub-problem with the objective of minimizing makespan, and a TSPTW. As the three sub-problems are classical problems, many approaches can be used to solve them very well. This research thus used genetic algorithms and heuristic rules to solve these sub-problems under the novel hybrid intelligent framework.

To evaluate the performance of the proposed novel hybrid intelligent framework to solve the investigated problem , we propose a nested genetic algorithm combined with the LPT rule (PGA) under this framework. A series of numerical experiments have been conducted to compare the performance of PGA and UGA. The experimental results showed that the PGA could find better optimization solutions and the PGA had a faster convergence rate.

Our future work will aims at investigating other integrated production and distribution scheduling problems and developing more efficient algorithms for these complex integrated scheduling problems.

6 ACKNOWLEDGMENT

The authors would like to thank the financial supports from the National Natural Science Foundation of China (Grant No. 71302134) and Sichuan University (Grant No. 2014SCU04A05).

7 REFERENCES

[1] **Potts, C. N. 1980.** Analysis of a heuristic for one machine sequencing with release dates and delivery times. Oper. Res. 28 1436-1441.

[2] **Toptal, A., Koc, U., & Sabuncuoglu, I. (2013).** A joint production and transportation planning problem with heterogeneous vehicles. Journal of the Operational Research Society, 65(2), 180-196.

[3] **Seyedhosseini, S. M., & Ghoreyshi, S. M. (2014).** An Integrated Model for Production and Distribution Planning of Perishable Products with Inventory and Routing Considerations. Mathematical Problems in Engineering, 2014.

[4] **Ullrich, C. A. (2013).** Integrated machine scheduling and vehicle routing with time windows. European Journal of Operational Research, 227(1), 152-165

[5] **Ross, G. T., & Soland, R. M. (1975).** A branch and bound algorithm for the generalized assignment problem. Mathematical programming, 8(1), 91-103.

[6] **Chu, P. C., & Beasley, J. E. (1997).** A genetic algorithm for the generalised assignment problem. Computers & Operations Research, 24(1), 17-23.

[7] **Romeijn, H. E., & Morales, D. R. (2000).** A class of greedy algorithms for the generalized assignment problem. Discrete Applied Mathematics, 103(1), 209-235.

[8] **Dıaz, J. A., & Fernández, E. (2001).** A tabu search heuristic for the generalized assignment problem. European Journal of Operational Research, 132(1), 22-38.

[9] **Pinedo, M. L. (2008).** Parallel Machine Models (Deterministic). In Scheduling (pp. 111-149). Springer New York.

[10] **Solomon, M. M. (1987).** Algorithms for the vehicle routing and scheduling problems with time window constraints. Operations research, 35(2), 254-265.

MANUFACTURING INFORMATION EVALUATION METHOD BASED ON VALUE IN NETWORKED MANUFACTURING ENVIRONMENT

Fei Li[1*]
[1] School of Mechatronics
Northwest Polytechnical University, Xi'an, China
lfxjzu@163.com

ABSTRACT

Manufacturing capacity (MC), the important information for describing manufacturing level, directly impacts on product quality and function. Improving MC has been one of the effectively ways to realize the manufacturing power. Particularly, cloud manufacturing and big data are put forward to provide the intelligent and service platform for improving MC. In the platform, analyzing and evaluating manufacturing information, especially to MC, is an indispensable part. The current studies of manufacturing information evaluation usually only aim for the importance of information, and seldom consider cost and value. In this paper, by analyzing the definition of MC by combining previous relative researches, a hierarchy tree of classes associated with properties based on weight is established for expressing the relation between MC and information. Based on this, MC is formally represented and done the multi-dimensions and multi-levels decomposition by set theory and hierarchical decomposition methods. Next, the new method of translating manufacturing information into value is constructed based on value engineering (VE), which can provide the supports for improving manufacturing level and product properties. Finally, the effectiveness of the method proposed in this paper is validated through case study.

Keywords: Integration, Manufacturing capacity, The relational model, Formal representation and decomposition, Value, Quantization and calculation

1 INTRODUCTION

With the development of information technology, the researches and applications of manufacturing informatization technology have made a series of achievements, which have played an important role in all areas of manufacturing industry and made a major contribution to promote manufacturing i informatization for our country. Particularly, cloud manufacturing and big data are put forward to provide the intelligent and service platform for improving MC. So many key technologies were worthy of study. In the platform, analyzing and evaluating manufacturing information, especially to MC, is an indispensable part.

What is the MC? Currently, scholars, both domestic and overseas, had been provided various definitions of MC. W.Skinner[1] , R. H. Hayes, and S. C. Wheelwright [2], pointed out that MC was comprised of cost, quality, flexibility, innovation, service, environment and the relationship among each element. A.Arafa([3] elaborated the relation between MC and enterprise performance for achieving the goal of less production cost and higher processing quality. Jiancheng GUAN[4] considered MC as the core of the innovation capacity. In the view of micro resource perspective,

[*] Ph.D. candidate for Manufacturing Systems Engineering

A .B. KASOGLU[5] stated that MC was a sets of resources for achieving the manufacturing goal. Luo yongliang[6] and other scholars perceived MC as a capacity of integrating and configuring manufacturing resources. Jinghui HAO[7] pointed out that MC was the evidence to choose the partners for enterprise. C. Liu[8] and some scholars defined MC as an effective integration of resources in the process of manufacturing work, and so on.

Based on these, the information is the basis of evaluation and analysis. The fuller the information is, the more accurate the result of evaluation and analysis is. Because the types and volume of capacity and information in cloud pool is various and huge, it is quite critical to clearly elaborate and represent the relation among MC, sub-capacities and information. For the moment, relative research is only a few in references retrieved. Though, Qiqi Zhu[9] has formally described the capacity of processing system, to provide more persuasive evidence to MC analysis and to analyze more accurate and faster, further information breakdown is needed.

Besides that, quantitative calculation of MC is an important part for analyzing. It is well known that dimension of each physical variable for mathematical calculation must be unified. Physical variable can add, subtract and equal to each other only in the condition of the same dimension. Regarding the diversity and the high volume of resources and information, it is difficult to calculate directly due to the non unified dimensions. So, unifying dimension is the key for calculation of MC.

In conclusion, this paper analyzed the definition of MC by combining previous relative research firstly. a hierarchy tree of classes associated with properties based on weight, the model for expressing the relation between MC and information, is established. Based on this, MC is formally represented and done the multi-dimensions and multi-levels decomposition by set theory and hierarchical decomposition methods. Next, the new method of translating manufacturing information into value is constructed based on value engineering, which can provide the supports for improving manufacturing level and product properties.Finally, the effectiveness of the method proposed in this paper is validated through case study.

2 THE METHOD OF DECOMPOSITION AND FORMAL REPRESENTATION FOR MC

According to the concept of broad manufacturing, MC can be perceived as a comprehensive system which includes market requirement analysis capacity, product design capacity, research development capacity, manufacturing, supportive activities capacity and environment treatment capacity and so on. It spans through the whole product life cycle.

Based on this, the types and volume of information related to the capacity is various and huge, so to decompose and to formally represent in detail are benefit to information usage, calculation and evaluation.

2.1 A Model for Expressing Relation Between MC and Information

In order to show the relationship between MC and information, hierarchy tree of classes associated with properties based on weight is established and done the multi-dimensional and hierarchical decomposition for MC. In the model, MC is mapped to class *A*, each sub capacity is mapped to class $A_i(i=1,2,...)$, and $A_i \in A$. Information is mapped to attribute set I_o, each sub-information is mapped to $I_j(j=1,2,...)$, and $I_j \in I_0$. The relation between class and attribute is importance of information for achieving capacity. It can be represented as weight *W*. Each relation can be represented as $w_k(k=1,2,...)$, and $w_k \in W$. So the relation among A, I_o and W can be denoted by W:A→ I_o.

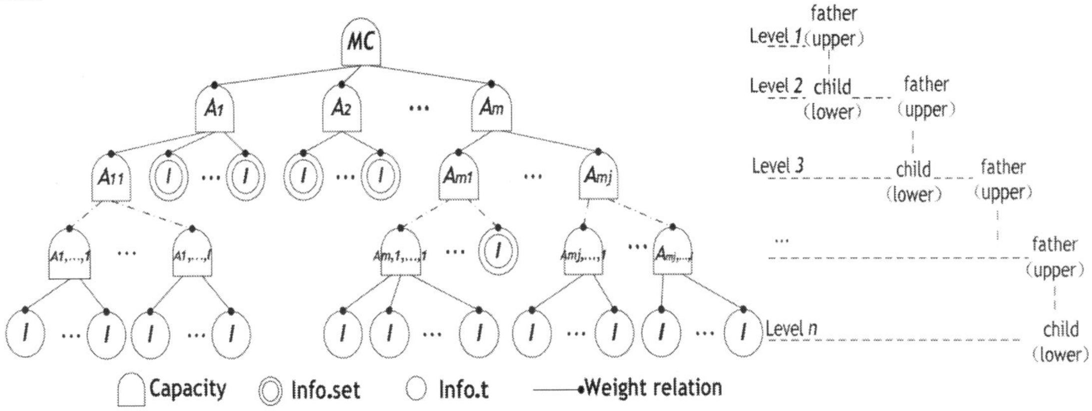

Figure 1: Hierarchy tree of classes associated with properties based on weight

In Figure 1, the relation includes upper-lower relation and coordinative relation. Upper-lower relation means relation of father and sub, and they can be regarded respectively as purposes and means.

2.2 The Formal Representation for MC and Information

The relationship among MC, sub-capacity and information can be expressed clearly based on the model. However, the more sufficient the decomposition of MC is, the huger the information volume is. So, it is difficult to express all the information only using the model. We do the multi-dimensional representation using set theory for MC and information in the paper to make up the shortage of the model. Where, $A_1, A_2, \ldots, A_m$ represents respectively sub-capacity, i, j, z represents respectively dimension.

$$MC = \{A_1, A_2, \ldots, A_i, \ldots, A_m\} \quad (1)$$

$$A_i = \{A_{i1}, A_{i2}, \ldots, A_{ij}, \ldots, A_{in}\} \; (i = 1, \ldots, m) \quad (2)$$

$$A_{ij} = \{A_{ij1}, A_{ij2}, \ldots, A_{ijk}, \ldots, I_O\}, (i = 1, \ldots, m; j = 1, \ldots, n) \quad (3)$$

....

$$A_{\underbrace{ijk,\ldots,l}_{levels}} = \{I_1, I_2, \ldots, I_m\} \quad (4)$$

3 THE METHOD OF TRANSLATING MANUFACTURING CAPACITY OR INFORMATION INTO VALUE

Based on connotation of MC, MC is the ability to produce the product for achieving the goal of the best quality, lower production cost, better function and excellent service at full speed, eventually to create new value. Thus, value can be seen as the embodiment of the MC. Here, we mainly study how to translate capacity or information into value to calculate and quantify MC.

3.1 The Value Coefficient of Information Based on Value Engineering

Value engineering (VE) has been presented firstly by L.D.Miles in the 1940s. It can be considered as a scientific method forachieving the essential function of products or serves. Especially, it can correctly handle the relationship between function and cost. In view of VE theory, capacity or information can be seen as the function. So, the value coefficient of capacity or information is shown as follows:

$$V = \frac{F}{C} \quad (5)$$

Where, F represents importance coefficient of capacity or information, C represents cost coefficient of capacity or information, V represents value coefficient of information.

3.2 The Value Coefficients of Information

Based on the definition of capacity, it can be seen as the function, so calculation of importance coefficients can refer to calculation of function coefficient. Because function evaluation methods we usually used, such as "01", "04" and so on, cannot express the relation of more or smaller function differences between two objectives.

In the paper, mutual evaluation weight method is established, then importance coefficient F are calculated.

Capacity or information in formula (1) to (4) is recorded as $\{S^{(1)}, S^{(2)}, \ldots, S^{(m)}\}$, m represents the number of elements in the set. The importance is recorded as y_{ij}, it denotes the scores by comparing capacity i and capacity j, or information i and information j.

- Step1: the evaluation standards are given according to Table1 and the importance is confirmed as shown in Table2.

Table1: Evaluation standards

importance	Standards
1	Comparing the same capacity or information
3	The slightly important capacity or information
5	The important capacity or information
7	The very important capacity or information
9	The excellent important capacity or information
2,4,6,8	Between adjacent standards above mentioned

Table2: The importance of information

I	$S^{(1)}$	$S^{(2)}$	$S^{(3)}$	…	$S^{(m)}$
$S^{(1)}$	y_{11}	y_{12}	y_{13}	…	y_{1m}
$S^{(2)}$	y_{21}	y_{22}	y_{23}	…	y_{2m}
$S^{(3)}$	y_{31}	y_{32}	y_{33}	…	y_{3m}
…	…	…	…	…	
$S^{(m)}$	y_{m1}	y_{m2}	y_{m3}	…	y_{mm}

In the Table2 , if $y_{ij}>0$,then $y_{ij} = \frac{1}{y_{ji}}$.

- Step2: Consistency of data is verified.

1. The product of every row in the matrix is calculated.

$$Y_i' = \prod_{j=1}^{m} y_{ij} \quad (6)$$

2. The m root of the product of every row is calculated.

$$\sqrt[m]{Y_i'} = \sqrt[m]{\prod_{j=1}^{m} y_{ij}} \quad (7)$$

3. Normalization processing is done:

$$W_0 = (w_1^0, w_2^0, \ldots w_m^0) \quad (8)$$

$$w_i^0 = \frac{\sqrt[m]{Y_i'}}{\sum_{i=1}^{m} \sqrt[m]{Y_i'}} \quad (9)$$

4. The maximal characteristic root is calculated.

$$\lambda_{max} = \sum_{i=1}^{m} \frac{(AW_0)^i}{nw_i^0} \quad (10)$$

5. The coincidence indicator CI is calculated.

$$CI = \frac{\lambda_{max} - m}{m-1} \quad (11)$$

6. The check value CR is calculated.

$$CR = \frac{CI}{RI} \quad (12)$$

Where, average random consistency index *RI* is shown in Table3.

Table 3: RI standard values

Matrix order	1	2	3	4	5
RI	0.00	0.00	0.58	0.90	1.12

If $CR < 0.1$, it illustrates data is effective.

- Step 3: W_1 is calculated as follows:

$$w_1^{(i)} = \frac{1}{m} \sum_{j=1}^{m} \frac{y_{ij}}{\sum_{i=1}^{m} y_{ij}} \quad (13)$$

$$F = W_1 = \{w_1^{(1)}, w_1^{(2)}, \ldots, w_1^{(m)}\} \quad (14)$$

4 CASE STUDY

4.1 A Model for Expressing Relation Between MC and Information

According to the above method, a model for expressing relation between MC and information is established and formally represented, as shown as follows.

MC={DA,PD,R&D,DE,SE,PR,SP,QU,WE,SR,EV,MA};
PR={E,C,P,T,I,O};

$$E = \{ EM, EB, ET, \dots, OT \};$$

$$EB = \{ p_{EB}, h_{EB}, f_{EB}, n_{EB}, b_{EB}, O_{EB} \};$$

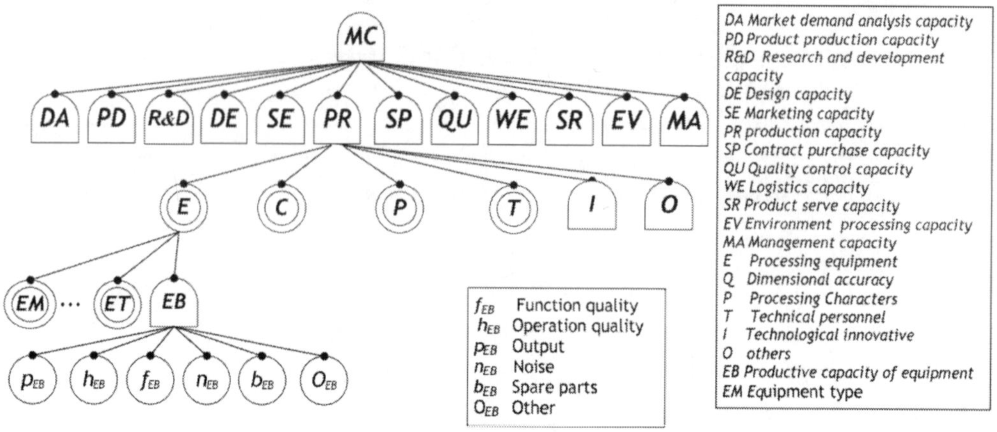

Figure2: A model for expressing relation between MC and information

4.2 The Calculation of Value Coefficients of EB

The method is confirmed by analyzing sub capacity EB and information. The importance of each information and evaluation scores between information is shown respectively in Table 4 and Table 5. w_{EB} represents motor power.

Table 4: The average scores of each sub information

EB	p_{EB}	h_{EB}	f_{EB}	n_{EB}	b_{EB}	w_{EB}
Average score	15.7	16.3	30	4	2.6	12.3
Sequence of importance	3	2	1	5	6	4

Table 5: The scores of importance evaluation

I	f_{EB}	h_{EB}	p_{EB}	n_{EB}	n_{EB}	b_{EB}
f_{EB}	1	3	3	4	5	6
h_{EB}	1/3	1	2	3	4	5
p_{EB}	1/3	1/2	1	3	4	5
w_{EB}	1/4	1/3	1/3	1	3	4
n_{EB}	1/5	1/4	1/4	1/3	1	2
b_{EB}	1/6	1/5	1/5	1/4	1/2	1

The cost coefficient C is provided by zhou[10] and then V is calculated. Besides that, we used other methods to compute F and V which are compared with the new method in this paper, as shown in Table6 and Table7.

Table 6: The importance coefficient of information in five methods

I	F(AHP)	F(01)	F(04)	F(DARE)	F(new)
f_{EB}	0.3900	0.3333	0.2833	0.3708	0.3830
h_{EB}	0.2225	0.2667	0.2333	0.2015	0.2148
p_{EB}	0.1822	0.2	0.1833	0.1907	0.1859
w_{EB}	0.1089	0.1333	0.15	0.1520	0.1274
n_{EB}	0.0574	0.0667	0.1	0.0494	0.0525
b_{EB}	0.0390	0	0.05	0.0321	0.0349

Table 7: The value coefficient of information in five methods

I	V(AHP)	V(01)	V(04)	V(DARE)	F(new)
f_{EB}	3.9	3.3	2.8	3.7	3.8
h_{EB}	0.89	1	0.93	0.81	0.86
p_{EB}	1.21	1.33	1.22	1.27	1.24
w_{EB}	0.726	0.889	1	1	0.85
n_{EB}	0.38	0.44	0.67	0.33	0.35
b_{EB}	0.195	0	0.25	0.16	0.17

In this table7, if V>1, it denotes that the shaving quality and output are important. If V<1, it denotes the cost is too high, so enterprise should take steps to reduce cost and improve value. If V=1, the information needn't to be studied.

However, we found the results are different for the same information in different methods. For example: the value of motor power in "04" method and DARE respectively equals 1, but the value is less than 1 in other methods. The value of operation quality in "01" equals 1, but the value is less than 1 in other methods, as shown in Figure 3. The reason is due to the computation of *F*. The more the levels of importance differences are, the better the results will be.

Furthermore, it is concluded that the value of shaving quality and output are higher with low costs. Thus, shaving quality and output can be perceived as advantage capacity, this is consistent with the actual situation.

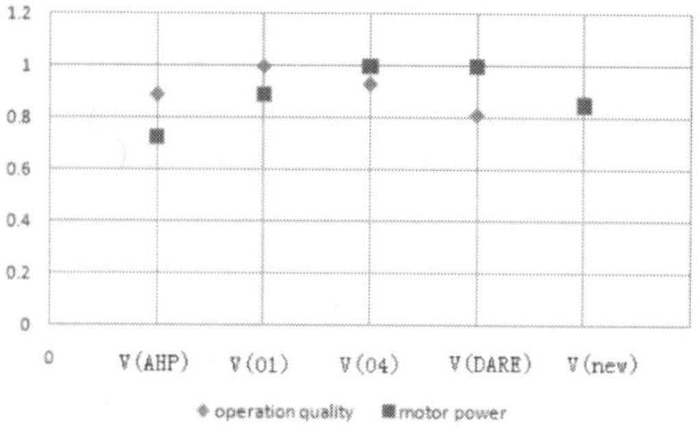

Figure 3: The result of value calculation for operation quality and motor power in different methods

5 CONCLUSION

In conclusion, we analyzed the definition of MC by combining previous relative research firstly. Then a hierarchy tree of classes associated with properties based on weight, the model for expressing the relation between MC and information, is established. Based on this, MC is formally represented and done the multi-dimensions and multi-levels decomposition by set theory and hierarchical decomposition methods. Next, the new method of translating manufacturing information into value is constructed based on value engineering. Finally, the effectiveness of the method proposed in this paper is validated through case study which can ensure essential information, remove unnecessary costs and provide the supports for improving manufacturing level and products properties.

6 REFERENCES

[1] **W.Skinner. 1969.** Manufacturing-missing link in corporate strategy, *Harvard Business Review*,pp136-145.

[2] **R. H. Hayes, S. C. Wheelwrigh. 1984.** Restoring our competitive edge: completing through manufacturing, New York: Wiley.

[3] **ARAFA , M. H. EIMARAGHY. 2011.** Manufacturing strategy and enterprise dynamic capability, CIRP Annals manufacturing Technology, 60(1), pp 656-660.

[4] **Jiancheng GUAN. 2004.** Study on the relationship between the enterprise manufacturing capability and innovation performance: empirical findings of Chinese, *scientific research management*, 25,pp 78-83.

[5] **A .B. KASOGLU. 2003.** Capability based distributed layout approach for virtual manufacturing cells, *Internanal of Production Research*, 41(11),pp 2597-2618.

[6] **Yongliang LUO, Lin ZHANG, etc. 2012.** Key technologies of manufacturing capability modeling in cloud manufacturing mode, *Computer Integrated Manufacturing System*,1(7), pp 1357-1366.

[7] **Jinghui HAO, Shudong SUN, Quanyou SHA. 2006.** Resource modeling oriented manufacturing capability, *Application research of computers*, (3), pp 60-63.

[8] **LIU,P.Y.JIANG,W.CAO. 2014.** Manufacturing capability match and evaluation foroutsourcing decision-making in one-of-a-kind production, Proceedings of the 2014 IEEE 18th In national Conference on Computer Supported Cooperative Work in Design, IEEE, pp 575-580.

[9] **Qiqi zhu, Pingyu JIANG. 2011.** Formal model of machining capability for a machine system, *Journal of Xi'an jiao tong university*, 45(11), pp63-67.

[10] **Cong zhou, Jian yang,etc. 2010.**Application analysis of the value engineering principle in product development,*Forestry machinery& woodworking equipment*, 28(3), pp 30-33.

A Hybrid Simplified Swarm Optimization with Local Search for Stochastic Function Optimization

James T. Lin[1*] and Chun-Hui Chao[1]

[1]Department of Industrial Engineering and Engineering Management

National Tsing-Hua University, Taiwan ROC

jtlin@ie.nthu.edu.tw

ABSTRACT

In this paper, a hybrid particle swarm optimization (PSO) and simplified swarm optimization (SSO) algorithm is developed for solving stochastic multimodal optimization problems. The most attractive features of PSO are its algorithmic easy implantation and fast convergence. However, most studies of PSO still focus on deterministic problem and mention that PSO has premature convergence in strongly multi-modal optimization problems. This paper proposes two methods (PSO and SSO) incorporating the optimal computing budget allocation (OCBA), hypothesis test (HT), and local search (LS) into the PSO or SSO algorithms, aimed at enhancing global search capability and searching efficiency in noisy environment. These two hybrid algorithms are called PSOOHL and SSOOHL, respectively. In these two hybrid algorithms, PSO and SSO are used to explore and exploit the solution space. The OCBA is applied to allocate computational budgets to provide reliable evaluation and identification in the swarm of PSO or SSO for updating next generation. Hypothesis test (HT) and local search are used for distinguishing the similarity of particles and generating the diversity particle to prevent premature convergence. Numerical experiments based on four benchmark functions are performed. The result shows that PSOOOHL is superior in terms of searching quality and efficiency.

Keywords: Stochastic multimodal optimization, Local search, Simplified Swarm Optimization, Particle Swarm Optimization, Optimal Computing Budget Allocation

1 INTRODUCTION

Combinatorial optimization problems are widely applied for engineering, science applications, and commercial analysis. Due to their mathematical models for real systems are usually nonlinear or non-differentiated, stochastic optimization algorithms have been important recently by their flexibility in searching the solutions. Therefore, the main goal which is committed by many researchers is whether the algorithm can find the global optimum solution or near global optimum solutions effectively [16]. Also, to develop a suitable algorithm becomes an important and urgent topic now.

Particle swarm optimization (PSO) algorithm is a useful and popular tool for solving some engineering optimization problems over the few years. It was firstly proposed by Kennedy and Eberhart in 1995 and has been proven to be a simple but powerful optimal algorithm among a lot of evolutionary algorithms (EAs)[1]. The main characteristics include the simplicity of updating formulas, easy execution on coding, and quick convergence, etc. According to these outstanding advantages, a variety of engineering implementations based on the PSO has been explored and developed successively, such as job-shop scheduling problems, digital filter designs, controller designs, power system designs, antenna array design...etc.[1]. PSO could memorize the best solution and performance that had been gone through, so the memory of good designs will be preserved by the population. Particles in PSO have the properties of competition and cooperation. Particles exchange their experience and information to each other, and the population will approach the best position like a flock of birds to search for foods. Simultaneously, they will abandon the previous place if there are better positions. Simplified swarm optimization (SSO) is an improved and simplified optimization technique of PSO, and it can be more efficient than PSO for discrete event problem proposed by Yeh, (2009) [18]. SSO also has the properties of competition and cooperation. However, the searching concept of SSO is different from PSO. SSO just remembers the previous position and go there but not be close to. Due to the simple concept, easy implementation, and quick convergence, nowadays PSO and SSO has gained much attention and wide applications in different fields.

Although PSO algorithms can quickly converge to the near-optimal solutions for many optimization problems, it has been reported that PSO has some difficulties in reaching the global optimum in some optimization problems and may trap into some stagnation, for example, suffer from premature convergence. It may cause the diversity of the swarm decreases when the swarm approaches the near-optimal solution [9]. This paper will focus on the issue of solving the multimodal function optimization problems using the modified PSO and SSO algorithm. It is well known that most of the function optimization problems with multiple designed variables are of multimodal, i.e., including multiple global or/and local optimal solutions, not a single optimum, over the search space. To solve this kind of optimization problems, various methods based on evolutionary algorithms have been employed and developed in recent years [1]. In previous studies, PSO was mainly applied to the deterministic optimization problem, while there are few type of researches about PSO for stochastic optimization problems [12]. However, in the real-world, there are many uncertainties that we have to take into account when solving optimization problems. Since the uncertainty is unpredictable, performance evaluation is usually estimated by multiple sampling with simulation. Because of the huge solution space, it is difficult to search the optimal solution in the global area. Accordingly, study of evolutionary algorithms for uncertain environments has been popular in the international academic fields, especially the research on designing effective and robust algorithms [3]. Pan et al. (2006) proposed an effective hybrid PSO approach named PSOOHT. The results indicate that the proposed PSOOHT can efficiently distinguish the performance in the process [12]. Lin and Chiu (2015) proposed a new approach for discrete resource allocation problem and to combine algorithm with OCBA and local search, called, PSO_{OTL}. The results indicate that the proposed PSO_{OTL} is very efficient in terms of quality and computational requirements [5].

For solving the stochastic multimodal optimization or optimization problem in complex system, PSO or SSO is rather improper and not applicable because these problems consist of several system optima including global optimal solutions or local optimal solutions. To deal with this problem, the hybrid version of the PSO and SSO algorithms are developed. In the proposed algorithm, the search algorithm of SSO is applied for well exploration and exploitation of the swarm in the solution space at first. Secondly, the optimal computing budget allocation (OCBA) technique is mainly used to reduce sampling error and enhance sampling efficiency by allocating limited sampling budgets to particles. Thirdly, Hypothesis test (HT) preserve good particles for the next generation to keep the diversity of the population, and local search generates new particle for the next generation. The numerical results and comparisons based on several well-known benchmarking demonstrate the advantages and properties of our proposed algorithm. The result shows that PSOOOHL is superior in terms of searching quality and efficiency compared to three algorithms.

2 PROPOSED APPROACH

• 2.1 Particle Swarm Optimization (PSO)

PSO is an evolutionary computation technique according to individual potentiality plus personal previous experience and cooperation, competition of the community, which is based on the simulation of simplified social model, such as bird flocking, fish schooling, and the swarm theory. The theoretical framework of PSO is very simple and easy to be coded and implemented by computer [4]. In addition, it is computationally inexpensive in terms of memory requirements and CPU times so nowadays PSO has gained much attention and wide applications in various fields [6].

PSO starts with the random initialization of a swarm of particles in the space and works on searching for the social behavior of the particles in the swarm. That is, PSO finds the best solution of all particles in searching generation, memorizes the position of the best and try to go near to it. For every particle, they will remember the best position they have gone through, and will have some experiences that the position may be better.

The position and the velocity of i-th particle in the search space of dimension d can be expressed as $X_i = [x_{i,1}, x_{i,2}, \dots, x_{i,d}]$ and $V_i = [v_{i,1}, v_{i,2}, \dots, v_{i,d}]$ respectively. Each particle has its own best position (P-best) $p_i = (p_{i,1}, p_{i,2}, \dots, p_{i,d})$ in searching procedure by comparing the objective value of historical position Global best position (G-best) is denoted by $p_g = (p_{g,1}, p_{g,2}, \dots, p_{g,d})$. The new velocity of each particle is calculated as:

$$n_{i,j}(t+1) = w \cdot v_{i,j}(t) + c_1 \cdot r_1 \cdot (p_{i,j} - x_{i,j}(t)) + c_2 \cdot r_2 \cdot (p_{g,j} - x_{i,j}(t)),\ j=1,2,\ldots,d. \quad (1)$$

where c_1, c_2 are acceleration coefficients of constants, w is the initial factor, r_1, r_2 are two independent random numbers followed uniform(0,1). Thus the position of each particle is updated by:

$$x_i(t+1) = x_i(t) + v_i(t+1),\ i=1,2,\ldots,N. \quad (2)$$

2.2 Simplified Swarm Optimization (SSO)

SSO is a simplified approach of particle swarm optimization. Inherit the advantage of PSO, SSO is also an evolutionary computation technique based on individual potentiality, personal previous experience, cooperation and competition. However, SSO doesn't have the concept of direction, it's evolution is leaping. Like PSO, SSO starts with random position of a swarm of particles in the space, but it's working behavior is different from PSO. Because of not having velocity, SSO updates the position by changing the coordinate of all dimensions. If the changing probability meets the setting parameter, particle will replace the original coordinate by P-best, G-best, or new position, or still be at the initial place. The position of each particle is updated by:

$$x_{id}^t = \begin{cases} g_{id}^{t-1} & if\ \rho_{[0,1]} \in [0, C_g) \\ p_{id}^{t-1} & if\ \rho_{[0,1]} \in [C_g, C_p) \\ x_{id}^{t-1} & if\ \rho_{[0,1]} \in [C_p, C_w) \\ x & if\ \rho_{[0,1]} \in [C_w, 1) \end{cases}, \quad C_g = c_g,\ C_p = C_g + c_p,\ C_w = C_p + c_w \quad (3)$$

2.3 Optimal Computing Budget Allocation (OCBA)

Because of the stochastic in the environment, we have to decide how many samples we want to evaluate the performance, however, it's impossible to have infinite resources, so we need a method to distribute the computing budgets intelligently. OCBA is a technique to optimally choose the number of simulations for all the solutions in order to maximize simulation efficiency with finite computing budget. In general, OCBA is used to solve the problem which is restricted by the total budget and expected to reach a desired level. In this paper, we should decide how many resources of the total budget should be allocated to different simulation alternatives until reaching the fixed probability of correct selection (P{CS}*). The objective functions and steps of OCBA is as follows:

$$\text{Minimize } n_1 + n_2 + \cdots\cdots + n_k = T \qquad \text{s.t. } \mathrm{P\{CS\}} \geq \mathrm{P\{CS\}} \quad (4)$$

2.4 Hybrid

With SSO and OCBA, it can be more efficiently allocating the computing budgets, however, it still has some disadvantages when two or more candidates are too similar. When the performance of some designs is close, OCBA will spend many resources to distinguish which is the best, but sometimes these consuming costs are unnecessary. Therefore, we need some methods to help us save the resources.

2.4.1 Hypothesis test

Hypothesis test is used to analyze the simulation performance and verify whether the means of the two alternatives are equal or different. The assumption for the test is that both alternatives are sampled from normal distributions with equal or unequal variances. The estimated mean value $\bar{J}$ and variance s_i^2 can be calculated as follows:

$$\bar{J} = \bar{L}(\theta_i) = \sum_{j=1}^{n_i} \frac{L(\theta_i, \omega_j)}{n_i}, \quad (5)$$

$$s_i^2 = \sum_{j=1}^{n_i} \left[L(\theta_i, \omega_j) - \bar{J}_i\right]^2 / (n_i - 1). \quad (6)$$

Before doing Hypothesis test, it needs a null hypothesis (H_0) and an alternative hypothesis (H_1) to describe what we want to verify. In this paper, we define H_0 as two designs' performance don't have significant differences, and H_1 as two designs' performance have significant difference. If the variance is unknown and the same, it should be used by pooled-t test as equation (7). In this paper, supposed the variance of those are unknown and different, so we use the formulation of two sample t-test as equation (8) and the confidence level of 95% to compute the critical value and p-value, then determine whether to reject or not reject H_0. If two designs have no significant difference, we skip the alternative that has worse performance, so it won't be preserved to the next generation, then continue doing Hypothesis test until the last designs. The critical region of H_0 is described as follows:

$$|\bar{J}_1 - \bar{J}_2| \geq t_{\frac{\alpha}{2}}(n_1 + n_2 - 2) \cdot \sqrt{\frac{n_1+n_2}{n_1 n_2}} \cdot \sqrt{\frac{(n_1-1){s_1}^2+(n_2-1){s_2}^2}{n_1+n_2-2}} = \tau. \quad (7)$$

$$|\bar{J}_1 - \bar{J}_2| \geq \cdot \sqrt{\frac{{s_1}^2}{n_1} + \frac{{s_2}^2}{n_2}} \cdot \frac{(\frac{{s_1}^2}{n_1}+\frac{{s_2}^2}{n_2})^2}{\frac{(\frac{{s_1}^2}{n_1})^2}{n_1-1}+\frac{(\frac{{s_2}^2}{n_2})^2}{n_2-1}} = \tau \quad (8)$$

If $|\bar{J}_1 - \bar{J}_2| < \tau$, it indicates that the null hypothesis holds, and it can be regarded that the performances of those two solutions have no significant difference in a statistical sense, otherwise they are significantly different. For a multi-modal stochastic optimization problem, repeat search can be avoided to some extent by applying HT to search process, but pure blind search with comparison under hypothesis test can often be trapped into local optima [13].

- **2.4.2 Local search**

The integration of SSO with local search method is to diversify the solution and avoid to stay in local optimal. After the Hypothesis test, there may be insufficient particle for SSO because of the skip rule. At that time, we should replenish the number of particles to n_0. There are two ways to add new designs, one is randomly selecting some new particles until n_0, other is using the local search method. When using local search, it has to randomly choose one of the preserved designs to do some fixed value of plus and minus for all positions. When the particle is not enough in each iteration, local search behavior generates new particles with diversity to prevent trapping in local optima.

- **2.4.3 Proposed SSOOHL**

This paper proposes the SSOOHL, and it's procedure is as follows:

Step 1: Initialization of SSO

1.1 Setting parameters for the c_g, c_p, c_w, c_r.

1.2 Randomly generate N feasible particles.

Step 2: Initialization of OCBA

2.1 Setting parameters including; $n_0, \Delta, T, P\{CS\}^*$.

Step3: All particles' performance evaluation is estimated using simulation by n_0 times sampling.

3.1: Calculate sample means and sample standard deviation of each particle.

3.2: Calculate $PCS \equiv 1 - \sum_{i=1,\, i\neq b}^{k} P\{\tilde{J}_b > \tilde{J}_i\}$ or $\prod_{i=1,\, i\neq b}^{k} P\{\tilde{J}_b - \tilde{J}_i\}$

3.3: If stopping criteria is not reached (P{CS}< P{CS}* or used total simulation budget <T) .

3.4: Increase the computing budget by Δ.

3.5: Repeat four step until the stopping criteria is satisfied.

Step4: If the stopping criteria are reached, go to step9; otherwise, go to Step5.

Step5: If SSO generation >=2, use hypothesis test (HT) to maintain the diversity of the swarm and distinguish between new particles and old particles; otherwise, go to Step 6.

5.1: Sort all particles of the old particle and the new particles from the best to worst and denote them sequentially $\theta_1, \theta_2, \ldots \theta_{2N}$. Let k=1, j=2, reserve θ_1 to next iteration and let $\theta_1 = \theta_k$.

5.2: Perform hypothesis test for θ_k with θ_j. If the null hypothesis is rejected, θ_j is reserved to the next swarm. Then k=k+1 and j=j+1 otherwise, let j=j+1.

5.3: If k < N and j ≤ 2N, go back to step 5.2; Otherwise, go to 5.4.

5.4: If k = N and j ≤ 2N, go to step 6, and continue the PSO procedure; Otherwise, go to Step8, use local search method to generate new particles.

Step6: Determines the P-best of each particle and the G-best of the swarm.

Step7: According to SSO procedure. Update the position.

Step8: Generate N-k new particle using the local search method of this study.

8.1 Repeat the duplication randomly from the Top-k of 2N particles and their all history information including; position, P-best and G-best until the new particle number is equal to N-k.

8.2 The dimension of each new duplicated particle's position will have the same probability of increase in one unit, decrease by one unit or maintain the same.

8.3 Loop the process of evaluating new particle by simulation with T/N times until all new particles are executed.

8.4 Go back to Step 6.

Step 9: Output the best solution and its performance.

3 Function optimization in noisy environment

Generally, uncertain function optimization problems can be described as follows:

$$\text{Min } J(\theta) \equiv E[L(\theta,\omega)],\ \theta = [x_1, x_2, \dots, x_d] \qquad \text{s.t. } x_i = [a_i, b_i], i = 1,2,\dots,d \tag{9}$$

where θ is the position with dimension d, ω is the noise of the function, $L(\theta,\omega)$ and $J(\theta)$ are the performance of one sample and the expected performance. In a noisy environment, it is expected that the algorithm won't be misled by the noise factor, so the expected value should be evaluated by multiple simulations, and can be calculated by

$$\overline{L}(\theta) = \frac{1}{n}\sum_{i=1}^{n} L(\theta,\omega), \tag{10}$$

By using $\overline{L}(\theta)$ as the objective function, the proposed approach is applied to the uncertain optimization benchmarking function as follows.

In this paper, we use four benchmark functions on experiment. First is Goldstein-Price (GP) Function, it has only one global optima at (0,-1) with the value of 3. The function is shown as follows:

$$f_G(x) = [1 + (x_1 + x_2 + 1)^2(19 - 14x_1 + 3{x_1}^2 - 14x_2 + 6x_1x_2 + 3{x_2}^2)] \times [30 + (2x_1 - 3x_2)^2(18 - 32x_1 + 12{x_1}^2 + 48x_2 - 36x_1x_2 + 27{x_2}^2)], \quad -2 \le x_i \le 2,\ \ i = 1,2 \tag{11}$$

Second is Branin (BR) Function, it has three global optima at (-3.142 , 12.275), (3.142 , 2.275), and (9.425 , 2.425) with the value of 0.397887, so it's also a multi-modal. The function is shown as follows:

$$f(x) = a(x_2 - b{x_1}^2 + cx_1 - r)^2 + s(1 - t)\cos(x_1) + s$$

$$a = 1, b = 5.1/4\pi^2, c = 5/\pi, r = 6, s = 10, t = 1/8\pi, \quad x_1 \in [-5,10], x_2 \in [0,15] \tag{12}$$

Third is Rastrigin (RA) Function, it has only one global optima at (0,0) with the value of -2 and has significant peak and about 50 local minima. The function is shown as follows:

$$f_R(x) = {x_1}^2 + {x_2}^2 - \cos 18x_1 - \cos 18x_2, \quad -1 \le x_i \le 1,\ i = 1,2 \tag{13}$$

Last is Shubert (SH) Function, it has eighteen global optima with the value of -186.7309 ant it's also a highly multi-modal. The function is shown as follows:

$$f(x) = \left(\sum_{i=1}^{5} i\cos((i+1)x_1 + i)\right)\left(\sum_{i=1}^{5} i\cos((i+1)x_2 + i)\right), \quad x_i \in [-5.12,5.12], i = 1,2 \tag{14}$$

4 Numerical simulation and comparisons

In this paper, we applied four algorithms, including PSOOHT, PSOOHL, SSOOHT, SSOOHL to conduct the experiment. Supposed the parameters are the best according to research, and the parameter setting is shown in Table 1 to conduct the experiment. We fix the noise magnitude at ω=0.5, and use four methods to run the experiments of four functions in 50 generations for 30 times and compare the performance of those four method. The statistical results are summarized in Table 2, where X* denote the optimal solution in a simulation run, $\overline{J}_a(X^*)$ and $\overline{L}_a(X^*)$ denote the average expected performance and the average estimated best performance respectively. The convergence graph of four hybrid methods for fifty generations are shown in figure 1 to figure 4.

Table 1: Parameter setting

PSO		SSO		OCBA	
Particle size	100	c_g	0.4	T	2000
C_1	2.05	c_p	0.3	n_0	10
C_2	2.05	c_w	0.2	$\triangle$	10
W	1	c_r	0.1		

Table 2: Simulation results by different methods with ω=0.5

Algorithm	Index	GP	BR	RA	SH
PSOOHT	$\overline{J}_a(X^*)$	2.537046	0.337559	-2.27441	-222.891
	$\overline{L}_a(X^*)$	1.893604	0.233611	-2.52226	-292.353
PSOOHL	$\overline{J}_a(X^*)$	2.487793	0.314614	-2.29355	-235.349
	$\overline{L}_a(X^*)$	1.645904	0.223472	-2.67295	-295.366
SSOOHT	$\overline{J}_a(X^*)$	2.682158	0.367618	-2.23043	-200.664
	$\overline{L}_a(X^*)$	2.284585	0.298812	-2.48378	-226.727
SSOOHL	$\overline{J}_a(X^*)$	2.652347	0.363388	-2.28126	-203.2
	$\overline{L}_a(X^*)$	2.20325	0.309405	-2.56497	-229.286

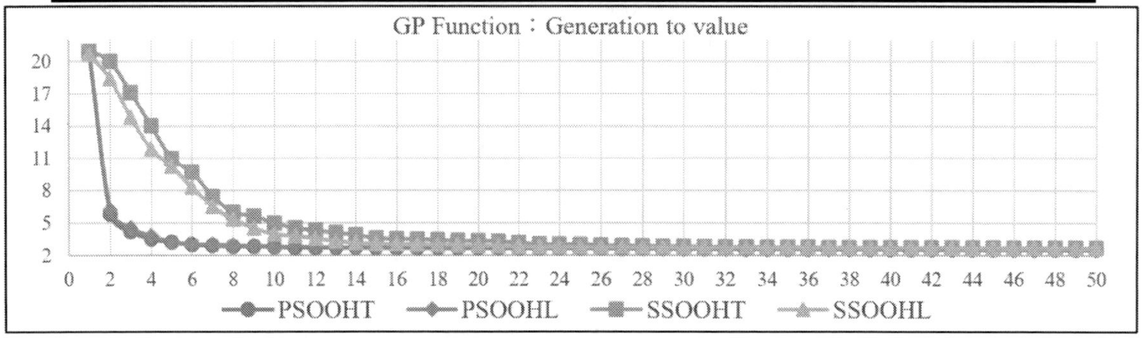

Figure 1: 50 generations of GP function

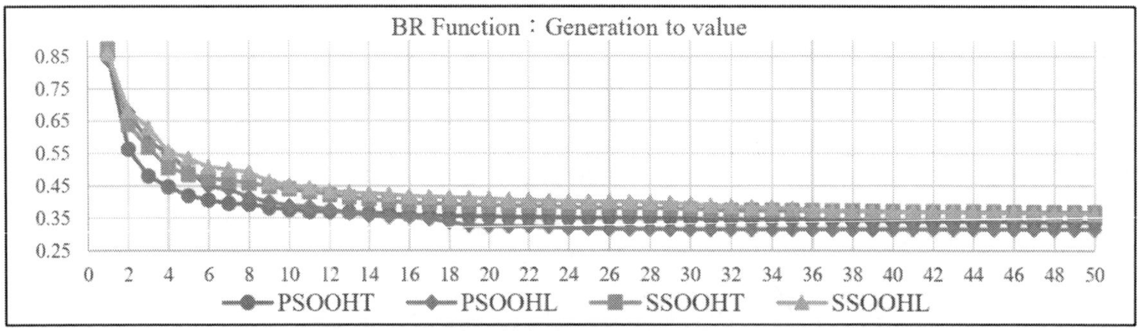

Figure 2: 50 generations of BR function

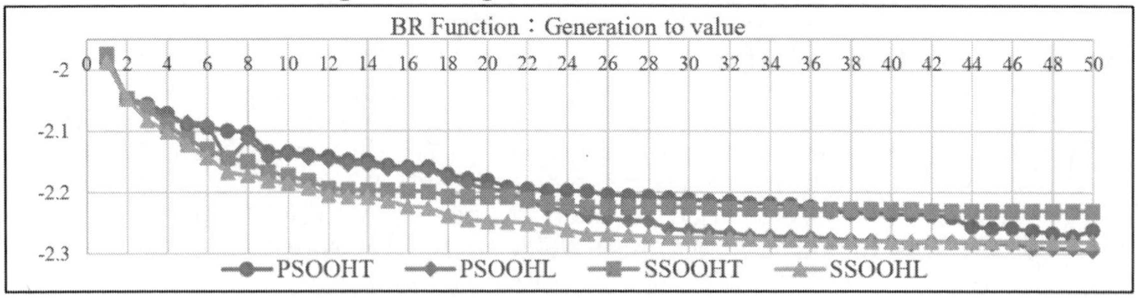

Figure 3: 50 generations of RA function

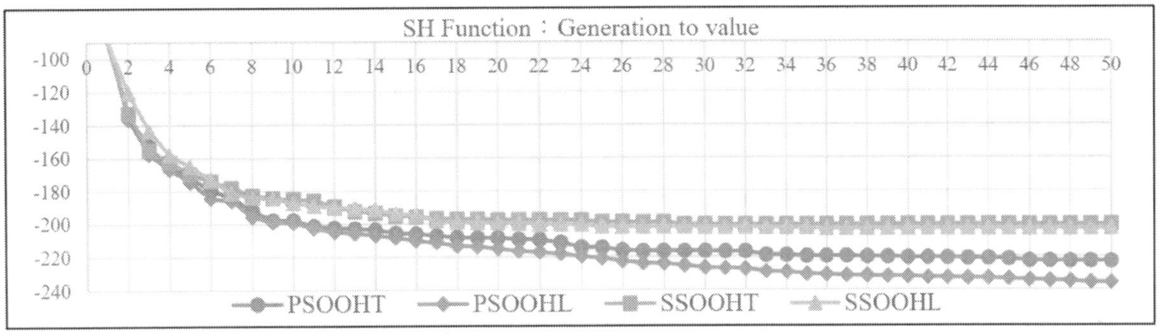

Figure 4: 50 generations of SH function

Comparing the performance of the methods which is hybrid local search or not using two sample t-test, supposed the variance is unknown and unequal, α=0.05, results are shown in Table 3. It can be concluded that the performance of methods has significant difference. With local search, the performance is better than without local search. According to this result, we can further compar the performance of methods used PSO or SSO hybrid OHL, using two sample t-test, also supposed the variance is unknown and unequal, α=0.05, results are shown in Table 5. It can be concluded that the performance of methods has significant difference. With PSO searching algorithm, the performance is better than with SSO searching algorithm.

Table 3: PSO/SSO OHT verse PSO/SSO OHL, PSO OHT/OHL verse SSO OHT/OHL

$H_0 : \mu_1 = \mu_2$ $H_1 : \mu_1 \neq \mu_2$						
Function	Algorithm	p-value	Significant	Algorithm	p-value	Significant
GP	PSO	0.0445	Yes	OHT	0.0216	Yes
	SSO	0.6329	No	OHL	0.0142	Yes
BR	PSO	0.0058	Yes	OHT	0.0071	Yes
	SSO	0.6921	No	OHL	0.0000	Yes
RA	PSO	0.5407	No	OHT	0.105	No
	SSO	0.0441	Yes	OHL	0.0478	Yes
SH	PSO	0.0484	Yes	OHT	0.0005	Yes
	SSO	0.6287	No	OHL	0.0000	Yes

In this paper, we can divide four functions into two analytical types, one is based on if the curve is smooth or not, and the other is based on numbers of global optimal. If the curve is smooth (GP, BR), although the method with local search has better performance than without local search on mean value, but their difference are not all significant among them. The reason is that there are no obvious peaks in the surface, when the position is altered by local search, the performance won't have obvious change. However, if the curve is non-smooth (RA, SH), the performance will obviously change by using local search. Since there are many peaks in the surface, the performance will alter significantly when the particle is moving.

If there is single global optimal (GP, RA), methods with local search have better performance than not, but there are not all have significant difference for all PSO and SSO. It is because particle moves with vector and directional path in PSO, but moves by leaping to another position in SSO. Therefore, if there are fewer local optima, SSO jumps to close to global optima exactly, and PSO also moves to there, then the performance will be similar and not have significant difference. Else, if there are more local optima, SSO has greater chance to trap into the local optima than PSO, so the performance will have significant difference between them. If there are multi global optima (BR, SH), PSO with local search will have significant difference, but SSO will not. Since there are many optimal solutions in the region, PSO will be close to one of the global optima through local search because of direction, but SSO may not approach to the global optima because it leaps randomly, accordingly, local search has less effect in SSO.

5 Conclusion

In this paper, we presented two effective hybrid algorithms (PSOOHL and SSOOHL) that integrated PSO or SSO with OCBA, HT and local search for stochastic multimodal optimization problems. It has been found that the effectiveness and efficient of OCBA, HT, and local search for PSO and SSO. OCBA intelligently allocate and save computing budgets in sampling procedure. With the help of HT and local search can reserve good particles and maintain the diversity of the swarm for preventing premature convergence. Through comparative works and convergence analysis based on four well-known functions, the proposed PSOOOHL algorithm showed the effectively and the efficiently in converging to optimal solution. The future work can consider applying these two approaches to other simulation optimization problems such as resource allocation problem or hybrid flow shop scheduling problem.

6 Reference

[1] Chang, W. D. (2015). A modified particle swarm optimization with multiple subpopulations for multimodal function optimization problems. Applied Soft Computing, 33, 170-182.

[2] Chen, C. H., Lin, J., Yücesan, E., & Chick, S. E. (2000). Simulation budget allocation for further enhancing the efficiency of ordinal optimization. Discrete Event Dynamic Systems, 10(3), 251-270.

[3] Jin, Y., & Branke, J. (2005). Evolutionary optimization in uncertain environments-a survey. IEEE Transactions on evolutionary computation, 9(3), 303-317.

[4] Kennedy, J., Eberhart, R.C. (1995, November). Particle swarm optimization. In Proceedings of 1995 IEEE International Conference on Neural Networks (pp. 1942-1948).

[5] Lin, J. T., & Chiu, C. C. (2015). A hybrid particle swarm optimization with local search for stochastic resource allocation problem. Journal of Intelligent Manufacturing, 1-15.

[6] Liu, B., Wang, L., Jin, Y. H., & Huang, D. X. (2005). Advances in particle swarm optimization algorithm. Control and Instruments in Chemical Industry, 32(3), 1.

[7] Liu, B., Wang, L., Jin, Y., & Huang, D. (2005, May). Designing neural networks using hybrid particle swarm optimization. In International Symposium on Neural Networks (pp. 391-397). Springer Berlin Heidelberg.

[8] Liu, Y., Chung, Y. Y., & Yeh, W. C. (2012, June). Simplified Swarm Optimization with Sorted Local Search for golf data classification. In 2012 IEEE Congress on Evolutionary Computation (pp. 1-8). IEEE.

[9] Lu, H., Sriyanyong, P., Song, Y. H., & Dillon, T. (2010). Experimental study of a new hybrid PSO with mutation for economic dispatch with non-smooth cost function. International Journal of Electrical Power & Energy Systems, 32(9), 921-935.

[10] Marinakis, Y. (2015). An improved particle swarm optimization algorithm for the capacitated location routing problem and for the location routing problem with stochastic demands. Applied Soft Computing, 37, 680-701.

[11] Netjinda, N., Achalakul, T., & Sirinaovakul, B. (2015). Particle Swarm Optimization inspired by starling flock behavior. Applied Soft Computing, 35, 411-422.

[12] Pan, H., Wang, L., & Liu, B. (2006). Particle swarm optimization for function optimization in noisy environment. Applied Mathematics and Computation, 181(2), 908-919.

[13] Pugachev, V. S. (2014). Probability theory and mathematical statistics for engineers. Elsevier.

[14] Selman, B., Kautz, H. A., & Cohen, B. (1994, October). Noise strategies for improving local search. In AAAI (Vol. 94, pp. 337-343).

[15] Sharifi, A., Kordestani, J. K., Mahdaviani, M., & Meybodi, M. R. (2015). A novel hybrid adaptive collaborative approach based on particle swarm optimization and local search for dynamic optimization problems. Applied Soft Computing, 32, 432-448.

[16] Shieh, H. L., Kuo, C. C., & Chiang, C. M. (2011). Modified particle swarm optimization algorithm with simulated annealing behavior and its numerical verification. Applied Mathematics and Computation, 218(8), 4365-4383.

[17] Wang, L. (2001). Intelligent optimization algorithms with applications. Tsinghua University & Springer Press, Beijing.

[18] Wang, L., Zhang, L., & Zheng, D. Z. (2005). A class of hypothesis-test-based genetic algorithms for flow shop scheduling with stochastic processing time. The International Journal of Advanced Manufacturing Technology, 25(11-12), 1157-1163.

[19] Yeh, W. C. (2012). Simplified swarm optimization in disassembly sequencing problems with learning effects. Computers & Operations Research, 39(9), 2168-2177.

AN ERGONOMICS STUDY IN REFUSE COLLECTION TASKS: USING PLASTIC BAG

Chien-Hsin Yang[1*] Swei-Pi Wu[2] Fong-Jung Yu[3]
[1]Department of Computer-Aided Industrial Design
Overseas Chinese University, Taiwan
chyang@ocu.edu.tw

[2]Department of Industrial Engineering and Management Information
Huafan University, Taiwan
spwu@huafan.hfu.edu.tw

[3]Department of Industrial Engineering & Management
Da-Yeh University, Taiwan
fischer@mail.dyu.edu.tw

ABSTRACT

Refuse collection is a key point to reach a cleaning and aesthetic of a city. However, a poor refuse collection task let collectors to suffer from much musculoskeletal disorders. In this study, we attempt to discover the effect of refuse collectors during refuse collection tasks. An ergonomics experiment designed as psychophysical approach to explore the effect of plastic bag type and carrying distances on maximum acceptable weight of carrying (MAWC), heart rate (HR), and rating of perceived exertion (RPE) on refuse collectors. The appropriate set was determined by ANOVA and Duncan multiple comparison test. The result showed that the short distance (5 meter) and handle plastic bag are benefit to refuse collection tasks.

Keywords: Refuse collection, musculoskeletal disorders, MAWC, HR, RPE.

1 INTRODUCTION

Manual material handling (MMH) tasks, especially heavy lifting and carrying are known to be risk factors for disorders in the low back (Hoogendoorn et al., 1999; Hoozemans et al., 2004; Window & O'keefe, 2006). Waste collection often includes manual handling of heavy containers and large plastic bags. To minimize body load during waste collection, wheeled containers have gradually replaced bins and others types of containers. Although using containers brings the benefit of reducing refuse collection loading, it is not a feasible collection task in poor spaces or narrow tunnels. Naturally, plastic bags are a convenient substitute. In practice, high frequency refuse collection tasks including pick up, lifting and carrying may result in musculoskeletal disorders, such as low back pain and upper limb injury (Wu, 2006). Previous ergonomic research has focused on the biomechanical study of refuse collection evaluation. However, the size of refuse collection equipment is not suitable for Chinese/Taiwanese people. Therefore, an investigation on the effects of refuse collection tasks among Chinese refuse collectors is necessary for physiological or occupation medical considerations.

* Corresponding Author

2 METHODS

2.1 Participants

A total of 12 participants were recruited for the ergonomic study. All participants are male and had more than 12 year experiences in classroom cleaning. The mean age was 20.67 years (SD = 2.67), mean height was 172 cm (SD = 4.26), and mean weight was 70.5 kg (SD = 5.4). All participants were right handed and none reported familial heredity diseases. Participants signed an informed consent form before the experiment.

2.2 Experiment

A two-factor randomized complete block design was used to evaluate plastic bag type and carrying distance effects on refuse collection tasks.

2.2.1 Independent variables

The two plastic bags investigated are illustrated in the following figure. The figure on the right is handled bag (HB); that on the left is a without handled bag (WOHB). The second factor investigated was carrying distance, which was divided to three levels: 5m, 12m, and 20m.

Without handled bag (WOHB)	Handled bag (HB)

Figure 1: Plastic bag types

2.2.2 Dependent variables

Refuse collection task was estimated by the maximum acceptable weight of carrying (MAWC), heart rate (HR), and rating of perceived exertion (RPE). The units of these variables were kg, beats per minute (bpm), and points (between 6 to 20 points).

2.3 Apparatus

The size of the plastic bags was 83 × 90 cm and contained approximately 75 l of refuse. The plastic bags were made from PET. In addition to the plastic bags, a heart rate trace instrument was used to collect participants' heart rates. To implement a psychophysical experiment, several irregular pieces of sand were added to the plastic bags to adjust carrying weight.

2.4 Procedure

Participants were required to finish three training tests before the formal experimental task. At the formal experiment, participants were instructed to adjust the amount of carrying weight for the different bags and distances without straining themselves or becoming unusually tried, weakened, overheated, or out of breath (Ciriello & Snook, 1983). During the carrying tasks, participants varied the weight in plastic bag using irregular pieces of sand. Each participant could adjust the weight as he felt necessary to arrive at the weight he felt he could carry for the tested distance. An experiment point included 20 runs with different types of plastic bags. Participants' heart rates were collected per minute, and the mean heart rates for the last five records were used for analysis. Each participant was asked to rate their perceived exertion in their fingers, wrist, elbows, shoulders, lower back, and waist.

2.5 Analysis

To examine the effects of MAWC, HR, and RPE, data were analyzed using ANOVA. Duncan's multiple range test was used for post hoc comparisons. An alpha level of 0.05 was selected as the minimum level of significance.

3 RESULTS

The experimental data were analyzed by ANOVA for each dependent variable: MAWC, HR, and RPE. The ANOVA results are summarized in Table 1.

Table 1 Summary ANOVA with MAWC, HR and Overall RPE as Dependent Variable (N = 12)

	MAWC	HR	Overall RPE
Plastic bag type	*	NS	*
Carrying distance	*	*	*
type × distance	NS	NS	NS

* Significant at 0.05; NS: not significant

3.1 Maximum acceptable weight of carrying (MAWC)

Table 2 Effect of MAWC on Refuse Collection Task

	Plastic bag type	
Distance	HB	WOHB
5 m	9.21 / 0.56*	8.63 / 0.68
12 m	8.63 / 0.83	8.13 / 0.62
20 m	8.33 / 0.90	7.81 / 0.75

Unit: kg; *(mean/stand deviation)

Table 3 Duncan Results for Carrying Distance on MAWC

Variable	level	Average	Duncan result	Significance
	5m	8.97	A	Alpha=.05
Carrying distance	12m	8.38	B	
	20m	8.07	C	

Unit: kg

As shown in Tables 1 and 2, we found that the effect of plastic bag type and carrying distance was significant. The Duncan test indicated that the MAWC at 5m was better than at 12m and 20m.

3.2 Heart rate (HR)

The experimental result indicated that the effect of carrying distance was significant. Carrying distance was divided into two groups based on the Duncan test (see Table 4).

Table 4 Duncan result for carrying distance on HR

Variable	level	Average	Duncan result	Significance
	5m	91.42	A	Alpha=.05
Carrying distance	12m	92.92	B	
	20m	93.12	B	

Unit: bpm

3.3 Rating of perceived exertion

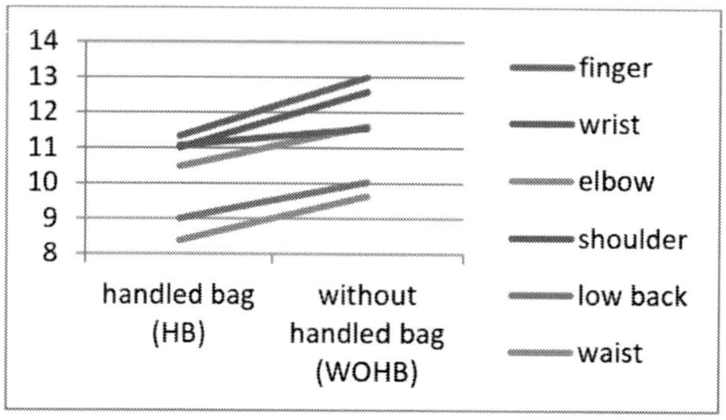

Figure 2: The effect of plastic bag types on RPE

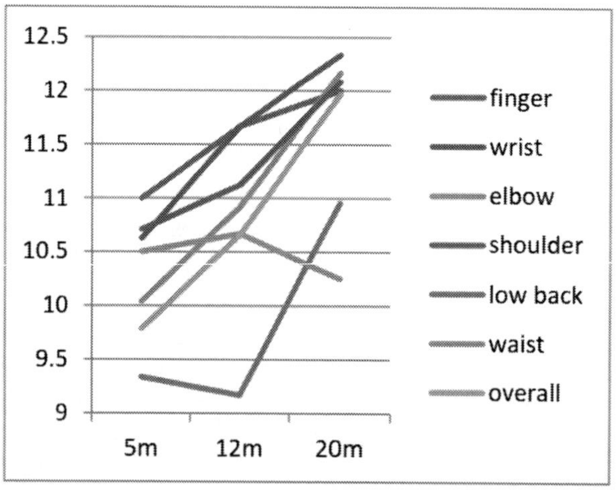

Figure 3: The effect of carrying distance on RPE

For plastic bag type and carrying distance, the effect on RPE on the fingers was greater than on the other positions. As shown in Figures 2 and 3, the loading of the lower back and waist was lower.

3.4 Summary

All factors (type and distance) were significant in this ergonomic experiment. Shorter distance yielded better responses for MAWC, HR and RPE. A reasonable inference of these findings is that the task end before the intersection of high physiological cost, such as aerobic capacity (O2 exchange). These results are similar to Snook (1978). As for plastic bag type, the handled bag had better carrying capacity in the current experiment. However, people often use the without handled bag to avoid refuse falling out, which is not good for refuse collection tasks and would include additional picking up, lifting, and carrying etc.

4 CONCLUSIONS

For this ergonomic study of refuse collection task, the researchers recruited 12 participants and used a psychophysical approach to evaluate the effects of plastic bag type and carrying distance on MAWC, HR, and RPE. The results showed that the effect of the two factors was significant on the response variables (in addition to HR). Higher MAWC and lower RPE appeared at shorter distance (5m). Although the effect of plastic bag type was not significant for HR, its effect on RPE was significant. Loading on the fingers was higher than the other positions; thus, RPE was also higher. The results of this brief ergonomics study showed that shorter carrying distance with handled plastic bags is a better setting for refuse collection tasks.

5 REFERENCES

[1] **Ciriello, V. M. and S. R. Snook.** 1983. A study of size, distance, height, and frequency effects on manual handling tasks, *Human Factors*, 25, pp 473-483.

[2] **Hoogendoorn, W.E., van Poppel, M.N.M., Koes, B.W., Bouter, L.M.** 1999. Physical load during work and leisure time as risk factors for back pain. *Scandinavian Journal of Work, Environment & Health*, 25(5), pp 387-403.

[3] **Hoozemans, M.J.M., Kuijer, P.P.F.M., Kingma, I., van Dieen, J.H., de Vries, W.H.K., ven der Woude, L.H.V., Veeger, H.E.J., van der Beek, A.J. and Fring-Dresen, M.H.W.** 2004. Mechanical loading of the low back and shoulders during pushing and pulling activities. *Ergonomics*, 47(1), pp 1-18.

[4] **Snook, S.** 1978. The design of manual handling tasks. *Ergonomics*, 21(12), pp 963-985.

[5] **Window, J. and O'Keefe, V.** 2006. Musculoskeletal disorders of refuse collectors emptying litter bins which are housed inside street furniture. 2006. Human Factors & Ergonomics Society of Australia 42nd Annual Conference.

[6] **Wu, S.P.** 2006. Psychophysically determined 1-hr load carrying capacity of Chinese females. *International Journal of Industrial Ergonomics*, 36, pp 891-899.

OPTIMIZATION OF RESERVATION STRATEGY IN MULTIPLE MEDICAL DIAGNOSIS E UIPMENT APPOINTMENT

ABSTRACT

J.F. Xu, . Ma, P. Xu, .W. Xu, .X. Liu and F. Liang*
Department of Management Science and Engineering
Nankai University, Tianjin, China
liangfeng79@nankai.edu.cn

Appointment scheduling of large medical diagnosis equipment, which aims at the needs of different patients, is an important issue in hospital management. In this paper, on the basis of different demand characteristics, the patients are divided into three basic types: outpatient, inpatient, emergency patient. Taking into account the situation, hospitals generally arrange two sets of equipment to service patients at the same time. Based on a given reasonable hypothesis, the system can be modelled as Markov Decision Process (MDP) and the corresponding Discrete Dynamic Programming Model is constructed. The optimizing decision of the model is the revenue of the diagnostic equipment considering the reasonable appointment scheduling. The solution is described in detail in the numerical examples. Finally, performance comparison indicates that the proposed policy is capable of sequencing patients' appointments more efficiently.

Keywords: Medical Diagnosis, Appointment Scheduling, Markov Decision Process (MDP), Dynamic Programming Model

1 INTRODUCTION

In a complex medical system, large diagnosis equipment, e.g. Nuclear magnetic resonance imaging equipment (MRI), Electronic computer tomography scanner (CT) and X-ray, belong to key components which decide efficiency of diagnosis and treatment. It is very important to assign limited medical equipment capacity rationally in different types of patients who compete with each other. Thus, the purpose of appointment scheduling meets the requirements of medical examination equipment allocation reducing the waiting time for the patients to check, making the best of the medical equipment, and allocating patient needs and equipment capacity reasonably. Appointment scheduling can be divided into present allocation strategy and previous service day's allocation strategy, according to the waiting time for patients to check. In addition, the waiting time can be divided into two parts, which are indirect delay time (The difference between the time of appointment and the actual service) and direct delay time (The difference between the time of sending service request and appointment).

Generally, demands of the medical diagnostic equipment come from different types of patients. These patients can be divided into three typical types: outpatient, inpatient and emergency patient, which have their own features. Outpatients usually make appointments a day or even a week in advance online or by telephone, and only a small part of the patients may cancel in advance' or break an appointment'. These patients complicate the appointment scheduling problem and bring troubles to the hospital managers; when some patients need to receive medical treatment on that day. After diagnosis and treatment, if patients need to further check, they need to pay fee at the paying office, then reserve diagnostic equipment on the artificial service office. Inpatients and

* Corresponding Author

emergency patients all inside the hospital generally need such equipment when they have physical examinations. The difference is that emergency patients need great timeliness, and require being met in a timely manner. Otherwise, it will cause irreparable consequences for the patients, which means that hospital refusal cost is high. Therefore, as long as emergency patients need diagnostic equipment, hospital won't refuse these patients. By contrast, the urgency of inpatients is relatively small. In a word, if set a priority for each type of patients, priority of the emergency patients is the highest. Because hospital managers regard medical equipment inspection cost as extra income and equipment maintenance cost as their fee, priorities of these two types of patients can be classified according to the specific circumstances.

The appointment system makes a great contribution to medical resources in recent years. However, it's a major problem to make use of the appointment system to improve the utilization of the medical equipment and reduce the patients' waiting time in the hospital. At present, after the online reserving, appointment system can arrange the patients who need diagnostic equipment on recent idle time, according to the principle of First Come First Served (FCFS). Only when all time periods are not available within the available periods, system will reject patients' request. However, as mentioned above, the expected waiting time (time allowed to wait) of different types of patients is also different. Some patients whose symptoms are serious need to receive checks as soon as possible. Once the hospital refuses these patients, it will cause the huge refusal cost. Some patients whose symptoms are mild can wait a little. Revenue from the patients can be less than a serious illness. However, due to the complexity of the appointment scheduling problem, most hospitals' appointment systems use first come first served strategy. These systems may lead that the hospital refuses higher yields of other types of patients, who appear later in the day or make huge refusal cost. Therefor, this appointment strategy is not an optimal control strategy.

This paper introduces Markov Decision Process to analyse the optimal appointment strategy and build the optimal model [13-14]. The rest of this paper is organized as follows: section 2 is the review of the related literature and section 3 proposes a MDP model to present the appointment policy. Performance evaluation of the proposed appointment policy is presented in section 4. Finally, conclusions and possible future research are discussed in section 5.

2 LITERATURE REVIEW

When it comes to the problem about reservation control decision of medical inspection equipment, some relevant literature has made the research. In the literature, for this type of appointment scheduling decision problem, there are various research methods, mainly including simulation, analytic method and so on. Among those methods, analytic method mainly includes MDP and queuing theory, etc. Each paper has different practical problems needed to be solved, and even slightly difference of assumptions can lead to different research methods.

References [1-2] mainly used the method of computer simulation, showed the actual effect according to the result of the simulation, and carried on the analysis. In reference [3], ueuing method was adopted. In view of the situation of a single set of MRI equipment, the relevant parameters were formulated when the patients were queuing for inspection, some important parameters like average queue length, average waiting time, and then they showed the main purpose of the study--achieving the maximization of system revenue. Reference [4] proposed MDP discrete dynamic model, and they used it to solve medical departments' patient scheduling problem with different priority classes, through the optimal solution of dynamic programming model, they determined in N day, how many patients with different priority classes were arranged to check, that was, the optimal capacity allocation scheduling of medical equipment. Reference [5] and [6] were also based on MDP method, in order to solve the appointment scheduling control problem.

References [7-9] mainly focused on stroke patients in need of diagnosis, and put forward the solution of Contracted Time Slots (CTS). The kind of stroke patients could be diagnosed in regular time slots, but also by the way of intentionally set CTS. However, other ordinary patients could only be diagnosed in regular time slots. These three papers were all based on the combination of MDP and stochastic programming, and they successively made joint optimization of stroke patients' dispatch strategy, CTS decision and reservation control strategy of the advance cancellation of CTS. References [10-12] also used the method of MDP, established the dynamic programming model of discrete time and studied the problem of appointment scheduling.

In summary, there are many references using MDP as the modelling approach of appointment policy. However, rare research considers the different patient types. In the paper, the patients are modelled as three types and MDP model is introduced to investigate the optimal appointment policy.

3 THE MDP MODEL

3.1 Foundation of MDP model

This paper aims at the situation that there are two sets of inspection equipment with same functions, and the patients who need to be examined are divided into three classes. In this case, the paper explores how to arrange the appointment scheduling of the two sets of equipment reasonably so that hospitals can obtain the expected total maximum revenue on the day of reservation in the medical inspection equipment system. When reservation requests of the patients arrive, hospitals need reservation system to make decisions to accept or reject the requests immediately.

This process can be specifically translated into a discrete MDP that contains the limited time domain. At any moment t of the day of the reservation period T(t=T,T-1, ,1),the reservation requests of two types of patients (outpatients and reservation patients) arrive randomly according to their arrival probabilities(λ_1 and λ_2) of respective reservation requests and the process is independent of each other. But for each reservation time t, due to the limitation of the number of equipment and equipment service ability, the maximum number of patients' reservation requests is equal to two. At the same time, there isn't interaction between the two reservation time slots. And the patients' reservation requests that each set of equipment corresponds to are also independent of each other. Whenever accepting the requests of different types of patients, the hospital will get the corresponding incomes (r1, r2 and r3); similarly, whenever refusing the reservation requests of different types of patients, hospital will have to receive corresponding penalty costs (c1, c2, and k). In addition, according to the reality, there are $c_1 < k$, $c_2 < k$, $r_1 + c_1 < r_3 + k$, $r_2 + c_2 < r_3 + k$ generally. It can be deduced that the hospital is better not to refuse but meet inspection demands from emergency patients as far as possible. According to the above brief description, the appointment scheduling problem of the medical inspection equipment can be translated into dynamic programming problem in the limited time domain.

3.2 Decision and system Stage of MDP

Accessible time range is t at per cycle in the decision stage. And S stands for the available capacity of inspection equipment A and B have, and S satisfies $S_j \in [\,0\,,\mathrm{N}\,]$ with j=A,B.

When reservation requests of patients arrive, reservation center will take action $a_t(S_A, S_B)$ based on the current system stage S=（S_A, S_B） ,which represents the decision of acceptance or refusal.

$$a_t(S_A, S_B) = \begin{cases} 0 & refusal \\ 1 & acceptance \end{cases}$$

3.3 Transition Probability of MDP

The arrival rates are λ_1 for out-patients and λ_2 for inpatients. Therefore, there are six situations and respective transition probabilities at per reservation period, which are shown as follows:

$P_1=\lambda_1^2$ (two reservation requests from out-patients arrive at the same time.)

$P_2=\lambda_2^2$ (two reservation requests from in-patients arrive at the same time.)

$P_3=2\lambda_1\lambda_2$ (two reservation requests from an in-patient and an out-patient arrive at the same time.)

$P_4=2\lambda_1(1-\lambda_1-\lambda_2)$ (one reservation request from an out-patient arrives.)

$P_5=2\lambda_2(1-\lambda_1-\lambda_2)$ (one reservation request from an in-patients arrives.)

$P_0=(1-\lambda_1-\lambda_2)^2$ (no reservation request arrives.)

As a result,

$$\sum_{i=1}^{5} P_i = 1 - P_0 \qquad (1)$$

3.4 Reward of MDP

Every time the hospital accepts request from patient i,the reward will increase by r_i ,and every time refuses request from patient i,the reward of hospital will reduce by c_i. Specifically, for emergency patients, penalty cost for being refused will increase by k. the unit punishment cost of available capacity is π . Furthermore, it's one of the crucial factors to achieve the optimal decision and decide the existence of optimal appointment scheduling that t penalty cost and reward between different kinds of patients are different. In this progress, reward from all these patients depends on the scale of hospital and inspection project. Hence, we do not initialize the magnitude relation betweenr_1 and r_2, c_1 andc_2. However, in order to reflect the priority of emergency patient, the magnitude relation between these parameters and reward of emergency patients are invariable as follows: $r_1 + c_1 < r_3 + k, r_2 + c_2 < r_3 + k$.

3.5 Value Function of MDP

The maximum expected reward of hospital of the period between t = t to t=0 is defined as $V_t(S_A, S_B)$ .The Value Function for t=T, T−1, ⋯, 2, 1 based on the MDP progress can be built as follows:

$$\boldsymbol{V_t(S_A, S_B) = \sum_{i=0}^{5} P_i H_{t-1}^i (S_A, S_B)} \qquad \textbf{(2)}$$

Iteration of the above function requires a marginal restriction to ensure the existence of optimal solution. The details are as follows: at the moment of t=0 when the reservation finishes and the equipment begins to work, emergency patients will show off that day. The assumed total number of arrival people obeys the random distribution of function y. If the total number of emergency patients Y surpasses the available capacity of two sets of equipment S ($S = S_A + S_B$) ,the hospital will reject the excess requests from emergency patients and afford penalty costs at the same time; If the available capacity of two sets of equipment S surpasses the number of emergency patients , equipment will be free. At the end of that day, the total available capacity is $(S-E(Y))^+$ and causes corresponding costs.

As a result,$V_0(S_A, S_B)$ can defined as:

$$\boldsymbol{V_0(S_A, S_B) = r_3 E((S_A + S_B)^\wedge Y) - \pi E\big((S_A + S_B) - Y\big)^+ - kE\big(Y - (S_A + S_B)\big)^+} \quad (3)$$

Where $(a \wedge b)$=min(a, b); $(a)^{+}$=max$(a, 0)$. In these formulas, $H_t^i(S_A, S_B)$ is defined as the probability of occurrence of situation i. is defined as the maximum expected revenue function at the end of reservation at situation i wherei $= 0,1,2,3,4,5$. i represent different situation of reservation arrival for two sets of equipment.

Specifically, when $i = 3$, which means there are two reservations from an outpatient and an inpatient arriving at the same time. Correspondingly, it can be the outpatient making a reservation to equipment A ,the in-patient making a reservation to equipment B, or the out-patient making a reservation to equipment B and the in-patient making a reservation to equipment A. Therefore the transition probability is $P_3 = 2\lambda_1\lambda_2$.Besides, because of the selecting priority of equipment A, the available capacity of equipment A may attain in advance. In this case, only one request can be accepted and equipment B will serve for it.

Through the MDP model, the practical problem of appointment scheduling of two sets of medical equipment has been transferred into mathematical model for quantitative analysis. A certain reservation decision from solving the optimal solution of the model can be made and the existence of optimal solution from equation (1) and (2) can be ensured. The relation between the summation of S_A and S_B and exception value of Y happens to be the marginal restriction of equation in dynamic programming.

When the series of random reservation requests that from different patients arrive, the critical value of the time-varying remaining capacities of two sets of equipment will be generated. Based on the comparison with the critical value, the system can make appropriate decisions for each period (i.e. accept the request or reject request) automatically, so that the final hospital revenue reaches the maximum. The proposed policy will be described in detail in the example of the next section.

4 NUMERICAL EXAMPLES

Considering the model assumptions, parameters are set as follows: The maximum daily capacity of each piece of equipment (N) is equal to 40. Each reservation period is divided into 50 equal parts averagely, which means that T=50 is the beginning of the reservation,T=1 is the last time of a reservation period and T=0 is the end of reservation, namely the start time of service. The arrival time of reservation requests of different patients is respectively: λ_1=0.7 and λ_2=0.2. So five transition probabilities corresponding to reservation request can be calculated as follows:$P_1=\lambda_1^2$=0.49; $P_2=\lambda_2^2$=0.04; $P_3=2\lambda_1\lambda_2$=0.28; $P_4=2\lambda_1(1-\lambda_1-\lambda_2)$ =0.14; $P_5=2\lambda_2(1-\lambda_1-\lambda_2)$ =0.04.

In addition, the possibility of no patient request arriving in the appointment time is P_0, P_0= $(1-\lambda_1-\lambda_2)2$=0.01. In the equipment's service day, the number of emergency patients is represented by the Poisson distribution and the parameter is 15. That means the excepted number of emergency patients arriving at hospital on the day is equal to 15. What's more, the parameters of rewards and punishment are set. According to the reality, it is assumed that the hospital receives the same benefit for checking different kinds of patients, that is $\gamma_1=\gamma_2=\gamma_3$=200. However, the penalty cost depends on the kinds of patients, generally speaking. Here,considering the real-world factors, two situations may happen: one is the penalty cost c_1 caused by rejecting an outpatient is less than c_2 caused by rejecting an inpatient; the other is that c_1 is higher than c_2. Limited by time and space,only the first situation is continued in detail. Therefore these parameters are set, c_1=100, c_2=200 and k =400. If hospital rejects an emergency patient's inspection requirement, the best diagnosis and treatment time will be delayed, which would influence the subsequent treatment and cause a more serious consequence compared to rejecting inpatients and outpatients. Finally, the available capacity of hospital inspection equipment (π) is assumed, which is equal to 100, π=100. According to the above parameters setting and the theory of value iteration algorithm, matlab is used to program. At first, a set of data are set, based on arriving of different patients' requests and the reservation request arriving rate of two kinds of patients. 0, 1 and 2 stand for corresponding situations, where 0 means that no patient reservation request arrives, 1 means that outpatients' requests arrive and 2 means that inpatients' requests arrive. Then, according to patients' reservation requests arrival, the system simulates the decisions for each reservation time, through multilevel iterations and comparisons. The decisions are receiving two patients' requests, receiving a patient' request or rejecting two requests, where 0 means 'rejection' and 1 means 'reception'. The process of patients' reservation and system decision are simulated by matlab. A surface chart is concluded showing (to show) the excepted total revenue of hospital, where S represents total available capacity of two sets of equipment, as shown in Figure 1. In the figure, when the total available capacity is equal to 40, excepted total revenue. In addition, the hospital reservation system would make the optimal decision for patients' requests at each time of reservation, as shown in Table 1. In the table, Accept(P) and Accept(T) represent the accept or not' decision in the proposed policy and FCFS policy respectively. 1 means accept' and 0 means reject'. AC(P) and AC(T) express the available capacity in the proposed policy and FCFS policy respectively.

Figure 1: Hospital e cepted revenue variation

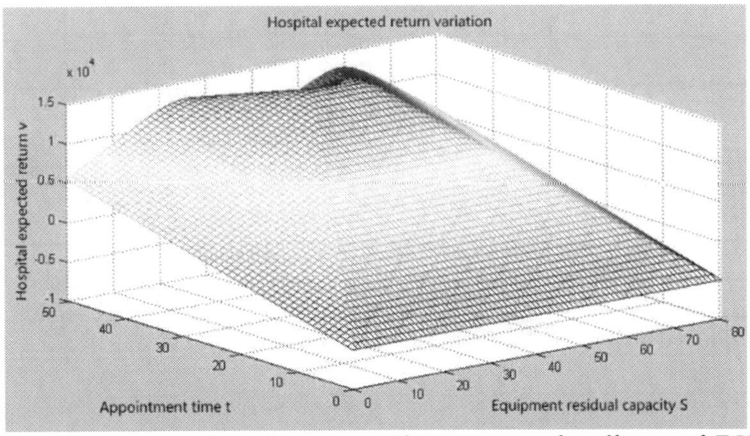

Table 1: The comparison between the proposed policy and FCFS policy

t	Patient type		Accept(P)		AC(P)		Accept(T)		AC(T)	
50	1	2	1	1	40	40	1	1	40	40
49	2	2	1	1	39	39	1	1	39	39
48	0	2	0	1	38	38	0	1	38	38
47	1	1	1	1	37	38	1	1	37	38
46	1	1	1	1	36	37	1	1	36	37
45	1	2	1	1	35	36	1	1	35	36
44	2	1	1	1	34	35	1	1	34	35
43	1	1	1	1	33	34	1	1	33	34
42	1	1	1	1	32	33	1	1	32	33
41	0	1	0	1	31	32	0	1	31	32
40	2	1	1	1	30	32	1	1	30	32
39	1	2	1	1	29	31	1	1	29	31
38	1	1	0	1	28	30	1	1	28	30
37	1	2	1	1	27	30	1	1	27	29
36	1	1	0	0	26	29	0	0	26	28
35	2	1	1	0	26	29	1	1	25	27

34	1	1	1	1	25	29	1	1	24	26
33	1	1	0	1	24	28	1	1	23	25
32	1	1	0	1	23	28	1	1	22	24
31	1	1	0	1	22	28	1	1	21	23
30	1	1	1	0	21	28	1	1	20	22
29	2	0	1	0	20	28	1	0	19	21
28	1	1	1	0	19	28	1	1	18	21
27	0	1	0	1	18	28	0	1	17	20
26	1	1	1	1	17	28	1	1	16	20
25	2	1	1	0	16	27	1	1	15	19
24	1	1	1	0	15	27	1	1	14	18
23	1	1	1	0	14	27	1	1	13	17
22	1	1	1	0	13	27	1	1	12	16
21	0	0	0	0	12	27	0	0	11	15
20	1	1	1	1	12	27	1	0	11	15
19	0	1	0	1	11	26	0	1	10	15
18	1	2	1	1	10	26	0	1	9	15
17	2	1	1	0	9	25	1	0	8	15
16	1	1	1	0	8	25	1	0	7	15
15	1	1	1	1	7	25	1	0	6	15
14	2	1	1	0	6	24	1	0	5	15
13	1	1	0	1	5	24	1	0	4	15
12	1	0	0	0	4	24	1	0	3	15
11	1	1	1	1	4	24	1	0	2	15
10	1	1	1	0	3	23	1	0	1	15
9	2	1	1	1	2	23	0	0	0	15
8	1	1	0	1	1	22	0	0	0	15
7	1	1	0	0	0	22	0	0	0	15
6	1	1	1	0	0	22	0	0	0	15
5	1	0	1	0	0	21	0	0	0	15
4	1	0	1	0	0	20	0	0	0	15
3	0	2	0	1	0	19	0	0	0	15
2	1	1	1	0	0	18	0	0	0	15
1	1	2	0	1	0	17	0	0	0	15
0	0	0	0	0	0	16	0	0	0	15

According to the above mentioned method, through making appointment scheduling decisions for two sets of inspection equipment, the total revenue of hospital is calculated, which is equal to 13600. The result increases 300 compared with the revenue 13300 computed by traditional method. This result proves the method proposed in this paper is efficient to help hospital administrators to solve this issue to a certain extent.

5 CONCLUSIONS

This paper focuses on the appointment reservation policy by analysing the situation of two sets of inspection equipment and three different needs of patients. MDP model is introduced to make an in-depth study of the problem and gives reasonable assumptions and constructs discrete dynamic programming model in accordance with the reservation problem. The dynamic programming model is further improved by reasonable setting of parameters in the model. Then, this paper obtains the optimal appointment scheduling strategy in correspondence with the instance, and also the maximum expected total profit the hospital can get after carrying out the decision. Compared with the traditional reservation strategy, the proposed policy can be observed to be more effective and efficient. In the future work, reservation policy taking into account of other patient types can be conducted.

ACKNOWLEDGE

This research was supported by the National Natural Science Foundation of China (Grant No. 71271122).

REFERENCES

[1]**Lev B, Revesz G, Shea F, Caltagirone R. 1 6.** Patient flow analysis and the delivery of radiology service, Socio - Economics Planning Sciences, 10: 159-166 .

[2]**Patrick , Puterman M L. 200 .** Improving resource utilization for diagnostic services through flexible inpatient scheduling: a method for improving resource utilization, Journal of Operational Research Society, 58: 235-245.

[3]**Den Boer A V, Koole G M, Van der Mei R D, Zwart B. 200 .** Capacity management for a diagnostic medical facility, Working paper.

[4]**Patrick , Puterman M L, ueyranne M. 2008.** Dynamic multi-priority patient scheduling for a diagnostic resource, Operations Research, 56(6) : 1507-1525.

[5]**Kolisch R, Sickinger S. 2008.** Providing radiology health care services to stochastic demand of different customer classes, O R Spectrum, 30: 375-395.

[6]**Saure A, Patrick , ueyranne M. 2012.** Dynamic Multi-appointment Patient Scheduling for radiation therapy, European Journal of Operational Research, 223(2): 573-584.

[7]**Geng N, Xie X, iang Z. 2011.** Capacity Reservation and Cancellation of Critical Resources, Automation Science and Engineering, IEEE Transactions on,8(3): 470-481.

[8]**Geng N, Xie X, Augusto V, et al. 2011.** A Monte Carlo optimization and dynamic programming approach for managing MRI examinations of stroke patients, Automatic Control, IEEE Transactions on, 56(11): 2515-2529.

[9]**Geng N, Xie X. 2012.** Optimizing contracted resource capacity with two advance cancellation modes, European Journal of Operational Research, 221(3): 501-512.

[10]**Green L, Savin S, Wang B . 2006.** Managing patient service in a diagnostic medical facility, Operations Research, 54 (1): 11-25.

[11]**Zhuang W F, Li M Z F . 2010.** A new method of proving structural properties for certain class of stochastic dynamic control problems, Operations Research Letters, 38: 462-467.

[12]**Gupta D,Wang L. 2008.** Revenue management for a primary-care clinic in the presence of patient choice, Operations Research, 56(3): 576-592.

[13] **Xiao Yongbo, Zhu, Yan.2016.** Value management of diagnostic equipment with cancelation, no-show, and emergency patients, Naval Research Logistics, 63(4):287-304.

[14] **G. Dobson, S. Hasi a, and E. . Pinker.2011.** Reserving capacity for urgent patients in primary care, Production and Operations Management, 20 (11):456 473.

ROBUSTNESS RESEARCH OF DATA-DRIVEN MEDICAL DECISIONS BASED ON MUTUAL INFORMATION FEATURE SELECTION MODEL

Man Xu[1], *Jiang Shen*[2*], *Haiyan Yu*[3], *Dan Gan*[2]
[1] *Business School, Nankai University, Tianjin, China*
td_xuman@nankai.edu.cn
[2] *College of Management and Economy, Tianjin University, Tianjin, China*
motoshen@163.com gandan@tju.edu.cn
[3] *School of Economics and Management, Chongqing University of Posts & Telecommunications, Chongqing, China*
yhy188@tju.edu.cn

ABSTRACT

In medical decision making, the feature dimensions of medical and health data are high and can easily be interfered with by the noisy attributes, thus causing random errors and frequent medical malpractice. To identify interesting relationships between pairs of variables in medical data resources, this paper presents a robust feature selection method based on relation-redundancy coefficient (R^2C), which captures a wide range of associations including the feature correlation, redundancy ratio, and the correlation of the features and their class labels. R^2C belongs to a larger class of mutual information-based feature selection methods for identifying and predicting relationships, in which robust decisions are obtained through a heuristic information retrieval algorithm in its feature space. Comparing with several conventional methods, the results on the benchmark data sets show that the proposed method effectively eliminates the redundant features, with its error decreasing to 14.28%±1.74% and the average time decreasing to 0.0019s for selecting each feature.

Keywords: *Data-driven decision-making, Relation-Redundancy coefficient, Feature selection criteria, Medical decision inference, Robustness*

1. INTRODUCTION

Since the New England Medical Center in America first put forward and applied the Clinical Pathway (CP) method in 1985, the health sector has been seeking a means which combines medical technology and industrial engineering standardized methodology to improve treatment efficiency and accuracy and regulate the behavior and process of physician diagnoses, to reduce systematic errors and to prevent medical accidents and mistakes. With the development of artificial intelligence, industrial engineering theory has taken advantage of artificial intelligence to create medical expert systems and medical artificial intelligence to provide assistance with medical diagnosis and treatment. In the big data environment, transforming linear decision models primarily consisting of CP to data-centric flat models of intelligent decision making can promote the tendency of parties involved in medical decisions and medical information flow to be pluralistic and interactive. According to the McKinsey report, the total amount of data produced in global medical service establishments was approximately 5000 PB in 2012. Furthermore, medical imaging archiving, electronic medical records, medical research information, hospital records and other documents were close to 10000 PB in 2014. As data have become a recognized resource, intelligent decision making driven by data has become a focus in managerial science research (Lee 2014). Data-driven medical decision making and intelligent medicine has also become an important theory in applied research. However, the implementation of intelligent decision making driven by data in medical practice has progressed very slowly and with difficultly. Although the progress is limited by constitution, stereotype, complexity of medical knowledge and other factors, the value of intelligent decision making driven by data in medical practice is expected to be significant (Cho et al, 2015). Harvard Business Review has claimed that the intelligent medicine industry will provide a comprehensive range of services for physicians, patients, and the community by consolidating evidence and value, and this process is also a subversion process of medical innovation (Porter and Lee, 2013). Traditional medical health systems are facing the impact of massive data, specifically, how to change data pressure into data advantage. Utilizing large sets of medical data as standardized evidences that can be responded at any time in a medical clinic has become a hot issue in the field of medical and health decisions. This approach helps to alleviate the medical resource deficiency and regional distribution inequality in China.

Medical decision making is a biochemical process that generates large quantities of heterogeneous data. The dimension of medical data is high and is easily influenced by environmental factors, which causes difficulties in improving the efficiency and accuracy of medical decision making, in acquiring high quality medical data, and in minimizing the impact of disturbing factors and redundant data in reasoning processes. Thus, the feature selection method has become an effective way to achieve robust medical decisions. Decision robustness was an important criterion in assessing candidate/feasible solutions under an uncertain environment (Lempert, 2003). Robust decision making systematically considers uncertainties and assumptions to achieve the insensitive decision scheme adapted to different decision-making environments and uncertainties (Lempert et al, 2006; NAGATOU et al, 2015). Hansen applied robust decision making to nonlinear system control problems and put forward a robust decision-making quantitative analysis method for modeling uncertainties (Hansen and Sargent, 2011). Robustness of medical decision systems was also defined from the perspective of system status (Xu, 2011; Xu et al, 2012). For a medical decision-making system (DMS), robustness requires that the DMS maintain high reasoning accuracy and sensitivity when the DMS data are uncertain or disorderly (Shen et al, 2015). Sargent and Hansen (2007), the 2011 Nobel Prize Winners, first proposed a mutual information method to measure the information differences between the specifications of the model and its true status. This paper studies the method to eliminate interference factors effectively from the perspective of feature selection. From the perspectives of system efficiency and generalization ability (Yao et al, 2012), the method selects the most effective features from the original feature space in the classification process to reduce the feature space dimension.

2. LITERATURE REVIEW

Exploring a large dataset has significant commercial value, social management and public service value and intelligent decision support value, thus the management and decision studies driven by large data sets have outstanding scientific and strategic significance and have been widely implemented in the fields of public administration, business management, financial management, manufacturing, and healthcare (Xu et al, 2014; Reshef et al, 2011). Nature and Science published special issues discussing processing of large data sets and the implications to internet technology, internet economics, supercomputing, environmental science, and biomedicine (Thomas, 2014; Fung et al, 2015). Feng and Chen also gradually investigated concepts, ideas, methods, academic routes and other issues associated with the application of large data sets in management and decision making (Feng et al, 2013; Chen et al, 2012). In healthcare, disease diagnosis and treatment standards based on big data analytics can investigate the trajectory and evolution phase of a disease, and data-driven disease prevention and lifestyle management can also reveal the treatment-related mechanisms to provide personalized service for healthcare providers (Manyika et al, 2011).

The uncertain information for medical decision making increases data redundancy (Pan et al, 2013), creating bottlenecks in improving system reasoning efficiency. Therefore, the research concerning efficient elimination of data redundancy causes widespread concern. To improve the case search performance, a genetic algorithm was adopted to eliminate the redundancy of high-dimensional data, but when the case library is in a disordered state, the case search easily falls into local optimum that leads to lower reasoning accuracy (Han et al, 2015). A hybrid genetic algorithm was proposed based on two-dimensional array coding to solve the problem of multiple levels of redundancy allocation, and the results showed that the two-dimensional array was resilient to solve such a problem (He et al,

2013). Moreover, the genetic algorithm was applied to a weighted linear combination of data fusion and attempted to eliminate redundant information in the integration process by scoring fusion correlation (Ghosh et al, 2015). Neural network was also implemented to eliminate the redundancy of the case library, which improved the matching efficiency in case-based reasoning (CBR). However, when the degree of case similarity represented by the global maximum of neurons is relatively low, the reasoning accuracy decreases (Das et al, 2009). Additionally, the joint feature selection algorithm based on the concepts of relevance and redundancy was designed to remove redundant features and effectively reduced the feature dimension (Tabakhi et al, 2015). To eliminate noisy features of the case library, the study exploited mutual information to select robust feature sets. This selection process can improve reasoning accuracy, but the reasoning efficiency is low because the information retrieval method (i.e., Nearest Neighbor method) requires traversal of the entire database node by node (Freeman et al, 2015; Xu et al, 2013).

Taking into account the impact that transmission path and channel type have on information transfer, the study needs to optimize information retrieval processes in the medical decision system to minimize transfer nodes and achieve efficient seamless information transfer. Information distance was used to study the cost of information transfer in the transmission path and node. The author studied the effect information transfer has on medical quality management by using information distance and an information weighting method (Chi, 2007). Reilly et al (2007) tried to improve health information transfer channels by eliminating the unreliability of clinical paths. To achieve a timely, accurate and consistent information transfer, Meloni et al. (2009) proposed a mutual information Bayesian Network by using mutual information pretreatment to improve the medical database.

The information transfer process for medical decision making is considered an information network system with multi-layer, multi-node, and interactive network structural features, in which different information sources are transferred through a separate evidence chain while the information is exchanged and coordinated continuously in the evidence chain (Shen et al, 2015; Yu et al, 2015). Archana et al. (2007) examined a multitasking medical emergency procedure under a single medical institution and studied the causes that lead to errors in medical decision systems. The results showed that the multitasking conversion process is the key factor leading to poor information flow convergence by analyzing the workflow of the medical emergency department.

Our motivation is to make an intelligent reasoning process efficient and robust by eliminating the redundant data and to infer robust decisions without reducing the accuracy and efficiency. This paper considered the statistic information of feature value, defined a new Relation-Redundancy Coefficient (R^2C) and proposed its corresponding heuristic feature selection algorithm. The method can capture the functional relationship of feature relevance and redundancy. Finally, we implement UCI datasets of heart disease diagnosis to evaluate its performance.

3. PROBLEM DESCRIPTION AND ASSUMPTIONS

3.1 Information Features of Medical Decision Support Systems

When the human body is digitized, the entire medical decision-making process constitutes a complex data system (Aliaga et al, 2015). Medical data sources from four resources: medical resources such as electronic medical records, medical imaging, clinical laboratory, and physician-patient behavior; industry-related data resources such as health insurance and medical literature; subject-related data such as life sciences, demography, and environmental science; and internet data such as internet social media. Medical data resources have four systematic features, namely, multi-dimension, polymorphism, dynamic and uncertainty.

(1) Multi-dimensional means multiple source data produced by a large of participants, multi-tasking and multi-devices in a medical DMS, and the heterogeneous information with different dimensions, types and structures in decision subspace, namely, the discrete, continuous, static and dynamic data have features of multiple heterogeneous complex systems. For example, multi-dimensional personnel data includes the data produced by interns, perfusionists, clinical engineers and nurses with different tasks and responsibilities; multi-dimensional patient vital sign data includes medical literature, expert clinical experience, medical records, test reports and pathology images; and multi-dimensional device parameter data includes defibrillator and single-lead ECG monitoring devices.

(2) Polymorphism means that medical decision data contain a limited number of cases that meet discrete and independent features, conflicting case features and different weights and dimensions. Polymorphism is a key factor that affects the medical decision-making process.

(3) Uncertainty means patient life sign errors and parameter errors of pathology. Uncertainty leads to interference(Fu et al, 2016), fault tolerance and redundancy in medical decision making such as the different levels of doctors' expertise,

patients' physiological and psychological changes, and data error of medical instruments.

As medical decision support resources, a variety of medical data, information and knowledge constitutes a multi-layer heterogeneous complex data space. Therefore, this paper uses a feature space to represent this unique medical data space.

3.2 Mutual Information Expression of Feature Space

Case data and diagnostic rule data exist in the medical DMS database. Let $A=\{f_i | i=1,2,...,n, j\leq n\}$, *where A denotes the feature space, and n denotes its size. Let* $U=\{u_i \mid i=1,2,\cdots,\mathrm{N}\}$, *where* U *denotes the field of medical case library, and N denotes its size. Let E denote the information uncertainty of the feature space, which can be measured by information entropy. When* P_{il} *denote the probability of the feature* $f_i=v_{il}$ *(*v_{il} *is an observation of* f_i*) and L denote the size of its values, we have*

$$\mathrm{E}(f_i)=-\sum_{l=1}^{L}P_{il}\log_2\frac{1}{P_{il}}. \tag{1}$$

If f_i, $f_j\in A$ *and* $f_j\neq f_i$, *, their conditional entropy (CE) can be inferred by the joint distribution, namely,*

$$E(f_i \mid f_j)=-\sum_{f_i,f_j\in A}\sum_{f_j\neq f_i}P(f_i,f_j)\log_2 P(f_i \mid f_j). \tag{2}$$

Let M denote mutual information (CM), which measures the amount of information between two features and describes the correlation and the degree of information redundancy of the features.

$$\mathrm{M}(f_i;f_j)=\sum_{f_i,f_j\in A}\sum_{f_j\neq f_i}\mathrm{P}(f_i,f_j)\log_2\frac{P(f_i \mid f_j)}{P(f_i)P(f_j)}. \tag{3}$$

The larger the mutual information value is, the more closely the two characteristic variables are coupling with each other. When the mutual information value approaches zero, the two characteristic variables are independent. The relationship between mutual information and information entropy can be expressed as

$$\mathrm{M}(f_i;f_j)=\mathrm{E}(f_i)-\mathrm{E}(f_i \mid f_j). \tag{4}$$

If $f_k\in A$, $f_k\neq f_i$ *and* $f_k\neq f_j$, *when* f_i *is a continuous variable and A is known, the conditional mutual information (CMI) between* f_k *and* f_i *can be expressed as*

$$\mathrm{M}(f_k;f_i \mid f_j)=\sum_{f_k\in A}\sum_{f_i\in A}\sum_{f_j\in A}P(f_k,f_i,f_j)\log_2\frac{P(f_k,f_i \mid f_j)}{P(f_k \mid f_i)P(f_k \mid f_j)}. \tag{5}$$

To calculate the information entropy and mutual information, this paper adapts the Boolean reasoning algorithm [18] to discrete the continuous features. It holds the assumption that this approach neither introduces information disturbances nor reduces the effect of feature selection. In the feature selection process, let $M(c;f_\lambda)$ denote the correlation between candidate features f_λ and the class label c.

3.3 Related - Redundant Coefficient

Let $S=\{f_s \mid s=1,2,\cdots,|S|\}$ denote the selected feature subsets from feature set A. Assuming F denotes the remaining feature subsets, F=A-S, c denotes the class label and T denotes the target feature subset. Given $f_\lambda \in F$, let $\mathrm{M}(c;S)$ denote the mutual information between S and c and $\mathrm{M}(c;f_\lambda \mid S)$ denote the conditional mutual information between S and c. Then, $\mathrm{M}(c;\{S,f_\lambda\})$ is constituted by $\mathrm{M}(c;f_\lambda \mid S)$ and $\mathrm{M}(c;S)$.

$$\mathrm{M}(c;\{S,f_\lambda\})=\mathrm{M}(c;S)+\mathrm{M}(c;f_\lambda \mid S). \qquad (6)$$

For any candidate feature of F, $\mathrm{M}(c;S)$ is a constant, and $\mathrm{M}(c;f_\lambda \mid S)$ denotes marginal information provided by f_λ, thus the maximum of $\mathrm{M}(c;\{S,f_\lambda\})$ can be transformed into the maximum of $\mathrm{M}(c;f_\lambda \mid S)$.

Definition 1. *In the feature selection process, the feature correlation, denoted as $\mathrm{M}(c;f_\lambda)$, measures the degree that candidate features act on the class label.*

Definition 2. *In the feature selection process, the feature redundancy, denoted as $\mathrm{M}(c;f_\lambda;S)$, measures the degree that the redundancy of f_λ, f_s and S acts on the class label. Then*

$$\mathrm{M}(c;f_\lambda \mid S)=\mathrm{M}(c;f_\lambda)-\mathrm{M}(c;f_\lambda;S). \qquad (7)$$

Therefore, f_λ provides marginal information by measuring feature correlation and redundancy. Since the mutual information of high dimensional features is very complex to calculate, this paper uses $\mathrm{M}(c;f_s;f_\lambda)$ to measure feature redundancy.

Definition 3. *The mutual information ratio (MIR), $\phi_s=\mathrm{M}(c;f_s)/\mathrm{E}(f_s)$, $0\le\phi_s\le 1$, which captures the functional relationship of information entropy and mutual information between f_s and c.*

MIR considers the feature value range and its distribution. Meanwhile, MIR reflects the importance of features acting on the class label, so it has the following properties: when the feature value range is the same, the more uniform the distribution of the values is, the smaller the importance is; when the feature value is evenly distributed, the larger the range is, the smaller the importance is.

For simplicity, assume that the features are uniformly distributed. When the class label c does not change, the ratio of unselected feature information entropy and mutual information between f_i and f_s is as follows.

$$\mathrm{M}(c;f_\lambda;S)=\frac{1}{|S|}\sum_{f_s\in S}\{\phi_s\cdot\mathrm{M}(f_s;f_\lambda)\}. \qquad (8)$$

Lemma 1(Kwak, 2002). *Given the relevance $\mathrm{M}(c;f_\lambda)$ between unselected features f_λ and c and the mutual information $\mathrm{M}(f_s;f_\lambda)$ unselected features f_λ and selected features f_s, the marginal information provided by f_λ is formulated as:*

$$\mathrm{M}(c;f_\lambda\,|\,S)=\mathrm{M}(c;f_\lambda)-\frac{1}{|S|}\sum_{f_s\in S}\{\phi_s\cdot\mathrm{M}(f_s;f_\lambda)\}. \qquad (9)$$

Where ϕ_s is the MIR, $|S|$ represent the size of set S, thus the second term in eq. (9) represents the weighted redundancy between f_λ and f_s.

Proof: *In the feature space, since*

$$\mathrm{M}(c;f_\lambda\,|\,f_s)=\mathrm{M}(c;f_\lambda)-\left\{\mathrm{M}(f_s;f_\lambda)-\mathrm{M}(f_s;f_\lambda|c)\right\},$$

the conditional mutual information of the set is approximated by the mean of the mutual information between f_λ and all selected features in S.

$$\mathrm{M}(c;f_\lambda\,|\,S)=\mathrm{M}(c;f_\lambda)-\frac{1}{|S|}\sum_{f_s\in S}\frac{\mathrm{M}(c;f_s)}{\mathrm{E}(f_s)}\mathrm{M}(f_s;f_\lambda). \qquad (10)$$

with MIR $\phi_s=\mathrm{M}(c;f_s)/\mathrm{E}(f_s)$, therefore, we obtain the eq. (9).

Definition 4. *Let a real non-negative number $\Re$ denote the relation-redundancy coefficient (R^2C), which measures the impact of the candidate feature f_λ to predict the class label.*

$$\Re=\begin{cases}\infty & S=\varnothing,\ or\sum_{f_s\in S}\{\phi_s\cdot\mathrm{M}(f_s;f_\lambda)\}=0\\ \gamma & \text{the others}\end{cases}. \qquad (11)$$

where $\gamma = \dfrac{\mathrm{M}(c\,;f_\lambda)}{\frac{1}{|S|}\sum_{f_s \in S}\{\phi_s \cdot \mathrm{M}(f_s;f_\lambda)\}}$.

Theorem 1. *When R^2C satisfies* $0<\Re<1$, $\mathrm{M}(c\,;f_\lambda) < \mathrm{M}(c;\{S,f_\lambda\})$

Proof: *With Lemma 1, we have*

$$\mathrm{M}(c;f_\lambda \mid S) = \mathrm{M}(c\,;f_\lambda)\left[1-\frac{\frac{1}{|S|}\sum_{f_s \in S}\{\phi_s \cdot \mathrm{M}(f_s;f_\lambda)\}}{\mathrm{M}(c\,;f_\lambda)}\right]. \tag{12}$$

Since $0<\Re<1$, *we take* $0<\gamma<1$.

When $\mathrm{M}(f_s;f_\lambda)\neq 0$, *we have* $\mathrm{M}(c\,;f_\lambda) < \mathrm{M}(c;\{S,f_\lambda\})$.

Therefore, $\mathrm{M}(c\,;f_\lambda \mid f_s)<0$.

Corollary 1. *When* $0<\Re<1$, *the redundancy of the candidate feature* f_λ *and other features* $f_s \in S$ *is stronger than the correlation between the feature and the class label.*

In this case, the candidate feature is called the redundancy feature.

Corollary 2. *When* $\Re>1$, *the correlation between the candidate feature* f_λ *and the class label is stronger than the redundancy of the candidate feature and other features.*

The candidate feature referred to a relative feature, which brings new information to the classification. The paper sets a threshold θ $(\theta>1)$ *based on the actual values for the correlation features. The features are strong correlation features when* $\Re \geq \theta$*; otherwise, they are weak correlation features.*

Corollary 3. *When* $\Re=0$ *or* $\gamma=0$, *the correlation of candidate features and the class label is zero, indicating that the corresponding features are irrelevant features.*

Since $\Re=0$, *we take* $\gamma=0$. *When* $\mathrm{M}(f_s;f_\lambda)\neq 0$, *we have* $\mathrm{M}(c\,;f_\lambda)=0$.

In the heart disease medical decision-making process, the correlation of candidate features (such as age, gender, chest pain type) and class label c(whether people have heart disease) is zero. In other words, candidate features are not related to heart disease.

Corollary 4. *When $\Re=\infty$, the decision maker needs only to analyze the correlation between the features and class label. Then he/she selects the corresponding features with a maximum $\Re$ into the target feature set.*

4. R2C BASED ROBUST FEATURE SELECTION

The selection process is named as R^2C mutual information based feature selection (R2CMI). When the data set contain n features, the complexity of this algorithm is $O(n^2)$, thus R2CMI can be implemented to select features of large data.

To achieve optimal feature sequences, the objective function of R2CMI is formulated as R2CMI Criterion,

$$\hbar=\max\left[\mathrm{M}(c;f_\lambda)\cdot(\Re-1)\right] \quad (13)$$

$$\hbar=\max\left\{\mathrm{M}(c;f_\lambda)\cdot\left[\frac{\mathrm{M}(c;f_\lambda)}{\frac{1}{|S|}\sum_{f_s\in S}\{\phi_s\cdot\mathrm{M}(f_s;f_\lambda)\}}-1\right]\right\}. \quad (14)$$

Among these functions, $\hbar(f_i,c)$ is a sort function, and it outputs a feature sequence about the optimal feature subset.

4.1 Feature Space Retrieval

Acquiring the optimal feature subsets is not only to construct the selection criterion of feature subset, but also to design the strategy of feature space retrieval. Thus, we adapted two heuristic algorithms to search the feature space.

(1) Forward search: when $f_\lambda\notin S$, $f_\lambda\in F$ (F=A-S), if $\Re\le\theta$, f_λ remains in F; otherwise, f_λ is moved from F to S.

(2) Backward search: when $f_\lambda\in S$, if $\Re<\theta$, move f_λ from S to F; otherwise, f_λ remains in S.

4.2 Reasoning Steps of R2CMI

(1) Initialization: assign the original input feature space A to F, set N, S and T empty, and set the threshold θ accordingly.

(2) Calculation: estimate $\mathrm{M}(c;f_\lambda)$ for each input feature f_λ.

(3) Forward search: Select the feature f_λ and set $\mathrm{M}(c;f_\lambda)=0$, subsequently, add f_λ to N and remove it from F, denote $\{f_\lambda\}:F\to N$.

(4) Backward search: Select the feature f_λ and set $\mathrm{M}(c;f_\lambda)=\mathrm{Max}$, subsequently, add f_λ to S and remove it from F, denote $\{f_\lambda\}:F\to S$.

(5) Repeat the steps (2)-(4), till $\Re < \theta$.

When $\forall f_s \in S$ *, calculate the new information entropy* $E(f_s)$ *. When* $\forall (f_s, f_\lambda), \forall f_s \in S, f_\lambda \in F$ *, calculate* $\mathrm{M}(c; f_s)$ *and* $\mathrm{M}(f_s; f_\lambda)$*. Select* f_λ *to satisfy*

$$\begin{cases} \max \mathrm{M}(c; f_\lambda) \cdot (\Re - 1) \\ \Re = \dfrac{|S| \cdot \mathrm{M}(c; f_\lambda)}{\sum_{f_s \in S} \left[\dfrac{\mathrm{M}(c; f_s)}{\mathrm{E}(f_s)} \cdot \mathrm{M}(f_s; f_\lambda) \right]} \end{cases} . \quad (15)$$

Finally, add f_λ *to S and remove it from F.*

(6) Adjustment: check and adjust feature space A based on R2CMI, and then assign the adjusted optimal feature subset as the solution.

5. NUMERICAL EXPERIMENT

Heart disease data sets provided by the Cleveland Clinic Foundation VA Medical Center were used to verify the effectiveness of the R2CMI model and compare with the MIFS method, the MIFS-U method and the mRMR method (Peng et al, 2006). The heart disease data sets source from many years of medical practice and cover a variety of heart disease cases. The data in the data sets are authentic and reliable (Sui Y, 2014). There are 303 samples in this data set, each sample has a class label $c = \{H, S\}$ *, and the basic information of the data set is shown in Table 1 and Table 2.*

Table 1. Data Description of Cleveland Heart Disease Cases.

Feature	Term	Multi-tasking	Multi-layer	Data Description
f_1	*Age*	*demography*	*nurses*	*age: age in years*
f_2	*Gender*	*demography*	*nurses*	*sex: sex (1 = male; 0 = female)*
f_3	*Cp*	*Subjective*	*nurses*	*(chest pain type)1: typical angina, 2: atypical angina, 3: non-anginal pain, 4: asymptomatic*
f_4	*Trestbps*	*General*	*nurses*	*resting blood pressure (in mm Hg on admission to the hospital)*
f_5	*Chol*	*test reports*	*clinical engineers*	*serum cholesterol in mg/dL*
f_6	*Fbs*	*test reports*	*clinical engineers*	*fasting blood sugar > 120 mg/dL (1 = true; 0 = false)*
f_7	*Restecg*	*ECG monitoring*	*Cardiac physician*	*Resting electrocardiographic results.0: normal, 1: having ST-T wave abnormality,2: showing probable or definite left ventricular hypertrophy by Estes' criteria*
f_8	*Thalach*	*ECG monitoring*	*Cardiac physician*	*maximum heart rate achieved*
f_9	*Exang*	*ECG monitoring*	*Cardiac physician*	*exercise induced angina (1 = yes; 0 = no)*
f_{10}	*Oldpeak*	*ECG monitoring*	*Cardiac physician*	*ST depression induced by exercise relative to rest*
f_{11}	*Slope*	*Pathology images*	*Cardiac physician*	*The slope of the peak exercise ST segment.* *1: upsloping, 2: flat, 3: downsloping*
f_{12}	*Ca*	*pathology images*	*perfusionists*	*number of major vessels (0-3) colored by fluoroscopy*
f_{13}	*Thal*	*pathology images*	*perfusionists*	*3 = normal; 6 = fixed defect; 7 = reversible defect*
d	*class*	*medical records*	*medical expert*	*No: Healthy; Yes: Sick*

Table 2. Information Table of the UCI Heart Disease Data Set.

Data		of Objects		Input Features		Class labels	
Name	Provider	Samples	Features	Discrete Features	Continuous Features	Number	Value
UCI-Heart	*Cleveland Clinic*	*303*	*13*	*7*	*6*	*1*	*2*

We use the Boolean reasoning algorithm for discrete, continuous and integer attributes, and obtain the division points of these attributes. The parameter of the Boolean reasoning algorithm is set to 0.97, and the discrete result is shown in Table 3.

Table 3. Discretion Processing of the Attributes.

Number	Feature	Discrete Intervals
f_1	*Age*	*(-∞, 55.5), [55.5, +∞)*
f_4	*Trestbps*	*(-∞, 122.5), [122.5, 139), [139, +∞)*
f_5	*Chol*	*(-∞, 236.5), [236.5, +∞)*
f_6	*Fbs*	*fasting blood sugar > 120 mg/dl (1 = true; 0 = false)*
f_8	*Thalach*	*(-∞, 147.5), [147.5, +∞)*
f_{10}	*Oldpeak*	*(-∞,0.7), [0.7, +∞)*

When the age range is $(-\infty, 55.5)$, *then it is mapped to 0; otherwise, it is mapped to 1. The slope of the peak exercise ST segment has three different forms, namely upsloped, flat and downsloped, which are mapped to 1, 2 and 3. When the person is healthy, then it is recorded as 0; otherwise, it is recorded as 1. After the above data pretreatment, the results of the first 10 medical record cases are shown in Table 4.*

Table 4. Preprocessing Results of the First Ten Medical Record Cases.

Age	Sex	Cp	Trestbp	Chol	Fbs	Restec	Thalacl	Exang	Oldpeal	Slope	Ca	Thal	Class
1	*1*	*1*	*2*	*0*	*1*	*2*	*1*	*0*	*1*	*3*	*0*	*6*	*0*

1	1	4	2	1	0	2	0	1	1	2	2	3	1
1	1	4	0	0	0	2	0	1	1	2	2	7	1
0	1	3	1	1	0	0	1	0	1	3	0	3	0
0	0	2	1	0	0	2	1	0	1	1	0	3	0
1	1	2	0	0	0	0	1	0	1	1	0	3	0
1	0	4	0	1	0	0	1	1	1	1	0	3	0
0	1	4	2	0	1	2	1	1	1	3	0	7	1
1	1	4	2	0	0	0	1	0	1	2	0	6	0
1	0	2	2	1	0	2	1	0	1	2	0	3	0

Use generalized float feature selection algorithm to eliminate redundant attributes in feature space and use equation (3) to calculate the mutual information of the normalized quadratic Renyi entropy $\mathrm{M}(c;f_l)$, *redundancy* $M(c;f_\lambda:S)$ *and the marginal value* $M(c;f_l\,|\,S)$. *The result is shown in Table 5.*

Table 5. Mutual Information Estimated by the Heuristic Feature Selection Algorithm.

Procedures	Feature	$\mathrm{M}(c;f_\lambda)$	$M(c;f_\lambda;S)$	$M(c;f_\lambda\,|\,S)$	Cumulative Time
1	f_{13}	0.217395	0	0.217395	0.040412
2	f_{12}	0.188451	0.039716	0.148735	0.068042
3	f_3	0.250012	0.104433	0.145579	0.096319
4	f_{10}	0.212886	0.116769	0.096116	0.127950
5	f_8	0.182942	0.118157	0.064785	0.152366
6	f_9	0.223176	0.170237	0.052939	0.151798
7	f_1	0.107515	0.055656	0.051859	0.206528
8	f_{11}	0.349667	0.301591	0.048076	0.237578

According to the feature selection process and floating search operator, we use equation (15) to achieve the target set, $\{f_{13}、f_3、f_{12}、f_{10}、f_8\}$.

In the feature selection process, the mutual information estimation results of $\mathrm{M}(f, c)$ *and* $\log_2(R^2C)$ *for all types of features are shown in Figure 1.*

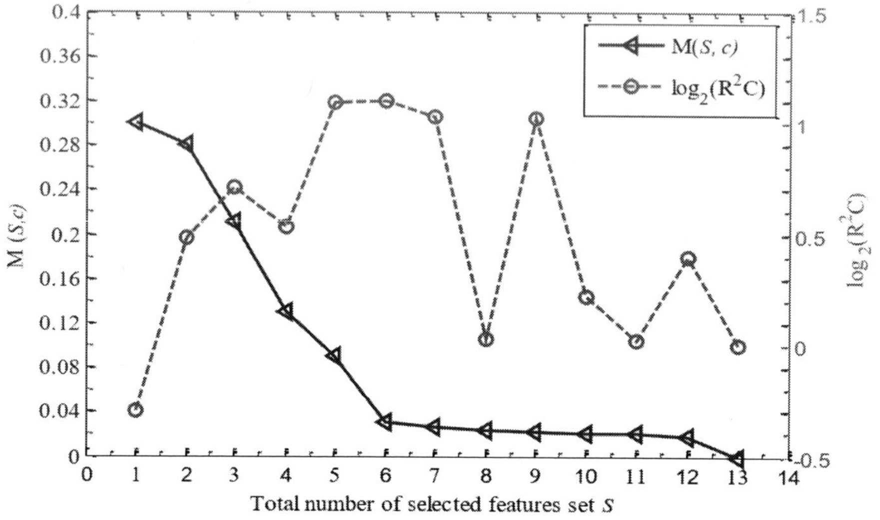

Fig. 1. Mutual Information Estimation of $\mathrm{M}(S, c)$ and $\log_2(R^2C)$.

We implemented the 1-NN classifier to verify the selected feature set, the size of the best features, the classification error and its average time for each feature selection method (Table 6).

Table 6. Comparison Results with Several Related Feature Selection Methods (1-NN Classifier).

Method	Feature Selection Sequence	Features	Error Rate (%)	Average Time for Each Selection (s)
MIFS ($\beta = 0.7$)	$f_{13}, f_{12}, f_3, f_{11}, f_6, f_7, f_2, f_9, f_5, f_4, f_1, f_{10}, f_8$	6	*15.44* $\pm$ *1.33*	*0.0021*
MIFSU ($\beta = 0.7$)	$f_{13}, f_3, f_{12}, f_{10}, f_9, f_{11}, f_2, f_8, f_7, f_6, f_1, f_5, f_4$	*7*	*14.52* $\pm$ *0.31*	*0.0021*
mRMR	$f_{13}, f_{12}, f_3, f_{10}, f_{11}, f_9, f_7$ $f_6, f_2, f_8, f_5, f_1, f_4$	6	*14.59* $\pm$ *2.11*	*0.0017*

R2CMI	$f_{13}, f_{12}, f_3, f_{10}, f_8, f_9, f_2$ $f_{11}, f_7, f_6, f_1, f_5, f_4$	*5*	*14.28 ± 1.74*	*0.0019*

The error changes are shown in Figure 2, illustrating that compared with other feature selection methods, R2CMI has the smallest errors. The results show that the selected feature set, { f_{13} *,* f_{12} *,* f_3 *,* f_{10} *,* f_8 *}, selected } is the best feature subset.*

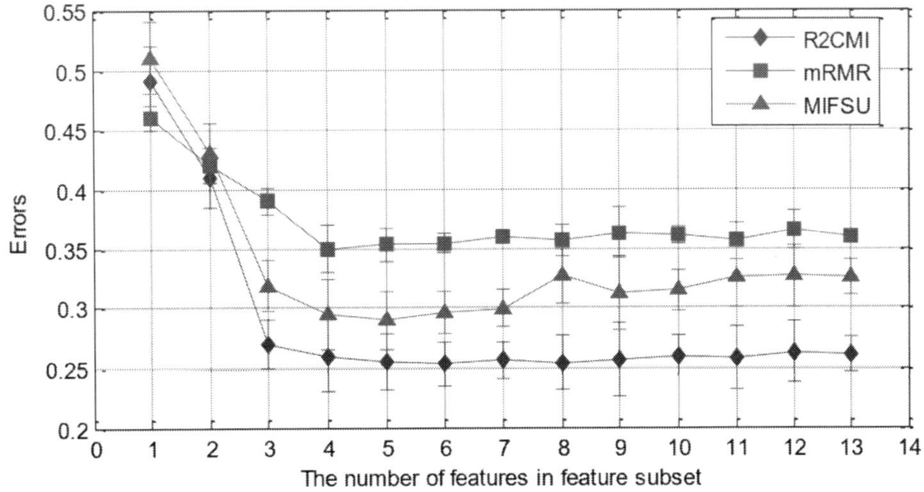

Fig. 2. Errors of Ten-fold Cross Validation by using 1-NN Classifier.

We also compares the results from three aspects: reduction rate of features, output information and system accuracy, shown in Table 7. The reduction rate of features of the R2CMI is between RST and HGA, but the output information and accuracy are higher than the other methods. In total, the R2CMI feature selection method is better.

Table 7. Comparison results with several related literature.

Feature Selection Criteria	Reference	Reduction rate of Features (%)	Output Information	Accuracy (%)
Rough Set (RST)	*Blake et al. (2010)*	*38.46*	*N/A*	*76.00*
Unfiltered RST rules	*Kumar et al. (2016)*			*81.5*
HGA	*Setiawan et al. (2009)*	*76.92*	*0.3266*	*82.59 ± 2.11*
R2CMI (1-NN)	*Wang et al. (2015)*	*61.54*	*0.3748*	*85.73±1.77*

6. CONCLUSIONS

The challenges of medical data-driven decision making are that medical data are huge and exist in a heterogeneous state. Intelligent decision-making methods gradually transform from pursuit of calculation speed into concerns regarding reasoning ability efficiency and accuracy of the heterogeneous data fusion process. With the development and innovation of theories and methods about large data, intelligent wearable, mobile and cloud computing, which are representative of information and communication technology, data-driven medical decisions have become the main trend in medical and health research in the field of decision support. The research about fast extracting high-quality decision-making knowledge from large-scale, multi-source heterogeneous medical data and improving the intelligence and accuracy of the decision-making process will become an important direction for the medical intelligent DMSs. Data-driven decision-making processes have characteristics of both medical systems and d DMSs. From the perspective of constitution of the data, the multi-layer, multi-node and networked interactive structure data sources of the health care system form their own independent data chain and transfer radioactively, while the transfer, feedback and coordination between different data chains is also ongoing. Health care systems not only link hardware systems such as medical testing equipment but also access to medical expertise, medical diagnostic rules and patient pathology data, which form a complex data environment. The knowledge and treatment services for patients are becoming increasingly complex.

To solve the redundancy data interference of large medical data sets on standardization and intelligent medical decision making effectively, this paper proposed a Related-Redundancy Coefficient (R^2C) method, the feature selection criterion (R2CMI) that considers feature correlation and the redundancy ratio simultaneously, and a heuristic search algorithm of feature space. The results showed that this method effectively eliminated the redundant UCI heart experimental data, that the 1-NN classifier effectively eliminated data interference and increased the error rate to 14.28% + 1.74%, and that the average time of selecting each feature fell to 0.0019 s. The R2CMI method used the histogram method to approximate the probability density of feature distribution. This approach has some estimation error, and the space needed by the storage histogram grows exponentially with the number of features in data sets. Improvement of the R2CMI method and its use in any feature distribution data sets still needs to be investigated.

ACKNOWLEDGMENTS

This research was supported by the National Natural Science Foundation of China (Grant No. 71571105, 71171143, 71201087).

REFERENCES

[1] Mckinsey Global Institute Analysis. 2012. *"North American Health care Provider Market Size and Forecast". ESG Research Report.*

[2] Lee G M. 2014. *System engineering approach to build an information system for emergency cesarean deliveries. International Journal of Industrial Engineering Theory Applications & Practice, 20(11-12):589-601.*

[3] CHO, Minsu, SONG, Minseok, YOO, Sooyoung. 2015. *A SYSTEMATIC METHODOLOGY FOR OUTPATIENT PROCESS ANALYSIS BASED ON PROCESS MINING. International Journal of Industrial Engineering: Theory, Applications and Practice, 22(4): 31-42.*

[4] Porter M E, Lee T H. 2013. *The Strategy That Will Fix Health Care. Harvard Business Review, 91(12):24-24.*

[5] Lempert R J. 2003. *Shaping the next one hundred years: new methods for quantitative, long-term policy analysis. Rand Corporation.*

[6] Xu Man. 2011. *Robustness of a Heart Attack First Aid Decision Based on Intelligent Fusion Model. Doctoral thesis of Tianjin University.*

[7] Xu M, Yu H, Shen J. 2012. *New algorithm for CBR-RBR fusion with robust thresholds. Chinese Journal of Mechanical Engineering, 25(6): 1255-1263.*

[8] Lempert R J, Groves D G, Popper S W, et al. 2006. *A general, analytic method for generating robust strategies and narrative scenarios. Management Science, 52 (4): 514-528.*

[9] Nagatou N, Watanabe T. 2015. *A Model-Checking Based Approach to Robustness Analysis of Procedures under Human-Made Faults. International Journal of Industrial Engineering: Theory, Applications and Practice, 22(4): 43-57.*

[10] Hansen L P, Sargent T J. 2011. *Robustness and ambiguity in continuous time. Journal of Economic Theory, 146(3): 1195-1223.*

[11] Shen J, Lin Y, Xu M. 2015. *Research on Quality and Efficiency of the Medical Decision-Making System Based on Robustness Theory//Proceedings of the 21st International Conference on Industrial Engineering and Engineering Management 2014. Atlantis Press,181-184.*

[12] Hansen L P, Sargent T J. 2007. *Robustness. Princeton University Press.*

[13] Yao Xu, Wang Xiaodan, Zhang Yuxi. 2012. *Summary of feature selection method. Control and Decision, 27(2): 161-166.*

[14] Xu Zongben, Feng Zhiyan, Guo Xunhua. 2014. *Leading subjects based on data-driven management and decision-making. Management World, 11: 158-163.*

[15] Reshef D N, Reshef Y A, Finucane H K, et al. 2011. *Detecting novel associations in large data sets. Science, 334(6062):1518-24.*

[16] Thomas H. 2014. *Data analysis: approximation aids handling of big data. Nature, 515(7526):198-198.*

[17] Fung I C, Tse Z T, Fu K W. 2015. *Converting Big Data into public health. Science, 347(6222).*

[18] Feng Zhiyan, Guo Xunhua, Zeng Dajun, et al. 2013. *Several leading subjects on business management in the big data background. Journal of Management Sciences in China, 16(1).*

[19] Chen H, Chiang R H L, Storey V C. 2012. *Business Intelligence and Analytics: From Big Data to Big Impact. MIS quarterly, 36(4): 1165-1188.*

[20] Manyika J, Chui M, Brown B, et al. 2011. *Big Data: The Next Frontier for Innovation, Competition, and Productivity. McKinsey Global Institute.*

[21] Freeman C, Kulić D, Basir O. 2015. *An evaluation of classifier-specific filter measure performance for feature selection. Pattern Recognition, 48(5):1812-1826*

[22] Xu M, Yu H, Shen J. 2013. *New approach to eliminate structural redundancy in case resource pools using a mutual information. Journal of Systems Engineering & Electronics, 24(4):625-633.*

[23] Pan R, Yang Q, Pan S J. 2007. *Mining competent case bases for case-based reasoning. Artificial Intelligence, 171(16-17):1039-1068.*

[24] Han M, Cao Z J. 2015. *An improved case-based reasoning method and its application in endpoint prediction of basic oxygen furnace. Neurocomputing, 149:1245-1252.*

[25] He P, Wu K, Xu J, et al. 2013. *Multilevel redundancy allocation using two dimensional arrays encoding and hybrid genetic algorithm. Computers & Industrial Engineering, 64(1): 69-83.*

[26] Ghosh K, Parui S K, Majumder P. 2015. *Learning combination weights in data fusion using Genetic Algorithms. Information Processing & Management, 51(3): 306-328.*

[27] Das R, Turkoglu I, Sengur A. 2009. *Effective diagnosis of heart disease through neural networks ensembles. Expert systems with applications, 36(4): 7675-7680.*

[28] Tabakhi S, Moradi P. 2015. *Relevance-redundancy feature selection based on ant colony optimization. Pattern Recognition, 48(9):2798-2811.*

[29] YU H Y, SHEN J, XU M. 2016. *Temporal case matching with information value maximization for predicting physiological states. Information Sciences, 10.1016/j.ins.2016.05.042.*

[30] Chi Z R, Wang H C. 2007. *DIT Measurement Model of Website's Ease of Use Based on User's Approach. Information Science, 25(1):139-145.*

[31] Reilly J, Newton R, Dowling R. 2007. *Implementation of a first presentation psychosis clinical pathway in an area mental health service: the trials of a continuing quality improvement process. Australas Psychiatry, 15(1):14-18.*

[32] Meloni A, Ripoli A, Positano V, et al. 2009. *Mutual Information Preconditioning Improves Structure Learning of Bayesian Networks From Medical Databases. IEEE Transactions on Information Technology in Biomedicine, 13(6): 984-989.*

[33] Shen Jiang, Yu Hai-Yan, Xu Man. 2015. *Heterogeneous evidence chains based fusion reasoning for multi-attribute group decision making. Acta Automatica Sinica, 41(4): 832–842*

[34] YU Hai-yan, SHEN Jiang, XU Man. 2015. *ECs-based reasoning for group decision analysis in the mislabeled classification context. Systems Engineering and Electronics, 37(11): 2546-2553.*

[35] Archana Laxmisan, Forogh Hakimzada Osman R. Sayan, Robert A. Green, et al. 2007. *The multitasking clinician: Decision-making and cognitive demand during and after team handoffs in emergency care. International journal of medical informatics, 76: 801-811.*

[36] Aliaga M, Forel J M, Bourmont S D, et al. 2015. *Diagnostic yield and safety of CT scans in ICU. Intensive Care Medicine, 41(3):436-443.*

[37] Fu Liping, Shen Meng, Yu Haiyan, et al. 2015. *Random disturbance reasoning model of decision-making system and its anti-interference capability. International Journal of Industrial Engineering: Theory, Applications and Practice, 22(5): 645-60.*

[38] **Kwak N, Choi C H. 2002.** *Input feature selection for classification problems. IEEE Transactions on Neural Networks, 13(1):143-59.*

[39] **Peng H, Long F, Ding C. 2005.** *Feature Selection Based on Mutual Information: Criteria of Max-Dependency, Max-Relevance, and Min-Redundancy. IEEE Transactions on Pattern Analysis & Machine Intelligence, 27(8):1226-38.*

[40] **Sui Y. 2014.** *A novel approach for discretization of continuous attributes in rough set theory. Knowledge-Based Systems, 73(1):223-238.*

[41] **Blake C L, Mertz C J. 2010.** *UCI Machine Learning Databases. [2010-03-20],*

http://mlearn.ics.uci.edu/databases/heartdisease/.

[42] **Kumar S U, Inbarani H H. 2016.** *Neighborhood rough set based ECG signal classification for diagnosis of cardiac diseases. Soft Computing: 1-13.*

[43] **Setiawan N A, Venkatachalam P A, M.H. 2009.** *A F. Rule Selection for Coronary Artery Disease Diagnosis Based on Rough Set. International Journal of Recent Trends in Engineering, 2(5): 198-202.*

[44] **Wang Z, Li M, Li J. 2015.** *A multi-objective evolutionary algorithm for feature selection based on mutual information with a new redundancy measure. Information Sciences, 307:73-88.*

Concurrent Control Chart Pattern Recognition Using a Novel Source Separation Method based on Sparse Representation

Akram Sadat Mozafari[1] and Majid Aminnayeri[2*]
[1]Department of Industrial Engineering and Management Systems
Amirkabir University of Technology, Tehran, Iran
Akram_mozafari@aut.ac.ir

[2]Department of Industrial Engineering and Management Systems
Amirkabir University of Technology, Tehran, Iran
mjnayeri@aut.ac.ir

ABSTRACT

Control chart has been widely utilized for statistical process control in industrial applications. Every abnormal pattern type in control chart originates from an assignable cause. Therefore, Control Chart Pattern Recognition (CCPR) is crucial to instantly detect abnormality cause(s) and get the process back into its normal state. Most methods in CCPR only focus on basic types of unnatural pattern; but in real world applications, processes can be influenced by several causes simultaneously. This leads to concurrent control chart patterns that are more complicated to recognize. In recent years, Blind Source Separation (BSS) methods such as Independent Component Analysis (ICA) and Singular Spectrum Analysis (SSA) have been used for un-mixing concurrent patterns to its components that are basic control chart patterns and can be more easily recognized.

In this paper, we propose a hybrid approach which consists of Source Separation and Sparse Representation based Classification (SS-SRC) for concurrent CCPR. SS-SRC has four major advantages: 1. unlike ICA and SSA which do not incorporate any prior information about sources, SS-SRC takes into account valid source patterns in separation step through a dictionary which has been learned using sources. 2. By sparse decomposition of the input pattern using the dictionary atoms, SS-SRC automatically detects Gaussian noise as the residual of the decomposition. 3. Unlike ICA, SS-SRC is a single channel source separation technique that needs only one concurrent observation, therefore it is more practical in real word applications. 4. Unlike BSS methods such as ICA and SSA that need a classification model, Sparse Representation Classification (SRC) in SS-SRC provides a simple classification technique based on source separation step. Finally, experimental results show the effectiveness of the proposed method for un-mixing, de-noising and recognition of concurrent Control Chart Patterns (CCPs) in comparison to ICA and SSA.

Keywords: control charts, concurrent patterns, blind source separation, sparse representation, dictionary learning

---

* Corresponding Author

1 INTRODUCTION

Monitoring the production process is a crucial task for quality improvement and timely fault detection. So, Statistical Process Control (SPC) has attracted a lot of attention in recent decades. Among all SPC tools, Control Chart -which shows whether a process is in control or not- is the most useful one. This chart has a center line, an upper control limit, and a lower control limit. If the points fall outside the control limits or non-randomly follow a certain pattern, the process is considered to be out of control. Generally, we have seven basic control chart patterns: normal, cyclic, systematic, up-ward trend, down-ward trend, up-ward shift, and down-ward shift [1]. Although many researchers [1], [2] have proposed useful methods for control chart patterns recognition, recognizing concurrent patterns has rarely been addressed. In real world applications, concurrent control chart patterns appears when two or more causes occur at the same time and it is vital to recognize them properly. Because of patterns interference, detection of concurrent control chart is more complicated and has been mentioned more often in recent years [3], [4].

In both single and concurrent CCPR, using raw data can be ineffective because of additive Gaussian noise (normal pattern), thus various methods have been proposed for pre-processing of raw CCP. This methods vary from simple ones such as statistical feature extraction [5], shape feature extraction [6] and wavelet transform [7] to advanced methods like Independent Component Analysis (ICA) [3], [8], [9], Singular Spectrum Analysis (SSA) [10], [11], and hybrid methods [2], [4].

In recent years, methods based on Blind Source Separation (BSS) for concurrent control chart pattern recognition have been investigated [4], [10]. In these methods, concurrent observation gets decomposed into its basic components in the first step. Then, various strategies such as Decision Tree [9] and Support Vector Machines (SVMs) [3], [8] are used for classification. One of well-known BSS algorithms is ICA, which is a powerful tool in source separation [12] . ICA has also been used for CCPR [3], [8], [9]. Some of studies have focused on the separation of noise (normal pattern) from a basic [3] or a concurrent [8] pattern using ICA. In other words, in this approaches one source is Gaussian noise and another source is the de-noised target pattern that must be recognized. Although this approaches [3], [8] can lead to an acceptable de-noising level and a high classification accuracy, they are very impractical due to modelling noise as a source in mixtures which is unrealistic. ICA has also been used for un-mixing concurrent pattern into basic types [9] that is a common strategy in BSS applications, but this is also impractical due to needing more than one mixture observation.

SSA is another useful BBS algorithm that needs only one observation and has been used by Gu et al. [11] and Xie et al. [10] for concurrent control chart pattern source separation. This method extracts periodic patterns but there is no guarantee that SSA components become equivalent to basic CCP types.

Recently, novel sparse representation based BSS techniques have gained a lot of attention, especially in speech domain [13], [14]. In this paper, we propose a single channel (with just one mixture observation) hybrid approach that consists of Source Separation and Sparse Representation based Classification (SS-SRC) stages for control chart pattern recognition. The rest of our paper is organized as follows. Section 2 describes the proposed method and its related algorithms. Section 3 presents experimental results. Finally, in section 4 we explain conclusions of our novel method.

2 THE PROPOSED SS-SRC SCHEME

Algorithm 1 shows proposed SS-SRC method for concurrent control chart pattern analysis. This method consists of two stages: 1. Steps 1-4 of Algorithm 1 provide an accurate source separation technique based on sparse coding. In this steps, using the dictionary atoms, the observed concurrent pattern $x_{mix} \in \mathbb{R}^{d\times 1}$ gets decomposed through non-negative sparse coding (Algorithm 2). The dictionary is composed of six (number of abnormal CCPs) sub-dictionaries ($D = [D^1, \dots, D^6]$, $D^i \in \mathbb{R}^{d\times K_i}$). Every sub-dictionary is obtained by applying k-means clustering algorithm on training samples of a basic CCP type. As Figure 1 shows in an example, every source in concurrent observation can be recovered using corresponding sub-dictionary. 2. Finally, in Step 5 of Algorithm 1, Sparse Representation based Classification (SRC) [15], [16] technique has been employed for detecting active basic CCP types.

In Algorithms 1 and 2, α denotes the coefficients vector (sparse representation) of x_{mix}, and α^s denotes the coefficients vector of the s-th source. Recovered component corresponding to the s-th sub-dictionary is denoted by $r^s \in \mathbb{R}^d$ where $r^s = D^s\alpha^s$. The contribution of s-th basic CCP type is defined as $c^s = ||r^s||_2 \in \mathbb{R}$, in which $||.||_p$ stands for $l_p - norm$.

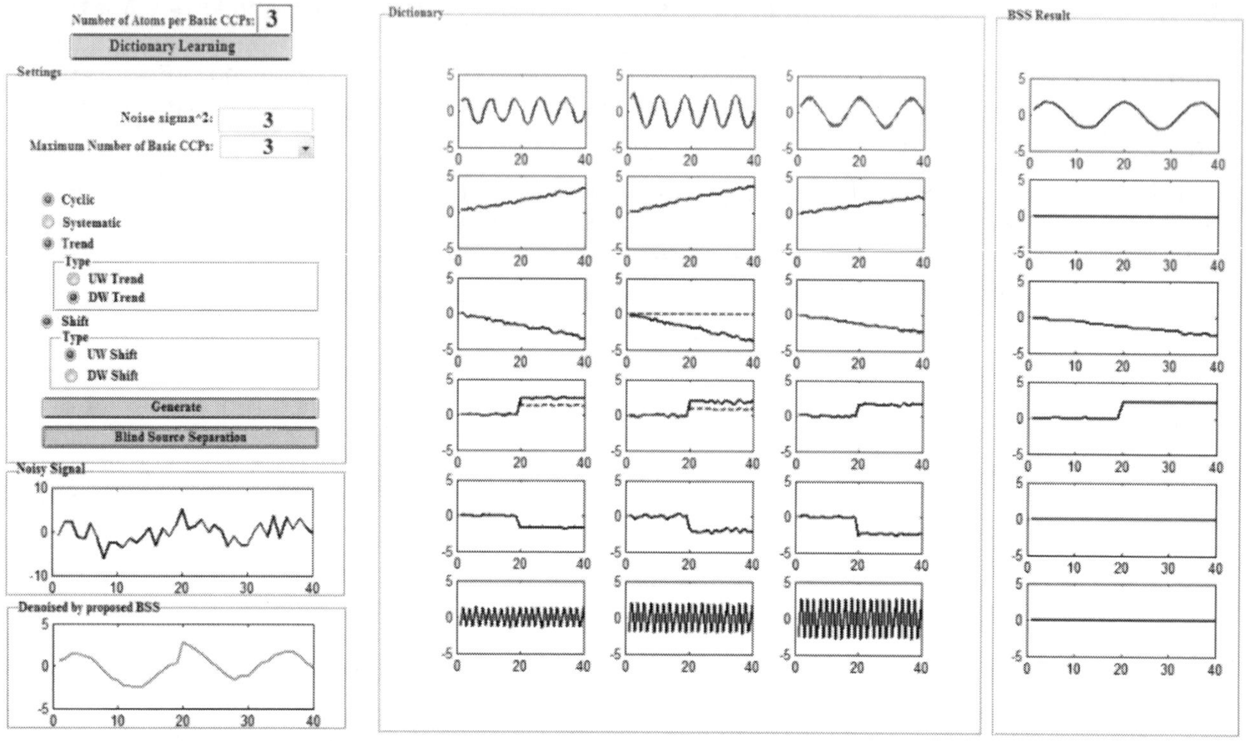

Figure 1: Learned Dictionary and Result of Proposed Source Separation for a Triplet Concurrent CCP (cyclic + down-ward trend + up-ward shift) with Noise Power = 3.

2.1 Clustering based Dictionary Learning

One of the main challenges in sparse representation based source separation approaches is dictionary learning. In applications with a few number of training data such as face recognition, the training data can be used directly as the dictionary [17]. This strategy is not suitable for control chart pattern applications due to additive Gaussian noise in the training samples. To overcome this issue, in this paper, k-means clustering algorithm has been utilized for learning sub-dictionaries. Every sub-dictionary corresponds to one of the basic CCP types and is obtained by applying k-means to its training samples [18]. Every cluster centre is considered to be a sub-dictionary atom (prototype). In k-means algorithm, every cluster centre is the mean of cluster members. Mean of additive Gaussian noise of the samples converges to zero or a constant. So, as shown in Figure 1, the resulting atoms (centres) converge to the desired noise-free prototypes. Figure 1 shows extracted sub-dictionaries (with three atoms) for six types of basic CCPs. As you can see in Figure 1, every single or concurrent control chart pattern can be estimated by sparse linear combination of the dictionary atoms. Since the Gaussian noise cannot be modelled by sparse selection of atoms, it will automatically get dropped as the residual of the sparse approximation. Components associated with each dictionary atom are shown by red colour in Figure 1 (for a triplet concurrent pattern with Gaussian noise power = 3). As Figure 1 shows, SS-SRC is able to un-mix and de-noise complicated concurrent patterns even with three basic types and in the presence of a Gaussian noise with high variance (power).

2.2 Sparse Representation based Source Separation

Sparse representation approaches reconstruct an input signal by a linear combination of atoms that has a sparse number of non-zero coefficients. Ideally, the number of non-zero coefficients ($l_0 - norm$) must be smaller than or equal to the maximum number of sources in the concurrent control chart pattern. This leads to following optimization problem:

$$\underset{\alpha}{argmin}\,||x_{mix} - \textstyle\sum_{s=1}^{S} D^s \alpha^s\,||_2^2 \;\; subject\ to\ \left|\left|\alpha\right|\right|_0 \leq maxS. \qquad (1)$$

This is a NP Hard optimization problem [18]. And, greedy approaches such as Orthogonal Matching Pursuit (OMP) have been proposed for finding its sub-optimal solution. But because of greedy atom selection in OMP, this approach is not robust to additive noise [18]. Therefore, a convex optimization objective function based on $l_1 - norm$ has been proposed in sparse coding literature [18] that guarantees global optimal solution. In this paper, we use the following $l_1 - norm$ based objective function:

$$\underset{\alpha}{argmin}\,||x_{mix} - \textstyle\sum_{s=1}^{S} D^s \alpha^s\,||_2^2 \;\; subject\ to\ \left|\left|\alpha\right|\right|_1 \leq maxS, \alpha_i \geq 0. \qquad (2)$$

where $maxS$ is the maximum number of sources (classes). This objective function seeks the (non-negative) minimum residual approximation in the convex set of $||\alpha||_1 \leq maxS$. In this paper, we use mexLasso function in spams-matlab package (available at http://spams-devel.gforge.inria.fr/) for solving EQ. 2.

to decrease ambiguity of classification in SRC approaches, the number of contributing classes (sub-dictionaries) in sparse coding should be minimized [15][16]. This is equivalent to a new sparsity constraint on the number of contributing classes. Step 4 of algorithm 1 makes sure that the maximum number of contributing sources is equal to $maxS$. This algorithm is a greedy approach for removing minimum contribution atoms until reaching the desired sparsity level ($maxS$) of sources and is inspired from the sparse coding algorithm that has been proposed by Xu et al. [17]. Precisely speaking, Steps 1-4 of Algorithm 1 seek a solution for the following objective function which guarantees the desired sparsity for both atoms and sources:

$$\underset{\alpha}{argmin}\left|\left|x_{mix} - \textstyle\sum_{s=1}^{S} D^s \alpha^s\right|\right|_2^2 \;\; s.t\ \left|\left|\alpha\right|\right|_1 \leq maxS, \alpha_i \geq 0, \left|\left|C\right|\right|_0 \leq maxS. \qquad (3)$$

2.3 Sparse Representation based Classification

In SRC techniques, the input signal is labelled with the most contributing class in signal decomposition [17]. In Step 5 of Algorithm 1, we use source-specific thresholds to detect the active sources. Thresholds are related to lower bound parameters such as minimum slope in trends and minimum magnitude in shift patterns, and they can be automatically learned by finding the minimum energy value of every basic type.

Algorithm 1: SS-SRC

Inputs:

- $D = [D^1, \ldots, D^S] \in \mathbb{R}^{d\times K}$: leaned dictionary (the algorithm is described in Section 2.1), $(S = 6)$.
- $x_{mix} \in \mathbb{R}^{d\times 1}$: observed concurrent CCP.
- $T = [t^1, \ldots, t^S]$: **activation** thresholds of sources.

Parameters:

- $maxS$: maximum number of sources

Outputs:

- $R = [r^1, \ldots, r^S]$: reconstructed sources' components.
- $C = [c^1, \ldots, c^S]$: sources' contributions (energies), while $||C||_0 \leq maxS$.
- $L = [l^1, \ldots, l^S]$, $L^s \in \{0, 1\}$: classification result, 1 for active sources, while $||L||_0 \leq maxS$.

Algorithm:

Step 1: Non-Negative Sparse Representation using Lasso

- $\alpha = \mathbf{Non-Negative-Sparse-Coding}(D, x_{mix}, maxS)$.
- $r^s = D^s\alpha^s \in \mathbb{R}^{d\times 1}$, $c^s = ||r^s||_2$ for $s = \{1, \ldots, S\}$.

Step 2: Correction of Up-Ward and Down-Ward Trends' Coefficients

- $\hat{x}_{mix-trend} = r^{UW-Trend} + r^{DW-Trend}$: reconstruction of x_{mix} by trend.
- $\alpha^{UW-Trend} = \mathbf{Non-Negative-Sparse-Coding}(D^{Up-Ward\,Trend}, \hat{x}_{mix-trend}, 1)$.
- $\alpha^{DW-Trend} = \mathbf{Non-Negative-Sparse-Coding}(D^{Down-Ward\,Trend}, \hat{x}_{mix-trend}, 1)$.
- $r^{UW-Trend} = D^{UW-Trend}\alpha^{UW-Trend}$, $r^{DW-Trend} = D^{DW-Trend}\alpha^{DW-Trend}$
- **IF** $(||\hat{x}_{mix-trend} - r^{UW-Trend}||_2 < ||\hat{x}_{mix-trend} - r^{DW-Trend}||_2)$
  $\alpha^{DW-Trend} = \vec{0}, r^{DW-Trend} = \vec{0}, c^{DW-Trend} = 0$.
  ELSE
  $\alpha^{UW-Trend} = \vec{0}, r^{UW-Trend} = \vec{0}, c^{UW-Trend} = 0$.

Step 3: Correction of Up-Ward and Down-Ward Shifts' Coefficients (same as Step 2)

Step 4: Constraining the Number of Sources in Sparse Coding ($||C||_0 \leq maxS$)

- D^r = atoms of D with non-zero coefficient.
- **REPEAT**
  Update sparse coding using remained atoms in D^r:
  $\alpha = \mathbf{Non-Negative-Sparse-Coding}(D^r, x_{mix}, maxS)$
- **Remove the atom (d_j) with minimum contribution from D^r:**
  $$j = \operatorname*{argmin}_j ||\alpha_j d_j||_2.$$
- **UNTILL** $||C||_0 \leq maxS$.

Step 5: Multi-Label SRC by putting Thresholds on the Contributions ($||L||_0 \leq maxS$)

- **For** $s = \{1, \ldots, S\}$,
- **IF** $(c^s \geq t^s)$ **THEN** $l^s = 1$, **ELSE** $l^s = 0$.

Algorithm 2: Non-Negative-Sparse-Coding

Inputs:

- $D' \in \mathbb{R}^{d\times K'}$: dictionary.
- $x \in \mathbb{R}^{d\times 1}$: input signal.

Parameters:

- $sparsity$: $l_1 - norm$ sparsity level.

Outputs:

- $\alpha \in \mathbb{R}^{K'}$: sparse representation of x over D'.

Algorithm:

$$\alpha = \operatorname*{argmin}_{\alpha} ||x_{mix} - D'\alpha||_2^2 \ \ subject\ to\ ||\alpha||_1 \leq sparsity, \alpha_i \geq 0.$$

3 EXPRIEMENTALS

Although SS-SRC can easily recognize single and concurrent CCPs simultaneously, generalization of ICA method for this purpose is not straightforward. Therefore, we evaluate the proposed SS-SRC approach in concurrent CCPR when input signals are originated from two basic CCPs.

3.1 Dataset and Evaluation Settings

To generate training and testing samples, we have used settings similar to [10]. The only difference in this paper is that we have taken into account all the combinations of the six abnormal CCP types. This leads to 15 mixture types, in which 2 combinations are ambiguous (up-ward/down-ward trend, up-ward/down-ward shift) and have been ignored. Linear combination of up-ward and down-ward trends (shifts) is invalid since it leads to another up-ward or down-ward trend (shift) depending on the slops (magnitudes). In our experiments, the number of dictionary atoms per basic CCP was selected to be 3. As we discussed in section 2, sparsity level must be selected depending on the maximum number of sources taken into consideration (in this settings 2).

3.2 Compared Methods

We compare our method with SSA-SVM [10] and ICA-SVM [8] methods. In this paper, we have used LIBSVM package [19] that is very fast and accurate. LIBSVM Software package is available at "http://www.csie.ntu.edu.tw/~cjlin/libsvm". We have used the RBF (Gaussian) kernel for SVM. Negative signs in the coefficients of recovered ICs is resolved by using the negative of the problematic ICs instead. FastICA [12] has been employed for ICA. The FastICA package can be downloaded from "http://research.ics.aalto.fi/ica/fastica/code/dlcode.shtml". Following Xie et al. [10], the parameters tuning for SVM have been done by a validation dataset. SS-SRC approach does not have any parameter tuning. Details of the SSA-SVM and ICA-SVM methods is as follows:

- ICA-SVM: In this method, two concurrent observations are required. Gaussian noise (normal pattern) and concurrent pattern are assumed to be sources [8]. We have used [0.4 0.6] and [0.6 0.4] as mixing coefficients. Based on [8], after IC extraction, the one with larger kurtosis value is used as target de-noised concurrent and the other IC is considered to be noise. A SVM classifier has been trained by using the target de-noised concurrent ICs of training samples.
- SSA-SVM: Only when one of the sources is periodic, SSA has the ability to un-mix accurately (in such a way that extracted components correspond to basic CCP types). Therefore, unlike [10], we have used same strategy as ICA-SVM method for SSA-SVM. After SSA decomposition, first components are considered as target concurrent pattern and remained components are assumed to be noise (normal pattern). This strategy leads to a higher accuracy than [10].

3.3 Recognition Results

From Table 1, it can be found that SS-SRC significantly improves classification accuracy rate in comparison to ICA-SVM and SSA-SVM methods.

Table 1: Accuracy rate (%) of the SS-SRC vs. ICA-SVM and SSA-SVM schemes (SVM with Gaussian kernel) for concurrent CCPR task.

Method	UTP+CP	USP+CP	DTP+CP	DSP+CP	SP+CP	UTP+SP	USP+SP	DTP+SP	DSP+SP	USP+UTP	DSP+UTP	DTP+DSP	DTP+USP	All (%)
SS-SRC	98.2	93	97.6	96.4	100	97.2	94.8	98	97.4	100	99.8	99.8	99.6	97.8
ICA-SVM	91.4	87.5	92.4	85.8	100	94.3	89.4	95.2	92.7	96.8	92.2	96.4	92.1	92.7
SSA-SVM	91.8	81.2	93.7	83.3	97.4	93.2	82.9	94.5	87.4	96.2	93.5	91.2	94.7	90.8

3.4 Evaluation in De-noising

In this section, we evaluate the performance of proposed approach in de-noising of concurrent CCPs by using Root Mean Squared Error (RMSE) of the recovered Gaussian noise (normal pattern). The last SSA component(s) and the extracted IC with smaller kurtosis value are considered to be Gaussian noise in SSA and ICA methods, respectively [8]. Table 2 shows that SS-SRC provides state-of-the-art de-noising technique for concurrent CCPs with a meaningful RMSE decrease. The RMSEs are calculated between 260,000 (6500 (number of test samples)*40 (dimensions)) recovered and real noise values.

Table 2: RMSE of the recovered noise in different methods.

Method	RMSE
ICA	0.7123
SSA, $I = \{2,3,4,5,6\}$ considered as noise components	1.1200
SSA, $I = \{3,4,5,6\}$ considered as noise components	0.5904
SSA, $I = \{4,5,6\}$ considered as noise components	0.6858
SSA, $I = \{5,6\}$ considered as noise components	0.8157
SSA, $I = \{6\}$ considered as noise components	0.9116
SS-SRC	0.3188

4 CONCLUSION

In production systems, concurrent CCPs appear when two or more problems occur simultaneously. Therefore, it is necessary to develop an effective system for concurrent control chart pattern recognition. In this paper, we proposed SS-SRC, which is a hybrid approach consisting of source separation and sparse representation based classification, for CCPR. SS-SRC utilizes a dictionary composed from sub-dictionaries (one per every CCP type). Sub-dictionaries are provided by applying k-means clustering on training samples of each basic CCP type. Unlike ICA and SSA, SS-SRC incorporates prior knowledge about sources in separation step using the mentioned dictionary. In the separation step, SS-SRC seeks a sparse (non-negative) linear combination of dictionary atoms to reconstruct the single/concurrent input CCP. Sparsity constraint on the reconstruction coefficients automatically leads to dropping noise as residual. So, unlike ICA and SSA, SS-SRC is able to separate sources and noise components carefully without any ambiguity. In addition, SS-SRC does not need a classification model such as SVM after the separation step. The SRC algorithm utilized here recognizes active CCPs based on the extracted components. SS-SRC is very practical in real word applications since it only needs one observation and can automatically detect the number of contributing sources. Experimental results show the effectiveness of the proposed method for un-mixing, de-noising and recognition of concurrent CCPs. Besides, we have designed a user-friendly system that can be easily monitored and detects contributing sources in a short time with high accuracy.

5 REFERENCES

[1] R.-S. Guh, F. Zorriassatine, J. D. T. Tannock, and C. O'Brien, "On-line control chart pattern detection and discrimination—a neural network approach," *Artif. Intell. Eng.*, vol. 13, no. 4, pp. 413-425, 1999.

[2] J. Addeh, A. Ebrahimzadeh, and V. Ranaee, "Application of the PSO-RBFNN model for recognition of control chart patterns," in *Proceedings 2011, 2nd International Conference on Control, Instrumentation and Automation, ICCIA*, 2011, vol. 49, no. 4, pp. 747-752.

[3] C.-J. Lu, Y. E. Shao, and P.-H. Li, "Mixture control chart patterns recognition using independent component analysis and support vector machine," *Neurocomputing*, vol. 74, no. 11, pp. 1908-1914, May 2011.

[4] G. D. Pelegrina, L. T. Duarte, and C. Jutten, "Blind source separation and feature extraction in concurrent control charts pattern recognition: Novel analyses and a comparison of different methods," *Comput. Ind. Eng.*, vol. 92, pp. 105-114, 2016.

[5] A. Hassan, M. S. N. Baksh, A. M. Shaharoun, and H. Jamaluddin, "Improved SPC chart pattern recognition using statistical features," *Int. J. Prod. Res.*, vol. 41, no. 7, pp. 1587-1603, 2003.

[6] M. Bag, S. Kumar Gauri, and S. Chakraborty, "Feature-based decision rules for control charts pattern recognition: A comparison between CART and QUEST algorithm," *Int. J. Ind. Eng. Comput.*, vol. 3, no. 2, pp. 199-210, 2012.

[7] S. Du, D. Huang, and J. Lv, "Recognition of concurrent control chart patterns using wavelet transform decomposition and multiclass support vector machines," *Comput. Ind. Eng.*, vol. 66, no. 4, pp. 683-695, 2013.

[8] C. J. Lu, Y. E. Shao, and C. C. Li, "Recognition of concurrent control chart patterns by integrating ICA and SVM," *Appl. Math. Inf. Sci.*, vol. 8, no. 2, pp. 681-689, 2014.

[9] C. H. Wang, T. P. Dong, and W. Kuo, "A hybrid approach for identification of concurrent control chart patterns," *J. Intell. Manuf.*, vol. 20, no. 4, pp. 409-419, 2009.

[10] L. Xie, N. Gu, D. Li, Z. Cao, M. Tan, and S. Nahavandi, "Concurrent control chart patterns recognition with singular spectrum analysis and support vector machine," *Comput. Ind. Eng.*, vol. 64, no. 1, pp. 280-289, Jan. 2013.

[11] N. Gu, Z. Q. Cao, L. J. Xie, D. Creighton, M. Tan, and S. Nahavandi, "Identification of concurrent control chart patterns with singular spectrum analysis and learning vector quantization," *J. Intell. Manuf.*, vol. 24, no. 1, pp. 1241-1252, 2013.

[12] A. Hyvärinen and E. Oja, "Independent component analysis: algorithms and applications," *Neural networks*, vol. 13, no. 4-5, pp. 411-430, 2000.

[13] T. Xu, W. Wang, and W. Dai, "Sparse coding with adaptive dictionary learning for underdetermined blind speech separation," *Speech Commun.*, vol. 55, no. 3, pp. 432-450, 2013.

[14] P. Parathai, W. L. Woo, S. S. Dlay, and B. Gao, "Single-channel blind separation using L1-sparse complex non-negative matrix factorization for acoustic signals," *J. Acoust. Soc. Am.*, vol. 137, no. 1, pp. EL124-EL129, 2015.

[15] G. Cheng and Z. Song, "Concentrative sparse representation based classification," *Opt. - Int. J. Light Electron Opt.*, vol. 124, no. 24, pp. 6846-6849, 2013.

[16] X. Song, Z. Liu, X. Yang, and S. Gao, "A new sparse representation-based classification algorithm using iterative class elimination," *Neural Comput. Appl.*, vol. 24, no. 7-8, pp. 1627-1637, 2014.

[17] Y. Xu, W. Zuo, and Z. Fan, "Supervised sparse representation method with a heuristic strategy and face recognition experiments," *Neurocomputing*, vol. 79, pp. 125-131, 2012.

[18] Y. Li, S. Amari, S. Shishkin, J. Cao, F. Gu, and A. S. Cichocki, "Sparse Representation and Its Applications in Blind Source Separation," *Adv. Neural Inf. Process. Syst. 16*, pp. 241-248, 2004.

[19] C. Chang and C. Lin, "LIBSVM : A Library for Support Vector Machines," *ACM Trans. Intell. Syst. Technol.*, vol. 2, pp. 1-39, 2011.

Distinguish of different maize variety based on bionic electronic nose

Mingtao Hou[1,2], Hongmei Zhang[1,2*], Miaosen Wang[1,2], Wanzhang Wang[1,2] and Liquan Yang[1,2]
[1] College of Mechanical and Electrical Engineering
Henan Agricultural University, Zhengzhou, China
mailto:14581264@sun.ac.za

[2] Collaborative Innovation Center of Henan Grain Crops
Zhengzhou China

ABSTRACT

The aim of this study is to explore methods of rapid identification in different maize varieties. This electronic nose (PEN3.5) comprising ten metal oxide semiconductor was used to generate a typical chemical fingerprint of the volatile compounds present in five different maize varieties samples. These different maize varieties were harvested in the same place and same time. The sensor array was optimized by the loading analysis and correlation coefficient. Principal component analysis (PCA) and linear discriminant analysis (LDA) were applied to achieve classification into the five groups of five different maize varieties, the result of PCA and LDA analysis did not achieve the ideal results. BP neural network analysis was used to analyze the response of the optimized sensor array and classify different maize varieties. In this work, the results obtained by BP neural network presented that: the correlation coefficient between measured and predicted value is 0.977, the average relative error is 0.2% and the maximum relative error is 23.89% for training set sample; the correlation coefficient between measured and predicted value is 0.985, the average relative error is 0.37% and the maximum relative error is 14.5% for test set sample. The results obtained indicated that the E-nose could discriminate successfully among maize of different varieties.

Keywords: Electronic nose, maize varieties, distinguish, neural networks

1 INTRODUCTION

Corn is one of the most important food crops in China and even in the world. China's corn production had exceeded rice and wheat. Maize varieties on the market are diverse. In China, the total number of maize varieties certification more than 6,000 from 1972 to 2013 [1]. With the

* Corresponding Author. : +86-371-63558040; fax: +86-371-63558040.

E-mail address:hmzh86022625@ sina.com (Hongmei Zhang)

transgenic technology rapid development in recent years, transgenic maize varieties are also increasing rapidly. The reason for the safety of genetically modified foods, there are still many consumers who doubt and resist transgenic technology [2]. Therefore, distinguish maize varieties is particularly important. Traditional detection methods of grain varieties include morphological identification, electrophoresis, gene detection, and many others. For these method is subjectivity, complex and long cycle [3], so, a fast and efficient method of identification of maize varieties is needed.

Electronic nose is a fast non-destructive detection technology, mainly used to distinguish and identify the smell of composite gas, which is widely used for farm and sideline products and food detection and has been paid attention increasingly [4,5]. It can collected and analyzed the odor information about rice, results showed that different varieties of rice can be distinguished [6-8]. Wu Lili et al. [9] have developed an electronic nose system and it can identification varieties of pepper rapidly. Fu Runze et al. [10] have used the electronic nose to discriminate different grade of fresh shrimp and scallop. Pei Gaopu et al. [11] have used the electronic nose to discriminate adulteration of honey rapidly. In addition, Zhang Hongmei et al. [12] also used electronic nose to distinguish three different grades of Xinyang Maojian tea better. There is a few literatures about electronic nose technology application on distinguish maize varieties.

The current study adopted the electronic nose technology to detect different varieties of maize and thus provide an effective detection method for the appraisal of maize quality. The electronic nose (PEN3.5) was used to test odor of five different maize varieties. The collected data were analyzed by PCA, LDA and BP neural network methods.

2 MATERIALS AND METHODS

2.1 Test materials

The experimented samples were five maize grain of different varieties, obtained from Xingyang city, Henan Province, China. Five maize varieties are labeled Y1, Y2, Y3, Y4, Y5, respectively. These maize ears with a moisture content of 20%, threshed by hand and preserved well. Each variety was divided into 10 (10 g) samples, i.e., tests were conducted in 10 replicates for each variety. Thus, there were 50 samples in all for detection by the electronic nose. The sample was placed in 250 ml flasks for analyses of responses to volatiles using electronic nose, respectively. The flask was closed tightly after introducing maize sample and was held at room temperature (25℃) for 40 min before static headspace sampling.

2.2 Electronic nose and test method

Analyses were performed with a portable electronic nose (PEN 3.5) operating with an enrichment and desorption unit (EDU). The portable electronic nose PEN 3.5 has an array of 10 different metal oxide sensors, they are W1C, W5S, W3C, W6S, W5C, W1S, W1W, W2S, W2W, W3S. These sensors were labeled S1, S2, S3, S4, S5, S6, S7, S8, S9, S10 in this paper. Sensors have different response values for different chemical components. These data were collected, measured and analyzed using the Winmuster software owned by electronic nose. The volatile odor of the maize sample to be tested was respond to sensor array in chamber of sensor array and produce an electronic signal. The data collection to the computer through the collection circuit. Sensors were cleaned before and after each measurement, this effectively ensure that the stability and accuracy of the data that electronic nose measurement. In this research, the detection time of electronic nose is set to 80 s and the cleaning time is set to 60 s, this can ensure that the electronic nose sensor to restore the initial state. When the measurement was completed, the acquired data were properly stored for later use.

3 CONCLUSION

3.1 Response of the sensor array to the odor of different maize variety

Type response curves of different maize variety samples are presented in figure 1. The gas response is expressed by G/G0. Where G and G0 are the resistance of a sensor in clean air and in a detecting gas, respectively. The result showed that after an initial period of low and stable conductivity, the conductivity increases sharply and then stabilizes after a collection time 60 s. The analysis of the maize samples by electronic nose showed that the sensor response curves stabilized after from 60 to 70 s. In this research, the signal of the sensors at 70 s was used in analysis of pattern.

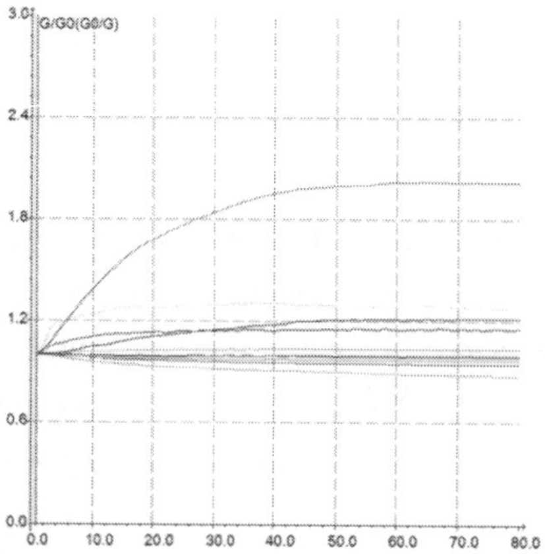

Figure 1: Type response curves of sensor array for different maize variety

Figure 2 is the average of the sensor array response of 5 different varieties of maize, the figure can be seen reaction of sensor S7 is relatively large for maize odor, the response value is also relatively large. So the sensor S7 is an important role in the detection result. Response value of sensor S1, S3, S4, S5, S6 and S8 changed little, it is shown that these sensors are not sensitive to the smell of maize. For the next analysis, need to remove redundant sensors.

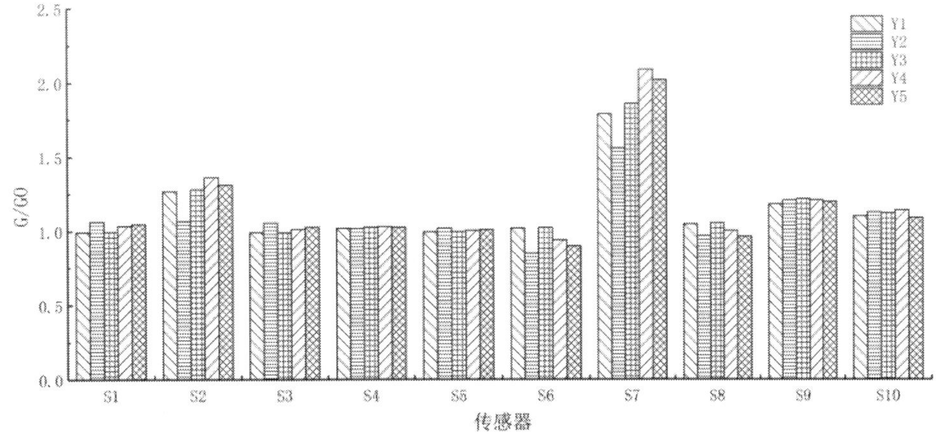

Figure 2: Average of the sensor array response of 5 different varieties of maize

3.2 Optimization of electronic nose sensor array

Figure 3 is a diagram of loading analysis. The factor load of some sensors is relatively close, it show that the signals collected by the sensors are very similar. For improve the efficiency of follow-up analysis, optimize the sensor array that signals are similar. Correlation coefficient reflects the degree of correlation between the sensors. The larger the correlation coefficient, the higher the degree of similarity between the sensors. These sensors contain redundant information, so can be remove [13]. Table 1 shows the correlation coefficient between the respective sensors. Therefore, considered the loading analysis and correlation coefficient, so kept Sensor S6, S7 and S10 as the new sensor array to identify maize varieties.

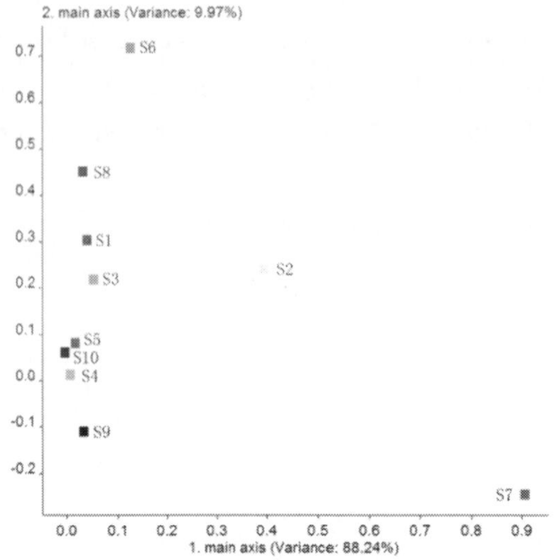

Figure 3: Loading analysis

Table 1: Correlation coefficient of respective sensors

correlation coefficient	S1	S2	S3	S4	S5	S6	S7	S8	S9	S10
S1	1.0000	0.2377	0.7669	0.0283	-0.0234	-0.4265	0.1771	0.1039	-0.1180	-0.0079
S2	-0.5384	1.0000	-0.3985	0.0581	0.0895	0.0732	0.8290	-0.0731	-0.6528	0.3753
S3	0.9675	-0.7262	1.0000	0.2151	0.2509	0.0202	-0.0975	-0.2281	0.1756	0.2108
S4	-0.1201	0.2723	-0.1664	1.0000	0.0423	-0.1056	0.1949	0.5349	-0.2785	-0.1676
S5	0.9391	-0.6231	0.9496	-0.0993	1.0000	0.0708	-0.0557	-0.1086	-0.0020	0.2214
S6	-0.9861	0.6111	-0.9770	0.1980	-0.9348	1.0000	0.1568	0.6035	-0.0355	-0.2501
S7	-0.2972	0.9435	-0.5138	0.2247	-0.4187	0.3830	1.0000	-0.2959	0.8561	-0.1459
S8	-0.9362	0.3743	-0.8668	0.2734	-0.8396	0.9393	0.1217	1.0000	0.2838	0.6626
S9	0.1954	0.1514	0.1231	0.0038	0.1344	-0.1310	0.3854	-0.1817	1.0000	0.1291
S10	-0.1128	0.0244	-0.0834	0.4493	-0.0003	0.1698	-0.0569	0.3637	0.0916	1.0000

3.3 Principal component analysis and linear discriminant analysis

PCA is a chemometric linear, unsupervised and pattern recognition technique used for analyzing, classifying and reducing the dimensionality of numerical datasets in a multivariate problem [14]. Figure 4 is the distinguished result of principal component analysis (PCA) on five different maize varieties. The contribution rate of the first and second principal component are 88.24% and 9.97%. It can be seen from the figure that the boundaries of the scatter plot of five maize varieties are not obvious. This indicating that PCA analysis method cannot distinguish five maize varieties better.

LDA is the most commonly used dimension reduction method for classification [15]. The method maximizes the variance between categories and minimizes the variance within categories. It merely look for a sensible rule to discriminate between them by forming linear functions of the data maximizing the ratio of the between-group sum of squares to the within-group sum of squares. After PCA the same data set using LDA. Figure 5 is the results of linear discriminant analysis method (LDA) to identify five different maize varieties. In this plot, about 72.49% of the total variance of the data is displayed. Functions 1 (LD1) and 2 (LD2) accounted for 42.08% and 30.41% of the variance, respectively. The data points are partially overlapped. This indicating that LDA analysis method also cannot distinguish five maize varieties.

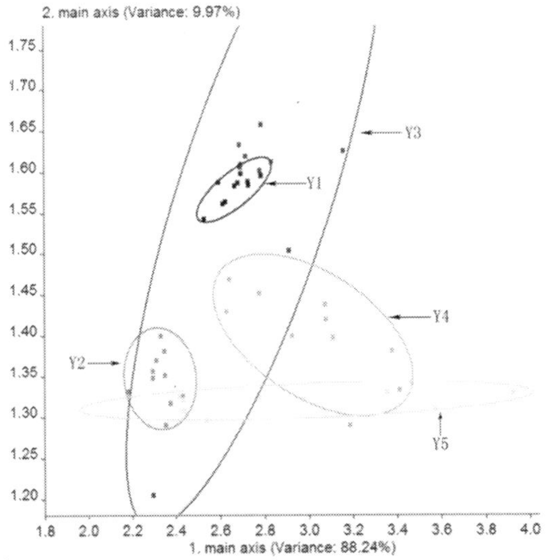

Figure 4: PCA analysis for maize variety

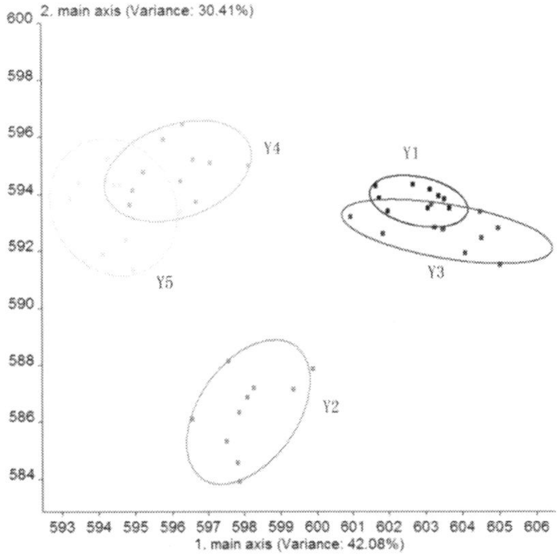

Figure 5: LDA analysis for maize variety

3.4 BP neural network analysis

BP neural network analysis is a non-linear classification methods of pattern recognition. The performance of the network can be improved by modifying its learning rate and additional momentum term and then good classification results are obtained in the gas analysis [7].

Optimization of electronic nose sensor array, retained sensors S6, S7 and S10 for BP neural network analysis. The total numbers of experiments carried out were 50 maize samples (10 maize samples for each variety). Eight samples were selected from each maize variety randomly and total numbers carried out were 40 for training set. Remaining maize samples of each group of two as a test set. In the process of identification, the desire output set to 1,2,3,4 and5 separately represent maize varieties of Y1, Y2, Y3, Y4 and Y5. The software parameters are set up before the analysis. The number of nodes in the input layer is set to 3, hidden layer is set to 1 and the number of nodes in the hidden layer is set to 7, the minimum training rate is set to 0.1, the dynamic parameter is set to 0.6, the sigmoid parameter is set to 0.6, the permissible error is set to 0.0001, the maximum iteration times is set to 5000, and the value of input node were standardized transformation.

Through BP network training to sensor array data and using the training and test sets to the network simulation. Simulation results of training set sample that the correlation coefficient between measured and predicted value is 0.977, the average relative error is 0.2% and the maximum relative error is 23.89%. Simulation results of test set sample that the correlation coefficient between measured and predicted value is 0.985, the average relative error is 0.37% and the maximum relative error is 14.5%. Figure 6 is the forecast result of BP network for maize variety. The figure shows that BP neural network analysis can be very good to distinguish five different maize varieties.

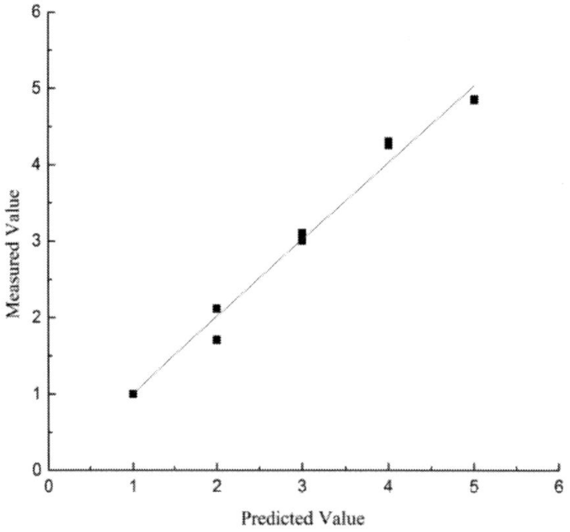

Figure 6: Forecast result of BP network for maize variety

4 CONCLUSION

This article described the research and analysis results of electronic nose for five different varieties of maize. This study shows that the sensor array response of five different maize varieties are different. The response of sensor S7 is maximum. Loading analysis was used to analyze the contribution rate of the electronic nose to detect the volatile matter of maize gas. The sensor of the electronic nose sensor array is optimized based on the eigenvalue and the correlation coefficient among the sensors. Using principal component analysis (PCA), linear discriminant method (LDA) and BP neural network analysis for classify the five different maize varieties. The results showed that the principal component analysis (PCA) and linear discriminant analysis (LDA) cannot distinguish different maize varieties. BP neural network analysis results show that correct rate of recognition for training set is 97.7% and for test set is 98.5%. Results indicate that electronic nose may be successfully applied as rapid and non-destructive method for distinguish different maize variety.

5 ACKNOWLEDGEMENTS

The authors acknowledge the financial support of the Chinese National Foundation of Nature and Science through Project 31501213 and the financial support Foundation for University Youth key Teacher of Henan Province through Project 2015GGJS-077.

6 REFERENCES

[1] **Yang Yang, Wang Fengge, Zhao Jiuran, et al. 2014.** Analysis of the Current Situation of Accredited, *Scientia Agricultura Sinica*, 47(22), pp 4360-4370.

[2] **Rui Yukui, Luo Yunbo, Huang Kunlun, et al. 2005.** Application of Near-Infrared Diffuse Reflectance Spectroscopy to the Detection and Identification of Transgenic Corn, *Spectroscopy and Spectral Analysis*, 25(10), pp 1581-1583.

[3] **Zha Jianwen, Chen Yongyan. 2004.** Recognition System for Corn Species by Exterior Parameters, *Transactions of the Chinese Society for Agricultural Machinery*, 35(6), pp 115-118.

[4] **Yu Yong, Wang Jun, Zhou Ming. 2003.** Research developments of electronic nose and its application in processing of agriculture products, *Journal of Zhejiang University*, 29(5), pp 579-584.

[5] **Wang Jun, Cui Shaoqing, Chen Xinwei, et al. 2013.** Advanced Technology and New Application in Electronic Nose, *Transactions of the Chinese Society for Agricultural Machinery*, 44(11), pp 160-167.

[6] **Xu Sai, Zhou Zhiyan, Luo Xiwen. 2014.** Classification and recognition of hybrid and inbred rough rice based on bionic electronic nose, *Transactions of the Chinese Society of Agricultural Engineering*, 30(9), pp 133-139.

[7] **Yu Huichun, Xiong Zuozhou, Yin Yong. 2012.** The Identification of Rice Varieties Based on Electronic Nose, *Journal of the Chinese Cereals and Oils Association*, (6), pp 105-109.

[8] **Hu Guixian, Wang Jun, Wang Jianjun, et al. 2011.** Detection for rice odors and identification of varieties based on electronic nose technique, *Journal of Zhejiang University*, 37(6), pp 670-676.

[9] **Wu Lili, Zheng Dan, Zheng Baozhou, et al. 2013.** Study of Zanthoxylurn Bungeanum Varieties Discriminating Method by Electronic Nose Technology, *Chinese Journal of Sensors and Actuators*, 26(11), pp 1473-1477.

[10] **Fu Runze, Shen Jian, Wang Xichang, et al. 2016.** Quality evaluation of live Yesso scallop and sensor selection based on artificial neural network and electronic nose, *Transactions of the Chinese Society of Agricultural Engineering*, 32(6), pp 268-275.

[11] **Pei Gaopu, Shi Bolin, Zhao Lei, et al. 2015.** Information variation feature and discriminant capabilities of electronic nose for typical adulteration honey identification, *Transactions of the Chinese Society of Agricultural Engineering*, 31(Supp.1), pp 325-331.

[12] **Zhang Hongmei, Gao Xiankun, Xu Guoqiang, et al. 2010.** Detection of tea grad e based on gas array sensor, *Journal of Henan Agricultural University*, 44(2), pp 177-179.

[13] **Yin Yong, Yu Huichun, Chu Bing, et al. 2014.** A sensor array optimization method of electronic nose based on elimination transform of Wilks statistic for discrimination of three kinds of vinegars, *Journal of Food Engineering*, 127(4), pp 43-48.

[14] **Taher Alizadeh. 2010.** Chemiresistor sensors array optimization by using the method of coupled statistical techniques and its application as an electronic nose for some organic vapors recognition, *Sensors and Actuators B*, 143(2), pp 740-749.

[15] **Sheng Wang, Jianfeng Lu, Xingjian Gu, et al. 2016.** Semi-supervised linear discriminant analysis for dimension reduction and classification, *Pattern Recognition*, 57(C), pp 179-189.

Study on Welding Operations of Seeder Rack Based on Ant Colony Algorithm

Xi-yang Li, Bin Cheng*, Hui Zhang, Cheng-song Li, Yu-cheng Wang
College of Mechanical and Electrical Engineering,
Shihezi University, Shihezi, China
ielixiyang@163.com

ABSTRACT

Welding and welding time path are important factors for 2 MBJ series seeder rack welding robot workstation to improve utilization and production efficiency. Aiming at the welding path planning problem, In-depth analyzed the spatial distribution and process requirement of the solder joints, Established the mathematical model of welding path planning, the shortest welding path is obtained by ant colony algorithms and Matlab software. Use MOD method sets the standard working hours of upper and lower parts in seeder rack welding processing, Analysis of man-machine operation time, in order to achieve to maximize the man-machine.

Keywords: Welding path, Welding time, Ant colony algorithm, MOD method

1 INTRODUCTION

Seeder are seasonal demand products, how to effectively use the welding robot, to reduce the welding time the cost of seeder rack, It is the seeder manufacturer urgent solved subject. Reasonable welding path to reduce the welding torch travel paths, to ensure the welding quality. Working time standards to determine its welding time before the seeder production, help rationalize production and efficient use of manpower, equipment and material resources. Therefore, Study on welding path and welding time has important significance for reduce welding time and welding costs.

The domestic and foreign scholars have done a lot of research for welding path and welding time. For welding path planning method, such as the branch and bound method, genetic algorithm, neural network, ant colony algorithm, etc[1, 2]. Lin Ju guang[3] et al, the ant colony algorithm is applied to find the optimal white underbody welding path. Wang Xuewu[4] et al. Used the improved particle swarm optimization algorithm to plan the solder joint path of the welding robot. In the study of welding time standard, Yu Xiaohong[5] used MOD method to accurately calculate the auxiliary time of machining process. Bai Lijie[6] will be divided into two parts of the mechanical processing time and manual time, using the MOD method to study the workers operating manual time. Due to the various enterprises are different from engaged in industry, the scale and the product , so companies need to find to match the product of welding path planning, and establish a set of corresponding to its own system of welding operation standard.

1.1 Seeder rack welding current situation

Current welding robot has gradually been applied to the machine rack, but due to the machine manufacturing enterprises are small miniature farm machinery manufacturing enterprises, considering the cost and the actual production situation of the enterprise, loading and unloading seeder rack, loading and unloading process is completed by the workers, for robot welding path planning, designers rely mainly on engineering experience, the lack of theoretical analysis, often appear planning unreasonable phenomenon, cause unnecessary waste.

* Bin Cheng: (1961-), professor, Research direction: production system and integration technology, supply chain management; email: 634992417@qq.com.

1.2 Seeder rack technological process

In this paper, regards the RT company 2 MBJ series machine rack as the research object, its main consists of have left beam assembly, right beam assembly, front beam, back beam and so on. The welding production beat is 576s. Seeder rack welding process is shown in figure1, specific welding process analysis is as follows:

Beam welding process, the left and right side beam assembly welding, each side beam is consists of 11 components. Both positive and negative side beam must be weld, the weld total of 34; Seeder rack assembly welding process, the main task is the first beam, back beam, retaining film rod connected to the left and the right beam assembly fabrication. Left, on the right side of the beam is divided two sides, it is need to clamped with which are front beam, back beam and retaining film vertical sides beam. Positive and negative assembly are weld, the weld total of 16, this requires welding need to flip.

Figure 1: Seeder rack welding process flow

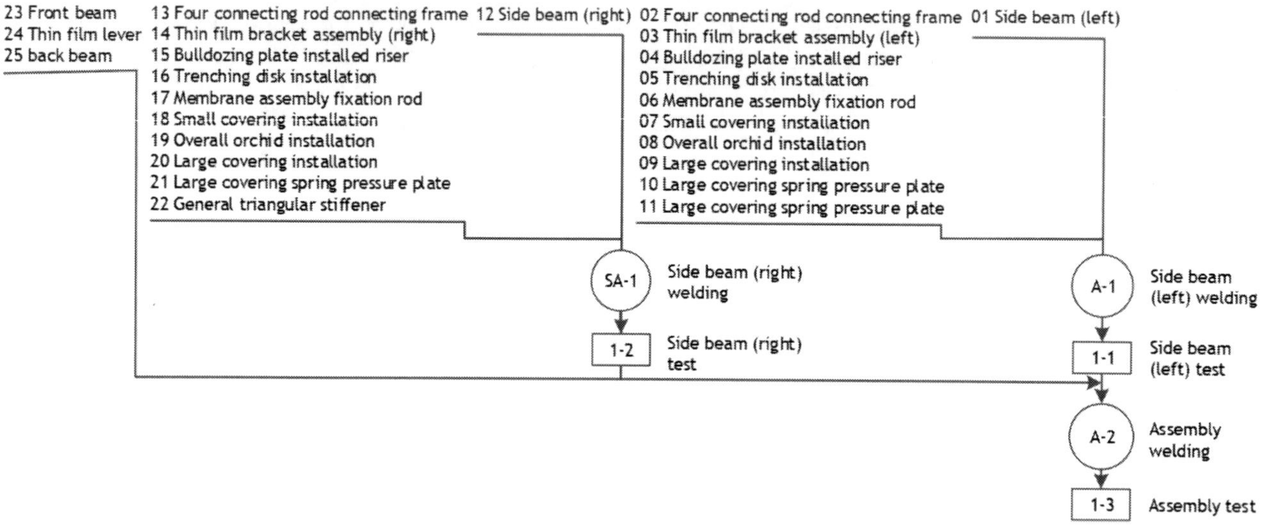

2 SEEDER RACK WELDING PATH PLANNING

In-depth analyze welding process of the seeder rack in this paper, presented regard the shortest welding path as the goal, set up mathematics model, and design of ant colony algorithm to solve the model, using the Matlab software get the shortest welding path.

2.1 Mathematical model of welding path

Welding path planning is the typical traveling salesman problem (TSP), the traveling salesman problem is a classic NP hard problem, the problem description is to the shortest path through all solder joints and may not be repeated.

Assuming the distance d_{ij} between the solder joints i and the solder joints j, the T for the path length collection $t=\{d_{12},d_{23},d_{34},\cdots,d_{ij}\}$, $f(t)$ represents the distance reciprocal between the all joints, the larger $f(t)$ the smaller the path of welding torch, the welding path optimization objective function:

$$d_{ij}=\sqrt{(x_i-x_j)^2+(y_i-y_j)^2} \quad (1)$$

$$f(t)=\max\left(\sum_{i,j=1}^{n} d_{ij}\right)^{-1} \quad (2)$$

Among them, $i,j \le n, i \ne j$.

2.2 Ant colony algorithm for welding path

(1)initialization

Initialization of the ant colony algorithm, because the number of iterations of the ant colony algorithm is more than 1000, the simulation results basically no difference, so the number of iterations of the algorithm set Nc_max=1000, when $t=0$, Cycle counter $NC=0$, Set the initial value of information on each path $\tau_{ij}(0)=C$ (C as constant), when $\Delta\tau_{ij}=0$ Place m ants on n welds. In order to solve the model simplification, set $m=n$. Set the attenuation coefficient of path pheromone rho = 0.5, Ants released pheromone quantity Q = 100. In the model, tabu_list is used to record welding seam which the ants has passed, The ants will not go through these weld seam in this cycles, city_remained represented haven't visited weld.

Variable initialization, using n to express the number of welds, that is the scale of the weld path planning, using d_{ij} to show the distance between the solder joints i and the solder joints j:

$$d_{ij}=\sqrt{(x_i-x_j)^2+(y_i-y_j)^2} \tag{3}$$

To initialize the shortest path of each cycle routh_best, the length of the shortest path of each cycle length_best and the average length of each cycle of all paths length_average.

(2)state transition

Initial time on each path of the initial value of the information amount $\tau_{ij}(0)=c$ (c as a constant), with the movement of ants, after t moments, the information amount on the path ij changed to:

$$\Delta\tau_{ij}^k=\begin{cases}\dfrac{Q}{L_k} & \text{when the } k \text{ ant passed the } i,j\\ 0 & \text{without}\end{cases} \tag{4}$$

$\Delta\tau_{ij}^k$ ——The residual information of the k ants in the weld ij this circulation;

$\Delta\tau_{ij}$ ——The residual information amount of the ant colony in the weld this circulation;

L_k——The path sum of the k ants this cycle.

the residual amount of information which is through the weld i, j, after t moments, in city_remained search and choose next weld, after the completion of the update tabu_list and city_remained, eventually make all weld don't repeat a walk. Calculate the path sum L_k of the k ant from the weld-line i movement to j, and updates the minimal path.

Calculation of ant path selection and the location transition probabilities:

$$p_{ij}^k=\begin{cases}\dfrac{\tau_{ij}^{\alpha}\eta_{ij}^{\beta}(t)}{\sum\limits_{s\in allowd_k}\tau_{is}^{\alpha}\eta_{is}^{\beta}(t)} & j\in allowed_k\\ 0 & \text{other}\end{cases} \tag{5}$$

p_{ij}^k ——The probability of the k ant move from position i to position j;

$\tau_{ij}(t)$——At t moment the pheromone concentration between weld i and j.

Define heuristic factor $\eta_{ij}=1/d_{ij}$, it indicates the visibility of weld between between weld i and j.The smaller d_{ij}, the larger η_{ij}, The more selection probability of weld i and j. The k ant in the movement process, according to the formula 4 to determine the transition probability, To determine the next step of the k ant to move the target weld j, the k ant will move from the weld i to weld j,

And j will be transferred from the collection city_remained to the collection tabu_list. Calculate the shortest path for each cycles routh_best and shortest path length for each cycle length_best, And comparing with last cycle, updates routh_best and length_best.

(3)Update the pheromone concentration

Update the residual information for each of the weld i, j, first calculate the total number of ants:

$$m = \sum_{i=1}^{n} b_i(t) \tag{6}$$

$$\Delta\tau_{ij} = \sum_{k=1}^{m} \Delta\tau_{ij}^{k} \tag{7}$$

$b_i(t)$——At t Moment the number of ants in the weld i;

m——the number of ants;

n——the number of welds.

At t Moment the sum of ants in the all weld, for each weld i to the weld j, set the total amount of the initial residual information $\Delta\tau_{ij} = 0$, according to the time of each iteration, update the $\Delta\tau_{ij}$ real-time

$$\Delta\tau_{ij} = \Delta\tau_{ij} + \Delta\tau_{ij}^{k} \tag{8}$$

$$\tau_{ij}(t+\lambda) = \rho \cdot \tau_{ij}(t) + \Delta\tau_{ij} \tag{9}$$

$\tau_{ij}(t+\lambda)$——At $t + \lambda$ monument the residual amount of information from the weld i to the weld j;

ρ——the retention of information.

After each circulation, updated the amount of information on each path, and the shortest path of this cycle is obtained. This process is repeated until the maximum number of iterations is reached Nc_max. When the maximum number of iterations, terminate the loop, output welding sequence and welding the shortest path.

2.3 welding path for MATLAB solution

(1)Determination the solder joint coordinates of seeder rack

Seeder rack for right Angle weld seam, and weld are parallel or perpendicular to the horizontal plane, the seeder rack weld projection in the two-dimensional plane, and determine the starting point of each welding line, The projection point of the welding seam which is perpendicular to the horizontal plane is coincident with the starting point or the end point of the horizontal plane, as a result, the projection point of the two-dimensional plane to replace three dimensional weld, simplified solution of the model, the solder joint coordinates, as shown in table 1, table 2 and table 3.

Table 1: Seam coordinates of seeder rack beam in 0° position

Weld serial number	**1**	**2**	**3**	**4**	**5**	**6**	**7**	**8**	**9**	**10**	**11**	**12**
X axis	0	60	60	100	100	220	260	585	615	720	780	720
Y axis	72	67	80	0	60	60	60	0	0	60	60	380
Weld serial number	**13**	**14**	**15**	**16**	**17**	**18**	**19**	**20**	**21**	**22**	**23**	**24**
X axis	780	890	930	122	122	108	108	138	142	153	159	153
Y axis	380	0	0	0	60	279	339	279	279	339	339	659
Weld serial number	**25**	**26**	**27**	**28**	**29**	**30**	**31**	**32**	**33**	**34**		
X axis	1593	169	172	205	209	221	221	225	225	225		
Y axis	659	279	279	339	339	279	339	359	290	351		

Table 2: Seam coordinates of seeder rack beam in 180°position

Weld serial number	**1**	**2**	**3**	**4**	**5**	**6**	**7**	**8**	**9**	**10**	**11**	**12**
X axis	0	60	60	220	585	780	780	890	1125	1088	1423	1533
Y axis	587	647	579	599	659	599	279	659	659	380	380	320
Weld serial number	**13**	**14**	**15**	**16**	**17**	**18**						
X axis	1533	1728	2093	2213	2313	2253						
Y axis	0	380	320	380	308	368						

Table 3: Seam coordinates of seeder rack assembly in 0° position

Weld serial number	**1**	**2**	**3**	**4**	**5**	**6**	**7**	**8**
X axis	140	140	100	100	890	890	930	930
Y axis	100	150	330	750	100	150	330	750

(2)The solution of welding shortest path

Seeder rack beam 0° and 180° position of solder joint coordinates were listed in the weld of ant colony algorithm model, through the Matlab programming tools to implement ant colony optimization algorithm, obtained the torch along the shortest path.

As shown in figure 2. The number of iterations for 1000 times, the number of ants is 34, obtained shortest welding path of seeder rack beam in 0 ° position shortest welding path of seeder rack beam order as follows: 31, 32, 34, 33, 30, 29, 28, 27, 26, 23, 22, 21, 20, 17, 16, 15, 14, 11, 10, 9, 8, 7, 6, 5, 2, 3, 1, 4, 12, 13, 18, 19, 24, 25.

As shown in figure 3. The number of iterations for 1000 times, the number of ants is 18, obtained shortest welding path of seeder rack beam order in 180° position as follows: 14, 12, 11, 9, 8, 6, 5, 4, 3, 2, 1, 7, 10, 13, 17, 18, 16, 15.

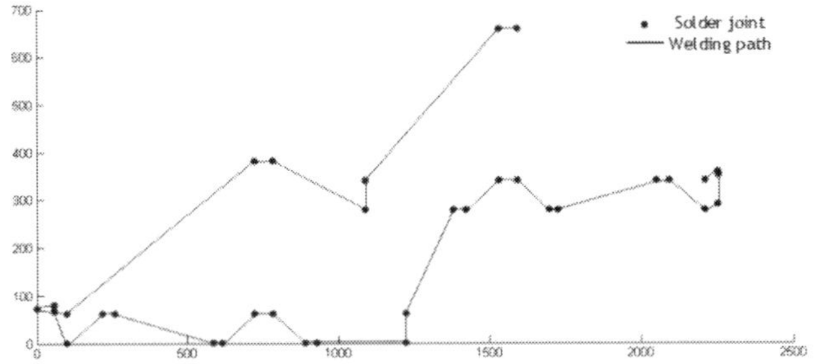

Figure 2: Shortest path of seeder rack side beam welding in 0° position

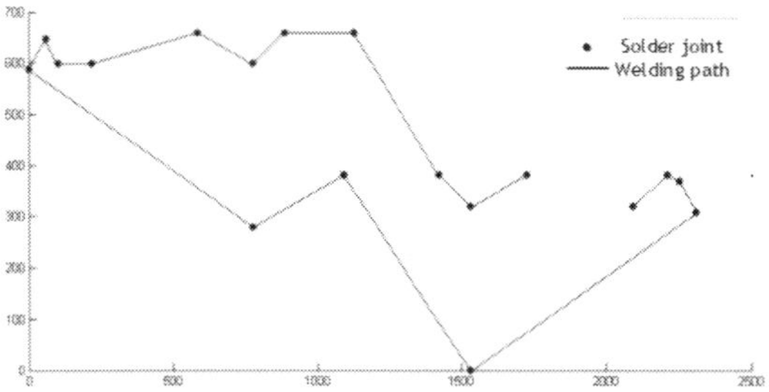

Figure 3: Shortest path of seeder rack side beam welding in 180° position

As shown in figure 4. The number of iterations for 1000 times, the number of ants is 8, obtained shortest welding path of seeder rack order in 0° position shortest welding path of seeder rack order as follows: 8, 7, 6, 5, 1, 2, 3, 4.

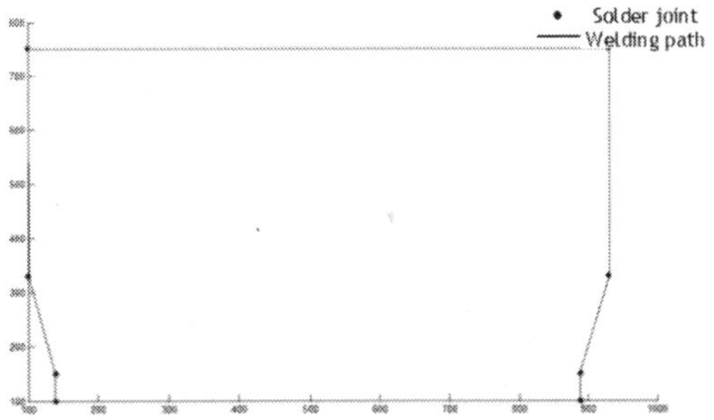

Figure 4: Shortest path seeder rack welding in 0° position

3 STUDY ON SEEDER RACK WELDING OPERATION TIME

The MOD method is a simple method which can ensure the high precision and easy to operate.

Seeder rack welding workstation predetermined motion time analysis, regard operatives as the research object, using the MOD method to record the left hand and right hand action of operatives, drawing hands operation factors analysis, to calculate the value of the mod. To select 1 MOD=0.129 seconds.

3.1 Study on the time of welding loading and unloading parts

Welding process involved in a variety of operations, can be considered by all sorts of grasp, put, walking, and others, such as through the MOD method to the decomposition of a standard work to action, and then to synthesize into standard operation time. From table 4, table 5 shows seeder rack assembly welding, values of loading is 400 MOD and and values of unloading is 447 MOD, respectively, namely the corresponding operation timing for 52 seconds and 58 seconds.

Table 4: Action factors analysis of loading and unloading materials for seeder rack assembly welding

Left hand movement		Time			Right hand movement	
Action description	Value	Times	MOD Value	Times	Value	Action description
Walk to the displacement machine	W5	3	35	3	W5	Walk to the displacement machine
Grab rack	M4G1		5		M4G1	Grab rack
……	……	……	……	……	……	
Return operation area	W5	5	25	5	W5	Return operation area
Rear loaded	M4P5		9		H	holding
Exit workstation	W5	3	15	3	W5	Exit workstation
Total	-	-	400	-	-	-

Table 5: Action factors analysis of loading and unloading materials for seeder rack side beam welding

Left hand movement		Time			Right hand movement	
Action description	Value	Times	MOD Value	Times	Value	Action description
Pull mobile shelves to the operation area	W5 2L1	6	42	6	W5 2L1	Pull mobile shelves to the operation area
Open the clamp	M4G1A4	9	81	9	M4G1A4	Open the clamp
……	……	…	……	…	……	……
Close the clamp 18	M3G1A4		8		H	holding
Pull mobile shelves to the queen	W5 2L1	6	42	6	W5 2L1	Pull mobile shelves to the queen
Total	-	-	447	-	-	-

Seeder rack welding production takt time is 576s. The the loading and unloading material time of seeder rack assembly and side beam welding less than takt time, so the workstation configuration one operatives.

3.2 man-machine operation analysis

Determined the welding mechanical time and operatives action time, based on the man-machine operation diagram to understand the relationship of operating personnel and the machine in seeder rack welding cycle, to excavate the machine or operatives idle time, and further combination and arrangement, to achieve reasonable process, increase productivity and maximize the man-machine. As shown in figure 5.

Machine rack welding robot utilization was 100%, the idle time of operatives is 466s, can undertake production preparation work, or other assigned work task in these time, to improve the utilization rate of workers and produce more value.

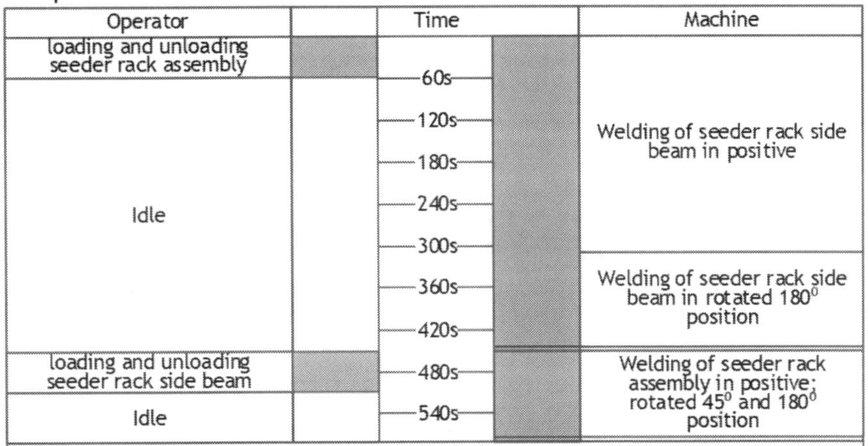

Statistical		
	Operator	Machine
Idle time/s	466	0
Working time/s	110	576
Cycle time/s	576	576
Utilization ratio	19.1%	100%

Figure 5: Man-machine analysis

4 CONCLUSION

With the progress of science and technology, automatic welding technology has been gradually used in small miniature farm machinery manufacturing enterprises, the welding path and the welding time research is an effective means to improve production efficiency and the utilization rate of equipment. This paper adopts the method of mathematical modeling, ant colony algorithm and Matlab software, the MOD analysis and man-machine operation analysis, planning machine rack welding robot welding path, standardized the working time for loading and unloading parts of seeder rack beam welding assembly welding, it has reference value for other similar products welding workstation to improve the efficiency.

5 REFERENCES

[1] **Guillaume Lavoue, Florent Dupont, Atilla Baskurt.** A new CAD mesh Segmentation method based on curvature tensor analysis [J]. Computer Aided Design, **2005**, 37(10): 975.

[2] **Garland M, Willmott A, Heckbert P.** Hierarchical face clustering on polygonal surfaces[J]. ACM Symp Interactive 3D Graph, **2001**(1): 49.

[3] **Lin Juguang, Chen Suxin, Dai Huaichu, Huang Wenjin.** Study on welding path planning for the white car body by Ant colony algorithm [J]. Journal of welding, **2015**, 01: 5-9+113.

[4] **Wang Xuewu, Shi Yingpan, Gu Xingsheng.** Based on partition particle swarm optimization algorithm for welding robot path planning [J]. East China Science Journal of University of Technology (NATURAL SCIENCE EDITION), **2014**, 05:95-101.

[5] **Xiao Hong Yu, Zhang Zhenming, Tian Xitian, Huang Lijiang.** Study on mechanical processing auxiliary time calculation method based on the MOD method [J]. Mechanical and electronic, **2009**(4): 68-70.

[6] **Bai Lijie.** Research and development of computer aided time quota management system based on MODAPTS. Dalian: Dalian Jiaotong University, **2007**:30-37.

COMPLETENESS ANALYSIS METHOD FOR INCOMPLETE MANUFACTURING INFORMATION IN THE INTELLIGENT MANUFACTURING SYSTEM

Fei Li[1*]
[1] School of Mechatronics
Northwest Polytechnical University, Xi'an, China
lfxjzu@163.com

ABSTRACT

With the development of computer and information technology, service-oriented manufacturing system, cloud manufacturing system and big data were put forward to provide the intelligent and service platform for collaborative manufacturing of enterprises. On this platform, whether they were the service providers or service consumers, Manufacturing capacity(MC) was one of the important information and resources which were the basis of functions achievement. Therefore, it is very essential to manage and analyze the MC effectively. But in this process, there are three problems: first, the types and volumes of manufacturing information and resources related to MC are various and huge; second, the relationships between MC and information are complex; third, not all the manufacturing information and resources we need can be obtained due to many reasons. Aiming these problems, in this paper, information and resources related to MC were classified firstly. Second of all, a model for expressing the relationships between information and MC was established. Based on this, the causes of information incomplete were analyzed. Then, the new Bayesian method based on relevant weight preference coefficient was presented in view of the independent and relative relations to process incomplete information, which can provide complete information for analysis and evaluation. Finally, the effectiveness of the method proposed in this paper was validated through case study.

Keywords: Intelligent manufacturing system, Manufacturing capacity, Incomplete manufacturing information, The relational model, The new Bayesian method

1 INTRODUCTION

With the development of information technology, the researches and applications of manufacturing informatization technology have made a series of achievements, which have played an important role in all areas of manufacturing industry. Particularly, service-oriented manufacturing system, cloud manufacturing system and big data were put forward to provide the intelligent and service platform for collaborate manufacturing and agile manufacturing.

On this platform, whether they were the service providers or service consumers, MC was one of the important information and resources which were the basis of functions achievement. Therefore, it is very essential to manage and analyze the MC effectively.

MC is a certain level achieved through integrating information and resources together. It is influenced by many factors, which can be summed up according to the connotation of MC. W.Skinner[1], R. H. Hayes and S. C. Wheelwright[2]pointed out that MC is comprised of cost, quality,

---

* Ph.D. candidate for Manufacturing Systems Engineering

flexibility, innovation, service, environment and the relationship among each element. K .HAFEEZ[3]and other scholars perceived MC as a capacity of integrating tangible resources and intangible resources, including machines, technicians, technology and so on; A.Arafa[4] stated relationship between the MC and enterprise performance to realize production of low cost and high quality. Jiancheng Guan[5]pointed out that the MC is the core of enterprise innovation ability. A.B.Kasoglu[6] pointed out that MC is the collection of tangible resources and intangible resources in order to achieve a task or goal. Liu[7] and other scholars perceived MC as a capacity of integrating and configuring manufacturing resources.

Based on this, the types and volumes of information related to MC are various and huge. Certainly, for analysis MC, The more the information is, the more accurate the analysis is. So to decompose and to formally represent in detail are benefit to information usage, calculation and evaluation. For the moment, relative research is only a few in references retrieved. Though, some capacity of process system has been formally described in order to provide more persuasive evidence for MC analysis. Further, to make the analysis more accurate and faster, information needs to be decomposed in detail and processed by computer easily.

Nevertheless, not all the information can be obtained in the process of actual manufacturing. So far, there are many information completeness processing methods. In these methods, How to choose a right method for different problem needs to be studied.

In conclusion, information and resources related to MC were classified firstly. Second of all, a model for expressing the relationships between information and MC was established. Based on this, the causes of information incomplete were analyzed. Then, the new Bayesian method based on relevant weight preference coefficient was presented in view of the independent and relative relations to process incomplete information, which can provide complete information for analysis and evaluation. Finally, the effectiveness of the method proposed in this paper was validated through case study.

2 THE METHOD OF DECOMPOSITION AND FORMAL REPRESENTATION FOR MC

According to the concept of broad manufacturing, MC can be perceived as a comprehensive system which includes market requirement analysis capacity, product design capacity, research development capacity, manufacturing, supportive activities capacity and environment treatment capacity and so on. It spans through the whole product life cycle. In the paper, information and resources related to MC were classified and the relationship was represented.

2.1 The Model for Expressing the Relationship between MC and Information

In order to show the relationship between MC and information, the hierarchy tree of classes associated with properties based on weight is established and done the multi-dimensional and hierarchical decomposition for MC, as shown in Figure 1. In the model, MC is mapped to class *A*, each sub capacity is mapped to class $A_i(i=1,2,...)$, and $A_i \in A$. Information is mapped to attribute set I_o, each sub-information is mapped to $I_j(j=1,2,...)$, and $I_j \in I_0$. The relation between class and attribute is importance of information for achieving capacity. It can be represented as weight *W*. Each relation can be represented as $w_k(k=1,2,...)$, and $w_k \in W$. So the relation among A , I_o and W can be denoted by W:A→ I_o

2.2 The Formal Representation for MC and Information

The relationship among MC, sub-capacity and information can be expressed clearly based on the model. However, the more sufficient the decomposition of MC is, the huger the information volume is. So, it is difficult to express all the information only using the model. We do the multi-dimensional representation using set theory for MC and information in the paper to make up the shortage of the model. Where, $A_1, A_2, ..., A_m$ represents respectively sub-capacity.

$$MC = \{A_1, A_2, ..., A_i, ..., A_m\} \quad (1)$$

$$A_i = \{A_{i1}, A_{i2}, ..., A_{ij}, ..., A_{in}\} \ (i = 1, ..., m) \quad (2)$$

$$A_{ij} = \{A_{ij1}, A_{ij2}, ..., A_{ijk}, ..., I_O\}, (i = 1, ..., m; j = 1, ..., n) \quad (3)$$

....

$$\underbrace{A_{ijk,...,l}}_{levels} = \{I_1, I_2, ..., I_m\} \quad (4)$$

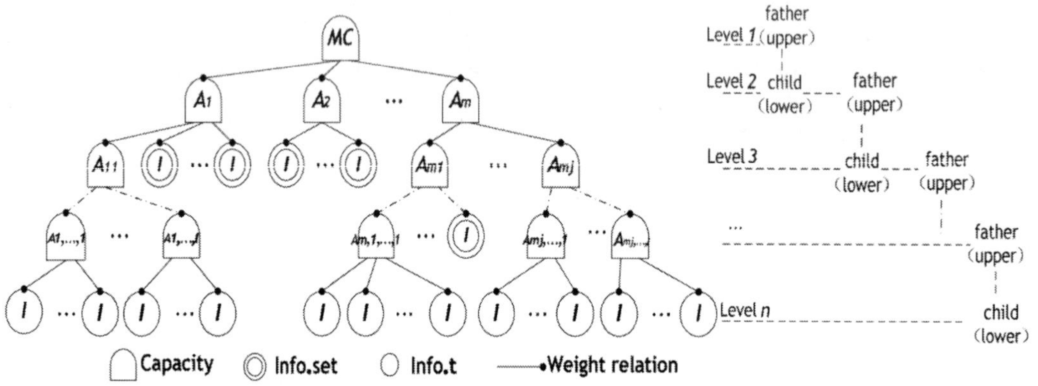

Figure 1: Hierarchy tree of classes associated with properties based on weight

3 COMPLETENESS PROCESSING METHOD FOR INCOMPLETE MANUFACTURING INFORMATION

3.1 The Causes Analysis of Incomplete Information

In the process of production, various uncertain factors were widespread existence, which can cause a lot of incomplete information. The main reasons are described as follows:

- Subjective judgments are not sure. Because human cognition is limited, translating the objective information into the subjective and available information can cause the incomplete information.
- Detailed history data or statistics data cannot be obtained.
- Information acquisition may be inaccuracy.
- Information is unknown.
- Non-uniform and non-standard information expression may be ascribed to the interdisciplinary study.
- Manufacturing information is random.

Thus, unknown, indistinguishability and uncertainty of information can cause information incompleteness. In the paper, incomplete information refers primarily to unknown information.

3.2 The Bayesian Method Based on Relative Weight Preference Coefficient

The types and volume of information are various and huge, but service consumers hope to find the suitable information within a certain range, rather than looking for a needle in a haystack. Hence, the essence of processing incomplete information is aiming at the small sample in large scale of data. Among these processing methods, directly processing on the incomplete data is the only way to make the original information no change. Naive Bayesian is a typical method. But this method also has its limitations, which requests that events are all independent.

For independence, maybe it is just a matter of each person having his or her own principles. From the perspective of physical variables, because the dimension and meanings of information were different, some people considered the relationship was independent. But from the perspective of logical relationship, the relationship was correlative. For example, there are three attributes in balloons data set of UCI, from the perspective of physical variables, the dimensions of attributes, size, act and age, are different, so they can be considered independent. From the perspective of logical relationship, size is related to act and age. Similarly, manufacturing information is probably with that. This will bring the uncertainty for choosing naive Bayesian method. Therefore, in this paper, the Bayesian method is improved.

3.2.1 The description of the problem

There are n data sample $X=\{x_1,x_2,...,x_n\}$, each data sample $x_i(i=1,2,...n)$ is constituted by m attributions $A=\{A_1,A_2,...,A_m\}$. Each attribution $A_j(j=1,2,...,m)$ is constituted by n dimensional vector, a_{ji} represents the value of attribute A_j in data sample i. Class T is divided into $T_1,T_2,...,T_n$, B is a set of attributes to be tested, and $A_k(k=1,2,...,n)\in B$.

- From the perspective of physical variables or physical significance, $\forall A_1 \cap A_2 \cap ,..., \cap A_m = \varphi$
- From the perspective of physical significance, $A_i \cup A_j \cup A_l = A$, $A_i \cap A_j = \varphi$, $A_i \cap A_l = \varphi$, $A_j \cap A_l \neq \varphi and A_i \cap \{A_j, A_l\} = \varphi$.

3.2.2 The Bayesian model based on relative weight preference coefficient

The Bayesian model based on relative weight preference coefficient is shown in formula (5).

$$P(T_i \mid B)=max[\delta(\frac{P(T_i)P(B \mid T_i)}{P'(B)} - \frac{P(T_i)\prod_{k=1}^{n} P(A_k \mid T_i)}{P(B)}) + \frac{P(T_i)\prod_{k=1}^{n} P(A_k \mid T_i)}{P(B)}] \quad \textbf{(5)}$$

Where: $P(T_i)$ represents the prior probability of T_i. $P(T_i \mid B)$ represents posterior probability of T_i when B occurs. $P(B)$ represents the prior probability of B. $P(B \mid T_i)$ represents posterior probability of B when T_i occurs.

Any two attributes (A_i, A_j) as a set are chosen for analyzing correlation, and then the number of correlation attributes is counted, denoted by n_{A_i,A_j}. Given that (A_i,A_j) is the same with (A_j,A_i), the number is denoted by 1. C_m^2 represents the number of all the attribute groups, and $\delta=\frac{2n_{A_i,A_j}}{C_m^2}$.

1. If $\forall A_1 \cap A_2 \cap ,..., \cap A_m = \varphi$, then $\delta=0$.

$$P(T_i \mid B)=max\frac{P(T_i)\prod_{k=1}^{n} P(A_k \mid T_i)}{P(B)}, i=1,2,...,n \quad \textit{(6)}$$

This expression is equivalent to Naive Bayesian model.

2. If $A_i \cup A_j \cup A_l = A$, $A_i \cap A_j = \varphi$, $A_i \cap A_l = \varphi$ $A_j \cap A_l \neq \varphi and A_i \cap \{A_j, A_l\} = \varphi$,then

$$P(T_i \mid B) = max[\delta(\frac{P(T_i)P(A_i \mid T_i)P(\{\ \} \mid T_i)}{P'(B)} - \frac{P(T_i)\prod_{k=1}^{n} P(A_k \mid T_i)}{P(B)}) + \frac{P(T_i)\prod_{k=1}^{n} P(A_k \mid T_i)}{P(B)}] \qquad (7)$$

Especially, if $P(B \mid T_i) = 0$,then

$$P(T_i \mid B) \approx max[P(T_i)P(A_i \mid T_i)P(\{\ \} \mid T_i)] \qquad (8)$$

3. If $\delta = 1$, then

$$P(T_i \mid B) = max\frac{P(T_i)P(B \mid T_i)}{P'(B)}, i = 1,2,...,n \qquad (9)$$

This expression is equivalent to the Bayesian model.

4 INSTANCE ANALYSIS

4.1 Decomposition of Capacity

In the paper, the manufacturing capacity is decomposed based on the process of production and is represented formally. $MC = \{A_1, A_2, A_3, A_4, A_5\}$. Then, only A_1 is decomposed continually considering the length of paper, as shown in Figure 2.

$$A_1 = \{A_{11}, A_{12}, A_{13}\} \text{ and } A_{13} = \{A_{131}, A_{132}, A_{133}, A_{134}\}$$

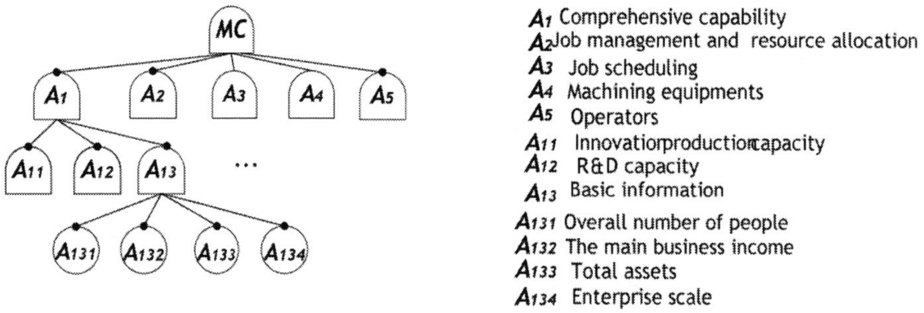

Figure 2: Hierarchical decomposition model for comprehensive capacity

Then, for illustrating and verified the methods, the information of last level in model we chosen is studied. Relative data are collected and the 11 samples are chosen to illustrate and verify the method presented in this paper.

4.2 The Completeness Processing Method for Incomplete Information.

Because the effectiveness of the method is associated with the data sample, so the method is verified with standard data set firstly, then it is verified again using the information set A_{13}.

1. Method verification

In standard data sets UCI, 4 complete data sets are chosen to test the effectiveness of method, data sets and basic attributes, as shown in Table 1.

Table 1: Basic information of test data sets

Data sets	Sample	The number of attributes	The number of incomplete attributes
adult-stretch	20	5	1
adult+stretch	20	5	1
small-yellow	20	5	1
small-yellow+adult-stretch	16	5	1

The data set "small-yellow+adult-stretch" is chosen and an attribute of any sample is extracted as missing data to verify, as shown in Figure 3. And the results compared with different method in the tool we developed are shown in Table2. By calculation, the conclusion is the same as the data of the original data set.

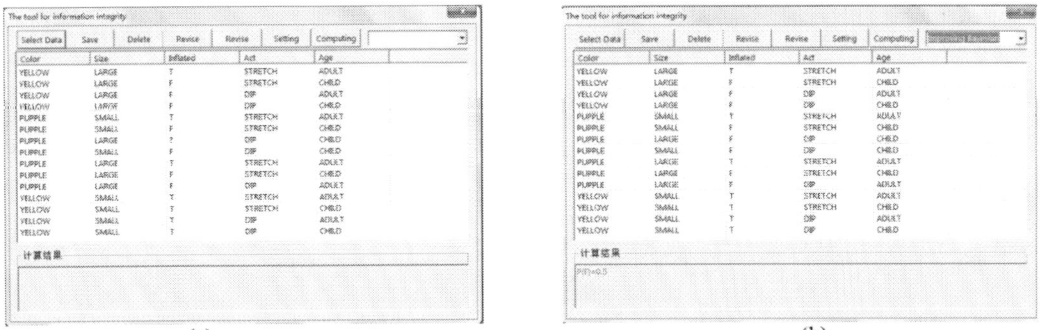

(a) (b)

Figure 3: (a)-(b) Data set and information completeness with the Bayesian method based on relative weight preference coefficient for standard data.

Table 2: The results with different methods for completing standard data

Methods	Max P	Result	Evaluation	Analysis
Bayesian	0	Unknown	improper	Inexistent the data in the known sample as the same as the data in test sample.
Naive Bayesian	0.98	F	improper	From the perspective of physical significance, attribute "size" is related to attribute "act" and "age".
Improving	0.05	F	proper	the independence and dependence considered simultaneously

2. Case study

The incomplete information table about A_{13} is shown in Table 3.

Table 3: The incomplete information about A_{13}

Enterprise names	Overall number of people	The main business income (ten thousand)	Total assets	Enterprise scale
Enterprise1	x>1000	y>4	300<z<500	big
Enterprise 2	x>1000	y>4	z>500	big
Enterprise 3	x>1000	y>4	z>500	small
Enterprise 4	300<x<1000	y>4	100<z<300	small
Enterprise 5	300<x<1000	y>4	z<100	small
Enterprise6	300<x<1000	y>4	100<z<300	big
Enterprise7	x>1000	y>4	100<z<300	big
Enterprise8	x>1000	2<y<4	z<100	middle
Enterprise9	x>1000	2<y<4	z<100	big
Enterprise10	x>1000	y>4	z<100	big
Enterprise11	300<x<1000	y>4	100<z<300	?

The result with improved Bayesian method is as shown as Figure 4. And the results compared with different method are shown in Table4. By calculation, the conclusion is consistent with actual.

In table 2 and table 4, although the result with Naïve Bayesian and improving Bayesian respectively is same, improving Bayesian considered independence and dependence simultaneously from the perspective of physical variables and physical significance, the results are more likely to be recognized.

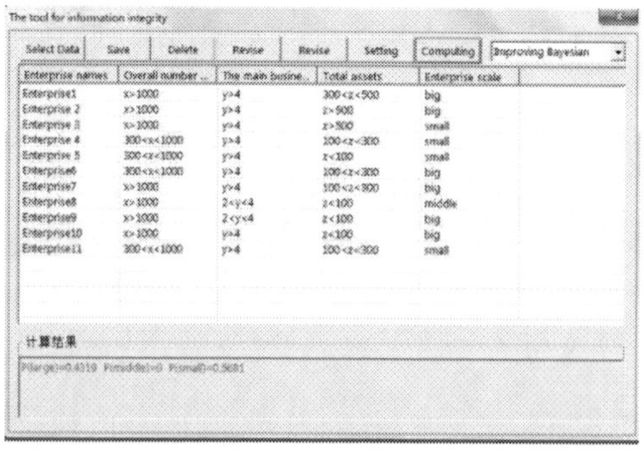

Figure 4: Information completeness with the Bayesian method based on relative weight preference coefficient for A_{13}.

Table 4: The results with different methods for completing information of ER

Methods	Max P	Result	Evaluation	Analysis
Bayesian	0.5	Unknown	improper	Inexistent the data in the known sample as the same as the data in test sample.
Naive Bayesian	0.704 2	Small	improper	From the perspective of physical significance, attribute "the number of people" is related to attribute "income" and "assets".
Improving	0.568 1	Small	proper	the independence and dependence considered simultaneously

5 CONCLUSION

The service-oriented manufacturing system, cloud manufacturing system and big data were put forward to provide the intelligent and service platform for collaborative manufacturing of enterprises. On this platform, manufacturing capacity was one of the important information and resources which were the basis of functions achievement. Therefore, it is very essential to manage and analyze the MC effectively. But in this process, there are three problems. Aiming these problems, in this paper, information and resources related to MC were classified firstly. Second of all, a model for expressing the relationships between information and MC was established. Based on this, the causes of information incomplete were analyzed. Then, the new Bayesian method was presented in view of the independent and relative relations to process incomplete information and the effectiveness of the method proposed in this paper was validated through case study, which can provide complete information for analysis and evaluation.

6 REFERENCES

[1] **W.Skinner.**1969. Manufacturing-missing link in corporate strategy, *Harvard Business Review*, pp 136-145.

[2] **R. H. Hayes and S. C. Wheelwrigh.** 1984. Restoring our competitive edge: completing through manufacturing, New York: Wiley.

[3] **K .HAFEEZ,ZHANG Yanbing, N .MALAK.** 2001. Determining key capabilities of a firm using analytic hierarchy process, *International Journal of Production Economics*, 76,(1),pp 39-51.

[4] **ARAFA , M. H. EIMARAGHY.**2011. Manufacturing strategy and enterprise dynamic capability, CIRP Annals manufacturing Technology, 60(1), pp 656-660.

[5] **Jiancheng GUAN.**2004. Study on the relationship between the enterprise manufacturing capability and innovation performance: empirical findings of Chinese, *scientific research management*, 25(ZK),pp 78-83.

[6] **A .B. KASOGLU.**2003. Capability based distributed layout approach for virtual manufacturing cells, *Internanal of Production Research*, 41(11),pp 2597-2618.

[7] **LIU,P.Y.JIANG,W.CAO.** 2014. Manufacturing capability match and evaluation foroutsourcing decision-making in one-of-a-kind production, Proceedings of the 2014 IEEE 18th In national Conference on Computer Supported Cooperative Work in Design, IEEE, pp 575-580.

ORDER ALLOCATION OPTIMIZATION OF DISTRIBUTED MULTI-PLANTS CONSIDERING MULTI-OB ECTIVES OPTIMIZATION

S.H. Wang, L.M. ang, .X. Zhou, .F. Zhang, A.W. an and F. Liang*
Department of Management Science and Engineering
Nankai University, Tianjin, China
Liangfeng79@nankai.edu.cn

ABSTRACT

This study faces the multi-objective multi-factory order allocation problem under the hypothesis that every order can be allocated to only one plant. A mathematical model is established based on the objective of cost minimization, total delay minimization and production load balancing. By providing a methodology, decision tree combined with heuristic rules, the problem can be solved efficiently and an optimized order allocation scheme can be inferred. Finally, the proposed allocation approach is proven in illustrative examples to perform more efficiently with multi-objective optimization.

Keywords: Multi-objective Optimization, Order Allocation, Decision Tree, Heuristic Rules

1 INTRODUCTION

With the gradual promotion of industrial 4.0 in the world, supply chain management model tends to be smart. Enterprises are more likely to be faced with a problem that orders from various districts are completed by distributed multi-plants. As an aspect of the industrial 4.0, smart order allocation is significant to the enterprise operation management, and it is also what this research focuses on. Due to the large demand for a product, enterprises can get massive orders in an allocation period, whereas order allocation mostly lacks of high-efficient quantitative optimization decision-making. A quick automatic allocating tool is required for a huge data computational task, which substitutes inefficient manual order allocation and improves allocation speed.

In this problem, one order only can be produced in one plant, and the cost as well as the time for each plant to produce the products differs. On the issue of order allocation, the short-term goal of the enterprise is to ensure finishing all the orders and to gain adequate profits. Taking a long view, plants are expected to coordinate with each other for the sake of achieving sustainable development. Therefore, the optimal solution to allocate a large number of orders is established based on the economic factors and capacity loads. In this paper, a kind of multi-plant order allocation problem is presented. The cost, time and load balancing are considered to optimize supply chain performance. Eventually, satisfactory order allocation programs are worked out. And then, numerical examples are given to show the feasibility and effectiveness.

The rest of this paper is organized as follows: a few similar problems have been discussed by previous scholars in Section 2. In section 3 a multi-objective 0-1 integer programming model is established with 5 assumptions. In section 4 and 5, Decision tree approach combined with heuristic rules is provided to describe and solve the problem briefly.

* Corresponding Author

2 LITERATURE REVIEW

The order allocation problem under multi-plant environment has been being discussed fiercely. A large number of papers treat the problem as a deterministic allocation optimization process based on the cost. Pan [1] proposes a mathematical model to minimize purchase cost, taking the constraints of product quality, service and delivery time into account. Based on the minimization of total manufacturing cost, distributing cost, setup cost and penalty cost for delay, ang [2] extends a more perfect mathematical model and solves it by genetic algorithms. Ji [3] offers a mixed integer linear programming (MILP) model for orders dispatching under the multi-product, multi-order, and multi-period hypothesis. The model focuses on minimizing the cost of the entire manufacturing / distribution network, which includes the production cost, transportation cost, Inventory cost and the penalty cost caused by the delay in delivery. A hybrid genetic algorithm that combines genetic algorithm with heuristic rules is designed, which can shorten the time for solving the model. Similarly, the model proposed by Zhou [4] aims to minimize the cost of early / delay penalty, production and transportation. In his paper, he describes the advanced / delayed production schedule according to the delivery period in distributed multiple plants. Then the mathematical model of the 0-1 programming is established, and the scheduling scheme is obtained by the branch and bound algorithm. Reference [3] and [4] are about one-off-manufacturing production. Compared with them, Ma's [5] research bases on the batch production. This paper holds a hypothesis: there may be a number of products in each order and an order only can be produced by one plant. To minimize the cost of production, inventory, adjustment of preparation and transportation, he constructs the mathematical model based on the limited production capacity and proposes the particle swarm optimization algorithm to solve the problem.

Korpela [6] proposes that the production capacity of the supplier is limited. This has brought inspiration that order allocation should not only be considered from the view of the purchaser but also from the angle of the supplier to consider the limits of its production capacity. Because ignoring the current production status of the plants will lead to the disadvantageous circumstances that some plants are too busy to complete the production task while others have no production plans. So some studies try to discuss afore allocation problems from the view of work-load balancing. Sanayei [7] disposes of the supplier selection and order allocation problem with the constraint of the supplier's production capacity. Cheng [8] focuses on order allocation problem in horizontal manufacturing collaboration alliance and establishes a multi-objective optimization model to minimize the integrated cost and production load balance. In order to provide the basis for the management of allocating orders in Horizontal Manufacturing Collaborative Alliance(HMCA), e [9] designs an algorithm named Multi-population Binary Ant Colony Algorithm with Hybrid Behaviors (MPBAHB) to minimize the integrated cost and balance the production loads among the selected manufacturing enterprises.

From above literatures, for the order allocation problem in multi-plant environment, the key point is to find an optimal allocation plan where the minimization of cost or punctual delivery or the balance of production load can be met [10-11]. But few researches have been conducted taking account of all those objectives. This research's major difference with others' is that the cost, delay time and load are considered at the same time, which can ensure the sustainable development of the plants while minimizing cost.

2.1 Description and Assumptions

An enterprise has n plants in different districts and produces only one product. Now it receives m orders from different customers in an allocation period. After the priority of these orders is determined through a specific strategy, all orders will be assigned to different plants.

Assumptions:

- The same order only can be allocated to one plant for production, but more than one order can be produced in the same plant.
- There have differences among the production capacity of those plants. The time and cost of producing the same product differ among those plants.
- If an order is delayed, the enterprise will pay a penalty cost. And the penalty cost is linear with time.
- Each plant can only process one order at the same time, but its waiting production list could have more than one order.
- Rush orders is not considered in this paper and orders received would not be cancelled.

2.2 Mathematical Model

The model of multi-plant orders allocation is demonstrated below and the notation is introduced:

i Order number, i =1,2,3, ,m

k Plant number, k=1,2,3, ,n

y_{ik} A binary variable to introduce the amount of production in order i to be produced at plant k, y_{ik}=1 if order i is produced at plant k, otherwise y_{ik}=0

x_i The amount of product in order i;

c_{ik} Unit transportation cost from order i to plant j

m_k Unit producing cost for order i in plant j

J_i Penalty coefficient

T_{Di} Delay time of order i

t'_{ik} ueueing time of order i to be produced in plant k, which consists of the time for waiting in queue based on priority and backlog of unfilled orders tasks in last production cycle （t_k）

t'_k The production time for plant k to produce one product

T_i The time from order i allocated to its plant to its completion

T_{ik} The time how long it will take to fulfill order i if order i is allocated to plant k (y_{ik}=1)

Di The time from order i allocated to plant k to the due time

t_k The time for other unfinished tasks in last production period to be produced in this period which are must be produced before the orders that are allocated now. t_k is a given number.

a Non-negative constant associated with penalty

Object function:

$$Min\ z1 = \sum_{i=1}^{m}\sum_{k=1}^{n} x_i\, y_{ik}(c_{ik} + m_k) + \sum_{i=1}^{m} J_i\, T_{Di} \quad (1)$$

$$Min\ z2 = \sum_{i=1}^{m} T_{Di} \quad (2)$$

$$Min\ z_3 = \frac{1}{n}\sum_{k=1}^{n}[(\sum_{i=1}^{m} t_k'\, x_i y_{ik} + t_k) - \frac{1}{n}\sum_{k=1}^{n}(\sum_{i=1}^{m} t_k'\, x_i y_{ik} + t_k)]^2 \quad (3)$$

$$J_i = a x_i \quad (4)$$

$$Ti = \sum_{k=1}^{n} T_{ik} y_{ik} \quad (5)$$

$$T_{ik} = t_{ik}' + x_i t_k' \quad (6)$$

$$t_{i_0 k}' = tk + \sum_{i=1}^{i_0 - 1} x_i t_k' y_{ik},\ i_0\ represent\ the\ priority\ (1 <= i_0 <= m) \quad (7)$$

$$T_{Di} = max\ \{Ti - Di,\ 0\} \quad (8)$$

Subject to:

$$\sum_{k=1}^{n} y_{ik} = 1\ \ i = 1,2,3,\dots,m \quad (9)$$

The objective function z1 in Eq.1 represents the total cost after order allocation. The first term includes total manufacturing and transport costs; the second term is total penalty cost for delay. The objective function z2 in Eq.2 is about to minimize the total delay time of the orders. The objective function z3 given in Eq.3 expresses the variance of production time of m plants. The punishing function for penalty coefficient is in Eq.4. In Eq.5, we present actually how long it will take to fulfil order i. The T_{ik} in Eq.6 is the time how long it will take to fulfill order i if order i is allocated to plant k (y_{ik}=1) and it consist of t_{ik}' and t_{ik}'. Equation 7 is to calculate the waiting time for plant k to finish all the tasks before order i. Equation 8 is to find out the delay time among of order i. Equation 9 is a constraint that every order is allocated to one and only one plant.

2.3 Methodology

In this section, a methodology incorporating a heuristic rule and a decision tree has been put forward, which aims to solve the problem and the model in a basic way. Initially, we get indefinite-quantity orders in a production cycle. Then a specific heuristic rule would be chosen to give a priority sequence for all batches of orders to determine which produce firstly. There are a lot of basic heuristic rules put forward by previous scholars, such as FIFO, SPT, LPT, EDD, ODD and etc. Nowadays most scholars are willing to use a combined heuristic rule that includes sever basic rules or propose a new heuristic rule based on other basic rules and their strategy towards some specific problems. In this research Earliest Due Date (EDD) is chosen to match the proposed decision tree. What needs more explanations is that the sequence of orders with same due date depends on the receiving time. The earlier the order receives, the higher priority it has when their due date is same.

Ultimately, a decision tree considering the multi-objective function has been established shown in Figure 1. According to the sequence of the heuristic rule, the order would be given to a definite plant through the decision tree one by one. It is assumed that the cost is most important and production load balance is least important. Hence, the cost is given priority and the first judgement is made in the first level. The plants after judgement have been divided into four situations in the first level. The plants which meet $m_k + c_{ik} \leqslant$0.7*(Non-profit price) belongs to the first group and they are regarded as the high profits plant. The plants which satisfied with 0.7*(Non-profit price)$< m_k + c_{ik} \leqslant$ 0.9*(Non-profit price) would be allocated to the second group and they are seen as the medium profits plant. The plants which subject to 0.9*(Non-profit price)$< m_k + c_{ik} \leqslant$ (Non-profit price) would be given to the third group and they are defined as the low profits plant. The plants mentioned in above three situations would go into second judgement in the second level. If there exist plants which is satisfied with $m_k + c_{ik} >$(non-profit price), the plant with lowest cost of them would enter the forth group and it never branch. In the second level it is judged by the completion date of each plants towards same order. In this level plants would be classified into three types under the three groups mentioned above. The plants whose completion date of the order is more than 3 days earlier than or equal its due date are seen as the high speed plant. The plants whose completion date of the order is less than 3 days earlier than its due date without delay are regarded as the low speed plant. The plant whose completion date of the order is later than its due date but is closest to it is the last type and it would never branch. The plant which balance the production load most would be selected from the first two types of plants in the third level. Finally, there are nine leaf nodes which would never branch and each node only has one specific plant at most. When dealing with an order, all plants should run in this decision tree and a few suitable plants would be chosen into the 9 leaf nodes. The closer to the left in figure 2 the node is, the higher priority it has because cost takes precedence over due date. For instance, when only the first leaf node (The number of leaf node is up to their location from left to right as Figure 1 shows) and the third leaf node has a plant, we prefer to select the plant in the first leaf node since it has higher priority. It is acceptable to determine which plant to get this order. In addition, it is worth mentioning that this decision tree has ability to feedback the key dynamic information of each plant like t'_{ik}.

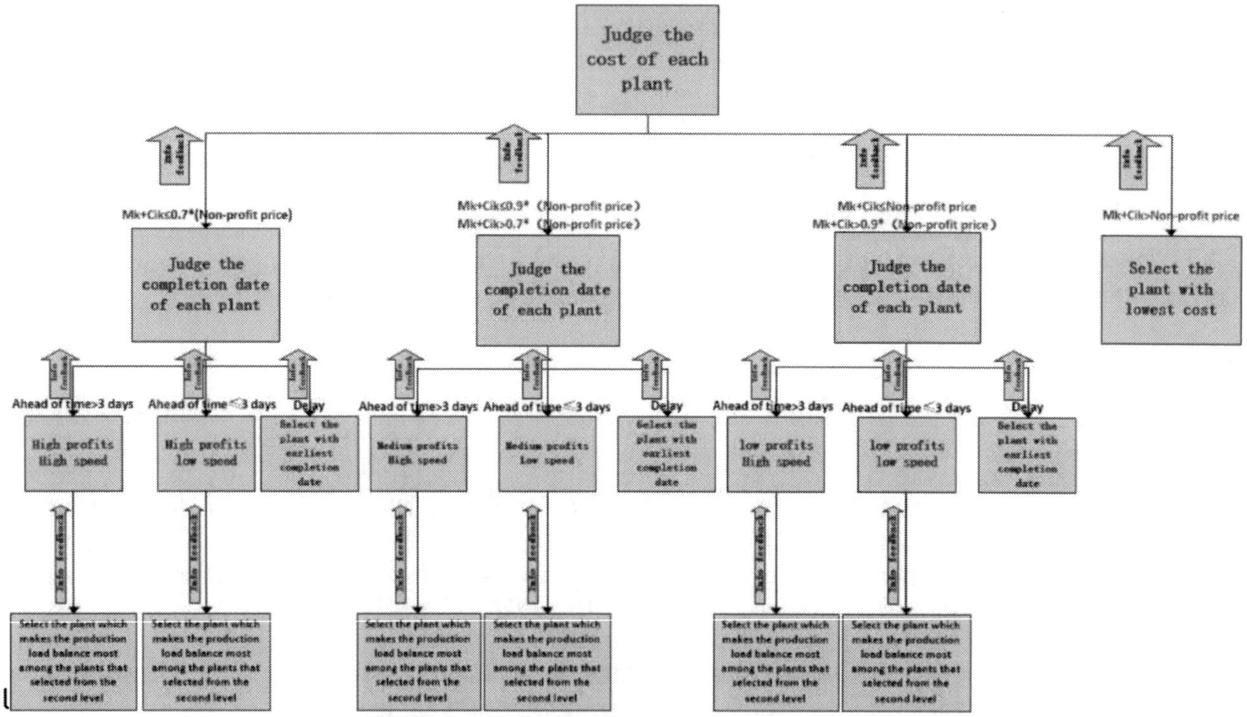

Figure 1: The Decision Tree

3 CASE ANALYSIS

In order to verify the validity of the model, following examples are constructed to numerical simulation analysis. It is assumed that a brand enterprise has received 9 different orders. The due time and the product quantity of each order list in the table below.

Table 1: Order Information

order number	due time Di (day)	amount of product xi (kg)	penalty cost Ji =3.5 xi
1	6	55	192.5

2	9	72	252
3	4	70	245
4	10	31	108.5
5	11	65	227.5
6	4	32	112
7	3	42	147
8	13	78	273
9	2	25	87.5

Now 5 plants with different cost and time are available to afford these orders. The unit producing cost, the unit production time and the unfinished time in last period of each plant are showed in the table 2. And the unit transportation cost from one plant to one order is given in Table 3.

Table 2: Plant Information

plant number	unit producing cost mk (uan/kg)	unit production timet$'_k$ (d/kg)	unfinished timet$_k$(d)
A	20	0.1	0.25
B	24	0.08	0.1
C	20	0.11	0.15
D	25	0.06	0.9
E	22	0.09	1

Table 3: Transportation Cost

order \ plant	A	B	C	D	E
1	4.5	1.2	1	5	2
2	2.5	0.3	5	1.2	1.2
3	16.5	8.2	13.6	8	10
4	9	3	5	5.3	7
5	12	13.5	14	12	11
6	8	8.2	10.3	13.5	3
7	7	2	5.5	8.5	12
8	13	8	17	6.9	11.5
9	16	16	15.5	13	15

Firstly, the heuristic rule EDD is used to give a priority sequence for the 9 orders:

Order9→Order7→Order6→Order3→Order1→Order2→Order4→Order5→Order8

Afterwards, the enterprise is able to allocate these orders through the decision tree one by one. Assume the Non-profit price is 35. Optimal order allocation based on this decision tree shows in Table 4 below that plant A produce order 4 and order 9, plant B produce order 2 and order 7, plant C produce order 5 and order 6, plant D produce order 3 and order 8, plant E produce order 1.

Table 4: Final Allocation of Multi-Ob ective

order \ plant	A	B	C	D	E

1	0	0	0	0	1
2	0	1	0	0	0
3	0	0	0	1	0
4	1	0	0	0	0
5	0	0	1	0	0
6	0	0	1	0	0
7	0	1	0	0	0
8	0	0	0	1	0
9	1	0	0	0	0

According to the order allocation above, object function of total cost (Z1=14396.585) is calculated. Object function of total delay time (Z2= 2.53 days) is calculated.

In order to find out the advantage of Multi-objective optimization model, a single optimization object (Minimizing the cost) is compared to the proposed allocation approach. The calculating result is given in Table 5.

Table 5: Final Allocation of Single-Ob ective

plant / order	A	B	C	D	E
1	0	0	1	0	0
2	1	0	0	0	0
3	0	0	0	0	1
4	0	0	1	0	0
5	0	1	0	0	0
6	0	0	0	0	1
7	0	1	0	0	0
8	0	0	0	1	0
9	1	0	0	0	0

Without thinking of issues related to time penalty, a new object function of cost is created as follows:

$$Z1' = \sum_{i=1}^{m}\sum_{k=1}^{n} x_i\, y_{ik}(c_{ik} + m_k)$$

Form the table, it can be calculated that: Single-Objective Z1'=13507.7 and Z2= 8.39 days; Multi-Objective Z1'=13938.4 and Z2= 2.53 days.

It is observed that although the cost of Single-Objective is lower, but the delay time is too long. The more an enterprise delays, the worse influence it obtains. The production of next period will also be affected.

By means of comparison and analysis, it witnessed the effectiveness and reasonability of the multi-objective model. And through it, enterprises are more possible to receive a satisfied order allocation.

4 CONCLUSION

In this paper, a kind of multi-plant order allocation problem is presented. A multi-objective 0-1 integer programming model is established with five assumptions. Decision tree approach combined with heuristic rules is provided to describe and solve the problem efficiently. Finally, the proposed allocation approach is proven in illustrative examples to perform more efficiently with multi-objective optimization. In the future work, more variables would be added to the model and more comprehensive situations that enterprises should face in actual operation management would be considered. A suitable solution like artificial intelligence, a specific algorithm would also be found and matched.

ACKNOWLEDGE

This research is supported by the National Natural Science Foundation of China (Grant No. 71271122).

REFERENCES

[1] **Pan A C. 1 8 .** Allocation of order quantity among suppliers, *Journal of Purchasing and Materials Management*, 25(3), pp 36-39.

[2] **Yang F C, Chen K, Wang M T, et al. 2010.** Mathematical modeling of multi-plant order allocation problem and solving by genetic algorithm with matrix representation, *International Journal of Advanced Manufacturing Technology*, 51(9), pp1251-1259.

[3] **I Xiao-li. 2005.** Order Allocation Model in Supply Chain Hybrid Genetic Algorithm, *Journal of Southwest Jiaotong University*, 40(6), pp811-815.

[4] **ZHOU in-hong, WANG Ding-wei. 2000.** Earliness and Tardiness Production Planning with Due-Date for Multi-Location OKP Manufacturing Systems, *Computer Integrated Manufacturing Systems*, 6(05), pp52-57.

[5] **MA Hui-min,YE Chun-ming,XU Sheng-liang,ZHANG Shuang. 200 .** Research on Multi-location Plants Capacitated Lot-sizing Problem Based on Particle Swarm Optimization Algorithm, *Operations Research and Management Science*, 18(03), pp123-127.

[6] **Korpela , Kylaheiko K, Lehmusvaara A, Tuominen M. 2002.** An analytic approach to production capacity allocation and supply chain design, *International Journal of Production Economics*, 78(2), pp187-195.

[7] **Sanayei A S, Mousavi F, Abdi M R, et a1. 2008.** An integrated group decision-making, process for supplier selection and order allocation using multi-attribute utility theory and linear programming, *Journal of the Franklin Institute (S0016-0032)*, 345(7), pp731-747.

[8] **CHENG Fang- i1,YE Fei-fan2,YANG ian-guo1. 2008.** Mutiobjective Order Allocation Optimization Model of Horizontal Manufacturing Collaborative Alliance ,*Computer Integrated Manufacturing Systems*, 14(12), pp2317-2321.

[9] **YE ing, XIONG Wei- ing, IANG Bao-chuan. 2011.** Solution of Multi-objective Order Allocation Problem Based on Binary Ant Colony Algorithm, *Computer Engineering*, 37(03), pp 175-177.

[10] **Moghaddam, Kamran S. 2015.** Supplier selection and order allocation in closed-loop supply chain systems using hybrid Monte Carlo simulation and goal programming, International Journal of Production Research, 53(20),pp6320-6338

[11] **Liu Yan- iu, Wang Hao. 2016.** Order allocation for service supply chain base on the customer best delivery time under the background of big data, International Journal of Computer Science and Applications, 13(1), pp84-92.

AN ADAPTIVE UNSCENTED KALMAN FILTER WITH TIME-VARIED NOISE IN INTEGRATED NAVIGATION

CUI Dan[1*], WU Nan[2], MENG Fankun[3]
Zhengzhou Information Science and Technology Institute, Zhengzhou, China
[1]cuidan1220@163.com, [2]wunan8471@nudt.edu.cn, [3]465744386@qq.com

ABSTRACT

The filtering accuracy easily declines or even filter may diverge because of the unknown statistics and time-correlated variation of noise in most practical applications. For these applications, an adaptive unscented Kalman filter was proposed. The covariance of measurement innovation may be used to simultaneously estimate the system noise covariance **Q** and the measurement noise covariance **R** in the adaptive filter. Simulations are performed in an INS/GPS integration. The results show that the adaptive algorithm can suppress the divergence caused by the time-correlated noise, and the positioning accuracies of three-dimensional space are respectively improved 10.3 times, 7.5 times and 2.3 times, and the filter is better in adaptability and stability.

Keywords: integrated navigation, adaptive unscented Kalman filter, innovation, covariance

1 INTRODUCTION

In order to solve the problem of a nonlinear measurement or system model, some scholars have proposed UKF (Unscented Kalman Filter) algorithm. The UKF algorithm uses a set of precise selection of sigma points, transformed by unscented transformation, to estimate the mean and covariance of the system state. In an UKF, it does not require linearization of nonlinear system, and the algorithm can be easily applied to estimate system state in nonlinear system. Therefore, UKF is widely used in many real systems[1].

During the application of conventional UKF, the system noise covariance matrix **Q** and the measurement noise covariance matrix **R** are accurately determined. However, there are some practical navigation systems which cannot be done. The filter may be easily divergent when the deviation between the assumed noise statistics and the actual noise statistics are very large. To solve this problem, domestic and foreign researchers had made a series of studies and proposed a variety of adaptive Kalman Filter algorithms, such as Sage-Husa Adaptive Filter algorithm, Memory Attenuation Kalman Filter algorithm, Fuzzy Adaptive Kalman Filter algorithm[2] and so on. Sage-Husa adaptive filter algorithm[3][4] can simultaneously estimate state and statistics of noise, but the algorithm needs to ensure the positive definiteness of measurement noise covariance matrix and the positive semi-definite of system noise covariance matrix, or the noise would potentially corrupt state estimates. Fagin and Sorenson et al had proposed fading memory filter algorithm[5][6] that increases the weight of the current measured value by making use of attenuation factor to suppress the divergence of filter, at the expense of filtering accuracy[7][8][9]. Combining the standard Kalman filter algorithm with fuzzy logic[10], the improved algorithm timely retuned the measurement noise variance. The disadvantage is that must have some knowledge of system noise statistics in advance[11][12].

* Corresponding Author

For the divergent phenomenon of filter caused by the time-correlated noise, there proposed an improved unscented Kalman filter algorithm. The method calculates both the system noise covariance **Q** and the measurement noise covariance **R** by the covariance of the last **N** measurement innovation of the measurement innovation statistics. The estimation of the time-correlated noise can suppress filter divergence. In an INS/GPS integration, it used the improved UKF algorithm to estimate and simulate the error states of aircraft. Its simulation results showed that the improved unscented Kalman filter could restrain filtering divergence and the accuracy was higher than classical Kalman filter, which had some certain reference significance for real application.

2 LOOSELY COUPLED INS/GPS INTEGRATION MODEL

The Inertial Navigation System, which can provide not only effective attitude, angular rate and accelerometer measurements, but also position and velocity. The system has strong autonomy and independence, and is not easy to be interfered by vibration environment and radio interference. The position accuracy of INS is very high and stable in a short time. However, the inertial navigation measurement error accumulates rapidly over time, as a result, the navigation accuracy declines. Consequently, INS cannot work alone for a long time, it must constantly be recalibrated. The Global Positioning Systems has a number of advantages. It can provide a high long-time accurate position with error just only a few meters, its navigation and positioning services can be global and all-weather. The error does not accumulate over time. On the contrary, compared to INS, the GPS signal is so easy to be interfered or blocked, and there is no attitude information. Therefore, the mobile carrier cannot rely on GPS to provide a continuous navigation solution.

The benefits and drawbacks of INS and GPS are complementary, so the advantages of both technologies are combined to give a continuous, robust and complete navigation solution with high long-time or short-time position accuracy by integrating them[13]. In loosely coupled INS/GPS integration, an error-state implementation is assumed with 15 states including attitudes, positions, velocities, accelerometer bias and gyroscope bias. In loosely coupled integration, no GPS states are normally estimated, so the choice of system state vector depends on INS. Generally, most INS/GPS integrations use both position and velocity as 6 measurement states to improve the observability of INS attitude and instrument errors.

The real dynamic system model is generally nonlinear, so the INS/GPS integrated system state equation and measurement equation can be written as followed:

$$\boldsymbol{X}_k = f(\boldsymbol{X}_{k-1}, \boldsymbol{u}_{k-1}) + \boldsymbol{V}_{k-1} \tag{1}$$

$$\boldsymbol{Z}_k = \begin{bmatrix} \boldsymbol{r}_{GPS} - \hat{\boldsymbol{r}}_{INS} \\ \boldsymbol{v}_{GPS} - \hat{\boldsymbol{v}}_{INS} \end{bmatrix} + \boldsymbol{W}_k = \boldsymbol{H}_k \boldsymbol{X}_k + \boldsymbol{W}_k \tag{2}$$

Where, $\boldsymbol{X}_k = [f_E\ f_N\ f_U\ \delta v_E\ \delta v_N\ \delta v_U\ \delta_L\ \delta_\lambda\ \delta_h\ \varepsilon_{bx}\ \varepsilon_{by}\ \varepsilon_{bz}\ \nabla_x\ \nabla_y\ \nabla_z]^T$ is the error-state vector for filter estimating attitude, velocity and position errors. The subscripts E, N, U are used to denote the east, north, down coordinates of the integration improved UKF filter; x, y, z respectively represent the dynamic carrier coordinate system, $\phi_E\ \phi_N\ \phi_U$ as attitude errors, $\delta_{v_E}\ \delta_{v_N}\ \delta_{v_U}$ as velocity errors, $\delta_L\ \delta_\lambda\ \delta_h$ as position errors, $\varepsilon_{bx}\ \varepsilon_{by}\ \varepsilon_{bz}$ as the gyroscope errors, $\nabla_x\ \nabla_y\ \nabla_z$ as accelerometer errors. $\boldsymbol{X}_k$, which is the error state estimated at time k with error covariance matrix $\boldsymbol{P}_k$. In the equation (2), the differences between measurements output by the GPS and predictions of those measurements from the INS solution are used to update the state vector. The measurements both of GPS and INS must be simultaneously efficacious.

3 ADAPTIVE UNSCENTED KALMAN FILTER WITH TIME-VARIED NOISE

3.1 Innovation-Based Noise Estimation

Both conventional unscented Kalman filter and Kalman filter are linear minimum variance estimation based on the model. The difference lies in the method of determining the optimum gain matrix[14]. UKF determines the optimum gain matrix from measurement and estimated covariance matrix. In UKF, the state vector is transformed from n dimensional to 2n-dimensions parallel vector, namely sigma point. These points calculate the state estimation and error covariance matrix according to the system equation and the measurement equation.

In a UKF, the noise statistics **Q** and **R** are preset Gaussian white noise, but the assumptions are not meet in applications[15]. This section looks at how the measurement innovations are used to handle the noise statistics.

Measurement innovations vector comprises the differences between GPS and corrected inertial position and velocity solution:

$$e_{Z_k} = Z_k - H_k X_{k|k-1} \tag{3}$$

Calculating the covariance of the last **N** measurement innovations, where **N** is the adaptive window with variable values:

$$C_k = \frac{1}{N} \sum_{j=k-N}^{k} e_{Z_j} e_{Z_j}^{\mathrm{T}} \tag{4}$$

Computing **Q** and **R** from (4):

$$\hat{Q}_k = K_k C_k K_k^{\mathrm{T}} \tag{5}$$

$$\hat{R}_k = C_k - H_k P_{k+1|k} H_k^{\mathrm{T}} \tag{6}$$

(5) and (6) are used to calculate the system state estimated covariance and measurement innovation covariance after giving initial values of **Q** and R, which can estimate the system state by the time-varied noise statistics. Improved UKF algorithm will be applied to the INS/GPS integrated system.

Figure 1 shows the calculation process of improved unscented Kalman filter algorithm.

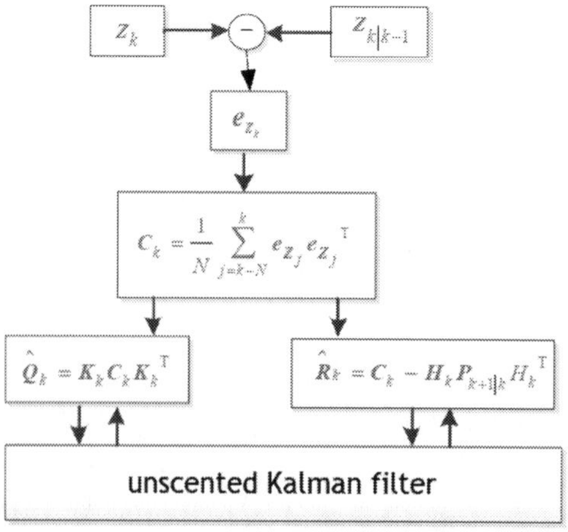

Figure 1: Improved unscented Kalman filter design process

3.2 Adaptive Unscented Kalman Filter Algorithm Based on the Time-Varied Noise Estimation

In loosely coupled INS/GPS integrated system, the INS error-state, estimated by the output of GPS position and velocity, is used to regulate the INS solution. The regulated INS solution is the output of integrated navigation system.

Firstly, determining the 2n+1 points sigma and the corresponding weights at time k. when the scaling factor λ take 0:

$$\begin{cases} \xi_0 = \hat{X}_k & i=0 \\ \xi_i = \hat{X}_k + (\sqrt{(n+\lambda)S_k})_i = \hat{X}_k + (\sqrt{15P_k})_i & i=1,\cdots,n \\ \xi_{i+n} = \hat{X}_k - (\sqrt{(n+\lambda)S_k})_i = \hat{X}_k - (\sqrt{15P_k})_i & i=1,\cdots,n \end{cases} \tag{7}$$

$$\begin{cases} W_0 = \lambda/(n+\lambda) = 0 & i=0 \\ W_i = 1/[2(n+\lambda)] = 1/(2\times 15) & i=1,\cdots,n \\ W_{i+n} = 1/[2(n+\lambda)] = 1/(2\times 15) & i=1,\cdots,n \end{cases} \tag{8}$$

where $(\sqrt{15P_k})_i$ is the i-th column of $\sqrt{15P_k}$,whose dimension is 15.

Each sigma point is propagated through system model using

$$X_{i,k+1|k} = \xi_i + f(\xi_i, u_k) \tag{9}$$

Calculating the state estimates and one step prediction error covariance:

$$\hat{X}_{k+1|k} = \frac{1}{2n+1}\sum_{i=0}^{2n} W_i X_{i,k+1|k} \tag{10}$$

$$\begin{aligned} P_{k+1|k} &= \frac{1}{2n+1}\sum_{i=0}^{2n} W_i (X_{i,k+1|k} - \hat{X}_{k+1|k})(X_{i,k+1|k} - \hat{X}_{k+1|k})^{\mathrm{T}} + \hat{Q}_k \\ &= \frac{1}{2n+1}\sum_{i=0}^{2n} W_i (X_{i,k+1|k} - \hat{X}_{k+1|k})(X_{i,k+1|k} - \hat{X}_{k+1|k})^{\mathrm{T}} + K_k C_k K_k^{\mathrm{T}} \end{aligned} \tag{11}$$

Substituting (11) into (7) updates the sigma points. Make one step prediction of measurement innovations by using the new sigma points. The new sigma points and the average measurement innovation are:

$$\begin{cases} e_{\hat{Z}_{i,k+1|k}} = Z_{k+1} - H_{k+1} X_{i,k+1|k} \\ e_{\hat{Z}_{k+1|k}} = \frac{1}{2n+1}\sum_{i=0}^{2n} e_{\hat{Z}_{i,k+1|k}} \end{cases} \tag{12}$$

This can be used to compute measurement innovation covariance:

$$\begin{aligned} C_{k+1|k} &= \frac{1}{2n+1}\sum_{i=0}^{2n} W_i (e_{\hat{Z}_{i,k+1|k}} - e_{\hat{Z}_{k+1|k}})(e_{\hat{Z}_{i,k+1|k}} - e_{\hat{Z}_{k+1|k}})^{\mathrm{T}} + \hat{R}_{k+1|k} \\ &= \frac{1}{2n+1}\sum_{i=0}^{2n} W_i (e_{\hat{Z}_{i,k+1|k}} - e_{\hat{Z}_{k+1|k}})(e_{\hat{Z}_{i,k+1|k}} - e_{\hat{Z}_{k+1|k}})^{\mathrm{T}} + C_k - H_k P_{k+1|k} H_k^{\mathrm{T}} \end{aligned} \tag{13}$$

Finally, the gain matrix, the state estimates and error covariance are updated using:

$$K_{k+1} = (C_{k+1|k})^{-1} \frac{1}{2n+1} \sum_{i=0}^{2n} W_i (X_{i,k+1|k} - \hat{X}_{k+1|k})(e_{\hat{z}_{i,k+1|k}} - e_{\hat{z}_{k+1|k}})^{\mathrm{T}} \quad (14)$$

$$\hat{X}_{k+1} = \hat{X}_{k+1|k} + K_{k+1} e_{\hat{z}_{k+1|k}} \quad (15)$$

$$P_{k+1} = P_{k+1|k} - K_{k+1} C_{k+1|k} K_{k+1}^{\mathrm{T}} \quad (16)$$

4 SIMULATION AND ANALYSIS

4.1 Simulation Conditions

To verify the performance of proposed innovation-based adaptive Unscented Kalman Filter algorithm in the INS/GPS loosely coupled integration, an application of the improved unscented Kalman filter algorithm was simulated in MATLAB. The initial dynamic positions are geodetic latitude 50.34388 °, longitude -3.31751 °, earth height as 10000m. The initial velocities are north -141.4m/s, east 141.4m/s, down 0m/s. The initial roll angle, pitch angle, yaw angle are respectively 0 °, 0 °, 135°. The dynamic time is 300s, at propagation intervals of 0.5s.

Under the condition of uncertain noise statistics, the initial parameters of innovation-based adaptive unscented Kalman filter are set as:

$X_0 = [0;0;0;0;0;0;0;0;0;0;0;0;0;0;0]$;

$P_0 = \mathrm{diag}(0.0003, 0.0003, 0.0003, 0.01, 0.01, 0.01, 100, 100, 100, 0.0001, 0.0001, 0.0001, 0, 0, 0)$;

$R_0 = \mathrm{diag}(6.25, 6.25, 6.25, 0.01, 0.01, 0.01)$;

$Q_0 = \mathrm{diag}(0, 0, 0, 0.1923, 0.1923, 0.1923, 0, 0, 0, 0.005, 0.005, 0.005, 0, 0, 0)$;

4.2 Simulation Results and Analysis

In the INS/GPS loosely coupled integration, the comparison experiment of improved unscented Kalman filter and Kalman filter was operated to plot the position, velocity and attitude error curves. Figure 2 shows the motion locus of the flight. The curves indicate position errors, velocity errors and attitude errors of improve filter and Kalman filter in figure 3, 4 and 5.

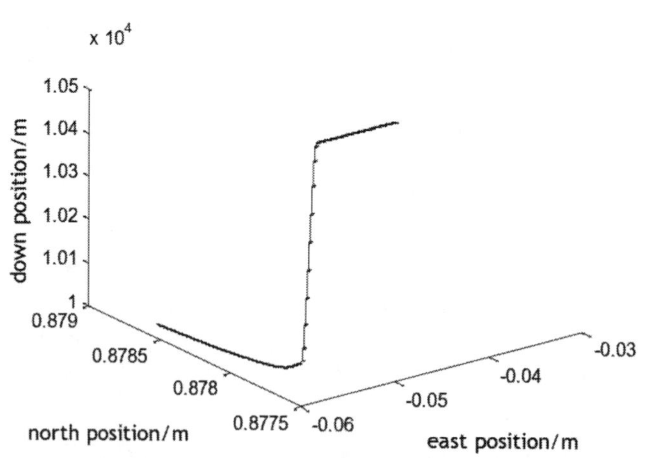

Figure 2: The Trajectory of Flight

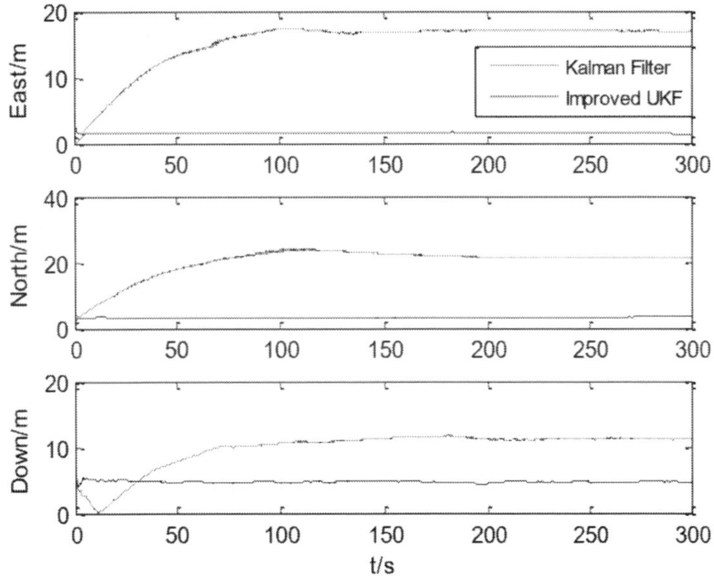

Figure 3: Comparison of East, North, Down Position Errors between Improved UKF and Kalman

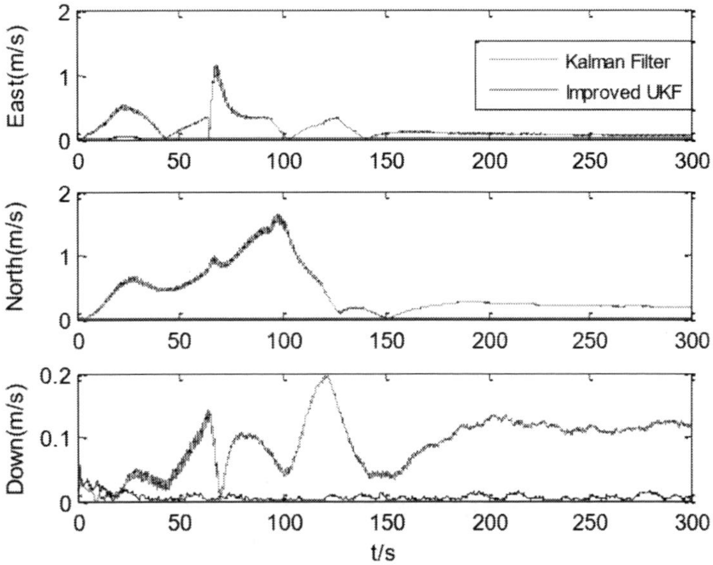

Figure 4: Comparison of East, North, Down Velocity Errors between Improved UKF and Kalman

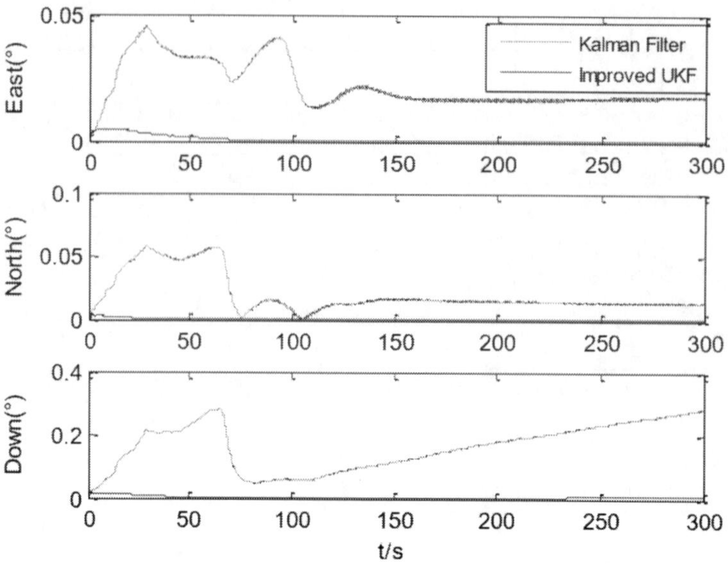

Figure 5: Comparison of East, North, Down Attitude Errors between Improved UKF and Kalman

It shows that the position, velocity and attitude error-state estimates of improved UKF are more stable than the Kalman filter in figure 3, 4 and 5. The east, north and down maximum position errors using Kalman filter reach up to 17.5m, 24.1m and 11.7m, while, the errors of improved UKF are only 1.7m, 3.2m and 5.2m. Thus, the improved UKF has higher position accuracies which is improved by 10.3 times, 7.5 times and 2.3 times, of three axes. The figure 4 shows that the east and north velocity errors significantly decline and the down velocity error also has a little decrease. The motion of flight is changed a lot at 64s, 94s and 127s, therefore, the velocity errors of Kalman filter, near the time of 64s, 94s and 127s, largely fluctuated. However, the improved UKF can well overcome the influence of trajectory mutations and the velocity error values are steady and close to 0. As can be seen from Kalman filtering curve in figure 5, the roll angle mutation make the north attitude error increased sharply at 64s, and the down attitude error has a divergent trend because of the accelerated climb during 94s to 127s. While, the attitude error values of the improved filter are close to 0 with no significant fluctuations and show good stability. Besides, the north attitude error is obviously restrained by the improved filter, and the east attitude error is only slightly modified. As for down attitude error, the output of the Kalman filter is divergent, while the improved UKF filter remains steady.

Compared to a standard Kalman filter, with the errors of position, velocity and attitude per axis in the loosely coupled INS/GPS integrated navigation system, it can be seen that the proposed UKF algorithm based on innovation is more stable and accurate when the system noise covariance matrix **Q** and the measurement noise covariance matrix **R** are uncertain.

5 CONCLUSION

In a standard Kalman filter, the statistics of system noise and measurement noise are known for the states estimated. This is not always the case for the actual physical process. Consequently, the adaptive UKF algorithm based on innovation is proposed. The improved filter uses the covariance of measurement innovation to estimate time-varied noise statistics. The paper describes INS/GPS loosely coupled integrated navigation system model, and also introduces the detailed process of the improved algorithm for the system state estimation. There are contrast simulation experiments between Kalman filter and improved unscented Kalman filter. The simulation results show that the innovation-based adaptive UKF can overcome the problems caused by unknown noise statistics, and have advantages of the stability and accuracy.

6 REFERENCES

[1] **HE You, XIU Jianjuan, GUAN Xin, et al.** Radar Data Processing with Applications [M] .Beijing: Electronic Industry Press, 2013.3.

[2] **H Guo, J Guo, M Yu, et al.** A Weighted Combination Filter with Nonholonomic Constrains for Integrated Navigation Systems [J]. Advances in Space Research, 2015, 55(5), pp 1470-1476.

[3] **Sage A P, Husa G W.** Adptive Filtering with Unknown Prior Statistics[C]//In Proceedings of Joint Automatic Control Conference.Boulder,USA:ASME,1969, pp 760-769.

[4] **QU Xinfen, LI Shiling, ZENG Chao, et al.** Application of Adaptive Fading Memory Adaptive Filter in Transferring Alignment [J]. Journal of Detection & Control,2014,36(5), pp 43-46.

[5] **TIAN Chong, WANG Xingliang, LU Yan'e.** An Adaptive Filter Algorithm for Exponentially-Weighted Multiple Attenuated Memory [J].Journal of Wuhan University of Technology ,2011,33(11), pp147-152

[6] **Long Zhao, Jing Liu.** An Improved Adaptive Filtering Algorithm with Applications in Integrated Navigation[C]//IEEE International Conference on Digital Manufacturing & Automation,2010, pp 182-185.

[7] **CH Wang, XU Yi-Bing, XU Ya-Wei.** Research on Memory Attenuation UKF Particle Filter Algorithm [J]. Microelectronics & Computer, 2012, 29(8), pp 50-52.

[8] **PK Tamboli, SP Duttagupta, K Roy.** Adaptive Fading Memory Filter Design for Compensation of Delayed Components in Self Powered Flux Detectors [J]. IEEE Transactions on Nuclear Science, 2015, 62(4), pp 1857-1864.

[9] **AM Al-Jumaily, Y Du.** Fading Memory Model for Airway Smooth Muscle Dynamic Response [J]. Journal of Theoretical Biology, 2011, 283(1), pp 10-13.

[10] **XU Tianlai, YOU Wenhu, CUI Pingyuan.** Research on GPS/INS Integrated Navigation System Based on Fuzzy Adaptive Kalman Filtering [J]. Journal of Astronautics,2005,26(5) ,pp 571-575.

[11] **A Amirzadeh, A Karimpour.** An Interacting Fuzzy-Fading-Memory-Based Augmented Kalman Filtering Method for Maneuvering Target Tracking [J]. Digital Signal Processing, 2013, 23(5), pp 1678-1685.

[12] **Enbo Shi.** An Improved Real-time Adaptive Kalman Filter For Low-Cost Integrated GPS/INS Navigation[C]//IEEE International Conference on Measurement, Information and Control,2012, pp 1093-1098.

[13] **D Dai, X Wang, D Zhan, et al.** An Improved Method for Gravity Disturbances Compensation in INS/GPS Integrated Navigation [C]//IEEE Signal Processing, 2014, PP 148-153.

[14] **QIN Yongyuan, ZHANG Hongyue, WANG Shuhua.** Theory of Kalman Filter and Integrated Navigation [M].Xi'an: Northwestern Polytechnical University Press,2015.2.

[15] **TL Xu.** Adaptive Kalman Filter for INS/GPS Integrated Navigation System [J]. Applied Mechanics & Materials, 2013, pp 332-338.

IMPROVING RELIABILITY USING EXPERIMENTAL DESIGN WITH A NON-CONSTANT SHAPE PARAMETER

Guodong Wang*
Department of Management Engineering
Zhengzhou University of Aeronautics, Zhengzhou, Henan, China
gdwang@tju.edu.cn

ABSTRACT

Reliability has always been considered to be one of the most important characteristics of a product. There is a strong need for design of experiments (DOE) that allow practitioners to identify the factors which affect lifetime significantly. In this article, we consider how to improve reliability through the use of designed experiments based on Weibull distribution with a non-constant shape parameter. We illustrate our proposed method through one real experiment contained censoring. Analysis results show that the misspecification of a constant shape parameter can lead to misidentify spurious effects.

Keywords: Design of experiments, reliability improvement, Weibull distribution, weighted least square method

1 INTRODUCTION

In today's increasingly competitive markets, leading practitioners have to find efficient and economical methods for improving product reliability. Many factors may affect product reliability, practitioners need to identify which factors are important and choose levels of these factors that lead to improved reliability. Design of experiments (DOE) has received a great deal of attention as quality improvement tool. Detailed description can be found in Box et al. [1], Wu and Hamada [2], and Montgomery [3]. DOE statistical techniques cannot, however, be directly applied to lifetime data analysis, because the distribution of lifetime data is usually non-normal. Additionally, the lifetime data may be censored. When censoring exists, DOE that based on least square method cannot be used to improve product reliability (Vining et al. [4]).

Maximum likelihood method, easily accounts for censored data and non-normal data, is generally recommended for calculating parameter estimates for lifetime models (Meeker and Escobar [5], Lawless [6], and Nelson [7]). Some statistical software packages have the capability to analyze lifetime data, such as SAS, R, MINITAB, and JMP. To test the significance of the estimates, they need to be compared with their standard errors. For example, in SAS software, the LIFEREG procedure estimates the parameters by maximum likelihood with a Newton-Raphson algorithm. The standard errors of the parameter estimates are computed from large sample normal approximations by using the observed information matrix. If convergence of the maximum likelihood estimates (MLEs) is attained, the Chi-square test statistic is computed for each effect, testing whether there is any contribution from any of the levels of the effect. Condra [8], Hamada [9], and Wu and

* Corresponding Author

Hamada [2] gave several examples to illustrated how to improve product reliability using design of experiments. Rigdon et al. [10] discussed a number of examples and presented results using these statistical software packages.

Examining the literature, we find few publications that use DOE method to improve product reliability considering the shape parameter is non-constant. Lawless [6] noted that DOE studies for reliability improvement often involve non-constant shape parameter. And he suggested using likelihood ratio tests to decide whether the scale parameter is constant. However, he did not present the method that can deal with non-constant shape parameter.

In this article, we identify important factors which affect lifetime significantly under the shape parameter is non-constant. The remainder of this article is organized as follows. Firstly, we introduce assumptions which will be used in this article. Secondly, we propose an analysis methodology of lifetime data obtained from DOE with a non-constant scale parameter. We illustrate our proposed method using a real example. Finally, we present the conclusions.

2 METHODOLOGY

In this section, we introduce our proposed methodology. Firstly, we present assumptions. Next we estimate the parameters of distribution using maximum likelihood method. Then we get the variance-covariance matrix by computing the inverse of Fisher information matrix. Finally, we identify important factors via weighted least square method and Wald confidence interval.

2.1 Assumptions

We consider an experiment that involves m treatment combinations; each treatment combination has n items. The lifetime data for individual items within *i*th treatment combination follows a Weibull distribution as below

$$F_i(t;\eta_i,\beta_i)=1-\exp\left(-\left(\frac{t}{\eta_i}\right)^{\beta_i}\right), \tag{1}$$

where $\eta_i>0,\beta_i>0$ are the scale and the shape parameters, respectively. $t>0$ is the time to failure. The experimental factors affect shape and scale parameters, i.e. $\beta_i\neq\beta_j$ and $\eta_i\neq\eta_j$ for $i\neq j$ and $i,j=1,2,\cdots,m$.

The scale parameter η_i depends on the experimental factors x_i through the model $\log\eta_i=x_i\gamma$, where **γ** are the corresponding coefficients.

2.2 Parameter Estimation

Let t_{ij} be the observed lifetime data for the *j*th item with the *i*th treatment combination. The likelihood function with right censoring present is

$$\mathcal{L}(t_{ij};\eta_i,\beta_i)=C\prod_{j=1}^{n_i}\left(f(t_{ij})\right)^{\delta_{ij}}\left(1-F(t_{ij})\right)^{1-\delta_{ij}} \tag{2}$$

where δ_{ij}=1 if the sample fails and δ_{ij}=0 if the sample is right censored. *C* is a constant dependent on censoring type.

The log likelihood function for the *i*th stress level with right censoring present is

$$\ell(t_{ij};\eta_i,\beta_i)=\log C+\log\beta_i\sum_{j=1}^{n_i}\delta_{ij}-\beta_i\sum_{j=1}^{n_i}\delta_{ij}\log\eta_i+(\beta_i-1)\sum_{j=1}^{n_i}\delta_{ij}\log t_{ij}-\sum_{j=1}^{n_i}\left(\frac{t_{ij}}{\eta_i}\right)^{\beta_i} \tag{3}$$

The parameter estimates can be obtained by solving likelihood equations.

From (3), it can be shown that

$$\begin{aligned}
-\frac{\partial^2 \ell}{\partial \beta_i^2} &= \frac{1}{\beta_i^2}\sum_{j=1}^{n}\delta_{ij} + \sum_{j=1}^{n}\left(\frac{t_{ij}}{\eta_i}\right)^{\beta_i}\log^2\left(\frac{t_{ij}}{\eta_i}\right) \\
-\frac{\partial^2 \ell}{\partial \eta_i^2} &= \frac{\beta_i(\beta_i+1)}{\eta_i^2}\sum_{j=1}^{n}\left(\frac{t_{ij}}{\eta_i}\right)^{\beta_i} - \frac{\beta_i}{\eta_i^2}\sum_{j=1}^{n}\delta_{ij} \\
-\frac{\partial^2 \ell}{\partial \beta_i \partial \eta_i} &= -\frac{\partial^2 \ell}{\partial \eta_i \partial \beta_i} = \frac{1}{\eta_i}\sum_{j=1}^{n}\delta_{ij} - \frac{\beta_i}{\eta_i}\sum_{j=1}^{n}\left(\frac{t_{ij}}{\eta_i}\right)^{\beta_i}\log\left(\frac{t_{ij}}{\eta_i}\right) - \frac{1}{\eta_i}\sum_{j=1}^{n}\left(\frac{t_{ij}}{\eta_i}\right)^{\beta_i}
\end{aligned} \qquad (4)$$

Rinne [11] pointed out if there does not exist censoring within the ith treatment combination, then the estimated asymptotic variance-covariance matrix of $\hat{\theta}_i = (\hat{\eta}_i, \hat{\beta}_i)$ is

$$\mathrm{Var}(\hat{\theta}_i) = \begin{bmatrix} -E\left(\frac{\partial^2 \ell}{\partial \eta_i^2}\right) & -E\left(\frac{\partial^2 \ell}{\partial \eta_i \partial \beta_i}\right) \\ -E\left(\frac{\partial^2 \ell}{\partial \beta_i \partial \eta_i}\right) & -E\left(\frac{\partial^2 \ell}{\partial \beta_i^2}\right) \end{bmatrix}^{-1}_{\theta_i = \hat{\theta}_i} = \frac{1}{n_i}\begin{bmatrix} 1.1087\frac{\hat{\eta}_i^2}{\hat{\beta}_i^2} & 0.2570\hat{\eta}_i \\ 0.2570\hat{\eta}_i & 0.6079\hat{\beta}_i^2 \end{bmatrix} \qquad (5)$$

If there exists censoring within the ith stress level, the estimated asymptotic variance-covariance matrix is then

$$\mathrm{Var}(\hat{\theta}_i) = \begin{bmatrix} -\frac{\partial^2 \ell}{\partial \eta_i^2} & -\frac{\partial^2 \ell}{\partial \eta_i \partial \beta_i} \\ -\frac{\partial^2 \ell}{\partial \beta_i \partial \eta_i} & -\frac{\partial^2 \ell}{\partial \beta_i^2} \end{bmatrix}^{-1}_{\theta_i = \hat{\theta}_i} \qquad (6)$$

2.3 Coefficient Estimation

The lifetime data may be censored and the variancesof error terms may be not constant for differen t treatment combinations, so there is aproblem of heteroscedasticity. Thus, we estimate model coefficients γ using weighted least square method. The coefficient estimates, $\hat{\gamma}$, can be computed by

$$\hat{\gamma} = \left(\boldsymbol{X}^T \boldsymbol{W}^{-1} \boldsymbol{X}\right)^{-1} \boldsymbol{X}^T \boldsymbol{W}^{-1} \log\hat{\boldsymbol{\eta}} \qquad (7)$$

where $\boldsymbol{X}$ is the matrix of treatment combinations, and

$$\log\hat{\boldsymbol{\eta}} = (\log\hat{\eta}_1, \ldots, \log\hat{\eta}_m)^T, \; \boldsymbol{W} = \mathrm{Var}(\log\hat{\eta}_i) = \begin{bmatrix} \mathrm{Var}(\log\hat{\eta}_1) & \cdots & 0 \\ \vdots & \ddots & \vdots \\ 0 & \cdots & \mathrm{Var}(\log\hat{\eta}_m) \end{bmatrix}.$$

We calculate the estimates of variances by $\mathrm{Var}(\log\hat{\eta}_i) = \mathrm{Var}(\hat{\eta}_i)/\hat{\eta}_i^2$.

2.4 Identify Important Factors

For large samples, it is known that $\hat{\gamma} - \gamma$ approximately distributed as $MN(\boldsymbol{0}, \mathrm{var}(\hat{\gamma}))$. Then the asymptotic normality result justifies the use of

$$\left(\hat{\gamma}_i - z_{\alpha/2} s.e.(\hat{\gamma}_i), \hat{\gamma}_i + z_{\alpha/2} s.e.(\hat{\gamma}_i)\right) \quad (8)$$

as a $100(1-\alpha)\%$ confidence interval. If this interval does not contain zero, the effect associated with a_i is declared to be significant at level α .

3 AN EXAMPLE

In this section, we demonstrate the application of our proposed method via an example from literature: industrial thermostat experiment. Wang and Kececioglu [12], and Mueller and Rigdon [13] reanalysed the data considering the scale parameter is non-constant and estimated the coefficients of models. However, they did not introduce the method that how to identify factors which affect lifetime significantly.

3.1 Experimental Design

Bullington et al. [14] provided a life cycle test to improve reliability of an industrial thermostat via Taguchi's methods. There were 10 observations for each of the 12 treatment combinations, and the 120 thermostats were cycled until failure, with testing stopped after 7,342,000 cycles. Thus, the data were right censored. There are 22 right censored data at trials 1, 6, 11. Table I shows the design and results of the test.

Table 1: Design and Results of Thermostat Experiment

	Factor																				
Trial	A	B	C	D	E	F	G	H	I	J	K	Order Lifetime Data (k-cycles)									
1	–	–	–	–	–	–	–	–	–	–	–	957	2846	7342	*	*	*	*	*	*	*
2	–	–	–	–	–	+	+	+	+	+	+	206	284	296	305	313	343	364	420	422	543
3	–	–	+	+	+	–	–	–	+	+	+	63	113	129	138	149	153	217	272	311	402
4	–	+	–	+	+	–	+	+	–	–	+	76	104	113	234	270	364	398	481	517	611
5	–	+	+	–	+	+	–	+	–	+	–	92	126	245	250	390	390	479	487	533	573
6	–	+	+	+	–	+	+	–	+	–	–	490	971	1615	6768	7342	*	*	*	*	*
7	+	–	+	+	–	–	+	+	–	+	–	232	326	326	351	372	446	459	590	597	732
8	+	–	+	–	+	+	+	–	–	–	+	56	71	92	104	126	156	161	167	216	263
9	+	–	–	+	+	+	–	+	+	–	–	142	142	238	247	310	318	420	482	663	672
10	+	+	+	–	–	–	–	+	+	–	+	259	266	306	337	347	368	372	426	451	510
11	+	+	–	+	–	+	–	–	–	+	+	381	420	7342	*	*	*	*	*	*	*
12	+	+	–	–	+	–	+	–	+	+	–	56	62	92	104	113	121	164	232	258	731

3.2 Data Analysis

If we assume that the shape parameter is constant, then we can get the conclusion that all the factors are important. The results can be found in Bullington et al. [14].

A type of factorial design, known as the fractional factorial design or the PB design, are often used to find the “vital few” significant factors out of a large group of potential factors. Lawless [6] suggested that practitioners need to decide whether the shape parameter is constant. Practitioners can test whether the scale parameter is constant or not and compute the *p*-value through MINITAB software. The estimates of Weibull Parameters are shown in Table III. We test the null hypothesis by likelihood ratio test. The result shows that the null hypothesis is rejected with *p*-value of 2.4725×10^{-12}. Thus we re-analyze the lifetime data under non-constant shape parameter.

Table 2: Parameter Estimates of Weibull Distributions

Group	Constant		Non-constant	
	Shape	Scale	Shape	Scale
1	1.8792	15547	0.722352	56500.1
2		359.367	4.10117	384.216
3		215.699	2.10769	220.974
4		358.468	1.84154	356.922
5		387.619	2.40902	402.083
6		98473	0.718205	18133.3
7		463.854	3.25951	495.069
8		152.562	2.48016	159.736
9		401.851	2.13918	412.446
10		371.106	5.13817	395.524
11		15362	0.372702	376500
12		261.112	1.22259	208.968

We obtain variance-covariance matrices of trials 1, 6, 11 by (6), and get the other variance-covariance matrices by (5). Table III shows the results under non-constant shape parameter. From Table III we find that the important factors are *C*, *E*, *G*, *H* and *I* at α =0.05 level.

Table 3: Analysis Results with Non-constant Shape Parameter

		Standard			95.0% Normal CI	
Predictor	Coef	Error	Z	P	Lower	Upper
Intercept	7.1194	0.3274	21.7451	0.0000	6.4777	7.7611
A	-0.2091	0.3274	-0.6388	0.5230	-0.8508	0.4326
B	0.5208	0.3274	1.5906	0.1117	-0.1209	1.1625
C	-0.7096	0.3274	-2.1673	0.0302	-1.3513	-0.0679
D	0.5717	0.3274	1.7463	0.0808	-0.0700	1.2134
E	-0.5010	0.3274	-4.5847	0.0000	-2.1427	-0.8593
F	0.4953	0.3274	1.5127	0.1304	-0.1465	1.1370
G	-0.7436	0.3274	-2.2712	0.0231	-1.3853	-0.1019
H	-1.1140	0.3274	-3.4024	0.0007	-1.7557	-0.4723
I	-0.7028	0.3274	-2.1467	0.0318	-1.3445	-0.0611
J	-0.1641	0.3274	-0.5013	0.6162	-0.8058	0.4776
K	-0.2662	0.3274	-0.8130	0.4162	-0.9079	0.3755

4 CONCLUSION

In this article, we analyze lifetime data from designed experiments for the target of improving product reliability. The proposed methodology is suitable for non-constant shape parameters as well as for a common shape parameter of Weibull distributions. The goal of this article is to illustrate the procedures of identify factors that affect lifetime significantly. In the example, we find different results if we consider the scale parameters are unequal. In fact, we can reject null hypothesis with a very small *p*-value using likelihood ratio test.

To identify the important factors which affect product lifetime significantly, we attempt at obtaining confidence intervals for model coefficients based on the asymptotic normality. It is very effective when the sample size is big. But it cannot be sure that our proposed can identify important factors correctly when the sample size is small. In this article, we do not evaluate sample size requirements for asymptotic properties to take effect. Kenett et al. [15] proposed applied bootstrap method to the analysis the data derived from designed experiments with censored data, heteroscedasticity and constraints related to the structure of the experiment. The next step in this research is to find out a smallest sample size such that our proposed method can identify important factors correctly by bootstrap method.

5 ACKNOWLEDGEMENT

This research was supported in part by grant 71402118 from the National Natural Science Foundation of China, grant 2013ZG55024 and 2014ZG55021 from the Aeronautical Science Foundation of China, grant 17A630072 from the Key Program for Scientific Research of Higher Education Institutions of Henan Province, China.

6 REFERENCES

[1] **Box, G.E.P., Hunter, W.G., Hunter, J.S.** 2005. *Statistics for Experimenters-Design, Innovation, and Discovery*, 2nd Edition, Wiley.

[2] **Wu, CFJ., Hamada, M.** 2009. *Experiments-Planning, Analysis, and Optimization*, 2nd Edition, Wiley.

[3] **Montgomery, D.C.** 2005. *Design and Analysis of Experiments*, 6th Edition, Wiley.

[4] **Vining, G.G., Murat, K., Søren, P. 2016.** Recent advances and future directions for quality engineering, *Quality and Reliability Engineering International*, 32(3), pp 863-875.

[5] **Meeker, W.Q., Escobar, L.A.** 1998. *Statistical Methods for Reliability Data*, Wiley.

[6] **Lawless, J.F.** 1982. *Statistical Models and Methods for Lifetime Data*, Wiley.

[7] **Nelson, W.B.** 1990. *Accelerated Testing: Statistical Models, Test Plans and Data Analysis*, Wiley.

[8] **Condra, L.W.** 2001. *Reliability Improvement with Design of Experiments*, Marcel Dekker.

[9] **Hamada, M. 1995.** Using statistically designed experiments to improve reliability and to achieve robust reliability, *IEEE Transactions on Reliability*, 44(2), pp. 206-215.

[10] **Rigdon, S.E., Englert, B.R.I., Lawson, A., Borror, C.M., Montgomery D.C., Pan, R. 2012.** Experiments for reliability achievement, *Quality Engineering*, 25(1), pp. 54-72.

[11] **Rinne, H.** 2008.*The Weibull Distribution: A Handbook*, CRC Press.

[12] **Wang, W.D., Kececioglu, D.B. 2000.** Fitting the Weibull log-linear model to accelerated life-test data, *IEEE Transactions on Reliability*, 49(2), pp. 217-223.

[13] **Mueller, G., Rigdon, S.E. 2015.** The constant shape parameter assumption in Weibull regression, *Quality Engineering*, 27(3), pp. 374-392.

[14] **Bullington, R. G., Lovin, S.G., Miller, D.M., Woodall, W.H. 1993.** Improvement of an industrial thermostat using designed experiments, *Journal of Quality Technology*, 25(4), pp. 262-270.

[15] Kenett, R.S., Rahav, E., Steinberg, D.M. 2006. Bootstrap analysis of designed experiments, *Quality and Reliability Engineering International*, 22(6), pp. 659-667.

SEGMENTATION OF MEDICAL IMAGES USING ENHANCED K-MEANS, WAVELET TRANSFORMATIONS AND MORPHOLOGICAL OPERATIONS

V RamakrishnaSajja [1] and N Gnaneswara Rao[2]
[1]Department of Computer Science & Engineering
Vignan's Foundation for Science, Technology & Research, India
bobysajja@gmail.com

[2]Department of Computer Science & Engineering
Vignan's Foundation for Science, Technology & Research, India
gnani.nitta@gmail.com

ABSTRACT

Image segmentation is a significant task in clinical investigative tools. Medical images typically include noise. As a result nowadays perfect segmentation of medical images is of great challenging issue. The proposed methodology for image segmentation depends on morphological operations and Enhanced K-Mean unsupervised clustering algorithm. As an initial step, the color image is transformed to gray image then the noise is removed by using default values for denoise & compression and wavelet denoise & compression. Global thresholding, symlet wavelet transform, soft or hard thresholding are used in wavelet denoise & compression. Later segmentation is performed using Enhanced K-Means unsupervised clustering algorithm on the medical image which leads to accurate results. Then Silhouette method is applied to know the quality of clustering. To get optimal number of clusters, the method evalclusters is used. Finally morphological operations are performed on the segmented images to get best results.

Keywords: Image segmentation, Enhanced k-Means, stationary wavelet transform, Silhouette method and morphological operations.

1 INTRODUCTION:

Image Segmentation[1,2] is the procedure of collecting similar picture elements together that have similar attributes in an image into non-intersecting part such that each part is uniform and the unification of no two neighbouring part should be uniform. Image segmentation is a critical step in clinical diagnoses. It is vital and decisive task in analysis of the image as well as determining the eminence of the ultimate result of analysis of the image. Enhanced K-Means an unsupervised clustering algorithm is useful for image segmentation and is compared with FCM [3-7], hard c-means [8], and it produces better results.

K-means clustering comes under the category of partitioning method. The k-means divides data into k reciprocally limited clusters, and gives the index of the cluster to which it has allocated each sample. K-means clustering is mostly appropriate than hierarchical clustering when data is huge.

The k-means method is regularly applied on unsupervised clustering procedure that aids to diminish the average squared distance between data points in the own cluster. Even though it does not provide substantiate accuracy, its ease in understanding the algorithm and stronghold are making it practice.

1.1 Standard k-means algorithm:

Step1: Choose k data points as the early centroids.

Step2: (Re) allocate each and every data points to its nearest centroids.

Step3: Re calculate the centroid for each and every newly assembled data points into the cluster.

Step4: Iterate step 2 and 3 until no change in the centroids.

2 PROPOSED METHOD:

Nowadays, a lot of scholars have gone through computer-aided methods for image segmentation. The proposed methodology is a fully automated system which is beneficiary to both the patients and clinicians. We take the clinical dataset and applied pre-processing techniques, such as converting RGB images of clinical dataset to gray images [9]. Then remove noise using stationary wavelet transforms [10-13]. Then we have given demised images to enhanced K-Means algorithm and we applied silhouette to know the quality of clusters. Finally apply morphological operations on segmented images for better analysis.

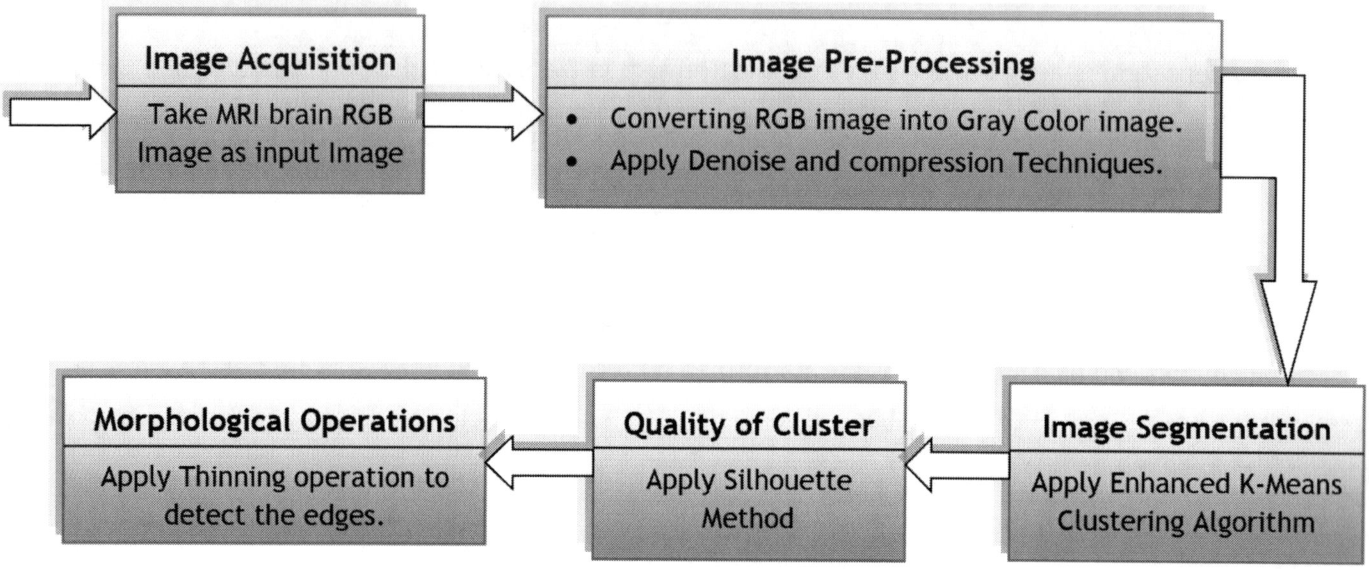

Figure 1: Block Diagram for proposed methodology

According to the proposed methodology, we represent a way of starting k-means by picking random initial centers with exacting probabilities that enhances the speed as well as the exactness of k-means. The process is explained in the following steps:

Step1: Consider the medical images that are true color images.

Step2: Convert the true color images into gray images.

Step3: Denoise and compression techniques and stationary wavelet transform are applied on the gray level images to remove noise.

Step4: The denoise images are given to enhanced k-means unsupervised clustering algorithm for segmentation.

Step5: Apply silhouette method to know the quality of segmentation.

Step6: Apply morphological operations for accurate results.

2.1 Transforming RGB images to GRAY images:

In the RGB color model each color will be displayed in its major components of Red, Green and Blue. RGB color space model is based on Cartesian coordinate system. It is used for Hardware oriented purpose. RGB model is not practical for Human Perception. So we transform rgbcolor image to gray image.

The outcome of computing the intensity of light at each and every pixel in a single band of the electromagnetic spectrum is the Gray scale image. Medical imaging or remote sensing often require more levels to make use of overall sensor informationrgb2gray () transforms the RGB image to the gray scale image by removing the hue and saturation information while sustaining the luminance.

2.2 Denoise and compression:

wdencmp is a 1D or 2D de-noised and compression-oriented method for a signal or for a image using wavelet transform. [XC,CXC,LXC,PERF0,PERFL2]=wdencmp('gbl',X,'wname',N,THR,SORH,KEEPAPP) gives a de-noised or compressed value XC of input image X given by stationary wavelet coefficients thresholding using global non negative threshold TH.

In addition to that output arguments [CXC, LXC] are the wavelet decomposition structure of XC. SORH ('s' or 'h') is for soft or hard thresholding.

Y = Wthresh(X, SORH, T) returns the soft (if SORH ='s') or hard (if SORH = 'h') thresholding of the input vector or matrix X. T is the threshold value.

2.2.1 Global Thresholding:

The compression characteristics of a wavelet basis are mainly connected to the comparative scarceness of the wavelet domain representation for the signal. The intention behind compression is depend on the perception that the normal signal component can be accurately approximated using the following elements: a small number of approximation coefficients and few of the detail coefficients.

2.3 Enhanced K-Means Algorithm:

The steps for Enhanced K-Means Algorithm are as shown below:

Step1: Select any data point randomly as the first centrec_1 for the first cluster from the dataset x.

Step2: Calculate the distance from each centrec_j to every data point in the dataset

$$m=d\left(\left(c_j, x_m\right)\right).$$

Step3: Specify the second centroid with the probabilityh

$$\frac{d^2(x_m, c_1))}{\sum_{j=1}^{n} d^2\left(x_j, c_1\right)}$$

Step4: Specify consequent centroid randomly from dataset with probability proportional to the distance from itself to the closest centre that you already chosen

$$\frac{\mathrm{d}^2(\mathrm{x_m}, \mathrm{c_p})}{\sum_{\{\mathrm{h}:\mathrm{x_h} \in c_p\}} \mathrm{d}^2\left(\mathrm{x_h}, c_p\right)}$$

Step5: Iterate step 4 until required number of clusters are formed and then follow the standard k-means algorithm.

Enhanced K-Means gives better results compared to K-Means and the process of identifying regions using enhanced k-means is shown in figure 1.

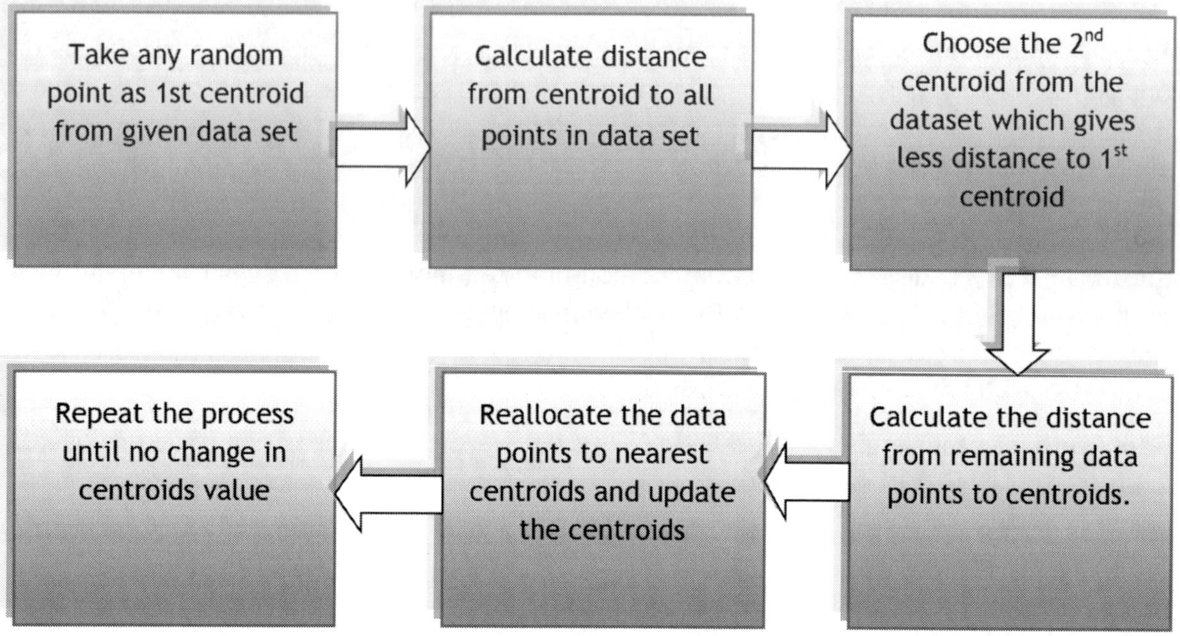

Figure 2:Block Diagram for enhanced K-Means

2.3.1 Avoid Local Minima:

Depending upon the initial seed points the enhanced k-means solutions may result in local minima. It is probable for enhanced k-means to attain a local minimum, where reassigning of any one data point to a new cluster would increase the overall sum of data point-to-centroid distances, but somewhere an improved way may be present on the other hand it can use the 'Replicates' name-value pair argument to prevent that difficulty. Replicate is defined as number of epochs to recur clustering by means of latest initial cluster centroid positions, specifically with the comma-separated couple containing of 'Replicates' and a non-negative integer.

2.3.2 Cluster Data Using Parallel Computing:

Clustering large data might take huge amount of time. If you have a Parallel Computing Toolbox then invoke a pool of workers that enhanced k-means runs each clustering job (or replicate) in parallel. So, if Replicates > 1, then the parallel computing reduces time to convergence.

2.4 Silhouette Method:

The silhouette value [14] for each and every data point is to calculate of how related that data point is to data points in its individual cluster, when compared to points in remaining clusters.

The silhouette value for the jth point, j, is termed as:

$$Sj = (qj - p_j) / \max(p_j, qj)$$

pj is the average distance from the j^{th}data pointto the remaining data points in the similar cluster as j

$qj$is the minimum average distance from the jth data point to data points in a different cluster, minimized over clusters.

The silhouette value ranges from -1 to +1. A greater silhouette value depicts that j is well-defined to its cluster itself.

2.5Morphological operations:

Morphology is a lane set of image processing operations that process images based on shapes. Morphological operations affect a structuring element to an input image, creating an output image of the same size. In a morphological process, the importance of each pixel in the output image is based on a comparison of the consequent pixel in the input image with its neighbours. By choosing the size and shape of the neighbourhood, you can make a morphological process that is perceptive to exact shapes in the input image.

After segmentation using enhanced k-means unsupervised clustering algorithm, binary image of each segmented one is obtained and the connected components of each and every binary image can be handled as well. To get rid of irrelevant connected components, Morphological filters are used for binary images for connected component. Furthermore we use a threshold to reduce little connected components.

3 EXPERIMENTAL RESULTS:

Brain tumour is one of the most detrimental diseases, and has exaggeratedbulk of people including children in the world. The possibility of continued existence can be improved if the tumour is detected at its impulsive stage. According to the process explained above we get the results by applying Enhanced K-Means Unsupervised clustering algorithm and silhouette method in gray image on various medical images.

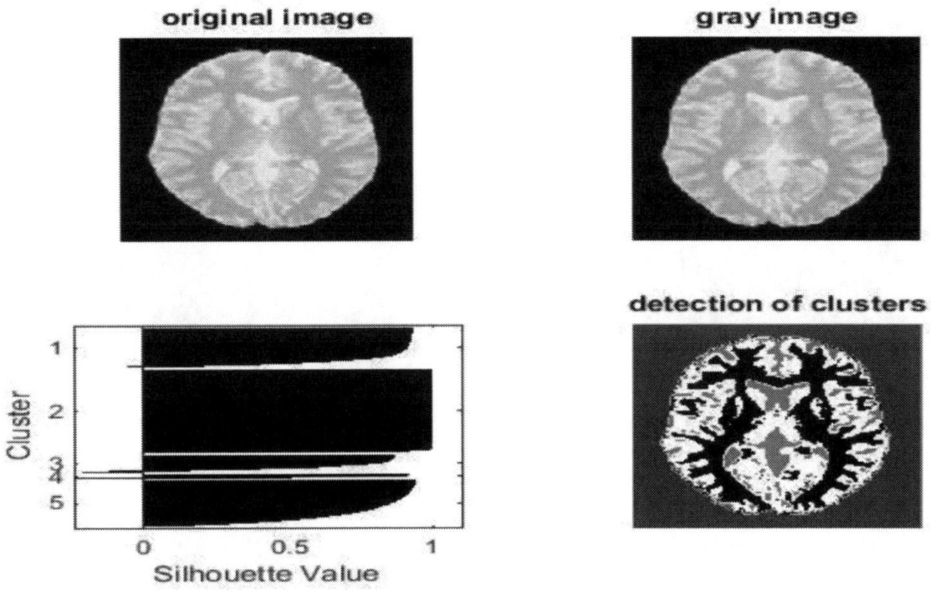

Figure 4: Detection of clusters from original image and silhouette plot for clusters.

In the below diagram we can observe the objects in an image and how the image after morphological operations have been done in their respective clusters.

Objects in Cluster1

Morphological thinning for cluster1

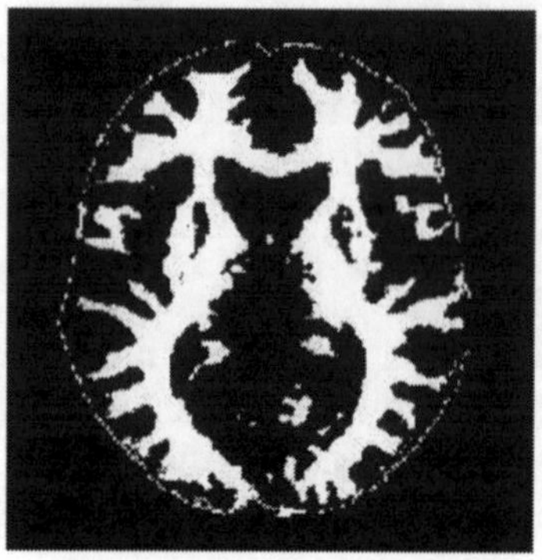

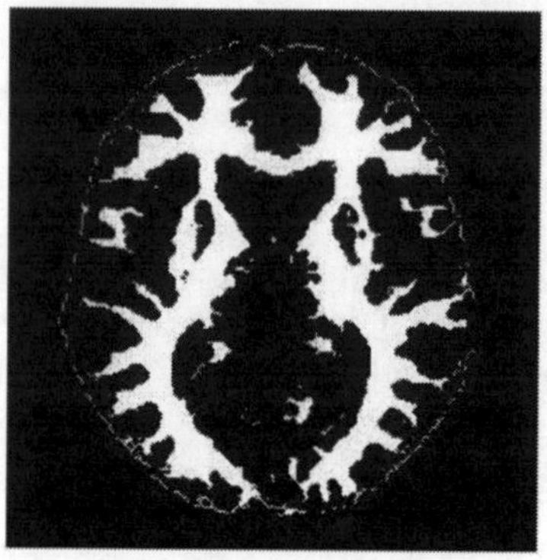

Objects in Cluster2

Morphological thinning for cluster2

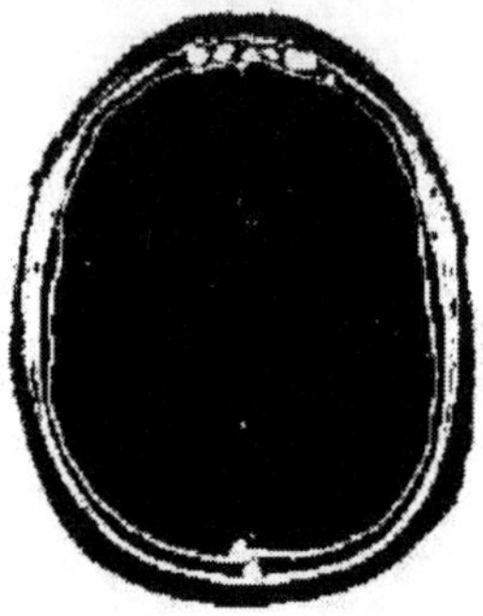

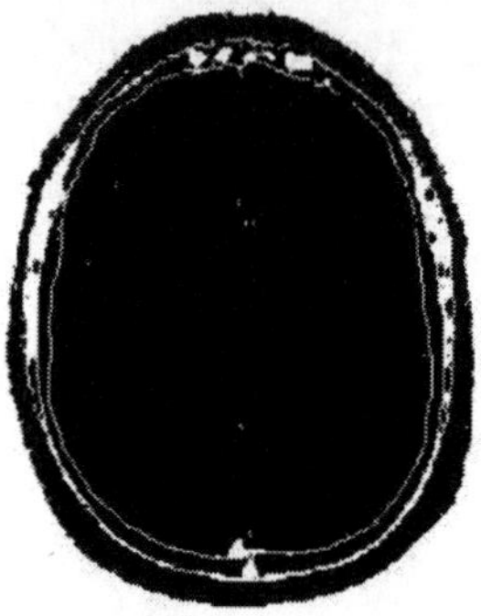

Objects in Cluster3

Morphological thinning for cluster3

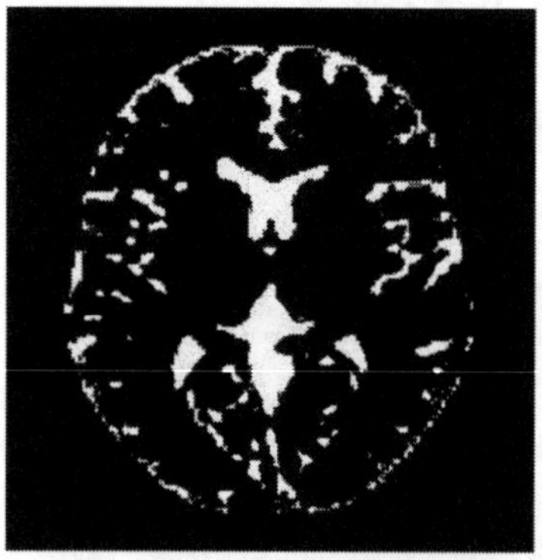

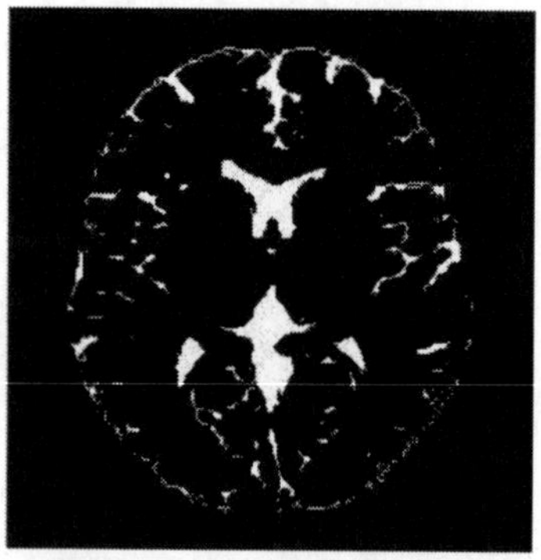

Objects in Cluster4

Morphological thinning for cluster4

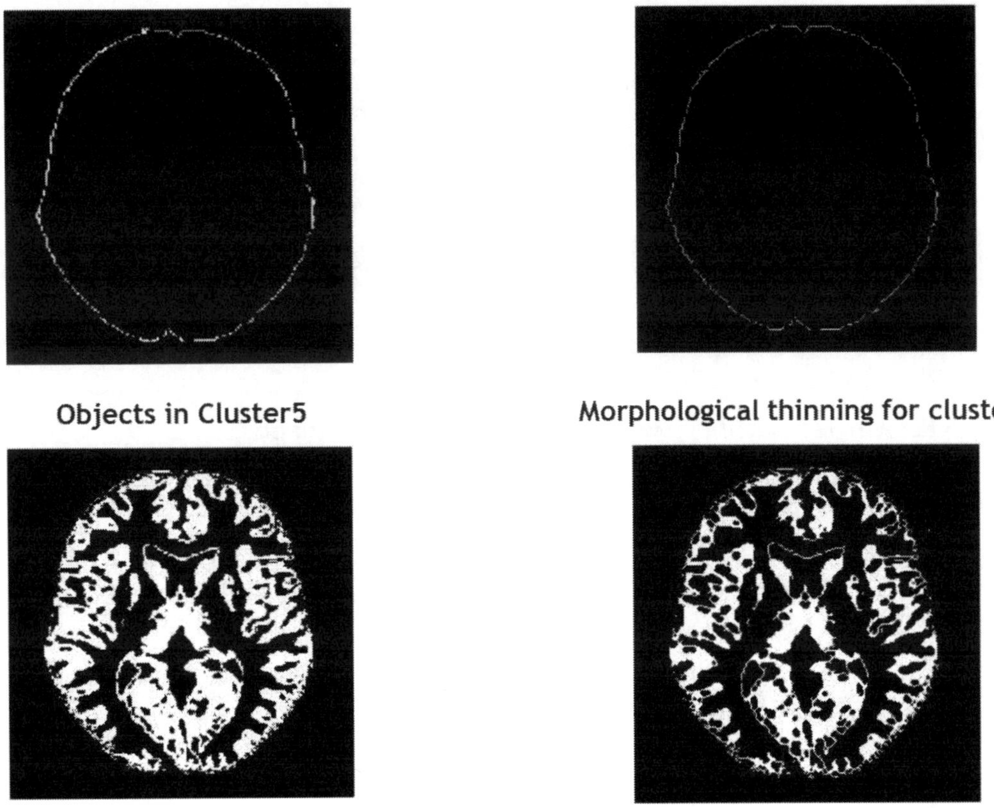

Figure 4: Objects in an image and image after morphological operations in respective clusters.

4 CONCLUSION:

This paper proposed a process for image segmentation of brain images using Enhanced K-Means unsupervised clustering algorithm and silhouette method in graycolor space model. The results are quantitative and are fully automated. So it is helpful for the Doctors in doing the diagnosis. In the future we can make changes to the proposed system by using any other optimization techniques and can improve the performance.

5 REFERENCES

[1] **H. D. Cheng, X. H. Jiang, Y. Sun et al**, "Color image segmentation: advances and prospects", Pattern recognition, Vol. 34, pp. 2259-2281, 2001.

[2] **C. G. Rafael, E. W. Richard, L. E. Steven,** Digital image processing using Matlab, Publishing House of Electronics Industry, Beijing, 2004

[3] **D. L. Pham, J. L. Prince,** "An adaptive fuzzy c-means algorithm for image segmentation in the presence of intensity in homogeneities", Pattern recognition letters, Vol. 20, pp. 57-68, 1999.

[4] **W. J. Chen, M. L. Giger, U. Bick,** "A fuzzy c-means (FCM)-based approach for computerized segmentation of breast lesions in dynamic contrast-enhanced MR images", Academic radiology, Vol. 13, No. 1, pp. 63-72, 2006.

[5] **J. Kang and Z. Ji,** "Dental plaque segmentation and quantification using histogram-aided fuzzy c-means algorithm", *the 29th Chinese Control Conference*, pp. 3068 - 3071, 2010.

[6] **D. L. Pham, J. L. Prince,** "An adaptive fuzzy c-means algorithm for image segmentation in the presence of intensity in homogeneities", Pattern recognition letters, Vol. 20, pp. 57-68, 1999.

[7] Medical Image Segmentation Using Fuzzy C-Mean (FCM) and Dominant Grey Levels of Image M.A. Balafar, ABD. Rahman ramli, M.Iqbalsaripan, Rozimahmud, Syamsiahmashohor

[8] **J. M. Gorriz, J. Ramirez, E. W. Lang et al,** "Hard c-means clustering for voice activity detection", Speech communication, Vol. 48, pp. 1638-1649, 2006.

[9] **Parveen, N. R.S., Sathik, M.M.,** 2010, Gray Scale Conversion Of X-Ray Image, ICISA 2010, and Chennai, India

[10] MR Brain Image Segmentation Based on Wavelet Transform andSOM Neural Network Dan Tian, Linan Fan *978-1-4244-5182-1/10/$26.00 _c 2010 IEEE*

[11] An Adaptive Logical Thresholding Methodfor Binarization of Blueprint Images

[12] **DeVore, R.A.; B. Jawerth, B.J. Lucier (1992),** "Image compressionthrough wavelet transform coding," IEEETrans. on Inf.

[13] Color Image Denoising Using Wavelet soft ThresholdingC Srisailam1, Parul Sharma, Sonali SuhaneInternational Journal of Emerging Technology and Advanced Engineering(ISSN 2250-2459, ISO 9001:2008 Certified Journal, Volume 4, Issue 7, July 2014)

[14] **Peter J. Rousseeuw (1987).** "Silhouettes: a Graphical Aid to the Interpretation and Validationof Cluster Analysis".*Computational and Applied Mathematics* **20**: 53-65.doi:10.1016/0377-0427(87)90125-7.

RESEARCH ON CHINESE PRIVATE MONEY SUPPLY AND AN EMPIRICAL ANALYSIS OF ITS IMPACT ON MONETARY POLICY - BASED ON THE BASIC DATA OF 1985-2015

Zhang Caixia[1]
[1]World Economy,School of business
China University of Political Science and Law, Beijing, China
zcx830322@163.com

ABSTRACT

In this paper, we firstly set forth the concept of private money supply and make an estimation for it in China between 1985 and 2014 on the basis of the unobserved financial calculation method put forward by Li Jianjun. And then, based on Vector Auto-Regression Model(SVAR), we conduct the empirical analysis on the influence of private money supply on monetary policy by using the annual data between 1985 and 2014. Theoretical and empirical analysis shows that the development of the private money supply affects the monetary policy transmission mechanism such as the conduction effect of credit and interest rate, which will make it more difficult for the central bank to make monetary policy than before. Also, the development of private money supply could result in the expansion of economy and arise a higher financial risk due to pro-cyclicality effects. Therefore, government should implement a prudent monetary policy and pay more attention to its implied risk.

KEY WORDS: Private Money Supply, Monetary Policy, SVAR Model

1 PRIVATE MONEY SUPPLY

There are two basic disputes in monetary economics. The first one is that whether money currency should be provided by government or by market; the other one is that whether an efficient monetary system should be completely monopolized by the government or public money and private money coexisted in the same time(azariadis et al., 2001)[1]. That will result in an important study of private money supply. In history, private money as a monetary phenomenon has existed in a long term, which is earlier than the national, government, central bank and so on.

During the period of barter, all currency can be seen as private money supply. By the end of 18th century, some countries began to monopoly the power to issue money, and public money supply gradually prevailed in most of countries. Before central bank, the Ministry of Finance issued currency directly, after then central bank issued currency and loaned to the Ministry of Finance. At this time, the right of issuing currency was completely monopolized by government and there is no private money supply. Also, there were some free banking system, such as Scotland's free banking(1831-1902), Sweden's free banking(1792-1845) and American free banking[2].

---

[1]Zhang Caixia, a doctor in China University of Political Science and Law.Email:zcx830322@163.com

After then, commercial banks became the main subject of money creation and the right of issuing currency was not monopolized by government either. Commercial bank, not as a government sector, its currency creation was strictly controlled by Government and Central Bank, and under the national financial monitoring; once run, government will provide more support and higher relief. Therefore credit money offered by commercial banks cannot can be seen as private money supply.

With the development of non bank financial institutions, modern financial system appears different levels of private money supply, such as some Off-balance-sheet currency given by commercial banks, credit money created by all kinds of non bank financial institutions or some other units of currency, such as community currency, virtual currency. These can be included in the scope of private money. They originate in the external commercial banks, while experience some core functions of bank, such as maturity transformation, liquidity transformation etc.. However, most of them are free from the formal financial system, and out of the control of modern monetary policy. All of them generally do not reserve for deposits, and regulate the money supply by open market operations and re-discount policy. That will affect the modern mechanism of money supply based on monetary multiplier theory in a large extent and make an important impact on the monetary policy implementation.

In this paper, we will study on private money supply and forecast the total amount of private money supply in China between 1985 and 2014, as well as the effect of the private money supply on the implementation of monetary policy.

2 THE MEASUREMENT OF PRIVATE MONEY SUPPLY

In our country, the financial system is dominated by banks, and the assets and liabilities of banks occupy the dominant position. While private money supply exists all the time. Maybe there are three factors, namely more demands of funds from private economy, the financial disinter-mediation phenomenon due to the strictly limited interest rate and the continual development of non-bank financial institutions, such as trust companies, securities and etc.. To some extent, private money supply can support the production and development of agriculture, private enterprise and some small commercial households, that is also to say, it can be helpful for the development of private economy. In this paper, the credit metrics of private money supply is chosen to reflect the increment and stock comprehensively.

2.1 The measurement of the scale of private money supply

Because of its strong concealment, there is no complete statistical data about private money supply, so a new suggestion for measuring is put forward from the perspective of the borrower, which was firstly put forward by Li Jianjun in 2010[3]. Some basic hypothesis are as follows: ①All of other credit money can be seen as private money except currency issued by formal financial institutions; ② Farmers, private enterprises and individual business are the main principal of private money supply; ③ Coefficient of per unit of GDP loans are unified among farmers, private enterprises and individual business, that is to say, unit of GDP needs to rely on some unified loans;④There is no difference in the organic composition of labor among different economic subjects;⑤Private money are mainly short-term loans. Specific prediction process is shown in Table 1.

Table 1: The measurement of Private money between 1985 and 2014

	Short loans	Loans	Private money supply	Proportion of total loans
1985	5153	6305.6	1629.7	25.8%
1986	6,548.0	8,159.3	2,025.9	24.8%
1987	7,714.0	9,803.9	2,341.8	23.9%
1988	9,085.7	11,459.9	2,637.0	23.0%
1989	10,780.4	14,360.1	3,369.8	23.5%
1990	13,816.7	17,680.7	5,470.2	30.9%
1991	16,323.2	21,337.8	6,118.2	28.7%
1992	19,682.3	26,322.9	6,556.1	24.9%
1993	24,251.9	32,943.1	8,121.7	24.7%
1994	27,879.5	39,976.0	12,609.6	31.5%
1995	32,072.9	50,544.1	18,593.5	36.8%
1996	41,230.8	61,156.6	21,703.5	35.5%
1997	55,418.3	74,914.1	27,502.2	36.7%
1998	60,613.2	86,524.1	33,752.0	39.0%
1999	63887.6	93734.3	37,301.6	39.8%
2000	65,748.1	99,371.1	35,961.0	36.2%
2001	67,327.2	112,314.7	40,844.5	36.4%
2002	74,247.9	131,293.9	50,828.8	38.7%
2003	83,661.2	158,996.2	64,713.2	40.7%
2004	86,836.8	177,363.5	73,878.8	41.7%
2005	87,449.2	194,690.4	81,869.2	42.1%
2006	98,509.5	225,285.3	97,387.1	43.2%
2007	114,477.9	261,690.9	116,013.4	44.3%
2008	125,181.7	303,394.6	144,846.1	47.7%
2009	146,611.3	399,684.8	226,343.8	56.6%
2010	166,233.4	478,976.5	286,306.0	59.8%
2011	203,132.6	546,398.3	338,938.8	62.0%
2012	248,272.8	628,100.5	400,744.1	63.8%
2013	290,237.8	717,087.7	466,111.8	65.0%
2014	314,795.7	814,780.3	582,561.7	71.5%

2.2 The basic situation of private money supply

In this paper, private money supply refers to all of credit currency issued by non-financial institutions, excluded formal currency issued by central bank, Ministry of Finance, commercial banks and so on. In our country, all of credit money results from banks off-balance business and kinds of non-bank financial institutions, also private lending, electronic money, community currency and virtual currency can be seen as private money. According to above calculation, there are some trends in our country (see Figure 1). Firstly, private money supply has been increasing year by year, and its ratio experiences the same trend, while the proportion reached to 70% in 2014. Secondly, the growth rate of private money supply is more faster than the rate of loans for central banks after 2008. So it is necessary to analyze how private money supply can be influence the efficiency and effectiveness of monetary policy.

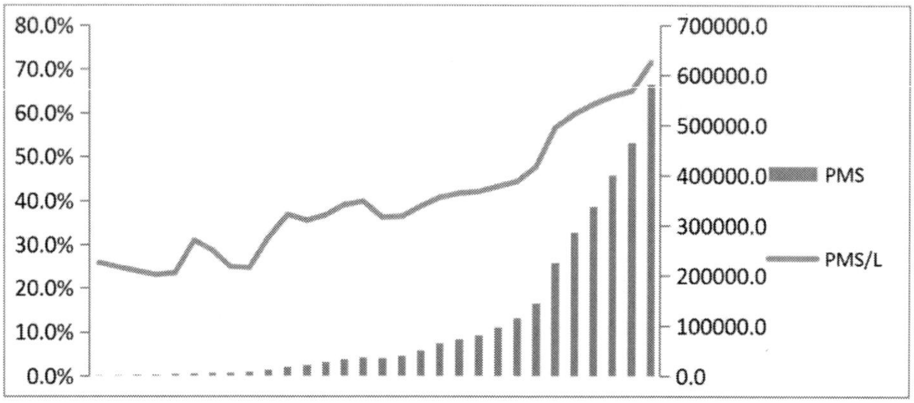

Figure 1 : The measurement of private money supply

3 EMPIRICAL ANALYSIS BASED ON VECTOR AUTO-REGRESSION MODEL(SVAR)

3.1 Variables and Model Selection

This paper is mainly to analyze the impact of private money supply on monetary policy. Monetary policy can be analyzed mainly through intermediate target and ultimate goal, so it is necessary to choose multiple variables as follows:

① Private money supply: Above, the private money supply has been predicted between 1985 and 2014 and we can get the annual growth rate of private money supply and make it as one variable, which can be represented by RPMS in short.

② Some intermediate target of monetary policy: Monetary policy can be measured mainly by money supply and interest rate. Because M2 is better to reflect the monetary environment, we choose the growth rate of M2 as one of the model variables, and RM2 in short. The other one is interest rate. Most of private money are short-term loans less than one year, so we make the rate of 6-12 month short term loans as the costs of borrowing, which can be represented by RATE[4].

③ The ultimate goal of monetary policy. How to maintain price stability and economic growth are the most important goals of macroeconomic regulation, this paper also focus on these two points. Economic growth can be measured by Gross Domestic Product (GDP) and the growth rate of GDP can be seen as useful variable. Consumer Price Index(CPI) is to measure the stability of economy and we can choose the growth rate of annual CPI as another variable, which is RCPI in short.

In this paper, empirical analysis is made on basis of annual data between 1985 and 2014, which are all from wind database and the website of The People's Bank of China[5] and National Bureau of Statistics of the people's Republic of China. The statistical software is used for Eviews6.0.

3.2 Model selection and correlation test

In quantitative economics, vector regression model (VAR) and structural vector regression (SVAR) are usually used to study the interaction between multi variables. Because of the limited parameters, the SVAR model is used in this paper.

3.2.1 ADF Stability Test

In order to establish SVAR model, all of variables should be tested by ADF method respectively, and the result shows that the five variables are all first-order single sequence and all of them meet the stability test under the significance level of1%, 5% and 10% (seen in Table 2).

Table 2: Stability Test

Variale	P	ADF	T(1%)	T 值(5%)	T 值(10%)	Result
RPMS	0.0001	-5.478951	-3.711457	-2.981038	-2.629906	Stable
RM2	0.0000	-9.313059	-3.689194	-2.971853	-2.625121	Stable
RGDP	0.0083	-3.802897	-3.724070	-2.986225	-2.632604	Stabile
RCPI	0.0005	-4.986206	-3.724070	-2.986225	-2.632604	Stabile
RATE	0.0078	-3.884725	-3.769597	-3.004861	-2.642242	Stabile

3.2.2 Cointegration Test

Cointegration relations is a long-term stable equilibrium relationship between two and more than two stationary time series. In this paper, we use the Johansen unrestricted cointegration rank test (trace) to test the relationship among RPMS,RM2, RGDP, RCPI and RATE.The result shows that there exists a long-term stable equilibrium status among these variables((see Table 3).

Table 3: Cointegration Test

协整变量	特征值	似然比	5%临界值	概率.**	假设的协整方程式

	0.311439	16.52358	15.49471	0.0349	None *
RPMS,RM2	0.212453	6.448484	3.841466	0.0111	At most 1 *
	0.644403	36.39623	15.49471	0.0000	None *
RPMS,RGDP	0.233491	7.445457	3.841466	0.0064	At most 1 *
	0.425873	23.07955	15.49471	0.0030	None *
RPMS,RCPI	0.259103	8.097143	3.841466	0.0044	At most 1 *
	0.512147	22.16173	15.49471	0.0043	None *
RPMS,RATE	0.286505	7.089177	3.841466	0.0078	At most 1 *

3.2.3 Stability Test

Stability test is the premise of impulse response function analysis in SVAR model. From analysis, it is found that the reciprocal of the characteristic unit roots are in the unit circle (see Figure 2). That indicates that SVAR model is stable, so impulse response function analysis can be used to the dynamic impact analysis.

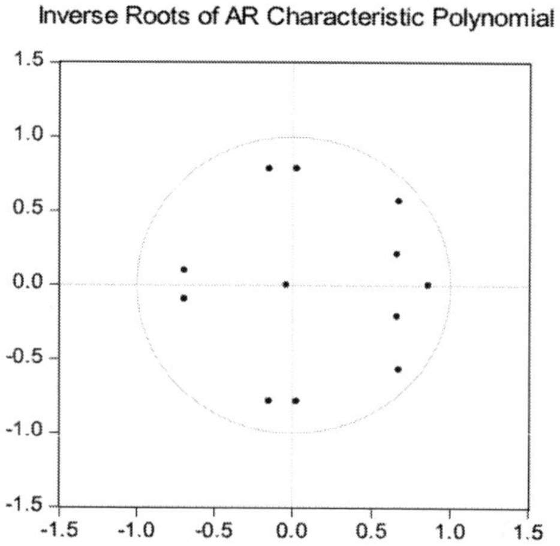

Figure 2 : Stability Test

3.3 Impulse Response Function and Variance Analysis

3.3.1 Impulse Response Function Analysis

The idea of Impulse Response Function Analysis is to analyze the mutual influence between variables,that is to say, how the disturbance term of single variable impacts the variation of every variable, and reveals the impact of the dynamic effect.The horizontal axis represents the number of the lagging period, and the vertical axis represents the rate of change.

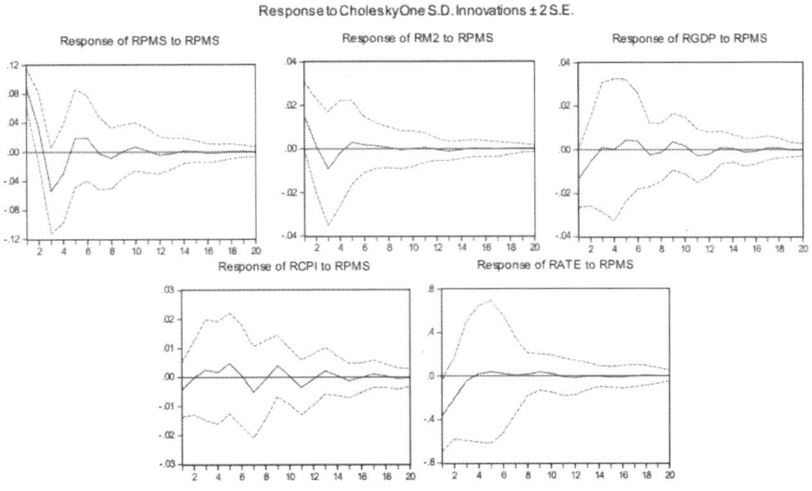

Figure 3: Impulse Response Function Analysis

Firstly the impact effect of RPMS, RM2, RGDP, RCPI, RATE on RPMS is analyzed,which is shown in figure 3. RM2 has an continual impact on RPMS for a long time, which will delay in the period of 15 or more. This is to say, an increase in the amount of currency in circulation, will increase the amount of GDP and upgrade the overall economic capacity to a certain extent, and then influence the scale of the private money supply through a certain production cycle. The impact of RCPI on RPSM is basically positive. This can be seen that when the price rises, according to the hedging considerations, residents will invest their idle funds into private money supply due to their higher revenue ratio. The impact of RATE on RPSM is negative mainly, and their impact experience a relatively short time. This is because when interest rates raise, residents will deposit the idle funds into commercial bank , which will has “crowding out effect” on RPMS.

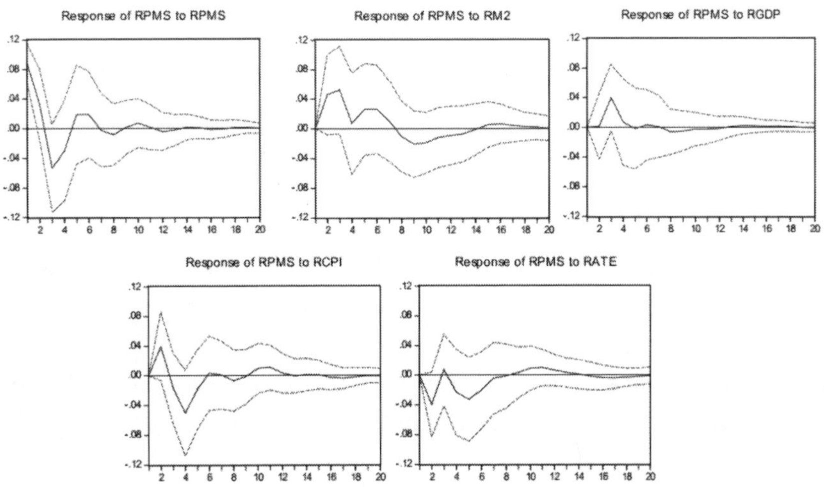

Figure 4: Impulse Response Function Analysis

Conversely, the impact effect of RPMS on RPMS, RM2, RGDP, RCPI, RATE is shown in figure 4. The impact of RPMS on RM2, RCPI and RATE are fading gradually, and the impact of their own delays very fast. The impact of RPMS on RM2 is positive, in the third period, the rate reached the peak, and as time goes on, the impact decreases slowly lower than that of the influence of RCPI. Those indicate that private money supply has direct impact on the currency especially M2, which will result in more difficulties for economic adjustment through monetary policy. The impact of RPMS on RGDP is positive, which indicates that private money supply will have positive effect on economic growth in short term. The impact of RPMS on RCPI also is positive, in its second period reaches the peak and decays gradually and tends to 0. This shows that, in the short term, the increase of private money supply may result in a slight rise for the price level. Private money supply have less impact on interest rates. This is because interest rate in our country is not completely decided by market, commercial banks just can adjust their interest rate in a limited floating range, which is not the real market interest rates.

3.3.2 Variance Decomposition Analysis

Decomposition Variance is mainly used to analyze the contribution of each structure impact on endogenous variables, and further to evaluate the importance of different structural shocks.

The first one is RPMS variance decomposition shown in Figure 5. We can see that the impact from itself is quite obvious,at the beginning of the period is 100%, the 10th period is 43.31%. This indicates the increases of private money in previous period will have an impact on investors' expectations,and then lead to another expanded increases,and vice verse. That is to say, private money supply exists pro-cyclicality effects. Secondly, the impact of RM2 on RPMS is significant, but its initial impact is very small, the impact increases gradually since the second period and reaches to 23.37% in the tenth period. With the increase of M2 , the investment and middle market are more active, but it will take a long time to transfer to real economy and private money supply. The impact of RCPI and RGDP on private money decrease gradually, in the 10th period reach to 15.69% or 5.6%. The impact of RATE on RPMS is slight, due to the special interest rate mechanism decided by government.

In the variance decomposition of RM2, RGDP, RCPI and RATE, RPMS has a largest contribution to RM2. The earlier in the stage, the larger in the impact. In the first period as high as 16.45%, the latter gradually reduces, while in the tenth period approaching 6.54%. The impact of RPMs on RGDP shows a decreasing trend, the initial impact is larger, in the first period as high as 17.65%, while the 3rd period rapidly decays to 4.28%. This result is consistent with the analysis in previous, which indicates that some increase in private money supply will boost the growth of GDP in a short term. The impact of RPMS on RCPI is very smaller, basically maintained between 1% and 3%, that is because the proportion of private money is still small.

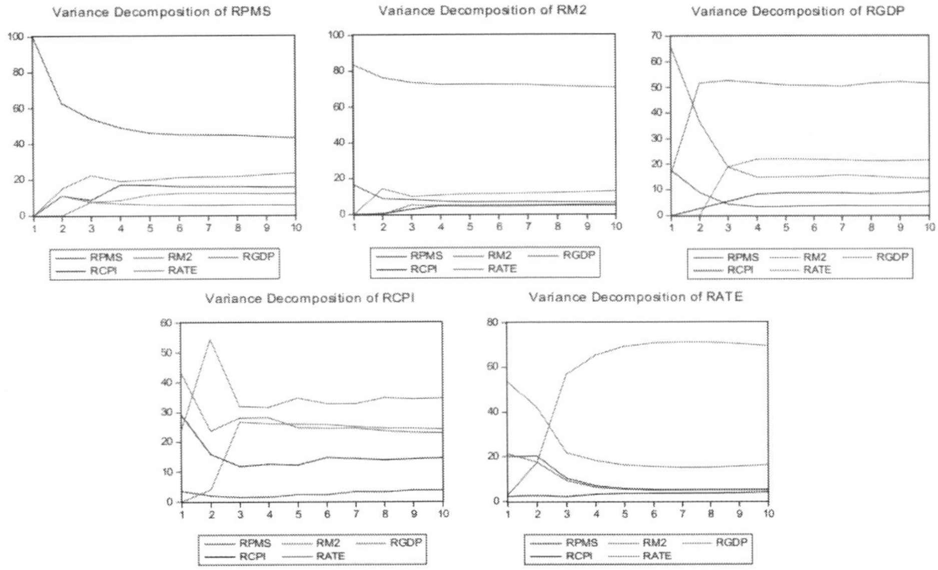

Figure 5: Variance Decomposition Analysis

4 CONCLUSIONS AND IMPLICATIONS

Firstly, the basic definition of private money supply is clarified in this paper, and an estimate for private money supply between 1985 and 2014 is analyzed. And then, an empirical analysis of private money supply on the monetary policy is put forward based on SVAR model. Some results show that there are long-term stable relationship among private money supply, M2, GDP, CPI and interest rates, and private money supply has a certain impact on the conduction mechanism of monetary policy. On one hand, M2, CPI, GDP and other economic indicators have direct impacts on private money supply, and there also exists Pro cyclical effect as strong as in the bank credit, which will hidden potential and high financial risk. On the other hand, private money supply also has impact on the policy transmission mechanism. Private money supply has direct impact on the currency especially M2, which will result in more difficulties for economic adjustment through monetary policy. Also, private money supply has positive effect on economic growth in short term, but more funds are invested into overheated industries and result in the lack of long-term effects that may lead to the negative effects of the economic crisis.

Therefore, central bank should adjust the original operating mechanism of monetary policy, and execute more prudent monetary policy (more lower monetary growth or more higher interest rate) to offset the expansion of private money supply. Otherwise, central bank staring at money supply as the intermediate target will not be able to realize expected, which is mainly used to adjust the economic growth and inflation. At the same time, private money supply should be understated objectively, more innovative more risks. It is the inevitable result of the financial development, which should be included in the scope of regulation to prevent the implied high risk。

5 REFERENCE

[1] **Azariadis,C., Bullard, j. And Smith, B..2011**.Private and Public Circulating Liabilities,Federal Reserve Bank of St. Louis Working Papers Series.

[2]**Zhou Liping.2013**. Shadow Banking System:The Return of Free Banking?,Social Sciences Academic Press.PP 22-25

[3] **Li Jianjun.2010**.The Change of Non-Observed Loan (NOL) in Our Country:1975~2008,Financial research,4,pp 40-49

[4]**Wang Zhen, Zeng Hui.2014**. Theoretical and empirical analysis of shadow banking system on the monetary policy, Study of International Finance, 12,pp 58-66

[5] **The People's Bank of China.1984-2015**.Statistics and Analysis Department.http://www.pbc.gov.cn/diaochatongjisi/116219/116319/index.html

[6]**Zhou Yingzhou, Jiang Zhendong. 2002**.Monetary channel and credit channel and monetary policy effectiveness, Journal of Financial Research, 12, pp 34-43

[7]**Zhao Jinwen, Gao Hui.2004**.The conduction behavior dynamic model test of monetary policy in China-An empirical analysis of the 1993-2002,3, pp:95-102

[8]**Financial Stability Board. 2011**.Shadow Banking: Scoping the Issues: A Background Note of the Financial Stability Board. FSB Working, Paper, pp:1-6

[9]**Nicola Gennaioli, Andrei Shleifer&Robert W.Vishny. 2013**. A Model of Shadow Banking, Journal of Finance,68 (4):1331-1363.

[10]**Pozsar Zoltan, Tobias Adrian, Adam Ashcraft and Hayley Boesky. 2010**. Shadow Banking. Federal Reserve Bank of New York Staff Report No. 458. pp:4-6

A SAP R/3 BASED SOLUTION FOR SUPPLY CHAIN PLANNING

Jun-Der Leu [1,*], Larry Jung-Hsing Lee[2], and i-Wei Huang[3]
[1,2,3] Department of Business Administration
National Central University, Taoyuan City, Taiwan
leujunder@mgt.ncu.edu.tw; larrylee@mgt.ncu.edu.tw; hiw@ey.gov.tw

[*,1] Department of Business Administration
Munich University of Applied Sciences, Munich, Germany
leu@hm.edu

ABSTRACT

Enterprise information systems, such as Enterprise Resource Planning (ERP) systems, have been applied to organization-wide coordination and the integration of business functions and processes. They do not, however, support the function of decision-making. ERP-based Decision Support Systems (DSS) have been developed in some industries, but there are few studies of application cases. In this research, we study the advanced planning system (APS) in decision logic design, and the system development of a master plan in a flat-panel display systems manufacturer. After the application of the ERP-based DSS, the case company is able to take efficient and high quality decisions. Improvements in the inventory level of final products also result, and shortages of Vendor Managed Inventory (VMI) hubs are identified. This successful case offers a reference to other enterprises.

Keywords: SAP R/3, Decision Support System (DSS), Advanced Planning System (APS)

1 INTRODUCTION

Many manufacturing enterprises use Enterprise Resource Planning (ERP) systems as operations platforms to support their businesses, integrating business data within the organization. Empirical studies have demonstrated the effectiveness of ERP application. Moharana and Sarmah [1] found that most IT firms have implemented ERP packages, and all transaction data related to consumable spare parts are then available to them. Torkul et al. [2] integrated ERP and CAD/CAM, which together provide a dynamic variant-generation mechanism that can be applied to any product and can use any CAD/CAM system data, thus enabling product variety to be automatically managed.

Logistics is defined as the integration of business processes regarding market demand fulfillment, transportation, and distribution, and production planning and manufacturing resources acquisition [3]. Logistical factors are critical for many high-tech manufacturing enterprises, who typically have low gross profits and short product life-cycles.

Many enterprises have in recent years reorganized their supply chains, creating new opportunities and increasing competitiveness. A supply chain is the set of entities involved in new product design, material procurement, production, distribution, and transportation, and the chain links two or more legally separated organizations through materials, information, and financial flow [4].

We develop an ERP-based decision support system for the advanced planning systems (APS) of manufacturing enterprises. Our paper is organized as follows: Section 2 reviews the research; Section 3 presents the development of ERP-based decision support systems of the APS in terms of theoretical background, systems framework, and planning logic. The industrial application of the developed system is presented in Section 4, and Section 5 gives the conclusion.

* Corresponding Author

2 LITERATURE REVIEW

The main components of the DSS are the database, the software system, and the user interface. The database is a collection of current or historical data and the software system contains several modules, including the decision models and the data analysis tools [5].

The Decision Support System (DSS) is a computer-based information system providing interactive information support to managers in their decision-making processes [6]. It is based on the information provided by an ERP system and can support collaboration and business decisions in a supply chain, but few studies examine the development and application of the DSS in the supply chain management (SCM) of actual industrial cases.

The APS is well-suited to simple planning methods for production scheduling. Chen, Huang, and Lai [7] combined agent technology with an APS system, and proposed a distributed production planning system for a multi-tier and multi-site production system. Vidoni and Vecchietti [8] developed a characterization of the APS, taking a systemic viewpoint of the general concept.

The system's design in the SCM is also the focus of study. Jolayemi and Olorunniwo [9] developed a deterministic model for production and distribution in a multi-plant and multi-warehouse environment. They also consider the problem of resource restriction and allow shortfalls to be met through subcontracting. Akyuz and Rehan [10] examine the requirements for creating an e-SCM from different perspectives, such as integration with legacy systems, ERP systems, business processes, and business/decision support needs. The APS of SCM has thus to some extent been developed.

3 DEVELOPMENT OF THE DECISION SUPPORT SYSTEM

3.1 Concept of the system (theoretical background)

Taking the perspective of the planning period, the three levels of decision hierarchy in SCM can be viewed as competitive strategy, tactical plans, and operational routines [11]. At the tactical planning level, the horizon of a master plan typically ranges from 6 to 24 months. This determines the outline of regular operations, such as master production scheduling, and mid-term sales, distribution, capacity, and material requirements planning synchronously within the scope of the strategic decisions [12]. Many constraints are considered, such as available resources and factory capacities, but these enterprise information systems neither include business partner information nor any decision support functions.

3.2 The System framework

Users can input their requirements through the interface, and the software system reads and calculates the historical data from the database and the real-time data from the ERP system, and presents the results to users through the interface (Figure 1).

3.3 The Planning models in the DSS

To address the operation problems, the case company develops a set of decision logic and system tools as the APS. The ERP and the Hub information system have been applied in internal operations and information sharing, but these do not include the functions of calculation and decision support found in SCM.

The case company decides to develop a DSS of the APS for its master planning. It defines 18 months as its mid-term plan period and selects week as the planning cycle. The scenarios of multi-period, multi-site, multi-hub, and multi-supplier are considered. Based on these scenarios, the case company designs the main decision logic of its master plan, as below (Figure 2).

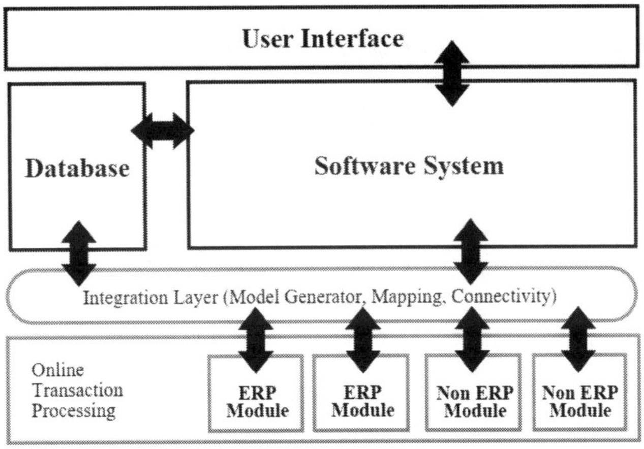

Figure 1: Main components of DSS.

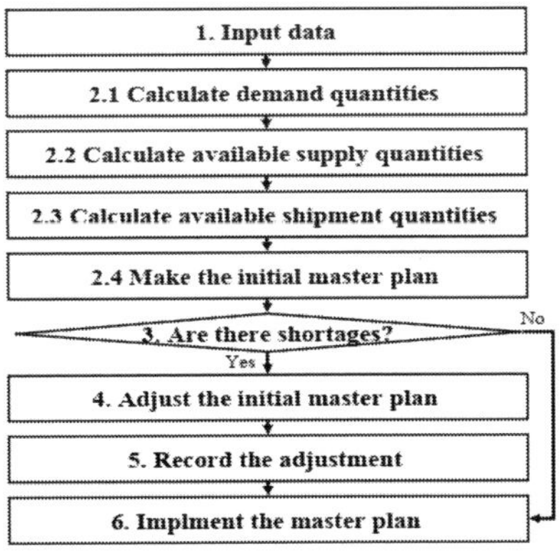

Figure 2: Decision logic of master planning

The notation used in this section is described below.

Indices/sets

I	Index of Product
T	Time of Planning
M	Material and Components
L	Path
K	Hubs
V	Vendor
J	Plant

Parameters

C_{mv}^{t}	Procurement cost of material m during t planning time
$Q_{mv}^{t,\max}$	Supply limit of material m during t planning time
$QI_{i}^{\min}$	Safety inventory level of product i
LT_{mj}^{t}	Lead time of material m during t planning time
CP_{ij}^{t}	Product processing costs of product i during t planning time

CI_{ij}^{t} Inventory costs of product i in plant j during t planning time

CI_{mj}^{t} Inventory costs of material m in plant j during t planning time

CIL_{ij}^{t} Inventory limit of product i in plant j during t planning time

CIL_{mj}^{t} Inventory limit of material m in plant j during t planning time

CAP_{ij}^{t} Capacity limit of product i in plant j during t planning time

PL_{il}^{t} Product price of product i for customer l during t planning time

CT_{ikl}^{t} Transportation costs of product i for customer l during t planning time

LT_{ikl}^{t} Lead time of product i from hub to customer l during t planning time

$QT^{t,\max}$ Maximum running limit during t planning time

QI_{k}^{t} Inventory level of hub k during t planning time

Decision variables

Q_{mj}^{t} Procurement quantity of material m in plant j during t planning time

Q_{m}^{t} Total procurement quantity of material m during t planning time

QI_{mj}^{t} Inventory quantity of material m in plant j during t planning time

Q_{i}^{t} Supply availability of product i during t planning time

Q_{ij}^{t} Supply availability of product i in plant j during t planning time

QI_{ij}^{t} Inventory quantity of product i in plant j during t planning time

QI_{i}^{t} Total inventory quantity of product i during t planning time

QT_{ijk}^{t} uantity of product i sent from plant j to hub during t planning time

QT_{ik}^{t} Product running quantity of product i in hub k during t planning time

QT^{t} Distribution availability during t planning time

QT_{ikl}^{t} Product distribution quantity of product i from hub to customer l during t planning time

D_{i}^{t} Total demand in t planning cycle

D_{ik}^{t} Demand of hub k in t planning cycle

X^{t} Demand shortage in t planning cycle

Stage 1 Input data. The input data includes the master data (e.g., bill of material, material item, bill of distribution, etc.), demand data (e.g., rolling forecast), inventory data, and constraint data (e.g. capacity constraint, material constraint, etc.).

Stage 2.1 Calculate demand quantities. Based on the rolling forecast, the demand quantities of the final products are calculated weekly.

$$D_{i}^{t} = \sum_{k} D_{ik}^{t} \tag{1}$$

Stage 2.2 Calculate supply availability quantities. Considering the demand quantity, the factory capacity, production line constraints, and supplier's capacity, the supply availability quantities are calculated weekly. If MIN ($\sum_j CAP_{ij}^t$, $Q_{mv}^{t,\max}$) $\geq$ D_i^t , use equation (2), otherwise use equation (3) to calculate the supply availability.

$$Q_i^t = D_i^t + (QI_i^{\min} - \sum_j QI_{ij}^{t-1}) \quad (2)$$

$$Q_i^t = \text{MIN} (\sum_j CAP_{ij}^t , Q_{mv}^{t,\max}) + \sum_j QI_{ij}^{t-1} \quad (3)$$

Stage 2.3 Calculate shipment availability quantities. According to the lot size, sailing, due-date, and the lead-time of shipment, the available shipment quantity is calculated weekly. If MIN ($\sum_j QT_{ijk}^t$, $\sum_k QT_{ikl}^t$) $\geq Q_i^t$, then use equation (4), otherwise use equation (5) for the shipment availability.

$$QT^t = Q_i^t \quad (4)$$

$$QT^t = \text{MIN} (\sum_j QT_{ijk}^t , \sum_k QT_{ikl}^t) \quad (5)$$

Stage 2.4 Create the initial master plan. Based on the planning results from Stage 2.1 to 2.3, the initial production, procurement, distribution, and inventory plans are decided weekly.

Stage 3 Check the initial master plan. If the demand cannot be satisfied, Stage 4 is executed. Otherwise, Stage 6 is executed.

Stage 4 Adjust the initial master plan. The three sub-stages in the adjustment are first to check if the shortage can be solved by overtime of the factory or the coordination of suppliers. Second, high property orders are produced first. The shortage of make-to-order in the t week then changes the master plan in the t week, and the shortage of make-to-stock in the t week changes the master plan in the t+2 week. Third, if the adjustment of the master plan does not solve the shortage, outsourcing is considered.

Stage 5 Record the adjustment. The adjustment is recorded as the input data of the next planning cycle.

Stage 6 Implement the master plan. The weekly plans of procurement, production, inventory, and distribution are decided. The decision logic of the APS stops at this stage.

4 THE APPLICATION CASE

The case company is a global flat-panel display systems manufacturer. It acquires the orders of flat computer monitors, flat TVs, etc. from TFT-LCD OEM (Original Equipment Manufacturer) companies in Asia, European, North America, and South America. Four factories and four Vendor Managed Inventory (VMI) hubs are shown on the logistic map (Figure 3). The orders that are made-to-order are transported to customers directly; those of that are made-to-stock are transported to the VMI hubs.

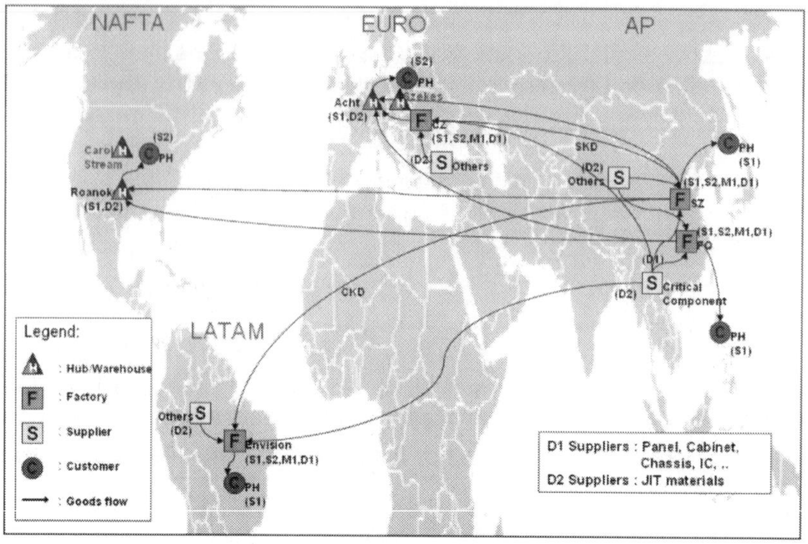

Figure 3: Global supply network of the case company

In the operation, the case company receives orders by rolling forecasting. Considering the due-date and capacity of factories and suppliers, it creates the production, procurement, and distribution plans. The transportation assignments are implemented by the logistic service providers assigned by the customers. According to the business model, there are problems of inventory, distribution, production, and procurement:

- Inventory. There is a requirement of safety stock to the orders of make-to-stock, and the inventory level of VMI hubs is controlled weekly.
- Distribution. The shipment is restricted by the lot size and sailing. The due-date and the lead-time of shipment are also considered.
- Production. Each factory has a capacity constraint. One production line can only process one size of product. Specific customers are prioritized.
- Procurement. The lead-time for procurement and the supplier capacity of critical components such as panels, molds, motherboards, hinges, bases, etc. affect the master plan.

The case company selects a software vendor to design its Master Planning Decision Support System (MP-DSS) of the APS, based on the decision logic of the master planning. The MP-DSS is expected to replace the original manpower planning approach and to enhance the accuracy and efficiency of master planning. After the calculation, the output data will be the goods movement report, the production plan, the vendor material supply plan, factory capacity loading, and the material constraint report.

Figure 4 illustrates the information flows of the MP-DSS of the APS. First, the MP-DSS reads the constraints of factory capacity and supplier capacity from the ERP system, the inventory level from the Hub information system, and the demand plan and rolling forecasting from the i2 SCP system of daily customers. After the MP-DSS is calculated, the distribution plan is sent to the customers' i2 SCP systems, and the production and procurement plans are sent to the ERP system of the case company. When the products are shipped, the shipment information is sent to the customers' ERP systems.

After the development of the MP-DSS of the APS, the case company inputs its data, which is the record over the past year, to verify the feasibility of the MP-DSS. These are the master, demand, inventory, and constraint data. These data cover 12 customers, 4 factories, 4 hubs, 181 kinds of product, 62 kinds of panel, and 169 kinds of model. From the calculation, the MP-DSS of the APS presents the capacity utilization rate and material procurement in the planned period (Figure 5), which helps to solve bottleneck and idle capacity problems. The results demonstrate the feasibility of the APS.

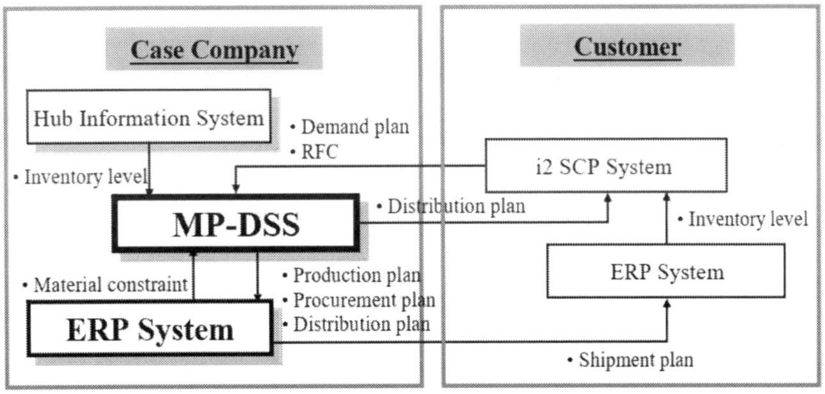

Figure 4: Information flow of the MP-DSS of the APS

Compared to the original planning approach, the calculation time of the master plan decreases from 10 to 3 hours, which improves the efficiency of decision-making. The average final product inventory also decreases from 20,656 to 12,084 units (about 41.2), and the shortage in VMI hubs decreases from 16,296 to 287 units (about 92.8). These findings demonstrate the decision quality of the MP-DSS of the APS.

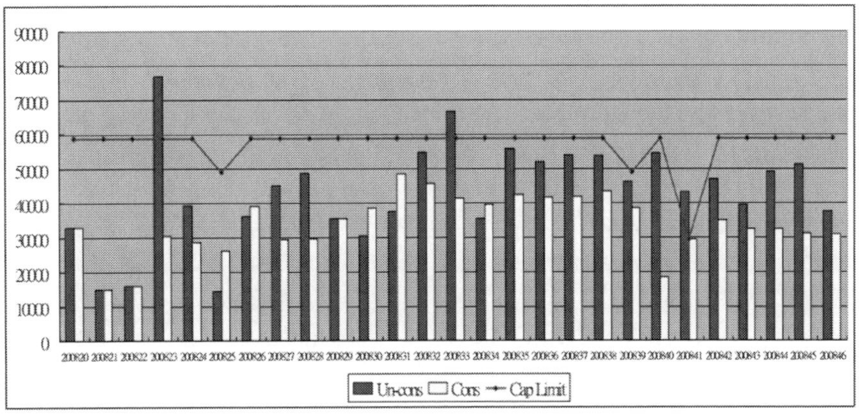

Figure 5: Capacity utilization of case company

5 CONCLUSIONS AND FURTHER RESEARCH

We examine the decision logic and APS development in the literature. In terms of decision logic, the literature suggests that procurement, production, and distribution should be synchronously planned in a master plan. The case company, however, forms its decision logic in three consecutive stages; calculating the demand quantities, the available supply quantities, and the available shipment quantities. The case company first considers the capacities of factories and suppliers, and after these constraints have been established, those of distribution are considered. The results from the flat-panel display company demonstrate that the TFT industry is material-oriented. The case company must consider the constraints of critical components, and the decision logic of master planning is found to change with the industry or business characteristics.

In the development of the MP-DSS of the APS, we know that the APS obtains operational data from the internal ERP and Hub information systems. It is an ERP-based APS according to its calculation functions. The APS also connects with its customer's systems, such as the i2 SCP system, which indicates that the APS is an inter-organizational system. However, the APS does not connect with the systems of its suppliers. The case company cannot obtain material information in real time, so its decision quality will be restricted.

In this research, we study the master plan and APS application in a global flat-panel display systems manufacturer. The case company gains improvements from the decision efficiency and quality. A successful case of ERP-based APS development is provided, which can inform other enterprises. A limitation of the research is that the case company is a material-oriented enterprise. Its master plan decision logic focuses on the collaboration between procurement and production. Its experience cannot apply to production- or market-oriented enterprises, and thus further research into these industries is required.

6 REFERENCES

[1] **Moharana, U.C., Sarmah, S.P. 2016.** Determination of optimal order-up to level quantities for dependent spare parts using data mining. *Computers & Industrial Engineering*, 95, pp 27 40

[2] **Torkul, O., Yilmaz, R., Selvi, I.H., Cesur, M.R. 2015.** Automatic generation of variants depending on changes of product properties in a flexible manufacturing environment. *Computers & Industrial Engineering*, 86, pp 22 28

[3] **Van Dyk, L. 2001.** The philosophy -tool continuum, *South African Journal of Industrial Engineering*, 12(1), pp 1-14.

[4] **Cooper, M.C., Lambert, D.M., Pagh, .D. 1** . Supply Chain Management: More than a New Name for Logistics. *The International Journal of Logistics Management*, 8(1), pp 1-13.

[5] **Laudon, K.C., Laudon, .P. 200** . *Essentials of Management Information Systems*, 8rd Edition, New Jersey: Pearson Education, Inc.

[6] **O Brien, .A., Marakas, G.M., 2006.** *Management Information Systems: Managing Information Technology in the Business Enterprise*, 8rd Edition, New ork: McGraw-Hill.

[7] **Chen, W.L., Huang, C.Y., Lai, Y.C. 200** . Multi-tier and multi-site collaborative production: Illustrated by a case example of TFT-LCD manufacturing. *Computers & Industrial Engineering*, 57(1), pp 61 72.

[8] **Vidoni, M.C., Vecchietti, A.R. 2016.** A systemic approach to define and characterize Advanced Planning Systems (APS). *Computers & Industrial Engineering*, 90, pp 326 338.

[9] **olayemi, .K., Olorunniwo, F.O. 2004.** A Deterministic Model for Planning Production uantities in a Multi-plant, Multi-warehouse environment with extensible capacities. *International Journal of Production Economics*, 87(1), pp 99-113.

[10] **Akyuz, G.A., Rehan, M. 200** . Requirements for Forming an e-Supply Chain'. *International Journal of Production Research*, 47(12), 3265-3287.

[11] **Min, H., Zhou, G. 2002.** Supply Chain Modelling: Past, Present and Future. *Computers & Industrial Engineering*, 43(1-2), pp 231-249.

[12] **Fleischmann, B., Meyr, H., Wagner, M. 2008.** *Supply Chain Management and Advanced Planning: Concepts, Models, Software, and Case Studies*, Berlin: Springer-Verlag.

CIE46 Proceedings, 29-31 October 2016, Tianjin / China

PARAMETER DEPLOYMENT AND SEPARATION FOR SOLVING PHYSICAL CONTRADICTIONS

D. Daniel Sheu and Li-Ching eh
[1] Dept. of Industrial Engineering Engineering Management, National Tsing Hua University, Hsinchu, Taiwan 300
dsheu@ie.nthu.edu.tw; dsheu@ssi.org.tw

ABSTRACT

Physical contradiction is at the heart of TRIZ contradiction problem solving. The essence of a physical contradiction is that for two objectives, we have two contradictory demands on the same parameter of the same system. As part one of the two-part Parameter Manipulation approach to solve physical contradiction, this paper proposes a systematic new method to solve physical contradictions using the parameter deployment and separation.

By defining the Local System to include the components directly at the immediate relevant components of the physical contradiction, the proposed parameter deployment systematically deploys the two objectives and the contradictory parameter into their respective constituent parameters. The essence of parameter separation is to assign the two contradictory requirements, either at the objective level or at the contradictory parameter level, to be satisfied separately by separate constituent parameters or distinct value ranges of a constituent parameter. In this way, the contradiction can be eliminated. An accompanying paper will describe the method of using Parameter Transfer to solve a physical contradiction by delegating the issue of satisfying one or both of the contradictory requirements to some parameters of a component seemingly irrelevant to the problem core. Out of the 25 preliminary cases, it was found that on average the number of solution ideas generated increased from 5.48 by all existing separation methods to 11.52 by this set of proposed methods - an increase of 110 .

The contributions of this research include: 1) Establishing the concept of Parameter deployment and strategy of using it to solve physical contradictions systematically, 2) Addressing the problem at both objective level and the contradictory parameter level producing 6 strategies for solving physical contradictions of which all existing separation methods to solve physical contradiction constitute just one of our 6 strategies, 3) Using parameter deployment to identify all possible combinations of separation methods to solve a physical contradiction, 4) Providing standard operational templates for problem-solving process, tool descriptions, and examples for ease of learning and applications, 5) Together with the Parameter Transfer to form a comprehensive 17 problem solving strategies in which all the disconnected existing methods are only one of the 17 strategies of this unified theory of parameter manipulation.

Keywords: Physical contradiction, Parameter deployment, parameter Separation, Parameter transfer, TRIZ, Systematic Innovation

1 INTRODUCTION:

1.1 Research Background and Ob ectives

Physical contradiction is at the heart of all problematic contradictions. At present, all methods to solve physical contradictions are based on either separation principles, by-passing contradictions, or satisfying contradictions. All of them at the end converted to inventive principles to solve the problems [1]. The deficiencies of the existing methods to solve physical contradictions include:

1) All existing methods appear to be independent and are lack of synergy among them.[2]
2) All existing methods at the end converted to the inventive principles. That means that the problem solver need only to examine the selected inventive principles. Many times, the 40

inventive principles [1] are not enough to inspire good specific solutions. There is no detail thinking process which can lead the problem solver to look at problems systematically.[3]

3) All the existing separating principles focus on the solving problem at the contradictory parameter level. Solving problem at the objective level has not been considered for separation principles.

This research established a comprehensive thinking process which generate 17 strategies to solving physical contradictions of which All the existing methods constitute only 4 of 17 strategies identified by this research. Furthermore, all the 17 problem solving strategies are under the same set of theory based on parameter manipulation.[4][5][6]

2 THEORETICAL BACKGROUND FOR PHYSICAL CONTRADICTION AND PARAMETER MANIPULATION

2.1 Formulation of Physical Contradiction

A physical contradiction occurs when one parameter cannot satisfy two demands in order to achieve two objectives at the same time. Model of physical contradiction can be expressed as:

- To O1, P should be +P. But,
- To O2, P should be -P.

Where O1 and O2 are two disparate objectives, P is the contradictory parameter which causes the problem. +P and -P represent the two incompatible demands which need to be satisfied in order to achieve O1 and O2. Taking desk area for example, in order to accommodate more stuffs on the desk, the area of the desk should be big. But, in order to occupy less space in a room, the area of the desk should be small. This can be shown in the following formulation:

- To accommodate more stuffs on the desk, the area of the desk should be big. But,
- To occupy less space in a room, the area of the desk should be small.

This paper proposes the systematic methods of:

1) parameter deployment: to assist users to identify all relevant parameters which contribute directly to the physical contradiction, and
2) parameter separation: To separate the contradictory demands either at the objective level or contradictory parameter level to solve the physical contradiction.

To solve physical contradiction, how those relevant parameters influencing the two objectives O1 and O2 or the contradictory parameter P needs to be investigated. By deploying the parameters representing O1, O2, and P into their corresponding constituent parameters, all the parameter separation strategies to solve the physical contradiction can be systematically and comprehensively identified thus ideas to solve the physical contradiction can be generated.

2.2 Parameter Deployment

2.2.1 *The Generic Form of Two Objectives*

The two objectives can be shown in the following generic forms:

- $O1=fn(P , \dots ; E_i^1 , \dots ; Z_k , \dots)$
- $O2=fn(P , \dots ; E_i^2 , \dots ; Z_k, \dots)$

Contradictory parameters can be shown in the following generic form:

- *P=fn(X1, X2, X3, ..., Xn)*

In the generic forms above, *P* represents contradictory parameter of two objectives, O1 and O2. E represents exclusive parameter of each objective, and Z represents compatible parameters of the same direction of the two objectives. The relationships between each parameter category and the definition of each category are shown in Table 1. In the contradictory parameter generic form above, X refers to the constituent parameter(s) of the contradictory parameter, *P*.

Some related symbols are defined in Table 1.

Table 1. Definitions and Descriptions of Each Parameter Categories

Symbol	Definition	Description
X_b^a	Constituent Parameter(s)	● All related parameters influencing the objective ● X_b^a represents the b-th constituent parameter of objective a . a being O1 or O2.
Z_k	Compatible parameter(s)of the same direction	● The k-th common constituent parameter of the two objectives having compatible demands. ● Z_k represents the k-th compatible parameter of the two objectives
P_j	Contradictory Parameter(s)	● A common constituent parameter of O1 and O2 with incompatible demands placing on the parameter. ● P_j represents the j-th contradictory parameter of the two objectives
E_c^a	Exclusive Parameter(s)	● Constituent parameters that are not common parameters of the two objectives ● E_c^a represents the c-th exclusive parameter of objective a

The existing separating principles, including separation in space, separation in time, separation in relations, separation in system levels, and separation in directions, all fall into one category of separation principle only with different parameters as the target of separation, such as space, time, relations, system levels, and directions to satisfy the two contradictory demands separately. For example, in the case of separation in time, time is used as the separating constituent parameter and so forth.

2.3 The Essence of Problem Solving Strategies in Parameter Manipulation

The method of parameter separation as a part of manipulation proposed in this paper is a systematically comprehensive theory for separation. The essence of the method is to separate the demand of contradictory parameter or to satisfy the two objectives separately . By manipulating parameters such as parameter separation , the two contradictory demands generated by the two objectives or the two contradictory parameters can be satisfied in separate parameters or in separate value range of the same parameter. There are various separation modes as follows:

- Satisfying the two contradictory demands with two different constituent parameters separately (as shown in Figure 1: COPE/COEP/COEE)
- Satisfying the two contradictory demands with distinct value ranges of a constituent parameter separately (as shown in Figure 1: IPV/IOV)
- Splitting the contradictory parameter into two parameters from component perspective and have these two parameters satisfy the two contradictory demands separately (as shown in Figure 1: CPS)

Parameter Separation can be divided into two modes: Separation within Parameters and Separation across Parameters. Separation within parameters satisfies two contradictory demands separately, either at objective level or at contradictory parameter level, with distinct value ranges of a constituent parameter. All the existing separation principles fall into this mode of parameter separation within different ranges of one parameter which may be space, time, system level, relationship, direction, etc. Separation across parameters is to satisfy two contradictory demands simultaneously, either at objective level or at contradictory parameter level, with two different constituent parameters. Furthermore, separation across parameters can be divided into Separation and Splitting . The former aims to satisfy two contradictory demands separately with two existing constituent parameters, while the latter aims to satisfy two contradictory demands by two different parameters which are split from a contradictory parameter using component perspectives.

2.4 Definition of Systems

It is the authors' observation that a problem of contradiction must be caused by some parameters of center component or some interactive components of the center component. Center component is either a component of contradictory parameters or a component of the two objectives. The interactive components are other components which have interactive functions with the center components. Often times, there are two or three center components corresponding to O1, O2, and P. The Local System (LS) of a contradiction is the combination of all center components and their interactive components with respective to the contradiction. A local system is the scope of parameter deployment which provides all related parameters influencing the two objectives or contradictory parameter for parameter separation that follows.

2.5 Strategies of Parameter Manipulation

Two aspects for parameter manipulation are discerned: manipulation targets and manipulation modes. In terms of manipulation target, it is observed that previous solutions by parameter separations to solve physical contradictions were mainly about separating demands of contradictory parameters. Separating parameters to satisfy the two objectives was not taken into consideration for separation methods. This paper proposes two levels of parameter manipulation targets to achieve: separation of parameter for problem solving at the contradictory parameter level, P, and at the objectives level, O1 and O2.

At the aspect of Manipulation Modes, there are Parameter Separation and Parameter Transfer. The essence of Parameter Transfer is to transfer the issue of satisfying one or two of the contradictory demands to some parameter(s) outside of the local system so as to solve the contradiction. Due to space limitation, this paper focuses on the Parameter Separation. With combinations of different Manipulation Modes and different Target Levels of Satisfaction, seventeen strategies of solving physical contradiction can be conceived as shown in Figure 1. The first letter of the strategies means the mode of manipulation (I: separation withIn parameter, C: separation aCross parameters, T: parameter Transfer); the second letter represents the target of manipulation (P: contradictory parameter as target, O: two objectives as target); the third and the fourth letter, if any, means manipulation methods (V: using different value ranges of a parameter, S: splitting parameters, P: using contradictory parameter to satisfy the demand, E: using

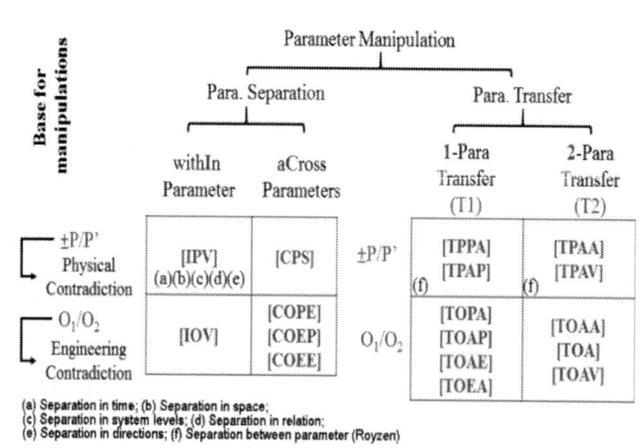

Figure 1. Overview of strategies of parameter manipulation

exclusive parameters to satisfy the demand, A: using non-constituent additional parameters). Problem solving strategies of parameter manipulation will be explained respectively in the following sections.

2.5.1 Parameter Separation

2.5.1.1 Contradictory Parameter Orientation

2.5.1.1.1 Separation within Parameter

The manipulation target in this section is set to be contradictory parameters and the mode of manipulation is separation within parameter . In this case, the problem solving strategy is IPV .

- [IPV]: I—separation within parameter; P—Separation of contradictory demands; V— separation of distinct value ranges of specific parameter +P/-P

It makes the contradictory parameter to satisfy different demands from two objectives separately in distinct value ranges of a specific parameter. That is, the specific parameter can satisfy +P in certain value range and satisfy -P in another value range. Specific parameters can be space, time, relation, system levels, and so on.

Take a pencil for example. To make it comfortable to hold (O1), the pencil shaft should have no angles (+P). To keep the pencil from rolling (O2), the pencil shaft should have angles (-P). With [IPV], separation in space is applicable to solve the problem with a pencil shaft having no angles (+P) on the part where people hold it and have angles (-P) at the ends of the pencil shaft. In this way, the pencil is easy to hold and will not roll on the table.

2.5.1.1.2 Separation between Parameters

Here, the manipulation target is set to be contradictory parameters and the mode of manipulation is separation across parameters . In this case, the problem solving strategy is CPS .

- [CPS]: C—separation across parameters; P—separation of contradictory parameters; S—splitting contradictory parameters.

CPS splits the contradictory parameter of the contradiction into two parameters so that the two contradictory demands can be assigned to the split parameters.

Take indoor temperature as an example. To prevent the machine from over-heating (O1), the indoor temperature should be low (+P). To prevent people from catching a cold (O2), the indoor temperature should be high (-P). With [CPS], the contradictory parameter, indoor temperature, can be split into two new parameters: indoor temperature around machine and indoor temperature around people . The former can satisfy the demand of keeping indoor temperature low (+P) while the latter can satisfy the demand of keeping indoor temperature high (-P).

2.5.2 Ob ective-Orientated

2.5.2.1 Separation within parameter

The manipulation target in this section is set to be two objectives and the mode of manipulation is separation within parameter . In this case, the problem solving strategy would be IOV .

- [IOV] : I—Separation within parameters; O—satisfy objectives separately; V— satisfy O1 and O2 separately with distinct value ranges of a specific parameter.

It makes the specific parameter satisfy the two objectives in distinct value ranges. That is, the specific 使 parameter can satisfy O1 in certain value range and satisfy O2 in another. Specific parameters include space, time, relation, system levels and so on.

Take a pencil for example. To make it comfortable to hold (O1), the pencil shaft should have no angles (+P). To keep the pencil from rolling (O2), the pencil shaft should have angles (-P). With [IOV], separation in space is applicable to solve the problem so that a pencil shaft should have no angles (+P) on the part where people can hold it comfortably (O1) and have angles (-P) at the end of the pencil shaft to create a rough and frictional part so the pencil will not roll that easily (O2).

2.5.2.2 Separation across Parameters

The manipulation target in this section is set to be two objectives and the mode of manipulation is separation across parameters . In this case, the problem solving strategy is COPE , COEP , and COEE .

- [COPE] : C—Separation across parameters; O—satisfy objectives separately; PE-P satisfies O1 、E2 satisfies O2

It makes contradictory parameter P to satisfy the demand of O1. The influence imposed on O2 by the contradictory parameter is offset by O2's exclusive parameter E2.

Take outdoor display panel for example. To prevent the panel from damage due to rain leakage (O1), the holes of front plate and rear plate should not exist (+P). To allow good heat dissipation for the panel (O2), the holes of front place and rear plate should exist (-P). With [COPE], there will be no holes on the plates which prevents the panel from damage due to rain leakage (O1). Moreover, by replacing the original material of the plates with metal that dissipates heat better, the panel will also dissipate heat better (O2).

- [COEP]: C—separation across parameters; O—satisfy objectives separately; EP—E1 satisfies O1, P satisfies O2

It makes contradictory parameter P to satisfy the demand of O2. The influence imposed on O1 by the contradictory parameter is offset by O1's exclusive parameter E1.

Take outdoor display panel for example. To prevent the panel from damage due to rain leakage (O1), the holes of front cover should not exist (+P). To allow good heat dissipation for the panel (O2), the holes of front cover should exist (-P). With [COEP], the holes will still be on the plates which allows good heat dissipation (O2). Plus, by enhancing waterproof capability of the front cover (ex: covering the plates with a waterproof but ventilative cloth such as Gortex), rain leakage can be avoided (O1) but heat can still dissipate.

- [COEE]: C—Separation across parameters; O—satisfy objectives separately; ignore P, E1 satisfies O1, E2 satisfies O2

It makes the contradictory parameter P to be an arbitrary value acceptable to the two objectives. Satisfy the demand of O1 with O1's exclusive parameter E1 and satisfy the demand of O2 with O2's exclusive parameter E2.

Take outdoor display panel as an example. To prevent the panel from damage due to rain leakage (O1), the holes of front plate and rear plate should not exist (+P). To allow good heat dissipation for the panel (O2), the holes of front place and rear plate should exist (-P). With [COEE], the material of the plates is changed to metal that dissipate heat better which helps heat dissipation for the panel (O2). Plus, by enhancing waterproof capability of the plates (ex: covering the plates with a waterproof cloth), rain leakage which damages the panel can be avoided (O1).

3 CASE STUDY

The physical contradiction of a sewing needle is taken as an example in this section.

A needle can be used to sew or repair clothes. To make the thread go through the hole on the needle eye easily, the diameter of the hole should be big. But, if the diameter of hole is big, the needle may damage the fabrics of the clothes when sewing [7].

3.1 Component Identification and Parameter Deployment

The objectives are to facilitate threading and to avoid damage the clothes respectively while the contradictory demand is that the same physical parameter (the diameter of the hole) should be big and small at the same time. By using Parameter Identification Table, the representative parameters of the two objectives and their center components as well as the center component of the contradictory parameter are filled in the table respectively. A contact component chart (as shown in Table 2) is drawn according to these center components and peripheral components.

Table 2. Parameter Identification Table of Sewing Needle Case

System : Needle				
To [O_1] facilitate threading, [P] hole diameter should be [+P] big. But, To [O_2] advoid damaing clothes, [P] hole diameter should be [-P] small.				
O_1	Representative parameter: length of the thread passing through the hole	Center Component: thread	P	Center Component: needle
O_2	Representative parameter: intactness of clothes	Center Component: clothes		
Contact Component Chart				

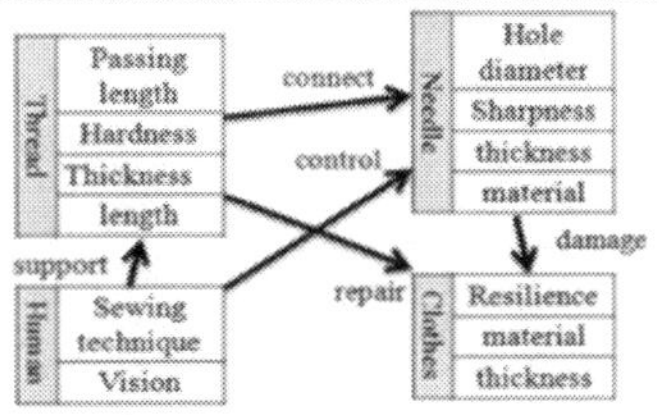

With the parameters identified via Parameter Identification Table, a Parameter Deployment Table can be used to examine whether or not the disturbances on these parameters impose influences on the two objectives or the contradictory parameter. By doing so, the objective functions of two objectives and the contradictory parameter function of the contradictory parameter can be generated as shown in Table 3.

Table 3. Parameter Deployment Table of Needle Case

System : Needle														
O_1	To facilitate threading						O_2	To avoid damaging clothes						
Center components/ Peripheral components														
Needle			Thread			Clothes			Human					
Parameter	O_1	O_2	Parameter	O_1	O_2	Parameter	O_1	O_2	Parameter	O_1	O_2	Para.	O_1	O_2
Hole diameter	↑	↓	Passing length	↑	X	Intactness	X	↑	Sewing technique	↑	↑			
Needle sharpness	X	↑	Hardness	↑	X	Resilience	X	↑	Vision	↑	X			
Thread thickness	X	↓	Thickness	↓	↓									
			Length	↑	X									
Parameter deployment	[O_1] To facilitate threading=fn(hole diameter ↑; passing length ↑, thread hardness ↑, thread length ↑, human vision ↑; thread thickness ↓, human sewing technique ↑) [O_2] To avoid damaging clothes=fn(hole diameter ↓; needle sharpness↑, clothes intactness ↑, clothes resilience ↑; →, thread thickness ↓, human sewing technique ↑) [±P] Hole diameter=fn(thread thickness)													

Lastly, use the formulas of two objectives and the contradictory parameter listed in the Parameter Integration Table to examine if any other parameters are left out and disturb them, so the functions of two objectives and the contradictory parameter can be completed more thoroughly.

Through Parameter Identification Table, Parameter Deployment Table, the functions of the two objectives and the contradictory parameter in the needle case are described below:

[O1] To facilitate threading=fn (hole diameter ↑; thread hardness ↑, thread material →, human vision ↑; thread thickness ↓, human sewing technique ↑, human carefulness ↑)

[O2] To avoid damage on the clothes=fn (hole diameter ↓; needle sharpness ↑, needle thickness ↓, clothes resilience ↑; thread thickness ↓, human sewing technique ↑, human carefulness ↑)

[P] hole diameter = fn (thread thickness)

Among them, hole diameter is the contradictory parameter of the two objectives while thread hardness, thread material, and human vision are exclusive parameters for the objective not to damage the clothes.

3.2 Parameter Manipulation and Contradiction Solution

The required time and the required objects of the contradictory demand can be figured out by using Parameter Manipulation Table-1 (IP). And the ideas generated accordingly can be put into the corresponding boxes as shown in Table 4.

Table 4. Parameter Manipulation Table-1 of Needle Case (IP)

System : Needle			
Separation within parameter -[IPV]			
[±P] Hole parameter=fn(thread thickness)			
Contradictory Parameter (P)	hole diameter	Decisional Factors	thread thickness
Questions	Answers	Ideas	
Where do we need +P? Where do we need –P?	Needle Needle		
When do we need +P? When do we need –P?	When threading When sewing	Choose soft and bendable metal to make needles and intensify the sharpness of the needle so as to satisfy both demands for a big hole when passing threads and a small hole when sewing	
Which object needs +P? Which object needs -P?	thread clothes	Choose soft and bendable metal to make needles and intensify the sharpness of the needle so as to satisfy both demands for a big hole when passing threads and a small hole when sewing	
+P/ -P can be satisfied in super system? +P/ -P can be satisfied in sub-system?	No No		
Other independent parameters	Value ranges	Ideas	

Through Parameter Manipulation Table-2 (IO), it is found that the required time and required object from contradictory demand are different. Since there is no other contradictory parameter in this case, it is impossible to do separation within parameter. However, separation of system levels can be adopted to solve the problem. Parameter Manipulation Table (IO) is shown in Table 5.

With Parameter Manipulation Table-3 (CPS/COPE/COEP/COEE), ideas resulted from manipulating contradictory parameter and each exclusive parameter are put in the corresponding boxes as shown in Table 6. Table list the Solution Compiled based Parameter Separations in the Needle case.

4 CONCLUSIONS AND CONTRIBUTIONS

The authors proposed a unified theory to encompass all existing and newly developed system of methods under the unified umbrella of Parameter Deployment and Manipulation. All existing methods constitute only four of the proposed 17 strategies for solving physical contradictions. This paper explains the integrated theory for parameter deployment and parameter separation. All existing separation theories constitute only one of the 6 strategies of the separation strategies proposed by this paper. Figure 2 shows a testing of the proposed methods on twenty five cases. It is found that the methods proposed can help locate many more solutions (11.52 on average) than traditional separation principles (1.28 on average) and traditional separation principles plus Royen's

Table 5. Parameter Manipulation Table-2 of Needle Case (OP)

System : Needle			
Separation within parameter -[IOV]			
[O_1] To facilitate threading=fn(hole diameter ↑; passing length ↑, thread hardness ↑, thread length ↑, human vision ↑; thread thickness ↓, human sewing technique ↑) [O_2] To avoid damaging clothes=fn(hole diameter ↓; needle sharpness↑, clothes intactness ↑, clothes resilience ↑; thread thickness ↓, human sewing technique ↑) [±P] Hole diameter=fn(thread thickness)			
Contradictory Parameter (P)	hole diameter	Parameter in same direction (Z)	thread thickness, human sewing techniques
Exclusive parameter of O_1 (E_1^1)	Passing length, thread hardness, length, human vision	Exclusive parameter of O_2 (E_1^2)	Needle sharpness, needle thickness, clothes intactness, clothes resilience
Questions	Answers	Ideas	
Where do we need O_1? Where do we need O_2?	Needle Needle	N/A	
When do we need O_1? When do we need O_2?	When threading When sewing	N/A	
Which object needs O_1? Which object needs O_2?	thread clothes	N/A	
O_1/O_2 can be satisfied in super system? O_1/O_2 can be satisfied in sub-system?	Super system –> O_1	Ask someone with good eyesight for help when threading (u)	
Other independent parameters	Value ranges	Ideas	

parameter separation [2](5.48 on average) do. The performance on solution search by methods proposed in this paper is about five times of that from using all existing methods.

The main contributions of this paper include:

1. Providing a systematic parameter identification and deployment method in which all problem related constituent parameters are exposed for possible resources of problem solving in each system. Problem solving can be achieved by manipulating all the parameters for systematic separation for problem solving at contradictory parameter level and objective level.
2. Modelling the expressions for problems and solutions to facilitate a systematic problem solving and the expressions thereof.
3. Proposing six parameter separation strategies of which all existing separation methods constitute only one of them. Together with Parameter Transfer methods to be explained in a separate paper, this unified theory of parameter manipulations include 17 strategies of solving physical contradictions which can greatly enhance the performance of identifying solutions for physical contradictions.
4. Providing a thorough list of systematic parameter separation strategies, along with all the processes, tables, explanations, and cases, making it convenient to learn and use.

Table 6. Parameter Manipulation Table-3 of Needle Case (CPS/COPE/COEP/COEE)

Separation across parameters -[CPS/COPE/COEP/COEE]			
[O_1] To facilitate threading=fn(hole diameter ↑; passing length ↑, thread hardness ↑, thread length ↑, human vision ↑; thread thickness ↓, human sewing technique ↑) [O_2] To avoid damaging clothes=fn(hole diameter ↓; needle sharpness↑, clothes intactness ↑, clothes resilience ↑; thread thickness ↓, human sewing technique ↑) [±P] Hole diameter=fn(thread thickness)			
Contradictory parameter (P)	hole diameter	Decisional factor	thread thickness, human sewing techniques
Exclusive parameter of O_1 (E_1^1)	Passing length, thread hardness, length, human vision	Exclusive parameter of O_2 (E_1^2)	Needle sharpness, needle thickness, clothes resilience, clothes intactness
Strategies	**Target**	**Ideas**	
[CPS] $P \Rightarrow E^1, E^2$	Contradictory parameter	Choose soft and foldable metal to make needles and intensify needle sharpness. Split hole diameter into "diameter for threading" and "diameter for sewing", satisfying the demands for big holes when threading as well as small holes when sewing.	
[COPE] $P \rightsquigarrow O_1$ $E^2 \rightsquigarrow O_2$	Objective	Satisfy the demand for big hole diameter (needle thickness is increased at the same time). Opt for fabrics of better resilience to make clothes (x).	
[COEP] $E^1 \rightsquigarrow O_1$ $P \rightsquigarrow O_2$	Objective	Satisfy the demand for small hole diameter. Opt for threads of greater hardness (x); ask someone with better eyesight for help when threading (u).	
[COEE] $E^1 \rightsquigarrow O_1$ $E^2 \rightsquigarrow O_2$	Objective	Opt for fabrics of better resilience to make clothes (x). Opt for threads of greater hardness (x); ask someone with better eyesight for help when threading (u).	

5 REFERENCES

[1] Altshuller, G. and Shulyak, L. 1988. *40 Principles: TRIZ Keys to Technical Innovation*, Genrich Altshuller and Lev Shulyak, Technical Innovation Center, Worcester, MA, USA

[2] Royzen, Z. 2010. *Solving Problems Using TOP-TRIZ,* TRIZCON 2010.

[3] Mann, D. 2007. *Hands on Systematic Innovation.* ISBN 90-77071-02-4, IFR.

[4] Yeh, Rachel, Parameter Deployment and Manipulation for Solving Physical Contradictions, Master's thesis, National Tsing Hua University, 2015. (In Chinese)

[5] Sheu, D. D. and Yeh, R. 2015. *Parameter Deployment and Separation: a New Way of Solving Physical Contradictions*, 2015 the 7th Systematic Innovation Conference and Annual Meeting of the Society of Systematic Innovation. Taipei, Taiwan, Outstanding paper award. (In Chinese)

[6] Li, H. C. 2013. *An Integrated Algorithm for TRIZ Inventive Problem Solving (Master's thesis).* National Tsing Hua University, Hsinchu, Taiwan.

[7] Sheu, D. D. 2015/9. *Mastering TRIZ Innovation Tools: Part I.* Agitek International Consulting, Inc. (4th Ed.) ISBN 978-986-85795-2-1. (In Chinese)

Table 7. Solution Compilation for Needle Case: Parameter Separations

Solutions		
Strategies	Modes	Ideas triggered
[IPV] [CPS] [TPAV]	Sol=(hole diameter: time-when threading/sewing→big/small; thread passing length~, thread hardness~, thread length~, human vision~; needle sharpness~, needle thickness~, clothes intactness~, clothes resilience~, clothes thickness~, human sewing techniques~; threader~, clothes self-cure capacity~, needle material →, sewing machine~)	Use soft and foldable metal to make needles and intensify the needle sharpness to pierce through the clothes. Satisfy the demands for big holes when threading and small holes when sewing.
[IOV] [COEP]	Sol=(hole diameter ↓; thread passing length~, thread hardness~, thread length~, human vision ↑; needle sharpness~, needle thickness~, clothes intactness~, clothes resilience~, thread thickness~, human sewing techniques~; threader~, clothes self-cure capacity~, needle material →, sewing machine~)	Satisfy the demand for small hole. Ask for someone with better eyesight for help when threading(u).
[COPE]	Sol=(hole diameter ↑; thread passing length~, thread hardness~, thread length~, human vision~; needle sharpness~, needle thickness~, clothes intactness~, clothes resilience ↑; thread thickness~, human sewing techniques~, clothes self-cure capacity~, needle material ~, sewing machine~)	Satisfy the demand for big hole (thicken the needle at the same time). Opt for fabrics of better resilience to prevent clothes from damage when sewing(u).
[COEP]	Sol=(hole diameter ↓; thread passing length~, thread hardness ↓, thread length~, human vision~; needle sharpness~, needle thickness~, clothes intactness~, clothes resilience~; thread thickness~, human sewing techniques~; threader~, clothes self-cure capacity~, needle material ~, sewing machine~)	Satisfy the demand for small hole. Opt for harder threads to facilitate threading(u).
[COEE]	Sol=(hole diameter~; thread passing length~, thread hardness ↑, thread length~, human vision~; needle sharpness~, needle thickness~, clothes intactness~, clothes resilience~; thread thickness~, human sewing techniques~; threader~, clothes self-cure capacity~, needle material ~, sewing machine~)	Opt for fabrics of better resilience to prevent clothes from damage when sewing (u). Opt for harder threads to facilitate threading (u).
[COEE]	Sol=(hole diameter~; thread passing length~, thread hardness~, thread length~, human vision ↑; needle sharpness~, needle thickness~, clothes intactness~, clothes resilience ↑; thread thickness~, human sewing techniques~; threader~, clothes self-cure capacity~, needle material ~, sewing machine~)	Opt for fabrics of better resilience to prevent clothes from damage when sewing. Ask for someone with better eyesight for help when threading (u).

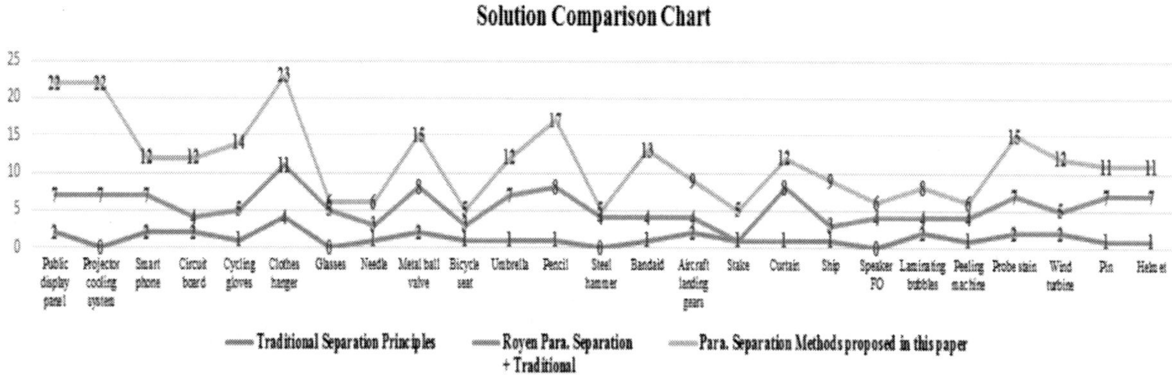

Figure 2. Comparison of Number of Solution Generated

A G-S CONCEPT BASED RECONFIGURABLE WIRELESS TELECOMMUNICATION SYSTEM

J.W. Wang[1*], . He[2], Z.C. Liang[1] and W.J. Zhang[2]
[1]Department of Industrial and Manufacturing Systems Engineering
The University of Hong Kong, Hong Kong, China
jwwang@hku.hk

[2]Department of Mechanical Engineering
University of Saskatchewan, Canada
chris.zhang@usask.ca

ABSTRACT

This paper presents an effective system reconfiguration model for the wireless system to achieve high survivability or service continuity in emergency situations. A new approach to reconfiguring the system is proposed based on the concept of general and specific function (G-S function) in engineering design proposed by Pahl and Beitz [1]. Further, a knowledge base which supports this new reconfiguration approach is introduced. This paper uses the third generation (3G) wireless telecommunication system, Wideband Code Division Multiple Access (WCDMA) system, as an example to explain the implementation of a reconfigurable service system.

Keywords: General-specific function, Reconfiguration, WCDMA, Domain model, Knowledge base

1 INTRODUCTION

The critical infrastructures were defined by [2] as physical and information technology facilities which have a serious impact on the health, safety, security, economic well-being of citizens, or the effective functioning of governments. Among all the critical infrastructures, the telecommunication infrastructure system plays an important role in the functioning of our society. Failure of telecommunication infrastructure will cause coordination capabilities disable and significantly hinder rescue and recovery operations [3]. In the absence of telecommunication infrastructure, it is hard to conduct the repair and recovery of other critical infrastructures which may dysfunction due to incidences or accidents. Harrop [4] examined the relationship of communications networks with other Canadian critical infrastructures and concluded that other critical infrastructures highly depend on the telecommunication network. Therefore, service continuity is a very important topic in telecommunication infrastructure systems.

Reconfiguration of a system means to change the connectivity of the system as well as its components for the purpose of changing the behaviors of the system. System reconfiguration in this paper is for enhancing the survivability and resilience of a service system [5, 6]. As such, there is always an assumption that a couple of components of a system are damaged and their functions are lost. The goal of system reconfiguration is then to find out ways upon the existing system (including those damaged components) to remedy the lost functions and maintain a satisfactory level of uality of Service (oS) for all the services or functions of the survivable service provider system [7].

* Corresponding Author

Generally, there are three ways to achieve system reconfiguration. First, reconfiguring the remaining system (excluding the damaged one). Second, exploring the damaged components. Third, exploring the capacity of the existing system under the assumption that the existing system usually does not run to its full capacity.

There are some concepts about operating system reconfiguration. The most general one is called general function and specific function (G-S concept) proposed by Pahl and Beitz [1]. A general function is defined as an abstract function of the system which plays its role independent of any particular environment. A specific function is defined as a concrete function with task-specific or service-specific environment where the system entities are to play their role. From these definitions, the general function can be concluded as the function without any context; while the specific function is the function with context.

The above three ways of system reconfiguration can be explained with the G-S concept. In the case of the first method, the two different configurations of original system and damaged system are considered to be two specific functions of the system, like a table with four legs being reconfigured with three legs. In the case of the second method, for one damaged component, its reuse is based on its general function, but the assembly of two damaged components into one is based on the fact that a new specific function is established. In the case of the third method, the general function of a system is to conduct a normal workload. The specific function of the system is to conduct abnormally larger workload after reconfiguration.

In the field of system reconfiguration, some other related and classic works are the work of Umeda, Tomiyama [8], the work of Ulrich and Seering [9] and the work of Wang et al.[6, 10-12]. Umeda, Tomiyama [8] illustrated a design methodology called function redundancy for highly reliable machines. They proposed that function redundancy is achieved by using the potential functions of the existing parts in a slightly different way from the original design. And they illustrated this idea using an emergency car engine stop example. This idea is a reflection of the third system reconfiguration method. Ulrich and Seering [9] proposed a concept called function sharing as a kind of design method for general product design. The function sharing is defined as a correspondence between several functional elements and one structural element or physical property of the design. In essence, whenever a single structural element can map to more than one functional element, then the device shows the function sharing. The function sharing concept is close to the function redundancy concept and seems to be restricted to a single component only in their original description. Wang et al. [6, 10-12] proposed reconfiguration of flow patterns to improve the resilience of transportation systems.

In comparison with these two concepts, the approach based on the G-S model is the most general one. The other two approaches are at most useful to the reliability of the system but not to the survivability of the system. This paper presents a reconfigurable WCDMA system which is based on the backbone (i.e. the domain model of the WCDMA) and G-S concept. The remainder of this paper is organized as follows. Section 2 illustrates a reconfigurable WCDMA system based on the G-S concept along with a procedure to perform reconfiguration in emergency situations. Section 3 presents a knowledge base which supports this new reconfiguration approach. Section 4 concludes the paper.

2 RECONFIGURABLE WCDMA SYSTEM BASED ON G-S CONCEPT

The overall architecture of a typical Universal Mobile Telecommunications System (UMTS) includes three basic blocks: User Equipment (UE), UMTS Terrestrial Radio Access Network (UTRAN), and Core Network (CN). These elements are connected among themselves and to external networks. As a kind of UMTS, a WCDMA system includes some detailed components. In the simplest infrastructure of a WCDMA system, a Node B serves hundreds of mobile users in a given area (Cell) by allocating radio resources for making new calls or for users to continue their calls. A RNC provides switching support for several neighboring base stations (Node B) and thus can serves thousands of users. The link between Node B and RNC and also between RNC and Media Gateway (MGw) are typically wireline, although the use of radio or microwave links to connect these components has been increasing. The MGw is a larger switch and can serve tens of thousands of users. The functions, behaviors, structural components of a WCDMA system are shown in Figure 1, and they are explained as follows.

The functions of a WCDMA system can be categorized into four service classes: conversational, streaming, interactive, and background. Each service class comprises a set of sub-functions. The four services are generalized and categorized into two overall functions: (1) function for voice transmission, and (2) function for data transmission. The overall function can be decomposed into more specific functions including the primary function and secondary function. This study focuses on the conversational service, so in the following, it is discussed on how to keep the voice function for a high continuity based on the G-S model.

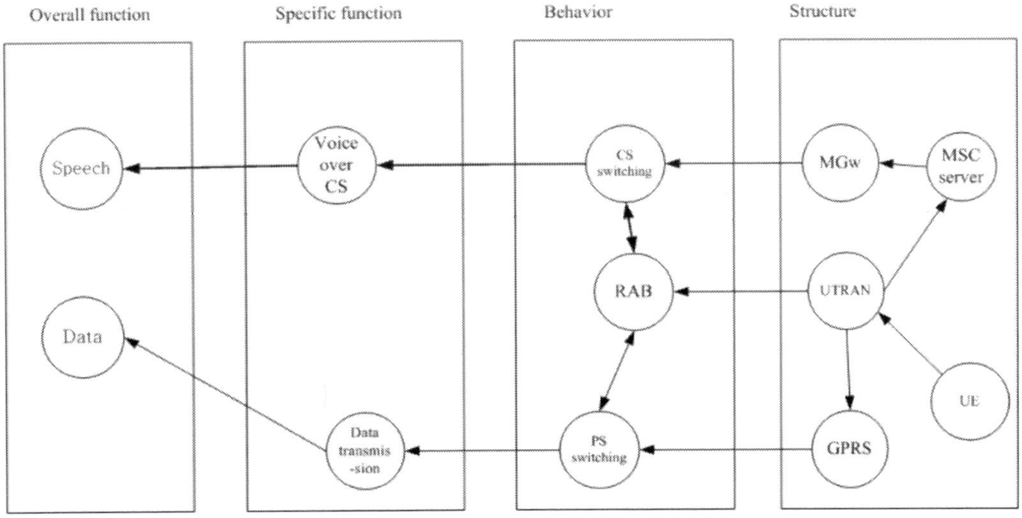

Figure 1: Typical overall function of a WCDMA system

The voice function is further divided into the sub-functions: (1) Speech function which is provided by the circuit switching system, (2) Voice over IP (VoIP) which is provided by the packet switching system, and (3) Videophone which is provided by both the circuit and packet switching systems. Further, since the videophone is at a higher service level, and it only works in the normal condition of the network and need more radio resources and bandwidths, it is thus out of the scope of this study. This study focuses on the speech function provided by circuit switch and the VoIP function by packet switch.

When the system is in failure, there are three levels of failure scenarios which are defined in this paper in accord with the telecommunication system: (1) Radio frequency access layer: failure of Nodes B, (2) Radio access control subsystem layer: failure of RNC, and (3) Transport layer: failure of MGw in CS domain, and failure of SGSN/GGSN in PS domain. In the following, Media Gateway (MGw) in the circuit switching domain is supposed to be failed in order to demonstrate the application of the G-S model for system reconfiguration to enhance the survivability of the system. The following steps are applied:

The first step is to analyze the lost function by decomposition it into a tree until to the component level. The function tree will include the particular function whose components (MGw) is damaged. The second step is to find a secondary function of any component including the one (MGw) which is damaged. For the example, the third method of system reconfiguration is considered; in particular, the PS route is considered for providing unused capacity, in the form of its secondary function to make this reconfiguration idea work. The third step is to explore the principle which is implicitly represented by the new-reconfigured structure.

Figure 2 shows this reconfiguration solution. The overall function speech , can be realized by primary function Voice over CS . The speech function is mainly realized by the behaviours of CS switching and Radio Access Bear (RAB) . Both behaviours are generated by the structural functional entities: MGw, MSC server, UTRAN and ME. Now, the lost primary function Voice over CS has be replaced by the secondary function through the component or subsystem such as: GPRS, UTRAN and ME. Thus, the original overall function speech' can be recovered.

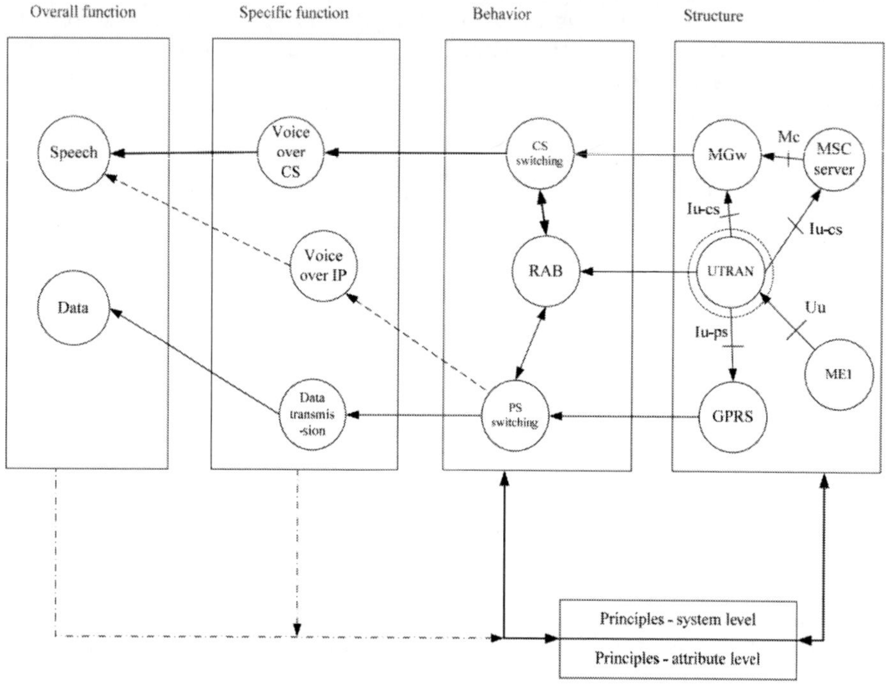

Figure 2: Generalized reconfigurable WCDMA

3 A KNOWLEDGE BASED SYSTEM

For the discussion above, it is clear that it is necessary to develop a knowledge base for supporting the G-S concept of reconfiguration. The knowledge base should contain function, structure, and principle of a particular system (WCDMA in this case). Therefore, this knowledge base consists of three parts: (1) function knowledge base, (2) structure knowledge base, and (3) principle knowledge base.

3.1 System function knowledge base

The system function knowledge base collects and describes the functions of the existing system. The function may be called function prototype. [13] shows the holistic architecture of the functions including the function decomposition and overall-specific function relations, G-S relationship.

There are two overall functions of the WCDMA system, the speech and data transmission. The speech' function can be realized mainly by three specific function services: Voice over Circuit

Switch, Vice over IP by Internet or Video conference service. The data' function is realized mainly by other three specific functions: Video conference for imagine, Audio/Video streaming, Web-browsing, etc. Such specific functions are also grouped into the four types of service class based on the oS standards: conversational, streaming, interactive, and background. For length limitation, a detailed description of the function decomposition for the example WCDMA system will be given in a following work.

3.2 System structure knowledge base

The structure of the WCDMA system is divided into two parts: the traffic transmission structure and the system control structure. The traffic transmission structure includes the entities that receive and transport the traffic data to destination, while the system control structure includes the entities that follow the rules or operation standards to manipulate the entities in the traffic transmission structure. These components are further categorized into five groups: (1) User interface, (2) Radio signal Access, (3) Signaling control (4) Data transportation, and (5) System database. For length limitation, a detailed description of the typical structure of a 3G WCDMA network and the states or properties of the components will be given in a following work.

The emergency manager searches the knowledge base by following the procedure (see Figure 3). It starts from the structures (either control structural part or traffic transmission structural), then the groups of each structure, then the components, and finally the attributes of the selected components.

Figure 4 shows a simple example of applying the above procedure for the voice function in the circuit switch domain. The manager starts with the transmission structure in the User Equipment (UE) layer. The UE supports the UMTS service with the Uu interface and within the radio coverage area of Node B, the Cell of Node B. The components in other groups such as Radio signal access, data transportation, are identified to provide the circuit switch function. All the selected components have their own state variables. The relation between the components is determined by the state variables (e.g. capacity and interfaces). When the state variables of the two different components can match each other, they can interact. For instance, the UE-1 is only related with the Node B because the Node B has the same Un interface. Their interfaces match each other. Meanwhile the capacity of the Node B also impacts the relation with the UE. Meanwhile, if the Node B-1 does not have enough capacity to process the service request from the UE, the UE has to request other node (e.g. Node B-2), which has enough capacity. The Node B-2 will transmit the traffic to the RNC-2 through the Iu-b interface between them. If the RNC-2 does not have enough capacity, UE-1 has to hand over to another Node B which can find an available traffic channel from RNC and CN. The RNC-2 is assumed to have enough capacity and MGw can allocate enough channels to the call services. Then the voice traffic can be transported from the RNC-2 to the MGw through the Iu-CS interface and forwarded to users in local network or other PLMN/PSTN network. Figure 10 shows the selected components and their provision of the WCDMA CS domain voice function. Similar to the Voice function realization, the VoIP function can be provided by the network components UE-1, Node B-2, RNC-2 and GPRS, etc. These components transport the packet Voice data to the IP network and relay to the users in other PSTN/PLMN networks.

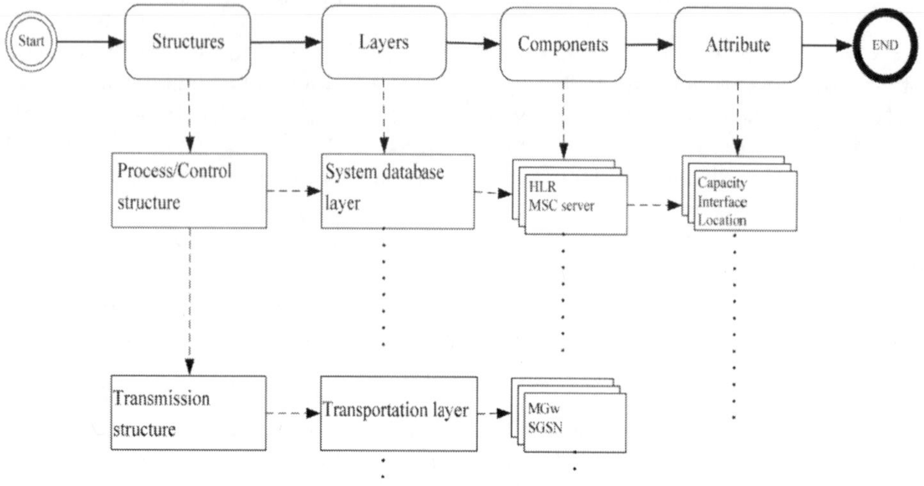

Figure 3: Search procedures for the structure knowledge base

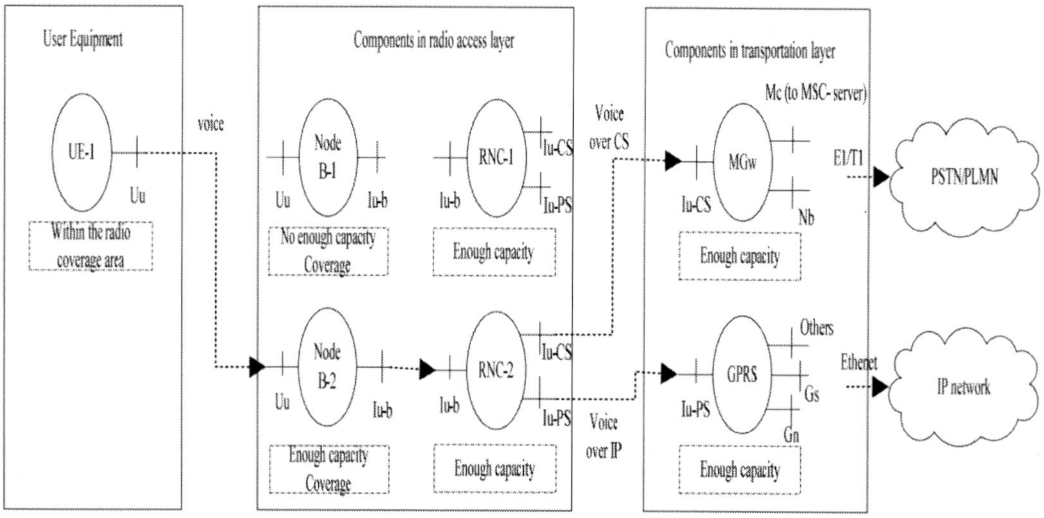

Figure 4: Configuration of network components

3.3 System principle knowledge base

The principle of telecommunication system (e.g. 3G WCDMA system) refers to standard interfaces or protocols. Protocols may be implemented by hardware, software, or a combination of the two. At the lowest level, a protocol defines the behavior of a hardware connection. The UMTS protocols are used to control the execution of network functions in a coordinated manner across system interfaces [14]. At a higher level, protocols are grouped in stacks, which are standard interfaces. There are three reference protocols and they are: (1) The Radio Interface Protocol Reference Model; (2) UTRAN Protocol Reference Model; and (3) The CN Protocol Reference Model. The architecture of each model consists of the protocol stacks [14]. In this figure, the protocols are grouped under the three reference models; the detailed knowledge about all these protocols is beyond the scope of this paper. Further, these protocols are classified into access stratum and non-access stratum. The non-access stratum concerns the user service request, while the access stratum concerns the processing of radio signals.

Table 1 is a schema of the principle knowledge base. In this table, the first column shows two operation levels, namely non-access stratum and access stratum. The second column shows the functions of the components which are governed by the protocols. The third column shows the components involved. The fourth column shows the protocols that are applied to the components (in the third column) to play the functions as shown in the second column.

Table 1: WCDMA system principle knowledge base

Operation level	Function	Components	Protocols
Non-access stratum	VoCS	UE, Node B, RNC, MSC server, MGw, HLR, EIR	Call Control (CC) protocol in non-access stratum; All the protocols in Access stratum
	VoIP	UE, Node B, RNC, SGSN, GGSN, HLR, EIR	Session Management (SM) protocol in non- access stratum; All the protocols in Access stratum
	MM	UE, Node B, RNC, MSC server, MGw, SGSN, GGSN, HLR, EIR	Mobility Management (MM) and GPRS Mobility Management (GMM) protocols in non- access stratum; All the protocols in Access stratum
Access stratum	PC	UE, Node B, RNC	RRC and PDCP protocols and other lower support protocols
	CAC	UE, Node B, RNC	RRC and PDCP protocols and other lower support protocols

4 CONCLUSION

This paper presented a novel approach to system reconfiguration. The approach is based on the General-Specific (G-S) concept. The essential point with the G-S model is that every component, subsystem, and system has two aspects: general function and specific function. To facilitate the modelling of the general and specific function, the general function can be thought of primary function, while the specific function can be thought of secondary function. It is considered that each system (or component or subsystem) has primary function and secondary function. A particular component or subsystem plays its primary function in one specific situation which may be called the named situation. When some components or subsystems are damaged, implying that they can no longer perform their primary functions, they will be replaced by other components or subsystem in the same system or in its relevant systems through their secondary functions which are not made available in their corresponding named situations.

To demonstrate of the application of this general and specific function concept to system reconfiguration, a reconfigurable WCDMA model was proposed. And then a knowledge base which supports this new reconfiguration approach was established in the field of telecommunication.

5 REFERENCES

[1] Pahl, G. and Beitz, W. 1996. *Engineering design: a systematic approach*, 3rd Edition, Springer-Verlag.

[2] Moore, A. 2002. *The vulnerability of mobile telecommunications to natural hazards*, Office of Critical Infrastructure Protection Emergency Preparedness.

[3] Samarajiva, R. 2001. Disaster-preparedness and recovery: a priority for telecom regulatory agencies in liberalized environments, *International Journal of Regulation and Governance*, 1(2), pp 181-196.

[4] Harrop, M. 2002. Creating Trust in Critical Network Infrastructures: The Canadian Case Study, *International Telecommunication Union Workshop on Creating Trust in Critical Network Infrastructures CN/07*, Seoul, Republic of Korea, May 20-22, pp 158-161.

[5] Wang, J. 2013. Towards a resilient networked service system, *Ph.D. dissertation, Dept. Mech. Eng., Univ. Saskatchewan, Saskatoon, SK, Canada.*

[6] Wang, J., Muddada, R.R., Wang, H., Ding, J., Lin, ., Liu, C., and Zhang, W. 2016. Toward a resilient holistic supply chain network system: Concept, review and future direction, *IEEE Systems Journal*, 10(2), pp 410-421.

[7] Wang, J., Wang, H., Zhang, W., Ip, W.-H., and Furuta, K. 2014. On a unified definition of the service system: What is its identity , *IEEE Systems Journal*, 8(3), pp 821-826.

[8] Umeda, ., Tomiyama, T., and oshikawa, H. 1995. FBS modeling: modeling scheme of function for conceptual design, *Proceedings of the 9th International Workshop on Qualitative Reasoning*, Amsterdam, May 11-19, pp 271-278.

[9] Ulrich, K.T. and Seering, W.P. 1988. Function Sharing in Mechanical Design, *The Seventh National Conference on Artificial Intelligence*, Saint Paul, America, August 21-26, pp 342-370.

[10] Wang, J., Ip, W., and Zhang, W. 2010. An integrated road construction and resource planning approach to the evacuation of victims from single source to multiple destinations, *IEEE Transactions on Intelligent Transportation Systems*, 11(2), pp 277-289.

[11] Wang, J., Wang, H., Zhang, W., Ip, W., and Furuta, K. 2013. Evacuation planning based on the contraflow technique with consideration of evacuation priorities and traffic setup time, *IEEE Transactions on Intelligent Transportation Systems*, 14(1), pp 480-485.

[12] Wang, J., Wang, H., Zhou, ., Wang, ., and Zhang, W. 2016. On an integrated approach to resilient transportation systems in emergency situations, *Natural computing, In press.*

[13] ITU-T. 1999. Network functional model for IMT-2000. Retrieved 09/20, 2016, from http://www.itu.int/rec/T-REC- .1711-199903-I/en.

[14] Kaaranen, H., Ahtiainen, A., Laitinen, L., Naghian, S., and Niemi, V. 2005. *Umts networks*, J. Wiley and Sons.

FACTORS INFLUENCING HUMAN-AUTOMATION INTERACTION IN SUPERVISOTY CONTROL: A PRELIMINARY REVIEW

R. Yang[1], H.W. Liu[2*], E.S. Qi[3] and P. Liu[4]
[1]College of Management and Economics
Tianjin University, Tianjin, China
Yangrun93@163.com

[2]College of Management and Economics
Tianjin University, Tianjin, China
hw_liu999@tju.edu.cn

[3]College of Management and Economics
Tianjin University, Tianjin, China
esqi@tju.edu.cn

[3]College of Management and Economics
Tianjin University, Tianjin, China
lp527965873@163.com

ABSTRACT

Automation is extensively used in complex systems (e.g., unmanned areial vehicles, aircrafts). The usage of automation is supposed to reduce operators' physical and mental workload, increase operation efficacy, and finally increase system performance. However, due to the breakdown of human-automation interaction (HAI) or improper HAI (e.g., overtrust), it could negatively affect operator performance, and damage and jeopardise the integrity of complex systems. This study will review the factors influencing HAI in supervisory control. The factors are classified into four types, automation, human, interface, and task and environment. This review will highlight the impact of these factors on the reliability of HAI. Automation-related factors to be reviewed include the level of automation, stage of automation, flexible automation, automation reliability. Human-related factors include trust, reliance, and compliance. Interface-related factors include appearance, ease of use, communication style, and transparency or feedback. Task and environment-related factors include task complexity, time pressure, task priorities, risk, interference, etc. Several potential directions for HAI research are suggested.

Keywords: Human-automation interaction, automation-related factors, human-related factors, interface, task and environment, reliability

1 INTRODUCTION

Automation means the whole or part work performed by human is completed by machines (especially by computers) [1]. On the one hand, automation is necessary. It can complete the tasks that operators can not complete, make up for the defects of human ability, improve work efficiency, and reduce operators' workload. On the other hand, automation also requires human cooperation and supervision. With the gradual development of automation, the role of the human is more important [2]. Therefore, it is required to investigate the relationship between human and automation and improve their interaction.

To address this issue, many studies on HAI have been carried out in the past years, which had different concerns. Some researchers focus on human elements. For instance, Chen & Barnes [3] reviewed human-related factors and discussed the user interface design for human-agent teaming.

Some researchers are more inclined to study the whole system of HAI. For example, Sheridan & Parasuraman [4] reviewed researches and challenges about HAI, and described studies of adaptive automation. In addition, there are several works evaluating the HAI by formal verification [5].

In this paper, some performance metrics which can reflect the reliability of HAI are described, including time and accuracy, mental workload, situation awareness. In addition, studies on the influencing factors of HAI are reviewed from four types: automation, human, interface, and task and environment. Especially, recent researches of flexible automation are summarized in detail. In the end, possible directions of future development are put forward. The paper is organized as follows. Section 2 introduces some performance metrics in the HAI realm; Section 3 reviews four different factors affecting the reliability of HAI, following by a brief conclusion in Section 4.

2 PERFORMANCE METRICS

Human performance is a meaningful indicator to demonstrate the effectiveness of HAI. The section discusses some metrics of performance, including: time and accuracy, mental workload, situation awareness.

2.1 Time and accuracy

Time is one of the important count variables in HAI. The mental workload can be measured by evaluating time-consuming [6]. In addition, the ratio between the time required in a task and the time available can help to assess operator workload [7]. Time indicators used in the field of HAI include operation modes time distribution [8], response time [9], and so on.

Accuracy is another indicator to measure the effectiveness of HAI. Accuracy is usually operationalized human error rate or human error probability. Signal Detection theory (SDT) is one of the measurement methods for detecting performance in many application domains such as HAI, which can provide the accuracy of decision outcomes. It has six metrics, hit, miss, false alarm rate, correct rejection, sensitivity (d'), and response criterion or observer bias (β)[10].

2.2 Mental Workload

The original purpose of the introduction of automation is to reduce operators' workload. Thus, mental workload is an important human performance index in HAI. Workload is assumed to be a multi-dimensional construct which has multiple explanatory factors. For example, Ref. [11] suggested that time load, mental effort load, and psychological stress load are three component factors. Workload measurement tools can be divided into four categories:

- Behavioural measurement, such as timeline analysis [6].
- Secondary task. Operator workload is evaluated according to operator performance in the secondary task in the case of completing the main task [12].
- Subjective measurement, such as Modified Cooper-Harper scale [13], NASA-Task Load Index (NASA-TLX) [14].
- Physiological measurement, such as electrophysiological (e.g., electroencephalogram (EEG), event-related potentials (ERPs), etc.), cerebral hemodynamics (e.g., positron emission tomography (PET), functional magnetic resonance imaging (fMRI), etc.), and autonomic nervous system (e.g., heart rate variability, pupil diameter, visual scanning, etc.) [10].

2.3 Situation awareness

Situation awareness (SA), which exists as a basis of performance, is widely used in many complex domains. The most common and widely applicable definition proposed by Endsley [15] is "the perception of the elements in the environment within a volume of time and space, the comprehension of their meaning and the projection of their status in the near future" (p. 97). SA has three levels: perception of elements in current situation, comprehension of current situation, and projection of future status. The analysis and methods of SA have been reviewed by Endsley and Garland [16]. For instance, SA global assessment technique (SAGAT) is a common method measuring the three levels of SA directly by a group of problems based on memory [???]. In addition, situation present assessment measure (SPAM) [17] has been used to assess the SA by recording the reaction time and the correct rate of the operator. In addition to the above explicit measures of SA, some implicit performance-based measures assess SA by imposing an abnormal or unexpected event in daily work [18].

3 INFLUENCING FACTORS OF PERFORMANCE

There are many factors influencing operator performance in HAI. The section summarizes the exist influencing factors categorised into four aspects (see Figure 1): automation, human, user interface design, and task and environment.

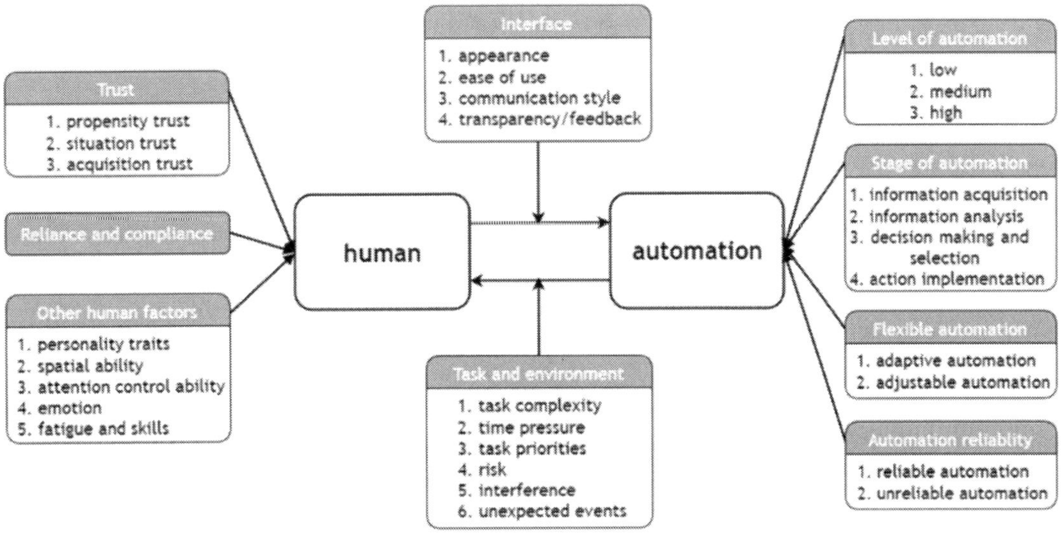

Figure 1: The Influencing Factors of HAI

3.1 Automation

3.1.1 Level of automation

Level of automation (LOA) describes the degree of automation, from a pure manual processing to fully automatic processing. LOA has multiple classification methods [6]. For instance, O'Hara & Higgins [19] divided LOA in the control systems of nuclear power plants into five levels. In the level of low, medium, or high, there are both have findings that automation can be used to improve the human performance [20-22]. In one meta-analysis, Onnasch et al. [23] found that, in automation failure-free condition, the increase of LOA could improve human performance and reduce the workload; however, in automation-failure condition, the reverse seems to be the case. The optimal LOA depends upon the task type, stage and reliability of automation in the specific context.

3.1.2 Stage of automation

The automation process can be divided into four stages: information acquisition, information analysis, decision making and selection, and action implementation, corresponding to the four primary stages of human information processing [6]. Automation at the later stage can improve more operators' performance and reduce more workload related to the early stage when it is reliable. However, automation at higher stage may have more catastrophic consequences when it is unreliable [23, 24].

3.1.3 Flexible automation

Flexible automation, compared with static automation, refers to that the function of human and automation is dynamically allocated. It can improve human performance and maintain situation awareness [25, 26]. Flexible automation can be divided into two categories, depending on whether the decision to automate is made by machine (adaptive automation) or human (adjustable automation). Adaptive automation is activated by different triggering mechanisms (e.g., operator performance, heart rate, electroencephalogram, and critical events). Adjustable automation requires human to decide whether to initiate automation or complete the task manually. At present, the research of adaptive automation is more than that of adjustable automation.

It is generally considered that adaptive automation is better than manual or full automation. Several experimental studies compared different trigger mechanisms. It has been found that the performance-based trigger mechanism is worse than heart rate-based and event-based [26]. The integration of multiple trigger mechanisms is better than single trigger mechanism [27].

Several studies suggested that adjustable automation does not improve human performance [28, 29], but more research supported it. Considering mental workload and diagnostic performance, adjustable automation is better than static automation at the level of low and medium [30]. Adjustable automation makes people clearly meet the role of "regulator", improve their performance, and keep their workload in a manageable range [25]. Currently, limited studies compared different techniques of adjustable automation. Valero-Gomez et al. [8] divided adjustable automation into static and flexible and found that flexible adjustable automation could produce higher performance.

Both of adaptive and adjustable automation have their own advantages and disadvantages. Adaptive automation can validly reduce human workload, however it will become more unpredictable to operators. Adjustable automation can ensure less unpredictability of systems, whereas human workload will become higher [25]. There are three comparative studies, which obtain different experimental results: one study supports the former [28], and the other two studies support the latter [26, 31]. Although studies have found that the integration of the two may be beneficial to maximize the strengths of both [32], but it has never been demonstrated by experiments. At present, the basic concept of complex system design is that people have the ultimate control power, so adaptive automation may be eventually unfeasible in practice [33]. Although adjustable automation may not be suitable for time critical or extremely complex conditions [3], it is a more practical option, for example, in the main control rooms of small modular reactors (SMRs) [22].

3.1.4 Automation reliability

Many complex system accidents relate to unreliable automation[34]. Manual operation will be better when the automation reliability is less than 70% [35]. Error behaviours of unreliable automation can be divided into two categories: omission error (such as miss-alarm) and execution error (such as false-alarm or executing a wrong operation). The performance cost of unreliable automation depends upon its type. Relevant results appear to be controversial. Dixon et al. [36] found that misstatement (i.e. one execution error) could cause more severe performance reduction than omission does. However, Rovira & Parasuraman [37] found that operator performance was more likely influenced by automation omission error.

3.2 Human

3.2.1 *Trust*

Trust is a core concept in HAI, which affects human willingness to rely on automation [38]. The inappropriate level of trust leads to abuse or abandon of automation [1]. Trust includes three types: dispositional trust, situational trust, and learned trust [38]. And different types of trust affect each other and a variety of influential factors (e.g., personality traits, system complexity, system performance, etc.) related to them have been discussed.

3.2.2 *Reliance and compliance*

Reliance and compliance are two kinds of mental states of automation dependence. Researchers found that the type of unreliability automation (omission error and execution error) has different effects on reliance and compliance. There are mix results. Some studies hold that execution error affects both reliance and compliance, whereas omission error only affects reliance [9]. There are also studies showing that omission and execution errors simultaneously affect reliance and compliance [39].

3.2.3 *Other human factors*

Some factors such as personality traits, spatial ability, attention control ability, emotion, fatigue and skills, etc., can also affect HAI [3, 40].

3.3 Interface

The key interface feature is observable or transparent [41]. On one hand, multi-mode feedback and ecological interface design are conducive to improve the transparency of automation and the quality of HAI [42]; on the other hand, transparent automation interface increases the complexity of the operators' tasks and reduces their performance [43]. This may relate to "the effect of attention tunnel" that it is difficult for operator to monitor the state of automation and other systems. In addition, the appearance, ease of use, communication style, and transparency or feedback of the automation can all affect human trust [38].

3.4 Task and environment

High complexity indicates that human need to devote more attention, resources, and effort to tasks [44]. Under complex conditions, human show greater dependence on automation and their performance are more vulnerable to the impact of automation reliability [1, 35]. Other tasks and environmental factors (time pressure, task priorities, risk, interference, and unexpected events) can also affect the reliability of HAI [45].

4 CONCLUSION

The objective of this paper is to review and synthesize findings on the factors related to HAI. Through a preliminary review process of literature on this topic, our review leads to some conclusions as following. Effective HAI does not only consider not only consider the automation-related factors, but also the human-related factors. In addition, excellent user interface can make people better understand the automation system, and task and environment-related factors have close relationship with human performance. Reliability is the foundation of HAI, that can be reflected by some performance metrics. Specifically, flexible automation meets the requirement of human greatly. With regard to the comparison of advantages and disadvantages of adjustable and adaptive automation, there are some controversies. More studies are needed to identify the best flexible automation strategy. Apart from the state of regulation, adjustable automation also can be divided by adjustable objects (such as automatic level / phase or subtask). It is necessary to further verify the advantages and optimal regulation mechanism of adjustable automation.

5 REFERENCES

[1] **Parasuraman, R. & Riley, V. 1997.** Humans and automation: use, misuse, disuse, abuse, *Human Factors*, 39(2), pp 230-253.

[2] **Parasuraman, R. & Wickens, C.D. 2008.** Humans: still vital after all these years of automation, *Human Factors*, 50(3), pp 511-520.

[3] **Chen, J.Y. & Barnes, M.J. 2014.** Human-agent teaming for multirobot control: a review of human factors issues, *IEEE Transactions on Human-Machine Systems*, 44(1), pp 13-29.

[4] **Sheridan, T.B. & Parasuraman, R. 2005.** Human-automation interaction, *Reviews of Human Factors and Ergonomics*, 1(1), pp 89-129.

[5] **Bolton, M.L., Bass, E.J., & Siminiceanu, R.I. 2013.** Using formal verification to evaluate human-automation interaction: a review, *IEEE Transactions on Systems, Man, and Cybernetics: Systems*, 43(3), pp 488-503.

[6] **Parasuraman, R., Sheridan, T.B., & Wickens, C.D. 2000.** A model for types and levels of human interaction with automation, *IEEE Transactions on Systems, Man, and Cybernetics-Part A: Systems and Humans*, 30(3), pp 286-297.

[7] **Kirwan, B. & Ainsworth, L.K.** 1992. *A Guide to Task Analysis: The Task Analysis Working Group*, London: Taylor & Francis.

[8] **Valero-Gomez, A., De La Puente, P., & Hernando, M. 2011.** Impact of two adjustable-autonomy models on the scalability of single-human/multiple-robot teams for exploration missions, *Human Factor*, 53(6), pp 703-716.

[9] **Meyer, J., Wiczorek, R., & Günzler, T. 2014.** Measures of reliance and compliance in aided visual scanning, *Human Factors*, 56(5), pp 840-849.

[10] **Wickens, C.D., Hollands, J.G., Banbury, S., *et al*.** 1992. *Engineering Psychology and Human Performance*, 2rd Edition, New York: Harper Collins.

[11] **Reid, G.B. & Nygren, T.E. 1988.** The subjective workload assessment technique: a scaling procedure for measuring mental workload, *Advances in Psychology*, 52, pp 185-218.

[12] **Raby, M. & Wickens, C.D. 1994.** Strategic workload management and decision biases in aviation, *The International Journal of Aviation Psychology*, 4(3), pp 211-240.

[13] **Wierwille, W.W. & Casali, J.G. 1983.** A validated rating scale for global mental workload measurement applications, Proceedings of the Human Factors and Ergonomics Society Annual Meeting, October, pp 129-133.

[14] **Hart, S.G. and Staveland, L.E. 1988.** Development of NASA-TLX (task load index): results of empirical and theoretical research, *Advances in Psychology*, 52, pp 139-183.

[15] **Endsley, M.R. 1988.** Design and Evaluation for Situation Awareness Enhancement, Proceedings of the Human Factors and Ergonomics Society Annual Meeting, October, pp 97-101.

[16] **Endsley, M.R.** 2000. Direct measurement of situation awareness: validity and use of SAGAT, In: *Situation Awareness Analysis and Measurement*, Mahwah, NJ: Erlbaum, pp 73-112.

[17] **Durso, F.T., Dattel, A.R., Banbury, S., *et al*. 2004.** SPAM: the real-time assessment of SA, *A Cognitive Approach to Situation Awareness: Theory and Application*, 1, pp 137-154.

[18] **Wickens, C. 1999.** Cognitive factors in aviation, *Handbook of Applied Cognition*, pp 247-282.

[19] **O'Hara, J. & Higgins, J. 2010.** Human-system interfaces to automatic systems: review guidance and technical basis, *Human Factors of Advanced Reactors (NRC JCN Y-6529) BNL Tech Report No BNL91017-2010*, pp 1-23.

[20] **Kaber, D.B. & Endsley, M.R. 2004.** The effects of level of automation and adaptive automation on human performance, situation awareness and workload in a dynamic control task, *Theoretical Issues in Ergonomics Science*, 5(2), pp 113-153.

[21] **Lin, C.J., Yenn, T.-C., & Yang, C.-W. 2010.** Automation design in advanced control rooms of the modernized nuclear power plants, *Safety Science*, 48(1), pp 63-71.

[22] **Oxstrand, J., & Le Blanc, K. L. L.** 2014. *Effects of Levels of Automation for Advanced Small Modular Reactors: Impacts on Performance, Workload, and Situation Awareness*, Idaho Falls, ID: Idaho National Laboratory.

[23] **Onnasch, L., Wickens, C.D., Li, H., *et al*. 2013.** Human performance consequences of stages and levels of automation: an Integrated meta-analysis, *Human Factors*, 56(3), pp 476-488.

[24] **Li, H., Wickens, C.D., Sarter, N., *et al*. 2014.** Stages and levels of automation in support of space teleoperations, *Human Factors*, 56(6), pp 1050-1061.

[25] **Miller, C.A. & Parasuraman, R. 2007.** Designing for flexible interaction between humans and automation: delegation interfaces for supervisory control, *Human Factors*, 49(1), pp 57-75.

[26] **Sauer, J., Kao, C.-S., & Wastell, D. 2012.** A comparison of adaptive and adaptable automation under different levels of environmental stress, *Ergonomics*, 55(8), pp 840-853.

[27] **Haarmann, A., Boucsein, W., & Schaefer, F. 2009.** Combining electrodermal responses and cardiovascular measures for probing adaptive automation during simulated flight, *Applied Ergonomics*, 40(6), pp 1026-1040.

[28] **Bailey, N.R., Scerbo, M.W., Freeman, F.G., *et al*. 2006.** Comparison of a brain-based adaptive system and a manual adaptable system for invoking automation, *Human Factors*, 48(4), pp 693-709.

[29] **Kaber, D.B. & Prinzel, L.J.** 2006. *Adaptive and Adaptable Automation Design: A Critical Review of the Literature and Recommendations for Future Research.* Hanover, NASA.

[30] **Sauer, J., Nickel, P., & Wastell, D. 2013.** Designing automation for complex work environments under different levels of stress, *Applied Ergonomics*, 44(1), pp 119-127.

[31] **Kidwell, B., Calhoun, G.L., Ruff, H.A., *et al*. 2012.** Adaptable and Adaptive Automation for Supervisory Control of Multiple Autonomous Vehicles, Proceedings of the Human Factors and Ergonomics Society Annual Meeting, Boston, MA, pp 428-432.

[32] **Miller, C.A., Funk, H., Goldman, R., *et al*. 2005.** Implications of Adaptive vs. Adaptable UIs on Decision Making: Why "Automated Adaptiveness" is not Always the Right Answer, Proceedings of the 1st International Conference on Augmented Cognition, Las Vegas, NV, pp 22-27.

[33] **Lee, J.D. and Seppelt, B.D. 2009.** Human factors in automation design, In: *Springer Handbook of Automation*, Springer, pp 417-436.

[34] **Woods, D.D., Dekker, S., Cook, R., Johannesen, L., & Sarter, N.** 2010. *Behind Human Error*, 2rd Edition, Burlington, VT: Ashgate.

[35] **Wickens, C.D. & Dixon, S.R.** 2005. *Is There a Magic Number 7 (to the Minus 1)?: The Benefits of Imperfect Diagnostic Automation: A Synthesis of the Literature*, Savoy, IL: University of Illinois at Urbana-Champaign.

[36] **Dixon, S.R., Wickens, C.D., & McCarley, J.S. 2007.** On the independence of compliance and reliance: are automation false alarms worse than misses?, *Human Factors*, 49(4), pp 564-572.

[37] **Rovira, E. & Parasuraman, R. 2010.** Transitioning to future air traffic management: Effects of imperfect automation on controller attention and performance, *Human Factors*, 52(3), pp 411-425.

[38] **Hoff, K.A. & Bashir, M. 2015.** Trust in automation integrating empirical evidence on factors that influence trust, *Human Factors*, 57(3), pp 407-434.

[39] **Rice, S. 2009.** Examining single-and multiple-process theories of trust in automation, *The Journal of General Psychology*, 136(3), pp 303-322.

[40] **Chen, J.Y., Procci, K., Boyce, M., *et al.*** 2014. *Situation Awareness-based Agent Transparency*, Aberdeen Proving Ground, MD: U.S. Army Research Laboratory.

[41] **Christoffersen, K. & Woods, D.D. 2002.** How to make automated systems team players, *Advances in Human Performance and Cognitive Engineering Research*, 2, pp 1-12.

[42] **Skjerve, A.B.M. & Skraaning, G. 2004.** The quality of human-automation cooperation in human-system interface for nuclear power plants, *International Journal of Human-Computer Studies*, 61(5), pp 649-677.

[43] **Skraaning Jr, G., Eitrheim, M.H., & Lau, N. 2010.** Coping with Automation in Future Plants, The 7th American Nuclear Society International Topical Meeting on Nuclear Plant Instrumentation, Control and Human-Machine Interface Technologies (NPIC&HMIT), Las Vegas, Nevada.

[44] **Liu, P. and Li, Z. 2012.** Task complexity: a review and conceptualization framework, *International Journal of Industrial Ergonomics*, 42(6), pp 553-568.

[45] **Mosier, K.L., Fischer, U., Morrow, D., *et al.* 2013.** Automation, task, and context features impacts on pilots' judgments of human-automation interaction, *Journal of Cognitive Engineering and Decision Making*, 7(4), pp 377-399.

METAHEURISTICS FOR AN OPERATING ROOM SCHEDULING PROBLEM

Mohamed Amine ABDELJAOUED[1], Nour El Houda SAADANI[1] and Zied BAHROUN[2]
[1]Université de Tunis El Manar, Faculté des Sciences de Tunis, LIP2-LR99ES18, 2092 Tunis, Tunisia
amine.abdeljawed@yahoo.fr; Houda.Saadani@ensi.rnu.tn

[2]Industrial Engineering, ESM Graduate Program, College of Engineering, American University of Sharjah, PO Box 26666, Sharjah, UAE
zbahroun@aus.edu

ABSTRACT

In this paper, we present two metaheuristic algorithms to solve an operating room scheduling problem. The problem is NP-hard and is characterized by a set of identical operating rooms on which groups of operations each one assigned to a single surgeon, have to be scheduled. The objective is to minimize the overall completion time of the operations. The first proposed metaheuristic is a simulated annealing and the second is a tabu search algorithm. We analyze and compare their performances by conducting an experimental study. These tests showed that both methods give very satisfying results, with a 0.55% relative gap compared to the lower bound and an improvement by 3% compared to the existing heuristics. We also notice that the two methods provide complementary solutions.

Keywords: scheduling algorithms, metaheuristics, open-shop, operating rooms

1 INTRODUCTION

Optimization in the health services field has become an increasingly important issue over the last decades. Hospitals attempt to reduce the costs related to the use of their resources by improving their management techniques. The resources that can be concerned by the optimization process are numerous; we can mention the beds [9], the operating rooms [7], or the transport equipments [22] to name a few. And of course, the improvement and the aimed benefits must not compromise the quality of services brought to patients.

In this paper, we study an operating room scheduling problem. Operating theatres are among the most important sectors in hospitals, both in terms of expenses related to it and the proportion of patients needing a surgical procedure among those admitted to the hospital [10]. The problem we tackle here has the particularity that it can be perceived as a well known production work-shop scheduling problem: Considering the operating rooms as machines and the groups of operations as jobs, the problem can be assimilated to an open-shop with parallel machines issue since each surgeon has a set of operations that can be performed in any order. The paper is organized as follows: In section 2, we introduce this problem and its constraints. Section 3 is dedicated to a literature review summarizing the main works on the fields of operating rooms scheduling problems and scheduling under resource constraints. We then present in section 4 and section 5 the two metaheuristics, a simulated annealing and a tabu search, that we propose to solve the problem. We compare and analyze the performances of these two methods by conducting an experimental study in section 6. Finally, we report our concluding remarks and future prospects in section 7.

2 PROBLEM DESCRIPTION

In this section we introduce the problem we are dealing with in this paper. This problem is the same as the one presented in [20]. It is characterized by a set of n operations that have to be scheduled on m identical operating rooms. The operations are divided into groups; each group is assigned to a single surgeon that can pursue the surgeries in any operating room. Operations of a same group can be performed in any order and waiting times between two of them are allowed. On the contrary, the preemption is banned: an operation cannot be interrupted once it has started. Of course, a surgeon, as well as an operating room, can perform at most one operation at a time. The objective is to find the best scheduling sequence of the operations on the rooms in order to minimize the makespan (i.e. the ending-date of the last completed operation). This issue is a part of the parallel machine scheduling problems under resources constraints (i.e. the surgeons are the resources needed by the operations to be performed) and is NP-hard for a number of operating rooms equal or exceeding 2 [3]. Using the three-field notation introduced by Graham et al [18], the problem is denoted *P|Res.11|Cmax*. *P* means that the machines are identical, *Res.11* that each operation requires at most one resource to be performed and that each resource type exists in only one copy, and *Cmax* refers to the objective function which is the minimization of the makespan.

If the number of operations by surgeons does not exceed the number of operating rooms, and due to the absence of defined order between the operations of a same surgeon, the problem can also be seen as an open-shop with parallel machines issue. In an open-shop problem, n sets of operations have to be scheduled on m machines. Operations of a same set can be processed in any order, but two of them cannot be processed at the same time. A machine can perform only an operation at a time. For the mathematical formulation of the problem, please refer to [1]. The authors provided a model inspired from the 2-dimensional strip-packing problems formulation and showed that it outperforms the open-shop based models, even if the sizes of the instances solved are still rather small, which is a natural thing for an NP-hard problem.

3 LITTERATURE REVIEW

Over the last twenty years, a large literature body on the operating rooms planning and scheduling problems has been developed. In [7], the authors provided a survey on this field where they classified the major works, depending on the performance measures considered and the constraints dealt with. These constraints can be very different, we can mention for instance the uncertainty parameters on patients arrival [4] or operations durations [24], or the patients nature: 'elective' (i.e. patients whose operations can be planned in advance) [6] or 'non-elective' (i.e. emergency cases taken into account) [23]. In most of the operating rooms scheduling problems, the objectives considered are financial. They usually consist in maximizing the rooms' utilization rate (i.e. the working time during the regular rooms' opening hours) [11] and minimizing the overtime work [15]. Minimizing the makespan [14] and the waiting times [2] are also commonly used objectives.

As mentioned, Guinet and Saadani [20] study the same problem we address in this work. They provide several two-step approaches and compare their performances. The first level of these heuristics arranges the operations according to certain sorting rules, and the second one assigns them to the operating rooms by respecting the precedence constraints between operations of a same group. Among the arrangement rules used for the first level we can mention SPT *(shortest processing time)* and LPT *(longest processing time)* and for the assignation level, the authors opted for the heuristics of Herrmann et al. [21] and Bräsel et al. [5], among others. In [1], the authors also tackle this same problem and present dichotomic algorithm that uses a lower and upper bounds and a feasibility test to solve it. Fei el al. [14] work on a planning and scheduling problem in a 2-rooms endoscopy center and provide a two-level approach; the first one consists in a mathematical model that determines the weekly plan by assigning a day and an operating room to each operation with the objective of maximizing the rooms utilization rate, and the second level regulates the scheduling sequence in each room using the polynomial algorithm of Gonzalez and Sahni [17] for the 2-machines open-shop problem. Some authors like Guinet and Chaabane [19] take account of the availability of recovery beds in their operating theatre scheduling process. They thus modeled their problem as a 2-stage hybrid flow-shop issue.

As mentioned, the problem we study is part of the resource-constrained scheduling issue. Considering this aspect, we also found interesting articles. Parallel machine scheduling is among the most studied subjects in the literature and in many works, the operations to schedule require the use of an additional resource, such as tools or workers. Edis et al. [12] presents a review on the parallel machine scheduling problems under resource constraints. Papers on the field are numerous and dealt with a wide variety of constraints, such as the machine type (parallel, uniform or unrelated), the number of resources required or specificities about the operations' durations [8]. Blazewicz et al. [3] provide a complexity classification of these parallel machine scheduling problems under resource constraints. In our literature review, we were also interested by the open-shop problems which are, as already mentioned, also characterized by the fact that the operations of a same job can be processed in any order. Indeed, the open-shop with parallel machines can be a particular case of the problem we are studying, when the number of operations by surgeon does not exceed the number of operating rooms. However, the published works on this issue are considerably less abundant than those related to the general open-shop. Among them, we can mention Peterkofsky and Daganzo [25]: the authors modeled a scheduling problem consisting in a set of cargo ships which have loading and unloading tasks to perform using cranes as an open-shop with parallel machines, and provided a branch-and-bound method to solve it.

4 SIMULATED ANNEALING ALGORITHM

The simulated annealing is a method inspired from a process used in metallurgy which consists in alternating heating and cooling cycles. Starting from an initial solution, it visits at every iteration a solution from a pre-defined neighborhood. If it is better than the previous one, it is accepted and the algorithm continues its process from it. Otherwise, the new solution is accepted or rejected according to a random rule. This is useful to avoid local optimums. More details about this method can be found in [13]. Data of this algorithm are the temperature T, the parameter $Kmax$ which controls the number of iterations at each temperature, and the cooling parameter r. As initial solution, we choose the one provided by the heuristic LPT-FAM (longest processing time - first available machine) used in [20] and classified as one of the most efficient. It sorts the operations according to the decreasing order of their durations and then assigns them, one by one, to the first available room, of course by respecting the precedence constraints between the operations of a same surgeon. The simulated annealing moves from one solution to another by randomly exchanging two operations in the assignation order. The operations, taken in this new order, are then assigned to the rooms as follows: If an operation of a given surgeon can be performed at the earliest at time t (in the case when the last assigned operation of the same surgeon ends at time t), the algorithm chooses the first room to be available at time $t_1 \geq t$. The acceptance probability of a worse solution is verified if the inequality $e^{-(\frac{S1-SM}{T})} > y$ is verified, with S1 referring to the solution whose acceptance is tested, SM to the best solution found so far and y a number randomly generated in the interval [0, 1]. Since the difference between S1 and SM gets greater as the solving process advances and the algorithm regularly improves SM, worse solutions are less and less accepted at the end. The solving continues until a stopping criterion eps is met. Below, we give the main steps of this simulated annealing algorithm:

```
Input: L[n]: List of operations; m: number of machines;
I-    Function
  Assign(List L, m): this procedure assigns each operation of the list L, to the
  first room to be available at a time t exceeding or equal to the ready-time of
  this operation.
```

```
II-   Initialization
   Data = (Temperature: T, iteration index: Kmax, cooling parameter: r, stopping
   criterion: eps)
   List O := classify the n operations according to LPT.
   Integer:
     Initial solution S := solution of the heuristic LPT-FAM
     Best solution SM := S;
     Iteration i := 0;
     Temperature step KT := 0;

III-  Loop
  1- i := i+1; KT := KT+1;
  2- O1 := order obtained by randomly exchanging two operations in O.
     S1 := Assign(O1, m).
  3- If(S1<SM)
   SM := S1; O := O1;
     Else
   Randomly generate y in [0, 1];
         If(e^(-(S1-SM)/T) > y) O := O1;
  4- If(KT ≤ Kmax) return to step 1.
  5- T := T*r;
  6- If (T>eps): KT := 0 and return to step 1;
     Else: return(SM); stop the algorithm;
```

5 TABU SEARCH METHOD

The second metaheuristic that we propose is a tabu search method. Starting from an initial solution, this procedure examines a set of solutions from a pre-defined neighborhood and moves to the best one, even if this solution is worse than the current one, to avoid local optimums. More details about the tabu search technique can be found in [16]. As initial solution, we used the LPT-FAM heuristic again. We defined three neighborhoods for a solution. Each neighborhood selects two operations to exchange in the assignation order (the initial assignation order is LPT). The first one is obtained by exchanging two non-successive operations of the same surgeon in the assignation order (the initial assignation order is LPT). We choose only the non-successive operations here because exchanging two successive operations of a same surgeon would not, in most cases, bring a change; they would be scheduled one after the other on a same room. The second (respectively the third) neighborhood is obtained by exchanging two successive (respectively non-successive) operations of different surgeons in this assignation order. In order to prevent cycling, a tabu list is defined to forbid the inverse moves during a certain number of iterations. The details of the algorithm's steps are given below. The tabu search algorithm stops if it performs Nmax successive iterations without improving the current solution.

```
Input: L[n]: List of operations; m: number of machines; Nmax: maximal number of
 iteration;
I-  Function
    Assign(List, m);
    Min(n1, n2, n3): return the minimal number between n1, n2 and n3;
II- Initialization
    List:
      Initial order O := classify the n operations according to LPT.
      Best order OM := O;
    Integer:
      Initial solution S := solution of the LPT-FAM heuristic
      Best solution SM := S;
```

```
    Iteration i := 0;
III- Loop
 1- i := i+1;
 2- Generate the neighborhood:
  - List V1: Randomly exchange two operations of the same Surgeon in the list O;
  - List V2: Randomly exchange two successive operations of different Surgeons in
    the list O;
  - List V3: Randomly exchange two non-successive operations of different Surgeons
    in the list O;
 3- Compute the solutions of the three neighborhood:
  - Solution SV1:= Assign(V1, m); Solution SV2:= Assign(V2, m); Solution SV3:=
    Assign(V3, m);
 4- S:= Min(SV1, SV2, SV3);
 O:= The list of the neighborhood that gave the minimal solution;
 5- If(S<SM) SM:=S;
 6- If(i>Nmax) return (SM); Stop the algorithm;
    Else return to step 1;
```

To ensure that the simulated annealing visits different solution areas and does not limit its search process to these three neighborhoods, we introduce two procedures that are periodically used; after a certain number of iterations without improving the solution. The first procedure is an intensification algorithm and it behaves as described below, and the second one is a diversification algorithm that allows switching solutions areas. Its process is simple: It exchanges the positions of three randomly chosen operations in the assignation order.

```
Intensification procedure:
Step 1: Choose, at random, a surgeon and remove its most long operation from the
        current assignation order.
Step 2: Consider the other operations in their assignation order, and assign each
        operation, to the first room to be available at a time t exceeding or equal
        to its ready-time.
Step 3: Choose the less occupied room and insert the removed operation in all
        positions (shift the other operations of the same surgeon, if necessary).
        Keep the position that gives the best makespan.
```

6 EXPERIMENTAL STUDY

6.1 Benchmark

In this section we will analyze and compare the performances of these two metaheuristics by conducting a battery of tests. We generated 10 instance sizes. Their details are given below:

- 5 operating rooms: with a number of surgeons included in {10, 15, 20, 25, 30}.
- 10 operating rooms: with a number of surgeons included in {30, 40, 45}.
- 15 operating rooms: with a number of surgeons included in {50, 60}.

The number of operation for each surgeon is generated in the interval [2, 4] using a uniform distribution. The durations of these operations are also generated with a uniform distribution within the interval [60, 180]. 10 tests are carried out for each instance size, so in total 100 tests have been performed. We restricted ourselves to these instance sizes since it is unusual to find bigger ones in operating room scheduling problems. Before the tests, the parameters of the simulated annealing have to be calibrated. So for each instance size, we run some preliminary tests with different parameters' values, and we kept those with which the algorithm returns close solutions when we execute many times the same test. From the (5 rooms, 10 surgeons) to the (10 rooms, 30 surgeons) instances, we thus fixed the temperature T, the iteration index $Kmax$ and the cooling parameter r of the simulated annealing respectively to 100, 20 and 0.8. We needed to raise T to 500 for the (10 rooms, 40 surgeons) instances and then to 800 for (15 rooms, 50 surgeons). We also increased $Kmax$ to 40 and then to 60 for the same instances. For the tabu search algorithm, we initially fixed the stopping criterion Nmax (i.e. the allowed number of successive iterations without improvement of the solution) to 200 and then we raised it to 250 from the (10 rooms, 45 surgeons) instances. For each test, we compute the relative gap between the solution found and the known lower bound of the problem, provided by Guinet and Saadani [20]. Out of these 100 tests, we also note how many times a metaheuristic gives a better solution than the other. As the tabu search visits 3 solutions at each iteration and the simulated annealing only one, a comparison of the total number of iterations each algorithm requires to perform to reach its final solution would not be fair. We choose instead to compare the total number of solutions each algorithm visits for each test.

6.2 Results of the tests

The tests carried out showed that both metaheuristics are high-performing. For all the instances, they return the solution instantly. In the table, 'OR' refers to operating rooms, 'Surg' to surgeons, 'SA' to the simulated annealing and 'TS' to the tabu search method. From Table 1 which reports for each instance size the relative gap between the solution provided by each algorithm and the lower bound, we can see that out from these 100 tests, both methods give exactly the same average gap: 0.55%. For each instance size, the relative gaps obtained are very close, even if the tabu search were better for 5 sizes from 10; compared with 3 instance sizes where the simulated annealing was better. Table 1 also compares the quality of the solutions and analyzes the number of iterations required by each method to reach them. As we can see, the simulated annealing provided a better solution in 49 tests, compared with 41 tests for the tabu search. In 10 tests both metaheuristics were equal. For the 49 tests where the simulated annealing was better, the average improvement obtained compared to the tabu search is about 0.27%. And for the 41 tests where the tabu search gave better solutions, the average improvement is about 0.30%. By analyzing these tests size by size, we notice that the simulated annealing gives a better total number of lower solutions for 5 instance sizes out of 10, while the tabu search was better only for 2 instance sizes (they were equal on the three remaining sizes).The only criterion in which a metaheuristic clearly makes the difference is the number of solutions each one visits before reaching its best one. The simulated annealing is clearly better here, reaching its solution in a lower number of 'visits' in 63 cases out of 100. This comparison makes more sense for the 10 tests where the two methods gave the same solution: the simulated annealing reaches it faster in 7 cases. Moreover, in the 49 tests where the simulated annealing gives a better solution, 23 times this solution were reached in a lower number of 'visits'; compared with 11 times from 41 for the tabu search. From this point, the simulated annealing is thus more satisfying, especially that it is easier to implement. From this experimental, an important remark can be made: Out from 100 tests, even if both metaheuristics are high-performing, they returned the same solution only 10 times. This would mean that the two algorithms do not explore the same solution regions and could be complementary. As their execution is instantly, we can run both of them and keep the best solution. This can be easily done for the operating rooms scheduling problem we are studying, as the realistic instances (even the bigger ones) for this problem rarely exceed those we tested.

Table 1: Comparing the performances of the simulated annealing and tabu search methods

Instance size	Relative gap		Number of better solutions		Cases of lower iterations	
	SA	TS	SA	TS	SA	TS
5 OR, 10 Surg	1.23%	1.23%	3	5	8	2
5 OR, 15 Surg	0.38%	0.37%	5	4	6	3

5 OR, 20 Surg	0.33%	0.20%	2	6	6	4
5 OR, 25 Surg	0.20%	0.16%	5	5	8	2
5 OR, 30 Surg	0.14%	0.14%	6	3	6	4
10 OR, 30 Surg	0.95%	1.05%	5	5	5	5
10 OR, 40 Surg	0.53%	0.57%	4	3	7	3
10 OR, 45 Surg	0.32%	0.44%	8	1	5	5
15 OR, 50 Surg	0.83%	0.81%	5	5	4	6
15 OR, 60 Surg	0.64%	0.61%	6	4	8	2
Average | Total	**0.55%**	**0.55%**	**49**	**41**	**63**	**36**

7 CONCLUSION

In this paper, we studied an operating room scheduling problem consisting in a set of operations that have to be scheduled on identical operating rooms. The operations are divided into groups and each group is assigned to a single surgeon. This problem is part of the parallel machines scheduling under resource constraints issue and is NP-hard for a number of rooms m equal or exceeding 2. We presented two metaheuristics to solve this problem: One is a simulated annealing algorithm and the second a tabu search method. The experimental study we conducted show that both methods give satisfying results. From the 100 tests carried out, an average gap of 0.55% compared to the lower bound emerges for both of them. We also noticed that the two metaheuristics can be complementary, as despite this very low gap, they returned the same solution only 10 times. Since their execution is nearly instantly, and that realistic instances for operating room scheduling are not that big, we can unite them and keep the best solution. Considering the operating rooms cost per working hour, which exceeds 700$ [26], such an improvement in the working time can constitute a financial gain for hospitals. As future work, we can think about a possible improvement of the tabu search method, maybe by integrating an aspiration function for the solutions of the neighborhood. We can also think about the addition of new constraints to the problem like the consideration of emergency cases or a possible uncertainty about the durations of the operations.

8 REFERENCES

[1] **Abdeljaouad M.A., Bahroun Z., Saadani N.E.H., 2014.** A dichotomic algorithm for an operating room scheduling problem, *in proceeding of international conference on control, decision and information technologies : CoDIT*, 134-139.

[2] **Arenas, M., Bilbao, A., Caballero, R., Gómez, T., Rodríguez, M. V., & Ruiz, F. 2002.** Analysis via goal programming of the minimum achievable stay in surgical waiting lists. *Journal of the Operational Research Society*, 53, 387-396.

[3] **Blazewicz, J., Lenstra, J.K., Kan A.H.G.R, 1983.** Scheduling subject to resource constraints: classification and complexity, *Discret applied mathematics* 5, 11-24

[4] **Bowers J., Mould G., 2004.** Managing uncertainty in orthopaedic trauma theatres, *European Journal of Operational Research* 154, 599-608.

[5] **Bräsel, H., Herms, A., Mörig, M., Tautenhahn, T., Tusch, J., & Werner, F. 2008.** Heuristic constructive algorithms for open shop scheduling to minimize mean flow time. *European Journal of Operational Research*, 189, 856-870.

[6] **Calichman M.V., 2005.** Creating an optimal operating room schedule, *AORN Journal* 81, 580-588.

[7] **Cardoen B., Demeulemeester E., Beliën J. 2010.** Operating room planning and scheduling: A literature review, *European Journal of Operational Research* 201, 921-932

[8] **Daniels, R.L., Hoopes, B.J., Mazzola, J.B., 1996.** Scheduling parallel manufacturing cells with resource flexibility. *Management Science* 42 (9), 1260-1276.

[9] **Demeester P., Souffriau W., Causmaecker D., Berghe G.V., 2009.** A hybrid tabu search algorithm for automatically assigning patients to beds, *Artificial Intelligence in Medecine: Article in Press.*

[10] **Denton, B., Viapiano, J., Vogl, A. 2007.** Optimization of surgery sequencing and scheduling decisions under uncertainty. *Health Care Management Science*, 10, 13-24.

[11] **Dexter F., 2003.** Operating room utilization: Information management systems, *Current Opinion in Anaesthesiology* 16, 619-622.

[12] **Edis E.B., Oguz C., Ozkarahan I., 2013.** Parallel machine scheduling with additional resources: Notation, classification, models and solution methods, *European Journal of Operational Research*, 230, 449-463

[13] **Eglese R.W., 1990.** Simulated annealing: A tool for operational research. *European Journal of Operational Research*, 46, 271-281.

[14] **Fei, H., Combes, C., Meskens, N., Chu, C. 2006.** Endoscopies scheduling problem: A case study, *Information Control Problems in Manufacturing*, 12, 635-640.

[15] **Fei H., Meskens N., Chu C., 2006.** An operating theatre planning and scheduling problem in the case of a block scheduling strategy, *in: Proceedings of the International Conference on Service Systems and Service Management.*

[16] **Glover, F. 1989.** Tabu Search - Part I, *ORSA Journal on Computing* 1, 190-206.

[17] **Gonzales T., Sahni S., 1976** Open shop scheduling to minimize finish time, *Journal of the Association for Computing Machinery* 23 (4), 665-679.

[18] **Graham, R.L., Lawler, E.L, Lenstra, J.K, & Kan A.H.G. R, 1979.** Optimization and approximation in deterministic sequencing and scheduling: a survey, *Annals of Discrete Mathematics*, 5, 287-326.

[19] **Guinet, A., Chaabane, S. 2003.** Operating theatre planning, *International Journal of Production Economics*, 85, 69-81

[20] **Guinet A, Saadani N.E.H, 2010.** Nomadisme chirurgical et ordonnancement de blocs opératoires, *in proceeding of Conférence Francophone en Gestion et Ingénierie des Systèmes Hospitaliers : GISEH.*

[21] **Herrmann, J., Proth, J., & Sauer, N. 1997.** Heuristics for unrelated machine scheduling with precedence constraints, *European Journal of Operational Research*, 102, 528-537.

[22] **Jacobson S.H., Hall S.N., Swisher J.R. 2006.** Patient Flow: Reducing delay in healthcare delivery, *International series in Operations Research and management science, Chapter: discrete-event simulation of health care systems*, 211-252.

[23] **Krempels K.H., Panchenko A., 2006.** An approach for automated surgery scheduling in *Proceeding of the sixth International Conference on the Practice and Theory of Automated Timetabling.*

[24] **Persson M., Persson J.A., 2007.** *Analyzing management policies for operating room planning using simulation*, Working Paper, Blekinge Institute of Technology, Sweden.

[25] **Peterkofsky, R. I., Daganzo, C. F. 1990.** A branch and bound solution method for the crane scheduling problem. *Transportation Research Part B: Methodological, 24*, 159-172.

[26] **Raft, J., Millet, F., & Meistelman, C. 2014.** Exemple de calcul du coût de fonctionnement d'un bloc opératoire avec la salle de surveillance post interventionnelle. *Annales Françaises d'Anesthésie et de Réanimation*, 33S, 360-364.

A FRACTAL-BASED COLLABORATIVE MODEL FOR DISASTER EMERGENCY ORGANIZATION

Annan li, Xiuquan Deng, Qiuhong Zhao

School of Economics and Management,
Beihang University, Beijing, China
qhzhao@buaa.edu.cn

ABSTRACT

The high degrees of uncertainty of the emergency environmental change make it difficult for the Emergency Organization (EO) to response quickly and efficiently. Coordination among the Emergency Organizations under the dynamic environments is getting more and more complex in the emergency management. The purpose of this paper is to propose a collaborative model including Local Emergency Command Organizations (LECOs), Rear Emergency Command Organizations (RECOs) and Rear Emergency Ability Organizations (REAOs), which supplies the emergency capability. And also this paper puts forward a Fractal Emergency Organization (FEO) framework based on the coordinate relationship. A fractal is consisted by some participants and its sub-fractals, where participants as suppliers form Demand-Supply (DS) relationship with sub-fractals as demands. Then, a fractal unit mathematical goal model is put forward with a parameter 'Collaborative Level' (CL). CL influences coordination cost in fractal and changes the task planning of fractal. Lastly, this paper uses a numerical example to illustrate the coordinate process of the FEO.

Keywords: Fractal theory; Emergency organization; Collaborative model

1 INTRODUCTION

Along with the uncertainty and complexity of social increases, unconventional emergencies such as the 9/11 terrorist attacks in USA, Fukishima nuclear power plant meltdown in Japan, Wenchuan and Yushu earthquake in China, MH370 flight incident in Malaysia, and recently 8.12 Tianjin Port explosion of China happened occasionally, with the characteristic of pressed for time, coordination difficulties, management process complex. The risk cognition capability and the routine response method of the conventional emergency organization cannot meet the increasing requirements of cross-time, cross-regional and cross-function response process. According to several large-scale emergency responses process, emergency organization shall dynamic integrate associated tasks and resources under the uncertainty conditions. It would be difficult to the mass organization management in dynamic environment (Kapucu [1]; Kapucu [2]).

To adapt to the dynamic and uncertain emergency environments, the cooperation emergency organizations are facing the following requirements, 1) it is necessary to shift from a centralized organizational structure to a more flexible decentralized organizational structure; 2) system must have a transparent mechanism promoting constructive cooperation between distributed emergency entities for their easy integration; 3) emergency organization should be flexible to adapt to endogenous and exogenous disturbances.

There are some advantageous features for emergency organization management adopting the fractal concept. Firstly, we make the complicated emergency organization system become understanding easily. We can use the hierarchical layers of fractals build by the similar fractal units instead of the traditional echelons structure. Secondly, emergency organization system has the characteristics of fractal including self-similarity, self-organization, goal-formation, and so on. The self-organizing characteristic, including self-optimization and dynamic restructuring process, optimizing the emergency goal and structure of fractals (Ryu [3]; Shin [4]; Shin [5]). Thirdly, the local optimization reduces the decision making time and encourages the rapid responses of the emergency organization. Since the variability of emergency environment make it difficult to find global optimum in emergency decision, the integration of the local optimized decision may give faster responsiveness to emergency events, and increase the emergency efficiency. Lastly, it makes the flexible management of EO enhancing because the parent fractals may increase the coordinate between child fractals (Mun [6]; Ryu [7]).

A fractal-based framework for emergency organization is proposed in this paper, in order to improve the coordination efficiency of organization members and quickly response in dealing with dynamically changing of emergency environment. Each set composed of participants have Demand and Supply, (DS) relationship in the fractal emergency organization is defined as a self-similar structure, named a Fractal Emergency Organizational Unit, (FEOU). It has the prominent features of openness, configurability and scalability.

This paper is composed as follows. In Section 2, we introduced the emergency organization research and the fractal theory. The basic concept and characteristics of a fractal are described in Section 3. And, in this section, a Fractal Emergency Organization, (FEO), framework formed by emergency task, (ET), Local Emergency Command Organizations, (LECOs), Rear Emergency Command Organizations, (RECOs) and Rear Emergency Ability Organizations, (REAOs), based on fractal collaborative mechanisms. In Section 4, a fractal mathematical goal model is established and Section 5 shows a case study. Finally, Section 6 concludes this paper and introduces further research.

2 EMERGENCY ORGANIZATION AND FRACTAL THEORY

With the change of time and environment, emergency response process is dynamic, and often involves multiple participants. 'Cooperation' has become a recognized response measures. The emergency organization constantly develops and evolves under dynamic demands, forms many organizations cooperative mode in the dynamic environment (Perry [8]; Kapucu [1]; Paulheim [9]; Kapucu [2]), such as centralized mode, authority mode, and collaborative mode, etc.

Koestler (1967) holds that the most of the complex systems are organized hierarchically, which the control flow is typically top-down and the feedback information is bottom-up. The Decision Makings, (DMs) at the upper level define tasks and coordinate lower level units, whereas DMs at the lowest level execute tasks (Yu [10]). There are many merits of a centralized hierarchical emergency organization structure, such as it provides centralized decision-making, which refers to that a sub unit should have one and only one superior to whom it is directly responsible. Because a special emergency organization needs fast decision-making in dynamic environment, some emergency organizations adopt the fixed vertically integrated hierarchy structure. But, studies indicate that the centralized structure has the following drawbacks: 1) there is limited ability to reconfiguration on unconventional situations; 2) the strict hierarchy limits the fast response to the occurrence of mutations of emergency situation; 3) the propagation of bottom-up information is hampered by a multilevel bureaucratic structure; 4) the construct of the structure which integrate all kinds of resources is difficult (Wang [11]).

Another emergency organization mode is the collaborative cooperation structure integrated by many of Emergency Command Units, (ECUs) and Emergency Ability Units, (EAUs) lived in different types and levels, which has the attributes of common decision-making, self-organization, and lateral accountability. Each Emergency Unit (EU) has equal rights of access to resources and the coordination among EUs is realized by using network electronic collaboration and coordination emergency platform. The collaborative model has the inherent capability of self-configuration, flexibility, fault tolerance, and reduced complexity. However, the collaborative model have the following drawbacks: 1) the decision process is delayed by the low efficiency negotiation mechanism; 2) achieve the quickly decision of individual EUs under the environment mutation is difficult (Wang [11]; Tang [12]).

In 1993, Germany Professor Warnecke presented the concept of fractal enterprise. He build dynamic corporate structure by combined with the characteristics of fractal theory of self-organization and adaptive, hoping to improve the dynamic adaptability to environmental change, which deal with various perturbation. And fractal enterprise should have 3 basic features: of self-similarity, self-optimization and self-organization. Subsequently, many scholars have given fractal theory method to different systems. Ryu ([13], [3], [7]) think that the fractal is a set of self-similar units; its objective can be achieved through coordination, consultation and cooperation. Each self-similar unit has its own objectives, and be able to be independent and autonomous decision-making, they were able to reconstruct to form a fractal system more efficient. Therefore, they put in fractal manufacturing systems (Fractal Manufacturing System, FMS) concept, and research based on fractal system goals conflict resolution mechanisms, goal coordination mechanism, as well as the dynamic reconfiguration process under the environment changes. Chen [14] and Ni [15] based on dynamic angle, applied fractal methodology to supply chain, proposed a reconfiguration Fractal Supply Chain, FSC). These suggest that the fractal theory is excellent methods to solve complex systems problems in dynamic environment. In recent years, Ryu [16] was also study the fractal supply chain based on trust and coordination, which is designed for supply chain coordination in dynamic environment, building mathematical models and simulation method for the appropriate conclusions.

Fractal emergency organization structure overcomes the drawbacks of both centralized hierarchy and collaborative structure and meets the key requirements of the emergency organization in dynamic and uncertain environments. 'Fractal' is derived from mathematical theory, which is a rough or fragmented geometric shape that can be split into parts, each of which is a reduced-size copy of the whole, a property called self-similarity (Mandelbrot [17]). The fractal organization structure is a hierarchy structure of self-similar and self-organization fractal organization units with the following advantages: 1) it make the complex system become simple because that the 'fractal' can represent an entire organization at the highest level or an elementary unit at the bottom level (Ryu [7]; Ryu [13]); 2) it is highly resilient to the internal and external disturbances of environments, attribute to each 'fractal' unit's attribute of self-optimize, self-organization can quickly response the mutation(Ryu [3]) .

3 FRACTAL-BASED COLLABORATIVE EMERGENCY ORGANIZATION FRAMEWORK

In some fractal organization structure, each unit becomes a fractal. However, in this paper, each interaction between participants becomes a fractal. Sets composed of participants have DS relationships are representing as a fractal. Participants who act as demand belong to the sub-fractals, and participants who act as suppliers belong to the fractal.

In order to build a goal model of FEO, some feature of EO must be considered.

Cross-region: The feature of cross-region makes the participants in the FEO classified into local EUs and cross-region EUs. This means that the ability of different participants to perform the same task start time is different. Emergency time is a key element of emergency response; therefore, delay time should be distinguished according to the different emergency participants.

Capability loss should be considered in the plan, because the environment of uncertainty and derivative disasters in the process of emergency often leads to loss of hundreds of emergency ability.

Collaborative level: CL of participants should be considered during the task planning. Because the quickly and effective emergency response emphasize the reliability of assessment to participants' EAUs and emergency cost.

To define fractal structure, we assume that various types of emergency organizational units can be specified in accordance with the abstraction level. When the event occurs, normal emergency organization change into non-normal emergency organization, non-normal organizations can be divided into Local Emergency Organizations, (LEO) and Rear Emergency Organizations, (REO). REO can be subdivided into the Rear Emergency Command Organizations, (RECO) and Rear Emergency Ability Organizations, (REAO). See in Fig. 1.

In order to complete emergency response tasks, all kinds of emergency ability were drawn from the normal emergency organization, build the LEO. For example, in Fig.1, LECO includes the participants of $JEO_1 - JEO_3$. $JEO_1 - JEO_3$ are the joint emergency organizations of cross-regions. On BECO level, $JEO_1 - JEO_3$ and local participants LEO_1, LEO_2 composed the module 1 of BECO; $JEO_1, JEO_2, LEO_3, LEO_4$ composed module 2 of BECO; and module 3 of BECO is composed only by LEO_5. $LEO_1 - LEO_5$ are rear local emergency organizations. On REAO level, $EA_1 - EA_8$ are the participants and composed $EAP_1 - EAP_5$. And the EAs' demand and supply flows throughout the entire structure. The structure of EO sees in Fig.1.

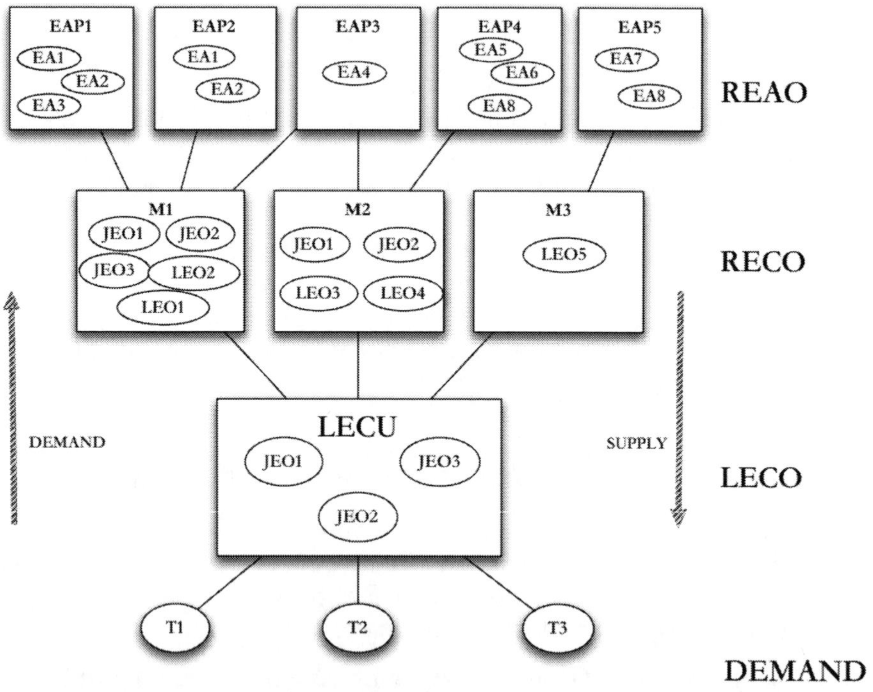

T1-T3: ET Demands JEO1-JEO3: Joint Emergency Organization
LEO1-LEO5: Local Emergency Organization
EA1-EA8: emergency ability unit

Fig. 1 EO structure

There are several fractals including f_1 to f_3. f_1 represents a fractal with DS relationship between participants-EAPs as a supplier and f_{B1} as a demander, which supply EAUs to BECUs. f_{B1} is a fractal formed by f_{L1} and participants of BECUs, in which f_{L1} act as demand and BECUs act as supplier. f_{L1} is composed of LECUs and ETs demand. According to the self-similarly of the fractal system, other fractal is form like this. See in Fig.2.

Each fractal has its own goal, and in this paper we transfer the goal for ETs and the task constrains. Through optimization of the goal model, task plans for each participant are optimized. Firstly, f_{L1} optimizes the task plans for the local emergency task based on emergency demand. According to the optimization results, f_{B1} generate task plans for the BECUs. And the task plans for EAUs are generated by f_1 optimization.

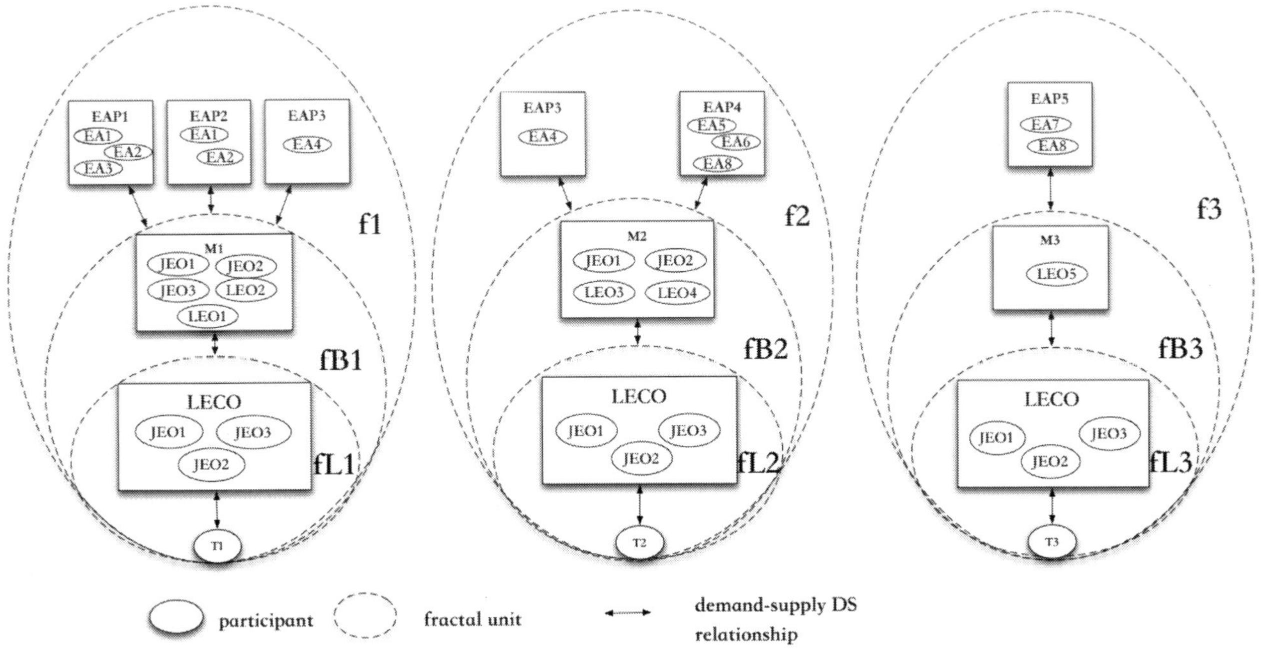

Fig.2 FEO structure

4 NUMERICAL GOAL MODELING OF A FRACTAL

In this paper, we give a parameter named 'collaborative level', which refers to the reliability of emergency ability assessment and the stipulated response time. In the FEO framework, CL will affect the accuracy of ET, processing time and task delay time, etc. In this paper, the CL is used for optimizing the task plans of participants in the fractal. The CL affects the task allocation and the quantities for DS relationship. The higher CL of a participant induces the participant to supply more EC and reduce the coordination workload. In this paper, the CL can be determined according to past information, and the range of it is 0 to 1.

In the model, firstly, the sub-fractals optimize task plans according the demand-supply relationship. Then, in the model of fractal, the optimized task plan make by sub-fractals are used as the demand. In this model, as an example, only one fractal is described. Other fractals' goal model has the same form, because of the characteristic of self-similarity.

Symbols:

g_f: Goal of fractal f;

M: Set of the emergency participants (EPs) in fractal f (m=1,2, ...,M);

I: Set of the ET (i=1,2,...,I);

K: Set of the Emergency Ability Units (EAUs) (k=1,2,...,K);

CL_{km}: Collaborative level of EP_m for EAU_k;

α : Coefficient for cost penalties;

p_{ik}: Unit profit of EAU_k for ET_i;

c_{ik}: Lost cost of EAU_k for ET_i;

ct_{ikm}: Time delay cost for unit EAU_k from EP_m to ET_i;

cp_{ikm}: Processing time cost for unit EAU_k from EP_m to ET_i;

R_{ik} : Demand of ET_i for EAU_k;

cap_{km}: Capacity of EP_m for EAU_k;

cap_m: Total capacity of EP_m;

Decision variables

x_{ikm}: Quantity of EAU_k supplied by participant m for ET_i;

The goal of a fractal is the maximization of emergency profit.

$$\max g_f = \sum_{i=1}^{I}\sum_{k=1}^{K}(p_{ik}\cdot R_{ik}-c_{ik})-\sum_{m=1}^{M}\sum_{k=1}^{K}\left(1+\frac{\alpha}{CL_{km}}\right)\cdot\sum_{i=1}^{I}x_{ikm}\cdot ct_{ikm}$$
$$-\sum_{m=1}^{M}\sum_{k=1}^{K}\left(1+\frac{\alpha}{CL_{km}}\right)\sum_{i=1}^{I}x_{ikm}\cdot cp_{ikm} \tag{1}$$

Subject to

$$\sum_{m=1}^{M}x_{ikm}\geq R_{ik},\forall k \text{ and i} \tag{2}$$

$$\sum_{m=1}^{M}(x_{ikm}\cdot CL_{km})\geq R_{ik},\forall k,\forall i \tag{3}$$

$$\sum_{i=1}^{I}x_{ikm}\leq cap_{km},\ \forall \text{i} \tag{4}$$

$$\sum_{i=1}^{I}\sum_{k=1}^{K}x_{ikm}\leq cap_m,\ \forall m \tag{5}$$

$$x_{ikm}\geq 0,\ \forall \text{i},\ \forall \text{k and } \forall \text{m} \tag{6}$$

Equation 1 is the objective function of the goal model. The main objective is maximizing the expected total profit of the fractal. The function includes profit of task disposal to control emergency event, EC loss cost, EC delay time cost and EC processing time cost. A coefficient of cost penalty is considered in the function according to the participants' CL. Equation 2,3 implies that the supply quantity of EA should be larger than the demand. Equation 4 and 5 are the constraints of the supply capacity. Equation 6 is the constraint for positive of x_{ikm}.

5 CASE STUDY

A FEO structure and each participant of fractal is listed in Table 1, we use the data listed in Table 2 to research task plan of f_{L1}.

Table 1 Fractal structure

fractal	Sub-fractal	participants	fractal	Sub-fractal	participants
f_1	f_{B1}	EAP_1-EAP_4	f_{B1}	f_{L1}	$LEO_1, LEO_2, JEO_1-JEO_3$
f_2	f_{B2}	EAP_1-EAP_4	f_{B2}	f_{L2}	$LEO_3, LEO_4, JEO_1-JEO_3$
f_3	f_{B3}	EAP_1-EAP_4	f_{B3}	f_{L3}	$LEO_5, LEO_6, JEO_1-JEO_3$
$f_{L1}-f_{L3}$	NONE	JEO_1-JEO_3			

As shown in Table 1, LECU is composed by JEO_1-JEO_3, f_{B1} could be supply by local participants JEO_1,LEO_1,LEO_2, and JEO_2,JEO_3 from other province.

Now, we explain the parameter CL uses the goal model of f_{L1}. First of all, we assume that the task plan of f_{L1} is known, that is, demand of task is known. The data of f_{L1} see in Table 2.

Table 2 data of participant of f_{L1}

Participants	EAU	cp_{ikm}	ct_{ikm}	c_{ik}	cap_{km}	p_{ik}	α
JEO1	LEA	7000	150	200	400	20000	2
JEO2	LEA	5000	150	200	400	20000	
JEO3	LEA	4000	150	200	400	20000	

We put the data into the target model, and obtained the solution. See in Table 3

Table 3 Results of two cases

	CL_{km}	R_{1k},R_{2k},R_{3k}	x_{ikm}	g_f		CL_{km}	R_{1k},R_{2k},R_{3k}	x_{ikm}	g_f
JEO_1	0.975	298,302,301	203,0, 0		JEO_1	0.985	298,302,301	200,0,0	
JEO_2	0.905		103,297,0		JEO_2	0.905		103,297,0	
JEO_3	0.830		0, 37,363		JEO_3	0.803		0,37363	
Case1				1318130	Case2				1413361

Figure 3 illustrates total goal profit of f_{L1} for different CL levels. As we see, case2 shows better performance than case 1, for we increase the CL level of participant JEO_1. And, we can find that the quantity of EA supplied by JEO_1 is decrease, but the total profit is increase. The comparative analysis of results of two case reflects a phenomenon, enhance the CL level can promote emergency efficiency and reduce the supply of emergency abilities.

6 CONCLUSION

In this paper, we have built a fractal-based EO framework to consider the CL in collaboration. In this structure, participants of EO by DS relationships form a fractal. Each fractal has a goal model, and generate task plan by optimizing the goal model. Fractal emergency organization has many advantages; through simple fractal unit we can know and understand the entire complex organizations. We can also simple reconfigure the fractal—add or delete participant, in response to the environmental changes. In the collaborative model of the fractal, we find that the CL is key factor, which helps to improve the emergency organization performance. In the coordination framework presented in this paper, the fractal emergency organization can expand the study of the dynamic reconfiguration.

7 ACKNOWLEDGEMENT

The work of this study is supported by National Natural Science Foundation of China under Project No. 91224007.

8 REFERENCES

[1] **Kapucu, N. 2008.** Collaborative emergency management: better community organising, better public preparedness and response. *Disasters*. 32(2), pp 39-62.

[2] **Kapucu, N., T. Arslan and F. Demiroz. 2010.** Collaborative emergency management and national emergency management network. *Disaster Prevention & Management*. 19(4), pp 452-468.

[3] **Ryu K, Jung E Y M. 2006.** Dynamic restructuring process for self-reconfiguration in the fractal manufacturing system. *International Journal of Production Research*. 44(15), pp 3105-3129.

[4] **Shin M, Mun J, Jung M. 2009.** Self-evolution framework of manufacturing systems based on fractal organization. *Computers & Industrial Engineering*. 56(3), pp 1029-1039.

[5] **Shin, M. 2013.** Employment Contract-Based Management Model of Production Resources on Relation-Driven Fractal Organization. *Journal of the Korean Institute of Industrial Engineers*. 39(4), pp 278-289.

[6] **Mun J, Shin M, Lee K, et al. 2009.** Manufacturing enterprise collaboration based on a goal-oriented fuzzy trust evaluation model in a virtual enterprise. *Computers & Industrial Engineering*. 56(3), pp 888-901.

[7] **Ryu K, Yücesan E. 2010.** A fuzzy newsvendor approach to supply chain coordination. *European Journal of Operational Research*. 200(2), pp 421-438.

[8] **Perry, R.W. 2003.** Preparedness for emergency response: guidelines for the emergency planning process. *Disasters*. 27(4): pp 336-350.

[9] **Paulheim, H., Döweling S, Tso-Sutter K.H.L., et al.** 2009. Improving Usability of Integrated Emergency Response Systems: The SoKNOS Approach. Informatik 2009: IM Focus Das Leben, Beiträge Der 39. Jahrestagung Der Gesellschaft Für Informatik E.v. 2009: pp 1435-1349.

[10] **Yu, F.L., Tu, F., & Pattipati, K. R. 2008.** Integration of a holonic organizational control architecture and multi-objective evolutionary algorithm for flexible distributed scheduling. *Systems Man & Cybernetics Part A--Systems & Humans IEEE Transactions on*, 38(5), pp 1001-1017.

[11] **Wang, H.M. 2014.** Research into Organization Structure of Floods Emergency Management under Bureaucracy- Cooperation System: A Case Study of Huaihe River Basin. *Journal of Hohai University (Philosophy and social sciences)*, (3), pp 42-48.

[12] **Tang, J. 2013.** Research on Emergency Response Organizing Structure and its Operational Model for Unconventional Emergencies. *Journal of Beijing Institute of Technology (Social Sciences Edition)*. 15(2), pp 82-89.

[13] **Ryu K, Son Y, Jung M. 2003.** Modelling and specifications of dynamic agents in fractal manufacturing systems. *Computers in Industry*. 52(2), pp 161-182.

[14] **Chen, J.F., Ni, S.B. 2004.** The application of fractal theory in supply chain constructing and operating models. *Modular Machine Tool & Automatic Manufacturing Technique*. (10), pp 40-43.

[15] **Ni, S.B., Chen, J.F. 2003.** The application of fractal theory to supply chain management. *Journal of Huazhong University of Science and Technology (Nature Science Edition)*. 31(11), pp 113-116.

[16] **Ryu K, Moon I, Oh S, et al. 2013.** A fractal echelon approach for inventory management in supply chain networks. *International Journal of Production Economics*. 143(2), pp 316-326.

[17] **Mandelbrot, B.B., 1982.** *The fractal geometry of nature*. New York, NY: Freeman.

APPLICATION OF INDUSTRIAL DEMILITARIZED ZONE OF FIREWALL IN INTEGRATED MANUFACTURING OR PRODUCTION SYSTEM

Ning Jiang[1*,2,4] ,Hu Lin[2,3] , Zhenyu Yin[2] and Yan Huang[3]
[1] University of Chinese Academy of Sciences, Beijing, China
jiangning@synu.edu.cn

[2] Shenyang Institute of Computing Technology
Chinese Academy of Sciences, Shenyang, China
linhu@sict.ac.cn, yinzhenyu@sict.ac.cn

[3] Shenyang Golding NC & Intelligence Tech. Co., Ltd.
Shenyang, China
huangyan@sict.ac.cn

[4]School of Educational Technology
Shenyang Normal University, Shenyang, China

ABSTRACT

The introduction of industrial Ethernet technology and the use of general standard protocols lead to the serious problem of information security in integrated manufacturing or production system. According to the reference model, the integrated manufacturing or production system is divided into different levels, which correspond to different security zone. According to the basic principle of defense in depth, the multi-layer defense-in-depth architecture of industrial control network is applied, and the corresponding protection is carried out at different levels. AS the key technologies of perimeter defense, industrial firewall is used between enterprise zone and manufacturing zone. By utilizing an IDMZ (industrial demilitarized zone), resources are shared between the enterprise and industrial networks without direct connections and under control of industrial firewalls. The security level for the IDMZ is higher than the enterprise zone but less than the manufacturing zone. Devices located in the IDMZ function as a bridge or buffer to provide higher level of controlled security and data management. To ensure no single point of failure exists, a standby firewall is deployed beside the active firewall.

Keywords: integrated manufacturing or production system, security, defense in depth, industrial firewall, industrial demilitarized zone

1 INTRODUCTION

A continuing trend in industrial networking is the convergence of technology, specifically industrial automation technology with information technology (IT). The trend of application convergence is being enabled by technology convergence. But some information security incidents have occurred

* Corresponding Author

frequently in recent years, such as Stuxnet attack. The rising incident count has increased focused on securing industrial control systems.

The issue of information security in industrial control system cannot depend only on a single security technology to achieve the overall system security. In integrated manufacturing or production system, it is not practical or necessary to apply the same level of security to all components. Every situation has a different acceptable level of security. Differences can be addressed by using the concept of a security zone. Deployment of security devices respectively according to security zones enhances the network defense capability. Integrated manufacturing or production system adopts defense-in-depth strategy. Firewall in the Defense-in-depth strategy applies Industrial demilitarized zone in integrated manufacturing or production system.

2 DEFENSE-IN-DEPTH STRATEGY APPLIED IN INTEGRATED MANUFACTURING OR PRODUCTION SYSTEM

2.1 A reference model

The openness of industrial control systems can be illustrated through the use of a reference model. The reference model describes a generic view of an integrated manufacturing or production system, expressed as a series of logical levels. The reference model used by the IEC 62443 series of standards appears in Fig. 1[1].

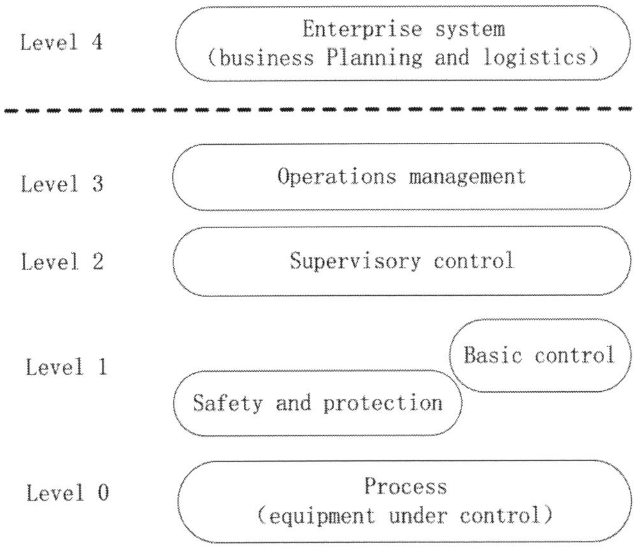

Figure 1: Reference Model for IEC 62443 Standards

Level 4 is Enterprise systems. This level is defined as including the functions involved in the business-related activities needed to manage a manufacturing organization. Functions include enterprise or regional financial systems and other enterprise infrastructure components such as production scheduling and operational management.

Level 3 is Operations management. This level includes the functions involved in managing the work flows to produce the desired end products. Examples include dispatching production, detailed production scheduling, reliability assurance, and site-wide control optimization.

Level 2 is Supervisory control. This level includes the functions involved in monitoring and controlling the physical process.

Level 1 is Local or basic control. This level includes the functions involved in sensing and manipulating the physical process. Level 1 includes continuous control, sequence control, batch control, and discrete control.

Level 0 is Process. This level is the actual physical process. The process includes a number of different types of production facilities in all sectors. Level 0 includes the sensors and actuators directly connected to the process and process equipment.

2.2 Defense-in-depth strategy

Defense in Depth is an information security practice adapted from a military defense strategy where an attacker is forced to overcome a great many obstacles that eventually expend the attacker's resources [2]. Defense-in-depth strategy provides multiple layers of protection between a potential attacker and critical data and systems. Defense in depth is the practice of implementing mutually supporting and overlapping defense measures designed to protect against intrusion and alert security staff in the event of a penetration. In terms of information security, information assets are layered in defensive measures. Layers of defense often overlap in order to ensure that traffic is processed multiple times by heterogeneous security technologies in hopes that the shortcomings of one security control are covered by another. A well-tuned defense in depth architecture can prevent a vast majority of attacks and alert an administrator to intrusions that pass through.

The defense-in-depth strategy bases on security zones [3]. A security zone is a logical grouping of physical, informational, and application assets sharing common security requirements. This concept applies to the electronic environment where some systems are included in the security zone and all others are outside the zone. There can also be zones within zones, or subzones. Security zones provide layered security, giving defense in depth and addressing multiple levels of security requirements. Defense in depth can also be accomplished by assigning different properties to security zones.

A security zone has a perimeter, which is the boundary between included and excluded elements. The concept of a zone also implies the need to access the assets in a zone from both within and without. Zones may be considered to be trusted or untrusted. Fortification of the perimeter is a key in a defense-in-depth architecture. Implementing anti-virus, intrusion detection system, and firewall is mandatory on defense of the host or the network perimeter.

2.3 Defense-in-depth architecture of integrated manufacturing or production system

Industrial security standards such as IEC-62443 use a defense-in-depth strategy. The defense-in-depth strategy uses security zones. Before a security zone is defined, the security requirements are first assessed. The security requirements involve communications access. According to the security and requirements and the reference model of integrated manufacturing system in Fig. 1, integrated manufacturing or production system network can be defined different security zones in Fig. 2. The external network is defined a security zone, called external zone. The level 4 is defined a security zone, called enterprise zone. Other levels are defined a security zone, called manufacturing zone [4].

After a security zone is defined, integrated manufacturing or production system networks are segmented through the use of some sort of a barrier device that has the ability to control what passes through the device. On Ethernet based networks running TCP/IP, the most common barrier devices in use are firewalls, routers and layer 3 switches. Frequently, integrated manufacturing or production system networks also employ barrier devices to separate and segment communications. The generally accepted practice is to use a barrier device such as an industrial firewall to manage the communication between the enterprise zone and the manufacturing zone.

Common filtering strategies at the barrier device are important. The base configuration of the barrier device should be to deny all communication by default and only allow communication by exception to meet a critical business need. Ports and services frequently used as attack vectors should not be opened through the barrier device. The fewer the number of ports and services open through the barrier device the better. The barrier device can serve as an automated tool to enforce security practices in the manufacturing zone.

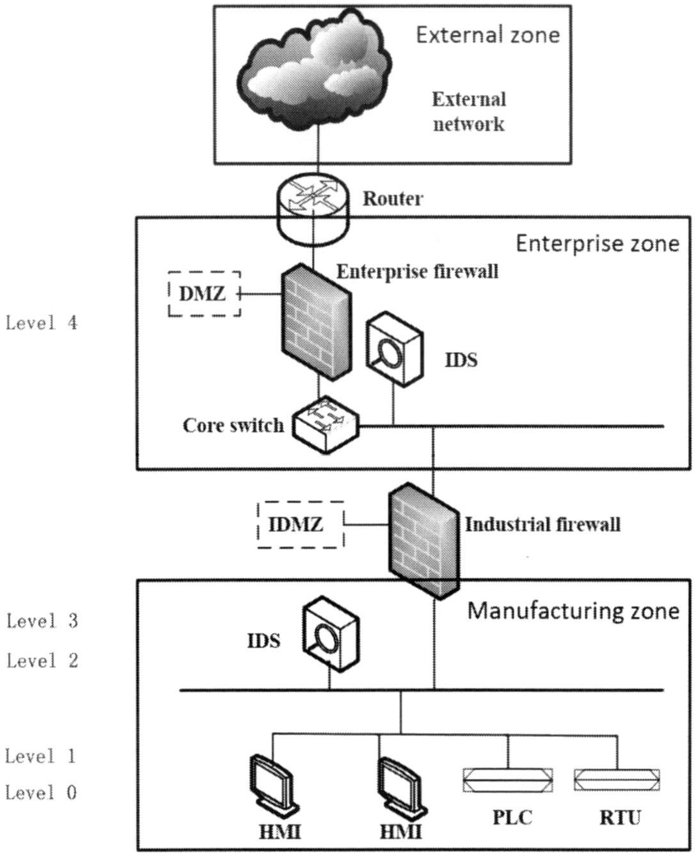

Figure 2: Defense-in-depth Architecture of Integrated Manufacturing or Production System

3 INDUSTRIAL DMZ (DEMILITARIZED ZONE)

Using a single firewall between a manufacturing zone and enterprise zone has a serious drawback. This is because the firewall must allow the data historian to have a wide range of access to the manufacturing zone. Essentially, each service needs a "hole" in the firewall to operate correctly. Configuring too many holes in the firewall reduces segregation between manufacturing zone and enterprise zone. It also opens the industrial network to a slew of attacks. This problem is typically addressed by creating an industrial DMZ [5].An architecture deploying an industrial firewall has three zones: an outside zone containing the enterprise zone, an inside zone containing the manufacturing zone, and an industrial DMZ.

Referred to as a perimeter network, the IDMZ (industrial DMZ) is a buffer that enforces data security policies between manufacturing zone and enterprise zone. The IDMZ is an additional layer of defense-in-depth to securely share industrial control system data and network services between the manufacturing and enterprise zones.

3.1 Concepts and application of IDMZ

The concepts of the IDMZ between the enterprise and manufacturing zones are very similar to those of the DMZs applied at the internet and enterprise interface. An IDMZ allows resources to be shared between the enterprise and industrial networks without direct connections and under control of firewalls placed at the boundaries [6]. Additionally, the IDMZ is also a point where access can easily be shut off if issues or threats arise in a zone that threatens operations in other zones.

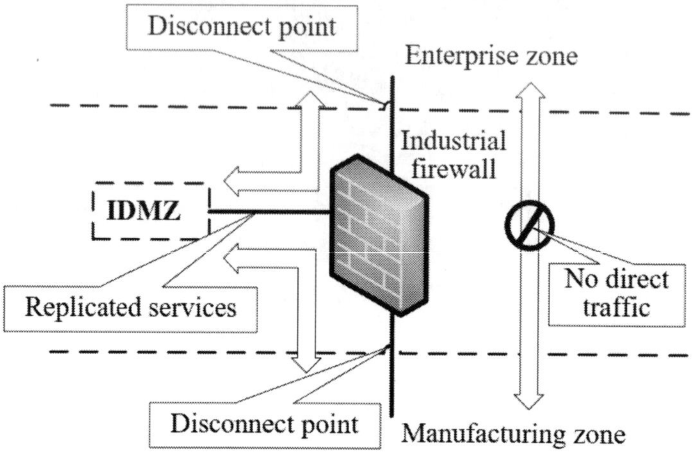

Figure 3: IDMZ Concepts

In Fig. 3, all network traffic from either side of the IDMZ terminates within the IDMZ. There is not network traffic which directly traverses the IDMZ [7]. There is not available path between the enterprise and manufacturing zones and there should are not common protocols in each logical firewall. There is not control traffic in the IDMZ and control traffic remains within the manufacturing zone. Additionally, there are not primary services or data stored in the IDMZ. There is an application data mirror to move data into and out of the manufacturing zone. Outbound connections from the IDMZ are limited.

3.2 Servers in IDMZ

The use of the IDMZ offers additional risk reduction opportunities between the low-security level zone and the high-security level zone. IN integrated manufacturing or production system, the security level for the IDMZ is higher than the enterprise zone but less than the manufacturing zone. The function of IDMZ is to eliminate or greatly reduce all direct communication between the enterprise zone and the manufacturing zone.

Devices located in the IDMZ function as a bridge or buffer between devices in the enterprise zone and manufacturing zone. Communication is setup between a device in the enterprise zone and the IDMZ. The device in the IDMZ then passes along the information to the recipient device in the manufacturing zone. Ideally the ports and services used between the enterprise zone and the IDMZ are different from those used between the IDMZ and the manufacturing zone. This reduces the likelihood of malicious code.

The IDMZ is a good place to locate devices like anti-virus servers and patch management servers. These devices can be used to manage deployment of security modules to the manufacturing zone and IDMZ devices in a more controlled manner without subjecting the high security level manufacturing zone to direct connection to servers that may be communicating to hundreds of devices.

Use of IDMZ can minimize the number of directly accessing manufacturing zone devices. For example, historian servers are often accessed from the enterprise zone. Rather than locating the historian server in the manufacturing zone and allowing direct access to this device from the enterprise zone by a large number of users, the security level of the manufacturing zone can be maintained at a higher level if the historian server is located in the IDMZ.

3.3 Deployment of IDMZ

In Fig. 4, an IDMZ is deployed. Servers, such as historian server, anti-virus server, patch management server, AV server, application mirror and remote desktop gateway server are in IDMZ.

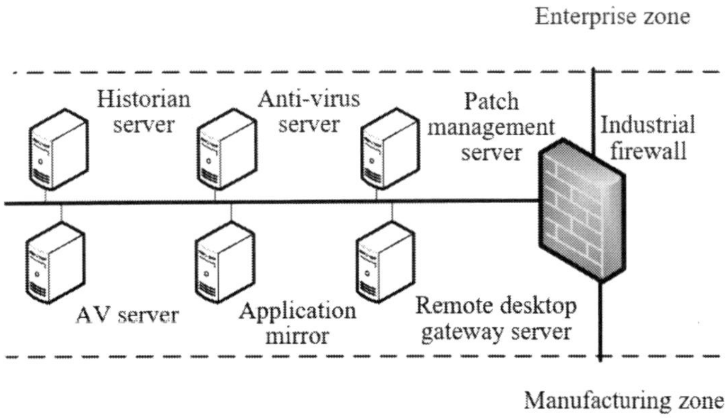

Figure 4: Deployment of IDMZ

It is necessary to provide fault-tolerant network services. Adding multiple components in a hot-standby configuration ensure no single point of failure exists [8]. For example, firewalls are implemented to provide failover. In Fig. 5, a standby firewall is deployed. The standby firewall is connected to active firewall by heartbeat. Many reliable hot-standby solutions make a firewall-induced outage very unlikely.

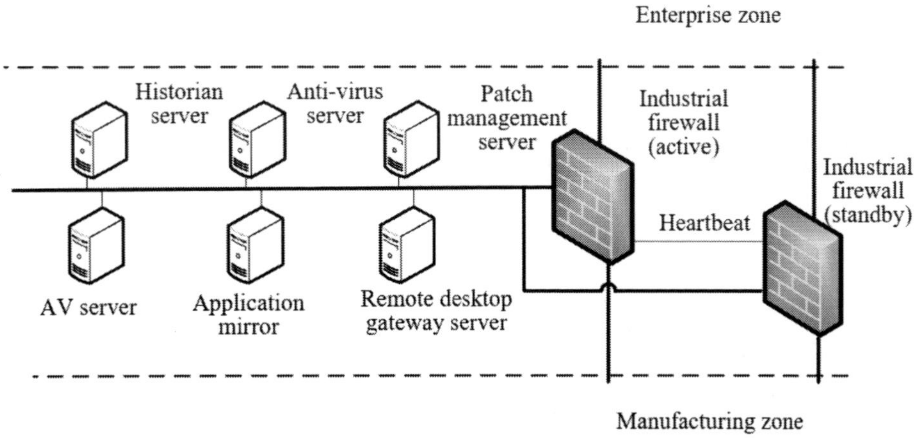

Figure 5: Deployment of IDMZ (two unit standby)

4 CONCLUSION

The information security issues of industrial control system are focused on. Integrated manufacturing or production system adopts defense-in-depth strategy. To implement defense-in-depth strategy, firewalls are deployed as barrier devices. A clear segmentation and classification of information assets, as either manufacturing or enterprise zone assets, is fundamental to a successful IDMZ implementation. The industrial control network is independent, needing nothing from the IDMZ to operate. A well-designed IDMZ facilitates the replication and control of data and only allows controlled access and communications from the manufacturing zone to the IDMZ and the IDMZ to the enterprise zone to provide higher level of controlled security and data management.

ACKNOWLEDGMENTS

This research was funded through Research on Functional Safety Technology of Numerical Control System (2014ZX04009031) supported by National Science and Technology Major Project of the Ministry of Science and Technology of China.

REFERENCES

[1] **IEC-62443-1-1,***Industrial communication networks -Network and system security -Part 1-1: Terminology, concepts and models.*

[2] **Cleghorn L. 2013.**Network Defense Methodology: A Comparison of Defense in Depth and Defense in Breadth,*Journal of Information Security*, 4(3),pp144-149.

[3] **DzungD, NaedeleM, Von HoffTP, et al. 2005.** Security for Industrial Communication Systems,*Proceedings of the IEEE*,93(6),pp1152-1177.

[4] **William Knowlesa,Daniel Princea,David Hutchisona, et al.2015.**A survey of cyber security management in industrial control systems,*International Journal of Critical Infrastructure Protection*,9(3),pp52-80.

[5] **Hamed Okhravi,David M. Nicol. 2009.**Application of trusted network technology to industrial control networks,*International Journal of Critical Infrastructure Protection*, 2(3),pp84-94.

[6] **Cheminod, M.,Durante, L.,Valenzano. 2013.**A Review of Security Issues in Industrial Networks, *IEEE Transactions on Industrial Informatics*, 9(1),pp277-293.

[7] **Mazur, D.C.,Kay, J.,Entzminger, R., et al. 2016.**Defining the Industrial Demilitarized Zone and its Benefits for Mining Applications, *IEEE Transactions on Industry Applications*,52(3),pp2731-2736.

[8] **Brooke, Paul. 2001.** Opening your e-business perimeter, *Network Computing*,12(1),pp38-46.

RESEARCH AND IMPLEMENTATION OF VISIBILITY EVALUATION TECHNOLOGY FOR ASSEMBLY WORK BASED ON SECONDARY DEVELOPMENT OF DELMIA

Chaozhong Guo[1], Shi An[2], Jihong an[1*], ingbao Hu[1], Binbin Ma[1]
[1]Department of Industrial Engineering
Harbin Institute of Technology, Harbin, China
jyan@hit.edu.cn

[2]School of Management
Harbin Institute of Technology, Harbin, China

ABSTRACT

As an important process to determine the quality of final products, assembly is of great significance in mechanical manufacturing. Virtual assembly simulation is an assembly inspection technology, which can find out the existing ergonomics and craft problems in advance to avoid unnecessary loss during the practical assembly. However, the current research of virtual assembly is basically confined to the stage of program control and physical simulation, and rarely consider human-computer interaction operations simulation, especially the visibility factor. In this paper, an automatic evaluation method for visibility is proposed for products assemblability based on the secondary development of DELMIA. First, a set of scoring criteria and evaluation algorithms are established based on the ergonomics theory, and an evaluation module is built based on the image processing and MATLAB programming. Then, the visibility evaluation module is embedded into DELMIA software by using the CAA based redevelopment technology. Finally, the assembly simulation of a mechanical product is adopted to validate the proposed technology. The results show that the automatic evaluation system provides feasible solutions for visibility assembly of complex products.

Keywords: Virtual assembly, visibility evaluation, secondary development of DELMIA, ergonomics

1 INTRODUCTION

With intelligent manufacturing time comes, how to improve the efficiency of product assembly process and the quality of the products is becoming one of the most important problems. The virtual assembly technology can simulate and optimize the process plan, and realize the rapid design of the products. It can evaluate assembly plans in the concept design stage of products, make scientific and reasonable, economical and practical assembly work plans, and effectively guide the workers [1]. Assembly visibility is one of the critical parameters in the evaluation of assembly works. It has great significance to analyze the visibility in the virtual assembly stage.

As a virtual assembly analysis platform, DELMIA software has powerful functions in assembly analysis, including push and pull analysis, transportation analysis, Rapid Upper Limb Assessment(RULA),

* Corresponding Author. Tel./fax: +86 45186402972

E-mail: jyan@hit.edu.cn (J. Yan)

biomechanical analysis, visibility analysis and so on [2]. Therefore, many scholars have finished assembly research based on the DELMIA. She and Fan [3] estiblished a virtual assembly modelof an automobile air condition by using the DELMIA software, and proposed a ergonomics based feasibility assessment method for product assembly. According to the assessement results of contact and collision simulation, the assembly sequence and path are adjusted, and the reasonable assembly process is obtained. Singh et al. [4] developed an ergonomics analysis module of the DELMIA software in the field of electronic industry in India, providing guidance for the actual workshop production. Liu et al. [5] studied the technology of complex products assembly evaluation based on DELMIA. By constructing a virtual assembly environment which is similar to the real environment, the simulation evaluation of an assembly process was carried out.

These researches mostly focused on the virtual assembly environment, assembly process flow, human fatigue analysis and so on. However, the visibility evaluation, a very important factor, is rarely to see. One of the reasons lies on the fact that the visibility of assembly is given by DELMIA through a simulation video, so it can not be directly used as a parameter for evaluation [6]. In our previous research [7], a visibility evaluation system was built by making the grade subjectively according to the images conversed from the simulation video provided by the DELMIA platform. However, the grading process is totally subjectively and manually. And for a real assembly task, the image library is very large which leads to time consuming assessment process and inaccurate results.

In this paper, an automatic evaluation system is estiblished with the secondary development technology of DELMIA. The grading of visibility is automatically calculated by using the distance between the assembly body and the vision filed center. The assessment module is developed based on the grading algorithm and embded into the DELMIA software. Finally, an real assembly task is used to verify the fucntion of the automatic evaluation sytem.

2 ESTABLISHMENT OF VISIBILITY AUTOMATIC EVALUATION MODULE

2.1 Establishment of visibility evaluation criteria

Visibility emphasizes the accessibility in vision, which is the extent of the visible field that people can see. It is of great significance to perform the visibility analysis in the virtual assembly to discover the problems in product design and improve the designs. However, DELMIA software only provides a simulation video for visibility analysis which cannot be directly used. The video screenshots of one assembly example is shown in Figure 1.

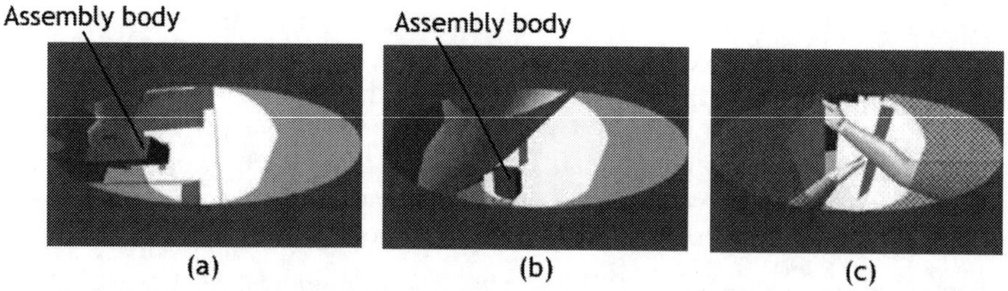

Figure 1: Video screenshots of one assembly e ample in the first view of assembly persons

As shown in Figure 1, the outer region of the ellipse is invisible, while the inner region is visible by human eyes. In the ellipse region, the left side with the red shadow is visible for the left eye but invisible for the right eye. Similarly, the right side with the red shadow is visible for the right eye but invisible for the left eye. The region with no shadow in the middle of picture is visible for both eyes. According to the actual assembly requirements, the evaluation criteria is given as follows: the assembly object is set as 0 point when it is invisible for both eyes, and the value is set as 1 point in the left shadow area (only the left eye can see) or the right shadow area (only the right eye can see). In the none shadow area (binocular vision), the evaluation value is given according to the distance from the picture center (the center of human visual field) to the assembly object. The farther the assembly object is away from the image center, the lower score is (the criteria is shown in Table 1). If the assembly body is located between the shadow free area and the shadow area, the score is also given according to the assembly part located in the shadow region.

2.2 Algorithm of visibility automatic evaluation

The visibility evaluation is based on the images derived from the simulation video in DELMIA. For each image, the image processing technology is used to identify the assembly body, and the single visibility is evaluated according to the relative distance between the body and the visual field centre. The overall visibility assessment of one assembly task is calculated using weighted sum of all the single visibility scores.

In order to implement image processing, the 3D assembly model in DELMIA is identified from the background by setting different characteristic grey values so that the position of the assembly part can be decided. In the image, each pixel is set as a 1x1 unit in order to quantize the distance between the assembly body and the centre of visual field. The boundary of the visible region is treated as a circle with a radius of 130 units.

Step 1: Obtain the RGB matrix A of the image. Because the image obtained from DELMIA is in a RGB24 encoding form, all the images are in the same size of 408 x 593 unit. The image can be converted to a multidimensional matrix A of RGB figure (408 593 3). The value of each unit in the matrix is 0-255, respectively representing the value of red, green and blue.

Step 2: Convert the RGB figure to the grey value form, obtaining a two-dimension matrix B which size is 408x593. The data in the matrix represents a certain range of grey level, and each value corresponds to a pixel point of the image, where the 0' represents the black colour and the 255' represents the white colour.

The characteristic matrix of the assembly body is derived by converting matrix B to a binary matrix C in the way of setting the values of the assembly body as 1' and the other values as 0', in which the outliers are eliminated and the matrix C is called the characteristic matrix.

Step 3: Calculate the distance between the assembly part and the visual field center. The cell coordinates with the value of 1 in the matrix are obtained by scanning the characteristic matrix C, the distance from the unit coordinates with value 1' to the center point of the visual field is calculated with two dimensional Euclidean distance formula [9]. The average distance is treated as the overall distance between the assembly body and the visual field center.

Step 4:

According to the calculation results, when the distance is less than 130, the assembly body is in the region of binocular vision. The interval [1, 5] is divided by 0.5 point as the evaluation scores in corresponding to distance of [0, 130], as shown in Table 1. When the distance value is greater than 130, then the assembly part is located in the shadow part or it is not in the visual field, which can be recognized by the matrix C. If C is an empty matrix, then the assembly part is not in the vision field and 0 point is given. Otherwise, the assembly part is in left or right shadow area. Hence, 1 point is given. By now the visibility evaluation of a single image is completed.

Table 1: uantitative relationship between the assembly part to the visual field center

Distance from the vision center to the assembly body (Unit)	Score
0-18	5

18-32	4.5
32-46	4
46-60	3.5
60-74	3
74-88	2.5
88-102	2
102-116	1.5
116-130	1

The final visibility evaluation formula for the whole assembly process is shown in equation 1:

$$X_{av} = \frac{X_1 + X_2 + X_3 + \cdots + X_n}{n} \tag{1}$$

where X_i represents the visibility value of the i image in the image library, n is the total number of images in this library, X_{av} is the overall visibility value for the assembly process. According to the above evaluation process and calculation equation, after the completion of all the images score, the overall visibility value is obtained.

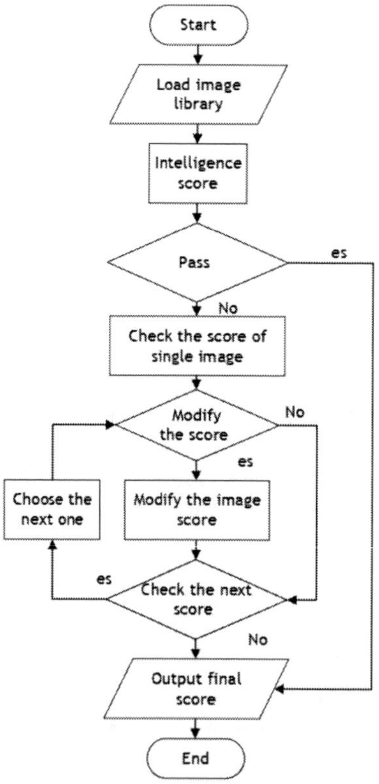

Figure 2: Flow chart of visibility evaluation module

In addition to the automatic scoring function, the visibility evaluation module also has the function of querying and modifying the deviation score. The integrated visibility evaluation flow chart for the assembly process is shown in Figure 2.

The GUI object is created by using GUIDE (Graphical User Interface Development Environment) in the MATLAB software, and an independent executable file (.exe) is compiled.

3 EVALUATION MODULE BASED ON SECONDARY DEVELOPMENT OF DELMIA

3.1 Secondary development environment estimation

As an independent executable program, the visibility evaluation module built in the MATLAB software cannot be directly linked to the DELMIA software. Therefore, it is necessary to carry out the secondary development of DELMIA to achieve the embedding of evaluation module so that it is more convenient to use in DELMIA software. Rapid Application Development Environment (RADE) is a visual integrated development environment offered by Dassault company, and provides a CAA framework program compiler [10]. Hence, the RADE integrated into Microsoft Visual Studio 2005 can be the secondary development tool of DELMIA.

A custom module designed by secondary development tools of DELMIA is used to connect the evaluation module and the DELMIA software. The embedding relationship among modules is shown in Figure 3.

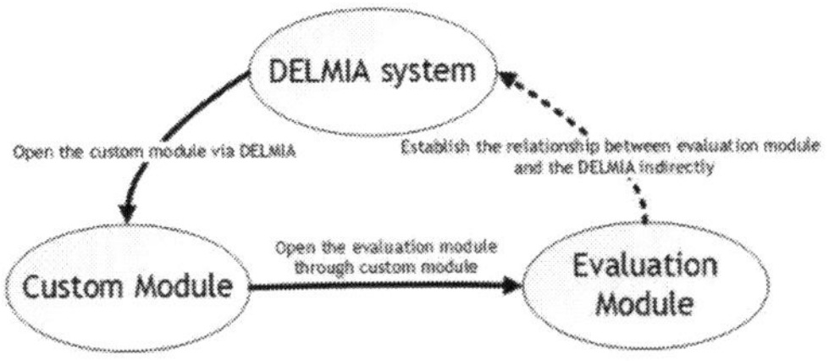

Figure 3: Embedding relationship among modules

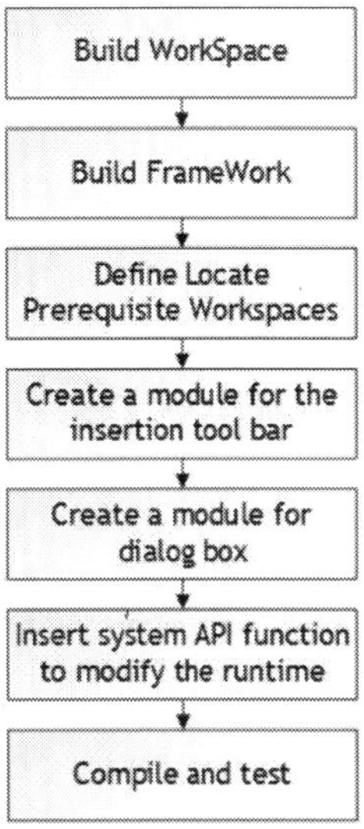

Figure 4: Flow chart based on the CAA

3.2 General steps of secondary development of DELMIA

The secondary development is based on the Component Application Architecture (CAA). The procedure is shown in Figure 4. Firstly, a WorkSpace is built in the RADE environment to generate a FrameWork. Then the custom module designed is completed under the FrameWork. On this basis, the required toolbar and dialog box module are added in this platform, as well as the necessary functions. After the final compilation, the custom module can be run in the DELMIA software.

3.3 Embedding the evaluation module

In the interface of DELMIA, there must be an entrance of the program to realize the embedding of evaluation module. Therefore, a tool bar should be created in the custom module by using the Addin’ wizard of the CAA custom module to create another WorkSpace, as shown in Figure 5. Then, the location of the resource is defined by the connection of the DELMIA system and the component interface, including the compilation of the header files, the shared library and resource files of building the running environment, etc. The Module files are edited according to the development purpose. A toolbar is added in the Module files as an entry to the external program.

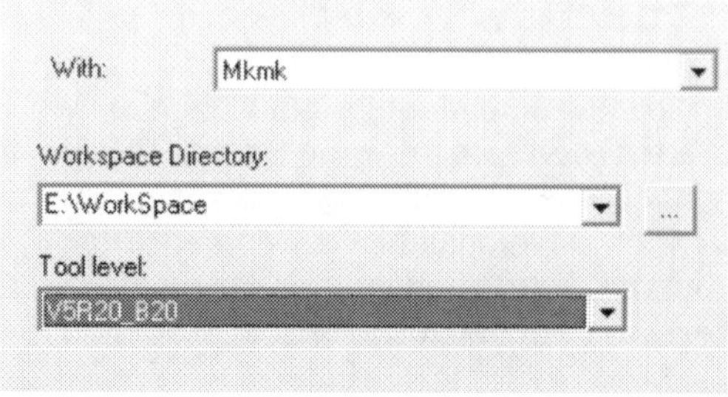

Figure 5: Building a WorkSpace

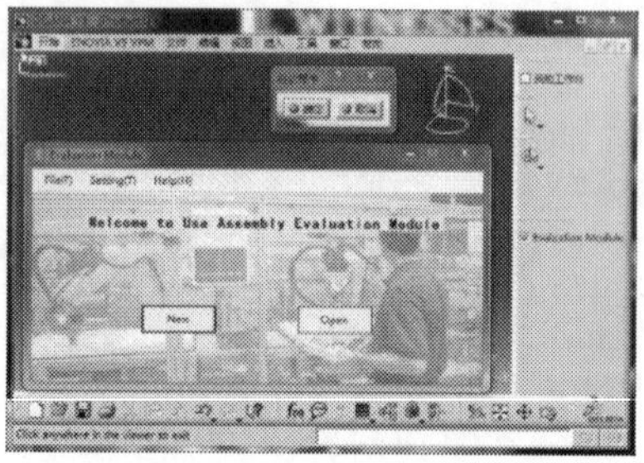

Figure 6: Access interface of automatic evaluation module

In order to complete the establishment of the toolbar, it also needs to add a necessary command to embed the visibility evaluation module into the tool bar. Herein, it is realized by using the state dialog’ command. Finally, the connection between the command and the dialog box is completed, and the embedding of visibility evaluation module is realized.

Compilation is indispensable for the developed modules. The RADE compiler is used to achieve the program compilation. Based on the configuration of program running environment, a user variable cnextoutput’ is created and set the value to console’. In the access interface, by clicking the open runtime window’ button and enter DELMIA’, the DELMIA can be called, and the automatic evaluation module developed is loaded on the right side of the toolbar, as shown in Figure 7.

4 CASE STUDY

In this section, the visibility evaluation of an assembly process case of electromechanical products is performed as a case study. To accomplish the task, one worker tightens the screws in the narrow environment with a lying down posture. The virtual assembly environment is built with using DELMIA software, and then the visibility analysis of key assembly process is carried out by loading the virtual human to evaluate whether the assembly process is feasible.

4.1 Correctness checking of visibility evaluation module

As described previously, the simulation video of the DELMIA software is decomposed into a series of image library, and then three images are selected to represent the typical spatial position of the assembly part (the red object in the images), as shown in Figure 7. The images (a), (b) and (c) represent the states that the assembly part in the shadow-free area, shadow area and out of the vision field.

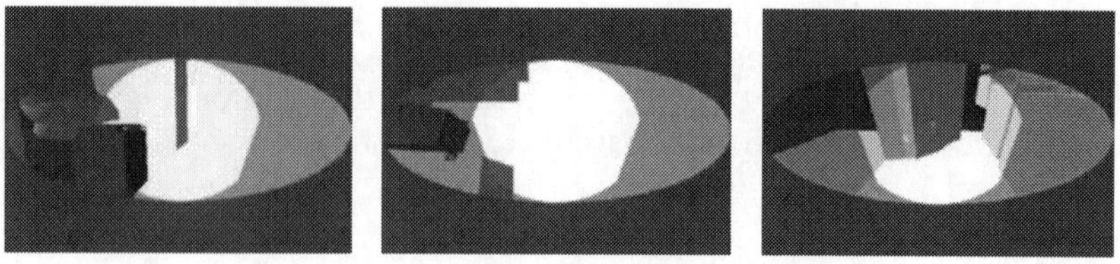

(a) (b) (c)

Figure : The typical spatial position of the assembly part

These three images are loaded in MATLAB and carried out the matrix operations in turn, then the corresponding eigenvalue matrixes C1, C2 and C3 are obtained, respectively. The characteristic value matrixes are plotted by using the function imshow ()', as shown in Figure 8.

(a) (b) (c)

Figure 8: Images corresponding to three characteristic matri es of Figure

After the characteristic value matrix is obtained, according to the score criterion and whether the characteristic matrix is null, the visibility score value of each image can be obtained. Scores of these three single images are 3.5, 1 and 0, respectively, conforming to the judgment standards of artificial formulation, which indicates that the evaluation module works well.

4.2 Application of visibility evaluation module

In order to assess the feasibility of the assembly process of an electrical components box, the developed evaluation system is used to analyze the assembly visibility. The evaluation system is quite friendly to use. First, start the DELMIA software, and click the button of the automatic visibility evaluation module embedded to access to the evaluation software interface. Then, load the image library obtained by decomposing the virtual assembly simulation video to start automatic scoring. In addition, in order to make it more flexible, an interface for modifying the score of some special actions, such as auxiliary operations (changing tools) is set. These actions are necessary, but their objects are not the assembly body. Therefore, the score is not calculated by the assembly evaluation module. Finally, the overall evaluation of the whole assembly can be calculated by Equ. 1 and shown in the interface. In Figure 9, the current image is about one tool changing action. If according to the evaluation algorithm, the score would be 0 and the overall score is 0.9 which is unreasonable. Therefore, it was modified as 2 according to the criteria, and the overall score is 1, which indicates that the assembly process is feasible. This result is concordant with the actual assembly process, proving that the DELMIA embedded evaluation module can be used for visibility evaluation for the assembly of electromechanical products.

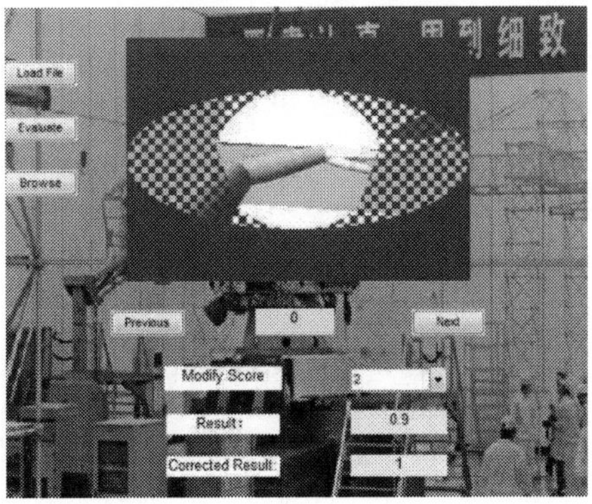

Figure : The interface of automatic visibility evaluation

5 CONCLUSION

In this paper, ergonomics theory is combined with assemblability evaluation technology in order to realize the automatic visibility evaluation, and a set of scoring criteria and scoring algorithms are established, and the visibility automatic evaluation module is completed by using the MATLAB programming. Meanwhile, the visibility evaluation module is embedded into DELMIA software menu based on the secondary development of DELMIA. Finally, on the basis of the assembly simulation of an actual example, the validity of the automatic evaluation system is verified, which provides feasible solutions for visibility assembly and ergonomics of complex products. Compared with artificial visibility evaluation, this automatic assembly visibility evaluation system has the function of the query evaluation process and modify the deviation, hence, it has higher reliability and rationality. Analysis shows that the assembly process is more complex, the automatic evaluation system is more efficient, accurate and objective. Through this method, designers can be more quickly and easily find the adverse effects in the assembly process during the products design, so as to improve the assembly performance and efficiency. The automatic visibility evaluation system has a wide range of applicability in practice.

ACKNOWLEDGMENTS

This work was funded by the National Natural Science Foundation of China (51505099).

REFERENCES

[1] **Seth A., Vance .M, Oliver .H. 2011.** Virtual reality for assembly methods prototyping: a review, Virtual reality, 15(1): 5-20.

[2] **Zhao L., Zhang Y., Wu X. 2016.** Virtual Assembly Simulation and Ergonomics Analysis for the Industrial Manipulator Based on DELMIA, Proceedings of the 6th International Asia Conference on Industrial Engineering and Management Innovation, Atlantis Press, 2016: 527-538.

[3] **She ., Fan X., Liu L., Meng X. 2012.** Assembly simulation of automotive air conditioner based on DELMIA, Journal of Jiangsu University of Science and Technology (Natural Science Edition, in Chinese), 3: 011.

[4] **Singh T., Singh . 2014.** Ergonomic evaluation of industrial tasks in Indian electronics industries, International Journal of Science and Research, 3(7): 1056-59.

[5] **Liu, X, Liu Y., Li, . 2011.** Virtual assembly and simulation of vibration sieve based on a human-interface environment, Procedia Engineering, 15: 2988-2992.

[6] **Zhao L., Zhang Y., Wu X. 2015.** Virtual Assembly Simulation and Ergonomics Analysis for the Industrial Manipulator Based on DELMIA, The 6th International Asia Conference on Industrial Engineering and Management Innovation (IEMI2015), 2015.

[7] **Zhou F., An L. 2012.** Assemblability Evaluation Technology for Products Based On Ergonomics, Machine Building Automation, 6: 005.

[8] **Liberti L., Lavor C., Maculan N. 2014.** Euclidean distance geometry and applications, Siam Review, 56(1): 3-69.

[9] **Zhang ., Yang B. 2015.** Extraction and Application of DELMIA Three-Dimensional Machining Model Information Oriented to Collaborative Process, Computer Systems Applications, 03: 246-250.

A CROSS-SYSTEM COLLABORATION ON ENGINEERING CHANGE ORDER EXECUTION

Wang Yan[1*]
[1]School of Mechanical Engineering and Automation
Beihang University, China
buaawangyan@buaa.edu.cn

ABSTRACT

To solve the process discontinuity and data asynchronism problems between product data management and computer aided quality systems which were used in engineering change order execution process, the engineering change management process in those two information systems were analysed. The objective difference was recognized and data exchange interfaces were identified. Then a cross-system collaboration model and a cross-system data exchange model were proposed. A case study in an aircraft manufacture company proved the validity and effectiveness of the proposed models.

Keywords: collaboration, engineering change order, execution process

1 INTRODUCTION:

Engineering Change (EC) plays a significant role in product quality improvement and has been proved inevitable in complex engineering product development. [1-3] Engineering Change Order (ECO) is the carrier of engineering change knowledge while implementing EC solutions within workshops of manufacturing companies.

In recent years, many companies have introduced information systems to improve their ability to control the engineering change process and improve the process efficiency of EC solution implementation. These information systems mainly include Product Data Management (PDM) system and Computer Aided Quality (CAQ) system. [4,5] However, the objective differs between PDM and CAQ.[6] On one hand, PDM mostly focus on the management of product-related data, including the engineering change data, but pays less attention to the ECO execution process.[7] On the other hand, CAQ helps employees in workshops to implement the EC solution. This means CAQ focuses more on the execution process of ECO. Each of those two information systems alone cannot cover the entire Engineering Change Management (ECM) process. This leads to the process discontinuity and data asynchronism when those two information systems running separately. Wasmer et al [8] have proposed a cross-company ECM framework that standardized product data communication and workflow task collaboration between enterprises and information systems. But their research requires a drastic change to information systems which is unacceptable for most software products. Li et al [9] have built a design change control system based on an architecture of design change control driven by workflow. Their work has realized information integration and process integration, but is not suitable for the situation of this research. Jin [10] have provided an engineering change management reference model to manage both document and bill of materials (BOM). But the model did not provide detailed description about controlling ECO execution process. To solve the process

* Corresponding Author

discontinuity and data asynchronism problems between PDM and CAQ with minimum cost, those two systems should work in a collaborative way to fulfil the ECM task.

This paper studies the ECM process both in PDM and CAQ, and proposes a cross-system process collaboration model and a cross-system data exchange model to help establish a collaborative management service based on the application of PDM and CAQ.

The structure of this paper is as follows: Section 2 presents the analysis on engineering change order execution process both in PDM and CAQ. Section 3 is the main section describing the cross-system collaboration mechanism, the cross-system process collaboration model, and the cross-system data exchange model. The exception handling mechanism is also given in this section. Section 4 applies these models to a study case and discusses the results. Finally, Section 5 concludes the paper.

2 ANALYSIS ON ENGINEERING CHANGE ORDER EXECUTION PROCESS

2.1 Engineering change order execution process

The objective of ECO execution process is to pass the engineering change knowledge from institutes of product design to workshops and finally change the work-in-pieces (WIP) and finished products. The flowchart of this process is shown in the middle layer of figure 1, which starts by receiving ECO from product design institute and ends with checking the implementation result of ECO execution plan. There are five stages altogether. After receiving ECO, process change tasks should be distributed to technologists to initiate the process change activities. After process change activities are completed, the production planners then make ECO execution plans according to production schedule of each product. The execution plans then be passed on to operators in workshops and guide the operators to deal with WIP or finished products. When ECO execution tasks finished, inspectors will check the execution result to ensure each product has been properly handled.

With the aim of PDM and CAQ, the ECO execution process has been carried out more efficiently in information systems. The top and bottom layer in figure 1 shows the activities in PDM and CAQ corresponding to each stage of the ECO execution process.

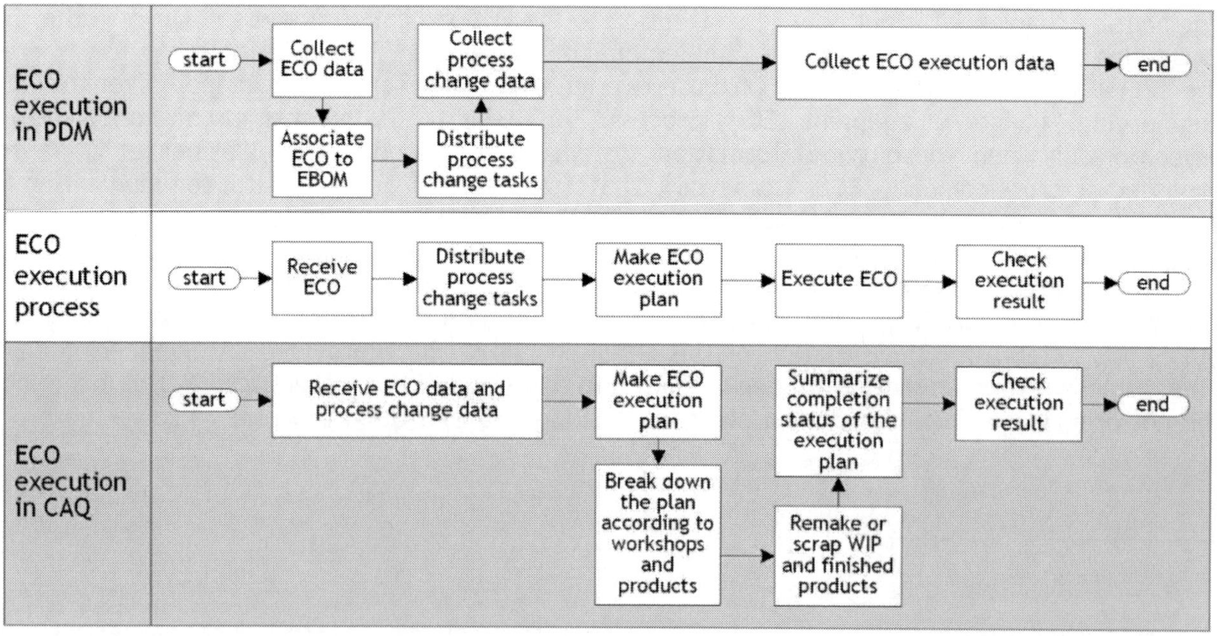

Figure 1: Flowchart of the ECO execution process and ECO execution in PDM&CAQ

2.2 Engineering change management in PDM

ECM in PDM efforts on the management of product data. From figure 1 we can see the purpose of most activities in the ECO execution process is to collect product-related data and associate them with Engineering Bill of Materials (EBOM). The workflow engine in PDM is capable for business process customization, but is mostly used to fulfil the data collection objective. In the process of engineering change management in PDM, three groups of data are collected and associated to EBOM, which are ECO data, process change data corresponding to the ECO, and ECO execution data. However, the ECO execution process is not involved in the process. Compared to ECM in CAQ, the ECM process in PDM chiefly orients to the employees in data management department.

2.3 Engineering change management in CAQ

ECM in CAQ efforts on the process regulation and automation in manufacturing process. In this ECM process, ECO execution plan is broken down to sub plans for each relevant workshop according to the production schedule of products. If the product has not been produced, then the latest process order that comes into force should be passed on to operators. As to WIP, repair or scrape treatment should be carried out. If the product is completely manufactured, remake or scrape procedure then be initiated. Compared to ECM in PDM, the ECM process in CAQ chiefly orients to the employees in workshops.

3 COLLABORATION METHOD ON ENGINEERING CHANGE ORDER EXECUTION

3.1 A cross-system process collaboration model

According to previous analysis on ECM processes in PDM and CAQ, those two processes concentrate on different stages in ECO execution process. In order to unite those two information systems as a collaborative management service, a cross-system process collaboration model (as shown in figure 2) is established to ensure a well-managed process flow.

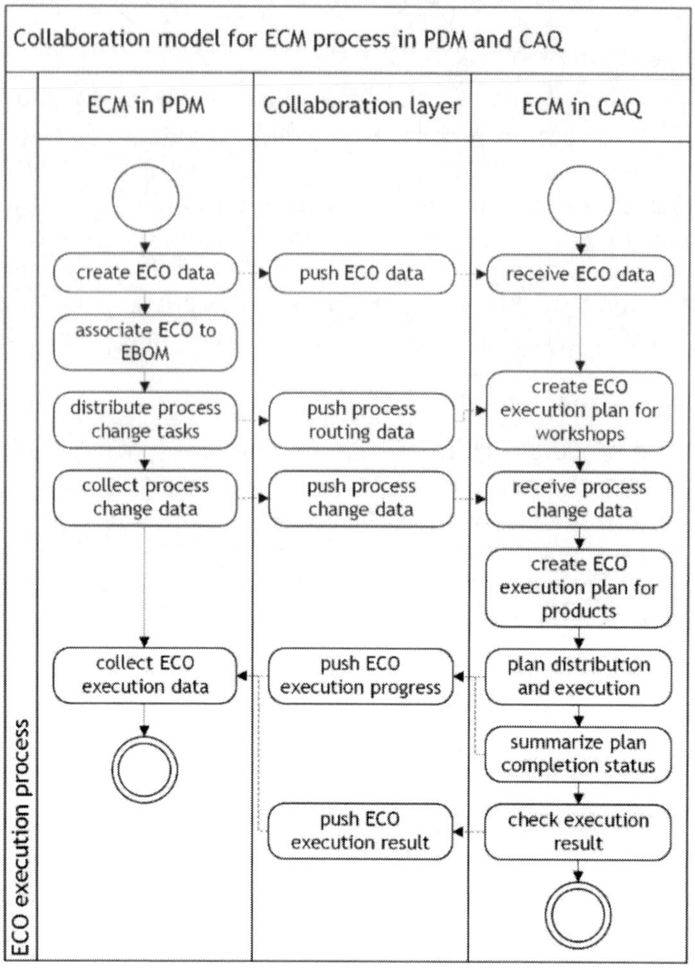

Figure 2: Collaboration model for ECM process in PDM and CAQ

In the collaboration layer of this model, five activities are involved to synchronize data between PDM and CAQ, meanwhile trigger activities in collaborative systems. In the former part of ECM process in CAQ, ECO data, process routing data, and process change data are derived from PDM. The process routing data is used to identify related workshops and help create ECO execution plan. In the latter part of ECM process in CAQ, ECO execution status and result data are pushed to PDM.

3.2 Identification of data exchange interfaces

From the process collaboration model in figure 2, data exchange interfaces can be identified. Table 1 gives the name, primary content, and use conditions of those interfaces. The server of the interface means which information system provide the interface, and the client means which one call the interface. For example, if PDM needs to push data to CAQ, PDM should call the interface and transport the data to CAQ, thus CAQ is the server and PDM is the client.

Table 1: Data exchange interfaces identified from figure 2

Interface name	Server	Client	Use condition	Primary content
Interface 1: Push ECO data	CAQ	PDM	Immediately after ECO data is created	<data identification number> <product drawing number>

			in PDM	<product name> <ECO data version> <EC code> <EC type> <effected products> <attachments>
Interface 2: Push process routing data	CAQ	PDM	Immediately after process change tasks are distributed to technologists	<data identification code> <EC code> <distributor's name> <distributor's ID number> <id of relevant workshop>
Interface 3: Push process change data	CAQ	PDM	Immediately after process change data is submitted by technologists	<data identification code> <ECO data version> <EC code> <technologist's name> <technologist's ID number> <auditor's name> <auditor's ID number> <if relevant> <treatment measure> <attachments>
Interface 4: Push ECO execution progress	PDM	CAQ	Immediately after any status change take place on the ECO execution plan	<data identification code> <ECO data version> <EC code> <execution plan code> <execution plan name> <execution result>
Interface 5: Push ECO execution result	PDM	CAQ	Immediately after the completion of ECO execution plan is confirmed	<data identification code> <ECO data version> <EC code> <execution result>

3.3 A cross-system data exchange model

After identifying data exchange interfaces, a cross-system data exchange model is established (as shown in figure 3). The directed arcs in this model imply the data transferring direction. The numbers next to those arcs are the order of data transmission. It should be noticed that ECO execution progress data that transferring from CAQ to PDM (marked by number 4) contains multiple arcs, which means the data transmission may happen many times during the ECM process. The transmission frequency and the total count for transportation depend on the number of sub plan that broken down from the main ECO execution plan.

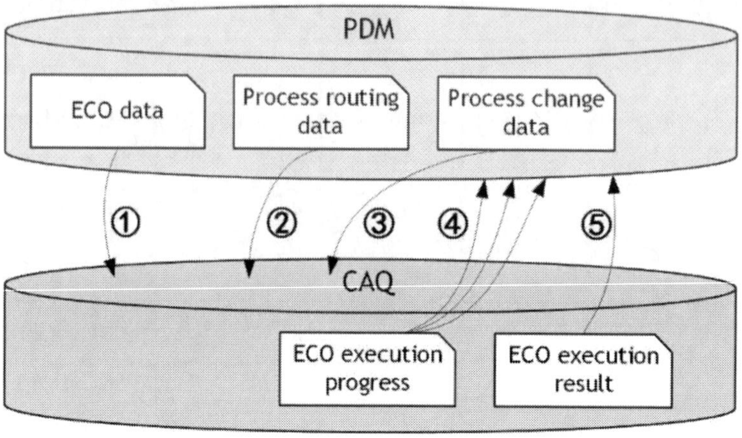

Figure 3: Data exchange model for PDM and CAQ

3.4 Exception handling mechanism

In practical environment, there are many unforeseen accidents that may disturb the data exchange procedure between two information systems, especially when those systems are running on different machines. Therefore, an exception handling mechanism is of vital important. In order to ensure the safety of data and the continuity of ECM process, several abnormal situations are considered and properly handled as follows.

- Data duplication: If the same data has been transferred more than one time, there may exist redundant data in the server and have negative impact on the ECM process. To solve this problem, a data identification method combining the data identification code, data receiving time, and data version information is introduced. When the server receives a data, the data identification procedure should take place first to ensure that this data has never been transmitted before.
- Transmission interruption: The whole procedure of data transmission usually takes no more than several milliseconds, but there still exists possibilities for interruption of the procedure. Once the procedure is interrupted, the client may wrongly assume that the data transmission was succeeded, but the server only receives incomplete information. To solve this problem, each interface should return a response information when data transmission is completed. The response should contain both information about whether the data transmits successfully and additional information if transmission fails.
- Fail of data transmission: There are many reasons that may lead to the failure of data transmission. These reasons include incomplete or error data, network disconnection, and accidental shutdown of server. The solution to this problem is the same as that to transmission interruption. If the client does not receive response information in a certain period of time after calling the server, the interface call should be regarded as failed. Then

a timer will be set on this data and the next calling procedure will take place after a certain period of time.

4 CASE STUDY

In this section, the cross-system collaboration methods are applied to an aircraft manufacture company. The PDM system in this company is Windchill^{TM} PDM System (v 8.0) from PTC Corporation. The CAQ system is a highly customized information system developed by Beihang University. The five data exchange interfaces in table 1 have been implemented using WebService technology. The PDM and CAQ systems have been running in a collaborative way and formed a collaborative management service. With the aid of this service, employees in product design institute, data management department, and workshops can share the same ECM process. The ECO data, process change data and ECO execution data created in the ECM process can be exchanged automatically and correctly between PDM and CAQ. These achievements have significantly improved the ability for the company to control the ECM process, and ensure an efficient execution of the ECO. Some typical screenshots of the system are shown in figure 4 and figure 5.

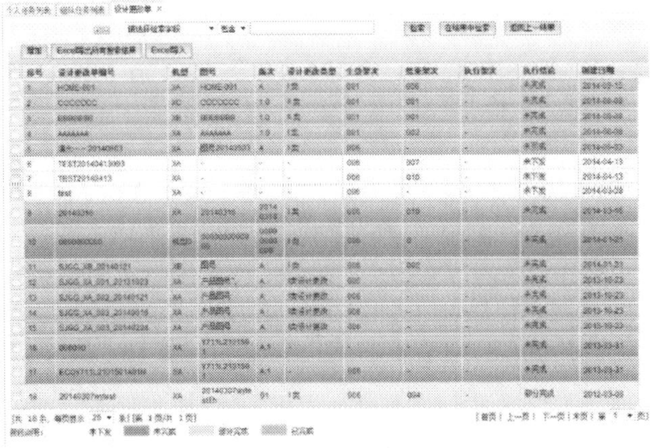

Figure 4: ECO received from PDM

Figure 5: ECO execution status in CAQ

5 CONCLUSION

The engineering change management processes in PDM and CAQ systems were analysed and the objective difference between those two information systems was recognized. To solve the process discontinuity and data asynchronism problems, a cross-system process collaboration model and a cross-system data exchange model were built. These models can help the company to design the total ECM process as a collaborative information system service, and identify possible data exchange interfaces between information systems. These methods were applied to an aircraft manufacture company and their validity and effectiveness have been proved.

6 ACKNOWLEDGEMENT

This research was supported by the Defence Industrial Technology Development Program, China (No. A0520132008).

7 REFERENCES

[1] **Hamraz, B., Clarkson, P.J. 2015.** Industrial evaluation of FBS Linkage - a method to support engineering change management, *Journal of Engineering Design*, 26(1-3), pp 24-47.

[2] **N. Do, I.J. Choi, and M. Song. 2008.** Propagation of engineering changes to multiple product data views using history of product structure changes, *International Journal of Computer Integrated Manufacturing*, 21(1), pp 19-32.

[3] **Fei Genyuan, James Gao, Oladele Owodunni, and Xiaoqing Tang. 2011.** A method for engineering design change analysis using system modelling and knowledge management techniques. *International Journal of Computer Integrated Manufacturing*, 24(6), pp 535-551.

[4] **Wang Zhonghao, Cai Ligang, Zhu Haiping, Gao Liang. 2003.** Design of engineering change management in PDM and its implementation. *Journal of Huazhong University of Science and technology (Nature Science Edition)*, 31(6), pp 38-40.

[5] **Qiao Yuliang. 2014.** Aircraft Design Change Implementation and System Implementation Study. *Equipment Manufacturing Technology*, 11, pp 186-188.

[6] **Zhu Hao, Yin Zeyong, Liu Jianwu, Chen Gaoyang, and Li Shengwen. 2004.** Research on engineering change control of collaborative product development for virtual enterprise. *Computer Integrated Manufacturing Systems*, 10(12), pp 1594-1599.

[7] **Shi Xiaojian, and Qiao Lihong. 2007.** Implementation of Process Change Management in PDM. *Acta Aeronautica Et Astronautica Sinica*, 23(1), pp 240-244.

[8] **Wasmer, A., Staub, G., and Vroom, R.W. 2011.** An industry approach to shared, cross-organisational engineering change handling - the road towards standards for product data processing. *Computer-Aided Design*, 43(5), pp 533-545.

[9] **Li Cheng, Gao Jianmin, and Li Qing. 2006.** Design change control systems based on process control. Computer Integrated Manufacturing Systems, 12(10), pp 1549-1555.

[10] **Jin Yulan. 2012.** Research on Product Engineering Change from the Perspective of Management Information System Framework. 2012 Second International Conference on Business Computing and Global Informatization. IEEE, Shanghai, China, Oct. 12-14, 2012, pp 710-712.

STUDY ON LIFE ESTIMATE METHOD FOR SECONDARY POWER SUPPLY OF SPACECRAFT BASED ON ACCELERATED TESTING

Hao Wang[1], Jinming Chen[2*], Yuege Zhou, Zemin Yao
[1]Beijing Institute of Spacecraft Environment Engineering
Beijing, China
shenhengwh@163.com

[2]Beijing Institute of Spacecraft Environment Engineering
Beijing, China
gfhchjm@163.com

ABSTRACT

The power subsystem of the spacecraft is of primary importance. However, the life testing is challenging because of its small sample size, long life, and highly reliability. In this work, a life estimation and reliability assessment method based on accelerated testing was proposed, which take advantages of the accelerated sensitive stress to improve the experiment efficiency. The multivariable modeling for performance degradation data was applied to improve the accuracy of the estimation. A certain model of secondary power supply of spacecraft was considered as an example. Then an accelerated degradation testing scheme was designed based on the failure mechanism analysis. Combined with the Wiener process,- the accelerated model was adopted to simulate the degradation process by using the copula function to build a multivariable model for the degradation data. Finally, the activation emery of the secondary power supply and the life estimation under the designed usage profile was obtained. This novel scheme provides a new approach for life estimation of the spacecraft power supply with small sample quantity and long life.

Keywords: spacecraft, secondary power supply, accelerated test, life estimate, small sample size

1 INTRODUCTION

As one of the spacecraft subsystems, the power subsystem has the simple missions but the primary importance. Based on statistics, 80% of the failures of the power subsystem result in severe negative impact on the spacecraft mission, and over half of the failures literally result in a failure of the mission [1].

In general, there are two types of failure of the power subsystems: one is early failure caused by potential defects originated in the components, material, or assembly process. This type of failure can be effectively reduced by environment stress screening and acceptance test. The other type is the real life time of product which is not as good as the designed life time. Therefore, it is necessary to assess the reliability and service life before using the power supply, especially for the newly developed products. As the designed life of the spacecraft and components has greatly increased over the years, the conventional life estimation method is invalid anymore. In the 1970s, NASA carried on the research and a series of applications of the accelerated test on the power

* Corresponding Author

supply for spacecraft [2]. However, because the testing is both time and money consuming, available samples for life estimation testing are very limited, which is way less than the conventional approaches needed [3].

In this work, a certain model of secondary power supply for spacecraft is studied, and a novel scheme is proposed for assessment of the reliability and life based on accelerated testing. First, we analyze the failure mechanism of the secondary power supple and use step-stress accelerated testing techniques to obtain degradation data of the key performance parameters under sensitive stress. Then the Wiener process and accelerated model are used to describe the degradation process of the secondary power supply. The multivariable accelerated degradation model is established using copula function. Finally, the evaluation results of the activation energy and life under the designed usage profile of the secondary power supply are analyzed.

2 PRINCIPLES OF EVALUATION METHOD

2.1 Accelerated degradation testing

Since the designed life of the spacecraft is getting longer and longer, the conventional environment simulation test is invalid for the mission. Accelerated testing which can significantly reduce the time and costs has attracted more and more attention. At present, common accelerated testing includes: Accelerated Life Testing (ALT) and Accelerated Degradation Testing (ADT).

ALT is based on the product failure data. Because of the random nature of the product failure, the accuracy of the assessment highly depends on the sample quantity. However, the spacecraft products are typically in small quantity, such that the available samples for testing are very limited as well. Thus it is difficult to ensure the accuracy of the assessment. Additionally, for some highly dependable components, it is difficult to obtain failure data even accelerated stress is adopted.

Under environment stress, some performance parameters will suffer degradation over time. These variations in data can provide valuable information about the life and reliability of the product. In the 1980s, Nelson introduced the method of valuation of ADT by studying data from ADT of the insulation material. This method which is suitable for the product with degradation on some performances, employed degradation data obtained from product under an accelerated stress to infer and predict the normal life of the product. Because the basis of the assessment is the performance degradation data other than the failure data, ADT requires much less samples than the ALT. Therefore, ADT is a proper tool to assess the spacecraft power supply products with degradation characteristics [4].

2.2 Performance degradation data modeling

Performance degradation data modeling can circumvent the study of the complicated mechanism of the degradation. It is a simple method which can be applied on various types of products. Modeling methods are mainly degradation paths modeling and random process modeling [5].

In degradation paths modeling, the degradation paths of all samples in a collection would be described using the same type of paths function [6]. However, in fact the degradation process of the product is affected by many random factors, such as the multiple stresses in the working environment, the failure mechanisms of the components and materials, etc. Even in stable testing environment, the stresses are usually fluctuating. Therefore, the random process modeling for degradation data is more proper than deterministic modeling in real applications [7].

In recent years, random process modeling is mainly used for random models with independent increment, such as Gamma process, geometric Brownian motion, Wiener process, etc. Gamma process is a non-negative and strictly monotone random process, which is suitable for describing the failure process involving wear-out, fatigue, and erosion [8,9]. Geometric Brownian motion is a random process model good for degradation process of the product with non-negative time [10]. Meanwhile, both monotone and non-monotone random processes can be well described by Wiener process. Hence, many products' performance degradation paths can be described by Wiener process [11]. If $\{B(t), t \geq 0\}$ is a standard Wiener process, this random process has stationary independent increment which equals $B(t+\Delta t)-B(t) \sim N(0, \Delta t)$. Drifted Wiener process $W(t) = \mu\tau(t)+\sigma B(\tau(t))$ is an important variant of the standard Wiener process, which has many merits in engineering. In the degradation data modeling, drift parameter μ indicates the product life characteristics and diffusion parameter σ indicates the instability and non-uniformness in the manufactory, testing, and measuring. Because the concordance of the failure mechanism of the failure process in ADT, σ can be regard as a constant. Moreover, $\tau(t)$ can be used to delineate the nonlinearity in the degradation process [12].

2.3 Evaluation method of the multivariable degradation data

In conventional accelerated testing, only the degradation data related to most important performance are selected by Failure Mode, Effects and Criticality (FMECA) to assess the life and reliability [13,14]. In fact, many kinds of products have two or more performance parameters characterized as indicators of the life and reliability. Moreover, the degradation information about each performance contains some information about the system reliability. It is an effective approach to improve the accuracy of the estimation by using multivariable performance degradation in the assessment, and also a good way of overcoming the small sample size issue of the spacecraft products.

By far, many useful results have developed in the area of multivariable data analysis. Crk assumed the degradation of all performance processes are independent, and analyzed the system reliability with series model [15]. Wang et al. proposed using multivariable Gaussian distribution to yield an accurate model of the multivariable performance degradation [16]. However, these methods have some limitations when applied to real applications. Copula function is an important mathematical tool in multivariable data analysis, which can describe the correlation among the random variables without the constrains on the marginal distribution. It has already been prevailing in many areas such as finance and risk assessment [17].

Assume $u_i = F_{X_i}(x_i)$ $(i = 1,2, \dots n)$ is a continuous one-dimensional distribution function. Then $H(x_1, x_2, \dots x_n)$ denotes the joint distribution function of the marginal distribution function $F_1(x_1), F_2(x_2), \dots F_n(x_n)$. There exist a Copula function C such that

$$H(x_1, x_2, \dots x_n) = C[F_1(x_1), F_2(x_2), \dots F_n(x_n)] \tag{1}$$

Because of all the good properties of Copula function, many researchers derived different types of Copula function. The most common ones are Gaussian Copula function, t-Copula function, and the Archimedean Copula functions. The Archimedean Copula function is simple in both construction and computation, which can build various and multivariate joint distributions with strong adaptivity. Archimedean Copula function is generated by generator function ϕ with a unified form as follows:

$$C(u_1, \dots u_i, \dots u_n) = \phi^{-1}(\phi(u_1), \dots \phi(u_i), \dots \phi(u_n)) \tag{2}$$

where ϕ satisfies $\sum_i^n \phi(u_i) \leq \phi(0)$, and for any $0 \leq u \leq 1$, we have $\phi(1) = 0$, $\dot{\phi} < 0$, $\ddot{\phi} > 0$, i.e. ϕ is a strictly monotone decreasing convex function.

The common Archimedean Copula functions include Clayton function, Frank function, and Gumbel function. Gumbel function and Clayton function are only applicable for variables with positive correlation. However, Frank function can be applied for both positive and negative correlated random variables, which has good symmetric property. Its generator function can be described as

$$\phi = -ln\frac{e^{-\theta u}-1}{e^{-\theta}-1} \quad (3)$$

3 EXAMPLE IN APPLICATION

Here a case of certain model of the secondary power supple was studied, whose exterior view is showed in Figure 1. The main function of this power product is to convert rated input voltage into adjustable DC output voltage and constant current. The normal operation temperature range is from -15°C to 50°C. In our experiments, we have three samples for life estimation, which have been screened and inspected by the manufacture to make sure working properly.

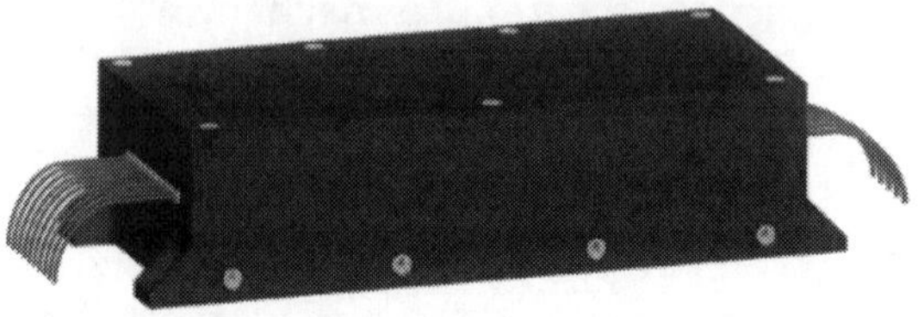

Figure 1: External View of the Secondary Power Supply

3.1 Test scheme design

According to the FMECA analysis, the thermal environment is the major environment stress that causes the secondary power supply to degrade. Because all components are designed according to derating requirements, the effect of the electrical stress is negligible compared to that of the thermal stress. In our testing the thermal environment is adopted as the accelerated stress. Referring to related standards, the high temperature limit is 65°C for acceptance test and 75°C for qualification test. Therefore, 70°C and 80°C -are chosen as the stress levels for the ADT to make sure the consistence of the failure mechanism. Additionally, because the secondary power supple is operated in vacuum, the heat exchange condition is quite different from environment of the ground. In order to assure the uniformness of the thermal distribution, we use the vacuum chamber whose vacuum degree is better than 6.65×10^{-3}Pa. And the test articles are placed on cold plate to simulate thermal boundary condition on orbit.

Based on the design requirements, several performances need to be monitored, such as, output voltage, output current, stability, and voltage ripple. In light of the importance of these indacators, only data of output voltage and output current are chosen for assessment. The time interval between adjacent data collecting is 6 hours.

3.2 Data modeling

Assume the measurement of the l-th performance of the i-th secondary power supply at time t_j under the s-th stress is Δx_{ijsl}. Then the value of degradation is $\Delta x_{ijsl} = x_{i,j+1,sl} - x_{ijsl}$. In the sequel, l=1 means the measurement of output voltage, l=2 means output current. Based on the assumption that the data follow a drifted Wiener process, we obtain

$$\Delta x_{ijsl} \sim N(\mu_{sl}\Delta t, \sigma_l^2 \Delta t_j) \quad (4)$$

where $\Delta t_j = t_{j+1} - t_j$, $j = 1,2,\dots$, μ_{sl} denotes life drift parameter of the secondary power supple under the stress S, μ_{sl} can be delineated by accelerated model.

Accelerated model describes the physical and chemistry relations between the life parameter μ and the stress level S. Utilizing the accelerated model and performance degradation data, we can extrapolate the life of the product under normal stress level. Since the accelerated stress in the test is thermal stress, Arrhenius model gives good estimation of the relation between the degradation rate of the performance and environment temperature, which is as follows:

$$\mu_{sl} = A_l exp\left(\frac{-Ea_l}{kT_s}\right) \tag{5}$$

where μ_{sl} is degradation rate of the l-th function at temperature T_s, and A_l is a positive parameter to be determined, Ea_l is activation energy, k is Boltzmann's constant ($8.617\times10^{-5}eV/K$), and T_s is the temperature of stress level S. The activation energy would represent the minimum amount of energy needed to allow a certain chemical reaction to occur. In engineering applications, activation energy indicates the rate of degradation of the product from normal state to failure state.

Sklar's Theorem gives the Copula function of the 2-dimensional degradation data as follows:

$$C(F_1(\Delta x_{ijs1};\theta_1), F_2(\Delta x_{ijs2};\theta_2);\alpha) \tag{6}$$

where $\theta_1 = (A_1, Ea_1, \sigma_1)$, $\theta_2 = (A_2, Ea_2, \sigma_2)$ and α are parameters to be determined.

3.3 Parameter estimation

The assessment model of multivariable accelerated degradation data has a rather complicated structure. The parameters to be determined are much more than one dimensional accelerated degradation model, so the conventional estimation methods are incapable to solve the problem. The characteristics of Copula function suggest using the inference of functions for margins (IFM) to solve the problem in steps, and using maximum likelihood estimation (MLE) for parameter estimation.

For the marginal distribution function of the l-th performance, whose density function is denoted as $f_{ijsl}(\Delta x_{ijsl};\theta_l)$, MLE gives

$$\hat{\theta}_l = arg\,max \sum_{i=1}^{3}\sum_{j=1}^{n}\sum_{s=1}^{2} ln\, f_{ijsl}(\Delta x_{ijsl};\theta_l) \tag{7}$$

Using the performance data obtained from ADT, estimated parameters values for output voltage and output current marginal distribution function are listed in Table 1.

Table 1. Parameters value estimation of the marginal distribution function

Output Voltage		**Output Current**	
A_1	1.315×10^{12}	A_2	1.132×10^{12}
Ea_1	1.072	Ea_2	1.2737
σ_1	1.367×10^{-2}	σ_2	3.233×10^{-5}

Then applying MLE again with $\hat{\theta}_l$, the estimation of parameter α in the Copula function is calculated as follows:

$$\hat{\alpha} = arg\,max \sum_{i=1}^{3}\sum_{j=1}^{n}\sum_{s=1}^{2} lnc[F_1(\Delta x_{ijs1};\hat{\theta}_1), F_2(\Delta x_{ijs2};\hat{\theta}_2);\alpha] \tag{8}$$

where c is the density function of the Copula function C. Substituting it into Frank generator function in equation 3, the density function becomes

$$c = \frac{\alpha(1-e^{-\alpha})e^{-\alpha(F_1+F_2)}}{[(1-e^{-\alpha})-(1-e^{-\alpha F_1})(1-e^{-\alpha F_2})]^2} \tag{9}$$

Using the accelerated degradation data and the parameters estimation of the marginal distribution function, we can get the parameter estimation of Frank-Copula function with a confidence of 0.95.

Table 2. Parameters estimation of copula function

Function	2.5%	$\hat{\alpha}$	97.5%
Frank-Copula	2.1478	3.0001	3.8523

3.4 Life estimation results

The life of secondary power supple is actually the time when the function parameter reaches the failure criteria. Let d_1 and d_2 denotes the thresholds of the degradation of output voltage and output current, respectively. The reliability of the secondary power supply under designed environment stress can be written as

$$R(t|T) = C\{F_1(d_1|T), F_2(d_2|T); \hat{\alpha}\} \tag{10}$$

Based on the design requirements, the environment temperature range of operation is -15°C to 50°C。 Take the harshest environment stress and set the environment temperature to 50°C(323K) as the stress level for the life estimation. The averaged measurements of functions before the experiments at 50°C are used as the initial value, and the designed failure criteria as the end. We could obtain d_1=3.73V and d_2=0.02A respectively. Substituting them in equation 10, the reliability curve can be derived based on accelerated performance degradation data of the secondary power supply (see Fig. 2). The assessment shows, if the designed service life is 5 years, the reliability of the product of proper operation for 5 years is greater than 0.82. We could also compute that if the temperature is set to 30°C, which is close to the mid-point of the designed temperature range, the reliability of the product is close to 0.9 for 5 years, and 0.81 for 10 years.

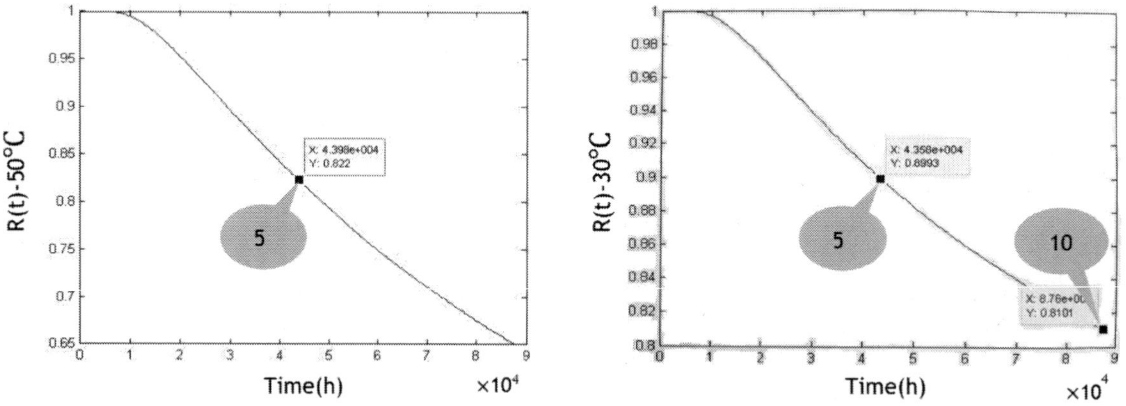

Figure 2: Reliability Curve at 50°C and 30°C

4 CONCLUSIONS

In this paper, a life estimation scheme is proposed based on ADT for spacecraft power product, and estimating a certain model of secondary power supply for spacecraft is used as a case study. The sensitive stress and key performance of the secondary power supply is determined using the failure mechanism analysis. Then the ADT is designed based on the analysis and obtain the 2-dimensional ADT data. The Wiener process and accelerated model is used to delineate the degradation process. The Copula function is adopted to establish the degradation model for the secondary power supply. In order to obtain activation energy and other parameters of the secondary power supply, the IMF and MLE are utilized. Then the life estimation for the product under the designed environmental stress is performed. The results show that the method proposed in this work improves the efficiency of the test and fully utilizes the information related to life and reliability in the multivariable degradation data. The proposed scheme expands the estimation method from a single performance indication to an estimation of the whole system, which provides a solution for life estimation of spacecraft power product with small sample size.

5 REFERENCES

[1] **Zhang, S. Shi, J. Wang, J.L.** 2010. Satellite On-board Failure Statistics and Analysis, *Spacecraft Engineering*, 19(4), pp 41-46.

[2] **Mccallum, J. Thomas, R.E. Waite, J. H.** Accelerated testing of space batteries, *Goddard Space Flight Center*, NASA-SP-323.

[3] **Dong, S. Zhang, L. Liu, C.L. et al.** 2013. Study on acceleration life test for electronic devices of satellite, *Quality and Reliability*, 5, pp 6-9.

[4] **Fu, H.Y. Dang, W. Li, K.** 2014. Study and Application of Accelerated Life Test for Space-Electronic Equipments Based on a Performance Degradation Model, *Manned Spaceflight*, 20(3), pp 267-272.

[5] **Chao, M.T. 1999.** Degradation Analysis and Related Topics: Some Thoughts and a Review, Proc-Natl. Sci. Counc. ROC, 23(5), pp 555-566.

[6] **Lu, C.J. Meeker, W.Q.** 1993. Using Degradation Measures to Estimate a Time to Failure Distribution, *Technometrics*, 35(2), pp 161-174.

[7] **Aalen, O.O. Borgan, O. Gjessing, H.K.** 2008. *Survival and Event History Analysis: A Process Point of View*, New York: Springer.

[8] **Wu, B.J. Jin G.** 2010. Reliability modeling and analyzing of momentum wheel based on gamma process, *Value Engineering*, pp 25-27.

[9] **Wang, H.W. Mi, Q.L.** 2015. Lifetime prediction based on Gamma processes from accelerated degradation data, *Chinese Journal of Aeronautics*, 28(1), pp 172-179.

[10] **Liao, H.T. Elsayed, E.A.** 2004. Reliability Prediction and Testing Plan Based on an Accelerated Degradation Rate Model, *International Journal of Materials and Product Technology*, 21(5), pp 402-422.

[11] **Tang, J. Su, T.S.** 2008. Estimating Failure Time Distribution and Its Parameters Based on Intermediate Data from a Wiener Degradation Model, *Naval Research Logistics*, 55(3), pp 265-276.

[12] **Hu, C.H. Lee, M.Y. Tang, J.** 2015. Optimum step-stress accelerated degradation test for Wiener degradation process under constraints, *European Journal of Operational Research*, 241(2), pp 412-421.

[13] **Liu, Y. Yang, Z. Ju, H.J. Gan, L.** 2012. Electronic Products Accelerated Testing System Design and Implementation, Prognostics & System Health Management Conference.

[14] **Chen, W.H. Liu, J. Gao L.** 2011. Accelerated Degradation Reliability Modeling and Test Data Statistical Analysis of Aerospace Electrical Connector, *Chinese Journal of Mechanical Engineering*, 24(6), pp 957-962.

[15] **Crk, V.** 2000. Reliability Assessment from Degradation Data, Annual Reliablity and Maintainability Symposium, pp 155-161.

[16] **Wang, P. Coit, D.W.** 2004. Reliability Prediction based on Degradation Modeling for Systems with Multiple Degradation Measures, Proceedings of Annual Reliability and Maintainability Symposium (RAMS), pp 302-307.

[17] **Frees, E.W. Wang, P.** 2006. Copula Credibility for Aggregate Loss Models, *Insurance: Mathematics and Economics*, 38, pp 360-373.

KEY TECHNOLOGIES FOR PRECISION MANAGEMENT AND INTEGRATED CONTROL IN LARGE SCALE DIGITAL ALIGNMENT

Fuzhou Du[1], Ke Wen[1*], and Jie Jin[2]
[1]School of Mechanical Engineering and Automation
Beihang University, Beijing, China
du_fuzhou@163.com

[2]Beijing institute of astronautical systems engineering
Beijing, China
jayking928@126.com

ABSTRACT

Alignment is one of the most critical stages during large scale assembly. The accuracy and setup time of the alignment process would significantly determine the quality and the cycle time of the final product. With the rapid development of technologies of large scale digital metrology and flexible tooling, the deployment of measurement assisted assembly technology becomes the new tendency in digital alignment of large scale products, such as aircrafts, aerospace and ships. However, there is still a lack of investigation on that, how to integrate those separate devices or systems (for example measurement devices, servo-controlled fixtures and data management systems) together to take their fully advantages. Nowadays many manufacturers have purchased plenty of advanced devices and systems, but still know little about how to use them effectively to improve assembly precision and product quality. Focusing on the process of large scale digital alignment using digital metrology and flexible fixtures, this paper studies the key technologies for precision management and control based on the integration of different systems. The methodology includes an integrated framework, automatically positioning planning algorithm and process simulation and optimization technology. Additionally, an implementation example is given to verify the feasibility of the proposed methodology. The research will benefit other stages of large scale assembly as well.

Keywords: Precision management, digital alignment, large scale assembly, positioning planning, assembly simulation

1 INTRODUCTION

Large scale alignment, which usually involves large size and heavy components, is one of the critical processes during the assembly of complex products, such as satellite, aircraft and ship. The precision and time of the alignment process would significantly contribute to the product quality and production cycle [1-2]. During the traditional large scale alignment process, the relative positions and postures between large size components are assured by large fixtures and jigs, and their precisions rely on the templates and are calibrated by optical instruments. However, those traditional approaches mentioned above are obsolete due to low accuracy and inefficiency [3-4].

* Corresponding Author: Ke Wen, School of Mechanical Engineering and Automation, Beihang University, Xueyuan Road 37, Haidian District, Beijing 100191, China. Email: rongyu_wen@163.com

With the rapid development of digital and flexible assembly technologies, digital alignment system based on large volume metrologies and flexible fixtures become the valid solution and new tendency for improving product quality and comprehensive competition, by assuring high-precision and reduce cost and time during large scale assembly [5-10].

The alignment process of large scale components includes many operations, and the most important three operations are described as: 1) data collection based on the integrated applications of large volume digital metrologies, 2) data management and analysis based on related algorithms, and 3) process simulation and control based on the integrated application of flexible fixtures and related control systems [11-12]. Those three operations are also called precision management and control during large scale alignment of complex products, and the researches on them are significantly important for developing and applying a digital alignment system.

Therefore, in this paper, the authors firstly propose a novel framework for precision management and integrated control during large scale digital alignment, which implemented the closed-loop of precision control based on an integrated platform. Then, the data analysis algorithms and applicative methodologies of this plat-form are illustrated in section 3, and the key technologies for data collection and components adjustment based on the integration of multi devices are detailed in section 4. Finally, a digital alignment case of satellite components is studied based on the implementation of the integrated platform, in order to verify the feasibility and validity of the proposed framework.

2 FRAMEWORK OF THE PRECISION MANAGEMENT AND CONTROL PROCESS

The process of the precision management and control include four parts: digital measuring, data analysis and process simulation, positioning planning, and flexible control. A closed loop has been achieved among the four parts in this paper, as shown in Figure 1.

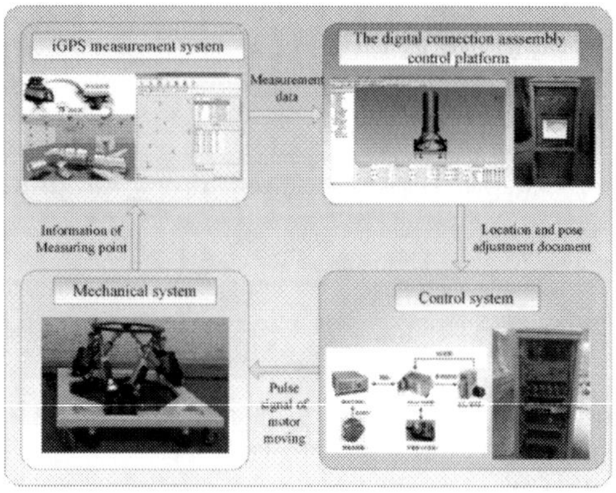

Figure 1: The closed loop process of precision management and control process

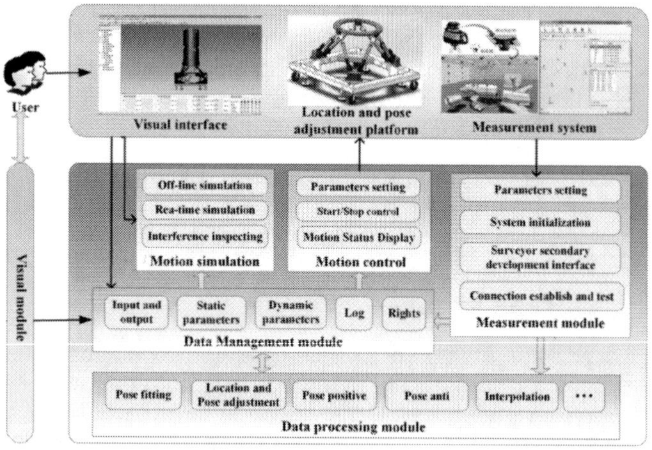

Figure 2: Functional structure of the integrated platform

As the core of the framework, the integrated plat-form acquires data from digital measurement system (Indoor Global Positioning System (iGPS) in this paper), and then analyzes those data based on built-in algorithms and knowledge to figure out the original and target position of large size components. After that, plans for adjusting the positions of the components are calculated and simulated in the three-dimensional virtual environment to verify their feasibility. Finally, the platform generates the plan documents of position and posture adjustment, as well as related movement simulation document which drive for the control system. Next, Control system sends pulse input, which is converted from displacement by processing interpolation, to drive the motors of the integrated control system for position adjustment. Meanwhile the digital measurement system instantly measures the positions of the target points on the components and sequential process repeat again and again, until the components are precisely aligned.

Figure 2 demonstrates the kernel functional structure of the software system for implementing the integrated platform. As the Figure 2 shown, the system firstly needs to implement the data transferring interfaces between digital measurement instrument and itself to collect required data. And then, built-in algorithms for position matching, coordination relocating, and position adjustment planning as well as movement simulation should be provided. Finally, it must have the functions for controlling the alignment process and provide a friendly user-interface to display actual and virtual process and critical data during those processes.

3 DATA ANALYSIS FOR PRECISION CONTROL

3.1 Coordination System Alignment

Multiple coordination systems are used during large scale digital alignment, such as assembly coordination system, geodetic coordination system, design coordination system, measurement coordination system and so on. The assembly process sets the measurement coordination system as alignment datum, and applies Rodriguez matrix based least square method to match position and location, also constructs transition matrix between measurement coordination system and other coordination systems. More specifically, the process establishes the transition matrix between geodetic coordination system and measurement coordination system by comparing actual points set data, retrieved from the actual value of each target point, with theoretical value of the corresponding standard point of earth. The reason for coordination matching is to provide a standard coordination system serving as a criterion for following processes.

3.2 Positioning Planning

Position Planning determines installed components' position and location for each moment based on its initial and target position and location. A reasonable plan requires smoothness during process, which, in other words, no sharp velocity and acceleration changes. The function of movements, velocities, accelerations and time in an acceptable solution is provided by inverse kinematics, further optimized for best time based on existed track.

The integrated platform exploits the mentioned planning and then generated position adjustment as well as movement simulation documents. The former records the information of the six-bar position adjustment mechanism at every moment which is used by UMAC controller to direct adjustment mechanism. And the latter stores position and location of alignment components at every moment for validating the plan by off-line simulation before actual aligning execution.

3.3 Alignment Process Simulation

Besides serving as the prerequisites of planning validating, adjustment track optimizing as well as alignment efficiency and quality enhancing, alignment process simulation also demonstrates, in the form of schematization, movement process of participated components more explicitly, which allows users to monitor components much more accurately and keeps safety during the process.

The proposed integrated platform of digital alignment provides off-line simulation and real-time simulation. The former simulates alignment process and inspect crash as well as interference between three-dimensional model of components prior to execution, which aims to validate scheme as well as modification. And its simulation documents, which records alignment track, are generated from adjustment planning on visualized UI of the integrated platform. And the latter aims to instantaneously solve the location and position of components, based on the information provided by iGPS measurement system during actual alignment. It illustrates the alignment process on its visualized UI and warns the potential crash as well as interference, i.e., a warning would jump out once the distance between participated components and environment is less than a given thresh-old. This method keeps safety of actual alignment, and offers real-time data from each sensor on its visualized UI [13-14].

3.4 Interference Inspection

The main purpose of alignment simulation is to inspect crashes or interferences between components. The crux of interference inspecting lies in real-time monitoring crash or overlapping events in geometrical enveloping area as well as determining specific region by rapid geometrical enveloping area calculation and displaying these events instantaneously during alignment process of three-dimensional components models. However with the fact that the models contains lots of sophisticated surfaces, Boolean intersection operation on visualized UI which is self-developed based on Open Cascade is unsatisfactory. In reality, it is the critical points on critical surfaces that may have crash or interference during vertical alignment, thus the author purposes, based on this fact, a monitoring and inspecting technology with respect to the critical points as well as critical measures, which is to pick some critical points-points, points-curve, as well as points-surface as objects according to the model and alignment process prior to simulation and then set up the threshold, i.e., the minimum distances.

The integrated platform monitors the distances between critical measures instantaneously during off-line alignment simulation, and it will warn and stop the simulation once the inspected distance reaches the threshold. Then it will adjust arguments and generates position adjusting documents. The interference inspecting has different function for off-line simulation or real-time simulation of components. As for former simulation the main purpose of interference inspecting is to validate the track documents for position adjusting based on initial and target position, which is, to determine the existence of crash or interference during alignment. While latter is to warn that the distance between large size components, or with environment, reaching the threshold during alignment process and then cease the process.

4 KEY TECHNOLOGIES FOR DATA COLLECTION AND COMPONENTS ADJUSTMENT

4.1 Data collecting Based on iGPS System

The integrated platform for digital alignment is mainly integrated with iGPS measurement system, control system and mechanical system, and it implements measurement data retrieval, mechanical system control as well as assembly process supervision.

Surveyor, the server software of iGPS indoor measurement system, is applied here to coordinate the function and configuration of the whole iGPS system. It provides five DLL-formed COM modules, which enable configuration, data visiting and modifying in iGPS system, in order to smooth data communication with clients. There are two methods to retrieve data from Surveyor server. One is to determine data port by the name of the frame, then retrieve real-time information about position and location from this port. The other utilizes COM ports can be connected to the receiver, which is used to inform applications that the position and location of the sensor has changed, an event previously registered by Surveyor server. Then it enables sensor's real-time coordination and location updating in event corresponding functions [15-16].

4.2 Component adjusting based on UMAC System

The integrated platform for digital alignment integrates control system mainly through secondary-development port Pcomm.dll and serial communication Mscom32.ocx provided by UMAC control. The integrated platform implements its direction on controller and obliquity sensor by DLL. It downloads position adjustment documents generated by adjustment planning to UMAC controller, then applies interpolation algorithm to transform pulse input into displacement and send it to driver for the purpose of directing motor to adjust the position and location of components. In the meanwhile, the integrated platform retrieves basic information of motor, absolute encoder, and obliquity sensor, etc., and then acquires necessary information, such as the length of rod , the speed of rod and pose and utilized for position and location positive solutions of simulation, through data collecting, analysing and processing.

5 CASE STUDY

Based on the proposed framework, algorithms and related key technologies, an integrated platform is developed using MFC technique. In this section, a case of digital alignment of satellite components is carried out, and the developed platform is applied into the process, in order to verify the feasibility of the framework, algorithms and technologies mentioned above. The entire process includes six phases, which are detailed below.

Initiation of the measurement system

Before the connection, by setting IP address and port on, the communication with iGPS measurement system is established; meanwhile through attaining name of sensor and its value, test for the connection is done.

Initiation of adjustment platform

After the communication between the integrated control platform and the paralleled adjustment platform, initialization and adjustment for the motion platform will be done, and related parameters of control system will be attained as well as the safety parameters of the motion platform. The statement parameters of the motion platform will be shown on the human machine interface real time.

Initial location and pose measurement

After the initiation of the integrated control platform, by attaining the value of the sensors which are to be tested, the location and pose of the component to be connected will be fitted and initialized. At the same time, the location and pose will be displayed on the floating panel on the bottom of the visual interface as shown in Figure 3.

Location and pose adjustment planning

After completion of initial location and pose measurement, both the initial and destination locations and poses could be attained, the operator could choose adjusted location and pose on the planning panel, and at the same time location and pose adjustment planning file and motion simulation file could be developed. Location and pose adjustment planning supports two adjustment modes. The first one is the six axis linkage quick motion mode. The second one is to adjust the pose on the initial location, and then to move to the destination location as shown in Figure 4.

Motion simulation

After finishing the location and pose adjustment planning, the integrated digital control platform will check the accuracy of the planning file by off line simulation and also check whether interference or collision will exist during the alignment process. The Figure 5 shows the process of off-line simulation of alignment components.

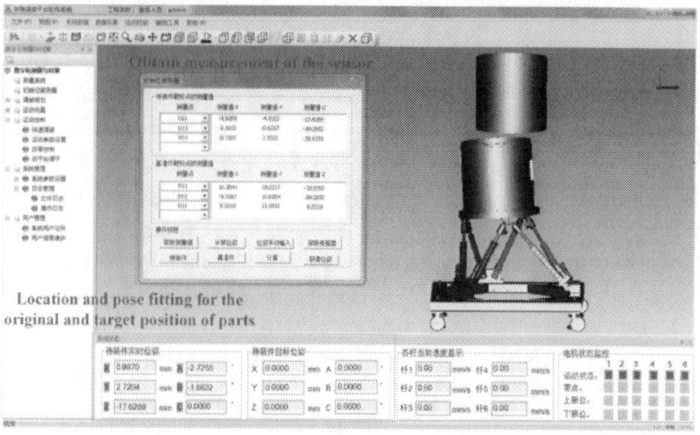

Figure 3: Initial location and pose measurement

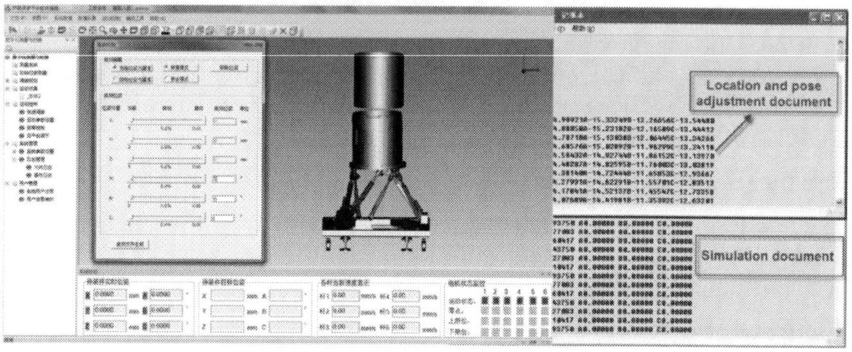

Figure 4: Location and pose adjustment planning

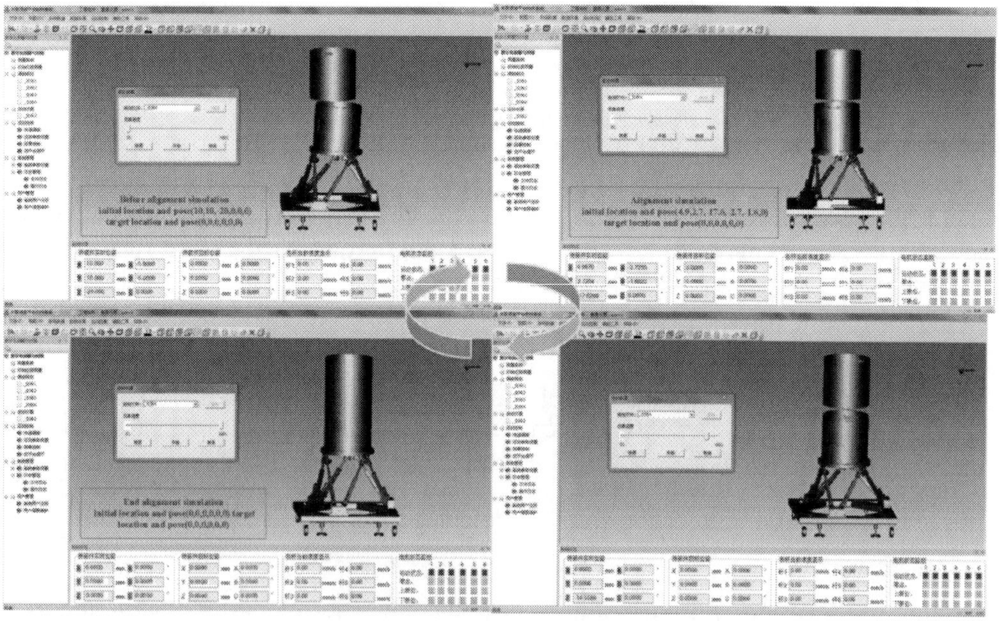

Figure 5: Off-line simulation of alignment components

Motion execution

This digital alignment software system provides an additional operation panel to quickly adjust the location and pose of the component to be connected, which is not based on measurement system as shown in Figure 6. It contains single axis quick motion, complex multiple axis motion, random adjustment of location and pose, which to high degree increase the flexibility of the adjustment.

Figure 6: Quickly adjustment of location and pose

6 CONCLUSION

In order to improve the precision and efficiency of large scale component alignment and assembly, large volume digital metrologies and flexible fixture technologies are more and more widely employed in the assembly of complex product such as satellite and aircraft. The integration of digital measurement devices, flexible fixture control system and related software become the tendency of digital and flexible large scale alignment; in this paper, a framework of the precision management and control process was proposed with an integrated platform, which aimed to implement the integration of iGPS system, data analysis software and fixture control system. The integrated platform was developed on the foundation of kernel algorithms for data analysing and key technologies for data collecting and components adjusting, which were detailed in the above sections.

The case of satellite components digital alignment which applied the developed integrated platform demonstrated that, the propose framework and its implementation were feasible for large scale digital alignment. The future work will focus on the improvement of related algorithms, and other possible applications of the proposed framework and technologies.

7 ACKNOWLEDGEMENT

This work is carried out under the support of Fund of National Engineering and Research Center for Commercial Aircraft Manufacturing (The project No. is SAMC14-JS-15-055) and the National Defense Basic Scientific Research, China (The project No. is A2120132007).

8 REFERENCES

[1] Maropoulos PG, Muelaner JE, Summers MD, et al. 2014. Paradigm in Large Scale Assembly-Research Priorities in Measurement Assisted Assembly, The International Journal of Advanced Manufacturing Technology, 70(1), pp 621-633.

[2] Maropoulos PG, Zhang D, Chapman P, et al. 2007. Key digital enterprise technology methods for large volume metrology and assembly integration, International Journal of Production Research, 45(7), pp 1539-59.

[3] Jamshidi J, Kayani A, Iravani P, et al. 2010. Manufacturing and assembly automation by integrated metrology systems for aircraft wing fabrication, Proceedings of IMechE Part B: J. Eng. Manuf, 224(1), pp 25-36.

[4] Jayaweera N, Webb P, Johnson C. 2010. Measurement assisted robotic assembly of fabricated aero-engine components, Assembly Automation, 30(1), pp 56-65.

[5] Mei ZY, Maropoulos PG. 2014. Review of the application of flexible, measurement-assisted assembly technology in aircraft manufacturing, Proc IMechE, Part B: J Engineering Manufacture, pp 1-13.

[6] Wen K, Du FZ. 2016. Key technologies for digital intelligent alignment of large-scale components, Computer Integrated Manufacturing Systems, 22(3), pp 686-694 [in Chinese].

[7] Mosqueira G, Apetz J, Santos KM, et al. 2012. Analysis of the indoor GPS system as feedback for the robotic alignment of fuselages using radar measurements as comparison, Robotics and Computer-Integrated Manufacturing, 28, pp 700-709.

[8] Peggs GN, Maropoulos PG, Hughes EB, et al. 2009. Recent developments in large-scale dimensional metrology, Proc IMechE, Part B: J Engineering Manufacture, 223(6), pp 571-595.

[9] Muelaner JE, Wang Z, Jamshidi J, et al. 2009. study of the uncertainty of angle measurement for a Rotary-Laser Automatic Theodo-lite(R-LAT). Proc IMechE, Part B: journal of Engineering Manufacture, 233(3), pp 217-229.

[10] Chen ZH, Du FZ, Tang XQ. 2013. Research on uncertainty in measurement assisted alignment in aircraft assembly, Chinese Journal of Aeronautics, 26(6), pp 1568-1576.

[11] Qiu BG, Jiang JX, Bi YB, et al. 2011. Study on posture alignment and joining test system for large aircraft fuselages, Acta Aeronautica et Astronautica Sinica, 32(5), pp 908-919 [Chinese].

[12] Zheng LY, Zhu XS, Liu RW, et al. 2013. A novel algorithm of posture best fit based on key characteristics for large components assembly, 12th CIRP Conference on Computer Aided Tolerancing, 10, pp 162-168 [Chinese].

[13] Muelaner JE, Maropoulos PG. 2008. Large Scale Metrology in Aerospace Assembly. Proceedings of DET2008, 5th International Conference on Digital Enterprise Technology, 2008.

[14] Du FZ, Chen ZH, Tang XQ. 2012. Precision analysis of iGPS measurement field and its application, Acta Aeronautica et Astronautica Sinica, 33(9), pp 101-109 [Chinese].

[15] Muelaner J, Kayani A, Martin O, et al. 2011. Measurement assisted assembly and the roadmap to part-to-part assembly, Proceedings of DET2011, 7th International Conference on Digital Enterprise Technology, 2011.

[16] Zhang B, Yao BG, Ke YL. 2009. A novel posture alignment system for aircraft wing assembly, J. Zhejiang Univ. Sci. A, 10(11), pp 1624-1630 [Chinese].

ROBUST MODULE PARTITION METHOD OF COMPLEX PRODUCT BASED ON IMPROVED EVOLUTIONARY ALGORITHM

Ming HUANG*
School of Mechanical Engineering and Automation
Beihang University, Beijing 100191, China
huangming@buaa.edu.cn

ABSTRACT

To improve the robustness of modular product to requirements change, a robust module partition method is proposed by introducing the robust design idea. Generalized and personalized or changeable requirements are identified as key noise factors. Components and modules change caused by requirements change is analyzed as influence of variation. For each module partition, the influence concentration degree is calculated as the robustness principle based on change-influence matrix between requirement noises and components, and the internal clustering and external independence are calculated simultaneously as the aggregation principle based on integrated interactions matrix. Bi-objective optimization model is established and an improved multi-objective evolutionary algorithm is used to address the problem and find the Pareto-optimal set. A selection method considering robustness is also used to select the best compromise Pareto solution. Finally, an industrial case study of coal mining machine is given to illustrate the practical application and rationality of the research.

Keywords: Module partition, Modularization principles, Robust design, Requirement change, Multi-objective optimization

1 INTRODUCTION

In today's highly competitive global business environment, modular design strategy has been proven to be beneficial for enterprises to reduce cost, increase product variety, decrease product lead time, cope with the emerging customized requirements, and improve manufacturing flexibility (Simpson [1]). Module partition is the basic of modular design. Much research has been carried out, from different viewpoints, in the module partition field. Modularization principles and partition mechanism are two major aspects.

Most of the modularization principles can be summarized as "Module aggregation + X". As a classic principle, the former focuses on multi-perspective interactions (i.e., energy flow, material flow, signal flow, behavior and structure) between components, and usually composed by degree of internal clustering and external independence (Wei [2]). The "X" represents a variety of supplementary principles, to emphasis some kind of preference of modular design, i.e., requirements from all stage of product life cycle related factors (Ji [3]). The aim of module partition mechanism is to divide all components into modules, and find a relative optimal solution. Classic mechanisms can be classified into three categories: 1) method based on the heuristic function flow analysis (Li [4]); 2) method based on homogeneous or heterogeneous fuzzy clustering analysis and multiple index evaluation (Zong [5]); 3) method based on multi-objective optimization model and intelligent algorithm (Wang [6]).

Fundation items: Project supported by the National Defense Technical Fundamental Research Foundation, China (No.JSZL2014204A002)

* Corresponding Author

Generalized requirements change inside and outside the enterprise will lead to a series of changes and fluctuations in the product itself, process and organization. After module partitioned, the customer's individual demand and technology evolution demand, as requirement noises, will lead to components change directly and indirectly, and the changes can be seen as influence of variation. It is necessary to make further study on module partition method considering requirements change (Wei [7]), nevertheless, relative studies are scant. This paper focuses on how to reduce the influence of requirements change, and guarantee the module partition as far as possible to meet the aggregation principle simultaneously. A strategy and method for robust module partitioning are proposed. Multi-objective optimization model is established and an improved multi-objective evolutionary algorithm is used to address the problem. Coal mining machine's module partition is performed as a case study, which proves the applicability and rationality of the proposed method.

2 ROBUST MODULE PARTITION STRATEGY

To improve the robustness of the module partition that is to reduce the influence of changes caused by the generalized requirement noises. According to the idea of robust design, three feasible strategies are proposed:

1. **Identify the key requirement noises.** Relative weight of requirement determines the priority and probability of its response. In the case of resource and time priority, the module partition should be firstly adaptive to the change of the key requirement noises.
2. **Improve the function structure of modular design process.** The former modular design process, as shown in figure 1(a), can be expressed as $Y = F(P(C,X),Z)$. The interaction process of Z lagged behind the forming process of M, as a result, the nonlinear effect determined by M is not well-targeted to reduce the influence caused by Z. The improved process, as shown in figure 1(b), has a sub process for analyzing change influence embedded within it, then can be expressed as $Y = F(P(C,X,R(Z)),Z)$. Since M becomes a function of Z, it will help improve the nonlinear effect better-targeted to reduce the influence caused by Z.

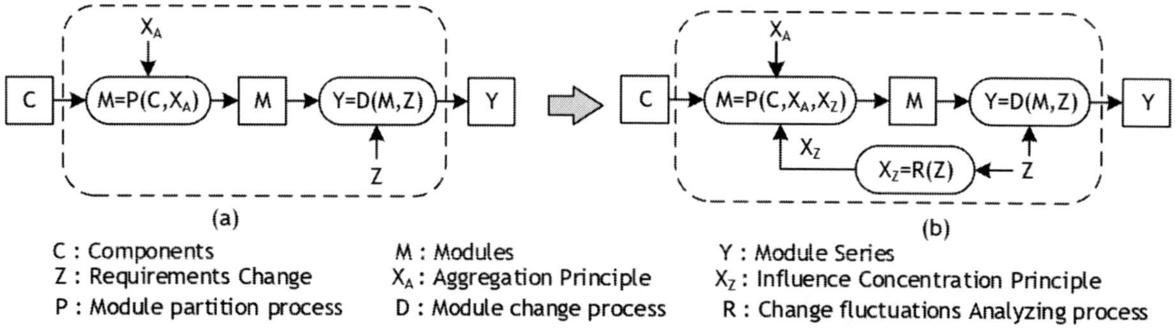

Figure 1: Improvement of modular design process

3. **Select an appropriate weight combination for modularization principles.** In the module partition process function, the weights of two principles are adjustable variables, which determines the best compromise selecting from the Pareto solution. Different weight combination corresponds to different compromise. The greater the weight of influence concentration principle, the better the nonlinear effect targets to reduce the influence. However, the aggregation principle may driver partition in another direction, so that an appropriate selection is important and necessary.

3 MATHEMATICAL MODELING OF BI-PRINCIPLE MODULARIZATION OPTIMIZATION PROBLEM

Suppose the system consists of a total of N components; M is the number of partitioned modules; and N_i is the number of components in the ith module $Moudle_i$, $i=1,2,\cdots,M$.

3.1 Influence concentration principle

3.1.1 Key Requirement noises identification

The generalized requirement noises include three categories: 1) the individual requirements or requirement change from the customers; 2) the requirement change caused by technology evolution; and 3) the requirement change caused by relative policy and regulation modified. In this paper, we focus on the first category, including individual requirements distributing in space and changing in time. Including the different needs of the proposed explicit for the product function, performance, structure, and hidden by the user and the environment are differences of performance requirements, although the expression is similar, especially the latter requires careful analysis and mining.

Based on customer requirements analysis, the key requirement noises set $RN=\{RN_1,RN_2,\cdots,RN_V\}$ are identified after rough selection and removal of redundancy. The weight vector $w=[w_1,w_2,\cdots,w_V]$ of RN is confirmed by compared with each other, where $0\le w_v \le 1$ and $\sum_{v=1}^{V} w_v =1$.

3.1.2 Change influence analysis

According to the key requirement noises identified, change influence can be analysed based on interactions between design requirements, product functions and physical components. A change-influence degree matrix C of every physical component caused by every requirement noise is built (Kreng [8])

$$C=\begin{bmatrix} c_{11} & c_{12} & \cdots & c_{1N} \\ c_{21} & c_{21} & \cdots & c_{2N} \\ \vdots & \vdots & & \vdots \\ c_{V1} & c_{V2} & \cdots & c_{VN} \end{bmatrix}, \tag{1}$$

where c_{vn} means the influence degree of the nth component caused by the vth requirement noise. According to the evaluation theory of fuzzy mathematics, the values of c_{vn} are subjectively assigned by the designer according to the scales of 9, 7, 4, 1, 0, with the scores of 9 to 0, each representing highest, higher, middle, lower, lowest probability, respectively. This specific scale was chosen as it has proven effective in various previous studies for similar settings (Wang [9]).

Requirements change will lead to components change directly and indirectly, and the influence of module can be expressed by the total influence of components change actively and passively, as a result, the components with high change frequency due to the same requirement noise should be combined together. The degree of the influence concentration F_R is

$$F_R=\sum_{v=1}^{V} w_v f(RN_v)=\sum_{v=1}^{V} w_v \left\{1-\frac{-1}{\ln(M)}\sum_{m=1}^{M}\left[\frac{\sum_{n=1}^{N_m} c_{vn}}{\sum_{n=1}^{N} c_{vn}} \ln\left(\frac{\sum_{n=1}^{N_m} c_{vn}}{\sum_{n=1}^{N} c_{vn}}\right)\right]\right\}, \tag{2}$$

where $f(RN_v)$ reflects the fuzzy entropy between all modules with the vth requirement noise. The higher is the value of F_R (the closer is to 1), the better the modules are partitioned.

3.2 Module aggregation principle

3.2.1 Integrated interactions analysis

According to multi-perspective interactions analysis (i.e., energy flow, material flow, signal flow, behavior and structure) between components, an integrated interactions matrix R is constructed

$$R = \begin{bmatrix} r_{11} & r_{12} & \cdots & r_{1N} \\ r_{21} & r_{21} & \cdots & r_{2N} \\ \vdots & \vdots & & \vdots \\ r_{N1} & r_{NN} & \cdots & r_{NN} \end{bmatrix}, \quad (3)$$

Where r_{ij} is the integrated interactions between ith component and jth component, and $0 \le r_{ij} \le 1$.

3.2.2 Internal clustering

The mth module's internal clustering degree F_{Im} can be calculated using Eq. (4), whereas all the modules' combined clustering F_I can be computed via Eq. (5).

$$F_{Im} = \sum_{i=1}^{N_m-1} \sum_{j=i+1}^{N_m} r_{ij} \Big/ \frac{N_m(N_m-1)}{2}, \quad (4)$$

$$F_I = \sum_{m=1}^{M} \left[\sum_{i=1}^{N_m-1} \sum_{j=i+1}^{N_m} r_{ij} \Big/ \frac{N_m(N_m-1)}{2} \right], \quad (5)$$

3.2.3 External independence

The external independence F_E can be measured by the interactions between components according to the integrated interactions matrix R. Eq. (6) is used to calculate the combined coupling degree R among all modules.

$$F_E = \sum_{p=1}^{M-1} \sum_{q=i+1}^{M} \left(\sum_{i=1}^{N_p} \sum_{j=1}^{N_q} r_{ij} \Big/ N_p N_q \right) \quad (6)$$

In general, the internal clustering and external independence of all modules are interrelated, an aggregate index form is used to reflect the module aggregation principle, as Eq. (7).

$$F_A = \frac{F_E}{M \bullet F_I} \quad (7)$$

3.3 BI-principle modularization optimization model

From a mathematical point of view, a bi-objective modularization optimization model for module partition can be expressed as the following Eq. (8).

$$\begin{cases}\min F_A \\ \max F_R\end{cases}, \quad \textit{subject to}$$

$$N = \sum_{m=1}^{M} N_m \tag{8}$$

$$M_{\min} \le M \le M_{\max}$$

4 OPTIMIZATION BASED ON IMPROVED EVOLUTIONARY ALGORITHM

As a new model of multi-objective genetic algorithm (MOGA), ISPEA2 can be regarded as a particular type of SPEA2 with three additional mechanisms (Kim [10]): (1) neighbourhood crossover that allows crossing over individuals located near each other in the objective space; (2) mating selection that reflects all good solutions within the archive; (3) application of two archives to maintain diverse solutions in both objective and variable spaces.

The algorithm flow of ISPEA2 is as follows (Wei [2]):

Procedure: ISPEA2
Parameters: N -population size
$\underline{N}$ -archive size
T -maximum number of generations
Begin
Initialization:
Generate an initial population P_0 and N random individuals
Create two empty archives: A_0^O and A_0^V
Repeat
Fitness assignment:
For each individual in P_t, A_0^O and A_0^V
Environmental selection:
From P_t, A_0^O and A_0^V creating new archives A_{t+1}^O , A_{t+1}^V
If
the number of individuals in A_{t+1}^O and $A_{t+1}^V \ge \underline{N}$
Then
Archive truncation in the objective space A_{t+1}^O, and archive truncation in the variable space A_{t+1}^V
End if
Neighbourhood crossover and mutation operation:
Generate P_{t+1} by copying A_{t+1}^V
$t = t + 1$
Until $t \ge T$
Print all non-dominated solutions in the final population and archive
End

5 CASE STUDY

Coal mining machine is a set of mechanical, electrical and hydraulic components as one of the large type complex system. Fig. 2 illustrates a structure of a typical coal mining machine, and Table 1 summarizes a list of its key components. According to the analysis of different operational environment, key requirement noises RN could be specified: (1) Cut, (2) Lift, (3) Load, (4) Tow, (5) Electronic control, (6) Auxiliary, and the weight vector is set as: $w = [0.2, 0.2, 0.2, 0.15, 0.15, 0.1]$.

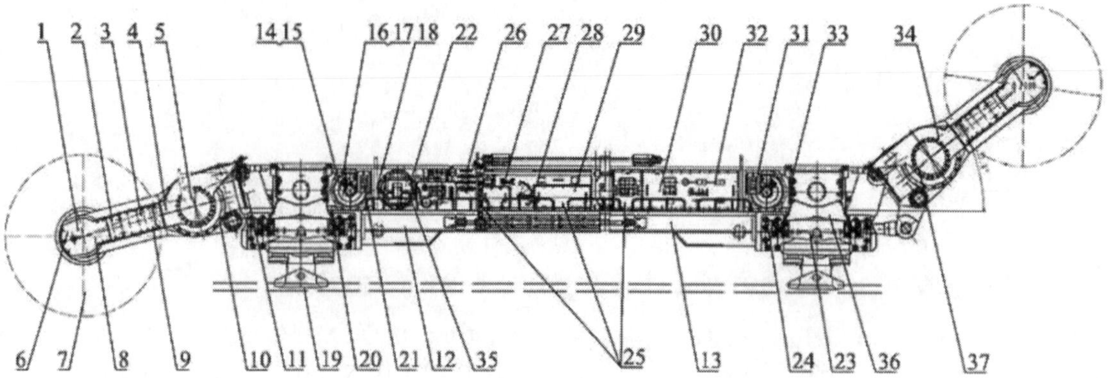

Figure 2: Structure of a typical coal mining machine

Table 1: List of key components of a typical coal mining machine

No.	Name	No.	Name	No.	Name
C1	Cutting motor	C14	Traction motor	C27	Pressure relay
C2	Rocker gear box	C15	Transmission shaft 3	C28	Oil level gauge
C3	Transmission shaft 1	C16	Transmission shaft 4	C29	Pressure gauge
C4	Double planet speed	C17	Double planet speed reducer 2	C30	Reversing valve handle
C5	Transmission shaft 2	C18	Transmission shaft 5	C31	Rings
C6	Roller seat	C19	Guide boot	C32	Cylinder body
C7	Cutting drum	C20	Outer traction shell	C33	hydraulic control one-
C8	Brake 1	C21	Joint	C34	Oil cooler
C9	Pick sprayer	C22	Pump motor	C35	Motor cover
C10	Clutch handle	C23	Adjustable high pump	C36	Removal tool
C11	Brake 2	C24	Oil tank	C37	Lubricating device
C12	Main frame 1st section	C25	Value group		
C13	Main frame 2nd section	C26	Energy storage device		

The results of evaluating the change influence are summarized in Table 2. And the results of developing the integrated interactions between components are illustrated in Table 3.

Table 2: Change influence from requirement noises to components

	C1	C2	C3	...	C35	C36	C37
RN1	9	4	4	...	0	0	4
RN2	0	7	7	...	0	0	1
RN3	0	0	0	...	0	0	0
RN4	0	0	0	...	0	0	1
RN5	1	0	0	...	0	0	0
RN6	0	0	0	...	1	1	1

Table 3: Integrated interactions between components

	C1	C2	C3	...	C35	C36	C37
C1	1	0.36	0.51	...	0	0	0.21
C2	0.36	1	0.380	...	0	0	0.18
C3	0.51	0.38	1	...	0	0	0.17
⋮	⋮	⋮	⋮		⋮	⋮	⋮
C35	0	0	0	...	1	0	0
C36	0	0	0	...	0	1	0
C37	0.21	0.18	0.17	...	0	0	1

According to the bi-principle modularization modelling method, the bi-objective modularization was carried out under the constraint conditions. Different solution will be selected when considering the aggregation principle and influence concentration principle of different weights. Based on the strategy explained in Section 2.2, the relative best and robust compromise solution is selected. The result of modules partitioned is shown in Table 4.

Table 4: The best compromise solution of module partition

No.	Module name	Components contained	No.	Module name	Components
1	Rocker arm module	C1,C2,C3,C4,C5,C6,C10	5	External traction module	C19,C20
2	Cutting drum module	C7,C8,C9	6	Pumping station module	C22,C23,C24,C25,C26,C27, C28,C29,C30
3	Main frame module	C12,C13	7	Auxiliary module	C34,C35,C36,C37
4	Traction and Drive module	C11,C14,C15,C16,C17,C18, C21	8	Cylinder height adjustment module	C31,C32,C33

6 CONCLUSION

This paper focuses on the strategy and method of module partition, to reduce the influence of requirements change in the following Configuration design based on modular product platform. Three strategies are proposed: (1) identify the key requirement noises, (2) improve the function structure of modular design process, and (3) select an appropriate weight combination for modularization principles. According to these strategies, a bi-principle modularization method is developed, which is transformed into a mathematical optimization model and solved as a MOOP. The case study of coal mining machine proves the applicability and rationality of the research.

The integrity of identifying generalized requirement noises, accuracy of analyzing change influence, and objectivity of evaluating interactions determine the reliability of final compromise solution. The significance of this study lies in proposing the strategies and establishing a process approach model to solve the problem, more efficient and accurate method is expected.

7 REFERENCES

[1] **Simpson, T.W. 2004**. Product platform design and customization: status and promise. AI EDAM: Artificial Intelligence for Engineering Design, Analysis and Manufacturing, 18(01):3-20.

[2] **Wei, W. 2015.** A multi-principle module identification method for product platform design. Journal of Zhejiang Universityence A, 16(1):1-10.

[3] **Ji, Y.J. 2013.** Modular design involving effectiveness of multiple phases for product life cycle. The International Journal of Advanced Manufacturing Technology, 66(9-12):1475-1488.

[4] **Li, Z.K., 2013.** An integrated method for flexible platform modular architecture design. Journal of Engineering Design, 24(1):25-44 (in Chinese).

[5] **Zong, M.Z. 2003.** A Multi-Perspective Hierarchical Division Approach for a Product in Its Modular Design. Transactions of Beijing Institute of Technology, 23(5):552-556 (in Chinese).

[6] **Wang, H.J. 2005.** Numerical programming approaches for the development of modular product family. Journal of Computer Aided Design & Computer Graphics, 17(3):473-478 (in Chinese)

[7] **Wei, W. 2015.** Module division method of robust product platform based on improved artificial immune algorithms. Computer Integrated Manufacturing Systems, 21(4):885-893 (in Chinese).

[8] **Kreng, V.B. 2003.** Product family design with grouping genetic algorithm-a case study. Journal of the Chinese Institute of Industrial Engineers, 20(4):373-388.

[9] **Wang, H.J. 2006.** Variant configuration design supporting personalization product customization. Chinese Journal of Mechanical Engineering, 42(1):90-97(in Chinese)

[10] **Kim, M. 2004.** SPEA2+: improving the performance of the strength Pareto evolutionary algorithm 2. In: Yao, X., Burke, E.K., Lozano, J.A., et al. (Eds.), Parallel Problem Solving from Nature-PPSN VIII. Springer Berlin Heidelberg, LNCS 3242:742-751.

MANAGEMENT OF AEROSPACE PRODUCT QUALITY DATA PACKAGE

SONG Shu-yuan[1], DUAN Gui-jiang [1*], LIU Tao-tao[1], YU Lin-yu[2]

[1]Department of Industry & Manufacturing System Engineering, Beihang University, China
[2]Beijinjing Jing-Hang Insistute of Cumputing and Communication, Box.7200-9, Beijing, China
emailssy@163.com, gjduan@buaa.edu.cn,hbliutaotao@126.com, yu_linyu@sina.com

ABSTRACT

Aerospace product quality data package is the objective record of product quality and reliability conditions, it's also the important basis of the implementation of process control, quality improvement and the acceptance of product delivery. The quality data package management system has great significance for enterprises to manage the data package documents more standardized and efficiently. Aiming at the quality management of the aerospace product, the function of the quality data package was analyzed, a multi-dimension division method of the quality data was presented, and management model of the quality data package was established. Aerospace product quality data package management is one of the important contents of the space model implementation of quality management, and one of the major basic works throughout the whole process of model development. It is important to improve the quality of aerospace products, enhance the quality and efficiency of aerospace model development and improve the capability of product assurance by carrying out the product assurance work with quality data package as the core and promoting the systematization, standardization and refinement of product assurance.

Keywords: Quality Data Package, Aerospace Products, Product Assurance, Management System

1 INTRODUCTION

With the increasing of aerospace product models and batches, the escalating of product complexity, and the rapidly growing of aerospace model quality data, facing of the massive product quality data, improving the reliability of product quality and ensuring the consistency of product quality data has become the urgent need for product quality management model. Aerospace product quality data package refers to both the requirements of the product quality and reliability in the development, production, inspection, delivery and other production processes of aerospace product, and the record of implementation process and results[1-2]. Product quality data package is not only an objective record of product quality and reliability characteristics, but also the objective evidence of the quality of product implementation and reliability, as well as the important basis for implementing process control, quality improvement and product delivery, in addition, an important sign of maturity stage of product.

2 QUALITY DATA PACKAGE FUNCTIONS AND MANAGEMENT REQUIREMENTS

* Corresponding Author: DUAN Gui-jiang, gjduan@buaa.edu.cn. SONG Shu-yuan, emailssy@163.com.

2.1 Analysis of the quality data package functions

As a new concept and management model proposed recently in the field of product quality data management, the production and usage of aerospace product quality data package has provided a new approach to manage various quality data in product development, which is more accurate, timely and standardized. The product quality data package contains the quality data information of each stage in the product life cycle. Based on the product quality control requirement and the quality data recording requirement from the whole production process of the product, data package is create from different forms of detailed record of quality data during all stages including design, manufacture, test validation and evaluation of delivery[3]. With the processing of the product development, more and more quality data are accumulated. This data is like product archive, which can be easily found and used whenever it's needed. The formation process of data package itself is the record of various types of quality data and status during the product development period. In addition, it's real and timely reflection of development staff's specific business. To a certain extent, the data can supervise the operation of the staff. With the processing of product testing, the data package becomes more and more perfect, which can be delivered to the user together with production[4].

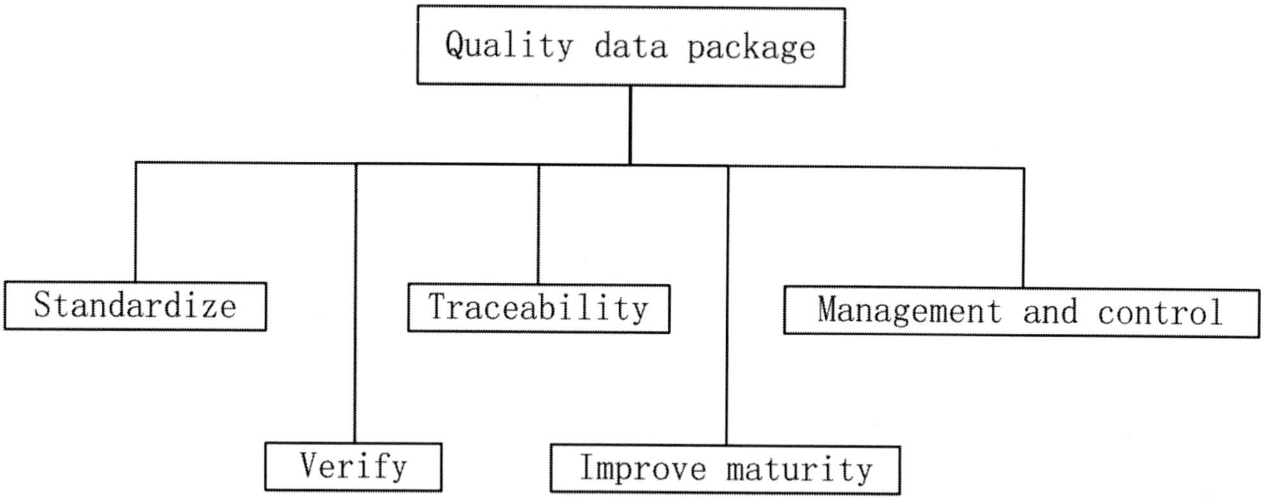

Figure 1: The Quality Data Package Functions

In order to collect, sort, statistic and analyze a large number of quality data of the model products, it is urgent to build a unified data structure around the quality data of the model products. At the same time, with the means of informatization, realize the collection, statistics and analysis of model product quality data. And completely control the quality information of components and parts of the product, which are in the sample and finalizing stages.

2.2 The purpose of quality data package management

The purpose of the construction of the quality data package management system is to provide a quick, convenient, uniform and standard approach for the aerospace enterprises to manage the product quality data packages so as to ensure the reliability of the quality data source of the products, the timely and standardized process, as well as the traceability of the results. The advantages of information technology provide a way for the management of complex and massive quality data, and clarify the clue. At the same time, the objectivity and standardization of the system eliminate the uncontrollable phenomenon of data acquisition distortion and data circulation process caused by subjective factors. Product quality data established through a quality data management system can be traced back to detailed reliability information for any component at any stage of development, as well as to manufacturer-related information for raw materials, also can be traced back to where it was installed on the product. This is conducive to aerospace products for quality analysis and control[5-7].

The quality of the data package management system is designed to achieve the quality of data packages integrity, correct, standardized, system management and control, as well as identify and control the key features of the product data. It ensures the traceability of product life cycle information. By Repeated iterations, gradually realize the meticulous management of entire aerospace products process. With the rapid development of information technology, the establishment of an effective integrated information resources platform and information management system has become an urgent need to adapt to enterprise's demands, while information management itself can greatly improve the quality and reliable data management efficiency, standardization and accuracy[8-10].

3 MULTI-DIMENSIONAL DIVISION OF QUALITY DATA

As the quality of data packages contain all the product life cycle, it involves multiple projects and the structure is so complex. The system will serves different users such as production workers, managerial personnel and production users and so on after the completion of the package management system. All the related users have their own requirements, so it is necessary to make comprehensive and systemic analysis in the early stages to make the system management tasks clear and to meet user's demand maximally. All the data item in the data package can be divided and categorized from multiple dimensions .In that case, we can emphasize the key points of different users and system function. It is better to meet the demands of users and manage the data package if the system was designed and planned by multiple dimensions.

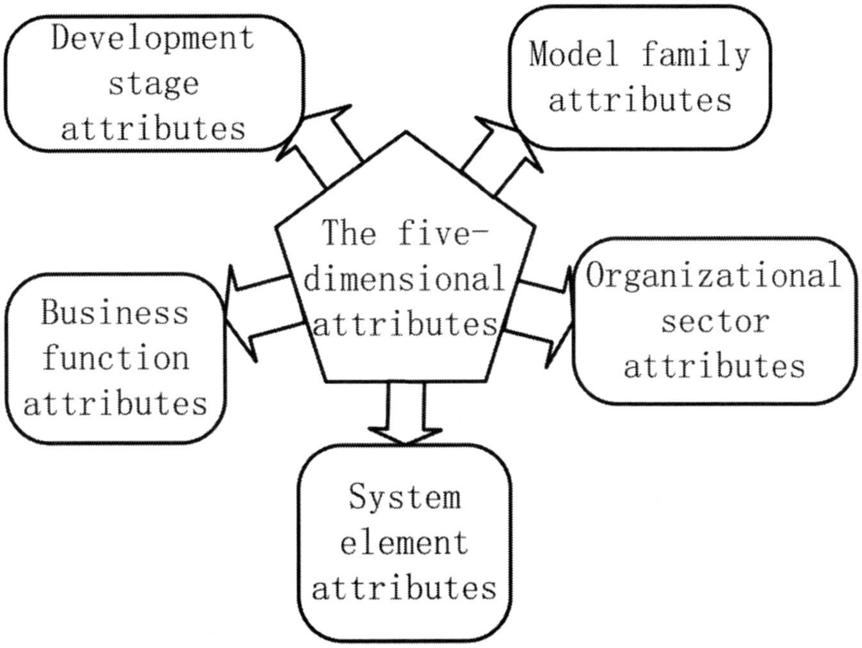

Figure 2: The Five-Dimensional Attributes of Quality Data

4 THE MANAGEMENT MODEL OF QUALITY DATA PACKAGE

Quality data package start with the product development and change in different product period. Finally, it will be delivered to users with product hardware as the basis for further use.

For each phase, the system user (employee) will operate, record and upload data as the quality data package required in this phase. First of all, the user select the data sample, fill in the sample after operation and then they upload data and submit to receiver. Secondly, the receiver will check whether the data and forms meet the requirements or not. If not, the receiver will inform the sender to fill in completely. Finally, the complete data package will be submitted with the product to store the quality data package and deliver.

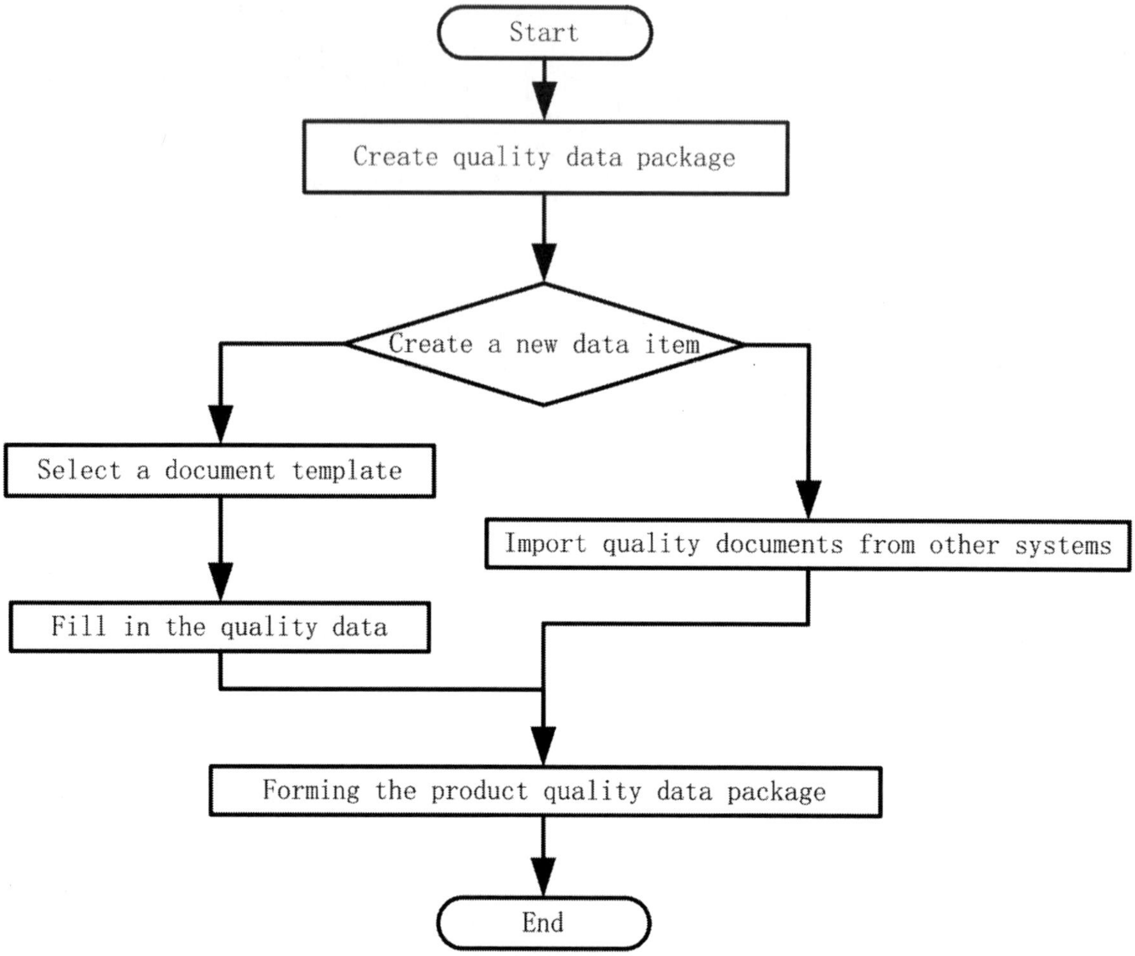

Figure 3: The Business Process Model of Quality Data Package Management

The quality and reliability data of aerospace product is too huge and the structure is very complex, involving the organizational structure, product structure, development stage and other dimensions. During the period of data collection, it is necessary to specify data collection item and content which should contain all the regulations and requirements of company. When it comes to integrate data, the quality data document should contain symbolic characters to help integrate data based by different dimensions. When it comes to make plan for system data structure, we should take the achievable and optimized system function into consideration.

5 CONCLUSION

The core idea of managing quality data package is to achieve informatization, standardization, convenience and timeliness. In summary, we analyze the content and function of aerospace product and clarify the goal of quality data management in first few paragraphs of this paper. Then, we put forward the multi-dimension partition method for data resources and provide solutions to solve the problem of data package management.

6 ACKNOWLEDGEMENT

This research is funded by the National High Technology Research and Development Program of China (863 Program) No(2015AA042101) and Natural Science Foundation of China (No.51175025).

7 REFERENCES

[1] **YUAN Jia-jun.** 2011. Aerospace Product Engineering, *China Astronautic Publishing House.*

[2] **YUAN Jia-jun.** 2009. Aerospace Product Quality and Reliability Data Package and Its Application, *China Quality*, 2009(4), pp 8-10.

[3] **TANG Xiao-qing, DUAN Gui-jiang, DU Fu-zhou.** 2009. Implementation Technology of Manufacturing Enterprise Quality Information Management System,*National Defense Industry Press.*

[4] **YANG Yang, LEI Zhi-feng, JI A-Qiang.** 2013(4). Construction and Management of Data Quality Package for Aerospace, *Aerospace Manufacturing Technology.*

[5] **ZHOU Xiu-bin, SU Qing, YANG Hai-cheng.** 2011. Content - oriented Product Quality Information Management of Aerospace Model, *Future and Development* , 2011(4), pp 50-53.

[6] **LIU Xu-yi.** 2012. The Idea and Method of Reliability Data Packet Management. *Aerospace Standardization*. 2012(1), pp 41-44.

[7] **TANG Xiao-qing, DUAN Gui-jiang.** 2001. The Informatization of Enterprise Quality Assurance System. *China Quality*, 2001(2), PP 35-37.

[8] **ZHANG Gen-bao, HE Zhen, LIU Ying.** 2006. Quality Management and Reliability. *China Science and Technology Press.*

[9] **Deitel P J, Deitel H M.** 2011. Java for Programmers[J]. *Prentice Hall.*

[10] **Duckett J.** 2008. Beginning Web programming with HTML, XHTML, and CSS[M]. *John Wiley & Sons.*

ENHANCE PERFORMANCE OF COORDINATE MEASURING MACHINE BASED ON UNIFORM DESIGN

GuijiangDuan[1*]and TaotaoLiu[1]
Zhuobing Hou[2]
[1]School of Mechanical Engineering and Automation, Beihang University, China
[2]AVIC Aircraft Co., Ltd. Xi'an Branch, China
gjduan@buaa.edu.cn , hbliutaotao@126.com
Ly6863366@163.com

ABSTRACT

Aiming at the low productivity of the CMM, a new measurement schemewhich could enhance the performance of the CMM is proposed in this paper. The uniform design is used in the measure points planning. The number of the measure points is reduced. The regression equation is used to analysis the result of this measurement scheme. This study indicates that the uniform design could increase the efficient of the CMM.

Keywords: Uniform Design, CMM, quality

1 INTRODUCTION

In order to improve the quality of products, it is necessary to inspect the products and its parts.[1]The advanced inspection equipment are widely used in quality control for several years, especially the CMM（Coordinate Measuring Machine）,owing to its high accuracy and precision. But the CMM is a time-consuming process with low productivity. Therefore, it has been critical to reduce measurement time for efficient performance of inspection process.

Syed HammadMian et al. introduced methodologies to generate most suitable measurement path resulting into minimum inspection time. These methodologies are based on different algorithms to reduce measurement cycle time for CMM[2]. JiandLi optimized measurement path for digitization of supercharger impeller by Ant Colony Optimization (ACO) algorithm[3].LiangshengQu et al. suggest genetic algorithms to optimize the measuring path on CMMs. The optimized operation can save nearly one-third of the total measuring time[4]. Zone-Ching Lin and Chein-Chung Chen proposed the nearest neighbour method and the refinement method to plan the measuring sequence[5]. C.G. LU et al. studied an approach of using an artificial neural network technique to carry out the inspection path management for a CMM, especially for multi-component inspection[6]. Measurement time can be determined by measurement path in the inspection process. Widespread research of measurement path sequence proved that suitable measurement path resulting into minimum inspection time. But limited research work has been found appropriate measure points can reduce measurement time.

In the inspection process, if the measuring points are appropriate, it will improve the inspection efficiency. Thus, the distribution of measuring points is important. In this paper, uniform design （UD） has been carried out to plan the measure points, then the regression analysis is used for the

[*] Corresponding Author:GuijiangDuan,gjduan@buaa.edu.cn.TaotaoLiu,hbliutaotao@126.com

measure results to find the maximum error point. We measure the maximum error point and compare with the theoretical point, In order to determine whether a workpiece is of tolerance.

2 UNIFORM DESIGN

In this section, we will start with a brief introduction on uniform design. Like orthogonal designs, uniform designs offer lots of experimental tables for users to conveniently utilize. Orthogonal design is the most popular fractional factorial designs, which can significantly reduce the number of experimental runs without losing the main information. However, the number of orthogonal experimental design sometimes is still too high for experimenters[7]. The uniform design method has been developed by Fang and Wang from number theory. In a uniform design of experiments, experimental trials are uniformly arranged within the designed scopes of investigated parameters, and the number of the experimental trials is equal to the maximum number of the designed grades for the investigated parameters[8].

The key idea for uniform design is to choose the most representative points in the experiment domain so as to obtain as much information as possible from the underlying model by as few experiments as possible. We use the uniform design in the inspection process.it will reduce the number of measuring points from n^2 to n.

3 THE DATA ANALYSIS OF UD

Since the number of experiments in UD is too small compared with the number of levels of factors in experiments, it is difficult to use the ANOVA model to fit the experiment data, because it is impossible to estimate all main effects of the factors. Thus, regression analysis is the main tool for data analysis[9].

The hypothesis of this paper is that the measuring points are on the same plane. The measured plane is shown in Fig1.

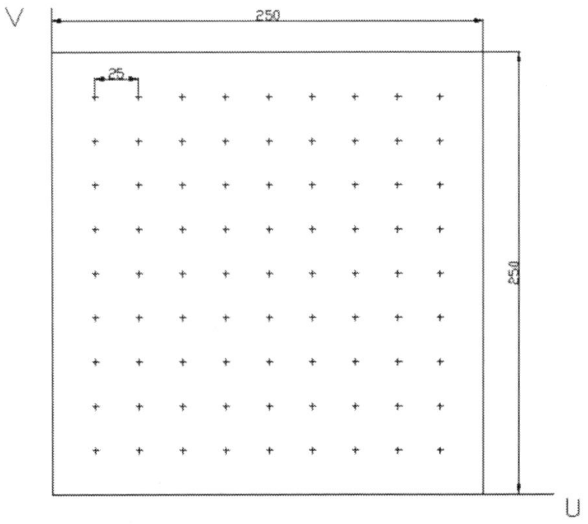

Figure 1 inspection plane

There are 2 factors influencing the value of the measuring points, the 2 factors are $X_1$and X_2.Using the second-order polynomial model for the regression equation.

$$y=\beta_0+\beta_1 x_1+\beta_2 x_2+\beta_3 x_1^2+\beta_4 x^2+\beta_5 x_1x_2+\varepsilon \quad (1)$$

$$E(\varepsilon)=0,\ D(\varepsilon)=\sigma^2<+\infty \quad (2)$$

Y stands for the normal direction coordinate values of the measuring points,$\beta_0,\beta_1,\beta_2,\beta_3,\beta_4,\beta_5$ are regression coefficients which are independent of X_1 and X_2.ε is Random error and $\varepsilon \sim N(0, \sigma^2)$.

4 CASE STUDY

According to the traditional distribution rule of measuring points, there are 81 measuring points in fig1. This paper use $U_9^*(9^4)$ for planning measuring points, then there are only 9 measuring points in the same plane.

From the uniform design tables, one can see that uniform design allows the largest possible amount of levels for each factor, that is, the number of levels could be equal to the number of experiment runs.

Table 1 Uniform Design $U_9^*(9^4)$

	1	**2**	**3**	**4**
1	1	3	7	9
2	2	6	4	8
3	3	9	1	7
4	4	2	8	6
5	5	5	5	5
6	6	8	2	4
7	7	1	9	3
8	8	4	6	2
9	9	7	3	1

Table 2 use table of $U_9^*(9^4)$

S	**Columns**			**D**
2	1	2		0.1574
3	2	3	4	0.1980

There are 2 factors X_1,X_2in this paper, according to Table 2 ,we use the columns1 and columns2 of Table 1.The coordinate values of measuring points are(1,3),(2,6),(3,9),(4,2),(5,5),(6,8),(7,1),(8,4)and(9,7), as the circled points show in Figure 2.

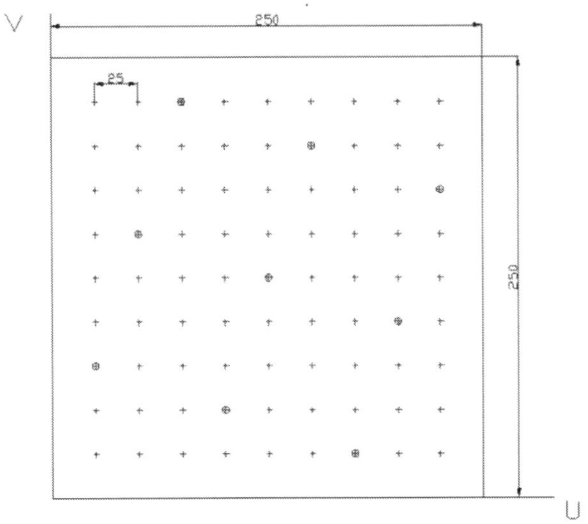

Figure 3 UD inspection plane

The mock data of the measuring points are shown in Table 3.

Table 3 The mock data of the measuring points

X_1 \ X_2	1	2	3	4	5	6	7	8	9
1	-0.029	-0.018	-0.007	-0.002	0.000	-0.002	-0.009	-0.016	-0.033
2	-0.011	0.000	0.011	0.018	0.016	0.014	0.007	-0.004	-0.019
3	0.001	0.011	0.025	0.031	0.028	0.026	0.020	0.006	-0.008
4	0.012	0.026	0.030	0.033	0.030	0.038	0.023	0.014	-0.002
5	0.018	0.026	0.034	0.033	0.039	0.030	0.030	0.018	0.000
6	0.020	0.024	0.035	0.044	0.043	0.031	0.024	0.012	-0.002
7	0.016	0.029	0.034	0.042	0.035	0.028	0.024	0.011	-0.006
8	0.008	0.022	0.022	0.030	0.028	0.020	0.011	0.000	-0.018
9	-0.002	0.010	0.017	0.016	0.018	0.010	0.001	-0.011	-0.026

Following $U_9^*(9^4)$,The normal direction coordinate values of the measuring points based on UD are shown in Table 4.

Table 4 The normal direction coordinate values of the measuring points based on UD

The No. of the measuring points	X_1	X_2	Value
1	1	3	-0.007
2	2	6	0.013
3	3	9	-0.009
4	4	2	0.023
5	5	5	0.040
6	6	8	0.015
7	7	1	0.015
8	8	4	0.029
9	9	7	0.001

According to the normal direction coordinate values of the measuring points based on UD in Table 4,the correlation coefficient R is demonstrated in Table 5 and the ANOVA in Table 6.

Table 5 Correlation coefficient R

Correlation coefficient R	0.995142
Standard deviation	0.002354
Number of samples	9

Table 6 ANOVA

	df	SS	MS	F	P
Regression	5	0.001699	0.00034	61.29974829	0.003211987
Error	3	1.66E-05	5.54E-06		
Total	8	0.001715601			

Regression coefficient and their P value of Eq. (1) are in Table 7.

Table 7 Regression coefficient

Regression coefficient	Regression coefficient value	P value

β_0	-0.06049	0.002593509
β_1	0.021068	0.000762433
β_2	0.01733	0.003481288
β_3	-0.00166	0.001741212
β_4	-0.00169	0.001635616
β_5	-0.00051	0.057440852

The regression equation:

$$y=-0.06049+0.021068x_1+0.01733x_2-0.00166x_1^2-0.00169x_2^2-0.000513x_1x_2 \quad (3)$$

The coordinate values of maximum error point by the regression equation is:

$x_1=6, x_2=4$.

The maximum error point of the plane is:

$x_1=6, x_2=4$.

Eq. (3) indicates that the regression equation can fitting the workpiece surface well.

5 CONCLUSION

In this paper, the uniform design have been introduced to plan the measuring points. A UD seeks measuring points that are uniformly scattered on the domain. Then the regression equation is built and achieves a satisfactory performance.

Through this preliminary investigation, a conclusion can be drawn that uniform design seems to be a very promising experimental design for model-building in planning the measuring points.

6 ACKNOWLEDGEMENT

This research is funded by the National High Technology Research and Development Program of China (863 Program) No(2015AA042101) and Natural Science Foundation of China (No.51175025).

7 REFERENCES

[1] Hongjun Yan, Xiaoqing Tang. 2002.Computer Aided Products Inspection Plan Design. China Quality. No.7 PP23-25.

[2] Syed Hammad Mian, Abdulrahman Al-Ahmari. 2014. Enhance performance of inspection process on CoordinateMeasuring Machine, Measurement 47.PP78-91.

[3] X. Ji, N. Li. 2010. ACO-Based Multiple Geometric Elements Measure PathPlanning for CMM, in: International Conference on ComputerApplication and System Modeling (ICCASM), China.PP 539-544.

[4] L. Qu, G. Xu, G. Wang.1998. Optimization of the measuring path on acoordinate measuring machine using genetic algorithms,Measurement 23 . 159-170.

[5] Z. Lin, C. Chen. 1997. Measuring-sequence planning by the nearestneighbor method and the refinement method. International Journalof Advance Manufacturing Technology. 13 PP271-281.

[6] C.G. Lu, D. Morton, P. Myler, M.H. Wu. 1995. An artificial intelligent (AI)inspection path management for multiple task measurement on coordinate measuring machine (CMM): an application of neuralnetwork technology, in: Proceedings of 31st internationalMATADOR conference. PP 353-357.

[7] Yi-zeng Liang, Kai-tai Fang, Qing-song Xu. 2001.Uniform design and its applications in chemistry andchemical engineering, Chemometrics and Intelligent Laboratory Systems. 58. PP43-57.

[8] J.F. Li, H.L. Liao, C.X. Ding, C. Coddet. 2005.Optimizing the plasma spray process parameters of yttria stabilizedzirconia coatings using a uniform design of experiments, Journal of Materials Processing Technology. 160. PP 34-42.

[9] Lin Zhang, Yi-Zeng Liang, Jian-Hui Jiang, Ru-Qin Yu, Kai-Tai Fang. 1998.Uniform design applied to nonlinear multivariate calibration by ANN, Analytica Chimica Acta 370 PP65-77.

THE IMPACT OF CONGESTION ON FUEL CONSUMPTION AND DELIVERY DRIVING SATISFACTION DEGREE

Hai-qing JIANG[1]
College of Modern Science & Technology, China JiLiang University , Hangzhou, P.R.China
Jhq191@cjlu.edu.cn

ABSTRACT

The activity of transport causes a high rate of negative effects on the environment, such as congestion. The traffic congestion increases extra fuel consumption, and at the same time, long-time frequent congestion causes negative mood of driving. In this paper, an idea that takes into account fuel consumption and drivers' driving satisfaction degree is proposed to control traffic congestion. Here, we get the situation of driving satisfaction about congestion time based on surveys of 240 valid questionnaire through net in China. Thus, the fuel consumption gallon per miles (GPM) is calculated by using fomulas considering driving speed. We draw the curves of GPM and satisfaction degree on congestion time. The upper limit of congestion time is obtained in the point of intersection of the two curves. Finally, we discuss the effect of different driving speed on drivers' satisfaction degree and fuel consumption.

Keywords： Driving mood, environmental protect- ion, fuel consumption, satisfaction degree

1 INTRODUCTION

The climate change due to emissions of greenhouse gases caused by human activity has been recognized as a threat to the global environment. According to the 19th annual report of overall US greenhouse gas emissions(GHGs) released by the US Environmental Protection Agency (EPA)， CO_2 from fossil fuel combu- stion is the largest source of anthropogenic greenhouse gas emissions in the US and has accounted for approx- imately 78% of global warming potential (GWP)-weighted emissions since 1990; it remained approximate- ly 78% of total GWP-weighted emissions in 2012. Emissions of CO_2 from fossil fuel combustion increased at an average annual rate of 0.3% from 1990 to 2012[1]. Globalization, with its increasing industrial trend towards outsourcing, has caused transportation to become the most visible sector that has increased GHGs emissions over the last two decades. Transportation activities are therefore one of the primary contributors to global environment change. The primary transportation-related emissions are CO_2 from fossil fuel combustion, which has increased by 16% from 1990 to 2012. This rise in CO_2 emissions, combined with an increase in HFCs from close to zero emissions in 1990 to 72.9 Tg CO_2 Eq. in 2012, led to an increase in overall emissions from transportation activities of 18%[1]. The increase is due to many factors such as increased demand for travel, lower fuel efficiency caused by long time congestion and so on. Due to the growing amount of traffic, when the traffic condition changes in the highway, such as traffic accidence, traffic information change etc. congestion occurs. On the condition of congestion,the vehicles often move in stop and go condition., which the average travel speed will be lower than smooth traffic condition. So the fuel consumption efficiency will increase following the congestion time. Besides that, congestion can cause heavy delays. The Dutch Organization for Transport and Logistics (TLN) estimated that over 10% of the truck drivers' working hours are lost due to delays as a result of traffic congestion[2], the drivers have to work overtime, and even violate the driving hours' regulations to finish work. the drivers' modes will be worse following the congestion time, as a result, their driving dissatisfaction degree will increase.
Existing traffic control systems do not cope well with unexpected traffic congested situations; essentially they were designed to respond to changes in moving traffic, instead of the buildup of stopped traffic; and they assume that the supply of road space remains fixed, congestion brings

[1]Corresponding Author :Hai-qing JIANG

about a reduction in supply. when the unexpected traffic congestion occurs, the traffic control is usually executed by the traffic police on experience[2].
Establishing a balance between fuel consumption efficiency and drivers' satisfaction degree when the congestion occurs can not only help the traffic police exactly implement traffic control to decrease drivers' bad mode, but also guide drivers to take measures to reduce fuel consumption and carbon emissions .
In this paper, an investigation is taken to show the relation between the drivers' driving satisfaction level and the waiting time caused by congestion. based on the results of the survey, a fitting curve is drawn. The fuel consumption is calculated by formulas and the corresponding GPM curve is drawn, and then the balance waiting time is got in the intersection of the two curves.
This article is organized as follows: Section 2 provides the necessary background and a literature review. Section 3 presents impact factors on vehicle fuel efficiency as well as an expression to calculate GPM and a survey about drivers' attitude to congestion time. Section 4 presents and analyzes experimental results. Section 5 ends with conclusions.

2. BACKGROUND AND LITERATURE REVIEW

The literature review for this paper covers two main areas of research: (a) the effects of congestion on vehicle fuel consumption; (b) the impact of congestion time on drivers' driving satisfaction degree.

2.1.Literature review for the fuel consumption by congestion

Fuel consumption cost is an important transport cost part. Several methods are avaiable to calculate vehicle fuel consumption. Article [3] models the fuel consumption not only taking loading weight into consideration but also satisfying the "non-passing" property in solving time-dependent vehicle routing problem (TDVRP).Article [4] introduces a formula of fuel consumption considering average speed and road conditions in a vehicle routing problem with time windows, which was studied based on minimizing total fuel consumption. Article [5]introduces a fuel consum- ption equation considering the impact of non-controllable fixed factors such as road conditions, altitude wind directions and controllable human factor like driving habits, vehicle speed and so on. These researches didn't consider the congestion factor. Congestion has a significant impact on fuel consumption and causes fuel consumption to rise sharply, partly because it takes a longer time to reach a destination, and partly because the rate of consumption is much higher at low average speeds. Direct and indirect impacts of congestion on fuel consumption have been studied and reported in the literature available.
The research shows in paper [6] there is a rapid fuel consumption as travel speed falls below 30 mph, and the result was obtained by using an emission model and freeway sensor data in California. Frequent changes in speed, i.e. stop-and-go traffic conditions, increase fuel consumption because fuel consumption is a function of not only speed but also acceleration rates [6].
Figliozzi analyzes the impact of congestion on urban tour based on modeling, numerical experiments, and real-world tour data. The analytical and numerical results indicate that travel time/distance between customers and depot is the crucial factor that exacerbates the negative impacts of congestion. Congestion increases the carriers' cost by worsening the relative weight of wages and overtime escalates[7].

2.2. Literature review for drviers satisfaction degree on congestion

The body of research devoted to investigating the impacts of congestion on drivers is relatively scant. There are some case studies on the impacts of waiting time to drivers.
Waiting time can evoke different emotions of the drivers. Paper [8] established emotional space of the driver. The emotional space consists of four factors, "Happy", "Angry", "Relief", and "Nervous". An emotion model of the driver is proposed based on fuzzy logic. example simulation results show that different waiting time and different road radius can evoke different emotions. the strong

emotion states, e.g. being angry, have potential to endanger driving safety, so the emotion of drivers should be considered when the signal timing is calculated at the intersection and road alignments in order to avoid traffic. Sometimes, waiting time is not the actual waiting time but perceived waiting time.Article [9] shows during the drivers' waiting at a traffic light, their perceived waiting time can differ from the actual waiting time. Through a comprehensive video survey, the perceived waiting time depends not only on the actual waiting time but also on other factors such as the number of stops in the queue and the presence of a red wave between adjacent intersections. the result shows that moving and stopping several times at the same intersection leads to lower perceived waiting time and compared to a long standstill waiting, very short or very long durations are likely to be overestimated. When passing two adjacent intersections, the drivers dislike stopping at both intersections, especially if the second stop is relatively short. models are proposed for estimating drivers' perception and their acceptance of waiting time based on the survey results and have been validated by a real-world experiment. congestion leads to waiting longer than expected for loads, which causes driver's fatigue. Article [10] develops a model of commercial vehicle driving environments to describe the driving environment effects on driver fatigue. Through a nationwide survey of 502 randomly selected over-the-road commercial truck drivers, data analysis revealed 12 driving-environment indicators to be good predictors of fatigue and crash outcomes. Implications for fatigue management are also discussed. these literatures show longer waiting time affects the emotion of the drivers, as a result, the satisfaction degree of the drivers to driving environment will decrease.

3. IMPACT FACTORS AND FORMULAS FOR VEHICLE FUEL EFFICIENCY

3.1.Impact factors

The factors affecting the fuel consumption can be summed up in three categories: vehicle parameter, the driving environmental factors and the driver's behavioral factors[5].Vehicle parameter is mainly composed by automotive design and related technology parameters, and the factors can be divided into several perspectives: the engine compression ratio and engine power, the rolling resistance of the car, the ramp resistance and the accelerated resistance and so on. For the specified models, the above factors are non-controllable fixed factors. The weight of the vehicle also affects fuel consumption, the weight of the vehicles are unchanged, so it is seen a fixed factor in this paper.

Different drivers have different driving habits and modes, which will impact on fuel consumption on the same condition. Paper[11] collected actual fuel consum- ption data in congested and smooth conditions obtained from four drivers, the changes in fuel consumption are almost perfectly explained by the changes in travel time, and the deviation from the regression line is only 2%. So there are no individual differences in these data. As a result, individual differences are very small and can be ignored when other conditions are the same.

3.2. Computing fuel consumption on congestion environme- nt

If miles per gallon(MPG) from node i to node j is MPGij and the correspongding transportation speed is Vij,according to [12],

$MPG_{ij} = \alpha_0 + \alpha_1 v_{ij}$ (1)

Where α_0 and α_1 are Parameters to be estimated. For formula(1) is based on long-term observation data, considering the inclination of the road and the weight(the sum of empty weight and the load weight),the equation (1)can be the following:

$MPG_{ij} = (\alpha_0 + \alpha_1 v_{ij}) \gamma_{ij} \pi_{ij}$ (2) Where γ_{ij} represents respectively the road inclinati-
on coefficient:

$$\begin{cases} \gamma_{ij}=1, & \text{road is flat} \\ \gamma_{ij}<1, & \text{road is uphill road} \end{cases}$$

γ ij>1, road is downhill road

when the vehicle is not empty, π ij >0. The fuel con- sumption (gallon) per travel distance (mile)from I to j is:

GPMij=1/ MPGij. (3)

when congestion occurs in the highway,the travel speed frequently changes. The speed is not easy to estimate, it can be replaced by the average speed when the distance from node i to node j (dij)is not too long, the vehicle when the average congestion time from i to j is tijcon, then the corresponding reach time is from tij to tijnew, where

tijnew= tij +tijcon (4)

So the average travel speed vijnew =dij/ tijnew.Accord -ing to formula (3) and (4),

GPMijnew = 1/(α 0+ α 1* dij/ (tij +tijcon)) γ ij π ij (5)

When the congestion occurs, the highway from i to j is the same as free flow, so α 0、 α 1、 γ ijand π ij can be seen constants. Table I is the actual data from US Department of Transportation[12]. The aim of this paper is to research the relation of the fuel consumption and drivers dissatisfaction degree caused by congestion, so we first let dij and vij be constants to get upper limited waiting time considering drivers' emotion and fuel consumption. the column 3 in Table I is the value we used to count GPMij and the GPMijnew.Fig.1 shows GPM following the congestion time.

3.3. Drivers satisfaction degree on congestion time

When driving in a congested route, the drivers' travel efficiency will decease. Their actual content of work or workload produces a certain deviation to expect, and the greater and more frequent it deviates, the more serious negative impact on the drivers' mood.

Table 1 e periement data

Parameters	Reference Value(or limit)	Actual value
α_0	2.819632	2.819632
α_1	0.065805	0.065805
γ_{ij}	0.75-1.25	1
π_{ij}	1	1
d_{ij}	5-50miles	30
v_{ij}	20-50mph	50
t_{ij}	d_{ij}/ v_{ij}*60 min	

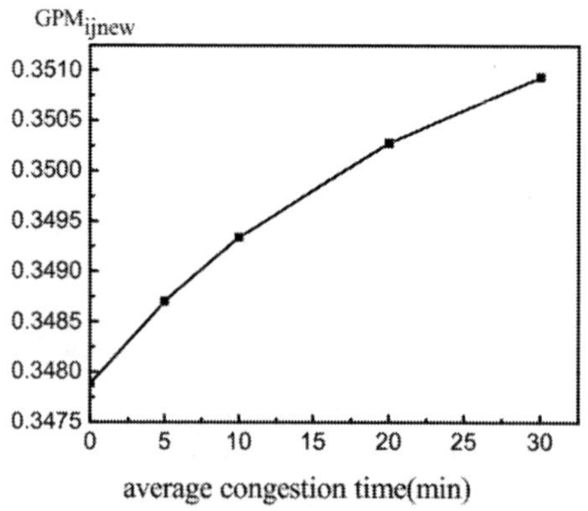

Figure.1 Congestion time and GPM

Due to the fact that differing drivers have different individual property, their endurance to the congestion time is different. Some ones will be very irritable when the congestion is in 3minutes, while some ones will be very ease when the congestion is over 10min.Then correspondingly, the satisfaction degree to the highway will be rather different. At the same time, different modes will affect the behaviors of drivers. We carried out an investigation to determine the fluency of congestion time to the drvers' satisfaction degree. A nationwide questionnaire survey of 210 drivers was conducted through site and web in China. The target population was defined as all individuals who have been a driver. According to the paper [13], in which the value of motorists' travel time varies with the traffic conditions as represented by the degree of congestion. The author categorized traffic conditions as six degrees that was free flowing, busy, light congestion, heavy congestion, stop start and gridlock. We design the range of average congestion time is from 3mins to 30 mins in our questionnaire.

The survey has four major sections:

1).Basic information of the drivers including driving period and gender;

2).Acceptable jam time in peak and flat peak traffic driving on the highway;

3).Fuel-related driving habits (such as whether flame-out waiting) when the traffic jams;

4).Acceptable jam time in peak and flat peak traffic driving in the general route.

3.4. Reliability Analysis

Using Cronbach 's Alpha(internal consistency) to examine the reliability. As can be seen in Table II, the total Table Cronbach ' s Alpha 0.886.

Scholars DeVellis (1991) believes when the Cronb- ach's Alpha is between 0.60 and 0.65, the questionnaire is bad and should be abandoned; acceptable value between 0.65 and 0.70; quite good between 0.70 and 0.80; very good between 0.80 and 0.90. Thus, reliability index of a good questionnaire should be more than 0.80 to the best. The reliability index of sub-scales should be more than 0.70 to the best. According to the calculated Crombach's Alpha, we got the conclusion that our questionnaire has a good internal consistency.

Table 2 cronbach' s alpha

Reliability index	Tolerance time in high-way Flat peak	Tolerance time in high-way peak	Tolerance time in Strai-ght road	Tolerance time in Cross-roads	Total table
Cronbach's Alpha	0.896	0.857	0.849	0.809	0.886

3.5. Survey results

Through data selected from the survey, 62.5% of drivers have their driver's license for less than 3 years, and a mean of 1.28 years, 54.17% are male and the others are female. 30.43% of drivers don't flame out and this can save fuel consumption. 60.87% of drivers will sometimes flame out,whose flameout follows the waiting time. Only 26.09% of drivers will flame out when the waiting time is in 5 mins. The drivers' satisfaction degree caused by congestion time on free-flow highway and busy highway is in Table III. The table shows when the average congestion time in free-flow highway is over 5 mins,the drivers' satisfaction degree sharply decreases, while the average congestion time is over 10 mins, the change of satisfaction degree is very little. But in busy highway, the satisfaction degree is different from that in the free-flow. Comparing to the free-flow highway, the drivers' tolerance to the congestion is higher. 10min is a key value. When the average congestion time is over 10mins. the satisfaction degree is rapid decrease. For the most congestion happen in the busy highway, this paper mainly concerns on the satisfaction degree in busy highway. The satisfaction degree at average congestion time in busy highway is shown in fig. 2.

4. ANALYSIS OF DATA

According to the results of fig.1 and fig.2,appropri-ate time value to control the travel can be got considering drivers' mode and fuel consumption. When the experiment data is shown in Table I, the joint of fig.1 and fig.2 is shown in fig.3. The average congestion time in the point of intersection of fig.4 is 11.89,so tijcon =11.89 mins ,which is consistent with our survey in general.

For the average speed of most vehicles in the route is from 20 to 50 miles per hour, we calculate the GPMijnew according to formula 5.Table IV shows GPMijnew of different speeds and congestion time. The optimal upper-limit average congestion time of different speeds considering on driving satisfaction degree is shown in fig. 4.

When the average speed changes from 20mph to 50mph, the tijcon is shown on the points of A、B、C、D in fig. 4, that is 5.8, 7.75, 9.38 and 11.89min, so the faster average speed is , the drivers' tolerance to congestion is higher.

We can find from fig.4, when the travel speed is 20mph, the curve is flatter than others.At the same time, fig.4 shows the lower of the speed ,the flatter of the curve, which means the fluency of congestion on fuel consumption is smaller when the average speed is lower .When the speed is 20mph,the curve is very flat. As a result, the speed impact on fuel consumption is little.

5. CONCLUSION

In this article, a problem on congestion is addressed. Congestion not only increases the fuel consumption, but

Table 3 waiting time and drivers satisfaction degree

Highway state	Free-flow					Busy				
Waiting time (min)	Wt<= 5	5<Wt<= 10	10<Wt <=20	20<Wt <=30	30<W t	Wt<=5	5<Wt< =10	10<Wt <=20	20<Wt<= 30	30<W t
Satisfaction degree(%)	100	78.26	4.35	4.35	4.35	100	95.65	43.48	17.39	4.35

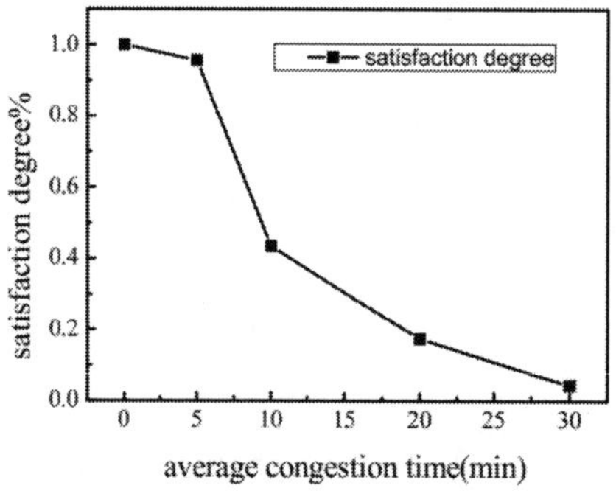

Figure2 Drivers satisfaction degree and congestion time in busy highway

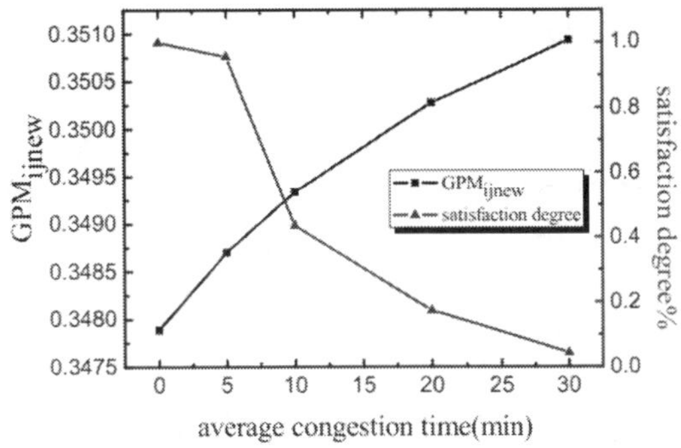

Figure 3 Optimal average congestion time on GPM and satisfaction degree

Table 4 gpmi new value of different speed

V_{ij}(mph)	Limit of congestion time(min)	GPM_{ijnew}	V_{ij}(mph)	Limit of congestion time(min)	GPM_{ijnew}
20	0	0.35192	40	0	0.34922
	5	0.35206		5	0.34976
	10	0.35219		10	0.35020
	20	0.35241		20	0.35088
	30	0.35260		30	0.35138
30	0	0.35057	50	0	0.34789
	5	0.35088		5	0.34870
	10	0.35114		10	0.34934
	20	0.35158		20	0.35028
	30	0.35192		30	0.35093

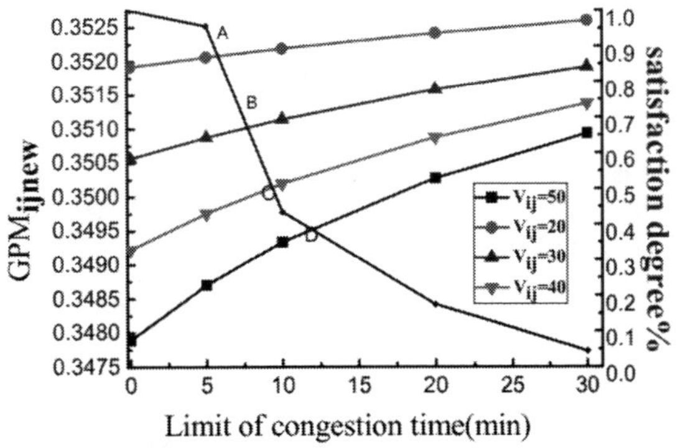

Figure4 Optimal average congestion time of different speed

causes the drivers' mode to be worse following the congestion time. Simultaneously considering the fuel consumption and drivers' satisfaction degree can help the transport managers establish reasonable measures to control traffic congestion time.

By analysis, the following conclusions are obtained:

1) The survey analysis shows the non-linear decline in the drives' satisfaction degree with congestion time increasing.

2) When the research drive speed is 50 mph, distance from node i to j is 30miles, the optimal average congestion is 10min.

3) When the driving speed is 20, 30 and 40 and the distance from node i to j is unchanged, the optimal average congestion time is 5.8, 7.75 and 9.38 mins respectively. At the same time, the fuel consumption (gallon) per travel distance increased to a lesser extent at 20mph.

In further studies, we will continue to study the impact of the transit environment on driving condition and driving behavior.

6. ACKNOWLEDGMENT

This work was supported, in part, by Science and Technology Plan Projects of Zhejiang Provincial Education Department(Y201431877)

. REFERENCES

[1] http://www.greecarcongress.com/2014/20140416-epaghg. html#more.

[2] **A.L. Kok a, E.W.Hans b , .M. .Schutten b**,"Vehicle routingunder time-dependent travel times: The impact of congestion av-oidance," Computers & operations Research, vol. 39 , pp. 910-918, 2012.

[3]**Yiyo Kuo**, "Using simulated annealing to minimize fuel con-sumption for the time-dependent vehicle routing problem,"Co-mputers&Industral Engineering, vol. 59 , pp. 157-165, 2010.

[4]**Li in,Fu Pei-hua**, "Model and simulation for vehicle routingproblem with time windows based on energy consumption," （inChina）,Journal of System Simulation, vol. 25, no. 6, pp. 1147 – 1157, 2013.

[5] **Yang Pei-ying, Tang ia-fu, Yu Yang, Pei in- iang**, "Mini- mizing carbon emissions through vehicle routing and scheduling in th eshuttle service of picking up and delivering customers to the airport,"Acta automatica sinaca, vol.39, no.4, pp.424-423,

2013.

[6] **Miguel Andres Figliozzi**, "The impacts of congestion on time-definitive urban freight distribution networks CO2 emis- sion levels: Results from a case study in Portland,Oregon," Transportation research part C,vol.19,pp. 766-778,2011.

[7] **Figliozzi, M.A.**, "The impact of congestion on commercial vehicle tours and costs," Transportation research part E: logis- tics and transportation,vol.46,no.4, pp.496-506,2009.

[8] **Mark Wardmana, . Nicol s Ib ezb** "The congestion mult- iplier: Variations in motorists' valuations of travel time with tra- ffic conditions ," transportation research part A: policy and practice, vol.46, no. 1, pp. 213-225,2012.

[9]**Van Der Bi l,Bas Vreeswi k, aap Bie, ing Van Berkum,Eric.** "Car drivers' perception and acceptance of waiting time at sign- alized insections," presented at the 14th International IEEE Co- nference on Intelligent Transportation Systems, Washington, DC, 2011.

[10]**Crum, Michael R. Morrow, Paula C. Olsgard, Patricia Roke, Philip** ., "Truck driving environments and their influ- ence on driver fatigue and crash rates, "Transportation Research Record, no 1779, p 125-133, 2001.

[11] **Kono, Toshiaki, Fushiki, Takumi Asada, Katsuya Nakano, Ken iro**, " Fuel consumption analysis and prediction model for "ECO" route search, "presented at the15th World Congress on Intelligent Transport Systems and ITS America Annual Meeting, New York, USA, 2008

[12] **Li in,Fu Pei-hua**, “Model and Simulation for Vehicle Rou- ting Problem with Time Windows Based on Energy Consumpti- on,”(in China),Journal of System Simulation,2013(25)6,1147-1154.

[13]**Mark Wardmana, . Nicol s Ib ezb** “The congestion mult- iplier: Variations in motorists’ valuations of travel time with traffic conditions,”transportation research part A: policy and practice,vol.46,no.1,pp.213-225,2012 .

[14] The Stationery Office, Committee on climate change-Annual report and accounts, (online),London,2011.

[15] **Barth, M., Boriboonsomsin, K.**, “Real-world carbon dioxi-de impacts of traffic congestion,” Transportation Research Rec- ord ,vol .2058, pp.163-171,2009.

Milk-Run Inventory Routing Optimization Problem in Automotive Inbound Logistics

F. CHEN* and Y. Wang
Department of Industrial Engineering and Management
Shanghai Jiao Tong University, Shanghai, 200240, China
fchen@sjtu.edu.cn

ABSTRACT

This paper studies milk-run inventory routing problem in automotive inbound logistics. The purpose of the problem is to minimize both transportation cost and inventory cost collaboratively. The problem is firstly formally described, and optimality conditions as well as valid inequalities are further investigated. A comprehensive mixed integer linear programming model is further built . Moreover, a heuristic algorithm is designed based on MILP model. Finally, computational experiments and case studies are given to show the efficiencies of proposed models and algorithms.

Keywords: Inbound Logistics, Milk-run, Inventory Routine

1 INTRODUCTION

Inventory routing problem(IRP) is a comprehensive issue in supplier chain management. This paper recognizes and formalizes a milk-run inventory-routing problem(MIRP) of automotive inbound logistics. In order to minimize both transportation and inventory cost and hence improve the efficiencies of supply chain, automotive companies prefer milk-run instead of direct delivery(DD) with vendor managed inventory(VMI) in inbound logistics. Such pattern will decide packing sizes, frequencies, and routines for every part.

A subset of suppliers is defined as a milk run line which will be served by a fleet. Each milk-run line consists of one or several milk-run routines, where each routine includes a group of components which generally picked by one truck.

Each truck will serve a subset of suppliers which is defined as a fixed line. The total transportation cost of each truck in each milk-run route is consisted of two parts. One is a travelling cost determined by the number and locations of suppliers visited by a truck during each period. The other is an idle cost determined by the empty volume. During each period, the inventory cost is considered only in the assembly plant according to the average volume of each component.

Figure 1 is an illustration of this problem. The line served by $Trcuck\ 1$ consists of three routes, $Route\ 1,2,3$, with the cycle time $T_1 = 14t$, which should be larger than the total travelling time required by the three routes, as the colored rectangle shown in the picture. The line served by $Truck\ 2$ consists of only one $Route4$ with the cycle time $T_2 = 13t$. Each route contains several suppliers and corresponding components with different picking quantities shown by sectors with different size. For example, in $Route\ 1$ the truck will pick 1 quantity of $Component\ 1$ and 2 quantity of $Component\ 2$ from $Supplier\ 1$, 1 quantity of $Component\ 4$ and 3 quantity of $Component\ 5$ from $Supplier\ 2$ and 2 quantity of $Component\ 6$ from $Supplier3$. The quantity of each component should

* Corresponding Author

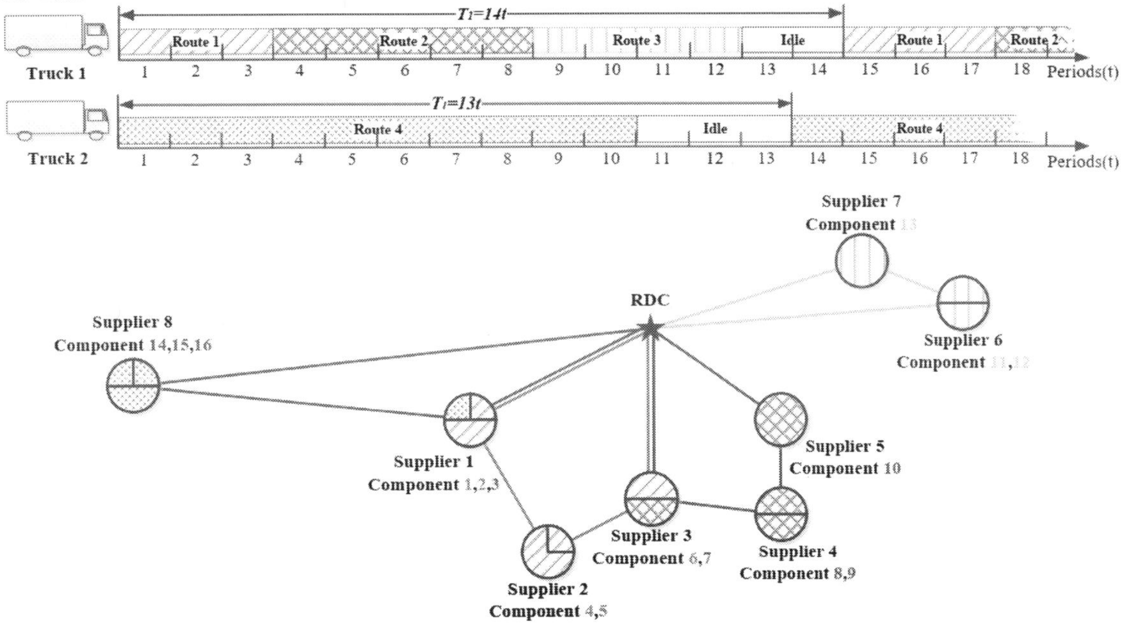

Figure 1. Overview of milk-run picking inventory-routing problem

be larger than the required quantity during T_1 while their total volume should be smaller than one truck volume.

Even though this paper is based on the real situation and data of an automotive enterprise in eastern China, it can be easily generalized to similar operations concerning both transportation and inventory in the situation of multi-vendor and single customer.

This paper is organized as follows. Section 2 is literature review. In section 3, formal description of this problem are given and spacial cases are discussed. Section 4 builds a mixed integer linear programming model and investigated the optimal conditions. Computational experiments and case studies are given in section 5 and section 6, respectively.

2 LITERATURE REVIEW

Early IRP with many-to-one structure can be traced to [13], but the problem is only about the collection of one special item, sewage. Recently more papers on multi-vendor single-buyer IRP problems arise[20, 15, 5], but they are lacking of either practicality or efficiency because of redundant assumptions or constraints. The majority of existing papers solve IRP problem on a finite time horizon, and usually divide them into several periods. Most of these papers assume that the demand of each period is deterministic[15, 5, 14, 10], while a few of them consider stochastic demands[3], or both deterministic and stochastic situations[11]. Research on infinite time horizon is limited[2, 19], and all of them assume that the demands are deterministic in order to simplify the problem. It is rather difficult to find exact solutions for NP-hard problems as IRP. [9] propose valid inequalities and an exact algorithm based on branch-and-cut for an IRP with multiple products, while most research solve IRP through approximation algorithms based on Lagrangian relaxation[7],reduced gradient[4] etc., heuristics[5, 14, 10] and strategy analysis[20].

Researches that addressing applications for automotive logistics are very limited[1, 18]. However, most existing IRP come from the application of outbound problems in energy[14], retail[6], maritime[17, 8] industries. Except focusing on these popular industries and a few special industries such as comkodity[16], sewage[13] and perishables[12].

3 PROBLEM DESCRIPTION AND PROPERTIES

The milk-run inventory routing problem(MIRP) can be stated as follows.

There are K components each of which needs to be picked up from a supplier $i \in \mathbb{S} = \{1, 2, ..., S\}$ to an automotive RDC. The distance between each supplier and the RDC, $d_{0i}(i \in \mathbb{S})$, and the distance between any two suppliers, $d_{ij}(i, j \in \mathbb{S})$, are given. The required quantity of each period, package quantity and package volume of each component $k \in \mathbb{K} = \{1, 2, ..., K\}$ is known respectively as r_{kt}, Q_k and v_k. Each truck $j \in \mathbb{J} = \{1, 2, ..., J\}$ used for picking these components has a limited capacity V. The problem is to decide fixed lines of suppliers and routines for milk-run inbound logistics with the objective of minimizing the total cost of both transportation and inventory in a finite horizon of T periods.

The following decisions will be taken into account.

Decision 1: Milk-run routes A subset of components $\mathbb{K}_e^{'} \subseteq \mathbb{K}$ will be picked from the corresponding subset of suppliers $\mathbb{S}_e^{'} \subset \mathbb{S}$, which is defined as a milk-run route MR_e, $e \in \mathbb{E} = \{1, 2, ..., E\}$. Each component can be assigned to only one route, and the total volume of components belongs to the same route should be within a truck volume.

Decision 2: Fixed lines All the suppliers served by a same truck j, $\mathbb{S}_j^* \subset \mathbb{S}$, is denoted as a fixed line FL_j. It is defined by the set of suppliers related to the milk-run routes MR_e^j, $e \in \mathbb{E}_j \subset \mathbb{E}$, assigned to truck j, thus the cycle time of this fixed line, T_j, should be larger than the total transportation time of MR_e^j.

Decision 3: Picking quantities Since the components cannot be unpacked during delivery, a proper number of packages of each component $k \in \mathbb{K}$, q_k, needs to be decided. The quantity of each component should cover the requirement of the production line during the cycle time of its related fixed line, T_j, and should be as fewer as possible to reduce the inventory costs.

Based on the three decisions, the objective is to minimize the total cost including three parts:

Cost 1: Travelling For each milk-run route e, to minimize the total travelling distance is a Travelling Salesman Problem(TSP). The travelling cost of each fixed line FL_j is defined by the sum of distance of each MR_e^j, D_{ej}, thus

$$CT = c_1 \cdot \sum_{j=1}^{J} \sum_{e \in \mathbb{E}_j} D_{ej}$$

Cost 2: Inventory The inventory cost depends on the sum of mean inventory quantity of each component $k \in \mathbb{K}$. Suppose $\mathbb{K}_e^j$ is the subset of components picked in route MR_e^j. The required quantity of component $k \in \mathbb{K}_e^j$ during each period is r_k. The number of packages picked in MR_e^j is p_k, each of which contains Q_k quantity of component k. Thus the inventory cost is

$$CI = c_2 \cdot \sum_{j=1}^{J} \sum_{e \in \mathbb{E}_j}^{E} \sum_{k \in \mathbb{K}_e^j} \frac{1}{2}(q_k \cdot Q_k - r_k \cdot T_j)$$

Cost 3: Tuck Usage Two parts of truck usage cost are considered. One is a fixed cost of each used truck, cf_j. The other is an empty cost of each MR_e^j, ce_{ej}. Suppose the capacity of each truck is V, and the package volume of component k is v_k, thus

$$CU = \sum_{j=1}^{J}(cf_j + c_3 \cdot \sum_{e \in \mathbb{E}_j}(V - \sum_{k} q_k \cdot v_k))$$

3.1 SPECIAL CASES

In several special cases, we can reduce our problem to other well-known problems to which many approaches have been proposed.

Case 1: Economic Order Quantity(EOQ) Model If $| \mathbb{S} |= 1$, the fixed line and milk-run routes are not necessary. Thus the MIRP can be solve through a classic EOQ model if we regard the transportation cost as the ordering cost. The question transforms into finding the trade-off between order cost and inventory cost of a single item.

Case 2: Joint Replenishment Model If $| \mathbb{S} |\geq 2$, and regarding the DD transportation cost of each component as its ordering cost, the MIRP becomes a joint replenishment problem. The objective is to find proper replenishing policy to minimize the ordering and inventory costs.

Case 3: Bin Packing Problem If there is only one fixed line FL_0, i.e., $| \mathbb{FL}_0 |=| \mathbb{S} |$, and the total cost is constant no matter how many suppliers will be visited by the truck during each milk-run route, i.e., $CE = CI = 0$, $CT =| \mathbb{E} |$, the MIRP is reduced to a bin packing problem. Thus the MIRP is a NP-hard problem since the well-know bin packing problem is proved to be a NP-hard problem.

4 MIXED INTEGER LINEAR PROGRAMMING

For building mixed integer linear programming model, parameters, decision variables, objectives, and constraints will be claimed.

4.1 PARAMETERS

c_{ki}	$c_{ki} = 1$ if component k belongs to supplier i, otherwise $c_{ki} = 0$, $k \in \mathbb{K} = \{1,2,...,K\}$, $i \in \mathbb{S} = \{1,2,...,S\}$. Each component k belongs to only one supplier, i.e., $\sum_{i=1}^{N} c_{ki} = 1$, $\forall k \in \mathbb{K}$
r_{kt}	required quantity of component k during period t, $k \in \mathbb{K}$, $t \in \mathbb{T} = \{1,2,...,T\}$. It is calculated based on production forecasting.
Q_k	quantity of component k in one package, $k \in \mathbb{K}$
v_k	package volume of component k, $k \in \mathbb{K}$
V	volume of truck
p	mean travelling speed of each truck
$d_{i_1 i_2}$	distance between i_1 and i_2, $d_{i_1 i_2} = 0$ if $i_1 = i_2$, $i_1, i_2 \in \mathbb{S}^* = \{0, \mathbb{S}\}$, where $i_1 = i_2 = 0$ represents for the RDC
α_i	$i = 1,2,3$ represent respectively the weight of the inventory cost, travelling cost, fixed cost of trucks
M	given constant number which is big enough

4.2 VARIABLES

q_{klrt}	nonnegative integer number of packages of component k picked in milk-run r of line l which will be finished at the beginning period t, $k \in \mathbb{K}$, $r \in \mathbb{R} = \{1, 2, ..., R\}$, $l \in \mathbb{L} = \{1, 2, ..., L\}$, $t \in \mathbb{T}$
I_{kt}	inentory quantity of component k at the beginning of period t, $k \in \mathbb{K}$, $t \in \mathbb{T}$
H_l	number of trucks required by line l, $l \in \mathbb{L}$
f_{lrt}	$f_{lrt} = 1$ if milk-run r of line l is finished at the beginning of period t, otherwise $f_{lrt} = 0$, $l \in \mathbb{L}$, $r \in \mathbb{R}$, $t \in \mathbb{T}$
e_{lrt}	$e_{lrt} = 1$ if milk-run r of line l begins at the beginning of period t, otherwise $f_{lrt} = 0$, $l \in \mathbb{L}$, $r \in \mathbb{R}$, $t \in \mathbb{T}$
x_{ilr}	$x_{ilr} = 1$ if supplier i will be visited during milk-run r of line l, otherwise $x_{ilr} = 0$, $i \in \mathbb{S}^*$, $l \in \mathbb{L}$, $r \in \mathbb{R}$
y_{il}	$y_{il} = 1$ if supplier i is assigned to line l, otherwise $y_{il} = 0$, $i \in \mathbb{S}$, $l \in \mathbb{L}$
$g_{i_1 i_2 lr}$	$g_{i_1 i_2 lr} = 1$ if there exists a travel from supplier i_1 to i_2 during milk-run r of line l, otherwise $g_{i_1 i_2 lr} = 0$, $i_1, i_2 \in \mathbb{S}^*$, $l \in \mathbb{L}$, $r \in \mathbb{R}$

4.3 MILP FORMULATION

$$\min \quad \alpha_1 \cdot \frac{1}{T} \cdot \sum_{k=1}^{K} \sum_{t=1}^{T-1} \frac{1}{2}(I_{kt+1} + I_{kt}) + \alpha_2 \cdot \sum_{l=1}^{L} \sum_{r=1}^{R} \sum_{i_1=0}^{S} \sum_{i_2=0}^{S} g_{i_1 i_2 lr} \cdot d_{i_1 i_2} + \alpha_3 \cdot \sum_{l=1}^{L} H_l$$

$$s.t. \quad \frac{1}{M} \cdot \sum_{k=1}^{K} \sum_{t=1}^{T} q_{klrt} \cdot c_{ki} \le x_{ilr} \le \sum_{k=1}^{K} \sum_{t=1}^{T} q_{klrt} \cdot c_{ki}, \ \forall i \in \mathbb{S}, l \in \mathbb{L}, r \in \mathbb{R} \quad (1)$$

$$\frac{1}{M} \cdot \sum_{r=1}^{R} x_{ilr} \le y_{il} \le \sum_{r=1}^{R} x_{ilr}, \ \forall i \in \mathbb{S}, l \in \mathbb{L} \quad (2)$$

$$\sum_{l=1}^{L} y_{il} = 1, \ \forall i \in \mathbb{S} \quad (3)$$

$$\sum_{k=1}^{K} \sum_{t=1}^{T} q_{klrt} \cdot v_k \le V, \ l \in \mathbb{L}, r \in \mathbb{R} \quad (4)$$

$$\frac{1}{M} \cdot \sum_{k=1}^{K} q_{klrt} \le f_{lrt} \le \sum_{k=1}^{K} q_{klrt}, \ l \in \mathbb{L}, r \in \mathbb{R}, t \in \mathbb{T} \quad (5)$$

$$\sum_{t=1}^{T} e_{lrt} = \sum_{t=1}^{T} f_{lrt} \le 1, \ l \in \mathbb{L}, r \in \mathbb{R} \quad (6)$$

$$I_{kt} + \sum_{l=1}^{L} \sum_{r=1}^{R} q_{klrt} \cdot Q_k - r_{kt} \ge 0, \ \forall k \in \mathbb{K}, t \in \mathbb{T} \quad (7)$$

$$\frac{1}{M} \sum_{i=1}^{S} x_{ilr} \le x_{0lr} \le \sum_{i=1}^{S} x_{ilr}, \ \forall l \in \mathbb{L}, r \in \mathbb{R} \quad (8)$$

$$2 g_{i_1 i_2 lr} \le x_{i_1 lr} + x_{i_2 lr}, \ \forall i_1 \neq i_2 \in \mathbb{S}^*, l \in \mathbb{L}, r \in \mathbb{R} \quad (9)$$

$$\sum_{i_2=0}^{S} g_{i_1 i_2 lr} = \sum_{i_2=0}^{S} g_{i_2 i_1 lr} = x_{i_1 lr}, \ \forall i_1 \in \mathbb{S}^*, l \in \mathbb{L}, r \in \mathbb{R} \quad (10)$$

$$g_{iilr} = 0, \ \forall i \in \mathbb{S}^*, l \in \mathbb{L}, r \in \mathbb{R} \quad (11)$$

$$g_{i_1 i_2 lr} + g_{i_2 i_1 lr} \le 1, \ \forall i_1 \neq i_2 \in \mathbb{S}, l \in \mathbb{L}, r \in \mathbb{R} \quad (12)$$

$$\sum_{t=1}^{T} f_{lrt} \cdot t - \sum_{t=1}^{T} e_{lrt} \cdot t \ge \left(\sum_{i_1=0}^{S} \sum_{i_2=0}^{S} g_{i_1 i_2 lr} \cdot d_{i_1 i_2} \right) / p, \ \forall l \in \mathbb{L}, r \in \mathbb{R} \quad (13)$$

$$H_l \ge \sum_{t'=1}^{t} \sum_{r=1}^{R} e_{lrt'} - \sum_{t'=1}^{t} f_{lrt'}, \ \forall t \in \mathbb{T}, l \in \mathbb{L} \quad (14)$$

f_{lrt}, e_{lrt}, x_{ilr}, y_{il}, $g_{i_1 i_2 lr}$ are binary; q_{klrt}, I_{kt}, H_l are nonnegative integers for all $k \in \mathbb{K}$, $i \in \mathbb{S}^*$, $t \in \mathbb{T}$, $l \in \mathbb{L}$, $r \in \mathbb{R}$.

MODEL EXPLANATIONS:

(1) Each component $k \in \mathbb{K}$ can be assigned to only one route $e \in \mathbb{E}$. (2) Component $k \in \mathbb{K}$ is assigned to route $e \in \mathbb{E}$ if there exists a positive picking quantity of the component in route e.

(3) The picking quantity of component $k \in \mathbb{K}$ should cover the requirement during the cycle time of the fixed line which it is assigned to.

(4) The trucks required by milk-run route $e \in \mathbb{E}$ in fixed line $j \in \mathbb{J}$ should be enough for the total volume of components assigned to that route.

(5) A supplier $i \in \mathbb{S}$ will be visited in route $e \in \mathbb{E}$ if and only if at least one component is assigned to the route.

(6) The RDC where $i = 0$ will be visited in route $e \in \mathbb{E}$ if and only if there exists at least one supplier assigned to the route.

(7) The travel from a point, namely a supplier or the RDC, a to another point b, $a, b \in \mathbb{S}^*$, exists if and only if both a and b belong to the same route $e \in \mathbb{E}$.

(8) For each point $a \in \mathbb{S}^*$, if point a is assigned to a route $e \in \mathbb{E}$, it will be visited once only, that is, the travel starts from a equals to the travel end to it, otherwise it will not be visited.

(9) Except the RDC, all the supplier can be visisted only once in each route e.

(10) Travel from point $a \in \mathbb{S}^*$ to a does not exist.

(11-12) A milk-run route $e \in \mathbb{E}$ is meaningful and should be assigned to at most one fixed line if and only if there are at least one component assigned to the route.

(13) The traveling distance of route $e \in \mathbb{E}$ counts as D_{ej} if it is assigned to fixed line $j \in \mathbb{J}$.

(14-15) If component k is assigned to milk-run route e which belongs to fixed line j, the cycle time of the component, T_{kej}, should be large than the required total transportation time of all the routes assigned to line j.

(16-18) The cycle time of a fixed line j, W_j equals to the cycle time of each component k, T_{kej}, if there exists component k assigned to milk-run route e which belongs to line j, otherwise $W_j = 0$.

(19) Number of trucks required by line j should be larger than that required by each route assigned to it.

5 OPTIMAL CONDITIONS

In this section, we will analyze the optimal conditions of the MIRP. Accordingly, valid inequalities will be presented to improve the MILP model in section 4. Numerical examples are given to show the effectiveness of model and how the MIRP problem can be reduced to the three special cases. *Moreover, sensitivity analysis of several key parameters is conducted to show the dynamic relation ship between transportation and inventory costs.*

This section gives some optimal conditions to the MIRP based on several special geometric structures of supplier locations observed from the practical environment.

Figure 2. Three suppliers on the radial line from RDC for theorem 1

Theorem 1. *As shown in figure 2, there are three suppliers, S_1, S_2 and S_3, locating on a radial line starting from the RDC S_0. S_1 is very far from the RDC, S_2 is near S_1 and S_3 is close to RDC. Milk-run performance of delivering the components of S_1 with S_2 is usually much better than with S_3, from the long-term planning prospective.*

Proof. For brevity, details are omitted.

6 ALGORITHM

In order to solve this MIRP more efficiently considering practical requirements, an algorithm is developed in this section based on the saving method of veihcle routing problem(VRP). There are generally five steps to conduct the algorithm.

(1) Data Preprocess:

As shown in figure 3 , we investigate the structure of the initial real data from an automotive company. The data need preprocess to three key inputs in order to conduct our algorithm as the following equations.

$$v_k = l_k \cdot w_k \cdot (h_k \cdot st_k + H \cdot (1 - st_k)) \tag{1}$$

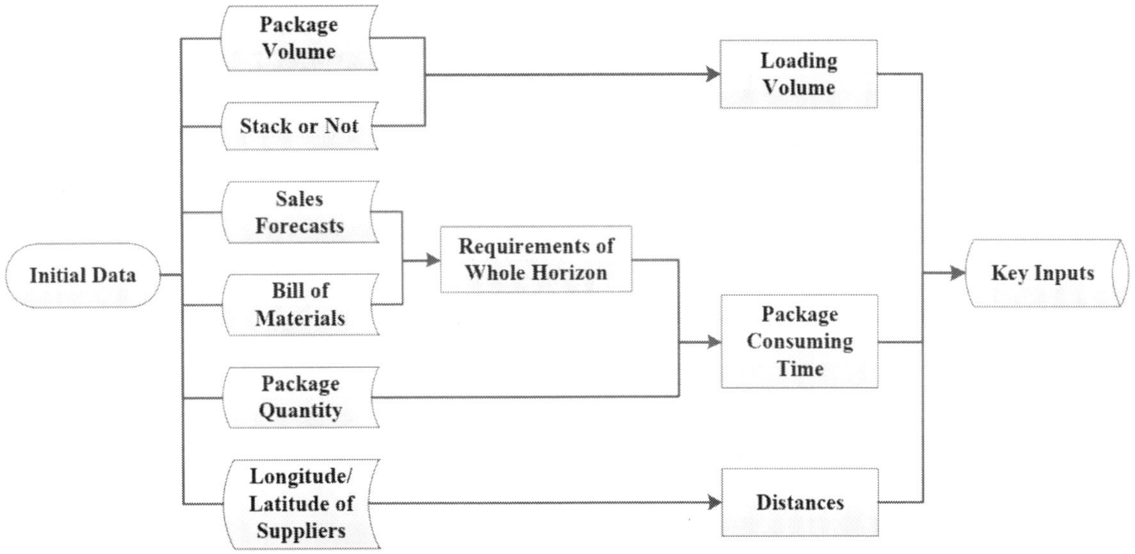

Figure 3. fig:data preprocess

$$t_k = T \cdot Q_k / \sum_{b=1}^{B} f_b \cdot r_{bk} \tag{2}$$

$$d_{ij} = \frac{RE \cdot \pi}{180} \cdot arccos(sin(la_i) \cdot sin(la_j) + cos(la_i) \cdot cos(la_j) \cdot cos(lo_i - lo_j)) \tag{3}$$

Equation 1 is the loading volume of one package of component $k \in \mathbb{K}$, v_k. It equals to the product of the package's length l_k, width w_k and height h_k, if it can be stacked with other packages, namely $st_k = 1$, otherwise the height of the package is considered to be the height of the truck H. Equation 2 estimates the time that one package of component $k \in \mathbb{K}$ would support the production procedure, which is denoted as a package consuming time t_k. The forecast of each product type $b \in \mathbb{B} = \{1, 2, ..., B\}$ is given as f_b for the finite time horizon T. r_{bk} is the quantity of component k required by one product b, and Q_k is the number of component k contained in one package. The distance between supplier i and j, d_{ij}, $i, j \in \mathbb{S}$, is calculated by equation 3, based on the latitude and longitude of the supplier location. Notice that west longitude should be negative for this equation.

(2) Component Integration: At the same supplier, a set of components will be regard as an integrated component if the total volume they need to cover the total requirement during their largest package consuming time $\max\{t_k\}$ is smaller than one truck volume V. Figure 4 is a three-component supplier illustration of this process. The reason of this integration is to simplify practical operations. Automotive company tends to order the components from the same supplier together. We regard each integrated component as a supply point to be assigned to a milk-run route.

(3) Calculate Distance Savings:

As shown in figure 5, comparing to directly delivery, the distance saving of milk-run of each pair of supply points $i_1 \neq i_2 \in \mathbb{S}' = \{1, 2, ..., S'\}$, where S' is the total number of integrated components,

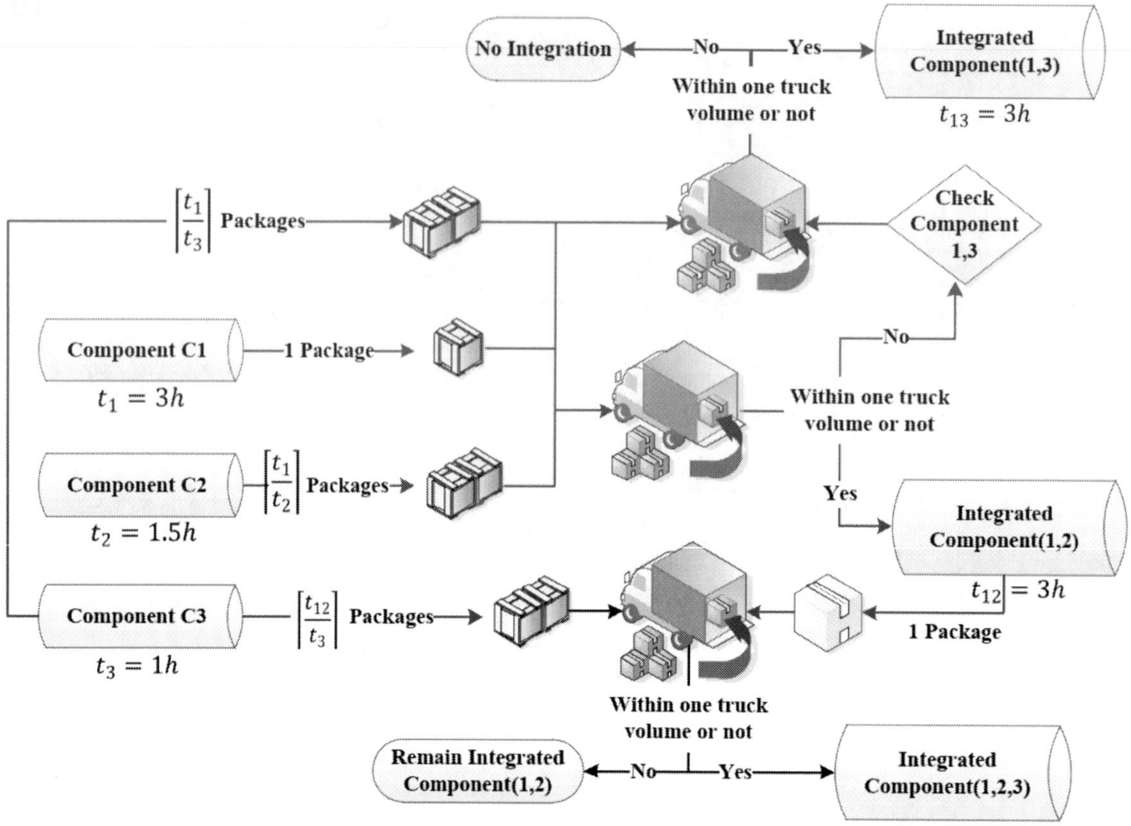

Figure 4. Illustration of 3 components integration at one supplier

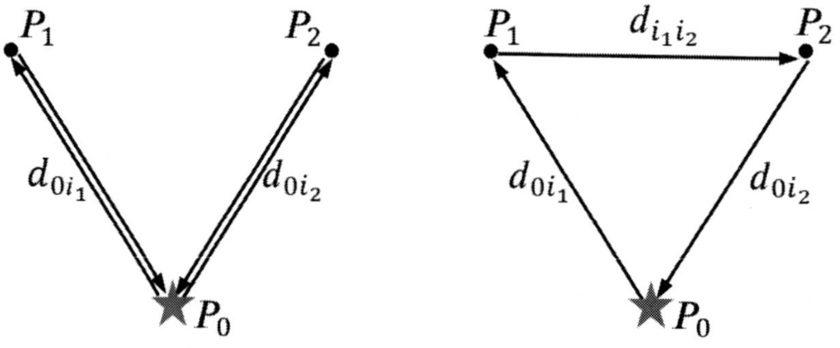

Figure 5. Distance saving of milk-run

can be calculated based on equation (5.4), denoted as $SD_{i_1i_2}$. Since $SD_{i_1i_2} = SD_{i_2i_1}$, only $M = S' \cdot (S' - 1)/2$ savings are taken into consideration.

$$SD_{i_1i_2} = d_{0i_1} + d_{0i_2} - d_{i_1i_2} \quad (4)$$

(4) Generate Milk-run Routes:

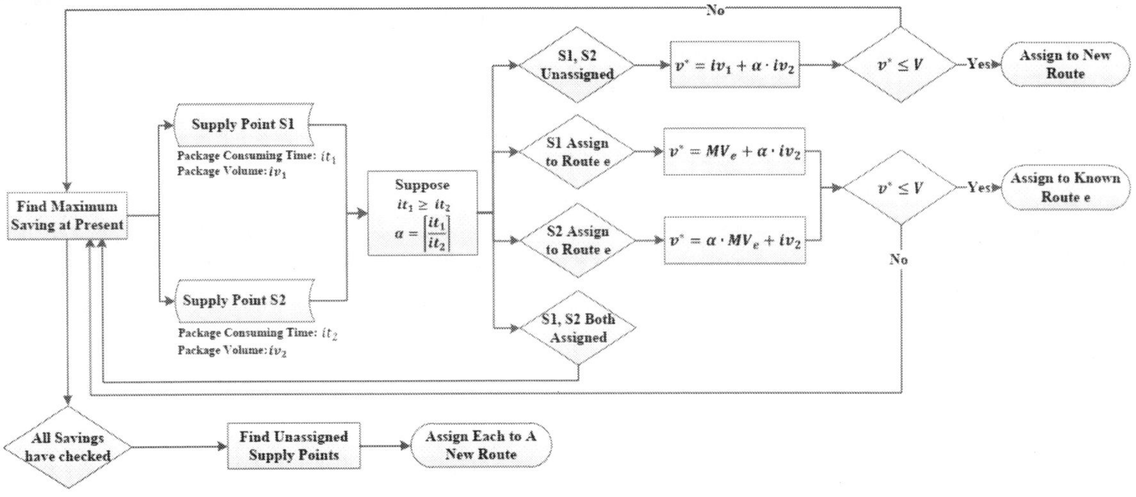

Figure 6. Assign integrated components to milk-run routes

As shown in figure 6, under the constraint of not exceeding one truck volume V, two supply points with larger saving will be assigned to the same milk-run route $e \in \mathbb{E}$ preferentially. Thus if a subset of intergrated components $\mathbb{K}_e^{'}$ is assigned to milk-run route e, the cycle time it is $MT_e = \max_{k \in \mathbb{K}_e^{'}}\{t_k\}$.

$$\min\{MT_e\} \geq \sum_{e \in \mathbb{E}_j} tr_e \tag{5}$$

(5) Generate Fixed lines:

Finally, if a set of milk-run routes $\mathbb{E}_j \subset \mathbb{E}$ satisfies inequation (5.5), they can be assigned to the same truck $j \in \mathbb{J}$ as a fixed line, and the the cycle time of truck j, $T_j^{'} = \min_{e \in \mathbb{E}_j}\{MT_e\}$. Namely, the minimum cycle time of these milk-run routes is larger than the total transportation time they need, where tr_e is the real transportation time of route e:

$$tr_e = D_e/p \tag{6}$$

where D_e is estimated distance of milk-run route e, which is defined with equation (5.7). p is the mean travelling speed of truks.

$$D_e = 2 \cdot \frac{\sum_{i \in \mathbb{E}_j} d_{0i}}{\mid \mathbb{E}_j \mid} + (\mid \mathbb{E}_j \mid -1) \cdot \frac{\sum_{i_1 \neq i_2 \in \mathbb{E}_j} d_{i_1 i_2}}{\mid \mathbb{E}_j \mid \cdot (\mid \mathbb{E}_j \mid -1)}$$

$$= \frac{1}{\mid \mathbb{E}_j \mid}(2 \sum_{i \in \mathbb{E}_j} d_{0i} + \sum_{i_1 \neq i_2 \in \mathbb{E}_j} d_{i_1 i_2}) \tag{7}$$

where $\frac{\sum_{i \in \mathbb{E}_j} d_{0i}}{\mid \mathbb{E}_j \mid}$ is the mean distance between each supplier and the RDC. $\frac{\sum_{i_1 \neq i_2 \in \mathbb{E}_j} d_{i_1 i_2}}{\mid \mathbb{E}_j \mid \cdot (\mid \mathbb{E}_j \mid -1)}$ is the mean distance between each pair of suppliers. We need 2 of the first part and $\mid \mathbb{E}_j \mid -1$ of the second part to generate the total mean traveling distance for a given milk-run route. Besides,notice that the cycle time of picking each component is changed while assigned milk-run routes to fixed lines.

The detailed steps of this algorithm is as follows:

Step 1: Data preprocess Generate three key inputs according to equations (5.1-5.3).

Step 2: Sort components in each supplier Suppose $\mathbb{K}^i = \{1, 2, ..., K^i\} \subset \mathbb{K}$ is the subset of components provided by it. Sort components $k \in \mathbb{K}^i$ such that package consuming time $t_1 \geq t_2 \geq ... \geq t_{K^i}$. For each $1 \leq k \leq K^j$, initialize all $IN_k = 0$.

Step 3: Generate integrated components Initialize integrated component number $n = 0$. For each supplier $i \in \mathbb{S}$, find the first $IN_{k*} = 0,\ k \in \mathbb{K}^i$.

Generat new integrated component $n = n + 1$, let $IN_{k^*} = n$. Then the volume of integrated component $iv_n = v_{k^*}$ and the package consuming time of integrated component $it_n = t_{k^*}$. Check each $k \in [k^* + 1, K^i]$:

If $IN_k = 0$, and $iv_n + \lceil \frac{it_n}{t_k} \rceil \cdot v_k \leq V$, then $iv_n = iv_n + \lceil \frac{it_n}{t_k} \rceil \cdot v_k,\ IN_k = k^*$.

Repeat step 3 until all $IN_k \neq 0,\ k \in \mathbb{K}$.

Step 4: Calculate and Sort Savings Calculate the distance saving between each pair of supply points $i_1 \neq i_2 \in \mathbb{S}'$ by equation (5.4). Sort the M non-redundant savings in descending order such that $SD_1 \geq SD_2 \geq ... \geq SD_M$.

Step 5: Generate Milk-run Routes Initialize $m = 1$, milk-run route number $e = 0$, consuming time of the component assigned to it is $MT_e = 0$, volume is $MV_e = 0$. The route number of each supply point $i \in \mathbb{S}'$, $SR_i = 0$.

Find the two supply point i_1, i_2 related to SD_m, suppose $it_{i_1} \geq it_{i_2}$(vice versa). Let $\alpha = \lceil \frac{it_{i_1}}{it_{i_2}} \rceil$.

If $SR_{i_1} = SR_{i_2} = 0$, and $iv_{i_1} + \alpha \cdot iv_{i_2} \leq V$, then $e = e + 1$; $SR_{i_1} = SR_{i_2} = e$, $MT_e = it_{i_1}$, $MV_e = iv_{i_1} + \alpha \cdot iv_{i_2}$. Otherwise $m = m + 1$;

If $SR_{i_1} = e^*$, $SR_{i_2} = 0$, and $MV_{e^*} + \alpha \cdot iv_{i_2} \leq V$, then $SR_{i_2} = e^*$, $MV_{e^*} = MV_{e^*} + \alpha \cdot iv_{i_2}$. Otherwise $m = m + 1$;

If $SR_{i_1} = 0$, $SR_{i_2} = e^*$, and $\alpha \cdot MV_{e^*} + iv_{i_1} \leq V$, then$SR_{i_1} = e^*$, $MV_{e^*} = \alpha \cdot MV_{e^*} + iv_{i_1}$, $MT_{e^*} = it_{i_1}$. Otherwise $m = m + 1$.

Repeat step 5 until $m = M$.

Step 6: Assign left supply point to DD routes If there exists a subset of supply points $i \in \mathbb{S}'' \subset \mathbb{S}'$ left with $SR_i = 0$ after step 5 iteration, for each $i \in \mathbb{S}''$, repeat $e = e + 1$, $SR_i = e$, until all $SR_i \neq 0$.

Step 7: Assign to Trucks Sort the cycle time of milk-run routes in descending order such that $MT_1 \leq MT_2 \leq ... \leq MT_E$. Initialize total truck number $J = 0$, the real traveling time of truck J is $TR_J = 0$, the fixed line number milk-run route $e \in \mathbb{E}$ assigned to is denoted as $FL_e = 0$.

Find the route e^* such that $MT_{e^*} = \min\{MT_e\}$ and $FL_e = 0$. $J = J + 1$, $FL_{e^*} = J$; $T_J' = T_{e^*}$; $TR_J = tr_{e^*}$; $e = e^* + 1$.

If $e \leq E$, $FL_e = 0$ and $T_J' \geq TR_J + tr_e$, then let $FL_e = J$, $TR_J = TR_J + tr_e$.

Repeat until all $FL_e \neq 0$, $e \in \mathbb{E}$.

Step 8: Adjust Picking Quantity If component k is assigned to milk-run route e, which is assigned to fixed line j, then the picking quantity of packages during each cycle time T_j' should be

$$q_k = \lceil \frac{T_j' \cdot r_k}{\delta t \cdot Q_k} \rceil \tag{8}$$

where r_k is the mean require quantity of component k during each period δt, Q_k is the package quantity of component k.

7 COMPUTATIONAL EXPERIMENTS

In this section numerical experiments are conducted to show the effectiveness of MILP model and algorithms.

Parameters about components and suppliers are required to be generated before the conducting of experiments. We divide the experiments into 10 groups based on the number of components(NC) and suppliers(NS), as shown in table 1. Notice that in each group, $NC >= NS$.

Table 1. Alternatives for experiment groups

Parameter	Alternatives
Component number(NC)	$\{2,3,4,5\}$
Supplier number(NS)	$\{2,3,4,5\}$

Then some parameters are constant numbers or generated based on NC and NS, as shown in table 2.

Table 2. Parameters based on NC and NS

Parameter	Generation
Milk-run routes(NE)	$NE = NC$
Fixed lines(NL)*	$NL = 2 \cdot NC$
Truck speed(SP)	$SP = 50(km/h)$
Component-supplier relations (CS_k)	$CS_k = rand(1,2,...,NS)$, $\forall k \in \{1,2,...,NC\}$

* Approximated.

Table 3. Parameters based on real data

Parameter	Real mean	Random Range
Require each period(CR_k)[1]	5.58	[0.3, 10.84]
Package quantity(PQ_k)	355	[11,699]
Package volume(PV_k)	0.606	[0.01,1.64] (m^3)
Truck volume (TV)[2]	$TV = 7.36 \cdot \max\{PV_k\}$	
Supplier location (LA, LO)	$LA \in [22.67853, 43.80996]$ $LO \in [104.1794, 125.24757]$	

[1] $k \in \{1,2,...,NC\}$
[2] Real truck volume is about 7.36 times of the maximum package volume.
[3] Real range of the latitude and longitude of the suppliers.

All the other parameters are generated according to the mean values of the real data. We have information of 2622 components with 217 suppliers as will be shown in a case study in section 7.

If the real mean value of one of the parameters shown in table 3 is u, the range limits of it are decided by $u \pm \sigma$, such that the number of real data in this range covers about 80% of the real data.

From table 4, it is not easy to see that MILP model is so complex that only small scale instance can be solved within reasonable running time. In contrast, the heuristic algorithm can performance competitive at running time but lose precisions.

Table 4. Experiment Results

NC	NS	Group	CPLEX Objective	CPLEX Time(sec)	Algorithm Objective	Algorithm Time(sec)	Objective Difference	NC	NS	Group	CPLEX Objective	CPLEX Time(sec)	Algorithm Objective	Algorithm Time(sec)	Objective Difference
		1	1886.62	0.0644	2088.35	0.0016	10.69 %			1	4093.45	6.0216	4770.34	0.0019	16.54%
		2	3644.86	0.0644	6303.92	0.0006	72.95%			2	2624.34	6.9995	4789.45	0.0017	82.50%
2	2	3	5029.32	0.0783	5130.05	0.0014	2.00%	6	3	3	4948.57	2.3760	5817.10	0.0025	17.55%
		4	2914.61	0.0646	2916.38	0.0013	0.06%			4	3937.89	5.4998	5003.22	0.0017	27.05%
		5	2002.66	0.0653	2576.40	0.0007	28.65%			5	4044.37	5.9388	8292.59	0.0031	105.04%
		1	5155.65	0.1352	5349.98	0.0015	3.77%			1	4122.77	0.5671	5035.20	0.0027	22.13%
		2	3121.21	0.0962	6235.70	0.0017	99.78%			2	3469.74	0.5309	6706.56	0.0039	93.29%
3	2	3	3018.51	0.0947	3193.48	0.0016	5.80%	4	4	3	4986.95	0.6263	6183.87	0.0031	24.00%
		4	3240.38	0.0941	3443.52	0.0027	6.27%			4	4286.45	0.6063	5767.72	0.0030	34.56%
		5	3774.62	0.0784	3794.42	0.0023	0.52%			5	4323.42	0.5258	6303.60	0.0025	45.80%
		1	4259.41	0.1688	4663.85	0.0012	9.50%			1	4358.78	1.4791	5921.80	0.0023	35.86%
		2	4822.72	0.6887	4983.23	0.0016	3.33%			2	5054.72	1.2612	7138.42	0.0021	41.22%
4	2	3	3921.43	0.1841	7782.58	0.0019	98.46%	5	4	3	5416.57	1.5249	6043.65	0.0028	11.58%
		4	3324.72	0.1909	3400.38	0.0014	2.28%			4	4795.22	1.2506	6503.81	0.0023	35.63%
		5	3133.59	0.5544	6485.64	0.0017	106.97%			5	3978.29	1.4724	8982.09	0.0032	125.78%
		1	2004.14	1.1974	3495.93	0.0016	74.44%			1	4634.59	24.8359	8555.88	0.0029	84.61%
		2	5095.45	1.4971	5178.07	0.0017	1.62%			2	4705.37	7.6989	7007.05	0.0022	48.92%
5	2	3	2843.36	1.7066	3694.15	0.0007	29.92%	6	4	3	4551.70	12.1462	7283.34	0.0035	60.01%
		4	2884.25	1.5908	3309.72	0.0006	14.75%			4	3617.53	13.4615	4921.33	0.0026	36.04%
		5	4144.85	1.6271	8960.87	0.0009	116.19%			5	4423.52	14.6167	9300.01	0.0022	110.24%
		1	1987.47	7.3739	4210.76	0.0028	111.87%			1	4366.32	90.1594	5138.75	0.0032	17.69%
		2	2885.71	9.2313	4507.04	0.0020	56.18%			2	6113.46	4.4343	7342.64	0.0035	20.11%
6	2	3	1865.98	4.0087	3914.38	0.0017	109.78%	5	5	3	4514.64	120.1119	5121.15	0.0078	13.43%
		4	3544.33	18.4510	8297.04	0.0025	134.09%			4	6092.20	62.9981	8586.12	0.0054	40.94%
		5	3979.86	5.5671	11520.38	0.0012	189.47%			5	3811.79	56.6709	6023.55	0.0029	58.02%
		1	4270.24	0.2656	5211.65	0.0020	22.05%			1	5223.33	>1800	8431.41	0.0036	61.42%
		2	4039.58	0.1981	6726.71	0.0013	66.52%			2	3871.36	10.2678	6454.62	0.0031	66.73%
3	3	3	3953.07	0.2024	5688.06	0.0021	43.89%	6	5	3	3451.97	>1800	5341.89	0.0035	54.75%
		4	5347.95	0.2463	5609.55	0.0023	4.89%			4	3668.67	>1800	5198.05	0.0032	41.69%
		5	3336.60	0.2020	4258.29	0.0022	27.62%			5	5175.88	>1800	7973.60	0.0033	54.05%
		1	2122.46	0.5934	2918.12	0.0016	37.49%			1	5567.23	39.7154	9691.53	0.0309	74.08%
		2	3490.55	0.5203	6283.04	0.0016	80.00%			2	5418.49	>1800	10283.47	0.0049	89.78%
4	3	3	4886.96	0.5478	7309.25	0.0014	49.57%	6	6	3	5416.45	>1800	8228.97	0.0046	51.93%
		4	3843.42	0.5241	7311.90	0.0024	90.24%			4	6806.55	45.3476	11429.45	0.0046	67.92%
		5	4078.59	0.6450	4730.97	0.0014	16.00%			5	5893.89	>1800	11275.32	0.0046	91.31%
		1	1244.83	1.4395	2229.18	0.0020	79.08%								
		2	2458.12	1.8679	3461.02	0.0018	40.80%								
5	3	3	4150.77	1.2567	6398.32	0.0026	54.15%								
		4	3304.20	1.5943	3748.09	0.0037	13.43%								
		5	5538.68	2.5396	7252.89	0.0021	30.95%								

8 CASE STUDY

In this section, a case study based on real data and requirements from an automotive company in south China is given to show the efficiency and practicability of our algorithm. According to the algorithm in section 5, the real MIRP is solved by using MATLAB R2013a on a PC with 1.7 GHz Intel Core i7 Processor, 8GB 1600 MHz DDR3 Memory. We summarize the results on milk-run routes, fixed lines, truck schedules, loading rates, and transportation and inventory costs, and compare the them with the traditional directly delivery(DD) operations and a planning service provided by a third-party logistics company(3PL).

There are totally 2622 components from 217 suppliers with different geographical locations taken into consideration. For instance, table 2 contains the first 30 components from 3 suppliers. Based on their initial information including package size, stack or not, sale forecasts, bill of materials, package quantities, we generate the three key inputs, loading volume, package consuming time and geographic coordinate.The truck volume is given as the most common one,

$16 \times 2.5 \times 2.4 \times 85\% = 81.6m^3$, since about 15% space is needed for loading and unloading in practice. After step 2 of the algorithm, 302 integrated component are generated.

Table 5. Integrated Components

Supply Point	Integrated to Component	Initial Supplier	Integrated Volume (m^3)	Integrated Package Consuming Time (hours)	North Latitude	East Longitude
1	19	1	65.09	262.80	39.67	116.36
2	6	1	18.90	31.54	39.67	116.36
3	23	2	14.19	175.20	40.36	116.75
4	33	3	67.98	219.00	31.35	121.38
5	35	3	20.12	15.00	31.35	121.38
6	44	4	61.73	70.08	31.36	121.43
7	54	4	7.64	9.56	31.36	121.43
8	56	5	0.08	273.75	31.30	121.37
9	96	6	70.38	671.69	31.27	121.47
10	89	6	31.11	32.85	31.27	121.47

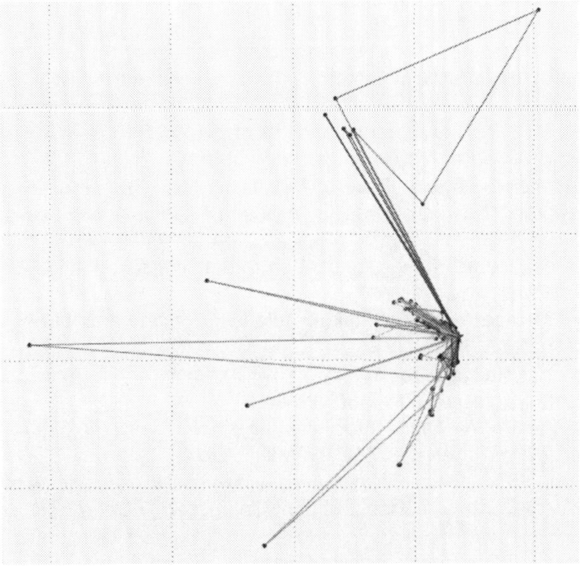

Figure 7. Milk-run Routes

As shown in figure 7, according to the distance savings and constraint of truck volume, we generate 123 milk-run routes. Table 5 is 10 illustration of the milk-run routes with the number of integrated components they contain, namely the number of supply points they will visit, the load rates, visiting cycle time, approximate distance and thus the required time of transportation considering mean speed of 50 km/h.

9 CONCLUSIONS

This paper presents milk-run picking inventory-routing problem(MIRP) for automotive inbound logistics. A mixed integer linear programming model is built . Several optimality conditions and valid inequalities are given to improve the model. Numerical examples and case studies show that proposed models and algorithms can be used to generate suitable lines and routines for inbound logistics.

10 ACKNOWLEDGEMENTS

The work is supported by National Science Foundation of China (No. 70972070; 71272115) and Shanghai Pujiang Program (No. 14PJC062).

11 REFERENCES

[1] Jesús Alegre, Manuel Laguna, and Joaquín Pacheco. Optimizing the periodic pick-up of raw materials for a manufacturer of auto parts. *European Journal of Operational Research*, 179(3):736-746, 2007.

[2] S Anily and A Federgruen. Two-echelon distribution systems with vehicle routing costs and central inventories. *Operations Research*, 41(1):37, 1993.

[3] Jonathan F Bard, Liu Huang, Patrick Jaillet, and Moshe Dror. A decomposition approach to the inventory routing problem with satellite facilities. *Transportation Science*, 32(2):189, 1998.

[4] Julian Benjamin. An analyst of inventory and transportation costs in a constrained network. *Transportation Science*, 23(3):177-183, 1989.

[5] Z Chen and B R Sarker. An integrated optimal inventory lot-sizing and vehicle-routing model for a multisupplier single-assembler system with jit delivery. *International Journal of Production Research*, 52(17):5086-5114, 2014.

[6] Z N Chen, C K M Lee, W H Ip, and G T S Ho. Design and evaluation of an integrated inventory and transportation system. *Transportation Planning & Technology*, 35(4):491-507, 2012.

[7] T William Chien, Anantaram Balakrishnan, and Richard T Wong. An tegrated inventory allocation and vehicle routing problem. *Transportation Science*, 23(2):67-76, 1989.

[8] Marielle Christiansen, Kjetil Fagerholt, Truls Flatberg, Oyvind Haugen, Oddvar Kloster, and Erik H Lund. Maritime inventory routing with multiple products: A case study from the cement industry. *European Journal of Operational Research*, 208(1):86-94, 2011.

[9] Leandro C Coelho and Gilbert Laporte. A branch-and-cut algorithm for the multi-product multi-vehicle inventory-routing problem. *International Journal of Production Research*, 51(23/24):7156-7169, 2013.

[10] Jean-François Cordeau, Demetrio Laganà, Roberto Musmanno, and Francesca Vocaturo. A decomposition-based heuristic for the multiple-product inventory-routing problem. *Computers & Operations Research*, 55(0):153-166, 2015.

[11] Moshe Dror and Pierre Trudeau. Cash flow optimization in delivery scheduling. *European Journal of Operational Research*, 88(3):504-515, 1996.

[12] Awi Federgruen, Gregory Prastacos, and Pual Zipkin. An allocation and distribution model for perishable products. *Operations Research*, 34(1):75-82, 1986.

[13] Richard C Larson. Transporting sludge to the 106-mile site: An inventory/routing model for fleet sizing and logistics system design. *Transportation Science*, 22(3):186, 1988.

[14] K Li, B Chen, A I Sivakumar, and Y Wu. An inventory-routing problem with the objective of travel time minimization. *European Journal of Operational Research*, 236(3):936-945, 2014.

[15] Anis Mjirda, Bassem Jarboui, Rita Macedo, Sa'id Hanafi, and Nenad Mladenović. A two phase variable neighborhood search for the multi-product inventory routing problem. *Computers & Operations Research*, 52, Part B(0):291-299, 2014.

[16] N Ramkumar, P Subramanian, T T Narendran, and K Ganesh. Mixed integer linear programming model for multi-commodity multi-depot inventory routing problem. *OPSEARCH*, 49(4):413-429, 2012.

[17] D Ronen. Marine inventory routing: Shipments planning. *The Journal of the Operational Research Society*, 53(1):108-114, 2002.

[18] Jennifer Stacey, Malini Natarajarathinam, and Charles Sox. The storage constrained, inbound inventory routing problem. *International Journal of Physical Distribution & Logistics Management*, 37(6):484-500, 2007.

[19] S Viswanathan and Kamlesh Mathur. Integrating routing and inventory decisions in one-warehouse multiretailer multiproduct. *Management Science*, 43(3):294, 1997.

[20] Thomas Volling, Martin Grunewald, and ThomasS Spengler. An integrated inventory-transportation system with periodic pick-ups and leveled replenishment. *BuR - Business Research*, 6(2):173-194, 2013.

PROMOTING INDUSTRIAL ENGINEERING AND TEACHING ANALYTICS AND OPTIMIZATION WITH LEGO ROBOTICS

oung Jae Jang[1*] and Geo Lee[2]

[1]Department of Industrial and Systems Engineering
KAIST, Daejeon, South Korea
yjang@kaist.ac.kr

[2]Department of Industrial and Systems Engineering
KAIST, Daejeon, South Korea
Acton-Boxborough Regional High School
Acton, MA, U.S.A.
17leeg@abschools.org

ABSTRACT

We present the development and implementation of a LEGO robotics-based project for an undergraduate-level class on manufacturing systems, Manufacturing Process Innovation (IE251), which is mandatory for undergraduate students majoring in industrial and systems engineering at KAIST. The proposed project was pilot tested during the Spring 2015 semester and proved to be effective in promoting industrial and systems engineering and delivering the concept of a large scale-complex system. We introduce how a curriculum in industrial engineering has been innovated and discuss the lessons learned from the pilot class.

Keywords: Education, Manufacturing systems, Project based learning, LEGO robotics

1 INTRODUCTION

The discipline of industrial and systems engineering (ISE) traditionally involves scientific decision making, based on data, to increase the efficiency and effectiveness of operations. ISE is concerned with the analysis, design, and control of *large-scale complex systems* that comprise people, materials, information, equipment, and money [1][2].

A large-scale complex system is one that consists of multiple subsystems or components. The term complex is used because good performance in the individual subsystems does not guarantee the good performance of the entire system. That is, the relationship between the performance of the subsystem and that of the system is not straightforward. Examples of such systems are manufacturing factories, transportation systems, and logistics systems.

It is well known that hands-on experience can facilitate the learning of an abstract concept. However, one of the most challenging issues in education for large-scale complex systems is the limited opportunities available for students to gain such hands-on experience. This limitation exists mainly because of the nature of the system. This issue can be illustrated by comparing a single robot system in mechanical or electrical engineering with an automated production line, which is

---

* Corresponding Author

the most investigated application of a large-scale complex system in ISE. In mechanical or electrical engineering, robot-building projects are commonly used to teach control theory, electromechanical system design, system dynamics, or similar topics. Students learn through the process of designing and building a single robot. However, in an automated manufacturing system, the production line is the system and the industrial robots are the components that support the system. The performance of the robots is measured through the performance of the system as a whole, rather than individually. That is, multiple robots must coordinate and interact with each other to build products and achieve the goals of the system, such as a target production rate within a cost constraint.

Ideally, to create a laboratory environment or project to design such an automated production system, a resource for building multiple robots should be provided. Accordingly, the physical scale of the laboratory or project must be larger than a project to build a single robot. This characteristic of scale also applies to other typical large-scale complex systems, such as supply chain networks, logistics, and transportation, in which multiple resources, components, and subsystems interact in a large-scale environment.

The Industrial and Systems Engineering at KAIST (Korea Advanced Institute of Science and Technology) recently created a laboratory and project using LEGO robotics kit, LEGO Mindstorms EV3, to teach the concept of the large-scale complex system as well as to promote the department. In particular, the promotion of the department is critical at KAIST since students declare their majors during the second year of their undergraduate studies. According to a recent survey conducted by the department, most undergraduate students admitted to KAIST did not know about the industrial engineering. This lack of understanding of the discipline may result in decreased rate of the major declaration in ISE. Therefore, the laboratory has been setup to demonstrate the issues and problems ISE engineers deal with.

In this paper, we introduce the large-scale system and project we developed and how the system and project were effectively used for the ISE education and used to successfully promoted ISE.

2 LEGO MANUFACTURING SYSTEM (LMS)

Figure 1: KAIST LEGO Manufacturing System

Figure 1 shows the production line built by an instructor and graduate teaching assistants using LEGO Mindstorms EV3. The production line consists of three processing machines in a serial configuration in which a part visits all three machines sequentially. The parts are transferred by automated conveyor systems that connect the machines. All machines and conveyor belts have sensors and can keep track of the part positions and machine states. Note that although the parts visit and travel through the line with a robotics system, no physical processing is done in the line. The parts stop at the processing machine for a while to imitate processing.

The students' mission is to improve the performance of the production line in a logical and scientific way. Specifically, they need to find a way to improve performance measures by increasing the production rate (number of parts produced per unit of time), reducing the cycle time (the total time taken to produce a product), reducing the amount of work-in-progress inventory, or any combination of these [3].

With this in mind, students should first define improvement in a specific way. This is a key learning process to translate an abstract meaning to a specific goal-oriented target in a logical manner. They need to understand the issue in the system, define the problems, and set the target. In an actual engineering or management situation, defining and structuring a problem from unstructured and abstract issues is critical and challenging. We expect students to first define their goals and set a target for what to improve and then outline their plan for how to modify the line to achieve the goals and target.

Note that there are some constraints in the modification. The processing times for the machine should be unchanged. In an actual production system, parts should be processed for a certain amount of time at each machine. Each machine is programmed to hold the part for a predefined time, and this process time should remain unchanged. In addition, each machine is preprogrammed to imitate machine failures and repairs. In an actual production line, unexpected failures are inevitable, and the LMS is programmed to imitate this behavior. The failure occurs in a random manner in the LMS. If a machine fails, it holds the part until it is repaired. The repair time is also random. The frequency of the failure and repair time are preprogrammed in each processing machine. These parameters are not known to the students. Therefore, it is the students' job to determine the frequency of failure and the repair time. The students cannot change the failure frequency or repair time in the machines. Instead, to improve the performance of the LMS they are able to modify the length of the conveyor systems, the sensor locations, the motor speed of the conveyor belts, and any conveyor logic software.

The students must first collect data to determine the machines' processing time, failure frequency, and repair time. They must also evaluate the initial performance measures of the original LMS from the data production rates, cycle time, and the WIP inventory level.

They then need to determine which performance measures show underperformance and propose a way to improve the performance. One of the key activities is to optimize the length of the conveyor system. Throughout the project, the students learn an optimization modeling and algorithm technique and apply optimization to allocate the optimal length of the conveyor system. In addition, the sensor location in the conveyor system and the speed of the conveyor belts can be modified.

Each team presents their final modification of the LMS, and the team with the greatest improvement in the line with proper analysis is credited. They may change the hardware configuration of the conveyor system, the software logics, and other parameters in the LMS, with the exception of the machines' processing time, failure frequency, and repair time.

In this project, a group of four students work as a team. The grading and performance of the activity in the project reflect their team effort.

Throughout the project, students

- explore some critical systems problems of uncertainty and dynamic behavior arising from the interaction of multiple subsystems (or components) using a large-scale complex system with LMS
- understand the problem-solving and system improvement processes for a large-scale comple x system by performing data analysis, system modeling, and optimization

- understand the concept of optimization and the circumstances in which optimization is nee ded for a physical system
- experience actual system improvement with data-based decision making

Table I: Class evaluations

	Fall 2012				**Spring 2015**			
	Undergraduate enrolment	Respondent	Average	Std. dev.	Undergraduate enrolment	Respondent	Average	**Std. dev.**
Overall	32286	27351	4.19	0.83	32137	25948	4.18	0.82
Department	934	795	4.12	0.89	777	633	4.24	0.83
Course	68	64	3.99	0.84	32	30	4.41	0.79

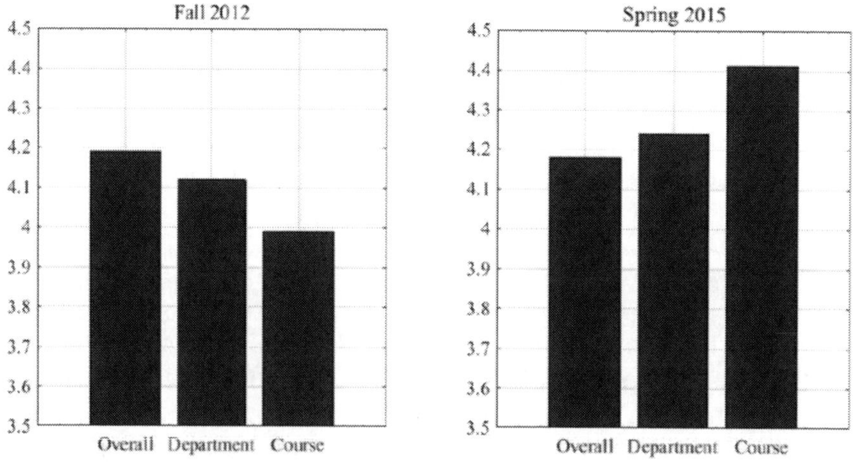

Figure 2: Class evaluation results

3 EVALUATION

Note that the detailed evaluation of the project and discussion are found in [1] and [2]. This section provides the summary in [1].

It is not easy to measure how the project impacts and effectiveness in improving the level of understanding of the students. The same class, Manufacturing Process Innovation, offered in Fall 2012 is used as a reference class to compare the effectiveness. The project has been developed since 2013 and during the development, it was tested in several different classes as a form of pilots. The class offered in Fall 2012 was the latest class where no LEGO project was introduced at all. The class in Fall 2012 was taught by the first author with the similar contents. There was no project in the Fall 2012 class. There were multiple field trips to factories and students had to write reports from the field trips. This field trips and reports were replaced with the LEGO project. Therefore, this class is a good reference to compare the effectiveness of the project.

Table I and Figure 2 show the overall course evaluation performed by the KAIST. For all the registered classes, KAIST conducts the class survey one week before the final exam period. Students take the survey and evaluate the class over the Internet class evaluation system. The results of the survey and evaluation are provided to the instructor after the final grades of the students are reported and finalized. The class evaluation report provides the overall class scores in the scale of 1 (no satisfied at all) to 5 (highly satisfied). In the report, the score of the class is also compared to the average scores of all the classes offered in KAIST and the classes offered in the department so that the instructors are able to understand the score level of their classes relative to other classes.

Table I shows the detailed metric of the satisfaction level for the Fall 2012 and Spring 2013 classes. Figure 8 presents the average satisfaction scores for the classes as bar graphs. Since 90 of the scores are within the 3.5 and 4.5 for undergraduate classes, the bar graph is drawn between these scores values. Note that in the Fall 2012 class, the satisfaction score is 3.99 which is lower than the average scores of the department and overall institution. In contrast, it is shown that the score of the class is significantly increased to 4.41 which is higher than those of the department and overall institution. A significant improvement is observed in the satisfaction score. Then what is the high scores contributed Is that caused by the LEGO project These questions are not easy to answer. However, given the survey results just for the project, it can be conjectured that the project did something positive in the overall scoring. More detailed analysis of the project evaluation such as evaluation analysis, survey analysis, test result analysis, and survey on the project are found in [1].

4 CONCLUSION

the LEGO-based project was effective in motivating students. The feedback obtained was positive and the students found the subject very interesting. Moreover, the test results indicated that the students were able to understand the issue of the dynamic flow behavior of manufacturing systems more effectively than with traditional lecture-based learning. The results of the survey and the class evaluation also showed that the students were highly satisfied with the class involving the LEGO-based project module. Although the module has only been used in one entry-level ISE major class, it can also be used in other ISE classes. In addition, LEGO robotics does not necessarily have to be used as a design project; it could also be used as an experimental device.

5 REFERENCES

[1] **ang, Y. . and Yosephine V. S. 2016.** LEGO Robotics Based Project for Industrial Engineering Education, *International Journal of Engineering Education*, 32(3A), pp. 1268 1278.

[2] **ang, Y. . 2014.** LEGO Project Report, KAIST Industrial and Systems Engineering Internal Report.

[3] **Gershwin. S. B.** 1994. *Manufacturing systems engineering*, PTR Prentice Hall.

HYBRID EVOLUTIONAL ALGORITHM FOR INDUSTRIAL AUTOMATION: AGV DISPATCHING AND ROBOT-BASED ASSENBLY LINE

Mitsuo Gen[1,2*], Lin Lin[1,3] and Hiroshi Kawakami[4]

[1] Fuzzy Logic Systems Institute, Iizuka, Japan
[2] Tokyo University of Science
gen@flsi.or.jp
[1,3] Dalian University of Technology, Dalian, China
lin@dlut.edu.cn
[4] Kyoto University, Kyoto, Japan
kawakami@design.kyoto-u.ac.jp

ABSTRACT

In real industrial manufacturing systems there are many combinatorial optimization problems (COP) imposing on more complex issues, such as complex structure, nonlinear constraints, and multiple objectives to be handled simultaneously. Scheduling optimization problem (SOP) is one of the important and complex COP models, where it can have a major impact on the productivity of a production process. Moreover, the COP models make the problem intractable to the traditional optimization techniques because most of scheduling problems fall into the class of NP-hard combinatorial problems.

Hybrid evolutionary algorithm (HEA) will be introduced for treating automation problems in factory, manufacturing, planning and scheduling, and logistics and transportation systems. HEA is the most popular metaheuristic method for solving NP-hard optimization problems. In the past few years, EA has been exploited to solve design automation problems. Concurrently, the field of HEA reveals a significant interest in evolvable hardware and problems such as routing, placement or test pattern generation. We propose a random key-based genetic algorithm with crossover and mutation operation to avoid premature convergence and to maintain diversity. Numerical experiments for case study show the effectiveness of proposed approach comparing with hybrid evolutionary algorithm.

Keywords: Genetic algorithm (GA), Random key-based GA, Hybrid evolutionary algorithm (HEA), Flexible Manufacturing System (FMS), Automated Guided vehicles (AGV), Robot-based assembly-line balancing (RALB) system,

1. INTRODUCTION

In real industrial manufacturing systems there are many combinatorial optimization problems (COP) imposing on more complex issues, such as complex structure, nonlinear constraints, and multiple objectives to be handled simultaneously. Scheduling optimization problem (SOP) is one of the important and complex COP models, where it can have a major impact on the productivity of a production process. Moreover, the COP models make the problem intractable to the traditional optimization techniques because most of scheduling problems fall into the class of NP-hard combinatorial problems (Yu & Gen 2010; Gen & Lin 2014; Gen *et al*, 2015).

The semiconductor based industry has grown rapidly and subsequently production planning problems have raised many important research issues. Because of short product lifecycles, it is crucial to rapidly respond to various customer needs and deliver products on time in high-tech manufacturing industries such as the

* Corresponding Author

various semiconductor devices including IC chips, LSI chips & microprocessors, hard disc device (HDD) and thin-film transistor-liquid crystal display (TFT-LCD). Manufacturing scheduling of TFT-LCD module assembly system is a key issue to enhance manufacture efficiency that could satisfy customer demand on time. By focusing on realistic settings, a module assembly process was formulated for use in the HDD and TFT-LCD industries as a generalization of the flexible job-shop scheduling problem (FJSP) and assembly line balancing (ALB) model, respectively. On a flexible job-shop floor, workstations employ non-identical parallel machines scheduling (PMS) model and re-entrant flowshop scheduling (RFS) model that exhibit distinct production velocities (Chou, Chien & Gen, 2014). An operation can be processed using an available machine from a given workstation. For example, the TFT-LCD module assembly scheduling problem can be divided into two subproblems: the routing (*i.e.*, assigning each operation to machines) and scheduling problems (*i.e.*, determining the start time of each operation to machines). However, to find the optimal solutions of manufacturing scheduling gives rise to complex combinatorial optimization, unfortunately, most of them fall into the class of NP-hard combinatorial problems (Chamnanlor *et al*, 2014).
In order to develop effective and efficient solution algorithms that are in a sense "good," i.e., whose computational time is small as within a few minutes, or at least reasonable for NP-hard combinatorial problems met in practice, we have to consider the following issues:

- Quality of solution,
- Computational time and
- Effectiveness of the nondominated solutions for multiobjective optimization problem (Yu & Gen, 2010).

This paper is organized as follows. First the background developments of hybrid evolutionary algorithms, in particular are described. Then basic schemes and working mechanism of genetic algorithm (GA) will be given, and fuzzy logic controller for enhancing evolutionary process. Lastly, automation and the challenges for applying evolutionary algorithms are specified. Next, the various applications based on EAs for solving factory automation (FA) problems will be surveyed, covering planning and scheduling problems, nonlinear optimization problems in manufacturing systems, and optimal design problems in logistics and transportation systems. Finally, among those applications based on EAs, detailed case studies will be introduced. The first case study covers dispatching of automated guided vehicles (AGV) and machine scheduling in a flexible manufacturing system (FMS). The second EA case study for treating automation problems is the robot-based assembly line balancing (RALB) problem. Numerical experiments for various scales of AGV dispatching problems and robot-based ALB problems will be described to show the effectiveness of the HEA approaches with greater search capability that improves the quality of solutions and enhances the rate of convergence over existing approaches.

2. HYBRID EVOLUTIONARY ALGORITHMS

Evolutionary algorithm (EA) is a subfield of artificial intelligence (AI), and refers to a synthesis of methodologies from fuzzy logic (FL), artificial neural networks (ANN), genetic algorithm (GA), and evolutionary programming (EP) and evolution strategy (ES). EA uses the evolutionary process within a computer to provide a means for addressing complex engineering problems involving chaotic disturbances, randomness, and complex nonlinear dynamics that traditional algorithms have been unable to conquer.

2.1 Genetic Algorithm

Among evolutionary algorithms (EAs), genetic algorithms (GA) is the most widely known type of EA today. GA includes the common essential elements of EA, and has wide real-world applications. The original form of GA was described by Goldberg (1989). GA is a stochastic search technique based on the mechanism of natural selection and natural genetics. The central theme of research on GA is to keep a balance between exploitation and exploration in its search to the optimal solution for survival in many different environments. Features for self-repair, self-guidance, and reproduction are the rules in biological systems, whereas they barely exist in the most sophisticated artificial systems. GA has been theoretically and empirically proven to provide a robust search in complex search spaces. GA, differing from conventional search techniques, starts with an initial set of random solutions called the population. Each individual in the population is called a chromosome, representing a solution to the problem at hand. A chromosome is a string of symbols, usually but not necessarily, a binary bit string. The chromosomes evolve through successive iterations, called generations. During each generation, the chromosomes are evaluated, using some measures of fitness. To create the next generation, new chromosomes, called offspring, are generated by either merging two chromosomes from the current generation using a crossover operator and/or modifying a chromosome using a mutation operator. A new generation is formed by selecting some of the parents, according to the fitness values, and offspring, and rejecting others so as to keep the population size constant. Fitter chromosomes have higher probabilities of being selected. After several generations, the algorithms converge to the best chromosome, which hopefully represents the optimum or suboptimal solution to the problem. In general, GA has five basic components, as summarized by Michalewicz (1994):

1. A genetic representation of potential solutions to the problem
2. A way to create a population (an initial set of potential solutions by encoding routine)
3. An evaluation function rating solutions in terms of their fitness (by decoding routine)
4. Genetic operators that alter the genetic composition of offspring (crossover, mutation, selection, etc.)
5. Parameter values that genetic algorithms use (population size, probabilities of applying genetic operators, etc.).

Major Advantages of GA: GA has received considerable attention regarding their potential as a novel optimization technique. Three major advantages exist when applying GA to various combinatorial optimization problems (Gen and Lin 2009a):

-**Adaptability**: GA does not have much mathematical requirements about the optimization problems. Because of the evolutionary nature, GA will search for solutions without regard to the specific inner workings of the problem. GA can handle any kind of objective functions and any kind of constraints, i.e., linear or nonlinear, defined on discrete, continuous, or mixed search spaces.

-**Robustness**: The use of evolution operators makes GA very effective in performing global search (in probability), whereas most of conventional heuristics usually perform local search. It has been proved by many studies that GA is more efficient and more robust in locating an optimal solution and reducing a computational effort than other conventional heuristics.

-**Flexibility**: GA provides us with a great flexibility to hybridize with domain-dependent heuristics to make an efficient implementation for a specific problem.

-**Applicability**: The practical software packages based on GA such as ERP and Simulation are standard usages in the real world.

Genetic Representation: In general, two ways exist to generate the initial population, *i.e.*, a set of solution candidates, heuristic initialization and random initialization, by using an encoding procedure satisfying system constraints and/or a boundary condition. How to present a solution of the scheduling problem into a chromosome is a key issue for GAs. For evaluating the effectiveness of the different chromosome representation, there are several critical issues are summarized by Gen and Lin (2009a).

- **Space**: Chromosomes should not require extravagant amounts of memory.
- **Time**: The time complexities of evaluating, recombining, and mutating chromosomes should be small.
- **Feasibility**: All chromosomes, particularly those generated by simple crossover (i.e., one-cut point crossover) and mutation, should represent feasible solutions.
- **Uniqueness**: The mapping from chromosomes to solutions (decoding) may belong to one of the following three cases: 1-to-1 mapping, n-to-1 mapping and 1-to-n mapping. The 1-to-1 mapping is the best one among three cases and 1-to-n mapping is the most undesired one.
- **Heritability**: Offspring of simple crossover (i.e., one-cut point crossover) should represent solutions that combine substructures of their parental solutions.
- **Locality**: A mutated chromosome should usually represent a solution similar to that of its parent.

We need to consider these critical issues carefully when designing an appropriate representation so as to build an effective GA. As known, scheduling problem is the implement of production plan, with considering production processes, lot-size, amount and customer requirements etc. And scheduling problem is how to decide the resources assignment to the production, with considering constrains of resources capabilities and capacities. There are two decision making parts for scheduling optimization:

1) operation sequencing and
2) resources assignment.

2.2 Fuzzy Logic Controller

Fuzzy logic is much closer in spirit to human thinking and natural language than the traditional logical systems. In essence, the fuzzy logic controller (FLC) provides an algorithm which can convert a linguistic control strategy based on expert knowledge into an automatic control strategy. In particular, this methodology appears very useful when the processes are too complex for analysis by conventional techniques or when the available sources of information are interpreted qualitatively, inexactly, or with uncertainty (Herrera and Lozano, 1996). The main idea is to use a FLC to compute new strategy parameter values that will be used by the genetic algorithms (Yun and Gen, 2003). A fuzzy logic controller is comprised of four principal components as shown in the generic structure of a FLC in Figure 1:

- Knowledge base,
- Fuzzification interface,
- Inference system,
- Defuzzification interface.

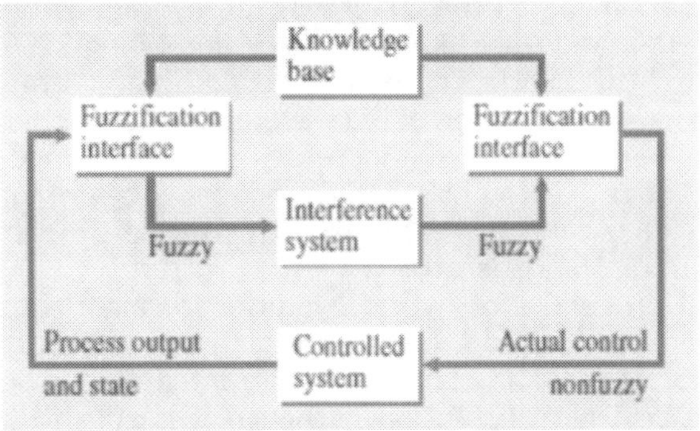

Figure 1: Generic structure of a fuzzy logic controller (FLC)

The experts' knowledge is stored in the knowledge base in the form of linguistic control rules. The inference system is the kernel of the controller, which provides an approximate reasoning based on the knowledge base. Yun and Gen (2003) proposed auto-tuning routine for the parameter crossover and mutation rates based on FLC and the change of the average fitness for enhancing the evolutionary process of the GA. Lin and Gen (2009) also proposed another enhanced auto-tuning strategy for GA for keeping a balance between exploration and exploitation during the evolutionary process.

One of the main problems related to GA is to find the optimal control parameter values required by the technique. Furthermore, different values may be necessary during the course of a run. There has been more work dedicated to finding the optimal parameters of genetic algorithms, which require different techniques for different problems. A hybrid genetic algorithms (HGA) has been built so that selected control parameters may be dynamically adjusted during the course of evolving a problem solution. In our paper, we use the basic concept of Yun and Gen (2003) to adaptively regulate GA operators using fuzzy logic control (FLC). The main scheme of this concept is to use two FLCs: the crossover FLC and mutation FLC as shown in Figure 2 and it is based on the coordinated strategy between the FLC and GA combined the change of the average fitness at each generation. These two FLCs are implemented independently to adaptively regulate the rates of crossover and mutation operators during genetic search process. For example, in minimization problem, we can set the change of the average fitness at generation t, $\Delta f_{avg}(t)$ as follows:

$$\Delta f_{avg}(t) = \frac{1}{popSize}\sum_{k=1}^{popSize} f_k(t) - \frac{1}{offSize}\sum_{k=1}^{offSize} f_k(t)$$

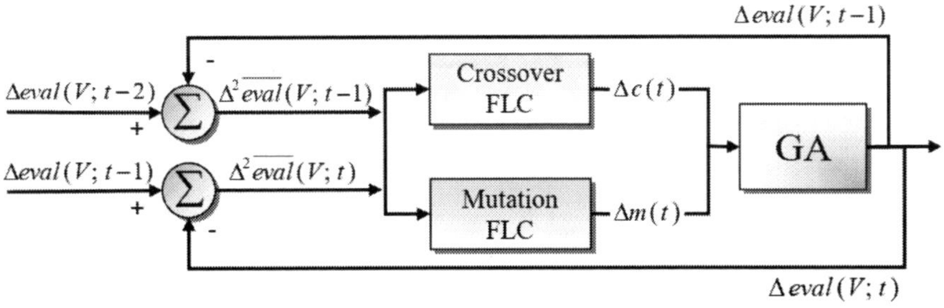

Figure 2 Coordinated strategy between the FLC and GA (Yun & Gen 2003)

2.3 Hybrid Evolutionary Algorithm

Adaptation techniques for designing hybrid evolutionary algorithm have been proposed and tested during the past few years. Problem adaptation and evolutionary processes adaptation are two kinds of different adaptations. To tune the parameters for changing configurations of evolutionary algorithms while finding the problem solution is an adaptation to evolutionary processes which was conducted. Parameters, such as population size, maximum generation, crossover probability, and mutation probability, strongly affect the balance between exploitation and exploration in the search space which characterize the EA behaviors.

Normally, a lot of applications of EA have fixed the parameters, so it is in contrast to the general determination of EA. Thus it is a natural idea to adjust the values of parameters during the run of the algorithm.

In this paper, an auto-tuning strategy is proposed by adopting FLC to tune the probabilities of the genetic operators depending on the change of the average fitness of parents and offspring which occurs at each generation (Lin and Gen 2009; Lin, Gen, and Wang 2009). By balancing between crossover and mutation probabilities, this is able to be done based on adjustment of the average fitness of the current and last generations. For the scheme, the average fitness change in parents and offspring populations during continuous generations of GA increases the occurrence probability of p_M and decreases the occurrence probability of p_C, if it consistently produces a better offspring. Otherwise, it decreases the occurrence probability of p_M and increases the occurrence probability of p_C, if it consistently produces a poorer offspring during the generations.

The overall procedure of hybrid evolutionary algorithm based on auto-tuning strategy FLC is proposed which combines GA and heuristics such as left-shift routine and traditional local search routine (Chamnanlor et al 2014). To solve reentrant flowshop scheduling problem by simple GA is difficult when finding the best global solution, although it is able to locate hopeful regions for global optimum in the search space. The hybrid EA, which is based on the concept of hybridization, might be better since it is possible to assist the local convergence.

```
procedure: Hybrid Evolutionary Algorithm with FLC
input: problem data, GA parameters (popSize, maxGen, p_M, p_C)
output: the best schedule
begin
    t ← 0;                                        // t: generation no.
    initialize P(t) by encoding routine;          // P(t): population
    check & repair P(t) the problem constraint for all chromosomes;
    evaluate P(t) by decoding routine and keep best solution;
    while (not terminating condition) do
        create C(t) from P(t) by crossover routine;    // C(t): offspring
        create C(t) from P(t) by mutation routine;
        create C(t) from P(t) by immigration routine;
        check & repair the precedence constraint for all offspring C(t);
        check & repair the time window constraint for all offspring C(t);
        improve offspring C(t) by left-shift routine;
        evaluate C(t) by decoding routine and update best solution;
        select P(t+1) from P(t) and C(t) by selection routine;
        tune parameters by auto-tuning strategy FLC routine;
        t ← t + 1;
    end;
    output the best schedule
end;
```

Figure 3: Overall procedure of hybrid evolutionary algorithm based on auto-tuning strategy FLC

3. METAHEURISTICS TECHNIQUES FOR INDUSTRIAL AUTOMATION

Currently, for manufacturing, the purpose of automation has shifted from increasing productivity and reducing costs to broader issues, such as increasing quality and flexibility in the manufacturing process. For example, automobile and truck pistons used to be installed into engines manually. This is rapidly being transitioned to automated machine installation, because the error rate for manual instalment was around 1-1.5%, but has been reduced to 0.00001% with automation. Hazardous operations, such as oil refining, manufacturing of industrial chemicals, and all forms of metal working, were always early contenders for automation (Gen & Lin 2009b).

However, many applications of automation, such as optimizations and automatic controls, are formulated with complex structures, complex constraints, and multiple objects simultaneously, which makes the problem intractable to traditional approaches. In recent years, the evolutionary techniques community has turned much of its attention toward applications in industrial automation.

3.1 Factory Automation

In a manufacturing system, layout design is concerned with the optimum arrangement of physical facilities, such as departments or machines, in a certain area. Usually the design criterion is considered to be minimizing material-handling costs. Because of the combinatorial nature of the facility layout problem, the heuristic technique is the most promising approach for solving practical-size layout problems. The interest in application of EAs to facility layout design has been growing rapidly (Gen & Lin 2009b). Tate and Smith (1995) applied GA to the shape-constrained unequal-area facility layout problem. Cohoon *et al* (1991) proposed a distributed GA for the floorplan design problem. Tam (1992) reported his experiences of applying genetic algorithms to the facility layout problem. Recently, Zheng et al (2015) proposed multiobjective multi-population biased random-key genetic algorithm for solving the 3-D container loading problem.

A flexible machining system is one of the forms of factory automation. The design of a flexible machining system (FMS) involves the layout of machine and workstations. Kusiak and Heragu (1987) reported a survey paper on the machine layout problem. The layout of machines in a FMS is typically determined by the type of material-handling devices used. The most used material-handling devices are: material handling robot, automated guided vehicle (AGV), and gantry robot. Gen and Cheng (1997) provide various GA approaches for the machine layout and facility layout problems.

3.2 Planning and Scheduling Automation

The planning and scheduling of manufacturing systems always require resource capacity constraints, disjunctive constraints, and precedence constraints, owing to the tight due dates, multiple customer-specific orders, and flexible process strategies. Here, some hot topics of applications of ETs in advanced planning and scheduling (APS) are introduced. These models mainly support the integrated, constraint-based planning of the manufacturing system to reduce lead times, lower inventories, increase throughput, etc.

The flexible jobshop problem (FJSP) is a generalization of the jobshop and parallel machine environment (Pinedo 2002), which provides a closer approximation to a wide range of real manufacturing systems. Kacem et al. proposed the operations machine-based GA approach (Kacem et al. 2002), which is based on a traditional representation called the schemata theorem representation. Zhang and Gen (2005) proposed a multistage operation-based encoding for FJSP. Gao et al (2006) proposed scheduling jobs and maintenance method in FJSP by using a hybrid genetic algorithm and Gao et al (2008) reported a hybrid genetic and variable neighbourhood descent algorithm for flexible job shop scheduling problems. Gen et al (2009) proposed multistage-based genetic algorithm for solving flexible Job-shop scheduling problem.

The objective of the resource-constrained project scheduling problem (rcPSP) is to schedule activities such that precedence and resource constraints are obeyed and the makespan of the project is to be minimized. Gen and Cheng (1997) adopted priority-based encoding for this rcPSP. In order to improve the effectiveness of priority-based GA approach for an extended resource constrained multiple project scheduling problem, Kim et al (2005) combined priority dispatching rules in priority-based encoding process.

The advanced planning and scheduling (APS) model includes a range of capabilities from finite capacity planning at the plant floor level through constraint-based planning to the latest applications of advanced logic for supply-chain planning and collaboration (Turbide 1998). Several related works by Moon et al (2004) and Moon & Seo (2005) have reported a GA approach especially for solving such kinds of APS problems. Recently Zhang et al (2014) proposed hybrid sampling strategy-based multiobjective evolutionary algorithm for solving process planning and scheduling problem. Chamnanlor et al (2015) reported embedding ant system in genetic algorithm for re-entrant hybrid flow shop scheduling problems with time window constraints.

3.3 Manufacturing Automation

Assembly-line balancing problems (ALB) consist of distributing work required to assemble a product in mass or series production on an assembly line among a set of workstations. Several constraints and different objectives may be considered. The simple assembly line balancing problem consists of assigning tasks to workstations such that precedence relations between tasks and zoning or other constraints are met. The objective is to make the work content at each station most balanced. GAs have been applied to solve various assembly-line balancing problems (Tsujimura et al 1995; Rubinovitz & Levitin 1995; Gen et al 1996). Gao et al (2009) proposed an innovative GA hybridized with local search for a robotic-based ALB problem. Based on different neighbourhood structures, five local search procedures are developed to enhance the search ability of GA.

An automated guided vehicle (AGV) is a mobile robot used widely in industrial applications to move materials from point to point. AGVs help to reduce costs of manufacturing and increase efficiency in a manufacturing system. For a recent review of AGV problems and issues the reader is referred to Hwang e al (2002), Qui et al (2002), Lim (2004), Vis (2006), Le-Anh & Koster (2006) and Lin et al (2006). Lin et al (2006) adopted priority-based encoding for solving the AGV dispatching problem in an FMS.

3.4 Logistics and Transportation Automation

Logistics is the last frontier for cost reduction and the third profit source of enterprises (Lau & Zhao 2008). The interest in developing effective logistics system design models and efficient optimization methods has been stimulated by high costs of logistics and the potential for securing considerable savings. One of the most important topics of logistics and transportation automation is automated container terminals.

Handling equipment scheduling (HES) is important for scheduling different types of handing equipment in order to improve the productivity of automated container terminals. Lau and Zhao reported a mixed integer programming (MIP) model, which considers various constraints related to the integrated operations between different types of handling equipment. This study proposes a multilayer GA to obtain the near optimal solution of the integrated scheduling problem (Lau & Zhao 2008).

Berth allocation planning (BAP) is to allocate space along the quayside to incoming ships at a container terminal in order to minimize some objective function. Preston and Kozan (2001) developed a container location model and proposed a GA approach with analyses of different resource levels and a comparison with current practice at the Port of Brisbane. Imai et al (2008) introduce a formulation for the simultaneous berth and crane allocation problem and employed GA to find an approximate solution for the problem. The aim of storage locations planning (SLP) is to determine the optimal storage strategy for various container-handling schedules.

4. AGV DISPATCHING IN MANUFACTURING SYSTEM

Automated material handling has been called the key to integrated manufacturing. An integrated system is useless without a fully integrated, automated material handling system. In the manufacturing environment, there are many automated material handling possibilities. Currently, automated guided vehicles systems (AGV systems), which include automated guided vehicles (AGVs), are the state of the art, and are often used to facilitate automatic storage and retrieval systems (AS/RS).

Traditionally, AGV systems were mostly used in manufacturing systems. In manufacturing areas, AGVs are used to transport all types of materials related to the manufacturing process. The transportation network connects all stationary installations (e.g., machines) in the center. At stations, pickup and delivery points are installed that operate as interfaces between the production/storage system and the transportation system of the center. At these points a load is transferred by, for example, a conveyor from the station to the AGV and vice versa. AGVs travel from one pickup and delivery point to another on fixed or free paths. Guide paths are determined by, for example, wires in the ground or markings on the floor. More recent technologies allow AGVs to operate without physical guide paths.

4.1 Network Modelling for AGV Dispatching

In this Subsection, we introduce simultaneous scheduling and routing of AGVs in a flexible manufacturing system (FMS) (Lin et al 2006; Lin & Gen 2009b). An FMS environment requires a flexible and adaptable material handling system. AGVs provide such a system. An AGV is a material handling equipment that travels on a network of guide paths. The FMS is composed of various cells, also called workstations (or machines), each with a specific operation such as milling, washing or assembly. Each cell is connected to the guide path network by a pickup/delivery (P/D) point where pallets are transferred from/to the AGVs. Pallets of products are moved between the cells by the AGVs. Assumptions considered are as follows:

1. AGVs only carry one kind of products at a time.
2. A network of guide paths is defined in advance, and the guide paths have to pass through all pickup/delivery points.
3. The vehicles are assumed to travel at a constant speed.
4. The vehicles can just travel forward, not backward.
5. As many vehicles travel on the guide path simultaneously, collisions are avoided by hardware and are not considered herein.
6. At each workstation, there is pickup space to store the operated material and delivery space to store the material for the next operation.
7. The operation can be started any time after an AGV took the material to come. And also the AGV can transport the operated material from the pickup point to the next delivery point any time.

Definition 1: A node is defined as task T_{ij}, which represents a transition task of the j-th process of job J_i for moving from the pickup point of machine $M_{i,j-1}$ to the delivering point of machine M_{ij}.

Definition 2: An arc can be defined as many decision variables, such as, capacity of AGVs, precedence constraints among the tasks, or costs of movement. Lin et al. defined an arc as a precedence constraint, and give a transition time $c_{jj'}$ from the delivery point of machine M_{ij} to the pickup point of machine $M_{ij'}$ on the arc.

Definition 3: We define the task precedence for each job; for example, task precedence for three jobs is shown as follows:

Job J_1 (4 tasks): $T_{11} \succ T_{12} \succ T_{13} \succ T_{14}$

Job J_2 (2 tasks): $T_{21} \succ T_{22}$

Job J_3 (3 tasks): $T_{31} \succ T_{32} \succ T_{33}$

The notation used for formulating a mathematical programming model is summarized as follows:

Indices:

i, i': index of jobs, i, i' = 1,..., n;

j, j': index of processes, j, j' = 1, ... ,n.

Parameters:

n: total number of jobs, m: total number of machines,

n_i : total number of operations of job i,

o_{ij} : the j-th operation of job i,

p_{ij} : processing time of operation o_{ij},

M_{ij} : machine assigned for operation o_{ij},

T_{ij} : transition task for operation o_{ij},

t_{ij} : transition time from $M_{i,j-1}$ to M_{ij}.

Decision variables:

x_{ij} : assigned AGV number for task T_{ij} ,

t^S_{ij} : starting time of task T_{ij},

c^S_{ij} : starting time of operation o_{ij}.

The objective functions are to minimize the time required to complete all jobs (*i.e.*, the makespan) t_{MS} and the number of AGVs, n_{AGV} and the problem can be formulated as a bicriteria mixed integer programming (BMIP) model as follows:

$$\min\ t_{\mathrm{MS}} = \max_{i} \left\{ t^{S}_{i,n_i} + t_{M_{i,n_i},0} \right\} \tag{1}$$

$$\min\ n_{\mathrm{AGV}} = \max_{i,j} \left\{ x_{ij} \right\} \tag{2}$$

$$\text{s.t.}\quad c^{S}_{ij} - c^{S}_{i,j-1} \geq p_{i,j-1} + t_{ij},\ \ \forall i, j = 2,\ldots,n_i \tag{3}$$

$$\left(c^{S}_{ij} - c^{S}_{i'j'} - p_{i'j'} + \Gamma\left|M_{ij} - M_{i'j'}\right| \geq 0\right) \vee \left(c^{S}_{i'j'} - c^{S}_{ij} - p_{ij} + \Gamma\left|M_{ij} - M_{i'j'}\right| \geq 0\right),\ \ \forall (i,j),(i',j') \tag{4}$$

$$\left(t^{S}_{ij} - t^{S}_{i'j'} - t_{i'j'} + \Gamma\left|x_{ij} - x_{i'j'}\right| \geq 0\right) \vee \left(t^{S}_{i'j'} - t^{S}_{ij} - t_{ij} + \Gamma\left|x_{ij} - x_{i'j'}\right| \geq 0\right),\ \ \forall (i,j),(i',j') \tag{5}$$

$$\left(t^{S}_{i,n_i} - t^{S}_{i'j'} - t_{i'j'} + \Gamma\left|x_{ij} - x_{i'j'}\right| \geq 0\right) \vee \left(t^{S}_{i'j'} - t^{S}_{i,n_i} - t_{i} + \Gamma\left|x_{ij} - x_{i'j'}\right| \geq 0\right),\ \ \forall (i,n_i),(i',j') \tag{6}$$

$$c^{S}_{ij} \geq t^{S}_{i,j+1} - p_{ij} \tag{7}$$

$$x_{ij} \geq 0,\ \ \forall i, j \tag{8}$$

$$t^{S}_{ij} \geq 0,\ \ \forall i, j \tag{9}$$

where Γ is a very large number, and t_i is the transition time for pickup point of machine to delivery point of Loading/Unloading. Inequality (3) describes the operation precedence constraints. In inequities (4)-(6), since one or the other constraint must hold, it is called disjunctive constraint.

4.2 Evolutionary Approach: Random Key-based Hybrid EA

How to encode a solution of the network problem into a chromosome is a key issue for GA approaches. In many GA applications, especially for the problems of network design, the simple approach of GA was difficult to apply directly (Michalewicz 1994; Gen and Cheng 2000; Deb 2001; Gen et al 2008). The AGV dispatching problem is a combination of AGV assignment and task sequencing. A solution can be described by the assignment of task on AGVs, and the task sequence that transport the material form a pickup point to a delivery point. In this paper, the chromosome is presented as a random key-based encoding, that is, the locus of each gene represent task ID and the gene are presented as real number. As we know, the real number can be separated with integer and decimal. We consider the part of integer to AGV assignment, and consider the part of decimal as giving the priority of each task (Lin & Gen 2009). The detailed encoding and decoding processes are separated as the following phases:

- Phase 1: random key-based encoding process.
- Phase 2: grouping process (AGV assignment).
  - 2-1 Extraction of integer part.
  - 2-2 AGV assignment.
- Phase 3: task sequencing growth process.
  - 3-1 Extraction of decimal part.
  - 3-2 Task sequencing.

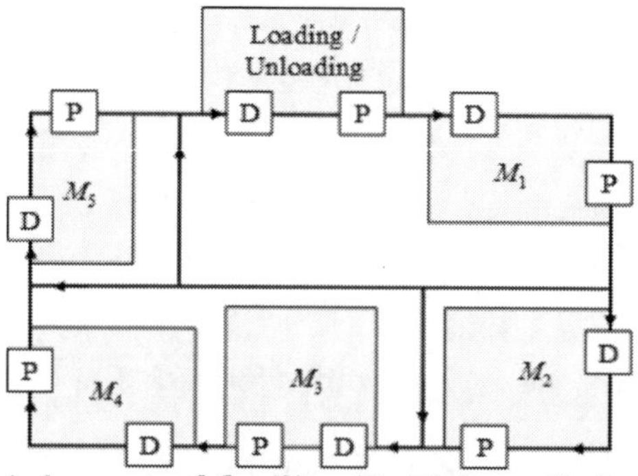

Figure 4: Layout of facility (P: Pickup, D: Delivery)

4.3 Numerical Experiment

For solving the AGV dispatching problem in FMS, the special difficulty arises from that the task sequencing is NP-hard problem and that a random sequence of AGV dispatching usually does not satisfy the operation precedence constraint and routing constraint. For evaluating the efficiency of the AGV dispatching algorithm suggested in the case study, a simulation program was developed using Java on a Pentium IV processor (3.2 GHz clock). We considered the AGV dispatching example in which a layout of facility is given for the experiment in Figure 4 with job requirements and transition time between pickup point u and delivery point v as shown in Tables 1-2, respectively. The detailed test data is given by Yang (2001) and Kim et al (2004). GA parameter settings were taken as follows: population size, popSize=20; crossover probability, $p_C = 0.70$; mutation probability, $p_M = 0.50$; immigration rate, $\mu = 0.15$. In an FMS case study, ten jobs are to be scheduled on five machines. The maximum number of processes for the operations is four. The detailed date sets are shown in (Kim & Hwang 1999).

The best machine schedule and AGV schedule in Gantt chart of the example for the AGV dispatching by the random key-based EA are shown in Figures 5 - 7, respectively in which the final time required to complete all jobs, *i.e.* makespan is 574 and 4 AGVs are used (Liang, Lin, Gen & Ohno, 2012).

Table 1: Job Requirements of the example

O_{ij}	M_{ij}				p_{ij}			
J_i \ P_j	P_1	P_2	P_3	P_4	P_1	P_2	P_3	P_4
J_1	1	2	1	-	80	120	60	-
J_2	2	1	-	-	100	60	-	-
J_3	5	3	3	-	70	100	70	-
J_4	5	3	2	2	70	100	100	40
J_5	4	2	-	-	90	40	-	-
J_6	4	4	1	2	90	70	60	40
J_7	1	3	-	-	80	70	-	-
J_8	5	4	5	4	70	70	70	80
J_9	5	4	1	-	70	70	60	-
J_{10}	5	1	3	-	70	60	70	-

Table 2: Transition time between pickup *u* and delivery *v*

t_{uv}/c_{uv}	Loading / Unloading	M_1	M_2	M_3	M_4	M_5
Loading / Unloading	1/ 1	1/ 7	8/13	14/18	16/23	18/20
M_1	13/18	3/ 3	2/ 9	8/14	10/19	13/18
M_2	18/22	22/28	2/ 2	2/ 7	4/12	12/18
M_3	13/11	17/22	24/29	1/ 1	1/ 6	7/11
M_4	8/14	12/20	18/26	24/29	3/ 3	2/10
M_5	5/ 7	9/12	15/18	19/23	23/28	2/ 2

$$\text{AGV1}: T_{11} \rightarrow T_{22} \rightarrow T_{41} \rightarrow T_{81} \rightarrow T_{91} \rightarrow T_{82} \rightarrow T_{92} \rightarrow T_{83} \rightarrow T_{84}$$
$$\text{AGV2}: T_{21} \rightarrow T_{41} \rightarrow T_{12} \rightarrow T_{15} \rightarrow T_{10,2} \rightarrow T_{52} \rightarrow T_{71} \rightarrow T_{44}$$
$$\text{AGV3}: T_{61} \rightarrow T_{62} \rightarrow T_{63} \rightarrow T_{64} \rightarrow T_{43} \rightarrow T_{72}$$
$$\text{AGV4}: T_{31} \rightarrow T_{32} \rightarrow T_{10,1} \rightarrow T_{33} \rightarrow T_{13} \rightarrow T_{10,3} \rightarrow T_{93}$$

Figure 5: The best machine schedule and AGV schedule

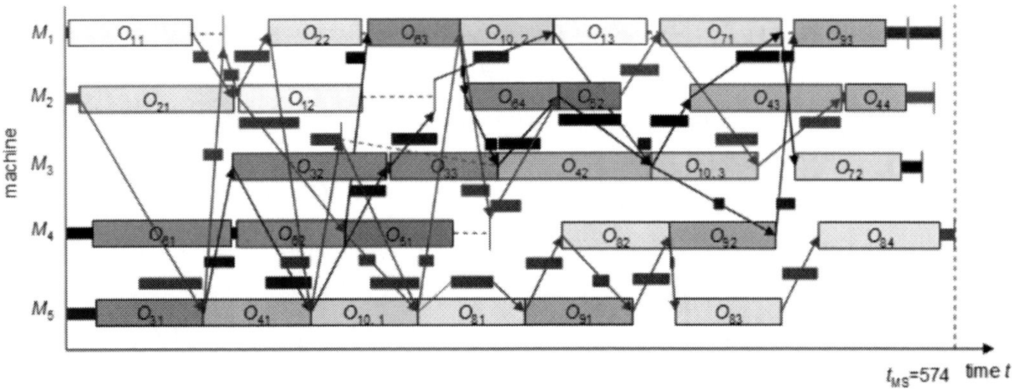

Figure 6: Gantt chart of the schedule of the example with five machines.

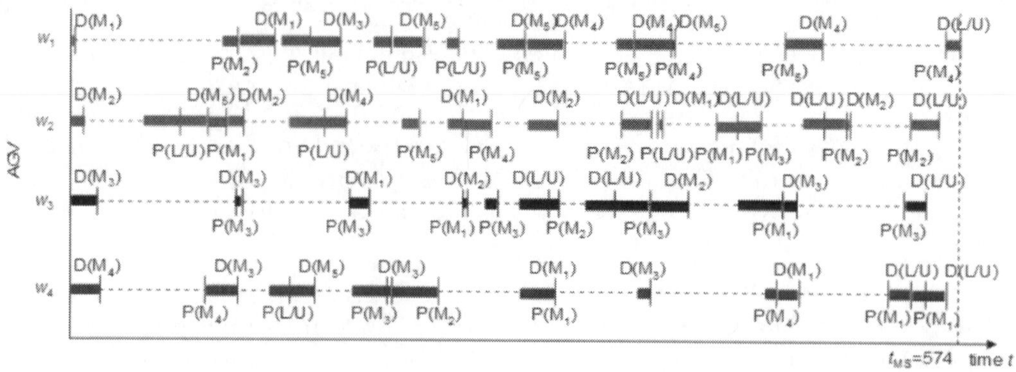

Figure 7: Gantt chart of the schedule of the example with considering AGVs routing.

As introduced above, the AGV dispatching problem is a difficult problem to solve by conventional heuristics. Adaptability, robustness, and flexibility make EA very effective for such automation problems.

5. ROBOT-BASED ASSEMBLY LINE SYSTEM

Assembly lines are flow-oriented production systems which are still typical in the industrial production of high-quantity standardized commodities and are even gaining importance in low-volume production of customized products. Usually, specific tooling is developed to perform the activities needed at each station. Such tooling is attached to the robot at the station. In order to avoid the wasted time required for tool change, the design of the tooling can take place only after the line has been balanced. Different robot types may exist at the assembly facility. Each robot type may have different capabilities and efficiencies for various elements of the assembly tasks. Hence, allocating the most appropriate robot for each station is critical for the performance of robotic assembly lines (Gen & Lin 2009a).

5.1 Assembly-Line Balancing Problems

The assembly-line balancing (ALB) problem concerns how to assign the tasks to stations and how to allocate the available robots for each station in order to minimize cycle time under the constraint of precedence relationships. Let us consider a simple example to describe the problem, in which ten tasks are to be assigned to four workstations, and four robots are to be equipped on the four stations. Figure 8 shows the precedence constraints for the ten tasks, and Table 3 gives the processing time for each of the tasks processed by each robot as the simple example of ALB problem. We show a feasible solution for this example in Figure 9. The balancing chart for the solution can be drawn to analyze the solution. Figure 9 shows that the idle time of stations 1-3 is very large, which means that this line is not balanced for production. In the real world, an assembly line is not just used for producing one unit of the product; it should produce several units. So we give the Gantt chart for three units to analyze the solution, as shown in Figure 10.

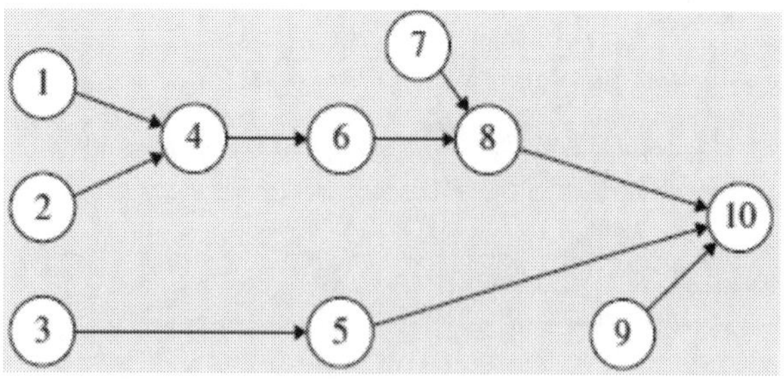

Figure 8: Precedence graph of the example ALB problem

Table 3: Data for the example

i	succ(i)	R_1	R_2	R_3	R_4
1	4	17	22	19	13
2	4	21	22	16	20
3	5	12	25	27	15
4	6	29	21	19	16
5	10	31	25	26	22
6	8	28	18	20	21
7	8	42	28	23	34
8	10	27	33	40	25
9	10	19	13	17	34
10	-	26	27	35	26

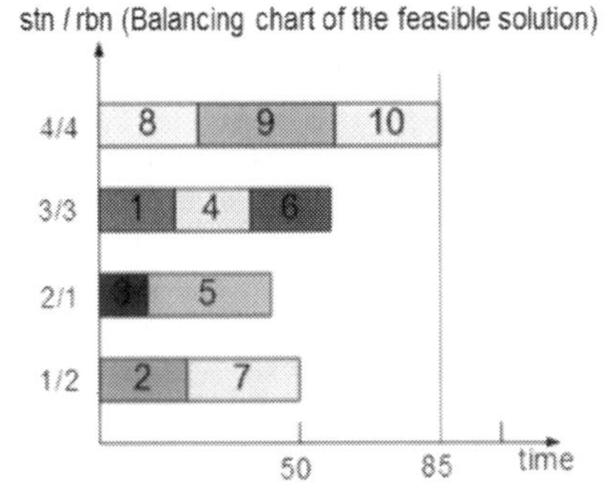

Figure: 9 Feasible solution for the example (stn: station #, rbn: robot #)

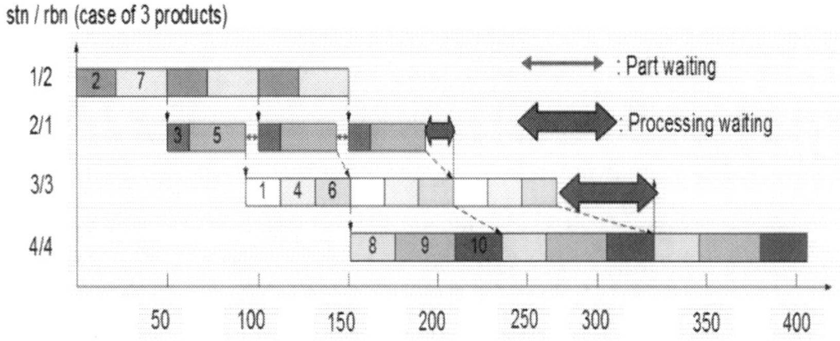

Figure: 10 Gantt chart example for producing three units

5.2 Robot-Based Assembly-Line Model

The following assumptions are stated to clarify the setting in which the problem arises:

1. The precedence relationship among assembly activities is known and invariable.
2. The duration of an activity is deterministic. Activities cannot be subdivided.
3. The duration of an activity depends on the assigned robot.
4. There are no limitations on the assignment of an activities or a robot to any station. If a task cannot be processed on a robot, the assembly time of the task on the robot is set to a very large number.
5. A single robot is assigned to each station.
6. Material handling, loading, and unloading times, as well as setup and tool changing times are negligible, or are included in the activity times. This assumption is realistic on a single model assembly line that works on the single product for which it is balanced. Tooling on such a robotic line is usually designed such that tool changes are minimized within a station. If tool change or other type of setup activity is necessary, it can be included in the activity time, since the transfer lot size on such line is of a single product.
7. The number of workstations is determined by the number of robots, since the problem aims to maximize the productivity by using all robots at hand.
8. The line is balanced for a single product.

The notation used for formulating a mathematical programming model can be summarized as follows:

Indices:

i, j: index of assembly tasks, $i, j = 1, 2, \ldots, n$;

k: index of workstations, $k = 1, 2, \ldots, m$;

l: index of robots, $l = 1, 2, \ldots, m$.

Parameters:

n: total number of assembly tasks,

m: total number of workstations (robots),

t_{il} : processing time of the i-th task by robot l,

pre(i): the set of predecessor of task i in the precedence diagram.

Decision variables:

x_{jk} = 1 if task j is assigned to workstation k, 0 otherwise.

y_{kl} = 1 if robot l is allocated to workstation k, 0 otherwise.

The mathematical model for the robot-based assembly-line balancing problem formulated as a 0-1 nonlinear programming (0-1NP) as follows:

$$\min \quad c_{\mathrm{T}} = \max_{1\le k\le m}\left\{\sum_{i=1}^{n}\sum_{l=1}^{m} t_{il} x_{ik} y_{kl}\right\} \tag{10}$$

$$\text{s. t.} \quad \sum_{k=1}^{m} k\,x_{jk} - \sum_{k=1}^{m} k\,x_{ik} \ge 0, \forall i,\ j \in \mathrm{pre}(j) \tag{11}$$

$$\sum_{k=1}^{m} x_{ik} = 1,\ \forall i \tag{12}$$

$$\sum_{l=1}^{m} y_{kl} = 1,\ \forall k \tag{13}$$

$$\sum_{k=1}^{m} y_{kl} = 1,\ \forall l \tag{14}$$

$$x_{ik} \in \{0,1\}, \forall i,k \tag{15}$$

$$y_{kl} \in \{0,1\}, \forall k,l \tag{16}$$

The objective (10) is to minimize the cycle time, C_T. Equation (11) represents the precedence constraints. It ensures that, for each pair of assembly activities, the precedent cannot be assigned to a station before the station of the successor if there is precedence between the two activities. Equation (12) ensures that each task has to be assigned to one station. Equation (13) ensures that each station is equipped with one robot. Equation (14) ensures that each robot can only be assigned to one station.

5.3 Numerical Experiment by Hybrid GA

Order encoding for task sequence: A GA's structure and parameter settings affect its performance. However, the primary determinants of a GA's success or failure are the coding by which its genotypes represent candidate solutions, and the interaction of the coding with the GA's recombination and mutation operators. A solution of the robot-based assembly line balancing (RALB) problem can be represented by two integer vectors: the task sequence vector, v_1, which contains a permutation of assembly tasks ordered according to their technological precedence sequence, and the robot assignment vector, v_2. The best solution representation method after solved by hybrid genetic algorithm is illustrated in Figure 11. The detailed processes of decoding the task sequence and assigning robots to workstations are shown in Qiu, et al (2002) and we can get the following best schedule of the RALB example:

S = {(t_1, R_1: 0-17), (t_2, R_1: 17-38), (t_3, R_1: 38-50), (t_4, R_2: 50-71), (t_5, R_3: 102-128), (t_6, R_2: 71-89), (t_7, R_3: 128-151), (t_8, R_4: 151-176), (t_9, R_2: 89-102), (t_{10}, R_4: 176-202)}

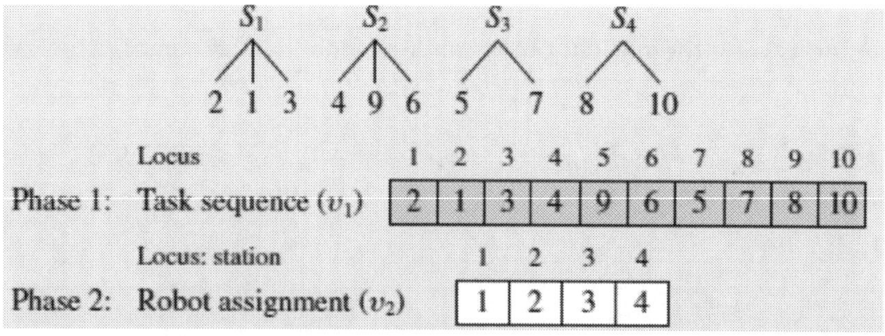

Figure 11: Best solution representation of the sample example

We show a best balance chart for this example in Figure 12. In the real world, an assembly line system is not just for producing one unit of the product; it should produce several units. So we give the Gantt chart with three units for analyzing the solution as in Figure 13. We can see the solution reduced the waiting time for the line by comparing with the feasible solution from Figure 13. This also means that the better solution improved the assembly-line balancing by using the hybrid genetic algorithm.

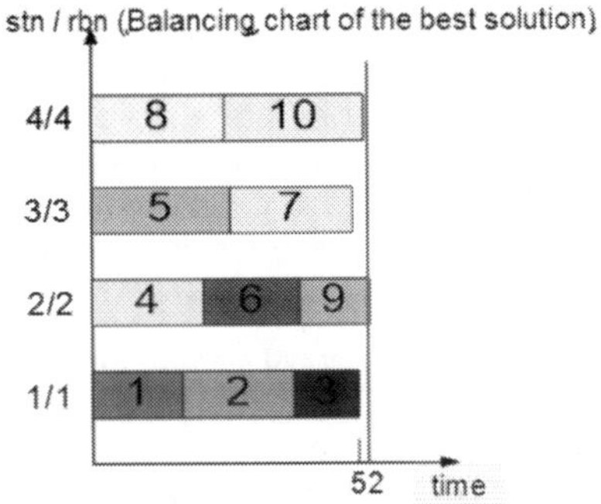

Figure 12: Best balance chart for the example

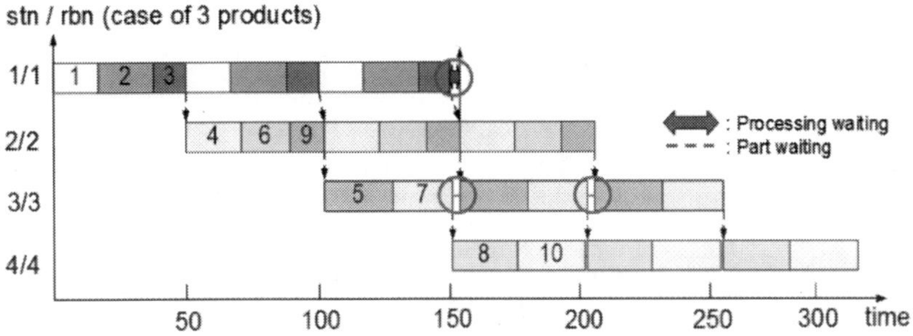

Figure 13: Gantt chart for producing three units

In order to evaluate the performance of the proposed method, a large set of problems were tested. In the literature, no benchmark data sets are available for RALB. There are eight representative precedence graphs (Scholl et al 2007) used in the simple assembly line balancing (sALB) literature (Scholl 2003). These precedence graphs contain 25 - 297 tasks.

Table 4 shows the experiment results of Levitin et al (2006)'s two genetic algorithms and proposed approach, hybrid GA for 32 different scale test problems. As depicted in Table 4, all of the results by the proposed approach are better than Levitin et al.'s recursive assignment method, and most results of the proposed approach are better than Levitin et al.'s consecutive assignment method, except for tests 25-3, 35-7, and 111-17. Depending on the experiment results, the way in which a solution of the automation problem is encoded into a chromosome is a key issue for applying the evolutionary algorithm.

Table 4: Performance of the proposed hybrid genetic algorithm

Test problem No. of tasks	No. of stations	WEST ratio	Cycle time (C_T) Levitin et al. recursive	Levitin et al. consecutive	Proposed approach
25	3	8.33	518	503	503
	4	6.25	351	330	327
	6	4.17	343	234	213
	9	2.78	138	125	123
35	4	8.75	551	450	449
	5	7.00	385	352	244
	7	5.00	250	222	222
	12	2.92	178	120	113
53	5	10.60	903	565	554
	7	7.57	390	342	320
	10	5.30	35	251	230
	14	3.79	243	166	162
70	7	10.00	546	490	449
	10	7.00	313	287	272
	14	5.00	231	213	204
	19	3.68	198	167	154
89	8	11.13	638	505	494
	12	7.42	455	371	370
	16	5.56	292	246	236
	21	4.24	277	209	205
111	9	12.33	695	586	557
	13	8.54	401	339	319
	17	6.53	322	257	257
	22	5.05	265	209	192
148	10	14.80	708	638	600
	14	10.57	537	441	427
	21	7.05	404	325	300
	29	5.10	249	210	202
297	19	15.63	1129	674	646
	29	10.24	571	444	430
	38	7.82	442	348	344
	50	5.94	363	275	256

6. CONCLUSION

In real industrial manufacturing systems there are many combinatorial optimization problems (COP) imposing on more complex issues with multiple objectives to be handled simultaneously. However it is very difficult for solving the COP problems intractable by the traditional solving approaches because of NP-hard combinatorial problems. For developing effective and efficient evolutionary algorithmic that are in a sense "good," *i.e.*, whose computational time is small as within a few minutes, or at least reasonable for the problems met in practice, we have to consider three issues: quality of solution, computational time and effectiveness of the nondominated solutions for multiobjective optimization problem (MOP).

We introduced hybrid evolutionary algorithm (HEA) for treating automation problems in factory, manufacturing, planning and scheduling, and logistics and transportation systems. HEA is the most popular metaheuristic method for solving NP-hard optimization problems. Numerical experiments based on hybrid evolutionary algorithm for case studies show the effectiveness of proposed approach comparing with other methods. The techniques presented in this paper are bioinspired and their influence on factory, planning and scheduling, manufacturing, logistics and transportation is discussed. Emerging trends will continue to follow bioinspired control and automation, such as the development of evolvable robots to better meet the needs of evolving and flexible lines.

Acknowledgement: This work is partly supported by JSPS: Grant-in-Aid for Scientific Research (C; No. 15K00357) and the Fundamental Research Funds for the Central Universities No. DUT15QY10.

REFERENCES

[1] Yu, X. and M. Gen. 2010. *Introduction to Evolutionary Algorithms*, 430pp, Springer, London.
[2] Gen, M. and L. Lin. 2014. Multiobjective evolutionary algorithm for manufacturing scheduling problems: State-of-the-art survey, *J. of Intelligent Manufacturing*, vol25(5), pp849-866.
[3] Gen, M., W.Q. Zhang and L. Lin. 2015. Multiobjective Hybrid Genetic Algorithms for Manufacturing Scheduling: Part II Case Studies of HDD and TFT-LCD, *Adv. in Intelligent Systems and Comp.* vol362, pp27-54..
[4] Chou, C-W, C-F Chien and M. Gen. 2014. A multiobjective hybrid genetic algorithm for TFT-LCD module assembly scheduling, *IEEE Trans. on Automation Science and Engineering*. vol10(3), pp692-705.
[5] Chamnanlor, C., K. Sethanan, C-F Chien and M. Gen. 2014. Re-entrant flow shop scheduling problem with time windows using hybrid genetic algorithm based on autotuning strategy. *Inter. Journal of Production Research*, vol52, pp2612-2629.
[6] Goldberg, D. E. 1989. *Genetic Algorithms in Search, Optimization, and Machine Learning*. Addison-Wiley, MA.
[7] Michalewicz, Z. 1994. *Genetic Algorithm + Data Structure = Evolution Programs*, 2nd ed., Springer, New York.
[8] Gen, M. and L. Lin. 2009a. Genetic Algorithms, in Benjamin Wah ed.: *Wiley Encyclopedia of Computer Science and Engineering*, pp1367-1381, John Wiley & Sons, Hoboken, N.J.
[9] Gen, M. and L. Lin. 2009b. Chapter C.29: Evolutionary Techniques for Automation, pp487-502, in Shimon Y. Nof ed.: Springer Handbook of Automation, Springer, London.
[10] Herrera F. and M. Lozano.1996. Adaptation of genetic algorithm parameters based on fuzzy logic controllers. In: Herrera F. and J. Verdegay (eds) *Genetic algorithms and soft computing*, Physica-Verlag, pp95-125.
[11] Yun, Y. and M. Gen. 2003: Performance analysis of adapted genetic algorithm with fuzzy logic and heuristics, *Fuzzy Optimization and Decision Making*, vol2, pp161-175.
[12] Lin, L., and M. Gen. 2009a. Auto-tuning Strategy for Evolutionary Algorithms: Balancing Between Exploration and Exploitation. *Soft Computing*, vol13, pp157-168.
[13] Tate, D. and A. Smith. 1995. Unequal-area facility layout by genetic search, *IIE Trans.* vol27, pp465-472.
[14] Cohoon, S. Hegde, and N. Martin. 1991. Distributed genetic algorithms for the floor-plan design problem, *IEEE Trans. Computer-Aided Design*. vol10, pp483-491.
[15] Tam, K. 1992. Genetic algorithms, function optimization, facility layout design, *Eur. J. Operational Res.* vol63, pp322-346.
[16] Kusiak, A. and S. Heragu. 1987. The facility layout problem, *Eur. J. Operational Res.* vol29, pp229-251.
[17] Gen, M. and R. Cheng. 1997. *Genetic Algorithms and Engineering Design*, John Wiley & Sons, New York.
[18] Pinedo, M. 2002. *Scheduling Theory, Algorithms and Systems*, Prentice-Hall, Upper Saddle River.
[19] Kacem, I., S. Hammavscheduling problems, *IEEE Trans. Systems, Man and Cyb.*, C, vol32, pp408-419.
[20] Zhang, H. and M. Gen. 2005. Multistage-based genetic algorithm for flexible job-shop scheduling problem, *J. of Complexity International*, vol11, pp223-232.
[21] Kim, K.W., Y.S. Yun, J.M. Yoon, M. Gen and G. Yamazaki. 2005. Hybrid genetic algorithm with adaptive abilities for resource-constrained multiple project scheduling, *Computer in Industry*, vol56(2), pp143-160.
[22] Turbide, D. 1998. *Advanced planning and scheduling* (APS) *systems*, Midrange ERP Mag.

[23] Moon, C., J.S. Kim and M. Gen. 2004. Advanced planning and scheduling based on precedence and resource constraints for e-Plant chains, Int. J. Production Res. vol42(15), pp2941-2955.
[24] Moon, C. and Y. Seo. 2005. Evolutionary algorithm for advanced process planning and scheduling in a multi-plant, *Computers & Industrial Eng.* vol48(2), pp311-325.
[25] Tsujimura, Y., M. Gen and E. Kubota. 1995. Solving fuzzy assembly-line balancing problem with genetic algorithms, *Computers & Industrial Eng.* vol.29, pp543-547.
[26] Rubinovitz, J. and G. Levitin. 1995. Genetic algorithm for line balancing, *Int. J. Prod. Econ.* vol41, pp343-354.
[27] Gen, M., Y. Tsujimura and Y. Li. 1996. Fuzzy assembly line balancing using genetic algorithms, *Computers & Industrial Eng.* vol.31, pp.631-634.
[28] Gao, J., L. Sun, L. Wang and M. Gen. 2009. An efficient approach for type II robotic assembly line balancing problems, *Computers & Industrial Engineering*, vol.56, no3, pp1065-1080.
[29] Hwang, H., S. Moon and M. Gen. 2002. An integrated model for the design of end-of-aisle order picking system and the determination of unit load sizes of AGVs, *Computers & Industrial Eng.*, vol42, pp249-258.
[30] Qiu, L., W. Hsu, S. Huang, H. Wang. 2002. Scheduling and routing algorithms for AGVs: a survey, *Int. J. Production Res.*, vol40(3), pp745-760.
[31] Lim, J.K. 2004. *Study on guide path design and path planning in automated guided vehicle system*. Ph.D. Thesis, Waseda University (In Japanese).
[32] Vis, I.F.A. 2006. Survey of research in the design and control of automated guided vehicle systems, *Eur. J. Operational Res.*, vol170(3), pp.677-709.
[33] Le-Anh, T. and D. Koster. 2006. A review of design and control of automated guided vehicle systems, *Eur. J. Operational Res.* vol171(1), pp.1-23.
[34] Lin, L. S.W. Shinn, M. Gen and H. Hwang. 2006. Network model and effective evolutionary approach for AGV dispatching in manufacturing system, *J. Intelligent Manuf.* vol17(4), pp465-477.
[35] Lau, Y.K. Y. Zhao. 2008. Integrated scheduling of handling equipment at automated container terminals, *Ann. Operations Res.* vol159(1), pp373-394 ()
[36] Preston, P. and E. Kozan. 2001. An approach to determine storage locations of containers at seaport terminals, Computers & Operations Res. vol28(10), pp983-995.
[37] Imai, A., H.C. Chen, E. Nishimura and S. Papadimitriou. 2008. T900-920.
[38] Lin, L. and M. Gen. 2009b. A random key-based genetic algorithm for AGV dispatching in FMS, *Inter. J. of Manufacturing Technology and Management*, 16, pp.58-75.
[39] Gen, M. and R. Cheng. 2000. Genetic Algorithms and Engineering Optimization, John Wiley & Sons, N.Y.
[40] Deb, K. 2001. *Multiobjective Optimization Using Evolutionary Algorithms*, John Wiley & Sons, Chichester.
[41] Gen, M., R. Cheng and L. Lin. 2008. *Network Models and Optimization: Multiple Objective Genetic Algorithms*, Springer, London.
[42] Yang, J.B. 2001. GA-based discrete dynamic programming approach for scheduling in FMS environment, *IEEE Trans. Systems, Man & Cybern.* B vol31(5), pp824-835.
[43] Kim, K., G. Yamazaki, L. Lin, M. Gen. 2004. Network-based hybrid genetic algorithm to the scheduling in FMS environments, J. Artificial Life & Robot. vol8(1), pp67-76.
[44] Kim, S.H. and H. Hwang. 1999. An adaptive dispatching algorithm for automated guided vehicles based on an evolutionary process, *Int. J. Production Econ.* vol60/61, pp465-472.
[45] Liang, Y., L. Lin, M. Gen and C-F Chien. 2012. A hybrid evolutionary algorithm for FMS optimization with AGV dispatching. *Proc. of the 42nd Inter*296.1-296.14.
[46] Scholl, A. 1993. Data of Assembly Line Balancing Problems. Schriften zur Quantitativen Betriebswirtschaftslehre 16/93, (TH Darmstadt, Darmstadt).
[47] Scholl, A., N. Boysen, M. Fliedner, R. Klein. 2007. Homepage for assembly line optimization research, http://www.assembly-line-balancing.de/
[48] Levitin, G., J. Rubinovitz, B. Shnits. 2006. A genetic algorithm for robotic assembly balancing, *Eur. J. Operational Res.* vol.168, pp.811-825.
[49] Gao, J., M. Gen, and L. Sun. 2006. "Scheduling Jobs and Maintenances in Flexible Job Shop with a Hybrid Genetic Algorithm." *Journal of Intelligent Manufacturing*, 17, pp.493-507.
[50] Gao, J., L. Sun and M. Gen. 2008. A hybrid genetic and variable neighborhood descent algorithm for flexible job shop scheduling problems, *Computers & Operations Research*, 35(9), pp.2892-2907.
[51] Gen, M., J. Gao and L. Lin. 2009. Multistage-based Genetic Algorithm for Flexible Job-shop Scheduling Problem. *Intelligent and Evolutionary Systems* 187, pp183-196.
[52] Zhang, Q.W., M. Gen and J.B. Jo. 2014. Hybrid sampling strategy-based multiobjective evolutionary algorithm for process planning and scheduling problem, *J. of Intelligent Manufacturing*, vol25(5), pp881-897.
[53] Chamnanlor, C., K. Sethanan, M. Gen and C-F Chien. 2015. Embedding ant system in genetic algorithm for re-entrant hybrid flow shop scheduling problems with time window constraints, *J. of Intelligent Manufacturing*, 17pp. DOI 10.1007/s10845-015-1078-9.
[54] Zheng, J-N, C-F Chien and M. Gen. 2015. Multi-objective multi-population biased random-key genetic algorithm for the 3-D container loading problem, *Computers & Industrial Engineering*, 8pp; DOI: 10.1016/j.cie.2014.07.012.

TWO-STEP TASK ASSIGNMENT TO MINIMIZE TOTAL COST

Jiayin Tan[1*] and Dakui Jiang[2]
[1]Department of Management Science and Engineering,
Nanjing Audit University, Nanjing, China
tanjiayin@nau.edu.cn

[2]College of Management and Economics,
Tianjin University, Tianjin
jiangdakui@tju.edu.cn

ABSTRACT

In recent years, humans have been creating more complicated systems than before due to the rapid development of technology. The tasks always require several steps to finish, which makes it more complex to assign them. The paper deals with an assignment problem to minimize total cost. In this problem, each task includes two steps, each of which will incurs cost for its assigned agent. The cost of each agent is dependent not only on the task assigned to it but also on the number of the tasks assigned to it. Firstly, a linear integer programming model for the problem was formulated. Then, a heuristic algorithm based on linear programs was developed to solve this problem and its worst-case ratio was also analyzed. Finally, experimental results demonstrated the developed heuristic can obtain better solution than the current best known algorithm.

Keywords: Assignment problem, Integer program, Heuristic, Worst-case analysis

1 INTRODUCTION

Assignment problem is a classical combination optimization problem, which examines the minimum cost assignment of n jobs to m agents under some constraints. The problem has many real-life applications, e.g., water resources allocation, the design of a communications network with capacity constraints at each network node, assigning software development tasks to programmers, and assigning jobs to computers in a network and storage space allocation. Over the past decades, many variations on the assignment problem have been proposed. Comprehensive surveys were presented by Loiola, de Abreu, Boaventura-Netto, et al.[1] and Pentico[2]. Also, a survey on algorithms for solving the problem was provided by Cattrysse and Van Wassenhove[3].

In traditional assignment models, it is always assumed that the task is considered as a whole (e.g., Fu et al.[4]; Benlic and Hao[5]; Dokeroglu[6]; Ghoniem, Flamand, and Haouari[7]; among others). However in some applications, a task will includes two steps. For example, to fulfill the water requirements of different regions in a water supply system, the decision maker needs to assign each requirement to the water suppliers according to their water-yielding capacities and water carrying costs; to minimize the total cost in a product manufacturing system, a manufacturing factory needs to consider the production and distribution of customer orders in a coordinated manner. Chen and

* Corresponding Author

Pundoor[8] ever studied an assignment model with two-step tasks. In their model, a customer order should be processed on one of the plants in a multi-plant supply chain and then batch delivered to the customer. They showed that the problem can be solved in polynomial time if the number of plants is fixed and is NP-hard in the strong sense if the number of plants is arbitrary. Then, they proposed a heuristic for the model and showed that the worst-case ratio of the proposed heuristic is $b + 1$, where b is the capacity of the delivery batch. After that, Jiang and Li[9] provided a tabu search algorithm for the problem. Similar models with different constraints were also studied by Chen and Li[10], Jiang and Li[11], and Jiang, Li, and Cao[12]. Chen and Li[10] considered that each task has a deadline constraint so that the agent should finish it under the deadline. Jiang and Li[11] and Jiang, Li, and Cao[12] considered that the agent can deal with the tasks in parallel. All of them formulated the problems as integer programs and proposed heuristics for them.

In recent years, humans have been creating more complicated systems, from the machines we live with to the informational systems. In these systems, the tasks always require several steps to finish, which makes it more complex to assign them. Hence, it is important to study the assignment problem with several steps in each task. In production systems, Chen[13] pointed out that production and distribution are two key operational functions in the supply chain and it is critical to integrate these two functions so as to achieve optimal operational performance. To finish a task, both order processing and order delivering are needed. Moreover, the task assigned to different agents will incurs different cost such as different raw material purchasing cost and different labor cost. Therefore, in this paper, we will study a two-step task assignment problem with the objective to minimize the total cost.

The remainder of this paper is organized as follows. We first formulate the problem in Section 2. Then, we design a heuristic for the problem in Section 3. In Section 4, we evaluate our designed algorithm by constructing a numerical example. Finally, we conclude the paper in Section 5.

2 FORMULATION

In this section, we first describe the problem mathematically. Then, we formulate the problem as a linear integer program.

The problem, represented by P hereafter, can be described as follows. Given a set of n independent tasks, $N = \{1, 2, ..., n\}$, and a set of m agents, $M = \{1, 2, ..., m\}$. All the tasks in N should be assigned and each task should be assigned to only one agent. Two steps are required to finish a task. In the first step, the cost c_{ij} incurs if task j is assigned to agent i, for $i \in M$ and $j \in N$. In the second step, the cost of each agent is dependent on the number of the tasks assigned to it in the first step. More precisely, if n_i tasks are assigned to agent i in the first step, then the cost of agent i in the second step is equal to $\lceil n_i/b \rceil f_i$, where $b \geq 1$ is an integer, for $i \in M$. $\lceil \cdot \rceil$ is a ceiling function and $\lceil n_i/b \rceil$ is the minimum integer no less than n_i/b. The objective is to minimize the total cost for assigning all the tasks. The notations of the problem is listed in Table 1.

Obviously, the mathematical model for problem P can be given as:

$$P : Z^* = \min \left\{ \sum_{i \in M} \left(\sum_{j \in N} c_{ij} x_{ij} + \left\lceil \frac{\sum_{j \in N} x_{ij}}{b} \right\rceil f_i \right) \right\} \tag{1}$$

$$s.t. \sum_{i \in M} x_{ij} = 1 \quad \forall j \in N, \tag{2}$$

$$x_{ij} \in \{0, 1\} \quad \forall i \in M, \forall j \in N \tag{3}$$

Table 1. Notations of The Problem

Notation	Description
M	The set of agents.
N	The set of customers
c_{ij}	The cost of task $j \in N$ in the first step if it is assigned to agent $i \in M$.
n_i	The number of tasks assigned to agent $i \in M$.
f_i	A constant associated with agent i for evaluating the cost of finishing the second step of tasks assigned to agent $i \in M$.
b	An integer for evaluating the cost of finishing the second step of tasks assigned to each agent, $b \geq 1$.

where $x_{ij} = 1$ if agent $i \in M$ is assigned to task $j \in N$, and 0 otherwise. The objective function (1) is to minimize the total cost, which includes both the total cost in the first step and in the second step. More precisely, if $x_{ij} = 1$, then the cost c_{ij} for agent i incurs in the first step; if $n_i = \sum_{j \in N} x_{ij}$ tasks are assigned to agent i in the first step, then the cost $\lceil n_i/b \rceil f_i$ for agent i incurs in the second step. Constraints (2) ensure that each task is assigned to only one agent.

The above model is a nonlinear integer program since the ceiling function is applied in (1). Thus, we will convert the model as the following one by adding integer parameters y_i, for $i \in M$.

$$P : Z^* = \min\left\{\sum_{i \in M}\Big(\sum_{j \in N} c_{ij}x_{ij} + y_i f_i\Big)\right\}$$

$$s.t. \sum_{i \in M} x_{ij} = 1 \quad \forall j \in N,$$

$$y_i \geq \frac{\sum_{j \in N} x_{ij}}{b} \qquad \forall i \in M,$$

$$x_{ij} \in \{0,1\}, y_i \geq 0 \quad \forall i \in M, \forall j \in N.$$

Due to y_i is an integer and $y_i = \lceil n_i/b \rceil = \lceil \sum_{j \in N} x_{ij}/b \rceil$, we can define it by using $y_i \geq \frac{\sum_{j \in N} x_{ij}}{b}$ for calculating the cost of agent $i \in M$ in the second step. The above converted model can obviously be solved by commercial softwares since it is a linear integer program.

3 HEURISTIC

In this section, we first provide an exact algorithm to get a lower bound of the optimal objective value of P.

In an optimal solution of P, there may exists some agents where no task is assigned due to the fact that the values of c_{ij} or f_i are extremely large associated with these agents. Hence,

Proposition 1. *In the optimal solution of P, there exists a non-empty subset Q of M, where each agent in Q should be assigned at least one task.*

Proposition 2. *In the optimal solution of P, if a task is assigned to agent $i \in M$, then $Z^* \geq f_i$, where Z^* is the optimal objective value of P.*

To get the lower bound, we construct the following linear program:

$$LP(Q) : F(Q) = \min\left\{\sum_{i\in Q}\sum_{j\in N}\left(c_{ij} + \frac{f_i}{b}\right)x_{ij}\right\}$$

$$s.t. \sum_{i\in M} x_{ij} = 1 \quad \forall j \in N,$$

$$\sum_{i\in Q}\left(c_{ij} + \frac{f_i}{b}\right)\sum_{j\in N} x_{ij} \geq f_h,$$

$$0 \leq x_{ij} \leq 1 \quad \forall i \in M, \forall j \in N$$

where Q is a non-empty subset of M, i.e., $Q \subseteq M$, $Q \neq \emptyset$. Compared with P, the integer variables x_{ij} and y_i are relaxed to real variables in the above linear program, for $i \in M$ and $j \in N$. Note that y_i is equal to $\frac{\sum_{j\in N} x_{ij}}{b}$ if it is relaxed to the real variable, for $i \in M$. Hence, y_i can be replaced by $\frac{\sum_{j\in N} x_{ij}}{b}$ in $LP(Q)$.

Lemma 1. *In an optimal solution of $LP(Q)$, if $n \geq |Q|$, then at least $(n - |Q|)x_{ij}$ variables take the value 1.*

The result of Lemma 1 has been proved and applied by Potts[14] and Jiang and Tan[15]. We thus omit the proof.

Algorithm EA

Step 0. Reindex the agents in the non-increasing order of f_i . Let Q_h be the set of the agents $\{h, \dots, m\}$, for $h \in M$.

Step 1. For $h = 1, 2, \dots, m$, solve $LP(Q_h)$ and get the optimal objective value $F(Q_h)$.

Step 2. Let U denote the subset of agents Q_h with the lowest objective value $F(Q_h)$, and let $u = |U|$.

In algorithm EA, only m linear programs need to be solved. Hence, algorithm EA is a polynomial-time algorithm.

Theorem 1. $F(U) \leq Z^*$, *where $F(U)$ and Z^* are the optimal objective values of $LP(U)$ and P, respectively.*

Proof. According to Proposition 2, if a task is assigned to agent $i \in M$ in an optimal solution of P, then $Z^* \geq f_i$. If a task is assigned to agent h in an optimal solution, then $f_h \leq F(Q_h) \leq Z^*$. Therefore, $F(U) = \min\{F(Q_h) | h \in M\} \leq Z^*$.

In the following, we develop the following heuristic for the problem.

Heuristic HA

Step 1. Run algorithm EA to obtain a subset of agents U and an optimal solution of $LP(U)$. Define a subset of tasks $J = \{j \in N | x_{ij} = 1$ for some $i \in U$ in this solution$\}$. According to Lemma 1, there are at least $n - u$ tasks in J and at most u tasks in $N \setminus J$.

Step 2. Assign each task $j \in J$ to agent $i \in U$ with $x_{ij} = 1$ and denote this partial assignment by σ_1. Enumerate all possible assignments of the no more than u tasks in $N \setminus J$ to the agents in M. Denote the partial assignment with lowest cost by σ_2.

Step 3. Concatenate σ_1 and σ_2, and Denote it by σ.

In the following, we analyze the worst-case performance of heuristic HA.

Theorem 2. $Z(\sigma) \leq (\min\{m, b\} + 1)Z^*$.

Proof. Define V to be the subset of agents in U where $\lceil \sum_{j \in J} x_{ij}/b \rceil \neq x_{ij}/b$. We have $Z(\sigma_1) \leq F(U) + \sum_{h \in V}(m-1)f_h$ and $Z(\sigma_1) \leq F(U) + \sum_{h \in V}(b-1)f_h/b$.

Since $mf_h \leq F(U)$, we have $Z(\sigma_1) \leq F(U) + (m-1)F(U) \leq mZ^*$. Similarly, since $\frac{b-1}{b}\sum_{i \in V} f_h \leq F(U)$, we have $Z(\sigma_1) \leq F(U) + (b-1)F(U) \leq bZ^*$. Also, σ_2 is an optimal solution for the jobs in $N \setminus J$. Because only a subset of jobs is involved in $N \setminus J$, $Z(\sigma_2) \leq Z^*$. Hence, $Z(\sigma) \leq Z(\sigma_1) + Z(\sigma_2) \leq (\min\{m, b\} + 1)Z^*$.

4 NUMERICAL EXAMPLE

In this section, we construct a small instance with two tasks for problem P as follows: There are two tasks to be assigned to two agents. For the agent 1, the cost of both tasks in the first step are 1, i.e., $c_{11} = c_{12} = 1$; the two parameters associated with the cost in the second step are $b = 100$ and $f_1 = 10$, respectively. For the agent 2, the cost of both tasks in the first step are also set as 1, i.e., $c_{21} = c_{22} = 1$; the two parameters associated with the cost in the second step are $b = 100$ and $f_2 = 100$, respectively.

We adopt the heuristic H3 proposed by Chen and Pundoor[8], our designed heuristic and IBM ILOG CPLEX 12.2 to solve this instance, respectively.Everything else in heuristic H3 are the same as in our heuristic except that only $LP(M)$ is solved in algorithm EA. Solve this instance by heuristic H3, a solution that the two tasks are assigned to agent 2 is obtained, the objective value of which is equal to 102. Further, solve it by our designed heuristic, another solution that the two tasks are assigned to agent 1 is obtained, the objective value of which is equal to 12. The solution obtained by CPLEX is same as that by ours, which is also the optimal one. Thus, our heuristic gets a better solution than heuristic H3.

Compared with our designed heuristic, only $LP(M)$ is solved to get a lower bound of the optimal objective value in heuristic H3. Hence, the computation result of our heuristic for all instances of problem P will no worse than that obtained by heuristic H3. However, the computation time of our heuristic will be longer since m linear programs need to be solved while only one linear program will be solved in heuristic H3.

5 CONCLUSION

In this paper, we have studied an assignment problem with two-step tasks to minimize total cost. The cost of each task incurs in each step. For each agent, the cost of the agent in the first step is dependent on the assigned task and the cost of the agent in the second step is associated with the number of the assigned tasks. The problem is NP-hard in the strong sense if the number of agent m is arbitrary. We developed a heuristic with the worst-case ratio of $\min\{m, b\} + 1$. However, the worst-case time complexity of the heuristic we developed is polynomial in the number of tasks n but exponential in the number of agents m. Also, there are only two steps for each task in our problem. For future research, it will be interested in developing more efficient and effective algorithms for the problem with more than two steps in each task.

6 ACKNOWLEDGEMENT

This research is supported by Natural Science Foundation of Jiangsu (no.BK20140762).

7 REFERENCES

[1] **Loiola, E. M., de Abreu, N. M. M., Boaventura-Netto, et al. 2007..** A survey for the quadratic assignment problem, *European Journal of Operational Research*, 176(2), pp 657-690.

[2] **Pentico, D. W. 2007.** Assignment problems: A golden anniversary survey, *European Journal of Operational Research*, 176(2), pp 774-793.

[3] **Cattrysse, D. G., Van Wassenhove, L. N. 1992.** A survey of algorithms for the generalized assignment problem, *European Journal of Operational Research*, 60(3), pp 260-272.

[4] **Fu, Y., Sun, J., Lai, K. K., Leung, J. W. 2015.** A robust optimization solution to bottleneck generalized assignment problem under uncertainty. *Annals of Operations Research*, 233(1), pp 123-133.

[5] **Benlic, U., Hao, J. K. 2015.** Memetic search for the quadratic assignment problem. *Expert Systems with Applications*, 42(1), pp 584-595.

[6] **Dokeroglu, T. 2015.** Hybrid teaching¨Clearning-based optimization algorithms for the Quadratic Assignment Problem. *Computers & Industrial Engineering*, 85(7), pp 86-101.

[7] **Ghoniem, A., Flamand, T., Haouari, M. 2016.** Exact Solution Methods for a Generalized Assignment Problem with Location/Allocation Considerations. *INFORMS Journal on Computing*, 28(3), pp 589-602.

[8] **Chen, Z. L., Pundoor, G. 2006.** Order assignment and scheduling in a supply chain, *Operations Research*, 54(3), pp 555-572.

[9] **Jiang, D., Li, B. 2011.** Supply chain scheduling based on hybrid taboo search algorithm, *Chinese Journal of Mechanical Engineering*, 47(20), pp 53-59.

[10] **Chen, Z. L., Li, C. L. 2008.** Scheduling with subcontracting options, *IIE Transactions*, 40(12), pp 1171-1184.

[11] **Jiang, D., Li, B. 2012.** Multi-plant supply chain scheduling with parallel machines based on taboo search algorithm, *China Mechanical Engineering*, 23(6): pp 688-693.

[12] **Jiang, D., Li, B., Cao, L. 2014.** Supply chain scheduling with subcontracting option under parallel machines, *Control and Decision*, 29(5), pp 951-955.

[13] **Chen, Z. L. 2010.** Integrated production and outbound distribution scheduling: review and extensions. *Operations Research*, 58(1), pp 130-148.

[14] **Potts, C. N. 1985**. Analysis of a linear programming heuristic for scheduling unrelated parallel machines, *Discrete Applied Mathematics*, 10(2), pp 155-164.

[15] **Jiang, D., Tan, J. 2016**. Scheduling with job rejection and nonsimultaneous machine available time on unrelated parallel machines, *Theoretical Computer Science*, 616, pp 94-99.

BUSINESS INTEROPERABILITY: THE RE-DESIGN AND MODELLING OF INTEROPERABLE BUYER-SUPPLIER DYADS

P. Espadinha-Cruz[1*], A. Grilo[1] and V. Cruz-Machado[1]
[1]UNIDEMI, Departamento de Engenharia Mecânica e Industrial
Faculdade de Ciências e Tecnologia, Universidade Nova de Lisboa, Portugal
pedroespadinhacruz@gmail.com

ABSTRACT

Buyer-supplier dyads are distinguished among other relationships of the supply chain as the simplest for of interaction, which allows achieving competitive synergy to compete with another dyads, networks or chains. Though, buyer-supplier activity is affected by business interoperability problems. Those hinder the dyad's performance and value created to the final customer. In an attempt to solve this problem, the present article proposes an integrated tool to identify and solve those problems following a comprehensive approach of the dyad's interaction at organizational, knowledge and technical perspectives of interoperability.

Keywords: Business Interoperability, buyer-supplier dyads, Axiomatic Design, BPMN

1 INTRODUCTION

Buyer-supplier dyads are distinguished as one valuable asset for effective management of the supply chain (SC) [1]. Collaboration between buyers and suppliers is emphasized under the concept of collaborative advantage [1], whereas the dyad struggles for win-win relationships with mutual benefits achieving competitive synergy [2]. In this context, business interoperability (BI), defined by Legner [3] as 'the organizational and operational ability of an enterprise to cooperate with its business partners and to efficiently establish, conduct and develop IT- supported business with the objective to create value', is an enabler that makes possible to execute the SC operations seamlessly, easing their alignment and the information flow, guaranteeing high performance and competitiveness [4]. However, lack of interoperability is an emerging issue in information technology (IT) based cooperation. BI affects organizational, knowledge and technical aspects of interaction. Still, most research in this area concentrates on identifying interoperability problems and barriers and forms to measure and remove them. According to Chen [5] a comprehensive interoperability methodology is missing. The problem with existing interoperability research is that is mostly single subject-focused rather than approaching the integration of several aspects that composes the business interoperability concept. Hence, the main research question (RQ) is: "How does the business interoperability problem identification and solving may be systemized in order to re-design buyer-supplier dyads, improving their performance and value?".

* Corresponding Author

The RQ aims at the development of a systematic methodology that allows to analyze and re-design buyer-supplier dyads, improving their performance and value. Hence, to tackle this RQ, research was breakdown in two functional areas: the identification and analysis of interoperability problems; and the design, modeling and performance measuring of buyer-supplier dyads. In the present article, we explore the assessment, design and modeling of the BI influence in buyer-supplier dyads.

In the next sections, is performed the literature revision in inter-firm relationships (section 2), and business interoperability (section 3). The proposed methodology is presented in section 4. In section 5 the proposed methodology is demonstrated in a case study performed in automotive industry. Last, the conclusions are presented in section 6.

2 INTER-FIRM RELATIONSHIPS: THE BUYER-SUPPLIER DYAD

In the context of supply chains (SC) and supply chain management (SCM), interoperability is being seen as a strong asset to achieve competitivity [6]. At a wide scope, SCM focuses on how a firm uses processes, technology and capability to enhance competitive advantage [7]. It relies on a precise allocation of resources that manage economies of scale, reduce redundant and duplicate operations, and increases the customer loyalty through a personalized service [8]. Harland [9] defends that, internally, SCM focuses in assets within an organization, and externally, in establishing relationships with first and second-tier suppliers, customers and the entire SC. This collaborative perspective of SCM looks at a set of approaches to efficiently integrate suppliers, manufacturers, warehouse, and stores, so that merchandise is produced and distributed at right quantities, to right locations, and at the right time, in order to minimize system wide cost while satisfying the service level requirements [10]. To achieve collaboration, two main approaches are followed: SC collaboration (SCC) and SC integration (SCI). While SCI relies on central control, ownership, or process integration governed by contract means, SCC puts more emphasis on governance through relational means [11]. In this sense, dyadic approaches are advocated by the relational view theory [2] and settled on the notion of 'collaborative advantage'[1]. This strategic aim settles a common ground between SCC and business interoperability (BI). More specifically, literature in buyer-supplier dyads set the strategic aim of these relationships towards win-win situations supported by partners collaboration and, ultimately, achieve synergies to compete with other chains [12]. Authors in SCC and buyer-supplier dyads defend constructs or practices that strive for the dyad unification, in principles similar to the ones defended in BI, as is explained in the next section.

3 BUSINESS INTEROPERABILITY STATE-OF-THE-ART

Business interoperability (BI) is a concept that evolved from the technical perspective of interoperability incorporating several aspects of organization interactions. Frameworks Levels of information systems interoperability (LISI), Organisational interoperability maturity model (OIM), Interoperability assessment methodology (IAM), Layers of coalition interoperability (LCI), IDEAS framework, European interoperability framework (EIF), Enterprise interoperability maturity model (EIMM), Business Interoperability Framework (BIF), Interoperability impact analysis model (IIAM), and INTEROP Framework, traced the evolutionary path that led to the existing notion of BI [13]-[19]. However, despite the growth in awareness for the subject, the BI body-of-knowledge (BoK) is considered dispersed in different perspectives of interoperability addresses in different levels of detail. In each framework a decomposition of interoperability is provided, either in interoperability types, as well as assessment criteria, requirements and interoperability metrics. The need for a cohesive framework is recognized in literature as a requirement not only to systemize the interoperability BoK, but to allow identifying and categorizing interoperability problems [5]. To achieve this, the aim was to address the existing distinct frameworks and models that attempt a classification of interoperability issues. The taxonomy of BI perspectives and types was the proposed approach to accomplish the systematization of the BI BoK. In Espadinha-Cruz [13] this BI decomposition is explained, whereas is proposed that BI can be achieved by 9 types of interoperability considered in two levels of detail.

Another issue in interoperability is the difficulty to correlate interoperability aspects with the interoperability measurements [5]. The BI decomposition defended on this article attempts to make the link between each interoperability perspective and the influence they have, in practice on the buyer-supplier dyad (see Figure 1). This decomposition represents the conceptual representation of the dyad's decisions for business relationship set up, as well as the BI criteria that characterize how interoperable firms are.

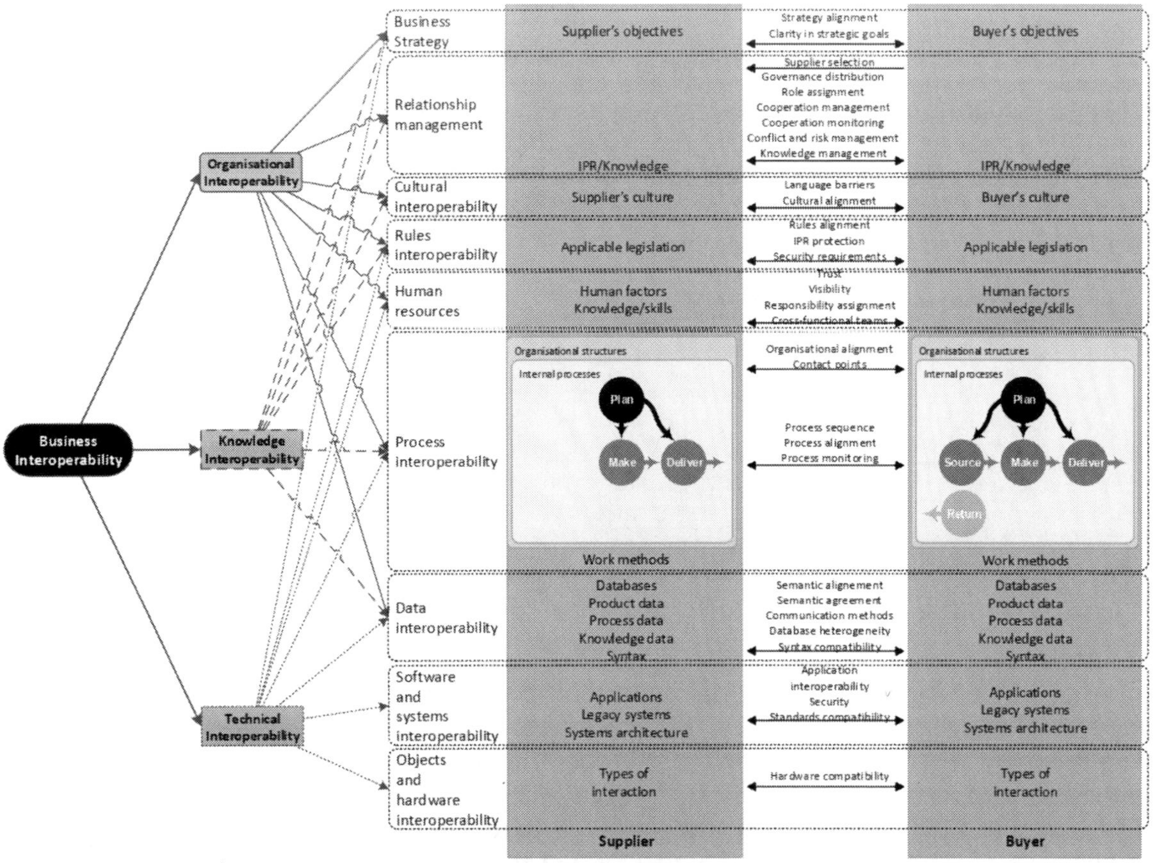

Figure 1: Representation of the buyer-supplied dyad in the main BI perspectives, BI criteria and SCM constructs

In subsequent propositions, the decomposition rationale is used to relate BI perspectives, business-context, criteria and performance metrics, which are represented in the methodology through an axiomatic design (AD) [20] framework.

4 METHODOLOGY TO ANALYSE AND RE-DESIGN INTEROPERABLE BUYER-SUPPLIER DYADS

Based on the notion of collaborative advantage, the dyad is focused on the final customer, whereas interoperability is seen as a service that delivers value added, distinguishing companies for their attractiveness and technology leadership, allowing cost reduction that, in turn, strengthens relationships helping them become competitive. The ultimate goal is the achievement of optimal interoperability, traduced in improved performance and increased value to customer (see Figure 2).

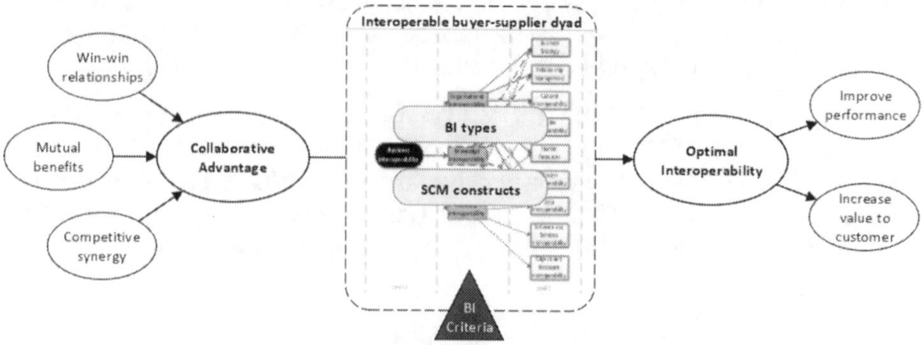

Figure 2: Framework for business interoperability impact in buyer-supplier dyads

The higher proposition of this research aims at interconnecting interoperability issues in a systematic manner so one can identify the issues that require improvement, and limit interaction, in order to accomplish a considered interoperable buyer-supplier dyad, which features higher performance and value created. To achieve this objective conceptually, an interoperability analysis and re-design method is proposed (see Figure 3).

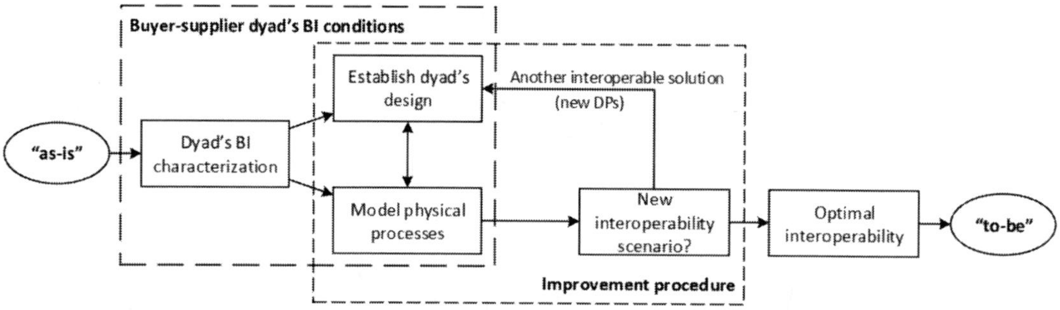

Figure 3: Proposed method to analyse and re-design interoperable buyer-supplier dyads

The method is accomplished in two stages: the determination of the buyer-supplier dyad's BI conditions - first stage - and the improvement procedure, allowing the improvement of the interoperable buyer-supplier dyad. The benchmark between the "as-is" to "to-be" states is accomplished by, firstly, characterizing interoperability in the buyer-supplier dyad (using the BI decomposition, criteria and SCM constructs represented in Figure 1); and, second, by implementing the first axiom of the Axiomatic design theory [20]. After determining the BI conditions of the dyad, these ones are registered in the AD framework and it is mapped each condition in interoperability requirements (or functional requirements - FRs) and interoperable solutions (or design parameters - DPs). Dependencies between each factor are registered in design matrices. According to the AD's first axiom, a called good design is the one that keeps FRs and DPs independent [20]. Upon the identification of interoperability problems (in this case, decoupled or coupled designs), providing new interoperable scenarios that comply with the 1st axiom, allows improving the dyad's interoperability. In the next section this method is demonstrated in an automotive buyer-supplier dyad.

5 CASE STUDY: INTEROPERABILITY IN THE EXCEPTIONS HANDLING OF AN ORDERING PROCESS

The present case study was implemented in a first tier supplier (FS) and a second tier supplier (SS) of an automotive industry. The context of interaction of the FS-SS dyad is an exceptional procedure the companies implemented to deal with orders where are missing components for FS's manufacturing. The customer need (CN) for the dyad's AD design is the achievement of higher interoperability in the design of the exceptions handling of the ordering process". The determination of interoperability conditions led to the breakdown of those ones into interoperability requirements (FRs) and interoperable solutions (DPs) (see Table 1 and "as-is" in **Figure 4**).

Table 1: The "as-is" design of the FS-SS dyad.

Interoperability requirements (FRs)	Interoperable solutions (DPs)
FR1: Establish liabilities and contingencies for failure to commitments. FR1.1: Establish liabilities for delivery failures. FR1.2: Reconcile liabilities for delivery failures with individual strategy. FR1.3: Ensure clarity in liabilities for both actors.	DP1: Firms' conditions regarding delays and order failures. DP1.1: Written contract specifying liabilities. DP1.2: The objectives are fully aligned. DP1.3: Occasional failures.
FR2: Establish preventive measures to deal with amount of orders less than ordered.	DP2: Standard procedure to identify faulty cases and exceptional procedure to deal with missing parts and contract obligations.
FR3: Align companies' internal processes to handle missing parts. FR3.1: Manage the complaint procedure. FP3.1.1: Assign employees to the interface for complaint reception. FR3.1.2: Manage the interface between systems (ICT and software) used to make and manage complaints. FR3.1.3: Manage the communication path to make complaints. FR3.2: Manage the confirmation procedure. FR3.2.1: Manage the communication path to answer complaints. FR3.2.2: Manage the interface between ICT's used to answer complaints.	DP3: The collaboration business process model (BPM) for missing parts solving (see Figure 5). DP3.1: Features of the complaint process. DP3.1.1: Contact points defined. DP3.1.2: Different systems used in each company. Information received from FS by e-mail/phone is unstructured due to a company personalized MS Access database to handle complaints. DP3.1.3: Standard procedure defined to communicate complaints. DP3.2: Features of complaint response. DP3.2.1: Standard procedure defined to communicate orders. DP3.2.2: Information exchanged by e-mail or phone.

The mapping of the conditions from BS to PI allowed identifying issues with regards to this exceptional procedure. BS clarity (see FR1.3) with regards the contractual liabilities presents a coupling on the matrix (see FR1.2.3 and DP1.1.1), which led to the occurrence of conflicts that require negotiation every time an order is incomplete.

The establishment of exceptional procedure to handle complaints is due to the alignment of individual objectives with (DP1.2) that resulted in the establishment of processes, in both companies, to handle missing parts and contract obligations (see FR2-DP2). As a result, FR3 deals with the need to align both procedures from the different companies. Problems in the PI perspective are evident on the design matrix (see "as-is" in **Figure 4**). Couplings on the FR3.1 and FR3.2 matrices are due to lack of interoperability between companies systems and the existence of a user-dependant procedure to treat the complain procedure by using e-mail or phone.

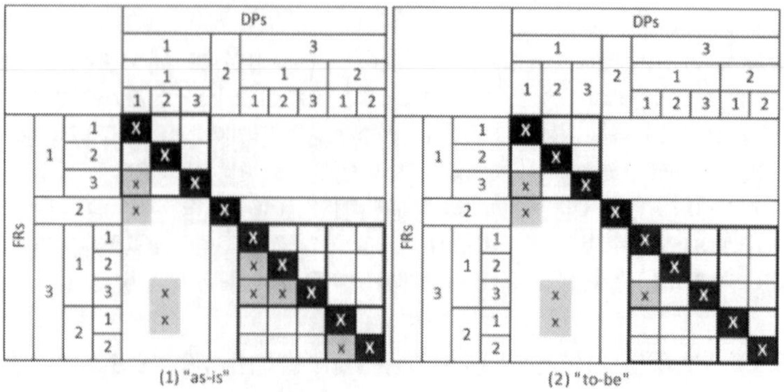

Figure 4: Design matrices for the "as-is" (1) and "to-be" (2) designs of the FS-SS dyad.

To solving the coupling on the matrices, the system interface (FR3.1.2) was provided with a new interoperable solution: the use of SAP system to handle complaints by issuing a credit memo and creating a new urgent order (ne DP3.1.2). Instead of using phone or e-mail, electronica data interchange (EDI) conveys the information to communicate the complaint and satisfy the order request.

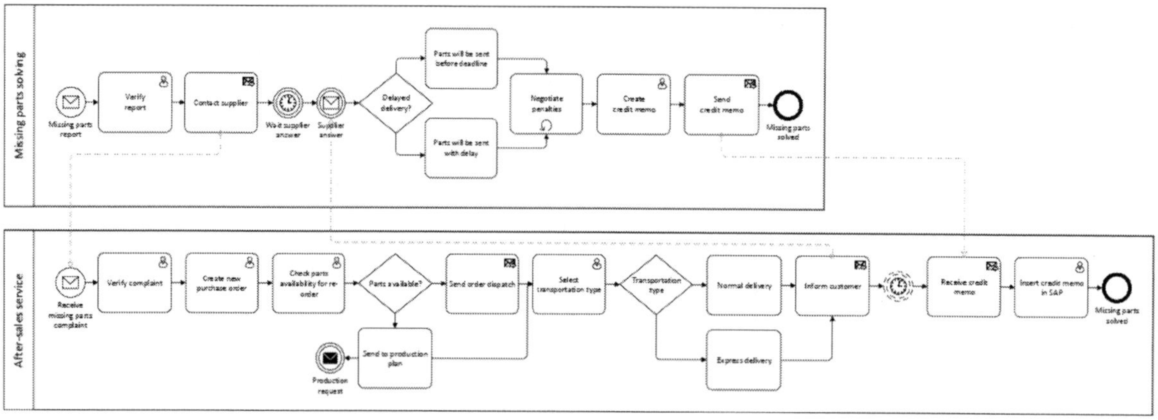

Figure 5: Collaboration BPM for missing parts solving ("as-is").

To deal with the coupling on the confirmation procedure (FR3.2), the use of the SAP system and the EDI allowed to send an advance shipment notice (ASN) to communicate the availability and expedition of the missing components.

The new DPs for the design are the following:

- DP3: New BPM for missing parts solving (see Figure 6).
- DP3.1.1: New contact points duly communicated.
- DP3.1.2: SAP and EDI implemented to process credit memo and re-ordering.
- DP3.1.3: New procedure to communicate complaints.
- DP3.2.1: ASN automatically sent upon expedition.
- DP3.2.2: Information exchanged by EDI.

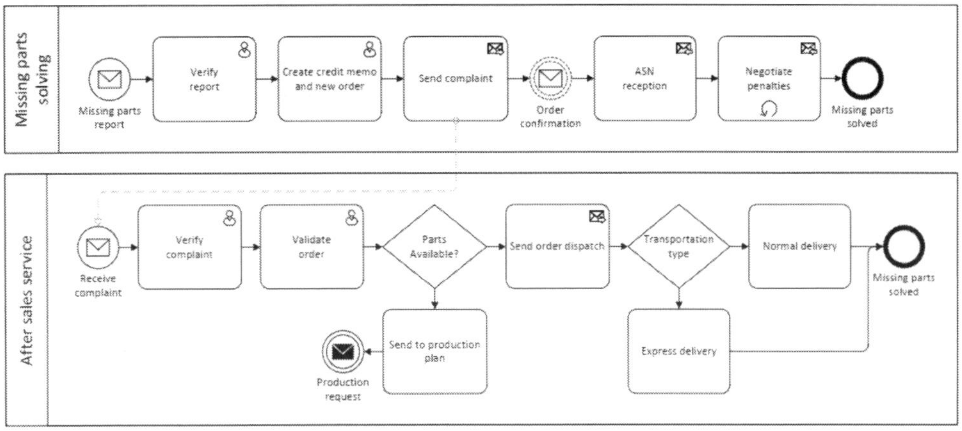

Figure 6: Proposed BPM for EDI and SAP implementation as a complaint management system.

6 CONCLUSION

The presented research contributes to the development of an integrated methodology to assess and re-design interoperable buyer-supplier dyads, by providing a systematic approach to identify interoperability problems, establish the dyad's design and map from conceptual (interoperability requirements) to physical (interoperability solutions) levels.

The decomposition of business interoperability into interaction perspectives allows looking and the buyer-supplier dyad from different points of view. That, combined with axiomatic design (AD) theory, allows mapping issues from the strategic foundations of the dyad to the technological interface that supports its processes. As so, in the present article was presented a case study in an automotive dyad where was possible to identify problems at strategic and process levels by implementing the 1st axiom. As a result, the proposal of a business process accompanied by the implementation of SAP and EDI information systems allows solving conceptually the compatibility problems between the two firms.

Future work will concentrate on linking the identified interoperability perspectives with the impact on the buyer-supplier dyad. As is illustrated in the theoretical framework in Figure 2, the aim of the buyer-supplier dyad is to achieve optimal interoperability. The presented procedure on this article shows, conceptually, how firms can be interoperability. By studying the influence of business interoperability on performance is possible to test the new scenarios using simulation, and determine which solutions provides higher performance.

7 REFERENCES

[1] **Chen I. and Paulraj A. 2004.** Towards a theory of supply chain management: the constructs and measurements, Journal of Operations Management, 22(2), pp. 119-150.

[2] **Mondini L. C., Machado D., and Scarpin M. 2014.** Dyad Buyer-Supplier and its Relation to Financial Performance, Journal of Operations and Supply Chain Management, 7(1), pp. 30-46.

[3] **Legner C. and Wende K. 2006.** Towards an Excellence Framework for Business Interoperability, Proceeding in 19th Bled eConference eValues, pp. 1-16.

[4] **Huhns M., Stephens L., and Ivezic N. 2002.** "Automating supply-chain management," Proceedings of the first international joint conference on Autonomous agents and multiagent systems. part 3 - AAMAS '02, no. 483, p. 1017.

[5] **Chen, D., Doumeingts G., and Vernadat F. 2008.** Architectures for enterprise integration and interoperability: Past, present and future, Computers in Industry, 59(7), pp. 647-659.

[6] **Blanc, S., Ducq, Y., and Vallespir B. 2007.** Evolution management towards interoperable supply chains using performance measurement, Computers in Industry, 58(7), pp. 720-732.

[7] **Choon Tan, K. 2001.** A framework of supply chain management literature, Europeran Journal of Purchashing & Supply Management, 7 pp. 39-48.

[8] **Bowersox, D., Closs D., and Stank T. 2003**, How to master cross-enterprise collaboration, Supply Chain Management Review, 7(4), pp. 18-26.

[9] **Harland, C. M.1996.** Supply Chain Management: Relationships, Chains and Networks, British Journal of Management, 7, pp. 63-80.

[10] **Simchi-Levi, D., Kaminsky, P., and Simchi-Levi E. 2008.** Designing and Managing the Supply Chain: Concepts, Strategies, and Case Studies, 3rd Edition, McGraw-Hill.

[11] **Cao, M. and Zhang, Q. 2011.** Supply chain collaboration: Impact on collaborative advantage and firm performance," J. Oper. Manag., vol. 29, no. 3, pp. 163-180, 2011.

[12] **Paulraj, A., Chen, I. J., and Flynn J.**, "Levels of strategic purchasing: Impact on supply integration and performance, Journal Purchashing and Supply Management, 12(3), pp. 107-122.

[13] **Espadinha-Cruz, P. and Grilo, A. 2014.** Methodology to analyse and re-design dyadic industrial cooperation, Proceeding of the 21st International Annual EurOMA Conference, EurOMA 2014, pp. 1-10.

[14] **Espadinha-Cruz, P. and Grilo, A. 2015.** A simulation approach to select interoperable solutions in supply chain dyads, Proceeding of the 7th International KES Conference on Itelligent Decision Technologies, KES-IDT 2015.

[15] **Espadinha-Cruz, P., Gonçalves-Coelho, A., Mourão, A., and Grilo, A. 2015.** Re-design of an Interoperable Buyer-seller Automotive Relationship Aided by Computer Simulation, Proceeding of the 9th International Conference on Axiomatic Design (ICAD 2015), vol. 34, pp. 98-105.

[16] **Espadinha-Cruz, P., Mourão, A., Gonçalves-Coelho, A., and Grilo, A. 2014.** Business Interoperability: Dyadic Supply Chain Process Decomposition Using Axiomatic Design, Proceeding of the 8th International Conference on Axiomatic Design (ICAD 2014), pp. 93-99.

[17] **Espadinha-Cruz, P., Cabral, I., and Grilo, A. 2013.** LARG Interoperable Supply Chains: from Cooperation Analysis to Design, Proceeding of the 5th International Conference of Intelligent Decision Technologies.

[18] **Grilo, A., Zutshi, A., Jardim-Goncalves, R., and Steiger-Garcao, A. 2013.** Construction collaborative networks, International Journal Computer Integrated Manufacturing, 26(1-2), pp. 152-165.

[19] **Grilo, A., Jardim-Goncalves, R., and Cruz-Machado, V. 2007.** A framework for measuring value in business interoperability, Proceeding of the 2007 IEEE International Conference Industrial Engineering and Engineering Management, pp. 520-524.

[20] **Suh, N. P. 1998.** Axiomatic Design Theory for Systems, *Research in Engineering Design*, 10(4), pp. 189-209.

ANALYSIS OF FINANCIAL PERFORMANCES AND INSOLVENCY RISKS OF MA OR COMMERCIAL BANKS IN CHINA

Hong Xiao*, Kapanya Theerawit
Department of Accounting,
Xiamen University, Xiamen, China
*xiaohong@xmu.edu.cn

ABSTRACT

The aim of this study was to compare financial performances and insolvency risks of large commercial banks and joint-stock commercial banks in China using data from 2010 to 2014. It was found that the joint-stock commercial banks played a more important role in the banking sector based on their proportions of total banking institution assets from 2010 to 2015. It was also found that banks with higher financial performances did not always mean that they were more stable and less risky. Results indicated that the average returns of on assets of large commercial bank were on a par with those of joint-stock commercial banks, but large commercial banks were less risky.

Keywords: Financial performance, Insolvency risk, Stability and security of commercial banks

1. INTRODUCTION

The banking sector is most important in the development of China's economic system as it has large assets, a variety of assets and liabilities, and a significant impact on the country's capital markets [1]. It also plays an increasingly important role in the Chinese industrialization process, and is a vital part of the national financial industry as banks are the only legal source of loans [2]. As a result, the stability and security of commercial banks are critical.

The Chinese government started to open its market-based economy in 1979. The period from 1979 to 1993 saw the emergence of the Peoples' Bank of China (PBC) that served as a central bank and four state-owned commercial banks that engaged in commercial bank lending, eight joint-stock commercial banks, and rural and urban credit cooperatives. At the end of 2014, China's banking institutions consisted of three policy banks, five large commercial banks, 12 joint-stock commercial banks, 133 city commercial banks, and a large number of other financial institutions, such as credit cooperatives, trust companies, and foreign banks. According to the China Banking Regulatory Commission (CBRC), the proportion of the total assets of banking institutions in large commercial banks decreased from 58.03% in 2003 to its lowest point of 41.21% in 2014, while on the other hand, the proportion of the total assets of banking institutions in joint-stock commercial banks increased from 10.70% in 2003 to its highest point of 18.21% in 2014. As a result, the joint-stock commercial banks are becoming an important part of the overall picture of Chinese banking institutions.

Four of the top ten of the largest banks in the world ranked by total assets are Chinese financial institutions [3] but the industry remains sensitive to mergers, restructures, and bankruptcies, and an analysis of financial performances and insolvency risks poses a challenge.

The aim of this study was to compare the financial performances and insolvency risks of large commercial banks and joint-stock commercial banks in China for the period from 2010 to 2014.The paper employed a z-score approach and probabilities of book-value insolvency banks for assessment of major commercial banks. Furthermore, the rank of financial performances and risks were divided into two groups, large commercial banks and joint-stock commercial banks.

* Corresponding Author: Hong Xiao

This paper presents a brief review of literature in section 2 and the study's methodology to calculate bank risks and probability of insolvency in section 3. Data and selected results are discussed in section 4, and section 5 provides a conclusion.

2. LITERATURE REVIEW

There is much literature related to insolvency risks in the banking industry, some focusing on ownership structure and some on bank regulations. Chun et al. (2010) [4] analyzed the effects of ownership structure and risk-taking Korean and Japanese behavior during the relatively regulated period of the late 1990s and early 2000s. It was found that managerial ownership alone did not affect either the risk or profit levels of Korean banks. In contrast, an increase in managerial ownership added to the total risk of Japanese banks.

Cornett et al. [5] examined the effects of government ownership and government involvement in countries' banking systems on the performances of Asian banks from 1989 to 2004. This study's examination of 16 Far East countries showed that state-owned banks exhibited more risk than private banks.

Haw et al. [6] compared the impact of foreign and state ownership on banking risk by using a sample of listed commercial banks in East Asia and Western Europe. This research found evidence that banks with concentrated ownership exhibited higher risk measured by return volatility and insolvency risk.

Lassoued et al. [7] analyzed the impact of foreign and state ownership on banking risk in the Middle East and North Africa (MENA) region from 2006 to 2012, and results showed that state ownership encouraged banks to take more risks while foreign ownership reduced risk-taking.

The effects of state ownership and bank regulation have been examined in many economies in Asia, Europe, and MENA, and these examinations showed that state-owned banks exhibited more risk than private banks. My study analyzed the overall financial performances and insolvency risks of two groups of Chinese banks, large commercial and joint-stock commercial.

Table 1 total assets of banking institutions (2010-2014)

Unit: RMB 100 Million

Year/bank	2010	2011	2012	2013	2014	Growth Rate
Large Commercial Banks (Proportion of Banking Institutions)	468,943 (49.20)	536,336 (47.34)	600,401 (44.93)	656,005 (43.34)	710,141 (41.21)	51.43% (-7.99)
Joint-Stock Commercial Banks (Proportion of Banking Institutions)	149,037 (15.64)	183,794 (16.22)	235,271 (17.61)	269,361 (17.80)	313,801 (18.21)	110.55% (2.57)
Other Financial Institutions (Proportion of Banking Institutions)	335,073 (35.16)	412,743 (36.44)	500,552 (37.46)	588,181 (38.86)	699,413 (40.58)	108.73% (5.42)
Total	953,053	1,132,873	1,336,224	1,513,547	1,723,355	80.82%

Source: China Banking Regulatory Commission 2014

3. METHODOLOGY

3.1. Measure of bank risk

The idea of risk measurement developed and used by Hannan and Hanweck (1988) The measure also has been used by Laeven and Levine (2009), Jin, Kanagaretnam, Lobo, and Mathieu (2013), and

Kanagaretnam, Lim, and Lobo (2014). Hannan and Hanweck originally developed their risk measure as probability of book-value insolvency. [8,9,10,11] was used to analyze the risks of commercial banks. The empirical form of their risk index (RI), which is labelled Z-score, is

$$Z-score=\frac{ROA+CAR}{\sigma(ROA)} \quad (1)$$

where ROA = the return on assets,

CAR = the capital-assets ratio (equity/assets),

σ(ROA) = the standard deviation of ROA.

The measure also has been used by Laeven and Levine (2009), Jin, Kanagaretnam, Lobo, and Mathieu (2013), and Kanagaretnam, Lim, and Lobo (2014). Hannan and Hanweck originally developed their risk measure as probability of book-value insolvency.

Five years of dada for each bank were used to calculate σ(ROA).The Z-score indicates the number of standard deviations a bank's return on assets has to drop below its expected value before equity is depleted and the bank becomes insolvent. Thus, a higher z-score indicates that the bank is more stable and less risky.

3.2. Probabilities of insolvency

Hannan and Hanweck (1988) [8] also analysed the probabilities of book-value insolvency banks as follows:

$$P(BVE<0)=\frac{1}{2(Z-\text{score})^2} \quad (2)$$

Where P(BVE<0) = Probabilities of book-value insolvency.

4. EMPIRICAL FINDINGS AND RESULTS

Data were collected from annual reports of the five largest commercial banks and the five largest joint-stock commercial banks from 2010 to 2014. According to statistics of the annual report from China Banking Regulatory Commission (CBRC) [12], at the end of 2014 the assets of the five largest commercial banks held 41.21% of the total assets of banking institutions, the five largest joint-stock commercial banks held 11.20% of the total assets of the banking institutions, and both groups held 52.41% of the total assets of banking institutions.

The probability of insolvency is more important for large banks than small banks, because, even without adverse situations such as the Global Financial Crisis, large banks are already scrutinised by regulators, investors, and the general public.

The total assets of China's banking institutions can be seen in the table 1. The table reveals that the proportion of the total assets of banking institutions in large commercial banks decreased from 49.20 % in 2010 to 41.21%, the lowest point, in 2014, while on the other hand, the proportion of the total assets of banking institutions in joint-stock commercial banks increased from 15.64% in 2010 to its highest point of 18.21% in 2014.

Table 2 shows the averages of financial information of the largest commercial banks and joint-stock commercial banks from 2010 to 2014, and that the largest commercial banks had the higher proportion of total assets (46.70), whereas the five joint-stock commercial banks' proportion of total assets was 11.20. The large commercial banks and joint-stock commercial banks held 57.90% of the total assets of banking institutions. Table 2 also shows the equity and net profit of each bank calculated under the Z-score model.

The probability of book-value insolvency banks model are given in Table 3. Consideration of the risk and the possibility of bankruptcy from table 3 shows that, except for the Agricultural Bank of China with a Z-score equal to 77.53 and P(BVE<0) of 0.00832%, all the large commercial banks had Z-scores higher than the joint-stock commercial banks. These results indicated that the large commercial banks were stable and had more regard for safety. The possible reasons for the high Z-score of large commercial banks were that they were state-owned and Chinese government injected much money to remove non-performing loans and non-performing assets.

Table 2 financial information:(average from 2010 to 2014)

Unit: RMB Million

Bank	Assets	Proportion of Total Assets	Equity	Net Profit
Large Commercial Banks				
Industrial and Commercial Bank of China	17,201,082	12.92	1,142,586	230,482
China Construction Bank	13,815,361	10.37	956,865	188,288
Agricultural Bank of China	13,159,116	9.88	764,107	141,543
Bank of China	12,819,190	9.63	887,947	145,331
Bank of Communications	5,213,077	3.91	354,596	55,392
Total	62,207,827	46.70	4,106,102	761,037
oint-Stock Commercial Banks				
China Merchants bank	3,316,468	2.49	218,048	42,993
Shanghai Pudong Development bank	3,156,111	2.37	182,484	33,546
China CITIC Bank	3,117,428	2.34	200,895	33,036
China Minsheng Bank	2,901,230	2.18	171,991	34,658
Industrial Bank	2,428,336	1.82	164,922	32,359
Total	14,919,573	11.20	938,339	176,591
Other Financial Institutions	56,053,640	42.09	3,794,999	529,314
Total Assets of Banking Institutions	133,181,040	100.00	8,839,440	1,466,942

Source: Annual report of each bank and China Banking Regulatory Commission 2014

Table 3 analyses of financial performances and risks

Bank	E(ROA)	CAP	SROA	Z-score	P(BVE<0)
Large Commercial Bank					
Industrial and Commercial Bank of China	1.34	0.0664	0.06%	130.8599	0.00292%
China Construction Bank	1.36	0.0693	0.06%	146.1903	0.00234%
Agricultural Bank of China	1.08	0.0581	0.09%	77.5296	0.00832%
Bank of China	1.13	0.0693	0.05%	153.1521	0.00213%
Bank of Communications	1.06	0.0680	0.05%	163.9828	0.00186%
oint-Stock Commercial banks					
China Merchants bank	1.30	0.0657	0.14%	46.9751	0.02266%
Shanghai Pudong Development bank	1.06	0.0578	0.10%	57.8299	0.01495%
China CITIC Bank	1.06	0.0644	0.05%	128.8958	0.00301%
China Minsheng Bank	1.19	0.0593	0.16%	37.0632	0.03640%
Industrial Bank	1.33	0.0679	0.14%	48.5244	0.02123%

It is possible to observe from the tables the Z-score corresponding to each probabilities of book-value insolvency. In contrast, except for China CITIC Bank with a Z-score equal to 128.90 and P(BVE<0) of 0.00301%, all the joint-stock commercial banks had Z-scores lower than the large commercial banks. Possible reasons for the lower Z-scores of joint-stock commercial banks have been their accelerated expansion of domestic and oversea markets, which led to the growth of non-performing loans and non-forming assets. The averages returns on assets of large commercial banks were on a par with the joint-stock commercial banks, but the large commercial banks were more stable and less risky. A summary of the classification of banks based on their ranking of financial performances and risks is given in Table 4.

The study separated the ranking of financial performances and risks into large commercial banks and joint-stock commercial banks. It can be seen that the China Construction Bank was first in ROA (1.36), followed by the Industrial and Commercial Bank of China (1.34) and the Bank of China (1.13) based on their risks. The Bank of Communications was last in ROA (1.06) despite being first in Z-score and P(BVE<0). Only the Agricultural Bank of China in fourth position in ROA (1.08) had a higher score in Z-score and P(BVE<0).

Table 4 ranking of financial performances and risks

Bank	E(ROA)	Z-score	P(BVE<0)
Large Commercial Banks			
Industrial and Commercial Bank of China	2	4	4
China Construction Bank	1	3	3
Agricultural Bank of China	4	5	5
Bank of China	3	2	2
Bank of Communications	5	1	1
oint-Stock Commercial banks			
China Merchants bank	2	4	4
Shanghai Pudong Development bank	4	2	2
China CITIC Bank	5	1	1
China Minsheng Bank	3	5	5
Industrial Bank	1	3	3

In the joint-stock commercial banks, the Industrial Bank was first in ROA (1.33), followed by the China Merchants Bank (1.30) and the China Minsheng Bank (1.16) based on their risks. The Shanghai Pudong Development Bank and the China CITIC Bank registered 1.06 despite being ranked first and second respectively in Z-score and P(BVE<0).

From the comparison of ranking of financial performances and risks between large commercial banks and joint-stock commercial banks it was also found that banks with higher financial performances dis not always mean that they were more stable and less risky.

5. CONCLUSION

This study examined the major commercial banks in China from 2010 to 2014. The study separated these banks into two groups, the five largest commercial banks and the five largest joint-stock commercial banks. The proportions of assets of each group in total assets of banking institutions were calculated, and it was found that the joint-stock commercial banks had a higher growth rate (110.55%) compared to commercial banks' growth rate of 51.43%. The proportion of assets of the large commercial banks in total assets of banking institutions decreased, while on the other hand, the proportion of joint-stock commercial banks in the total assets of banking institutions increased

The study showed that the joint-stock commercial banks play a more important role in the banking sector. Further, the application of a risk index (Z-score) to measure the risks of commercial bank and the possibility of bankruptcy found evidence that large commercial banks were more stable and less risky than joint-stock commercial banks. It was also found that banks with higher financial performances dis not always mean that they were more stable and less risky, while averages of returns on assets of large commercial bank were on a par with joint-stock commercial banks, but the large commercial banks were less risky.

6. ACKNOWLEDGMENT

This paper was supported by Social Science Fund of the Ministry of Education under grant No.15JJD630012 and The national accounting research of China under grand No.2015KJB008, which authors gratefully acknowledge.

. REFERENCES

[1] **Y. Tan,** "The Impacts of Risk and Competition on Bank Profitability in China", Journal of International Financial Markets, Institutions and Money, vol.40, pp. 85-110, 2016.

[2] **H. Zhao and S.** "Kang, Banking Performance Evaluation in China Based on Non-Radial Super-Efficiency Data Envelopment Analysis", Procedia Economics and Finance, Prague, Czech Republic, 2015.

[3] **The banker,** "The Banker Top 1000 World Banks 2015 ranking", Retrieved January 25, (2016).

[4] **S. E. Chan, M. Nagano and M. H. Lee**, "Ownership Structure and Risk-taking Behavior: Evidence from Bank in Korea and Japan", Asian Economic Journal, vol.25, No.2. Pp.151-175, 2011.

[5] **M. M. Cornett, L. Guo, S. Khaksari, and H. Tehranian**, "The Impact of State Ownership on Performance Differences in Privately-Owned Versus State-Owned Banks: An International Comparison", Journal of financial Intermediation, vol.19, pp. 74-94, 2010.

[6] **I. M. Haw, S. S. M. Ho, B. Hu and D. Wu,** "Concentrated Control, Institutions, and Banking Sector: an International Study", Journal of Bank and Finance, Vol.34,pp. 485-497, 2010.

[7] **N. Lassoued, H. Sassi and M. B. R. Attia**, "The impact of state and foreign ownership on banking risk: Evidence from the MENA countries", Research in International Business and Finance, vol.36. pp. 167-178, 2016.

[8] T. **H. Hannan and A. Gearal,** "Bank Insolvency Risk and Market for Large Certificates of Deposit", Journal of Money, Deposit and Banking, 1988, May, pp. 203-211, 1988.

[9] **L. Laeven and R. Levine,** "Bank Governance, Regulation and Risk-Taking", Journal of Finance Economics, vol.2. pp. 259-275, 2009.

[10] **. Y. in, K. Kanagaretnam, G. . Lobo and R. Mathieu**, "Impact of FDICIA Internal Controls on Bank Risk-Taking", Journal of Banking and Finance, 2013, vol.37. pp. 614-624, 2013.

[11] **K. Kanagaretnam, C. Y. Lim, and G. . Lobo,** "Influence of National Culture on Accounting Conservatism and Risk-Taking in the Banking Industry", Journal Accounting Review, vol.89, No.3. pp. 1115-1149, 2014.

[12] China Banking Regulatory Commission. "2014 Annual Report". Retrieved May 25, 2015.

A Comprehensive Evaluation Methodology for Assembly Manufacturing Industry Based on Virtual Assembly and Ergonomics Analysis

Ji-hong an**, Xu-wei Lou, Wei-gang Li
Department of Industrial Engineering,
Harbin Institute of Technology, Harbin, China
*jyan@hit.edu.cn

ABSTRACT

Nowadays science has gradually become a kind of link. Assembly manufacturing is also not out of this trend. The analysis and evaluation of the assembly operations need to not only meet the requirements of assimilability, but also take into more consideration the human factors. This paper will make a deep analysis and research on the comprehensive evaluation method in the assembly operation, which can adapt to the characteristics of the assembly and meet the basic needs of assembly workers. Moreover, a comprehensive evaluation method is built to achieve the goal that multiple factors are investigated comprehensively. Furthermore, this research developed a comprehensive evaluation software and discussed a practical application to show the operating effect and application value of the method.

Keywords： Virtual Assembly, Ergonomics, Analytic Hierarchy Process, Fuzzy Comprehensive Evaluation

I. INTRODUCTION

With the rise of the new science and technology revolution recently, more and more assembly manufacturing enterprises have introduced the virtual assembly simulation and Evaluation Technology in order to solve the problems of product design, staff training, etc. Virtual assembly is a kind of technology, which is used to simulate the process of products assembly in the computer to directly display assembly design and planning [1]. It makes up for the deficiency of traditional design and development of mechanical products, and provides a convenient means for the design of products. At present, there are many experts and scholars to have in-depth research in the application of virtual assembly technology. Germany's Fraunhofer developed a virtual assembly planning system, which can interact in a virtual environment to assemble and disassemble parts. The Carnegie-Mellon University and University of Maryland in USA jointly developed the IAMS assembly system, which can be used to check the interference of assembly path [2]. Some research institutions in China have also done some research to develop an evaluation system for assembly process inspection and simulation [3-6].

It is not difficult to find out from the above research, the focus of the virtual assembly simulation and evaluation technology is the verification of the assembly of the product. However, it shouldn't be ignored that the full realization of the potential productivity of mechanical equipment depends on good Human-Computer interaction and cooperation [7]. Moreover, the level of assembly is mainly dependent on the operational quality of the staff for the assembly manufacturing industry [8]. At present, most researches on virtual assembly evaluation technology ignore the investigation of assembly operators, which is one of the main reasons leading to the reason why current assembly simulation systems cannot achieve the anticipated results.

---

* Corresponding Author
†This work is funded by National Natural Science Foundation of China (Grant No. 51561125002) and CAST-BISEE Foundation (Grant No. CAST-BISEE-2014-05).

Viewing of the current shortcoming, this research will establish a comprehensive evaluation method considering the assimilability and ergonomic. DELMIA will be used as the virtual simulation platform, which can simulate and analyze the whole manufacturing and maintenance process [9]. In this paper, the evaluation indexes are selected based on the depth study of the assembly process. Secondly, the hierarchical structure and index weight system are established by using AHP (Analytic Hierarchy Process). Furthermore, the fuzzy comprehensive evaluation method is used to set up the evaluation system of each index. Finally, based on the establishment of comprehensive evaluation method, a comprehensive evaluation software is built by using of the programming language. Moreover, the comprehensive evaluation method is applied and tested through an actual case.

2. THE EVALUATION INDEX SYSTEM

2.1. Methodology

For a comprehensive evaluation problem, the establishment of evaluation index system is the basis of the final analysis and evaluation. The accuracy of the final evaluation results will be determined by whether the index system is reasonable. In the process of setting up the index system, this research will adopt the system analysis method, namely decomposing the problem first and then consideration comprehensively. In addition, the selection of index should follow the principles of scientific, practical, comprehensive and concise [10]. Firstly, the goal need to be decomposed into all the indicators related to the decision problem according to the objectives and the nature of the problem in the selection process. Then, based on their mutual relations, these indexes can be combined to form a final multi-level evaluation model [11].

2.2. The Selection of Inde and the Construction of Hierarchy System

According to the index selection method and principle, this research divides the index system into three layers, namely the target layer, the criterion layer and the index layer. The target layer is the specific goal to be achieved. The criterion layer is subordinate to the target layer, which reflects the general index of the evaluation problems from different aspects. The index layer is the further decomposition of the criterion layer, and it reflects the various more specific aspects of the evaluation and convenient to evaluate and assessment indexes [12]. Based on the characteristics of the analytical tools and analysis methods, this research will mainly adopt the five aspects of the assimilability, biomechanics, comfort, psychological load and operating arrangement as the criterion layer of index system. The criterion layer can be further decomposed to get the index layer: accessibility, visibility, work piece placement, operating space, lifting force, push/pull force, carry force, lumbar load, RULA, metabolic rate, operating posture, work repetition, noise, color, lighting, process rationality, operation rationality and operation economy.

3. THE EVALUATION INDEX WEIGHT SYSTEM

3.1. Methodology

In the process of comprehensive evaluation, the important degree of the indexes at the same level is different, so the weight of index system need to be built in the comprehensive evaluation methods. There are some common methods of determining weights, such as SMART (Simple Multi-Attribute Ranking Technique) [13], DEA (Data Envelopment Analysis) [14], Grey Related Degree Analysis [15] and so on. However, these methods cannot reflect the preference of decision makers and optimality and fairness of the evaluation. So in this research, AHP (Analytic Hierarchy Process) will be used, which is a more mature method in the multi-attribute decision-making method. AHP is a combination of the qualitative and quantitative analysis method, which is proposed by the famous American operation research expert Satty [16].

AHP is not only applicable to the situation of uncertainty and subjective information, but also allows the use of e experience, insight and intuition to measure the relative weight of indexes in a logical way [17].

There are several steps in the analytic hierarchy process. Firstly, the hierarchy model is established according to the index system. Secondly, the judgment matrix of each hierarchy is built by using the method of nine level evaluation scale and pairwise comparison. Then, the weight of the same level can be got by solving the characteristic value of the judgment matrix. Finally, the consistency checking method is used to modify the index weight system.

3.2. The Construction of udgment Matri

The hierarchy of indicators has been established in the last section, the following step is to build the judgment matrix. In the establishment of judgment matrix, this research adopts the method of pairwise comparison, which is to get the relative important degree judgment matrix by comparing between every two indexes of the same hierarchy. In the process of the implementation of the pairwise comparison method, the data can be obtained by means of the questionnaire. Thus, the design of questionnaire will directly influence whether the judgment matrix reflects the important degree faithfully. The actual focus is the expression of the times compared to the relative factors, so that people who do not learn the detail of AHP can understand the contents of the questionnaire.

3.3. The Weight Calculation and Consistency Check

After the judgment matrix is established, the next step is to calculate weight and consistency check. A large amount of calculation is required to calculate eigenvalues of judgment matrix, because there are plenty of indexes in the index system. Based on this situation, this research will use the Choice Expert software to make calculations. Choice Expert is an Expert-Decision system software, which is developed by the United States DSS (Decision Support Software) company. The software has the basic function of advanced AHP, including the construction criteria, pairwise comparison, consistency check, sensitivity analysis, output report and so on [18].

By using Expert Choice software, it is only needed to input the corresponding judgment matrix and the remaining work can be completed by the software. As well as that, the consistency check will be done after the input judgment matrix. The result will become the basis of the comprehensive evaluation score. The specific use of the process is as shown in Figure. 1. After the above steps, every index can get the final weight.

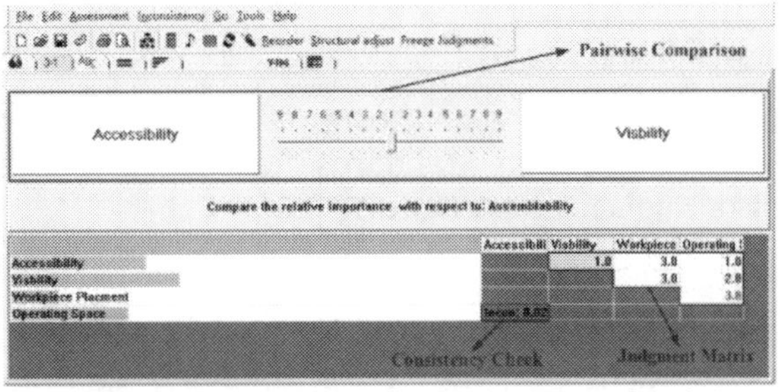

Figure.1. Input judgment matrix and consistency check in expert choice

4. THE EVALUATION INDEX SCORE SYSTEM

4.1. Methodology

The method of AHP is used to decompose the decision problem into a hierarchy framework and make qualitative and quantitative analysis. However, it is not enough to use AHP to carry out comprehensive evaluation. AHP can only determine the weight value of each index and cannot give the quantitative evaluation score. Therefore, in the comprehensive evaluation method the introduction of the evaluation index score system is also needed. At present, the main methods used in the construction of scoring system are Principal Component Analysis [19], Entropy-weighted Method [20] and Fuzzy Comprehensive Evaluation Method [21]. Due to the indexes characteristic of fuzzy and uncertainty, Fuzzy Comprehensive Evaluation Method will be used to construct the index score system. Using this quantitative analysis method, decision makers can make different degrees of fuzzy evaluation to the various factors of complex problems with reference to some evaluation criteria. Then, with the help of fuzzy mathematics, a quantitative comprehensive evaluation result can be obtained.

4.2. The Construction of udgment Matri

Based on the fuzzy comprehensive evaluation, this study uses the 5 point system as the evaluation set of the evaluation criteria. The corresponding relationship between score and evaluation grade is listed in TABLE I.

Table 1 the corresponding relationship between score and evaluation grade

Score	1	2	3	4	5
Grade	Unqualified	ualified	Fair	Good	Perfect

The following will be the introduction of the evaluation criteria for the evaluation of indexes at all levels:

1) Evaluation criteria of assimilability

For accessibility, visibility, work piece placement and operating space can use human envelope analysis of DELMIA, virtual human visual area analysis, work piece interference test and measuring tool respectively.

2) Evaluation criteria of biomechanics

For lifting force, push/pull force, carry force and lumbar load, the data of each index should be normalized according to the biomechanical evaluation criteria provided by DELMIA software firstly. Then the linear cross membership function is used to give the score of each indexes. The linear cross membership function can reflect the fuzziness of the index perfectly [22]. Its specific function form is as shown in Figure. 2.

3) Evaluation criteria of comfort

The score of RULA and the operating posture will be evaluated according to the analytical data provided by DELMIA software. The index of metabolic rate will be analyzed and evaluated by the evaluation standard of national work intensity grade.

4) Evaluation criteria of psychological load

It can refer to the relevant national and industry standards such as Hygienic Standard for Noise in Industrial Enterprises , Lighting Design Standards for Industrial Enterprises for analysis and evaluation of work repetition, noise, color, lighting [23].

5) Evaluation criteria of operating arrangement

For process rationality, operation rationality and operation economy, these indexes will be analyzed and evaluated according to product working procedure chart, man-machine relationship diagram and assembly motion image analysis separately .

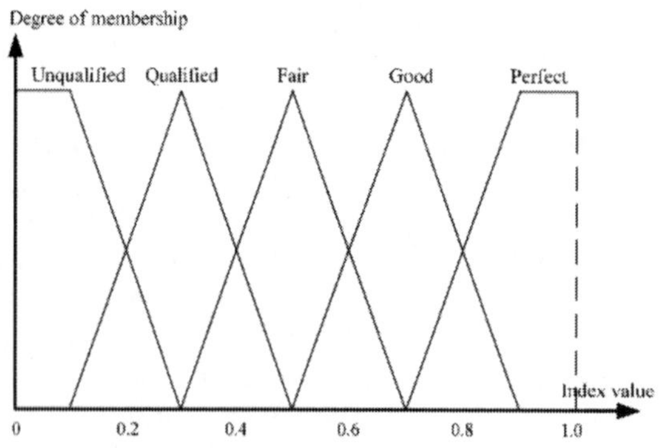

Figure.2. The graph of linear cross membership function

5. SOFTWARE DEVELOPMENT AND APPLICATION

5.1. Software Development

Based on the above comprehensive evaluation method, this study built a comprehensive evaluation software for assembly work by the C language, which make the comprehensive evaluation present intuitively. The main interface of software is as shown in Figure. 3

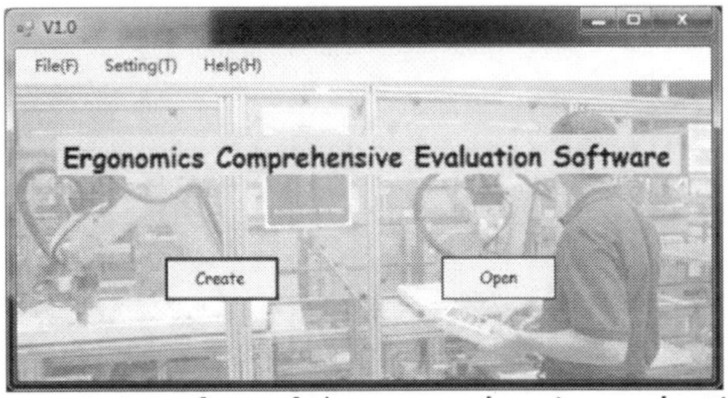

Figure.3. The main interface of the comprehensive evaluation software

In order to make the software used more conveniently, this software provides many useful functions such as personalized selection function for evaluation index, weight system and membership function. In the software, the user only need input the analysis report from DELMIA and related environmental parameters to realize the automation evaluation of assembly tasks. The specific use of the software process is as shown in Figure. 4.

5.2. Application

In this section, an assembly case will be implemented to illustrate the application of the comprehensive evaluation method. The case of this research is the installation process of the supporting plate of a certain type of manipulator, which is located in the top of dynamic transmission components. In this case, the operator need carry the support plate from the workbench to the workstation for installation. With the use of DELMIA software, the final virtual assembly environment is as shown in Figure. 5.

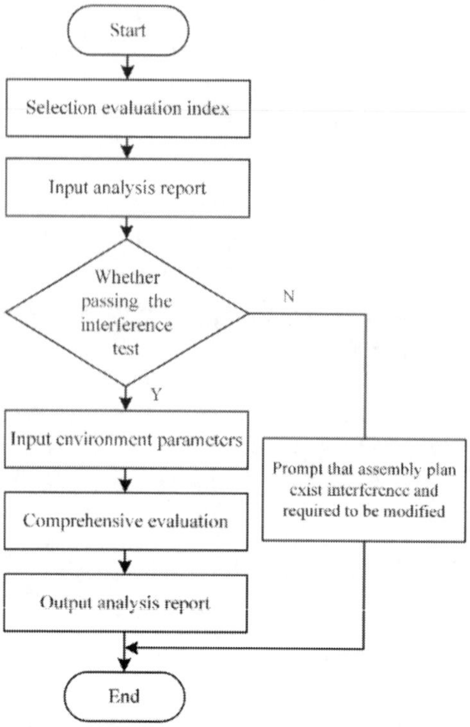

Figure.4. The flow diagram of the comprehensive evaluation software

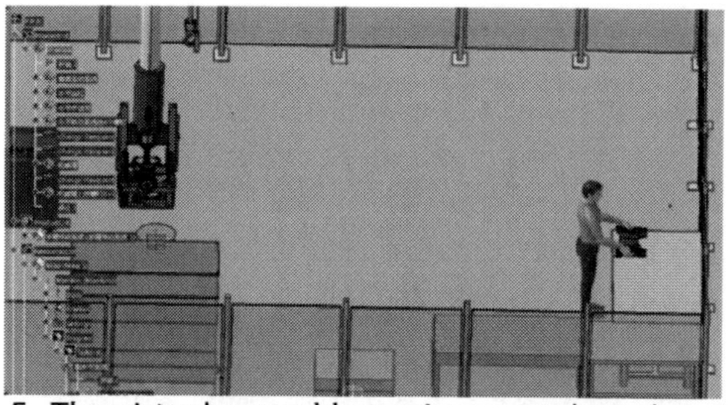

Figure.5. The virtual assembly environment based on DELMIA

Based on the characteristics of this case and the data obtained from the virtual simulation, the indexes of accessibility, visibility, workpiece placement, operating space, lifting force, carry force, lumbar load, RULA, and metabolic rate will be chosen in the comprehensive evaluation. Then, the final comprehensive evaluation results can be obtained according to the software application process. The specific score results of comprehensive evaluation is listed in TABLE II.

It is not difficult to establish that there are still some places to be improved in the assembly plan of this case, though the comprehensive score is 3.77 points. First, the lifting force and carry force scores of the operator were lower than 3 points. This situation is due to the heavyweight of the support plate. In addition, the force of operator in lifting and carrying operation is close to or more than the limit of the operator force at that time. On the other hand, because the distance of handling the support plate is short, the carrying operation in short time is still feasible. If it is a long time and long distance transportation, the assembly plan will be required to be improved. In that assembly process, the heavier parts should be transported by using the electric lifting arm in order to reduce the body's stress fatigue. Moreover, because the supporting plate assembly station is low, the virtual human in the assembly process always keep the bent posture. It leads to the fact that some scores in indexes of RULA and lumbar stress are deducted in the final evaluation result.

The specific working environment is as shown in Figure.6. The solution for this problem is to lift the station to a reasonable height, so that the operator can maintain the best working attitude.

Table 2 the comprehensive evaluation result of support plate assembly

Criterion Layer	Score	Inde Layer	Score
Assemblability	4.17	Accessibility,	4
		Visibility	4
		Workpiece Placement	4
		Operating Space	5
Biomechanics	2.91	Lifting force	2.58
		Carry force	1.23
		Lumbar load	4.58
Comfort	3.91	RULA	3.72
		Metabolic Rate	4
Comprehensive score		3.77	

Figure.6. The virtual human install the support plate

Through the results of the above cases, it can be shown that the comprehensive evaluation method established in this study can reflect the level of the integrated design of the assembly operation perfectly. Therefore, the investigation of this research has a significant theoretical and practical meaning.

6. CONCLUSIONS

In order to improve the design level of assembly plan and the comfort of the assembly workers, a comprehensive evaluation methodology is studied to modify and optimize the assembly plan in this paper. Moreover, this research has developed the comprehensive evaluation software and carried out the verification of the evaluation method. The main research conclusions are as follows:

(1) In the formulation of evaluation methods, firstly, this research selected 18 evaluation indexes and the evaluation index system is established, which are related to the evaluation problem based on the idea of problem decomposition. After that, through the method of AHP, the hierarchical model of evaluation problem is established and the weight system of evaluation index is built based on the pairwise comparison method. At last, the fuzzy comprehensive evaluation method is used to determine the evaluation criterions of each evaluation index.

(2) A comprehensive evaluation software is developed to realize automatic evaluation of assembly plan. In addition, the other functions such as selection of individual index, the automatic matching of index weight, the formulation of individual membership function, the comprehensive evaluation of multi index score and the result analysis report are realized in the software. Finally, the comprehensive evaluation method is applied and verified based on the virtual simulation model of manipulator. From the result, it can be concluded that the comprehensive analysis and evaluation can help in achieving safer and more productive workplaces for workers.

. ACKNOWLEDGMENT

Our deepest gratitude goes first and foremost to Dr. Li-zhong Zhao, who gave us a lot of advice on the Ergonomics. Besides, we'd like to express our sincere thank to Mr. Kun Wang and Ms. ong-hui Zhang, who offered us plenty of references and information in the construction of the virtual assembly model.

8. REFERENCE

[1] **G. Peng, G. Chen, X. Liu.** Modular Fixture Assembly Model for Virtual Assembly Design, Journal of Donghua University (English Edition), vol. 01, pp.10-15, 2009.

[2] **S. K. Gupta, C. . Paredis, R. Sinha.** Intelligent assembly modeling and simulation, Assembly Automation, vol. 21, no.03, pp. 215-235, 1980 J. Tan, Z. Liu, S. Zhang. Intelligent assembly modeling based on semantics knowledge in virtual environment, Computer Supported Cooperative Work in Design, The Sixth International Conference on, pp. 568-571, 2001.

[3] **Y. Yao, P. Xia, . Liu, et al.** A pragmatic system to support interactive assembly planning and training in an immersive virtual environment (I-VAPTS) The International Journal of Advanced Manufacturing Technology, vol. 30, no.9-10, pp.959-967, 2006.

[4] **X. Chen, A. Sui, M. Wang,** VEADAM:A Collaborative Virtual Design and Virtual Manufacture System, Acta Simulata Systematica Sinica, vol.17, no.2, pp.371-374, 2005.

[5] **G. Peng, G. Wang, W. Liu.** A desktop virtual reality-based interactive modular fixture conFigureuration design system, Computer-Aided Design, vol. 42, no. 5, pp.432-444, 2010.

[6] **L. Zeng, L. Zhang, T. Xiao.** Implementation of a virtual assembly support system, Acta Simulata Systematica Sinica, vol.14, no.9, pp.1149-1153, 2002.

[7] **. Posada, C. Toro, I. Barandiaran.** Visual computing as a key enabling technology for industrie 4.0 and industrial internet, IEEE Computer Graphics Applications, vol. 35, no.2, pp.26-40, 2015.

[8] **X. Li, Z. Zhang, . Gao.** Assembly Oriented Control Algorithm of Collaborative Disassembly and Assembly Operation in Collaborative Virtual Maintenance Process, Journal of Manufacturing Systems, vol. 36, pp. 95-108, 2015.

[9] **G. Chen, C. Lv, . Geng.** Virtual human control technology for immersed virtual maintenance system in DELMIA environment, Audio Language and Image Processing (ICALIP), pp. 1352-1355, 2010.

[10] **M. Gan.** Maintainability Design and Test (in Chinese). Beijing: National Defence Industry Press, 1995, pp. 132-137.

[11] **Y. Ye, L. Ke, D.** Huang. System Comprehensive Evaluation Technology and its Application (in Chinese). Beijing: Metallurgical Industry Press, 2006.

[12] S. **Gupta, G. S. Dangayach, A. K. Singh.** Analytic Hierarchy Process (AHP) Model for Evaluating Sustainable Manufacturing Practices in Indian Electrical Panel Industries, Procedia - Social and Behavioral Sciences, vol. 189, pp. 208-216, 2015.

[13] **G. Valiris, P. Chytas, M. Glykas.** Making decisions using the balanced scorecard and the simple multi-attribute rating technique, Performance Measurement Metrics, vol. 6, no. 6, pp. 159-171, 2000.

[14] **P. Zhou, B. W. Ang, K. L. Poh.** A survey of data envelopment analysis in energy and environmental studies, European Journal of Operational Research, vol. 189, no. 1, pp.1-18, 2008.

[15] . N. K. Liu, Y. Hu. Application of feature-weighted Support Vector regression using grey correlation degree to stock price forecasting, Neural Computing Applications, vol. 22, no.1, pp. 143-152, 2012.

[16] T. L. Saaty. How to make a decision: The analytic hierarchy process, European Journal of Operational Research, vol. 48, no.1, pp.9-26, 1990.

[17] W. Pedrycz, M. Song. Analytic Hierarchy Process (AHP) in Group Decision Making and its Optimization With an Allocation of Information Granularity, IEEE Transactions on Fuzzy Systems, vol.19, no.3, pp.527-539.2011.

[18] Y. Chou, C. Sun, H. Yen. Evaluating the Criteria for Human Resource for Science and Technology (Hrst) Based on an Integrated Fuzzy Ahp and Fuzzy Dematel Approach . Applied Soft Computing, vol.12, no.1 pp.64-71, 2012.

[19] I. T. olliffe. Principal Component Analysis, Springer Berlin, vol. 87, no. 100, pp.41-64, 2010.

[20] H. Zhi, N. ing. Entropy method for determination of weight of evaluating indicators in fuzzy synthetic evaluation for water quality assessment, Journal of Environmental Sciences, vol. 18, no.5, pp.1020-1023, 2006.

[21] A. Kumar, S. K. Choi, L. Goksel. Tolerance allocation of assemblies using fuzzy comprehensive evaluation and decision support process, International Journal of Advanced Manufacturing Technology, vol. 55, no.1-4, pp.379-391, 2011.

[22] C. Su. Manufacturing System Modeling and Simulation (in Chinese), Beijing: China Machine Press, 2008.

[23] X. Zhu. Industrial Psychology (in Chinese), Hangzhou: Zhejiang Education Press, 2003.

COST MANAGEMENT SYSTEM IN THE CONTAINER TERMINAL: A CASE ON SHANGHAI PORT

DING Yi [a], WEI Xu-Jun [a,*], Liang Cheng-Ji [a],ZHOU Wei-Wen [e], GU Tian-Yi [c], HUANG Heng [b], ZHU Chang-ming [d]

a Logistics Research Center, Shanghai Maritime University, Shanghai 200136, China

b Shanghai International Port Group, Shanghai 200136, China

c Department of Computer Science, University of New Hampshire, Durham, NH 03824, United States

d College of Information Engineering, Shanghai Maritime University, Shanghai 201306, PR China

e Department of Industrial and System Engineering, Texas A&M University, College Station, TX 77843, United States

ABSTRACT

Cost management has become a key factor in determining the competitiveness of container terminals. This study examines the application of a Time-Driven Activity-Based Costing (TDABC) modeling to a container terminal and elucidates how to use TDABC in the cost management of the operations at a container terminal. Moreover, a case study about Shanghai Shengdong International Container Terminal (SSICT) is performed. A large amount of raw activity data is extracted from the Terminal Operation System (TOS) data base to the data warehouse of Container Terminal TDABC System. The system can be used for calculations and statistical analyses. The data analysis clearly demonstrates that this system provides accurate cost information and is an effective analysis tool based on TDABC. The case study explains how the containers are transferred at the container terminal in a three-stage material handling process firstly. In this case study performed on the operation of a given vessel, TDABC model reveals that Non-Value-Added Cost (NVAC) in SSICT accounts for 42.87% of the quay crane cost, 50.12% of the yard truck cost, and 28.89% of the yard crane cost. A study performed by shipping lines reveals that more than 50% of the yard truck cost of shipping lines except North America line is non-value-added cost, so container terminals must effectively allocate yard truck resources. The artificial neutral network model is integrated with TDABC model to increase the general using of TDABC in cost budgeting. This is the first time to apply TDABC to container terminal for cost management systematically, and the research findings provide deep insight into the causes of excessive logistical operating cost and improve the procedure of cost management in SSICT.

Keyword : Container terminal, Complex logistical process, Time-driven activity-based costing, Cost management, Terminal operation system

1 INTRODUCTION

In current maritime industry, there are two categories of port performance indicators [1]. One is macro one quantifying aggregate port impacts on economic activity, the other is micro one evaluating input/output ratio measurements of port operations. In this paper, we focus on the micro level and its cost. Various references, particularly UNCTAD monographs, provide a range of port indicators by ratio type and category of operation [2-5]. There are many ways of measuring port efficiency or productivity, although reducible to three broad categories: physical indicators, factor productivity indicators, and economic and financial indicators [6]. Frontier models have evaluated port efficiency both within a single country [7] and across different countries [8]. Recent models propose or use Data Envelopment Analysis (DEA) to evaluate port efficiency [9-12]. A detailed review of such techniques for measuring port efficiency is provided by [13]. Those techniques do not always provide consistent results. For example, Cullinane et al [14] suggest that the size of a port or a terminal is positively correlated with its efficiency, whereas Coto-Milla´n et al. found that larger ports are more likely to be economically inefficient [15]. In this paper, the authors focus on evaluating the port efficiency from the perspective of the operating cost in container terminal. In a global container terminal today, owing to high investment costs [16], the productivity of a material handling system must be accurately estimated before various logistical, industrial, or transportation systems could be implemented. The overall profitability of a container terminal depends upon whether the sales income covers operating costs. Therefore, for container terminals, the management of operating cost management is an important way to ensure competitiveness. Hence, accurate cost information is in great demand. At a container terminal, a material handling system that consists of Quay Cranes (QC), Yard Trucks (YT) and Yard Cranes (YC) is used to handle the containers. Accordingly, the transfer of containers at a container terminal has a multi-stage logistical process [17]. The direct handling cost includes the energy cost (such as diesel oil cost and electric power cost), the maintenance cost (including costs for preventing breakdown and for lubricant), and the salaries of drivers. A lack of knowledge of how to distribute these costs in each step of the multi-stage logistics process makes us cannot ensure the profitability of a container terminal in a highly competitive environment. Thus, the cost model should be accurate and sufficiently detailed to capture the various factors that influence the handling costs. Logistical processes must be cost-efficient and the accuracy and specifications of cost models are also especially important. Although it has been argued that Activity-Based Costing (ABC) information can be used to support strategic and operational decisions [18-20], there is still little empirical evidence that adoption of ABC has an impact on performance (with the notable exception of Banker et al., 2008) [21]. Many companies have abandoned ABC since it does not capture the complexity of their operations, is time consuming in implement, and requires building and maintaining expensive systems [22]. Time-Driven Activity-Based Costing (TDABC) can solve these problems. Although all investigations concern logistics companies, this work develops an appropriate cost management method that is applicable to container terminals. One claim made TDABC in consulting research is that it enables operational improvements [23]. For example, in their descriptions of the Kemps LLC case, Kaplan and Anderson (2007) illustrated that by tracing the costs of changeovers in producing, packaging, picking, loading, and delivering products, the implementation of TDABC system enabled the detection of process inefficiencies [24]. Since TDABC system reveals the high costs of these frequent changeovers, customers are encouraged to place their orders less often. Hence, using TDABC information allowed Kemp LLC and other companies described by Kaplan and Anderson (2007) to realize certain process improvements [24].

The competition between container terminals has become fierce due to large growth rates on major seaborne container routes. Terminals are faced with more and more containers to be handled in short time period at low cost. Most literature focus on enlarging handling capacities and achieving gains in productivity [17], what has been missing from the literature is systematically analysis on the operational cost of container terminal. This paper describes the handling process of Shanghai Shengdong International Container Terminal (SSICT) with TDABC, which is a cost modeling technique that was developed by Kaplan and Anderson (2004) [22]. In SSICT, container throughput has been increasing for the last five years, but the growth of operating cost has exceeded that of revenue. Consequently, the profit rate has been falling. Detailed and accurate cost information is vital to the survival of the terminal. All container terminals in Shanghai port regard Cost Per Container (CPC) as the key performance indicator in cost management, but the operators of Shanghai port have already realized that an average cost index of CPC fails to provide useful information about profitability. The operating characteristics associated with different containers, vessels and shipping lines, and the bottleneck in the allocation and scheduling of container terminal resources cannot be captured by CPC alone. TDABC is a revised version of ABC that is already being used in other fields. This study discusses the experience of the SSICT with using TDABC in modeling its complex handling operations. The objectives are given as below.

- To trace how the container terminal models complex handling processes using the TDABC approach;
- To discover how the cost information that is provided by the TDABC model is used to improve container terminal resource allocation.

In our study, the TDABC method herein captures the time of each operating activities with TDABC information system descript by a general formula at the SSICT, called time equations. In the time equation, time spent to handle a given order which consists of several operating activities, as will be discussed in details in Section 3.

The paper is organized as follows. First, TDABC is used to build a model of the complex handling operation process in SSICT to capture the cost distribution of each handling of container, ship, or shipping line and the bottleneck of the allocation and scheduling plan through Container Terminal TDABC System, which also enables the value-added cost (VAC) and the non-value added cost (NVAC) of the handling processes to be captured. Second, the artificial neural network (ANN) is utilized to budget quay crane resource hours, yard truck resource hours, and yard crane resource hours for the handling of vessels that will arrive in the future, and to provide guidance for improving resource utilization and formulating a better berthing price strategy.

2. Methodology

Kaplan and Anderson (2007) argued that “ABC systems were expensive to build, complex to sustain, and difficult to modify” and hence they made ABC’s weakness as TDABC’s primary strengths [24]. The breakthrough of TDABC lies in time equations, which greatly simplify the process of estimating and generate a far more accurate cost model than can be generated obtained using traditional ABC techniques. Time equation can be used as a Taylorized control device to enforce employee compliance, they may also enhance staff members’ understanding of their work process, which, in turn, may result in operational improvements.

As such, TDABC is very suitable for a service company such as a container terminal or other logistics enterprises, because the operational activities are primarily measured on the basis of handling time used for performing a given activity. The usage of traditional cost system to determine CPC not only causes inflexible but also fails to trace the detail capacity consumption information for each activity.

In addition, CPC cannot reflect the variation environment in the container terminal. The processing time will vary according to the specific characteristics of a particular task and order. To solve this problem, TDABC accommodates the complexity of real-world operations with time equations, a feature that enables the model to reflect how particular order and activity characteristics cause processing times to vary. With using TDABC, the cost calculations can be more precise and also more flexible with the changing environmental circumstances [22].

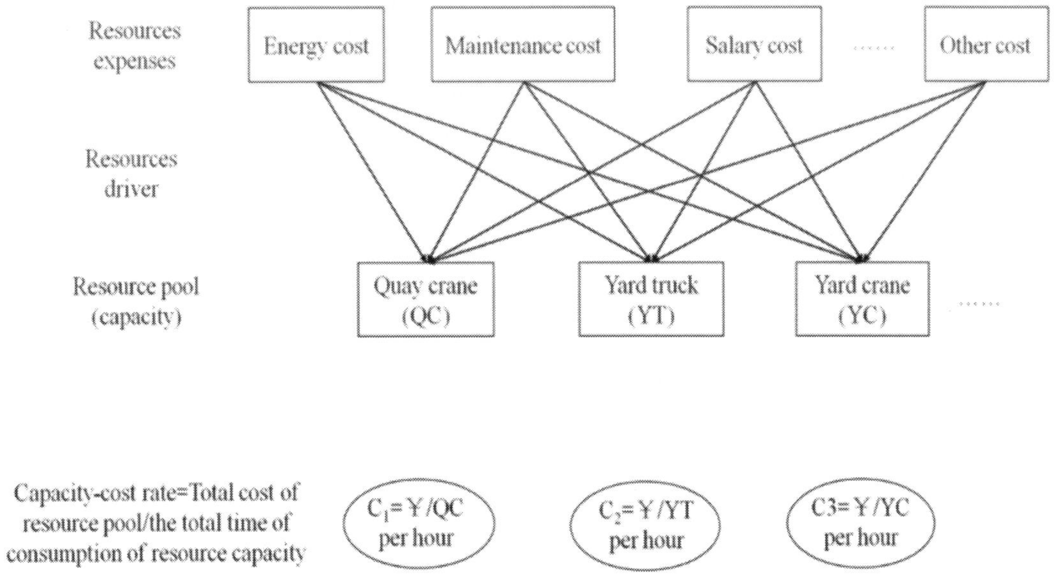

Fig. 1. Capacity-cost rate

The following elements are required to construct a TDABC model: (1) the cost per time unit of resource capacity (capacity-cost rate) (Fig.1.), (2) the units of time of consumption for resource capacity by products, services, or customers. The implementation of the model consists of the following steps in a container terminal.

Step 1. Identify the various resource groups (QC, YT or YC).

Step 2. Calculate the cost of each resource group (including energy cost, maintenance cost, salary cost etc.) based on data derived from financial data base.

Step 3. Obtain the time equations for each order of a container terminal machine operations from TDABC data warehouse, including QC, YC, and YT operations based on the time consumption for each activity within these orders.

Step 4. Calculate the practical capacity of each resource group (including available handling hours, and maintenance, waiting, and moving hours) basing on the time equations. (Fig.2.)

Step 5. Calculate the capacity-cost rate with dividing the total cost of the resource group by the total time of resource capacity consumption.

Step 6. Multiply the capacity-cost rate by the estimated time for the cost object such as containers, vessels or shipping lines in loading and discharging process.

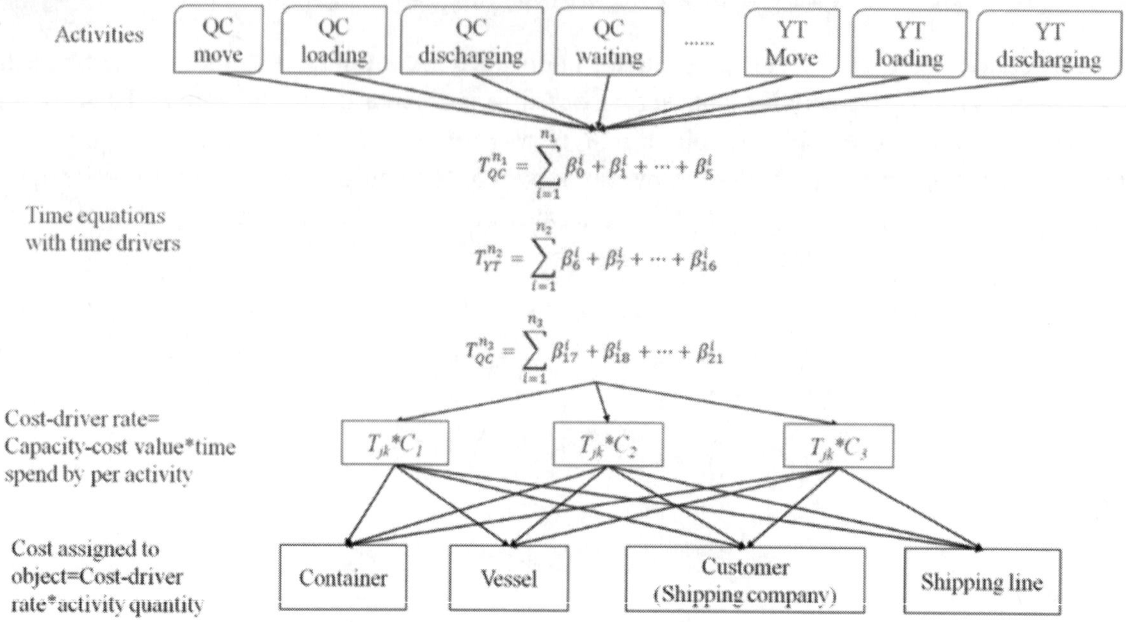

Fig. 2. Time equations

With TDABC, a flexible model for estimating operational time is established, which captures the activities in a complex logistics process. With the collection of relevant data, the manager can use the flexible model for estimating operation time to calculate the time and cost consumption of each cost object, such as a vessel, a shipping line, or a shipping company, as displayed in Fig.2.

: Preparation time for handling operation order i

: Time spent on various activities in handling operation order i

: Resource capacity of machine k (quay crane, yard truck or yard crane) that is consumed by product j (container, vessel, shipping line), which can be measured in resource hours

: Total time spent by quay crane, yard truck and yard crane in handling operation orders, respectively. (represent the quay crane order quantity, yard truck order quantity and yard crane order quantity respectively)

The inclusion of multiple time drivers enables complex activities to be modeled without increasing the number of activities [22]. Therefore, TDABC potentially provides many opportunities to design cost models of complex activities, as in logistics and distribution companies, hospitals, and servicing companies. However, in the container terminal context, some operation managers indicated that an exclusive accounting information system should be developed to assist the traditional operation system to attain a stabilized behavior of management accounting practice. They are complemented by specialized Information Technology (IT) infrastructure (including data warehouses and Extract Transform & Load (ETL) tools) which is necessary for resource deployment and usage efficiency. The following contents will further elucidate the experiences of SSICT Shanghai Port in using TDABC to cost management by a TDABC information system in the container terminal.

3. Application of TDABC in container terminal cost management

3.1. Shanghai Shengdong International Container Terminal Co. Ltd (SSICT)

Shanghai International Ports Group (SPIG) is one of the world's largest container terminal operators on account of its activities at its home port, where the total throughput of its 11 terminals was 35.7 million TEU in 2015.

Shanghai Shengdong International Container Terminal Co. Ltd, located on Yangshan Island, was capitalized by SIPG with a registered capital of 5 billion dollars. The major responsibilities of the company are to operate and manage the Phase-1 and Phase-2 terminals in the Yangshan Deepwater Port Area as well as to fund and run the auxiliary and supporting project: the International Logistics Park in the Port Area. Currently, the port has a 3,000m-long deep-water quay length, 34 of the world's most up-to-date container quay cranes, 120 yard cranes and other handling and transportation facilities.

Fig. 3. Shanghai Shengdong International container terminal (SSICT)

3.2 Container terminal handling process

Container terminal perform various handling operation by utilizing such resource as quays or berthes, quay cranes (QCs), storage yards (SYs), yard cranes (YCs), traveling areas (TAs), and yard trucks (YTs) [25]. The handling process in the container terminals consists of three stages. The loading-discharging processes are performed by quay cranes, yard trucks, and yard cranes. At the container terminal, quay cranes process the seaside handling operations; yard trucks carry out horizontal transport operations, and yard cranes perform the yard handling operations. In this three-stage material handling system, there is no buffer exists between adjacent handling equipment. In some situations, an equipment A must wait for another one B before A can pass a container to B or receive a container from B. Seaside handling operations, horizontal transport operations, and yard handling operations involve various activities.

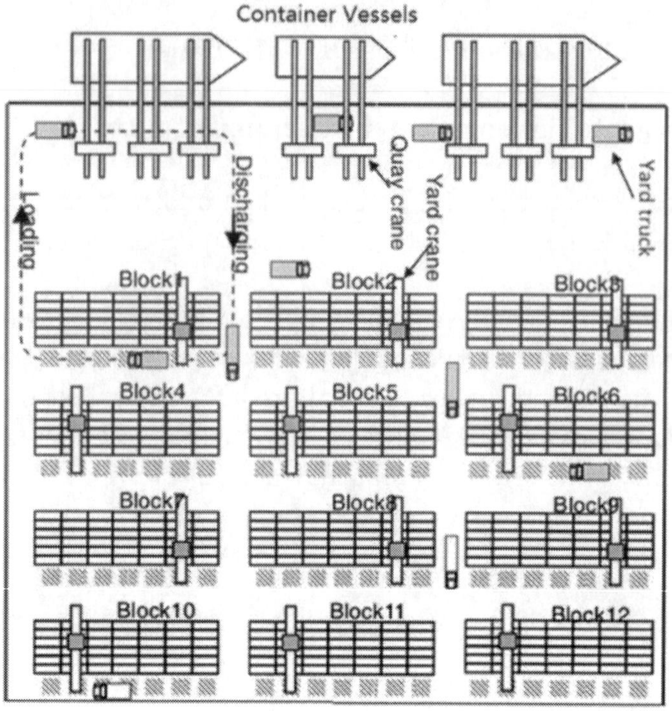

Fig.4. Handling process of container terminal

In the discharging operation (Vessel to Yard, VY), a QC picks up a container from a vessel and waits for the arrival of a YT. After a YT has received a container from a QC, it delivers that container to a particular block in yard area. The YC in turn picks up the container from the YT and puts it into a slot in a yard-bay. The process of loading operation (Yard to Vessel, YV) is in reverse. The description can also be shown in Fig.4.

The time of performing an operation order is estimated for each specific instance of the activities, based on its particular characteristics. These characteristics are called "time-drivers" because they "drive" the time spent on a given activity. Time equations formulate how time drivers drive the time spent on activities. In complex environments in which the time needed to perform an activity is driven by many drivers, TDABC can include these multiple drivers. The Table 1 displays activities of QC, YT, and YC associated with handling operations.

Table 1 Activities of QC, YT, and TC associated with handling operations (order i=1,2,..., 1,...)

QC activities	YT activities	YC activities
QC waiting for order i	YT waiting for order i	YC waiting for order i
QC moving	YT moving to loading point	YC move
QC waiting for YT	YT waiting for YC(YV)	YC waiting for YT
QC loading container on YT	YT waiting for QC(VY)	YC loading container on YT
QC unloading container from YT	QC loading the container on YT(YV)	YC unloading container from YT
	YC loading the container on YT(VY)	
	YT moving to unloading point	
	YT waiting for QC (YV)	

YT waiting for YC (VY)
QC unloading the container from YT (YV)
YC unloading the container from YT (VY)

: Time spent by the activity *n* in processing order *i (minute)*

3.3 Non-value-added activities

The activities of handling operation can be divided into two types: value-added ones and non-value-added ones. A value-added activity can be identified by considering the following criteria. (1) The activity can change the state of an operation object (container); (2) only this kind of activities can change the state of the container whereas the other cannot; (3) this activity enables other activities to perform. All the characteristics are necessary for value-added activities, otherwise, the activities are non-value-added ones.

Table 2 Non-value-added activities of handling operations (order *i=1, 2... n, n+1...*)

QC non-value added activities		YT non-value added activities		YC non-value added activities	
QC waiting for order *i*		YT waiting for order *i*		YC waiting for order *i*	
QC waiting for YT		YT waiting for QC(YV)		YC waiting for YT	
		YT waiting for YC(VY)			
		YT waiting for YC (YV)			
		YT waiting for QC(VY)			

: Time spent by the activity *n* in processing order *i (minute)*

(1)

In container terminal handling operation, all waiting activities are non-value-added, as shown in Table 2 and the non-value added cost rate is defined by formulation (1), where is the number of quay crane handling operation orders; is the number of yard handling operation orders, and is the number of yard crane handling operation orders. For years, SIPG has been considering how to improve its cost management strategy, our team sought to develop a cost management information system, practically applying the TDABC theory to one of the container terminals of SIPG.

3.4. TDABC information system

Some operation managers indicate that the relationship between IT and container terminal operation performance appears to be the most complex, penetrating, and unpredictable in the field of accounting information systems. Nicolaou and Bhattacharya reported a positive relationship between Enterprise Resource Planning (ERP) implementation and firm efficiencies [26-27]. Similarly, Ruivo et al. (2014) argued that the benefits of ERPs reside in how firms use and exploit the integration capabilities (both data and process) of these systems after the implementation phase [28].

SSICT adopts TOS (Terminal Operation System) as the IT system for container terminal's ERP, which enhances the operational efficiency of container terminal and coordinate and optimize the planning and management of container and equipment moves. The TOS connects directly to the Operational Data Base (ODB) and provides necessary information to enable various departments to implement business processes.

To support the implementation of TDABC in SSICT, an information system called the Container Terminal TDABC System (CTTS) is developed; it embeds different statistical methods to calculate capacity-cost rate, incorporates the time consumption of various activities in the time functions for three types of handling operational equipment, derives the time and cost consumptions of different cost objects, and analyzes the value-added cost and non-value-added cost of these objects (such as vessels and shipping lines).

Fig.5. presents the integration framework of CTTS and the data flows among the related module components. All input data are extracted from the ODB, the financial system data base to the TDABC data warehouse. The ODB stores the general terminal configurations (including the numbers of available QCs, YTs, and YCs, their attributes and etc.), the operational data (information about operating containers, time spent on different activities, order type (loading operation or discharging operation), vessel structure, and etc.), which are filtered and organized by the TOS modules. The financial system data base stores the terminal cost data, including energy cost, maintenance cost and salary of the drivers of each type of machine. Historical operational data and cost data are extracted from the ODB and periodically (every night) transferred to the oracle data warehouse by the ETL tool. Using raw operation activity's data that are stored in the oracle data warehouse, the computing servers perform computing that are related to the time equations. The report servers generate a series of analytical reports, which help the container terminal operator evaluate the operational performance and resource management, and adjust the operation strategy accordingly.

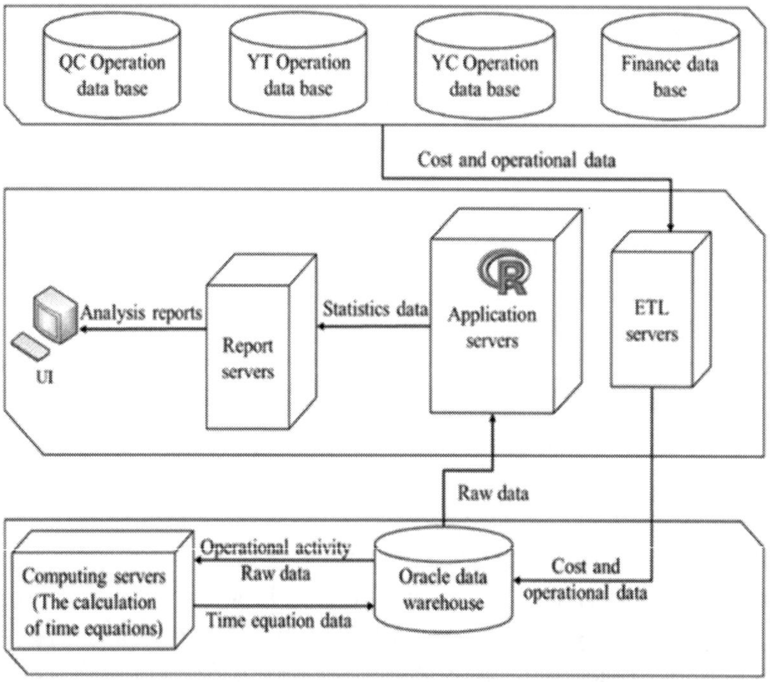

Fig. 5. Framework of Container Terminal TDABC System

4. Case study

4.1 Capacity-cost rate

During the terminal operations, the utilization of the terminal equipment (QC, YT, and YC) produces direct operating costs, including the salary of drivers, the energy (diesel oil and electricity) and the maintenance of these equipment. The total cost consumptions of each resource pool could be achieved from financial data base. From the data of time equations in the TDABC data warehouse, the total available time for consuming resource capacity (Resource Hour, RH) can be obtained, the capacity-cost rate of each resource pool can be calculated with dividing the total cost of the resource by the total time of consumption of the corresponding resource capacity for each resource pool. Table 3 presents the cost data and resource capacity data that corresponded to SSICT from June 2015 to December 2015.

Table 3 Related cost data and capacity-cost rate for quay crane, yard truck and yard crane of SSICT from June 2015 to December 2015 Data sources: SSICT finance system

Machine	Energy cost (￥)	Maintenance cost (￥)	Salary cost (￥)	Total cost (￥)	Available capacity (Hour)	Capacity-cost rate (￥/ Hour)
QC	12,869,204	4,484,192	3,663,908	21,017,304	15,368	1,367
YT	2,972,970	939,120	3,351,790	7,263,880	54,470	133
YC	8,674,862	4,489,513	2,914,905	16,079,280	52,360	307

4.2. Cost per vessel

TDABC system was utilized to collect the data on handling operations for a vessel that was serviced at SSICT. Then the TDABC model was used to analyze the costs associated with the activities of this vessel. Table 4 presents the loading and discharging order quantities for this ship, the Non-Value-Added Resource Hours (NVARH) and the Total Resource Hours (TRH) that were consumed by these orders during handling operations.

Table 4 Quantity of orders and capacity supplied for the vessel MAERSEK KLEVEN

Machine	Operation type	Order quantity	NVARH (hours)	TRH(hours)
QC	Loading (YV) Discharging (VY)	481 634	29.55	68.92
YT	Loading (YV) Discharging (VY)	523 672	305.10	612.24
YC	Loading (YV) Discharging (VY)	649 719	29.20	100.56

The capacity-cost rates of the resource pools for C, YT, and YC are ￥1,367/hour, ￥133/hour and ￥307/hour respectively, as indicated in Table 3. The following indicators are thus calculated; non-value-added cost, non-value-added cost rate, the total cost per vessel:

=68.92*1,367+612.24*133+100.56*307=206,513.48￥

=29.55*1,367+305.10*133+29.20*307=89,937.55￥= = =43.6%

The non-value-added cost rate is 43.6 %, revealing that plenty of room exists for improving the resource utilization in the handling of this vessel. The non-value-added cost rates of QC, YT, and YC are 42.88%, 49.83%, and 29.04% respectively, indicating that the operation of QC and YT have more room to improve. Most of non-value-added cost of QCs for this vessel is caused by waiting for YTs. Fig. 6 shows the resource hours for each activity of the vessel.

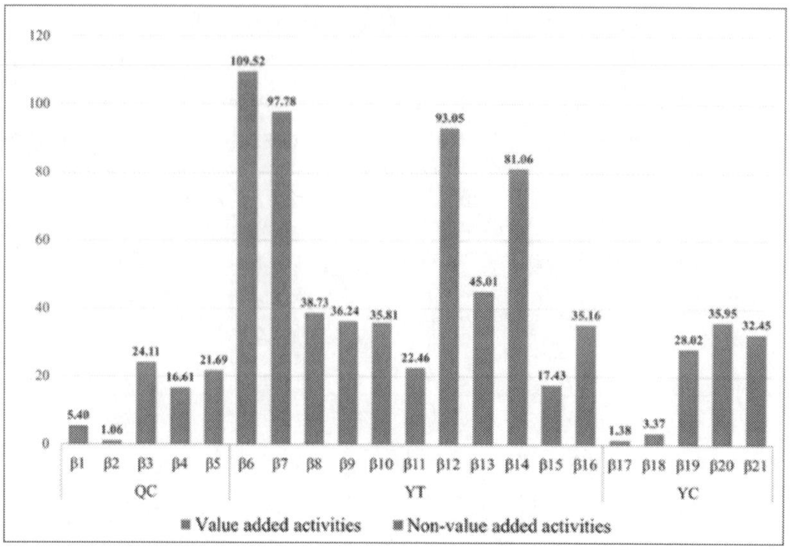

Fig.6. Resource hours consumed by different activities

QC: ()
YT:
YC:)

The time consumed by YTs' waiting for orders is 109.52 hours, indicating that much yard truck capacity is wasted and the allocation of yard truck capacity substantially exceeds that requirement of this vessel's handling operations.

With respect to seaside operations, the cost of waiting by yard truck accounts for 13.28% of the total cost of yard truck; the corresponding percentage for yard operations is 19.57%. Three possible causes are as follows.

1. Traffic congestion that is caused by an overuse of yard trucks that enter the yard blocks for a period.
2. A lack of available yard crane resources at a particular time.
3. Many yard trucks were waiting to be operated under the quay cranes, which were being operated for the same vessel.

Yard truck (YT) is not the only resource that consumes waiting time: quay crane and yard crane also consume 24.11 and 28.02 resource hours during the handling process of this vessel, mostly waiting for YTs. Although YT resources were overused overall, the overuse of YT resources frequently alternated with a shortage of YTs in both the seaside area and the yard area, revealing that collaborative operations in this three-stage logistics system are not very effective at SSICT.

4.3. Cost to shipping lines

According to the historical operational data of vessels, operational activity cost varies dramatically even for a certain given amount of operations. It reveals that the difficulty of operations changes among the categories of vessels associated with the different shipping lines. Table 5 presents the main shipping lines whose vessels are served by the SSICT, along with the number of handling operation orders for each shipping line. The time equation data (from June 2015 to December 2015) are extracted from the Oracle data warehouse to analyze the operational cost performance for vessels that belong to different shipping lines at SSICT.

Table 5 Quantity of operational orders of shipping lines serviced by SSICT from June 2015 to December 2015

Shipping lines	QC orders		YT orders		YC orders	
	YV	VY	YV	VY	YV	VY
Northwest Europe (NWE)	125208	144414	173420	175034	182132	179942
North America (NA)	82212	71450	111972	83618	112028	88332
South America (SA)	34468	37730	48600	46934	52150	47830
Mediterranean (MED)	51688	72514	71578	80836	78428	89976
Africa (AFR)	16416	29584	23542	35622	25346	35936
Australia (AUS)	4308	5212	6830	7426	6904	7068
CHJ	61072	49578	84716	70996	89486	85100
SHD	2012	1554	2802	2148	3038	3110
YHA	3172	628	4328	942	4670	1160
YSK	16996	21076	22914	30026	23870	43970

From June 2015 to December 2015, the vessels of ten shipping lines were served at SSICT.

Among these, CHJ, SHD, YHA, YSK are domestic shipping lines and the others are international ones. In this research, the international shipping lines NA and AFR and the domestic shipping lines SHD and YHA are used as examples. First, the average time consumptions for the various activities (from) for these four shipping lines are calculated, where m is the quantity of quay crane (QC) handling operation orders; n is the quantity of yard truck (YT) operation orders and o is the quantity of yard crane (YC) operation orders during the handling process for a shipping line.

For instance, the average waiting time of domestic shipping lines YHA and SHD exceeded that of vessels of international shipping lines, which reflects the fact that the terminal is likely to put more focused on the international shipping lines. Vessels of African shipping lines wasted a large amount of YT capacity in waiting for QC () - nine point nine minutes per order. This result is explained by the fact that the congestion of YTs inhibited the trucks' passing through the seaside area. A similar situation arises in operations for shipping lines YHA and SHD, on account of the operator's failing to maintain the pass-through capacity of the seaside areas. The mean time spent by YTs that were transporting containers for lines SHD and YHA in waiting for each order () exceeded seven minutes, and the waiting time of YTs that transported containers for the Africa lines also exceeded five minutes. These facts considerably affect the utilization of YT resource capacity, and the efficiency of the generation of orders for YTs must be improved.

Next, the data concerning resource capacity and the costs for these ten shipping lines from June 2015 to December 2015 at SSICT are analyzed. The value-added cost, non-value-added cost, total cost, and non-value-added cost rate of each shipping line were calculated. Meanwhile, the results will also bring some elevations.

a. Non-value-added cost rate and total cost of shipping lines

The non-value-added cost rate for international shipping line ranges from 42% to 47%, while that for domestic shipping lines ranges from 51% to 59%. The operational performance of international shipping lines is generally better than that of domestic shipping lines. The non-value-added cost rate of shipping line AUS is 46.48%, indicating that it is the most difficult international shipping line to serve. YHA is the most difficult domestic line to serve, as its non-value-added cost accounts for 58.83%. Several factors contribute to this situation: the first is the poor reliability of the vessel schedule since delayed vessel arrivals increase operational difficulty for YT (the non-value-added cost rate of the YT for shipping lines AUS and YHA were 57.32% and 66.00%, respectively); the second is the intensive arrival of vessels, which leads to an operation peak. The container port can negotiate with shipping companies to improve the reliability of vessel schedules.

Terminals may, through better synchronization and resource planning, utilize all available terminal capacities, and thereby increase productivity and profitability. Fernando Alvarez et al. (2010) proposed a simulation model to solve the vessel speed optimization problem. The planner seeks to minimize the weighted sum of fuel costs and delay penalties while generating feasible berth-vessel-equipment assignments, and respecting the contractual agreements of all parties. SSICT should make efforts to minimize both the probability that an operation peak appears and delay penalties. The following approaches can be applied to estimate the time of arrival.

- The ship determines the earliest expected time of arrival. The vessel could set the appropriate speed with synchronization of terminals.
- The terminal determines the most realistic time to start the cargo operations, taking into account the LAYCAN and the speed range of the vessel.
- A proposed estimated arrival time and a queue number are determined and conveyed to the vessel. Slack time may be introduced to improve robustness.

b. Non-value added cost rate and total cost of QCs, YTs, and YCs

Resources are undoubtedly overused to various degrees in the operations of QC, YT, and YC. QC has the highest operating cost owing to its high capacity-cost rate. Improving the operation efficiency and minimizing the waiting time of QCs should be treated as important tasks for the terminal. The wasting of YT resources is the most serious, except in the case of the NA shipping line (whose non-value-added cost rate is 48.14%); the non-value-added cost rates of other shipping lines are greater than 50% (and the non-value-added cost rate of shipping line YHA even reaches approximately 70%). Although the capacity-cost rate of YT is not very high (￥133/hour), the non-value-added cost is high (￥30,787,964.28 for all YTs). Traffic congestion is one of the most important reason leads to the high waiting RHs of YTs. Since the YTs provide the link between the seaside area and the yard area, traffic congestion caused by high rates of loading and unloading containers from a particular block is a significant concern. Clearly, there is huge scope for improvement in YT resource use. The amelioration of YT operations can reduce the operational cost of the container terminal and improve its ability of profit control.

c. Actions based on previous result

The poor utilization of YT resources capacity cannot be solely attributed to the YT scheduling since SIPG's operation strategy affects the YTs, and thus container terminals are inclined to allocate more YTs than needed which waste YT resources. Usually, SIPG allocates YTs to the terminal company, such as SSICT and pays drivers' salaries on a piecework basis for each container and thus it does not care about the details of the operating cost of the YTs, as described above. Accordingly, only the number of containers influences the cost of YTs for a terminal company while the number of YTs does not. To reduce the difficulty of decision-making and protect against uncertainty, the container terminal company will request more than enough YTs from SIPG, resulting in surplus YT resource capacity. The waiting YTs generate unnecessary maintenance and fuel costs which reduces the profit of the SIPG. SIPG should think prudently about how to utilize this surplus resource capacity. One way is to lease out YTs for revenue, such as by renting them to logistics companies that lack transport capability. Limiting the quantity of allocated YTs (without affecting the operating efficiency of the terminal), will necessarily improve the scheduling by the dispatchers, preventing casual scheduling. Compared with supporting the decision-making, the more widespread usage of cost information is in budgeting, which is a management accounting tool that projects or forecasts costs for various levels of production and operation activity. Budgets are important in planning, which set the organization's direction for the budget period. Budgets provide the basis for earnings forecasting that senior executives issue to the market. Under the changing environmental circumstances in container terminals, the fact that capacity consumption increases with the operating quantity could be found, but the inter-relationships between them are still unavailable. Over the past few decades, a nature-inspired computational tool, the Artificial Neural Network (ANN), has been used to model real-world problems because of its immense ability to capture inter-relationships among input and output data pairs that are unknown, nonlinear or too difficult to formulate [29]. In next section, we will illustrate how to adopt ANN with the principles of TDABC to predict the capacity consumption for dealing with the handling operations in Container Terminal TDABC System's budget module.

4.4. Application of artificial neural network to generate cost budget of North West Europe shipping line

4.4.1. Artificial Neural Network (ANN) model

ANN exhibits two characteristics of the brain as primary features, one is the ability to 'learn' and the other is the ability to generalize from limited information. ANN models can recognize trends and patterns, and learn from their interactions with the environment [30]. In most of ANN applications as seen in the literature, the feed forward error back propagation neural network is the most popular and widely used ANN architecture.

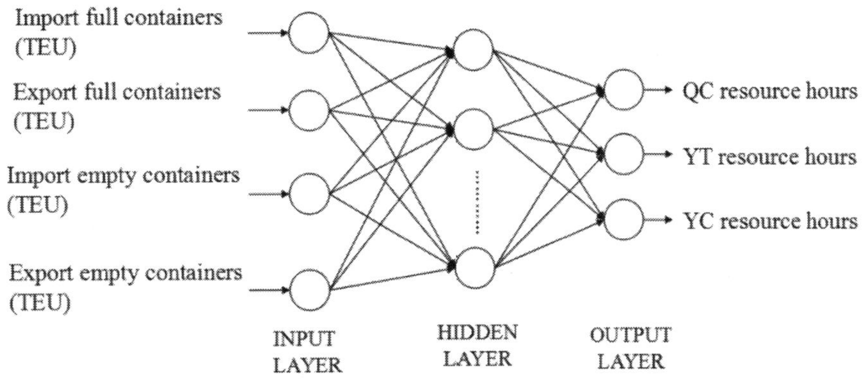

Fig. 7. Network architecture of North West Europe shipping line. (4-20-3)

Fig.7. displays the network model of the North West Europe shipping line. The number of hidden layers of neurons is 20 and the architecture of neural network model is 4-20-3. A learning rate used in the training of the neural networks is 0.1. Ten network models are established for each shipping line and the network parameters (learning rate, number of hidden layers of neurons and number of iterations) are adjusted to optimize the performance of the network models.

4.4.2. Measures of accuracy of models

To observe the modeling performance of ANN, different vessels of North West Europe shipping line are considered. All operational data of North West Europe shipping line are obtained from the SSICT. The data from June 2015 to December 2015 are selected to train the networks and the data from January 2016 to March 2016 are selected to test the performance of the networks. The neural networks adopted to predict the resource hours of QC, YT, and YC are trained with the Back Propagation (BP) algorithm [30].

The ANN model takes 300 iterations to reduce the biases of observed values and predicts values to under 2.5 (Fig.8 (A)). Finally, the BP algorithm takes 3000 iterations to reach the desired performance (Fig.8 (B)).

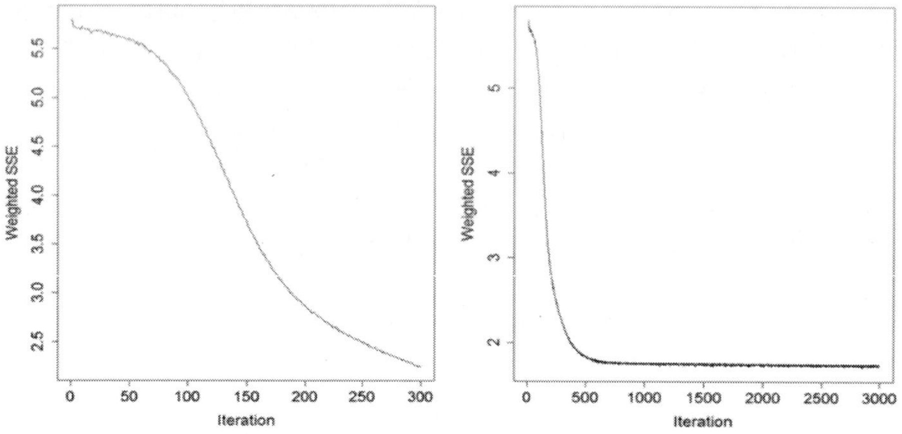

(A) (B)

FIG. 8. TRAINING OF ANN MODEL (NORTH WEST EUROPE SHIPPING LINE)

Fig. 9 (a) and (b) display regression plots of the prediction of training ANN models and the error histogram (normalization), respectively. In Fig. 9(a), the darker region has the more plots than the light region. The regression function is.

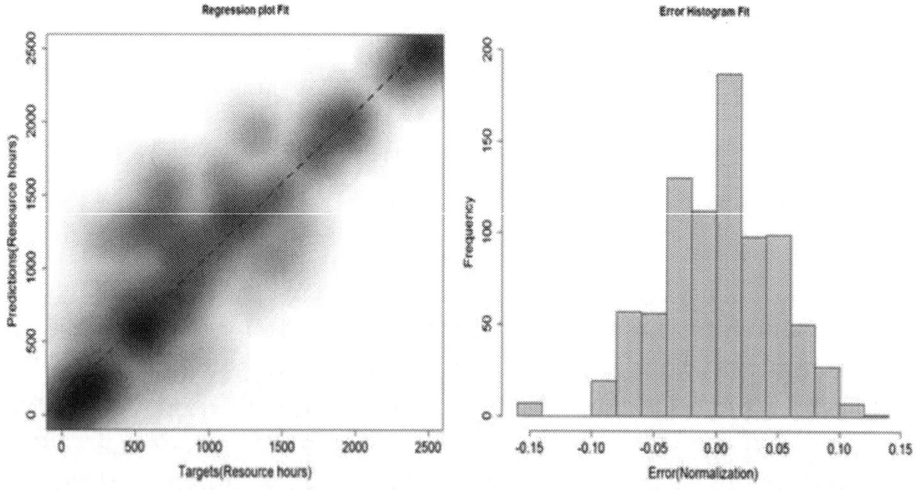

(a) (b)

Fig. 9. Regression plot and error histogram of ANN model (NWE shipping line)

The ANN model generates a satisfactory result, and most of the models also produced relatively favourable results, with some outperformed others. ANNs have the advantage of simplicity over more sophisticated models. Therefore, when the relationship between inputs and outputs is difficult to describe and information is lacking or difficult to obtain, the ANN method provides the most viable option for cost-budgeting.

5. Conclusions and future work

This paper describes how logistics operations in SSICT can be modeled to provide detailed and accurate cost information. In SSICT, QCs, YTs, and YCs perform loading and discharging operations. Indeed, SSICT was facing various forms of complexity in its operations. The TDABC can model this complexity by introducing an appropriate time driver for each activity. The TDABC model also captures the variable property of each cost object by obtaining corresponding activities in the time equations. In addition, data warehouse technology and the rapidly increased supply of analytical package software have brought many enhancements in multidimensional analytical power and the efficiency of accounting processes, which enable the TDABC model to be fed with the "actual" time spent on each activity. In the case study, the operating cost of a vessel was computed and the details of the cost were obtained. Then, the bottlenecks in handling process of the vessel—surplus YT resource capacity, caused by a poor allocation plan or the excess of YTs in the yard area or seaside are identified. The research is then extended from the vessels to the shipping lines. Some important findings were obtained: paying by piece works and YT outsourcing lead to the over-allocation of YTs, which should be changed or abolished. Finally, the ANN model attempts to predict the resource hours of QC, YT, and YC in budget, and its feasibility is verified.

In future, indirect costs such as the managers' salaries, the cost of security, vessel drainage cost and the extra operating costs associated with special containers must also be taken into account to analyze the operating cost based on some certain cost distribution strategy. In this paper, TDABC is only used to model the general loading and discharging process ("Terminal's three-stage material handling system"). Our future work will involve in the loading and discharging of dangerous goods containers (which may not be stored in the yard area and the transfer of the general container between blocks). Other research topics include the number of runs that are required before the TDABC model provides generally accepted time estimates, and the association of dynamic pricing with the TDABC model, which would probably increase the general level of acceptance of the model and improve the satisfaction of customers.

Acknowledgments

We would like to express our gratitude to all the employees at SSICT for their full support in this project. In particular, we thank Mrs. Shen Lian-Hong, IT manager of the SSICT, for her valuable suggestions concerning system implementation and data collection. This project was partially supported by the National Natural Science Foundation of China [71301101] and Shanghai Natural Science Foundation [16ZR1414500].

References

1. UNCTAD, 1999, Technical note: the fourth generation port. UNCTAD Ports Newsletter, 19, 9-12.
2. UNCTAD, 1976, Port Performance Indicators (Geneva: UNCTAD).
3. UNCTAD, 1983, Manual on a Uniform System of Port Statistics and Performance Indicators (Geneva:

UNCTAD).
4. De Monie, G., 1987, Measuring and Evaluating Port Performance and Productivity. Monograph no. 6 on Port Management, (Geneva: UNCTAD).
5. ISL, 1990, General Aspects of Port Management: Port Management Text Book Series Vol. 1 (Bremen, Germany: Institute of Shipping Economics and Logistics (ISL)).
6. Trujillo, L. and Nombela, G., 1999, Privatisation and Regulation of the Seaport Industry. Policy Research Working Paper 2181. Washington, DC: The World Bank.
7. Liu, Z. and Zhuang, J., 1998, Evaluating partial reforms in the Chinese state industrial sector: a stochastic frontier cost function approach. Review of Applied Economics, 12(1), 9-35.
8. Song, D. K, Cullinane, K. and Roe, M. S., 2001, The Productive Efficiency of Container Terminals: An Application to Korea and the UK (Aldershot: Ashgate).
9. Roll Y. and Hayuth, Y., 1993, Port performance comparison applying data envelopment analysis (DEA). Maritime Policy & Management, 20(2), 153-161.
10. Notteboom, T., Coeck, C. and Van Den Broeck, J., 2000, Measuring and explaining the relative efficiency of container terminals by means of Bayesian stochastic frontier models. International Journal of Maritime Economics, 2(2), 83-106.
11. Tongzon, J. L., 1995, Determinants of port performance and efficiency. Transportation Research Part A: Policy and Practice, 29A(3), 245-252.
12. Valentine V. F. and Gray, R., 2001, The measurement of port efficiency using data envelopment analysis, World Conference on Transport Research, Seoul, South Korea, July.
13. Estache, A., Gonzales, M. and Trujillo, L., 2002, Efficiency Gains from Port Reform and the Potential for Yardstick Competition: Lessons from Mexico. World Bank, Washington, http://www.worldbank.org/wbi/regulation/pdfs/portreform_mexico. pdf.
14. Cullinane, K., Song, D. W. and Gray, R., 2002, A stochastic frontier model of the efficiency of major container terminals in Asia: assessing the influence of administrative and ownership structures. Transportation Research, 36(A), 743-762.
15. Coto-Milla¤ n, P., Banos-Pino, J. and Rodriguez-A ´ lvarez, A., 2000, Economic efficiency in Spanish ports: some empirical evidence, Maritime Policy & Management, 27(2), 169-174.
16. Milorad Vidovic and Kap Hwan Kim., 2006. Estimating the cycle time of three-stage material handling systems, Ann Oper Res, 144:181-200.
17. Robert Stahlbock · Stefan Voß, 2008.Operations research at container terminals: a literature update. OR Spectrum (2008) 30:1-52
18. Cooper, R., Kaplan, R.S., 1992. From ABC to ABM. Management Accounting 74 (5), 54-57.
19. Narayanan, V.G., Sarkar, R.G., 2002. The impact of activity-based costing on managerial decisions at Insteel Industries: a field study. Journal of Economics and Management Strategy[J] 11 (2), 257-288.
20. Swenson,D., 1995. The benefits of activity-based cost management to the manufacturing industry.Journal of Management Accounting Research 7, 167-180.
21. Banker, R.D., Bardhan, I.R., Chen, T.Y., 2008. The role of manufacturing practices in mediating the impact of activity-based costing on plant performance. Accounting, Organizations and Society 33 (1), 1-19.
22. Kaplan, R. and Anderson, S., 2004. Time-driven activity-based costing, Harvard Business Review [J], November, pp, 131-8.

23. Lukka, K., Granlund, M., 2002. The fragmented communication structure within the accounting academia: the case of activity-based costing research genres. Accounting, Organizations and Society 27 (1-2), 165-190.
24. Kaplan, R. and Anderson, S., 2007. Time-driven Activity-based Costing, a Simpler and more Powerful Path to Higher Profits, Harvard Business School Press, Boston, MA.
25. S. H. Won, K. H. KiM., 2009. An integrated framework for various operation plans in container terminals. Polish Maritime Research 3(61) Vol 16; pp. 51-61.
26. Nicolaou AI, Bhattacharya S, 2006. Organizational performance effects of ERP systems usage: the impact of post-implementation changes. Int J Account Inf Syst,7:18-35.
27. Nicolaou AI, Bhattacharya S, 2008. Sustainability of ERPs performance outcomes: the role of post-implementation review quality. Int J Account Inf Syst,9:43-60.
28. Pedro Ruivo, Tiago Oliveira and Miguel Neto, 2014. Examine ERP post-implementation stages of use and value: Empirical evidence from Portuguese SMEs. Int J Account Inf Syst,15: 166-184.
29. Archana Sarkar, Prashant Pandey, 2015. River Water Quality Modelling using Artificial Neural Network Technique. Aquatic Procedia 4 (2015) 1070 - 1077
30. Rumelhart, De.E., Hinton, G. E. and Williams, R.J,1986. Learning representation by back propagating errors. Nature, 323(9), 533-536.

HUMAN RELIABILITY STUDY ON CONTROL ROOM OPERATORS AND DECISION MAKERS, PART 1: THE LITERATURE REVIEW FOR HRA METHODS

an Zhan[a, 1], Pandu R. Tadikamalla[b], Jiansha Lu[a], Zhi Pei[a]

a College of Mechanical Engineering,

Zhejiang University of Technology, Hangzhou, 310014, PR China

yuanyan608@hotmail.com, ljs@zjut.edu.cn, peizhi82@163.com

b Katz Graduate School of Business,

University of Pittsburgh, Pittsburgh, PA 15260, United States

pandu@katz.pitt.edu

ABSTRACT

Despite the advances made in developing human reliability analysis (HRA) methods, there has not previously been an attempt to systematically synthesize the knowledge in the span of cognition from perception to action to inform HRA. Automatic control and control room operators supervision are of critical importance in major process industries. The focus is on incorporating operator's operational and cognitive errors in the process reliability analysis. The goal of this work is to develop a hybrid of existing methods of current HRA models and in particular issues related to those methods variabilities and results. Those methods under development are referred to as the Integrated Decision-tree Human Event Analysis System (IDHEAS) and Standardized Plant Analysis Risk-HRA (SPAR-H).

This is the first of two papers that detail the literature review (Part 1) and the results (Part 2). This paper describes the literature review and the high-level architecture used to organize the literature review; the second paper describes the procedural framework.

Keywords: Human reliability analysis (HRA), CREAM, IDHEAS, SPAR-H

1 INTRODUCTION

Nowadays leading organizations in the industry sector acknowledge the fact that human errors and system failures can never be totally eliminated. Technology development has witnessed the increasing reliability and durability of instruments in modern complex systems, while people as the most flexible and the least understandable part of those systems [1], continue to be the main incidents contribution. Human Reliability Analysis (HRA) is the aspect of Probabilistic Risk Assessment (PRA) that is concerned with systematically identifying and analyzing the causes and consequences of human errors, which started to be studied as a part of a Probabilistic Safety Assessment (PSA) in 1960s [2]. More and more researchers in HRA tend to think HRA needs combination of different disciplines such as human factors, social psychology,

---

[1] Corresponding author. E-mail address: yuanyan608@hotmail.com (Y. Zhan).
(Yan Zhan was) on leave from Zhejiang University of Technology as a visiting scholar in Katz Graduate School of Business, University of Pittsburgh, Pittsburgh, PA 15260, United States.

behavioral science and organizational management, and so on [3,4]. HRA can offer a comprehensive description of the contribution on human error analysis and reliability assessment in qualitative and quantitative ways [5]. However, the discussion of independent, coherent and complex behavior can be applied to physical, biological or social systems; the complexity profile of human beings consists of the distribution of scales at which behavior is manifested.

In the last few decades, different HRA methods have been proposed to analyze, predict and minimize human errors in nuclear power plants, and other process industries, such as THERP (Technique for Human Error Rate Prediction), CREAM (Cognitive Reliability and Error Analysis Method), and ASP/SPAR (Accident Sequence Precursor/Standardized Plant Analysis Risk Model) [6-10]. Recent HRA methods emphasize potential cognitive mechanisms and causes of human errors from the perspective of psychology, cognition and neuroscience, which make a far more reasonable and convincing qualitative analysis possible, and provide a fairly vibrant structure to facilitate quantitative analysis. However, qualitative analysis is a subjective bias which inevitably used task analysis and expert participation remains open. Different methods of HRA quantitative methods are far from convincing, suffering lack of sources of data, experts estimate bias, suspension performance shaping factors (PSFs), and so on.

In a typical oil/electric facility, time lack, system complexity and low likelihood of occurrence are the main features of emergency events in the major hazard industries which contribute to human error. Several studies have been completed in order to enhance the quality of Detailed Action Plans (DAPs), by developing a quantitative or semi-quantitative method for evaluating the effectiveness and efficiency of DAPs or emergency operating procedures (EOPs). Reer analyzed the reliability of emergency procedures based on probabilistic and statistical modeling, considering time and organizational factors [11]. Park et al. developed the method for quantifying the level of complexity of each task in an EOP from task complexity point of view [12]. Carvalho et al. proposed methods to evaluate the logical sequence of the tasks provided in an EOP, considering the cognitive concepts of human beings as well as the compatibility of the time required with the available time to perform an action by an operator [13]. Other studies found researches on contextual factors affecting complex decision-making process are either limited or have produced contradictory results, especially studies relating decision's familiarity, magnitude and significance of impact, cognitive diversity, uncertainty aversion, and manager's need for achievement to decision-making processes[14-18]. There is an extensive literature on decision making heuristics [19] and the existence of systematic and robust cognitive biases [20]. Therefore, human error probability (HEP) of complex control room operators and decision makers should be considered in the security and reliability studies.

2 OVERVIEW OF THE LITERATURE REVIEW

A systematic literature review was conducted with the goals to identify psychological and cognitive mechanisms for human performance failure, identify proximate causes of failure (based on psychological literature), and identify factors that influence human performance.

The focus of the literature review was on identifying information that is relevant to the metro industry, which has unique characteristics that separate it from other domains. In particular, metros are operated by crews groups who work together, are highly trained, supervised, and who work in a regulated, procedure-driven environment. Therefore, the literature review focused on individual/team aspect of human performance, performance when working with procedures, and research related to working with highly trained or expert personnel.

The literature review was conducted in two main stages. Stage 1 was the initial, exploratory review of the literature to identify psychological failure mechanisms. The literature search results then underwent an internal peer review to ensure adequate coverage of the relevant psychological literature. The conclusion of this internal review was that the literature review had adequate breadth; it had covered the appropriate areas, but it needed more depth within those areas. Following the internal review, Stage 2 of the literature review commenced, focusing expressly on supporting the existing cognitive HRA.

The literature review resulted in an extensive list of articles, book chapters, books, and other sources, focusing on answering three key questions: What kind of cognitive failures do operators make in metro control rooms Why do operators make failures under various situations How do the performance influencing factors (PIFs) affect the chances of failures In other words, the literature that was reviewed and used in the development of the cognitive framework had to provide information about human tasks in response to control room events, and describe the basic human functions to achieve the tasks.

3 COMPARISON OF CREAM/IDHEAS

This part will compare the analytical process of two cognitive-based methods: CREAM and IDHEAS (Integrated Decision-tree Human Event Analysis System), a state-of-the-art HRA method newly proposed. Then we will compare them among other HRA methods.

CREAM emphasizes the importance of cognitive factors in HRA, which distinguishes it from those earlier HRA methods. There are two versions of CREAM: basic CREAM and extended CREAM. The former is used to analyze operators' control modes that further define the probability interval to human error, which is required for the latter.

On the assumption that human behavior is governed by two basic principles: the cyclical nature of human cognition and the dependence of cognitive processes from context and working environment, the distinction between competence and control is based upon Contextual Control Model (COCOM), in which a classification scheme clearly separates genotypes (causes) and phenotypes (manifestations) [7]. Several aspects of the context are identified as Common Performance Conditions (CPCs). Genotypes are separated out into three categories. Basic phenotypes (error modes) are divided into four sub-groups, as action at the wrong time, type, object and place. Phenotypes and genotypes are classified into general consequents and for each general consequent there are numerous general and specific antecedents. For the purpose of HRA (basic CREAM), the first step is task analysis and one of the nine CPCs analysis is carried out based on a list of operator activities. Basic CREAM method can be used as a screening process to decide whether or not to continue with HRA. The next stage of extended analysis requires a cognitive demands profile to be built. This involves describing each cognitive activity in terms of observation, interpretation, planning and execution (i.e. COCOM functions) and plotting this in graphical form.

In order to address the strengths and limitations of existing HRA methods, under the support of the U.S. Nuclear Regulatory Commission (USNRC), researchers carried out detailed review in psychology, cognition, team performance and other related areas, and developed the latest HRA method IDHEAS[21]. There are two important concepts in IDHEAS: macrocognition and proximate cause. Macrocognition is a high level description of what humans do with their brains, which is comprised of five functions (detecting/noticing, sensemaking/understanding, decision making, action, and team coordination). The concept of proximate cause is developed to explain the cause of the malfunction of a macrocognition function. IDHEAS develops guidance for defining human failure event (HFE) led by crew response tree (CRT), and further constructs a cause-based quantification model. Fourteen Crew Failure Modes (CFMs) are categorized into three phases of response model (plant status assessment, response planning, and execution). et, IDHEAS is currently not mature enough to be followed in actual practices. The comparison would focus on their qualification processes, rather than quantification results.

3.1 Cognitive failures analysis

Basic CREAM links between control modes and task failure probabilities, and HEP range for each control mode, based on expert judgment. Extended CREAM combines basic cognitive function failure probabilities with adjustment factors for CPCs. Failure probabilities for each type of cognitive function and CPC-specific multipliers are based on empirical generic values as well as expert judgments compiled from the literature, maintaining division between logical causes and consequences of human error.

In IDHEAS, crew failure scenarios are informed by and consistent with the work done to identify cognitive failure mechanisms, the consequences of those mechanisms (proximate causes of failure), and the characteristics of those performance influencing factors (PIFs) that enable those mechanisms to result in errors, which sounds more reasonable than CREAM. However, further research should be undertaken to more precisely integrate HRA with contextual factors, influences of main factors on both cognitive and action performance in the process industries. Since there is no apparent dispute between diagnosis and execution, which can be taken as two phase response model as SPAR-H, might well bypass the issues mentioned above.

3.2 Sub-task probability

CREAM methodology relies on a task analysis to identify the opportunities for decision failures and/or execution failures. Because task analyses are relatively resource-intensive, an approach would need to be developed to identify the scenarios (and sub-scenarios) that should be examined in task analyses [22]. IDHEAS is essential to perform an analysis to identify in detail the tasks required of the crew to perform the function whose failure is being addressed, and identification of the contextual factors that may influence whether the crew succeeds or fails in performing the required function.

3.3 Final probability

The application of CREAM for quantification in predictive analyses requires a very moderate level of effort. On the other hand, the quality of the qualitative and quantitative results is highly dependent on a comprehensive set of detailed task analyses. In practice, task analyses will have to be performed for the nominal response to all PSA initiating events and a substantial number of additional analyses will probably have to be performed for the scenario variants with additional hardware/operator failures. In IDHEAS it has been found useful to document the results of this analysis in the form of CRT, which is the result of a procedural task analysis that defines what has to happen for a human action for which an HFE is defined to succeed and provides a graphical representation of the entire task analysis. The dependency between two HFEs or between subtasks of one HFE, events and/or sub-tasks can be sorted by time and space, making dependency model based on time or space possible.

Unfortunately, as mentioned earlier, deficiencies in these task analyses will systematically lead to inaccurate qualitative findings (critical performance factors and scenario conditions) and to under-estimation of the HEPs. Moreover, the CREAM method does not support the screening of tasks or sub-tasks (in order to reduce the effort devoted to task analyses). IDHEAS is so far developing and the quantification model still needs more work, dependency remains as an unsolved problem, and several decision trees still do not have HEP numbers[23].

4 COMPARISONS AMONG OTHER HRA METHODS

Researchers have shown that risky human behaviors can be explained as being economically rational [24-26]. An operator/manager will break security rules if doing so is expected to show positive results in terms of physical or emotional rewards. For example, meeting production goals results in a bonus for the manager and cutting through in the standard operating procedures makes the task easier for the operator. The whole area of behavioral safety has become extremely popular, and resources have made a significant contribution to the migration of behavioral safety ideas from research into practice at companies all over the world [27-30]. Petersen believed system-caused human error derived from management, culture, design, overload and decision, including logical decision, proneness and perception of risk, especially roles of line management and staff[31].

Bell and Holroyd [32] considered 17 methods out of 35 potentially relevant HRA to be of potential use to major hazard directorates, such as CREAM and SPAR-H; developing method like IDHEAS is not included for its unsolved dependency model and Decision Trees (DTs) for internal events[23]. Boring proposed that external PSFs should be focused on, representing those factors external to the individual that affect their performance, such as the Human Machine Interface (HMI), procedures, or the environment, especially when generalizing the PSFs to new interface technologies[33]. Rasmussen and Laumman also suggested PSF level descriptions should be clearly offered when evaluating if the ergonomics/HMI are good, nominal or bad [34].

The required failure rate values could be obtained from available data banks specific to the industry. The most useful data for oil and gas industries are OREDA, IAEA-TECDOC-478 and Guidelines for Process Equipment Reliability Data [35-37]. After studying several HRA techniques, the following seven methods have been reviewed for detail based on five selection criteria: domain, validity, availability of the methodology details, whether or not the method has the HEP data and PSFs. The results of this investigation are given in Table 1.

- Technique for Human Error Rate Prediction (THERP) [6]

- Human Error Assessment and Reduction Technique (HEART) [38]

- A Technique for Human Event Analysis (ATHEANA) [39]

- Cognitive Reliability and Error Analysis Method (CREAM) [7]

- Standardized Plant Analysis Risk-HRA (SPAR-H) [8]

- Nuclear Action Reliability Assessment (NARA) [40]

- Integrated Decision-tree Human Event Analysis System (IDHEAS) [21,23]

Table 1: Comparisons between some HRA methods.

Method	Domain	Publicly available	Include HEP	Validity	Contains PSFs
THERP	Nuclear with wider application	es	es	es	es
HEART	Generic	es	es	es	es
ATHEANA	Nuclear with wider application	es	es	No empirical validation	es
CREAM	Nuclear with wider application	es	Not required	Ongoing	es
SPAR-H	Nuclear other major hazard industries	es	es	es	es
NARA	Nuclear specific	No (A proprietary tool)	es	es	Consists 18 EPCs(Error-Producing Conditions)
IDHEAS	Not clear	Partly available	Not required	Ongoing	Contains PIFs(Performance Influencing Factors)

Among these methods, SPAR-H is utilized in this research for HRA for the following reasons:

- It is one of the newest techniques developed for HRA and its base error rates are calibrated against other HRA methods such as THERP and HEART.
- Time has been regarded as an important factor in the calculation.
- It is validated formally, as a comprehensive view on human probability analysis, i.e. this approach provides for sufficient focus on the details of a task without having to go into in-depth analysis.
- It provides basic data for human error probability. Human performance choices are tabulated based on domain experts' opinions

5 CONCLUSIONS

The authors performed a systematic literature review to identify the cognitive mechanisms underlying human failures. The literature review produced more than just a summarization of the major cognitive models of macrocognition and other specific literature; cognitive mechanisms were identified for each element in the framework. While the reality of human cognition is that there is some conceptual overlap among macrocognitive functions, this paper provides more discussion of the cognitive framework produced from this literature review, which is a tool designed to inform HRA and SPAR-H method development with the PIFs via the mechanisms that show how context can lead to failure.

6 ACKNOWLEDGEMENTS

The authors would like to acknowledge the contributions of Dr. Shiyun Li and the late Prof. James A. Craft to this work. Their inspiration, guidance, and support throughout this effort have been invaluable. The authors are very grateful to the anonymous referees for their insightful and constructive comments and suggestions that have led to an improved version of this paper. The authors are also grateful for the financial support from the National Natural Science Foundation of Zhejiang Province (No. L 14G010003), the National Natural Science Foundation of China (No. 71201144), the Scientific Research Project of Science and Technology Department of Zhejiang Province (No. 2013C33054), and the Scientific Research Startup Foundation of Zhejiang University of Technology (No. 102006429).

REFERENCES

[1] **Fuld, R. B. 1 3.** The fiction of function allocation, *Ergonomics in Design: The Quarterly of Human Factors Applications*, 1(1), pp 20-24.

[2] **Nuclear Regulatory Commission. 1 5.** Reactor safety study. An assessment of accident risks in US commercial nuclear power plants. Appendix XI. Analysis of comments on the draft WASH-1400 report [R]. Nuclear Regulatory Commission, Washington, DC (USA),

[3] **LaSala K P. 1 8.** Human performance reliability: a historical perspective [J]. *IEEE Transactions on Reliability*, 47(3), pp 365-371.

[4] **He X H, Huang X R. 200 .** *Human reliability analysis in industrial systems: Theory, methods and practice* [M], Beijing: Tsinghua Press

[5] **Boring R L. 200 .** Dynamic human reliability analysis: Benefits and challenges of simulating human performance [J]. *Risk, Reliability and Societal Safety*, 2, pp 1043-1049.

[6] **Swain A D, Guttmann H E. 1 83.** Handbook of human reliability analysis with emphasis on nuclear power plant applications, Final Report, NUREG/CR-1278, Washington, DC: US Nuclear Regulatory Commission

[7] **Hollnagel E. 1 8.** *Cognitive reliability and error analysis method (CREAM)* [M]. UK Oxford: Elsevier Science Ltd.

[8] **Gertman D, Blackman H, Marble , Byers and Smith C. 2004.** The SPAR-H Human Reliability Analysis Method (NUREG/CR-6883) [R], Washington D.C: US Nuclear Regulatory Commission

[9] **Chang , Lois E. 2012.** Overview of the NRC's HRA data program and current activities [C]//Proceedings of 11th probabilistic safety assessment and management/European safety and reliability, pp 25-29.

[10] **Groth K M, Smith C L, Swiler L P. 2014.** A Bayesian method for using simulator data to enhance human error probabilities assigned by existing HRA methods [J]. *Reliability Engineering & System Safety*, 128, pp 32-40.

[11] **Reer B. 1 4.** A probabilistic method for analyzing the reliability effect of time and organizational factors [J]. *European journal of operational research*, 75(3), pp 521-539.

[12] **Park , ung W. 200 .** A study on the development of a task complexity measure for emergency operating procedures of nuclear power plants [J]. *Reliability Engineering & System Safety*, 92(8), pp 1102-1116.

[13] **Carvalho P V R, de Oliveira M V, dos Santos I L. 200 .** A computerized tool to evaluate the cognitive compatibility of the emergency operational procedures task flow [J]. *Progress in Nuclear Energy*, 51(3), pp 409-419.

[14] **Cheng M M, Luckett P F, Schulz A K D. 2003.** The effects of cognitive style diversity on decision-making dyads, pp an empirical analysis in the context of a complex task [J]. *Behavioral research in accounting*, 15(1), pp 39-62.

[15] **French S, Bedford T, Pollard S T, et al. 2011.** Human reliability analysis, pp A critique and review for managers [J]. *Safety science*, 49(6), pp 753-763.

[16] **Nooraie M. 2012.** Factors influencing strategic decision-making processes [J]. *International Journal of Academic Research in Business and Social Sciences*, 2(7), pp 405-429.

[17] **Zhan Y, Lu S, Li S Y. 2013.** A hybrid GA-TS algorithm for optimizing networked manufacturing resources configuration [J]. *Applied Mathematics & Information Sciences*, 7(5), pp 2045-2053.

[18] **Borgonovo E, Marinacci M. 2015.** Decision analysis under ambiguity [J]. *European Journal of*

Operational Research, 244(3), pp 823-836.

[19] **French S, Maule , Papamichail N, 200 .** *Decision behaviour, analysis and support* [M]. Cambridge University Press,

[20] **Bazerman M H, Moore D A. 2012.** *Judgment in Managerial Decision Making* [M], 8th. Wiley Sons,

[21] **Whaley A M, Hendrickson S M L, Boring R L, et al. 2012.** Bridging Human Reliability Analysis and Psychology, Part 2: A Cognitive Framework to Support HRA [C]//Proceedings of the 11th International Conference on Probabilistic Safety Assessment and Management (PSAM11), Helsinki, Finland.

[22] **He X, Wang Y, Shen Z, et al. 2008.** A simplified CREAM prospective quantification process and its application [J]. *Reliability Engineering & System Safety*, 93(2), pp 298-306.

[23] **. Xing.** SRM-061020 on HRA methods: Progress and status in 2013 [EB/OL]. http://pbadupws.nrc.gov/docs/ML1400/ML14009A355.pdf, 2013-12-15/2015-01-14.

[24] **Battmann W, Klumb P. 1 3.** Behavioural economics and compliance with safety regulations [J]. *Safety science*, 16(1), pp 35-46.

[25] **Reason , Parker D, Lawton R. 1 8.** Organizational controls and safety: The varieties of rule-related behavior [J]. *Journal of occupational and organizational psychology*, 71, pp 289.

[26] **Mosleh A, Chang Y H. 2004.** Model-based human reliability analysis: prospects and requirements [J]. *Reliability Engineering & System Safety*, 83(2), pp 241-253.

[27] **Krause T R. 1 6.** *The behavior-based safety process: Managing involvement for an injury-free culture* [M]. 2nd ed. John Wiley Sons.

[28] **Geller E S. 2000.** *The psychology of safety handbook* [M]. 2nd ed. Lewis Publishers Inc

[29] **Zhan Y, Lu , Li S. 2012.** Research on Organizational Culture and Resources Integration Model Based on MICK-4FI [J]. *Sensors & Transducers*, 16, pp 295-303.

[30] **Zhan Y, Lu , Solvang W D. 2012.** An Empirical Study on the Impact of Cultural Types on Resources Integration Model [J]. *Journal of Service Science and Management*, 5(4), pp 331-338.

[31] **Petersen D. 1 6.** *Human-error reduction and safety management* [J]. 3rd ed. John Wiley Sons.

[32] **Bell , Holroyd . 200 .** Review of human reliability assessment methods [J]. *Health and Safety Laboratory*, United Kingdom

[33] **Boring R L. 2014.** Human Reliability Analysis for Digital Human-Machine Interfaces: A Wish List for Future Research [C] //Proceedings of the Probabilistic Safety Assessment and Management (PSAM 12) Conference.

[34] **Rasmussen M, Laumman K. 2014.** The suitability of the SPAR-H Ergonomics/HMI PSF in a computerized control room in the petroleum industry [C]//Proceedings of the Probabilistic Safety Assessment and Management (PSAM 12) Conference.

[35] **International Atomic Energy Agency. 1 88.** Component Reliability Data for use in Probabilistic Safety Assessment [J]. IAEA TECDOC 478.

[36] **American Institute of Chemical Engineers Center for Chemical Process Safety (CCPS). 1 8 .** Guidelines for Process Equipment Reliability Data (PERD) with Data Tables. Wiley, John Sons

[37] **OREDA-Offshore Reliability Data Handbook** [M], 4th Ed, Dnv Technica, 2002

[38] **Williams C. 1 85.** HEART A proposed method for achieving high reliability in process operation by means of human factors engineering technology [C]//Proceedings of a Symposium on the Achievement of Reliability in Operating Plant, Safety and Reliability Society. 16.

[39] **Cooper S E, Ramey-Smith A M, Wreathall , Parry G. W, Bley D C, Luckas W , Taylor H, Barriere M T. 1 6.** A Technique for Human Error Analysis (ATHEANA) [R], NUREG/CR-6350. USNRC, Washington D.C.

[40] **Kirwan B, Gibson H, Kennedy R, et al. 2005.** Nuclear action reliability assessment (NARA): a data-based HRA tool [J]. *Safety & Reliability*, 25(2), pp 38-45.

HUMAN RELIABILITY STUDY ON CONTROL ROOM OPERATORS AND DECISION MAKERS, PART 2: DOOR OPERATION OF GMC USING THE SPAR-H METHOD

an Zhan [a, 1], Pandu R. Tadikamalla [b], James A. Craft [b], Jiansha Lu [a], Jijun uan [c], Zhi Pei [a]

a College of Mechanical Engineering,

Zhejiang University of Technology, Hangzhou, 310014, PR China

yuanyan608@hotmail.com, ljs@zjut.edu.cn, peizhi82@163.com

b Katz Graduate School of Business,

University of Pittsburgh, Pittsburgh, PA 15260, United States

pandu@katz.pitt.edu, craft@katz.pitt.edu

c School of Finance,

Guangdong University of Finance Economics, Guangzhou, 510320, PR China

yuanjijun113@163.com

ABSTRACT

This is the second of two papers that discuss the literature review to develop a hybrid human reliability analysis (HRA) method. This review was conducted with the goal of strengthening the technical basis within psychology, cognitive science and human factors for the hybrid HRA method being proposed. The first paper provides an overview of the literature review approach and high-level structure, whereas this paper presents the results of the review.

Resulting from a comparison of some current HRA methods, the human reliability analysis method Standardized Plant Analysis Risk-HRA (SPAR-H) is utilized to evaluate human error probabilities (HEPs) of action and diagnosis in the Door Operation of Guangzhou Metro Corporation (GMC) Line 1. The results show that raising the experience/training level of staff will enhance the human reliability of door operation. Taking diagnosis complexity and situation awareness into consideration, different decision makers have different impacts on the human reliability. It is also argued that those should be evaluated and takes reference points on ergonomics/Human Machine Interface (HMI) quality from the standards used in the industry, and existing ergonomics/HMI evaluation tools on usability.

Keywords: Human reliability analysis (HRA); ualitative screening; SPAR-H; Human error probability (HEP)

1 INTRODUCTION

Probabilistic Risk Assessment (PRA) helps identify potential areas of significant risk and suggest improvements, and quantify overall risk from plant operation, but not accounting for reducing the risk of human operations. HRA is an important tool when dealing with human factors in control

[1] Corresponding author. E-mail address: yuanyan608@hotmail.com (Y. Zhan).
(Yan Zhan was) on leave from Zhejiang University of Technology as a visiting scholar in Katz Graduate School of Business, University of Pittsburgh, Pittsburgh, PA 15260, United States.

room design or modernizations, especially the reliability of complex systems. In general, the human contribution to overall performance of obstacles and improvement possibilities is at least as important as that of hardware reliability. To get accurate and precise measure of system reliability, human error should be taken into consideration.

2 HUMAN RELIABILITY ANALYSIS

Various classification models of human error behavior have been proposed based on different theoretical foundations, such as Swain's error of commission and omission classification, Rouse's concept model of operating procedure error [1], and Rasmussen's skill-based, rule-based, and knowledge-based behaviors [2].

2.1 Classification of human action

To perform an HRA, one has to perform a number of tasks in the right order. During normal situations, the reliability of safety systems remains within the acceptable range, which is subjected to maintenance and test. If manual human action (HA) error in the maintenance process, which may lead to the unavailability of the systems, is not discovered, the systems will remain in unavailable state until the occurrences of abnormal situation, i.e. an initiating event (IE). The component/system unavailability due to human error prior to the IE is known as Type A HA. Mistake or error in some manipulations, performed manually by operators, or in coincidence with the unavailability of other component/system may lead to an IE, which is known as Type B HA. Contextually, operator performances are degraded and the probability that operator will perform the right action would be less. Error in operator action in this situation is known as Type C HA (as dynamic HA or post-initiator type human error).

2.2 Identification of human action

In order to carefully search for the potential Type A and B HAs during normal situation, dedicated procedures and other documents are reviewed, which address the maintenance steps for the major system/component that are important to safety. There is a general guide for operators in the main control room (MCR), considering that the operators have experienced well about the general component and system. To get the possible dynamic HAs for certain accident sequences, event reporting documents that briefly describe events occurred in the past, are reviewed and panel discussion which includes the HRA analysts, accident sequences analysts, systems analysts, and control operators are iteratively held. The operators offer their possible response actions to the postulated accident sequences and scenarios. For this approach, the qualitative screening phase is not performed on the Type C HAs.

2.3 Qualitative screening

Following identification of HAs, qualitative screening is carried out on these HAs. Each HA is analyzed for its steps, possible human error which could reduce component/system reliability and induce possible consequences or initiating event (IE), type of HA and applicability to the current PSA scope. The screening phase provides a reduced number of HAs to be considered (Fig. 1).

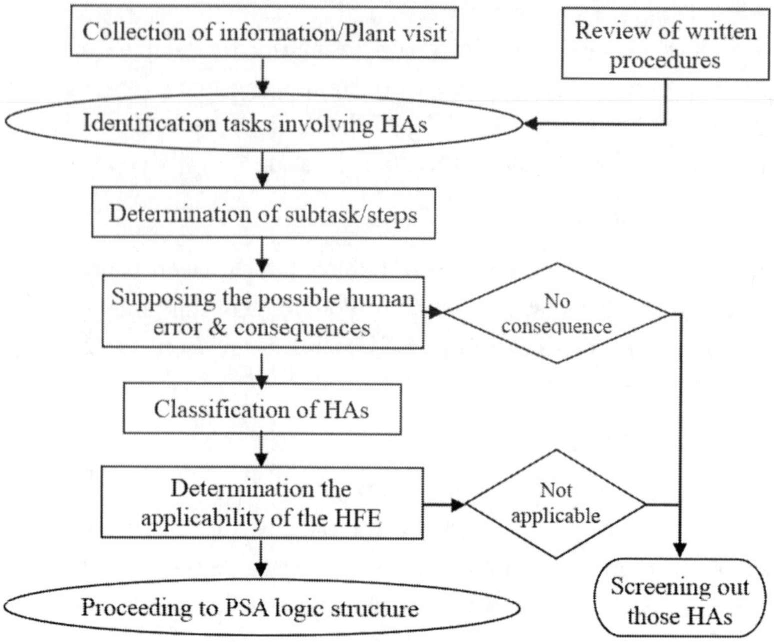

Fig. 1 Procedural framework of qualitative screening phase

Through the qualitative screening phase, the significance of the identified HAs to the safety issues of interest is recognized and then the effect of these HAs to the safety is represented by the incorporation of the HFEs into the PSA logic structure.

2.4 SPAR-H method

More recently, the U.S. Nuclear Regulatory Commission (NRC) has published the Standardized Plant Analysis Risk-HRA (SPAR-H) method to take into account of the potential for human error, which has been applied to more than seventy US nuclear power plants as a further simplification and generalization of those approaches[3].

SPAR-H tried to address the practical need the NRC had, which was to do risk assessments in the accident precursor program and to see events from the point of view of the implementation process. SPAR-H method intended to make the characterization of human performance more realistic and better representation of uncertainty and dependency assignment for HEP calculations. When using SPAR-H, the analysis makes decisions regarding the assignment of a particular post- or pre-initiator activity to either diagnosis or action. The method is based on eight performance shaping factors (PSFs), which are available time, stress/stressors, complexity, experience/training, procedures, ergonomics/Human Machine Interface (HMI), fitness for duty, and work processes. They cover most of the contributors to human error. Within different task categories, each PSF features a list of levels and associated multipliers (Table 1).

3 CASE STUDY

In this section, this study collected information about management of Door Operation of GMC (Guangzhou Metro Corporation) Line 1. Through statistical analysis of failure data on the subway doors systems, combined with recent maintenance records of GMC Line 1, a number of key factors causing door operation failures were confirmed. Among those factors, this paper focused on door keeping open or no locking that had higher probability of occurrence and more serious consequences, compared with signals showing wrongly , and foreign body blocking , frequently caused by passengers.

Table 1 SPAR-H PSF levels and multipliers for diagnosis/action portion

PSFs	PSF levels	Multipliers for diagnosis	Multipliers for action
Available time	Inadequate time	P (failure) = 1.0	P (failure) = 1.0
	Barely adequate time(2/3 average time)	10	10
	Nominal time	1	1
	Extra time(1-2 nominal time & >30 mins①)	0.1	0.1
	Expansive time(>2 nominal time & >30 mins)	0.1-0.01③	0.01④
	Insufficient information	1	1
Stress/ stressors	Extreme(disruptive stress)	5	5
	High(higher than nominal level)	2	2
	Nominal	1	1
	Insufficient information	1	1
Complexity	Highly complex(very difficult to perform)	5	5
	Moderately complex(somewhat difficult to perform)	2	2
	Nominal(not difficult to perform)	1	1
	Obvious diagnosis	0.1	
	Insufficient information	1	1
Experience/ training	Low(< 6 mos②)	10	3
	Nominal(> 6 mos)	1	1
	High(extensive experience)	0.5	0.5
	Insufficient information	1	1
Procedures	Not available	50	50
	Incomplete	20	20
	Available, but poor	5	5
	Nominal	1	1
	Diagnostic/symptom oriented(function-oriented)	0.5	
	Insufficient information	1	1
Ergonomics / HMI	Missing/misleading	50	50
	Poor	10	10
	Nominal	1	1
	Good	0.5	0.5
	Insufficient information	1	1
Fitness for duty	Unfit	P (failure) = 1.0	P (failure) = 1.0
	Degraded fitness	5	5
	Nominal	1	1
	Insufficient information	1	1
Work processes	Poor	2	5
	Normal	1	1
	Good	0.8	0.5
	Insufficient information	1	1

Notes:

①mins: minutes *②mos: months*

③Analyst's choice, depending on complexity of diagnosis, such as available help and cues.

④Analyst's choice, depending on complexity, work environment and ease of checking.

Based on those results, the performance of subway doors systems in similar accidents was evaluated. Strictly speaking the PSF levels are scalars and the mean HEP is a probability. It has proved difficult to account for the influence of multiple shaping factors that had different levels of influence without imposing a high degree of expert consensus judgment of the HRA process.

SPAR-H makes the assignment of human error probability a more repeatable function and less reliant on the analyst performing the HRA. SPAR-H has clear criteria for determining the multipliers on most PSFs excluding PSFs of experience/training and ergonomics/HMI which need experts to make judgments. The requirements of experts in this study include a professional background in human factors and operating knowledge of Communication-Based Train Control (CBTC) System [4].

3.1 PSF overview for door operation of GMC Line 1

This paper used SPAR-H to evaluate HEPs of diagnosis and action for door operation of GMC Line 1 management model. The features of door operation of GMC Line 1 were gathered and categorized for each PSF as in Table 2.

Table 2 The summary of each PSF for door operation of GMC Line 1

PSFs	Situations
Available time	In most cases, there is sufficient time.
Stress/stressors	Such problem could endanger public security. Facing the problem of public security or property loss.
Complexity	There is little ambiguity and not difficult to perform.
Experience/training	OHSMS (Occupation Health and Safety Management System) methods in late 1980s and has been use in GMC since 2004.
Procedures	Following OHSMS and ERP (Emergency Response Plan).
Ergonomics/ HMI	The interface is adequate.
Fitness for duty	The fitness for duty is corporation and team specific. This analysis was determined from a general model and therefore nominal is an appropriate choice.
Work processes	Determined on the basis of analyst evaluation of corporation specific information.

3.2 Calculating the composite PSF

The composite PSF is calculated as the product of the analysts ratings of all PSFs provided in SPAR-H method. The ratings are multiplied by one another, despite whether the PSF influence is positive or negative. The adjustment factor covers a great variety of combinations where the combined PSF influence would be a probability greater than one, if no adjustments were applied.

The final HEP values are calculated by multiplying the nominal HEP by the weighing factors derived from SPAR-H tables. In the formula (1), nominal human error probability (NHEP) is 1.0E-2 on diagnosis analysis and 1.0E-3 on action analysis.

$$HEP = NHEP \cdot PSF_{composite} \tag{1}$$

When three or more negative multiple PSFs are present, SPAR-H offer a composite PSF score used in conjunction with the adjustment factor, which is a simple modification of the nominal human error probability that meets mathematical requirements(2).

$$HEP = \frac{NHEP \cdot PSF_{composite}}{NHEP \cdot (PSF_{composite} - 1) + 1} \tag{2}$$

3.3 Quantification of each task

Based on what is stated in SPAR-H method, the procedure of task quantification is given as follows:

Step 1. Rate PSFs for the diagnosis (cognitive processing) portion of the task; identify the salient performance drivers, both positive and negative, if any, making use of the "obvious diagnosis" category in order to reduce conservatisms used in employing the nominal rate.

Step 2. Calculate the diagnosis failure probability.

Step 3. Calculate the adjustment factor and modify HEP if there are three or more negative PSF influences.

Step 4. Record the final diagnosis HEP.

Step 5. Evaluate PSFs for the action (execution) portion of the task, if any, the diagnosis and action elements can be modeled separately on the basic event level.

Step 6. Calculate the action failure probability.

Step 7. Calculate the adjustment factor and modify HEP if there are three or more negative PSF influences.

Step 8. Record the final action HEP.

Step 9. Calculate the task failure probability without formal dependence by adding the diagnosis failure probability to the action failure probability.

In the event that a combined Diagnosis and Action HEP approaches or exceeds 1.0, the equation (3) should be applied:

$$HEP(Diagnosis+Action) = HEP(Diagnosis) + HEP(Action) - [HEP(Diagnosis) \times HEP(Action)]$$

Step 10. Perform dependency analysis, aiding the analyst in mapping tasks, PSFs, and other aspects of context to the appropriate dependency level.

3.4 Examination the performance of door operation

To examine the performance of door operation, this paper uses SPAR-H to evaluate HEPs of action and diagnosis in door keeping open or no locking procedures. The scenario evaluated in this study makes use of a risk model based on door keeping open or no locking event analysis. The model assumes that the door of metro kept open or no locking, several tasks such as Activation of emergency braking, Implementation delays safety, Closure line should be performed. A door keeping open or no locking event tree is presented as in Fig. 2.

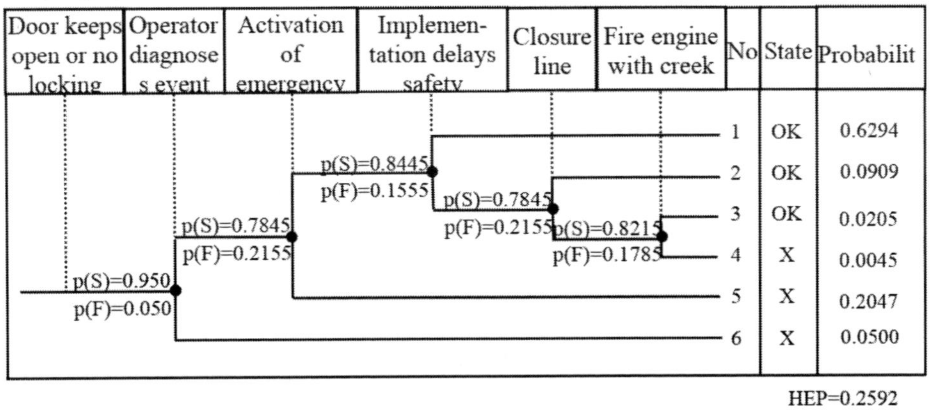

Fig. 2 The HEP in door keeping open or no locking event tree taking vice president as decision maker with low training level

In this model, the end state OK is defined as one in which emergency braking has been activated and delays safety (closure line, or fire engine with creek) was implemented to prevent metro accident. This study evaluated the effect of different levels of experience/training of staffs and different decision makers (the vice president or the control center director). Three experience/training levels in this study are defined as follows:

Low not provide adequate practice required to adequately perform the required tasks.

Normal more than 6 months of formal training to ensure that individuals are skilled in routine operations and have been exposed to abnormal situations.

High this level of training provides operators with comprehensive knowledge and practice in a broad range of possible scenarios.

With this event analysis, the HEP determined by the SPAR-H method can indicate the human error contribution to the event frequency.

4 RESULTS AND DISCUSSION

4.1 HEP Values in SPAR-H

It is widely accepted that human functions follow the lognormal probability distribution. Several HRA techniques utilize this assumption for calculating their HEP. However, SPAR-H uses Constrained Non-Informative (CNI) distribution to calculate the uncertainty in the level of HEP estimation, which is a form of beta distribution with the mean equal to HEP. According to the SPAR-H manual, expert judgments were performed to determine whether the task contain a significant amount of diagnosis activity, to discover the appropriate error and assign the correct level of influence (i.e., multiplier for diagnosis or action of each PSF). Hence, the PSF levels are: stress/stressors is high, ergonomics/HMI is good, procedures are available, and other factors are normal. The weight multipliers for PSFs are presented in Table 3. The adjustment factor is only calculated when three or more negative PSFs are present, the result of the diagnosis failure probability is calculated using formula (1). The final diagnosis HEP of this task is: $\mathbf{1.0}\boldsymbol{E}-\mathbf{2}\times\mathbf{1}\times\mathbf{2}\times\mathbf{1}\times\mathbf{1}\times\mathbf{5}\times\mathbf{0.5}\times\mathbf{1}\times\mathbf{1}=\mathbf{0.05}$.

Table 3. Evaluate PSFs for diagnosis portion of the operator diagnoses event task

PSFs	PSF Levels	Multiplier	Note
Available time	Nominal time	1	Based on operating characteristics, it is assumed that the crew had sufficient time.
Stress/stressors	High	2	Such problem could endanger public security or property loss. It is assumed that stress will be higher than normal.
Complexity	Normal	1	There is little ambiguity and not difficult to perform.
Experience/training	Normal	1	Lots of OHS training and preparation is provided.
Procedures	Available	5	Procedure is available but there is a lack of adequate exposure to consistent instruction.
Ergonomics/ HMI	Good	0.5	The interface is adequate and learnable but not easy to use.
Fitness for duty	Normal	1	The fitness for duty is corporation and team specific. This analysis was determined from a generic model and therefore nominal is an appropriate choice.
Work processes	Normal	1	Determined on the basis of analyst evaluation of corporation specific information.

4.2 Experience/training level

After calculation of the probability of each task in the door keeping open or no locking procedures determined by the SPAR-H method, the net probability of state X is the HEP of the procedures. An example of HEP calculation in door keeping open or no locking procedures is presented in Fig. 2. The results of event analysis with different situations are presented in Table 4, Fig. 3 and Fig. 4.

Table 4 The probability of each subsequent HEP & human reliability with different levels of experience/training of staffs and decision maker

Experience/training level		Low		Normal		High	
Decision maker		The vice president	The control center director	The vice president	The control center director	The vice president	The control center director
No	State						
1	OK	0.6294	0.6598	0.7828	0.8195	0.8469	0.8678
2	OK	0.0909	0.0953	0.0157	0.0165	0.0101	0.0090
3	OK	0.0205	0.0215	0.0035	0.0038	0.0023	0.0020
4	X	0.0045	0.0047	0.0010	0.0008	0.0005	0.0004
5	X	0.2047	0.1687	0.1470	0.1094	0.0902	0.0708
6	X	0.0500	0.0500	0.0500	0.0500	0.0500	0.0500
HEP		0.2592	0.2234	0.1980	0.1602	0.1407	0.1212
Human reliability		0.7408	0.7766	0.8020	0.8398	0.8593	0.8788

Fig. 3 show that raising the experience/training level of staffs from low to normal and from normal to high would increase the human reliability about 6 and 4 , respectively. The PSF in SPAR-H captures the years of experience of the operator or crew, whether the operator/crew has been trained on this type of accident, the amount of time since training, and whether the scenario is novel or unique. These measures for experience/training can be easily extracted from data at each corporation or group, which would be part of the employee records or typically gathered for use in evaluating the training programs through training coverage and frequency. Lots of OHS training and preparation is provided in Guangzhou Metro Line 1. In general, a licensed operator has at least reached the normal level of training, considering a repetitive task that an operator performs in the normal course of events.

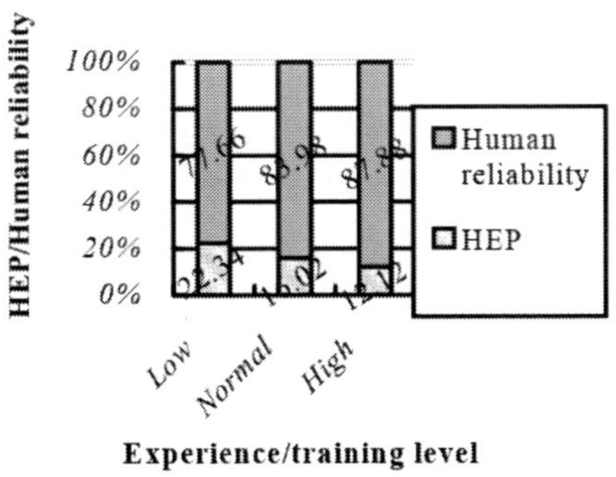

Fig. 3 The HEP & human reliability with different levels of experience/training of staffs.

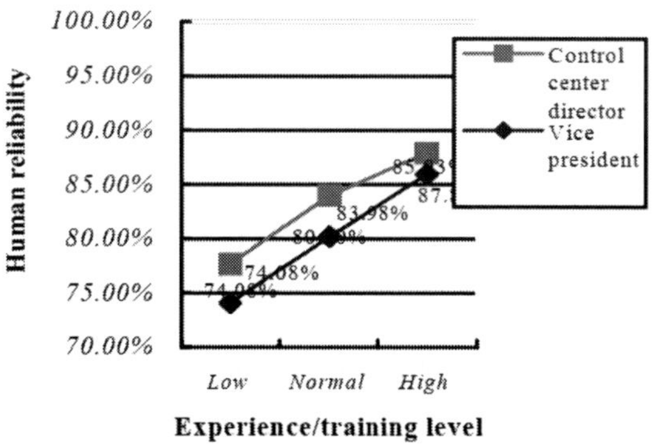

Fig. 4. The human reliability with different levels of experience/training of staffs & decision maker.

Another consideration is whether the needed experience/training is explicit or tacit in nature (i.e., whether or not the crew/individual has been involved in a teaching or learning scenario, in either a traditional training or an operational setting). Specific examples where experience/training might be deficient are those learning experience/training are expressed in the ways that humans live and deal with ambiguity and uncertainty originating in the environment, or self generated for innovative purposes, and those teaching experience/training needs to be transferred to the next group/generation, characterized by their certainty or explicit know-ability, which can be observed, documented and measured in a fairly straightforward manner. Such situation should be involved in future research. For example, we may use age as a supporting surrogate for experience/training. Using data collected from our national labor statistics, according to the age distribution for the electric power industry workforce, we may consider the logical possibilities for persons aged 18 25 to low experience/training", persons aged 26 40 to nominal experience/training and persons over age 41 to high experience/training .

4.3 Different decision makers

Another observation is the impact caused by different decision makers performing door open operation. As illustrated in Fig. 4, the decision by the control center director would increase human reliability 1 3 over the decision by the vice president. Based on the above-mentioned analysis, the major differences between the control center director and the vice president are the diagnosis complexity and situation awareness. Environmental and situational factors are contributors to the success or failure of human performance via their impact upon perception, processing, and response [5]. For example, system relationships may be relatively complex, and the configuration and flow of events are not well recognized. The operator may not be able to recognize the true nature of the problem and, thus, is challenged to determine a solution from memory. In this situation, it would be more difficult to determine what was occurring and to take the correct course of action.

The control center director's scope of review is limited to the corporation, whose understanding of the situation can influence his decision. Considering how difficult the task is to perform in the given context, the vice president's diagnosis will take much more mental effort required, such as performing mental calculations, memory requirements, understanding the underlying model of how the system works, and relying on knowledge instead of training or practice[6]. As for situation awareness, defined as the perception of the elements in the environment within a volume of time and space, the comprehension of their meaning, and the projection of their status in the near future [7], in many cases, an on-site manager has more capability to take single-step retrieval of appropriate actions than an off-site one, after rapid retrieval of comprehension relevant to the recognized situation.

4.4 Ergonomics/HMI PSF level in the control room

The human reliability quantification method used in this study was applied in a case study, which was carried out in cooperation with GMC. Due to the specificity of the organization and the nature of the tasks performed under accident conditions (multiple actors, complex chain of command, and complex diagnosis), SPAR-H methodology was chosen and applied to accident management [8]. It provided simplified assessment of the failure mode, action vs. diagnosis failures and led to a simplified but conservative HEP assessment via the PSF rating. However current SPAR-H does offer guidance on the ergonomics/HMI PSF, there are difficulties in reducing everything related to ergonomics/HMI into the choice of just one PSF level. For example, analyst may experience where he evaluated the general HMI in the control room to be good, while one aspect of the information presented to the operator that was in the specific scenario he was investigating could have been clearer. The analyst may find it challenging to choose between good ergonomics/HMI, based on the general evaluation and poor one, based on the small aspect of the HMI that could have conveyed the information clearer. Such situation should be involved in future research and development of HRA.

Another issue was on reference points. When evaluating if the ergonomics/HMI are good, nominal or bad, a basis for comparison (a frame of reference) is required. Today large physical distances between indicators are rare but as the operators have a massive amount of information available, the HMI quality will greatly influence how much of the relevant information that reaches the operator. This could be a significant HEP contributor in cases where task relevant information is poorly conveyed through monitor interface or HMI. Further work is currently being carried out by authors in order to guide on what should be evaluated and take reference points on ergonomics/HMI quality from the standards used in the industry, and existing ergonomics/HMI evaluation tools on usability.

5 CONCLUSIONS

It is especially important for HRA models to be capable of leveraging data from specific simulator, experimental data collection activities, and documentation when possible, including event and near-miss databases and procedures [9]. Adding the data enhances the technical basis of the SPAR-H HEPs, and this enhanced technical basis increases the credibility of the SPAR-H method. With the intent of forcing consistency, SPAR-H requires that the same set of PSFs always be evaluated for every HEP under consideration. However, a qualitative screening process is needed to identify which PSFs might be important for a particular human error evaluation, if any PSFs that are not defined will dominate. In this study, the qualitative aspects of the operator reliability have been evaluated by performing the procedural framework of qualitative screening phase.

Current practices in and uses of HRA are insufficient for the complexities of modern society. The approach taken in this study generates reasonable results despite the fact that there was no directly applicable database. Uncertainty was expressed through probability distributions that were accepted by the group. Experts from many different organizations and backgrounds work together and reach consensus that different types and/or degrees of uncertainty exist between the nominal condition and having insufficient information. The results/discussion sections learned from performing this example analysis were documented and demonstrate SPAR-H, as an HRA method, can be used to evaluate HEP and human reliability with limited information in future HRA studies, both related to metro and to other industries.

6 ACKNOWLEDGMENT

The authors are very grateful to the anonymous referees for their insightful and constructive comments and suggestions that have led to an improved version of this paper. The authors are also grateful for the financial support from the National Natural Science Foundation of Zhejiang Province (No. L 14G010003), the National Natural Science Foundation of China (Nos. 71201144 and 71272084), the Scientific Research Project of Science and Technology Department of Zhejiang Province (No. 2013C33054), and the Scientific Research Startup Foundation of Zhejiang University of Technology (No. 102006429).

REFERENCES

[1] **Rouse W B, Rouse S H. 1 83.** Analysis and classification of human error [J]. Systems, Man and Cybernetics, IEEE Transactions on, (4), pp 539-549.

[2] **Rasmussen . 1 83.** Skills, rules, and knowledge; signals, signs, and symbols, and other distinctions in human performance models [J]. Systems, Man and Cybernetics, IEEE Transactions on, (3), pp 257-266

[3] **Kamyab S, Nematollahi M. 2012.** Performance evaluating of the AP1000 passive safety systems for mitigation of small break loss of coolant accident using risk assessment tool-II software [J]. Nuclear Engineering and Design, 253, pp 32-40.

[4] **Li G. 2013.** Research on the maintenance strategy and fault diagnosis of the on-board equipments in metro train based on RAMS [D]. Beijing Jiaotong University doctoral dissertation.

[5] **D s M, Backstr m T, Sundstr m-Frisk C. 2004.** Human actions and errors in risk handling an empirically grounded discussion of cognitive action-regulation levels [J]. Safety science, 42(3), pp 185-204.

[6] **Boring R L, Hendrickson S M L, Forester A, et al. 2010.** Issues in benchmarking human reliability analysis methods: A literature review[J]. Reliability Engineering System Safety, 95(6): 591-605.

[7] **Endsley M R. 1 88.** Design and evaluation for situation awareness enhancement [C]//Proceedings of the Human Factors and Ergonomics Society Annual Meeting. SAGE Publications, 32(2), pp 97-101.

[8] **Whaley A M, Kelly D L, Boring R L, et al. 2011.** SPAR-H Step-by-step Guidance [J]. Idaho National Laboratory Risk, Reliability, and NRC Programs Department Idaho Falls, Idaho.

[9] **Groth K M, Swiler L P. 2013.** Bridging the gap between HRA research and HRA practice: A Bayesian network version of SPAR-H [J]. Reliability Engineering System Safety, 115, pp 33-42.

AUTO PARTS INBOUND LOGISTICS RESEARCH ON PRODUCTION LEVELING

Sheng ZHONG[1], Ji-long LI[2*]
[1] Department of Business,
University of Sichuan, Chengdu, China
[2] Department of logistics engineering,
University of Sichuan, Chengdu, China
*18328008943@163.com

ABSTRACT

This paper researches the theory of production leveling and think, as the leveling production assistance, inbound logistics need to be leveling. Inbound logistics leveling should consist of orders leveling, goods collection leveling and receiving leveling. The paper is based on the demand of production leveling to build inbound logistic system and use improved C-W saving mileage algorithm and penalty function to solve the specific case to prove that inbound logistics leveling is feasible.

Keywords: Production leveling, penalty function, inbound logistics, improved C-W saving mileage algorithm

1. INTRODUCTION

The logistic of car plant consist of inbound logistic, production logistic and outbound logistic. Inbound logistic is the important link of linking car plant and suppliers. It serves car production. Production leveling is a mode of pull-production and is based on market-oriented. In the production line, it is based on custom orders to organize small multi-bath production in the established order and proportion [1]. The target of production leveling is reducing unnecessary waste, like unnecessary inventory and unnecessary production [2]. In order to better serve the production leveling, inbound logistic also needs to be leveling. It needs to organize small multi-bath distribution according to the demand of production in the shortest time.

At present, the researches of inbound logistic are mainly about route majorization. Jinlian Wang has built mathematical model by the limit of maximum inventory, the minimum transportation cost and inventory cost and used tabu search algorithm to solve the problem [3]. Wang Xu has built time window distribution mathematical model and used improved heuristic algorithm to solve it [4]. Jiwu Ding has researched the meaning of production leveling and built a new production leveling model [5]. ating Wang has studied the production leveling in FAW [6]. Mingang Zeng has researched inbound logistic leveling [7]. Oded Berman has built mathematical model to choose suitable route [8]. Kaneko and Nojiri has researched the JIT parts distribution in Japan [9]. Daughery and Pittman has researched business logistic, viewed from business resource [10].

From what has been said, we know the research that inbound logistic how to better serve production leveling is scarce. This paper researches production leveling and think inbound logistic should be leveling to better serve production leveling. So, first, we study the theory of production leveling and build the system of inbound logistic leveling. Second, we build mathematical mold of penalty

*Corresponding Author: Ji-long LI

function that is about three interest-concerned groups of third-party logistic service provider, suppliers and RDC. Third, we put forward improved C-W saving mileage algorithm to solve it. The last, we give specific case.

2. THE THEORY OF PRODUCTION LEVELING AND ITS DEMAND FOR INBOUND LOGISTICS LEVELING

2.1. The theory of production leveling

Production leveling is put forward by Taiichi Ohno who is the founder of Toyota production system [11].It is a pull-production. It organize small multi-bath production in the established order and proportion according to the custom orders [12]. For example, a car plant product three kinds of cars of A, B and C, according to the market demand, the demand quantity of A, B and C is 100, 200 and 300 in April. According to production leveling principle, daily production plan could be 10, 20 and 30 and we can product the cars in the order of C, B, C, A, C, B. It demand capacity equilibrium, variety equilibrium, working house equilibrium and production load equilibrium [13]. It requires inbound logistic to reduce unnecessary waste in the time of meeting production leveling, like unnecessary inventory and production [14].

2.2. Inbound logistic leveling

Through production leveling research, we can find production leveling is a small multi-bath recycle production. So I think inbound logistic leveling is according to the order of production to complete the plan of the parts inbound logistic. According to the demand of production leveling, I think inbound logistic leveling consist of three parts: order leveling, goods collection leveling and receiving leveling.

Because production leveling is a recycle production in established proportion and order, order leveling is according to the order and proportion of production leveling to complete the parts distribution. We could take the parts demand quantity of finishing a recycle production as an order. And we could take an order or multiple orders as a batch of a recycle goods collection. We could according to the number of the batches to confirm the times of recycle goods collection and finish the small multi-bath distribution. We also need to consider the demand of production time and the limit of inventory to confirm the suitable time of distribution.

Because production leveling has strict requirement for inbound logistic in time and inventory. So, the accuracy of time and quantity is the basic requirement for good collection leveling, that is to say goods collection leveling is giving suppliers accurate time of collecting goods. And suppliers can have enough time to stock up and reduce unnecessary inventory.

Receiving leveling require the distribution cars complete the discharging within the stipulated time. After the cars finish the plan of goods collection, it will go to the RDC to unload the cargo. If the crossing of discharging was enough, the cars could complete the discharging in any time. But, in reality, the crossing of discharging is limited. If many cars come to the RDC to unload cargo, the phenomenon of waiting in line will happen, which would be a waste of time and destroy the plan of goods collection. In order to avoid the phenomenon, we need to level receiving cargo. That is to say we should make reasonable planning of the car unloading time and unloading crossing.

According to order leveling, goods collections leveling and receiving leveling, we can make the flow chart as the Figure 1.

2.3. Penalty function

Inbound logistic leveling need to carry out strict time window. So, we introduce penalty system to punish the behavior of destroying time window and use the penalty function count the cost of penalty [9]. And we make the hypothesis that if the cars arrive in time or in advance, we will don't give penalty, or we will give penalty. Penalty function is as follows.

$$P=P_1T_1+\sum_{i=1}^{n}P_2T_{2i}+P_3T_3 \qquad (1)$$

P is the total penalty cost in one day. T_1 is the loss of time that due to TPL in one day and P_1 is its cost of loss one minute. T_{2i} is the loss of time that due to the supplier of i and P_2 is its cost of loss one minute. T_3 is the loss of time that due to the RDC and P_3 is its cost of loss one minute.

2.4. The improved c-w saving mileage algorithm

Inbound logistic leveling has strict requirement for transportation time. In reality, the shortest distance route is not likely the shortest time route. According to the requirement of inbound logistic leveling, we introduce the improved C-W saving mileage algorithm [15]. We introduce the formula of (2). In the formula, V stand for the normal speed of the car, stand for the actual average speed in the route between I supplier and j supplier and it can be obtained through statistic. Zij stands for the relative distance between I supplier and j supplier, it can reflect the travel time between i supplier and j supplier. We change the saving mileage to the saving relative mileage. Other step is same as C-W saving mileage algorithm

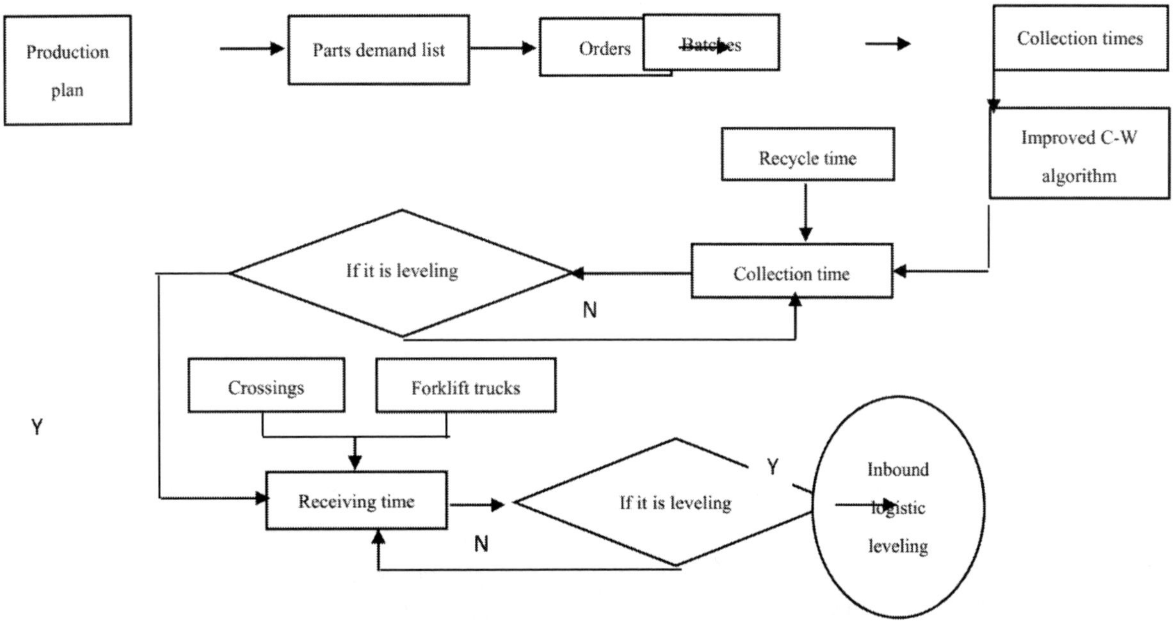

Figure. 1 Inbound logistic leveling flow chart

That is improved C-W saving mileage algorithm. And we use it to make the schedule of collection route.

$$Z_{ij}=S_{ij}*\frac{V}{V_{ij}}=V*T_{ij} \qquad (2)$$

3. THE CASE ANANLYSISI

3.1 The background of this case

VL is an automobile plant and product three types cars of A, B and C. the model of production is production leveling. The production order is A-B-A-A-B-A-A-C-B, it finish one circulation in 30 minutes. Around it, there have 10 suppliers to support 13 types of parts. The parts inventory keep half of work day. One day, the demand of the parts is as the following table Ⅰ. The relative distance of Zij between the 10 suppliers is as the following table Ⅱ and the distance between suppliers and RDC is as the table Ⅲ. We assume the normal speed of the cars is 60 km/h and the plant work time is from 8:00 to 12:00 and 13:00 to 17:00. The receiving time is from 8:00 to 17:00.The car's maximum number of miles is 100km and the biggest cargo capacity is 50m3.

3.2 The process of calculation

3.2.1 Order leveling

We know the plant work time is 8 hours one day and the time of finishing a circulation production is 0.5 hours, so we put the demand for parts in 0.5 hours as an order. Then we can produce the table Ⅳ.

3.2.2 Vehicle routing

According to the table Ⅱ, we can Figureure out the saving relative mileage table, then we expend it from big to small and base on maximum number of miles and the biggest cargo capacity to plan the route. We can Figureure the route is P-J-A-B-C-P, P-D-E-F-G-P and P-H-I-P. The cars traveling mileage and loading rate is as table Ⅳ.

3.2.3 Collection leveling and receiving leveling

We assume that the time of loading in one circulation is 28 minutes and the time of loading for every suppliers in the route is the same. And the time of receiving one carload is 30 minutes. There is one crossing to receive the cargo. According to the mileage of route, normal speed, the time of loading and receiving and the rule of collection and receiving leveling, we can Figureure out the schedule of collection and receiving, that is as the table Ⅴ and table Ⅵ

3.3 Result analysis.

The leveling collection schedule give supplier accurate collection time, so they have enough time to stock. The leveling receiving schedule ensure the RDC complete the task of receiving and could avoid the phenomenon of waiting in line. For the behavior of destroy inbound logistic leveling, we give punishment, which can ensure effectively carry out inbound logistic leveling. Inbound logistic leveling could meet the normal production and keep minimum inventory.

4. CONCLUSION

We research the production leveling and think inbound logistic also need to be leveling to meet production and reduce the level of inventory. We give the way of inbound logistic leveling and use the specific case to prove it is feasible. It can give inbound logistic of automobile plant some help. But, in reality, some uncontrollable factors may destroy inbound logistic leveling. So the pre-arranged plans of solving the problem need further research.

Table 1 the demand of parts

supplier	A	A	B	C	C	D	E
part	1	2	3	4	5	6	7
quantity m^3	32	32	64	16	16	128	32
supplier	F	G	H	I	J	J	
part	8	9	10	11	12	13	
quantity m^3	64	80	80	96	16	32	

Table 2 the distance between suppliers and rdc

A	B	C	D	E	F	G	H	I	J
30	27	21	24	24	24	9	12	30	21

TABLE 3 THE DISTANCE BETWEEN SUPPLIERS

	B	C	D	E	F	G	H	I	J
A	12	27	42	54	54	39	42	33	12
B		15	30	42	51	36	42	45	24
C			15	27	45	30	33	51	39
D				18	39	33	36	54	45
E					21	30	36	54	45
F						18	24	51	45
G							6	33	30
H								27	33
I									24

Table 4 order leveling

supplier	A	A	B	C	C	D	E
part	1	2	3	4	5	6	7
quantity m^3	8	32	64	32	16	80	32
order m^3	0.5	2	4	2	1	5	2
batch m^3	2	8	16	8	4	20	8
times	4	4	4	4	4	4	4
supplier	F	G	H	I	J	J	
part	8	9	10	11	12	13	
quantity m^3	64	8	96	96	32	4	
order m^3	4	0.5	6	6	2	0.3	
batch m^3	16	2	24	24	8	1	
times	4	4	4	4	4	4	

Table 5 the list of collecting time

Supplier	Part	Batch				Collecting time			
		1	2	3	4	1	2	3	4
J	12	8	8	8	8	07:01-07:08	09:30-09:37	11:58-12:05	14:26-14:33

A	1	2	2	2	2	07:21-07:28	09:49-09:56	12:17-12:24	14:45-14:52
A	2	8	8	8	8	07:21-07:28	09:49-09:56	12:17-12:24	14:45-14:52
B	3	16	16	16	16	07:40-07:47	10:08-10:15	12:36-12:43	15:04-15:11
C	4	8	8	8	8	08:02-08:09	10:30-10:37	12:58-13:05	15:26-15:33
C	5	4	4	4	4	08:02-08:09	10:30-10:37	12:58-13:05	15:26-15:33
total		46	46	46	46				
D	6	20	20	20	20	06:26-06:33	08:54-09:01	11:22-11:29	13:50-13:57
E	7	8	8	8	8	06:51-06:58	09:19-09:26	11:47-11:54	14:15-14:22
F	8	16	16	16	16	07:19-07:26	09:47-09:54	12:15-12:22	14:43-14:50
G	9	2	2	2	2	07:44-07:51	10:12-10:19	12:40-12:47	15:08-15:15
total		46	46	46	46				
H	10	24	24	24	24	07:43-07:53	10:11-10:21	12:39-12:49	15:07-15:17
I	11	24	24	24	24	08:20-08:30	10:48-10:58	13:16-13:26	15:44-15:54
total		48	48	48	48				

Table 6 the list of receiving time

Route	Receiving time			
	1	2	3	4
P-J-A-B-C-P	08:30-09:00	10:58-11:28	13:26-13:56	15:54-16:24
P-D-E-F-G-P	08:00-08:30	10:28-10:58	12:56-13:26	15:24-15:54
P-H-I-P	09:00-09:30	11:28-11:58	13:56-14:26	16:24-16:54

5. REFERENCE

[1] **Suo Zhang, Li-fang Wang,** The Application of E-order Segmentation in Automobile Plant [J], Logistic Material Handing, 2009,14(04): 88-92. (Chinese)

[2] **Er-shi i,** Toyota Production System and the Analysis of Its Application in China [J], Industrial Engineering and Management, 1998, 4(4):37-39. (Chinese)

[3] **in-lian Wang,** Study on Milk-run Routing Models and Optimization Algorithms for Parts Inbound Logistic in Automobile Manufacturing [D], Shanghai Jiao Tong University, 2009. (Chinese)

[4] **Xu Wang, Dong Chen, Zhen-feng Chen,** A Research on Leveling Production Model and Its Application [J], Journal of Computer Application, 2011,31(04):1125-1129. (Chinese)

[5] **i-wu Ding, Hong-wei Liu, Dan Xu,** A Research on Leveling Production Model and Its Application [J], Industrial Engineering Journal, 2015, 18(01):31-36. (Chinese)

[6] **Ya-ting Wang,** Leveling in Car Production Organization Application Research and Innovation [D], Tianjin University,2012. (Chinese)

[7]**Min-gang Zeng, Shui- ing Su, Yong-lin Li,** Application of Heijunka for Parts Inbound Logistics with Milk-run [J], Industrial Engineering and Management, 2011.16(2):40-45. (Chinese)

[8] **Berman O, Wang ian.** Inbound Logistic Planning Minimizing Transportation and Inventory Cost [J]. SAE International Magazine. 2001(1):2600.

[9] **Kaneko , No iri.** The Logistics of Just-in-Time between Parts Suppliers and Car Assemblers in Japan [J]. Journal of Trans-port Geography 2008, 16(3):155-173. [10] Daughery P J, Pittman P H.

Utilization of Time-based Strategies: Creating Distribution Flexibility [J]. International Joural of Operation Management 1995,15(2):54-60.

[11] **Monden Y.** Toyota Production System: an Integrated Approach to Just-in-time [M].US: CRC Press, 2011.

[12] **Bo iang,** Study of Vehicle Routing Problem with Time Windows Based on Genetic Algorithm [D], Beijing Jiao Tong University, 2010. (Chinese)

[13] **An- ing Wang,** Study on Inbound Logistic of Automobile Parts on Milk Run and Optimization the Route [D], Anhui Polytechnic University, 2010. (Chinese)

[14] **Ee-shi i, Zi- ian Liu**, Toyota Production System and Application , Business Management, 1998.23(4):55-57. (Chinese)

[15] **Clarke G, Wright . w.** Scheduling of Vehicle Routing Problem [J]. European of Operations Research.1998, 34:336-334.

Implementing Lean Improvement in A Machine Tool Plant

Xing-yu Jiang , Bo-xue Song*, Ming-ming Dai, Hao-yin Zhang, u-fei Li
School of Mechanical Engineering,
Shenyang University of Technology, Shenyang, China
*1097451013@qq.com

ABSTRACT

Lean Production can improve the productivity in many aspects. Aiming to the problem that the production cycle is too long in a machine tool plant, Lean analysis is implemented and problems are found according to the casualty diagram and the current workshop layout. Then, different strategies are proposed respectively to improve the mentioned aspects. Besides, the workshop layout is improved by the discrete event simulation applying Plant Simulation software. The current layout DES model of the plant is established. Then the optimized layout DES model can be obtained by applying genetic algorithm integrated in the software. By implementing these strategies, the production cycle is reduced by the decrease of WIP, the improvement of the manufacturing processes, the more effective management and the optimized workshop layout obtained by the simulation. Besides, the strategies proposed can be the reference to other plants that have the similar problems

Keywords: Lean improvement, discrete event simulation, layout improvements

1. INTRODUCTION

Lean Production(LP), which was proposed by Ohno [1], is a multi-varieties and batch-variable production type to meet the individual or various customer demand. The core of lean production is to eliminate the waste. The waste is the material, machine or manpower that is beyond the quantity needed to add value to the product. LP is one of the best ways to deal with it, which is proved to be an effective way of production management. LP helps companies to find and eliminate seven types of waste[2]: overproduction, waiting, transportation, inappropriate processing, unnecessary inventory, unnecessary motion and defects.

The application of Lean principles, derived from the Toyota Production System, has enabled many organizations to simultaneously improve productivity, quality and customer service[3]. Now it is worldwide used in various fields such as manufacturing and services by various tools such as Kanban[4] and poka yoke[5]. Uwe et al.[6] applied a new approach based on the principle of Toyota Production System at Mercedes-Benz do Brazil, San Paulo. The new approach eliminated redundant actions by documenting standard assembly operation and provide a high degree of understanding and information accessibility to production workers compared with early main-frame based process planning systems. Carminati[7] presented a structured approach for the development of a lean management project highlighting what are the characteristics of this particular sector with a case study in a major Italian company in the fashion industry. Besides, there is a study that indicates employees' participative decision-making and propensity for participative decision-making significantly differ by the duration of lean production in operation[8]. Meanwhile, LP is continuous improved to be more practical by summaries in production practice.

As can be seen, transportation is one of the waste that needs to be eliminated and one of the efficient way is to change the current layout. Many researchers have analyzed the layout problems. Abhishek et al.[9] proposed several advantages of layout improvement that an ideal plant layout should provide the optimum relationship among output, floor area and manufacturing process. Wang et al.[10] applied fundamental industrial engineering methods to increase the output efficiency of the down line area. Hosseini et al.[11] used SLP to improve the multi-floor layout of

* Corresponding Author: Bo-xue Song

a card and packet production company. Reiner et al.[12] proposed a new method that is intended for further performance improvement after initial layout with a standard heliostat field layout tool.

Implementing simulation has many advantages, which allows reusing data, supports an effective cooperation between different departments, and provides up-to-date and relevant data to every user who needs it[13]. For example, Mohd et al.[14] used arena simulation software to analyses the production line layout of the manufacturing system design to identify the key performance and to improve the production performance of a bicarbonate beverage production company.

In this paper, a machine tool plant is taken as the object. Aiming to the problems found during the production processes, several leans improvement strategies are proposed. Besides, DES is applied to analyze the current layout and obtain the optimized layout. With these methods, the productivity can be raised so that the production cycle can be reduced and the wastes among the processes can be eliminated.

2. PROBLEM ANALYSIS

2.1. Background

The machine tool plant in this paper is a professional manufacturer with many types of machine tools. Machine tool protection is one of the products with stamping parts, cutting parts, welding parts and sheet metal parts, which mainly support the protection of VMC850B, VMC850T, VMC 850E and TC series. Some of the products are shown in Figure 1. As can be seen, the machine tool protection is a type of hood which can protect external damages or corrosion to ensure the safety when the machine tool is operating.

At present, the plant owns two laser cutting machines. The cutting power is 1800W and the cutting thickness is 16mm for carbon steel and 6mm for stainless steel. Metal tubes less than 300mm can be processed by the rotating chuck in the machine, which can greatly improve welding quality. Besides, a set of FMS with stereoscopic warehouse and a numerical control bending machine are introduced from Belgium.

Production management problems arise with the increase of sales volume such as the delay of delivery period, accumulation of WIP, chaos in field management, occlusion of semi-manufactures flow in manufacturing process and low productivity.

VMC series are the main products of the enterprise with mature, stable processes and large market demand. In this article, the manufacturing process of VMC series machine tool protection is selected as the research objective. The manufacturing process mainly contains the following five procedures: stock preparation, bending, riveting, flame plating and assembly. Every procedure has 3-5 stations. The detailed data of the procedures is shown in Table I.

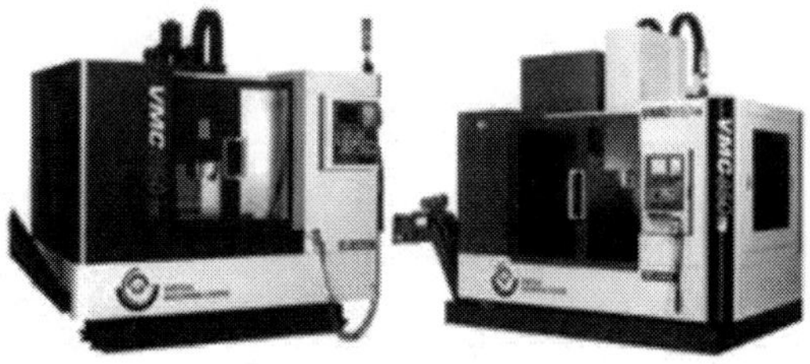

Figure.1. VMC850E and VMC850B

2.2. Problem Analysis

Aiming to find the problems among the production processed, the order V1304008 is tracked as an example to make value stream analysis. The processing procedure and data information are shown in Table II.

From table II, it can be found that nearly 8 days are needed to wait for a set of machine tool protection to be finished. Besides, the active time accounts little in a day, or the value-added time is too short, which means most of time is wasted. Besides the manufacturing or management problems, the main reason causing long production cycle is the unreasonable layout.

Table 1 process data sheet of vmc series

Process name	Shifts System	Numbers (people/station)	Devices
Stock preparation	2	1	Laser cutting machine
Bending	3	2	Bending machine
Riveting	1	2	CO2 protection welding
Flame plating	2	2	Automatic flame plating line
Assembly	1	2	Assemble manually

Then, a casualty diagram is drawn to illustrate the problems, shown in figure 2 by the field investigation.

From figure 2, it is found that there are seraral problems in every process. For example, wrong parts are always delivered to the bending process due to less efficiency communications in stock preparation process. Problems shown in figure 2 can be integrated into the following problems:

1) WIP overstock is severe. From figure 2. Most of the WIP problems are raised from unreasonable manufacturing processes. For example, the failure rate is high in the stock preparation process; bending machines are not enough to meet the bending process; and the riveting process and the flame plating process are much complex. All these problems will restrict the productivity of the machine tool protecton.

2) The production information flow jam is everywhere. For example, there is no synchronization in the tock preparation of shears, which will delay the stock preparation efficiency and the delivery to the bending process, and also influence the whole processes, finally causing long production cycle.

3) The management confusion problems in assembly processes. As can be seen, workers are assembling parts with undisciplined attitude and lack of standard assembling methods. This phenomenon also exists in other processes. Another problem is that the orders are not classified that it is hard to identify whether the order type is rush or normal.

Therefore, the improvements should be implemented into these aspects: WIP overstock, production information flow jam, management confusion and unreasonable workshop layout.

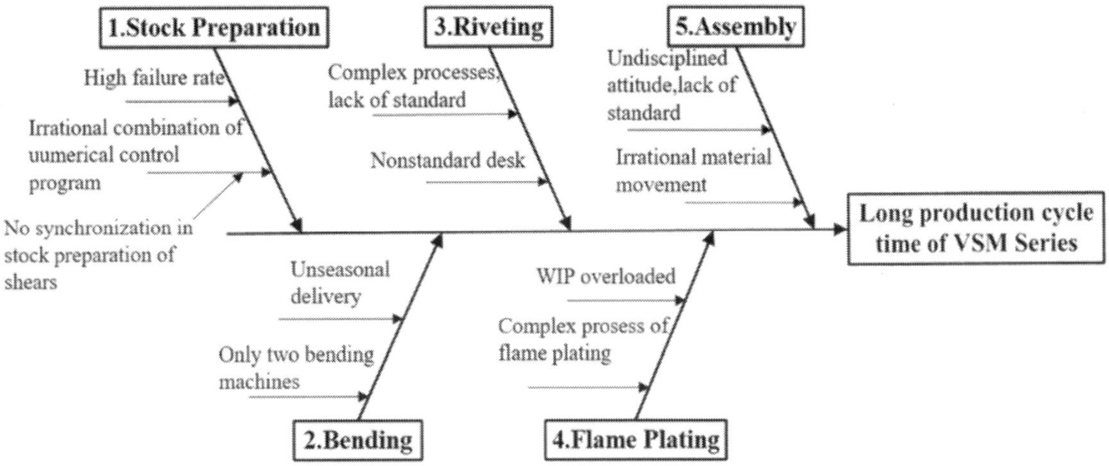

Figure.2, Causality diagram of the manufacturing process

III. LEAN IMPROVEMETS

3.1. Layout Improvements

By the field investigation, the current workshop layout can be concluded and shown in figure

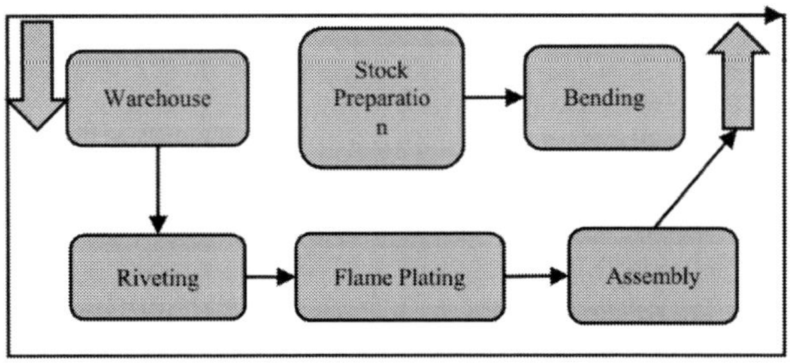

Figure.3. The current workshop layout of the plant

Table 2 data of order v1304008

Time (Day)	Process	Active time (h)			Available time per day(h)	WIP(set)
20-22		Stock preparation	1.0	16	7	
22-24		Bending	1.2	24	7	
24-25		Riveting	1.1	8	7	
25-26		Flame plating	0.8	16	7	
27-28		Assembly	2.0	8	7	

The principle of the effective layout is to ensure the minimum logistic scale. To find the optimized layout, a simulation model is established in Plant Simulation software, shown in figure 4. Two tables named W From To Chart and D From To Chart are logistic from-to-chart and distance from-to-chart respectively and both charts are shown in table III and table IV. The method named InitPartsTable controls the initial state of the manufacturing process. The variety named

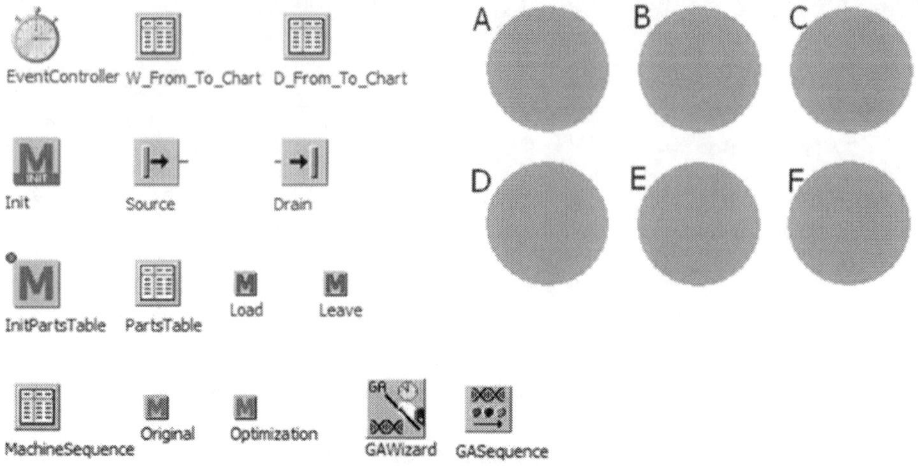

Figure.4. The simulation model of the layout of process workshops

HandlingCost shows the current total logistic quantity.

After running the method named Init in the model and starting EventController, the original layout of the plant and the corresponding HandlingCost can be shown in figure 5. As can be seen, this is a typical sequence problem. To find out the optimistic layout to ensure the minimum logistic quantity, Genetic Algorithm is applied to optimize the layout. GAWizard in the model controls the whole process of Genetic Algorithm and GASequence defines how to optimize the chromosome. In GASequence, the operator is defined as Order Crossover. Two crossover positions are randomly verified and the fragment between two crossover positions is exchanged, and the genetic fragment exchanging from the other father individual will be deleted from the second crossover position, and then the remaining genetic will be placed after the second crossover position.

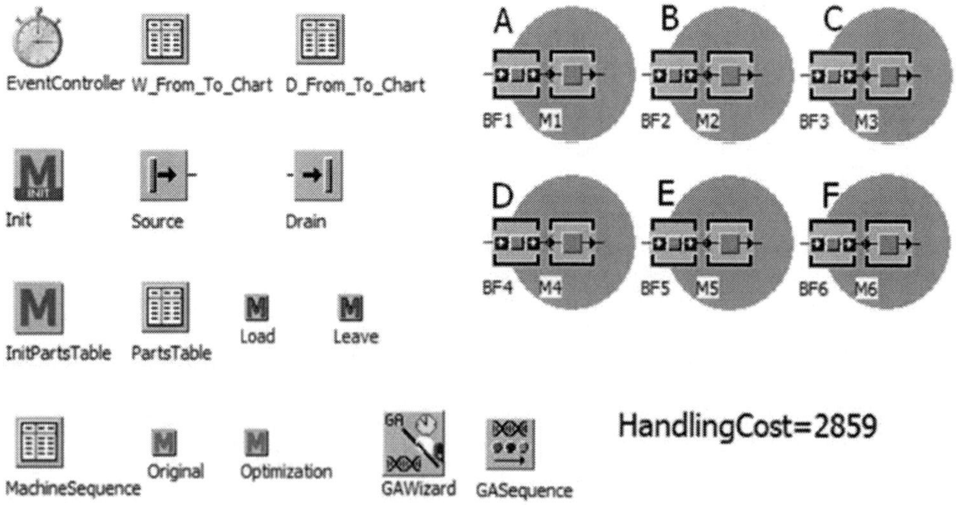

Figure.5. The original layout and the corresponding HandlingCost

By clicking start button in GAWizard, the simulation process begins and the optimized result is shown in the model that the minimum HandlingCost is 2719. Besides, the optimized layout of the process workshops is also shown in the model, which is shown in figure 6.

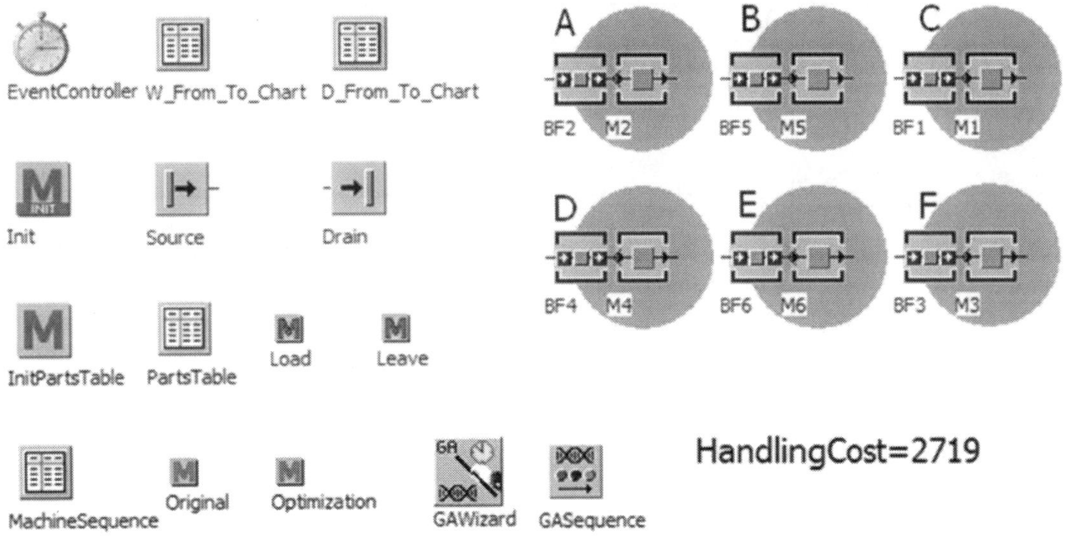

Figure.6. The optimized layout and the corresponding Handling Cost

3.2. WIP Overstock Improvements

To improve WIP overstock problems, strategies can be divided into two aspects: the manufacturing problems and the production type. As for manufacturing problems mentioned in the problem analysis section, the following strategies will be proposed. Firstly, regular overhauling should be implemented to ensure the normal operation of the machine tools. Secondly, one bending machine should be added to the bending process to increase the bending productivity. Finally, the original riveting and flame plating process should be improved and re-organized. This involves the technique strategies and should be implemented with technicians.

As for the production type, the original production type is push-type, which may easily cause WIP overstock problems. If the production type is changed to pull-type, volume production will be ensured by withdrawal Kanban and production Kanban of pull-type production, with which the whole production will be pulled and the inventory waste will be reduced.

Besides, production scheduling will be implemented applying withdrawal Kanban and production Kanban to control the quantity of WIP among processes. After implementing pull-type production and Kanban technique, the quantity of WIP decreased from 7 sets to 3 sets and the time of bending and waiting for assembly also decreased 1 day.

Table 3 logistic from-to-chart

	Warehouse	Stock Preparetion	Bending	Riveting	Flame Plating	Assembly
Warehouse	0	1	1	2	3	3
Stock Preparation		0	2	1	3	3
Bending			0	1	1	2
Riveting				0	2	1
Flame Plating					0	1
Assembly						0

Table 4 distance from-to-chart

	Warehouse	Stock Preparetion	Bending	Riveting	Flame Plating	Assembly
Warehouse	0	192	177	168	150	130
Stock Preparation		0	150	80	67	30
Bending			0	98	0	87
Riveting				0	77	105
Flame Plating					0	112
Assembly						0

3.3. Information Flow am Improvements

To solve poor communication among processes, an effective information management system should be established. The consummation of information management system is the prerequisite to realize pull-type production. The information management platform of every process should be established and the information subsystem among processes should be consummated to ensure the integrity of information stream. Laser cutting database of common parts and interchangeable parts should be founded to avoid the waste of over-programming. The stock of shears and cutting is synchronized to shorten the time to assort by 1 day. Thus, the preparation time of waiting for stock will become just 1 day.

3.4. Management confusion improvements

For the management problems, order forms should be regularly managed according to working load of the equipment. Casual delivery of order forms causes WIP backlog, which will further impact financing cost. When an order form is received, it should be classified by rush orders or normal orders and production series. Then, the production will be arranged according to the rhythm of production after verifying the information of the order form. Finally, the flow condition of orders will be tracked to ensure the accomplishment. The coordination time of orders will be reduced by the processes above. The assembly time of stocking decreased to 1 day compared with before.

In addition, to implement pull-type production, the thinking patterns transforming and the supporting system should also be established. Educational activities should be carried out to the employees to spread lean thinking among them. Every production element such as production files, assistant tools and quality records should be standardized.

The new supporting system includes three sections: field management section, quality management section and equipment management section. The first section aims to standardize the production field such as 5S to ensure implementing pull-type production. The second section aims to find and improve quality defects in the products. With continuous recording the quality and implementing ISO9001 and S9000 quality standard in the process, problems causing quality defects will be found to prevent the quality accidents from happening. The last section aims to maintain the operation of the machine tools. By preventing failures rather than fixing after failures in the machine tools, the manufacturing plan will be ensured.

4. CONCLUSION

Aiming to the problems that the production cycle is too long in a machine tool plant, several lean improvement strategies are proposed to improve the problems. The WIP overstock problems are improved by implementing regular overhauling, adding an extra bending machine, improving the original processed and implementing pull-type production. These strategies reduce the quantity of WIP from 7 sets to 3 sets and the time of bending and waiting for assembly also decrease 1 day.

The information flow jam problems are improved by establishing an effective information management system. Thus, the stock of shears and cutting is synchronized and the stock preparation time is reduced to just one day.

The management confusion problems are improved by setting a series of standards for workers' action that the assembly time can be decreased. Besides, the order types are classified that it is easy to arrange production plans. Besides, the thinking patterns transforming and the product supporting system are both established to fit the new production type and ensure the product quality.

5. ACKNOWLEDGMENT

This research is supported by National Natural Science Foundation of China, No. 51305279.

6. REFERENCES

[1] **Ohno, T.,** Toyota Production System: Beyond Large-scale Production . Portland, OR: Productivity Press, 1988.

[2] **Peter Hines, Nick Rich,** The seven value stream mapping tools . International Journal of Operations Production Management, vol.17, no.1, pp.46-64, 1997.

[3] **Riezebos ., Klingenberg W., Hicks C,** Lean Production and information technology: Connection or contradiction . Computers in industry, vol.60, no.4, pp.237-247, 2009.

[4] **Monden, Y.,** Adaptable Kanban System Helps Toyota Maintain Just-in-time Production. Industrial Engineering, vol.13, no.5, pp.29 46, 1981.

[5] **Shingo, S.,** Zero uality Control: Source Inspection and the Poka- oke System . Portland, OR: Productivity Press, 1986.

[6] **Bracht U., Meyer O., Oelt enbruns H., Clausthal Z.,** Standardized assembly process planning based on The Toyota production system . SAE Technical Papers, 7th International Mobility Technology Conference and Exhibit, 1999.

[7] **Gionata C.,** Lean Supply Chain Model and Application in an Italian Fashion Luxury Company . Understanding the Lean Enterprise, pp.203-220, 2015.

[8] **Dharmasri W., Vathsala W.,** Differences in Organizational Factors by Lean Duration . Operations Management Research, vol.4, no.3, pp.111-126, 2011.

[9] **Jain A., Bhatti R., Singh H.,** Improving Employee Manpower Productivity by Plant Layout Improvement . 2014 Recent Advances in Engineering and Computational Sciences, 2014.

[10] **Wang H.Y., Zhuo Y. ., in .Y.** Layout Improvement for Mixed-Model Production Offline Area Based on Fundamental Industrial Engineering and Simulation . Applied Mechanics and Materials, vol.328, pp.154-157, 2013.

[11] **Hosseini.S.S., Wong K.Y.,** Mirzpour S.A., Ahmadi R. Multi-Floor Facility Layout Improvement Using Systematic Layout Planning . Advanced Materials Research, vol.845, pp532-537, 2014.

[12] **Buck R.** Heliostat Field Layout Improvement by Nonrestricted refinement . Journal of Solar Energy Engineering, vol.136, no.2, 2014.

[13] **Steffen Bangsow.** Tecnomatix Plant Simulation . Springer International Publishing AG Switzerland, preface, 2015.

[14] **Said M.A.B.M.,** Ismail B. Improvement of Production Line Layout Using Arena Simulation Software . Applied Mechanics and Materials, vol.446-447, pp.1340-1346, 2014.

DESIGN AND SIMULATION ANALYSIS OF THE REEL OF WHEAT COMBINE HARVESTER

Ling-he Yuan[1,2],Wan-zhang Wang[1,2],Ya-dong Tang[1], Jing-zhao Shi[1,*]
[1]College of Mechanical and Electrical Engineering,
Henan Agricultural University, Zhengzhou, China
[2]Collaborative Innovation Center of Henan Grain Crops.
Zhengzhou, China
*13838384792@163.com

ABSTRACT

In view of the problem of grain loss caused by the reel of wheat combine harvester, the rotation speed and the radius of eccentric reel were rationally designed by analyzing the motion of the cam-action reel and the selection of relevant parameters. The trajectory of tube shaft and spring tooth was get by Establishing the 3D model of cam-action reel in three dimensional modeling software and importing it into ADAMS software for motion simulation. The blank area was reduced by optimizing the radius of the cam-action reel. Combined with the mechanical properties of wheat stalk, its finite element model was set up in the ADAMS software, then done a co-simulation with the cam-action reel, getting the displacement curve of wheat stalk and the maximum moving distance Bmax.

Keywords: Cam-action reel, simulation analysis, wheat combine harvester, wheat stalk.

1. INTRODUCTION

In recent years, with improvement of wheat Mechanized sowing and harvesting mechanization level, planting area and output of wheat increased year by year in china. In the year of 2015, national grain area was about 415 million mu, increasing by 1 million 660 thousand mu over the previous year. The total output of summer grain was 1410 billion 50 million kg, the output of wheat was 123billion 600 million kg, increasing by 3 billion 550 million kg [1]. However, the grain loss of mechanical harvest was seriously affected the yield of wheat. At the time of wheat harvest, grain loss caused by reaping is one of the important reasons, especially the reel hitting and pushing the wheat stalk its installation height can cause grain loss of wheat. Therefore, this paper had a design and simulation analysis of real for wheat combine harvester, in order to reduce the loss caused by the harvest process.

2. STRUCTURE AND FUNCTION OF THE REEL

Generally, the reel of the harvester can be divided into plate type, cam type and cam-action type. The plate reel has the advantages of simple structure, low manufacturing cost, but its adaptability for lodging crop is poor and causing a serious blow to the crops. The cam reel is of low quality and easy maintenance, but it is mainly used in forage harvesting machinery. The cam-action reel is of strong lifting capability and small blow to crop corn, mostly for rice and wheat combine harvester [2]. Therefore, the paper chose the cam-action reel, its structure is shown in Figureure 1.

* Corresponding Author: Jing-zhao Shi
Earmarked Fund for Henan Agriculture Research System (S10-02-G07/2016)

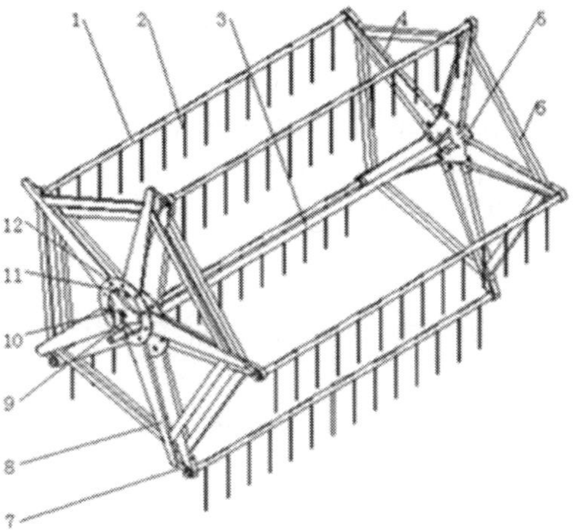

1. Tube shaft; 2. Spring tooth; 3. Wheel spindle; 4. Main spoke; 5.Reel spider; 6.Stiffener; 7. Crank; 8. Eccentric spoke; 9.Eccentric boom; 10.Supporting plate; 11. Supporting roller; 12. Eccentric reel spider

Figure.1 structure of the cam-action reel

The cam-action reel is in front of the header of wheat combine harvester table in front, it is a key component of header. Its main function is lifting the wheat stalk and leading to cutter. In operation process , the cam-action reel lifts wheat stalk, then pushes the wheat stalk that be cut off to the rear of the header, to prevent stalk accumulating on the cutter, so as not to affect the next stage of cutting stalk.

3. KINEMATIC ANALYSIS AND DESIGN OF THE CAM-ACTION REEL

The motion trajectory shape of the cam-action reel is determined by λ, which is the ratio of the peripheral speed of the reel vs and the forward speed of combine harvester vm. The change regulation of motion trajectory with different values of λ is shown in Figureure 2. Only tube shaft and spring tooth have backward horizontal velocity, the reel can guide and lift wheat stalk, then push it. From Figureure 2, we can know that only when λ is bigger than one, the trajectory is a trochoid, and forming a sway ring in the motion process [3]. Tube shaft and spring tooth have backward horizontal velocity in the lower of swing ring.

3.1. Motion Analysis of the Tube Shaft

In the operation, the tube shaft of the cam-action reel makes the linear motion with the harvester, and at the same time, goes around the wheel spindle, so the motion trajectory of the tube shaft relative to the ground is a synthetic motion. The motion equation of the tube shaft

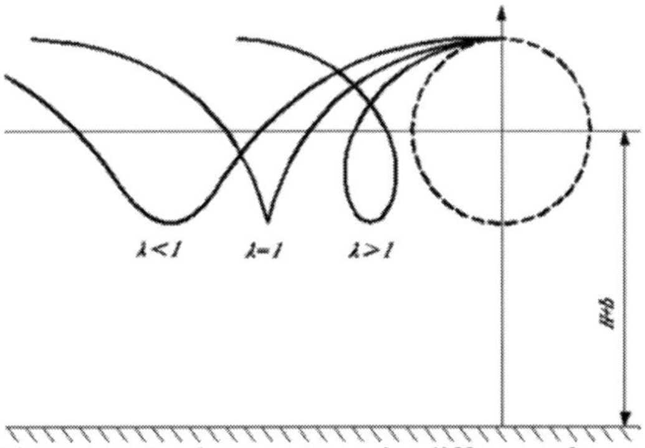

Figure.2.Reel tra ectory in different λ

is as follows.

$$\begin{cases} x = v_m t + R\cos\omega t \\ y = H + h - R\sin\omega t \end{cases} \tag{1}$$

In the equation, x is the horizontal coordinate of any point on the tube shaft. y is the longitudinal coordinates of any point on the tube shaft. vm is the forward speed of the harvester. t is the working time. R is the radius of the cam-action reel. ω is the angular velocity of the cam-action reel. H is the installation height of wheel spindle. h is the height of the cutting knife from the ground.

The horizontal and vertical velocity equation of the tube shaft is as follows.

$$\begin{cases} v_x = \dfrac{dx}{dt} = v_m t - \omega R\sin\omega t \\ v_y = \dfrac{dy}{dt} - \omega R\cos\omega t \end{cases} \tag{2}$$

3.2. Motion Analysis of the Spring Tooth

In the process of movement, eccentric reel spider goes around the center of supporting plate, the crank that is fixed on the tube shaft goes around the hinge joint with the eccentric spoke, and the spring tooth keeps a certain Angle with the ground. Adjusting Angle range of the spring tooth is generally from vertical downward to leaning forward or backward of 30 °. Harvesting the upright wheat, the spring tooth is vertical to the ground. Harvesting the lodging wheat, the spring tooth is adjusted to leaning forward or backward of 15°to 30 °, and harvesting along the direction of the wheat lodging. When harvesting the upright wheat, the motion equation of spring tooth is as follows:

$$\begin{cases} x = v_m t + R\cos\omega t \\ y = H + h - R\sin\omega t - d \end{cases} \tag{3}$$

In the equation, d is the distance between any point on the spring tooth and the tube shaft.

3.3. Design of the Cam-action Reel

In order to decrease tube shaft hitting wheat stalk in the process, the horizontal velocity of tube shaft is zero when it inserts wheat stalks, as shown in Figureure 3. The condition that should be meet is as follows.

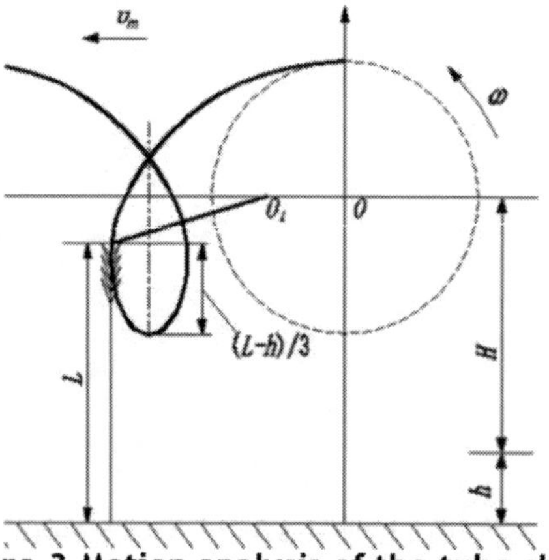

Figure.3.Motion analysis of the tube shaft

$$\begin{cases} v_x = \frac{dx}{dt} = v_m - \omega R \sin \omega t_1 = 0 \\ \frac{\omega R}{v_m} = \lambda \\ L + R \sin \omega t_1 = H + h \end{cases} \quad (4)$$

In the equation, ωt1 is the angle that tube shaft enters between wheat stalks. L is the height of wheat stalk.

As shown in Figureure 3, in order to make the reel lay wheat stalks on the header and prevent spring tooth pushing wheat stalks off the header, avoiding grain loss, it should be effect on a third position of the wheat stalks that are cut off (the position of above the centre of gravity) when the tube shaft moves to the lowest position [4]. The condition is as follow.

$$H \geq R + \frac{2(L-h)}{3} \quad (5)$$

In practice, the two conditions cannot be satisfied at the same time, generally quantifying radius of the reel by solving the two equations.

$$R = \frac{\lambda(L-h)}{3(\lambda-1)} \quad (6)$$

The range of λ is from 1.5 to 1.7. The range of vm is from 1.2 to 1.5 m/s [5]. The operation quality of wheat harvest machinery requires that the height of stubble is less than or equal to 18 cm [6]. The height of Wheat stalks are from 0.6 m to 0.9 m. Selecting that λ was 1.5 and vm was 1.2 m/s. So R was equal to from 0.42 m to 0.72 m. In this paper, selecting R to be 0.5 m and solving the speed of the cam-action reel was 34.4 RPM, then turning it to 35 RPM. At last, vs is equal to1.832 m/s, λ is equal to 1.526.

4. SIMULATION AND ANALYSIS OF THE CAM-ACTION REEL

The 3D model of the cam-action reel was established by software, and was saved as a file format that could be identified by ADAMS software, then importing it to ADAMS software for dynamics simulation. Defining material properties of the cam-action reel and add joints in

Table.1. situation of constraint

joint name	joint object	joint type
Joint_1	Ground and Frame	Translational
Joint_2	Wheel spindle and Frame	Revolute
Joint_3	Wheel spindle and Reel spider	Fixed
Joint_4	Eccentric boom and Frame	Fixed
Joint_5	Eccentric boom and Supporting plate	Fixed
Joint_6	Supporting plate and roller	Revolute
Joint_7	Eccentric spoke and Eccentric reel spider	Revolute
Joint_8	Eccentric spoke and Crank	Revolute
Joint_9	Crank and tube shaft	Fixed
Joint_10	tube shaft and spring tooth	Fixed
Joint_11	tube shaft and main spoke	Revolute
Joint_12	main spoke and reel spider	Fixed

the interface of ADAMS/View .The constraint conditions are as shown in table I.

In order to realize the spring tooth that is fixed on the shaft tube keeps vertical, adding contact force between the supporting roller and the eccentric reel spider. Adding the translational joint motion in 'joint_1', which speed is 1.2 m/s. Adding the rotational joint motion in 'joint_2', which speed is 35 RPM. Setting simulation time for 5 s and the simulation step for 100 steps. After the simulation, the motion trajectory and the velocity curve in forward direction of tube shaft and spring tooth were drawn from post processer module, as shown in Figureure 4 and Figureure 5. In Figureure 4, the motion trajectory of the tube shaft and the spring tooth were trochoids. In Figureure 5, along the velocity in forward direction of tube shaft was zero at 0.623 s and 1.091 s. The tube shaft inserted wheat talks at 0.623 s, and had a backward horizontal velocity from 0.623 s to 1.091 s, so it pushed wheat stalks during this time.

The trajectory curves of adjacent tube shafts are shown in Figureure 6. b that is the maximum transverse chord of sway rings is equal to 194.7 mm, bk that is the maximum transverse chord of the blank area of adjacent sway rings is 250.8 mm. If there is a blank between adjacent swing rings, wheat stalks in the blank are hit by the tube shaft indirectly, leading to that stalks cannot be pushed and pile on the cutter, causing grain loss. If there is overlap between adjacent sway rings, the force that the tube shaft hits wheat stalks and the grain loss will increase. For avoid the overlap and blank, the radius of cam-action reel must be changed to make bk to be zero. Establishing the parameterized model of cam-action reel in ADAMS software to optimize its radius, R is equal to 0.62 meter. The trajectory curves of adjacent tube shafts after optimization are shown in Figureure 7.

5. SIMULATION BETWEEN WHEAT STALK AND CAM-ACTION REEL

The rotary motion of cam-action reel effects on wheat stalks to push it. The cam-action reel pushes to cutter directly in the area that is under the transverse chord ΔX and a (ΔX is half of b, a is the horizontal distance between the cutting line and the lowest point of sway ring). The wheat stalks still cannot be cut off when the tube shaft separates from them, which leads to stalk rebound and cutting leakage, so the maximum forward distance of reel is Bmax, as shown in Figureure 8. In the past ,the wheat stalk was been as a rigid body to analyse the maximum forward distance of reel[7-8], but it is a typical cantilever beam in natural growth state, it will be bend by external force. The interaction between reel and wheat stalks must be considered in order to get actual motion situation of the reel. Only by establishing the virtual model of wheat stalk and building joint force with the reel in ADAMS software, the accurate performance parameters of the reel can be got.

5.1. Establishing the Model of Wheat Stalk

Establishing the flexible model of wheat stalk by finite element method in ADAMS software to ensure that it is in keep with the real one. The model of wheat stalk is actually composed of beam with nodes, and has inertial properties. The property of the cross section at the node can be defined by selecting the section type in ADAMS/View database. Wheat stalk was regarded as a variable cross-section cantilever beam with 5 internodes, the fixed end is on the base of stalk. The elastic modulus of mature wheat stalks is from 2 to 3.16 GPa, its average value is 2.47GPa. The density of mature wheat stalks is from 1007 to 2349 kg/m3, its average value is 1510 kg/m3. The Poisson's ratio is 0.27[9-10]. Establishing the flexible model of wheat stalk according to geometric dimension parameter in the literature [9].

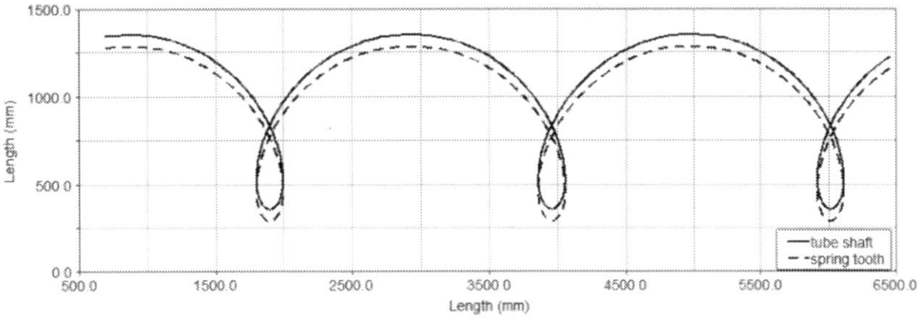

Figure.4.Motion tra ectory of tube shaft and spring tooth

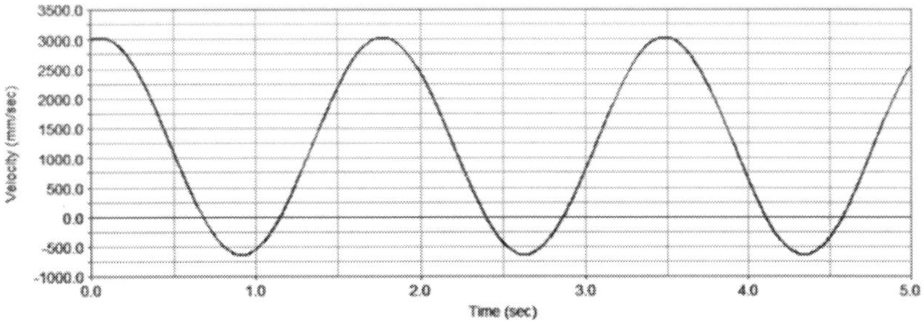

Figure.5.Velocity curve in forward direction of tube shaft and spring tooth

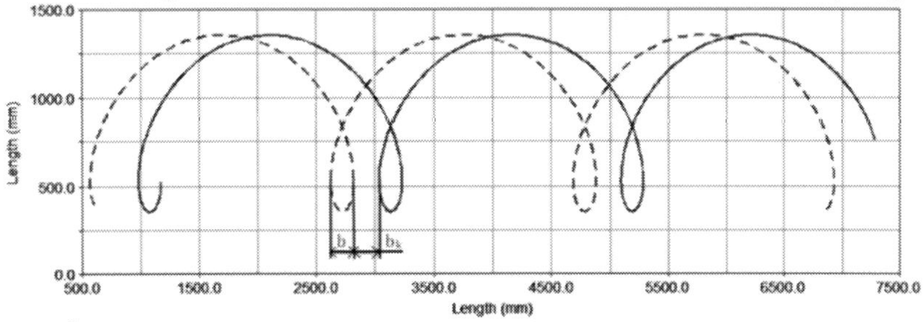

Figure.6.Tra ectory curves of ad acent tube shafts

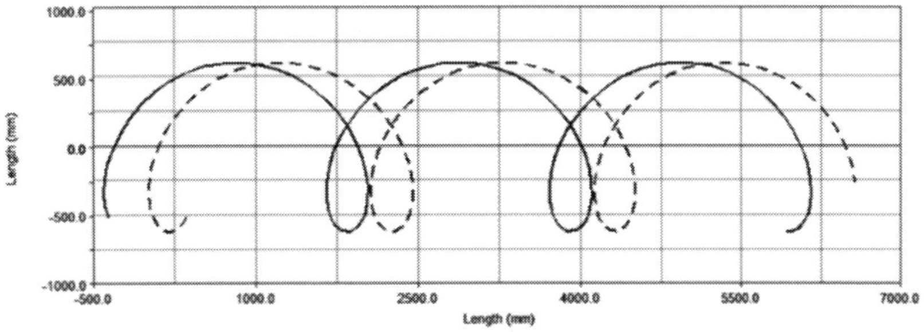

Figure. .Tra ectory curves of ad acent tube shafts after optimization

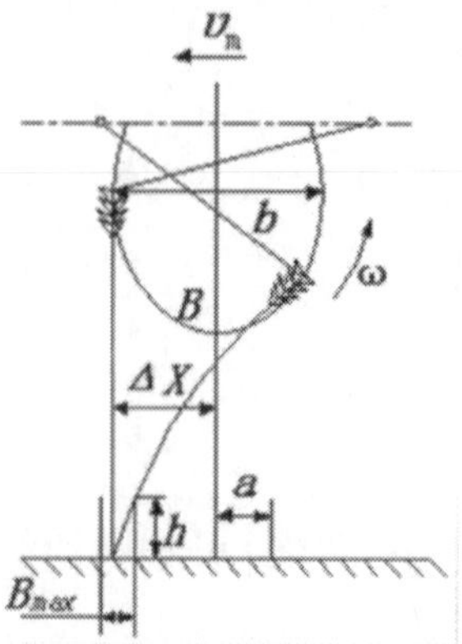

Figure.8. Effect on stalk from the tube shaft

5.2. Simulation and Analysis of Reel and Wheat Stalk

The tube shaft should insert stalks to simulate the work process of reel. Adding contact force and distance sensor between the tube axis and the wheat stalk, when the tube shaft separated from the wheat stalk, stopping simulation. The displacement curves of the top of wheat stalk, the being cut point of stalk, the wheel spindle and the tube shaft in the forward direction are as shown in Figureure 9.

In Figureure 9, the tube shaft inserted stalks vertically at 0.5804 s, then began to hit and bend wheat; simulation stops at 1.075 s, X1, the displacement of wheel spindle in the forward direction, is 1289 mm. X2, the displacement of the being cut point of stalk, is 1182 mm, so Bmax is 107 mm, which from X1 subtracts X2.

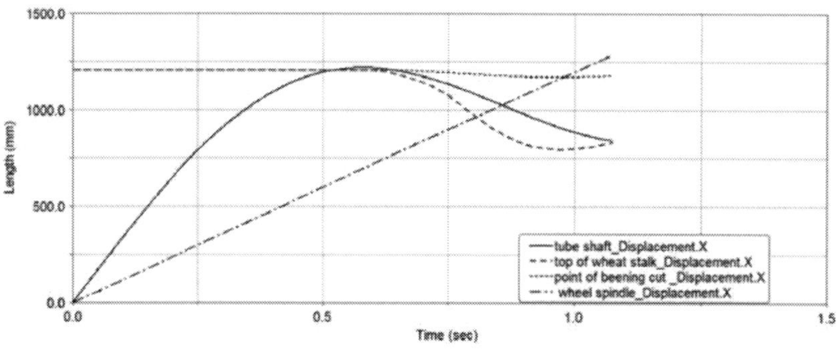

Figure. . Displacement in the forward direction

6. CONCLUSIONS

(1) The Simulation Analysis of the cam-action reel of wheat combine harvester had been done by selecting appropriate parameters to achieve lifting and pushing function of the reel. The ratio of the peripheral speed of the reel vs and the forward speed of combine harvester vm is 1.5. The speed of the reel is 35 RPM, and its radius is 0.5 m.

(2) The motion trajectories tube shaft and spring tooth were obtained by establishing the 3D model of the cam-action reel in software and importing to ADAMS software for dynamics simulation. The blank area between adjacent swing rings was decreased by optimizing the radius of reel, which was 0.62 m at last.

(3) The finite element model of wheat stalk was established in ADAMS software and done a joint motion simulation with the reel, determining that the maximum forward distance of the reel was 107 mm through analyzing the displacement curve of wheat stalk and wheel spindle in forward direction.

. REFERENCES

[1] Agricultural Economy Department of the National Development and Reform Commission, "The primary, secondary and tertiary industrial convergence development becoming a new trend in the countryside," China Economic & Trade Herald, no.24, pp.20-22, 2015. (CHINESE)

[2] i-chun Dai, Xin-min Pei, Cong-wen Song, Yan-bin Zhang, "Design on dial reel of alfalfa flattening harvester," Agricultural Egineering, vol.4, no.3, pp.113-116, May. 2014. (CHINESE)

[3] Shu-chuan Yang, Shu-ming Yang, Yong-wei She, "Computer simulation on reel trajectory," Journal of Agricultural Mechanization Research, vol.32, no.12, pp.141-145, Dec.2012. (CHINESE)

[4] Bao-fa Li, Agricultural Mechanics. Beijing: China Agriculture Press, pp.157-158, 2003. (CHINESE)

[5] Chinese Academy of Agricultural Mechanization Sciences, Agricultural Machinery Design Manual. Beijing: China Agricultural Science and Technology Press, pp.902-906, 2007. (CHINESE)

[6] Chinese Agricultural Mechanization Management Department of the Ministry of Agriculture, Chinese Ministry of agriculture machinery station identification of agricultural test, Outline of Agricultural Mechanization Standard, Beijing: China Agricultural Science and Technology Press, pp.246, 2013. (CHINESE)

[7] Wen-tao Ren, Hong-lai Wang, Hong-guang Cui, Zhan-yong Zhang, Zu-lin Li, "Simulation Analysis of Plant Rebounding in Combine Harvesting," Journal of Shenyang Agricultural University,vol.34, no.2, pp.121-123, Apr.2003.

[8] Lin Wan, Yong-cai Ma, Bing-yi Zhang. "Computer simulation of straw rebound in combining machine," Journal of Agricultural Mechanization Research, vol.26, no.3, pp.125-26, Mar.2012. (CHINESE)

[9] Zhi-hua Yuan, Ai- in Liang, Xiu-li Zhang, Hui- in Li, Dong Wang, "Simulation and analysis on ovalization of Wheat Stem Section," Hubei Agricultural Sciences, vol.51, no.18, pp.4113-4115, Sep.2012. (CHINESE)

[10] Hui-min Fang, Chang-ying i, ing-yi Zhang, F.A. Chandio, "Research on mechanical properties of wheat straw at home and abroad," Journal of Chinese Agricultural Mechanization, vol.35, no.6, pp.304-308,Vol.2014. (CHINESE)

A REVIEW OF CLOUD SERVICE RECOMMENDATION BASED ON DYNAMIC TRUST MODELING UNDER A SOCIAL NETWORK ENVIRONMENT

Jian Xu*, Shuai Ding, Feng Niu
School of Management,
Hefei University of Technology, Hefei, China
*hfut_xujian@sina.com

ABSTRACT

The advancement of Cloud computing, and the rapid development of intelligent terminal and mobile Internet application make cloud services as important technical support and impetus of people's lives. The open innovation of cloud service has set off a new round of industry change. This paper describes the research status about trust modeling and evolution, virtual communities discovery and recommendation system of cloud services.

Keywords: cloud service, social network, trust, virtual community, recommendation

1. INTRODUCTION

The interaction and penetration of cloud computing, Internet of Things, big data, mobile Internet and other new generation in information technology accelerate the cross-border integration of manufacturing, service, financial, business and other related industries. Google, Apple, Facebook, Twitter, Amazon and many other internet companies actively build the ecological system which is supported by cloud platform and is at the core of cloud service, greatly facilitating the publishing of innovative service and personalized options of global users. Cloud service is defined as applications and services which are deployed at the distal end and all the service access is through the Internet or private network. In the cloud services market, the homogenization of service function, the uncertainty of QoS, the diversity of supplier's reputation and the dynamic nature of user demands and preferences will bring about the overloaded information problems for users during the cloud service selectional process. In order to support decision-making for the users of cloud service, a growing number of open operators begin to use recommendation technology based on model or collaborative filtering to build cloud service recommendation system. After analyzing the data of demographic characteristics, behavior (browsing, purchasing, evaluation, etc.) and cloud service, the recommendation system will provide accurate and personalized recommendation information for the target users.

In recent years, the social network has been changing the cloud service users' ways to communicate and collaborate. Making friends, contacts, labels, groups and parties among users would make impact on decision-making behavior of the target users, especially the data of users who are trusted by others will play a key role in determining the target users' choice.

2. THE RESEARCH STATUS AND DEVELOPMENT TREND OF TRUST MODELING AND EVOLUTION

Social network platform provides users with an open and equal communication environment, and broadens the channels for users to collect cloud services information of technology or product, and also reduces the perceived risk for users to select cloud product or cloud service. Trust relationship

* Corresponding Author: Jian Xu

between users of cloud services directly affects the target user's decision-making intent under the social network environment. The suggestions or recommendations of trusted users tend to be more likely to be adopted by the target users. In order to analyze the trust relationship between users and their impact on recommending service under the social network environment, more and more scholars begin to focus on the key issues of trust metric, trust transmission, trust evolution and so on.

2.1Trust measures under the social network environment

Trust is a subjective feeling among users under the constraints of a particular context[1]. As the premise work of trust modeling under the social network environment, trust metrics need to rely on the specific scene to choose more reasonable measures. Skopic et al[2] note that the foundation trust model among online virtual community users is important for building SOA framework in a collaborative network, and proposed rule-based approach to build interactive and experience-oriented trust metrics. Liu et al[3] analyzed the diversity of trust metrics path between entities under the social network environment, which also established the trust metrics based on multi-QoT constraints and built the most suitable model on choosing the trust path. Shekarpour et al[4] summarized methods of trust modeling in the semantic Web framework as four kinds of models, which are the centralized model, distributed model, global model and localized model, and moreover they also analyzed the advantages ,disadvantages and scope of the models. Hang[5] put forward two trust modeling and choosing methods of cloud services which were based on Bayesian network and hybridßmodel, focused on solving the problems about traditional trust model which lacks attention to the dynamic relationship between the service and assumes the operating environment blindly. Gorner et al[6] studied the method of enhancing trust modeling which was based on Multi-Agent Systems and advisor Network under the social network environment, and from the size restrictions of advisor network, trusted threshold settings and advisor recommended system integration, they gave a comprehensive upgraded program.

Trust relationship is an essential feature of cloud services among users under the social network environment, which is transitive, dynamic and usually measured by the direct trust and indirect trust. It is necessary to consider the strength of relationship, the degree of neighborhood overlap and the similarity of interest between cloud service users under the social networking environment, on basis of that to build the direct trust estimation method. In order to promote the indirect trust estimation method which is limited by distance, we need to build trust transfer model which is based on the complex networks of propagation dynamics. Finally, we will study the multiple trust path in order to build the comprehensive trust measure from the cloud service users.

2.2Trust transitivity under the social network environment

Trust transitivity means that subject1 takes advantage of the trust relationship between the subject2 and subject3. The research about trust transitivity originates from the network security, focuses on solving the identity authentication and identification problems between the nodes which are not directly related in the distributed network. The trust transitivity problem between users who are not directly related under the social network environment has always been the core issue of academia, and how to get a more accurate trust degree is an important focus of the trust transitivity model under the social network environment. Kim and Song[7] believes trust transitivity model is an important tool in the social network research. By analysing the trust path from original user to unknown target user and integrating the trust paths, we could achieve the accurate characterization to trust relationships between any users who are not directly related under the social network environment. Hamdi[8] et al study the trust transitivity model TISoN among users under the social network environment. They design the trust path searching algorithm and trust inference measurement algorithm TIM to build trust network. Lesani and Montazeri[9] proposed the method to gather and deliver blur trust under the virtual social network., and they proved that the transitivity efficiency of long trust path which has high trust degree is higher than short trust path's by experiments. Guha[10] et al used subjective logic TNA-SL to research the problem about how to analyze the trust network. They provided the semantic notation of trust transitivity and the parallel rust path. In addition, they establish trust measurement and transitivity model based on the subjective logic. What's more, some new methods or algorithms about trust link, trust paths gathering, and attenuation of trust transitivity in the field of network security have been proposed, and these methods or algorithms get a better validation result[11].

2.3Trust evolution under the social network environment

Trust relationship between users of the social network environment is not enduring, instead it will change dynamically along with the periodic oscillation of interactive strength, the update of intensity interactive mode and the adjustment of preferences[12,13]. Huang and Wilkinson[14] studied the dynamics, evolvability and generation mechanism of trust under a social network environment, proposing trust evolution model based on the complex adaptive systems, of which the trust measurement changes dynamically by the experience and interactive strength of time dimension. Zolfaghar and Aghaiel[15] researched trust forecasting techniques focusing on time, and they established a trust network evolution model based on the supervised learning method. Schall et al[16] analyzed dynamic trust of the collaborative service, studying elastic composite service framework and trust assessment method which play attention to the cross integration of human services and software services. Yahyaoui[17] used the game theory based on trust to research the service combination problem theory, and took advantage of the promotion or demotion decision obtained by game analysis to proceed trust evolution. Long Jun et al[18] built trust reasoning and evolution system which is composed of public trust evolution system and entity evolution trust evolution system, proposing the trust evolution approach which is limited by trusted entity, collection of entities and entities gradually approaching the evaluation, etc. Yin Gui-sheng et al[19] proposed the trust evolution model based on the global entity revenue function, evolutionary game of WrightFisher multi-strategy trust and incentive mechanism in order to enhance the adaptability and predictive accuracy of trust evolutionary model.

3. STATUS AND DEVELOPMENT TRENDS OF VIRTUAL COMMUNITY DISCOVERY METHOD

Social networking is a complex network structure, which is composed by different kinds of virtual communities. Depending on the difference of interests or preferences, users of social network environment will spontaneously form a large number of virtual communities, and intersection between the virtual communities will come into being, which could continue to adjust dynamically as the migration of interest or preference. Virtual communities discovery is the premise of cloud services recommendation system research under the social network environment. Identifying the virtual community of similar interests or preferences in a social network will help enhance the efficiency and accuracy of the recommendation system.

3.1Virtual community discovery

Virtual community reflects partial features and relationship of individual behavior in social network, and doing research on virtual community of social network plays a crucial role in understanding the structure and function of network. Chen and Dinh[20,21] studied non-overlapped community discovery according to the random walk method and optimization of modularity. Nikolaev[22] et al discussed the significance of virtual community discovery under the social network environment, and study the network entropy-centric approach based on the random Markov chain in order to enhance the computational efficiency of traditional methods in large-scale centralized network data. Lin et al[23] pointed out that virtual community discovery could be applied to predict social behavior of social network, and proposed a new integer programming method for analyzing the hierarchical community structure of social network. Meo[24] used the pretreatment strategies of central empowerment to enhance existing virtual community discovery method of social network that the center of any side reflects its transmission efficiency in network topology.

There are a lot of research results about virtual community discovery and structural evolution. However we still lack virtual community discovery method based on dynamic trust modeling. The technology of virtual community discovery have been applied to improve intelligent recommendation system of e-commerce by some scholars, yet little of them get involved in cloud service recommendation problem under the social network environment.

3.2Virtual community evolution

The dynamics of the social network environment and the diversity of user interest will make structure of virtual community unfixed, and thus it is necessary to analyze the evolution of this kind of phenomenon. Palla[25] noted that the stability of small community is a prerequisite to ensure its existence and the dynamics of large community is the foundation of its existence by analyzing the essential characteristic of community which is time-dependent overlapping. Folino and Pizzuti[26] adopted the evolutionary clustering method to construct multi-object optimization model on virtual community discovery, and respectively set community structures' quality of current timestamp and communities' distance of adjacent timestamp as the optimize target, and then studied some basic features about the evolution of virtual community. Fan and Yeung[27] studied the structure evolution of virtual community caused by the adjustment of users' demand under the social network environment, and analyzed the similarity of the community structure under different social networking environments. At the same time, in order to deal with the community discovery and community evolution under the dynamic social network, domestic universities have done a lot of research work and get some pretty research results. Wu Bin et al[28] studied evolution rule and time characteristics of the dynamic social networks, and proposed an efficient algorithm for community discovery to support community structures discovery and dynamic characteristics analysis of community evolution under the dynamic network. Zhang Chun-ying et al[29] utilized the concept of relevant community which is given a threshold value α to design a relevant community mining algorithm under dynamic social network in order to deal with the certain - uncertain relationship which exists in Web social network.

At present, there are more research findings of virtual community discovery and structure evolution, but less virtual community discovery methods basd on dynamic trust modeling. Scholars have already begun to apply the technology of virtual community discovery to improve the intelligent recommendation system in e-commerce, but little attention is paid to the cloud service recommendation under the social network environment. The large scale application of the open platform makes higher requirements for the cloud service recommendation system, while the partial virtual community discovery and evolution method which are oriented to the cloud service dynamic recommendation have not come into being, so it is urgent to do some related research work.

4. REASEARCH STATUS AND DEVELOPMENT DYNAMIC ANALYSIS OF CLOUD SERVICE RECOMMENDATION METHOD

Cloud service recommendation is not only an important way to solve the problem of information overload, but also a critical path for the operators of open platform to carry out automated mass customization. According to the users' interests and preferences, cloud service recommendation system could initiatively response and recommend candidate item of cloud service users may be interested in, and help users make selection decisions.

4.1Semantic Modeling and discovery of cloud service

In order to support intelligent cloud services discovery, selection and recommendation driven by demand, we need to give functional features, QoS features, using rules and restrictions of cloud services grammatical definition and logical description for building semantics specification of cloud service which has a clear structure and could be understood by computer, thus we can achieve intelligent cloud service discovery through semantic analysis and demand matching. Taking into account users' demand for quality to cloud service applications, some scholars also blended security, availability, and users' personalization preference or expectations in descriptive model in order to build a patulous cloud service semantic model which could describe users' credible constraint or QoS requirement[30].

4.2Recommendation system of cloud service

Cloud service owns a clear service object but lacks explicit resources to support the recommendation of which, so we should follow the user's demands or preferences and reach a predetermined target in the process of cloud service recommendation through multi-attribute match, related search or collaborative filtering. In earlier research, the functional demand of user is usually the key point of cloud service recommendation system[31]. With the gradual increase of homogeneous cloud service, the recommended method which takes into account QoS attributes has become a hot topic of related fields. Zhao et al[32] inspected service response time, trustrank and monetary cost, and proposed a recommendation algorithm of cloud service which could meet user's QOS demand and the constraints of fixing a price. Han[33] designed a cloud service recommendation system which could make QoS network match platforms factor of VMvare perfectly by analyzing user's demands and the actual performance of cloud service. In order to ensure the effectiveness of cloud service recommended under the background of scarce data, collaborative filtering (Collaborative Filtering, CF) technology has been widely used to study cloud service recommendation system of multiple QoS attributes, then a number of important achievements such as cloud service model based on user's CF recommendation algorithm[34], cloud service QoS prediction model based on hybrid CF algorithm[35], cloud service recommendation model based on user distribution[36] come into being. On this basis, we also put forward a set of cloud service recommendation methods of multiple QoS attributes in the information-missing background with the techniques of multi-attribute match, credible evaluation and collaborative filtering recommendation.

4.3Cloud service Recommendation based on trust

In face of massive homogenized cloud service products and continuous adjustment of the service level agreements and service standards, a user is difficult to complete cloud service selection decisions independently, who need to rely on other users, especially the advice or recommendation information from users he trusts. Cho et al[37] studied the influence of other users' selection behavior on the target users' service decision under the social network environment. Deng et al[38] studied the recommendation method involving trust relationship in cloud services under the social network environment, and applied service information, user information, interaction logs, and trust relationship of social network to the process of cloud service selection and recommendation in order to enhance the accuracy and customer satisfaction of cloud service recommendation. Liu Yingchun[39] put forward a discovery method of trusted service based on trust and recommendation relationship, and proposed a formalized definition and calculation method of the service network model. Though cloud service recommendation system based on trust under the social network environment has not come into being in academia, a large number of recommended methods based on trust about goods or service have been proposed. Chen et al[40] proposed a personalized recommendation system involving trust relationship between users to deal with cold start problems of the service or goods recommendation. Shambour and Lu[41] applied social trust network and semantic knowledge to the process of recommendation and then established a collaborative filtering framework based on trust semantic integration in order to overcome the problem of poor accuracy and low coverage of the e-business.

Social network environment brings many new challenges for structure design and algorithm research, but the process model and architecture design of the cloud service recommendation system are still insufficient. The research about cloud service recommendation based on trust is still in its infancy, and QoS missing values prediction model involving the trust relationship between users, collaborative filtering method , sort algorithms and other issues need more time to study. In addition, the existing system lacks consideration for the dynamic nature of cloud service recommendation, so it is difficult to form a cloud service dynamic recommendation mechanism which could adapt to two-dimensional evolution of time and demands.

5. CONCLUSION

Although there are some achievements in cloud service recommendation based on dynamic trust modeling in a social network environment, we still own wide prospects in researching this field, and our future work could focus on:1) About comprehensive measure of trust between the cloud service users under a social network environment, we could explore the relationship between validity and robustness of cloud service recommendation system based on trust relationship modeling and the functional requirements, QoS requirements, browsing behavior, attention comment, preferences for users;2) About local object-oriented virtual community discovery of cloud service, in order to ensure the accuracy of cloud service recommendation, we could find local virtual community which makes the target user as the core node and contain other trusted users within a certain range by the construction of trust relationship network and fitness function analysis;3) About cloud service recommendation based on dynamic trust modeling and QoS collaborative filtering, we should focuse on the effective integration of other trusted users' behavior information, service information, and trust metrics under the social network environment.

6. REFERENCES

[1] **Chang E, Hussain F, Dillon T.** Trust and reputation for service-oriented environments: technologies for building business intelligence and consumer confidence[M]. John Wiley & Sons, 2006.

[2] **Skopik F, Schall D, Dustdar S.** Modeling and mining of dynamic trust in complex service-oriented systems[J]. Information Systems, 2010, 35(7): 735-757.

[3] **Liu G, Wang Y, Orgun M A, et al.** Finding the optimal social trust path for the selection of trustworthy service providers in complex social networks[J]. Services Computing, IEEE Transactions on, 2013, 6(2): 152-167.

[4] **Shekarpour S, Katebi S D.** Modeling and evaluation of trust with an extension in semantic web[J]. Web Semantics: Science, Services and Agents on the World Wide Web, 2010, 8(1): 26-36.

[5] **Hang C W, Singh M P.** Trustworthy service selection and composition[J]. ACM Transactions on Autonomous and Adaptive Systems (TAAS), 2011, 6(1): 5.

[6] **Gorner , Zhang , Cohen R.** Improving trust modeling through the limit of advisor network size and use of referrals[J]. Electronic Commerce Research and Applications, 2013, 12(2): 112-123.

[7] **Kim Y A, Song H S.** Strategies for predicting local trust based on trust propagation in social networks[J]. Knowledge-Based Systems, 2011, 24(8): 1360-1371.

[8] **Hamdi S, Bouzeghoub A, Gancarski A L, et al.** Trust inference computation for online social networks[C]//Trust, Security and Privacy in Computing and Communications (TrustCom), 2013 12th IEEE International Conference on. IEEE, 2013: 210-217.

[9] **Lesani M, Montazeri N.** Fuzzy trust aggregation and personalized trust inference in virtual social networks[J]. Computational Intelligence, 2009, 25(2): 51-83.

[10] **sang A, Gray E, Kinateder M.** Simplification and analysis of transitive trust networks[J]. Web Intelligence and Agent Systems, 2006, 4(2): 139-161.

[11] **De Mello E R, Van Moorsel A, da Silva Fraga .** Evaluation of P2P search algorithms for discovering trust paths[M]. Springer Berlin Heidelberg, 2007.

[12] **Chen S, Wang G, ia W.** Cluster-group based trusted computing for mobile social networks using implicit social behavioral graph[J]. Future Generation Computer Systems, 2014.

[13] **De Meo P, Ferrara E, Rosaci D, et al.** Trust and compactness in social network groups[J]. Cybernetics, IEEE Transactions on, 2015, 45(2): 205-216.

[14] **Huang Y, Wilkinson I F.** The dynamics and evolution of trust in business relationships[J]. Industrial Marketing Management, 2013, 42(3): 455-465.

[15] **Zolfaghar K, Aghaie A.** Evolution of trust networks in social web applications using supervised learning[J]. Procedia Computer Science, 2011, 3: 833-839.

[16] **Schall D, Skopik F, Dustdar S.** Expert discovery and interactions in mixed service-oriented systems[J]. Services Computing, IEEE Transactions on, 2012, 5(2): 233-245.

[17] **Yahyaoui H.** A trust-based game theoretical model for Web services collaboration[J]. Knowledge-Based Systems, 2012, 27: 162-169.

[18] **Long un, Liu Xin Min, Yuan Xin climbing, etc.** A Web service composition strategy based on trust reasoning and evolution[J]. Journal of Computers, 2012, 35 (2): 298-314.

[19] **Yin Gui-sheng, Wang Ying ie, Dong Yu in.** Wright-Fisher multi-strategy trust evolution model of network configuration software[J]. Journal of Software, 2012, 23 (8): 1978-1991.

[20] **Chen , Saad Y.** Dense subgraph extraction with application to community detection[J]. Knowledge and Data Engineering, IEEE Transactions on, 2012, 24(7): 1216-1230.

[21] **Dinh T N, Thai M T.** Community Detection in Scale-Free Networks: Approximation Algorithms for Maximizing Modularity[J]. Selected Areas in Communications, IEEE Journal on, 2013, 31(6): 997-1006.

[22] **Nikolaev A G, Razib R, Kucheriya A.** On efficient use of entropy centrality for social network analysis and community detection[J]. Social Networks, 2015, 40: 154-162.

[23] **Lin C C, Kang R, Chen Y.** An integer programming approach and visual analysis for detecting hierarchical community structures in social networks[J]. Information Sciences, 2015, 299: 296-311.

[24] **De Meo P, Ferrara E, Fiumara G, et al.** Enhancing community detection using a network weighting strategy[J]. Information Sciences, 2013, 222: 648-668.

[25] **Palla G, Barab si A L, Vicsek T.** Quantifying social group evolution[J]. Nature, 2007, 446(7136): 664-667.

[26] **Folino F, Pizzuti C.** A multiobjective and evolutionary clustering method for dynamic networks[C]//Advances in Social Networks Analysis and Mining (ASONAM), 2010 International Conference on. IEEE, 2010: 256-263.

[27] **Fan W, Yeung K H.** Similarity between community structures of different online social networks and its impact on underlying community detection[J]. Communications in Nonlinear Science and Numerical Simulation, 2015, 20(3): 1015-1025.

[28] **Wu Bin, Wang Bai, Yang Sheng i.** A social network evolution analysis framework based on event [J] Journal of Software, 2011, 22 (7): 1488-1502.

[29] **Zhang Chun-ying, Guo ingfeng.** Set of social networking α community relations and dynamic mining algorithm [J]. Journal of Computers, 2013, 36 (8): 1682-1692.

[30] **egadeesan H, Balasubramaniam S.** A method to support variability of enterprise services on the cloud[C]//Cloud Computing, 2009. CLOUD'09. IEEE International Conference on. IEEE, 2009: 117-124.

[31] **Blake M B, Nowlan M F.** A web service recommender system using enhanced syntactical matching[C]//Web Services, 2007. ICWS 2007. IEEE International Conference on. IEEE, 2007: 575-582.

[32] **Zhao L, Ren Y, Li M, et al.** Flexible service selection with user-specific QoS support in service-oriented architecture[J]. Journal of Network and Computer Applications, 2012, 35(3): 962-973.

[33] **Han S M, Hassan M M, Yoon C W, et al.** Efficient service recommendation system for cloud computing market[C]//Proceedings of the 2nd international conference on interaction sciences: information technology, culture and human. ACM, 2009: 839-845.

[34] **Shao L, Zhang , Wei Y, et al.** Personalized qos prediction forweb services via collaborative filtering[C]//Web Services, 2007. ICWS 2007. IEEE International Conference on. IEEE, 2007: 439-446.

[35] **Zheng Z, Ma H, Lyu M R, et al.** Qos-aware web service recommendation by collaborative filtering[J]. Services Computing, IEEE Transactions on, 2011, 4(2): 140-152.

[36] **Chen X, Zheng Z, Yu , et al.** Web service recommendation via exploiting location and qos information[J]. Parallel and Distributed Systems, IEEE Transactions on, 2014, 25(7): 1913-1924.

[37] **Cho Y S, Steeg G V, Galstyan A.** Co-evolution of selection and influence in social networks[J]. arXiv preprint arXiv:1106.2788, 2011.

[38] **Deng S, Huang L, Xu G.** Social network-based service recommendation with trust enhancement[J]. Expert Systems with Applications, 2014, 41(18): 8075-8084.

[39] **Liu Yingchun, Zheng Xiaolin, Chen Deren.** The trusted service discovery based on trust relationship and recommendation relationship[J]. Systems Engineering-theory & Practice, 2012, 32 (12): 2789-2795.

[40] **Chen C C, Wan Y H, Chung M C, et al.** An effective recommendation method for cold start new users using trust and distrust networks[J]. Information Sciences, 2013, 224: 19-36.

[41] **Shambour , Lu .** A trust-semantic fusion-based recommendation approach for e-business applications[J]. Decision Support Systems, 2012, 54(1): 768-780.

EVALUATION FOR NODE IMPORTANCE OF MARITIME COAL SUPPLY CHAIN IN CHINA BASED ON NODE CONTRACTION

Xu-mei uan[*1,*], Xu Zhang[1], Ji-ge uan[1,2]
[1]Department of Industrial Engineering,
School of Management and Economics, anshan University, China
[2]Hebei Port Group Company Limited
[*]yxm@ysu.edu.cn

ABSTRACT

The inverse distribution of coal in China makes its supply chain has a network structure with many nodes which will trigger strong uncertainties and risks. To prevent these problems, the paper identifies the structure and node importance of maritime coal supply chain in China with node contraction method. A realistic maritime coal supply chain with mines, land transportation enterprises, load-out stations, railway administration, loading ports, marine enterprises, discharging ports, land transportation enterprises and coal owners is introduced to anchor the evaluation scheme, we calculate the node importance degree order and compare the evaluation results of different methods like connectivity (CT), betweenness (BT), node deletion (ND), and node contraction (NC). The computational results show that node contraction is more suitable for this research and through this method, the downstream land transportation enterprises and loading ports have a relatively high importance. The evaluation scheme proposed in this article can be extended to other similar supply chains.

Keywords: Maritime coal supply chain; node contraction; node importance

1. INTRODUCTION

Coal, as China's major energy, is an important component of the national economy and at the core of any other resources. According to the China National Bureau of Statistics, the resource production of coal, crude oil and natural gas are 36.8Mt, 2.15Mt and 1350CBM respectively which feature is rich coal, lack oil and miss gas. This condition determines coal accounts for 70 of energy consumption in China and it will not be 50 until 2050 according to the forecast by authority organizations.

Supply chain management based on mutual trust and open sharing is necessary among coal transportation companies to avoid uncoordinated activities, improve scheduling efficiency and meet customer personalized demand. However, because of its uniqueness, maritime coal supply chain in China has different characteristics with traditional manufacturing industries. i) Maritime coal supply chain has a long-distance transportation. About 90.29 of the coal is mined in the western provinces of Shanxi and Shaanxi and the northwestern region of Inner Mongolia. But many consumption groups are concentrated in the industrial developed areas of the southeast and central coastal provinces. Due to the inverse distribution of supply and demand, western coal eastward transporting and northern coal southward transporting have become the two major sea-rail combined transport pattern in China which make it difficult to respond quickly and supply steadily. ii) Maritime coal supply chain is greatly affected by the natural environment. The distribution features of coal resources make the transportation extensive and long-distance which needs several processes like highway, railway and sea transport. These processes and even the overall maritime coal supply chain may disrupt by natural factors of storm, rain and snow. In addition, coal in China is greatly affected by seasonal fluctuations, the demand phenomena of peak load of summer is one of the challenges facing by coal supply chain. iii) Maritime coal supply chain is obviously influenced by government policies. As already stated, coal is the major energy in China, its production, transportation and transformation can not be developed without the management and control of

[*] Corresponding Author: Xu-mei Yuan

government. For example, coal output and demand are both decline recently with the increasing intensification of environmental constraints, energy-saving emission reduction efforts and new alternative energies. The factors above have made strong uncertainties and risks for the maritime coal supply chain.

Maritime coal supply chain has a network structure with many nodes whose normal running are the precondition and guarantee for a well functioning supply chain. So, it is the key to protect the nodes to cope with supply chain risks. But, the paradox of more nodes and less invests is significant which intensifies the necessity of protected object selection based on node importance.

The research goal underlying the work described in this paper is to identify the structure and importance of nodes in maritime coal supply chain in China. We develop an evaluation scheme in the context of nodes in a real world maritime coal supply chain using node contraction method and the scheme can also be extended to other similar supply chains. In addition, we compare the results of different approaches like node deletion, betweenness to prove the efficiency and applicability of the method in this paper. Section 2 provides a summary of related work. Section 3 introduces the method of node contraction. Section 4 describes the problem and its features in China. In Sections 5 we present the evaluation scheme of node importance based on a realistic maritime coal supply chain, and calculate the importance degree order. Section 6 summarizes detailed results and comparisons of the different approaches.

2. RELATED WORK

As a typical industry supply chain, maritime coal supply chain has a late starter and the vast literature available in this topic highlights its nodes composing and transport characteristics as the research foundation to study the problems of supply chain optimization (see Peng et al[1]; Chen[2]), logistic scheduling (see Conradie et al[3]; uan et al[4]), capacity expansion (see Singh et al[5]; Thomas et al[6]), network design(see Fan[7]; Fan[8]; uan et al[9]), customer relationship management (see Peng et al[10]) and so on. Specific to nodes and their evaluation in coal chains, Peng et al (2003) emphasized the opening functional system structure based on the analysis of the traditional pyramidal structure, characteristic and material, information, and capital flows of coal supply chain[11]; Thomas et al (2014) divided the coal chain into four nodes: mines, load-out stations, ports and customers[6]; uan et al (2016) studied a coal supply chain which includes four nodes, such as mines, load points, ports and customers[9]; uan et al (2013) evaluated the importance of coal suppliers from ports perspective in maritime coal supply chain[12].

Node importance evaluation in complex networks are often solved by the methods of social network (see Poulin et al[13]) on the basis of the idea that "importance is equivalent to significance", systems science (see Chen et al[14]) on the basis of the thinking that "destruction is equivalent to importance", and information search area (see Kleinberg[15]) which take into account more factors. Herein, the typical evaluation methods include maximum likelihood estimation, F statistics and other statistical models (see Kazumi et al[16]; Kermarrec et al[17]; Kazumi et al[18]; Xiao[19]), node deletion (see Li[20]), betweenness, node contraction(see Tan et al[21]; Wang et al[22]) and so on. As to node contraction method, it is proposed by Tan et al (2006) assumes that the most important node is the one whose contraction results leads to the largest increase of the networks agglomeration by contracting the edges connected with it[21]. Compared with node degree, node deletion, and betweenness, the contraction method allows for node connectivity and position simultaneously and comprehensively which is more intuitionistic and effective.

To the best of our knowledge, this paper introduces a maritime coal supply chain in detail which includes mines, load-out stations, loading ports, discharging ports, coal owners and functional services providers as companies for rail transport, highway transport and maritime transport. According to the material flows of this chain, the importance of each node is evaluated by node contraction method and the results of different methods are compared.

3. METHODOLOGY

Node Contraction (NC) is a method of node importance evaluation in the complex network which can be described as follows in Figure.1. Suppose that v_i is a node in a network and other k_i nodes connected with v_i work well. The so-called node contraction can synchronize v_i with k_i into a new node v_i' to take place of the primary k_i+1 nodes, and then the connections with k_i+1 nodes are turned to the node v_i'.

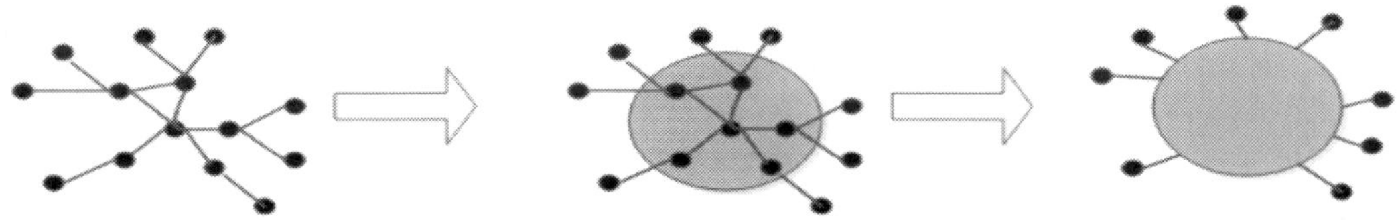

Figure.1. node contraction

The research of node contraction bases on a node-centered perspective. Take maritime coal for example, it supplies by mines and transports to resource customers by the coordination of functional services providers and transportation enterprises, and forms a complete supply chain finally. The importance of functional services providers in maritime coal supply chain can be calculated with node contraction.

The core of node contraction can be objectively translated into agglomeration which means to agglomerate v_i with its surrounding neighborhood nodes when contracting or merging nodes connected with v_i. The agglomeration of a network depends on the number of nodes and the connectivity capacity among nodes. And practically, the network agglomeration will be higher when the fewer nodes it has and the easier contacting the nodes conduct.

Supposing that a complex network $G(V,E)$ has N nodes, M edges, and the distance between nodes v_i and v_j is d_{ij}, then the connectivity capacity can be measured by the average path length $L(G)$ of the network.

$$L(G)=\frac{\sum_{i\neq j\in V} d_{ij}}{N(N-1)} \tag{1}$$

And the agglomeration of the network $\partial(G)$ is defined as follow:

$$\partial(G)=\frac{1}{N\cdot L(G)}=\frac{1}{N\cdot\frac{\sum_{i\neq j\in V} d_{ij}}{N(N-1)}}=\frac{N-1}{\sum_{i\neq j\in V} d_{ij}} \tag{2}$$

When $N=1$, the agglomeration of the network reaches the maximum value 1, and $0<\partial(G)\leq 1$.

According to formula (2), the importance of the node v_i can be defined as follow:

$$\begin{aligned}IMC(v_i)&=1-\frac{\partial(G)}{\partial(G'(v_i))}=1-\frac{\frac{1}{N\cdot L(G)}}{\frac{1}{(N-k_i)\cdot L(G'(v_i))}}\\&=\frac{N\cdot L(G)-(N-k_i)\cdot L(G'(v_i))}{N\cdot L(G)}\end{aligned} \tag{3}$$

Where $\partial(G'(v_i))$ and $L(G'(v_i))$ donate the agglomeration of network G' and its average path length after contracting node v_i respectively.

In addition, from $1 \le L(G(v_i)) < L(G)$ and $k_i \ge 1$,

$$0 < \frac{1}{N} < IMC(v_i) \le 1 - \frac{1}{N \cdot L(G)} < 1 \tag{4}$$

Iff v_i is a node in star network, the equality holds in formula (4) and $k_i = N - 1$, $L(G) = \frac{2(N-1)}{N}$, $L(G(v_i)) = 1$, $IMC(v_i) = 1 - \frac{1}{N \cdot L(G)} = 1 - \frac{1}{2(N-1)}$.)

From (4) we can see that the importance of the node v_i in the network lies on two factors: one is its connectivity degree k_i, the other is its position. The higher the degree is and the more critical the position of the node is, the greater the agglomeration is, and the more important the node is accordingly.

4. PROBLEM DESCRIPTIONS

Consider an industry supply chain where a set of independent members are connected with a single shared resource and affected by many external factors. The nodes and network structure are stable and the relationship is complicate with serve uncertainty and risks, so, the optimization of nodes and relations is rather important than the selection of partners in these supply chains to guarantee operation and defense risks with less invests. Thereinto, the evaluation of node importance is the foundation to those studies. Maritime coal supply chain is a typical industry supply chain whose operation relies on specific practices guidance and particular industry laws. To understand and solve the research problem better, we detail the problem by clearing the node composing and relationship of maritime coal supply chain.

4.1. Node composing

According to the distributing pattern and demand status, the shipping process of maritime coal in China including five stages: i) the coal is loaded by mining excavators and trucks in mine sites to the load-out stations; ii) different destinations of coal are transported by rail from mines to loading ports in the north such as inhuangdao, Tianjin, Jingtang, Huanghua, Caofeidian and so on; iii) the coal is unloaded and stacked to stockpiles at coal ports by dumpers and stackers; iv) the coal is loaded onto ships by conveyors and reclaimers to discharging ports like Ningbo, Rizhao, Penglai, Lianyungang, Jingjiang, Huizhou and so on for temporary storage or turnaround; v) the coal is transported to the ultimate customers such as the plants, cement plants and coal dealers by sea or land.

In conclusion, a typical maritime coal supply chain shown in Figure.2 includes the following node components: mines, load-out stations, loading ports, discharging ports, coal owners and functional services providers as companies for rail transport, highway transport and maritime transport.

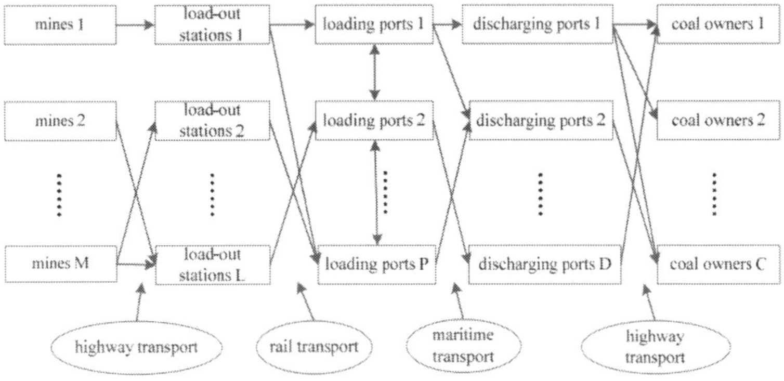

Figure. 2. nodes and structure of maritime coal supply chain

4.2. Node relationship

As a complex network of systems, it is the key to establish relationship among nodes properly for supply chain management. As already stated in our previous letter, the relationship among maritime coal supply chain is complicate which not only includes the cooperation for the overall interests and competition for own interests among nodes, but also the co-opetition relation represented by specialized cooperation and dislocation competition within nodes. Take loading ports in maritime coal supply chain as an example, they share important information with other nodes about transport paths, lay time and so on and schedule logistics synergistically to provide high-quality services for customers. Meanwhile, the ports will play competition games between cooperation and noncooperation with other nodes from their own interests. In addition, loading ports in China mainly located along bohai bay coastline and form Jin-Ji ports group including Tianjin Port, inhuangdao Port, Tangshan Port and Huanghua Port. The economic hinterland among Jin-Ji ports is obliqueness and as shown in Figure.3, there are coordinated cooperation and serve competition for information, resources and customers simultaneously.

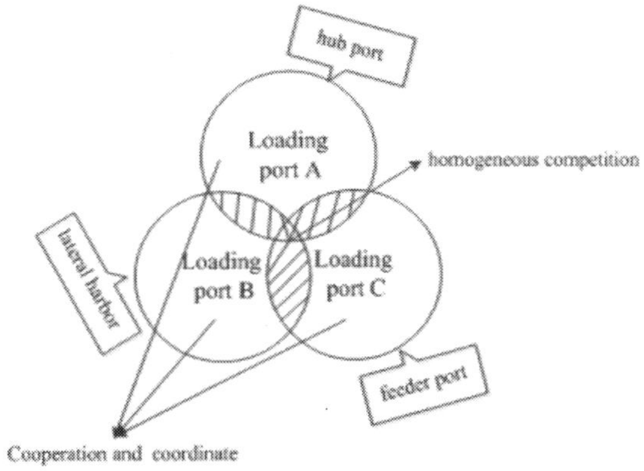

Figure.3. relations among in- i ports group

The maritime coal supply chain includes three flows: materials flows, information flows, and financial flows with no exception (see Figure.4): Because of the inverse distribution of mines and customers in China, materials flows from upstream to downstream in turn by land transportation enterprises, load-out stations, railway administration, loading ports, marine enterprises, discharging ports, land transportation enterprises ; Finance which is the motivation and guarantee for the supply chain flows from downstream to upstream; The information enables two-way flow and there could be information transfer and exchange between nodes directly in maritime coal supply chain.

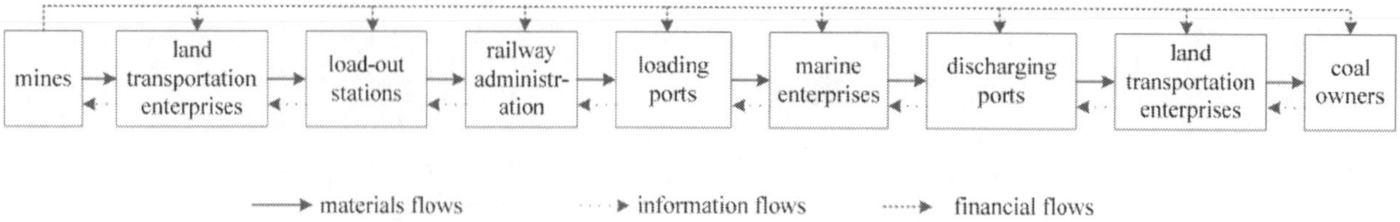

Figure.4. flows of maritime coal supply chain

The core objective of this paper is to identify the node importance based on the basic situation of maritime coal supply chain in China. To ensure the functionality and effectiveness of this research, we build the supply chain structure and relationship with logistics circumstance.

V. EVALUATION SCHEME AND CASE STUDY

This section focus on a certain maritime coal supply chain in reality, see Figure.5. The concrete planes and steps for calculating node importance in this coal supply chain using node contraction method can be concluded as follows though the analysis above.

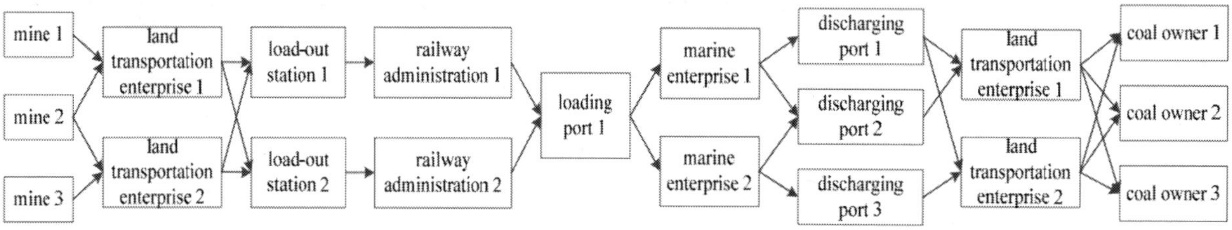

Figure.5. a certain maritime coal supply chain

Step 1 Construct the complicated network. The topological structure based on maritime coal supply chain structure is shown in Figure.6.

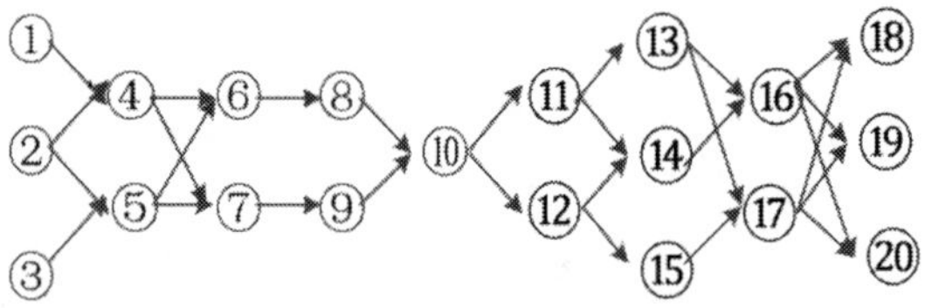

Figure.6. the topological structure of maritime coal supply chain

Therein, the corresponding relation between nodes in Figure.5 and symbols in Figure.6 can be expressed in Table I. And then the complex network can be constructed.

Table 1 corresponding relation between nodes and symbols

symbols in Figure.6	nodes in Figure.5	symbols in Figure.6	nodes in Figure.5
①, ②, ③	mines	⑪, ⑫	marine enterprises
④, ⑤	land transportation enterprises	⑬, ⑭, ⑮	discharging ports
⑥, ⑦	load-out stations	⑯, ⑰	land transportation enterprises
⑧, ⑨	railway administration	⑱, ⑲, ⑳	coal owners

⑩	loading ports

Step 2 Calculate $L(G)$. The average path length of the network can be calculated by Floyd as formula (1). Then

$$L(G) = \frac{1368}{19*20} = 3.6$$

Step 3 Calculate $L(G'(v_i))$. By contracting node v_i and drawing up the contracted topological structure accordingly, the network G' can be constructed and the corresponding average path length $L(G')$ can be obtained. For node ⑩, the contracted topological structure is shown in Figure.7.

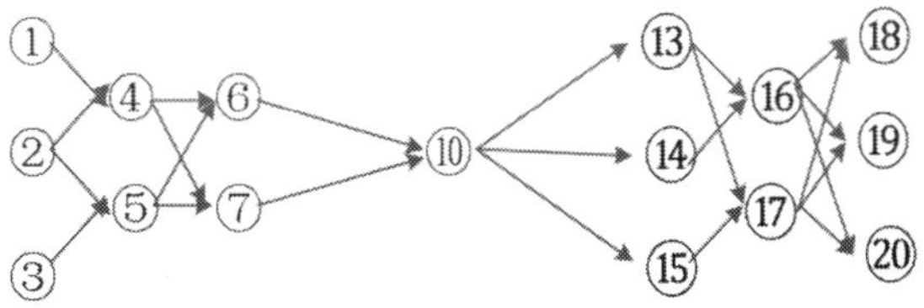

Figure. . the topological structure by contracting node ⑩

And

$$L(G'(v_{10})) = \frac{698}{16*15} = 2.9083$$

Similarly, the results of $L(G')$ for each node v_i can be reached (see Table II).

Table 2 $L(G')$ for each node v_i

node	$L(G'(v))$	node	$L(G'(v))$	node	$L(G'(v))$	node	$L(G'(v))$
①	3.4737	⑥	2.9853	⑫	3.0441	⑯	3.0755
②	3.4183	⑦	2.9853	⑬	3.0294	⑰	3.0667
③	3.4737	⑧	2.9869	⑭	3.1177	⑱	3.4837
④	3.1750	⑨	2.9869	⑮	2.2206	⑲	3.4837
⑤	3.1750	⑩	2.9083	⑯	3.1830	⑳	3.4837

Step 4 Calculate $IMC(v_i)$. Take node ⑩ as an example still, then

$$IMC(v_{10}) = \frac{3.6*20-16*2.9083}{20*3.6} = 0.353$$

based on formula (3). Table.III gives the results of $IMC(v_i)$ for each node v_i

Table 3 $IMC(v_i)$ for each node v_i

node	$IMC(v)$	node	$IMC(v)$	node	$IMC(v)$	node	$IMC(v)$
①	0.083	⑥	0.295	⑪	0.281	⑯	0.359
②	0.145	⑦	0.295	⑫	0.285	⑰	0.361
③	0.083	⑧	0.253	⑬	0.264	⑱	0.129
④	0.294	⑨	0.253	⑭	0.240	⑲	0.129
⑤	0.294	⑩	0.354	⑮	0.204	⑳	0.129

VI. RESULTS AND DISCUSSIONS

6.1. Node importance analysis of maritime coal supply chain

The following summarized features are provided by the analysis the results given in Table III: (i) the order of the node importance in the coal chain is as follows: ①③<⑱⑲⑳<②<⑮<⑭<⑧⑨<⑬<⑪<⑫< ④ ⑤ < ⑥ ⑦ < ⑩ <⑯<⑰, i.e. mines<coal owners<discharging ports<railway administration<marine enterprises < upstream land transportation enterprises < load-out stations < loading ports < downstream land transportation enterprises which intuitively shows the difference of node importance; (ii) the downstream land transportation enterprises and loading ports have a high importance because of their high connectivity. Meanwhile, the loading ports are not only the inevitable part of maritime coal transportation but also the gathering center of various information. Comparatively speaking, the importance of apex (mine nodes) and terminal (coal owner nodes) of the coal chain are relatively low by their low connectivity; (iii) the importance of nodes can be impacted by many factors such as node connectivity, node locations, importance of edges connected with nodes and so on. That is, node importance lies not only on its own importance, but also the characteristics of node connections. However, node contraction just consider the node itself that is why some nodes in this chain have the same importance in one level. To solve this problem, improved methods based on node contraction like weighted complex network (see u et al[23]; Wang et al[24]) are introduced. Because the focus of this article is not method innovation but evaluating node importance of one maritime coal supply chain to finding out the important nodes and withstanding risks effectively in the real world, we take node contraction for the article considering algorithm complexity and research operability.

6.2. Comparison of node importance of different methods

Table IV compares the node importance results calculated by the method of connectivity (CT), betweenness (BT), node deletion (ND), and node contraction (NC).

Table 4 results comparison of ct, bt, nd and nc

node	CT	BT	ND	NC	node	CT	BT	ND	NC
①	1	0.10	0	0.08	⑪	3	0.37	1	0.28
②	2	0.12	0	0.15	⑫	3	0.37	1	0.28
③	1	0.10	0	0.08	⑬	3	0.35	0.75	0.26
④	4	0.39	1	0.29	⑭	3	0.31	0.75	0.24
⑤	4	0.39	1	0.29	⑮	2	0.27	0.75	0.20
⑥	3	0.41	0.75	0.30	⑯	5	0.57	1	0.36
⑦	3	0.41	0.75	0.30	⑰	5	0.57	1	0.36
⑧	2	0.32	1	0.25	⑱	2	0.10	0	0.13
⑨	2	0.32	1	0.25	⑲	2	0.10	0	0.13
⑩	4	0.55	1	0.35	⑳	2	0.10	0	0.13

The observations are summarized below:

(1)There are many same values in the results of connectivity and node deletion method which are decided by the attribute of the algorithm itself. For example, the importance of ④⑤ and ⑯⑰ all equal to one based on node deletion method. That is because the network is turned into unconnected curves by deleting ④⑤ and ⑯⑰ which shows an inaccurate and unexpected results according to the objective reality.

(2)The results of betweenness and node contraction method have nearly the same importance ordering with node contraction which can illustrate the importance of nodes in each level and demonstrate the effectiveness of each other. However, based on the research of Tan et al (2006) based on large scale operation with many nodes, the CPU time of node contraction is significantly less than the method of betweenness.

. CONCLUSION

This paper studies a node importance evaluation problem in a real world maritime coal supply chain. We firstly propose a systemic structure of the maritime coal supply chain which includes mines, land transportation enterprises, load-out stations, railway administration, loading ports, marine enterprises, discharging ports, land transportation enterprises and coal owners. Then the relationship among the nodes which is the premise of research problem is provided. The evaluation problem is solved by the typical node contraction method based on agglomeration. Computational results show that node contraction is more suitable and effective compared with the method of betweenness, node deletion and connectivity. Although the research method is tested with the context of a maritime coal supply chain, it can be extended to other general RCSPs in contexts such as airline, wine, automobile manufacturing and the service industry.

The purpose of the article is to evaluating node importance of one maritime coal supply chain to finding out the important nodes and withstanding risks effectively in the real world. We find the downstream land transportation enterprises and loading ports are the most important nodes in our proposed coal chain and the importance of mine nodes and coal owner nodes is relatively low which is mainly because of the connectivity. Although node contraction can reach our research object, it really has some limitations. Further research will need to introduce new and feasible methods to make the results of node importance evaluation in real word supply chain more fascinating and practical.

8. ACKNOWLEDGMENT

The authors would like to thank National Social Science Foundation (NO. 13BGL006), Social Science Fund of Hebei Province (NO. HB15GL002) and Hebei Province Department of Education (NO. 2016SJBS022) for supporting this work.

. REFERENCE

[1] **Peng, H.- ., Zhou, M.-h., Liu, M.-z.,** Integrated decision model of large-scale coal supply chain and its application , Computer Integrated Manufacturing Systems, vol. 15, no. 9, pp. 1738-1743, 2009. (Chinese)

[2] **Chen, H.-m.,** The research on collecting distributing integrated collaborative management model of maritime coal logistics in China , Journal of ShanXi Finance and Economics University, no. S3, pp. 137-138, 2011. (Chinese)

[3] **Conradie, D., Morison, L., oubert, .,** Scheduling at coal handling facilities using simulated annealing , Mathematical Methods of Operations Research, vol. 68, no. 68, pp. 277-293, 2008.

[4] **Yuan .-g., Yuan, X.-m.,** Sell order-liner transport model of inhuangdao port , Business Management, no. 11, pp. 71-73, 2013. (Chinese)

[5] **Singh, G., Sier, D., Ernst, A. T., Gavriliouk, O., Oyston, R., Giles, T.,** A mixed integer programming model for long term capacity expansion planning: A case study from the hunter valley coal chain , European Journal of Operational Research, vol. 220, no. 1, pp. 210-224, 2012.

[6] **Anu Thomas, ayendran Venkateswaran, Gaurav Singh,.** A resource constrained scheduling problem with multiple independent producers and a single linking constraint: A coal supply chain example . European Journal of Operational Research, vol. 236, no. 3, pp. 946-956, 2014.

[7] **Fan, Z.- .,** Genetic algorithm for multi-stage coal supply chain network design . Computer Engineering and Applications, vol. 48, no. 24, pp. 206-211, 2012. (Chinese)

[8] **Fan, Z.- .,** Optimal design for multi-echelon coal supply chain network with complex demands , Computer Engineering and Applications, vol. 50, no. 5, pp. 21-28, 2014. (Chinese)

[9] **Yuan, X.-m., Zhang, X.,** Multi-objective seaborne coal supply chain network optimization model under port logistics capability , Mathematics in Practice and Theory, in press. (Chinese)

[10] **Peng, H.- ., Zhou, M.-h., Liu, M.-z.,** Integrated decision model of large-scale coal supply chain and its application , Computer Integrated Manufacturing Systems, vol. 15, no. 9, pp. 1738-1743, 2009. (Chinese)

[11] **Peng, C., Yue, D,** The research of opening functional structure of coal supply chain , Journal of China Coal Society, vol. 28, no. 3, 326-331, 2003. (Chinese)

[12] **Yuan, X.-m., Zhang, X., Yuan .-g.,** Research on coal suppliers evaluation from ports perspective . China Coal, vol. 46, no. 12, pp. 5-10, 2015. (Chinese)

[13] **Poulin R, Boily M C, Masse B R,** Dynamical systems to define centrality in social networks , Social Networks, vol. 22, no. 3, pp. 187-220, 2000.

[14] **Chen Y, Hu A- , Hu ,** A method for finding the most vital node in communication networks , High Technology Letters, no.1, pp. 573-575, 2004. (Chinese)

[15] **Kleinberg M,** Authoritative sources in a hyperlinked environment , Journal of the ACM, vol. 46, no. 5, pp. 604-632, 1999.

[16] **Kazumi S, Masahiro K, Kouzou O,** Super mediator - A new centrality measure of node importance for information diffusion over social network , Information Sciences, vol. 329, no.6, pp. 985-1000, 2016.

[17] **Kermarrec A, Merrer E L,** Second order centrality: distributed assessment of nodes critical in complex networks , Computer Communications, vol. 34, no. 5, pp. 619-628, 2011.

[18] **Kazumi S, Masahiro K, Kouzou O,** Discovery of super-mediators of information diffusion in social networks , presented at Proceedings of the Thirteenth International Conference of Discovery Science, vol. 6332, no. 2, pp. 144-158, 2010.

[19] **Xiao Y-m,** Identification analysis on core enterprise of supply chain from the resource based view , Science and Technology Management Research, no. 6, pp. 202-205, 2015. (Chinese)

[20] **Li X-y, Luo Z, Yan B-p.,** Research and application of importance of habitats of migratory bird based on node deletion , Application Research of Computers, vol. 32, no. 6, pp. 409-412, 2015. (Chinese)

[21] **Tan Y- , Wu , Deng H-z,** Evaluation method for node importance based on node contraction in complex networks , Systems Engineering-Theory Practice, no. 11, pp. 79-83, 2006. (Chinese)

[22] **Wang -s, Wu X-p, Yan B, Guo -w,** Improved method of node importance evaluation based on node contraction in complex networks , Procedia Engineering, vol. 11, no.1, pp. 1600-1604, 2011.

[23] **Yu, K.-p., Yang, Y., Liu, N.,** Vulnerability of complicated supply chain network based on weighted improved nodes contraction method . Computer Integrated Manufacturing Systems, vol. 20, no. 4, pp. 963-970, 2014. (Chinese)

[24] **Wang, B., Ma, R.-n, Wang G.,** Evaluation method for node importance based on improved node contraction in weighted networks , Application Research of Computers, vol. 33, no.5, pp. 1-5, 2015. (Chinese)

SIMULATION AND OPTIMIZATION OF THE PRODUCTION LOGISTICS SYSTEM IN FLUORINE ROLLER WORKSHOP

Jing-ying Zhao*
College of Mechanical Electrical Engineering,
Guangdong University of Petrochemical Technology, Maoming, Guangdong, China
mmczjy@163.com

ABSTRACT

With simulation model of the manufacturing process in Fluorine Roller Workshop by Witness, which was run on computer, we can obtain such information of the production logistics system as manufacturing outcome, rate of equipment utilization, as well as some issues which were related to material handling and production bottleneck. By the analysis of the simulation results, had pointed out each kinds of question existed in the production process and came up with the improvement plan, simulation model's running results show that the improvement plan can make the production capacity increase by 30 .†

Keywords: Improvement plan, Production logistics system, TCPN model, Witness

1. INTRODUCTION

As part of manufacturing process, production logistics is becoming important in modern manufacturing to improve their competitiveness [1]. Manufacturing practices both at home and abroad have showed fully that optimizing the production logistics plays a crucial role in cutting cost and improving both economical benefits and product quality [2].

2. BASIC DATA AND ITS BACKGROUND

This Fluorine Roller Workshop is an important part of one instrumentation company, large-scale state-owned enterprise with electrostatic duplicator as main leading product. Founded in 1958, this company is the key enterprise of duplicator industry in china. Major production of the Fluorine Roller Workshop is fluorine roller applied to printers and duplicators. Production process and material handling of all types of products are similar, thus, the optimization of production logistics is based on the main product of Fluorine Roller Workshop.

Some simple assemblies of the final products are done in this workshop. After assembly, having been lacquered, the product will be sent directly to the inspection area. Then, they will reach to packaging area while the unqualified product will be sent back for lacquering again. As the following Figure.1, the initial layout and material handling plan, the workshop has to finish assembly of 120 products each day according to the production program. We can also see from TABLE I about the specific conFigureuration of man, equipment and the time used.

* Corresponding Author: Jing-ying Zhao
† Foundation item: Guangdong provincial science and technology projects(2015A030401102)

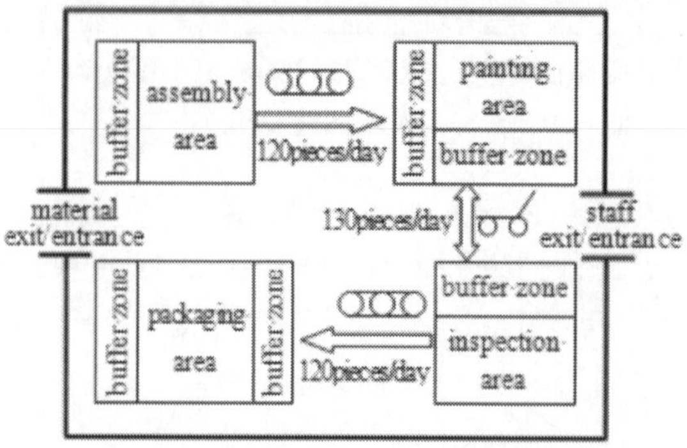

Figure.1. Initial layout and material handling plan

Table 1 configureuration of man, e uipment and the time

work or handling	equipment	personnel	time(min)
Assembly	1 assembly machine	1 operator	TNORMAL (3.2,0.5,2.0,4.5)
Painting	1 painting equipment	1 painter	UNIFORM (3.0,4.2)
Inspection		1 inspector	TNORMAL (2.5,0.2,2.2,3.0)
Packaging	1packaging equipment	1 operator	UNIFORM (3.5,5.2)
Assembly to painting(one way)	1 electric conveyor		0.25 10
Painting to inspection(two way)	1 cart	1 porter	No-load: 8 2.5; load: 8 2
Inspection to packaging(one way)	1 electric conveyor		0.25 10
Malfunction of assembly and packaging		1 repairer	ERLANG (4.0,3) ERLANG (3.2,3)

3. MODELING BY PETRI NET

The initial plan of layout and handling, together with configureuration of man and equipment [3], express the static model of Fluorine Roller Workshop [4]. We need to know how it runs and the time information by building up a dynamic model in order to describe the production logistics system in more detail if we want to simulate the whole production process [5]. Then, we established a TCPN (Timed Colored Petri net) model, as the following Figure.2, that according to some related data on technological process [6]. What do place and transition mean in Figure.2 can be seen in TABLE II.

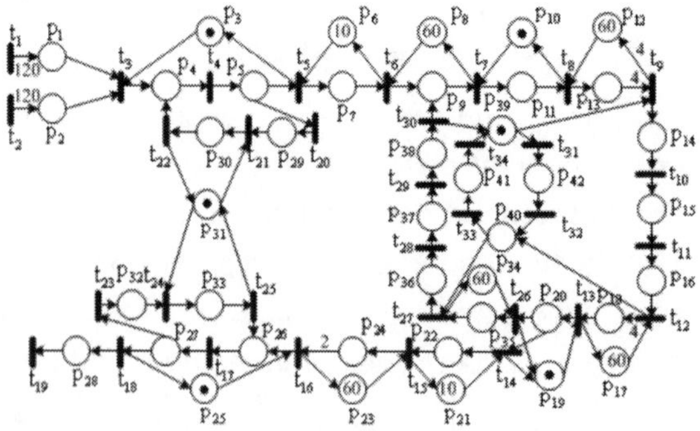

Figure.2. TCPN model of production logistics system

Table 2 meaning of place and transition

p_1	buffer zone for part1	p_{20}	inspecting	p_{39}	available cart in painting	t_{16}	start loading
p_2	buffer zone for part2	p_{21}	remained capacity of conveyor	p_{40}	available cart in inspection	t_{17}	start packing
p_3	assembly machine operator idle	p_{22}	used capacity of conveyor	p_{41}	from inspection to painting	t_{18}	finish packing
p_4	loading on assembly machine	p_{23}	remained capacity of buffer	p_{42}	from inspection to painting	t_{19}	leaving workshop
p_5	assembly	p_{24}	used capacity of buffer	t_1	arrival of part1	t_{20}	breakdown
p_6	remained capacity of conveyor	p_{25}	packaging equipment operator idle	t_2	arrival of part2	t_{21}	start repairing
p_7	used capacity of conveyor	p_{26}	loading on packaging equipment	t_3	start loading	t_{22}	finish repairing
p_8	remained capacity of buffer	p_{27}	packing	t_4	start processing	t_{23}	breakdown
p_9	used capacity of buffer	p_{28}	buffer after packing	t_5	start carrying	t_{24}	start repairing
p_{10}	painter idle	p_{29}	assembly machine breakdown	t_6	enter buffer zone	t_{25}	finish repairing
p_{11}	painting	p_{30}	repairing	t_7	start painting	t_{26}	failed products
p_{12}	remained capacity of buffer	p_{31}	repairer idle	t_8	enter buffer zone	t_{27}	start loading
p_{13}	used capacity of buffer	p_{32}	packaging equipment breakdown	t_9	start loading	t_{28}	start carrying
p_{14}	loading on cart	p_{33}	repairing	t_{10}	start carrying	t_{29}	start unloading
p_{15}	carrying	p_{34}	remained capacity of buffer	t_{11}	start unloading	t_{30}	enter buffer zone
p_{16}	unloading	p_{35}	used capacity of buffer	t_{12}	enter buffer zone	t_{31}	start off painting
p_{17}	remained capacity of buffer	p_{36}	loading unqualified products	t_{13}	start inspecting	t_{32}	reach to inspection
p_{18}	used capacity of buffer	p_{37}	carrying	t_{14}	passed products	t_{33}	start off inspection
p_{19}	Inspector idle	p_{38}	unloading	t_{15}	enter buffer	t_{34}	reach to painting

D={0,0,0,0.5,2.7,0,2.5,0,0,0,3.6,0,0,2.5,4.0,0.8,0,0,0,2.5,0,2.5,0,0,0,0.5,3.0,0,0,4.0,0,0,3.2,0, 0,1.0,4.0,0.5,0,0,3.2,3.2}. Due to the random numbers from certain distribution, the delay aggregate D is averaged out. Both the initial icons and input/output functions are marked on Figure.2.

4. SIMULATION MODEL OF PRODUCTION LOGISTICS SYSTEM IN FLUORINE ROLLER WORKSHOP BY WITNESS

Details of simulation required vary with different simulated target, which means that simulation model should be built on simulated target and simplified in order to achieve the target [7]. It not only reduces the workload but also helps avoid inaccuracy caused by unnecessary interference.

4.1. Defining Elements

Elements definition is that we define names and types after creating elements according to simulated target as well as the component of simulated system [8]. Elements of production logistics system in this Fluorine Roller Workshop are defined in TABLE III as following.

Table 3 elements definition

Name	Defined Name	Type	Amount
part1	PART1	Part	1
part2	PART2	Part	1
Assembly Machine	ASSMEBL	Machine	1
Painter	CORLORED	Machine	1
Inspector	INSPECT	Machine	1
Packaging Machine	PACKING	Machine	1
Buffer Zone 1	BUFFER1	Buffer	2
Buffer Zone 2	BUFFER2	Buffer	3
Buffer Zone 3	BUFFER3	Buffer	3
Conveyor1	CONVE OR1	Conveyor	1
Conveyor2	CONVE OR2	Conveyor	1
Track1	TRACK1	Track	1
Track2	TRACK2	Track	1
Cart	CAR	Vehicle	1
Repairer	LABOR1	Labor	1

4.2. Designing in detail

Designing in detail is to describe the dynamic mechanism and all kinds of parameters, and to make sure the running process of simulation model [9]. By defining elements' properties, rules and activities, we can keep the simulation process run as what the mathematical model describes.

Take assembly machine for example, the analysis of operating rules and the specific design are as follow. In Petri Net model, assembly machine consists of transition $t_3, t_4, t_5, t_{20}, t_{21}, t_{22}$ and place $p_3, p_4, p_5, p_{29}, p_{30}$. Transition t_3 represents loading part on assembly machine. We are required to take out Tokens respectively from p_1, p_2, p_3 to stimulate t_3, which shows that we need to take out one part1 and one part2 from buffer zone respectively so as to start assembly as long as the assembly machine and the operator are idle. Transition t_{21} represents starting to repair the machine. We are required to take out Tokens respectively from p_{29}, p_{31} to stimulate t_{21}, which shows that repair takes place on the condition that machine is malfunction and repairer is idle. Times of respective operation of the system are defined by delay function D.

5. OPERATION, ANALYSIS AND OPTIMIZATION OF SIMULATION MODEL

5.1. Operation and analysis of simulation model

After simulation testing and modifying the simulation model, run the simulation system according to simulation target, which is showed on the screen, for certain range of time, observing changes of all relevant data. We can analyze several points such as rate of machine utilization, productive time of each product, in-process inventory etc [10]. by the statistics report of Witness at an appropriate running time. Due to the production program that we have to finish 120 pieces each day in this Fluorine Roller Workshop, we defined the unit of time for simulation as minute and the time that part1, part2 takes to reach as 120 per 480 minutes. After having simulated the production process of one work day for 480 minutes, we then analyzed respectively the relevant elements.

We are required to analyze Machine statistics first since equipment utilization, quantity of production and breakdown condition through the production process reflect greatly any problems that might exist in the production logistics system. As long as operation of simulation model ended, statistically analyzing the Machine with Witness, we got the bar chart as following Figure.3.

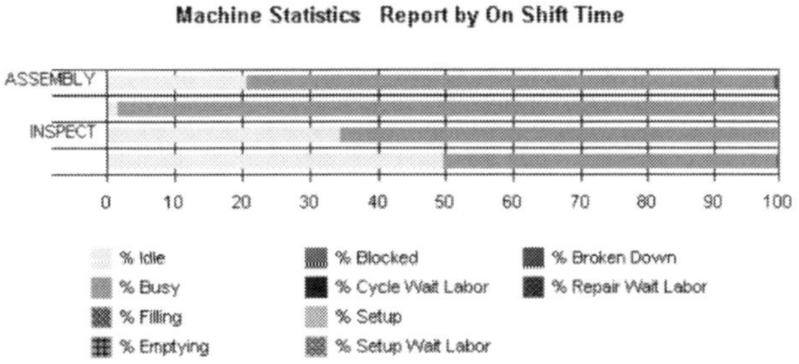

Figure.3. Bar chart of machine statistics

Bar chart above reflects directly all the condition of Machine element, through which it is easy to find out the highly or lowly utilized machine by comparing all machines horizontally. Rate of painter utilization is highest since painters are at work most of time of simulation process, while rate of packaging utilization is lowest because the previous procedure is unable to provide sufficient finished product.

Bar chart above also reflects conditions of all kinds of machine, which makes it easy to compare with, yet without providing detailed statistical data. Thus, statistical report by Witness is still required in order to analyze Machine elements further. As can be seen from Figure.4, No.of packing operation was 54 and the quantity of finished package was only 108, which didn't reach production program. The time painter used was the longest accounting for 98.49 of the whole time throughout the simulation and the amount of products he handled was the most, which were 131. The reason why this happened was that CORLORED process took too much time, what's more, some products need repainting. The rate of packing equipment utilization was low, which accounted for 49.94 of the whole time throughout the simulation, leading to too much waste of labor, this was caused by unbalance among production processes.

Name	% Idle	% Busy	% Broken Down	No. Of Operations
ASSEMBLY	20.76	78.70	0.54	120
COLORED	1.51	98.49	0.00	131
INSPECT	34.64	65.36	0.00	125
PACKING	49.80	49.94	0.26	54

Figure.4. Statistical report on machine

5.2. Improvement and optimization

We can find out relevant solutions by adding one painter. At the same time, considering very few staff enter or exit the workshop, we plan to shut down the entrance for staff, letting them as well as materials use the same entrance. Besides, place one conveyor between painting zone and inspection zone, the cart will be only responsible for carrying the unqualified products. The following Figure.5 is the statistical reports after carrying out improvement plan.

Name	% Idle	% Busy	% Blocked	% Broken Down	No. Of Operations
ASSEMBLY	20.76	78.70	0.00	0.54	120
COLORED(1)	50.28	49.68	0.04	0.00	67
COLORED(2)	49.17	50.83	0.00	0.00	67
INSPECT	29.92	70.08	0.00	0.00	134
PACKING	44.48	55.27	0.00	0.26	60

Figure.5. Statistical report on machine after carrying out improvement plan

As we can see from statistical reports above, efficiency of the production logistics system has been increased obviously by changing the layout of workshop and the way material being carried. Rate of all equipments utilization and the amount of in-process products remain at an acceptable level. In fact, if parts for study are sufficient, production capacity can reach more than 150 products per day after being improved, which is higher than before by about 30 .

6. CONCLUSIONS

Considering the studied object simulation model of production logistics system in Fluorine Roller Workshop, simulation model was modified appropriately according to the results of running simulation model which shows productivity can be increased by 30 . The limitation of this study is that studied object we chose is too simple in the structure and the scale is comparatively small. However, the idea and methods of the study still contain references on solving similar problems. Furthermore, to some big scale workshops of complicated structure, we still need to abstract and simplification according to the studied issue in order to build up relevant simulation model.

.REFERENCES

[1] **Ming Dong, Yi-fan Wu and Chang Wu,** Simulation optimization for semiconductor manufacturing logistics systems under stochastic demand, Journal of Shanghai Jiao tong University, vol. 44, no. 3, pp. 306-310, 2010(in Chinese).

[2] **Drira A, Pierreval H and Ha ri-Gabou S,** Facility layout problems: A survey, Annual Reviews in Control, vol. 31, no. 6, pp. 255-267, 2007.

[3] **Xia iang, Rong Fang and Xiu- u Lan,** Hot research areas of simulation technique in the country review of journal of system simulation in recent years, Journal of System Simulation, vol. 16, no. 9, pp. 1910-1913, 2004(in Chinese).

[4] **Y D Cui and L Huang,** Dynamic programming algorithms for generating optimal strip layouts, Computational Optimization and Applications, vol. 33, no. 23, pp. 287-301, 2006.

[5] **Fung. S. H,** A virtual warehouse system for production logistics, Production Planning and Control, no. 6, pp. 597-607, 2005.

[6] **URBAN T L,** "Inventory models with inventory-level-dependent demand: a comprehensive review and unifying theory," European Journal of Operational Research, vol. 162, no. 3, pp. 792-804, 2005.

[7] **MISHRA S, MISHRA S P, MISHRA N,** "Inventory management for deteriorating items with salvage value under time varying demand condition," American Journal of Applied Mathematics and Statistics, vol. 2, no. 3, pp. 160-162, 2014.

[8] **SONI H N,** "Optimal replenishment policies for non-instantaneous deteriorating items with price and stock sensitive demand under permissible delay in payment," International Journal of Production Economics, vol. 146, no. 1, pp. 259-268, 2013.

[9] **SANA S S,** "Price-sensitive demand for perishable items an EO model," Applied Mathematics and Computation, vol. 217, no. 13, pp. 6248-6259, 2011.

[10] **Sa adieh M S, THORSTENSON A, OKAR M R A,** "An integrated vendor buyer model with stock-dependent demand," Transportation Research Part E: Logistics and Transportation Review, vol. 46, no. 6, pp. 963-974, 2010.

THE LEARNING BEHAVIOR AND ORGANIZATIONAL INNOVATION IN BUSINESS

Xiao-bing ang*
Arts School in Xuzhou Institute of Technology,
Xuzhou, China
*xbyang_711@sina.com

†ABSTRACT

The purpose of this paper is to enhance the innovation ability of the enterprise organization. For this purpose, firstly, the enterprises organizational innovation is divided into three stages from the perspective of individuals, groups and organizational learning in enterprise, that is, decision-making learning, work groups learning, and organizational members learning. Secondly, the factors affecting enterprise organizational innovation are analyzed. Thirdly, the suggestions enhancing innovation capability are proposed based on the above points. The research of this paper could provide decision-making reference for successful implementation of enterprise organizational innovation.

Keywords: Learning behavior, innovation process, organizational innovation, suggestions

1. INTRODUCTION

In the market competition, only continuous innovating can the organization enhance the environmental adaptability, improve the efficiency of business activities and have a broader space for survival and development. With the growing global competition, more and more enterprises have realized the organization innovation is not only an important condition of enterprise growth and wealth creation, but also is the key factor for the enterprise to long-time survival and success [1].

Early people realized organized innovation from the result of innovation. They thought that organized innovation just occurs when the enterprise designed and manufactured new products and obtained better market results or benefits[2]. Though the meaning of organized innovation was expand to adopting new ideas or new behavior successfully, but the process of organizational innovation was all ignored. In fact, organizational innovation is not only a result, but also a process[3]. Amabile considered organization innovation of the enterprise as working process of using knowledge and related information for innovation[4]. Obstfeld also viewed organizational innovation as a process of creating new social connections between people and their ideas and resources[5]. Damanpour summarized above well. He thought that business organization innovation can be new products, new services, new technologies, new management and the emergence, adoption and implementation of new ideas[6].

In this paper, the innovation of enterprise organization is divided into three different stages. The paper puts forward some suggestions on strengthening the innovation ability of enterprises on the basis of analyzing the influencing factors of enterprises' organizational innovation, providing theoretical reference and decision-making reference for enterprises to implement organizational innovation.

2. THE PROCESS OF LEARNING AND ORGANIZATION INNOVATION IN ENTERPRISES

Western scholars have already revealed the process of organizational innovation from the perspective of organizational change. For example, Lewin divided organizational innovation into three stages of unfreezing, freezing and refreezing. Custer divided organizational change into six

* Corresponding Author: Xiao-bing Yang
†This research was supported by Jiangsu Province Social Science Fund of China under Grant 15JD038.

parts, such as the review and reflection, the awareness of problems, the analysis of problems, the proposed scheme, the implementation of change, the effect of feedback, etc[7,8]. However, these studies are rarely considered together with organizational learning and organizational learning. And these scholars pay little attention to the different roles of decision makers and employees in organizational innovation[9]. Next, we will research on the process of organization innovation based on decision maker from the perspective of individual, group and organizational learning in the enterprise. From this point of view, enterprise organizational innovation can be divided into three stages.

First of all, the decision makers have the idea of organizational innovation by learning, and make consensus in the decision-making layer.

The presuppose of organizational innovation is the decision makers have the ideas of organizational innovation by individual learning. Organization innovation ideas of enterprise decision-makers generally come from two aspects: one aspect is organization innovation ideas came from decision maker's own learning. Such as the decision makers formed their own organization innovation ideas based on the learning and absorbing by observing the organization and management activities of other enterprises. The other aspect is organization innovation ideas came from the others advices. Such as employees who can't make decisions persuade the decision makers to adopt their own organizational innovation by telling the decision makers their organizational innovation ideas. The above two are based on the individual learning of decision makers. But the former can be classified as direct learning, and the latter can be classified as indirect learning based on the learning of others.

It is not enough for a decision maker to have innovative ideas, and sometimes he needs to convince other leaders to have this kind of innovation. This process of persuading is actually a group learning process. Only when the enterprise decision-makers accept an organizational innovation idea by learning, can the enterprises decision-makers reach a consensus on the innovation idea. And this innovative idea is easy to implement in the enterprise. Sometimes a decision maker has sufficient power to put the idea of organization innovation into practice without the consensus of the decision-making level. In this case, the decision makers should accept this kind of innovation sooner or later. Otherwise, the organizational innovation will not be successful for an organizational innovation means that the organizational behavior, including all members of the organization, changes.

So, the first stage of the enterprise organizational innovation is the learning stage for the decision-making individuals and the decision-making groups. The learning of other members can be effective when the decision-making individuals and the decision-making groups have learned effectively. Of course, the study of this stage is mainly the study of thinking.

Secondly, the working group can formalize organization innovation ideas by learning to form a new organization management mode.

Organizational innovation ideas are often very abstract, and it is impossible to apply it directly to the organization and management of the enterprise. At this time, the decision makers need to consider how to formalize the innovative ideas, and then translate organization innovation ideas into modes and methods of organization management. This kind of organization management mode or method should make the appropriate arrangement for the group rules and individual rules, and create new forms of individual, group and organization.

The formalization of the organization innovation ideas is usually completed by a number of the enterprise members under the leadership of the decision-making level. This is a kind of deep group learning of enterprise organization management, under the guidance of organizational innovation idea. These members find the original organization management problems by group learning. Led by organizational innovation ideas, they put forward a new method of organization management which can guide the following work of organization innovation. It is very important for the working group to learn how to understand the innovation idea of decision-maker, how to combine the practice and the ideas of enterprise organization and management, and what kind of group learning and formal method should be used.

Thirdly, the members of the organization put the new organization management mode into the actual behavior by learning.

Enterprise has only obtained the behavior rules of various actors by formal learning when translating organizational innovation ideas into organization management mode. Then specific organizational behavior should be innovated according to this model to achieve the unification of form and behavior of organizational innovation. It is difficult for an enterprise to complete the change from the behavior rules to specific behavior, a long time is needed and a lot of effort must be made. After all, behavior change involved both attitudes and habits.

At this stage, behavior change occurs in three levels: the individual, the group, and the organization. All enterprises staffs including decision makers should change their behavior according to the new rules and gradually develop the habit of doing things according to the rules of innovation. In addition to the behavior of the individual needs to be changed, the behavior of groups and organizations also needs to change and behavior patterns need to be cured. If the work on the stage of formalizing organization innovation ideas is not good enough, formal management mode should be modified and completed at this stage.

Various problems encountered in the implementation of innovative models of organization management need to be solved at this stage in order to make the innovation model can be carried out smoothly. And original innovation model should be rethought and perfected from time to time to ensure its rationality. If the work at this stage cannot be done well, organizational innovation cannot be carried out smoothly.

For a larger enterprise, experimental innovation should be carried out in the part of the individuals and groups before the entire organization is implemented, because the organization innovation work involves many factors and the process is complex. Enterprise management personnel can discover problems by experiments, and then apply the experience of success or failure to the organization innovation work of the whole enterprise, improving success rate of organizational innovation.

In summary, the enterprise organization innovation generally needs to experience three stages: the innovation thought formation, the innovation thought forms and the innovation mode implementation. Any improper disposal of the three stages is likely to lead to innovation failure. Enterprises achieve one organizational innovation after another in the innovation cycle which is composed of these three stages, constantly changing their organizational management model, improving the adaptability of the organization.

3. INFLUENCING FACTORS OF ENTERPRISE ORGANIZATIONAL INNOVATION

The premise of the enterprise organization innovation is that the decision makers have the idea of organizational innovation. But in order to ensure the success of the enterprise organization innovation, the following factors should be paid attention to.

3.1. The effectiveness of organizational innovation

Innovative idea is the essence of the implementation of organizational innovation. It not only guides the organization innovation process, but also guides the following enterprise practice. The effectiveness of organization and management of enterprises directly affects the success or failure of organizational innovation. In general, the organization innovative ideas cannot be too far ahead, so that it will be unable to implement because of the lack of practice; cannot be too conservative, so that it will not miss the opportunity to innovate. The organization innovation ideas should be find the balance point between advancement and practicability. During the early stage of reform and opening-up, many China enterprises blindly learn and introduce some advanced management ideas and management modes of Western Enterprises. The consequence of this is that most of the organizational innovation is forced to stop because of their poor environmental adaptability, and few enterprises have successfully completed their organizational innovation. If we trace the cause of this phenomenon, we can find that one of the important reasons is the organizations management mode is too advanced, but Chinese enterprises lacking the basis of practice.

3.2. The effectiveness of the organization innovation ideas converting into the organization management mode

The good idea of innovation must be translated into an effective management model, for the transformation between thought and action requires rules or models as a medium. Organizational innovation is often related to the various levels and aspects of the enterprise, related to the responsibility, power and interests of many people in the enterprise. If innovative ideas are just some ideas and have not been translated into a systematic, scientific, effective management model, it is hard to imagine that the organization members will have the innovation behavior. It can be said that the effect of translating organizational innovation ideas into organization management mode directly determines the difficulty of its innovation practice, and then affecting the success or failure of the whole innovation activities. The more scientific, practical and easily-operational the organization management mode is, stronger its ability to guide the practice of enterprise innovation is. The resistance encountered in the implementation will be smaller, and success rate of organizational innovation will be higher.

3.3. The ability and determination of the decision makers to control the process of organizational innovation

Innovation means change, changing the past and reshaping the future. Organizational innovation will always encounter a variety of problems. The effectiveness of organizational innovation needs to be assessed from time to time, the communication of organizations needs to be strengthened, extra rewards and punishments need to be executed. To deal with the above problems will increase the burden on the staff. Organizational innovation will increase people's psychological burden and anxiety, bringing inconvenience to the staff and bringing conflicts of interests. Decision makers need to have enough ability and patience to deal with many problems, such as individual behavior defects, management behavior rigidity and organizational behavior inertia. In addition, it is difficult and time-consuming for the enterprise organizational that includes a large number of people and groups to complete the whole process from ideas to action, and then to the new organization behavior model. This requires decision makers have enough determination to persist in innovation. It can be said that the ability and determination of decision makers to manage organizational innovation directly determines the success or failure of organizational innovation activities.

4. STRATEGIES FOR ENTERPRISE TO PROMOTE LEARNING AND ORGANIZATIONAL INNOVATION

In order to promote the innovation of enterprise organization, it is required to develop strategies by the decision makers and organization members on the basis of understanding the process of organizational innovation and its influencing factors[10]. Specifically, promoting enterprise organizational innovation needs to do the following works.

4.1 To strengthen the innovative consciousness of the decision makers and improve their ability to manage organizational innovation

Organizational innovation is a kind of organizational learning behavior to improve the adaptability of the whole enterprise. It can't be separated from the innovation consciousness and strong leadership of the decision makers. If the decision maker does not have the innovation consciousness, good innovative ideas cannot be transformed into innovative practice of enterprises; if the decision maker is not a strong leader, organizational innovation activities involving the entire enterprise cannot be carried out.

According to psychoanalytic theory, human brain has both the tendency of mental fantasy, such as positive thinking, denying present situation, having dreams; and the tendency of mental depression, such as escaping thinking, rigid thinking, following the rules. The human brain has a strong creative mind only when a person can control abstract sign, and can associate thinking with the rules of dealing with external things. That is to say, in order to strengthen the management of innovation consciousness, it is necessary for the managers to work in the creative space of the combination of spiritual fantasy and external reality, the dependent mentality and exploiting ideas[11]. Therefore, managers should study and explore, constantly grasping new concepts, new theories and new methods of organization and management. At the same time, they have to dare to deny himself or the status quo, associating new knowledge with the practice of organization management and constructing the new model of enterprise organization and management. We can use some measures to strengthen the innovative consciousness of managers, such as learning the experience and lessons from other organizations, participating in the organization and management activities, listening to all kinds of employees carefully, treating the comments and suggestions made by employees correctly, daring to take responsibility of innovations fail, etc.

At the same time, managers should fully understand the complexity and difficulty of organizational innovation. They should be actively involved in the various innovative activities of the enterprise on the basis of full knowledge reserve. They should be willing to take responsibility for the failure of organizational innovation, be good at summarizing the experience and lessons of the failure, accumulate innovative management experience in practical innovation continually and improve their ability to manage organizational innovation.

4.2. Enhance learning consciousness of the staff and improving their enthusiasm for innovation

Enterprise organization innovation cannot be separated from the change of enterprise staff thought and behavior mode. The change of behavior mode is the employee's individual behavior innovation, which is realized by the employee learning.

Employee learning is an analysis of employee behavior, done by employee himself. Then the behavior in favor of goal realization will be repeated, the behavior which is not conducive to the realization of the goal will be avoided or corrected. Employee learning in an enterprise is a continuous process. By learning, enterprise employees gain new knowledge and experience continuously, adjust their own behavior rules constantly, change their behavior, and enhance their organizational adaptability. Innovation is the inevitable result of study. The change of employee behavior mode based on learning is the behavior innovation of employees. In the process of organizational innovation, the effectiveness of the employees accepting and adapting to new standards mostly depends on learning ability of the employees. It's hard to imagine that inflexible and conventional employees can easily change their behavior.

Employee participation is helpful to the organization innovation of the enterprise. Employee participation is not only helpful for the production of excellent innovative ideas which is suitable for the actual, but also helpful for translating organizational innovation ideas into organization management mode, and has positive significance to the implementation of organizational innovation. Employee participation can improve their learning awareness, enhancing their ability to innovate, reducing the resistance of group innovation and organization innovation, and ensuring the success of enterprise organization innovation. In addition, employee participation can also promote the communication between the staff at all levels and help the development of organizational innovation.

4.3. Promoting the organizational learning, and creating a suitable organizational environment for innovation

Organizational learning is the process of constantly developing new rules. The change of organization and management modal of enterprise which based on organizational learning is enterprise organizational innovation. Many scholars believe that organizational learning is organizational innovation. Argyris divides the learning in the organization into single-loop learning and double-loop learning[12]. The double-loop learning is actually a kind of organizational innovation, which improves the adaptability of organizational management by changing organization management rules. Senge looks organizational learning as a cycle which consists of four links[13]. The four links are generating new insights, creating new choices, taking new action and observing the action result. This cycle is in fact an organizational innovation cycle. To promote the enterprise organizational learning, not only can make the enterprise full of new ideas, but also can help employees accept all kinds of new ideas, new rules and new management mode. Organizational learning can also improve the problem-solving capacity of employees and managers, and enhance their ability to manage organizational innovation.

In order to promote the innovation of enterprise organization, a suitable environment of organization innovation should be created. Therefore, the decision makers should not pursue stability of the organization and management excessively, because the pursuit of stability may stifle innovation capacity of enterprises; On the contrary, decision makers should stimulate diversity and differences of thinking and behavior modal of enterprise members, encourage business members to maintain the vitality of organizational innovation, encourage them to break the normal procedure, challenge the traditional and conservative forces. There are a lot of differences, contradictions and imbalances in the process of organizational innovation. Decision makers should constantly improve and promote the learning of business organizations; constantly initiate and induce the innovation consciousness of the enterprise members, and prepare the required resources, energy and vitality for the organization of the enterprise.

V. CONCLUSION

Organizational innovation is crucial to the success of the enterprise market competition. Organizational innovation requires the decision makers to have innovative ideas, and to achieve consensus in the decision-making layer firstly. Next, they need to form a new common organization model in a group way. The final performance of the innovation is the change of employees' thoughts and behaviors

The factors that influence the success of the enterprise organization innovation mainly includes three aspects, namely, whether an innovative ideas is good or bad, how difficult is to transform the ideas into the organizational result, and whether the decision-maker have the adequate abilities to solve the problems that may encountered during the process of organizational innovation.

In order to promote organizational innovation, the decision makers should strengthen their consciousness of innovation and improve their ability to control the organization innovation, the employees should enhance their awareness of learning and improve their enthusiasm for participation in innovation. Besides. an support organizational environment or atmosphere should be promoted and formed.

6. REFERENCES

[1] **R. D. Ireland, M. A. Hitt, S. M. Camo, and D. L. Se ton,** "Integrating entrepreneurship actions and strategic management actions to create firm wealth", academy of Management Executive, vol. 15, no. 1, pp. 49 63, 2001.

[2] **D. Giovanni, M. Luigi, B. Andrea, and V. Marco,** "Norms as emergent properties of adaptive learning: The case of economic routines . Journal of Evolutionary Economics, Vol. 9, no. 1, pp. 5-26, 1999.

[3] **H. Wang, X. Xu, H. Wu, and . M. Sun,** "A Review of Studies on Organizational Innovation Measurement", Lanzhou Academic Journal, Vol. 11, no. 1, pp. 83-91, 2009.

[4] **T. M. Amabile,** "A model of creativity and innovation in organizations", Research in organizational behavior, Vol.22, no. 10, pp.123-167, 1988.

[5] **D. Obstfeld,** "Social networks, the Tetris Lunges orientation, and involvement in innovation", Administrative Science uarterly, Vol. 46, no. 3, pp. 100-130, 2005.

[6] **F. Damanpou,** "Organizational Size and Innovation", Organization Studies, Vol. 16, no.13, pp. 375-402, 1991

[7] **A. Kieser,** "The rules and origination learning: behaviors theory models"(in Chinese), in Organizational learning and knowledge creation, M. Dierkes Ed. SH: Shanghai People's Publishing House, pp. 125-134, 2001.

[8] **D. A. Kolb,** "Experiential learning: Experience as the source of learning and development". Englewood Cliffs, NJ: Prentice-Hall, 1984.

[9] **B. Levitt, and . G. March,** "Organizational learning", Annual Review of Sociology, Vol. 14, no.4, pp. 319-340, 1988.

[10] **P. C. Du, and X. F. GAO,** "Research on the Influence Factors of Enterprise Independent Innovation Ability and Suggestions", Science Technology and Economy, Vol. 8, no. 4, pp.21-24, 2010.

[11] **R. D. Stacey,** "Complexity and Creativity in Organizations", San Francisco: Berett-Koehler Publishers, 1996, pp. 123-132.

[12] **C. Argyris,** "On organization learning" (in Chinese), Beijing: Renmin University Press, 2004, pp. 35-51.

[13] **P. M. Senge**, "The Fifth Discipline: The art and practice of the learning organization", New ork: Doubleday Currency, 1990, pp. 53-86.

VARIOUS PROBLEMS STATISTICS IN ENGINEERING PRO ECT MANAGEMENT

Xueliang Hou, Wang uan*
College of economy and management,
North China Electric Power University, Beijing, China
*wang yuan93113@163.com

ABSTRACT

It is sure that engineering projects exist many problems which always produce disadvantage to projects and often appear suddenly. Obviously, these problems bring much bother to engineering project managers. For the purpose of researching and solving these problems, many managers have been engaged in investigating the problems in engineering projects. Some direct relationships of all problems with their stage are discovered by analyzing the statistics to various problems. The results show that it is construction stage where the amount of problems is the maximum in engineering project; the increasing of the problems' amount is outstanding from decision-making stage to arranged stage. However, the problems' amount from construction stage to conclusion stage will decrease. The result warns that it is significant to project managers to make effective ready for solving some problems in advance. So, what is very important for project managers is to improve engineering management level .

Keywords: Problems, classification, engineering project management, regulation

1. INTRODUCTION

In the implementing process of construction engineering projects (CEP), there are many indeterminate factors which always bring disadvantage to projects. They often show in various forms in construction projects just like illness phenomena in medicine. Because of these problems' randomization and variety, project managers could not be easy to find out their characteristics and distribution regulations, and it is more difficult to forecast and control them. Once these problems happen in construction projects, they might result in unqualified engineering, delay in schedule or higher engineering cost, and project failure in the worse case[1]. So, how to find out these problems is very important for construction project managers.

2. STATISTICS AND ANALYSIS OF INVESTIGATION RESULTS

If these problems in CEP are named as morbidity phenomena, the most important task is investigation to these morbidity phenomena before analyzing them. In order to make the investigation result have universality and representative, 575 project managers in 35 building companies at different districts in china have been visited, and 6782 samples with a large amount of confirmatory information which can represent and describe engineering morbidity phenomena are obtained[2]. It can be seen from the investigation results obviously that these morbidity phenomena not only have many categories and large amounts, but also are very disorder on the surface. This shows that these morbidity phenomena have a characteristic of variety. So, these

* Corresponding Author: Wang Yuan

morbidity phenomena should be analyzed from the macro point of view firstly by classifiable method.

Generally, construction project is made up of four stages, which are decision-making stage, arranged stage, construction stage and Conclusion stage. All projects' plans and objects will be implemented step by step from first stage to last stage, so, it can be sure that all problem phenomena in engineering projects would correlate with their corresponding plans and objects tightly. This means all problem phenomena can be compartmentalized into the four stages. Based upon which, some problem phenomena with same characteristic in one stage can also be united by apriority[3].

According to the method, it can be known that representational problem phenomena in CEP have 22 categories 260 subcategories, which are 4 categories and 22 subcategories in decision-making stage, 6 categories and 65 subcategories in arranged stage, 7 categories and 144 subcategories in construction stage and 5 categories 29 subcategories in the last stage[4]. All problem categories and their accumulated frequency reflected in investigation are shown in table 1.

3. DISTRIBUTION ANALYSIS OF ALL PROBLEM PHENOMENA IN EVERY STAGE

It can be seen from table 1 that the subcategory amounts and frequency of all kinds of problem phenomena in every stage are very different, this means that different problem would bring different degree's affection to project. It can be sure that the problems which often happen will bring more disadvantage and restriction to project development, and must consume more project resources, so, these problems are just major problems [5]. This characteristic shows there is a kind of primary and secondary relationship among these problem phenomena. If we want to reveal these primary and secondary relationships, the distribution regulations of all kinds of problem phenomena in every stage should be analyzed and described[6].

In table 1, all problem phenomena in decision-making stage are divided into three kinds, which are project plan, feasibility analysis and project's approval process. In the three categories, there are 7 subcategories in project plan, their accumulated frequencies are 97; there are 5 subcategories in feasibility analysis, their accumulated frequencies are 134; there are 8 subcategories in project approval process, their accumulated frequencies are 187. According to these data, we can describe them as figure 1 by statistics and Luos paslik model[7].

Table 1 statistics of all problem phennomena categories and their cumulating fre uency

Stage	Category	Subcategory amount	Frequency
Decision making stage	Project plan	7	95
	Feasibility analysis	6	136
	Approval process	7	189
Arranged stage	Organization establishment	6	123
	Engineering design	19	440
	Bid	15	551
	Engineering preparation	11	310
	Mutual cooperation	15	430
Construction stage	risk	14	278
	quality	26	534
	schedule	27	540
	cost	13	537
	safety	30	488
	Environment protection	8	283
	Resource supply	11	332
	Other tasks	13	354
Conclusion	Checking and accepting	12	356

stage	settlement	9	322
	handover	6	163
	Other tasks	5	321

Supported by Beijing Natural Science Foundation(9162014)

Similarly, according to the same method and combine with table 1, the distribution states of problem phenomena in arranged stage, construction stage and conclusion stage can also be obtained. The three stage's problem distribution figures are shown as follow from figure 2 to 4.

Meanwhile, according to above analysis results together with investigation results, some conclusions can be draw.

1)In decision-making stage, occurred problem phenomena mostly concentrate on the face of inefficient decision-making. They are often shown in such forms as overtime making-decision, delay of project plan or taking too long times in project approval process[8].

2) In arranged stage, occurred problem phenomena often concentrate on the face of unfair competition in bid, disobeying standards in design, unsuccessful project financing, complicated project procedures, etc.

3) In construction stage, although a large amount of problem phenomena are shown in such forms as bad quality, delay in schedule, high cost or safety accidents and so on, in fact, they prove that there are many such problems as inconsistent action, disordered management, unconscionable individual behaviors, breach of duty and so on in the stage.

4) In conclusion stage, the main problems concentrate on project settlement or engineering checking and accepting. But it can be known from the investigation results that the default in fund payment has become the most troublesome problem at present.

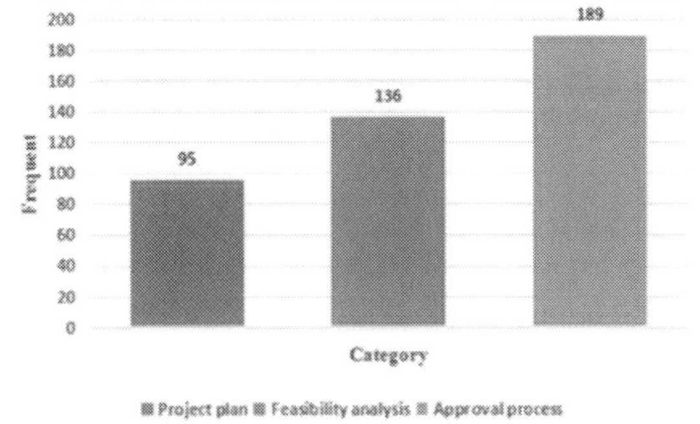

Figure 1. Distribution of problem in decision-making stage

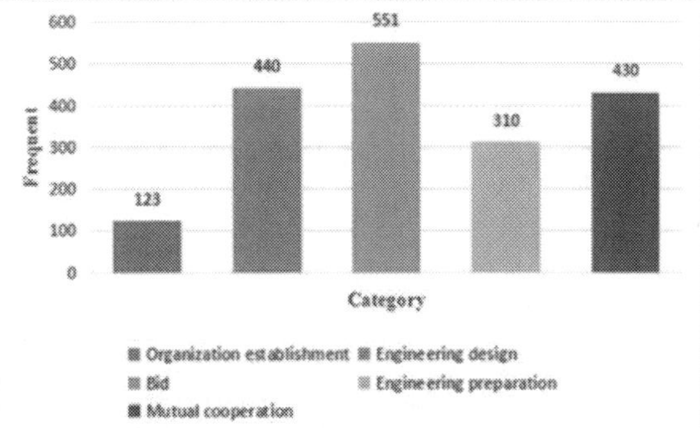

Figure 2. Distribution of problem in arranged stage

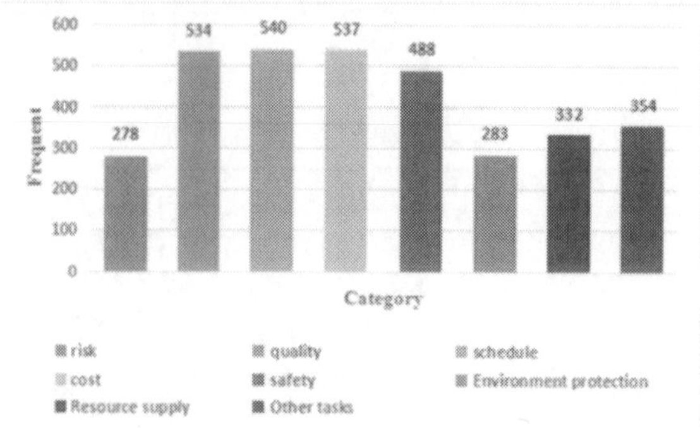

Figure 3. Distribution of problem in construction stage

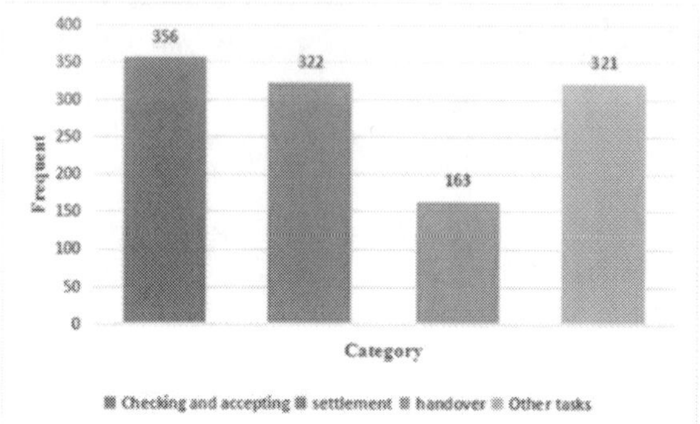

Figure 4. Distribution of problem in conclusion stage

Meanwhile, seeing from the whole of problem phenomena, two following problems occur all the time in project[9].

1) Unharmonious cooperation among many departments and the inefficiency in cooperation go though whole project, and often appear in different forms in different stages. The investigation results show the success of the project is closely connected with the two points.

2) Unconscionable individual behaviors produced by project managers often happen during the project. They usually appear in such forms as project examining and approving in decision-making stage, public bidding in arranged stage, purchasing materials in construction stage, project settlement or engineering checking as well as payment in conclusion stage. The investigation result shows project managers' unconscionable behaviors have been another main problem that closely connected with the success of the project.

3.1Total Distribution of Problem Phenomena

It can be known from above analysis of problem phenomena in four stages that the frequencies of problem phenomena in every stage are very different. If there are many classifications of problem phenomena in one stage, their accumulated frequencies are also higher, the total representation degree must be greater, and more project resource for solving them would be consumed too. So, finding out and master their distributions of problem phenomena in construction project is very important for managing project problems and optimizing project resources.

Here, let all problem phenomena in projects as a parent population. If we compartmentalize the parent population by four stages in construction project, the four stages are just subsets of the parent population[10]. According to table 1, the four subsets' sub-sample capacitance of cumulative frequencies are respectively 418, 1825, 3346 and 1162. Similarly, according to above same method and the four sub-sample capacitance, the total distribution state of problem phenomena in construction can be obtained, and it can be shown as table 2 and figure 5.

Table 2 total distributions of problem phenomena in cep

Number	Stage	Accumulated Frequencies	Frequencies rate
A	Decision-making stage	420	6
B	Arranged stage	1854	28
C	Construction stage	3346	49
D	Conclusion stage	1162	17
total		6782	100

Seeing from table 2, the cumulative frequencies of problem phenomena in construction stage come to a head in CEP. The amount in construction stage is almost a half of quantity of all problem phenomena. Although the value could not be sure as absolute rate which compare problem amount in construction stage with problem amount in other stages, at least,

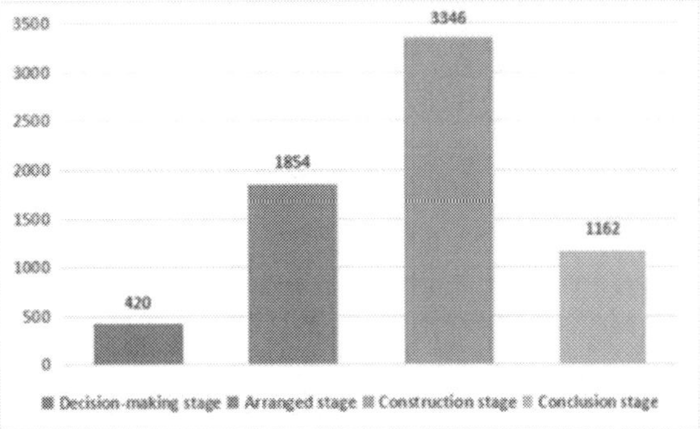

Figure 5. Total distribution of problem phenomena in CEP

It reflects the real distribution state of problem and a real relative rate in construction projects as a whole. The most important of all is that the result provides a reliable and confirmatory evidence for project managers to arrange their jobs reasonably, forecasting project problems scientifically, controlling whole project as well as planning and optimizing project resources efficiently, so, the research result has very high value in project management.

3.2. Mathematic Characteristic of Total Distribution

1)Mean value and median

The mean value and median represent a sample character of position in statistics. According to table 2, if the mean value of four stages is μ , then,

$$\mu = \frac{1}{n}\sum_{i=1}^{n}\mu_i = \frac{1}{4}(420+1854+3346+1162) = 1695.5$$

Because the sample number is even number, so, the median is 3346. Thus, it can be known by this value that problem phenomena amounts will obviously increase in the term from decision-making stage to arranged stage. The result warns that project managers should make sufficient ready for solving some problems in advance, because project is gradually going into a high peak of problems[11].

2)Range R , variance σ^2 and standard deviation $\sqrt{\sigma^2}$

In table 2, $Max(\mu_i)=3346, Min(\mu_i)=420$,

then , $R = Max(\mu_i) - Min(\mu_i) = 2926$

$$\sigma^2 = \frac{1}{n}\sum_{i=1}^{n}(\mu_i - \mu)^2$$
$$= \frac{1}{4}[(420-1695.5)^2 + (1854-1695.5)^2$$
$$+ (3346-1695.5)^2 + (1162-1695.5)^2]$$
$$= 11663167.75$$
$$\sqrt{\sigma^2} = 1079.96$$

Range, variance and standard deviation describe a sample character of decentralization in statistics[11]. Therefore, the result shows problem phenomena run though the whole life cycle of construction engineering projects.

4. CONCLUSION

It can be known from above research results that the distributions of problem phenomena in CEP are regular although they have a characteristic of variety on the surface. Especially, some following regulations are very important to project managers.

1) All problem phenomena correlate with their stage characteristics, their corresponding plans and objects tightly. This result can offer science evidences for project managers forecasting problem phenomena in projects.

2) Because unconscionable individual behaviors often happen in construction project, and they closely connected with the success of the project, project managers must forbid such behaviors if they want to realize project targets smoothly.

3) The amount of problem phenomena in construction stage is the maximum in CEP. This means they would consume lots of project resources, so, project managers should make resources scheme in advance in order to avoid resource conflict.

4) The problem phenomena amount will obviously increase from decision-making stage to arranged stage. The result warns that project managers should make sufficient ready for solving some problems in advance.

5. ACKNOWLEDGMENT

The research has been sponsored by Supported by Beijing Natural Science Foundation(9162014). So, the author is grateful their financial support in the study and investigation.

6. REFERENCES

[1] **XueLiang Hou, Mei Lu, Kai Wang.** Reduced analysis and confirmatory research on co-adaptability theoretical solution to conflicting events in construction engineering projects[J].Science China Technological Sciences,2010,532:.

[2] **F. Amato,R. Canonico,A. Mazzeo,A. Picariello.** Statistical and Lexical Analysis for Semi-automatic Extraction of Relevant Information from Legal Documents[J]. Journal of Applied Sciences,2011,114:.

[3] **Bon-Gang Hwang,Xianbo Zhao,Lene Lay Ghim Tan.** Green building projects: schedule performance, influential factors and solutions[J]. Engineering, Construction and Architectural Management,2015,223:.

[4] **Christina Kl ver.** Solving problems of project management with a self enforcing network (SEN)[J]. Computational and Mathematical Organization Theory,2012,182:.[5] Remon F. Aziz,Asmaa A. Abdel-Hakam. Exploring Delay Causes of Road Construction Projects in Egypt[J]. Alexandria Engineering Journal,2016,:.

[6] **Bon-Gang Hwang,Lay Peng Leong,Young-Ki Huh.** Sustainable green construction management: schedule performance and improvement[J]. Technological and Economic Development of Economy,2013,19sup1:.

[7] **Ma Muozhuang.** Management statistics. Beijing:Science press, 2002.

[8] **i-Man Shao,Kan Zhang,Wen-Xin Zhou.** Stein's method for nonlinear statistics: A brief survey and recent progress[J]. Journal of Statistical Planning and Inference,2016,168:.

[9] **A. V. Kolesnichenko.** On construction of the entropy transport model based on the formalism of nonextensive statistics[J]. Mathematical Models and Computer Simulations,2014,66:.

[10] **Ann T.W. Yu,Geoffrey .P. Shen.** Problems and solutions of requirements management for construction projects under the traditional procurement systems[J]. Facilities,2013,315:.

[11] **V. N. Burkov,I. V. Burkova.** Network programming technique in project management problems[J]. Automation and Remote Control,2012,737:.

FOREIGN OWNERSHIP AND OPERATING PERFORMANCE OF LISTED FIRMS: EVIDENCE FROM CHINESE A-SHARE MARKET

Crystal Xiaobei CHEN, Huimin FAN[1], Donglian MA
School of Urban Planning and Management,
Harbin Institute of Technology Shenzhen Graduate School, Shenzhen, China
*fanhuiminhit@foxmail.com

ABSTRACT

This study examines the relationship between foreign ownership or shareholding and operating performance of listed firms in Chinese stock market. Considering the influence of stock market cycles, further investigation is conducted to inspect how the relationship change in different stock market stages. Our findings prove that operating performance is positively and significantly correlated with foreign ownership in Chinese A-share market as that in many other stock markets. Further empirical analysis indicates that correlative directions are different for different stock market stages.

Keywords: A-share, foreign ownership, operating performance, stock market stage

1. INTRODUCTION

The impact of foreign ownership or shareholding on the operating performance has been well documented. With regard to researches on this relationship, the most popular method is accounting research which employs operating performance indicators as evaluation criteria based on the financial statements. Basically, there are three different perspectives on the relationship between foreign ownership and firm operating performance, which means positive correlation (Aitken Harrison 1999; Chhibber Majumdar 1999; Harris Robinson 2003; Gurbuz Asli 2010; Greenaway u 2014) [1,2,3,4], negative correlation (Xu et al. 2006; Masulis et al. 2011) [5,6] and no correlation (Barbosa Louri; Abraham 2013) [7,8]. However, an investigation by Guo Tsafack (2011) [9] proves that there exists an inverse U-shaped relation between foreign ownership and firms accounting performance represented by ROA or ROE from Chinese companies.

In this paper, we focus on listed firms in Chinese A-share market to which is paid less attention than mature stock markets in developed countries. As an emerging stock market in a developing country, Chinese stock market with a history of only 25 years has become one of the most important stock markets in the world. At the end of 2015, there's more than 2800 listed firms in both Shanghai and Shenzhen Stock Exchange accompanied by total market capitalization of about 10 trillion just after US. Due to the impressive performance of the Chinese economy and Chinese stock market, international capital flows in through FDI, FII, underground channel, etc. Meanwhile, foreign-invested firms in China continue to increase during last three decades. It's of value to conduct this research in Chinese stock market.

While a large body of literature has explore the relationship between foreign ownership and firm operating performance in US, Europe and emerging countries including China, these researches neglect the influence of stock market cycle on the relationship. International capital flows are sensitive to state of the economy, especially international indirect investment. Our study attempts to shed lights on how firm operating performance correlates with foreign ownership during different stock market stages.

Our results indicate that foreign ownership is positively and significantly correlated with firm operating performance of listed firms in Chinese A-share market. The relation between foreign ownership and operating performance is proved to be different in different stock market stages.

[1]Corresponding Author: Huimin FAN

And there are asymmetric changing amplitude of operating performance in the bear markets and the bull markets when foreign ownership is given.

2. RESEARCH DESIGN

2.1. Hypothesis

According to previous literature related on this topic and our research aim, we develop three hypothesis in this paper.

As part of emerging economies, China's capital market attracts a considerable involvement of foreign investment from countries with developed market economy. Due to the participation of foreign capital, foreign-invested firms which are supervised by the ultimate controllers and their government regulation will also be constrained by foreign investors. As part of the ownership structure, foreign ownership has its own advantages over others. On one side, foreign shareholders presented by trust and investment firms, financial companies and large international banks from the developed countries generally have professional management teams which can bring advanced management concept and management modes for listed companies to provide more business resources and management talent. As a result, operating performance of foreign-invested firms are possible to benefit from the experience and knowledge spillover. On the other side, firms with foreign ownership should be subject to more perfect legal system and stricter monitoring measures of foreign countries whose information disclosure quality are higher. Empirical study by Chen (2014) [10] suggests that listed firms are more likely to improve the quality of financial information disclosure and reduce non-standard audit opinions issued by external audits when the proportions of foreign ownership in listed firms increase. Based on the above, Hypothesis 1 is proposed.

Hypothesis 1: Foreign ownership is positively related to the operating performance of the listed firms;

In the bear market, a lot of stock prices will substantially decline, which results in inconvenience to the circulation of funds and great loss of interest. When it comes to the stock market slump or economic crisis, foreign investors lose faith in the market environment and consider the benefits of participation in corporate governance less than the cost. Therefore, most foreign investors will give up to participate in corporate governance. At the same time, foreign investors will choose to transfer funds out of the domestic stock market to other countries capital markets for less risk and higher profit. However, this kind of behavior itself will often cause great harm to the domestic stock market and expand the economic crisis, which is also the dispute point that whether foreign capital should be introduced into domestic market. So, for listed firms with foreign ownership, withdrawal of foreign capital will tend to cause greater loss when there's a downturn in the economy. According, we propose Hypothesis 2.

Hypothesis 2: In the bear market, foreign ownership is negatively related to the operating performance of listed firms.

In the bull market, rising stock price increases the enthusiasm and confidence of foreign shareholders in domestic listed firms. Foreign shareholders generally increase capital investment to meet the needs of the funds for expansion. With foreign stake climbing, foreign shareholders become major participants in firm governance. And there is a convergence between the interests of foreign shareholders and listed firms development. Owning to the great controlling stake, large foreign shareholders have incentive to motivate the managers. They have enough ability to take on the cost of incentive which can effectively avoid the phenomenon of "free riding". In addition, foreign shareholders may directly take part in firm governance with advanced management method and technological innovation in order to improve the value of listed firms to obtain long-term benefits. For these reason, we propose Hypothesis 3.

Hypothesis 3: In the bull market, foreign ownership is positively related to the performance of the listed firms.

2.2. Data and Estimated Model

Monthly market data and financial data of Chinese A-share stock listed in both Shanghai and Shenzhen Stock Exchange is adopted from January 2011 to September 2015, which is drawn from Wind. Stocks with ST or *ST are eliminated in data selection process. In this paper, we employ return-on-asset (ROA) and earnings-per-share (EPS) as indicators of firm operating performance. And all variables are defined in Table I.

Table 1 variable definition

Variable properties	Variable	Variable definition
Dependent variable	PF	Operating performance expressed in ROA and EPS.
Independent variable	FS	Proportion of foreign capital of top ten shareholders.
Control variable	SIZE	Natural logarithm of total assets of enterprises.
	HER	The sum of the proportion of the top three shareholder.
	LEV	Ratio of total liabilities to total assets.
	RPS	The current ratio of the total amount of undistributed profit divided by total equity.
	GROW	The difference between the current operating profit and previous operating profit divided by previous operating profit.

In order to examine the influence of foreign ownership on the operating performance of listed firms, SIZE, HER, LEV, RPS and GROW are used to control the influence of other specific features to make sure that the regression results will not be impacted by the market capitalization, ownership concentration, liability ratio, open distributed profit per share and the growth speed of different firms. The estimated model is designed as follows.

$$PF = \beta_0 + \beta_1 FS + \beta_2 SIZE + \beta_3 LEV + \beta_4 HER + \beta_5 RPS + \beta_6 GROW + \varepsilon \quad (1)$$

3. EMPIRICAL RESULT

In this paper, we use nonparametric method of Pagan and Sossounov (2003) [11] to divide stock market cycle into the bull market and bear market during based on the Shanghai Composite Index. The average closing price of the Shanghai Composite Index is calculated as Pt at month t. Then Pt is compared with both average closing price of the previous three months and the three months following. If the value of Pt is the maximum, then month t is peak of stock market cycle; if value of Pt is the minimum, then month t is the bottom. Additionally, the peaks and bottoms should appear alternately and repeatedly according to period of stock market. When the peak or trough shows up continuously, then the smaller successive peaks and the larger continuous bottom should be deleted. What's more, cycle time of a single bull market or bear market is required to be no less than 4 months. However, according to Edward (2003) [12], if the fluctuation range is more than 20 , then 4 months rule will be neglected. Besides, financial information disclosure of listed firm are released every quarter. Considering the financial statement release frequency, stock market cycles are rearranged as quarter data. According to the above judgment method, the sample range can be partitioned as bull market stages and bear market stages illustrated in Table II.

Table 2 Stock market cycles of sample period

Cycle	Bull		Bear	
1	2011.01 2011.04	2011 1	2011.05 2011.12	2011 2 2011 4
2	2012.01 2012.05	2012 1 2012 2	2012.06 2012.11	2012 3 2012 4
3	2012.12 2013.10	2013 1 2013 3	2013.11 2014.05	2013 4 2014 2
4	2014.06 2015.06	2014 3 2015 2	2015.07 2015.09	2015 3

The relationship between operating performance and foreign ownership is drawn from the panel data regression results obtained from fixed effect model corrected for heteroscedasticity. Empirical analysis in this paper begins with the whole sample. Table III shows the regression results of all sample range.

Table 3 regression resuils of all sample range

	ROA		EPS	
	Coef.	t-value	Coef.	t-value
FS	1.320***	3.248	6.814***	3.221
SIZE	0.339**	5.027	-3.971***	-4.562
HER	6.520***	15.936	31.780***	7.730
LEV	-3.745***	-13.636	-11.070**	-3.4668
RPS	0.731***	20.105	16.300***	12.1077
GROW	0.000	4.182	0.000	1.0503
N	42332		42332	
F-statistic	209.233		928.598	
Adj R^2	0.029		0.115	

*: p<0.10; **: p<0.05; ***: p<0.001

As reported in Table 3, the regression results of all sample range states that operating performance is positively and significantly related to foreign ownership at 1 confidence level, which is consistent with our Hypothesis 1. This means that listed firms with large foreign shareholders perform well than general listed firms. And if foreign ownership rises by 1 percent, then ROE will increase 1.32 percent and ROE will increase 6.814 percent.

Then, in order to inspect how foreign ownership and operating performance correlate during different stock market stages, we conduct the same regressions separately. Regression results are showed in Table IV and Table V.

Table 4 regression results of the bear markets

	ROE		EPS	
	Coef.	t-value	Coef.	t-value
FS	-1.727***	-2.970	-9.104***	-3.248

SIZE	-0.848***	-3.251	-12.929***	-8.693
HER	1.118***	9.324	61.213***	9.302
LEV	-5.480***	-4.663	-15.935***	-3.099
RPS	1.091***	4.146	19.855***	8.948
GROW	0.002*	1.726	0.017	1.568
N	42332		42332	
F-statistic	182.522		562.488	
Adj R2	0.052		0.144	

*: p<0.10; **: p<0.05; ***: p<0.001

Table 5 regression results of the bull markets

	ROE		EPS	
	Coef.	t-value	Coef.	t-value
FS	3.779***	6.496	18.879***	7.831
SIZE	1.718***	13.033	5.648***	8.463
HER	0.718	0.879	-3.952	-1.154
LEV	-2.063***	-3.978	-4.578*	-1.812
RPS	0.641***	5.239	14.866***	18.139
GROW	0.000	1.004	0.001	0.671
N	42332		42332	
F-statistic	200.189		613.171	
Adj R2	0.050		0.137	

*: p<0.10; **: p<0.05; ***: p<0.001

According to the results of regression analysis, when all control variable are given, foreign ownership is still positively and significantly related to operating performance during the bull market. While in bear market, operating performance is also significantly correlated with foreign ownership, but they tend to move in opposite direction. As FS is significant at 1 confidence level in all regressions, so we can infer from analysis results that foreign ownership and operating performance are positively correlated in the bull market and negatively correlated in the bear market, which is consistent with Hypothesis 2 and Hypothesis 3.

Comparing the regression results in the bear markets and the bull markets, the absolute value of slopes of FS are quite different, which imply that the declining range of operating performance during the bear market is smaller than the increasing range of operating performance during the bull market. The correlation coefficients of the bull market are approximately twice as that in the bear market.

4. CONCLUSIONS

This study used panel data model to capture the relationship between foreign ownership or shareholding and operating performance of listed firms in Chinese stock market. ROE and EPS are employed as proxies of operating performance from quarterly released financial statements. And the proportion of foreign ownership in top ten investors represents the foreign ownership in our research. It is proved that foreign ownership is positively and significantly related to operating performance in Chinese A-share market, which means that an increase in foreign ownership of listed firms will be accompanied by better operating performance.

Further analysis shows that foreign ownership and operating performance are positively correlated in the bull market, and negatively correlated in the bear market when all control variables are presumed. And the decline range and increase range are asymmetric. The cause may be that the confidence of foreign investors varies from different stock market stages, which results in the level change of firm governance.

5. REFERENCE

[1] **B. . Aitken, and A. E. Harrison,** Do domestic firms benefit from direct foreign investment evidence from Venezuela, American Economic Review, vol. 89, no. 3, pp. 605-618 , 1999.

[2] **R. Harris, and C. Robinson,** Foreign ownership and productivity in the United Kingdom estimates for U.K. manufacturing using the ARD, Review of Industrial Organization, vol. 22, no. 3, pp. 207-223, 2003.

[3] **A.O. Gurbuz, and A. Asli,** The impact of foreign ownership on firm performance, evidence from an emerging market: Turkey, American Journal of Economics Business Administration, vol. 2, no. 4, pp. 350-359, 2010.

[4] **D. Greenaway, A. Guariglia, and Z. Yu,** The more the better foreign ownership and corporate performance in China, Discussion paper, vol. 20, pp. 7-9, 2009.

[5] **D. Xu, Y. Pan, and C. Wu,** Performance of domestic and foreign-invested enterprises in China, Journal of World Business, vol. 41, no. 3, pp. 261-274, 2005.

[6] **R. W. Masulis, C. Wang, and F. Xie,** Globalizing the boardroom The effects of foreign directors on corporate governance and firm performance, Journal of Accounting Economics, vol. 53, no. 3, pp. 527-554, 2011.

[7] **N. Barbosa, and H. Louri,** Corporate performance: Does ownership matter A comparison of foreign- and domestic-owned firms in Greece and Portugal, Review of Industrial Organization, vol. 27, no. 1, pp. 73-102, 2005.

[8] **A. Abraham,** Foreign ownership and bank performance metrics in Saudi Arabia, International Journal of Islamic and Middle Eastern Finance and Management, vol. 66, no. 6, pp. 43-50, 2013.

[9] **L. Guo, and G. Tsafack,** Foreign shareholding, corporate governance and firm performance: evidence from Chinese companies, Working paper, 2011

[10] **B. B. Chen,** Research on the influence of foreign shareholders to China's Commercial Bank, Journal of Regional Financial Research, vol. 6, pp. 33- , 2014.

[11] **A. R. Pagan, and K. A. Sossounov,** A simple framework for analyzing bull and bear markets, Journal of Applied Econometrics, vol. 18, no. 1, pp. 23-46, 2001.

[12] **S. Edvards, . G. Biscarri, and F. P. D. Gracia,** Stock market cycles, financial liberalization and volatility, Journal of International Money and Finance, vol. 22, no. 7, pp. 925-955, 2003.

Research on the Practical Teaching System for Business Administration Based Upon 3C Educational Concepts

Dan-xia Song [1,*], Philip D. Taylor[2], Long-yi in[1]
[1]School of Business Administration,
Guangzhou University, Guangzhou, China
[2]Department of Business Economy,
Wesleyan College, Macon, U.S.A.
*dannasong@163.com

ABSTRACT

Practical teaching plays a significant role in the training of business administration students in universities. This paper takes the undergraduate students whose major is business administration in Guangzhou University as an example, through the questionnaire investigation to find the major problems exist in the current practical teaching system and reconstruct the practical teaching system for business administration based upon the 3C educational concepts. Finally, some improvements are proposed to ensure the system operates smoothly.

Keywords: Business Administration; practical teaching system; 3C educational concepts

1. INTRODUCTION

With the rapid increase in the demand of business administration talents, both the quantity and quality of graduates are raised gradually. How to train the applicative and innovative talents with high quality becomes an urgent problem to be solved by the major of business administration of universities and colleges. Business administration is a subject with high applicability and practicability. Students of business administration should not only master the theories of management, but also know how to apply this knowledge to analyze and solve the problems facing the company. A professor from the Business School of Harvard University has written that the business school should give the students a key which includes three qualities that the students should possess after graduation: managerial knowledge, ethics and practical skill of management. [1] How do students internalize the knowledge and transform it into ability Practical teaching may play a very important role in this process. In this paper, a questionnaire survey was given to the students whose major is business administration in Guangzhou University. The purpose of the investigation was to gather the information about the current status and students' satisfaction with practical teaching. Analysis of the data found that there are some problems existing in our current practical teaching system and tried to reconfigure the practical teaching system for students in business administration based upon the 3C concepts. Finally, some improvements are proposed to ensure the system operates effectively and efficiently.

2. ANALYSIS OF CURRENT SITUATION OF PRACTICAL TEACHING FOR BUSINESS ADMINISTRATION

In order to know the current situation and the students' satisfaction with practical teaching in business administration, our research faculty conducted a questionnaire survey to the junior and senior students whose major is business administration in Guangzhou University. We gave out 136 questionnaires and retrieved 136 questionnaires, all of them are valid, and the valid questionnaire retrieval rate is 100 .

The questionnaire consisted of two parts, one is an objective part which maintained multiple choices referring to the importance of practical teaching, the current implementation of practical

* Corresponding Author: Dan-xia Song

teaching and the satisfaction of the students. The other is a subjective part which asked the students to give some suggestions of how to improve the practical teaching. According to the results of the survey, although 83.82 students think that the practices for business administration is very important, their participation in practices is not satisfactory. As to the frequency of participation in practices, only10.29 students choose very often and often, 52.21 students choose rarely and almost none. With regard to the satisfaction of practical teaching, only 8.09 students replied they are satisfied, 47.06 replied just so-so and 44.85 replied unsatisfied.

The results showed that students have a high cognition of the significance of practical teaching, but their actual participation and satisfaction are not high. It is worthy to think deeply and improve the practical teaching system to increase the students' enthusiasm and satisfaction. Through the data analysis from questionnaire survey, several problems were found in the current practical teaching system.

1) The practical teaching goals are not clear or well-defined: In the environment of a market economy, the graduates of business administration must possess good professional qualities and personal comprehensive capabilities to adapt to the demand of society. These qualities and capabilities need to be trained by the practices. The goals of practical teaching for business administration are to train the students to become more applicative and innovative, to become competent to manage a business. The training goals should be set according to the demands derived from the competitiveness of the market and the unique characteristics of the enterprises. Also, the students' self-development demand should be given some consideration. However, through the questionnaires survey, we can find that the goals of practical teaching for business administration are neither clear nor well-defined. Both the students and the teachers paid insufficient attention to practices. They all conducted the practical activities as their teaching or studying tasks. The practices for business administration became a mere formality in some degree, and thus the results were unsatisfactory.

2) The time design of practices is not reasonable: The time design of practices is not reasonable in two primary ways. One is that the proportion of practical teaching is insufficient. The other is that the time schedule of professional practices is not workable. According to the document issued by the Educational Ministry of China, for those majors with regard to humanities and social sciences, the accumulated credits of practical teaching should be no less than 15 of the total credits. Some researchers even suggested that proportion should be increased to 20 -25 . [2] In the undergraduates' training plan for business administration of Guangzhou University which was newly revised in 2013, the total credits of practical teaching activities in business administration is up to 26 credits, about 15.9 of the total credits. Although it reached the basic requirement of Educational Ministry of China, but the proportion should be increased to be in line with the percentage advocated by the researchers. In addition, the time schedule of professional practices is not workable. As an important practical means, the professional practices are arranged from the second year in Guangzhou University. The corresponding professional practices are cognitive practice, productive practice and graduation practice, and the duration is two to four weeks at the end of each semester. The students generally complained that the practice duration was too short and they had to come back to school to have their final examinations during the practice, so the practice was not continual, both aspects contributing to their not being able to achieve their desired objective of professional practices.

3) The forms of practical teaching activities are not diversified: Normally, there are two categories of practical teaching: one is inside-campus practices and another is outside-campus practices. As for the inside-campus practices, the questionnaire survey showed that the students' participation rates are higher in case study and scenario simulation, and the participation rates are lower in academic competition, course practical training and subject research. As for the outside-campus practices, the students mainly participate in the part-time job and professional practice which are initiated and organized by the students themselves. According to the survey, 74.27 students think that their outside-campus practices are not closely related to their major. In terms of usefulness,

51.47 of students replied that they get benefits from their practices while 48.53 of students do not benefit. The investigation data also showed that the current practical teaching forms in Guangzhou University are simple and monotonous. They are mainly concentrated in case study, course experiments and professional practices. Due to the lack of practical bases and shortage of practical teaching instructors, the professional practices are conducted by the form of scattered practice, but the beneficial effect of practices cannot be guaranteed due to lack of effective supervision and inspection in this scattered practice.

4) The contents of outside-campus practices are not specific: Normally, we should design different practical teaching syllabi and define specific practice contents aimed at different forms of practices after we determine our practical teaching goals and forms. Actually in Guangzhou University, we have some specific practical teaching programs for some inside-campus practices. For example, we have syllabi for case study, course experiments, comprehensive training and academic competitions, but we do not have specific instruction for the outside-campus practices. For instance, we did not have specific instructions and documents about the cognitive practice, productive practice and graduation practice. The students reflected that they did not know what to see, what to learn, how to apply the knowledge to their internship work and how to solve the actual problems. So, short of definitive practice goals and specific practice instructions we end up with inefficiencies of practical teaching.

5) The performance evaluation of professional practices is not scientific: The performance evaluation is a very important part of practical teaching system. It helps to supervise and control the whole procedure of practical teaching and thus ensures that practical teaching can be implemented as expected. Now, we have a comparatively good and scientific evaluation system for the inside-campus practices, but the evaluation system for outside-campus professional practices still has some problems. For example, the scores of the students' professional practice were normally given by the instructors by means of a five-level system, but the standards of each level are not uniform as they can be determined by the teachers themselves. Obviously, the evaluation results are subjective because they are just according to the practice diaries and reports submitted by the students after they finished the practices. So it is absolutely necessary to set up a scientific evaluation system for the practical teaching with unified standards, scientific methods and multi-dimensions.

3. PRACTICAL TEACHING SYSTEM IN BUSINESS ADMINISTRATION BASED UPON 3C CONCEPS

According to the questionnaire survey, the significance of practices in business administration is indubitable, but the participation enthusiasm and satisfaction of students are not good. This research is aimed to solve the problems which exist in the current practical teaching system for business administration in Guangzhou University. Through the investigation and analysis, we try to constitute a newly practical teaching system in business administration which is based upon the 3C educational concepts.

The 3C educational concepts is an integrated teaching concept including three features described with three words with the beginning of letter C --they are Cubed, Customized and Clinical. [1] Cubed means the teaching environment of business administration should not be limited in class, it needs to be extended to enterprises, and it should be multidimensional interactive teaching and learning environment. Customized means the practical teaching forms and contents should be designed according to the different needs of enterprises and the career development of students. Clinical refers to conduct the practical teaching is similar to the medical clinical teaching which is opened in the teaching space and both the teachers and entrepreneurs can be the teaching subjects. Based upon the 3C concepts, we reconfigure the practical teaching system in business administration by the five dimensions which are objects, space, means, time and evaluation dimensions. The detailed structure is depicted in Figure 1.

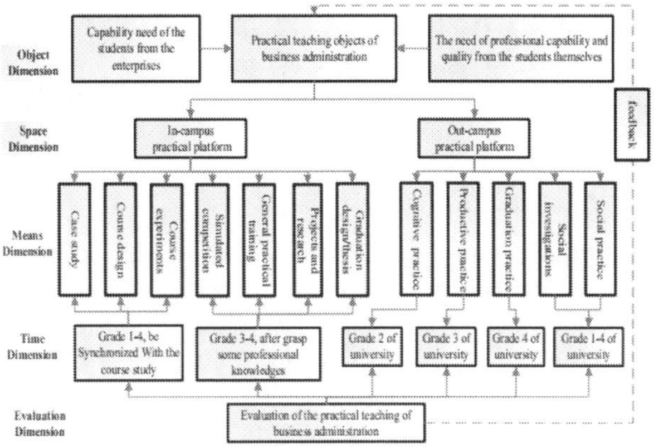

Figure.1. Reconfiguration of practical system of business administration

The five dimensions which mentioned above are illustrated as follows. The Objective Dimension means the practical system in business administration should combine the needs from the talents market, both of the enterprises and the students. The training objects of this practical system should be made comprehensively upon the above needs and should be divided into sub goals according to the different stage and different level. The Space Dimension means that the practical teaching should not be limited to the classroom; it should be extended outside of the class in order to cultivate the students' capability of solving the practical problems of the enterprises. The space dimension of practical system mainly be classified into two types of practical platforms, one is the inside-campus platform and another is the outside-campus platform. The space dimension is closely and inseparably related to the Means Dimension. Different practical space requires different practical means and contents to correspond to. The practical teaching activities during the four years of university should be conducted in stages and progressively. The space dimension and the means dimension should be integrated with Time Dimension. Different practical teaching activities should be designed in different periods in order to cultivate the student's professional capabilities progressively and systematically. The Evaluation Dimension is the evaluation of the process control and the effectiveness of the whole practical teaching system. The evaluation results need to be fed back in time, and the corrective actions should be made in time if the deviation is occur, thus the expected practical teaching objects can be satisfied eventually.

4. IMPROVEMENTS OF PRACTICAL TEACHING IN BUSINESS ADMINISTRATION

Some measures need to be taken to assure the well-functioning of the above practical teaching system for business administration. Through the questionnaire-investigation and problem analysis outlined above, the improvements of practical teaching in business administration are as follows.

1) Setting up definite goals of practical teaching: A definite goal of practical teaching should be set up based on the considering of both the needs of enterprises and the needs of the students themselves. According to the related survey, the sequencing of the capabilities which should be possessed by the business management undergraduates from the most important to the least important is like this: the capability of analysis and solving problems, communication and organization, manipulative ability, the ability of collecting and sorting out the information, the ability of combining the enterprise management reality with the advanced management thoughts, independent logical thinking and judgments, team work and etc. [3] However, the investigation from the students showed that they think the order of the capabilities they should have are sequenced in significance like this: coordination and communication, analysis of problems, learning ability, managerial ability, adaptability, professional skills and innovation. Based on the above

analysis, a feasible goal of the practical teaching system should be set up based upon the comprehensive consideration of the needs both from the enterprises and from the students themselves. The different sub goals should be set up according to the students' level and the professional knowledge they have grasped. The sub goals of each level are to build on each other one by one. Thus the capability of students can be improved gradually and the training goals can be realized eventually. [4]

2) Designing and improving the knowledge system of practices: With regard to the unsatisfied effects of the current practical teaching system in business administration, the design and improvements of practical knowledge system should be enhanced. The practical teaching syllabi should be composed according to the major core courses, and the practice manual or instruction also should be compiled in order to clarify which knowledge points the students should apply to or which ability the students should be trained during the practices. If the university has outside practice bases, the practice instructions should clarify what role the practice enterprise plays and what condition and cooperation it should provide to the internships. Thus, the university-enterprise cooperation can really have rules to follow. After having the specific practice contents as the instruction, with the addition of problem-oriented practice evaluation which consists of designing several different problems before practices, dividing the students into different project groups, where each group will be instructed by the university tutor and enterprise tutor together. The students are required to analyze and solve the problem during their practice and show the results in their practice report or make a presentation. Thus, the students will practice with problems and are instructed in specific practice contents, so they can have a definite objective in view and improve the effectiveness of scattered practices. [5]

3) Arranging the practical teaching activities systematically: As to the design of the time dimension, the time schedule arrangement of practical teaching activities should be theory first and practice second, basic practice first and general practice second with the consideration of the professional knowledge the students have learned. For example, the case study as well as the course experiment and design which are mainly designed for the certain knowledge points or a certain course can be conducted in the process of course study or just after the course learning. However, some other means of practices should be conducted after the students have finished some professional courses and processed certain professional qualities, such as general practice training, simulated competition, projects and research and graduation design and thesis. The time of graduation design and thesis can be moved up appropriately, thus the students can collect some related information from the enterprises during their graduation practice to prepare for their graduation thesis.

As to the time schedule of outside-campus practices, a survey of corporate managers shows that it will take at least two to three months for a student in the major of business administration to become familiar with the managerial procedures of a company, so two-to-four weeks of practice is obviously far from enough. [3] So, it is difficult for the students to find such a short term internship. We can shift the final examination to an earlier date, thus the practices can be extended to winter or summer holidays and this will help the students to find internship job more easily and have a long enough period to finish their practices.

4) Adopting diverse practical teaching means: With respect to practical teaching means, the traditional single practical mode needs to be replaced by diverse practical forms in order to create learning enthusiasm of students. In addition to the currently used means such as case study, course experiments and professional practices, other forms such as scenario simulation and practical rehearsal can also be used in the classroom to help the students apply the theory to practice. Students should be encouraged to participate in some competitions such as challenge cup , venture contest and ERP simulation game to practice their applicative abilities. As to the professional practices, different practices can be adopted to help the students to complete their practical activities. For example, the cognitive practice and productive practice can be conducted by means of concentrated practice which is taking the students to the local famous enterprises to visit and observe. It can also be conducted by dividing the students into several groups, designing

problems in advance, organizing the students to the out-campus practice bases or enterprises to make the post training and solving the problem during their practice. The graduation practice can be conducted by means of scattered practice. Students need to choose proper position and enterprise according to their career preference to complete their practice while collecting related information to prepare for their graduation thesis. Teachers and practical instructors should supervise the entire process. Besides this, students should be encouraged to work on part-time jobs or start self-employed businesses after class under the premise of not influencing their study. These activities are very good practices which should be encouraged and guided by university.

5) Setting up a scientific and effective evaluation system of practical teaching: The evaluation system is a very important part of the practical teaching system of business administration. The realization of the practical teaching goals needs an effective evaluation system of supervision and feedback. A scientific evaluation system can not only test the practice activity to see whether it meets the expected objects, but also help to find the deviation and correct it timely. At present, the evaluation of in-class and in-campus practical teaching in business administration is relatively perfect, but the evaluation of out-campus practices is subjective and become formalistic due to the scattered form. So it is highly necessary to work out a scientific and effective evaluation system of practical teaching in business administration. There are four types of evaluation methods of practical teaching which are commonly used. They are evaluation by process, results, obtaining evidence and comprehensive consideration. [6] In the improvement of evaluation system, these methods can be adopted integrative. Meanwhile, a 360 degrees evaluation method used in employee performance assessment can be used for reference. The evaluation from instructors both in university and in practice enterprise, customers, coworkers and students themselves all need to be considered. The students can also evaluate the instructor's performance. The improved evaluation system should be more objective and scientific.

5. CONCLUSION

The training of applicative and innovative students in business administration cannot do without the practical teaching. In this paper, through the questionnaire investigation in business administration, found the major problems existed in the current practical system. Based this analysis, reconfiguring the practical teaching system which includes five dimensions. The paper also provides some improvements to ensure the smooth operation of practical teaching system of business administration. The design of practical knowledge system and evaluation method of practical teaching system need to be further researched.

6. ACKNOWLEDGMENT

This paper was supported in part by the educational research project of Guangdong Province under Grant 2013JK136 and also supported partly by China Scholarship Council and Guangzhou University.

. APPENDICES

uestionnaire of Current Practical Teaching for Business Administration

Dear students:

We are the teaching research project group of business administration school in Guangzhou University. We need to communicate with you through this questionnaire survey in order to know the current status and satisfaction of practical teaching for undergraduates of business administration, please take a few minutes to answer the following questions. Thank you for your help

Class ________ Gender:

1、Do you think the practical teaching is important in your majors learning

A. Very important B. Somewhat important C. Medium D. Less important E. No importance

2、What is the participation frequency of practices

A. Very high B. High C. Medium D. Seldom E. Almost not

3、Which capability do you think should be stressed to train during the practical teaching

A. Coordination and communication B. Professional skills C. Capability of Management D. Learning ability E. Market insight F. Adaptability G. Innovation H. Analyzing ability

4、What kind of practices have you participated in during your study (Can be multiple choice)

A. Case study B. Scenario simulation C. Academic competition D. Course practical training E. Project research F. Professional practices G. Graduation practice H. Others

5、What kind of practice have you attend outside the campus (Can be multiple choice)

A. Professional practices B. Social study C. Part-time job D. Observation of enterprises E. Others

6、Who is the major organizer of the above practical teaching activities (No more than two choices)

A. Teachers B. The university C. The class
D. Oneself E. Classmates F. Others

7、Is the practice which you participated in related to your major

A. Closely related to B. Related to C. Medium
D. A little related to E. Have nothing to do with

8、Do you think you get benefits from your practices

A. Absolutely yes B. Get some benefits C. Medium
D. Not so much benefits E. Get no benefits

9、How do you think the atmosphere about the practical teaching in your school

A. Both the teachers and students attach importance to it; B. The teachers devote much attention to it but the students do not; C. The teachers do not care about it but the students have a high enthusiasm; D. Neither the teachers nor the students pay attention to it;

10、Are you satisfied with the practical teaching activities arranged for your major study

A. Very satisfied B. Satisfied C. Just so-so D. Not very satisfied E. Not satisfied

11、If the practices arranged in enterprises will occupy your winter or summer holidays, do you still want to strive for this chance

A. es. So long as I can improve my working ability and enhance knowledge learning through the practices. B. es, but the premise is that I have some allowance of the practices, otherwise I will not attend. C. No, I don't want to waste time since the practices have no benefits. D. It depend on the specific situation.

12、What are the major problems do you think in our practical teaching (Can be multiple choice)

A. The supports from the government to build some practical bases are not enough; B. The leaders of our university pay insufficient attention to practical teaching; C. The teachers ignore the practical teaching and make it become a mere formality; D. The students attach no importance on practices; E. Lack of cooperation between enterprises and university and the capital support is insufficient; F. There are too much theory curriculums to preempt the practical teaching; G. The practical teaching form is unitary; H. The practices are not related to my major and have no benefit for me; I. The evaluation system of practical teaching are not good. J. Some other problems

13、Which forms of practical teaching do you want to participate in (Can be multiple choice)

A. Case study B. Course practical training C. Project research D. Academic competition E. Scenario simulation F. Cooperation of enterprises and university G. Observing and visiting enterprises H. Social survey I. Professional practices J. Others

14、What are your suggestions about the practical teaching

8. REFERENCES

[1] **Chen Zhi iang.** New teaching mode for training practical management personnel--3C teaching mode [J]. Research in Higher Education of Engineering, 2010(2):134-137

[2] **Zhu Bo ing.** Construction of practical teaching system for business administration of applicative undergraduates [J]. Helongjiang Research on Higher Education, 2009(1):163-165

[3] **Xia Wenli.** Investigation of practical teaching for business administration of undergraduates and discussion of teaching reform [J]. Shanxi Jiaou (Higher Education Version), 2008(12):44-45

[4] **u Li, Liu Yu, Ge Xin uan, in Chunhua.** Design and Practice of Practical Teaching System for Business Administration [J]. Research and Exploration in Laboratory. 2009, 28(11):228-230

[5] **Xie ian, Wu Lei, Zhang iming.** Interaction between practical instructors and student project groups: Probe of out-campus practical teaching organization of business administration [J]. Journal of Xianning University, 2012, 32(5):7-9

[6] **Niu Honglei.** Construction of practical teaching system for business administration based upon employment-oriented [J]. Market Modernization, 2008(28):398-399

The analysis on constraints of human resource and financial resource in the independent innovation capacity of manufacturing enterprise in Liuzhou

Min Wei*, Lu Ma
Mangement School,
Guangxi University of Science and Technology, Liuzhou, P.R.China 545006
*weimin7201@163.com

ABSTRACT

The restriction of resources has been the biggest obstacle in the human economic development. It is the essence of economic development to overcome the scarcity of resources and approve as many as products to meet the needs of human beings. With the abundant of labor force and natural resources Liuzhou has occupied a place in Guangxi manufacturing industry chain, but due to lack of human resources and financial resources that the innovation ability of manufacturing is limited. How to enhance the innovation ability of enterprises become the focus of the task during the transition period of manufacturing enterprises in Liuzhou. Through the investigation and literature reading we start to analyze the restriction of manufacturing industry enterprise independent innovation capacity in Liuzhou from human resources and financial resources, and puts forward strategies and measures on how to enhance the innovation capability of manufacturing enterprises in Liuzhou.

Keywords: Financial resources, human resources, innovation ability, manufacturing enterprises

I. INTRODUCTION

The 21st century is the the era of knowledge economy characterized by knowledge innovation, technological innovation and high-tech. If the enterprises wants to achieve sustained and rapid growth, it must continue to innovate. However, only rely on their own efforts are not enough, the support and guarantee of external innovation resources and environment of enterprise is needed, especially the support of human resources and financial resources. Hart (1998), Capello (1999), who found that the merits of human resources will determine the success or failure of innovative performance and innovative activities. Technological innovation is inseparable from the support of human resources, and it has a serious dependence.

II. THEORETICAL BASIS

2.1. Introduction to Resource Constraints

Resource is refers to the sum of a certain place investment as economic activity or potential input factors in a certain period of time. Affected by the scarcity of resources, theory of constraints become to reach the constraints on economic development of resources factors and some of the resources is considered to be restricting the regional economic development of the "bottleneck". Resource constraint is a processes and phenomena of constraints formation since the amount of supply of resources to reduce and the decline of the quality, the increase of difficulty on development and utilization resources and favorable resource endowment in the process of social-economic development. Mainly reflected in: on the one hand, due to the restriction of the natural conditions, natural resources are at a disadvantage, that is the structural imbalance of resources; on the other hand, with the rapid growth of Chinese economy and growing population, the demand for resources is rising, coupled with the mode of extensive economic growth has not been effectively changed, specially the part of the high consumption of resources industry blind investment and low-level redundant construction is serious, inefficient use of resources, serious waste of resources insufficient to make more acute contradictions[1].

* Corresponding Author: Min Wei

Resource constraints theory believes that the enterprise is a collection of heterogeneous resources. Due to the various reasons, enterprises have the resources that they are not identical, with heterogeneity[2]. This heterogeneity determines the difference of the competitiveness of enterprises. Enterprise resources can be divided into tangible resources and intangible resources, which tangible resources including physical resources and financial resources, and intangible resources including social capital, organizational capital and human resources.

2.2. Summary of Human Resources

The innovation of human resources is the person who able to participate in science and technology innovation and promote technological development of the city or country. It represents a situation of talent supply. Human resources is the personification of knowledge and technology, It is very active as the innovative resources in the innovation activities of enterprise. Human resources plays a more important role especially in the high-tech enterprises[3]. Of course, the level of technological innovation of the enterprises don't depend on the level of one or several research and development personnel, and is usually caused by the overall level of corporate research and development team. Because of the characteristics of human resources with initiative and creativity in the process of enterprise innovation, and creativity of human resources has a body and infinite features, so a large number of outstanding talents can accelerate the pace of business innovation, the success of innovation activities rate will be greatly enhanced[4].

Zhou Guohong, Lu Lijun, Rong Fei, Li Rongping, Daniele and other studies suggest that the measurement of human resources mainly have the following indicators: talent availability, the development of talent market, talent policy, talent incentive, the talent density of science and technology, the utilization of technical talent, scientific research reserves, returned overseas students, entrepreneurs and so on[5]. In this study, combined with the availability of indicators and great influence factor on enterprise innovation, we think about the conduction of talent resources in Liuzhou from the three aspects that existing human resources of innovation, the existing senior human resources of innovation, the reserve of human resources. The indicators used are as follows: the full-time staff equivalents of study and experimental development, students per million in the number of colleges and universities, colleges and universities average number of students per teacher burden, science and technology activities in the university bachelor degree or above[6].

2.3. Summary of Financial Resources

Financial resources mainly refers to the enterprise funds used for innovation activities (including science and technology funds, subsidies, etc.)[7]. It is an important resource and material security for the activities of enterprise innovation. Financial resources have a significant role in the technology innovation activities of enterprise[8]. In this study, we consider about the availability of indicators and the greater impact factors of innovation that we use the government funds for research and experimental development expenditures and the intensity of input to measure the financial resources of Liuzhou.

Table 1 summary of elements of human resource and financial

Resource Element	Two level Index	Measurement Content
Human Resources	The full-time equivalent staff of research and experimental development	The number of full-time staff plus part-time staff translated into full-time work
	The number of students per10,000 people in colleges and universities	The number of students per100,000 people in colleges and universities
	Colleges and universities average number of students per teacher burden	Colleges and universities average number of students per teacher burden
	The number of bachelor degree or above in Science and technology activities	The number of bachelor degree or above in Science and technology activities

	Research and experimental development expenditures of government funds	It refers to R D Expenditure for the various types of funding from all levels of
Financial Resources	The total funding of Research and experimental development(the intensity of input)	Refers to the actual expenditures for basic research, applied research and experimental development in statistical year

3. ANALYSIS

The element of talent is the main body of enterprise innovation. It mainly refers to the research and development staff, its size reflects the companies emphasis on R D activities and reflects the human capital resources that businesses invest in technological innovation and R D activities to some extent[9]. At the same time, the capital is the most important factor affecting on enterprise innovation. The technological level of developed countries is greatly leading our country, the reasons is that the innovation investment funds is far more than our country. During the 1990s, the rapid growth in investment in the technology business become the main driving force to promote the technological innovation and economic growth in America[10].

3.1. Full-time E uivalent Staff of Research and E perimental Development

The full-time equivalent staff of research and experimental development refers to personnel who involved in research and experimental development research projects, administrative and support work, including items (topics) crews, technology administrative staff and support staff directly providing services for the project (task) activities. This indicator reflects the scale of human investment in activities of research and development with independent intellectual property rights.

Table 2 full-time e uivalent staff of research and development of liuzhou and the ma or cites of guang i(person years)

	2010	2011	2012	2013	2014
Guangxi	15044	20045.1	21582	20699.6	22792.8
Nanning	3501	5104.5	5279	4230.6	3508.1
Liuzhou	4555	6308.5	6401	9014.7	10303.5
Guilin	3150	3122.1	3168	2481.9	3677.9
Wuzhou	715	894.1	1021	845.5	864.1

As can be seen from Table II, from 2010 to 2014, full-time equivalent personnel of research and experimental development manufacturing enterprises in Liuzhou respectively were 4555 person-years, 6308.5 person-years, 6401 person-years, 9014.7 person-years, 10,303.5 person-years. It was a increasing trend year by year and the growth rate was 27 , 1.4 , 29 , 13 . As can be seen from the comparison of data, compared to

Nanning, Guilin and Wuzhou that Liuzhou is in the lead, and much higher than the average level of the Guangxi region, but the growth rate is not stable.

3.2. The number of students who are enrollment in colleges and universities per 100,000 people

Colleges and universities is the main source and reserves cradle of human resources to foster research activities. The following TABLE III is a compared between Liuzhou and Guangxi major cities in the average number of students per one hundred thousand and the average number of students per teacher burden

Table 3 the enrollment student and teacher burden situation in liuzhou and the ma or cities of guang i (person)

Time (year)	2010		2011		2012		2013		2014	
	The average number of students per one hundred thousand	The average number of students per teacher burden	The average number of students per one hundred thousand	The average number of students per teacher burden	The average number of students per one hundred thousand	The average number of students per teacher burden	The average number of students per one hundred thousand	The average number of students per teacher burden	The average number of students per one hundred thousand	The average number of students per teacher burden
Nanning	3844	17.34	4090	17.70	4349	18.16	4597	18.94	4882	19.32
Liuzhou	1660	17.67	1871	19.69	1897	21.29	1751	19.52	2318	26.12
Guilin	2566	22.20	2887	23.29	3021	24.09	3063	24.15	3518	26.00
Wuzhou	333	18.54	413	20.09	565	24.57	458	34.15	491	23.03
Guangxi	1100	17.93	1154	17.94	1201	17.96	1220	17.21	1282	18.63

Deadline by the end of 2014, the total population of Liuzhou is 3.779 million, far below the Nanning 7.997 million, Guilin 5.2648. From TABLE III can be seen that from 2010 to 2014 years colleges and universities average per 100000 people, the number of students in the school are far lower than those of Nanning and Guilin, but the average number of students per teacher burden is higher than the Nanning level. It shows that the popularity of higher education in Liuzhou is not high and the scientific research personnel reserve level is low, heavy burden of teachers, the strength is weak, the scientific research personnel cultivation heavier tasks, will also affect the strength and quality of research activity of teachers participate.

3.3. The Number of Bachelor Degree or Above in Science and Technology Activities

To better illustrate the level of scientific research strength, We use the number of college degree or above in science and technology activities to measure.

As can be seen from the following TABLE IV, from 2010 to 2014 the number of bachelor degree or above in Science and technology activities in Liuzhou and Guangxi major city is increasing year by year, and the annual number of Liuzhou is higher than the average of Guangxi region, Nanning, Guilin and Wuzhou. It suggests that Liuzhou in the Guangxi in terms of scientific research and high-level talent is an ideal size, but the overall growth rate of Liuzhou is slower and the growth potential is not enough optimism to be further improved.

Table 4 the number of bachelor degree or above in science and technology activities in liuzhou and guang i ma or city (person)

	2010	2011	2012	2013	2014
Guangxi	1210	1805	1998	2147	2241
Nanning	313	462	557	565	621
Liuzhou	404	532	568	643	839

Guilin	198	279	251	338	325
Wuzhou	25	72	70	96	69

3.4. The Overall Situation of Liuzhou R D Funds

According to the science and technology statistics index system of the United Nations education, scientific and Cultural Organization (UNESCO), there are two main indicators to describe the scale of investment in R D activities: one is the absolute R D funding; the other is R D expenditure as a percentage of gross domestic product (GDP). As

From the Table we can see that The sources of R D funding mainly come from the government (including institutions), enterprises, foreign and other funds, which the government and enterprises are the main channels for R D funding. The R D expenditure of Liuzhou manufacturing business financed by the government from 67.65 million yuan in 2010 increased to 108.804 million yuan in 2014 and the rate of average annual growth is 10.4 , with an annual increment of funds reached above 10.289 million yuan, accounted for 3.47 about the average of the total R D funding. The R D expenditure of Liuzhou manufacturing business financed by the enterprise from 1.85681 billion yuan in 2010 increased to 3.27288 billion yuan in 2014 and the rate of average annual growth is 13.01 percent, higher than the growth rate of government funds. The average annual increment the total R D expenditure. It indicates that Liuzhou R D activities of enterprises have occupied the absolute dominant position. But there are very few capital about the foreign and the total funds in the five years is only 105,000 yuan, negligible proportion of the total appropriations.financing reached 354.018 million yuan, far above the government's; accounted for 88.59 about the average of shown in TABLE V, the R D funding of Liuzhou from the 2.08261 billion yuan in 2010 to 3.593013 billion yuan in 2014, an average annual growth rate of 14.8 , with an annual increment of funding reached 359.6655 million yuan or more. With the growing of R D funding the overall strength trend of Liuzhou is rising, indicating that Liuzhou has recognized the role of science and technology to promote investment in the economy.

Table 5 the constitutes of liuzhou r d e penditure

Time (year)	The total funding of R D	Fund input intensity of R D	The source of funds(ten thousand yuan)				The composition of funds(ten thousand yuan)		
			Government funding	Enterprise funding	Foreign funding	Other funding	Basic research	Applied research	Experimental development
2010	208261	0.86	6765.0	185681.0	0.0	959.0	0	3468	189937
2011	239576.2	0.88	8578.9	210708.8	10.5	1448.4	0	336.2	220410.4
2012	267370	0.86	10717.0	228707.0	0.0	464.0	0	8269	231618
2013	334685.5	0.95	11920.4	296004.0	0.0	462.9	0	343.1	308044.2
2014	359301.3	0.949	10880.4	327288.0	0	1693.8	0	6059	333803.2

3.5. The Comparison of R D funding between Liuzhou and the ma or city of Guang i

First, it is a comparison of R D Expenditure of the total amount of funding between Liuzhou and the major city of Guangxi form 2010 to 2014. According to the relevant statistics, we summarize the total expenditure of R D about Liuzhou and the major city of Guangxi form 2010 to 2014 in the TABLE VI.

Table 6 the r d funds of liuzhou and guang i main city

	2010	2011	2012	2013	2014

	Total investment(ten thousand yuan)	Investment intensity( )	Total investment(ten thousand yuan)	Investment intensity( )	Total investment(ten thousand yuan)	Investment intensity( )	Total investment(ten thousand yuan)	Investment intensity( )	Total investment(ten thousand yuan)	Investment intensity( )
Guangxi	464653	0.52	624677.9	0.51	746793	0.50	854252.6	0.49	884194.8	0.47
Nanning	88227	0.73	114852	0.54	125193	0.49	117389.1	0.39	126841.1	0.38
Liuzhou	208261	0.86	239576.2	0.88	267370	0.86	334685.5	0.94	359301.3	0.95
Guilin	65019	0.76	87011.8	0.77	102678	0.70	114912.3	0.66	124846.3	0.64
Wuzhou	11246	0.16	25246.1	0.27	31190	0.24	32785.1	0.20	19767	0.11

As can be seen from TABLE VI, from 2010 to 2014, the total R D expenditure of Liuzhou and Guangxi major cities are showing increasing trend, but the total R D funding of Liuzhou is far greater than Nanning, Guilin and Wuzhou, and the rate of growth is more obvious. Meanwhile, the intensity of Liuzhou R D funding also increased year by year, Nanning, Guilin, Wuzhou have declining signs. But the rate of growth of Liuzhou R D funding is very slow. The resources of Liuzhou research funding is no advantage that does not match with the goals.

4. POLICY AND RECOMMENDATIONS

First of all, intensify the cultivation of talents. We should strengthen the overall planning and guidance of the manufacturing talent and to organize and implement the training program of manufacturing talent, According the current development of Liuzhou manufacturing industry at the present stage. It is necessary to build a number of training center of engineering innovation in colleges and universities.

Secondly, we must focus on training high-level innovative talents; Secondly, promote the cooperation between enterprises and universities. To encourage enterprises to cooperate with schools and to train the scientific research personnel, technical skills and talent compound talents that Liuzhou manufacturing is much-needed, and actively promote the combination.

Thirdly, to improve the industrial talent management system. Strengthening the forecasting of industry talent demand and improve all kinds of personnel information database, to establish talent incentive mechanism, increasing talent recognition and reward efforts.

Fourthly, to innovate the investment way of financial capital. We should integrate the related financial resources and science and technology plans. At the same time we must give full play to the guiding role and multiplier effect of government funds to support the building of business innovation capacity. To set up R D institutions and technology innovation capacity-building projects of industrial.

Fifthly, to promote financial services of innovation and technology actively. We should promote the pilot projects of science and finance and build financial system to support for the science and technology. To guide the banking financial institutions especially specializing in science and technology branch increase credit support for small and medium sized enterprises of science and technology. To encourage state-owned commercial banks to support enterprise innovation especially the small micro enterprise.

5. CONCLUSION

Generally speaking, the overall level of Liuzhou manufacturing innovation ability of enterprises is not high that the space of the progress and development is large. The improve of Liuzhou manufacturing innovation ability of enterprises has a great significance to the process of industrialization and economic, science and technology building in Liuzhou. It needs to work together with Liuzhou government, business, academia and other parties. At the same time we should build a multi-level human resources training system and strengthen overall planning and manufacturing talent development guidance and organize and implement the personnel training programs, and increase efforts to train personnel, improve the personnel training system.

6. ACKNOWLEDGMENT

This research was financially supported by the Liuzhou Science and Technology Department's Soft Science Program of 2013, NO2013J040404.

. REFERENCES

[1] **Kryukov, Valery, Moe,l.** How the Industry Structure Is Changing in the Former.U.S.S.R.[J].Oil Gas Joumah 1992, 90(20):36-38.

[2] **ayakar, Krishna Prasad.** Industry structure, regulatory choices and the diffusion of telecommunications services what governments can do to further universal service [D].Indiana University. l 999. 80-110.

[3] **Tony Proctor.** Corporate restructuring: the pitfalls of changing industry structure [J].Management Decision 2001,39 (3):197-201.

[4] **Eswar S Prasad.** Wage inequality in the United Kingdom, 1975-99[J].IMF Staff Papers.2002,49(3):339-363.

[5] **Hotelling, Harold,** The Economics of Exhaustible Resources[J]. Journal of Political Economy, 1931(4):137-175.

[6] **Solow.** Inter-generational Equity and Exhaustible Resources [J].The Review of Economic Studies, 1974 (41): 29-45.

[7] **Hartwick. .M.** Non-renewable Resource extraction programme and exhaustible resource[M]. New ork: Cambridge University Press, 1989.

[8] 2010--2014 Liuzhou statistical yearbook [M] Beijing: China Statistics Press, 2014.

[9] **Wang Li ing,** An inquiry into study Chinese Manufacturing Enterprise Independent Innovation Mechanism,[J]. Technology and Economy, 2011 (1).

[10] Bin Jiancheng in international division of labor in the manufacturing background-Chinese manufacturing development direction and countermeasures new international division system [J]. APEC, 2013, 01.

AN ENERGY-EFFICIENT SCHEDULING METHOD FOR A PARALLEL-MACHINE SCHEDULING PROBLEM IN AUTOMOBILE MOULD MACHINING

Shijin Wang*, Shuan Ma
Department of Management Science Engineering,
School of Economics Management, Tongji University, Shanghai, China, 200092
*shijinwang@tongji.edu.cn

ABSTRACT

Recently, production scheduling with energy consideration has been attracting more and more attentions in both academia and industry. In this paper, we discuss an energy-efficient scheduling method for a parallel machine scheduling problem in an automobile mould manufacturing shop floor. The scheduling problem considers two objectives, namely makespan for production aspects and total energy consumption for energy aspects. A constructive heuristic method is proposed to solve the problem. Finally, a case study is applied to demonstrate the effectiveness of the proposed heuristic method.

Keywords: Automobile mould, constructive heuristic method, energy-efficient scheduling, parallel machine

1. INTRODUCTION

Energy consumption in manufacturing leads to the increase of cost, for instance, an unit car manufacturing may consume up to 700kwh electricity and the energy cost takes around 9-12 of the total manufacturing cost [1]. However, more importantly, energy consumption is one of the main reasons for the environment pollution due to the greenhouse gas emissions, in particular, the carbon dioxide (CO2)[2]. For instance, at least 26 of the total CO2 emissions in China [3], are from the manufacturing energy consumption. Therefore, faced with the rise in energy prices, increasing energy demand, and rigorous requirements of carbon emissions reduction for environment protection, there has been a growing interests for energy consumption minimization in industry, especially for energy-intensive industries, like automobile, die and mould, glass manufacturing, and chemical industry.

The production operation plays an important role in the reduction of both production costs and energy consumption. As one of key parts of production operations, scheduling is the potential way to reduce energy consumption. However, many practical scheduling problems, even those without energy consideration, are already strongly NP-hard. It is more difficult to solve large-scale problems in real-world. Therefore, energy-efficient scheduling becomes a challenge task for both researchers and practitioners [4,5,6].

In this paper, motivated by the scheduling challenge in a numerical control (NC) machining for automobile parts moulds, we propose an energy-efficient scheduling method, in which the scheduling problem in mould machining shop floor is transformed into a parallel-machine scheduling problem with energy consideration. The possible methods for exact and suboptimal solutions are proposed. Then, we focus on a constructive heuristic method. The remainder of the paper is organized as follows. Section II reviews the relevant literature, especially on moulds manufacturing with energy considerations, and parallel-machine scheduling with energy consumptions. Section III describes the manufacturing process of automobile moulds and the scheduling problem in NC machining center. Section IV proposes the scheduling method. Section V gives a case study. Finally, conclusions and future works are given in Section VI.

* Corresponding Author: Shijin Wang

2. LITERATURE REVIEW

By far, there are many different approaches for energy consumption analysis at the industry, factory, machine and process levels [7]. In the following, we review the energy-efficient scheduling and management in mould industry and in the parallel machine setting, which are most highly related to the paper.

Spiering, Kohlitz, Sundmaeker and Herrmann (2015) [8] presented a systematic approach to energy efficiency benchmarking in injection moulding, more specifically, they developed methods for energy consumption forecasting and controlling in injection moulding. Madan, Mani, Lee and Lyons (2015) [7] proposed a science-based guideline for energy consumption estimate, energy performance evaluation and improvement of unit-manufacturing processes, and especially, they focused on the injection mould case. The guideline they proposed is more about early estimation of manufacturing energy and continuous improvement of energy performance, rather than the operations management methods. Motivated by the mould heat-treatment processing in dynamic environment, Liu, Li, Chen and Mao (2016) [9] investigated the dynamic batch control of parallel processors with objectives related to delivery and energy. They developed an event-driven look-ahead batch control strategy to solve the problem.

In mould manufacturing shop floor, there are many parallel processors with same functions but with different energy consumption rates since some processors employ green technology while others employ regular technology. Hence, in many cases, scheduling and operations are optimized in parallel machine setting. By far, there have been many researches on parallel-machine scheduling with energy considerations. For instance, Fang and Lin (2013) [10] addressed a parallel machine scheduling problem to minimize the total weighed job tardiness and power cost. They proposed an integer programming model formulation and two objectives are added up as a single objective function: weighted tardiness and power consumption. Two constructive heuristic methods and a particle swarm optimization algorithm are developed to obtain the solutions. Rager, Gahm and Denz (2015) [11] considered an energy-oriented scheduling problem in parallel machine to minimize the demand of final energy sources. A genetic algorithm and two memetic algorithms are developed. Li, ang, Zhang, and Liu (2015) [12] studied an unrelated parallel machine scheduling problem to minimize the energy and tardiness cost. A mathematical model is developed and the two objectives are added together to come up with a single objective. Ten heuristic methods are proposed. Li, Zhang, and Leung (2016) [13] studied a parallel machine scheduling problem to minimize the makespan or the total completion time, with the constraints that the total cost (the sum of the energy cost and the clean up cost) is within a given range. They proposed efficient heuristics.

3. MANUFACTURING PROCESS AND PROBLEM DESCRIPTION

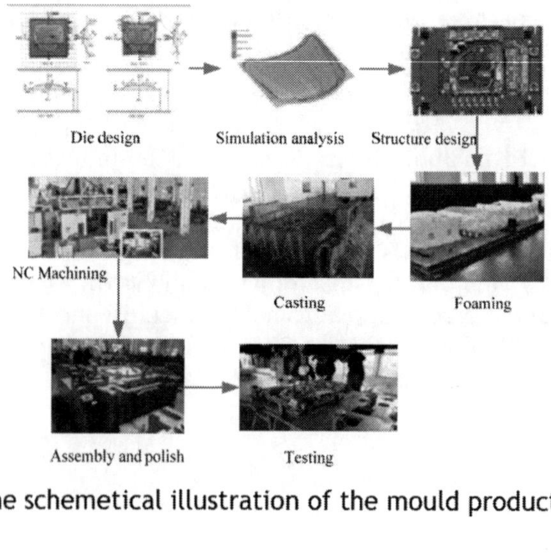

Figure 1. The schemetical illustration of the mould production process

3.1 MANUFACTURING PROCESS

The mould manufacturing is of high importance for the automobile industry, since many automobile parts including hood, fender, doors, body side and roof are made by moulds. The general production process of automobile parts moulds is briefly described as follows:

(1) Design. Once the orders from customers are received, manufacturer uses various design software like UG, CATIA to design moulds. Then, process analysis of flexibility is conducted, followed by the mould structure design, which are all realized through 3D simulations.

(2) Prototype. The plastics foaming is used to manufacture the moulds and all parts of moulds, by filling the cavity with plastics. Then, the moulds are casted using the molten iron.

(3) Machining. In the NC machining center, bases and inserts of moulds are machined through 3-axle machining center, surface is machined through 5-axle machining center to ensure the quality.

(4) Assembly, polish and testing. Inserts and other components of a mould are assembled into the base with bench work. The surfaces of the mould are grinded, and finally the mould is tested with press.

In this paper, we focus on the energy-efficient scheduling in NC machining center with many parallel machines, since it is the bottleneck in the mould manufacturing as we know it from the interviewed company. When these resources are used to process moulds, a large amount of energy, especially electricity, are used. How to finish processing of mould jobs with the least energy consumption and at the same time meet some productive objective is a big challenge for production operational decisions.

3.2 SCHEDULING PROBLEM DESCRIPTION OF NC MACHINING

During the process of mould manufacturing, a lot of energy is consumed when melting iron to cast the mould. However, the energy for melting iron is dependent on the number of moulds that are delivered to customers, which are usually fixed. Therefore, in this paper, we focus on the scheduling with energy consideration in the NC machining center with parallel machines.

The scheduling problem in the NC machining center is described as follows:

There are n independent, non-preemptive jobs to be processed on m parallel identical machines in a given time periods. At time 0, all jobs are available for processing. Each machine can only process one job at a time and one job can only be processed on one machine. Though the processing time of a job on any alternative machine is same, the energy consumption rate of machines are different, $e_k, k = 1,2,\dots,m$. Also, due to time-of-use (TOU) policy, the energy (especially electricity) price of each time period is not necessary same, c_t, $t = 1,2,\dots,T$, where T is the length of the planning horizon. Energy consumption is occurred on each time period when a machine is processing a job and the energy consumption when machine is idle can be ignored. The scheduling objective is to assign the jobs to the machines, to sequence the jobs on each machine and to determine the starting time of each job so that the makespan and the total energy consumptions (TEC) are minimized.

4. AN ENERGY-EFFICIENT SCHEDULING FRAMEWORK

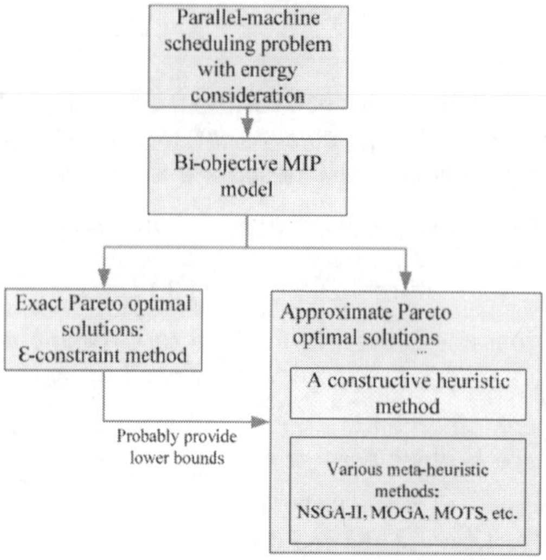

Figure 2. The scheduling optimization framework for Parallel-machine shop floor of automobile mould maching

4.1 THE ENERGY-EFFICIENT SCHEDULING METHODS

The scheduling framework is shown in Figure.2. The scheduling problem can be formulated as a mixed integer programming (MIP) model, in which two objectives, one with production aim and another with energy aim, are considered. Then, in order to solve the bi-objective optimization model, different approaches can be used, which includes aggregation of objectives, hierarchical structuring of objectives, replacement of objectives by constraints, analysis of Pareto optimal solutions, and approximation of multiobjectives [14].

The ε-constraint method can be used to find the exact non-dominated solutions. For a bi-objective optimization problem, the ε-constraint method is to solve a set of single-objective auxiliary problems in which one objective is kept as main objective while another objective is transferred to a constraint. However, on the one hand, the number of iterations of the ε-constraint method is dependent on the number of partitions for each objective, on the other hand, to solve a single-objective auxiliary problem optimally with commercial software like CPLEX is still time consuming. Therefore, in most cases, the ε-constraint method can only be used to solve the small scale problems. However, the practical problems in mould machining shop floor are usually large scales.

For the approximate multiobjective solutions, there are many meta-heuristic methods: NSGA-II, multiobjective evolutionary algorithm (MOEA), multiobjective genetic algorithm (MOGA), multiobjective tabu search (MOTS), and so on. These methods are widely used in multiobjective optimization and their efficiency and effectiveness are demonstrated in the literature. However, usually, there are many parameters in these algorithms, and the values of parameters affect the performance of the algorithms significantly, which needs tremendous efforts of parameter tuning.

4.2 A CONSTRUCTIVE HEURISTIC METHOD

To overcome the shortage of high computation burden of the ε-constraint method and the disadvantages of parameter tuning of meta-heuristic algorithms, a constructive heuristic method is proposed. When makespan is limited by a certain value, an idea is intuitively to avoid the job processing that occurred on machines with high energy consumption rates and during periods with high electricity prices. Therefore, the key point is how to avoid the high energy consumption job production. For any given makespan, the constructive heuristic method could obtain the near-optimal TEC fast. The detailed procedure is showed as follows:

Step 1: Sort the jobs in non-increasing order of processing times $p_i, i = 1, 2, \ldots, n$.

Step 2: Sort the machines in non-decreasing order of energy consumption rates $e_k, k = 1,2,\ldots,m$.

Step 3: Define $h_k^t = e_k c_t$, $k = 1,2,\ldots,m$, $t = 1,2,\ldots,T$.

Step 4: For the assignment of job i, select certain time interval with minimum h_k^t from all idle and continuous time intervals whose length is p_i. The selected time interval is the best option for job i which means it causes the least energy consumption for processing job i.

Step 5: Update the status of periods of the machine that are assigned to process job i.

Step 6: $i = i+1$, back to Step 4 until all jobs are assigned.

In these steps, step 1 is aim to guarantee that all jobs could be assigned. Step 2 could reduce the time of Step 3. Step 3 is conducted to compute the energy consumption for each time period of each machine.

5. A CASE STUDY

5.1 DATA OF THE CASE STUDY

The following case study is originated from a real-world scheduling problem in an automobile mould machining shop floor with 17 parallel machines. The powers of machines are listed in Table I (The data are preprocessed for some commercial confidentiality reasons). We consider two cases.

In the first case, all 17 machines can process each job (i.e., full flexibility of machines). In the coming $|T| = 120$ time periods, there are 50 jobs needed to be processed in the shop floor, and their processing times p_i are listed in Table II. The automobile parts mould machining usually takes quite long time, since the moulds are very large and heavy, and many operations on the machines are needed.

In the second case, there are requirements of surface

Table 1.the power of 17 machines

Machine k	Power (kW/h)	Machine k	Power (kW/h)
1	48	10	52
2	48	11	40
3	48	12	68
4	60	13	68
5	48	14	50
6	48	15	40
7	64	16	40
8	52	17	48
9	80	/	/

Table 2.the processing times of jobs(unit: hour)

Job i	p_i	i	p_i	i	p_i	i	p_i	i	p_i
1	45	11	42	21	15	31	11	41	18
2	8	12	17	22	5	32	11	42	70
3	31	13	58	23	5	33	25	43	19
4	13	14	32	24	12	34	42	44	28
5	11	15	18	25	31	35	86	45	36
6	35	16	74	26	14	36	17	46	17

7	36	17	54	27	5	37	17	47	93
8	14	18	47	28	4	38	19	48	47
9	18	19	79	29	15	39	19	49	43
10	36	20	32	30	44	40	20	50	91

Table 3.the processing requirements of jobs

Job i	Normal direction	Accuracy level	Job i	Normal direction	Accuracy level
1	N	B	10		A
2	N	B	11	N	A
3		B	12	N	C
4	N	C	13	N	C
5		B	14	N	A
6	N	B	15		B
7	N	C	16	N	B
8		B	17		B
9		A	/	/	/

Table 4.the processing capabilities of machines

Job i	Accuracy level	Normal direction	Job i	Accuracy level	Normal direction
1	B	N	26	C	N
2	B		27	C	N
3	B		28	C	N
4	C	N	29	C	N
5	C	N	30	B	N
6	B	N	31	C	N
7	B	N	32	C	N
8	C	N	33	B	N
9	C	N	34	B	N
10	B	N	35	A	N
11	B	N	36	C	N
12	C	N	37	C	N
13	B	N	38	C	N
14	B	N	39	C	N
15	C	N	40	C	N
16	A	N	41	C	N
17	B	N	42	A	N
18	B	N	43	C	N
19	A	N	44	B	N
20	B	N	45	B	N
21	C	N	46	C	N
22	C	N	47	A	N
23	C	N	48	B	N
24	C	N	49	B	N
25	B	N	50	A	N

processing accuracy of moulds and there are capabilities of machines for processing accuracy. The machine with accuracy level A can process jobs with accuracy requirements A or B, but it cannot process the jobs with C. Similarly, the jobs with accuracy requirements A can only be processed by machines with A. The jobs with accuracy requirements B can be processed by machines with either A or B. In addition, some machines support normal direction processing. Jobs with the requirements of normal direction processing can only be processed by machines with normal direction functions. Jobs without the normal direction requirements can be processed by any machine. The information of the jobs and machines are shown in Table III and Table IV. In this case, machines have eligibility of processing, i.e., not full flexibility. $|T|=210$ is used for this case.

The main energy considered here is electricity and the electricity price for industry is dynamic due to the peak and valley periods. To simulate the dynamic electricity price, we set the price c_t be a random value from {1,2,3} for each period. Since p_i is long, the changes of electricity prices influence the total energy consumption largely.

5.2 SCHEDULING RESULTS

The computational results with the constructive heuristic method are shown in Figure. 3 and Figure. 4. The approximate Pareto fronts of the method include 16 non-dominated solutions for both the cases of full flexibility and eligibility. It takes the heuristic method less than 60 seconds to obtain all these non-dominated solutions, while it is impossible to obtain one non-dominated solution within 60 seconds if we use CPLEX to solve the problem with the ε-constraint method. For the case of full flexibility, the solutions along the front range from 166048 TEC and makespan $C_{\max}=99$, to 155590 TEC and makespan $C_{\max}=120$. From the solutions (99, 166048) to (120, 155590), there is 6.3 potential of energy cost saving at the cost of 17.5 longer makespan. For the case of eligibility, the solutions along the front range from 147296 TEC and makespan $C_{\max}=187$, to 143680 TEC and makespan $C_{\max}=206$. From the solutions (187, 147296) to (206, 143680), there is 2.5 potential of energy cost saving at the cost of 9.2 longer makespan.

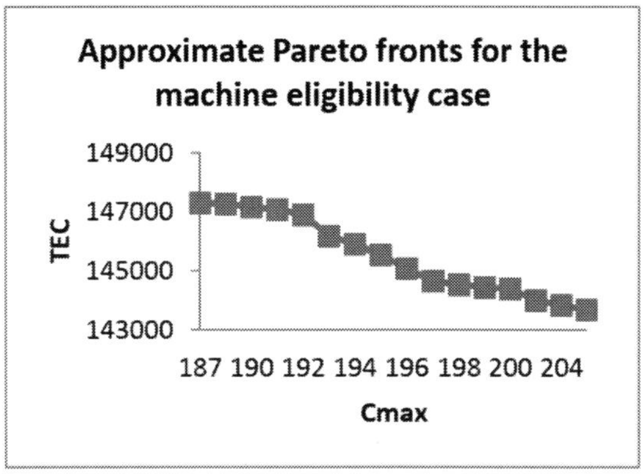

Figure 3. The computational results for the case of full flexibility of machines

Figure 4. The computational results for the case of machine eligibility

6. CONCLUSION

With the increasing concern on the energy efficiency of manufacturing industry, more and more researches are on energy-efficient scheduling in various manufacturing setting. This paper studies the scheduling problem with energy consideration in automobile mould manufacturing shop floor with parallel machines. A scheduling framework is proposed, in which the problem is modelled as a bi-objective optimization model, and a constructive heuristic method is proposed to obtain the approximate Pareto fronts. A case study from a real-world automobile mould manufacturing shop floor is used to demonstrate the effectiveness of the proposed constructive method. For the future work, various meta-heuristic algorithms, like NSGA-II, could be used to solve the problem to obtain better solutions in relatively less computation times.

. ACKNOWLEDGMENT

This work was supported by the National Science Foundation of China (NSFC) under Grants 71171149, 71428002, 71571134 and 71571135. This work was also supported by the Fundamental Research Funds for the Central Universities.

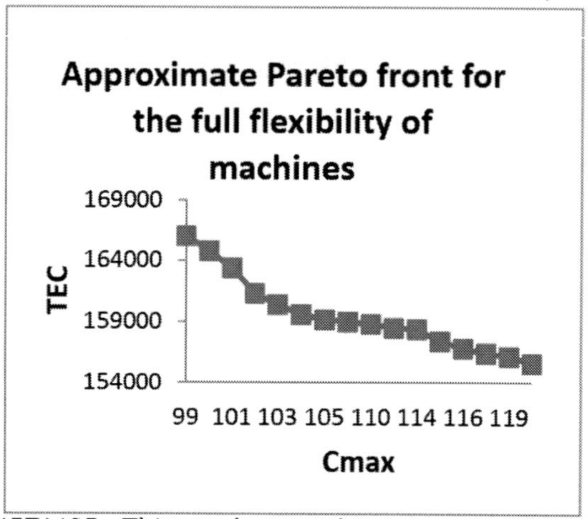

8. REFERENCES

[1] **A. Fysikopoulos, D. Anagnostakis, K. Salonitis, and G. Chryssolouris**, An empirical study of the energy consumption in automotive assembly, Procedia CIRP 3, vol. 3, pp. 477 482, 2012.

[2] **.Y. Ding, S. . Song, and C. Wu**, Carbon-efficient scheduling of flow shops by multi-objective optimization, European Journal of Operational Research, vol. 248, no. 3, pp. 758 771, 2016.

[3] **Y. Liu, H.B. Dong, N. Lohse, S. Petrovic, and N. Gindy**, An investigation into minimising total energy consumption and total weighted tardiness in job shops, Journal of Cleaner Production, vol. 65, pp. 87 89, 2014.

[4] **C. Gahm, F. Denz, M. Dirr, and A. Tuma**, Energy-efficient scheduling in manufacturing companies: a review and research framework, European Journal of Operational Research, vol. 248, no. 3, pp. 744 757, 2016.

[5] **A. Giret, D. Trentesau , and V. Prabhu,** Sustainability in manufacturing operations scheduling: a state of the art review, Journal of Manufacturing Systems, vol. 37, no. 1, pp. 126 140, 2015.

[6] **L. Merkert, I. Har unkoski, A. Isaksson, S. Saynevirta, A. Saarela, and G. Sand,** Scheduling and energy: industrial challenges and opportunities, Computers and Chemical Engineering, vol. 72, pp. 183 198, 2015.

[7] **. Madan, M. Mani, .H. Lee, and K.W. Lyons,** Energy performance evaluation and improvement of unit-manufacturing processes: injection molding case study, Journal of Cleaner Production, vol. 105, pp. 157 170, 2015.

[8] **T. Spiering, S. Kohlitz, H. Sundmaeker, amd C. Herrmann,** Energy efficiency benchmarking for injection moulding processes, Robotics and Computer-Integrated Manufacturing, vol. 36, pp. 45 59, 2015.

[9] **. . Liu, Z.T. Li, .X. Chen, and N. Mao,** Controlling delivery and energy performance of parallel batch processors in dynamic mould manufacturing, Computers Operations Research, vol. 66, pp. 116 129, 2016.

[10] **K.T. Fang, and B.M. Lin,** Parallel-machine scheduling to minimize tardiness penalty and power cost, Computers Industrial Engineering, vol. 64, no. 1, pp. 224 234, 2013.

[11] **M. Rager, C. Gahm, and F. Denz,** Energy-oriented scheduling based on Evolutionary Algorithms Computers Operations Research, vol. 54, pp. 218 231, 2015.

[12] **Z. Li, H. Yang, S. Zhang, and G. Liu,** Unrelated parallel machine scheduling problem with energy and tardiness cost, The International Journal of Advanced Manufacturing Technology, in press, pp. 1 14, 2015.

[13] **K. Li, X. Zhang, .Y.T. Leung, and S.L. Yang,** Parallel machine scheduling problems in green manufacturing industry, Journal of Manufacturing Systems, vol. 38, pp. 98 106, 2016.

[14] **.Y.T. Leung, K. Lee, and M.L. Pinedo,** Bi-criteria scheduling with machine assignment costs, International Journal of Production Economics, vol. 139, no. 1, pp. 321 329, 2012.

ANALYSIS ON THE EFFICIENCY OF R D FUNDS IN MANUFACTURING ENTERPRISES IN LIUZHOU-- BASED ON DEA

Min i*, Lu Ma
School of Management,
Guangxi University of Science and Technology, Liuzhou, China
(*e-mail: qimin0802@163. com)

ABSTRACT

Manufacturing industry has been the focus of the process of economic development in China. It has gradually embarked on the road of innovation with the change of the economic environment. Liuzhou is a city with heavy industrial city, so the research on the R D funds efficiency of manufacturing enterprises in Liuzhou has a certain guiding and reference significance to the development of Liuzhou's manufacturing industry. The comparison of R D funds investment efficiency of four representative cities in Guangxi Provence in recent 5 years which is based on DEA method is an attempt to detect Liuzhou manufacturing enterprises' own problems. Finally, the corresponding countermeasures and suggestions are put forward according to the results obtained.

Keywords: DEA method, Liuzhou manufacturing enterprise, R D funds efficiency

1. INTRODUCTION

Innovation is one of the most concerned topics in the business and academic circles in recent years. For the enterprise, sustainable innovation is the key to win the market seat and manufacturing is no exception[1-2]. Industry 4 and "Made in China 2025" put forward the innovation of manufacturing industry to a new height, and pointed out the direction for the development of the manufacturing industry.

Investment in science and technology activities of manufacturing enterprises is an important symbol to evaluate the overall level of scientific and technological development and the importance of science and technology. Generally they can be reflected in the two aspects of investment in human resources and economic resources from scientific and technological activities[3]. Manufacturing enterprise technical personnel refers to personnel directly engaged in scientific and technological activities and to provide direct services for scientific and technological activities[4]. Manufacturing enterprise science and technology activity economic resources mainly refers to the enterprise engaged in scientific and technological activities of funds. It is not an important part of the manufacturing enterprise technology investment, but a strong support to promote the innovation capability of manufacturing enterprises[5].This paper analyzes the economic resources input to the research and development(R D) input and output efficiency.

2. RELATIVE RESEARCH REVIEW

* Corresponding Author: Min Qi
† Work partially supported by grant 2013J040404

At present, there are a lot of research on R D input and output efficiency at home and abroad. Chinese scholars have also conducted a lot of research, but scholars on the R D efficiency and the application of the method is not the same.

V.J. Thomas Seema, Sharma Sudhir and K. Jain analyzed and compared the R D efficiency of America's fifty states and the District of Columbia. Then they put forward the policy recommendations for the United States which can ensure their enterprises high R D efficiency[6]. Pilar Beneito and other scholars studied the Spanish manufacturing enterprise panel data during 1990-2006 to find different models of R D efficiency over time through the number of innovative products the enterprises obtained[7].

Cao Xianzhong and other scholars compared and analyzed the angtze River Delta urban agglomeration efficiency of different degrees through R D resource input and output efficiency by CRS,VRS model in the Data Envelopment Analysis(DEA) and Malmquist index method[8]. Ma Xiaojun and Diao Xiaoqun use the DEA cross evaluation model to evaluate the input and output efficiency of China's enterprise R D[9]. Zhao Xicang and Ren ang explored the relationship between the R D input and the strength of economic growth using PVAR model, impulse response function and variance separately[10]. Cheng Shixiong and Liu Jianping empirically researched on the output efficiency of China's R D input, which uses the knowledge production function model and stochastic production function method[11]. Dong Jie and Liu Hang used the Malmquist method to study the R D efficiency of Chinese traditional industry[12]. Although there are many researches on R D efficiency in manufacturing industry, DEA as the research method and the research on capital investment are few. Based on DEA method, this paper analyzes the R D capital input and output efficiency of Liuzhou(LZ) manufacturing enterprises, and compares it with the three regions of Nanning(NN), Guilin(GL) and Wuzhou(WZ), which are representative of Guangxi province.

3. EMPIRICAL RESEARCH

3.1SELECTION OF EVALUATION INDEX

This article is focused on the R D funds. Economic resources of science and technology activities not only represents a company's technological strength, but also reflects the advantages and disadvantages of a corporate innovation external environment, but also reflects the degree of support for innovative activities of enterprises. Through the collation of the literature, we found that more literatures select R D funds as the financial input for the calculation of indicators.

Patent is one of the important output of scientific and technological activities in manufacturing enterprises. The number of patents, as well as the extent to which it can be accepted by the market, can directly or indirectly reflect an enterprise's ability to innovate[13-17]. The ultimate goal of enterprise R D investment is to enable the business to obtain economic benefits, meanwhile new product is the secret weapon in an invincible position in the fierce market competition of enterprises, but also the contribution of the innovation of science and technology to the economy reflects the new products on directly[18]. Therefore, this paper uses the amount of patent applications and new product sales revenue as a measure of R D capital output indicators.

Due to the absence of Liuzhou manufacturing enterprises specialized data sources, we visited the Guangxi Bureau of Statistics and Liuzhou Bureau of statistics, and through communication with relevant departments responsible person, eventually we choose the data of Industrial Enterprises above Designated Size instead of the data of manufacturing industry enterprise.

The R D funds of Industrial Enterprises above Designated Size is set to X, the amount of patent application to 1 and new product sales volume to 2. They are summarized in TABLE I.

3.2EXPERIMENTAL RESULTS

We manipulate the data gained from Technical earbook of Industrial Enterprises above Designated Size which is a source that contains detailed science and technology activities record of industrial enterprises above designated size meeting the needs of our research by MaxDEA and gain the results obtained which are shown in TABLE II.

Table 1 evalutation index variables of r d efficiency in manufacturing industry

Variable	Variable Name	Variable Declaration
Input variable	X (ten thousand yuan)	R D funds of Industrial Enterprises above Designated Size
Output variable	1 (item)	The amount of patent application
	2（ten thousand yuan）	New product sales

Table 2 the results of envelopment model

Time	DMU	DMU Order in Data	Score	Benchmark(Lambda)
2010	NN	1	0.268578	WZ5(4.463348)
2010	LZ	2	0.981321	WZ5(10.535792)
2010	GL	3	0.598596	WZ5(3.289270)
2010	WZ	4	0.459814	WZ5(0.568928)
2011	NN	5	0.338872	WZ5(5.810290)
2011	LZ	6	0.963376	WZ5(12.120008)
2011	GL	7	0.634964	WZ5(4.401872)
2011	WZ	8	0.316107	WZ5(1.277184)
2012	NN	9	0.558017	WZ5(6.333435)

2012	LZ	10	0.819052	WZ5(13.526079)
2012	GL	11	0.534497	WZ5(5.194415)
2012	WZ	12	0.669188	WZ5(1.577882)
2013	NN	13	0.678784	WZ5(5.938640)
2013	LZ	14	0.785208	WZ5(16.931527)
2013	GL	15	0.658156	WZ5(5.813340)
2013	WZ	16	0.895283	WZ5(1.658577)
2014	NN	17	0.966986	WZ5(6.416811)
2014	LZ	18	0.667471	WZ5(18.176825)
2014	GL	19	0.539898	WZ5(6.315895)
2014	WZ	20	1	WZ5(1.000000)

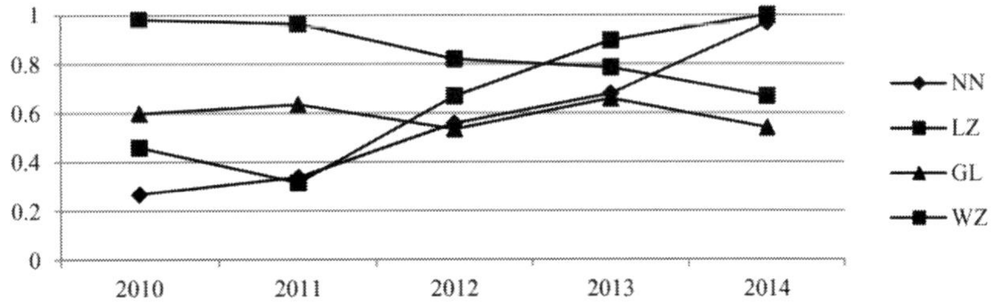

Fig.1.R D funds input and output efficiency of four cities

According to experimental results(Table II), we made an intuitive chart which is shown in the Fig.1.. We can see that Liuzhou R D funds input and output efficiency has been the first of the four cities from 2010 to 2012 period, while it was overtake by Wuzhou in 2013. Liuzhou was caught up by Nanning as three of the four till 2014 and it has been showing a downward trend during the five years. On the other hand, the R D funds input and output efficiency of Nanning has gained successive incremental in the past five years. Wuzhou except for 2010 that its R D funds input and output efficiency decreases, the remaining four years it keeps rapid increasing. Eventually Nanning and Wuzhou are both catching up with Liuzhou becoming the top two among cities of Guangxi Rank by the R D funds input and output efficiency.

4. DISCUSSION

Manufacturing enterprises have always been the object of attention of the government of Liuzhou and Liuzhou manufacturing enterprises in the input of the R D is the largest in the four provinces, while the output of the innovation is not ideal. Manufacturing enterprises in Nanning and Wuzhou started later than Liuzhou, but the output efficiency is increasing year by year. Through comprehensive analysis, we can draw the following three reasons: firstly, the construction of enterprise independent innovation system is not perfect. The construction of Liuzhou enterprise information platform for independent innovation are sadly lacking and the development of the scientific and technological intermediary institutions urgent needs to be strengthened. Secondly, the enterprise innovation power is insufficient, and the innovation foundation is not firm. In addition to leading enterprises, other manufacturing enterprises in Liuzhou began to pay attention to innovation in recent years, and the amount of innovative investment accounted for the proportion of the amount of the main business income is small. Finally, they are lack of talent and innovating force. The number of colleges and universities in Liuzhou is not large and part of college graduates choose their working place outside the city due to the geographical location or policy. At the same time, government talent introduction policy did not achieve the desired effect.

In order to enhance the Liuzhou manufacturing enterprise innovation ability, they can considerate from the following several aspects: firstly, cooperate and learn from units with strong innovation ability. Actively learning the experience from advanced units can shorten the cycle of innovation, reduce the waste of resources, improve the quality of innovation. Secondly, introduce and cultivate innovative talents. Talent is the key asset of economic development, which is a necessary condition for sustainable and long-term innovation to gain competitive advantage. Thirdly, improve the exiting innovation system. Innovation system plays a role of bridge in the process of innovation. Perfect innovative system is favor of brainstorming, intelligence resources conformity in order to increase the innovative vitality and create a benign atmosphere for innovation.

Due to the limitations of time and funds, this paper is still insufficient. First, this paper analyzes Liuzhou manufacturing industry enterprise input-output efficiency only considering R D funds investment as the independent variable, patent applications and the amount of new product sales as the dependent variable. However, there are other factors which can effect and reflect enterprise innovation that this paper has not been discussed. Second, the use of data from industrial enterprises above designated size in Liuzhou to replace the Liuzhou manufacturing enterprise data may cause some certain errors. In this regard, we will make further research in the future so as to refine our research.

5. REFERENCES

[1] **Craig Lee, Rob Hallak, Shruti R. Sardeshmukh.** Innovation, entrepreneurship, and restaurant performance: A higher-order structural model[J]. Tourism Management, 2016,53: 215-228.

[2] **os Ram n Cobo-Benita, Enri ue Rodr guez-Segura, Isabel Ortiz-Marcos, Luis Ballesteros-S nchez.** Innovation projects performance: Analyzing the impact of organizational characteristics[J]. Journal of Business Research,2016,69:1357-1360.

[3]**Schankerman, M..** The effects of double-counting and expensing on the measured returns to R D[J]. Rev. Econ. Stat. 1981, 64: 454 458.

[4]**Guan , He Y.** Comparison and evaluation of domestic and international outputs in information science and technology research of China[J].Scientometrics,2005,65:215 44.

[5]**Wallsten S.** .. The Effects of Government-Industry R D programs on private R D: the case of the small business innovation research program[J]. Rand Journal of Economics,2000,31:82-100.

[6]**V. . Thomas, Seema Sharma, Sudhir K. ain.** Using patents and publications to assess R D ef ciency in the states of the USA[J]. World Patent Information.2011,33: 4 10.

[7]**PilarBeneito, Mar a Engracia Rochina-Barrachina, Amparo Sanchis.** The path of R D efficiency over time[J]. International Journal of Industrial Organization,2015,42: 57-69.

[8]**Cao Xianzhong, Zeng Gang, Zou Lin.** Spatial differentiation of input-output efficiency of R D resources in case of angtze River Delta Urban agglomeration[J]. Economic Geography,2015,01:104-111.(China)

[9]**Ma Xiao un, Diao Xiao un.** The evaluation and analysis of R D efficiency of firms in China[J]. Mathematics in Practice and Therory,2015,05:35-43.(China)

[10]**Zhao Xicang, Ren Yang.** Study on the dynamic relationship among R D input, patent output efficiency and economic growth strength based on 13 cities' panel VAR Model in Jiangsu Province[J]. Soft Science,2014,10:18-21+26. (China)

[11]**Cheng Shi iong, Liu ianping.** R D production efficiency and influencing factors on China's industrial sector[J]. The Journal of uantitative Technical Economics,2014,02: 36-51+85.(China)

[12]**Dong ie, Liu Hang.** R D efficiency measurement and influencing factors analysis of China's traditional industry[J]. Science and Technology Management Research, 2015,23:64-68.(China)

[13]**Griliches Z.** Patent statistics as economic indicators: a survey[J]. Journal of Economic Literature, 1990,28: 1661-1707.

[14]**Hausman, erry A., Gregory K. Leonard.** Estimation of patent licensing value using a flexible Dem and specification[J]. Journal of Econometrics,2007,139:242- 258.

[15 **Teitel S.** Patents, R D expenditures, country size, and per capita income: an international comparison[J]. Scientometrics,1994,29:137 59.

[16]**Beneito, P.**. The innovative performance of in-house and contracted R D in terms of patents and utility models[J]. Res. Policy,2006,35:502 517.

[17]**Hausman, ., Hall, B.H., Griliches, Z.**. Econometric models for count data with and application to the patents-R D relationship[J]. Econometrica,1984,52:354 383.

[18]**Griliches Z.** R D and the productivity slow down[J]. American Economic Review,1980,70:343-348.

DETERMINANTS OF TRADE CREDIT SUPPLY: THE CASE OF TEXTILE SECTOR OF PAKISTAN

Adnan Abbas[1], Xiao-guang Luo [2*], Jaleel Ahmed[3]
[1,2]School of Economy,
Harbin University of Science and Technology, Harbin, China

[3] School of Economy and Management,
Harbin Institute of Technology, Harbin, China
* luoxg@126.com

ABSTRACT

The basic objective of this research is to examine the determinants of trade credit supply in textile sector of Pakistan. Trade credit can be affected by firm specifics characteristics. After collecting seven year data from 2006 to 2012 from textile firms of Pakistan, we have applied fixed effect model to estimate regression equation to obtain results. Research found that the following firm specifics independent variables such as current asset, cash bank balance, gross profit, current liabilities, shareholders equity, inventory and tax expense significantly related to trade credit supply.

Keywords: Trade credit, Textile sector, Determinants of trade credit, KSE

1. INTRODUCTION

In different countries many Organizations seek various means to finance their capital need in diverse societies. There are two major types of financing which include short term financing and long term financing. Trade credit is one of major source of short term funding in all over the world (Petersen and Rajan, 1995; Van Horen, 2007) [1],[2]. Trade credit and bank loans play an important rule during tight monetary conditions (Guariglia and Mateut, 2006) [3].

Studies found that the companies working in countries having inefficient legal and financial system depend more on trade credit (Rajan and Zinglase, 1995,;Johnson, McMillian and Woodruff, 2002) [4][5], for the contingency effect are higher in this case (La Porta et al. 1998; Pike et al , 2005)[6] [7].

Trade credit, as explained by Martinez-Sola et al. (2012) is an arrangement between buyer and seller allowing an exchange of goods with some deferred payments terms and condition [8]. Trade credit is an important source of financing for firms, especially for small and medium size firms (Boyery and Gobert 2007) [9]. Trade credit is general term that is used for either supply or demand for credit goods. Trade credit increases benefits for both sellers and buyers part by smoothing or stable out the uncertainties in delivery cycle and simplifying cash management (Schwartz 1974) [10].

Wilson and summers (2002) argue that trade credit is an important marketing tools. The logic is that trade credit could be used to build business relationships in a new market and new entrance could be more easily as compared to building their reputation after year of hard work and marketing their products [11].

According to the study of Rajan Zingales, (1995) aggregate volume of trade credit of American firms was 17.8 and was gradually increasing before 1990[12]. But in United Kingdom it was about 55 of total credit of the firms based on the trade credit as studied by (Kohler et al. 2000; Guariglia et al. 2006) [13][14].

---

* Corresponding Author Xiao-guang Luo

Trade credit represents more than a quarter of total corporate assets in France, Germany and Italy and it is also vital for an emerging countries like China, where organizations acquire limited support from banks (Ge and iu 2007) [15]. The gap of research remains when the trade credit literature is sought in Pakistan. There is no study till date had carried out for detailed about determinates of trade credit analysis and effecting trade credit in non-financial firms listed at Karachi Stock exchange (KSE). KSE is a Pakistan largest stock exchange.

The remainder of this paper is planned as following: section 2 reviews of literature that leads us to the gap in the literature. Section 3 discusses the data and empirical methodology. Section 4 reports the results with analysis and at lastly section 5 conclude the study discussion the implications of the findings.

2. LITERATURE REVIEW

In every financing policy trade credit always plays a vital role. Funds includes in trade credit supply have always been considered as a profit in current assets. According to Giannetti et al. (2003), account of trade credit supply stands as a quarter of total assets in balance sheets in European firms [16]. Trade credit is a major or important source of financing in underdeveloped countries because these countries have less access to standardized capital market (Beck et al. 2008; Ge and iu, 2007) [17], [18].

Jain (2001) argued that trade credit play a significant role between the relationship of financial institutions and business [19]. In case of shortage of financial market, it is beneficial for buyers and suppliers to use this to meet their external financing needs (Frank Maksimovic, 2004) [20].

France Delannay and Weill (2004): Studied on the determinants of trade credit in transition countries. They used regression, equity, size, profit, and leverage as a variable. They consider regression of the ratio of trade payable to total assets and of the ratio of trade receivable to turnover on a set of variables as methodology. At the end it has been determined that survival of differences between countries concerning theses determinants [21].

Love and Zaidi (2010: Worked on trade bank credit and financial crises in four East Asian countries Thailand,

Philippines, Korea and Indonesia. They used size, export, import as variables. It is determined that after the financial crises, firms constrained in bank finance obtain less trade credit in terms of percent of inputs acquired on credit and short time payments [22].

Rajendra R. Vaidya (2011): Studied on determinants of trade credit: Evidence from Indian manufacturing firms. They used dummy variables, explanatory, inventories, fixed assets, profit and size as a variable. The experiential evidence recommends that strong evidence exists to support management of inventory object for the existence of trade credit [23].

Kestens et al. (2012): Investigated trade credit and company performance during the 2008 financial crises in Belgium. They used variables like, size, liquidity and inventory as a variable dummy as variables. They found that financial crises had a negative impact on firm profitability, but this negative impact was lower for companies that reported an increase in trade receivables during the crises [24].

Martinez- Sola et al. (2012): Studied on trade credit policy and firm value in Spain. They used variables like dummy, economic, Control, Growth and Size as Variables. They found that a quadratic relationship between accounts receivable and firm value, furthermore divergence from the required level of account receivables significantly decrease the firm value [25].

Coulibaly et al. (2013): Investigated the financial frictions, trade credit and the 2008-09 global financial crises in USA. They used explanatory, sale and size as Variables. They found that the effectiveness of financial friction in genera and trade credit in particular, on economic activity in their sample of EMEs during the global financial crises. Their focus on trade credit as an alternative source of economic financing for the period of crises is guided by the finding of existing studies, which documents the function of trade credit in reducing financing constriction during past in emerging market economies (EMES) [26].

Rene Guy and Mazra (2012): Studied the determinants of trade credit demand: An empirical study from Cameroonian Firms. By using transactional element, financial component and socio-cultural variables, they concluded that financial objectives are typically correlated with the organizational behavior [27].

Ahmed et al. (2014): Worked on determinants of trade credit: The case of a developing economy in Pakistan. They use size, liquidity, price, and inventory and sale growth as variable. They have applied three models named, pooled ordinary least square method, fixed effects method and random affects method. They found out that the study adds to the literature by analyzing the determinants of trade credit in context of Pakistan [28].

Many other researchers also work on determinates of trade credit such as Ahmed et al. (2014): Can trade credit serve as a cushion against the financial setbacks in a developing economy. They have found that large firms are willing to grant trade credit but they are not interested in receiving goods on credit [29].

In light of the above mentioned discussion about the determinants of the trade credit supply, we have developed the following hypotheses.

H1: There is a positive and significant relationship between current assets and trade credit supply.

H2: There is a negative and significant relationship between cash and bank balance and trade credit supply.

H3: There is a significant relationship between gross profit and trade credit supply.

H4: There is an inverse and significant relationship between inventory and trade credit supply.

H5: There is a positive and significant relationship between current liabilities and trade credit supply.

H6: There is a significant relationship between shareholder equity and trade credit supply.

H7: There is a positive and significant relationship between tax expense and trade credit supply.

3. METHODOLOGY AND DATA

To test the hypotheses raised in above paragraph, by following Ahmed et al. (2014), we have applied fixed effects model. Rationality behind using this model is to check more suitable explanation of the results. As our simple regression model is

$$Y_{it} = \alpha + \beta_i \sum_{i=1}^{n} X_{it} + \varepsilon_{it} \qquad (1)$$

By introducing fixed effects in the Equation 1

$$Y_{it} = \alpha_i + \beta_i \sum_{i=1}^{n} X_{it} + \varepsilon_{it} \qquad (2)$$

Where, X is a vector of explanatory variables. After the selection of appropriate explanatory variables we have developed the Equation 3 as

$$TCS_{it} = \alpha_i + \beta_1 CA_{it} + \beta_2 CAB_{it} + \beta_3 GP_{it} + \beta_4 INV_{it} + \beta_5 CL_{it} + \beta_6 SRE_{it} + \beta_7 TE_{it} + \varepsilon_{it} \qquad (3)$$

Where, i" is representing ith firm at time t. ai is different for all cross sections. TCS stands for trade credit supply CA stands for current assets. CAB stand for cash and bank balance of a company during the year, where GP stands for gross profit of a firms, INV stands for inventory, CL stands for current liabilities, SRE stand for shareholder equity and TE stands for trade expense of a firm during the year. E epsilon is error term.

To use the equation (3), we used seven year data from 2006 to 2012 for analysis of determinants of trade credit supply for textile sectors of Pakistan. Relevant data has been taken from Balance Sheet analysis (BSA) published by The State Bank of Pakistan (SBP). Secondly, data have been taken from financial statements of non-financial firms listed at Karachi Stock Exchange (KSE). Only those firms are included that provides the complete data in context to the analysis. Firms which have missing values of assets and liabilities were also excluded.

By looking at minimum and maximum values of the other variables, they are describing that extreme values do not exist in the ratios of these variables. Moreover we can observe that there is not a pronounced variation in data. TE (tax expensive) has a greater variation than any other variables. Inventory, cash and bank balance lesser variation in their data sets.

Table I providing the data of Descriptive statistics and also contains the values of mean, median, maximum, minimum and Std. Dev and we used variables Trade Credit Supply (TCS), Current Asset (CA), Cash and Bank Balance (CAB), Gross Profit(GP), Current Liabilities (CL), Shareholder Equity (SRE) and Tax expenses (TE).

Table 1 descriptive statistics

	TCS	CA	CAB	GP	CL	SRE	INV	TE
Mean	0.10	0.42	0.02	0.03	0.71	0.29	0.21	0.01
Median	0.07	0.41	0.01	0.09	0.73	0.29	0.19	0.01
Max	0.87	1.00	0.69	0.84	1.00	0.97	0.67	4.99
Min	0.0	0.02	0.0	-8.0	0.08	-1.0	0.00	- 0.3
Std. Dev.	0.09	0.17	0.03	0.54	0.21	0.26	0.12	0.17
Sum	92.600	403	13.9	26.6	673	280	194	13.3
Sums. Dev.	9.11	28.697	1.0180	280.78	38.603	61.482	12.945	26.228
Observation	955	955	955	955	955	955	955	955

To check the correlation among variables a Pearson correlation test has been conducted. Table II is providing the correlation matrix for variables.

Trade debt (TCS) is a ratio of account receivable to total assets and it has a positive correlation with (CA), (GP), (SRE), (TE) and (INV).

Table 2 correlation matrix

	TCS	CA	CAB	GP	CL	SRE	TE	INV
TCS	1.00							
CA	0.51	1.00						
CAB	-0.03	0.11	1.00					
GP	0.11	0.19	0.04	1.00				
CL	0.27	0.52	0.02	0.09	1.00			
SRE	0.04	0.17	0.01	0.02	0.15	1.00		
TE	0.01	0.05	-0.01	0.02	0.02	-0.01	1.00	
INV	0.01	0.61	0.07	0.21	0.37	0.15	0.02	1.00

TCS have negative correlation with CAB. CA is the ratio of account receivable to sales, it has a positive correlation with all variable.CAB is a ratio of account receivable to asset. It has a negative correlation with TE and positive relation with GP, CL, SRE and INV. GP is an account receivable and it is a ratio of gross profit to sale. It is a positive correlation with CL, SRE, TE and INV. CL stand for current liabilities and it is account payable and calculated as current liabilities to total liabilities. It has positive correlation with all variables SR, TE and INV. SRE stand for shareholder equity and can be calculated as total assets minus total liabilities. It has positive correlation with INV but negative correlation with TE. TE stands for tax expense and be calculated as tax expense to sale.it has positive correlation with INV. TE is positively correlated with inventory. INV can be calculated as inventory to total sales.

By looking at the magnitude of the correlations among variables, we can observe that no variables have greater correlation with other variables, which means variables do not have the problem of multi co-linearity.

To ensure the contravention of ordinary least square (OLS) assumptions, we apply various tests to identify the problem of multi co-linearity, autocorrelation, and heteroscedasticity. The value of VIF indicates that there is no problem of multi co-linearity because all variables have VIF less than 5.

4. RESULTS

Table III is showing the effects of different firm specific variables on trade credit supply. In table III, we can see the results of fixed effect method. The results are in line with the prior conclusion of various studies on trade credit determinants.

In Table III the results of fixed effects models are use; the current ratio of CA is calculated by dividing total current asset by total current liabilities. It is highly significant with positive sign of coefficient 0.1371. Positive sign of coefficient is indicating that one unit increase in current asset would leads 13.71percent increase in trade credit supply. The value of t-statistic is significant because its value is more than 2 .value of t-statistic is 7.3709 and p value is 0.0000 therefore at last current assets is significant variables in our results. Results of CA prove our first hypothesis regarding the positive and significant relationship between trade credit supply and CA.

In Table III the ratio of CAB is calculated by dividing cash and bank balance by total asset. Negative sign of coefficient is indicating that one unit increase would lead towards 11.06 percent decrease in trade credit supply. The value of t-stat is -2.0018 which more than 2 and p value of CAB is 0.045 thus we consider CAB variables is significant. Results of CAB also prove that our finding hypothesis regarding the relationship CAB. Results of CAB prove our second hypothesis regarding the negative and significant relationship between trade credit supply and CAB.

GP is calculated by dividing gross profit by sale. A company total revenue (equivalence to total sales) minus the cost of goods sold. Positive sign of coefficient is indicating that one unit increase in gross profit would be 5.3 percent increase in trade credit supply. In order to increase sales non-financial firms in Pakistan tries to grant more credit sales which lead to increase gross profit. The value of t-stat in GP variables are 1.2744 which is less than 2 therefore GP is insignificant. It does not prove our third hypothesis.

Table 3 fixed effect results

Variable	Coefficient	Std. Error	t-Statistic	Prob.
C	0.0660	0.0094	6.9783	0.0000
CA	0.1371	0.0186	7.3709	0.0000
CAB	-0.1106	0.0552	-2.0018	0.0456
GP	0.0053	0.0041	1.2744	0.2029
INV	-0.2264	0.0271	-8.3270	0.0000
CL	0.0274	0.0112	2.4321	0.0152

SRE	0.0027	0.0128	0.2166	0.8286
TE	0.0163	0.0094	1.7328	0.0835
R-squared	0.8277			
Adjusted R-squared	0.7930			
F-statistic	23.843			
Prob. (F-statistic)	0.0000			

In Table III the result of fixed effects model for inventory turn out to be highly significant with negative mark of coefficient. According to the results of fixed effects model, inventory to sale ratio significantly related to trade credit supply. Negative sign of coefficient is showing that one unit increase in inventory would lead towards 22.64 percent decrease in trade credit supply. The value of t-stat is -8.3270 is more than 2 therefore it is significant and the value of p is 0.0000 therefore it is significant. In order to increase sales of non-financial firms in Pakistan and tries to grant more credit on sale which leads towards the decrease in inventory (Ahmed et al. 2014). This statement turned our direction towards the important motive of inventory management. Results of INV prove our fourth hypothesis regarding the negative and significant relationship between trade credit supply and INV.

CL is calculated by dividing current liabilities to total liabilities. In Table III our variables is significant because its t value is more than 2 therefore we can consider that our variable is significant. Results of CL prove our fifth hypothesis regarding the positive and significant relationship between trade credit supply and CL.

The ratio of SRE used to help determine that how much shareholder receives in the event of company liquidation. Positive sign of coefficient is indicating that one unit increase in shareholder would leads toward 2.7 percent increase in trade credit supply. In Table III the results of fixed effect model for shareholder variables turn out to be insignificant because the value of t-statistics in this variables is 0.2166 which is less than 2 and the p value is 0.82 is also insignificant therefore we consider this variables said to be insignificant. SRE does not prove our six hypotheses because SRE is insignificant.

The ratio of TE is calculated dividing tax expense by total sale. In table III the positive sign of coefficient is indicating that one unit increase in tax expense leads towards 0.16 percent to increase in trade credit supply. The value of t-stat is significant because its vale is 1.7328 which is near to 2 so we can observe it is significant but the value of p is 0.835 which is insignificant. Results of TE prove our seventh hypothesis regarding the positive and significant relationship between trade credit supply and TE.

The coefficient of determining, indicated by R-squared (R2) that described how much data is fit for further analysis and also the model is in line or not. And R2 of 1 show that the regression line is entirely fit to the data; while an R2 of 0 describes that the line does not fit on the data. It happens because the data is highly non-linear than the curve permit or for the reason that it is random. In the model the value of R-squared is 82.77 .

5. CONCLUSION

The main focus of this research paper is to define the trade credit patterns in the textile sector of Pakistan. The study has selected the textile sector as it has a major share in the economy of Pakistan. This study adds the literature by finding the determinants of trade credit supply in context of textile firms of Pakistan. Many studies have been called that trade credit practices should be studied in different cultures in order to have insight into the different patterns used across the nations. This study has tried to analyze the determinants of account receivable of balance sheet in Pakistani textile sector. To find the appropriate results, we used fixed effect model. The results showed the current assets, cash bank balance, gross profit, inventory and current liabilities are the significant determinants of trade credit supply. For future research, it is need to explore more factors affecting trade credit supply and trade credit demand in the textile sector of Pakistan.

6. REFERENCES

[1] **Petersen M. A. and Ra an R. G** The effect of credit market competition on lending relationships , uarterly Journal of Economics, Vol.110, pp.407-443, 1995.

[2] **Van Horen , N:** Customer market power and the provision of trade credit : evidence from Eastern Europe and Central Asia , World Bank Policy Research Working Paper, No.4284, 2007.

[3] **Guariglia and Mateut:** Credit channel, trade credit channel, and inventory investment: Evidence from a panel of UK firms , Journal of banking and finance, Elsevier, Vol. 30, No.10, pp.2835-2856, 2006.

[4] **Ra an, R., and Zingales, L:** What do I know about capital structure Some evidence from international data , Journal of Finance, Vol.50, No pp.1421 1460, 1995.

[5] **ohnson, S., McMillian, ., and Woodruff, C:** Courts and Relational Contracts , Journal of Law, Economics and Organization, Vol.18, pp.221-277, 2002.

[6] **La Porta, R., Lopez- De-Silanes, F., Dhleifer, A., and Vishny, R.W:** Law and Finance , Journal of Political Economy, Vol.106, pp.1113-1155, 1998.

[7] **Pike, R CHENG, N.S Cravens, K and Lamminmaki, D:** Trade credits terms asymmetric information and price discrimination evidence from three continents , Journal of Business Finance Accounting, vol.32, No.5/6, pp.1197 1236, 2005.

[8] **Mart nez-Sola, C., Garc a-Teruel, P. . and Mart nez-Solano, P.:** Trade credit policy and firm value , Accounting and Finance, Vol.53, No.3, pp.791-808, 2012.

[9] **Boyery M. M. and Gobert k:** The impact of switching costs on vendor financing , Finance Research Letters, Vol.6, No.4, pp.569-241, 2007.

[10] **Schwartz R. A:** An economic model of trade credit , The Journal of Financial and uantitative Analysis, Vol. 9, No. 4, pp.643-657, 1974.

[11] **Wilson N. and Summers B:** Trade credit terms offered by small firms: Survey evidence and empirical analysis , Journal of Business Finance Accounting, Vol.29, No.3, 4, pp.317-351. 2002.

[12] **Ra an, R., and Zingales, L:** What do I know about capital structure Some evidence from international data , Journal of Finance, Vol.50, No.5, pp.1421 1460, 1995.

[13] **Kohler M., Britton E. and Yates T:** Trade credit and the monetary transmission mechanism , Bank of England, London, Working paper EC2R 8AH. 2000.

[14] **Guariglia and Mateut:** Credit channel, trade credit channel, and inventory investment: Evidence from a panel of UK firms , Journal of banking and finance, Elsevier, Vol. 30, No.10, pp.2835-2856, 2006.

[15] **Ge, Y. and iu, :** Financial development, bank discrimination and trade credit , Journal of Banking and Finance, Vol. 31, pp. 513 530, 2007.

[16] **Giannetti, M:** Do better institutions mitigate agency problems Evidence from corporate finance choices , Journal of Financial and uantitative Analysis, Vol. 38, pp.185 212, 2003.

[17] **Beck T, Demirg c-Kunt A., and Maksimovic V:** Financing patterns around the world: Are small firms different , Journal of Financial Economics, Vol. 89, No.3, pp. 467 487, 2008.

[18] **Ge, Y. and iu, :** Financial development, bank discrimination and trade credit , Journal of Banking and Finance, Vol. 31, pp. 513 530, 2007.

[19] **ain N:** Monitoring costs and trade credit , The uarterly Review of Economics and Finance, Vol.41, No. 1, pp. 89-110, 2001.

[20] **Frank M.Z and Maksimovic:** Trade credit collateral and adverse selection , University of Maryland, working paper, 2071, 2004.

[21] **France Delannay and Weill:** The Determinants of Trade Credit in Transition Countries 9 Central and Eastern European Countries , Economics of planning, Vol.3, No.7, pp.173-193, 2004.

[22] **Love and Zaidi:** Trade Credit, Bank Credit and Financial Crises , International Review of Finance, Vol.10, Vol.1, pp.125-147, 2010.

[23] **Ra endra R. Vaidya:** The Determinants of Trade credit: Evidence from Indian Manufacturing Firms , Modern Economy, Vol.2, No pp.707-716, 2011

[24] **Kestens.K, Cauwenberge, P.V and Bauwhede, H.V:** Trade Credit and company performance during the 2008 financial crises , Accounting and Finance, Vol. 52, No.4, pp.1125-1151, 2012.

[25] **Mart nez-Sola, C., Garc a-Teruel, P. . and Mart nez- Solano, P.:** Trade credit policy and firm value , Accounting and Finance, Vol.53, No.3. pp.791-808, 2012

[26] **Coulibaly. B, Sapriza.H and Zlate.A:** Financial frictions, trade credit, and the 2008-09 global financial crises , International Review of Economics and Finance, Vol.26, pp. 25-48, 2013.

[27] **Rene Guy and Mazra:** The Determinants of Trade Credit Demand: An Empirical Study from Cameroonian Firms , International Journal of Business and Management, Vol.7, No. pp. 2012

[28] **Ahmed, . Xiaofeng, H., and Khalid, :** Determinants of trade credit: The case of a developing economy , European Researcher, Vol.83, pp.1694-1706, 2014a.

[29] **Ahmed. , Xiaofeng.H, and Khalid, :** Can trade credit serve as a cushion against the financial setbacks in a developing economy , Journal of Convergence Information Technology, Vol. 9, No. 6, pp. 143-150, 2014b.

RESEARCH ON AGGLOMERATION EFFECT AND ITS INFLUENCING FACTORS OF RENEWABLE RESOURCE INDUSTRY FOR THREE MA OR ECONOMIC CIRCLES IN CHINA

an Tang[1*], Jian Li[1,2], Weisha Hu[1], u-xi Chen[1]
[1]School of Management,
Tianjin University of Technology, Tianjin, China

[2]School of Management,
Tianjin University, Tianjin, China
(*sugaryan@yeah.net)

ABSTRACT

As three major economic circles in China are located in the economically developed coastal areas, industrial and economic development of renewable resources and distribution situation has some similarities. On the basis of three major economic circles renewable resource industry agglomeration level of the measure, using the traditional fixed effect panel regression model analysis, comparing the results of the estimates of the spatial panel data model, selecting the time fixed effects model for spatial analysis panel, the results indicates:(1)The angtze river delta industry development level is higher than other two economic circle, having a better aggregation effect;(2)The industry has shown a strong spatial clustering feature, and there is a positive spatial correlation between regions;(3)For industry gathering in the region, labor factor, the level of manufacturing industry, and the enterprise scale have significant positive influence, and technology-intensive has a significant negative effects;(4)For inter-regional industry gathering, labor factor, manufacturing level and opening show a significant positive impact, and enterprise scale has a significant negative effect.

Keywords: Renewable resource industry; spatial agglomeration; spatial Durban model; spatial correlation

1. INTRODUCTION

Economic circle is the inevitable result of the market development and urbanization. China's three major economic circles are based on the rapid development of economic characteristics and on their own regional industrial advantages which gradually show convergence trend aggregation in the leading industry [1]. Renewable resources industry is an important industry in the industrial development of China's reforms which take a large proportion in the total national economy and have a significant value in use in the comprehensive utilization of resources, recycling economy and other fields [2]. In recent years, three major economic circles of industrial have took a rapid development of renewable resources ,but in the influence of the traditional industrial structure and mode there are still significant limitations to the industrial development which show in the performance of the recycling of resources, enterprise spatial distribution structure and other

* Corresponding Author: Yan Tang

† The Ministry of education of Humanities and Social Science project, Number: 14YJC630117

aspects there insufficient. It's necessary to take further study of its key factors for the efforts of building a conservation-oriented society.

Domestic and foreign experts and scholars have given full attention to the study of the factors of Industry gathering. There are many methods applied to deeper research for different regions and industries, such as factor analysis, Super - efficiency DEA, Malmquist, panel data, aggregate fractal dimension [3-10]. However, in the context of the increasing recession of the urban ecological environment, the renewable resource industry of China's three major economic circles is relatively weak and vulnerable to economic and social factors and so on. This essay selects the angtze River Delta, Pearl River Delta, the Bohai Sea region as object of study to discuss the three representatives of the area of renewable resource industries' tendency to aggregate and its influencing factors, which has a certain reference value to the three economic circles renewable resource industries' distribution planning and development.

2. ECONOMETRIC MODEL OF PANEL DATA

2.1. CALCULATION FORMULA OF INDUSTRY AGGREGATION

To prevent the distortion of Gini Coefficient, this essay selects EG Coefficient as calculation formula of the three major economic circle industry aggregation [11].

$$\mathrm{EG}=\frac{\sum_{j=1}^{r}(x_j-s_{ij})-\left(1-\sum_{j=1}^{r}\mathrm{x}_j^2\right)\sum_{j=1}^{r}\frac{1}{n_{ij}}s_{ij}^2}{\left(1-\sum_{j=1}^{r}\mathrm{x}_j^2\right)\left(1-\sum_{j=1}^{r}\frac{1}{n_{ij}}s_{ij}^2\right)} \quad (1)$$

Among them, x_j is the proportion of the Region J's total industrial output value of the three economic circles in Chinese industrial output value, r is the total number of regions, n_{ij} is the number of enterprises of the Industry I in Region J, s_{ij} is the national market share of Industry i in Region j.

2.2. PANEL DATA MODEL

To test the combined effect of various factors on the renewable resource industries in Beijing, Tianjin Region, by construct the panel model to the data in the decade of these three provinces, this essay test the effect of the stability, the independent impact and the combined effect of the nine factors in renewable resource industries [12].

If the sectional individual intercept α_j is different, econometric model follows.

$$EQ_{jt}=\bar{\alpha}+\alpha_j^*+\beta_1 X_{1jt}+\beta_{2j}X_{2jt}+\cdots+\beta_k X_{kjt}+\varepsilon_{jt} \quad (2)$$

Among them, EQ_{jt} is the aggregation of renewable resource industries, X_{it} is the vector of various factors, t is sample time.

2.3. MODEL OF SPATIAL PANEL DATA

Basic spatial panel models include the spatial lag model (SLM) and the spatial error model (SEM) as indicated by Formula (4) and (5) respectively. Considering importance of the economy-space matrix's influences on analysis of factors affecting the renewable resources industry in the three major economic circles, the quantitative analysis of the extensive spatial panel models should be carried out with the Spatial Dubin Model (SDM), which contains spatial lag items of dependent and independent variables as well[13]. It is showed in Formula (6).

$$y_{it} = \rho \sum_{j=1}^{N} w_{ij} y_{jt} + \partial + x_{it}\beta + \mu_i + \lambda_t + \varphi_{it} \tag{4}$$

$$y_{it} = \alpha + x_{it}\beta + \mu_i + \lambda_t + \varphi_{it}, \quad \varphi_{it} = \rho \sum_{j=1}^{n} w_{ij}\varphi_{it} + \varepsilon_{it} \tag{5}$$

$$y_{it} = \delta \sum_{j=1}^{N} w_{ij} y_{jt} + \alpha + x_{it}\beta + \sum_{j=1}^{N} w_{ij} x_{jt}\gamma + \mu_i + \lambda_t + \varphi_{it} \tag{6}$$

Among them, y_{it} is dependent variable in Sectional unit i at time t; w_{ij} is component of N N dimensional nonnegative spatial weight matrix W; ∂ is endogenous parameters of interactions size between variable; x_{it} is exogenous variable of 1 K dimension; β is the corresponding coefficient vector of K 1 dimension; μ_i represents a specific space effect; λ_t represents the specific time effect; φ_{it} indicates spatial autocorrelation errors; ρ indicates spatial autocorrelation coefficient between the error terms; γ indicates parameter vector of K 1 dimension. Spatial Durbin model can be simplified as spatial lag model or spatial error model under certain conditions.

3. VARIABLE SELECTION AND DATA COLLECTION

3.1. VARIABLE SELECTION

By referring to theories of the classical economics and the new geographical economics, influencing factors have been analyzed by choosing variables from the following aspects, as showed in Table I.

(1) Resource endowment. According to the Heckscher Ohlin model and the renewable resources industry features, the raw resources will be transformed into re-manufactured products after they are manufactured and used for certain time, so as to bring the huge value-added effect to utilization of renewable resources, by fully utilizing the sufficient labor resources, introducing the senior science and technology talents and reclaiming the waste products and the waster metal resources from the manufacture industry.

(2)Regional economy. By referring to the concept of radiation in the regional economics theory, the added value of re-manufactured products will be increased, through technical innovation for comprehensive utilization of renewable resources, the potential renewable resources reserve will be increased, and the industrial scale of leading enterprises will be expanded to drive development of the peripheral micro and small enterprises.

(3) Policy instruments. It has been an important research field in the managerial economics. Related government policies of the renewable resources industry are considered as the mixed instruments, the opening degree is considered as the voluntary instrument, and the environmental regulation intensity is considered as the compulsory instrument, so as to achieve the policy goal of the industry's sustainable development.

Table 1 variable factors description

category	variable	method
resource endowment	Labor(LW)	average wage level
	equipment manufacturing (EM)	Industrial output
	human capital (HC)	number of higher education
regional economy	technology intensive (TI)	R D expenditure
	consumer (CL)	purchase level per capita
	enterprise size (ES)	industries industrial output / corporate
policy instruments	government investment (GS)	nontransfer payments
	opening (OW)	Imports / GDP
	environmental regulation (ER)	pollution completed investment / GDP

3.2. DATA COLLECTION

The three major economic circles of China, namely the angtze River Delta Region, the Pan Pearl River Delta Region and the Bohai Rim Region have been taken as the research targets. The Bohai Rim Region includes Beijing, Tianjin and Hebei, the angtze River Delta Region includes Shanghai, Jiangsu and Zhejiang, and the Pan Pearl River Delta Region includes Guangdong, Fujian, Jiangxi, Hunan, unnan, Guizhou, Guangxi and Sichuan. All original data required in this paper are from China Statistical earbook, different local statistical yearbooks, China Statistical earbook of Science and Technology, China Labor Statistical earbook, and China Statistical earbook on Environment between 2004 and 2014[14].

4. THE EMPIRICAL RESULT ANALYSIS

4.1. INDUSTRIAL CONCENTRATION TREND

With the EG factors, the cluster development trend of the renewable resources industry in these three major economic circles can be calculated. Chronologically, there have been not significant changes with the oscillating and ascending increase trend of EG indexes between 2004 -2014, which indicates that the renewable resources industry structure has to be adjusted further; on the other hand, the EG indexes have turned out to be large values generally, which indicates clustering of enterprises in the industry, preliminary formation of the scale economy and rapid development of the industry.

Spatially, although these three economic circles do not adjoin each other, they are all located on the eastern coast of China. The renewable resources industry development tends to be fluctuating, which is consistent with their general economic levels. In the ten years, the renewable industry has achieved a high development level generally in these economic circles, accounting for 65 of the total GDP of the national renewable resources enterprises; it has showed a trend of rapid ascending first and slow decrease later, and then tended to develop rationally. Specifically, the angtze River Delta Region has kept the highest industrial cluster degree, followed by the Pan Pearl River Delta Region at the beginning, the Bohai Rim in the middle term and both regions basically in the late period [15]. However, there has not been a co-built and co-existing integrated resources network among them, so key factors promoting the renewable resources industry development have to be found from the results of the influencing factors analysis and test.

4.2. LSDV ESTIMATION

With four panel data unit root test methods of LLC, IPS, ADF and PP, and with the LSDV method to estimate the comprehensive recession effect of the fixed effect model, test results of 9 initial variables (Comprehensive Model 1) and 8 variables excluding the colinearity variables (Comprehensive Model 2) have been listed in Table II.

Table 2 the combined effects of the test results

Variable	Coefficient	t statics
lnLW	-0.0011(-0.0011)	-4.5757*** (-4.6547***)
lnEM	0.0007(0.0007)	3.0581*** (3.6869***)
lnTI	-0.0432(-0.0448)	-3.8637*** (-4.1920***)
lnHC	-0.0004(-0.0004)	-2.3303**(-2.2540**)
lnCL	0.0009(0.0009)	4.8603*** (5.1309***)
lnES	0.0018(00018)	3.4842*** (3.5205***)
lnGS	-0.0021(-0.0020)	-2.7441*** (-2.8312***)
lnOW	-0.0003(-0.0003)	-2.0841**(-2.4170**)
lnER	0.0000(-0.0007)	0.0421
c	0.0075(0.0077)	4.0772*** (4.3469***)
LLC	--	-26.6707***
IPS	--	-7.38633***
ADF	--	292.136***
PP	--	325.425***

Note: correlation value of comprehensive regression1 is outside the parentheses , correlation value of comprehensive regression2 is in brackets.

Table 3 cross-sectional fixed effect coefficients of areas

region	province	fixed	parameter	statics
jingjinji economic band	Beijing	0.0050	R^2	0.4400
	Tianjin	0.0018	A+R^2	0.3400
	Hebei	-0.0004	SER	0.0023
yangtze river delta region	Shanghai	0.0004	LOG	666.1588
	Jiangsu	0.0006	F-statistic	4.4064
	Zhejiang	0.0063	P(F)	0.0000
pearl river delta region	Guangdong	0.0005	S.D.	0.0028
	Fujian	-0.0028	DW	1.8529

Jiangxi	-0.0021	AIC	-9.2023
Hunan	0.0012	SC	-8.7400
Guangxi	-0.0030	H C	-9.0144
Sichuan	-0.0002	SSR	0.0006
Guizhou	-0.0019	MD	0.0013
unnan	-0.0020		

It can be known that the higher equipment manufacturing industry development level, the higher household consumption level, the larger renewable resource enterprise scale and the larger environment investment intensity mean a better development trend of the renewable resources industry in these three economic circles, while the science and technology development level, the human resources level, the government investment level, and the opening degree haven't played their due roles, taking no account of spatial relevancy, which indicates that the industry is still labor intensive, so its production model should be adjusted further and the cluster development strategy should be carried out.

From Table III, it is known that DW=1.8529, which has the rather weak self-correlation. To verify whether the self-correlation belongs to the spatial correlation, the spatial recession model test should be carried out to study the characteristics of its influencing factors further.

4.3. SPATIAL PANEL MODEL ESTIMATION

In order to select the comfortable space panel model for provinces of renewable resources in three regions, the result of four non space panel estimation were estimated as shown in Table IV.

Table 4 nonspace panel estimation

	Pooled OLS	Spatial fixed	Time fixed	Spatial and time fixed
Ln(lw)	-0.1608*** (-3.0839)	-0.0160 (-0.1562)	-0.0885*** (4.8225)	-0.2176*** (-4.0604)
Ln(em)	-9.7700*** (-3.7070)	-12.3029 (-1.3212)	-9.0984*** (-3.9966)	-8.0921*** (-3.1254)
Ln(ti)	-0.0591 (-0.6867)	-0.0934 (-1.2541)	-0.2416*** (-2.7455)	-0.1004 (-1.0045)
Ln(hc)	0.1902*** (2.7825)	0.0392 (0.4307)	-0.1319** (-2.3954)	0.1133* (2.0554)
Ln(cl)	-0.0200 (-1.1273)	-0.0129 (-0.6562)	-0.0160** (-2.8444)	-0.0352* (-1.9651)
Ln(es)	-0.1367 (-0.3622)	2.3979*** (3.1682)	1.6912 (1.2116)	-0.2733 (-0.7457)
Ln(ow)	0.3005 (1.7281)	-1.1367 (-1.5680)	-1.2876* (-1.9006)	0.2019 (1.1761)
Ln(gs)	-2.5241 (-0.6787)	-0.5170 (-0.1625)	-0.2613 (-0.0839)	-1.7939 (0.4844)
c	0.7955*** (3.7277)	--	--	--
σ^2	0.0592	0.0389	0.0345	0.0538
R^2	0.1955	0.1896	0.1547	0.1833
LogL	3.8554	32.7103	41.0773	10.0274
LM-Lag	12.0327***	3.0321*	0.0059	20.7651***

LM-Error	10.0699***	3.0051*	3.2475*	14.7508***
Moran I	3.0443	2.5120*	1.0834	3.1453*

Note:*, **, *** is significance on 10 , 5 , 1 level respectively, the amount of t statistics is in parentheses .The same below.

Table 5 estimation of sdm with time fixed effects

Variable	coefficient	Variable	coefficient
Ln(lw)	0.3075***	W*Ln(lw)	0.4994***
Ln(em)	1.4800***	W*Ln(em)	1.1801***
Ln(ti)	-0.1511*	W*Ln(ti)	-0.0510
Ln(hc)	-0.0725	W*Ln(hc)	-0.0506
Ln(cl)	-0.0152	W*Ln(cl)	0.0320
Ln(es)	0.5409*	W*Ln(es)	-2.2172**
Ln(ow)	-0.0899	W*Ln(ow)	1.2941***
Ln(gs)	-1.8251	W*Ln(gs)	0.3531
σ^2	0.0109	Wald (SAR)	40.6252***
R^2	0.6489	LR (SAR)	37.2651***
Corrected R^2	0.5347	Wald (SEM)	37.5041***
LogL	329.0442	LR (SEM)	35.6623***

From the results, the time fixed effects significant presences. Thus, the space panel model should be established at the time fixed effects. The statistics have passed the significance test of LM test at the 10 level. SDM model needs to be constructed for further development. And the final form of the model should be determined by Wald and LR test, as shown in Table V. By simplifying SAR model, the statistical value of Wald and LR is 40.6252 and 37.2651 respectively for SDM. And by simplifying SEM model, the statistical value of Wald and LR is 37.5041 and 35.6623 respectively for SDM.

The estimation results of SDM with time fixed effects are as follows:

(1) The angle of impact in promoting types, local labor factors, the level of manufacturing industries, firm size has a significant positive influence; labor factors adjacent areas, the level of manufacturing, opening also showing significant positive impact effects. This shows that the three major economic circles adequate labor force, a higher degree of manufacturing economy, renewable resource industry gathering area more developed, renewable resource mobility between neighboring regions stronger, a considerable portion of renewable raw materials from import input, which can be good for following up development of renewable resources industry, to promote a sustained effect.

(2) In the perspective of inhibitory effect, the local technology intensity has a significant negative impact, and the size of the adjacent region also has a significant negative impact. This is mainly due to the improvement of local technology intensive is not very good to promote the development of the industry, weaken the monopoly of the local high technology industry. Three major economic circles of the renewable resources industrial park initially formed the scale of the economy, the impact on the adjacent areas of greater competition; competition and cooperation are not closely linked.

(3) In not significant impact angle, local human capital, consumer purchase, degree of openness, government investment effect was not significant, the area adjacent to the technology intensive degree, human capital, consumer purchase level, government investment is not significant. Renewable resources industry is a traditional resource waste dissembling processing industry development, and remains a labor-intensive economy characteristic, the demand for highly educated talents with the local export less, affected by the consumers to purchase level and direct government investment is small.

5. CONCLUSIONS AND RECOMMENDATIONS

Renewable resource industries as the realization of "three development" and the construction of ecological civilization specific carrier have gone beyond the scope of its own. It is showing a new vitality of development, degree of aggregation and the level of development of the industry which determines the ability of sustainable economic development. Selection of more developed coastal three big economic circles the relevant data from 2004 to 2014, industry concentration is obtained by using the index of EG, and builds influence the aggregation level of index system. First of all, based on the traditional panel data model, including LLC, IPS, ADF and PP unit root test four methods, estimated combined effect of regression fixed effects model using LSDV method, the regression model includes eight variables. Secondly, combining space-related economic effects, by comparing the non spatial panel estimates four models, model selection panel established under the time fixed effects model SDM empirical analysis. Results Explanation:(1) Three Economic Circles of renewable resources, industrial cluster development there is a positive spatial correlation between the regions and cities should be enhanced interactivity, collaboration, integration development, promoting technological, information, cross-regional flow of personnel, to improve inter-regional regeneration resource industry cluster effect. (2) For the industry gathering in the region, labor factor, the level of manufacturing industry, firm size to have a significant positive impact on technology-intensive have a significant negative effect to influence; For inter-regional industry gathering, labor factor, manufacturing level, opening up showing significant positive impact on firm size has a significant negative effect on the impact.

Therefore, the industry should focus on renewable resources to attract talent, and strive to build the domestic first-class brand of the leading enterprises in manufacturing and other related industries and to increase cooperation and docking, increased interregional introduction and utilization of waste resources, improve the system of regional infrastructure services. Better achieve "Internet +" renewable resources, industrial transformation and ecological protection in the context of the same total fertility, improve the economic circle of resources and environment carrying capacity.

6. REFERENCE

[1] **Ma un, Guan Yifan**. Influencing factors of industrial structure of three major economic circles [J]. Science and Technology Management Research, 2015,(7):63-66.(in Chinese)

[2] **He Yi ing, Wang Yaoyu.** Financial agglomeration and regional industrial structure and ecological efficiency coupling coordination cmpirical research three big economic circle, for example [J]. Inquiry into Economic Issues, 2015,(5):131-137.(in Chinese)

[3] **Liu ia, Zhao in in, Zhang Guanghai**. Spatial econometric research on the relationship between tourism industry agglomeration and tourism economic grow in China [J]. Economic Geography, 2013, (4):186-192. (in Chinese)

[4] **Ma Hailiang, Huang Dechun, Yao Huize**. Total-factor energy efficiency analysis of three major economic regions in China: based on super-DEA and Malmquist [J].China Population, Resources and Environment, 2011, 21(11):38-43. (in Chinese)

[5] **Yao Fengge, Zhong Shen, Zhou Zhongyuan**. The influence of service industry on regional economic growth empirically based on 1990-2008 China's three big economic circle city panel data [J]. Urban Problems, 2011, (1):57-61.(in Chinese)

[6] **Zhang Yun, Tang Yan**. International industrial transfer and the regional industrial structure, based on the research on China's three big economic circles [J]. Journal of Suzhou University of Science and Technology (Social Science), 2011, (2):23-28.(in Chinese)

[7] **Yin Desheng, Tang Haiyan**. Determinants of China are manufacturing industry agglomeration and the change trend, based on an empirical analysis of the three big economic circles [J]. World Economy Study, 2007, (12):3-9.(in Chinese)

[8] **Yang Yan, Zhou Wenyuan, Zhang Wanchun**. A fractal study on the spatial structure of urban system in the three economic circles of China [J]. Journal of Subtropical Resources and Environment, 2008, 3(4):66-71.(in Chinese)

[9] **Rosenthal, S., Strange, W. C. Geography**, Industrial Organization, and Agglomeration [J]. Review of Economics and Statistics, 2003, 85(2):377-393.

[10] **Puga, D**. The magnitude and causes of agglomeration economies [J]. Joumal of Regional Science, 2010, 50(1):203-219.

[11] **Syverson, C**. Product substitutability and productivity dispersion [J]. Review of Economics and Statistics, 2004, 86(2):534-550.

[12] **Chen Ying, Zhang Tunzhen**. Industry selection and configuration of the new city district: taking Tianjin Binhai New Area as an example [J]. urban studies, 2014, 21(5): 17-20.(in Chinese)

[13] **Chen Zhen, Yan Liang, Xie Xiongbao**. Analysis of external environmental factors of the evolution of resource-based industrial clusters [J]. China population, resources and environment, 2011, 21(04): 153-157. (in Chinese)

[14] **Teemu Lyytik inen**. Tax competition among local governments: evidence from a property tax reform in Finland [J]. Journal of Public Economics, 2012, 96(7):584-595.

[15] **Anselin L, varga, A., Z. . Acs**. Geographic spillovers and university research: a spatial econometric perspective [J]. Growth and Change, 2000, (31):501-516.

Knowledge Workers' Job Stress and Work Engagement: The Mediating Effect of Psychological Capital

Fu Yao[1,2,a],Cheng Shuo*[1,b], Yang Chang[1] , Yan Shi-bin[2]

[1] School of Management, Guangdong University of Technology, Guangzhou

[2] Guangzhou Academy of Social Sciences, Guangzhou

([a] yoyo@gzass.net, [b]cs_gdut@163.com,)

Abstract-**This study, from the perspective of positive psychology, is based on the psychological capital four-dimensional theory that was put forward by Luthans, grounded theory which was raised by Linjiang Ke, and the theory of resources conservation proposed by Hobfoll. We chose a sample of 592 knowledge-based employees from 11 companies, using both questionnaire survey and multiple regression analysis method, to investigate the relationship between job stress and work engagement, especially the mediating role of psychological capital. The results indicated that there is an inverted U shape relationship between job stress and work engagement, which means the job involvement of knowledge-based employees enable to reached the highest condition with a certain level of job stress.**

Keywords-**Job stress, psychological capital, work engagement, mediating effect, knowledge worker**

I.INTRODUCTION

In the background of economic globalization and knowledge economy, the economy in various regions has rapidly developed and industry competition is becoming fiercer and fiercer. At the same time, the rapid development of science and technology requires enterprises to carry out technological innovation, mergers, restructuring and other changes. It's a big challenge for enterprises to break through the fierce competition in the market and cater to the development of science and technology. Also, the market risk brought by the various uncertain factors leads to the increasing pressure of enterprise's employees. While knowledge worker is the backbone of enterprise development, their job stress is increasingly heavy.

At present, many enterprises try their best to attract and manage the knowledge workers to fulfill the most of these workers' potential, such as more generous pay and benefits, growth space, work design, balance of work and life. Nevertheless, the ceaseless change of economic globalization, political, social ethics and technology have brought some challenges to the knowledge workers and enhanced the intensity of competition for talent and market share, which took great pressure to the employees. If they are not able to deal with the pressure appropriately, these pressure will have a negative impact on the individual and the organization.

II.LITERATURE REVIEW

It's an important task for enterprise human resource management to find out how to effectively manage knowledge workers and release the most of their potential under the objective pressure. The traditional theories pay more attention to their weakness and try to find out why this negative result happened[1]. However, the positive psychology helps employees prevent from the negative result by finding positive methods and strategies, and helps to improve employees' performance, playing a positive role in the employees' interpersonal relationships and health (including psychological and physical health)[2]. Now a great number of mature companies from western developed economies actively carry out the research on employee well-being, and pay attention to the employee strength, and explore what aspects of person is right or good instead of focusing on the problems which may be occurred to employees.

In recent years, entrepreneurs are becoming increasingly aware of the importance of intangible capital—psychologicalcapital[2]. Based on the two aspects of organizational behavior and psychology, the famous organization behavior expert Luthans put forward the thory of positive organizational behavior and psychological capital[2]. Avey(2006)pointed out that psychological capital has become another important resource for enterprises to build competitive power[4]. It is known, from "conservation of resources theory" proposed by Hobfoll, that valuable resources have become the focus. They not only relieve job stress, but also ameliorate employee's work engagement. As one of the main sources of individual power, psychological capital is this kind of resource. However, numerous current studies have ignored the personality of the individual in the organization because not all individuals will be influenced by the factors, and impact on each different individual will vary from person to person. Based on the psychological capital four-dimensional theory proposed by Luthans, Chinese scholar Ke (2008)used grounded theory to propose that the psychological capital that fits China's national situation consists of two factors: transactional psychological capital and interpersonal psychological capital. [10] This study corroborates the main effect model of psychological capital--psychological capital has direct effect on job stress, work engagement and other variables[13]. In previous domestic and foreign research, more attention was paid to the negative psychology, such as job burnout, and less to the positive psychology. Kaln first proposed the concept of work engagement. [7]On the basis of that, Schaufeli (2002) particularly analyzed work engagement, and gave the definition of work engagement [8].

At present, studies on working pressure of employees lay stress on the working pressure source. While as to the organization operation and individual development, the employee's involvement in the work period is more important than the job satisfaction, job burnout and other factors. Therefore, from the empiricism point of view, this paper divides the job stress of knowledge worker into transaction stress and interpersonal stress, and analyzes the effect of interaction among the affairs, interpersonal stress and work engagement of the knowledge workers. Also, this paper considers the psychological capital as moderating variables, studying its moderating effect on the work engagement, analysing methods to reduce stress of knowledge workers ,improving their work engagement.

III. THEORY AND HYPOTHESIS

There are many definitions of job stress, but French's(1982)[1] is widely accepted. When the subjective intention dose not match the condition provided by objective environment, it will bring the individual job stress. For example, salary and status are not equal to subjective expectations. Based on the Cooper job stress scale[6] ,job stress questionnaire(OSI-2) and Chinese revised version(OSI-PRC) drawn up by the Institute of Psychology of Chinese Academy of Sciences, we combined the actual situation with specific characteristics of Chinese knowledge workers' job stress, and then modified several times according to expert's advice and on-site questionnaire pre test results. Consequently, we identified the job stress measurement index system. From four dimensions of the transaction stress(work load, occupation development) and interpersonal stress(social support, interpersonal relationships), etc, we measured the impact that job stress has on work engagement. Then, a questionnaire survey was conducted on the knowledge workers in Guangzhou area. Moreover, we collected the data on this field. The results showed that job stress of knowledge workers in Guangzhou area has high reliability and validity after analysis and test.

Mainly referring to Schaufeli's work scale about work engagement and relevant literature in China and abroad, the paper sets up the work scale including 3 dimensions of energy, dedication and focus, a total of 12 measurement issues. In view of the fact that work engagement is affected by many factors, this paper mainly studied the relationship between the job stress and work engagement. But, how the employee's work engagement changes while the job stress increases? So we put forward the hypothesis:

H1 The relationship between the job stress of knowledge workers (transactional stress / interpersonal stress) and work engagement is that the work engagement increases at the beginning and decreases afterward.

Psychological capital is a habit or preference of individual positive psychology. For example, with strong confidence and sufficient energy to work in a challenging job, people show a strong resilience ability, which make them quickly recover from difficulties in order to win. Based on Luthans' theory(PCQ-24)[6], the questionnaire of psychological capital includes four dimension-- self efficacy, hope, optimism and resilience. From the view of the characteristics of psychological capital, the greater the psychological capital of employees, the better their working attitude, and the stronger their resistance to job stress, the higher efficiency to work. Based on the research results of the existing literature, this paper made the following speculation: Psychological capital has a certain effect on job stress and work engagement. Thus, we put forward the following hypothesis:

H2 Psychological capital plays a mediating role on the knowledge workers' job stress (transaction stress/interpersonal stress) and work engagement (transaction stress/interpersonal stress), knowledge workers' job stress (transaction stress/interpersonal stress) has a significant positive influence on work engagement through the partial mediating effect of psychological capital.

H3 Psychological capital can reduce the impact that knowledge workers' job stress (transaction stress / interpersonal stress) has on work engagement.

IV. RESEARCH OBJECTS AND METHODS

This paper took knowledge workers of Guangzhou manufacturing enterprise as the research sample, and we distribute questionnaire to 800 knowledge employees of 11 companies from various industries of Guangzhou by random sampling method. In all 800 questionnaires, we recovered 667 questionnaires and acquired 592 valid questionnaires, and the effective recovery rate was 74%. In this survey, men accounted for 51.86%, and their age mainly concentrated in the 40 years old(83.61%) and their level of education is undergraduate(above 81.25%). Their working years mainly concentrated in 2 to 10 years (79.73%) and job mainly concentrated in the middle and senior staff (75.51%).

This paper used job stress survey scale and questionnaire(OSI-2), Chinese revised version(OSI-PRC), which included the transaction stress (workload, career development) and inter- personal stress (social support, interpersonal relationships)[18]. When it comes to the work engagement scale, we used simplified scale which drawn up by Schaufeli[8], including 3 dimensions: vigor, dedication and focus. And we used Luthans psychological capital questionnaire(PCQ-24)[6], including 4 dimensions: self efficacy, hope, optimism and resilience. All scales were used Likert 5 point scoring method to count, and SPSS21.0 software to analyse statistically.

V. RESEARCH RESULTS

4.1 Analysis of reliability and validity of scale

First we sorted out the questionnaire's data, and then analyzed and dealt with the data of the questionnaire, finally calculated the reliability and validity, which is shown in the table 1. The results from the test analysis showed that the cumulative variance interpretation rate of each quantity is more than 60%, and the overall reliability coefficient is higher than 0.80, which can prove that this table has a good reliability.

Table 1 Numerical value of reliability and validity of scale

	transaction stress	interpersonal stress	psychological capital	work engagement
1.umulative variance interpretation rate	70.308%	63.134%	66.329%	66.358%
2.integral reliability coefficient	0.895	0.872	0.887*	0.849

Analyze the correlation of the 4 factors of the scale, the results was shown in table 2 after statistics and analysis. It can be concluded from the analysis results that there is significant correction relationship among these 4 factors.

Table 2 Correlation analysis of variables

factor	mean value	standard deviation	1	2	3	4
1.transaction stress	3.927	0.510	1.000			
2.interpersonal stress	3.890	0.506	0.762**	1.000		
3.Psychological capital	3.801	0.465	0.673**	0.663**	1.000	
4.work engagement	3.879	0.552	0.752**	0.738**	0.683**	1.000

Note: * * means $p<0.01$, * means $p<0.05$.

4.2 Mediating effect test

In order to test the mediating effect of psychological capital that job stress has on the work engagement, the regression analysis method is used to test.

The multiple regression analysis to the demographic variables of knowledge workers showed that psychological capital has a positive effect on the work engagement, this is the positive predictive effect (β=0.753, p<0.01; $R2$=0.570, $\triangle R2$=0.565, F=110.568, p<0.01). The results from Table 3 showed: (1)The results of model 1 and model 4 show that there is a significant relationship of the regression coefficient among knowledge worker's job stress, job stress square and work engagement(β_1=0.738, p<0.01;β_1=0.756, p<0.01, β_2= -0.242, p<0.01). From the point of overall trend, work engagement will increase with the transaction stress increasing, but after reaching a critical value, it will decrease, that is to say, work engagement increases at the beginning and then decreases when transaction stress increasing, which supports the hypothesis H1; (2)The results of model 2 and model 5 show that there is a significant relationship of the regression coefficient among transaction stress, transaction stress (square) and psychological capital (β_1=0.758, p<0.01; β_1=0.632, p<0.01, β_2=-0.066, p<0.01); (3)We can know from the test results of model 3 that regression coefficient of knowledge worker's transaction stress to work engagement decreased from 0.738 to 0.397, but the regression coefficient relationship was still obvious (β_1=0.397, p<0.01). And the regression coefficient of the knowledge worker's psychological capital on the work engagement increased to 0.450(p<0.01). Therefore, it was concluded that the transaction stress of knowledge workers through the partial mediating effect of psychological capital has significant positive effects on work engagement, which supports the hypothesis H2; (4)We can also know from the results of model 6 that regression coefficient of knowledge worker's transaction stress to work engagement decreased from 0.756 to 0.447(p<0.01), and regression coefficient of transaction stress square to work engagement increased from -0.242 to -0.110(p<0.05), at the same time, regression coefficient of psychological capital to work engagement increased to 0.490(p<0.01), which proved that psychological capital can relieve the impact that transaction stress has on work engagement. To sum up, we can conclude that knowledge worker's transaction stress and transaction stress square have significant positive effects on work engagement, which supports the hypothesis H3.

Table 3 The test results of the mediating effect of psychological capital on transaction stress and work engagement

Variable		Work engagement	Psychological capital	Work engagement	Work engagement	Psychological capital	Work engagement
		Model 1	Model 2	Model 3	Model 4	Model 5	Model 6
Independet variable	Transaction stress	0.738**	0.758**	0.397**	0.756**	0.632**	0.447**
	Transaction stress square				-0.242**	0.066**	-0.110*
Mediator variable	Psychological capital			0.450**			0.490**
	R^2	0.552	0.585	0.636	0.551	0.611	0.646
	$\triangle R^2$	0.547**	0.580**	0.631**	0.546**	0.606**	0.640**
	F	102.743**	117.601**	127.261**	89.947**	114.621**	117.91**

Note: * * means p<0.01, * means p<0.05.

In the same way, after the multiple regression analysis to the demographic variables of knowledge workers, we can conclude these following theories from Table 4: (1)The results of model 1 and model 4 show that there is a significant relationship of the regression coefficient among the knowledge worker's interpersonal stress and work engagement (β_1=0.686,p<0.01;β_1=0.714,p<0.01, β_2=-0.351,p<0.01). Work engagement increases at the beginning and then decreases with interpersonal stress increasing, which supports the hypothesis H1; (2)The results of model 2 and model 5 show that there is a significant relationship of the regression coefficient between interpersonal stress, interpersonal stress (square) and psychological capital (β_1=0.671, p<0.01;β_1=0.524, p<0.01,β_2=-0.066, p<0.01); (3)We can know from the test results of model 3 that regression coefficient of knowledge worker's interpersonal stress to work engagement decreased from

0.686 to 0.327, but the regression coefficient relationship was still obvious (β_1=0.0.327, p<0.01), and the regression coefficient of the knowledge worker's psychological capital on the work engagement increased to 0.535(p<0.01). Therefore, it was concluded that the interpersonal stress of knowledge workers through the partial mediating effect of psychological capital has significant positive effects on work engagement, which supports the hypothesis H2; (4)We can know from the results of model 6 that regression coefficient of knowledge worker's interpersonal stress to work engagement decreased from 0.714 to 0.405(p<0.01), and regression coefficient of interpersonal stress square to work engagement increased from -0.351 to -0.121(p<0.05), at the same time, regression coefficient of psychological capital to work engagement increased to 0.590 (p<0.01), which proved that psychological capital can relieve the impact that interpersonal stress has on work engagement. Therefore, we can conclude that knowledge worker's interpersonal stress, interpersonal stress square have significant positive effects on work engagement, which supports the hypothesis H3.

Table 4 The test results of the mediating effect of psychological capital on interpersonal stress and work engagement

Variable		Work engagement	Psychological capital	Work engagement	Work engagement	Psychological capital	Work engagement
		Model 1	Model 2	Model 3	Model 4	Model 5	Model 6
Independet variable	Interpersonal stress	0.686**	0.671**	0.327**	0.714**	0.524**	0.405**
	interpersonal stress square				-0.351**	0.066**	-0.121*
Mediator variable	Psychological capital			0.535**			0.590**
	R^2	0.472	0.456	0.628	0.473	0.494	0.649
	ΔR^2	0.465**	0.450**	0.622**	0.466**	0.487**	0.644**
	F	74.449**	69.951**	122.77**	65.387**	71.162**	119.550**

Note: * * means p<0.01, * means p<0.05.

VI. CONCLUSION

This paper mainly studied the relationship among knowledge worker's job stress (transaction stress, interpersonal stress), work engagement and psychological capital, we can conclude these following theories after the analysis to the data: (1)From the point of overall trend, work engagement increases at the beginning and then decreases with the knowledge worker's job stress (transaction stress/interpersonal stress) increasing; (2)Knowledge worker's job stress (transaction stress, interpersonal stress) has positive effects on the work engagement with the influence of the partial mediating effect of psychological capital. That is to say, knowledge workers can increase their work engagement by strengthening psychological capital with the effect of the job stress (transaction stress/interpersonal stress); (3)The job stress of knowledge workers has a positive effect on work engagement, so does the psychological capital. Therefore, in the view of the problem of knowledge workers' job stress and work engagement, we can strengthen the management and take the corresponding measure through these following aspects: 1.Enterprise should focus on the management of knowledge workers' job stress, improving the ability to deal with the job stress, designing the corresponding work mainly based on the personal characteristics; 2.Focus on strengthening the role of psychological capital among job stress and work engagement, and improve the work efficiency of knowledge workers through the aspects of self-efficacy, hope, optimism, resilience, and so on; 3.Inspire the initiative of knowledge workers, then form a positive working attitude, and fulfill the most of their self potential, finally, it will bring the vitality and power for the development of enterprise.

Moreover, this paper is based on the Chinese situation, and the main data comes from knowledge workers of domestic enterprises. However, in order to ensure the reliability and validity of the strategy, the scales all refer to authoritative international journal did not verify the mechanism\; At the same time, the source of the sample is relatively single because of the restriction of research conditions, which may influence the stability and effectiveness of the regression model. In the future study, we should further research the restriction issues above, and make a more comprehensive and in-depth analysis.

ACKNOWLEDGMENTS

The authors wish to express their sincerest thanks to the editors and anonymous referees for their constructive comments and suggestions on the paper. We acknowledge the support of The research base of world cultural city construction and cultural industry, and National Natural Science Fund of China (No.71271060).

REFERENCES

[1]Hunter J,Schmidt F. Quantifying the effects of psychological interventions on employee job performance and work-force productivity[J]. American Psychologist,1983,38(4):473- 478.

[2]Luthans F,Avey J B, Avolio B J, et al. Psychological capital development: Toward a micro-intervention[J]. Journal of Organizational Behavior, 2002(27):387-393.

[3]Luthans F, Avolio B J. Authentic leadership: A positive developmental approach [M] // Cameron K S,Dutton J E, Quinn R E. Positive Organizational Scholarship. San Francisco: Berrett-Koehler, 2003.

[4]Avey J B, Patera J, West B. The implications of positive psychological capital on employee absenteeism[J]. Journal of Leadership and Organizational Studies, 2006,13(2): 42-60.

[5]Luthans F, Luthans K, Luthans B. Positive psychological capital: Going beyond human and social capital[J]. Business Horizons, 2004 (47):45-50.

[6]Luthans F, Youssef C. Human, social, and now positive psychological capital management: Investing in people for competitive advantage[J]. Organizational Dynamics, 2004 (33):151-152.

[7]Kahn W A. Psychological conditions of personal engagement and disengagement at work[J]. Academy of Management Journal, 1990, 33(4): 692-724 .

[8]Schaufeli,Salanova,Gonzale-Roma,etal.The measurement of engagement and burn-out: a two sample confirmatory factor analytic approach[J]. Journal of Happiness Studies,2002,3(1):71 -92.

[9]Hobfoll, Stevan E. The Influence of Culture, Community, and the Nested - Self in the Stress Process: Advancing Conservation of Resources Theory[J].Applied Psychology An International Review,2001,50(3): 337-421.

[10]Luthans F, Norman S M, Avolio B J, et al. The Mediating Role of Psychological Capital in the Supportive Organizational Climate-Employee Performance Relation- ship[J].Journal of Organizational Behavior, 2008,29 (2): 219-318.

[11]French JRP,Caplan RD,Van Harrison.The mechanisms of job stress and strain[J].Chichester, England:Wiley,1982: 5-11.

STUDY ON THE INNOVATION CAPABILITY OF MANUFACTURING ENTERPRISES IN LIUZHOU BASED ON FACTOR ANALYSIS

Chao LIN[1], i-yi WEI[2], Lu MA[1,*]
[1]Department of Management,
Guangxi University of Science and Technology, Liuzhou, China
*malu6655@163.com
[2]School of Management,
University of Liverpool, Liverpool, The United Kingdom

ABSTRACT

This paper aims to analyze the academic achievement's research of the domestic and foreign related innovation ability evaluating indicator system as well as evaluation method from three aspects: the innovation input, activity and output. By using the factor analytic method and the empirical data from Compilation of scientific and technical statistics data of large scaled industrial enterprises in Guangxi 2014' to generate a quantitative estimate and empirical analysis concerning to the innovation ability of Liuzhou's manufacturing enterprises.

Keywords: Evaluation, factor analysis, innovation capability, manufacturing enterprises

1. INTRODUCTION

Liuzhou as an industrial city, the development level of its manufacturing enterprises reflects the city's independent innovation ability and the status of industrial competitiveness. The manufacturing enterprises in Liuzhou have been developed for many years and have become an important pillar of the economic progress [1]. However, due to the obvious characteristics of Liuzhou manufacturing industry "big but not strong", especially in the development of the scale, the technological level and economic benefits compared with the developed industrial cities, there are still some issues regarding to capacity of independent innovation, uncontrollable core technology [2]. Furthermore, lack of innovative capacity has become the bottleneck factor of Liuzhou manufacturing enterprises to upgrade. This paper adopts the method of empirical analysis, starting from the construction of Liuzhou manufacturing enterprise innovation ability index system, objectively and effectively evaluate the existing situation and potential of the innovation and development of manufacturing enterprises in Liuzhou, which has important significance to improve the level of independent innovation of manufacturing enterprises in Liuzhou [3]. This paper conducts a comprehensive evaluation of innovation capacity from 3 aspects, including technological innovation investment, innovation activities and innovation output.

2. CONSTRUCTED THE EVALUATION INDEX SYSTEM OF INNOVATION CAPABILITY

At present, the domestic research on enterprise innovation ability, Dong Feng (2008) selected 24 original indicators, such as R D investment strength, R D personnel input strength, etc. constructed the evaluation index system of enterprise innovation capability[4]. Jiang ang (2011) designed the evaluation index system of innovation capability of large and medium enterprises in Sichuan Province from four aspects of potential innovation resources, innovation activities, innovation output capability and innovation environment. Xia Zhiyong (2008) constructed the enterprise innovation ability evaluation index system, from the research and development ability, production ability, management ability, marketing ability, market support ability, and finance and tax support ability and so on. Cao Hongjun and other (2009) set up the evaluation index system of enterprise innovation ability from six angles, for instance innovation consciousness, innovation

* Corresponding Author: Lu MA

investment ability, innovation output ability, innovation activity, management ability, innovation way [5]. Moreover, Jing Runtian, Chen Rui (2009) constructed the innovation ability evaluation index system from four aspects of innovation resources, innovation activities, innovation environment, and innovation output capacity. Ning Lianju, Li Meng (2011) built an evaluation index system of innovation ability from innovative input, innovation activities carried out and innovation output [6].

Although the selection of evaluation indicators from the scholars stated above is not identical. However, from a comprehensive point of view, the core content and the common point of the evaluation of enterprise innovation capability is the innovation investment, the ability of innovation activities and the output capacity of innovation [7]. Such as manufacturing capacity factors is able to incorporate into the input indexes of innovation; research and development ability, the ability to digest and absorb, internal management capacity and other factors can be incorporated into the innovation activity index [8]; marketing capability and new product sales income and factors, which can be incorporated into the innovation output indicators. Therefore, this research constructs the evaluation index system of enterprise innovation ability mainly from three aspects of innovation investment, innovation activity, innovation output, and two eight level indexes are selected in the first level index, as shown in Table I.

Table 1 evaluation inde system of innovation capbility of manuefacturing enterprises

First level index	Two level index	Variable
Innovation input	R D personnel	X1
	R D expenditure	X2
Innovation activities	Number of R D projects	X3
	R D project funds	X4
	Expenditure on technology transformation	X5
Innovation Output	The amount of patent application	X6
	New product sales revenue	X7
	New product research and development project	X8

3. EMPIRICAL ANALYSIS ON THE INNOVATION CAPABILITY OF MANUFACTURING ENTERPRISES IN LIUZHOU

3.1. Data sources and Processing

In this paper, we select eight indicators that will reflect the innovation capability of manufacturing enterprises as the original variables. In order to make the innovation capability of manufacturing enterprises in Liuzhou city have a more intuitive comparison, this paper also analyses and evaluates the innovation capability of manufacturing enterprises in four major cities in Guangxi. The key is to compare the differences in the innovation ability of the four major cities in Guangxi, so as to give a better demonstration of Liuzhou's manufacturing enterprise innovation ability. The original indicator data used in this paper is mainly from the compilation of scientific and technical statistics data of Industrial Enterprises above the scale of Guangxi 2014. In order to avoid the difference between the dimensions of the data, this paper uses the SPSS 22 statistical analysis software to collect the original data for the same direction and standardization.

3.2. Factor analysis condition udgment

Before doing the factor analysis, we must carry out statistical test at first; otherwise the analysis may be blind, resulting in lack of credibility. The KMO test is generally used to compare variables between simple correlation coefficient and partial correlation coefficient index, when the value is approximately to 1, which means it is more suitable for factor analysis [9]. When this value is less than 0.6, it is not suitable for factor analysis. Bartlett spherical test is aim to detect the correlation coefficient matrix and determine whether it is a standard unit array, when the significant probability of the statistical value of Bartlett is less than or equal to 0.05, the variables are independent of each other, and can be used as factor analysis. In this paper, we use SPSS 22 statistical analysis software to analyze the original indicators of the four major cities in Guangxi City, and get the results of KMO and Bartlett ball test as shown in Table II.

Table 2 kmo and bartlett inspection

Index	value
Kaiser-Meyer-Olkin metric for sampling sufficient degree	0.676
Bartlett of the spherical degree test approximate chi square	307.700
Df	28
Sig.	0.000

From Table II, the KMO value is 0.676. As the value more than 0.6, it is therefore suitable for factor analysis. Bartlett degree test statistic is 307.700 and corresponding probability p = 0.000, which is less than the significant level of 0.05. Thus, this set of data for factor analysis is in line with the conditions.

3.3. Solving eigenvalue, Contribution rate and Cumulative Contribution Rate

The contribution rate of public factors indicates that the common factor reflects the information of the original index; the cumulative contribution rate indicates that the corresponding several common factors reflect the information of the original index. From table III we can find that the first two common factors of the cumulative contribution rate reached 88.635 , which means that the two common factors will reflect the rate of 88.635 of the original information. It is manifested that these two factors can gather and sum up most of the original variable information and is able to facilitate the analysis of the manufacturing enterprise innovation capacity.

Table 3 eigenvalue and cumulative contribution rate

Ingredients	Initial eigenvalue	Extraction square and loading	Rotating square and loading
	Total; variance ;accumulate	Total; variance ;accumulate	Total; variance ;accumulate
1	5.689;71.116; 71.116	5.689; 71.116; 71.116	4.837; 60.46; 60.46
2	1.614; 20.172; 91.288	1.614; 20.172; 91.288	2.466; 30.82; 91.28

3.4. Factor analysis

Through the analysis of the initial factor loading table and factor model, it is found that the typical representative each factor's variables of is not outstanding and the interpretation of factor of evaluation index is not obvious. In order to illustrate the factor of practical economic meaning, it needs to rotate the factor loading matrix and further analysis is required. The factor load matrix after the rotation is shown in Table IV.

Table 4 factor load matri

Variable	Ingredients

	1	2
New product sales revenue	0.977	-0.070
R D expenditure	0.973	0.215
R D project funds	0.972	0.210
The amount of patent application	0.856	0.286
R D personnel	0.826	0.520
Expenditure on technology transformation	0.707	0.415
Number of R D projects	0.101	0.983
New product research and development project	0.261	0.938

From the Table IV, it can be seen that the rotation factor load matrix is divided into two levels, which is suggested to be a good explanation of the original indicators. The load value is greater means the more closely the relationship between the variables and factors can be. According to each factors' relative principal component score coefficient, the new product sales revenue and other eight indicators can be classified into two categories of main factors. Factor 1 in new product sales revenue (X_7), R D expenditures (X_2) research and development project funding (X_4), the amount of patent applications (X_8), R D personnel (X_1) , technical transformation expenditures (X_5) have a greater load. These indicators reflex the development activities within the enterprise. And these activities can improve enterprise's innovation. So, the factor 1 can be called innovation internal development factor. Factor 2 in number of R D projects (X_3), number of new product development projects (X_8) have a greater load. Due to the manufacturing enterprises products R D and new product development project is conducted with external partners, so factor 2 is called external innovation development factor.

3.5. Main factor score and ranking

In order to sort the factor scores, this paper uses regression method to get the factor score matrix, as shown in Table V.

Table 5 component score coefficient matri

Variable	Ingredients	
	1	2
R D personnel	0.128	0.125
R D expenditure	0.222	-0.062
R D project funds	-0.150	0.500
The amount of patent application	0.223	-0.065
New product sales revenue	0.115	0.091
Expenditure on technology transformation	0.178	-0.004
New product research and development project	0.275	-0.213
Number of R D projects	-0.099	0.447

According to Table V, we can obtain the factor score after rotation.

$F1=0.128X1+0.222X2-0.150X3+0.223X4+0.115X5+0.178X6+0.275X7-0.099X8$;

$F2=0.125X1-0.062X2+0.500X3-0.065X4+0.091X5-0.004X6-0.213X7+0.447 X8$.

Weighted synthesis was carried out by two main factors, and the main factor accounted for the proportion of the cumulative variance contribution rate, we can get comprehensive score calculation formula : $K = 0.60469F1+0.30819F2$.

3.6. Comprehensive score analysis on factors of manufacturing enterprises in Liuzhou and Guang i other ma or city

Liuzhou manufacturing enterprises and other major prefecture level city manufacturing enterprise factor comprehensive score as shown in Table VI. From the comprehensive score of main factor, Liuzhou being one of the largest industrial cities in Guangxi ranked the top. The enterprises of manufacturing in Nanning and Guilin ranked the second and third. The ranking of these two cities is in accordance with their level of economic development. Wuzhou manufacturing enterprises ranked the fourth. It is largely related with its West River Economic Belt leading city and its traffic fortress.

Table 6 factor scores and comprehansive score

City	F1		F2		Innovation ability	
	score	rank	score	rank	comprehensive	rank
Nanning	0.32520	2	0.95010	1	0.48946	2
Liuzhou	2.03071	1	-0.90313	4	0.94961	1
Guilin	-0.13424	3	0.37942	2	0.03576	3
Wuzhou	-1.23771	4	-0.79292	3	-0.99280	4

4. DISCUSSION

(1) According to table VI shows, the variance contribution rate of factor 1 is 60.469 , turn out to be the most impacting facotor on the innovation ability. From the four major prefecture level city manufacturing industry factor 1 scores, internal innovation ability of Liuzhou is the strongest, the score reached 2.03071. Nanning manufacturing enterprises came in second place, which reflects Nanning have increase the strength of the internal innovation in recent years. Guilin's manufacturing enterprises' scored more than Wuzhou, which is ranked third in the region. The ranking of Guilin is mainly because of Guilin in recent years has been more emphasise on science and technology research and development and technical innovation of enterprises. Wuzhou's manufacturing enterprises ranked fourth, but with respect to Liuzhou, its manufacturing enterprise innovation ability is still at the lower stage.

(2) Factor 2's variance contribution rate is 30.819 . It is the second principal component, influenced the innovation capability. From the four major cities' scores, the innovation capability of manufacturing enterprises in Nanning topped the list, Guilin manufacturing enterprises innovation ability ranked second, Wuzhou manufacturing enterprise innovation ability ranked third, and the innovation capability of manufacturing enterprises in Liuzhou is ranked the bottom. The rank manifests that the external development of the innovation ability in Liuzhou requires for further attention and concern.

(3) As is known to all, Liuzhou is the largest industrial city in Guangxi, and its manufacturing enterprises are the key to the prosperity of the city economy. The importance of manufacturing enterprises in Liuzhou is self-evident to the whole Liuzhou city. Liuzhou filled with abundant labor force and natural resources in Guangxi manufacturing industry. Although Liuzhou manufacturing enterprise development relatively fast, but it still have some questions to resolve, such as R D investment accounted for the proportion of sales income is too low, some important products are mainly imported from abroad, enterprises lack of external innovation development activities [10]. For manufacturing enterprises in Liuzhou, it need further increase the external innovation development capacity. Nanning's manufacturing enterprises ranked first in the overall ranking, but factor 1 ranked the second, so Nanning manufacturing enterprises need further enhance the internal strength of enterprise innovation in the future. For Nanning, it is effective to increase investment in R D human resources and economic resources, and increase sales of new products.

Guilin and Wuzhou manufacturing enterprises is ranking at last, so the two cities also need increase manufacturing enterprise R D funding and R D personnel input.

5. CONCLUSION

Above all, for one thing , Liuzhou municipal government need to further create a good policy environment and increase financial support effectively. First, Liuzhou should conduct tax policies to encourage innovation. Manufacturing enterprises, which focus on science, technology development and scientific research, shall be exempted from import tax in accordance with the law. Second, Liuzhou should innovate financial investment way. Government should integrate the relevant financial capital and various kinds of science and technology plan, give full play to the guiding role of government funds, and to support the construction of enterprise innovation capacity. Give financial support to the outstanding enterprises identified by research institutions. Support small and micro enterprises in technological innovation activities and increase innovation capital. Fund those enterprises, which breakthrough innovation. Third, to promote the combination of technology and finance, established financial support system of science and technology. To guide the banking financial institutions to increase SME credit support. Encourage state-owned commercial banks to support enterprise innovation, especially small and micro enterprises in technological innovation. To develop the science and technology financial services platform. Establish and use the venture capital fund. Develop science and technology innovation of the insurance products positively. Promote intellectual property rights and equity pledge loan. Promote the implementation of direct debt financing mode.

For another, in the light of absence of the external innovation development in Liuzhou manufacturing enterprises, Liuzhou should increase government funding for independent innovation, through effective incentive mechanism to encourage both engineering and technical personnel in order to conduct innovation. Furthermore, Liuzhou should establish independent innovation platform, strengthen cooperation with institutions of higher learning and scientific research institutes. Liuzhou manufacturing enterprises should take the external innovation as a booster for the independent innovation and improve the technological content of products, enhance the market competitiveness of products.

6.ACKNOWLEDGMENT

This research was financially supported by the Liuzhou Science and Technology Department's Soft Science Program of 2013, NO2013J040404.

.REFERENCES

[1] **GUAN , MA N.** Innovative capability and export performance of Chinese firms [J].Technovation,2003

[2] **BROWN S, FAI F.** Strategic resonance between and capabilities in the Technological organizational innovation process within firms [J]. Technovation,2006,26 (1):6075.

[3] **GUAN C, YAM R C M, MOK C K, et al.** A study of the relationship between competitiveness and innovation capability based on DEA models Technological[J]. European Journal of Operational Research, 2006, 170 (3):971986.

[4] **BURGELMAN R A, CHRISTENSEN C M.** Strategic Management of Technology and Innovation[M] . New ork: McGraw Hill, 2004

[5] **Hu haibo.** Review on the research of industrial independent innovation capability [J]. Options Derivatives Traders, 2011,33(7).

[6] **Liu Guo in.** Review of research on independent innovation at home and abroad [J].Scientific and Technological progress and Countermeasures, 2012.

[7] **Du Kai**. Research on the development path of China's manufacturing industry based on resource constraints [J].Jiangsu social sciences, 2013.

[8] **Hao Xiaoyan**. Evaluation of technological innovation efficiency of manufacturing industry in Inner Mongolia [J].Scientific management research, 2013.

[9] **ELTON , GRUBER , DAS S, et al**. Efficiency with costly information: a reinterpretation of evidence for managed portfolios [J].Review of Financial Studies, 1993 (60):122. 9

[10] **WERMERS R**. Mutual fund performance: an empirical into Stock-picking talent, style, transactions costs, and expenses [J].Journal of Finance, 2000 (8):16551703.

STRATEGY BASED ON GAME THEORY IN THE DEVELOPMENT OF CHINA LOGISTICS PARKS

Shi-Zheng Huang[12*]

[1] School of Economics and Management, Guangdong University of Petrochemical Technology, Maoming 525000, China
[2] Institute of Management Sciences, Lanzhou University, ShenZhen 518057, China
szhuang06@gmail.com

ABSTRACT

The logistics industry plays an important role in the developments of the national economy. Logistics parks promote the developments of the modern logistics industry through the integration of resources and the scale effect. However, there occur some phenomena such as blind constructions, homogeneity in the logistics parks in operation, the high vacancy and low efficiency of parks, which lead the developments into prisoner's dilemma . This article analyses the local governments' selection of development strategies for logistics parks by applying game theory, and reveals that local government's interest is the key factor in the strategy selections. Moreover, the interference acts of the central government is an effective approach to change the developments of logistics parks from prisoner's dilemma to the cooperative game.

Keywords: Logistics Park, Prisoner's Dilemma, Game Theory

1. INTRODUCTION

The logistics park plays an important part in logistics development. Peter F. Drunck (1962) first proposed that logistics is last border in reducing cost in the economic domain. The operation under the global environment, logistics has become a critical industry in promoting national economy. In order to find the solutions regarding the logistical distribution and the cost reduction, logistics parks have been built in most cities all over the world. In China, the logistics industry was taken as a new industry in 2009, and logistics park project was listed as one of the nine key projects. According to the research, the constructions of logistics parks could not catch up with the economic development; moreover, the homogeneity of the logistics parks has appeared to become a serious problem. Logistics parks have been overbuilt under the blind constructions, the lack of overall planning in some small and medium-sized cities as well. Meanwhile, some logistics parks lacked the differential orientations and the core competence in their function designs and layouts. The radiation areas of the parks overlap and the homogeneity in service quality is getting worse. This article discovers that local governments develop and rebuild logistics parks without taking accounting of the cost and benefit. Behind the vicious competition, except for the achievement promotion systems of the local government officials, the protection policy of the Chinese local government's interest will be the key problem.

An analysis from the developments of the logistics parks of developed countries shows that the developments of the logistics parks in Germany and Japan benefited from the guidance of the scientific general plans and policy (Wang, 2006; ang and Sun, 2008). The constructions of the logistics parks in China lacked scientific general plans, without the explicit purpose as well. The scale competitions and enclosures exist as serious phenomena. (Li, 2007; Feng, 2008; Fu, 2008; Wang, 2009). Wang et al. (2006) proposed that the falseness in motivations of constructing logistics parks indicates government's blind constructions in pursue of achievements. Wang and

* Corresponding Author: Shi-Zheng Huang

Hong (2008) put forward that logistics parks should reflect economic effect of integrated logistics nodes and proposed that the planning and constructions of logistics parks should keep pace with local economic developments. In the policy of Adjustment and Revitalization Planning for Logistics Industry released by Chinese government in 2009, announced nine key projects. Among the policy, the general plan dominates the primary position, which indicates that logistics parks should be planned in important logistics node cities, manufacturing industry bases and comprehensive traffic hubs. Kevin O'Connor (2009) proposed that logistics parks should be constructed on the basis of the global city regional frameworks with the traffic and nodes. Zhao et al. (2006) pointed out that central government should plan spatial layouts for various sectors and types from the national macroscopic overall levels and provide the guidance for local governments to develop the logistics park planning, and avoid the duplication of investment and the unnecessary waste of resources. Moreover, the local governments should cooperate with the superior planning actively and get rid of the barriers between the reigns and industries, in order to create an open, fair, and equitable park environment. Meanwhile, strengthen the communications and coordination between the regions as well.

Currently there are a lot of research literatures regarding the development of Chinese logistics parks mainly concentrate on the planning and layouts, demand estimations, park scale determinations, the node locations, explore models, profit models, etc. Most literatures and researches focus on the single logistics park. Rare researches concentrate on the competitions among the logistics parks developed by governments. Game theory provides a framework to study the strategic act by single decision-maker and develops the solutions which can be wildly accepted. Game theory can solve the strategy decisions not only among the nations, local governments, but also between the central government and the local governments (Kaveh Madani, 2010). Anderson et al. (2012) made decisions on the government infrastructure constructions by applying game theory. This article elaborates the game theory in the development competitions among logistics parks in China and its solutions, analyses strategy options of developing logistics parks for local governments, discusses the decisions made by local governments according to the resource advantage, and proposes the initial ideas of the solving problems.

II. CURRENT DEVELOPMENT SITUATIONS OF LOGISTICS PARKS IN CHINA

Logistics parks can also be named as distribution parks or logistics bases. According to China National Standard logistics term in 2006, the logistics park is a variety of logistics facilities and various types of logistics enterprises concentrated in space on the layouts of the locations, a specific area with a certain scale and comprehensive service functions. In the rule of Logistics park classification and basic requirements , logistics parks would be built to fit in with the freight forwarding, industrial developments, and the needs of the logistics residents and consumers. Relying on the local traffic conditions, an industry system phase with the city's integrated functionality and the public characters of the logistics foundation facility groups, the logistics function area will gather the layouts of the concentrated planning, land conservations, enterprises gathering and intensive operations. Currently, the national definition of the logistics park is still indistinct although there's the explicit provision in the national standard in China. The main reason lies in the different development level of logistics parks, the academic only views a part of the logistics park growth while the society has diverse understandings of the logistics parks.

China is relatively behind in the developing of the logistics parks compared in the whole process, but it has considerable achievements and the rapid growths during the last decade, with the emergence of the first logistics park in Pinghu, Shenzhen in 1998. According to the survey in 2012 conducted by China Logistics Parks and Purchasing Federation and China Logistics Association of National Logistics Park after the comprehensive research of three times . There will be a total of 754 logistics parks with all types, compared with 475 in 2008, a significant increase of 58.7 , far exceeding the growth rate of foreign logistics parks in terms of the quantity and scale. Third National Logistics Park (Base) Survey Report data show that the number of logistics parks which are in operation is 348, accounting for 46 , 32 of the whole are under-construction, and 22 have been planned. According to the local distribution, northern coastal economic zone parks have the faster pace of the development, exact region distributions is as below in table I.

The development of the logistics parks in China is not mature whether from the theory research or in practice. The local governments lack the scientific planning and connections with the local economic development in the process of the logistics park developments. They plan logistics parks without the consideration of the local actual situation or the future market demands, but only focus on the short-term profit, which to some extent results in runaway of parks' scale and quantity. The government's exploitations are lack of the whole planning and management systems. Moreover, the competition among parks results in the park's relatively high vacancy rate, the planning lacked of rationality and serious "enclosure movement." It's estimated that vacancy rate in most logistics parks is as high as 30 , causing a great deal of waste in the material, financial, land and human resources.

Table 1 The construction statuses of the logistics parks In 8 economic zones

Region	In Operation	Under Construction	Plan	Total
Northern costal economic zone	88	35	5	128
Middle reaches of angtze economic zone	40	31	68	139
North-east economic zone	26	23	7	56
Eastern costal economic zone	65	14	14	93
Middle reaches of ellow River economic zone	35	47	11	93
Southern costal economic zone	52	25	6	83
North-west economic zone	14	25	25	64
South-west economic zone	28	41	29	98
Total	**348**	**241**	**165**	**54**

1) Disequilibrium in logistics park development: Current logistics parks in China are mainly located in economically developed coastal cities, few in the central and western cities. Under the adjustment and upgrade of Chinese industrial structures, some industries will transfer from the coastal developed cities to the mid-west cities, which will definitely bring new opportunities and challenges to logistics parks in the mid-west areas, accompanied with surplus and vacancy in supply and demand in the coastal areas as well. Meanwhile, many important industrial and agricultural materials concentrate in western China, which will bound to material handling difficulties and the high cost of collecting and distributing contradictions.

2) Irrational planning and blind construction of logistics park: It is more and more apparent that logistics parks in some areas are over constructions in the scale and quantity. The construction of logistics parks should refer to the number of logistics parks per billion GDP (number of logistics parks/total GDP). Japan, which ranks the third in GDP in 2010 possesses only more than 20 logistics parks; while Germany who ranks the fourth only has 30 logistics parks, the numbers of logistics parks that Germany and Japan possess per billion GDP are respectively 0.0014 and 0.0018, compared with the number of China 0.0077. Moreover, some cities with both two key economic factors, which are overall GDP and population rank in the middle or lower levels, their traffic positions don't belong to nodes, but own a number of logistics parks even more than the average

number of national level node cities. Therefore, the over- constructions and over-scales of logistics parks cause the two main reasons of the logistics park vacancy.

3. LOGISTICS PARK DEVELOPMENTS FROM PRISONER S DILEMMA TO CORPORATE GAME THEORY

3.1. Game Theory

Game theory is in essence applied mathematics research of the competition and corporation (Stanford Encyclopedia of Philosophy, 2006), to study the behavior of the decision-making body in the direct interaction between people how to make decision and let the decisions reach the equilibrium. In the mid-1980s, game theory has been widely applied in economics and other disciplines, its ideas penetrated to almost all fields of economic analysis, and developed as part of mainstream economics, an important discipline of operations research as well.

A game consists four basic elements: (1) Participants (Player), also called as decision-makers, each participant is based as rational. (2) Action, participants can choose all decision-making. (3)Payoffs, participants get paid remuneration in the end of the game in accordance with the combination of strategies each participant selected. (4) Information, what every participant knows when making decisions.

The core concept of non-cooperative game is Nash Equilibrium, which the equilibrium itself is an estimation of competitor's strategy in a given environment. Proposed by Nash (1950), the definition of Nash Equilibrium is a combination of strategies, in which one will never take the best strategies and assure the one's own strategy will not be poorer than others', regardless of what the other party do.

3.2. Logistics park developments from prisoner s dilemma to corporate game theory

Logistics parks in China mainly adopt the model that government planning with enterprises exploitation. In the corporation and competition under the local economy, local government stakeholders in different areas, we personify and take it as a rational economic man. In the process of economic development, local governments usually implement the competition and protection approaches due to the consideration of the priority of local government's interests over local public's interests in the decision process. Meanwhile, the local government owns different competitive capabilities according to its capability to acquire resources. Logistics parks feature the operating function rather than the public welfare-oriented, thus it is exclusive and competitive. And in the process of development, when it occurs that multiple logistics parks coexist in one area, each park can be taken as chess square, and each chess strategy space can be chosen and the value of which is continuous. Consequently the game parties have infinite variety of optional strategies. This article concentrates on non-cooperative game theory and pure strategy, mixed strategy is out of extent of this research.

3.2.1Prisoner s Dilemma

Prisoner's dilemma was proposed by A.W. Tucker (1940) to discuss whether cooperative equilibrium exists. The cooperation of both opponent sides allows both sides to achieve optimal results. Betrayal the other side allows him to get more benefits, so the two sides had been planted betrayal incentives in the beginning. Each behavior can take the cooperation strategy at the expense of short-term self-interest for the common interests or betray at the expense of long-term common interests in order to pursue individual self-interest. Each behavior's reward depends on the strategy taken by the other side, all in the obedience to the compensation structure.

Model 1: The competitions between the local governments with the same level of the competitiveness in developing logistics parks

The current management system of the logistics parks in China is chaotic. Local governments develop parks in their own ways. In the game, assume that local government A and B both intend to invest logistics parks, the game parties both make decisions in the pursuit of the profit maximization as their game strategy. When they cooperate; both sides will get a profit of 60 units with fair competition. When they don't cooperate, under the implement of the protectionism and both can keep their own profit of 100 units each player. If one side chooses to cooperate while the other side chooses no, then the competitive side will lose 10 units due to the other side's protectionism policy. If both sides don't implement the local protectionism, each will gain 50 units of profit. Resulting in the combinations of various strategies payoff matrix as shown in Table II, among which column represent strategy space of the local government B, the first number stands for its profit; the row represents strategy space of the local government A, and the second number stands for its profit.

Table 2The strategy of the developing logistics parks between the local governments of the same level of the competitiveness

Local Government A / Local Government B	Competition	Local Protection
Competition	60, 60	10, 100
Local Protection	100, 10	50, 50

Apparently, there exist two Nash Equilibriums in this game: (Competition, Competition) and (Protection, Protection), which means that under these two circumstances the two local governments won't change their strategy. They will pay respectively (60, 60) and (50, 50), but the level of payment they gain makes a big difference. According to Nash Equilibrium, both sides will make strategy on the assumption that the opponent side takes strategic action at its optimal. If they choose to compete, they will face the risk of losing 10 units' profit if the opponent doesn't corporate. If they don't adopt local protection, the worst outcome they may have can still remain the current level of profit. The local government's implement of the protection strategy always has the advantage over the competitiveness. (100>60, or 50>10). The local protection strategy is at non-pare to optimal equilibrium state, the optimal strategy for game parties. Thus (Protection, Protection) means balanced gains for both sides, who are not motivated to change this selection, so that the two sides stuck into a game "Prisoner's dilemma".

Model 2: The competitions between the local governments with different competitivenesseses in developing logistics parks

Various regions of China are at different stages of development and different levels of the developments. Differences in locations, resources, management systems are all notable, thus the competitiveness between the local governments is various as well. Assume the local government A possesses absolute advantage in the economy scale, management systems, the cost and innovation capabilities, resulting in stronger competitivenesses in developing logistics parks which enables to gain more profit in cooperation. Meanwhile, the local government B is at a disadvantage in the competition, associated with impaired interests. Assume even the two local government adopt (Competition, Competition) for same level of capability, each will gain 60 units of profit. But with economic development, disequilibrium will occur in different regions' economic development, game payoff matrix will have relevant changes seen in Table III.

Table 3 The strategy of the developing logistics parks between the local governments of the different competitivenesses

Local Government A / Local Government B	Competition	Local Protection
Competition	120, 180	60, 100
Local Protection	100, 60	80, 80

There's only one Nash Equilibrium in the game (Protection, Protection), the local government strategy (Competition, Competition) is no longer a Nash Equilibrium. We can indicate that local government A has comparatively more advantage in reducing the cost, showing competitive advantages with innovations, attracting capitals and occupying the market. While local government B doesn't have the competitive advantage which will lead to run of resource and loss of market, eventually results in local government's interest loss (120<180). Thus, the local governments will consider their own interest primarily and tend to implement the local protection strategy before the emergence of the comparative advantage in cost, management system and innovation capability. And when the local government adopts the local protection strategy, the optimal strategy for other local governments will be the same. However it is not an effective strategy for the development of logistics parks, thus stuck into prisoner's dilemma again.

If the game only performs once, both local governments will select the ineffective disadvantage strategy. The only equilibrium is to implement of local protection for both sides, which will be hard to break without external force. If, in another case, games are repeated and communications are allowed, with the mutual trust, in the period time of t, sub-game perfection are the last one and are non-cooperative game, the last one will be betrayed into a prisoner's dilemma, with which infer that the previous one must also be a non-cooperative game. For infinite games, the existence of equilibrium strategy is more difficult to solve. Prisoner's dilemma is a typical case of a non-cooperative game, which reveals the local government to select the optimum strategy in the game strategy for the development of logistics parks, maximizing their own interests. The result is not the ideal state to achieve "win-win".

3.2.2 Cooperative game

Local protection strategy is the initial important approach to achieve the interests of the local government. But to break the low efficiency of the competitive strategy of the local government (Protection, Protection) and to resolve the "prisoner's dilemma" require the central government's involvement in the game, in order to change the local governments' profit strategy in the game.

From the analysis of table II, the initial selection of the local government is not in favor of the healthy development of logistics parks, thus the external interference by the central government is the key to break this path dependence. The central government can change the game gains between the local governments through unified planning of the logistics parks, the establishment and improvement of the market rules, and the punishment for violations of the local protection. According to table II, if the central government gets the punishment of 100 units for the local government implementing the local protection strategy, then the payoff will change to table IV. Meanwhile the Local Protection is no longer equilibrium; the only Nash Equilibrium is (Competition, Competition).

Table 4the strategy of the developing logistics parks between the local governments of the same level of the competitiveness

Local Government A / Local Government B	Competition	Local Protection
Competition	150, 150	80, 0
Local Protection	0, 80	0, 0

If the game repeats, it should work out with joint developments of the cooperation between the local governments. Since a good market order is public goods, and creating a good market order will face the constraints of transaction costs, the stronger the local government is, will greater benefit it can obtain in a good market. And it will be more motivated to promote the perfection of the market order, thus a party with the stronger competitiveness can pay the larger share of the transaction costs. For example, under the circumstance of table III, the local government A with stronger competitiveness pays the other side 20 units as long as it adopts competition strategy. Meanwhile, the payoff matrix of two parties has following changes as shown in table V, (Competition, Competition) also becomes the only Nash Equilibrium.

Table 5 The strategy of the developing logistics parks between the local governments of the different competitivenesses

Local Government A / Local Government B	Competition	Local Protection
Competition	140, 160	60，100
Local Protection	100, 60	80，80

The central government can play the role as the agent in coordinating behaviors between the local governments. On the other hand, urge and help the local governments to establish a coordination mechanism and reduce the cost of coordination between the local governments. And provide the compensation to support backward areas to maintain a favorable market environment at the same time.

In reality, the inadequacy in punishment of violations in the market performed by the central government is one of the important causes of the local protection. The infinite game of the both sides emphasis on the future profits, constrained the efficient agreement. Thus, it should be large enough, with fewer profits of betrayal. Moreover, the cost of punishment should be greater than profits of betrayal. The final solution may become balanced with the cooperation (Competition, Competition) optimal strategy.

4. CONCLUSION

As an important composing part of Chinese modern logistics industry, the development of logistic park enhances national strategic height. However, local governments take more consideration to the interests of their own in the process of logistics parks decisions, resulting in the blind construction of the logistics parks and failure to adapt to the regional economic development demands for those parks in operation. This article analyze the competition strategy of the local government in developing logistics parks, applying the game theory and discover that local governments with the same or different levels of competitiveness will all of the initially implement local protection strategies, which is equilibrium of the low efficiency and not favorable for long-term developments. The market competition will reduce the cost and enhance the efficiency, thus more favorable for the labor division and cooperation. As supreme governing body, the central government's maintenance of the market order enables the development of logistics parks change from prisoner's dilemma to the cooperative game.

Although the assumption of this article is based on two local governments, constructing and the analyzing model under conditions with respectively the same and different competitivenesseses, the conclusion won't change even with several participants. On account that the achievements of the local government officials are closely bound to local economic interests, thus how the local government officials influence the decision and constraints issues regarding the information cost and the implementation when the central government participate in the game both deserve the further research.

5.REFERENCES

[1] **He Li Ming et al.**, The investigation report of the Third National Logistics Park (base), China Logistics Purchasing, No.18, pp.70-71.2012.60-65.

[2] **Tang Yu Sun and You Wang**, Revelation of the foreign experience of the Chinese logistics park construction, China Storage Transport, No.4, pp.26-30.2008.

[3] **Hong Mei**, Construction of China logistics park, Science and Technology Innovation Herald, No.13, pp.115.2008.

[4] **Zhao Hong Bo.**et al., The development of China's logistics park, Academic Exchange, No.7, pp.86-89.2006.

[5] **Wang Ming**, Foreign logistics park operator mode and learn, China Storage Transport, No.5, pp.14-16.2003.

[6] **Fu Lan**, Problems of China urban logistics park construction, China Logistics Purchasing, No.7, pp.70-71.2006.

[7] **Xiang Xian Wen.**et al., Product attributes of China logistics park, Railway Transport and Economy, No.7, pp.66-69.2009.

[8] **Li Guo** i, Dissect the logistics park planning and building, Transport Business China, Z1,PP.54-55,2008.

[9] Notice of the State Council on the issuance of the logistics industry restructuring and revitalization plan, http://www.gov.cn/zwgk/2009-03/13/content 1259194.htm, 2009.

[10] **Zhang Wei ying**, Game Theory and Information Economics, Shanghai: Shanghai People's Publishing House, 2004.

[11] Stanford Encyclopedia of Philosophy, http://plato.stanford.edu/entries/game-theory, 2006.

[12] **Paul. R.** Doommel, Intergovernmental relations in Mangaging Local Government, Sage Publication,Inc., pp118-125, 1991.

[13] **Porter, M. E.**, How Competitive Forces Shape Strategy. Harvard Business Review, March-April 1979.

[14] **Kaveh Madani**, Game theory and water resources. Journal of Hydrology 381 (2010), pp 225 238, 2010.

[15] **Christopher M. Anderson, et al**, A game-theoretic analysis of competition among container port hubs: the case of Busan and Shanghai, Maritime Policy Management, vol. 35, No. 1, pp.5-26, 2012

[16] **Kevin O Connor. (2010)**, Global city regions and the location of logistics activity, Journal of Transport Geography, vol.18,Issue 3,No.5, pp.354-362, 2010.

CHROMATIC CONTRAST SENSITIVITY TEST AND ITS APPLICATION ON THE EVALUATION OF DISPLAY E UIPMENT

Hong-qiang YU , Ting JIANG , Chun-hui WANG*
National Key Laboratory of Human Factors Engineering,
China Astronaut Research and Training Center, Beijing, China
chunhui_89@163.com*

ABSTRACT

In order to test people's ability to distinguish tiny differences of color and use it to evaluate display equipment, a chromatic contrast sensitivity test was designed. Firstly, we measured twelve normal subjects' chromatic contrast sensitivity twice and verified reliability of the test. Then, we also measured sixteen subjects with abnormal intraocular pressure. Finally, a comparison was made between normal and abnormal subjects. Results showed that all subjects' chromatic contrast sensitivity varied similarly with spatial frequency and a downward trend of discrimination threshold was seen which means chromatic contrast sensitivity enhanced as spatial frequency increased. Chromatic contrast sensitivity of subjects with abnormal intraocular pressure was significantly lower than normal subjects. Design of display equipment should be based on characteristics of the human eye and different people with different visual ability need different display equipment corresponding.

Keywords: Chromatic, contrast sensitivity, display

1. INTRODUCTION

Traditional contrast sensitivity test was black and white contrast sensitivity [1-4]. Rarely study about chromatic contrast sensitivity was seen because of the complexity of color [5-8]. And contrast sensitivity characteristic data was an important basis for a chromatic display interface design in ergonomics research. However, black and white contrast sensitivity which was extensively studied in the past could not reflect two aspects of both color and contrast. For example, the ergonomic requirements of display contrast between characters and background was from 8:1 to 12:1 was also based on the black and white contrast sensitivity of human eye. But display contrast between color characters and background was lower than 5:1 which means the original ergonomic requirements was improper in this situation. Reference [9-11] measured visual contrast sensitivity function and used it in image compression and color image evaluation, but they mainly used test methods and results of traditional black and white contrast sensitivity, rarely involved the application of chromatic contrast sensitivity.

In recent years, with the development of computer technology, the display equipment has become an important information presentity for almost everyone. The contrast sensitivity test based on display characteristics attracted the attention of researchers and much significant progress was made [12-16]. A few researchers have proposed several definitions of chromatic contrast sensitivity and measurements have been done to get contrast sensitivity function curve of basic colors, but its application was not satisfactory. One important reason was that the standard color space which they used to define chromatic contrast sensitivity was not RGB color space which was currently widely used by display equipment.

Therefore, in order to test people's ability to distinguish tiny differences of color and use it to evaluate display equipment, a chromatic contrast sensitivity test was designed based on RGB color space and corresponding contrast sensitivity function curves were measured.

* Corresponding Author: Chun-hui WANG

2. METHODOLOGY

2.1. Test method

There were two parts in generating test method of chromatic contrast sensitivity. One was generating image of color grating used for discrimination and the other one was measuring threshold of discrimination using method of limits in psychology.

The core color of color grating used recommended standard values for color vision research from CIE [17] (see Table I and Figure. 1). Choose R, G, or B as single color variable respectively for red, green, or blue while ensuring the other two parameters unchanged. Meanwhile, the color variable changed according to a sinusoidal variation and the difference between the peaks and valleys of the sinusoidal variation was defined as chromatic contrast sensitivity. The cycle of the sine curve which related to grating width was reflected by spatial frequency which meant number of sine cycles in 1 degree angle of view. Two kinds of color grating images with different orientation (horizontal or vertical) were made to reduce the subjective effects.

Table 1 parameters of three core colors

Color	L	a	b	R	G	B
Red	44	37	23	165	76	68
Green	56	-32	0	55	149	133
Blue	36	5	-31	67	84	135

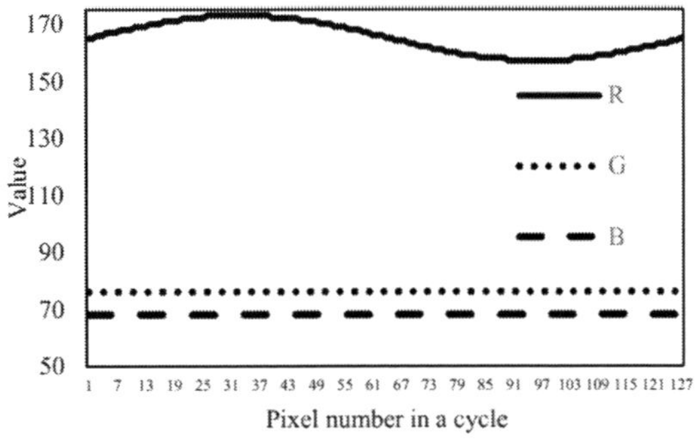

Figure. 1. Parameter changes of red in a cycle

Testing process was controlled by E-prime 2.0 program to realize method of limits in psychology. The program automatically showed color images and recorded subjects' feedback. In each spatial frequency, two kinds of contrast threshold was recorded. One was contrast changing from big to small and the threshold was the smallest contrast which subjects could distinguished. The other one was contrast changing from small to big and the threshold was the contrast of first color image which subjects could tell its orientation. The final threshold was average value of two kinds of threshold abovementioned.

2.2. Sub ects

Twelve normal subjects with an average age of 24 years (ranging from 21 to 28) participated in the experiment. None of the participants suffer from eye diseases and their visual acuity or corrected visual acuity was 1.0 or more.

Sixteen abnormal subjects whose intraocular pressure of both eyes were more than 20 mmHg with an average age of 24.2 years (ranging from 19 to 30) also participated in the experiment.

All subjects were gave informed consent prior to their participant in the study and received reward after participation. The experiment obtained approval of ethics committee.

2.3. E uipment

GT 740M VGA Card with 2GB DDR3 memory and 64bit memory interface was used. DELL P2415Q display was used to presented color images of gratings. It was professional equipment whose sRGB + ΔE <3 and gamma value was 2.2 which meant stable and accurate color display. Display size was 23.8 inches (60.4 cm) with a resolution of 3840 × 2160 and 185 dots per inch. Chin rest which was fixed on experiment table was used to make sure the test distance was 30cm and also make sure color images presented at center of subject's vision field. Subjects gave feedback through a wireless keyboard. Experiments was done in a standard darkroom.

2.4. E perimental design

Before formal testing, subjects had enough time to be familiar with the test content and methods of operation and dark adaptation for 2 minutes and light adaptation for 20 minutes was conducted.

E-prime test program described previously sequentially measured chromatic contrast sensitivity threshold of three core colors in 10 spatial frequency. Subject was required to give feedback of grating direction (horizontal, vertical or no gratings) by keyboard and the program automatically record the results. Testing process lasted about 30 minutes to avoid the impact of visual fatigue on the results.

Every normal subject completed same measurement twice separated by at least 48 hours. And subjects with abnormal intraocular pressure completed the measurement once.

III. RESULTS AND DISCUSSION

3.1. Reliability verification

Twelve normal subjects tested twice and chromatic contrast sensitivity threshold of two test in three core colors with ten spatial frequencies was obtained. Pearson correlation test was conducted between two threshold test of three colors and significant correlations were found of corresponding results (see Table II). What's more, the correlation coefficient were greater than 0.8.

Correlation analysis results showed that chromatic contrast sensitivity test method based on RGB color space and LED display was stable and reliability of the test method was verified.

Table 2 results of test method reliability test

Color	Red	Green	Blue
Coefficient	0.946	0.874	0.806
Significance	$P < 0.01$	$P < 0.01$	$P < 0.01$

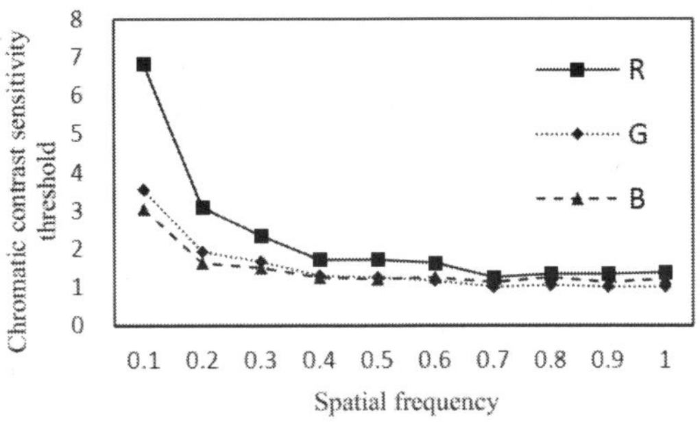

Figure. 2. Chromatic contrast sensitivity threshold of normal sub ects

Table 3 significance analysis of difference between red and green or red and blue at same fre uency (p value)

Spatial frequency	0.1	0.2	0.3	0.4	0.5	0.6	0.7
Green	<0.01	<0.01	0.01	0.002	0.020	0.014	0.053
Blue	<0.01	<0.01	<0.01	0.005	0.020	0.032	0.082

3.2. Chromatic contrast sensitivity threshold curve of normal sub ects

Chromatic contrast sensitivity threshold was obtained by averaging the two test results. Contrast sensitivity threshold curves of three colors were drawn using ten spatial frequencies as abscissa and threshold as ordinate (see Figure. 2).

There are similar trends in twelve subjects' threshold curve and the differences between subjects was small on the whole. The curve was flat at spatial frequency of 0.3-1.0 c/d but steep at spatial frequency of 0.1 c/d and 0.2 c/d. The threshold increased significantly at 0.1 c/d and 0.2 c/d indicating weakening of the ability to distinguish small color differences.

Contrast sensitivity threshold of three color can be seen in Figure. 2. Green and blue curve were approached to each other and red curve was higher than them at same frequency. The difference was significant (see Table III)when spatial frequency was between 0.1 c/d and 0.6 c/d indicating that the subjects were weak in distinguishing small differences of red compared to green and blue. In conventional applications, the striking red color which often cause people's attention was usually used as a reminder, while people were harder to find small changes in the red color presented by electronic display. Relatively speaking, small changes of green and blue aroused people's attention more.

3.3. Chromatic contrast sensitivity threshold curve of sub ects with abnormal intraocular pressure

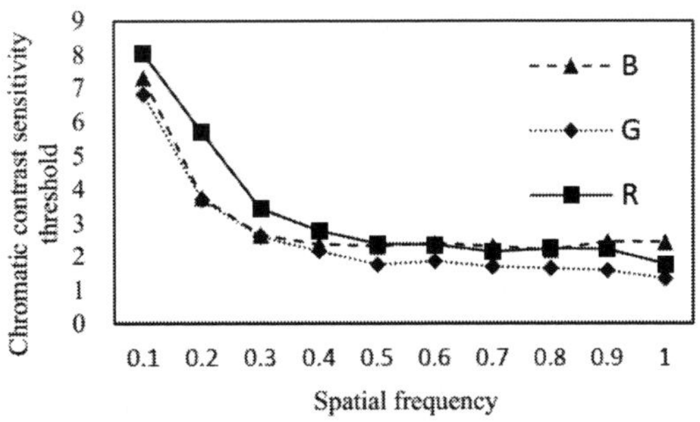

Figure. 3. Chromatic contrast sensitivity threshold of normal sub ects

Sixteen subjects with abnormal intraocular pressure test chromatic contrast sensitivity threshold. Their contrast sensitivity threshold curves of three colors were shown in Figure. 3.

Threshold curve trend of subjects with abnormal intraocular pressure was similar with normal subjects. With increasing spatial frequency, chromatic contrast sensitivity threshold was decreased indicating ability of distinguish small color differences upgrade. The curve trend of three colors was also similar and the threshold significantly increased when spatial frequency was 0.1-0.3 c/d which meant that similarly to the normal subjects, ability to distinguish slight color differences of subjects with abnormal intraocular pressure decline rapidly at extreme low spatial frequencies. In addition, threshold of red was higher than green and blue when the spatial frequency was 0.1-0.3 c/d and threshold of green was lower than red and blue when the spatial frequency was 0.4-1.0 c/d. However, color difference of subjects with abnormal intraocular was small compared to that of normal subjects.

3.4. Comparison between normal and abnormal sub ects

A comparison was made between normal and abnormal subjects. Three colors' chromatic contrast sensitivity threshold mean value of twelve normal subjects and mean value of sixteen subjects with abnormal intraocular pressure at ten spatial frequencies were shown in Figure. 4. Threshold of abnormal subjects at all spatial frequencies were higher compared to normal subjects obviously and the difference of red color was relatively small.

Chromatic contrast sensitivity threshold test results could be used to guide image compression, industrial design and other engineering applications directly, and also it's the foundation for ergonomic evaluation of image quality presented by display equipment. The test results from subjects with abnormal intraocular pressure indicated that ergonomics standard should not only follow normal people's demand but also pay close attention to features of many abnormal people.

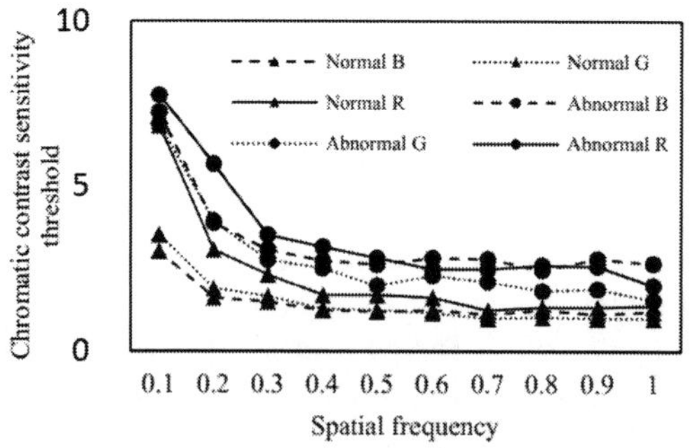

Figure. 4. Chromatic contrast sensitivity threshold of normal sub ects and sub ects with abnormal intraocular pressure

4. CONCLUSION

Chromatic contrast sensitivity test based on RGB color space and LED display was stable and effective. Ability to distinguish slight color differences decline rapidly at extreme low spatial frequencies and small changes of green and blue aroused people's attention more than red. Discrimination threshold of abnormal subjects at all spatial frequencies were higher compared to normal subjects indicated that ergonomics standard for display equipment should pay close attention to both people's demand and features of many abnormal people.

5. ACKNOWLEDGMENT

The study was supported by the Foundation of National Key Laboratory of Human Factors Engineering (SYFD140051806).

6. REFERENCES

[1] **E. Peli,** "Contrast sensitivity function and image discrimination," JOSA A, vol. 18, no. 2, pp. 283-293, 2001.

[2] **A. D. Straw, T. Rainsford, and D. C. O Carroll.** "Contrast sensitivity of insect motion detectors to natural images," Journal of Vision, vol. 8, no. 3, pp. 32-32, 2008.

[3] **I Yuanyuan, ZHANG Feng u, and YU Fanglei.** "Analysis of influential factors to the contrast sensitivity on human eyes" (in Chinese), Chinese Journal of Optometry & Ophthalmology, vol. 9, no. 5, pp. 328-331, 2007.

[4] **F. T. Oen, T. H. Lim, and M. P. Chung.** "Contrast sensitivity in a large adult population," Annals of the Academy of Medicine, vol. 23, no. 3, 322-326, 1994.

[5] **K. T. Mullen.** "The contrast sensitivity of human colour vision to red-green and blue-yellow chromatic gratings," The Journal of Physiology, vol. 359, no. 1, pp. 381-400, 1985.

[6] **. M. Rovamo, M. I. Kankaanp , and H. Kukkonen.** "Modelling spatial contrast sensitivity functions for chromatic and luminance-modulated gratings," Vision Research, vol. 39, no. 14, pp. 2387-2398, 1999.

[7] **H. C. Owens, S. Westland, and Van de Velde K.** "Contrast sensitivity for lime-purple and cyan-orange gratings," Society for Imaging Science and Technology, vol. 2002, no. 1, pp. 145-148, 2002.

[8] **Xu H, and Yaguchi H.** "Visual evaluation at scale of threshold to suprathreshold color difference," Color Research & Application, vol. 23, no. 3, pp. 198-208, 2005.

[9] **S. Westland, H. Owens, and V. Cheung.** "Model of luminance contrast-sensitivity function for application to image assessment," Color Research & Application, vol. 31, no. 4, pp. 315-319, 2006.

[10] **YAO uncai.** "Color Image Compression Technology Based on Contrast Sensitivity Characteristics of Human Vision System" (in Chinese), Chinese Journal of Liquid Crystals & Displays, vol. 27, no. 6, pp. 800-807, 2012.

[11] **YAO uncai.** "Image Quality Assessment Method Based on Contrast Sensitivity Characteristics of Human Vision System" (in Chinese), Chinese Journal of Liquid Crystals & Displays, vol. 26, no. 3, pp. 390-396, 2011.

[12] **LU Weige, XU Haisong, and WANG Zhehong.** "Investigation of Chromatic Contrast Sensitivity Based on Different Color Directions and Spatial Frequencies" (in Chinese), Acta Optica Sinica, vol. 31, no. 1, pp. 293-298, 2011.

[13] **ZENG Wang, XU Haisong, and WANG Zhehong.** "Investigation of Color Discrimination Threshold Characteristics Under Different Chromatic Backgrounds" (in Chinese), Acta Optica Sinica, vol. 31, no. 1, pp. 287-292, 2011.

[14] **SHI unsheng, YAO uncai, and YU Hongfei.** "Measurement of Luminance Contrast Sensitivity Function of Human Visual System on Cathode Ray Tube Display" (in Chinese), Acta Optica Sinica, vol. 27, no. 4, pp. 744-748, 2007.

[15] **YAO uncai, SHEN ing, and HE unfeng.** "Measuring luminance contrast sensitivity characteristics of human visual system using LCD display" (in Chinese), Chinese Journal of Liquid Crystals & Displays, vol. 30, no. 6, pp. 1063-1069, 2015.

[16] **WANG Pengfei.** "Study on the color spatial frequency response of the human visual system" (in Chinese), Beijing: Beijing Institute Of Graphic Communication, 2011.

[17] **A. Robertson.** "CIE guidelines for coordinated research on color-difference evaluation," Color Research & Application, vol. 3, no. 3, pp. 149-151, 1978.

SOLVING THE ASSEMBLY LINE BALANCING PROBLEM IN A GARMENT FACTORY BASED ON GENETIC ALGORITHM

Lin Li, Lian-qing Yu*
Department of Industrial Engineering,
Wuhan Textile University, Wuhan, China
*yulq@wtu.edu.cn

ABSTRACT

A kind of genetic algorithm is used to solve the assembly line balancing problem in a garment factory in this paper. Firstly, a genetic algorithm for assembly line balancing problem is introduced briefly. Secondly, a sewing assembly line which is used for the production of XFM jacket is analyzed in the workshop of Xtep Company. Then the assembly line balancing problem is optimized based on the genetic algorithm. After optimization, the station number is reduced by 3 and the equilibrium rate is increased by 32%. Finally, the simulation with Flexsim and the workshop experiment are conducted, the results show that GA is an appropriate tool to solve the assembly line balancing problem in a garment factory.

Keywords: Assembly line balancing (ALB), Flexsim simulation, garment, genetic algorithm(GA)

1. INTRODUCTION

The purpose of assembly line balancing is to minimize the idle time of the assembly lines, improve the utilization rate of workers and equipment. The assembly line balancing problem is a typical combinatorial optimization problem which has great importance in engineering applications. It generally be divided into three categories. Class Ⅰ: to find out the minimum location number with a given assembly line cycle time; Class Ⅱ: to find out the assembly line cycle time with a given the minimum location number; ClassⅢ: to find out the optimal location balanced load with the minimum location number and assembly line cycle time[1]. Amount of researches on this issue have been carried out, and many kinds of algorithms have been put forward including industrial engineering method, mathematical analysis, computer simulation method, random algorithm, heuristic algorithm and so on[2-7]. Facing the large scale assembly line balancing, heuristic algorithms are effective methods for this problem. But the searching efficiency is very low, because some heuristic search methods in the search process of approaching a optimal solution or a sub optimal solution always have a reverse process, and be apt to produce the phenomenon of "combinatorial explosion" too[8]. Genetic algorithm (GA) is a global optimization algorithm in the field of artificial intelligence proposed by Jhon Holland, which simulates the process of biologic evolution to search and study[9]. Its coding technology and genetic operation is simpler, optimization is not bound by restrictive conditions, and there are two most significant characteristic is the implicit parallelism and global search.

2. GANETIC ALGORITHM FOR ASSEMBLY LINE BALANCING

In this paper, we intend to solve the first assembly

line balancing problem(Class Ⅰ). The purpose is to find the minimized assembly line smooth index and location satisfying for the constraints with given production cycle time. The objective function is defined as:

* Corresponding Author: Lian-qing Yu

$$\min P = \alpha \cdot n + \beta \cdot SI$$
$$SI = \sqrt{\sum_{i=1}^{m}\left(\max T(S_k) - T(S_k)\right)^2} \quad (1)$$

n stands for the workstation number, α, β as the weight coefficient, SI for the assembly line smooth index, workstations time are described by T(Sk) and practical production cycle times are specified by max T(Sk). When balance goal focused on workstation for minimum or workstations load more evenly, we can control the optimization results of the algorithm with adjusting the value of α and β according to the circumstance.

Assembly line balance which is based on genetic algorithm is a process of optimizing the order of the work unit. The process is illustrated in Fig 1.

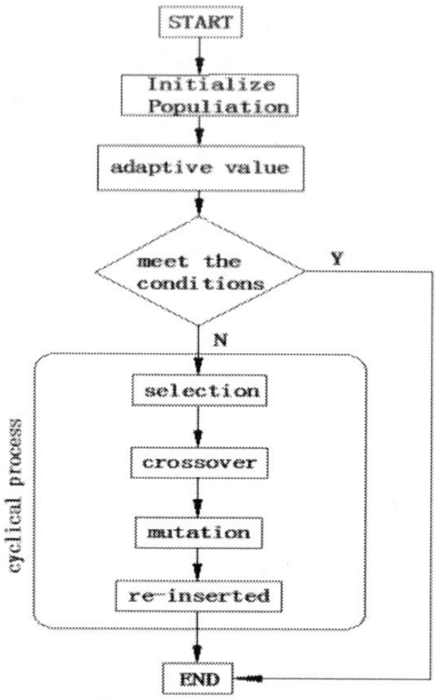

Figure.1 Flow chart of genetic algorithm

2.1. Parameter input

Including coding and decoding. Coding establishes a corresponding relationship between the order of the manufacturing units and chromosomes.[10] Due to the particularity of apparel industry, coding needs special design. Here we choose sequence encoding: allocating the workstation according to the operation element order and each operation element in a row corresponds to a gene. A good order corresponds to a chromosome. Decoding is a process of assigning work elements to workstation under the condition of meeting the minimum number of workstations. For example, there are n elements, so each chromosome with n genes. n operation elements are originally encoded and arranged in a chromosome. The sequence of each operation element assigned as A1, A2, A3...An.

2.2. Initialize the population

We adopt the method of random and loop search, until generate that N (p) individuals.

2.3. Fitness function

As the fitness function is the demand function of the maximum problem, however, our objective function is the demand function of the minimize problem. So the suitable fitness function is designed which is aim at individual i.

$$F(i) = 0.95^{\alpha \cdot n + \beta \cdot SI} \quad (2)$$

Therefore the smaller assembly line balancing smooth index and location number of chromosomes, the greater its fitness.

2.4. Selection, crossover, and mutation operator

Selection is used to determine the crossover individuals which should be reorganized, as well as how many individuals produced by the selected individuals. Firstly, calculating the fitness value, and then using the expected value method choice, which eliminated the individual with bigger fitness, and speeded up the convergence speed. crossing is a process of combing with information from parent mating populations to produce new individual. Choosing randomly two individuals in a population selected as cross subject, and randomly generated the same binary string as the chromosome gene digits. Corresponding to the selected binary string of 1 gene, and order them according to the arranges corresponding the other one's parents. The rest of the order stays the same. The process or result of crossover are illustrated by the following.

Binary string	0	1	0	1	1	...	1
Parent 1	A1	A2	A4	A6	A3	...	An
Parent 2	A1	A3	A2	A4	A5	...	An
Child 1	A1	A3	A4	A2	A6	...	An
Child 2	A1	A4	A2	A3	A5	...	An

Mutation operation should be used on some individual selected randomly, in order to improve the diversity of the individuals in the population, and avoid falling into local optimum. Using individual species consist of binary genetic code to realize the small probability reversal which is 0 and 1 at the swap.

3. SEWING ASSEMBLY LINE BALANCING BASED ON GENETIC ALGORITHM

Assembly line of Xtep Company, LTD. Garment sewing workshop exist 12 assembly lines, and its average production time is 9.5 hours a day. We chose a sewing assembly line which is producing XFM jacket (Fig 2) as the research object. According to their actual state of low assembly line balancing rate, the genetic algorithm is adopted to make the assembly line balancing studied. Through field investigation and analysis, we get the sequence of sewing assembly line and tasks required in XFM jacket manufacturing which is shown in figure 3 and table I.

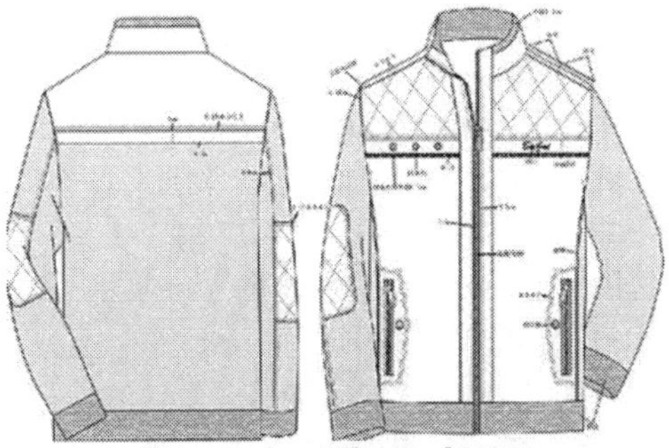

Figure 2 acket of XFM

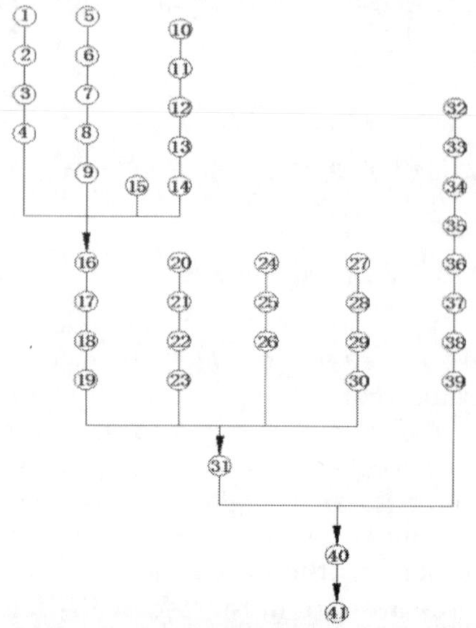

Figure 3 Se uence of sewing assembly line

Figure 4 The sewing line in workshop(before optimazation)

Table 1 the 41 tasks re uired in fm acket manufacturing

Task No.	Task Name	SAM	Preceding Task
1	set main label *1	26	
2	back-stitch neckline *1	10	1
3	stitch collar band*1	49	2
4	join under collar and top collar*1	16	3
5	five thread overlock the front facing	48	1
6	join sleeve seam *2	34	5
7	join shoulder seam*2	29	6
8	match and stitch sleeve lining seam *1	60	7
9	joining front edg*1	62	8
10	double needle at front pocket opening	24	1
11	inner stitch of the front pocket	93	10
12	match flap facing*2	139	11
13	stitch ends of pocket mouth*2	31	12
14	attach pocket to garment*2	41	13

15	five thread overlock the side sleeve seam*2	42	1
16	set back yoke*2	22	4、9、14、15
17	set sleeve*2	104	16
18	slip-stitch sleeve lining to garment *2	38	17
19	double needle at cuff*2	67	18
20	stitch collar band *1	24	16
21	sew collar on and down	28	20
22	fuse interlining *1	18	21
23	attach collar to band *1	72	22
24	hot shrink rib cuff *2	5	16
25	sew rib cuff *2	10	24
26	mark elastic band position	96	25
27	hot shrink rib hem*1	5	16
28	double needle overlocking the back*1	32	27
29	stitch elastic band	74	28
30	join 1/16 side seam*2	35	29
31	join front edge *2	120	19、23、26、30
32	join under collar and top collar	101	31
33	join side seam *1	50	32
34	sew in a zipper *1	96	33
35	mark position	43	34
36	join sleeve bottom*1	72	35
37	blind stitching hem*1	26	36
38	turning over lining	41	37
39	attaching lining to garment	19	38
40	join under collar and top collar	56	31、39
41	bartack	39	40

If the active working time of assembly line is averaged out to 8 hour days, plan period is expected to output qualified production 250 per day. And then we can calculate the theoretical minimum work stations N, daily output CT, line balancing rate P of the sewing assembly line respectively.

$$N=\left[\frac{T}{R}\right]=\left[\frac{T}{Fe/n}\right]=\left[\frac{2030}{8\times3600\div250}\right]=[17.62]=18 \qquad (3)$$

$$CT=Fe/Ft\ =8\times60\times60/170=170 \qquad (4)$$

$$\mathrm{P}\ =\left[\frac{\sum Ti}{N*CT}\right]*100\%=\frac{2030}{24*170}\%=49.87\% \qquad (5)$$

Among them: [a] - the minimum integer which is greater than or equal to a; T - all processes total operation time of single items; R - assembly line average tempo; Fe - operational use time of planning period; n - the expected output of qualified products in plan period; Ti - the sum in each process, N - station number; CT - bottleneck time.

We put the original data to genetic algorithm, the GA population scale of sewing assembly line select 50, crossover probability choose 0.8, mutation probability select 0.2, setting the value of the weight of a and b is 0.8 and 0.2 respectively. It is concluded that the assembly line need 21 points by Matlab programming arithmetic finally. The results are shown in table II.

Table 2 the optimal solution of worker's assignment from task 1-41

Task No	Task	SAM
1	1、2、3	85
2	5、6	82
3	8、9	122
4	10、11	117
5	12	139
6	4、7、13	76
7	14、18、37	105
8	15、20、21	94
9	16、19	89
10	17	104
11	22、23	90
12	24、27、28、41	102
13	25、26	106
14	29、30	109
15	31	120
16	32	101
17	33、35	93
18	34	96
19	36、38、39	132
20	40	101
21	41	29

4. RESULTS ANALYSIS

4.1. Results analysis based on Fle sim

According to a series of optimization scheme, we implement the optimal concrete action. Using the Flexsim software, the simulation model(figure 5) was established and analyzed based on the efficiency of optimization, and state simulation diagram including the state characteristic data such as free, run, jams and so on. Each trial run was carried out for about 1 day, or 54000 seconds. As shown in Fig 6.

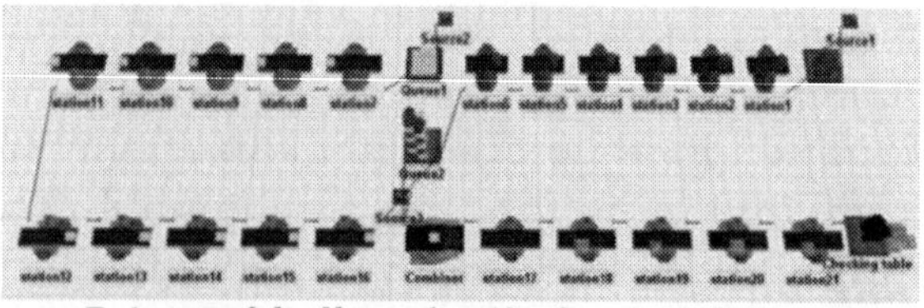

Figure 5 Assembly line simulation model in Fle sim

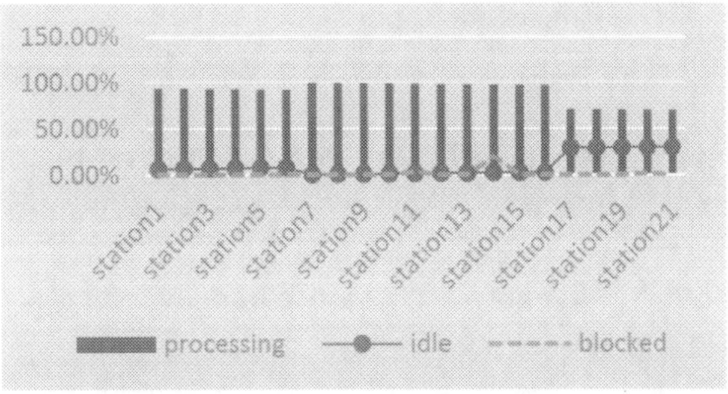

Figure 6 The diagram of simulation status(optimized)

Figure The sewing line in workshop(optimized)

4.2. The e perimental results

The comparison of computation results before and after optimization is made(see table III). This assembly line balance rate increased by 32.5%, from the current level 50.0% to 82.5%; and production capacity per shift increased by 17.6%, from the original 170 pieces to 200 pieces; number of employees reduce the three, from 24 down to 21. The optimized balance effect is good, and each workstation load more balanced. The effect of site application shown in Fig.7 is better than before(Fig.4).

Table 3 alb effect analysis

Index	Equilibrium Rate	Capacity	Station
Before optimization	50%	170 piece	24
Optimized	82%	200 piece	21
Variation	32%	30 piece	-3
Change percentage	65.0%	17.6%	12.5%

5. CONCLUSION

In this paper, genetic algorithm is employed to solve the garment sewing assembly line balancing problem based on minimize assembly line smooth index and location number. After the optimization, the station numbers is deceased 3 and the working load is evened better than before. The balancing result showed the method is has strong application value and practical meaning.

6. ACKNOWLEDGMENT

This paper is supported by the National Natural Science Foundation of China (Grant No. 51275363) The authors are grateful for the financial support.

. REFERENCES

[1] **ing Hu, Ming-zhou Liu, ing-fa Lei, Mao-gen Ge, Cheng-he Chen,** "Method for dynamic balancing of assembly line in workshop," (in Chinese) Journal of Agricultural machinery, vol. 40, no. 3, pp. 194-198, 2009.

[2] **Watanabe T, Hashimoto Y, Nishikawa I,** "Line balancing using a genetic model[J] ," Control Eng Practice, vol. 3, no. 1, pp. 69- 76,2006.

[3] **Ciobanu G.A,** "Branch and bound algorithm to solve an assembly line balancing problem[J],"Economic Computation and Economic Cybernetics Studies and Research, no. 4, pp. 69- 79,1977.

[4] **Pang Chuan- Chen,** "An heuristic method for manufacturing assembly line balancing[J]," Western Elecric Engineer, vol.19, no. 4, pp. 36- 45, 1975.

[5] **Xiu Li, Wen-huang Liu, Cheng-yu iang, Ning-sheng Wang,**"The solving of job-shop scheduling problem based on genetic algorithm," (in Chinese) Journal of System Simulation, vol. 13, no. 6, pp. 736-739, 2001.

[6] **Bhattachar ee T K,** "An heuristic app roach to general assembly line balancing [J],"International Journal of Operations & Production Management, vol.8, no. 6, pp. 67- 77,1988.

[7] **Rubinovitz , Levitin G,** "Genetic algorithm for assembly line balancing[J] ,"Int J Production Economics, vol.41, pp. 343- 353,1995.

[8] **Fu-lin Wang, i- uan Wang, Chang-you Wu, iu-feng Wu,** "The improved research on actual number genetic algorithms," (in Chinese) Journal of Biomathematics", vol. 21, no. 1, pp. 153-158,2006.

[9] **Hong- un Wang, ian-hui Zhao,** "Study of assembly line balancing system based on genetic algorithm,"(in Chinese) Computer Engineering and Applications, vol.44, no. 10, pp. 195-197,2008.

[10] **Man, K.F., Tang, K.S. and K wong, S,** "Genetic algorithms: concepts and applications," IEEE Trans. on Industrial Electronics, Vol. 43 No. 5, October, pp. 519-33.,1996.

EVALUATION ON THE CURRENT SITUATION OF PPP DEVELOPMENT IN CHINA BASED ON THE MATURITY MODEL

Ziqiang uan*[1], Xing Bi
College of Management and Economics,
TianJin University, TianJin, China
*1874581029@qq.com

ABSTRACT

The infrastructural construction of our country has long been mainly supported by the fiscal investment of the government and state-owned enterprises. But this model is gradually on the decline. Introducing and applying the PPP model to make the private capital involved in the infrastructural construction and urbanization has been the prevailing trend to implement a win-win result. Firstly, this paper combines with China's PPP practice for the PPP definition, analyses the type of China's PPP project and its three stages in the development. Then, according to its developmental maturity model and analyses the three dimensions of China's PPP model in maturity from the laws and regulations, management organizations and the public participation, concluding that China's PPP model is still in the primary level of maturity, but in the continuous development. Finally, the paper gives suggestions for the development of China's PPP model.

Keywords: Public-private partnership; maturity; the growing trend

1. INTRODUCTION

1.1. Background

For many years, the local government mainly relies on establishing financing platform and other forms to finance in the performance of the infrastructure and public services. Although at that time, it got the money to improve people's livelihood and the development of social programs, which brought the disadvantages of the local debt scale. So, it does not have sustainability. By the end of June 2013, the national government entities at all levels are responsible for repayment of the debt of 20.698865 trillion. Guaranteed responsibility for the debt of 2.925649 trillion, and they may be liable of the debt of 6.650456 trillion. In addition, by the end of 2012, there are three provincial and 99 municipal, 195 country, 3465 villages of the government at the county level responsible for repaying debt, which is higher than 100 . The heavy financial burden of local government has seriously hindered the development of infrastructure. Because of the introduction of social capital in public-private partnership (PPP), it has the property of innovation and efficiency, and become an important mode in China's further development of infrastructure, which provides a feasible framework for public infrastructure development. The PPP model relative to the completely by the traditional model of the public sector to provide infrastructure services have many advantages, such as a more reasonable risk management, lower cost in life cycle, more convenient contract management, more innovative public services and more efficient project management[1], etc. So, it wins worldwide attention. In developed countries, the PPP model has covered the application scope of water supply, public transportation, energy, health care and other fields[2]. The PPP project accounts for roughly a year from 10 to 14 of the investment of public service, and has expanded into areas such as national defense, police and prison. In current innovative investment and financing system of urbanization, focusing on local financing platform to resolve the debt risk, and actively promote the enterprise under the background of "going out", promoting the use of the PPP model, not only upgrades the micro operation, but also changes the institutional mechanisms at the macro level.

[1] Corresponding Author: Ziqiang Yuan

1.2. The definition of China s PPP model

There isn't a recognized concept of the PPP currently, including the United Nations developmental program, the European commission, the PPP national committee in the United States as well as the different experts and scholars all have their own understandings [3][4] .The main definition shows in the TABLE I.

Table 1 the definition of ppp model

Origin	Definition
UK Department of the Treasury	PPP is an arrangement characterized by the cooperation of the public sector and the corporate sector. Broadly speaking, the PPP can contain all cooperation of the public sector and the corporate sector in formulating policies, providing services and the infrastructure.
The PPP commission of the United States	The PPP (federal, state or local) is an agreement between the public sector and the corporate sector. Through this agreement, the public sector and the corporate sector can share resources of technology, assets in the process of providing service for the public. In addition, the two sides can also share the risks and gain interests by providing the services.
The PPP commission of the Canada	The public and the corporate sector based on their respective expertise invest cooperatively, through the reasonable distribution of the resource, risk and return to the greatest extent to meet the public demand determined in advance.
The team of promotion in efficiency of Hong Kong	The PPP provides complementary technologies respectively among the public sector and the corporate sector to supply public service or project in the contracted program. Both the degree of involvement and responsibility varies with the project
The Development Administration of the United States	PPP refers to the government、for-profit and non-profit enterprises based on a project and the formation of the mutual relations of cooperation. Through this form of cooperation, the parties can achieve more favorable results than acting alone expected.

This paper defines the PPP as the public sector to provide public facilities or services, which establishes the institutionalization of cooperation with adequate capital and financing ability of for-profit companies. Cooperation is the essence of the PPP, the purpose of cooperation is to fully take their advantages, by sharing the risks and the interests to achieve the optimal allocation of resources.

2. MODEL AND ANALYSIS

2.1The PPP s stage of development in China

The PPP model has experienced three rounds of boom when it was introduced into China since the 1980s. In the first round of the PPP boom since the mid of 1980s to 1990s, which relieved the pressure on infrastructural investment funds[5], China's central government and local government implemented in the represent of a series of the PPP projects: Sha Jiao B power plant, Lai bin B power plant and Chen du Shui Liu Chang B power plant factories, which is mainly by introducing foreign investment to solve China's shortage of construction funds. Ten years after entering the 21st century, China ushered in the second round of the PPP boom. China's economy continues to develop rapidly, the bottlenecks effects of infrastructure on economic growth highlights again, and the government's fiscal power alone can't completely meet the huge required investment[6], resulting a representative PPP projects: such as metro line 4 in Beijing, the first phase of project in Beijing LuGou iao's waste water treatment, GaoAnTun waste incinerators and the franchised gas projects in the east new district of i-town road in Beijing.

The PPP has transformed from the original way of financing tools, technological updates to achieve the transformation of government functions, innovating budget system, integrating social resources, the mechanism of reform to eliminate hidden barriers in China. The top design of legal policy, the ability of construction, the environment of investment and so on all progress rapidly. China's PPP steps from the childhood to youth and will soon celebrate the PPP's development of the third wave.

2.2. The maturity of PPP in the development of China

The PPP model grows continuously in China, meaning that the public sector puts many expectations on this model. However, compared with Britain and Australia, the PPP model has a long way to go if it wants to play a better role in China.

Deloitte accounting affairs puts forward the PPP model maturity model in 2011, dividing it into three stages[7], as is shown in TABLE 2 .

Table 2 the PPP developmental phase of the criteria

The first stage	The second stage	The third stage
Determining the framework of the policy; Testing the applicability of the law; Determining the channel of the project; Developing the basic conception (such as public comparative benchmarking) ; Drawing the experience from previous projects or other departments; Beginning to marketing	Promoting the reform of the laws; Publishing policies and practical guides; Setting up a dedicated institution of the PPP; Improving the delivery of PPP model; Continuing to marketing; Expanding the channel of the project to expense to other industries; Driving the new financing	Defining comprehensively and establishing a comprehensive system; Eliminating legal barriers; Redefining and reproducing PPP model; Allocating the risk elaborately; Having a solid deal flow; Having a long-term political consensus; Using money comprehensively; Formulating the prosperity of infrastructure investment market , including pensions and the shares of the enterprise; Using the experience of PPP for municipal services training

After the comprehensive analysis of the economical intelligence unit and the standard of classification of the Deloitte, an accounting firm[7], we can divide the PPP maturity into three dimensions: the laws and regulations, managemental organizations and the public participation, and analyze it respectively.

2.2.1 The PPP maturity of laws and regulations

China's PPP practice led the world and constructed numerous PPP projects since the PPP mode adopted in China in the 1990s. According to the World Bank's public and private infrastructure consultancy (PPIAF) database statistics, China implemented 614 PPP projects during 2000-2009, which being the highest in the world. But the progress is slow in the laws and regulations, and it's construction gradually becomes an obstacle of the development of China's PPP projects[8].

The Asian development bank investigated and graded the PPP maturity in laws and regulations to 15 countries in the Asia-pacific region in 2011. China's legal construction closes to Thailand、Kazakhstan, etc, ranking 11th, as is shown in TABLE III , Australia and Britain as a benchmark to compare.

Table 3 the ranking of legal maturity in the asia-pacific region (and benchmarking country)

Ranking	Country	Goals
1	Australia	100.0
2	UK	96.9
3	Korea, Rep.	78.1
4	Gujarat State	65.6
5	India	59.4
6	Japan	50.0
7	Philippines	43.8
8	Bangladesh	40.6
9	Indonesia	40.6
10	Pakistan	34.4
11	China	31.3
12	Thailand	28.1
13	Kazakhstan	25.0
14	Mongolia	25.0
15	Vietnam	18.8
16	Papua New Guinea	15.6

By the analysis above, China's existing laws and regulations of the PPP maturity is low, still in its infancy, and exists serious problem of shortage, vagueness, unstableness.

2.2.2The maturity in management institutions of PPP

China's national government entrusted the Treasury to set up the first PPP work leading group, fully responsible for the management of PPP in May 2014. Before this, the national government has not designated a ministry to govern the PPP project uniformly. According to The several opinions concerning encouraging and guiding the healthy development of private investment, the PPP projects are both governed by the several ministries jointly in all works of life. However, the division of labor in liability between the various ministries is not clear, which can lead to problems, such as mutual shuffle or fragment between ministries. These problems once happen, which can cause the administrative chaos, seriously affecting the efficiency of the PPP project.

Local governments in the PPP project are in a specific operating level, but when the local government initiated the infrastructure of PPP project, they didn’t generally organize a full-time administrative supervision department to take the responsibility for the administrative examination and approval and the supervision of the PPP project, nor the professional consulting agency for the PPP project providing services, consulting, training, summary work and so on. It often appears that the local government intervenes the operations of the corporate sector excessively, or appears the extreme phenomenon that it allows the corporate sector to operate freely.

The above analysis reveals that the maturity of current administration in China's PPP is low, and governments at all levels haven’t yet set up a corresponding management institutions of PPP.

2.2.3The maturity in public participation of PPP

The participant in the PPP model is not only the government and enterprise, but also the public as consumers and regulators. Strengthening the supervision and the management of the public to the PPP project is an important factor in the success of the project. Now, it requires the information to be transparent, otherwise, problems may occur. Public participation in the supervision of the PPP projects gives play to the role of the supervision of the media, which can prevent a project with no conditions but also uses the PPP model, and prevents the PPP model that brings to the decrease in the quality of service. To prosper the PPP market, there must be more participants to join, through public participation to create transparent procedures, which can also be more attractive to the private sector. Thus, the government can choose more capable and technological partners, and provide consumers with more excellent service, which is a virtuous cycle.

The decision-making stage in the infrastructure of PPP project rarely invited public participation in China. At the same time, local governments have not established the feedback channels of the public opinion, making the real demand of the public difficult to reflect. In terms of disclosure of information, the disclosure of information level in China's infrastructure PPP project is generally low, and has not set up the official website of the PPP project to the operation of the public project information to accept the supervision of the public.

Thus, the public participation is in low maturity in China's PPP, and the mechanism of PPP public participation has not yet been established, and the transparency of information remains to be improved.

3. RESULTS AND SUGGESTIONS

3.1. Results

Through the analysis of the economist intelligence unit and Deloitte, an accounting firm to classification standard of the development of the Asia-pacific countries of the PPP model, the PPP model in our country maturity model can be divided into laws and regulations, the PPP management maturity, the PPP public participation and maturity in three dimensions. By the analysis of the present situation in three dimensions one by one, it has concluded that our country is still in the primary level and existed the problems of shortage, obscurity, instability in the laws and regulations. Governments at all levels have not yet set up a corresponding PPP management institution, and the PPP public participation mechanism has not yet been established and the transparency of the information needs to be improved.

3.2. Suggestions

Along with the development of Chinese PPP model of the third wave, China will make an influence in the following aspects, which makes the development of China's PPP model to be mature.

Through setting perfectly the PPP management institution, formulating and improving the laws and regulations of the PPP, strengthening the guidance of the policy and improving the mechanism of the public participation in four aspects, it provided guidance to the development of China's PPP model and established Chinese PPP model on the basis of the system frame structure, an essential part of the planning and design of system frame in advance, which is driven by the experiments of PPP projects to promote the whole PPP system gradually.

Predictably, the elements of various systems will not orderly appear and being set up on the same day. But based on the analysis of the PPP about present situation and trend in our country, and the international experience [9], China will gradually set up the PPP system framework by the exploration and practice. Structure is shown in Fig. 1

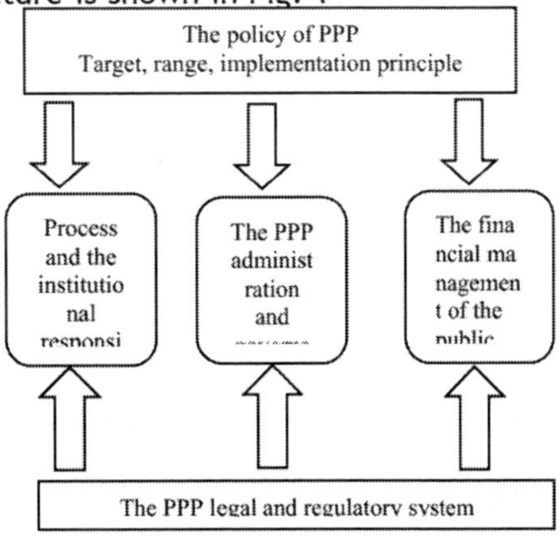

Fig.1.PPP institutional framework structure

Among them, the PPP special legal and regulatory system forms the foundation of the PPP institutional framework, it shall be formulated by the legislature, establishing a clear value orientation for the long term. PPP policy reflects the government's tenure policy goals, including the private sectors' participation in infrastructure and public service in the middle of policy options, such as the target, scope, implementation principle and so on elements, which is likely to change with the change of government or a priority and adjust periodically. The operational process and related responsibilities of PPP project, the PPP governance structure and management of public finance, constituting the public management system of continuous implementation of PPP projects.

4. CONCLUSION

In the past 20 years, the PPP model has received the widespread attention in the world and made great success in some countries. Faced with the huge demand for the infrastructural construction in urbanization, the PPP model has obtained widespread application in China, but the effect is poor, which has exposed the insufficient knowledge of the PPP mode. This paper combines with China's PPP practice to define the PPP's concept and analyses the types of the China's PPP project and the three stages of it's development concretely, then, it evaluates the maturity of the PPP model in China from the laws and regulations, management organizations and the public participation in three dimensions, concluding that China's PPP model is still in the primary level of maturity but in the continuous development. Finally, the thesis gives suggestions for the development of China's PPP model.

5.REFERENCES

[1] **NewSouth Wales Treasury**, Working with the government: Guidelines for privately financed projects.2006.

[2] **TiemanR.**Arevolution in public procurement:UK's private finance initiative. Finance Times, London, 2003:17-23.

[3] **Chen Shuang, Xia Zhi ian**,Analyses the the development of the PPP mode in China [J].Economic forum, 2010 (11): 39-42.

[4] **Michael , Garvin.Ph.D.,M.ASCE(200)** Enabling Development of the Transportation Public-Private Partnership Market in the United States . Journal of Construction Engineering and Management, Vol. 136, No.4, April 1, 2010.

[5] **Wang Shou ing**, The financing of the franchised project. [M]. The press of the Beijing: Tsinghua university, 2006.

[6] **Wang Shou ing,Ke Yong ian**, China's BOT/PPP project practice experience and investment in Beijing,2008:82-83.

[7] United Nations. Guidebook on Promoting Good Governance in Public Private Partnerships [M]. New ork and Geneva: United Nations, 2008: 7-8.

[8] **Zhao, X. B. (200)**, Research on PPP Projects Performance Evaluation. Beijing, Tsinghua University. M.c.

[9] **Yuan, ., et al. (200)**, "Selection of performance objectives and key performance indicators in public private partnership projects to achieve value for money." Construction Management and Economics 27(3): 253-270.

RESEARCH ON DYNAMIC RECONFIGURATION OF VIRTUAL MANUFACTURING CELL CONSIDERING NEW ORDER INTRODUCED

Wen-min Han [1], Ling Jin[1,*1], Zheng-yi Li[1], Jia-shang Jiang[2]
[1] College of Economic and Management,
Jiangsu University of Science and Technology, Zhenjiang, Jiangsu, China

[2] College of Mathematical and Physical Science,
Jiangsu University of Science and Technology, Zhenjiang, Jiangsu, China
15195891128@163.com

ABSTRACT

In order to make real-time response to the dynamic change of product demand, introducing new parts into virtual cell manufacturing system during processing of the current parts is needed to be considered. For the virtual cell reconfiguration problem, a method which combine neural network with heuristic is used to design the two-phase program. In the first phase, in order to make full use of the existing cell structure, neural network is used to achieve similarity matching results between new parts and existing cells; in the second phase, after new orders were added into cell system, adjusting cell structure based on the goals of minimizing cross-cell times and cell load balancing to optimize the overall system. The results of case study verifies the feasibility and effectiveness of the program, responding to changes in time while ensuring the smooth operation of the system.

Keywords: Cell reconfiguration, new order, neural network, virtual manufacturing cell

1. INTRODUCTION

With economic development, the product demand of consumers is growing to be diverse and personalized, product manufacturing showing a feature of multi-species and variable-volume. To meet the fierce market competition and rapid product replacement requirements, make-to-order manufacturing become the development direction of the manufacturing industry, how to make a manufacturing plan that balance between flexibility and efficiency have caused widespread concern of scholars and companies. In this circumstance, virtual cell manufacturing system (VCMS) which is based on group technology have provided an effective way to solve the above problems. The advantage of VCMS lies in fixed physical location of machines in the system, and it forms manufacturing cells only in a logical way, therefore, when the tasks change and the manufacturing cells need to be rebuilt, VCMS can achieve rapid reconfiguration of manufacturing systems by the way of forming virtual cell, this effectively reduce the reconfiguration costs due to changes in the physical location of machines.

Cell formation is the premise and basis of the implementation of VCMS, scholars across the world have made a lot of research around this issue. Chen used a two-phase method to solve the formation problem of virtual manufacturing cell, with the optimization objectives of machine load balancing and minimum cross-cell processing times[1]. Jia et al proposed the idea of modular process which combining parts clustering with processes clustering to deal with modular decomposition, it effectively improve the speed of cell formation[2]. Wang provided a method combining fuzzy neural network with heuristic algorithm to form cells[3]. Elbenani B used genetic

[1] Corresponding Author: Ling Jin

algorithms to solve the problem of cell formation[4]. Rafiee et al proposed a mathematical model which integrated cell formation with inventory problems[5]. N.Safaei presented a mixed integer model to form manufacturing cell under a dynamic environment[6]. Nima Safaei considered outsourcing production plan while designing the dynamic multi-period manufacturing cell system[7]. Ahkioon established a multi-period dynamic cell system, taking into account scheduling plan and optional operation path[8]. Jayakumar studied the adaptive strategy of dynamic cell formation method and proposed an adaptive model in which cell structure changed periodically[9]. R.Jayachitra used fuzzy programming method to deal with uncertain information in the cell formation process[10].

Through analysis, we can see these literature are mainly focus on either static cell formation or periodically cell reconfiguration, studies on event-driven and real-time reconfiguration of cells are less. But in the actual production process, new orders are normally randomly arriving, and this uncertainty requires manufacturing system to make real-time response, however, traditional periodically cell reconfiguration such as quarterly or monthly obviously does not meet such requirements. Han and Garbie IH proposed a similarity coefficient method to introduce new parts into existing cells[11-12], but they did not consider adjustment and reconfiguration of cells after new parts were added. For the insignificancy of above research, this paper studies cell reconfiguration when new orders were introduced into the production process, considering that reconfiguration program should minimize the impact on the existing scheduling plan, we propose the idea that VCMS should be reconstructed on the basis of existing cells, so that it can response to changes in time while ensuring the smooth operation of the system. In this paper, a method which combine neural network with heuristic is used to design the two-phase virtual cell reconfiguration program. The program has considered the similarity matching between new parts and existing cells, the optimization goals of minimizing times of cross-cell processing and cell load balancing have also been considered.

2. ANALYSIS AND DESIGN OF VIRTUAL CELL RECONFIGURATION

2.1. Analysis of cell reconfiguration

The objects of traditional periodically cell reconfiguration are accumulated production tasks over a period of time, rather than randomly arrived orders, and it need not consider cell structure in the last period while restructuring the cell system, so it belongs to an independent process of forming an equivalent new cell. However, under conditions of new order arriving, cell reconfiguration require manufacturing system to make real-time response, and while taking into account the existing production process, reconfiguration should be emphasized on the basis of existing cells, so that to ensure the efficiency and stable operation of overall system. Therefore, cell reconfiguration need to solve problems of two aspects under conditions of new orders adding:

First of all, the designed program should be required to maximize the continuity of the existing cell configurations, thereby reducing cell costs of reconfiguration and reduce the impact on the existing scheduling plan;

Second, apart from ensuring the stable operation of the system, program should also focus on system consistency and efficiency after new orders are added, so adjustment of cell structure is needed to achieve overall optimization of the system.

2.2. Design flow of cell reconfiguration program

According to data information of new orders and existing cell system, this paper propose a two-phase method to solve virtual cell reconfiguration problem, as shown in Figure 1. In the first phase, in order to make full use of the existing cell structure, we use neural network algorithm to achieve similarity matching results between new parts and existing cells. It is prior to assign new part into existing cell, but if there are too much transport produced between cells after new parts are added, then we should create a new cell for production of new parts. In the second phase, after new orders are added into cell system, it is needed to make adjustment of cell structure based on the goals of minimizing cross-cell times and cell load balancing, as to optimize the overall system and complete the process of cell reconfiguration.

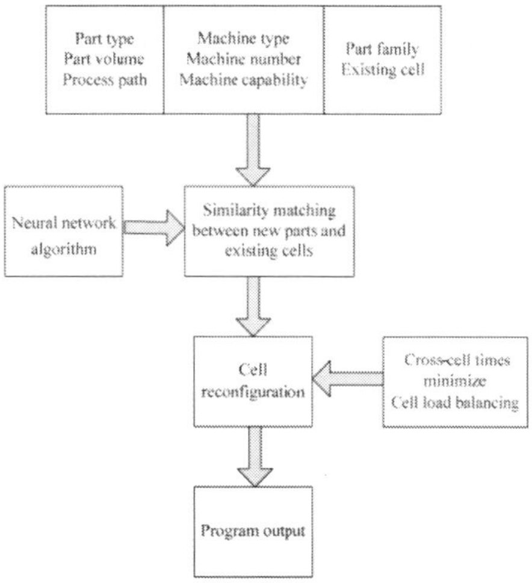

Figure.1. Overall framework of virtual cell reconfiguration program

3. SIMILARITY MATCHING PROCESS BETWEEN NEW PARTS AND EXISTING CELLS

Fuzzy adaptive resonance theory (FART) neural network has the features of self-learning and self-organization, and relying on its own learning rules, it can store the mode learned before and identify new mode. In conjunction with the features, this paper select FART neural network to complete the process of matching new parts with existing cells based on the similarity degree. The process of cell reconfiguration is as follows: in order to assign new part into part family of existing cell, FART network first calculates the similarity values between them, then it select the greatest value and compared to the threshold value of the network, if the value meet the threshold requirement, that means the new parts can be introduced into existing cell, otherwise a new cell should be reconstructed.

3.1. FART Neural Network Model

The neural network in this paper comprises two layers: input layer and output layer, as shown in Figure 2. The neurons number of input layer is the dimension of input vector. The input vector of this paper is the processing path of parts, so the neurons number is the kinds of required machine. Output layer neurons represents a class category, and its number equals to the cell number to be formed. Connection weights corresponding to the cluster center, when the difference between input vector and cluster center is small, then the vector can be regarded as belonging to known classes, otherwise a new category is created.

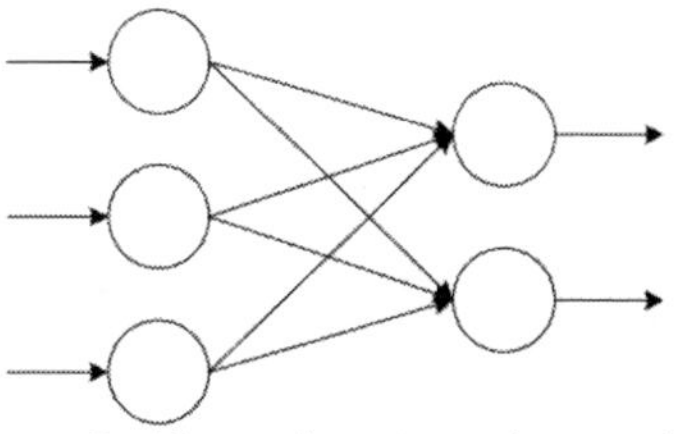

Figure.2. Neural network topology

3.2. Concrete steps of FART Neural Network algorithm

Specific steps of improved FART neural network algorithm which matching new parts with existing cells based on similarity are as follows:

1) Parameter setting. Setting the initial value of connection weights as one, set threshold ρ ,the role of ρ is to control the thickness of classification, a larger value of ρ indicates the requirement of the similarity degree between new parts and existing cells is higher.

2) Input mode initialization. First select randomly for parts one of its multiple processing paths as an input vector of the network:

$$I_p = \left[i_{1p}, i_{2p}, \ldots, i_{mp} \right] \quad (1)$$

Where p∈[1,P], m∈[1,m], P is the kinds of parts, M is the types of machines, imp is the product of processing time and quantity while part p is processed on machine m.

Second, initialize cluster center mode vector based on the existing cell formation program:

$$\mathbf{I}_c = \left[i_{1c}, i_{2c}, \ldots, i_{mc} \right] \quad (2)$$

Where Ic indicates an existing cell, imc is average operation property values of existing part family which is processed on machine m within cell c.

$$i_{mc} = \frac{\sum_{x=1}^{X} i_{mp}}{X} \quad (3)$$

Where X is the existing number of parts processed on machine m, X∈[1,P].

3) Input Mode normalized. Due to large magnitude of difference between initialized input mode, clustering performance of the network will be affected, therefore we need to adopt normalization method to deal with elements of input mode:

$$r_{jp} = \frac{i_{jp} - i_{jp}^{\min}}{i_{jp}^{\max} - i_{jp}^{\min}} \quad (4)$$

Where ijp is the jth element of Ip (j=1,2,...,M), ijpmin is the min value of the jth element of all input mode, ijpmax is the max value of the jth element of all input mode. The input mode after normalized is:

$$\mathbf{R}_p = \left[r_{1p}, r_{2p}, \ldots, r_{mp} \right] \quad (5)$$

4) Similarity between input mode and cluster center. The Euclidean distance between input vector and cluster center is used to judge the similarity degree of the two, the smaller the distance the greater the similarity. The distance between input mode and each cluster center is calculated as:

$$D_{pc} = \left\| \mathbf{R}_p - \mathbf{I}_c \right\| = \sqrt{\sum_{i=1}^{M} (r_{ip} - i_{ic})^2} \quad (6)$$

According to the principle of competitive learning, the output layer neurons with minimum Dpc will win, set as Dpcmin.

5) Threshold value test. If Dpcmin < ρ, then the current input mode is considered belonging to the category that output neurons represents, which means new part can be assigned to existing cell. Adjust the connection weights according to the following formula:

$$w_{ic,j+1} = w_{ic,j} + ku_{cj}(r_{ij} - w_{ic,j}) / n_c \qquad (7)$$

Where k is the adjustment factor , turn to step 4.

6) If Dpcmin > ρ, which indicates although output neurons compete to win, but it exceeding the threshold requirements, so the input mode can not belong to the category represented by the output neurons, then increasing an output layer neurons which represents a new cell. The input mode will be the initial value of newly added weight, turn to step 4.

7) Loop until matching process of all parts is completed, the final network model represents the clustering and distribution results.

4. AD USTMENT OF MANUFACTURING CELL BASED ON HEURISTIC METHODS

In order to ensure the efficient operation of the system as a whole, after the allotment of new parts by neural network algorithm, adjustment for cells must be made considering the goal of minimizing cross-cell times and cell load balancing, as to complete the reconfiguration process.

4.1. Cross-cell times minimize

After clustering distribution of new parts by improved FART neural network, there are cross-cell operations of parts exist generally, while too much cross-cell times can cause an increase cost of logistics, thus, it is needed to deal with the shared machine which related to cross-cell times when reconstructing cells. The total cross-cell times of all parts are:

$$s = \sum_{c=1}^{C}\sum_{d=1}^{C}\sum_{p=1}^{P} v_p \cdot s_{cdp} \qquad (8)$$

Where scdp represents the total cross-cell times of part p between cell c and d, vp is the processing volume of part p.

If machine m is shared by cell c and d, assume cross-cell times are sc when the shared machine is adjusted to cell c, and cross-cell times are sd when the shared machine is adjusted to cell d, then adjustment steps of sharing machine are as follows: 1) Read cell reconfiguration results of last phase; 2) Finding shared machine and stating the number of cells to which they belong. If the number is two, adjusting machine to the two cells c and d respectively, and calculate cross-cell times sc and sd after adjustment respectively. If sc<sd , then adjust machine to cell c, if not, adjust machine to cell d. 3)When machine is shared by more than two cells, then adjusted it to each cell in turn and calculated cross-cell times after each adjustment. By comparing these numbers, select the cell that have the minimum cross-cell times as the result of adjustments. Store cell information and algorithm transfer to cell load balancing phase.

4.2. Cell load balancing

The purpose of cell load balancing is to balance production load of machines in the cell, avoiding individual machine become overloading or idle. Cell load balancing index is:

$$\sqrt{\sum_{c=1}^{C}\left(K_{\max}-K_{c}\right)^{2}/C} \qquad (9)$$

Kc represents load of cell c, C represents total number of virtual cells; Kmax is maximum of all cell load. Load balancing adjustment is mainly to matching machine load with its capacity, therefore, load capacity ratio is used as a basis for the adjustment, defined as follows:

1) Machine load capacity ratio: the ratio of machines load and its production capacity, which measures utilization rate of each machine, the formula is:

$$U_{mc}=\frac{K_{mc}}{C_{mc}}=\frac{1}{C_{mc}}\sum_{p=1}^{P}t_{pmc}v_{p} \qquad (10)$$

Where Umc represents load capacity ratio of machine m in virtual cell c.

2) Cell load capacity ratio: The average value of machine load capacity ratio in each cell, measures the average utilization of machines in the cell, calculated as:

$$U_{c}=\frac{1}{M_{c}}\sum_{m=1}^{M_{c}}U_{mc}=\frac{1}{M_{c}}\sum_{m=1}^{M_{c}}\left[\frac{1}{C_{mc}}\sum_{p=1}^{P}t_{pmc}v_{p}\right] \qquad (11)$$

Load balancing will be handled following the process from whole to part, and adhering to the principle of transferring load of overload machines to other machines. Adjustment steps are as follows: 1) Calculate load capacity ratio Umc of each machine; 2) calculate load capacity ratio Uc of each cell; 3) Identify cell q with the maximum load capacity ratio; 4) Identify the machine n with maximum load capacity ratio in the cell; 5) If Unq>1, turn to step 6), else turn to step 7); 6)There may be multiple parts of current operation on the machine, select part that needed to be transferred of the procedure which should satisfy Unq≤1 after transferred to cell q', it is also need to ensure that the cell is working within its capacity range after cell q' receiving the current operation. If there are a plurality of cells satisfy the conditions of receiving part, cell with the minimum cell load capacity ratio is preferred; 7) loop until the adjustments of all machines and cells are completed.

5. CASE ANALYSIS

Shipbuilding industry is under a typical market environment with multi-species and small-quantities, products operations are more and complex, a shipyard in China adopts cell manufacturing methods in outfitting operations, this has reduced production costs and enhanced business efficiency.

Now we use a particular pipe processing operations as an example to carry out virtual cell reconfiguration, the current production task contains seven kinds of parts, available machine information as shown in Table I, processing information of parts as shown in Table II, the company's existing cell construction scheme as shown in Table III. Suppose the production capacity of each machine can provide for is 6000min. After post-trial confirm the neural network algorithm to build a good effect parameter value ρ=0.6.

Table 1 machine information

Machine Types	Operation Name	Machine ID
A	Opening	A1，A2
B	Cutting	B3，B4，B5

C	Welding	C6, C7, C8
D	Pressure test	D9, D10
E	Surface treatment	E11
F	Bending	F12

Table 2 processing information of parts

Task	Process type	Task	Process type
P1 (15)	Cutting (63) Welding (84) Bending (56) Pressure test (72)	P4 (18)	Cutting (50) Welding (67) Pressure test (57)
		P5 (46)	Cutting (32) Welding (43) Pressure test (53)
P2 (12)	Cutting (82) Welding (109) Bending (68) Pressure test (93) Surface treatment (42)	P6 (25)	Openings (35) Cutting (38) Welding (50)
P3 (13)	Openings (50) Cutting (74) Welding (92) Pressure test (69) Surface treatment (42)	P7 (48)	Openings (25) Cutting (35) Welding (46) Pressure test (50)

Note: the number in brackets after parts is quantity of products, the number in brackets after operation type indicates operation time and the unit is minute.

Table 3 e isting manufacturing cellular formation

Cell program	C1	C2
Part Family	P1,P2,P4,P5	P3,P6,P7
Machine cell	B4,C6,D9,F12	A1,B3,C7,D10,E11

Now the company receives a new order, needing to produce two new parts. Information of processing operations as shown in Table IV.

Table 4 processing information of new parts

Part	Operation	Operation type
P8 (26)	1	Openings (45)
	2	Cutting (59)
	3	Welding (79)
	4	Bending (47)
	5	Surface treatment (30)
P9 (32)	1	Cutting (35)
	2	Welding (46)
	3	Pressure test (53)

For the adding of new orders, the program proposed in this paper is used to reconstruct the cell, the reconfiguration results shown in Table V.

Table 5 cellular reconfiguration program after new parts added

	C1	C2	C3
Original	B4,C6,D9,F12 P1,P2,P4, P5	A1,B3,C7,D10, E11 P3,P6,P7	
Phase One	B4,C6,D9,F12	A1,B3,C7,D10,	A2,B5,C8

	P1,P2,P4, P5	E11 P6,P7,P9	P8
Phase Two	B4,C6,D9,F12 P1,P2,P4, P5	A1,B3,C7,D10 P3,P7,P9	A2,B5,C8,E11 P6,P8

6. DISCUSSION

Seen from the results, in the first phase, after the similarity matching of new parts and the existing cells by neural network, new part P9 has a greater similarity with C2 and meets the threshold requirement, so it can be directly assigned into the cell, and without any cross-cell processing; For new part P8, because of its similarity with the existing cells is low, it cannot be directly added to the existing cells, so needing to reconstruct new cell. Taking into account the priority of using more idle machine of the system, so reconstruct new cell C3 with machine A2, B5 and C8, also cross-cell operation is needed for P8 to complete all the process; In the second phase, firstly, adjusting the results of the last phase combined with cross-cell goal, finding that E11 is shared by P2, P3 and P8 in three cells, calculate the total number of cross-cell one by one while putting E11 into each cell, it can be seen the number of total cross-cell is least when it is put into cell C3. Similarly, F12 is shared by P1, P2 and P8 in cell C1 and C3, in accordance with the same method to know the machine does not need to get adjusted. Combined with load balancing target it can be seen that machine C7 in cell C2 is overload, while newly constructed cell C2 is free and satisfy the condition to receive part P6, so adjust P6 into C3, to give a final reconfiguration program.

. CONCLUSION

Dynamic virtual cell reconfiguration is the basis for rapid response manufacturing. This paper studies the virtual cell reconfiguration under conditions of introducing new order for process, In order to reduce the impact of cell reconfiguration on existing production scheduling, match new part with existing cell based on similarity firstly, making full use of the original cell configuration form, increasing the possibility of assigning new parts into original cell. Then adjusting cell system based on the goals of minimizing times of cross-cell processing and cell load balancing, for the purpose of achieving overall system optimization. Program in this paper has implemented cell reconfiguration on the basis of existing cells, achieving real-time response to changes while ensuring the smooth operation of the system. Direction of further studies may be on the resource conflicts after new orders were added.

8. ACKNOWLEDGMENT

This paper is sponsored by National Natural Science Foundation of China (GN:71271105) and Humanities and Social Sciences Planning Fund of Ministry of Education(GN:12YJA630036).

. REFERENCES

[1] **Ya-rong Chen, Yu- ing Zhou, Hong-ming Zhou, Pei Li**, "Virtual manufacturing cell formation method for reconfigurable manufacturing based on two-stage solving"(in Chinese), Journal of China Mechanical Engineering, vol. 24, no. 22, pp. 3024-3029, 2013.

[2] **Guo-zhu ia, Shi-chao Luan, Liang Chen**, "Cell formation problem for virtual cellular manufacturing based on the process modularization"(in Chinese), System Engineering Theory& Practice, vol. 32, no. 10, pp. 2111-2118, 2012.

[3] **Guo- in Wang, Yan Yan, Ru- in Ning, Ai-min Wang**, "Manufacturing cell formation approach combing fuzzy neural network with heuristic algorithm"(in Chinese), Computer Integrated Manufacturing Systems, vol. 14, no. 11, pp. 2175-2183, 2008.

[4] **Elbenani B, Ferland A, Bellemare** , "Genetic algorithm and large neighbourhood search to solve the cell formation problem", Expert Systems with Applications, vol. 39, no. 3, pp. 2408-2414, 2012.

[5] **K.Rafiee, M.Rabbani, H.Rafiei, A.Rahimi-Vahed**, "A new approach towards integrated cell formation and inventory lot sizing in an unreliable cellular manufacturing system", Applied Mathematical Modeling, no. 35, pp. 1810-1819, 2011.

[6] **N. Safaei, M. Saidi-Mehrabad, M.S**, "Jabal-Ameli. A hybrid simulated annealing for solving an extended model of dynamic cellular manufacturing system". European Journal of Operational Research, no. 185, pp. 563-592, 2008.

[7] **Nima Safaei, RezaTavakkoli-Moghaddam**, "Integrated multi-period cell formation and subcontracting production planning in dynamic cellular manufacturing systems", Int. J. Production Economics, no. 120, pp. 301-314, 2009.

[8] **S. Ahkioon, A.A. Bulgak, T. Bektas**, "Integrated cellular manufacturing systems design with production planning and dynamic system reconfiguration", European Journal of Operational Research, no. 192, pp. 414-428, 2009.

[9] **V. ayakumar, R. Ra u**, "An adaptive cellular manufacturing system design with routing flexibility and dynamic system reconfiguration", European Journal of Scientific Research, no. 136, pp. 595-611, 2010.

[10] **R. ayachitra et al**, "A fuzzy programming approach for formation of virtual cells under dynamic and uncertain conditions", International Journal of Engineering Science and Technology, vol. 2, no. 6, pp. 1708-1724, 2010.

[11] **Wen-min Han, Li-yi Zeng, Yun-de Wei**, "Study on dynamic reconfiguration approach of manufacturing cell for make-to-order"(in Chinese), Process Automation Instrumentation, vol. 32, no. 3, pp. 29-35, 2011.

[12] **Garbie IH, ParsaeiH R, Leep H R**, "Introducing new parts into existing cellular manufacturing systems based on a novel similarity coefficient", International Journal of Production Research, vol. 43, no. 5, pp. 1007-1037, 2005.

THE COMPREHENSIVE EVALUATE METHODS AND SOFTWARE IMPLEMENTATION OF INTEGRATED SUPPLY CHAIN PERFORMANCE

Li-zhen Du*[1], un-fa Wang, Kai Liu
Department of Industrial Engineering,
Wuhan Textile University, Wuhan, China
dlzay@163.com

ABSTRACT

The ultimate goal to valuate the integrated supply chain performance is to carry out the overall optimization. It is the key that a set of perfect integrated supply chain performance evaluation index system must be established. Because of time-varying characteristics of the evaluation object, multi-level dynamic fuzzy evaluation method and improved grey relational method are adopted to comprehensively evaluate the overall performance of the integrated supply chain. Based on this, the Performance Evaluation System (PES) of the integrated supply chain is developed on MATLAB software platform. In the end, the validity and feasibility of the system is validated by an example.

Keywords: Integrated supply chain management, performance evaluation, multi-level dynamic fuzzy evaluation method, improved grey relational method, software development

1. INTRODUCTION

Decision-making of each node enterprise will have effect on the upstream and downstream enterprises of the integrated supply chain[1]. The overall performance is the most important evaluation criterion to evaluate the operating effectiveness of the supply chain. The integrated supply chain performance evaluation has attracted more and more concern and discussion, and plays a significant and increasing role in the modern corporation management[2].

The ultimate goal of the integrated supply chain performance evaluation is not only to achieve the whole high-efficient operating status, but beyond that to optimize the operation flow which can provide a scientific and objective decision basis for manufacturers to optimize the integrated supply chain[3-4].

2. INTEGRATED SUPPLY CHAIN PERFORMANCE EVALUATION INDEX SYSTEM

To evaluate the overall performance of the integrated supply chain, a set of perfect integrated supply chain performance evaluation index system must be established, the key of supply chain performance evaluation. According to the definition of the integrated supply chain performance, which includes the value of end-user and the value of the integrated supply chain[5], a four-level evaluation index system for the integrated supply chain performance is established as shown in Fig.1.

The first level indexes:

$U = \{U_1(\text{users value}), U_2(\text{supply chain value})\}$

The second level indexes:

[1] Corresponding Author: Li-zhen Du

$U_1 = \{u_{11}(\text{quality}), u_{12}(\text{price}), u_{13}(\text{reliablility})\}$,

$$U_2 = \left\{ \begin{array}{l} u_{21}(\text{business process value}), u_{22}(\text{partnership value}), \\ u_{23}(\text{economic value}), u_{24}(\text{innovative value}) \end{array} \right\}$$

The third level indexes:

$$u_{11} = \left\{ \begin{array}{l} u_{111}(\text{return and repair rate}), \\ u_{112}(\text{response time to users complain}) \end{array} \right\}$$

$$u_{12} = \left\{ \begin{array}{l} u_{121}(\text{the average price advantage}), \\ u_{122}(\text{commodities promotional frequency}) \end{array} \right\}$$

$$u_{13} = \left\{ \begin{array}{l} u_{131}(\text{sale wastagerate}), u_{132}(\text{deliveryreliability}), \\ u_{133}(\text{userscomplainrate}) \end{array} \right\}$$

$$u_{21} = \left\{ \begin{array}{l} u_{211}(\text{production} - \text{sale rate}), u_{212}(\text{production cycle time}), \\ u_{213}(\text{order fulfillment cycle time}) \end{array} \right\}$$

$$u_{22} = \left\{ \begin{array}{l} u_{221}(\text{on} - \text{time delivery rate}), u_{222}(\text{production qualified rate}), \\ u_{223}(\text{information sharing rate}) \end{array} \right\}$$

$$u_{23} = \{u_{231}(\text{finance income}), u_{232}(\text{asset operating conditions})\}$$

$$u_{24} = \left\{ \begin{array}{l} u_{241}(\text{new technology adoption rate}), u_{242}(\text{investment rate}), \\ u_{243}(\text{new} - \text{product revenue rate}) \end{array} \right\}$$

The fourth level indexes:

$$u_{211} = \left\{ \begin{array}{l} u_{2111}(\text{node} - \text{enterprise production} - \text{sale rate}), \\ u_{2112}(\text{core} - \text{enterprise production sale rate}) \end{array} \right\}$$

$$u_{212} = \left\{ \begin{array}{l} u_{2121}(\text{node} - \text{enterprise production cycle time}), \\ u_{2122}(\text{core} - \text{enterprise production cycle time}) \end{array} \right\}$$

$$u_{231} = \{u_{2311}(\text{market share}), u_{2312}(\text{capital yield})\}$$

$$u_{232} = \{u_{2321}(\text{capital turnover rate}), u_{2322}(\text{inventory turnover rate})\}$$

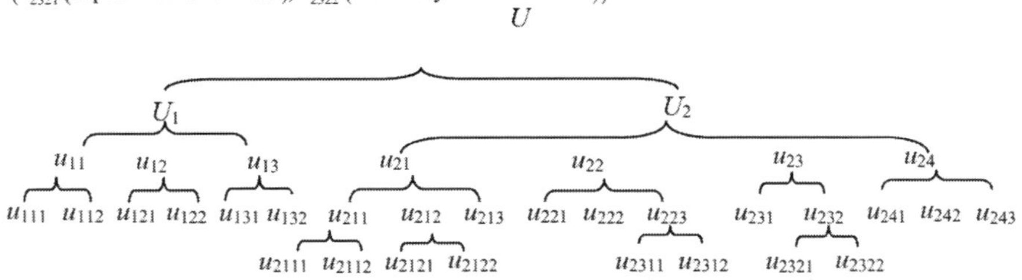

Figure. 1. An evaluation inde system for the integrated supply chain performance

3. THE COMPREHENSIVE EVALUATION METHODS OF INTEGRATED SUPPLY CHAIN PERFORMANCE

With scientific and systematic evaluation indexes system of integrated supply chain performance, it is necessary to choose suitable comprehensive evaluation methods to evaluate the performance of integrated supply chain and to realize the whole optimization of the integrated supply chain.

Two methods are used to comprehensive evaluate[6], which are the multi-level dynamic fuzzy comprehensive evaluation method, the improved method on grey relational evaluation.

3.1. Improved Method on Grey Relational Evaluation

A method which uses the cosines vector included angle to determine the index weight is proposed in this paper. It establishes the best and worst sequence according to index-values of several evaluation schemes; calculates good matrix and poor matrix of deviation values; and then it calculates the cosine values of the angle between relevant vectors of the two matrixes in respect of each indicator. The weights of all indicators are obtained after normalization. Grey relation is not mentioned in the cosine method, yet the calculation process is similar to the method on grey relational evaluation[7]. An improved grey relational method is obtained by combining the cosine method and grey relational evaluation method, which can determine the indicator weight and evaluation result more effectively. This method is divided into 9 steps.

3.1.1 Generation of evaluation matri

Suppose there are m indicators and n schemes in the indicator system, then the evaluation matrix is $A=(a_{ij})_{m\times n}$ and $(i=1,2,\cdots,m; j=1,2,\cdots,n)$, whereas a_{ij} means the indictor of the i evaluation indicator on j schemes.

3.1.2 Select the reference se uence (optimal se uence is U and worst one is L)

Since the given evaluation matrix is non-dimensionally treated, so all of them are positive indicators (the bigger, the better).

$$U=(u_i)_{1\times m}=(u_1,u_2,\cdots,u_m)^T$$

$$L=(l_i)_{1\times m}=(l_1,l_2,\cdots,l_m)^T$$

3.1.3 Determine the matri of deviation value

Good matrix of deviation value $UA=(ua_{ij})_{m\times n}$, poor matrix of deviation value $LA=(la_{ij})_{m\times n}$.

Whereas $ua_{ij}=|u_i-a_{ij}|$, $la_{ij}=|l_i-a_{ij}|$ (1)

3.1.4 Determine the matri of deviation rate

Good matrix of deviation rate $R=(r_{ij})_{m\times n}$, poor matrix of deviation rate $S=(s_{ij})_{m\times n}$.

Whereas

$$r_{ij}=\frac{ua_{ij}}{\max_j\{a_{ij}\}-\min_j\{a_{ij}\}} \tag{2}$$

$$s_{ij}=\frac{la_{ij}}{\max_j\{a_{ij}\}-\min_j\{a_{ij}\}} \tag{3}$$

3.1.5 Determine the deviation e tremum at two poles

Maximum value of good deviation at two poles $\Delta^u_{\max}=\max_j\max_i ua_{ij}$

Minimum value of good deviation at two poles $\Delta^u_{\min}=\min_j\min_i ua_{ij}$

Maximum value of poor deviation at two poles $\Delta^u_{\max}=\max_j\max_i ua_{ij}$

Minimum value of poor deviation at two poles $\Delta^l_{\min}=\min_j\min_i la_{ij}$

3.1.6 Calculate the weight

Regarding the indicator i, select the corresponding row vector r_i from R and corresponding row vector s_i in S, and then calculate the included angle cosine of two vectors;

$$c_i=\frac{\sum_{j=1}^{n} r_{ij}\cdot s_{ij}}{\sqrt{\sum_{j=1}^{n} r_{ij}^2}\sqrt{\sum_{j=1}^{n} s_{ij}^2}} \tag{4}$$

Ultimately normalize c_i and obtain the weight vector of indicator $\bar{\omega}=(\omega_1,\omega_2,\cdots,\omega_m)$.

Whereas $$\omega_i=\frac{c_i}{\sum_{i=1}^{m}c_i} \quad (5)$$

3.1. Calculate the matri of correlation coefficient

Good matrix of correlation coefficient $\xi_u=(\xi_{uj}(i))_{m\times n}$, poor matrix of correlation coefficient $\xi_l=(\xi_{lj}(i))_{m\times n}$, whereas $\xi_{uj}(i)$ and $\xi_{lj}(i)$ are the indicators x_{ij} of vector $\bar{x}_j$ in scheme j and correlation coefficient l_i of indicator u_i of reference vector $\bar{u}$ in $\bar{l}$, respectively.

$$\xi_{uj}(i)=\frac{\Delta^u_{\min}+\rho\Delta^u_{\max}}{ua_{ij}+\rho\Delta^u_{\max}} \quad (6)$$

$$\xi_{lj}(i)=\frac{\Delta^l_{\min}+\rho\Delta^l_{\max}}{la_{ij}+\rho\Delta^l_{\max}} \quad (7)$$

Whereas $\rho\in(0,\infty)$ is the identification coefficient, which plays the role in increasing the difference among values. The smaller it is, the bigger resolution it will get. Normally 0.5 is selected.

3.1.8 Calculate the relational degree

The good relation of scheme j is (relational with $\bar{u}$):

$$D(u,j)=\sum_{k=1}^{n}\xi_{uj}(k)\omega(k) \quad (8)$$

The poor relation of scheme j is (relational with $\bar{l}$):

$$D(l,j)=\sum_{k=1}^{n}\xi_{lj}(k)\omega(k) \quad (9)$$

3.1. Comprehensive relation

The good relation (the bigger, the better) or poor relation (the smaller, the better) are used to sequence the scheme. The comprehensive judgment method can also be used to sequence the comprehensive relation.

Comprehensive relation of scheme j is:

$$V_j=\frac{1}{1+[D(l,j)/D(u,j)]} \quad (10)$$

This method considers the good and poor relation, and improves the accuracy of evaluation at the same time.

3.1.10 Se uencing and preference

The sequence of good and poor relation or comprehensive relation in a size down can be used to make consequence. Select the maximum relation as the preferential scheme.

3.2. The Multi-level Dynamic Fuzzy Comprehensive Evaluation Method

The fuzzy comprehensive evaluation method is a systematic evaluation method that combines fuzzy mathematics with Analytic Hierarchy Process (AHP)[8]. It is better to solve the problem that to evaluate comprehensively multi-objective system[9-11]. However, when using the fuzzy comprehensive evaluation method, we often regarded little for characteristic values of the evaluation object that can change over time, but rather regard these values as constants, or evaluate only according to the values of a certain point in time, and then regard the evaluation results as final results at a whole time period. Obviously, this evaluation method is not suitable to deal with some special problems with time change.

Dynamic fuzzy evaluation method is an improved fuzzy comprehensive evaluation method, time parameter k is introduced into evaluation matrixes[12-13]. Then, the improved fuzzy comprehensive evaluation method take into account these index values change over time, namely, it is a kind of real-time dynamic evaluation.

The multi-level dynamic fuzzy comprehensive evaluation method has several main steps, as follows.

3.2.1 To establish inde es set and evaluation set

According to the automotive supply chain performance evaluation index system, a four-level evaluation index set can be established. The evaluation set have four grades, respectively can be labeled as excellent, good, medium, and poor.

3.2.2 To calculate single-factor evaluation matri

On the time period k, evaluation matrix R_k of quantitative index can be determined by using fuzzy membership function $r_{ijk} = \mu(X)$. Index values can adopt the recent three-years values, that is, $k = 1,2,3$. With the premise of information-sharing, rapid data acquisition become possible. Therefore, the final evaluation result takes into account these evaluation indexes values change at different time point in 3 years.

It is more difficult to quantitatively evaluate these qualitative indexes. The fuzzy statistical method is usually used. Concrete realization method as follows: let several experts involved in the evaluation, according to designated evaluation criteria to divided these evaluation factors into different grades; and then to statistic the frequency N_{ij} that evaluation factor U_i belongs to grade V_j, finally to calculate membership function $\mu(U_i) = N_{ij} / n$, that is, the membership degree of the index U_i belongs to grade V_j. In the formula, n is the number of experts involved in the evaluation.

Combining the evaluation results of these qualitative indexes with that of quantitative indexes, single-factor evaluation matrix R can be obtained.

3.2.3 To determine weights of inde es by using AHP

Many methods can be used to determine weights, such as empirical method, regression analysis method, direct scoring method, entropy weight method and analytic hierarchy process (AHP) and so on[14]. Because the automotive supply chain performance evaluation is actually a typical multi-level and multi-objective decision-making issue containing quantitative and qualitative factors[15]. AHP is a systematic, hierarchy and comprehensive evaluation method that combines quantitative and qualitative factors. Therefore, based on analyzing the automotive supply chain performance evaluation index system, AHP is suitable to determine weights of these indexes.

3.2.4 Comprehensive evaluation

The process of comprehensive evaluation is to fuzzy calculate by using (11) from the fourth-level indexes to the first-level ones step by step, then to obtain the comprehensive evaluation result B_k in each year.

$$B_k = W \circ R = (b_{1k}, b_{2k}, b_{3k}, b_{4k}) \tag{11}$$

Finally, taking into account that evaluation result in each year is different, and the impact of the result on the final comprehensive evaluation result is different, the overall comprehensive evaluation result can be obtained by using (12).

$$B = \sum_{k=1}^{3} \lambda_k \cdot B_k = (b_1, b_2, b_3, b_4) \tag{12}$$

In this formula, λ_k is the weight at different period, and where $k = 1,2,3$, and $\sum \lambda_k = 1$.

3.2.5 Fuzzy evaluation result

According to the principle of the largest membership degree, $b_j = \max\{b_1, b_2, b_3, b_4\}$ then, the evaluation grade of the program should be b_j. That is, through multi-level dynamic fuzzy comprehensive evaluation, the evaluation results show the performance grade of the automotive supply chain should be b_j.

4. SOFTWARE DEVELOPMENT AND APPLICATION

The steps to evaluate the performance of integrated supply chain by using the improved method on grey relational evaluation and multi-level dynamic fuzzy comprehensive evaluation model is very complicated, and the computation is very large and complicated. For managers, these methods and the evaluation process are cumbersome. For this reason, we have developed the Performance Evaluation System (PES) of integrated supply chain based on MATLAB software platform, which can help auto manufacturers to evaluate conveniently the performance of the integrated supply chain. The system mainly includes two function modules: IGRE (Improved method on Grey Relational Evaluation) and DFCE (multi-level Dynamic Fuzzy Comprehensive Evaluation). Fig.2 is the main interface of the Performance Evaluation System (PES).

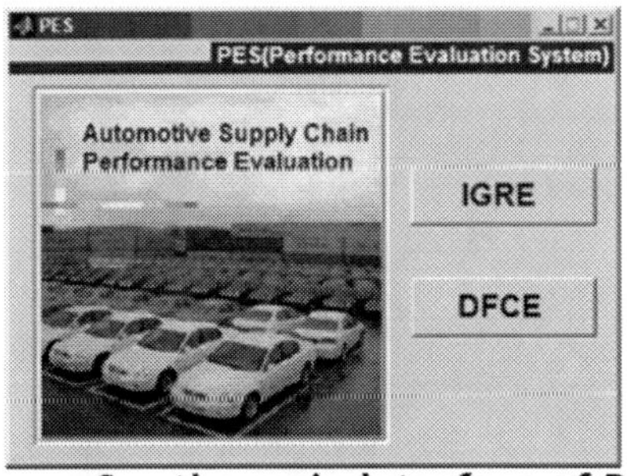

Figure.2. the main interface of PES

4.1. The Function Module of the IGRE

By clicking on the IGRE button to active the function module of IGRE, as shown in Fig.3.

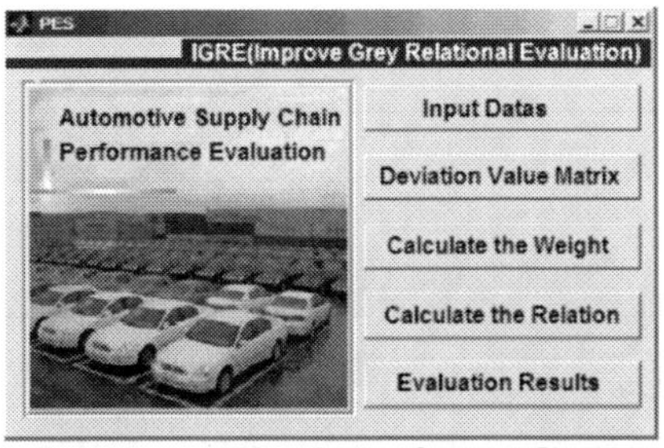

Figure.3. The interface of IGRE

4.2 The Function Module of the DFCE

By clicking on the DFCE button to active the function module of DFCE, as shown in Fig.4.

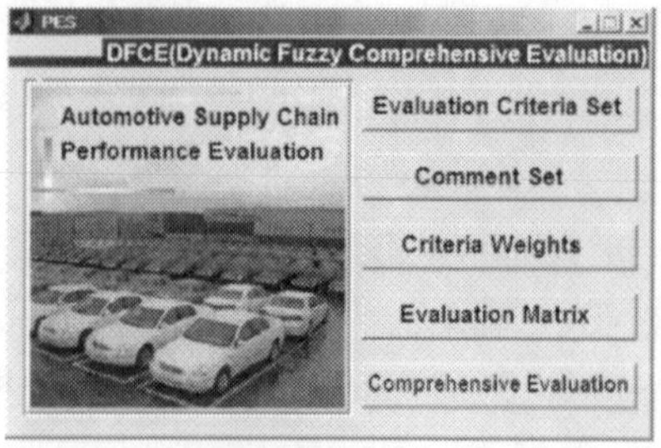

Figure.4. The interface of DFCE

4.3 A application case

A certain integrated assembly plant has gradually formed an integrated supply chain which around the vehicle manufacturer as the core, upward integrated with auto parts suppliers, downward with vehicle distributors.

According to the above-mentioned integrated supply chain performance evaluation index system, statistical data of the value of each evaluation index in the last three-years is shown in TABLE I. Most of original data directly come from the production management database, sales management database and customer relationship management database and so on.

Table 1 the inde values of the integrated supply chain performance in three years

Evaluation index	2009	2010	2011
u_{111}	0.8	0.5	0.2
u_{112} (hours/time)	24	20	12
u_{121}	70	80	90
u_{122} (times/quarter)	0.5	0.5	1
u_{131}	0.9	0.6	0.2
u_{132}	95	98	99
u_{133}	0.5	0.4	0.2
u_{2111}	96	97	98
u_{2112}	99	100	100
u_{2121} (hours)	80	75	60
u_{2122} (hours)	42	38	31
u_{213} (days)	45	40	30
u_{221}	85	90	95
u_{222}	90	93	98
u_{223}	80	90	95
u_{2311}	2	3	6
u_{2312}	10	20	35
u_{2321} (times/year)	1	2	4
u_{2322} (times/year)	1	2	4
u_{241}	10	25	40
u_{242}	5	8	12
u_{243}	98	99	99

First to click the Input Data button in Fig.3 to input all initial dates, these data can be non-dimension treatment by the system automatically. Then, click the Deviation Value Matrix , Calculate the Weight to obtain the deviation value matrixes and the weight of index successively. Click the Calculate the Relation to get annual good relation, poor relation and comprehensive relation. Finally, click the last button to show the order of annual relation, as shown in the Fig.5.

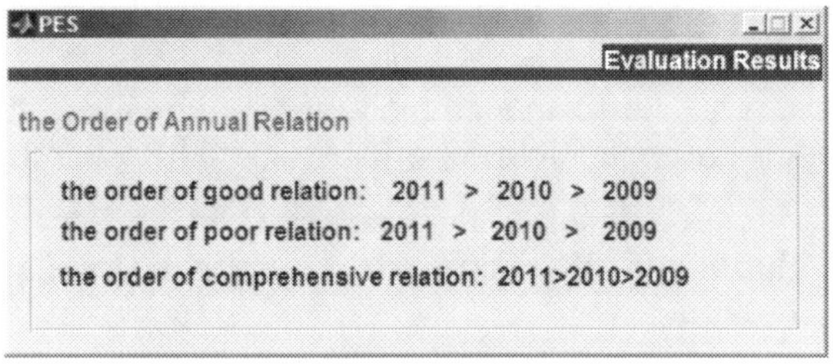

Figure.5. The evaluation results

In DFCE function module, the former four buttons must be input dates. After inputting the first-year comparative matrix pairs to the Evaluation Matrix module, to click the Comprehensive Evaluation button to obtain the evaluation result of the first-year is medium . In the same way, by inputting the second-year and the third-year evaluation matrix pairs to get the evaluation result of the second-year is good and the third-year is excellent sequentially. Finally, input the weight of each year (0.2, 0.3, 0.5) to the Weight of Each ear text box, and click the Comprehension button to obtain the three-year' weighted comprehensive evaluation is excellent , as shown in the Fig.6.

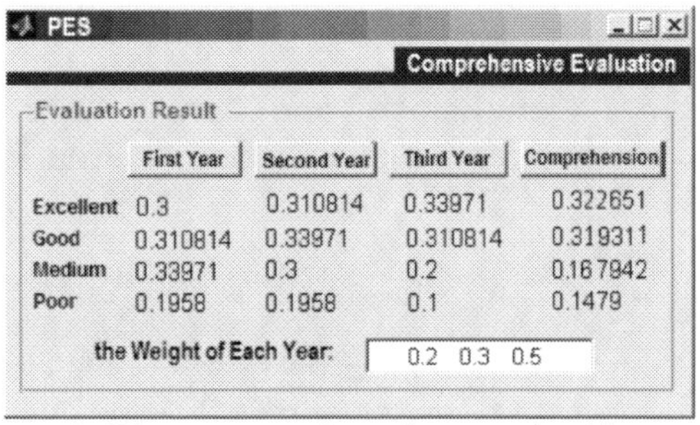

Figure.6. The comprehensive evaluation result

5. CONCLUSION

A multi-level performance evaluation index system of integrated supply chain has been established according to the definition of the integrated supply chain performance, then improved method of grey relational evaluation and multi-level dynamic fuzzy comprehensive evaluation method have been adopt to analyze and evaluate systematically the performance of whole integrated supply chain. Based on these studies, we have explored a set of integrated supply chain performance evaluation system (PES) on the Matalab7.0 software platform. Even if the user is not familiar with the algorithms or does not have strong computer skills, he can also undertake the performance evaluation of the integrated supply chain by using the system under a simple prompt or guidance. Through the example which has been tested on a real case, it is obviously that the system is very effective and feasible, and it can become a useful tool to evaluate dynamically the integrated supply chain performance.

6. REFERENCES

[1] **Peter Trkman, Kevin McCormack, Marcos Paulo Valadares de Oliveira**, The impact of business analytics on supply chain performance, Decision Support Systems, vol.49, pp. 318-327, 2010.

[2] **ian Cai, Xiangdong Liu, Zhihu Xiao**. Improving supply chain performance management: A systematic approach to analyzing iterative KPI accomplishment, Decision Support Systems, vol.46, pp. 512-521,2009.

[3] **Ming Dong, F. Frank Chen**. Performance modeling and analysis of integrated logistic chains: An analytic framework, European Journal of Operational Research, vol.162, pp.83-98, 2005.

[4] **Haritha Saaranga, Roger Moser**. Performance evaluation of purchasing and supply management using value chain DEA approach, European Journal of Operational Research, vol.207, pp.197-205, 2009.

[5] **G. Tomas M. Hult, David . Ketchen r., Brian R. Chabowski**. Leadership, the buying center, and supply chain performance: A study of linked users, buyers, and suppliers, Industrial Marketing Management, vol.36, pp. 393-403, 2007.

[6] **osef ablonsky**. Measuring the efficiency of production units by AHP models, Mathematical and Computer Modeling, vol. 46, pp.1091-1098, 2010.

[7] **iuping Xu, Bin Li, Desheng Wu**. Rough data envelopment analysis and its application to supply chain performance evaluation, International Journal of Production Economics, vol.122, pp. 628-638, 2009.

[8] **Dani ela Tadic, Alev Taskin Gumus, Slavko Arsovski, atc**. An evaluation of quality goals by fuzzy AHP and fuzzy TOPSIS methodology, Journal of Intelligent Fuzzy Systems, vol.25, pp. 547-556, 2013.

[9] **Ahmet Can Kutlu, Mehmet Ekmekcioglu and Cengiz Kahraman**. A fuzzy multi-criteria approach to point-factor method for job evaluation, Journal of Intelligent Fuzzy Systems, vol.25, pp. 659-671, 2013.

[10] **Chin-Nung Liao**. An evaluation model using fuzzy TOPSIS and goal programming for T M consultant selection, Journal of Testing and Evaluation, Vol.41, no.1, pp.1-9, 2013.

[11] **Xin Zhang, Fang in, Peide Liu**. A grey relation projection method for multi-attribute decision making based on intuitionist trapezoidal fuzzy number, Applied Mathematical Modelling, vol.37, pp.3467-3477, 2013.

[12] **Ali Azadeh, Maryam Seifoory, Morteza Abbasi**. Integration of simulation and fuzzy multi-attribute decision making for modeling and assessment of fuzzy parameters, International Journal of Industrial and Systems Engineering, vol.6, no.4, pp.483-502, 2010.

[13] **Li Cunfang, Zhou De un**. Evaluation and demonstration of the synthesis competitive capacity of enterprises basing on fuzzy maths (in Chinese), Control and Decision, vol.22, pp.337-340, 2007.

[14] **Morteza Pakdin Amiri**. Project selection for oil-fields development by using the AHP and fuzzy TOPSIS methods, Expert Systems with Applications, vol.37, pp. 6218-6224, 2010.

[15] **Darshan Kumar, agdev Singh, Om Pal Singh, Seema**. A fuzzy logic based decision support system for evaluation of suppliers in supply chain management practices, Mathematical and Computer Modelling, Vol.57, pp.2945-2960, 2013.

THE IMPACT OF INTER-FIRM RELATIONSHIP NETWORK ON COLLABORATIVE INNOVATION: A CASE ANALYSIS

Lan Wang[1]
School of Business Beijing Language and Culture University Beijing, China
im2003@163.com

ABSTRACT

Alliance and partner-like relationship have continued to become more and more popular across a broader boundary spanning of firms, industries and nationalities in the environment of increasing globalization and the demands of a networked, fast-paced economy. More and more companies depend on the complimentary resources and capability in the supply chain network to enhance their innovation competence. This article summarized different view inter-firm relationship and its impact on performance. Based on the case analysis of Company A, this article reveals how to integrate recourses in the business ecosystem to realize localized researching of production equipment.

Keywords : inter-firm relationship; collaborative innovation; network resource;

1.INTRODUCTION

In the dynamic environment, continuous innovation is the necessity of an enterprise's survival and development. Companies that can integrate global resources and innovate rapidly can become the winners of competitions. Under the background of knowledge economy, the innovation actions of companies also have higher requirements. These requirements mainly embody in the complexity trend in the technology innovation process. On the one hand, the exploration of new products or the improvement of skills depends on the diversified heterogeneity knowledge and technology. Many technology innovations emerge in cross and integrated area of knowledge. On the other hand, as the technology develops, the technology itself becomes more complex. Under the background of knowledge economy, knowledge division is more detailed. The companies cannot equip all necessary knowledge. The challenge is how to find complementary knowledge.

Companies' strategies constantly change according to changes of the competition environment. For Chinese companies, in the first period of economy transition, the changes of companies' strategies are concentrated on business strategies, mainly on the transferring from production-oriented to market-oriented. Meanwhile, products offered by companies are changing from one type to multi-types, and the characteristics of products emphasized by companies changed from only quantity to quality, function and costs etc. One of the specific behaviors of strategy changes is that market orientation becomes an important trait of companies' strategy. Marketing function was built rapidly and gradually occupied the hosting position. In order to better understand and predict market needs, to provide excellent after-sale services, and to expand product channels, companies build their own sales and marketing departments, and the leaders of sales departments gradually become companies' senior managers. Companies that built sales departments early developed rapidly, typically the development of the township and village enterprises. Although township and

[1] Corresponding Author: Lan Wang

village enterprises' technology and production capability are lower than medium and large state owned companies, township enterprises obtain constant development depending on their flexible methods of sales and marketing operation. The second characteristic of strategy changes is that the company gradually focuses on technology orientation. Companies concentrate more on research and exploration, as well as technology innovation. Companies' strategy adjustments on technology problems have important effects for companies to guarantee their constant competitive advantages. The third phenomenon of strategy changes is the change of the responsibilities of companies' production system. In the period of the planned economy, product quantity was the only indicator that measures the company and its production system performance, so the production system emphasized efficiency. Along with economy transition, the goals and responsibilities of production system are diversified. On one hand, the production system emphasizes efficiency; on the other hand, it emphasizes indicators such as production costs, types, quality and deadline, as well as undertaking new responsibilities such as technology innovation. The production system's traits have radical changes because of the production management system imports, such as agile manufacturing, overall quality management, and computer integrated manufacturing system. The trait of this generation is that technology accumulation is more important, the combination of technology and innovation is more attended, flexibility and reaction speed are emphasized, and the diversity of products and their quality are more focused on, so companies must have the capability to respond to diversified customer needs (Gerwin, 1987)[1], and ensure that environment adaptation gradually becomes an important indicator of strategy.

The improvement of the products' function and quality must depend on the enhancement of the company's research ability and the increase of its technology innovation speed. In order to realize innovation, companies must quickly accumulate the knowledge, technology and ability. The increase of the research and technology not only embody within the company, but also demonstrates in technology collaboration and other external approaches that can improve the company's products and technology ability. Companies import product lines or equipment directly, purchase technology patents, and establish joint venture with oversea companies. These activities are the economy phenomena hotspot in the initial stage of transition economy. The import of technology brings the improvements of companies' competence, and the rapid development of some emerging industries also enhances the overall economic benefits effectively. However, when facing the global competition, domestic companies have the problem of lack of experience and product competence, and these companies' technology innovations are limited in scales, capitals and human resources etc., therefore these companies can hardly implement effective heterogeneity strategies. In order to acquire competence in international competitions and build constant competitive advantages, companies' innovation ability can be effectively enhanced by the access of knowledge learning and exploration collaborations among companies. Domestic companies such as Haier and Lenovo experienced the development from imitation to innovation and win international competitive advantages by domestic and foreign cooperation as well as constant learning and absorbing of new knowledge. Under the complex trend of innovation, companies constantly adjust their strategies to adapt the environment changes. From the evolvement of the human gene plan of the world, to the forming of industry technology alliance, to the collaboration of upstream and downstream companies, this trend demonstrates the network changes of innovation. It can be seen that the innovation mode experiences a linear innovation dyadic relationship mode integration mode network mode. Therefore, in the network environment, how to promote collaborative innovation performance by relationship driven competence and how the company relationship affect innovation performance are important issues for companies to establish sustainable competitive advantages.

2.LITERATURE REVIEW

2.1Inter-firm relationship theory

Traditional resource view believes that the heterogeneity resource that brings competitive advantages comes from the inside of companies (Barney, 1991)[2]. Along with the speed up of globalization and networking, and the form of global network, companies not only depend on the occupation of scarce resource, but also obtain competitive advantages by continuous learning and knowledge creation with the help of abundant social capital (Lane Lubatkin, 1998)[3]. Therefore, traditional resource theory has new extension and development about the growth of network research. Company network theory believes that the heterogeneity resource of companies is not only from the inside of companies, but also comes from the outside of companies, which is the network that companies embedded in. Except for the resource that companies hold, their key resource can be obtained from the various relationships with other external entities. The multiple relationships among companies can generate significant relational rent and competitive advantages (Dyer Hatch, 2006)[4]. Hence, companies' competitive advantages not only depend on the internal resource, but also depend on the various resource and ability that are hard to be imitated by competitors in the network (Gualti, 1999)[5]. The boundary of organizations becomes unclear, and the research about inter-relationship is more focused by scholars. The behaviors and performance can be explained by the network (Gulati, Nohria, Zaheer, 2000)[6]. Under the globalization, researches about firm's resources, behaviors and capability have already extended their views from singe firm to their external networks, as well as the effect of inter-firm relationship on firm performance. From the view of network structure, the characteristics of companies in the network affect their acquisition and utilization of resources. These characteristics influence strategic behaviors and the influence will generate difference on firm performance and competitive advantages .

In recent years, long-term oriented collaborative inter-firm relationship becomes hotspot among practitioners and academic managers. Effective supplier management and the establishment of cooperative/partnership relationship with suppliers can significantly improve performances. The performance improvements are shown as the reduction of costs, the improvement of companies' flexibility and agility, the enhancement of companies' dynamic ability, and the improvement of innovative performance etc. (Tiwana, 2008)[7]. Such relationship among companies can be seen as their heterogeneity resource that is valuable, scarce, inimitable and insubstitutable. This resource can bring constant competitive advantages to companies (Kogut, 2000)[8].

2.2Type of inter-firm relationship

2.2.1Strong tie and weak tie

The resource of relational network among companies can bring advantages to resource and actions to movers in the network. Three major theories in social network are the representative theories that explain the nature of the advantage of network characteristics : Granovetter's (1973, 1985) [9][10] strong tie advantages theory and weak tie advantages theory , social capital theory represented by Coleman (1988) [11]and structure hole theory represented by Burt (1992)[12].

As the leader of new economic sociology, Granovetter made great contribution to social networking theory. In 1973, Granovetter firstly came up with the definition of the strength of tie, and divided ties to strong connection and weak connection. He believes that strong ties maintain the internal relationships of groups and organizations, and that weak ties link groups and organizations. He measured the strengths of ties from four dimensions: the frequency of interaction, emotional strength, closeness and reciprocal exchange.

2.2.2The strength of weak ties

Granovetter (1973) believes that in the process of information and knowledge diffusion. Weak ties have more effects. He claims that in this network, the subjects of strong ties understand each other, and their knowledge structure, experiences and backgrounds are fairly similar. Therefore, in the communications of strong ties, the information is repeated highly, and the knowledge and information increased by frequent interactions are mainly the same, so they are redundant. However, weak ties develop among individuals that have different economy traits, and they are distributed to a wide area. Weak ties can bring important information from other groups to individuals who do not belong to these groups, and these heterogeneous information and knowledge have more values to the subjects of activities in the network, so the judgment that weak ties are used as information bridges is brought up.

2.2.3The advantages of strong ties

Another group of scholars believe that strong ties have important influence in network relationships, especially in business behaviors that are supported by organizational relationships. For example, organizations and individuals who are at uncertain positions can obtain connector's protection by developing strong ties to lower the uncertainty of environment. Gulati (1995) also believes that strong ties are more advantageous, because when the interaction number of both acting subjects increases, familiarity exists. Familiarity can lead to mutual trust and can generate stronger mutual support. Similarly, Rowley et al. (2000)[13] believes that when an organization has long-term close relationship crosses organizational boundaries (strong ties), this relationship can help the organization to deal with the changes of environment and the pressure of various uncertainties, and so strong ties are beneficial for the organization to cope with some risks. Moreover, the flow and transfer of some complex knowledge usually show proximity effect and viscous effect , meaning that the efficiency of information and knowledge delivery increases, especially face to face sharing and frequent communications. Uzzi's (1997)[14] research showed that those advantages of strong ties could help enhance trust and cooperation. By the function mechanism such as information sharing, trust and joint problems solvings, strong ties can help organizations and individuals obtain more fine-grained, dependent knowledge and tacit knowledge.

2.2.4Relational embeddedness parado

Based on the definition of strong tie and weak tie, Granovetter(1985) states that all economic activities are embedded in social relationships. Pure economic relations hypothesized in new classical economics do not exist in real life, and that real economic relations contain a large amount of embedded social relationship factors. Among these, relational embeddedness refers to relationships' strength, Familiarity, consistency and directions etc. Structure embeddedness is more focused on the factors of the system structure that reflect network relationships. Structure embeddedness is how the mover's structure and location have influences on its behaviors and results. It can be seen that the definition of relational embeddedness originates from Granovetter's (1973) point on the ties between individuals and organizations, and the strength of relationships shows the depth of relational embeddedness. Overall, scholars obtain various research results on relational embeddedness's influences on companies, some conclusions are even opposite, forming the relational embeddedness paradox .

As it is written in above, from the availability of resource, some scholars claim that weak ties in the network play the role of bridges that can obtain information and resource across organizations, and are helpful for companies to obtain useful external information and knowledge, promoting company's ability and performance more effectively. The most representative scholar is Granovetter (1973). Strong ties generated from individuals and organizations have similar traits, and the individuals in that organization have similar understanding of objects and experiences, so the information acquired by strong ties is highly repeated and has redundancy.

Some other scholars' opinionis different from the opinion of weak connection's power, believing that the level of embeddedness positively affects company's performance, and that strong ties are advantages for companies to communicate, and thus bringing the sharing of sticky knowledge and the transferring of high quality information. Movery and Oxley's (1996) [15]research discovered that strong network relationships could promote knowledge transferring among organizations and enhance company's absorbing ability. McEvily and Marcus (2005)[16] testified relational embeddedness's influence on competence by a empirical research of 234 manufacturing companies, pointing out that the trust, the sharing of high quality information, and the joint problem solvings among companies, suppliers and customers are beneficial for companies to increase their competence, proving that embeddedness positively influences the competence.

2.2.5Other typology of inter-firm relationship

Cannon and Perreault (1999) classify relationship types according to five traits: basic purchasing and selling relationship, contract-trading relationship, customer and supplier relationship, cooperation relationship, collaboration relationship, adaptation adjusting relationship and customer-centered relationship. previous research divides relationships into seven types according to different standards: non-strategic trading relationship, management relationship, contract relationship, special contract relationship, partner relationship, joint venture and strategy alliance. Some other scholars believe that the different type of cooperative relationship can be understood as the difference of management level of company relationships. However, although in a same type of relationship, such as alliance, the cooperation levels are at different phases during the procedure of relationship upgrading. Santoro's research (2000) showed that the cooperation relationships between schools and companies have horizontal differentiation. Gulati (1998) believes that the alliance of companies can be described as high level relational embedding. Speckman, Kamauff and Myhr (1998) believe that companies are closer from trading relationship to cooperation relationship. Both trading sides can conduct collaborative behaviors, but the two sides cannot be stated as partnership relationship, because close trading partners should have top level of trust and commitment. Some organization theories researched and discussed the difference of different relationship types on their closeness, indirectly described the differences of the same type of relationship. strategic alliances show various forms because of the differences in commitments, symmetry of returns and the level of trust. Lambert, Emmelhainz and Garder (1996) classified three types of partnership relationships according to the difference of trading dependence and closeness: coordination between the partners, integration, and significant integration.

2.3Collaborative innovation

It is because the increasing unpredictability of industry integration and market environment, companies cannot completely depend on their own resource and ability to improve their innovation performance. Meanwhile, supplier management is more emphasized, and company cooperation becomes a key and effective innovation strategy. In recent years, more scholars connect collaboration with innovation. Specifically, companies develop new products by technology cooperation, and improve their technology ability. Companies create new products quickly by collaborating with downstream customers and understanding customer needs. Especially in the high technology industry, technology collaboration is significantly important, and collaborative innovation almost becomes a model. Many of recent new products are created by collaborative innovation, such as the Ipod from Apple and the Xbox from Microsoft. These products demonstrate that technology collaboration is an important method of organizational changes and competitive advantages acquisition. However, it is undeniable that there are many examples of failed collaborative innovation.

Strategic management literatures refer most that the factors of collaborative performance are the relationship factors among companies, including the similarity of partners, management mode, experience, and relational embedding etc. Although these research are based on different theories, many results show that the relationship between companies have positive effects on innovation performance, especially on the development and improvement of companies' technology (Zahra, 1996)[17]. Companies can also learn new management techniques to develop creative products and expand new markets. Companies can obtain sufficient knowledge and experience from suppliers and customers, and can acquire multiple resources, including management and financial resource. Previous research reveals that coherent motivation, mutual trust between companies and current routines are beneficial for risk reduction and problem solving, and thus generate successful collaborative innovation.

Innovation is an important factor for companies to remain their vitality. As Jarillo (1988)[18] said, realizing an innovative solution is the key problem faced by entrepreneurs, so they have to accumulate and obtain necessary resource and abilities. For the research of company innovation, some scholars focus on the degree of innovation classified as incremental innovation and radical innovation. Some other scholars focus more on the objects of innovation, distinguishing product innovation and procedure innovation. However, in order to adapt to external environment and changes of market needs, the problems companies face are not how to realize a certain type of innovation, but how to require the innovation that can reflect the environment and market needs timely. It is because the increasing unpredictability of industry integration and market environment, companies cannot completely depend on their own resource and ability to improve their innovation performance. Meanwhile, supplier management is more emphasized, and company cooperation becomes a key and effective innovation strategy. In recent years, more scholars connect collaboration to creation. Specifically, companies develop new products by technology cooperation, and improve their technology ability. Companies create new products quickly by collaborating with downstream customers and understanding customer needs. Collaborative innovation is defined as companies' research, new products, new services, new procedures, and the capability to adapt to market needs with several organizations by integrating supplier chain value activities (Vickery et al., 199)[19]. Because of the constraints of their resource and efficiency, the relationship driven competence among companies is more concentrated. Companies obtain diversified and complementary abilities and resource by establishing strategic alliance and partnership, promoting the collaborative innovation.

3.CASE ANALYSES

3.1Company overview

Company A is a big resource company. Company A builds multi-industry business modules based on coal, forming a specialized technology innovation system that the technology transferring platform is technology companies and the postdoctoral research working stations, and the method is to completely utilize external technology resource. An important reason that company A can realize its quick development is to stick to and depend on technology innovation.

3.2Systematic supplier evaluation and management regulations

The supplier management is more important. Key equipment purchasing, service purchasing and high value creation determined that the focal firm should maintain close relationship with suppliers. In fact, the purchases based on cooperation will lead to supply chain management and collaborative logistics management, because purchasing management is realized on the basis of realizing organization efficiency. Company A starts from internal departments to strengthen the collaborative efficiency among departments, improves the relational performance with suppliers, and extends to the whole supply chain network. At the same time, company A should realize that the coordinate supply relationship is not for all suppliers. It is impossible that all suppliers can offer strong strategic support to downstream companies because the support level is decided by the core competence of suppliers and the characteristics of product. Company A bases on realizing above questions and establishes long-term dynamic supplier evaluation mechanism to evaluate suppliers. The evaluation standards include technology, services, logistics and pricing etc., establishing a supplier certificate system.

It is worth saying that company A's supplier segmentation management and evaluation systems are just built. So far, these systems have not yet established supplier elimination and renewing mechanism, and this is company A's next direction of development. It is also the same under the background of economy globalization. If a company wants to obtain advantages in market competitions, it has to use R D advantages to improve products' quality, or using advanced equipment and procedure advantages to create low cost products, or using the brand advantage to distribute the products. No matter in what way, technology innovation is the key. In 21st century, technology innovation has become the hotspot between countries and companies, becoming an effective way to improve core competence. Company A, as a big national company, is the leader of coal industry. Its competitors are not only domestic companies, but are transferring to international companies. These international companies have sufficient capital, advanced technology and equipment, as well as strong R D capability, controlling the top levels of technology in most industries. Practices show that without core technology and technology innovation capability, domestic companies can hardly break the technology monopoly of developed countries and their international companies, and can hardly move to international labor division, so company development faces heavier international pressure and impulsive force. Under this situation, it is a must for company A to increase market competence and participate in international competitions by increasing technology innovation ability as well as grasping core technology and proprietary intellectual property rights. Company A promotes technology innovation and conduct behavior improvements as well as improving its competence. Company A starts the establishment of coal exploitation and utilization technology innovation strategic alliance , leading the transferring of the coal industry from resource scaling and rough increasing to resource saving and connotation development type. Company A has many original innovation results in coal, electricity, road, harbor and chemistry, significantly leading the development of the coal industry and other related industries.

3.3 A Company collaborative innovation

Domestic equipment manufacturing obtained significant improvements, but the localization of heavy equipment still face many difficulties. The main reason of the situation is that the research of the localization of heavy equipment is occupied by manufacturing companies and lacks the participation of users. Therefore, these companies have not yet formed a mechanism that is guided by market needs, motivated by technology support and product risks shared jointly. In order to promote the localization of heavy equipment, equipment-using companies must take actions. In the research and production of heavy equipment, first performance is a difficulty faced by equipment production companies. In order to motivate the positive strength to join the research of digging equipment localization and overcome the difficulty of first performance , company A actively implements strategic collaboration with excellent equipment production companies such as Aerospace Science and Technology Group and Shanghai Electric Power Group, creatively bringing up with a method of research replacing purchasing. Equipment-using companies come up with goals and main technical parameters and standards of localization production, and manufacturing companies input capital to overcome difficulties, design and produce products according to the requirements. When the first sample is proved to be successful after experiments and tryouts, users purchase the product at the market price, and admit that the producer has the production performance of that series of products, as well as having the permission to bid in the future market-based purchases. If the sample is unsuccessful, the manufacturer pays the research fee. In this way, the first performance problem is solved, and the research input can be largely increased to guarantee the product quality. In 2006, the research of hydraulic support in company saved the equipment investments to more than 3 billion RMB. It is proved that as long as we positively coordinate every aspect, strengthen the cooperation with users, manufacturers and research institutions, and overcome difficulties, we can surely make great breakthroughs in the technology of equipment, to promote the overall level of domestic equipment manufacturing.

Lead the development of the implementation plan. A company, as the largest user of high-enddigging equipment, integrates its needs and organizes all aspects of power to promote local research concentrated on hydraulic support. In the research and development of hydraulic support, A company firstly made the localization implementation plans of hydraulic support, cleared the localization implementation principle, overall planning and progress arrangement. A company established a technology discussion group consisted of academician and industry professionals, deeply researched the technology possibility of local exploitation, economic rationality and technology plans. Meanwhile, A company set main technical parameter and standards of localized equipment as a leader. In the process of digging equipment localization, A company exerts an important linking, leading and promoting effect.

Organize researching . In research strategies, A company develops diversified and various forms of key technology researching mode. In the localization research of hydraulic support, A company firstly absorbed and deeply researched the design and concept of imported equipment, and then innovated when technical problems occur. Secondly, A company went through specialized research of the welding process of high strength steel independently. Thirdly, A company entrusted Aerospace Science and Technology Group to research and develop electro-hydraulic control system. Electro-hydraulic control system is the key part and core of hydraulic support. Domestic coal industry started the research of this system during the ninth five-year plan , but never succeeded. The electro-hydraulic control technology developed by Aerospace Science and Technology Group is the leader in domestic industry. By completely using the advantages of core technology, the complete localization of high strength hydraulic support will be largely improved. Fourthly, A company demanded that the manufacturers of hydraulic support strengthen the establishment of quality assurance system by contracts, and set supervising regulations through the entire design and production process of hydraulic support to guarantee the quality to the most.

Selecting research institutes by strategy collaborations and bidding. In the selection of digging equipment localization research, A company applied the method combined by strategy collaborations and inviting bids. First of all, by bidding, A company selected Zhengzhou coal machine factory and Beijing coal machine factory to produce hydraulic supports of 4.5 meters and 5.5 meters for mining area respectively. Experienced coal machine manufacturers can not only exert professional technologies and mature techniques to accelerate localization process, but also force foreign suppliers to lower their prices under pressure. At the same time when A company supported coal machine companies to develop high-end hydraulic supports, it started the strategy collaborations with Shanghai and China Aerospace Science and Technology Group in technology innovation A company invited Aerospace Science and Technology Group, Shanghai Electric Group and other leading equipment manufacturers to participate in the localization research. Meanwhile, A company uses the way of producing rather than purchasing to obtain digging equipment, which set the basis of localization research. By combining strategy collaborations and inviting bidding, A company not only created a scene where several equipment manufacturers compete, but also broke the monopoly in the coal machine manufacturing industry, and thus reduced the costs of equipment purchases.

4. CONCLUSION AND DISCUSSION

Apparently, the emergence of company A's original innovative results not only depends on the current internal resource. Companies should creatively utilize the external resource that they need for self-dependent innovation. The realization of companies' technology innovation requires that companies to integrate current internal resource as well as absorbing valuable and scarce external resource. Therefore, in order to promote companies' self-dependent innovation capability and increase companies' competitive advantages, companies have to integrate internal and external innovative resource effectively. Self-dependent technology innovation mainly contains three aspects: the first is to strengthen original innovation and to obtain more scientific discoveries and inventions; the second is to strengthen integrated innovation and form competitive products and industries by accumulating various related technologies; the third is to absorb thoroughly and recreate on the basis of learning and importing international advanced technologies. These three ways have important and positive influence on the self-dependent innovation for companies. Company A is an integrated energy company, and its main products (coal, electricity, gas and transportation) are all key and basic products of the national energy guarantee. The products lie in various industries, wide area, and high threshold and require advanced technologies. However, company A is different from other national companies, because it does not have many research entities. These factors make company A stick to its innovation key parts and technology, as well as exploring a specialized road that combines university-industry cooperation, potential market, technology needs, cooperation, strategic alliance, and integrating company A's external resource to the most. In company A's unusual and leaping development, it has formed an innovation mode of chain innovation, service integration innovation and agglomeration innovation.

After China's economy transition passed its primary phrase, companies' external environment was basically market-oriented. In this situation, business operations in companies nearly adapt to the requirements of market competitions. Companies should not only emphasize production efficiency, but also focus on the flexibility and integration ability. So far, market competitions are fiercer, and there are many products and brands compete in any industry or any down market. Where the opportunities are for new comers and laggards in this seemed full and excessive-competing market How to compete with industry leaders How to win over competitions, to survive, to develop and to expand These are important strategy questions that many companies are faced with. Currently, there are several successful challenge takers or followers in the domestic markets of IT, beverage, vehicle and household electrical appliances, their paths of success embody sufficient business methods, management tools and management thoughts. The changes of companies' business strategy are not the necessary reflect of economic transition, but an ordinary behavior in market competitions, and these strategy changes and interaction behaviors lift to a level among companies. Especially under the environment where Chinese medium and small private companies are wholly market-oriented, there is nothing permanent except change . Companies have to face different external competition pressure, and the cultivation of mutual strategy orientation among companies have significant meanings for companies to adapt to changes, to overcome risks and to increase performances. To enhance the innovation capability and create new business model, the company they have to embed in the business ecosystem.

5. ACKNOWLEDGEMENTS

This research is supported by Humanity and Social Science foundation of Ministry of Education of China (Project No.15 JC630123)

6. REFERENCE

[1]**Gerwin, D**. An agenda for research on the flexibility of manufacturing processes. International Journal of Operations and Production Management, 1987, 7(1): 38-49.

[2]**Barney, .B**. Firm resource and sustained competitive advantage. Journal of management, 1991, 17(l): 99-120.

[3]**Lane, P. ., Lubatkin, M**. Relative absorptive capacity and inter-organizational learning. Strategic Management Journal, 1998, 19: 461-477.

[4]**Dyer, . H., Hatch, N**. Relation-Specific Capabilities and Barriers to Knowledge Transfers: Creating Advantage through Network Relationships. Strategic Management Journal, 2006, 27: 701-719.

[5]**Gulati, R**. Network location and learning: the influences of network Resources and firm capabilities on alliance formation. Strategic Management Journal, 1999, 20: 397-420.

[6]**Gulati, R., Nohira, N., Zaheer, A**. Strategic Networks. Strategic Management Journal, 2000, 21(3): 203-215.

[7]**Tiwana, A**. Do bridging ties complement strong ties An empirical examination of alliance ambedexiterity. Strategic Management Journal, 2008, 29: 251 272.

[8]**Kogut, B., Zander, U**. Knowledge of the firm, combinative capabilities, and the replication of technology. Organization Science, 1992, 3: 383-397.

[9]**Granovetter, M**. Getting a Job: A Study of Contacts and Careers .Cambridge, MA: Harvard University Press, 1985.

[10]**Granovetter, M**. The strength of weak ties. American Journal of Sociology, 1973, 78: 1360-1 380.

[11]**Coleman, .S**. Social capital in the creation of human capital, American Journal of Sociology, 1988, 94: 95-120.

[12]**Burt, R.S**. Structural Holes: The Social Structure of Competition. Cambridge, MA: Harvard University Press, 1992.

[13]**Rowley, T., Behrens, D., Krackhardt, D**. Redundant governance structures: An Analysis of structural and relational embeddedness in the steel and Semi-conduetor industries. Strategic Management Joumal, 2000, 21:369 386.

[14 **Uzzi, B**. Social structure and competition in interfirm networks. Administrative Science uarterly, 1997, 42: 35-6.

[15]**McKee, D. O., Varadara an, P.R., . Pride, W. M**. Strategic Adaptability and Firm Performance: A Market Contingent Perspective, Journal of Marketing, 1989, 53 (3): 21.

[16]**Mowery, D.C., O ley, .E**. Strategic alliances and inter-firm knowledge and Inter-firm knowledge transfer. Strategy Management Journal, 1996, 17:77-91.

[17]**Zahra, S.A.** Technology strategy and financial performance: Examining the moderating role of the firm's competitive environment. Journal of Business Venturing, 1996, 11: 189-219.

[18] **arilli, C. .** On strategic networks. Strategic Management Journal, 1988, 9(1): 31-41.

[19]**Vickery, S., Calantone, R. Dr oge, C.** Supply chain flexibility: an empirical study. The Journal of Supply Chain Management, 1999, 35 (1): 16-24.

THE RESEARCH OF DATA PREPROCESS OF 3D RAPID PROTOTYPING

Xia Han[1]
[1]Beijing Institute Of Fashion Technology Institute of art and design,
hepingjie Street, Beijing, 100029, China
hanxia_vip@126.com

ABSTRACT

3D Rapid Prototyping technology is the technology using a process of increasing the material layer by layer to form three-dimensional solid model via RP processing equipment produced. Therefore, before rapid prototyping, geometry data should be obtained first from the CAD system, reverse engineering, CT, etc. Then pretreat the data that matching with 3D rapid prototyping layered software and output to the corresponding 3D rapid prototyping system, and then to complete the 3D printing.

Keywords :3D Rapid Prototyping technology; data pre-processing; data output

1. INTRODUCTION

3D Rapid Prototyping technology is the technology using a process of increasing the material layer by layer to form three-dimensional solid model via RP processing equipment produced. Therefore, before rapid prototyping, geometry data should be obtained first from the CAD system, reverse engineering, CT, etc. and save it as a matching format with a 3D rapid prototyping layered software; then, two-dimensionally process the three-dimensional data, including process on technical, STL files and other documents, etc., that is, transforming the complex three-dimensional data into a series of two-dimensional level information; then output to the corresponding 3D rapid Prototyping system; 3D rapid prototyping system then accumulate the material layer by layer according to the two-dimensional plies information to make the final product or three-dimensional solid model[1]. The data processes of 3D rapid prototyping technology are shown in Figure 1[2].

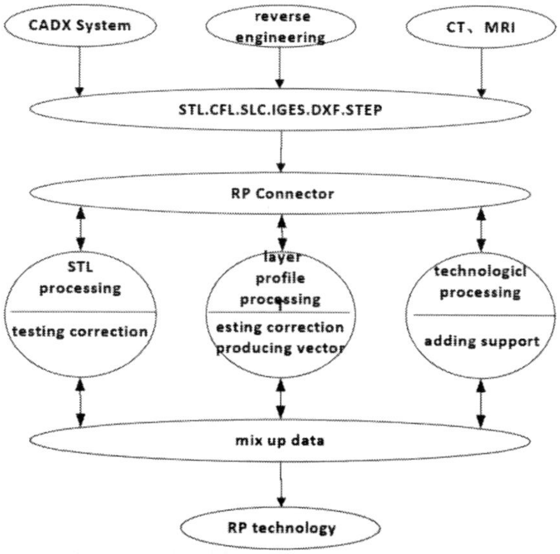

Figure.1 the data processing of 3D rapid prototyping

[1] Corresponding Author: Xia Han

2. THE DATA SOURCE OF RP TECHNOLOGY

2.1CAD data

This is the most important and widely used sources of data. Products generated by the three-dimensional solid modeling software (PROE, Solidworks, Autocad, etc.) and then the three-dimensional CAD data model will be separated by layers to form precise cross-sectional profile. The most common method is to transform the three-dimensional CAD data model to triangular mesh data and to separated by layers. Then we will get processing path of RP system.

2.2. Reverse engineering data

The main source of such data is using reverse engineering software and measuring equipment (such as 3D scanner) to scan the existing parts three-dimensionally to obtain point cloud data; and then process the point cloud data and generate triangular mesh data to form STL file, and then process the data by layers or processing 3D point cloud data by layers directly.

The main data interface format used by the current RP systems is shown in FIG. 2. As can be seen from the figure, at present, common data interface of 3D rapid prototyping system has two formats: one is based on the geometric model data, such as STL, IGES, STEP and other types, which can accept data from the outside (such as RE, CAD systems, etc.) directly; the other is based on RP systems slice data, such as CLI, HP / L, CT and other data formats. Since the forming way of RP system is accumulating layer by layer, so when we use geometric model as an interface format, RP system molding required for "slicing" process of the geometric model to generate NC code for 3D rapid prototyping system. The structure of STL file format is relatively simple and easy to implement, thus has become the current RP systems interface format[3].

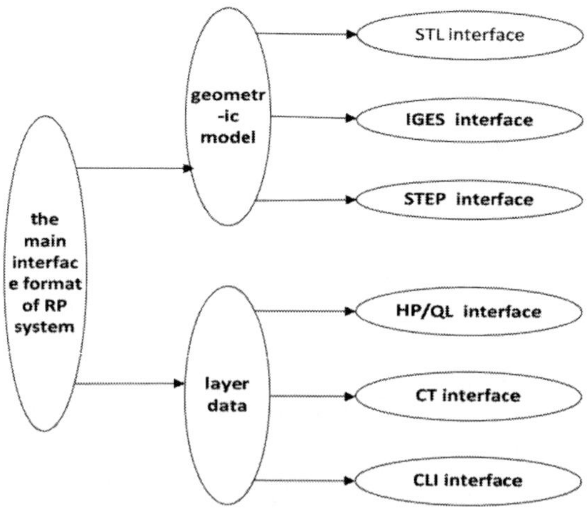

Fig.2 the main interfaces format of rapid prototyping system

Apart from the main interface format described in FIG. 2, there are other common data formats such as DXF, LMI, SLC, HPGL, etc. Some data preprocessing methods of common RP technology system formats will be described in detail below.

3 THE PREPROCESSING OF STL DATA 4 5

STL data format is three-dimensional surface of the Table type data format. Currently, most CAD software on the international market are equipped with an interface for STL data files. And STL files are almost the most popular interface type of most RP system and it has become a recognized industry standard in this field. The effect of STL data file in RP system is shown in Figure.3.

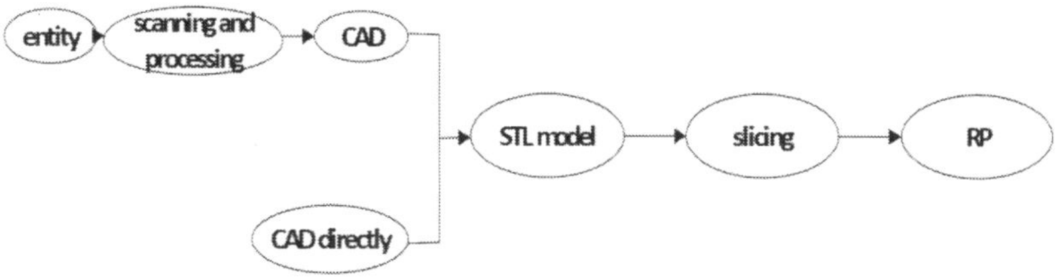

Figure.3 The role of STL data profile in 3D rapid prototyping

The principal of STL data format is using the form of small triangular dough piece to approximate three-dimensional solid free-form surfaces. That is a collection of obtained triangular grid from three-dimensional CAD solid model. In each triangular face Table, it can be made up of three vertices of the triangle, point normal vector model outside triangular facets, namely, STL data format is based on the three components and three triangle vertex coordinates of the triangle. FIG 4 shows all triangle surfaces of solid model of STL file. CAD models described by STL format is showed in FIG. 5.

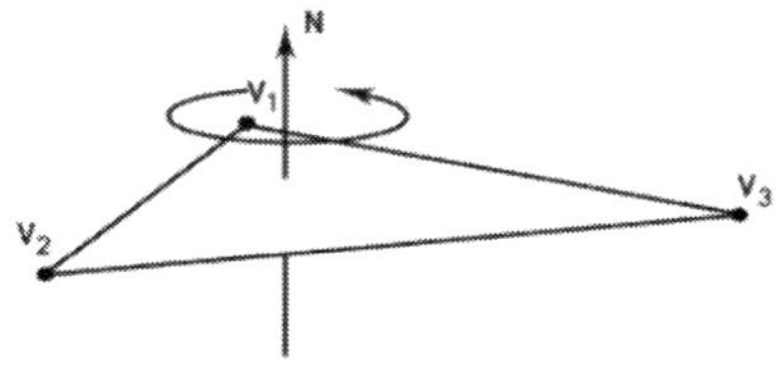

Figure.4 triangular surface

Figure.5 CAD model described by STL data

STL file is the most widely used data interface format of RP technology, which has the following significant advantages: a wide range of files can be input, and almost all three-dimensional geometric data model can be transformed to STL file by triangular facets. The process is very simple. Most of the three-dimensional CAD software has a function of outputting STL file directly and controlling the initial model accuracy of the STL. The slicing algorithm is relatively simple. It is easy to split the model when it is too large to be produced a shape.

But the STL file format also has its own shortcomings. It has a great amount of data and there is a phenomenon of redundancy; Errors sometimes occur in the conversion process of data; There is approximation error because it is using the format of the triangular facets to approximate the entire entity, etc. Thus, STL file format of data need to be improved in processing speed, accuracy and stability.

3.1The accuracy of STL file

As mentioned above, the data format STL file is using the form of small triangular facets to approximate the outer surface of the three-dimensional solid model. Therefore, the number of small triangles will directly affect the accuracy of models. The more triangular facets, the higher accuracy of products will have. However, too small triangular facets will cause large STL file hence increasing the storage of the computer, thereby increasing the slices processing time for the RP system. Thus, the complexity and accuracy of the model should be considered when outputting STL file format from 3D CAD / CAM software so as to select the most appropriate accuracy indication and control parameters.

Precision control parameter of the output STL file format is closely related to the quality of the finished RP article. The accuracy indication of CAD model that STL file approaching is the number of small triangles. Or we can use the size of the chord height of triangle plane when approaching the surface to describe the accuracy indication.

Chord height is described as the radial distance between the contour edge and surface of triangle. Using small triangular surfaces to approximate CAD model surface is just a first-order approximation of the original model and it does not contain adjacency information so it can not express the intent of the original design fully and completely. So there are errors from a real surface. FIG. 6 (a), (b) shows the triangle partition while output a sphere model into STL file. As can be seen from the figure, the size of the chord height determines the number of triangle and directly affects the surface quality of the output model.

(a)

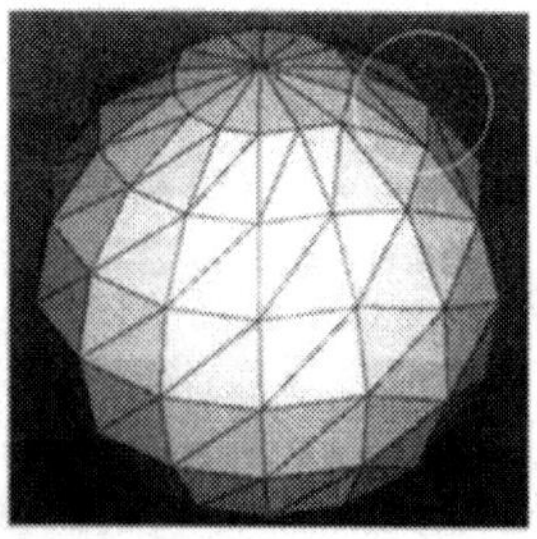

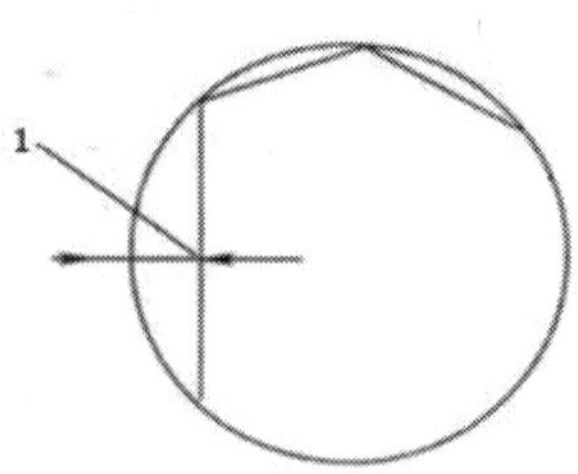

(b)

Figure.6 the triangle partition
(a) surfaces of triangle of STL profile outputting from sphere
(b) chord height

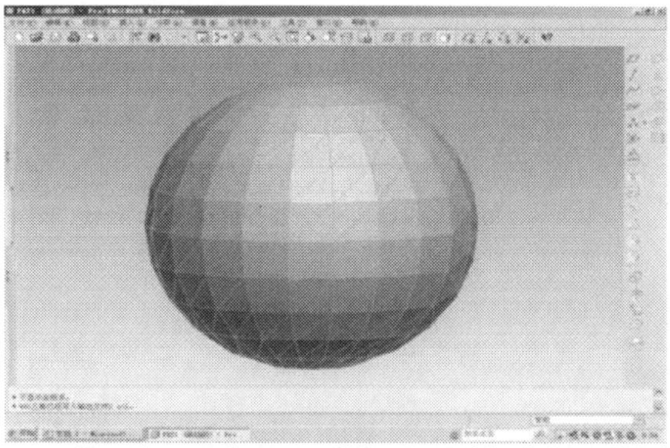

Figure. triangle surfaces with chord height of 2 inches

Figure.8 triangle surfaces with chord height of 1 inch

Figure. triangle surfaces with chord height of 0.05 inch

3.2. the basic principles of STL file

(1) Common vertex principle after making triangular mesh for three-dimensional solid mode, every little triangle-shaped faces must share two vertices with adjacent small triangular-shaped faces, that is, a vertex of a small triangular-shaped surface must fall on a point of adjacent triangular shaped faces, rather than on the edge.

For example, four triangles share one vertex A in FIG. 10 (a), this is a correct expression; and Figure 10 (b) shows that two triangles share vertex A but it falls on the edge of the another triangle, which shows that this expression is wrong.

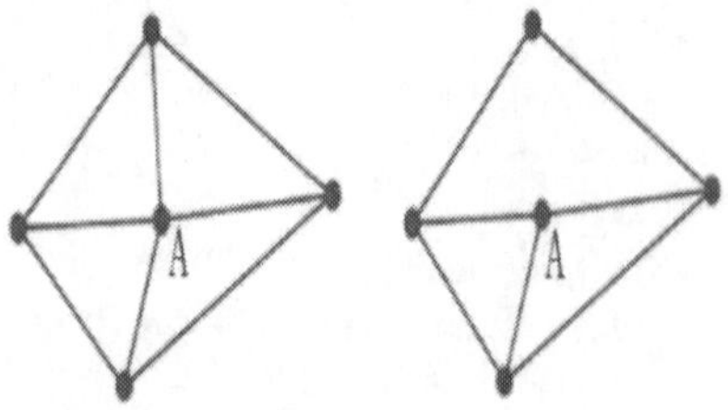

(a) (b)

Figure.10 two triangles share a verte

(2) Orientation principle For each of the small triangles STL file format, order the vertex to represent edge and using right-hand rule to determine vector direction which connected by three vertices. FIG. 11 (a) shows that the direction of the vector arrow of the triangle is towards the outside of the paper so we called it the outer surface; otherwise we call it the inner surface; As shown in FIG. 11 (b), it is correct that for two adjacent triangles faces, vector direction of the arrow is toward the inside of the paper; And as shown in Fig. 11 (c), it is uncorrect that for the two adjacent triangular faces, the vector direction of the arrow is in contradiction.

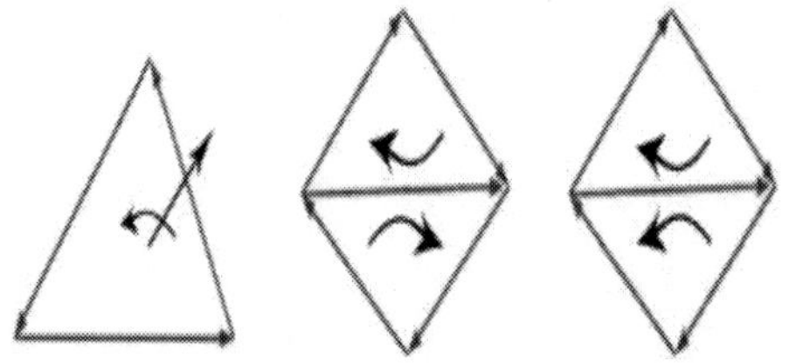

(a) (b) (c)

Figure.11 （a）、（b）、（c）Orientation principle

(3) Value principle the vertex coordinate values for each small triangular faces of STL file format should be positive and shouldn't be zero and negative.

(4) Filling principle all surfaces of model in STL file format should be covered with small triangular facets and must not have omissions. The gap shown in FIG.12 should be complemented with the small triangles and eliminate errors, otherwise broken surface will occur during RP3D rapid prototyping.

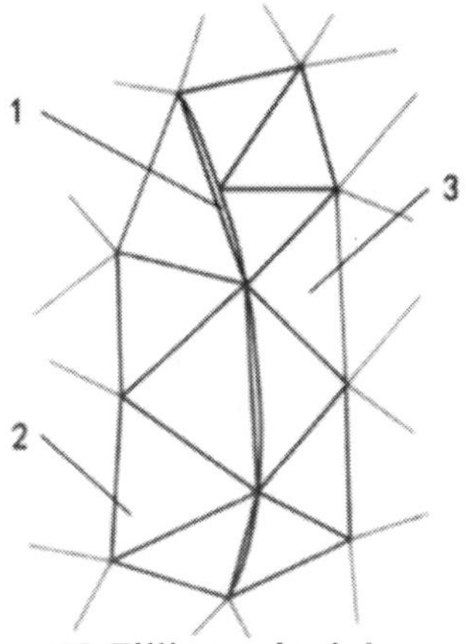

Figure.12 Filling principle
1-interval 2-surface1 3-surface2

4. THE DATA INTERFACE FORMAT OF RP TECHNOLOGY SYSTEM [6]

In addition to the STL file format, the data interface format used by RP technology system including other two-dimensional, such as dimensional layer data format files have SLC, CLI, HPGL, etc and three-dimensional plies data formats, such as 3-D ply data file format there IGES, STEP, DXF, DXF, etc.

4.1Two-dimensional layer data file format

SLC, CLI, HPGL, etc. belong to two-dimensional layer data file format. These files are basically independent of RP technology and equipment and necessarily complement the STL file so as to establish better contact between three-dimensional CAD model and RP technology and equipment. These documents can be obtained from reverse engineering, for which these documents have impact on RP technology and equipment and reverse engineering.

Two-dimensional layer format has the following advantages: It can be layered directly in the CAD modeling system so as to omit processing time and improve manufacturing accuracy of the model and reduce the amount of storage. Its biggest advantage is that there will be less wrong and easy to fix.

Two-dimensional format has the following disadvantages: Because it is a two-dimensional layer Table file format, the layer thickness can be changed. In addition, the model can not be added support and can not be rotated or re-oriented.

Currently, the two-dimensional format is mainly used after the three-dimensional data model stratified treatment and assisting the conversion from STL file to data files that can be identified by RP technology and equipment. FIG.13 shows the main role of its role.

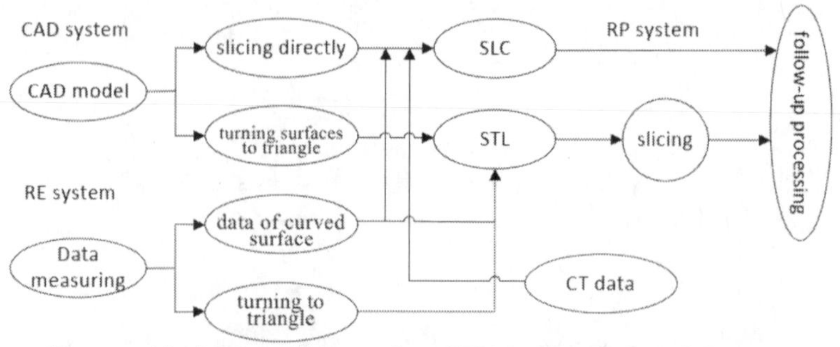

Figure.13 the au iliary role of Two-dimensional format

(1) SLC file format SLC (Stereo lithography Contour) file format is a format developed by the American company 3D System, which is the description for a three-dimensional CAD data model of the two-dimensional semi-contour. That is, the file is formed by a series of cross-sectional composition in the Z-axis direction and in each layer the three-dimensional solid model is described by inner and outer boundary lines.

There are many ways to obtain SLC data files. For example, the SLC data files can be transformed from three-dimensional CAD data model and surface model, CT scan data, etc. It is similar with cut layer data of RP technology system. STL file uses SLC data as a interface between CAD and RP systems to slice directly by three-dimensional CAD system so as to avoid partial loss of the accuracy of the model, while the CAD system is also easier to work. In addition, SLC data file as input data of RP technology has better adaptability for most of the current RP processes and equipments. And there is no choice constraint on parameter molding of RP technology system in the processing parameters and generating support and processing methods, etc.

The entity elements in SLC file format includes contoured layer, contour boundary line, multi-line and so on. Contour layer represents a layer of contour data on the Z-axis of three-dimensional CAD data. The cross-sectional slice layer lies on the cross-section parallel to the layers of X- plane and have a certain thickness. It is used to describe the closed boundary line in and out contour model. Multi-line within the boundaries are arranged in a clockwise direction and material portion entity is out of multi-line; Multi-line out of the boundaries are arranged in a counterclockwise direction and material portion entity is in the multi-line. Segment refers to the line connecting two vertices in two-dimensional. Multi-line refers to the sequence of a plurality of successive segments connected together, which is closed.

SLC data has following shortcomings: Each layer of data is approximate expression for three-dimensional solid and the accuracy is not high. In addition, it takes longer time to calculate the large amount of data.

(2)CLI file CLI (Common Layer Interface) file format is a two-dimensional data format developed by the European Automobile Manufacturers supported by Brite Euram project. The objective of this research is to provide another form of data which is 2.5-dimension for RP technology system. The geometry form of parts is described by these 2.5-dimension data and each layer of it is composed of a series of contour cross-sectional line which have a certain thickness.

The entity elements used by CLI data format include filling line and multi-line. Filling line consists of a series of parallel straight lines; multi-line is a series vertices which arranged in good order connected by a section of straight lines. By setting the fill lines and multi-lines, we can design and process the supporting structure.

CLI file is a widely accepted two-dimensional format for RP technology and equipment. The most important feature of CLI data files is simple and efficient, and easy to slice and correct model errors in the layers of the model, and has automatic repair function. It is widely used in the medical

CT (Computed Tomography, CT), layered manufacturing technology (Laminated Manufacturing Technology, LMT) and very suitable for the resin layer processing, optical processing, extrusion or fuse the powder sintering of RP system of three-dimensional solid model processing and manufacturing.

CLI file also has its shortcomings. The contour lines which describing CLI data file layers have a certain thickness and it just define the boundaries of regional entities. That is, it is an approximate expression Thus, the contour accuracy is lower and the size of the data files is large so the processing time is longer.

(3)HPGL (Hewlett packard Graphics Language) file format is a standard data format for plotter. It also belongs to the two-dimensional data type, including splines, text, curves, circles. The advantages of the file format are that the general three-dimensional CAD software has an interface to output HPGL files without additional development; In addition, we can use HPGL file format to transfer data to RP process system directly to manufacture without slicing.

Table.1 shows the advantages and disadvantages of three above formats.

Table.1 the advantages and disadvantages of three above formats

property	SLC	CLI	HPGL
veracity	ordinary	ordinary	ordinary
form	ordinary	ordinary	ordinary
repairability	good	good	good
redundancy	null	null	null
completeness	yes	yes	yes
storage	ordinary	ordinary	ordinary
neutral	null	yes	yes
source of data	geometry	geometry	geometry

4.2. The data formats of three-dimensional(STEP、DXF、IGES、LMI)

(1)STEP Files STEP (Standard for the Exchange of Product) file is a standard interface file developed this year for engineering data exchange. Currently, it covers all areas of industry and has a perfect industry standard, archiving as ISO 10303.

The advantages of STEP file are: large amount of information; STEP file format is now one of international standard interface formats for product data exchange, so the STEP file is an interface file between 3D CAD data and RP processing. The disadvantages of STEP data format are: the files contain many redundant additional data of RP process. Therefore, during the conversion we must first remove some unnecessary redundancy information while compressing data and add the topology information.

(2)DXF format DXF (Drawing Exchange Format) is a vector file format in AutoCAD. It is stored in ASCII files and is accurate in terms of the size of graphics performance. Currently many software supports input and output DXF format. Currently, DXF file format is the de facto standard for industrial applications.

The data volume of DXF file format is large and the structure is complex. The information may loss while describing the complex three-dimensional products.

(3)IGES format IGES (Initial Graphics Exchange Specification) file format, became the ANSI standard in 1982 and it has become standard file commercially available CAD systems for graphics information interchange which is widely used in the industrial fields. Most commercial CAD systems can converse data by means of IGES file. There are many PR process system accept IGES format.

FIG.14 shows the basic principle of data transmission and conversion in CAD system. IGES format file defines the unit and the cell properties, mainly used to describe the language, geometry information, topology and structure of the model. It can accurately describe the three-dimensional CAD data model. In addition, we can defined three-dimensional structure of a specific entity by using structural units and attributes of IGES file formats.

The advantages of IGES format are that it can provide entity information including lines, curves, arcs, surface, body, etc. and can accurately show the three-dimensional CAD data model information. The disadvantages of IGES format are that though it is a common standard, it contains a large number of unnecessary redundant information and it doesn't support to describe patch format; its slicing algorithm is more complicated than slicing algorithm of STL file format; The support structure should be first established by three-dimensional system and then converted to IGES format. Otherwise it cannot be achieved. Therefore, IGES format process is more complicated compared with the STL file.

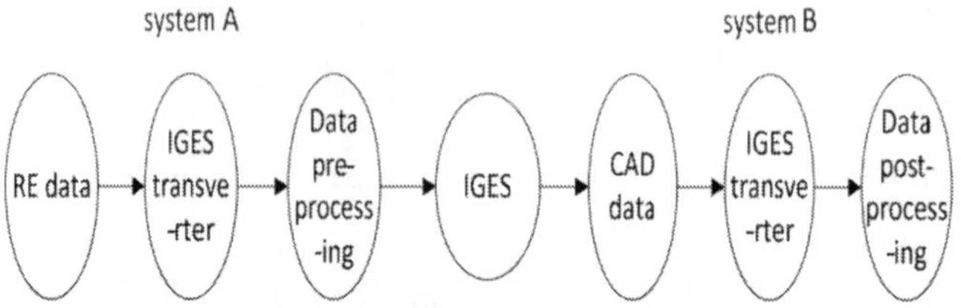

Figure.14 the data transmission and transform between IGES and CAD

Although there are still some problems in using IGES, STEP file format, the use of IGES, STEP file is basically independent of CAD software and the universality of algorithm is better. So the performance is more complete compared to STL format.

(4)LMI format LMI (Layer Manufacturing Interface) file format supports patch model and the most important feature of LMI file format is not dependent on any hardware and software environment and RP processes including topology information triangular facets; There is no redundant data; It supports accurate model and has certain flexibility and scalability and there was no ambiguity in the description.

(5)RPI format RPI (Rapid Prototyping Interface) file format, which can be derived from the STL file format, that is, RPI format information can describe patch models and defined entities such as rim, surface. It not only list vertex coordinates but also add the topology information. RPI format is a combination of multiple entities, within each entity it contains data information.

The advantages of RPI format are that it has great flexibility and easy to be extended. The file structure is compact and there is no redundant information due to its data format of topology information and plug-in data; It can describe CSG (Constructive solid Geometry) entity so as to not only describing facets of entities but also expressing the model in STL file format. However, due to the great flexibility of RPI file format, the slicing process is more cumbersome and complex.

Table.2 shows the figures of three above three-dimensional data format.

Table.2 the figures of three above three-dimensional data formats

property	STEP	STL	CFL	RPI
veracity	good	ordinary	ordinary	ordinary
form	B-rep CSG	triangle model	surfaces	surfaces CSG
repairability	ordinary	ordinary	ordinary	ordinary
redundancy	good	ordinary	good	good
completeness	yes	null	null	yes
storage	good	bad	ordinary	good

neutral	yes	bad	null	yes
source of data	Geometry/RP technology /topology	geometry	Geometry /topology	Geometry /topology RP technology

In summary, the data preprocessing of technology 3D rapid prototyping technology is critical. It relates to the quality of 3D rapid prototyping process and post-molding product. Therefore, it is necessary to attach great importance to and continue to improve and develop this technology in the future 3D rapid prototyping technology research and development.

5. REFERENCES

[1] **Liu Wei un**. Technology and application of rapid prototyping[M]. Machinery Industry Press, 2016,pp.23-25.

[2] **Chen Xuefang, Sun Chunhua**. Technology and application of reverse engineering and rapid prototyping[M]. Machinery Industry Press,2009,pp.67-82.

[3] **Wang Xuerang, Yang Zhanyao**. Rapid prototyping and rapid mould manufacturing technology[M]. Tsinghua University Press,2006,pp.89-90.

[4] **Zhang Shu, Chen Chao iang**. Product innovation and rapid development[M]. Machinery Industry Press, 2008,pp.124-126.

[5] **Wang guangchun**. Add material manufacturing technology and application example[M]. Machinery Industry Press, 2014,pp.133-135.

[6] **Andrea Gebhard**. Rapid prototyping technology[M]. Chemical Industry Press,2005,pp.221-223.

[7] **Du Zhizhong, Lu unhua**.3D printing technology[M]. Zhejiang University Press,2015,pp.122-125.

MODELING OF ISOMETRIC MUSCULAR STRENGTHS RELATED TO MANUAL MATERIALS HANDLING TASKS

Lu Peng[1], Cai-jun Zhao[1], Ai-qun Liu[1], Kai-Way Li[1, 2, 1]
[1] School of Safety Environmental Engineering
Hunan Institute of Technology, Hengyang, PRC
[2] Department of Industrial Management
Chung Hua University, Hsin-Chu, Taiwan
* kaiwayli@qq.com

ABSTRACT

A fundamental concept in ergonomics is that a job should be designed so that job demand does not exceed worker's capability. Isometric muscular strengths have been proposed as predictors of physical capabilities for human. Determinations of isometric strength are essential in the prevention of work-related musculoskeletal injuries. China encompasses the largest work force in the globe, huge amount of manual materials handling (MMH) tasks are performed in industry daily. Investigations of isometric strengths are essential for MMH task designs. In this paper, body dimensions and three isometric strengths related to the MMH tasks of 73 subjects are measured. Linear regression models for the three isometric strengths were established. The results showed that functional overhead reach and shoulder width were significant dependent variables for all three isometric strengths. Subjects of higher BMI had greater isometric strengths and each pair of isometric strengths was of significant positive correlation ($p<0.0001$).

Keywords : Anthropometry, employee screening, musculoskeletal disorders, physical work

1. INTRODUCTION

Even though robotics and automation devices are widely used in industrial settings today, manual materials handling (MMH) tasks are still common [1]. The MMH tasks have been recognized as major contributors of musculoskeletal disorders at work places [2]. The concept of designing jobs not to exceed workers' capability has been introduced decades ago [3]. However, job requirements exceeding workers' capacity is still common worldwide.

Muscular strength is the capability of a person to exert force, generated by muscular contraction, on an external object. It is the maximum force output that a muscle, or muscle groups can develop when contract voluntarily and is normally term maximum voluntary contraction (MVC). The force exertion may or may not result in movement of body segments or external object. Even though the MVC is the maximum force a person can generate, it is well below the failure limit of the musculoskeletal system. Since the MVC is the upper limit of force that a person can generate safely in a specific testing condition, it may be adopted as an indicator of the capability of the person to perform physical work [4-5].

In designing MMH tasks, human strength are used to ensure that job demands do not exceed the MVC of the person under similar conditions. If the strength requirements exceed the MVC of the worker, risk of musculoskeletal injuries increases substantially [6]. Isometric strengths have been proposed as predictors for human lifting capability [7]. They are also adopted to measure the maximum endurance time (MET) for manual tasks [8-10]. Muscular strength measurements are, therefore, essential to prevent incidences and severity of musculoskeletal injuries [11]. The literature has indicated the significance of muscular strength measurements as a part of pre-employment screening for musculoskeletal injury prevention [2, 12].

In addition to job design, muscular strength is also important in analyzing hazards associated with product usage. It is important to predict what the effect of specific amounts of muscular strength

[1] Corresponding Author: Kai-Way Li

exertion will be on the products being used. A product safety assessment might include determining how much grip, push or pull strength is required for children to play with certain toys in a playground or estimating the push strength required to operate a snow blower. It is not possible to predict the risk of injury associated with product use or misuse without comprehensive and accurate muscular strength data [13].

As both the factory for the world and one of the leading commercial markets in the world, China encompasses the largest work force in the globe. Huge amount of MMH tasks are handled daily. However, the concept of designing jobs to meet the workers' capacity is still not popular, not to mention the measurement of muscular strength as a pre-employment tool for workers allocation. The objects of this study were i) to study the correlations of anthropometric measures of Chinese adult samples and their isometric strength data; and ii) to establish predictive models for the isometric muscular strength.

2. METHODOLOGY

2.1. Sub ects

A total of 73 human subjects, including 26 females and 47 males, were recruited. All the subjects were free of musculoskeletal injury. They signed an informed consent before join the study. The age, stature, body weight, and body mass index for the females were 20.4 (1.1) yrs, 158.4 (4.7) cm, 51.3 (6.9) kg, and 20.4 (2.4) kg/m2, respectively. The age, stature, body weight, and body mass index for the males were 20.3 (1.0) yrs, 172.0 (5.2) cm, 63.3 (10.6) kg, and 21.4 (3.1) kg/m2, respectively.

2.2. Apparatus

An anthropometric measurement kit was used to measure body dimensions of the subjects. In addition, a testing apparatus to measure the isometric strength was adopted. This apparatus included a steel platform, a chain, an S-shape loadcell (Lutron Inc., FG-5100), and a handle. The handle was a steel one with a diameter of 3 cm. There was a hook on the platform. One end of the chain was anchored to this hook and the other end was connected to the handle. The loadcell was connected in the middle of the chain. When the subject pulled the handle, the peak pulling force was shown in a digital display.

2.3. Procedure

The isometric arm strength measurement requires that the subject stands erect on the platform of the strength measurement unit, with back and legs straight with the feet flat. His upper arms straight down and the forearms flexed at 90°. The subject grasps the handle of the strength measurement unit and applies a vertical, upward force. The erected force is generated only by the arm muscles (see Fig. 1). The procedure follows those in the literature [2, 14-18].

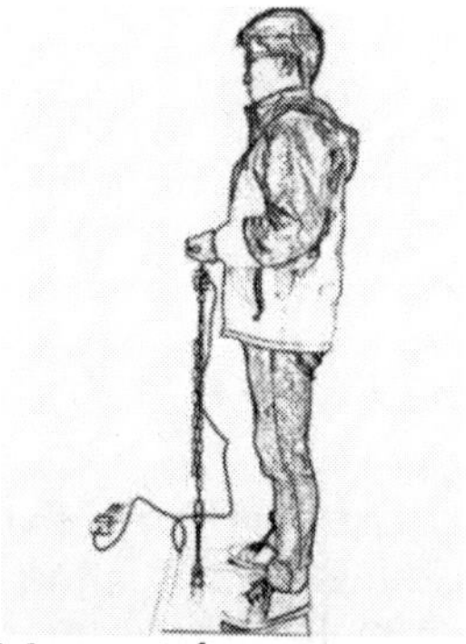

Figure. 1 Isometric arm measurement

In the isometric composite strength measurement, or leg lifting strength measurement, the subject takes a semi-squat posture. The subject's elbow must be extended. The feet were flat on the platform. The handle of the testing device was adjusted approximately 38 cm above the platform and was between the legs of the subject. An upward force is generated by simultaneously extending the knees and the trunk.

Figure. 2 Isometric composite strength measurement

In the isometric back strength measurement, or torso lifting strength measurement, the subject's feet are separated at shoulder width on the platform. The elbow and knees are fully extended. The handle of the testing device was 38 cm above the platform. The subject flexed the trunk in order to grip the handle. The vertical upward force is exerted by torso extension.

Figure. 3 Isometric back strength measurement

2.4. Data processing Analyses

Descriptive statistics, analyses of variance (ANOVA), correlation analyses, and regression analyses were performed. SAS 9.0 was adopted for statistical analyses.

3. RESULTS AND DICUSSION

3.1. *Descriptive statistics*

The body dimensions and three isometric strengths of 73 young Chinese adults were collected. The descriptive statistics of anthropometric measures for the female and male subjects are shown in TABLE I and TABLE II, respectively. For males, the isometric arm, back, and composite strengths were 35.0 (5.7) kgf, 95.9 (13.6) kgf, and 89.0 (15.2) kgf, respectively. For females, the isometric arm, back, and composite strengths were 20.3 (2.7) kgf, 57.8 (11.8) kgf, and 52.5 (10.3) kgf, respectively.

Fig. 4 depicts the comparison of isometric strengths between female subjects and male subjects, where we can get the isometric strengths of male adults are overwhelmingly higher than female adults. Anyway, Fig. 4 complies with the nature of human physiology.

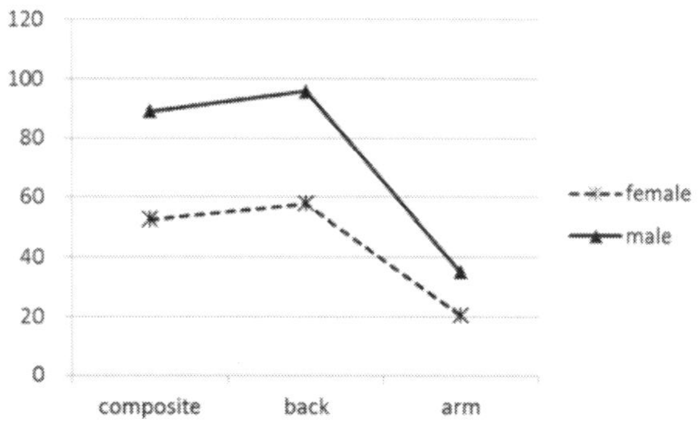

Figure. 4 Comparison of isometric strengths (kgf)

Table 1 body dimensions (cm) for female sub ects

Variable	Mean	Std	Min	Max
BMI (kg/m²)	20.4	2.4	16.6	26.5
thigh length	69.2	21.3	38.3	93.0
shank length	34.8	2.6	31.9	42.5
eye height	147.7	10.6	137.5	193.0
shoulder height	127.7	4.6	120.0	144.0
elbow height	98.4	4.1	91.0	108.0
functional overhead reach	188.1	7.4	177.0	212.1
both arm abduction width	153.8	6.9	144.6	171.0
shoulder width	36.9	2.7	29.0	43.0

Table 2 body dimensions (cm) for male sub ects

Variable	Mean	Std	Min	Max
BMI (kg/m²)	21.4	3.1	15.4	30.9
thigh length	88.0	16.0	46.0	106.0
shank length	40.5	4.2	33.0	51.0
eye height	160.1	5.8	147.0	174.0
shoulder height	140.4	5.4	129.2	153.0
elbow height	106.8	5.3	99.0	121.0
functional overhead reach	209.2	9.8	190.0	232.0
both arm abduction width	168.4	14.7	78.3	183.0
shoulder width	42.5	2.4	34.5	49.5

3.2. Analyses of variance

The BMI, an index of body stature was shown in TABLE 3.

Table 3 descriptive statistics of bmi (kg/m²)

Mean	Median	Std Dev	Min	Max	5th Pctl	95th Pctl
21.0	20.5	2.9	15.4	30.9	16.6	26.0

The BMI might be grouped into two levels. The BMI values less than the median (20.5) were grouped into level 1, denoted as K1. Conversely, the second was denoted as K2.

TABLE IV indicates that the difference of BMI level accounts for statistically significant effects on isometric strengths. Specifically, subjects of higher BMI were equipped with greater isometric strengths. It is noteworthy that the result of Table IV was obtained against a specific BMI range:15.4 to 30.9.

Table 4 analyses of variance for bmi level

	Composite strength[a]	Back strength[a]	Arm strength[a]
Mean for K1	69.72	73.26	27.51
Mean for K2	82.41	91.56	32.12
p-value	0.0136	0.0003	0.0208
Homogeneity of variance (Pr>F)	0.873	0.152	0.165

[a]kgf

3.3. Regression analyses

Equations (1) to (3) present the results of regression analyses among isometric composite strength (ICT), isometric back strength (IBT) and isometric arm strength (IAT). We found that functional overhead reach and shoulder width are significant contributing variables for all the three isometric strengths. The coefficients of determination, or simply R2, for the three equations were 0.58, 0.68, and 0.69, respectively.

$$ICT=-169.9+0.8x_1+2.3x_4 \quad (1)$$

$$IBT=-217.8+1.7BMI+0.9x_1+0.25x_2+1.04x_4 \quad (2)$$

$$IAT=-74+0.3x_1+0.3x_3+0.6x_4 \quad (3)$$

where x_1 is the functional overhead reach, x_2 is the width with both arm abduction, x_3 is the shank length, and x_4 the shoulder width.

Fig. 5, 6, and 7 show the actual values versus predicted values for the isometric composite, back, and arm strength, respectively. All these figures show positive linear correlations between the predicted values and actuals values.

Composite strength (kgf)

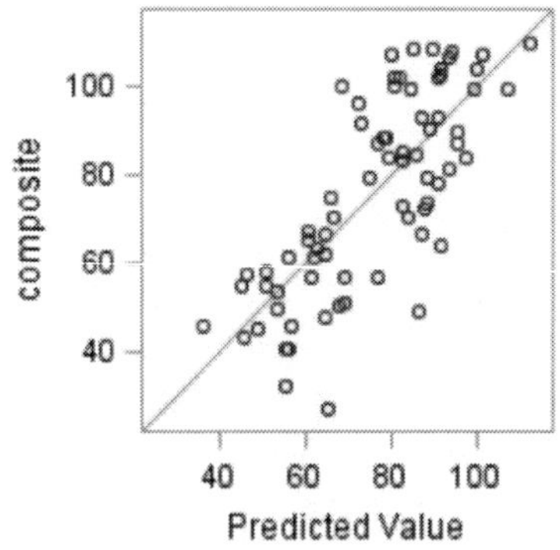

Predicted value (kgf)

Figure. 5 Actual values versus predicted values for isometric composite strength

Back strength (kgf)

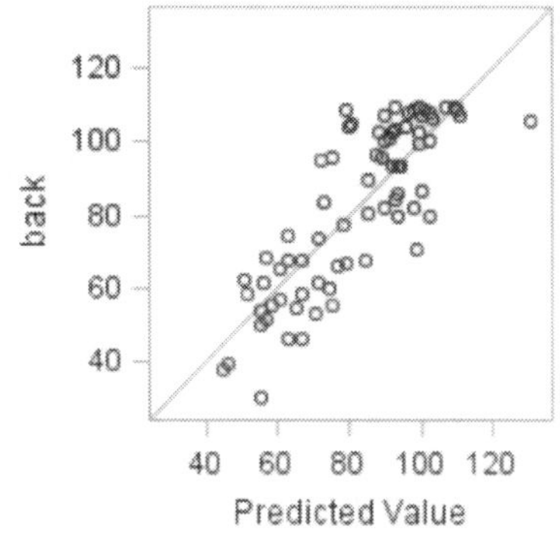

Predicted value (kgf)

Figure. 6 Actual values versus predicted values for isometric back strength

Arm strength (kgf)

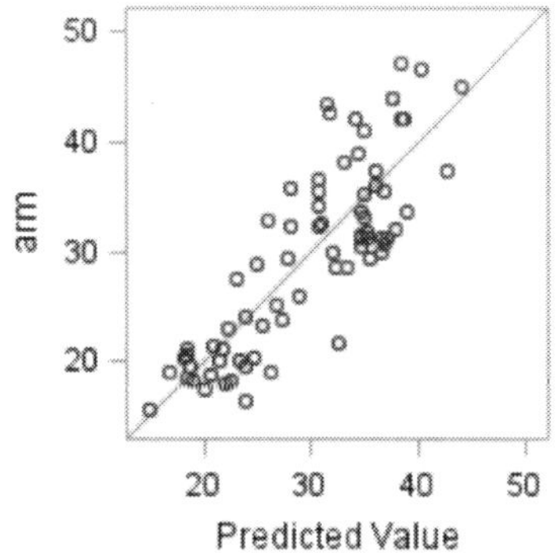

Predicted value (kgf)

Figure. Actual values versus predicted values for isometric arm strength

3.4. correlation analyses

Regarding to different body parts, the measured isometric strengths vary a lot. In this paper, the correlation analyses among isometric composite strength, isometric back strength and isometric arm strength was performed. TABLE V shows correlation coefficients among these three isometric strengths are high. Linear regression relationships between isometric composite strength and back strength, between isometric composite strength and arm strength, and between isometric back strength and arm strength are shown in Fig. 8, 9, and 10, respectively.

Table 5 the pearson analyses of isometric strengths

	composite	back	arm
composite	1	0.79 <0.0001	0.80 <0.0001
back	-	1	0.83 <0.0001

Composite strength (kgf)

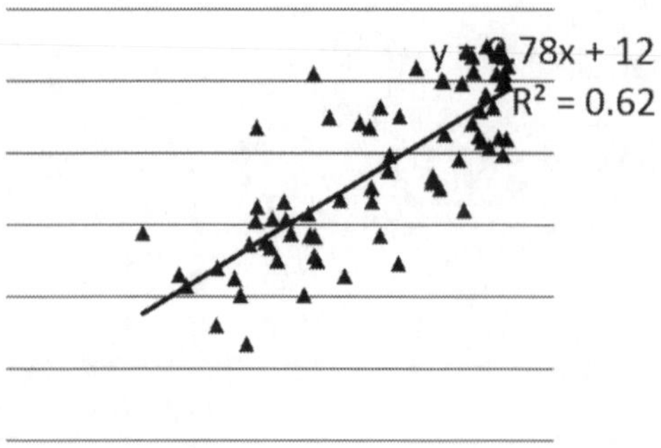

Back strength (kgf)

Figure 8. Regression relationship between isometric compositestrength and isometric back strength

Composite strength (kgf)

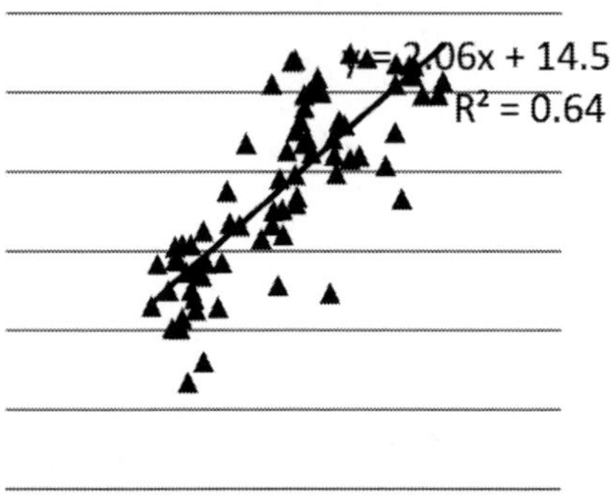

Arm strength (kgf)

Figure . Regression relationship between isometric compositestrength and isometric arm strength

Back strength (kgf)

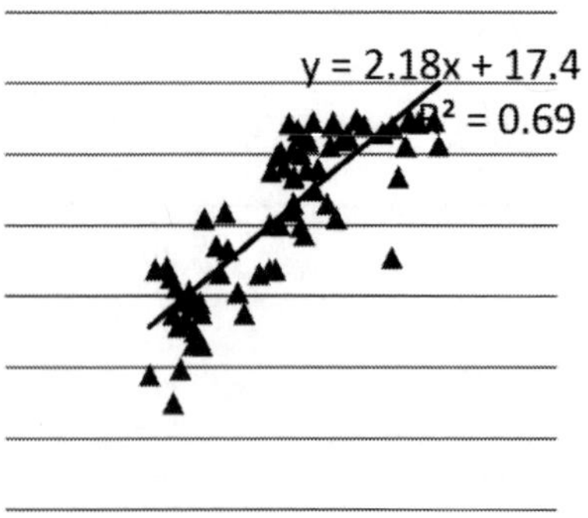

Arm strength (kgf)

Figure 10. Regression relationship between isometric backstrength and isometric arm strength

4. CONCLUSION

With the comprehensive measurement and statistical analyses for body dimensions and isometric strengths regarding 73 young Chinese adults, the statistics of body dimensions and three isometric strength measures were obtained. Furthermore, linear regression models for these three isometric strengths were established. It was found that both the functional overhead reach and shoulder width were significant contributing variables for all the three isometric strengths. The subjects with higher BMI had greater isometric strengths. The isometric strengths had significant positive correlation with one another.

Measurement and analysis of isometric strengths are very important for optimizing MMH tasks design, or being as a pre-employment tool for workers allocation. All the subjects were young adults. Future researches may be conducted for workers of different age levels.

5.ACKNOWLEDGMENT

The authors thank the subjects for participating in the experiment.

6.REFERENCES

[1] **K. N. Agrawal, R. K. P. Singh, K. K. Satapthy**, Isometric strength of agricultural workers of Meghalaya: A case study of an Indian population, International Journal of Industrial Ergonomics, vol. 39, no. 6, pp. 919-923, Nov. 2009.

[2] **M. M. Ayoub, A. Mital, Manual Materials Handling**. London, Taylor Francis, 1989, pp. 9-35.

[3] **V. M. Ciriello, S. H. Snook, L. Hashemi, . Cotnam**, Distribution of manual materials handling task parameters, International Journal of Industrial Ergonomics, vol. 24, no. 4, pp. 379-388, Aug. 1999.

[4] **S. H. Snook, V. M. Ciriello**, The design of manual tasks: revised tables of maximum acceptable weights and forces, Ergonomics, vol. 34, no. 9, pp. 1197-1213, 1991.

[5] **V. M. Ciriello, S. H. Snook**, Survey of manual materials handling tasks, International Journal of Industrial Ergonomics, vol. 23, pp. 149-156, 1999.

[6] **Y. H. Lee, Y. L. Chen**, An Isometric Predictor for Maximum Acceptable Weight of Lift for Chinese Men, Human Factors, vol. 38, no. 4, pp. 646-653, 1996.

[7] **A. Mital, M. M. Ayoub**, Modeling of Isometric Strength and Lifting Capacity, Human Factors, vol. 22, no. 3, pp. 285-291, June 1980.

[8] **Z. Zhang, K. W. Li, W. Zhang, L. Ma, Z. Chen**, Muscular fatigue and maximum endurance time assessment for male and female industrial workers, International Journal of Industrial Ergonomics, vol. 44, no. 2, pp. 292-297, Mar. 2014.

[9] **K. W. Li, W. S. Chiu**, Isometric Arm Strength and Subjective Rating of Upper Limb Fatigue in Two-Handed Carrying Tasks, PLOS One 10(3): e0119550. doi:10.1371/journal.pone.0119550, 2015.

[10] **K. W. Li, C. W. Wang, R. Yu**, Modeling of Predictive Muscular Strength for Sustained One-Handed Carrying Task, Work, vol. 52, no. 4, pp. 911-919, Sep. 2015.

[11] **T. Rikkonen, . Sirola, K. Salovaara, M. Tuppurainen, . urvelin, R. Honkanen, H. Kr ger**, Muscle Strength and Body Composition Are Clinical Indicators of Osteoporosis, Calcified Tissue International, vol. 91, no. 2, pp. 131-138, Aug. 2012.

[12] **D. B. Chaffin, G. D. Herrin, W. M. Keyserling**, Pre-employment strength testing, Journal of Occupational Medicine, vol. 20, pp. 403-408, 1978.

[13] **M. D. Boulette**, Strength Considerations for Product Safety Assessments, Proceedings of the Human Factors and Ergonomics Society Annual Meeting, vol. 29, no. 3: pp. 255-259, October 1985.

[14] **L. S. Caldwell, D. B. Chaffin, F. N. Dukes-Dobos, K. H. E. Kromer, L. L. Laubach, S. N. Snook, D. E. Wasserman**, A proposed standard procedure for static muscle strength testing, American Industrial Hygiene Association Journal, vol. 35, no. 4, pp. 201-206, 1974.

[15] **G. B. Xiao, L. Ling, P. G. Dempsey, B. B. Lu, Y. X. Liang,** Isometric muscle strength and anthropometric characteristic of a Chinese sample, International Journal of Industrial Ergonomics, vol. 35, no. 7, pp. 674-679, July 2005.

[16] **D. B. Chaffin**, Ergonomics guide for assessment of human static strength, American Industrial Hygiene Association Journal, vol. 36, pp. 505-511, 1975.

[17] **L. S. Caldwell, D. B. Chaffin, F. N. Dukes-Dobos, K. H. E. Kromer, L. L. Laubach, S. N. Snook, D. E. Wasserman** , A proposed standard procedure for static muscle strength testing, American Industrial Hygiene Association Journal, vol. 35, pp. 201-206, 1974.

[18] **K. N. Agrawal, R. K. P. Singh, K. K. Satapthy**, Isometric strength of agricultural workers of Meghalaya: A case study of an Indian population, International Journal of Industrial Ergonomics, vol. 39, no. 6, pp. 919-923, 2009.

MUSCULAR STRENGTH AND PERCEIVED PHYSICAL EXERTION IN A PULLING TASK

Fan Tang[1], Can-nan i[1], Kai-Way Li[1,2,*1]

[1]School of Safety and Environmental Engineering, Hunan Institute of Technology, Hengyang, PRC
[2]Department of Industrial Management, Chung Hua University, Hsin-Chu, Taiwan
*kaiwayli@qq.com

ABSTRACT

A simulated pulling experiment was performed in this study. Eight male subjects participated in the study. The subjects pulled a handle using their dominant hand, under two loading conditions, for five minutes. Their single arm pulling strengths and the subjective rating of perceived physical exertion were collected. The results indicated that the pulling strengths after the trial were significantly lower than those of the maximum voluntary contraction (MVC) of the subjects. The pulling strength for the 30 kg loading condition was lower than that of the 20 kg condition. The difference, however, did not reach statistical significance level. The subjective rating scores for the 20 kg and 30 kg conditions were 5.87 and 7.12, respectively. The Pearson's correlation coefficient between the pulling strength and the subjective rating was -0.67 ($p<0.0001$). A linear regression model describing the relationship between the pulling strength and the subjective rating score was established.

Keywords: Borg CR-10, manual materials handling, muscular fatigue, regression model

1. INTRODUCTION

Musculoskeletal disorders (MSDs), such as low back pain and carpal tunnel syndrome, at work are common. The literature has pointed out risk factors contributing to the MSDs. Unnatural postures, repetitive motion, forceful exertion, vibration, low temperature, and so on, are typical risk factors for the MSDs [1-4]. The MSDs may develop if those so-called ergonomic risk factors persisted for a period of time. There are bodily symptoms before the MSDs do occur. Muscular pain and fatigue are the most typical ones.

Muscular fatigue has been one of the major issues in the study of job design [5-7]. The symptoms associated with muscular fatigue including pain, swelling, weakness, and so on, in the body part. However, weakness of the body parts is adopted most by scientists in assessing muscular fatigue. For example, Vøllestad [8] defined muscular fatigue as any exercise-induced reduction in the capacity to generate force or power output. Chaffin et al. [2], on the other hand, defined muscular fatigue as the reduction in the ability to exert force in response to voluntary effort. Similar definitions may be found in the literature [9].

Measurement of muscular strength has been adopted to assess muscular fatigue in terms of muscle weakness. In this approach, the maximum voluntary contraction (MVC) is measured. Muscular strength is measured after a certain task has been performed for a period of time. The decrease of the muscular strength may, therefore, be examined to indicate the level of muscular fatigue. Mathematic equations have been adopted to describe the decrease of muscular strength. Non-linear

[1] Corresponding Author: Kai-Way Li

functions, such as exponential function, logarithm function, and polynomial functions have been employed for this purpose [10, 11].

Isometric muscular strength measurements have been studied very common because the measurements are relatively simple both in the instrumentation and control of bodily posture. Many examples of muscular fatigue studies employing isometric strength measurements may be found in the literature [8, 10, 12- 15].

The developing of MSDs is task dependent. Manual materials handing (MMH) tasks are the primary one to be blamed for the MSDs. Assessments of muscular strength were normally conducted simulating the performance of a certain MMH task. Lifting and lowering, as has been justified as the most risky tasks contributing to low back pain, are probably the industrial activities been studied most by scientists globally. The famous NIOSH lifting equation was, for example, developed specifically for designing lifting tasks [16]. Other MMH tasks, such as pushing, pulling, and carrying have also been addressed. For examples, Ma et al. [17] studied the muscular fatigue phenomenon of upper limbs via studying drilling tasks in aircraft assembly. Li and Chiu al. [14] assessed the upper arm fatigue of human subjects in box holding tasks. Li et al. [15] assessed muscular strength of adult males while carrying a weight using single arm.

Pulling is a one of the most common MMH tasks in industrial sectors. Pulling requires grasping of a handle and hand/arm exertion to drag a cart or a load to move toward the puller. The forceful exertion of the hand/arm system could result in muscular fatigue when the task is performed for a period of time. Such phenomenon may be assessed by studying the muscular strength before and after the task is performed. In addition, muscular strength may also be assessed by having the human subject report his/her subjective rating of perceived exertion. The Borg rating scale [18-20] has been widely adopted for this purpose. There are two subjecting ratings developed by Borg: the rating of perceived exertion (RPE) ranging from 6 to 20 and the CR-10 rating. The former has been widely adopted to assess muscular exertion, perceived loading, pain, and bodily discomfort for whole body condition. The latter has been used for the same purposes for specific body segment or local area of the body.

The objective of this study was to assess muscular fatigue in performing a pulling task. Specifically, the objectives were to i) compared the muscular strength of pulling before and after performing a pulling task; and ii) study the correlation and mathematic relationship between the subjective rating of human perceived exertion and the muscular strength in performing the pulling task.

2. METHODOLOGY

An experiment was conducted in the laboratory to simulate the pulling of a pallet truck using single arm. The temperature and humidity were 9.6 C (1.3) and 78.3 (5.8), respectively.

2.1. Human Sub ects

Eight male adults volunteered as human subjects in the pulling experiment without payment. All the subjects were right-handlers. All of them did not have a history of musculoskeletal injuries. They read and signed an informed consent for participating in the study. The age, stature, body weight, body mass index were 22.13 (1.05) yrs, 57.94 (9.03) kg, 163.78 (2.62) cm, and 21.60 (3.35) kg/m^2, respectively. All the subjects were requested to refrain from strenuous physical exercise a day before joining the experiment.

2.2. Apparatus

An apparatus to measure the pull strength was employed. This apparatus encompassed a chain, a loadcell (Lutron Inc., FG-5100), and a handle. The handle was made from steel. The diameter of the handle was 3 cm. One end of the chain was anchored to a hook which was mounted on a wall 37 cm above the ground in the laboratory. The other end of the chain was connected to the handle.

The S-shape loadcell was connected between the anchor and the handle to sense the tension of the chain when it was been pulled. The peak force of the pulling was displayed in a digital window. A Borg CR-10 rating scale [19] was employed to measure perceived physical exertion of the subject at the end of each trial.

2.3. Stick of a Pallet Truck

A stick with a handle, simulating the stick unit of a commercially available pallet truck, was fabricated and installed (see Fig. 1). This steel stick has a length of 84.5 cm and a mass of 1.5 kg. The diameter of the handle was 3 cm. The stick was a T-bar and was suspended by two wires at one and two thirds of the length from the end. The stick was swing to a side so that it was tilted with an angle of 42 with the horizon and the lower end of the stick was approximately 37 cm above the ground as were normally observed in pulling of a real truck. The angles between the two wires and the vertical were 18 and 21 , respectively. A load (cast irons) with a mass of 20 kg or 30 kg was suspended in the middle of the stick to generate a back swing force. Based on a kinetic analysis, the back swing forces for the 20 kg and 30 kg conditions were 76.1 N and 111.65 N, respectively. These forces were the forces that the subject applied in order to balance the load when simulating a truck pulling task.

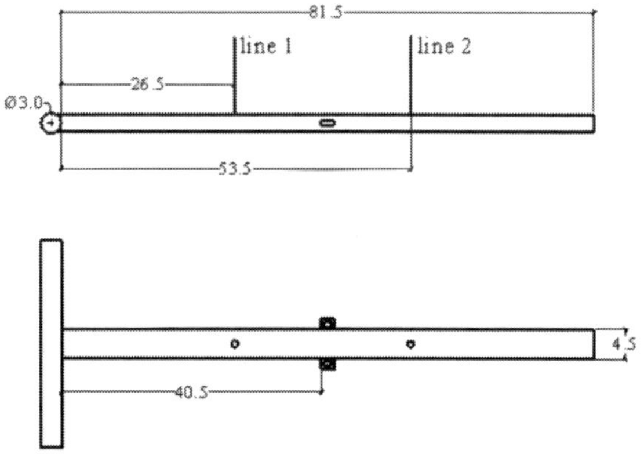

Figure.1. Simulated stick/handle unit of a truck

2.4. E perimental Procedure

Before each trial, the subject was requested to do warm up exercise for five minutes. The isometric pulling strength was, then, measured. In this measurement, the subject pulled the handle of the strength testing unit as hard as he could using his right arm for 4 to 6 seconds (see Fig. 2). The elbow of the subject was flexed at approximately 15 . The posture was similar to that when pulling a truck. The subject stood with his left foot in the front and two feet apart. This is a typical posture for truck pulling. The test protocol followed those in the literature [21-23]. The peak force was recorded as the maximum voluntary contraction (MVC) for the trial.

For the simulated pulling task, the subject stood and grasped the handle to simulate a truck pulling posture using his right arm and maintained this pull for a period of 5 min (see Fig. 3). He was instructed to maintain the same posture during the trial. At the end of the trial, the subject released the handle and reported his perceived physical exertion or physical burden on the Borg CR-10 scale for the pull. He then went through the pulling strength measurement procedure described in the previous paragraph. The subject took a break for ten minutes or longer until next test.

Figure.2. Pulling strength measurement

Figure.3. Simulated pulling task

2.5. E periment Design Data Analysis

Each subject needs to complete both the 20 kg and 30 kg loading conditions. A total of 32 (2 loading conditions 2 strength testing 8 subjects) pulling strength measures were collected. There were 16 Borg CR-10 ratings (2 loading conditions 8 subjects). Microsoft Excel was employed for data analysis and figure preparation. The SAS 9.4 was used for statistical analyses.

3. RESULTS

3.1. Isometric Pulling Strength

The isometric pulling strength for the subjects was 155.02 (25.13) N. Their pulling strengths after pulled the simulated handle for 5 minutes under the 20 kg and 30 kg conditions were 121.46 (24.65) N and 97.51 (21.84) N, respectively. The ANOVA results indicated that the pulling strength before and after the pull test was significantly different ($p<0.0001$). The pulling strengths between the 20 kg and 30 kg loading conditions after the trial were, however, insignificant. Fig. 4 and 5 show the box plots of the pulling strength (N).

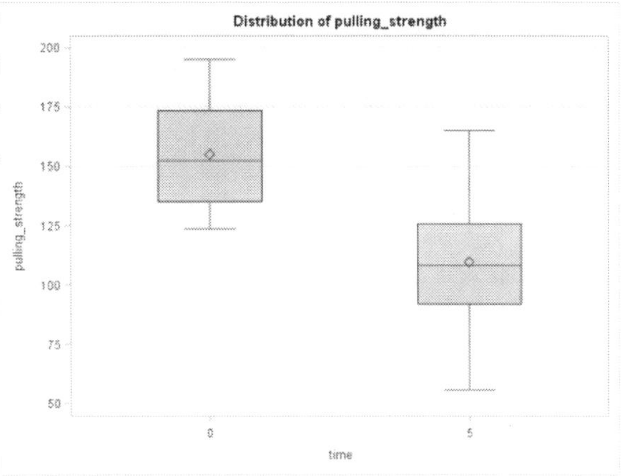

Figure.4. Bo plot of pulling strength before and after the trial

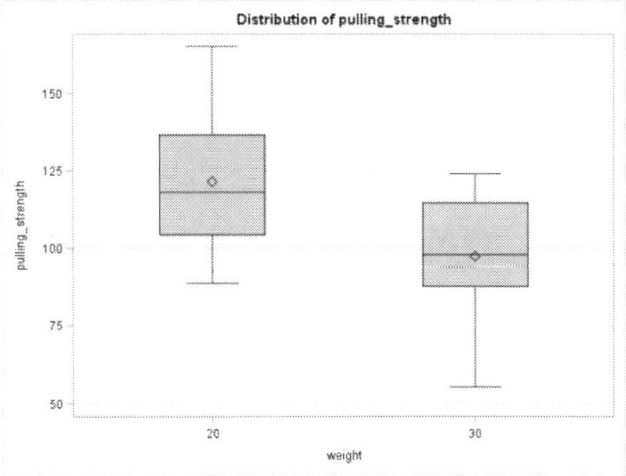

Figure.5. Bo plot of pulling strength after the trial

3.2. CR-10 Ratings after the trial

The CR-10 ratings for the 20 and 30 kg loading conditions were 5.87 (1.73) and 7.12 (2.03), respectively (see Fig. 6). The difference between the two loading conditions did not reach the 0.05 significance level.

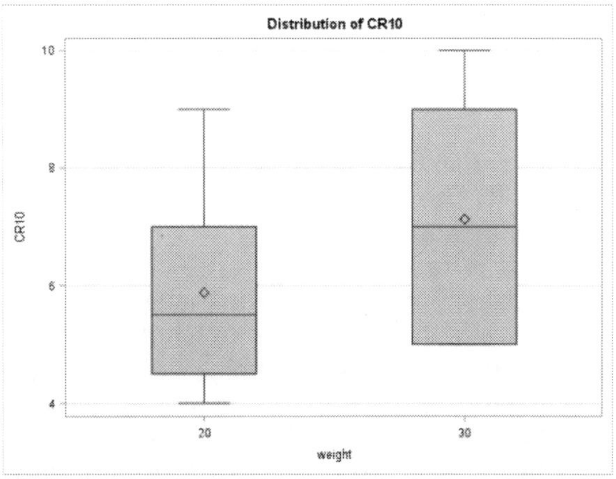

Figure.6. CR-10 scores for the loading conditions

3.3. Correlation   Regression Analysis

The Pearson's correlation coefficient between the pulling strength and the CR-10 score was -0.67 ($p<0.0001$) indicating the negative correlation between the two variables.
A regression analysis was performed using the pulling strength as the dependent variable and the CR-10 score as the independent variable. The following equation was obtained:

Pulling strength (N) = 152.95   6.37   *CR-10 score*   (1)

The two-tailed *t*-tests for both regression coefficients were statistically significant at $p<0.0001$. The root mean square error (MSE) of the model is 25.39 and the R^2 is 0.45. The regression equation and the corresponding 95   confidence and prediction intervals are shown in Fig. 7.

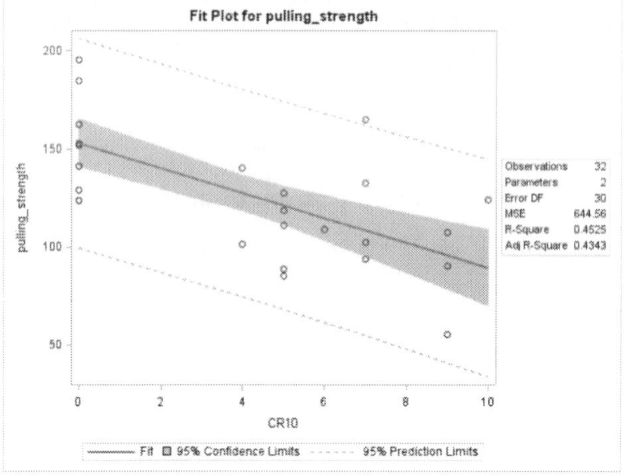

Figure. . Plot of regression line and confidence intervals in (1)

4. DISCUSSION

The experiment was conducted in the laboratory in the winter in Hengyany where there is no heating system in the room. The temperature for the experiment was 9.6 C (1.3). This temperature was lower than those in many other studies [13-15, 17]. However, such temperature is common at workplaces in Hunan during the winter.

A suspended stick was installed to simulate the stick/handle unit of a commercially available pallet truck. The cast irons suspended on the stick generated a back swing force which needed to be balance by the pulling force of the subjects. The pulling force for the 20 kg and 30 kg loading conditions were 76.1 N and 111.65 N, respectively. These pulling force were equivalent to 50.2 and 73.65 of the MVC of the subjects. Such loads were higher than those recommended as the load for a fair day's work [1] and muscular fatigue might be expected after the loads were handled for a few minutes.

In theory, a subject pull a load for a period of time, the pulling strength should be decreasing upon the developing of muscular fatigue. The major hypotheses of the study were that pulling strength reduced significantly after the subject pulled for a period of time and the after-trial pulling strength for the 30 kg condition was lower than that of the 20 kg condition as the load is a significant factor affecting the exertion of the pull. The first hypothesis was supported by our data as the pulling strengths after the trial were significantly ($p<0.0001$) lower than those before the trials. The second hypothesis was, however, not supported as the difference of the pulling strength between the two loading conditions was not statistically significant even though the averaged pulling strength (97.51 N) for the 30 kg condition was lower than that (121.46 N) of the 20 kg condition. The reason for this insignificance may be of two folds. First, the sample size of the experiment may be not enough. It is very likely to obtain significant results if more samples are tested. Second, variations caused by the posture of the subjects could introduce noise in the data. It is known that body posture affects muscular strength. The subjects were requested to pull with a flexion of the elbow for approximately 15 . However, it was extremely difficultly difficult to have the subjects maintained the same posture during the trials. It was believed that elbow posture could induce variation for both inter- and intra-subjects.

The CR-10 scores between the two loading conditions were not statistically significant even though the CR-10 score for the 30 kg condition (7.12) was higher than that of the 20 kg condition (5.87). The same reasons explained in the previous paragraph may also be applied.

The correlation coefficient between the pulling strength and the CR-10 score was -0.67 ($p<0.0001$). This is as was expected. The pulling effort resulted in pulling strength decrease which then increased the CR-10 score. The regression equation in (1) indicating the CR-10 score may be adopted to predict the decrease of puling strength.

5. CONCLUSION

A simulated pallet truck pulling experiment was performed. The single-arm pulling strength before and after the pulling was measured. The subjective ratings of the simulated pulling task were also recorded. The single-arm pulling strength decrease significantly after the subject performed the pulling task due to muscular fatigue. Such decrease corresponded to the increase of the subjective rating scores. The regression equation in (1) may be adopted to describe and predict the pulling strength under a similar task scenario. The suspended stick unit may be adopted for future studies involving pallet truck pulling.

6.ACKNOWLEDGMENT

The authors thank all the subjects for their participating in the study.

.REFERENCES

[1] **M. M. Ayoub, A. Mital**, Manual Materials Handling, Taylor Francis, London, 1989.

[2] **D. B. Chaffin, G. B. . Andersson, B. . Martin**, Occupational biomechanics, 3rd Edition, Wiley-Interscience, 1999.

[3] **V. M. Ciriello, S. H. Snook**, Survey of manual materials handling tasks, International Journal of Industrial Ergonomics, vol. 23, pp. 149-156, 1999.

[4] **D. B. Chaffin, M. M. Ayoub**, The problems of manual materials handling, Industrial Engineering, no. 7, pp. 24-29.

[5] **W. Rohmert, M. Wangenheim, . Mainzer, P. Zipp, W. Lesser**, A study stressing the need for a static postural force model for work analysis, Ergonomics, vol. 29, no. 10, pp. 1235-1249, 1986.

[6] **W. Rohmert**, Problems in determining rest allowances Part 1: use of modern methods to evaluate stress and strain in static muscular work, Appl Ergon, vol. 4, no. 2, pp. 91-95, 1973.

[7] **D. Roman-Liu, T. Tokarski, K. Wo cik**, uantitative assessment of upper limb muscle fatigue depending on the conditions of repetitive task load, Journal of Electromyography and Kinesiology, vol. 14, no. 6, pp. 671 682, 2004.

[8] **N. V llestad**, Measurement of human muscle fatigue, Journal of Neuroscience Methods, vol. 74, no. 2, pp. 219 227, 1997.

[9] **R. M. Enoka, . Duchateau**, Muscle fatigue: what, why and how it influences muscle function, J Physiol, vol. 586, no. 1, pp. 11-23, 2008.

[10] **L. Ma, D. Chablatt, F. Bennis, W. Zhang**, A new simple dynamic muscle fatigue model and its validation, International Journal of Industrial Ergonomics, vol. 39, no. 1, pp. 211-220, 2009.

[11] **H. Sato, . Ohashi, K. Iwanaga, R. Yoshitake, K. Shimada**, Endurance time and fatigue in static contractions, Journal of Human ergology, no. 13, pp. 147-154, 1984.

[12] **L. Ma, D. Chablatt, F. Bennis, W. Zhang, B. Hu, F. Guillaume**, A novel approach for determining fatigue resistances of different muscle groups in static cases, International Journal of Industrial Ergonomics, vol. 41, no. 1, pp. 10-18, 2011.

[13] **Z. Zhang, K. W. Li, W. Zhang, L. Ma, Z. Chen**, Muscular fatigue and maximum endurance time assessment for male and female industrial workers, International Journal of Industrial Ergonomics, no. 44, pp. 292-297, 2014.

[14] **K. W. Li, W. S. Chiu**, Isometric Arm Strength and Subjective Rating of Upper Limb Fatigue in Two-Handed Carrying Tasks, PLOS One, vol. 10, no. 3, e0119550. doi:10.1371/journal.pone.0119550 , 2015.

[15] **K. W. Li, C. W. Wang, R. Yu**, Modeling of Predictive Muscular Strength for Sustained One-Handed Carrying Task, Work , no. 52, pp. 911-919, 2015.

[16] **P. G. Dempsey**, Usability of the revised NIOSH lifting equation, Ergonomics, vol. 45, no. 12, pp. 817-828, 2002.

[17] **L. Ma, W. Zhang, B. Hu, D. Chablet, F. Bennis, F. Guillaume**, Determination of subject-speci c muscle fatigue rates under static fatiguing operations, Ergonomics, vol. 56, no. 12, pp. 1889 1900, 2013.

[18] **A. Borg**, Psychophysical bases of perceived exertion, Medicine and Science in Sports and Exercise, vol. 14, no. 5, pp. 377-381, 1982.

[19]**A. Borg**, Psychophysical scaling with applications in physical work and the perception of exertion, Scand Journal of Work Environmental Health, vol. 16, no. 10, pp. 55-58, 1990. [20] A. Borg, H. Sj berg, The variation in hand steadiness with physical stress, Journal of Behavior, vol. 13, pp. 110-116, 1981.

[21] **D. B. Chaffin**, Ergonomics guide for assessment of human static strength, American Industrial Hygiene Association Journal, vol. 36, pp. 505-511, 1975.

[22] **L. S. Caldwell, D. B. Chaffin, F. N. Dukes-Dobos, K. H. E. Kromer, L. L. Laubach, S. N. Snook, D. E. Wasserman** , A proposed standard procedure for static muscle strength testing, American Industrial Hygiene Association Journal, vol. 35, pp. 201-206, 1974.

[23] **K. N. Agrawal, R. K. P. Singh, K. K. Satapthy,** Isometric strength of agricultural workers of Meghalaya: A case study of an Indian population, International Journal of Industrial Ergonomics, vol. 39, no. 6, pp. 919-923. 2009.

.

AUTHOR INDEX

AUTHOR INDEX

AUTHOR INDEX

AUTHOR INDEX